WORLD *of* INVENTION

2ND EDITION

WORLD *of* INVENTION

Kimberley A. McGrath, *Editor*

Bridget Travers, *Contributing Editor*

GALE

DETROIT • LONDON

STAFF

The paper used in this publication meets the minimum requirements of American National Standard for Information Sciences—Permanence Paper for Printed Library Materials, ANSI Z39.48-1984. ∞™

ISBN 0-7876-2759-3
ISSN 1071-0973

Printed in the United States of America
10 9 8 7 6 5 4 3 2 1

I(T)P™

The trademark **ITP** is used under license.

Contents

Introduction

It is staggering to contemplate just how many inventions have come into being since the beginning of recorded time. Attempts to improve the lot of human beings by improving on the state of nature have been practically incessant, if not always successful. Almost as staggering is the amount that has been written on the subject. Any large public library is likely to contain more than 300 volumes devoted to inventions and inventors, and new books on the topic appear each year. Given the sometimes comic, sometimes awesome nature of the things people have invented—from nineteenth-century failsafe coffins equipped with ladders and alarm bells to twentieth-century wonders like the particle accelerator—the myriad strange ways in which they have invented them, and the colorful nature of the inventors themselves—of every stripe from brilliant, mad, eccentric to absent-minded or lucky, from humanitarian to charlatan—the fascination with the subject is not surprising, and many authors and editors have done it full justice.

With so much material already available on inventions and inventors, why, then, yet another on the topic? This one, written for high school students and other lay readers, has a combination of features that we believe distinguishes it from others in the field.

Scope of the Book

Comprehensive, one-volume coverage of well-known, as well as lesser-known, inventions and inventors.

This second edition of *World of Invention* comprises 907 nonbiographical and 397 biographical entries covering everything from the wheel and axle to genetic engineering, skateboards to chewing gum, razors to lasers, hydrogen bombs to disposable diapers, and people as different in background, method, and interests as Charles Babbage, the brilliant, irascible nineteenth-century mathematician who spent his life and his fortune using careful scientific methods to develop the prototype of today's digital computer and Charles Goodyear, the hardware salesman and would-be chemist who, in his dogged attempts to create vulcanized rubber, randomly mixed crude rubber with everything from castor oil to cottage cheese. The connecting thread among these dissimilar topics is that all have had some kind of impact on our world, often revolutionary, but sometimes blessedly fleeting (as in the case of the hula hoop, which before its demise was the cause of many an aching lower back). The reader will also find information on noteworthy contributions of inventors from groups often overlooked in the past, specifically, women and minority members.

No one volume could, of course, contain useful, substantive information on all inventions and inventors. While we have included the most outstanding ones of prehistory and antiquity, we have put our main focus on significant developments from the start of the Industrial Revolution to the present (needless to say, a tall enough order in itself); thus, although readers will not find information on the Roman chariot, the will find information on the steam engine and the space shuttle. We have also limited coverage to the following categories in order to be as comprehensive as possible in covering important people and topics within these areas:

Agriculture
Automotive engineering
Aviation and aerospace
Chemistry
Civil engineering and construction
Clothing, textiles, and their manufacture
Communications/Graphic arts
Computer science and mathematical devices
Electrical engineering/Electrical devices
Electronics
Everyday items
Food/Food science
Geology
Household appliances
Lighting and illumination

Materials
Mechanical engineering
Medicine and dentistry
Metallurgy
Meteorology
Musical instruments
Navigation
Oceanography
Optics
Personal care items
Physics
Security systems and related items
Sports, games, toys, and fads
Timepieces, measuring devices, and related items
Transportation
Weapons and related items

Readers seeking information not included in this volume may fine it in the book's companion volume, *World of Scientific Discovery*, 2nd edition, which covers topics and people in the fields of agriculture, astronomy, biochemistry, environmental sciences, geology, mathematics, medicine/health sciences, meteorology, oceanography, pharmacology, and physics. Most books on inventions or on scientific discoveries do no attempt to distinguish between an invention and a discovery, for the very good reason that it is often a difficult, if not impossible, distinction to make. We have done so in the interest of producing two stand-alone one-volume works, rather than a multi-volume series. In making the distinction, we defined an *invention* as something that would not exist unless some person had made it happen, and a *discovery* as something inherent in nature, always there, just waiting for someone with the insight to shout "Eureka!"

There are, to be sure, difficulties with these definitions. Nylon, for instance, can be thought of as both a discovery and an invention: its components are inherent in nature, and at the moment Carothers put them together, it certainly was a discovery; nonetheless, nylon is human made and synthetic, and for that reason, we have included it in *World of Invention*, 2nd edition. Similarly, inoculation can be thought of as both invention and discovery, but the readers's convenience, we have included it in *World of Scientific Discovery*, 2nd edition, where the related diseases and scientists are discussed.

Concise but substantive information presented in a non-technical fashion

The entries focus not on how things work, but on how they came to be. In as few words as possible and in as much depth as the documentation allows, they related the story behind the invention and the forces that motivated the inventor. Our aim has been not only to provide the skeletal facts of what, when, where, and by whom; we have also tried to describe the how and why of the invention, as well as its social impact, and in a way easily understood by the lay reader.

Some inventions—aircraft and patent medicines, for example—have gone through so many stages and involved so many people that we have presented these stories in what might be called *umbrella entries*. Longer than most, briefly sketching the highlights of the evolution while alerting the reader to important individuals and topics that are covered in separate entries.

How to Use the Book

World of Invention, 2nd edition, has been designed with ready reference in mind.

Entries are arranged alphabetically, rather than by chronology or scientific field.

Bold-faced terms direct reader to related entries.

Cross-references at the end of entries alert the reader to related entries not specifically mentioned in the body of the text.

A **Sources Consulted** section lists the most worthwhile print material we encountered in the compilation of this volume. It is there for the inspired reader who wants more information on the people and discoveries covered in this volume.

A **Subject Index** lists inventions by scientific category.

A **comprehensive General Index** guides the reader to all topics and persons mentioned in the book. Bolded page references refer the reader to the term's full entry.

Special Thanks

Once again, in compiling this edition, we have been fortunate in also being able to call upon the following people—our panel of advisors—who not only contributed to the accuracy of the information is this book and in *World of Scientific Discovery*, 2nd edition, but who also helped us mold both volumes as well, and to them we would like to express sincere appreciation:

Nancy Bard
Librarian
Thomas Jefferson High School for Science and Technology
Alexandria, Virginia

Leonard C. Bruno
Science Curator, Manuscripts
Library of Congress
Washington, D.C.

Josephine Davies
Library Manager
Whitmore Library
Salt Lake City, Utah

Doug Fodeman
Technology Director
Brookwood School
Manchester, Massachusetts

Jenny Grogg
Faculty Associate
University High School
Normal, Illinois

Joseph Hoffman
Faculty Associate
West Bloomfield High School
West Bloomfield, Michigan

Richard Jeryan
Principal Staff Engineer
Ford Motor Company
Dearborn, Michigan

David Newton
Educator
Ashland, Oregon

ACKNOWLEDGMENTS

The editors wish to thank the copyright holders of the illustrations and photographs included in this volume and the permissions managers of the institutions and companies listed below for assisting us in securing reproduction rights. Following is a list of the copyright holders who have granted us permission to reproduce material in *WORLD OF INVENTION*, 2nd Edition. Every effort has been made to trace copyright, but if omissions have been made, please let us know.

PHOTOGRAPHS AND ILLUSTRATIONS AP-PEARING IN *WORLD OF INVENTION*, 2nd EDITION, WERE RECEIVED FROM THE FOLLOWING SOURCES:

AP/Wide World Photos. Reproduced by permission: **Artificial heart.** Jarvik, Robert, photograph; **Bomb**. Pipe bomb, photograph; **Bus.** New York City bus with passengers, 1914, photograph; **Communications satellite.** Jupiter C rocket, lauching 1958, photograph; **Dialysis machine.** Artifical kidney (unit for home use), photograph; **Drew, Charles Richard** (with microscope), photograph; **Earthquake measurement scale.** Charles Richter and Clarence Allen looking at an apparatus showing seismic activity, photograph; **Edgerton, Harold E.** in his laboratory at the Massachusetts Institute of Technology, photograph; **Escalator,** photograph; **Faget, Maxime A.**, photograph; **Fat substitute.** Shown by John Pepper, Procter & Gamble CEO, photograph by Steven M. Herppich; **Fiber, synthetic.** Synthetic fiber clothing, worn by three models, 1940, photograph by M. Becker; **Gates, Bill** (holding Pentach), photograph; **Haber, Fritz** (elbows on bookshelf), photograph; **High-speed flash photagraphy.** Harold E. Edgerton demonstrating high-speed flash of photography; **Kettering, Charles**, photograph; **Laser.** Theodore H. Maiman holding the original laser, photograph by Wally Fong; **Lawn mower.** First patented lawn mower, photograph; **Lemelson, Jerome**, with a loop race car track, photograph by Mark Elias; **Margarine.** Man putting boxes of Parkay margarine, photograph; **Mass production.** Ford Motor Company Model T Assembly line, photograph; **Morgan, Garrett A.**, 1945, photograph; **Noyce, Robert** (semiconductor in hand), 1989, photograph; **Reber, Grote** (with radio telescope), photograph; **Shortwave radio**. Two men, one is listening and typing out the conversations, while the other is tuning in on a station, photograph; **Transplant, surgical.** Eye transfer, photograph; **Van Straten, Florence W.** (writing a diagram), photograph; **Walton, Ernest T. S.**, photograph; **Wang, Ang**, photograph.

Archive Photos, Inc. Reproduced by permission: **Air conditioning.** Advertisement outside of a restaurant, photograph; **Barbed wire**. Cutting barbed wire fences, in Brighton Ranch, County County, Nebraska, 1885, photograph**; Bottle tops.** Assorted vintage bottle caps; **Breakfast cereal.** Toasted Corn Flakes ad, photograph; **Corset and girdle.** 1950s model with old metal corset, photograph; **Dome.** Dome of San Sophia in Istambul, photograph; **Iron ship.** *Great Eastern* sitting at dock; **Howe, Elias,** photograph; **Jet engine.** Whittle, Sir Frank, standing next to jet engine he built, photograph; **Korolyov, Sergei P.**, photograph; **Land mine.** Soldiers putting unexploded mines into the ground, Kabul, Afghanistan, 1997, photograph by Ed Grazda; **Lie detector.** Being given to a man, photograph; **Locomotive.** Trevithick's locomotive, invention from 1804, illustration; **Maiman, Theodore H.**, photograph; **Otis, Elisha Graves** (full beard and mustache), photograph; **Parachute**. WWII soldier parachuting, photograph; **Polyester**. Man wearing a leisure suit, early 1970s, photograph by Lambert; **Postage stamp**. Penny Black, the first postage stamps to be produced in Britain being held by Simon Browne of England, photograph by Mike Bramley; **Pullman car.** Passengers sitting and standing in the standard sleeping car, on the Chicago, Milwaukee, St. Paul & Pacific Railroad, photograph; **Robot.** Designed by Honda, first humaniod robot, photograph by Honda; **Space station.** Russia's *Mir* complex, during approach for docking, photograph; **Shoe and shoemaking.** Illustration; **Stephenson, Robert** (seated), print; **Steel production.** The Bessemer-steel converting process, photograph; **Submachine gun**. "Tommy" Thompson Machine Gun sitting on a stand, photograph; **Submarine.** *U.S.S. Nautilus* in New York, photograph; **Subway.** London Tube Station, photograph by Neil Strassberg; **Train and railroad.** The Golden Spike Ceremony, railroad, photograph; **Weather balloon.** Being filled with helium, photograph.

Archive Photos/Svenskt. Reproduced by permission: **Gabor, Dennis,** photograph.

Corbis-Bettmann. Reproduced by permission: **Armored vehicles.** First completed American tank, World War I, photograph; **Automobile, gasoline.** Ford Model T (man and woman riding), 1920, photograph; **Bridge.** Golden Gate bridge under construction, 1936, photograph; **Carlson, Chester** (standing with first Xerox copier), photograph; **Chemical warfare.** American troops during World War I gas attack, 1918, France, photograph; **Dam and dike.** Hoover dam (after completion), photograph; **ENIAC** (first electronic calculator, six people working it), photograph; **Helicopter.** Berlinger helicopter just taking off, 1924, photograph; **Henry, Joseph,** photograph; **Hypatia of Alexandria,** conte crayon drawing; **Linde, Karl Paul Gottfried,** photograph; **Ocean liner**. Front page of the New York Times, 1912, photograph; **Particle accelerator**. Lawrence, Ernest (with atom smasher), photograph; **Photography**. The first known photograph, of a courtyard outside Niepce's room, photograph; **Power supply.** Faraday's magneto-electric induction apparatus, line drawing; **Revolver.** The original Colt's Revolver, photograph; **Siemens, Werner,** reading a book, photograph; **Teddy bear**. Advertisement for Teddy Bears, photograph; **Television**. Man watching television, 1930s, photograph; **Tesla, Nikola,** photograph; **Torricelli, Evangelista,** engraving; **Typewriter**. Daughter of Christopher Latham Sholes, typing on her father's experimental typewriting machine, 1862, photograph.

Corbis Corporation. Reproduced by permission: **Alphabet.** Inscription showing decree of Emperor of Rome, photographed; **Baekeland, Leo H.** (wearing dark suit and vest), photograph; **Nylon.** Man displaying to crowd nylon stockings, photograph; **Stained glass.** Chartres Cathedral, rosette and stained glass window, photograph; **Tunnel.** Channel tunnel (Chunnel), (workers in hard-hats), photograph; **Warship.** First battle between iron ships of war, photograph; **Wozniak, Steve** (mouth open, dark zip jacket), 1983, San Bernardino, CA, photograph.

Federal Aviation Administration. Reproduced by permission: **Seaplane**. *Spruce Goose*, largest plane flown, photograph.

The Granger Collection. Reproduced by permission: **Edison, Thomas Alva**, photograph; **Wright, Orville and Wilbur**, photograph.

Hulton-Deutsch Collection/Corbis-Bettmann. Reproduced by permission: **Watson-Watt, Robert A.** (seated), photograph; **Whittle, Frank** (seated, in uniform), photograph.

IBM Corporation, Research Division, Almaden Research Center. Reproduced by permission: **WORM optical drives**. WORM optical disc containing permanent data bits, photograph.

JLM Visuals. Reproduced by permission: **Windmill and wind turbine**. Wind generators in a line, Whitewater, California, photograph.

The Library of Congress: **Abbe, Cleveland**, portrait; **Abbe, Ernst**, photograph; **Alvarez, Luis** (working on an experiment), photograph; **Amputation**. First picture ever made of an amputation, photograph; **Arkwright, Richard** Spinning machine designed by Sir Richard Arkwright, illustration; **Armstrong, Edwin Howard**, photograph; **Babbage, Charles**, engraving; **Baeyer, Johann Friedrich Wilhelm Adolf von**, portrait; **Baldwin, Matthias**, portrait; **Barnard, Christiaan N.** (wearing plaid suit, striped shirt), photograph; **Battery.** First battery, diagram, illustration; **Benz, Karl Friedrich** (seated in first Mercedes Benz automobile), photograph; **Berliner, Emile**, portrait; **Berthelot, Pierre Eugene Marcellin**, engraving; **Bessemer, Henry**, engraving; **Birdseye, Clarence**, photograph; **Birkeland, Kristian Olaf Bernhard**, portrait; **Blanchard, Thomas**, portrait; **Bleriot, Louis**, portrait; **Blood transfusion**, 1882, drawing by Dr. Roussel; **Boyle, Robert**, painting; **Bragg, William Henry**, portrait; **Brattain, Walter Houser**, portrait; **Braun, Wernher von**, photograph; **Brown, Rachel Fuller** (right) and Elizabeth Lee Hazen (working with chemicals), photograph; **Bunsen burner**. 19th century Bunsen burner; **Bunsen, Robert**, portrait; **Bush, Vannevar**, photograph; **Calculating machine**, Blaise Pascal, illustration; **Camera obscura**, diagram, illustration; **Canal and canal lock**. *U.S.S. Ancon* passing through the Panama Canal (Culebra Cut), August 23, 1914, photograph; **Carrel, Alexis**, photograph; **Cartwright, Edmund**, portrait; **Carver, George Washington** (blowing into tube that is inserted into a beaker), photograph; **Clock and watch**. Pendulum clocks, Christian Huygen, illustration; **Cockcroft, John D.**, photograph; **Compound microscope** invented by Robert Hook, 1665, illustration; **Cooper, Peter**, portrait; **Corliss, George Henry**, portrait; **Cotton gin.** First cotton gin, men working, illustration; **Cousteau, Jacques** (wearing swimsuit and scuba gear), photograph; **Crick, Francis** (hand on chin), photograph; **Crompton, Samuel**, portrait; **Daguerre, Louis** (seated), engraving; **Daimler, Gottlieb** (seated in old car with unidentified man), photograph; **DDT** being sprayed on women at the beach, photograph; **De Forest, Lee** (seated, speaking into microphone), photograph; **Diesel, Rudolf Christian Karl**, portrait; **Dobzhansky, Theodosius**, photograph; **Draper, Charles Stark**, portrait; **Eastman, George** Kodak Brownie Camera, two men, photograph; **Edison, Thomas Alva** (pouring substance), photograph; **Ericsson, John**, portrait; **Evans, Oliver** (turned left, facing front, hands crossed in front on cane), drawing; **Faraday,**

Michael, sketch; **Fitch, John**, portrait; **Francis, Joseph**, portrait; **Franklin, Benjamin** (wearing dark coat with fur collar), painting; **Fraunhofer, Joseph von**, portrait; **Fuller, Buckminster**, portrait; **Fulton, Robert** (seated at drawing table), painting; **Galileo** (with telescope, pointed toward sky), illustration; **Gaslight**. Gas lighting on street, photograph; **Gatling, Richard Gordon**, photograph; **Gauss, Carl Friedrich**, painting; **Glaser, Donald**, portrait; **Goddard, Robert H.**, photograph; **Goodyear, Charles**, sketch; **Gray, Elisha**, portrait; **Greek fire**. 13th century Greek machines, portrait; **Guericke, Otto von**, drawing; **Gutenberg, Johannes**, sketch; **Haeckel, Ernst** (standing, hand on hip), photograph; **Hancock, Thomas**, portrait; **Harrison, John**, portrait; **Hofmann, August Wilhelm von** (standing, right hand on hip), print; **Holland, John Philip**, portrait; **Hollerith, Herman**, portrait; **Hunter, John**, portrait; **Julian, Percy Lavon**, photograph; **Kelly, William**, portrait; **Kinetoscope**, man sneezing, photograph; **Land, Edwin H.**, photograph; **Langley, Samuel**, painting; **Langmuir, Irving**, photograph; **Lawrence, Ernest Orlando**, photograph; **Leonardo Da Vinici** (head and shoulders, facing left), engraved by Cosomo Colombini; **Liebig, Jushes von**, portrait; **Lilienthal, Otto**, portrait; **Lincoln, Abraham** (sitting in chair), postcard; **Linotype**, newspaper publishing, photograph; **Lodge, Oliver Joseph**, portrait; **Lumiere, Louis** (arms folded), photograph; **Marconi, Guglielmo**, photograph; **Maxim, Hiram Stevens**, portrait; **McCormick, Cyrus H.** (oval format, dark hair, white beard), engraving; McMillan, Edwin M., portrait; **Mercer, John,** portrait; **Mergenthaler, Ottmar**, portrait; **Montgolfier, Jacques and Joseph**, portrait; **Morse, Samuel Finley Breese** (standing by invention), painting; **Morse code** alphabet, illustration; **Murdock, William**, portrait; **Muybridge, Eadweard James**, portrait; **Nasmyth, James** (turning to left, lock of graying hair hangs on forehead), painting; **Natta, Giulio** (seated, writing), illustration; **Nobel, Alfred** (wearing suit, checked bow tie), drawing; **Pacemaker**, photograph; **Papin, Denis**, portrait; **Pare, Ambroise**, portrait; **Parsons, Charles A.** (in light jacket, dark tie, gray mustache), drawing; Photovoltaic cell. Solar panels on one side of building, photograph; **Plows**. Roman farmer plowing his field, illustration; **Popov, Aleksandr Stepanovich**, engraving; **Printing technology**. Printing press, illustration; **Rennie, John**, portrait; **Schobein, Christian**, portrait; **Sextant**. Hadley's sextant, photograph; **Shockley, William**, photograph; **Slater, Samuel** (lock of hair in center of forehead), engraving; **Smeaton, John** (seated, left hand inside vest), engraving; **Sperry, Elmer** (seated at desk), photograph; **Stealth aircraft**. Stealth bomber under construction, photograph; **Steam engine**, diagram, illustration; **Steam-powered road vehicles**. Steam engine wagon, illustration; **Steinmetz, Charles P.** (knee on chair, cigar in hand), photograph; Stephenson, George (standing, left hand on coat lapel), engraving; **Svedberg, Theodor**, portrait; **Szilard, Leo** (seated, speaking), photograph; **Telegraph cable, undersea** Transatlantic cable, photograph; **Teller, Edward**, photograph; **Thomson, William,** portrait; **Transistor**, by paperclip; **Tsiolkovsky, Konstantin,** painting; **UNIVAC** mainframe computer, photograph; **Watt, James**, painting; **Westinghouse, George**, portrait; **Wheatstone, Charles**, portrait; **Whitney, Eli,** (sitting near table with books), engravings; **Wright, Wilbur, Orville Wright** (in their airplane), photograph; **Yalow, Rosalyn S.**, photograph; **Ziegler, Karl**, portrait; **Zsigmondy, Richard**, portrait; **Zworykin, Vladimir**, photograph.

Metropolitan Museum of Art/Corbis Corporation. Reproduced by permission: **Winemaking.** Ancient Egyptian winemaking, Thebes, photograph of an illustration.

Photo Researchers, Inc. Reproduced by permission: **Aqueduct.** Pont du Gard, a Roman aqueduct, Nimes, photograph by John Moss. National Audubon Society Collection/Photo Researchers, Inc.; **Astrolabe**. Galileo's astrolabe, photograph by AKG. Photo Researchers, Inc.; **Biodegradable plastics**. SEM of biodegradable plastic, large sphere with top opening, photograph. National Audubon Society Collection/Photo Researchers, Inc.; **Cloud chamber.** Cosmic ray tracks, Leningrad, Russia, 1927, photograph by Dmitry Skobeltzn; **Cyclotron**, photograph. Lawrence Berkeley Laboratory/ Science Photo Library, National Audubon Society Collection/Photo Researchers, Inc.; **Fiber optics** bundle, photograph. (c) Erich Schrempp, The National Audubon Society Collection/Photo Researchers, Inc.; **Genetic fingerprinting**. DNA blueprint (gloved hand holding blueprint), photograph by Phillippe Plailly. Science Photo Library, National Audubon Society Collection/Photo Researchers, Inc.; **Open heart surgery**, photograph by Martin Dohrn. National Audubon Society Collection/Photo Researchers, Inc.

Stock Market. Reproduced by permission. **Metric system**. Standard scale, copper, photograph by Thomas Ernsting.

Underwood & Underwood/Corbis-Bettmann. Reproduced by permission: **Skyscraper.** Home Insurance Building, Chicago, photograph.

UPI/Corbis-Bettmann. Reproduced by permission: **Ambulance**. Ernest Hemmingway next to ambulance, World War I; **Amplifier**. Electric amplifier for radio, sitting between two men, 1900, engraving; **Artificial limb and joint.** Two crippled veterans with artificial legs, 1648, woodcut; **Basketball.** Naismith, James A. (holding ball and basket), photograph; **Bath and shower.** Thermae of Caracalla, interior of ornate Roman bath house, Rome, Italy, drawing; **Boyle, Robert**, standing in laboratory with Denis Papin, engraving; **Caisson.** Men wearing hip boots, holding poles in a caisson, New York City, illustration; **Colt, Samuel**, photograph; **Foucault, Jean-Bernard-Leon**, photograph; **Gas mask**. Israeli soldier demonstrating to two children how to use gas masks, Haifa, Israel, photograph; **In-vitro fertilization**. Brown, Louise Lesley ("Test Tube Baby," being held my mother), Chicago, Illinois, 1979, photograph; **Incandescent light bulb**, photograph; **Iron lung and other respirators**. Hospital ward with iron lungs, photograph; **Looms.** Jacquard loom with strings and punch cards hanging in middle, photograph; **Mercator, Gerhardus** (octigon border has his name printed inside), woodcut; **Milking machines**. Electric Dairy Milker, photograph;

Napalm. Napalm burned child, mother attempting to give water, Cam Che, Viet Nam, 1966, photograph; **Neon light**. Las Vegas, Nevada, 1953, photograph; **Neumann, John von**, photograph by United Press Photo; **Elevator.** Otis, Elisha Graves (standing on elevator), New York City, drawing; **Paper.** Paper making (Chinese men drying paper), China, 1634, drawing; **Radar.** Radar room on World War II ship, China Sea, 1944, photograph; **Radio.** Radio receiving set, London, England, 1926, photograph; **Refridgeration.** Frigidaire refrigerator (wooden appliance with metal hinges), photograph; **Rumford, Benjamin,** engraving, photograph by Kellenhofer; **Sandwich.** Montagu, John (Earl Of Sandwich, oval format, in right profile), illustration; **Scuba diving.** Cousteau, Captain Jacques-Yves, with Janney Ethel and Commander Douglas Fane, both demonstrators are wearing air tanks and swim gear, October 3, 1950, photograph; **Jobs, Steven P.** With John Sculley and Steve Wozniak, photograph; **Silkmaking.** Silk making by four Chinese women, China, engraving by J. Davies after a drawing by T. Allom; **Slot machine and vending machine.** Charlot, Pierre, chef, shaking hands with a hand that is in the vending machine, photograph; **Soda pop.** Coca Cola sign with scrolled border, illustration; **Spinning jenny.** 1765, engraving; **Sputnik 1.** (metal sphere sitting in three legged stand), Moscow, Russia, 1957, photograph; **Steamboat.** Fulton's steamboat ("Clermont" painted at the stern), drawing; **Stove.** 1800, illustration by H. Humphrey; **Streetcar and trolley.** Cable streetcar (writing at top reads "Market St...Park,"), San Francisco California, photograph; **Thermometer.** Fahrenheit's thermometer, engraving; **Tobacco products.** Tobacco pipes being handed by women in center to man on the right, 1747, engraving; **Townes, Charles** (on the telephone), photograph; **Van De Graaf, Robert J.**, and Dr. Karl Taylor (standing next to round objects), photograph; **Zeppelin, Ferdinand von** (white handlebar mustache), photograph.

U.S. National Aeronautics and Space Administration (NASA): **Bell, Alexander Graham** (sitting, testing new telephone invention), photograph; **Communications satellites**. *Echo I*, photograph; **Space probe**. *Voyager*, photograph; **Space shuttle**. *Challenger*, satellite view, photograph; **Spacecraft, manned**. *Apollo 11* launch, photograph; **Wind tunnel**, photograph.

U.S. National Archives and Records Administration. Reproduced by permission. **Atomic bomb**. Mushroom cloud rising from atomic bomb, 1945, Nagasaki, Japan, photograph.

U.S. National Library of Medicine. Reproduced by permission: **Acupunture**, photograph; **Cataract surgery**, photograph; **Cesarean section**, 1533 woodcut, photograph; **Dental drill**, 19th century, photograph; **False teeth**, early 19th century, french carved, photograph; **Fluoride treatment**, dental, photograph; **Forceps**, photograph; **Fractures, treatments and devices**, photograph; **Incubator**. Baby in incubator, photograph; **Opthalmoscope**, 1800s drawing; **Patent medicine**. Man in Indian headress hawking medicine, photograph; **Stethoscope**, 1844, photograph; **Surveying instruments**, photograph; **Tooth-extraction devices**, photograph; **Toothbrush and toothpaste**, photograph; **Wheelchair**, photograph; **X-ray machine**, photograph.

Public Domain: **Ammunition**. Women making bullets, photograph; **Cartoon**. "Hogan's Alley", illustration by R.F. Outcault; **Combine harvester**. Walla Walla, Washington, 24-horse drawn combine harvester. Source unknown. **Ferris wheel**. First ferris wheel, photograph; **Ford, Henry** (wearing tweed suit and patterned tie), photograph; **Lovelace, Ada Augusta Countess of**, portrait; **Seeding device**, line drawing; **Singer, Isaac Merritt** (jutting grey beard, tuxedo), illustration. **Vacuum cleaner**. Advertisement for The Hoover vacuum, illustration; **Volta, Alessandro** (head and shoulders), portrait.

A

Abacus

Although the exact time of its discovery is not truly known, the abacus is believed to have originated in the Middle East (in the region of modern-day Iraq) around 3000 B.C. Later, the abacus was used by the Hindus, and later still by the ancient Greeks and Romans of the Classical Period. It is thought that Arabic traders brought the abacus to the Orient, where its use spread widely.

Around 1000 A.D., Pope Sylvester II (930-1003) led a re-introduction of the abacus to Europe, but its use died out once again beginning in the 1200s, when Arabic numerals gained popular use. By the 1800s, when Frenchman Jean Victor Poncelet (1788-1867) brought an abacus from Russia to France, it was looked on as a quaint amusement. In some areas of China, Japan, and the Middle East, the abacus is still frequently used today.

The abacus was intended to assist shopkeepers, tax collectors, and merchants who had previously kept track of their accounts using small pebbles or stones. In fact, the word "calculate" comes from the Latin word *calculi*, meaning pebbles. The earliest abacus was probably a wooden tablet sprinkled with sand. Marks made in the sand could be wiped clean and re-used as often as required. Modern day versions of the abacus usually contain thirteen columns of beads strung on rods or wires set in a rectangular frame. Beginning with the right-most column and working left, the columns represent ones, tens, hundreds and so forth, up to the trillions.

Although the abacus is generally used for quick addition and subtraction operations only, experienced users are able to multiply, divide, and calculate square or cube roots about as quickly as electronic calculators. In 1946 in Tokyo, Japan, a contest was held between a Japanese abacus and the latest model electric **calculating machine** and surprisingly enough, the abacus decisively won the match. More recently, the abacus has been used in schools to help students learn place values of the numbering system and in some cases has been a valuable aid in teaching blind children arithmetic.

Abbe, Cleveland (1838-1916)

American meteorologist

Often referred to as America's first weatherman, Cleveland Abbe organized and promoted a nationwide network of weather reporting services that used new rapid modes of communication to notify various locations of approaching weather conditions well in advance of their arrival.

Born in New York City, Abbe studied atmospheric physics at the City College of New York from 1861 to 1864, then traveled to Russia to study astronomy at the Pulkova Observatory near Leningrad until 1866. After his return to the United States, he went to the Cincinnati Observatory, serving as director from 1868 to 1870. There he set about establishing a system of weather observations and warnings using the **telegraph**. Under the auspices of the Cincinnati Chamber of Commerce, he operated a weather forecasting service on a trial basis from August through October 1869. Beginning on September 1, 1869, his rudimentary forecasts were posted daily at the Chamber office and sent to about thirty subscribers.

Spurred by the popularity of those forecasts, Congress established the Signal Corps in 1870. Abbe helped guide the initial plans and was named the Service's meteorologist the following year, issuing forecasts nationally three times daily. In 1891, the service was reorganized under civilian control as the United States Weather Bureau, and Abbe remained as its meteorologist until he retired at seventy-seven. He is credited with the creation of the first weather **maps**, which served a more informational than analytical need.

In 1902, Abbe started a training course for Bureau personnel at George Washington University in Washington, D.C. As a result of his contributions to the field of meteorology, the United States became the world leader in weather reporting and forecasting.

Abbe is also credited with establishing the four **time zones** that span the continental United States. Before the zones' inception, every town and county in America had its

Chinese abacus.

own local time. The nearest things to a unified time system were train timetables. Time zones were essential for a meaningful national weather reporting system as well as for railroad travel.

By the time of Abbe's death in 1916 and during the years following, European meteorologists surpassed Americans with their innovations. However, the United States Weather Bureau and its successor, the National Weather Service, never lost their role as the most important weather reporting force in the world.

See also Weather forecasting methods

Abbe, Ernst (1840-1905)

German optical engineer

Ernst Abbe was probably the first optical engineer, designing and perfecting methods for manufacturing **microscope**s and lens systems of very high quality. Though he was a great scientist in his own right, he might have remained anonymous but for the foresight of his employer, Carl Zeiss (1816-1888). In his early twenties, Abbe was working as a lecturer in Jena, Germany. He was recognized as being intelligent and industrious, particularly in mathematics, but he was unable to secure a professorial position at the university. In 1855 Zeiss, the owner and operator of a local company which built optical instruments, approached him. Zeiss had realized that the dramatic rise in scientific interest and research in Europe would create a demand for precision instruments—instruments his shop could easily provide. However, neither Zeiss nor his employees possessed the scientific knowledge to design such instruments. Abbe was hired as a consultant to mathematically design **lens**es of unrivaled excellence.

The science of lenscrafting had stalled since the time of Anton van Leeuwenhoek (1632-1723), chiefly due to certain seemingly insurmountable flaws in man-made lenses. Foremost among these was the problem of chromatic aberration, which manifested itself as colored circles around the subject. Scientists were also frustrated with the poor quality of the **glass** used to make lenses. During the following decade, Abbe worked on new grinding procedures that might correct chromatic aberration; by combining his efforts with Zeiss's glassmaker, Otto Schott, he eventually succeeded in producing near-flawless scientific lenses of very high power.

These same ten years were profitable ones for Abbe. With the increasing success of the Zeiss Works, Abbe was recognized as a brilliant scientist and was given a professorship at Jena University in 1875. Zeiss, who realized that the success of his business was in no small part due to Abbe's efforts, made the young professor a partner in 1876.

Abbe's work on theoretical optics earned him international notoriety, and he was offered a position at the more prestigious University of Berlin (a position he declined in order to continue his research at Zeiss). During their collaboration, Abbe and Zeiss produced thousands of scientific optical instruments. Their innovations set important standards for the development of **telescope**s and photographic equipment. Carl Zeiss died in 1888, leaving the entire Zeiss Works to Abbe. In addition to running the company, Abbe used his own considerable funds to set up the Carl Zeiss Foundation, an organization for the advancement of science and social improvement.

Abel, Frederick (1827-1902)

English chemist and inventor

A distinguished chemist and explosives specialist, Abel is remembered as one of the inventors of **cordite**, a smokeless **gunpowder** still used today. Born in Woolwich, near London, Abel studied under August Wilhelm von Hofmann (1818-1892), an enormously influential experimental and industrial scientist, at the Royal College of Chemistry. During a lengthy career he served as a researcher, scholar, and lecturer, and became the leading British authority on explosives. One of his most significant early discoveries was that **guncotton** could be chemically stabilized through thorough washing with water to remove all traces of acid and impurities.

His most important work, however, came after the British government's establishment in 1888 of an Explosives Commission, dedicated in particular to the military uses of new discoveries in the field. As a member of this Commission, Abel kept in close contact with **Alfred Nobel**, an acquaintance from previous years whose latest invention, ballistite, was received with skepticism due to the volatility of camphor as an ingredient. In 1889 Abel, together with **James Dewar**, invented cordite, a versatile **smokeless powder** which purportedly improved upon ballistite through the introduction of acetone and petroleum jelly.

Although Nobel contested the cordite patent, the English rights of which Abel and Dewar handed over to the government, his efforts were unsuccessful and widespread production of the propellant continued. Abel was knighted in 1891 for his invention and was made a baronet two years later.

Abrasive

One method of working various materials is to rub or grind them to achieve a desired shape, sharpness, or smoothness. Grinding or polishing is often one of the final stages in creating a product. The corrosive agents used to wear the surface of the product to the desired condition are called abrasives. In the past, many materials, from sand or dust to vegetable fibers, were used as abrasives. In more recent times, grindstones, grinding wheels and tools, and rubbing cloths and papers have all been employed.

Natural abrasives include sandstone and solid quartz, emery, corundum, diamonds, and garnet. Manufactured abrasives include silicon carbide (Carborundum), aluminum oxide, boron carbide, synthetic diamonds, and boron nitride.

Silicon carbide was first manufactured in 1891 when Edward G. Acheson (1856-1931) heated a mixture of clay and coke. A former employee of **Thomas Edison** (1847-1931), in 1884 Acheson had struck out on his own in search of a method of creating artificial diamonds. By 1893 he turned to abrasives and patented Carborundum, which was manufactured in trough-like furnaces. For fifty years it was the second hardest known substance after diamonds. A few years after the invention of Carborundum, Charles B. Jacobs developed aluminum oxide, which is commercially known as activated alumina. It is slightly softer but tougher than silicon carbide and is used in the manufacture of most grinding wheels. Before 1900 aluminum oxide was identified as the abrasive element in emery. Emery is a natural rock that, when crushed, can be applied to cloth or paper (hence the term emery cloth). Many so-called sandpapers are actually coated with aluminum oxide or Carborundum particles. Natural industrial-grade diamonds are used in tool sharpeners and on drill bits. Natural diamonds were used extensively before World War II, but synthetic diamonds were developed in the 1950s. In the 1960s, cubic boron nitride (CBN) was developed by General Electric Corporation as an abrasive. Both synthetic diamonds and CBN were originally considered too expensive to produce to be practical, since they required such extremes of temperature and pressure to create. However, the efficiency created by their use soon justified their high cost.

Abrasives have become a major part of many manufacturing processes. Grinders are used for smoothing, polishing, and buffing. Narrow grinding wheels are capable of cutting nearly any type of material and have often replaced saws for cutting jobs. Abrasives are used in barrel finishing, or tumbling. Objects having burrs or rough edges can be polished in a rotating drum. Rock collectors use this method to polish rocks, accomplishing in a matter of days what might take hundreds of years in nature. Hand-applied abrasives are used for paint removal and in the removal of surface blemishes and imperfections.

Cleveland Abbe

See also Aluminum production

Accelerator • See Particle accelerator

Accelerometer

The accelerometer is a device used to measure different kinds of acceleration, the rate of change of velocity. Its initial use was to validate the principles of Newtonian physics, including those of universal gravity.

The first accelerometer, originally known as the Atwood machine, was invented by the English physicist George Atwood (1746-1807) in 1783. There are two types of accelerometers. The instrument constructed by Atwood measures linear acceleration, such as that experienced by a falling object. A spring system is used to measure the accelerating force, which provides the acceleration via Newton's famous second law, force equals mass times acceleration (F=ma). Later accelerometers were designed to measure circular or twisting acceleration, such as that experienced by a weight spun at the end of a string. Here the acceleration depends on the radius of the circle of the spinning object.

Ernst Abbe

There was little need for the Atwood accelerometer until the rise of the **automobile** industry. As demand increased, automakers found it necessary to make their vehicles safer and more efficient. By placing many accelerometers in a test vehicle, researchers are able to determine where the engine's power is being dispersed. Simple forward acceleration can be measured, of course, but so can the sideways and up-and-down shiftings within the car's frame. Automotive researchers often place a human-sized dummy containing several accelerometers inside a moving vehicle in order to determine the effect of the car's motion upon a passenger.

Acheson, Edward Goodrich • See Abrasive

Acrylic plastic

Acrylic resins are any thermoplastic polymer or copolymer of acrylic acid, methacrylic acid, or acrylonitrile.

In 1843 Ferdinand Redtenbacher (1809-1895) oxidized acrolein with aqueous silver oxide and isolated acrylic acid. Friedrich Beilstein (1838-1883) produced acrylic acid by distilling hydroacrylic acids in 1862. Research in the field continued with the efforts of Edward Frankland (1825-1899), Duppon, Schneider, Richard Erlenmeyer (1825-1909), Engelhorn, Carpary and Tollens and accelerated after French chemist Charles Moureu (1803-1929) discovered acrylonitrile in 1893. He demonstrated that it was a nitrile of acrylic acid.

During World War I, acrylonitrile was put to work in the manufacture of a synthetic rubber. With the restoration of trade after the war, the supply of natural rubber increased and made the synthetic less profitable, so companies began researching other uses for acrylonitrile. The synthetic fiber industry was one of the first options investigated. Early developments in acrylonitrile fibers were hampered until appropriate solvents were discovered that allowed the fibers to be formed by wet or dry spinning. The relatively high melting temperatures made melt extrusion impractical. Finding a suitable dyeing method also delayed the debut of the fibers. In 1942, Du Pont introduced polyacrylonitrile fibers under the name Orlon, and large-scale Orlon production was underway by the early 1950s. The first use of acrylonitrile-butadiene-styrene (ABS) copolymer in the manufacture of luggage occurred in 1948. In 1966, ABS was used for the first time on the exterior surfaces of helicopters.

Meanwhile, Otto Rohm had been studying methyl acrylate compounds, trying to create an elastomer. His research eventually led to patents on several synthetic drying oils. He spent the next 15 years searching for a substitute for methyl acrylate. Eventually he and Otto Haas discovered a new leather tanning process and founded the Rohm and Haas Company. In 1927 they attempted to prepare sheets of polymethyl acrylate by pressing the polymer between two sheets of glass. The new safety glass proved superior to the older variety, which used cellulose nitrate in the middle layer. In 1935, Rohm & Haas marketed another acrylic, polymethyl methacrylate (which had been discovered 60 years earlier), as Plexiglas. Plexiglas proved transparent, strong, and tough enough to be used in the cockpits of military aircraft. In 1940, Plexiglas was employed in the bomber noses of war planes, and three years later acrylic aircraft canapies were being produced. By 1946, acrylics had been introduced for dentures and for automobile tailight lenses. In 1974, acrylic sheets stiffened with reinforced plastic were used for the first time in all exterior body panels of an automobile. Today Plexiglas is manufactured in forms ranging from clear to opaque; it is nearly unbreakable and is used in place of glass in airplanes, automobiles, light fixtures, signs, and household appliances.

In 1937, Du Pont introduced its own polymethyl methacrylate product under the name Lucite, and in 1956 introduced Lucite acrylic lacquers. Like Plexiglas, Lucite may be machined in a variety of ways. Varieties range from transparent to opaque to colored. Clear Lucite has been used in place of glass in laboratory instruments, and cameras. Formed into a rod, Lucite can direct light; because of this it is used in medical applications. Lucite paints drip less than other paints, dry quickly, and rarely blister, even when used outdoors over bare wood. Lucite paints and coatings are used in a variety of industrial applications, as automotive and fabric finishes, and in lacquers and inks.

See also Fiber, synthetic; Plastics; Rubber, synthetic

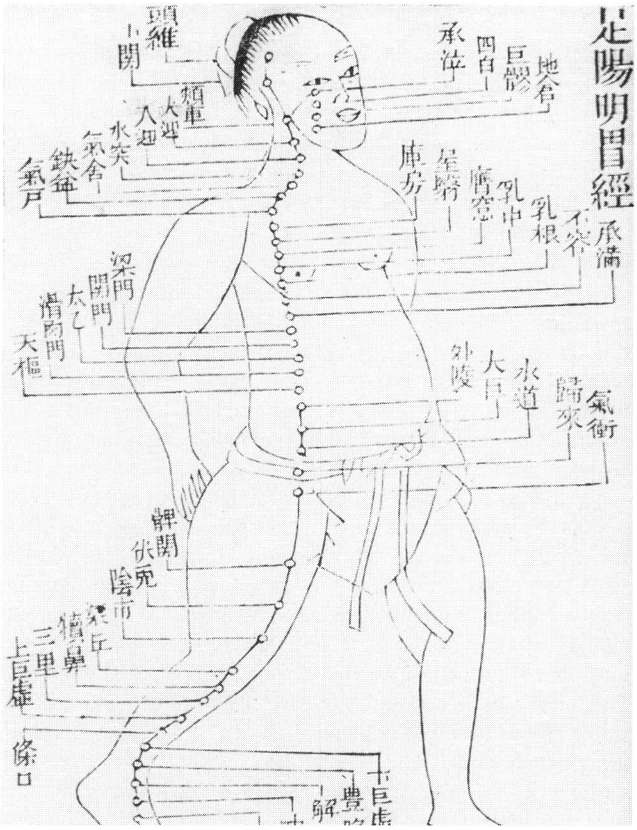

Illustration from a 1875 Chinese manuscript demonstrating major acupuncture points.

Acupuncture

Acupuncture is an ancient Chinese method of relieving pain and treating disease by inserting fine metal needles into particular parts of the body.

The invention of acupuncture cannot be dated. The procedure of needle insertion was practiced in the Stone Age and apparently evolved into the formal system of acupuncture over time. Ancient Chinese medical practitioners learned that certain areas of the skin showed sensitivity during illness or organ malfunction, and that the particular areas of sensitivity depended on the particular disease or dysfunction. The points were found to be part of a pattern rather than being randomly distributed. The lines linking the patterns of points were called *meridians*; each meridian was linked with certain body organs and physical conditions.

These points of skin sensitivity were then related to Taoist philosophy, which asserts that the life force, or energy, called *Qi*, circulates throughout the body. Balance within the body depends on the measured interplay between two forms of energy called *yin* and *yang*. When these forces are in harmony, the body is healthy; when either force becomes dominant, disease or pain occurs. Acupuncture restores the balance between yin and yang by tapping into the body channels, or meridians, through which these energy forms flow.

The basic reference book on acupuncture is the *Nei Ching*, or *Yellow Emperor's Classic of Internal Medicine*, which is ascribed to the legendary Yellow Emperor, Huang Ti, believed to have lived from 2697 to 2596 B.C. The *Nei Ching* is divided into two parts. The first, *Su Wen*, explains the theoretical basis of Chinese medicine. The second, *Ling Shu*, tells exactly how to use acupuncture to treat and prevent every known disease, and gives detailed needle insertion points. An edition of the original *Nei Ching* was compiled by Wang Ping in 762 A.D. and revised around 1200 A.D. This edition is the basis for the modern *Nei Ching* and remains the foundation of acupuncture practice which is still an important element of Chinese medicine. It was supported by Mao Ze-dong and is often used as the only anesthetic when performing surgery.

The earliest acupuncture needles were made of stone, fish bones, and bamboo. These materials gave way to metals such as copper, brass, silver, and gold. Today, most acupuncture needles are made of stainless steel, gold, or silver, varying in length up to several inches. Acupuncture needles are inserted to various depths then twirled or vibrated. A tiny electric charge may be added. Insertion is painless or, at the most, mildly uncomfortable for a moment. Acupuncture students practice needle insertion on themselves thousands of times before perfecting their technique. The *Nei Ching* prescribed 365 insertion points; modern acupuncturists use 650 to 800.

Knowledge of acupuncture was brought to the West by Jesuit missionaries in the seventeenth century, although detailed descriptions of acupuncture theory and practice were not available for Westerners until Soulié de Morant's writings in the 1940s. Western interest in acupuncture has been growing steadily since then, and scientific explanations for the proven effectiveness of acupuncture note that the skin does, in fact, have different levels of electrical resistance at the ancient acupuncture points. Researchers speculate that acupuncture may stimulate production of the body's natural pain relievers (endorphins), or it may interrupt nervous system pain messages.

Regardless of its mechanism of action, the use of acupuncture in the United States has risen dramatically in the last decade. The fact that, in 1996, the federal court required insurance companies to cover treatment with acupuncture, and the National Institutes of Health began funding acupuncture research across the nation, is evidence of its effectiveness and the demand for its use.

See also Surgical instruments

Adams, Thomas, Jr. • See Chewing gum

Adhesives and adhesive tape

Adhesives are substances that hold materials together by surface attachment. Natural adhesives such as beeswax, resin, and bitumen have been used since earliest times. Ancient Egyptians used glue made from animal skin and bones for woodworking. Medieval monks used egg white to bind gold leaf to their illuminated manuscripts. The nineteenth century saw the advent of rubber and pyroxylin cements.

Although adhesive tape was first patented in 1848 by the American Henry Day, and an adhesive bandage was invented by a German pharmacist, Paul Beiersdorf, in 1882, the real advances in adhesive technology came in the twentieth century when synthetic materials were developed. While natural adhesives—such as glue made from animal and plant proteins, pastes of dextrin, starch or latex, and natural rubber, resin, and bitumen—remain in wide use, synthetics now dominate the industry. Synthetic resins are either thermoplastic (softened by heating) or thermosetting (hardened by heating).

The most familiar adhesive tape, Scotch tape, was the invention of Richard Drew of the Minnesota Mining and Manufacturing Company (3M). At the time, two-toned paint jobs were all the rage on automobiles. Painters at auto body shops used gummed tape to make a sharp edge between contrasting colors. When they removed the tape, paint often came with it. Drew, a 3M lab researcher then taking samples of waterproof sandpaper to auto body shops, responded to mechanics' complaints. He developed a pressure-sensitive masking tape in the mid-1920s that did not mar the paint. To make the two-inch wide tape more affordable, only the edges of the tape had adhesive applied. When the car painters used the lightly treated tape, it fell off. They told the 3M salesmen to take "this Scotch tape" back to their stingy bosses (or "this tape back to your Scotch bosses") and put adhesive all over it. Like the improved tape, the Scotch name stuck. In 1930 Drew invented a way to coat Du Pont's newly invented (1924) cellophane with adhesive. The resulting pressure-sensitive transparent tape was also given the Scotch name. Sales were slow, though, until a sales manager invented a dispenser for the tape. Scotch tape became a huge seller for 3M, with hundreds of varieties; it was largely supplanted in the 1960s by Magic Tape, which could be written on and did not yellow or become brittle with age. Removable, restickable transparent tape arrived in the 1980s.

Other modern adhesives include epoxy resins, developed in the 1950s, which allow bonding of previously difficult materials, such as glass and metal. Superglue is a cyanoacrylate ester—based adhesive that sets in seconds and works on many different materials—including (to the dismay of careless users) human skin.

Throughout history, a wide range of materials has been used to close deep cuts and other wounds, including cobwebs, the jaws of leaf-cutting insects and, in modern times, stitches and staples. By 1997 another material was showing promise: surgical glue. Doctors found that instead of sewing patient's wounds and having to remove the stitches a week or so later, they could simply glue the edges together and send their patients home. The adhesive, Dermabond, is an octylcyanoacrylate polymer manufactured by Closure Medical Corporation of Raleigh, North Carolina. The reactive ingredient in Dermabond, 2-octyl acrylate, reacts with moisture on the skin's surface to form a strong plastic coating.

State-of-the-art adhesives continue to be widely used in industry—notably on aircraft—in place of such traditional fastening methods as rivets, bolts, and welding.

Aerogels

The first aerogels were prepared in 1931 by Steven S. Kistler of the College of the Pacific in Stockton, California. Kistler was trying to prove that a gel contains a continuous solid network of the same size and shape as the wet gel. Kistler observed that a wet gel, if allowed to dry on its own, would shrink and become severely cracked. He correctly surmised that the solid component of the gel was microporous, and that the pore structure collapsed due to the liquid-vapor surface tension forces of the evaporating liquid. Kistler noted that in order to produce an aerogel, it would be necessary to replace the liquid of a wet gel with air by some means in which the surface of the liquid does not recede within the gel. The gels studied by Kistler were silica gels prepared by the acidic condensation of aqueous sodium silicate. Kistler succeeded in creating aerogels by washing the silica gels with water, exchanging the water for alcohol, and finally converting the alcohol to a supercritical fluid and allowing it to escape. Kistler's aerogels were very similar to those prepared today. They were transparent, of low density, and highly porous. Kistler later prepared aerogels from many other materials, including alumina, tungsten oxide, ferric oxide, tin oxide, nickel tartarate, **cellulose**, cellulose nitrate, gelatin, agar, egg albumen, and rubber. Upon leaving the College of the Pacific, Kistler joined Monsanto, which shortly thereafter began marketing an aerogel for use as a thixotropic agent in **cosmetics** and **toothpaste**. In the 1960s, the development of inexpensive fumed silica undercut the market for aerogel, and Monsanto ceased production of its aerogel.

In the late 1970s, the French government approached Professor Stanislaus Teichner at Universite Claud Bernard in Lyon, France, about developing a method for storing oxygen and **rocket** fuels in porous materials. Teichner's laboratory replaced the sodium silcate used by Kistler with tetramethylorthosilicate (TMOS). Hydrolyzing TMOS with methanol produced a gel in a single step. Drying these gels under supercritical alcohol conditions produced high quality silica aerogels.

In the 1980s, elementary particle physics researchers began using silica aerogels as a medium for the production and detection of Cherenkov radiation.

In 1983, researchers at Lawrence Berkeley Laboratory found that TMOS, which is highly toxic, could be replaced with tetraethylorthosilicate (TEOS), without sacrificing quality in the aerogel produced. They also found that they could replace the alcohol in the gel with liquid carbon dioxide, thereby easing the severity of the processing conditions.

In 1984 silica gel technologists in Sweden suffered a setback when the first pilot plant for the production of silica aerogel monoliths using TMOS exploded. The plant was subsequently rebuilt and is now operated by Airglass Corporation.

Also in the 1980s, workers at Lawrence Livermore National Laboratory prepared the world's least dense silica aerogel (and the world's least dense solid material). This aerogel had a density of 0.003 g/cm^3, about 3 times the density of air. Subsequent work at Lawrence Livermore led to the preparation of aerogels of organic **polymers**, including resorcinilformaldehyde and melamine-formaldehyde aerogels. It was found that the resorcinol-formaldehyde gels could be pyrolyzed to give aerogels of pure carbon.

Silica aerogel prepared at the Jet Propulsion Laboratory in Pasadena, California, has flown on several **space shuttle** missions, where it has been used to collect samples of high velocity cosmic dust.

The 1990s saw several new commercializations of aerogel technology, including operations by Nanopore and Aerojet Corporation in the United States, and Hoechst Corporation in Germany.

Aerosol spray

The aerosol spray can—used for many years to apply everything from deodorant to paint to insecticides—dates back to 1926, when Norwegian inventor Eric Rotheim discovered that a product could be contained in and sprayed from an aluminum can injected with gas or liquid to build pressure.

In 1939, American Julian S. Kahn received a patent for a disposable spray can, but the product remained largely undeveloped,even after Lyle David Goodhue, credited as the inventor of the spray can, earned a patent for a refillable aerosol spray can in 1941. A breakthrough came in 1949 when Bronx machine shop proprietor Robert H. Abplanalp developed a cheap, efficient valve that facilitated widespread use of the can.

The propellants in aerosol spray cans—usually liquified gases such as hydrocarbons,carbon dioxide,or nitrous oxide—are mixed with the contents of the can and pressure-sealed for eventual release through a valve. Although products like shaving cream or whipped cream also come in pressurized cans, they are not actually aerosols; they contain gases suspended in liquid bases rather than solids or liquids suspended in gaseous bases. **Chloroflourocarbons** (CFCs) were once used extensively in aerosol spray cans, but during the 1970s controversy developed over the use of these freon-type gas propellants. Scientists began to fear that chloroflourocarbons released in large quantities posed a significant danger to the environment because they could break down and react with chlorine. Such a reaction could destroy some of the protective ozone layer in the upper atmosphere, which shields the earth from harmful solar radiation. In 1978, the United States government banned the use of certain chloroflourocarbons in connection with the first global environmental agreement, the Montreal Protocol, and manufacturers were forced to find other propellants for their product. A complete phaseout of CFCs by the year 2005 in developed countries is included in the Montreal agreement. In the mid-1980s as public concern grew on the issue of an Antarctic Ozone Hole clearly detected in NASA photographs, substitutions were made for propellants using HCFCs including Liquefied Petroleum Gas (LPG),a flammable mixture which requires careful disposal; di methyl ether largely present in air fresheners and personal care products; and nonsoluble gasses, an alternative to LPG and soluble compressed gases. In response to continuing concerns about environmental and fire safety hazards of aerosol can recycling, a 1996 study was conducted by the Factory Mutual Research Corporation (FMRC), an independent fire engineering research group. The study concluded that recycling empty aerosol cans does not increase fire risk at the recycling station, although some risk is present when the cans are not emptied.

Agent Orange

Agent Orange is one of several herbicide preparations used by the U.S. Defense Department to destroy forests and enemy crops in Vietnam in the 1960s. Agent Orange contains a 50:50 mixture of two herbicides: 2,4-D (2,4-dichlorophenoxyacetic acid) and 2,4,5-T (2,4,5-trichlorophenoxyacetic acid). A contaminant of 2,4,5-T is TCDD (2,3,7,8-tetrachlorodibenzo-*p*-dioxin), which is highly toxic to animals.

During the Vietnam conflict, North Vietnamese guerrillas found cover in the lush jungles of South Vietnam. To deprive their opponents of cover and food crops, the U.S. military sprayed Agent Orange and other herbicides on about 6,600 sq mi (1,709,000 ha), an area equivalent to one tenth of the land mass of South Vietnam. The bulk of the herbicides were applied by "Operation Ranch Hand," a tactical program for spraying herbicides from airplanes by the Air Force.

Developed and tested during World War II, 2,4-D and 2,4,5-T were commonly used agricultural herbicides in the United States. In 1950, more than 10 million lbs (4.5 million kg) of these herbicides were applied annually for weed control in the United States. By 1960, their use had increased to 35 million lbs (16 million kg).

Between 1962 and 1971, over 10.5 million gal (40.5 million l) of Agent Orange were sprayed in Vietnam. Peak use occurred between 1967 and 1969. Agent Orange was sprayed in Vietnam at a rate that was approximately ten times that normally used in U.S. forestry. U.S. forces sprayed approximately 90% of Agent Orange on tropical forests in Vietnam; the remainder fell on crops and mangrove swamps. The herbicide was sprayed on the boundaries of military bases and river banks from sprayers on boats and trucks and backpack sprayers carried by men.

In 1969, a scientific study found that 2,4,5-T, one of the chemicals in Agent Orange, causes birth defects when given to mice in high doses. A later study revealed that 2,3,7,8 TCDD (tetrachlorodibenzo- *p*-dioxin), a contaminant of 2,4,5-T, actually caused the birth defects. In 1970, the American Association for the Advancement of Science (AAAS) established a commission to study the effects of herbicides on the ecology and population of South Vietnam. The commission reported that herbicides had not only destroyed vegetation and food, but 2,4,5-T and its associated dioxin contaminant might possibly have caused birth defects in South Vietnamese people.

A second study sponsored by the South Vietnamese Ministry of Health found no difference in the number of stillbirths and birth defects in the years during and after the herbicide spraying. The National Academy of Sciences also found no evidence for a higher rate of birth defects resulting from herbicide spraying in South Vietnam. However, a North Vietnamese study published in 1971 reported a high frequency of retarded and malformed children born to more than 900 South Vietnamese who had been exposed to herbicide spraying and later hospitalized in Hanoi.

On April 15, 1970, federal agencies responsible for regulating health and agriculture suspended all use of 2,4,5-T in the U.S, except for killing weeds and brush on non-crop land. On May 9, 1970, Operation Ranch Hand flew its last mission in Vietnam. U.S. forces ceased herbicide spraying in Vietnam in 1971.

Like **aspirin** and caffeine, 2,4-D and 2,4,5-T are moderately toxic: they can alter the body's normal chemical activities, but substantial quantities must be absorbed into the body to produce visible harmful effects. A substantial dose—0.0045 oz per lb of body weight (300 milligrams per kg body weight)—of these herbicides produces death in laboratory animals. An acute reaction occurs only at high doses because animals rapidly excrete these chemicals in the urine. Long-term exposure to 2,4-D and 2,4,5-T does not cause accumulated tissue damage.

The toxicity of TCDD, a contaminant of 2,4,5-T, is the subject of continuing controversy and study. The TCDD content of 2,4,5-T samples used in the 1969 animal studies was 30 parts per million, or 0.003%. In later studies, scientists found that when the level of the dioxin contaminant in 2,4,5-T samples was much lower, no birth defects occurred in test animals.

Since the early 1970s, TCDD has been known as the most toxic of all chlorodibenzodioxins. However, while some animals are very sensitive to TCDD, others are moderately tolerant to it. As little as 0.25 micrograms (a millionth of a gram) per lb body weight (0.6 microgram per kg) can kill 50% of the guinea pigs exposed to TCDD, but it takes a dose of TCDD that is 10,000 times larger to cause the same number of deaths in hamsters.

Workers involved in accidents or spills at factories where 2,4,5-T is manufactured have developed a condition called *chloracne*, a rash of skin lesions on the face, neck, and back. This condition is an easily observed toxic effect resulting from exposure to TCDD.

TCDD is formed during the synthesis of trichlorophenol, an ingredient needed to produce 2,4,5-T. After the TCDD contaminant was identified, researchers developed a method to produce trichlorophenol with a reduced level of TCDD in the final product. This change substantially decreased the number of chloracne cases among factory workers in the herbicide industry.

Since the liver is the body's specialized organ for handling poisons, it is not surprising that chemical plant workers exposed to TCDD in occupational accidents or activities suffered from enlarged livers and impaired liver functions. They also reported having neurological problems.

In 1977, veterans of the Vietnam War began to report serious health symptoms, which they believed to result from exposure to Agent Orange. In response to veterans' concerns, the White House established an Interagency Work Group to study possible long-term health effects of phenoxy herbicides and contaminants in 1979.

The Veterans Administration (VA) sponsored studies of Vietnam veterans and their families to evaluate any relationship between health problems and the spraying of Agent Orange. The agency also endorsed studies of exposure of workers in the herbicide industry to TCDD.

Inadequate records of herbicide spraying and troop movements have made it difficult to determine to what degree individuals were exposed to herbicides and TCDD in Vietnam. The blood and tissue levels of TCDD in Vietnam veterans suggest that, with the exception of Operation Ranch Hand personnel, troops in Vietnam received substantially lower exposures than did workers in chemical factories. Interpreting the measurements is difficult because TCDD comes from widespread sources in the United States: cigarette smoke, incinerator emissions, automobile emissions, charcoal grilled steaks, and paper products. The average level of TCDD in fatty tissue of Americans is four to seven parts per trillion (ppt).

Studies of workers at herbicide factories and Vietnam veterans have determined the following facts about exposure to TCDD: TCDD concentrations in the blood serum of chemical plant workers were strongly correlated with the length of time that the workers had been exposed to TCDD; median blood levels of TCDD in over 800 Operation Ranch Hand veterans who were heavily exposed to Agent Orange were significantly higher (12.8 ppt) than in a comparable group of veterans (4.2 ppt); the average TCDD concentrations in fatty tissues and blood sera of Vietnam veterans who, based on records of herbicide spraying and troop positions, might have been exposed to Agent Orange on the ground, did not differ significantly from that of other veterans and civilians; the number of deaths or birth defects among people who suffered from chloracne after accidental exposure to TCDD at manufacturing plants did not differ significantly from those of other industrial workers.

As Vietnam veterans continued to express concern about the possible health effects of Agent Orange, the U.S. Congress passed Public Law 102-4, the "Agent Orange Act of 1991." This law ordered a comprehensive review and evaluation of information regarding the health effects of exposure to Agent Orange and its components by the National Academy of Sciences (NAS).

The NAS found that, except for individuals directly involved in the spraying of Agent Orange, information on the degree of herbicide exposure among Vietnam veterans was insufficient to make any conclusions about health effects. Therefore, the NAS evaluated studies of people who were actually exposed to herbicides-often at high levels and for long periods of time. After an extensive review of all available information, the NAS classified possible health effects in four categories, according to their likelihood of being caused by TCDD exposure:

I. Sufficient evidence of an association: Soft tissue sarcoma, Non-Hodgkin's lymphoma, Hodgkin's disease, chloracne, and porphyria cutanea tarda (a skin condition in genetically susceptible individuals). The CDC (Centers for Disease Control) found that service in Vietnam was associated with an increased risk of non-Hodgkin's lymphoma, but the evidence suggests that 2,4-D, rather than TCDD, is responsible.

II. Limited/suggestive evidence of an association: Respiratory cancers (lung, larynx, trachea), prostate cancer, multiple myeloma (cancer of bone marrow cells). Respiratory cancers were found to be related to TCDD only when workers were exposed to high levels of TCDD for prolonged periods of time. No conclusions could be made from studies of Vietnam veterans.

III. Insufficient evidence to determine whether an association exists: Liver, nasal, bone, female reproductive, kidney, testicular cancers, leukemia, spontaneous abortion, birth defects, stillbirths, low birthweight, childhood cancer in offspring, infertility, cognitive, neurological disorders, motor/coordination dysfunction, nervous system, metabolic and digestive, immune system, circulatory, and respiratory disorders.

IV. Limited/suggestive evidence of no association: Skin cancer, gastrointestinal tumors, bladder cancer, and brain tumors.

Evidence gathered to date from animal studies and occupational exposures in the herbicide industry is still inadequate to resolve many of the veterans' concerns over possible health effects of Agent Orange. Consequently, the NAS recommends continuing the health study of Air Force Ranch Hand personnel and including in that study Army Chemical Corps veterans who also applied Agent Orange. In addition, the NAS concludes that if scientists can adequately estimate herbicide exposures by reconstructing the timing of Agent Orange applications and troop movements in Vietnam, they might be able to link the level of exposure to health effects—or lack of them—in veterans.

The damage to the flora of Indochina caused by the spraying of Agent Orange is still visible today. Although Agent Orange applications affected foliage over the diverse areas of Vietnam, the most severe damage occurred in the mangrove forests of coastal areas. About 306,280 acres (124,000 ha/1,240 sq km) of coastal mangrove, comprising about 40% of the mangrove swamps in South Vietnam, was sprayed at least once. The spraying killed extensive thickets of the dominant mangrove species, *Rhizophora apiculata*, which is very sensitive to Agent Orange herbicides, leaving barren, badly eroded coastlines. A 1970 National Academy of Sciences report described the harvesting of dead mangroves for fuel, an occupation that would sustain fewer people than it did before the war, as the supply of mangrove wood was not being renewed. The report warned of a future economic loss when the forest would be stripped, unless a vigorous replanting program was undertaken. The destruction of mature seed-bearing trees has made regeneration of mangroves slow and sporadic, and "weed" species have become dominant. The Academy of Sciences estimates that full recovery of the mangrove forest to its former state will take at least 100 years.

After the mangrove trees were destroyed, the number of coastal birds declined dramatically. The web of channels winding beneath the mangrove trees provide important breeding and nursery grounds for a rich variety of fish and crustaceans. The wartime spraying of mangrove swamps is thought to be the cause of a post-war decline in South Vietnam's offshore fishery.

Aerial spraying of Agent Orange over 14 million acres (5.8 million ha) of dense inland forests killed about 10% of the tall trees comprising the forest canopy. Defoliation of the trees was extensive, but some defoliated trees continued to live. Smaller shrubs, protected from the herbicides by the high canopy, or those which replaced dying trees-now comprise the majority of vegetation in affected areas. The total loss of commercially useful timber caused by the military application of herbicides in South Vietnam is estimated to be 26-61 million cu yards (20-47 million cu m). In areas that were sprayed repeatedly (approximately 34% of the sprayed land), all woody plants, except for a few resistant, commercially unimportant species (*Irvingia malayana* and *Parinari annamense*), were replaced by tussock grasses and bamboo. In the dry season, the stands of grass easily catch fire, and, if burned repeatedly, the lands are less likely to return to forests. The beginnings of natural recovery can be seen, but if the lands are left to natural processes, it will be many years before the forests will approach their former productivity.

As the rich, diverse tropical forests disappeared, so did animal habitat. A uniform grassland has poor habitat diversity compared to the complex, multilayered tropical forest. As a result, the number of bird and mammal species living in the areas that were sprayed declined dramatically. Most of the forest animals are adapted to living in one specific type of habitat and are unable to adjust to survival in the grassland. Wild boar, wild goat, water buffalo, tiger, and various species of deer became less common once the cover and food resources of the forest were removed. Domestic animals such as water buffalo, zebus, pigs, chickens, and ducks were also reported to become ill after the spraying of Agent Orange.

Because Agent Orange herbicides might remain in the soil, there is concern that these residues might inhibit the growth of crops and other plants. Soil bacteria break down the herbicides into smaller molecules, but complete decomposition of 2,4,5-T requires that two types of bacteria be present in the soil. Studies performed 15 years after the spraying in South Vietnam found degradation products of Agent Orange in the soil. These byproducts, which can be toxic, can be passed through the food web.

How much Agent Orange actually reached the soil is subject to question. A large proportion of the herbicides falling onto the forest are trapped by the canopy; few drops reach the soil directly. In open areas, however, virtually the entire application reached the soil.

The contaminant TCDD is quite persistent in soil, with a half-life of three to five years. (In that period of time, one half of the dioxin originally applied would still be present in the soil.) In studies conducted in the United States, samples from inland soils and sediments in mangrove areas treated with TCDD had substantial levels of the chemical after 10 years.

An indirect effect of the Agent Orange spraying is the poor fertility of soils in many areas, due to erosion following the destruction of soil-binding vegetation.

Aiken, Howard H. (1900-1973)

American computer scientist

Aiken designed and built the Automatic Sequence Control Calculator (ASCC), the first digital computer in the United States to work from a program and produce reliable results. The machine was also called **Mark I**.

Howard Aiken was born in Hoboken, New Jersey, and grew up in Indianapolis. In 1923 he received a degree in electrical engineering from the University of Wisconsin, having supported himself through school by working at the local power company. After working as an industrial engineer, he entered graduate school at Harvard University, earning an M.A. (1937) and Ph.D. (1939) in physics. He then joined the Harvard faculty, becoming a full professor in 1946.

While in graduate school, Aiken began to design a machine to speed up the calculating of differential equations. He based it on the Analytical Engine designed by the English computer pioneer **Charles Babbage**. Several calculating machine manufacturers and even the president of Harvard turned down his design. Finally in 1939 IBM's president, **Thomas J. Watson**, agreed to build it. Construction continued during World War II, while Aiken was a Navy Commander and then head of the Navy's Computing Project.

When the machine was completed, in January 1943, it was moved to Harvard, where it performed its first calculations in May 1944. Mark I was used for various military calculations during its fifteen year life.

Mark I was an electromechanical calculator, composed of seventy-eight linked calculators and adding machines. It could perform arithmetic and look up data on tables. Operations were controlled by punched paper tape. Input was entered from punched cards or tape or from hand-set switches; output was punched on cards or printed on paper.

Aiken also built more advanced versions, called Mark II, Mark III, and Mark IV. In January 1947, he became head of Harvard's new Computation Laboratory, which pioneered in advanced computer research, including use of natural languages, computer commands and processing, computer circuit theory, data processing, and magnetic storage devices.

After retiring from Harvard in 1961, he became Distinguished Professor of Information Technology at the University of Miami, in Florida.

Air compressor

Compressed air is used in a variety of ways—from stoking a fire to propelling a jet **aircraft**—and is produced by many types of machines. Although such mechanisms as pile drivers, **fans**, or **drills** use air under pressure to perform some task, machines whose sole purpose is to compress air are generically referred to as *air compressors.*

The earliest type of air compressor was the bellows. It was, and still is, used for building fires. At first bellows consisted of nothing more than leather bags. Later, handles and intake **valves** were added to facilitate continuous operation.

Hydraulic compressors were used in early Roman times in the metal-forging process. Water falling down a pipe under the influence of gravity forced air out of the bottom and into the furnace. In 1650, German scientist **Otto von Guericke** devised an air pump consisting of a single **piston** and cylinder. With it he was able to experiment with compressed air and vacuums.

George Medhurst of England designed the first motorized air compressor in 1799 and used it in mining. In 1852, Isambard K. Brunel designed a pressurized **caisson** for workers in the Chepstow railway bridge project that prevented water and mud from entering the work space. Englishman Thomas Cochrane invented a compressed air rock drill in 1857 for use in tunneling. In the late 1800s, Amedee Bollee used compressed air to improve the hydraulic ram.

Whether applied directly or placed behind a movable object, compressed air can supply a powerful force that cannot be conveniently produced by other means. The modern compressors that produce this force can be categorized in three types. The reciprocating compressor uses a piston that draws air through an intake valve and then pushes the air out through an outlet valve. These devices are used for general industrial purposes, such as in blast furnaces. Rotary compressors consist of enclosed fans that take air in at the center of a bladed wheel, or impeller, and blow the air out through a shaft. These compressors are used primarily in cooling systems. Finally, jet compressors use either pressurized water or highly pressurized gas or vapor to force air outward from a cylinder called a diffuser. Jet compressors have no moving parts and are capable of producing great force.

Air conditioning

Throughout the centuries, people have devised many methods of manipulating the heat and humidity of the air surrounding them. Ancient Egyptians, Indians, Greeks, and Romans hung wet mats over doors and windows (wind blowing through the mats was cooled by evaporation), while their rulers were cooled by slave-operated **fans**. **Leonardo da Vinci** invented a water-driven fan in the middle ages. In the nineteenth century, the textile industry used various devices to humidify factory air; South Carolina physician John Gorrie (1803-1855) invented an air-conditioning and **refrigeration** machine; and in New York City, both Carnegie Hall and the Stock Exchange building had air-cooling systems around the turn of the century.

None of these systems was truly effective, however. The breakthrough came in 1902 from a young mechanical engineer named **Willis H. Carrier** (1876-1950). Carrier had grown up on a farm in New York and then attended Cornell University on scholarship, where he studied engineering. While working for the Buffalo Forge Company, Carrier was assigned to solve the problem of a Brooklyn, New York, printing company. Humid weather seriously compromised the quality of color printing. Noting that very cold air would absorb humidity from very warm air, Carrier devised a cold-water-spray apparatus that dehumidified and cleaned interior air, and added a *dew-point control* that automatically regulated interior air humidity by adjusting the water-spray temperature. Carrier patented his system in 1904. By 1911 he had codified the laws of psychrometry —the science measuring the water-vapor content of air—and presented a paper outlining the basis of a complete air-conditioning system. For this, he became known as ''the Father of Air Conditioning.'' The term *air conditioning* itself was first used by engineer Stuart W. Cramer in 1906. In 1919, the first air-conditioned movie theater opened in Chicago, Illinois, along with the first air-conditioned department store, Abraham & Straus, in New York City. The first fully air-conditioned high-rise structure, the Milam Building in San Antonio, Texas, was erected in 1929. The Baltimore & Ohio railway line introduced air conditioning to that industry in 1931; Packard did the same for **automobiles** in 1939, followed by Greyhound **bus**es in 1940.

Individual room air conditioners were developed in the 1930s by the Americans H. H. Schutz and J. Q. Sherman, with compact versions becoming available in the 1950s. Early air conditioners used ammonia, an irritating and corrosive substance, as a refrigerant. In 1930, Thomas Midgley, Jr. (1889-1944) of the Du Pont company discovered *freon*, a nonflammable, highly efficient refrigerant gas which, in various forms, was the industry standard. Freon is a halogenated hydrocarbon, also known as a chlorofluorocarbon. When these compounds escape from freon production facilities or refrigeration systems and find their way into the upper atmosphere of the earth, they break down and release chlorine. This contributes to the destruction of the **ozone** layer, which screens dangerous ultraviolet radiation from sunlight, preventing possible skin cancers. The production of freon was therefore banned as of 1996.

Alternatives to freon are being developed that are less environmentally destructive, such as the chemical R134a for the air-conditioning (A/C) systems of automobiles. Cars built prior to 1992 must convert their A/C system once their existing freon system empties.

Scientists are exploring other ways to cool air, especially for smaller systems like refrigerators. Acoustic cooling uses a loudspeaker to create pressure fluctuations in a gas, decompressing it, and therefore cooling it. Another experimental freon-free refrigerator uses a **thermocouple** device that cools by passing electric current between two semiconductors, though it requires larger amounts of power than conventional refrigerators.

AIR PUMP • See Vacuum pump

AIR TRAFFIC CONTROL

A horrible air disaster occurred in the skies over the Grand Canyon in 1956, when a two commercial **aircraft**, operated by

Before air conditioning became commonplace, restaurants and movie theaters would use their cool environment as an enticement to customers.

Trans World Airlines and United Airlines and flying in clear skies, collided and crashed, killing more than 100 people. As a result of this tragedy, Congress in 1958 established the Federal Aviation Administration (FAA), to regulate and oversee the operation of aircraft in the skies over the United States, which were becoming quite congested. The resulting structure of air traffic control has greatly increased the safety of flight in the United States, and similar air traffic control procedures are in place over much of the rest of the world as well.

Rudimentary air traffic control (ATC) existed well before the Grand Canyon disaster. As early as the 1920s, the earliest air traffic controllers guided aircraft in the vicinity of airports manually, using lights and flags, while beacons and flashing lights were placed along cross-country routes to establish the earliest airways. This purely visual system was useless in bad weather, however, and by the 1930s, radio communication was coming into use for ATC. The first region having something approximating today's ATC was New York City, with other major metropolitan areas soon getting ATC centers as well.

After World War II, ATC centers could and did take advantage of the newly developed **radar** and improved **radio** communication brought about by the war, but the system remained rudimentary. It was only after the creation of the FAA that full-scale regulation of America's airspace took place, and this was fortuitous, for the advent of the jet engine suddenly resulted in a large number of very fast planes, reducing pilots' margins of error and practically demanding some set of rules to keep everyone well-separated and operating safely in the air.

Many people think that ATC consists of a row of controllers sitting in front of their radar screens at the nation's airports, telling arriving and departing traffic what to do. This is only a ver incomplete part of the picture. The FAA realized that the airspace over the United States would at any time have many different kinds of planes, flying for many different purposes, in a variety of weather conditions, and the some kind of structure was need to accommodate all of them. The following elements were put into effect.

First, ATC extends over virtually the entire United States. In general, from 1,200 feet (365.76 m) above the ground and higher, the entire country is blanketed by *controlled airspace*. In certain areas, mainly near airports, controlled airspace extends down to 700 feet (213.36 m) above the ground, and, in the immediate vicinity of an airport, all the way to the surface. Controlled airspace is that airspace in which FAA regulations apply. Elsewhere, in uncontrolled airspace, pilots are bound by far fewer regulations. In this way, the recreational pilot who simply wished to go flying for a while without all the strictures imposed by the FAA had only to stay in uncontrolled airspace, below 1,200 feet (365.76 m), while the pilot who did want the protection afforded by ATC could easily get it.

The FAA then recognized two types of operating environments. Good weather conditions would permit flying under visual flight rules (VFR), which, as the name implies, suggests a strong reliance on visual cues to maintain an acceptable level of safety. Poor weather conditions (as well as some other conditions, described below), necessitated a set of instrument flight rules (IFR), under which the pilot relied on attitude and navigational information provided by the plane's instrument panel to fly safely. On a clear day, a pilot in controlled airspace can choose a VFR or IFR flight plan, and the FAA regulations were devised in a way to accommodate both VFR and IFR operations could in the same airspace. A pilot can only choose to fly IFR if he or she has possesses an instrument rating, however, which is above and beyond the basic pilot's license that must also be held.

Controlled airspace is divided into several different types, designated by letters of the alphabet. Uncontrolled airspace is designated Class G, while controlled airspace below 18,000 feet (5,486.4 m) above sea level and not in the vicinity of and airport is Class E. All airspace above 18,000 feet is designated Class A. The reason for the division of Class E and Class A airspace stems from the type of planes operating in them. Generally, Class E airspace is where one finds general aviation aircraft (few of which can climb above 18,000 feet [5,486.4 m] anyway) and commercial turboprop aircraft. Above 18,000 feet is the realm of the heavy jets, since jet engines operate more efficiently at higher altitudes. The difference between Class E and A airspace is that in class A, all operations are IFR, and pilots must be instrument-rated. This is because the aircraft at these altitudes operate at very high speeds, with little margin for error, and ATC control of the entire airspace is essential.

Three other types of airspace, Classes D, C, and B, govern the vicinity of airports. These correspond roughly to small municipal, medium-size metropolitan, and major metropolitan airports, respectively, and encompass an increasingly rigorous set of regulations. For example, all a VFR pilot has to do to enter Class C airspace (such as that around Norfolk, Virginia and comparably-sized cities), is establish two-way radio contact with ATC. No explicit permission from ATC to enter is needed, although the pilot must continue to obey all regulations governing VFR flight. To enter Class B airspace, such as on approach to Chicago O'Hare, an explicit ATC clearance is required. The private pilot who cruises without permission into the airspace over O'Hare risks losing his license.

In addition to the airspace classes, the FAA created a network of airways and navigational aids to assist pilots. Two sets of airways exist, the low-level network, in Class E airspace, and the upper-level IFR airways in Class A airspace. A pilot uses these airways to devise an IFR flight plan, and can confidently fly them even in a solid mass of clouds, since no VFR traffic can operate amidst clouds and since ATC will ensure that no other IFR traffic threatens the safety of the flight.

The men and women who staff air traffic control centers work at airports, handling traffic on the ground, as well as departing and arriving traffic, traffic entering an airport's control area, and traffic en route between airports, at a variety of altitudes. Traffic is ''handed off'' from one control center to another by ATC direction, requiring only a change of radio frequency from the pilot. We hear frequent comments that the ATC hardware in the United States is antiquated, even dangerously so. While the system could probably benefit from equipment upgrades, we should also recognize that it has developed since 1960 into a network of staggering complexity, and that it works very well. The remarkable safety of commercial air travel results not only from the quality of the aircraft now being produced, but from the protective umbrella, stemming from the needless tragedy over the Grand Canyon, that governs to movement of aircraft in our very crowded skies.

AIRBAG, AUTOMOBILE

Seat belts have been proven to be effective in saving lives and preventing or lessening injuries in automobile accidents. The key to their success, however, is that occupants must use these safety devices. Because seat belts require active use by the wearer, inventors developed passive restraints to ensure driver and passenger safety. The first passive restraints were modifications of seat belts themselves; the belts were coordinated with operations of opening the car doors and starting the automobile, which caused belts built into tracks in the doors to wrap around the driver or passenger when the seat was occupied. Concurrently, the airbag was devised as a secondary form of passive restraint during impact.

In 1961, Arnold Kent of England considered the use of inflatable cushions. He suggested they could be stowed in front of the occupants and would inflate automatically from a container of compressed gas in the event of a collision. In the United States, automotive engineers developed Kent's idea further. Air cushions made of nylon-reinforced plastic were stored in the hub of the steering wheel on the driver's side and in the dashboard on the passenger's side of the front seat. In serious front-end collisions (the equivalent of hitting a brick wall at a speed greater than 12 miles per hour), crash sensors activate the airbags. Within 1/25th of a second after impact, the bag inflates to cushion the driver or passenger from hurtling forward into the steering wheel, dashboard, or **windshield**. The bag then deflates rapidly so the occupant can get out of the vehicle. A solid chemical, sodium azide, triggers the nitrogen gas used to inflate the bag.

Unfortunately, airbags themselves were proven to cause death and injury. The National Highway Safety Administration (NHSA) found that more than 50 drivers and passengers lost their lives due to impact from airbags in the period from 1994 through 1996, and a number of those injured or killed were children. Airbags had been designed based on car occupants of average size, so the force and direction of impact of the bag itself could hurt children or small adults. Lobbyists and others proposed that cutoff or deactivation switches be installed in automobiles so the driver would have the option of preventing the passenger-side airbag from working. Fear that riders would now stop using both seat belts and airbags prompted the NHSA to ask car manufacturers to develop ''smart'' airbags that would sense more accurately the sizes of the car's occupants and adjust the amount of inflation accordingly.

''Smart'' airbags are being developed. A switch under the seat measures the occupant's position in the seat, and a piezoelectric crystal (like those used in electronic bathroom scales) weighs the occupant. These data tell a microprocessor whether or not to inflate the bag, by how much, and how quickly. A still more sophisticated measuring system in development uses a camera (standard video cameras, infra-red devices, and ultrasonic transceivers are being considered) to detect the rider's size and position in the seat. Ultimately, a radar sensor in the front of the car may anticipate impact and deploy the bag before the accident occurs.

The ''smart'' airbag must also inflate safely. Compressed helium (instead of nitrogen) is being studied, and dual-stage inflators may be used in which the first stage only inflates for small-sized occupants and a second stage is triggered if the rider is large. Widespread use of ''smart'' airbags is expected by the year 2000; in 1996, Mercedes introduced a system that disables the airbag when a child safety seat is locked into position, and side impact bags also became available in some car models in the mid-1990s.

Honda is exploring a new system that will eliminate the airbag controversy altogether; it has developed an ''Air Belt'', which is a safety belt with an inflatable shoulder portion that inflates on impact to prevent head movement and to cushion the body. Air Belts can be used to protect back-seat as well as front-seat occupants, and they inflate away from the wearer instead of creating a point of impact by inflating toward the occupant. The parts of the Air Belt are a control unit that signals an inflator to inflate the seatbelt shoulder. The seatbelt tenses during the process so the belt itself is not slack, and the belt deflates rapidly as it absorbs the impact. The Air Belt is also less expensive to produce and install than airbags embedded in dashboards and steering columns. Honda expects to begin marketing the Air Belt in 1998-1999. Regardless of future developments in airbags, the automobile passenger's first line of defense is the seat belt, and other devices should be used in conjunction with seat belts.

AIR-COOLED ENGINE • See Engine cooling systems

AIRCRAFT

The human fascination with flight is older than recorded history. An ancient Chinese legend tells of a prince who builds a wind-powered flying chariot, and in the famed Greek myth, Daedalus devises wings of feathers and wax so he and his son Icarus can escape their island prison. Icarus flies too close to the sun, his wings melt, and he plunges to his death.

Early history records a number of other failed attempts at imitating the flight of birds. Over nine hundred years ago, in what surely must have been a leap of faith, an English monk with a wing attached to each arm jumped off a church roof; somewhat remarkably, he suffered only two broken legs. A Frenchman attempting a similar feat some four centuries later was not so lucky; he fell to his death from a high church steeple. Around 1500, Leonardo da Vinci was busily observing birds in flight and sketching designs for flying machines that employed flapping wings; he even made some models. Wisely, given the fate of others before him, he did not attempt to use them.

It could be said that the Chinese were the first to successfully put human beings in the air. In the 1200s, they were

Orville Wright makes the first successful powered flight on December 17, 1903, as his bother Wilbur watches.

using manned **kite**s not only for reconnaissance during wartime but also as punishment for criminals. The kites, however, were tethered to the ground, and so the dream of a human being flying as free as a bird was still not realized, nor would it be until 1783.

The first successful free flight of human beings relied not on wings but on the fact that hot air is lighter than cool air. In France in the early 1780s, **Joseph-Michel Montgolfier** and **Jacques-Étienne Montgolfier**, sons of a wealthy paper manufacturer, noticed that bags grew lighter and lifted into the air when held above a fire. In November 1783, after a series of experiments, the Montgolfiers successfully launched a hot-air **balloon** containing two human beings—the first manned aircraft. Other enthusiasts, such as the French physicist Jacques Charles, began filling balloons with hydrogen—a gas lighter than air—to create buoyancy. This procedure not only eliminated the necessity of keeping a fire burning beneath the craft; it also meant that the balloon did not sink as the hot air within it cooled.

Balloons were still dependent on the wind for direction, and by 1852, the desire for better control had led Henri Giffard to design a balloon that had a propeller driven by a steam engine. Known as a dirigible, or **airship** (as engine-powered lighter-than-air craft came to be called), it was perfected by **Ferdinand von Zeppelin** in Germany in the late nineteenth century. Zeppelin's design, based on **David Schwarz**'s, used an enormous metal frame with a cloth skin stretched over it. The hydrogen-filled *zeppelins* could be steered with complete control and by 1914 had transported more than 10,000 passengers.

Despite the success of balloons and airships, winged flight never lost its allure. As early as 1804, an Englishman by the name of **George Cayley** had built a small, operable winged **glider**. He wanted to use some kind of power to run a propeller, but no suitable engine was then available. By the 1890s, **Otto Lilienthal** of Germany had made over two thousand glider flights, sometimes traveling over 700 feet (213.5 m). His goal, too, was to motorize a glider, but he died in a crash before he could attempt it.

An American, Octave Chanute, improved on Lilienthal's glider design by placing another set of wings above the lower ones, thus creating a biplane. Although some would-be aviators in the nineteenth century did succeed in attaching steam engines to aircraft, their machines produced only uncontrolled jumps. Even with a head start down a ramp in 1874, Frenchman Felix Du Temple's steam-powered craft stayed airborne for only two seconds. In 1884, a similar attempt by a Russian, Alexander Mozhaiski, had similar results. In 1890, a French engineer, Clement Ader, managed a precarious hop of 150 feet (46 km), but he had no control of the flight and could not stay aloft. **Hiram Maxim**, a prolific American-born inventor, somehow managed to get his 3.5 ton (3.2 metric ton) biplane off the ground for a few seconds in 1894.

But **Samuel Langley**, an American scientist, had the most promising early results. His unmanned *Aerodrome* flew half a mile (0.8045 km) in ninety seconds in 1896, but he could not follow up with a piloted flight.

The triumph of the first controlled and sustained flight in engine-powered, heavier-than-air craft belongs to the Americans **Orville and Wilbur Wright**. Using knowledge gained by the early glider pioneers and their own engineering skill, they constructed a steerable biplane with a gasoline engine and two propellers. Orville flew it for the first time for 120 feet (37 m) in 1903, and by 1905 the Wright brothers were making flights of more than 20 miles (32 km).

A number of other firsts quickly followed the Wrights' exploits. In 1906, Alberto Santos-Dumont, a Brazilian, became the first man to fly a plane in Europe. In 1908, Henri Farman, an Englishman living in France, flew almost 17 miles (27 km) directly across the French countryside and because the Wright brothers' flights were circular, made the world's first long-distance cross-country flight.

In 1909, a London newspaper offered a reward to the first person who could cross the English Channel in an airplane. A daring Frenchman, **Louis Blériot**, took the challenge and completed his rough flight from France to England in thirty-seven minutes. Three years later, Harriet Quimby, an American, became the first woman to fly across the channel. Other innovations during these years included Henri Fabre's **seaplane** with floats for landing and Armand Deperdussin's sleek, single-winged plane, which in 1912 became the first craft to fly over 100 miles (161 km) per hour.

World War I (1914-1918) created a demand for a different kind of aircraft. At first, planes were used only for scouting out enemy positions, but it wasn't long before **machine guns** were being mounted on the front of planes to rid the skies of enemy craft. The problem in the ensuing dogfights was that the pilot, who was also the gunner, was as likely to hit his own propeller as he was an enemy plane. One solution on the Allied side was to attach steel deflector plates to the propeller. But Anthony Fokker gave the Germans a better system by designing an interrupter gear that prevented the machine gun from firing when the propeller blade was in front of the gun muzzle.

By 1916, many French and British planes were using a revolutionary engine: the rotary, which unlike conventional engines, had a stationary crankshaft around which the cylinders and crankcase rotated with the propeller bolted to their front. The rotary had both power and mechanical simplicity. The increased torque (or tendency to twist) that it produced allowed experienced pilots to turn abruptly, an advantage in dogfights.

The typical plane at the start of the war was a biplane that could fly 60 to 70 miles (97 to 113 km) per hour. By 1918, planes were stronger, more maneuverable, and capable of speeds in excess of 130 miles (209 km) per hour. Hugo Junkers of Germany had introduced the Junkers J-1, the first all-metal plane with cantilevered wings (i.e., wings supported by the body of the plane rather than by outside bracing). World War I also produced the first aerial bomber planes, which carried out raids over London, Paris, and several German cities. *Zeppelins*, too, were put into service as bombers and created havoc during their dozen attacks on London. Ultimately, however, the aircraft of this time were ineffective as weapons of war. No airplane or airship ever managed to sink or even seriously damage a **warship**, nor were any land battles decided mainly on the basis of air power.

When World War I ended, enthusiasm for flying was high. Many former air force pilots found an outlet for their passion in barnstorming—performing dramatic, risky stunts at state fairs and carnivals. Others set out to break distance or endurance records, such as crossing the Atlantic nonstop. In 1919, two British airmen, Alcock and Brown, took off from Newfoundland in a bomber and headed for Europe. The weather was so terrible they had to take turns crawling out on the wings to chip the ice away; after sixteen hours of flying, they landed in Ireland. In 1926, two Americans, Floyd Bennett and the explorer Richard Byrd, became the first people to fly over the North Pole.

The most celebrated feat of the era was the first solo crossing of the Atlantic Ocean by Charles Lindbergh, a mail pilot, in 1927. Using an extra fuel tank on his specially rebuilt plane, *The Spirit of St. Louis,* Lindbergh flew without a radio or parachute from New York to Paris in just under 34 hours. Four years later, Hugh Herndon and Clyde Pangborn completed the first nonstop flight across the Pacific. One year after that, in 1932, Amelia Earhart became the first woman to cross the Atlantic solo, only to disappear over the Pacific five years later as she attempted to circle the globe. By 1933, Wiley Post had made the first solo flight around the world; with eleven stops, it took him seven days and eight hours. In 1938, Howard Hughes reduced the time for a transworld flight to three days and nineteen hours.

The sale of surplus military planes after the war helped bolster the enthusiasm for flight. Just before 1920, the world's first scheduled airlines went into business, many of them using rebuilt military craft on runs between European cities. They had plenty of competition from airships. By 1929, a huge German airship, the *Graf Zeppelin*, had flown around the world in just over twenty-one days, and airships were making regular runs across the Atlantic. Many believed these craft would always be the best means of air travel. However, the dramatic explosion of the *Hindenburg* over New Jersey in 1937, which killed thirty-six people aboard, put an end to the use of airships as passenger craft.

Heavier-than-air craft improved significantly during the 1920s and 1930s. Planes became larger, stronger, and faster, achieving speeds of over 300 miles (483 km) per hour, and workhorse planes, such as the all-metal Douglas DC-3, began providing reliable transport for cargo and passengers.

Helicopters, craft that can take off and land vertically and hover in mid-air, also came into their own at this time. Although the idea of vertical flight goes at least as far back as Leonardo da Vinci, the technical difficulties involved were not overcome until the late 1930s, when Heinrich Focke built a two-rotor helicopter and **Igor Sikorsky** built the first workable single-rotor machine. Helicopters were well on their way to useful service by 1940.

At the start of World War II (1939-1945), many countries still used biplanes that performed little better than those used in the earlier war. Germany held the technological lead with excellent fighter planes like the Messerschmidt Bf 109, an all-metal monoplane with one to three cannons and two machine guns, and good twin-engine bombers like the Junkers Ju-88. The Japanese also began the war with a superior fighter, the Zero. As the war progressed, the United States and Britain produced aircraft that matched and then surpassed the enemy planes they met in battle. By the last year of the war, the United States' Boeing B-29 *Superfortress*, an immense bomber, had made its predecessors obsolete. It could fly for distances of 5,000 miles (8,045 km) carrying enormous loads of **bombs**. It was a B-29 bomber that dropped the **atomic bomb** at Hiroshima.

Perhaps the most important technological advance of World War II was the development of **jet engines** for aircraft. Rather than using fuel to turn a propeller, jet engines mix fuel and air in a combustion chamber, igniting it and blowing it out of the engine as high-speed exhaust. This creates an immense thrust forward. Both the Allied and Axis scientists knew that these engines would make airplanes capable of speeds from 500 to 600 miles (804.5 to 965 km) per hour, and so a frenzied race to develop them began.

In 1939, a jet engine developed by the German Hans von Ohain was tested on a Heinkel plane, and the Allies scrambled to match this advance. In 1941, the jet engine developed by the British scientist **Frank Whittle** successfully propelled a plane at speeds matching the best British fighter then in use. The Americans quickly borrowed Whittle's technology, hoping to create their own jet engines. But the more immediate need for conventional planes precluded the introduction of jets by either the Axis or the Allies, and by the time Germany introduced its awesome jet-powered Messerschmidt Me-262 late in 1944, the outcome of the war was already decided.

The jet opened new frontiers for aviation. Scientists and fliers could now turn their attention to the challenge of breaking the sound barrier. Though some German rocket-powered planes had exceeded 600 miles (965 km) per hour, many felt that would be the absolute speed limit, because at speeds approaching 740 miles (1,191 km) per hour—the speed of sound—air piled up in front of a plane, threatening to shake it apart if the pilot tried to push through the resistance. In the United States, Bell Aircraft built the X-1, an experimental plane with thin wings and tail designed to slice through the air smoothly, and four powerful rocket engines to drive the plane through piled-up air. In October 1947, a B-29 carried the X-1 and its pilot, Chuck Yeager, to 30,000 feet (9,150 m), where the plane was released, and Yeager ignited his rocket engines. With only a little shaking, Yeager reached Mach 1, the speed of sound, and those on the ground heard the first sonic boom, resulting from a shock wave produced by the world's first supersonic plane.

With the sound barrier broken, new speed records were made regularly. By the 1960s, the American-made experimental X-15 had traveled at the speed of Mach 5, equal to five times the speed of sound. It had also flown to an amazing altitude of 50 miles (80 km), qualifying it as the first airplane that was also a **spacecraft**.

Highly efficient jet engines came into wide use during the Korean War (1950-1953), powering the first successful jet fighters. The jet technology developed for military use was also applied to civilian transport and in the 1950s began supplanting propeller-driven systems on commercial and passenger craft.

The first passenger jet, Britain's De Havilland Comet, made its debut in 1952, carrying passengers at 500 miles (804 km) per hour. The Boeing 707 introduced jet flight to American passengers six years later, and Boeing followed up with other successful jets, culminating in 1970 with the gargantuan Boeing 747, which still carries huge loads of passengers for thousands of miles. Supersonic passenger jets have also been introduced, though their poor cost-efficiency precludes their wide use at present. The Soviet-made Tupolev Tu-144 was the first in 1968, followed by Great Britain and France's collaborative Concorde in 1969.

Demands for military superiority continue to drive advances in aircraft technology. In 1993, the first B-2 Stealth bomber was delivered to the United States Air Force, the culmination of decades of development of the concept of the ''flying wing.'' The Stealth's lack of right angles cause it to present a vanishingly small radar signal, rendering it very difficult to detect. The B-2's sister plane, the F-117 Stealth fighter, played an important role in the 1991 Gulf war against Iraq, and in 1997, the newest generation Stealth, the F-22 Raptor, took to the skies for the first time.

Similar demands in the realm of commerical aircraft have led to increasingly advanced and safe passenger jets. The two giant builders of commercial aircraft, the Boeing Aircraft Corporation and the European consortium Airbus Industrie, spurred by customer demands for reliability and cost-efficiency, now produce aircraft of remarkable durability and superior performance, as do numerous other commercial and general aviation aircraft manufacturers.

Less, than one century after the Wright brothers made their first tentative flight over Kill Devil Hill at Kitty Hawk, aircraft have evolved into the some of the most complex and succesful machines ever devised by the human mind. And although supersonic transports, radar-evading planes, and even **space shuttles** now seem almost commonplace, none of these advances in aviation would have been possible without the same kind of curiosity, imagination, and faith displayed by that English monk over nine hundred years ago.

See also Parachute; Piccard, Auguste; Rockets; Rutan, Bert; Stealth aircraft; Vertical takeoff and landing craft (VTOL); Wallis, Barnes; Wind tunnel

Aircraft carrier

World War I was instrumental in spurring the development of the airplane. The tentative unstable Wright Flyer, which **Orville and Wilbur Wright** had barely gotten off the ground in 1903, had developed into a variety of sleek fighting machines. Amidst the general euphoria following the end of World War I, one man saw both the danger of repeated escalation of hostilities, and the role that the newfangled airplanes might play.

His name was Billy Mitchell, a decorated hero of the 1915-1917 air war in France. A passionate aviator, he had seen firsthand the advantage that air superiority had conferred in the skies over France, and upon his return to the United States after the war, he brought these opinions forcibly to military planning tables. Here the United States was expending vast sums on battleships and other ships of the line, he argued, when it should be spending money on airplanes. These opinions encountered stubborn resistance from the Navy, and the 1920s saw years of inter-service squabbling.

Mitchell threw himself recklessly into the fray. Finally permitted to carry out test bombing runs against obsolete warships, he proved that airplanes could sink ships with **bomb**s. He attacked opponents of his views right and left, perhaps more sharply than was wise, but never without notice. In 1926, after a particularly vicious attack on his superiors, he was court-martialed and found guilty of insubordination. He resigned from the service, but continued his campaign as a private citizen, warning the public about the short-sightedness of the military and, ominously, about the air power being amassed by the Japanese empire.

Despite Mitchell's ousting from the service, the United States did begin a limited program of **aircraft** carrier construction. The earliest bona fide aircraft carrier was the U.S.S. *Langley*, commissioned in 1922. It was an ungainly ship, capable of only limited air operations. Over the next decade, the United States put only a few other carriers in service. The *Lexington* and *Saratoga* were converted battlecruisers, and the small *Ranger,*, a true aircraft carrier, was slow and hard to manage. The more modern *Yorktown*, and *Enterprise* joined the fleet in the late 1930s.

By the time these ships entered service, the world situation had deteriorated drastically. Adolf Hitler was rapidly annexing parts of Europe, while Japan had invaded Manchuria in 1936 and was amassing naval striking power that greatly outmatched that of the United States. In 1939, Japan proved Mitchell right once and for all, sinking the British battlecruisers *Prince of Wales* and *Repulse* with airplanes.

On December 7, 1941, the Japanese attacked the United States navy in Pearl Harbor, Oahu, Hawaii. By chance all three of the United States's major carriers were not in the harbor that day, a fact grimly noted by Japan's brilliant naval strategist, Isoroku Yamamoto. The war in the Pacific was joined, and the scattered islands around which the United States strategy revolved were the perfect demonstration of the necessity of air superiority. For six months the Japanese, with their enormous aircraft carrier superiority, stormed across the Pacific. Then, in early June, 1942 the battle of Midway turned the tide. In return for losing only the *Yorktown*, United States aircraft destroyed the core of the Japanese carrier fleet: the mighty quartet consisting of the *Akagi, Kaga, Hiryu*, and *Soryu*. In one stroke nearly half of Japan's carrier capacity slipped beneath the waves, and though the war dragged on for another three years, the outcome was no longer in doubt. The great carrier battles of the Pacific were the tactical conflicts on which the greater strategic victory won by the United States hinged.

With America's war machine in full gear, the 1940s saw an explosion in the number and types of aircraft carriers. The slanted deck was introduced, which effectively gave carriers two runways from which to conduct operations. **Catapults** at the bow of the ships enabled efficient launch operations. Some fifty fleet carriers and a host of light and escort carriers joined the service during the war.

Following World War II, the United States continued a vigorous program of aircraft carrier construction. Fast attack carriers constructed during the late 1940s and early 1950s saw service through the 1990s, including during the Korean and Vietnam wars.

The aircraft carrier entered a truly new era in 1961, with the commissioning of the *Enterprise* as the first nuclear-powered carrier. Literally a floating city, this giant ship carried a crew of more than 5,000 sailors and could accommodate nearly 100 jet aircraft. Its descendents, the *Nimitz*-class carriers, have been continuously under construction since the early 1970s. The latest of these, the U. S. S. *Ronald Reagan*, is an appropriately-named testament to the ex-President, whose commitment to aircraft carrier superiority in the 1980s led to substantial upgrading of the fleet.

The presence of long-range nuclear **missiles** might at first glance seem to make aircraft carriers obsolete in the way that carriers themselves made the old battleships obsolete. Aircraft carriers have maintained a vital presence in modern conflicts, however, serving crucial roles in the 1980s British campaign in the Falkand Islands, and in the Gulf war of 1991 against Iraq. The great ships stemming from Billy Mitchell's vision still play an essential role in military policy and operations, and the vital role of air superiority in regional conflicts has been dramatically demonstrated even in the most recent conflicts.

AIRPLANE • See Aircraft

AIRSHIPS

Shortly after the first **balloon** flights in the late 1700s, people began working on ways to steer the new invention. Some suggested sails, others advocated flappers, while still others wanted to try hand-cranked propellers. None proved practical because they could not generate sufficient power.

In 1852, however, the Frenchman Henri Giffard built and flew the first powered aircraft. He hung a three-horsepower **steam engine** with a propeller below the balloon, which he shaped like a cigar to move more easily through the air. His control, however, was very poor, and he could only go aloft on the most windless days.

Despite the experimentation of Giffard and others, steerable balloons did not become truly successful until the arrival of compact **internal combustion engines** in the late 1800s. German engineer Paul Haenlein was the first to fly an airship powered by such an engine, and in 1883 Albert and Gaston Tissandier were the first to fly an aircraft with an electrical engine.

In addition, the introduction of aluminum made rigid frames a possibility. In 1897, the Austrian **David Schwarz** designed an airship with aluminum, and although it crashed on its maiden flight, he had proved the usefulness of the new metal. Alberto Santos-Dumont, an Argentinean living in France, built several successful airships between 1897 and 1904 and won a prize for circumnavigating the Eiffel Tower.

The man who truly popularized airships, however, was **Count Ferdinand von Zeppelin,** a retired German army officer. Obtaining the patent rights from the widow of David Schwarz, he built a series of rigid airships in hope of interesting the German government in the military potential of the craft. After several mishaps, the government abandoned the project, but public donations financed the building of other Zeppelins, which eventually succeeded in meeting the government's requirement that the airship remain aloft for twenty-four hours. Zeppelins were eventually used in bombing raids against the British during World War I. Although the actual damages were slight, the massive airships were an effective tool of intimidation.

After the war, airships became a popular mode of transportation. While the British and the Germans both worked on better designs, the Germans retained the lead in airship technology, due largely to the fact that they had been restricted in their manufacture of airplanes by the Treaty of Versailles. The British, however, were the first to cross the Atlantic in an airship, taking 108 hours in 1918. In 1928, the Germans constructed the most luxurious airship, the *Graf Zeppelin*, and its size was spectacular: 775 feet (236 m) long and 100 feet (30.5 m) wide. It pampered its twenty passengers, who paid between $1,000-3,000 for the transatlantic round trip. It eventually became the first aircraft to fly over 1,000,000 miles (1,609,000 km), it made 144 crossings of the Atlantic, and it carried more than 13,000 passengers.

In 1923, the United States launched the *Shenandoah*, which was filled with helium rather than hydrogen, which was extremely dangerous due to its flammability. Helium was rare, more difficult to produce, and therefore more expensive. The United States also refused to export helium to Germany following the rise of Nazism.

However, this did not keep the Germans from completing the *Hindenburg* in 1936. This airship was even bigger and better than the *Graf Zeppelin*, holding fifty passengers. Its only drawback was its use of hydrogen gas, and this proved fatal: on May 6, 1937, it burst into flames over Lakehurst, New Jersey, a disaster that killed thirty-six people. No one is sure to this day what happened. Some believe there was sabotage, but most think it was a natural electrical spark setting off hydrogen gas flowing from a leak. This disaster, which followed the crashes of numerous other airships (including the *Shenandoah* in a thunderstorm), and the advances of airplane technology effectively ended the use of airships for regular passenger service.

For a long time, airships seemed to be relics of the past, but in recent years there has been a revival of interest. The Goodyear Tire Company has flown airships over major sporting events and gained widespread fame. They are also used for continuous reconnaissance and patrol.

ALARM CLOCK • See Clock and watch

ALCOHOL, DISTILLING OF

Alcohol distillation is a heat-dependent process used to purify or separate alcohol mixtures. This process is used in the manufacture of such alcoholic beverages as brandy, whiskey, and rum. These beverages, also known as spirits, are made by first preparing a mash of grains or fruit juices. The mash is fermented, that is, natural sugars in the grain or fruit are converted to alcohol by microscopic yeast. Then the fermented mash is heated in a boiler, causing the alcohol to evaporate. The alcohol vapors are collected and then cooled in a condenser to produce the alcoholic beverage.

Distilling is done with a still, either a pot still or a column still. A pot still produces one batch of liquid at a time and is a simple still that can be set up and used at home. A column still permits distillation to occur continuously. Most commercial distilleries use column stills.

For thousands of years it was known that heating a liquid to boiling and then condensing it could rid the liquid of impurities. For example, distillation could convert salt water into fresh water. Both the Romans and Egyptians distilled plant oils using air-cooled outlets from a boiling vessel to condense the

The *Santos Dumont* flies over Paris.

liquid product. Most likely the Arabians were the first to use a water-cooled condenser to collect the more easily evaporated components which would have escaped with the earlier method. However, distilling alcohol, first for medicinal purposes, probably began in the early twelfth century in Italy when wine was distilled to produce brandy. To this mixture herbs and spices were added, and the first liqueurs were developed, such as Benedictine and Chartreuse.

During the 1400s in Ireland and Scotland, whiskey was distilled from a fermented mash of grains, primarily barley. In the 1600s gin was invented, a mixture of alcohol and water flavored with juniper berries. Later barley, corn, rye or sometimes even potatoes were mashed and distilled to make vodka. In the United States, the very first distillery was established in New York City in 1640. In 1789 the first bourbon was distilled in Kentucky using corn mash. In Barbados sugar cane juice was fermented and distilled into rum and in Mexico fermented juice from the maguey plant was distilled into tequila.

By the 1800s a large-scale distilled spirits industry had been established in Europe and North America. Distilled spirits are usually clear in color but if they are aged in kegs to develop flavor as many whiskeys are, they may also take on an amber hue. The term proof is used to describe the alcohol content of a beverage. In the United States, the proof of a distilled beverage is twice the amount of its alcohol content. For example 100 proof whiskey is 50 percent alcohol. Proof is slightly different in Great Britain and Canada. Today in the United States distilled whiskey consumption has dropped as consumption of fermented beverages such as beer and wine have increased.

Alcorn, George E. (? - ?)

American physicist

A strong student and athlete during high school, George Alcorn attended Los Angeles's Occidental College. He graduated in the top of his class with a degree in physics and eight sports letters. In 1963, after a summer working on trajectories and orbital mechanics for *Titan 1* and *Titan 2*, *Saturn IV*, and *Nova* at the space division of North American Rockwell, he completed his M.S. in nuclear physics from Howard University. Two years later, he obtained a doctorate in atomic and molecular physics from the same university. His next project was a NASA grant to research negative ion formation. Alcorn enjoys teaching science and math, but his greatest contributions have been in the production of **semiconductor**s, for which he holds eight patents. He has adapted chemical ionization mass spectrometers to detect amino acids, developed experimental methods to study planetary life, perfected systems to enable missiles to reenter the atmosphere, masterminded secret defense projects, designed instruments for space travel, created devices to test atmospheric contamination, built magnetic mass spectrometers, devised mass analyzers, taken new directions in **magnet** design, and invented a high x-ray spectrometer.

Alexanderson, Ernst Frederik Werner (1878-1975)

American engineer and inventor

Several important inventors—among them **Vladimir Zworykin**, **Philo Farnsworth**, and **John Logie Baird**—played key roles in the development of **television**. After these pioneers laid the foundation, new inventors arrived to expand the field of television, some by making it more practical, and some by refining the new medium's gadgetry. A strong competition emerged between Columbia Broadcasting System (CBS) and the Radio Corporation of America (RCA), which were vying to debut the first practical color television system. With the help of a Swedish immigrant, Ernst Alexanderson, RCA developed the modern system of color broadcasting in 1955. After completing his graduate studies at the Royal Institute of Technology in Stockholm, Sweden, Alexanderson came to the United States in 1901. The following year he secured a job at General Electric (GE), where he designed a high-frequency **alternator** to be used in broadcasting experiments. Using Alexanderson's alternator, Reginald Aubrey Fessenden made the first transmission of voice over radio waves. Impressed with Alexanderson, the Radio Corporation of America (RCA) hired him as their chief engineer in 1919. Over the next ten years he transmitted the first transatlantic **facsimile**, invented a home television receiver, and demonstrated a working television system which included a camera and transmitter as well as a projector that showed the picture on a theater screen. When Alexanderson retired from GE in 1948, he had intended to serve only as a consultant. However, in 1952 RCA approached him once again. Their goal was to develop a practical system for broadcasting color television before CBS and their electrical wizard **Peter Goldmark** could finish their own. Alexanderson agreed to take on the challenge. After several failures, the RCA team introduced a system that not only produced a good color image on new sets, but was compatible with existing black-and-white televisions. CBS's system, on the other hand, could not be received at all by existing sets. Alexanderson continued to work as a television consultant until his death in 1975. During his life, he was awarded more than 300 patents.

Alidade • See Astrolabe

al-Jazari, Ismaeel (1150-1220)

Arab engineer

Little is known about the life of Ismaeel al-Jazari, an Arab who began serving the sultans in northern Mesopotamia in 1181. However, through his writings we can learn about the engineering craftsmanship of the medieval Arab society.

In 1206 al-Jazari presented his sultan with the *Book of Knowledge of Ingenious Mechanical Devices*, devised for both educational and entertainment value. In this singularly important work, he described contemporary labor-saving devices and unusual clocks, including some of his own designs. Even though the book does not contain descriptions of the **astrolabe** or **balance**, for which Arab scientists were famous, it plainly displays stronger and weaker aspects of Arabic engineering.

Most of the machines, with varying degrees of utility, used age-old principles of mechanics, including systems of weights, pulleys, gears, **cams**, and **levers**. The crankshaft was first described by al-Jazari, if not invented by him.

One water-raising device described in his book had a **water wheel** which brought a series of scoops to the surface through cogs along its shaft. These types of water machines comprise the majority of the useful gadgets al-Jazari notes, and consequently, he has been called a hydraulic engineer.

He also mentioned many amusing, yet practical, devices, including several unusual clocks, which were much more elaborate than those developed elsewhere. They were sometimes driven by a weighted cord on a pulley-wheel that also operated the time mechanism. As the weight fell, it encountered resistance from a float riding in a water tank. Others involved a bucket that tipped over when filled with water, hitting a ratchet that moved the clock ahead. These extraordinary clocks struck the hour with a cacophony of noise from moving figures with musical instruments.

Al-Jazari noted a number of practical joke devices in his text. Some were trick drinking vessels that appeared to contain water but could not be emptied. Others looked empty but produced water when tipped over.

In producing these not-so-useful inventions, al-Jazari was typical of his age. Arab engineers had a reputation for inventing frivolous machines that dated back to the writings of the ancient Greeks.

Alloys

A mixture of two or more metals is called an *alloy.* Alloys are distinguished from composite metals in that alloys are thoroughly mixed, creating, in effect, a synthetic metal. In metal composites, the introduced metal retains its identity within the matrix in the form of fibers, beads, or other shapes.

Alloys can be created by mixing the metals while in a molten state or by bonding metal powders. Various alloys have different desired properties such as strength, visual attractiveness, or malleability. The number of possible alloy combinations is almost endless since any metal can be alloyed in pairs or in multiples. An entire period of human prehistoryis named for the earliest known alloy—bronze.

During the Bronze Age (c. 3500-1000 B.C.) humans first fashioned tools and weapons from something other than basic materials found in nature. Humans combined copper and tin to form a strong metal that was still easily malleable. Modern bronze contains a 25:75 ratio of tin to copper. The use of bronze in early times was greatest in nations where tin deposits were most plentiful, like Asia Minor, and among countries that traded with tin-mining nations.

Brass is an alloy of copper and zinc. It is valued for its light weight and rigid strength. It has a ratio of about one-third zinc to two-thirds copper. The exact ratio of metals determines the qualities of the alloy. For example, brass having less than 63 percent copper must be heated to be worked. Brass is noted for its beauty when polished. Brass was perhaps first produced in Palestine from 1400-1200 B.C. It was later used by the Romans for coins. Many references to brass in the Bible and other ancient documents are really mistranslations of mentions of bronze.

Pewter is an alloy of copper, tin, and antimony. It is a very soft mixture that can be worked when cold and beat repeatedly without becoming brittle. It was used in Roman times, but its greatest period of popularity began in England in the fourteenth century and continued into the eighteenth. Colonial American metalworks produced some notable pewter work. As a cheaper version of silver, it was used in plates, cups, pitchers, and candelabra.

The various types of steel and iron are all alloys classifiable by their content of other materials. For instance, wrought iron has a very small carbon content, while cast iron has at least two percent carbon. Steels contain varying amounts of carbon and metals such as tungsten, molybdenum, vanadium, and cobalt, giving them the strength, durability, and anticorrosion capabilities required by their different uses. Stainless steel, which has eighteen percent chromium and eight percent nickel alloyed to it, is valued for its anti-corrosive qualities. Duraluminum contains one-third steel and two-thirds aluminum. It was developed during World War I for the superstructures of the Zeppelin **airship**s built in Germany.

Many alloys add function to physical beauty. For example, sterling silver is made with eight percent copper to add strength so that it can be made into chalices and silverware. All American coins are made from copper alloy, sometimes sandwiched between layers of silver. Alloys greatly enhance the versatility of metals. Without them there would be total dependency on pure metals, which would affect their cost and availability. Alloys are a very important part of humankind's past and future.

See also Aluminum production; Construction, cast-iron and wrought-iron; Iron production; Steel alloy; Steel production

Alphabet

The twenty-six-letter alphabet used by English-speaking cultures is called the Roman or Latin alphabet, for it was during the Roman Empire that it came to resemble the alphabet we use today. However, the Roman alphabet actually has its roots in ancient Egyptian picture-writing, in which a symbol would be used to denote an entire word (just as the symbols "+" and "$" stand for "plus" and "dollar" in the modern alphabet). Eventually the Egyptians found their hieroglyphics too limiting, and they added new symbols to represent the syllables within words: for example, the word "water" has two syllables, and would thus be represented by two symbols ("wa" and "ter"). By combining the syllabic symbols, words and sentences could be formed.

Sometime before 1000 B.C., the Semitic people of Syria and Palestine developed their own alphabet, borrowing largely from the Egyptians. By this time, though, the Egyptian alphabet had grown to several hundred symbols and was difficult to learn. The Semites replaced all of the Egyptian pictures with symbols of their own, simultaneously eliminating all syllabic symbols containing more than one consonant. In the Semitic alphabet, "water" would be written using the symbols WA, TE, and RE (one symbol for each consonant). This made the alphabet, which was composed of approximately thirty symbols, much less cumbersome. The problem with the Semitic alphabet, however, was that it did not distinguish between different vowels; the same symbol was used for WA, WE, and WO, instead of three different symbols.

The Semitic alphabet was used for only a few generations before it, too, was modified—this time by the Phoenicians. By eliminating a few more consonants, they reduced the number of symbols to twenty-two. While the Phoenician alphabet was not too different from that of the Semites, it had a greater impact upon the world; this was due, primarily, to the importance of the Phoenician city of Byblos, a major trading port. Thousands of merchants would pass through Byblos, spreading the new language throughout the Mediterranean. The name of this city was later immortalized as the root for the words *biblia* (Latin, meaning "book") and *bible.*

Though many cultures adopted (and adapted) the Phoenician alphabet, none did so with such lasting impact as the Greeks. At first, the Greek alphabet was nearly identical to the Phoenician, with only minor changes. However, sometime around 800 B.C., the Greeks dropped three of the Phoenician symbols and replaced the remaining letters with symbols of their own. They also changed the direction of their **writing**, so that words and sentences were read from left-to-right.

The most important change made by the Greeks was the addition of six new symbols. In the Phoenician alphabet, these symbols had been considered "weak consonants" that were occasionally used to distinguish between similar-sounding consonants. The Greek language, however, did not utilize these particular sounds, and so the symbols were made into the vowels *a*, *e*, *u*, *long e*, and *i*. Once the vowels had been established, it was no longer necessary to have the other symbols represent consonant-vowel pairs: instead of using WA TE RE to spell "water," the Greeks could use one symbol for each consonant and one for each vowel. This twenty-four-letter alphabet was the first to contain individual vowels and consonants.

The Greek alphabet was adopted by the Etruscans, who moved to Italy around 700 B.C. They taught the alphabet to the Romans, who modified it even further, dropping the symbols for *z*, *x*, and *th* (*x* and *z* were added again at a later time, along with *y*, and placed at the end of the alphabet). Many historians consider the Roman alphabet to have been completed by 114 A.D., though the letters *j* and *w* were not added until the middle ages.

The Romans were also instrumental in the development of the style in which the alphabet was written. At first, all letters were capitalized and squarish; however, Roman stonecutters who were required to carve long inscriptions into marble, began to round the edges, making them easier to carve. They also added *serifs*—little curling strokes at the top and bottom of many letters—so that each letter did not end quite so abruptly (serifs also enabled stonecutters to more easily conceal their mistakes). Eventually, a smaller version of capital letters called *unicals* began to be used in less formal situations, and eventually evolved into a run-together style called *cursive.* Meanwhile, scribes, in an effort to conserve paper, had developed the smaller version of capital letters now known as *lower case.*

This Latin inscription from the third century exemplifies some of the later developments in the style of the Roman alphabet, including serifs and uncials.

Although the Roman alphabet has been used to represent the English language for hundreds of years, it is not the ideal system for that language. For example, the letter *e* can be short, as in *pet*; long, as in *deep*; or silent, as in *late*. The Roman alphabet also does not contain symbols for the *th*, *sh*, or *qu* sounds commonly used in the English language.

In addition to the Roman, there are several other major groups of alphabet systems used in the world. The Arabic, Hebrew, and Sanskrit alphabets all evolved from the Aramaic alphabet, which was, in turn, developed from the early Phoenician system. The alphabets used by Russians, Serbs, Poles, and other Slavic cultures are derived from the Cyrillic alphabet developed by Saint Cyril (c.827-869) and Saint Methodius (c. 825-884), who were both Christian missionaries. The Chinese language—used by more people than any other—is still written using thousands of pictures, and is the only major system that does not use an alphabetical system; Japanese, based upon the Chinese alphabet, uses both pictures and syllabic symbols.

Alternating current

Alternating current (AC) is one of two types of electrical flow. The other, **Direct current** (DC), was the predecessor of alternating current. Following the invention of the electric battery by **Alessandro Volta** in 1800, the next eighty years saw the invention and development of numerous devices that utilized DC. It was not until 1884 that AC posed a real challenge to DC.

Unlike direct current, which flows in one direction, alternating current oscillates forward and backward at a specific frequency, usually 50 or 60 cycles per second (a cycle per second is referred to as Hertz, abbreviated Hz). In other words, the current peaks first in one direction, drops to zero, peaks in the other direction, drops to zero, and then repeats the cycle. AC in North America oscillates at 60 Hz, so it takes only one-sixtieth of a second for a single cycle. In other parts of the world 50 Hz is common, and 400 Hz is common on aircraft, to make possible lighter electrical machines.

Alternating current was the brainchild of **Nikola Tesla**, a brilliant Croatian electrical engineer initially employed by Continental Edison in Paris. In 1884 Tesla came to the United States to work with the chief proponent of DC, Thomas Alva Edison, and to convince him of the benefits of AC.

Alternating current has a number of advantages over DC. **Alternators** (**generators** designed for AC operation) did not require the slip-rings and commutators (brushes) upon which their DC cousins depended. AC operates on the process of electrical induction, which was discovered by **Michael Faraday** in 1831. Inducing the flow of electricity from one coil of **wire** into another eliminated the arcing that plagued DC generators. As mentioned above, an even greater advantage to AC is that its voltage can be stepped-up to higher levels with a **transformer**, sent great distances through high tension wires, and stepped-down at its destination.

Alternators at power stations produce "three-phase electricity;" they have three coils equally spaced around their primary coil, each of which is induced to produce a 60 Hz alternating current for three circuits. Three-phase electricity can supply as much current through three thin wires as it would normally take two thick wires to carry. The advantage in using a thinner wire is to minimize the electrical resistance a thick wire would produce.

Since the current of an alternating current peaks in opposite directions over one cycle, the average voltage for the cycle is zero. The voltage of the current intensity, or amplitude, as displayed on an **oscilloscope**, is that of a sine wave. In three phase power, as the voltage in one wire peaks, the voltage in the other two are halfway to peak (one increasing, the other decreasing).

Unfortunately for him, Edison rejected alternating current. First, he thought, the world was geared to DC; there were no AC applications so there was no need to change. Secondly, Edison considered alternating electricity to be "killer current," and would not accept arguments to the contrary. Tesla remained with Edison for only one year before quitting in disgust.

The opportunity for AC to prove itself came in 1893. Tesla, with his backer **George Westinghouse**, underbid Edison for the contract to provide power for the Columbian Exposition in Chicago. This was the first electrical fair in history, and Tesla's polyphase (two-phase) AC system was a spectacular success. The consummate showman, Tesla put on impressive performances; at one point he sent a potential of 200,000 volts of AC through his body and challenged Edison to do the same with DC. (Edison may have had the last laugh; after the state of New York began using AC to electrocute prisoners, Tesla became convinced that Edison had helped establish the system to prove to the public that alternating current was indeed deadly.) Also during the fair, Westinghouse demonstrated a rotary converter that, ironically, changed the polyphase AC into direct current to operate a DC motor and run a railway car.

The case in favor of alternating current was further advanced by **Charles Steinmetz**. He established the law governing hysteresis,the residual magnetism that occurs in generators and motors. Hysteresis causes a loss of power and was little understood at the time. His theoretical studies of AC in 1893 resulted in making what was then a very complex field understandable to the average engineer and electrician.

ALTERNATING-CURRENT MOTOR • See Electric motor

ALTERNATOR

The alternator is related to its cousin the dynamo in that both convert mechanical energy into electricity. Unlike the dynamo, which creates direct current (DC), the alternator produces **alternating current** (AC) by rotating an armature through a magnetic field. It is essentially an **electromagnet** rotating within coils of **wire**.

The alternator makes use of the principal of electrical induction, discovered by **Michael Faraday** and **Joseph Henry** working independently in 1831. They found that moving an electrical conductor, such as a wire, across lines of magnetic force caused an electric current to flow in the wire. Alternatively, moving a **magnet** within a coil of wire induced electricity to flow through the wire. In addition, when the movement of the magnet was reversed, the electrical flow also reversed—alternating the movement of the magnet created an alternating current.

Faraday showed that voltage could be produced by magnetism provided three conditions existed: there had to be a conductor in which to induce the voltage, a magnetic field had to be close to the conductor, and a relative motion had to occur between the magnetic field and the conductor. Either the conductor (the rotor) had to be moved so as to cut across the magnetic field, or the magnet (the stator) had to be moved so that the magnetic field was cut by the conductor. The movement of one or the other induced electrons to flow within the conductor, creating the voltage.

Faraday's discovery did not cause an immediate impact in the world of electricity. At the time, electricity was basically a scientific curiosity. **Alessandro Volta** had just invented the DC battery in 1800, and it showed great potential for the future. Most scientists were occupied in the study of the applications of DC electricity, and the next step in developing the alternator had to wait for thirty-six years.

It was the Belgian-French inventor **Zénobe Gramme** who rescued the alternator from obscurity. In 1867 he improved the AC generator (alternator) making it a more practical device for the production of alternating current; two years later he improved the DC dynamo. In 1871 he and his associate

Hippolyte Fontaine opened a factory, and the electrical generating equipment they produced marked a turning point, for it is upon their work that the electrical industry was established.

Still, there was not a great flurry of activity leading to the overthrow of direct current. That was not until the arrival of **Nikola Tesla**, a brilliant Croatian electrical engineer, who appeared on the doorstep of **Thomas Alva Edison** in 1884.

Edison was one of the world's chief proponents of DC electricity. Tesla began working with Edison and spent a year in the futile attempt to convince Edison of the advantages of AC. Because alternators operated under the process of induction, they did not require the slip rings and commutators (metal brushes) which caused arcing in the DC dynamos. In addition, alternators produced voltages that could be increased with a **transformer** and carried great distances with high tension wires. Edison refused to accept Tesla's arguments; the world was wired for DC and he had no inclination to change the status quo. When Tesla obtained a patent for his electric motor in 1888, he sold the rights to **George Westinghouse**. Ironically, Westinghouse underbid Edison for the contract to provide electricity at the Columbia Exposition in Chicago in 1893. He immediately contacted Tesla, who manufactured a polyphase (two-phase) AC system, wired the fairgrounds, and made this first electrical exposition in history a spectacular success. By 1895 the water power of Niagara Falls was being used to generate alternating current, and this led to an increasing preference for alternating current.

The current produced by an alternator oscillates forward and backward at a specific frequency, depending on the speed at which the magnet (or loop of wire) rotates. This is usually 50 or 60 cycles per second; a cycle per second is called a Hertz (Hz). The current peaks first in one direction, drops to zero, peaks in the other direction, drops to zero then repeats the cycle. Alternators in North America produce electricity at 60 Hz, and so it takes only one-sixtieth of a second for a single cycle.

Modern alternators produce three-phase electricity. There are three coils equally spaced around a primary coil, each of which is induced to produce a 60 Hz alternating current for three **electric circuit**s. The voltage of the electromagnetic wave, as seen on an oscilloscope, is a sine wave. With three-phase power, as the voltage in one wire peaks, the voltage in the other two are halfway to peak (one increasing, the other decreasing). The benefit of three-phase electricity is that it can supply as much current through three thinner wires as it would otherwise take two thick wires to carry. Using a thin wire instead of a thick one minimizes the electrical resistance a thick wire would produce.

ALTIMETER

An altimeter is an instrument that measures altitude (the distance above sea level or some other chosen point). **Aircraft** use two different types: the aneroid (barometric) altimeter and the radio altimeter.

The aneroid altimeter derives a quantity for altitude by measuring the decreasing air pressure that occurs with increase in altitude, while a radio altimeter bounces a **radio** signal from an object aloft to the ground and determines the distance aloft as a function of the time it takes for the signal to return.

The history of the altimeter begins with the invention of the mercury **barometer**, the first device to measure air pressure. In 1643, Italian physicist **Evangelista Torricelli**, a pupil of **Galileo**, filled a tube with mercury. One end of the tube was closed; the other open end was turned upside down and inserted in a cup of mercury. Because the air exerted pressure on the mercury in the cup, about 30 in (76.2 cm) of mercury remained in the tube.

After Torricelli performed this experiment to prove that air exerted an intrinsic pressure, others wondered if there was a way to use it to show the difference in pressure at sea level and the tops of mountains. Since air thins as altitude increases, the mercury in a barometer should fall as less pressure is exerted on the cup's contents. Once Blaise Pascal (1623-1662), the French mathematician and physicist, proved this was the case in the 1648, the barometer came to be used to measure altitude. The device's earliest use in flight was by French balloonist **Jacques Charles**, who took one aloft in 1783.

Airplanes use a variation of the barometer called the aneroid altimeter, which does not depend on any liquid to determine the altitude. Virtually the same as the aneroid barometer, invented by Lucius Vidi in 1843, it records the pressure exerted on an sealed compartment from which some of the air has been removed. The surface of the compartment bulges or contracts slightly as the outside air pressure changes, and the movements of the compartment are indicated on a needle which is connected to the box by a chain, lever, and springs. Aneroid altimeters are surprisingly accurate (giving readings within 1.5 ft [.5 m] at 11,000 ft [3,353 m]) and are more sensitive to changes than mercury-filled altimeters. An additional benefit of the aneroid altimeter is its ease of operation since it does not need liquids and stands up well to jarring.

A drawback of the aneroid altimeter is its sensitivity to weather conditions. If the barometric pressure changes during the course of a flight, the altimeter will not indicate the correct altitude at the arrival airport. For example, if the pilot flies into an area of lower barometric pressure, the altimeter will read slightly too high. For this reason, pilots must constantly monitor local weather reports. Additionally, at or above 18,000 ft (5,486.4 m), all aircraft must set their aneroid altimeters to 29.91 in (76 cm) of mercury; this ensures that everyone using the high-altitude flight corridors have altimeters that read identically, reducing the chance of mid-air collisions.

The second type of altimeter uses **radar**, developed by **Sir Robert Watson-Watt** in the 1930s. It sends out a radio wave which bounces off the surface of the earth. The reflected signal returns to the source, and a device measures how long the waves took to make the trip back and forth. Since electromagnetic waves always travel at the speed of light (186,000 mi [299,274 km] per second), an absolute measurement of altitude is quickly computed by the device. Unlike aneroid altimeters, radar altimeters do not require adjustment to weather conditions or frequent calibrations. Additionally, such radar capability can be used to warn of terrain proximity, aiding pilots in avoiding instances of controlled flight into terrain (CFIT), a common cause of aviation fatalities.

ALUMINUM PRODUCTION

Although aluminum makes up about 8 percent of the earth's crust (it is the third most plentiful crustal element), it does not occur in its pure form, but is found associated with other elements in compounds, such as aluminum chloride (cryolite) and aluminum hydroxide (bauxite). It was identified as a possible new metal prior to 1800, but it was not until 1825 that it was isolated in powder form by Danish chemist Hans Christian Oersted (1777-1857). Two years later German chemist Friedrich Wöhler produced it in solid ingot form (1880-82). These discoveries, as essential as they were, did not make it possible to produce aluminum in large quantities. Thus it was considered a precious metal for most of the nineteenth century.

Mass production of aluminum was made possible with the simultaneous invention of the electrolytic reduction method by American chemist **Charles Martin Hall** and French chemist **Paul-Louis-Toussaint Héroult** in 1886. Both men used electrodes to isolate aluminum within a batch of molten cryolite, creating a method by which the metal could be produced in large quantities over a short period of time. With the *Hall-Heroult process*, aluminum became much less expensive, passing from a precious substance to one that was practical for a variety of everyday applications. Production in 1887 totaled 28.6t (26 t).

Hall and a consortium of investors and supporters went to Pittsburgh financier Andrew Mellon (1855-1937) in 1889 and made a cash-for-stock swap. With the cash, Hall founded the Pittsburgh Reduction Company, which was renamed the Aluminum Company of America (Alcoa) in 1907.

The Pittsburgh Reduction Company produced large quantities of aluminum, which was quickly adopted as material for buildings, boats, kitchen utensils, electrical wiring, automobiles and appliances. It was even used by the Wright brothers in their original 1903 airplane. In 1906 the Germans began building their World War I Zeppelin airships with a structural material called duraluminum. It was aluminum with an alloy of one-third steel, making it slightly heavier than aluminum but also making it nearly as strong as steel.

Aluminum foil was invented in 1910. Nine years later Virginian R. S. Reynolds, founded the United States Foil Company, which made foil lining for his uncle R. J.'s cigarette packages. In 1930 the company became the Reynolds Metal Company.

Bauxite became the primary ore from which aluminum was extracted. This earthy compound exists in great abundance in Arkansas, Haiti, and Jamaica. Alcoa and Reynolds Metal Company own most of these reserves. Development of the *Bayer process* of refining aluminum from bauxite and the proximity of the reserves to the North American aluminum market helped boost the popularity of bauxite and aluminum. Invented by Karl Wilhelm Bayer in 1888, the process refined bauxite to a stage where it could be accommodated by the Hall-Heroult process.

The aluminum industry greatly expanded with military applications during World War II, especially in the airplane industry. During the postwar 1940s and 1950s, when record numbers of Americans were raising families and purchasing middle-class single-family homes, household goods made with aluminum, from coffee pots to lawn chairs, became a common part of daily life, often replacing wood where weathering was a concern.

Aluminum has been at the center of the recycling movement of the 1980s and 1990s. Its widespread use in beer and soft drink containers makes it one of the most readily recyclable commodities on the market. In this same time period, the idea of using aluminum in car components, even replacing steel in car bodies, came to fruition. Aluminum can make cars lighter so they would consume less gasoline and reduce exhaust-gas emissions. Potentially, automobile fuel efficiency could triple, if structural safety issues can be successfully addressed. The first all-aluminum car, the Audi A8, was produced in 1994, though it was only available in Europe. The Reynolds Company also promoted the use of aluminum in bridges and their in the mid-1990s. Though aluminum is more expensive than other materials used in bridges, it has other advantages, including a higher strength-to-weight ratio than, for example, concrete.

See also Aircraft; Alloy; Recycling; Steel alloy; Wire

ALVAREZ, LUIS WALTER (1911-1988)

American physicist

Alvarez was born in San Francisco, California, on June 13, 1911. He attended the University of Chicago, where he received his bachelor's degree (1932), master's degree (1934) and doctorate (1936). Upon graduation, he joined the faculty at the University of California at Berkeley. He became professor of physics there in 1945 and served in that capacity until 1978. Because of his wide-ranging interests, Alvarez was called the ''prize wild-idea man'' by colleagues at Berkeley. His earliest research was in the area of nuclear physics and cosmic rays. He was involved in research on the ''east-west'' effect in cosmic rays, the radioactivity of tritium (hydrogen-3) and other isotopes, the nature of nuclear fission, and the magnetic properties of the neutron. Alvarez received the Nobel Prize in physics in 1968 for his work on liquid-hydrogen **bubble chamber**s. After **Donald Glaser**'s invention of the bubble chamber in 1952, scientists attempted to construct larger, more efficient models of the detecting device. Alvarez and his colleagues were at the forefront of this research. In 1959, his research team completed a 72-inch (183 cm) chamber that was, for five years, the largest of its kind in the world. At the same time, Alvarez worked on improving methods for analyzing **photograph**s obtained in the bubble chamber. He realized that existing methods were too slow to observe and record particles with very short lives. As a result of improved computing and analytical techniques, Alvarez and his colleagues were eventually able to discover a number of very short-lived particles, known as *resonances*, with lifetimes of less than 10-20 second.

During World War II, Alvarez was actively involved in research on the **atomic bomb** and **radar**. Working with Lawrence Johnston, he developed a narrow-beam ground-based radar system that permitted airplanes to land in fog and bad weather. He also developed a system for "blind" bombing of targets. In 1980, Alvarez and his son proposed a theory regarding the extinction of dinosaurs that occurred 65 million years ago. They had discovered a layer of sedimentary rock in Italy that is unusually rich in iridium. Similar iridium-rich strata were later found in other parts of the world. Alvarez and his son suggested that the iridium came from a giant meteorite that collided with Earth 65 million years ago. The dust produced by the collision, they said, blocked out solar radiation for three years and destroyed so much vegetation that dinosaurs became extinct.

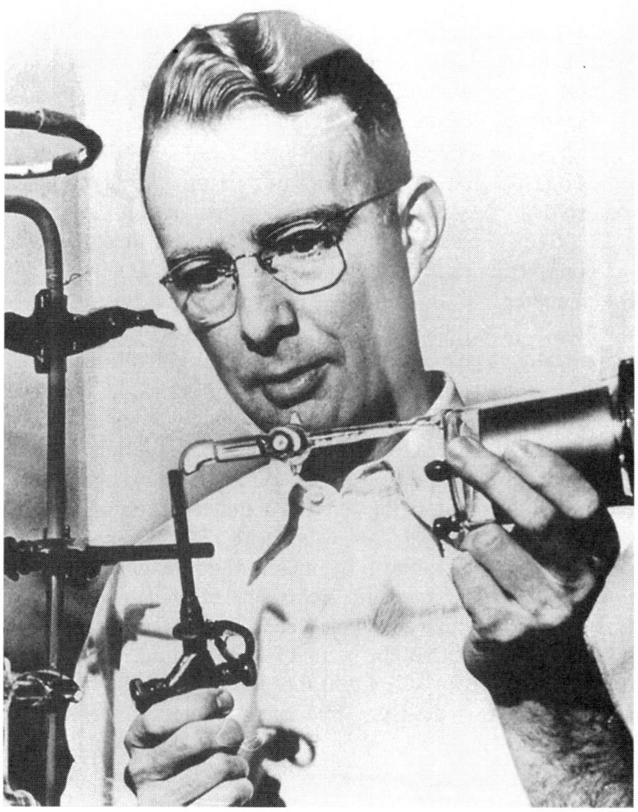

Luis Walter Alvarez

Ambulance

Ambulances were used as long ago as the eleventh and twelfth centuries, when men wounded in battle during the Crusades were carried by horse-drawn wagons behind the front lines for treatment. A similar system was used in 1487 by the Spanish forces during the siege of Málaga. The modern ambulance, providing swift and efficient transportation for medical treatment, evolved from a creation by Dominique-Jean Larrey (1766-1842) in 1792. Larrey, Napoleon's private surgeon, strove to improve battlefield treatment of wounded soldiers. Larrey's horse-drawn "flying ambulance" carried surgeons and medical supplies onto the battlefields during the Rhine campaign in 1792. For the Italian campaign of 1794, Larrey employed light ambulance carriages with stretchers to carry the wounded, and camels powered his ambulances in Egypt in 1799. With fellow surgeon Pierre Percy (1754-1825), Larrey formed a battalion of ambulance soldiers, including stretcher bearers and surgeons. Larrey's flying ambulances, and the swift medical attention they brought, greatly impressed Napoleon's troops and significantly boosted morale.

Ambulance service was expanded from the military to the civilian world in 1869 by Bellevue Hospital in New York City. Larrey's flying ambulance remained standard until the first motorized ambulances appeared around the turn of the century, pioneered by Panhard-Levassor of France. Until the mid-1960s, ambulances were mostly hearses, because these vehicles could transport a lying-down patient. Hearses afforded little room for supplies or attendants, however, let alone treatment en route. A National Academy of Science/National Research Council report in 1966 focused attention on the need for both professional training of emergency care technicians and standardization of ambulance design. This resulted in today's modern ambulance, with working space and sophisticated supplies and equipment operated by medical paraprofessionals, so that the ambulance is no longer simply a transportation vehicle—it is also a moving treatment center.

The first airborne ambulances were hot-air **balloon**s used to evacuate the wounded from Paris during the Prussian siege in 1870. **Helicopter**s began to transport wounded soldiers during World War II and became vital evacuation vehicles in the Korean and Vietnam wars. Today, air ambulances—both fixed-wing and helicopters—are increasingly used for quick transportation of civilian patients. Independent air ambulance companies dispatch professional medical teams nationally and internationally on fully-equipped, twin- engine jets, coordinating ground support for the critically-ill patient and their families.

Ammeter

An ammeter is an instrument that measures electric current, usually in units of amperes (or milliamps or microamps when very small currents are involved). Both the magnitude of direct current (DC) and **alternating current** (AC) can be measured, though with ammeters of different types. Due to differences in design, an ammeter is not as sensitive as a **galvanometer**.

There are two basic types of ammeters. The *moving coil ammeter* has a scale with equally spaced divisions; the *moving iron* or *moving magnet ammeter* has a nonlinear scale. Both versions use a principle discovered by Hans Christian Oersted in 1820, which holds that an electric current in a **wire** produces a magnetic field. The magnetic field exerts a force on a coil, making it turn. A pointer that is attached to the coil or magnet also moves, allowing the measurement to be read off the scale.

The moving coil ammeter has three basic parts: a permanent **magnet**, a coil, and a coil spring. When current is passed through the coil it creates a magnetic field that is opposite that of the permanent magnet, causing the coil, and its attached pointer, to pivot. The coil spring not only limits the distance the coil can rotate, but the amount of tension it exerts depends on the strength of the force turning the coil.

There is a fixed maximum of DC, called a full scale deflection (fsd), that an ammeter can measure. To obtain high values, a shunt **resistor** is attached to each end of the coil in a DC ammeter, causing a part of the current to bypass the meter movement and preventing an overload. To determine the true amperage of the current, it is then necessary to multiply the amperage indicated on the linear scale by the quantity of one plus the ratio of the coil resistance to the shunt resistance.

In order to measure alternating current (AC), rectifiers that convert AC to DC are incorporated in the ammeter's shunt resistors. A simpler and less expensive option, with its own set of limitations, involves a moving iron (moving magnet) ammeter with its pointer attached to a counterbalance weight and a piece of soft iron; another piece of iron is located nearby. A coil surrounds the entire movement; when the current passes through the coil, magnetic fields are created around both pieces of iron which then repel each other. This results in a nonlinear movement of the pointer that corresponds to the nonlinear scale. When AC passes through the coil the magnetic fields alternate at the same frequency as the current, but the force remains one of repulsion. AC with frequencies up to 300 Hz can be measured just as well as DC. Nonetheless the moving iron ammeter has fallen into disuse, because its accuracy is limited and it has high resistance.

Modern **electronics**, **integrated circuit**s, and **semiconductor**s have ushered in a new era for measuring current. A digital display ammeter, functioning as a **voltmeter** (which is an integral part of the **multimeter**), has no moving parts, but instead indicates current by measuring the proportional voltage drop across an internal resistor. Digital displays are extremely accurate and easy to operate, but they are more expensive than their electro-mechanical cousins.

Ammonia synthesis

Ammonia (NH_3) is one of the most familiar compounds of **nitrogen** and **hydrogen** as well as one of the world's most valuable industrial and agricultural chemicals. Under ordinary conditions, ammonia is a colorless gas that is lighter than air. It has a sharp odor that stings the nostrils and can cause suffocation if inhaled in great enough concentrations. Because of its physical properties, liquefied ammonia has been used as a refrigerant; when the liquid vaporizes, it absorbs a large amount of heat from its surroundings. Ammonia is used

Motorized ambulances were first put to widespread use during World War I. Above, one of the war's most famous ambulance drivers, Ernest Hemingway, stands in front of his vehicle.

throughout the industrial world as a valuable fertilizer. Ammonia and its compounds, primarily ammonium nitrate and other ammonium salts, replenish nitrogen in depleted soils. This greatly increases yields of agricultural crops, especially those that cannot obtain nitrogen from the atmosphere. Some farmers apply anhydrous (dry) ammonia directly to their fields from pressurized tanks. Ammonia is also oxidized to produce large quantities of nitric acid (HNO3) for making explosives such as **TNT (trinitrotoluene)** and **nitroglycerin**. The ammonia is converted to nitric acid, an essential ingredient in explosives, via the **Ostwald-Bauer process** invented by German chemist Wilhelm Ostwald (1853-1932) in 1901. During World War I, synthetic ammonia was used to manufacture **ammunition** after Germany's nitrate supplies had been cut off. The textile industry also uses ammonia to produce **nylon**, **rayon**, and other synthetic **fiber**s. Some ammonium salts are used extensively in galvanizing, tinning, and soldering processes and in fireproofing fabrics, **paper**, and wood. Ammonia is also used in dyeing and scouring fibers and in manufacturing many chemicals, **plastics**, vitamins, and drugs.

In ancient times, ammonia was derived from organic material such as manure and urine. During the Middle Ages, alchemists produced a weak solution of ammonia in water by distilling the horns and hoofs of oxen. They called this ammonia solution ''spirits of hartshorn.'' Later, more substantial amounts of ammonia were recovered from the vapors of natural steam vents in Italy, and large quantities of ammonia were produced from Chile saltpeter (sodium nitrate, or $NaNO_3$), which is mined in South American deserts. Around the turn of the twentieth century, Sir **William Crookes** predicted that the world would soon run short of Chile saltpeter. Scientists feared that without nitrogen-based fertilizers to increase crop yields, the growing world population would starve. Thus began the search for an alternative source of ammonia. Although a practically inexhaustible source of nitrogen exists in the atmosphere, which consists of about 80 percent nitrogen, no one knew how to convert the element into ammonia on a large scale. In Germany, **Fritz Haber** began investigating the possibility of combining nitrogen from the atmosphere with hydrogen to form ammonia. In principle the chemistry of ammonia synthesis was simple but the execution was difficult because high temperatures and very high pressures were needed to produce enough quantities of ammonia to make the process economical. Haber succeeded in developing an ammonia synthesis process that worked in the laboratory in 1909. Initially, Haber used osmium or uranium as a catalyst to speed up the synthesis reactions, which took place at a temperature of 1,025°F (550°C) and at pressures up to two hundred times greater than normal atmospheric pressure. In 1913 **Carl Bosch** developed the process further, making it practical on a larger scale. Early industrial processes, which were operated at slightly lower temperatures and higher pressures to maximize ammonia production, used finely divided iron as a catalyst.

In 1917 French engineer **Georges Claude** independently developed an ammonia synthesis process similar to that of Haber and Bosch. Claude's process was based on earlier work by fellow chemist Henry-Louis Le Châtelier (1850-1936). Essentially, the Haber-Bosch process has remained unchanged since the early 1900s. It is used today to manufacture thousands of tons of ammonia worldwide, accounting for more than 85 percent of total ammonia production.

Commercial ammonia synthesis takes place in large steel reactors that are designed to withstand very high pressures (up to one thousand atmospheres) and high temperatures (about 700°C or 1300°F). The reactors also resist attack by hydrogen gases which can corrode many metals, especially at high temperature and pressure. The catalyst in commercial ammonia synthesis is usually an iron oxide mixed with a small amount of additional ''promoter'' material. When the process begins, the hydrogen gas converts the catalyst to pure iron. Gradually, as the iron catalyst is poisoned by carbon and sulfur compounds, it becomes less effective and must be replaced. Nitrogen gas for the synthesis process comes from the atmosphere, while most of the hydrogen needed is extracted from natural gas sources. Once ammonia gas has been formed, it is liquefied by cooling it with water. Nitrogen and hydrogen that have not yet been converted to ammonia are recycled through the system. In addition to Haber-Bosch synthesis, other industrial sources of ammonia also exist; it is created as a by-product, for example, during the production of coal and coke-oven gas.

In the early 1990s the M.W. Kellogg Co. developed a new reduced energy process for manufacturing ammonia. This technology was successfully implemented by Pacific Ammonia Inc at its Kitimat, British Columbia plant in 1994. This new technique can increase ammonia production by 40% by using two innovative processes. The first innovation, known as the Kellogg Advanced Ammonia Process, involves a ruthenium-based catalyst instead of a traditional iron catalyst. The second innovation, the Kellogg Reforming Exchanger System, uses a new open-tube design in the reforming exchanger.

See also Birkeland, Kristian Olaf Bernhard; Dye; Electroplating; Galvanized iron; Nitrogen fixation; Refrigeration; Solder and soldering iron

Ammunition

Ammunition is a general term used to describe **rocket**s, **bombs**, **grenades**, **torpedoes**, mines, guided missiles, smoke bombs, flares, gun cartridges, shells, and other projectile materials used in warfare, hunting, and target shooting.

Primitive peoples first used stones and sticks as ammunition. They later discovered that, if a stone was wrapped in a thong made of animal hide or vine, it could be sent great distances with deadly accuracy, depending upon its velocity. The sling soon followed, often used with sharpened sticks and stones, which were later also employed as arrowheads. The Greeks and Romans used various types of catapults to throw stones. Wooden engines were also used in the ancient world to throw showers of darts or to launch flaming matter.

Gunpowder was probably discovered in China during the tenth century and was used primarily in **fireworks**. Its military use was not realized until the early 1300s, when the Arabs are believed to have invented the first gun and **cannon**. The first guns appear to have been simple tubes rested on the ground or on wooden trestles, and they probably fired arrows or conveniently sized stones. The earliest cannon consisted of a deep wooden bowl which held the gunpowder and a cannon ball balanced on the rim. The cannon ball popped off when the powder exploded. There were also experiments with arrows fired by gunpowder (the so-called ''pot de fer'') and with larger cannon, called bombards, which shot stone balls wrapped with iron hoops to keep them from shattering in the explosion.

During the fifteenth century, the Germans started casting iron balls that fit the bore of cannon more tightly than stones did, so that less of the explosive force was lost. The Germans also invented the bomb, which was initially no more than an iron ball filled with gunpowder. Bombs were first tossed by hand, as the modern grenade still is, but later they were fired from guns.

In the early sixteenth century, soldiers discovered that gunpowder could be ground to a grainy consistency, which allowed small air pockets to remain when the powder was rammed into the weapon. When the powder was ignited, the air pockets allowed fire to travel more quickly and uniformly for a more powerful explosion.

In the late sixteenth century, the Dutch began using bombs as ammunition for mortars, short cannons with thick walls. A hollow metal ball filled with powder had a small hole in it for a fuse. They tried putting this bomb into a mortar with the fuse down toward the powder, but the explosion often drove the fuse into the bomb, causing it to blow up in the mortar. Then they tried turning the bomb over with the fuse sticking up toward the muzzle. The gunner was to light the fuse and the powder at the same time, which proved too difficult. It was not until 1850 that someone realized that the heat of firing would set off the fuse no matter which way it faced in the mortar.

Until the early sixteenth century, all guns and cannons had to be ignited by hand, an often dangerous and cumbersome procedure. With the invention of the wheellock—an ignition mechanism that produced an internal spark—firearms of all shapes and sizes started proliferating, and there was widespread experimentation with different forms of ammunition. Handheld guns and **rifles** became more popular, using both single bullets and shot. Cartridges, packages that held both the explosive charge and the bullet or shot in a cloth or paper casing, came into use. Loading became less cumbersome as a consequence, also because the shooter did not have to measure the amount of gunpowder himself anymore. Gustavus Adolphus (1594-1632), the great Swedish leader, was apparently the first to use cartridges with shot for light cannon to speed up the loading of the gun. He also popularized the use of case shot, often called canister, against enemy infantry. A canister is a

Workers in a munitions factory during World War II sort through bullets.

tin can filled with musket balls or scrap metal designed to scatter among the enemy when fired.

The Prussian ruler Frederick the Great (1740-1786) advanced artillery in the eighteenth century. He depended heavily on grapeshot, which was much like canister. A charge of grape consisted of 50 to 60 iron balls, each about one inch in diameter. They were bunched around a wooden rod attached at one end to a wooden disk. The entire charge was wrapped in a cotton bag and loaded into the gun, where the heat of firing burned the bag away and sent the balls spraying out of the muzzle.

The British started using rockets as part of their military arsenal in the late eighteenth century. They had picked up the idea for these weapons in India and produced iron-headed rockets that carried an explosive charge in the nose for up to two miles. A crude time fuse exploded the **rocket** after it landed. It was not until World War II, though, that rockets, notably the German V1 and V2 rockets, became truly effective weapons that could be sent over great distances with a reasonable degree of accuracy.

By the late eighteenth century, virtually all ammunition for large guns was prepared in advance. We can distinguish between fixed ammunition, in which all elements are put into one waterproof unit or cartridge; semi-fixed ammunition, which is made in two pieces and allows for on-site adjustments in the amount of propellant charge according to the required range; and separate loading ammunition for rounds so large they would be difficult to handle in one piece.

Advances in the first half of the nineteenth century focused on the discovery of *percussion*. Alexander Forsyth (1769-1843), a Scottish clergyman, found that certain chemicals exploded when struck a sharp blow. As a consequence, it was no longer necessary to rely exclusively on gunpowder to set off the primary explosion. Both guns and cartridges soon incorporated the principles of percussion, using detonating mixtures rather than powder for the primer. Another area of experimentation concerned the design of bullets. A variety of shapes and sizes were tried out, until a cylindri-conical shape was settled on as the most effective. There were also attempts to design bullets with raised belts or lugs to run through spiral rifling grooves in the gun barrel. The object was to make the bullet fit the gun barrel more tightly to enhance velocity and to put spin on the bullet to improve accuracy. At the same time, the entire cartridge had to be easy to load.

A Frenchman, Claude-Etienne Minié (1804-1879), designed a cartridge much like a modern one, known as the Mineé ball, with a pointed nose and a deep hollow in the base where an iron cup was fitted. When the powder exploded, it shoved the cup into the base, flaring the outer edges of the base into the rifling grooves. The Minié ball fit tightly into the gun barrel when it expanded on explosion, creating an air-tight seal behind it, and so it fired much more forcefully than earlier cartridges. An unknown American mechanic discovered later that the cup was not necessary, since the explosion expanded the hollowed-out base anyway.

In the 1880s **Alfred Nobel** invented smokeless powder, which represented a great improvement over the traditional black powder, made from charcoal, sulphur and saltpeter. This powder created a great deal of smoke, which, particularly with repeater rifles, could seriously limit visibility in large-scale battles. It also caused a great deal of fouling in the gun barrel, necessitating frequent scrubbing with warm water to retain accuracy of aim. Nobel's smokeless powder was made out of guncotton (cotton dipped in nitric and sulfuric acid) and **nitroglycerine**. The new mixture gave much higher pressure and greater velocity and had the added advantage that it fired even when damp.

Ammunition for handguns and rifles has remained virtually unchanged in the twentieth century. In the late nineteenth and early twentieth centuries, bullets were developed that expanded or exploded upon impact, such as the dum-dum bullet. These bullets had such atrocious effects on human targets that international conventions outlawed them for use against personnel. Expanding bullets are commonly used in hunting, however, to ensure a quick kill. Another early twentieth-century development was shrapnel, named after its original inventor, Henry Shrapnel (1761-1842). Essentially an improvement of the canister, shrapnel consists of a blunt-nosed **steel** can with up to 1,200 lead balls. As it approaches the target a pre-set fuse fires another charge of powder that discharges the pellets much like shot fired from a shotgun.

The most significant developments, however, have come with the increasing variety and increasingly lethal nature of shells, missiles, and bombs. Modern artillery shells are either thin-walled to hold a maximum bursting charge or thick-walled to scatter lethal fragments upon exploding. Shell fuses have been created either to cause an explosion after a set time or upon impact. Armor-piercing shells are made with the usual lead casing and a blunt, solid steel center hidden under a streamlined exterior. Chemical or incendiary shells have only enough bursting charge to break open the shell and spread the contents of poison gases, incendiary chemicals, defoliants, and other harmful materials. A similar variety of charges is possible with rockets, which are self-powered using the principles of jet propulsion. A new class of ammunition is represented by guided missiles, whose course can be altered in flight by such mechanisms as a target-seeking **radar** device or a heat-seeking device. The accuracy of modern ammunition has also been greatly enhanced with the use of computers. Computer programmed ''smart'' missiles may contain tiny cameras allowing remote steering, infra-red sensing devices or laser homing systems.

The most destructive innovation of the twentieth century, of course, has come with the ability of scientists to set off nuclear reactions. The development of the nuclear bomb during World War II, which made possible such intense explosions with such devastating after-effects as to enable the destruction of entire countries with a single charge, has permanently altered global politics. Although some countries are trying to reduce or dismantle their nuclear arsenals, more and more others are acquiring the capacity to produce one. It is not unimaginable that, with nuclear weaponry, humankind has reached the limits of its power to destroy itself.

Many countries have developed and tested nuclear weapons since World War II, but they have not been deployed in battle since the bombs dropped on Hiroshima and Nagasaki.

Ammunition made with recycled nuclear waste, called depleted uranium or DU, was developed by the U.S. military and used during the 1991 Persian Gulf War. Bullets made of depleted uranium are twice as dense as lead and are able to pierce steel-armored tanks with ease. Depleted uranium is only slightly radioactive, however its dust or shrapnel is believed to be highly toxic and have unknown health effects. The use of depleted uranium in ammunition and tank armor during the Persian Gulf War was not made known until several years after the warÕs end. While still under investigation, depleted uranium may prove to be a type of ammunition almost as deadly for its users as for its targets.

Amniocentesis

Amniocentesis is the process of removing from a pregnant woman's uterus a sample of the amniotic fluid which surrounds the fetus. This fluid contains fetal cells which are analyzed to identify genetic defects of the fetus. Until the advent of amniocentesis, **prenatal diagnostic techniques** were severely limited. In February 1952, the English physician Douglas Bevis published an article in the journal *The Lancet* describing his use of amniocentesis to identify risk factors in the fetuses of Rh-negative women whose partner had the Rh-positive blood type. Bevis, who conducted his study at St. Mary's Hospital in Manchester, England, chemically analyzed the iron and urobilinogen content of the fluid to determine the possibility of hemolytic (blood) disease in the unborn child. While the technique of using a needle to obtain samples of amniotic fluid was available in the late 1920s or early 1930s, the Bevis study is considered a landmark that led to the wide use of amniocentesis for fetal diagnosis. Bevis's technique was later (1961) refined by measuring amounts of bilirubin in the amniotic fluid of Rh-sensitized women.

During the 1950s the ability to determine fetal sex was developed, based on the 1949 observations of doctors Murray Barr and Ewart Bartram who noted that all female cells—but no male cells—contain a chromatin mass on the edge of the nucleus. If fetal cells in the amniotic fluid contain this Barr body, the fetus is female. Knowing the fetal sex is important in assessing the risk of a child being born with a sex-linked (affecting males only) disease such as hemophilia. During the mid-1960s it became possible to grow human cells in the laboratory. By 1966, fetal cells from amniotic fluid were cultured and their chromosomes analyzed. It was now possible to determine whether a fetus was affected by Down Syndrome, which would cause the child to be mentally retarded. The first such diagnosis was made in 1968 by Dr. Carlo Valenti in New York. Testing of the fetus for genetic disease is now widespread, particularly for women over age 35 (who have a heightened risk of carrying a child with Down syndrome) and parents with a family history of genetic problems. Today, hundreds of hereditary diseases can be diagnosed through amniocentesis.

At first, amniocentesis was done ''blind.'' The practitioner relied only on external feel to guide the needle into the uterus while avoiding the placenta, the fetus, and the umbilical cord. By the 1980s, **ultrasound devices** were routinely used to guide the needle visually within the womb. The development of amniocentesis has made it possible for prospective parents to opt for therapeutic abortion in cases where they know the fetus is afflicted by severe mental or physical defects. Diagnosis of fetal disorders has also made it possible to treat some of those disorders in the uterus, before the baby is born. Amniocentesis can also reveal the maturity of a fetus, especially important when early delivery may be necessary. For example, when amniocentesis shows that the fetal lungs are not mature enough to function properly after birth, a hormone can be injected into the fetus to help the lungs mature.

Amniocentesis carries with it the risk spontaneous abortion, variously estimated to be from 0.5 to 1%. In most instances, amniocentesis is conducted during the sixteenth or eighteenth week of gestation. Analysis takes 10 to 21 days, so diagnosis of fetal abnormalities isn't available until the twentieth or twenty-first week of pregnancy. *Chorionic villus sampling*, an alternative method of fetal diagnosis, can be done much earlier, but carries a higher risk of causing spontaneous abortion (miscarriage). Earlier amniocentesis also carries a higher risk to the fetus: A three-year study of amniocentesis performed at 11 or 12 weeks showed an increase in fetal loss of 7.6%, and a higher incidence of ''club foot'' deformity. This study resulted in wide-spread cessation of early testing across the United States and Canada

In August 1997, *The Lancet* reported that researchers from Britain, Italy, and Hong Kong discovered samples of fetal DNA in maternal blood plasma, holding promise of a safer and more effective method of diagnosing prenatal genetic abnormalities. *Pre-implantation genetic diagnosis* (PGD), a procedure in which a woman's eggs are fertilized *in vitro* and checked for genetic defect before implantation into the uterus, allows for the earliest detection of genetic fetal defects.

See also Prenatal diagnostic techniques; Ultrasound devices

Amontons, Guillaume (1666-1705)
French physicist

Amontons was one of the earliest scientists to develop improved scientific instruments for measuring temperature and pressure. Born in Paris in 1663, Amontons became deaf at a very early age. This apparent tragedy served to steer his interests toward books and academia, and later in life he was said to have been thankful for the concentration his deafness provided him. As a youth Amontons attempted to construct a perpetual motion machine, a fruitless attempt that nevertheless solidified his interest in science and mechanics. After working on several public works projects, Amontons applied his skills to inventing.

One of his first major projects was the invention in 1687 of an improved **hygrometer**, a device used to measure humidity and which consisted of a mercury-filled ball that expanded or contracted according to the air's water content. Just a year later he constructed a **barometer**.

Beginning in 1695 Amontons worked on several instruments to be used on ships. Many devices of the age relied upon alcohol, mercury, or other liquids to provide a reading; unfortunately, these liquid-based instruments were thrown off by a ship's constant pitching. Several authors have attributed the invention of a fixed-volume air **thermometer** to Amontons. However, there is some doubt as to whether Amontons truly constructed such a device, and the invention is credited to Grand Duke Ferdinand II of Tuscany (1610-1670). What *is* clear is that Amontons did invent a pressure-independent air thermometer as well as a cisternless barometer, both designed for shipboard use.

Another area of controversy concerning Amonton's life is centered on the theory of an *absolute zero temperature*. In 1699 Amontons published a series of papers in which he discussed the effect of low temperature upon gas volumes. While he may have considered the possibility of a temperature so low that gas would contract into nothingness, there is little evidence within his papers that Amontons authored the concept of absolute zero. It is quite possible, though, that his work inspired the research of German physicist and mathematician Johann Heinrich Lambert (1728-1777) and William Thomson, a nineteenth-century Irish mathematician and physicist who invented the absolute scale, or Kelvin scale, in 1848.

Amplifiers

Amplifiers are not only a fundamental component of **radio**s, **television**s and **telephone**s, they are essential to all modern **electronics**. Amplifiers differ considerably in design and in the amount of amplification (called gain) they produce, but they all work in much the same way.

The evolution of the amplifier began with **Thomas Alva Edison** who, in the process of studying his **direct current light bulb**s, inserted a metal plate near the filament. He discovered that electricity would flow from the positive side to the plate, **diode**, but treated it as curiosity.

John Ambrose Fleming, who worked for Edison, modified the diode in 1904 to detect radio waves. The Englishman called his invention the thermionic valve because it controlled the flow of water, but in the United States, it was called a vacuum tube, which better described its construction.

Illustration from the early twentieth century of two men using a crude amplifier.

In 1906, American scientist Lee de Forest added a third element, called a *grid*, to Fleming's invention. De Forest discovered the device, which he called an ''audion'' (now known as a triode), made a superior radio wave detector. What he did not realize was his audion set up an electrical current that could be amplified considerably.

In 1912, **Edwin Howard Armstrong** discovered the amplifying capabilities of the audions by linking several of them together. He applied for a patent for his regenerative circuit and was sued by de Forest, who claimed the invention was his. The case dragged on for twenty years; the courts initially found in favor of Armstrong, but de Forest convinced the Supreme Court to ooverturn the findings.

Basically, the amplifier uses two circuits: a weak circuit in one part of the tube and a stronger current in another part. The flow of the weak current induces the flow of the strong current. When the weak current is modulated, it passes the modulation to the stronger current, which reproduces it at a much higher power. The stronger current is then sent into a loudspeaker where the signal is converted into sound waves.

Today transistors serve the same function as vacuum tubes. In a **transistor** the weak current flows between the emitter and the base, which induces a strong current to flow between the emitter and the collector. Because the emitter is shared by both the weak and strong current, transistors have only three terminals, as oppposed to four in a vacuum tube.

Triodes have a major flaw; they amplify distortion as well as audio. In 1923 Harold S. Black discovered that he could subtract the amplitude of the output signal from that of the input signal and cancel both signals. That left just the distortion which could be amplified, fed back into the system, and used to cancel out the original distortion. This ''feedback-feedforward'' system did not completely eliminate distortion, but it reduced it considerably. Three years later he discovered that by taking the the output signal of the amplifier and feeding it back into the system out of phase (negative feedback), he could obtain nearly any amount of distortion reduction.

Amplifiers today are used in devices never imagined by the early inventors: compact disk players, tape recorders, radar, digital computers, servomechanisms, and electronic musical instruments are only a few of the inventions that depend upon amplifiers for their operation.

See also Electric circuit

Amplitude modulation

Amplitude modulation (AM) was the invention that made **radio** broadcasting of voice and music possible.

In its infancy, radio waves were used to send nothing more than **Morse code** signals. In fact, the waves themselves have nothing to do with sound. Radio waves are a form of electromagnetic radiation that make an excellent ''carrier'' of audio signals, first used for Morse code.

Guglielmo Marconi, using an intermittent spark-pulse radio wave **generator**, amazed the world in 1901 by sending Morse code 2,137 miles (3,440 km) from England to Newfoundland. In that same year, **Reginald Aubrey Fessenden** invented a high-frequency **alternator** which produced a continuous radio wave. It was a great improvement over Marconi's intermittent generator and was essential to the development of amplitude modulation.

Fessenden invented the way to modulate the amplitude (the distance from ''peak'' to ''trough'') of radio waves. A **microphone** was used to convert ordinary sound into an electric signal which was superimposed on the continuous radio waves. The modulated radio waves, now matching the amplitude of the electric signal, were sent out over the transmitting **antenna**.

At the receiving antenna, the radio waves are converted back into electric signals and demodulated; the radio wave carrier is removed, leaving behind the audio signal that is amplified and sent to a **loudspeaker**.

On Christmas Eve, 1906, Fessenden made the world's first radio broadcast. He transmitted his voice, as well as violin and recorded music a distance of several hundred miles to ships on the Atlantic Ocean.

Amplitude modulation was a great improvement in radio transmission, but it has limitations; interference caused by the earth's ionosphere is a big problem. This layer of charged particles, theorized in 1902 by Arthur Kennelly (1861-1939) and Oliver Heaviside (1850-1925), and discovered in 1924 by Edward Appleton (1892-1965), was a double-edged sword; it reflected Marconi's signal, making his trans-Atlantic message possible, but it also interfered with the modulated signals. Changes in the ionosphere at night alter its reflecting characteristics; some AM signals fade, while others travel greater distances to interfere with other AM signals. In addition, lightning, a strong source of electromagnetic radiation, breaks up the modulated radio signal, causing a great deal of static. These problems were solved with **Edwin Howard Armstrong**'s invention of **frequency modulation** (FM) broadcasting in 1924, but he traded one set of problems for another.

Amputation

Amputation—cutting off all or part of a leg or arm—has been practiced since earliest times, but only out of desperation in cases of crushed or mangled limbs. It was a desperate measure because the patient was highly likely to die from bleeding after the amputation was performed. From the time of Hippocrates (460-370 B.C.) until the 1500s, the reluctant amputations that were performed usually cut through dead rather than living tissue because the dead tissue did not hemorrhage. Stumps were then cauterized with red-hot irons or boiling oil or tar, which stopped most bleeding and was also considered to help prevent rotting. The German surgeon Fabricius Hildanus (Wilhelm Fabry; 1560-1634) used a red-hot **knife** for amputations, which accomplished both removal of the limb and control of bleeding at the same time.

Cauterization was, of course, terribly painful to the unfortunate patient. A giant step forward in amputation was made by the French surgeon **Ambroise Paré** in the 1500s. Paré, an unschooled provincial, gained his surgical knowledge in service on the battlefield. By Paré's time, **gunpowder** had made battlefield injuries so devastating that amputation was now commonplace, at least for soldiers. Even amputation at the thigh, which previously had been very rare because of the extremely heavy (usually fatal) bleeding, was now often necessary.

Paré's great improvement in amputation surgery was ligature—tying off of the blood vessels rather than cauterizing. Earlier surgeons, such as Celsus (30 B.C. -50 A.D.), Avicenna (980-1037), Guy de Chauliac (1300-1368), and Giovanni de Vigo, had advocated ligature, but it was Paré who developed a successful technique to carry it out. He also devised a curved instrument he called a crow's beak to draw out the severed blood vessels. Paré began using ligatures in 1552 and described the technique in his *Ten Books* in 1564.

Although Paré's method was effective, it was impractical because of the large number of blood vessels involved in major amputations. A method of controlling bleeding until the surgeon could tie off all the vessels was needed. This control was finally provided by the effective tourniquet designed by J. L. Petit (1647-1750) in 1718. Earlier versions of the tourniquet had been used by, among others, Fabricius and then Morel in the 1600s. Petit's screw tourniquet was fixed to the lower abdomen and put direct pressure on the main artery.

With bleeding controlled by Petit's tourniquet, Paré's ligatures were now practical. Amputations on the battlefield were carried out swiftly and in great number. The French surgeon Dominique-Jean Larrey performed 200 amputations in one day during the Battle of Borodino in 1812. Unfortunately, while patients no longer died routinely of bleeding during an amputation, many—often most—died of infections afterward. It remained for Joseph Lister (1811-1886) to introduce antiseptics for amputation to become a successful procedure, and as modern physicians learned new, effective ways to treat illnesses and infections, amputation steadily became less necessary.

See also Surgical instruments

Anchor

An anchor is a device attached to a boat or ship and cast overboard that keeps the vessel in a particular place by digging into the bottom. The word anchor comes from the Greek word meaning hook or crook. Prehistoric anchors held vessels in place simply by their great weight and bottom friction, and therefore consisted of a single, large stone or a basketful of stones. Sacks of sand also were sometimes used as were logs of wood loaded with lead. All of these primitive types are called deadweight anchors.

During ancient or Graeco-Roman times, anchors were made of iron and were given teeth or "flukes" to fasten themselves to the bottom. It is known that during Caesar's time a variety of anchor types and sizes were carried when extensive seafaring voyages were made. Each had its own particular use for certain conditions. During the Middle Ages, little or no improvements on anchors were made, and some knowledge was actually lost. Until the early 1700s, anchors were poorly made, mainly because of bad welding and poor quality iron that was brittle. By 1770, the entire anchor was made of iron, with the old wooden stock replaced by iron. The stock is the crossbar perpendicular to the anchor's shank or vertical bar. In 1813, an Englishman named Pering introduced curved arms, and by 1852, the British Admiralty used what is recognized as the traditionally-shaped, curved anchors on all its ships. This became the classic stock anchor that we all know today and which sometimes serves as the symbol of hope. These hook-type anchors were used on all types of sea floors, but they were not very efficient since little of the anchor itself actually penetrated into the bottom soil. Also, as vessels, especially warships, became progressively larger and heavier, anchors grew proportionally, and they soon became difficult both to handle and to stow on board.

Woodcut dating from 1517 depicting a leg amputation.

The Admiralty eventually switched from hemp to chain cable to attach its anchors and soon adopted the new stockless anchor called Trotman's or Rodger's anchor. This type had arms that could pivot and thus would dig very well into the bottom at the most effective angle when dragged. Stockless anchors eventually came to be used on most large ships, and they soon developed pivoting flukes (arrowhead-like projections on the ends of the curved arms). A great advantage of the stockless anchor was that it could be pulled up into a hole in the ship (the hawse pipe) until only its arms were protruding, and remain there always ready to be used. Other specialized anchor shapes that have become popular are the CQR or plow anchor with a single fluke; the grapnel anchor which resembles four or five fishhooks clustered together; and the mushroom anchor shaped like an upside-down mushroom. The effectiveness of an anchor depends mainly on the nature of the ocean bottom to which its attaches itself. For a mud bottom, a mushroom anchor is best; for a rocky bottom, a grapnel serves the purpose. Finally, the size of the vessel being held determines the size and type of anchor. A vessel the size of an aircraft carrier would need an anchor weighing more than 60,000 pounds (27,200 kilograms).

Anemometer

Anemometers are devices which measure the velocity of the wind. The type most commonly in use today consists of three or four cups radiating from the top of a spindle; the cups catch the wind from any direction. The wind speed is either displayed on a dial at the base of the spindle or is recorded at a remote location on a **clock**-driven drum (anemograph).

The earliest anemometers date from the 1400s and consisted of a swinging plate suspended from an axis. The plate would swing vertically in a quarter-circle across a fixed indicator as the wind would blow against it.

Other types of anemometers developed were the normal-plate anemometer which used a spring-balanced plate. There was also the pressure-tube anemometer—wind blowing into the lower open end of the tube would move mercury upward in the closed end, which had indicating marks much like a thermometer. A type of anemometer employing this principle but using water was made by Dr. James Lind of England in 1776.

Robert Hooke, an English scientist, developed the first practical recording anemometer with a rotating drum in 1644. It was part of a ''weather clock'' which also recorded the temperature, rainfall, humidity, and barometric pressure.

All anemometers, old and current, share a common principle: an indicator reacting physically or electronically to a mechanical device subjected to pressure from air movement.

See also Weather forecasting methods

Angioplasty, balloon

Balloon angioplasty is an invasive surgical procedure to relieve a condition called *atherosclerosis* in which the arteries become clogged by deposits of fatty material called *plaque*, severely restricting blood circulation to the heart, brain, or extremities. In severe cases, the end result is heart attack, stroke, or the amputation of a limb due to gangrene. By inserting a balloon-tipped **catheter** into a blocked artery and inflating the balloon, plaque is compressed against the inner vessel walls allowing blood to pass freely.

In 1929, German physician Werner Forssmann (1904-1979) devised a way to thread a catheter—a long, thin, flexible tube—through a vein from an arm into the heart. In 1958, Mason Sones invented a way to visualize individual coronary arteries, a technique called selective **coronary arteriography**. In 1964 Charles T. Dotter and Melvin Judkins of the University of Oregon combined these advances to successfully perform *transluminal* (along the lumen, or cavity, of a blood vessel) *angioplasty* (blood vessel repair) using a fluoroscope to guide a catheter to the artery and dilate the blocked area. At the University Hospital in Zurich, Switzerland, Andreas Gruentzig began investigating ways of adapting transluminal angioplasty to clear blockage in the relatively small coronary arteries by adding a tiny inflatable **balloon** to the tip of the catheter. On September 16, 1977, he performed the first coronary balloon angioplasty on a human patient, a thirty-six-year-old insurance salesman, avoiding the need for coronary **bypass surgery**.

With refinements in technique and technology, *Percutaneous Transluminal Coronary Angioplasty* (PTCA), a procedure in which the catheter is inserted via the arteries and the aorta into a coronary artery, rapidly came into widespread use around the world as a relatively simple, inexpensive, and safe alternative to **open-heart surgery** for coronary bypass, dramatically improving the quality of life for many patients suffering from angina (heart pain). However, results are temporary due to *restenosis*, or renarrowing of the arteries, which occurs in 20-50% of cases, often within weeks or months of the procedure. *Stents* are often inserted to hold the vessel open.

A technique pioneered by several groups of researchers in the early 1980s uses **laser**s, introduced by catheter, to vaporize plaque in arteries followed by balloon angioplasty. Laser angioplasty is currently approved for use in leg arteries only, as it carries a significant risk of perforating the blood vessels being treated. **Ultrasound devices** are often used to remove plaque, and device called a ''Rotoblator''—an extremely thin catheter wire tipped with a miniature diamond-coated drill which revolves at 190,000 rpm—is used in severe cases. Phase II clinical trials scheduled to begin in 1998 will test the AngioRad(TM) Gamma System, a procedure designed to reduce the risk restenosis. During this procedure—the first to focus on in-stent restenosis—a tiny radioactive wire is inserted into the artery or stent, which appears to inhibit growth of smooth muscle cells, lowering the incidence of restenosis.

Animal breeding

Selectivity is one of nature's ways of ensuring the health and survival of plant and animal species. For example, when an unhealthy chick is evicted from its nest or when a healthy pine tree crowds out others that are less vigorous in the competition for light and soil, natural selection is taking place for the good of the species. Human beings, whether knowingly or unknowingly, have been a part of this process for ten thousand years. By selecting the plumpest seeds for planting and the healthiest animals from a litter of puppies, people involve themselves in the natural selection process.

The domestication and breeding of animals began around 9000 B.C. It is even possible that humanity's close relationship with dogs began at the end of the Ice Age. Animal breeding represents a deliberate effort to induce specific traits beneficial to man. Breeding weeds out undesirable characteristics and channels the desirable genes into future generations. The intent may be to keep a genetic strain purebred, as with pedigrees, or it may be to crossbreed, producing trait combinations which, if left to chance, may never have occurred. Crossbreeding is practiced among different species and among groups within species. For instance, a male donkey can be crossbred with a female horse, or mare, to produce a mule. Mules possess the strength of the donkey and the agility and temperament of the horse. Mules, like most hybrids, are sterile, incapable of reproducing. Selection is done to mass-distribute a desired trait among a stock of animals. The greatest example of successful selectivity is in the dairy industry. The high demands placed on dairymen for cleanliness and product quality and consistency have resulted in high standards in testing and record-keeping, both of which are essential for higher gains in selection efforts. This was made possible by the establishment of strong dairy associations in the early 1900s.

One of the earliest methods of human intervention in animal reproduction was **artificial insemination**. The Arabs practiced it as early as the 1320s. In 1420, French monk Dom Pinchon attempted the artificial fertilization of fish eggs, and in 1780, Italian physiologist Lazzaro Spallanzani (1729-1799) experimented with artificial insemination to obtain puppies. English biologist Robert Bakewell (1725-1795), the founder of the science of animal breeding, developed several new breeds of livestock in the late 1700s. Only within the last two hundred years has science developed an understanding of the hereditary process. The result of this new knowledge has been the direct involvement, especially in the last twenty years, with gene structures, the mechanics of heredity. An accurate understanding of the principles of heredity began with the research of Austrian monk and botanist Gregor Mendel (1822-1884), considered the father of genetics. His experiments with garden peas established the existence of the paired hereditary units he called genes. His findings were relevant to plants, animals and humans alike. Mendel's work coupled with the theories of Charles Darwin (1809-1882) formed the basis of genetics research during the twentieth century; however, Mendel's laws went unnoticed until they were rediscovered in 1901. Research during the 1920s and 1930s by American Jay L. Lush applied Mendel's work directly to animal breeding. Statistical research was conducted by American geneticist Sewall Wright on evolution theory and by the team of C. R. Henderson and Alan Robertson.

The discovery in 1953 of deoxyribonucleic acid (DNA) by American biologist James D. Watson (1928-) and English biologist Francis Clark has led to the new field of **genetic engineering**. DNA, the ''stuff of life,'' is the spiral-ladder-shaped structure within every living cell that determines the genetic makeup of every individual. Genetic engineering of the 1980s and 1990s was successful enough in plants to allow for the preliminary marketing of gene-altered tomatoes and potatoes. Desired traits are achieved by the direct injection of genes into plant and animal cell nuclei.

Animal genetics has been marked by a low success rate although the birth in early 1997 of Dolly, a sheep cloned at the Roslin Institute near Edinburgh, Scotland, may change that outlook. Dolly was cloned from adult cells, a follow up to an earlier experiment in which a pair of lambs were cloned from fetal fibroblast cells in 1996. Ethical debates concerning this process will undoubtedly continue for a long time to come, particularly as issues of human cloning proliferate.

Animation

Animation is a series of still drawings that, when viewed in rapid succession, gives the impression of a moving picture. The word animation derives from the Latin words *anima* meaning life, and *animare* meaning to breathe life into. Throughout history, people have employed various techniques to give the impression of moving pictures. Cave drawings depicted animals with their legs overlapping so that they appeared to be running. The properties of animation can be seen in Asian puppet shows, Greek bas-relief, Egyptian funeral paintings, medieval stained glass, and modern comic strips.

In 1640, a Jesuit monk named Althanasius Kircher invented a ''magic lantern'' that projected enlarged drawings on a wall. A fellow Jesuit, Gaspar Schott, developed this idea further by creating a straight strip of pictures, a sort of early filmstrip, that could be pulled across the lantern's lens. Schott further modified the lantern until it became a revolving disk. A century later, in 1736, a Dutch scientist named Pieter Van Musschenbroek created a series of drawings of windmill vanes that, when projected in rapid succession, gave the illusion of the windmill circling around and around.

The magic lantern became a popular form of entertainment. Traveling entertainers, visiting the villages and towns of Europe, included it in their shows. In London, the Swiss-born physician and scholar Peter Mark Roget, most famous for compiling the *Thesaurus of English Words and Phrases,* was fascinated by the scientific phenomenon at play and wrote an essay entitled ''Persistence of Vision with Regard to Moving Objects'' that was widely read and used as a basis for subsequent inventions. One of the first was the thaumatrope, developed in the 1820s by John Paris, also an English doctor. The thaumatrope was simply a small disk with a different image drawn on either side. Strings were knotted onto two edges so that the disk could be spun. As the disk twirled around, the two images appeared to blend. For example, a monkey on one side appeared to sit inside the cage on the opposite side.

The next major innovation was the phenakistoscope, created by Joseph Plateau, a Belgian physicist and doctor. Plateau's contribution was a flat disk perforated with evenly spaced slots. Figures were drawn around the edges, depicting successive movements. A stick attached to the back allowed the disk to be held at eye level in front of a mirror. The viewer then spun the disk and watched the reflection of the figures pass through the slits, once again giving the illusion of movement.

In Austria, Simon Ritter von Stampfer was toying with the same idea and called his invention a stroboscope. A number of other scopes followed, culminating in the zoetrope, created by William Horner. The zoetrope was a drum-shaped cylinder that was open at the top with slits placed at regularly spaced intervals. A paper strip with a series of drawings could be inserted inside the drum, so that when it was spun the images appeared to move.

By 1845, Baron Franz von Uchatius invented the first movie projector. Images painted on glass were passed in front of the projected light. Forty-three years later, **George Eastman** introduced celluloid film, a strip of cellulose acetate coated with a light-sensitive emulsion that retained and projected images better than those painted on glass. The first animated cartoon *Humorous Phases of Funny Faces* by J. Stuart Blackton, of the *New York Evening World,* was shown in the United States in 1906. Two years later, French animator Emile Cohl followed suit with *Phantasmagorie*. Winsor McCay introduced *Gertie the Dinosaur* in 1911. Other cartoonists who brought their characters to the screen included George McManus (Maggie and Jiggs) and Max Fleischer (Betty Boop and Popeye). By 1923, Walt Disney, the world's most famous animator, began turning children's stories into animated cartoons. Mickey Mouse was introduced in *Steamboat Willie* in 1928. Disney's first animated full-length film, *Snow White and the Seven Dwarfs*, debuted in 1937.

Yellow Submarine, a 1968 animated film starring the Beatles, featured the process of pixilation in which live people are photographed in stop-motion to give the illusion of humanly-impossible movements. In the film *The Lord of the Rings*, directed in 1978 by Ralph Bakshi using rotoscoping, live action was filmed first. Then each frame was traced and colored to create a series of animation cels. By the late twentieth century, many in the industry were experimenting with computer technology to create animation. In 1995, John Lassiter directed *Toy Story*, the first feature film created entirely with **computer** animation.

Antenna

An antenna is one of the fundamental parts of a **radio**; it is not only needed to receive radio waves, but to transmit them as well.

In 1887 Heinrich Hertz (1857-1894) discovered that electromagnetic radio waves were produced by the oscillations of alternating current in a **wire**. Using a simple dipole antenna (two metal plates connected to a rod), he was able to detect what he called ''Hertzian waves.'' His Hertzian dipole antenna allowed him to measure both the shape and intensity of the invisible waves. The rate of the oscillation of the current that produces radio waves, known as the frequency, is now measured in kilohertz or megahertz in honor of Hertz's discovery.

A transmitting antenna converts the oscillating electric current into radio waves, whose frequency is identical to that of the original electrical oscillations. The waves radiate outward from the antenna in all directions like ripples on water. When intercepted by a receiving antenna, the oscillation of the waves sets up a weak electric current which exactly matches the frequency of the waves transmitted. This weak electric signal can be increased with an **amplifier**.

In 1889 Édouard Branly (1844-1940) discovered that metal filings in a container cohered in the presence of Hertzian waves. While not a true antenna, this was an early radio wave detector. Immediately on his heels came **Oliver Lodge** who improved the ''coherer,'' making it a more efficient detector. Lodge was an early proponent of ''tuning'' frequencies to obtain better results and devised a resonant antenna circuit. Russian physicist **Aleksandr Popov** added a wire to the coherer, thereby inventing the first ''modern'' antenna, and was able to detect waves over the ''immense'' distance of 262 ft (82 m).

The next advancement in antenna technology was invented by **Guglielmo Marconi**. He discovered that adding a ground wire to his transmitter and receiver allowed him to send radio waves over greater distances. He also discovered he could focus radio waves into beams by putting sheets of metal around his antenna. On December 11, 1901, Marconi amazed the world when he successfully transmitted a **Morse code** signal 2,137 mi (3,440 km) from England to Newfoundland. His receiver used a single wire (a monopole antenna) 515 ft. (157 m) long, flown aloft by a kite. After Marconi's success, radio (and later **television**) technology advanced rapidly.

Depending on the type of signal being broadcast, different types of antennae are needed. A directional antenna has a horizontal element on which short cross beams are attached. When aimed at the transmitter, the cross beams receive and radiate the signal. A reflector at the end of the antenna directs the radiated signal back into the cross beams, strengthening the original signal. This helps eliminate interference caused by other nearby sources. In Japan, Dr. Hidetsugu Yagi and Dr. S. Uda invented a Very High Frequency (VHF) antenna in 1926 that was eventually put into use for television reception.

In 1928 Karl Jansky, a radio engineer at Bell Telephone Laboratories, built an antenna designed to detect radio signals that were causing interference with radio-**telephone** calls over the Atlantic Ocean. His discovery set the foundation of radio astronomy. **Grote Reber** built the world's first **radio telescope** in 1937.

Andre G. Clavier established a microwave link between two stations in New Jersey in 1930 using a 10 ft (3 m) parabolic dish antenna. In the following year, he began transmitting across the English Channel.

With the advent of direct **satellite** broadcasting, parabolic ''dish'' antennae have sprouted in many suburban and rural areas. They function very much like radio telescopes. Developments such as Digital satellite systems and Satellite Master Antenna Television have built on this technology.

Mobile technology has necessitated a significantly increased dependence on antennae; cellular phone antennae now dot much of the nation's open landscape and devices such as beepers utilize satellite signals. The preponderance of cellular phone antennae in populated areas has generated some controversy, as some observers fear the microwaves radiating from cellular phone antennae may be cancer-causing. Thus, residents sometimes oppose the installation of a mobile phone antennae near their homes.

Antiballistic missile • See Rocket and missile

Apgar score

Until the early 1950s, physicians had no reliable means of assessing the health of newborns in the critical first minutes of life. As a result of delays in diagnosis, conditions that might have been corrected sometimes proved fatal.

In 1952, Virginia Apgar (1909-1974), a physician then affiliated with Columbia-Presbyterian Medical Center in New York, introduced a scoring system that became the standard instrument for immediate evaluation of newborns. Known as the Apgar score, the test is administered one minute after birth and again five minutes after birth. A rating of 0, 1, or 2 is given in each of these five categories: ''A''—Activity (muscle tone), ''P''—Pulse; ''G''—Grimace (reflex irritability or response to stimulation); ''A''— Appearance (skin color); and ''R''— Respiration. For example, in the Appearance (color) category, a baby who possesses a healthy skin tone receives 2 points, whereas a bluish infant receives only 1 or 0 points. The highest possible total score is 10. It is not unusual for infants to score 7 at one minute of age and 9 or 10 at five minutes of age; by this time, they generally have a healthier skin tone and are breathing better.

Armed with information provided by the Apgar score, medical personnel can take immediate measures to ensure a newborn's survival. An initial score of 3 or lower is a signal that the baby's condition is critical and requires urgent attention; a score of 7 or higher signifies that all is well. Studies of the extended Apgar score (the five-minute recheck) have shown it to be an accurate indicator of an infant's chances for survival and normalcy.

Apgar, one of the first female graduates of Columbia University's College of Physicians and Surgeons and the first woman ever to hold a full professorship there, invented this system after years of studying the effects of anesthesia in childbirth. Anesthesia is one of the factors that can suppress the Apgar score; others include lack of oxygen, inhalation of amniotic fluid, and maternal drug use.

Appleby, John (1840-1917)

American inventor

John Appleby was a clever inventor who, with patience and perseverance, turned an idea he conceived as a young man into a profitable business later in life. He was born in 1840 in Westmoreland, New York, the son of English immigrants. In 1845 his family moved to Walworth County, Wisconsin, where they established a farm homestead. It was in this frontier environment that Appleby, as a boy, had to rely upon his inventiveness to make his own toys. After his father's death in 1849, he shared the responsibility of supporting his family by working on neighbors' farms as well as on their own. In doing so, he showed a great acumen for repairing farm tools and equipment.

When he was only fifteen, Appleby witnessed a demonstration of a recently invented grain-cutting machine. He immediately saw a major shortcoming in its function. It could mechanically cut grain, but the binding operation—tying the grain into bundles—still required slow, backbreaking manual labor. He suggested that a device was needed to mechanically tie knots and automatically bind the grain. Those attending the demonstration scoffed at him as much for his idea as for his age.

Appleby knew that such a device could be created. In 1859, before he was nineteen, the concept for such a knot-tying device struck him while he was working in the field. So vivid was his mental picture of it that he stopped his chores, got out his penknife, and fashioned a model of the hooked device from a piece of apple wood.

He was aware that he needed capital to convert his idea into a working model. To go around simply showing his idea to people could lead to its falling into the hands of dishonest entrepreneurs, so he put his wood carving in the attic, where it remained for many years. In the intervening time, many other new farm machines were invented, yet no one was able to invent a workable twine binder.

During the Civil War, while serving in the Union Army trenches at Vicksburg, Mississippi, Appleby invented a magazine and automatic **ammunition** feeder for **rifles**. In 1864, he sold this invention for $500; it was resold for $7,000, but Appleby still made enough money to launch his farm machine venture.

During the 1870s, machines were being introduced that bound grain with **wire**. However, small bits of wire fell into the sheaves of grain, causing animals to die from eating it and farm machinery to break down when wire became caught in their mechanisms. Appleby knew that twine was the answer. He organized the Appleby Reaper Works in 1874 at Mazomanie, Wisconsin. After persistent experimenting, he patented the Appleby Knotter in 1878 and a binder in 1879.

In 1881, Appleby sold his invention to **Cyrus McCormick** for $35,000. **Twine binders** completely replaced wire binders by 1882, and twine binders using Appleby's principle continue to be used on 90 percent of the grain harvested today.

See also Reaper and binder; Combine harvester

Aqueduct

Aqueducts are structures used to carry water from a supply source to distant areas in need of water. The word aqueduct comes from two Latin words, aqua (water) and ducere (to lead). The first aqueducts were built as early as the tenth century B.C. by ancient communities. As people moved inland and away from direct water supplies, they created systems of water retrieval. Wells were dug to reach underground water supplies. Also, cisterns, underground collecting tanks, were used to store water. Eventually, dams were constructed to block water flow, allowing water pressure to increase, and run-off channels were constructed to guide water to specific regions. Early aqueducts redirected water by use of conduits (covered canals, or pipes, usually made of stone) that were often buried a few inches below ground for protection. Aqueducts were driven by the force of gravity pulling water downhill and extended for many miles. The use of aqueducts for drinking water, agriculture, and other uses is a part of Greek, Mexican, Roman, and Asian history. The city of ancient Rome had 11 active aqueducts traversing roughly 300 mi (485 km). Modern aqueducts use electrical power to elevate water that travels through many miles of pipes. Many modern pipes are deep beneath the ground and supply cities with water for personal and industrial use.

The first record of an aqueduct appeared in 691 B.C. in Assyria. This 34-mi (55-km) long aqueduct was fairly simple, consisting of a single arch over one valley. At that time, Greeks were using wells to retrieve water from underground pools. Certain plants, such as fig trees, marked water sources because their roots grow in water. The first Greek aqueduct followed in 530 B.C.. on the island of Samos. This aqueduct was built by an engineer named Eupalinus, who was told to supply the city with water by tunneling a pathway through a mountain. The Samos aqueduct extended for about 1 mi (1.6 km) underground, and had a diameter of 8 ft (2.4 m). These first aqueducts demonstrated an understanding of siphons and other basic hydraulic principles.

While ancient Roman aqueducts evolved into an extensive network of canals supplying the city, the first one, the Aqua Appia, was not built until 312 B.C. This aqueduct was a simple subterranean covered ditch. Roman aqueducts were usually built as open troughs, covered with a top, and then covered with soil. They were made from a variety of materials including masonry, lead, terra cotta, and wood. The Appia was about 50 ft (15 m) underground to make it inaccessible to Roman enemies on the city's outskirts. The Anio Vetus, built in 272 B.C., brought more water to the city, but both the Appia and the Vetus had sewerlike designs. The Aqua Marcia, built in 140 B.C., was made of stone and had lofty arches. The Aqua Tepula of 125 B.C. was made from poured concrete. Later Roman aqueducts were mainly built to meet the needs of the people or the desires of the rulers of the time. The average Roman aqueduct was 10-50 mi (16-80 km) long with a 7-15 sq ft (0.7-1.4 sq m) cross-section. Aqueducts were generally wide enough for a man to enter and clean.

Around the world, communities made advances in irrigation and water management. In the Mexican Tehuacan Valley, evidence of irrigation dates back to around 700 B.C. in the remains of the Purron Dam. The dam was used to direct water to domestic and crop regions for several hundred years. In the same valley, the Xiquila Aqueduct was built around 400 A.D. Early North American aqueducts include the Potomac aqueduct in Washington, D.C. This aqueduct, which was built in 1830, extends over the Potomac River at the Key Bridge which joins Northern Virginia and the Georgetown area of the city. It was built with support from eight piers and two stone abutments to carry water from the upper Potomac to the city.

Later aqueducts of the United States include the Colorado River Aqueduct that supplies Los Angeles and the Delaware River Aqueduct that carries water into New York. In addition, aqueducts carry excess water from northern to southern California. The southwestern region of the United States is particularly dry and requires water import. Water can be collected from aquifers (underground water tables), rivers, lakes, or man-made reservoirs.

While modern water pipes are much wider (20-30 ft or 6.1-9.1 m in diameter) and significantly longer (hundreds of miles long) than the first aqueducts, the hydraulic principles governing water carriage remain essentially the same. Water flows along gradients, and its velocity depends on a number of factors. Water flows more quickly along steeper gradients, but wear and tear on such pipes is greater, resulting in the need for more frequent repair. More gradual sloping pipes result in slower-flowing water with greater sludge deposits; hence, these pipes require more cleaning with less repair.

Water velocity along conduits is also greater in larger, smoother pipes. Pipes or canals that have rough surfaces disrupt water flow, slowing it down. In addition, larger diameter passageways provide less resistance, because a smaller percentage of the flowing water is retarded by the surface friction of the conduit. Thus, smaller diameter pipes slow the flow of water compared to larger diameter pipes.

The use of water to generate other forms of power is not new at all. The ancient Greeks and Romans both used water mills for work in such places as flour factories. In such mills, aqueducts were used to supply water on a relatively continuous basis. A modern application of water power is hydroelectric power.

ARCH

Arches were originally developed as a means of providing interior support to buildings, and creating doors and windows. Their use in construction permitted longer single spans built from smaller, more easily handled units, rather than from huge separate beams.

The earliest forerunner of the arch consisted of two vertical pillars with an interlocking beam. Both pillar and beam were made of wood or stone. The Babylonians used arches as early as the sixth century B.C., as did the Egyptians, Chinese, and Greeks.

As urban centers grew and the need for larger buildings increased, larger pillars and beams were used to build temples, courts, and public baths. When it became more and more difficult to work with the enormous pillars and beams, true arches, curved overhead and made of small units stacked on top of one another, came into use. The stacking technique made massive arches possible, since blocks could be hauled in one at a time and handled more easily.

The Roman Empire initiated large projects using sophisticated archwork. Its quest to create an elaborate system of public works demanded large numbers of buildings, highways, and aqueducts. Efficiency was a key component of a project of that scale, and the Roman arch, with its characteristic keystone at the top locking the arch into place, answered that demand, holding loads firmly yet remaining simple to construct. The Romans also used **concrete** extensively, bracing arches with cross vaults where immense loads demanded it. Reliance on cross-vaulting led to the construction of **domes**—essentially three-dimensional arches.

The Roman arch, an example of form following function, was massive and perhaps a bit monotonous in appearance. By the twelfth century the Gothic arch, characterized by a point and often relying on a joint at the apex rather than a keystone, began to appear in Europe, primarily in the design of the continent's great cathedrals. The Gothic arch was part of a trend toward a greater diversity in design which employed vaulted ceilings, flying buttresses, large stained glass windows, and ledges for statuary.

Arches were used extensively in **bridge** construction, with stone arches used during the Roman Empire and again in Europe from medieval times to the present. During the 1700s and 1800s, iron replaced stone in bridge-building, and by the mid-1800s, steel arches came into use to support the increased loads created by trains and longer spans.

While modern architects seldom use arches, a few examples can be found such as the unique Gateway Arch in St. Louis. Designed by Eero Saarinen (1910-1961) and completed in 1965, the stainless steel arch stands at a height of 630 ft (220 m) alongside the Mississippi River, attracting tourists and sightseers. The world's longest arch spans are the Sidney Harbor Bridge in Australia and the Bayone Bridge over the Kill van Kull in Bayone, New Jersey, with spans of 1650 ft (502.9 m) and 1652 ft (503.5 m) respectively.

See also Concrete and cement; Construction, cast-iron and wrought-iron; Steel alloy

ARCHER, FREDERICK SCOTT • See Collodion

ARCHIMEDEAN SCREW • See Water pump

ARCHIMEDES (ca. 287 B.C.-212)

Greek mathematician and inventor

Archimedes was an ancient Greek mathematician, philosopher, and inventor. It seems, however, that he did not think as much of his numerous inventions—important and fundamental as they were—as he did of his work in the field of mathematics. He felt his mechanical *toys* were not the most important pursuit of a mathematician-philosopher.

Archimedes was born around 287 B.C. in Syracuse, a town in the Greek colony of Sicily. His father was the astronomer Phidias, and he was related to the tyrant Hieron II (308 B.C.?-216 or 215 B.C.). Archimedes went to Alexandria about 250 B.C. to study under Conon and other mathematicians who had studied under Euclid (ca. 300 B.C.). He later returned to Syracuse where he apparently stayed the rest of his life. Archimedes performed countless experiments on **screws**, **levers**, and **pulleys**. The Archimedean screw, also called a water snail, is still used in certain parts of the world to raise and move water. This screw enclosed in a cylinder created, in essence, the first **water pump**, and is perhaps his most remembered invention.

The Archimedean screw has been the basis for the creation of many other tools, such as the combine harvester and auger **drills**. His work with levers and pulleys led to the inventions of compound pulley systems and **cranes**. His compound pulleys are highlighted in a story that reports that Archimedes moved a fully-loaded ship single-handedly while seated at a distance. His crane was reportedly used in warfare during the Roman siege of his home, Syracuse.

Other wartime inventions attributed to Archimedes include rock-throwing **catapults**, grappling hooks, and **lenses** or **mirrors** that could allegedly reflect the sun's rays and cause ships to catch on fire. He invented a self-moving celestial model representing the sun, moon, and constellations, so accurate that it even showed eclipses in a time-lapse manner. This invention utilized a system of screws and pulleys that moved the globes in their various courses and speeds.

Many of Archimedes's inventions were spawned by the experiments he conducted to prove his theories. He earned the honorary title *father of experimental science* because he not only discussed and explained many basic scientific principles, but he also tested them in a process of trial and experimentation which was based upon three essential principles. The first of these principles is the idea that natural laws continue to work even with large changes in size. The second principle proposes that mechanical power can be transferred from models used in laboratory work to practical applications. The third principle states that a rational, step-by-step logic is involved in solving mechanical problems and designing equipment. Adherence to these principles led Archimedes to such inventions as block and tackle systems, the *water snail* screw, and devices for *driving* objects using axles and drums. Even today, inventors and scientists assume Archimedean principles to be basic to their fields.

Archimedes did more than create a number of useful inventions. For instance, he dealt with mathematical principles such as calculating the value of pi to figure the areas and volumes of curved surfaces and circular forms. During this process, Archimedes used methods similar to calculus, which was not to be created for almost another two thousand years. He also created a form of exponential notation to allow him to prove that nothing exists that is too large to measure. His theories in the realm of statics, particularly in the studies of gravity, balance, and equilibrium were based on experiments with levers. ("Give me a place to stand and I will move the Earth," he also is often reported to have said about the lever.)

The famous story of Archimedes death, though again apocryphal, is that a Roman soldier killed him during an invasion because Archimedes refused to leave the contemplation of the mathematical diagrams he had drawn in the dirt.

See also Balance and scale

ARC LAMP

Long before the incandescent electric light bulb was invented, arc lamps had given birth to the science of electric lighting. In the early 1800s, when the first large batteries were being built, researchers noticed that electric current would leap across a gap in a circuit, from one electrode to the other, creating a brilliant light. Sir **Humphry Davy** is credited with discovering this electric arc and inventing the first arc lamp, which used carbon electrodes.

Yet the electric arc lamp remained a curiosity for decades. Many scientists gave public demonstrations of arc lighting, and the invention of automatic controls in the 1840s made it possible for arc lamps to be used in special applications such as **lighthouses**, theaters, and **microscopes**. But arc lamps still relied on expensive batteries or **generators** as their source of power. Then a flurry of inventions brought arc lighting into widespread use. First came the development in 1871 of a relatively cheap source of electricity, the dynamo, a type of generator which produces **direct current** power. Public interest quickly reawakened, and people began installing arc lighting in factories, mills, and railway stations—any place light was required over a large, open space. France pioneered in this field, though great Britain and America soon followed.

The next step forward was the Yablochkov candle, a type of arc lamp invented in 1876 by Pavel Yablochkov (1847-1894), a Russian engineer who later moved to Paris. This device, which could burn for two hours without adjustment, eliminated the need for expensive automatic controls. Although defects soon led to its downfall, this arc lamp greatly stimulated development of electric lighting and increased the demand for better generating equipment. By this time, American scientists were active in improving and installing arc lighting systems. In 1877, a dynamo invented earlier by William Wallace and American inventor and electrician Moses Farmer (1820-1893) was adapted for arc lighting by Wallace. This was probably the first commercial arc lamp made in the United States. Around the same time, American Charles Brush (1849-1929) introduced arc lamps for street lighting. Brush's arc lamp, which used **magnets** to move the electrodes, could be lit by remote control. He also invented a way to operate multiple arc lamps from a single dynamo, which greatly improved upon the European method. In 1879, Brush demonstrated his first streetlight system in Cleveland, a success that led many other American and European cities to install Brush arc lighting. Finally, a team of two American electrical engineers, Edwin Houston (1847-1937) and **Elihu Thomson** (1853-1937), introduced an arc lighting system that wasted less electricity by maintaining a constant current. Two years later, in 1881, they patented automatic controls for the system.

At the turn of the century, after other improvements to arc lamps, spin-off technologies began to spring from the original concept. Scientists knew that electricity, when passed through certain gases at very low pressures, would discharge light, producing a glow instead of an arc. Although high voltage was needed to start the process, a much lower voltage would sustain it. An American engineer, Peter Hewitt (1861-1921), invented a starting device and developed the first *discharge light*, which used mercury vapor in a glass tube. Soon, higher pressure lamps began to be developed, using either mercury or sodium vapor. In contrast to arc lamps and incandescent light bulbs, discharge lamps deliver nearly all their energy in the form of visible light or ultraviolet rays, rather than producing large amounts of useless heat. The color of the light varies depending on the gas. Mercury gives a bluish light, which can be made to look more natural by coating the tube with phosphors, while sodium vapor light is distinctly yellow. Both types provide excellent illumination for large areas such as roadways, shopping malls, parking lots, and exhibit halls. Mercury lamps are used where the quality of light is an aesthetic concern—in a city's downtown area, for example—while sodium lights work well where visibility is more important than appearance. Metal halide lamps, a fairly recent development, produce a spectrum that is ideal for color television pickup, so they are often used in sport stadiums and athletic fields. **Fluorescent lamp**s and **neon light**s are also variations of discharge lamps.

Meanwhile, the original arc lamp has come full circle. Ironically, today's extremely powerful versions make use of the lamp's heat, rather than its light. These high-tech arc lamps, which can simulate the heat of the sun, have proved useful in testing aerospace materials and hardening metal surfaces.

See also Battery, electric

Arc searchlight

Searchlights depend on special **lenses** and reflectors to focus electric light into a pinpoint beam that can illuminate objects thousands of feet away. Since about 1870, carbon **arc lamp**s have been used as the light source for searchlights. During World War I, **Elmer Sperry**, an American engineer, invented a high-intensity arc searchlight. The United States Navy and other armed forces quickly adopted Sperry's light for military purposes.

In today's large searchlights, chemicals are added to the carbon to increase the arc light's brilliance. Like a car's headlights, searchlights focus their beam with a parabolic reflector, a curved metal cup which has the special property that it directs the light scattering from the source into a narrow stream of parallel rays. Parabolic reflectors came into use in the late 1800s. Before then, searchlights had used a special **mirror** invented by Colonel Alphonse Mangin for the French army in 1877. Some searchlights also use a **Fresnel lens** to concentrate the light beam. This type of lens, which has a surface divided into concentric rings, was originally developed in 1820 for **lighthouse**s by the French physicist **Augustin-Jean Fresnel** (1788-1827).

Argand, Aime • See Oil lamp

Arkwright, Richard (1732-1792)

English inventor

Sir Richard Arkwright is generally considered the inventor of the first automated process for spinning cotton yarn. He began with the construction of a machine that used rollers to separate the individual fibers, allowing for the spinning of sturdier yarn; this enabled weavers to construct cloth that was, for the first time, entirely made of cotton.

Arkwright was born in Lancashire, England, in 1732. As a boy he was apprenticed to a Preston barber, and at the age of 18 he opened his own shop. After several years of mild prosperity he decided to abandon his business, choosing instead to travel the country to purchase human hair for the construction of wigs. Although he had developed a secret formula for dyeing hair, the demand for wigs gradually declined, and Arkwright was forced to change professions once again.

During his travels, Arkwright had come in frequent contact with weavers and spinners. The recent invention of the

spinning jenny by **James Hargreaves** had revitalized the industry, and Arkwright recognized that other such inventions could be a source of tremendous profits. He consulted with two engineers, John Kay (1704-1764) and Thomas Highs, and in the late 1760s began experimenting with spinning machines of his own design. Arkwright and Kay rented a workshop in a Preston schoolhouse, where they constructed their first spinning machine. The doors of the shop were kept closed and the activities within were a guarded secret; this, along with the strange noises emanating from within, led the local population to believe that sorcery or witchcraft was being practiced there.

In 1768 Arkwright completed the construction of his first spinning machine. Unlike the spinning jenny—which was simply a modification of the traditional **spinning wheel**—this machine applied new technology to produce yarn of a much higher quality. Arkwright's machine used four pairs of rollers, through which the cotton was drawn. The top roller of each pair was **leather**, which enabled it to grip the cotton. The bottom roller was made of metal or wood and was fluted, allowing the cotton fibers to pass through. Each pair of rollers turned at a slightly different speed: the first set was the slowest, with the final set turning the fastest. By increasing the speed as it passed through, the cotton's fibers were drafted, or drawn apart; the fibers could then be twisted into stronger, smoother yarn.

Arkwright's machine was a vast improvement over existing spinners. However, it was driven by horses—a cumbersome and expensive method that was not easily applied on a large scale. Lacking the funds to make improvements to his design, Arkwright enlisted the partnership of Samuel Need and Jedediah Strutt, two hosiers from Derby. With their financial support he was able to convert his spinning machine to water power, and in 1771 he was granted a patent for the **water frame** spinner. The cotton yarn produced by the spinning jenny had been too weak to be used as *warp* (the strong threads used to hold a fabric together); instead, expensive **linen** threads were used as warp, with cotton threads useful only as *weft* (the threads that cross the warp).

With Arkwright's machine, cotton threads could be made strong enough to serve as both warp and weft, producing for the first time in England cloth made entirely of cotton. Despite the great demand for such material, the sale of cotton cloth was hindered by a high tax on all-cotton fabrics; this tax had been designed to restrict the importing of Indian calicoes, but Arkwright's competitors quickly applied it to his new cottons.

In 1774 Arkwright was granted an exception to the calico tax, making the production of cotton fabrics (available exclusively through the use of the water frame spinner) northern England's leading industry. At the same time, Arkwright succeeded in inventing and patenting machines responsible for every process associated with the production of yarn, including carding, drawing, and spinning. While immensely profitable, Arkwright's machines angered his competitors, who suddenly found that most of their technology was owned by Arkwright. Nevertheless they continued to use their own machines without his permission, and in 1781 Arkwright brought legal action against them.

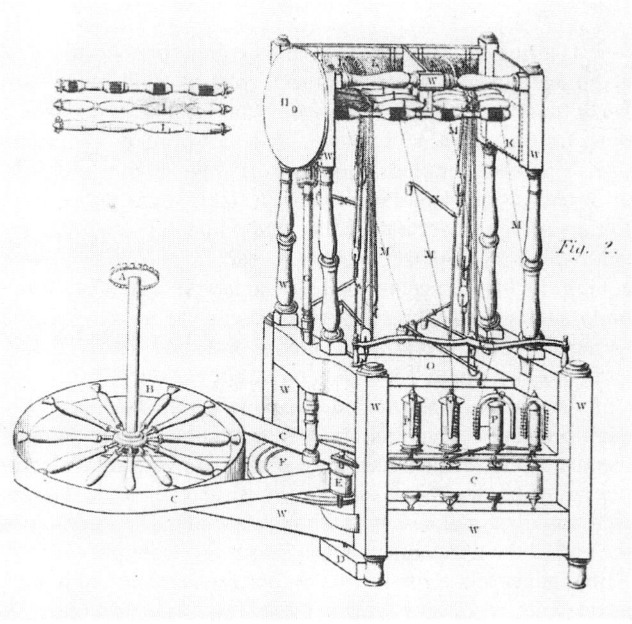

Richard Arwkright's spinning machine.

The nine firms sued by Arkwright argued that the language of his patents was too vague and claimed credit for technology that had existed for years; Arkwright, who had arranged for the language to do that very thing, was found to have no case.

In 1785 he again took legal action to enforce his patents; by that time, however, more than thirty thousand workers were using Arkwright's patented machinery. The combined support of all these manufacturers was too much for Arkwright to contend with, and in November of 1785 his patents were cancelled. Despite the loss of legal authority, Arkwright's business prospered. He was still the most powerful individual producer in the yarn industry, and for years he fixed the price of that commodity. He set up mills all over England and Scotland, and financed other inventions that would benefit the spinning industry. He was knighted by King George III in 1786.

See also Textiles

Armor

Since antiquity, soldiers have worn protective clothing while in combat. The earliest type of body armor was usually a cloak layered with leather, wood, or shells. With their advanced ability to work with metals, the Greeks made large bronze plates to cover parts of the body. The plates for the chest and back were joined by suspenders over the shoulders. In addition, they created springy metal shin guards and fancy helmets. Around 300 B.C. the Romans created the first known examples of mail, a series of interlocked metal rings. They later crafted the *lorica seamentata*, metal plates shaped to encircle the body while allowing freedom for arm movements. By the first century A.D., Roman soldiers wore iron helmets equipped with large cheek pieces and a neck guard.

During the Middle Ages, armor grew in sophistication and usage. Knights wore long metal coats of mail, called *hauberks* (the Norman term) or *byrnies* (the English term), which hung from the neck to the knees. The coat of mail often came in varying configurations: woven rings, rings sewn upon leather or cotton, overlapping scales of leather, small plates sewn to canvas, or thick cotton padded and quilted in squares. They were so heavy it took two men to carry one. Helmets were equipped with nose guards that hid the face so well it was often impossible to tell who the leader was unless he took off the helmet in battle. Other metal pieces protected the head, but, surprisingly enough, there was no armor for the legs or feet.

As more sophisticated weapons came into existence, extra protection from them became essential. During the thirteenth century, metal plates were added to cover such vulnerable points as the knees, elbows, shins, and forearms. Horses were also armored, either with mail or with quilted fabric. By the end of the thirteenth century, horses too wore plate armor. Armor advanced in Europe in the fourteenth century to protect against the increasing power of the bow. Gloves made of metal, called gauntlets, became standard, replacing earlier chain mail mittens. The well-equipped fighter now protected his entire body with plate armor. In Germany and Austria another type of body armor appeared; it had raised ridges to deflect enemy weapons and weighed nearly 60 lb (27.24 kg).

The effectiveness of body armor became obsolete following the introduction of **gunpowder** and firearms. Soldiers soon discovered that even crude guns could fire bullets capable of penetrating most armor plates. Thicker armor was not the answer, for it reduced the wearer to nearly total immobility. By the end of the seventeenth century, most European armies had abandoned the use of armor. However, it remained popular among the Indians, Persians, Turks, and Japanese.

With the advent of trench warfare during World War I, the necessity of wearing body protection in battle returned. Soldiers vulnerable to head wounds due to flying shrapnel were equipped with tin hats or steel helmets. Machine gunners and snipers wore back and breast plates in addition to helmets. Tank crews were issued metal plates for the upper face and a short curtain of mail for the mouth and chin.

Since then, soldiers have increased their reliance upon armor. Shell fragments caused enough injuries in World War II to create a need for metal reinforced garments made of steel, aluminum, or **glass** fiber plates combined with heavy **nylon**. Bomber crews wore flak jackets and bulletproof vests to protect them from enemy anti-aircraft fire. Today, body armor is manufactured with aluminum oxide and boron carbide **ceramics** in order to lessen the weight. **Kevlar** has been a successful addition to armored protection. Demolition experts are

Armor of the Middle Ages.

equipped with a special type of body armor, weighing about 48 lb (21.8 kg) and complete with a cooling system.

In the late 1990s researchers at American chemical companies made significant advances in bullet-resistant materials. They produced a highly flexible, breathable material out of aramid fiber layered with polyethylene yarn that was considered ten times stronger than steel, and so lightweight that a bulletproof vest weighed only 3 lb (1.36 kg). This material advance led to the creation of ordinary garments, including slacks and brasiers, that were highly bullet-proof. While the bulk of armored clothing is used by military and law enforcement personnel, it is available to civilians, who form a growing market.

Armored vehicles

In 1896, E. J. Pennington, an American, designed (but never built) a steam-powered armored car equipped with metal skirts and two **machine guns**, one facing forward and the other facing to the rear. Three years later, English engineer F. R. Simms took a small steam-powered, four-wheeled vehicle and added a bulletproof shield as well as a machine gun. In 1902, Simms introduced the first armored car powered by an **internal combustion engine**. His "War Car," which carried three guns, weighed over five tons, rode on four steel-tired wheels, had a boat-shaped hull of thin armor, and was capable of a top speed of 9 mi (14.48 km) per hour.

The British army began experimenting with armored car designs during World War I. They took existing cars and chassis and added open-topped bodies equipped with a machine gun that could shoot over the side armor. Trucks were fitted with **armor**, thus turning them into the first armored personnel carriers. By 1914, the British had manufactured three fully-armored vehicles (built on the chassis of Rolls Royce touring cars), complete with machine guns in turrets that could sweep around in a full circle. The future of armored vehicles appeared to be bleak, however, since wheeled armored cars could not cross trenches effectively. In October 1914, E. D. Swinton, a lieutenant-colonel in the British army, suggested tracked vehicles, an adaptation of an American tractor using the British patent of a caterpillar track. To keep the new weapon a secret from any hostile sources, the British referred to the new weapon as a tank and the name stuck. A revised tank, based on the designs of Lieutenant Wilson and William Tritton, pulled its twenty-eight ton mass along at little more than marching speed, but it carried several machine guns and two six-pound guns.

The English tank was first used in battle in September 1916. The British generals, who did not understand how to use the new weapon, sent it out onto a battlefield that had already

One of the first tanks, developed for use in World War I.

endured a three-day bombardment from 1,000 guns. It is not surprising that the tanks bogged down in the churned-up ground. At the battle of Cambrai in 1917 they were used much more effectively: 400 massed tanks penetrated the German line to a depth of four miles.

The tank's performance in battle led to further experimentation with its design. The British came out with a larger and heavier model, capable of crossing a ten-foot trench while carrying twenty soldiers. The French created a tank with a fully rotatable gun turret. The Germans, who had lagged in tank production, came up with a vehicle that weighed 148 tons and required twenty-two men to run all the equipment and weapons inside.

By World War II, armored vehicles were capable of speeds of over 30 mi (48.27 km) per hour. The Germans created a miniature underwater tank that carried two **torpedoes**. Newer tanks have developed accurate laser-control fire systems. In addition, there are now very high velocity guns with barrel sizes ranging from 4-5 in (105-125 mm).

As more and more effective anti-tank weapons were developed in the late twentieth century, including "smart" weapons with sophisticated computer devices that allowed them to probe a tank's surface for weakness, tank armor became more elaborate. Rather than becoming thicker and heavier, some tanks relied on "smart" armor using an array of high-technology defenses. American tanks developed in the late 1990s used so-called integrated defense systems, or IDS. The IDS used electronic scanners to survey the electromagnetic spectrum above and around the tank. Sensors could thus detect when weapons had been launched against the tank. Software guided by artificial intelligence then activated a variety of responses, from radar jamming to launching smoke screens and anti-missile weapons.

Lighter weight tanks were also developed in the late 1990s, using advanced materials. Tanks so light they could be air-lifted and dropped by parachute were developed for the U.S. Army, beginning in the 1980s, culminating in a 1990s tank made of aluminum, titanium and ceramics. The light tank was designed so that troops could attach additional modular armor to it once it landed.

Portable armor similar to body armor was also developed in the late 1990s for use in transport airplanes. Peacekeeping crews landing transport planes during the war in the former Yugoslavia found themselves vulnerable to rifle fire when flying at low altitudes. In response, military engineers developed armored tiles made of ceramic-metal composites bonded to Kevlar and then attached to Velcro-backed fabric. The armored tiles could be rapidly attached to the floor and walls of the plane. The advanced material was stronger than steel and considerably lighter, an issue crucial to air transport.

Armstrong, Edwin Howard (1890-1954)

American electrical engineer

An extremely innovative inventor, Edwin Armstrong unfortunately spent as much time in court fighting lawsuits as he did in the laboratory.

Born on December 18, 1890, Armstrong decided he was going to be an inventor by age fourteen. He had read of **Guglielmo Marconi** and became very interested in wireless **radio** communication. Before he was twenty he had built his own transmitter and was broadcasting radio signals. He enrolled at Columbia University, studied under Michael Pupin (1858-1935), and received a degree in electrical engineering in 1913.

In 1912 Armstrong created one of his three major inventions, the regenerative circuit. Working with the *audion*, a **vacuum tube** invented by **Lee de Forest** to detect radio signals, he discovered that feeding the tube's current back into itself enhanced the sensitivity of the tube greatly. He was able to amplify distant radio signals loudly enough to be heard without the use of headphones. He also noted that a suitably overloaded audion could be used to transmit radio waves.

Armstrong applied for a patent, and promptly wound up in the midst of a four-way lawsuit. De Forest, **Irving Langmuir** at General Electric, and Alexander Meissner in Germany all claimed priority. Langmuir and Meissner were soon eliminated from the suit, but the complex case with de Forest dragged on for twenty years. De Forest had essentially invented a **telephone** circuit that had the *potential* for radio amplification, whereas Armstrong had invented the circuit specifically for radio. The courts initially upheld Armstrong's patent, but de Forest, backed by AT&T, was able to convince the Supreme Court to overturn the original ruling. Technically knowledgeable scientists continued to credit Armstrong for the discovery and gave him numerous awards.

Before Armstrong's invention could be put to wide use, World War I intervened. Armstrong became an officer in the U. S. Signal Corps and was sent to Paris, France, where he made his second important invention. In 1917 he improved **Reginald Aubrey Fessenden**'s heterodyne technology to amplify weak signals, and created the superheterodyne circuit. Although his original hope of using it to detect enemy **aircraft** did not materialize, the circuit became the basis for 98 percent of the receivers used in radio and **television** to make them tunable with the twist of a dial.

After the war Armstrong returned to Columbia and developed the supergenerative circuit in 1920, which became widely used in police, aircraft, and amateur radio receivers. He then addressed the problem of static that was plaguing radio reception. Fessenden had invented **amplitude modulation** (AM) of radio waves in 1906, which made transmission of voice and music possible, but electromagnetic interference caused by lightning storms created bursts of static that broke up the signal. In 1933 Armstrong devised a method of modulating the frequency of the wavelengths, rather than the amplitude. **Frequency modulation** (FM), Armstrong's third major invention, produced extremely clear sound and was immune to electromagnetic interference. It also opened the door for multiplex and **stereo** broadcasting, as well as microwave relay links.

But frequency modulation had a very difficult birth. First, the depressed economy of the 1930s did not encourage major investments in new technology. The Federal Communications Commission (FCC) didn't grant his permit to initiate FM broadcasting until 1940. Then World War II came along and delayed development. Following the war, the major broadcasters did all they could to hamper development, seeing FM's potential as a competitor with their established AM stations and new television technology. The FCC was influenced to arbitrarily change the frequencies allotted to FM and limit FM signal power. When the major corporations finally got involved with FM, they began using Armstrong's patents. Armstrong filed suit against RCA and its subsidiaries for infringing on his patents. But he could bear no more. On January 31, 1954, demoralized and his health failing, Armstrong jumped to his death from his apartment in New York City. His widow continued to press 21 lawsuits, eventually won every one of them, and received a $10,000,000 settlement.

Edwin Howard Armstrong

Arteriography, coronary

Coronary arteriography, or coronary angiography, is used to detect coronary artery disease. In this procedure a special x-ray opaque dye is injected into the bloodstream through a thin plastic tube (**catheter**) which has been inserted into an artery in either the arm or groin and threaded through a vein to the arteries located inside the aorta—the body's main blood vessel which originates in the heart. Moving pictures are then taken of the heart with a special **x-ray machine**. If there are no obstructions, the dye flows smoothly through the arteries to all branches. Partial blockages appear as tiny breaks in the dye stream; complete blockages prevent the dye flowing into branches that otherwise would be visible with x-ray. Coronary arteriography is a valuable tool used by physicians to assess the extent of damage or disease in coronary arteries and determine the most appropriate treatment. Although there are certain risks associated with the procedure, when performed by specialists in a catheterization laboratory (''cath lab'') it is relatively safe.

The coronary arteriography was a technique researchers had sought to develop, with very limited success, since the 1930s. The ability to view coronary arteries was considered fundamental to the development of effective diagnosis and treatment of coronary artery disease. Early attempts at coronary arteriography were hindered by two major problems: massive amounts of *contrast agent*, or dye, had to be injected, which often caused serious side effects, and only a single radiographic plate was obtained. Significant improvements came with the introduction of serial film changers and the image intensifier in 1949, allowing true motion cinematography. The technique remained less than ideal, however, until 1958, when an accident led Mason Sones of the Cleveland Clinic to the development of selective coronary arteriography.

Sones was working in his laboratory with a huge, heavy image amplifier that required the physician to stand in a pit beneath the patient's table while an assistant on the platform above injected the contrast medium via a catheter into the aorta, a major blood vessel, rather than into the heart itself due to the danger of the heart going into ventricular fibrillation (irregular contractions that frequently cause cardiac arrest). As chance would have it, the catheter slipped, and the injection traveled into the coronary artery itself. Sones was surprised—and overwhelmingly relieved—when the patient's heart did not go into fibrillation. This episode prompted Sones to conclude that smaller, more dilute amounts of contrast could be introduced directly into individual coronary arteries, providing consistently clear imaging of selected coronary arteries. Melvin Judkins of the University of Oregon improved the technique and catheter design, and Sones's coronary arteriography became widely used.

See also Catheter; X-ray machine

Artesian well • See Water supply and drainage

Artificial bone

Bone grafts are second only to blood transfusions on America's list of transplants. Artificial bone is becoming an increasing alternative to the two most common types of bone grafts or repairs: the *autografts*, in which a piece of bone is taken from elsewhere in the patient's body (usually the pelvis), or the *allograft* bone transplanted from cadavers; the Red Cross maintains a bone bank for this purpose. A third type of bone-replacement surgery involves titanium and cobalt chromium **alloys**. Bone is very alive and constantly rebuilds itself. Its porus framework is a composition of collagen protein fibers running through *hydroxyapatite*, a mineral that makes up about

70% of living bone. Due to the fact that there is little ''spare'' bone in the body for use in autografts, possible rejection by the recipient's immune system and the risk of transmitted diseases from cadaver bones, and to find a substance that more closely resembled real bone than the metal alloys, researchers turned to the artificial production of hydroxyapatite.

In the 1960s, Marshall R. Urist, a professor of orthopedic surgery at the University of California, Los Angeles, uncovered the process of bone formation. Improved diagnostic techniques and research into the structure and composition of bone, has led to the development and FDA approval, or pending approval, of a number of synthetic calcium-based bone repair products. The first FDA approval of a synthetic bone implant (or bone void filler) was granted in 1992 to a product called *Pro Osteon*, a calcium phosphate material that mimics hydroxyapatite manufactured from coral through a thermochemical process developed in the 1970s. Because this material remains porus, it acts as a framework through which cells and blood vessels from the natural bone intertwine. This new bone growth connects both ends of the fracture and the body ultimately reabsorbs the synthetic framework. In over 20,000 procedures using this product, there have been zero rejections. Because of its porosity, however, this material lacks the strength needed for weight-bearing bones.

Richard J. Lagow, a chemist at the University of Texas at Austin, developed a way to synthesize hydroxyapatite into a porous and much stronger form suitable for bone replacement, and a denser form similar to tooth enamel. The finished product, a bioceramic called *Megagraft 1000*, is processed by calcium metal, calcium hydroxide, and phosphoric acid reaction at 1,292-1,562°F (700-850°C). Not only does it's porosity invite invasion by blood vessels and cells that gradually break down the implant as new, natural bone grows, the number of ''pores'' can be adjusted to match that of the graft recipient, encouraging faster regrowth of natural bone. In collaboration with professor Joel W. Barlow, a chemical engineer at the same university, Lagow devised a computer-guided **laser** to generate different bone shapes and sizes with the intention of providing bone ''blanks'' customized for individual patient needs. Hydroxyapatite is also used as a coating for **artificial joints** and prostheses to encourage bone to grow and bind tightly to the implant.

1993 saw FDA approval of *Collagraft*, a hydroxyapatite/tricalcium phosphate and bovine collagen which is mixed with the patients bone marrow. In 1998, Dr. Jay Lieberman, of the University of California, Los Angeles began clinical trials to treat *osteonecrosis* (bone death) of the hip, by inserting a capsule containing *bone-morphogenetic protein* (BMP), into an allograft. BMP, discovered by Urist in 1965, occurs naturally in small quantities in the bone matrix and imitates the bone development process which occurs during fetal development. John M. Wozney with Genetics Institute in Massachusetts, first cloned BMP in 1988—in 1998, more than 30 BMPs have been cloned. Over the next decade, researchers look to tissue-engineered implants for reconstructive skeletal deformities and cell-based therapies for osteoarthritis and osteoporosis. Carbon fiber composites—materials used in **skis** and **tennis** rackets—are also being tested for use because the composites resemble bone in both stiffness and flexibility.

Artificial chromosome • See Genetic engineering

Artificial gene • See Genetic engineering

Artificial heart

Because the heart functions primarily as a **pump** to keep blood circulating through the body, medical researchers have long attempted to develop a mechanical pump to take over its job in the event of damage or disease. In 1935 the French-born surgeon **Alexis Carrel**, and famed American aviator Charles Lindbergh (1902-1974), designed a perfusion pump that kept excised organs, including the heart, alive by circulating blood through them. News reports called this device an ''artificial'' or ''robot'' heart. The first total artificial heart (TAH) was implanted in 1957 in a dog at the Cleveland Clinic by Willem Kolff, a Dutch-born surgeon, and Tetsuzo Akutsu. Kolff later led a medical team at the University of Utah at Salt Lake City in developing the artificial heart. At the urging of another TAH pioneer, Michael DeBakey (1908-), the United States government, through the National Institutes of Health, established an Artificial Heart Program in 1964 to develop both partial and total artificial heart devices. By 1966, DeBakey had designed and implanted a pneumatically driven component called a Left Ventricular Assist Device LVAD) to serve the chamber of the heart that pumps blood out into the arteries. This was an important development, for the great majority of severe heart disease is caused by left ventricle failure.

The first human artificial heart implant was carried out by Denton Cooley (1920-) and his surgical team at the Texas Heart Institute in 1969. The pneumatically driven **Dacron**-lined **plastic** heart, designed by Argentine-born Domingo Liotta, was a temporary measure to keep a patient alive until a heart transplant could be performed. Artificial heart implantation captured worldwide headlines in 1982 when the first TAH intended for permanent use was implanted in the chest of a patient on the verge of death—dentist Barney Clark. The procedure was done at the University of Utah by a surgical team headed by William DeVries. The device, called a *Jarvik-7*, was designed by American physician Robert Jarvik. The plastic and titanium pump was powered by compressed air delivered by a large external **air compressor** through two tubes that passed into the body via incisions in the abdomen. Clark survived for 112 days. DeVries then joined the staff at Humana Hospital in Louisville, Kentucky, where he carried out four other Jarvik-7 implants during 1984 and 1985. Each of these patients also died, including William Schroeder, who suffered a repeated series of debilitating setbacks during his 620-day struggle to survive. The results of permanent implantation of the Jarvik-7 revealed its insurmountable limitations: it caused blood clots to form, which traveled to the brain and precipitated stroke; the abdominal incisions provided a pathway for infection-causing bacteria; and the patient's mobility was severely restricted by the cumbersome compressor. These problems and the concurrent development of successful heart transplantation, permanent installation of the Jarvik-7 and any other TAHs rapidly fell out of use, especially after Schroeder's death in 1986.

At the same time as heart transplantation became an established procedure for terminal heart disease, demand for donor hearts outstripped supply. In 1998, more than 50,000 Americans could benefit from a heart transplant while the available number of donor hearts remains steady at approximately 2,000. The mechanical heart thus became a "bridge for transplantation"—a last resort to keep a patient alive until a donor heart became available. Meanwhile, research focuses on a new generation of electrically powered artificial hearts, both TAHs and LVADs which use portable **battery** packs to transmit power via **radio** signals through unbroken skin to an implanted mechanical heart pump. The first of these devices was experimentally implanted in a human subject in 1991. While no significant progress has been made, two groups of researchers are working under the National Heart Lung and Blood Institute contract program. One group—The Texas Heart Institute and ABIOMED, Inc., anticipates clinical trials of a new TAH by the year 2000. The second—Pennsylvania State University College of Medicine and 3M Health Care—foresees formal testing of a long-term (two-year) LVAD beginning in 1998. A third study at the University of Pittsburgh Artificial Heart Program aims at improving fluid dynamics in artificial hearts by minimizing reverse blood flow within the pump which causes blood clots.

See also Barnard, Christiaan; Surgical transplants

Artificial heart valve

Heart valves are flaps of tissue within the heart that open to allow blood to flow from one of the heart's four chambers to the next and then close to prevent any blood from leaking back. When any one of the heart's four valves becomes too diseased or damaged to function adequately, the only effective treatment is valve replacement. This was not possible until the advent of **open-heart surgery** in the 1950s. Researchers then set out to design a valve that could be easily implanted and tolerated by surrounding tissue, would not promote clot formation, and would be durable.

A precursor of the artificial heart valve was developed by American surgeon Charles A. Hufnagel (1916-), who inserted a tube-and-float device in a patient's descending *aorta* in 1952, to prevent aortic backflow. An artificial cardiac aortic valve (a caged-ball device) was implanted into a human being in March 1960 by Dwight Harken at Peter Bent Brigham Hospital in Boston. This was followed shortly by a total artificial mitral valve replacement (a flexible-leaflet) performed by Nina Braunwald at the National Institutes of Health.

The first completely successful artificial heart valve, an invention that became, and remains, a standard in the field, was designed and implanted in a fifty-two-year-old man by surgeons Albert Starr (1926-) and M. L. Edwards in Portland, Oregon, in 1961. This Starr-Edwards valve consisted of a **silicone-rubber** ball enclosed in a **stainless steel** cage; today, the valve has a hollow metal ball, an **alloy** cage, and a **Teflon** base.

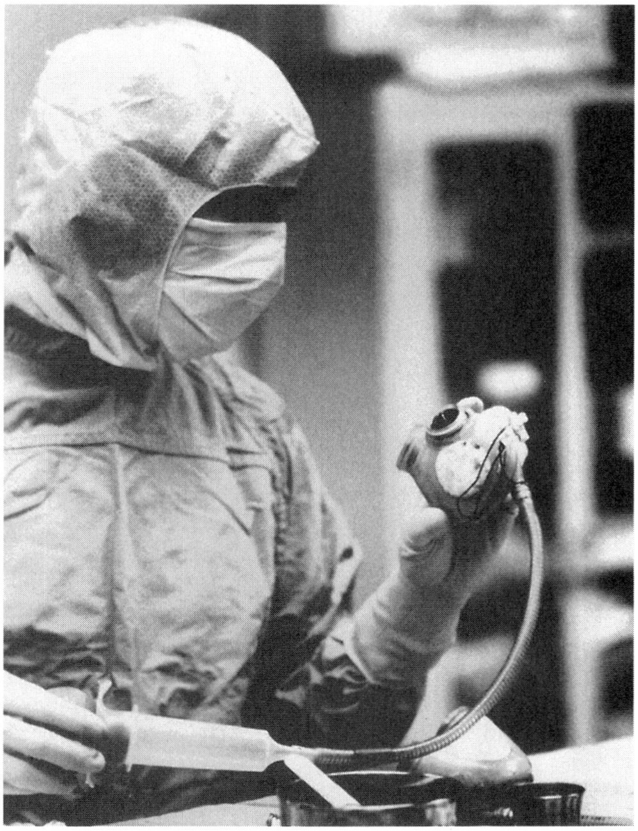
Inventor Robert Jarvik holds the Jarvik-7, the first total artificial heart to be permanently implanted in a human.

Later in the 1960s, the tilting-disc valve was developed. Valve design was improved after the space program produced pyrolite carbon, a strong and durable new material. The St. Jude Medical pivoting bileaflet valve, consisting of two leaflets rotating within a ring, was introduced in 1976. Mechanical heart valves made the headlines in 1990 when Shiley Laboratories recalled all its convexo-concave tilting-disk valves from the market because of the risk of strut fractures that could cause sudden cardiac failure; many patients with other types of Shiley valves wrongly believed their implants were also likely to fail suddenly.

Because of the danger of blood clot formation, implantation of a mechanical heart valve requires that the patient take *anticoagulation* medication for life. Alternatives to artificial valves carry a much lower risk of blood clots and reduce the need for anticoagulation therapy. These include both human and animal valve transplants: Human tissue valves are either *allografts* (or homografts)—valves transplanted from a deceased person whose organs have been donated for transplantation, or *autografts*—in which the pulmonic valve is removed from the patient to replace the abnormal aortic valve and the pulmonic valve is then replaced with an allograft. Animal valve transplants are either *porcine* (pig) or *bovine* (cow) tissue, traditinoally held in place by a stent. In 1997, the United States Food and Drug Administration approved the "Stentless Toronto SPV Valve," a porcine tissue valve manufactured by St. Jude Medical, Inc, which is entirely supported by the patient's aorta. With no stent to take up space, a larger valve can be implanted, improving blood flow.

ARTIFICIAL HIP • See Artificial limb and joint

ARTIFICIAL INTELLIGENCE (AI)

The invention of the steam engine and industrial **mass production** astonished a society familiar only with manual labor, generating visions of a leisurely future in which human toil was no longer necessary and machines provided all of the necessities of life. Similarly, the power of **digital computer** technology to perform amazing feats of calculation has raised the hope of scientists that computers might be built which have the ability to perform complex tasks, make sound decisions in novel situations, initiate activity and learn from mistakes—in a word, to *think.*

One of the most important figures in the early study of artificial intelligence was **Alan Turing** (1912-1954), an English mathematician and logician. Turing proposed a means by which one could determine whether a computer is "intelligent." Place a computer, Turing said, in one room and place a keyboard attached to the computer in another room. Then allow an individual to communicate with the computer via the keyboard. If the individual cannot determine whether there is a computer or a human in the other room, then the computer could be termed "intelligent." He argued that intelligent computers could eventually be created with the ability to play games like chess and checkers, learn and translate languages, create and break codes, and tackle tough mathematical problems.

In the late 1930s Turing considered what mathematical procedures a computer would need to use in order to perform any calculation. Using a binary method, he illustrated on paper the operations by which a hypothetical computer could carry out any algorithm (a finite set of mathematical instructions). Turing's theorem was dubbed the "Universal Turing Machine" because it clearly demonstrated the properties that any genuine calculating machine would need to possess. Turing went on to help create a machine called *Enigma* which was capable of deciphering German codes during World War II.

In 1956 an important conference on artificial intelligence was held at Dartmouth College. The Dartmouth symposium was the meeting place of a number of individuals who would shape the future of artificial intelligence including John McCarthy, a mathematics professor, Marvin Minsky (1916-), a Harvard mathematician interested in how the brain worked, Nathaniel Rochester (1919-), an IBM researcher, and Claude Shannon (1916-), a mathematician at Bell Laboratories. It was at this conference that Minsky began writing an influential paper eventually titled "Steps Toward Artificial Intelligence." In it he described several machines capable of proving mathematical theorems or playing chess.

About the same time researchers at RAND Corporation completed work on an intelligent program. Allen Newell (1927-1992), Herbert Simon (1916-) and J. C. Shaw unveiled the Logic Theorist, capable of proving theorems in mathematics. It could carry out operations no one though a machine capable of performing, and it even provided some novel solutions to long-standing computer programming dilemmas. The Logic Theorist was the first device that really gave credence to the claim that computers could be made "intelligent."

It was in the area of games that artificial intelligence gained international fame. IBM was responsible for two of the early game-playing programs. One of IBM's experts in **vacuum tubes**, Arthur Samuel (1901-1990), became fascinated with the idea of creating a checkers-playing program that would learn from its mistakes. He wrote a computer program that could recognize a position it had already come across in a previous game and alter its moves based on the results of the completed game. In the late 1950s another IBM employee, Alex Bernstein, became one of the first to design a chess-playing computer program, but had limited success, as he encountered difficulty defining a clear set of principles that could guide the computer without testing every possible move.

A team from Northwestern University eventually created a more successful chess program called CHESS. Their strategy was to "crunch" as many moves as possible; at any board position their program analyzes millions of possibilities. An early version of the program, CHESS 4.5, earned an expert rating from the United States Chess Federation. By 1978, their program could compete at the master level, though CHESS 4.7 lost to British master David Levy that year.

Another chess-playing program called DEEP THOUGHT was developed by Feng-Hsuing Hsu at Carnegie-Mellon University. Capable of analyzing 34 million board positions per minute, DEEP THOUGHT beat several players at the master level before it was defeated by the world champion. In 1993 the next incarnation of DEEP THOUGHT was unveiled: DEEP BLUE. DEEP BLUE was capable of analyzing over 100 million board positions per second. In February of 1996, DEEP BLUE played chess champion Gary Kasparov. Kasparov beat the computer four games to two. However, in a rematch the following year, DEEP BLUE beat Kasparov 3 1/2 games to 2 1/2. Although DEEP THOUGHT, and later DEEP BLUE, are highly successful at winning chess matches, a number of AI experts have become disillusioned with such programs, contending that they resemble a massive primitive calculators more than "thinking" machines.

One of the most promising avenues of recent research in AI has been in language processing. One of the foremost innovators in the field is Raymond Kurzweil (1948-). After graduating from the Massachusetts Institute of Technology in 1974, Kurzweil began work on a reading machine for the blind that could translate any printed material into understandable speech. Kurzweil succeeded in producing such a device two years later. The Kurzweil Reader caught the attention of blind singer Stevie Wonder (1950-) who contacted Kurzweil, expressing interest in a keyboard synthesizer that could reproduce the sounds of acoustical instruments. In 1983 Kurzweil completed work on a groundbreaking synthesizer that could reproduce the sound of more than 100 musical instruments with surprising accuracy. He also began work on a voice-activated **typewriter**, a machine that IBM and other companies had been attempting to perfect since the 1940s.

In the late 1980s Kurzweil produced Voice Works, a computer that can take dictation. He plans to apply the technology to create a writing machine that enables a deaf person to see a real-time readout of conversations and discussions.

Other AI machines and programs have been designed to carry out foreign language translations, robot control, automatic programming, and natural language interaction. Recently developed **expert systems** can diagnose system malfunctions, develop manufacturing plant schedules, and even analyze chemical structures or disease symptoms.

ARTIFICIAL KINDEY • See Dialysis machine

ARTIFICIAL LIGAMENTS

Ligaments are bands of tough, elastic tissue that bind bones together at a joint, such as the knee. When a ligament is torn, it can either be repaired or replaced. Repair is the first choice; however, severe injuries usually necessitate replacement. Perhaps the most common ligament replacement surgery is for the *Anterior Cruciate Ligament* (ACL) in the knee. *Autografts* (using tendons from elsewhere in the patient's body) were instituted around the 1900s with varying results, and are still the most popular and successful. In Germany in 1912, surgeons began using tendons from Kangaroo tails in *xenografts*. These replacements were relatively unsuccessful due to rejection and infection. Rehabilitation in both instances could take up to two years; however, advances in autografts and rehabilitation techniques have considerably reduced recovery time.

Use of artificial ligaments began when the W. L. Gore Company developed a ligament made from **Gortex**, the porous, expanded **Teflon** invented in 1969 and best known for its use in **waterproof materials**. The Food and Drug Administration (FDA) approved use of this synthetic ligament in humans in 1988, but only for patients in whom an autograft had failed. The six-inch-long Gortex ligament, constructed from about 1,000 fibers braided together for strength and attached to the bones above and below the knee with **stainless steel** screws, soon becomes naturally anchored as the bone grows into and through the Gortex, assimilating it as if it were a natural body part. Rehabilitation time with the Gortex ligament was shortened to six weeks. Surgery is done as an outpatient arthroscopic procedure. An unacceptably high failure rate of artificial ligaments after two years still exists in 1998, primarily due to mechanical failure or inflammation caused by products shed from the graft. Dacron and synthetic **polymers** are also used in artificial ligaments, tendons, and joints; and Japanese researchers working in the emerging field of *fiber- reinforced elastomers* (FRE) see potential in this biomaterial for artificial ligaments.

Before the French surgeon Ambroise Paré developed a feasible working joint in the 1500s, peg legs were the only means of replacing a lost limb.

In 1998, the FDA approved Phase I clinical trials of a porcine (pig) implant manufactured from the skin of a pig's small intestines. This material, called *small-intestinal submucosa* (SIS)—in essence, sausage casing— was developed by Purdue University and DePuy Orthopaedics, Inc. In March, 1998, James McDonalds, a 38-year-old ranch caretaker in Colorado, became the first person to receive this implant. After being processed into long, paper-thin but extremely strong strands and inserted, the SIS tissue acts as a framework for the body's natural healing process, creating new blood vessels in the surrounding area which nourish the graft and attract molecules necessary for tissue regeneration. SIS implants begin weaker than artificial ligaments, but grow stronger as the natural tissue regrows, the reverse of the trend in synthetic-fiber implants. Researchers believe there is no apparent rejection by the body, and that the SIS eventually dissolves and is eliminated.

See also Artificial limb and joint

ARTIFICIAL LIMB AND JOINT

Almost four million Americans use artificial limbs, the majority necessitated by amputation resulting from poor circulation caused by diabetes or atheroscleroses. Crude artificial limbs have been used since the earliest loss of an extremity; Greek historian Herodotus mentioned a wooden foot in 500 B.C.; a Roman mosaic depicts a peg-leg; and medieval knights had artificial limbs to improve their appearance. Today, prostheses rival natural limbs both in function and appearance.

The modern era of artificial limbs began with the famous French surgeon **Ambroise Paré** (1517-1590), a barber-surgeon who, in 1536, became a battlefield surgeon. After devising safer, more effective methods of **amputation**, Paré turned his attention to the design of artificial limbs, exercising great ingenuity and striving to simulate some degree of natural movement. An artificial leg pictured in Paré's *Works of 1575* featured a movable knee joint controlled by a string, a flexible foot operated with a strong **spring**, and an artificial hand with fingers that moved individually by means of tiny internal cogs and **levers**. As the many nineteenth-century wars created a larger demand, more sophisticated devices were invented. The most significant impetus for improvement in prosthetic design was the birth in the early 1960s of babies with vestiges of arms, a birth defect caused by a drug called *thalidomide* often prescribed to pregnant women. Artificial arms powered by carbon dioxide were developed for these children. Modern limbs include electric arms which use a small battery, others controlled by a tiny switch, and the *myoelectric prosthesses*which uses electrical impulses detected by small electrodes placed on the skin over the remaining arm muscles. Artificial legs can be fitted with spring-loaded feet, artificial feet constructed with toes, synthetic coverings made to match skin tone and hair patterns, and electrodes in the artificial limb leading to the natural skin which allows the brain to register ''feeling'' in the prosthesis. Artificial limbs take advantage of **plastics** and **fiberglass** for enhanced strength and comfort.

Joints are the movable points where two bones come together—as at the knee or shoulder. Surgical joint replacement--*joint arthroplasty*—began in the 1950s. Today, more than 500,000 joint repairs and replacements are performed in the United States each year, including 20,000 new cases of hip *osteonecrosis* (bone death). The first total knee arthroplasty was performed in 1951; 10 years later the first total hip replacement occurred, and shoulder replacements began in the 1960s. Artificial joints are secured either by cement or ''bone ingrowth,'' a process in which the natural bone grows into the porous surface of the a prosthesis. In 1998, Dr. Jay Lieberman, of the University of California, Los Angeles began clinical trials to treat osteonecrosis of the hip by inserting a capsule containing *bone-morphogenetic protein* (BMP) into an allograft (implant of purified human bone). Total hip replacements are metal and have a limited life span—BMP, which occurs naturally in small quantities in the bone matrix, imitates the bone development process which occurs during fetal development. Researchers believe BMP may reducing the need for total hip replacement. John M. Wozney with Genetics Institute in Massachusetts, first cloned BMP in 1988—in early 1998, more than 30 BMPs were cloned. 1998 also saw results of a seven-year study in which cells from a patient's healthy cartilage are transplanted into the damaged portion of the knee joint, regenerating cartilage and reducing the need for knee replacement. Researchers look ahead to tissue-engineered implants and cell-based therapies for degenerative diseases such as osteoarthritis and osteoporosis. Research is ongoing to improve prosthetic materials, surgical techniques, ways of securing joints, and postoperative mobility.

Artificial skin

Artificial skin, a synthetic equivalent to human skin, can dramatically increase the chance of survival of severely burned patients. The first synthetic skin was invented by John F. Burke, chief of Trauma Services at Massachusetts General Hospital, and Ioannis V. Yannas, chemistry professor at Massachusetts Institute of Technology.

Seeing so many burn victims during his career, Burke had long been seeking a replacement for human skin that would prevent infection and dehydration. Meanwhile, Ioannis Yannas had been studying collagen, a protein found in human skin. Teaming up during the 1970s, the two found that collagen fibers and a long sugar molecule (called a **polymer**) could be combined to form a porous material that resembled skin and, when placed on wounds of lab animals, seemed to encourage the growth of new skin cells around it. The pair then created a kind of artificial skin using polymers from shark cartilage and collagen from cowhide.

Using their synthetic material, called *Silastic*, Burke and Yannas continued experimenting and found that artificial skin acts like a framework onto which new skin tissue and blood vessels grow, although these new cells are unable to produce hair follicles or sweat glands normally formed in the dermis. As the new skin grows, the cowhide and shark substances from the artificial skin are broken down and absorbed by the body. In 1979 Burke and Yannas used their artificial skin on their first patient, a woman who had suffered burns over half her body. After peeling away her burned skin, Burke applied a layer of artificial skin and, where possible, grafted to it some of her own unburned skin. Three weeks later, the woman's new skin, the same color as her unburned skin, was growing at an amazingly healthy rate.

Meanwhile, at nearby Harvard University, Howard Green had begun culturing human skin cells under sterile conditions and growing a sheet of human epidermis cells from a tiny piece of a person's skin. However, when the cultured skin was placed on a wound it was rejected by the body's immune system. Green later began work with Eugene Bell of MIT who founded a research group called Organogenesis. The goal of this Boston-based firm was to make artificial skin that would include an epidermis layer and solve the problem of rejection. This research has produced a product called *Graftskin*—a living skin equivalent made of purified bovine collagen into which dermal cells from infant boys' foreskin have been ''seeded.'' On top of that layer is an epidermal layer of the cultured human skin cells. It is formed into 4 × 8 in (10 × 20 cm) sheets that can be sutured or stapled onto a patient during surgery. No rejection was seen in clinical trials. Hospital trials included burn victims, patients needing skin grafts after cancer surgery, and those with chronic nonhealing wounds. In January, 1998, this multilayered, tissue-engineered skin trade marked *Apligraf* received unconditional recommendation from the FDA advisory panel for treating venous leg ulcers, making it the first living manufactured organ ever recommended for approval by the FDA. (A welcome side benefit of this research is that it can be used to test dermatological products without testing on animals.)

Another company in California, Advanced Tissue Sciences, engineers a product called *Dermagraft* using a similar method, producing 250,000 square feet of final product from one foreskin. In a study of 85 burn patients, Dermagraft, used as a temporary skin until permanent skin grafts could be done, cut the need for surgery in half and reduced hospital stays my months. Hy-Gene, Inc. in New Jersey, clones a patient's own skin cells producing an epidermis about 10 layers thick. In three weeks, a small patch of cloned skin can grow enough to cover an entire human from head to toe. This product was patented and went into clinical trials in November, 1997. A fourth company, Integra Life Sciences, does not clone skin but manufactures a matrix from collagen and shark cartilage with appropriate pore size and absorption rate for a wide range of human body parts. The environment of this lattice-like structure "tricks" the body into growing replacement tissue.

Artificial sweetener

Artificial sweeteners are synthetic substances used to replace sugar in foods and beverages. Typically, they are much sweeter than sugar and have few or no calories. They are widely used by dieters and people afflicted with diabetes. Saccharin, cyclamates, and aspartame are the three widely used artificial sweeteners.

Saccharin was the first true artificial sweetener invented. It was discovered in 1879 by two chemists at Johns Hopkins University, the German Constantin Fahlberg (1850-1910) and the American Ira Remsen (1846-1927). While investigating the derivatives of *toluene*, an ingredient found in coal tar, the scientists noticed a sweet taste on their hands. They then traced toluene back to a new compound that was 300 to 500 times sweeter than sugar. Saccharin came on the market in the early 1900s. Doubts about its safety soon arose however, and it was banned in 1912. It reappeared during the sugar shortages of World War I. In the early 1970s, tests indicated a possible link between saccharin and bladder cancer in rats. When the FDA began making preparations to ban the sweetener in 1977, congress halted their efforts. At that point, the FDA required manufacturers to publish warning labels on all foods containing saccharin the following year. Canada outlawed the use of saccharin in prepared foods beginning in 1977.

In 1937, a new synthetic sweetener was accidentally discovered by Ph.D. candidate Michael Sveda (1912-) at a University of Illinois chemistry laboratory. Like Remsen and Fahlberg, Sveda noticed a sweet taste on his fingers and found a chemical responsible, which was eventually called cyclamate. Ludwig Audrieth (1901-) joined Sveda's investigations during the 1940s, and in June 1950 Abbott Laboratories marketed the substance as the sweetener Sucaryl. It was 30 to 50 times as sweet as sugar, did not have saccharin's bitter aftertaste, and signaled an explosion of artificially sweetened low-calorie products, especially diet soft drinks. As with saccharin, however, cyclamates were linked with the possibility of bladder cancer in rats. Amid widespread media coverage, cyclamates were banned in the United States in 1970. Scientific opinion remains split on the safety of cyclamates. They are banned in some countries, but allowed in others worldwide.

Today, aspartame is the most widely used artificial sweetener. Made from two amino acids, aspartame was discovered by an American chemist, James M. Schlatter (1942-), who—like his two predecessors—noticed during a 1965 experiment that his fingers tasted sweet. Approximately 200 times as sweet as sugar and having no bitter aftertaste, aspartame was approved by the FDA in 1974 for use in dry foods (such as cereal or chewing gum) and as a tabletop sweetener. The approval was withdrawn in 1975, then reinstated in 1981; in 1983 aspartame was approved for use in carbonated beverages.

Research into other synthetic sugar substitutes continues. One of these, *acesulfame K*, was discovered in 1967 by Karl Clauss and Harald Jensen of Germany and was approved for use in the United States in 1988. Another sugar substitute, *N-(4-nitrophenylcarbamoyl)-L-aspartyl-L-phenylalanine methyl ester*, which is comparable to aspartame, was discovered in 1982 by Frenchmen Claude Noffre and Jean-Marie Tinti. An incredibly sweet substance, it is 17,000 to 52,000 times sweeter than sugar.

In 1998, the FDA approved the use of sucralose, another high-intensity sweetener. It is derived from sucrose but is 600 times as sweet. To produce sucralose, the manufacturers start with sucrose and chemically modify it by replacing three hydroxyl groups with chlorine. This results in a compound that tastes sweeter than sucrose, but can not be digested by the body. It is anticipated that this material will become the standard in artificial sugars.

Artillery

The term *artillery* refers to any mounted **gun** or weapon that discharges **missile**s. The first example of artillery can be traced back to the Stone Age when there is evidence that prehistoric people used slings (probably made of animal hides or vines) to hurl rocks at prey. Another form of primitive artillery, the **bow and arrow**, was first used widely by the ancient Egyptians around 5000 B.C.

During the Middle Ages, **catapult**s were used to attack walled cities and castles. Catapults operate on the same basic premise as bows and slings; each consists of a large arm that is restrained with a trigger mechanism. When the mechanism is released, the arm flings forward, hurling its **ammunition** at the target. Early catapults were capable of projecting a 180 lb (82 kg) shot up to 650 ft (198 m) away.

Probably the best known type of artillery is the cannon, or "firetube," which first appeared in the fourteenth century. These early cannons were made of circular bundles of iron bars held together with rings or hoops. They shot arrows, iron bolts, or round stones. Cannon made of cast iron or bronze appeared by the end of the fourteenth century. The fine **gunpowder** used in early cannons tended to burn slowly, but around 1425 it was discovered that wet powder, passed through a sieve, formed granules, which, when dried before igniting, provided a more powerful explosion. By the fifteenth century, huge cannon called bombards were built, which were capable of devastating castle and city walls. One famous bombard used against Constantinople in 1453 was so heavy it took sixty oxen and 200 men to move it. It took two hours to load it with a 1,600-pound (726 kg) stone ball, and could hit a target about a mile (1.6 km) distant.

Cannon design improved over the centuries, leading to smaller but more powerful, mobile and accurate weapons. Cannons of a Swiss design became prevalent in the eighteenth century, made from a bored-out piece of solid cast metal. The barrel of this kind of cannon was more uniform than earlier models, which were cast around a hollow core. Cannon accuracy also improved as military leaders gained greater understanding of the scientific principles underlying trajectory, projectile flight, and air resistance. The early nineteenth century also saw many innovations in cannon shot. Heavy iron balls were in many cases replaced by exploding shells filled with small shot. In 1851, John Dahlgren (1809-1870), an American naval officer, created an 11 inch (28 cm) gun (called the Dahlgren gun) which gained a well-deserved reputation during the American Civil War. He came up with a design in which the barrel wall was no thicker than necessary to contain the pressure of the expanding gases within the barrel. To do this, Dahlgren trimmed a great deal of the weight from the barrel of the cannon, giving the weapon a bottle shape.

Another type of artillery is the howitzer, a device that fires shells in a high arc. It was first used near the end of the seventeenth century by Anglo-Dutch armies. The name itself comes from the Dutch word, houwitser, meaning ''slingshot.'' The mortar is another artillery piece that fires high-arcing projectiles. Early mortars were metal pots secured to a timber base used to fire large stone balls that, having fallen from a great height, were capable of inflicting extensive damage on their targets. Wilfred Stokes developed the modern 3 in (8 cm) mortar for the English Army during World War I.

Another type of artillery to emerge in World War I was the infamous ''Paris Gun,'' a giant weapon with a firing barrel that was over 100 feet long. It was capable of firing its 250 lb (114 kg) shells a record-breaking distance of 75 mi (120 km). This much-feared piece of artillery was used by the German army to shell Paris over a 140-day period in 1918.

During World War II, the United States military introduced the 2.36 in (5.6 cm) rocket launcher, nicknamed the **bazooka** by American GIs. The bazooka weighed about 10 lbs (4.5 kg) and was operated by two people; one positioned it on his shoulder and fired while the other loaded the ammunition. Primarily used as an antitank weapon, the bazooka could hit a target up to several hundred yards away with great accuracy.

A lesser-known artillery piece is the recoilless rifle, first developed by the Germans around 1840 and used extensively during World War II. It could fire shells the size of small-caliber artillery shells, but it was much lighter than other artillery pieces. Its 4 in (105 mm) shell could reach over 7,000 yd (6,398 m), yet the gun, mounted on a gun carriage, weighed only 855 lbs (388 kg). With the rise of small missiles, the recoilless rifle has nearly disappeared.

After World War II, the most important artillery pieces became principally rockets and guided missiles. These are vastly more powerful and accurate than any earlier artillery. Modern field artillery uses radar, computer guidance systems, acoustic locators, and other sophisticated technology to locate targets. These precision-guided munitions can home in on enemy signals from a great distance, and some can be reprogrammed in flight. The pin-point accuracy of such weapons makes it possible to remove enemy artillery with minimal damage to surrounding, non-military targets, though this has not always been the case in actual battle practice.

The advent of nuclear weapons at the close of World War II threatened to make all other weapons obsolete. American official policy was that of ''Mutual Assured Destruction,'' that is, the development of weapons with such intolerable destructive force that no one would dare to use them. The United States, Soviet Union, and several other countries concentrated their weapons programs on massive nuclear bombs. Nevertheless, during the 1950s and 1960s, American military scientists experimented with small nuclear warheads that could be deployed on the battlefield. Scientists in other countries also apparently developed similar weapons. Nuclear field artillery is assumed to exist in the late 1990s. These battlefield warheads are based on nuclear fusion using tritium, and are thought to be able to kill troops with radiation at several hundred meters distant, yet not contaminate the side that uses them.

ASPARTAME • See Artificial sweetener

ASPHALT • See Pitch

ASPIRIN

Aspirin is one of the safest and least expensive pain relievers on the marketplace. While other pain relievers were discovered and manufactured before aspirin, they only gained acceptance as over-the-counter drugs in Europe and the United States after aspirin's success at the turn of the twentieth century.

Today, Americans alone consume 16,000 tons of aspirin tablets a year, equaling 80 million pills, and we spend about $2 billion a year for non-prescription pain relievers, many of which contain aspirin or similar drugs.

Currently, the drug is available in several dosage forms in various concentrations from .0021 to .00227 ounces (60 to 650 milligrams), but the drug is most widely used in tablet form. Other dosage forms include capsules, caplets, suppositories, and liquid elixir.

Aspirin can be used to fight a host of health problems: cerebral thromboses (with less than one tablet a day); general pain or fever (two to six tablets a day); and diseases such as rheumatic fever, gout, and rheumatoid arthritis. The drug is also beneficial in helping to ward off heart attacks. In addition, biologists use aspirin to interfere with white blood cell action, and molecular biologists use the drug to activate genes.

The wide range of effects that aspirin can produce made it difficult to pinpoint how it actually works, and it wasn't until the 1970s that biologists hypothesized that aspirin and related drugs (such as ibuprofen) work by inhibiting the synthesis of certain hormones that cause pain and inflammation. Since then, scientists have made further progress in understanding how aspirin works. They now know, for instance, that aspirin and its relatives actually prevent the growth of cells that cause inflammation.

The compound from which the active ingredient in aspirin was first derived, salicylic acid, was found in the bark of a willow tree in 1763 by Reverend Edmund Stone of Chipping-Norton, England. (The bark from the willow tree—Salix Alba—contains high levels of salicin, the glycoside of salicylic acid.) Earlier accounts indicate that Hippocrates of ancient Greece used willow leaves for the same purpose—to reduce fever and relieve the aches of a variety of illnesses.

During the 1800s, various scientists extracted salicylic acid from willow bark and produced the compound synthetically. Then, in 1853, French chemist Charles F. Gerhardt synthesized a primitive form of aspirin, a derivative of salicylic acid. In 1897 Felix Hoffmann, a German chemist working at the Bayer division of I.G. Farber, discovered a better method for synthesizing the drug. Though sometimes Hoffmann is improperly given credit for the discovery of aspirin, he did understand that aspirin was an effective pain reliever that did not have the side effects of salicylic acid (it burned throats and upset stomachs).

Bayer marketed aspirin beginning in 1899 and dominated the production of pain relievers until after World War I, when Sterling Drug bought German-owned Bayer's New York operations. Today, "Aspirin" is a registered trademark of Bayer in many countries around the world, but in the United States and the United Kingdom aspirin is simply the common name for acetylsalicylic acid.

The manufacture of aspirin has paralleled advancements in pharmaceutical manufacturing as a whole, with significant mechanization occurring during the early twentieth century. Now, the manufacture of aspirin is highly automated and, in certain pharmaceutical companies, completely computerized.

While the aspirin production process varies between pharmaceutical companies, dosage forms and amounts, the process is not as complex as the process for many other drugs. In particular, the production of hard aspirin tablets requires only four ingredients: the active ingredient (acetylsalicylic acid), corn starch, water, and a lubricant.

ASSEMBLY LINE • See Mass production

ASTROLABE

The astrolabe is an astronomical instrument used to observe the positions of the stars. With modifications it has also been used for timekeeping, navigation, and surveying. The most common type, the planispheric astrolabe, consisted of a star **map** (the rete) engraved on a round sheet of metal. Only the angular relationship of the stars was accurate, since this was all that was needed. A metal ring could be moved across the map to represent the position of the local horizon. An outer ring could be adjusted to allow for the apparent rotation of the stars around the North Star, using prominent stars as reference points.

Galileo's astrolabe.

Astrolabes were forerunners of mechanical **clocks**, and looked somewhat like watches. With a set of tables, the observer could determine the day and hour for a fixed location by the position of the stars. With the addition of a sighting-rule, called an alidade, an astrolabe could be used as a **surveying instrument**. The rule could be moved across a scale to measure elevation.

It is thought that the Greeks had astrolabes, but it is known that the Arabs at least perfected them and made regular use of them. With the clear desert sky at their constant disposal, the Arab people excelled in astronomy and used the stars to navigate across the seas of sand. Regular use of astrolabes continued into the 1800s. The newer prismatic astrolabe continues to be used for precision surveying.

Atanasoff, John (1903-1995)

American mathematician and computer inventor

John Atanasoff is recognized as the first designer of an electronic digital computer. He was born in Hamilton, New York, into a family of mathematicians. His father was an electrical engineer, his mother a mathematics teacher. Atanasoff was raised in Florida, and he loved science and engineering as a child. He earned an engineering degree from the University of Florida in 1925, then did graduate studies in mathematics, receiving a master's degree from Iowa State College (now University) in 1926 and a Ph.D. from the University of Wisconsin in 1930, after which he joined Iowa State's mathematics and physics faculties.

Atanasoff determined to improve computing machines so it would be easier for his graduate students to perform mathematical calculations. From the first, he wanted a digital device, but was unable to develop the concept. Then one winter night in 1937, while on a long drive to clear his mind, he stopped off at a roadside tavern. Suddenly, the entire concept came to him. He could use **vacuum tube**s as the computing medium, binary (base 2) numbers, serial calculation, and condensers that placed a charge on Bakelite (plastic) drums as the computer's memory. Each drum held thirty fifty-digit numbers. Punched cards provided input and output, but each step had to be hand-started. Atanasoff and a graduate student, Clifford Berry, built working portions of the A-B-C (Atanasoff-Berry Computer) over the next two years, but the university refused to patent it for him.

During World War II, **John Mauchly** and **J. Presper Eckert** designed and patented the **ENIAC** computer, later selling their patents to the manufacturer Sperry Rand. After a bitter court battle, in 1973 Atanasoff was awarded the patent instead, on the ground that Mauchly and Eckert had knowledge of the A-B-C and used its design as the basis for theirs. Atanasoff's other interests have included acoustics detection and package-handling automation. In 1990, Atanasoff received the National Medal of Technology.

Atomic bomb

The invention of the atomic bomb was an historical inevitability. The scientific discovery necessary for construction of such a bomb occurred just as World War II was about to begin. It seems impossible that an enormous new source of energy (nuclear fission) would *not* eventually be applied in the construction of a bomb.

The scientific discovery that made the bomb possible occurred in 1934. Enrico Fermi (1901-1954), an Italian physicist, was trying to produce element number 93 by bombarding uranium with neutrons. Fermi thought he had been successful, but his results were not clear-cut. Other scientists attempted to repeat his work to clarify Fermi's findings. In 1938, Otto Hahn (1879-1968) and Fritz Strassman showed that the products of Fermi's reaction were nuclei from the middle of the periodic table.

The significance of these findings was made clear by Lise Meitner (1878-1968) and Otto Frisch (1904-1979), who demonstrated that the uranium nuclei in Fermi's experiment had actually been fissioned by neutrons. The word *fission* means to break apart. In the case of nuclear fission, a neutron causes a large nucleus, like that of uranium, to break apart into two roughly equal pieces, the products identified by Hahn and Strassman.

Three other important discoveries followed quickly. First, Niels Bohr (1885-1962) predicted—and the American physicist John Dunning (1937-) soon confirmed—that only one isotope of uranium, uranium-235, actually undergoes fission. Second, **Leo Szilard** and Walter Zinn (1906-), in rediscovering the fission reaction, found that each time a uranium nucleus fissions, an average of about two neutrons are also released.

The production of these neutrons creates the possibility of a *chain reaction*. A chain reaction is any reaction in which the substance needed to start the reaction is also produced as a result of the reaction. In this case, the neutrons needed to initiate nuclear fission are also one of the products of fission.

A nuclear chain reaction will occur in any piece of fissionable material that is large enough to prevent the loss of neutrons to the surrounding environment. The size of the material needed is called its critical size or critical mass.

Finally, the discovery was made that enormous amounts of energy are released during fission. Each fission of a uranium-235 nucleus releases 100 million times more energy than is released in a chemical reaction.

Many scientists recognized the significance of this series of discoveries. The first rumblings of World War II were just being heard in Europe. The possibility of an explosive weapon that used nuclear fission was obvious to at least some scientists. In a fortunate twist of history, most of the German and Italian scientists working on nuclear fission were forced to flee, or chose to flee, their native lands, thus denying Adolph Hitler (1889-1945) the possible use of such a weapon.

By 1939, a number of nuclear scientists were convinced that the United States should begin construction of a fission weapon, an atomic bomb. They persuaded Albert Einstein (1879-1955) to present a letter (actually written by Szilard) to President Franklin D. Roosevelt (1882-1945), describing the bomb and encouraging a national effort for its development.

The U.S. government moved slowly on Einstein's letter. As late as 1943, less than $300,000 had been spent on the development of a fission bomb. It was not until mid-1942 that a program began in earnest to build a bomb. At that point, President Roosevelt authorized the creation of the Manhattan Project. The Manhattan Project was the name given the new Manhattan Engineering District created within the Army Corps of Engineers. Brigadier General Leslie R. Groves (1896-1970) of the Corps was put in charge of the Project.

One of the major tasks facing the project was the production of fissionable material for a bomb. Uranium-235, the only naturally occurring fissionable isotope of the elements, makes up only 0.7 percent of the element. In order to achieve a self-sustaining chain reaction, a much larger concentration of uranium-235 was needed. Since uranium-235 and the far more common uranium-238 are chemically similar, separation of the two isotopes presents a difficult challenge.

The project explored three methods of separating uranium-235 from uranium-238, gaseous diffusion, centrifuging, and electromagnetic separation. Most of the research on separation techniques took place at a new laboratory, built specially for the purpose at Oak Ridge, Tennessee. Directing research on the three separation techniques were Harold Urey (1893-1981) (diffusion), Eger Murphree (1898-1962) (centrifuging), and **E. O. Lawrence** (electromagnetic).

In addition, research was begun on the production of plutonium-239, a synthetic transuranium element that is also fissionable. This research was conducted in Hanford, Washington, under the direction of Eugene P. Wigner (1901-1995), an emigre from Hungary.

Two other major research centers were established at the University of Chicago and at Los Alamos, New Mexico. Fermi directed research on the first atomic pile and the first chain reaction at Chicago, while J. Robert Oppenheimer (1904-1967) was in charge of basic research and final assembly at Los Alamos.

The first atomic bombs were assembled and tested in mid-1945. Their design was simple. Two pieces of fissionable material, each less than critical size, were placed at opposite ends of the bomb. When the bomb was dropped, one piece was propelled into the second piece by means of a conventional explosive. A mass larger than critical size was produced, and a self-sustaining chain reaction was initiated.

The first test of a nuclear weapon was conducted on July 16, 1945, near Alamogordo, New Mexico. The weapon was a success, releasing an amount of energy equivalent to 20,000 tons of **TNT**. Within a month, two bombs were dropped on Japan, a uranium-235 weapon over Hiroshima, and a plutonium weapon over Nagasaki. An estimated 110,000-150,000 people were killed by the two bombs and another 200,000 or more were injured by them. Research on the atomic bomb continued after the war, soon leading to the invention of an immensely more powerful weapon, the nuclear fusion, or hydrogen, bomb.

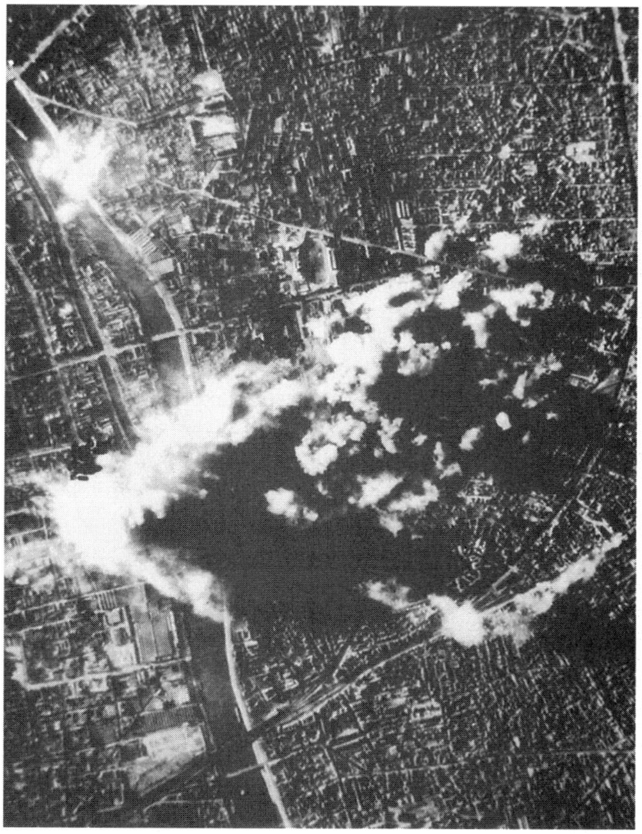

Explosion of the atomic bomb over Nagasaki, Japan, on August 9, 1945.

Atomic clock

The atomic clock is a timekeeping device of unparalleled precision that uses the vibration frequencies within certain atoms and molecules. The time kept by atomic clocks is now accepted as the international standard. American physicist William Frank Libby (1908-1980) formulated the atomic clock theory in 1946. He envisioned a clock whose timing was controlled by an **oscillator**, but one whose frequency did not drift. This was accomplished by comparing the oscillator's frequency to that of an electron moving from one energy level to another. Since the electron's frequency is unfaltering for any particular element, Libby's clock was very accurate—the model built at the National Bureau of Measures in Washington, D.C., was precise to about 21 picoseconds per year.

The name most commonly associated with the invention of the atomic clock is that of **Charles Hard Townes**. The research Townes was conducting on microwave oscillation closely paralleled that of Libby. Townes, however, took the invention a step further: by using an ammonia molecule—the electrons of which vibrate within the microwave range—he constructed an atomic clock whose timing deviated by less than one second for every 30,000 years of operation—far more accurate than Libby's device. The work Townes did on the atomic clock led directly to his invention of the **maser** and, eventually, the **laser**. The atomic clock most often found in laboratories was designed by Norman F. Ramsey, Jr. (1915-) of Harvard University and is driven by the atomic frequency of the element cesium's electrons.

There are two basic types of atomic clocks: active and passive. Active clocks generate an oscillation signal directly from their atoms using the process called stimulated emission. Passive clocks use atoms that have been irradiated with electromagnetic energy to provide a signal. In addition to the cesium and ammonia varieties, clocks using rubidium, hydrogen, thallium, or ions of mercury, barium, and magnesium have been developed. Clocks that produce optical and infrared oscillation frequencies are being developed.

In a world of speed-of-light communications and hyperdistant exploration, the need for a device that accurately and reliably measures very small increments of time is great. For example, atomic clocks are often used in navigation systems: the distance and position of a vessel can be precisely defined by measuring the difference between the time it takes for two signals to reach it; this difference is usually very small, and so a very accurate timing device is essential. The ability to measure small deviations in time has made the atomic clock a useful tool for testing Albert Einstein's (1879-1955) theory of relativity, which claims that time is slowed at speeds approaching that of light. Since manmade technology can achieve only a fraction of the speed of light, atomic clocks are vital to recording these very small shifts in time. Atomic clocks are also valued for their extremely stable frequencies, which are used in communicating with deep-**space probes**.

The most accurate atomic clock yet constructed was put into operation in April, 1993, at the National Institute of Standards and Technology in Boulder, Colorado. It is designed neither to gain nor lose a second in the next one million years. As of the late 1990s, research was under way to shift the standard of timekeeping from atomic clocks to devices that measure time by millisecond pulsars, the extremely fast speed of neutron stars.

Atomic force microscope

In recent years, tremendous advances have been made in the field of microscopy. The electron microscope, once considered state-of-the-art, has been made obsolete by the **field ion microscope** and the powerful **scanning tunneling microscope** (STM). In 1985 a new member was added to this list: the atomic force microscope (AFM). Invented by **Gerd Binnig** (co-inventor of the STM), Christoph Gerber in Zurich, Switzerland, and Calvin Quate (1923-) in California, the AFM represents the technological pinnacle of microscopy.

The AFM uses a tiny needle made of diamond, tungsten, or silicon, much like those used in the STM. While the STM relies upon a subject's ability to conduct electricity through its needle, the AFM scans its subjects by actually lightly touching them with the needle. Like that of a **phonograph** record, the AFM's needle reads the bumps on the subject's surface, rising as it hits the peaks and dipping as it traces the valleys. Of course, the topography read by the AFM varies by only a few molecules up or down, so a very sensitive device must be used to detect the needle's rising and falling. In the original model, Binnig and Gerber used an STM to sense these movements. Other AFM's use a fine-tuned **laser**.

The AFM has already been used to study the supermicroscopic structures of living cells, objects that could not be viewed with the STM. American physicist Paul Hansma (1946-) and his colleagues at the University of California, Santa Barbara, are quickly becoming experts in AFM research. In 1989, this team succeeded in observing the blood-clotting process within blood cells. Hansma's team presented their findings in a thirty-three-minute movie, assembled from AFM pictures taken every ten seconds.

Other scientists are utilizing the AFM's ability to remove samples of cells without harming the cell structure. By adding a bit more force to the scanning needle the AFM can scrape cells, making it the world's most delicate dissecting tool. Scientists hope to apply this method to the study of living cells, particularly floppy protein cells, whose fragility makes them nearly impossible to view without distortion.

Atom smasher • See Particle accelerator

Audiocassette

In 1963 the Philips Company introduced the audiocassette, a device that used magnetic tape to record and replay sound. This convenient form of recording was developed to offer a simple, cost effective alternative to its forerunner, audiotape.

Magnetic recording was originally introduced in the form of audiotape in 1929 by Fritz Pfleumer, a German engineer. Prior to the use of audiotape, sound was recorded and stored on **record**s made of wax or **vinyl**, which proved to be a cumbersome and damage-prone system.

In 1935, a German electronics firm, AEG, began widely marketing a record/playback device that used audiotape. Although the electronics industry refined the clarity and range of AEG's machine, its open-reel format proved to be inconvenient and complicated, since the user had to thread the tape through the machine and onto a take-up reel to record or play back. The technology was invaluable for business use, but because of the complexities of the system, it failed to spread into extensive personal use until the introduction of the audiocassette.

The audiocassette solved many of the problems presented by the reel-to-reel system. For example, instead of threading the tape, the user simply inserted a cassette into the tape player and pressed a button to record or produce the sound. The new format also reduced the time and effort needed to mount, search, advance, and rewind a tape. In addition, the cassette could be stopped and ejected from the machine at any point in its cycle and its hard **plastic** shell protected the tape from damage.

The introduction of the audiocassette was part of a series of interrelated developments within the field of sound recording. Technological innovations created highly dynamic sound reproduction capabilities, as well as an increase in the storage capacity of the magnetic tape itself. Through a significant reduction in width and running time, and the ability to record on both sides of the tape, the recording capacity of the cassettes more than doubled that of audiotape. The popularity of the cassette format was such that within only a few years of its debut, more than sixty companies were manufacturing tapes based on the original Philips design.

In 1966, three years after monophonic cassettes were made available, Philips began marketing **stereo** cassettes. Perhaps the most significant factor in popularizing the use of the medium was that the convenience and compact size of cassettes allowed the development of smaller, more portable tape players and recorders. With the eventual development of personal cassette players such as the Sony Walkman, anyone could listen to anything they chose anywhere they chose.

Continual improvements in recording technology have provided serious competition for the audiocassette. Digital technologies have made it possible to record, store, and play back sound with greater clarity and integrity. **Digital audio tape** (DAT) and its recorders, which became widely available in the United States in 1990, and digital compact cassettes (DCC), introduced for consumer use in 1992, provide the amateur audiophile with the ability to make a tape copy that is an exact reproduction of original sounds rather than just an approximation. Using a code of binary numbers (a series of 0s and 1s) rather than the wave pattern used in standard tape technology, digital technology virtually eliminates the deterioration of subsequent copies that is present in the standard analog method.

Despite these digital developments, however, the audiocassette is dying a much slower death than the **LP records** did in the face of compact discs. In 1992, the Dolby S-type noise reduction audiocassette was introduced. With the right taping equipment, the sound of this type of audiocassette is comparable to digital tape or compact discs. Though sales of music compact discs surpassed those of audiocassettes in 1992, as of 1997, audiocassette players are far more prevalent than compact disc players in cars. It is predicted that while audiocassettes will continue to lose their market share in the music medium, there will be place for them for at least several more years in books on tape. Beginning in 1990, books on tape, an audio recording of a book, usually abridged, were pushed for non-blind consumer use and quickly caught on. Books on tapes remain popular in the late 1990s, and are often used in automobile cassette players and portable cassette players. However, this segment of the market for audiocassettes is also threatened by new technologies including books recorded on compact discs as well as CD-ROM and CD-ROM games. CD-ROM games can feature a level of interactiveness with the text that audiocassettes cannot match.

See also Camras, Marvin; Optical disc; Compact disk player; Phonograph; Poulsen, Valdemar

Audiometer

An audiometer is an instrument used to measure how well a person hears during a basic hearing test, called an *audiogram*, which tests the ability to hear pure tones in each ear.The ear, a complex organ, receives sound in the form of vibrations that strike the eardrum. These vibrations move from the eardrum through the bones of the middle ear to the cochlea, a spiral-shaped organ filled with fluid. The vibration sets the fluid in motion and sensory cells along the cochlea's basilar membrane send messages of the sound to the brain. A complete audiogram tests *bone conduction*—the ability to hear sound transmitted through bone, and *air conduction*—sound transmitted through air. Comparing these two can help determine whether hearing loss is occurring within the middle ear which conducts sound from the ear canal to the inner ear (conductive hearing loss), or in the inner ear or the nerve which transmits sound signals to the brain (sensorineural hearing loss).

The brain distinguishes many distinct sounds: *Pitch*, or *frequency* is a measurement of how high or low the sound is. Frequency is measured in units called *hertz* (Hz), the number of vibrations per second. Each sound has a degree of loudness which is measured in units called *decibels* (dB). The human ear hears a range of sounds from 15 to 100 dB. Sounds over 140 dB, such as those made by a jet **aircraft**, can damage hearing. The audiometer may consist of a **oscillator** to change the frequency of sounds, an audio **amplifier**, and an atenuator to control the volume. The person being tested wears a pair of headphones through which the amplitude of a tone is slowly increased until the person hears the sound. The lowest decibel level at which a sound is heard is called the *threshold*. Using the audiometer, frequency is varied from 64 Hz to over 8,000 Hz. Amplitude can be varied by 5 dB increments. (Newer digital audiometers range from 20 Hz to 20,000 Hz with amplitude resolution at 1 dB and are important for new generation digital hearing aids.) The oscillator is used to change pitch so a range of sounds can be tested. The test produces a graph showing the lowest decibel level at which each frequency is heard and tracing a profile of the person's threshold of hearing, which is then compared to a line on the graph representing normal hearing levels. Hearing is considered good if every tone sounded between 64 and 8,192 Hz is heard at a volume of 20 dB. Losses over 20 dB are considered abnormal. Hearing loss is generally greatest at the high frequencies, which seems to occur in many people over fifty.

The pure-tone audiometer was invented by Georg von Békésy (1899-1972), a Hungarian-American physicist. His was a patient-operated instrument that came out in 1946. Békésy, a Nobel Prize winner who explained the acoustics of the ear, had for many years studied the transmission of sound for the **telephone** company in Hungary. Testing the telephone lines was routinely carried out and often done with pure tones (tones of one frequency). He listened to everything he heard over the telephone lines—even the clicks when phones were being connected and disconnected. He started using the clicks as test signals. The clicks themselves were a combination of many pure tones that came along the telephone lines in a single short pulse. His early experiences helped him study hearing in great detail and arrive at his design for the audiometer.

Auer, Karl (Baron von Welsbach) (1858-1929)

Austrian chemist

Unlike many other notable scientists of the past, Karl Auer had the benefit of a good education, thanks to his father's well-paid position as director of Austria's Imperial Printing Press. After studying at the Vienna Polytechnic, Auer trained in Heidelberg under **Robert Bunsen** (who later invented the famous **Bunsen burner**). There Auer grew interested in the rare earth elements, which led to his discovery of two new metals, neodymium and praseodymium.

Auer's devotion to this class of elements had many practical results. By the late 1800s, gas made from coal had become a common source of light in towns and cities. Although **gaslight** was an enormous improvement over **candle**s and **lamp**s, Auer came up with an idea that would enable gas flames to give off much more light. He noticed that certain rare earth metals would glow brilliantly when heated by a gas flame. In 1885, after testing several substances, he invented the incandescent gas mantle—a cylindrical shell that glows with a bright white light when placed over a gas flame. These mantles were made by soaking a piece of cotton in a metal salt solution (99% thorium and 1% cerium), then burning off the fabric to leave a shell of incandescent metal oxide.

Auer's invention could not have been better timed for the gaslight industry. Electric lights were just being introduced, but the early models produced a relatively feeble light, comparable to old-fashioned gaslights. Auer's mantle gave gas lighting the edge it needed to compete with electric lights for many years.

However, Auer was a true scientist and did not take sides in the war between gaslight and incandescent electric light. In 1898, to improve the electric bulb, Auer invented a filament made of osmium—one of his beloved rare earth metals—to replace the carbon filaments then in use. Although this metal was too expensive for commercial bulbs, Auer had paved the way for development of metallic filaments. Within a decade, an American chemist, **Irving Langmuir**, had invented the tungsten filament, which is still used in today's bulbs.

Also, around 1900, Auer invented an alloy called misch metal to replace flint in sparking devices such as cigarette lighters. Auer's misch metal contained, of course, rare earth elements, notably cerium. At the time, cerium was being dumped in factory yards and given away free with other elements, so Auer was trying to find a practical use for it. His invention represented the first improvement on the prehistoric method of sparking with flint and steel. Modern ''flint-ignited'' lighters actually use misch metal instead of true flint.

Auer's interest in lighting and, specifically, in rare earth metals, dominated his scientific career. When he was elevated to the Austrian aristocracy in 1901, he chose the words ''more light'' as his baronial motto.

Automata

The ancient Greeks were among the first people to create *automata*, machines that imitate the motions of humans and animals. **Hero of Alexandria** wrote extensively about automata and invented several special devices for the miniature stage during his experiments on mechanics and pneumatics. Over the centuries, automata moved beyond one-piece wooden characters on miniature stages.

In the twelfth century, Islamic craftsman and hydraulic engineer **Ismaeel al-Jazari** spent much of his spare time incorporating cogs, **gears**, **cams**, ratchets, **pulleys**, **levers**, and for the first mention in history, the crankshaft, into designing unusual timepieces. One of his designs marked the hour with an orchestra of figures that beat drums, clashed cymbals, and blew trumpets. This piece was much more elaborate and complicated than the European timepieces of the period.

Automata eventually developed into larger, more complex, and often very finely crafted works of art. Some of these became known as androids, a term that originated in 1727. In fact, by the 1700s, very complex and realistic pieces were being made. In 1738, **Jacques de Vaucanson** entertained all of Europe with his duck that swam, quacked, flapped its wings, ate, drank, and defecated.

One remarkable Swiss family became known around the world for their creations. Pierre Jacquet-Droz and his two sons made a living constructing clocks, but they also enjoyed experimenting. In 1768, the family finished a four-year project named ''The Scribe and Charles.'' This automaton could write messages of forty characters or less and motion like a human: he could dip his **pen** into **ink**, move to a new line on the **paper**, and follow his writing with his eyes and head. He wrote with his right hand and moved the paper across the table top with his left hand. Brass cams and removable metal disks in the figure's back controlled his movements and the content of the messages. (The Scribe is now exhibited at an art museum in Switzerland.) The Jacquet-Droz family created another automaton named ''Henry, the Draftsman,'' who could draw four different pre-programmed **pencil** sketches. If any graphite remained on the paper, Henry would even blow it away. The family also created ''Marianne, the Musician.'' This ''young woman'' could play a small organ with all ten fingers. While playing, she would flirt and glance at the audience, lean forward occasionally as if to read her sheet music, and stand up to curtsy at the end of her performance. In time, people began accusing the Jacquet-Droz family of sorcery, as they began creating other objects such as a bleating lamb and a dog that guarded a fruit basket—if someone removed a banana or an apple, the dog would bark until the fruit was put back into place.

Of course, not all automata was as complex and amazingly crafted as these examples. Wolfgang Von Kempelen (1734-1804), for instance, allegedly built an automaton, dressed like a Turk, who could play chess. Von Kempelen toured Europe playing at fairs and in courts, amazing people with the Turk's abilities. One day it was discovered that instead of a machine, the solid-bottomed table contained a small man who actually played the game.

Today, automata abound in all sectors of life. There are toy dolls that move, cry, walk, and talk. Cars that ''go by themselves'' and electric trains with switches, lights, and smoke are enjoyed by people of all ages. The ''Tiki Room'' and other animatronic programs at Disneyland are other examples of automata, conceptually not very different from the creations of the Greeks thousands of years ago.

Industrial robots are a growing aspect of modern automata. Industrial robots are usually equipped with only a single arm that is programmed to perform repetitive tasks in an assembly line. These robots can perform five basic movements that are controlled by small motors or pneumatics: extension or retraction of the arm, vertical movement of the arm, rotation of the arm, rotation of the hand, and pivoting the hand. These ''Hands'' can also be made to grip objects and place components. Two fairly common models now in use are the Versatran and Unimate robots.

With the creation of the computer, the theory of automata has become a highly-developed field. In general automata are designed to respond to changes in external conditions or other input—a category that describes a large number of modern electronic devices, sometimes with the input of sensors, humans, or another automation. Automata theorists have considered the vast class of such systems by idealizing the situation, considering a machine that reads one symbol at a time off a ticker tape, and performs an action based on what it has read. Such a theoretical machine, called a *Turing machine,* was the prototype for the modern digital computer, and has proved useful in the mathematical study of the foundations of mathematics, and for the study of languages, pattern recognition, and artificial intelligence.

Another area of automata research has been in nerve-like or brain-like models, called *neural networks,* which can receive, interpret, store, and use information, analagous to the neurons, nerve endings, and synapses in the brain. They can be used to extract patterns from large amounts of data, and, analagous to biological systems, can be ''trained'' to adapt themselves to future situations based on configurations they have already encountered. With the aid of modern computers, neural networks are being applied to an increasing number of real-world problems of considerable complexity. Besides recognizing patterns, they have the ability to help make decisions with imprecise data. They offer solutions to a variety of classification problems such as speech, character, and signal recognition, as well as system modeling where the underlying processes are not well understood.

See also Robotics

Automated highway system

After World War II, construction of an interstate freeway system seemed like a concept for the distant future—but it quickly turned to reality, thanks to major investments of money, technology, and agency coordination. In the 1990s, development of an automated highway system or AHS seems like a similarly futuristic vision that is surprisingly close to reality.

In 1991, the United States Congress mandated the Intelligent Transportation Systems program and supported its directive to the Department of Transportation with more than $1.2 billion dollars. The Department of Transportation was instructed to create prototype vehicles and an automated highway system by 1997, and, at the National Automated Highway System Consortium in San Diego in 1997, these developments were demonstrated for the first time. The automated highway system blends vehicles and highway sections that both have specialized electronic equipment. Each ''intelligent'' **automobile** has an on-board **computer**, **radar** units on the front and sides, laser rangefinders on the rear, video cameras, and motors to power the steering, throttle, and brakes. Displays feed information to the driver who can choose to activate or deactiviate the automated system and press a button for lane changes. The highway is also equipped with magnets, radio transceivers, or sensor strips that the vehicle can detect. With the automated system operating, the vehicle adjusts its speed to traffic, avoids obstacles, changes lanes to pass slower cars, and stops at the programmed destination.

The Automated Highway System has two primary objectives. Safety is the first goal driving the merging of science and industry. Despite all the safety features in automobiles, human error causes 40,000 fatalities on American roads annually. Five million injuries result, and the cost of these accidents (many resulting from driver fatigue) is about $150 billion dollars. The automated highway system can be activated by the driver, and automated and nonautomated vehicles can drive on the same roads. The second goal of the program is to make better use of the capacity of our existing system of freeways. Irregularities in driving patterns and reaction time lead to congestion, and the automated system can feed vehicles into less traveled lanes or other roadways.

Additional funds for the automated highway system were made available in 1998 to accelerate the availability of the first part of this technology. A collision avoidance system that consists of radar to detect other vehicles and obstacles in the street will be the first advance sold to the public. It will either apply the brakes automatically or warn the driver of the obstacle. Adaptive cruise control, which adjusts speed automatically to surrounding traffic, is also nearing commercial release. Sensors that detect magnets in lane boundaries are also expected to become standard equipment as a result of the research leading to the automated highway system. Automated cars will move in groups toward a shared destination in a concept called ''platooning.'' The driver who wants to maintain control will have so-called ''free-agent control'' with the comfort of safety devices that provide warnings.

In the United States, automobile manufacturers, government agencies, and high-tech firms have formed the National Automated Highway System Consortium (NAHSC) to explore different technologies for intelligent vehicles. Similar alliances have been forged among governments and manufacturers in Europe and Japan. Sensor and radar systems are being studied, as are methods that rely on video cameras and image readouts; some combination of these systems is likely to prove both safe and cost-effective.

The navigational computers that get their bearings from satellite signals and that are available in some car models are one of the first technologies to emerge on a large scale from AHS research. Some states also have ''EZ Pass'' systems at tollbooths that debit the driver's account as a car passes through. ''Incremental deployment'' of these pieces of the system is expected to have lengths of highway and vehicles equipped with some of the companion technology available across the United States by 2008. Computers may completely manage the steering wheel by 2015. While some conflicts may arise as different technologies become commercially available, the phased introduction of components of the system will let drivers adjust to technology with time.

Automatic drip coffee maker

Coffee was first cultivated in Ethiopia in the sixth century A.D. The coffee berries were consumed whole, or a wine was made out of the fermented fruits. Coffee as we know it, made from ground, roasted beans, dates to the thirteenth century, and by the fifteenth century, coffee was popular all across the Islamic world. The drink was introduced to Europe around 1615. The ancient method of preparing coffee was to boil the crushed roasted beans in water until the liquid reached the desired strength. The typical coffee pot was a long-handled brass pot with a narrow throat. This kind of pot is still used throughout the Arab world, and is known in the West as a Turkish coffee pot.

In England and America, boiling coffee in a sauce pan was for a long time the standard method. Sometimes the coffee was boiled for several hours; other classic recipes called for additions to the pot such as egg white, salt, and even mustard.

More sophisticated methods of brewing coffee evolved in France. The coffee bag, similar to the familiar tea bag, appeared in France in 1711. Ground coffee was placed in a cloth bag, the bag into a pot, and boiling water poured on top. Nearly a hundred years later, Jean Baptiste de Belloy, who was Archbishop of Paris, invented a three-part drip coffee pot. The top part of the pot held inside it a filter section made of perforated metal or china. Boiling water was poured through the filter section, and it slowly dripped down to fill the pot below. The percolator was invented in 1825. In a percolator, the pot full of water is placed directly on the stove burner. When the water boils, it condenses in the top of the pot, and then drips through a strainer basket filled with coffee. The Melitta filter—a plastic cone with several openings in the bottom, that holds a paper filter of finely ground coffee—appeared around 1910, as did the glass Silex, an hourglass-shaped filter pot.

The automatic drip coffee maker operates on the same principle as the Melitta and Silex, by dripping boiled water through finely ground coffee in a paper filter. This machine debuted in the United States in 1972 as the well-known Mr. Coffee™. Mr. Coffee™ was an immediate success, and popularized the automatic drip method. As of 1996, some 73% of American households report owning an automatic drip coffee maker. In an automatic drip coffee maker, a measured amount of cold water is poured into a reservoir. Inside the reservoir, a heating element heats the water to boiling. The steam rises through a tube and condenses. The condensed water is distributed over the ground coffee in the filter through a device like a shower head. The water flows through the filter, infusing with the coffee, and falls into a carafe. The carafe sits on a metal plate which has another heating element inside it. This keeps the coffee warm. Some models have timing features, so that they can be pre-filled at night to make coffee at dawn. Other units have a temporary shut-off function, so the carafe can be removed from the warmer plate while the coffee is filtering. Others pulse the water over the filter at intervals, for a slower drip and more concentrated brew.

AUTOMATIC FLIGHT CONTROL • See Automatic pilot

AUTOMATIC PILOT

The automatic pilot has its roots in the **gyroscope**, a weighted, balanced wheel mounted in bearings and spinning at high velocity. As early as 1852 the French scientist **Jean-Bernard-Léon Foucault** had experimented with the gyroscope and found that it tended to stay aligned with its original position and also tended to orient itself parallel to the earth's axis in a north-south direction. Thus, he reasoned, the gyroscope could be used as a **compass** because it designated true, or geographic, north rather than magnetic north, as traditional compasses did, which varied according to their location.

By the early 1900s the **gyrocompass** was a crucial part of navigation. A German manufacturer, Hermann Anschutz-Kaempfe, and an American, inventor, **Elmer Sperry**, had produced two gyrocompasses for use on board ships. Sperry also invented the first automatic pilot for ships, named "Metal Mike," which used the information from the ship's gyrocompass to steer the vessel.

Soon interest arose in applying this method to control **aircraft**. Sperry again led the way when one of his devices was used aboard a Curtiss flying boat in 1912. It used a single gyroscope which, like all spinning masses, tended to resist any change in the plane's axis of rotation. Whenever the airplane departed from its original altitude, a small force was applied to a spring connected to one end of the gyro axis, and this, magnified mechanically, was used to restore movement of the aircraft controls. In 1914, Sperry's son, Lawrence, competed in Paris with fifty-three other entrants to win a prize of fifty thousand francs for the most stable airplane. He demonstrated his plane's stability by flying low over the judges while he took his hands off the controls and a mechanic walked out on the wing.

All simple autopilots since that time have used similar principles. However, for airplane control in three directions (left/right, up/down, wing up/wing down) three gyros are needed. By 1930 the British Royal Aircraft Establishment and several private companies had developed refined autopilot systems that were gradually introduced to both military and civilian aircraft. During World War II much more complicated autopilots were produced. By a simple movement of a control knob, the aircraft could be held on a steady heading at a constant altitude, made to turn at a steady rate in either direction, or even change altitude in a precise manner.

A German, Irmgard Flugge-Lotz (1903-1974) played a key role in developing these automatic controls. As a child in Germany she spent her time at the movies watching engineering documentaries rather than Charlie Chaplin comedies. During World War II she developed methods of controlling aircraft during acceleration and in flying curves; with faster planes, pilots had no chance to correct for any miscalculations. She worked on what she called "discontinuous automatic control," which laid the foundation for automatic on-off aircraft control systems in jets.

After World War II, the United States armed forces became interested in developing an inertial guidance system that used autopilots for **rockets**, **submarines** and manned aircraft. Such a system would not rely on any information from the outside, such as radio waves or celestial bodies, for "fixes." Instead, it would plot its course from information via gyroscopes and from calculations accounting for the turn of the Earth. The United States Air Force turned to a Massachusetts Institute of Technology professor, **Charles Stark Draper**, to develop such a system. He worked on improving gyro units, specific force receivers, amplifiers, servodrives, and other elements. On February 8, 1953, his equipment was aboard a B-29 bomber that left Bedford, Massachusetts, for Los Angeles, California, on a twelve-hour flight. It kept course without deviation, corrected for wind and currents, rose to clear the Rocky Mountains—all with no input from the ground or from the pilot and co-pilot. The inertial guidance system sensed every change in the forward velocity, every move right and left and up and down. By continually digesting these changes, and remembering the plane's take-off point, the system was able to determine in-flight position with great accuracy.

Today, much of the burden of flight has been transferred to autopilots. They can control attitude and altitude, speed, heading, and flight path selection. One great advance has been in landings: signals from an instrument landing system control an aircraft automatically during approach, to land the plane down the glide slope beam, keep it on the runway, and even turn it onto the taxiway, all in totally blind conditions.

Automation

The introduction of automation in the workplace, particularly among industrial settings, has escalated at an astonishing pace. For the factory worker, it has been both a blessing and a curse. While automation has made many dangerous and tedious tasks obsolete, its implementation has also eliminated some jobs entirely.

One of the earliest examples of industrial automation was introduced with the invention of mechanized weaving machines in the 1700s. The many intricate designs and patterns popular at the time caused weavers to spend many frustrating hours creating and often re-working designs. With the use of mechanical weaving machines, weavers increased their output and saved themselves from many tedious hours of labor. In 1801, Joseph Jacquard took the idea a step further, developing a weaving loom programmed and operated entirely by punched cards, an important concept adapted for use in early digital computers.

One group that protested the loss of their jobs to automation was the self-named "Ludds," or *Luddites*, as we know them today. The Luddites were organized bands of English craftsman who rioted around 1811 over the loss of their jobs to textile automation. They destroyed textile machinery around Nottingham, England, but abstained from violence against local citizens, and often received local support. After several deaths, they were repressed by hangings and deportations after a mass trial at York in 1813. Today the word Luddite is used to refer to anyone who is thought to fear automation, technology, and its consequences, and, usually by implication, progress.

A second important forerunner of modern automation was Scottish engineer **James Watt**'s device for mechanical self-regulation. Wishing to improve the design and operation of the modern **steam engine**, Watt developed a centrifugal **governor**—a self-adjusting, automatically controlled component which regulated engine speed. This device was important because it was one of the earliest mechanical devices to use *feedback*, the modification of performance based on input from the environment. The introduction of the governor in 1789 proved an important aid to engine efficiency and enjoyed widespread use in factories across the country.

During the next one hundred years, mechanization in the workplace reached unprecedented levels. By the 1900s, an increasing number of labor-saving machines were being utilized, including electromechanical devices. The application of computer technology to mechanization made completely automated systems possible. In 1946, Delmar S. Harder, a system designer for the Ford Motor Company, created a completely automatic, self-regulating manufacturing system which made use of many of the concepts employed by **robotics** to this day. His system was capable of producing a car engine every fourteen minutes, compared to the previous time of twenty-one hours. Harder introduced the term *automation* to describe automatic industrial processes like his.

Realizing that the feedback laws governing living nervous systems, computers, and industrial automation systems had a great deal in common, Norbert Wiener (1894-1964) launched a new science he called **cybernetics** in the early 1950s. This science elucidated the relationship between stimuli (or inputs) and responses (or outputs) in self-regulating systems, helping to clarify the function of such systems for scientists and engineers working in robotics.

Industrial robots and the use of programmable robot arms made their first appearance around 1954. By 1980, there were about five thousand industrial robots in use in the United States; by the year 2000, industrial experts say there could be half a million. Automation is most widespread in the automotive realm, although it is also prominent in **paper**, steel, and oil manufacturing, as well as the brewing, mining, traffic control, and postal sorting industries.

See also Computer, digital; Computer, industrial uses of; Computer input and output devices

Automobile, electric

The first electric car was perhaps the one built by Robert Davidson of Scotland in 1837. Later, during the 1880s and 1890s, many electric cars were built and sold in Europe and the United States. In fact, there were more electric cars than **gasoline**-powered **automobile**s in America during the late 1890s.

One of the first Americans to market a successful electric car was William Morrison of Des Moines, Iowa. Morrison began his production about 1890. In 1898, the carriage makers Bailey and Company of Amesbury, Massachusetts, fitted one of their carriages with an **electric motor**. Unfortunately, the batteries they used made the carriage too heavy to be moved by the motor. S. R. Bailey and his son Edwin continued to refine their electric carriage, and, in 1908, they introduced their first practical model which could travel about 50 miles (80 km) before it needed to be recharged. About five years later, the American inventor **Thomas Alva Edison** turned to the Bailey carriage firm to test a new storage **battery** that he had invented. A Bailey carriage fitted with the new Edison storage batteries was subjected to severe endurance tests and passed with flying colors. The Bailey-Edison team devised several public demonstrations to promote the electric carriage. Electric cars were popular because they were easy to operate, ran quietly, and ran without emitting any pollution; but the success of electric cars was short-lived. The Bailey Company stopped production of the electric carriage in 1915; and, among the more than 100 electric car companies in business early in the century, only one survived to the 1940s—the Detroit Electric Vehicle Manufacturing Company, which closed its doors in 1941.

Two factors led to the demise of early versions of the electric car. First, **Charles Kettering**'s electric starter, invented around 1912, made gasoline-powered automobiles much more practical. Second, **Henry Ford**'s assembly-line methods reduced the cost of gasoline-powered cars. By the late 1920s, electric vehicles were almost totally supplanted by gasoline-powered ones; delivery vehicles that made frequent stops and electric **trolley**s used in public transportation were among the few uses of electric vehicles to survive beyond the 1920s.

Due to the rising cost and diminishing supply of oil and the pollution problems associated with **internal combustion engine**s, the electric car has enjoyed a popular resurgence since the late 1960s. California's law, which mandates that 2 percent of new cars sold in the state be powered by zero-emission engines by 1998, has led to even greater interest in electric vehicles. In response to the California regulation, General Motors introduced an electric-powered car called the Impact in January 1990. A far cry from the Bailey electric of the 1910s, the *Impact* is a two-passenger sports car that can accelerate from 0 to 60 mph (96 km) in 8 seconds with a top speed of 110 mph (176 kph). It can travel for about 120 miles (193 km) at 55 mph (88 kph) before its batteries need recharging. The major drawback to the *Impact* and other current electric-powered prototypes is that the batteries will need to be replaced about every two years, bringing the operating cost to about twice that of existing gasoline-powered cars. Batteries also require frequent recharging, and, unlike the corner gas station, recharging stations are not readily available.

The lithium-ion battery has been touted as a future hope for electric automobiles. This battery is constructed with an anode made of lithium and cobalt and a carbon cathode. The lithium-ion battery has an energy density three times that of a lead-acid battery, so, in theory, a car equipped with the lithium-ion power source should have three times the range. It also produces greater acceleration and does not need to be fully discharged to be recharged. Cost of the lithium-ion battery remains high, but Nissan plans to introduce its ''Altra'' model, with a lithium-ion power pack, in 1998 and to emphasize the durability and range of this electric car.

Other developments that may make the electric automobile feasible include the fuel cell and the hybrid automobile. Fuel cells use hydrogen, methanol, ethanol, or gasoline as the chemical source for hydrogen which supplies electrons to the fuel cell to generate electricity. Cells that require hydrogen or other gases are not practical because the gases are dangerous and require storage reservoirs too large for the family automobile. But cells that use gasoline produce water vapor as their major waste product, consume the gasoline more efficiently than internal combustion engines, and can benefit from the gasoline stations that are readily available. The hybrid automobile has both electric and internal combustion power plants that function only when they are most efficient; the electric motor runs during low speed driving conditions with frequent stops, and the gas engine works during highway travel. The electric portion helps conserve gasoline and reduce pollution while the internal combustion engine limits the need to recharge the electric battery so frequently. Prototype vehicles equipped with fuel cells are in limited use, and mass production of hybrid cars has proven successful in Japan as of 1998.

Automobile, gasoline

Gasoline was first used as a fuel for vehicles in the early to mid-1820s. Inventors had experimented with vehicles powered by other means almost two centuries earlier; the first horseless carriage was a toy built in 1680 by Sir Isaac Newton (1642-1727), and a road vehicle using a **steam engine** was exhibited as early as 1771. A claim exists that an English mechanical engineer named Samuel Brown drove a self-powered vehicle in London. Brown's engine apparently had separate combustion and working cylinders and used hydrogen gas as a fuel. The patent specifications for his vehicle and eyewitness accounts of his drive both exist, but details are sketchy.

Setting aside the claim made for Brown, the first well-documented instance of an automobile using gas as a fuel dates from around 1862, when French inventor **Jean-Joseph-Étienne Lenoir** adapted a one-cylinder engine to a vehicle that traveled about six miles in two hours. For fuel, Lenoir's engine used the same gas that was used in streetlights, which was called illuminating gas. In 1864, Austrian inventor Siegfried Marcus was ordered to remove his noisy *Kraftswagon* from the streets of Vienna. His engine was similar to Lenoir's but used gasoline as a fuel; consequently, Marcus is credited with developing the first gasoline-powered automobile. Neither Lenoir nor Marcus pursued their early success; these inventors could not envision any commercial value for a ''horseless carriage.''

Others were not so quick to abandon the idea of the automobile, and the 1880s saw several inventors develop automobiles with an eye toward the commercial market. Foremost among these were the Germans **Carl Benz** and **Gottlieb Daimler**. Although the two men apparently never met, they were both employed by the Karlsruhe Maschinenbaugesellschaft—Benz first and Daimler somewhat later. Benz ran his first car early in 1885; it was a three-wheeler powered by a two-cycle, one-cylinder engine. Later that same year, Daimler and his associate **Wilhelm Maybach** mounted a high-speed engine of their own design on a wooden **bicycle**, creating the first **motorcycle**. Both Daimler and Maybach had earlier worked for the firm created by **Nikolaus August Otto**, to develop a four-stroke **internal combustion engine**.

The engine that Otto invented is in large part responsible for making the automobile possible, but Daimler and Maybach believed that Otto poorly understood the engine's potential. For this reason, Daimler and Maybach left Otto's firm. A year after they built their motorcycle, Daimler and Maybach mounted their engine on a modified four-wheel horse carriage, apparently believing that the public would demand that automobiles be essentially converted carriages. Benz felt differently and, from the start, created a vehicle that was designed as a ''motorcar,'' distinct from the horse carriages of the day. By 1880, Benz was employing fifty workmen to build his three-wheeled car. In 1890, he began production of a four-wheeled vehicle. Both of these early German automotive firms were successful, and after the deaths of both Daimler and Benz, their firms merged to form the Daimler-Benz company in 1926, selling the products of the combined enterprise under the *Mercedes-Benz* nameplate.

The French were also innovators in the early European auto industry. In 1891, the French firm of Panhard-Levassor was the first to mount a gasoline engine in the front of a vehicle, and, in 1898, **Louis Renault** was the first to abandon the chain-drive, which had been the universal method of transferring power from the engine to the drive wheels, and adapt a drive shaft to transfer engine power.

The Americans were not sitting idly by while the Europeans developed their automobiles. In fact, while Daimler and Benz both claim the invention of the automobile, their claim is disputed by American backers of George Baldwin Selden (1846-1922). Selden received U.S. patent 549,160 for the invention of the automobile. Selden filed his patent claim on May 8, 1879, several years before Daimler or Benz exhibited their automobiles. Despite Selden's patent claim, there is a dispute as to who actually created the first American gasoline-powered automobile. Some feel the credit should go to the brothers Charles Edgar Duryea (1861-1938) and James Frank Duryea (1869-1967) of Chicopee, Massachusetts, who drove their first automobile on September 21, 1893. The Duryeas were very early automobile success stories (in 1895, they won the first automobile race held in the United States), but their "first American gasoline auto" claim is challenged by supporters of John William Lambert of Ohio City, Ohio, who some maintain ran a gasoline auto in 1891.

Disregarding the claim of originality, other automobile manufacturers and would-be auto manufacturers soon followed. Charles Brady King of Detroit, built a car in 1896—the first of the millions of automobiles that have been produced in the city that has become synonymous with the automobile industry. In 1896, **Henry Ford** introduced his first automobile in Dearborn, Michigan. Alexander Winton (1860-1932) of Cleveland, Ohio, and Ransom Eli Olds (1864-1950) of Lansing, Michigan, also produced their gasoline cars that year. Introduced in 1901, Olds' three-horsepower, curved-dash Oldsmobile was America's first commercially successful automobile. That first year, 425 were built, and, in 1904, five thousand Oldsmobiles were built. The success of the Oldsmobile was quickly challenged, and from 1904 to 1908, 241 companies joined the burgeoning auto industry.

Henry Ford's Model T (his ninth model) was put into **mass production** in 1908. The Model T was powered by a four-cylinder, four-stroke, in-line, water-cooled engine and

Henry Ford's Model T, the first mass-produced automobile. Ford and the relatively inexpensive Model T revolutionized transportation by making automobiles affordable for the general public.

was a huge success. Between 1908 and 1927, Ford's assembly line production methods turned out fifteen million Model Ts. By 1926, the Ford Motor Company was producing one-half of the world's automobiles. Ford's Model T led a revolution in the American auto industry by offering the public an automobile that was relatively inexpensive, versatile, and easy to maintain. The Model T insured that the automobile would no longer be a toy for the well-to-do—its price in 1926 was only $290.

This same period saw the emergence in Europe of the automotive giants Austin, Morris, Singer, Fiat, Peugeot, and Citroën. Innovations in the auto industry were widespread and continuous. Peugeot began experimenting with **pneumatic tires** as early as 1895, replacing the solid rubber tires generally in use. Tubeless tires were introduced in 1948. Another French firm, Renault, introduced the manual transmission in 1899. It had one reverse and three forward gears. The German inventor Fottinger invented the automatic transmission in 1910. The honeycomb radiator for engine cooling was invented by his countryman Maybach, and the Renault company of France introduced a sealed cooling system in 1960. General Motors, founded in 1908 by carriage maker William C. Durant (1861-1947) of Flint, Michigan, introduced the electric starter (developed by **Charles Kettering**) in 1912. Safety glass was first used in 1926, the same year that the Dodge Brothers introduced the first all steel-bodied car. The synchromesh gearbox was introduced by General Motors in 1929, and Citroën produced a front-wheel drive vehicle in 1934, as well as a car with "unibody" construction.

In braking, the first automobiles used a wooden shoe that pressed against the solid rubber tires, but drum brakes, applied by a foot pedal, were developed quite early, though for the rear wheels only at first. Although the Dutch firm Spyker was building cars with four-wheel **braking systems** as early as 1903, four-wheel brakes were not widely adapted until the twenties. Disc brakes were not common until the 1950s, although they too were in experimental use early in the century.

Power windows could be found on autos in 1946, and 1950 saw the introduction of seat belts and power steering. Also in 1950, the British auto manufacturer Rover became the first to build an automobile powered by a **gas turbine** engine. Many other manufacturers followed suit, and, in 1956, Renault's gas turbine-powered "Etoile Filante" (Shooting Star) attained a speed of 192 mph (308 kph).

After nearly a century-long love affair with the ease, speed, and mobility of the "horseless carriage," exhaust emissions from the millions and millions of gasoline-powered vehicles on the road became a major concern. This concern resulted in legislation that mandated pollution-controlling exhaust systems for most U.S. cars by 1968. These exhaust-system controls have done little to reverse the effects of auto-source pollution. What is more, recent environmental studies have suggested that the combustion of fossil fuels—in automobiles and other devices—may be causing global warming, a phenomenon that may alter our planet's climate with results that cannot yet be foreseen. While it is true that gasoline-powered vehicles unquestionably revolutionized human existence, it is possible that the automobile's convenience and mobility have placed an excessive burden on our environment.

The next step in the evolution of the automobile will be to create an environmentally harmless power source. Electric automobiles have been explored as the solution and may yet be proven to be convenient and inexpensive. Engineers believe that some kind of hybrid may be the best answer and are experimenting with fuel cells using different types of fuel. A fuel cell is a type of battery that continuously recharges itself with a fuel source; however, those sources are usually hydrogen and methane, which are dangerous and hard to store. Other possible fuel sources include methanol, which has been used to run an experimental bus in Chicago since March 1998. As engineers come full circle in their study of various fuels, gasoline may again be the answer. Gasoline stores hydrogen as part of its chemical composition. By breaking down the gasoline into its components, the fuel cell obtains the electrons it needs to generate electricity from the hydrogen in the gasoline and more of the energy in the fuel is released than in conventional ignition. Harmful waste products are also eliminated—water vapor is the chief waste from operation of a fuel cell.

Other than the fuel cell, the most eco-friendly solution to the gasoline-powered automobile may be a hybrid that combines the advantages of the electric and internal combustion engines. Toyota's "Prius" went on sale in Japan in 1997; the car works by using its electric power source in stop-and-go driving conditions then switching to its gasoline engine at speed on the highway. Battery use is conserved when the gasoline engine operates, and some braking energy is used to recharge the battery. High speeds and excellent acceleration—advantages of the gasoline engine—are provided by the hybrid while the availability of electric power helps increase fuel economy. Improvements in fuel injection systems and weight reduction by manufacturing auto bodies from recycled plastic are other advances that may meld with fuel cells or hybrid engines to create a more ideal automobile.

See also Automobile, electric; Oil engine; Steam-powered road vehicle; Wankel engine

Automotive electronics

Automotive engineers were slow to adopt electronics applications because the cost was high compared to the benefits to consumers. Electrical systems were also very limited in the beginning. A starter on a 1912 model was the first electrical device used on an automobile. Electric **headlights** and horns followed quickly and replaced dangerous lanterns and bulb-type horns. By 1930, the six-volt electrical system was standard to most American automobile makes. The key electrical components at that time were the battery, the starter or cranking motor, the generator (also called a dynamo), the **ignition system**, the lighting system, and power for accessories and controls.

By the mid-1950s, 12-volt systems replaced the six-volt standard. After World War II, engine speeds and cylinder pressures increased, so the voltage demands for ignition were much greater. Manufacturers also provided a wider range of accessories such as multispeed windshield wipers, radios, and power windows. To ignite fuel, a much higher voltage of about 20,000 volts is needed. The ignition system forms a circuit that routes from the battery through the ignition switch, the ignition coil (an induction coil that increases voltage), a circuit breaker, and back to ground. The circuit breaker or interrupter switch develops high voltage across the secondary coil of the ignition; the coil transfers power to the distributor which connects the coil to each of the **spark plugs** to cause combustion.

The generator feeds electricity to the lighting system, controls, and accessories. The engine crankshaft drives the generator through a belt system. The generator produces electric power, but a voltage regulator balances the electrical needs of the whole system including the requirements for keeping the battery charged. The regulator also provides this balance regardless of the engine speed. When the generator (a direct-current device) can't carry the electrical load, an alternator (using alternating current) furnishes the added charge. Alternators were made available in 1958 when the number of electrically powered accessories began to tax the generator's capacity. The battery stores excess energy so power can be provided to start the engine or operate accessories when the engine is turned off. Power is also drawn from the battery during operations when the generator speed is too low.

The first electronic devices in automobiles were radio receivers that became standard equipment in the 1930s and 1940s. Manufacturers attempted to introduce electronic ignition and **fuel injection** systems in the late 1950s, but costs were prohibitive and the public saw little advantage to these options. In the 1970s, however, the regulations restricting exhaust emissions made ignition and fuel injection systems necessary. Electronics applications were ideal for these systems because their controls were so precise.

By the 1990s, almost all automotive systems are equipped with sensors and actuators that are linked to electronic control units (ECUs) and an onboard computer. Sensors detect manifold pressure, crankshaft angular position (which controls the timing of both the ignition and fuel injection systems), the ratio of air to fuel that must be optimal iffuel injection and emissions controls are to be effective, and coolant temperature. Other sensors measure exhaust gas oxygen (EGO), which is a direct indicator of other emissions. A "knock" sensor evaluates the rise in cylinder pressure during combustion, which causes an annoying knocking sound in the engine, and retards ignition in order to stop the knocking.

Critical actuators that are parts of the car's electronics system include the fuel metering actuator that controls fuel metering rods in electronic **carburetors**, and the exhaust gas recirculator actuator or EGR that links the intake and exhaust manifolds so that exhaust that is high in emissions is recycled through the system until the emissions are reduced. An actuator with functions the driver can observe directly is the cruise control throttle actuator, which adjusts acceleration and throttle motion to maintain the selected cruise speed.

Electronics are being adapted to many other automotive systems. Occupant safety devices like **safety belts** and **airbag**s are connected to electronic warning signals and actuators. Soon, the driver will be able to monitor low tire pressures, thanks to electronics. Collision avoidance and electronic navigation systems are past initial development and introduction stages and show great promise for linking drivers on the intelligent highway. Speech synthesis devices are being tested so that gauges on the instrument panel can deliver verbal reports to the driver, and "sensor multiplexing" will allow improved coordination of the automobile's devices.

Axe

Axes, perhaps the first all-purpose tools, date back some 400,000 years, before the discovery of fire. Made originally from crudely chipped stones, the axe eventually evolved into the flint hand variety of the Old Stone Age, or Paleolithic period. Primitive axes of this type were commonly used for killing animals, cutting meat, and fashioning other tools from flint—a hard, yet versatile, siliceous rock.

By 30,000 B.C., wooden hafts, or handles, were added to flint axes. Around 4000 B.C., axe heads were being molded of copper, which was replaced over time by bronze, **iron**, and steel. Heads were originally attached to handles by a variety of means, none of which ensured lasting durability; eventually, axe heads were specially formed with large openings called eyes to allow custom fitting of handles.

During the Medieval Age the axe became a valuable agricultural tool, indispensable for the clearing of forests. It was also used for military purposes in the form of the battle-axe, for hand-to-hand combat.

More recently, particularly in the United States, the axe was prized for its many uses in settling frontier wilderness. Because it was employed to fell trees and build log homes, the axe became a prominent symbol of westward expansion.

Early pioneers arriving in America brought with them heavy English and German axes, carryovers from older styles used in war. American conditions brought a need for a newer, lighter axe shape and better manufacturing procedures.

Two New England hardware store owners who took advantage of the situation were brothers Samuel and David Collins. During the 1830s they bought an old gristmill and converted its wheel, once used for grinding, to drive huge triphammers and furnace bellows. Large grindstones, six feet (1.8 m) in diameter, were used to shape the blades. Instead of the old bellows system, air for the **forge**s was carried by a series of hollow chestnut pipes from a huge wooden pump driven by a **water wheel**. Pioneers began using charcoal rather than coal to fire their furnaces because the coal provided higher temperatures. From iron ore imported from Sweden they made their own steel, which was used for the bit (cutting edge) of the axe. The steel was then fitted between two iron strips that formed an eye for the handle.

Pioneering manufacturing procedures were advanced for their time. A Collins workman could temper and forge eight axes per day due to improved conditions in the plant and the ingenuity of **Elisha King Root**, who invented and refined special machines that could stamp and sharpen the axes. One of his machines could forge the axe blade in a standardized shape so that it could be fitted to prepunched holes in the axe handle.

The new American tools were so good that a British company tried to imitate them, but lost to the Collins Brothers in a court case. This litigation represented a great victory for the fledgling American manufacturing industry. Although several varieties of American axes existed at the time, nearly all featured curved, lightweight handles and heavier heads for easier swinging and increased power. The design of the head was the true innovation of the American axe; because of the nearly equal weight of the bit and the poll (flat edge), the new axe offered superior balance and accuracy.

By the 1880s the crosscut saw had all but replaced the axe for tree-cutting. During the twentieth century, chainsaws and automatic log-splitters have become the preferred tools. Nonetheless, specialty axes, such as the fireman's axe and the logging axe, are still used today, as are the descendants of the early American axes, which remain effective tools for chopping and splitting wood. In addition, specialized axes designed for ice and mountain climbing are sold.

See also Alloy; Blast furnace; Blower; Heating systems

AXLE • See Wheel and axle

B

Babbage, Charles (1791-1871)
English inventor

Charles Babbage, the son of a wealthy English banker, was born in the early years of the Industrial Revolution. As that revolution progressed, he became one of its foremost—and most controversial—spokesmen. Because of his views on what the scientific method could do for industry and commerce (and a few other eccentricities), most of his contemporaries regarded his as a crank. Today, he is remembered as the brilliant mathematician who invented the prototype of the **digital computer**.

Babbage entered Cambridge University at the age of 19, and it was there that he thought of a computer first crossed his mind. One evening in 1812, as he sat gazing at a table of mathematical data, it occurred to him that a machine could calculate such data faster than humans and without human error. In the increasingly complex world in which he lived, errors in mathematical tables were becoming a matter of real concern; particularly serious—since they were the cause of frequent shipwrecks—were inaccuracies in navigational tables. Babbage had occasion to return to his idea of a computing machine about 1820, when as a member of a learned society, he was given the task of verifying tables of astronomical data. The numerous errors he found in them led him to exclaim, "I wish God these calculations had been accomplished by steam." Babbage never did see numbers calculated by a steam-powered machine, though he spent the rest of his life and much of his fortune trying to build one.

In 1822 he completed a small working model of the Difference Engine, a machine that could compile and print mathematical tables. The next year, after convincing the British government to fund the project, he began building a full-scale version. Progress was slow and expensive, since **machine tools** for making the parts had to be custom-crafted. But Babbage's efforts, which included a study of all mechanical devices that could be used in building his machine, had a profound effect on mechanical engineering. The ensuing improvements to machine tools and techniques were worth far more to the British industry than the £17,000 the government ultimately spent on the project. The government withdrew its support of the Difference Engine in 1834, but by then Babbage had already conceived the idea for his Analytical Engine. A programmable automatic machine, the Analytical Engine was the direct ancestor of the modern digital computer.

Babbage's constant improvements and redrafting of plans may be one reason the Analytical Engine was never finished. Another was the technology available to him. It would have been far simpler for him to implement his ideas with electromechanical devices than with mechanical ones, but it would be many years before electrical technology was reliable enough for Babbage's purposes. Babbage himself would have blamed the shortsightedness of the British government for the failure of his machines ever to see practical application. But the real wonder is how he ever managed to get any government funding at all. The idea of a machine that seemed to perform human thought processes struck many of his contemporaries as ridiculous, if not sacrilegious.

Babbage tossed off ideas and designs for other devices with great abandon—among them, a **cowcatcher** for **locomotives**, a meter to reduce water waste, a device for recording earthquake shock waves, and a skeleton **key**. All these seem to have been of fleeting interest to him, obsessed as he was with his computing engines. Typically, credit for the **ophthalmoscope** that he invented in 1848 went to **Hermann von Helmholtz**, who invented the same device in 1851; his multipurpose machine tool, which would have been very useful in small factories, was never built; and his **lighthouse** signalling system was pirated by a Russian naval officer (to whom he had gladly shown it) and later patented by an English admiral (who profited from it).

Had Babbage been born a few generations later, when technology was more advanced, the world might not have had to wait until the mid-twentieth century for the digital computer, and Babbage might have died content. As it was, a friend who visited him before his death reported that he sounded as

Charles Babbage

if he hated ''mankind in general, Englishmen in particular, and the English government and organ grinders most of all'' (organ grinders and other street musicians have been among the few things that could distract him from his work). His hatred was, however, abstract; it melted away in the face of anyone with a hard-luck story. He was, as one of his biographers characterized him, a lovable but ''irascible genius.''

See also Aiken, Howard; Heliograph; Lovelace, Ada Augusta, Countess of; Speedometer; Stage lighting

Baby bottle

Bottles to feed infants have been in use for many centuries. The first consisted of urns with two openings: one for pouring the liquid into the bottle and the other to be put in the baby's mouth. Predictably, the baby bottle has changed markedly over the last five centuries. The sixteenth-century bottle resembled a duck, with the baby being fed from the beak. Modern baby bottles are cylindrically shaped and are made of **plastic** or **glass**. Some modern bottles have disposable plastic liners which eliminate the sterilizing process that sometimes makes glass and plastic bottles inconvenient. A bottle is available that stores formula and water in separate chambers. Unlocking the mechanism mixes the two, allowing convenient transport without refrigeration. Nipples for baby bottles have also changed. Until the end of the eighteenth century nipples were made from a piece of rolled linen, inserted into the liquid at one end and into the baby's mouth at the other. Sponge leather, wood and dried cow's udder were also used. When **rubber** appeared on the market during the nineteenth century, it became the favored material. Strangely, the rubber was used primarily to cushion a metal nipple, sometimes including a spring mechanism to increase the flow. Modern nipples are either **silicone** or a synthetic rubber-like material with holes of varying sizes to match the child's age and bottle contents.

Baby carriers/pouches

Ann Moore, the mother of three grown children, is the force behind the *Snugli*, a cloth baby-carrier she devised after returning to the United States following a Peace Corps stint in Togo. After her first child was born in the early sixties, Moore longed for a baby carrier similar to those the Africans used, which dispensed with the cumbersome bulk of the traditional Western stroller and permitted parent-child contact. Moore's first attempt to fashion a model like the African carrier she admired was a lengthy strip of fabric she used to bind her child to her chest. The next effort was a joint project with her grandmother, Lucy Aukerman. The pair made a pouch from an old sheet, added crisscrossed backstraps and holes for the infant's hands and feet, and the *Snugli* was born. By 1983 their creation was earning Moore and her husband $6 million a year.

As popular as the *Snugli* was, one mother found it uncomfortable. Andrea Proudfoot of Eugene, Oregon, designed a baby carrier that shifted the weight of the child to the back rather than the chest. Her carrier, *Andrea's Baby Pack*, was then patented, and Proudfoot's cottage industry was launched. Infant car seats with a removable carrier allow sleeping babies to continue to sleep when being lifted from the car. Modern baby carriers are made in styles that cart babies in front, in back, or in a sling or pouch arrangement. Baby may face out or face in, and sometimes a single carrier offers both choices. Selection of a carrier depends on the age and size of the child, the size and possibly the fitness level of the person doing the carrying, and personal preference, both parent and child's. For some mothers, the ability to conveniently nurse while the baby is in the carrier is important. As with the earliest *Snugli*, the inspiration for many carrier designs is an ethnic group such as Inuit, African, or Native American.

Baekeland, Leo (1863-1944)
Belgian chemist

Leo Hendrik Baekeland was born November 14, 1863, in Ghent, Belgium. He graduated from high school first in his class when he was 16 and earned a scholarship to the University of Ghent. At 21 he received his doctorate degree with highest honors. He arrived in the United States by way of a traveling fellowship, settled in the state of New York and went to work for a photographic firm.

In 1891, Baekeland perfected the manufacturing process for ''Velox,'' a gelatine silver chloride **paper** invented by Josef Eder (1885-1944). The paper made it possible to develop photographic prints under artificial light. He sold his invention to **George Eastman**, owner of *Kodak*, for one million dollars in 1899.

Baekeland bought a home in Yonkers, New York, with his fortune and built a laboratory where he began experiments in electrochemistry. He was granted patents for his work with electrolytic cells. Baekeland then began searching for a substitute for **shellac** which at the time was an entirely natural product. Baekeland felt the market was ready for a cheaper substitute.

Baekeland centered his research on finding a solvent that would dissolve a resinous substance formed by a condensation reaction of formaldehyde with phenol. As noted by Baeyer in the 1800s, this tacky residue was nearly impossible to remove from laboratory glassware. Baekeland felt that a solvent capable of breaking down this residue would have to possess the shellac-like properties he was looking for.

After long research turned up no appropriate solvent, it occurred to Baekeland that a residue impervious to solvents might be more than a nuisance after all. He began attempting to *create* an impervious resin. He built a reaction vessel, which he called a Bakelizer, and began experimenting with the phenol-formaldehyde reaction. By controlling the chemical proportions, catalysts, pressure, and temperature, he eventually succeeded in forming a clear solid that was heat, water, and solvent resistant, and nonconductive. Baekeland patented the solid in 1907, naming it **Bakelite**. It could be easily machined and could be dyed any color with no adverse effects on its physical properties. Bakelite was first used in automotive applications. Soon it replaced hard **rubber** and amber for electrical uses and is still used in industrial arts where thermoplastics are unsuitable. Though Baekeland did not fully understand the chemical structure of his invention his careful record keeping and observation during his experiments made his search a success.

Bakelite was the first totally synthetic **plastic** and the first thermoset plastic. After Bakelite was successfully developed, the search was on for other artificial substitutes for natural materials such as rubber and silk. By the 1940s, this research began to pay dividends—and today, petrochemical plastics and fabrics are important in almost every aspect of our daily lives.

Leo Hendrik Baekeland

Baeyer, Adolf von (1835-1917)

German chemist

Adolf von Baeyer was born in 1835 in Berlin. He studied at the Friedrich-Wilhelms Gymnasium. He attended the University of Berlin where he studied physics and mathematics. He studied physical chemistry at Heidelberg under **Robert Bunsen**. Unsatisfied with the field of study, he transferred to the private laboratory of organic chemist August von Stradonitz Kekule, also in Heidelberg. While working with Kekule, Baeyer compiled enough data to support his thesis on organic arsenic compounds; based on that thesis he received his doctoral degree in 1858. Later that same year, Kekule move the lab activities to Ghent; Baeyer followed. En route to the new site, Baeyer met Adolf Schlieper, who had once worked on a uric acid research team and gave Baeyer one of his remaining samples of the substance. Through further investigation, Baeyer discovered barbituric acid, the basic compound of the barbiturate family. While in Ghent he earned the basic requisites to qualify as university instructor and lecturer in the field. Returning to Berlin two years later, he began intensive research in uncovering the parent compound from which indigo is derived. In 1841 Auguste Laurent introduced oxygen into the indigo molecule, which broke it into two equal molecules—this resultant compound he called isatin. In 1870, Baeyer further broke down the substance (in an effort to discover the parent compound) by distilling the isatin over hot zinc dust. This action removed both oxygen atoms leaving the substance he called indole—the parent compound of indigo. Eventually he developed a new technique to create isatin from phenylacetic acid. Thus, Baeyer had succeeded in producing indigo from common chemicals. The distillation process had a great impact on the **dye** industry and became widely used to produce dyes from other materials, such as madder (used to dye British army and fox-hunting coats). Although Baeyer did not participate in the fine tuning of the industrial production of indigo, by 1897 synthetic indigo was readily available to the open market. In 1875 he moved to Munich to accept a professorship. It was at Munich that he began work with acetylene and polyacetylene compounds. Eventually, this research led to his study of cyclic compounds. Cyclic (ring) compounds are compounds of carbon atoms which are arranged as rings. Baeyer theorized the reason that naturally occurring ring compounds have five and six carbon atoms was due to the bond angles. According to Baeyer's strain theory, the more a bond angle deviates from a tetrahedral angle the more unstable the bond is. The smaller the angle (less three or four member rings) the more strain was put on the bond; larger angles (more than six member rings) would also result in more strain. He also made the presupposition that all the rings were planar. His theory came under fire in 1890 and was disproved in 1969. Baeyer also conducted research in the areas of polyacetylenes, oxonium salts, and rosaniline dyes. In 1905, he was awarded the Nobel Prize in chemistry. He died in 1917.

Baird, John Logie (1888-1946)

Scottish inventor

John Logie Baird was one of the principal players in the early days of **television**. His invention, the photomechanical television, was the first to broadcast a live transmission. Born in Scotland in 1888, Baird received his education at the Royal Technical College and the University of Glasgow. Plagued by poor health, he was unable to serve in World War I and was ultimately forced to resign his position as an electrical engineer. He then decided to become a ''professional amateur,'' and pursued many different interests and enterprises. However, after exhaustion led to a nervous breakdown, Baird chose to concentrate on **electronics**, especially following **Guglielmo Marconi**'s demonstration of how **radio** waves could be used to carry an audio signal. Baird was certain that a similar process could transmit a visual signal, and he began working upon a design that would do so. At the heart of Baird's design was a device called a Nipkow disk, a scanning disk invented in 1884 by the German scientist **Paul Nipkow**. Basically, this device was comprised of a cardboard disk with a series of square holes, situated in a spiral. When coupled with a **photoelectric cell** and spun, the Nipkow disk is able to scan areas of lightness and darkness and convert that information into an electrical signal. By using a second disk, synchronized with the first, Nipkow was able to retranslate that signal into a primitive visual image. Baird took Nipkow's idea one step farther, developing a system by which the signal could be sent via electromagnetic waves, rather than cables. While still in the developmental stages, Baird's invention found little financial support, since most investors considered it a mere novelty. During this time Baird worked as a shoe shiner and a razor blade salesman, earning just enough money to pay for food, shelter, and mechanical supplies. Much of the prototype for his invention was built out of household items such as a cake tin, knitting needles, a bicycle lamp, and string. On October 2, 1925, Baird succeeded in sending the image of a ventriloquist's dummy from one end of his attic to the other. Exhilarated, he ran to the shop downstairs and persuaded a young boy to become the first person to have his image transmitted by television. Baird became famous nearly overnight, and soon investors were giving him enough money to pursue even more ambitious goals. In 1927 he sent a television signal from London to Glasgow and in 1928 from London to New York. Unfortunately, the Nipkow disk and the photomechanical design produced an image of very poor resolution—a flaw inherent to the mechanical design. Soon, Baird's invention would be replaced by the **cathode-ray tube** design of **Vladimir Zworykin**. Still, Baird continued to strive for better television designs. He helped to develop natural color television as well as large-screen projection, which, he envisioned, would ultimately allow the public to watch television on a movie screen. Baird died in obscurity, his early contributions nearly forgotten, in 1946, at the age of 58.

Bakelite

Belgian-born chemist **Leo Hendrik Baekeland**, working independently, patented Bakelite in 1909. For several years Baekeland had been searching for a solvent to dissolve the product of a condensation reaction between formaldehyde and phenol. His intention was to use the solvent as a substitute for **shellac**. Because the residue was resistant to so many solvents, Baekeland became intrigued with it and began to search for a way to produce it in large quantities. He began to experiment with the substance, using a reaction vessel he built and named for himself, the Bakelizer. Baekeland found that by controlling the temperature and pressure of the reaction, he was able to produce a clear solid in large amounts. He called it Bakelite after himself.

For the first 10 years or so following its introduction, Bakelite was used primarily for electrical and automobile insulators, and in heavy industrial products. Eventually, Bakelite spread to the consumer market in the form of castings that could be sculpted by designers using fast **wheels** and razor-like tools. After World War II, many companies began using patterned molds instead of hand-carving.

When the Bakelite patent expired in 1927, it was acquired by the Catalin Corporation. Catalin began mass production using the cast resin formula, and succeeded in adding 15 new colors (including a marbleized effect) to the 5 originally introduced by Baekeland.

In 1942, production of novelty items was suspended in order to concentrate on wartime needs. These wartime products, which included defense phones and aviator goggles, numbered in the thousands. By the end of the war, a new generation of **plastics**, including **fiberglass**, **vinyl**, and **acrylics** had been developed; and Bakelite and Catalin had largely become obsolete.

Nevertheless, because Bakelite is resistant to solvents, water, and heat, and is nonconducting, Bakelite retained a specialty market niche. It was the first fully synthetic thermoset polymer and one of the first artificial substitutes for a natural product. (A thermoset polymer is one that does not soften when heated.) It replaced hard **rubber** and amber as an electrical insulator, and is still widely used in automotive applications and in **radios**.

The discovery of Bakelite prompted a concerted effort to produce plastics and polymers unlike anything found in nature (like polystyrene and low density polyethylene). When

Baekeland died in 1944, more than 125,000 tons of Bakelite were being produced annually and research into plastics and polymers showed no signs of diminishing.

A resurgence of interest in Bakelite occurred in the mid-1970's when artist, photographer Andy Warhol (1930-1987) started collecting Bakelite carvings, eventually amassing one of the world's largest collections. At the time of his death in 1987, Bakelite had reached the high prices that it never could command during the Depression, with deeply carved bracelets and designed pins selling for between $900 and $1500.

See also Artificial silk; Insulating technique; Polymer and polymerization; Rubber, synthetic

Baker, Sara Josephine (1873-1945)

American physician

Sara Josephine Baker was a pioneer in public health care and preventive medicine in the early part of the twentieth century. Josephine Baker (as she preferred to be called) was born into a wealthy New York family. When her father died, she decided to become a physician, an unheard of ambition for girls of that time. Finding herself ill-prepared for medical school, Baker studied biology and chemistry at home for a year before applying to the Women's Medical College of the New York Infirmary for Women and Children, one of the few colleges that offered women work in medical clinics. She entered school at 18 and received her M.D. in 1898. Baker first served as an intern at the New England Hospital for Women and Children in Boston where she learned the harsh realities of sickness and death in turn-of-the-century slums. Once, while taking care of a sick woman, Baker had to defend herself by kicking a drunk husband down the stairs.

Baker returned to New York and opened a medical practice with a classmate, Dr. Florence M. Laighton. Unfortunately, women doctors were such a rarity at that time that they had few patients and had to close the practice. Baker then became a medical inspector for the City of New York Department of Health. Unlike other medical examiners who never left their offices, Baker went to the people, climbing stairs into slum apartments and entering schools to check on sick children. In the poverty-stricken parts of New York City, she saw first-hand how filthy conditions led to the spread of disease. She went to the Bowery in the middle of the night to vaccinate against smallpox, during the meningitis epidemic in 1905 she worked tirelessly diagnosing and treating the disease, and also helped the Department of Health track down "Typhoid Mary" Mallon, the infamous carrier of that deadly disease. Baker was a crusader in encouraging school children to report and treat common problems of the time such as head lice and eye and skin infections. With Lina Rogers (who may have been the first public health nurse in the country) she started the school nurse program in New York City.

Baker was familiar with the statistics: one-third of the deaths reported in New York City were of children under five years old, and one-fifth were babies less than a year old. Knowing from experience that poor people had little education and knew little about disease prevention, she started a pioneer project in the summer of 1908, sending 30 school nurses into homes of newborns educating mothers on caring for their infants. The results were dramatic—1200 fewer deaths that year—which led the city to form the Division of Child Hygiene, later called the Bureau of Child Health. This was the first taxpayer-supported agency in the world devoted to the health of children. As director of this bureau, Baker spent the next 15 years instituting sound practices that improved the lives of thousands of children. Among her observations, Baker noticed that there were hundreds of children suffering from preventable blindness. She helped design and distribute a dispenser for administering the compound, *silver nitrate* to the eyes of all newborns. This compound helps prevent infection and blindness that results from the mother's infection with the sexually-transmitted disease *gonorrhea.* Baker also created a design for newborn clothing so that babies could be dressed in easily managed layers: underwear and outerwear could be laid out flat so that a baby's arm could be pulled through all layers at once. Sewing companies produced patterns for these clothes which became very popular and affordable.

Sara Josephine Baker

Despite all of Baker's achievements, some bureaucrats sought to have her dismissed; however, mothers whose children had benefitted marched in protest and her position was saved. Baker was no stranger to gender bias in her professional life. Members of one medical organization narrow-mindedly made comments like "it (her program) was ruining medical practice by keeping babies well.'' When a group of Brooklyn physicians sought to eliminate the Bureau of Child Hygiene, Baker took it as a compliment and continued about her work. She often wrote using just her initials—Dr. S. J. Baker. When she was invited to present a paper in Philadelphia, the audience of male physicians were shocked and astounded to discover her gender. Despite this lack of support from the medical profession, she wrote numerous respected books on the health of children. She was also an activist for women's right to vote, and a successful fund-raiser for all causes in which she believed.

Baking powder

Baking powder is a leavening agent, which is a substance that helps baked goods rise. It is based on baking soda, also known as sodium bicarbonate, which has the ability to neutralize acids. When baking soda is mixed with an acidic ingredient, such as buttermilk, chocolate, or molasses, it neutralizes the acid and releases carbon dioxide bubbles. These bubbles, in turn, cause the dough to swell.

Baking powder also contains an acid salt, such as cream of tartar, sodium aluminum sulfate, or calcium acid phosphate, with which it reacts when added to a liquid. This allows baking powder to work in recipes that do not include other acidic ingredients. Modern double-acting baking powder contains two types of acid, one activated by liquid and the other by heat. The extra, tiny air holes thus created during baking give baked goods a lighter texture. In addition, baking powder may contain starch, which serves as a filler and drying agent.

In the 1790s, people in the United States first began using a crude baking powder based on potassium carbonate, which came from perlash obtained from wood ashes. This powder was packaged and sold as *American Saleratus*, the latter word meaning ''aerated salt,'' because the powder seemed to add air and lift to **bread** dough.

In the 1830s, a New York doctor named Austin Church began looking for a way to produce a cheaper, cleaner *salteratus*. Church heated soda ash, also known as sodium carbonate, over a wood-burning fire and found that in time it changed into sodium bicarbonate. In 1846, Church and his brother-in-law, John Dwight, began large-scale distribution of baking soda. Eventually the two men split to form competing companies, and Church introduced the *Arm & Hammer* brand name. In 1896, descendants of the two men rejoined the companies under the name Church & Dwight. This company still controls about 90 percent of the U.S. consumer market for **baking soda**.

Around 1835, the first modern-style baking powder was developed. It was a mixture of baking soda and cream of tartar, an acidic substance obtained from the residue in wine barrels. A commercial version of this mixture was introduced in 1850. By the end of the nineteenth century, other acid salts began to replace cream of tartar. Today baking powder for home use typically consists of baking soda, sodium aluminum sulfate, calcium acid phosphate, and cornstarch.

Baking soda

Baking soda is a white crystalline powder ($NaHCO_3$) better known to chemists as sodium bicarbonate, bicarbonate of soda, sodium hydrogen carbonate, or sodium acid carbonate. It is classified as an acid salt, formed by combining an acid (carbonic) and a base (sodium hydroxide), and it reacts with other chemicals as a mild alkali. At temperatures above 300 degrees Fahrenheit (149 degrees Celsius), baking soda decomposes into sodium carbonate (a more stable substance), water, and carbon dioxide.

The native chemical and physical properties of baking soda account for its wide range of applications, including cleaning, deodorizing, buffering, and fire extinguishing. Baking soda neutralizes odors chemically, rather than masking or absorbing them. Consequently, it is used in bath salts and deodorant body powders. Baking soda tends to maintain a pH of 8.1 (7 is neutral) even when acids, which lower pH, or bases, which raise pH, are added to the solution. Its ability to tabletize makes it a good effervescent ingredient in antacids and denture cleaning products. Sodium bicarbonate is also found in some anti-plaque **mouthwash** products and **toothpaste**. When baking soda is used as a cleaner in paste form or dry on a damp sponge, its crystalline structure provides a gentle abrasion that helps to remove dirt without scratching sensitive surfaces. Its mild alkalinity works to turn up fatty acids contained in dirt and grease into a form of soap that can be dissolved in water and rinsed easily. Baking soda is also used as a leavening agent in making baked goods such as **bread** or pancakes. When combined with an acidic agent (such as lemon juice), carbon dioxide gas is released and is absorbed by the product's cells. As the gas expands during baking, the cell walls expand as well, creating a leavened product.

In addition to its many home uses, baking soda also has many industrial applications. For instance, baking soda releases carbon dioxide when heated. Since carbon dioxide is heavier than air, it can smother flames by keeping oxygen out, making sodium bicarbonate a useful agent in **fire extinguishers**. Other applications include air pollution control (because it absorbs sulfur dioxide and other acid gas emissions), abrasive blastings for removal of surface coatings, chemical manufacturing, **leather** tanning, oil well drilling fluids (because it precipitates calcium and acts as a lubricant), **rubber** and **plastic** manufacturing, **paper** manufacturing, textile processing, and water treatment (because it reduces the level of lead and other heavy metals).

Imported from England, baking soda was first used in America during colonial times, but it was not produced in the United States until 1839. In 1846, Austin Church, a Connecticut physician, and John Dwight, a farmer from Massachusetts, established a factory in New York to manufacture baking soda. Dr. Church's son, John, owned a mill called the Vulcan Spice Mills. Vulcan, the Roman god of forge and fire, was represented by an arm and hammer, and the new baking soda company adopted the arm and hammer logo as its own. Today, the Arm & Hammer brand of baking soda is among the most widely recognized brand names.

Named after Nicolas Leblanc, the French chemist who invented it, the Leblanc process was the earliest means of manufacturing soda ash (Na_2C0_3), from which sodium bicarbonate is made. Sodium chloride (table salt) was heated with sulfuric acid, producing sodium sulfate and hydrochloric acid. The sodium sulfate was then heated with coal and limestone to form sodium carbonate, or soda ash.

In the late 1800s, another method of producing soda ash was devised by **Ernest Solvay**, a Belgian chemical engineer. The Solvay method was soon adapted in the United States, where it replaced the Leblanc process. In the Solvay process, carbon dioxide and ammonia are passed into a concentrated solution of sodium chloride. Crude sodium bicarbonate precipitates out and is heated to form soda ash, which is then further treated and refined to form sodium bicarbonate of *United States Pharmacopoeia* (U.S.P.) purity.

Although this method of producing baking soda ash is widely used, it is also problematic because the chemicals used in the process are pollutants and cause disposal problems. An alternative is to refine soda ash from trona ore, a natural deposit.

BALANCE AND SCALE

One of the earliest known measuring devices, the equal-arm balance was used in Egypt during the fourth millennium B.C. to determine exact and relative weights. The Egyptians used the scale for measuring goods to be traded, as well as for weighing gold dust received in payment. This ancient balance consisted of a horizontal, pivoting beam from which were suspended a pan at each end, equidistant from the pivot. One pan held the item to be weighed; standard weights were added to the other. The weight of the item could be determined by adding or removing weights from the second pan until the two were equal. Though primitive, these two-pan balances were capable of accuracy as high as 99%. Probably the most significant improvement to the balance design was made by the Romans. They added a knife-edge, called a fulcrum, to the pivot point, which made the device much more sensitive and accurate, particularly when determining very small weights. For years agate was used as the fulcrum-piece, but many balances have since replaced these with fulcrums made of corundum (synthetic sapphire).

Most modern balances are of the *single-pan* variety. Instead of using two pans, this device replaces one pan with a large, heavy counterweight. The remaining pan is equipped with a number of smaller weights; when the item to be weighed is placed within the pan, these small weights make the pan heavier than the counterweight. The weight of the unknown can then be determined by removing enough small weights for the two sides to balance. Another balance, called the deflection balance, does not require the two sides to be balanced exactly. This type uses a vertical needle to make a more accurate measurement of the unknown's weight. When the two sides are nearly balanced the needle stands almost straight up, tilting slightly to one side or the other. The amount the needle is tilting can be measured to determine the exact weight of the item to several decimal points. The most accurate balance yet invented is the microbalance. Just a few inches long and weighing about 50 mg, the microbalance is ideal for determining extremely small weights to within a few millionths of a gram.

Matthias William Baldwin

BALANCE SPRING • See Spring

BALDWIN, MATTHIAS (1795-1866)

American engineer and manufacturer

The youngest of five children, Baldwin was born in Elizabethtown, New Jersey. His father, a carriagemaker, died when Baldwin was five, and the fortune he left to his wife was squandered by lawyers. Baldwin's mother, however, saw to it that her youngest son received an education and apprenticed him to a jeweler in Philadelphia.

Baldwin soon tired of jewelry making and in 1825 decided to open a manufacturing business along with a partner, David Mason. Baldwin demonstrated a great deal of talent in manufacturing, producing a flurry of products in succession, from bookmaking tools to **textile** and printing processes. Mason, unnerved by Baldwin's engineering prowess, quit the business.

In 1828 Baldwin had constructed a **steam engine** to run equipment for his firm and when he had the opportunity to examine a **locomotive** that had been imported to the United States from England, he decided to try his hand at building locomotives. This was a timely decision, for the crude locomotives being built at that time were ripe for improvement. Baldwin, with his keen engineering skills, succeeded immediately. Within two years Baldwin constructed *Old Ironsides*, a six-ton locomotive, for the Philadelphia and Germantown Railroad. *Old Ironsides* became one of the first practical locomotives in America, making daily trips between the two Pennsylvania cities. Over the next ten years, Baldwin built ten more locomotives, and, by the time of his death in 1866, his company had built 1,500. The efforts of Baldwin, who epitomized the entrepreneurial engineer of nineteenth-century America, helped lay the foundation for the **railroad** revolution that later swept through the United States.

The company that Baldwin founded, the Baldwin Locomotive Works, became one of the world's great industrial giants. Yet despite its experimentation with **steam turbines** and diesel engines after 1939, the company never was able to make a satisfactory tranisition from steam and was eventually put out of business by two newcomers, General Motors and General Electric.

BALLISTIC MISSILE • See Rocket and missile

BALLOON

The first successful balloon was designed in the late eighteenth century by **Joseph-Michel Montgolfier** and **Jacques-Étienne Montgolfier**, the sons of a successful paper manufacturer. Aware of Joseph Priestley's discovery in 1766 of hydrogen, a gas that weighed just one fourteenth as much as air, Joseph Montgolfier attempted to use this gas to lift small bags, but it escaped too easily. He turned next to smoke, thinking that it contained a similar, though unknown, gas, not realizing that it was the heat in smoke that made air expand, and therefore rise.

In 1783, he and his brother built several small, experimental balloons, which eventually soared to 1,000 feet (305 m). In June of that year they constructed a balloon 110 feet (33.5 m) in circumference with a paper liner inside a strong **linen** skin. It ascended to 6,000 feet (1,830 m) floated serenely for ten minutes, and finally settled back to earth over a mile (1.6 km) downwind. The first manned flight of the Montgolfier's balloon occurred on November 21, 1783, when physician Jean-François Pilatre de Rozier and the Marquis d'Arlandes, an infantry major, sailed aloft for twenty-five minutes.

While the Montgolfier brothers were experimenting with their hot-air balloon, physicist **Jacques Charles** had designed a balloon inflated by hydrogen. At this time collecting hydrogen from diluted sulfuric acid, iron, or zinc was an expensive, time-consuming process. **Benjamin Franklin** was one of the spectators to witness the unmanned flight of the Charles's balloon on August 27, 1783. Someone asked him what good it was, to which he replied: "What good is a new born baby?" The balloon met an unfortunate end, however. After it landed in a village outside Paris, peasants attacked it with stones and pitchforks, thinking that it was a monster. Charles eventually modified his balloon until it had many of today's features: a **valve** where expanded gas can be vented, sand to be used as ballast, and a wicker car suspended from a cord net that covered the upper half of the balloon to distribute the weight more evenly.

The achievements of Charles and the Montgolfiers sparked widespread interest in balloons. Jean Pierre Blanchard, first to sail across the English Channel in a balloon, helped introduce the wonders of lighter-than-air flight throughout Europe and the United States. Balloons were used for coronations, fairs, and other gatherings where the balloonist could get paid. During the early 1800s balloonists turned from the Montgolfier type, with burning materials to those designs using hydrogen or coal gas, made from hydrogen and methane. One successful balloon called Le Grand Ballon Captif was exhibited at the 1878 Paris World's Fair. It was taller than the Arc de Triumphe and held fifty-two people who were lifted 2,000 feet (610 m) into the air for a panoramic view of the city.

During this same time, balloons, due to their ability to see great distances, performed military missions. They were used by both sides during the American Civil War. They helped the French as they fought in the Franco-Prussian War of 1870 and 1871. In World War I they were used by both sides for **artillery** spotting and general observation of troop movements. London had tethered barrage balloons floating over the city to protect it from German planes in World War II.

Since their invention, balloons have also been used for scientific purposes, particularly in meteorology. English balloonists Henry Coxwell and James Glaisher embarked on tandem flights in 1862 to better understand the atmosphere. They ascended to nearly 37,000 feet (11,285 m) and learned a great deal about wind currents, temperature variations, humidity, the earth's magnetism, and a warm current of air blowing from the southwest, responsible for the relatively mild English winters.

By 1930, balloonists were exploring the stratosphere, a region between seven and twelve miles (11.26 km and 19.20 km) above the earth. Survival at these heights required a sealed cabin that would maintain the atmospheric pressure equal to that on the ground. In 1930, a Swiss physicist, **Auguste Piccard**, devised the first such cabin to reach altitudes where he could study cosmic rays. A flight in 1932 took Piccard to 51,961 feet (15,848 m). A record height of 113,000 feet (34,465 m) was reached in 1961 by Malcolm Ross and Victor Prather, United States Navy personnel who used the ascent to test space suits designed for the Mercury astronauts.

Balloons have also become popular for the sheer sporting aspects. James Gordon Bennett, publisher of the *New York Herald*, established an international balloon race in 1906. The excitement lay in the fact that no one could predict where or when the race would end. However, as the twentieth century unfolded, sporting balloons were fast being eclipsed by airplanes and, more importantly, the cost of helium for lift soon became prohibitively expensive. The United States Navy, however, revived interest by returning to the old Montgolfier idea of hot-air ballooning, but used a propane burner instead of a straw fire. Soon balloon clubs and meets were back in style. Albuquerque, New Mexico, has become the ballooning capital of the world with more than 100 aeronauts in residence. One inhabitant, Max Anderson, became the first person to cross the Atlantic in a balloon in 1978, appropriately landing outside Paris, France.

The 1990s witnessed a resurgence of interest in attempts to circumnavigate the world in a balloon, a feat that has eluded aeronauts since the Montgolfiers. This interest spanned the globe: European teams attempted the feat from Switzerland, a British team launched from Morocco, and American teams initiated attempts from Nebraska and New Mexico. Twice the Americans reached Asia from the United States, but more frequently the attempts ended in failure considerably sooner. One ill-fated attempt lasted only several hours, as the balloon began leaking helium over the New Mexico desert, and the balloonists were forced to parachute to safety, one of them landing ignominiously on a cactus.

BALLOON FRAME

During the period of time when the European pioneers were moving across the United States and Canada, they adapted a Scandinavian method of building houses called mortise-and-tenon construction, in which the projecting tenon at the end of one log was fitted into the mortise, or notch, of another to form a corner joint. This technique required more lumber than other methods that used one log to make several boards, but it provided a simple way to construct homes—the familiar log cabins—in the forests of timber-rich North America.

By the 1830s, however, city dwellers were exploring new a type of construction, the balloon frame. The balloon frame was developed at a time when North American cities were growing and as the **mass production** of **nail**s to standard specifications allowed carpenters and architects to specify the type of nail for each particular job required. The balloon frame borrowed its essential elements from the log cabin. Most other structures required an internal skeleton for support, but the log cabin's integrity was based on the fact that its interlocking logs created a ''shell'' that could support itself and the roof. Like the mortar-and-tenon cabin, the balloon frame structure required only its wooden ''exoskeleton'' for its support, hence its name—internal walls were nothing more than room partitions.

The new technique used milled timbers instead of the traditional heavy timbers and raw logs. In the 1820s, sawmills began creating lumber in standard sizes, which made it easier to transport, and also to design houses ahead of their construction. Instead of having to be interlocked in the mortar-and-tenon way or fastened by large wooden pegs, the milled timbers were small enough to receive nails. Joining them required little skill and only a few hand tools, and was also inexpensive. The job of joining them became significantly easier. The exterior of the balloon frame building could be of any veneer, or covering: wood, **brick**, stone or stucco for walls, shingle, **ceramic** tile or tin for roofs.

The person generally credited with the invention of the balloon frame was Chicagoan George W. Snow. Snow built a balloon-frame warehouse in Chicago in 1832. This building employed the now classic two-by-four (inch) vertical building stud, and established the standard 16-inch spacing between them. The studs supported two-by-ten inch roof and floor joists. A year later, carpenter Augustus D. Taylor used the balloon frame in the construction of Chicago's St. Mary's Church. The church was taken down and reassembled three times during its existence.

The frame-support system was later adopted by **James Bogardus** and the architects that followed him in the late 1800s for the construction of office buildings. The frames were of iron and steel beams and columns and were fastened with steel bolts and rivets. Balloon frame construction of residential and commercial buildings increased in the twentieth century and continues to be the standard building form for ranch houses, ''garden'' apartments and other small-building construction. The early 1900s marked the height of popularity for balloon frame construction when Sears, Roebuck and Company offered frame house kits in its mail order catalogs. The offer has long been discontinued, but thousands of the neat, white, two-story Sears and Roebuck houses remain in towns and cities throughout Illinois, Iowa, and other Midwestern states.

See also Jenney, William

BALLPOINT PEN • See Pen

BANDAGES AND DRESSINGS

In one form or another, bandages and dressings have likely been in use since prehistoric times, with plant materials and strips of animal hide serving the purpose initially and, later, fabrics. Early writings from Mesopotamia, Egypt, China, Greece, and Rome describe wound ointments and dressings, and Homer (c. 900-800 B.C.) mentions bandages for battle wounds, as do Hippocrates (c. 460 B.C.) and the Bible. Ancient Egyptian embalmers were highly skilled in the art of bandaging. The great French surgeon **Ambroise Paré** revived and modernized the treatment of wounds by abandoning cauterization in favor of ointments covered with carefully applied bandages. Three hundred years later, English surgeon **Joseph Lister** (1827-1912) pioneered the use of bandages and dressings soaked in carbolic acid as an antiseptic. Adhesive plasters, the precursors of today's adhesive bandages, were mentioned in an 1830 Philadelphia medical journal, patented in 1845 by Drs. William Shecut and Horace Day of New Jersey, and marketed as *Allcock's Porous Plaster* by Dr. Thomas Allcock. A German pharmacist, Paul Beiersdorf, patented a plaster-covered bandage called *Hansaplast* in 1882.

The adhesive bandage as we know it was the invention of Earl Dickson, an employee of the Johnson & Johnson medical supply company. Dickson's young bride continually cut and burned herself in the kitchen, and the concerned husband repeatedly bandaged her with pieces of gauze and surgical tape. Dickson saw that his wife needed a prepared supply of these dressings she could apply herself, and he began experimenting. He laid out a strip of Johnson & Johnson's surgical tape sticky side up on a table and placed a folded-up gauze pad in the middle of the tape. To keep the gauze clean and the tape sticky, Dickson covered the strip with crinoline. Mrs. Dickson appreciated her husband's invention, and so did Dickson's co-workers and bosses. Johnson & Johnson quickly put the bandages on the market, and, in 1920, they became Band-Aids, a name suggested by a Johnson & Johnson mill superintendent, W. Johnson.

A modern wound care dressing widely used in health care facilities has no absorbent pad beneath it's adhesive surface yet does not adhere to the wound. ''SureSkin'' is very thin, transparent *hydrocolloid* dressing in which absorbent materials are built into the adhesive materials. The dressing absorbs excretion from the wound, forming a gel which creates a healing environment for tissue regeneration, and its transparency allows monitoring of the wound without removal. The dressing can remain in place for long periods and, although absorbent and allows penetration of oxygen and water vapor, its protective **polyurethane** film protects against external bacteria and water, so patients can wear it in the shower or bath without the wound getting wet.

Banneker, Benjamin (1731-1806)

American mathematician and inventor

Benjamin Banneker is credited with being America's first black scientist. A native of Ellicott's Lower Mills, Maryland, he was born the first of three children November 9, 1731, to Robert, a slave from Guinea, West Africa, and his free wife Mary Banneky, of English-African descent. Benjamin's grandfather, Banneka, was an African prince; his grandmother, Molly Welsh Banneker, was an indentured servant who read and discussed the Bible with him. His father purchased his own freedom. The family lived ten miles from Baltimore on a 120-acre farm which Benjamin inherited following his father's death in 1757.

Despite having to work hard to support his family, Banneker received eight years of schooling from a Quaker teacher at an integrated private academy. He read borrowed books by Addison, Pope, Shakespeare, Milton, and Dryden, studied the stars, and created and solved math puzzles as entertainment and a means of self-education. He owned no books until the age of thirty-two.

Banneker was a mechanical genius. Famous as a mathematician and inventor, in 1753 he completed the hand-carving of America's first **clock**, a faithful wooden mechanism studded with **iron** and **brass** that took two years to make. The clock kept time and struck the hour for over twenty years. The astounding part of this achievement is the fact that he had seen only one timepiece, a watch, and memorized its workings so that he could create his own model.

To improve his grasp of agriculture and obtain a better yield from tobacco, Banneker used his expertise in math and astronomy to study the heavens. He predicted the solar eclipse of 1789. Three years later he launched the *Pennsylvania, Delaware, Maryland, and Virginia Almanac and Ephemeris*, an almanac which contained data concerning tides, eclipses, formulas, history, literature, astrology, and medicines. With the support of abolitionist societies in Maryland and Pennsylvania, the popular volume ran for six editions and remained in publication for ten years. Banneker sent a copy to **Thomas Jefferson** as a means of proving that black people, if given better living conditions and a proper education, were capable of intellectual accomplishments. Jefferson, who championed Banneker's efforts, passed the almanac along to the French Academy of Sciences.

As a result of Banneker's correspondence with Jefferson, in 1791 President George Washington appointed Banneker as assistant to Major Andrew Ellicott as a part of the six-man team who surveyed the Territory of Columbia and planned Washington, D. C. Here he achieved his most notable work. After the abrupt resignation and departure of Pierre-Charles L'Enfant, the initiator of the project, Banneker drew on his memory for the meticulous details which made the plan a success.

Banneker remained a bachelor and depended on his two sisters, Minta and Molly, to keep house for him. Famed as a recluse, he kept late hours in his laboratory and slept by day. He also involved himself in antiwar and antislavery movements by writing pamphlets and essays, the most significant being ''A Plan of Peace-Office for the United States.'' His hobbies included grafting fruit trees in his orchard and teaching himself to play the violin. Suffering ill health, he leased his land to tenant farmers and later sold off parcels, maintaining only enough money to finance his scientific experiments. He died in poverty at his farm on October 9, 1806. As his body was being interred, his house caught fire. His books and possessions, including his prized wooden clock, were consumed by flames.

Bar code

Almost everyone is familiar with the striped bars found on grocery and retail store items. These are bar codes, or more specifically, the Universal Product Code (UPC). UPC codes first appeared in stores in 1973 and have since revolutionized the sales industry. The UPC code consists of ten pairs of thick and thin vertical bars that represent the manufacturer's identity, product size and name. Price information, which is not part of the bar code, is determined by the store. Bar codes are read by hand-held wand readers or fixed scanners linked to point of sale (POS) terminals. Bar codes are also used for non-retail purposes. One of the earliest uses for bar codes was as an identifier on railroad cars. Organizers of sporting events also take advantage of bar code technology. For example, as runners of the Boston Marathon complete the 26-mi (41.86-km) course, they turn over a bar code tag that allows race officials to quickly tabulate results. From 1965 through 1982, the United States Post Office experimented with optical character recognition (OCR) and bar code technology to speed up mail delivery. The Post Office now utilizes another type of bar code called the

POSTNET bar code. Consisting of full and half height bars representing the zip-code and delivery address, the bar code allows mail to be sorted automatically at speeds of up to 700 pieces per minute.

See also Computer pattern and character recognition; Computer, industrial uses of; Computer input and output devices

Barbecue spit

The origin of the barbecue spit has been traced to the smokejack that was popular in Europe during the Renaissance. It is uncertain whether the smokejack was invented in Switzerland or Italy. A famous Renaissance cookbook contained sketches of smokejacks in the papal kitchens, and Montaigne (1533-1592) made note of the machine when he visited Switzerland in 1581. Smokejacks were like inverted **windmills** that were built into chimneys. Panes of wood above the flames of the fire were rotated by the smoke and steam of the fire below. The vanes rotated a spit on which a roast was skewered. These machines were first to channel heat into mechanical power. There were no significant changes in the design of the smokejack, except for the introduction of bevel **gears**. Some great houses in Europe still have their nineteenth-century smokejacks. **Benjamin Franklin**, ever a source of innovation, applied electricity conveyed through a Leyden jar to revolve a disc attached to a shaft on which a fowl was spitted. In April 1749, Franklin held a famous electrical picnic at which he cooked a turkey on his electrical smokejack. This machine is closer to the barbecue spits of today. Barbecues now range from standard portable charcoal grills for home or camping use, to built in gas grills connected to residential gas lines. Portable grills may be powered by charcoal, propane, or electricity. Elaborate accessories are available including fish cookers, smokers, fryers, and manual or electrically powered spits.In many areas of the country, grills are seen as part of an extension of indoor living, even to the extent of creating complete outdoor kitchens if the climate permits.

Barbed wire

A key weapon in the struggle to settle America's plains, barbed wire was patented in 1873. Years of armed range fights and lawsuits ensued as ranchers and farmers separated livestock and grain crops, ending open grazing and encouraging small-scale farming in the West. This fencing innovation consisted of two or more pieces of twisted **wire** bearing thorn-like barbs at regular intervals. Called "Devil's Rope" by Native Americans, it replaced more conventional and familiar types of fencing materials. Barbed wire provided a means of immediate enclosure, compared with slow-growing shrubs like the osage orange used to enclose land in the midwest, which tended to die during the droughts and high winds common to the plains. Likewise, the wire material was quicker, easier, and cheaper to install than was wood fencing, which often had to be imported from the east; it was more practical than rocks for stone walls, since these had to be shipped from the east or south and required great amounts of labor to install.

Benjamin Banneker

Barbed wire was also stronger than regular wire fencing, which frequently snapped during the plains' drastically changing temperatures, or broke when livestock leaned into it. Legendary and vicious "range wars," actually a series of small, unrelated fights, constitute a major chapter in the social history of the American plains. Before barbed wire was introduced, cattle roamed and grazed freely on the open range, then were driven to slaughterhouses once a year along well-established stock routes. Trouble arose when homesteading farmers enclosed their land, sometimes blocking well-traveled driving routes and access to water. Though most of the wire was probably painted black or bright red, many cattle ran into the newly installed barriers and were injured. Brief but violent fights ensued in Texas, Wyoming, New Mexico, and throughout the West, dividing the combatants into "fencemen" and "no-fencemen." Some of the fighting involved fence-cutters who were known to destroy miles of newly hung wire by cutting pieces from fences.

Henry M. Rose was the actual inventor of handmade barbed wire and demonstrated his creation at the 1873 county fair in DeKalb, Illinois. Friends and neighbors Joseph Farwell Glidden, Jacob Haish, and Isaac L. Ellwood all left the exhibit with ideas for improving Rose's brainchild. Glidden and Haish both submitted patent applications, Glidden in October 1873 and Haish in December. Glidden's improvement consisted of a means to hold the wire's barbs in place, an idea that has outlasted hundreds of other improvements and changes since then. After developing a machine to produce the wire in large quantities, Glidden asked hardware-store owner Ellwood to invest $265 in a partnership. The invention was an immediate success, and Glidden eventually sold his half-interest for more than $60,000 to the Washburn and Moen Manufacturing Company of Worcester, Massachusetts. Meanwhile, building contractor and lumberman Haish designed and patented an S-shaped barbed wire. He proceeded to manufacture it, but Washburn and Moen bought Haish's patents and tried to monopolize the market. Haish fought the company for 18 years until the Supreme Court returned the rights to him in 1892. He eventually realized a profit not only from the production of the wire itself, but also from building and leasing barbed wire-making machinery to other manufacturers, including Washburn and Moen.

Barbed wire was used during World War I to protect front lines from surprise infantry attacks, and today different types are sometimes used to safeguard construction sites, storage yards, and government outposts, as well as continuing to divide land in the western part of the United States. As a testimony to its utility, the Barbed Wire Museum at Canyon, Texas, displays over 200 specimens of the fencing material. Some people collect antique barbed wire as a symbol of our country's westward expansion.

Bardeen, John (1908-1991)

American physicist

The only person ever to win two Nobel Prizes in physics, Bardeen deserves special regard not only from the scientific community but also from consumers who use products arising from his work, including ''boom-box'' **radios** and desktop **microcomputer**s. Bardeen's invention of the electrical **transistor** opened the door to today's electronic age, and his research on the phenomenon of superconductivity is now being used to develop more powerful computers and **artificial intelligence**. Born in Madison, Wisconsin, Bardeen earned his bachelor's degree in electrical engineering at the University of Wisconsin. Although he did some graduate research, he left school to work for the oil industry as a geophysicist in Pittsburgh. After a few years, however, the attraction of pure science lured him to Princeton University, where he was awarded a Ph.D. in physics and mathematics in 1936. There he was introduced to the rapidly developing field of solid-state physics—the study of electronic devices that control electrical current without using moving parts or **vacuum tubes**. By the 1930s vacuum tubes were standard in electronic equipment, but the limitations of these tubes restricted technological advances. Vacuum tubes were bulky and fragile, required large amounts of power,

Range wars erupted in the American West when homesteading farmers, using barbed wire to fence off their land, blocked cattlemen's access to well-established driving routes and water sources. Above, masked men destroy a line of barbed wire.

and needed cooling systems to protect them from overheating. Scientists who wanted to develop bigger computers were frustrated by the huge, complex arrays of tubes and cooling systems that were required. To overcome this limitation, Bell Telephone Laboratories formed a research team to develop a solid-state device that could replace vacuum tubes. The most promising replacement appeared to be **semiconductors**—materials such as **silicon** and germanium that can either conduct or resist an electrical current. After teaching at the University of Minnesota for two years and working for the U.S. Navy during World War II, (developing new technology for the detection of enemy **submarines**), Bardeen joined the research team at Bell Laboratories in 1945. Another team member, **William Shockley**, had suggested that electrical current could be increased or amplified using semiconductors and metals. When Shockley's tests failed, it was Bardeen who explained theoretically how a semiconductor device works. With the assistance of another researcher, American physicist **Walter Houser Brattain**, the team conducted groundbreaking experiments on semiconductor technology. By 1947, the team had succeeded in amplifying electrical current with a ''transfer **resistor**,'' or transistor, that used a germanium semiconductor placed between metallic contacts. The invention of the transistor revolutionized electronic technology, making it possible for scientists to develop much more powerful computers. As a result, Bardeen and his co-workers shared the 1956 Nobel Prize in physics. In 1951, Bardeen left Bell Labs to become professor of physics and electrical engineering at the University of Illinois. There he performed the research on superconductivity that won him a second Nobel Prize in physics, awarded in 1972. The phenomenon of superconductivity occurs at extremely low temperatures, just a few degrees above absolute zero. At that critical point, some metals completely lose all resistance to electrical current and become ''super'' conductive. This effect had been discovered in 1911 by Dutch physicist Heike Kamerlingh Onnes (1853-1926), but scientists had been unable to explain it. Bardeen, working with Leon Cooper (1930-) and John R. Schrieffer (1931-), developed an interpretation of superconductivity which was named the BCS theory (for the three scientists' initials). According to this theory, electrons can attract one another and form pairs under certain conditions. Zero resistivity, they showed, occurs when there is not enough thermal energy to break the pairs of electrons apart. Using the BCS theory, scientists have greatly increased the operating speed of computers and produced exceptionally small and powerful **electromagnets**. Bardeen remained on the faculty at the University of Illinois until his death in 1991.

Barnard, Christiaan (1922-)

South African surgeon

Christiaan Barnard

Dr. Christiaan Barnard became internationally famous on December 3, 1967, when he performed the first human-to-human heart **transplant**. Many other surgeons had been struggling to prepare for this same historic operation, but Barnard's personal drive and fascination with his field propelled him onto the world stage as the first to undertake such ground-breaking surgery. Barnard was born and raised in the arid South African countryside known for its sheep farms. His father was a Dutch Reformed missionary and the family of six lived very humbly. Barnard, known for his excellent academic performance and photographic memory, graduated from the University of Cape Town medical school in 1946. During his residency, he devoted most of his studies to tubercular meningitis, writing his doctoral thesis on the subject in 1953. When he was transferred to Groote Schuur Hospital—the site of his historic operation—Barnard became interested in surgery. A grant in 1955 to study cardiothoracic surgery at the University of Minnesota enabled him to work under the guidance of the prominent surgeon C. Walton Lillehei, one of the many researchers around the world attempting to develop techniques that would lead to the first human heart transplant. Together, Barnard and Lillehei performed experimental open-heart surgery in animals.

When Barnard returned to South Africa, he and his surgeon brother, Marius, began a rigorous series of heart transplant experiments in dogs using a method called the Shumway technique. By the end of 1967, Barnard felt ready to perform the transplant operation on humans, and it was only a matter of time before the right donor and the right patient appeared. Barnard performed the innovative and, to many, shocking surgery on Louis Washkansky, a fifty-three-year-old grocer with debilitating heart disease. Washkansky received the heart of twenty-five-year-old Denise Darvall, who had been killed by an **automobile**. Almost immediately, controversy spread, and people worldwide voiced moral, legal, and ethical objections to this operation. Of particular concern was the question of how to define the death of a potential donor, since comatose patients can be maintained by artificial means for an indefinite period. Barnard's reaction to the uproar was one of steely determination to hold his ground; he had no second thoughts and intended to continue with more transplants. When Washkansky first awoke from his surgery, he is reported to have said, "I am the new Frankenstein." He survived with his new heart for eighteen days, at which point the infections that ravaged his body became lethal. It became evident that the surgical methods for accomplishing transplantation had been achieved, but in order to suppress the body's fundamental mechanism of rejecting foreign tissue, powerful antirejection medications had to be administered. These drugs, called immunosuppressants, lower the body's resistance to foreign tissue but simultaneously suppress its overall immune response, or natural ability to fight viral and bacterial infection. Patients are prone to severe infections as a result.

After Barnard's initial heart transplant operation, many other surgeons undertook the same operation in the following weeks, but with very poor results. Survival rates were unacceptably low— doctors still could not control infections which flourished in their patients. Barnard's second heart transplant recipient, Philip Blaiberg, survived a remarkable 593 days, but Barnard recognized that this case was an aberration rather than the rule. After a total of four transplant attempts, he decided to stop until more research could be done. On November 25, 1974, Barnard tried a new technique, the first double-heart transplant, in which he implanted the heart of a ten-year-old girl into a fifty-eight-year-old man without removing the patient's diseased heart. He performed this operation again in 1975, but both patients lived only a few months. Only a handful of centers continued cardiac transplantation after 1969. Many experts felt that research should be directed toward developing the **artificial heart**, since the heart is a simple pumping mechanism and relatively easy to replicate as opposed to other organs, such as the kidney. It remained imperative, however, to develop a successful method for controlling post-transplantation infections. The development in the 1970s of **cyclosporin**, a specific immunosuppressive agent, once again turned the tide toward organ transplantation. Cyclosporin, which suppresses only certain aspects of the immune system allowing the body to receive a new organ but also fight off the more virulent infections, is by no means a wonder drug but it has clearly reduced the rate of mortality associated with transplants. In the 1990s, investigations into *tolerance induction* and *chimerism* (the coexistence of donor and recipient cells), particularly through simultaneous donor bone marrow infusion with the organ transplant and the role of dendritic cells, hold promise for transplants without the need for immunosuppressive drugs. Barnard's landmark surgery forced many issues surrounding transplantation, legal, ethical, and medical, and he remains a symbol of tenacity and daring in pursuing his goal.

See also Transplant surgical

Barnes, Albert C. • See Patent medicine

Barometer

The invention of the barometer, a device to measure air pressure, was a long and arduous process. It required the creation and understanding of a vacuum. Unfortunately one notable philosopher had said a vacuum couldn't exist, and that stalled advancement for centuries. When a physicist accidentally succeeded in creating a barometer, he didn't realize what he had invented. Eventually a third scientist built a barometer and put it to proper use, although he didn't give it its name. Aristotle (384 B.C.-322 B.C.) was the philosopher who said a vacuum could not exist. He correctly believed that the atmosphere had weight, but had no method of measuring it. **Galileo** disproved a number of Aristotle's claims. In 1638 he published *Two New Sciences*, in which he stated his belief that a vacuum could exist, but discounted the idea that air had weight and could exert a pressure.

About two years later, physicist Gasparo Berti attached a long lead pipe up the side of his house. At the bottom he placed a vessel partially filled with water; at the top, a sealed **glass** vessel filled with water. Berti opened a **valve** at the bottom of his pipe and a portion of the water drained out. The evacuated glass vessel at the top now contained a vacuum and Berti was able to experiment with striking the bell. His sole purpose was to discover if sound carried in a vacuum. In the process he had essentially created a water barometer, the water level of which would fluctuate with atmospheric pressure. Surprisingly, neither he nor his colleagues realized what he had invented. Meanwhile, Galileo's former assistant, **Evangelista Torricelli**, decided to undertake additional experiments following his master's death. In 1643, he filled a tube with mercury, upended it in a bowl filled with mercury, and noticed only a portion of the tube emptied. This was the first barometer. He correctly surmised that the column of mercury was being supported by the atmosphere which was pressing down on the open bowl. Torricelli noted small changes in the level of the mercury from day to day and realized that these deviations represented changes in the pressure of the air, though he never built a permanent instrument to measure these changes.

Mathematician Blaise Pascal (1623-1662) followed along in Torricelli's footsteps. He reasoned that if the atmosphere had weight it should decrease as the altitude increased. The best way to find out if this were so would be to make measurements at various altitudes on a mountain. Pascal was a sickly individual and not up to mountain-climbing, so he convinced his strong brother-in-law to carry two barometers one mile (1.5 kilometers) up a mountain in 1648 and make measurements—not once, but five times. His theory was proven correct.

Otto von Guericke's claim to fame was the invention of an air pump in 1647, by which he was able to create a vacuum. He gave considerable thought to air pressure and density and, in 1672, built a water barometer. Guericke, the P. T. Barnum of the seventeenth century, made a brass tube over 34 feet (10m) high, attached a glass ball at the top, and filled it with water. Within the ball was a little figure of a man which floated high in the ball during good weather; in bad weather the water level dropped and the figure floated lower. In 1660, Guericke became the first to use a barometer to forecast weather. **Robert Boyle** improved on Guericke's air pump and duplicated Torricelli's experiments. It was Boyle who gave the barometer its name in 1665. Boyle and Pascal independently suggested that a portable barometer could be created by bending the bottom of the tube 180 degrees, thus creating a siphon barometer and eliminating the need for a bowl to contain the mercury.

About this time, improvements in barometers came fast and furious. In 1664 **Robert Hooke**, Boyle's assistant, designed a *wheel barometer* that used a mechanical linkage to magnify the movement of the mercury, which was an improvement on a device made by Christopher Wren (1632-1723). French physicist **Guillaume Amontons** also made improvements to numerous instruments; he designed a barometer that did not use mercury and could be used on board a ship at sea. Several different types of barometers have been invented since then. A typical mercury barometer is made of a glass tube more than 30 inches (760 mm) long. (''One atmosphere'' holds mercury at about the 30 inch level, hence the need for a tube greater than that length.) The *Fortin barometer*, invented by Jean Fortin (1750-1831) in the nineteenth century, uses an adjusting **screw** and flexible **leather** bag to raise or lower the mercury level. The *English Kew barometer*, often used on ships, uses a simple mercury cistern, a contracted scale, and a restriction in the tube to lessen oscillations caused by the sea. Italian scientist Lucius Vidi invented the *aneroid barometer* in 1843. This ingenious device uses no liquid. Instead, a bellows expands and contracts with changes in the atmosphere. The movement is amplified with a rack and pinion arrangement that moves a pointer on a dial. It can also be used as an **altimeter**. Another version, called an *aneroid barograph*, traces a continuous record of air pressure changes on paper.

See also Vacuum pump

Barton, Otis • See Bathysphere

Baseball

Baseball, America's favorite pastime, is rooted in the English games of cricket and rounders. In colonial times, children in Boston, Massachusetts played variations of these games, including ''one-o-cat,'' (using one base and three players) and ''two-o-cat,'' (using two bases and three players). When there were enough people to form sides, the game was called rounders, town ball, goal ball, or base ball. Each variation involved hitting balls and running for goals. United States Army General Abner Doubleday (1819-1893) never claimed to have invented the game, but he is credited with developing the first baseball diamond. In 1839, at the age of twenty, he drew a diamond-shaped field with four bases 60 feet (18.3 m) apart in his hometown of Cooperstown, New York, and dubbed it a ''baseball diamond.'' One hundred years later, Cooperstown became the home of the National Baseball Hall of Fame. In 1845, Alexander J. Cartwright (1820-1892), a New York City surveyor, established standard rules for the game of baseball and organized the New York Knickerbockers, one of the sport's first teams. The first professional team was the Cincinnati Red Stockings, which formed in 1869. Players earned anywhere between $600 and $1,400 annually. By 1871, nine teams composed the National Association of Professional Baseball Players. The now well-known National League was organized in 1876, while the American League wasn't established until 1900. How did the equipment used in baseball come into being? The catcher's mask, first used by Winthrop Thayer in 1877, was a welcomed invention for bruised and battered catchers used to taking wallops from overhand pitching. William ''Gunner'' McGunnigle is credited with the invention of the catcher's mitt. Thin lead sheets inserted into the palms of bricklayer's gloves evolved into the catcher's mitt used today.

James Naismith, the inventor of basketball, with the peach baskets he nailed to the walls of a YMCA gymnasium in1891 for the sport's first game.

Basic oxygen process • See Steel production

Basketball

Baseball is America's favorite pastime, but it's basketball that is wholly American in origin. While baseball evolved from cricket and rounders and **football** from **soccer** and rugby, basketball has no origins in the foreign sports world. In 1891, James A. Naismith (1861-1939), a physical education professor at the International YMCA Training School in Springfield, Massachusetts, was asked to develop an indoor game that could be played during winter at night. He nailed two peach baskets to opposite walls of the gymnasium and published the rules of this new game in the school paper on January 15, 1892. Five days later, the first basketball game was played. There were no backboards, and, when a shot went into the basket, a student was obliged to climb a ladder and fish out the ball. Soon, the peach baskets were replaced with wire-like wastebaskets with pullcords. In 1893, bottomless cord nets were standard fare. Other additions to the game included regulation-size basketballs in 1894, backboards in 1895, and dribbling in 1900. Within four years of its invention, basketball was a country-wide craze. The first collegiate game was played between the University of Iowa and the University of Chicago. Now, colleges and universities across the nation take part in the National Collegiate Athletic Association (NCAA). In 1936, basketball was included in the Olympic Games for the first time.

Basov, Nicolai G. • See Maser

Bath and shower

People's bathing habits have varied greatly over the centuries and among varying civilizations. Ancient Egyptians bathed for religious purposes, as did the Hindus of India. The city of Mohenjo-daro in the Indus Valley had a number of baths, complete with drainage system, in 2500 B.C. The Cretan Palace of Knossos and some ancient Greek palaces had separate bathrooms with individual tubs, piped-in water, and drainage systems. The Greeks considered bathing as a way to tone up the body, so they used cold water. The Romans turned bathing into a social occasion. Wealthy Romans had their own bathrooms. Soon enormous, elaborate public baths were constructed. These featured cold, warm, and hot bathing rooms, plus theaters libraries, exercise yards, and more. The Baths of Diocletian, built in A.D. 302, could hold 3,200 people at once; the Baths of Caracalla, built in 217, held 1,600 people. The Roman baths were originally intended, as Greek baths had been, for exercise and health, but over time they became centers for luxurious leisure and dissipation.

After the fall of the Roman Empire, bathing all but disappeared in Western civilization. Clergymen frowned on bathing because they disapproved of nakedness; bathing was even considered to be unhealthy. So the early medieval population lived in filth and, therefore, sickness. Some public baths appeared in the thirteenth and fourteenth centuries, but they became known as centers of immorality and were outlawed in many localities.The general disinterest in bathing among all levels of the population is highlighted by the fact that Louis XIV's (1638-1715) sumptuous royal palace of Versailles, built in 1661, had absolutely no bathing facilities (or **toilets**).

Bathing finally came back into favor during the 1700s. **Benjamin Franklin** brought the first bathtub to the United States from France in 1790. It was a ''slipper bath''—a tub shaped like a shoe, with a water-heating device in the heel and a drain in the toe. As long as running water remained a rarity, though, bathtubs were generally just wooden tubs filled by hand and put away when not in use. The Virginia stool shower became popular in the 1830s. The bather worked a **lever** on one side to pump water up through a hose while simultaneously working another lever or pedal to move an attached scrub brush up and down his or her back.

Steam baths were popular during the 1840s, after home boilers for heating water became more widespread. Steam baths, also called vapor or Russian baths, were considered to have medical benefits.

Closet tubs had a vogue in the 1880s; a comfortable tub folded out of a closet tucked into a corner of the kitchen. Closet as well as free-standing tubs were usually wooden boxes lined with tin, copper, or galvanized iron. The Glamor Tub of 1888 was one of the first to have its own needle spray and shower. Most showers, however, consisted of a hand-held spray nozzle connected to the tub's faucet with a hose. Built-in shower baths only became commonplace after World War I, boosted by the introduction around 1920 of the mass-produced, double-shelled, porcelain-enamel lead tub that could be installed flush to the wall, unlike the old claw-foot tubs.

A variation on bathing is the sauna, developed into a national tradition by the Finnish people over 1,000 years ago. Inside wooden huts, the Finns would pile stones over a fire; water poured over the stones created steam in which to bathe. After sweating in the hut, bathers would jump into an adjacent icy lake, or roll outdoors in the snow. In the United States today, saunas are generally found in health clubs and resorts.

Group bathing reappeared in the United States in the 1970s when the hot tub made its appearance in California. Luxury fixtures and equipment are the hallmark of many modern bathrooms. In place of standard porcelain bathroom tubs and showers, materials now range from one piece plastic surrounds to marble or other opulent materials, the latter often featuring faucets available not only in chrome, but also in brass or gold plate. Luxury whirlpools that accommodate one or two people, bidets, and personal spas or hot tubs that seat varying numbers of persons are part of this trend.

Bathyscaphe

Since the early 1700s when Edmund Halley invented the first **diving bell**, oceanographers and engineers had been putting forth a series of efforts to overcome the obstacles associated with deep sea exploration.

The diving bell and diving helmet extended a person's time in the water by increasing the duration of the air supply. The **bathysphere**, invented by Americans William Beebe and Otis Barton in 1930, extended greatly a diver's depth range by protecting the diver from deep water pressure. However, to more fully explore the oceans, researchers still needed lateral mobility and independence from surface support vessels.

During the late 1940s and early 1950s, the Swiss physicist, Professor **Auguste Piccard**, worked on various prototypes of a vessel intended to give deep sea explorers the mobility they were after; he called it the *bathyscaphe* from the Greek words for ''deep'' and ''boat.''

His success came with two vessels, the French Navy *FNRS 3*, manned by Georges Houot and Pierre-Henri Willm, and the *Trieste*, manned by Piccard and his son, Jacques. The latter vessel was named for the Italian port at the north end of the Adriatic Sea which hosted the latter part of his research project.

Piccard's bathyscaphes were built along the same concept as his lighter-than-air atmospheric research **balloon**s. The round bathysphere-like diving chamber, six feet seven in (2.0 m) in diameter, was suspended under the belly of a fifty foot (15.2 m) compartmentalized float. The compartments were filled with lighter-than-water **gasoline**. The gasoline was used for buoyancy, not as fuel. To descend, the occupants would release measured the ascent, iron ballast pellets were released. The vessel could be maneuvered by releasing a combination of gasoline and ballast. A pair of **battery**-driven **screw propellers** gave the bathyscaphe lateral mobility.

The *Trieste* made a record-setting dive of 10,335 feet (3,150 m) in the Mediterranean Sea, off Capri, in 1953, after several years of work and unsuccessful test dives.

The French took the record in the *FNRS 3* with a dive of 13,287 feet (4,063 m) in 1954 in the Atlantic Ocean off Dakar, West Africa. Jacques Piccard and the U.S. Navy Lt. Don Walsh regained the title in the *Trieste* in 1960 with a descent to the bottom of the Marianas Trench in the Pacific Ocean 250 miles (402.25 km) southeast of Guam. This dive went to 35,800 feet (10,911 m), a mile deeper than the height of Mt. Everest.

Today, sophisticated methods of remote sensing have enabled humans to explore the ocean in greater detail. As with space exploration, unmanned missions using semi-autonomous vehicles can yield more information at less cost and risk. Unmanned submersibles are being used for locating and recovering items from deep shipwrecks (the *Titanic*, for instance). Marine geologists remotely map the ocean bottom with increasing detail, as well as observe and photograph activities of plate tectonics along mid-ocean ridges and rifts.

Despite this, manned deep sea expeditions will continue as long as direct participation is required.

See also Diving Apparatus, Submarine

BATHYSPHERE

Throughout human history, the sea has yielded an abundance of resources for man's existence and has provided efficient routes for exploration and transportation. In return, it has exacted a toll in terms of human life and property. The fear and respect that it earned from those who ventured out upon its surface was itself a deterrent to learning more about its mysteries. The physical restrictions of penetrating the sea made subsurface exploration nearly impossible. The most immediate restriction was air supply. Only the most disciplined divers could stay under for more than a few minutes. As scientists began to recognize the sea as a realm to be explored, this was their first hurdle.

Massive, elaborate baths were a common part of daily life in ancient Rome. For example, the Baths of Caracalla, above, could hold 1,600 people and was a social center for the city.

In 1716, the English astronomer, Edmund Halley, whose interests in the universe included Earth, invented a wooden **diving bell** which was open at the bottom. The significance of Halley's bell was the system developed with it to provide its occupants air. The trapped air in the bell was resupplied by sending down weighted barrels of fresh air. Divers could also venture from the bell with the aid of **leather** helmets and leather air hoses. The next major restriction to be surmounted was the tremendous water pressure which increases with depth. While some sea creatures have adapted to this environment, an unprotected human can only dive to about 330 feet (100 m) even with an air supply. The diving suits, bells, and **submarine**s invented in the eighteenth and nineteenth centuries were not constructed well enough to protect divers very far below the surface.

The collaboration in the late 1920s of two scientists, William Beebe, a naturalist from Columbia University in New York, and Otis Barton, an engineer at Harvard University in Boston, led to the invention of the bathysphere, the first ever deep sea exploration vessel. The sphere was nearly five feet in diameter, had steel walls and weighed 5400 pounds (2,451 kg). It had a circular manhole at the top and three windows made of thick fused quartz. Air was released from two tanks and chemicals were used to absorb moisture and carbon dioxide. It was equipped with a **telephone** and searchlight. The sphere was tethered to the surface vessel by a steel cable. After two unmanned test dives, Beebe and Barton made the first manned descent on June 6, 1930, to a world-record depth of 798 ft (244 m). Continued dives led to a record of 3,024 ft (923 m) in the Atlantic Ocean off Bermuda.

The primary purpose of the bathysphere was to explore rather than to set records. The explorations led to discovery of deep sea plant and animal species and observation of already known species. It also gave scientists new knowledge of submarine topography, geology, and geomorphology. After some improvements to his diving vessel, Barton set a final diving record of 4,500 ft (1372 m) in the Pacific Ocean off of Southern California in 1949. But by this time efforts led by Swiss physicist **Auguste Piccard** were being made to develop the successor to the bathysphere, the self-propelled **bathyscaphe**. The bathysphere was limited to vertical travel and was constantly dependent on its umbilical connection to its surface support ship. Besides visual observation, other exploration activities, such as specimen collection, were difficult with the sphere. Increased water pressures and potential problems with the support line, including the shear weight of the steel cable, made the deeper dives ever more risky. The development of the bathyscaphe in the early 1950s would allow scientists to overcome the third, and perhaps final, obstacle to deep sea exploration, the need for freedom and independence for meaningful research.

Battery, electric

The electric battery makes use of a chemical reaction to produce direct current. There are two basic types: the primary battery, known as a dry cell, which is not rechargeable and is discarded when used up, and the secondary (or storage) battery, which can be recharged. Without batteries modern civilization would come to a grinding halt, yet the origin of the battery dates back at least two hundred years; some sources even say two thousand years.

In the third century B.C. the Parthians, living in what is now Iran, left behind artifacts indicating their knowledge of **electroplating**.

However, the modern history of the battery begins with an erroneous conclusion made by Luigi Galvani (1737-1798) in 1771. Experimenting with dissected frogs, Galvani noticed that their leg muscles twitched when touched with metal probes, and he decided that ''animal electricity'' was responsible. Not everyone accepted Galvani's theory, however—one of his fiercest critics was **Alessandro Volta**.

Volta, who began investigating the phenomena in 1794, strongly believed that a chemical interaction of the probes produced Galvani's current. To Galvani's irritation, the evidence began to lean in Volta's favor. To establish definitive proof, Volta set out to build a device that would produce a steady flow of electricity. After much trial and error, he was successful in 1800.

Volta filled a number of bowls with a saline solution and then connected them with strips of metal; one end of the strip was copper, the other end was either tin or zinc. The reaction between the salt and metals produced an electric current. This was the world's first electric battery, and it vindicated Volta. He improved his invention by using small round discs of copper and zinc with cardboard discs that had been soaked in a salt solution. The result was a more compact battery which became known as the Voltaic pile. The more discs Volta used, the more current that was produced. Some of his piles were sixty cells high, held in place by a metal rod.

The chemical reaction that creates electricity is very straightforward: the salt solution (which is the electrolyte, a liquid that transmits charged ions) causes the molecules of copper and zinc (which comprise the electrodes) to break up, producing electrically-charged ions that are sent through an **electric circuit** and put to work. The drawback is that a thin film of hydrogen bubbles forms on the positive electrode (anode) and causes the internal resistance of the battery to increase, resulting in a decrease of current output. This process is known as polarization. When the circuit is turned off the bubbles dissolve, but they return when the circuit is operated again.

The problem of polarization was lessened by **John Frederic Daniell** in 1836. He incorporated a depolarizer into his battery design; the negative electrode (cathode), which was zinc, was dipped into a weak sulfuric acid electrolyte, and the copper anode was placed in a saturated copper sulphate solution. A porous membrane kept the two liquids separated. This resulted in a battery that had a longer life and was more reliable.

In 1839, William Grove (1811-1896) invented an electric cell that produced power from the interaction of hydrogen and oxygen gases. Known as a **fuel cell**, it is not a true battery; its chief disadvantage is in having to replenish the gases. The fuel cell was fairly ignored until the advent of the space age, when it is used to provide electricity on **spacecraft**.

The biggest disadvantage with the early batteries was that they eventually died and had to be replaced. French physicist Gaston Planté (1834-1889) made a great improvement with his ''storage battery.'' Built in 1859, this battery had lead plates immersed in sulfuric acid and was completely rechargeable. It is essentially the same battery that is used in today's **automobile**s.

In 1868 Georges Leclanché introduced the battery with which people are most familiar. The original Leclanché cell was as ''wet'' as all the other batteries and was originally used to recharge storage batteries, but a modification that made the liquid electrolyte a paste resulted in the type of battery known as a ''dry'' cell.

It is the dry cell battery that is used in **flashlights**, **radios**, **calculators**, and many other devices. Dry cells are often used in series, where the positive electrode of one touches the negative electrode of the other. The number of cells depends on the voltage required to operate the device. Although the Leclanché cell has a depolarizer, it is not immune to polarization and is best used in devices that operate intermittently.

The alkaline zinc-mercuric oxide cell (*mercury battery*), which is considered to be quite modern, was invented in 1884, although it did not become widely available until after World War II. It is best suited for devices that operate continuously, such as a **hearing aid** or wrist **watch**.

Thomas Alva Edison invented a nickel-iron cell in 1900, and in the same year the nickel-cadmium (''ni-cad'') battery was invented in Sweden by Junger and Berg. Both of these batteries can withstand considerable abuse; they can be overcharged, over-discharged, or left idle without suffering any ill effects.

Batteries are still undergoing development today, and are essential in the areas of laptop computers, **cellular telephones**, and automobiles. Nickel metal hybride (NiMH) batteries have been used in laptops for years. Lithium-ion batteries are used in many portable electronics as well, but their cathode contains the toxic metal cobalt, and they cannot be recharged if charged too many times. Manganese oxyiodide cathodes have recently shown promise as a way to avoid these problems. Lithium batteries made from flexible sheets of plastic, developed in 1998, show promise for someday making laptop computers even smaller.

The same manganese oxyiodide compound shows promise for the batteries in electric cars. **Electric automobiles**, which are very clean compared to **gasoline automobiles**, await a powerful, safe, cheap battery before they will be widely adopted. Again, lithium-ion batteries would be preferred, since lithium is a very reactive metal. But lithium reacts violently when immersed in water, a cheap, abundant electrolyte. New prototype batteries developed in 1995 and improved in 1997 mix lithium with the manganese oxide for the anode, and mix dissolved lithium salts with the water to prevent reactions. Meanwhile, using the older lithium-cobalt batteries, the car company Nissan Altra plans to introduce an electric car, with a range of 120 miles (192 km).

BATTLESHIP • See Warship

BAYONET

Bayonets may have been made for hunters in Europe sometime in the sixteenth century. They were introduced as military weapons in France in the next century, with their earliest recorded use being1642. The bayonet at first was simply a dagger with a tapered handle that could be shoved into the barrel of a gun. It replaced the pike—a long wooden shaft with a sharp metal tip—typically used by foot soldiers. These early bayonets were known as plug bayonets. While still a formidable weapon, the plug bayonet was imperfect, as the gun could not be fired with the **knife** in it, and it was often difficult to remove the sharp blade. Various versions that fit the blade to the outside of the gun barrel followed. A French colonel, Jean Martinet, is credited with inventing the socket bayonet in

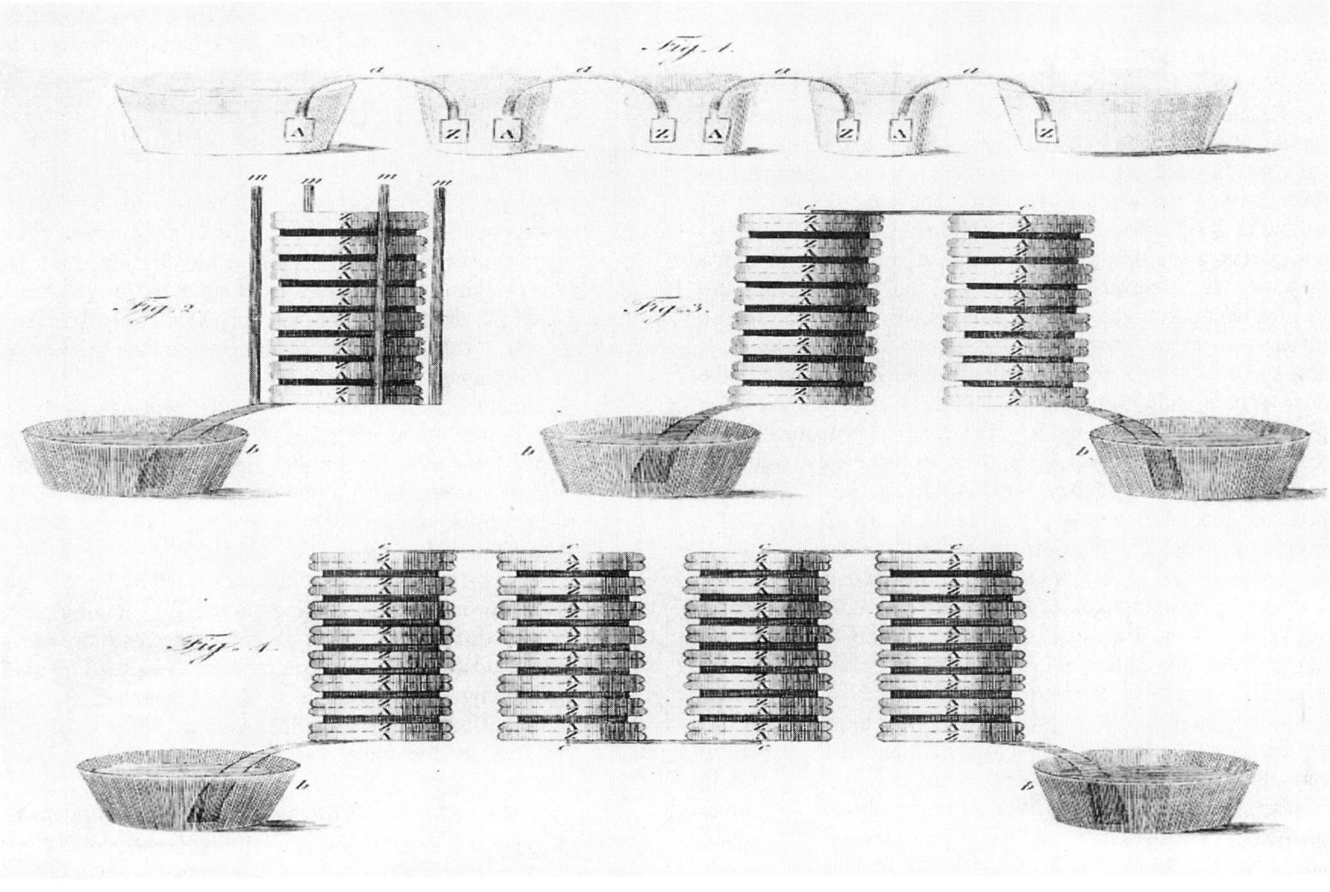

Voltaic piles. Invented in 1800 by Alessandro Volta, the voltaic pile is considered the first electrical battery.

1687, and this was introduced to the French army around 1700. The socket bayonet consisted of a short tube linked by a curved neck to a blade. The tube slipped over the muzzle of the firearm and was held in place by a protruding stud which engaged a Z-shaped slot. With the socket bayonet, the gun could be fired with the blade in place. Another change came in 1781 when John Waters was granted a British patent for a hinged bayonet that was permanently fixed to the gun and could be folded back along the barrel when not in use. The success of the bayonet in hand-to-hand combat dramatically transformed battlefield tactics. Infantrymen could reload their firearms while being covered by their comrades' bayonets, fight cavalry soldier with better odds, and have a reliable weapon at their disposal even if rain soaked their **gunpowder**. Advances in gun technology eventually made the bayonet obsolete. However bayonets with smaller, detachable blades were still used throughout World War I and World War II.

Bazooka

At the beginning of World War II, there was a need for an effective lightweight anti-tank weapon that a single soldier could carry. The United States military introduced the most famous one—the bazooka, named after a musical horn used by American comedian Bob Burns. The bazooka was developed in 1941 and first used by American troops in late 1942. The principle behind the bazooka goes back to English physicist Sir Isaac Newton (1642-1727) and his three laws of motion, in which he stated that for every force there is an equal and opposite force. The bazooka had a warhead with a rocket motor that produced pressurized gas which escaped through a nozzle; this force heading out of the nozzle produced an equal and opposite force driving the warhead forward. The bazooka was fairly simple in design. There was a 3.5-pound (1.6 kg), 21-inch (53 cm) rocket made up of a fused warhead at the front, a sturdy tube with a solid propellant, a nozzle for the gas to escape, and fins for accurate flight. The muzzle velocity was 265 feet (80 m) per second and the warhead could penetrate five inches of armor at a range of 700 yards (640 m). The launcher was a tube open at both ends, weighing 13 pounds (6 kg) and measuring 54.5 inches (138 cm) long. Before the rocket was loaded into the tube, its safety pin was removed, and the tube's circuits were checked to see if they were working properly. The tube was then placed on a soldier's shoulder while another soldier inserted the rocket. When the trigger on the tube was pulled, the bazooka fired its rocket motor. The key here was to ensure that the engine completed its firing before the tail of the rocket emerged from the tube, since the soldier holding the tube could be burned badly if the engine was still firing as it exited the tube. The only other problem was the use of batteries to fire the rocket engine: cold weather caused them to produce insufficient power to fire the rocket. Soldiers ended up having to carry the batteries in their pockets until they were needed in combat. During the Korean War a bigger, aluminum "Super Bazooka" was used. But the bazooka was essentially abandoned after this point in favor of lighter and more accurate anti-tank and anti-aircraft weapons. These include shoulder-fired **rocket and missile** launchers. These modern weapons are much more sophisticated than the bazooka. For example a shoulder-fired anti-aircraft missile used extensively by the United States in the Persian Gulf War contained a computer to calculate an intercepting course and a heat-seeking device to home in on airplane jet trails.

Beard, Andrew J. (1849-1941)

American inventor

Born in Alabama, Andrew J. Beard was set free from slavery at the age of 15. He remained in Alabama, taking up farming for several years. In addition to farming, Beard experimented with mechanical devices, and had soon built his own flour mill. During the years he operated the mill, he began to channel his profits into other inventions. His first success was a **plow** he patented and sold for four thousand dollars. Other inventions such as a second type of plow and a rotary **steam engine** followed with similar results.

In the 1890s, Beard took a job at a **railroad** yard in Eastlake, Alabama that led to his best-known invention. He observed the danger of serious injury or death workers faced when coupling **train** cars by securing the cars with a metal pin as they came together. The accidents he saw and stories he heard motivated him to create a device, called the Jenny coupler, which automatically locked the two cars together when they were forced against each other. In 1897, he sold his patent rights to the device for fifty thousand dollars, a considerable sum at the time.

Bearings

In ancient times, bearings may have taken the form of roller (cylindrical) bearings in the **wheel** hubs of wagons or, more probably, of a conveyor-belt system of logs designed by the Egyptians to transport large sections of stone for the construction of their pyramids. The first ball-shaped bearings, made of bronze, were invented by the Romans to assist the movement of metal turntables on **ship** decks.

In none of these instances were the bearings likely to have moved freely and without skidding. This development came in 1543 when Italian sculptor Benvenuto Cellini (1500-1571) placed his statue of Jupiter atop a platform in which were housed four wooden bearings that permitted free lateral as well as rotational movement of the statue.

The key mechanical role of all bearings, as Cellini knew, is to serve as a contact and support for moving parts through the reduction of friction. Bearings became more prevalent in later centuries as greater attention was placed on axle design in all types of inventions. With the advent of fast-moving metal machine parts, which produce high levels of heat and friction, the importance of bearings became even more pronounced.

A host of manufacturing industries rely on bearings to enhance machine performance and prolong the stability of moving parts. The addition of a liquid film such as oil or grease to the surface of bearings further enhances machining capabilities. The most common bearings, of either the rolling or the ball type, are retained in a *cage*, which in turn is placed within two cylinders. Among the early inventors who obtained patents for ball bearings were French mechanic Jules Pierre Suriray and Welsh ironmaster Philip Vaughn.

Bushings are used to fill cylindrical holes, to preserve their dimensions and to provide a smooth surface for whatever object will fill the hole. In a larger sense bushings are removable metal linings, usually in the form of a bearing to carry a shaft. Generally a bushing is a small bearing in the form of a cylinder, made of soft metal.

See also Wheel and axle

Beau de Rochas, Alphonse • See Internal combustion engine

Beaufort wind scale

In 1806, to standardize nautical observations, Sir Francis Beaufort, an Irish hydrographer and member of the British Admiralty, created a scale for judging the strength of wind at sea. His scale is still a useful standard for the determination of wind force.

Each of the Beaufort Scale's 12 wind-force levels, ranging from calm to hurricane force, includes a description of the effect of the wind on readily observable, common objects. Thus, the scale gives even an unskilled observer a means of estimating wind force. Originally limited to a description of the effects of wind on a sailing vessel's canvas (force 12, for instance, was "that which no canvas could withstand"), the scale was revised in 1939 by the International Meteorological Committee to include the effect of wind on land features. The numbers from the Beaufort Scale were used on weather **maps** until 1955, when a system of wind feathers, which show wind direction and intensity, was adopted.

See also Anemometer; Douglas sea scale; Petersen Scale; Wind vane

Becquerel, Alexandre-Edmond (1820-1891)

French physicist

Alexandre-Edmond Becquerel was the second of three generations of eminent French physicists. His father, Antoine-César Becquerel (1788-1878), was an experimental professor at the Paris Museum of Natural History; his son was Antoine-Henri Becquerel (1852-1908), the discoverer of radioactivity. Alexandre also served at the Museum of Natural History, turning down opportunities at the Ecole Polytechnique and the Ecole Normale Superieure in order to work with his father. He later received his doctorate from the University of Paris, and eventually took a professorial position at the Agronomic Institute of Versailles.

Becquerel conducted extensive research in the fields of magnetism, electricity, and optics. He was especially interested in phosphorescence and luminescence, chemical reactions caused by exposing certain substances to light. In the 1840s he found that these reactions could produce an electric current in both liquids and metals. He devised an instrument, called an actinometer, that could measure the intensity of a light source by observing the amount of electrical current it elicited in these substances. The connection between light energy and chemical energy was seized upon by many scientists in the following years, and research has led to the development of the **photoelectric cell**.

During this time, Becquerel was considered the foremost authority on luminescent phenomena and published several important papers on the subject. He conducted three pioneering experiments between the years 1857 and 1859 that formed the foundation of all later studies of luminescence.

Beebe, Charles William • See Bathysphere

Beekeeping

Honey has been a prized commodity from ancient times. For many, it was the sole source of sugar in the diet. It was often used for religious purposes, as medicinal ointment and fermented to make an alcoholic drink.

Wild honey was once the only kind available. Bee cultivation, or apiculture, became a well-developed craft by the second millennium B.C. in Egypt. Bees were kept in pottery or in bark. The honey was often used in bartering as well as for local consumption. Bee culture was also practiced in Greece and Asia Minor.

The first true manmade beehive was invented in 1789 by Swiss naturalist François Huber. It contained movable frames that opened in book-like fashion. A primer comb was added; the bees did the rest of the work. Further improvements were made during the 1800s, including hives that used a synthetic comb foundation. By the 1880s, enough had been learned about managing bee colonies to make it commercially viable.

Today there are an estimated five million manmade beehives in the United States alone. Each colony, or apiary, has 40 to 75 hives. There must be sufficient numbers of flowering plants to support the apiary and to produce the desired amount of honey. For this reason, apiaries must be kept some distance apart from each other, though bees usually stay within two miles of their hives. It is estimated that bees must travel a total distance of 13,000 miles from flower to flower to collect enough nectar for one pound of honey. In addition to making honey, the bees also pollinate the farmers' crops, a valuable double service.

The nectar is stored in the bees' stomachs for the journey to the hive. The bee will travel in a serpentine path until it reaches its capacity of nectar. It then makes a direct flight, or beeline, to the hive.

At the hive, the nectar is transferred to other bees or directly to the hive cells. The nectar undergoes chemical changes both in the bee's stomach and in the hive that convert it to honey. The honey serves as food for the bees. When a beekeeper collects the honey, some must be left in the hive for the bees.

Collecting honey requires patience and skill. A professional keeper may shun protective gear and move slowly and deliberately while receiving only a few stings, having built an immunity to the stings' poison.

In recent years, concerns have been raised about so called killer bees, Africanized bees (*Apis mellifica scutellata*) that are significantly more agressive than current domestic varieties, an effect that increases with each subsequent generation. Through a cooperative program with Mexican beekeepers, the United States government has been tracking killer bee development in that country and in the southern United States since the late 1980s. Although not quite as dangerous as some early reports suggested, these bees, which have been in Mexico since the early 1990s, are now resident in hives in Texas and other states, and have been known to kill nearby cattle. Some beekeepers have experimented with changing the queen bee of a hive each year in an attempt to slow the advent of the ferocious characteristics of the bees. Beekeeping continues to be a popular hobby for many people despite a continuing increase in the number of Africanized bees in this country.

See also Ceramics

Békésy, Georg von • See Audiometer

Bell, Alexander Graham (1847-1922)
American inventor and educator

Alexander Graham Bell is remembered today as the inventor of the **telephone**, but he was also an outstanding teacher of the deaf and a prolific inventor of other devices.

Bell was born on March 3, 1847, in Edinburgh, Scotland, to a family of speech educators. His father, Melville Bell, had invented *Visible Speech*, a code of symbols for all spoken sounds that was used in teaching deaf people to speak. Aleck Bell studied at Edinburgh University in 1864 and assisted his father at University College, London, from 1868-70. During these years he became deeply interested in the study of sound and the mechanics of speech, inspired in part by the acoustic experiments of German physicist Hermann von Helmholtz (1821-1894), which gave Bell the idea of telegraphing speech.

When young Bell's two brothers died of tuberculosis, Melville Bell took his remaining family to the healthier climate of Canada in 1870. From there, Aleck Bell journeyed to Boston, Massachusetts, in 1871 and joined the staff of the Boston School for the Deaf. The following year, Bell opened his own school in Boston for training teachers of the deaf; in 1873 he became a professor of vocal physiology at Boston University, and he also tutored private pupils.

Bell's interest in speech and communication led him to investigate the transmission of sound over **wires**. In particular, he experimented with development of the harmonic **telegraph**—a device that could send multiple messages at the same time over a single wire. Bell also worked with the possibility of transmitting the human voice, experimenting with vibrating membranes and an actual human ear. Bell was backed financially in his investigations by Gardiner Hubbard (1822-1897) and Thomas Sanders, fathers of two of his deaf pupils.

Early in 1874, Bell met Thomas A. Watson (1854-1934), a young machinist at a Boston electrical shop. Watson became Bell's indispensable assistant, bringing to Bell's experiments the crucial ingredient that had been lacking—his technical expertise in electrical engineering. Together the two men spent endless hours experimenting. Although Bell formed the basic concept of the telephone—using a varying but unbroken electric current to transmit the varying sound waves of human speech—in the summer of 1874, Hubbard insisted that the young inventor focus his efforts on the harmonic telegraph instead. Bell complied, but when he patented one of his telegraph designs in February 1875, he found that **Elisha Gray** had patented a multiple telegraph two days earlier. Greatly discouraged, Bell consulted in Washington with the elderly **Joseph Henry**, who urged Bell to pursue his "germ of a great invention"—speech transmission.

Back in Boston, Bell and Watson continued to work on the harmonic telegraph, but still with the telephone in mind. By accident on a June day in 1875, an intermittent transmitter produced a steady current and transmitted sound. Bell had proof of his 1874 idea; he quickly sketched a design for an electric telephone, and Watson built it. The partners experimented all summer, but failed actually to transmit voice sounds. That fall, Bell began to write the patent specifications, but delayed application; Hubbard finally filed for the patent on February 14, 1876, just hours before Gray appeared at the same patent office to file an intent to patent his telephone design. Bell's patent was granted on March 7, 1876, and on March 10, the first message transmitted by telephone passed from Bell to Watson in their workshop: "Mr. Watson, come here, I want you!"

After a year of refining the new device, Watson and Bell, along with Hubbard and Sanders, formed the Bell Telephone Company in 1877. Bell immediately married Mabel Hubbard, daughter of his new partner, and sailed to England to promote his telephone. The phone company grew rapidly, and Bell became a wealthy man. He turned to other interests on his return to the United States in 1879, while also defending his patents (which were upheld in 1888) against numerous lawsuits.

With money from the Volta Prize, awarded to him in 1880 by the French government, Bell established the Volta Laboratory. Among the new devices he invented there were the graphophone for recording sound on wax cylinders or disks; the photophone, for transmitting speech on a beam of light; an **audiometer**; a telephone probe, used in surgery until the discovery of the X-ray; and an induction balance for detecting metal within the human body.

Bell founded several organizations to support teaching of the deaf. He helped to establish *Science* magazine and the National Geographic Society. He also worked on **air conditioning**, an improved strain of sheep (to bear multiple lambs), an early **iron lung**, solar distillation of water, and **sonar** detection of icebergs.

The possibility of flight fascinated Bell. He built tetrahedral **kites** capable of carrying a human being. He supported **Samuel Langley**'s pioneering experiments in aviation, and helped found the Aerial Experiment Association in 1907. He also designed a **hydrofoil** boat that set the world water-speed record in 1918.

Bell became a United States citizen in 1882. He died at his summer home on Cape Breton Island, Nova Scotia, on August 2, 1922.

Alexander Graham Bell

Bennett, Willard Harrison (1903-1987)

American physicist

Willard Bennett was an American physicist noted for his contributions to several areas of physics, including plasma physics, astrophysics, geophysics, surface physics, and physical chemistry.

Bennett was born in Ohio in 1903, and studied at the Carnegie Institute of Technology and the University of Wisconsin before receiving his Ph.D. from the University of Michigan in 1928. He was elected to a National Research Fellowship in Physics, and in 1928-29 studied at the California Institute of Technology. In 1930 Bennett returned to his native state to join the Physics faculty at Ohio State University.

In the 1930s Bennett did pioneering research in plasma physics—the study of gases ionized by high-frequency electricity. In 1934 he discovered the pinch effect, a electromagnetic phenomena that has important implications for plasmas at very high temperatures.

André Ampère demonstrated that parallel electric currents in a **wire** having the same direction attract one another through the magnetic fields they generate. Imagining a current through a wire of a certain thickness as being composed of smaller individual components of that current all traveling within the wire, there will be a net inward pressure on the surface of the wire. This force is negligible at common current values, but can be quite significant for large values of the current.

The pinch effect, named by C. Hering in 1907, first showed up in early types of induction furnaces in which large, low-frequency **alternating current**s of about 100,000 amps were induced. At high enough currents, the metal, ring-shaped conductors in the furnace could actually be closed off, or pinched, by the pressure. (Something similar has been seen to affect tube-shaped conductors in the presence of a lightning strike.)

Bennett discovered the pinch effect in the context of plasma physics, where an extremely-high temperature plasma is confined to a region of space in an attempt to extract energy from a thermonuclear reaction in the plasma. For a plasma confined in the shape of a torus, and a transformer core passing through the axis of the torus, Bennett discovered the conditions for the pinch effect to occur, and derived a fundamental equation that described the current needed in order that a certain plasma temperature be reached. In this configuration, the current in the plasma has the double function of heating the plasma and also compressing it toward the center of the torus tube. These studies and later research have been extensively used as scientists attempt to magnetically confine a plasma at temperatures high enough for controlled nuclear reactions to take place, such as at the Tokamak Fusion Test Reactor at Princeton University.

In 1936 Bennett proposed the tandem van de Graaff accelerator, which later became widely used in nuclear research. In 1950 he developed the radio frequency mass spectrometer, a nonmagnetic spectrometer that could determine the masses of atoms. Because it did not rely on a heavy magnet, it was the first launched into space, carried by **Sputnik III**. At the time it was the only space instrument used by the Russians that was credited to an American inventor in their own Russian-language publications. Also in the 1950s, Bennett's experimental tube called the Stormertron predicted and modeled the Van Allen radiation belts surrounding the earth six years before they were discovered by satellite.

Bennett worked on **aircraft** equipment development as a U.S. Army officer during World War II, and from 1946-1950 served as chief of the physical **electronics** section of the National Bureau of Standards. He was a consultant for the U.S. Naval Research Laboratory, and went to North Carolina State University in 1961, where he held the Burlington professor-

Carl Friedrich Benz

ship. Bennett held 67 patents before his death in 1987. In 1991 he was inducted into the National Inventors Hall of Fame.

Bentz, Melitta (? - ?)

German inventor

In 1908, Bentz, a German housewife, improvised a means of drip-brewing coffee using a piece of **paper** from a school notebook. Tired of the bitter and granular brew yielded by the coffee-making methods of the day (either wrapping coffee in cloth and boiling it or simply dumping loose coffee into boiling water), Bentz cut out a circle of paper and fit it into the bottom of a pot she had perforated, hoping that by setting the coffee on top of this crude filter and pouring boiling water over it, she could brew a far better tasting beverage. She did, and Bentz and her husband, Hugo, commissioned a tinsmith to fashion a supply of their new drip **coffeemakers**, eventually selling them at a trade fair. From this venture grew the Melitta Company, which went on to manufacture coffee filters and replace the metal pots with **porcelain** and **plastic** ones.In addition, Melitta Corporation has opened retail coffee shops,and sells a variety of coffee related products including flavored syrups. An entrepreneurial award for women, the ''Melitta Bentz Woman of Innovation'' award offers $5000 to the winner and $1000 to her favorite charity.

Benz, Karl Friedrich (1844-1929)

German engineer

Karl Friedrich Benz was born in Karlsruhe, Germany, on November 26, 1844. He attended the local gymnasium and polytechnic school. At age 21, Benz went to work in a local machine shop. He continued to work in the machine tool industry and, in 1871, opened a small engineering works with August Ritter as his partner. His first efforts to power a horseless carriage were not very fruitful, but he persisted. Some said he was obsessed with the idea. Shut out of the race to develop a four-stroke engine by **Nikolaus August Otto**'s patented engine, Benz continued to make improvements to a one-horsepower, two-stroke engine of his own design at his factory in Mannheim. Eventually, this new engine attracted enough financial support for Benz to open a new company (without Ritter) in 1883.

It was this company, called Benz & Company, which built the world's first **automobile** powered by an **internal combustion engine**. Benz and another German, **Gottlieb Daimler**, feuded over who should be the deserving recipient of this honor. In fact, the two men never met and developed their automobiles independently, but it appears that Benz was about two years ahead of Daimler.

Benz also saw the motor car differently than did Daimler. Daimler believed the public would demand that the automobile closely resemble a converted horse carriage; Benz maintained that a completely different vehicle was called for. His first self-propelled vehicle was a three-wheeled, horseshoe-shaped contrivance that resembled a giant baby carriage. Benz may also have been the casualty of an automobile accident, having driven one of his first autos into a wall at his factory, while apparently forgetting to steer the vehicle during a demonstration.

Benz made his first commercial sale in 1887 in Paris, and, within one year, he was employing about fifty men to produce his three-wheeled vehicle. By this time, the Otto four-stroke engine patent had been revoked, and Benz (as well as others) was using four-cycle engines of a design that had been refined by his rival, Daimler. The Benz Company's first four-wheel model, introduced in 1893, was powered by a 1.5 HP engine and was capable of 15 mph (24 kph).

In 1899, the successful Benz Company reorganized as a limited company. Four years later, Benz left the company after a lingering dispute with its board of directors. He returned to the company briefly in 1904 but quickly retired to Ladenburg, where he died on April 4, 1929.

The Daimler and Benz companies merged in 1926 to form the Daimler-Benz company, which marketed their products under the Mercedes-Benz nameplate.

Berliner, Emile (1851-1929)

German American inventor

With the invention of his flat disk record (which, in 1888, replaced **Thomas Edison**'s more expensive and more fragile cylinder) Emile Berliner elevated the record industry to prominence in home entertainment. Born on May 20, 1851 in Hanover, Germany, Berliner studied the printing trade before immigrating to the United States at age nineteen, where he studied sound and electricity at Cooper Union in New York City. In 1877, he invented an improved voice transmitter with a variable-pressure contact for the **telephone**. This device, which came to be called a **microphone**, won him a job as chief inspector for Bell Telephone the following year. The sale of his patent made him a wealthy man, but it brought him fifteen years of court battles against Edison, who patented the same device two weeks after Berliner.

Ten years later, Berliner patented a gramophone that played a flat record. He produced records for the new gramophone in the following way: a moving stylus recorded a musical performance onto a seven inch zinc disk covered with a fatty film. The stylus scratched through the acid-proof, film coating the disc as it moved. When this ''master'' disc was dipped into acid, the acid was able to eat away only where the wiggling of the stylus exposed the zinc beneath the coating, thereby leaving a groove in the zinc matching the original movement of the stylus. Unfortunately, copies made from these masters had a ''fuzzy'' sound quality because the grooves etched by the acid on the master did not always retrace the movement of the original stylus with perfect accuracy.

New Jersey mechanic Eldridge R. Johnson, seeing the potential of Berliner's discs, went to work on an improved duplication method. Rather than starting with a zinc disc, he began with a hard wax disc (similar to the wax Edison was using for his cylinders), into which a recording stylus cut grooves directly. Then the disc was dusted with gold powder and slowly *electroplated* (coated with layer after layer of metal) to form a ''negative'' made of metal. This negative, or *matrix*, was separated from the wax original and used to stamp out many copies of the recording in a malleable substance like **shellac**.

Via this simple method of duplication, a speech, musical performance, or other aural experience could be made available to the public in quantity at a reasonably low cost. Berliner's discs had two advantages over Edison's cumbersome ''cylinders'': first, they were more easily stored, and second, although the duplication process developed by Johnson was similar to the procedure Edison used on his cylinders, reproducing the flat discs was far easier and more reliable.

Using his own patents as well as those of Berliner, Johnson founded the Victor Talking Machine Company in 1901, which later became Radio Corporation of America or RCA. Berliner's recording equipment became the record industry's standard. As increasing numbers of musical performers made presses of their works for sale on Berliner's disks, sales of the Berliner gramophones and discs began to rise. At fifty cents a disc, records quickly gained popularity—and when the great tenor Enrico Caruso (1873-1921) recorded a series of performances for Victor in 1903, any doubt about the future of the Berliner gramophone was demolished as legions of excited fans bought his recordings. In the process, Caruso became the first performer to sell a million records.

Berliner continued creating and patenting ideas, including an airplane engine in 1908 and acoustical tiles to enhance soundproofing in 1925. In his last years, he worked to promote the compulsory pasteurization of milk as a means of improving infant health and nutrition. He died in Washington, D. C. on August 3, 1929.

Emile Berliner

See also Jet engine; Milk, pasteurized; Phonograph

Berthelot, Pierre Eugène Marcellin (1827-1907)

French chemist

Berthelot was not only an outstanding scientist but also an accomplished historian, philosopher, and public servant. He lived all his life in Paris, France, where his father, a doctor, treated patients in the city's poor neighborhoods. While still a student, Berthelot showed that organic compounds such as phenol could be synthesized from acetic acid which, in turn, could be prepared from other chemicals including carbon tetrachloride. Berthelot was one of the first scientists to use the word *synthesis* to describe the production of organic compounds from their elements. In 1854 Berthelot earned his doctorate with a thesis on the synthesis of natural fats, which he created by combining glycerol with fatty acids.

Berthelot continued his studies, graduated as a pharmacist in 1858, and became a professor of organic chemistry. In addition to his research on fats, Berthelot is known for synthe-

Pierre Eugéne Marcellin Berthelot

sizing alcohols, which he defined as neutral compounds containing carbon, hydrogen, and oxygen. Traditionally, ethyl alcohol (ethanol) was prepared by fermenting sugars with yeast, but Berthelot showed that it could be obtained from ethylene, an organic gas. Berthelot also synthesized acetylene gas, which he named although it had been discovered earlier, by passing hydrogen through an electric arc produced with carbon poles.

He published a definitive work on the synthesis of organic chemicals in 1860, concluding that an almost infinite number of organic compounds could be synthesized. Although other chemists had prepared natural organic substances, Berthelot was one of the first to synthesize organic compounds that did not occur in nature. Some scientists still believed that only a living organism could produce organic chemicals, and that vegetable and animal substances were essentially different from chemicals made in the laboratory. Berthelot's work helped disprove this theory and showed that the same physical forces operated in both organic and inorganic chemistry.

Berthelot also developed new experimental techniques in the field of thermochemistry, which studies the amount of heat that is released or absorbed during chemical reactions. His most notable contribution was the invention of the bomb **calorimeter**, a device in which gas is mixed with excess oxygen, compressed, and then sparked. Berthelot's calorimeter measured the heat of combustion much more accurately than previous methods. Berthelot also introduced the terms exothermic to describe a reaction that releases heat and endothermic for one that requires heat.

Although Berthelot devoted more time to research than to teaching, he continued to give lectures through the 1860s. During France's war with Prussia in 1870-71, Berthelot was put in charge of the city's defense committee. When the new French Republic was established, he was elected to the Senate and began an active career in public affairs. Late in his career, Berthelot interpreted ancient and medieval manuscripts on alchemy and analyzed metal objects from ancient Egypt and Mesopotamia. He also published a book on Antoine-Laurent Lavoisier (1743-1794), which included excerpts from the renowned chemist's laboratory notebooks. In 1883 Berthelot established a research farm near Paris for studying agricultural chemistry. In experiments there, he discovered that some plants can absorb nitrogen from the air in the same way that nitrogen compounds are formed in the laboratory. Although Berthelot had been raised as a Catholic, his exposure to philosophy had led him to question his religion. Throughout his life, he resisted clerical influence in education and promoted a greater emphasis on science in the classroom. Berthelot was devoted to his wife, who came from a Protestant family. They were married for 45 years and had six children together. When she became ill, he tended her night and day. Less than an hour after her death, he also died. A special law was passed to allow them to be buried together.

BERTHOLLET, CLAUDE LOUIS • See Bleach

BESSEMER, HENRY (1813-1898)

English engineer

Born in Charlton, Hertfordshire, England on January 19, 1813, Sir Henry Bessemer was the son of an engineer who was a French immigrant. His inventiveness and keen business sense led him to a life of fame and wealth.

At an early age, he showed an interest in inventing things. His involvement with steel production began in the 1850s, during the Crimean War, when he experimented with new types of cannon projectiles. His most significant innovation at the time was a *spinning projectile*. The spinning motion helped stabilize the trajectory of the object. Ignored by the British military, Bessemer took his spinning projectile idea to France, where it was better received. The main problem was that the projectile tended to explode before leaving the cannon chamber. Bessemer needed a higher-strength material for the cannons.

Steel was scarce for there was no efficient method of removing carbon from iron ore to the specifications needed for steel. It was thought that cast iron had to be converted to wrought iron by removing most of the carbon, then converted to steel by re-adding carbon. Bessemer contended that cast iron could be converted directly into steel by applying blasts of cool air to the molten iron. Although many believed that the cool air would only solidify the iron prematurely, Bessemer's demonstrations proved the opposite to be true. The cool air caused the carbonic impurities to ignite and burn off more readily.

Bessemer's process, also called the pneumatic conversion process, was patented in 1856. Ironically, American inventor **William Kelly** had developed virtually the same process in 1851. Kelly, however, had chosen to protect his discovery through secrecy instead of through the patent office, so Bessemer had no knowledge of Kelly's work. Not until a year after Bessemer announced his process did Kelly go public and secure an American patent. The Bessemer interests eventually overtook Kelly's steelmaking firm, and most of the profit and notoriety went to Bessemer.

In 1878 **Sidney Gilchrist Thomas** adapted the Bessemer process to the removal of phosphorus in the steel. Bessemer had produced his steel from phosphorus-free ores, and was therefore limited by the selection of ores. The Bessemer process was the most important development in the steel industry because it allowed steel to become, with iron, the foundation of modern industry. For his accomplishments, Bessemer was admitted to the Royal Society in 1877 and was knighted in 1879.

The open-hearth steelmaking process developed by **Charles William Siemens** and his brother, **Ernst Werner von Siemens**, and **Pierre-Émile Martin** surpassed the Bessemer process at the turn of the twentieth century in terms of total steel output. The last Bessemer converters were phased out during the 1970s.

Bessemer went on to invent other things, such as a steamship cabin that remained stable while the **ship** rolled, preventing seasickness. He also invented a solar furnace, an astronomical **telescope**, and diamond polishing machines.

None of Bessemer's other inventions approached the importance of the steelmaking process that bears his name. The process did not become fully successful until Bessemer was in his seventies. Nonetheless, both he and the industrialized world benefited greatly from it. Bessermer died in London, England, on March 15, 1898.

See also Artillery; Blast furnace; Solar heating ; Steel production

Henry Bessemer

Betatron • See Cyclotron

Bevan, Edward John (1856-1921)

English chemist

Bevan studied chemistry at Owens College, Manchester, England, where he met his future collaborator **Charles F. Cross**. After completing his studies, Bevan worked for a **paper** mill in Scotland. The main ingredient in paper is cellulose, primarily from trees. Cellulose does not dissolve in water, although it will swell when wet, and will react to a few chemicals, namely strong acids and bases.

In 1885 Bevan and Cross began a consulting firm in London, concentrating on the investigation of industrial applications of cellulose. Together with Beadle they patented a process that resulted in viscose **rayon**, dissolving cellulose in carbon disulfide and sodium xanthate and then passing it through fine holes in a metal plate into a dilute acid. The acid regenerated the cellulose into fibers that can be spun and dried.

In 1892 Bevan became the public analyst to the Middlesex County Council. He died in October of 1921, near London.

See also Cellulose, chemical uses of; Fiber, synthetic; Textiles

Bicycle

Forerunners of the bicycle are depicted on ancient Greek tombstones and in ancient Greek and Egyptian drawings. The first real bicycle is considered to be the *célérifère*, a two-wheeled wooden horse designed by Comte Mede de Sivrac of France around 1790. The rider straddled a horizontal bar attached to a **wheel** at either end, grasped the fixed handlebar, and propelled forward by pushing with alternate feet. Authorities disagree as to whether Sivrac ever actually built his célérifère; in any event, it was no more than a toy. A major improvement in two-wheeler design, perfected in 1818, came from the German Baron Karl von Drais de Sauerbrun. This ''Draisienne'' or ''Draisine'' featured a steering bar attached to a spindle on the front wheel. A craze for the machine developed among fashionable people in France and then in England, where it was called a ''dandy horse'' or ''hobby horse.''

The first bicycle that could be propelled without the rider pushing against the ground with the feet was invented by the Scots blacksmith Kirkpatrick Macmillan in 1839. He connected foot treadles to cranks attached to the rear wheel. Macmillan's design included an improved handlebar and a more

Examples of early bicycles, clockwise from top, left: early nineteenth-century célélfère; velocipede; early chain-driven "safety" bicycle; "reclining" bicycle; 1870s tricycle.

comfortable seat. This machine never became popular. The next important development of the bicycle came from two Frenchmen, Pierre Michaux and his son Ernest, in 1861. Approached by a customer to repair his dandy horse, the Michaux changed its design instead. They fitted rotating pedals to the front axle. This machine came to be known as the *velocipede* and was quite popular; by 1865 the Michaux were manufacturing 400 of them a year. The smoothness of the velocipede's ride can be guessed at by its English nickname, ''boneshaker.'' A Michaux employee, Pierre Lallement, departed for the United States in 1866 when he was denied equal credit for the design of the velocipede. With James Carrol of Connecticut, Lallement took out the first United States bicycle patent that year.

The tubular metal bicycle frame was patented by American Madison in 1867. The French bicycle industry was stalled by the Franco-Prussian War of 1870, and England took the lead in bicycle development. Hard rubber tires appeared in 1868, followed by the cycle **lamp** and crude brakes. To improve speed, the front wheel gradually became larger. In 1870, James Starley (1830-1881) of England designed a bicycle with a mammoth front wheel and very small rear wheel. Formally named the Ariel, Starley's machine was commonly called a ''penny-farthing,'' after two very different-sized British coins of the time. Also called a ''high'' or ''ordinary,'' the bicycle was both easier to pedal and faster but was very unstable because of its high center of gravity. Applying the brakes was likely to propel the rider head-first over the handlebars—hence its nickname, ''bonebreaker.''

The hazards of the penny-farthing led to two new developments. One was the tricycle, which became quite popular with female cyclists; in 1881 even Queen Victoria (1819-1901) ordered one. The other innovation was the ''safety'' bicycle built by Englishman H. J. Lawson in 1876. The safety had two medium-sized wheels of approximately equal diameter and a chain-driven rear wheel. Mass production of the Rover safety bicycle began in 1884 or 1885 in Coventry by John Kemp Starley, nephew of the penny-farthing inventor. With the advent of the diamond metal frame in 1893, the modern bicycle design was essentially complete. All that remained was the reinvention of the **pneumatic tire**, accomplished by Scots veterinarian **John Boyd Dunlop** in 1888.

The ensuing wave of popularity for bicycle riding had several important social effects. Women for the first time had an acceptable, healthful form of exercise that also promoted sensible changes in restrictive female clothing. The public gained an inexpensive means of increased mobility. And in the United States, bicycling gave birth to a demand for greatly improved roads—which, in turn, helped pave the way for the **automobile**.

By the beginning of the 20th century, bicycles had effective **braking systems**, electric lights, and multiple speed **gearing** mechanisms and cost less than a month's wages. Improvements continued to be made between the wars, especially in Europe where the bicycle was very popular, with advances being made in tube technology and in the development of alloy components. In America however, bicycle popularity declined until it was regarded as only a child's toy. In 1933, Schwinn produced an impractical, single-speed bike whose wide tires made it resemble a **motorcycle**. After World War II, bike sales plummeted in Europe and began to rise in America as the lightweight, derailleur-gear bikes began to be imported from Europe. By the 1960s, the road bike, built for speed with its thin tires and drop handlebars, became very popular. In the mid-1970s, designers combined new lightweight technology with old-style balloon tires and invented the mountain bike. In 1986, bicycle sales in the United States increased by 80 percent, with mountain bikes accounting for a 35 percent share. Bicycles at the start of the next millennium will have superlight but strong carbon-fiber composite frames, motorcycle-style brakes, an electronic transmission, and even electric-assist pedaling. Shock absorbers will become standard as will titanium pedals and Kevlar tires. As the world total of bicycles approaches nearly one billion, people are using bicycles for sport, touring, transport, fitness, and even a way to get back to nature.

BIFOCALS • See Eyeglasses

BIGELOW, ERASTUS BRIGHAM (1814-1879)

American inventor

One of the keys to the English Industrial Revolution was the development of the **water frame** spinning machine. This device, invented in 1769 by Sir **Richard Arkwright**, accelerated the spinning process to such a degree that the entire textile industry was forced to modernize itself in order to keep up with yarn production. In the United States, a similar revolution was facilitated by the invention of the power **loom** for the weaving of **carpets**. Like the water frame, the power loom forced the entire American textile industry to operate more efficiently. The inventor of the power loom was Erastus Bigelow.

Bigelow was born in West Boylston, Massachusetts. The small farm where he was raised could not provide enough to feed and clothe him and his brother, and at a fairly young age the boys were forced to earn their own keep. For many years Bigelow labored on neighboring farms, played violin at the local church, taught penmanship to the town's children, and performed other odd jobs. At the age of eighteen he wrote and published a stenography textbook, earning a small sum of money in the process; he invested his profits in a second, larger volume that he attempted to market across New England. His attempts were unsuccessful, and he soon found himself several hundred dollars in debt.

During this time Bigelow had been studying medicine at the Leicester Academy, with the intention of continuing at Harvard. Forced by his financial situation to leave school, he turned his attention toward becoming an inventor. It is not clear what inspired Bigelow to concentrate on loom technology; whatever the reason, he developed in 1837 a power loom for the weaving of lace. It was immensely successful, and Bigelow soon realized that the basic principles behind his loom could be adapted to produce ginghams and other fabrics. In 1838 he, along with his older brother, founded the Clinton Company to build and operate his new looms. The company became so successful that the town in which it was based became known as Clintonville and, eventually, Clinton, Massachusetts.

The next thirty years were very productive for Bigelow. His power looms were adapted to make the very popular Brussels, Wilton, tapestry, and velvet carpets, as well as counterpanes and pile fabrics. He patented almost every device used in the production of carpeting in the United States and England, most of which are still used today (with some modifications).

Outside of the textile industry, Bigelow was very active in social and political forums. He ran for Congress in 1860, but was defeated by a handful of votes. He was a respected economist, publishing several volumes on English and American tariff policies (he was a staunch advocate of American protectionism). In 1861 he was part of the twenty-one member committee that established the Massachusetts Institute of Technology.

BINAC • See Stored program

Binnig, Gerd (1947-)

German physicist

Along with his research colleague **Heinrich Rohrer**, Gerd Binnig invented the first **microscope** that opened the individual atom to view. The Royal Swedish Academy of Sciences found this **scanning tunneling microscope** (STM) so important that it awarded the device's inventors half of the 1986 Nobel Prize in physics just five years after the first successful test of the STM. The academy declared that even though development of the STM was in its infancy, it was already clear that "entirely new fields are opening up for the study of the structure of matter." Binnig was only 39 years old when he received the honor.

Binnig was born in Frankfurt am Main, then West Germany, on July 20, 1947, the son of Ruth Bracke Binnig, a drafter, and Karl Franz Binnig, a machine engineer. Binnig earned both a diploma and a Ph.D. in physics from Johann Wolfgang Goethe University in Frankfurt. Immediately after receiving his doctorate for work on superconductivity in 1978, Binnig joined the staff of the research laboratory operated by International Business Machines (IBM) in Zurich, Switzerland, and began his collaboration with Rohrer on the development of the STM.

Rohrer had been at the IBM lab since 1963 and also had a background in superconductivity. Together Binnig and Rohrer became interested in exploring the characteristics of the surface of materials. The atomic structure of the surface of a solid differs from the atomic structure of the solid's interior in that atoms on the surface can interact only with atoms at, on, and immediately below the surface, making surface structures frustratingly complex.

To accomplish their goal, Binnig and Rohrer turned to a phenomenon of quantum mechanics known as tunneling. Quantum mechanics had earlier revealed that the wavelike nature of electrons permits them to escape the surface boundary of a solid—they "smear out" beyond the surface and form an electron cloud around the solid. Electrons can "tunnel" through touching and overlapping clouds between two surfaces. Ivar Giaever of General Electric verified this experimentally in 1960. Binnig had investigated tunneling in superconductors during his graduate studies. Now he and Rohrer decided to make electrons tunnel through a vacuum from a sample solid surface to a sharp, needlelike probe. As the needle tip approaches within a nanometer (one billionth of a meter) of the sample, their electron clouds touch and a tunneling current starts to flow. The probe's tip follows this current at a constant height above the surface atoms, producing a three-dimensional map of the solid's surface, atom by atom.

In order to insulate their microscope against the serious problem of distorting vibration and noise, Binnig and Rohrer made a series of technical advances that included the creation of a probe tip consisting of a single atom. The colleagues and their research team soon demonstrated practical uses of the STM, revealing the surface structure of crystals, observing chemical interactions, and scanning the surface of DNA (deoxyribonucleic acid) chains. Using the STM, Binnig became the first person to observe a virus escape from a living cell. The tremendous importance of the STM lies in its many applications—for basic research in chemistry, physics, and biology and for applied research in semiconductor physics, microelectronics, metallurgy, and bioengineering.

While he was on leave at Stanford University in California in 1985, Binnig examined the use of the atomic force between atoms, rather than tunneling current, to move the scanning tip over a solid's surface. Binnig shared his ideas with Christoph Gerber of IBM Zurich and Calvin Quate of Stanford, and soon they had produced a prototype of a new type of scanner, the **atomic force microscope** (AFM), which started a new field of microscopy. The AFM made it possible for the first time to image materials that are not electrically conductive.

Binnig became group leader at IBM's Zurich lab in 1984, and also an IBM Fellow. He was visiting professor at Stanford from 1986 to 1988. He was the author of a popular German book on human creativity and chaos titled *Aus dem Nichts* (*Out of Nothing*), which argued that creativity arises from disordered thoughts. He and Rohrer shared a number of prestigious international awards for their pioneering research in microscopy.

Binocular

In 1823, a new optical instrument began to appear in French opera houses that allowed patrons in the distant (and less expensive) seats to view the opera as if they were in the front row. Called opera glasses, the device combined **telescope** lenses with stereoscopic **prism**s to provide a magnified, three-dimensional view. After many years (but relatively few modifications), opera glasses have evolved into the binocular.

In their simplest form, *binocular*s are a pair of small refracting telescope **lens**es, one for each eye. The brain assembles the two views, one from each lens, into a single picture. Because each eye sees its own view, the final image has depth; this is not so with conventional telescopes, which possess only one eyepiece and, therefore, a two-dimensional image.

While some simple binoculars can be found, most quality binoculars possess a more intricate design. In more complex binoculars, there is a system of **prism**s between the large front lens, called the objective lens, and the smaller eyepiece. These prisms serve two important functions. First, they bend the light so that the final image is both upright and nonreversed. In a common telescopic view, the image is both reversed and upside-down. Second, the bending produced by the prisms lengthens the overall light path, which allows for much greater magnification while staying within the binocular's short tube. Without prisms, average-powered binoculars would need to be more than one foot (0.305 m) in length.

In order to enhance the stereoscopic, or three-dimensional, effect, the binocular's two objective lenses are placed further apart than the viewer's eyes. When the two views are then assembled by the brain, a greater impression of depth and clarity results.

Many different factors influence the quality of a binocular. For the majority of users, the most important of these is magnifying power. Binocular magnification usually ranges from six to twenty times—that is, the object appears six to twenty times larger in the binoculars than it would with the unaided eye. Magnification is usually expressed as "X," whereby six times magnification would be written 6 X.

Another factor governing a binocular's quality is the size of the two object lenses, called the aperture. Larger aperture sizes are valued, because they collect a greater amount of light. This is crucial, because an image becomes fainter as the magnification increases. Thus, high magnifications are usually coupled with wide apertures. While object lenses of thirty to eighty millimeters are common, apertures as large as 150 millimeters have been designed; these, however, are used chiefly in military reconnaissance. Binocular makers generally express the quality of their instruments in terms of both magnification and aperture size. A common rating is 7 × 50, which describes a magnification of seven power and an aperture of fifty millimeters.

Professional users of binoculars, such as astronomers and military personnel, also consider the size of the light beam that exits the eyepiece, called the exit pupil. The closer this light beam is to the width of the viewer's own pupil, the more efficient the binocular. This is a tricky factor because the size of a human pupil varies: in bright light (such as daylight) the pupil is only 0.078 in (2 mm) across, while in dim light (such as moonlight) it opens to almost 0.273 in (7 mm). Thus, binoculars with a small exit pupil are best for daytime use, while those with a wider exit pupil are essential for nighttime observing.

To appeal to sports enthusiasts who do not want to carry the heavy, bulky high-performance binoculars used for military and hunting applications, manufacturers have developed compact, lightweight, economical binoculars. Such binoculars feature cases made of a hard, durable **plastic** called polycarbonate, and lightweight optics that feature antireflection coatings to maximize the amount of light that passes through the system.

Yet another factor affecting the quality of a binocular is the straightness of its beams. In a perfectly adjusted instrument, the two beams entering each eye will be parallel. If these beams are offset even slightly, a doubled image will be produced. Such a poorly adjusted view is uncomfortable and bad for the eyes. In order to fix the image, the binoculars should be collimated so that the beams are parallel.

Although they are inferior to telescopes in magnification, binoculars are often better devices for viewing the heavens. Because of their wider object lenses, binoculars can collect more light than telescopes; this makes objects such as distant stars or planetary satellites appear much brighter than they would in many telescopes. Even household binoculars are sufficient for viewing the Moon and the visible planets, and they are usually much less expensive than a telescope.

Biodegradable plastics

Most of the world's **plastics** are made from hydrocarbons, specifically oil and petroleum products that take hundreds of years to disintegrate once they are thrown away. However, there are indications that tomorrow's plastics will be made from materials that are more quickly and completely biodegradable than oil-based plastics. Plastics of this sort may one day be discarded in backyard compost piles.

During the 1980s, compostable bags made of **polyethylene** plus starch first came upon the market. These early biodegradable bags caused disappointment when it was discovered they did not totally degrade, but left polyethylene residues that were invisible to the naked eye and difficult to find by normal chemical analysis. Soil scientists expressed concern that the polyethylene fragments might accumulate in soils and hinder both microbial and plant growth. The test of true biodegradability, they argued, is whether in a composting environment 100 percent of the carbon of the material will mineralize and convert to carbon dioxide.

In 1990 the British-based company Imperial Chemical Industries (ICI) released a material called Biopol, used to manufacture shampoo **bottles**. Biopol is made from a plastic called PHB-V, which is produced from bacteria and glucose. When thrown away, this material is broken down by the microorganisms present in waste and decomposed completely within a couple of months.

Other versions of biodegradable plastics have been developed in the United States. In 1991 Procter and Gamble, Du Pont, and Exxon funded bacteria-based plastic research at a number of academic institutions, including the University of Massachusetts at Amherst. In addition, Battelle, a private research company, produced a completely biodegradable plastic from vegetable oils. Other research and development efforts have shown that plastics can be made from other glucose-intensive materials such as potato scraps, corn, molasses, and beets.

Even though the general concept of truly biodegradable plastics and the specific development of Biopol was exciting, the cost of manufacturing these plastics proved to be much higher than the cost of making oil-based plastics. In 1990 biodegradable plastics were seven times more costly to produce than standard oil-based plastics.

By 1997, compostable bags still sold for more than nondegradable bags, but this did not take into account the high cost of removing nondegradable plastic at composting facilities. The outlook for biodegradable plastics thus appeared to depend on success in developing applications that favored composting versus landfilling.

Again in 1997, examples of both natural and synthetic biodegradable polymers were known. One synthetic polymer used in biodegradable plastic bags was an aliphatic **polyester** called polycaprolactone, marketed by Union Carbide. Another biodegradable plastic film called MaterBi contained corn starch and other proprietary ingredients. Corn was also used to make polylactic acid for Cargill's EcoPLA biopolymer. Research was underway at the University of Michigan on a material called Envar consisting of an alloy of caprolactone and a thermoplastic starch.

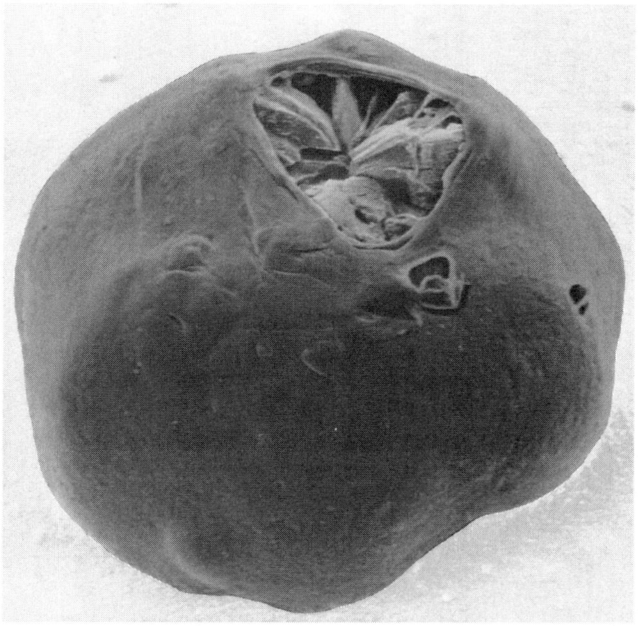

A scanning electron micrograph (SEM) of the surface of a sheet of biodegradable plastic. The spherical object that dominates the image is one of the many granules of starch embedded in the surface of the plastic. When the plastic is buried, the starch takes up water and expands, thereby breaking up the plastic so that more of it is exposed to the bacteria that feeds on it.

BIOLOGICAL WARFARE

Long ago humans discovered that certain animals and plants were capable of causing severe illness or death if anyone consumed them. Some were clever enough to extract liquids from these poisonous plants and tip their arrows with the poison before hunting. It was inevitable that these natural poisons would be used on the battlefield. In 600 B.C. Greek soldiers fighting against Greeks from another city became mysteriously ill after drinking water that their enemies had poisoned with rotting animal carcasses. In the 1400s, Tartars captured a town by catapulting the bodies of plague victims over the walled city. In eighteenth-century North America, British military leaders sent blankets containing smallpox germs to Indian camps in the hopes of causing outbreaks of the dreaded disease.

The twentieth century has seen an increase in the sophistication of biological warfare. During World War I, the Germans were accused of inoculating horses and mules with glanders, a very infectious disease that occurs among animals. They also were charged with infecting cattle with anthrax, an infectious disease that can be transmitted to humans. There are also stories of German spies caught attempting to spread plague bacteria in Russia.

The fear and revulsion of biological weapons were so great that several nations outlawed their use. However these were often ignored. For example, the Japanese built the first known major biological warfare installation in 1937, where they experimented with such bacteria as typhus, typhoid, anthrax, cholera, botulism, and smallpox as potential weapons. And, since bacteria become effective by the chemical poisons they secrete, technically, the ''soldier'' or ''assassin'' is neither alive (biological) nor chemical (like chlorine). As a result, they are not covered by treaties dealing with biological and chemical weapons. The Japanese may have put their research to use on the battlefield; spies supposedly carried biological agents to Chinese towns.

Since the Allies feared large-scale biological warfare during World War II, they set about creating their own weapons. The British developed and tested anthrax bombs, 4 lb. (1.8 kg) small **bombs** loaded into 500 lb. (227 kg) cluster bombs.

Clarence Birdseye

They proved that fragile living organisms could withstand being produced, transported, loaded into munitions, and exploded. In the United States, a secret operation was set up to research, test, and produce biological weapons. After the war, experiments in biological warfare continued. In the United States, for example, the CIA (Central Intelligence Agency) wanted to find a better suicide pill for its agents to swallow if captured. Researchers came up with saxitoxin, a poison given off by a tiny marine plankton and able to kill a human within ten seconds. The Soviets developed a fungal poison that was apparently used in Laos, Yemen, and Afghanistan.

An international treaty banning the production of biological weapons was signed by 140 countries in 1972. The United States apparently ceased investigating offensive biological weapons at this time. However, the peak of the Soviet Unionòs biological weapons research is said to have come in the twenty years after the treaty was signed. Hundreds of biological agents were stockpiled for use in missile attacks. Soviet scientists also developed genetically altered strains of plague, anthrax, glanders, smallpox and other bacteria to make them resistant to antibiotics. They also cultivated rare but particularly virulent viruses such as Marburg virus, which causes uncontrolled bleeding.

The actual feasibility of a battlefield biological weapon attack is uncertain, since many biological agents are vulnerable to weather conditions, and their dispersion may be difficult to control. But a number of countries, including Iran, Iraq, Israel and Egypt, are known to have biological weapons capabilities in the 1990s. Biological weapons programs are feared to be proliferating because they may be cheaper to obtain and more lethal than conventional weapons. Defense against a biological attack in a densely populated area is extremely problematic. Vaccines usually take months to develop, and so would be no use after the event of an attack. Another difficulty would be gathering enough antibiotics to treat a massive outbreak of infection caused by biological weapons. American scientists began work in 1998 on advanced vaccine manufacturing methods to counteract possible biological attacks. The technique involves attaching antigens to harmful bacteria to the common bacteria present in the nose or mouth, thus triggering the body's immune response. This method could speed up the time needed to develop a vaccine from several months to just a week. But even assuming the success of this research, possible biological weapons attacks still inspire terror. In the late 1990s there is increasing pressure to enforce the existing biological weapons treaty, which has essentially been ignored since 1972.

BIPLANE • See Aircraft

BIRDSEYE, CLARENCE (1886-1956)

American inventor and industrialist

Born in Brooklyn, New York, Clarence ''Bob'' Birdseye attended Amherst College for two years before leaving in 1912 in order to indulge his spirit of adventure by fur-trading and trapping in Labrador, Canada. Birdseye returned to Labrador in 1916 with his new wife and infant. In order to preserve the few fresh vegetables that found their way to Labrador by **ship**, Birdseye began experimenting with the Eskimo method of quick-freezing foods. He stored fresh cabbages in a barrel with sea water which froze quickly in the subzero Arctic climate. Birdseye also experimented with quick-freezing fish and caribou meat. When thawed, these foods remained tender and fresh-flavored, unlike previous methods involving slow cold storage.

Birdseye returned to the United States in 1917 determined to develop commercial methods of rapid freezing, experimenting with an electric fan, cakes of ice, and salt brine. In 1923 he invested everything he had in Birdseye Seafoods, marketing frozen fish. In 1924 he and three partners founded General Seafoods in Gloucester, Massachusetts, which became the first company to use the technique of rapid dry freezing of foods in compact, packageable blocks.

The Postum Company bought Birdseye's business and 168 patents in 1929 for $22 million. The company renamed itself General Foods and marketed its **frozen foods** under the *Birds Eye* trademark.

After the sale, Clarence Birdseye continued as a consultant to General Foods and promoted the development of the frozen foods industry by lecturing and writing. He devoted himself to more inventing, obtaining over 300 patents, including ones for an infrared heat lamp, a whale-fishing **harpoon**, a method of dehydrating foods, and a spotlight for store window displays. When Birdseye died in New York City, he was recognized as the father of the frozen food industry. His name lives on in the *Birds Eye* brand of frozen vegetables.

Birkeland, Kristian Olaf Bernhard (1867-1917)

Norwegian physicist

Birkeland's scientific career might have turned out differently had he not grown up in a country of mighty rivers. Electricity produced from Norway's hydropower plants was very cheap, and the process that Birkeland invented for synthesizing ammonia depended on low-cost power. Unfortunately, similar conditions did not apply throughout the rest of the world, and other processes soon eclipsed Birkeland's local success.

As a student, Birkeland concentrated on mathematical physics, studying primarily in Paris and Geneva. For a short time, he studied under Heinrich Hertz (1857-1894) in Bonn, Germany.

During the 1890s, Birkeland presented solutions to Scottish physicist James Clerk Maxwell's (1831-1879) equations describing magnetic and electrical forces and fields. He also led three research expeditions that studied the northern lights (aurora borealis), and he founded an observatory for geophysical studies at an extreme northern latitude (70°N).

During the late 1800s and early 1900s, scientists became interested in how "free" (elemental) nitrogen from the atmosphere is "fixed," or combined with other elements. Normally, nitrogen gas is chemically inert, yet nitrogen compounds are found throughout nature. **Fertilizers**, explosives, and other products manufactured from nitrogen compounds were made from limited natural resources, mainly Chile saltpeter (sodium nitrate, or $NaNO_3$), which was imported from the deserts of South America. As these resources began to dwindle, scientists became concerned that without enough nitrogen-rich fertilizers to improve crop yields, agricultural production might fall behind the rate of global population growth.

In the early 1900s, Birkeland and his co-worker Samuel Eyde (1866-1940) invented the first commercially successful process for artificially fixing atmospheric nitrogen. Eyde, whose education was in construction engineering, had been involved in building railway stations and harbors throughout Scandinavia. He was also one of the founders of Norway's electrochemical industry. With such a background, Eyde became interested in entrepreneurial ways to make use of the electrochemical industry's expertise. In 1901 Birkeland and Eyde set up a small laboratory for studying electric arc methods of combining atmospheric nitrogen and oxygen into nitrogenoxides.

Their experiments were based on a technique pioneered in 1784 by Henry Cavendish (1731-1810) who had produced nitric acid by passing sparks through a jar of air confined over water. The electricity forced the nitrogen and oxygen in the air to combine, forming nitrogen dioxide gas, which then dissolved in the water to create nitric acid.

While Eyde secured power supplies and additional financing for the venture from Swedish and French sources,

Kristian Olaf Bernhard Birkeland

Birkeland focused on the technical aspects of the process. They succeeded in developing a practical process and building a small plant. In the Birkeland-Eyde process, a magnetic field is used to spread the electric arc into a disk of flame. Air is blown into the disk, then mixed immediately with cold air. The process is able to maintain a high-temperature equilibrium.

In 1908, a larger plant using the Birkeland-Eyde arc process was built at Notodden, Norway. Several process modifications were developed, and more plants were built in other countries as well. But the arc process uses energy inefficiently, making it impractical for most applications.

By 1913, **Fritz Haber** and **Carl Bosch** had established a method of fixing nitrogen as ammonia instead of nitrogen oxide. Their **ammonia synthesis** process finally overshadowed the Birkeland-Eyde method.

Later, Birkeland moved to Egypt, to improve his health and also to pursue his interest in astronomy. He attempted to return to Norway during World War I but died en route. Eyde continued to be successful in both business and politics and was appointed Norway's minister to Poland in 1920.

See also Electromagnet; Hydroelectric plant; Nitrogen fixation; Ostwald-Bosch process

Biro, Ladislao and Biro, George • See Pen

Birth control

Birth control, or contraception, is the use of physical barriers, timing, chemicals, or a combination of these methods to prevent pregnancy. While many methods have been designed for women's use, few are available to men. Abortion and the controversial antigestation drug **RU 486** (Roussel-Uclaf), are not considered routine birth control methods in the United States. The principal birth control methods in order of effectiveness are: surgical procedures, oral contraceptives, long-lasting hormonal contraceptives, intrauterine device (IUD), diaphragm, spermicides, and selective abstinence (or ''rhythm method'') and withdrawal.

The surgical procedure of tubal ligation (''tying the tubes'') involves severing or tying the fallopian tubes which prevents eggs from passing from the ovary into the uterus. For men, the vasectomy, in which the vas deferens is severed preventing sperm from leaving the testes, is a simple, reliable method. Both procedures can be done on an outpatient basis and—although generally considered permanent—can often be reversed.

Oral contraceptive include birth control pills contain various amounts of the female hormones estrogen and progesterone and mimic the natural condition of pregnancy when a woman normally cannot become pregnant. The first pill was developed in Massachusetts by endocrinologist **Gregory Pincus**, biologist Min-Chueh Chang (1908-), and physician John Rock (1890-1984). It contained progestin, a synthetic progesterone developed in Mexico by the American chemist Carl Djerassi (1923-). In 1960, it was approved by the U.S. Food and Drug Administration (FDA) and first became available by prescription as Enovid (G.D. Searle and Co.). Today, many brands of oral contraceptives are available and appropriate for most women when prescribed under medical supervision.

Long-lasting hormonal contraceptives include Norplant (Wyeth-Ayerst Laboratories), a device containing a form of progesterone, is implanted under the skin for timed release over five years. Approved by the FDA in 1990, questions have been raised about possible links to cancer, side effects of heavy menstruation and ''break-through bleeding,'' and difficulty removing the matchstick-sized implants. Depo-Provera, produced by Upjohn Co., is a synthetic form of progesterone and is used in almost 100 countries. One injection works for three months. Approval by the FDA was withheld in the 1970s because of possible links to cancer and osteoporosis. In 1992, the FDA once again began the process of approving Depo-Provera.

The intrauterine device (IUD) is placed in the uterus for long-term use. It is not known precisely how an IUD prevents pregnancy. It may cause an egg to be released prematurely, before the uterus is prepared to accommodate it, or metals used in the loop form of the device may disable enzymes necessary for embryo implantation. Modern IUDs date back approximately 100 years. Today's devices, made of **plastic**, copper, or **steel** are formed into loops, coils, and T-shapes. An IUD must be inserted and removed by a physician. Some types have been associated with perforation of the uterus and pelvic inflammatory disease, in some cases resulting in death. The most successful IUD is the plastic Lippes loop, developed in the 1960s by Dr. Jack Lippes (1924-). Most contemporary IUDs are variants of the Lippes loop, containing either copper or progesterone.

A condom is a latex rubber or lambskin sheath placed over an erect penis to prevent sperm from entering the vagina during intercourse. It is also the most effective method of preventing transmission of venereal diseases and AIDS/HIV. The practice of using condoms during intercourse dates back to at least the sixteenth century. A condom can be used alone or in conjunction with a spermicidal foam or gel inserted into the vagina.

Developed in Germany in the late nineteenth century, the diaphragm is a flexible rubber barrier that a woman inserts vaginally before intercourse to cover the cervix and is most effective when used with a spermicide.

In the 1970s, chemicals such as nonoxynol-9 that kill sperm became available in suppositories, foams, creams, jellies, and sponges (which have been taken off the market). Inserted into the vagina before intercourse, spermicides can be used alone or, more effectively, with condoms and diaphragms.

The rhythm method requires a woman to track her monthly cycle and avoid intercourse near the time of ovulation. The temperature method involves monitoring body temperature variation during the monthly cycle. Temperature falls below normal in the days before ovulation, drops further during ovulation, and then rises above normal until menstruation. Monitoring differences in vaginal secretions can also help pinpoint the ovulation time. The calendar method assumes that ovulation occurs on the 15th day of the monthly cycle. Because this differs between women and even for the same woman each month, this is the least reliable of the rhythm methods. Although withdrawal—the man withdrawing before ejaculation—is also unreliable, when no other contraception is used, the risk of pregnancy can be greatly reduced.

In May, 1998, a new oral contraceptive for women—the first to use a shortened ''hormone-free interval'' and lower daily doses of estrogen—was approved by the FDA. In 1996, the World Health Organization hailed a contraceptive injection for men of testosterone that drastically reduces the sperm count and which is 99 percent effective. This contraceptive will not be available for five to ten years. Two other studies in male birth control are ongoing—one focuses on preventing sperm from breaking the egg's gel-like protective coating; the other on blocking protein receptors on the sperm so it cannot ''dock'' on the egg's exterior.

Bissell, Melville and Bissell, Anna •

See Carpet sweeper

Bitumen • See Pitch

BLAIR, HENRY (ca. 1804-1860)
American inventor

Little is known of the life of Henry Blair, except that he was probably a free farmer of Montgomery County, Maryland. He was one of the first African-Americans to receive a patent. His invention, a corn-planting machine, was able to seed corn in a checkerboard fashion, a system which provided good weed control. Two years later, he designed a similar device suitable for planting cotton.

BLALOCK, ALFRED • See Blue baby operation

BLANCHARD, THOMAS (1788-1864)
American inventor

Blanchard is considered one of the founders of the American **machine tool** industry. His contributions anticipated the development of **mass production** in the late nineteenth and early twentieth centuries.

Born in Sutton, Massachusetts, in 1788, Blanchard worked in his brother's factory while still a boy, designing and producing such inventions as an apple parer and a tack-making machine. His most important work, however, came with his employment at the U.S. arsenal in Springfield, which sought his services after his construction of a **lathe** that could produce irregular gunstocks. Around 1818, while focusing on the problem of manufacturing identical gunstocks in rapid succession, Blanchard conceived of a lathe capable of endlessly duplicating any pre-set machine pattern. Integral to Blanchard's invention was the movement of a friction **wheel** over the pattern and the transmission of the movement to a cutting wheel. In addition to gunstocks, among the model patterns that Blanchard experimented with were **shoe** lasts, wheel spokes, and hatblocks.

Because of his machine's ability to produce not only exact duplicates but replicas in various sizes as well, Blanchard is considered an inventor of the first rank. Lathes based on his original became the foundation of many modern manufacturing processes. Unfortunately, pirating of his idea, complicated by weak patents, was common until Congress eventually reaffirmed the primacy of his work.

Blanchard is also respected for his visionary, though failed, promotion of railroads and for his work with steam carriages and **steamboat**s.

Thomas Blanchard

BLAST FURNACE

Furnaces have been used in the metal-smelting process for thousands of years. The walls of a furnace trap heat to achieve a temperature great enough to melt metal. For millennia, people melted metal simply by building huge fires in furnaces of various types and letting the natural build-up of heat occur. However, in more recent history, metallurgists have developed various techniques to make the furnaces cleaner, hotter, and more efficient.

The blast furnace is one of the greatest improvements to the smelting process. By allowing blasts of air into the furnace, greater heat can be produced. The individual credited with the invention of the blast furnace is **Abraham Darby**. Although blast furnaces had already existed for a long time, Darby introduced coke as fuel for the furnace. Coke is made from coal heated between 9,000°C (16,232°F) and 11,500°C (20,732°F). It burns slower and hotter than coal, with little or no flame. Darby's process made it possible to produce iron more efficiently.

Blast furnaces were further improved in the 1850s by the simultaneous invention of the pneumatic conversion process by **William Kelly** and **Henry Bessemer**. This process directed the blasts of air into the molten metal, either overhead or directly into the metal through holes in the bottom of the kettle. It allowed for the efficient elimination of carbonic impurities from the iron, making **steel production** possible.

In 1856 brothers **Charles William Siemens** and **Ernst Werner von Siemens** invented their regenerative metal-processing chamber, which trapped escaping heat and returned it to the furnace. Improvements were made by **Pierre-Emile Martin** as well.

A newer generation of blast furnace employs the basic oxygen method, which produces supersonic blasts of oxygen. An automated blast-furnace system was introduced in 1997. Developed in Austria, the furnace is regulated by computers and specific computer models. These models control each phase of the operation of the blast furnace for optimal output.

See also Iron production

Bleach

Although ancient methods of bleaching remain unknown, historians have evidence that early civilizations must have known how to bleach fabrics. White cloth was produced by the ancient Egyptians, Babylonians, Phoenicians, and Hebrews, as well as by the Greeks and Romans. After the Crusades of the 1100s and 1200s, the practice of bleaching fabric spread throughout Europe. In the old days, people simply spread wet cloth on the ground outdoors and left it to dry in the sunlight until it turned white, which could take weeks or even months. This process came to be called crofting, after the Scottish word for a small meadow (*croft*). As early as 1322, crofting was practiced on bleaching grounds in England near Manchester. In Scotland and Ireland, some people still bleach their cloth on the grass in this way. High-quality **linen** that was dried on plots of grass became known as *lawn*.

By the 1700s, Dutch weavers had improved the bleaching process and emerged as the leaders of Europe's bleaching industry. They discovered that linen, which was still the most common type of cloth, could be bleached more efficiently by first soaking it in lye (a concentrated alkaline solution of potassium or sodium hydroxide). After the lye was washed out, the linen was spread on the ground as usual. After repeating this step a few times the Dutch soaked the linen in buttermilk, or soured milk, then washed it and dried it outdoors again. Although major bleaching operations were known outside Holland, the Dutch enjoyed a near-monopoly on bleaching linen through the 1700s. Fabric produced by the Dutch process was called *holland* cloth. However, this process was problematic in that it could take several months, especially in northern countries with limited sunlight. Furthermore, it used up large amounts of valuable space.

In 1756, scientists found that dilute sulfuric acid would work better than buttermilk and the time required for the bleaching process was greatly reduced. An even more dramatic improvement in bleaching technology resulted from the discovery of chlorine in 1774 by Swedish chemist Karl Wilhelm Scheele (1742-1786). French chemist Claude Louis Berthollet (1748-1822) discovered that this gas is a very effective bleaching agent. Berthollet, who was director of a French tapestry factory, developed a method of using chlorine to bleach **textiles**. In 1785, he introduced a bleaching liquid called lye de Javelle and publicized his technique without patenting it. When **James Watt** learned of the method, he passed the information on to Scottish chemist and manufacturer Charles Tennant, who began using the bleaching liquid in Glasgow. But the chlorine gas needed for the liquid bleaching process was not readily available, so Tennant invented a more convenient bleaching powder and introduced it in 1799. The solid powder, which was made by combining chlorine with slaked lime (calcium hydroxide), was much easier to handle and ship to other fabric manufacturers. When added to a little dilute acid, the powder released the chlorine gas which bleached the cloth very quickly. By the 1830s, factories were churning out huge quantities of bleaching powder for textile use. This abundant supply of chlorine bleach helped stimulate the cotton industry.

The fabric industry found hydrogen peroxide to be another useful bleaching chemical. Peroxide is used on some fabrics, such as wool and **silk**, which can not tolerate exposure to chlorine. About 95 percent of textiles that require whitening are now bleached with hydrogen peroxide which, like chlorine, is a powerful oxidizing agent. Pure hydrogen peroxide was first prepared by Louis Jacques Thenard (1777-1857) in 1818. Even though it costs more than chlorine, hydrogen peroxide is often preferred because its action on fibers is milder, it leaves no undesirable residues, and the bleaching process is faster. Several sulfur compounds, which act as reducing agents rather than oxidants, are also used in bleaching wool and silk fabrics, as well as feathers and straw.

Historically a variety of other industries have used similar bleaching technology. For example, in the **paper** industry, wood pulp is usually treated with chlorine compounds first followed with hydrogen peroxide as a final bleaching step. This two-step process improves the pulp's ability to stay white through subsequent paper manufacturing steps. Other industries which extensively use bleaching technology include the flour industry, which uses chlorine compounds to bleach over 80 percent of all wheat flour produced in the United States, and the petroleum industry which uses chlorine bleach to deodorize gas filters. Chlorine compounds are also known for their antibacterial properties and are used in certain water treatment processes.

Through the 1990's chlorine-based products have remained the preferred method of fabric bleaching for the consumer. However, perborate bleaches are becoming increasingly popular because they can be safely used on a variety of fabrics. Improved bleaching performance is expected to come from other oxygen-containing bleaches such as percarbonates which are expected to play an important role in the future of commercial bleaching.

Blériot, Louis (1872-1936)

French aviator

Born in Cambrai, France, Blériot was a successful manufacturer of **automobile** accessories, including head lamps, foot warmers for car passengers, and luminous license plates. He spent much of his money on aviation research, but he was considered a terrible pilot—uncoordinated, impatient and a faulty designer. Out of thirteen different **aircraft** configurations he built or tested, more than half either would not fly or crashed with him at the controls.

In 1909, the London *Daily Mail* offered a prize of $5,000 for the first flight across the English channel by a heavier-than-air machine. Blériot entered enthusiastically, but his plane appeared terribly inadequate. It was small, underpowered, with only 150 square feet (45.75 sq. m) of wing area. The fuselage was only partly covered (the rest was a trellis design), and the pilot sat on a wooden seat with only a leather strap for a back rest and had no instruments to help guide him. The craft's engine was a crude, three-cylinder with holes punched in the bottoms of the cylinders to let hot gases escape. It could run for approximately a half hour, enough time, Bléroit thought, to fly across the channel.

On Sunday, July 25, 1909, Blériot beat his competition into the air. His engine ran relatively smoothly and a light rain kept it from overheating. Winds drove him past his intended landing place in Dover, England, but a French newsman, who had been assigned to cover Blériot's arrival, waved him to a safe landing, using a French flag. Blériot won the competition, completing the twenty-two mile (35 m) trip in thirty-seven minutes.

Blériot eventually formed an aircraft company that produced several well-known fighters in World War I, including the famous S.P. A.D. fighter which German pilot Eddie Rickenbacker used for many of his aerial victories. Blériot remained active in flight research until his death in 1936.

Blissboard

A blissboard is a means of communication using Blissymbolics, a language based on symbols that was invented by Charles Bliss in the mid-twentieth century. Blissymbolics has become a language used by people who cannot speak, including the severely handicapped, though Bliss had different intentions when he devised it. Born in Austria-Hungary in 1897, Bliss wanted to create an international system of symbols that could be understood by everyone. After being imprisoned in Nazi concentration camps, Bliss eventually came to live in Australia where he developed his language. He based it on symbols and pictographs, and published his work under the title *semantography* in 1949. Bliss's language went largely unnoticed, though he continued to revise and refine it. He published another edition of this book in 1965, but added his new name for his language, Blissymbolics, to avoid confusion with the academic discussion of other semantographies. Eventually, Bliss found support in Canada where in 1982, three years before his death, he granted the exclusive rights to Blissymbolics to Blissymbolics Communication International (BCI) in Canada for the use of handicapped people as well as others with language, communication, and learning problems.

While Blissymbolics can be a simple form of communication, with symbols representing common things and ideas, the language is flexible enough that complicated ideas and complex emotions can be expressed as well. There is grammar and tense forms in Blissymbolics, and room for innovation of new symbols by combining other Blissymbols. The symbols can be arranged into sentences with an intricate structure. But

Louis Blériot

Blissymbolics is easy to learn, for there are a core of about 100 symbols which are the basis to the language, and can be combined in different ways to communicated different ideas. Because of the nature of Blissymbolics, it is widely used by non-speaking people as well as those of limited literacy. Physicist Stephen Hawking is one of the most famous users of Blissymbolics.

One way handicapped people use Blissymbolics is with a blissboard, which features the key symbols on it. A ''speaker'' points to the symbols to communicate with his or her listener. Blissboards can be mounted on wheelchairs, but can only be used to communicate by two people in the same room. A Canadian woman, Rachel Zimmerman, invented the first computer program to use Blissymbols when she was 12 years old. Her program, called Blissymbol Printer, allows the user to communicate by tapping a touch-sensitive blissboard which is translated into a written language by a computer, then sent or stored. As technology has developed, similar blissboards can be input devices with the output being synthetic speech instead of text. Another software package, Bliss Write, is a Blissymbolics word processor. With the advent of the Internet, another Blissymbolics computer program has been developed, BlissNet. BlissNet translates messages between Blissymbolics and text so users can send and receive e-mail and participate in other on-line forms of communication. Though there are no exact numbers of people who use Blissymbolics, it is safe to assume that the language will continue to become more widespread.

Block and tackle

A *block and tackle* is a configuration of ropes and **pulley**s designed to gain a mechanical advantage when lifting objects.

Lifting a heavy object with a single rope wrapped around one pulley wheel requires a pulling force equal to the weight of the object. But by letting a second pulley hang below the first, and wrapping the rope around this second pulley, then back around the top pulley, the force required to lift the object is only one-half the weight of the object. The *mechanical advantage,* the ratio of weight to force applied, is two.

The rope can be looped around the bottom pulley yet again, then back around the top pulley, and the required force will be one-forth the weight, for a mechanical advantage of 4. Still more complicated arrangements can decrease the force even further, making it possible to lift heavy objects that would not be otherwise possible, such as a car engine out of the chassis of a car.

No less *work* is required, though, because the less force required to do the lifting, the greater the distance the rope must be pulled to lift the object. The block and tackle provides a mechanical advantage, but not an energy advantage.

Historically, the block and tackle has proven useful anywhere a force needed to be provided only occasionally and with no great urgency, such as farms and on **ships**. On ships the block and tackle was used with gears to move equipment such as sails and spars. Block and tackle using **steel** cable are commonly used in **crane**s for hoisting equipment.

See also Energy efficiency

Blodgett, Katherine Burr (1898-1979)

American physicist and chemist

Katherine Burr Blodgett is credited with the invention of the color gauge and non-reflecting or ''invisible'' **glass**. Born in Schenectady, New York, in 1898, Blodgett had an unusual early education. Her mother, widowed just months before Katherine was born, made education a priority for her children. Blodgett attended school in France and Germany and was privately tutored in New York City before attending Bryn Mawr College. She went on to earn a master's degree in physics from the University of Chicago. Upon graduation, Blodgett returned to Schenectady and applied for a position at General Electric Laboratories, where her father had worked as a patent attorney.

At the GE laboratory she was fortunate to work with the well-respected Dr. **Irving Langmuir**, who would later receive a Nobel Prize for his work in chemistry. Dr. Langmuir saw much promise in his young colleague and upon his recommendation Blodgett decided to pursue a doctorate degree in physics. She was the first woman ever awarded a Ph.D. in physics from Cambridge University. Blodgett's most important work came from her independent research on an oily substance that Dr. Langmuir had developed in the lab. This unusual material was unique in that it formed a film of exactly one molecule in thickness on the surface of water. In 1933 Blodgett made her first breakthrough with the film. She lowered a metal plate into the liquid and discovered that the film would move toward the plate and adhere to it as she raised the plate from the water. In fact, each time she lowered and raised the plate another layer of film would adhere to it. This was the first time a scientist had ever been able to build up layers of molecules one at a time.

Blodgett knew there were practical applications for these built-up layers of film. She observed that the number of layers of molecules accumulated determined the amount of light reflected by the film and concluded that it would be possible to measure the thickness of the film by using the color of the light it reflected. By sealing this film in varied thicknesses within a glass tube, Blodgett could match the color with a corresponding layer of thickness. This device, called the color gauge, now enables scientists in the fields of chemistry, biochemistry, physics, and metallurgy to measure the thickness of transparent and semitransparent substances within millionths of an inch. Blodgett continued working with what has come to be known as the Langmuir-Blodgett film and in 1938 created non-reflecting glass by applying a thin layer of it to transparent glass. The light reflected by the film canceled out the light reflected by the glass itself, thus rendering the glass invisible. This ''invisible glass'' has been used in many consumer products from picture frames to **camera** lenses.

Blodgett's invisible glass has also been extremely useful in optics. Because the film eliminates all reflection, it allows 100 percent of the light that falls on a **lens** to pass through it. Previously, eight to ten percent of light falling on a lens would be lost due to the glass lens's reflection of it. Many of today's camera lenses, created using Blodgett's invisible glass, allow photographers to capture all of the available light. The Langmuir-Blodgett film was also used briefly in the process of artificial rainmaking. Though Blodgett found no other practical use for the film before her retirement in 1963, her invention of the color gauge and her method of creating ''invisible glass'' have assisted scientists in countless endeavors which require the measurement of extremely thin substances. Blodgett died in 1979 at the age of eighty-one.

Blood, artificial

In 1957, Thomas Chang, an undergraduate at McGill University, decided his research project would be to make the first artificial cell and from this project reported the preparation of artificial red blood cells. That project grew into today's dynamic field of biomedical research in which ''designer cells'' are a major focus. Chang is credited with inventing *microencapsulation*, which allows biochemicals to be contained inside artificial membranes and, ever since that development, he and many other researchers have been on a quest to develop viable artificial blood to replace whole blood in transfusions. However, success has remained elusive. A safe, effective blood substitute is highly desirable because it would eliminate several problems of using fresh blood, including supply shortages in the face of increasing demand; short shelf life even under **refrigeration**; transmission of hepatitis, the AIDS virus, and other viral diseases, and the need for careful blood typing. The primary job of artificial blood is to duplicate the oxygen-carrying function of *hemoglobin*, an iron-containing protein within the red cells of natural blood. Researchers and manufacturers have taken two approaches: so-called ''white blood'' using *perfluorochemicals* (or fluorocarbons), and ''red blood'' made from *modified hemoglobin*.

The potential of fluorocarbons, a family of organic molecules, was dramatically demonstrated in 1966 by American physician Leland C. Clark, Jr., of the University of Cincinnati when he completely immersed a mouse in liquid fluorocarbon and discovered it could continue to breathe. Researchers subsequently determined that liquid fluorocarbon has the ability to dissolve large amounts of oxygen. The early fluorocarbons could not be used in human medicine, however, because they concentrated in the liver and spleen causing toxic side-effects. In 1973 Clark found that *perfluorodecalin* was completely eliminated from the body through exhalation; however, the drawback of this substance was that it could form large droplets capable of blocking capillaries. Ryoichi Naito, a chemist at the Japanese pharmaceutical firm of Green Cross, found he could overcome this problem by adding a second fluorocarbon, *perfluoropropylamine*, to the first. The result was a milky white solution called Fluosol-DA. In 1989, Fluosol was approved by the U.S. Food and Drug Administration (FDA) for use during **balloon angioplasty**. During the time the inflated balloon cuts off blood supply to some tissues, injected Fluosol can carry oxygen to the deprived tissue cells. While much has been done to improve perfluorochemicals, they are limited to low doses because of toxic side-effects. Also, because they lack the oxygen binding characteristics of hemoglobin, sufficient oxygen carriage occurs only when patients are breathing 100% oxygen. This means only small quantities (amounting to one to two units of whole blood) can be used—sufficient only to offset the need for this amount of blood during surgery.

A different approach to blood substitution has centered on hemoglobin isolated from the red blood cell and, by 1998, six North American companies were conducting modified hemoglobin clinical trials in humans, some progressing to Phase III clinical trials, in the continuing search for an appropriate substitute for blood. Use of free hemoglobin was suggested back in the nineteenth century, but modern researchers several decades ago discovered that such use had several severe problems. Outside the red blood cell, hemoglobin holds on tightly to oxygen and does not release enough to the tissue cells. Free hemoglobin also breaks down into two halves that are filtered out by the kidneys, which often causes severe kidney damage.

To eliminate these difficulties, researchers have worked on various methods of effectively modifying hemoglobin. One approach has been to chemically link the hemoglobin subunits together, forming a bigger molecule (polymer) that will not break down. Although researchers were concerned that the giant molecules could damage body organs, animal tests were encouraging. Surgeon Gerald Moss of the University of Chicago licensed his multi-molecule linkage technique to Northfield Laboratories of Illinois and, in 1987, Northfield began clinical trials of modified hemoglobin in humans. The first round of tests were successful, but during the second round in 1989 several trauma patients suffered allergic reactions to the hemoglobin product. An FDA Advisory Committee investigated and learned of a German trial conducted by physician Konrad Messmer at Heidelberg University in the early 1980s in which the two volunteers suffered kidney failure after receiving a modified hemoglobin product. The FDA concluded that many organs of the body could be damaged by hemoglobin-based blood substitutes, and abruptly halted Northfield's human trials.

Somatogen Inc. of Boulder, Colorado, used the understanding of hemoglobin's atomic structure discovered by Max Perutz to produce a genetically engineered modified hemoglobin that gave up its oxygen more easily and did not break down quickly. Somatogen produced its hemoglobin in yeast or bacteria rather than human or animal substances and sought approval for human testing in the early 1990s. Biopure of Boston began working with modified hemoglobin from cows in 1984 and began human trials of its product in Guatemala in 1990. DNX, a biotechnology company in Princeton, New Jersey, announced in 1991 that it had produced genetically-engineered pigs that made normal human hemoglobin. Some components of plasma, too, could be synthesized by the early 1990s. Genentech of California produced a plasma that promotes coagulation in hemophiliacs.

All these approaches hold promise, but researchers remain uncertain about the cause of the toxic side effects produced by hemoglobin blood substitutes.

See also Blood transfusion; medical technology

BLOOD PRESSURE MEASURING DEVICES

Hypertension, or high blood pressure—also known as the "silent killer"—affects more than 60 million Americans. Blood pressure measuring devises are the only means through which high blood pressure can be detected.

In the course of his pioneering work on blood circulation in the 1600s, William Harvey (1578-1657) noted that blood pulsated out of a severed artery as if it were under rhythmic pressure. Nearly a century later, Stephen Hales (1677-1761), an English clergyman and physiologist, devised a technique to measure the pressure exerted on the vessels as blood was pumped through them. Hales inserted a brass pipe into an animal's blood vessel and used the windpipe of a goose (for its flexibility) to connect the pipe to a long glass tube. The height to which the animal's blood spurted up into the tube gave a measure of the force propelling the blood. One of Hales' most dramatic experiments using this simple **manometer** involved a white mare, tied flat on the ground to a stable door; the glass tube in this instance was 12 ft. 9 in. (3.8 m) long, and the horse's blood rose in it to a height of 9 ft. 6 in. (2.9 m). Hales began his blood pressure measurement experimentation around 1706, continued around 1712-13, and finally reported his technique in his 1733 book *Haemastaticks.*

Another century passed before the Hales manometer was improved upon. In 1828 French physician Jean Leonard Marie Poiseuille (1797-1869) replaced the long glass tube with a U-shaped tube filled with mercury and calibrated to record pressure levels in millimeters of mercury. The German physiologist **Karl Friedrich Wilhelm Ludwig** modified Poiseuille's manometer in 1847, adding a revolving cylinder and float with a revolving drum on which the blood pressure was recorded. This device was called a *kymograph.* It was further refined by **Etienne-Jules Marey** with his 1863 *sphygmograph*, another blood pressure recorder.

The first blood pressure measuring device that required no skin penetration was the *sphygmomanometer* pioneered by Samuel Siegfried von Basch (1837-1905), a German physician, in 1876. This rather inaccurate device was replaced in 1896 by the sphygmomanometer of the Italian physician Scipione Riva-Rocci (1863-1937), the prototype of today's standard instrument which uses an arm band which is inflated until the blood flow through the arteries can no longer be detected. Air is then released from the band, and blood pressure measured on a mercury manometer at the moment when the pulse reappears. While Riva-Rocci's instrument was accurate, it measured only *systolic* pressure—pressure within the artery when the heart is contracting. A Russian physician, Nikolai Korotkoff, added the missing element when he suggested in 1905 that a **stethoscope** be used to listen to the blood flow in the brachial artery of the elbow. Heard through the stethoscope, the tapping that begins when air is released from the band is the systolic pressure; the moment the tapping sound disappears is the *diastolic* (between contractions) pressure.

Wide clinical use of blood pressure measurement using the sphygmomanometer was promoted by American surgeon **Harvey Williams Cushing**. Standard readings were soon established and became basic indicators of heart and lung health or problems. Today, virtually every visit by an adult to a doctor includes a blood pressure test. Many types of blood pressure measuring devises are now available, even for home use, for which *automated meters* are preferable because of their simplicity. Automated meters are battery-operated electronic or digital devices which measure either sound or force and eliminate the need for a stethoscope, using instead a microphone device. There is even a cuffless *oscillometric* meter, which reads blood pressure from a device in which the patient's finger is placed. Home blood pressure measuring devices offer obvious advantages, even though they provide less accurate readings than those obtained at a physician's office.

BLOOD TRANSFUSION

Blood transfusion is the process of transferring blood from one person into the body of another person. Folk medicine and ancient practice long considered blood to have beneficial, curative properties when swallowed. Actual transfusion was seldom attempted, however. Perhaps the earliest recorded case was that of Pope Innocent VIII (1432-1492), who was transfused in April 1492 with the blood of three young boys. The outcome indicated why transfusion attempts were rare and dangerous: the boys died.

After William Harvey (1578-1657) explained the mechanism of blood circulation in 1628, interest in transfusion was rekindled. An Italian physician, Giovanni Colle, gave the first concise description of a blood transfusion in 1628. An English clergyman, Francis Potter, seems to have experimented with transfusions in the 1650s. In the 1660s, the Royal Society of London sponsored a series of transfusion trials after Sir Christopher Wren (1632-1723), the famous architect, used a quill-and-bladder **syringe** to inject fluid into the vein of a dog owned by **Robert Boyle** in 1659, to demonstrate a new method of administering medications. Richard Lower (1631-1691) continued the experiments at Oxford University and performed the first direct blood transfusion from one dog to another in February 1665 by connecting an artery to a vein via a silver tube.

A French physician, Jean Baptiste Denis (1643-1704), used Lower's technique in June 1667 to perform a transfusion from a lamb to an ill young man. Several months later both Denis and Lower transfused blood from a sheep to a man. The promising new technique was abruptly halted in 1668, when one of Denis's transfused patients died. Although the patient was found to have been poisoned by his wife and Denis was acquitted of wrongdoing, transfusions were banned in France and did not become medically established in England.

In 1818 James Blundell, an English physician at Guy's Hospital in London, revived the possibility of transfusion by using a syringe to inject blood from human donors. At first Blundell transfused only hopeless cases, but in 1829 he used blood transfusion successfully to treat a woman with postpartum hemorrhage. Both Blundell and James N. Aveling improved the apparatus for carrying out transfusions, and the technique was widely used during the Franco-Prussian War of 1870-1871.

Blood transfusion remained a risky procedure, however: the donor's blood tended to coagulate, and recipients were likely to suffer a fatal transfusion reaction. The discovery of blood groups in 1900 and of sodium citrate as an anticoagulant in 1914 solved both of these critical problems. Austrian-American pathologist Karl Landsteiner (1868-1943) showed the existence of three distinct blood types (which became four in 1902) in an ABO system. Antigens in some types reacted adversely to antibodies in other types, causing the clumping of red cells that could fatally block blood vessels. Landsteiner's findings made it possible to identify donor and recipient blood types and thus avoid the deadly transfusion reaction in most cases. Typing of blood for transfusion began in 1907. Transfusion reaction was more fully overcome in 1940 when Landsteiner and Walter Weiner (1899-) discovered the Rhesus factor, which typically causes the antigen/antibody reaction.

At first, blood transfusion was done via direct connection between donor and recipient (although occasionally via syringe). George Washington Crile (1864-1943), an American surgeon, developed a standard surgical method of blood transfusion. After surgically exposing a recipient's vein and a donor's artery, a physician clamped shut the vessels and attached a small tube as a conduit between them. When the clamps were opened, blood flowed from donor to recipient. Edward Lindeman took the procedure out of the operating room in 1913 with a simple needle puncture technique that also allowed exact measurement of the amounts of blood being transfused. With all these advances in place, blood transfusion spread rapidly and became firmly established during World War I.

Once blood transfusion was in wide use, storage of donated blood became a problem. The first ''blood bank'' was set up by Dr. Bernard Fantus in 1937 at Cook County Hospital in Chicago, Illinois. A method of preserving red blood cells for up to 21 days with acid citrate dextrose was developed in the 1940s. The African-American surgeon **Charles Richard Drew** studied in depth a way to preserve and store blood ready for instant use. He discovered that plasma could be processed and reserved for a long time, and transfused in place of whole blood without regard to blood type or matching. Drew established blood banks in England and the United States during World War II, which saved thousands of lives by making blood transfusion available to the wounded.

Today blood transfusion remains a widely used and critical medical procedure. After World War II, methods were developed for separating the various constituents of blood, so in addition to whole blood, a patient may receive ''packed'' red cells, granulocytes (white cells), platelets, fresh-frozen or liquid plasma, plasma components, and **artificial blood**. Perhaps most serious remaining risks of blood transfusion is the possibility of transmitting disease via the donor's blood, particularly the HIV virus and hepatitis. Although donated blood is

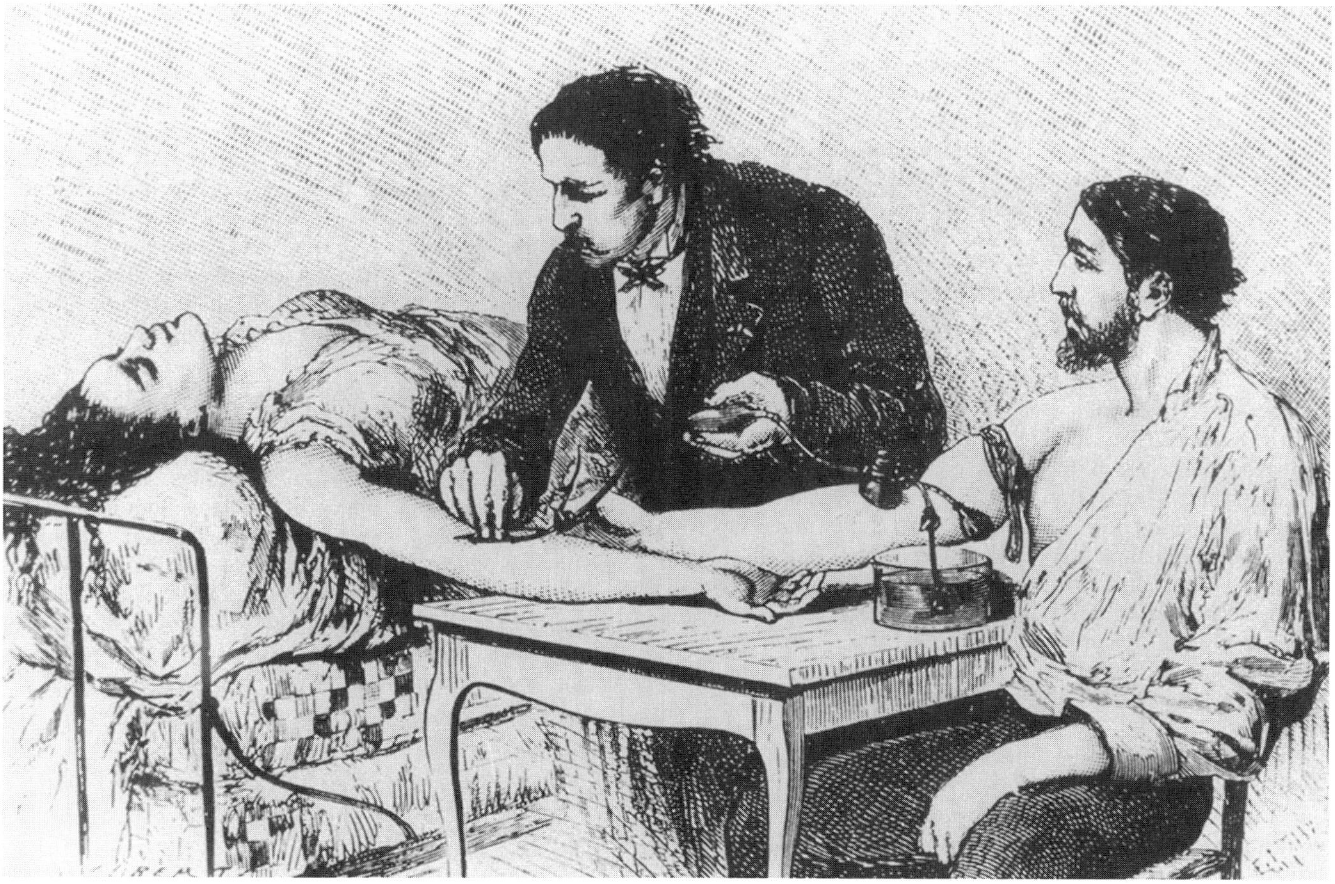

Illustration of an early blood transfusion.

carefully screened, patients concerned about transmitted diseases have the option of *autologous* transfusions—in which the patient donates his or her own blood which is banked until the surgery, or *direct donation*—donation by friends or family members.

Blood vessels, artificial

More than 430,000 Americans with heart disease require coronary artery bypass surgery each year, and up to 200,000 suffer from *critical limb ischemia*, or clogged arteries in the leg—one of the major causes of **amputation**. In the 1940s and 1950s, techniques were developed to surgically **transplant** blood vessels by grafting sections of arteries or veins to replace diseased or damaged portions of other vessels. These transplants were frequently unsuccessful—donor arteries often were rejected by the recipient or quickly developed arteriosclerosis, while transplanting vessels from the patient's own body required two surgeries and many patients had no suitable vessels for transplantation.

To overcome these problems, researchers began to experiment with synthetic blood vessel materials. During World War I, French-born surgeon **Alexis Carrel** had tried using tubes of **glass** and aluminum, and later attempts were made with **polyethylene** and siliconized **rubber**. Results from all these experiments indicated that synthetic fabric was most likely to be successful. A porous material called *vinyon* was tried on dogs, and A. B. Voorhees used this for the first time on humans in 1953. Many other synthetic fabrics were subsequently experimented with; Teflon (Gortex is an example of a Teflon-based fabric) and **Dacron** proved to be best. Blood vessels made from these synthetics are not rejected by the body's immune system, and they are easily available and extremely durable.

While large Dacron blood vessels work very well, small ones have a tendency to become blocked by clots. Researchers are working on ways to make the interior walls of these small synthetic vessels smoother so clots won't form. Chemist Donald Lyman of the University of Utah synthesized a polymer in the early 1980s that had both a high affinity for albumin, which reduced clot formation in synthetic blood vessels, and elasticity, which reduced strain at the juncture of the natural and artificial vessels. Research Industries of Salt Lake City began testing Lyman's vessels on humans in 1988. Surgeon David Annis of the University of Liverpool produced a similar flexible, smooth-walled plastic vessel and also began human trials in the late 1980s. In 1990 Organogenesis of Cambridge, Massachusetts, began animal testing of its *living blood vessel equivalent* which featured a smooth inner layer grown in the laboratory from human cadaver artery cells and tubules strengthened with Dacron mesh. Another approach worked out by Stuart Williams at Jefferson Medical College, Philadelphia, uses cells from the patient's own inner blood vessel lining to grow a lining on the inside of Dacron synthetic vessels.

In 1997, Jeffrey Isner in Boston announced successful **gene therapy** for generating new blood vessels. Injecting a growth-factor-producing DNA into a patient's leg muscles instructs existing vessel cells to create new blood vessels which bypass the clogged vessel. In 1998, Thomas-Joseph Stegmann of Germany announced a similar development through protein therapy by **genetic engineering** of human growth factor FGF-I which, when injected near a blocked vessel in the heart, created a new vascular system. On a three-month follow-up of patients in his clinical trial, no blockages had formed in the new vessels. Also in 1998, Nicholas L'Heureux and his team in San Diego succesfully engineered human tissue without synthetic scaffolding, creating an entirely new replacement blood vessel from human cells able to withstand 20 times the normal human blood pressure.

Blowpipe

It is a generally known fact that fire can be made to burn hotter by fanning the flames with air. This is important in metallurgy, where extremely high temperatures are required to convert metals to their molten state. The blowpipe, which consists of a small tapered tube, is used to direct a stream of compressed air at a heat source. There are seven different types of blowtypes, and are usually made of metal, brass or silver commonly. It has played a significant role discovering at least fifteen elements and the composition of numerous minerals from the end of the eighteeth century to about 1860. Blowpipes were also used in smithies.

Blowpipes were probably invented by the ancient Egyptians, for a depiction can be found in tomb walling paintings dating from 2,400 B.C.. Blowpipes were used by metal artisans and glass blowers. Swedish mineralogist Axel Fredrik Cronstedt (1722-1765) was the first to use the blowpipe, which was operated by mouth or with a bellows, during metallurgical experiments, specifically mineral analysis, in 1738. Metals display different identifying characteristics when placed under high heat, including fusibility, the color of the metal, the color of the flame, the types of vapors emitted, and the general nature of the metal itself. Cronstedt was able to use these characteristics to systematize his analysis of metals. He was also able to isolate a previously unknown element—nickel—all with the aid of his blowpipe.

Another Swede, Johan Gottlieb Gahn (1745-1818), made proficient use of the blowpipe in the late 1700s, particularly in the isolation of manganese in 1774. In 1801, Robert Hare (1781-1858), a chemist from Pennsylvania, invented the oxyhydrogen blowpipe, which incorporates both oxygen and hydrogen at the blowpipe tip. It is the most efficient type of blowpipe, producing the hottest flame possible. Blowpipes were used primarily until the 1860s and 1870s when other apparatus and methodologies replaced them in identifying metals. However, blowpipes are still used commercially in **welding** and metal cutting.

See also Blast furnace

Blowtorch

The blowtorch, a portable device that shoots out an extremely hot flame of pressurized burning alcohol or **gasoline**, can be used in plumbing and metalworking to melt metals, **solder** pipes, or thaw out frozen pipes. In 1862 James Rhodes, an American, developed and patented a design for a vapor lamp much like modern ones, but he never built a working model. In 1880, several different Swedish inventors claimed credit for the development of the modern blowtorch. Max Sievert, a manufacturer and exporter, patented a design for a tool with a small fuel reservoir connected by a metal tube to a small nozzle located in a perforated combustion chamber. Fuel, which was heated to a gaseous state issued from a nozzle into the combustion chamber, was mixed with air. The fuel was then ignited to form a clean, hot flame that burned strongly because it was pressurized. Although Sievert filed for the patent, it seems more likely that his engineer, C.R. Nyberg, actually devised this blowtorch. Propane is the most common fuel of choice since it is safer than the oxyacetylene gases used in earlier torches. A new tool, a multilayer foil, patented by Troy Barbee and Timothy Weihs uses hair-thin metal foils which, when exposed to heat from a match releases sufficient energy to melt metal and form a welded joint. As production costs for this process drop, it may become a safe alternative to a blowtorch for many applications although it's use is currently limited to such low oxygen environments of space and underwater.

Blue baby operation

Before 1944, babies born with *cyanosis*—bluish skin caused by lack of oxygen in the blood—either died or lived with painful physical defects. The plight of these ''blue babies'' aroused the interest of Dr. **Helen Taussig** of Johns Hopkins after she became head of that hospital's Children's Heart Clinic in the 1930s. After much pioneering fluoroscopy study, Taussig developed a theory that the cyanosis was due to constriction of the pulmonary artery, the vessel that carries oxygen-depleted bluish blood from the heart to the lungs, where the blood absorbs oxygen and becomes red once again. Next, Taussig visited heart surgeon Robert Gross of Boston, Massachusetts, who had developed an operation to close a baby's blood vessel. This convinced Taussig that a reverse operation—to open a blocked blood vessel—should be possible.

In 1941 Dr. Alfred Blalock (1899-1964) became chief surgeon at Johns Hopkins. Taussig, who knew of his reputation as a vascular surgeon and research in blood vessel bypasses, interested Blalock in her theory about cyanosis. Together they experimented on hundreds of dogs to perfect an operation in which a branch of the aorta is joined to the pulmonary artery to create a bypass of the defective portion, assuring adequate flow of blood to the lungs. On November 29, 1944, Blalock performed the first ''blue baby'' operation on a 15-month-old girl, assisted by Taussig. Two more successful operations followed. A paper by Taussig and Blalock reported the procedure in the May 19, 1945, issue of the *Journal of the American Medical Association.*

The ''Blalock-Taussig Shunt,'' as the operation came to be called, was soon widely adopted and saved thousands of babies' lives. Surgeons came to Johns Hopkins from around the world to learn the new procedure, and Blalock traveled abroad to further spread knowledge of the operation. This operative technique is still used today for very young children, keeping them alive until they are old enough for **open-heart surgery**. A modified procedure using man-made material for the shunt was first performed in 1963. The Blalock-Taussig procedure was the beginning of the modern era of heart surgery, paving the way for open-heart surgery and surgical correction of many congenital heart defects.

Blue jeans

The denim pants that began as a working garment in the 1849 San Francisco Gold Rush and evolved in the 1980s to a designer status item were invented by peddler Levi Strauss (1829-1902). Strauss was only seventeen years old when he left his native Bavaria, Germany, speaking little English. He became a peddler of housewares in and around New York, New York, but found much competition. Strauss heard of the boom in San Francisco and in 1850 boarded a clipper ship with his wares.

By the time he reached his destination, he had sold everything except a great amount of brown canvas, which he assumed miners would buy to make tents and wagon covers. Strauss soon discovered that his customers desired comfortable but rugged pants that could survive the rigors of mining. Strauss contracted with a local tailor to make his ''waist high overalls.'' He soon began making them of a cotton fabric called ''serge de Nimes,'' a French word that was Americanized to denim.

The next improvement to his pants actually came from tailor Jacob Davis in Carson City, Nevada. Strauss had been manufacturing Davis's design of canvas pants that featured rivets at stress points to strengthen the pants. Davis and Strauss applied for and received the patent on their riveted pants, and Davis began overseeing manufacturing of the product in San Francisco. Davis began cutting the fabric himself and delivering it to seamstresses, who sewed the garments that were picked up by Davis in the evening. Eventually this method, called the ''putting out'' system, was replaced by **mass production** in a factory.

Sweet-Orr, a New York-based producer of the canvas pants, was founded in 1871 and exhibited its product at the 1876 Centennial Exposition in Philadelphia. The design boasted extra room in the seat, and Sweet-Orr claimed they were the most comfortable work pants. But the company's most important contribution was to standardize basic sizes for jeans, since ready-to-wear clothing became the norm as a result of the Civil War demand for off-the-rack speed and convenience. Up to that point, most other garments had been produced on a custom basis.

Lee, a Kansas-based company, entered the denim pants market in 1886. Like other jeans companies, Lee stressed its product's durability, claiming that its pants were ''tough as mule skin.'' Sweet-Orr advertised that their pants could withstand a tug-o-war. The garment was soon worn by railroad workers, cowboys, and any other worker who needed durable pants. In 1926 Lee replaced the button fly with a **zipper**, and Levi's did the same in 1955.

By the 1930s the pants came to be called ''jeans.'' Although the origin of the name is unclear, it might be from the French Revolutionaries called ''Jeans'' (pronounced ''johns'') of the late 1700s who wore a heavy cotton fabric, or from the city of Genoa, Italy, where such fabric was made. At any rate, the 1930s saw the transformation of jeans from strictly work clothes to the realm of fashion. Jeans were touted as the perfect garment on dude ranches and at rodeos. During World War II, jeans again became work clothes. In fact, the country was encouraged to wear old jeans so defense workers would have them to wear in the factories. The suspender buttons and back cinch were eliminated, too, as the government saw these as a waste of valuable wartime materials.

In the 1950s jeans remained a work garment, but by the 1960s jeans became both a fashion item (in a faded, torn, and worn version) and a symbol of the working class. Jeans began to enter the designer phase during the late 1970s. The influence of the movie Urban Cowboy sent American sales alone skyrocketing to 600 million pairs in the peak year of 1981. Blue jeans became an expensive status symbol, and such trade names as ''Guess,'' ''Jordache,'' ''Sassoon,'' and ''Calvin'' were sought after by fashion-conscious consumers. Prewashed, acid-washed, stone-washed, pre-worn, and pre-ripped became part of the clothing vocabulary. Basic jeans made a comeback in 1985, and throughout the 1990s, the trend was still popular.

BOCAGE, ANDRE • See Computerized axial tomography

BOGARDUS, JAMES (1800-1874)
American inventor

Architecture underwent a significant transformation during the 1800s. At the beginning of the century, the heavy walls of a building were designed to bear the load of the entire structure. By the end of the century, facades of brick and masonry were built upon skeletal frames which supported the weight of the structure. The modern architecture movement had begun.

An early leader in this movement was James Bogardus. Bogardus was born in Catskill, New York in 1800. As a young man, he became a prolific inventor and a lecturer on technical subjects. He became interested in architecture on a trip to Europe in 1836, where he was influenced by the form and function of Europe's classic structures. Upon returning to the United States in 1940, he established a foundry in New York City which produced prefabricated **cast-iron** beams and columns that could be shipped anywhere and then easily assembled or disassembled at the building site. This new method was fully demonstrated in the construction of Bogardus's own five-story factory in New York in 1848.

Bogardus preferred cast iron as a building material because of its malleability. It could be easily molded into the classic shapes he had seen in Europe, and it was both functional and decorative. A cast-iron shot tower which he built with a cast-iron framework is thought to have influenced architect William Jenney. Bogardus is credited with several other inventions, including a refillable lead **pencil**, a new engraving process for **postage stamps**, a dry-gas meter, and deep-sea sounding and drilling machines.

BOIVIN, MARIE GILLAIN (1773-1841)
French midwife

Marie Gillain Boivin was considered to be the most outstanding obstetrician of the nineteenth century. Born in Montreuil, a suburb of Paris, she was educated by nuns whose order ran a hospital at Etampes, and married Louis Boivin at the age of twenty-four. After being widowed, she became a midwife in 1800, practicing in Versailles. When her young daughter was killed in an accident, Boivin returned to Paris where she worked at the Hospice de la Maternité under Maria Louise Dugès La Chapelle (1769-1821), another renowned midwife. Boivin soon became known for her obstetrical skill and knowledge, especially in difficult cases; the leading surgeon of the time said she had an eye at the tip of each finger. She was appointed codirector of the General Hospital for Seine and Oise in 1814, directed a temporary military hospital in 1815, and later directed the Hospice de la Maternité and the Maison Royale de Santé. The king of Prussia invested Boivin with the Order of Merit in 1814, and in 1827 she received an honorary M.D. from the University of Marburg in Germany, one of the few women so honored at the time.

Following her break with Mme. La Chapelle, Boivin turned down lucrative offers and worked instead for minimal pay at a hospital for prostitutes. Her pension was so small she died in severe poverty after one year of retirement. Boivin's contributions to the science of obstetrics included the invention of a new pelvimeter and a vaginal speculum, the use of a **stethoscope** to listen to the fetal heartbeat, and discoveries about causes of miscarriage and diseases of the placenta and uterus. She published a number of widely read treatises on obstetrics, including *Mémorial de l'art des Accouchements* in 1812, which became a textbook for medical students and midwives. Boivin's work on diseases of the uterus, published in 1833, was said to be as modern as was possible at the time.

BOMB

A bomb is conventionally an encased explosive or chemical, triggered by a fuse or by impact, that is used in warfare. Bombs can be flung by hand or through a barrel, propelled by a **rocket**, carried within a missile, dropped from an **aircraft**, or left in the ground or water as mines. In addition, bombs take effect in various ways: through fragmentation, which scatters bullet-like fragments such as shrapnel; through the heat and fire of incendiary bombs; through the destructive surge of air called the ''blast''; through the suction after-effect of this displaced air; and through shock, which can convulse the earth, buildings, and water.

The history of bombs, intertwined with the history of war, extends far back in time. **Catapults**, which projected stones, arrows, or incendiary mixtures generally based on tar, date back to antiquity. The relatively late, petroleum-based incendiary mixture known as ''**Greek fire**'' was apparently first used by the Byzantines in the seventh century. The medieval trebuchet, which appeared around 1100, introduced the most important technical changes to the catapult.

The widespread exploitation in the 1300s of **gunpowder**, which had been developed much earlier by the Chinese, opened a new era in the history of bombs. As gunpowder burns explosively when heated, it can be detonated to hurl projectiles from wooden or metal barrels. Western use of the cannon dates from the 1320s, and soon became standard to European armies; the term ''bomb'' derives from the short-barreled cannons of this period known as bombarda. By the late 1500s, the Dutch began using bombs fired in high arcs through the use of short-barrelled mortars. These bombs were hollow metal balls filled with gunpowder; a fuse set into a small hole burned its way down to set off the explosion.

Aerial bombing was first attempted in 1849, when the Austrians deployed hot-air **balloons** loaded with small bombs against the city of Venice. These ''bombers,'' carried by the wind, caused little actual damage but proved psychologically effective. They also foreshadowed the twentieth century use of aerial bombing, made possible by the invention of the airplane in 1903 and spurred by the advent of the two World Wars.

In a 1911 experiment, an American soldier hand-dropped a homemade, two-pound bomb from a Wright brothers airplane onto a test target on the ground. Later that year, an Italian army officer dropped four converted hand **grenade**s onto enemy forces from the air. By the end of the war, the Germans had used zeppelins to drop more than 200 tons of bombs on London, England. To do so, the Germans adapted conventional bombshells with better fuses and with fins for improved navigation.

Projectile bombs were also developed during World War I. The Germans wanted a weapon that could drop straight down into narrow trenches, which were a key arena of this war. They developed the Minewerfer, which was loaded manually and had both the barrel and the bomb rifled (grooved) to achieve the desired trajectory. Eventually, several countries designed automatic mortars; the bombs for these mortars had built-in percussion caps that, at impact with the bottom of the barrel, ignited the bomb's propellant.

Chemical warfare bombs, in addition, were primarily deployed during the first World War. The Germans began using poisonous chlorine gasbombs in 1915; soon both sides of the war had resorted to chlorine, mustard gas, and phosgene bombs—releasing horrendous gasses that destroyed human skin and lungs. Chemical warfare, though still intermittently deployed, is now largely proscribed. However chemical bombs were set to have been deployed by Iraq in the Persian Gulf War, and several other countries are known to possess chemical warheads.

Another type of bomb developed in World War I was the depth charge, designed by the British to protect their navy against the German U-boat (**submarine**), whose torpedoes could carry up to 220 pounds of explosives. Ironically, the **torpedo** had been invented and perfected by the British themselves in the second half of the nineteenth century. Depth charge bombs consisted of canisters filled with explosives and fitted with a valve triggered by water pressure against a spring; the tension of the spring could be adjusted to set off the explosive at the desired depth. The depth charge created a small lethal blast radius of about twenty-five feet for 300 pounds of **dynamite** (which was first manufactured in 1875); damaging shock waves extended much farther. Underwater mine bombs, though used by the Dutch as early as 1585, also became significant strategic weapons during the first World War.

Simple pipe bombs have been used throughout the twentieth century by groups and individuals without access to more sophisticated weaponry.

It was World War II, nonetheless, that saw extensive advances in and deployment of bombs, which now weighed in at up to 2000 pounds. New fragmentation bombs were designed using metal wire wound around an explosive charge. **Armor**-piercing bombs were created with a thick case and a pointed tip to penetrate and explode inside of **ships**, tanks, and bunkers. Thinly-cased blockbuster bombs, which had little penetrating power but generated a powerful blast, were also developed. A new form of incendiary bomb generated intense heat through the burning of thermite, a mixture of aluminum powder and iron oxide, which in turn ignited a bomb casing made of magnesium.

One bomb created by the Germans during the second World War, the V-1, is often classified as a rocket. It was actually a jet-propelled, pilotless bomb with a powerplant known as a pulsejet. Its air-breathing engine operated intermittently, admitting air in pulses like modern turbo jets. This system saved on fuel, leaving the V-1 capable of carrying a one-ton warhead 300 miles from Germany to London. However, the relatively slow speed of this weapon, about 450 miles per hour, made it vulnerable to attack or neutralization by fighter planes. In several documented cases, Allied fighter pilots flipped the V-1 off its course by flying alongside the bomb and gently deflecting it with their wings.

Since the end of the second World War in 1945, bombs have undergone even further elaboration. Cluster bombs were developed for aerial use. These consist of an outer casing enclosing dozens of small bombs; the casing splits open in the air, releasing a shower of bomblets to explode upon contact. Fuel-air bombs release a cloud of volatile and explosive vapor close to the ground. Napalm bombs, a mixture of napalm and aviation fuel, scatter flaming gasoline in all directions. ''Smart bombs,'' fitted with small wings, adjustable fins, and compact laser beam or video-remote guidance systems, have been designed for long-distance precision bombing. Some ''smart bombs'' are equipped with a ''smart fuzes'' that contains elaborate sensory equipment. For example such a bomb can sense when it has penetrated a buried structure such as a bunker, and then detonate itself. This type of bomb was used against Iraq in the Persian Gulf War, with mixed accuracy.

The most significant era in the history of bombs dates from the end of World War II, when the United States dropped the **atomic bomb** on the Japanese cities of Hiroshima and Nagasaki, starting the nuclear age. Since then, so-called hydrogen or nuclear fusion bombs have been created with hundreds of times the explosive power of the bombs dropped on Japan. Long-range delivery systems have also been developed so that targets thousands of miles distant can be hit. Though with the end of the Cold War, many people perceive that the danger of nuclear war has lessened, in fact nuclear bomb technology has spread to more countries, making disarmament a more complex and daunting process in the post-Cold War era.

Even relatively small and conventional bombs are capable of inflicting great damage, especially in a crowded urban area. Two deadly bomb attacks in the United States in the 1990s were attributed to small groups using easy to obtain materials. A bomb set off in the World Trade Center in New York City in 1993 killed six people and injured more than a thousand. The bomb was made in an apartment and conveyed in a van. A similar bomb, made of ammonium nitrate and fuel oil and ignited from a car, destroyed the Alfred P. Murrah Federal Building in Oklahoma City in 1995. Hundreds of people were killed and the nine-story building was destroyed in the attack, carried out by a pair of men. The relative simplicity of the technology and the availability of materials makes this type of bombing difficult to prevent.

BOMB CALORIMETER • See Calorimeter

BOMBER PLANE

The first use of **aircraft** as a method for dropping **bombs** on an enemy was the largely unsuccessful use of aircraft by Germany in World War I. The appearance of these behemoths over Britain was intimidating, but the giant airships were highly susceptible to attack by nimble **fighter planes**, and although they did do some damage they had little effect on the outcome of the war.

Following World War I, it was already obvious that airplanes, well-established as fighting machines, couldalso be use to transport bombs. All that was needed was sturdier planes with more powerful engines. Bombs are deliberately designed to be heavy, to give them greater ability to punch through defensive armor possessed by and enemy, and planes would need to have respectable payload capacity to be effective as bombers.

Spurred by the tireless efforts of World War I aviation hero Billy Mitchell, the Boeing aircraft corporation developed the immortal B-17 ''Flying Fortress,'' first put into commission in 1935. The redoubtable B-17s participated in early attacks against Germany in 1941, and continued to perform missions throughout the war. Far from the flimsy plane devised less than 40 years previously by **Orville and Wilbur Wright**, the B-17 richly deserved its nickname. It was capable of flying far past the range of escorting fighter planes, and so was on its own as it flew over German airspace and into hordes of waiting German fighters. Capable of absorbing an enormous amount of damage, the four-engine, ten-man B-17s delivered load after load of bombs on German industrial and military targets, often with frightful losses to the B-17 squadrons, but equally often with devastating effect against the enemy.

While the B-17 carried out their missions typically during the day, the British Royal Air Force executed night attacks against Germany employing the new, massive, four-engine Avro Lancaster. Mercilessly the night attacks came against Germany, with major industrial centers such as Hamburg and Dresden being almost literally razed to the ground.

Over the course of World War II, many different bomber types appeared, including the B-24 Liberator, the B-25 Mitchell, and, toward the latter part of the war, the huge Boeing B-29 Superfortress. Able to carry up to eight tons of bombs, the B-29s saw action primarily over Japan and Asia. On August 6, 1945, the B-29 *Enola Gay* dropped the newly-invented atomic bomb on Hiroshima, Japan. Three days later, Nagasaki was obliterated under an immense mushroom cloud as well, and Japan immediately surrendered.

The advent of the **jet engine** following World War II pushed all previous bombers into obsolescence, and the countries embroiled in the cold war developed a suite of new planes. The United States' B-52 Statofortress was powered by eight engines and could fly halfway around the world without refuelling, while Great Britain and the Soviet Union also added long-range bombers to their fleets. All these jet-powered bombers were capable of delivering nuclear bombs against targets, creating an uneasy balance of power in which the great distances separating the involved nations were no longer a cushion against attack. A large variety of shorter-range bomber were developed as well, seeing actions in numerous regional conflicts in east Asia and the Middle East during the 1950s and 1960s. Particular concern developed in 1970 when the Soviet Union was found to have developed a long-range bomber, capable of flying more than twice the speed of sound, known as the ''Backfire.'' Soviet protestations that their creation was a purely ''defensive'' bomber produced grim smiles in other countries. This gave the Soviets a clear advantage in bomber technology, and the analogous United States aircraft was the B-1B, a subsonic bomber developed during the 1980s.

One difficulty with the early jet-powered bombers was that they returned a positively enormous **radar** signal. This led to long and expensive attempts to devise low-radar-return bombers. In the United States, the culmination of these efforts is the B-2 **Stealth** bomber, costing about two billion dollars each, but virtually invisible to radar. As of the late 1990s, the B-2 remains the most advanced bomber plane ever designed.

BONE, ARTIFICIAL • See Artificial bone

BONE SETTING • See Fractures, treatments and devices for

BOOMERANG

There is no agreement on who first invented the boomerang. Many scientists believe that boomerangs were created independently by a number of prehistoric hunting peoples because they have been found at ancient sites in Africa, Asia, Europe, Indonesia, and North America. However, the boomerang is associated most closely with the Aborigines, the original people of Australia, who use them for hunting and warfare. The boomerang is usually made from hardwood and shaped into a simple curve. It is flat on one side and convex on the other, much like airplane wings, so that it can fly far. The ones most people are familiar with, the returning boomerangs, are generally used more for sport because they are difficult to aim and catch. The nonreturning variety have less of a curve and are used more for hunting or warfare, since their spinning action causes a greater impact than that of a thrown rock or stick. A nonreturning boomerang is capable of nearly piercing a small animal in two from a distance of 400 feet (122 m).

BOOTH, HERBERT CECIL • See Vacuum cleaner

BORING MACHINE

Boring machines produce smooth and accurate holes in metal or other material by enlarging existing holes with a bore, a single cutting tip or a grinding **wheel**. Although boring mills existed for centuries prior to the Industrial Revolution for the purposes of gun and cannon production, they were crude operations that, despite the rotation of a cutting tool, necessitated the regular pivoting of the mold being cut. The first true boring machine was constructed by **John Smeaton** in 1765. Powered by a **waterwheel** and distinguished by a traveling carriage, the machine produced an uneven bore hole, but nonetheless provided the basis for **John Wilkinson**'s more stable, cylinder-supported machine that was produced a decade later.

With his machine, Wilkinson became the foremost expert on casting and boring iron cannons and engine cylinders. More importantly, he became the first ironmaster to fulfill the stringent design specifications of **James Watt**'s **steam engine**, and then employ the steam engine as the heart of what was to become the English Industrial Revolution. Viewed in this context, his cylinder boring mill may be considered the first industrial **machine tool**.

Although Wilkinson's invention was designed for horizontal boring, most boring machines prior to 1800 were restricted to vertical boring, which accommodated heavier operations, and were, in effect, **lathe**s in which the cutting tool, rather than the piece to be cut, was rotated. Johann Georg Bodmer (1786-1864), a Swiss mechanic and inventor, is credited with designing the earliest modern boring machine, which features variable-speed drilling. Boring is distinct from drilling in that it is a finishing or smoothing process executed after drilling and it is accomplished with only a single-edged cutting tool.

One of the most impressive use of boring machines was the construction of the ''Chunnel,'' the **tunnel** underneath the English Channel. A total of 11 boring machines were used over the three and a half years of digging the tunnel. Like huge, carnivorous earthworms, each tunnel boring machine (TBM) stretched about 600 ft (180 m) and included the cutterhead, its drive motor, excavation units, fans and other machinery, and operated as a self-contained tunnel-processing plant. The TBM was was studded with tungsten teeth on its cutterhead on the front, and could chew through 15 feet (4.6 m) of rock an hour. Precast concrete segments were put in place to line the hole as the TBM moved forward. One TBM worked from each end to create the tunnel, and the excavation produced 10 million cubic yards (7.7 million cubic meters) of dirt.

BOSCH, CARL (1874-1940)

German chemist

As a chemist working for a large German company, Carl Bosch became responsible for transforming a laboratory demonstration project into a practical, large-scale operation. Through Bosch's efforts, **Fritz Haber**'s method of **ammonia synthesis** became one of the twentieth century's most important industrial processes.

The son of an engineer, Bosch worked as an apprentice metalworker in a machine shop after graduating from high school. He later attended college and earned a doctorate in 1898 from the University of Leipzig. The following year, Bosch was hired by Badische Anilin und Sodafabrik (BASF). When Fritz Haber approached BASF with his proposal for commercial production of ammonia in 1908, Bosch was assigned to solve the task of bringing Haber's process up to industrial scale. Bosch solved two of these problems: the supply of raw materials and the provision of suitable catalysts fairly easily. His most difficult challenge lay in the design of equipment that could withstand the extremely high pressures required for ammonia synthesis and survive the corrosive effects of hydrogen gas.

During several years of round-the-clock operation, Bosch conducted more than 20,000 separate tests in some two dozen experimental reactors. He substituted **alloy**ed **steel** in place of Haber's carbon steel reactor, and he invented an ingenious method of shielding the inside of the reactor with a thin, perforated iron liner. Bosch also developed instruments to control and monitor gas purity, temperature, and heat flow within the reactors. In 1913, a huge ammonia synthesis plant in Oppau was opened for production.

During World War I, Bosch arranged with Haber to begin producing nitric acid from ammonia via the **Ostwald-Bauer process**. Nitric acid was needed for Germany to continue manufacturing ammunition after its supply of raw material for explosives had been cut off by Allied ships. In 1931, Bosch shared the Nobel Prize in chemistry with Friedrich Bergius (1884-1949) for developing the high-pressure reactor technology needed for ammonia synthesis. Bosch had also been involved in research on manufacturing synthetic fuels such as methanol and **gasoline** from coal and oil. In 1935, Bosch reached the highest position in executive management in German industry when he succeeded Max Planck (1858-1947) as head of the Kaiser Wilhelm Society. Bosch continued to work in Germany until his death in 1940, but managed to avoid working for the Nazi regime.

See also Nitrogen fixation

Bosch, Robert August (1861-1942)

German inventor and industrialist

Robert August Bosch was born near Ulm (Württemberg), Germany on September 23, 1861. Bosch gained experience and honed his technical ingenuity in the United States, where he worked in **Thomas Edison**'s laboratories.

In 1886, Bosch and German electrical engineer Siegmund Bergmann founded the Bosch GmbH in Stuttgart, Germany. This company manufactured Bosch's most significant invention, the electrical magneto. Bosch's magneto was first used only on stationary **internal combustion engines**, but when his co-worker Gottlieb Honold invented the high-tension **spark plug** in 1902 to complement Bosch's high-tension magneto, the Bosch magneto-spark plug combination became the standard for the world's **automobile** industry.

By 1920, the Bosch company had sold over a million magnetos. Bosch's United States operation was taken over by the Alien Property Custodian in 1918 as a consequence of the U.S. government's declaration of war against Germany. After the war, Bosch reentered the U.S. market under his own name, and a ten-year legal battle ensued. In 1930, Bosch entered into an agreement with the U.S. Bosch company, which would market the products of the German Bosch company in the United States.

After the war, Bosch also expanded his electrical ignition manufacturing empire by adding brake components and his famed horn. At this time he was already involved in manufacturing his best-selling magnetos and spark plugs, starters, **generators**, wiper motors, and **radios**, as well as diesel engine accessories and gas appliances. As an industrialist, Robert Bosch was considered to hold rather advanced social views for his time. He introduced an eight-hour day in 1906 and advocated industrial arbitration and free trade. He remained active in his industrial enterprises well into his 70s; he died in Stuttgart on March 9, 1942.

See also Ignition systems

Bottle and bottle machinery

Bottles are narrow-necked containers for holding liquids, usually made of **glass** or **plastic**,and closed with corks, stoppers, or caps. *Glass bottle*s were first made in about 1500 B.C. by the Egyptians, who formed molten glass around a core of sand and clay; when the glass cooled, the core was dug out. Because this process was so difficult and time-consuming, glass bottles were a luxury item for the ancient Egyptians.

By 200 B.C., glass bottles were made in China, Persia, and Egypt by blowing molten glass into molds, a method also used somewhat later by the Romans. This was a cheaper and quicker way to produce bottles. Glassmaking by this method spread through Europe during the 1400s and 1500s. Skilled workers set up the New World's first bottle- and glass-making factory in Virginia in 1608.

Various well-known types of glass containers appeared in the United States during the 1800s. The **baby bottle** was patented in 1841. John L. Mason, a glass-blower, introduced his Mason jar, the first glass jar with a screw-on cap, in 1858. The milk bottle was developed by Dr. Hervey Thatcher also during the mid-1800s. The twentieth century saw the introduction of the renowned Coca-cola design in 1915. Other **soda pop** brands soon followed suit with distinctive individual bottle shapes, but soda bottle shapes were standardized after 1934, when fired-on permanent color could be used for brand identification.

Bottles were a common item in the eighteenth century, and glassmaking was one of America's earliest industries. However, as late as 1890, bottles were still handmade by individual glassblowers working with a crew of assistants, and production was limited to 250 bottles a day at most. In 1865 a pressing and blowing machine was invented to automate the bottle-making process; another bottle-making machine was invented by Astley in 1881.

The first fully automatic machine for making all types of glass bottles and jars was invented by Michael J. Owens, an employee of a Toledo, Ohio, lamp-chimney company. The Owens Bottle Machine was put into commercial operation about 1903. This was the first major advance in the manufacture of glass containers since the ancient discovery of glass-blowing. The Owens machine made possible inexpensive, large-scale, commercial production of glass bottles. Together with the Crown bottle cap, it set the stage for the rapid growth of the bottled carbonated beverage industry. By 1920, most bottles in the United States were made on Owens-type machines.

Blow molding machines to produce *bottles* appeared in the early 1940s. The first plastic bottles were **polyethylene** squeeze types. A plastic bottle strong enough to hold carbonated beverages without bursting was invented by engineer **Nat Wyeth** (a relation of American artist Andrew Wyeth) for the Du Pont Corporation.

Plastic bottles are strong, lightweight, and resist dents and breakage. Glass bottles are easily recyclable, reusable, and made from renewable natural resources. Particularly popular are plastic bottles in brand specific shapes.

See also Bottle tops; Soda water

Bottle tops

Cork has been used since antiquity as a stopper for bottles because of its compressive abilities. During the Renaissance, cork stoppers were commonplace, and cork-oak trees were grown and processed in the Pyrenees Mountains especially for this purpose. Wine bottles were commonly sealed with oiled hemp. When Pierre Pérignon (1638-1715) invented champagne in 1688, he found that the gaseous pressure inside his bottles blew out the hemp stoppers. To solve the problem, he invented corks held in place by **wire**. During the 1800s, Hiram Codd's **glass** ball stopper was widely used in England. It was held against a **rubber** gasket by gas pressure within the bottle. The stopper of choice in the United States was invented by Charles G. Hutchinson in 1879. In this design, a rubber gasket inside the bottle was attached to a wire loop that extended outside the bottle; a blow to the loop released the gasket, while a pull on the loop closed it. The modern metal bottle cap was developed by the prolific Maryland inventor William Painter, who patented his first stopper in 1885. By 1891, his definitive design, a cork-lined metal cap with a corrugated edge that is crimped around the bottle lip, appeared. Painter called his invention the "crown cap," founded the Crown Cork and Seal Company to market it, and became very wealthy from it. The crown cap was the industry standard for nearly 80 years. In 1955 the crown cap's cork liner was replaced by **plastic**, and a high-speed machine to inspect crown seals was introduced in 1958. In the 1960s, the **Coca-Cola** company offered lift-top crown caps. The push-on, twist-off cap was first developed for **baby food**. Screw caps for carbonated beverages appeared in the 1960s and 1970s and are the standard today.

Boulton, Matthew (1728-1809)

English manufacturer and engineer

Matthew Boulton had a reputation as a restless man who constantly strove for new innovations. He was born in Birmingham, England, on September 3, 1728, and became a partner in his father's buckle manufacturing business in 1750, which he eventually inherited in 1759. In 1761, he launched a new factory in Soho; after meeting **Benjamin Franklin**, however, he became more interested in the scientific and technological advancements of the day, particularly the development of **steam engine**s.

In 1764, Erasmus Darwin (1731-1802; grandfather of naturalist Charles Darwin, 1809-1882) proposed to Boulton

Invented by American William Painter, crown caps have been an integral part of the bottling industry since the turn of the century.

the idea of building a steam-powered carriage. Three years later the two met Scottish engineer **James Watt**, and Boulton decided to join forces with him to further the development of steam power, a field still ripe with promise after the introduction in 1712 of **Thomas Newcomen**'s problematic steam engine.

His chance came years later when, in 1772, inventor and close friend John Roebuck (1718-1794) went bankrupt and lost patents on his own, mostly unsuccessful, steam engine developments. Boulton acquired Roebuck's patents in 1775 and then entered into a partnership with Watt. During the next decade, their engines were to become prime motivators of Europe's Industrial Revolution. The task Boulton and Watt set themselves was monstrous and fraught with financial difficulties, legal battles, and mechanical delays. Yet they succeeded perhaps beyond anyone's expectations, for the new Boulton & Watt engines boasted an energy efficiency three times that of the most contemporary Newcomen engines.

During the 1780s Boulton's fame spread as he became the leading English manufacturer, the man who brought Watt's genius to fruition by means of his optimism and manufacturing expertise. He was quoted by James Boswell as saying, "I sell here what all the world desires—power." In his later years, Boulton patented a steam-driven coin press that restored public confidence in the English monetary system and forced counterfeiters out of the business. A founding member of the prestigious Lunar Society of Birmingham, Boulton ranked at the time of his death, on August 17, 1809, as one of the most prominent figures among scientific circles of his time.

Bow and arrow

The bow and arrow is an early weapon that provides greater power to its user than a hand-thrown instrument by slowly storing energy which is quickly expended when the bowstring is released. The bow and arrow is thought to have been invented in northern Africa and southern Europe around 15,000 B.C. It was widespread across Europe by around 9000 B.C. The bow and arrow was an important hunting tool and weapon across Asia and the Americas as well. Most early bows were simple devices, constructed of slightly curved springy wood, and strung with gut. But even some very early bows have been found in Siberia made of a composite of wood and antler, revealing sophisticated craftsmanship. Arrow-making technique is believed to have been similar to the way some modern aboriginal peoples still make them. Shafts of wood were passed through holes bored in bone or antler and bent until they were straight. Then the shaft was passed through another hole, which shaved the stick to a precise diameter. Arrows needed to be consistent in weight and diameter for the bow to shoot well, and even prehistoric peoples must have possessed considerable expertise to manufacture reliable arrows. Most arrows were weighted with feathers. The sharpened end would have been hardened in fire by early peoples, and later tipped with metal.

Simple bows could be deadly weapons, and they were small and light enough to be used from horseback or even from a moving chariot. Arrows could also be wrapped in incendiary material and lit just before being shot, to set fire to distant targets. The Greeks are credited with developing a crossbow around 400 B.C., a more complex bow that could send arrows off with a higher velocity than simple bows. Called a gastraphete, it was a bow with one end that could be rested on the ground while the other end had a wooden semicircle which pressed against the archer's stomach. A soldier used both hands to pull the bow back, the bowstring was caught in a hook, and the arrow or bolt was laid into a trough. After aiming, the soldier pulled a trigger which released the bowstring. The next step was to increase the user's ability to draw back the bow. Medieval European artisans designed crossbows with sophisticated mechanisms for loading. Since there was a limit on how much even two hands could pull it back, crossbows were designed with a small winch called a cranequin controlled by a **wheel** with notches, and a ratchet to keep the wheel turning in one direction. A hook engaged the bowstring, and by cranking the cranequin, the bowman pulled the string back. By the fourteenth century, crossbows were being made of **steel**, as well as of horn and wood. Medieval crossbows could fire heavy bolts that were shorter than arrows but equipped with heavier points or heads. The big drawback with all crossbows, however, was the time it took for the user to load and fire.

Another type of bow that became famous was the English longbow, which came into use around the thirteenth century. Originally discovered in Wales, the longbow was five to six feet (1.5 to 1.8 m) long with a yard-long (.92 m) arrow that could be sent a maximum distance of 500 yards (457 m). A good archer could fire six of these arrows per minute. These longbows were responsible for the English victory over France at Crecy in 1346, where the French knights were struck before they could reach the English lines. English longbows were usually strung with **linen** or hemp. Though simpler and actually less powerful than the crossbow of the same era, because they could be aimed and fired quickly, they gave England a distinct military advantage.

The use of the bow and arrow persisted in Europe well after the introduction of in the thirteenth century. Composite weapons, such as cannons that shot arrows or crossbows that shot bullets, were manufactured even up until the nineteenth century. But by the end of the nineteenth century the increasing power of the gun had made the bow and arrow obsolete as all but an archaic hunting weapon. Bowhunting was kept alive in the United States by a few dedicated sportmen. After the American Civil War, ex-Confederate soldiers were not allowed to own firearms, and some turned to bowhunting. The book *The Witchery of Archery*, written by ex-Confederate Maurice Thompson in 1878, is credited with sparking the formation of hundreds of American bowhunting clubs. Another American proponent of bowhunting, Dr. Saxton Pope, learned bow craft from Ishi, the last member of a tribe of Stone Age Indians discovered in San Francisco in the early 1900s. Pope's writings were also influential in keeping interest in bowhunting alive.

Bowhunting resurged in the United States after the invention of the modern compound crossbow by H.W. Allen in the 1960s. The compound crossbow uses internal wheels in a pulley system to draw back the string. Both powerful and compact, this modern crossbow gained great popularity. Crossbows in the 1990s are increasingly sophisticated, made from new materials such as aluminum/graphite composites or from a fiberglass-like material called CarbonGlass. Many use electronic sights or fiber-optic sights. Bow design is becoming simpler, with the number of wheels reduced from the four to six standard in the 1970s to only two or even one. Arrows in the 1990s are made from aluminum or graphite, and are lightweight as well as strong.

BOYKIN, OTIS (1920-1982)

American engineer

Born in Dallas, Texas, on August 29, 1920, Boykin invented dozens of electronic and mechanical devices including a control unit used in heart **pacemaker**s, components for guided **missile**s and **computer**s, an electronic air filter, and a theft-resistant **cash register**. Boykin attended Fisk University in Nashville, Tennessee, from 1938 to 1941 and subsequently worked for a number of electronics firms in the Chicago, Illinois area. From 1947-1949, he pursued graduate studies at the Illinois Institute of Technology. For the next thirty-three years, Boykin contributed his expertise to the fields of chemistry and **electronics** as a consultant to a number of firms in Chicago as well as in Paris, France. He died of heart failure in Chicago in 1982.

BOYLE, ROBERT (1627-1691)

Irish English physicist and chemist

Before the 1700s alchemists and some scientists believed that one substance could be changed into another, and they tried to prove it by transforming such common metals as lead into gold and silver. Robert Boyle, one of the first modern chemists, knew that these assertions were incorrect and believed that theories must be proved by careful experiments, the experimental results must be reported publicly, and other scientists must be able to repeat the experiments and confirm the results. Thus, Boyle became one of the pioneers of modern scientific methods.

Boyle was fortunate enough to be born into one of the wealthiest aristocratic families of his day. He was a gifted and industrious student, and after attending an excellent school in England, he and his brother studied in Europe for several years, accompanied by a personal tutor. In Italy Boyle became acquainted with **Galileo**'s experimental work, which was a strong influence on his scientific career. He also became a devout Christian after experiencing a frightening thunderstorm in Switzerland.

In the 1640s Boyle returned to England and began meeting regularly with other scholars in London to discuss the philosophy of experimentation and other innovative ideas. At first, they called themselves the Invisible College, but they later became known as the Royal Society. Boyle's interest in physics and chemistry grew when he moved to Oxford in 1654 and became acquainted with the scientists there. To pursue his interests, Boyle set up an elaborate research laboratory and hired skilled assistants to conduct experiments. Although Boyle is best known for Boyle's law and other discoveries demonstrating the relationship between the pressure and volume of gases, he and his co-workers also developed many innovative scientific techniques and research devices.

One of Boyle's most brilliant research assistants was **Robert Hooke**, who built an improved air pump based on an earlier one designed by German engineer **Otto von Guericke**. Air pumps were used in early laboratories to create vacuums inside cylinders. Although vacuums were poorly understood at the time, they were the subject of much interest and experimentation. Boyle and Hooke developed a better method of supporting the air pump's cylinder and cranking its **piston**; they also improved the design and placement of the pump's **valves**. Boyle also became the first scientist to collect a gas by means of an ingenious experimental set-up. He filled a flask with sulfuric acid and dropped iron **nails** into it, then inverted the flask and submerged it in more acid. As the iron reacted with the sulfur, hydrogen bubbles rose upward to collect at the top of the inverted flask. This apparatus represented a primitive form of the pneumatic trough, which is now a familiar laboratory device.

In 1668 Boyle returned to London to live with his sister. In his later years, he devoted greater effort to promoting his Christian ideals, but he continued to correspond with many scientists and scholars. In 1680 Boyle invented the first **match** although it would be many years before matches became widely used. He and an assistant coated a piece of coarse **paper** with phosphorus, then produced a flame by drawing a sulfur-tipped wooden splint through a fold in the paper. Boyle is also credited with coining many new scientific terms, such as *analysis* and *element* (in their modern meanings) and **barometer**.

Robert Boyle

See also Vacuum pump

Bragg, William Henry
Bragg, William Lawrence (1890-1971)
English physicists

The team of William Henry and William Lawrence Bragg is certainly one of the most scientifically productive in history. Combining their intellect and research skills, they succeeded in constructing the first X-ray **spectroscope**, establishing the science of **X-ray crystallography**. They were jointly awarded the 1915 Nobel Prize for Physics and remain the only father-and-son team ever so honored.

William Henry Bragg was educated on the Isle of Man, at King William's College. Always at the top of his classes, he was particularly talented in mathematics. He entered Cambridge at age nineteen to study physics under John William Strutt, Lord Rayleigh (1842-1919) and Sir Joseph J. Thomson (1856-1940). Thomson steered Bragg to an opening in the physics and mathematics department at the University of Adelaide, Australia, and Bragg undertook the long sea voyage, becoming a professor in 1886. During the next eighteen years, he would establish a reputation as a masterful lecturer; however, he published almost nothing and until he was forty-one conducted no original research.

William Henry Bragg

The turning point in Bragg's career came in 1906. As co-president of the Australasian Association for the Advancement of Science, he was expected to deliver an address. For his topic, Bragg chose the recent breakthroughs in radioactivity that had been discovered by Antoine-Henri Becquerel (1852-1908), Marie Curie, and Pierre Curie (1859-1906). While researching the subject, Bragg became quite interested in it himself; he found certain flaws in reasoning in the accounts he read and decided to conduct his own research in the field. For the next two years he made his own not inconsequential contributions to radiation physics, particularly in the study of alpha particle emission. He also published his findings regularly, achieving worldwide respect.

Meanwhile, his son, William Lawrence Bragg, was walking firmly in his father's footsteps: recognized as a child prodigy, the younger Bragg entered the University of Adelaide when he was fifteen. He spent much time observing and assisting with his father's research, and the two would often spend long nights discussing their findings. William Lawrence Bragg was, in fact, the recipient of Australia's first medical X-ray examination, when his father used a home-built **X-ray machine** to examine his fractured elbow.

In 1909 the Braggs returned to England, William Henry to teach at Leeds, William Lawrence to attend graduate courses at Cambridge (he had graduated from Adelaide when he was eighteen). At that time, the British scientific community was excited about Max von Laue's (1879-1960) discovery of X-ray diffraction. Laue had used a crystal to create a diffraction pattern, proving that X-rays were transverse electromagnetic waves, like those of light. Both Braggs immediately became intrigued by this discovery and discussed between them the ramifications. William Lawrence developed a system of equations based upon the theory that crystals were arranged in planes of molecules; using these equations (now known as Bragg's law), they began the series of experiments that culminated in the invention of the X-ray spectroscope in 1913. They published their work in 1913 and that year were awarded the Nobel Prize; then twenty-five, William Lawrence was the youngest person to win the prize.

The first application of the X-ray spectroscope was to examine the structure of certain crystals. The Braggs discovered that sodium chloride crystals are not made of molecules at all, but rather patterns of sodium ions and chloride ions, providing more support for the Bohr model of the atom. This early experiment also served as the foundation for the science of X-ray crystallography; crystallography has since become an important tool for chemists and mineralogists and was the key process in the research of DNA structure.

Though their accomplishments as experimental physicists gained them a place in the highest ranks of the scientific community, the Braggs each possessed the ability to convey the wonders of science to the common man; William Henry was a sought-after public speaker in Europe, and William Lawrence enjoyed writing science books for children. William Henry Bragg was knighted in 1920, William Lawrence Bragg in 1941.

Braille, Louis (1809-1852)

French teacher

Braille designed a coding system, based on patterns of raised dots, which the blind could read by touch. Born on January 4, 1809, Coupvray, France, Braille was accidentally blinded in one eye at the age of three. Within two years, a disease in his other eye left him completely blind.

In 1819, Braille received a scholarship to the Institut National des Jeunes Aveugles (National Institute of Blind Youth), founded by Valentin Haüy (1745-1822). The same year Braille entered the school, Captain Charles Barbier invented sonography, or nightwriting, a system of embossed symbols used by soldiers to communicate silently at night on the battlefield. Inspired by a lecture Barbier gave at the Institute a few years later, the fifteen-year-old Braille adapted Barbier's system to replace Haüy's awkward embossed type, which he and his classmates had been obliged to learn.

In his initial study, Braille had experimented with geometric shapes cut from **leather** as well as with **nails** and tacks hammered into boards. He finally settled on a fingertip-sized six-dot code, based on the twenty-five letters of the alphabet, which could be recognized with a single contact of one digit. By varying the number and placement of dots, he coded letters, punctuation, numbers, diphthongs, familiar words, scientific symbols, mathematical and musical notation, and capitalization. With the right hand, the reader touched individual dots and, with the left, moved on toward the next line, comprehending as smoothly and rapidly as sighted readers. Using the Braille system, students were also able to take notes and write themes by punching dots into paper with a pointed stylus which was aligned with a metal guide.

At the age of twenty, Braille published a monograph describing the use of his coded system. In 1837, he issued a second publication featuring an expanded system of coding text. Despite the students' favorable response to the Braille code, sighted instructors and school board members, fearing for their jobs should the number of well-educated blind individuals increase, opposed his system.

Braille grew seriously ill with incurable tuberculosis in 1835 and was forced to resign his teaching post. He died in Paris on January 6, 1852. The Braille writing system—though demonstrated at the Paris Exposition of Industry in 1834 and praised by King Louis-Philippe—was not fully accepted until 1854, two years after the inventor's death. The system underwent periodic alteration; the standardized system employed today was first used in the United States in 1860 at the Missouri School for the Blind.

See also Blind, communications systems for; Kurzweil reading machine; Musical instruments, electric

Braking systems

Many variations of braking systems have evolved over the years. Some have been more successful than others depending on the distance required to stop, the speed of the vehicle, and the person operating the brake.

The first truly effective and reliable brakes were made for **trains** and were patented by **George Westinghouse** in 1869. He was met with great resistance, however, from railroad brakemen who made their living by listening for whistle signals from the engineer. Westinghouse's brake was originally activated by opening **valves** to release compressed air into tubes that ran the length of the train. He soon realized, however, that a leak or break in the lines rendered the entire system ineffective. In 1872, he improved the mechanism so that air pressure held the brake open until it was released.

While steam, compressed air, vacuum, and hydraulic brakes were all in use in various kinds of transportation at the turn of the century, early **automobiles** reverted to an older brake type known as band brakes. Band brakes simply consist of a flexible band that loops around a drum that revolves with the engine. In early automobiles, a **leather** band was connected to a **wheel**. When the vehicle needed to be stopped, a **lever** was pulled or a pedal pushed to activate the brake, the band then tightened, and the wheel slowed through friction. Variations of band brakes are still used in large, slow-moving machinery such as **crane**s, hoists, and **tractor**s.

An interesting phenomenon related to band brakes is the principle of self-energization. This is the tendency of the band to tighten itself, without more pressure being applied, as it is pulled forward by the drum. Self-energization works only if the band is allowed to tighten in the same direction as the drum is rotating, so unless the drum changes rotation directions, the brake takes much more pressure to stop reverse movement.

Louis Renault's 1902 invention of the standard drum brake provided more consistently reliable braking in both the forward and reverse directions. The drum is an iron casting that rotates with the wheel. Two curved brake shoes are hinged together and fastened to a backing plate that doesn't turn. When the brakes are applied, the shoes are pushed against the inside of the spinning drum, and friction slows the drum and the wheel. Renault's brake relied on two hinged shoes that were forced apart by a **cam**. In 1917, R. Stevens patented adjustable shoes to compensate for wear on the shoes or brake drum. Many other patents over the years have improved the drum brake system. **Pistons** and cylinders have replaced cams, for example, but the same basic design is still used on many cars, particularly on rear wheels.

Hydraulic, or fluid-activated, brakes came into use in the mid-1930s. A master cylinder is the primary component of a hydraulic system. This master cylinder serves as a central reservoir that contains a thick brake fluid. It is connected to drum brakes by tubes called lines which are also filled with brake fluid. When pressure is applied to the brakes, the master cylinder pressurizes the fluid trapped in the system, which in turn applies pressure to close shoes against the brake drum. The emergency brake is a manual method of bypassing the hydraulic system so the brakes can be engaged. A wire cable from the parking brake lever or pedal is attached to the brakes (usually on the rear wheels) and sets the brakes when it is pulled.

Disc brakes were first patented for use on sports cars where frequent, high-pressure braking overheated the shoes on drum brakes, reducing their effectiveness. Disc brakes consist of a caliper, or clamp, that fits over the thin, vertical edge of a wheel, known as a rotor. **Bicycle** brakes often operate with calipers that squeeze bicycle rims in much the same way. When pressure is applied to engage disc brakes, the caliper (which does not rotate) grips each side of the rotor (which moves as the wheel turns). As the caliper closes, it pushes pads against the rotor disc, and friction again causes the wheel to slow. These brakes apply more equal pressure to wheels for safer stops regardless of vehicle speed or amount of pressure applied. Because these brakes are not self-energizing, more initial pressure is required to squeeze the calipers; therefore, servo-assist (relay) mechanisms are often integrated into the brake systems.

Antilock braking systems (ABS systems) have evolved during the 1970s and 1980s in **aircraft**. On land, they also provide safer stops, especially in front-wheel-drive cars with heavy front loads and on wet or icy pavements. The main component of an antilock system is a small mechanism that senses and regulates the fluid pressure to each wheel and maintains a steady pressure to prevent wheel lockup and skidding. During the braking process, sensors compare pressure in the hydraulic system to wheel movement. If a sensor detects that a wheel is beginning to lockup or skid, the computer activates a small, electric valve that reduces the fluid pressure to a particular wheel and allows it to continue rotating alone. Pressure is blocked until the tire regains traction or the brake pedal is released. These measurements and changes are calculated as often as fifteen times each second in some antilock systems.

In the 1990s, ABS brakes have been improved through the development of the so-called ''self-steering'' ABS. This system is equipped with a yaw sensor that evaluates the rate at which the car is turning; when oversteering or understeering (spinning or loss of steering control) begin, the system modifies the braking force applied to the individual wheels. The driver can slow the vehicle while steering to avoid an object without losing control of the car. Electronic sensors in new ABS systems also assist the driver by measuring the speed at which the brake pedal is depressed. If the driver stabs at the brake, the vehicle is fully stopped by the ABS system that recognizes an emergency situation even if the driver has backed off the brake.

Other braking systems in use include regenerative brakes on electric trains and air brakes on aircraft. Regenerative brakes involve an **electric motor**, an accumulator, and a **generator**; the motor ''steals'' the energy supply and slows a vehicle through energy loss, the accumulator stores the stolen electricity, and the generator releases the electrical power after braking is completed. Airplanes are slowed in-flight by air brakes or flaps that extend at right angles above and below their wings to create drag. On the ground, wheel brakes are aided by reverse thrust of the engines, which, in effect, uses the engine exhaust stream to push the aircraft in an opposite direction.

Bramah, Joseph (1748-1814)

English engineer and inventor

Joseph Bramah, son of a Yorkshire farmer, was one of the fathers of the **machine tool** industry whose inventions greatly contributed to the development of the Industrial Revolution. Born Joe Brammer in Stainborough, Yorkshire, England, Bramah was seriously injured in an accident at the age of sixteen, and so turned from his family profession of farming to become a cabinetmaker's apprentice. After completing his apprenticeship, Bramah set up his own carpentry and cabinet-making shop in London. Alexander Cumming had recently patented a water-closet **valve** system, which Bramah found unsatisfactory while installing water closets for his customers.

In 1778 Bramah patented his own, improved flushing system, changing his name on the patent to the more elegant sounding Bramah. In 1784 the inventor patented his burglar-proof Bramah **lock**. He exhibited the lock in his Piccadilly shop window with a notice offering a two hundred-guinea award to anyone who could pick it. All attempts failed until, sixty-seven years later in 1851, an American mechanic named Alfred Hobbs succeeded in opening the lock—after fifty-one hours.

Bramah's lock was effective because it was intricate. In order to manufacture it economically, Bramah realized he needed finely-designed machine tools capable of turning out precisely-made parts. To help with this, he hired a young blacksmith named **Henry Maudslay**, who became superintendent of Bramah's shop the following year at the age of nineteen. Together, around 1794, the two men made a crucial improvement to the crude **lathes** of the day: the slide rest. Instead of holding the cutting tool by hand against the metal to be cut, the iron fist of the slide rest held the tool firmly and rigidly against the metal and moved the tool uniformly along a carriage. The slide rest permitted much greater accuracy and output in metal working.

Another very important Bramah invention was the *hydraulic press* of 1795. This was the first practical application of hydraulic principles and opened a tremendous new source of power to the manufacturers and builders of the Industrial Revolution.

Bramah was a marvelously inventive man who secured a total of 18 patents. His other inventions included a machine for numbering bank notes, a wood-planing machine, a device to make quill nibs for **pens**, a beer pump, and **paper**-making and soda-water machines. He was the first to suggest using a **screw propeller** to drive ships rather than a paddle wheel. Bramah died in London in 1814.

Brandenberger, Jacques E. • See Cellophane

Brassiere

Methods of supporting breasts have been known since ancient times. Greek women wore breast bands, apparently to flatten or minimize the bust. A fourth-century Roman mosaic in Sicily shows a female athlete wearing bikini pants and a bra. In modern times, ''patent bust improvers'' similar to the bra were advertised in Great Britain in 1902, and a ''brassiere'' was illustrated in the magazine *Vogue* in 1907. A German immigrant to America, Otto Titzling, claimed to have invented the bra in 1912, but he never patented the design.

The origin of the word *brassiere* is uncertain. The French use the term *soutien-gorge* (''throat-supporter'') for the bra. The claims of the Frenchman Philippe de Brassiere to be the origin of the name were rejected in court. Apparently the French word for ''arm,'' *bras*, is the basis of the term. The word ''bra'' came into use around 1937.

The first person to patent a design for the brassiere was the American socialite Mary Phelps Jacob (later known as Caresse Crosby). Dressing for a debutante ball, Jacob became fed up by, in her own words, ''being encased in a sort of boxlike armor of whalebone and pink cordage''—her **corset**. With her maid, Jacob fashioned two silk handkerchiefs and pink ribbon into the first modern bra—short and soft.

Jacob showed her creation secretly to her female friends, who asked her to make copies for them. When a total stranger sent a dollar asking for one of the contraptions, Jacob realized she had a money-maker. She patented her ''Backless Brassiere'' in November 1914. After having several hundred brassieres produced, Jacob lost interest in her product and then sold her patent to Warner Brothers Corset Company for $15,000.

Jacob's bra design went on to earn Warners $15 to $20 million in the next few decades. The company introduced bra cup sizes—A, B, C, and D—in the 1930s. The thirties also saw the introduction of the padded, strapless, and wired bras. An odd development in the 1950s was the ''sweater girl'' bra that was designed to produce high, pointed breasts.

In the 1990s, the pushup bra has reemerged in the form of the Wonderbra. The lowcut Wonderbra incorporates cups that are tightly shaped across the bottom. The these cups push the breasts upward and inward, an effect that is enhanced by small pads place in the outer bottom area of the cup. The result is the appearance of highly sought-after cleavage. Like the new ''foundation garments'' that are merely updates of the traditional girdle, the Wonderbra puts a new interpretation on an old idea, and has been wildly successful.

Brattain, Walter Houser (1902-1987)

American physicist

Brattain was born in China, and he grew up on his parents' cattle ranch in Washington state. In 1924 he graduated from Whitman College in Walla Walla, Washington, with a degree in physics and mathematics. He later received his Masters from the University of Oregon and his Ph.D. from the University of Minnesota. After working for the United States National Bureau of Standards while in graduate school, Brattain joined Bell Telephone Laboratories in 1929. At Bell Labs he began studying semiconducting substances such as germanium and silicon, which had been used for years to rectify *electrical currents*, forcing them to flow in only one direction. In 1936 **William Shockley** was hired by Bell Labs and joined Brattain in his semiconductor research. Their efforts, however, were interrupted by World War II, during which time Brattain conducted research on the magnetic detection of **submarines**.

Soon after the war was over, another scientist, **John Bardeen**, was added to Bell Labs research team, and together the three scientists studied how semiconductors could amplify, or increase, electrical current as well as rectify it. When Shockley's idea for a semiconducting device failed in tests, Bardeen explained that a layer of electrons between surfaces prevented the flow of electricity. With Bardeen's help, Brattain performed experiments in 1947 that demonstrated this concept for the first time, and, in the course of their research, Brattain, Bardeen, and Shockley developed the world's first *point contact resistor*, nicknamed the **transistor** by another Bell Labs researcher. The experimental device was placed between metallic contacts, and used germanium oxide as a semiconducting material. By making tiny changes in the current, Brattain was able to induce great variations in the power output. Brattain's experiments showed that the point-contact transistor could amplify electrical current by a factor of fifty. This soon led to the development of an improved transistor that was more versatile and easier to mass-produce.

Although it took a few years to perfect the device, the transistor created a revolution in electronic and communications technology by serving as a replacement for the bulky **vacuum tube**s that had been used in early computers. With these smaller, more efficient transistors, scientists were able to design more complex equipment that used much less power to operate. Technology has since developed **integrated circuit**s,

Walter Houser Brattain

which contain an entire electronic circuit on a single piece of semiconducting material such as a *silicon wafer*, or chip.

Brattain, Bardeen, and Shockley received the Nobel Prize in Physics in 1956 for their investigation of the properties of semiconductors. All three researchers were widely honored for their invention of the transistor, and Brattain was elected to the National Academy of Sciences in 1959. During the 1950s and 1960s, he continued working on semiconductors and the transistor effect. After Brattain retired from Bell Labs in 1967, he accepted a post as physics professor at his alma mater, Whitman College.

See also Resistor

Braun, Karl Ferdinand (1850-1918)

German physicist

The physicist Karl Ferdinand Braun was an important contributor to the fields of **television**, **radio**, and **electronics**. For his work he shared the 1909 Nobel Prize for Physics with **Guglielmo Marconi**.

Braun was born in Fulda, Germany, and later attended the Universities of Marburg and Berlin. After receiving his doctorate in 1872, he held several academic positions at the Technical University of Karlsruhe, the University of Tübingen, and the University of Strasbourg. He began his first important experimental work in 1874. That year, while studying mineral metal sulfide, he noted that certain crystals transmitted electricity in one direction more easily than in the other. These crystals, called rectifiers, were useful in converting **alternating current** (which travels in two directions) into **direct current** (which travels in one). The first use of **crystal rectifier**s was in crystal-set radios. These rectifiers were replaced eventually by **triodes** and other valve circuits, but returned later in the twentieth century—in an improved form—in solid-state circuitry.

The 1890s marked a period of intense activity for Braun. It had been known for some time that the path of electrons that coursed through a **cathode-ray tube** could be altered by applying a magnetic current. Braun devised a system by which the magnetic current alternated, causing the beam to ''scan'' slowly; as the scanning beam hit the side of the tube, a glowing dot appeared, moving up and down. When a high-frequency alternating current was applied, the dot would scan at a faster rate, creating observable patterns. What resulted in 1897 was Braun's **oscilloscope**, an important laboratory tool and the precursor to the television tube. Braun also experimented with ways to improve the power and range of radio transmissions. At the time, radio **antenna** were connected directly to the power supply, resulting in a maximum transmitter range of about nine miles (15 km). Braun's idea was to remove the antenna from the power circuit and use magnetic coupling to boost the output. He patented his system in 1899, and since then magnetic coupling has been applied to radio, television, and **radar**. In addition to magnetic coupling Braun also invented the directional antenna. In the early twentieth century considerable controversy developed over the ownership of radio technology patents. In order to help, Braun came to the United States in 1917 to testify in a series of patent hearings. During his stay, America entered World War I. Braun, a German citizen, was taken prisoner as an enemy alien and died a year later while still imprisoned.

von Braun, Wernher (1912-1977)
German engineer

The son of a German nobleman and his wife, an amateur astronomer, von Braun was fascinated early on with astronomy and the idea of interplanetary travel. Upon graduating from high school, von Braun joined the Verein für Raumschiffahrt (Society for Space Travel), a group of talented amateur scientists who built several experimental **rocket**s that achieved heights of up to a mile (1.6 km).

When Adolf Hitler came to power in the early 1930s, the German military took control of the group and banned further private research in rocket technology. In 1932 von Braun became a civilian employee of the German army while he worked on his doctorate in physics from the University of Berlin, which he earned in 1934. Although he dreamed of using rockets for space exploration, von Braun worked with indefatigable energy throughout World War II at Peenemünde, an army research center on the Baltic coast, developing the long range **missile**s needed by Germany to achieve air superiority. At one point the Gestapo arrested and jailed von Braun for allegedly focusing on exploratory spaceflight to the detriment of Germany's war effort.

The most important achievement of von Braun and his colleagues was the A-4 rocket. Successfully launched in 1942, the A-4 weighed 5.5 tons (4.9 metric tons) and travelled at a speed 3,000 miles per hour (4,800 kph) and a height of 60 miles (96.5 km). The rocket was later renamed the V-2 (Vengeance Weapon 2) and used against Great Britain, beginning on September 7, 1944. ''When I heard about this,'' von Braun later recounted, ''it was the darkest hour of my life.'' By the end of the war, over 1,000 rockets carrying 2,500 pounds (1,130 kg) of explosives were fired at Great Britain. However, the V-2 had been developed too late in the war and did little to reverse the tide against Nazi Germany. Near the end of the war von Braun was ordered to move to a camp in Bavaria. Because he suspected that he and his co-workers would be killed rather than allowed to surrender, they hid in small villages until they were able to turn themselves—and their equipment—over to American forces.

Von Braun and 126 other German scientists were hired and brought to the United States under the code name Project Paperclip. They continued their rocket research at the White Sands Proving Grounds in New Mexico and soon transferred to the U.S. Army base in Huntsville, Alabama, where they constructed a new rocket called the Redstone. Twice as large as the V-2, it stood 70 feet (21.35 m) tall and was capable of a thrust of 78,000 pounds (35,412 kg). In addition to his research, von Braun wrote articles for periodicals and spoke with missionary zeal about future space travel. In 1954 he suggested using the new rocket to launch a satellite into orbit, but the government decided to go instead with a Navy proposal, called Project Vanguard.

Undeterred, von Braun developed the Jupiter-C, a four-stage jet capable of flights 700 miles (1,126 km) high and 3,300 miles (5,309 km) in distance. When the Russians launched **Sputnik** in 1957, the U.S. space program shifted into high gear. The Vanguard had one miserable failure after another, so the Americans turned to von Braun and his Jupiter-C. On January 31, 1958, it launched **Explorer 1** into orbit.

With this success and enthusiastic public support, von Braun pursued more projects for his rockets. He proposed boosting a small capsule with a human passenger into a short flight through outer space. It would be only a suborbital flight since the Redstone did not have the thrust to boost such a large object into orbit. Von Braun's Redstone was used for two of these short flights: Alan Shepard, Jr. (1923-) went up for 15 minutes in May, 1961; Virgil (''Gus'') Grissom (1926-1967) made a similar flight the following July. When von Braun moved to NASA (National Aeronautical and Space Administration), he worked on the Saturn rocket. This rocket was to be America's answer to the giant boosters used by the Russians. The Saturn was a monster: it was 150 feet (45.75 m) tall and 21 feet (6.4 m) thick at the base. A variation, called Saturn 5, was used to send the first men to the moon, starting with the *Apollo 8*, launched in December, 1968.

Von Braun stayed with NASA until the Apollo missions halted in the early 1970s. He wanted to work on a manned mission to Mars, but the public had lost interest in the space program. He did, however, help in the early stages of the space shuttle program before retiring in 1972.

Breakfast cereals

While it may seem odd to us today, accustomed as we are to charges that many dry breakfast cereals lack food value and to advertising attempts to convince us otherwise, breakfast cereals actually began as part of the health-food craze of the 1890s. The first dry, flaked breakfast cereal was developed by the Kellogg brothers, two of sixteen Michigan siblings. They came from a family of Seventh Day Adventists, a religious group headquartered in Battle Creek, Michigan, that emphasized attention to health and a simple, vegetarian diet. **John Harvey Kellogg** (1852-1943) was a young medical doctor when he took over the Adventist Health Reform Institute in Battle Creek in 1876. He renamed that Institute the Battle Creek Sanitarium—deliberately changing the common term *sanitorium*—and promoted his patients' health through good diet. Dr. Kellogg constantly experimented with ways to concoct tasty meals for his patrons using nuts and grains. He invented Granola in 1877, as well as peanut butter and a number of meat-substitute foods such as Protose (for beef) and Nuttose (for veal). He also invented Caramel Coffee, a grain substitute for real coffee.

Dr. Kellogg's younger brother, Will Keith (1860-1951), worked as the Sanitarium's administrator and assisted John Harvey in the development of his foodstuffs, although the brothers didn't get along well. In 1894 the Kelloggs accidentally produced flaked breakfast cereal when they left an experimental boiled wheat dough in the kitchen for a few days. When they pushed the dried dough through rollers, it produced flakes that were tasty when baked. Sanitarium patients loved the new concoction when served with milk, and after they returned home, ordered the flakes by mail. The Kelloggs introduced corn flakes in 1898. In 1903, Will Keith split with his brother John and started the Kellogg Toasted Corn Flake Company, marketing the corn flakes with added malt, sugar, and salt. Through massive advertising and free sample give-aways, Will Keith made his product the world's most popular breakfast cereal and revolutionized American breakfast habits.

The Kelloggs weren't the first to market dry breakfast cereal, however. The inventor Henry D. Perky of Denver, Colorado, had made a machine in 1893 that shredded wheat and formed it into little pillow-shaped biscuits. He marketed it as Shredded Wheat.

Another breakfast cereal pioneer who got his start in 1891 as a patient at the Kelloggs' Sanitarium was Charles W. Post (1854-1914). When Dr. Kellogg refused Post's offer to promote the Kellogg cereal coffee substitute, Post set up business in Battle Creek in 1895 and developed his own product, Postum Cereal Food Coffee. Postum was an instant success and is still on the market today. Two years later, Post introduced his first dry breakfast cereal, Grape-Nuts, baked in the form of bread sticks. The name came from Post's belief that grape sugar was formed during the baking, and from the cereal's nutty flavor. In 1904 Post came out with his own version of corn flakes, which he originally called Elijah's Manna. Vigorously attacked from the pulpit for blasphemy, Post quickly changed the cereal's name to Post Toasties. Post's business, which began in a little white barn, eventually grew into the giant General Foods Corporation, which merged with Kraft in 1989 to become the largest food company in the United States.

Wernher von Braun

Inspired by the Kelloggs' and Post's successes, a breakfast-cereal rush took place in Battle Creek around the turn of the century. Dozens of factories sprang up, producing cereals like Tryabita, Strengtho, Corno, Malta Vita, and Maple-Flakes. Most did not last long. Those that did emulated the Post and Kellogg formula for success: intensive advertising and promotion. Breakfast cereal manufacture, which did not exist before the 1890s, is today a billion-dollar industry marketing over 100 different brands. As of 1998, it was estimated that 95 percent of American households purchase cold, ready-to-eat breakfast cereals. Hot, cooked cereals, such as oatmeal, cream of wheat, and grits, are popular as well. Most now come with microwave instructions for quick preparation, and many are sold in instant varieties.

Modern cold cereals are often made by a process that starts with cooking a mixture of grain, flavorings, and vitamins and minerals under steam pressure. The cooked mixture is then put in dryers until it reaches the desired moisture level. Next large flaking mills exert tons of pressure on the grain to flatten it into thin flakes. The flakes are toasted in ovens until they achieve the right color, flavor, and crispness. Finally the cereal is sprayed with more vitamins before being packaged in cartons lined with moisture-resistant material. Other cereals are exploded into puffs, shredded and formed into little biscuits, or extruded in fanciful shapes. In addition to flavorings and nu-

Corn flakes, the first cold breakfast cereal, was invented in 1894 by W. K. Kellogg and his elder bother John Harvey Kellogg, the founder of the Battle Creek Sanitarium. While the flakes were originally intended only for "San" guests, W. K. Kellogg eventually broke with his brother to form the Kellogg Toasted Corn Flake Company, which, with free samples and advertisements such as the one above, succeeded in revolutionizing American breakfast habits.

trients, cereal makers often add sugar, salt, and preservatives to their products. Some cereals also have extra ingredients such as nuts, raisins, and marshmallows.

Breech-loader • See Rifle

Breeder reactor • See Nuclear reactor

Brewster, David (1781-1868)
Scottish physicist

During the early 1800s, a scientific revolution was underway in Britain and France as physicists fiercely debated the truth or falsity of two optical theories: that light is composed of either waves or particles. Many famous scientists, including Sir David Brewster, conducted independent research in an attempt to ascertain the true nature of light. Out of this research came the basis of modern optical study, but what made Brewster remarkable was his ability to apply his research toward practical ends, inventing such devices as the **kaleidoscope** and the improved **stereoscope**.

Brewster entered the University of Edinburgh in 1794 as a divinity student, and although he completed his studies in 1800 and was licensed to preach in 1804, he never was ordained a minister. Instead, he developed a knack for building precise scientific instruments while at college and performed some fledgling experiments with light.

His curiosity about light and the new controversy surrounding it grew, and once out of school, he began his research in earnest, concentrating on the phenomenon of polarized light—light in which all waves lie in the same plane. The best evidence to support this new hypothesis was the fact that light, when passed through a piece of Iceland spar, was split into two rays bearing different properties. Brewster took this thinking even further, showing that the two new rays were completely polarized and turned at right angles to each other. This discovery was dubbed Brewster's law. The law states that a beam of light can be split into two beams at right angles to each other, and the reflected beam will be completely polarized, while the refracted beam will be partly polarized. This formulation earned him the Royal Society of London's Rumford Medal in 1819 and was the primary proof for the transverse wave light theory.

While using **mirrors** to reflect light, Brewster noticed that two mirrors placed in a tube could create an image. By placing colored beads at the bottom of the tube, he discovered that an ever-changing picture was formed. Thus the kaleidoscope was born. Brewster received the patent for his device in 1816, and though it became immensely popular, it was also very easy to duplicate, and so Brewster's earnings were meager.

In addition to the kaleidoscope, Brewster perfected the stereoscope, an ingenious device that was the first "three-dimensional" viewing apparatus. Brewster also persuaded the British government to adopt a new **lighthouse** technology developed by **Augustin-Jean Fresnel** (1788-1827) in France, utilizing lightweight, flat lenses. Brewster helped found the British Association for the Advancement of Science in 1821 and was knighted in 1862, at the age of 51.

Bricks

Bricks are among the oldest construction materials known. The raw materials needed to make them are found almost everywhere, and the method of making them is very simple. The Sumerians were using plano-convex bricks, flat on one side and curved on the other, as early as 4000 B.C.. Brickwork at Ur showed traits of sophistication, included corbeled **vaulting.** At the time bricks were often stamped with the names of their makers.

Brick-making spread to Persia, India, China, Greece, and eventually to most of the world. In some areas, such as Egypt, stone was reserved for official buildings and the homes of the privileged, while sun-dried brick was used in common construction. In other areas, brick was preferred for the officials while wood and other less permanent materials were left to the commoners. Sun-dried brick was adequate as long as there was no danger of destructive rains. Even so, the Egyptians and many other ancient peoples often found it desirable to strengthen their sun-dried bricks by scattering chopped straw around the brick pits and then walking about the pits to mix the straw with the clay, literally "treading the mortar" as references in the *Bible* suggest.

In wetter climates, bricks required kiln-firing to made them hard enough to fend off moisture. Brick kilns, or ovens, have been used for millennia, and the decision to adopt baked or unbaked bricks was sometimes as much a matter of cultural choice as of technological necessity. By the seventh century B.C., straw was no longer mixed in the clay, but the clay was still "trod" to ensure an even distribution of moisture and proper baking in the kiln.

Most of the natural clay used for bricks are complex mixtures of clay minerals with salts of common elements like iron or calcium mixed in. Bricks are held together with mortar, an adhesive usually made of cement, lime putty, and sand in various proportions. Brick-making in Europe, like many other technical advances, died with the Roman Empire. There was a lengthy period in which only the most basic construction techniques were practiced during the Middle Ages. Not until the 1500s, when King Henry VIII promoted it, was the use of brick revived in England. In 1619, a clay working machine was patented, and after London burned in 1666, brick-making making flourished.

European brick construction was introduced to North America in Virginia in 1611 and in Massachusetts in 1629. However, the adobe cultures of what is now the Southwest United States and Mexico had already been building their cliff dwellings with brick for centuries. (Adobe is the Spanish word for unbaked brick.) Advances in brick-making continued when Apollos Kinsley of Connecticut patented a brick-making machine in 1793. It consisted of a mill with an auger that forced clay into molds. A type of reinforced brick was introduced in the early 1800s, and iron and **steel** were used to strengthen the clay instead of straw. In 1825, **Marc Isambard Brunel** used bricks reinforced with wrought-iron bolts in his Thames River Tunnel project.

The revolution in construction that occurred in the late 1800s shifted structural support from building walls to iron or steel frameworks. Brick became a facade material, still of great importance, but no longer the main ingredient in a building's support. Yet, for several decades, brick was nearly the only facade material in use in some cities. Brickyards were as numerous as breweries at the turn of the twentieth century. It was also used to pave city streets and river wharves (cobblestones). In some areas, bricks were being made from granite blocks. Brick was used extensively in residential construction after World War II. Commercially, however, it took a back seat to **glass** and **steel**. After some thirty years of futuristic modern architecture based on glass and metal, brick is making a comeback as an attractive embellishment to high-rise buildings and shopping centers. It is also being used once again to pave sidewalks. After several millennia, the principles of brick-making have remained virtually unchanged. The use of brick has varied, but its role in the construction industry seems assured.

Bridge

For ages humans have built bridges to allow transport over streams, rivers, gorges, and other transportation routes. The primitive clapper bridge—a stone slab-and-pier construction—and the basic beam bridge in which a stone or tree breached a gap, are among the earliest bridges.

The Romans used the semi-circular masonry arch in their bridge-building and also used timber, while in China segmental arch bridges were constructed to span greater distances. As commerce escalated, the load requirements of roads and bridges became greater, and complex bridge engineering was crucial to ensure against collapses. The late eighteenth century saw the introduction of iron and **steel** into bridge-building, with the stronger **cast-iron** eventually replacing **wrought iron**. The bridge-builders of Great Britain were particularly innovative and showed a keenness for competition. Such names as **Thomas Telford** and **Isambard K. Brunel** are associated with magnificent early suspension bridges.

Bridges are of three basic types that encompass a range of variations: the beam, the arch, and the suspension. Beam or girder bridges press downward on supports, while arch bridges are compressed and push outward against the supports, placing the burden on the endpoints or abutments. The suspension bridge relies on tension between cables and exerts a pull on its end anchorages; its earliest form was constructed of vines, bamboo, or cane tied to tree trunks. The need to span greater distances and deeper channels has led to the reliance on suspension bridges rather than on piers and arches.

Truss bridges—bridges with rigid frameworks for support—were simple to build and were typically American. Cantilever bridges, a variation on the beam or girder, were also developed in the later part of the 1800s and could extend three to four times farther than the diagonal truss. The first cantilever bridge was built across the Main River in Germany in 1867. The renowned Firth of Forth Bridge was built in Scotland in 1889 and was the first long-span railway bridge made of steel.

In 1874, American James Eads (1820-1887) designed a magnificent steel arch bridge with two decks across the Mississippi River at St. Louis, marking the beginning of the replacement of iron bridges with steel. Iron bridges tended to collapse after ten years or so of continued stress, and steel was far more durable. **John Roebling** pioneered the use of steel cable in the construction of the Brooklyn Bridge. He died in 1869, but his son, Washington Augustus Roebling (1837-1926), engineered the project and saw its completion in 1883.

Variations on these types of bridges include the covered bridge, first built by Timothy Palmer 1806 across Philadelphia's Schuylkill River. The covering over the framework protected the beams from rot and the deck from snow and ice. Palmer's innovation was popular, and covered bridges were common sights in the southern and midwestern United States, and especially in New England.

Pontoon bridges are generally used during wartime, replacing destroyed or blocked bridges or facilitating crossings where no bridges exist. The earliest pontoon bridges were used by Persian armies in the fifth century B.C. Primitive pontoons consisted of boats placed side by side and overlaid with planks.

Modern pontoon bridges consist of a deck laid across airtight floats tethered or anchored to the bottom and were used extensively during World War II.

Another type of bridge used in military campaigns is the Bailey bridge. Invented in 1941 by England's Sir Donald Bailey (1901-1985), it consists of interchangeable steel panels held together with steel pins that can be quickly assembled on shore with hand tools, then unrolled across a river on temporary piers. An entire class of bridge is the movable deck bridge, used most commonly to span narrow shipping channels where the height of passing boats is greater than that of the deck.

The most common type of movable bridge is the drawbridge, or bascule bridge, which is either hinged at one end, permitting the entire deck to swing upward like a gate, or split at the middle, allowing both halves to swing upward. The vertical lift bridge uses weights to raise the entire center deck and its truss above the shipping channel. One of the most well known of these is the Aerial Bridge at Duluth harbor. Other verticals are used in the St. Lawrence Seaway and in many harbors of the world. Swinging bridges pivot the bridge structure on its pier and are useful in lock and dam situations. Only the bridge span over the lock is moved to allow towboats to pass.

Early suspension bridges used iron rods or chains. In the later nineteenth century, steel cables came into use. Some bridge decks have been suspended from arches, like the Sydney Harbour Bridge (1932) in Australia. Several classic bridges, like the George Washington (1931) in New York and the Golden Gate (1937) in San Francisco, were built during the early half of the twentieth century. Suspension bridge construction reached its peak between the late 1950s and early 1970s.

Modern construction materials are making it possible to build bridges of simpler design and across greater expanses. Reinforced and prestressed concrete and high-strength steel are allowing for the return to simple self-supporting bridge decks reminiscent of the early beam bridge. The Interstate Highway system of the United States has made extensive use of the continuous box-and-plate girder bridge as a result. Suspension bridges are now being built at record lengths, crossing channels previously considered unbridgeable. The longest spanning suspension bride in the United States (and sixth in the world) is the Verrazano-Narrows bridge in New York City, completed in 1964 and spanning 4,259 ft (1,298 m). The longest suspension bridge in the world is the Akashi Kaikyo Bridge, the final link in a chain of spans that connects Japan's four main islands. Completed in 1998 after ten years of work, it measures 12,828 ft (3,910 m), or 2.43 mi (3.91 km). Its center span alone measures 6,529 ft (1,990 m). With the nearest seismic fault 90 miles (145 km) away, the bridge was designed

San Francisco's Golden Gate Bridge under construction.

to withstand an **earthquake** that measures 8.5 on the Richter scale.

The longest bridge in the world is the over-water highway bridge over the Lake Pontchartrain Causeway in Louisiana, spanning 34.3 mi (55.2 km). The twin spans of the Causeway are made of pre-stressed panels supported by over 9,000 concrete pilings. The first span opened to the public in 1956, and the second in 1969. The result was one-way driving safety on straight, two-lane railed surfaces 80-feet (24 m) apart connected by seven crossovers that function as pull-over areas for auto emergencies.

The future for bridge design includes "smart bridges," which have **fiber optic** sensors embedded into their structure to monitor their health. The sensors are linked to a computer that monitors the bridge for cracks, strains, and road-salt corrosion, minimizing the need for on-site maintenance. (An estimated 230,000 bridges in the United States alone are in need of serious repair.) Efforts are also underway to construct bridges of composite materials that require virtually no maintenance, such as a 36-ton bridge in Butler County, Ohio. It is made of lightweight, corrosive-resistant polymers, and is expected to last 150 years, about three times longer than conventional bridges. Such bridges cost about twice that of a normal bridge, which has hindered their widespread use. In other prototypes, high-performance concrete (HPC) allows longer beams to be used, and is more resistive to the salt erosion which corrodes the reinforcing steel bars inside the concrete beam.

See also Concrete and cement; Cast-iron and wrought-iron construction; Steel alloys

Bridge, electrical

One problem facing scientists experimenting with electricity in the mid-1800s was taking accurate measurements of electromotive force (emf), current strength, and resistance. A **galvanometer** was fine for indicating current flow, but it was an unstable device for making an actual measurement, and there was no standard way to calibrate it. As far as determining the unknown resistance in an electric circuit, a solution was invented by English physicist **Charles Wheatstone** in 1843. Wheatstone had come across a neglected German publication written in 1827 by Georg Ohm (1787-1854) which set forth the theory that emf, current, and resistance were all mathematically related. This finding put Wheatstone on the right track.

Wheatstone created an electrical bridge circuit which included four **resistor**s, three having a known value, with the fourth unknown. When an emf is applied to the inputs of this four terminal network, the output terminals of the circuit in question are "bridged" with a detector such as a voltmeter. Using a calibrated variable resistance, which Wheatstone called a rheostat, he could adjust the components in the network, "balancing" the voltage across the bridge (bringing it to zero as detected on the voltmeter). Since the components in the network have specific relations to each other, the resistance of the fourth component could be determined from the resistances of the other three.

The Wheatstone bridge is one of the best known electrical bridges, but there are a variety of other bridges used to convert physical quantities like temperature or pressure into electrical quantities like voltage or current. Examples are the Carey-Foster bridge, which also measures resistance; the Kelvin double bridge, which has eight arms instead of four; the Wien bridge which measures **alternating current** capacitance, as does the Nernst high-frequency capacitance bridge; and the Heaviside mutual inductance bridge and the Maxwell bridge, which compare inductance with capacitance.

Bridgman, Percy (1882-1961)

American physicist

Percy Williams Bridgman was born in Cambridge, Massachusetts, on April 21, 1882. He attended public schools in Newton, Massachusetts, and then entered Harvard University in 1900. He earned his bachelors, masters, and finally, his doctorate (1908) at Harvard. Upon graduation, Bridgman was appointed to the faculty at Harvard, where he remained for the rest of his life. Even after his retirement in 1954, he continued doing research at the University's Jefferson Physical Laboratory.

Bridgman devoted his academic career to a single topic, high pressure physics. In the 1880s French physicists Louis Paul Cailletet (1832-1913) and Emile Hilaire Amagat (1841-1915) invented devices for studying the effects of pressures as high as 3,000 atmospheres (44,000 pounds per square inch) on materials. At pressures greater than 3,000 atmospheres, however, the seals on their devices failed. It seemed that a limit on high pressure research had been reached. In 1905, Bridgman decided to study the effects of high pressure on certain optical phenomena. When an explosion damaged the equipment he was using, he turned his attention to the machine used to produce high pressure, rather than the optical phenomena. He set about to develop a high pressure device in which the seal would not fail and which, therefore, could produce pressures greater than those achieved by Cailletet and Amagat. The approach he developed involved a seal that became tighter with increased pressures. The external support on the device was also constructed so that it became stronger as the pressure increased. Bridgman's device was so successful that he was soon able to produce pressures of 20,000 atmospheres (300,000 pounds per square inch). Further developments over the years made possible devices that could produce pressures of 100,000 atmospheres (1,500,000 pounds per square inch).

With these devices, Bridgman and other scientists were able to explore a great variety of high pressure phenomena. He discovered a new form of ice, for example, that has a melting point of 200° C at 40,000 atmospheres of pressure. His devices have also been used by geologists to study the behavior of matter at pressures similar to those at the earth's core. In 1955, a team of General Electric researchers, with Bridgman as consultant, used a high pressure device to create the first synthetic diamonds.

See also Diamond anvil cell; Synthetic gemstones

BROCKHOUSE, BERTRAM NEVILLE (1918-)

Canadian physicist

Neville Brockhouse was born in Lethbridge, Alberta, Canada, on July 15, 1918, to homestead farmers. Brockhouse had one sister, Alice Evelyn, and two brothers, one of whom died in infancy, and another who later became a railroad civil engineer. In 1926 the Brockhouse family moved to Vancouver, British Columbia, where Bertram and his siblings grew up.

Financial difficulties during the Great Depression caused Brockhouse's family to move to Chicago, Illinois, in 1935 in search of better opportunities. Brockhouse began attending Central YMCA College (later renamed Roosevelt University), taking technical courses. He learned to repair and design radios and worked for a short time in the laboratory of a small electronics firm. In 1938 the family decided to return to Vancouver, where Brockhouse continued to repair radios for a living.

Anti-totalitarian and anti-Communist feelings inspired Brockhouse to enlist in the Royal Canadian Navy in 1939. He was in the navy for six years, during which he spent most of his time repairing equipment at a shore base. In 1944 the navy enrolled him in electrical engineering courses at Nova Scotia Technical College and assigned him to the test facilities at the National Research Council in Ottawa. It was during this time that Brockhouse met his future wife, Doris Miller. They would lose contact for a while after the war was over and Brockhouse was released from the navy in 1945.

In September 1945 Brockhouse began studying at the University of British Columbia. He felt that engineering or physics were obvious choices for study considering his background, but decided the combination of physics and mathematics was the best choice. In the summer of 1946 Brockhouse took a motorcycle trip to Ottawa. He graduated with a B.S. from the University of British Columbia in April of 1947 and took a summer job at the National Research Council Laboratory in Ottawa.

Brockhouse entered the University of Toronto at the age of 29 and began studying the effects of stress and temperature on ferromagnetism. He ran into some difficulties, though, when three faculty members left to take other positions, and he was left unsupervised.

Brockhouse earned his Ph.D. in 1950 and then joined the Atomic Energy Project of the National Research Council of Canada (later renamed the Atomic Energy of Canada Limited), in Chalk River, Ontario. It was during this period that Brockhouse conducted his ground-breaking research in neutron physics. He had been reading a paper about neutrons, whose existence had only been verified 20 years before. He was supposed to be working on another project, but could not stop thinking about testing the theories he had read about in the laboratory. He began working on an apparatus that would focus a neutron beam on solids like minerals, metals, and gems and would therefore reveal their structure. Brockhouse would go on to improve this neutron beam, and his apparatus, the triple-axis neutron spectrometer, is now used all over the world to examine the structure of crystals.

In 1962, he accepted a position as professor of physics at McMaster University in Hamilton, Ontario, which had a ''swimming pool'' reactor that would allow him to continue his research. He continued to conduct preliminary research for experiments that would be conducted at the atomic research lab in Chalk River, but now his work included teaching and training students to work with the reactor. In 1962 he won the Buckley Prize of the American Physical Society and in 1963 the British Institute of Physics and Physical Society's prize. In 1971 he supervised a group of talented students in the building of the university's own triple-axis spectrometer.

Brockhouse retired from McMaster University as a professor emeritus in 1984. In retirement he pursued interests in the philosophy of physics. Exactly ten years after his retirement, Brockhouse awoke on a Wednesday morning to find a message on his answering machine. The message said that B.N. Brockhouse and C.G. Shull had been chosen for the 1994 Nobel Prize for physics. Brockhouse was stunned and called his wife in to listen to the message. Because of the award, he was promptly called out of retirement for a year of lectures, awards, and travel, all for work he had completed over 40 years ago.

BRONCHOSCOPE • See Endoscope

BROWNING, JOHN (1855-1926)

American inventor

Born in Ogden, Utah, John Browning spent most of his formative years in his father's gunsmith shop, where he made his first rifle out of scrap iron at the age of thirteen.

In 1879, Browning received his first patent for a single-shot rifle that was loaded from the rear rather than from the muzzle. The Winchester Repeating Arms Company was so impressed with this weapon that they paid Browning large royalties to produce it. He then invented a lever-action repeating **rifle** and shotgun for Winchester, introduced in 1884. In 1888, Browning designed a pump-action **shotgun** that was more advantageous for the hunter.

Browning soon turned his attention to automatic weapons. The Gatling **machine gun** (patented in 1862 by **Richard Jordan Gatling**), was the dominant automatic weapon of the era; it was a rapid-fire gun powered by a hand-operated crank. Browning attempted to find a way of using expanding gases and recoil from exploding **ammunition** to eject, reload, and fire weapons automatically. From 1889 to 1892, he experimented on the machine gun, finally creating a crude weapon that captured the gases at the muzzle. In 1895, Browning improved the gun by diverting the gases from a hole drilled in the back of the barrel. The resulting weapon was used in the Spanish-American War as well as in the Boxer Rebellion in China. Browning was also able to employ the same principle for a semi-automatic **pistol** that became the standard military sidearm for the United States in 1911. A later variation of a heavy machine gun designed by Browning was used aboard United States military **aircraft**, having the ability of firing at the rate of 1,200 shots per minute. In 1918, Browning's machine gun was pronounced the finest in the world: it was water-cooled, mounted on a tripod, and weighed thirty-six pounds (sixteen kg). During one test, this weapon was fired continuously for forty-eight minutes. Browning's success with automatic weapons carried over to the sporting arms industry. For example, he developed an automatic shotgun and rifle, as well as a double-barreled shotgun with the two barrels arranged vertically.

No design of Browning's had ever proved to be a failure. He invented more successful firearms than any other American: today, there are over thirty million modern weapons which are based on his designs. One measure of Browning's phenomenal success is the fact that no fundamental changes have appeared in the firearms industry since his death in 1926.

Brunel, Isambard Kingdom (1806-1859)

English engineer

Isambard Kingdom Brunel was the son of **Marc Isambard Brunel**, a French engineer who escaped the revolutionary turmoil of France in 1793 by immigrating to the United States, where he became an American citizen and engineered several projects, including the Hudson-Champlain Canal. In 1799 Marc Brunel moved to England and married Sophia Kingdom, an Englishwoman who had begun corresponding with Brunel while she was imprisoned during the French Revolution.

Isambard Brunel was born in 1806 in Portsmouth, England. Demonstrating an early aptitude for mathematics, he was sent as a 14-year-old to college in France, where he studied and apprenticed in the design of mechanical instruments. At the age of 20 Brunel was appointed resident engineer of his father's Thames River Tunnel project, and it was during this early experience that Brunel's legendary resilience and flamboyance were first publicly realized. Brunel tenaciously overcame many obstacles, but the project, overwhelmed with mishaps, was aborted when an undetected low section in the river bed caused a collapse and flooded part of the **tunnel** in 1828, nearly drowning Brunel. The tunnel was not completed until nearly fifteen years later.

While Brunel was recuperating from a broken leg suffered during the collapse of the tunnel, he submitted four designs to the River Avon **Bridge**, which were rejected by **Thomas Telford**. Telford, who was famous for his Holyhead Road and Menai Strait Bridge projects, rejected all submissions and entered his own design. However, the Bridge Committee rejected Telford as well, and in 1830, Brunel finally succeeded in having his design accepted for the bridge project, also known as Clifton Gorge. Although not completed until 1864, five years after his death, Brunel's elegant suspension bridge is still in use today.

One of Brunel's greatest achievements was the design and construction of the London-to-Bristol rail line, which became know as the Great Western Railway (GWR). It was noted for its low grades, low-arch bridges and its two-mile tunnel. He gave the tracks a 7-foot gauge to give the **train**s better stability. Although the rail system was later converted to conform to England's standard 4 feet, 8 1/2 inch rail gauge, the "gauge war" Brunel prompted helped spur innovations in England's **locomotive** industry.

Brunel went on to build 1,600 miles (2,574 km) of rail lines in England; he also served as an advisor on the construction of rail lines in Australia and India. His construction techniques were as significant as his designs, including his use of compressed-air **caisson**s for underwater construction of bridgepiers. One notable and extremely costly failure, an "atmospheric railway," which ran on atmospheric pressure generated by steam pumps, fell victim to unforeseen technical difficulties but did operate briefly between Exeter and Newton Abbot, achieving noteworthy speeds of 64 miles per hour (103 kph). That failure was overshadowed by his many successes, among them the design of the railway stations at Paddington and Temple Meads, Bristol; a number of tunnels and bridges; and a prefabricated hospital for wartime field use.

Brunel's work for GWR led to another triumph for the civil engineer—pioneering efforts in steam navigation. In 1836 he was challenged by the notion that the GWR could "extend" its terminus from Bristol—the British gateway to westbound shipping—across the Atlantic to New York via transatlantic **steamship**. He set to work immediately on the *Great Western*, which became the first steam-powered **ship** to make a complete trans-Atlantic crossing, arriving in New York in 1838. While it had previously been thought that no ship could carry enough coal to make the journey, the *Great Western* encountered no such difficulties, thanks to Brunel's design.

While the *Great Western* had a wooden hull, the hull of Brunel's next ship, the Great Britain, was made of iron. Three years after its 1843 launching, its hull survived a grounding on the Irish coast. However, Brunel was not yet satisfied. He designed his next ship, the *Great Eastern*, to extend steam service to Australia and carry a year's exports to India. The effort, though, ruined Brunel's finances, reputation, health, due in large part to the questionable business dealings of his partner in the venture, John Scott Russell.

Brunel, Marc Isambard (1769-1849)

French-English engineer

Marc Isambard Brunel was born in Hacqueville, Normandy, France. He served as an officer in the French navy from 1786 to 1792. In 1793, he left revolutionary France for the United States because of his royalist sympathies. He practiced as a civil engineer and architect and soon became New York City's chief engineer. In that post, he constructed many buildings, an arsenal, and a cannon foundry, and advised the city on ways to improve the defenses of the channel between Staten and Long Islands.

While in New York, Brunel designed a method for manufacturing ship's blocks, or pulleys, mechanically rather than by hand. Since a single large warship of that time could use fourteen hundred blocks, Brunel sailed to England in 1799 to present his ideas to the British navy. In London, Brunel and **Henry Maudslay** constructed models of the machines. The proposal was accepted, and Maudslay constructed the block-making machinery at the Portsmouth Dockyard between 1803 and 1807. As completed, a system of forty-three machines run by ten unskilled workers turned elm logs into finished blocks of higher quality and consistency, and in much greater numbers, than the handmade blocks previously made by 110 skilled workmen. This was the first example of large-scale mass production by a series of specialized machine tools, replacing a series of steps performed by hand.

Brunel was a prolific inventor but a poor businessman. A fire at his London sawmills plus financial mismanagement by his partners put Brunel deeply in debt. This condition worsened when the Napoleonic wars ended and the British government refused to honor an agreement to purchase the output of Brunel's second mass-production line, nailed-boot-making machinery. Brunel spent several months in debtor's prison in 1821, until his friends secured his release.

In 1825, Brunel began work on his best-known project, a **tunnel** under the Thames River. He had patented a tunneling shield in 1818, which won him the appointment as engineer of the tunnel project. The shield was a metal cylinder pushed forward mechanically as tunneling progressed, and with separate cells for individual workmen. Although water above the tunnel burst through the floor of the riverbed five times, the shield always held. These delays, however, caused financial problems, and at one point the tunnel project was halted entirely for seven years. It was finally completed in 1843, the first successful resolution of the underwater tunneling challenge. The Thames Tunnel immediately became a popular public attraction, and it is now part of the London Underground (subway) system. Brunel was knighted for his achievement in 1841.

Brunel also designed a suspension bridge in France and floating piers in Liverpool. His other inventions include machines for sawing and bending timber, knitting stockings, printing, and making **nails**. He died in London in 1849.

Bubble chamber

For nearly half a century, the **cloud chamber** invented by **C. T. R. Wilson** in 1911 was about the only method for observing the tracks of very small particles such as alpha and beta particles, protons and deuterons. The cloud chamber operates on the principle that a particle passing through a supersaturated vapor will cause condensation on ions formed along its path. The condensed liquid can then be observed directly or photographed for further study. For all its power as a detection device, however, the cloud chamber has some serious limitations. Most important is the fact that the gaseous atoms and molecules within the chamber are relatively far apart. If a nuclear event occurs in which a particle travels only a very short distance—less than the distance between two atoms or molecules—that event will not be recorded by the cloud chamber. It was this problem to which the American physicist, **Donald Glaser**, turned his attention in the early 1950s. He was working on the "strange" particles that are produced in cosmic ray showers and in **particle accelerator** reactions. Some of these particles are produced and decay within such a short distance that they are unlikely to be recorded by a cloud chamber.

Glaser's idea was to use a liquid as a detecting medium since the particles in a liquid are much closer together than are those in a gas. Glaser's bubble chamber is essentially the opposite of a cloud chamber. It contains a liquid that is heated beyond its normal boiling point. However, if the liquid is kept under pressure and if all foreign material is removed from the liquid, it will not boil. Instead, it will remain in a superheated state. Imagine, now, that the bubble chamber is exposed to a source of radiation. Particles released from the radioactive source will travel through the bubble chamber and interact with atoms and molecules in the liquid. This interaction will result in the formation of ions, atoms, or molecules that carry an electrical charge. The ions act as nuclei on which the liquid can begin to boil. The path taken by the particle as it moves through the bubble chamber is marked by the formation of many very tiny bubbles, formed where the liquid has changed into a gas. Bubble chambers are widely used today in the study of nuclear events. The target area in a particle accelerator may be surrounded by banks of bubble chambers in many different orientations. Computer analysis of the photographs taken by these bubble chambers provides a three-dimensional view of particle tracks. These tracks can then be used to identify particles produced in the accelerator.

Bubble gum • See Chewing gum

Bubble memory • See Computer memory

Bullet • See Ammunition

Bulletproof vest • See Armor

Bungee cord

Bungee cord is made of one or more strands of an elastic material, usually **rubber**, bound together by a fabric covering. It is commonly used as a tie-down for luggage or equipment carried on the outside of a vehicle. Bungee cord is also used by the military to absorb the opening shock of the large cargo **parachutes** when dropping heavy loads such as tanks. The development of long, heavy-duty bungee cord for the military has led to the recreational sport of bungee jumping. In this sport, the participant jumps from an elevated structure while wearing a harness attached to one end of a long bungee cord with the other end attached to the structure.

The term "bungee" or "bungie" is thought to be British slang for india-rubber. Some references to the india-rubber originally used for erasing pencil marks on paper call it "india-bungie." Another source claims the term was derived from the Anglo-Indian word "bangy" referring to the colloquial term for a yoke carried on the shoulder with two equal loads suspended by cords front and rear. In either case, the concepts of an elastic material and load-bearing cords both apply to the modern bungee cord.

The history of bungee jumping as a sport or test of courage is believed to date back 1500 years to Pentecost Island in what is now the Republic of Vanuatu in the South Pacific. According to local legend, a wife felt she was being mistreated by her husband and fled, taking refuge in a tall tree. As her husband was climbing the tree in pursuit, she secretly tied vines around her ankles. When he tried to grab her, she jumped. He jumped after her and fell to his death. The springy vines broke her fall and she lived. After that, the men of the island decided that no woman should ever trick them again, and they began to practice jumping with vines. In time, it became a test of courage, with the bravest men jumping from a height of 80 feet (24 m) to have their heads brush the ground before the vines completely stopped their fall.

Modern bungee jumping using elastic bungee cords started in the late 1970s. On April Fools' Day in 1979, the Oxford Dangerous Sports Club of Britain caught the public's attention when members bungeed off the 245-foot (75 m) Clifton Bridge in Bristol, England. Since that time, bungee jumping has become a commercialized sport with thousands of participants.

Bunsen burner

A very common instrument in most school and research laboratories, the Bunsen gas burner was first used by **Robert Bunsen** and **Gustav Kirchhoff** in their experiments with spectroscopy, a means of analyzing the elemental constituents of chemicals. Because of its ability to create a very hot and nearly invisible flame, the Bunsen burner was ideal for heating elements to incandescence; samples could be heated until they glowed, allowing Bunsen and Kirchhoff to observe the characteristic spectral pattern emitted by the element. Actually a modification of a device used by **Michael Faraday**, the Bunsen burner is essentially a metal tube into which a gas line is fed. At the base of the tube is a series of openings that allow air into the system. Because air is added prior to burning, the flame produces very little light and smoke. The height and heat of the flame are very easy to control, making it a useful tool for student scientists. Torches using a design similar to Bunsen's are often used for **soldering**.

Bunsen, Robert (1811-1899)

German chemist

Robert Bunsen is best known for his invention of the improved gas flame device which bears his name: the **Bunsen burner**. However, Robert Wilhelm Eberhard Bunsen's contributions to science extend far beyond this one invention; he developed a number of other common laboratory instruments, as well as a new device and process for analyzing the elemental constituents of chemicals called spectroscopy. This new science, developed with **Gustav Kirchhoff**, is probably his most important legacy to science.

Bunsen received his doctorate in 1830 from the University of Göttingen, in his hometown. His high marks earned him a grant from the Hanoverian government and with those funds he traveled to Berlin, Bonn, Paris, and Vienna during the next three years. At each stop he met with the cities' great thinkers and toured their centers of industry. Upon his return to Göttingen he was made an instructor at the university; his stay there was brief, however, and he taught at several other German universities until 1852, when he settled into a professorship at the University of Heidelberg.

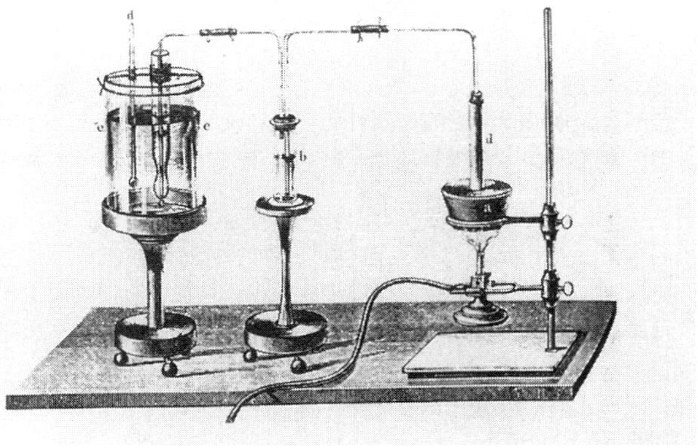

Robert Bunsen and Gustav Kirchoff invented the bunsen burner in 1854 to aid in their spectroscopy experiments.

During the years before Heidelberg, Bunsen's research efforts were concentrated in the field of organic chemistry, particularly in the study of arsenic and its compounds. Though his studies yielded some advances, such as the discovery of an antidote for cyanide poisoning, he abandoned this research when an explosion of cacodyl cyanide in 1843 cost him his right eye.

A determined scientist, Bunsen applied his talents toward inorganic chemistry and the behavior of inorganic gases. He observed heated gases in geysers and **furnace**s and published his only book as a compilation of his gas research findings. At this time he began the work that would ultimately lead him to the discovery of spectroscopy. Bunsen had become interested in the chemical properties of alkali metals, such as barium and sodium. In order to isolate these elements, he invented new types of galvanic and carbon-zinc batteries (many of which are still called Bunsen batteries). To properly analyze the elements, he constructed a very sensitive ice **calorimeter**, measuring the volume of melted ice, rather than the mass.

In 1851, while visiting the University at Breslau, Bunsen developed a working friendship with Gustav Kirchhoff, then an instructor at Breslau. Kirchhoff shared his interest in the properties of chemicals, and in 1854 Bunsen persuaded Kirchhoff to transfer to the University of Heidelberg. Together, they began research on the spectral emission of elements. In their experiments, an element was superheated or burned, so that the color of its flame could be observed. Unfortunately, even the best gas burner of the time imparted a glow from its flame that crept into the element's spectrum, skewing the scientists' findings. Bunsen was again forced to invent equipment whose precision would match that of his own research. Working with a design used by **Michael Faraday**, Bunsen made improvements that resulted in the modern laboratory burner, a device which produces a very hot, nearly invisible flame. Using this tool, Bunsen and Kirchhoff went on to pioneer the science of spectroscopy.

Over the course of his career, Bunsen also invented a grease-spot photometer (used for measuring light), a process for mass-producing magnesium, a laboratory filter pump for washing precipitate samples, and a steam calorimeter. Bunsen died in 1899.

Buoy

The use of buoys—in the form of casks or iron-banded wood staves—dates back to the time of the Norman invasion of England in 1066. From that time, development of the English buoy was slow. It was not until the reign of Henry VIII (1491-1547) that the importance of maritime trade to England and, in turn, the need for a means to direct **ships** through navigable waters was realized. By royal charter in 1514, Henry founded Trinity House, an organization that provided essential assistance in shipping for 450 years. By the mid-1500s, use of buoys had spread and most of the principal ports and harbors used them to mark channel entrances. Beginning in the 1600s, the first different colored and shaped buoys were used by the Germans along the Baltic coast. Each color and shape of these wooden buoys had a different meaning to the local sailors.

In 1845 George William Lennox, whose company supplied the old wooden buoys to Trinity House, designed the first iron buoy. The prototype was barrel-shaped with riveted iron construction. The iron buoys were also painted to provide directives to sailing crews. The shape and color of buoys remained of vital importance. In 1889 the Trinity House convened a conference in an effort to standardize the meanings and uses of buoy types, thus facilitating international use. It was agreed at this conference, for example, that black, conical buoys mark the starboard side of a channel, while red (or checkered) and can (or truncated cone) buoys mark the port side. There are also buoys with country-specific uses.

Before the use of lights, buoys could be detected by ship crews by the sound of a bell or shrill whistle. Early buoys were equipped with bells at the top. The bells chimed as the entire buoy moved with the motion of the water. In the late 1800s, based on the creations of American J. M. Courtenay, the bells were replaced by whistles. His automatic whistle used the motion of the waves to compress air into a tube fitted to the bottom of the buoy. This motion caused a **piston** in the buoy to move in a cylinder, expelling air through a whistle attached to the top of the buoy. Both of these provided for a fair means of buoy detection in times of storms, intense fog, or at night. Eventually, anchored and lighted vessels called *lightships* were developed to guide ships through dangerous waters.

Many of these crafts have since been replaced by the more cost-effective Lanbys (Large Automatic Navigational Buoys). Lanbys have a mast and use a light (the beam of which can extend up to sixteen miles); Lanbys also use fog signals, **radar** beacons, and **radio** beacons. Other more up-to-date and complex buoys use cylinders of compressed gas to power their lights, many have reflectors to facilitate detection by radar, and still others are made of newly developed **plastics**. To standardize buoy meanings worldwide, the International Association of Lighthouse Authorities devised the IALA system B, which was put in place in the United States by 1989. In addition to important navigational purposes, buoys are also used for such purposes as markers for cables or mining grounds and as reference points, ensuring that lightships don't drift from their station.

See also Polymer and polymerization

Burbank, Luther • See Plant breeding

Buret

A buret (also spelled burette) is a long glass tube open at both ends, that is used to measure out precise volumes of liquids or gases. Most burets are about 1 mm in diameter and 75 cm long. The bottom of a buret is tapered so that its diameter is only about 0.1 mm in diameter. Burets are most commonly designed to hold volumes of 1 mL or less.

Fluid is dispensed form a buret through a **glass** stopcock at the lower end of the glass tube. The stopcock consists of an inner piece of ground glass that fits tightly into the glass tube and that can be rotated in a tightly fitting casing. The stopcock allows a fluid to be released in very small, precise amounts. Commercially available burets can usually be read with an accuracy of +/-0.01 mL.

Probably the most familiar application of a buret is in the process known as titration. In this process, accurately measured amounts of two solutions are allowed to react with each other in order to determine the concentration of one. Burets have far more uses, however. A single buret can be used, for example, to release a known volume of a solution of known concentration in order to determine the mass of an unknown solid. Oxidation-reduction reactions can also be studied quantitatively using burets.

Gas-dispensing burets consist of arrangements in which some gas is contained by and forced out of a graduated cylindrical tube by means of some liquid, such as mercury. In their appearance, gas burets look something like an upside-down version of their liquid counterparts.

Burets became necessary in chemical research only with the development of relatively precise analytical techniques in the eighteenth century. Credit for their invention is usually given to the French chemist Joseph Louis Gay-Lussac, who first developed them for the purpose of assaying silver.

Burglar alarm

To protect their business, homes, automobiles, and the possessions therein, owners enlist the help of burglar alarms. Functioning as a type of invisible guardian, devices to scare away thieves or alert authorities have been around since antiquity. The first modern burglar alarm was one of thousands of innovations developed at the outset of the Industrial Revolution in eighteenth-century England. The English inventor Tildesley is credited with inventing the burglar alarm. His model mechanically linked a set of bells to a door **lock**. When an intruder attempted to open the lock with a skeleton key or other device, the chimes would sound. The commotion caused by the ringing bells would attract attention and cause the intruder to flee. Similar alarms were developed in the American colonies around this time.

The next major innovation in the burglar alarm was the addition of electricity. In the 1850s, Boston inventor Augustus Pope linked his alarm bells to the point of contact (usually a door or window) by wires carrying electricity. In addition, Pope used magnets to form contact points. When an intruder opened a door or window the magnetic contacts would close, forming an electrical circuit. The closed circuit sent electricity to the bells, which sounded the alarm.

Pope sold his patented invention to Edwin Holmes in 1857. Holmes eventually took his business to New York City in search of a larger market for his product. Many potential customers misunderstood electricity and therefore were skeptical of the burglar alarm. The electrification of city lights in 1880 relieved doubts and gave birth to numerous new alarm companies.

The invention of the telegraph and telephone gave birth to a new function for burglar alarms. In addition to sounding bells to chase away intruders, alarms could be used to call for help. Consequently, the American Telephone and Telegraph Company (AT&T) bought Holmes's burglar alarm business in 1905. Burglar alarms and fire alarms essentially merged at this time, becoming emergency call systems that were linked directly to police and fire stations.

World War II brought further innovations, such as alarms that would sound and call the police automatically. This around-the-clock protection was made available to homeowners and businesses after the war. As burglar alarm technology became less expensive and more versatile in the 1970s and 80s, it found wider application. For example, by the mid-1990s, the alarms became virtually a standard feature of **automobiles**. Burglar alarms have also continued to adopt military technology. Although the electric contact systems of the past are still used, most advanced burglar alarms now rely heavily on **motion detectors**, **surveillance equipment**, and **electronic tracking devices**.

Burroughs, William Seward (1855-1898)

American inventor

Born in Auburn, New York, Burroughs began tinkering in his father's machine shop early in his childhood—a hobby that he continued throughout most of his life. While he was clerking in a bank at age 15, Burroughs conceived the idea of a device that would perform arithmetical calculation. Although the Comptometer developed by Eugene Felt (1862-1930) was successfully utilized by most businesses of that time, Burroughs wished to design a **calculator** capable of printing figures as they were entered, permitting greater permanency and accuracy.

In 1881 he moved to St. Louis, Missouri, and began developing such a device in his spare time. Three years later he had a working model along with his first patent. Once in production, the design of his machine proved impractical. Aided by three partners, Burroughs sold $100,000 of stock in the American Arithmetic Company and continued with his experimentation. His second design was also unsuccessful because it was unable to stand up to heavy use.

Finally in 1891, Burroughs produced a well designed, fully functional adding machine. Each figure entered, as well as the final results of the computation, was printed on paper. Burroughs enlisted the Boyer Machine Company to aid in production, and the adding machine proved a great success, particularly in the banking and insurance industries.

Burroughs died in 1898 before he could see the impact his machine had on improving the efficiency and accuracy of the business world. In 1905 the American Arithmetic Company was renamed the Burroughs Adding Machine Company (later shortened to Burroughs Corporation).

See also Calculating machine

Bus

The first bus, like the first **taxicab**, originated in Paris, when **Blaise Pascal** inaugurated a short-lived free bus service in 1662. Its popularity declined once a fare was charged, however, bringing public transportation to a halt. In 1827, Colonel Stanislas Baudry, another Frenchman, owned a bathhouse in the suburbs of Paris and needed to ferry his customers to and from the city. He realized if he adapted the long-distance **stagecoach** for shorter runs, he could improve his business. Baudry's coaches carried fifteen passengers and a conductor. In that same year Sir Goldsworthy Gurney (1793-1875), an Engish inventor, developed an 18-passenger carriage powered by a **steam engine** that could travel at 15 miles (24 km) per hour. Within four years he had other carriages running a 9-mile (14 km) suburban route around London. In 1829 George Shilibeer, a coach maker, established urban bus service in London with his 22-seat buses that were drawn by three horses. The conductors wore naval midshipmen's uniforms to add a touch of class. Shilibeer called his new transportation an *omnibus*, a Latin word that means "for everyone." The idea spread, and models of Shilibeer's buses soon appeared in other major cities. Walter Hancock, another Englishman, designed steam carriages that could travel as fast as 20 miles (32 km) per hour; they logged 4,000 miles (6,436 km) in suburban London. One of them, the *Enterprise*, began passenger routes in 1833. Modern buses arose from the growth of the **automobile** and interstate highway system. Until 1920, buses were little more than long cars set on a truck chassis. By 1931, the Greyhound Bus Company in the United States had built a 40,000-mile (65,000 km) bus network that linked cities throughout America, Mexico, and Canada and offered reasonably-priced, comfortable transportation. Today's buses are large and powerful, driven by **diesel engine**s. Most are air-conditioned, with comfortable seats, air suspension systems, and plenty of room for cargo.

Designed from the beginning as buses rather than trucks in disguise, the new buses of the 1990s are aerodynamically crafted and have wide windows with shaded glass. Using air pressure to open and shut doors, many city buses can ''kneel'' for handicapped access or lower their steps for wheelchair or walker access. Others are articulated or separated into sections. In the late 1990s, a prototype city transit bus was produced by Northrop Grumman that used laminated fiberglass and weighed tons less than its metal cousin. Combining this light but strong frame with a hybrid propulsion system in which a compressed natural gas engine drives an electric generator that in turn, powers two electric motors, this new 72-passenger bus reduces assembly time, eliminates corrosion problems, slashes fuel costs by 20 percent, and extends its useful life to 25 years, all without compromising safety.

Bush, Vannevar (1890-1974)

American engineer

Vannevar Bush was the inventor of one of the first computers, the differential analyzer. He was also a key figure in directing United States scientific research during World War II.

Bush was born in Everett, Massachusetts, where his father was a Universalist minister. After receiving B.S. and M.S. degrees in 1913 from Tufts College (now University), in 1916 he earned a doctorate in engineering at MIT and Harvard University. After performing acoustic research for the Navy during World War I, Bush joined the MIT faculty, becoming dean of the school of engineering in 1932.

Bush patented his first engineering invention, a surveying device, while still in college. In the 1920s and 1930s he began to develop a series of differential analyzers, for solving differential equations with as many as eighteen independent variables. They were **analog computer**s, which used continuous values, such as distance and movement, rather than the binary values of today's digital computers. In this work, Bush was one of the first to construct a machine based on the computer concepts developed in the nineteenth century by **Charles Babbage**. First intended for use in calculating electrical current movement in power grids, the analyzers eventually became general equation solvers. Later versions could store data. The largest of them used thousands of **vacuum tube**s and relays and hundreds of miles of wiring.

In the late 1930s, Bush also designed a **digital computer**. Bush's other inventions include a rapid selector for retrieval of specific information on specially-coded microfilm and a typewriter that can produce even margins on both the left and right (justification). Bush's science policy activities also began in the 1930s. From 1938 to 1955 Bush was president of the Carnegie Institution in Washington, one of the first privately-financed scientific research organizations in the United States. During World War II he became director of the Office of Sci-

A 1914 model of the open-top buses that were a fixture on New York City's Fifth Avenue at the turn of the century.

entific Research and Development, which created science policy and carried it out. This included development of **radar** and the **atomic bomb**. After World War II, Bush played an important role in the Federal government's financial support of basic research, including establishment of the National Science Foundation.

Vannevar Bush

Bushnell, David (1742-1824)
American inventor

David Bushnell grew up on a family farm in a rural area near Saybrook, Connecticut. After his father's death, Bushnell sold the farm and decided to pursue an education at Yale, starting in 1771. While at Yale he discovered that, contrary to popular belief, **gunpowder** would explode underwater and that the density of water actually increases the effectiveness of an explosion. After graduating in 1775 Bushnell returned to Saybrook, where he began building a one-man **submarine** designed to deliver an explosive charge to the hull of a **warship**.

Bushnell's submarine became known as the *Turtle* because the contour of its hull vaguely resembled a turtle floating head up and tail down in the water. Thought to have been constructed mainly of wood, the *Turtle* had room for one operator and held enough air to remain submerged for up to thirty minutes. Horizontal and vertical propulsion was provided by two hand-operated screw propellers projecting from the top and front of the vessel. Steering was controlled by a rudder at the rear, and a system of foot-operated pumps regulated the amount of water in the ballast tank for ascents and descents. The *Turtle's* navigation instruments included a depth gauge and a **compass** lit by phosphorus. After maneuvering into position, the operator would attach a gunpowder charge below the waterline of a ship by driving a large screw into the hull with a crank inside the sub. As the submarine moved away from the ship, a time-delay detonator would be activated, allowing the operator to leave the area safely before the gunpowder charge exploded.

In 1776 Ezra Lee, a colonial soldier, made the first attempt to use the *Turtle* in actual warfare. Lee successfully maneuvered the submarine through New York Harbor to a position below the *H.M.S Eagle*, a British warship. He was unable, however, to drive the mine attachment screw into the copper-clad hull of the *Eagle* and ultimately released the charge, which floated away and exploded harmlessly. Attempts to sink other ships were similarly unsuccessful.

Bushnell later joined the Continental army as an engineer and briefly served as the commander of the Corps of Engineers at West Point. After retiring from military service, Bushnell travelled to France and then returned to America to teach at a school in Georgia. Although his submarine ultimately failed to carry out its intended purpose, the *Turtle* is nevertheless recognized as the first submarine to be used in warfare.

Butterfat test

The late 1800s and early 1900s saw many advances in the dairying industry. One such development was the butterfat test invented by American Stephen M. Babcock (1843-1931) in 1890 and marketed in 1891. Before Babcock invented the test, dairy farmers had to wait for the cream to rise in a container of milk to estimate its percentage of butterfat. This method was slow and inaccurate. The Babcock test employed a combination of acid, heat and centrifugal force to quickly precipitate a measurable amount of butterfat. Since many farmers were not scientifically qualified to conduct the tests themselves, they usually had it done by an expert for a small fee. Dairy farmers ultimately profited from the test because they could learn which cows gave high-fat milk and what farming methods and environmental conditions helped induce high-fat milk production. Additionally, butter manufacturers could make advance payments to farmers who knew the percentage of butterfat that their cows would produce. The Babcock test, which removed the element of chance in the production of butterfat, eventually led to the organization of dairying associations in Wisconsin, New York and other dairying regions. The presence of such associations and testing methods led to standardization in dairy production. Mechanization and improved transportation, especially the **railroads**, moved butter and **cheese** making from the farm to the factory.

Butterick, Ellen • See Dressmaking

BUTTONS AND OTHER FASTENERS

Although buttons were used in ancient times, they generally served as decoration rather than as garment fasteners. The word *button* comes from the French *bouton*, meaning something that sticks or butts out, implying a decorative purpose. Prior to the thirteenth century, most people wore loose, flowing clothing that was secured where necessary with clasps, buckles, girdles, sashes, strings, and similar items. In thirteenth-century Europe, however, garments became more fitted; secure fasteners became necessary, and the buttonhole was invented.

By the 1300s buttons had become important as both fasteners and decorations, adorning garments from wrist to elbow and from throat to waist. A profusion of buttons made from precious metals and gems became a symbol of wealth and status. As with many other products, the popularity of buttons increased once they became affordable.

The mass production of buttons, first centered in the city of Birmingham, England, became firmly established throughout that nation in the eighteenth century, supplanting hand production of buttons. The Dane Bertel Sanders invented the *two-shell* manufacturing method in England in 1897: two metal shells locked together, securing any appropriate covering material between them.

Button manufacturing came to the United States early in the 1700s. Casper Wistar (1696-1752), a German immigrant, was making *brass buttons* in Philadelphia, Pennsylvania, by 1720. Other early button-making centers in America were located in Waterbury, Connecticut, and Attleboro, Massachusetts. The War of 1812 provided the impetus needed to make button manufacturing one of America's most mechanized industries of the time.

In addition to metals and precious gems, early buttons were made of bone, wood, **leather**, and sometimes shell. With the advent of the **plastics**, manufacturers began to mold buttons of celluloid, and later of other types of plastic. Today, although buttons made of wood, abalone shell, leather, and metals can be found, plastic buttons are by far the most common and offer the greatest variety of styles and colors. In fact, it is often easier to find a plastic button that simulates a leather-covered button than to find a leather-covered button itself. As an interesting adjunct to buttons, clothing stores sell decorative metal button covers that slip over existing shirt buttons to add drama and variety to the fasteners in minutes.

Hook-and-eye fasteners had been known in Europe during the Renaissance; their manufacture was a recognized trade by the mid-sixteenth century. The first American hook-and-eye patent, however, was taken out in 1831 by Scottish inventor James Stewart, and more patents for these fasteners were granted in 1841.

Pin-back buttons bearing pictures or slogans are ubiquitous in the United States today. They made their first appearance in the presidential election of 1896, promoting Democrat William Jennings Bryan (1860-1925) and Republican William McKinley (1843-1901). Now they carry everything from social commentary to jokes to pop art.

See also Pin; Zipper

BYPASS SURGERY • See Open-heart surgery

C

Cable television

Unlike broadcast **television**, cable television's origins were unportentous: known as CATV (Community Antenna Television), cable was first used to deliver a clear signal to rural communities. At the time, a CATV system generally consisted of a single large **antenna** mounted in a high, clear area to receive signals from distant broadcasters. Cables were fed to the houses in the community, usually delivering two or three channels. In the mid-1960s, new technology allowed for up to twelve channels to be carried through a single cable. In order to fill these new channels, cable operators began to import television signals from more distant sources, allowing viewers to watch stations from large cities and neighboring states. With access to a wider variety of stations, the demand for cable increased.

Apart from dramatically improved reception, cable television held the capacity for educational, cultural, and community service, since it was feasible for cable systems to be bidirectional: viewers could use their cable television to answer questions or participate in polls, among other things, all from their armchairs. Still, though cable penetration increased steadily, it did so very slowly.

In the early 1970s several small companies in California and on the East Coast began offering pay-per-view broadcasting: first-run films and major sporting events delivered by cable to a viewer's home for a monthly fee. This caused demand to skyrocket, and by 1975 the first nationwide pay-per-view cable station—Home Box Office (HBO)—was in service.

What made cable transmission practical is its use of coaxial cable, which consists of an insulated tube of conducting material that contains a central conductor at its core. The thick, layered cable allows transmission of a wide band of frequencies and rejects interference from automobiles, electrical appliances, and the like. As coaxial technology improved, the number of stations available to cable operators rose from twelve to more than fifty; with the help of digital video compression, that number can be increased to almost 150. The antennas once used to deliver a signal to a cable system are long since gone, replaced by microwave dishes often fed by **communication satellites**. Once a signal is delivered to a cable company in this manner, it is distributed over cable lines to customers. Broadcasts are often scrambled to prevent nonsubscribers from splicing into a cable line without paying for the service.

In the mid-1990s, coaxial cable began being replaced by fiber optic cable, or a combination of both called hybrid fiber-coaxial (HFC). Fiber optic cable, which consists of thin strands of glass, and the combination are superior because they can carry more channels and improve the quality of the signal. In 1998, about 50% of all subscribers get their cable via fiber optic technology, and continues to expand throughout the country. With digital compression, in which the video signal is compressed digitally, many more viewing options are available to consumers. This also reflects the proliferation of cable television network which now can find space on cable systems.

Another area of innovation in the cable television industry is the combination of television with computers. The twenty-first century should bring widespread use of these infant technologies. Set-top computer boxes will sit on most new digital televisions within five to ten years. These boxes have computer chips to manage a variety of information and entertainment possibilities, including the number of increasing number of available networks. The boxes also transmit better picture and sound. Cable modems are superior to computer modems using telephone lines because they have a high capacity and high speed connection already available. Cable modems will have many of the same functions as telephone modems, including Internet access and telecommuting. Interactive television will come to fruition, allowing viewers to do their banking, grocery shopping, and perhaps even voting via their cable systems. However, cable remains threatened by direct-broadcast satellite (DBS) services, which can provide many more channels than cable currently can, wireless cable,

Engraving of workers in a caisson during the construction of the Brooklyn Bridge.

a lower-cost version of DBS, and the telephone company's interest in developing their own cable.

CAE (computer aided engineering) • See Computer, industrial uses of

Caisson

In the construction of **bridge** piers and deep-water docks, it is necessary to provide the workers a water-free environment to excavate the foundations. In shallow water and on dry land, coffer **dam**s and open-air caissons are sufficient. For deep water, an enclosed caisson is necessary.

The caisson is a water tight compartment that is open at the bottom. (The earliest caissons were shaped like cylinders, open at both ends.) Workers descend to the caisson through a shaft, or man-lock. As they excavate the bottom, the material is transported to the surface through the excavation shaft, or lock. As the excavating continues, the massive caisson sinks deeper into the soft bed until the bedrock, or any other desired depth, is reached. The downward edges of the caisson walls are wedge-shaped to aid in its downward penetration of the bottom.

Pressurized caissons were used by **Isambard K. Brunel** in the 1838 construction of the London docks. Compressed air was used to keep water and soil out of the work space. Without it, pressure at depths of just 40 ft (12 m) were great enough to push the water and soil into the bottom opening of the caisson. But it also put the same pressure on the workers; pressures of about 52 lb/in^2 (360 kilopascals), or about 3.7 times atmospheric pressure, is about all the human body can stand. This corresponds to a caisson depth of 120 ft (37 m).

In 1839, a French researcher observed that miners were emerging from mine shafts with medical complaints that included incoherence, nosebleeds, muscle cramps, and itching. The symptoms sometimes were not experienced until long after the miners returned home, and several even died. These same symptoms were also experienced by caisson workers. Hence, the condition became known as caisson disease, and later by deep sea divers as "the bends." The symptoms are the result of too rapid decompression as a person ascends to the surface from a great depth or pressure, due to the expansion of bubbles of air trapped in joints, muscles, or blood.

One of the most renowned cases of caisson disease is that of Washington Roebling (1837-1926). In 1869, Roebling replaced his father John Roebling as chief engineer of the Brooklyn Bridge building project after the elder Roebling's death. Shortly thereafter, Washington fell victim to caisson disease due to repeated descents to the bases of the bridge piers. Seriously weakened, he was forced to oversee the bridge's completion from his home near the Brooklyn end of the bridge. He never fully recovered. Caisson disease today is averted by slower ascents and by use of decompression chambers. Modern caissons are incorporated directly into the foundation of the structure, sometimes as openings in the massive **concrete** foundations which are allowed to remain after the completion of the project.

Calculating machine

Although the **abacus**, the first tool of calculation, has existed since ancient times, advanced calculating machines did not appear until the early 1600s. Scientists and mathematicians were determined to simplify complex astronomical and navigational calculations and realized that mechanical machines were the means to achieve this. In 1617 John Napier (1550-1617), a Scottish mathematician, originated the concept of logarithms. By using logarithms, multiplication and division could be performed by repeated addition and subtraction. Napier mechanized the calculating process by placing his logarithms on wooden cylinders, the surface of which contained numbers that could be manipulated to perform the accurate calculation. the cylinders, enventually made of bone or ivory, came to be known as **Napier's bones**. For years following its invention, other mathematicians used Napier's bones as the basis for their calculating devices.

In 1623 German mathematician **Wilhelm Schickard** etched the logarithms on cylinders turned by a dial, so that the results were viewed through small windows. Capable of six-digit calculations involving all four mathematical operations, Schickard called his invention a *calculator-clock*. Unfortunately, both the first and second models of his invention were lost, and scientists did not realize the full significance of his calculator-clock until 1935.

In 1642, at the age of nineteen, French mathematician Blaise Pascal (1623-1662) developed the *Pascaline*, a machine capable of adding and subtracting nine-digit numbers. Figures were entered moving numbered **wheel**s linked to each other by gear, similar to a car's **odometer**. Reportedly, Pascal invented the machine to help his father, a tax collector who spent many hours doing complex calculations. A second mathematician inspired by his tax collector father was Gottfried Wilhelm von Leibniz (1646-1716) of Germany. Leinbniz introduced the idea of stored information and improved Pascal's design to perform multiplication and division as well. Developed in 1673, the Leibniz calculator, also known as the stepped reckoner, was so advanced in design, variations were used by other inventors for the next three hundred years.

Another inventor from the same period was Rene Grillet, a French clockmaker who presented his calculator at local fairs and charged admission to see it. It is unclear whether Grillet's machine was capable of performing complex multiplication and division, but it is known that his machine could handle up to eighteen digits. Perhaps the most interesting aspect of Grillet's machine was its size—small enough to be carried in a pocket. A German scientist, Gaspar Schott (1608-1666), developed several methods for inscribing logarithms on mechanical devices. During the 1650s he created a calculator containing ten sets of tables to perform arithmetic, geometry, astronomy, sundial, calendar and music calculations. Although very similar in design to Schikard's calculator, Schott's device was not at all accurate and people soon abandoned its use.

In the 1670s Samuel Morland (1625-1695), an English diplomat and inventor, experimented with simple adding machines and more complex models using Napier's logarithms. Morland recognized the usefulness of a machine that could count money and designed a device specifically for this purpose. Although it was not a financial success, his attempt represented the first commercially sold calculator in Europe, as all earlier devices had been strictly for experimental purposes. In 1820, a commercially successful machine was manufactured by Charles Thomas from France. A well constructed machine, the *Arithmometer* as it was dubbed, operated with the same stepped wheel design of earlier models. The machine enjoyed great success and dominated the market until the twentieth century when made obsolete by the introduction of keyboards.

Calculator design remained basically unchanged through the next fifty years. Limitations in machinery, physical tolerances in the metals and wood used for production restricted designs of that time. In 1822, a prototype for a *difference engine* was built by **Charles Babbage**. The next year, he proposed a steam powered *analytical engine*, but like the difference engine, it was never completed. In 1874, the first alternative to Leibniz's wheel mechanism (in use since 1623) was introduced. A Swedish inventor, Willgodt T. Odhner invented the *Odhner*, a calculating machine based on a variable toothed **gear**—the pinwheel, later referred to as "barrel machines." This design proved reliable, easy to use and reduced the size of previous models. Two years earlier, Frank Steven Baldwin from the United States, independently developed the same type of pinwheel calculator, called the *Monroe*. Although both machines proved profitable to their inventors, the Monroe calculator was able to convert to a motor drive in 1925.

In 1880, **Herman Hollerith**, of the United States, set about tackling the immense task of counting the country's population. He built the first electro-mechanical **punched card** tabulating machine used in the 1890 census. Soon the need for faster, more accurate recordkeeping, particularly for population studies, insurance and business use became apparent. In

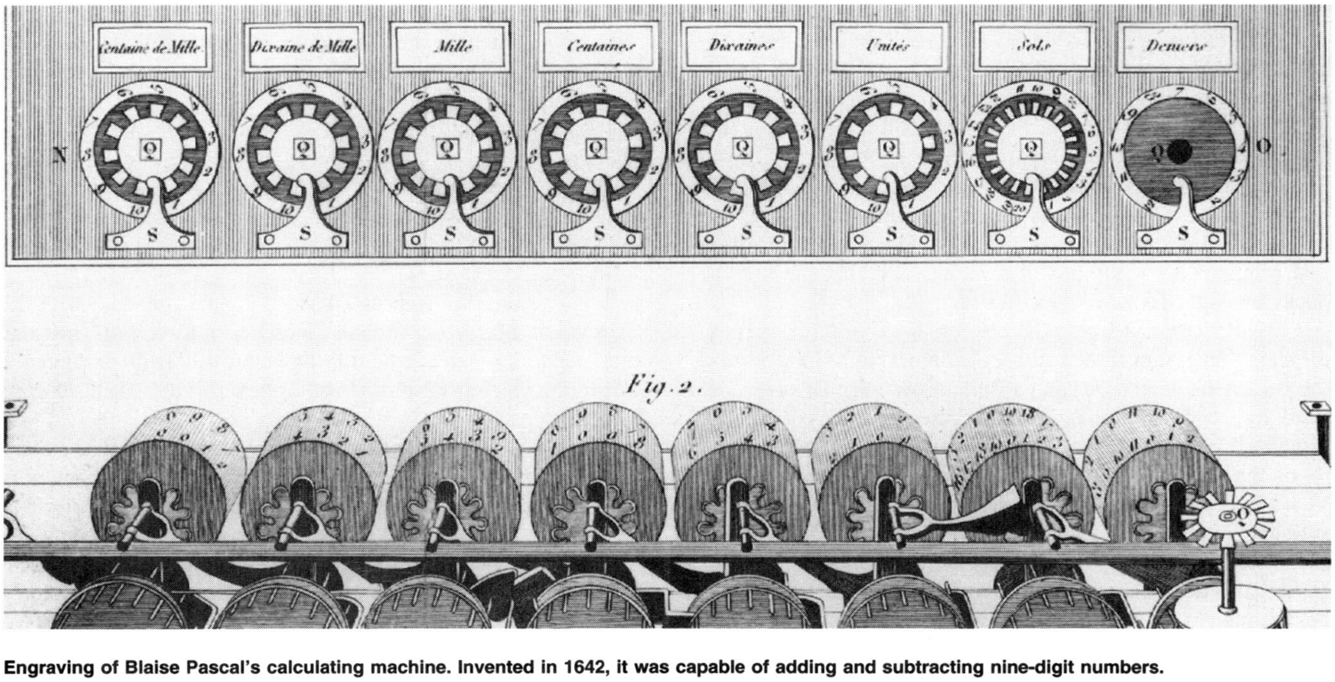

Engraving of Blaise Pascal's calculating machine. Invented in 1642, it was capable of adding and subtracting nine-digit numbers.

1884, Eugene Felt introduced the *Comptometer*—a time saving device designed to keep from ''turning men in veritable machines.'' His first model was built with a macaroni box, metal staples, rubber bands and meat skewers. While the Comptometer could do all the mathematical operations, it was not able to record the results of those calculations. **William Seward Burroughs** recognized the usefulness of such a feature, and in 1891 produced a machine capable of printing numbers and results in a grand total. The machines proved very successful, particularly in the banking and insurance industries. In 1905, Robert A. Pelham (1859-1943), an African-American employed by the Census Bureau, devised the first tabulating machine used in the Census of Manufacturers and in 1913 also developed a similar tallying machine used by the Population Division. In 1972, Texas Instruments introduced the first completely electronic **calculator**, later versions added memory storage, paper print-outs and programmable features.

See also Cash register; Microcomputer

Calculator, pocket

Within a few years of its invention in 1959, the **integrated circuit** (microchip)—tiny, complex electronic circuits on a single chip of silicon—had become reliable and inexpensive. The president of Texas Instruments, Patrick Haggerty, wanted to demonstrate the potentially pervasive uses of the integrated circuit, which was mostly confined to military and industrial functions.

In October 1965, Haggerty challenged Jack Kilby, a TI engineer and coinventor of the microchip, to design a miniature calculator that would be as powerful as desk models but small enough to fit into a coat pocket. Kilby was a man who thrived on solving difficult technical problems. He assembled a three-man team of himself and two fellow TI engineers that produced a prototype within a year. Jerry Merryman, a self-taught electrical engineer, designed the logic circuits to fit within the power and space limitations. James Van Tassel, an expert on semiconductor components, developed a small, power-efficient keyboard for the input. Kilby found a suitable rechargeable battery. Displaying the output remained a problem. **Light-emitting diode** (LED [light-emitting diode]) technology, which became the standard for calculator display, was not yet advanced enough to use. So Kilby invented a new thermal printer with a low-power printing head that pressed the paper readout against a heated digit.

Kilby, Merryman, and Van Tassel applied for a patent on their ''Miniature Electronic Calculator'' in 1967. Because the pocket calculator was an entirely new device, it took some years to get it into production. TI formed a joint venture with Canon of Tokyo and placed the Pocketronic Printing Calculator on the market in the United States in 1971. It was not nearly as portable and cheap as today's models: the pocketronic weighed 2 1/2 pounds and cost $150. But it was an immediate success.

Like the other electronic pocket calculators that soon began appearing, it surprised its inventors by appealing not just to businesspeople, engineers, and scientists, but also to average consumers, who used the devices to total grocery bills, figure square footage of rooms, prepare income-tax forms, and other common mathematical tasks. Improved models soon followed.

In 1972 Hewlett-Packard introduced the HP-35. The first hand-held scientific calculator, it featured LED display. Also in 1972, TI marketed the Datamath, which used a single chip and had a full-floating decimal point, LED display, and limited memory.

Today, pocket calculators with a wide range of functions are available, including programmable calculators which are in effect miniature computers. More than fifty million portable calculators are now sold in the United States each year, many for less than $10. In 1975 the Smithsonian Institution made the original pocket calculator part of its permanent collection. In 1976 Keuffel & Esser, manufacturer of millions of slide rules over the years, presented its last slide rule to the Smithsonian.

Calendar

A calendar is a system of measuring the passage of time. Early peoples used the lunar month—the interval between a complete sequence of phases of the moon—to calculate time. Because a lunar month contains about 29 1/2 days, a lunar calendar of 12 months results in a 354-day year. This is about 11 days less than the solar year—the length of time it takes the earth to make one complete orbit around the sun—of approximately 365 1/4 days. This discrepancy confounded calendar makers for thousands of years.

Early Mesopotamian cultures used a lunar year to calculate time. In order to align their 354-day lunar year with the solar year, the Babylonians added a month (a procedure called intercalation) whenever their calendar got out of balance. Because intercalation was done irregularly, the Babylonian calendar became extremely complicated and confusing. To correct for this deficiency, the Babylonian lunisolar cycle was standardized in the fourth century B.C. to a calendar of 19 years and 235 months.

The ancient Egyptians developed a solar calendar based on their observations of the annual appearance of the Dog Star, Sirius, and the nearly simultaneous flooding of the Nile. Sometime between 4000 and 3000 B.C. the Egyptians adopted a 365-day year, with 12 months of 30 days each plus five extra end-of-the-year feast days.

The ancient Hebrews used a lunar calendar, with 12 months of alternating 30 and 29 days, adjusted with irregular intercalations. After A.D. 300, the Hebrew calendar adopted a fixed intercalation system of adding seven extra months during each 19-year cycle.

The Chinese calendar, allegedly invented in 2637 B.C. by the legendary emperor Huangdi, contained 12 lunar months with the same seven month-19 year intercalation schedule. The sequence of years in the Chinese calendar is traced in cycles of 60, with each year in the cycle designated by two terms, one of which is the name of one of 12 animals that recur in order.

In Mesomerica, the Maya Indians, who were excellent astronomers, developed a very accurate calendar. It consisted of 18 months of 20 days each, for a 360-day year, with five extra days (considered to be unlucky) added. The Mayans also observed a parallel 260-day cycle, which was probably used in religious ritual.

The ancient Romans, like the Greeks before them, at first based their calendar on the lunar month. This 10-month year was supposedly established by Romulus, the legendary founder of Rome. The months of January and February were added, according to tradition, by the ruler Numa Pompilius around 700 B.C. This 355-day year was supposed to be periodically adjusted by insertion of an extra month, but the intercalation was erratic. By 46 B.C. it was evident to Julius Caesar (100-44 B.C.) that the calendar needed revision. He called on the Greek astronomer Sosigenes, whose suggestions Caesar adopted.

The new Roman calendar year was divided into 12 months of 30 and 31 days, except that February had 29 days; every four years an extra day was added to the year to account for the 365 1/4-day solar year. (When the Roman emperor Augustus (63 B.C. -14 A.D.) renamed one of the months after himself, he took a day away from February and added it to August so that his month would be as long as July, the month named for Julius Caesar.) This revision, called the Julian calendar, remained in use for 1,600 years.

Although the Julian calendar was very accurate, it was 11 minutes and 14 seconds too long per year. By 1582 the discrepancy had caused the spring equinox to occur 10 days too early. Pope Gregory XIII (1502-1585) reformed the calendar again so it would conform to the now-known actual solar year. He ordained that 10 full days be cut from the current year, so that October 4, 1582, was followed by October 15, 1582. He also devised a system whereby 3.12 days are dropped every 4 centuries.

The new Gregorian calendar was extremely accurate and gradually was adopted worldwide, with Protestant countries lagging behind Roman Catholic nations. England waited until 1752 to switch to the Gregorian calendar. Because the year 1700 had been a leap year according to the Julian calendar, England now had to subtract 11 days from its calendar in order to adopt the new system. The disappearance of thse 11 days led to riots in that country. China adopted the Gregorian calendar in 1912, the Soviet Union in 1918. Many Islamic countries continue to use the 354-day Muslim calendar, which has 12 lunar months of 29 and 30 days each, with an extra day added in 11 years of a 30-year cycle. The Bahai calendar, introduced in 1844 by Seyyèd Ali Muhammad, operates on a 19-week year within a 361-year cycle. The Hebrew, Hindu, and Bahai calendars remain in religious use.

With the growth of computerization in the late twentieth century, increasing amounts of calendrically based data began to be stored in computer software fixed-file formats. The inaccurate programming of the software binary dating systems led to the potential misreporting of the digit "00" in the year "2000" as "1900." If left uncorrected, at the turn of the millennium this discrepancy would result in a massive miscalculation of records worldwide. In addition, the Gregorian rule that centennial years divisible by 400 are leap years raised the possibility that two-digit binary systems not recognizing the year 2000 would be thrown one day off reckoning after February 28, 2000. As of 1998, businesses and governments across the world were working together to bring about a year 2000 software conversion.

CALNE, ROY (1930-)
English surgeon

A pioneer in the field of organ transplantation, Roy Calne was born in London, England, in 1930. He studied at Lancing College and then received his M.B. and B.S. from Guy's Hospital Medical School in London. He practiced at Guy's Hospital for one year after qualifying there in 1953, served in the Royal Army Medical Corps from 1954 to 1956, and was then an orthopedic surgeon in Oxford for two years. In 1958 he became a surgeon at the Royal Free Hospital in London.

While doing postgraduate surgical research in London, Calne conducted experimental kidney transplants on dogs, using one of the earliest immunosuppression drugs—drugs that suppress the body's natural immune reaction against transplanted tissue, which causes rejection of the graft. The results were encouraging, and Calne decided to go to the United States to learn more about immunosuppressants and work with other transplant researchers.From 1960 to 1961 Calne was associated with the Harvard Medical School and Peter Bent Brigham Hospital in Boston, Massachusetts, and with the team there headed by Dr. Joseph Murray that had carried out the first successful human kidney transplant in 1954. Calne and Murray collaborated with future Nobel Laureate George Hitchings (1905-) of Burroughs-Wellcome drug company to develop and test improved immunosuppressants. The result was Imuran, which Calne and Murray found allowed extended survival of dog kidney transplants.

Back in England, Calne was appointed lecturer in surgery first at St. Mary's Hospital in London and then, from 1962 to 1965, at Westminster Hospital. He joined Cambridge University in 1965 as professor of surgery, and later became department head. During these years, he improved the techniques of kidney transplantation and developed ways of transplanting the liver, a complicated organ. He published a standard text on kidney transplantation in 1963. Continuing his interest in immunosuppression, Calne and his junior associate, Dr. David White, carried out animal and then human trials on a new drug supplied to them by the Swiss scientist Jean-Francois Borel of the Sandoz pharmaceutical firm. The trials proved the effectiveness of what came to be the standard drug for suppressing rejection of organ transplants: cyclosporin. Calne was knighted for his achievements in 1981.

See also Kountz, Samuel L.; Surgical transplant

CALORIMETER

A calorimeter is a scientific instrument used to measure the heat generated or absorbed in a chemical reaction or physical process. We owe the invention of this device to an observation made just before the turn of the nineteenth century by the preeminent scientist **Benjamin Thompson** (**Count Rumford**). While supervising the construction of cannons, Rumford noticed that as the fire chamber was bored out, the metal cannons would heat up. He observed that the more work the drill exerted in the boring process, the greater the temperature increase. To measure the amount of heat generated by this process, Count Rumford placed the warm cannon into a tub of water and measured the increase in the water's temperature. In doing so, he simultaneously invented the science of calorimetry and the first primitive calorimeter. In simplest terms, a modern calorimeter is a water-filled insulated chamber. When a hot object is set within the water, the system's temperature increases. By measuring the increase in the calorimeter's temperature, a scientist can calculate such factors as the specific heat (the amount of heat lost per gram) of a substance. Another application of calorimetry is the determination of the calorific value of certain fuels—that is, the amount of energy obtained when fuel is burned. Engineers burn the fuel completely within a calorimeter system and then measure the temperature increase within the device. The amount of heat generated by this burning is indicative of the fuel's calorific value.

Later scientists continued Count Rumford's work with calorimeters. During the 1860s, the French chemist **Pierre Berthelot** constructed a calorimeter specifically designed to measure the heat exchanged during chemical reactions, providing a wealth of insight into the nature of exothermic and endothermic reactions. At about this same time, the German physiologist Karl von Voit (1831-1908) constructed a calorimeter large enough to contain a human being. By creating a closed, measured system, Voit was able to observe the consumption of oxygen, the release of carbon dioxide, and the overall heat given off by a human body. Voit's research became the foundation for the modern understanding of human metabolism.

Calorimeters continue to be useful in modern science and in 1995 two major advances were made in calorimetery. A Swedish company developed an ultra-sensitive microcalorimeter, also known as a Thermal Activity, which is capable measuring changes in chemical reactions down to the millionths of a watt. The device is being used by the paints and coating system to detect corrosion of metallic surfaces. The same year, the American Association for the Advancement of Science reported that a newly designed single-crystal surface calorimeter has been used to measure the energy difference between two solid surface structures. This technology could lead to exciting new discoveries in surface chemistry.

See also Artillery; Boring machine

CALOTYPE PROCESS • See Photography

CAMCORDER • See Video recording

CAMERA OBSCURA

Aristotle (384-322 B.C.) noted the ability of small holes to project images. As he viewed a partial eclipse of the sun, he noticed that an image of the eclipse could be seen on the ground before him through one of the holes of a strainer he held. Eventually, someone noticed that a small hole in the wall of a dark room produced the same effect—an inverted image of the view outside the room appeared on the wall opposite the hole. Thus, the camera obscura (''dark chamber''), which is the direct ancestor of the modern photographic camera, was born.

Probably the earliest account of the camera obscura came from an Arabian scholar named Ibn Al-Haitham, known as Alhazen (c. 900). Alhazen not only documented the concept in detail, but also demonstrated the significance of the relationship between the size of aperture and the sharpness of image. For some reason, his writings were overlooked for many centuries. Roger Bacon (1220-1292), however, was familiar with Alhazen's observations as shown by writings he made in 1267.

Although **Leonardo da Vinci**, the great Renaissance artist, also gave clear descriptions of the camera obscura in his manuscripts, it was his student, Cesare Cesariano who provided a published account of da Vinci's observations in 1521 (Da Vinci's own manuscripts, which date from about 1491, were not discovered until 1797). By 1550, mathematician Girolamo Cardano (1501-1576) had suggested replacing the hole with a biconvex **lens** to improve the image.

The Italian scientist Giovanni Battista Della Porta is credited with bringing the knowledge of the camera obscura to the general public. Della Porta was the first to suggest that it be used in drawing and painting. His account, published in 1558, gave instruction on how images could be traced and thus reproduced on drawing-boards, and how images could be righted using a system of **mirror**s and lenses.

To this point, any camera obscura had to remain stationary, as only a darkened room had been used to produce images. It was not until 1572 that the idea of the camera obscura as a portable apparatus was developed. This idea came from Friedrich Risner, a German mathematician. He suggested that a light wooden hut be constructed in place of the darkened room so the apparatus could be moved. This ''miniaturization'' brought the camera obscura one step closer to becoming the photographic camera.

Following Aristotle, another German mathematician, Christopher Scheiner, used the camera obscura for solar observation in 1617. By combining it with a refracting **telescope**, he was able to project magnified images of sunspots onto a portable screen covered by a cloth.

In 1685, the camera obscura began to take on the appearance of the modern day camera at the hand of Johann Zahn. Zahn, also a German, produced illustrations of many types of small box cameras that could be taken virtually anywhere. These prototypes showed a strong resemblance to box and reflex cameras of the nineteenth century. Now the camera obscura was ready for **photography**, but its development would stall until the mid-nineteenth century, when a light-sensitive film would be placed at the back of a box camera to capture images permanently. Camera obscura were also used as telescopes in the nineteenth century. These types of telescopes were often find that era's English seaside resorts, to view faraway terrestrial images, and continue to be used on both sides of the Atlantic today.

See also Camera lucida; Photography; Telescope.

Camouflage

Methods of concealment and deception in hunting and in warfare has been practiced since antiquity. Camouflage has become a complex military tool that involves concealment and deception. We readily think of concealment as the merging of troops, weapons, and equipment. However, camouflage also evokes deception in which decoys are created to mislead an enemy as to intention, to give a false idea of strength, or to draw the enemy's attention from a real attack.

Pre-twentieth-century warfare involved camouflage. There is the famous Greek story of the Trojan horse. In 212 B.C., Greeks constructed a false beachfront with straw and supported by a light structure. When Roman forces assaulted the "beach," they floundered and were cut down by the defending Greeks. The Scottish King Robert I (1274-1329) had his troops dig hidden pits with which to trap English cavalry at the battle of Bannockburn in 1314. Dummy forts and artillery were frequently employed to deceive enemies. As guns became more accurate, armies attempted to merge into the landscape by abandoning their distinctive uniforms. The first troops to wear khaki uniforms were the Indian Guides, an English army unit who dyed their white uniforms a mud color (*khaki* means "dust" in Urdu) in 1846.

Camouflage was used extensively during World War I. In 1914, a French portrait painter, Guirand de Scevola, painted canvas sheets which were used to cover and hide 75mm guns. He eventually became the head of the first camouflage section in the history of war. An English artist, Solomon J. Solomon, also headed up a similar section for the British; he created artificial trees for observers to hide in while checking out the German lines. Solomon also painted the first British tanks with a disruptive pattern of browns and greens to blend in with the landscape. Nets were a favorite way to hide equipment and artillery during the war. After the United States entered the war, military leaders set up a camouflage unit with Homer Saint-Gaudens (the son of the celebrated American sculptor Augustus Saint-Gaudens, 1848-1907) in charge.

Two camouflage measures were used extensively at sea during World War I: the decoy ship and dazzle painting. The British used decoy vessels, called Q-ships, to lure German U-boats close where they could be dispatched with gunfire. The Q-ships looked like ordinary merchant vessels, but they had hidden guns that could be quickly exposed and fired at the unsuspecting submarine floating on the surface nearby. Dazzle

Illustration depicting how to use a camera obscura.

painting was first suggested by an English zoologist, John Kerr (1869-1957). Kerr and a painter, P. Tudor Hart, argued that disruptive paint schemes of strongly contrasting colors could break up the outline of a ship's structure and confuse enemy observers. Norman Wilkinson, a naval lieutenant who shared these ideas, oversaw the dazzle painting on British ships.

Airplanes were also first camouflaged during World War I. An American physicist, M. Luckiesh, pointed out that an airplane needed two paint schemes: one to help hide it when seen from above, and one to hide it when seen from the ground. The Germans spent much more time on airplane camouflage: they painted the upper surfaces a dull green, purple, and earth tones. Under surfaces were either painted white, pale or turquoise blue, or pale blue-green.

Since World War I, armies have continued developing camouflage skills. Leslie Watson, a British horticulturalist, devised the use of small holes bored into concrete airplane runways. These holes were filled in with soil and tufts of grass, causing the runway to resemble grassland rather than a strategic landing place for aircraft. Before the D-Day landings of World War II, the Allies used fake radio messages as well as dummy equipment to confuse the Germans. Today's camouflage specialists are scientists rather than painters because modern weapons can operate in various wavelengths: ultraviolet, infrared, and microwave. Camouflage nets are still in use, particularly those made from polyvinyl chloride (PVC) but most are rendered ineffective against radar. Aircraft and ships now use radar-absorbing paint.

Camouflage clothing became more sophisticated after World War II, for both hunting and military use. Camouflage used in the Vietnam War used tiger stripes and woodland patterns to blend in to the vegetation there, while camouflage developed for the Persian Gulf War used desert patterns. A treebark pattern camouflage for hunters was developed in the 1980s, with gray and black shapes and vertical lines which mimic a wooded environment. In the late 1990s, camouflage designers began working on patterns that vanish into the grids of night-vision scopes. High-tech camouflage clothing that actually changes colors is under development, but not perfected.

See also Stealth aircraft

Can opener

Curiously, the first patent for a can opener wasn't issued until forty-eight years after Peter Durand secured his 1810 patent for the tin can. It wasn't as though the can opener was an unneeded invention. A tin of roast veal taken on William Parry's voyage to the Arctic in 1824 instructed: "Cut round on the top near to the outer edge with a chisel and hammer." The hammer-and-chisel or-screwdriver method of liberating a can's contents was finally made obsolete in 1858 when Ezra J. Warner of Waterbury, Connecticut, patented his can opener. The device used a long blade or spike to pierce the can and a shorter blade to grasp the container's rim. During the Civil War, Union troops received Warner's can opener along with their rations of canned food. Many can openers were patented during the 1860s through the turn of the century, some designed for multiple uses. A notable 1878 patent design by J. Cox was a "combined can opener, knife and scissor sharpener, tack drawer, putty knife, tack hammer; complete with finger rests." The modern type of can opener, with a cutting wheel instead of a spike or blade turned by a crank, was introduced by Lyman in 1878. However, can openers didn't become popular until wall-mounted models appeared in 1930. The electric can opener which debuted in 1957, now appears in under the counter mounts and countertop models as well as standard wall-mounted units.Aluminum is sometimes used for the tops of tin cans. These tops, which usually include a plastic cover,have a tab,which may be pulled open without the need for a can opener.

Canal and canal lock

Canals—humanmade waterways—divert water from natural sources to facilitate transportation, irrigate or drain land, and supply water. **Irrigation** canals were a feature of most ancient civilizations. The Nahrwan Canal, 185 mi. (300 km) long, was built between the Tigris and Euphrates rivers circa 2400 B.C., and Egypt's pharoahs linked the Mediterranean and Red seas with a canal that the Romans later restored and used for shipping. China built the initial stretch of its Grand Canal in 610 A.D., a waterway 600 mi. (1,000 km) in length at its completion. Canal systems for transportation were not widespread in Europe until the 1600s and 1700s. The first network came into use in the Netherlands, which had adapted drainage canals to handle boats by the thirteenth and fourteenth centuries. France's *Canal du Midi* linked the Atlantic and Mediterranean by 1681. Cargo boats used in the canal systems were designed with flat bottoms to accommodate the shallow depths.

Making allowances for changes in land levels is a primary concern in canal construction. Lake Erie, for example, is roughly 325 ft. (99.13 m) higher than Lake Ontario. An early means of addressing such variations was to construct slipways between various levels, up or down which boats could be pulled. China relied on these as early as the fourth century A.D. A more sophisticated means of compensating for level variations is the lock—a chamber linking different water levels. Boats enter and leave the lock through a pair of gates. When both gates are closed, the water level in the chamber is raised or lowered, taking the boats to the desired level. The forerunner of the contemporary lock was the *navigation weir* or *flash lock*—an opening in a masonry dam that was closed with a wooden gate by a sluice or other moving device. These were in use in China in 50 B.C. The first pound lock probably grew from a discovery by China's Chiao Wei-Yo, who constructed two flash-locks within 250 ft. (76.25 m) and found that the stretch of river between them functioned as an equalizer. The first such lock in the West was constructed in Belgium in 1396. **Leonardo da Vinci** (1452-1519) is believed to have invented a pound lock featuring mitre gates that swung vertically while he was an engineer for the Duke of Milan.

Many of Western Europe's major rivers are today linked by an extensive canal system used for shipping, and the opening of the *Erie Canal* in 1825 signalled the onset of a national water transportation network in America. Its success spurred construction of canals in cities like Philadelphia and Washington, D.C., although the proliferation of railroads dramatically curtailed America's reliance on canals for shipping and transportation. Canal construction reached its technological peak when canals began accommodating oceangoing vessels. These require depths of 30 to 40 ft. (9.15-12.2 m). To justify the cost and effort involved, deep-water canals either must offer an alternative to lengthy shipping routes, as do the Suez (1869) and Panama (1914) canals, or permit access to inland ports. The Panama Canal, currently administered by the United States, will be turned over to Panama in December 1999, although a separate treaty guarantees international access in perpetuity. Completion of the *St. Lawrence Seaway*, for example, allowed ocean access to the Great Lakes ports. Modern canal systems are often intricately engineered affairs. Canada's *Trent-Severn Waterway* linking Lake Huron's Georgian Bay with Lake Ontario is 242 mi. (387 km) long; its main course alone features 33 mi. (53 km) of manmade channels, marine railways, and two hydraulic lift locks, one of which is the world's highest—65 ft. (20 m).

CANDLE

Peace, comfort, reverence, gaiety—all of these emotions and more are symbolized by the flame of a candle. But for thousands of years, candles also served as humankind's main source of lighting. Although experts differ on just who invented the candle and when, it is believed that primitive forms of candles were used by ancient civilizations in Egypt, Crete, and Italy. The candle may have begun to evolve when hunters, cooking meat over an open fire, noticed that the burning fat created a bright light. Early candles were made of this animal fat, called tallow, and of beeswax, which comes from honeycombs.Initially, reeds were utilized as a means of keeping the candle burning; eventually,the burning properties of candles were improved with the invention of fiber wicks. The modern braided wick, which burns itself up instead of having to be trimmed, was not introduced until 1824.

By the Middle Ages, candlemaking had become a thriving trade, supporting two guilds—one for makers of tallow candles and one for makers of beeswax candles. Wax candles smelled better and gave off less smoke than tallow candles, but only wealthy people could afford them. Servants who worked for kings and nobles often received candles as part of their wages. This candle allowance was a status symbol, and the number allotted was jealously eyed by workers of lower rank.

Ship passing through the Panama Canal in 1914.

Although less prestigious, tallow candles held the advantage of being edible. Once, when British lighthouse owners investigated the cause of high candle consumption, they found that the lighthouse keepers were eating them.

In the American colonies, women discovered that boiling bayberries would produce a sweet-smelling, clean-burning wax, though the process of extracting the wax was tedious. Toward the end of the 1700s, as the fishing industry grew, people began making candles from a by-product of whale oil called spermaceti, which solidifies as a wax. Then when paraffin wax was crystallized from petroleum in the 1800s, it too became a raw material for candles. Around the same time, French chemists Michel-Eugène Chevreul (1786-1889) and Joseph-Louis Gay-Lussac (1788-1850) patented yet another, improved candle material—stearic acid, which Chevreul had isolated during his studies of animal fat. He had found that the unpleasant odor of burning tallow was due to glycerin, not to the fatty acid itself. Often, candles were blended from more than one raw material to maximize stiffness, brightness, and cleanliness. Today, candles are made mostly of paraffin.

Just as candle materials evolved, so did processes for making them. Early tapered candles were made simply by dipping the wick into molten fat, letting it cool, and then repeating the process until the candle reached the desired thickness. Beeswax candles were made by pouring melted wax onto the wick repeatedly, after which the candle maker would roll the wax into the desired shape. Candles were molded by hand as early as the seventeenth century in France. In 1790 the second patent ever granted in the newly created United States was for a candlemaking process invented by Joseph Sampson. Then in 1834, Joseph Morgan invented a machine for molding candles, which led to today's automated,continuous candle production.

Despite the introduction of electric light in the late 1800s, candles have remained popular, and their production has grown steadily. Candles are now inexpensive and plentiful,and people use them for emergency lighting,for atmosphere and to add pleasant scents to a room.

See also Lamp

CANNON • See Artillery

CANNON-BORING MACHINE • See Boring machine

CANTILEVER

A cantilever is any structure supported at one point which carries a load at the other end. Examples are a **dome** crane, an overhanging roof, and an unsupported balcony.

The cantilever's support need not be at one end, but can be in the middle of its horizontal span. For example, a seesaw is a cantilever. Horizontal cantilevers often have a counterweight on the their other end, such as with a crane. An **aircraft** wing is an example of a horizontal cantilever, where the support is on the *end* of the wing, supported by the wind, and the plane is the weight supported by the beam (that is, the wing).

Cantilevers need not be horizontal. A flagpole stuck in the ground is a cantilever, where the **force** of the wind against both flag and pole is the load causing stress. (Notice that the force against a cantilever's beam need not be applied only to its end, but can be along its entire length.) The ground acts as the point of support. Other examples of vertical cantilevers are **skyscrapers**, chimneys, and trees, the largest vertical cantilevers found in nature.

Some of the most important applications of the cantilever are in **bridge**s. A steel truss is constructed horizontally, supported by a foundational support set in the river or gorge to be crossed. The truss then ''juts out'' over and past the support, meeting a truss from the other side. Sometimes a central connecting section, or ''spanning section,'' is needed to complete the bridge between the trusses.

The earliest cantilever bridges were made of timber, and primitive cantilever bridges that are still standing can be found across the world. The Forth Railway Bridge in Scotland was the first cantilever bridge made entirely of steel, completed in 1892. It consisted of two connected suspended spans, each 1,710 ft (521 m) long. Another early steel cantilever bridge was over the East River in New York City, completed in 1901, with two main spans of 1,182 ft (360 m) connected by a central anchor span of 630 ft (192 m) on Blackwell Island.

The Quebec Bridge across the St. Lawrence River provided tragic lessons in the engineering of steel cantilever bridges. Begun in 1904, it was a railroad bridge designed to have a 1,800 ft (550 m) total span, which would have made it the world's longest. There were reports of a piece of metal on a tower buckling, and not long after an entire 9,000 ton (80 million newtons) section of the bridge fell into the river with 86 men on it, killing 75. Analysis showed that the collapse was a result of errors in calculating the weights and stresses and in assembling the parts.

After recover and redesign, the spanning section was again assembled and attached to the ends of the cantilever arms, only to break when a lifting stirrup broke away, killing 13 men. The new span was finally erected a year later, and the bridge opened in 1918, and has been in use ever since—the longest cantilever span in the world. (Today's steel strength would allow, in theory, a span of 2,500 ft (760 m)). The Golden Gate bridge in San Francisco is part cantilever bridge and part suspension bridge, with a cantilever span of 1,400 ft (430 m).

CAPACITOR

A capacitor is an electrical device that can store electrical energy. In this way it is like a **battery**; however, unlike a battery, a capacitor has the property of capacitance, which is the ability to store electric charge and let it out all at once, and it can resist changes of voltage across its terminals. Capicatance is measures in units of farads, though typical capacitors have values of picofarads, one trillionth of a farad.

A capacitor consists of two conductors that are separated by a dielectric (insulator). The larger the surface area of the conductors and the smaller the separation between them, the larger the capacitance will be. There are two main types of capacitors—variable (which are tunable) and fixed (which have a single, fixed capacitance). Fixed capacitors, which comprise the lion's share of the market, come in two types: electrolytic and non-electrolytic.

The origin of the capacitor goes back more than 200 years. In October 1745, Ewald von Kleist, Dean of the Cathedral of Camin (in Germany), made the discovery of capacitance. Three months later, in January 1746, Professor Peter von Muschenbrock at the University of Leyden, made the same discovery and built the first working model of a capacitor. The device consisted of a **glass** jar partially filled with water, with an electrode dipping into the water, and is called a Leyden jar. A student named Cuneus, experimenting in Leyden, was using a Hawkesbee machine to generate static electricity which was conducted by a chain into water in the jar—a simple type of Leyden jar. When Cuneus touched the chain he received such an intense electric shock that it nearly killed him. The news of this ability to store a large electrical current spread rapidly, and demonstrations in which the public willingly offered themselves as guinea pigs to be shocked took place.

It was one such demonstration that caught the attention of **Benjamin Franklin**. The electricity he conducted out of the sky with his famous kite experiment was stored in a Leyden jar. Franklin ultimately discovered that the electrical charge was stored within the glass itself, not the water, and determined that the electrostatic action was due to the material that insulated the conductor. He placed two sheets of lead on his jar (one on the inside and one on the outside), electrified them, then removed them and found that the glass insulator held the charge. Franklin's invention became known as the electrical "condenser," a name which is often still applied to capacitors.

The Leyden jar became obsolete in 1775 when **Alessandro Volta** (1745-1827) invented a device he called the electrophorous. It consisted of one metal plate covered with ebonite and a second plate that had an insulated handle. When the ebonite plate was rubbed, it built up a negative charge. When the second plate was placed above the ebonite, a positive charge was attracted to the lower surface, and a negative charge was repelled to the upper surface. This negative charge was drawn off with a grounding wire, leaving the positive charge behind. Repeating the process caused a strong positive charge to be built up. This "charge-accumulating" device became the basis for today's electrical capacitors.

The use of more efficient dielectric (insulating) material improved the performance of capacitors. In the 1850s mica (a naturally occurring dielectric) was made into sheets for use as insulators, although they didn't come into commercial use until World War I. They could withstand shock better than glass insulators, and their efficiency permitted the size of the capacitor to be reduced. Rolled **paper** and ceramic insulators came into use in 1876 and 1900 respectively; today plastic film, electrolytes, and even air are used as insulators. Ceramic capacitors are similar to the paper type but they have several advantages; they have larger capacitances, are smaller in size, and operate at voltages up to 1,000 volts. Mica capacitors can be used with voltages as high as 35,000 volts. Plastic film capacitors, using polystyrene as an insulator, are more expensive than the paper capacitor, but they are smaller and have a very high insulation resistance.

The ability to store and control electric charges and currents is essential to modern electronics. The dielectric inhibits the flow of direct current (DC), but the constant charging and discharging allows alternating current (AC) to pass. This type of capacitor is used to couple amplifiers and is known as a blocking, or coupling, capacitor. Electrolytic capacitors have a very thin dielectric and a high capacitance. Because of their small size they are very well-suited for modern electronics, but they do have a disadvantage in that they have a high leakage of current. They are used as electronic filters to rectify AC (convert AC to DC). Electrolytic capacitors are also used in electronic photographic strobe units. There is a great variety of uses for capacitors. They are used in ignition systems to reduce arcing. Trimmer capacitors are used where just a small change in value is needed. Variable (tunable) capacitors are found in old-style radio tuners; two sets of interleaved metal plates, one set fixed and one movable, are used to tune in specific frequencies. The leaves are insulated by the air between them, which serves as the dielectric.

CAR • See Automobile, gasoline

CAR WASH, AUTOMATIC

The first American car-washing facilities were very expensive to operate because they were little more than conveyors that pulled the car past groups of workers who performed all the tasks associated with cleaning a car. In the late 1930s the first fully automatic car wash went into operation in the United States. It had many of the typical features we recognize today, including large rotating rollers that have flexible strands to flick away the dirt. One horizontal roller is used to clean the upper surfaces, while vertical rollers are used to clean the sides. Modern car washes incorporate a number of other features. Many of the rollers now have limit switches, operated by simple levers that control the motors. When the roller meets an object such as a **mirror**, the roller will slow down and move around it. Many car washes pass cars under blowers for drying and many are also equipped to apply liquid wax after the washing cycle is complete. With water conservation becoming more important, car washes often recycle much of their water. Many car washes feature brushless, no-cloth washing, relying on water pressure and chemicals to clean the car. Detailing, a service for fastidious customers is a thorough, minute, immaculate cleaning, vacuuming and polishing of a car, inside and out, usually by hand.

See also Soap

CARBINE • See Rifle

CARBONATED BEVERAGE • See Soda water

Carborundum • See Abrasive

Carburetor

The carburetor was a critical invention in the advancement of the gas engine and its adaptation to the motor car. The carburetor allowed the use of gasoline as a fuel, which replaced vaporous gas (the "illuminating gas" that was used for street lights) as the source of power for **internal combustion engine**s.

The modern carburetor was invented by German engineer, **Wilhelm Maybach**, an associate of **Gottlieb Daimler**. Together these two men developed one of Europe's first **automobile**s, powered by a two-cylinder engine fitted with Maybach's carburetor. Their car was first driven in 1885.

Maybach's design consisted of a spray nozzle that released a fine mist of gasoline which was then mixed with about 15 times its weight in air to produce a combustible mixture. This fuel-air mixture was drawn into the engine by the downward action of the **piston** and admitted into the combustion chamber by the opening of the intake valve at precisely the proper moment. Maybach's early carburetor was regulated by two screws. The first regulated the amount of gasoline, the second, the amount of air. This design, while functional, did not allow for easy variations in the fuel-air mixture; hence, engine speed was hard to regulate. **Carl Benz**, another German inventor, solved this problem in 1893. Benz invented the carburetor butterfly **valve** which is situated below the spray nozzle in the tube that connects the carburetor and the engine. This butterfly valve can be turned to vary the fuel-air mixture. Consequently, engine speed and power can be easily regulated.

Other refinements of the carburetor include the choke valve, which is closed to start a cold engine. The choke is essentially another butterfly valve, but this valve is situated *above* the spray nozzle mechanism. Closing the choke restricts the flow of air to the engine and results in a richer mixture of gasoline that is easier for the **spark plug** to ignite in a cold engine. Chokes today are automatically controlled by engine temperature.

More than one spray nozzle-butterfly valve combination can be incorporated into a single carburetor, resulting in a two-barrel or a four-barrel carburetor. Cars designed for racing may be outfitted with more than one of these multi-barreled carburetors. Modern carburetors are built with several innovations designed to burn the fuel-air mixture more efficiently so as to reduce polluting exhaust emissions.

Carding machine

In all types of spinning, it is important that the textile be first disentangled and cleaned of seeds or debris. This was done by hand for thousands of years. Sometime during the Middle Ages, spinners began pulling the raw fibers through thistles or teazles; the tiny hooks on the teazle heads would comb the fibers, removing any dirt and arranging the fibers so that they would be roughly parallel. The Latin name for the teazles was *cardus*, and the process became known as *carding*.

Soon the teazles were mounted side by side on two wooden slats, one for each hand. The carder would then clean the material by drawing one slat against the other. The carded material, called *sliver* or *roving*, would be removed by hand from one of the carding slats and then could be sent to the spinner. By the early 1700s, the natural teazles were replaced by metal wires, bent slightly and mounted on **leather** strips, which were in turn fixed on larger wooden slats. These wire carders were more efficient than those using teazles because their hooks were stronger and arranged more evenly.

Carding continued to be done by hand until the mid-1700s. About that time, many improvements were being made in the spinning and weaving industries; as these industries became more and more mechanized, they demanded a great deal more carded fiber than could be produced by hand. Several inventors attempted to speed up the carding process, but it was not until 1748 that David Bourne developed and patented a cylinder carding machine.

In Bourne's machine, the wire hooks were mounted onto a set of turning cylinders. As the material passed through, it was cleaned and straightened by each successive cylinder. Although the sliver still had to be manually removed from the final cylinder, Bourne's carding machine was the first successful machine of its kind.

The next man responsible for the development of a continuous carding machine was the English inventor **Richard Arkwright**. In the 1760s and 1770s, Arkwright built and patented the **water frame** spinning machine, a water-powered continuous spinner. Because the water frame was so efficient, Arkwright was forced to mechanize every other sequence in the spinning process. In 1775 he patented his carding engine and completed the first textile factory, wherein all of the procedures were automated.

More improvements to carding technology were made in the 1800s. In 1834, a self-cleaning carding engine was invented eliminating the need to clean the cylinders twice a day. A few years later the bent wires were replaced with a length of steel ribbon, which was wrapped around the cylinder. This ribbon could clean the fiber material much faster than wires and thus sped up the entire process.

Cargo ship

Cargo ships are necessary for world commerce. There are two basic kinds of cargo ships: freighters and tankers. Freighters come in all shapes and sizes in order to accommodate various loads. Dry cargo vessels comprise one category and are used to transport coal, grain, iron ore, and similar products that can be loaded in bulk. The first of these hauled iron ore on the Great Lakes during the late 1800s; it resembled a long steel box, with crew quarters at the front and engine at the back. Modern freighters, by contrast, have a double hull for strength and safety, and include storage areas for fuel oil, water ballast, or fresh water. The largest of these in use today, with the bridge and engine room near the tail, can haul over 100,000 tons of cargo. Barges are smaller forms of these bulk carriers. Formerly powered by sail, they now have diesel engines or are towed by tugboats.

There are also general cargo ships used for hauling packaged items. Around the turn of the century these ships had three islands, or three structures, that stood out above the main deck: the crew quarters at the front, the bridge near the middle, and the cabins for officers and passengers at the stern. Cargo was loaded between these three structures.

One-island vessels eventually replaced these early designs because they provided room for additional and larger hatches, thus simplifying loading and unloading. During World War II the United States built many of these one-island vessels, called Liberty and Victory ships, which transported troops and supplies throughout the world. Modern cargo ships house powerful, electrically driven **crane**s and derricks. These ships can be loaded from the side and stern as well as from the hatches and they feature automatic engine and navigational controls.

The need to improve efficiency has led to at least three specialized cargo ship designs: containers, roll on/roll off, and LASH (Lighter Aboard Ship). Container vessels reduce the time that is spent in port and are filled with prepackaged aluminum boxes of various sizes. Money is saved because there is less loading and unloading time as well as less breakage and theft. Some of these ships can hold around 1,000 containers; each one equals the cargo-carrying capacity of 17 standard World War II freighters. Roll on/roll off ships also utilize containers, but these boxes have wheels on them which allow them to be hauled like truck trailers both on and off the ship. A LASH is a huge freighter which can stow small loaded barges (lighters).

The second kind of cargo ship is the tanker. In 1878 Ludwig Nobel revolutionized the way oil was carried. Previous to his time, all liquids were transported in barrels or tanks, yet Nobel launched a ship that was one huge tank itself. The early tankers were 300 feet long with a capacity of 2,300 tons of oil. Today, there are tankers 1,000 feet long that can carry 300,000 tons of oil. Their hulls are divided into tanks, which are loaded or unloaded by **pump**s. Because of recent well-known oil spills from some of these vessels, there are plans for double-hulled construction to prevent such disasters. Other tankers can carry liquefied natural gases in specially designed tanks. Space between the gas tanks and the hull is filled with an inert gas to prevent oxygen in the air and any leak of gas from combining to produce an explosive mixture.

Finally, there are multipurpose ships that combine the features of a freighter and a tanker. Some have refrigerated space for foods that spoil easily, tank space to haul liquids, and a deck for roll on/roll off containers. In 1968, the United States launched its first multipurpose vessel with the capability of carrying containers, roll on/roll off cargo, various general cargo, and refrigerated items.

Rising fuel prices during the 1970s prompted the development of a new technology that used sails shaped like aircraft wings turned on end to take some of the burden off the engines and save fuel. Slightly curved to form a wing shape, these sails were attached to a mast that could pivot and locate the best angle for the sail to catch the wind. Once the computers set the mast at the best angle to the wind, the sail created the same ''lifting'' force that an airplane's wing generates, except that the force pushed the ship along the water. However, this system did not always prove to be efficient for extremely large vessels.

By the late 1990s, the trend was toward building even larger vessels for container freight, and in 1997, the largest such ship was the 1,043-foot (316.3-m) long Dutch ship, *Regina Maersk.* Capable of handling about 6,000 steel or aluminum containers (each 20-feet [6-m] long), it requires a crew of only 15. One of the most revolutionary designs for cargo ships is the planned waterjet-propelled cargo vessel proposed to travel between the United States and Europe. Called the *FastShip Atlantic*, it is designed to cruise at 42 knots (48 mph). Since this is twice the speed of conventional freighters, it will cut transatlantic travel time in half — to three and a half days. Powered by six to eight gas turbine engines which drive five water jets (which use no moving parts and operate on the same principle as a jet ski), the new freighter will range between 560-860 ft (186.6-286.6 m) long. The key to attaining such high speed with such a huge vessel is the ship's radically new ''semi-planing'' hull. Flatter and wider than a conventional hull at the stern (front), it allows the vessel to skim over the water rather than push through it. This means it need not slow at all for waves as high as 50 feet. Once it reaches port with its cargo, transfer is accomplished rapidly by gliding them off or onto the ship using large metal pallets buoyed by a cushion of air. *FastShip* promises to transform all aspects of transatlantic delivery since it will not only shorten door-to-door delivery from the traditional 14-35 days to an astounding five days, but will make cargo ships dependable, on time freight delivery systems since they will no longer be subject to heavy seas.

Carlson, Chester Floyd (1906-1968)

American inventor

Chester F. Carlson's life was one of hard work and diligence that paid off. At age 14 Carlson supported his ailing parents by working odd jobs. He attended college at night, graduating in 1930 from California Institute of Technology with a B.S. in physics.

After working with Bell Telephone Laboratories, Carlson labored in the patents department of a New York electronics firm and began studying patent law at night. To save money, he copied longhand from borrowed textbooks. By day Carlson spent much time obtaining copies of blueprints and patent descriptions. Because of his experiences of hand copying, Carlson set out to develop a quick and inexpensive duplicating process.

Within three years, Carlson received his own ''electrophotography'' patent. He and a young German physicist and engineer, Otto Kornei, Carlson worked on his system to electrostatically produce copies. Using a mirror Carlson projected an image onto a electrically charged drum coated with selenium. Because selenium will only hold an electric charge in the dark, the charge remained exclusively on the areas where the dark sections of the image were reflected. Carlson then coated the drum with a powder made of carbon and thermoplastic resin. Rolling an electrically charged sheet of paper along the drum released the image onto the paper, where it was

Chester Floyd Carlson

fused into a permanent impression by infrared light. In his laboratory on October 22, 1938, Carlson made the first copy of an original document using the process he called xerography, from the Greek for "dry writing."

After a six-year search for a backer, in 1947 the Haloid Company purchased the rights to Carlson's photocopier. Haloid, which was renamed Xerox Corporation, hit its stride 12 years later when on September 16, 1959 it introduced its first automatic plain-paper copier, the Xerox 914.

Chester Carlson became a very wealthy man and philanthropist, who during his lifetime he donated over $150 million to worthy causes. On September 19, 1968, Carlson died of a heart attack in a New York movie theater.

See also Duplicating machine

Caro, Heinrich (1834-1910)
German chemist

Caro was born in Posen, Prussia (now Poznan, Poland) in 1834. He moved to Berlin in 1834 and eventually studied dyeing at the *Gewerbeinstitut* and chemistry at the University of Berlin. In 1855 a calico factory hired Caro as a colorist. There he was able to apply his chemical training to solve production problems, and he was soon sent to England for training in more advanced dyeing processes. He studied there for two years, returning to Germany to work with synthetic **dyes**, which were growing in popularity, for Roberts, Dale and Company.

Caro was able to build on the work of others to discover improved processes. He enhanced **William Henry Perkin**'s technique for deriving mauve dye from aniline and was made a partner in the firm. His work on induline was built on the Johann Peter Griess's (1829-1888) work with aniline dye. Caro also worked with C. A. Martius to produce Bismarck brown and Martius yellow. Caro's research helped to define the structure of triphenylmethane, the parent compound of rosaniline dyes.

In 1866 Caro moved back to Berlin, where he soon became the director of the Badische Anilin und Soda Fabrik (BASF) laboratory. Caro improved the technique of Carl Graebe (1841-1927) and Liebermann to produce alizarin. He added bromine to **Adolf von Baeyer**'s flourescein and produced flourescent red. In collaboration with Baeyer, Caro developed the dye known as methylene blue. In the mid 1870s numerous azo dyes were developed and discovered. Caro, with others, developed chrysoidine, orange and fast red. He went on to discover naphthol yellow, persulfuric acid, also called Caro's acid. Caro resigned from BASF in 1889. Heinrich Caro was responsible for more of Germany's successes in the dye industry than any other individual. He died in Dresden in 1910.

See also Textiles

Carothers, Wallace Hume (1896-1937)
American chemist

Wallace H. Carothers was born on April 27, 1896 in Berlington, Iowa. He graduated from Terkin College with a bachelor of science in 1920, and earned his Ph.D. in 1924 from the University of Illinois. For a short time Carothers taught at the University of Illinois and at Harvard University. Because he preferred research to teaching, Carothers accepted a position with Du Pont in 1928 to work on a basic research program.

Until then, Du Pont had simply bought the technologies developed by others. In order to diversify, Du Pont decided to make a commitment to fundamental research. Carothers was asked to research polymer chemistry. It was felt that by discovering the underlying properties and structures of polymers, Du Pont could develop synthetic substitutes for natural products.

There were various theories on the structures of polymers. Carothers supported the theory that polymers were large molecules made up of repeating units. He developed a research plan and in 1930 succeeded in proving that polymers are macromolecules, thus laying the foundation for modern polymer science.

With his research team, Carothers went on to develop a substitute for rubber called neoprene. Natural rubber is a polymer of isoprene. Neoprene was made with a chemical of a similar structure called chlorprene. Although it was more expensive than natural rubber, neoprene proved more resistant to sun, weather, ozone, oil, and gasoline. First marketed in 1930, it is still used in hoses, shoe soles, gaskets, drive belts, wet suits, and cable jacketing.

Carothers's research team then began to experiment with polymer fibers to replace silk. They started with polyamides but soon went on to work with **polyester**. During experimentation with the polyesters, one of the researchers drew the polymer out into strands and noticed its silky appearance. Because of their earlier research the team realized that stretching the material oriented the polymer molecules. There was one problem, however. Polyesters melt at temperatures too low for use in the textile industry. **Nylon**, a polyamide, has a higher melting point. The team repeated the experiment with nylon and had greater success forming fibers. Industrially, nylon is forced through small holes to form fibers and then stretched by passing it through a pair of rollers rotating at different speeds. This process, called cold drawing, stretches the fiber several hundred percent and increases the strength of nylon by more than 90 percent.

Nylon is used in guitar strings and fabric blends where, with proper treatment, it resembles wool, silk, or cotton. Debuting as stockings in 1938, nylon was the first commercially successful synthetic fiber. Nylon is resistant to heat, oils, grease, and water, and is used in apparel, **carpet**s, and tire cords.

As substitutes for rubber and silk, both neoprene and nylon became very important during World War II, when these natural products were in short supply. Carothers contributed more to our basic understanding of polymer science than anyone else. Sadly, Carothers did not live to see the global impact of his discoveries; he committed suicide in 1937.

See also Polymer and polymerization; Rubber, synthetic; Textiles; Tire, pneumatic

Carpet

When the word carpet is used today, it usually refers to the machine-woven wall-to-wall carpets found in homes and offices. However, the modern carpet has its origins in the hand-woven carpets of England and, before that, the intricate rugs of Persia and the Far East.

The first rugs were probably woven by nomadic shepherds. These thick and durable rugs were used as ground coverings, blankets, walls, and doors. Since these nomads had little in the way of music or literature, the patterns displayed upon their rugs became their principal art form. As trade between Europe and Asia increased, the rugs of the Far East became very popular. In most cases they were deemed too beautiful to be trod upon, and were instead hung on walls or over furniture, as decoration.

Until the mid-1800s, all rugs were woven by hand. This was a time-consuming process, for the wool that made up the rug's pattern was knotted through the backing one piece at a time. Generally, rugs with more knots were considered more valuable, and some priceless Oriental rugs contained more than two thousand knots per square inch. Such rugs were coveted in England, where they were considered a status symbol, as well as in Islamic nations, where they were valued as prayer rugs.

In 1839 Scottish inventor James Templeton developed a power **loom** for the weaving of Asian-style rugs. The British began to use this device, attempting with some success to duplicate the rugs of the Far East. Just a year later, American inventor **Erastus Bigelow** perfected the power loom for carpets. Carpets then began to replace rugs as floor covering.

Originally, power looms were only twenty-seven inches wide. However, the broad loom was developed in the early 1900s, allowing for the construction of carpets nine to eighteen feet wide; later looms could produce carpets almost thirty feet wide. Using a bank of needles, the power carpet looms accurately simulated the action of hand-weaving. Depending upon the way the yarn was stitched into the *warp* and *weft*, different types of carpet could be woven; these included Axminster, Wilton, chenille, and velvet. Rugs and carpets woven on these power looms were popular until just after World War II when they, too, were replaced—this time by *tufted carpet.*

Most of the carpeting found in homes today is tufted. It is made by stitching yarn through a pre-woven fabric backing. Using hundreds of hollow needles (sometimes more than a thousand), the tufting machine fills the carpet with *loops*—or *tufts*—that make up the pile. The yarn can then be left in a *loop pile* or can be snipped at each loop, which is called *cut pile.* By varying the height of the tufts, a wide range of patterns can be created. The carpet's backing is coated with an adhesive, usually *latex*, in order to hold the tufts in place. The yarn itself, originally spun from wool, is now usually composed of such synthetic fibers as **nylon** and **polyester**.

See also Fiber, synthetic

Carpet sweeper

Carpet cleaning before the advent of the mechanical carpet sweeper was a laborious task. Carpets were swept with hand-held whisk brooms and, once or twice a year, were taken outside, hung over a clothesline, and pounded with a carpet beater.

The first floor-sweeping machine was patented by Jane (or James) Hume of England in 1811. It consisted of a box containing a brush turned by a **pulley**—a design that didn't work very well. Lucius Bigelow invented an improved version in 1858; across the ocean, Hiram H. Herrick of Massachusetts patented a mechanical carpet sweeper that same year. All of these devices tended to stir up more dust than they collected and didn't work equally well on floors and carpets.

A carpet sweeper patented in 1876 by Melville R. Bissell (1843-1889) of Grand Rapids, Michigan, solved these problems. Bissell owned a china shop and had developed an allergy to the dust from the straw that china was packed in. To eliminate the dust, Bissell designed a sweeper with cylindrical brushes that swept dust into a box and a knob that adjusted the brushes to the surface being swept.

Bissell and his wife Anna formed the Bissell Carpet Sweeper Company to market their new product. Women working in their homes made various parts of the sweepers; Anna collected the parts, and she and Melville then assembled them into whole machines. The Bissells' vigorous promotion of their sweepers, coupled with the machine's effectiveness and the new public awareness of the existence of germs, secured the widespread popularity of their carpet sweeper. People

Alexis Carrel

spoke of ''Bisselling'' their carpets; the machine was often called a ''bissell'' rather than a carpet sweeper.

After Melville Bissell died in 1888, Anna Bissell took over as president of the firm, later becoming chairman of the board and remaining so until her death in 1934. Only the advent of the portable electric home **vacuum cleaner** blunted the popularity of the Bissells' invention.

Carrel, Alexis (1873-1944)

French surgeon and physiologist

Alexis Carrel was born in Lyons, France, and brought up by his devout Roman Catholic mother after his father, a **textile** manufacturer, died when Alexis was five. Carrel expressed an early interest in science by dissecting birds and conducting chemistry experiments. He received university degrees in both letters (1890) and science (1891) and then studied medicine at the University of Lyons, earning his medical degree in 1900.

As a medical student working at hospitals in Lyons, Carrel displayed a deft talent for dissection and surgery. His interest in blood vessel surgery was aroused in 1894 when the French president, Marie Francois Carnot (1837-1894), was shot. The assassin's bullet severed a major artery, and Carnot bled to death because no techniques existed at the time to repair severed blood vessels. Carrel set out to develop such methods. He learned through embroidery lessons how to use very fine needles and silk thread. He used strict asepsis to avoid infections. To prevent clotting—the major cause of failure in blood vessel suturing—Carrel coated needles, other instruments, and thread with paraffin. To expose blood only to the smooth inner walls of the vessels—thereby further reducing the risk of clotting—Carrel invented the technique of rolling back the vessel ends like cuffs and then stitching the turned-back ends together. Carrel's suturing technique was successfully implemented in 1902. The ability to stitch blood vessels together opened the door to far more sophisticated surgery than had previously been possible, including organ transplantation.

Unable to advance professionally at Lyons, Carrel furthered his study of advanced medicine in Paris, France, in 1903 and then moved to Canada, intending to become a cattle rancher. Instead, he became an assistant in physiology at the University of Chicago from 1904 to 1906 and, from 1906 to 1938, was a research member of the Rockefeller Institute for Medical Research in New York City. At both Chicago and the Rockefeller Institute, Carrel expanded his work with blood vessel surgery into the field of organ transplantation, transferring kidneys and other organs in animals. His successful grafting of veins to arteries laid the basis for today's common coronary artery bypass surgery. For his work in suturing and transplantation, Carrel received the 1912 Nobel Prize for medicine or physiology.

Carrel also experimented with tissue cultivation. Expanding on the earlier work of Ross Harrison, Carrel kept a piece of tissue from a chick embryo's heart alive and reproducing in his lab for thirty-four years; the tissue culture outlived Carrel! During service for the French army in World War I, Carrel developed a very effective means of irrigating deep wounds with a disinfectant solution.

After the war, Carrel collaborated with the famous aviator Charles Lindbergh (1902-1974) to develop a device that would keep entire organs alive outside the body. The Carrel-Lindbergh perfusion pump of the 1930s circulated blood or a nutrient fluid through the organ via the organ's blood vessels. This perfusion pump was also called an artificial heart and was an important early step in the development of methods to maintain circulation when major organs of the body are undergoing surgical intervention.

In 1935 Carrel published a best-selling book, *Man, the Unknown*, that promoted an ideal world ruled by an intellectual elite. He returned to Paris in 1939, where he remained during the German occupation, establishing an Institute for the Study of Human Problems. Because Carrel accepted support from the Vichy government and dealt with the Germans in connection with his institute, his reputation was maligned by charges of collaboration at the time of this death in 1944 from heart failure in Paris.

See also Open-heart surgery

Carrier, Willis (1876-1950)

American engineer and inventor

Willis Carrier is widely regarded as the father of modern **refrigeration** and **air conditioning**. Carrier received more than 80 patents over 50 years in the field of air conditioning. He was the founder and chief executive officer of the giant Carrier Corporation, one of the largest manufacturing companies in the United States, and his name has become almost synonymous with air conditioning.

Willis Haviland Carrier, an only child, was born on November 26, 1876, to Duane and Elizabeth Carrier, in Angola, New York. He received his secondary education in the local school system and also attended the Buffalo General High School. Carrier won a competitive scholarship to study mechanical engineering at Cornell University in Ithaca, New York.

Carrier graduated from Cornell in 1901, and accepted a position as an engineer with the Buffalo Forge Company in Buffalo, New York. From the beginning of his career, he was determined to put mechanical engineering on a more rational basis then what was the practice at that time. Because engineers did not really understand the reasons their machinery functioned the way they did, they were forced to build large "factors of safety" into their designs, which led to inefficiencies. To investigate the principals behind the functioning of the company's products, Carrier founded the world's first industrial laboratory at the Buffalo Forge Company. With the better understanding that he gave them, the Buffalo Forge designers were able to design better, safer, more efficient products.

Although the term air conditioning would not be coined for five years, Carrier's first design job as a young engineer was to develop an air conditioning system for the Sackett-Wilhelm lithographing company in Brooklyn, New York. The problem faced by the Sackett-Wilhelm Company was not heat, but humidity. The paper in their printing plant would shrink or expand depending on the amount of water it absorbed from the air. Carrier met this problem by designing a system for the company that cooled the air in its plant to a constant temperature and reduced the air's humidity.

Work on the Sackett-Wilhelm problem with humidity sparked Carrier's interest in the physical properties of air. His first major contribution to the field of air conditioning and refrigeration came in 1904 when he discovered that air could be dehumidified by spraying water through it. Carrier came to this conceptual breakthrough while waiting for a train on a fog-shrouded railroad platform. He realized that fog is nothing more than water vapor that has condensed out of air. From his research, Carrier knew that the amount of water vapor air can contain is dependent on the temperature. As air's temperature drops, the amount of water vapor it can contain also drops. So, if humid air is cooled (by spraying cold water through it, for instance) the amount of water vapor it can contain drops. Eventually, despite the extra water being sprayed through it, the amount of water vapor in the air becomes greater than the maximum amount of water vapor the air can contain. When this happens, water vapor becomes water and fog or morning dew forms. The temperature where this occurs is called the air's "dew point."

Within a few years the system was in use in industries as widespread as textiles and brewing. The fact that cool air was produced by Carrier's invention was only a side effect of the dehumidifying process. The original purchasers of his dehumidifier weren't interested in comfort, only in increasing the productivity of their facilities by getting rid of excess moisture in the air.

Five years after joining the Buffalo Forge Company, Carrier had risen to become the company's chief engineer. In 1906 he convinced management that the promise of air conditioning was so great that they allowed him to found the Carrier Air Conditioning Corporation of America as a wholly-owned subsidiary of the Buffalo Forge Company. As Carrier had predicted, air conditioning soon became more important for the comfort it provided than for the industrial processes that it made possible.

Air conditioning began to appear in theaters and concert halls and in the skyscrapers that were beginning to appear in New York City and Chicago. In 1911 Carrier was invited to give a lecture to the American Society of Mechanical Engineers, a high honor for someone so young. Carrier's lecture, published with the unpromising title of "Rational Psychometric Formulae," became the basis upon which the air conditioning industry was founded. Carrier's psychometric formulas describe the different combinations of temperature and humidity that are possible for air at a given pressure, and are still in use today. Using the formulas, an engineer can design an efficient air conditioner that is capable of converting air at one temperature and humidity into air at a second, presumably lower, temperature and humidity.

Carrier and six of his subordinates left the Buffalo Forge Company in 1915 to found the Carrier Engineering Company. Started with $35,000, mainly the savings of its founders, the company prospered. The net worth of the company had risen to $4,000,000 in 1930 at the start of the Great Depression. Despite the Depression, air conditioning was an invention that's time had come. Carrier and his staff designed the first centrifugal compressor for use in air conditioners. The centrifugal compressor, besides being much smaller than its predecessors, was also much safer and efficient. Using it in combination with lightweight finned coils, he was able to design the portable air conditioners that became ubiquitous in the 1940s and 1950s. He had made air conditioning available for everyone.

Carrier was also a pioneer in the refrigeration industry. He was elected president of the American Society of Refrigeration Engineers in 1927 in recognition of his work in that field. He also served as the president of the American Society of Heating and Ventilating Engineers in 1931. He received an honorary Doctor of Engineering degree from Lehigh University and an honorary Doctor of Science from Alfred University. In 1949, the prestigious Newcomen Society of Great Britain honored him as the father of air conditioning. Carrier retired as chairman of the board of the Carrier Corporation in 1943. He devoted his retirement to spreading the benefits of air conditioning to the rest of the world. He died on October 10, 1950.

Edmund Cartwright

Cartwright, Edmund (1743-1823)

English inventor

The English **textile** industry was revolutionized by **Richard Arkwright**'s **water frame** spinning machine. Using the water frame, wool and cotton yarn could be spun faster and more efficiently than with a **spinning wheel** or **spinning jenny**—all without the need for manual labor. However, the new spinning machine created a new problem, for hand **looms** could not match the pace of the water frame, causing a surplus of yarn to accrue. This problem was partially solved by the invention of Edmund Cartwright's mechanical loom.

Cartwright was born in Nottingham, England. In 1784 Cartwright took a holiday to Matlock, not far from Arkwright's water frame cotton mills. It is said that, while there, he engaged Arkwright in a conversation as to the difficulty of creating a powered weaving machine. Inspired, Cartwright began the construction of just such a machine the moment he returned home.

Cartwright's first loom was clumsy and ineffective—primarily due to the fact that he had built it without first actually seeing a hand loom. Although the device was barely functional, Cartwright took out a patent and began to make improvements. Employing several local manufacturers as advisors he constructed two more prototypes, and by 1790 had completed a mechanical loom able to weave wide cloth, such as calico.

Cartwright established a factory for his looms in 1786. While his earlier models did not attract much attention, the power loom completed in 179O worried the local weavers, who feared (correctly) the machine would replace them. Thus, in 1791 the factory burned down under mysterious circumstances, and Cartwright found himself in debt. He attempted to offset his losses by inventing an ingenious wool-combing machine; however, this, too, was opposed by local workers, and Cartwright collected little from its patent. In 1793 he was forced by his financial situation to sell his factory and patents.

He soon moved to London, where he applied his inventive abilities toward agriculture. He invented a **reaper**, and won prizes from the Board of Agriculture for his essays on husbandry and manure use. He also acquired the patent for a rope making machine and for an alcohol fueled engine. When the patent for his mechanical loom expired in 1804, Cartwright petitioned the House of Commons for restitution, and was awarded a sum of 10,000 pounds. He continued to develop new agricultural devices until his death in 1823.

Carver, George Washington (c. 1864-1943)

American botanist and inventor

George Washington Carver, an undersized, softspoken genius, achieved fame as an agricultural chemist, botanist, educator, and inventor, though he refused to patent or capitalize on most of his innovations. His study of peanuts and **peanut products**, sweet potatoes, and soybeans led to an economic bonanza for the southern farmer by providing an alternative to cotton and tobacco as staple crops. A native of Diamond Grove, Missouri, he was born a slave. His birthdate was never recorded. When Carver was an infant he was kidnapped along with his brother Jim and their mother Mary by slave rustlers. His mother was sold, but George suffered from whooping cough and was left to die. When Carver's original master sought to find his stolen slaves, the price for the sickly child immediately increased. The thieves required Carver's master to trade a three-hundred-dollar race horse for the boy's ransom.

Following the Emancipation Proclamation in 1865, George was adopted by his former owners, Moses and Sue Carver, who gave him their surname, and tried to obtain an education for him at African-American schools. From the age of six, he studied on his own, focusing on woods and wildflowers, which he enjoyed cultivating and using as subjects for his oil and water color paintings. At age ten, he moved to Neosho, Missouri, to attend a one-room school. He formed strong family ties with a African-American couple, Mariah and Andy Watkins, who became his foster parents. By age thirteen, Carver migrated to Minneapolis, Kansas, obtained a high school diploma, and worked as an independent field laborer. In an effort to further his education, he applied for and received a scholarship to Highland University, but the offer was rescinded when the university president realized that Carver was African-American.

To avoid further constraints on his efforts, Carver had to move farther north to seek an education. As the first African-American to attend Simpson College in Indianola, Iowa, he worked as a cook and earned the $12 annual tuition. At his teachers' suggestion, he advanced to Iowa State College of Agriculture and Mechanic Arts (Iowa State University). There he concentrated his efforts in the natural sciences. He obtained his bachelor's degree in agriculture in 1894 and joined the school's faculty as an assistant professor of botany and overseer of the greenhouse. In this capacity, he became the first African-American ever to teach at the university. Carver received his master's degree in 1896.

Carver achieved so notable a name that Booker T. Washington offered him an annual salary of $1,500 to head the new agriculture program at Tuskegee Institute in Tuskegee, Alabama. In the school's primitive, understaffed laboratory, he made the bulk of his discoveries, such as the hybridization of short-and tall-stalk cotton. He also devoted himself to educating farmers on crop rotation and diversification to avoid overworking their fields. Carver's dedication to the school and to agriculture was so great that he refused lucrative offers to join the laboratories of **Henry Ford** and **Thomas Alva Edison**.

To bolster farm income, Carver focused on humble plants. He created over 300 products from the peanut, including **dye**, **shoe** polish, **soap**, **plastic**, wood stain, flour, and milk and cheese substitutes. In 1921, he testified before the Congressional Ways and Means Committee concerning the need to protect American peanut growers from foreign competition. He derived 118 by-products from the sweet potato, such as syrup, starch, wood stain, and flour. Experiments with other plants produced **cosmetics**, **breakfast cereals**, fertilizer, oil, food additives, dye, **paint**, and medicine prototypes. He promoted uses for okra fiber and native clays; also, three fungi he discovered were named for him.

Along with laboratory work, Carver involved himself in community outreach. He took his "school on wheels" into the countryside to educate illiterate rural families about scientific farming methods and **food preservation**. One of his most beneficial lessons was support for the tomato as versatile food for home and sale. He became a champion of **recycling** and waste control, lectured on horticulture at leading universities, and wrote numerous pamphlets explaining to farmers how improved techniques could raise their standard of living. As unofficial spokesperson for African-Americans, he influenced newspaper publishers, liberal congressmen, and other notable people. His example became the hallmark of African-American achievement.

For his contributions, he received honorary doctoral degrees, the Theodore Roosevelt medal, commendation from the Edison Foundation and London's Royal Society of Arts, and the NAACP's Spingarn Medal. In 1936, Tuskegee honored him in his fortieth year of teaching as the school's most productive and prestigious staff member. Notables, including the Prince of Wales and President Theodore Roosevelt, visited him at the Institute.

Unaffected by fame, Carver remained unmarried, lived frugally, and refused to commercialize his success. He never collected his teaching salary and, in fact, donated $30,000 to the George Washington Carver Foundation. As a gesture of love, he willed his estate to Tuskegee for the preservation and continuance of his work. After suffering declining health in the last months of his life, he died in his campus quarters on January 5, 1943. His epitaph characterizes his humanistic attitude toward scientific discovery: "He could have added fortune to fame, but caring for neither, he found happiness and honor in being helpful to the world." His birthplace became a national monument. In 1940, the Carver Foundation established the Carver Memorial Museum and preserved the Tuskegee laboratory in honor of his service to humanity. In 1973, Carver was elected to the Hall of Fame for Great Americans; Congress designated January 5 as George Washington Carver Day.

George Washington Carver

Cash register

We owe the invention of our modern day cash register to a Dayton, Ohio saloonkeeper, James J. Ritty. Frustrated by dishonest bartenders who helped themselves to saloons profits, Ritty also desired a more efficient method to track sales than haphazard scribbles on scraps of paper.

While vacationing in 1879, Ritty was touring the engine room of a transatlantic steamer and noticed a device which counted revolutions of the ships propeller. He realized a similar device could benefit his business. Working along with his brother John, Ritty developed "Ritty's Incorruptible Cash Register." A large device consisting of two rows of keys, a clocklike face with a "minute" hand to record cents, an "hour" hand to record dollars and a bell which signalled the completion of a transaction. The first model did not include any cash drawer controls or any method to maintain permanent records. In the following years, James Ritty and his brother experimented with several improved models, including one which used a paper roll to record transactions. Unable to raise additional capital to continue, Ritty sold the company and patent for $1,000 to the National Manufacturing Company in 1881. In 1884, the name was changed to National Cash Register Company when John Henry Patterson of Ohio became owner. Patterson added an internal printing mechanism which kept a running record of sales and printed a paper sales slip. By 1910, National Cash Register monopolized the market to such an extent, an antitrust suit was filed, although the sentence was later reversed. In 1974, under owner William Anderson, the name was shortened to NCR. With the advent of electronic components and integrated circuits in the 1970s, NCR became completely electronic, branching into computers and 24-hour automated bank teller machines.

See also Calculating machine; Digital computer; Screw propeller

Cast-iron and wrought-iron construction

Iron can be obtained in two forms through the smelting process. *Wrought iron* is more pure, having less than 0.3% carbon and 1% or 2% slag. *Cast iron* contains 2% to 4% carbon, plus a mixture of other impurities. Wrought iron is considered superior in strength over cast iron; it never does reach a truly molten state and must be hammered into the desired shape. Cast iron can be poured into casting molds and so is easier and cheaper to use.

Wrought iron was first used in Asia Minor before 1000 B.C. and was used in making tools and weapons. Both cast and wrought iron were in use at about the same time in China, about 300 and 600 B.C. respectively. During the Middle Ages, wrought iron was used in Europe for detailed grille work, balconies, and other appointments. Cast iron was produced but not widely used until the 1600s, when casting methods were improved. From then on, it became an increasingly important building material.

During the 1700s and 1800s, both types of iron were popular for building. Cast iron was gradually singled out for structural purposes as attention shifted from supporting walls to load-bearing frameworks. Cast iron could be mass produced and was therefore much cheaper. In 1848, inventor **James Bogardus** went into full production with cast iron by establishing a foundry that made factory beams and columns that could be shipped anywhere and assembled at the construction site. The foundry stayed in business for over thirty years.

Architect **William Jenney** used cast iron as a structural material for a Chicago office building in 1885, and continued to use it in all types of buildings. His iron framework designs were a response to a call for **fireproofing techniques** resulting from the Chicago fire of 1871. He later switched to steel because iron succumbed to fatigue over time and buckled in extremely hot fires.

The 984-foot (300 m) Eiffel Tower in Paris, completed in 1889, was the last major wrought-iron structure built. Countless **bridge** structures were made of cast iron during the 1700s and 1800s, many of them very graceful in design. Then the coming of the railroads from 1830 on placed new load demands on bridges, and there were numerous bridge failures during the 1860s and 1870s, especially in the United States. Fatigue was claiming the bridges after about ten years, and this led to the adoption of steel for bridge construction.

Wrought iron remains a decorative material in the twentieth century, but is rarely used in building. Cast iron is not used widely for construction but is used for **automobile** parts, tools, and furnaces.

See also Iron production; Train and railroad

Catalytic converter

Automotive developments late in the twentieth century often seem to pit government regulations against technology and the consumer's pocketbook. A case in point is the catalytic converter, but, surprisingly, technical advances have met increasingly stringent standards at reasonable costs to automobile owners. In the 1970s, the U.S. Environmental Protection Agency, along with other federal and state agencies, developed a set of strict emission-control standards (the Clean Air Act and others) to be met by car manufacturers by 1980.

As of the early 1980s, all new cars sold in the U.S. have had catalytic converters. The catalytic converter reduces the amount of pollutants in exhaust gases. The exhaust flows through the converter that contains beads coated with metals that act as catalysts by reacting with hydrocarbons and carbon monoxide and converting them to water and carbon dioxide that can be eliminated with less harm to the environment. Platinum and palladium are the metals typically used in catalytic converters, although other metals and alloys are used or are in development to improve effectiveness and reduce cost. Widespread use of catalytic converters also necessitated production of lead-free gasoline because lead coats the beads in the converter and makes them useless.

The process of improving the catalytic converter to meet emission standards is complicated. Many engine systems are involved including the ratio of air to fuel in **carburetor**s or fuel injectors, combustion timing, valve timing, and cylinder compression, among many others. Fuel chemistry is also crucial. A combined system is needed to reduce emissions of carbon monoxide, hydrogen, and nitrogen oxide; this system also has to use fuel conservatively, be easy to manufacture, cost as little as possible, and be simple to maintain. Other standards require that catalytic converters in cars built since 1996 last for at least 100,000 mi (160,900 km).

In the 1990s, amendments to the Clean Air Act imposed demands that a single catalytic converter cannot meet. Researchers have returned to the drawing board to develop new catalytic materials and approaches for converting exhaust gases including coupling several conversion devices. According to the regulations, ultralow emissions vehicles (ULEVs) must reach the marketplace by 2003. This requirement has not only stimulated new technologies for **internal combustion engines** but has pushed development of electric cars, which are zero-emissions vehicles (ZEVs), and hybrid vehicles, which will probably have their own standard between the ULEV and ZEV groups.

The amendments also include a requirement for reducing emissions during engine startup. Conventional catalyst materials don't work during the first few minutes of engine operation when temperatures are low, and unburned hydrocarbons are released directly into the atmosphere. Several designs are being considered. Converters that use an electrically heated catalyst (EHC) contain electrical honeycombs or coils that preheat the catalyst to cause quick reactions. Hydrocarbon trap (HTC) systems work by storing those first hydrocarbon emissions until the catalyst is hot and able to react. Successful development of HTC systems involves the thermal and adsorptive (adhering) properties of materials. A third and most promising cold-start device is the close-coupled catalyst (CCC) that uses two catalytic converters in line, with one located next to the exhaust manifold so it heats quickly. Emissions then pass to the second catalyst where more reactions occur, and remaining gases are emitted at levels below the ULEV standard.

Research for improving catalytic converters requires a variety of engineering specialties including ceramics, thermodynamics, metal properties, ionic and fuel chemistry, and manufacturing efficiency. Copper, cobalt, iron, and titanium are used experimentally to cause catalysis. Series or step-type burner systems for combusting exhaust gases are also being studied. Fuel additives like ethanol reduce emissions but are hard to produce, so biocatalysts are being developed to generate ethanol from sugars with the help of bacteria. It is hoped that the complex considerations in catalytic converter technology will produce an invisible benefit—cleaner air.

See also Fuel injection

Catapult

The catapult is an ancient **artillery** device that used the power of recovery of twisted fibers to hurl projectiles at long range. The invention of torsion artillery and siege engines has been credited to the Greeks. They were using torsion catapults from about 300 B.C.. These catapults used vast quantities of human hair twisted into skeins. There are several recorded instances of gifts between Greek cities of tons of human hair, evidence of how much was needed for these war machines.

The Romans made improvements to the catapult's basic design and perfected its use in battle. The simplest of the early catapults used a thick bundle of human hair or animal sinew, with one wooden beam inserted through it. Geared winches were used to twist the hair or sinew without letting it unwind. To load the catapult, soldiers manned a windlass, which pulled the beam down until it was horizontal. A stone was attached to the end of the beam. When a soldier pulled a rope, releasing the beam from its mooring, the beam sprang upward with the energy of the released tension of the twisted hair, flinging the stone with great force. The basic Roman catapult had a single arm that flew up to hurl a stone toward a distant enemy.

In medieval Europe, a catapult that operated like a crossbow came into usage. It had two short arms that moved horizontally instead of one large beam that moved vertically like the earlier Roman catapult. A heavy bowstring connected the ends of the arms. The usual **ammunition** was a short, heavy javelin.

The ballista closely resembled the crossbow-catapult but differed in size; it could weigh up to four tons (3.63 t) and send a 60-pound (27.2 kg) rock as far as 500 yards (457.2 m). Both machines were capable of throwing ammunition at the same target continuously. However, these early devices had their drawbacks. They worked fine in dry weather, but when wet, the fibers lost their resilience and power.

Around A.D. 1000, a new type of catapult, called the *huo-pa'o*, appeared in China; its design was probably based on an irrigation device used in Egypt around 1500 B.C. An unequal lever had a heavy weight attached to the shorter arm. The longer arm was attached to a rope that held a bucket with the ammunition. The arm was then pulled down to engage a hook, the ammunition (stones, most likely) was loaded in the bucket or sling, and the hook was released. The heavy weight on the short arm fell, sending the bucket skyward, thus launching the ammunition. These weapons, called *trebuchets* by Europeans, appeared throughout the continent in the eleventh and twelfth centuries. The largest trebuchets had 50-foot (15 m) arms and used a 10-ton (9 t) weight, were capable of throwing a 250-pound (114 kg) rock or ball about 900 feet (275 m).

Surprisingly enough, catapults were seen much later in history. In 1849 Richard Hodges received an English patent for an airgun in which a piston was forced up the cylinder by a twisted India-rubber spring. He also invented an elastic-rubber catapult mounted on a gun-shaped stock.

Catapult devices were even used in recreational sports, such as trap-shooting, in which elastic catapults hurled inanimate targets in the air for hunters to shoot. An American, Captain Adam Bongardus, created an effective ball trap device for this sport in 1876.

The catapult is used now to launch heavy jet planes from aircraft carriers. Without enough space for a long take-off path, jet planes are hurled into the air from a steel carriage mounted on a catapult arm on the carrieròs deck. Developed in World War II, the first airplane catapults were hydraulic. The British developed a steam-driven catapult in 1951, using steam from the carrieròs boilers. The underbody of the plane rests in a truss mounted on a rolling carriage. This whole assembly is fixed to a turntable so that the it can be rotated to face the airplane into the wind. When the plane is ready for takeoff, it starts its engines, and the catapult arm is released, sending the carriage forward and vaulting the plane safely away from the ship.

Cataract surgery

Cataract surgery involves removing cataracts from the lens of an eye or removing the lens altogether and replacing it with new lens. Cataracts are spots on the lens that may cloud the lens, causing a blurring of vision and eventually blindness. The lens of the eye helps to focus light. Light from an object first strikes the transparent covering of the eyeball, the cornea. Then the light passes through the lens, which bends the light rays enough so that they focus on the back of the eye, the retina. Sometimes with age or due to other conditions, cataracts form and light can no longer pass through the affected part of the lens. To correct the problem, several alternative procedures may be used.

A surgical procedure called an *introcapsular extraction* involves removing the entire lens through a cut made along the top edge of the cornea. In this procedure, invented by an American named Kelman in 1976, an ultrasonic device actually emulsifies or breaks the lens into tiny fragments so that it can be aspirated from the eye. A new intraocular lens made of **plastic** is inserted and the incision closed with tiny sutures. Plastic lenses were first invented in 1952 by an English physician, Harold Ridley. After the plastic lens is transplanted, the patient is able to see. The patient may be fitted for special glasses or contact lenses.

Cataract surgery has a long history. It was mentioned in the code of Hammurabi, the Babylonian king who lived 4,000 years ago. The first known cataract operation to extract a clouded lens was performed by J. Daviel, a Frenchman, in 1748. Another well-known English surgeon, William Cheselden (1688-1752), restored sight to a man born blind.

In 1905 Austrian physician Eduard Zirm performed the first known cornea transplant by transplanting the cornea of one person into the eye of a blind person. Basing his work on Zirm, Doctor Elschwig of Prague also successfully performed a cornea transplant in 1914. Since 1944, eye banks have been established in many places around the world where donated eyes may be stored in order to be used in such transplants.

In 1961, an American physician, Irving S. Cooper, began using a freezing technique known as cryosurgery to freeze and destroy damaged tissue. He first used cryosurgery on damaged brain tissue of Parkinson's disease patients. Now it is successfully used to remove cataracts from the lens of the eye. In 1979, the first **laser** eye surgery was performed using an ultra-rapid pulsated Yag laser. Done by Professor Daniele Aron-Rosa, laser eye surgery allows surgery without having to cut the eye. Since that time laser surgery has successfully been used on corneas and detached retinas. For cataracts, it is used primarily to make a clear opening in the posterior capsule if it turns cloudy after the original cataract operation.

Further advances in cataract surgery have made it a routine and, for the most part, outpatient therapy. For example, anesthetic eye drops have been developed to replace the standard injection method, thus eliminating complications sometimes associated with injections. Advanced surgical techniques have also led to smaller and smaller incisions that no long re-

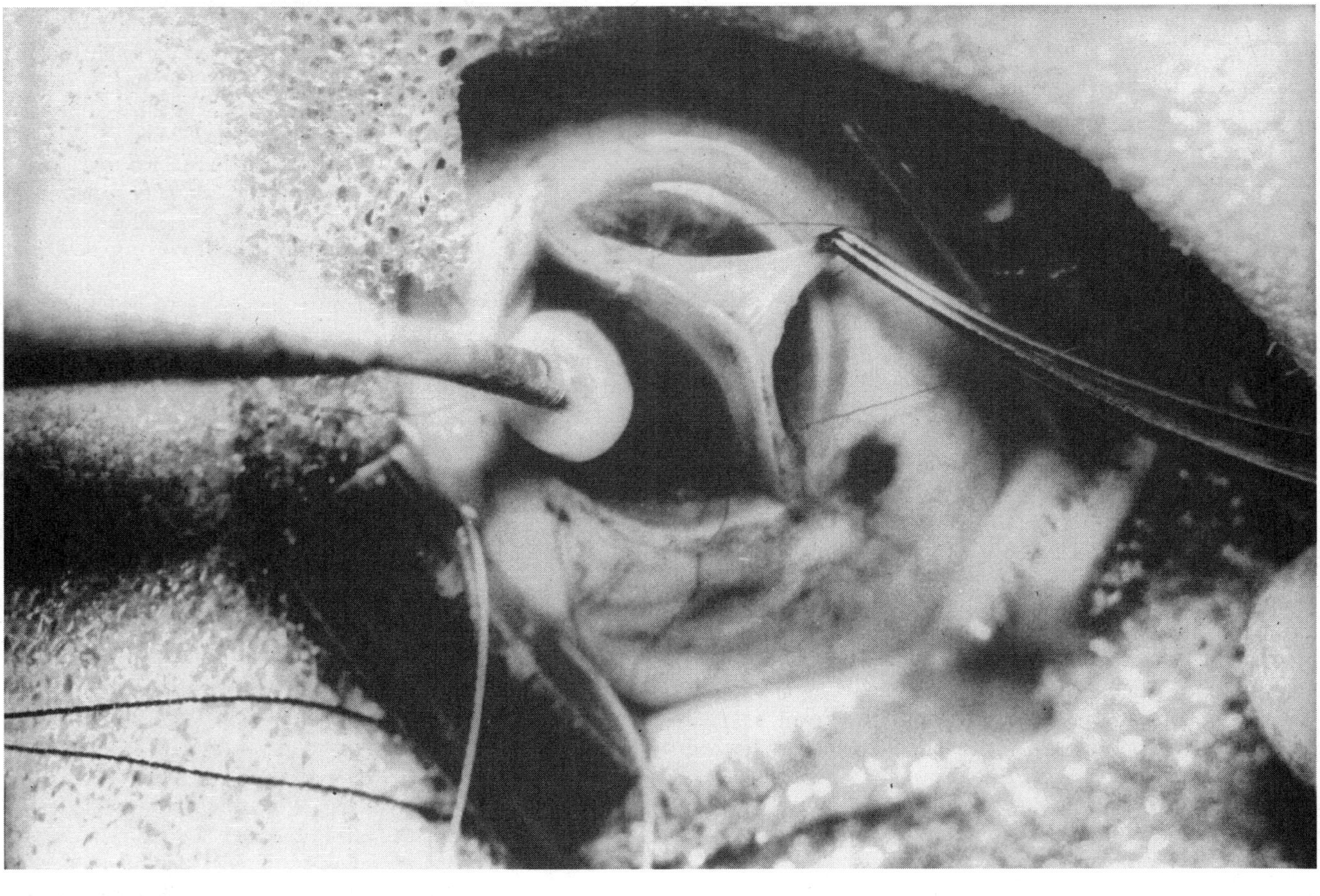

Cryosurgical removal of cataract.

quire sutures. These small incisions, in turn, are leading surgeons to concentrate more on surgical techniques that will restore multifocal vision in cataract patients, thus reducing their dependency on eye glasses after surgery.

Caterpillar vehicles

Since about 2500 B.C., draught animals have been used to perform such strenuous tasks as plowing fields and hauling heavy loads. Animals long remained essential for agriculture, road building, logging, and other major industries. It was not until the 1800s that **steam-powered road vehicle**s began to replace animals for heavy work.

The work that machines were beginning to perform was not usually done under ideal conditions. Rough or muddy fields and roads often bogged down wheeled vehicles. In 1770 an Englishman, Richard Edgeworth, patented his idea to equip a vehicle with a set of large tracks, which he called an ''artificial road.'' But the idea was never developed.

Alvin Orlando Lombard, a farm boy from Springfield, Maine, developed the first hauler with caterpillar treads. Instead of cutting into the surface as conventional wheels do, the treads rode over the surface, using grips to push the vehicle forward. Lombard's hauler was used for 30 years to tow logs in Maine. In 1904, Benjamin Holt developed the first crawler for road building. When David Roberts' sophisticated design failed to attract notice in England, he sold his patents to Holt as well. Holt's success led to the emergence of the Caterpillar Tractor Company.

Caterpillar treads were not widely accepted until after World War I, after tanks had proven the **tractor** tread's usefulness. Treads enabled tanks to ride through all types of terrain, pushing small trees aside and riding over trenches. By 1925, there were a half million crawlers in the United States, mostly farm tractors.

The Caterpillar Company later introduced improvements, such as diesel-powered equipment in the 1930s. Caterpillar and its competitors added treads to a multitude of heavy work machines: graders, haulers, bulldozers, and backhoes, for example. Between the world wars, farm tractors abandoned the caterpillar tread for large-treaded rear wheels that facilitated passing between crop rows.

Today few think of the modern snowmobile as a tracked vehicle, but this useful vehicle that was first called a ''motor sled'' is propelled over snow and ice by a small gasoline engine that turns a wide track or belt at its rear. Developed in the early fifties, snowmobiles can carry up to two people and are used for recreation, delivery, and rescue in snowy environments. As with all tracked vehicles, the tread of its tracks is rolled around an endless belt that its small wheels ride on. The tracks allow it to ride on and through snow as the driver steers the front runners or skis with the handlebars.

In 1992, tracks also were used in the new stair-climbing ACCESS wheelchair. This battery-powered, microprocessor-controlled wheelchair travels over flat, level surfaces in its wheel mode, but when switched to track mode, its dual tank tracks enable it to climb or come down safely staircases as steep as 36 degrees.

Catheter, cardiac

The first practical system for cardiac catheterization—passing a tube through a vein into the heart—was devised by physician Werner Forssmann (1904-1979) in 1929. Some earlier investigators had experimented with cardiac catheterization in animals and human cadavers. Forssmann, working at the Eberswalde Surgical Clinic near Berlin, Germany, set out to prove that the procedure was safe for live patients. He exposed a vein in his own arm, inserted a catheter (a long, thin, flexible tube), and advanced it about two feet through the vein to the right side (atrium) of his heart, confirming the position by X-ray. After eight more experimental procedures, Forssmann published a report describing the technique and suggesting its usefulness in examining and diagnosing diseased hearts and poor blood circulation. Forssmann was unable to develop the technique further due to financial strictures at the clinic.

In the United States, two physicians at Columbia University read about Forssmann's experiments—André F. Cournand (1895-1988) and his associate, Dickinson W. Richards (1895-1973). They thought cardiac catheterization would provide a reliable method of determining blood flow through the lungs, blood pressure in the heart, and oxygen content of the blood, all important indications of cardiac and circulatory health. Cournand and Richards began experimenting with catheterization in the 1930s, progressing to dogs and chimpanzees in 1936. In 1941 Cournand performed the first human heart catheterization since Forssmann's, assisted by Dr. Hilmert Ranges. This and subsequent catheterizations showed that the procedure was safe and yielded the valuable measurements the colleagues had hoped for, providing an effective nonsurgical means of diagnosing cardiovascular problems. Cardiac catheterization rapidly spread through the world and became the single most important diagnostic tool for heart disease.

In 1956, Forssmann, Cournand, and Richards shared the Nobel Prize in medicine or physiology ''for their discoveries concerning heart catheterization and pathological changes in the circulatory system.''

At the end of the 1980s, an intracoronary prosthetic device called a stent was introduced. Used to hold vessels open, stents have revolutionized cardiac catheterization and interventional cardiology. In 1997 more than one-million stents were used worldwide. In a study of stent usage in **angioplasty** procedures, investigators found that when stents were utilized the percentage of patients requiring another angioplasty a year later was reduced from 25 percent to 16 percent.

Cathode ray tube

In the mid-to late-1800s, the world was experiencing a scientific revolution. Phenomena that had never before been truly understood, such as light, heat, and radiation, were systematically unravelled, and by such great scientists as Henri Becquerel, Marie Curie and Thomas Young. Among these phenomena was the nature of electricity: how it worked, and why. The early experiments to solve the riddle of electricity often included the use of anode-cathode tubes—glass tubes containing an

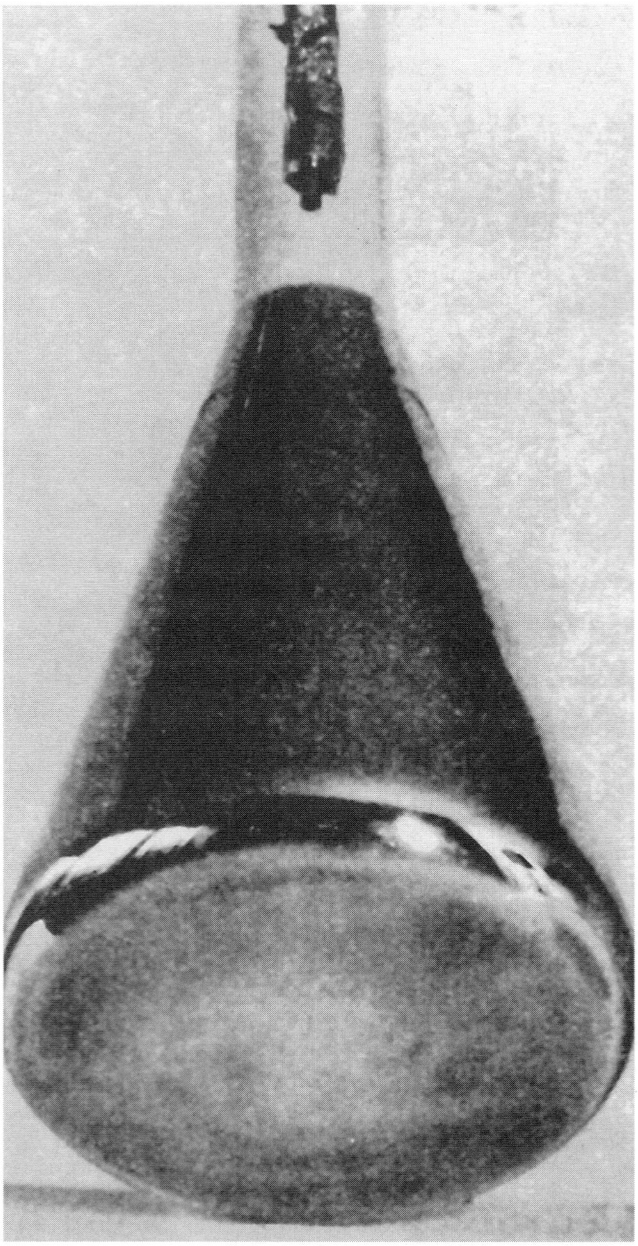

German physicists Heinrich Geissler and Julius Plücker developed the first cathode ray tubes around 1855.

anode at one end and a cathode at the other. When most of the air was evacuated from this tube, an electrical charge could be observed jumping across the gap between the two electrodes. one scientist who performed such an experiment was **Michael Faraday**; he noticed that, as the amount of air within the tube decreased, a faint glow could be detected between the anode and the cathode. However, the technology of the time was not sufficient to produce a high vacuum within the glass tube, and so Faraday was unable to further explore this effect.

The pioneers in the study of cathode-ray tubes were the German team of **Heinrich Geissler** and Julius Plucker. Geissler, a skilled glassworker, was employed by the University of Bonn as a maker of scientific instruments. While at the university he met Plucker, then a young professor. Some time around 1855, Plucker convinced Geissler to design an apparatus for evacuating a glass tube. Geissler did just that, constructing a hand-crank mercury pump that could (after a laborious session of pumping) remove most of the air from a tube. The new vacuum tubes were very popular, and became known as Geissler's tubes. Using the improved vacuum tube, Plucker made some startling discoveries. First, he was able to produce a bright stream-like glow between the electrodes—much brighter than any achieved in previous experiments. Second, he found that the glow responded to a magnetic field, and that it could be moved by a powerful **magnet**. This discovery was monumental, for it indicated that the stream crossing the vacuum was composed of particles rather than rays.

The next scientist to conduct important research using vacuum tubes was conducted in 1869 by Johann Hittorf. A student of Plucker's, Hittorf further improved the method for creating a vacuum within glass tubes of his own design. He observed that the luminescent glow increased dramatically as the pressure within the tube continued to decrease. He also placed tiny obstacles inside the tube, in the path between the two electrodes. When a current was applied, the glow would by partially obscured by these obstacles, casting shadows. This further confirmed the idea that the glow was caused by a particle emission.

Probably the most important research using cathode-ray tubes was performed in 1875 by the English physicist **William Crookes**. In order to confirm the experiments of Plucker and Hittorf, Crookes designed his own vacuum tube from which the air could be almost completely removed. So great an improvement over Geissler's tubes were these that the Crookes tube quickly became the standard vacuum tube for use in scientific experiments. Crookes continued Plucker's experiments with magnetic fields, finding the glow easily deflected. He also installed tiny vanes within his tubes; as the current was applied the vanes would turn slightly, as if they were blown by a gust of wind. These experiments incontrovertibly showed that a stream of particles travelled through the tube. Crookes believed that cathode rays were a ''fourth state of matter,'' possibly associated with an invisible aether.

German scientist Eugen Goldstein first dubbed Crookes's rays cathode rays in 1876. In 1892, Phillip Lenard, following up on Heinrich Hertz's discovery that under certain conditions cathode rays could penetrate metal, succeeded in passing cathode rays through a ''window'' of thin metal set into the side of a Crookes tube. The rays exited the tube through the window into the air, showing that cathode rays were not a phenomenon exclusive to a vacuum. While performing a similar experiment in 1895, the German physicist Wilhelm Roentgen accidentally discovered an even more penetrating form of radiation, which he called X-ray radiation.

While many scientists were busy trying to unlock the secrets of cathode rays, others were searching for ways to apply them toward practical ends. The first such application came in 1897 in the form of **Karl Ferdinand Braun**'s oscilloscope. This device used a cathode ray tube to produce luminescence on a chemically treated screen. The cathode rays were allowed to pass through a narrow aperture, effectively focusing them into a beam which appeared on the screen as a dot. The dot was then made to "scan" across the screen according to the frequency of an incoming signal. An observer viewing the oscilloscope's screen would then see a visual representation of an electrical pulse. About the same time, the final proof of the particulate nature of cathode rays was provided by the great British physicist J. J. Thomson. Thomson also succeeded in measuring the mass and charge of the particles, which were shown to be smaller than an atom. The cathode-ray particles became known as electrons, and the cathode ray tube as an electron gun. For his discovery of the first subatomic particle, Thomson was awarded the 1906 Nobel Prize in Physics.

During the first three decades of the twentieth century, inventors continued to devise uses for cathode ray technology. Inspired by Braun's oscilloscope, A. A. Campbell-Swinton suggested that a cathode ray tube could be used to project a video image upon a screen; unfortunately, the technology of the time was unable to match Campbell-Swinton's vision. It was not until 1922 that **Philo T. Farnsworth** used a magnet to focus a stream of electrons onto a screen, producing a crude image. Though the first of its kind, Farnsworth's invention was quickly superseded by **Vladimir Zworykin**'s kinescope, the ancestor of the modern **television**. Today, almost every form of image-viewing device is based upon cathode-ray technology.

Electron guns are used widely in scientific and medical applications, and also based on the principals of cathode ray tubes. One application of particular importance has been the **electron microscope**, invented in 1928 by **Ernst Ruska**. The electron microscope uses a stream of electrons, rather than light, to magnify an image. Because electrons have a much smaller wavelength, they can be used to magnify objects, such as ultramicroscopic viruses, that are too small to be resolved by visible light. Just as Plucker and Crookes did, Ruska used a strong magnetic field to focus the electron stream into an image.

CAVITRON • See Ultrasound devices

CAYLEY, GEORGE (1773-1857)
English aviator

Cayley was an individual of varied interests. He liked mechanical things, he kept a notebook full of sketches of plants and animals, and he had a journal with a wide range of entries.

Due to the success of the **balloon** ascents by brothers **Jacques-Étienne Montgolfier** and **Joseph-Michel Montgolfier** in the 1780s, Cayley was interested in human flight. Instead of concentrating on balloons like so many did, he concentrated on heavier-than-air flight. In 1796, he built a **helicopter** model with feather propellers, applying the idea of using an airscrew for mechanical flight.

In 1799 he envisioned a fixed-wing **aircraft**. It was a feed-wing **glider** whose wings were stretched cloth. There was a boat-shaped fuselage for the pilot and a modern tail unit for control. His sketch looks very modern because he understood the major concerns of lift, propulsion, and control. However, his power source—a paddle—would prove useless.

In 1804, he designed and built the world's first flyable model aircraft. It used a paper kite for a wing resting on a slender wooden pole. At the back, he attached a tail assembly like today's planes have: a cruciform assembly for horizontal and vertical control.

Between 1809 and 1810, Cayley published his important article "On Aerial Navigation," in which he outlined his carefully thought-out views on flying. These theories would later become the principles and practical applications of aeronautics. The article also points to the possibility that Cayley had at that time constructed a full-size glider capable of carrying a man briefly into the air. However, no other evidence pointing to its existence has been found.

Unfortunately, readers greeted this remarkable paper with little enthusiasm, but Cayley remained undaunted. He adopted the use of two and three wings placed one above the other, a forerunner of the designs conceived by **Orville** and **Wilbur Wright** and other twentieth-century aviators. In 1849 he designed a triplane glider, which was the first inherently stable, full-size heavier-than-aircraft. However, this machine could only carry the weight of a child, leading Cayley to construct another, more powerful one. In 1853, this glider carried Cayley's unwilling coachman 900 ft (274.5 m) before crashing. Although emerging unhurt, the shaken coachman quit his job and Cayley stopped building gliders. While he died in 1857 without receiving acclaim for his feats, Cayley is today hailed as the founder of the science of aeronautics.

CCD

The CCD (Charge-Coupled Device) is a type of semiconductor that can store packets of electrical charge in tiny surface regions called *potential wells*. Because of the way the device is constructed, the packets of charge can be moved from place to place within the semiconductor without rearranging their order; thus, the electronic "information" is stored until it is needed.

There are three basic ways in which information can be stored in a CCD. First, since the amount of charge in an individual well can be anywhere from zero to maximum, the CCD can be used as an analog storage device. The potential wells can also be thought of as being either empty or full (off or on); in this way the CCD can be thought of as a digital device. Finally, if a light source is focused upon the CCD's surface, the wells will fill according to the amount of light and shadow, converting the light into an electrical signal; thus, the CCD can also be an optical storage device.

Because of the variety of storage capabilities, CCDs have found a vast number of applications. The simplest is as a device for transferring analog signals to digital and vice versa. They are also used as scramblers, particularly to code and decode the video signals transmitted by cable and satellite broadcasters. As an optical device, CCDs are often found in facsimile devices, bar code readers, photocopiers, and aerial mapping cameras. Astronomers have discovered that CCDs are almost fifty times more sensitive than normal film, allowing them to photograph more distant objects and over a much longer period of time. Since the CCD yields an electrical signal, it is much easier to input its image into a computer for enhancement. CCDs have been used to capture images of the cosmos and in attempts to discover the source of ionized hydrogen. CCDs are also used for optical diagnostics of intense ion beams.

Probably the most practical use for CCDs is in **television** cameras. When first invented, TV cameras used an electron gun to scan a photoelectric array; this scanning would produce an electronic equivalent of the image. In a solid-state camera, the image is focused onto an array of CCDs; a strong signal will be generated where the light is bright, a weak signal where it is dim. The signal can then be transferred without the use of an electron gun. CCDs are now also used in many high-end **video cameras**. CCD video cameras are even being utilized in some ''smart'' air bags which can determine the presence of a child-safety seat. The CCD also is also the primary component of the digital camera. Rather than recording images onto film, digital cameras record images on a CCD and allow the transfer of those images to **television** or the **computer**.

See also Communication satellite; Computer input and output devices; Digitizer; Duplicating machine; Fax machine; Telescope

Celanese • See Rayon

Cellophane

Cellophane, a plastic film commonly used in packaging, wasdeveloped from cellulose acetate by Jacques E. Brandenburger in 1912. Cellophane was the first clear plastic packaging material. The importance of this discovery as a marketing tool should not be underestimated; sales of some products increased over ten times once they were wrapped in cellophane. During the depression, cellophane production actually increased while many other industries failed.

Brandenburger was disgusted with the sloppy tablecloths at his favorite café and he began to search for a way to protect tablecloths from stains. After researching cellulose nitrate solutions, which produced brittle fibers too rigid for tablecloths, Brandenburger began working with a material developed by **Charles Frederick Cross** and **Edward John Bevan**. The material was produced by treating the cellulose from wood with sodium hydroxide in water. This caused the fibers to swell, a process called mercerization. The mixture was allowed to ageseveral days and then was treated with carbon disulfide to produce a cellulose xanthate solution, called viscose. After another several days of aging, the viscose was forced through a thin holeand converted back to a nonsoluble cellulose by treating it in a bath of acid-containing salts. Brandenburger kept working with the viscose, attempting to form it into thinfilms. In 1908, he found that by using a bath of aqueous ammonium sulfate and an aqueous acid bath with glycerol, the cellophane he produced was more flexible. He also forced the viscose through thin slots to form sheets. It took four more years for Brandenburger to perfect the manufacturing process. He eventually was successful in producing transparent sheets, which he called cellophane from the French words cellulose and diaphane, meaning diaphanous or transparent. He patented his process for producing continuous cellophane films in 1912.

He originally sold his invention to the motion-picture industry as a replacement for cellulose nitrate, a highly flammable film. But cellophane proved unsuitable for this application because it distorted under heat. So Brandenberger marketed it as a wrap, selling it from a fancy Paris boutique. Prices were so high that at first cellophane was used only to swathe the finest French perfumes—and was kept locked in store safes at night. Cellophane was later used for eyepieces in **gas masks** made during World War I. The North American rights were bought by Du Pont in 1920 and four years later production of cellophane began in the United States. Cellophane was eventually replaced by other plastics which had better water-vapor impermeability and heat sealability, but it remains unique for creating a new niche in the packaging industry.

See also Cellulose, chemical uses of; Mercer, John

Cellular telephone

Until 1984, mobile **telephone** service was severely limited. Each city had a single **antenna** to transmit signals to and from a car phone's antenna. The Federal Communications Commission (FCC) assigned only 12-24 frequencies to an urban area, so only one or two dozen car-phone calls could take place in the entire city at one time. Users had to wait up to 30 minutes to get a dial tone, and potential mobile phone customers were put on 5- or 10-year waiting lists.

Cellular phone technology changed all this. In a cellular system, each metropolitan area is divided into broadcasting zones, or ''cells.'' Each 6-10-mi^2 (15.5-26 km^2) cell has its own antenna. As a car phone moves through the city, a computer automatically passes its frequency from one cell to the next. A single frequency can be used for multiple nonadjacent cells, and as the number of users increases, cells can be subdivided into any number of smaller cells, so the cellular system is capable of far greater usage than the old mobile service.

Rudimentary cellular technology was known as early as 1947. The increasing overcrowding of mobile phone service in the 1960s and 1970s gave impetus to cellular development. Bell Laboratories took the lead, proposing a system in 1971 and putting the first experimental cellular service—called Advanced Mobile Phone Service (AMPS)—into operation in 1978 in the Chicago, Illinois, area. After the Bell Telephone monopoly was disbanded in 1978, the seven new regional phone companies began to pursue the cellular phone market vigorously. In 1981 the FCC issued cellular phone regulations and in October 1983 Bell's Ameritech Mobile Communications subsidiary introduced the first American commercial cellular system in Chicago. (Cellular service was also available by then in a number of other countries.) The FCC also allowed one non-Bell (''nonwire-line'') service in each metropolitan area. In December 1983, Cellular One began transmitting in Washington, D.C. As the number of cellular phone systems and subscribers increased, the costs for equipment and service decreased, and the technology has improved.

Until the early 1990s, the primary type of cellular telephone was analog, that is it transmitted calls via a series of soundwaves. In the mid-1990s, digital cellular service began to emerge, and is expected to be the dominant in the early twenty-first century. With digital cellular telephones, the calls are sent as series of ones and zeroes. Digital is superior to analog in that they are less prone to static, have improved voice quality, and are less vulnerable to cellular fraud. (Analog calls can be intercepted with the right equipment since the signal is transmitted as sound waves; digital calls can be encrypted and is thus protected.) Also, many cellular telephone companies offer many more service options for free on digital telephones that are extra costs for analog, including pager service, voice mail, caller ID, three-way calling, and call waiting. Digital cellular telephones also have text messaging and data connectivity abilities. In the late 1990s, as the transition to digital takes place, many companies offer a dual-mode phone that can handle both digital and analog technology.

By 1998, there were 43 million cellular telephone subscribers in the United States. As more people sign up for the service, there has been an increased problem with drivers talking on their cellular phones involved in car accidents. According the *New England Journal of Medicine*, there is an increased risk of a car accident when driving and talking at the same time, similar in size to driving under the influence. Using a cellular phone while driving drastically increases response time to changing traffic conditions. It is not the only problem with cellular telephones. Digital phones also can interfere with pacemakers.

Cellophane, being produced above, was the first packaging material that allowed customers to view the contents of a product before they bought it.

Celluloid

Celluloid, the first **plastic**, was developed from **cellulose** in 1855 by an English inventor named **Alexander Parkes**. When Parkes dissolved pyroxylin, a partially nitrated cellulose, in alcohol and ether-containing camphor it yielded a hard solid which could be heated and molded. Subsequently, Parkes tried to market his product, which he called parkesine, but was unsuccessful because the substance was so revolutionary that no one knew how to use it. Furthermore, it required such large amounts of solvent to make that it was extremely expensive.

Celluloid resurfaced in 1869 with help from a contest sponsored by Phelanand Collender to find a substitute for ivory used in billiard balls. At the time depletion of the herds of elephants in Africa was leading to a shortage in natural ivory. Inventor John Wesley Hyatt and his brother sought to win the $10,000 prize by improving on Parkes' process. They dissolved cotton in nitric acid, and added camphor. They then applied heat and pressure to mold the celluloid into billiard balls. This resulted in a thermoplastic product that would not break when bent and which could be made into thin sheets or molded into a variety of shapes. It could be treated to look like amber, onyx, tortoise shell and many other natural products. It soon substituted for ivory and ebony on piano keyboards and was successfully marketed in many forms, including shirt collars, knife handles, dice, buttons, fountain pens, and baby rattles. For a while the celluloid business flourished, although its growth was impeded somewhat when rumors spread that celluloid billiard balls burst into flame when they struck each other. John Hyatt received the Perkin Medal of Honor for his developments in celluloid.

In addition, celluloid revolutionized the **photography** industry. Until this time **glass** plates smeared with an emulsion or **paper** treated with a similar emulsion were exposed to form photos. In 1887 Hannibal Goodwin first proposed the use of rolls of celluloid film. George Eastman used Hyatt's celluloid to develop sheet film for still photography in 1889. Both of these developments led to cheaper and faster developing. Still photography was revolutionized and **motion pictures** were born. Unfortunately celluloid was highly flammable and films often burst into flames when the celluloid jammed in front of the projector light bulb. By 1924 celluloid film was set aside in favor of safety film, which is made from cellulose acetate. In safety film, acetic acid replaces the nitric acid used to form the nitrocellulose and forms instead cellulose acetate, a much more stable compound.

Celluloid continued to be an important chemical for many industrial applications until 1897 when the next major advance in plastics occurred with the invention of Galalith, which is a plastic made from casein and formaldehyde.

Cellulose, chemical uses of

Cellulose is the major component of the cell walls in plants strengthens the roots, leaves, and stems, making them rigid. It is a polysaccharide, a long molecule composed of oxygen, hydrogen, and sugar, which is very similar to sugars and starches. The hydroxide groups in cellulose are isotactic—that is, they are placed on alternating sides of the main chain. These pendant groups bind with others on long cellulose molecules to form microfibrils.

Henri Braconnet (1780-1855), a French chemist, verified the existence of cellulose in 1819. Braconnet was a leading scientist in the field of animal and plant chemistry; he was apprenticed to an apothecary in 1793 and became a pharmacist in 1795 in Strasbourg, Germany. He discovered cellulose in mushrooms and called it fungine, then began experimenting on the effects of sulfuric acid on wood. Braconnet eventually succeeded in creating a nitrated cellulose, which he called xyloidine, by adding wood or potato starch to nitric acid.

The next major development for cellulose came in 1834 when **Anselme Payen**, a French chemist, isolated cellu from wood. Four years later, Théophile Jules Pelouze (1807-1867), another French chemist, also developed nitrocellulose. This discovery was just one achievement in a long and significant career in chemistry which began when Pelouze encountered Joseph-Louis Gay-Lussac (1775-1850) on an omnibus. He impressed Gay-Lussac enough that the famous scientist made Pelouze his laboratory assistant. Pelouze became a professor at Lille in 1830 and then moved on to teach at France's major scientific academy, the École Polytechnique. He conducted many successful experiments in collaboration with Gay-Lussac as well as on his own and eventually established a private laboratory school for chemistry in Paris; there he trained students and allowed foreign chemists to conduct private research.

In 1845 the German chemist **Christian Schönbein** developed **guncotton** by accident. He was experimenting with nitric and sulfuric acids in his kitchen and spilled them. After he used his wife's cotton apron to clean up the mess and hung it up to dry over the oven, it burst into flames. Schönbein realized that the nitric acids had bonded with the cotton fibers to form nitrocellulose, which he marketed as guncotton. Guncotton flourished briefly as a smokeless **gunpowder**, but it was highly unstable. Schönbein eventually developed a gel he called **collodion**, made from guncotton mixed with ether, which helped render the substance much more stable than was guncotton alone. Alexander Parkes developed the first **plastic**, made from cellulose, in 1855. He dissolved pyroxylin (a nitrated cellulose) in alcohol and camphor containing ether to produce a moldable solid. Parkes was not successful in marketing his invention. Fifteen years later **John Wesley Hyatt**, in an attempt to win a contest held to discover a substitute for ivory, improved on Parkes's technique and developed celluloid. Hyatt did not win the prize but was successful in marketing celluloid as a material for rattles, collars, **photographic film**, and dental plates. It was, however, highly flammable, which limited its utility in range of ways. During the same year the **rayon** industry began. The term *rayon* refers to any fiber developed from cellulose. George Audemars, a Swiss chemist, was granted the first patent for synthetic fibers in 1855. Sir **Joseph Swan**, an English chemist, produced fibers from nitrocellulose in 1880.

Hilaire Chardonnet was the man who accelerated the production of nitrocellulose fibers. He developed a process in which the highly flammable nitrocellulose could be denitrated using an acid sulfide solutions. Previous attempts to denitrate fibers had left them weak and unable to compete with the strength of natural fibers. Chardonnet's technique left the cellulose xanthate fibers with strength comparable to silk. In fact rayon, as the fibers were called, was sometimes called **artificial silk**. Another production techniques for rayon was introduced by Hermann Pauly in 1897. The process yields rayon known as Pauly silk, Bemberg rayon, or cuprammonium rayon. The process yields a finer thread but is not widely used.

The most common process for the production of viscose rayon was developed by **Charles F. Cross** and **Edward J. Bevan**, both of England, in 1894. They also received a patent for the production of cellulose acetate. The cellulose for acetate rayon comes from cotton, rather than wood pulp as in viscose rayon.

It is interesting that with all of the uses developed for cellulose, the method of cellulose production in plants was not discovered until the late 1930s. Wanda Farr, an accomplish biologist and microscopist, made the discovery. She established that in each cell of every cellulose-containing plant there are plastids in the protoplasm that produce cellulose. Even then, scientists did not understand the exact mechanism of cellulose production. Then, in 1997 researchers finally discovered a gene in the plant Arabidopsis that is responsible for the synthesis of cellulose. Although there is much yet to be learned about polysaccharide synthesis in plants, this breakthrough discovery is anticipated to lead to in an understanding of how plants generate their cell walls.

Cellulose is widely used today in photographic film, fibers, tapes and many other products. The largest applications for cellulose polymers are in the production of inks and paint systems. One increasingly important use of cellulose is in low fat non-dairy foods that closely resemble real dairy products. For example, cellulose derivatives are being used in low fat frozen desserts, processed cheese and sour cream. NutraSweet Kelco Co. of San Diego, California, is one unveiled developers of cellulose food products.

See also Cellophane; Cordite; Lacquer; Varnish

Celsius, Anders (1701-1744)

Swedish astronomer

Celsius is a familiar name to much of the world since it represents the most widely accepted scale of temperature. It is ironic that its inventor, Anders Celsius, the inventor of the Celsius scale, was primarily an astronomer and did not conceive of his temperature scale until shortly before his death.

The son of an astronomy professor and grandson of a mathematician, Celsius chose a life within academia. He studied at the University of Uppsala where his father taught, and in 1730 he, too, was given a professorship there. His earliest research concerned the aurora borealis (northern lights), and he was the first to suggest a connection between these lights and changes in the earth's magnetic field.

Celsius traveled for several years, including an expedition into Lapland with French astronomer Pierre-Louis Maupertuis (1698-1759) to measure a degree of longitude. Upon his return he was appointed steward to Uppsala's new observatory. He began a series of observations using colored glass plates to record the magnitude of certain stars. This constituted the first attempt to measure the intensity of starlight with a tool other than the human eye.

The work for which Celsius is best known is his creation of a hundred-point scale for temperature, although he was not the first to have done so since several hundred-point scales existed at that time. Celsius's unique and lasting contribution was the modification of assigning the freezing and boiling points of water as the constant temperatures at either end of the scale. When the Celsius scale debuted in 1747 it was the reverse of today's scale, with zero degrees being the boiling point of water and one hundred degrees being the freezing point. A year later the two constants were exchanged, creating the temperature scale we use today. Celsius originally called his scale *centigrade* (from the Latin for "hundred steps"), and for years it was simply referred to as the Swedish **thermometer**. In 1948 most of the world adopted the hundred-point scale, calling it the Celsius scale.

CEMENT • See Concrete and cement

CENTRIFUGE

A centrifuge is a device used to separate liquids from solids by spinning. Long ago, people saw that gravity could eventually separate a sediment from a liquid or separate two liquids which do not mix. The heavier element within a container would descend, while the lighter element would rise to the surface. This process was extremely slow if left up to nature alone and was also wasteful, as evidenced by the way farmers used to separate cream from milk. They would let whole milk stand for several hours until the lighter cream rose to the top. They then skimmed off the cream with a wooden spoon, but as much as 40 percent of the cream was left in the milk. Later, small strainer dishes were used to extract the cream, yet this too was a slow and tedious process.

In 1877, Swedish inventor **Carl Gustaf Patrik de Laval** introduced a high-speed centrifugal cream separator. Milk was placed in a chamber where it was heated and then sent through tubes to a container that was spun at 4,000 revolutions per minute by a **steam engine**. The centrifugal force separated the lighter cream, causing it to settle in the center of the container. The heavier milk was pushed to the outer part and forced up to a discharge pipe. Thus, only the cream was left in the container. Several years later an improved **cream separator** was introduced with the capability for self-skimming and self-emptying. Other separators can extract impurities from lubricating oils, beer and wine, and numerous other substances. Other types of centrifuges were created in which spin dryers were used for filtering solids: a perforated drum was spun, driving any separated liquids to the outside where they were collected. These spin dryers can now develop accelerations of up to 2,000 times the force of gravity. They are used in the food, chemical, and mineral industries to separate water from all sorts of solids. Other centrifuges remove blood serum (plasma) from the heavier blood cells.

However, some scientists needed faster rotations for separating smaller particles. Such particles, like DNA (deoxyribonucleic acid), proteins, and viruses, are too small to settle out with normal gravity; the banging of water molecules is enough to keep the particles from separating. The key was to build an ultracentrifuge that could spin fast enough to cause these small particles to settle out. In 1923 a Swedish chemist, **Theodor Svedberg**, developed a device that could spin fast enough to create gravity over 100,000 times normal. It could take small samples in **glass** containers, balance them on a cushion of air, and send jets of compressed air that touched the outer surface. By 1936 Svedberg had produced an ultracentrifuge that spun at 120,000 times per minute and created a centrifugal force equal to 525,000 times that of normal gravity. Newer models can accelerate samples to 2,000,000 times the force of gravity. This machine enabled biologists, biochemists, physicians, and other life scientists to examine viruses; cell nuclei; small parts within cells; and individual protein and nucleic acid molecules. Thus, **genetic engineering** became a field ripe with possibility. Centrifuges are also commonly used in the petroleum industry to separate oil components. In 1997 state of the art centrifuges, such as those built by U.S. Centrifuge, use a finite-element-analysis software made by Algor Inc. to design advanced centrifuge systems.

CERAMIC

Linked to synthetic metals and **plastics** in their versatility and durability, ceramics comprise a broad category of such specially shaped and heated products as **pottery**, porcelain, china, and ceramic tile; ceramics may also be found in **cement**, plumbing and construction materials, and **spacecraft** components.

The basic ingredient in all forms of ceramics are silicates, minerals which are so common and widespread that it is virtually impossible to completely trace their history. When silicates (feldspar and silica, for example) are combined with a liquid such as water, they form a mixture that can be kneaded and shaped into any form. After shaping, the object is dried and fired in a high-temperature oven called a **kiln**. A glaze (a **glass**-like substance that makes the surface glossy and nonporous) may be added between drying and firing.

From ancient days to the present, this process has survived almost unchanged, save for the addition of mechanical aids. The oldest examples of pottery, found in Moravia and dating back to 25,000 B.C., are animal shapes made of fired clay. Similar figures, dating from around 5500 B.C., have been found in the Near East, and Japanese pots, made as early as 9000 B.C., have also been discovered. Pottery was originally dried in the sun or fired in bonfires, which produced articles that were ultimately too porous. Only kilns (which first appeared in Mesopotamia, as did the first potter's wheels, around 3000 B.C.), which allowed for articles to be fired at high temperatures followed by slow cooling, could harden clay enough that it became airtight. Decorative techniques, which developed concurrently with the making of pottery, included the use of mineral-based glazes to produce various colors; etching with sticks, fingernails, or shells; and painting before glazing. Some of the most fascinating pottery in history was made by the ancient Greeks, whose vases were skillfully decorated in the ''black figure'' (black paint applied to red clay) or ''red figure'' (black paint covering all but the design, which stood out in red clay) methods. The early Islamic potters of the Middle East produced colorful, imaginatively glazed tiles and other items. Their elaborate pictorial designs have provided archeologists with many clues to their daily lives. Perhaps the most renowned potters of all time, however, are the Chinese, who developed the finest form of pottery known as porcelain.

Porcelain is made of a clay called kaolin mixed with a petuntse, which is comprised of feldspar, aluminum silicate, and potash or soda. It is fired at an extremely high temperature to achieve its smooth finish. The result is a milky white ceramic material that is uniformly translucent and glasslike. Porcelain was first made in T'ang Dynasty China (A.D. 618-906) and by A.D. 1000 it was being mass-produced in the city of Ching-te Chen. Much later, increased trade with China brought porcelain to the attention of Europeans. At the same time, Middle Eastern methods of glazing were introduced to Europe through Spanish Moors and adapted into such forms as the Italian majolica and the French faience. These were vibrantly colored, glazed floor tiles. Europeans had seen porcelain as early as the fifteenth century but it remained quite rare until the English, French, and Dutch East India Companies began the widespread, economical importation of chinaware. As the use of china cups gained in popularity, potters in Italy, France, Germany, England, and the United States searched for ways to reproduce fine Chinese porcelain. Their efforts produced an imitative material called soft porcelain, for which a factory at Sèvres, France, became famous during the eighteenth century. Johann Friedrich Böttger (1682-1719), a German apothecary's apprentice, discovered how to make true or hard-paste porcelain in 1709 and established a factory in Dresden the next year.

One of the best-known contributors to the development of ceramics was English potter Josiah Wedgwood (1730-1795). In his early twenties he began developing a highly scientific approach to his craft, and after going into business for himself succeeded in developing a cream-colored earthenware that, because of Queen Charlotte's patronage, became known as Queen's ware. Simple, elegant, and durable, this product eventually became popular worldwide. In 1768, Wedgwood entered into a partnership with Thomas Bentley and began making jasperware, unglazed stoneware of various colors with white relief figures for which he became renowned. Wedgwood also produced items in finely textured, unglazed black basalt decorated with red painting to imitate the red figure vases of ancient Greece, and he developed bone china, which is made of clay mixed with bone ash. Finally, Wedgwood is remembered as the inventor of the **pyrometer** (a high-temperature gauge) and as the first pottery craftsman to utilize a steam-powered **engine** in his factory.

Ceramics began to be used for industrial purposes with the development of other technologies during the early twentieth century. A high demand for military materials during World War II hastened the science's evolution, and ceramics are now commonly found in a wide variety of products, including **abrasives**, bathroom fixtures, and electrical insulation. During the 1960s and 1970s, the burgeoning fields of atomic energy, **electronics**, **communication**, and space travel increased demands for more sophisticated ceramic products—for instance, spacecraft requiring ''skins'' and engine linings that are highly heat-resistant.

In the 1980s, research to develop ceramic engine parts that would better withstand heat than metal and thus burn fuel more efficiently was undertaken. As recently as 1990, a team of scientists at Japan's Government Industrial Research Institute found further applications for ceramics by developing stretchable compounds made from silicon carbide, silicon nitride, and other materials. When made into strips and heated, this special ceramic material can be stretched to two-and-a-half times its original length without losing its hardness and durability. Such ceramic fibers have revolutionary applications in the automotive and aerospace industries. One company, Thermal Ceramics Inc., has introduced a series of ceramic fiber products for use in transportation equipments such as automotive air bags, turbochargers, plastic gasoline tanks and fuel lines. In addition, they have developed ceramic fiber papers which are lightweight, durable, cost-effective alternatives to traditional insulating, friction, and filtering media.

Two other exciting developments have been made in the ceramic industry in the late 1990s. The first is a new ceramic gelcasting process developed by scientists at Oak Ridge National Laboratory. Their procedure is a simple process that allows the formation of complete ceramic components without machining. In 1997 General Motors Corp. engineers began to explore this process as a new way to manufacture improved ceramic components for turbines. Turbine engineers hope this technology can be used in the future to produce high quality ceramic nozzles, blades, rotors and other parts economically. The second breakthrough came from chemist Andrew R. Barron, an inorganic chemist at Rice University in Houston, and his colleagues C. Jeff Harlan and Rhonda L. Callender who found a way to break down the aluminum-oxide particles found in ceramics. Their process eliminates potentially hazardous waste by creating small sub-units that easily and safely dissolve in water. The impact of these two discoveries on the ceramics remains to be seen.

See also Steam Engine

Cermet

Although examples of metal-ceramic **composites** in the **electronics** industry date to the 1950s, true cermets consisting of

ceramic (cer) reinforcements interspersed in a metal (met) matrix comprise a relatively new class of composite material. The primary support for cermets has come from the aerospace industry for airframe and spacecraft structures. More recently, the automotive, electronic and recreation industries have started to employ these composites. The driving force behind the development of most of the existing composites has been their ability to provide selected types of material behavior. Many combinations of matrices and reinforcing ceramics have been investigated since the late 1950s.

The ceramic reinforcements of the 1990s can be divided into four categories: continuous fibers (boron, graphite, alumina, silicon carbide), discontinuous fibers (alumina, alumina-silica), whiskers (silicon carbide), and particulates (silicon carbide, boron carbide). Numerous metals have been used as matrices; the most important have been aluminum, titanium, magnesium, and copper alloys and superalloys.

The decades that began in the 1960s saw rapid development in the area of composites that provide high strength, toughness, and low weight, as needed, for example, in automobiles and **aircraft**. In the cermets of the 1960s, boron and graphite were used to reinforce light **alloys**. The high-strength, lightweight boron filaments found use in advanced aerospace structures. The 1970s saw the introduction of oxide fibers (especially alumina-based) in cermets. Alumina had been used for ages as a temperature-resistant material, and was especially attractive owing to its relatively low price. In the 1980s, cermets began to make use of silicon carbide and silicon nitride fibers. By the 1990s the driving force for composite development had become the production of components that approximated very closely the required final shape.

By the end of the 1990s, cermets were either in use or in prototype for the **space shuttle**, commercial airliners, electronic substrates, **bicycles**, automobiles, lightweight body **armor**, components for hard disk drives, bearings and other wear parts, sporting goods, cutting tools, construction materials for fusion applications, and a variety of other applications. Titanium reinforced with silicon carbide fibers was under development as skin material for the national aerospace plane. While the vast majority were aluminum matrix composites, a growing number of applications were beginning to require the matrix properties of superalloys, titanium, copper, magnesium, or iron. A boron carbide-aluminum cermet developed at Lawrence Livermore National Laboratory for military armor employed boron carbide as the dispersed phase and aluminum as the matrix. The cermet was stronger for its weight than steel, lighter than aluminum, and five times more resistant to fracture than conventional structural ceramics.

See also Composite materials; Ceramics

Cesarean section

Cesarean section is the removal of an unborn child from the uterus by means of surgical incision through the abdominal wall. Originally practiced only on dead women, cesarean section today is a common and relatively safe birth experience.

Surgical removal of a fetus from a dead or dying mother was mandated for religious purposes by several ancient cultures—in Egypt in 3000 B.C. and in India in 1500 B.C. in order to provide separate burial for the two individuals. The ancient Roman law code, known as *lex caesaria*, mandated this procedure in an attempt to save the baby and is the probable source of the operation's name (not the legend about the unlikely surgical birth of Julius Caesar [100-44 B.C.], whose mother lived for many years after his delivery).

Sporadic attempts to perform cesarean section as a means of saving both mother and baby seem to have occurred in medieval Europe. Records from Frankfurt-am-Main, Germany, claim seven caesareans were performed there before 1411; a French physician reported fifteen cesarean operations by 1581. In this era of incredibly crude surgery, however, it is unlikely that many of these cases could have had nonfatal consequences for the mother.

One of the earliest reports of a successful cesarean operation dates to the year 1500, when a Swiss pork butcher or sow gelder named Jacob Nufer used his practiced skills to deliver his own wife of their child. The first reliably documented cesarean section was performed by Jeremiah Trautman in 1610 in Wittenberg, Germany. A renowned Dutch physician, Hendrik van Roonhuyze, championed the procedure and included illustrations of his method of cesarean incision in his 1663 book on operative gynecology. Cesarean section came to the British Isles in 1738, when an Irish midwife named Mary Donally performed a successful emergency operation. Cesarean delivery was practiced successfully in the United States by John Lambert of Ohio in 1827 and Francois Prevost in Louisiana before 1832. A patient of William Gibson of Baltimore, Maryland, lived for fifty years after her first delivery of two cesarean births in 1835.

Although these and other cases demonstrated that cesarean delivery could be successful, the operation was largely avoided throughout the eighteenth and most of the nineteenth century because of the dreadful maternal mortality rate—between 50 and 75 percent. Before the discovery of anesthesia in 1847, a cesarean was an agonizing procedure for the mother, strictly a last-resort option. Massive infection was an extremely likely outcome until the adoption of antiseptic principles for surgery after 1867. Internal bleeding, too, killed many mothers until two German physicians, Max Sanger and Adolf Kehrer, began to practice uterine suture—previously ignored—in 1882.

Once anesthesia, antisepsis, and uterine suture became standard, cesarean delivery became a viable and sensible option. During the early 1900s cesarean section gradually replaced other alternatives such as high **forceps** delivery, cutting of the pubic bone, and destruction of the fetus. As the birthplace moved from home to hospital, and the cesarean mortality rate dropped to near zero by 1960, the rate of cesarean delivery rose dramatically. This rate was spurred on by the dictum "Once a cesarean, always a cesarean" and by the fear of today's physicians that they will be sued for malpractice if they fail to perform a cesarean. Cesarean section is the most common major operation performed in the United States. However, in the 1990s, the percentage of cesarean sections in the U.S. has steadily declined from a high of 25 percent in 1990 to approximately 20 percent at the end of the decade. The

World Health Organization has set a goal of reducing the use of cesarean sections to 10 to 15 percent, citing that the procedure is costly, often unnecessary, and though largely safe still increases the chances of maternal death by two to four times compared to vaginal births.

CFCs • See Chlorofluorocarbon

Chain

Decorative small chains fashioned from gold and silver were made back in the time before Christ. By 200 B.C. Greeks used chains for water-raising machines; later, Romans used bronze chains in their galleys. Other uses of chains were explored during the Renaissance. **Leonardo da Vinci** sketched both hinged link chains and continuous chain drives with sprocket wheels in his manuscripts, but its not known whether the chains he drew were ever made. Iron chains and buckets were used in the 1500s as part of pumping machinery to replace ropes and earthenware pots. In a 1588 drawing of a water-raising machine, there is a chain with square links that fit over projecting teeth on wooden wheels; each square link is connected to the next one by three oval links.

Due to the high cost and scarcity of suitable metals, as well as the absence of good tools to create chains, there was a very limited use of chains before the Industrial Revolution before Thomas Bunton invented chain links with a central stud, which added strength to the links, neutralizing their tendency to stretch under strain. About the same time, in 1808, Samuel Brown patented a design for an improved iron chain. This was followed in 1820 by the first ship's cable, created by Noah Hingley, an English blacksmith who used only a forge, hammer, and anvil.

Because England had a rich source of iron ore, it became the center for chain making; its iron was superior to steel, which tended to corrode too badly to be useful over a long period of time. The chain drive was perfected about this same time. In 1864 Englishman James Slater patented a driving chain precise enough to be useful for driving **bicycle**s and various machines. His factory was later acquired by a Swiss, Hans Renold, who devised a better chain in 1880—the bush roller chain which allowed more weight to be added to the chain. In France Andre Guilmet invented a bicyclette that had the chain drive attached to the rear wheel. This early bicycle was then manufactured by Meyer et Cie in 1868. The chain drive in use then was still comparatively weak until another Frenchman, G. Juzan, developed what he called bicyclette moderne in 1885:

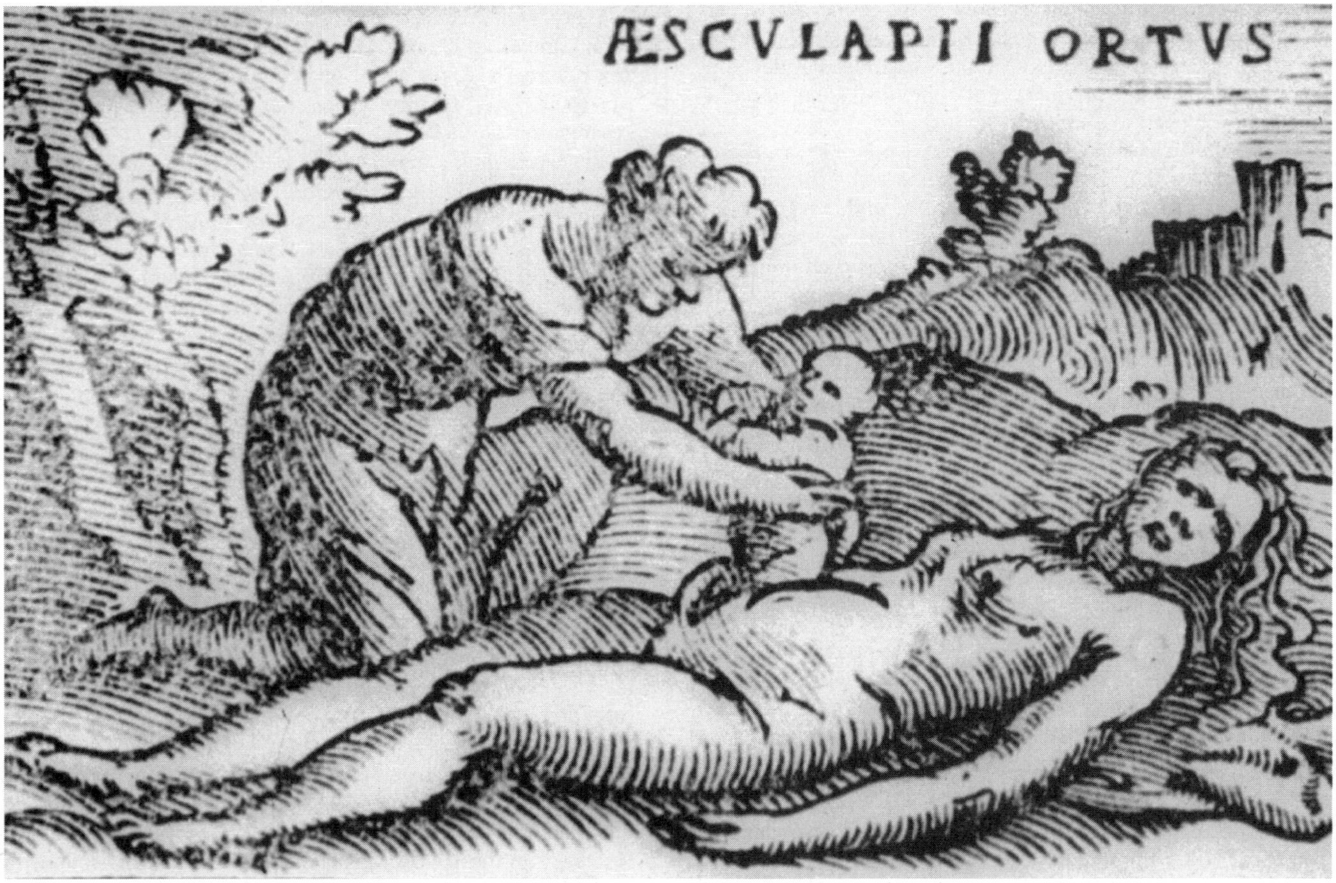

A woodcut dated 1533 depicting a cesarean section. Until the advent of antisepsis and more advanced surgical techniques, such a procedure was only done as a last resort, often when the mother was dying or already dead.

it had same-size wheels with a better chain drive to the back wheel. Finally, John Kemp Starley created a commercially successful "safety" bicycle in 1885 with the same features as the French model. From the bicycle came the use of chain drives in **motorcycles**.

Today's large chains are made by forge **welding**. With this process, metal studs can be inserted into the links while they are still hot if added strength is necessary. Special hydraulic machines can then test the finished product.

Chalk

Chalk used in school classrooms comes in slender sticks approximately 0.35 of an inch (9 mm) in diameter and 3.15 in (80 mm) long. Lessons are often presented to entire classes on chalkboards (or blackboards, as they were originally called) using sticks of chalk because this method has proven cheap and easy.

As found in nature, chalk has been used for drawing since prehistoric times, when, according to archaeologists, it helped to create some of the earliest cave drawings. Later, artists of different countries and styles used chalk mainly for sketches, and some such drawings, protected with shellac or a similar substance, have survived. Chalk was first formed into sticks for the convenience of artists. The method was to grind natural chalk to a fine powder, then add water, clay as a binder, and various dry colors. The resultant putty was then rolled into cylinders and dried. Although impurities produce natural chalk in many colors, when artists made their own chalk they usually added pigments to render these colors more vivid. Carbon, for example, was used to enhance black, and ferric oxide (Fe_2O_3) created a more vivid red.

Chalk did not become standard in schoolrooms until the nineteenth century, when class sizes began to increase and teachers needed a convenient way of conveying information to many students at one time. Not only did instructors use large blackboards, but students also worked with individual chalkboards, complete with chalk sticks and a sponge or cloth to use as an eraser. These small chalkboards were used for practice, especially among the younger students. Pens dipped in ink wells were the preferred tool for writing final copy, but these were reserved for older students who could be trusted not to make a mess: **paper**—made solely from rags at this time—was expensive.

An important change in the nature of classroom chalk paralleled a change in chalkboards. Blackboards used to be black, because they were made from true slate. While some experts advocated a change to yellow chalkboards and dark blue or purple chalk to simulate writing on paper, when manufacturers began to fashion chalkboards from synthetic materials during the twentieth century, they chose the color green, arguing that it was easier on the eyes. Yellow became the preferred color for chalk.

Almost all chalk produced today is dustless. Earlier, softer chalk tended to produce a cloud of dust that some feared might contribute to respiratory problems. Dustless chalk still produces dust; it's just that the dust settles faster. Manufacturers accomplish this by baking their chalk longer to harden it more. Another method, used by a French company, is to dip 80% of each dustless chalk stick in **shellac** to prevent the chalk from rubbing off onto the hands.

Chang Min Chueh • See Birth control

Chappe, Claude • See Semaphore

Chardonnet, Hilaire, Comte de (1839-1924)

French chemist

Chardonnet's career in science began with engineering studies at the École Polytechnique; he also assisted Louis Pasteur (1822-1845) in his efforts to save the French silk industry from a devastating silkworm epidemic. Realizing there was a market for an artificial silk, Chardonnet built upon the work of the Swiss chemist George Audemars and Sir **Joseph Swan** of England to develop **cellulose**-based **fibers**. Audemars had received a patent in 1855 for the manufacture of synthetic fibers; by 1880 Swan had developed threads from nitrocellulose.

Chardonnet first treated cotton with nitric and sulfuric acids and then dissolved the mixture in alcohol and ether. He then passed the solution through glass tubes, forming fibers, and allowed them to dry. These fibers, called **rayon** (the term used in referring to any fiber developed from cellulose) were highly flammable until they were denitrated. Reportedly, some garments made of early rayon burst into flames when lit cigarettes were nearby, Unfortunately the techniques that existed at that time to denitrate the material weakened it and made it unsuitable for the textile industry. Chardonnet used ammonium sulfide to denitrate these fibers thus reducing the flammability and retaining fiber strength comparable to that of silk. He received the first patent for his work in 1884 and began manufacturing rayon in 1891. The material was displayed at the 1891 Paris Exposition where it won the grand prize. He was also awarded the Perkin medal in 1914 for his development of rayon.

After his work with fibers, Chardonnet went on to study a number of other subjects including ultraviolet light, telephony, and the movements of birds' eyes. He died in 1924 in Paris, France.

Charnley, John (1911-1982)

English surgeon

A skilled and innovative orthopedic surgeon, John Charnley was born in Bury, Lancashire, England. He was an outstanding medical student at Manchester University, receiving his degree there in 1936. He served in the British army medical corps during World War II, supervising production of splints to be used for wounded soldiers suffering from bone fractures. After the war he joined the orthopedic department of the Manchester

Royal Infirmary, leaving in the mid-1960s to develop the Centre for Hip Surgery at Wrightington Hospital in Lancashire. Under Charnley's direction, the Centre became the world's major, state-of-the-art center for hip replacement surgery.

In the earlier part of his career, Charnley worked out a method, called arthrodesis, of surgically fusing the surfaces of joints in patients suffering from rheumatoid arthritis. Although this made the joint immobile, it did eliminate pain in the joint. Charnley also wrote a notable book on orthopedic surgery, *The Closed Treatment of Common Fractures*, published in 1950.

Charnley's major contribution to orthopedic surgery was his development of successful methods of replacing hip joints. Earlier attempts at hip replacement had not achieved satisfactory results. Charnley realized that more than medical skill was needed to solve the problem. He made very careful, detailed studies of the engineering principals involved, and he investigated and tested new synthetic materials. He achieved initial success with low-friction Teflon to produce a smoothly moving joint. During the 1960s he had increasingly satisfactory results with high-density plastics and had perfected the procedure and materials by 1972. His studies on the control of infection after surgery, especially the use of air tents to maintain sterile conditions during the operation, also contributed to the success of his hip replacement procedure, which has become almost standard treatment for severe hip joint degeneration.

Charnley was knighted for his achievements in 1977 and became a Fellow of the Royal Society in 1975. He died suddenly in 1982.

Charpak, Georges (1924-)

Polish French physicist

Georges Charpak received the Nobel Prize in physics in 1992 for his invention and development of particle detectors, most notably the multiwire proportional chamber. A number of his colleagues, who received the Nobel Prize before him, had used his invention to make important discoveries in physics. Charpak is credited with creating instrumentation that is used by thousands of other scientists at CERN, the European laboratory for particle physics located in Geneva, Switzerland, as well as by researchers in other prominent laboratories involved in the study of the nature of matter.

Born on August 1, 1924, in Dabrovica, Poland, to Maurice Charpak and Anna Szapiro, Charpak moved to France with his family in 1929. In 1943 the French Vichy government accused the young Charpak of being a terrorist and sentenced him to the concentration camp in Dachau, West Germany (now Germany). Charpak remained in the camp until its liberation in 1945. Upon his return to France, he completed a degree in civil engineering from the Ecole des Mines in Paris, and in 1946 he became a French citizen.

Two years later, as a graduate student in nuclear physics at the Collège de France in Paris, Charpak went to work in the laboratory of physicist Frédéric Joliot-Curie. In this laboratory Charpak began building the equipment he needed to perform his experiments (he constructed his own equipment out of necessity, as the laboratory had none). In 1955 he received his Ph.D. from the Collège de France.

In awarding the Nobel Prize to Charpak, the Swedish Academy of Sciences traced the history of the development of detector devices in physics. The cloud chamber and the bubble chamber were two earlier inventions that had received recognition from the academy. Both relied on photographic techniques to capture particle events. In 1958 Charpak was invited by experimental physicist Leon Lederman, who had heard Charpak lecture in Padua, to come to CERN to work on sparking devices to detect particles. Though these devices were an improvement over existing techniques, they, too, relied on photographic recording, which was slow and cumbersome to analyze. Charpak turned to the problem of a spark chamber reading without photographic film, and built his first multiwire proportional tracking chamber in 1968. The multiwire proportional chamber extended the technology of the Geiger-Muller tube, bubble chamber, and cloud chamber in two ways: The multiwire proportional chamber replaced the single positively charged wire of the Geiger-Muller tube, which attracts electrons in a chamber of ionized gas, with a multiwire device; it also replaced photographic analysis of a trail of bubbles with computerized electronic analysis of current produced in the wires as they attract electrons. Charpak credited his background in nuclear physics with the success of his invention.

Since liberating physics from dependence on film readings, Charpak has turned his interests to medicine and aerospace problems. Work he has done in the latter area makes it possible to produce an X-ray radiograph of turbine blades as they spin. In the field of medicine, his chamber is able to analyze the structure of a protein with X rays a thousand times faster than was previously possible. He also is working on imaging problems to identify receptors in the brain.

Charpak's particle work mimics the state of the universe as it was a fraction of a second after the ''big bang.'' It is believed that some of the particles have not existed in nature since that time, and the ability of physicists to study them will reveal and increase the understanding of the relationships among the forces of nature. Whereas Charpak built on the work of his predecessors with the bubble and spark chambers, others have used his invention to make their own contributions to the field of physics. A group led by Samuel Ting discovered the first manifestation of charmed quarks at Brookhaven in 1974, and Carlo Rubbia led a group to discover the W and Z particles. (Charmed quarks and W and Z particles are subatomic particles.)

Charpak has worked on behalf of other scientists imprisoned by repressive governments. He is the founder of the SOS committee at CERN. This association worked diligently on the part of Soviet dissidents, such as Andrei Sakharov, Yuri Orlov, and Anatoly Sharansky, when they were deprived of their civil rights under the former Soviet Union.

Charpak has been a member of the French Academy of Sciences since 1985. In addition to his continued association with CERN, Charpak is also the Joliot-Curie Professor at the Ecole Supérieure de Physique et Chimie in Paris, a position he has held since 1984. In 1989 he received the High Energy and

Particle Physics Prize from the European Physical Society. He has published numerous papers in scientific journals.

Cheese

Cheese is a fermented food derived from the milk of various mammals. Since humans began to domesticate milk-producing animals around 10,000 B.C., they have known about the propensity of milk to separate into curds and whey. As milk sours, it breaks down into curds, lumps of phosphoprotein, and whey, a watery, grey fluid that contains lactose, minerals, vitamins, and traces of fat. It is the curds that are used to make cheese, and practically every culture on Earth has developed its own methods, the only major exceptions being China and the ancient Americas.

The first cheeses were "fresh," that is, not fermented. They consisted solely of salted white curds drained of whey, similar to today's cottage cheese. The next step was to develop ways of accelerating the natural separation process. This was achieved by adding rennet to the milk. Rennet is an enzyme from the stomachs of young ruminants—a ruminant is an animal that chews its food very thoroughly and possesses a complex digestive system with three or four stomach chambers; in the United States, cows are the best known creatures of this kind. Rennet remains the most popular way of "starting" cheese, though other starting agents such as lactic acid and various plant extracts are also used.

By 100 cheese makers in various countries knew how to press, ripen, and cure fresh cheeses, thereby creating a product that could be stored for long periods. Each country or region developed different types of cheese that reflected local ingredients and conditions. The number of cheeses thus developed is staggering. France, famous for the quality and variety of its cheeses, is home to about 400 commercially available cheeses.

The next significant step to affect the manufacture of cheese occurred in the 1860s, when Louis Pasteur introduced the process that bears his name. Pasteurization entails heating milk to partially sterilize it without altering its basic chemical structure. Because the process destroys dangerous microorganisms, pasteurized milk is considered more healthful, and most cheese is made from pasteurized milk today.

The first and simplest way of extending the length cheese would keep without spoiling was simply ageing it. Aged cheese was popular from the start because it kept well for domestic use. In the 1300s, the Dutch began to seal cheese intended for export in hard rinds to maintain its freshness, and, in the early 1800s, the Swiss became the first to process cheese. Frustrated by the speed with which their cheese went bad in the days before refrigeration, they developed a method of grinding old cheese, adding filler ingredients, and heating the mixture to produce a sterile, uniform, long-lasting product. Another advantage of processing cheese was that it permitted the makers to recycle edible, second-grade cheeses in a palatable form.

Prior to the twentieth century, most people considered cheese a specialty food, produced in individual households and eaten rarely. However, with the advent of mass production, both the supply of and the demand for cheese have increased. In 1955, 13 percent of milk was made into cheese. By 1984, this percentage had grown to 31 percent, and it continues to increase. Interestingly, though processed cheese is now widely available, it represents only one-third of the cheese being made today. Despite the fact that most cheeses are produced in large factories, a majority are still made using natural methods. In fact, small, "farmhouse" cheese making has made a comeback in recent years. Many Americans now own their own small cheese- making businesses, and their products have become quite popular, particularly among connoisseurs.

Chemical warfare

Throughout history there has been continued use of chemicals in warfare. Ancient armies burned painted wood to create smoke screens or to force enemies from hiding places. Two thousand years before Christ, Indian soldiers used smoke screens, incendiary weapons, and toxic fumes in battle. The Spartans used sulfur dioxide against their rivals, the Athenians, during the Peloponnesian War in 429 B.C. by burning pitch and sulfur on wood to create poisonous sulfur smoke near an enemy city. The Byzantine's were able to destroy enemy ships by using **Greek fire**, a chemical mixture that burst into flames when it came into contact with water. During the Middle Ages, a group of Christians saved Belgrade from advancing Turks by dipping rags in poison, lighting them, and fanning the fumes at the enemy.

Chemical warfare was put to greater use during World War I. When England set up a naval blockade that prevented Germany from importing nitrates to use in manufacturing explosives, Germany turned to its chemical industry. **Fritz Haber**, a chemist who had devised a way to make nitrogen into ammonia for use as fertilizers or explosives, was appointed the head of Germany's chemical warfare service. The naval blockade convinced many Germans that they had to turn to chemicals to continue the war. Haber devised a way to use chlorine, released from cylinders, to form a gas cloud that would blow onto the Allied front lines. The German army introduced this weapon on April 22, 1915, by releasing 160 tons of liquid chlorine from nearly 6,000 pressurized cylinders over Ypres, Belgium. The Allied soldiers were totally unprepared, and the casualties were horrendous: 5,000 dead and 10,000 injured. Both sides worked on developing better delivery systems, which included **artillery** shells. Haber created another chemical called phosgene, an asphyxiating gas that had delayed effects on those who inhaled it, and later created mustard gas, an agent that produces severe blisters on all body surfaces. In addition, mustard gas remained on the ground and equipment where it could cause casualties long after the original attack. Unlike chlorine and phosgene, mustard gas was so vicious that there was no way to effectively defend against it.

During World War II newer and more powerful chemicals were created for use on battlefields. In Germany, Dr. Gerhardt Schrader, a chemist, was developing ways to destroy insects when by accident he came upon a compound that proved highly toxic to mammals. It was called tabun, the first of the nerve gases. When tabun was inhaled or absorbed through the skin, it affected the human nervous system. This substance was very effective; it killed its victim in minutes, while phosgene and mustard gas took hours. Soon Schrader developed a second nerve agent, sarin. Because the Germans had erroneously assumed that the Allies had access to such weapons, the military never did use it during the war. The Japanese, however, had assembled a huge arsenal of both chemical and biological weapons between 1930 and 1945, and these were used in Manchuria.

After World War II, the arms race between the United States and the Soviet Union included the development of deadlier chemical weapons. The United States developed and used Agent Orange, a defoliant, in the Vietnam War in hopes of depriving the enemy of hiding places in forests and jungles. Both Iran and Iraq built up chemical weapons arsenals in the 1980s. These weapons of mass destruction are presumably held as a check to the nuclear threat of Israel and the United States. Iraq used its chemical weapons against the Iraqi Kurdish minority in the 1980s, and threatened chemical attack during the Persian Gulf War. Other small nations, including Libya and Syria, are known to have chemical weapons arsenals. The growing reliance on chemical weapons in the 1990s may be because they are cheaper and easier to procure than nuclear weapons.

In Japan in 1995 a religious cult released the chemical agent sarin in the Tokyo subway system, killing twelve people and injuring another 3,000. This was the most damaging terrorist use of chemical weapons to date, inspiring fear worldwide that other splinter groups might make similar attacks. Beginning in 1997, the Domestic Preparedness Program of the United States Defense Department began spending millions of dollars to stage mock chemical attack exercises in more than a hundred U.S. cities, to prepare emergency workers for chemical or biological attack. Many terrorism experts worldwide agree that the knowledge and materials to build chemical weapons is becoming easier to obtain, legitimizing fears that more chemical attacks may take place in the future.

See also Gas mask; Napalm

Chewing gum

People have been chewing naturally gummy substances since earliest times. The ancient Greeks chewed resin from the mastic tree, and the Maya of Central America chewed chicle, the latex sap of the wild sapodilla tree. The North American Indi-

American troops advancing on a World War I battlefield while wearing gas masks. The soldier at left, unable to don his mask, clutches his throat as he breaths in poisonous gas.

ans chewed spruce tree sap and taught the New England colonists to do the same. In the mid-1800s, sweetened paraffin replaced spruce resin as the preferred substance for chewing gum.

Chicle, the base for chewing gum, is presumed to have been brought to the United States by the Mexican general Antonio López de Santa Anna (1794-1876) in the 1860s. When he returned to Mexico, the general left a chunk of chicle behind with an acquaintance, inventor Thomas Adams, Sr. Adams experimented with uses for the chicle; while it did not make a good rubber substitute, Adams found that it did make an excellent chewing gum. The inventor set up a Jersey City factory and began marketing his tasteless "Adams' New York Gum—Snapping and Stretching."This was soon followed by licorice-flavored Black Jack.

Salesman Thomas Adams, Jr., promoted the gum, and sales and competitors soon proliferated. William F. Semple of Ohio added more flavors to chewing gum. Adams responded by dispensing gum in vending machines. William Wrigley, Jr. (1861-1932) made his gum the world's most advertised product; he began by giving away sticks of "chewing candy" to promote sales of his father's baking powder. William White added peppermint flavoring to corn syrup, blended this with chicle, and became a millionaire on the sales of his Yucatan gum.

Workers slit the bark of chicle-producing trees and collect in cups the latex that runs out. The small quantities are later combined, boiled, and formed into blocks for shipment. At the chewing gum factory, the chicle is ground, melted, and purified. After sweetners and flavors are added, the gum is rolled into balls or flattened and sliced into the sticks we find in stores.

The popularity of chewing gum was not universal. Some people, including parents and teachers, considered it a vice to chew gum. Doctors warned that chewing gum would exhaust the salivary glands and that a piece of gum accidentally swallowed would cause the intestines to stick together. Nevertheless, people continued to chew gum, especially soldiers in the two world wars.

Bubble gum was first produced in the early 1900s by Frank Fleer of the Fleer Company. Called Blibber-Blubber, it was unsatisfactorily wet and sticky. Fleer's much-improved Dubble Bubble Gum appeared in 1928 and was an instant success. Frank's brother, Henry, came up with the candy-coated chicle tablets which the Fleers named Chiclets.

Chicle remained the basic ingredient of chewing gum until World War II. Wartime shortages spurred the development of synthetic gum and plastic substitutes for chicle. Sugar-free gum came on the market in the mid-1960s and is a top seller today. Some gum manufacturers produce chewing gum that is advertised not to stick to or damage dentures or fillings.

Gum, currently a mix of 40 to 50 compounds, has included additives that claim to promote weight loss, stop smoking, and provide vitamin supplements and relieve motion sickness. The most popular choices continue to be the standard chewing gums found in retail outlets.

Chipboard

Chipboard is heavy paperboard made from recycled **paper**. Examples of chipboard include the backing of pads of paper, egg cartons, and stiffeners for photographs in mailing; it is chiefly used as boxboard and containerboard for corrugated packaging. Chipboard makes up the largest class of paper products consumed in the United States. While most cardboards are designed for strength and attractiveness for packaging and other purposes, chipboard is used when these qualities are not essential. When treated with wax, it is useful as a grease-proof packaging material used for bacon and motor oil containers.

A patent for the first machine for pasting sheets of paper together, forming chipboard, was granted in 1824 to an Englishman by the name of John Dickinson (1782-1869), inventor of the cylinder papermaking machine (1809). The cylinder papermaking machine has a wire screen fitted like a jacket around a cylinder that is immersed in a vat of pulp. Multiple layers of pulp are then worked together in the couching, or laying out, stage to form the chipboard. This type of machine is still used for the production of chipboard. American Thomas Gilpin (1776-1853) erected the first paper-machine in America in 1817; this was a cylinder machine suitable for the production of heavy paper from rags. The first patent for corrugated paper of the type used later in boxes was granted to American Albert L. Jones in 1871.

The early 1900s saw the development of a variety of cardboards and chipboards. In 1903, commercial use of corrugated containers began. By 1936, the use of chipboard packaging for milk containers was widespread. Consumer acceptance of frozen orange juice in 1948 led to the expansion of the **frozen foods** industry with its chipboard packaging. Wastepaper, consisting chiefly of recycled newspaper, is an important input material for paper mills in their production of chipboard. In 1990, chipboard consumed 51 percent of the wastepaper remaining in the United States after export. In Britain, the term chipboard refers to a wood product made from chips that are compressed and veneered. Used in construction and furniture manufacturing, this material is called particle board or hardboard in the United States.

Chlorofluorocarbon

A chlorofluorocarbon (CFC) is an organic compound typically consisting of chlorine, fluorine, carbon, and hydrogen. Freon, a trade name, is often used to refer to CFCs, which were invented in the 1930s and have been used widely as aerosol propellants, refrigerants, and solvents. Odorless, colorless, nontoxic, and nonflammable, CFCs are considered valuable industrial products and have proven an especially safe and reliable aid in food preservation. However, the accumulation of CFCs in the stratosphere, which may be linked to possible ozone depletion, has generated considerable public debate and has led to legislation and international agreements banning the production of CFCs by the year 2000.

In the late 1920s, researchers had been trying to develop a coolant that was both nontoxic and nonflammable. At that time, methyl chloride was used, but if it leaked from the refrigerator, it could explode. This danger was demonstrated in one case when methyl chloride gas escaped, causing a disastrous explosion in a Cleveland hospital. Sulfur dioxide was sometimes used as an alternative coolant because its unpleasant odor could be easily noticed in the event of a leak. The problem was brought to the attention of Thomas Midgley Jr., a mechanical engineer at the research laboratory of General Motors. He was asked by his superiors to try to manufacture a safe, workable coolant. (At that time, General Motors was the parent company to Frigidaire.) Midgley and his associate chemists thought that fluorine might work because they had read that carbon tetrafluoride had a boiling point of-15 degrees Celsius. The compound, as it turns out, had accidentally been referenced. Its actual boiling point is 133.96°F (92.2°C), not nearly the level necessary to produce refrigeration. Nevertheless, the incident proved useful because it prompted Midgley to look at other carbon compounds containing both fluorine and chlorine. Within three days, Midgley's team discovered the right mix: dichlorodifluoromethane, a compound whose molecules contain one carbon, two chlorine, and two fluorine atoms. It is now referred to as CFC-12 or F-12 and marketed as Freon—as are a number of other compounds, including trichlorofluoromethane, dichlorotetrafluoroethane, and chlorodifluoromethane.

Midgley and his colleagues had been correct in guessing that CFCs would have the desired thermal properties and boiling points to serve as refrigerant gases. Because they remained unreactive, and therefore safe, CFCs were seen as ideal for many applications. Through the 1960s, the widespread manufacture of CFCs allowed for accelerated production of refrigerators and **air conditioners**. Other applications for CFCs were discovered as well, including their use as blowing agents in polystyrene foam. Despite their popularity, CFCs became the target of growing environmental concern by certain groups of researchers. In 1972, two scientists from the University of California, F. Sherwood Rowland (1927-) and Mario Jose Molina, conducted tests to determine if the persistent characteristics of CFCs could pose a problem by remaining indefinitely in the atmosphere. Soon after, their tests confirmed that CFCs do indeed persist, until they gradually ascend into the stratosphere, break down due to ultraviolet radiation, and release chlorine, which in turn affects ozone production. Their discovery set the stage for vehement public debate about the continued use of CFCs. By the mid-1970s, the United States government banned the use of CFCs as aerosol propellants but it resisted a total ban for all industries. Instead, countries and industries began negotiating the process of phasing out CFCs. As CFC use is allowed in fewer and fewer applications, a black market has been growing for the chemical. In 1997, the Environmental Protection Agency, Customs Service, State Department, Justice Department, IRS, and the FBI initiated enforcement actions to prevent (CFC) smuggling in the US.

Many CFCs have been replaced with related materials known as hydrochlorofluorocarbons (HCFCs) which reportedly do not interfere with the ozone layer. However, these materials are not free from controversy either. A 1997 Belgian health officials report indicate hydrochlorofluorocarbon-123 may be responsible for a rise in acute hepatitis among workers of a smelting plant. Liver damage found in nine workers is believed to be caused by a leaking refrigerant pipe in the workers' cabin. The U.S. Environmental Protection Agency, disputes these findings.

Chromatography

Chromatography is a technique used in analytical chemistry to separate and identify components of mixtures. The name comes from the Greek term for ''color writing'' because this method was originally used to separate colored samples. Today a variety of chromatography methods work by separating the individual components of mixtures so that each one can be analyzed and identified. Instruments used to perform chromatography are known as chromatographs, which have become essential pieces of equipment in chemical laboratories. With these devices, scientists can tell what chemical compounds are present in complex mixtures such as smog, cigarette smoke, petroleum products, or even coffee aroma.

The first chromatograph was invented just after the turn of the twentieth century by Russian botanist Mikhail Semenovich Tsvett (1872-1919), whose name (sometimes rendered as ''Tswett'') coincidentally means ''color'' in Russian. Tsvett was born in Italy of a Russian father and Italian mother. After studying in Switzerland and doing some research in St. Petersburg, Tsvett settled in Warsaw, which at that time belonged to the Russian empire. During his years in Poland, Tsvett's botanical research led to his great moment of insight. Tsvett was looking for a method of separating a mixture of plant pigments, which are chemically very similar to each other. To isolate different types of chlorophyll, Tsvett trickled a mixture of dissolved pigments through a **glass** tube packed with calcium carbonate powder. As the solution washed downward, each pigment stuck to the powder with a different degree of strength, creating a series of colored bands. Each band of color represented a different substance, and Tsvett referred to the colored bands as a chromatogram. He also suggested that the technique (now called adsorption chromatography) could be used to separate colorless substances. Although Tsvett published a report of this work in the early 1900s, chemists paid very little attention to it. For one thing, the report was written in Russian, which few Western chemists of the time read. Also, the technique of chromatography may have seemed too simplistic to chemists, who were used to relying on lengthy extraction, crystallization, or distillation to separate mixtures. But within a few years, Tsvett's technique was rediscovered by German organic chemist Richard Martin Willstatter (1872-1942), who was also studying chlorophyll. By introducing chromatography to Western European scientists, Willstatter helped establish one of the most versatile analytical techniques known to chemistry.

Willstatter went on to define the major types of chlorophyll and discover the importance of magnesium in the chlorophyll molecule. Soon chromatography was found to work on almost all kinds of mixtures, including colorless ones, as Tsvett had predicted. Absorbing powders were discovered that perform better than calcium carbonate for separating ordinary molecules. Also, compounds known as *zeolites* were introduced to separate individual ions, or electrically charged particles, in a process called ion-exchange chromatography. American chemist Frank Harold Spedding adapted this technique to the separation of rare-earth metals. In the 1930s, synthetic resins were developed for complex ion-exchange processes. During World War II, life rafts were equipped with survival kits that contained resins for removing most salts from seawater.

The most dramatic advance in the history of chromatography took place in 1944, when scientists discovered that a strip of porous filter **paper** could substitute for the column of absorbing powder. In this technique, called paper chromatography, a drop of the mixture to be separated is placed on the paper, then one edge is dipped into a solvent. The solvent spreads across the paper, carrying the mixture's components with it.

When the components are finished spreading, the paper is dried and sprayed with a reagent that reveals a change in color. Because the components move at different speeds, they show up as distinct, physically separated spots that can be cut out with scissors and further analyzed. The paper method is a type of partition chromatography, which is based on differences in solubility rather than differences in adsorption. One of its advantages is that it requires only a small sample of material.

Paper chromatography was invented by two British biochemists, Archer John Porter Martin (1910-) and Richard Laurence Millington Synge (1914-). Martin was the son of a physician, while Synge's father was a stockbroker. Both scientists studied at Cambridge University, where Martin earned his Ph.D. in 1936 and Synge in 1941. Martin began working in the university's nutritional laboratory, where he investigated problems related to vitamin E. For a few years, he also became involved in a study of the felting of wool.

In 1941, Martin and Synge began working together on proteins, which are made up of chains of amino acids. Martin and Synge were trying to characterize a particular protein by determining the precise numbers of each amino acid present. Amino acids are so similar to each other, however, that the problem of separating them had defeated a whole generation of biochemists. Martin and Synge's development of paper chromatography to solve this problem was an instant success, not only on amino acids but also on various other mixtures. The two scientists were awarded the Nobel Prize in chemistry in 1952 for their work.

Martin and Synge's research led to a number of other important scientific advances. After Synge determined the structure of an antibiotic peptide called *Gramicidin-S*, Frederick Sanger (1918-) used paper chromatography to figure out the structure of the insulin molecule—not only the number of particular amino acids in it, but also the order in which they occurred. Insulin is now used to control blood sugar levels in people afflicted with diabetes.

The same technique was used by Melvin Calvin (1911-1997) during the 1950s to discover the complex series of reactions that enable green plants to convert solar energy into the chemical energy stored in food. Working with green algal cells, Calvin interrupted the photosynthetic process at different stages by plunging the cells into alcohol. Then he crushed them and separated their components via paper chromatography. Calvin was thus able to identify at least ten different intermediate products that had been created within a few seconds. Paper chromatography was also used by Austrian-American biochemist Erwin Chargaff (1905-), who modified the technique to study the components of the nucleic acid molecule. His research revealed four components, or nitrogenous bases, that occur in pairs. British biochemists Watson and Crick later used these results to work out the structure of DNA (deoxyribonucleic acid). The genetic material of humans, other animals, and plants is made of DNA, which is passed on from generation to generation and is responsible for all inherited traits.

In addition to inventing paper chromatography, Archer Martin developed another technique called gas chromatography, which enables chemists to separate mixtures of gases, or substances that can be vaporized or gasified by heat. Instead of a liquid solvent, helium gas is usually used to force the mixture through a column and separate the gaseous components. Martin and his colleague A. T. James first used gas chromatography to microanalyze fatty acids. The widespread acceptance of gas chromatography is unique in the laboratory instrumentation field. Today it is used in almost every branch of the chemical industry, particularly in the production of petrochemicals from oil and natural gas. One of the most common fixtures in biochemical laboratories is ''GCMS'' analytical equipment, which uses gas chromatography to separate individual components from complex organic mixtures, then uses mass spectrometry to identify each component.

Recently, chromatography has evolved into even more sophisticated analytical techniques. In thin-layer chromatography, for example, an alumina gel, silica gel, or other finely divided solid is spread onto a glass plate in a thin, uniform layer that takes the place of filter paper in the chromatographic process. This technique is not only faster than paper chromatography, but it can also separate smaller quantities of pure components. It is often used in the pharmaceutical industry to isolate penicillin and other antibiotics. A number of new products for chromatography were featured at the 1998 Pittsburgh Conference and Exposition on Analystical Chemistry and Applied Spectroscopy. For example, the Dionex company showed its latest ion exchange technology, the EG40, which can separate charged molecules. This device promises to be a valuable tool in the analysis of water quality and environmental contaminants and on-line monitoring of raw materials and production processes.

Chronometer

When Christopher Columbus arrived in America in 1492, he thought he had reached the shores of Asia. He probably would have been able to avoid the error and anticipate the achievement if he only had an accurate timepiece and an accurate star chart. With the two combined, he could have reliably charted his course.

Sailors had long been able to determine latitude (north-south position). By using a quadrant to measure the altitude of the sun or the polestar—the brightest star that appears nearest to either celestial pole at a certain latitude. But latitude is only half the data needed. Longitude and latitude intersect to pinpoint position.

While on land longitude could be measured and checked against reference points, at sea navigators had to relate the positions of the stars and other objects in the sky with time to determine where they were on the ocean. The necessary celestial measurements were made and eventually published in the early 1700s by such astronomers as John Flamsteed (1646-1719), England's astronomer royal, and the Frenchman Nicolas Louis de Lacaille (1713-1762), among others. Yet since early **clocks** did not work accurately at sea, navigation remained treacherous.

As far back as 1533, the German geographer Rainer (Regnier) Gemma Frisius (1508-1555) suggested that a navigator could determine his location with an accurate timepiece. In 1656, **Christiaan Huygens** (1629-1695) invented a clock that used a weight-and-pendulum system. But a clock that relied upon gravity as its driving force could not be very portable and certainly would not work on a ship being tossed about at sea.

Development of an accurate timepiece for navigators, or marine chronometer, was of such importance to the seafaring English that the British Parliament posted a 20,000 pound award for whomever could invent one of sufficient accuracy.

A carpenter's son and a self-taught carpenter, **John Harrison** (1693-1776) invented a chronometer that was successfully employed on a voyage to Lisbon in 1736. The English Board of Longitude did not give Harrison the recognition due him, however, until he developed his fourth chronometer, which proved to be accurate to within five seconds on a voyage to Jamaica in 1761-62. Still, it took a fifth improvement and a personal appeal to King George III before Harrison received his prize.

Also in 1761, another Englishman, Nevil Maskelyne (1732-1811), drew upon the work of Lacaille and others to determine his location on a voyage to St. Helena by relating the position of the moon with lunar navigation tables. These tables were later distributed to mariners. But this method was too complex to become a substitute for the chronometer.

The Frenchman Pierre Le Roy (1717-1785) created a chronometer in 1763 that used an adjustable gyration radius to compensate for temperature changes, which affect the accuracy of timepieces. Appointed Horologer de la Marine in 1764, Ferdinand Berthoud (1727-1807) improved on Harrison's and Le Roy's devices and wrote 10 volumes on the subject of marine chronometers.

In the late 1700s, John Arnold (1736?-1799) improved and simplified the chronometer by combining a helical balance spring with a pivoted detent escapement and was able to achieve amazing accuracy. About the same time, Thomas Mudge (1717-1794) devised a detached leaver escapement which allowed for uniform movement independent of external influences. His design became the basis for watch making during the nineteenth century. Also in the late eighteenth century, Thomas Earnshaw (1749-1829) constructed a spring-dent escapement and a temperature-compensated or ''bimetallic'' balance that became the pattern of marine chronometer construction for more than a century.

Despite its increasing accuracy, the chronometer could not be truly accurate until a standard system of marking off meridians of longitude was set. This took place in 1884, when an international convention agreed that the observatory at Greenwich, England, should become the Prime Meridian, or starting place, with a longitude of 0 degrees.

All timepieces originally were referred to as chronometers. As time passed the term acquired its current meaning: any precise time-keeping instrument used in navigation. The need for chronometers became less important in the 1920s when broadcasting time by radio signals became widespread. Later satellite locating systems further lessened their importance, though many navigators keep them on board for backup use. By the late twentieth century, quartz digital chronometers were gradually replacing their mechanical ancestors.

While at sea, beyond the sight of land, a marine navigator will take star sightings through a **sextant**. With a chronometer to time the readings, the ship's location can be calculated and charted.

CHRONOSCOPE • See Wheatstone, Charles

CHU, STEVEN (1948-)

American atomic physicist

Steven Chu was born in St. Louis, Missouri, on February 28, 1948. He attended the University of Rochester, from which he received a B.S. in physics and A.B. degree in mathematics. He then continued his education at the University of California at Berkeley, which awarded him a Ph.D. in physics in 1976. He remained at Berkeley for two more years, then accepted a position at Bell Laboratories in New Jersey.

In 1983, he was appointed head of the quantum electronics research department at the AT&T Bell Laboratories in Holmdel, New Jersey. It was during this time that Chu conducted the experiments for which he was awarded the 1997 Nobel Prize for Physics.

This research was motivated by a growing interest on the part of scientists to study the properties of individual particles, such as atoms, molecules, and ions. Such an objective would have seemed absurd only a few decades ago. In the first place, there are almost no mechanisms for viewing or otherwise measuring individual particles as small as atoms and molecules. In the second place, even if such instruments were available, they would be difficult to use with individual particles since such particles normally move at very great speeds. For example, the molecules of which air is composed move at speeds of about 2,480 miles (4,000 km) per hour at room temperature.

Chu developed a method for slowing down the motion of atoms and molecules so that their properties could be studied more carefully. The technique that he developed has been termed ''atom traps'' or ''optical molasses'' because it restrains atoms and molecules to a relatively small space and prevents them from moving out of that space for a period of time. The principle behind this technique is that a moving particle can be slowed down if it can be bombarded by a pulse of radiation. For example, suppose that an atom is moving from left to right across this page with a speed of 2,480 miles (4,000 km) per hour. Then suppose that one was to fire a pulse of **laser** radiation from the right to the left, aimed directly at the oncoming atom. When the laser pulse strikes the atom, the atom will absorb energy from the laser pulse and slow down (that is, have an increased energy in the backward direction). In theory, one might hope to stop the atom simply by hitting it with repeated bursts of radiation.

In fact, the process is not quite that simple. Very shortly after the atom absorbs energy from the laser beam, it emits a photon that contains as much energy as was absorbed. That energy may be given off in any direction. As it does so, the atom recoils in some random direction. Its left-to-right speed has been reduced, but a second laser pulse cannot be used to reduce its speed further in the same way.

Chu solved this problem by designing a ''box'' whose ''walls'' consisted of six laser beams arranged in three pairs. The beams were arranged along three parallel axes in such a way that an atom placed in the middle of the system was trapped. No matter which direction the atom tried to move, it was struck with one of the laser beams and was suspended in space. Even with this arrangement, atoms are not brought to a complete standstill. The system is able, however, to reduce the speed of particles to about 30 centimeters per second, a speed that corresponds to a temperature of about 240 micro kelvin, or 240 millionths of a degree kelvin. Calculations suggest that this temperature may be the lowest that can be obtained by the process described here.

If the above experiment is continued over a period of time, a second effect is observed. Instead of trapping a single particle, groups of particles are collected, all with similar energies and properties. This collection of particles acts like a very thin gas consisting of slowly moving particles. It corresponds to a new state of matter predicted in 1927 by Albert Einstein and now known as a Bose-Einstein condensate (BSC). The BSC has properties distinct from those of the other four states of matter (solid, liquid, gas, and plasma) and appears to behave in some respects like a single ''super-atom.''

In 1987, Chu accepted a professorship at Stanford University. There, Chu has continued his work on atom traps and optical molasses, but has also found new applications of these techniques. In these applications, laser beams are used to trap the ends of molecules and hold them in suspension long enough for their properties to be studied. The laser beams used in this way are sometimes referred to as ''optical tweezers.''

One of the interesting aspects of Chu's work with optical tweezers has been the study of deoxyribonucleic acid (DNA) molecules. As with protein molecules, DNA molecules consist of very long polymeric chains that have complex levels of organization. For example, they form spiral staircase-like coils whose structures, in that configuration, are sometimes difficult to study. Chu has found ways to use his ''optical tweezers'' laser beams to stretch DNA molecules out to a straight, spaghetti-like string, and then to study their properties in that form. One of the interesting discoveries he has made is that unraveled DNA molecules tend to vibrate, when agitated, in much the same way as a guitar string. His team had expected, Chu explains, that the vibration of a single DNA molecule might be more complex or, at least, different from that of a plucked string. Instead, the team has discovered that the behavior of a plucked strand of DNA can be described completely, simply by applying well known and traditional laws of simple harmonics.

In another series of experiments, Chu and his co-workers have found that identical DNA strands unravel when exposed to mild perturbations in completely different ways. This discovery is remarkable since one would assume that two molecules with identical chemical structures would behave in the same way. These results indicate that some factors are responsible for differences in the unraveling process that DNA molecules experience.

CIM (COMPUTER INTEGRATED MANUFACTURING) • See Computer, industrial uses of

CIRCUIT BREAKER AND FUSE

A circuit breaker does exactly what its name implies—it is designed to automatically break, or cut off, an electric circuit if the current in that circuit gets dangerously high.

The resistance electricity encounters as it flows through a circuit creates heat. If the electric current is great enough, the resulting heat can generate a fire. While circuits are designed very carefully, a circuit breaker or fuse provides a fail-safe condition and eliminates risk.

The first circuit breaker was based on the electric relay, which was invented by American physicist **Joseph Henry** in 1829. Turning electricity on and off activated a relay which controlled another circuit. In the late 1890s circuit breakers using **spring**s and **electromagnet**s came into use. If the current in a circuit became too high it would ''trip'' a relay, thus breaking the connection. When a circuit breaker trips, the fault in the circuit must be corrected before turning the breaker back on, otherwise the breaker will simply trip again.

A fuse performs the same task as a circuit breaker, but in a different manner. It is composed of a metal that melts at a low temperature, and in doing so breaking the overheated circuit. It is based on a discovery made by English physicist James Joule (1818-1889). In 1840 he devised a formula relating the development of heat due to an electric current. He found that the heat generated was proportional to the square of the intensity of the current multiplied by the resistance of the circuit. It was thus possible to determine how thick, or thin, a fuse could be before melting, because the resistance of a certain piece of metal could be measured beforehand.

Fuses are located in older homes and buildings. As in the case of a circuit breaker, when a fuse ''blows'' the fault in the circuit must be determined before replacing the fuse.

Unfortunately, unlike a circuit breaker, it is easy to bypass a fuse by inserting another piece of a conducting metal, like a copper penny, in its place. This is a very dangerous practice and can lead to disaster; the penny allows current to flow but does not have any mechanism for limiting the amount of current. Overloaded circuits can cause fires leading to property damage and loss of life.

CIRCUIT, ELECTRIC • See Electric circuit

CISTERN • See Water supply and drainage

CLAUDE, GEORGES (1870-1960)

French engineer

Although Georges Claude made his fortune by inventing the **neon light**, his other scientific advancements were just as important as his lucky contribution to the advertising business. Claude began his career in the municipal electricity works of Paris, where he experienced a near-fatal accident with a high-tension wire that led him to develop better safety precautions. In 1897, he invented a novel way to transport and store the dangerously flammable gas acetylene. He dissolved the gas in a liquid, acetone, which made it much easier and safer to handle. This idea greatly facilitated the industrial use of acetylene as an organic chemical. Then in 1902, Claude developed a process to make liquid air in commercial quantities, around the same time that German chemist **Karl Paul Gottfried von Linde** invented a similar process. Claude also improved methods of generating power from the energy released when liquid oxygen is re-gasified. When the renowned Scottish chemist Sir William Ramsay (1852-1916) needed liquid oxygen for his research on inert gases, it was Georges Claude who supplied it. As a result, Claude himself became interested in the inert gases, a group of gases that is relatively non-reactive.

In 1910, Claude showed that neon gas would glow with colored light when electricity was discharged in a neon-filled tube. He also invented a means of purifying the gas in the tube using a charcoal filter. Claude's development of neon tubes, which could be twisted to form letters and pictures, soon created quite a stir in the advertising industry and made him a rich man. With the threat of World War I, Claude undertook more serious ventures. First he produced liquid chlorine, which was used in poison gas attacks. Then in 1917, he developed a higher pressure, less expensive process to synthesize ammonia, another important industrial chemical and fertilizer. Based on earlier work by fellow French chemist Henri Le Châtelier (1850-1936), Claude's process was similar to one developed independently by **Fritz Haber**, a German chemist.

Georges Claude's intimate understanding of thermodynamics—the conversion of heat into mechanical work—also resulted in a visionary project that is just now being reexamined. Claude was one of the first scientists to realize that electric energy could be produced using the difference in temperature between the ocean's warm surface water and its colder depths. Although Claude's project ended in dismal failure in 1933, researchers today are evaluating the possibilities of this alternative approach to power generation.

Whatever research Claude undertook after World War II was not recorded. Because of certain statements he made during the war, Claude was convicted of collaborating with Nazi Germany and spent some years in prison before he was freed by the efforts of his friends. He died in 1960.

See also Ammonia synthesis; Liquid gas, commercial production of

CLIPPER SHIPS

The term ''clipper'' is used loosely to describe types of very fast sailing ships that ''clip'' along. The term was first applied to speedy schooners called Baltimore clippers (built in Virginia and Maryland) which became famous for escaping the British blockade and acting as privateers during the War of 1812. These early clippers, which actually had come from previous French designs, had a long and low hull design with weight distributed more toward the rear than other ships. They also had a stern that sloped sharply back toward the bow, thus making for more deck space but less hull. Of course, this design limited cargo space, but it increased speed. The lack of carrying capacity was to eventually bring an end, at least commercially, to these beautiful ships.

The first true clipper was built by John Griffiths (1809-1882), an American naval architect. By the age of nineteen, he had already designed a major vessel and had proposed numerous innovations in ship design. In 1842, Griffiths delivered the first formal lectures on naval architecture in the United States and opened a free school in shipbuilding. In 1845, he built the *Rainbow*, a clipper designed for the tea trade with China and, unfortunately, for the slave trade between Africa and the United States.

Griffiths revolutionized ship design with his clippers. Until his time, all large American ships were built with their greatest width near the front, which made the bows fairly blunt. He advocated a sharply pointed bow and stern with an overhanging deck space. The *Rainbow*, which resulted from Griffiths' studies, was extremely fast: she sailed to China in 92 days and raced home in 88 days, much faster than other large ships of the day.

Another American shipbuilder, Donald McKay (1810-1880), has been associated with clippers for his successful designs. The California gold rush of 1848 and the discovery of gold in Australia in 1850 spurred the rise of these clippers. McKay launched his first clipper, a vessel of 1,500 tons (1,365 t) named the *Stag Hound* in 1850. His most famous ship was probably the *Flying Cloud*. It was 229 feet (69.4 m) long, 40 feet (12 m) wide, 21 feet (6.4 m) deep, weighed 1,700 tons (1,547 t), and was able to reach San Francisco from New York in 89 days. McKay faced great skepticism in 1852 when he built *Sovereign of the Seas*, a clipper weighing 2,400 tons (2,637 t), but it proved to be successful when it set the all-time record for a sailing ship for the voyage from New York to Liverpool in 13 days. At times it reached speeds of 22 knots per hour; this is little less than the cruising speeds of the great Atlantic liners built just before the first world war. McKay went on to build the *Great Republic*, a mammoth clipper of 3,300 tons (3,003 t), and the *James Baines*, which sailed the 14,000 nautical miles from London to Melbourne, Australia, in 1854 in 63.5 days while carrying 700 passengers and 1,400 tons (1,274 t) of cargo.

The clippers were unusual in their material as well as their shape. Unlike other ships built of oak or other hard woods, clippers were usually built of the soft woods abundant on the American coast. These soon became soggy and warped, and the hull might only last five years. Because they were so much cheaper to build, they still were able to earn a big profit for their owners.

Several events helped to lessen the need for clippers during the 1860s and the 1870s. **Steam engine**s improved, and **railroad**s crossed the vast American West. When the Suez Canal opened in 1869, the need to circle the southern cape of Africa at record speeds had diminished. **Cargo ship**s with a greater holding capacity began to dominate the seas. However, the clipper and its lore were firmly established in American culture. The beauty and grace of these vessels continues to inspire and delight many today.

Although clippers had virtually disappeared as working vessels during the twentieth century, they reappeared by the 1990s to serve another purpose—specialized luxury cruises. By the late 1990s, two much larger but authentic copies of Yankee clippers were built in Belgium by a Swedish-owned company which sailed them to the Caribbean from New York. Although 100 ft (30.3 m) longer than the original clipper ships, and equipped with the latest in seafaring technology as well as an engine, these new clippers are outfitted with 36,000 sq ft (3,240 sq ft) of barkentine rig sails which give the 160 passengers the true experience of a tall sailing ship with all the amenities of a yacht.

See also Train and railroad

Clock and watch

In the early days of humanity, only three divisions of time existed: days (the interval between successive sunrises), months (the interval between complete lunar orbits), and years (the interval between the start of one planting season and the start of the next). The first artificial division of time was the hour, probably established by the Egyptians during the fourth millennium B.C. Beginning at dawn and dusk, twelve hours each were given to night and day. Unfortunately, since the changing seasons cause the length of night and day to vary by several hours, the Egyptian hour was not really a fixed unit. In winter, for example, because nights are longer than days, twelve night "hours" would last longer than twelve day "hours." What was needed was a device that could measure time in regular, unvarying amounts. Toward this end, early scientists began the evolution of the modern clock.

Ancient observers noticed that as the sun traveled across the sky, shadows on the ground would move and vary in size. This led to the invention of the **sundial**, most likely beginning as a simple stick in the ground and eventually leading to the construction of large *obelisks*. Ancient writers credit the Greek scientist Anaximander of Miletus (610-c. 547 B.C.) with the invention of the sundial during the sixth century B.C., but it is presently considered almost certain that it appeared in the Chinese and Egyptians civilizations many years earlier. When properly read, sundials served as a fairly accurate method for marking the passage of time; however, they proved to be difficult for many to interpret. In addition, the markings for each sundial had to be adjusted according to its latitude, and the readings differed as the seasons progressed.

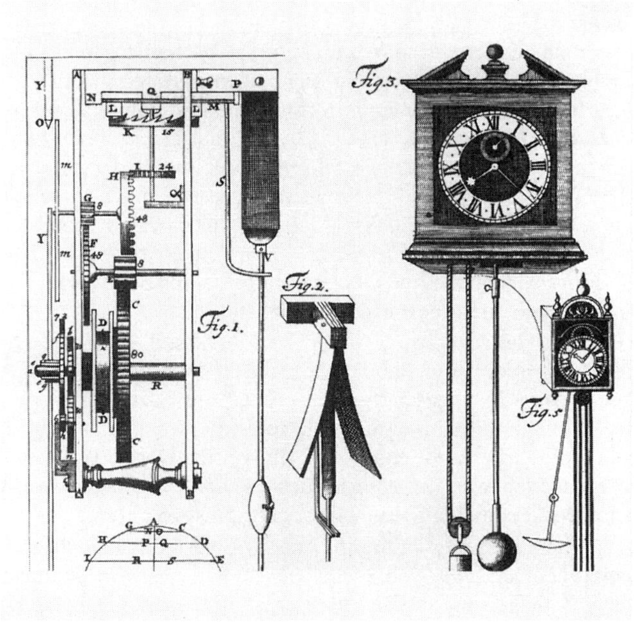

Dutch scientist Christiaan Huygens is credited with inventing the first working pendulum clock in 1656.

The first mechanical timekeeping device was a water clock called a clepsydra. It operated by pouring a steady stream of water into a vessel; after a certain period the vessel would fill, then tip itself empty, and be ready for refilling. The amount of time this took could be regulated by changing the size of the vessel. Later versions of the clepsydra began with a filled vessel, which would release the water over time. Such clepsydras were first used from about 1500 B.C. through the Middle Ages. During this time some rather elaborate water clocks were constructed. One built for the Emperor Charlemagne in 800 A.D. dropped a metal ball into a bowl to mark the arrival of a new hour, while others, which were used by astronomers, regulated **astrolabes** and other equipment.

There were many problems with water clocks. Depending upon the climate, the water in the clock would often evaporate, causing the device to lose time, or freeze solid, stopping the machine entirely. Even under optimal temperatures the continual flow of water through an aperture would cause the opening to erode and widen, thus making the clock increasingly inaccurate. It became evident to clockmakers of the time that a completely mechanical clock was necessary.

During the Middle Ages the two professionals most skilled in the construction of clocks were astronomers, who used the devices to plot the motions of the heavens, and monks, who needed them to determine when to toll the monastery bell. A monk probably invented the first completely mechanical clock around A.D. 1275. This first clock, which was driven by the slow pull of a falling weight that had to be reset to its starting position after several hours, was much more accurate than the water clocks of the past. The clocks in monasteries were also among the first alarm clocks; fitted with a striking mechanism, the clock could be set to sound when the monastery's great bell needed to be rung.

As the accuracy of timepieces increased, society came to certain realizations about the nature of the world. First, it became readily apparent that days (that is, daylight hours) varied in length throughout the year. Second, it was found that the sun did not rise at the same time all over the world. This latter phenomenon was not addressed until 1884, when the world adopted Greenwich Mean Time, giving us **time zones**.

The next step in the evolution of the clock was the development of improved *escapements*, which are mechanical devices that ensure regular motion within the clock. Often pictured as a tiny hammer falling into the teeth of a gear, the escapement allows the minute hand to move once each minute. The first escapement was designed around A.D. 1300 and was fitted into a *weight-driven* clock. Called the *verge-and-foliot* escapement, it used a rotating bar with foliots to alternately halt and release the teeth of a ratchet wheel. Improvements in escapement design allowed the Italian **Giovanni dé Dondi** to build an elaborate astronomical clock over a span of sixteen years in the 1300s.

Near the beginning of the fifteenth century, engineers were using coiled springs in door locks and handguns. Borrowing from this technology, clockmakers developed the first *spring-driven clocks* around 1430. By replacing the heavy, long-corded weights, horologists were able to build timepieces small enough to be carried on one's person. The main drawback of coiled springs was that they unwound quickly at first and then more slowly. Clockmakers soon added the *fusee*, a trumpet-shaped pulley that increased mechanical leverage as a spring wound down, allowing the watch to run at a constant rate. First suggested by **Leonardo da Vinci** in 1407, the fusee is so efficient that it is often used in many of today's clocks.

Even with the addition of springs, clocks before the mid-1600s were notoriously inaccurate. About that time, history tells us, **Galileo** was in the Tower of Pisa during an earthquake. As the ground shook, Galileo watched the motion of swinging chandeliers with fascination; by timing their swing against his own pulse, Galileo found that the amount of time it took a chandelier to swing from one side to another was constant, no matter what the distance. This supposedly was the inspiration for his swinging pendulum, an invention he designed but never actually built. The first working pendulum clock was constructed in 1656 by the brilliant Dutch scientist **Christiaan Huygens**. With the pendulum imparting a steady motion and the addition in 1675 of the anchor escapement by William Clement, the *weight-driven pendulum clock* became the most precise yet.

During this same time scientists became occupied with a new puzzle—inventing a timekeeping device that could be used aboard sailing ships for navigation. Because of the turbulence at sea, both weight-driven and pendulum clocks were unsuitable. In 1674 Huygens introduced a watch that featured a *balance spring* as a regulator, acting in place of a pendulum. Besides sparking a tremendous controversy with the English scientist **Robert Hooke**, who claimed that Huygens had stolen his idea, the introduction of the balance spring made an immediate impact upon the world of clockmakers. However, it was not until 1761 that an English inventor, **John Harrison**, joined a balance spring with a *mainspring-driven clock* to produce a precise and completely portable watch, suitable for ships as well as a person's wrist. Harrison's design forms the basis for most modern spring-driven clocks.

The common wristwatch is among the most precise mechanical instruments. If a watch loses 20 seconds every day it is still operating at an error rate of only 0.023 percent—all the more remarkable since it is expected to run 24 hours a day, 365 days a year, a task required of no other measuring device. Still, clockmakers at the turn of the twentieth century were not yet satisfied.

Clocks powered by electricity had been in existence since the late 1800s, but most required large and ungainly machinery in order to function. These early *battery clocks* used tiny motors to wind the mainspring when it ran down. The real revolution in battery operated timepieces came during the 1950s when the Swiss put a tiny electric tuning fork inside a watch. When the battery applies a small charge to the tuning fork, it will vibrate continuously at a very specific rate, and that vibration can be used in place of a clock's slowly unwinding mainspring. Since the invention of the *Swiss-movement* watch, most tuning forks have been replaced with tiny pieces of quartz crystal, a natural substance which can vibrate with much greater precision than any man-made tuning fork. Even the most affordable *quartz clock* is accurate to within one minute per year.

Currently the apogee of human efforts to monitor time is the **atomic clock**. Technically not a timepiece since it does not indicate time, the atomic clock is used as a reference standard for absolute time. First constructed in 1948, the atomic clock measures the unvarying frequencies at which molecules vibrate. By knowing how many times a molecule will vibrate within a unit of time, the atomic clock can be used to regulate the accuracy of other clocks. Such absolute precision is essential for navigation, particularly in space, as well as research on the atomic level. Chronographs regulated by atomic clocks will lose less than one second every one thousand years.

See also Chronometer; Hourglass

Clothes dryer

The development of the clothes dryer followed that of the **washing machine**. An early design was patented by the African-American inventor G.T. Sampson in 1892.

The first successful home drying machines were designed in the 1930s by J. Ross Moore, originally of North Dakota and then of Minneapolis, Minnesota. Moore sold his designs to the Hamilton Manufacturing Company of Two Rivers, Wisconsin, who then called in industrial designer Brooks Stevens to help redesign the machine. It was Stevens who came up with the idea for the window in the dryer's door so consumers would know what the machine was for—Stevens advised Hamilton to display the window-doored dryer in stores with a pair of "boxer shorts flying around in there."

Hamilton made the only dryers marketed in the United States before World War II. They were relatively simple machines that operated at fixed temperatures. Modern dryer improvements began in 1960 with Maytag's introduction of **electronics**. Today's dryers have automatic cycles with varying temperatures and can shut themselves off when the clothes reach the correct stage of dryness.

Energy-saving dryers have been developed that do not vent to the outside; instead they employ a condensing process that relies on a heat exchanger and a small fan pulling in room air to dry the clothes, thereby avoiding the heat loss associated with a more traditional clothes dryer. Dryers come in varying sizes and colors, and are available to stand independently or stacked with a washer for small spaces.

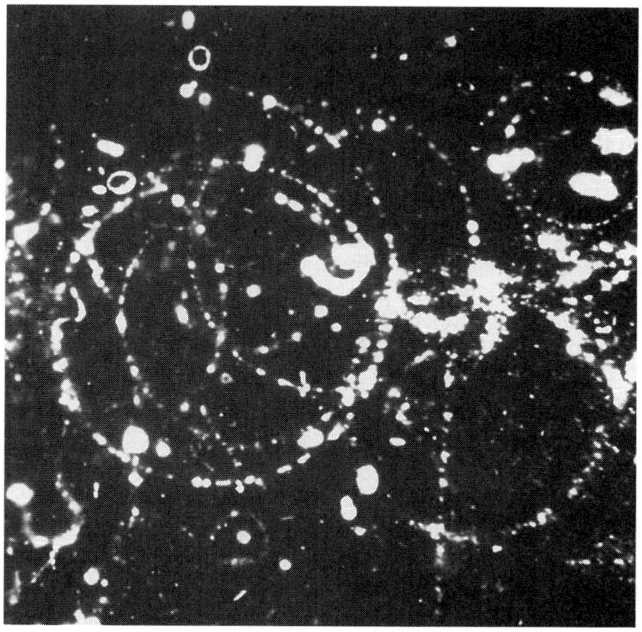

One of the first cloud chamber photographs showing the track of a cosmic ray, taken in 1927.

Cloud chamber

C. T. R. Wilson, inventor of the cloud chamber, claimed that the earliest ideas for this device came to him in 1894. While still a young student, he stood at the summit of Ben Novis, highest peak in Scotland, and was amazed by the optical phenomena produced when sun shone through the clouds. He resolved to build a "cloud machine" of his own in the laboratory and to further study these phenomena.

The principle for building such a machine was well known at the time. Imagine an air-tight container filled with a **piston** that can be moved up and down. Then suppose that the air inside this container is saturated with some vapor, such as that of water or alcohol. Then, if the piston is suddenly withdrawn, the air inside the container expands and is cooled. The cooler air is able to hold less vapor than could the original warmer air. One might expect the excess vapor to condense in the form of droplets.

Indeed, condensation does occur if dust or other foreign particles are present in the air-vapor mixture. Such is exactly the process by which clouds are formed in the atmosphere. But if the air-vapor mixture is completely free of foreign materials, condensation often does not occur. The vapor remains in a supersaturated state.

As early as 1887, the German physicist H. von Helmholtz had learned that an electrical charge can cause condensation in this kind of supersaturated state. Wilson was eager to find out if he could produce a similar effect with x rays. In 1896, he performed this experiment and found that, like electricity, x rays could induce condensation in the supersaturated vapor.

In 1911, Wilson adapted these ideas in the design of his first cloud chamber. He found that radiation from a radioactive material left an easily observable track when it passed through the cloud chamber. The track results from the fact that alpha particles, beta particles, and other charged particles interact with air and vapor atoms and molecules within the container. This interaction results in the formation of ions, atoms or molecules that carry an electrical charge. The ions act as nuclei on which condensation can occur.

In use, the cloud chamber is placed between the poles of a **magnet**. The magnetic field causes particles to bend in one direction or another, depending on the electrical charge they carry.

The events that occur in a cloud chamber can be viewed by taking a photograph of them. In the presence of a radioactive source, the piston is released and, after a short delay, the camera shutter is opened. The kinds of particles that have passed through the chamber can be determined by the types of tracks they leave.

Although the cloud chamber has many useful applications, it has been supplanted in many instances by the **bubble chamber** invented in 1953 by **Donald Glaser**.

See also Photograph

Cloud seeding

Mark Twain once said that everyone talks about the weather, but no one ever does anything about it. Mr. Twain may have been correct in his day. However, since the 1940s researchers have been at least partially successful in modifying one aspect of the weather—precipitation.

After about three years of investigative work at the General Electric Research Laboratory in Schenectady, New York, researchers **Irving Langmuir** and his assistant, Vincent Joseph Schaefer, created the first human-made rainfall. Their work had originated as war-influenced research on airplane wing icing. On November 13, 1946, Schaefer sprinkled several pounds of dry ice (frozen carbon dioxide) from an airplane into a supercooled cloud, a cloud in which the water droplets remain liquid in sub-zero temperatures. He then flew under the cloud to experience a self-induced snowfall. The snow changed to rain by the time it reached Langmuir, who was observing the experiment on the ground.

Langmuir and Schaefer selected dry ice as cloud "seed" for its quick cooling ability. As the dry ice travels through the cloud, the water vapor behind it condenses into rain-producing crystals. As the crystals gain weight, they begin to fall and grow larger as they collide with other droplets.

Another General Electric (GE) scientist who had worked with Langmuir and Schaefer, Bernard Vonnegut, developed a different cloud-seeding strategy. The formation of water droplets requires microscopic nuclei. Under natural conditions, these nuclei can consist of dust, smoke, or sea salt particles. Instead of using dry ice as a catalyst, Vonnegut decided to use a substitute nuclei around which the water droplets in the cloud could condense. He chose silver iodide as this substitute because the shape of its crystals resembled the shape of the ice crystals he was attempting to create.

The silver iodide was not only successful, it had practical advantages over dry ice. It could be distributed from the ground through the use of cannons, smoke generators, and natural cumulonimbus cloud updrafts. Also, it could be stored indefinitely at room temperature.

There is general disagreement over the success and practicality of cloud seeding. Opponents of cloud seeding contend that there is no real proof that the precipitation experienced by the seeders is actually of their own making. Proponents, on the other hand, declare that the effect of seeding may be more than local.

Over the years, cloud seeding has become an accepted part of the strategy to combat drought. It may indeed bring crop-saving relief to a dry field or may help reinforce subsurface water tables. However, the practice has not begun to eliminate deserts or devastating droughts, for researchers have yet to reproduce the general ground-soaking effects of a well-organized natural storm system so necessary for agriculture and replenishment of water reserves. And today there are environmental concerns over any activity that threatens to change or destroy a bio-community such as the desert.

As researchers collect and analyze more information about the weather, other attempts to modify it are bound to be developed. Eventually, it may become easier to change the weather than to predict it.

Coaxial Cable • See Cable television; Microwave transmission

Coca Cola • See Soda pop

Cockcroft, John Douglas (1897-1967)

English physicist

Cockcroft was born in Todmorden, England, on May 27, 1897, the son of a textile manufacturer. After completing secondary school at Todmorden in 1914, he entered Manchester University. He left the university after only one year to join the British army and enter World War I. He survived some of the worst battles of the war and returned to take a job at the Metropolitan Vickers Electrical Company. Officials at Vickers encouraged him to return to college and, in 1924, he earned his bachelors degree from St. John's College, Cambridge. After graduation, he joined Ernest Rutherford's (1871-1937) research team at the Cavendish Laboratory.

While at the Cavendish, Cockcroft met the Russian physicist George Gamow (1904-1968) who outlined a new theory about particle bombardment. Gamow explained that his calculations showed that subatomic particles with a relatively modest amount of kinetic energy had a small, but significant, probability of entering atomic nuclei and causing their disintegration.

To test this theory, Cockcroft, working with **E. T. S. Walton**, designed a machine to accelerate protons. The machine had the ability to generate voltages greater than 500,000 volts. Protons introduced at one end of the machine were accelerated by this potential difference and then directed at a lithium target at the opposite end of the machine.

Cockcroft and Walton found that the product of the reaction between protons and lithium nuclei was alpha particles (helium nuclei):

$$_1H^1 + {}_3Li^7 = 2\,{}_2He^4$$

Evidently the protons had entered the lithium nucleus and then caused it to break apart into two alpha particles. This reaction was the first nuclear change brought about by artificially accelerated particles.

Cockcroft and Walton repeated this experiment using other targets, such as boron, carbon, fluorine, and other elements. They found similar results with these elements. With boron as a target, for example, three alpha particles were produced:

$$_1H^1 + {}_5B^{11} = 3\,{}_2He^4$$

Shortly after deuterium was discovered, the two researchers began using deuterons as projectiles in their accelerator.

When World War II began, Cockcroft once more put his talents to work for Great Britain. He was appointed Chief Superintendent of the Government's Air Defence Research and Establishment. In that position, he contributed to the development of the nation's **radar** defense system. He was also involved in the development of nuclear energy as head of the Canadian Atomic Energy Project. After the war, he became director of the first atomic energy research laboratory at Harwell, England.

Cockcroft and Walton were awarded the Nobel Prize in Physics in 1951 for the development of their **particle accelerator** and for the transmutation of atomic nuclei. In 1961, Cockcroft also received the Atoms for Peace award. He died in Cambridge on September 18, 1967.

See also Linear accelerator

John Douglas Cockcroft

COFFEE, INSTANT • See Instant coffee

COHEN-TANNOUDJI, CLAUDE (1933-)
French physicist

Cohen-Tannoudji was born April 1, 1933, in Constantine, Algeria. He studied at Ecole Normale Superieure from 1953 until 1957, when he received his *agrégation.* He earned his D.Sc. (Ph.D.) from the University of Paris in physics in 1962. While a graduate student, he studied under 1966 Nobel Prize winner Alfred Kastler. His thesis advisor was Jean Brossel. Following in both of his mentors' theoretical footsteps, Cohen-Tannoudji has arguably become their most distinguished student. Also, while still a student as well as shortly after graduation, from 1960-64, Cohen-Tannoudji worked as a researcher at the Centre National La Recherche Scientifique in Paris. He then served as a professor at the University of Paris from 1964-73.

In his early research, Cohen-Tannoudji studied the phenomenon that formed the foundation of his later work. Cohen-Tannoudji worked methodically, predicting theoretically, then demonstrating his theories through experiments. His investigations concerned atoms and their energy levels. He found that energy levels shifted a small amount when, in a field intensely radiated with light, they absorbed and emitted photons of light. (Photons are the smallest pieces of electromagnetic energy.) This is Cohen-Tannoudji's so-called light shift discovery.

It was in the 1970s that Cohen-Tannoudji came up with another theory that eventually was applied specifically to lasers. His theory explained the interactions between high-intensity electromagnetic fields and the atoms within them. When an atom is surrounded by photons, it is constantly absorbing and re-emitting them. He described the atom as being ''dressed'' by the photons. This observation became important when it was applied to lasers in the late 1970s and beyond.

By the 1980s, Cohen-Tannoudji theorized specifically about the light (of a **laser**) and its effect on the motion of an atom. His experiments in laser cooling were a response to other scientists' work with laser beams, cooling atoms, and trapping atoms. Cohen-Tannoudji, leading his own team of experts, focused on the modification of atomic motion when it interacted with photons. He had his first breakthrough in 1985, when he saw that an atom in motion slows in a standing light wave. As the laser beam intensifies, the atom slows increasingly. Because a material's temperature measures how fast its atoms are moving, this effect came to be known as ''Sisyphus cooling,'' named for the Greek mythological personage who had to roll a stone up a hill for eternity. However, this method was limited in its cooling effects.

By 1988 Cohen-Tannoudji brought together his ''dressed'' atom ideas with the problems raised by the cooling and trapping atoms theories in order to try to bring the temperature of atoms closer to absolute zero. A few years later, Cohen-Tannoudji had an unexpected breakthrough when studying an atom trapped in a standing light wave. He found that it was dependent on optical pumping, an advanced laser technology in which emits light energy in pulses. This approach keeps the atoms from rebounding, in effect ''trapping'' them. This type of Sisyphus cooling brought the temperature to 0.18 millionths of degree above absolute zero. The phenomenon had been anticipated by Albert Einstein and others.

In the 1990s, Cohen received important recognition for his work: the Julius Edgar Lilienfeld Prize from the American Physics Society (1992), the Charles Hard Townes medal from the Optical Society of America (1993), the Harvey prize in science and technology from Technion, Israel (1996). Also in 1996, Cohen-Tannoudji won the Médaille d'Or of the National Center for Scientific Research, the highest honor a scientist can receive in France. He won the Médaille specifically for his research in atoms and related cooling methods. In 1997 Cohen-Tannoudji shared the Nobel Prize in physics for his explanation of the phenomenon and his ''trapping'' discovery.

Cohen-Tannoudji's work has opened up long range research possibilities, but affects basic physics as well. It can be used to study atomic structure, as well as help bridge the frontier between classical and quantum physics. Cohen-Tannoudji's theories also have proven useful for practical technology now and in the future. Ultracold atoms can be used to create more precise atomic clocks. (**Atomic clocks** are devices which use the constant frequency associated with some atomic and subatomic phenomena to define an accurate and reproducible time scale.) His theories are also influential in the development of measuring instruments of extreme precision. Indeed, an atomic laser might be possible. (An atomic laser would be like an ordinary one, save atoms would take the place of photons used in ordinary lasers.)

Collodion

Collodion, a clear, viscous solution of cellulose nitrate, was discovered in 1846 by **Christian Schönbein**. Experimenting with nitric and sulfuric acids in his kitchen, he accidentally spilled the acids and cleaned up the mess with his wife's cotton apron. When the apron dried, it burst into flames. He had discovered cellulose nitrate or **guncotton**. Later he dissolved the guncotton in ether and alcohol to form collodion, a more stable mixture that could be handled with greater safety. It was used as a protective dressing on wounds, in **photography**, and in blasting gelatin. When it is spread on a surface, the solvents evaporate, leaving behind an airtight film which has a multitude of uses.

Frederick Scott Archer, an English inventor, is credited with developing the wet collodion process in 1851. This photographic method was the first to create fine-grain negatives, but it was not convenient because it required a portable darkroom. A **glass** plate coated with collodion containing potassium iodide was dipped into a silver nitrate in solution. This process had to take place in the dark. The plate was then exposed, developed, and printed. **William Henry Fox Talbot**, the inventor of a process using silver iodide paper to make paper negatives, sued Archer for patent infringement. Talbot lost the case, and Archer went on to market his technique. Unfortunately, he went the way of many inventors and died penniless in 1857. For twenty years the wet collodion process was the main photographic process and was used, for example, to chronicle the American Civil War. The dry plate method replaced the wet process in 1871.

Alfred Nobel, the Swedish inventor of **dynamite** and the founder of the Nobel prize, also used collodion. By combining it with **nitroglycerin**, he produced an explosive more powerful than either component and much more stable. He patented blasting gelatin in 1875. Although collodion has since been replaced by more advanced polymeric materials, it is considered to be the forerunner of the lacquer paints that are now widely used in the automobile industry.

See also Cellulose, chemical uses of

Color photography

As **photography** evolved in the nineteenth century, attaining a full-colored rather than black-and-white or monotone image became an obvious goal.

The English astronomer Sir **John Frederick William Herschel** and the French scientist **Alexandre-Edmond Becquerel** made the pioneering attempts at color photography. Both were able to photographically reproduce color, but neither succeeded in creating permanent color images.

In the late 1850s, the Scottish physicist James Clerk Maxwell (1831-1879), assisted by Thomas Sutton, developed what became a standard method of color photography: the additive three-color process. With this process, three distinct negatives were taken through color filters. These filters—alternately red, green, and blue-violet—screened out all other color. Synthesizing these separate monochrome negatives resulted in an image with a full range of color.

In 1868-69, the French chemist Louis Ducos du Hauron patented and then published his *Les Couluers en Photographie, Solution du Probleme* (''Colors in photography, the solution to the problem''). In this work, du Hauron outlined both the additive and a subtractive process for color photography. The subtractive process, which also relied on superimposing monochrome negatives, used pigments to ''subtract'' color from full-spectrum light. On February 25, 1869, two days after du Hauron had been granted his patent, the French inventor and poet Charles Cros (1842-1888) published an independent work that also outlined the subtractive method.

Two decades later, the French physics professor **Gabriel Lippmann** arrived at color photography by a wholly original route. At the Sorbonne, France's leading university, Lippmann lectured on photographic methods based on the principle that light rays can overlap to create a multihued ''interference.'' Such interference accounts, for example, for the color phenomena produced by soap bubbles and pools of oil. While this phenomenon was well-known, exploiting it in photography was a new idea. Lippmann's method was to apply a light-sensitive emulsion to a **glass** plate on the side opposite to the camera lens. The **mirror** effect that resulted would create the interference. By 1893, Lippmann was able to producing weak color photographs through this technique. Although practical drawbacks limited the application of his discovery and precluded commercial success, Lippmann was awarded the Nobel Prize in physics in 1908 for having devised the first direct process of color photography.

Employing the conventional additive three-color process, the American inventor **Frederic Eugene Ives** first made color photography practical for professional photographers. With his 1893 Photochromoscope camera, three ''separation'' negatives were taken in succession—using the red, green, and blue-violet filters—onto a single photographic plate. The professional photographer could use a customized viewing machine developed by Ives to derive a positive color image.

The French inventors and manufacturers **Auguste and Louis Lumiere**, in addition to launching cinematography, developed the first method of color photography accessible to amateurs. In 1904, Louis Lumiere published this discovery, the autochrome method. While relying on the additive process, the Lumieres' method did not depend on the superimposition of monochrome negatives. Instead, a single glass plate was coated with fine potato starch that contained a mixture of red, green, and blue grains. Thus, these grains functioned as color filters. The colored grains were set in a panchromatic emulsion commercially available after 1907. The final autochrome image consisted of small flecks of color; these blended visually in an effect analogous to that of an Impressionistic painting.

Although the Lumieres technique remained a popular method of color photography until the 1930s, the autochrome had its drawbacks. Like the modern photographic slide, the autochrome yielded a transparency that had to be projected onto a screen or viewed though a hand-held device. In addition, due to its light-screening layer of potato starch grains, the autochrome plate required an exposure period up to sixty times longer than was feasible with the best black-and-white plates available. The longer exposure time, of course, limited the photographer's choice of subject.

In 1912, the German scientist Hans Fischer (1181-1945) made a breakthrough by proposing that color photography could be achieved chemically, rather than optically, through oxidation or "coupling" among chemicals in a multilayered film. Although Fischer's own attempts to devise a color film failed, his concept is the basis for modern color film.

In 1935, Leopold Godowsky and Leopold Mannes, both professional musicians, declared their invention of a practical color film that exploited this oxidizing process. The method they developed was in some ways a stopgap: rather than incorporating the oxidizing chemicals into the film itself (the challenge Fischer had raised), the chemicals were added during a development-stage bath. The film, nonetheless, was a success. Godowsky and Mannes ultimately collaborated with the American photographic inventor **George Eastman**, the founder of Eastman-Kodak, who marketed the film as *Kodachrome*.

Within a few years, various companies had achieved Fischer's goal of incorporating color-forming chemicals into the emulsion layers of the film itself. These first color films produced transparencies, as had the Lumieres' autochrome plate, and the procedures for satisfactorily transferring a color photograph onto paper were gradually perfected.

In 1963, the American founder of the Polaroid Company, **Edwin Herbert Land**, marketed a product that made Fischer's goal seem modest. This was polacolor, a color film for the **instant camera** that produced photographs instantly and automatically. With this innovation, the century-old goal of achieving color photography had clearly been reached. However, with the advent of the digital camera, and related technology, this kind of color photography is gradually being displaced. Digital color photography requires only a computer, no darkroom, chemicals, processing, or even film is needed. The resulting image can be edited in ways unimagined by the inventors of color photography.

See also Computer input and output devices; Digital camera.

Color television • See Television

Colossus

Colossus was the name of a series of electronic, special-purpose computers, built by the British during World War II to break German military codes. The first Colossus was put into operation in 1943, and was thought to be the first electronic, programmable machine. Later it was learned that **Konrad Zuse** had built and operated such a machine, his Z3, several years earlier. In all, ten Colossus machines were built.

Message-code wheels were among the earliest decoding devices, including one invented by Thomas Jefferson in the late eighteenth century. The first machines that both decoded and printed their results were developed in France in the 1870s. In the 1920s, German code experts developed the first of a series of machines called Enigma that worked like teleprinters, sending messages over telephone lines. But when plain text was typed on the keyboard, it was translated into code before being transmitted. When a coded message was received, the machine automatically decoded it.

The first Enigma machines used straight substitutions, such as E for H, F for I, and so on. Polish scientists had invented electromechanical machines called Bombas to decipher German messages that they intercepted between 1928 and 1938. When World War II started in 1939, British scientists could decode advanced Enigma messages.

But by 1942, the Germans had built an even more advanced machine that the British called the Fish. It had a set of twelve interlocking code wheels, each with a different diameter. Each wheel also had movable cams that allowed a somewhat random variation in the code-to-key relationship. This made the code difficult and often impossible to break.

The British government had established a top-secret decoding operation called Bletchley Park. There, in 1943, a team of British scientists led by Thomas W. Flowers designed a machine to handle Fish's messages. Flowers was a London University-trained electrical engineer. His specialty was the electronics of telephone exchanges, such as the development of electronic switches that were designed to replace the electromechanical switches then used for direct-distance dialing.

The first decoding machine used vacuum tube processors for binary numbers. It was faster than previous British machines, able to read 2,000 characters per second and completely automatic. The design was influenced by the British mathematician and computer theorist **Alan Turing**, whose scientific papers on machine processes were cornerstones of modern computer design and also **artificial intelligence**. The scientists named it Heath Robinson, after a British cartoonist who drew outrageous machines.

The coded German messages were intercepted and each code character, represented by a five-digit binary number, was punched into paper tape. The ends of the message tape were pasted together, forming a loop. A key tape was also punched and made into a loop, but with a different circumference. When the two tape loops were run at the same time for thousands of passes, every possible code-key combination occurred. As the tapes revolved, the patterns of punched holes were "read" by photoelectric cells and translated into electrical impulses. Patterns detected by the machine were sent to code specialists, who provided the final decoded message.

But the Heath Robinson design was quickly outdated, as the volume of German messages increased. The scientists began building a new decoding machine, named Colossus.

Samuel Colt

Colossus differed from Heath Robinson in several ways. It was entirely electronic. It was digital and programmable, with hand-set switches.

The Colossus **Mark I** was a 1,500 vacuum-tube machine that could handle 5,000 characters per second. The Mark II model, built in 1944, had 2,400 vacuum tubes. It could read 25,000 characters per second because it used parallel processing of five operations at once. An electronic signal generator replaced the key tape. And it had a memory consisting of thyratrons—a type of gas-filled anode-cathode tube that produces a current and holds it until the power is cut off.

In all, ten versions of Colossus were built, and some of the scientists involved became the leaders of British computer research in the late 1940s. Flowers later received an honorary degree from the University of Newcastle upon Tyne for his work on Colossus.

See also Code and cipher

Colt, Samuel (1814-1862)
American inventor

Samuel Colt, born near Hartford, Connecticut, in 1814, Colt was shipped off to sea by his father. While on board, Colt noticed that the helmsman's wheel always lined up with one of its spokes through the use of a simple clutch and locking device. In 1830 his ship sailed to India, where Colt saw a repeating **pistol** invented by Bostonian Elisha Collier. On his return voyage, Colt combined the mechanism of the helmsman's wheel with the idea of a repeating pistol and developed his Colt "**revolver**."

Colt used a pawl mechanism to rotate the cylinder that held the bullets. The cylinder was locked into place by cocking the hammer. Colt was in his teens at the time, and his design was so simple that he whittled a wooden model while still on board ship. In 1832 Colt returned to the United States with plans to manufacture his pistol.

In 1836 Colt received a patent for his revolver and opened the Patent Arms Manufacturing Company in Paterson, New Jersey. (He had received patents from both France and Britain a year earlier). The country was at peace at this time, and there was little enthusiasm for a rapid-fire weapon. The U.S. Army bought a few of Colt's weapons for use in its war against the Seminoles in Florida; the Texas Rangers also purchased several of Colt's pistols. But due to poor sales, the company was unable to purchase the sophisticated machine tools necessary to produce the weapons cheaply, and the company went bankrupt in 1842. Colt abandoned the arms business and turned his ingenuity to such pursuits as perfecting underwater mines and laying an underwater telegraph line from New York City to Fire Island and Coney Island.

In 1846 the Mexican-American War changed Colt's fortunes. Colt's revolvers saved the life of the lone survivor of an early battle in the war. When the survivor, Captain Thornton, told his superiors of the value of Colt's weapon, the Army ordered a thousand of the revolvers. Although Colt had no factory, he accepted the order and obtained the services of **Eli Whitney** and Whitney's arms factory at Whitneyville, Connecticut, to help him meet the Army's order. Two years later, with the help of New England's finest mechanic, **Elisha King Root**, Colt opened his own factory—the Colt Patent Arms plant in Hartford, Connecticut. Together Colt and Root created the first modern assembly line, utilizing parts machined so accurately that they were interchangeable and required little hand finishing.

Colt vigorously defended his patents against competition from other arms manufacturers and, due to his success, he enjoyed a virtual monopoly on arms sales. By 1855 Colt was the largest arms manufacturer in the world—his Hartford factory alone housed 1,400 machine tools. Colt also established a factory in Great Britain to supply the European market.

Colt's early repeating pistol—often called the "Gun that Won the West"—became only one of myriad other weapons being produced at the Hartford armory. At the outbreak of the American Civil War, Colt was in the enviable position of supplying the United States government with thousands of weapons. Yet he did not live to see the end of the war; he died suddenly at the age of forty-eight in Hartford on January 10, 1862. The firearm company he founded was later integrated into a large multinational industrial conglomerate called Colt Industries which still manufactures weapons at its armory in Hartford.

See also Ammunition; Mass production

Column, architectural

Early construction methods relied on pillars and beams to support roofs and wall openings, and wood and stone were the most common building materials. As temples and public buildings became more sophisticated in their construction, however, pillars became columns, beams became cornices, and detailed reliefs and paintings embellished buildings. The column grew out of a desire to build higher structures. Stone pillars were extremely heavy and difficult to obtain and maneuver, while the column could be assembled in pieces. Though its segments were not light, moving an 8-ton portion of a column was easier than maneuvering an 88-ton pillar. Adding artistic detail to column segments was also less difficult.

Though columns are found at the Great Pyramid of Zoser in Egypt, built around 2600 B.C., the ancient Greeks were the first to make extensive use of them. The Parthenon at Athens is the best example of a columned Greek structure. Its columns each boast 11 marble segments of about 9 tons (8 tonnes), reaching a total height of 33 ft. (10 m). The Greeks designed three distinct styles of column: Doric, Ionic, and Corinthian.

The Romans, too, relied heavily on columns, in some cases adopting Greek styles with little modification, and in others creating additional styles, specifically Tuscan and Composite. The style of a column can often be identified by its capital, the section at the top that tapers outward. The Corinthian capital features detailed leafwork, while the Ionic column ends in a scroll and the Doric has a flat slab. Some columns feature smooth shafts, while others are channeled.

Columns reappeared on the architectural scene during the Greek revival period in the newly independent United States. Many official buildings in Washington, D.C. mirror the grand constructions of ancient Greece. Southern plantation owners also favored columns, generally constructing them of wood or cement, while columns in government buildings were made of stone, true to the original. Office buildings of the late 1800s and early 1900s often featured ornamental columns on upper stories and facades.

Combination lock • See Lock and key

Combine harvester

Traditional grain harvesting required separate cutting, binding, and threshing operations. The mechanical **reaper**, invented in 1831 by **Cyrus McCormick**, performed the cutting, but it required that laborers follow the machine in the field to bind the

Holt combine harvester being drawn through a wheat field by 24 horse in turn-of-the-century Walla Walla, Washington.

Created by Richard Outcault in 1895, "Hogan's Alley" was the first comic strip to achieve widespread popularity in the United States. Featuring the Kid in his trademark yellow nightshirt (lower right), the comic inspired the phrase "yellow journalism" when it became the focus of an intense bidding war between two sensational papers, William Randolph Hearst's *New York Journal* and Joseph Pulitzer's *New York World*.

sheaves of grain. After binding, the grain had to be loaded onto wagons and delivered to stationary threshing machines, which separated the grain kernels from the rest of the plant. Binding had to be completed before the reaper could turn around to start a new row. The combine harvester combined the cutting and threshing procedures, eliminating the need to bind the stalks. Combine harvesters were developed almost at the same time as the mechanical reapers. The first patent on a combine was issued to **Samuel Lane** of Hallowell, Maine, in 1828. His machine was complex and no record exists of it having been actually used. The first successful combine, a horse-drawn machine requiring a team of twenty horses, was invented in 1836 by Hiram Moore and J. Hascall in Kalamazoo, Michigan. The cutter bars and thresher were powered by ground **wheel**s, which meant that the machine had to be moving to perform both functions. In 1854 one of their combines was shipped around South America's Cape Horn to San Francisco. This machine and copies of it were used in the wheat fields of the San Joaquin Valley in California.

During this same period, Australian H. V. McKay built an identical machine, called a stripper, and later made many improvements to it. McCormick's reapers, though less technically advanced, possessed certain advantages over these early combines. While these combines were successful in the dry climates of Australia and California, the threshing machinery often bogged down in the humidity of the midwestern climate in the United States or in wet conditions.

Improvements continued to be made to the combine in the late 1800s. Levelers were added to prevent toppling. Wheels were introduced that could be raised and lowered on rough ground. In 1886 Henry Holt added a link chain that was designed to break in order to prevent a runaway horse from stripping the machine's gears. Most importantly, the thresher was provided with an independent steam power source so it could function while the combine was idle.

A prototype self-propelled combine was invented by Daniel Best in 1888. It was not until 1922 that a motorized combine was introduced to the market by Canadian firm Massey-Harris (later Massey-Ferguson), with research assistance provided by the same Australians who had built the earlier combines. The Massey-Harris combine was horse-drawn, but its machinery was driven by a motor.

In 1935 Allis-Chalmers, an American firm, introduced the first totally self-propelled combine. Massey-Harris responded with its own self-propelled combine in 1938, which was widely used in Argentina. In 1944 the first harvest brigades were organized in the United States by Massey-Harris. Men who worked for these brigades migrated during the cutting season from Texas northward across the Great Plains to Canada, harvesting wheat as it ripened. In 1975 the axial combine harvester was introduced. Developed by American firms New Holland and International Harvester, the axial combine threshed the grain with blades rotating in a cylindrical cage. The compact and highly effective axial combine surpassed other combines that only shook the grain loose. This was the most significant improvement to be made to the combine since the addition of the **internal combustion engine**.

See also Corn picker, mechanical; Threshing machine

Comic strip and comic book

The late 1800s and early 1900s saw the rise of the comic strip, a panel of drawings that form a narrative. England's W. F. Thomas created one of the first regularly featured characters, Ally Sloper, who appeared from 1884 to 1920. In the United States Richard F. Outcault (1863-1928) created ''Hogan's Alley'' (later renamed ''The Yellow Kid'') in 1895, a popular strip starring Mickey Dugan, a kind of aged baby attired in a trademark yellow sack bearing printed comments. Originally published in the *New York World*, it also appeared in the *New York Journal* and spurred a tug-of-war between the two newspapers, giving rise to the term yellow journalism.

In 1897, Rudolph Dirks's (1877-1968) ''Katzenjammer Kids,'' based on an earlier cartoon called ''Max und Moritz'' by Wilhelm Busch (1832-1908), appeared in the *New York Journal*. The year 1907 brought readers the first successful daily comic strip, Bud Fisher's ''Mr. Mutt'' (later ''Mutt and Jeff'') drawn for the *San Francisco Chronicle*.

By the 1930s adventure strips had gained popularity in the United States, among them ''Dick Tracy'' created by Chester Gould (1900-1985), featuring a square-jawed police detective who tangled with creatively named, odd-looking villains. Other early adventure strips included ''Superman,'' ''Tarzan,'' ''Terry and the Pirates,'' and ''Prince Valiant.'' Comic strip fare today includes offerings such as ''Mary Worth,'' featuring a middle-aged, maternal advisor, and ''Peanuts,'' a creation of Charles Schultz (1922-), whose dog Snoopy and round-headed child Charlie Brown have helped make his strip one of the most successful.

Beginning in 1924 with ''Little Orphan Annie,'' by Harold Gray (1894-1968), strips have often blended entertainment with social commentary or political satire. Other early examples include ''L'il Abner,'' originated in 1934 by Al Capp (1909-1979), and ''Pogo'' in 1949, by Walt Kelly (1913-1973). ''Doonesbury'' by Garry Trudeau chronicles the progress of its college-friend characters from the turbulent era of the 1960s through the present, and ''Outland'' and its predecessor ''Bloom County'' by Berke Breathed (1957-) use animal and human characters to render satirical commentary.

The first comic book was an 1897 collection of reprints of the ''Yellow Kid'' strip. A collection of ''Mutt and Jeff'' reprints appeared in 1911, and Japan published the first comic book featuring *original* material in 1920. Comic books became especially popular during World War II, providing welcome diversion for readers stationed away from home. While the rise of television in the late 1940s somewhat lessened public interest in comic books, their popularity rose in the 1960s, and they still attract ardent readers and collectors. An assortment of ''alternative press'' comic books has sprung up, including such standouts as Peter Bagge's ''Hate,'' Julie Doucet's ''Dirty Plotte,'' and Neil Gaiman's fanciful ''Sandman.''

Communications satellites

Prior to 1956 it was not possible for people to speak to each other across the Atlantic Ocean except by radiotelephone. If atmospheric conditions were poor, the connection was often interrupted. In 1956 the first transatlantic **telephone** cable went into operation, providing 36 static-free telephone circuits. By the early 1960s, there were still only about forty channels of

Launched in 1960, *Echo 1* was the first U.S.-built communications satellite.

trans-Atlantic communication, with little promise that the ocean-bottom cables could handle future demand.

As early as 1945, however, Arthur C. Clarke, a British science fiction writer, had described the possibility of establishing a communications network covering the entire globe with three strategically orbiting satellites. Obstacles encountered in realizing these ideas immediately were many and various, however. A satellite would be exposed to extreme conditions of heat and cold. It needs an independent **power supply** to last for months or years. Once in orbit, it can only be reached by **radio** signal, so that it must be virtually independently operational. Finally, rocket science was in its infancy, and launching a satellite into orbit posed serious problems.

Several experiments were carried out in the late 1950s, in the United States and the Soviet Union. *Sputnik I* was the first artificial satellite to be launched, in 1957. In 1958, the U.S. established a satellite connection between New Jersey and California using the moon as a reflecting surface for radio signals. The *Echo I*, a giant gas-filled balloon with a shiny surface to reflect radio beams directed at it, was launched in 1960 to serve a similar purpose. The most successful early advances were made by AT&T, which began developing a satellite with receiving and transmitting equipment early in 1961. In July of 1962, the company paid NASA to launch *Telstar I*, the first transatlantic relay satellite. It was in operation until February 1963 and transmitted telephone calls, facsimiles and **television** broadcasts between groundstations in Andover, Maine, England, and France. NASA was also working on satellite technology and achieved the first communications link between North and South America and the first transpacific television transmission with its Relay satellites. These early satellites orbited around the earth at low and medium altitudes and moved in and out of range of the fixed groundstations. *Telstar I*, in fact, had been in range of the Andover station for no more than 60 to 260 minutes a day. Connections, then, were intermittent.

The Syncom series of satellites, developed by Hughes Aircraft, were the first to be launched into high altitude, geosynchronous orbit, so as to stay in the same position relative to the earth and maintain continuous contact with the groundstations. The first *Syncom* failed as soon as it was placed in orbit in 1963, but *Syncom II* and *III*, launched later the same year, were completely successful and inspired most of the satellites put up subsequently. The Soviet Union also developed satellites, though more slowly. Its Molnya series, the first of which was launched in 1965, was mainly notable for its mixed-use design. Besides its equipment for communications, it had cameras for meteorological and other scientific observation. Meteorological satellite pictures are now widely used by television weather forecasters.

Since then, so many satellites have been launched that an unexpected problem has arisen—crowding of the synchronous orbit. To preserve high transmission quality while keeping the groundstations small, a distance of 7° between satellites has to be maintained to eliminate interference. The U.S. and Canada, for instance, are competing for the same spots in the synchronous orbit for their domestic communications satellites. This problem may have been solved by the announcement of the Teledisc project in 1994. In order to reach areas where wiring would be too costly or just impossible, the people behind Teledisc, including Bill Gates of Microsoft, proposed using a system of about 300 low Earth-orbiting (LEO) satellites. These satellites would be smaller and smarter than those in synchronous orbit, and circle only several hundred miles about Earth. LEOs would be capable of transmitting the same kinds of data that a fiber optic network would—sound, video, and data—but with increased speed and image resolution. Since the announcement of Teledisc, several other companies announced similar plans, including Motorola's Celestri and Iridium systems. These systems are expected to be fully operational by 2005.

COMPACT DISC • See Optical disk

COMPACT DISC PLAYER

Until the 1980s most popular music had come into homes in the form of records. But traditional sound recording had been done in analog, a continuous waveform which is theoretically capable of producing an exact replica of the sound recorded, but which loses fidelity when it is sonically "compressed" onto a record. In addition, analog records are prone to high levels of noise from dust or scratches in the record groove, which are reproduced by the needle as sound. Records also lose their quality of sound over years of use, as the needle gradually wears the playing surface away. With digital recording, instantaneous "samples" of an analog wave are taken at set intervals; the resulting information is stored as binary code (a system of 1s and 0s). Such binary code is then replayed from a disc using a laser beam, which never touches the playing surface and is less prone to transform most slight imperfections into sound.

The first to suggest the digital approach for information storage was the French mathematician, Jean Joseph Fourier (1768-1830). Fourier postulated that if samples of a continuous wave (such as a musical passage or voice) could be taken, the original wave could be precisely reproduced as long as a sufficient number of samples was obtained.

Acoustic technicians worked to determine how many samples would be necessary to duplicate a waveform to the ear. In 1928, Harry Nyquist (1889-1976), a Bell Laboratories mathematician working on formulas for the improvement of telegraph signaling, found that maximizing the accuracy of such signals required twice as many samples as the length of the actual wave involved. In other words, each wave had to be sampled at least twice to obtain accurate reproduction. Applying the principle to sound, a sampling rate of at least 40,000 times per second should then produce flawless sound, since a human can hear no higher than a frequency of 20 kilohertz (20,000 hz).

Not long after, a second Bell Laboratories researcher, mathematician Claude Shannon (b. 1916), developed a new science, called information theory, that would provide an additional theoretical background for digital data storage. In 1948, Shannon clarified how units of information carried along communication lines can be measured, transmitted, and tested, and he suggested ways to reduce or eliminate errors that distort the flow of information. Using his work, two Massachusetts Institute of Technology researchers (Irving Reed and Gustave Solomon) were able to develop procedures to correct information errors, which were later used in compact disc players.

Even though the conditions required to store information digitally were understood by this time, the devices capable of carrying out such storage were yet to be invented, and many technological advances had to take place before compact discs could become a reality. For example, computer and **electronics** technology had to be developed capable of handling the large number of samples involved in digital recording. The introduction of the **laser** in 1960 by **Theodore Maiman** would make the compact disc player a reality. Engineers at a number of audio firms realized that a laser beam could be used to retrieve digital data from a recording without physically touching it, and set about developing a practical **optical disk**.

A number of digital recording and playing systems came out of this effort. First, a system had to be devised to *record* data in a digital form. NHK and the Sony Corporation were the first to develop working digital recorders in the late 1960s, and within a few years a number of recording studios were using digital recorders, which store sounds from a recording session more faithfully than analog equipment does. Producing an economical digital *player* took longer, and it was not until 1978 that Magnavox introduced the first digital compact disc player for the consumer. This player, called the "videodisc" or "laserdisc," was for movies rather than for music. Around this time, nearly a dozen companies began to unveil competing digital music systems. Recalling the marketing nightmare associated with the conflicting **audiocassette** and eight-track systems, as well as the VHS and Beta video tape disaster, which confused consumers and cost companies millions, Philips and Sony Corporation began a joint effort to create a standard, superior digital playback system which they could license to any other company wishing to produce a CD player. Within about a year, they had succeeded, and beginning in the early 1980, CD players came onto the market in earnest.

The system adopted by Sony and Philips works as follows: digital impulses coded from music, pictures or voice are embossed as peaks and valleys on a hard plastic disk. The plastic is coated with a thin microscopic layer of metal such as **gold** or **aluminum**. The metal is then coated with a resinous protective layer. The finished disk, called a CD (compact disc), is about 4.7 inches (12 cm) in diameter. In a compact disc player, a small **laser** beam shines upon the peaks and valleys on the metalized portion of the disc while the disc spins. A **mirror** or prism between the laser and the disc picks up light reflected from the disc and bounces it onto a photosensitive **diode**, which transduces the impulses into electrical current. The current is then converted into an analog waveform for playback through stereo speakers.

Compact disc technology continued to develop throughout the 1980s and 1990s. In 1988, the first rewritable compact disc (CD-R) was introduced, though it could only be rewritten once. This technology was not commonly employed in compact disc players. The year 1992 saw the introduction of Sony's MiniDisc system. MiniDiscs were miniaturized versions of compact discs, and had similarly small players. They were intended to be the compact disc equivalent of Sony's Walkman, and had early erasing and recording technology, but remained an anomaly through 1998. In 1996, an erasable compact disc, called the CD-E, became available. CD-Es can be read, erased, and rewritten repeatedly in special compact disc players, which became available that same year, yet can still be played on standard compact disc players. CD-E discs have the same dimension as compact discs, but have extra layers of different materials. They read and write data using the technology of phase-change. In 1997, Paras Prassad, a chemist at SUNY-Buffalo, announced his discovery of a technology that allows one thousand times more data to be squeezed on to one disc, by layering the data and using a precise laser to read it. Despite these innovations, experts predict that compact disc and compact disc players will be eventually replaced by DVD technology.

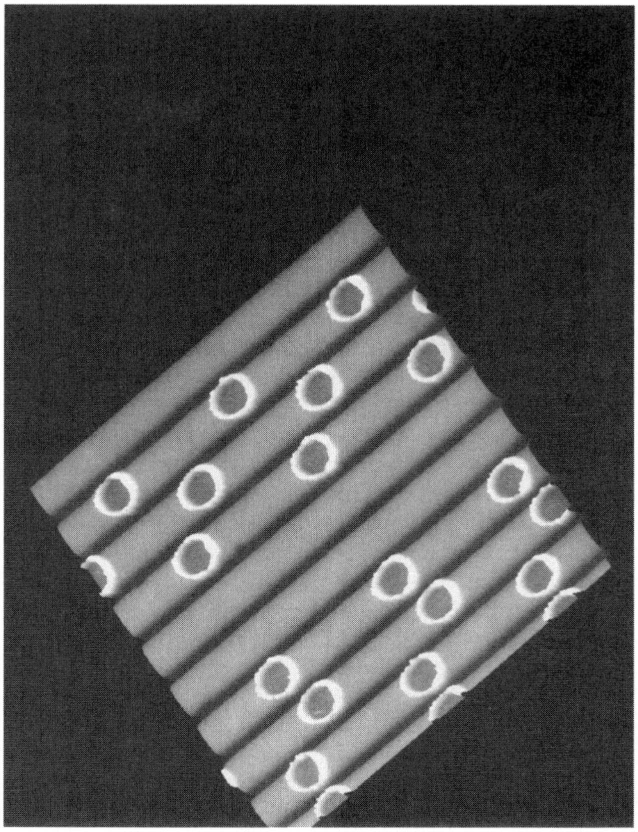

An atomic force microscope image of marks (data) burned into a write-once optical disc (WORM). The write laser burns holes in an ablative material that is part of the disc. The marks cannot be erased, hence "write-once."

Compass

The compass is a tool used by land and sea explorers to help them navigate their journeys. Before compasses were developed, people used the sun, wind, and stars as their guides, but the compass allowed them to calculate their location and direction with greater accuracy.

During the first century B.C., the Chinese observed that pieces of lodestone, an iron mineral, always pointed north when they were placed on a surface; this discovery led to the development of the compass. The first Chinese compass was a spoon made of lodestone that rested on a smooth surface with markings indicating the four directions. The next step in the advancement of compasses was to enclose the lodestone in a decorative casing with a projecting needle to indicate which direction was north. One problem compass makers encountered was that the iron they used for the pointer lost its magnetism easily. After experimenting with different metals, the Chinese combined carbon and iron to make **steel**. The steel was stronger than iron and held its magnetic charge for a long time.

Meanwhile, other navigators around the world were also discovering the compass. Evidence suggests that Arab sailors were using compasses as early as A.D. 600. As the Arab influence spread into North Africa, Spain and France, it brought an extensive knowledge of the most advanced navigation techniques known at the time. Thus, the compass was introduced to Europe. The European to write about the use of the compass was Alexander Neckham (1157-1217) in A.D. 1187. He described two compasses, one floating and one dry. By the fourteenth century, European ships carried maps on which compass readings to reach different destinations were charted. These charts used a wind rose to indicate directions of north, south, east, and west with each quarter of the rose divided into ninety degrees.

An innovation in compass design followed the development of the windrose. Compass makers glued a **magnet**ic needle to the bottom of the wind rose with the north point of the rose and the north end of the needle aligned. When it was attached to a pivot the whole card rotated to point north rather than just the needle which made the compass much easier to read than previous styles.

During the fifteenth century, Prince Henry of Portugal (1394-1460) had a great influence on the development of sailing and navigation. He established a school for navigators in Portugal, and encouraged sailors and map makers to coordinate their information to make more accurate maps of the seas. Because of his interest in sea travel and exploration, Prince Henry fostered many improvements to the compass. One was the development of the binnacle—a glass case to protect compasses on board ship decks. The binnacle also had a place for a lamp to allow ship's pilots to read the compass at night. A stand for the compass was also developed at this time. Two brass rings were attached one inside the other so that they could move up and down. The compass was set inside the rings and attached to a stand which enabled the compass to remain level even though the ship swayed in rough waters.

Another of the innovations that came in the fifteenth century was the discovery of magnetic declination. When Christopher Columbus (1451-1506) sailed from Spain to the New World, he noticed that his compass did not align directly with the North Star. The difference between magnetic north and true north was called declination. Although sailors had known that the needle of the compass pointed slightly to the east of true north, they had assumed that the readings were consistent around the world. Columbus discovered that the degree to which the compass readings varied changed as he sailed across the ocean. In 1581, an English navigator named Robert Norman began investigating what made the compass readings vary at different points. Norman noticed that the angle of the compass needle varied according to its location as well. He conducted experiments to find out what was causing this ''dip.'' In the course of his investigations, Norman discovered what he called the ''attractive power on the earth.'' He deduced that it was this ''attractive power'' that pulled the magnetic compass toward the earth's surface.

This was consistent because the angle at which compasses dipped corresponded exactly to the angle of the earth's surface. Through further work in the area, Norman hypothesized that the earth had magnetic fields that run parallel to its surface. He made diagrams of these fields and explained the difference between the geographical axis of the earth and its magnetic axis. Charts similar to his still appear on modern maps. Norman's diagrams made it possible for navigators to determine the magnetic declination for any given position on the map, thus their calculations were much more accurate than ever before. Further improvements were made to maps and compasses when it was discovered that declination was affected not only by location but by time. Over a period of years, the earth's magnetic field shifts, changing the declination for any given place.

In the late 1500s, another Englishman, William Gilbert (1544-1603), experimented with magnets to find out why they behaved as they did. After carving a chunk of lodestone into a ball and then holding a needle near the lodestone. From this, he was able to determine north and south poles on this ball and deduced that the Earth itself must attract magnets through its magnetic poles in a similar way.

As maps and ships improved, new compasses and new problems developed. In 1789 a British doctor, Gowin Knight, rubbed a bundle of magnets on an iron bar to create super magnets, which in turn magnetized compass needles for longer periods of time. A great success, Knight's super magnets were used by the British navy for over eighty years. However, by 1850 all ships in the British Navy were iron, leading to magnetic deviation in which compasses were attracted to the small amounts of magnetism in the ships and thus gave false readings. **Lord Kelvin** corrected this by using small corrector magnets that surrounded the compass and prevented deviation.

Navigators next turned to the liquid compass rather than the old, dry card compasses. Scientists mixed water with alcohol to prevent the liquid from freezing, they sealed the compass bowl with rubber to prevent leaks, and they invented strong paints so that the markings would not flake away in alcohol. By the early twentieth century, liquid compasses were commonplace. In World War I, the vertical card compass was created to replace flat compasses which were too difficult to read during battle maneuvers. It was shaped like a tube and spun around on a pivot. Similar ones can sometimes be found on the dashboard of cars.

An American, **Elmer Sperry**, built the first **gyrocompass**, a device that worked day or night, anywhere on Earth even at the poles where lines of force are too close together for magnetic uses to function properly. When the gyrocompass is pointed north, it holds that position because it uses a **gyroscope**, a spinning wheel set in gimbals. The wheel spins due to an electric motor, and this wheel will turn in the same direction unless the position of the gimbals is changed. By 1935, many pilots used gyrocompasses because they were steady in all types of weather and they never spun wildly, even in the sharpest turns. The U.S. **submarine** *Nautilus* used a gyrocompass when it crossed under the North Pole because it was not affected by the powerful magnetic forces there. **Rocket**s are dependent on gyrocompasses because in space magnetic poles cannot pull needles in their directions.

The future of compasses lies in the electronic and satellite-based arenas. Electronic compasses are not inclined to the same errors magnetic compasses are like extraneous magnetic fields and vibration. Electronic compasses are more accurate and easier to calibrate. **Global Positioning System**s (GPS) are not compasses in the traditional sense, but provide precise information about an object's immediate location from 24 satellites orbiting Earth. A GPS receiver process radio signals put out by the satellites to determine their position. This information can be used to steer ships or planes around other vessels or other potential path-blockers. When combined with other technologies, such as computerized **map**s, GPS can help guide motorists through roadways. Indeed, one company, Silva, manufactured a GPS Compass, which combines an electronic compass with a GPS receiver.

Composite materials

There are benefits to compositing two or more materials. Composite materials take advantage of the unique properties of each component and are designed to optimize the strength, density, electrical properties, and cost of the materials.

Many conventional materials such as plywood, reinforced rubber, and reinforced concrete are composites since they consist of combinations of materials. Concrete, with its cement matrix and rock aggregate, is a composite. To many, however, the term composite material refers to a specific category of advanced composites.

Advanced composites have emerged primarily as a result of the **aircraft** and spacecraft industries, in which there are extremes in strength and weight requirements. New metals have been created that not only meet this demand, but also are capable of resisting fatigue, corrosion, and vibration.

Fiber-reinforced *metal-matrix composites* (MMCs) consist of high-performance metallic reinforcements—**glass**, boron or graphite filaments—in matrices of aluminum, titanium, magnesium, copper, or another material. MMCs are created by diffusion bonding, casting, hot molding, or by plasma spraying the matrix around the reinforcement, which can be in the shape of particles, whiskers, or fibers.

MMCs were first used in 1974 in the United States **space shuttle** project. A boron-aluminum composite was used in tubing that went into the craft. This same technology was passed on commercially and has come to be used in the manufacture of many products, from **pistons** and connecting rods to tennis racquets and fishing rods.

Boron and silicon carbide are the most common of the very high modulus reinforcements. They were developed for the United States Air Force Materials Laboratory by Texaco and United Technology corporations.

Composites can be found in the textile industry. Felt is a type of composite consisting of interlocking fibers created by the use of heat, moisture, and pressure. Wool, or wool combined with other fibers, is mechanically or chemically meshed together while the fibrous mass is kept warm and moist. The three types of felts are pressed, needled, and woven. Felt is a popular fabric because of its similarity to fur.

Composites are becoming common in **plastics**. The same reinforcement materials—glass, boron, and graphite—are used to increase the strength and stiffness of plastics. Polyvinyl chloride is used as a laminate for metal. It protects metal from corrosion and allows colors to be added. Plastic composites can be molded into shapes, needed for kitchenware and electrical insulation, uses that require light weight, strength, and flexibility.

Super-strength composites are employed in the construction industry. Bridge and building designs that demand that larger spaces be spanned require synthetic materials to meet the new challenges. This requirement extends to the machinery and equipment used in building or roadway construction. New types of reinforced concretes containing fibers are being used for highway overpasses, especially in areas that are earthquake prone. High-strength metal girders are used to support bridge decks and in stress-skin highrise buildings.

A main concern in the development of composites is that an adequate bond be achieved between the matrix and the introduced material. There must also be an equitable distribution of the composited materials. Fibers can be created by spinning, drawing (pulling), extrusion, and deposition. They are then combined with the host matrix by controlled mixing.

Material compositing probably began when the Egyptians mixed straw in clay to strengthen their unbaked bricks. Anticipating modern construction practices, the Romans began to mix volcanic ash into cement to create a subaqueous building material in about 200 B.C. Today laboratories with multimillion dollar budgets are continuing the tradition as they attempt to bring composites into the age of high technology.

See also Aluminum production; Concrete and cement; Road building; Textiles; Wire

Composting

As farmers and gardeners grow their crops, nutrients are removed from the soil as the plants grow and produce fruit. To keep the soil in condition for future crops, nutrients must be reintroduced to the soil. Natural and man-made **fertilizer**s are the most common source of introduced nutrients.

Field composting is a major supplement to fertilizing. Farmers achieve composting by simply ''plowing under'' the crop's remains, or trash, allowing it to decompose in the field during the fallow months.

Compost heaps are especially beneficial to dedicated gardeners, greenhouse farmers, and nursery operators. For the gardener who must continually reuse the same plot of ground, the compost heap can be a major source of soil nutrients. For the greenhouse grower, introduction of soils from outside sources every year is comparable to moving to a different farm annually. Since gardeners cannot develop their soil the way that a farmer can, composting helps ensure them a consistent soil balance.

Composting was a standard practice for centuries before commercial fertilizers became available. It was only natural for a farmer to return biodegradable material back to the earth. Composting has had a resurgence in popularity since the 1960s because of increased awareness of ecological concerns. The compost heap has become a common feature of urban, suburban and rural garden systems. More recently, in the 1980s and 1990s, as concerns were raised in the United States over landfill capacities, the compost heap became popular with suburban lawn owners as **recycling** systems for their leaf and grass litter.

Grass, leaves, twigs, garden trash, table scraps, and even small amounts of meat scraps can be added to the heap. When interspersed with layers of soil, manure and fertilizer, and with frequent water applications and turning over of the collected material, the proper conditions are created for the breakdown of the material by microorganisms, primarily bacteria. Some bacteria are chemical decomposers, which produce enzymes that digest the material. Other organisms digest the material directly.

The chemical decomposers produce heat as they convert the compounds such as phosphorus, carbon, and nitrogen into energy for their growth and development. This can raise the temperature at the center of the heap to 103° F (39° C) or more. The heat not only accelerates decomposition, but also kills pests and weed seeds. Larger organisms, especially earthworms, create tunnel systems in the heap, aerating the material and leaving behind feces rich in nitrogen, an element that would otherwise be lacking in the composition.

Decomposition generally takes six months before the composted material can be used. There has been some debate over the practicality of compost heaps, especially in large-scale farming. It is a labor intensive operation. It has also been found that plowed-in organics deliver a higher nutrient yield than compost. On the other hand, material plowed in to soil that is already well-drained can accelerate moisture loss. Nevertheless, for small-scale farming, composting offers the double benefit of recycling waste and keeping the soil in good condition.

Computer, analog

Unlike a digital computer, which performs calculations strictly upon numbers or symbols, an analog computer translates continuously varying quantities such as temperature, pressure, weight, or speed into corresponding voltages or gear movements. It then performs "calculations" by comparing, adding or subtracting voltages or gear motions in various ways, finally directing the result to an output device such as a **cathode-ray tube** or pen plotter on a roll of paper. Common devices like thermostats and bathroom scales are actually simple analog computers: they "compute" one thing by measuring another; they do not count.

The earliest known analog computer is an **astrolabe**. Built in Greece during the first century B.C., the device used pointers and scales on its face and a complex arrangement of bronze gears to predict the motions of the sun, planets, and stars.

Other early measuring devices were also analog computers. **Sundials** traced a shadow's path to show the time of day. Springweight scales, which have been used for centuries, convert the pull on a stretched **spring** to avoirdupois. The **slide rule** was invented about 1620 and is still used, although it has been almost completely superseded by the electronic **calculator**.

In 1905 Rollin Harris and E. G. Fisher of the U.S. Coast and Geodetic Survey started work on a calculating device that would forecast tides. Dubbed the "Great Brass Brain," it was 11 feet (3.3 m) long, 7 feet (2.1 m) high, and weighed 2,500 pounds (1,125 kg). It contained a maze of **cams**, **gears**, and rotating shafts. Completed in 1910, the machine worked as follows: an operator set 37 dials (each representing a particular geological or astronomical variable), turned a crank, and the computer drew up tidal charts for as far into the future as the operator wished. It made accurate predictions and was used for 56 years before being retired in 1966.

Vannevar Bush, an electrical engineer at the Massachusetts Institute of Technology, created–what is considered to be the first modern computer in the 1930s. He and a team from MIT's electrical engineering staff, discouraged by the time-consuming mathematical computations, called differential equations, required to solve certain engineering problems, began work on a device to solve these equations automatically. In 1935, the incredible second version of their device, dubbed the "differential analyzer," was unveiled: it weighed 100 tons (91 t), contained 150 motors, and hundreds of miles of wires connecting relays and **vacuum tube**s. Three copies of the machine were built for military and research use. Over the next 15 years, MIT built several new versions of the computer. By present standards the machine was slow, only about 100 times faster than a human operator using a desk calculator.

In the 1950s RCA produced the first reliable design for a fully electronic analog computer, but by this time, many of the most complex functions of analog computers were being assumed by faster and more accurate digital computers. Analog computers are still used today for some applications such as scientific calculation, engineering design, industrial process control, and **spacecraft** navigation.

See also Computer, digital

Computer applications

Applications are computer programs that allow a user to manipulate data in specified ways. Databases, spreadsheets, word processing, page layout programs, and drawing programs are electronic versions of older business systems. Communications and networking programs are extensions of the computer as an electronic medium. Some applications even include their own languages, allowing the user to write related routines or customize use of the application.

Applications were written for **mainframe computer** systems beginning in the 1950s. The wide variety of user-oriented applications began with the personal computer, beginning in the mid-1970s, developing along with faster processing chips, large computer memory and disk storage, and more flexible operating systems.

The modern software industry began in 1975, when **William Gates** and Paul Allen wrote an interpreted version of the language BASIC for the first personal computer, the Altair (MITS). This made it easy and even fun to write applications. As other computers were invented, Gates licensed versions of BASIC to run on them as well. By the mid-1980s, Gates's company, Microsoft, had become the leading software publisher and he was a billionaire.

Besides Microsoft, other prominent software companies are Lotus Development Corp. (founded by Mitchell Kapor), Novell Inc. (founded by Ray Noorda), Borland International Inc. (now Inprise Corporation), headed by Philippe Kahn, and WordPerfect Corp.

Database Management Systems.

A database is an electronic version of a set of index cards, in which each set of information is a record and each type of information is a field. A database management system (or DBMS) allows the user to find specific words and names, to sort the records by specific fields, such as last name or ZIP code, and otherwise specify which information is presented and how it is displayed. Programs exist for single desktop computers, mainframes, and computer networks. Databases are now being designed to include illustrations, video, and music and other sounds, as well as words and numbers.

The two main types of databases are the flat file and the relational database. A flat file is like a single set of index cards. A relational database is composed of one or more flat files and provides user-specified links among fields that allow complex searches and associations. On-line information systems are examples.

Many relational databases are searched with a special computer language called the Structured Query Language, or SQL. This in turn is often made easier to use with various graphical interfaces or front ends, such as those supplied by Oracle Corp. or Sybase.

For many years the most popular database has been dBASE, introduced by Ashton-Tate in 1983 and now owned by Inprise Corporation. Other widely-used databases are Borland's Paradox, and FoxPro and Access, which are both produced by Microsoft.

Specialized database management systems include information management systems, such as Microsoft Project, which are used for scheduling projects and assembling the necessary ideas and data. Going a step farther, expert systems combine databases with routines that allow them to make decisions. They are widely used in the banking and credit industries.

Word Processing.

Word processing programs begin with the basic features of a typewriter and add abilities available only with a computer: quickly moving text from one position to another (cutting and pasting), producing individually addressed copies of the same letter (mail merge), checking spelling, and replacing one word or phrase with another throughout a document (global search and replace).

The first word processors were single-purpose machines, such as those produced in the 1970s by Wang Laboratories. The first popular word processing software for personal computers was WordStar, published by Micro Pro in 1978. Today the most widely used programs are WordPerfect (WordPerfect Corp.) and Microsoft Word, both dating from the early 1980s.

Many word processing programs also provide features for professional-quality desktop publishing, such as page design and layout and insertion of illustrations. The finished product is printed on a high-resolution laser printer. There are also specialized page layout programs, such as PageMaker (Aldus Corp.) and Ventura Publisher (Xerox).

Spreadsheets.

Spreadsheets are electronic versions of the traditional accounting spreadsheet written on columnar paper. Like the paper version, an electronic spreadsheet categorizes information in columns and rows, which intersect in *cells*. Each cell can contain a number or a formula that can be automatically calculated, totaled, and linked to other totals and cells. Some programs provide a ''third dimension'' that allows information on several spreadsheets to be considered together.

Spreadsheets are also used to project future values in ''what if'' situations, such as increasing sales or opening a new factory. Today's spreadsheets can also provide charts and other visualization of the data.

The first spreadsheet for personal computers was VisiCalc, created in 1979 by Dan Bricklin and Bob Frankston. Complex spreadsheets include Lotus 1-2-3 (Lotus Development Corp.) and Microsoft's Excel, both introduced in the mid-1980s.

Accounting programs are spreadsheets that are already formatted for business management, such as payrolls, inventories, accounts payable, and accounts receivable.

Graphics.

Computerized graphics programs perform the jobs that used to be the work of drafters and graphic artists. Two-dimensional programs are used for line drawings or for black-and-white or color paintings. Popular software includes Adobe PhotoShop, Adobe Illustrator, and CorelDRAW (Corel).

Three-dimensional figures, both outlined (or wireframed) and solid are also available with some graphics programs. However, there are dedicated 3D graphics programs. These 3D programs such as Lightwave 3D (NewTek), Bryce

3D (MetaCreations), and SoftImage (Microsoft) are used to produce three-dimensional graphics and animation sequences for television shows and feature films. These are also used by industry as **CAD/CAM** (Computer-Aided Design). When combined with a database of parts, specifications, and costs, someone with a computer can not only design a new product but also specify the materials and its expected performance. AutoCAD (AutoDesk) is a widely-used example.

Graphics programs are used in business and research to visualize data produced by statistical calculations and research observations. Graphics is also the technique that makes possible the games played on computers, game systems, and arcade game machines. It is also the basis for many educational applications.

Communications.

Communications programs allow computers to communicate with each other either over telephone lines by modem or over the direct cabling of computer networks. Transferring files, sending and receiving data, using data stored on another computer, and electronic mail (e-mail) systems that allow people to receive messages in their own ''mailboxes'' are some common uses of communications software. Numerous communications programs are available, one of the more popular being ProComm (Datastorm Technologies).

A network operating system, such as NetWare (Novell Inc.) and Lantastic (Artisoft), handles the basic functions of network communications, such as directing data to the correct computer address and maintaining the flow of data.

Networking has led to the formation of workgroups, in which people work together by using the same computer programs rather than being in the same place physically. Workgroup management programs include Lotus Notes (Lotus Development Corp.) and Outlook/Exchange (Microsoft).

Computer, digital

The digital computer is a programmable electronic device that processes numbers and words accurately and at enormous speed. It comes in a variety of shapes and sizes, ranging from the familiar desktop **microcomputer** to the **minicomputer**, **mainframe**, and **supercomputer**. The supercomputer is the most powerful in this hierarchy and is used by organizations such as NASA to process upwards of 100 million instructions per second. The impact of the digital computer on society has been tremendous; in its various forms, it is used to run everything from spacecraft to factories, healthcare systems to telecommunications, banks to household budgets.

The story of how the digital computer evolved is largely the story of an unending search for labor-saving devices. Its roots go back beyond the **calculating machines** of the 1600s to the pebbles (in Latin, *calculi*) that the merchants of Rome used for counting, to the **abacus** of the fifth century B.C. Although none of these early devices were automatic, they were useful in a world where mathematical calculations, laboriously performed by human beings, were riddled with human error.

By the early 1800s, with the Industrial Revolution well underway, errors in mathematical data had assumed new importance; faulty navigational tables, for example, were the cause of frequent shipwrecks. Such errors were also the source of irritation to **Charles Babbage**, a brilliant young English mathematician. Convinced that a machine could do mathematical calculations faster and more accurately than humans, Babbage in 1822 produced a small working model of his difference engine. The difference engine's arithmetic functioning was limited, but it could compile and print mathematical table with no more human intervention needed than a hand to turn the handles at the top of the model. Although the British government was impressed enough to invest £17,000 in construction of a full-scale difference engine, it was never built; the project came to a halt in 1833 in a dispute over payments between Babbage and his workmen.

By that time, Babbage had already started to work on an improved version—the analytical engine, an automated programmable machine that could perform all types of arithmetic functions. The Analytical Engine had all the essential parts of the modern computer: an input device, a memory, a central processing unit, and a printer. For input and programming, Babbage used **punched cards**, an idea borrowed from **Joseph Jacquard**, who had used them in his revolutionary weaving **loom** in 1801.

Although the analytical engine has gone down in history as the prototype of the modern computer, a full-scale version was never built. Among the deterrents were lack of funding and a technology that lagged well behind Babbage's vision. Even if the analytical engine had been built, it would have been powered by a steam engine, and given its purely mechanical components, its computing speed would not have been great. Less than twenty years after Babbage's death in 1871, an American by the name of **Herman Hollerith** was able to make use of a new technology—electricity—when he submitted to the United States government a plan for a machine that could compute census data. Hollerith's electromechanical device tabulated the results of the 1890 United States census in less than six weeks, something of an improvement over the seven years it had taken to tabulate the results of the 1880 census. Hollerith went on to found the company that ultimately emerged as IBM.

World War II was the movitation for the next significant stage in the evolution of the digital computer. Out of it came the **Colossus**, a special-purpose electronic computer built by the British to decipher German codes; the **Mark I**, a gigantic electromechanical device constructed at Harvard University under the direction of **Howard Aiken**; and the **ENIAC**, another huge machine, but one that was fully electronic and thus much faster that the Mark I. Built at the University of Pennsylvania under the direction of **John Mauchly** and **J. Presper Eckert**, the ENIAC operated on some 18,000 **vacuum tubes**. If its electronic components had been laid side by side two inches apart, they would have covered a football field.

The ENIAC was a general-purpose computer in theory, but to switch form one program to another meant that a part of the machine had to be disassembled and rewired. To circumvent this tedious process, **John von Neumann**, a Hungarian-born American mathematician, proposed the concept of the **stored program**—that is, coding the program in the same way as the stored data and keeping it in the computer for as long as needed. The computer could then be instructed to change programs, and the programs themselves could even be written to interact with each other. For coding, Neumann proposed using the **binary numbering system**—0 and 1—rather than the 0 to 9 of the decimal system. Because 0 and 1 correspond to the on or off states of electric current, computer design was greatly simplified.

Neumann's concepts were incorporated in the British-built EDSAC and the University of Pennsylvania's EDVAC in 1949, and in the **UNIVAC** and other first-generation computers that followed in the 1950s. All these machines were large, plodding dinosaurs by today's standards. Since then, advances in **programming languages** and electronics—among them, the **transistor**, the integrated circuit, and the **microprocessor**—have brought about computing power in the forms we know it today, ranging from the supercomputer to far more compact models.

See also Computer, analog

Computer discs and tape

The first commercial computers used **punched card**s and paper tape to store information. During the late 1940s, however, computer engineers began to explore other means of storage because the cards and paper were bulky, prone to damage and difficult to access. When magnetic recording technology emerged after World War II, computer designers were quick to see the possibilities such a technology could offer them.

In 1949 **J. Presper Eckert** and **John Mauchly** introduced BINAC (Binary Automatic Computer), the first computer to employ magnetic tape for storage. The plastic-based tape it used was able to hold far more information per unit size than punched cards. BINAC, however, was merely a stripped-down temporary predecessor of a more advanced machine called **UNIVAC** (Universal Automatic Computer), completed by Eckert and Mauchly two years later. UNIVAC replaced BINAC's plastic tape with a stronger metal tape. Once better **plastic**s were developed, however, they returned to use as they were safer and caused less wear to tape reading components.

Tapes, however, were still not fast enough for computers that could perform thousands of calculations per second, so other proposals were advanced. For a time, rapidly rotating magnetic drums were used as a storage medium, but drums could not be engineered to meet the need for greater speed of access that emerged as computers continued to improve.

Reynold Johnson, an engineer at IBM, inspired by the work of inventor **Jacob Rabinow**, began work on the concept of using spinning discs as a data storage method in 1952. By 1955 he and his staff had developed a massive hard disk unit, which consisted of a set of fifty platters, each two feet wide, coated with a magnetic material and mounted on spindles rotating 1,200 times per minute. A read/write head moved back and forth between discs to obtain information. Later models added separate heads for each disc. Dubbed "jukeboxes," the machines were quite effective despite their bulk.

The development of the floppy disk followed the hard disc drive of the 1950s. The floppy was able to hold less data than the hard disc and it was slower in operation; but Alan Shugart's design for the floppy, devised in the late 1960s, had two advantages over the hard disc drive system: portability and less susceptibility to damage. His creation, initially conceived as a device to aid restoration of the operating system of **mainframe computer**s after power failures, was an 8-inch (20.3-cm) plastic disc touched by a read-write head. The floppy was less prone to damage from the head than that a hard disc (which had a head that floated a fraction of an inch above the disc, occasionally striking the disc and erasing information when it encountered dust or vibration). The flexible plastic discs were also extremely durable, light and compact. These advantages rapidly made floppy discs a mainstay in computing, and floppy discs were included with personal computers when they came on the market in 1981. Newer floppy discs are much smaller (common are the 5.25-inch (13.3-cm) and 3.5-inch (8.9-cm) sizes) and can hold far more data—up to 1.44 megabytes (1.44 million bytes, equivalent to 720 typed pages of information).

The success of personal **microcomputer**s led to further innovations. Once again, the need of users for more data storage outstripped the capacity of floppy discs. In the early 1980s, a team of IBM engineers headed by Philip D. Estridge was assigned the task of improving the IBM-PC by adding a better storage system. Estridge and his team developed the fixed hard drive, a miniaturized throwback to the Johnson hard disc of the 1950s. Also called a Winchester disc, the hard drive was a unit mounted permanently within the chassis of the computer and coated with a magnetic material. It was capable of holding thirty times the data of the early floppy discs—ten megabytes (five thousand pages) of information.

The hard drive, in fact, made so much information available to the user at once that IBM had to modify its PC design to accommodate the disc. First, a new **computer operating system** capable of handling and organizing large numbers of files was introduced. In addition, the computer was given the ability to "boot" or load its operating system directly from the hard disc when the computer is turned on. This meant that a user no longer needed to insert a series of floppy discs when switching the computer on.

In the late 1980s, Plus Development of California introduced a HardCard that contains a fixed disc which fits directly into the expansion slots of IBM's personal computers for those who wish to add a hard drive. Small hard disc drives are available for in laptop computers as well. In 1987 IBM came out with a new 3.5 inch hard disc capable of holding the equivalent of over 120 megabytes (60,000 pages of data). In addition, a new medium, the **optical disk** was introduced. Systems such as CD-ROM and **WORM** surpassed the floppies and hard discs in performance by delivering portability, reliability, greater storage capacity, and lower cost. During the mid- to late-1990s, hard drive technology advanced to create 3.5 inch hard discs capable of holding over 12 gigabytes (6,000,000 pages of data). Removable high-capacity disk drives became popular. These rigid floppy disk-like cartridges could hold nearly 2 gigabytes (1,000,000 pages of data).

See also Computer, digital; Stored program

Computer graphics

Computer graphics is a pictorial communication between humans and computers. While almost any problem and solution can be reduced to alphanumeric form, it is often an unnatural and time-consuming way to think and work. Introduced in the 1970s, graphic screen displays and plotters bridge the gap between computers and humans, allowing for creation and output of any form imaginable. Graphic screen displays differ from alphanumeric displays, as they are capable of showing the usual letters, numbers and symbols plus graphs, diagrams, lines, curves, circles and other shapes. Input can be accomplished using a keyboard, but other input devices—such as a light pen, mouse, track ball, or **digitizer**—are frequently used to speed up the process. Such input devices used on a graphic workstation are an essential part of **CAD/CAM** (computer-aided design and manufacturing) applications, enabling complex drawings to be created or changed at will on-screen.

The development of various **microcomputer** graphics software, particularly that from Lotus Development Corporation in 1982, has been gladly hailed by the business community. Graphics software converts numerical data into high resolution bar, pie or line graphs, giving greater impact to business documents, reports and presentations. Draw and paint programs, another type of graphics software, first appeared in the mid-to late 1980s. Such programs, called interactive software, are used by designers, artists and engineers to create or modify illustrations directly on screen. Specialized CAD/CAM software is also available, however, it often requires powerful computer workstations to operate to its greatest potential.

Although graphic screens can generate high quality drawings, they cannot produce hard copy output. The graphic plotter, first conceived in 1833 by **Charles Babbage** for his analytical engine, is most useful for this purpose. Plotters produce high resolution three-dimensional output on single sheets or continuous rolls of paper. Capable of large, intricate drawings of exacting precision, plotters utilize computer-controlled **pens** positionable at up to 45,000 different points in each square inch. One type of plotter, the drum plotter, uses a continuous roll of paper that moves forward and backward over a drum. A number of pens suspended over the paper move from side to side and are raised or lowered to make contact with the paper. Drum plotters are often used as the output device on CAD/CAM stations to produce immense drawings of electronic circuitry. Commonly used by the business community, the flatbed plotter looks much like a drafting table, with various colored pens held in place by a frame. Although flatbed plotters are limited to using single sheets of paper, they are much less expensive than drum plotters, which may cost upwards of $250,000.

See also Computer, industrial uses of; Computer input and output devices

Computer, industrial uses of

Labor saving ideas involving the computer, particularly for industrial and mechanical purposes, is a never-ending process. Computers in industry are drastically changing the way things are made and profoundly changing the jobs of the people who make them.

In 1947 the California Institute of Technology devised a method for using computers to aid in designing **aircraft**. By 1950, the Massachusetts Institute of Technology (MIT) had developed an automatically controlled **milling machine** used in cutting metal parts. In 1963, Sketchpad, an early forerunner of the Computer Aided Design (CAD) system, was developed by Lincoln Laboratories. A similar system, DAC-1 from General Motors, was developed at the same time. In 1968 MIT and the U.S. Air Force jointly developed a **CAD/CAM** (computer aided manufacturing) system to drive **lathes** and tooling machines used in the aeronautics industry.

By the 1970s a number of new methods were developed to remove much of the tedium and manual processes from design and manufacturing work. Now an essential part of industry, CAD/CAM, as it is more commonly called, is the process of using the computer in design and manufacturing functions. Thousands or tens of thousands of highly technical and accurate drawings and charts are required for the many design specifications, blueprints, material lists and other documents used to build complex machines. If the engineers decide structural components need to be changed, all the plans and drawings must also be changed. Prior to CAD/CAM, human designers and draftspersons had to change them manually, a time-consuming and error-prone process.

When a CAD system is used, the computer can automatically evaluate and change all corresponding documents instantly. Using interactive graphics workstations, designers, engineers and architects can create models or drawings, increase or decrease sizes, rotate or change them at will, and see results instantly on screen. CAD use is particularly valuable in space programs, where many unknown design variables are involved. Previously, engineers depended upon trial-and-error testing and modification, a time consuming and possibly life threatening process. However, when aided by **computer simulation** and testing, a great deal of time, money and possibly lives can be saved. Besides its use in the military, CAD is also used in civil aeronautics, automotive and data processing industries.

CAM, commonly utilized in conjunction with CAD, uses the computer to communicate instructions to automated machinery. CAM techniques are especially suited for manufacturing plants, where tasks are repetitive, tedious or dangerous for human workers. While the use of CAD/CAM systems enables the production of better, less expensive products, workers stand to lose their livelihoods due to the increased acceptance of automated systems. Computer Integrated Manufacturing (CIM), a term popularized by Joseph Harrington in 1975, is also known as Autofacturing. CIM is a programmable manufacturing method designed to link CAD, CAM, industrial **robotics**, and machine manufacturing using unattended processing workstations. CIM offers uninterrupted operation from raw materials to finished product, with the added benefits of quality assurance and automated assembly. Unfortunately, this means less direct human attention is required as well as fewer jobs in the workplace.

CAE (computer aided engineering), which appeared in the late 1970s, combines software, hardware, graphics, automated analysis, simulated operation and physical testing to improve accuracy, effectiveness and productivity.

See also Computer graphics; Computer input and output devices; Computer vision; Cybernetics

Computer input and output devices

When computers were introduced into wide usage after World War II, they revolutionized the way modern societies function. Primarily, they have allowed for the more efficient processing and storing of vast amounts of information. The material processed by a computer, however, must still be entered into its data banks in some way, and must be output to be used. The development of ''user friendly'' devices to enter and extract information has been essential to the computer's success because these are the parts of a computer system with which most people regularly interact.

Many methods of input have been developed, each presenting certain advantages and disadvantages. **Punched cards** were the first output devices used by modern computer designers. They were developed by **Joseph-Marie Jacquard**, a French weaver, who designed them in 1801 as a means of automating his weaving machines. In 1888 **Herman Hollerith**, an American businessman, devised a machine that used punched cards in conjunction with electrically charged **nails** that connected wherever the card had a hole. This machine was used to help process information from the census of 1890.

The use of computers expanded rapidly after the end of World War II, and by the 1950s, many businesses were incorporating them into their essential activities. This gave rise to the development of special input devices for different needs. The modem was one of the first input devices to help create world-wide linking capabilities among computer systems. The word MODEM is an acronym for MODulator/DEModulator which describes the process of changing digital (computer) signals to analog (**telephone**) signals and vice versa. Modems came into use in the 1950s and allowed computers to hook up to each other through telephone lines.

Magnetic tape, developed and used during the late 1950s, is now used more as a storage medium than an input method. Magnetic ink and magnetic ink character recognition (MICR) were also introduced in the 1950s. The banking industry is the primary user of magnetic ink. Checks and deposit slips are printed with the ink so that they can be scanned through MICR readers which translate the characters into electronic signals.

The terminal was first introduced as an input device in the mid-1960s. Known by several names (VDTs—video display terminals, CRTs— **cathode-ray tubes**), a terminal usually consists of a keyboard and display screen, though keyboard/printer combinations are available. Keyboards produce letters by using codes and electronic signals to represent characters.

In 1963, I. E. Sutherland (1938-), introduced the light pen which he had developed while working at the Massachusetts Institute of Technology. When the pen makes contact with the screen, light-sensitive cells in the pen tip transmit an electronic signal, allowing users to write or modify on-screen images. Touch-sensitive monitors work in a similar manner except the user simply touches the appropriate place on the monitor rather than using a light pen. Digitizing tablets, invented by the Rand Corporation in 1964, allow a graphic image to be scanned, converted into digital data and displayed or revised on screen. These systems are widely used in graphics, engineering, and design work.

Specialized Point-of-Sale (POS) terminals began appearing in retail outlets in 1973. POS terminals perform typical **cash register** functions, but also track inventory, transmit data, and check credit inquiries. Often paired with optical scanners or wand readers, POS terminals have helped decrease operating costs for many retailers.

A mouse or ''pet'' peripheral, is a fairly new input device, first included with the Apple Computer Corporation's LISA microcomputer in 1983. So named because its shape somewhat resembles a small rodent, it contains a rolling ball and a panel with one or more buttons. When rolled on a smooth, flat surface, several small, light-sensing diodes track the mouse's movements and the computer responds by moving the cursor on-screen accordingly. Mice are a particular necessity when using graphics, painting and design programs.

Printers are the most common peripherals used purely for output, although technically, video displays and modems may also be considered output devices as well. One of the earliest printers was a line printer, developed in 1953 by the Remington Rand Company. Line printers produce entire lines of characters at a time using an engraved band or chain to strike the ribbon. Printers which produce characters by striking against a ribbon are called impact printers. Another impact printer is the daisy wheel printer, so named for its print mechanism, a spoked wheel containing raised characters. Daisy wheels produce excellent print quality, but print slowly.

Dot matrix printers produce letters in the same fashion as bank temperature signs, whereby light patterns are turned on or off to produce numbers. In the same way, different pin patterns strike the ribbon to produce dot-formed characters. Ink jet printers, a non-impact printer introduced by IBM in 1976, improved upon dot matrix appearance by squirting dots of ink to form letters. In 1982, IBM also introduced the laser printer, an extremely fast printer which utilizes laser beams moving across a rotating drum to create near typeset quality print.

Pen plotters, available since the mid 1950s, convert graphs, charts and line drawings into large, high resolution color output. One or several **pens** move across the paper to produce a 3-D effect image. On the downside, plotters are often bulky (some as large as pool tables), expensive and not useful for text output.

With Computer Output Microfilm (COM), developed in 1968, printed output is photographed as very small images on sheets or rolls of film. COM output produces amazing space savings; a 450-page book can be stored on three 4 x 6 inch (10 x 15 cm) microfiche pages.

See also Computer disk and tape; Computer network; Computer pattern and character recognition; Computer speech recognition; Computer vision; Digitizer

Computer memory

A typical computer is organized into four essential components: a central processing unit (CPU) to manipulate numbers; a primary memory that feeds data and instructions of immediate importance into the CPU; *secondary* or peripheral memory that stores data and programs until they are fed into the main memory; and **input-output devices**, which link the CPU and the information the computer will work with.

Primary Memory.

Early computers stored instructions and data in bulky mechanical devices, **vacuum tubes**, or transistorized elements. These devices were most often fed by **punched cards** or paper tapes. In the 1950s several memory systems were developed. The most important of these was the magnetic core memory devised by **Jay Forrester**, an engineer at the Massachusetts Institute of Technology. His system used tiny magnetic "doughnuts" suspended on a gridwork of wires inside the computer. These discs could be electrically "flipped" with an electric current to represent binary ones or zeros. Magnetic core memory could calculate at high speeds and had the added ability to retain its information even when the power to the computer was shut off. Even more important, magnetic memory could access information *randomly*—in other words, data from one portion of the memory was accessible as data from any other, unlike a stack of cards or roll of magnetic tape that had to be rewound or shuffled to find the information. Random-access memory (RAM) remains the heart of computer memory today.

The final step in developing primary memory came with the creation of the computer chip. When **Jack Kilby** and **Robert Noyce** invented the **integrated circuit** around 1960, Noyce realized that these tiny circuits could be perfected to produce a *RAM chip* for computers. By 1970, he had devised a chip that could be made in quantity, and he cofounded a company, Intel, to manufacture it. These chips were so small and cost-efficient that they began to be used in a host of electronic devices from watches to pocket **calculators**.

Around the same time, Marian "Ted" Hoff Jr., an employee at Intel, designed a chip called the **microprocessor** that contained a CPU. Now the information that once had to be stored in a jumble of vacuum tubes and stacks of punched cards could be kept on three tiny chips: one containing the CPU, one containing the program or instructions, and one with the data.

Peripheral or Auxiliary Memory.

As mentioned previously, most computers before the 1950s used punched cards or tapes to store data. Furthermore, until the construction of the EDVAC computer by **John von Neumann** in the mid-1940s, no computer was able to hold stored programs in its memory. For this reason, computers such as **J. Presper Eckert** and **John Mauchly**'s **ENIAC** depended on manipulation of dials or wires and a constant flow of cards to function. When Eckert and Mauchly built **UNIVAC** in 1951, they remedied this problem by providing the computer with internal program storage and by using magnetic tape to add data to the computer. Although more rapid and compact than cards, magnetic tape was still inconvenient: the fragile reels of tape had to be constantly wound and rewound to retrieve information.

Once primary high-speed RAM memories had been developed, a replacement for the sluggish punched cards and reels of magnetic tape had to be found; with the advent of magnetic drums and discs, cards and tapes were relegated to marginal use. Punched cards eventually fell from use entirely, but magnetic tapes (often variations of the cassette) are still used because they can store large amounts of information.

Most peripheral memory systems developed since the 1950s rely on magnetism to store data. Bubble memory, introduced in 1969 by Andrew H. Bobeck, a scientist a Bell Telephone Laboratories, consists of a thin card lined with ferromagnetic materials and placed next to a permanent magnet of the same shape. Data is stored and manipulated on "bubbles" or tiny magnetized regions within the card. This system's advantages include its lack of moving parts, its rapid access time, and its ability to perform logical operations that other magnetic systems cannot. But bubble memory was hampered by high production costs and reliability problems. Once some of the difficulties had been worked out, it began to find specialized uses in the 1980s in "intelligent" terminals and other electronic equipment.

At a low cost per unit of information, computer discs and tapes have come to dominate the market for peripheral memory devices. With the advent of **audiocassette** technology, magnetic tapes, with extremely high capacity for information, have been used for "archival" or bulk storage, though they are too slow for more immediate use because, like their predecessors, they must be wound and rewound to access information.

The latest audiocassette innovation is the Digital Audio Tape, or DAT. DAT tapes are very high-density and can store several gigabytes of information on a single cassette. High-density floppy disks can store the equivalent of a thousand pages of information, are accessed randomly, and are portable. Hard disks must be built into a computer but have over one thousand times the capacity of floppy discs.

The newest storage system, the **optical disk**, better known as a CD-ROM, reads data with a **laser** beam. CD-ROMs can store 275,000 pages of information on a single five-inch disc. An improvement to CD-ROMs was introduced in 1997. The Digital Versatile Disk, or DVD, can store nearly 5 gigabytes of information. This is approximately seven times more storage than on a CD-ROM.

Computer network

Computer networks have been a boon for businesses because they connect different makes and models of **microcomputers**,

minicomputers or even **mainframe computers**, allowing for shared communication, files and equipment.

The first networks appeared in the 1960s when multi-user networks were introduced. In multi-user systems, dumb terminals, which do not have processing capabilities of their own, are connected to a central host computer, which shares processing time with these dumb terminals. Now, much more popular are Local Area Networks, or LANS, which appeared in the early 1970s. A LAN is a communication network privately owned by the organization using it. LANs utilize stand-alone microcomputers rather than dumb terminals and can vary greatly in size, range, and complexity. The actual distance and number of computers that can be connected is highly dependent upon the type of LAN and communication line used. **Telephone** lines are sometimes used and are most convenient; however, coaxial cable connections permit faster, higher quality transmissions. **Fiber optics**, tiny tubes of glass half the diameter of a human hair, has become the preferred technology of the 1990s, allowing faster and less expensive data transmissions than wire cabling. All networks are different in size and complexity. However, they all share one of three similar shapes. These shapes are known as network topologies.

There are three basic topologies (networks) to which most networks conform: Bus, Ring, and Star. The Bus topology uses a single communication line to connect equipment, allowing contact between all stations, or nodes. Failure of a single microcomputer may interfere with the rest of the network. A Star topology consists of several microcomputers connected to a centralized networking device called a hub. This acts as a crossing guard for network traffic. Star topologies may also allow the computers to connect to a centralized host computer, called the file server, which shares its files among all users of the network. A variation of this, the Multi-star network, has several file servers connected to each other, yet each also has a Star network of its own. Star networks are more costly to install due to the extra cabling involved, but are frequently used among retail stores having several branch outlets. The third topology, the Ring, is somewhat similar in design to the Bus topology. Each computer is connected to two other computers and can communicate and share files with any station in the network. However, the two ends of the Bus are connected together, making the Bus into a Ring. A Ring network is slower and more expensive to install than the Star, but it is also more reliable, since alternative paths exist if a communications line is interrupted. A modification on the Ring network is the Complex or Distributed network. Communication lines connect all stations in the network, but on the downside, installation is complicated and requires high maintainance.

LANs offer many benefits within companies, industrial sites, college campuses and hospitals, allowing for more efficient and convenient computer services. Future predictions call for worldwide interconnected networks resembling the telephone systems of today. Accessing these networks will likely become just as commonplace as making a telephone call. One application of networking, electronic mail (E-Mail), introduced in the early 1970s, is very popular among businesses. E-mail service allows letters to be sent across the United States for less than the cost of mail, with the added advantage of arriving within seconds. Another popular offshoot of networking are Information Networks, such as America Online, CompuServe, or Prodigy, wherein computer users may access a wide variety of computerized services, including E-Mail, investment advice, reference and travel information, shopping and much more.

See also Computer input and output devices; Computer operating systems

Computer operating systems

In early computer systems, a human operator monitored operations, determined the order of programs to run, and handled all input and output processes. However, much time was wasted when a program would end its run, and the entire system sat idle while the operator prepared the next job. As computer processing speeds increased, the use of human operators became unrealistic. The need for more efficient use of computer resources resulted in the development of operating systems, which act somewhat like traffic directors for the computer. Operating systems are a collection of programs designed to permit a computer to manage its own operations, including peripherals (printers and modems), programs, and data. In 1954 Gene Amdahl (1922-) developed the first operating system, which was used on an IBM 704. In 1961 Frederick Brookes (1931-), an IBM engineer, began developing an operating system for the IBM S/360. By 1965 Brookes's development team consisted of 2,000 programmers with a budget of $60 million. The resultant operating system, known as OS/360, was a well designed, efficient program that became an industry standard for the next several decades. It was the largest software development project ever attempted and remained so through the mid-1980s. OS/360 also served as the basis for other **mainframe computer** software developed in the 1980s, such as DOS/VS, MVS/XA, and OS/VS.

Generally, systems software is developed for a particular type of computer processor chip or to meet a particular function. UNIX, developed in 1971 by Ken Thompson (1917-) and Denis Ritchie at Bell Laboratories, has become the dominant operating system of the late 1980s and early 1990s. Usable on mainframes, **minicomputers** and even **microcomputers**, UNIX is often the preferred operating system for network use. Another mainframe system, PICK, developed by Dick Pick, also goes by the name REALITY, PRIME-INFORMATION or ULTIMATE, depending upon the type of computer it is used on. When microcomputers were introduced in the early 1970s, almost all ran under the CP/M operating system, written in 1973 by Gary Kildall. Soon after its introduction, the CP/M system was overtaken completely by MS-DOS, an operating system from Microsoft Corporation and its founder William Gates. The earliest version of MS-DOS was based on a product from Seattle Computer Products known as 86-DOS, which was basically an enhanced version of CP/M. Many powerful features have been added to MS-DOS and current versions bear little resemblance to the first product. Some newer operating systems, including OS/2 introduced in 1987 by Microsoft and IBM, have been hailed as the next industry standard; however, users were slow to accept a replacement for DOS. Microsoft enhanced MS-DOS over the years and eventually created a graphical user interface called Windows. Windows was a success in that it hid the complexities of MS-DOS behind on-screen graphics and the use of a mouse. But acceptance of Windows was slow until 1995 when Microsoft released Windows 95. Windows 95 was more than a thin shell over MS-DOS, it was in its own right an operating system. Windows software from Microsoft is now used on over 90% of the computers on the planet.

See also Computer, digital; Computer, industrial uses of; Computer input and output devices; Computer network

Computer pattern and character recognition

Character recognition is the technology of using machines to identify symbols (usually alphanumeric characters) in order to express them as machine readable codes. The first commercial attempts at character recognition appeared in the 1950s when magnetic ink character recognition (MICR) was introduced into the banking industry. With the MICR system, characters and symbols are printed on checks or deposit and withdrawal slips using magnetic ink and are then put through a magnetizing process. Although MICR characters are readable by humans, often special MICR reader/sorters are used to interpret and sort the checks more rapidly.

In other areas of industry during the 1950s, the keypunch, keytape and key-to-disk systems were being used for most data input. Although popular at that time, these devices were expensive, prone to errors, and not well suited for large quantities of data entry. The desire to improve data entry methods paved the way for the development of character recognition and optical recognition machines.

Character recognition systems read numbers, letters, special characters and marks. An electronic scanning device converts the data into electrical signals and sends the signals to the computer for processing. Data may be represented in a variety of ways, including optical marks, optical characters, handwritten characters or **bar codes**.

Optical mark recognition (OMR), which was introduced in the early 1960s, is the simplest form of character recognition. OMR consists of a series of marks, lines or bar-codes, which are detected and decoded by fixed optical scanners or hand-held wand readers. College entrance exams are commonly scored using OMR technology.

Optical character recognition (OCR), a more advanced concept, is the process of optically reading typed, printed, or handwritten characters. OCR has been hailed as the solution to computer input bottlenecks because it may help to reduce or eliminate manual keystroke operations. Early OCR machines were restricted to recognizing just one or two standardized fonts, a long-standing problem which American engineer Raymond Kurzweil (1948-) resolved to tackle. In 1974 he began developing a machine that would read to blind people. Two years later the Kurzweil Reading Machine, which could scan any type of printed page and read aloud everything from utility bills to full length novels, was enthusiastically received upon its release.

Today modern OCR systems can be ''trained'' to recognize almost any font, as well as handwritten materials. OCR is utilized by the U.S. Post Office. OCR readers/sorters and bar-code readers began appearing in post offices in the 1960s. Using OCR technology to read and sort zip codes saves much time and effort.

Computer pattern recognition developed as a specialized form of character recognition and is used by manufacturers who utilize industrial robots. When a robot picks up a part such as a screw, nut or other component, it must be able to sense the precise shape and alignment of the part in order to grasp it properly. Optical sensing gives the robot this pattern recognition ability. Although a robot can be taught to sense a physical part or object, when symbols or handwriting are the materials, the problem of pattern recognition becomes much more complex.

Since the 1970s researchers have been attempting to merge pattern recognition with **artificial intelligence** systems. Artificial intelligence is the ability to solve thinking problems; problems which require judgment, reasoning, imagination, intuition and other human characteristics, such as emotion and loyalty. To install this kind of human intelligence in computers is the ultimate goal of pattern recognition and artificial intelligence research.

See also Computer, industrial uses of; Computer input and output devices; Computer vision; Robotics; Voice synthesizer

Computer, personal • See Microcomputer

Computer plotter • See Computer graphics

Computer printer • See Computer input and output devices

Computer simulation

Computer simulation involves designing a model of a real system for the purpose of training, teaching, predicting or entertaining.

A forerunner of computer simulators was the famous Link trainer used to teach people how to fly in the 1930s. Edwin A. Link, originally a designer of pipe organs and air-driven player pianos, used a device resembling a bellows to twist and turn a cockpit mounted on a movable platform. The ''pilot'' maneuvered the device by manipulating a mock set of cockpit controls.

More complex flight simulators began using computer graphics starting in 1968. Two men, David Evans and Ivan Sutherland, spearheaded a computer graphics program for ARPA, the Advanced Research Projects Agency. ARPA was designed to short-circuit the traditional research funding process, directly funding the creative projects that might help the United States maintain its technological vigor. A key member of ARPA was J.C.R. Licklider, a researcher and professor at Massachusetts Institute of Technology, who wanted to try using computers to interact with people rather than just to crunch numbers. He found Ivan Sutherland at Lincoln Laboratory; it was Sutherland who single-handedly created the field of interactive computer graphics. He and David Evans helped create the graphics used by flight simulators ever since.

Computer simulation was also used in the field of psychology. Joseph Weizenbaum, who had worked for General Electric and Bank of America in data processing, began a friendship with Kenneth Colby, a psychiatrist who had grown disenchanted with traditional one-to-one psychotherapy. They believed computers could offer a way of gaining new insights into neurotic behavior and perhaps even develop new therapeutic methods. Several question-answer machines had been developed already; this led Weizenbaum to try to create a more sophisticated one. The result was ELIZA (named after the character in *My Fair Lady* who learned to speak properly), a program that was intended to simulate the conversation between a psychoanalyst and a patient, with the machine in the role of the analyst. The program might start out with the following: "What brought you here to see me today?" If the patient typed in the response "I'm feeling tired," the machine responded with the question "Why do you think you're feeling tired?" ELIZA took its clues from whatever the patient said in an attempt to create useful questions. The program was such a success that many people spent a great deal of time telling their troubles to the computer even though they knew it was just a computer program. The key to ELIZA was its natural responses to the statements it received. If it was puzzled by a statement it didn't understand, it would fall back on "I see" or "That's very interesting," much like humans do when confused in similar situations.

Probably the most exciting current frontier in the area of computer simulation is virtual reality. This is the most interactive form of simulation. Unlike passive viewing of computer graphics, virtual reality is created by a display and control technology that can surround its user with an artificial environment that attempts to mimic reality.

Important contributors to virtual reality technology include Jaron Lanier, a computer programmer who developed successful video games for Atari in the early 1980s, then founded VPL Research, a company that has begun producing a virtual reality headset. Equipped with goggles that contain tiny liquid crystal screens, the headset projects a computer-generated landscape before the user's eyes. Since the image projection program is sensitive to the user's movements, it alters the perspective of the viewed scene as the user's head moves, producing stunningly realistic images. Lanier also created a glove called DataGlove with optical fibers and sensors that can measure the position of the wearer's hand and the movement of the fingers. The glove allows the wearer to move or grab things in the artificial world presented on the headset. A simplified version of the glove, known as the PowerGlove, is sold by Mattel for use with the popular Nintendo video games.

Another pathfinder is John Walker, president of Autodesk. Starting as a programmer, Walker founded Autodesk in the belief that architects and designers who had been using personal computers in their businesses would find useful a type of computer-aided design (CAD) software. His company's AutoCAD became one of the best-selling personal computer software packages. Autodesk is working now on virtual reality CAD.

Thomas Furness, a designer of visual displays for the military since 1966, developed a headset in 1982 for use by pilots in the Air Force similar to the one produced by VPL. He then worked on voice-operated commands that allowed a pilot to look toward a symbol (as the computer tracked the pilot's eye motion), determine what it was, and, if he wanted to shoot at it, utter a word or two to open fire on it.

Finally, Myron Krueger created an artificial reality laboratory at the University of Connecticut in the 1970s, where he has developed virtual reality programs such as CRITTER, a virtual reality cartoon featuring a four-legged yellow creature that interacts with the user.

The tremendous practical potential of computer simulation is just beginning to unfold. Matsushita Corporation of Japan is using virtual reality to help customers shop for custom-built kitchens. The customers can "tour" three-dimensional images of their proposed kitchens, using headsets and gloves, suggesting changes before actual installation takes place. Architects and aerospace engineers will be using more simulations in their work as well. Boeing has already developed a new airliner designed and engineered entirely within a computer. Researchers in drug companies are generating models of molecules to tailor specific drugs for specific conditions. Computer simulation may even become the newest medium for entertainment. VPL Research has formed a joint venture with MCA to build a series of test theaters employing virtual reality on a large scale.

Computer speech recognition

The advent of machines capable of recognizing human speech has been anxiously anticipated for many years. Prior to 1950, experiments were based upon the idea that language could be analyzed by syntax (the formal structure of a language) and semantics (the meanings of the individual words). Early speech recognition researchers were convinced that if all the proper word meanings and rules of grammar were stored inside a computer, the machine would be able to effectively translate languages. Unfortunately, more than $20 million dollars was spent by the military and various other government agencies on this type of research—all of which failed miserably. In the 1950s the Bell System began experimenting with a system that would allow **telephone** numbers to be spoken into receivers instead of being dialed, but that project also failed. Many of the problems encountered involved handling variances in accent, pronunciation, background noise or even speakers with head colds, all of which easily confused the computer.

It began to look as if language translation would never be possible, since interpretation is based to a great extent on understanding, and no one knew what that was, much less how to endow a computer with it. Finally, in 1950 K. H. Davis, an American scientist at Bell Telephone Laboratories built the first machine able to successfully recognize speech. His machine could distinguish ten spoken numbers from a series of acoustic signals. Later in 1959, L. S. Green, Edmund Berkely, and Calvin Gotlieb constructed something they called The Conversation Machine, which could figure out simple questions about the weather or time and surprisingly enough, make sensible answers in reply to them. Several more sophisticated question/answer machines appeared during the early 1960s, although these also were limited in the number and types of words which could be used.

During the mid-1960s the Project MAC computer research laboratory at the Massachusetts Institute of Technology, headed by Marvin Minsky (1927-), was the source of many speech recognition ideas, although none of the experiments proved highly successful. Between 1971 and 1976, an agency known as the Advanced Research Project Agency for Speech Understanding Research (ARPASUR), set aside $15 million dollars for speech recognition research. Based on these investigations, HEARSAY-II and Harpy were developed in 1976, by Raj Reddy (1937-), V. R. Lesser, and Lee Daniel Erman (1944-) at the Carnegie-Mellon University in Pittsburgh, Pennsylvania. HEARSAY-II was capable of understanding connected speech, limited by a 1,000 word vocabulary. A few years later a similar experiment known as Hwin, was developed by Woods and Wolf and their research team. Industry giants IBM and Nippon Electric Corporation and Bell Labs have also funded research projects into voice recognition.

One particularly beneficial application of computer speech recognition is in the education of the handicapped as well as people unable to operate keyboards. A speech recognition system can help deaf students to speak by providing them with visual feedback when they attempt to form words. Conversely, they can receive verbal instructions from physically handicapped pupils unable to enter information on a keyboard. Speech recognition has also been used in factories to control machinery, enter data, inspect parts, and take inventory. In some hospitals, doctors and nurses wear **microphone**s to describe their actions to a computer that interprets and logs instructions and records. As more advanced and lower-cost speech recognition chips are being developed, scores of valuable applications and uses for speech recognition systems are being realized.

See also Computer, digital; Computer, industrial uses of; Computer input and output devices

COMPUTER STORAGE • See Computer memory

COMPUTER VISION

Computer vision derives complex information from two-or three-dimensional images or objects through computer analysis. The technology, also called machine vision, grew from specialized research into pattern recognition. Enabling a computer to ''see'' is a complex process, involving many factors. In general, images to be analyzed must have any special patterns identified, then precise measurements of the characteristics of the patterns are taken and finally, a comparison of each part or pattern takes place.

Jerome Lemelson was a pioneer in this field. During the 1950s, he experimented with programmed industrial robots. By equipping them with special sensors, they were able to see and could thus inspect products being passed before them. Lemelson's prototypes used light beams and **photoelectric cell**s to scan objects, but with rapid advances in hardware and computer technology, his designs were soon improved. By 1960, he used **television** cameras linked to **digitizer**-equipped computers to measure object dimensions, detect surface imperfections, and compare colors and tolerances.

While most of early vision systems dealt with two-dimensional images such as documents, Larry Roberts, a MIT graduate student, began work in the early 1960s on what was called visual scene analysis. This led to important breakthroughs in computer vision, particularly in recognizing three-dimensional scenes, a type of technology required in **robotics** vision.

By the late 1960s, MIT, Stanford, and other universities began experimenting with robots capable of visual perception. In 1969, the Stanford Institute developed a mobile robot, which they affectionately named Shakey for its haphazard way of moving. Outfitted with **television** cameras coupled to sensory feedback systems, Shakey was an excellent example of early **artificial intelligence** and specialized pattern-recognition technology.

Another development in machine vision was the testudo (*turtle* in Latin), by William G. Walter. This device was a small turtle-like object equipped with a photoelectric cell for an eye, a sensing mechanism to detect contact and motors that enabled it to turn or move forward and backwards. It could detect light sources and automatically moved toward illumination when it required recharging.

One interesting tangent that arose from computer vision is computer art. An early example was produced in 1966 by Bell Lab engineers Leon D. Harmon and Kenneth C. Knowlton. Using a computer linked to a TV camera to scan pictures or paintings, each row in the picture is separated into points, and assigned a number representing brightness. By specifying an appropriate symbol for each number, a fairly accurate rendering of the painting is seen by the computer.

The ability of electronic circuits to ''see'' an image has also been put to use in Israeli police work. A program called PATREC (short for pattern recognition) has been used to analyze a police artist's sketch of a criminal suspect. Running it through the main police file, similar photographs are matched and selected for further scrutiny.

Computer vision has also been used in the military to provide long-distance eyes on the battlefield for military commanders and has proved valuable in analyzing **satellite** pictures. But perhaps the largest application of machine vision is in the **automobile** manufacturing field. Companies such as General Motors employ thousands of machine vision systems.

As **microprocessors** continue to gain in performance, speed, and capacity, many innovative uses for computer vision are sure to emerge.

Computer-Aided Design (CAD)/ Computer-Aided Manufacturing (CAM)

In the area of computer software, the trend has been to make difficult and tedious manual tasks easier to perform. One such area where this is most prevalent is computer-aided design and computer-aided manufacturing. More commonly referred to as CAD/CAM, these two areas of computer software have made life easier for thousands of designers, architects, and draftspeople all over the globe.

Thousands or tens of thousands of highly technical and accurate drawings and charts are required for the many design specifications, blueprints, material lists and other documents used to build complex machines. If the engineers decide structural components need to be changed, all the plans and drawings must also be changed. Prior to CAD/CAM, human designers and draftspersons had to change them manually, a time consuming and error-prone process. When a CAD system is used, the computer can automatically evaluate and change all corresponding documents instantly. Using interactive graphics workstations, designers, engineers and architects can create models or drawings, increase or decrease sizes, rotate or change them at will, and see results instantly on screen.

CAD use is particularly valuable in space programs, where many unknown design variables are involved. Previously, engineers depended upon trial-and-error testing and modification, a time consuming and possibly life threatening process. However, when aided by **computer simulation** and testing, a great deal of time, money and possibly lives can be saved. Besides its use in the military, CAD is also used in civil aeronautics, automotive, and data processing industries.

CAM, commonly utilized in conjunction with CAD, uses the computer to communicate instructions to automated machinery. CAM techniques are especially suited for manufacturing plants, where tasks are repetitive, tedious or dangerous for human workers. While the use of CAD/CAM systems enables the production of better, less expensive products, positions held within companies and factories are, in some cases, likely to be phased out as these automated systems become more readily accepted.

See also Computer graphics; Computer input and output devices; Computer vision; Cybernetics

Computerized axial tomography (CAT)

Since the invention of the **X-ray machine**, it has been apparent to physicians that an efficient and precise method for viewing patients' internal structures was essential. Fluoroscopes could outline bone structure and help locate foreign objects, but were not sensitive enough to show in detail organs such as the brain. During the late 1960s, Alan Cormack (1924-), an American, and Godfrey Hounsfield (1919-), an Englishman, independently developed a method called *computerized axial tomography* (CAT) scanning by which the internal structure of the body could be seen by assembling X-ray cross-sections, taken along a body axis, into a three-dimensional picture.

Cormack was the first to construct a tomographic device. His initial model used a thin beam of X rays aimed at one section of the body but repeated from many different angles. This method allowed him to combine the many X-ray pictures into one complete view. Though Cormack published his findings and theories, he received little recognition, chiefly because he lacked a system to process the volumes of information that went into a single CAT scan.

The solution to this problem was the **computer**. Hounsfield began working on his own CAT scanner in 1967, using a system remarkably similar to Cormack's with one important difference: Hounsfield used computers to collate the X-ray data and create a tomographic picture. The prototype CAT scanner used gamma rays to view inside the body; they required very long exposures, and the first CAT scans (of a preserved human brain) took nine days to complete. This was far too long for patients to endure, so Hounsfield began to improve his design. Later models required nine hours, then nine minutes. His final apparatus could complete its X-ray scan in 4.5 minutes, with an additional 20 minutes for the computer to assemble the information.

The design of most CAT scanners is relatively simple. A series of X-ray scanners are placed around the periphery of a tube-like cavity (large enough for a patient to insert his head, for example). Each X-ray scanner has a detector placed exactly opposite it, and both scanner and detector are built to move together so that, wherever the scanner moves, it is always aimed precisely at its detector. During the CAT scan, the X-ray scanners emit short pulses, usually lasting no more than a few milliseconds. Once this X-ray snapshot has been taken, the scanners and detectors rotate slightly (providing the same view but from a different angle) and another pulse is emitted. After all of the snapshots have been taken, a computer collates the information and constructs from it a complete picture. This can be a monumental task, since many CAT machines use up to 300 X-ray scanners taking 300 snapshots each, resulting in almost 90,000 X-ray "slices." Newer machines have also been designed to scan a patient's entire body.

In the 1970s, a more advanced tomography technique was developed by Michael Phelps and Edward Hoffman, a pair of biophysicists from the UCLA School of Medicine. Their technique, called positron emission tomography (PET) uses radioactive tracers injected into a patient that emit positrons and gamma rays. The scanner "reads" the gamma rays in much the same way that X-rays are scanned in a CAT machine, but a much finer image can be obtained by tailoring the radioactive tracers used to the organs being viewed. For instance, certain tracers bond to neurotransmitters in the brain, so a detailed image of brain functions can be obtained.

CAT and PET scanners have significantly reduced the need for dangerous exploratory surgery, particularly on the brain. They also view soft tissue in unprecedented detail. However, tomographic technology is prohibitively expensive: the machines cost hospitals up to one million dollars each, and for this reason they have been criticized as expensive toys for wealthy establishments. However, CAT scanners are nearly one hundred times more efficient than X-ray machines, since they require less energy per view and use all of the information gathered, rather than the estimated one percent recorded by conventional X-ray machines.

Cormack and Hounsfield shared a 1979 Nobel Prize for physiology or medicine for their work on computerized tomography. It is interesting to note that neither scientist held a doctoral degree at the time of their ground-breaking work. Research continues into various uses for CAT scans. One of the most recently developed uses for CAT scans is to help create a virtual three-dimensional reality via the computer that allow doctors to "fly" through areas of the body, in effect, doing a reconnaissance of diseased areas. The technique will also allow surgeons to "practice" procedures in a virtual reality setting.

Concentrated fruit juice

The juice squeezed from raw fruit is marketed today mostly as juice concentrate. Even canned and bottled ready-to-drink fruit juice is usually reconstituted from concentrate. **Gail Borden** experimented with producing juice concentrates after the Civil War. However, these products were not commercially successful until just after World War II, when frozen concentrated orange juice was developed. A group of citrus growers in Lake Wales, Florida, experimented with alternatives to canned juice, which didn't taste fresh. They developed a method of evaporating most of the water from fresh orange juice in an airless tank at temperatures below 80°F (26.4°C).

This same basic method is still used today. To make up for the loss of some flavor and aroma during evaporation, a small amount of fresh, unconcentrated juice is added back to the concentrate. The resulting mix is then frozen and pasteurized. Finally, the processed concentrate may be shipped to a juice packager, who adds filtered water to the mix until it reaches the right brix level, a measure of soluble sugar content. For cans of frozen orange concentrate, packagers aim for a brix level of 42°, which is three times more concentrated than fresh juice. For ready-to-drink reconstituted juice, packagers aim for a brix level around 12° which is about the same as fresh juice.

Frozen orange juice concentrate was immediately popular with consumers; even in Florida, it far outsells freshly-squeezed juice. Other frozen juice concentrates followed, especially apple and grape juices, although they are not nearly as popular as orange juice. Research scientist **George Speri Sperti** obtained a patent for freeze-drying orange juice concentrate.

See also Freeze-dried food

Concorde • See Aircraft

Concrete and cement

In geology, cementations and concretions occur when a process called lithification takes place, which means that loose particles of rock are bonded together by a mineral such as calcium carbonate (calcite) or iron oxide (limonite). Humans have recognized this natural phenomenon and been able to duplicate it from early times for construction purposes.

In approximately 200 B.C. the Romans used lime and gypsum to bond sand particles together. The resulting cement was used as mortar and paving material. By adding stones and pottery shards to the cement, they were able to create a material that was adequately hard, but because it required less cement, it was cheaper to make. The Romans also created a forerunner of Portland cement. They mixed volcanic ash with cement to create a variety of cement that could be used under water for aqueducts, drains, and bridges.

Concrete and cement have been used virtually throughout history; yet, their physical properties have been fully understood for less than 300 years. The consequence of not knowing about it was a trial and error process that often had disastrous results. In 1758 **John Smeaton** rebuilt the Eddystone lighthouse in England. It had originally been built from rock quarried at Portland. After much experimentation, Smeaton developed a hydraulic cement to seal the lighthouse from the rough conditions of the English Channel. In 1824 Joseph Aspdin (1799-1855) patented Portland cement, named for the gray Portland limestone—such as that used in the Eddystone lighthouse—that it resembled. It was developed mainly in response to the growth of canal systems in Europe and, especially, in the United States.

Portland cement has become the most widely used construction material in the world. Portland cement is made of calcium derived from limestone, chalk, marl, or shells and of silica derived from clay or shale. It is manufactured by quarrying rock and crushing it to a high fineness. It is then subjected to baking in a rotary kiln. The "klinkers" (remaining large pieces) are then reground into a fine powder that binds when mixed with water.

The two main keys to the strength of concrete is its proportion of aggregate to matrix and water to matrix. Generally, the bond increases with the amount of cement matrix used. Water is the key bonding agent. The bond is not established from drying. On the contrary, the cement gets stronger the longer it sets in water. This is the reason that construction workers hose down concrete as it sets. This process is referred to as curing.

The first reinforced concrete using iron rods was developed by Frenchman Jean Louis Lambot and patented in 1867 by Joseph Monier, who used Portland cement instead of concrete. The purpose of the iron rods was to increase the tensile strength of the material, a property that cement and concrete lack. Later, steel rods replaced iron as reinforcement.

Ferrocement is similar to reinforced concrete, though instead of separate reinforcing bars, a wire mesh is used to form a cement cast. The steel mesh can be formed into nearly any shape desired, making it strong and versatile. Italian engineer Pier Luigi Nerri used ferrocement in 1942 and 1943 to build a fleet of boats. Nerri referred to ferrocement as ''melted stone,'' for it could be easily formed into statues and building adornments.

Fiber reinforcement is increasingly used in concrete. Asbestos cement was developed in 1899 by Ludwig Hatschek for its insulating ability. Steel fiber cement has been used in England in the construction of motorways. In other parts of the world, such natural fibers as sisal, sugar cane, coconut, and bamboo are used because they are cheap and readily available.

There has been some debate over the cost-effectiveness of concrete reinforcement. On the one hand, its greater strength reduces construction costs. This is countered by increased materials cost, especially when steel is used. It cannot be denied that it has made many previously impossible, ambitious construction projects feasible. The entire Interstate Highway System was made possible by reinforced concrete.

The most recent development, which is taking construction to new heights, is *prestressed* concrete. First introduced by Eugene Freyssinet (1879-1962) in 1927, steel rods are placed under tension while the concrete sets around them. This creates pressure within the beam. Instead of passively carrying a load, the beam actively resists the load placed upon it. It is commonly used in bridges and **dome**s.

Despite its simple utility, concrete and cement continue to evolve to meet demands of modern times. Special cements are produced for mortars and architectural or engineering applications, such as masonry cement or plastic cement. Cement made from the compound supercritical carbon dioxide may have a host of applications—metals could be introduced to make concretes conduct electricity, or make them radar-invisible. Bacteria have been used to make cement, as a by-product of their breakdown of urea. This breakdown makes nearby calcium precipitate out as limestone, which can fill in the cracks in fissures in cement, from the inside out. (The bacteria die off when their nutrient urea die out.) Such cement would be useful for sealing up cracks in buildings and other structures.

Smart cement is cement embedded with sensors that can provide nondestructive damage assessment. By placing sensors like **fiber optic** strain gauges directly in the concrete of a structure like a bridge, a **computer** can directly and instantly monitor the bridge for cracks, strain, and road-salt corrosion, eliminating expensive maintenance checks. Prototype concrete-containing short carbon fibers may be able to do the same thing, without embedded sensors, by monitoring its electrical response.

See also Composite material; Road building; Water supply and drainage

CONDENSED MILK • See Milk, condensed

CONDENSER, ELECTRIC • See Capacitor

CONDOM • See Birth control

CONTACT LENSES

Modern contact **lens**es are thin corrective pieces of eyewear made of **plastic**. Because they float on a thin film of tears and cover the cornea—the transparent tissue covering the pupil and the iris—they offer the benefits of near invisibility, unimpaired side vision, and ease of wear.

The first contact lenses were made by Adolf Fick in 1887 for the purpose of correcting irregular astigmatism. Like **eyeglasses**, they were comprised of **glass**. In 1912, Carl Zeiss (1816-1888) also designed glass lenses. Then, in 1938, Obrig and Muller developed the first hard plastic lenses (*scleral*, or *haptic*, lenses that virtually covered the front of the eyeball). Although optically inferior to glass, plastic proved considerably lighter and more comfortable to wear. Such contact lenses were prevalent until the 1950s; they required an impression of the eyeball to mold the lenses.

In 1948, Kevin Touhy designed plastic *corneal* lenses, which were decidedly smaller than scleral lenses and allowed for increased circulation of tears and oxygen. The development of the keratometer, which takes measurements of the cornea reflected from a light source, eliminated the need for eyeball impressions.

Bifocal contact lenses were first produced in 1958. In 1971, soft or *hydrophilic* lenses—made from a water-absorbent plastic gel—were developed. Although more comfortable to wear than hard-plastic lenses (though gas-permeable hard lenses have been offered as a compromise), soft lenses are subject to easy damage and must be sterilized before each insertion to avoid any bacterial infection.

Despite their widespread use for cosmetic reasons, contact lenses do offer some medical advantages over eyeglasses. For example, contacts are more effective for restoring sight in people who have had cataracts surgically removed. Contacts are also effective when the refracting ability of each eye is different and are used successfully on people with keratoconus, a condition that causes the cornea to bulge at its center and cannot be corrected by ordinary glasses. However, contact lenses cannot be used by everyone and often are not recommended for patients with chronic glaucoma, diabetes, allergies, or eye inflammations.

Contacts can be made in different colors and can be shaded as sunglasses. In addition, they can incorporate magnifying lenses to help people read small print. Because of their popularity, advances continue to be made in developing new types of contact lenses. Recent advances included extended-wear lenses, which may be worn for several days before cleaning, and disposable lenses, which can be safely worn for up to one week without removing.

See also Magnifying glass

CONTRACEPTIVE • See Birth control

CONVECTION OVEN • See Stove

Peter Cooper

CONVEYOR BELT • See Mass production

COOKE, WILLIAM FOTHERGILL • See Wheatstone, Charles

COOLIDGE, WILLIAM (1873-1975)
American physical chemist

William David Coolidge was born in Hudson, Massachusetts, the son of a farmer and a dressmaker. As a young boy, he worked in a shoe factory to help support his family. After attending public schools, Coolidge funded his own college education by borrowing money and earning scholarships and fellowships. With a degree from the Massachusetts Institute of Technology (MIT), Coolidge went to Germany to study physics. After earning his Ph.D. with high honors, he returned to MIT to do research. Although Coolidge was content there, he was lured in 1905 to General Electric Company's research laboratory, which offered to double his MIT salary. Coolidge had avoided a career in industry after experiencing factory work, but General Electric (GE) promised him freedom to pursue his own interests as well as the company's commercial research goals. As it turned out, he remained at the lab for his entire career, like his GE colleague **Irving Langmuir**.

In just a few years, Coolidge solved one of the greatest technological problems of the time—developing a superior filament for incandescent light bulbs. Early electric light bulbs used carbon filaments, which were not only delicate to handle but were also limited in the amount of light they could produce. Scientists knew that tungsten, the metal with the highest melting point, would perform better than carbon; but because tungsten is brittle, no one could figure out a way to make filaments from it. Coolidge invented a process for making tungsten ductile, or capable of being drawn into fine wire for filaments. Modern electric light bulbs are still manufactured this way.

Continuing his research with tungsten, Coolidge invented an X-ray tube that is still used by doctors and dentists today. His revolutionary tube was based on a tungsten "target," which is bombarded under high vacuum by a stream of electrons to produce X rays. Coolidge's tube allowed much more precise control over the X-ray wave length and could also accommodate much higher voltages. Patented in 1913, the Coolidge tube introduced X-ray technology into the worlds of industry and medicine. During World War I, Coolidge collaborated with Langmuir on the development of the first successful submarine detection system. Coolidge frequently incorporated many techniques used in nuclear research to produce a total of eighty-three patents.

In 1932 he became director of GE's lab, earning respect for his modesty and calm authority. He postponed his retirement until 1945 in order to work throughout World War II. Coolidge lived to the age of 102, continuing to enjoy hobbies such as travel and photography.

See also X-ray machine

COOPER, PETER (1791-1883)
American inventor, manufacturer, and philanthropist

Best remembered as a philanthropist, Peter Cooper was a prolific inventive genius and a highly successful manufacturer. Cooper was born in New York City, the son a Revolutionary army soldier who was active in numerous enterprises and involved young Peter in all of them. Although Cooper had only one year of formal education, his early experiences with his father prepared him for success in his varied business career. Apprenticed to a coachmaker at the age of 17, Cooper did so well that his employer paid him a salary and offered to back him in his own enterprise. Instead, Cooper went into the cloth-shearing business, in which he prospered. He then bought the rights to a glue-making process, improved it with his own invention, began operating a glue factory, and secured a virtual monopoly of the American glue business.

In 1828 Cooper moved into iron manufacturing, building the Canton Iron Works in Baltimore, Maryland, intending to supply the Baltimore & Ohio Railroad. The railroad was on the verge of failure, however, because of the twisting and hilly route its tracks followed. Most engineers at that time held that **locomotive**s couldn't run on such terrain. Cooper promptly built America's first steam locomotive, which was small but powerful. In 1830 this "Tom Thumb" pulled 40 passengers at a speed of 10 mph (16 kph) and proved that railroads could run on track that curved.

Cooper's business enterprises grew rapidly after this success. His iron business expanded into mines, foundries, **wire** manufactories, and rolling mills. In 1854 Cooper's Trenton factory produced the first iron structural beams for use in erecting fireproof buildings. Cooper became a principal backer and unwavering supporter of Cyrus Field's (1819-1892) project for laying the Atlantic **telegraph** cable. As president of the North American Telegraph Company, Cooper owned and controlled half of the telegraph lines in the United States. As an inventor, Cooper designed an early **washing machine** and various engines for powering watercraft.

As a member of the Board of Aldermen of New York, Cooper advocated free public schools, sanitary water supplies, and paid police and firefighting forces. In 1859 he founded Cooper Union in New York City, a college offering free courses in science, technical subjects, and art, and in 1876 Cooper was the presidential candidate of the Greenback party. Upon his death in New York City in 1883, Cooper was widely eulogized.

See also Adhesive and adhesive tape

Cordite

Patented by Sir Frederick Abel and Sir **James Dewar** in 1889, cordite is a **smokeless powder** derived from **nitroglycerine** and nitrocellulose (**guncotton**). Although not the first of the explosive mixtures to supersede **gunpowder**, cordite nonetheless represented an important advance due to its plasticity, its ability to be molded into cord-like shapes and then be divided precisely into various sizes. Like all smokeless powders, cordite enabled armies to conceal their battle positions more effectively and also brought about improvements in gun and bullet technology. Variants of cordite and Paul Vieille's (1833-1896) *Poudre B*, invented in 1884, formed the basic ammunition during World War I and are still in use today. **Alfred Nobel**'s development of ballistite in 1888 made possible Abel and Dewar's invention, which was essentially ballistite with the addition of acetone and petroleum jelly for increased stability.

Corliss, George Henry (1817-1888)

American inventor

Prominent inventor and entrepreneur George Henry Corliss was born in Easton, New York, on June 2, 1817, the only son of a physician. He worked for a textile manufacturer for four years and finished school in Vermont. By age twenty-five he had returned to New York, opened a boot store, and patented a boot-stitching machine. Two years later he began work as a draftsman for the Providence, Rhode Island, engineering firm who had helped him with that patent. Corliss devoted himself thereafter to improving the steam engines the firm manufactured.

Corliss was given free reign at the firm, and within four years had started his own business to manufacture and promote his earliest ideas—to create a more efficient and economical system of rocking valves and **governor**s to control steam and

George Henry Corliss

exhaust valves. He received patents in 1849 and 1851 and opened the Corliss Engine Company in 1856, and within three years Corliss engines were being exported to Scotland for use in cotton mills. By 1864, valves for the engines were being made at Bolton, England. Corliss directed both the business and research sides of this company, and over the years invented many assembly line improvements such as a bevel-gear cutter.

Corliss's company gained international acclaim at both the Paris World's Fair of 1867 and the Centennial Exhibition in Philadelphia in 1876. A 700-ton (637-t), 1400-1600 horsepower Corliss engine provided power to all the exhibits in Machinery Hall for six months continuously in Philadelphia. While there, historian Henry Adams (1838-1918) proclaimed the engine the symbol of the era. Corliss directed his company until his death on February 21, 1888.

Cormack, Alan • See Computerized axial tomography

Corn picker, mechanical

The introduction of machinery to the farmer's field has served to make farming operations more efficient and increase productivity. In the midst of this trend toward mechanization is the corn picker which today harvests the majority of corn produced in developed countries. The origins of the mechanical corn picker can be traced to the 1st century A.D. The Roman naturalist Pliny the Elder (23-79 A.D.) referred to an iron-toothed cart that was pushed by oxen to strip corn from the stalks. A Scottish inventor, Patrick Bell, invented a corn reaper in 1826 with reciprocating clipping **knives** driven by ground **wheels**. He refused to obtain a patent on it insisting that his invention was a humanitarian gesture, a decision which left him basically penniless for the remainder of his life.

Many patents were issued for corn harvesters in Europe and America during the 1800s, few of which were accepted for use in the field. The first machines to be marketed in the United States were sold on a small scale. Interest in the machines lagged into the late 1920s due to the invention and popularity of **Cyrus McCormick**'s corn binder, an adaptation of **John Appleby**'s twine binder, which cut and bound the corn stalks for later shelling by field laborers. The William Deering Company of Chicago was also involved in development of a corn binder and was McCormick's chief competitor during this period. In 1928, two-row mechanical corn pickers were mounted to power-take-off (PTO) **tractors** for the first time.

The first picker-shellers were developed in the 1930s. The pickers stripped the ears of corn from the stalk while the shellers stripped the kernels from the ears while in the field. Acceptance for these machines was delayed until after World War II due to the absence of corn drying equipment. The first self-propelled corn picker was introduced in 1946. Mechanized corn picking increased rapidly during the 1950s and 1960s. Today, corn pickers fall into five categories. Picker-huskers pick and husk the corn in-field. Picker-shellers pick the ears and strip the kernels. Corn harvesters remove the entire stalk from the field. Corn snappers remove the ears only. Finally, **combine harvesters** can be adapted for corn picking by adding corn heads, pointed devices at the front of the machine that snap the corn. Corn picking machines are also classified by the number of rows they can harvest in one sweep and by the type of mount. Field loss by mechanical picking in North America is about 12 percent in August. The loss increases with lateness of the harvest season with a 50 percent loss in October mainly due to the increase of moisture in the crop. Efficiency of mechanical over hand picking offsets the lower early-harvest losses.

Cornell, Ezra (1807-1874)

American businessman

Ezra Cornell was born at Westchester Landing, New York, the oldest of the eleven children of a Quaker family. As a boy, Cornell attended village schools, helped with his father's farm and pottery works, and learned carpentry. He left home at 18, worked as a machinist, and settled in Ithaca, New York, in 1828, where he became manager of the Beebe flour and plaster mills.

When the mills were sold in 1841, Cornell secured the rights to promote a patented **plow**. This led to a meeting with F. O. J. Smith, congressman from Maine and a partner of **Samuel Morse**. Smith had contracted with Morse to lay the underground cable for an experimental **telegraph** line from Baltimore, Maryland, to Washington and was struggling to design a plow that could do the job. Cornell stopped by Smith's office in 1843 and promptly sketched a machine that would dig the trench and lay and bury the cable in one operation. Smith immediately hired Cornell to supervise construction of the line. Cornell's machine was a success, but the underground line was a failure due to faulty insulation. Cornell convinced Morse to string his lines overhead on poles and designed a practical way (variously reported to use broken bottle necks or glass doorknobs) to insulate the **wire**.

Cornell now made development of the telegraph his career. He built, demonstrated, and promoted numerous lines between major cities in the Northeast and Midwest, included the one over which Morse sent his first message.

By 1855, the ruthless competition among rival telegraph companies led Cornell and other industry leaders to combine their businesses into a single entity, the Western Union Telegraph Company. Cornell was the largest stockholder in the company for the next 15 years, and was a director for the rest of his life.

Western Union made Cornell a wealthy man. He turned to public and educational interests, serving in the state legislature, establishing a model farm near Ithaca and, in 1863, building a free public library for the city. He was a principal founder of Cornell University, which opened in 1868 and was designed, according to Cornell's vision, to offer university-level instruction in technical and agricultural subjects as well as liberal arts. Cornell endowed his namesake school liberally, and his public-land transactions also benefitted the university handsomely. The economic panic of 1873 nearly bankrupted Cornell, however, and he died in Ithaca in 1874.

Corrugated iron

Iron by itself is a very strong metal and was used especially during the nineteenth century for construction and railroad manufacturing. Yet, several methods have been invented to add to its strength, including alloying, compositing, and reinforcing. Corrugating is an efficient method of reinforcing sheet iron—marks, irregularities, or (usually) parallel, curved marks or ridges and hollows are shaped on the surface of a sheet of iron, to which the cement sticks better than on a flat surface. The corrugate waves, or ribs, add significantly to the strength and rigidity of sheet iron while not adding much to its weight. The ribbing has virtually the same effect as reinforcement framing.

Corrugated metal was first used in the Mediterranean area as early as 500 B.C. Bronze was corrugated by hammering it into the form of buckets. One of the first attempts to mass produce corrugated metal was that of the Englishman Richard Walker whose Patent Corrugated Iron Factory was established in 1833. All that is known about Walker's corrugating process is that it was developed about 1828.

The individual most credited with the advancement of corrugating iron and other metals is Joseph Francis. This Bostonian invented a hydraulic press during the 1840s to form corrugated boat hulls for lifeboats. His boats were very popular during the late 1800s because of their light weight, strength, and watertight qualities, and were much more reliable than traditional wooden boats. Francis also invented the lifecar, an enclosed lifeboat that was reeled to shore on a lifeline launched to offshore shipwrecks. His boats saved hundreds of people in America and Europe, and earned him many awards, including a gold Congressional Medal of Honor in 1890. Francis was also known for inventing the first amphibious duck, a watertight military wagon with a corrugated iron hull designed for crossing rivers.

Francis's hydraulic press pushed, or crimped, the iron sheets into the desired shape. The press dies had to be highly detailed, and care had to be taken not to allow the metal to crease, because creasing created weak points in the corrugation. Frenchman Pierre Carpentier patented a corrugating machine in 1851 that added ribbing to galvanized metal. Galvanizing added a layer of zinc to iron as an anti-corrosion measure. During the Australian gold rush of 1851, corrugated galvanized iron sheets were shipped as construction material. Today corrugated roofs and water tanks are a typical feature of cattle and sheep stations in the Australian ''outback''. The material is cheap, strong, easy to transport, and easy to construct and repair. Corrugated iron and steel have become popular construction materials in the United States as well. They are used as complete building material for warehouses, farm buildings, and public buildings. Some building material companies have developed entire prefabricated building systems for low-cost construction consisting of corrugated materials.

Corrugating has been applied to materials other than metal. Fiberglass is often corrugated for use as covering, as awning material, and in fencing. Corrugating is also used extensively in the paper box industry. Corrugated cardboard is used as a sturdy packaging material. This simple idea is one that has had far-reaching implications.

See also Iron production; Steel production

Corset and girdle

Corsets and girdles are undergarments that control the shape of the body, usually the female body. The earliest recorded corset is shown on a figurine of a snake goddess from Crete, dated about 2,000 B.C. Ancient Greek and Roman women, too, wore bands around the torso to shape and control their figures. But since loose-fitting garments were the fashion norm until the mid-fourteenth century, corsets were not necessary and seldom worn.

The first modern corsets appeared in France in the fourteenth century, as outer garments. Spanish fashion of the late 1500s featured tight bodices stiffened with wood or metal; some of these figure-constrictors were made entirely of metal. European men as well as women wore these, including Sir Walter Raleigh (1552-1618). The waist in its normal position then disappeared from female fashion for a while.

When it reappeared around 1820, the corset also came back, and in the next decades, reached its heights of restrictiveness and discomfort. Made of heavy canvas reinforced with whalebone or steel, the corset held a woman in a rigid frame that eventually reached from the bust all the way down to the mid-thigh. Waist measurements of 17-21 in (43-53 cm) were aimed for. These corsets were considered a medical necessity; women were supposed to suffer from delicate health, and their muscles, it was believed in Victorian times, were too fragile to support their frames unassisted. It was a self-fulfilling prophecy. Corsets deformed women's internal organs and constricted breathing; unused back muscles atrophied. Women in fact became weak and delicate.

An undergarment model from the 1950s next to a Victorian-era corset constructed from steel.

The tightly laced corset began to die out in the early 1900s as the fashionable female silhouette became slim, straight, and waistless. The advent of elastic created the first girdle in 1911. Instead of being encased in a whaleboned, laced corset, women's bodies would now be held firmly in place by the tension of elastic girdles. The girdle also performed the vital function of holding up a woman's stockings. At first the girdle was a simple tube. The panty girdle, which became standard wear for almost all women, appeared in 1935. The advent of the mini-skirt in the late 1960s necessitated the invention of panty hose, which was the death knell for girdles. Freed of the need to wear a garment to hold up their stockings, women by the millions happily gave up their girdles, and have kept them off ever since.

Cort, Henry (1740-1800)
English inventor

Born in Lancaster, England, in 1740, Henry Cort was one of many individuals who were central to the development of the iron and steel industries before and during the Industrial Revolution.

From 1765 Cort developed an interest in iron while he was in the Royal Navy, where he was in charge of improving wrought-iron ordnance. During this time, he was able to accumulate funds needed to start his own business.

Iron working had been a laborious process that required hammering to shape and purify the metals. After Cort established a foundry at Fareham, England, he developed a method of creating iron bars with grooved rollers, a process he patented in 1783.

The following year, he patented a puddling process in which the carbon in the iron was separated by stirring molten pig iron in a reverberating furnace. The action of the air in the decarburization causes pure iron to form. Cort had used to his advantage the high-grade ores mined at Dannemara, near Uppsala, Sweden. After the slag was removed, the molten iron was then applied to Cort's grooved rollers.

Both inventions were little more than improvements over what already existed. Cort's original contribution came with the combination of the two processes, which made iron more readily available throughout Britain and quadrupled production in 20 years.

Although insightful in metallurgy, Cort had little success in business. He made a fateful mistake when he established a partnership with Samuel Jellicoe at Gosport, just south of Fareham. Jellicoe's father was a dishonest naval official who embezzled public funds for use in the partnership. Although Cort had nothing to do with the crime, he took the brunt of the blame for it. The British Admiralty forced Cort to repay the monies and denied him the rights to his patents. Cort's competitors made full use of the patents, leaving Cort with nothing but a small pension. Cort died in Hampstead, London, England, in 1800.

See also Construction, cast-iron and wrought-iron; Iron production; Steel production

Cosmetics

Cosmetics are personal care products used to improve one's appearance by cleansing, conditioning, or coloring the body or face. Today cosmetics include soaps, shampoos, conditioning products for hair and skin as well as lipstick, blush, eyeshadow, and other make-up for the face.

Archaeological evidence found at paleolithic sites indicates that prehistoric peoples used various pigments mixed with grease to adorn not only the walls of their caves but also their bodies. Body painting, in addition to being considered ornamental, was thought to afford supernatural protection, and at some burial sites, large quantities of paints have been found interred with the dead. Among the seventeen different colors that archaeologists have identified as in use in prehistoric times, white, black, orange, red, and yellow seem to have been the most popular.

Both historical records and archaeological evidence show that the ancient peoples of the Middle East, like prehistoric peoples, used cosmetics for aesthetic as well as religious purposes. Their lavish use of eye cosmetics seems to have served the additional purpose of warding off the glare of the sun. Eye cosmetics included an eyeliner made from ground ants' eggs and kohl, a paste made from soot, antimony, or galena (a type of lead ore) that was applied to the eyelashes, lids, and brows. Two famous Egyptian queens who adorned themselves in this manner were Nefertiti (c. 1365 B.C.) and, much later, Cleopatra (c. 50 B.C.). Also in common use among the upper classes of Egypt were rouges, henna for dying hair and fingernails, white powders, bath oils, and abrasives for cleaning teeth. The oldest cosmetic item that archaeologists have found in the Middle East is one in common use today—lipstick. Dating from about 4000 B.C., it was found in a Babylonian tomb and in all likelihood belonged to a man.

Although the word *cosmetics* is derived from the Greek *kosmetikos*, meaning "skilled in decorating," the classical Greeks apparently frowned on the use of cosmetics. Similarly, the early Romans regarded cosmetics as a sign of decadence. With the rise of the Roman Empire, however, and the ensuing abundance of luxurious goods imported from the conquered lands of the Middle East, cosmetics became status symbols for Romans of both sexes. In his castigating comments to one of his female friends, Martial, a Roman of the late first century A.D., gives some flavor of the use of cosmetics in his day: "While you, Galla, are at home, your hair is being dressed at the hairdressers; though you lay aside your teeth at night with your silk garments...though even your face does not sleep with you, and you ogle me from under eyebrows which are brought to you in the morning... nevertheless, you offer me delights."

Throughout Europe, Roman concepts of personal cleanliness and beauty waned as the Empire went into decline, and they did not re-emerge until the Crusaders began returning from the Holy Land in the Middle Ages. The medieval ideal of feminine beauty—skin white as a lily and cheeks of rosy red—was pursued by nobility and commoners alike. But while the ruling class could afford expensive cosmetic preparations, their subjects had to whiten their skin with wheat flour and rouge their cheeks with beet juice. By the 1500s, the French had become famous for their skill in applying cosmetics. Highborn men and women powdered their hair with saffron, or flower pollen, and painted their faces with "supernatural luster," a preparation of gold leaf and hot lemon juice. Venice, however, was the major producer of cosmetics. A skin whitener known as Venetian ceruse was very popular, even though it was known that the white lead it contained could damage the skin and result in baldness or even death. Other dangerous concoctions included red mercuric sulfide, used for lipstick, and sulfuric acid, used for bleaching hair.

Cosmetics enjoyed heightened popularity in eighteenth-century Europe, where members of the French court whitened their faces and etched the veins of their faces in blue. Beauty patches of black silk or velvet, originally invented as small dots or crescents to hide the disfiguring marks of smallpox, became larger and took the form of flowers, stars, birds, and symbols of personal occupation and politics. In England powdered wigs and reddened lips were in fashion.

With the French Revolution and the dawn of the Victorian Age, cosmetics again went into decline. Men eschewed them, and respectable women adhered to the Victorian ideal of a "natural" beauty; anything more than a dab of rice powder or scent was the province of the streetwalker.

But by the late 1800s, advertising was coming into its own, and with it printed testimonials of cosmetic preparations by famous women—such as the endorsement of Pear's Soap by the English actress Lillie Langtry—respectable women again began experimenting with cosmetics. It was not until the 1920s, however, when **motion picture**s and movie stars were becoming the rage, that cosmetics started growing into the multibillion dollar industry that it is today.

Two entrepreneurs who contributed significantly to the rise of the cosmetics trade were Elizabeth Arden (1884-1966) and Helena Rubenstein (1870-1965), whose rivalry became as well known to Americans as the Hatfield-McCoy feud. Born in Poland, Helena Rubenstein emigrated to Australia in 1902 with a fair complexion and 12 pots of her mother's face cream, the invention of a European chemist. Australian women were so impressed with Rubenstein's skin that she was finally persuaded to open a small beauty shop in Melbourne. By 1908, she had fashionable salons in London and Paris. In 1915, with World War I underway, she fled to the United States, where she opened salons in New York, Boston, Massachusetts, Philadelphia, Pennsylvania, and San Francisco, California. In the "flapper" era that followed the war, she created the vamp look for the movie actress Theda Bara (1890-1955). Madame, as Rubenstein was known, died in 1965 at the age of 94, leaving an estate valued at more than $130 million.

Elizabeth Arden's career was as meteoric as that of Rubenstein. Born into a family of British immigrants in rural Ontario, Canada, probably in 1878, Arden (nee Florence Nightingale Graham) rose from her impoverished childhood to become one of the wealthiest women in the world. At one time, she operated more than 100 salons on three continents. Although early in her career she tutored herself by having a facial at almost every salon in Paris, even at that stage she would not deign to patronize the salon of her archrival Rubenstein. In 1938, in one memorable episode, Arden hired away a dozen members of Rubenstein's New York salon. Not to be outdone, Madame retaliated by hiring Arden's former husband and business partner, Thomas Jenkins Lewis. Despite—or perhaps helped by—such well-publicized skirmishes, the modern cosmetics industry continued to grow.

Over the last 50 years, cosmetic scientists have developed advanced techniques for caring for and decorating the skin. For example, state of the art skin cleansers and moisturizers can now help minimize external signs of aging while color cosmetics are longer lasting and available in more shades than ever before. Further aided by improvements in **mass production**, packaging, and advertising, the cosmetic industry has become a multibillion dollars enterprise in the 1990s. Nonetheless, the industry continues to be challenged by environmental and ethical concerns. One such challenge came as a result of laws passed by certain state legislatures which limited the solvents (Volatile Organic Compounds) that can be used in hair sprays. Another came from the animal rights movement which has pressured the industry to produce cruelty free cosmetics which do not involve animal testing.

See also Haircare; Toothbrush and toothpaste

Cotton candy

Cotton candy is a fluffy, cottony confection made from long, thin strands of spun sugar. It was invented in 1900 by Thomas Patton, who called it Fairy Floss. By 1905, the confection was made by several companies, one of which had adopted the name Kotton Kandy. Some of the machines used at this time appear to have been electric; others were hand-driven. Today cotton candy is still a popular treat at circuses, county fairs, and amusement parks. It is known as *barbe de grandpere* ("grandfather's beard") in France and *zuckerwolle* ("sugar wool") in Germany.

A cotton candy machine is composed of a spinning drum inside a larger bowl. Sucrose, better known as ordinary table sugar, is placed in the inner drum, where it is heated until it melts into syrup. This syrup starts streaming up the inside walls of the spinning drum and coming out the top in fine strands. These strands of molten sugar cool and solidify very quickly in the air. They begin to build up into cottony fluff inside the outer bowl. As a paper cone is rolled around inside the bowl, it picks up the fluff for easier eating. Cotton candy is often tinted with food coloring, most commonly pink. Sometimes flavorings are added as well.

In 1988, pediatrician Richard Fuisz founded Fuisz Technologies, a company built on an unusual use for cotton candy technology. Fuisz put powdered medicine along with sugar or similar ingredients in a cotton candy-type machine to create a rapidly dissolving tablet.

Cotton fabric

Perhaps no other natural product has influenced the destiny of men as has cotton. It has clothed nations, enslaved men, monopolized labor, and given direction to entire industries.

The first historical mention of cotton was in the writings of the Greek historian Herodotus (484?-425? B.C.), who described having seen trees with fleece growing in them in India. Archeological discoveries have traced the use of cotton in India to 3000 B.C. or earlier. Cotton spread by trade to the Middle East, particularly Egypt, and in the 7th and 8th centuries, was brought to Spain by the Moors.

New World explorers found cotton fabrics being manufactured in Peru, Mexico, and what is now the southwestern U.S. Carbon 14 tests have dated the use of cotton in Peru as far back as 2500 B.C.

Europeans first planted cotton in the New World in Virginia, using seed from the West Indies. The need to harvest cotton when the weather is perfectly dry at first meant that the European colonists had to spend long days working in the hot sun. They eventually got around their dislike for this type of labor by importing slaves to do the work for them.

In 1793, a young American inventor named Eli Whitney (1765-1825) developed the **cotton gin**, which successfully separated cotton seed from the fibers by mechanical means. This single invention raised cotton exports from 400 bales a year in 1791 to 30,000 bales in 1800 and 180,000 bales in 1810. As a corollary, in the decade between 1790 and 1800, the slave population of the United states increased 33%. By 1810, there were more than a million slaves in the Southern states; by 1860, the number had risen to more than 4 million.

Since the end of World War II, the demand for cotton has been largely supplanted by one for synthetic fibers, particularly polyester and nylon. Incursions into the cotton market have in part been due to the dwindling availability of land to raise cotton.

The cultivation of cotton requires a warm, humid climate and sandy soil. From planting to flowering takes from 80 to 110 days; another 55 to 80 days are required for the flower to produce the cotton ball.

Today mechanical harvesters are most often used to gather cotton from the plants. A mechanical harvester can pick up to 650 pounds (295 kg) per hour, compared to the 15 pounds (about 7 kg) of so that a hand picker can gather in the same time. Even so, hand pickers are still sometimes preferred because mechanical harvesters tend to gather a great deal of waste matter along with the fiber.

After the cotton has been harvested, the seeds must be removed. This is done by what is essentially the same method devised by Whitney in 1793. The seed cotton is fed into a gin consisting of a series of circular saws that separate the fiber from the seeds. The fiber is then compressed into bales weighing about 478 pounds. A second ginning separates out the short fibers, leaving the more desirable longer fibers.

Turning the raw cotton into yarn requires many steps. First, any heavy impurities such as dirt or seeds are removed from the opened bales of cotton at the mill. Then the fibers are drawn into wide thin webs, which are gathered together into ropes or strands. The finer quality fibers are combed until all short lengths have been removed. Both combed and uncombed fibers are drawn and twisted. Finally the twisted fiber is further drawn and twisted to produced the finished yarn. Weaving the cotton fabric consists of interlacing lengthwise yarns with crosswise yarns to produce cotton cloth.

See also Linen fabric; Textiles

Cotton gin

As is the case with many revolutionary ideas, the origin of the cotton gin (*gin* is short for *engine*) has been in hot dispute. Cotton gins of various designs had been used in the British colonies since the seventeenth century, most notably the gin designed by the Philadelphia, Pennsylvania-born inventor, Joseph Eve (1760-1835), for use by him and others in the West Indies. His gin was in use there from 1787 onward. When Southern planters took an interest in Eve's gin, he applied for a patent, returned to Charleston, South Carolina, and set up a factory to produce them. While somewhat successful, his gins could not compete with the more sophisticated machines later developed by **Eli Whitney**; consequently, the invention of the cotton gin has been widely attributed to Whitney. On a trip south from New England, Eli Whitney—a recent graduate of Yale University—made the acquaintance of another traveler, Catherine Littlefield Greene, widow of the American Revolutionary General Nathanael Greene (1742-1786). It was Catherine Greene who apprised Whitney of the need among Southern cotton merchants for a device to clean cotton, and Whitney used her basement as a shop to develop his ideas. Slaves who worked the cotton on Catherine Greene's plantation were in the practice of using a simple comblike device to separate the seeds and other debris from the cotton fibers. Whitney mechanized and expanded upon this simple comb (with design assistance from Catherine Greene, some maintain) and within six months he had developed his first cotton gin.

This early cotton gin consisted of a wooden cylinder that contained a row of slender spikes set one-half inch apart. As the machine was turned, these spikes meshed with a wire grid. The cotton fibers were pulled through the grid, but the seeds were too big and were ejected. A revolving brush cleaned the cotton off the spikes. This early hand-cranked machine could clean fifty pounds (25.5 kg) of cotton per day. This output was a fifty-fold increase in output per worker and revolutionized the cotton industry. The new, efficient method of cleaning cotton led to increased demand for more agricultural land, which in turn led to the forced removal of thousands of Native Americans from the South to land west of the Mississippi River. This agricultural expansion also led to the importation of more slaves to work the newly opened land. Some claim that due to these factors the cotton gin contributed to the North-South antagonisms that led to the American Civil War.

Whitney formed a partnership to manufacture his new machines with Phineas Miller, who three years later married Catherine Greene. Although Whitney obtained a patent on the machine in March 1794, court battles to assert his patent rights and the setting up of a large-scale manufacturing plant to build cotton gins proved to be very costly. As a result, Whitney never profited financially from his invention. In addition, Greene and Miller had to sell off large parts of the Greene plantation to pay the expenses. The courts eventually vindicated Whitney's ownership of the cotton gin patent but not in time to allow the partnership to gain control of the market. While Whitney and Miller fought in the courts, blacksmiths all over the South were constructing their own machines—the machine was very simple to duplicate—with no regard for the patent owned by Whitney. By the late 1790s, Whitney had given up on the cotton gin and turned his attention to the manufacture of firearms for the United States government; but others continued to improve on the cotton gin design. The most fundamental change was the substitution of a series of circular toothed "saws" in place of Whitney's rows of spikes. This refined machine came to be known as the "saw gin" in some areas. To give some idea of the revolutionary nature of the cotton gin, in 1793—one year before Whitney received his patent—the United States exported about 500,000 pounds (227,000 kg) of cotton; by 1803 that figure had risen to over 40 million pounds (18 million kg).

Cotton picker, mechanical

The growing and harvesting of cotton was central to the social and economic structure of the southern United States during its early history. The cotton-based economy consisted of large plantations primarily built and sustained through the labor of slaves who had been brought from Africa. Because of this continuing supply of human labor, the development of mechanical cotton harvesters was slower than other types of farm equipment. Although some early cotton pickers were invented, they were commercial failures. In 1850 S. S. Rembert and J. Prescott of Memphis, Tennessee, patented a mechanical cotton picker. In 1895 August Campbell patented the spindle, which plucked the cotton from the boll, the basis for today's pickers. After the end of the American Civil War in 1865, many newly freed slaves migrated to cities in the North. However, most remained as poorly paid field laborers, continuing to make machinery unnecessary.

The first mechanical **tractor**-pulled cotton strippers, which removed the entire cotton boll from the plant, did not appear until 1914. Consisting of nothing more than a section of picket fence dragged across the field, the process was referred to as "sledding cotton." During the 1920s International Harvester purchased Campbell's patents, and, after nearly twenty years of research, produced and marketed the barbed-spindle cotton picker, beginning in 1942. Texans Mack and John Rust patented a smooth spindle picker in 1932. However, the recurrence of cheap labor during the Great Depression of the 1930s again delayed acceptance of cotton pickers by farmers, and the pair did not market their machine until 1949. After 1950 mechanical pickers were rapidly brought into the cotton fields. In 1953, 25 percent of United States cotton was mechanically harvested, and by 1962, that figure had risen to 60 percent. The increased used of mechanical pickers led in part to the second wave of migration of black laborers to northern cities, some of whom went to work in the factories that made the machines. Hand-picked cotton, is rarely available as most picking is currently mechanical. Cotton fibers are now routinely mixed with synthetics adding strength to the cotton and moisture absorption and wrinkle resistance to the synthetic material.While there is a 5 to 10 percent field loss with mechanical picking, the cost-effectiveness of machine harvesting makes it preferred for the mass market.

See also Tractor

Cottrell, Frederick Gardner (1877-1948)

American chemist

Frederick Cottrell was the inventor of the electrostatic percipitator, an important device that removes dust and smoke particles from gases. It is mainly used to collect polluting emissions from industrial plants—in the United States alone, more than 100 million tons of otherwise-polluting particles are captured by this device each year.

The concept of the electrostatic precipitator was first described in 1824 by a German mathematics teacher, M. Hohlfeld. Cottrell began his work to actually build such a device in 1906, while a professor of chemistry at the University of California at Berkeley. He succeeded four years later, and his device was quickly adopted by heavy industries confronted with strict new regulations for air pollution control at the time.

The invention of the cotton gin in the late eighteenth century revolutionized agriculture in the American South.

The electrostatic precipitator uses electric forces to remove the harmful particles from a gas. The dust and smoke particles become charged, or ionized, after the "dirty" gas is passed through a high voltage discharge. A charged electrode or electrified plate of the opposite charge as the particles then attracts the particles out of the gas and onto the plate. The deposited particles are removed from the plate by washing or scraping.

The electrostatic precipitator was a quick success and was adopted by smelting and sulfuric acid industries, and cement plants. It was used to remove tar from coke-oven gases, to clean the air in air-conditioning systems, to remove acid fumes in refineries and chemical plants, to remove fly ash from the released gases of large coal-burning power plants, to remove smoke from air (as in some household air purifiers), and to recover specks of valuable metals like copper and tin.

Cottrell believed that the proceeds from inventions should be used to fund further research, and so in 1912 he created the nonprofit Research Corporation and gave his electrostatic precipitator patents to the company. The significant earnings from these patents were eventually used to fund research in rockets, particle accelerators, and a very early form of the laser called a **maser**.

In 1911 Cottrell joined the United States Bureau of Mines, where he supervised work on the recovery of helium from gas wells. In 1922 he worked for the Department of Agriculture, contributing to the production of **ammonia** for fertilizers. Later in his life he became a consultant. Cottrell was an early advocate of the cooperation of international scientists, as well as the universal language Esperanto.

See also Electrostatic devices

Cough drops

Nineteenth-century **patent medicine** hucksters peddled their fair share of cough syrups—Buffalo Bill Cody marketed one called Cough Cream. But the world-famous cough remedy in drop form was first advertised by the Smith family in 1852.

James Smith moved his family from Scotland to St. Armand, Quebec, then to Poughkeepsie, New York, in 1847, where he established a restaurant. There, according to legend, a traveling peddler called Sly Hawkins gave Smith the formula for a tasty and effective cough candy. Smith cooked up a batch on the kitchen stove and began advertising the "James Smith and Sons Compound of Wild Cherry Cough Candy" in 1852 as a remedy for coughs, colds, hoarseness, and sore throats. Smith's sons William and Andrew soon joined their father in the new enterprise, energetically peddling the cough candy around Poughkeepsie. Production moved from the kitchen to a nearby building, which became the world's first cough-drop factory. After James Smith's death, William and Andrew renamed their business Smith Brothers.

Because the popularity of the cough drops gave birth to a rash of imitators who used similar names, the brothers designed a distinctive trademark for their product. They put their own pictures on retailers' dispensing bowls and on the customers' envelopes, with the word *Trade* under William's visage and *Mark* under Andrew's. After the brothers began producing factory-filled packages in 1877 (one of the first ever on the market) carrying their portrait trademark, William and Andrew became evermore known by the aliases Trade and Mark.

Today, there are many brands and manufacturers of cough drops. Recently cough drops and throat lozenges have been manufactured containing the vitamin zinc, which some researchers have found may reduce the length of the average cold.

Coulomb, Charles Augustin de • See Torsion balance

Cousteau, Jacques Yves (1910-1997)
French oceanographer

Jacques-Yves Cousteau is perhaps the most well-known modern scuba diver and undersea explorer. He brought the world of undersea diving within the capabilities of ordinary people by inventing (with Emile Gagnan) the aqualung in 1942. This self-contained underwater breathing apparatus (SCUBA) now enables divers to remain under water for an hour or more instead of minutes. Also, Cousteau helped design other diving equipment such as the two-person diving saucer. His explorations, conducted from his famous oceanic ship *Calypso*, were documented and earned him numerous honors and awards.

Cousteau was born in 1910 in St. Andre-de-Cubzak, France. He attended the Brest Naval Academy and became a naval gunnery officer in 1933. At the age of 47, Cousteau retired from the navy with the rank of corvette captain.

During World War II Cousteau co-invented the aqualung, a device that provides pressurized air to the diver while underwater. Before the aqualung, divers had to wear heavy suits and fishbowl-like helmets that made swimming nearly impossible. But Cousteau's invention helped popularize **scuba diving** by providing far greater mobility for underwater exploration and leading to the development of modern scuba gear.

After the war, Cousteau involved himself in a series of projects, exploring the oceans, and designing and developing equipment for marine exploration. He developed the *Calypso* underwater camera, a forerunner to today's *Nikonos* camera. He participated in testing the **bathyscaphe**, a deep-diving vessel invented by Swiss physicist **Auguste Piccard**. In 1959, he co-invented the first exploration mini-submarine, the two person diving saucer, *Soucoupe*.

Diving research was another of Cousteau's many interests. Cousteau and Philippe Taillez established the Undersea Research Group at Toulon which became Europe's leading center for studying diving techniques and undersea living. Cousteau is also known for developing early underwater habitats—the Conshelf I, II, and III—located between 33 ft (10 m) and 330 ft (100 m) below the ocean surface, and capable of sheltering people for prolonged periods of time. Cousteau's documentary film, *World Without Sun*, which recorded man's first prolonged (30 days) stay in an undersea habitat, the Conshelf II, received an Oscar for Best Documentary in 1964.

Cousteau did not consider himself a scientist, but rather, a filmmaker, explorer, and environmentalist. He produced 65 films and lead over 80 expeditions. He won awards at both the Cannes Film Festival and the Academy Awards for his 1956 feature film *The Silent World*. From the late 1950s on, he produced numerous television programs, including *The World of Jacques-Yves Cousteau* and the series called *The Undersea World of Jacques Cousteau*. He wrote, co-wrote, or edited 80 books about the oceans and his adventures in them. The last of these, his memoir *The Man, The Octopus, and the Orchid*, was published just six days after his death. In 1974, Cousteau founded the Cousteau Society, one of the world's largest conservation groups. He spent much of his life trying to educate the public about the importance of the marine environment, its vulnerability and beauty.

See also Diving apparatus

Cramer, S. W. • See Air conditioning

Crane

An invention of ancient origin, the crane is used for the loading and unloading of heavy objects and as an aid in the construction of tall buildings. It is distinct in that it moves objects both horizontally and vertically, unlike hoists or **elevators**. One common forerunner of the crane was the *shaduf*, prevalent in Egypt and India around 1500 B.C. Employed by a single person for lifting water, the shaduf consisted of a vertical support, a long, pivoting beam, and a counterweight.

The first true cranes, founded on principles governing the use of **lever**s and counterweights, employed a **pulley** system affixed to a single mast (or boom). Lifting power was provided by humans or draft animals operating a treadmill or large wheel. Eventually, a second mast and guy wires were added to increase the strength and stability of this early form of crane.

One of the most significant developments in crane design, which probably occurred during Medieval times with the advent of Gothic architecture, was the jib-crane, which featured a pivoting horizontal arm that projected outward from the top of the boom. The addition of hinged movement to the outermost section of the jib allowed for even further versatility and movement. Jib-cranes are also known as derrick cranes, derrick being the term originally applied to gallows structures at the time when Englishman Godfrey Derrick became a prominent hangman.

In modern nomenclature, the derrick is a large hoisting machine similar in most respects to the crane, save for its typically stationary foundation. Oil derricks, for example, are specialized steel towers used for raising and lowering equipment for drilling oil wells. One of the most powerful cranes, a barge derrick, is a double-boomed structure capable of lifting and moving ships weighing up to 3,000 tons (2,730 t). Other cranes with specialized uses include the cantilever crane (featuring a suspended horizontal boom and used in shipyards), the overhead traveling crane (guided by rails and a trolley-suspended **pulley** system and used for indoor work), and the tractor-mounted crawler crane (a hydraulic-powered crane with a telescoping boom). A simple example of a small-scale crane is the fork-lift truck. Like its much larger relatives, the fork-lift is limited not so much by the size of its hoisting apparatus as by the force of its rear counterweight.

CRASH TEST DUMMY

The standard procedure for safety-testing **automobiles** and other vehicles is the crash test in which the car is rammed into a brick wall. The driver and passengers are models of humans called crash test dummies that are fitted with sensors. The sensors provide data on acceleration, rapid deceleration on impact, force of impact, and other motions of each dummy's torso and limbs that can be related to potential injuries for real human occupants. Although this testing method is widely accepted, instrumented, analyzed, and standardized, it has several major flaws. The vehicle is sacrificed, of course, and, although the "human" counterparts can be reused, the process of getting them to respond with all the complications of human bodies has been long and imperfect.

Early testing of vehicles used cadavers, but this practice had serious drawbacks including a shortage of cadavers. Damage could be readily observed but did not necessarily show what would happen to a live, flexible being. Crash test dummies were developed in 1949 under contract to the United States Air Force. "Sierra Sam," the first dummy, was used to test aircraft ejection seats that were mounted on sleds on rails, with the sleds propelled by rockets. The same type of dummy was used later by the automotive industry to develop lapsaftey belts and shoulder harnesses. Sierra Sam resembled a man on the surface, had the body weight of a statistically average adult male, and had articulated limbs. His spine and neck were rigid, however, and only the acceleration of his head as it followed the path of ejection could be measured.

The state of the art for dummies improved considerably by 1952 when "Mark 1" was created from a plaster cast of a live subject. His skull was cast from two pieces of aluminum that housed instruments for measuring acceleration and force of impact. The neck and back were made of a series of ball-and-socket joints with spacers to simulate the range of motion of a real spine, and the dummy had a set of steel tubing ribs. Mark 1 had vinyl skin and foam "flesh," but the limbs were quite stiff. A series of dummies developed through 1956 had special modifications to measure a selected range of motions. Different sizes and weights were made for the first time, and these post-Mark 1 dummies were used to test tractor safety, frogmen's suits for underwater escapes, and aspects of the Apollo Program for the National Aeronautics and Space Administration (NASA).

When American and European space programs needed to evaluate the effects of rocket thrust on astronauts strapped in seats, the Grumman-Alderson Research Dummy (GARD) was invented. The man in the seat had to be perfectly aligned with the direction of rocket thrust, or the man and seat would twist out of place. GARD was an important technological step because the process of locating the center of gravity and evaluating the moment of inertia (the tendency to rotate with acceleration) in the dummy were perfected and well instrumented. Scientists could measure rotations, acceleration history, and the stresses between the man and the seat. Motion ranges in this dummy are simple because the seat limits them, and the manikin is durable. GARD is still used to test ejection seats.

By 1966, the emphasis in developing crash test dummies shifted from aviation to the automotive industry. Improvements were needed in the rib cage, spine, pelvis, and abdominal cavity to evaluate restraint systems. Mathematical models showed that motions could be simulated with models made of separate body ranges connected by springs. Not only did the springs move on impact, but they rebounded so effects like whiplash could be studied. Development of the shoulder harness required the dummy to have a breastbone (sternum), clavicles, and shoulder blades. To mimic the actions of internal organs, a visceral sac was installed and instrumented. Cavities in the thighs, chest, and head were designed to hold instruments to measure more complicated movements and force deflections.

Improvements to test dummies were based on careful analysis of data from real automobile crashes; by 1970, injuries to women, children, large men, and all persons with smaller bodies were demonstrating that the statistically correct adult male dummy didn't represent these groups. A large male model and the first female dummy (complete with a ladylike wig) were manufactured in 1970; named Sierra Stan and Sierra

Susie, their family was completed with the addition of Sierra Sammy (a 6-year-old) and Sierra Toddler (a 3-year-old) that same year. Neither child had the correct weight distribution, and children complicated the problem of studying the effects of impacts if they were standing or not sitting conventionally during accidents. The child dummies were redesigned using improved data to have more flexible skeletons and supple limb joints than adults and to be adaptable to various positions in the car; instrumentation was also improved to measure the broader range of variables. Improved child dummies were essential to development and testing of child safety seats and, later, to **airbags**.

By the mid-1970s, improvements in **computer** analysis allowed measurements of almost forty different parameters in the behavior of test dummies and materials were similarly improved to make construction of realistic, fully articulated dummies possible. For example, the dummy's bones could be constructed of fiberglass to behave like human bones except with greater breaking strength so the dummy could be used again. Designs became so specific, however, that models like the Supermorphic Dummy were made exclusively for car crash tests—they were too fragile for ejection testing. The Limb Restraint Evaluator (LRE) Anthropomorphic Manikin was created especially for testing restraint devices to prevent injuries due to flailing during ejection from military jets. Other specialized dummies tested experimental parachutes and helicopter crashes.

Although the crash test dummy has proven its life-saving capabilities in countless tests, **computer** simulations may prove fatal to the dummy family. Computer software that can simulate car crashes, damages to the vehicle itself, and damages to the passengers are in development. Simulations are attractive to auto manufacturers because they eliminate some, if not all, destructive testing. The human body can be ''examined'' after simulated impact to better understand the details of injuries to the head and spine particularly. The time from design to production of a vehicle may be reduced as much as three months, and new designs may be safer for a broader range of physical types.

Cray, Seymour • See Supercomputer

Cream separator

Cream is the yellowish part of milk, rich in fat, that naturally rises to the surface if milk is allowed to stand. Before mechanical separators came into practical use, cream had been separated from milk with small strainer dishes—obviously a slow and tedious process. In 1877, the Swedish inventor **Carl Gustaf Patrik de Laval** introduced a high-speed centrifugal cream separator. Milk was placed in a chamber heated to 86°F (29.7°C), which is the best temperature for separating cream. From there the milk went through tubes to a container that was spun at 4,000 revolutions per minute by a **steam engine**. The centrifugal force created by this spinning action separated the cream, which was lighter, causing it to settle in the center of the container. The milk, which was heavier, was pushed to the outer part and forced up to a discharge pipe, leaving only the cream remaining in the container. This machine was successfully marketed and used in large dairies throughout the world.

Over the years, improvements were made to this basic concept. In 1888, an advanced cream separator was introduced in Scotland; it was self-skimming, emptied itself completely, and was easily cleaned. Manual machines soon appeared for those unable to afford a steam engine as a power source. In 1896, a new machine called the butter accumulator appeared with the ability to separate and make butter in one continuous operation. Today the most efficient cream separators leave less than 0.01% fat in the skim milk.

See also Centrifuge

Crompton, Samuel (1753-1827)
English inventor

The evolution of the spinning machine came to its fruition in 1779 with the invention of the **spinning mule**. This device borrowed from **James Hargreaves**'s **spinning jenny** and **Richard Arkwright**'s **water frame**. **Textile** industries worldwide used the mule for almost two hundred years, during which time the machine's design was modified only slightly. Despite the overwhelming success of the spinning mule, however, its inventor, Samuel Crompton, enjoyed little prosperity. In fact, the last half-century of his life was spent fighting the industry that had effectively stolen his machine.

Crompton was born near Bolton, England, in 1753. His father died when he was five, and he learned at an early age to help his mother with all the household tasks. Among these was working with a spinning jenny, the yarn from which was sold at a local market. Crompton's mother was a stern woman with a short temper, a temper which flared every time the jenny broke its yarn. Realizing this was the fault of a flaw in the jenny's design, Crompton set out to construct his own spinning machine.

The building of the spinning mule took five years and all the money Crompton had earned as a fiddler at a local theater. He worked in secrecy, usually at night; the noises coming from his workshop led many neighbors to believe the building was haunted. The final product—crude, but very efficient—was finished in 1779, just after Crompton's twenty-seventh birthday.

The spinning mule used the two most important elements of the water-frame and the spinning jenny. From Arkwright's water frame it borrowed a set of rollers to draw out the cotton fibers; from the jenny, a moving carriage that gently stretched the roving. Added to this was Crompton's own spindle carriage, which insured that no tension was applied to the yarn before it had been completely spun. The yarn spun by this machine was strong and smooth, able to be used in materials such as muslin, and did not break as easily as that spun by the jenny. Though it was originally called the Hall-in-the-Wood wheel (after Crompton's birthplace), it soon became known as the spinning mule because of its hybrid nature.

For a short time, Crompton's family took the mule's yarn to the local market, selling it at a considerable profit.

Soon, however, the demand for muslin yarn rose dramatically. Fortuneseekers hounded Crompton to reveal his secret, some resorting to attempted burglary (one story claims that Arkwright himself tried to break into Crompton's workshop). After a few months Crompton could bear no more: to preserve his own sanity he decided to sell the spinning mule to the public. Still too poor to afford the price of a patent, he agreed to sell the machine's design for a subscription of seventy pounds each year.

As soon as the textile industry gained possession of the spinning mule they reneged on their agreement. Crompton collected just six pounds in the ten years following the invention of the mule. He was able to successfully appeal to a number of manufacturers, who granted him a subscription of 400 pounds—a nominal amount, considering that the mule had replaced both the jenny and the water frame, and was responsible for the employment of more than 200,000 people. Crompton received a grant in the amount of 5,000 pounds in 1812 from the House of Commons; still, this was not enough to rescue him from poverty. He attempted twice to establish his own business, failing each time. In 1824 a group of friends anonymously donated to Crompton an annuity of 63 pounds—enough to live on, but little else. He died in Bolton in 1927.

In contrast to Crompton's financial decline, the spinning mule became the most important machine in British manufacturing. In 1827 the British engineer Richard Roberts introduced the self-acting mule, which no longer required any manual labor.

Samuel Crompton

Crookes, William (1832-1919)

English physicist

Proficient in the fields of chemistry and physics, Crookes is best remembered for his invention of the Crookes tube, a **cathode-ray tube** that was the precursor to modern **television** and video tubes.

Crookes was born in London in 1832, the oldest of 16 children of a wealthy tailor and real estate investor. At age 16 he entered the Royal College of Chemistry in the hopes of studying organic chemistry. While there he became the assistant to August Wilhelm von Hofmann, a position which allowed him to attend meetings at the Royal Institution. It was at one such meeting that Crookes met the eminent physicist **Michael Faraday** who convinced him to change his area of concentration from chemistry to physics, and particularly to optics.

After graduation from the Royal College of Chemistry, Crookes briefly attempted two positions in academia: superintendent of the meteorological department at the Radcliffe Observatory in 1854, and Lecturer in Chemistry at Chester Training College in 1855. Upon his father's death he received a substantial inheritance and was able to open his own laboratory and concentrate on physics.

It was not long before this investment paid off. While experimenting with spectroscopy (physics pertaining to the theory and interpretation of interactions between matter and radiation) Crookes discovered a lime green band in the spectrum of selenium, a band that belongs to no known element at that time. After several years he succeeded in isolating the element, which he named thallium (from the Greek word *thallos*, meaning "green twig"). At the same time, the French scientist C. A. Lamy had also isolated a sample of thallium, sparking a small controversy over who would be attributed with the discovery. Today, the discovery is usually credited to Crookes. In 1861 he published his discovery.

While trying to determine the precise atomic weight of thallium, Crookes became interested in the use of vacuum tubes, which had recently been improved upon by **Johann Heinrich Geissler**. By placing a tiny amount of thallium within a vacuum tube and weighing it there, a very accurate measurement could be obtained. However, he noticed that the arms of the scale would occasionally jerk unexpectedly, even though no visible force was acting upon them. Crookes later discovered that very small amounts of air had remained within the tube, and when struck by sunlight or lamplight it became disrupted, thus striking and moving the sensitive scale. In order to demonstrate this phenomenon, Crookes invented a device called a radiometer, consisting of a series of four small vanes balanced upon a pin. The two sides of each vane were painted different colors—one side black, the other silver. The entire assembly was then sealed in a vacuum bulb. When light struck the vanes the black sides would heat up, causing them to turn as the excited air molecules struck them. This device was (and still is) mostly a science toy, but it was also used by Scottish physicist James Clerk Maxwell (1831-1879) to prove the kinetic theory of gases.

Perhaps Crookes' most important work also stemmed from his experiments with vacuum tubes, particularly with **cathode ray tubes**. At that time a great deal of research was being done on **cathode rays**; Heinrich Geissler (1815-1879) had invented an improved vacuum tube in which two electrodes could be placed, one at each end. An electrical current was sent through the system, and as the air in the tube was evacuated, it would begin to glow. Crookes made further improvements to the system and in the late 1870s the first practical cathode ray tube was designed. It was comprised of a vacuum tube with two electrodes, a cathode and an anode, one at each end of the tube. When an electric current was introduced and the tube evacuated, a green glow would appear. While this effect had been observed before, it was Crookes who first noted the dark area closest to the cathode that he called Crookes' dark space. More importantly, he noted that when he placed a pivoting vane within the tube, the vane would turn slightly, as if within the current of a stream when the tube was evacuated.

At the time it was not known just what made up this invisible stream. Crookes himself postulated an ultragaseous fourth state of matter that manifested itself within the tube. It was not until 1897 that English physicist Joseph J. Thomson (1856-1940) announced that a stream of electrons was created by the Crookes tube. Today, the same kind of technology exists in the electron guns used in **televisions** and **electron microscopes**.

In his later years, Crookes was fascinated by the emerging field of radioactivity. After researching uranium for several years he invented an instrument designed to detect alpha-rays called a spinthariscope, which means "spark-viewer." When an alpha particle strikes the instrument's zinc sulphide covered screen, a microscopic flash of light can be observed. During his career Crookes was awarded numerous accolades, including the Royal Medal, and was knighted in 1897. For some, however, his reputation is tarnished by the fact that he was a spiritualist, publishing several papers on the validity of psychic phonomena and the occult.

See also Video recording

Crop rotation

As crops grow, they remove nutrients from the soil; however, some crops return certain nutrients over the seasons and between species. Unlike a natural ecosystem, in which there is a balanced exchange of nutrients between old growth and new growth, a crop system is totally dependent on the farmer's techniques. Farmers have three alternatives for keeping their fields productive. They can add natural or chemical fertilizers; they can let the field remain fallow, or uncropped, for one or more growing seasons; or they can rotate their crops on a regular basis.

Crop rotation requires a knowledge of how different crops interact with the soil. One of the most fundamental and earliest known facets of crop rotation was the use of legumes, which include peas, beans, and lentils. Legumes, as well as other crops like oats, return vital nitrogen to the soil. The Greeks and Romans rotated crops on a three-field system as early as 200 B.C., but knowledge of that practice, like so much other technical expertise, was lost with the decline of Classical civilization. Europe's rediscovery of crop rotation was in the form of the four-course Norfolk system in which four fields were alternately planted in wheat or rye, root crops, legumes, and nitrogen-rich clover. This system provided a balanced nutrient exchange. It also provided the people with a more diverse diet. Farm animals were allowed to graze in the clover, further enriching the soil with their droppings. The Norfolk system was responsible in part for the use of horses as draft animals in Europe and provided horses with oats. An improved horse collar and the iron-edged moldboard **plow**, gave Europe the means to feed its growing population and assert its influence over the rest of the world.

It was not until the 1880s that Herman Hellriegel (1831-1895), a German chemist, learned that legumes extracted nitrogen from the atmosphere and introduced it into the soil. Prior to Hellriegel's discovery, farmers were simply aware that rotating legumes with other crops increased production and diminished the level of plant diseases. During the mid-twentieth century, chemical fertilizers became more readily available, and farmers, intent on mass production and increased profits, shifted from crop rotation to single-crop production. Wheat, corn, and soybeans became the major cash crops. Over time, constant application of fertilizers, both natural and manmade, along with pesticides and herbicides, led to pollution of the soil and streams. Intensive agriculture also depletes the water table and damages the soil structure. More recently, although agribusiness still utilizes fertilizers, there has been a return to crop rotation and other natural farming techniques in conjunction with fertilizer usage.

See also Composting; Farming, mass-production; Fertilizer, synthetic

Cross, Charles Frederick (1855-1935)

English chemist

Charles Cross was born in 1855 in Brentford, England. Upon his graduation from King's College of London in 1878, he set up an analytical and consulting chemical firm with **Edward John Bevan** in London.

In 1892 Cross, Bevan, and Beadle developed viscous rayon which is formed by first dissolving cellulose from wood fibers in a sodium hydroxide solution and then treating it with carbon disulfide after drying. The resulting solution—called viscose—is extruded through a small hole into an acid bath to form. After drying these fibers can be spun into fabric.

Cross wrote several technical books, including *Cellulose*, before his death in Sussex in 1935.

See also Cellulose, chemical uses of; Fiber, synthetic

Crosthwait, David N., Jr. (1898-1976)

American engineer

Crosthwait was born on May 27, 1898, in Nashville, Tennessee. His father was a biology and chemistry teacher, and the principal of Nashville's first high school for African-Americans. The family moved to Kansas City, Missouri, where Crosthwait received his high school diploma in 1916. He went on to study at Purdue University where he earned both his B.S. and Master's degrees.

After receiving his advanced degree, Crosthwait became a consultant to engineering firms and public utilities, specializing in heating, cooling, and refrigeration technology. He designed and tested heating and ventilation systems, and introduced numerous electronic and mechanical inventions. Among his patents were an automobile turn signal, a component for **thermostat**s, a **vacuum pump**, and several other devices related to heating systems. Crosthwait also designed a steam system that was used to heat the towers of Radio City in New York, New York.

Cruise control, automobile

Ralph Teeter was blinded in an accident at the age of eight, but this mishap did not prevent him from building his own **gasoline**-powered car at the age of 12 and going on later to earn an engineering degree from the University of Pennsylvania.

During a drive to the university in the 1930s, Teeter realized that few motorists maintained a constant speed on the long stretches of the Pennsylvania Turnpike. Relying on the sound of their **automobile**'s engine and the feel of the vibrations, he sensed that drivers tend to move at uneven and inconsistent speeds, a problem that could lead to accidents.

Near the end of World War II, Teeter completed his system of cruise control, called a Speedostat. This device relied on manifold pressure from the exhaust system to control engine speed by arranging a valve to maintain an automatic throttle. This essentially allowed the driver to select a speed by dialing a number; when the car reached that speed, the accelerator would resist further pressure. The driver did have the ability to increase the speed by exerting extra pressure on the gas pedal. This early mechanism only prevented drivers from moving at speeds higher than those they had selected; it did not allow them to maintain a steady speed.

To create the first true cruise control, Teeter added a **magnet** to his invention; this allowed a simple lock-in device to permit the driver to maintain the same speed without pressing the accelerator, regardless of the terrain or wind resistance. This version of cruise control soon become more viable than the first device, which was eventually dropped altogether.

Cruise control was not an immediate success, because motorists were often unsure of its benefits. Some felt that relying on cruise control was not safe because it might increase the monotony of the ride and lead to less driver awareness. Others, however, acknowledged the control's advantages to the driver, who was free to devote greater attention to the road. By 1956, cruise control systems were manufactured by a larger company that improved the design and marketed the product more aggressively. Today, 70% of all new American cars are equipped with cruise control. Teeter himself went on to invent more than twenty other products and was granted some fifty patents.

As car designs approach the year 2000, adaptive cruise control systems are being devised, among many other safety systems collectively called intelligent transportation systems (ITS). Adaptive cruise control allows the car to sense the vehicle in the driving lane ahead of it and speed up or slow down (including coming to a full stop) with traffic. It helps avoid rear-end collisions and facilitates "vehicle platooning" in which a group of vehicles (commuters, for example) move smoothly as a unit toward a shared general destination.

The car is fitted with a small radar **antenna** behind its front grille or bumper, which detects vehicles in its lane up to 400 feet (150 meters) ahead. The driver can adjust the distance to the leading vehicle to provide a comfort zone; interior readouts tell the driver whether the car is cruising at the selected speed or following the speed of the leading vehicle. If the car has to slow down, adaptive cruise control also activates the rear brake lights. If a third vehicle moves into the lane between the leading and cruising cars, the cruising car decelerates and assumes a speed appropriate to the new leader. When the driver accelerates above the selected cruise speed, the readout displays "driver." The actions are all determined by the driver's settings and the radar sensor readings that feed to a computer that activates braking or accelerating. Adaptive cruise control allows commuters in heavy traffic to rely on cruise control without continuously overriding or resetting the system.

Cryobiology

Cryobiology is the study of the effects of very low temperatures on living things. Temperatures ranging from the freezing point of water down to absolute zero are used to freeze living matter, often to preserve the cells for the future.

Usually liquid nitrogen is used to get temperatures down quickly so that cells can be placed in "suspended animation" and then, if quickly thawed, resume their normal activity. When cells are frozen slowly, a great deal of damage occurs because water within the cell crystallizes first, leaving great concentrations of other compounds such as salts, which throws off the delicate osmotic balance. The cells dehydrate and the acid-base balance is destroyed. To avoid cell death, the freezing process is controlled and aided by certain antifreeze agents.

Cryobiology techniques are used to freeze and store eye corneas, blood, sperm, and pollen in "banks" for future use. Although currently applicable in human beings only in the realm of science fiction, cryobiologists have conducted studies of freezing entire animals and then "bringing them back to life" an hour or so later. For humans, such cryopreservation could be used, for example, to freeze a person who has a terminal illness in hopes of "unfreezing" them in the future when a cure would be found.

In 1961, an American physician, Irving S. Cooper, began using a freezing technique known as cryosurgery to freeze and destroy damaged tissue within the brains of Parkinson's disease patients. His technique destroyed cells that caused some of the symptoms of Parkinson's such as tremors and rigidity. Since that time cryosurgery has been used successfully in delicate eye surgery to remove cataracts from the lens of the eye and repair detached retinas. It has more recently been used in the treatment of liver tumors and prostate and cervix cancer.

Cryosurgery is relatively bloodless because the reduced temperatures cause the constriction of blood vessels so there is little bleeding. Special instruments are used with freezing tips which kill unwanted tissue while the healthy tissue around it is shielded.

Crystal rectifier

A crystal rectifier is a device used to convert **alternating current** into direct current. Alternating current is easier to generate and distribute using **transformers**, but direct current is often more appropriate for use in certain electrical devices, particularly in radios and televisions.

The crystal rectifier was discovered by the German physicist **Karl Ferdinand Braun**, who patented his device in 1874. While experimenting with metal sulfides, he noticed that some crystals allowed current to pass through in only one direction. Since alternating current travels in two directions, moving forward to a maximum value and then doubling back on itself to the negative of that value (known as sine-wave motion), when it is passed through a rectifier the forward direction only is allowed to pass, while the doubling-back motion is eliminated. The alternating current is therefore converted into direct current. Such crystal rectifiers are also known as *diodes.*

When first introduced in the late nineteenth century, crystal rectifiers were used almost exclusively in crystal set radios, which are still available at hobby shops in do-it-yourself kits. After 1906 they were replaced by more efficient vacuum tubes. In 1948, the American physicist **William Shockley** revived the crystal rectifier and the transistor in new solid-state communications equipment. Today, rectifiers are used in numerous processes, including anodizing, battery charging, and metal refining, as well as such devices as **amplifiers** and measuring equipment.

See also Battery, electric

Crystal set radio • See Braun, Karl Ferdinand

Cugnot, Nicholas-Joseph (1725-1804)
French engineer

Cugnot is credited with developing the world's first self-propelled vehicle. Powered by a **steam engine** refined by Cugnot, his three-wheeled vehicle could carry four passengers and moved at a walking pace. Built in 1769 and first used the following year, Cugnot's vehicle was originally designed to haul heavy **artillery** pieces and should more properly be called a **tractor** and not a carriage.

Cugnot was born at Void, in the Meuse province of France. As a young man, he joined the French army and while in the service in Germany and Belgium, he invented a new kind of **rifle** for use by French troops. He was also encouraged to work on a steam-powered gun-carriage. Cugnot was aware of the improvements in steam power developed by **Thomas Savery**, an English inventor, and **Denis Papin**, a French physicist.

Cugnot added further improvements, which employed steam power to move **piston**s without condensation, greatly improving engine efficiency. His engine consisted of two, 13-inch (33-cm), 1.75-cubic-foot (50-liter) pistons connected by a rocking beam which were synchronized so that when atmospheric pressure forced one piston up, high pressure steam forced the other piston down. The reciprocating motion was transferred to the **axle**, where it produced the rotary motion that turned the **wheel**. This arrangement is considered to be the first successful device for converting reciprocating motion into rotary motion.

Cugnot's first carriage had serious limitations. Its three-wheel design, with the boiler well out in front, was inherently unstable, and the whole heavy boiler-drive-wheel mechanism had to be turned to steer the carriage. It carried no reserves of water or fuel and required that the driver stop periodically to refire the furnace and add water to the boiler.

Despite these obvious drawbacks, the French Minister of War, the Duc de Choiseul, was encouraged by Cugnot's first demonstration, and he commissioned Cugnot to build a second larger, more powerful, and faster vehicle. This second carriage was completed by Cugnot in 1771 at a cost of 20,000 livres. Unfortunately, de Choiseul fell from power before this second vehicle could be fully tested, and his successor showed no interest in Cugnot's steam gun-carriage. It sat in a military shed for 30 years until it was moved to the Conservatoire National des Arts et Métiers in Paris, where it has remained on exhibit ever since.

Cugnot was granted a pension by the War Ministry in 1779, and he moved to Brussels. The French revolutionaries eliminated Cugnot's pension in 1789, but it was restored during the Consulate by Napoleon and continued until Cugnot's death, which came on October 2, 1804, in Paris.

See also Steam-powered road vehicle

Cultivator, mechanical

Cultivators, or tillers, prepare the soil for growing crops. Whereas the **plow** breaks the soil and turns it over, the cultivator further pulverizes the soil to facilitate planting. Cultivating helps promote plant growth by eradicating weeds, preparing the land for irrigation, and facilitating the incorporation of **fertilizer** and pesticide into the soil. For some crops, the cultivator can also prepare the soil for harvesting. The hoe, for manual cultivating, has dates to the earliest of humankind's tool-making periods and remains an important gardening tool.

In the early 1700s, Jethro Tull of England invented a horse-drawn mechanical hoe with three coulters, or hoes, and seed funnels for planting. In 1856 George Esterly patented a straddle row cultivator drawn by two horses. Horse-drawn riding cultivators were introduced in the late 1880s. Two-row cultivators appeared after 1900. In 1912 Australian A. C. Howard invented a rotary cultivator with revolving blades. It was later adapted to operate on **tractor** power.

A tractor-mounted cultivator was developed in 1918 by the B. F. Avery Company. International Harvester developed an integral mounted cultivator in 1925. In these devices, the gangs, or rows of rotors that perform the tilling, had to be manually lifted. The first cultivators with power-lift gangs were developed around 1937.

The modern rotary cultivator works through the plowed soil with running blades, or tines, to give the soil a powdery consistency ideal for planting and crop growth. The curved blades project from a central **axle** made of high-strength steel that can withstand the stress of moving through dense soil or over obstructions. Another version of the machine, the spiked-rotor cultivator, works the soil with spikes that rotate around a long axle. Baloon tires are a common addition, to add stability. Modern cultivators can also be modified to allow precision sowing of seed that otherwise would have been broadcasted, or randomly distributed.

Small pedestrian controlled rotary cultivators, or rototillers, have become increasingly popular with home gardeners. The **gasoline**-powered tillers spare the gardener hours of back breaking hoeing and raking.

See also Combine harvester; Reaper and binder; Seeding devices

Cuneiform • See Writing

Curie, Marie • See Geiger counter

Cybernetics

Cybernetics is the study of communication and feedback control in machines and humans. Cybernetics analyzes the ability of humans, animals and some machines to respond to or make adjustments based upon sensory input from the environment. This process of response or adjustment is called feedback or automatic control. For example, the household **thermostat** uses feedback when it turns a **furnace** on or off based on its measurements of temperature.

The earliest known feedback control mechanism, the centrifugal **governor**, was developed by Scotsman **James Watt** in 1788. Watt's **steam engine** governor contained two weighted arms that were hurled outward by centrifugal force as engine speed increased. Once the arms reached a certain point, they triggered a mechanical link which closed a **valve**, thus preventing engine from exceeding a certain speed, keeping it at a constant rate.

The principles of feedback control were first clearly defined by Norbert Wiener (1894-1964), an American mathematician. Wiener was a child prodigy who could read and write by the time he was three years old. Wiener earned his Ph.D. in mathematics from Harvard at the age of nineteen. Wiener was particularly intrigued by the parallels between the behavior of computers and the functioning of the brain and nervous system in higher organisms. Turned down for military duty due to poor eyesight, Wiener and his colleague, Julian Bigelow, worked for the government during World War II, developing **radar** and missile guidance systems using automatic information processing and machine controls.

After the war Wiener continued to work in machine and human feedback research along with another associate, Arturo Rosenblueth (1900-1970), a Mexican physiologist. In 1948, Wiener summarized his findings in *Cybernetics, or Control and Communication in the Animal and Machine*. The word ''cybernetics,'' coined by Wiener, comes from a Greek word *kybernetes*, meaning steersman. Wiener's book, a popular one even outside of the scientific community, enumerated the principles of feedback systems. Wiener continued to lecture and teach the many uses and possibilities for cybernetics, but also warned of the possible dangers in his second book, *The Human Use of Human Beings: Cybernetics and Society*, published in 1950. In this book, Wiener cautioned that an increased reliance on machines might initiate a decline in human intellectual capabilities.

With the advent of the digital **computer**, cybernetic principles such as those illuminated by Wiener could be applied to increasingly complex tasks, resulting in machines with the practical ability to carry out meaningful work. In 1946, Delmar S. Harder devised one of the earliest such systems to automate the manufacture of car engines at the Ford Motor Company. The system involved an element of thinking—the machines regulated themselves, without human supervision, to produce the desired results. Harder's assembly-line **automation** produced one car engine every 14 minutes—compared with the 21 hours it had taken humans previously.

By the 1960s and 1970s, the fields of cybernetics, **robotics** and artificial intelligence began to skyrocket. A large number of industrial and manufacturing plants devised and installed cybernetic systems such as robots in the workplace. In 1980 there were roughly about 5,000 industrial robots in the United States; by the year 2000, researchers estimate there could be half a million.

See also Computer, digital; Computer, industrial use of; Computer operating systems

Cyclamate • See Artificial sweetener

Cyclotron

The first particle accelerators built were linear accelerators (linacs). These machines were constructed on the notion of producing particles in one part of a machine and then using a strong electric field to accelerate them to high energies across the machine. At the opposite end of the machine, the particles were caused to collide with a target.

The first successful cyclotron, built in 1930 by Ernest Lawrence at the University of California, Berkeley. Only 5 in (13 cm) in diameter, it accelerated protons to 80 kV. The protons originated from a radioactive a radioactive source at the center of the cyclotron and then circle round within the hollow chamber.

Linacs possess one inherent disadvantage, however. To increase the energy of particles, the machine must be made longer and longer. For example, in order to build the 2-mile (3.2-km) linac at Stanford University in an exactly straight line, the engineers who designed it had to make sure it did not follow the Earth's curvature.

Scientists were aware of this problem early on and looked for ways to deal with it. The most successful solution was that proposed by **E. O. Lawrence** in the early 1930s. Lawrence suggested that particles be subjected to electric and magnetic fields that would force them to travel in curved paths rather than straight lines.

This type of cyclotron, or *circular accelerator*, consists of two hollow D-shaped compartments with a gap between them. One compartment (called a *dee*) is charged positively and the other, negatively. Particles to be accelerated are introduced in the center of the gap between the two dees. They are attracted to the dee whose charge is opposite their own and are repelled by the dee whose charge is the same as their own.

As the particles move toward and then into the oppositely-charged dee, the charge on the dees is reversed. The particles are now repelled by the dee into which they had traveled and are attracted to the opposite dee. They reverse their path and head back to the opposite dee.

Their path is not, however, a straight, back and forth line between the two dees. Large **magnets** above and below the dees cause the particles to move in a curved path, a spiral. Each time the electric field changes polarity, the particles change direction, pick up energy, and move through the machine in a larger and larger spiral.

At some point, the particles attain maximum energy, reach the circumference of the **cyclotron**, are directed out of the machine, and strike a target.

The first cyclotron was built by Lawrence at the University of California. It was made out of coffee cans, sealing wax, and left-over laboratory equipment. The machine was 4.3 inches (11 cm) in diameter and produced particles with energies of 80 keV (kiloelectron, or thousand-electron, volts).

The next year, Lawrence and M. Stanley Livingston (1905-1986) built a larger machine, 27 inches (69 cm) in diameter that they upgraded to 37 inches (94 cm) in 1937. Within a short time, dozens of cyclotrons were being constructed in research centers all over the world. The two largest of these machines were an 85.8-inch (218-cm) cyclotron at the Oak Ridge National Laboratory and a 88.6-inch (225-cm) machine at the Nobel Institute in Stockholm, Sweden. Both of these cyclotrons could accelerate particles to energies of 22 MeV (megaelectron, or million-electron, volts).

The way to build more powerful cyclotrons would seem to be obvious: simply make them larger and larger. That solution eventually breaks down, however. Albert Einstein (1879-1955) had showed as early as 1905 that mass and energy are closely related. As a particle moves through an accelerator and picks up energy, some of that energy is converted to mass. The more energy the particle receives, the more massive it becomes.

At low velocities, this mass increase is modest. A particle moving at one-fifth the speed of light, for example, gains only two percent over its rest mass. But at velocities that are easily attained in large cyclotrons, the mass increase becomes very large.

The consequence of increasing mass is that particles slow down as they spiral through the machine. With each revolution, they tend to fall a little behind the alternating electric field, which is changing at a regular rate. Eventually, they are completely ''out of synch'' with the electric field, and they become lost within the machine.

The solution to this problem was devised independently by **Vladmir I. Veksler** in what was then the Soviet Union and **Edwin McMillan** at the University of California in the 1940s. The solution is to gradually reduce the speed at which the electric field alternates. When that happens, particles within the machine will automatically adjust to the field and will stay ''in synch'' with it. The term *phase stability* is used to describe the principle discovered by Veksler and McMillan.

The first machine to employ this principle, called a synchrocyclotron, was put into operation at the University of California at Berkeley in November 1946. It was 184 inches (4.67 m) in diameter and soon produced protons with energies of 720 MeV.

Synchrocyclotrons are capable of accelerating protons to energies up to 1 GeV (gigaelectron, or billion-electron, volts). The practical upper limit for such machines is the size and cost of the magnets they require. A synchrocyclotron four times larger than the 5-yd (4.67-m) machine requires magnets fifty times as large to keep particles in their proper path. At this larger size, it becomes less expensive to adopt another cyclotron design than it is to keep making bigger synchrocyclotrons.

Fortunately, a second solution exists for the problem of mass increase at high energies. That solution is the sector-focusing cyclotron. In a sector-focusing cyclotron, it is the magnetic field, rather than the electric field, that is changed. Recall that the path taken by particles in a cyclotron is determined by the magnetic field imposed on them. By increasing the strength of the magnetic field at a regular rate, it is possible to make particles move in smaller circles each time they go around the machine. In this way, they can be kept ''in synch'' with the alternating electric field.

The term *sector-focusing* refers to the technique by which the above process is accomplished. In 1938 Llewellyn H. Thomas, at Ohio State University, showed that the most efficient design for controlling the magnetic field was to divide the two dees into pie-shaped wedges, known as *sectors*. The magnetic fields in each sector alternate between strong and weak.

The next stage in the development of circular accelerators was the betatron, invented by Donald Kerst (1911-), at the University of Illinois, in 1939. The betatron uses a doughnut-shaped hollow ring, rather than dees or pie-shaped sectors. It uses two sets of magnets. The first set acts as do magnets in other accelerators, guiding the path of the particles. The second magnet, placed in the center of the ring, is used to accelerate particles. Because of technical limitations, the betatron is used to accelerate electrons only.

The final stage in the development of Lawrence's original cyclotron concept has been the synchrotron. The synchrotron, like the betatron, consists of a doughnut-shaped hollow ring. Particles are first accelerated by a linear accelerator and then injected into the synchrotron. Once inside the synchrotron, they are given additional energy when they pass through *accelerating cavities* placed at various positions along the ring. The particles' paths are carefully controlled by magnets that are also placed at various locations along the ring. Instead of giant magnets that surround the whole machine, much smaller magnets can be used in specific locations around the ring.

The first electron synchrotron came into full operation at the University of California at Berkeley in 1949. It produced electron beams of 320 MeV energy. The first proton synchrotron was the 3-GeV Cosmotron opened at the Brookhaven National Laboratory in 1960. The largest synchrotron currently in use is the machine at the Fermi National Laboratory near Chicago, Illinois. It produces protons with a maximum energy of 1,000 GeV, or 1 TeV (teraelectron, or trillion-electron volts). Because of this energy output, the Fermi machine is also known as the Tevatron.

See also Nuclear reactor

CYTOSCOPE • See Endoscope

D

Dacron

Dacron is one of the names the Du Pont Company uses for a polymer fiber—polyethylene terephthalate—that **John Rex Whinfield** (1901-1966) and J. T. Dickinson developed in Great Britain in 1941. Whinfield and Dickinson called their new synthetic fiber *Terylene.*

At the time that Whinfield and Dickinson were working on *Terylene,* the race to develop substitutes for natural fibers was on. **Nylon**, invented by **Wallace Hume Carothers** (1896-1937), an American chemist at Du Pont, had been doing a booming business since its introduction in 1935. **Rayon**, invented in 1892, was another synthetic fiber in high demand. Any similar product stood a good chance of being successful.

Despite intensive research at Du Pont, Carothers had found the fibers of most **polyesters** unsuitable for use in **textiles** because of their low melting temperatures. One reaction Carothers and his team had not tried was the one Whinfield and Dickinson used to create polyethylene terephthalate—that is, reacting ethylene glycol with terephthalic acid to produce long polymer chains.

Dacron fibers are made by heating the polymer and forcing it through a metal plate with small holes. The fibers are drawn out and passed through a pair of rollers rotating at different speeds. This orients the molecules into long linear chains and allows cross-linking to occur. Cross-linking connects the linear chains to each other, forming a three-dimensional network that strengthens the macromolecules. The high melting temperature of polyethylene terephthalate, 496°F (256°C), is due to this cross-linking. The discovery of Dacron revolutionized the textile industry. As one of the first artificial fibers, it could be used both in wool blends or by itself. Hundreds of tons of Dacron are produced each year.

And although originally desired as a textile fiber, polyethylene terephthalate is now used in film and tape (virtually all magnetic recording tape is based on this material), as a molding material, and as the matrix for glass-filled composites. One its largest uses is in bottles, especially **soda pop bottles**. It is also used in blends with other polymers, such as polycarbonate.

Recycling of polyethylene terephthalate can be accomplished by chemical depolymerization to polyethylene terephthalate and other products. Such processes are currently in use for products that come in contact with food.

See also Fiber, synthetic; Polymer and polymerization

Daguerre, Louis Jacque Mandé (1789-1851)

French inventor

Born on November 18, 1789, Louis J. M. Daguerre began his career as an artist, serving first as an apprentice to the chief designer of the Paris Opera and later working for nine years as an assistant to the Opera's panorama painter. In 1814 he exhibited his work at the Salon and later worked as an independent stage designer. He cofounded the Diorama, a means of entertainment based on enormous paintings on semi-transparent linen through which light was transmitted and reflected. The lighting could be manipulated to represent changes of seasons or times of day, and the realistic effects captivated audiences.

The simple **camera obscura** (''dark room'') or pinhole camera, through which light falling through a small opening could be directed onto a screen to create a precise image of something outside the box or room, helped spur Daguerre's interest in chemically fixing images. How, he wondered, could the *camera obscura* image be preserved?

In 1826 Daguerre learned that **Joseph-Nicéphore Niepce** also was researching means of chemically fixing images. Daguerre had not yet produced an image, unfixed or otherwise, while Niépce had succeeded in fixing the outlines of some objects on light-sensitive plates. In 1829 the two formed a partnership. Though Niepce died in 1833, Daguerre continued the work they had begun, experimenting with copper plates and

Louis-Jacque-Mandé Daguerre

silver iodide, which he discovered was light-sensitive. Quite by accident he then discovered that mercury vapor could develop images. Daguerre had placed a silver iodide plate in a cabinet containing various chemicals, later discovering a clear picture on the plate; by process of elimination he determined that the miracle he had been seeking was wrought by mercury vapor from a broken **thermometer**.

Exposure time for the photographs was about twenty minutes. In 1837 Daguerre fixed photographs permanently with sodium chloride, and after 1839, using **J. F. W. Herschel**'s discovery, sodium thiosulfate. The process produced a shiny, inverted, and very clear image. By 1839 Daguerre had improved upon Niepce's initial discovery so much that he believed the process should rightly be named the daguerreotype. Daguerre published his findings and gave the rights to the *daguerreotype* to the French government in exchange for annuities to be paid both to himself and to Niepce's son. He was concerned primarily with the recognition of being the inventor, however, as he still agreed that the profits from the new process should be divided equally.

Daguerre also patented his process in England. In 1839 France presented Daguerre's findings to the world. In the same year Daguerre received a reward from the French government and his process was presented to the Institut de France.

The excitement the *daguerreotype* caused has been likened to that generated in 1969 when man first walked on the moon. The *daguerreotype* was lauded throughout the world, particularly in the United States, where it had its greatest and longest-lived success. Daguerre himself produced thirty-two editions, in eight languages, of a manual describing his process. He also gave weekly demonstrations of the process at the Conservatory of Arts and Letters in Paris.

Nonetheless, the *daguerreotype* was not without shortcomings. Though its image was permanent, it was damaged by the slightest friction, since the imprinted layer was very thin. Furthermore, no copies could be made from a *daguerreotype* since the image was comprised of a single direct positive. The equipment required to produce *daguerreotypes* weighed about 110 lb. (50 kg), and all sensitizing and developing operations had to be done on the spot. For these reasons, open-air **photography** was cumbersome and rarely practiced. An additional drawback was the cost: a new metal plate was required for each exposure.

By the 1850s the *daguerreotype* had been improved to its practical limit. In 1851, just four months before Daguerre's death, the wet **collodion** process invented by Frederick Scott Archer (1813-1857) replaced the *daguerreotype* as the most popular and most convenient method of photography. Daguerre died July 12, 1851.

Daimler, Gottlieb (1834-1900)

German engineer

Gottlieb Daimler, the son of a master baker, was born in Schorndorf, Württemburg, on March 17, 1834. His father wanted his son to become a municipal employee, but the young, mechanically inclined Daimler instead apprenticed himself to a gunsmith. After four years of his apprenticeship (during which he attended a technical school), Daimler worked in a **steam-engine** factory and eventually completed his schooling at the Stuttgart Polytechnic. He spent the next three decades working as an engineer and technical director of engine development for several companies. It was during this period that he worked with **Nikolaus August Otto**, the inventor of the four-cycle internal combustion engine, and **Wilhelm Maybach**, who become Daimler's lifelong collaborator.

In 1882 Daimler and Maybach set up a factory to produce a light, high-speed, gasoline-powered **internal combustion engine**. They intended to design an engine that could be used to power a vehicle. Despite early disappointments, Daimler's invention of a reliable self-firing **ignition systems** (a red-hot, porcelain tube that protruded into the cylinder), and Maybach's invention of a device similar to the **carburetor** for their smaller gas engine, helped push them ahead of other inventors who were emerging as competitors.

In an attempt to find a commercial use for his engine, Daimler fitted it to a boat in 1882. Three years later, Daimler and Maybach fitted their engine to a crude wooden **bicycle**—the first **motorcycle**—and drove it on the streets of Mannheim, Baden. In 1887 their sophisticated gasoline engine was adapted to power a four-wheeled vehicle, creating one of the first true automobiles (its unique features included a belt-drive mechanism to turn the wheels, a ''tiller'' for steering, and a four-speed gearbox). At the 1889 Paris Exposition, the Daimler company exhibited a two-cylinder V-shaped engine, perhaps the first engine to use the "V" design.

In 1890 Daimler and Maybach formed the Daimler Motoren Gesellschaft in Stuttgart, but they left the company only a year later in order to concentrate on various technical and commercial development projects.

A Daimler-powered car won the first international car race—the 1894 Paris-to-Rouen race. Of the 102 cars that started the competition, only fifteen completed it, and all finishers were powered by a Daimler engine. The race focused attention on the Daimler Motor Company and helped to promote the concept of motoring in general. The success of the Paris-to-Rouen race may also have been a factor in Daimler's and Maybach's decision to rejoin the Daimler Motor Company in 1895. In the following year, the Daimler company produced the first road **truck**, and in 1900 the company produced the first Mercedes automobile (named for the daughter of the financier backing Daimler). Daimler did not live to see this first Mercedes; he died of heart disease on March 6, 1900, in Stuttgart, Germany, after a lifetime as an inventor in the forefront of automobile development. Daimler's auto company merged with the Benz Company (also of Germany) in 1926, forming the Mercedes-Benz **automobile** company.

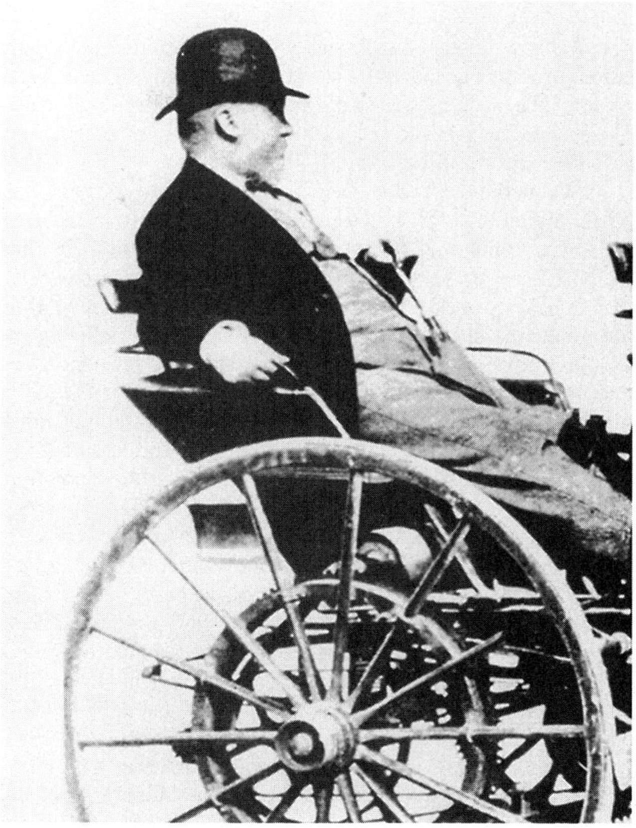
Gottlieb Daimler

Dalén, Nils (1869-1937)
Swedish engineer

As an agriculture student, Dalén appeared ready to follow his father into farming. However, at school he designed and constructed several improved variations of farm equipment, prompting a businessman to encourage him to pursue a technical education. After graduating with a degree in mechanical engineering, Dalén joined the Swedish Carbide and Acetylene Company as technical chief in 1901. This company was working on a way to improve a device called a gas accumulator, used to produce a safe and effective fuel for the illumination of lighthouses and buoys. It was comprised of acetylene (a hydrocarbon gas that burns with an extremely bright, white light), acetone (a highly flammable organic liquid), a porous mass used to absorb the solution of acetylene and acetone, and a metallic container. The problem was that no porous mass had yet been manufactured that adequately absorbed the acetylene gas. For that reason the gas tended to accumulate, creating the possibility of an explosion. Dalén solved the problem when he built a better gas accumulator. He produced a substance he called aga, put it in a steel container half-filled with acetone, and forced acetylene into the container under pressure of ten atmospheres. The risk of explosion was eliminated. He then invented a regulator to control the pressure of the gas inside the container. In 1905 he designed a reliable device that increased the number of brief flashes emitted by a **lighthouse**. In 1907 he created a special **valve** that prevented the gas accumulator from providing gas during bright days when the flashes were not necessary, thus saving a large amount of acetylene and acetone. Dalén was blinded by an explosion in 1912 while conducting an experiment with acetylene. He was awarded the Nobel prize for physics the same year for his aga lighting, which was not only used in lighthouses but also for **trains, automobile** headlights, welding, smelting, and cutting metals. He also invented an extremely efficient **stove**. In spite of his blindness Dalén continued to conduct research until his death.

Damadian, Raymond Vahan (1936-)
American physicist and entrepreneur

Raymond Damadian was the first to take a nuclear **magnetic resonance image (MIR)** image of a human body, and went on to develop the MRI as an indispensable tool for medicine.

At the age of 16 Damadian entered the University of Wisconsin, studying mathematics. He then went to medical school, believing it would afford him greater options than a traditional science education. He earned a medical degree at the Albert Einstein College of Medicine in Bronx, New York, and undertook postgraduate studies in nephrology (the study of the kidneys), biophysics, and physiological chemistry. In 1967 Damadian began a professorship at the State University of New York's Downstate Medical Center.

Damadian began investigating **sodium** and **potassium** in living **cells**, and was able to borrow time on a **nuclear magnetic resonance** (NMR) machine for his research. Nuclear magnetic resonances were discovered in 1938 by physicist Isidor I. Rabi. When nuclei are placed in a strong, unchanging magnetic field, they will absorb **electromagnetic waves** of radio frequency in a way characteristic of each nucleus. They then re-emit waves at these frequencies, and the resulting spectrum is a clear indicator of the nucleus's identity. NMR provides a nondestructive, nonradioactive technique that can be used to measure the radiation spectra of cells and living organisms.

Damadian asked a crucial question: how would the NMR signal change between healthy cells and cancerous cells? Experimenting on mice, Damadian found in 1971 that the NMR signals persisted for much longer in cancerous cells than in healthy ones. The next year he filed a patent application for a machine that would use magnetic fields and radio waves to allow physicians to screen the human body for cancerous tumors. The patent was awarded to Damadian in 1974.

Damadian had grand claims for his machine, but most of his colleagues thought it to be outlandish, and he had difficulty obtaining research funds to build it. After a disappointing rejection from the National Institute of Health, Damadian bypassed bureaucratic channels and wrote a letter directly to President Richard M. Nixon. Damadian's letter received attention, and he subsequently received a modest grant. When those funds disappeared Damadian was able to find private backers. A crucial contribution was made by Paul C. Lauterbur, who in 1971 conceived the idea of using a changing magnetic field to produce information on the location of nuclei, allowing NMR information to be displayed as a pictorial map. In 1977 Damadian's group's first machine designed to make a magnetic resonate scan and viewable image of the human chest, *Indomitable*, was finished. On July 3, 1977 one of Damadian's skinnier graduate students, Larry Minkoff, sat in the wrap-around machine for several hours, and the first crude images were obtained, showing Minkoff's heart, lungs, and chest wall. Damadian's machine began to receive popular attention.

Damadian began his own company, FONAR (Field fOcused Nuclear mAgnetic Resonance), to build MRI machines, and in 1980 the first commercial scanner was sold. The Food and Drug Administration approved the device in 1984. MRI imaging developed to its current point where doctors can see the interior of the body in minute detail, including not just tumors but many other small structures that do not register on x-rays.

Indomitable now sits in the Smithsonian Institute. Damadian and Lauterbur were awarded the National Medal of Technology in 1988, and Damadian was inducted into the National Inventors Hall of Fame in 1989.

Dam and dike

During the early stages of civilization—when humans followed food resources, hunting and foraging in seasonal movements—the main concern about water was *finding* it. When people started building permanent settlements, it became necessary in many places to *manage* water resources.

Management systems can be divided into two basic types: systems built to ensure a sufficient and continuous water supply and systems built to protect against flooding—roughly, dams and dikes. Many cultures, including the Mayans and Southeast Asians, have impounded, diverted and transported water for agricultural and urban needs. Ancient ruins in the Middle East attest to early efforts in water resources management on a grand scale. For millenia, the Egyptians maintained extensive **irrigation** systems consisting of low earthen dikes along the Nile to control the water level for agricultural purposes. Large dam projects were also carried out in Egypt, though not always successfully. Some time during the Old Kingdom, a dam 361 feet (110 m) wide and 39 feet (12 m) high was built near present Helwan to impound the flash floods coming down the wadi in winter. Apparently, the barrage collapsed the first winter it was in use, but the abutments of the dam still remain. The Orontes dam, 500 feet (150 m) long and 20 feet (6 m) high, was built in 1300 B.C. with greater success. Both the dam and the 3-by-6-mile (5-by-10 km) lake it created still exist. Remains of similar age are also to be found in the Tigris river valley. Water diversion for agriculture and direct consumption also formed a major component of the infrastructure of the Roman Empire. A number of dams constructed in that era in Italy, Spain and North Africa are still in use. The next big project that has survived the test of time is the Tibi dam near Alicante in Spain, which was completed in 1594 and provides water for irrigation to this day.

A little earlier, in the second half of the fifteenth century, a major land reclamation project was completed in Holland. Although different societies, including the Dutch, had built dike systems in river areas to contain seasonal rises in water levels and had diked in tidal areas to keep out seawater at high tide, the Holland project gave a new dimension to land reclamation. For the first time, land below sea level was reclaimed, requiring pumping systems to be in constant operation after the so-called polders were created. For centuries, **windmills** maintained groundwater in these polders at levels suitable for agriculture, until they were replaced by **steam engine**s.

With the advent of industrialization dam building once again became a major focus of interest in the Western world. The growth of urban centers overtaxed the naturally available supply of water in many places. Great Britain was the first country to start building great numbers of dams in the nineteenth century. Unfortunately, these efforts proceeded largely by trial and error, causing catastrophic flooding on a number of occasions. Most of these early nineteenth-century installations were earthen or rockfill dams that did not provide watertight barriers. The leakage would undermine the stability of the dam, and the weight of the water bearing down on it would eventually cause a collapse. It is estimated that about 150,000 dams across the world pose a potential hazard to lives or property.

There have been about 200 dam failures since 1900. One of the worst took place in Pennsylvania, near Johnstown, in 1889, when a dam built to impound water for the Pennsylvania Canal failed, sending a 70-foot (21 m) wall of water over Johnstown and seven other towns. About 2,200 people died in the disaster. With these failures in mind, the French, taking a more scientific approach, started developing masonry dams of different designs that proved much more reliable and laid the foundation of modern dam design.

The development of technology to generate hydroelectric power gave new impetus to dam projects, stimulating such gigantic multi-purpose achievements as the Hoover Dam in the Colorado river in Nevada. The Hoover Dam, which was finished in 1936, impounds water for irrigation purposes and supply and can generate 1.8 million horsepower per day. At 725 ft (221 m) it is the 17th highest dam in the world. Another modern giant is the High Dam at Aswan, Egypt, which was built with aid, not always very efficient, of the Soviet Union. The High Dam can impound 212 billion cubic yards (162 billion cubic meters) of water in a lake 310 miles (500 km) long and on average 6 miles (10 k) wide. With the exception of the Aswan High Dam, which is an earth and rockfill barrage, the modern dams are more like buildings, with galleries at different heights to enable continuous monitoring of the performance and stability of the structure. China has by far the largest number of dams, over 18,000 that are at least 50 ft (15 m) high. The United States has about 5,300.

While flood prevention and irrigation were often the reasons behind the construction of many early dams, the development of hydroelectricity factored into the building of massive projects, such as the Hoover Dam, in the twentieth century.

Daniell, John Frederic (1790-1845)
English chemist

John Frederic Daniell was a scientist and inventor whose widespread interests reflected the relatively unified nature of the science of his day.

His contributions to technology started with the introduction of improvements to the sugar refining industry when he was a young man. Later, he brought improvements in the lighting industry to Europe and America by producing a new gas from a distillate of resin in turpentine.

In 1820, he made a contribution to meteorology by inventing the dew-point **hygrometer**, an instrument designed to measure relative humidity. It consisted of two thin glass bulbs suspended from a base and connected with a glass tube. One bulb contained liquid ether and a **thermometer**. As the other bulb was slowly cooled and reheated, dew would appear and disappear at the end with the thermometer The mean temperature of this action was taken as the dew point. A description of this was published in 1820.

He was rewarded by horticulturalists with a medal for his suggestion that the humidity and temperature in hothouses should be regulated. In 1836 he invented a new **battery**, the Daniell cell. Unlike the recently invented zinc-copper voltaic battery, its current did not decline rapidly. By introducing a barrier between the zinc and copper, he was able to stop the formation of hydrogen, which was impairing battery function.

Daniell was active as a teacher, writer, and illustrator, and made himself known as a social philosopher. His early death occurred while attending a Royal Society meeting in London in 1845.

Darby, Abraham (ca. 1678-1717)
English engineer

The work of Abraham Darby is considered by many to be the cornerstone of the Industrial Revolution. His development of the coke-burning blast furnace in 1709 made possible the mass production of commercial grade iron. It, along with later improvements made by others, led to the evolution of the iron and steel industries, and the many industries they spawned in turn—**aircraft**, automobiles, shipbuilding, and construction.

Darby was born near Dudley, Worcestershire, England in 1678. The iron industry at this time was hampered by its inability to produce sustained amounts of heat at high temperatures necessary for continuous smelting operations. In the seventeenth century, charcoal was the leading fuel for stoking (feeding) furnaces. As demand for iron grew, so did the demand for charcoal. This drove the price for charcoal higher. Also, soft charcoal was incapable of physically supporting large amounts of iron ore inside the furnaces.

Darby had been employed in the copper-smelting industry in Bristol, England, where coke was used as fuel. Coke is a derivative of coal, produced by heating the coal and removing the sulfur and combustible impurities. Coke delivers a hotter, more sustained heat without flame. Coke became central to Darby's smelting process. When he established his Bristol Iron Works Company in 1708, he wisely chose the village of Coalbrookdale, in the upper Severn River valley in the west of England, where coal and coke were readily available.

Darby's first iron products were primarily small implements and cooking utensils. His business was greatly bolstered by an order from Thomas Newcomen for 6t (5.4t) cylinders for his steam powered mine-pumping engines. The engines, in turn, proved to be useful in the **blast furnace** industry.

Darby managed to keep the coking process within his family. His son, Abraham Darby II, continued making the

Newcomen cylinders well after Darby's death on March 8, 1717. By 1758, 100 of the cylinders had been delivered by the Darby foundry.

Darby's grandson, Abraham Darby III, incorporated iron in construction when he collaborated with architect **John Wilkinson** on the Severn River bridge at Iron Bridge in 1779. Well after Abraham III's death in 1791, the Darby foundry was commissioned by **Richard Trevithick** in 1802 to produce the first **locomotive** engine, which required a high-pressure steam boiler.

The area around Coalbrookdale grew into an iron-producing district. In time, however, it became a victim of its own success, succumbing to the depletion of its coal reserves, pollution, and changing markets; however, the iron and steel-making industries of North America, and Asia can trace their origins to Coalbrookdale and Abraham Darby.

See also Cast-iron and wrought-iron construction; Iron production; Steel production

Davenport, Thomas (1802-1851)

American inventor

Born to a poor family on July 9, 1802, at Williamstown, Vermont, Thomas Davenport was the fourth of eleven children. He received only three years of schooling and was apprenticed to a blacksmith when he was fourteen. The hard work was agreeable to the boy, and seven years later, he set up his own blacksmith business in Brandon, Vermont.

In 1831 Davenport went to the Penfield Iron Works, where **Joseph Henry** had installed the first commercial **electromagnet**. Davenport watched as the electromagnet lifted 750 lbs. (340 kg) of iron. Amazed and intrigued, he decided to acquire an electromagnet of his own and experiment. In order to raise money for the purchase, Davenport sold his brother's horse. Once he had the magnet in hand to study, he was able to build a larger version on his own, using his wife's silk wedding dress to insulate the wires.

Davenport eventually ended up with four electromagnets and conceived a way to convert the electromagnetic force into mechanical power. In 1834 he attached two electromagnets to a pivot, mounting that between the other two magnets which were fixed in place as poles. He attached a **battery** to the magnets via a "commutator" (a crude switching device), threw the switch, and the pivot rotated. Davenport had just invented the **electric motor**.

Joseph Henry had done much the same thing a few years earlier, but his effort was seen as more of a curiosity, because it had an inefficient oscillatory movement that was not very practical. Davenport's motor, however, had a rotary movement, and it became the prototype for every electric motor in use today. Realizing the importance of his invention, Davenport quit blacksmithing to promote his device.

However, Davenport waited until 1836 before deciding to patent his invention, and ran into nothing but trouble. He decided to personally make a trip to the Patent Office in Washington D.C., and set off on foot. By the time he made it to Washington, he had spent all the money for the patent application, so back to Vermont he went. He came up with more money and this time decided to mail the application. But then the Patent Office was destroyed by a fire.

Eventually Davenport received his patent, on February 5, 1837, though renown and financial reward never followed. He built several miniature motor-driven models, including a working electric trolley car, but most people treated the devices as mere curiosities. The probable reason why Davenport's inventions failed to receive enthusiastic acclaim was that the motor upon which his devices depended was heavy, fragile, and expensive. In addition to his trolley, he invented an electric railway, electric printing press (upon which he printed his newsletter *Electro-Magnet and Mechanics Intelligencer*), **electric telegraph**, and electric piano.

On July 6, 1851, three days short of his 49th birthday, Davenport died in Salisbury, Vermont. His sons claimed he had died from a broken heart after **Samuel Morse** was given credit for inventing the telegraph.

Davidson, Shelby (1868-1931)

American inventor

Shelby Davidson was born in Lexington, Kentucky, on May 10, 1868. He graduated from Howard University in 1893 with a Bachelor of Arts degree. Davidson went on to practice law, but is most remembered as an inventor.

Davidson journeyed to Washington, D.C. in 1887. Working in the auditing department of the United States Post Office, the young African-American became interested in finding a more efficient way to handle government auditing functions. In 1908, after studying adding machines for two years, Davidson found his answer in the form of his first invention: a rewind device for adding machines. Davidson was convinced his device would greatly improve efficiency by reducing both **paper** usage and the time clerks spent in paperwork. The government agreed with him and officially began utilizing the invention that same year.

Davidson was also keenly interested in mechanical tabulation and in 1911 patented a device designed to help postal clerks. This "automatic fee device," which Davidson continued to improve, paved the way for more efficient processing of postal fees.

See also Burroughs, William; Calculating machine

Davies-Hinton test • See Syphilis test

da Vinci, Leonardo • See Leonardo da Vinci

Davy, Humphry (1778-1829)

English chemist

Humphry Davy grew up poor, helping his mother pay off debts left by his father, a woodcarver who had lost his earnings in speculative investments. As a result, Davy's education was haphazard, and he disliked being a student. The schools in his part of the country (Cornwall, the southwest tip of England) were far from outstanding at that time; still, Davy managed to absorb knowledge of classic literature and science. In later life, he said that he was happy he didn't have to study too hard in school so that he had more time to think on his own. Without money for further education, Davy began at age seventeen to serve as an apprentice to a pharmacist/surgeon. During this time, he took it upon himself to learn more about whatever interested him, such as geography, languages, and philosophy, as well as science. When he was nineteen, Davy read a book on chemistry by the French scientist Antoine-Laurent Lavoisier (1743–1794) that convinced him to concentrate on that subject. For the rest of his life, Davy's career was marked by brilliant, if impetuous, scientific explorations in chemistry and electrochemistry that led to the inventions for which he is known today.

Davy's style in the laboratory was to work quickly and intensely, pursuing one new idea after another. He aimed at originality and creativity, rather than tediously repeating tests and confirming results. Stimulated by the Italian physicist **Alessandro Volta**'s invention of the electric **battery**, Davy rushed into the new field of electrochemistry and, in 1808, invented the carbon **arc lamp**. Although scientists knew that sparks were created between electrodes in a battery, Davy proposed using carbon as the electrode material instead of metal. With carbon electrodes, Davy made a strong electric current leap from one electrode to the other, creating an intense white light that was practical for illumination. Davy's invention thus gave birth to the entire science of electric lighting. Arc lamps are still used today; so is arc welding, another practical outcome of Davy's electric arc research. Luckily for Davy's reputation, the arc lamp made a perfect subject for public display. Davy had been hired in the 1800s to lecture for the Royal Institution, a new scientific institution that was having financial problems. Davy's charm as a speaker, along with his spectacular demonstrations of electric arc lighting, drew enthusiastic crowds from London's high society and soon reversed the institution's fortunes. (Some historians have also referred to Davy's good looks, which probably contributed to his popularity with the fashionable women in the audience.)

In his early thirties, after being knighted in 1812, Davy married a wealthy Scottish widow and began to travel extensively, enjoying his fame wherever he went. He was accompanied on some of these tours by his assistant/valet **Michael Faraday**, who was destined to eclipse his mentor's reputation in the realm of science. Upon his return to England, Davy was called upon to study coal-mine explosions, which in those days were killing hundreds of miners each year. In less than three months, he invented the miner's safety lamp, also called the *Davy lamp*. When Davy tested samples of the "fire-damp" gas that caused the explosions, he confirmed that it was mainly methane and that it would ignite only at high temperatures. In Davy's safety lamp, the flame is surrounded by wire gauze to dissipate heat and prevent ignition of flammable gases. This invention was the first major step toward safety in the coal mining industry.

Davy's career, which included several discoveries as well as inventions, was rewarded by many honors and medals. In addition to his knighthood, he was made a baronet in 1818 and was elected president of the prestigious Royal Society in 1820. In his conflicts with other scientists, however, Davy made some enemies who thought he was arrogant, as he well may have been. He even tried to prevent his associate Faraday from being elected to the Royal Society. While still in his thirties, Davy began to be plagued by ill health. The same curiosity that drove him to discover and invent with such success had also taken its toll on his body. By sniffing and tasting unknown chemicals, he had poisoned his system, and his eyes had been damaged in a laboratory explosion. Although Davy continued to pursue scientific interests, he suffered a stroke when he was only forty-nine and died abroad just two years later. One of Davy's last inventions was a method for protecting metal from corrosion. Called cathodic protection, it was used to prevent the corrosion of copper-bottomed ships by seawater. Although only partially successful, Davy's method represents the first application of cathodic protection in scientific history. Today similar methods are used to protect metal pipelines and other equipment from corrosion.

DDT

DDT (*dichloro-diphenyl-trichloroethane*) is perhaps the most recognized of all **insecticide**s because it revealed the many hazards associated with using synthetic pesticides. This colorless, odorless, insoluble toxic pesticide contains up to fourteen chemical compounds and is known for its ability to eradicate pesky insects such as flies, lice, mosquitoes, and agricultural pests. Although first synthesized in 1874 by German chemist Othmar Zeidler, DDT was not used as an insecticide until 1939, when Swiss scientist Paul Hermann Müller (1899-1965) discovered its insect-killing properties. The benefits of DDT were demonstrated in the 1940s, however, when it was used in World War II to clear out mosquito-infested areas prior to invasion. Even after the war, the use of DDT in the United States almost completely wiped out malaria and yellow fever. And in tropical areas, the use of DDT has helped save millions of lives that would otherwise have been lost to disease. DDT was also routinely applied as a crop dust or water spray on orchards, gardens, fields, and forests. At one point it was registered for use on 334 agricultural crops.

Unfortunately, DDT is too durable; in some applications, it is effective for up to 12 years. Water cannot wash it away and it resists breakdown by light and air. This strength and persistence has resulted in DDT's transfer to non-target living organisms. Once in an ecosystem, it can pass on from crops to birds and from water to fish, eventually affecting the whole food chain. When ingested by humans, DDT is stored in body fats and can be passed on to nursing babies. Low levels of DDT in humans are harmless but large concentrations can cause severe health problems such as liver cancer. When applied to an insect, DDT is easily absorbed through the body surface and after attacking the nervous system, causes paralysis. Some insects, however, have a resistance to DDT, thereby making the insecticide ineffective. These resistant insects are able to reproduce and pass this trait on to their offspring. Many problems arise when larger animals are exposed to DDT or eat smaller animals that have ingested the toxin. For example, while DDT is more toxic to fish than birds, it still causes widespread bird deaths. With high levels of exposure, DDT causes convulsions and paralyzes the birds' nerve centers. In smaller concentrations, it can weaken their egg shells and can cause sharp declines in the species' reproductive rate. DDT ingestion by peregrine falcons is thought to have caused their almost complete extinction in most regions of the United States. Evidence documenting DDT's adverse impact on the environment and human health was detailed in Rachel Carson's landmark book, *Silent Spring* (1962) which exposed the dangers of unregulated pesticide use. Spurred by public pressure, state and federal governments turned their attention to the regulation of pesticides, and, in 1972, the United States Environmental Protection Agency banned the use of DDT. Today, DDT is restricted in the United States, Europe, and Japan. However, many other countries still use DDT widely for malaria control, delousing, and the eradication of other disease-spreading insects.

The debate on the adverse health effects of DDT continues today. A 1997 report in the New England Journal of Medicine indicates that, despite all previous studies to the contrary, DDT and its derivative DDE can not be conclusively linked to breast cancer.

DEADBOLT • See Lock and key

DECOMPRESSION, STAGE • See Scuba diving

DEERE, JOHN (1804-1886)

American inventor and manufacturer

Wooden **plows** were no match for the sticky, thick clay soils encountered by settlers of the Central Plains states of Wisconsin, Illinois, and Indiana. Even the iron plows invented by Charles Newbold of New Jersey in 1703 and by Jethro Wood of Massachusetts in 1819 caused the soil to bunch up in large sections rather than scour into neat furrows.

Deere developed a steel plow that could cut through the plains mud with the speed and efficiency demanded by the farmers. Deere's implements played a major role in the transformation of the continent's wild prairies into fertile farmland.

Deere was born in Rutland, Vermont. At the age of seventeen, he began a four-year blacksmith apprenticeship in

Before the dangers of DDT were known, crops and people alike were sprayed with the chemical to prevent bothersome insects.

Middlebury, Vermont. After mastering his trade, he spent twelve years working as a blacksmith in various towns throughout Vermont.

In 1837 Deere moved to the small northern Illinois town of Grand Detour where he became aware of the iron plow's inability to penetrate the clay soil. Fashioning a plow from a discarded steel circular saw blade, he was able to successfully plow a dozen rows nonstop. The blade was curved to one side, eliminating the need to push the dirt aside by hand. Establishing a partnership with Leonard Andrus, Deere built three steel-plows in 1838. By 1842 over 100 plows were being made.

In 1846 the Pittsburgh firm of Jones and Quigg shipped a large quantity of steel to Deere and Andrus. That year they produced a total of 1,000 plows. Deere also contracted with the English firm of Naylor and Company for the finest steel. Although the shipment of steel arrived pitted from exposure to the salty sea air, Deere was able to manufacture fifty plows from it in 1847.

It was about this time that Deere decided that Grand Detour was not ideally situated for a growing business. He sold his interest in the partnership to Andrus and moved to Moline, Illinois, which had the advantage of access to major rail and river routes. It was here that he organized his own company.

By 1857 Deere's company had produced over 10,000 plows, most of which were being carried by nearly every covered wagon that headed across the western prairie. Deere's plow became known as the "singing plow" because of the distinctive humming sound it made while sliding through the dirt. In 1858 Deere made his son, Charles Deere, his partner, and in 1868 the company became Deere and Company.

Deere and Company steadily expanded its output and diversified into production of **cultivators**, hay balers, and other agricultural equipment. The company is especially well known for its line of **tractors**. Also recognized for its generosity, the company helped numerous farmers during the Great Depression of the 1930s by extending credit and forgiving debts.

See also Haying machines and tools; Steel production

Defibrillator and cardioverter

Ventricular fibrillation is a state of cardiac arrhythmia in which the individual heart muscles contract in a random, uncoordinated way. The heart appears to shiver, and blood circulation ceases. Ventricular fibrillation is fatal unless an electric shock is applied within minutes to restore normal heart contraction. As early as 1899, French physiologists, Jean Louis Prevost and Frederic Battelli were able to stop ventricular fibrillation in a dog by applying an electric shock to the animal's exposed heart. In 1930, William B. Kouwenhoven, an American electrical engineer at Johns Hopkins University, developed with colleagues a closed-chest defibrillator that sent **alternating current** (AC) electrical shocks to the heart through electrodes placed on a dog's chest. In 1947, Claude Beck, professor of surgery at Case Western Reserve University, first successfully resuscitated a human patient by internal cardiac massage and electrical defibrillation; American cardiologist Paul Zoll applied AC defibrillator to human patients in 1961; and the **direct current** (DC) defibrillator introduced by Lown and Neuman in 1962 provided greater reliability and safety.

Defibrillators greatly improved the ability of patients to survive heart surgery, invasive cardiac diagnostic and treatment techniques, and heart attacks, all of which can send the heart into ventricular fibrillation. Since the 1970s, most hospital emergency rooms have been equipped with electric defibrillators, and portable devices have becoming standard equipment for **ambulances**. Most recently, automated external defibrillators (AEDs), light-weight, portable, user-friendly devices about the size of a lunch box, can be found at sports stadiums, hotels, and casinos, in coast guard marine vehicles and police cars, and many other places where people gather. These devices can be used by operators with much less training than paramedics, and are intended for use by on-site personnel like fire fighters until paramedics arrive. Sometimes called the "smart" defibrillator, they provide audio instructions and visual prompts to walk the operator through the defibrillating process, as well as recording the sequences of events such as when the operator connected the analyzer and pushed the shock button.

An implantable device, called the *automatic implantable cardioverter defibrillator* (AICD) was invented by Mieczyslaw Mirowski of the Johns Hopkins University Medical School to stop heart arrhythmias. Approved for use by the Food and Drug Administration in October 1985, Mirowski's AICD senses two kinds of abnormal heart rhythms—ventricular fibrillation and ventricular tachycardia—automatically sending an electric shock to the heart to correct the disturbance. As a defibrillator, the device jolts the heart out of ventricular fibrillation; as a cardioverter, it shocks the heart out of an abnormally fast heartbeat called ventricular tachycardia, restoring normal heart rate. Because the AICD requires a heftier power pack than the standard cardiac **pacemaker**, the **battery** pack for the AICD is separately implanted in the patient's abdomen. The lithium batteries can deliver 100 to 150 shocks during their three-year lifetime. The AICD is a potential lifesaver for the 700,000 people in the United States who survive heart attacks each year and are therefore at risk for potentially fatal arrhythmias. The AICD is also routinely used for patients whose arrhythmias cannot be treated with medication or surgery.

De Forest, Lee (1873-1961)

American inventor

Like many inventors, Lee De Forest was a brilliant innovator but a poor businessman; he spent much of his life fighting lawsuits and lost as many fortunes as he made. A prolific inventor, De Forest had more than 300 patents to his credit. He worked in a host of areas including **radar**, diathermy (application of **alternating current** to the body in medicine), **telephone**, and color **television**. But his most important achievement was his audion, an electronic device that helped launch a century of innovation in electronics.

De Forest was born at Council Bluffs, Iowa, on August 26, 1873. His father was a minister whose interest in astrono-

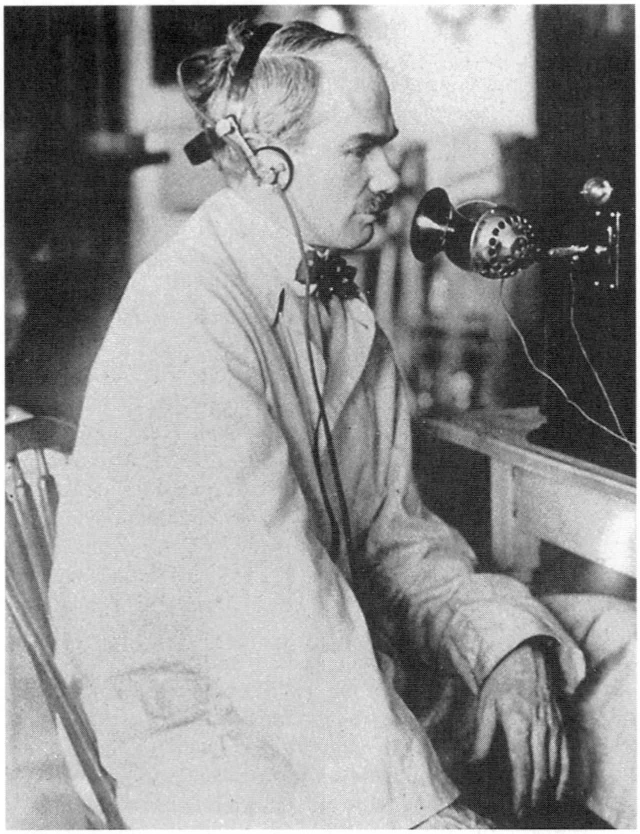

Lee De Forest

my may have sparked his son's love for science. De Forest attended Yale University, studying under Josiah Willard Gibbs (1839-1903), a pioneer in the disciplines of physical chemistry and advanced thermodynamics. De Forest's 1899 Ph.D. dissertation, on the reflection of ''Hertzian'' (or radio) waves, was possibly the first scholarly treatise in the United States to deal with Heinrich Hertz's (1857-1894) discovery.

When he heard of **Guglielmo Marconi**'s early success with wireless, or **radio**, De Forest began building his own wireless set ups. De Forest's first major invention was the responder, which he introduced in 1901. A device for detecting radio waves, the responder was superior to the coherer detectors then in use, in that it contained a liquid electrolyte instead of metal filings like the coherer did. These filings tended to stick together, impairing the coherer's sensitivity. De Forest found a financial backer and established the American De Forest Wireless Telegraph Company. Competing with Marconi, De Forest began selling equipment to the United States Army and Navy, and aggressively marketed his devices to the public by carrying out wireless demonstrations around the country.

In 1903, De Forest created an improved version of his responder which was very similar to a design patented by **Reginald Fessenden**. Fessenden sued, and in 1905 court injunctions were issued to keep De Forest's company from manufacturing the detector. Disillusioned by the defeat and angered by the shady activities of his board of directors (who were embezzling funds), De Forest took his patents and quit the company in 1907.

By this time, De Forest had already made a new invention of far greater importance than his responder. He had discovered that by adding a third element (a wire grid) to the **diode** invented by **John Ambrose Fleming**, he could amplify weak signals far more effectively than available detectors could. He obtained a patent for his device, which he called an *audion*, in 1906.

In order to make the first wireless voice transmission on Christmas Day, 1906, Reginald Fessenden had to have specialized equipment constructed, but De Forest's audion or triode soon made radio voice communication commonplace. The audion could perform three functions which would revolutionize electronics. First, it could amplify weak signals including radio waves, telephone signals, or any signal that required boosting. Second, it could *modulate* a signal—that is, it could impose voice, music or other sounds upon a radio ''carrier'' wave for broadcast. Finally, when overloaded with a signal, the audion became an *oscillator* with the ability to generate radio waves, though De Forest did not make this discovery until 1912. Until **Walter H. Brattain**, **John Bardeen**, and **William B. Shockley** invented the **transistor** in 1948, De Forest's triode would remain the most important component in virtually any electronic device.

The success of the audion enabled De Forest to raise capital for another company he called the De Forest Radio Telephone Company, and again he began staging broadcasts around the country. On January 20, 1910, in a spectacular demonstration, he used his broadcasting system to transmit the singing of Enrico Caruso to listeners on the east coast of the United States. He had won more Navy contracts and appeared to be flourishing at last. But once more, De Forest found himself in trouble when he and a number of company officials were arrested for mail fraud. Apparently, his backers were again involved in fraudulent activity, and prosecutors, bent on putting a stop to an epidemic of swindling in the radio industry, charged De Forest and his associates with trying to exploit ''a strange device like an electric lamp (the audion), which... had proven to be useless.'' While out on bail, a desperate De Forest offered to sell the broadcating rights to his audion to the American Telephone and Telegraph Company. Feigning disinterest, AT&T sent an anonymous agent to offer De Forest a paltry $50,000 for the device, swearing ''on his word of honor as a gentleman'' that he did not work for AT&T. De Forest accepted. Shortly after, he made up some of his losses when he sold AT&T the radio-signalling rights to the audion for a better price of $340,000. In 1913, De Forest was acquitted of fraud, but the president of his company was found guilty and sentenced to prison.

De Forest continued to find new ways to use the audion. In the 1910s, he discovered that he could increase the sensitivity of his audion detector by sending back some energy from the diode circuit into the grid, creating a feedback loop. When he applied for a patent for this ''regenerative'' circuit, he promptly embroiled himself in yet another lawsuit, this time with **Edwin H. Armstrong**, **Irving Langmuir** at General Electric, and Alexander Meissner in Germany. Armstrong had been issued a patent for such a circuit in 1914 and won initially, but after a 20 year battle, De Forest was victorious in his appeal to the Supreme Court, though most radio engineers believe that the idea belonged to Armstrong.

This was enough for De Forest. In 1919 he stopped working on radio and took up the challenge of synchronizing sound to **motion pictures**. He devised an electrical-optical method of directly recording a sound track on film and, on April 12, 1923, gave a demonstration of a talking motion picture. Incredibly, movie executives expressed indifference to this innovation, thinking that public interest was lacking. When "talkies" became an instant sensation years later, De Forest gained nothing.

De Forest died on June 30, 1961, in Hollywood, California. He had received his last patent, for an automatic telephone dialing device, at the age of 84. Despite the misfortunes that De Forest's work in wireless brought him, he always expressed a special fondness for radio, which he considered his "child." Though his reputation as an originator seems to have diminished over time, his audion undoubtedly was more important in the development of electronics over the first half of the twentieth century than any other single invention.

See also Amplitude modulation

Dehmelt, Hans (1922-)

German American physicist

Hans Dehmelt has devoted his career as a physicist to developing means of performing highly precise, accurate atomic measurements. In pursuit of this goal, Dehmelt devised a trap that for the first time isolated a single electron, which in turn made possible the precise experimental measurements that verified the theory of quantum electrodynamics. For his work, Dehmelt was honored with a share in the 1989 Nobel Prize in physics.

The son of Georg Karl and Asta Ella Klemmt Dehmelt, Hans Georg Dehmelt was born on September 9, 1922, in Görlitz, Germany. An early interest in the workings of **radios** led Dehmelt to the study of physics, which was interrupted by service in the German army and a period as a prisoner of war in France during World War II. After the war, Dehmelt continued his studies in physics at the University of Göttingen, financing himself in part by repairing broken radios. During one college course, Dehmelt became interested in the possibility of isolating the electron, something theorists said couldn't be done.

Dehmelt earned his Ph.D. from Göttingen in 1950 and stayed on for postdoctoral work until 1952, concentrating on nuclear magnetic resonance, and using his skill in electronics to make up for a postwar lack of conventional experimental equipment. He continued these studies from 1952 to 1955 at Duke University in the United States, and became a professor at the University of Washington in 1955, remaining in that position into the 1990s. While at Washington, Dehmelt returned to his original interest in isolating the electron in order to accurately measure it. Although the electron's mass and charge could be measured precisely with existing methods, its spin and magnetism could not be. Quantum electrodynamics (QED) did predict the electron's spin and magnetism, as part of QED's explanation of all interactions between electrically charged particles, but until spin and magnetism were actually measured on existing electrons, QED remained an unproven theory.

In 1959, Dehmelt's electronics background once again came into play. He used an electronic component called a Penning discharge tube to build what he called a Penning trap: a **vacuum tube** with a strong magnetic field and a weak electric field that trapped and held electrons. Throughout the 1960s Dehmelt and his colleagues tried to devise ways of measuring the gyromagnetic ratio, or *g*-factor, of an electron, which had been calculated theoretically via quantum mechanics. Dehmelt and his team found that by manipulating voltage, they could drive electrons out of the trap until only a single one remained. This breakthrough was accomplished in 1973, and soon the team was able to capture and hold single electrons for extended periods of time. Two years later, Dehmelt worked out a way to cool the trapped electron, which slowed its motion and made accurate measurement easier. In 1976 Dehmelt and his coworkers used their trap to observe the quantum jump of a single ion, again confirming a prediction of quantum mechanics.

Soon Dehmelt and his team were able to measure the *g*-factor to ten times its previously calculated value. Driven by his goal for ever more accurate atomic measurements, Dehmelt applied a number of refinements to his trap so the *g*-factor could be measured to 12 decimal places by 1991. Dehmelt and his colleagues have also trapped positrons (antielectrons). One of these positrons, which he named Priscilla, was held for three months, performing its quantum jumps and thus allowing the first experimental measurement of the *g*-factor for antimatter. Dehmelt also initiated and guided an experiment at the University of Heidelberg in 1974 to trap and observe a single barium ion. This eventually produced the first photograph, in black and white, of a charged atom. With his long-time collaborator Robert Van Dyck, Dehmelt later was able to take a color photograph of a blue barium ion they named Astrid.

For his particle-trapping work, Dehmelt shared half of the 1989 Nobel Prize in physics with **Wolfgang Pauli** of the University of Bonn, who also devised an ion trap. The work of **Norman Foster Ramsey** of Harvard University, the other 1989 Nobel physics laureate, led to the development of the **atomic clock**. The precise measurement methods made possible by the work of these three men made it possible "to conduct experiments that might force us to reconsider some basic physical laws," according the the Nobel Committee.

Late in his career, Dehmelt noted the discrepancies between the theoretical *g*-factor predicted by QED and the actual *g*-factor he had measured. Although the discrepancy was minuscule, Dehmelt speculated that it might indicate that the electron could be broken down into smaller and smaller particles, or subquarks, finally ending in the "cosmon." A linked cosmon and anticosmon, Dehmelt theorized, spontaneously came into being out of nothingness in a quantum jump and then instantly decayed, creating the Big Bang that began the universe.

Dehmelt became a naturalized U.S. citizen in 1962. He acted as a consultant to Varian Associates of Palo Alto, California, from 1956 through 1976, and has contributed numerous articles to professional journals. He was elected to the National Academy of Sciences in 1978 and has won several prestigious scientific awards.

DEHYDRATED FOOD • See Food preservation

DEMPSTER, GEORGE (1887-1964)

American inventor

George Dempster revolutionized refuse collection with his invention of a loading mechanism which allows a **truck** to maneuver a covered trash bin over a receptacle, empty it, and replace the container for later use.

Born September 12, 1887, in Knoxville, Tennessee, Dempster was the third of eleven children. At 14, he worked as a day laborer on the C & O Railroad. Two years later, he traveled with the W. J. Oliver Construction Company and worked as a track laborer in Elkwood, Virginia, and a locomotive fireman in Guilford, Indiana. He graduated from high school in 1906 and studied until 1907 at Knoxville's John R. Neal Law School. He failed the bar exam and worked in construction, on a fruit farm, and as a stable hand.

Dempster was employed as a heavy equipment operator and engineer on the Panama Canal for five years. He returned to Knoxville in 1912, when the city became the headquarters for the TennesseeValley Authority and a center for experiments on the atomic power industry. He joined his four brothers in a contracting partnership called Dempster Construction Company and constructed railroads, highways, **dams**, and **bridge**s during a flurry of highway construction in the Southern states.

At the age of 39, Dempster made his contribution to American technology by inventing the Dempster-Dumpster, a multi-purpose storage device capable of holding liquids or solids that could be emptied by movable lifts into a truck manned by a single operator. The machine was applied to over 3,000 types of containers, although it was originally intended to load stone from quarries. Dempster patented his device and sales of the Dempster-Dumpster soared.

Dempster gave up construction work to develop an equipment line which included a hydraulic press for baling scrap metal, hydraulic scoop and front-end loader suitable for road construction and coal mining, and compactor. He developed the 500-employee Dempster Motor Company in Knoxville, which produced Maxwell and Chalmers **automobiles** and trucks.

DENTAL DRILL

When a tooth develops a cavity, the decayed tissue must be removed. The earliest devices for doing this were picks and enamel **scissors**. Then two-edged cutting instruments were designed; they were twirled in both directions between the fingers. The father of modern dentistry, the Frenchman Pierre Fauchard (1678-1761), described an improved drill in 1728. Its rotary movement was powered by catgut twisted around a cylinder, or by jewelers' bowstrings. A hand-cranked dental drill bit was patented by John Lewis in 1838. George Washington's dentist, John Greenwood (1760-1819), invented the first known "dental foot engine" in 1790 when he adapted his mother's foot-treadle spinning wheel to rotate a drill. Greenwood's dentist son continued to use the drill, but the idea went no further. The Scottish inventor **James Nasmyth** used a coiled wire spring to drive a drill in 1829. Charles Merry of St. Louis, Missouri, adapted Nasmyth's drill, adding a flexible cable, in 1858. The first "motor-driven" drill appeared in 1864, the design of Englishman George F. Harrington; hand-held, it was powered by the spring action of a **clock** movement. In 1868 the American George F. Green introduced a pneumatic drill powered by a pedal bellows. Fellow American James B. Morrison patented a pedal bur drill in 1871. A further improvement of the Nasmyth-Merry design, it featured a flexible arm with a "hand" piece to hold the drill, plus a foot treadle and **pulleys**. Each of these advances increased the speed at which the drill operated. In 1874 Green added electricity to the dental drill; powered by electromagnetic motors, it worked well but was heavy and expensive. Plug-in electric drills became available in 1908; by then most dental offices were electrified. Once efficient, mechanically-driven drills became widely available, teeth could be properly and accurately prepared for well-fitting crowns and fillings, and American teeth blossomed with gold. Modern dental drills are turbine-powered, rotating at speeds of 300,000 to 400,000 revolutions per minute. The Morrison drill, by comparison, operated at 600 to 800 r.p.m.

Re-emerging on the scene is the *air abrasion* drill, first developed in the 1940s as an alternative to the slow-speed, belt driven handpieces used at the time but which fell from favor with the arrival of the Borden air rotor, the first air turbine handpiece. Air abrasion drills, which use a pressurized stream of microscopic abrasive powder to remove decay, enamel, and previous dental work, reduce the amount of heat, vibrations and bone-conducted noise typical with conventional drills. Patients find this technology less traumatic and rarely request anesthetics. In May, 1997, the Food and Drug Administration cleared for marketing in the United States the first *erbium:yttrium-aluminum-garnet*, or Er:YAG **laser**, for use in "hard tissue." Clinical studies proved its equivalence to the high-speed drill for making cavity preparations in enamel and removing caries, in most cases without the need for anesthetic. The laser is also used for cavity prevention and etching. This approval concludes a search begun by Dr. Leo Goldman in 1964 and many subsequent searches for an appropriate laser wavelength.

See also Dental filling, crown, and bridge

DENTAL FILLING, CROWN AND BRIDGE

Fillings

Cavities in teeth have been filled since earliest times with a variety of materials: stone chips, turpentine resin, gum, metals.

Arculanus (Giovanni d'Arcoli) recommended gold-leaf fillings in 1484. The renowned physician **Ambroise Paré** used lead or cork; in the 1700s, Pierre Fauchard (1678-1761), the father of modern dentistry, favored **tin foil** or lead cylinders; and Philip Pfaff (1715-1767), dentist to Frederick the Great of Prussia (1712-1786), used gold foil. Gold leaf as a filling became popular in the United States in the early nineteenth century; Marcus Bull of Hartford, Connecticut, began producing beaten gold for dental use in 1812 which was replaced in 1853 in the United States and England by sponge gold. This was followed by cohesive, or adhesive, gold introduced by American dentist Robert A. Arthur in 1855 and, by 1847, gutta percha was being used.

The invention of the power-driven **dental drill** led to increased demand for fillings and so for an inexpensive filling material. In 1826, Auguste Taveau of Paris, France, used silver coins to developed what was probably the first dental amalgam—a solution of one or more metals mixed with mercury. The French Crawcour brothers emigrated to the United States in 1833, introducing Taveau's amalgam; however, the poor quality of the amalgam led to its condemnation by many dentists, kicking off the so-called "amalgam war," a 10-year period from 1840 to 1850 of bitter controversy about the merits and deficiencies of mercury amalgam. Numerous experiments were made from the 1860s through the 1890s to develop improved amalgam filling materials, and Chicago, Illinois, dentist G. V. Black (1836-1915) finally standardized both cavity preparation and amalgam manufacture in 1895.

After truly effective dental cement was developed, baked **porcelain** inlays became popular for filling large cavities. These were first described by B. Wood in 1862. In 1897 an Iowa dentist, B. F. Philbrook, described his method of casting metallic fillings from a wax impression that matched the shape of the cavity perfectly. Dr. William H. Taggart of Chicago described a similar method for casting gold inlays in 1907. This technique made possible the modern era of accurate filling and inlay fitting. Amalgams—"silver fillings"—are still used, even though periodic controversies arise regarding the possibility of mercury toxicity. "White fillings," however, are gaining in popularity as they give a more aesthetic and natural look. Several types of cosmetic fillings are available today: *Direct composites* consist of quartz **resin** usually containing a light sensitive agent which allows them to be hardened, or bonded, into place by shining an intense light for about 40 seconds; *indirect composite/porcelain inlays*, specifically designed for strength as well as aesthetics, are usually fabricated in a laboratory then bonded in the dentist's office. Available in a variety of shades to match natural teeth, composites are extremely strong and can withstand the 40,000 pounds of pressure per square inch exerted when chewing with the back teeth.

Crowns and Bridges

Crowns (used to replace and cover missing portions of teeth) and bridges (mountings for artificial teeth anchored to and bridging the gap between the natural tooth at either side) were made of gold and used by the Etruscans at least 2,500 years before modern dentistry. Crowns and bridges fell from popularity during the Middle Ages and were only gradually rediscovered. The gold shell crown, described by Pierre Mouton of Paris, France, in 1746, was not patented until 1873 by Beers. The Logan crown, patented in 1885, used porcelain fused to a platinum post, replacing the unsatisfactory wooden posts previously used. In 1907 the detached-post crown was introduced, which was more easily adjustable. Bridge work developed along with crowns; dentists would add extra facing to a crown to hold a replacement for an adjacent missing tooth. The major advance came with the detachable facings patented by Dr. Walter Mason of New Jersey in 1890 and the improved interchangeable facings introduced by Mason's associate Dr. Thomas Steele in 1904. Today, crowns and bridges can be made from full porcelain, porcelain-fused-to-metal, **polymer** glass, or all metal. Cement and bonding materials for metal-free restorations—called *adhesive dentistry*—utilize zinc phosphate or glass ionometer cements. Cosmetic dentistry—reproducing a tooth for aesthetic reasons—is now popular and widely practiced: a dental implant is a metal screw placed into the jaw bone providing an anchor for a permanent false tooth or a set of false teeth; veneers are porcelain laminates which cover the natural tooth to mask and strengthen discolored or fractured teeth; and full caps (similar to crowns) are permanently placed over the original tooth to give an attractive appearance to a misshaped tooth or to narrow gaps between natural teeth.

Hand-cranked dental drill from the nineteenth century.

DENTURES • See False teeth

DERRICK • See Crane

DESALINATION TECHNIQUES

Desalination is the process of removing salt from sea water. This process is also known as desalinization, desalting or saline water reclamation. Approximately 97 percent of the Earth's water is either sea water or brackish (salt water contained in inland bodies), both of which are undrinkable by humans. Sea water contains 35,000 parts per million (ppm) of dissolved solids, mostly sodium chloride and calcium and magnesium salts. Brackish water contains 5000 to 10,000 ppm dissolved solids. To be consumable, or potable, water must contain less than 500 ppm dissolved solids. The method used to reach this level depends on the local water supply, the water needs of the community, and economics. Growing populations in arid or desert lands, contaminated groundwater, and sailors at sea all created the need for desalting techniques.

In the fourth century B.C. Aristotle (384 B.C.-322 B.C.) told of Greek sailors desalting water using evaporation techniques. Sand filters were also used. Another technique used a wool wick to siphon the water. The salts were trapped in the wool. During the first century A.D. the Romans employed clay filters to trap salt. Distillation was widely used from the fourth century on—salt water was boiled and the steam collected in sponges. The first scientific paper on desalting was published by Arab chemists in the eighth century. By the 1500s methods included filtering water through sand, distillation, and the use of white wax bowls to absorb the salt. The techniques have become more sophisticated, but distillation and filtering are still the primary methods of desalination for most of the world. The first desalination patent was granted in 1869 and in that same year the first land-based steam distillation plant was established in Britain, to replenish the fresh water supplies of the ships at anchor in the harbor. A constant problem in such a process is scaling. When the water is heated over 160°F (71°C), the dissolved solids in water will precipitate as a crusty residue known as scale. The scale interferes with the transfer of heat in desalting machinery, greatly reducing the effectiveness. The majority of desalting plants today use a procedure known as multistage flash distillation to avoid scale. Lowering the pressure on the sea water allows it to boil at temperatures below 160°F, avoiding scaling. Some of the water evaporates, or flashes, during this low pressure boiling. The remaining water is now at a lower temperature, having lost some energy during the flashing. It is passed to the next stage at a lower temperature and pressure, where it flashes again. The condensate of the previous stage is piped through the water at the following stage to heat the water. The process is repeated many times. The water vapor is filtered to remove any remaining brine, then condensed and stored. The entire process takes advantage of thermodynamic laws to get the maximum use of any energy. It is the reuse of heat that makes these plants economical. Over eighty percent of land-based desalting plants are multistage flash distillation facilities.

A host of other desalinization processes have been developed. An increasingly popular process, reverse osmosis, essentially filters water at the molecular level, by forcing it through a membrane. The pressures required for brackish water range from 250-400 pounds per square inch (psi), while those for sea water are between 800-1200 psi. The pressure required depends on the type of membrane used. Membranes have been steadily improving with the introduction of polymers. Membranes were formerly made of cellulose acetate, but today they are made from polyamide **plastics**. The polyamide membranes are more durable than those of cellulose acetate and require about half the pressure. Solar distillation is used in the subtropical regions of the world. Sea water is placed in a black tray and covered by a sloping sheet of **glass** or plastic. Sunlight passes through the cover. Water evaporates and then condenses on the cover. It runs down the cover and is collected. The salts are left behind in the trays. This method has been used successfully in the Greek islands.

In the 1990s desalination technology is advancing in several directions. One increasingly popular approach is to use desalinated water to supplement regular drinking water. The state of Florida, for example, is using dozens of reverse-osmotic plants to treat undrinkable brackish and then mixing the treated water with the regular water supply. The intent is to extend the local water supply. Another approach is to make traditional methods, like distillation, more economically feasible. In 1998, engineers at California's Metropolitan Water District reported they have designed a new, improved distillation plant which uses cheaper materials (e.g., aluminium instead of titanium, **concrete** instead of steel) to keep construction costs down.

See also Cellulose, chemical uses of; Polymer and polymerization

Desktop publishing

Desktop publishing involves the production of printed matter performed on personal **computers** for much less cost and effort than traditional printing methods. Desktop publishing became possible in 1985 with the introduction of several page-layout programs, including Aldus's PageMaker which remains a market leader in 1998. Using a personal computer (usually a Macintosh), a laser printer with PostScript page description language, and this software, users could design page layouts, including fonts and graphics, right on their computer. There have been two areas of competitive controversy in desktop publishing. One is over the preferential software for page layout design. PageMaker has been challenged by Adobe's version, QuarkXPress. The latter is especially favored by magazine designers. Another contention concerns Macintosh or IBM-type personal computers (PCs). The Macintosh has been favored from the beginning and is usually considered the most desktop publishing friendly. However, PCs have improved in this area throughout the 1990s.

In the late 1980s, software was released that could output complex drawings, such as Adobe's Illustrator and Aldus's FreeHand. In these programs, a user could create art that could be placed in the page layout program. In 1990, Adobe Photoshop was introduced. With Photoshop or other similar software, photographs, both color and black and white, are scanned into the computer for image editing using a scanner. A flat-bed desktop scanner became another piece of essential hardware in desktop publishing, a favored input device. By 1991, Apple introduced TrueType fonts that allowed the fonts themselves to be manipulated by designers. One drawback to desktop publishing in this area is the massive amounts of memory these programs and their documents take up. As the technology has improved, more memory and ways of storing it have become available.

In its beginning, desktop publishing was primarily used for publications such as newsletters and other smaller publications. As the technology has improved and become less expensive, more and more magazines, newspapers, even books go through the pre-printing process on desktop publishing software. Desktop publishing is an attractive option for publications because of the flexible control over design, and the less time, money, and labor it takes to produce a publication. The only area in desktop publishing that still can require outside service is high-quality color photos which have to be taken to a service bureau to be scanned on a certain type of high-end scanner. Throughout the 1990s, the quality of desktop scanners keeps improving and with the advent of **digital cameras**, a totally digital, totally in-house publishing experience is a real possibility early in the next century.

Desktop publishing software is also used to create online documents for the World Wide Web. Specific programs like FrameMaker allows desktop publishers to manipulate images and text for electronic publication much like PageMaker does for printed matter. In the mid to late 1990s, desktop publishing software has been developed that can morph images, and edit sound and videotape, which also can be posted on the Internet. With this technology, the tools of publishing are literally available to anyone with access to a computer.

See also Computer applications; Computer graphics; Computer input and output devices; Computer, industrial uses of; Digital camera; Printing technology.

DESTROYER • See Warship

DETERGENT • See Soap and detergent

DEWAR, JAMES (1842-1923)

Scottish chemist and physicist

Born on September 20, 1852, in Kincardine, Fife, Scotland, James Dewar became famous for his work in cryogenics—the study of objects at extremely low temperatures. He was educated at the University of Edinburgh and obtained professorial posts at both Cambridge and the Royal Institution in London. Dewar's interests covered a variety of fields. He was involved with spectroscopy, organic chemistry, the effect of light upon the retina, electricity, and the measurement of high temperatures. Dewar is best known for his work with the liquification of gases at extremely low temperatures.

In 1877, Louis Cailletet and Raoul Pictet independently were able to create small amounts of oxygen and nitrogen in liquid form at temperatures less than 80° above absolute zero, a feat even **Michael Faraday**, who had liquified most of the known gases by 1845, had been unable to carry out. Dewar decided to pursue this line of research beginning in the late 1870s. However, Dewar was not so much concerned with the mechanism of liquefying the gases as he was with studying their properties as they approached absolute zero.

The greatest stumbling block he encountered in his work with liquification was keeping the gases cold long enough to study them. Liquid oxygen kept in a flask absorbed heat from the surrounding air and returned to its gaseous phase. To eliminate the effect of the warm air, Dewar put the flask of liquid gas inside a larger flask and created a vacuum between them. A vacuum would prevent the transfer of energy that occurred through conduction or convection; heat would not penetrate and cold would not escape. To eliminate the transfer of radiant energy, Dewar silvered the walls of the flasks so they would reflect, rather than absorb, energy. He also invented a technique to create a more efficient vacuum. In 1902, Dewar discovered that when charcoal was cooled, it became very efficient at absorbing molecules. He was able to create a better vacuum by using charcoal to absorb any molecules that had not been pumped out when the vacuum was created. The final result of Dewar's efforts was the **vacuum bottle** or *Dewar flask* which, strangely, Dewar did not bother to patent. Since a vacuum prevented the transmission of both heat and cold, the Dewar flask was found to be an excellent device for storing hot and cold liquids. It is best known for its use as in insulated beverage container, better known as a Thermos ™ brand bottle.

Dewar was the first person to produce a large quantity of liquid oxygen in 1885 but it was not until 1898 that he managed to liquify hydrogen gas. In further efforts he developed methods of lowering the temperature as low as -258°C (-433°F or 13 ° above absolute zero). At this temperature, every gas but helium could be cooled to a liquid or a solid. Although Dewar attempted to liquify helium he failed because his helium source was contaminated with neon gas which froze at a higher temperature, plugging his apparatus with ice. It was left to another cryogenics specialist, Heike Kamerlingh Onnes (1853-1926), to accomplish the task in 1908 using Dewar's methods.

Dewar also collaborated with two inventors famous in their own right. The first of these collaborations landed Dewar in the midst of a major controversy. While on a government committee studying explosives at the end of the 1880s, Dewar and a colleague **Frederick Abel** invented a smokeless **gunpowder** called **cordite** and were granted a patent for it. Unfortunately, the invention had come following long discussions with **Alfred Nobel**, who was the creator of **TNT**, and Nobel sued to overturn the patent. The court ruled against Nobel, allowing Dewar and Abel to keep their lucrative patent. Dewar also worked with **John Ambrose Fleming** (who later invented the **vacuum tube**) from 1892 to 1895 to conduct a study of the electrical properties of supercooled gases.

Dewar was knighted in 1904. While many of his discoveries came about through the collaboration of colleagues, Dewar did not work well in a team situation; he could be brusque and individualistic at times. He was indifferent as a teacher, but excellent as a public lecturer. At the time of his death, on March 27, 1923, the 80-year-old was still experimenting, using a charcoal-gas thermoscope to measure **infrared radiation** from the sky.

See also Liquid gas, commercial production of

DIALYSIS MACHINE

The kidneys perform the vital function of filtering waste materials out of the blood. When the kidneys stop functioning,

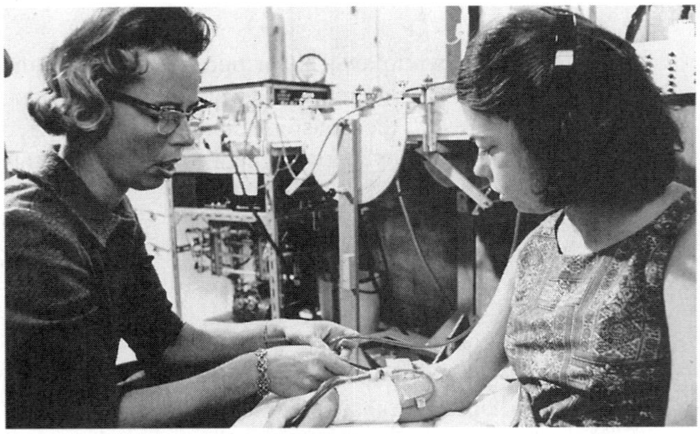

Introduced in the mid-1960s, the first portable kidney dialysis machines freed patients from constant trips to hospitals or dialysis centers for treatment.

death due to waste buildup occurs quickly. As early as 1861 a Scottish chemist, Thomas Graham (1748-1843), described a procedure he called *dialysis* to purify the blood in cases of kidney failure. The blood would be diffused across a membrane that allowed wastes to pass into a balanced fluid, while replenishing substances would pass from the fluid into the blood. Practical application of dialysis was developed by **John Jacob Abel**, the first professor of pharmacology at Johns Hopkins University School of Medicine. In 1912 Abel was investigating byproducts in the blood, and needed a device to filter these substances out. With colleagues Benjamin Turner and Leonard Rowntree, he built a machine that circulated blood through celloidin tubing immersed in a saline-dextrose solution and wrapped around a rotating drum. Urea and other toxins passed out into the solution, and oxygen passed into the blood. Abel tested this process, which he called *vividiffusion*, on rabbits and dogs, and published the findings in 1914. A major problem, however, was the tendency of the blood to clot while circulating through the tubes. Abel had used *hirudin*, an anticoagulant obtained from leeches, to prevent clotting. Once the effective anticlotting agent *heparin* became widely available, dialysis was ready for clinical use.

Several pioneers developed early versions of dialysis machines during World War II when many injured soldiers and civilians suffered kidney damage and died. In 1937, a young Dutch physician, Willem Kolff, working in Groningen, Holland, had already put together a crude dialyzing machine and worked to refine it. After the Germans occupied the Netherlands in 1941, Kolff moved to Kampen where, in spite of wartime shortages, he constructed a dialysis machine using cellophane tubing and beer cans. He first used his device on a human patient in March 1943 and, although all but one of the 15 patients he treated from 1943 to 1944 died, he persevered. By the end of the war, Kolff had refined his machine and began to promote its use, bringing dialyzers to The Hague, Amsterdam, and London. Meanwhile, with no knowledge of Kolff's work, Nils Alwall of Sweden and **G. Murray** of Canada were also developing a dialysis machine. In 1947 Kolff brought blueprints for his latest machine to doctors at Peter Bent Brigham Hospital at Harvard Medical School in Boston. These doctors, along with John Merrell, Karl Walter, and George Thorn, developed kidney dialysis into a standard treatment, using it to support patients in their pioneering kidney transplantations in 1954. Kidney transplants, the first organs ever to be transplanted, were made possible because of dialysis, which kept the patient alive until the transplanted kidney began functioning or by maintaining patients awaiting a donor organ. Long-term dialysis was not possible, however, until 1960, because each time a patient was attached to a dialysis machine, both an artery and a vein had to be punctured, leading to eventual vessel deterioration. Dr. Belding Scribner of Seattle overcame this problem when he designed a **Teflon** and Silastic shunt (two parallel tubes with a U-connection) that could be inserted into a patient's artery and vein and left in place for months or even years.

Today, *hemodialysis*, a refined version of this technique, allows the patient the option of home treatment with the aid of a family member or friend, or by specialists at a dialysis center. Patients also have the option of using *Peritoneal* dialysis, in which the abdomen lining (peritoneal membrane) filters waste from the blood into a cleansing solution called *dialystate*. *Continuous Ambulatory Peritoneal Dialysis*, the most common of three types, requires no machine and can be done by the patient. The dialystate, contained in a plastic bag, is transported through a permanent catheter inserted into the abdomen. The catheter is then plugged and, after four to six hours, the patient removes the plug, draining the solution containing waste matter back into the bag, which is then disposed of. This process is repeated continuously. *Continuous Cyclic Peritoneal Dialysis* is a similar function done by a machine connected to the catheter and performed at night while the patient sleeps. This procedure lasts from 10 to 12 hours every night. *Intermittent Peritoneal Dialysis* can also be done at home with a similar machine, but is usually done in hospital several times a week for a total of 36 to 42 hours. Some sessions may last 24 hours.

Diamond anvil cell

Interest in high pressure phenomena peaked during the 1880s. The French scientists Louis Paul Cailletet and Emile Hillaire built devices for studying the behavior of materials under pressures up to three thousand atmospheres (44,000 pounds per square inch). Their devices failed at about three thousand atmospheres, however, and a limit to such research seemed to have been reached. Two decades later, however, a new approach was tried. **Percy Bridgman**, then a doctoral student at Harvard, designed a new device for creating high pressures. Bridgman's first diamond anvil cell was able to produce pressures four times greater than those achieved by Cailletet and Hillaire. Further refinements over the next fifty years resulted in cells that could produce pressures of thirty thousand, fifty thousand, and eventually one hundred thousand atmospheres. In some specialized situations, the cell was modified to produce pressures of 425,000 atmospheres (6,250,000 pounds per square inch). The diamond anvil cell is a small device that can easily be held in the hand. It consists of two **piston**s with very different surface areas, A1 and A2. A1 can be as large as a few square centimeters while A2 is about 0.1 square millimeters. When a force is applied on the large surface area by turning a **screw**, that force is multiplied on the smaller surface area by a factor of A1/A2. The sample to be studied is placed between two flat parallel plates made of diamond in the smaller piston. Diamond is chosen for the plates because it is the hardest substance known and it is transparent to light. Thus, the experimenter can watch what happens to the sample as pressure on it is increased. One application of the diamond anvil cell is simply to observe changes in the properties of a substance at high pressures. For example, the cell has been used to convert hydrogen to a metallic solid at pressures of fifty-seven thousand atmospheres (840,000 pounds per square inch). Geologists also use the cell to simulate pressures within the Earth's core and to test hypotheses about the deepest regions of the Earth. Finally, the cell has been used to convert graphite to synthetic diamond at pressures of sixty-five thousand atmospheres (960,000 pounds per square inch) and temperatures of about 2552°F (1400°C).

Diamond, synthetic • See Synthetic gemstones

Diaper, disposable

Before the advent of the disposable diaper in the 1960s, babies had to resign themselves to the discomfort of soggy cloth nappies, while their parents or nannies were saddled with the unpleasant tasks of washing, bleaching, and drying. Tired of diapers that had to be washed or sent to an expensive laundry service, New York homemaker Marion Donovan used some absorbent padding and a piece of shower curtain to invent the first disposable diaper, called *the Boater*, in 1951. Manufacturers were skeptical of the snap-on, throwaway diaper, so Donovan marketed her product herself. Her diaper grew more and more popular with both babies and adults and Donovan eventually sold her interest in the product for one million dollars. The first mass-produced disposable diapers were crude in construction; pulpy, uncomfortable, and often messy. Nevertheless, they were deemed superior to the cloth variety. By the late 1950s, Proctor and Gamble had begun research and development for a disposable diaper; it is said that P&G spent more research money on diapers than **Henry Ford** spent on the first **automobile**. The company's initial incarnation was a pair of **plastic** pants with an elasticized waistband and leg openings, but this product never gained much popularity. In 1961, Proctor and Gamble introduced Pampers, the **rayon**-plastic-fluff model that would set the standard and lead the market for diapers for the next several decades. By 1975, the company unveiled another brand called Luvs (designed by Kenneth Buell, whose resume included working on the Gemini space program). Improvements to the basic disposable diaper design have included adhesive tabs to replace the awkward and potentially dangerous diaper pin, and separate models specially designed to comfortably accommodate the anatomy of male and female infants. By the 1980s, most babies wore disposable diapers, despite a return to cloth diapers by some families concerned about the environmental implications of disposable diapers. Although manufacturers and devotees claim that disposable diapers comprise only a minute percentage of landfill waste, a number of babies again sport the cloth nappies their diaperers had earlier abandoned.

See also Polymer and polymerization

Dick test

The Dick test is a laboratory test designed to indicate whether or not a person is immune to scarlet fever. Scarlet fever, at one time fairly common in children, is a disease named for the flushed face, red rash, and fever it causes. The Dick test is named after George and Gladys Dick, American bacteriologists who worked on the diagnosis and treatment of scarlet fever in the 1920s. The test involves administering two different injections, one into each arm of a patient. In one arm, toxin taken from a culture of scarlet fever bacteria is injected; in the other arm is injected neutralized toxins which act as a control. If the toxin causes redness, tenderness and swelling after 24 hours, the person is not immune to scarlet fever. The control normally shows no swelling for comparison.

For isolating the streptococcus bacteria that causes scarlet fever, and for the preparation of an antitoxin for the treatment of the disease, George and Gladys Dick obtained a British patent in 1924 and a U.S. patent in 1925. Armed with this test and the antitoxin, doctors could diagnose and treat the disease much more efficiently, resulting in a decline in the number of cases. As the age of antibiotics arrived, scarlet fever was successfully treated with antibiotics and thus the number of cases dwindled sufficiently to make it fairly uncommon today.

Dielectric materials

A perfect dielectric is a material in which there are no free electrical charges, and which therefore does not conduct electricity. Real dielectrics, also known as insulators, are never perfectly insulating, with the result that there has been a 150-year search for ever better dielectric materials.

The first practical need for dielectric materials arose out of the discovery of electricity, which in turn led to the invention of the light bulb, telegraph, telephone, and other electricity-oriented innovations. Electric line wires in service must be kept as dry as possible in order to function efficiently, and to cut down on loss of current. As electric technology developed, insulators were needed for telephone lines, electric power lines, and other applications. Also, as railroads began crossing the continent, there came the need for signal devices. Electricity had to be moved economically from one place to another to meet the increasing demands generated by these new inventions.

An early challenge was the possibility of running electrical cables under the sea. The only materials for insulating and waterproofing wire in the mid-1800s, however, were such materials as oakite and asphaltum. These materials did not last long for even short, shallow ocean crossings. The discovery of a close relative of **rubber** called **gutta-percha** (an inelastic latex) in 1843, which English physicist and chemist **Michael Faraday** (1791-1867) determined to be an electrical insulator and that proved reliable in salt-water, led to electricians and telegraphers developing a suitable means of extending telegraph lines undersea, the first line being placed between England and France in 1850. A screw extruder was developed to coat the copper conductor, thus laying the foundations for today's **plastics** extrusion industry.

In 1848, the first U.S. factory to insulate wire with gutta-percha started operation. In Germany, **Ernst Werner von Siemens** (1816-1892) developed a gutta-percha insulating machine in 1847, and succeeded in laying a gutta-percha cable in Kiel Harbor a year later. In 1921, the Key West-Havana telephone cable began service, using a blend of rubber and gutta-percha for cable insulation. By 1947, polyethylene had replaced rubber and gutta-percha as the preferred underwater insulator. In 1949, AT&T developed a submarine cable insulated by **polyethylene** and 5% butyl rubber. Modern undersea cables are comprised of steel strength members protecting multiple hair thin glass communication fibers, with the whole package surrounded by a waterproof electrical insulator such as polyethylene.

Meanwhile, across land surfaces, communication and electric line wires were kept off of the ground by being strung between poles. But something was needed to keep the wires and poles apart. This device was the **glass** or porcelain insulator. Glass insulators were first produced in the 1850s for use with telegraph lines. Porcelain insulators had their start when local potteries began making telegraph insulators in the 1850s and 1860s. With the advent of electric power distribution in the 1880s, larger and more reliable insulators were needed to carry the higher voltages of power lines. Glass at that time was not sufficient. Glaze-welding in which insulators were produced in two or three pieces and cemented together became popular in the late 1890s and early 1900s and was used by most major insulator manufacturers. From around 1915 on, the porcelain insulator virtually replaced glass on all electrical distribution, even at low voltages, as the superiority was demonstrated in both insulation quality and strength. Other types of insulators such as rubber, and a very dense, oily wood called lignum vitae were used extensively from the 1930s to the 1970s.

The addition of sulphur under heat to natural **rubber** to produce a hard, rigid, thermoplastically moldable product was patented by **Thomas Hancock** in 1843. Termed vulcanite, ebonite, or hard rubber, the material soon came to be used as an insulating solid in the electrical industry. Rubber provided the electrical and communications industries with the effective insulation they badly needed. In the years since **Thomas Edison's** revolutionary inventions, several different wiring styles have come and gone. The oldest wiring system, the knob and tube type, was made up of individual conductors with a cloth insulation. Following knob and tube insulation came multi-conductor cable. The earliest multi-conductor cable used cloth and varnish insulation. It was employed in the 1940s and 1950s in the U.S. Afterwards, the more modern asphalt-impregnated romex cloth was invented. Despite its stickiness, romex was as safe as its modern replacement, thermoplastic insulation wire.

For the first 10 years or so following **Bakelite**'s introduction in 1909, it was used primarily for electrical and **automobile** insulators, and in heavy industrial products.

High pressure **laminates** of **paper** and phenol formaldehyde **resins** for electrical insulation date back to 1913.

Cables connecting to oil-filled units that are not completely sealed, such as transformers, condensers, oil coolers, and high-voltage switches, require liquid or gas insulation. **Polychlorinated biphenyls (PCBs)** were commonly used as liquid dielectrics in the 1930s. Colorless oily compounds, PCBs are resistant to dielectric breakdown. They are chemically stable and resistant to heat and burning, making them especially useful in industry. However, adults who have been exposed to high concentrations of PCBs have suffered debilitating stomach pain and disfiguring skin eruptions, while their children have suffered physical, behavior, and cognitive problems. Consequently PCBs have been banned by the Environmental Protection Agency for most uses. Alternatives to PCB insulators today include hydrocarbon oils, askarel, and silicone oils. Gas insulation, such as SF_6, is another alternative.

See also Polymer and polymerization; Plastics; Ceramic; Plunkett, Roy J.; Semon, Waldo

DIESEL ENGINE • See Engine, oil

DIESEL, RUDOLF CHRISTIAN KARL (1858-1913)

German engineer

Rudolf Diesel was born in Paris, France, on March 18, 1858, the son of Bavarian Germans from Augsburg. At the outbreak of the Franco-Prussian War in 1870, the family was expelled from France and they moved to London. His father sent the young Diesel back to Augsburg to continue the education that he had begun in France. Diesel eventually found his way to the technical high school in Munich, where he excelled in engineering, passing his examinations with the highest marks on record.

In 1880, he went to work for the firm of his **refrigeration** professor, **Karl Paul Gottfried von Linde**. While an employee of the Linde firm, Diesel became fascinated with the theoretical work of the French physicist Nicholas Carnot (1796-1832), which presented the principles of the modern **internal combustion engine**. Diesel was convinced that an engine four times as efficient as a **steam engine** could be built by injecting fuel into an engine in which the **piston** compresses air in a ratio as great as 25 to 1. Such high compression causes the air to reach temperatures of nearly 1000°F (537°C), a temperature that is high enough to ignite the fuel without the need for complex spark **ignition systems**.

Rudolf Diesel worked on various designs for his engines for over a decade, and he was granted a patent in 1892 for an engine designed to burn the cheapest fuel then available—powdered coal. By 1897, Diesel abandoned powdered coal, substituting **kerosene** as the fuel. Diesel's engine proved to be very efficient, as his calculations showed it would be, but due to the restrictions he placed upon the manufacture of engines built under his patent, Diesel engines were very heavy and not suited to anything but stationary applications. Hence, early Diesel engines were not adapted to power the burgeoning **automobile** industry. Diesel engines did, however, find wide acceptance in the shipping and locomotive industries.

The diesel engine's high efficiency and comparatively simple design made it commercially successful, and its manufacture made Rudolf Diesel a millionaire. However, while crossing the English Channel on the way to consult with the British Admiralty, Diesel disappeared at sea on September 30, 1913. His body was never recovered and suicide is considered to have been a possible reason for his death as he was known to be emotionally unstable and given to occasional breakdowns.

See also Oil engine

Rudolf Christian Karl Diesel

Differential

As soon as power-driven vehicles were capable of cornering, a problem became apparent. When the vehicle went around a curve, the **wheel** on the inside of the curve traveled a shorter distance than the other and thus turned more slowly. The differing speeds of the wheels on either side of the car caused it to skid, as well as stressing the axle assembly. As a result, there was a need for a **gear** system able to drive two shafts at different speeds.

The first successful split axle and gear, or differential, was designed by O. Pecquer in 1828. The drive shaft joined with a large housing that contained the ends of two rear axles and was located between the rear wheels. Here, the drive shaft ended with a drive pinion gear that matched up with a large crown wheel. This crown wheel had two planet pinion gears mounted permanently to it, and it was these planet pinions that contacted the ends of the two axles, which also had toothed gears, and spun them for power to the rear wheels. When the vehicle traveled in a straight line, the planet pinions did not spin, so the two half-shafts were driven at the same speed. As the vehicle cornered, the planet pinions did spin, allowing the half-shafts to revolve at different speeds.

Differentials were improved greatly over the years. Charles Mott and Henry Timken (1831-1909) were important creators of better differentials for American cars. Early differentials were composed of pinion and crown gears with straight teeth, which caused a great deal of noise and wore out quickly.

Modern differentials have *helical* gears; the toothed surfaces are bevelled and the teeth themselves are curved, thus preventing excess wear and cutting down on noise. The limited-slip differential was developed to improve traction by allowing the axles to spin at independent speeds while cornering and applying equal torque for better traction while accelerating on the straights. Holding power within the differential may be provided by clutches, sets of interlocking cones, ratchets, or air pressure; street cars typically rely on clutches, but other, stronger devices are used in off-road vehicles and race cars that have more powerful engines producing greater torque.

Systems of gears may not comprise the definitive differential; the Daf car, built in Holland, doesn't use any gears at all and relies instead on a belt-drive system that allows slippage of the belt on **pulley**s to adjust power applied to the wheels.

See also Automobile; Wheel and axle automotive

Diffraction grating

An important part of the study of light has been the ability to disperse light, splitting it into a band of component colors called a spectrum. By doing so, scientists are able to examine the bright and dark emission and absorption lines that cross the spectrum, each one indicating the presence of an element. Early physicists such as Isaac Newton (1642-1727) used **prisms** to disperse light; later, beams of light were passed through a narrow slit, bending them slightly to produce a spectrum. Today, the most widely used instrument for obtaining the spectra of light is the diffraction grating.

Very crude diffraction gratings were used by American astronomer David Rittenhouse (1732-1796) and English physician and physicist Thomas Young (1773-1829); however, the first person to take an analytical approach toward the construction of gratings was the German optician **Joseph von Fraunhofer**, and it is he who is generally credited with their invention.

Fraunhofer, like many others, had been experimenting with light, shining it through a narrow aperture to produce its spectrum. In 1821, he tried passing the light through a wire mesh and was rewarded with a much brighter and more well-defined band of color. He altered the spacing of the wires, finding that narrower openings further improved the spectrum's resolution. Fraunhofer published his findings, and soon every respectable laboratory possessed a set of diffraction gratings.

More than fifty years after its invention, the diffraction grating was once again modified, this time by the English physicist John William Strutt (Baron Rayleigh) (1842-1919). Baron Rayleigh was the first to determine the precise resolving power of a diffraction grating, making them easier to use and construct for specific purposes. This also led directly to modifications in another important optical instrument, the **spectroscope**.

Diffraction gratings illustrate a theory proposed by the Dutch physicist **Christiaan Huygens** in the seventeenth century. Huygens suggested that light travels in a wave; if that wave should encounter an obstacle or an aperture, a new wave will be created at that point. A diffraction grating is essentially made up of thousands of tiny apertures—as the light encounters the mesh, it is split up into thousands of new waves. As the waves overlap they create interference: in some places the waves amplify each other, while in other places they negate each other. The resulting pattern of light and dark areas is called the *interference pattern*, and a spectrum is really just one example of such a pattern.

Modern diffraction gratings are, of course, much more intricate than those used by Fraunhofer and Baron Rayleigh, but they still obey the same optical laws. They are usually constructed from **glass**, and they must be ground precisely using a diamond bit. There are two basic types of diffraction gratings. Transmission gratings, like those used by Fraunhofer, are a perforated mesh that allows light to pass all the way through. They are used infrequently, since they are only useful for diffracting visible light, and they are most often found in small spectroscopes.

The more popular variety is the reflection grating. It is either a flat or concave glass into which thousands of tiny grooves have been cut. The surface is then given a thin coating of metal and is polished for maximum reflectivity; aluminum is commonly used, as are gold, silver, copper, and tin. The spacing between grooves varies depending upon the type of radiation to be diffracted and the distance at which the spectrum will be projected. The grooves must be etched very precisely—within a fraction of the smallest incident wavelength—and must be both perfectly straight and parallel; a flawed grating will result in ghosts, false lines, and visual ''satellites.''

On the cutting edge are holographic gratings. Instead of a mesh or a series of grooves, these transmission devices bear a **hologram** that diffracts light to produce the same effect as conventional gratings. The devices are lightweight and precise. Holographic gratings can either be produced by conventional **holography**, a complex laser process, or by etching a computer-designed hologram into a substrate with an electron beam. Once a master has been produced, however, large volumes of replicas can be produced at relatively low cost.

Diffraction gratings have been designed to examine the spectra of many different radiation varieties-including infrared, ultraviolet, and microwaves—and has a wide range of applications. One of the most prominent has been in astronomy, where diffraction gratings are used to disperse the light from distant stars. By observing the dark lines that cross their spectra, astronomers have been able to determine the elements that make up the universe. These lines also show a distinct shift toward one end of the spectrum; this shift was shown to be caused by a star moving away from the earth and is viewed as proof for the expanding universe theory.

Digital Audiotape

Digital audiotape (DAT) records audio signals digitally, as a coded series of zeroes and ones. DAT also uses helical-scan technology, that is, it records diagonally across the width of the tape. Introduced in Japan and Europe in the late 1980s, DAT became available to consumers in the United States in 1990. The DAT cartridge itself is much smaller than its analog counterpart, about half its size, and the tape itself is 0.4-0.5 mm (about 0.15-0.2 inches) thick. It looks like a smaller version of a VHS videocassette. DAT has only become a standard in the music recording industry. As a commercial format for music and computer data back-up, DAT has faced intense competition and prevailed only in the latter.

A Sony product, DAT is primarily thought of as a music recording and playback format, comparable to the **compact disc** (CD) in sound quality. Indeed, DAT tapes feature high fidelity, crystal clear sound and hold about two minutes of music per meter (1.1 yd) of tape. DAT has been used in the recording studio since in 1987, but did not hit the consumer market until 1990 because of legal action threatened by the music industry. DAT recorders can make perfect bit-for-bit copies of compact discs, leading to concerns that people would copy CDS, make pristine DAT copies of DAT copies, and sell them, undercutting profits. This issue was resolved by DAT manufacturers agreeing to insert a computer chip that prevents such copies of copies from being made. Indeed, DAT is primarily used as a recording medium, whether it be musicians making recordings at home or in the studio—where DAT is used to make digital master tapes and demo tapes— or fans taping concerts or their favorite compact discs. There are limited numbers of prerecorded DATs available. DATs can be played back on a DAT deck, or a device similar to a **walkman**, called the DATman, manufactured by Sony and introduced in 1991. Despite DAT's superior sound quality, it has faced stiff competition on the consumer market from rival technologies like the digital compact cassette (DCC), introduced by Philips in 1992. DCCs look like analog **audiocassette**s, and unlike DAT, DCC players can play both analog and digital cassettes. The sound quality of DCC is necessarily as good as DAT. However, neither format has caught on in the mainstream consumer market of the late 1990s, and remain in the realm of musicians and high-end audiophiles.

The use of DAT as a data archive for computers is much more widespread because of its low cost, high capacity, and small size. An uncompressed 60-m (about 20 ft) DAT cartridge can hold up to 1.3 gigabytes of data. When compressed, the amount jumps to 2.5 gigabytes. To use DAT as a tape back-up system, a special drive is attached to the computer and files are copied onto it. Competitors have included 8-mm videotape, sometimes known under the brand name Exabyte; quarter-inch tape, also called QIC or "quick;" and, more recently, digital linear tape (DLT). But DAT speed and capacity, as well as the quietness and ease of use of DAT drives have made it the preferred tape back-up system as of 1996. A future threat to DAT's supremacy in this area is the advent recordable compact disc.

See also Sound mixing and recording

Digital cameras

Though they often look like traditional film cameras, digital cameras take pictures electronically. The image is captured by charge-coupled devices (or **CCD**s, light-sensitive semiconductors) and generated as pixels, then stored as a file. After the picture is taken, it can be immediately seen—there is no photo-processing involved. Many digital cameras have **LCD**s (liquid-crystal display) in the back of the camera for image viewing. The image can also be seen on a computer monitor or **television** screen, again right off the camera if the camera has a video-out port. Digital camera technology has been a boon to photojournalists and other professional photographers, and continues to make headway in the consumer market of the late 1990s. Indeed, more than 2 million digital cameras were sold in 1997.

Introduced to consumers in 1994, digital cameras have several advantages over film cameras. There are no film or processing costs, which eventually offsets the higher camera price. There are no chemicals involved, like the silver nitrate essential to film processing, which makes the process better for the environment. For professional photographers, it is often advantageous to use digital cameras because their work ends up being processed electronically anyway. But the most important advantage may be the flexibility of image management. Digital cameras with LCD viewers allow the photographer to preview the picture before it is taken. Because pictures taken by digital cameras are electronic files, they do not have to be permanent. If the image is imperfect for whatever reason, it can be erased immediately. It can also be edited: internal camera software like Photo Studio allows the photographer to adjust contrast and sharpness.

Images are stored inside of a digital camera in one of several ways. Some cameras have a flash memory inside or a similar kind of internal memory chip. Others have external memory cards, which work like floppy disks: they are put in the camera before use. After the images are loaded on to a computer as files, a photographer can use computer software such as Adobe's PhotoShop to further edit the image. The photo can then be printed out or sent over the Internet. However, when printed the resulting images often do not look as sharp as film prints, mostly due to the lesser quality of printers than the images themselves. This is one of the major drawbacks to using a digital camera, but it should be solved as printer technology advances.

While the bodies of digital camera often resemble their film counterparts, there are several key differences. Digital cameras rely on batteries, which are often easily drained, especially if the LCD viewer is used. Most digital cameras do not yet feature the same kind of zoom, shutter speed, and similar controls—they are still primarily point-and-shoot instruments. As of 1998, the resolution of image produced by the digital camera is not quite as good as that of film. Film is finely grained and can hold much more detail than the best digital camera can, though this again is something that should change as digital camera technology improves. The future of digital cameras should bring other such improvements—including many features available on film cameras like telephoto capabilities. In 1997, digital camcorders became available on the consumer market that work in much the same way that digital cameras do, and some, like Canon's Optura, feature both products in one.

See also Computer applications, Computer discs and tape, Computer graphics, Computer input and output devices, Photography.

Digital clock • See Clock and watch

Digital computer • See Computer, digital

Digitizer

Digitizers are computer hardware devices used to convert analog (continuously varying) signals into a digital or electrical signal. This process allows many common measurements—such as temperature, air pressure, or engine speed which are usually output in analog form—to be converted and used by **digital computer**s. Perhaps the most component of digitizers today is the digitizing tablet (also known as area digitizers), first developed by the Rand Corporation in 1964. Graphic images can be traced on the digitizer tablet using a stylus, and the corresponding image is converted and displayed on the computer screen in digital format. Digitizing tablets are commonly used in the **CAD/CAM** and graphics fields allowing drawings, charts, or graphs to be input in pictorial form onto the computer screen. Another example of digitizer use occurs in the conversion of speech into a digital format. A **microphone** is used to break speech pressure waves into equivalent varying voltages, and the electronic digitizer converts the voltages into digital code. While this type of digitizer depends mainly on electronics, there are also linear and rotary digitizers, which combine optical and electronic sensing devices. Linear and rotary digitizers are most commonly used in engineering or scientific applications to provide digital information on the position of a rotating shaft.

See also Computer, industrial uses of; Computer input and output devices; Computer graphics; Computer speech recognition

Diode

Many people have heard of and encountered light-emitting diodes, also called **LEDs**; they are used often in electronic toys and readouts. The first diodes were invented to serve as rectifiers—that is, they were used to convert **alternating current** (AC) into **direct current** (DC). AC is very easy to generate using a **transformer**; however, AC moves in a sine-wave pattern, moving first in one direction and then doubling back in the opposite direction. Many electronic devices require a current that moves in only one direction, such as DC. A device was needed that would eliminate the sine-wave motion of AC.

Karl Ferdinand Braun, a German physicist, noticed that certain crystalline substances served as rectifiers, allowing current to pass very easily in one direction but inhibiting its motion in the opposite direction. When an AC current was passed through these crystals, it was easily converted into DC. Braun patented his **crystal rectifier** in 1874, but its use was limited to crystal set **radio**s. The crystal rectifier was replaced as an electronic tool in 1906 when the vacuum-tube **transistor** became available.

The **vacuum tube** (invented by **John Ambrose Fleming**) was essentially the first true diode. It consisted of two metal poles, an anode and a cathode (thus the word di-ode), encased in a glass vacuum tube. The cathode was usually made from cesium (or some other alkali metal) that released electrons when heated, while the anode was made from a metal that did not boil off electrons when hot. When a current was applied through the diode so that the cathode was negatively charged, it would heat up and release electrons; these electrons would flow across the vacuum to the anode, thus completing the circuit. This is called the forward current direction and can be thought of as the diode's ''on'' position. However, when a current was applied so that the anode was negatively charged, it, too, would heat up, but would not release any electrons. Consequently, no electrons would flow to the cathode and the circuit would be incomplete or in the ''off'' position. The vacuum tube, or valve diode, would thus allow current to flow in one direction only, just as the crystal rectifier did.

If a third type of electrode—the control grid—is added to the system, the device is called a triode. Triodes are used to control the passage of electrons and can often be used to amplify the current as it passes through.

Today, most valve diodes have been replaced by semiconductor diodes. A semiconductor, such as silicon or germanium, is a material, usually crystalline, that does not carry current as well as a true conductor. In order to make a semiconductor diode, the crystal is sliced and coated at each end with different substances, so that each end of the crystal has different electronic properties. One end is called the n-type, because its material gives it a greater number of free, negatively-charged electrons. The other end is called the p-type, because it has a greater number of positively-charged ''holes.'' Both the electrons and the holes are allowed to move within the crystal, but they are strongest near the ends.

The middle of the crystal is called the p-n junction; this is where electrons and holes exist in equal amounts. When a current is applied to the crystal such that the n-type side is stronger (more negative), more electrons will flow past the p-n junction. However, when the voltage is reversed, the p-type side will become stronger and the p-n junction will act as a barrier. Just like the crystal rectifier and the valve diode, the semiconductor diode only allows current to flow freely in one direction.

In order to make an LED, gallium arsenide or a similar substance must be used as the semiconductor. When treated with phosphorus, the LED will glow in the ''on'' position. If no phosphorus is used the LED will glow in the invisible infrared spectrum.

An evolving development is that of molecular diodes, whereby data could be stored, transmitted, and retrieved via molecules. Such a process is expected to allow for the decrease in size and increase in speed of certain electronic instruments.

Direct current

One of two types of electrical flow, direct current (DC) predates **alternating current**. Unlike alternating current, direct current flows in only one direction. Current is measured in amperes, frequently shortened to ''amps,'' which is the amount of electrical charge that flows over a given time span.

Benjamin Franklin, who first suggested the terms ''positive'' and ''negative'' electrical charges, believed the flow of current went from positive to negative. In fact the opposite is true, because current is carried by **electrons**, which are negatively-charged and so attracted to positive polarities. The electric battery that **Alessandro Volta** invented in 1800 produced direct current and was an immediate success. It, its successors, and direct current **generator**s were used to operate a variety of electrical devices during the next eighty years.

Then, in 1881, serious competition arose in the form of a brilliant Croatian electrical engineer named **Nikola Tesla**. Employed by the Continental Edison Company in Paris, Tesla was a firm believer in the advantages of alternating current, but he had been unable to convince others. In 1884 he came to the United States to work with the great inventor **Thomas Alva Edison**, a major proponent of direct current.

The fact that direct current was prone to numerous problems was not in question. Direct current generators had carbon brushes that touched a rotating commutator to draw electrical power. This rotating contact point caused arcing, and the higher the voltage that was being generated, the greater the arcing. This effectively limited direct current to low levels. Another problem was due to resistance within the wires conducting the current. Resistance causes heat, and that resulted in reducing the current further. Raising the level of the current to compensate was ineffective; doubling the current caused heating loss to be squared, plus it worsened generator arcing. This meant that Edison's generating plants were limited to delivering power to customers within a few blocks. In addition, Tesla and Edison did not get along; each described the other's mentality as low, for example. Edison believed direct current was far safer than alternating current, which he thought was deadly, though Tesla disagreed. Edison argued that no one could switch from direct current because there were no successful alternating current motors. Tesla asserted the benefits of his alternating current polyphase system. Each man was determined his system was better and the arguments became more emotional. Eventually, Tesla left Edison in disgust in 1885.

The nail in direct current's coffin was delivered eight years later in 1893, and it was Tesla who held the hammer. He and George Westinghouse had underbid Edison for the contract to provide electricity to the Columbian Exposition in Chicago, and their alternating current was a spectacular triumph.

Direct current may have lost out as the primary source of the world's electric power, but it is still extremely vital. Batteries, both large and small, produce direct current for a variety of applications. Modern **electronics**, **telephone**s, computers, and a host of other devices require direct current to operate. Ironically it is often Tesla's alternating current that is transformed into direct current to provide the power.

DIRECT CURRENT MOTOR • See Electric motor

DIRECTIONAL SIGNALS, AUTOMOBILE

As long ago as 1889, E. Martin worked on a rudimentary signal system that would enhance automobile and road safety. He thought of using a board mounted to the back of his vehicle with a set of flaps that would conceal the word STOP. When the brakes were applied, the flaps would open to reveal the message to those following him. With the development of four-wheel **braking systems** and electrical systems, a switch was connected to a brake rod in such a way that, when the brakes were applied, a red lamp at the rear would light. Later, the use of hydraulic brakes changed the system so that the switch was made sensitive to pressure in the brake line.

Features were added to this system to enable the driver to denote turns as well as stops. In 1893, J. B. Freeman employed a roller at the back of the car which was attached to a cord and could indicate STOP, LEFT, and RIGHT. Other inventors created mechanical arms that swung out on either side of the car when the driver operated a control. Just after the turn of the century, such signals were pneumatically operated and self-canceling, thanks to F. Berger. In 1908, Alfredo Barrachini used separate **semaphore** arms that were lit by small electric bulbs.

The next step was the use of flashing lights to indicate turns. Several designers used flashing lights in semaphores and other early indicators. H. G. Wheeler devised a small motor to make the lights blink, while a Vermont company used a pneumatic device that worked from engine suction to flash the lights. Inventors at a Delaware corporation realized, in 1935, that a device called a thermal interrupter switch could also be used to trigger flashing car lights. In this switch, a tension wire heats up as electricity flows through it; it then expands and allows current to pass to the bulb. This short-circuits the wire, which then cools off enough to shrink and break off the contact to the bulb. In 1962, John Ridout and Frank Hill made a flasher circuit that used **transistors** for the first time. Microprocessors activate brake lights and directional signals in cars of the 1990s; sizes and locations of these safety devices have been modified to include a centrally located and elevated brake light and turn, reverse, and brake lights that are generally larger than those of earlier models.

See also Automobile; Automotive electronics

DISCHARGE LAMP • See Arc lamp

DISHWASHER

The first patent for a mechanical dishwashing device was granted in 1850 to a man named Houghton from Ogden, New York. His wooden contraption splashed water on dishes. In 1866, Josephine Cochrane of Shelbyville, Indiana, built a dishwasher in her home. Her interest in developing tools that would ease the labor of household chores extended beyond dishwashing machines, and she established one of the first kitchen equipment companies in America. Cochrane worked diligently on her dishwashing machine but was unable to invest enough time and money into its commercial production. After her husband's death, she began commercial production of her machine with financial help from friends. Cochrane's first two models were both hand-operated and ultimately proved as tiring as washing dishes by hand. However, Cochrane built improved models for home and hotel use. Her larger machines were powered by a **steam engine** and were capable of washing hundreds of dishes in two minutes. The modern Kitchenaid dishwasher is the descendant of Cochrane's early machines. In 1911, the first motor-powered dishwashing appliance was developed. The first practical dishwasher for home use was marketed in 1932. Despite the fact that the average daily time spent washing dishes was sixty-eight minutes, the machine was not in great demand. Early machines, designed for restaurants and hotels, passed dirty dishes under jets of hot water on a **conveyor belt**. Modern machines operate by mechanical arms that rotate and spray water on dishes placed in stationary **wire** baskets. Machines draw in cold water and heat it; the average home dishwasher uses approximately 3.5 gallons (13.2 l) of water heated to 140° F (60° C). An entire day's dishes can be washed in one load. The dishwasher dries the dishes by circulating hot air, thus condensing moisture. In Seattle, Washington, in 1962 Kelvinator demonstrated a dishwasher that used neither **soap** nor water. The machine relied on high-frequency sound waves to clean dishes, but proved too costly to develop on a mass scale. Newer dishwashers remove soil without pre-rinsing by hand, an energy saving boon.

DISPOSABLE DIAPER • See Diaper, disposable

DISTILLING • See Alcohol, distilling of

DISTILLING OF ALCOHOL • See Alcohol, distilling of

DISTRIBUTOR, AUTOMOBILE • See Kettering, Charles

DIVING APPARATUS

The first divers to use any type of diving apparatus were probably breathhold, or free divers. An Assyrian artifact from about 900 B.C. is among the earliest records of free diving gear. A frescoe depicts a swimmer breathing underwater from a goat skin. By the 5th century B.C. Greek divers used breathing tubes while in the water. The earliest record of swimming goggles is from a second century A.D. vase of Peruvian origin. Tortoiseshell goggles and breathing tubes were routinely used by free divers by the Middle Ages.

The diving bell, another early device that allowed man to descend and observe the underwater environment, is open to the water at the bottom. From its earliest use, this allowed divers to go on short forays out into the open water and return to the bell for air. However, the air inside would quickly become stale. Edmond Halley (1656-1742) finally made diving bells practical in 1716 by introducing the idea of air replenishment while underwater. More recent developments related to the diving bell include the **bathysphere** and **bathyscaphe**. These twentieth century inventions provided scientists their first glimpse of the deepest areas of the ocean. The **submarine** has been instrumental both in marine warfare and marine research.

Today's oceanographers use a wide variety of personal diving apparatuses for underwater exploration. In modern free diving, or skin diving, the diver carries no air supply and is equipped only with mask, fins, snorkel, and when in cold water, a wet suit. The mask allows the diver to see clearly, while the snorkel, a J-shaped tube, lets the diver breathe while swimming just below the surface of the water. Fins, flat rubber shoes that resemble duck feet, provide much better propulsion than bare feet with less effort.

In **scuba diving** (an acronym for Self-Contained Underwater Breathing Apparatus) cylinders of pressurized gas (usually compressed air) are strapped onto a diver's back. This air supply is connected via hoses to the diver's head gear, which includes a mask, a pressure regulator, and a mouthpiece.

There are three major types of scuba equipment. In closed and semi-closed circuit systems, known as rebreathers, exhaled air is recirculated and purified by passing through a canister of carbon dioxide absorbent, which extracts the carbon dioxide from exhaled air, making it safe to recycle. With these types of systems, divers can remain underwater for longer periods of time, but their complexity makes rebreathers potentially more dangerous. Open-circuit systems simply vent exhaled air directly into the water. They are preferred by most divers because they are relatively safe, easy to use, and less costly than other systems.

Divers use weighted belts and inflatable vests to adjust their buoyancy so they can descend in a gradual, controlled fashion, neither paddling furiously to get underwater nor sinking like a stone to the bottom. When water is cold, divers wear wet suits to withstand the chill. These full-length suits are made from a spongy rubber material that insulates against the cold water. While clinging to the body, the wet suit retains an inner layer of water, the temperature of which is regulated by the diver's body heat. A warmer alternative for very cold water is the dry suit — so-named because it keeps the diver dry even when fully submerged. The first practical diving suit was devised by a German named Klingert around 1797. His combination of metal parts for the main body shield and leather for the sleeves and breeches was innovative, but it was extremely heavy, and the leather restricted the diver's movements.

Far more complex than snorkeling or scuba diving, is tethered diving (hard-hat diving), in which air is pumped through a hose from the surface or an immersed structure. This approach, although more expensive, is commonly used for deep-sea excavations or to repair underwater structures like oil rigs. There are many different designs available, but the classic hard-hat assembly is heavy and awkward, usually consisting of a 60-pound (27-kg) helmet connected to a rubberized body suit, an 80-pound (36-kg) weighted belt, and a 40-pound (18-kg) pair of weighted shoes. The diver controls the air supply by using valves on the belt or inside the helmet. Air is important not only for breathing, but aids in ascent and descent as well. A non-return valve allows air pressure to build in the suit so the diver can rise. The chin valve releases air into the water, allowing the diver to vent air and sink deeper.

Another type of tethered-diving apparatus is the JIM suit, a magnesium **alloy** and **fiberglass** shell invented by Joseph Peress in 1922. It is named for Jim Jarrett, who made the first experimental dives using this suit in the 1920s. The JIM suit maintains the diver at a constant 1-atmosphere (sea-level) pressure and has its own 20-hour air supply. It weighs in at a massive 910 pounds (413 kg), which helps the diver withstand the strong ocean pressure encountered at its maximum working depth—about 2,000 ft (610 m).

See also Cousteau, Jacques

DIVING BELL

The first device that allowed man to descend underwater and observe the marine environment for an extended period was the diving bell. Named for their shape, these chambers began as open-ended, metal-rimmed wooden barrels and have existed since antiquity. Legend claims that Alexander the Great (356-323 B.C.) used a diving bell in 332 B.C. to take a trip beneath the Bosporus. Whether this tale is true or not, **Aristotle**, a contemporary, told of a simple but functional diving bell. By 1620 Sir Francis Bacon (1561-1626) knew of a ''sort of metal barrel'' that could take men underwater. And in 1665, a diving bell was used to salvage guns from a shipwrecked Armada vessel. Although many eighteenth-century inventors worked on diving bells, the noted English astronomer Edmund Halley (1656-1743) is credited with the first modern version. Although still open at the bottom, his wooden bell had a capacity of 60 cubic feet (1.7 cu m) and contained a bench and platform for its occupants. In 1716, he demonstrated a way to replenish the air inside so it could remain submerged for even more extended periods of time. While divers remained underwater inside the bell, lead containers filled with fresh air were lowered from the surface. The containers were encased in wooden barrels to which hoses were attached. When the barrel reached the bell, it was tipped in such a way that the air was forced upward and entered the bell. As the new air entered, the occupants released the spent air through a tap at the top of the chamber. Halley later developed a device he called the ''cap of Maintenance,'' a kind of headgear that allowed divers to leave the bell yet remain attached to its air-supply system. By the 1720s developers added an air pump to the diving bell, ensuring a constant supply of fresh air. In 1788, British engineer **John Smeaton** built the first truly modern diving bell, which was both safe and well designed. Modern bells can hold up to four passengers and travel to depths of about 1,500 feet (460 m). These are primarily used for construction, maintenance, and repair of offshore oil platforms and for naval operations.

See also Bathyscaphe, Bathysphere, Diving Apparatus, Submarine

DIVING SUIT • See Diving apparatus

DNA FINGERPRINTING • See Genetic fingerprinting

DOG BISCUIT

Perhaps as many dog owners as dogs are grateful to F. H. Bennett, who invented the first dog biscuit in 1908. The F. H. Bennett Company created the small bone-shaped biscuit that is regarded by many today as a way to keep a dog's breath fresh and his attention focused on appropriate behavior. The biscuit began as a novelty item made from a combination of minerals, meat products, and milk. It was the only product of the company that remained in production after the National Biscuit Company took over the Bennett bakery in 1931. The Nabisco company named the product Milk-Bone and marketed it as a ''dog's dessert.'' The *Milk-Bone* dog biscuit endures as one of the most popular dog treats sold in America.Owners of today's pampered pets offer homemade dog buscuits or purchase gourmet and natural food treats for their animals.

DOME

A dome—essentially a three-dimensional arch—is a special class of vaulted roof. As opposed to other forms of **vaulting**, the dome is round, like an inverted bowl.

The earliest known domes are found in Cretan tombs dating to about 2500 B.C. The pantheons, or temples, of the Roman Empire were round structures capped by domes. The

Constructed in the sixth century, St. Sophia Church, or the Hagia Sophia, was the first building to feature a dome atop a rectangular base.

Romans made extensive use of concrete, bracing arches with cross-vaults when the load demanded it. The dome of St. Sophia church in Istanbul was the first to be placed atop a rectangular building.

Early domes were constructed of masonry. Just as an **arch** is held together by its keystone and interlocking pieces, a dome's stability relies upon the center piece at the top. The weight of the structure is then dispersed outward and downward through the walls of the dome.

The value of a dome is that it can span large spaces with no internal support. Domes are used as the focal point for public structures such as capitol buildings, churches, and mosques, and are often as ornamental as they are functional.

During the 1800s, domes began to be built with iron and steel frameworks. In the mid-1900s, R. Buckminster Fuller invented the **geodesic dome**, which made maximum use of minimum materials and consisted of many small interlocking tetrahedrons (pyramid-like forms with a base and three sides). Made of aluminum and glass or plastic, the geodesic dome is lightweight, resistant to wind, and capable of covering acres of ground with no internal support.

While the dome is seldom used in modern architectural designs, it has enjoyed renewed popularity with the construction of enclosed stadiums. Relying more on the sheer strength of new building materials, stadium roofs must be high enough and must cover a large enough playing area so as not to interfere with the game being played. The Houston, Texas *Astrodome*, built in 1966, was the first domed stadium. Its roof is 208 ft (63.4 m) high and 642 ft (195 m) across. Some stadium roofs are retractable. The roof at the *Metrodome* in Minneapolis, Minnesota is composed of lightweight materials supported by compressed air.

Dondi, Giovanni de' (1318-1389)

Italian astronomer and physician

Giovanni de' Dondi (1318-1389) was the second in a line of Italian clockmakers. His father, Jacopo de' Dondi (1298-1355), designed a **clock** for the Palazzo del Capitanio at Padua. In 1359 or 1352, Giovanni became a professor of medicine at the University of Padua; several years later he was appointed to that university's four faculties of medicine, logic, philosophy, and astrology. He taught astronomy at the University of Padua, later lecturing on medicine in Florence, Italy. He also served as ambassador to Venice, Italy, and held several other political offices.

Dondi is remembered chiefly for his design and construction of an astronomical clock known as an astrarium. The astrarium (also referred to in contemporary accounts as a "planetarium" or a "clock") as it was called, took 16 years to complete. The astrarium connected the Ptolemaic model of the universe with contemporary horological knowledge. It also set a model for the construction of astronomical clocks on European buildings.

At the astrarium's unveiling in 1364, it proved to be one of the most elaborate astronomical devices yet created. The mechanism was set upon a two-part heptagonal framework. The top frame had seven dials to plot the movements of the sun and moon, as well as the planets Mercury, Venus, Mars, Jupiter, and Saturn—the only known planets in the fourteenth century. The bottom frame comprised a 24-hour dial, upon which were recorded ecclesiastical feasts as well as the lunar nodes (used to compute eclipses). At either side of this dial, the winter and summer solstices were recorded in keeping with the Julian **calendar**.

Dondi also provided a detailed treatise describing the device's workings, now considered by historians as one of the earliest recorded descriptions of any clock. Twenty years after the astrarium's completion, it was purchased by Duke Gran Galazzo Visconti, who housed it in his library at Pavia. The astrarium was prized as one of the most remarkable and beautiful artifaces of its time and was said to have been studied by Leonardo da Vinci, Regiomontanus, and Petrarch. The clock changed hands several times before falling into the possession of Charles V in 1529 or 1530, when it was considered beyond repair and was retired.

Modern clockmakers constructed replicas of the astrarium by following the design in Dondi's treatise. One of these replicas resides in the Museum of History and Technology in Washington, D.C.

Donkin, Bryan (1768-1855)

English engineer

Bryan Donkin was born in Northumberland, England, in 1768. Apprenticed to papermaker John Hall in Kent, England, Donkin was given the task of setting up and improving the automated papermaking machine designed in 1798 by the Frenchman Nicolas Robert (1761-1828). Donkin established plants in 1803 and 1804 at Bermondsey and Frogmore with the first practical Robert machines and continued to build better models—191 of them—through 1851.

In 1811, Donkin and Hall purchased Peter Durand's 1810 patent for using tin cans to preserve food by the method of heat sterilization invented by **Nicolas Francois Appert**. Donkin and his partners set up England's first cannery in 1812 at Bermondsey, supplying tinned meats and soups to the Royal Navy as well as to the royal family. The cannery's success was assured after its products were used on John Ross's (1777-1856) expedition to the Arctic in 1814 and on Otto Kotzebue's (1787-1846) voyage in search of the Northwest Passage in 1815.

Donkin's involvement in papermaking fostered his subsequent interest in **printing technology**. In 1813 he patented one of the first rotary presses. Although the machine itself was a failure, the composition rollers made of glue and treacle were an important innovation widely adopted by the industry.

Donkin was a founder in 1818 of the Institution of Civil Engineers. An avid amateur astronomer, he served for a time as president of the Royal Astronomical Society. Donkin died in London, England, in 1855.

See also Can and canned food

Doppler Radar

The changing frequency of sound waves emitted from a moving object was documented by Austrian physicist Christian Doppler in 1842 and termed the "Doppler effect." This concept was applied to **radar** in the 1970s, spurring the development of Doppler radar. Doppler radar is based on the concept that an approaching storm emits sound waves at a higher frequency than does a retreating storm. Doppler radar allowed for more advanced methods of weather forecasting because, unlike previous radar, it could not only determine whether precipitation and other severe weather conditions were in an area, it could determine the direction and wind speed of the system. Doppler radar also allows for earlier detection of a severe weather system. For example, conventional radar can detect a tornado only two minutes before its arrival; Doppler radar allows for a 20-minute warning.

The National Weather Service began replacing its conventional radar system with Doppler radar in the mid-1990s. The Service now operates 156 high-powered Doppler radar stations across the United States (including Alaska and Hawaii), Guam, and Korea. Each station can detect precipitation and severe weather systems from approximately 285 mi (456 km) away and measure winds from approximately 150 mi (240 km) away.

Double boiler

The double boiler—water-bath or bain-marie—was invented by a woman known as Maria the Jewess, Mary the Jewess, Maria Prophetissa, or Miriam the Prophetess. An alchemist of either the first or second century A.D. who lived in Alexandria, she wrote under the title "Miriam, sister of Moses." Only fragments of her writings on alchemy, the Maria Practica, survive. The double boiler, still in use today, maintains a constant temperature for preparing experimental substances or food and heating reagents slowly and steadily. Substances undergoing phase changes—in the case of the double boiler, liquid turning to vapor—usually remain at a constant temperature, so the temperature of a pan immersed in boiling water would hold steady at about 212° F (100° C) while excess energy boils off as steam. The modern-day double boiler is small and compact, usually consisting of a saucepan nesting in a slightly larger one. Invention of the military **microwave** oven has decrease use of double boilers, since the same effect can be achieved without the inconvenience of boiling water. Other inventions attributed to Maria the Jewess are still used today. The tribikos, or three-part still, may have been the first distillation device and was constructed of pottery and copper tubing. The kerotakis, a device used for the condensation and reflux of vapors, is still a fixture in the field of chemistry.

Doughnut

The invention of the doughnut, which evolved from the round, fried Dutch cakes brought to colonial America, is usually attributed to Hanson Crockett Gregory, a sea captain born in Rockport, Maine. In 1947, a plaque commemorating the captain's achievement was affixed to his birthplace. While it is generally agreed that the doughnut dates back to 1847, storytellers relate several different versions of Gregory's discovery. One tale holds that Gregory was commanding a vessel called the *Frypan* when six men who fell overboard drowned because their bellies were full of fried cakes. Distressed by this tragedy and determined to avert it in the future, Gregory pondered the dilemma of the too-heavy fried cake. His solution was to jab a hole in the middle of the cake (thus, appropriately, making it resemble a life preserver) to lessen its danger to hungry sailors. One of the most frequently told stories about Gregory and the doughnut also takes place on a **ship**. During a sea voyage, Gregory and his crew struggled long and hard to guide their vessel through a ferocious storm. A thoughtful cook brought the valiant, exhausted captain a snack of fried cakes to eat as he stood at the ship's helm. When the ship suddenly encountered a huge wave, Gregory impaled his fried cake on a spoke of the wheel to free both of his hands for steering (in a slightly different version of this story, the power of the wave's impact knocked the cake onto the spoke without Gregory's help). The proud captain publicized his invention after his return to Maine, and the doughnut became a favorite treat for seamen.

Despite the colorful appeal of these stories, their veracity was questioned by Captain Gregory's descendent, Fred E. Crockett of Camden, Maine, during the Great Doughnut Debate. Held in October 1941 at the Hotel Astor in New York, the event was sponsored by the National Dunking Association, founded in 1938 and reportedly comprised of three million members (including Martha Graham and Bob Hope). The debate featured appearances not only by Crockett but by Chief High Eagle of the Wampanoag tribe, who claimed that his own people invented the doughnut when a wayward arrow missed a Pilgrim homemaker and pierced her fried cake instead. Crockett, however, dismissed both the chief's story and those popularly told about his ancestor. He noted that the captain would have been only fifteen years old in 1847, too young to have command of a ship. Crockett related instead how young Hanson Gregory had instructed his mother to poke out the middle of her fried cakes to avoid the sogginess that frequently lingered there. Mrs. Gregory shared her doughnuts with her neighbors and their fame gradually spread.

The first doughnut-hole machine, which featured a spring-loaded tube to push the dough out of the cake's middle, was patented in 1872 by John F. Blondel of Thomaston, Maine. During the First World War, United States soldiers stationed in France received doughnuts from the Salvation Army, and the national fondness for the hole-less fried cakes grew. In 1921, a Bulgarian immigrant named Arnold Levitt invented a machine that could mass-produce doughnuts. His Donut Corporation of America, founded just after World War II, helped bring the doughnut worldwide acclaim. By the end of the 1980s, America's two most famous doughnut makers, Dunkin' Donuts and Mister Donut, had 1,878 and 558 franchises, respectively. By the late 1990s, Americans were eating an estimated 10 billion doughnuts every year.

There are two main types of doughnut today. Raised doughnuts are made with yeast and allowed to rise at least once before being fried. Cake doughnuts are made with baking powder and chilled before frying to keep the dough from absorbing too much oil when being cooked. Raised doughnuts can be filled with jelly or cream, while the dough for cake doughnuts is often flavored with spices, fruit, or chocolate.

Douglas sea scale

The Douglas Sea Scale was devised by the English Admiral H. P. Douglas in 1921. Its purpose is to estimate the sea's roughness for navigation. The Douglas Scale consists of two codes, one for estimating the state of the sea (fresh waves attributable to local wind conditions), the other for describing sea swell (large rolling waves attributable to previous or distant winds).

It was difficult to relate the **Beaufort wind scale** to a ship's features, especially as sails were replaced with the rigid structures of powered ships. The Douglas Sea Scale standardized the many variations being used by **ship** captains from many nations.

See also Petersen scale

Dow, Herbert Henry (1866-1930)

American chemist

Dow was born in Belleville, Ontario, Canada, on February 26, 1866. His parents were Americans who had moved to Canada when his father could no longer find work in their native New England. Only a few months after Herbert's birth, the Dow family moved back to the United States. They settled first in Derby, Connecticut, then moved on to Cleveland, where Herbert's father got a job at the Chisholm Steel Shovel works. In 1884, Dow entered the Case School of Applied Science (now Case Western Reserve University). He majored in chemistry and became especially interested in the study of brines, a form of salt water. During his senior year at Case, he presented a paper on brines before the American Association for the Advancement of Science.

After graduation in 1888, Dow began to work on methods for extracting bromine from brine by electrolytic methods. He received his first patent on the topic in 1889 and decided to open a plant based on the process. The plant, located in Canton, Ohio, had an impressive name—The Canton Chemical Company—but consisted of little more than an ill-equipped shed. Less than a year after it opened the business failed and Dow looked for a new location where he could begin again.

The site he selected was Midland, Michigan at the base of the state's "thumb" region. One reason for this choice was the presence of enormous underground reserves of brine beneath the eastern part of the state. The brine, a remnant of an ancient sea, constituted an essentially limitless supply of raw material for Dow's new process.

In August 1890, Dow moved to Midland and opened his new company, The Midland Chemical Company. At first his neighbors were puzzled by the newcomer's business, and they referred to him behind his back as "Crazy Dow." However, less than six months later Dow's plant was in operation producing bromine from brine by using electric current.

Dow rapidly proved that he was a chemical genius, soon developing methods for producing chlorine, magnesium, and other commercially valuable chemicals from brine.

He was not as successful at first as a businessman, however. At one point, he lost control of his own patents and of the company he had founded. By 1897, however, his business skills had improved and he founded another new company, the Dow Chemical Company. Dow Chemical has since grown to become one of the half dozen largest chemical companies in the United States. The business is now a multimillion dollar operation that produces hundreds of different chemicals, including drugs, agricultural chemicals, **plastics**, **dyes**, caustic soda, hydrochloric acid, **rubber**, industrial solvents, and the chlorine-and bromine-based products with which the company started.

Dow became ill in September 1930 and entered the Mayo Brothers Clinic at Rochester, Minnesota. He was diagnosed with cirrhosis of the liver and failed rapidly. He died at the Clinic on October 15, 1930.

See also Polymer and polymerization

Drainage • See Water supply and drainage

Draper, Charles Stark (1901-1987)

American engineer

Charles Stark Draper is best known for his work on gyroscopic instruments and other navigation systems for sea, air, and **spacecraft**. He was born on October 2, 1901, in Windsor, Missouri. After high school, he attended the University of Missouri for two years before transferring to the psychology program at Stanford University in Palo Alto, California, in 1919. After his graduation from Stanford with a bachelor's degree in psychology, Draper made a decision that changed the course of his life: he agreed to drive to Boston with a friend who planned to enroll at Harvard. As the two young men drove through Cambridge, Massachusetts, the scenic view from the road near the Massachusetts Institute of Technology (MIT) caught Draper's attention, and he convinced his friend to stop for a better look. Several hours later, Draper emerged from the school's campus having enrolled in its electrochemistry program.

Draper continued as a student at MIT through 1938, receiving a bachelor's degree in electrochemical engineering in 1926, a master's degree in 1928, and doctorate in physics in 1938. In 1939, he became a Professor of Aeronautical Engineering at MIT. Draper's interest in aviation greatly influenced his research. As an **aircraft** owner and pilot, he was directly affected by airplane engine performance and navigational capabilities. This practical approach always informed his theoretical research. Among his accomplishments during his early years are improvements to the magnetic **compass**, the rate of turn indicator, the rate of speed indicator, the rate of climb indicator, and other gyroscopic instruments.

Developments in Draper's work eventually led to the 1940 founding of MIT's Instrumentation Laboratory. Through the Instrumentation Laboratory (originally named the Confidential Instrument Development Laboratory) Draper played a crucial role in improving navigational technology during World War II. His work on a more accurate and reliable gunsight for use onboard United States Navy vessels was prompted by the sinking of two **warships** in 1941. The problem with the existing antiaircraft guns was that their **gyroscopes** did not give enough tracking data to allow for computer-aided targeting. Draper addressed this difficulty in his Mark 14 gunsight. A "rated" gyroscope floating in viscous fluid was the basis of Draper's device. The gyroscope could compute the data necessary to achieve a direct hit of a designated object including the precise target range, wind velocity, and angle of the weapon. This allowed the gunner to simply bring the marked plane within a circle of dots on a reflecting glass to aim his gun. The Mark 14 proved its worth at the 1942 battle of Santa Cruz on board the U.S.S. *South Dakota*, and thereafter became standard in all United States Navy antiaircraft guns.

Draper also helped develop a gunsight system for aircraft. Called the A-1, it also was an immediate success. Advanced models of this later gave the F-86 Sabre an advantage in the Korean War against Russian-built planes. Although the war ended, Draper continued to work on research for the United States military. His areas of research included inertial navigation and guidance systems for sea and air that calculated the course of the craft without input from external sources. The first project he took on was developing a partially inertial navigating system including a **gyrocompass** and Marine Stable Element (from which the project, MAST, took its name) for use on **ships** and **submarines**.

In 1951, Draper began a project known as SINS—Ship's Inertial Navigation System—which made the theory of an entirely inertial sea navigation system a reality. The problem SINS overcame was the tendency of gyros to drift off course over several months of operation unless they are corrected. Not many people took notice of Draper's initial demonstration of SINS, but among those who did was Vice Admiral Hyman Rickover—an instrumental figure in the development of the nuclear submarine. Rickover immediately saw the potential use of SINS with nuclear submarines and agreed to sponsor the further development of the system. Once the success of SINS was proved, United States military leaders asked Draper to develop navigational systems for Polaris ballistic missiles that would interface with the SINS unit on the submarine from which the missile was launched.

The success of his work on the navigation system of the Polaris missiles resulted in a Distinguished Public Service Award from the Navy. Draper was subsequently involved in many ground-breaking **rocket**, aviation, and space projects. Project SPIRE, for example, was an inertial autopilot system that succeeded in keeping the plane on course throughout a flight across the United States, correcting for winds and currents, and elevating the plane to clear the Rocky Mountains. Draper was instrumental in designing navigational systems for the mid-range *Thor* and long-range *Titan* rockets. During the 1960s, the Instrumental Laboratory under Draper's supervision designed navigational systems for the *Apollo* manned missions to the moon. Draper's chief technical work was clearly the inertial guidance system. His influence, however, spread beyond the bounds of his technical contributions to the work of his fellow researchers and students. Draper's philosophies, techniques, and methods of examining scientific results informed an entire generation of scientists and engineers who followed him. Draper died in Cambridge, Massachusetts, on July 25, 1987.

Charles Stark Draper

Drebbel, Cornelius Van (1572-1633)

Dutch inventor

As an adolescent Drebbel was apprenticed to an engraver but soon developed an interest in alchemy and mechanical inventions. In 1598 he was granted a patent for a perpetual motion machine which reportedly used changes in atmospheric pressure to power a **clock**. This "invention" is thought to have established Drebbel's fame in scientific and aristocratic circles in Europe. Around 1604 he journeyed to England, where he was awarded an annuity by King James I (1566-1625) to continue his scientific work. Drebbel has been credited with constructing the first compound **microscope** using two sets of convex **lenses**, and he employed his skill as lens grinder to manufacture a variety of optical instruments.

During the early 1620s Drebbel designed and built his most famous invention, the **submarine**. Although a similar design had been described some fifty years earlier, Drebbel's is the first known to have been constructed. Consisting mainly of greased **leather** stretched over a wooden frame, Drebbel's submarine was propelled by oars projecting through the sides and sealed with leather flaps. The vessel was capable of traveling twelve to fifteen feet (3.6 to 4.5 meters) below the surface, and fresh air was supplied by tubes running to the surface with floats at the top. Drebbel successfully tested his submarine several times in the Thames River in England.

Drebbel invented the first **thermostat**, which used a column of mercury and a system of floats and levers to maintain a steady temperature within a furnace. He later invented an **incubator** for hatching eggs which used the same principle for temperature regulation. Drebbel also discovered the first permanent scarlet fabric **dye**, which became popular throughout Europe, and developed a process for manufacturing sulfuric acid from sulfur and saltpeter.

Dredger

River channels and estuaries sometimes become obstructed by accumulations of silt or by submerged objects. This can result in the blockage of transportation routes or even in flooding. A dredger is a self-powered barge-like vessel that is used primarily to clear these waterways. It is also used to construct **dams** and **canals** for drainage projects, for the recovery of submerged objects, or to extract valuable minerals, such as gold, from silt deposits.

The first known dredgers were in use in the Netherlands during the twelfth century. For many centuries, the Dutch have used bottom material to build dikes. The seawater behind the dikes is drained using **windmills**, forming polders. Then the new land can be desalinized and used for agriculture. The first steam-powered dredger was invented in England in 1796. Scottish engineer, **John Rennie**, made improvements to it, and used them in the construction of the London Docks in 1804 and 1805.

Dredgers can be secured to the bottom with the use of pinions, called spuds. The material at the bottom is removed with either a scoop shovel, grab bucket, or bucket ladder, and is deposited either on an adjacent barge or directly on the shore. Some hydraulic dredgers even use hoses to suck matter from the bottom, similar to a **vacuum cleaner**. Modern dredgers are powered by diesel fuel, though steam is still used in some areas. They are commonly seen in river systems, harbors and estuaries in the United States and other parts of the world.

Dressmaking

Ready-made clothing available in standard sizes in retail shops is a relatively recent phenomenon. The modern garment industry that produces these goods did not evolve until after the refinement of assembly-line **mass production** techniques in the late nineteenth century. Until that time, dressmaking was generally one of the numerous and time-consuming chores of the housewife. Those who could afford it hired professional seamstresses to do the work.

One enterprising professional seamstress of the nineteenth century who achieved a certain renown was Elizabeth Keckley (1818-1907). An African-American born as a slave in Virginia, Keckley—through her efforts as a seamstress and dressmaker—managed to buy her freedom in 1855. After moving to Washington, D.C., in 1860, she instructed other seamstresses in a system she had invented for cutting and fitting dresses. Her fame evidently spread quickly, for she soon had a select group of clients, among them Varina Howell Davis, whose husband was about to become president of the Confederacy, and Mary Todd Lincoln (1818-1882), wife of the newly-elected president of the United States.

Just after moving into the White House in 1861, Mrs. Lincoln hired Keckley to make her a gown. Keckley's work and personality were apparently very satisfactory, for she was almost immediately not only the First Lady's dressmaker, but also her maid, traveling companion, and a confidante to whom Mrs. Lincoln entrusted her private views of public figures and issues. In 1868, three years after Lincoln's assassination, a ghost-written book called *Behind the Scenes* appeared under Keckley's name. Although purporting to improve Mrs. Lincoln's public image, it had quite the opposite effect. It also had unfortunate results for Keckley: her relationship with Mrs. Lincoln ended abruptly, her dressmaking business declined, and other African-Americans shunned her as a traitor to the much-loved Lincolns.

At the time Keckley was introducing her cutting and fitting system to the seamstresses of Washington, dressmaking generally involved laboriously picking apart an old garment to use as a pattern for a new one. **Paper** patterns—full-scale models for cutting pieces of garments designed in standard sizes—did not come into general use until after 1863. This labor-saving invention was, however, not unknown. When the women's magazines that became so popular in the course of the nineteenth century first appeared, they often included patterns for lace caps, embroidery, and the like. The patterns must have been well received by homemakers, for by the 1840s some of these magazines were including inserts of full-scale paper patterns of various garments with each issue.

Meanwhile, professional seamstresses and tailors had been able to avail themselves of several books on cutting, the first of these having appeared in the late 1700s. By 1834, professionals in Great Britain could purchase a set of "models make-up in paper" for the sum of ten shillings. It was, however, a Yankee, Ebenezer Butterick (1826-1903), a native of Sterling, Massachusetts, and a tailor by trade, who, with the help of his wife Eleanor, first streamlined the process of patternmaking and made the paper pattern available to professionals and amateurs alike.

It is said that Butterick brought fashion within the reach of any woman who owned a **sewing machine**, cloth, and the fifty cents that his patterns cost. Instead of drawing individual patterns onto the cloth with wax chalk, Butterick created an assembly-line operation by cutting a standard-sized pattern out of stiff paper and then using it as a template for cutting other patterns. In 1863, he closed his tailor shop and went into the business of patternmaking full time. By 1870, Butterick had competition, for in that year McCall's Patterns came on the market. Meanwhile, the women's magazines had continued to include patterns with each of their issues. In 1892, *Vogue* magazine came into existence, and its patterns proved so popular that a separate publication, the *Vogue Pattern Book*, appeared.

None of this seems to have hurt Butterick's business. When he died in 1903, his company was still selling millions of patterns each year. Despite the efficiency of the modern garment industry and the disappearance of patterns from women's magazines, home dressmaking is apparently not yet a lost art. The labor-saving paper pattern is still available from Butterick's, McCall's, Vogue, and other companies.

Drew, Charles Richard (1904-1950)

American physician

Charles Drew, noted authority on hematology and human **blood transfusion**, developed a method for storing blood plasma and thus became known as the ''Father of the Blood Bank.'' Drew also exerted his influence on the American Medical Association to rid its affiliates of racism. Born the eldest of five children on June 3, 1904 in Washington, D. C., he was the son of Richard Thomas Drew, a **carpet** installer, and Nora Rosella Burrell Drew, a teacher. As a child, Drew earned money with a **paper** route, expanding his territory by hiring other boys to help him deliver papers. He won swimming medals in elementary school and graduated from Paul Laurence Dunbar High School in 1922 as star halfback with honors in football, basketball, track, and baseball. On an athletic scholarship, Drew waited tables so that he could complete undergraduate work at Amherst College, where he earned the Mossman and Thomas W. Ashley trophies, Canadian Championship, and Pentathlon Award for athletic prowess in track and football.

An indifferent student in his early years, he was so taken by his studies in biology that he resolved to become a doctor, although he lacked the funds for medical training. Following graduation in 1926, he taught chemistry and biology and directed the sports program at Baltimore's Morgan State College. In 1933, paying his way by working as a referee, he enrolled at McGill University in Montreal, where he earned medical and surgical degrees, a Rosenwald Fellowship, the Williams Prize, a prize in neuroanatomy, and membership in the medical honor society. Drew completed an internship at the Royal Victoria Hospital, followed by a year of surgical residency at Montreal General, where he concentrated on blood typing, surgery, and transfusion. He taught pathology at Freedmen's Hospital and was a professor of surgery at Howard University in Washington, D. C. In 1938 he received a Rockefeller fellowship at New York's Presbyterian Hospital. While serving as a General Education Board Fellow at Columbia-Presbyterian Medical Center, Drew married teacher Minnie Lenore Robbins and began a family. Professionally, Drew concentrated on surgical shock, blood preservation, and the use of plasma in transfusion. He also opened Columbia's first blood bank. Drew completed a Sc.D. in medical science in 1940. His dissertation described the preservation of banked blood, which he learned about from work done at the Cook County Hospital in Chicago. He published his findings in ''Plasma Potassium Content of Cardiac Blood at Death'' (1939) and ''Studies in Blood Preservation: Some Effects of Carbon Dioxide'' (1940).

Charles Richard Drew

At the beginning of World War II, Drew took a leave of absence from Howard University to perfect plasma use and blood processing and storage methods for both England and France. During the height of Hitler's assault on England, he directed the ''Blood for Britain'' drive and organized a local civil defense team. His interest in trauma medicine led to the creation of lifesaving blood banks and an appointment with the American Red Cross. In 1941, he directed Red Cross blood collection and the use of dried plasma for the U. S. military. Because of a regulation requiring that blood from non-white donors be stored separately from Caucasian blood and not be administered to military wounded, he spoke out against the absurdity of blood segregation. He was forced to resign his post after he took a firm stand on the fact that blood cannot be identified by race in the laboratory. Returning to Howard University, Drew, humiliated and despairing, gave up research and returned to the classroom, where he distinguished himself through warm, personable relationships with students. He rose to head of surgery at Freedmen's Hospital and, in 1942, became the first African-American to serve as examiner for the American Board of Surgery. This honor was followed by the NAACP's Spingarn Medal, honorary doctorates from Virginia State College and Amherst College, and a consultancy with the army surgeon general, during which he upgraded European medical facilities. He performed community service at local hospitals, spoke against racial bias in medical hiring, and published articles on hematology.

On April 1, 1950, while traveling with three associates to deliver a speech at the Andrew Memorial Clinic of Tuskegee Institute, the car Drew was driving crashed when he fell asleep at the wheel and ran off the road outside Burlington, North Carolina; Drew was killed. His colleagues honored him with the Charles R. Drew Memorial Fund, which maintains scholarships and lectures. Schools in eight states have been named for him, a Drew clinic operates in Brooklyn, and Howard University boasts a Drew Hall.

DRIED FOOD • See Food preservation

DRILL

The drill, which dates back to the Stone Age, was first used to start fires. A slender piece of wood was rotated between the hands to create a flame, through friction. This device became the bow drill by the addition of a bow-shaped stick and a simple cord attachment that allowed for increased rotation.

As early as the eighth century B.C., drills were used by the Greeks for hole-cutting, and this remains their primary purpose today. By the third century B.C., the Chinese found that drills could be used to bore through the earth to tap oil deposits. Heavy drilling through rock was accomplished by means of metal or weighted attachments to systematically pound and penetrate.

Drills developed little until well into the nineteenth century. They were most extensively used in clockmaking, an application which required only that the drill bit be sharp enough to puncture thin metal. The clockmaker's drill press was operated either by hand crank or foot treadle. In 1835, Joseph Whitworth, a British mechanical engineer, introduced a heavy-duty vertical drill that represented a huge advance over previous presses in both speed and stability. Innovations that followed included the portable electric drill, the radial drill (for aligned drilling over large surfaces), and the pneumatic drill (especially useful for tunneling work).

Lightweight cordless drills have severed the link to the electrical outlet and offer easy use in almost any situation. Improvements in the nickel-cadmium **battery** technology used in these tools, and other design refinements have made the cordless drill the most popular of cordless tools.

Drill bits typically contain spiral-shaped grooves, though some bits have straight cutting edges. These grooves enable waste material to be channeled away from the cutting surface. Closely related to the drill is the **boring machine**. Together these two machine tools facilitate the manufacture and assembly of countless objects of varying shape, size, and use.

See also Lathe

DRIVING CHAIN • See Chain

DRUMS, ELECTRONIC • See Synthesizer, music

DRY CLEANING

Dry cleaning is a process which removes dirt and stains from garments using a solvent other than water. As a profession, dry cleaning dates back to Mycenean civilization, around 1600 B.C. Archaeologists surmise that these ancient dry cleaners used absorbent earth or powdered meal to draw sweat, odors, and soil from clothing. Later innovations depended more on liquid solvents than on absorbent powders. In the seventeenth century, for instance, one adviser noted that turpentine would remove rosin by evaporating stains from the weave. Since the eighteenth century, people have protected fabrics from shrinkage and warpage by replacing water-based cleaners with chemical substitutes such as naphtha, benzine, and benzol. One technique involved the vigorous application of grated potatoes as a means of cleaning unwashable materials, furniture, and even oil paintings. Other products, such as pinene and camphene, successfully removed spots. A common home dry cleaning method called for the application of **gasoline** to dirty clothes to dissolve protein and fatty or oily substances, such as blood or gravy stains, but the flammability of gasoline proved hazardous.

In 1821, Thomas L. Jennings (1791-1859), an African-American tailor and abolitionist, patented a dry-cleaning process, but it was not until the mid-1800s that entrepreneurs instituted formal dry cleaning. Various legends describe how chemical solvents came to light. Some credit a maid who spilled turpentine on a soiled dress; another version describes a French sailor who fell into a barrel of turpentine and came out with a spotless uniform. In 1845 Parisian dyer Jean-Baptiste Jolly spilled **kerosene** on a soiled tablecloth and discovered that the substance cleaned the spot. He coined the term "dry cleaning" to differentiate it from regular **soap** and water wash; his firm, Jolly-Belin, became Europe's first professional dry cleaner. By 1897, Ludwig Anthelin of Leipzig, Germany, discovered that application of carbon tetrachloride gave acceptable results without creating a fire hazard. Because the solution was an irritant, fans to ventilate solvent storage chambers were necessary to protect workers from severe respiratory distress. Around 1910, American dry cleaners were capitalizing on European successes. Valet shops and pressing clubs expanded into adjuncts for laundries, where a variety of services, including alterations, reweaving, and repairs led to full-time clothing care.

By 1919, dry cleaning had grown into a $55 million industry and cleaners were searching for improved, nonflammable solvents. Around 1930, dry cleaners introduced a less harmful chemical named Stoddard solvent to honor W.J. Stoddard, president of the National Institute of Dry Cleaning. Other nonflammable synthetic solvents, such as trichloroethylene and perchloroethylene, were developed for use in both professional cleaning plants and coin-operated machines. The usual procedure for dry cleaning required that clothing be inspected, tagged, then sorted as to color, **fiber**, and weave. Delicate fabrics—such as cashmere, angora, ornate lace, **leather**, suede, fur, and other fragile materials—and seriously stained garments were spot-cleaned by hand. Ornamentation, belts, lace collars and cuffs, fur pieces, and shoulder pads were removed for individualized care. The soiled items were churned in a large drum similar to a **washing machine** through mists of solvent and detergent, which loosened and suspended stains and grime. Then the bundle underwent a second cycle to agitate and remove the chemical cleaning agents. Used solvent was filtered or distilled to remove soil and to allow the reuse of the chemicals. To complete the job, a spotter brushed and steam-gunned stubborn spots. Finally, the clothing was dried. To remove wrinkles and restore shape and texture to the fabric at the conclusion of the cleaning process, pressers shaped and steamed garments with a professional-sized iron or over a basket-shaped hoop.

A half century later the industry still uses the same basic methods and has grown to be a $2.8 billion enterprise. However, the industry has not been without its difficulties. In 1960, the advent of permanent-press materials introduced the public to the ease and low cost of wash-and-wear clothing, thereby lessening dependence on professional cleaning. A second blow to the industry was the environmental movement, which urged city councils to regulate or ban the release of noxious solvents into the atmosphere.

In the last few decades several advances have been made in dry cleaning technology. One advance involves the creation of a sensor which helps optimize the amount of moisture used in the dry cleaning process. Despite the name, dry cleaning does involve some small amounts of water. Too much water causes fabrics to wrinkle and shrink and too little leads to residual spotting. A relative humidity transmitter developed in the mid 1990s has been used to help control the exact amount of moisture.

In 1997, Micell Technologies, Inc. developed technology which uses liquid carbon dioxide, instead of harmful chlorinated compounds, to remove dirt and grease from fabric. Due to the cost of the high pressure equipment used in this process, it may take a while for this technology to replace existing equipment. Micell's key to this technology was the discovery of a surfactant, or detergent, which was soluble in liquid carbon dioxide.

The latest development in dry cleaning is not really a dry cleaning process at all. In 1998 both Procter & Gamble's and Creative Products Resources introduced home dry cleaning kits which are designed to allow consumers to treat dry cleanable garments at home. Both systems consist of a non-woven fabric pad which is saturated with a cleaning solution. This pad is used to pre-treat the stained area of the garment and is placed in a large resealable plastic bag along with the garment. This bag is tightly sealed, placed in a dryer, and tumbled with mimimal heat. This heat helps loosen the pre-treated stain while the fragrance and other chemicals in the pad penetrate throughout the fabric leaving it refreshed. While this technology is not anticipated to completely replace existing dry cleaning operations, it may have a serious impact on the industry.

DRYER • See Clothes dryer

DUNLOP, JOHN BOYD (1840-1921)

Scottish inventor

John B. Dunlop was born in Aryshire, Scotland. After completing his schooling, Dunlop practiced as a veterinary surgeon first in Edinburgh, Scotland, and then in Belfast, Ireland. In 1887 Dunlop was asked by his young son to think of some way to make the boy's tricycle ride more comfortably on Belfast's cobblestone streets. Dunlop began experimenting with improvements to the tricycle's solid **rubber** tires. He cut up an old garden hose, made it into a tube, pumped it up with air, and fitted it to the rear **wheel**s of his son's tricycle. Dunlop had reinvented the **pneumatic tire**, which, unknown to Dunlop, had first been invented by **Robert Thomson** in 1845.

Dunlop tested and patented his pneumatic tire in Great Britain in 1888, and secured a U.S. patent in 1890. An Irish industrialist named W. H. Du Cros became interested in Dunlop's invention after a cyclist won a Belfast race with his pneumatic-tired **bicycle**. Du Cros organized a company with Dunlop, which became the Dunlop Rubber Company, eventually a worldwide concern.

Dunlop did not profit greatly from his invention. He sold his patent and interest in the company to Du Cros in 1896 and retired to Dublin, where his only business activity until his death there in 1921 was an interest in a drapery firm.

After its introduction in 1888, Dunlop's pneumatic tires became standard equipment for bicycles since they so well absorbed the direct road shock that bicyclists experienced. The same year as Dunlop's patent, the first **automobile** was produced by **Karl Benz** of Mannheim, Germany. The availability of Dunlop's new tires led to experiments by others to produce a thicker-tread tire for motor vehicles, and in 1895 Michelin's first automobile tire appeared. With the growing popularity of the automobile after 1900, the demand for increasingly durable rubber compounds for both bicycles and autos became enormous. Dunlop's invention gave rise to two major industries.

DVD

The term ''DVD'' is not an acronym, although it is popularly understood to stand for digital video disc or digital versitle disc. This product grew out of **laser disc**, or compact disc (CD), technology and offers high-density optical and audio storage for both entertainment and computing purposes. The storage capacity of DVD is greater than that of the CD, and is available with a double-layer design for twice the capacity. Like the laser disc, DVD offers video, audio and, data storage capabilities. DVD offers enhancements that a conventional CD does not, however. The audio DVD offers Dolby digital surround-sound. In the realm of video, DVD offers higher-quality video, the ability to contain longer films and the ability to move to a specific scene instantaneously. For computing, the DVD-ROM can store 20 times the information of a CD-ROM and offers enhanced graphics and faster movement.

DYES, SYNTHETIC

A dye is a material used to impart color to a substrate such as fabric or hair. Natural dyes have been used for this purpose for thousands of years. Egyptian clothing dating from 3000 B.C. was dyed with indigo, yellow, red, and green colorants. These dyes were derived from from plant and occasionally animal sources. Dyes from plants included blue dye from the indigo plant; blue dye from woad; red and brown dyes from the madder plant; yellow, orange, brown, and black dyes from various trees; orange or red dye from henna; and yellow from safflower and weld. Animal dyes included red dyes from the cochineal insect and Tyrian purple from shellfish. Dye baths were made by mixing the dye with an alkali, perhaps wood ash, and stale urine, a process noteworthy for the odors it produced. (Despite the smell, urine was a convenient source of ammonia for both dyeing and disinfecting for numerous ancient and medieval cultures.)

Dyes were dissolved and applied to the cloth, and the color change would take place either because the dye molecules became bonded to the fiber molecules or because the dye was unable to pass out of the fibers' fine capillaries once it is inside of them. Dyes that do not wash out easily are called *fast dyes*. Some dyes are not easily absorbed by fibers or are easily washed out. These dyes require the addition of a mordant which breaks down the fibers and helps the dye molecules to be absorbed more easily. For example, in 1844 **John Mercer**, the inventor of mercerized cotton, realized that wool dipped in chlorine was more easily dyed.

The synthetic dye industry began in Britain when **August Wilhelm von Hofmann**, a German chemist, was invited by Prince Albert (1819-1861) to become the director of the Royal College of Chemistry. Hofmann had received his Ph.D. for his work with coal tar and aniline. Hofmann, unfortunately, was somewhat clumsy in the laboratory and so hired good laboratory assistants, among them **William Henry Perkin**. While trying to synthesize quinine, Perkin created the first synthetic dye in 1856 when he added potassium dichromate and alcohol to aniline. Perkin and his family founded a dye factory to produce aniline purple dye for the silk industry. The color became known as mauve, from the French word for the plant that had previously been used to produce violet. Perkin followed up these inventions with aniline red (1859) and aniline black (1863) dyes. At first, Hofmann disapproved of Perkin's commercial activities, but he soon followed the path blazed by his student. Hofmann began researching dyes and eventually reacted carbon tetrachloride and aniline to form a dye called rosaniline. The reaction would not have been possible if the aniline had been pure. The presence of the impurities orthotoluidine and paratoluidine were necessary. In 1863 Hofmann succeeded in replacing the hydrogen in rosaniline with aniline to form aniline blue.

The next breakthrough in dye technology came when **Heinrich Caro**, building on the work of Peter Greiss (1829-1888), produced induline, and later Bismarck brown and Martius yellow; the two latter compounds were derived with the help of Martius. There followed a flurry of discoveries in the industry. Carl Graebe (1841-1927), a German chemist, developed aizarine, a red dye, in 1869. **Adolf von Baeyer** synthesized indigotin in 1880. He distilled isatin over hot zinc dust, a process that had a great impact on the development of synthetic dyes. Baeyer left the commercial development of indigotin to Karl Huemann, who is credited with the production of indigotin on an industrial scale. Huemann's process was improved by accident when a chemist named Sapper broke a **thermometer** while heating the naphthalene dictated by Huemann's method. The mercury combined with the sulfuric acid used in the process and formed mercury sulfate. The mercury sulfate converted the naphthalene to phthalic anhydride, which in turn was converted to indigo.

The first vat dye was developed by Rene Bohn in 1901. Bohn had received his doctorate from the University of Zurich with Huemann. He then went to work for BASF Corporation. Bohn tried to formulate a dye that would have the best traits of both alizarin and indigo. Instead he produced indanthrone, a blue dye. He soon discovered that when the reaction was run at a high temperature, it resulted in flavanthrone, a yellow dye. Today more than 200 dyes have been developed from indanthrone or are related to it chemically. These synthetic dyes are brighter than their natural counterparts but tend to wash out or fade when used in cotton or linen. Thus these fabrics require a mordant to set the dye. Johann Peter Griess (1829-1888) discovered that compounds could be bonded together using nitrogen compounds to form a chemical bond called azo bond. Numerous dyes contain this bond. Some, called acid dyes, dissolve in water and react easily with animal derived fibers, wool, and silk. Metal complex dyes are azo dyes that contain a metal molecule. Azo dyes used on cotton or cellulose fibers, such as **rayon**, must be formed inside the fiber itself. Many cellulose fabrics are dyed using vat dyes. Vat dyes and sulfur dyes are similar in that they are both embedded in the cellulose fiber rather than bonding to the fiber molecules. Reactive dyes bond with the fiber molecules and are used on cellulose fibers.

See also Cellulose, chemical uses of; Fiber, synthetic; Textiles

Dynamite

Dynamite is an explosive that was invented in 1866 by Swedish physicist **Alfred Nobel** and patented by him a year later. Nobel is most familiar to us today as the founder of the Nobel Prize. Many have observed the irony in the fact that he left his multimillion dollar fortune, made by the patenting and manufacture of dynamite and other inventions, to establish prizes awarded "to those, who during the preceding year, shall have conferred the greatest benefit on mankind."

Actually, Nobel invented dynamite to make the dangerous explosive, **nitroglycerin**, a safer substance. As a pacifist, he did not intend his invention to be used for war; indeed, there are indications he believed its use could bring war to an end more swiftly, or that the horrors of such an explosive would prevent warfare in the first place. Nobel also saw the need for explosives in mining, engineering, industry, transport, and other peaceful applications. His intention was to meet this need and in this he succeeded, making many of the great engineering projects of the nineteenth and twentieth centuries possible.

For centuries prior to the introduction of dynamite, **gunpowder** was the only explosive available; it was useful but limited in its applications. Then in 1846, an Italian chemist named

Ascanio Sobrero (1812-1888) invented nitroglycerine. The pale-yellow oil was not detonated by flame or spark like gunpowder, but by impact or percussion. Unfortunately, it was so highly combustible that if even a small bottle was dropped, it could easily blow up a building.

In the mid-nineteenth century, Nobel and his father Immanuel, also a noted inventor, became convinced of nitroglycerine's promise and began manufacturing it. Numerous accidents due to the volatility of the compound limited its usefulness. In fact, Nobel's own factory blew up in 1864, killing five people, including his younger brother. Many unfortunate experiences involving the inappropriate use of nitroglycerine were reported. Not realizing that even a slight shock or temperature change could cause the oil to explode, people reportedly used nitroglycerine for lamp oil, boot polish, or for greasing wagon wheels—often with fatal consequences. Despite the enormous risks associated with transporting nitroglycerine, miners were particularly anxious to get their hands on it. The resulting effort to meet demand for the liquid increased the occurrence of explosions during the transportation process. Many warehouses, factories, and **ships** around the globe were damaged or destroyed, which created a worldwide stir. Governments, pressured by public outcry, began to ban the transport and/or possession of the material.

The need for a safe but powerful explosive prompted Nobel to find a way to make nitroglycerine safer without significantly diminishing its power. He had already invented (in 1865) a detonating cap to give more control over the timing of an explosion once the substance was in place, but that did not solve the problem of handling or transportation. Nobel then discovered that he could turn the oil into a manageable solid by soaking it up into a porous material. Many nonexplosive inert substances were tried: paper, wood, waste, brick dust, and dry clay. Finally in 1864, he discovered that a mineral called kieselguhr, found mainly in northern Germany, seemed to do the job. It soaked up the nitroglycerine without changing its chemical makeup, and the resulting doughlike substance could be made into hard cakes or sticks. Though it was 25 percent less powerful than pure nitroglycerine, it was still far more powerful than gunpowder, and easily manageable. Nobel named this new explosive *dynamite* (after the Greek word *dynamis*, meaning power). He patented it under two names: Dynamite and Nobel's Safety Powder.

The story about Nobel's invention was often changed to one in which the discovery was accidental—this version claimed that some nitroglycerine had leaked into kieselguhr used in packing some containers. Nobel strenuously denied this account throughout his life. Although he readily admitted the accidental nature of some later inventions, he remained proud of the systematic scientific research that had been involved in the invention of dynamite.

Even in this safer form, dynamite was at first regarded suspiciously by some governments, which refused to allow its importation. The major gunpowder manufacturers tried to stop it from being patented. Later, when the military usages of dynamite became apparent, governments reversed their decision. Soon Nobel had factories worldwide. Within 20 years of the patent, 66,500 tons of dynamite were produced worldwide; 90 years after its invention, 400,000 tons a year were produced in the United States alone. Well into the twentieth century, dynamite was essential in warfare; the explosive power of bombs used in World War II was made possible by Nobel's inventions. Thus Nobel's peaceful intentions were overshadowed by the violent use of his product.

Dynamo • See Generator

Dynamometer

Deriving its name from the dyne, the fundamental metric unit of force, the dynamometer is an apparatus designed to measure the power, force, or energy of any machine that has a spinning shaft. Dynamometers are frequently used in determining the horsepower of **internal combustion engines**, and are often employed by the **automobile** industry in the testing and development of car engines.

Though there are many types of dynamometers, most fall into two general varieties: absorption and transmission. The earliest dynamometers were of the absorption variety, measuring the horsepower of a machine by finding out how much force had to be absorbed in order to impede the shaft's motion. In 1821 the French engineer Baron Gaspard de Prony (1755-1839) designed a rope-and-scale dynamometer called the Prony brake; when the rope was wrapped tightly around an engine's flywheel, the turning shaft would slow. At the same time, the weighted rope would be pulled slightly, and this increased tension could be measured upon the attached scale. This type of system is also called a rope brake.

The Prony brake was used extensively until the development of the water brake by **William Froude**, an English engineer. Froude had spent much of his career studying the ability of fluids to resist motion (such as water's resistance to the action of a boat moving through it). The water brake he designed consisted of a rotor encased within a water-filled compartment. The engine being tested would turn the rotor, and the horsepower could be found by measuring the amount of resistance exerted by the water.

A third type of brake dynamometer has since been invented that uses a large fan; using a principle similar to that of the water brake, the fan brake measures horsepower by measuring air resistance. There are also brakes that measure electrical resistance called electromagnetic brakes, but their efficiency is limited to relatively low horsepowers. Most automakers use brake dynamometers in their research, and the horsepower of their vehicles are usually expressed in brake horsepower (bhp).

The second, less commonly used type of dynamometer is the transmission variety. The transmission dynamometer allows researchers to measure an engine's power without slowing its motion. Because this is nearly impossible in any resistance-type device, transmission dynamometers are used principally in electric engines; here, the power is measured by calculating the current and voltage necessary to turn the shaft at a constant speed and is then translated into horsepower.

See also Braking systems; Electric motor; Steam engine

E

Earth survey satellite

Before the advent of the space age, scientists generally believed that very little detail of Earth could be seen from altitudes above 100 miles (161 km). Astronaut Gordon Cooper shattered that idea when he reported seeing roads, buildings, and even smoke from chimneys during his flight in a *Mercury* capsule in May 1963. Later observations used multi-spectral imagery in visible light and infra-red cameras, which made it possible to detect small variations of energy that features on the Earth's surface emit or reflect from the Sun. Vegetation could be distinguished from rock, soil, or water. It was easy to tell the difference between healthy and diseased crops, healthy or poor soil conditions, wet or dry soil, and even the difference between various crops under cultivation.

The first satellites to carry out detailed observations of the earth were the *Landsats* developed by General Electric for NASA. First launched in 1972, they orbited about 500 miles (805 km) high from north to south so the entire globe would rotate underneath them. One satellite was able to report on nearly every area of the world every 18 days. The early satellites could discriminate features no smaller than one acre, but *Landsat 4* was able to narrow this down to 1/5 acre. This satellite also was equipped to be retrieved by the **space shuttle** for possible repair and re-use.

Landsat satellites have yielded valuable data for a wide variety of fields. Cartographers have used them to map normally inaccessible areas, and environmentalists have used them to monitor air pollution and oil spills as well as to study the ozone layer. The agriculture industry has also benefited from *Landsat*, which improved management of crop and timber resources by determining when to plant and harvest for maximum yield and giving advanced warning of drought. The satellites also aid shipping companies in determining the flow of ice in arctic regions and hydroelectric engineers in predicting snow-melts in mountain ranges. However, the largest purchasers of Landsat data have been the oil and mining industries. Large folds and ruptures of Earth's surface, as well as certain colors, can indicate the location of mineral and shale deposits.

In addition to *Landsat*, oceanographers benefited greatly from *Seasat I.* Launched in 1978, this survey satellite was able to obtain data on sea conditions by using radar microwave techniques. It could tell sea-surface temperature, wind speed, wind direction and the amount of water in the atmosphere. These satellites also helped determine locations of sea currents, useful for fishing fleets and shipping companies which can route their tankers accordingly for fuel savings.

The successor to the Landsat program is the Earth Observing Satellite (EOS), scheduled for launch in mid-1998. Designed to carry out a comprehensive series of observations of the Earth's energy balance, atmospheric conditions, sea surface temperatures, the mission stems from our growing awareness of the complexity of manmade and natural influences on climate change, and the need for comprehensive data sets to help us understand the many effects observed.

Earthquake detection • See Earthquake measurement scale

Earthquake measurement scale

The earliest earthquake measurements were simple descriptions of seismic events and their consequences. These results were unreliable depending on the distance between the earthquake's source (epicenter), and the people evaluating the event.

A more systematic approach was developed by an Italian seismologist, Guiseppe Mercalli in 1902. He gauged earthquake intensity by qualitatively rating observer's experiences and the damage done to structures. The United States Coast and Geodetic Survey adapted his method, which they called the modified Mercalli Scale, dividing the measurements into

Charles Richter and colleague with seismograph.

12 categories: level II was "felt by persons at rest," but at level VII it was "difficult to stand." Level X caused most buildings to collapse, and level XII, the most intense, combined ground fissures with tsunamis (tidal waves) and almost total destruction. Despite the specific detail of descriptions, Mercalli intensity, like the earlier descriptions, was influenced by the measurement's distance from the earthquake's epicenter, as well as building design, and the character of the bedrock and soils.

Seismologists needed a way to determine the size, or magnitude, of an earthquake. They needed a quantitative, numerical measurement that would compare the strength of earthquakes in a meaningful way, not merely catalog damage or record perceptions as Mercalli's qualitative method did. This critical factor was finally developed in 1935 by American seismologist Charles F. Richter, a professor of seismology at the California Institute of Technology. His system of measurement, called the Richter magnitude scale, was based on his studies of earthquakes in southern California. It has become the most widely used assessment of earthquake severity in the world.

Richter measured ground movement with a **seismograph** and calculated the magnitude based on the height of the waves on the seismogram (the paper record). He then compared the reading to others taken at various distances from the epicenter and calculated an average magnitude after correcting for distance. The results are plotted on a logarithmic scale, in whole numbers and tenths. The scale is open-ended, meaning there is no upper limit to the magnitude of an earthquake. However, earthquake magnitude, as measured by the Richter magnitude scale, probably cannot exceed 9. Although rarely reported, very small earthquakes may have negative magnitude values. There are several different methods for determining earthquake magnitude and seismologists in different areas use different methods. This is why, in part, you may hear different magnitudes for the same quake, depending on the source of the information.

Each whole number increase in Richter magnitude means that the ground motion of the quake is ten times greater than the previous whole number. Thus, an earthquake with a magnitude of 6.5 has ten times the ground motion of one with a magnitude of 5.5; an earthquake of 7.5 has 100 times the ground motion of the 5.5 earthquake, and so on.

The relative difference in the amount of energy from different earthquakes follows a different pattern. Instead of tenfold jumps with each increase in magnitude, energy released differs in roughly thirtyfold increments. Thus, an earthquake with a value of 7 releases about 30 times the amount of energy as an earthquake measured at 6, while an earthquake of 8 would release about 900 times the energy of a Richter magnitude 6 earthquake.

Today the modified Mercalli intensity scale is still used in combination with the Richter magnitude scale because both methods are helpful in gauging the total impact of an earthquake. Building destruction and people's perceptions are considered important means of measurement for several reasons. Mercalli intensity allows geologists to better understand the role that bedrock and soil type play in the degree of earthquake damage. Deep earthquakes are much less damaging than shallow earthquakes. The type of fault movement (horizontal, vertical, oblique) can be related to earthquake damage as well. Finally, engineers can study how various construction methods and building materials respond to shaking, in hopes of developing more earthquake-resistant structures in the future.

For very damaging earthquakes, those above magnitude 7, Richter magnitude does not accurately express the amount of energy released during an earthquake. Therefore, a new scale, called moment magnitude, has been implemented by seismologists. It is generally the scale used when seismologists report seismic events to news agencies and, therefore, is the scale that is most often reported to the public. The moment magnitude scale takes into consideration the amount of slip on the fault, the area of rupture, and the strength of the rocks that failed, and thereby more correctly expresses the size of the largest earthquakes. Unlike Richter magnitude, moment magnitudes may exceed 9. For the 1960 Chilean earthquake, the strongest earthquake recorded by a seismograph, the Richter magnitude was about 8.5 and the moment magnitude about 9.5. Magnitudes for less powerful earthquakes are very similar to those based on Richter magnitude.

Earthquake-proofing techniques

Earthquakes occur in many parts of the world, sometimes with great regularity. The magnitude of an earthquake is measured by the Richter seismic scale, while the severity of a earthquake is measured in terms of damage and lives lost. The damage and casualties are related to more than the strength of the tremor, but also to population density and the quality of building construction.

In the San Francisco earthquake of 1906, a majority of the buildings withstood the tremors but were destroyed by the fire that followed. It not only spawned demands for **fireproofing techniques**, but also brought about earthquake-proof primary and secondary water supply systems for fighting fires.

During later earthquakes, many overpasses of the California interstate highway system collapsed or were damaged because of their inflexible design. In the 1950s, the concept of ductility, or pliancy, was formulated. It called for the use of energy-absorbing features and reinforced building materials.

The major thrust of earthquake-proofing by architects is to prevent the collapse of buildings. The ability of a building to withstand the stress of an earthquake depends upon its type of construction, shape, mass distribution, and rigidity. Various combinations of techniques are used. Square, rectangular, or shell-shaped buildings, and buildings with few stories, can better resist vibrations than L-shaped structures or skyscrapers. To reduce stress, a building's ground floor can be supported by very rigid, hollow columns, while the rest of the building is supported by flexible columns located inside the hollow columns. Another method is to use rollers or rubber pads to separate the foundation columns from the ground, allowing the columns to shake horizontally during an earthquake.

To help prevent collapse, roofs should be made of lightweight materials. Exterior walls can be made more durable by fortifying them with steel or wooden beams, or with reinforced **concrete**. Interior walls can bolster exterior walls, and a continuous collar can cap a rectangular shaped structure, aiding its stability. If nonstructural walls (not used for support) are attached only to the floor *or* only to the ceiling, they can move sideways as the building sways. Flexible window frames can hold windows in place without breaking during tremors.

Some architectural ideas are theoretical and, even after thorough laboratory testing, are not proven until an earthquake occurs. The San Fernando earthquake of 1971 taught engineers a valuable lesson, as the so-called soft story concept failed completely. It was thought that the upper stories of a high-rise building would suffer less damage if the first story was allowed to flex, having windows and facades instead of rigid walls and columns. However, many of these buildings collapsed.

The collapse of the Nimitz Freeway in Oakland, California, during the San Francisco earthquake of 1989 made it clear that despite extensive research and building codes for resistant construction, not enough had been done to prevent damage. The success of architects in dealing with the destructive force of earthquakes will no doubt take many years of trial and error.

See also Earthquake measurement scale

After inventing photographic film in 1884, George Eastman introduced a series of cameras that greatly simplified the formerly cumbersome photographic process and made taking pictures a favorite pastime of the general public. Above, a customer holds Eastman's Kodak Brownie camera, which sold for one dollar in 1900.

Eastman, George (1854-1932)

American inventor

Born in Waterville, New York, George Eastman quit school at the age of fourteen to help support his family, working first as a messenger for an insurance company and then as a bank clerk. In 1878 as he planned a vacation, a friend suggested he take along a camera. Eastman took the suggestion to heart, equipping himself with the paraphernalia then required in the wet-plate process of **photography**: a sizable camera, a heavy tripod, a plate holder and a number of the fragile and cumbersome glass plates, and the developing necessities, among them chemicals and a portable tent-like darkroom—a "packhorse load," as Eastman himself called it. That firsthand encounter with the complexities and inconvenience of the photographic method of the day launched Eastman on his quest to simplify the process and make it accessible and enjoyable to the general public.

Eastman learned soon after that an English photographer, Dr. Richard Leach Maddox, had invented dry photographic plates in 1871 to replace the glass plates, which had to be smeared with an emulsion of wet chemicals before a photo could be taken. Experimenting in his mother's kitchen after work hours, Eastman devised a means of coating glass plates with a gelatin emulsion. After the emulsion dried, the plate would last for long periods of time. In 1879 Eastman sailed to England and obtained the first patent for his invention and received the corresponding American patent within the next year.

The following year Eastman started a company to produce and market dry photographic plates, and by the end of 1881 he had a business partner and six employees. He quit his position at the bank to devote his full attention to his photographic business. In 1884 Eastman patented photographic film, on which the emulsion was smeared on paper. That same year Eastman and an associate invented a container for rolls of negative paper.

The inventor's love of the letter K led him to create the name *Kodak* for his company. In 1888, the *Number One Kodak Camera* was introduced at a cost of $25, which covered film for a hundred exposures, a shoulder strap, and a case. After the film was shot, the camera's owner sent the camera back to Kodak, which then developed and printed the film, inserted new film in the camera, and returned everything to the owner. Kodak's slogan in those days was apt: *You press the button—we do the rest.*

In 1889, Eastman abandoned paper and turned to a tougher material, celluloid nitrite—**celluloid**—to produce flexible transparent film with the help of a staff research scientist. Soon thereafter, **Thomas Alva Edison** used this type of film to realize his dream of a practical **motion picture** process. (Celluloid proved to be flammable, however. When Kodak came out with a home movie camera and projector in 1923, it used 16-mm movie film on a nonflammable base of cellulose acetate.)

Eastman's company produced new products in rapid succession: the first folding camera in 1890, film that could be loaded in daylight in 1891, a pocket camera with a window that showed the number of exposures in the camera in 1895, and a folding pocket camera in 1898. In 1900 Kodak introduced the Brownie camera at a cost of one dollar; a roll of film was fifteen cents.

In his lifetime, George Eastman gave away nearly $100 million, including $20 million to Massachusetts Institute of Technology and $51 million to the University of Rochester. He also established the Rochester Dental Dispensary and dental clinics in several large European cities.

By the time he was in his 70s, he had given away the majority of his wealth. He never married and had no close relationships, and on March 14, 1932, after leaving a simple note—*My work is done. Why wait?* —Eastman committed suicide.

ECHOCARDIOGRAM • See Ultrasound devices

ECHOENCEPHALOGRAPHY • See Ultrasound devices

ECKERT, J(OHN) PRESPER, JR. (1913-1995)

American engineer

Presper Eckert, with John Mauchly, designed and built several significant computers in the 1940s— **ENIAC**, EDVAC, BINAC, and **UNIVAC**. Eckert was born in Philadelphia, where his father was a real estate developer. He received a B.S. in 1941 and an M.S. in 1943 from the University of Pennsylvania's Moore School of Electrical Engineering. Eckert began working with Mauchly, a faculty member, because they were both interested in electronic computer design. World War II was in progress, and the university had a United States Army contract to develop a calculating machine known as ENIAC (Electronic Numerical Integrator And Computer). Eckert and Mauchly designed and patented ENIAC. It was much faster than earlier computers and featured electronic processing with **vacuum tubes**. However, it used punched cards for the program and intermediate processing results, which greatly slowed its speed. Also, each processing sequence had to be set up by hand. The school's next computer was EDVAC (Electronic Discrete Variable Automatic Computer), a computer milestone, featuring a stored program in which the program and data are both in the computer's memory and treated alike. Eckert and Mauchly designed the memory, which used mercury-filled delay lines to retain the incoming electronic signal as a much slower sound wave. In 1948, the University wanted control of all patents for equipment produced by its faculty. Eckert and Mauchly resigned and formed their own firm, the Electronic Control Company, to produce stored-program computers based on the patents. BINAC (BINary Automatic Computer), a fast and relatively small computer to be used on a guided missile, was their first computer, in 1949. It was the first computer to use magnetic tape for input and output, and featured two processing units that performed the same calculations for accuracy. At the same time, Eckert and Mauchly were building the much larger UNIVAC (UNIVersal Automatic Computer), which they completed in 1950. It was the first widely-available commercial computer to use stored programs. After financial losses in 1950, Eckert and Mauchly sold their company and patents to the typewriter and calculator manufacturer Remington Rand. Eckert remained with Remington Rand to continue development of UNIVAC, eventually becoming a vice-president of the successor company, Sperry Univac. During the 1960s, Eckert and Mauchly were sued by the computer manufacturer Honeywell and **John Atanasoff**, a physics professor at Iowa State University. They claimed that the patents were based on Atanasoff's A-B-C computer, which he had demonstrated for Mauchly in 1941, before ENIAC was built. Eckert and Mauchly maintained that much of their design work was completed before Mauchly met Atanasoff, and that the two machines were different. However, in 1973, the judge ruled in Atanasoff's favor, declaring Eckert and Mauchly's patents invalid.

EDSAC • See Stored program

EDVAC • See Stored program

Edgerton, Harold E. (1903-1990)

American engineer

Harold Edgerton was born in Fremont, Nebraska, on April 6, 1903. Although his family moved to Washington, D.C., due to his father's career with the federal government, the family eventually returned to their native Nebraska. After earning an electrical engineering degree from the University of Nebraska, Edgerton worked for General Electric in Schenectady, New York, then entered the Massachusetts Institute of Technology (MIT). At MIT Edgerton earned both his masters and doctorate degrees in electrical engineering. He received a faculty appointment at MIT and eventually was awarded the highest rank at the university, that of Institute Professor. While conducting a study of power **generator**s at MIT, Edgerton noticed that flashes of light from mercury rectifiers made a generator's rotors appear to stand still. He recognized the opportunity to invent a lamp that could ''freeze'' motion with rapid pulses of intense light. By 1931, he had described such a device in the journal *Electrical Engineering*. This **stroboscope** was the first high-powered reusable electronic flash. Combined with a **camera**, it was able to photograph a particular microsecond of history. Not only could the stroboscopic flash make an object appear to stand ''still,'' but objects in cyclic motion could be made to appear slower, in reverse, or simply juggled by adjusting the speed of the stroboscopic light. Though taking a ''still'' picture of a rapidly moving object had been accomplished by **William Henry Fox Talbot** in 1851 and had been improved upon by many others, Edgerton was not only the first to substitute a control circuit for a tube or lamp, but the first to make **high-speed flash photography** practical. With his two students, Kenneth Germeshausen and Herbert E. Grier, Edgerton continued to develop other appliances and circuits for stroboscopy. Around 1940, Edgerton and Germeshausen were invited to take their experimental stroboscopic high-speed **movie camera** to the MGM studio. They produced a ''short'' entitled *Quicker Than a Wink.* It received an Oscar for the best short

Harold Edgerton with the stroboscopic lights he used to create his pioneering high-speed flash photographs.

Thomas Alva Edison

film. In later years, Edgerton, who came to be known as "Papa Flash," took up oceanographic photography. He eventually made ten voyages on **Jacques Cousteau**'s research vessel, *Calypso*, before his death on January 4, 1990.

See also Photography.

Edison, Thomas Alva (1847-1931)

American inventor

During his lifetime, Thomas Edison patented more than 1,000 inventions—a number that no one else has ever approached. Among these were crucial innovations such as the **phonograph**, the **motion picture**, and the **incandescent light bulb**. Yet despite Edison's reputation as a prodigy, his success resulted as much from hard work as from natural intelligence. Edison's self-confidence and determination helped him to overcome poverty, physical handicap, and disastrous financial setbacks. Like many strong-willed people, Edison therefore stepped on some toes along the way, especially by aggressively patenting improvements to other people's work. One longstanding conflict cost him the Nobel Prize.

Edison, who was born on February 11, 1847, in Milan, Ohio, but grew up in Port Huron, Michigan, went to school for a total of only three months. His teacher, failing to relate to the way Edison's mind worked, dismissed him as being "addled," or retarded. Luckily, Edison's mother was a teacher too, and she was happy to supervise his education at home, especially since his health was delicate. Under the guidance of his mother for the next four years, Edison flourished. He began to read voraciously, and became interested in science, especially chemistry and electricity. Edison even put together a makeshift laboratory in the basement of the family house.

Edison was also fascinated by **locomotive**s and railroads, and began to spend a great deal of time at the machine shop operated by the Grand Trunk Railroad in Port Huron. When a job selling newspapers and candy on the Detroit-Port Huron train became available, he eagerly snatched it up. He worked hard, often earning ten dollars a day, and spent the long afternoon stopovers in Detroit devouring book after book in the large public library there.

With his earnings, Edison bought some used printing equipment so that he could publish and sell his own newspaper on the train. By the age of twelve, he was employing other boys. He bought more laboratory equipment and received permission to construct a lab in the baggage car of the train, where he performed experiments until, some time later, a jar of phosphorus fell from a shelf and set off a fire there and the angry conductor "evicted" Edison from the baggage car.

Though Edison developed a broad range of knowledge, he was never able to master mathematics. He later said that one major mathematical work, Isaac Newton's *Principia*, "gave me a distaste for mathematics from which I have never recovered." Around this time, Edison began to grow deaf—a condition that would worsen through his life until he could hear only a loud shout.

After Edison bravely pulled a stationmaster's son from certain death before an oncoming train in 1862, the boy's grateful father offered to teach Edison telegraphy in appreciation. At the time, the **telegraph** was the principal means of communication and the demand for skilled operators was high. He picked up telegraphy so well that, for the next several years, Edison was able to roam the country, living very roughly but making good money while building a reputation as one of the best and fastest telegraphers in the country. He also continued to study; he bought the collected works of Michael Faraday, an English physicist and pioneer in electricity, who greatly impressed Edison because "he used no mathematics."

Edison's first invention, an electric vote-recording machine patented in 1869 for use in legislative chambers, was unsuccessful because politicians were not interested in speeding up the voting process. From then on, Edison swore, he would invent only those products that he knew people would want. He formed a small company in Boston to produce an improved stock ticker he had invented, but was soon bought out by a telegraph company, with little to show for his efforts. After his attempt to demonstrate a new telegraphy system failed, Edison, with not a dime in his pocket, departed for New York to seek his fortune.

The gold-market scandals of the day caught Edison's attention when he took a job at the Gold Indicator Company soon after coming to New York. As he observed the havoc of the gold panic on "Black Friday," September 24, 1869, Edison realized that he could capitalize on the gold traders' need for information, so he used what he had learned about stock tickers to build an improved ''gold printer,'' which he and his partners sold to Western Union for $15,000.

Though his share was only $5,000, the invention yielded Edison a position on Western Union's technical staff, and within a few weeks, Edison had devised a revolutionary stock ticker, for which the president of Western Union paid Edison $40,000. Edison, who had always been uninterested in accumulating wealth, didn't even know how to cash the check and was fooled by a mischievous bank teller into accepting payment in huge stacks of small bills.

Edison parlayed the sum into his projects. At the age of twenty-three, he set up his own engineering firm that, besides manufacturing his stock tickers, began churning out numerous inventions including the mimeograph, improvements to the **typewriter**, and an improved telegraph that could send four messages at once on a single wire. He married Mary Stilwell, who had assisted him with the invention of paraffin paper, in 1871.

Thanks to his flourishing business, Edison was now in a position to realize his dream of establishing an ''invention factory'' that would focus on practical products. Abandoning the monotonous task of manufacturing, he moved to Menlo Park, New Jersey, in 1876, where he set up a huge industrial research laboratory, the first of its kind. Edison made it his goal to produce a new invention every ten days, and during one four-year period, he obtained an average of one new patent every five days, earning him the nickname ''the Wizard of Menlo Park.''

Just a few of the inventions Edison's lab produced during this time were major improvements to Alexander Graham Bell's **telephone**, including a ''carbon button'' **microphone**, which enhanced its sound quality. **Emile Berliner** had invented a similar device in Germany, and Edison was drawn into a lawsuit over primacy for the invention. Fourteen years later, Berliner won his case.

Edison's favorite project, and perhaps his most original invention, was the phonograph, or record player. The idea for the phonograph came to Edison while he was studying a telephone receiver. He attached a point to the diaphragm of the receiver so he could feel the sound vibrations with his finger as they were emitted. It dawned on him that the vibrations were so strong that the point might ''etch'' them onto a piece of moving tinfoil. He reasoned that a similar point could then trace the grooves left on the foil and pass the vibrations onto another diaphragm to produce sound.

His original phonograph used a tin-foil covered cylinder that was hand-cranked while a needle traced a groove on it. The ''talking machine'' created a sensation, and others quickly began introducing improvements to it. Although the phonograph underwent a century of technological improvement after Edison, its essential principle remains the same today.

By now Edison had earned a reputation for being able to do just about anything he put his mind to. When he announced that he intended to produce an electric light that would compete with gaslight, the stock prices of gaslight companies tumbled as their executives panicked. Many people, most notably Sir **Joseph Swan**, had tried to invent an electric light using an incandescent filament, or wire, enclosed in a glass bulb, but had not been able to create a filament that could withstand intense heat over long enough periods of time to be practical. Even Edison had a tough time of it, going through a long, trial-and-error process in which he tested thousands of materials. Undaunted by failures, Edison finally found that a scorched cotton thread would work best. When heated in a vacuum, it produced a white glow without melting, evaporating, or breaking.

Although Swan came up with a similar light bulb around the same time, Edison patented his idea more aggressively, promoted his product more effectively, and sketched out a practical system of power supply which could support its use on a large scale. On New Year's Eve of 1879, Edison gave a public demonstration of the new bulb, lighting up his laboratory and a half mile of streets in Menlo Park before thousands of spectators. Edison had not only invented an economical light source, but developed an entire system for generating and distributing electricity from a central power station. By 1881, Edison's Pearl Street station in New York was supplying about 400 outlets for eighty-five customers using a parallel wiring system that made it possible to switch off individual bulbs without turning off others. Eventually, Edison's electric business became the General Electric Company.

Though he never considered himself a theoretical scientist, while he tinkered with his incandescent light bulb in 1883, Edison inadvertently made a tremendously important scientific discovery. He inserted a small metal plate near the filament of a light bulb, and found that the plate drew a current when he connected it to the positive terminal of the light bulb circuit—even though the plate was not touching the filament. Puzzled by the phenomenon, he put it aside. When J. J. Thomson discovered the electron over a decade later, one of Edison's assistants, **John Ambrose Fleming**, realized that the long-forgotten ''Edison Effect'' was caused by electrons boiling off of the filament and streaming onto the positively- charged plate. This led him to invent the **vacuum tube**, a device that would become essential to **radio**, **television**, and electronics.

In 1884, Edison's wife died of typhoid fever. Two years later, he married twenty-year-old Mina Miller, the daughter of a successful Ohio inventor. Edison had now accumulated millions of dollars to pursue new ideas. In 1887, he moved to a new, much larger lab in West Orange, New Jersey. Through the 1880s and '90s, Edison continued to produce major innovations such as the motion picture camera and a viewer which he called the **kinetoscope**. Although the invention of motion pictures cannot be credited to Edison alone, Edison (with the help of his assistant William Dickson) made the crucial discovery that the images could be advanced by sprocketed wheels inserted in holes along the sides of the film. Edison opened his first kinetoscope parlor in 1894, and then was persuaded to team with Thomas Armat, who had invented a movie projector in 1895 that made it possible to show films before groups of people. Everywhere, vacant shops were converted to five-cent theaters, or ''nickelodeons'' using the ''Edison Projection Kinetoscope,'' as it was called.

Edison built a film studio to produce movies for the nickelodeons and kinetoscope parlors. He also experimented with adding sound to his films by combining his kinetoscope with a phonograph to make the **kinetophone**, which produced moving images with sound—a kind of crude "talking picture." But Edison never fully appreciated the importance of sound to the future of motion pictures, and left the creation of "talkies" to others.

Though his ability to create seemed almost magical by this time, Edison would soon be humbled by his worst financial disaster. In 1890, he had sunk two million dollars—an incredible sum at the time—into developing a process for producing iron from ore using magnetic separation. Edison's process was a technical success, and he created some of the most awesome machinery yet seen by civilization to carry it out. But in 1900, a tremendous reserve of cheap iron ore was discovered in Minnesota, making Edison's technique uncompetitive overnight. Even though he was left deeply in debt, he simply closed the project, made sure his creditors were paid off, and went on to tackle other problems.

His next development was a new type of storage battery, which Edison hoped would replace conventional batteries in the rapidly growing automobile industry. It is typical of him that even after his first 8,000 experiments failed, he said, "Well, at least we know 8,000 things that don't work." Although Edison's battery turned out to be unsuitable for cars, it did succeed in railroad and marine shipping applications, which required batteries with longer life and greater durability. During this period he also worked on an electric railroad, the manufacture and use of **concrete**, and various kinds of office machinery.

In 1912, Edison was proposed as a co-winner of the Nobel Prize with **Nikola Tesla**. Tesla had been employed in Edison's lab in the late 1880s, but soon quit after a disagreement with Edison, and went on to achieve great success on his own. The root of their hostility was their disagreement about how to supply electricity to the public. Tesla favored using a system of **alternating current** (AC) he had helped to develop, while Edison favored **direct current** (DC). Even when the superior AC system began to take hold around the country, Edison stubbornly resisted, something the temperamental Tesla never forgot. When he refused to share the Nobel Prize with Edison, the committee awarded the prize to someone else.

By the time a fire destroyed most of Edison's laboratory complex in 1914, his stamina had been running thin and he was devoting less time to inventing. He spent time camping each summer with his friends **Henry Ford**, Harvey Firestone, and John Burroughs. Though he felt repulsed by the idea of war, Edison, out of a sense of patriotic obligation, turned his attention to naval research on torpedoes, periscopes, and flame throwers during World War I. He suggested that the Navy create a permanent scientific laboratory, which it has since established.

Edison's final project, begun in 1927, was a far-sighted attempt to find a cheap domestic source of rubber. Although Edison tested 17,000 plants, he never found an ideal source. Edison lived to take part in the "golden jubilee of light"—a celebration commemorating the fiftieth anniversary of his invention of the light bulb. Two years later, on October 18, 1931, he died peacefully at his home in West Orange, New Jersey, at the age of 84.

See also Automobile, electric; Battery, electric; Duplicating machine; Langmuir, Irving; Latimer, Lewis Howard; Movie camera; Tire, pneumatic

EGG PRODUCTION • See Incubator

EIGHT-TRACK TAPE

The eight-track cartridge tape system was devised by William Powell Lear, the founder of the LearJet corporation, in the 1960s. Lear's invention is similar to its predecessor, the four-track tape. The main difference is that the eight-track tape is divided into eight channels, or tracks, where the four-track is divided into only four. Doubling the number of tracks doubled the amount of playing time on the tape, from about 40 minutes to about one hour and 20 minutes. The narrower channels of the eight-track tape, however, sometimes resulted in one track "bleeding" into another so that strains of one track could be heard over another. The eight-track tape was a popular audio format from the mid-1960s through the early 1980s, when their popularity finally almost completely gave way to **audiocasette** tapes and vinyl **LP records**.

EINTHOVEN, WILLEM (1860-1927)

Dutch physiologist

Willem Einthoven was born in Java, Dutch East Indies (now Indonesia). When he was six years old, his father, Jacob, (a physician) died and, in 1870, the widowed Louise de Vogel Einthoven returned with her six children to Utrecht, The Netherlands. Einthoven studied medicine at the University of Utrecht and earned his medical degree *cum laude* in 1885. Later that year he was appointed professor of physiology at the University of Leiden in Holland, a post he held for the rest of his life.

An interest in physics as well as physiology soon led Einthoven into a study of the electrical impulses produced by movement of the heart muscle. Since existing instruments were inadequate for measuring and recording these electric currents, Einthoven set about inventing a sufficiently sensitive device for this purpose. The result was the string **galvanometer** of 1903, which used a fine quartz **wire** that reacted to changes in cardiac currents. Einthoven called his invention the **electrocardiograph**, and the recordings of heart rhythms it made electrocardiograms. Using the machine, Einthoven identified the normal heart-current waves and set the standards for interpretation of electrocardiograms. He also pioneered the diagnosis of different types of heart disease and irregularity through electrocardiography. In recognition of the profound importance of the electrocardiograph in the diagnosis of heart ailments, Einthoven was awarded the Nobel Prize in medicine or physiology in 1924.

Einthoven was an open, humorous, and generous man who often invited scientists to work with him. He was elected a foreign member of the Royal Society of London, and he traveled and lectured widely in both Europe and in the United States. In his last major experiment, conducted with his son, Willem, Einthoven received radiotelegrams from Java using a string galvanometer. His final work, a treatise on the action current of the heart, was published posthumously. Einthoven died in Leiden in 1927.

ELASTIC • See Rubber, vulcanized

ELECTRET

The electret is an insulating material (a dielectric) that has the ability to hold an electrostatic charge, making it the electrostatic equivalent of a permanent **magnet**.

In 1919 Japanese scientist Mototaro Eguchi applied a high voltage to two electrodes that were immersed in a molten mixture of wax and resin. Keeping the voltage on, he allowed the mixture to solidify, creating a ''frozen'' polarization in the mixture between the electrodes. The dielectrics that are the electret come in two types. In a homocharge electret one side of the dielectric becomes positively charged, while the other side acquires a negative charge. In a heterocharge electret the positive and negative charges are aligned in the same direction within and throughout the material, but do not reach the electret's surface.

Electrets were first used by the Japanese in field **telephone**s during World War II. In the United States and Great Britain, patents were taken out for their use in **microphone**s, **loudspeaker**s, **voltmeter**s and electrometers. Most recently, homocharged electrets, which hold their charge longer, have been made with polymers in the form of a flexible film. The electret film forms a flexible diaphragm in a condenser microphone and can produce a wide, flat frequency response with low distortion without the need of an external polarizing voltage. These microphones are small and lightweight, insensitive to shock, and inexpensive.

The electret **transducer** is a device for the conversion of sound or motion into electrical energy using an electret. It has a flexible diaphragm that has been coated on one side with a thin metal film. A metal plate faces the uncoated side. Electrical leads are connected to both the film and plate. When the diaphragm moves it induces a charge on the metal plate, creating a flowing current. The current goes through the leads and into the **amplifier** input circuit where it produces a voltage which can be amplified.

Electret foil transducers have been used in a wide range of frequencies, from ultrasonic (up to 10^8 Hz) to seismic (down to 10^{-3} Hz) detectors. They are also used in record players, hearing aids, earphones, and push-button keys on telephones and calculators. About 100 million electret microphones are produced annually for studio and home casette recorders.

See also Electric circuit; Electrostatic devices

ELECTRIC ARC PROCESS FOR STEEL • See Steel production

ELECTRIC BATTERY • See Battery, electric

ELECTRIC BLANKET

The forerunner of the modern electric blanket was designed in 1912 by American physician Sidney Russell. In trying to devise an effective heating pad for his patients, Russell found that he could produce the desired heat by passing an electrical current through insulated metal tape that had been secured inside a blanket covering. It was not until the 1930s, however, that the electric blanket was commercially developed in the United States. Sometime earlier, in 1926, the British Ex-Services Mental Welfare Society acquired the rights to the Thermega Underblanket, but did not begin producing it until about the same time American manufacturers introduced electric blankets.

Safety was a major concern in electric blanket use. Research conducted during World War II on electrically heated uniforms for airplane pilots led to some safety improvements to electric blankets—most notably the addition of protective **vinyl** coverings for the electrical elements. In 1967, after a series of accidents, a U. S. company called Dreamland incorporated a monitoring system that cuts off electricity to the blanket if it overheats. In 1970, Britain's Thermega marketed a blanket that replaced the electrical elements with tubes carrying hot water.

Current (1998) research results on a possible relationship between electromagnetic fields and various cancers are uncertain, although prudence dictates taking reasonable precautions when using electric blankets. Although newer blankets have better shielding and use less electricity, heating the bed then unplugging the blanket for most of the night seems a sensible approach.

ELECTRIC CIRCUIT

An electrical circuit controls the flow of electricity. It is composed of various components, depending on its purpose. The most common components in a circuit are switches, **capacitor**s, **resistor**s, inductors, and **potentiometer**s. Adding **vacuum tube**s (also known as valves) or semiconductors (such as **transistor**s) changes the nature of the circuit, giving rise to the name *electronic* circuit. Circuits often include a fuse or circuit breaker to prevent a power overload.

The first electric circuit was invented by **Alessandro Volta** in 1800. He discovered he could produce a steady flow of electricity using bowls of salt solution that were connected by metal strips. Next, he used alternating discs of copper, zinc, and cardboard that had been soaked in a salt solution to create his voltaic pile (an early battery). By attaching a **wire** running from the top to the bottom, he caused an electric current to flow through his circuit. The first practical use of the circuit was in electrolysis, which is a process where electric current passed through a substance induces a chemical change. This led to the discovery of several new chemical elements.

The flow of electricity through a circuit is measured in amperes, named for André Marie Ampère (1775-1836), is the amount of charge flowing past a specific point in the circuit each second. Electric charge is measured in Coulombs, for Charles Coulomb (1736-1806). One ampere equals one Coulomb per second. (The electrical charge of one electron is 1.602×10^{-19} Coulomb). Voltage, named for Volta, is the driving force (electrostatic field) that "pushes" electrons through the circuit. The resistance of a circuit opposes the flow of electrons, and often causes heat. Georg Ohm (1787-1854) discovered that some conductors have more resistance than others, which affects their efficiency in a circuit. His famous law, known as *Ohm's Law*, states that the resistance in a circuit equals the voltage across a conductor divided by the current, or R=V/I. In its more common form the equation reads V=IR.

See also Electronics

ELECTRIC CONDENSER • See Capacitor

ELECTRIC HIGH-TENSION LINES • See Alternating current

ELECTRIC MOTOR

The modern electric motor dates back to 1831, when American physicist **Joseph Henry** published a paper describing an electric motor which, he showed, was basically the reverse of the electric **generator**. Instead of converting mechanical movement into an electric current, like the generator, his motor used electric current to produce mechanical movement. Henry's motor was the first to be constructed, although inefficiency limited its potential.

In 1834 American blacksmith Thomas Davenport improved the motor's operating principles, using four **magnet**s, two fixed and two revolving. Davenport used his motor to operate his own **drill**s and wood-turning **lathe**s. He went on to incorporate his motor in the electric railway, electric trolley, electric piano, and electric printing press.

However, across the Atlantic Ocean, English physicist **Michael Faraday** had been making advances of his own. Some of Faraday's discoveries were anticipated by Henry, but Faraday received credit by publishing the results of his experiments first. Faraday, having learned of Hans Christian Oersted's (1777-1851) 1821 discovery that an electric current created a magnetic field which could deflect a **compass** needle, set out to reverse the results and create an electric current from a magnetic field.

Faraday built a device that consisted of rods, wires, and magnets, some of which were fixed in position, others mounted on pivots. When he sent an electric current through his device, the moveable wire pivoted around the fixed magnet, and the movable magnet pivoted around the fixed wire. Although it did not perform work tasks, it did convert electrical and magnetic force into mechanical movement, the essence of a electric motor. Ten years later, Faraday constructed the first electric generator, making a continuous supply of electricity available for use.

Hippolyte Pixii built a hand-driven generator that produced **alternating current** (AC), but, on the suggestion of André Marie Ampère (1775-1836), he added a commutator to convert the power into direct current (DC). The scientific world had concentrated on DC since 1800, when **Alessandro Volta** invented his battery. Although AC had a number of advantages over DC, there was little interest in it, and early electric motors operated on the principle of direct current.

Alternating current took a step forward in 1867 when Belgian-French inventor **Zénobe Gramme** built an improved dynamo for producing AC; two years later he improved the DC dynamo. Using the principles of Henry's and Faraday's discoveries, Gramme, with his associate Hippolyte Fontaine (1833-1917), opened a factory that manufactured electrical devices, setting the standards for the industry.

In 1884, **Nikola Tesla** went to work for **Thomas Edison**, and tried to convince him of the advantages of AC power. Edison, whose electric company was already established with DC, could not be convinced. A year later, Tesla took his AC technology and induction motor to industrialist **George Westinghouse**, initiating a conflict between Edison and Westinghouse that was only resolved in 1893, when AC generators were successfully used to provide power for the World Columbia Exposition in Chicago. Westinghouse made a fortune building AC motors, and Tesla, ironically, was awarded the Edison medal in 1917 for his work in electricity.

All motors have two basic parts: a *rotor* and a *stator*. The stator is usually stationary, and the rotor revolves around it. The stator is a magnet that shapes a magnetic field. The rotor is a conductor connected to the electric circuit that interacts with this magnetic field, and in turn produces a magnetic field that acts on the stator.

Modern motors can be classified into two groups: electromagnetic motors and magnetic motors. Electromagnetic motors improve in performance as they are enlarged; magnetic machines improve as they are scaled-down. Electromagnetic motors include the induction, AC polyphase commutator, AC single-phase commutator, DC, synchronous, and repulsion motors. The magnetic motors include the solenoid, relay, reluctance, and hysteresis motors.

An induction motor uses AC and a ring of fixed electromagnets (the stator) to produce a rotating magnetic field. The moving electromagnetic field causes the rotor to spin, producing mechanical energy. More than 90% of the world's motors are of this type. A synchronous motor uses either permanent magnets or DC-fed **electromagnet**s to produce a magnetic field. Unlike the induction motor, whose rotor "chases" after the rotating magnetic field, the synchronous motor has a magnetized rotor. The rotor's magnetic field matches the rotating magnetic field, resulting in a synchronized mechanical motion that has very little slippage.

There are two main disadvantages to induction and synchronous motors: without a variable power supply, they cannot provide efficient speed variation over a wide range. They are also limited to speeds under 3,000 revolutions per minute (rpm) when they draw power directly from power mains. When these factors are a problem, commutator motors are used.

The commutator motor has insulated coils whose ends are connected to a pair of conducting segments. Carbon blocks or copper brushes make direct contact with the pairs of commutator segments from which the current flows. The chief problem with this type of motor is the sparking and arcing that occurs between the brushes and the commutator segments.

The reluctance motor, used in clocks, is a synchronous motor whose magnetized rotor has been replaced by a piece of metal that is shaped so that it fits into a number of preferred positions, where the resistance is minimized.

The hysteresis motor is similar to the reluctance motor, except the shaped piece of metal is replaced by a smooth cylinder. The magnetic field passes over the cylinder, leaving it permanently magnetized. The motor operates more and more like a synchronous motor as it speeds up, until it locks on the rotating magnetic field.

The linear induction motor is a hybrid of its rotary cousin. In this type of motor, the rotor, which does not rotate, and the stator move past each other, separated by a small gap of air, resulting in a magnetic field that moves in a straight line instead of in a circle. A liner induction motor is superbly suited for use in high-speed magnetic levitation **train**s, whose upper speed is limited only by wind resistance and safety considerations. There are more than 200 different types of linear induction motors in use today.

Electric piano • See Musical instruments, electric

Electric range • See Stove

Electric resistor • See Resistor

Electric train • See Train and railroad

Electrocardiograph (ECG)

In the late 1700s medical researchers learned that muscles produce tiny electric impulses now known as "action potentials." Italian biophysicist, Carlo Matteucci (1811-1868), identified action potentials in a pigeon's heart in 1843 and, in 1856, German scientists Rudolf Albert von Kölliker (1817-1905) and Heinrich Müller (1820-1864) recorded these electric currents from a frog's heart.

Reasoning that these recordings could reveal irregularities and, hence, heart disease, researchers attempted to develop accurate measuring devices. The French physiologist, Augustus Waller (1856-1922), found that cardiac currents could be recorded by placing surface electrodes on the body. In 1887, Waller developed a capillary electrometer—tubes of mercury that rose and fell with the changes in heart muscle current—which, however, was imprecise and difficult to use. So Dutch physiologist **Willem Einthoven** set out to design an improved apparatus. In 1903 he described the result, his string **galvanometer** consisting of a thin, silver-coated quartz **wire** stretched between the poles of a **magnet**. As the heart's electric impulse flowed through, the wire was deflected and motion was magnified and projected onto moving photographic film. The extreme sensitivity of the device, which weighed 600 lb (272.4 kg), allowed it to detect the tiny cardiAC currents very accurately.

Einthoven called his machine the electrocardiograph and the recorded electrical impulses an electrocardiogram. He devised the standard positioning of the electrodes and described the regular heart waves and the triangle used to interpret electrocardiograms. Through clinical studies, Einthoven identified a number of heart problems with his galvanometer. Einthoven won the Nobel Prize in 1924 for inventing the ECG. In 1942, Emanuel Goldberger added three augmented limb leads to Einthoven's three limb leads and the six chest leads making the 12-lead electrocardiogram that is used today, and English physician Sir Thomas Lewis (1881-1945) established the electrocardiogram as a standard clinical tool.

With refinements in instrumentation and technique, electrocardiography became one of the most useful diagnostic tools in medicine. This highly accurate, easy to interpret, and relatively inexpensive device permits diagnosis of heart conditions without needle or incision and even portable devices are now available: the "HeartMirror," weighing only 3.3 lb (1.5 kg) operates on four AA NiCad batteries and will record 12 selectable leads. Housed in a carrying case, it is ideal for a physician's office. The "HeartVision" portable emergency device weighing only 13.4 oz (380 gr.) and using four AA batteries, fits into a pocket. Four built-in emergency electrodes eliminate contact leads—the device is simply placed on a patient's chest and displays the reading on a liquid crystal display (LCD). It also stores the reading, which can be printed out later on an analog ECG recorder or personal computer.

Electroencephalogram (EEG)

An electroencephalogram (EEG) is a graphic picture of the electrical activity of the brain. This visual image is created when electrodes are placed on a subject's scalp and connected by wires to an *electroencephalograph*, a devise which records the patterns of brain waves—the rhythmic changes in electric potentials in the brain—by tracing them with a fine, inked needle or nib onto a sheet of paper. EEGs are useful in diagnosing epilepsy, brain tumors, strokes, brain damage caused through head trauma, and other neurological conditions characterized by distinctive, abnormal brain wave patterns. They are also used in investigating psychiatric disorders such as schizophrenia, and in determining brain death—an important process for use in conjunction with organ donation for **surgical transplant** recipients.

The founder of electroencephalography was Hans Berger (1873-1941), a German psychiatrist who made the first human EEG in 1924. Interested primarily in psychophysiology—that is, the relationship between the mind and the brain—Berger set about measuring the brain's electrical activity in the hope that a physiological record of this kind would provide insight into mental processes. He found inspiration for his work in the **electrocardiograph (ECG)** invented by **Willem Einthoven** in 1900, and in work done earlier on the brain waves of animals. In 1875, Richard Caton (1842-1926), an English physiologist and surgeon, had measured electrical activity in the exposed brains of rabbits and monkeys but had been unable to make a graphic recording; the first recording of this kind was made in 1913 by a scientist named Vladimir Pravdich-Neminskii, who used the Einthoven string **galvanometer** to record from the intact skulls of dogs.

Using a galvonometer much like the one Pravdich-Neminskii had used a decade earlier, Berger began his search for the human EEG by experimenting with the exposed brains of dogs. He then started placing needle electrodes under the scalp of patients who had lost some of their skull bones in surgery. It was while working with one of these patients—a seventeen-year-old who had been operated on because of a suspected brain tumor—that Berger recorded the first human EEG in 1924. He was initially uncertain whether the electrical oscillations he recorded originated in the brain. It was not until after conducting many other experiments—including experiments on the intact skulls of healthy people and of people with brain disorders—that he published his first paper on the human electroencephalogram in 1929.

The initial reaction of other scientists to Berger's work was one of disbelief; like Berger himself, the scientific world at first doubted whether the workings of an organ as complex as the brain could be recorded through the skull. Berger did not achieve an international reputation until 1934, when Edgar Douglas Adrian (1889-1977), a renowned English neurophysiologist, confirmed his findings. Even then, however, Berger remained unappreciated in his own country. In the late 1930s, the Nazis forced him to retire from the University of Jena, where he had been professor and director of psychiatry since 1919. With his laboratory dismantled and no facilities to carry on his work, Berger fell into a depression and committed suicide in 1941.

Despite his reputation as a reserved and inflexible man, Berger would no doubt be pleased to know that, over the years, research scientists have used the EEG to identify the parts of the brain involved in the mental processes of reasoning, memory, and feeling. He would also, no doubt, be interested in a system that has simplified EEG interpretation. Known as BEAM (brain electrical activity mapping), this system was invented by Frank Duffy of the Harvard Medical School in the early 1980s. It uses computer technology to combine the signals from the individual electrodes into a overall, color-coded map of the brain's electrical activity. BEAM can store large amounts of EEG data, compare healthy profiles with abnormal ones, and provide detailed analyses that have been used to accurately diagnose such conditions as dyslexia and schizophrenia, which are usually difficult to detect. Efforts are currently underway to use BEAM in matching EEG patterns to specific brain functions. For example, research scientists at Johns Hopkins University have used BEAM to map the electrical activity involved in the movement of a monkey's arm; their studies have shown that when the monkey anticipates moving its arm, the pattern of electrical activity in its brain changes. If efforts like these are successful, it may one day be possible to use computers and the electrical activity of the brain not only to control artificial limbs but in many other revolutionary applications as well.

The use of EEGs in neurofeedback (or biofeedback) is becoming increasingly popular; for example, in helping children cope with Attention Deficit/Hyperactivity Disorders. In 1997, Neuropathways EEG Imaging system became the first U.S. patented digital EEG neurofeedback device, displaying brain wave voltages and frequencies in *real time*, (there is less than one thousandth of a second delay between when the brain wave begins and its display on a computer screen). This allows interventional feedback when the event is occurring, not after it has passed, and has produced improvement in concentration and other cognitive functions of both ADD clients and head trauma victims. Meanwhile, Mind Media in The Netherlands is marketing *The BrainTracer*, a non-medical device which allows individuals to see their real time brain wave activity on standard home computer screens and print them on standard printers.

Electromagnet

An electromagnet is a device that converts the flow of an electric current into a magnetic force. Unlike a permanent **magnet**, an electromagnet can be turned on and off.

The first hint that the flow of electricity created a magnetic field came in 1819 by the Danish physicist Hans Christian Oersted (1777-1851), who discovered that a **compass** needle was deflected when brought near a wire that had electricity flowing through it. (The same discovery was made almost 20 years earlier by the Italian Gian Domenico Romagnosi, but was completely ignored.) This unexpected discovery encouraged others, notably **Michael Faraday** and André Marie Ampère (1775-1836), to investigate further.

In 1820 English physicist William Sturgeon (1783-1850) built upon Oersted's discovery and Ampère's concept of the **solenoid** to invent the electromagnet. He took a soft-iron bar, wrapped it with eighteen turns of **wire**, and sent electricity through the wire. The electricity set up a magnetic field which was concentrated in the iron core. (''Soft-iron'' is iron which retains this magnetism only when current flows in the coil.) Sturgeon's second electromagnet, the core of which was bent into a horseshoe, could lift a remarkable 9 lb (4 kg) of another magnetic material, twenty times its own weight.

American physicist Joseph Henry learned of Sturgeon's invention in 1829 and decided he could do better. The secret was to increase the number of turns of wire. The difficulty was to insulate the wire from itself as well as the iron core. Sturgeon had put a coat of shellac on his core to insulate it from the wire; his eighteen turns were spaced far enough apart to prevent their touching each other. Henry, however, was going to use many more turns of wire, so they *would* touch. He laboriously hand-wrapped all his wire in silk cloth, which he obtained by ripping up his wife's silk petticoat. Henry's electromagnet was a great success. By 1831 he had built one that could lift 750 lb (340 kg); another, using the current from an ordinary 1830s-style battery, lifted more than one ton (900 kg) of iron.

In addition to moving great weights, Henry also used his electromagnet to move a small iron bar. In this case, the small bar was located about one mile (1.6 km) away from the switch that activated the electricity. Because of electrical resistance in the wire, there was a limit to how far Henry could remotely operate his electromagnet. He found, however, that he could use one electromagnet to control another and linked them together, thus inventing the relay in 1835. In fact, Henry had just invented the **telegraph**, but **Samuel Morse** promoted the device and got the patent in 1844.

Meanwhile, Faraday and Ampère had been busy as well. In 1821 Faraday established the concept of "magnetic lines of force"; the principle of the electric motor came to him ten years later. Ampère discovered the laws that relate electric current and magnetic force in 1827, suggesting that magnetic fields and electric fields were just different forms of the same thing. Scientists today call this "electromagnetism."

Today electromagnets are used in a variety of electrical devices with moving parts, including **generator**s, motors, **brake**s, and clutches. Small relays and solenoids are used in doorbells, circuit breakers, and **telephone** receivers, while gigantic electromagnets are used to control **particle accelerator**s as well as other scientific devices.

ELECTRON MICROSCOPE

Described by the Nobel Society as "one of the most important inventions of the century," the electron **microscope** is a valuable and versatile research tool. The first working models were constructed by German engineers **Ernst Ruska** (1906-) and Max Knoll in 1932, and since that time the electron microscope has found numerous applications in chemistry, engineering, and medicine.

At the turn of the twentieth century, the science of microscopy had reached an impasse: because all microscopes relied upon visible light, even the most powerful could not detect an image smaller than one wavelength of light. This was tremendously frustrating for physicists, who were anxious to study the structure of matter on an atomic level. Around this time, French scientist Louis de Broglie (1892-1987) theorized that subatomic particles sometimes act like waves, but with much shorter wavelengths. Ruska, then a student at the University of Berlin, wondered why a microscope couldn't be designed that was similar in function to a normal microscope but used a beam of electrons instead of a beam of light. Such a microscope could resolve images thousands of times smaller than a wavelength.

There was one major obstacle to Ruska's plan, however. In a compound microscope, a series of lenses are used to focus, magnify, and refocus the image. In order for an electron-based instrument to perform as a microscope, some device was required to focus the electron beam. Ruska knew that electrons could be manipulated within a magnetic field, and in the late 1920s he designed a magnetic coil that acted as an electron lens. With this breakthrough, Ruska and Knoll constructed their first electron microscope. Though the prototype model was capable of magnification of only a few hundred power (about that of an average laboratory microscope), it proved that electrons could indeed be used in microscopy.

The microscope built by Ruska and Knoll is very similar in principle to a compound microscope. A beam of electrons is directed at a specimen sliced thin enough to allow the beam to pass through. As they travel through, the electrons are deflected according to the atomic structure of the specimen. The beam is then focused by the magnetic coil onto a photographic plate; when developed, the image on the plate shows the specimen at very high magnification.

Scientists worldwide immediately embraced Ruska's invention as the breakthrough in optical research, and they directed their own efforts toward improving upon its precision and flexibility. A Canadian-American physicist, James Hillier (1915-), constructed a microscope from Ruska's design that was nearly 20 times more powerful. In 1939, modifications made by Vladimir Kosma Zworykin (1889-1982) enabled the electron microscope to be used for studying viruses and protein molecules. Eventually, electron microscopy was greatly improved, with microscopes able to magnify an image 2,000,000 times. One particularly interesting outcome of such research was the invention of holography and the **hologram** by Hungarian-born engineer **Dennis Gabor** (1900-1979) in 1947. Gabor's work with this three-dimensional photography found numerous applications upon development of the **laser** in 1960.

There are now two distinct types of electron microscopes: the transmission variety (such as Ruska's), and the scanning variety. Scanning electron microscopes, instead of being focused by the scanner to peer through the specimen, are used to observe electrons that are scattered from the surface of the specimen as the beam contacts it. The beam is moved along the surface, scanning for any irregularities. The scanning electron microscope yields an extremely detailed three-dimensional image of a specimen but can only be used at low resolution; used in tandem, the scanning and transmission electron microscopes are powerful research tools.

Today, electron microscopes can be found in most hospital and medical research laboratories. One of the more interesting applications for the invention is that by petroleum companies, who use electron microscopy to study the molecular links in petrochemicals. The results of their studies help in the search for petroleum substitutes and synthetic fuels, especially those derived from coal.

The advances made by Ruska, Knoll, and Hillier have contributed directly to the development of the **field ion microscope** (invented by Erwin Wilhelm Muller) and the **scanning tunneling microscope** (invented by Heinrich Rohrer and **Gerd Binnig**), now considered the most powerful optical tools in the world. For his work, Ruska shared the 1986 Nobel Prize for physics with Binnig and Rohrer.

ELECTRON TUBE • See Vacuum tube

Electronic computer • See Computer, analog

Electronic mail

Electronic mail, or the sending of messages via computer, was the very first application for networked computers. Invented in 1971, electronic mail, or email as it is commonly called, was the offspring of a project started in 1969 by the U.S. Department of Defense. This project was designed to connect vital organizations together in case of any normal communications failure. The Advanced Research Projects Agency, or ARPA, used telephone lines to link the computers of defense contractors, university researchers, and military personnel. This ''network'' of computers was called ARPANET. ARPANET slowly grew and grew to become the Internet. The application these scientists developed to communicate with each other was email.

The technology behind email has changed little since its invention. The content on the other hand is radically different. In the early years of email, transmissions were limited to small squibs of text that could be sent quickly between two computers. With today's computer technology, entire multimedia presentations can be sent to your next-door neighbor, or someone around the globe, with little difficulty.

In order to utilize email, one must be connected to a computer network like the Internet. Once connected, you are assigned a uniquely identifying user name. This name, in conjunction with the name of the network you are using, is called an email address. This is analogous to your own home address. It uniquely identifies you within the system. Email addresses are in the form of *username@networkname*. An example of a real-world email address is *prez@whitehouse.gov*. This is the email address of the President of the United States. In this example, *prez* is the user name and *whitehouse.gov* is the network name. These are separated by the ''at'' character. This address can be used to send and receive email from almost anywhere in the entire world. The use of email is so widespread in America that the U.S. Postal Service in 1997 began testing electronic postmarks and email storage services.

Email has become the communication method of choice within the corporate world. This has allowed for better communication between employers and employees. In addition, email has drastically reduced the amount of time it takes to relay business and personal information between two parties. It has become one of the most widely used computer applications on the planet. However, this ease of communication has found its way into many households. What once was just available to larger businesses, universities, and governments, now allows your grandmother to receive birthday pictures of your child over her computer.

See also Internet/Worldwide Web

Electronic tracking devices

Electronic devices can be used to track people, animals, vehicles, and virtually anything else that moves. Electronic tracking can be distinguished from electronic **surveillance equipment** in that the location of the subject is the primary goal of tracking. Surveillance involves actually seeing or hearing the subject. Prior to electronics, tracking meant following the trail of evidence left behind by the subject: his or her scent, fingerprints, footprints, etc.

Modern electronic tracking began in earnest with the advent of radar during World War II. American **Hedy Lamarr**, better known as a movie star, patented a device to track and guide torpedoes. The torpedoes emitted a radio signal that was received back at the submarine that launched it. At the time, Lamarr's invention was ignored, but gained a great deal of interest after the war.

During the 1950s, the first manmade satellites were launched into space, allowing the expansion of electronic tracking devices. With this technology, the subject transmits its signal to a receiver in the satellite. The satellite, maintaining a geosynchronous orbit with a fixed coordinate on the ground, can track the subject anywhere on the planet. In the 1960s, the U.S. military proposed a vast global positioning system (GPS) network. This technology was initially designed to track nuclear missiles, but soon found numerous civilian applications.

By the mid-1980s electronic tracking devices were being used in a variety of ways. Convicts serving their sentences under house arrest in some jurisdictions have been fitted with monitoring anklets that contain transmitters. A receiver is planted inside the subject's telephone, and the signal is forwarded to a central computer. If the wearer strays too far from home, the signal is interrupted and the authorities are notified. More advanced systems allow the prisoner to travel only to particular destinations (such as a job). Rental car companies use electronic tracking devices in their vehicles as well. In addition to security, drivers can use the devices with onboard computerized maps to navigate unfamiliar areas. Electronic tracking devices have many other such uses, but have also raised concerns regarding privacy rights since continuous monitoring from space is now possible.

Electronics

Though they had been experimenting with electricity for many years, it was not until the beginning of the nineteenth century that scientists began to comprehend how it worked. This increased awareness was probably a result of the invention of the first battery—**Alessandro Volta**'s voltaic pile—in 1800. By using batteries, scientists had a reliable supply of electricity to work with. They began to test the effect of electricity upon chemicals, metals, and other substances in an effort to better understand its nature and, eventually, harness its power.

As worldwide trade increased so did the need for instant long-distance communication, and it was the science of electronics that addressed this need. By 1870 **Alexander Graham Bell** was experimenting with a means for sending a voice signal over a wire. While this was a viable idea, the signal was generally too weak to travel over long distances; a similar problem existed with **Guglielmo Marconi**'s **radio**, which couldn't put out a signal powerful enough to clearly transmit a human voice. These problems were solved by the invention of the amplifying triode in 1911 and the high-frequency **alternator** in 1900. Stimulated by the two World Wars, electronic innovations came to dominate technology and industry in the twentieth century.

Most electronic devices are based upon combinations of **electric circuits**. There are many possible components to such a circuit, including **resistors** (to hinder the current of electricity), **capacitors** (to store a charge), **potentiometers** (to control the amount of current), and switches (to prevent or allow a circuit being completed). A circuit can contain one or more of these components, in series or in parallel.

As the understanding of electronics grew, the devices used became more and more sophisticated. By the late 1800s scientists were experimenting with Crookes tubes (named after the British scientist **William Crookes**). One such scientist, Sir **John Ambrose Fleming**, reconfigured the Crookes tube into a device which became known as the **vacuum tube** or thermionic valve. The vacuum tube consists of a hollow, oblong tube containing two electrodes—an anode and a cathode—at each end. In order for a circuit to be completed, most of the air must be pumped out of the tube, creating a vacuum. When a current is applied to the negatively-charged cathode it releases electrons; these electrons cross the gap to the positively-charged anode, completing the circuit.

Because this particular vacuum tube contains just two electrodes it is called a **diode**. One useful feature of a diode is that it allows current to flow through it in one direction only: from the cathode to the anode. (Electrons are not released by the positive terminal, nor would they jump to the negative terminal.) Diodes are often used to convert **alternating current** (which travels in a back-and-forth motion) into **direct current** (which travels in only one direction). Such circuits are often called *rectifiers.*

Sometimes a metal mesh is placed within the vacuum tube, between the cathode and the anode. This type of device, invented by **Lee De Forest**, is called a triode. In a triode, the current passing from the cathode to the anode (called the *plate*) varies in direct proportion to the voltage applied to the mesh (called the *grid*). Therefore, a strong current passed through the plate can be made to ''mimic'' a feeble voltage fed into the grid (for example, a weak radio signal), thereby providing amplification. For many years, triodes were an indispensable component of radios and amplifying devices.

Another type of vacuum tube is the **cathode ray tube** (CRT). In a CRT, a tiny aperture is placed in the anode, allowing some of the electron stream to escape from the tube. Using magnetic fields and deflection plates, the escaping stream is directed toward a fluorescent screen. Where they strike the screen, the electrons create a brief glow. The first invention to utilize this phenomenon was **Karl Ferdinand Braun**'s **oscilloscope**, used to observe wave patterns in electrical signals. Today, televisions use similar electron beams to produce a visual image upon a screen.

For the most part, vacuum tubes and thermionic valves have been replaced by semiconductors. A very primitive semiconductor called a ''cat's whisker'' diode (first discovered by K. F. Braun) was used in the early 1900s in crystal radio sets, but it was not until after World War II that the real importance of semiconductors began to be realized.

A semiconductor is any material that is not quite a conductor and not quite an insulator—that is, it can act as either under the proper conditions. A good example of semiconducting material is germanium, which forms the base of many modern semiconductors. Generally, a small amount of impurity—such as antimony—is added, increasing the electrical effects.

The key to using semiconductors is to place two or more different semiconducting surfaces together, for the semiconducting effects manifest themselves most dramatically at the place where two materials meet. For example, if two pieces of germanium—one coated with antimony, the other with indium—are placed next to each other, and area called a p-n junction is formed; at this junction, electrons pass back and forth easily. If a current is applied in one direction the electrons will flow along, completing the circuit; however, if the current is applied in the other direction the electrons will build up, impeding the flow and preventing the circuit from being completed. In this way, semiconductors serve the same function as vacuum tube diodes.

The most important kind of semiconductor is the **transistor**—a device that performs all of the functions of a triode. The first transistor was built in 1948 by a team at Bell Telephone Laboratories in New Jersey headed by **William Shockley**. Shockley realized immediately that transistors could almost completely replace the vacuum tubes used in most electronic devices; they were much smaller, allowing for the designing of hand-held and portable appliances, and did not give off heat as tubes did. The first transistor radios were mass produced in 1955; Shockley's team was awarded the Nobel Prize in physics the next year.

Today, the transistor is itself becoming obsolete, as electronics manufacturers perfect tiny components called **integrated circuits** (IC). Made from semiconducting materials, an IC less than 10 millimeters square can hold thousands of different electronic components. Because they are so small, ICs have virtually eliminated the problem of overheating while dramatically increasing the speed by which functions can be performed. In 1971 a microprocessor chip—containing a complete central processing unit for a computer—was introduced by the Intel Corporation.

The generation of electronic components that will replace the integrated circuit can be seen on the horizon. Some scientists believe that metal-oxide semiconductors (MOS); just a few millimeters in diameter, will replace the IC. These devices perform the same functions as an IC, only faster—and require less energy to do so.

Another important development in electronics has been that of the electronic countermeasure, which emits signals that block reception of radar and radio signals. Such a device is commonly used in military functions to steer radar-guided missiles from their targets.

Electroplating

Some metals are recognized for their strength and durability, while others are better known for their visual beauty. When both qualities are desired, it is possible to coat a stronger metal, such as iron, with a more attractive one, such as gold or silver. For example, jewelry is commonly gold-plated, and silverwear is usually a cheaper metal coated with silver.

The coating process, or gilding, was achieved in earlier times by direct pounding of the exterior metal to the base metal. Later a bonding method was developed in which both metals were joined while in a molten state. However, such coatings are often thicker than was necessary, wasting the valuable coating material.

Establishing a bond by use of an electrical current, or *electroplating*, was first developed in 1800 by the German physicist **Johann Wilhelm Ritter** (1776-1810). Within months of the invention of the electric battery by **Alessandro Volta**, Ritter was able to demonstrate how copper plates could be created by running an electrical current through copper sulfate. Positively charged copper ions in the copper sulfate would be attracted to the negatively-charged plate, which served as the electrode of the battery. These ions would then cover the plate. It would be another 40 years before this process would be used on a large scale.

In 1839 Germans Carl Jacobi and **Ernst Werner von Siemens** were granted patents for the electroplating process for gold- and silver-plating. In England the Elkingtons of Birmingham developed their own electroplating system in 1840. Gold or silver dissolved in cyanide would adhere to the item being plated as an electric current was passed through the solution. (Other metals commonly applied by electroplating are cadmium, chromium, and nickel.) By 1900 the Germans had established primacy in the electroplating industry.

Electroscope • See Electrostatic devices

Electrostatic devices

Electrostatics is the branch of science that studies the behavior of electric charges that are at rest, or "static." The phenomena of static electricity has been known for well over two thousand years, and a variety of electrostatic devices have been created over the centuries.

The ancient Greek philosopher Thales (624-546 B.C.) discovered that when a piece of amber was rubbed, it could pick up light objects, a process known as triboelectrification. (The Greek name for amber, *elektron*, gave rise to many of the words we use today in connection with electricity.) It was also noted that lodestone had the natural ability to pick up iron objects, although the early Greeks did not know that electricity and magnetism were linked.

In the late sixteenth century, William Gilbert (1544-1603) began experimenting with static electricity, pointing out the difference between static electric attraction and magnetic attraction. Later, in the mid-1600s, Otto von Guericke built the first electrostatic machine. His device consisted of a sulfur globe that was rotated by a crank and stroked by hand. It released a considerable static electric charge with a large spark. A similar device was invented by Francis Hawkesbee in 1706. In his design, an iron chain contacted a spinning globe and conducted the electric charge to a suspended gun barrel; at the other end of the barrel another chain conducted the charge.

In 1745 the first electrostatic storage device was invented, created nearly simultaneously by two scientists unaware of each other's work. Peter von Muschenbrock, a Professor at the University of Leyden, and Ewald von Kleist of the Cathedral of Camin, Germany, devised a water-filled glass jar with two electrodes. Also, a Leyden student named Cuneus, who had been using a Hawkesbee machine to electrify the water, touched the chain to remove it, and nearly died from the electric shock.

The Leyden jar could accumulate a considerable electric charge, and audiences willingly received electric shocks in public displays. One of these displays aroused the curiosity of Benjamin Franklin, who obtained a Leyden Jar for study. He determined that it was not the water that held the electric charge, but the **glass** insulator. This is the principle behind the electrical condenser (**capacitor**), one of the most important electrical components in use today.

Charles F. DuFay (1698-1739) discovered that suspended bits of cork, electrified with a statically charged glass rod, repelled each other. DuFay concluded that any two objects which had the same charge repelled each other, while unlike charges attracted. The science of electrostatics, so named by André Marie Ampère (1775-1836) is based on this fact.

French physicist Charles Coulomb (1736-1806) became interested in the work of Joseph Priestley (1733-1804), who had built an electrostatic generator in 1769 and studied electrical repulsion. Coulomb used his torsion balance to make precise measurements of the force of attraction between two electrically charged spheres, and found they obeyed an inverse square law where the force decreased with the square of the distance, analagous to Newton's law of gravity. The mathematical relationship between the forces is known as Coulomb's law, and the unit of electric charge is named the Coulomb in his honor.

Alessandro Volta invented a device in 1775 that could create and store an electrostatic charge. Called an electrophorus, it used two plates to accumulate a strong positive charge. The device replaced the Leyden Jar, and the two-plate principle is behind the capacitors in use today.

Several other electrostatic machines have been devised. In 1765 John Reid, an instrument maker in London, built a portable static electric generating machine to treat medical problems. John Cuthbertson built a huge device in 1783, which could produce electrical discharges two feet (61 cm) in length. The gold leaf electroscope, invented in 1787, consists of two leaves which repel when they receive an electric charge. In 1881, British engineer James Wimshurst invented his Wimshurst machine, two glass discs with metal segments spinning opposite to each other. Brushes touching the metal segments removed the charge created and conducted it to a pair of Leyden jars, to be stored for later use.

The most famous of all the electrostatic devices is the van de Graff **generator**. Invented in 1929 by Robert J. van de Graff, it uses a conveyor belt to carry an electric charge from a high-voltage supply to a hollow ball. It had various applications. For his experiments on properties of atoms, van de Graff needed to accelerate subatomic particles to very high velocity, and he knew that storing an electrostatic charge could result in a high potential. Another generator was modified to produce x rays for use in the treatment of internal tumors, and was installed in a hospital in Boston in 1937. Van de Graff's first generator operated at 80,000 volts, but was eventually improved to five million volts. It remains one of the most widely used experimental exhibits in schools and museums today.

ELECTROSTATIC PRECIPITATOR • See Electrostatic devices

ELEVATED RAILWAY • See Streetcar and trolley

ELEVATOR

The elevator, also called a lift in Great Britain, is a device consisting of a car or platform that moves passengers and freight up and down between the floors of a building. The car, enclosed in a steel frame, glides quickly and smoothly between steel rails inside a vertical shaft. Although the earliest elevators were frustratingly slow and dangerous, today's electric elevators can transport passengers at speeds up to 1,800 feet (550 m) per minute. Such speed is especially important in the hugely-tall modern skyscrapers like New York's World Trade Center or Chicago's John Hancock Center whose elevators move tens of thousands of people to and from their offices every day. In the 1980s, glass elevators put on the outside of tall buildings were thought by architects to add elegance to their structures, in addition to giving passengers a spectacular view of the surrounding area. The elevator has had a tremendous impact on our modern urban landscape: it made the **skyscraper** practical.

The concept of a device to raise and lower heavy loads mechanically reaches back to the earliest days of humanity, a part of the human will to stretch accomplishment past the confines of mere muscle power. The ancient Greeks and Romans knew how to hoist loads using **winch**es and **pulley**s powered by human labor, animals, or water. Similar lifting devices were used in the Middle Ages. An early attempt at transporting people to the tops of buildings, the seventeenth-century ''flying chair'' was thwarted by the knotty problem of passenger safety.

Freight elevators powered by steam or hydraulic systems were in common use in England and the United States by the early 1800s. Both types had serious drawbacks. The hydraulic elevators were very slow, and passengers wisely refused to ride in the steam elevators—the ropes used to lift their cabs quite often broke, plunging the cab and its contents to the building's basement.

The passenger-safety problem, which had severely hindered elevator use and development, was solved in 1852 by **Elisha Graves Otis**. As a master mechanic for a New York bedstead factory, Otis invented a safety device for his employer's freight hoists that stopped the elevator cab from falling when the rope broke. Teeth along the sides of the cab were held back when the elevator's hoist rope was taut. When tension on the rope was released, the cab teeth sprang out and clamped onto the elevator's guide rails, which held the cab securely in place.

Elisha Graves Otis demonstrating his "safety elevator" at New York's Crystal Palace Exhibition in 1853. He repeatedly climbed aboard the cab and, after being raised above the crowd, ordered the hoisting rope to be cut. To onlookers' amazement, the cab would lock safely into place.

Initial orders for this new ''safety elevator'' were slow, so Otis hit upon a dramatic way to publicize his invention. He set up an open elevator cab and shaft at New York's Crystal Palace Exhibition in 1853-54. He repeatedly climbed aboard the cab and had himself hoisted above gathered onlookers. The crowd, gasping as Otis, with a flourish, ordered the hoisting rope cut, was amazed to see the cab lock safely into place.

The first passenger elevator was installed in the five-story Haughwout department store in New York City in 1857. Hotel and office installations soon followed, and public acceptance of elevator safety grew. Building owners discovered they could rent rooms and offices on the top floor for as much as, or even more than, similar space on the lower floors. City buildings, previously limited to five stories (the extent of most people's stair-climbing ability), suddenly grew to ten or twelve stories. Structural steel had made the building of skyscrapers technically possible; elevator safety made such buildings physically and economically feasible.

Improved versions of Otis's steam-powered elevator soon followed. A modified hydraulic elevator permitted safer stops at higher speeds. The electric passenger elevator did away with the furnaces and boilers required for steam operation and replaced the pipes and tanks needed for hydraulic systems. An early electric Otis installation in 1889 carried delighted visitors up the newly erected 984-foot (300 m) Eiffel Tower in Paris, France. Push-button controls appeared in 1894. The basic modern gearless-traction elevator system was introduced about 1903. After this, efforts to improve the elevator focused on such convenience and efficiency features as automatic leveling, power control of doors, automatic operation, and increased speed.

Most of today's elevators are powered by electric traction systems, with an **electric motor** turning the sheave (pulley) around which the hoisting cables run. No doubt Elisha Otis, whose factory only employed eight or ten men at the time of his death in 1861, would be astonished to know that over 2 million elevators are now in service around the world, and that the average elevator in New York City travels about 10,000 miles each year.

Many very large buildings have elevators dedicated to certain floor levels or to odd or even-numbered floors. Newer buildings and malls often have observation elevators whose glass sides are thought to add elegance. They also allow passengers to view the surrounding areas. Some of these elevators are placed on the exterior walls of a building, offering an occasionally spectacular view.

Perhaps the ultimate modern elevator is the concept of the ''space elevator.'' Proposed during the 1980s as a serious method of transporting people and materials into outer space, this ''elevator'' would essentially be a tether or cable that connected a stationary satellite to the earth's surface. The elevator itself would be some sort of cab connected to the tether that would literally go up and down the tether between the satellite and the earth. Although still only a concept, this potentially simple and inexpensive transport system has already been tested on the United States Space Shuttle which experimented with tethered satellites.

Elster, Jules • See Photoelectric cell

Embalming • See Hunter, John

Embryonic transfer

Embryonic transfer is the transfer of an embryo from the uterus of one female to the uterus of another. English doctor Walter Heape first performed successful embryo transfer in 1890 when he removed embryos from a female rabbit and placed them into a female hare which subsequently gave birth. Further animal experiments were carried out on 15 different species. Embryo transfer became a valuable technique in the cattle industry; superior animals could produce large numbers of embryos, which could then be carried by less valuable cows.

Human embryo transfer got its start in 1972 when a Chilean team led by Horacio Croxatto described their technique for flushing an unimplanted, unfertilized egg out of the uterus. Brothers Randolph Seed, a surgeon, and Richard Seed, head of a cattle-breeding company, emulated Croxatto's technique in 1980 to recover a fertilized egg. They founded a company called ''Fertility and Genetics Research,'' which funded a medical team at Harbor-UCLA Medical Center in Torrance, California, headed by John E. Buster and Maria Bustillo, which began treating a group of infertile women and advertised for egg donors. Healthy donors were matched with recipient couples for blood type, Rh, and hair and eye color. As each donor reached readiness for ovulation, she was inseminated with sperm from the recipient's husband. Since a fertilized egg does not implant in the uterus immediately, uterine lavage (washing) was performed on each donor five days after insemination. Recovered embryos, or blastocysts, were then transferred to the uterus of the recipient. Two transfers performed by the Bustillo-Buster team in 1983 resulted in the birth of two babies early in 1984.

Embryo transfer is a valuable technique for overcoming infertility in women who are unable to produce viable eggs or who are at risk for transmitting genetic disease. It carries risks of infection for both donor and recipient and of unwanted pregnancy for the donor if the embryo implants in the uterine wall before washing occurs. Legal and ethical issues are involved, since the embryo recipient—the woman who carries and gives birth to the child—is not the child's genetic mother. Medical ethics also became involved when Fertility and Genetics Research tried to patent the transfer technique for profit.

See also *In vitro* fertilization

Encryption devices

Encryption is the process of converting text into code. Encryption devices are the machines that perform the actual encrypting. Such machines have existed throughout history. Though encryption began more as a need in war than commerce, in the late twentieth century, the latter has taken precedent, especially as more and more business is conducted electronically and digitally.

One of the first encryption devices, the skytale, was devised by the Spartans in the fifth century B.C.. The skytale was a wooden staff, and text was written length-wise on papyrus or the like wrapped tightly around the staff. When the papyrus was unwrapped and delivered, it looked nonsensical until re-wrapped around a skytale of the similar thickness. By the late nineteenth century, **Thomas Jefferson** invented a cipher wheel, a simple encryption device which featured two disks. The **alphabet** was on one, and the other featured its scrambled counterpart.

By the early twentieth century, rotor-based encryption devices became popular. The Enigma machine was invented in 1918 by Arthur Scherbius, a German electrical engineer. On the outside, Enigma looked somewhat like a typewriter with a display. When typing in text, the display shows the cipher. When decoding, the display features the decoded text. Inside, Enigma featured an electromechanical rotor system for converting the typed text into code, and vice versa. Each time a letter was typed on the Enigma keyboard, each rotor would turn and the letter would undergo up to six individual transcriptions. If the same letter was typed again, it would not be coded the same because of the individual rotor turns. Though initially unwanted by German business or government, the latter began to find military applications for Enigma by 1926. Then Germany also continued to improve Enigma over the years, making it both smaller and more powerful. Thus, Enigma's code was not solved by the Allies in World War II until 1940, but only with the help of information bought from a German involved with the machine. The Americans also had several encryption devices of their own at this time, including SIGABA, a United States Navy machine that eventually became the standard for all branches of the American military in the early 1940s. SIGABA, also known as Converter M-134-C, was also an electrical rotor-based device.

These encryption devices are primitive when compared with possibilities generated by a computer and computer software, and business has become the primary customer as more and more transactions are handled over electronically. Cryptographic software is used to encode these transactions, from ATM machines and credit card verifications to illegal activities such as insider trading and drug sales, as well as communication on the Internet and over **cellular telephone**s. The programs, such as Pretty Good Privacy and SynCrypt, have become nearly impossible to crack. These programs have their roots in a late 1970s invention by several scholars at the Massachusetts Institute of Technology. With their public-key encryption, a message is sent with both its encoding and decoding keys. The keys themselves are based on complex mathematics, so the just knowing the encoding key would not allow an interceptor to read the message. The United States government has placed limitations on the export of the strongest (i.e., most mathematically complex) encryption software, and there has been debate in the late 1990s as to if the government should hold the keys to such software or if it should be banned if it is uncrackable. Rulings have generally been against such a key, but this is still an on-going discussion.

See also Code and cipher

Endoscope

Sometimes called the fiberscope, the endoscope is an optical instrument that allows doctors to view the inner workings of the human body without having to perform surgery. The endoscope is a flexible narrow tube containing several bundles of hair-thin **glass** fibers that are covered with a reflective coating. A highly intense light source, usually a **halogen lamp**, is used to transmit light along one bundle of fibers toward the target area inside the body. Another bundle of fibers carries an image of the target area back up the tube where it is viewed through an eyepiece.

Endoscopes are primarily used in the health care field, but they may also be used for industrial purposes examining such inaccessible places as the inside of fuel tanks and nuclear reactors. Crude versions of the endoscope were used as early as the nineteenth century and included long, rigid tubes illuminated by candles. In 1854, Manuel Patricio Rodriguez Garcia, a Spanish-born vocal teacher, designed the forerunner of the laryngoscope; it allowed a clear view of the glottis (the vocal cords and the opening between them) and made it possible to see obstructions in the larynx.

The first efforts to develop the kind of glass fibers that would eventually be used in endoscopes were made by the Atomic Energy Authority and by the Rank Organization in Britain. By 1965, a 25-micron (a micron is one-thousandth of a millimeter) fiber had been produced. An American company, Bausch and Lomb, subsequently developed a 15-micron fiber used in their "Flexiscope," which could be used for industrial inspections because it gave off a "cold" light that was safe even in fuel tanks. When the American Cytoscopic Company succeeded in sterilizing glass fibers, the possibilities for medical uses of the endoscope increased greatly.

The modern endoscope can perform an amazing variety of medical procedures. In addition to the fiber optic bundles that transmit light, the endoscopic tube also contains air and water channels for flushing water through or inflating targeted areas. Tiny forceps can be placed at the tip of the endoscope to take specimen samples for laboratory analysis and to perform simple operations such as removing colon polyps or gallstones. Endoscopes can also be used to stop hemorrhaging by delivering laser beams directly to the point of bleeding; the **blood** coagulates and the bleeding is stopped.

Different types of endoscopes are specially designed to examine specific parts of the body. The angioscope passes through the arteries that carry blood to the heart; the arthroscope is used to explore the interiors of joints; the bronchoscope is used with a special dye and fluorescent light to detect lung malignancies; the gastroscope probes the stomach and upper intestinal tract; and the laparoscope is used for diagnosis and treatment of abdominal conditions.

Endoscopes can be used for a variety of remote sensing applications. They can be connected to spectrometers to perform sophisticated chemical analysis optically. The endfaces of the fibers can be impregnated with molecules that are sensitive to certain compounds, allowing researchers to detect minute concentrations of pollutants and toxins.

Energy efficiency

Energy efficiency is a measure of how much of the input **energy** is useful as output energy. An automobile, for example, typically has a low efficiency, around 30%, as much of the consumed fuel is converted to heat instead of power in the driveshaft. A conventional electric **generator** converts about 95-99% of the mechanical energy input to the rotor shaft into electrical energy at the generator¤s terminals.

The first law of thermodynamics is the statement of conservation of energy: an isolated system has the same amount of energy at all times. Energy is neither created or destroyed. (A nonisolated system has external forces acting on it, or energy added to or deleted from it, which disrupts the energy balance. But almost any system is isolated if viewed from a considerable distance.) However, energy can be converted from one form into another—for example, in the case of a coin sliding across a tabletop, some of its kinetic energy is invariably lost and converted to heat energy (and perhaps a small amount to sound energy).

Energy efficiency, then, is a measure of how much energy is lost as a system changes. In the case above where some kinetic energy is lost to heat, the system (consisting of the coin) loses energy to friction and sound. But the coin plus the tabletop plus the atoms underneath the tabletop plus the air around it is isolated, and so energy is conserved in that system. Energy efficiency will be 100% only in extreme examples like collisions between a small number of elementary particles like **electron**s and **photon**s. In more everyday systems, energy efficiency will never be 100%, because in all real systems some energy is lost to the production of heat, light, sound, chemical reactions, potential energy, atomic reactions, etc.

Engineers of all types often spend great effort in the design of an instrument or machine to increase its efficiency. Doing so not only can save money, it often leads to reduced weight and size of the instrument (important for **aircraft** and space vehicles), makes it more reliable or provides a longer lifetime.

See also Energy; Heat and thermodynamics; Otto cycle; Wankel engine

Engine, air-cooled • See Engine cooling systems

Engine cooling systems

Engines that produce their energy by heat and combustion have always presented engineers and inventors with the problem of maintaining safe operating temperatures. In 1823, Samuel Brown, an English inventor, came up with a way of cooling an engine by using water. In his design, the water, contained within a casing or cylinder lining, circulated around the cylinders. The water was constantly kept moving through the action of a **pump** and was recooled by contact with outside air. Two years later, Brown founded a company to build his water-cooled engine. In 1859, **Jean-Joseph-Étienne Lenoir**, a Belgian-French inventor, created the first usable **internal combustion engine**; his device employed the same water-cooling system.

It was not until 1897 that the radiator was first invented. A German engineer, **Wilhelm Maybach**, worked for a long time to perfect a successful radiator. Finally, he came up with a honeycomb design composed of a network of short straight ducts through which air, blown in by a small **fan** located in front of the ductwork, could circulate. These ducts also allowed the system's water to contact as much air as possible and thus cool down before returning to the heat of the engine. The company Maybach worked for, Daimler Motoren Gesellschaft, was the first to build these radiators.

Engines may also be cooled by air alone. In 1875, a Frenchman, Alexis de Bischop, created an engine with a cylinder surrounded by vertical blades. These blades afforded a large surface area to disperse heat better. The main advantage of the air-cooled system is simpler engine casting; there is no need for water ducts, a radiator or water pump, hoses, or antifreeze. Water has proven to be a much better coolant than air, however, and a far better insulator against noise as well.

The components of an engine cooling system include cavities called water jackets around the hottest parts of the engine, such as the exhaust valves and cylinders, a pump that circulates water through these jackets and the radiator, the radiator components including tubes filled with coolant (a mixture of water and antifreeze) and the fins providing surface area to transfer heat to the surrounding air, and the pressure relief valve in the radiator cap that maintains the balance between air pressure and fluid temperature. The water pump and fan are driven by pulleys and belts (or a camshaft in some car models), and a thermostat regulates the engine's operating temperature. Oil also cools the engine block and crankshaft while providing lubrication for the moving parts of the engine. An oil pump pressures oil through specific oil passages and a filter and returns it to the oil pan.

In the 1990s, developments in cooling system technologies have included design of a truck radiator with multiple modules instead of a single core. Damaged modules can be replaced in minutes, and the shock-mounted modules resist thermal expansion and impact better than the standard single core. Aluminum and copper-brass, the materials of choice for radiator construction, may also be replaced by brazed copper. Treatment of the copper in a brazing furnace makes it tolerant of a wider range of operating temperatures, and copper radiators are lighter in weight, smaller in size and require less energy for manufacture. Radiators of the future may also reduce pollution by applying the same principles used in the **catalytic converter**. As smog-bearing air moves through radiator coils that are treated with a catalytic coating of platinum and other additives, the ozone in the smog is converted to oxygen and carbon monoxide becomes carbon dioxide.

See also Automobile, gasoline

Engine, hot-air • See Stirling engine

ENGINE, INTERNAL COMBUSTION • See Internal combustion engine

ENGINE, JET • See Jet engine

ENGINE, OIL

The first **internal combustion engine** was developed in 1680 by the Dutch physicist **Christiaan Huygens**, who experimented with **gunpowder** as a fuel. The engine was far too dangerous to be practical, but the idea of an internal combustion engine continued to fascinate inventors.

Huygens's use of gunpowder as fuel for internal combustion engines gave way to the use of coal gas, which was extensively used to light city streets. Frenchman **Jean-Joseph Étienne Lenoir** is usually credited with developing in 1859 the world's first practical internal combustion engine, which was fueled by coal gas. Although Lenoir's two-cycle engine wasted fuel, his design was made considerably more efficient by the development of the four-stroke engine, an accomplishment generally credited to the German inventor **Nikolaus August Otto**.

Also in 1859 Edwin Drake bore through 69 feet (21 m) of rock and struck oil in Pennsylvania. It was an event that changed forever how the world generated power. The oil was first refined to yield **kerosene**, a fuel that immediately began to replace illuminating gas as the fuel of choice for internal combustion engines. Many innovators altered the Otto four-cycle engine to burn the gasoline made possible by Drake's discovery.

The most notable of these innovators was the German **Rudolf Diesel** who saw fuel oil as a cheaper and thus more desirable fuel. Diesel was a keen student of thermodynamics, a science whose development was greatly aided by the French physicist Nicolas Carnot (1796-1832). Carnot's theories prompted Rudolf Diesel to develop an engine designed to convert the greatest amount of a fuel's energy to usable power. Steam engines convert about seven percent of the energy contained in the burnt fuel into mechanical energy; Diesel was convinced that he could design an engine that would capture over 70% of the fuel's energy. His theory, based on Carnot's principles, led to an engine that radically compressed air to a pressure of about 4,000 pounds per square inch (28 million pascals). (The idea to compress the air drawn into the cylinder was one of Otto's contributions to the development of the internal combustion engine.) By compressing the air in a ratio of about 25 to 1, its temperature rises to as much as 1,000°F (537°C). When injected into this super-heated air, fuel will spontaneously ignite, providing the power to push the engine's **piston**. Diesel first used powdered coal as a fuel because it was the cheapest fuel available, but soon abandoned the powder in favor of heavy fuel oil. Diesel's engine never reached the theoretical efficiencies that Carnot's principles suggested, but it did prove to be an engine vastly more efficient than any Otto engine then in use. Diesel received a patent on his engine in 1892.

In addition to its improved efficiency, the diesel engine has other advantages over the more conventional gasoline-powered four-stroke ''Otto-cycle'' engine. Because it is self-igniting, it needs no complex **ignition systems**, and because the power output is regulated by the amount of fuel injected directly into the cylinder, it requires no **carburetor** to mix air with the engine's fuel. Finally, diesel fuel is cheaper than gasoline and, because it is less volatile, is a safer fuel to use.

Diesel engines possess certain disadvantages as well. Due to the necessity of containing high compression ratios, a diesel engine must be considerably heavier than a gasoline engine that produces the same amount of usable power. Consequently, they initially cost more to build. Diesel engines run roughly at low speeds, they have high levels of pollutants in their exhaust, and the exhaust odor is considered to be more obnoxious than that emitted by conventional gasoline engines. Despite these disadvantages, the diesel engine's efficiency has led to its widespread use, and the engine has undergone several refinements since the death of Rudolf Diesel in 1913. It was first installed in a ship in 1910, and although its weight precludes its use in **aircraft**, it was adapted for use in an automobile in 1922. The diesel-electric **locomotive** has all but replaced the steam engine for rail power. The diesel engine can also be found in trucks, buses, tractors and other agricultural machinery. It is almost the exclusive choice for industrial power throughout the world due to its ability to burn relatively unrefined fuels and deliver an efficient amount of the fuel's energy as usable power.

ENGINE, ROTARY • See Wankel engine

ENGINE, STEAM • See Steam engine

ENGINE, TURBOPROP • See Jet engine

ENGRAVING AND ETCHING

The practice of carving designs, or engraving, on stone and metal objects has been used by crafts workers since early antiquity; the surviving monuments, vessels, and jewelry of Greek, Sumerian, Chinese, and Egyptian civilizations, among others, provide ample examples. The practice of etching, using acid to burn designs into metal or **glass**, was originally used for decorative purposes by medieval armorers. The Chinese were using wood carvings to create reproducible artwork in the first millennium A.D., but it was not until the fifteenth century that engraving and etching designs on metal developed into art forms unto themselves. The first known artistic engravings in Europe were created by goldsmiths in Germany and northern Italy; later, engravers employed other soft metals, including copper, which is the primary medium used for plates today.

There are two principal types of engraving. In *intaglio*, a design is carved with a pointed tool, called a *burin*, into a metal plate or wood block, which is then rolled with **ink**. The ink is wiped from the plate, but remains inside the grooves;

paper is then pressed onto the plate. In *relief* engraving, the artists used the burin to cut a design that is raised above, rather than carved below, the surface of the plate. Since the Renaissance, the essential mechanics of engraving has changed little, although in the twentieth century, many engravers, particularly commercial engravers, have begun to use burin-like power tools.

In etching, which began in the early sixteenth century, the plate is covered with a compound of beeswax, bitumen, and resin, in which the artist carves the design with a sharp tool; when the plate is treated with acid, the acid eats through the lines cut into the wax to engrave the design (the word *etching* is a derivative of the Dutch for ''eat''). Like engraving, etching has undergone few changes since its introduction, an exception being the development of soft-ground etching (*vernis mou*) in the late eighteenth century. In this type of etching, the plate is coated with an acid-resistant substance that is softer and stickier than the compound normally used, and the artist traces on it with a **pencil** rather than a pointed tool. The result is a finished, reproducible plate with soft lines reminiscent of those of a pencil or charcoal sketch. Both etching and engraving lost prevalence in the nineteenth and twentieth centuries, when **printing technology** left the forms without an edge in reproducibly; nonetheless, both have still been employed by major 20th-century artists, including Pablo Picasso and Henri Matisse.

See also Woodcut

ENIAC

ENIAC (Electronic Numerical Integrator And Computer) was designed during World War II by **J. Presper Eckert** and **John Mauchly** at the University of Pennsylvania for the United States Army. It was the forerunner of Eckert and Mauchly's **UNIVAC**, which was the first widely-available commercial computer.

Because of the war, the University of Pennsylvania's Moore School of Engineering had a contract with the United States Army to design an advanced machine that could perform military-related calculations, such as cannon trajectories. The machine, which came to be called ENIAC, was needed quickly, and the design team decided to use available materials and technology, such as vacuum tube processors and punched cards to store the program and data. In 1941 when the project began, John Mauchly was a physicist who had recently joined the faculty and Presper Eckert was an engineering student. Also on the team was Arthur Burks (1915-), a mathematician and logic expert, who was responsible for ENIAC's circuitry.

ENIAC had 18,000 vacuum tube processors and required 150,000 watts of power. It weighed thirty tons, and took up over 1,500 square feet. Using ten-digit decimal numbers, it could perform 5,000 additions or subtractions per second. It also multiplied, divided, and performed square roots. Programming was provided by interconnected wiring, like a telephone switchboard. Later, however, it was refitted to use punched card programs.

The computer was dedicated in February 1946, and was used to compute the trajectories of artillery shells for the Army, Navy, and Air Force. It also performed various calculations related to nuclear weapons research, including various radiation studies and predicting the weather.

ENIAC continued to operate until 1955. During its lifetime it demonstrated that electronic systems were reliable as well as accurate, paving the way for future computers. But its design became the center of a legal controversy.

In 1937, **John Atanasoff**, a physicist at Iowa State College (now University) designed a computer with vacuum tube processors, binary numbers, serial calculation, punched cards for input and output, and memory consisting of condensers placing charges on Bakelite (a type of plastic) drums. Under Atanasoff's faculty contract, the patent on anything that he invented would be held by the school—a common practice. However, when he asked Iowa State to patent his computer, the school refused. By 1942, Atanasoff and Clifford Berry, a graduate student, built enough of the A-B-C computer to calculate differential equations. Atanasoff then took up other World War II-related research and did not go back to the computer project.

However, in December 1940, Atanasoff had met John Mauchly at a scientific meeting and invited him to Iowa to see the computer. The visit took place in June 1941 and Atanasoff demonstrated the A-B-C. Mauchly, who already had computer design experience, and Eckert went on to design and build ENIAC and to take out a patent on it.

The ENIAC patent served as the basis for their later computers, EDVAC, BINAC, and UNIVAC, built by their own firm. In 1950, Remington Rand bought their company, including all UNIVAC-related patents. By the 1960s, Remington Rand had become Sperry Rand. During that time a competitor, Honeywell, contacted Atanasoff and they decided to challenge the Sperry Rand patent.

In court, Atanasoff's case was based on his original design and on Mauchly's visit to Iowa. Atanasoff's machine was strictly a differential analyzer, and Mauchly countered that he had not stolen Atanasoff's design. He contended that he had the main elements of his computer design before meeting with Atanasoff and seeing the A-B-C. Mauchly believed the design was flawed, though he was impressed with the performance, and he intended to build a general-purpose machine.

Several scientists who had worked on ENIAC supported Atanasoff, while others supported Eckert and Mauchly. In 1973, the judge ruled in Atanasoff's favor, declaring the ENIAC-UNIVAC patent invalid.

Enzymatic engineering

An *enzyme* is a protein which accelerates, or (*catalyzes*), the chemical reactions of living cells. Most biochemical reactions would be too slow to support life without enzymes. Enzymes can be manipulated through the science of enzymatic engineering which alters existing natural substances or builds new ones from natural or artificial materials. Enzyme research and enzymatic engineering include methods and concepts from protein chemistry, molecular biophysics, and molecular biology.

The term enzyme, coined in the late nineteenth century, means leavening and people have recognized the importance of enzyme activity for thousands of years. One of the earliest and most well known of all enzyme-catalyzed reactions is the fermentation of sugar to alcohol. The first enzyme to be isolated was diastase, which catalyzes the hydrolysis of starches into sugars, in 1833. It was also the first enzyme to be patented, in 1894 by Jokichi Takemine (1854-1922).

Pepsin, an important enzyme in the digestive process, was identified in 1839 through efforts by many scientists, including those of the American William Beaumont (1785-1853). The isolation of pepsin in 1930 proved that enzymes are proteins. The first enzymatic engineering also began in the 1930s, when scientists modified pepsin and other enzymes by removing or replacing groups.

Since that time six classes of enzymes have been identified: *oxidoreductases*carry out redox (electron transfer) reactions; *transferases* transfer a group (such as a phosphate) from one substance to another; *hydrolases* break bonds by hydrolysis, such as carbon-oxygen or carbon-carbon; *lyases* break bonds by elimination; *isomerases* rearrange molecules structurally; and *ligases* join molecules, along with hydrolysis of a triphosphate bond.

All these abilities are being used in modern enzymatic engineering, which began in the late 1950s by such pioneers as the three winners of the 1987 Nobel prize in chemistry—Donald J. Cram (1919-), of the University of California, Los Angeles; Jean-Marie Lehm (1939-), of Louis Pasteur University (Strasbourg) and College de France (Paris); and Charles Pederson (1904-), now retired from E.I. duPont de Nemours & Co. Their work involved discovering the structure and function of enzymes, and attempting to create synthetic counterparts of the enzymes, their receptors (called hosts), and substrates, inhibitors, or cofactors (guests).

Enzymes are now used in science and medicine for cancer therapy, treatment of wounds, and genetic engineering. They are also part of chemical production, pulp and paper processing, and municipal solid waste disposal. Other industrial uses include production of detergents, leather, and fuel alcohol, as well as manufacturing of syrups and sweeteners, starch processing and baking, meat processing, cheesemaking, and wine production.

Enzyme engineers work to adapt or mimic or adapt these processes in several ways. An existing enzyme can be modified by amino acid substitution, without disturbing the structure, for instance, so it will work at high temperatures instead of body temperature.

An enzyme can be combined with other substances to carry out a reaction. Scientists at Rockefeller University have combined the enzyme *papain*, which breaks down proteins (it is used as a meat tenderizer), with nitrogen compounds called flavins. The resulting compound is more versatile.

ENIAC, the first electronic computer.

Substances can also be engineered to work like enzymes. Scientists at California Institute of Technology have combined an oxygen-binding protein called *myoglobin* with the metal ruthenium, which can start chemical reactions. The resulting enzyme can carry oxygen to a site, then perform an oxidation reaction.

Genetic engineering techniques can be used to change the genes, adding, deleting, or substituting DNA nucleotides. Once these changes are made, the cell assembles different amino acids, making a new enzyme. Mutations are being studied by many scientists for use in medical treatment and industry. For example, research in 1997 have revealed how the enzyme human topoisomerase I removes the additional twists that accumulate in DNA. This research suggests a new mechanism for enzyme functionality.

Eraser

The first erasers, for wiping out lead **pencil** marks, were pieces of bread. The modern eraser made of **rubber** appeared in the eighteenth century. The first suggestion to use rubber—a vegetable gum from South America called *caoutchouc*—as an eraser was recorded in 1752, probably from a Frenchman named Jean de Magellan (1723-1790). In 1770 the English scientist Joseph Priestley (1733-1804) noted that he could use caoutchouc to rub out lead pencil marks in a manuscript. From this, caoutchouc got its familiar name of rubber, and in Great Britain erasers are still called rubbers.

The idea for attaching a rubber eraser to the end of a pencil was patented in 1858 by both Hyman Lipman of Philadelphia, Pennsylvania, and Joseph Rechendorfer of New York City. The eraser with a hollowed-out end into which a pencil could be inserted was invented by J. B. Blair of Philadelphia in 1867; earlier versions also existed.

Modern erasers are a mixture of rubber, vegetable oil, sulfur, and pumice. Erasable **ink** is used on some pens. **Plastic** and synthetic rubber are also used to produce erasers.

Ericsson, John (1803-1889)

Swedish American engineer

Primarily remembered for designing the first iron-clad **warship**, Ericsson was born in the Värmland region of Sweden and displayed an early talent for science and mathematics. As an adolescent he worked as a draftsman on the Swedish Göta Canal project and later served as a land surveyor in the Swedish Army.

Ericsson went to London, England, in 1826 to pursue a career in engineering. During the next thirteen years, he developed an interest in propulsion systems, making several improvements to **steam engine** design, experimenting with the use of compressed air for power, and working on a device he termed the caloric engine. In 1829, Ericsson and John Braithwaite designed and built a locomotive which they entered in the Rainhill Trials, a competition to find the best new locomotive design. The competition was won **George Stephenson**'s famous locomotive, the Rocket. Ericsson later turned his attentions to naval engineering. His innovations in ship design include placing the engines below the waterline and replacing the commonly used paddle wheel with a **screw propeller**. These design modifications considerably reduced the vulnerability of ship propulsion systems to damage from hostile fire. In 1837 a ship incorporating these design elements, the *Francis B. Ogden*, was successfully launched.

In 1839 Robert Field Stockton, a captain in the U.S. Navy, brought Ericsson to the United States to build the engines and propulsion system for the *U.S.S. Princeton*, the first propeller-driven, steam-powered, iron-hulled warship. During an 1844 demonstration attended by President John Tyler, one of the *Princeton*'s guns exploded, killing the U.S. secretary of the Navy and several others. No blame, however, was attached to any of the designers. Ericsson became a naturalized U.S. citizen in 1848.

During the Civil War, Ericsson presented the United States government with a design for a new type of heavily armored warship. Built in 1861 and launched in January of the following year, Ericsson's vessel, dubbed the *Monitor,* was the first completely iron-clad warship. Driven by a steam-powered screw propeller, the *Monitor* had a low box-like shape, 172-feet long, and was armored with five inches of iron plate on the sides and one inch on the deck. The *Monitor*'s two eleven-inch guns were enclosed in a deck-mounted turret covered with eight inches of iron plate and rotated by steam power. Soon after the *Monitor* was launched, Confederate forces salvaged a ship called the *Merrimack* (also known as the *Virginia*) and covered it with iron railroad track.

On March 8, 1862, the *Merrimack* sailed into the harbor at the mouth of the James River in Virginia and used its ten guns to sink two wooden-hulled Union warships. While the *Merrimack* attacked a third ship on the following day, the *Monitor* arrived on the scene. The two iron-clad vessels exchanged numerous rounds in a heated battle. When the *Merrimack* scored a direct hit on the pilothouse of the *Monitor,* the captain of the Union vessel was blinded by flying iron fragments and let his ship wander into shallow water. The crew of the *Merrimack* concluded that they had won and returned to their home base leaking water and low on **ammunition**. The *Merrimack*'s success in sinking wooden ships served as a propaganda victory for the Confederacy, but most commentators either interpret the battle between the two iron-clad ships as a victory for the *Monitor* or find the results inconclusive. Never particularly seaworthy, the *Monitor* sank during a storm in December 1862 with a loss of sixteen lives.

Ericsson went on to design and build other *Monitor*-type vessels for the U.S. government. He also experimented with **torpedo**es and investigated uses for solar energy. After his death in 1889, Ericsson's body was returned to his native land at the request of the Swedish government. *Monitors* of Civil-War era design were used by various navies until the early part of the twentieth century, last serving as **submarine** tenders during World War I before being scrapped. *Monitor*-class vessels of modern design have been used by the United States, the Soviet Union, Britain, and Romania as river gunboats and landing craft.

Few seem to know that Ericsson was a true mechanical genius who, as a very young boy, constructed models of machines he had never seen and that had only been described to him. In one instance, despite never having seen a windmill, he correctly sketched the mechanism that connected the windmill crank with the power lever. Once Ericsson's prodigious talents were brought to the attention of the president of the Gotha ship canal project, the twelve year old boy was made a member of the canal surveying team and was soon directing the work of some 600 soldier/laborers. So respected was the twelve year old that the army assigned one soldier to do nothing but carry a stool for the young engineer to stand upon while he used the tall surveying instruments.

Ernst, Richard R. (1933-)

Swiss chemist

Richard R. Ernst was the 1991 recipient of the Nobel and Wolf Prizes in chemistry and the Louisa Gross Horwitz Prize at Columbia University, a joint award to Ernst and his colleague Kurt Wüthrich for their development of the methodology of high-resolution nuclear magnetic resonance (NMR) spectroscopy, the most important instrumental measuring technique within chemistry. Ernst's contributions in this field led to the development of **magnetic resonance imaging** (MRI), the biomedical instrument widely used today to perform noninvasive diagnosis of the human body.

Born on August 14, 1933, to Robert Ernst and Irma Brunner in Winterthur, Switzerland, Richard Robert Ernst was educated at the Eidgenössiche Technische Hochschule (ETH) in Zurich, where he received his doctorate in 1962. In 1976 he became a full professor at the ETH. After receiving his Ph.D., Ernst moved to Palo Alto, California, to become a research scientist at Varian Associates. There he worked with Weston A. Anderson on efforts to make NMR spectroscopy more sensitive.

Ernst's contribution built on the NMR experiments reported in 1945 by Felix Bloch at Stanford University and Edward Mills Purcell at Harvard. Bloch and Purcell shared the Nobel Prize in physics for this work. They demonstrated that some atomic nuclei can be knocked out of alignment in a strong magnetic field when exposed to a slow sweep of radio frequencies. The nuclei realign in response to "resonant" frequencies, emitting signals that are like a chemical signature. When he received his prize, Purcell predicted NMR would become a tool for chemical analysis. For that to happen, however, it was necessary to overcome limited sensitivity of the early NMR method to the chemical signature of the substance being analyzed. Only a few substances—hydrogen, fluorine, and phosphorus—had spectra strong enough to identify reliably.

In 1966 Ernst and Anderson were able to enhance NMR spectra by replacing the slow sweep of radio frequencies with short, intense pulses. Spectra too weak to identify previously became discernable. The spectra resulting from exposure to the pulse of radio frequencies were complex, and to analyze them, Ernst made use of a Fourier transformation, which computers of the mid–1960s could use to interpret the small fluctuations in brightness of the NMR spectra. NMR equipment became widely available to the chemical research community by the early 1970s.

John Ericsson

Ernst later developed an even more sophisticated variant, two-dimensional NMR spectroscopy, which replaced single pulses of radio frequencies with a sequence of pulses, just as the pulse had replaced a sweep of radio frequencies. Building further on the concept underlying this advance, subsequent multidimensional techniques (COSY, for example, which stands for correlation spectroscopy; others are SECSY, NOESY, and ROESY) made it possible to analyze the three-dimensional structures of proteins and other large biological molecules. They were the basis for the development of magnetic resonance imaging (MRI), which is used in medical diagnosis. The methods also enable researchers to gather information about the chemical environment of the molecules under study.

In 1991 Ernst received the Nobel Prize in chemistry and the Louisa Gross Horwitz Prize at Columbia University. The Horwitz Prize was a joint award to Ernst and his colleague Kurt Wüthrich for their work in developing NMR methods that could show both the behavior and structure of complex biological molecules. Ernst also received the Wolf Prize in chemistry that year.

Ernst and his group are engaged in studying how molecules interact with one another over time and how they change shape. His work is considered cross-disciplinary since it falls

One of the first two publicly installed escalators. Built by the Otis Elevator Company in 1900, it carried passengers to New York City's Third Avenue "L" station at 59th Street. Otis built its other escalator in Paris.

between chemistry and physics and also involves problems in quantum mechanics (a theory that assumes that energy exists in discrete units). His book on NMR is considered a classic in the field, invaluable for its cross-references to related literature.

Ernst holds a number of patents for his inventions and has received recognition from many scientific societies for his work. He serves on the editorial boards of several journals dealing with magnetic resonance.

Escalator

The escalator is a continuous moving staircase that carries people from one level or floor to another. A series of steps are pulled by chains along two sets of tracks, powered by an **electric motor**.

The first United States patent for an escalator was issued in 1859, but the invention wasn't used. The practical escalator was the result of independent inventions in the early 1890s by two Americans, Jesse Reno and Charles Seeberger. Reno invented an inclined belt with a grooved tread for steady footing. Seeberger produced a flat-step design with side entrance. A problem with both designs, however, was the stationary handrail.

During the 1890s, the Otis Elevator Company acquired rights to these two designs. Otis put several escalators into operation at its factory in Yonkers, New York, in 1899. The first public installations of an escalator were at the 1900 Paris Exposition and in the New York City elevated railroad. Otis made other early installations in Chicago, Illinois, and Philadelphia, Pennsylvania, department stores.

Around 1920 the Reno and Seeberger designs were combined and the result was the modern escalator. The word *escalator*, derived from a Latin word meaning ladder, was originally an Otis Elevator Company trademark. The term lost trademark status in 1949 because of its widely popular use.

Modern escalators are powered by electric motors that move their steps at a fairly slow 100 feet per minute. All standard-use escalators carry their passengers upwards at a 30-degree angle. They are used in many multistoried public buildings, especially in malls and department stores where people move continuously between connecting floors. Viewed in cross-section or from the side, an escalator is seen as an endless belt. When its stairs reach the top, they fold flat and go around a large wheel to make a return trip to the stairway bottom. There, the stairs or steps go around another wheel that rolls them back up, and they reappear. Many of the moving sidewalks or walkways appearing in increasingly spread-out airports are essentially a flat type of escalator. The world's tallest bank of escalators are in London, England. In 1993 in Hong Kong, a 2,600-ft (800-meter) Hillside Escalator Link opened and was able to move about 27,000 people a day. The Link is a series of moving sidewalks and escalators — all covered — that move residents from one part of town to another. It has 20 exits and entrances, takes about 20 minutes to run end-to-end, and reverses itself for the afternoon rush hour.

By the end of the 1990s, the newest escalator designs stress slimmer, sleeker profiles and major safety enhancements such as a gap of only 1/16th of an inch between the escalator steps and the skirt (side) panels. This provides a greater measure of user safety, especially for small childern and slow-moving seniors. The newest models will also feature a totally enclosed and quieter drive system instead of a chain drive.

Esperanto • See Language, universal

Evans, Oliver (1755-1819)

American inventor

Oliver Evans was born in Newport, Delaware, on September 13, 1755, and as a young man was apprenticed to a wheelwright. At 22 he invented a machine for making the carding teeth used in the **textiles** industry. Two years later he went into the flour-milling business with two of his brothers. While working at the flour mill, he invented the grain elevator, conveyor, drill, hopper boy, and descender. These inventions essentially automated the flour-milling process to the point that the mill could be run by one person. In the late 1780s the legislatures of Maryland and Pennsylvania granted Evans the exclusive right to the application of these improvements, and the

U.S. Congress granted Evans U.S. patents for his flour-milling inventions in 1790. His was only the third patent granted by the U.S. government.

Around 1800 Evans refined the **steam engines** of his day, developing perhaps the first steam engine constructed on the high-pressure principle. Earlier he had sent copies of his plans to England, where **Richard Trevithick**, who is often credited with the invention, had access to them. Although Evans had been working on plans for a steam-powered carriage, he adapted his high-pressure steam engine to further improve the milling process, which previously had been for the most part powered by **waterwheels**.

In 1803 and 1804 the Philadelphia Board of Health commissioned Evans to build a steam-powered **dredger**, the first to be used in the United States. The dredge consisted of a small steam engine and the machinery to raise the mud from the Schuylkill River. It was powered to move on land over wheels and in the river by means of a **paddle wheel**. Evans named his craft the *Orukter Amphibolos*, or amphibious vehicle, and it is considered to be the first instance in the United States in which steam power was used to propel a land carriage. He urged that his idea be adapted to move vehicles on rails of wood or iron and, although he lobbied for a railroad to be built between Philadelphia and New York, the country's first commercial railroad track was not laid until the early 1830s, years after Evans's death.

Throughout his life, Evans refined the steam engine and initiated innovative manufacturing techniques. While he failed to develop his *Orukter Amphibolos* into a true steam-carriage, or a true paddle wheel boat, Evans long maintained that he very well could have created these conveyances. He grieved that the credit had gone to other inventors.

In 1797 Evans published the details of his early inventions in a book entitled *The Young Millwright and Miller's Guide*. Another Evans book, *The Young Engineer's Guide*, was published in 1805. Both works were translated into French and published in Paris in the early 1800s. Oliver Evans died in New York City on April 21, 1819.

See also Mass production; Steam-powered road vehicles

Evinrude, Ole (1877-1934)

Norwegian American inventor

Ole Evinrude was born in Norway on April 19, 1877; five years later, his family emigrated to the United States and settled near Cambridge, Wisconsin. Interested in mechanics from an early age, Evinrude became an apprentice machinist at age 16 and eventually a master patternmaker as well.

Along with a growing number of people at the turn of the century, Ole Evinrude was fascinated by the potential of the newly developed **internal combustion engine**, and at the turn of the century, he set up a firm to build small engines.

While Evinrude concentrated on the mechanical and engineering aspects of the new firm, he entrusted the bookkeeping and business end of the firm to his assistant, Bessie Cary. The story surrounding Evinrude's invention of the outboard boat engine revolves around a picnic that Cary and Evinrude enjoyed on an island in Lake Michigan two and one-half miles from shore. Cary expressed a desire for a dish of ice cream and Evinrude rowed back to shore for it. Of course, the ice cream was melted by the time he returned, but Evinrude, inspired by the incident, was determined to design an engine that would replace the oar as a means of boat propulsion.

Cary and Evinrude were married in 1906. Their firm immediately began to develop its first outboard motor, a one-cylinder, 1.5 hp model, which became an instant success upon its introduction in 1909. A year later Evinrude founded Evinrude Motors in Milwaukee, Wisconsin, to build his new engines.

Due to the poor health of Bessie, the Evinrudes sold their company in 1913, and Ole agreed to not re-enter the outboard motor business for five years. His inventive mind kept busy, however, and during his "retirement," he devised a much improved, two-cylinder outboard engine. In 1921 he and Bessie formed the ELTO Outboard Motor Company (ELTO standing for Evinrude's Light Twin Outboard). This new outboard engine was also very successful, and in 1929 the ELTO company merged with the original Evinrude company (since renamed the Outboard Motor Corporation) and the Lockwood Motor Company with Evinrude as president.

Bessie, who had retired in 1928 for health reasons, died in 1933. Ole Evinrude died the following year on July 12 in Milwaukee, and the company was taken over by their son, Ralph. In 1936 the Evinrude company merged with the Johnson Motor Company to form the Outboard Marine Corporation, which has enjoyed continuing success in the outboard motor business.

Evinrude's name continues today not only on many an outboard motor but also by the presentation of the Ole Evinrude Award. Given annually by the New York Boat Show, this award is presented in recognition of an individual's contributions to the growth of recreational boating.

See also Powerboat

Expert system

An expert system is a computer program that combines a knowledge base of information about a particular field with a system of rules that applies the knowledge to specific situations. Such a system can analyze a problem in a given discipline and provide solutions for it. The system of rules, often called an inference mechanism, consists of a complicated "tree" of logical operations that is carried out by the program. An expert system is considered a form of **artificial intelligence**.

Edward Feigenbaum, a computer specialist, and **Joshua Lederberg**, a professor of genetics, are generally credited with developing the first expert system at Stanford University in the 1960s. Their program, DENDRAL, determines the structures of organic chemicals using only complex spectroscopic or instrument readings.

Other researchers realized that expert systems would be useful in medical diagnosis and began developing programs to assist physicians in investigating disease. An early program of this type, also developed at Stanford University in the mid-1970s, diagnoses blood-related infections. The program, called MYCIN, analyzes data input by the physician, asks relevant questions about the patient's test results and condition, then uses this information to provide a diagnosis and recommendations for treatment. Although its accuracy seems to be quite good, it has not come into use in medical practice because physicians, fearing serious errors, seem hesitant to rely on a computer to diagnose serious illness. A training program based on MYCIN called GUIDON is, however, being used in medical school classrooms.

Another medical expert system, PUFF, has found more use in real-life situations. When a patient undergoes lung tests, PUFF receives the data from the tests as they are carried out and, like MYCIN, provides the physician with a diagnosis and suggested treatment.

Expert systems have become very useful in the field of finance and accounting as well. For example, an expert system called TAXMAN has been developed to help the many people bewildered by the complexity of their annual tax returns. Operable on a personal **computer** (PC), TAXMAN creates a return by prompting the user to input data, performing necessary calculations, and neatly printing the results.

A host of other expert systems has been devised, including programs to perform geological assessment of various land sites, determine the cause of diesel locomotive malfunctions, solve difficult mathematical equations, and help sales representatives tailor merchandise orders to a customer's needs, especially in technically complex markets such as that for computer equipment. Finally, there are even expert systems to aid computer programmers in designing and "debugging" other computer systems.

See also Computer, digital; Microcomputer

Explorer 1

On October 4, 1957, the Soviets launched the first **satellite** to reach Earth orbit: **Sputnik 1**. The American government tried to downplay the event, but in actuality began aggressively pursuing their own satellite program.

The German-born American scientist **Wernher von Braun** had developed a Jupiter-C **rocket** for the Army and asked permission to launch a satellite with it. The government turned him down; however, having already decided on a Navy project, called Vanguard. On December 6, 1957, officials gathered the press to witness the first launch of the Vanguard rocket, which, after hovering a few feet off the ground, exploded into flames. After other disasters with Vanguard, von Braun was given the green light. On January 31, 1958, a Jupiter-C lifted off. Two and a half minutes later, the first stage shut down and the next fired. Orbit was attained. The satellite itself orbited with the top stage of the rocket, ranging in distance from the earth between 218 miles (352 km) and 1,586 miles (2,554 km).

This satellite, called *Explorer 1*, was small in comparison to the *Sputniks*. It weighed only 10.5 pounds (4.7 kg). The Russians had larger rockets to launch their mammoth satellites, but the American representative had its advantages. Unlike *Sputnik 1*, *Explorer 1* contained more than just instruments to measure the temperature and density of the upper atmosphere. It had a **micrometer** to measure debris in space and a **radiation detector** that found intense rings of radiation surrounding the earth at great altitude. Later, these rings were called the Van Allen radiation belts, named after the scientist who designed the experiment. *Explorer 1* thus proved the great scientific value of satellites.

Extraterrestrial lander and rover

Planners of space exploration face an interesting trade-off. On one hand is financial reality: manned missions are much more expensive than unmanned missions, as the huge cost of the *Apollo* and **space shuttle** programs attests. One the other hand is the desire for knowledge: it is very difficult or even impossible to teach a machine to do the variety of tasks a human could do.

Since the end of the *Apollo* missions, and aggravated by a wane in public interest in space exploration, budgetary restrictions have limited planners of missions to other worlds in the solar system to unmanned craft. Since the last *Apollo* mission in 1983, advances in technology have allowed mankind to reach to point where we can vicariously explore the surface of another world — even if only on a small scale so far.

In 1976, the *Viking* landers arrived on Mars. They were static craft; once they had touched down, they would never budge again. They did, however, incorporate experiments employing robot arms, extendable appendages that could scoop a small amount of soil form the Martian surface and drop it into a small receptacle for analysis. The picture of the small hole in the Martian soil left by *Viking*'s robot arm led the late Carl Sagan, in his book *Cosmos*, to point out that for the first time, humans had begun to alter the face of another planet.

Interplanetary missions of the 1980s and early 1990s were fly-by missions. The *Pioneer* and *Voyager* missions revolutionized our view of the outer solar system, particularly *Voyager 2*, with its "Grand Tour" of all four giant planets. But it was not until 1997 that the next major stride in extraterrestrial lander technology occurred.

On December 4, 1996, the *Mars Pathfinder* mission was launched aboard a Delta II rocket, with a scheduled arrival date at mars of July 4, 1997.

Unlike the doomed 1993 *Mars Observer* and 1996 *Mars96* missions, *Mars Pathfinder* arrived right on schedule, touching down and deploying safely on the Ares Vallis flood basin on Mars. Most significantly, the craft contained a robot rover, called *Sojourner*, a six-wheeled crawler capable of straying from the fixed station, taking close-up pictures of what it encountered. Scientists renamed the area the Sagan Memorial Station on July 5, 1997, in honor of the late astronomer who had lobbied so effectively for exploration of the planet.

Pathfinder returned almost 18,000 pictures of the area around it, while also providing new information about the Martian atmosphere, geology, and magnetic field.

But the real hit was *Sojourner*. Designed to operate for a week, *Sojourner* resolutely crept around the Martian surface for more than a month, analyzing the content of rocks and soil it encountered and photographing them. Although it covered only the area in the immediate vicinity of the lander, at the leisurely pace of 0.7 centimeters per second, it was by any definition a smashing success.

Sojourner became a huge public hit, helped by easy access on the Internet (the *Pathfinder* site eventually registered more than half a billion hits) and by a toy *Sojourner* released by Mattel. Behind all the hype, however, was an immensely sophisticated machine capable of analyzing its position and avoiding rocks that might tip it over without explicit human control to tell it to do so. *Sojourner* was therefore a ''smart'' robot, and although its capacities were limited, they represented a major step forard in rover technology, and a sure precursor to future missions, where **robotic** craft may perform far more extensive analysis. The *Pathfinder* mission will certainly be long regarding as a milestone in the exploration of other worlds.

Launch of the Jupiter C rocket that carried *Explorer 1* into space.

Eyeglasses

Eyeglasses are corrective lenses mounted in frames that help those with vision problems see clearly. The lenses are shaped in order to bend light rays so that they will focus at the back of the eye, the retina. Some people who can see distant objects clearly but to whom near objects look blurry suffer from hyperopia or farsightedness. As explained by Franciscus Donders (1818-1889), a Dutch physiologist, the cause of farsightedness is that the eyeball is too shallow and that the image actually focuses beyond the eye. To correct hyperopia, convex corrective lenses are used to make the light rays converge or come together on the retina. Some people suffer from myopia or nearsightedness, in which the image is focused in front of the retina so that only near objects can be seen clearly. Concave lenses can be worn to diverge the light rays and permit light from far away objects to focus directly on the retina. A condition called presbyopia occurs when the lens of the eye loses it elasticity and it can no longer change shape. The condition is usually associated with age and becomes evident after 40. Presbyopia causes people to be somewhat farsighted. Sometimes this is corrected by wearing bifocals, or eyeglasses that have a second lens below the top lens. A person with presbyopia can look through the bottom lens while reading and use the top lens for distant objects.

The invention of eyeglasses has a long and colorful history. It is said that during the days of the Roman Empire, the emperor Nero watched exhibitions in the Coliseum holding a jewel with curved facets up to one eye, but this cannot be verified. However, Roger Bacon, an English scholar, is said to have suggested the use of eyeglasses in the 1200s. An Italian physicist, Salvino degli Armati probably invented eyeglasses in around 1285. He shared the design of his new device with an Italian monk, Allesandro della Spina, who made public the invention and is often given credit for inventing eyeglasses.

In the fourteenth century Venetian craftsmen, known for their work in glass, were making ''disks for the eyes.'' The finely ground glass disks were given the name lenses by the Italians because of their similarity in shape to lentils. For hundreds of years thereafter, **lenses** were called glass lentils. The earliest lenses were convex—that is they bulged outward in the middle and aided people who were far-sighted. Wearing spectacles became common enough so that a 1352 portrait of St. Hugh showed him wearing them, although he had died some hundred years before. By the fifteenth century, eyeglasses had found their way to China. But long before, Chinese judges had worn smoky quartz spectacles, but it is thought they were worn so the judges could remain impartial and not show expression in their eyes when they heard cases rather than correct vision.

In 1451, Nicholas of Cusa in Germany invented eyeglasses to correct nearsightedness using concave lenses. Rather than bulging in the middle like convex lenses, concave lenses are thinner at the center and thicker at the ends. Pope Leo X was one of the first to wear them.

Early eyeglasses had glass lenses mounted on heavy frames of wood, lead or copper. Natural materials of **leather**, bone and horn were later used and then lighter frames of steel were made by the early seventeenth century. Tortoiseshell frames came into use in the eighteenth century. In 1746, a French optician named Thomin invented actual eyeglass frames that could be placed over the ears and nose.

In the United States, **Benjamin Franklin**, statesman and scientist, designed the first bifocals in 1760. In this way he could use the top lens to see distant objects and peer down into the bottom lens when he read without needing two pair of glasses. The two lenses were joined in a metal frame. In England in 1827, Sir George Biddle Airy (1801-1892), an English astronomer and mathematician made the first glasses to correct astigmatism, a condition he himself had. Astigmatism is blurry vision caused by irregular curves in the cornea, the transparent covering of the eye. The irregular curvature makes it impossible for light rays to focus on a single point. To correct this, the exact area of the irregularity of the cornea is located, and a corresponding area on the eyeglass lens is ground to bring light rays passing through that area into proper focus.

Today eyeglasses come in a wide array of styles and designs. Frames are generally made of metal or **plastic**, and lenses are made of glass or plastic. Polycarbonate is a type of very hard plastic used for lenses. Robust and lightweight, it is scratch-resistant and provides 100% protection from ultraviolet radiation that can contribute to cataract formation. Along with lightweight frames, such lenses are commonly used in sports glasses and childrens' glasses.

In 1955 the first unbreakable lenses were made and in 1971 a new lens came out which combined the properties of plastic with glass. During the 1950s the Varilux was invented, corrective lenses of variable strength that can be used in place of bifocals. Testing the eyes for visual acuity and examining the eye with a retinoscope are routine before determining the strength, or refractive index, of the lenses prescribed.

Eyewear has been revolutionized with the invention of the contact lens, corrective lenses without the frames, which put a tiny corrective lens directly on the cornea of the eye. However the idea dates back to Leonardo da Vinci who described a way of correcting vision using a water-filled tube, and to a number of scientists who experimented with layers of gelatin to correct vision during the seventeenth and eighteenth centuries. Contact lenses were first made in Europe near the turn of the twentieth century using glass. In 1936 IG Farben, a German company, made the first contact lens from Plexiglas—still used today for "hard" contact lenses. An American inventor named Tuohy began use of a lens that covers only the cornea and in 1964 a Czech named Wichterle made the first flexible or "soft" lens. Today there is an array of hard and soft lenses, extended wear lenses, tinted lenses, lenses that can correct astigmatism, and lenses to correct color blindness.

F

FACSIMILE MACHINE

The facsimile, or fax, machine is both a transmitting and receiving device that "reads" text, **maps**, **photographs**, fingerprints, and graphics and communicates via **telephone** line. Since the 1980s, fax machines have undergone rapid development and refinement, and are now indispensable communication aids for news services, businesses, government agencies, and individuals.

The fax was invented by Alexander Bain of Scotland in 1842. His crude device, along with scanning systems invented by Frederick Bakewell in 1848, evolved into several modern versions. In 1869, a Frenchman, Ludovic d'Arlincourt, synchronized transmitters and receivers with tuning forks and thus aided further developments. In 1924, faxes were first used to transmit wire photos from Cleveland to New York, a boon to the newspaper industry. Two years later, RCA inaugurated a trans-Atlantic **radio** photo service for businesses. Models either required an electrolytic or photo-sensitive **paper**, which changed color when current passed through it; or thermal paper, a material coated with colorless dye which became visible upon contact with a toner. The use of faxes, and fax technology itself, remained comparatively limited until the mid-1980s.

To used a fax machine, the user inserts the material to be transmitted into a slot, then makes a telephone connection with another facsimile machine. When the number is dialed, the two machines make electronic connection. A rotating drum advances the original before an optical **scanner**. The scanner reads the original document either in horizontal rows or vertical columns and converts the printed image into a pattern of several million tiny electronic signals, or pixels, per page. The facsimile machine can adjust the number of pixels so that the sender can control the sharpness and quality of the transmission. Within seconds, the encoded pattern is converted into electric current by a **photoelectric cell**, then travels via **telegraph** or telephone wires to the receiving fax, which is synchronized to accept the signal and produce an exact replica of the original by reverse process.

In the 1990s, fax machine technology has improved as they have become more common and less expensive. Indeed, most than 600 different fax models were available in this time period. Almost all faxes now employ plain paper (which, unlike thermal paper, avoids curling) and are preferred for their superior reproduction. Another improvement is the invention of a scrambler, an encoder which allows the sender to secure secrecy for documents. While most fax machines still stand alone or are part of a telephone unit, using a computer to fax has become more widespread. Computers can be equipped with a fax board, an electronic circuit that allows the computer to receive messages, thus saving on paper and telephone charges. In the late 1990s, fax machines have also been incorporated into multifunction printers. These printers often contain a laser or ink-jet printer, scanner, copier and fax machine in one. Stand-alone faxes have also become more compact in the late 1990s as problems with roller and chip design have been solved. Hitachi introduced the first compact fax, and since then, faxes have become small enough to be portable.

See also Code and cipher; Computer, digital; Computer input and output devices; Electronics

FAGET, MAXIME (1921-)

American engineer

Maxime Faget holds a position few can claim—spacecraft design engineer. Among his creations are the *Mercury* capsule, the *Apollo* command and service modules, and the **space shuttle** orbiter.

Faget was born in British Honduras. As a child, he read airplane magazines and *Astounding Science Fiction*, and built model airplanes. Faget earned his B.S. from Louisiana State University in 1943.

Three years later he went to work for the National Advisory Committee for Aeronautics, later renamed National Aero-

Maxime Faget

nautics and Space Administration (NASA). At Langley Aeronautical Laboratory in Hampton, Virginia, he designed ramjets and began a fruitful collaboration with Caldwell Johnson. Faget eventually was assigned to the propulsion-and-performance team that developed the design for the X-l5, the experimental plane that flew Mach 6. The flight of the Russian satellite **Sputnik 1** in 1957 challenged Americans to pursue manned space flight. A space craft must protect its occupant from high G (gravity) forces and atmospheric friction upon re-entry. Faget argued for a blunt bodied capsule because it could slow down high in the atmosphere where the friction and heat were less. Not only was Faget a good designer, but he was equally effective as a debater, winning over those opposed to his idea. In the late 1950s Faget headed the flight systems division that designed the *Mercury* capsule.

From 1962 to 1981, Faget was the director of engineering and development at the Manned Spacecraft Center, later renamed the Johnson Space Center, in Houston, Texas. Faget and Johnson designed both the *Apollo* capsule and service module. For the lunar missions, Faget liked the direct approach method in which a huge rocket launched the command-and-service module that would land on the moon as a single unit and return to the earth. A serious drawback was the huge **rocket** needed for takeoff and lunar landing, so Faget converted it to the final form in which the command-and-service module went into lunar orbit with a small lunar-landing craft attached. This system required much less weight for earth launch.

Even before the lunar landings, NASA wanted plans for a reusable space shuttle. Though Faget envisioned a straight-wing plane for greater maneuverability, NASA opted for the delta-shaped design. Faget also argued unsuccessfully for engines that could swing into the payload bay for easier re-entry and one piece solid-rocket boosters, designs that, if used, might have prevented the *Challenger* disaster and eliminated the hydrogen-leak problem that grounded the fleet for six months.

In 1981 Faget left NASA to work for the aerospace firm Eagle Engineering, and in 1983 he founded Space Industries, Inc. to design an industrial space facility.

See also Braun, Wernher von

Fahrenheit, Daniel Gabriel (1686-1736)

German Dutch physicist

Fahrenheit invented the first truly accurate **thermometer** using mercury instead of alcohol and water mixtures. In the laboratory, he used his invention to develop the first temperature scale precise enough to become a worldwide standard.

The eldest of five children born to a wealthy merchant, Fahrenheit was in Danzig (Gdansk), Poland. When he was fifteen his parents died suddenly, and he was sent to Amsterdam to study business. Instead of pursuing this trade, Fahrenheit became interested in the growing field of scientific instruments and their construction. Sometime around 1707 he began to wander the European countryside, visiting instrument makers in Germany, Denmark, and elsewhere, learning their skills. He began constructing his own thermometers in 1714, and it was in these that he used mercury for the first time.

Previous thermometers, such as those constructed by **Galileo** and **Guillaume Amontons**, used combinations of alcohol and water; as the temperature rose, the alcohol would expand and the level within the thermometer would increase. These thermometers were not particularly accurate, however, since they were too easily thrown off by changing air pressure. The key to Fahrenheit's thermometer was a new method for cleaning mercury that enabled it to rise and fall within the tube without sticking to the sides. Mercury was an ideal substance for reading temperatures since it expanded at a more constant rate than alcohol and is able to be read at much higher and lower temperatures.

The next important step in the development of a standard temperature scale was the choosing of fixed high and low points. It was common in the early eighteenth century to choose as the high point the temperature of the body, and as the low point the freezing temperature of an ice-and-salt mixture—then believed to be the coldest temperature achievable in the laboratory. These were the points chosen by Claus Roemer, a German scientist whom Fahrenheit visited in 1701. Roemer's scale placed blood temperature at 22.5° and the freezing point of pure water at 7.5 °. When Fahrenheit graduated his own scale he emulated Roemer's fixed points; however, with the improved accuracy of a mercury thermometer, he was able to split each degree into four, making the freezing point of water 30° and the temperature of the human body 90°. In 1717 he moved his points to 32° and 96° in order to eliminate fractions.

These points remained fixed for several years, during which time Fahrenheit performed extensive research on the freezing and boiling points of water. He found that the boiling point was constant, but that it could be changed as atmospheric pressure was decreased (such as by increasing elevation to many thousand feet above sea level). He placed the boiling point of water at 212°, a figure that was actually several degrees too low. After Fahrenheit's death scientists chose to adopt this temperature as the boiling point of water and to shift the scale slightly to accommodate the change. With 212° as the boiling point of water and 32° as the freezing point, the new normal temperature for the human body became 98.6°.

In 1742 Fahrenheit was admitted to the British Royal Society despite having had no formal scientific training and having published just one collection of research papers.

See also Barometer

False teeth

Replacements for decayed or lost teeth have been produced for millennia. The Etruscans made skillfully designed false teeth out of ivory and bone, secured by gold bridgework, as early as 700 B.C. Unfortunately, this level of sophistication for false teeth was not regained until the 1800s.

During medieval times, the practice of dentistry was largely confined to **tooth extraction**; replacement was seldom considered. Gaps between teeth were expected, even among the rich and powerful. Queen Elizabeth I (1533-1603) filled the holes in her mouth with cloth to improve her appearance in public.

When false teeth were installed, they were hand-carved and tied in place with silk threads. If not enough natural teeth remained, anchoring false ones was difficult. People who wore full sets of dentures had to remove them when they wanted to eat. Upper and lower plates fit poorly and were held together with steel **springs**; disconcertingly, the set of teeth could spring suddenly out of the wearer's mouth. Even George Washington (1732-1799) suffered terribly from tooth loss and ill-fitting dentures. The major obstacles to progress were finding suitable materials for false teeth, making accurate measurements of a patient's mouth, and getting the teeth to stay in place. These problems began to be solved during the 1700s.

Since antiquity, the most common material for false teeth was animal bone or ivory, especially from elephants or hippopotami. Human teeth were also used, pulled from the dead or sold by poor people from their own mouths. These kinds of false teeth soon rotted, turning brown and rancid. Rich people preferred teeth of silver, gold, mother of pearl, or agate.

In 1774 the French pharmacist Duchateau enlisted the help of the prominent dentist Dubois de Chemant to design hard-baked, rot-proof porcelain dentures. De Chemant patented his improved version of these "Mineral Paste Teeth" in 1789 and took them with him when he emigrated to England shortly afterward. The single porcelain tooth held in place by an imbedded platinum pin was invented in 1808 by the Italian dentist Giuseppangelo Fonzi. Inspired by his dislike of handling dead people's teeth, Claudius Ash of London, England, invented an improved porcelain tooth around 1837.

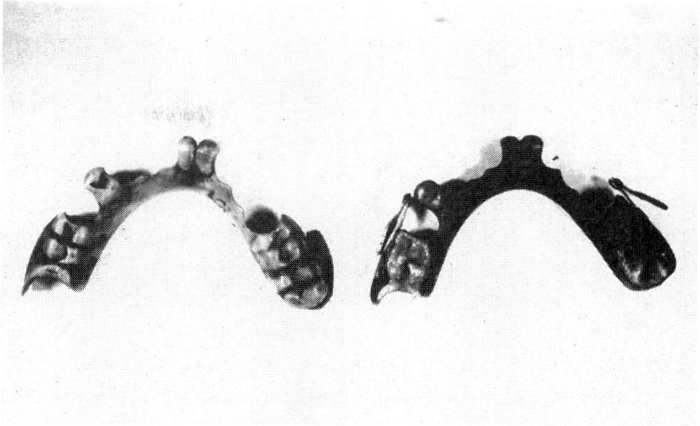

A pair of hand-carved, ivory false teeth from nineteenth-century France.

Porcelain teeth came to the United States in 1817 via the French dentist A. A. Planteau. The famous artist Charles Peale (1741-1847) began baking mineral teeth in Philadelphia, Pennsylvania, in 1822. Commercial manufacture of porcelain teeth in the United States was begun, also in Philadelphia, around 1825 by Samuel Stockton. In 1844 Stockton's nephew founded the S. S. White Company, which greatly improved the design of artificial teeth and marketed them on a large scale.

Fit and comfort, too, gradually improved. The German Philip Pfaff (1715-1767) introduced plaster of paris impressions of the patient's mouth in 1756. Daniel Evans of Philadelphia also devised a method of accurate mouth measurement in 1836. The real breakthrough came with **Charles Goodyear**'s discovery of **vulcanized rubber** in 1839. This cheap, easy-to-work material could be molded to fit the mouth and made a good base to hold false teeth. Well-mounted dentures could now be made cheaply. The timing was fortuitous. Horace Wells (1815-1848) had just introduced painless tooth extraction using nitrous oxide. The number of people having teeth removed skyrocketed, creating a great demand for good, affordable dentures, which Goodyear's invention made possible. After 1870, another cheap base, **celluloid**, was tried in place of rubber, but it too had drawbacks.

Today dentures are primarily made of **plastic** or **ceramic**. However, research continues into new and better materials to make dentures longer lasting and more resistant to stain. One such material is cobalt chromium, which is a hard metal that does not rust or change shape. A more recent development in dentistry is the dental implant. Dental implants are constructed from biocompatible materials, such as titanium, which are not recognized by the body's immune system as foreign. Endosteal implant are placed into the bone to replace the root portion of a tooth. A subperiosteal implant—used when there is not enough bone left—fits in a framework fashion over the remaining bone. These implants are then used to support a natural looking artificial tooth, which is usually attached with a dental adhesive.

Fan

Fans have been used since the days of antiquity. Egyptians pharaohs were fanned by their slaves with huge lotus leaves; ancient Greeks and Romans used their own versions, often trimming their fans with peacock feathers. The folding or pleated fan is thought to have been invented by the Japanese in about A.D. 700 and may have been modeled after the way a bat folds its wings. The fan served both practical and ceremonial functions in China and Japan. In China, it was especially popular during the Ming Dynasty (1368-1644) and was carried by both sexes as well as by members of many different social classes, from courtiers to warriors, during tea ceremonies and on stage. Some of the most talented Asian painters applied their skill to the exquisite decoration of fans, an art that was not developed in Europe until the nineteenth century.

During the Middle Ages in Europe, rigid fans were used; metal disks on long handles shooed flies during church ceremonies, and ornamental fans were fashioned from parchment mounted on ivory, gold, or silver handles. Beginning in the fifteenth century, Portuguese traders brought large quantities of folding fans from Asia, and by the seventeenth century they were highly popular. Although the fans exported to Europe were of much poorer quality than those used in Asian countries, they were much admired by their European purchasers. During the reign of King Louis XV in France, even men carried dainty fans. A French favorite was the briséa fan composed of wide, overlapping, blade-like sticks connected with a ribbon at the top.

Fans were decorated in accordance with the styles of the day and ranged from simple to ornate, some bearing reproductions of famous paintings. Their sizes ranged from the eight-inch version popular in the early 1800s to as large as twenty inches during the Victorian era. The most expensive fans were made of such materials as ass's skin, parchment, or silk, with handles of carved ivory, tortoise shell, horn, bone, or sandalwood.

The use of handheld fans for cooling purposes and as decorative accessories died out, for the most part, after the nineteenth century. The electric fan—first produced commercially (in a two-bladed desk version) by Dr. Schuyler Skaats Wheeler for the Crocker & Curtis Electric Motor Company in 1882—proved highly effective and less physically taxing as a cooling aid. The first oscillating, **gear**-driven electric fan was produced by the Eck Dynamo & Electric Company in 1908. This fan swiveled back and forth on its stand, blowing air over a large area. The first extractor or self-contained window unit fan was introduced in 1934; designed by an English company, Vent-Axia, it was made of **plastic** instead of metal. In 1937 a plastic laminate called Micarta came into use as a coating for electric fan blades, rendering them quieter and less prone to corrosion or warping. Today, most inexpensive household fans have blades composed entirely of plastic. Aside from being cheaper to produce, they are not as likely to injure fingers accidentally inserted in the path of the blade. Ceiling fans have had a resurgence in popularity for both decorative and practical purposes.

Faraday, Michael (1791-1867)

English physicist and chemist

Michael Faraday was one of the greatest scientists of the 19th century. His early life closely paralleled that of **Benjamin Franklin**. Both were part of a large family; both were apprenticed in the printing trade; both read voraciously and became self-educated; and both loved science.

Faraday was born in Newington, Surrey, England, on September 22, 1791. His father, a blacksmith, could not afford a formal education for Michael, and so the boy received just the bare essentials and was apprenticed to a bookbinder. In some ways this apprenticeship was a stroke of good fortune for Michael because it gave him the opportunity to read all that he desired. He studied the articles about electricity in the *Encyclopaedia Britannica,* read a chemistry textbook, and was very interested in magnetism. In 1812 Faraday obtained tickets to attend the lectures of **Humphry Davy** at the Royal Institution. Faraday took 386 pages of notes and had them bound in leather and sent to Sir Joseph Banks (1743-1820), who was president of the Royal Society of London, with the hope of making a favorable impression. Unfortunately, Banks never responded. No matter—Faraday then sent a copy directly to Davy along with a job application to be Davy's assistant. Davy was very impressed, but he already had an assistant. However, shortly thereafter, Davy fired his assistant for brawling, contacted Faraday, and offered him the job of "washing bottles." This was not exactly what Faraday had in mind, but it was a step in the right direction and he accepted.

In 1813 Davy resigned his post at the Royal Institution, married a wealthy widow, and began an extended trip through Europe. The trip afforded Faraday the opportunity to meet such famous men as Italian physicist **Alessandro Volta** and French chemist Louis-Nicolas Vauquelin (1763-1829). In 1820, Danish physicist Hans Christian Oersted (1777-1851) had discovered that an electric current produced a magnetic field. This had set off a flurry of investigation by other scientists, among them Faraday, who was now back in England. Within a year of Oersted's discovery, Faraday had built a device which essentially consisted of a hinged wire, a **magnet** and a chemical **battery**. When the current was turned on, a magnetic field was set up in the wire, and it began to spin around the magnet. Faraday had just invented the **electric motor**.

Faraday's motor was certainly an interesting device, but it was treated as a toy. But Faraday had a greater goal in sight. Oersted had converted electric current into a magnetic force; Faraday intended to reverse the process and create electricity from magnetism. Taking an iron ring, Faraday wrapped half of it with a coil of **wire** that was attached to a battery and switch. **André Marie Ampère** (1775-1836) had shown that electricity would set up a magnetic field in the coil. The other half of the ring was wrapped with a wire that led to a **galvanometer**. In theory, the first coil would set up a magnetic field that the second coil would intercept and convert back to electric current which the galvanometer would register. Faraday threw the switch and received instant gratification: the experiment worked, resulting in a device that became known as the **trans-**

former. However, the result was not exactly what he expected. Instead of registering a continuous current, the galvanometer moved only when the circuit was opened or closed. Ampère had observed the same effect a decade earlier but ignored it because it did not fit his theories. Deciding to make the theory fit the observation, instead of the other way around, Faraday concluded that when the current was turned on or off, it caused magnetic "lines of force" from the first coil to expand or contract across the second coil, inducing a momentary flow of current in the second coil. In this way Faraday discovered the principle of electrical induction. Meanwhile, in the United States, physicist **Joseph Henry** had independently made the same discovery.

Faraday's affiliation with Davy had been suffering because Davy was extremely jealous of his former assistant, who was now eclipsing him. The situation escalated following Faraday's invention of the transformer; Davy claimed the idea for the experiment had been his. When Faraday was nominated to become a member of the Royal Society in 1824, Davy cast the only negative vote.

Having shown that magnetism could produce electricity, Faraday's next goal was to produce a continuous current instead of just a momentary spurt. This time he decided to reverse an experiment made by Dominique Arago (1786-1853). In 1824 Arago had discovered that a rotating copper disk deflected a magnetic needle. This, explained Faraday, was an example of magnetic induction. Faraday planned to use a magnetic field to set up an electric current. In 1831 Faraday took a copper disk and spun it between the poles of a permanent magnet. This set up an electric current in the disk which could be passed through a wire and put to work. So long as the wheel spun, current was produced. This simple experiment produced the greatest electrical invention in history: the electric **generator**. It took five decades and other inventions to make generators practical, but Faraday had pointed the way.

Faraday is especially remembered for his use of intuition in his scientific discoveries, making minimal use of mathematics. Unfortunately, he suffered a mental breakdown in 1839 from which he never fully recovered, and he was forced to leave the laboratory work to others. In addition to his inventions, he had compiled a number of notable discoveries: "magnetic lines of force," the compound benzene, how to liquify various gasses, and the laws of **electrolysis**. He also developed the concept of a "field"—a force, like magnetism or electric fields or gravity, that extends throughout space and is produced by magnets or electric charge or, in the case of gravity, mass. James Clerk Maxwell later developed his famous equations describing electromagnetism using this concept, acknowledging his debt to Faraday.

On August 25, 1867, Faraday died at Hampton Court, Middlesex, England. His accomplishments were all the more remarkable considering he had had no formal training in science or mathematics, yet was able to establish the fundamental nature of electricity and magnetism.

Michael Faraday

FARMING, MASS PRODUCTION

The great irony of agriculture is that more efficient farming practices have invariably led to population increases. The very effort of meeting the demand for food also creates it. Some anthropologists speculate that the practice of agriculture was in itself a response to increasing population in approximately 10000 B.C. The simplest and earliest method of obtaining food, hunting and gathering, could support only a limited number of people over a given geographic area. Although, according to scholarly thought, people may have grown small numbers of vegetables or penned young animals to fatten in earlier times, reliance on agriculture as the main source of food may have come about largely when the population of a given region rose beyond the critical maximum that hunting alone could support. Agriculture provided a more reliable food supply. It also meant a higher yield and better quality of meat and produce, which in turn meant that people could live longer, remain healthier, and successfully raise more children.

Most scholars agree that Western agriculture had its origins in the Fertile Crescent of Mesopotamia and Palestine. However, other early centers of civilization in Asia, the Americas, and elsewhere also developed farming. Animal domestication and **animal breeding** began with the dog in prehistoric times. Dogs were raised for food, for hunting, and for pest control and only later as companions. Other animals—cattle, hogs, sheep, horses—were raised for their meat, for field work, for their manure, and for the consumption of crop waste. All these animals were bred from their wild counterparts, favor being given to those that possessed traits farmers found beneficial.

Plants were also domesticated. Pesticides, **herbicides**, and **fertilizers** were developed. The first fertilizers came from natural sources—manure, compost, and crop litter, but chemical fertilizers became popular after World War II and lured farmers away from crop rotation and into mass production of single crops, or monoculture. Perhaps the most significant technological development in farming was the **plow**. Presumably, early farming the world over utilized the technique practiced by native North Americans of making holes with sharp sticks for one or two seeds each, or else of simply throwing seed onto bare soil with little or no preparation. Although this method is practical for small gardens, planting large fields is laborious and time-consuming when done in this way.

With the development of the plow, which appeared in Mesopotamia in approximately 4000 B.C., planting could be done in rows, or furrows, instead of in holes, allowing more crops to be planted in less time and more food to be produced. The time saved in planting allowed more leisure for the development of other activities, such as industry and commerce. The Mesopotamian plow was a simple device with a wooden, stone, or bone protrusion to scratch furrows into the soil. Eventually, development of the plowshare, or blade, usually made of metal, increased the plow's efficiency. The Celtic tribes used iron plowshares by about 400 B.C. The invention of the iron plowshare in Central Europe around 600 A.D. led to a significant population increase.

During the Middle Ages and Renaissance, the gradual replacement of oxen with horses as plow animals also improved the efficiency of plowing. Mechanization of the farm began in earnest during the 1800s, coinciding with the Industrial Revolution. The invention of **reapers**, **threshing machines**, **combine harvesters**, and **mowing machines** greatly reduced the amount of manual labor farming required. The **internal combustion engine** gradually eliminated the need for animal labor during the 1900s.

One of the most controversial developments in agriculture is the development of the factory farming of animals. Practitioners view it as a useful means of producing more food at less cost, while animal rights activists protest the disregard for the animals' comfort and welfare it embodies. Although factory farming of a sort was practiced in Hellenistic and Roman times, when farmers, for instance, kept fowl caged or broke their legs and wings so they would not exercise and therefore fatten quickly, modern factory farming was first practiced in 1771 by an English farmer named Moody who kept his cattle tightly confined in a dark, stuffy room. The profuse heat from their breath and warm bodies caused them to fatten faster and yield better meat.

Such intensive animal production has been widely adopted during the twentieth century. Hogs are kept in so-called sweat boxes, calves are bled to produce whiter veal, and chickens are mass-produced in large buildings, never being allowed contact with the ground or fresh air. Growth hormones in fodder make the animals mature at unnatural rates. Mass animal production in such close quarters creates a potential for the rapid spread of animal diseases that could devastate a farmer's operation. To combat this, antibiotics were developed during the 1950s to kill disease-carrying bacteria, through injection or by mixing it with the animals' feed. Recently, much concern has developed over the possible effects of hormone- and antibiotic-laden fodder on the human consumers of animal products.

Another concern over chemicals in agriculture involves the widespread use of fertilizers, herbicides, and pesticides around the world. Chemical agents, used mainly since World War II, have lead to serious pollution crises in many areas. Even the application of natural manure may have detrimental effects. For instance, recent fish kills in the Chesapeake Bay have been attributed to generations of manure applications by Amish farmers in Pennsylvania, which have overloaded the bay waters with nitrogen and deprived the fish of oxygen. Accumulations of chemicals in farm products have become a concern for the consumer. The long-term health effects of chemicals, in food have yet to be fully understood. Poor farming practices can lead to wind and water erosion and depletion of the topsoil. The plowing and overgrazing of unstable prairie soils by American settlers, in the West led to the devastating dust storms of the 1930s. Similar consequences have been felt in the famine-plagued Sahel of sub-Saharan Africa.

Modernization of farming has enabled highly efficient production. The savings, however, have been offset by inflation and increased operating costs. Only those farmers with keen business skills and luck are surviving. During the last fifty years, the number of farms has decreased while the size of farms has increased. The small family farm is rapidly being replaced by the large corporate farm. As farm production has greatly increased, so has the world's population. The resulting stress on food production systems spurs perennial fear that failure of any of the crucial links in the mass production puzzle, such as water supply, could undermine the entire scheme.

Today's farms are really agribusinesses, the most successful of which rely on sophisticated computerization and global satellite weather forecasting to analyze national and international trends and needs. More farms use specialized services instead of relying on integrated farming in which all operations are conducted by the farmer and his employees; outside operators and consultants are hired to provide advice and services. The 1996 Freedom to Farm bill, which decreases subsidies over a period of years, forces farmers to respond to global market pressures, for example by changing the amount and variety of crops they plant, or by understanding the relationship of grain exports to global beef markets. Closer to home, public concerns about water quality including agricultural runoff in rivers and streams is a continuing worry.

See also Animal breeding; Composting; Crop rotation ; Grain elevator; Harrow; Haying machines and tools; Insecticide; Plant breeding; Plant reproduction; Threshing machine; Tractor

Farnsworth, Philo Taylor (1906-1971)

American inventor and engineer

In the late 1800s several European inventors were demonstrating photomechanical **television** systems, systems that relied upon a spinning disc that would scan an image and then play it back. Though these systems enjoyed moderate success, it was clear that an all-electronic device would provide a clearer image while using less energy. The eventual inventor of the all-electronic television was a Russian immigrant named **Vladimir Zworykin**, although a young man from Utah named Philo Farnsworth is credited with the prototypes.

As young boy, Farnsworth had heard about the European advances in television technology. While still in high school he drew out a schematic for an electronic television, but did not as yet possess the skill to build the device. After two years at Brigham Young University he finally found financial support for his television, and he began working on its development. Three years later he debuted the first phase: an electronic camera called the ''image dissector.''

The costs to complete Farnsworth's television were becoming too great for his financiers to carry. In the early 1930s the project was handed to the Philco Corporation, whose greater resources allowed Farnsworth the freedom to devote his full attention to his invention's completion. Within a few years he had demonstrated a working model to the Franklin Institute. In 1938, however, Philco withdrew its financing, and Farnsworth was forced to seek other support. When World War II began, Farnsworth's television project was suspended indefinitely. By the time the war had ended Zworykin—with the help of the Radio Corporation of America (RCA)—had completed his own electronic television, the iconoscope. Though Farnsworth's design was still considered viable, the RCA system was quickly made the industry standard. The Farnsworth Television and RCA soon became the electronic research arm of the International Telephone and Telegraph Company (ITT), with Farnsworth remaining as a research consultant. All told, Farnsworth was awarded more than 300 domestic and foreign patents in fields ranging from television to atomic energy.

Fat substitute

Fat contributes 35 to 40 percent of the daily caloric intake of Americans. Most physicians agree that a far healthier level would be no more than 30 percent. As public awareness of the need to cut fat intake increased, food manufacturers saw the developing possibility of a large, new market: fat substitutes.

Proctor & Gamble Chairman John Pepper introduces Olean, the company's brand name for its fat substitute, olestra, which is a synthetic chemical derived from sugar and vegetable oil.

Starting in the 1960s, some companies developed carbohydrate-based fat replacers. For example, *Hellman's Light* mayonnaise used emulsified starch. These kinds of fat substitutes have only 1 to 4 calories per gram, compared to 9 calories per gram of fat. Still the most widely used replacements for fat, carbohydrate-based substitutes are identified on food labels as dextrins, maltodextrins, modified food starches, polydextrose, cellulose, and various gums.

In the 1980s, the NutraSweet Company (a subsidiary of Monsanto headquartered in Deerfield, Illinois) became especially interested in fat substitutes because the patents on its main money-maker, the **artificial sweetener** *NutraSweet*, would start expiring in 1992, and it needed a new high-profit product. A food scientist at NutraSweet, Norman Singer, developed an emulsified protein substance. In this, proteins from milk and egg whites are mixed and heated then made into round microscopic particles which, as they roll on the tongue, simulate the creamy feel of fat—but with only 15 percent of the calories of fat. In February 1990, NutraSweet's product, called *Simplesse*, became the first fat substitute approved by the Food and Drug Administration (FDA), for use in frozen desserts. NutraSweet immediately launched its *Simplesse*-based reduced-calorie *Simple Pleasures* frozen dessert.

While *Simplesse* works well for some products, it has a large drawback: it can not be used for cooking because heat causes it to stiffen. One solution is Proctor & Gamble's *Olean*, the brand name for olestra, a fat-free oil. Unlike *Simplesse*, which is made from natural ingredients, olestra is a synthetic compound of sugar and fatty acids. It has no cholesterol or calories, and is able to pass through the body without being absorbed. Olestra can be used to cook foods.

Olestra was discovered by chance. In the 1960's, Proctor & Gamble scientists were looking for a more easily digested fat, when ironically they stumbled upon a fat that isn't digested at all. Development of olestra took over 20 years, and Proctor & Gamble applied for FDA approval of the imitation fat in 1987. Because olestra is a synthetic food additive, Proctor & Gamble's application triggered lengthy safety studies. Research on the product included more than 150 studies involving over 20,000 people. Finally, in 1996, snacks made with *Olean* cooking oil began selling in test markets. Widespread sale of potato and tortilla chips made with *Olean* started in 1998.

Although the FDA approved olestra's use in snack foods, controversy lingers over the fat substitute's safety. One concern is that, because olestra isn't digested, it may cause abdominal cramps and diarrhea. Another concern is that olestra inhibits the absorption of certain vitamins and other nutrients. It is still unclear how great the risks may be.

Many other companies are developing fat substitutes, some with gelatin-water mixtures, and are currently test-marketing them in butter and **margarine** products. More artificial fats are certain to appear on the market in the near future, as Americans continue to search for new ways to cut the fat from their diet.

Favaloro, Rene • See Open-heart surgery

Felt

A legend concerning the invention of felt begins with a group of merchants traveling on foot. After many hours their feet became sore and chafed and, seeking relief, they padded their shoes with soft fur plucked from their camels' coats. When they removed their shoes that evening the moisture, heat, and pressure had compressed the fur into a new material—felt. While the authenticity of this tale is dubious, it demonstrates the basic method for preparing felt.

For thousands of years felt has been made by cultures that domesticate wool-bearing animals, such as sheep. The natural wool fibers are scaly, so that they will become interlocked when meshed together. In order to make felt, a quantity of shorn wool is dampened and pressed; heat is often applied as well. The thickness of the felt depends upon the quantity of wool and the amount of pressure used; the resulting felt can be soft and thick, or very thin and rigid (such as the felt used to cover piano hammers). Felt is a versatile material: it can be used for clothing, upholstery, **carpet**ing, and hats.

While felt, unlike woven fabrics, cannot fray, its meshing will begin to unravel if it is not stitched or quilted in some way. Felt can also be easily shaped, just by pressing the loose wool against a mold. Because it is created through a single-step process, felt is an inexpensive material to manufacture, and simple felt ''planking'' was carried on until the 1970s. Recently, attempts have been made to replace wool felt with felt made from **synthetic fiber**s, called bonded-fiber fabrics. Because the artificial fibers (usually **polyester** or **acrylic**) do not possess scales, an adhesive must be used to make them mesh together.

Ferris wheel

A Ferris wheel is a carnival ride consisting of a large, vertical **wheel** rotating around an axle, invented by engineer George Washington Gale Ferris for the World's Columbian Exposition in Chicago, Illinois, in 1893. Ferris was born February 14, 1859, in Galesburg, Illinois. As a child he lived in Carson City, Nevada. He graduated from Rensselaer Polytechnic Institute in 1881 and set to work designing railroads, bridges and tunnels. In 1885 he began inspecting structural steel for the Kentucky and Indiana Bridge Company. He took what he learned on the job and started his own business, the G. W. C. Ferris & Company, which acted as a consultant to steel users. In 1892 Daniel Burnham (1846-1912), planner for the World's Columbian Exposition in Chicago, announced he was looking for a visual focal point for the site that would rival the Eiffel Tower built for the Paris Exposition three years earlier. Ferris reacted by producing a design for a giant upright wheel. It was both an engineering and an aesthetic marvel. The original Ferris wheel towered 250 feet over the exposition and could hold 60 people in each of its 36 cars. It weighed 2100 tons, boasted a forged steel axle 33 inches by 45 feet, could carry 150 tons of riders, and was driven by two, 1000-horsepower reversible **steam engines**. The combination of its lights and the rhythm of its whirling motion was similar to the visual impact of a steam-driven **paddle wheel**. To finance the venture, Ferris raised $250,000 through a stock issue. His wheel made a profit of more than a million dollars. Ferris died three years later, in 1896.

See also Train and railroad

Fertilizer, synthetic

Soon after crops were domesticated, the need for fertilizers was realized. The first widely used fertilizer was manure, but bonemeal, fishmeal, dried blood, sewage, and seaweed all have served as fertilizers as well. In some cases green manuring is used; this takes place when crops such as clover or alfalfa are grown and then plowed under and left to rot. The first scientific study of fertilizers was conducted in 1550 when Frenchman Bernard Palissy (c.1499 or 1510-1589), voiced the opinion that since plants absorbed minerals—which he called vegetable salts—from the soil, the minerals needed to be replenished. Olivier de Serres (1539-1619), also of France, suggested crop rotation as a way of preserving soil nutrients. A 1669 paper published in England, however, suggested the use of fertilizers to replace soil minerals—phosphorus, potassium, nitrogen, sulfur, calcium, iron, and magnesium. Trace elements plants use include copper, born, zinc, manganese, and cobalt. Plants also require carbon, hydrogen, and oxygen, usually supplied through the air and water. Depending on the plant and the soil, some or all of these may need to be added. If the minerals are not replenished, the soil will not support the crop growth.

This was not such a problem in the United States or Canada during the early years of westward expansion because farmers would simply move to more fertile land when they had depleted the soil's mineral supply. However, replenishing these minerals was critical in Britain, Europe, and Asia, where arable land was in shorter supply. In general, the most commonly needed nutrients are potassium, nitrogen, and phosphorus. Natural reserves of potassium are ample. Originally the largest supplies were in Germany, but today potassium chloride mines in New Mexico and California supply much of the potassium used in synthetic fertilizers. Sulfur is present in the phosphate or ammonium fertilizers, usually as an impurity of the production process rather than as a deliberate additive. Calcium is currently supplied by dolomite, limestone, burned lime, or wood ash. Iron and magnesium are also added to fertilizers, as are the trace elements. Manure, sodium nitrate, urea, and ammonia sulphate have long been used as fertilizers for crops. Manures have a high percentage of nitrogen, as do sodium nitrate and sulphate of ammonia. Manure also improves the porosity of the soil. Unfortunately, processed manure was not in great supply while sodium nitrate, and sulphate of ammonia were available only from Chile and also were in short supply.

What was needed was a way of utilizing the nitrogen from the air, a way of fixing atmospheric nitrogen. Phosphorus is also needed in large quantities for plant growth. It was supplied as bonemeal, produced by pulverizing bones. This also became scarce. James Murray, an Irish physician, suggested that mineral phosphates from rocks could be used in place of bonemeal; however, he was not able to market his idea successfully. John Bennett Lawes (1814-1900), a English chemist, founded a research station in 1834 after inheriting his father's estate. He patented the idea of using phosphate rocks in superphosphates in 1841 and founded a superphosphate factory in 1843. Today most phosphorus is supplied in superphosphates or triple superphosphates. These are produced by applying sulphuric acid to natural phosphorus found in rocks.

The first Ferris wheel, which was erected at the 1893 World's Columbian Exposition in Chicago. The carnival ride was 250-ft (76-m) wheel with a 45-ft (13.7-m) axle.

In 1876 **Sidney Gilchrist Thomas**, also of England, used slag to develop a fertilizer for soil lacking phosphorus. A by-product of steel manufacturing, slag is also rich in phosphorus and is sometimes used. In 1802 Daniel Thaer investigated the effects of fertilizer in Saxony. At the same time other experimental farms began operation in France and Germany. **Humphry Davy** 's lecture in 1813 on agricultural chemistry signaled the beginning of intense study into fertilizers and the beginning of a new era in agriculture. In the 1840s a French chemist, Jean Baptiste Boussingault (1802-1887), began studying the effects of nitrogen on plant growth. He was puzzled by the fact that plants could grow in nitrogen-poor soil. At the time he presumed that the nitrogen required for their growth came from the air, but he did not know how the plants utilized the gaseous nitrogen. Actually the plants contained nitrogen-fixing bacteria at their roots. Ten years later **Justus von Liebig**, a German chemist, published Chemistry in its Application to Agriculture and Physiology.

While fertilizers containing phosphorus were being produced, nitrogen containing fertilizers were slower in developing. Not until the early 1900s did a German chemist, **Fritz Haber**, develop the technique for producing ammonia synthetically. He combined hydrogen with nitrogen and a catalyst at a high temperature and pressure to form synthetic nitrogen. By 1913 **Carl Bosch**, a German chemist working for BASF, developed the industrial ammonia production process. The first ammonia factory produced nine thousand tons of ammonia in its first year. The production techniques developed for the fertilizer were later applied to the petroleum industry.

Synthetic fertilizers have dramatically increased crop yields since the early twentieth century. Today, however, the amount of synthetic fertilizers in use, coupled with increased emissions from fossil fuels has resulted in a rate of excess nitrogen deposits on fields and through runoff in streams and rivers that has caused an increase in growth of algae in estuaries, oceans, and lakes. As the algae die, they deplete the oxygen supply in the water resulting in massive fish kills. Clearly, this state of affairs must be addressed, as it will have a direct effect on biodiversity, and on the health of the planet and its inhabitants.

See also Ammonia synthesis; Nitrogen fixation

Fessenden, Reginald Aubrey (1866-1932)

Canadian American physicist

It is unfortunate that Reginald Fessenden is not better remembered for his greatest accomplishment, which was eclipsed by the achievements of his contemporaries. Born on October 6, 1866, in Milton, Quebec, Fessenden was the son of a minister. He chose not to follow in those footsteps because his true calling was in science.

In 1886, Fessenden moved to New York City and started working at the Edison Machine Works. The following year, **Thomas Alva Edison** recognized Fessenden's talents and promoted him to the position of chief chemist at his laboratory in West Orange, New Jersey. Fessenden stayed with Edison for only three years. In 1890, he moved to Pittsfield, Massachusetts, where he was chief electrician for Edison's competitor, Westinghouse Electric Corporation. Two years later, he was a professor of engineering at Purdue University, and from 1893 to 1900 he taught at the Western University of Pennsylvania (now University of Pittsburgh) where he held the same post.

During those 14 years Fessenden went through five jobs. This migrant electrician had gained enough experience to set the stage for his greatest accomplishment. While at Western University, he studied some problems with wireless **radio** communication, which the U. S. Weather Bureau wanted to use in forecasting.

Fessenden invented a liquid barretter detector which gradually replaced the coherer that was used by **Guglielmo Marconi**. But Fessenden had a more lofty goal: he wanted to send actual voice messages through the air. To this end he patented, in 1901, a high-frequency **alternator** (a **generator** that produces **alternating current**). This invention produced a continuous radio wave, instead of the intermittent spark-generated pulse with which Marconi was experimenting in **Morse code** transmission. This wave could be modulated to encode the ''shape'' of a voice or musical sound.

Fessenden was not alone in his attempt to surpass Marconi's achievement. **Lee de Forest**, who would soon invent the ''audion'' **vacuum tube**, was busily building wireless systems which put him in competition with Fessenden. But Fessenden was a step ahead of de Forest.

Fessenden's plan was to limit the signal to one frequency for better reception over long distances, and to provide modulation which would be decoded at the receiver into voice and musical sound. He approached General Electric for help in constructing a more powerful generator. The task went to **Charles Steinmetz**, the genius who had used mathematics to work out the details of alternating current circuits. Steinmetz designed an alternator that generated current at 10,000 cycles per second, but it was not good enough for Fessenden.

The job next went to **Ernst Alexanderson**. He built a large 80-kilohertz generator that generated 50,000 cycles per second, though not without some disagreements with Fessenden, who thought he knew how to do it better. It took three years to build this alternator, which was then installed at a transmitter at Brant Rock, Massachusetts. The transmitter had a range of about 100 miles (160 km).

By 1905, Marconi, de Forest and Fessenden were at each others throats, embroiled in patent infringement lawsuits. Fessenden eventually won a suit against de Forest for using his detector patent.

The next year, Fessenden's system was finally ready, using **amplitude modulation** (AM) and an ordinary **telephone** microphone to impress sound on the radio waves produced by his alternator. After a great deal of promotion, on Christmas Eve, 1906, Fessenden broadcast his voice, violin, and recorded music. This transmission, the first radio broadcast in history, was received by the stunned owners of Fessenden radio sets along the Eastern seaboard of the United States and by ships several hundred miles away in the Atlantic Ocean. Later that year, he made the first trans-Atlantic voice communication and announced the first trans-Atlantic two-way connection between the United States and Scotland.

Fessenden made other improvements to his receiver. In 1912, he blended the incoming signal with a locally produced but slightly different signal, creating a beat-frequency tone that was audible. This was the first heterodyne circuit which later became a standard part of radio technology.

Although he was an extremely creative engineer, Fessenden lacked the abilities of a businessman. As impressive as his broadcasting device was, it was too massive and cumbersome to market for permanent use, and de Forest's invention of the audion in 1906 soon made it obsolete. He had founded the National Electric Signaling Company in 1902 with the help of wealthy financiers, but his abrasive personality alienated them, and the financial panic of 1907 wiped out his chance to sell the business to AT&T at a great profit.

Deserted by his backers, Fessenden spent more and more of his time in court where he fought infringements of his patents, which were ultimately sold to the Westinghouse Corporation. This period marked the end of the era of independent inventors; afterwards, most technological advances were made under the firm control of corporations. Yet the experiments of ''lone wolves'' like Fessenden excited the public and encouraged a generation of enthusiastic radio amateurs.

At the time of his death, on July 22, 1932, Fessenden had over 500 patents to his credit, making him one of America's most prolific inventors. Most of his inventions were related to radio, such as a radio **compass**, but some of his other inventions include a sonic fathometer, a signaling device for **submarines**, and a turboelectric drive for battleships.

Fiber optics

Since the late 1950s, optical fibers have emerged as revolutionary tools in the fields of medicine and telecommunications. They are capable of transmitting light pulses containing data up to 13,000 miles (20,917 km) without significant distortion. They also permit the ''piping'' of light into otherwise inaccessible locations, making them useful in diagnostic procedures which would have previously required invasive surgery.

Optical fibers operate on the principle of total internal reflection. Every medium through which light can pass possesses a certain refractive index, the amount by which a beam of light is bent as it enters the medium. As the angle at which the light strikes the medium is decreased from the perpendicular, a point is reached at which the light is bent so much at the surface that it reflects completely back into the medium from which it originated; thus, the light will bounce back rather than escape. In an optical fiber, total internal reflection is accomplished by a layer of material, known as cladding, with a lower refractive index. Once light enters the fiber it is internally reflected by the cladding; this prevents light loss by keeping the beam of light zigzagging inside the **glass** core. The manufacturing of optical fibers consists of coating the inner wall of a silica glass tube with a hundred or more successive layers of thin glass. The tube is then heated to 2,000 degrees Celsius and stretched into a strand of thin, flexible fiber. The result is a clad fiber, approximately 0.0005 in. (0.0127 mm) in diameter. By comparison, a human hair measures 0.002 in. (0.0508 mm).

Fiber optics received its first application in medicine. In the late 1950s, Dr. Narinder S. Kapany (b. 1927) hit upon the idea of building an **endoscope** capable of seeing around twists and turns in a patient's body by using fiber optic bundles. His device, which came to be called the fiberscope, consists of two bundles of fiber: one incoherent bundle, in which there is no relationship between the order of fibers from one end of the bundle to the other, to transmit light into the body of the patient, and one coherent bundle, in which the individual fibers have the same position at both ends of the bundle, to carry a color image back to the physician. Because of its small size and flexibility, the fiberscope can be used to view many areas inside the body, such as the cardiovascular and digestive sys-

Fiber optics were first conceived of by Dr. Narinder S. Kapany, who built a flexible "fiberscope" in the late 1950s to see around the twists and turns in the human body. This revolutionary technology has since become an integral part of the telecommunications industry.

tems, that physicians cannot otherwise see without performing surgery.

Optical fibers were first used in the field of telecommunications in 1966, when it became apparent that data transmitted by a **laser**, however bright, could be broken up and absorbed by uncontrollable elements such as fog and snow. The first optical fibers produced contained flaws that resulted in significant amounts of light loss. To boost the range of the light signal, energized atoms of the rare element erbium were used to amplify the signal at 1.54 micrometers, the wavelength at which the fibers are able to transmit light the farthest. This replaced the more costly method of converting weak light signals to electronic form and back to light again before sending them through the next segment. A **telephone** conversation is carried over optical fibers by a method called digital transmission. This is achieved by first converting sound waves into electrical signals, each of which are then assigned a digital code of 1 or 0. The light carries the digitally encoded information by emitting a series of pulses: a 1 would be represented by a light pulse, while a 0 would be represented by the absence of a pulse. Upon reception, the light waves are converted back into electronic data, which are then converted back into sound waves. By utilizing digital transmission, telecommunications systems carry more information farther, over a smaller cable system than its copper wire predecessor. A typical copper bundle measuring 3 in. (7.62 cm) in diameter can be replaced by a 0.25 in.(0.635 cm) wide optical fiber carrying the same amount of data. This improvement becomes important in areas where telephone cables must be placed underground, in which space is so highly limited.

The minute size of optical fibers also allows for a significant reduction in the weight of a particular system. The reduced weight is beneficial in systems that require rapid deployment of information, such as in military communications and in **aircraft** instrument wiring. By replacing the copper wiring on a jet aircraft, up to 1,000 lbs. (454 kg) may be saved, allowing for more economical fuel consumption. Optical fibers are also immune to electromagnetic interference, making them roughly one hundred times more accurate than copper; they typically allow only one error in one hundred million bits of data transmitted.

Optical fibers have been demonstrated as an ideal method of transmitting high-definition **television** (HDTV) signals. Because its transmissions contain twice as much information as those of conventional television, HDTV allows for much greater clarity and definition in its picture; however, standard transmission technology is not capable of transmitting so much information at once. Using optical fibers, the HDTV signal can be transmitted as a digital light-pulse, providing a near-flawless image reproduction that is far superior to broadcast transmission, just as music from a digital CD is superior to that broadcast over FM **radio**. More commonly in the late 1990s though, **cable television** providers as well as telecommunications companies are laying hybrid cable combining fiber optic with coaxial, as a cost effective approach of bringing fiber optic technology within service of their customers. As of 1997, 25 million miles of fiber optic strands had been laid in the United States.

Fiber, synthetic

Uses for synthetic fibers range from nylon stockings and clothing to cables and tire reinforcement. Synthetic fibers are made from **polymers** that are either melted into a solution by heat or dissolved by a solvent. The solution is then passed through a metal plate with fine holes, called spinnerets. This process forms the polymers into strands. The fibers are then either cooled or passed through a jet of air to allow the solvent to evaporate. Most fibers at this point are subjected to cold drawing, a strengthening technique developed by **Wallace Carothers**'s team in its search for artificial silk. After drawing, the fibers are washed, dried, dyed, and woven.

The first patent for synthetic fiber was granted to George Audemars in 1855. A related patent was granted to Sir **Joseph Swan** in 1880. Both of these men produced fibers from cellulose, which, unfortunately, were not very strong. Louis Comte de Chardonnet later found that by denitrating the fibers he could strengthen them until they were as durable as silk. **Edward John Bevan** and **Charles F. Cross** developed the industrial production process for this material, which is called **rayon**.

Acrylonitrile, which was discovered by Moureau in 1893, is used in the production of nitrile rubber, acrylic fibers, **insecticide**s, and **plastics**. The fibers formed by acrylonitrile are unstable at their melting point, so melt extrusion is impractical. Solution spinning was not possible for many years because no appropriate solvent had been found. Today cold drawing is used to strengthen the fibers by orienting the molecules and allowing hydrogen bonding to occur. The fibers are then dried and woven into a fabric that resembles soft wool, which is used for sails, cords, blankets, and clothing.

Nylon was developed by DuPont Company researchers as a substitute for silk. Carothers and his team of assistants had been researching long chain **polyester**s and polyamides. In 1938 they had almost given up on finding a suitable fiber when two members of the team discovered cold drawing. Nylon is used in clothing, laces, tooth brushes, sails, fish nets, and carpets.

John Whinfield and J. Dickson continued Carothers's research and made a polyester with terephalic acid in 1941. The fiber was christened terylene and marketed as **Dacron** in the U.S. by the DuPont Company. Acrilan, produced by the Chemstrand Corporation, is an acrylic fiber used in fabrics and may be blended with wool or cotton to form clothing, carpeting, linens, draperies and upholstery. Fabrics made from acrilan resist mildew, moths, and wrinkling. They also tend to dry quickly.

Orlon is a class of synthetic fibers first produced commercially by the DuPont Company in 1950. The fibers vary in size, texture, and ability to hold dyes. They can be woven or knitted, usually into bulky garments. Orlon is used in upholstery and carpets.

Vinyon filaments and fibers were developed by the Carbide and Carbon Chemicals Corporation, which licensed American Viscose Corporation to produce them in 1939. The fiber is a copolymer of 88 percent **vinyl** chloride and 12 percent vinyl acetate. It was the first plastic fiber produced on a large scale in the United States. The fibers are stretched in a process similar to cold drawing. The stretching increases the strength of the fiber but lowers its elasticity. The fiber does not take dyes and becomes sticky if heated to over 149° F (65° C). At 167° F (75° C). garments made of the fiber will shrink. Further research led to the development of a fiber Vinyon N, which is a copolymer of vinyl chloride with acrylonitrile. For fibers, the copolymer ranges from fifty-six to sixty percent copolymerized vinyl chloride. It was patented in 1947.

Kevlar is a polyamide fiber developed by Stephanie Kwolek of DuPont in 1965. It is incredibly strong due to its molecular structure of alternating aromatic rings and amide groups crosslinked by hydrogen bonding. Kevlar's light weight and high strength make it very marketable. It is commonly used to make bulletproof vests.

See also Acrylic, plastic; Toothbrush and toothpaste

Models from the 1940 meeting of the National Association of Manufacturers in Synthetic Clothing display fashions made of (l-r) casein, vinyon, and nylon.

FIBERGLASS

Fiberglass consists of very fine threads of **glass**, sometimes combined with other materials, loosely bunched together in woolly masses. Flexible and strong, fiberglass resists burning, and will not decay, stretch or fade. It is also an excellent insulator, as the minute spaces between the fibers trap air, preventing the flow of heat. Such properties make it desirable for weaving into cloth for curtains and tablecloths, for combining with **plastics** for **automobiles**, boat bodies and fishing rods, and for packing into woolly bulk form for air filters as well as **insulation**.

The ancient Egyptians used glass fibers, incorporating them into vases and containers as decorative trim by winding the glass around a core of glazed clay. But fiberglass manufacture on a large scale did not take place until the twentieth century. The first modern technique for making fiberglass was patented in 1836 by Dubus-Bonnel of France, who wove hot glass strands on a Jacquard-type loom. During World War I, the Germans, facing a shortage of asbestos, began manufacturing fiberglass as an excellent substitute insulator.

In the United States, The Owens Illinois Glass Company and the Corning Glass Works conducted experiments between 1931 and 1939 resulting in the development of successful commercial manufacture of fiberglass. In 1941, a flexible fiberglass cloth suitable for use as reinforcement in plastic laminates was developed. Owens Corning developed the first fiberglass-reinforced boat hull in 1944 for the Navy, followed two years later by the production of fiberglass-reinforced fishing rods, serving trays, and pleasure boats. In 1953, Owens Corning and General Motors announced the first production automobile made entirely of fiberglass-reinforced plastic, the Chevrolet Corvette. During the 1960s, fiberglass applications grew to include underground tanks and pipe. In 1975, the Pontiac Silverdome, with its 10-acre fiberglass fiber roof, opened. In 1990, fiberglass windows were introduced.

Fiberglass is made of the same materials as regular glass—sand, soda, and lime—and can be produced by three methods. The first involves forming glass into marbles that are examined for impurities, then melted down; the liquified glass runs through tiny holes at the bottom of the furnace and is caught by a spinning drum that winds the strands on a bobbin while simultaneously pulling them into even finer fibers. One ⅝-inch marble can yield up to 95 miles of fiberglass. Fiberglass *yarns* are then formed by twisting these very fine fibers together. The second process, the *direct melt process*, eliminates the use of marbles. In the third process, bulk fiberglass (fiberglass wool) is produced by melting the raw materials in a furnace and allowing the liquified glass to flow through tiny holes at the bottom; high-pressure steam jets catch the resulting strands and force them into fibers 8 to 15 inches (20 to 38 cm) long. The fibers are gathered on a conveyor belt into a white, woolly mass.

See also Composite material

FIBERSCOPE • See Endoscope

FIELD EMISSION MICROSCOPE • See Field ion microscope

FIELD ION MICROSCOPE

The field ion microscope is a remarkably powerful optical device. It represents several generations of scientific evolution of the electron microscope and is probably the most powerful magnifying instrument yet invented.

Before German electrical engineer **Ernst Ruska** and Max Knoll invented the **electron microscope** in 1932, the science of microscopy had reached a standstill. Though scientists were theorizing about the structure and composition of the atom, no microscope yet invented had the ability to resolve an image smaller than a wavelength of light; thus, there was no way to actually observe atomic and subatomic particles. Ruska and Knoll overcame this barrier by substituting a beam of electrons for the microscope's light source. With wavelengths far shorter than those of visible light, electrons could be used to view specimens at magnification levels thousands of times higher than those possible with simple microscopes. The electron microscope provided the first views of viruses and bacteria, as well as large molecules.

Though the power of the electron microscope was unsurpassed at the time, many scientists were unsatisfied with its results. Ruska's invention had revealed a new world of submicroscopic particles, but it was not powerful enough to view individual atoms. Though this seems like a purely academic goal, it was of crucial importance to a scientific community struggling to understand the properties of these universal building blocks.

Erwin Wilhelm Müller (1911-1977) began work on a new type of particle microscope while still attending the Technical University at Berlin. In 1936—just a year after he graduated—he developed the field emission microscope. Based on J. Robert Oppenheimer 's (1904-1967) theories of quantum mechanics, as well as on Ruska's groundbreaking research, the field emission microscope allowed magnification up to two million times.

In the field emission microscope, a very thin metal needle is placed within a **vacuum tube**. When a powerful electric current (up to 30,000 volts) is applied, electrons are emitted from the sharpened tip of the negatively charged needle. The electrons, spreading as they move away from the needle, strike the surface of a (positively charged) fluorescent screen. The image produced is an enlarged view of the atomic lattice structure of the tip of the needle.

The field emission microscope provided unparalleled images of atomic structure. However, there remained several shortcomings: first, the process could only observe the atomic structure of metals, and sturdy ones at that, since the intense electric current necessary to liberate the electrons would destroy any other substance (organic and biological matter could not be observed); second, the random nature of electron movement created images of high magnification that often were too blurry to be useful. Most frustrating to Müller was his invention's inability to achieve its primary goal: the imaging of individual atoms.

Müller remained in Germany until 1952, when he moved to the United States to work at the Pennsylvania State University Field Emission Laboratory. Here he tried a new design, utilizing ion emission rather than electrons. This involved switching the polarity of the system, so that the needle was positively charged and the screen negatively charged. The random travel of ions was much easier to contain by supercooling the needle tip (usually with liquid nitrogen or hydrogen). Lastly, he added a small amount of inert gas to the vacuum tube. Müller's new design, called the field ion microscope, proved to be even more powerful than the emission microscope—and without the latter's resolution problems. Using this device, Müller finally was able to resolve individual atoms.

The field ion microscope is the most powerful optical tool available and is the only instrument capable of examining the structure of metals on an atomic level. It is used primarily in atomic research, particularly in the study of how metals crystallize. It was still impossible to examine biological specimens until the development of field evaporation technology; this allows scientists to evaporate metal surrounding a needle of organic material. While this technique has provided some insights into the structure of biological molecules, the field ion microscope is still not flexible enough to be used on living cells; this was later made possible by the invention of the **scanning tunneling microscope**.

FIGHTER JET

Aircraft powered by **jet engines** first flew during the early part of World War II, with the United States, Britain, and Germany all testing initial models. However, the technology was too new, and the production capacity too limited, for these aircraft to play a significant role in the war. Only one jet, the British Gloster Meteor, particpated in World War II operations, and then only toward the end of the conflict.

By the time of the 1948-1952 Korean War, jet aircraft were becoming commonplace, and this conflict was the first time that jets were used to any great extent in military action. The Lockheed P-80 Shooting Star became the first jet that saw significant use, but more common sights in the Korean skies were the first planes that resembled truly modern fighters.

Throughout the Korean War, there was direct military action between American and Chinese pilots. The Americans flew the new F-86 Sabres, while the Chinese had been supplied with Russian Built MiG-15s. These two aircraft dueled throughout the war, with the F-86's playing a critical role in deflecting the veritable horde of MiGs available to the Chinese. The Sabres were also the first fighters to exceed the speed of sound, though they were not specifically designed to sustain such speed.

The renewed demonstration of the importance of air superiority that emerged from the Korean War, combined with rapidly advancing jet airplane technology and designs, led to an amazing proliferation of jet fighters. Before long they were the workhorses of most nations' air forces, whether designed by the country itself (as with the United States, the Soviet Union, and a number of European countries) or purchased from suppliers (as many Third World countries did). Literally hundreds of designs emerged, and they are impossible to enumerate in a short space.

Major developments in United States jet fighter technology appeared rapidly after the Korean War. The F-100 Super Sabre improved on its earlier cousin by achieving level supersonic flight. By the time of the Vietnam War, from 1965 to

1975, the McDonnell Douglas F-4 Phantom had become the United States' jet workhorse, followed closely by the 1970s Grumman F-14 Tomcat and McDonnell Douglas F-15 Eagle. The Soviets kept pace with the increasingly advanced MiG-23S and MiG-25. The more advanced of these fighters and flew well in excess of Mach 2 (twice the speed of sound), could fly up to 50,000 feet (and in the case of the MiG-25, an astonishing 80,000 feet). Increasingly advanced planes also meant increasingly expensive planes, and in the late 1970s, the United States undertook development of a lighter, less expensive fighter, eventually to enter service as the F-18 Hornet.

In the 1970s and 1980s, aircraft designers came under increasing demand to create fighters and **bomber planes** that were nearly invisible to **radar**. During the 1980s, the United States developed the F-117A ''Stealth'' aircraft, which is designed to minimize its radar return by using radar-absorbent material as part of its ''skin,'' as well as incorporating numerous unusual angles to deflect radar signals hitting it. Despite some well-publicized incidents, including the breakup and crash of an F117-A during an air show near Baltimore in 1998, the futuristic-looking Stealth fighter fought successfully in the 1991 Guld war against Iraw, and remains the state-of-the-art in jet fighter technology.

Fingerprinting

Fingerprinting is probably the oldest form of biometrics, which refers to the process of identifying people through their unique physical characteristics. A fingerprint is made up of the pattern of ridges (papillary ridges) on a person's fingertips. Each fingerprint is unique, its pattern never changes, and fingerprints can be classified according to general types. These three characteristics make fingerprinting an infallible method of identification. By studying the number and sequence of the ridges in the fingerprint patterns, fingerprint specialists can positively match an individual to a set of prints.

With over 200 million fingerprints on record at the Federal Bureau of Investigation (FBI), matching fingerprints is not an easy job. Categorizing of fingerprint patterns makes the matching process easier. There are three general types: *whorl*, *arch*, and *loop*. Eight subcategories have also been developed to define the different combinations these general patterns can create. Fingerprints are especially helpful in crime detection work. Criminals can change their names and their appearance but not their fingerprints, despite some attempts to do so with everything from sandpaper to strong acids. Their best efforts have produced temporary discomfort or permanent injury.

Detectives look for three types of prints at a crime scene: *visible*, *molded*, or *latent*. Visible fingerprints are left by fingertips coated with paint, grease, or some other visible substance. Molded fingerprints are left in soft items like wax or putty. Latent or hidden prints are left by the skin's natural oils and secretions. These may not be visible to the naked eye but can be brought out by special powders and chemicals. In addition to their use in criminal investigations, fingerprints are also useful in other areas. For example, they can be used to identify accident victims, armed forces personnel, government employees, and amnesia victims.

Fingerprints are recorded by rolling the fingertips on an **ink**-covered surface. Then each finger is pressed onto a standard card divided into ten squares, one print per square. As a check for the proper sequence, all five fingers are simultaneously pressed onto another large square.

Establishing exactly how long fingerprinting has been around is difficult. Before fingerprinting was introduced, people had to be identified by tattoos, brands, **photography**, body measurements, and other inefficient means that resulted in many misidentifications. Prehistoric carvings resembling fingerprint patterns have been found in cliffs, caves, and even ancient clay tablets; their significance is still uncertain. The Chinese began to use thumbprints to sign documents long before the birth of Christ, but probably as a legal signature rather than a means of identification. In 1823, Dr. J. E. Purkinje (1787-1869), anatomy professor at the University of Breslau, published a paper noting the diverse ridge patterns on human fingertips, but made no mention of adapting his observation to personal identification. In 1858, Sir William J. Herschel devised a workable fingerprint identification system thought to be the first of its kind. In the 1880s, Francis Galton (1822-1911) obtained the first extensive collection of fingerprints for his studies on heredity. He also established a bureau for the registration of civilians by means of fingerprints and measurements. In 1891, Juan Vucetich, an Argentine police officer, was believed to have created the first usable fingerprint identification system that could be applied to criminal investigation.

A few years later Sir Edward R. Henry of Great Britain developed a more simplified fingerprint classification system. All subsequent systems have built upon Henry's invention. Throughout the twentieth century, advances have been made in the detection of latent fingerprints and processing fingerprint data. In the 1970s, police began to use **lasers** to detect fingerprints. Lasers were found to be much more sensitive to ''weak'' fingerprints and therefore extremely useful to detectives. In addition, computerization of fingerprint records has a become standard procedure. This serves two purposes. The digitized print records are much easier to store than large numbers of the old fingerprint cards. Second, prints can be searched, scanned, and matched quickly using special software. Future developments to this procedure will include computer enhancement to obtain more detailed fingerprints.

Finsen light

The finsen light, named for its inventor, Niels Tyberg Finsen, was a powerful light used to cure people of a skin disorder. Niels Finsen was born in 1860 in the Faroe Islands, Danish islands in the North Atlantic Ocean. His parents were of Icelandic origin, so young Niels spent his early schooling in the capital of Iceland. He was a frail and sickly youth. Living so close to the Arctic Circle where the days are short in winter, Niels became aware of the effects of sunlight on his disposition and health. Later Finsen went to Denmark to study for his medical degree at the University of Copenhagen. While at the university he was incorrectly diagnosed as having heart disease, but actually he suffered from Pick's disease, a condition that affects the liver and the lining around the heart. (He later developed ascites, fluid accumulation in the abdominal cavity, and was confined to a wheelchair.)

After receiving his medical degree in 1891, Finsen became very interested in how light affects disease as he himself found a benefit from exposure to sunlight. He was familiar with the work of a Swedish researcher who in 1889 had discovered that ultraviolet light (short waves) irritates biological tissue more than infrared light (longer waves such as heat waves). So Finsen began recording the effects of sunlight on insects and amphibians, becoming ever convinced that light could be used to treat human disease. He found that ultraviolet light from the sun or from electric lights could kill bacteria. Finsen was convinced that it was the effect of the light and not the heat it caused and wrote several papers in 1893 and 1894 on the beneficial use of phototherapy. He also thought that red light was helpful in curing smallpox but this idea was later abandoned.

In 1895 Finsen made an arrangement with Copenhagen Electric Light Works to treat patients two hours each day with ultraviolet light. His patients were diagnosed with lupus vulgaris, a skin disease associated with the tuberculosis bacteria. He designed a powerful lamp (the finsen light) for this purpose, a bright artificial light generated by electrical carbon arcs. In 1896 Finsen founded the Finsen Institute for Phototherapy in Copenhagen dedicated to studying effects of light and curing people of disease. At the Institute 800 lupus patients were treated. Half were cured of the disease and nearly the rest showed improvement in their conditions. For this achievement he was awarded the 1903 Nobel Prize in medicine and physiology. Finsen donated half the prize money to the Institute. In failing health, he died the next year at only 43 years old.

During Finsen's era both X rays and gamma rays were discovered by Wilhelm Röntgen (1845-1923) and Antoine-Henri Becquerel (1852-1908) respectively. With Finsen's success with light therapy leading the way, the idea of **radiotherapy** was born. Since his time, X rays and gamma rays have been used for the diagnoses and treatment of disease. Even today some foods are irradiated with UV light to kill bacteria (although this practice is sometimes controversial). Finsen was also ahead of his time in his concept of the effect of sunlight on disposition and health. In is only fairly recently that Seasonal Adjustment Disorder (SAD) has been recognized as a type of depression caused by a lack of sunlight in winter. People diagnosed with SAD can be treated by sitting under lights to extend exposure to light on short days.

FIRE ENGINE • See Fire-fighting equipment

FIRE EXTINGUISHER

George William Manby (1765-1854) invented the first fire extinguisher in 1813. Manby, who had been a member of the British militia, had observed the inability of firemen in Edinburgh to reach the upper floors of burning buildings and was inspired to create a means of remedying that difficulty. Manby's extinguisher consisted of a four-gallon copper cylinder which held three gallons of water; the remainder contained compressed air. When the stopcock at the top of the cylinder was engaged, the compressed water would be forced out through a tube running from the **valve** to the inside base of the cylinder and directed toward the fire. In 1866, Frenchman Francois Carlier invented an extinguisher which had a cylinder containing a mixture of water and bicarbonate of soda and a separate **bottle** filled with sulfuric acid. When the bottle was punctured, the acid mixed with the bicarbonate, producing carbonic acid which, when it bubbled, forced the water out. A Russian, Alexander Laurent, developed a solution of aluminum sulfate and bicarbonate of soda in 1905. The carbonic acid bubbles themselves, rather than water, were forced out of the cylinder, smothering the fire. Laurent's device was meant to be used on oil-based and electrical fires, against which water is useless. During World War II, a commercial product called Aero Foam, a derivative of soy protein, was invented by **Percy L. Julian** and used by the United States' military. Modern fire extinguishers have an inner cartridge containing carbon dioxide that acts as a pressurizing agent. When the operating valve is pressed, the gas is released into the main cylinder and forces the extinguishing agent, whether water, foam, or powder, through a nozzle. Entire buildings are now required to have fire extinguishing sprinkler systems; these are installed in the ceilings and are heat-activated.

See also Fire-fighting equipment

FIRE-FIGHTING EQUIPMENT

Among the earliest attempts to organize fire-fighting were the ancient Egyptians' gathering of volunteer fire-fighters and the Romans' use of slaves stationed in strategic locations to spot and douse fires. The Greeks and Romans even devised primitive fire engines—small human-powered **water pumps** mounted on **wheels** or skids. But for many centuries fire fighting essentially consisted of little more than bucket brigades.

The great London fire of 1666, which decimated nearly thirteen thousand buildings, drew attention to the need for preparedness in the face of fire emergencies. Hand-operated pumps on wheels drawn by humans came into use, replaced in the 1800s by horse-drawn wagons. These early pumps could produce streams of water of no more than 50 ft (15.25 m). In 1830, John Giraud of Baltimore invented a pump with a chamber of compressed air that ultimately was capable of boosting the stream of water delivered through the **leather** hoses to about 200 ft (62 m).

The first fire alarms were invented by Ithiel Richardson in 1830. He ran strings through rooms of a building that, when burned through, would set off a central bell. Rufus Porter used a metal bar that would drop onto a bell apparatus when heated. Welshman George William Manby (1765-1854) invented the first portable **fire extinguisher** in 1813. The four-gallon copper vessel contained water and compressed air. Later extinguisher designs were chemically activated and were capable of putting out electrical and oil fires.

A rescue ladder attached to a fire wagon on a revolving base was first introduced in 1840. That same year P. R. Hodge developed a prototype steam-powered fire engine. However, it was not until 1852, when A. B. Little made improvements to it, that the steam-powered engine was accepted for use by firemen. Gasoline-powered engines came into use in the 1900s. No longer was it necessary to "fire up" the engines to build up steam before responding to the fire. The entire set of mechanical equipment could be operated automatically by electricity.

Modern fire trucks are fully equipped with life-support systems and a range of tools for battling fires, including pumps capable of throwing 750 to 2,000 gal. (2,838.75 to 7,570 l) of water per minute, steel water towers, and ladders that can extend over a hundred ft. (30.5 m) into the air. Sophisticated alarm systems are activated directly from commercial and industrial sites. The computerized *Enhanced 9-1-1 Telephone System* allows individuals quick-dial access to emergency services. Fire fighting chemicals as well as water are delivered as needed.

Special measures are often required for other types of fires. The main course of action in fighting a forest fire for example, is to contain it rather than extinguish it. Fire fighters are flown or driven in to create a fire line. The line is formed by cutting, dousing, or burning a break in the trees in the hope of stopping the fire's advance. **Ships** at sea must have their own fire-fighting capabilities. In harbors, fireboats can be called into action.

See also Fire extinguisher; Internal combustion engine

Fireproofing techniques

The threat of fire has been among humankind's constant concerns, and fireproofing is one of the most common means of prevention. The earliest fireproofing materials were tiles and clay, installed to protect timbers and walls from stray sparks from cookstoves and fireplaces. Yet little could be done to prevent a spark from landing on a thatched roof, or to keep lightning from striking.

Wood was, and continues to be, one of the cheapest types of building material, but where wood buildings are crowded together at the center of cities, a small accident involving fire can decimate an entire city. The Great Chicago Fire of 1871 prompted a call to architects to devise fireproof buildings. In response to this, **William Jenney** designed the Home Insurance Building in 1884, featuring an iron framework that held up the external "skin" of the building. Jenney's construction launched a revolution in structural design.

Today buildings are classified according to how fire resistant they are; that designation is now preferred over the term *fireproof* and refers to the number of hours a structure can bear exposure to heat before it is no longer safe—as when, for example, its metal begins to weaken and threatens to collapse. Materials favored in the construction of fire-resistant buildings include **brick**, tile, concrete, cement, gypsum, and plaster.

While steel itself is noncombustible, high temperatures will weaken it, and it must be encased in fire-resistant building materials. Another architectural tactic to increase a structure's fire resistance involves eliminating spaces within frames where fires can spread. While some substances are inherently fireproof—asbestos, **glass**, and vermiculite, among others—combustible materials can be treated with special paints or chemicals that render them less susceptible to fire. Wood, for example, can be treated with a mixture of ammonium sulfate, ammonium carbonate, alum, borax, and boric acid in water. Common households items like curtains and clothing also can be fireproofed with similar types of mixtures. Technology even may provide for proactive fire defense if a fire is detected—a prototype fire shield for homes is a large flame-retardant blanket that self-deploys from the attic through a door in a home's roof, if infrared and ultraviolet sensors detect fire. The shield, deployed by the force of compressed-air cannisters, covers the structure and nearby landscaping, and an elastic lip secures the blanket around the base of the house. At the same time an alarm sounds to warn occupants.

Some of the deadliest fires occur in buildings made of noncombustible materials but filled with furniture, carpeting, paneling, and, now, **plastics** that, when ignited, can turn a structure made of brick or corrugated metal into a toxic disaster. The fumes produced, especially from some of the new synthetic flame-retardant fibers manufactured from polyvinyl chloride (PVC) and other synthetics, can prove deadlier than flames themselves. Natural and manmade fibers are now being modified chemically to alter their reactions to heat, either by modifying the burning chemistry of the material or causing them to break down under heat to release a fire-extinguishing substance.

See also Construction, cast-iron and wrought-iron

Fireworks

The first fireworks were most likely created in China during the tenth century and employed for ceremonial purposes. Dependent on black powder (**gunpowder**) for their pyrotechnical displays, fireworks typically consist of a fuel source, an oxidizer, a fuse, and color-producing compounds. The gunpowder itself, a ground-up mixture of potassium nitrate, sulfur, and charcoal, is the most important variable and determines the speed, height, and bursting power of the charge. The large-scale manufacture of fireworks is carried on today in a number of countries, including Japan, France, England, Spain, Italy, and the United States. Major fireworks displays are enjoyed the world over and have become requisite centerpieces of such festivals, observances, and holidays as the Fourth of July, New Year's Eve, Mardi Gras and the Chinese New Year.

The history of fireworks, a peaceful invention, is intertwined with that of early instruments of war. From the tenth until the late seventeenth century, gunpowder was employed primarily with projectiles, first arrows and, later, gunshot and cannon balls. Similarly, the first firework was also a projectile, resembling the modern-day Roman candle and spewing balls of fire from a bamboo tube. Although the Chinese were the originators of this device, the Arabs hold claim to developing, in 1353, the first gun, a bamboo tube reinforced with iron to withstand the explosive pressure of the compacted powder. However, it was an Englishman, Roger Bacon (c. 1214-1292), who wrote the first specific instructions for the preparation of black powder in the form of a Latin anagram printed in 1242. Bacon's recipe has since been altered and improved upon for both military and industrial uses. For fireworks manufacturers—nearly all longstanding, family-owned businesses—black powder recipes are highly prized and carefully guarded trade secrets.

Perhaps the first person to ardently promote and cultivate the stunning visual and dramatic possibilities of fireworks was King Louis XIV of France, who reigned from 1643 to 1715 and whose palace at Versailles formed the perfect backdrop for his lavish, fireworks-punctuated galas. In much the same Baroque spirit today, the best fireworks displays emphasize pageantry, ornateness, and surprise, all with the assistance of computerized choreography, split-second electrical firing, elaborate one-of-a-kind set pieces, and massive lines of mortars from which the large shells (stars, chrysanthemums, comets, peonies, salutes, etc.) are launched. The six basic fireworks colors (along with their key ingredients) are white (magnesium or aluminum), yellow (sodium salts), red (strontium nitrate or carbonate), green (barium nitrate or chlorate), blue (copper salts), and orange (charcoal or iron). Fireworks, which come in three general classifications, are widely restricted or prohibited for private use, and range in size from the 3/4 inch-long (2 cm) ladyfinger firecracker to the world record Fat Man II, a 720-pound (326 kg), 40.5 inch-diameter (102 cm) shell fired near Titusville, Florida, on October 22, 1977.

Firing mechanism, gun

The firing mechanisms of primitive firearms applied one of two methods: the slow match or the quick match. The slow match was a rope or cord boiled in lye and **gunpowder** and then dried, which would smolder indefinitely in the touchhole before igniting. For the quick match, the cord was moistened and then rolled in finely-ground gunpowder. Both methods often rendered uncertain results and were quite dangerous.

In the fifteenth century, the Spanish invented the harquebus or arquebus, a portable but heavy gun equipped with a matchlock, which was a slow-burning match lowered into the breech of the barrel to ignite the gunpowder. This weapon was equipped with a movable clamp, known as a *serpentine*, that held the match on the gun. The serpentine was hooked to the trigger so that as it was pulled, the match dipped into the powder in the pan. The matchlock did not work like modern triggers, which snap the hammer forward quickly; instead, it moved only as fast as the soldier pulled on the trigger. The great advantage of this firing mechanism was that the gun could be held up to one shoulder and aimed without the hunter or soldier having to take his eye off the target to look for the touchhole.

The next step in the evolution of firing mechanisms was the development of a gun that could self-ignite. One solution that later proved to be impractical was a German gun called the monk's gun. Its firing mechanism was copied from a fire-lighting device already in general use. A serpentine that held a piece of iron pyrites close to the flash pan was constructed so that it pressed down on the roughened surface of a flat **steel** bar. The shooter pulled a handle that was connected to the bar, thus causing the pyrites to create a shower of sparks that fell on the powder in the pan.

In the early 1500s, a wheel-lock mechanism was produced in Germany. It was simpler in design and more steady for the shooter using it. It also had a serpentine with pyrites, but it used a steel **wheel** instead of a roughened bar. The wheel was wound up like a clock with a key. When the shooter pulled the trigger, the wheel rotated rapidly against the stone, thus producing sparks. This wheel lock device was very dependable but expensive.

The expensive manufacturing of the wheel lock led some to devise a better system of firing guns. The snaphance, or *flintlock*, which was developed in Scandinavia and the Netherlands around 1550, used the principle of the flint-and-steel method to start household fires. The flint was held in a vise-like arm called the cock, which was drawn back by hand and locked into a ready position. Pulling the trigger caused the cock to snap forward and strike its flint against a steel plate. This produced a shower of sparks that fell into the priming pan.

Both the French and the English claim to have improved the flintlock in the early 1600s, and in any case the flintlock was in general use by the 1650s. The main advantage of the flintlock was that the soldier or hunter no longer had to open the pan cover by hand. Instead, the flint struck the steel plate and flipped open the cover all in the same movement. For the next two hundred years the simplicity of the flintlock held center stage in firearm technology. One of the most famous flintlocks was the Kentucky **rifle**.

The flintlock had its disadvantages, however. It failed to fire an average of three times in ten shots, and it often failed to explode the main charge in the breech. In 1805 Alexander Forsyth (1768-1843), a Scottish minister and inventor, began experiments to devise the first percussion lock. Realizing that certain chemicals would explode if they were hit sharply with a blunt object, he came up with a potassium-chlorate compound. To make it work, Forsyth installed a small **bottle** of it alongside the barrel of the gun. When the bottle was turned upside down, it poured a small amount of the compound into the flash pan. The hammer then landed on a firing pin after the trigger was pulled, and this set off the charge.

An important variation of the percussion lock, the percussion cap, was created by an English artist living in the United States. In 1816 Joshua Shaw (1776-1860) fashioned a gun with a small steel nipple (replacing the flash pan) that jutted up from the barrel. A small hole in the nipple led to the powder chamber. He designed copper caps, which contained a small charge, that fit over the nipple. When the hammer struck these caps, their charge went off, thus setting off the main powder.

It was a small step from these early percussion caps to the modern bullet. In the early 1800s, Swiss gunmaker Johannes Samuel Pauly (1766-1820?) created a cartridge with a soft brass head containing a depression to hold a pinch of explosive priming mixture. The hammer ignited this mixture. The pin-fire cartridge was developed in France in the 1830s. The cartridge was ignited by the firing mechanism's hammer falling on a pin jutting up out of the end of the cartridge. This was replaced by the center-fire cartridge in the 1860s, in which the pin became part of the gun, and not the cartridge. The primer is placed in the center of the base of the cartridge. This ignites when struck by the firing pin of the hammer. This is essentially the modern bullet.

FISCHER, EMIL HERMANN (1852-1919)

German chemist

Emil Hermann Fischer was born on October 9, 1852 in Euskirchen, Germany. With five older sisters, he was the only son of Laurenz Fischer and Julie Poensgen Fischer. Fischer described his youth as happy in his unfinished autobiography, *Aus meinem Leben (Out of My Life)*. Fischer graduated at the top of his class from the Gymnasium (high school) of Bonn in 1869. He entered the University of Bonn in the spring of 1871.

After less than a year at the University of Bonn, Fischer transferred to Strasbourg where he studied under the noted chemist, **Adolf von Baeyer**. For his doctorate Fischer did research on fluorescein, a coal tar dye that shows a fine yellow-green **fluorescence** in solution, and is used to trace water through systems. Fischer's researches into **coal**, coal tar, and the synthesis of organic chemicals, did much to build the German dye industry.

Fischer received his doctoral degree in 1874 from Strasbourg, but he continued his research on coal tar dyes with a cousin, Otto Philipp Fischer, until 1878. Ultimately he acquired a number of patents for industrially useful chemicals. In 1875 Fischer was invited to follow Baeyer to the University of Munich where Fischer became associate professor of analytical chemistry in 1879. His researches included the discovery of a new compound, phenylhydrazine, a chemical he later used extensively in research on sugars. By 1878 he figured out the chemical formula for phenylhydrazine, and this discovery stimulated other researches leading to the development of such synthetic drugs as **novocaine**. In 1881 Fischer began investigations into a new field, purine chemistry (part of a group of nucleic acids), identifying three amino acids and synthesizing many more. This research resulted in many more advances in the German drug industry.

Fischer left in 1882 to accept the position of professor of chemistry at the University of Erlangen, near Nuremberg. At Erlangen, Fischer continued his work on purines and began to study carbohydrates in 1884. His subsequent work with phenylhydrazine in an unventilated laboratory caused him to suffer the effects of phenylhydrazine poisoning which attacks the kidney, liver, and respiratory system. Upon his recovery in 1885 he accepted a chair in Würzburg. In 1888 Fischer married Agnes Gerlach. They had three sons before she died in 1895. While Fischer was at Würzburg he was honored with a Bavarian medal.

In 1892 Fischer accepted the position of professor in charge of the chemistry department at the University of Berlin, the most prestigious position for an academic chemist in Germany at that time. He was offered full freedom in the construction of a new building at the chemical institute of Berlin, and his subsequent design of a well-ventilated laboratory became a model for university laboratories all over the world. In addition, his teaching methods led to the formation of small groups of students involved in basic scientific research. With the help of his cooperative teams of students, and fellow researchers from many countries, he designed a careful plan for each research project. As the work progressed he always looked for deviations from the expected results. Each unusual occurrence was researched systematically to its conclusion. This strategy permanently influenced both graduate education in chemistry and the expectations of universities for research and publication from their professors worldwide.

Fischer's researches into sugar and purines had proven especially successful. He synthesized about one hundred and thirty purines, which included caffeine, theophylline (used in the preparation of the motion sickness drug Dramamine), and uric acid. In addition, after studying the three-dimensional shapes of sugar molecules, Fischer synthesized glucose as well as about thirty other sugars. By 1899 Fischer finished most of his work on sugars and purines and began research on proteins and enzymes in an effort to identify their chemical nature. Fischer was elected to membership in the Academy of Sciences, and, in 1902, he received the Nobel Prize "for his synthesis in the groups of sugars and purine," as quoted by Eduard Farber in *Nobel Prize Winners in Chemistry*. In 1909 he received the Helmholtz Medal for his work on sugar and protein chemistry.

Determined to keep the preeminent position of world leader in chemical research for Germany, a position he did much to create, he gathered support from industry, government, and other scientists to establish a number of research institutes—the Kaiser Wilhelm Society for the Advancement of Sciences, the Kaiser Wilhelm Institute for Chemistry in Berlin-Dahlem, and the Kaiser Wilhelm Institute for Coal Research in Mulheim-Ruhr. As director of the University of Berlin laboratories he started a radiochemistry laboratory where, years after his death, scientists **Otto Hahn** and **Lise Meitner** worked on research that led to the fission of uranium and the ultimate development of the **atomic bomb**.

World War I took Fischer away from most of his experimental investigations as he redirected his research concentrations toward the war effort. The British blockade would have brought the defeat of Germany by 1915 had Fischer and his colleagues not succeeded in using the resources they had to synthesize much of what they could no longer get on the world market. He led the development of synthetic saltpeter (potassium nitrate) and nitric acid, both used in the manufacture of explosives. As food became in short supply he coordinated research and production of synthetic **fertilizers**. Fischer directed research to replace diminishing supplies of camphor (used to stabilize gunpowder) and pyrites which supplied sulfur for explosives.

Research alone could not win the war, and not all of Fischer's projects were successful. It was obvious to Fischer that Germany would be defeated. In an effort to organize the re-

John Fitch

building of chemical research and industry in Germany to gain back the leadership it had before the war, Fischer and a friend made plans to form the German Society for the Advancement of Chemical Instruction.

Fischer lost his two younger sons in the war, which left him depressed, and he also was suffering from cancer. Emil Fischer died in Berlin, July 15, 1919. His remaining son, Hermann Otto Laurenz Fischer (1888–1960) went on to become a Professor of Biochemistry at the University of California in Berkeley. On October 9, 1952, Fischer's son dedicated the Emil Fischer Library at the University of California which is the repository of the collected works of Fischer, including the manuscript for his autobiography, research files, and Fischer's correspondences in World War I.

Fitch, John (1743-1798)

American inventor

As a child Fitch had few years of schooling and suffered from a harsh father. He apprenticed himself to a **clock** maker, learned **brass** working, and later opened his own brass foundry. However, due to inefficient business practices he was unsuccessful in both this endeavor and as a silversmith. He did make some money managing a gun factory during the Revolutionary War, but the colonial currency became worthless, and he passed the last part of the war as a British prisoner.

Settling in Pennsylvania after the war, Fitch concentrated on inventing a steam-powered boat. Using a working model, he was granted a fourteen-year monopoly from five state governments to operate his craft on their waters and secured the backing of a group of Philadelphia investors. On August 22, 1787, he successfully demonstrated a 45-foot (13.7 m) boat on the Delaware River before delegates to the Constitutional Convention. This craft was propelled by six paddles on a side like an Indian canoe but was driven by a **steam engine**. The next year he built a 60-foot (18 m) paddle wheeler with paddles that moved like ducks' feet, and in 1790 an even larger boat was launched, one which for a time maintained a regular schedule of trips between Philadelphia and Trenton. However, the number of passengers, never great in the first place, dwindled, making the route a money loser. His backers then quit, and a fourth boat was wrecked in a storm in 1792.

Fitch later attempted to spread his ideas in Europe. He secured a French patent in 1795, but the French Revolution prevented him from continuing business in that country. The project failed, and he had to return home as a seaman to pay for his passage. Although in poor health, Fitch continued to pursue backers but to no avail. He died in 1798, nine years before **Robert Fulton** repeated his work and received credit for the invention of the **steamboat**.

See also Morey, Samuel

Flame thrower

As a weapon, fire was limited in its effectiveness by the difficulty of controlling and delivering it. The Chinese are credited with the first controlled wartime use of fire about 500 B.C. Other ancient peoples used pots full of burning oil to defend city walls against invaders attempting to scale them. Around 671 A.D., the Byzantines successfully defended Constantinople by using **Greek fire** against the Arabs. For centuries, the ingredients of Greek fire were kept a closely guarded secret, but eventually the Arabs were able to acquire it. The Middle Ages also saw the use of fire arrows, which were shot into beseiged castles and towns.

Like many other devices, fire had advanced as a weapon in the twentieth century. The first modern flame thrower was invented in 1901, and was used by the Germans during World War I. It consisted of a pack containing a liquid agent worn by the soldier, a hose connected to the liquid agent, and a nozzle that measured the amount of liquid released. The liquid was squirted out of the hose and ignited by an attached spark device. The British army in World War II carried an effective flame thrower known as the Ack Pak that was capable of burning continuously for ten seconds. It was equipped with a cylinder of inert gas (usually nitrogen) that provided the pressure to propel petroleum gel a distance of 150 feet (46 m). A magnesium cartridge lit the fuel to produce the jet of flame. In addition, this device could shoot unlit fuel onto a target, which could be ignited later. The British also developed an **armored vehicle** with flame-throwing capability—the Crocodile. Its main gun remained intact, but the **machine gun** was replaced by a flame thrower that could shoot out to about 450 feet (137 m).

The flame-thrower does not seem to have been significant as a weapon after World War II. Its corollary might be **napalm**, a petroleum gel with an aluminum-based additive, that ignites when dropped from planes and burns even under water. It was used by the United States armed forces against North Korea and Vietnam.

See also Napalm

FLAMSTEED, JOHN • See Chronometer

FLASH BULB • See Photography

FLASHLIGHT

Now a common household item, the lowly flashlight was once considered a novel toy. When it was introduced in 1898 at an electrical show in New York, the flashlight weighed more than six pounds, and its **battery** alone was half a foot long. Although patents for the device were issued to American Electrical Novelty and Manufacturing Co. in the 1890s, no single person has laid claim to its invention. The inventor of the toy electric train, American Joshua Cowen (1880-1965), built a prototype flashlight, but used it merely to illuminate potted flowers. When flashlights were first sold to the public, they were regarded as amazing toys rather than practical appliances. Today's lightweight, powerful flashlights are considerably more convenient and useful, thanks to improvements in **light bulb**s, batteries, and controls. In most flashlights, which use incandescent electric light bulbs, the light is focused into a narrow beam by a reflector and a **lens**. Small fluorescent models are also available, and some flashlights use extremely brilliant **arc lamp**s that can illuminate objects in darkness half a mile (.8 km) away. Flashlight bodies can be twisted to fit around corners, worn on bands around the forehead, or hung from tent poles as they become more versatile, lighter and brighter. At Georgia Tech Research Institute scientists are making in **radar** flashlights that use in **microwave** technology to detect a human presence through walls that are as much as four feet away.

See also Battery, electric

FLEMING, JOHN AMBROSE (1849-1945)

English engineer

John Ambrose Fleming was the common thread that linked the work of three individual geniuses, yet every one of those three now overshadow him. Fleming, born on November 29, 1849, in Lancaster, England, was the son of a Congregational minister. He attended University College in London, England, graduated in 1870, and taught science for seven years. In 1877, Fleming entered Cambridge University to work for the brilliant Scottish physicist James Clerk Maxwell (1831-1879), who had established equations describing the behavior of electricity and magnetism. Unfortunately, Fleming's association with Maxwell was a brief two years; Maxwell died at age 48 from cancer.

Three years later Fleming became a consultant for the Edison Electric Light Company in London. In 1885, he was appointed professor of electrical technology at University College, where he remained for forty-one years. There he devised the "right-hand rule" which became an easy way to remember the relationship between the direction of a magnetic field, the motion of the conductor, and the resulting electromotive force. During his long tenure at University College, Fleming experimented a great deal with wireless telegraphy.

Having worked with **Thomas Alva Edison**, Fleming became intrigued by a discovery that had been made in 1884. In the process of studying his light bulbs, Edison inserted a metal plate near the filament. He discovered that electricity would flow to the plate when it was hooked to the positive terminal of the bulb, but not to the negative terminal. This "Edison effect" was a curiosity for which he had no explanation; in reality he had unwittingly invented the first **vacuum tube**, which eventually came to be called the **diode**. Its ability to convert **alternating current** to **direct current** was ignored by Edison, who instead patented the device for use in controlling electric **generator**s.

Meanwhile, Fleming had become consultant to the Marconi Wireless Telegraph Company, and helped to design the transmitter that **Guglielmo Marconi** used in his 1901 trans-Atlantic broadcast. Later Marconi expressed interest in devising a more efficient method of amplifying **radio** signals. **Karl Ferdinand Braun** discovered in 1874 that some crystals had the ability to transmit electricity better in one direction than another. These **crystal rectifier**s could be used to convert the alternating current generated by radio waves into direct current for amplification, but the crystals were inefficient at higher frequencies. Following the 1896 discovery of the electron in 1896 by English physicist Joseph J. Thomson (1856-1940), it became clear that the metal plate had the ability to absorb hot electrons in a vacuum. Fleming saw that this tube could more efficiently do what Braun's rectifier did. In 1904, Fleming designed a receiver for Marconi. At first he essentially used Edison's patented device, but Fleming's circuit had an entirely different purpose. Fleming applied for his own patent on November 16, 1904.

Fleming named his invention the thermionic valve, the ancestor of all electronic tubes, because it controlled the flow of electricity just as a valve controls the flow of water. In the United States the invention was called a vacuum tube, which better described its construction. In 1906, American scientist **Lee de Forest** improved on Fleming's invention, but ran afoul of the Marconi Company which owned Fleming's patent. The recipient of many honors, Fleming had an unusually long and active life. He died on April 18, 1945, at the age of ninety-five, after having witnessed radio's growth from infancy to maturity as the world's major form of communication.

FLIGHT DATA RECORDER

Flight Data Recorders (FDRs) were developed in the 1950s, shorly after the creation of the Federal Aviation Administration (FAA). The formation of the FAA was spurred in large part by the disastrous collision of two commercial **aircraft** over the Grand Canyon in 1956, in response to the obvious need for greater control and regulation of safety measures in the nation's increasingly crowded skies. One of the requirements developed by the FAA was that planes be equipped with FDRs, in crash-proof boxes, that would record the history of the flight and the various control inputs made to the plane during the flight. Following a crash, the FDR information could be analyzed to determine what had gone wrong, and to allow recommendations for ensuring the accident did not recur. The FDR quickly acquired the somewhat ominous nickname ''black box,'' which is still in use today.

Several variants on the original FDR have been developed, mostly in response to FAA rule changes. As planes increased in sophistication, the amount of data needed to properly analyze an accident increased as well, soon outstripping the capability of older FDRs. The new digital FDRs (DFDRs) and Flight Data Acquisition Units (DFDAUs) are standard issue on modern commercial airliners, recording a dizzying variety of data. Combined with the transcripts recovered from the cockpit voice recorders, the National Traffic Safety Board (NTSB) can now conduct exhaustive analyses of accident circumstances. Even so, so accidents go withou adequate explanation, such as in the crash of a Boeing 737 outside Pittsburgh in 1994.

Modern FDRs record flight data information digitally on a repeating tape. Up to about 25 hours of data can be stored, more than enough for even the longest individual flights. Once the tape is full, recording simply beings again at the start of the tape, overwriting the old information. There is therefore a continuous 25-hour history of the plane's performance for investigators to sift through. This history contains such essential flight characteristics as the plane's airspeed and altitude, its vertical speed (rate of ascent or descent), its heading, its position along each flight axis (roll, pitch, and yaw). It also includes the status of the plane's hardware at each instant, e.g., the position of its control surfaces (ailerons, elevator, and rudder, which control the movements along the three flight axes), and engine thrust. All events are recorded along with the exact time at which they occurred. The result is a superb performance record that has repeatedly been used to institute both safety alterations to existing airplanes, maintenance practices, and design modifications to new planes.

Flight simulator

Flight simulators play an essential role in maintaining the exceptional level of safety that commercial airlines offer the traveling public. Modern flight simulators are available for all types of commercial **aircraft**, with mock cockpits that include functional analogs of every control in the real cockpit, and simulator time is an required component of every airline transport pilot's training.

Flight simulators developed from the obvious desire to provide a training environment for student pilots that did not necessitate actually being in a plane. Early airline pilot training, in the 1940s, did require the use of actual planes, which was both expensive and provided a less-than-ideal practice environment.

The simplest type of simulator is a cockpit procedure trainer, which is simply a replica of the cockpit of a particular type of plane, in which a student can be taught the location and use of the instruments and controls. These CPT's are still in use today for general training.

Full-scale flight simulation had to await the advent of **digital computers** powerful enough to simulate the many events that occur during flight. Since the 1970s, these machines have developed in complexity until they were effectively able to simulate every aspect of normal and abnormal flight. Especially important was the ability to place pilots in dire emergencies—situations that would likely result in fatal accidents without proper response from the cockpit—and this emergency training was probably the single most beneficial aspect of modern flight simulation. In addition, the flight instructor could tailor the simulator's programs to suit the identified weaknesses of the student. This would be impossible in a real plane, since the instructor would have little control over real-world situations, as would certainly not wish to create emergencies while actually airborne.

The modern flight simulator is an enclosed space that is roughly the size of the cockpit it is meant to simulate. It is mounted several feet off the floor on movable supports, and these supports can move the enclosure to simulate the actual motion of the airplane in response to the pilot's control inputs. Images are generated outside the ''window'' of the simulated cockpit, simulating different types of aviation situations from takeoff to landing. A pilot inside the simulator sees the flight situation projected outside the window exactly as it would appear from the cockpit of the real plane. The simulator can create any type of emergency situation, from failed engines or landing gear, to instrument failures, to weather-related problems such as severe turbulence or icing. Simulators can be programmed to replicate operations in the vicinity of particular cities that the given airline serves. Pilots in the simulator are closely monitored by trainers and their responses to different situations evaluated. All flight simulators used by airlines are FAA-approved, and pilots are required to use them. The expense is substantial—a new simulator can cost as much as $20,000,000—but the steady improvement in airline safety records can be attributed at least in part to the realism of the ground-based training pilots now receive.

Float glass • See Glass

Flowers, Thomas H. • See Colossus

Fluorescent lamp

You might think that the everyday light bulb is the most common type of lighting today, but actually—if you count all the offices and factories in the United States—the fluorescent lamp outshines the **incandescent light bulb** as our main source of electric lighting. The introduction of the fluorescent lamp in 1936 was staged with great drama at the U.S. Patent Office's 100th anniversary celebration, where more than 1,000 people were gathered in a hotel ballroom. Just when the names of America's twelve greatest inventors were being announced, the room was flooded with brilliant fluorescent light. The invention of the lamp resulted from the work of many different scientists over a long period of time, beginning in the 1600s with the discovery of a substance that would glow in the dark after exposure to the sun. The visible glow produced by materials exposed to ultraviolet rays was later named fluorescence by Sir George Gabriel Stokes, who studied various fluorescent substances.

In 1855 a German glassblower, **Johann Heinrich Wilhelm Geissler**, invented a **vacuum pump** that could extract most of the air from a tube. Once these **vacuum tube**s became available, scientists were able to pass electricity through different gases at very low pressures inside the tubes. The electric current would cause certain gases to discharge light, producing a glow. This led to the invention of the mercury vapor discharge lamp in 1901 by Peter Hewitt, an American engineer.

Although the early mercury vapor lamps were quite successful, they produced a high proportion of ultraviolet light, which is invisible to the human eye. Scientists already knew that fluorescent chemicals called phosphors would convert ultraviolet rays to visible light. As early as 1859, **Antoine-Henri Becquerel** had built a primitive fluorescent lamp by coating the inside of a vacuum tube with a phosphor. During the 1920s and 1930s, laboratories began formulating more durable, efficient phosphor coatings. Engineers also improved the mercury vapor lamp by reducing its operating voltage. The first practical fluorescent lamp was developed by a team of American scientists at General Electric's laboratory around 1934. Soon after its dramatic introduction, the fluorescent light was promoted at the 1939 New York World's Fair and at an exposition in San Francisco. Various colors and sizes were offered, and sales were healthy. Then when America joined World War II, factories began working at full capacity to supply troops, and sales of fluorescent lighting multiplied 100-fold.

Since then, many technical advances have been made, leading to longer life and higher efficiency. Fluorescent lamps now supply two-thirds of lighting demand worldwide. By using different phosphor-coating formulas, a greater variety of shades and colors have been created for different uses. For example, the *warmest* shades of white, which contain the most red light, are desirable for grocery meat counters, where they make the meat look more appetizing. However, fluorescent light is still somewhat glaring, and the incandescent lamp will continue to reign in people's homes until fluorescent models can produce a cozier, more flattering light.

Recent years have seen great interest in the health effects of fluorescent lighting. In the 1950s John Ott, a banker and photographer, noticed while shooting time-lapse photography for Walt Disney that pumpkin seed sprouts would not fully mature under fluorescent lights. After studies of various laboratory animals, he initiated careful studies in the 1970s of first-grade school children in Florida. Ott determined that students who worked under ordinary cool-white fluorescent lighting demonstrated fatigue and attention problems, but that their behavior noticeably improved if they worked under full-spectrum fluorescent lights, which give off more ultraviolet light than the standard cool-white fluorescent lights. It was even found that students in classrooms with full-spectrum fluorescent lights had fewer dental cavities than those working under cool-white fluorescent lights. Some schools and businesses have begun installing full-spectrum fluorescent lighting in their rooms in the belief that it leaves workers less tired and able to pay better attention to their work.

See also Arc lamp

Fluoride treatment, dental

Fluoride is a chemical found in many substances. In the human body, fluoride acts to prevent tooth decay by strengthening tooth enamel and inhibiting the growth of plaque-forming bacteria. After researchers discovered this characteristic of fluoride, *fluoridation*—the process of adding the fluoride to public water supplies—began.

It all started with Frederick S. McKay, a Colorado Springs, Colorado, dentist, in the early 1900s. McKay noticed that many of his patients had brown stains, called "mottled enamel," on their teeth. McKay set out to find the cause, helped by researcher Greene V. Black (1836-1915) of Northwestern University and other dentists. By 1916, McKay believed the mottling was caused by something in the patients' drinking water. By 1928, he concluded that mottling was linked to reduced tooth decay.

In 1931, at the suggestion of an Alcoa chemist in Bauxite, Arkansas, McKay verified that drinking water from places with a high degree of tooth mottling contained unusually high levels of naturally occurring fluoride. H. Trendley Dean, a dentist with the United States Public Health Service, also studied the connection between mottling and fluoride in the 1930s. By the early 1940s, Dean and his research team had established that one part per million was the ideal level of fluoride in drinking water, substantially reducing decay while not causing mottling.

Following safety tests on animals, the Public Health Service conducted field tests. In 1945 the public water systems of Newburgh, New York, and Grand Rapids, Michigan, became the first ever to be artificially fluoridated with sodium fluoride. Simultaneously, a group of Wisconsin dentists led by John G. Frisch inaugurated fluoridation in their state. Results of these tests seemed to show that fluoridation reduced dental cavities by as much as two thirds. Based on those results, the United States Public Health Service recommended in 1950 that all United States communities with public water systems fluoridate. Later that year the American Dental Association (ADA) followed suit, and the American Medical Association added its endorsement in 1951.

Even though virtually the entire dental, medical, and public health establishment favored fluoridation, the recommendation was immediately controversial and has remained so. Opponents objected to fluoridation because of possible health risks (fluoride is toxic in large amounts) and concerns about being deprived of the choice whether or not to consume a chemical. While referenda have blocked fluoridation in a number of communities, nearly 60 percent of people in the United States now drink fluoridated water. Fluoridation is also practiced in about thirty other countries.

The initial claims that fluoridation of drinking water produced two-thirds less tooth decay have been modified to about 20 to 25 percent reduction. Other ways of applying fluoride have been developed. In the 1950s Procter and Gamble had the idea of adding the chemical to toothpaste. First, researchers at Indiana University had to find a way to keep stannous fluoride from bonding with toothpaste abrasives. Once this problem was overcome, Procter & Gamble introduced its new ''Crest—with Fluoristan'' in 1956, launched with an advertising blitz that included the popular line ''Look, Mom—no cavities!'' Four years later, P&G scored a coup when the ADA Council on Dental Therapeutic gave Crest its seal of approval as ''an effective decay-preventive dentifrice.'' The ADA now estimates that brushing with fluoride-containing toothpaste reduces tooth decay by as much as 20 or 30 percent. In addition to toothpaste, fluoride can also be taken in tablet form and as a solution either ''painted'' directly onto the teeth or swished around as a mouthwash. Recently, a team of British scientists have been working on developing a vaccine to prevent tooth decay, which could one day serve as an alternative to fluoridation. A plant-based substance, the vaccine is ''painted'' on the teeth and prevents tooth decay by eliminating bacteria from the mouth.

See also Toothbrush and toothpaste

Flying buttress

A buttress is an arrangement of stone or **bricks** projecting from the face of a wall to provide it with sideways support. A *flying buttress* is one in which the buttress does not make contact all the way up the wall.

In the construction of cathedrals, large tower-like ''piers'' (sometimes also referred to as buttresses) were erected next to the cathedral walls. Flying buttresses then extended from these piers to the cathedral itself—thick, sloping, half-arch-like structures. With this design, the flying buttresses provide a upward and inward force on the columns of the cathedral. Inside the cathedral, the roof is a vault, with ribs extended along the vault for support. The upward, inward force of the flying buttress supports the downward force of the ribbed vaults.

Flying buttresses are part of the Gothic style of architecture, born in France during the twelfth century. It was a time

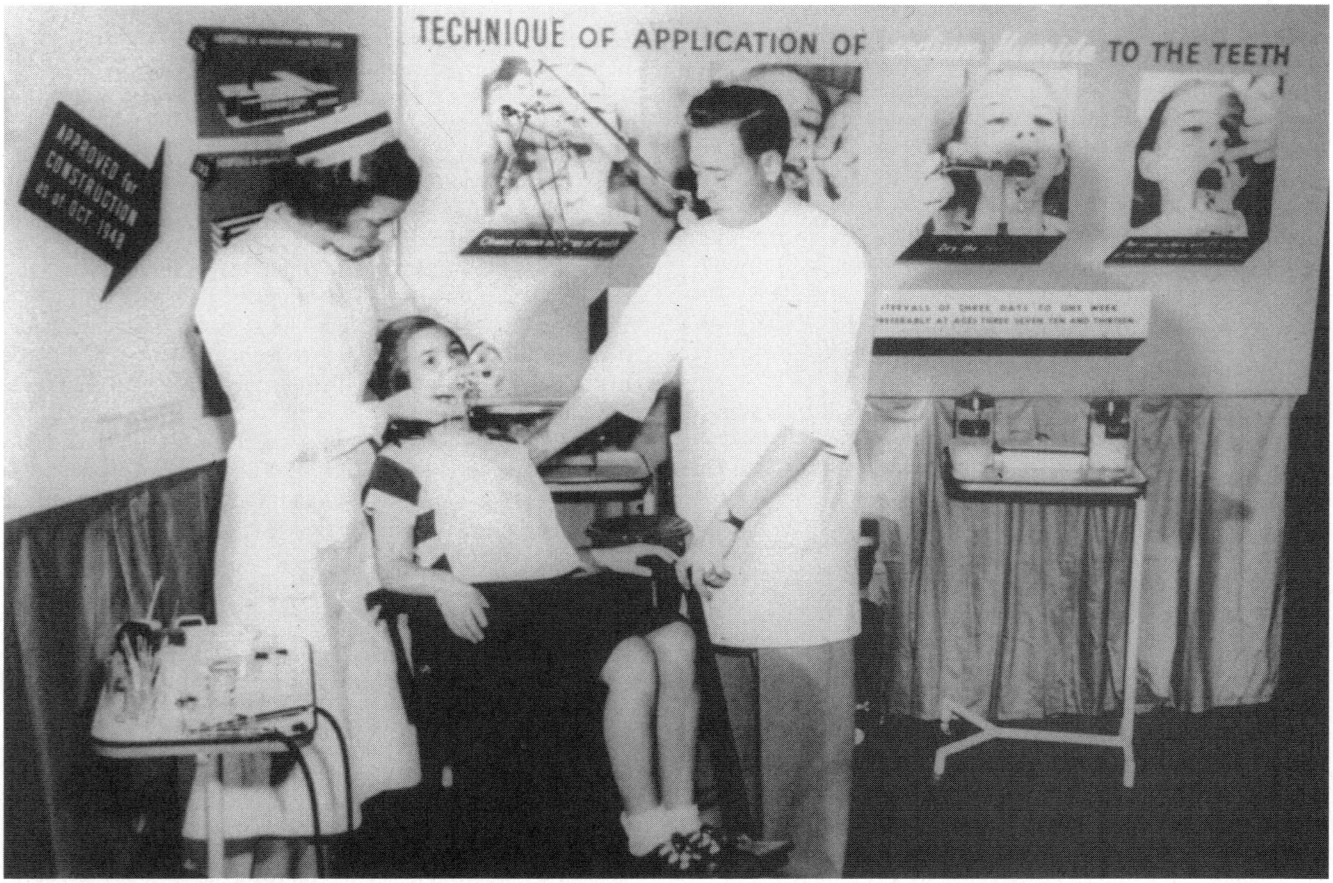

Fluoride treatments became standard in the 1950s.

in which churches and cathedrals were centerpieces of society, often spectacular and beautiful and at the forefront of architectural development. At first the support piers were internal or inside the structure's walls. But competition for the highest vaulting lead to piers of great size, which not only had to support the weight of the vaulting but also support the tremendous sideways forces coming from such great heights. This required the piers to be thick at the bottom, and eventually they were simply placed outside the walls, to which the flying buttresses were attached. In this way the spherical vaulting was kept from pushing apart from its own weight.

The first appearance of rows of fully-developed flying buttresses was at the cathedral known as Cluny III in France around 1130 A.D. Flying buttresses were not only functional, but, as the Gothic architectural style progressed, were often elaborate and majestic when set in rows alongside a cathedral, conveying an image of delicacy. When it was found that if a pier was weighted on top it did not have to be as massive at the base, elaborate pinnacles came to top the piers, ornamental in appearance but mostly serving as additional weight.

Flying shuttle

It is generally considered that the world's first Industrial Revolution was begun in the British **textile** industry, and was stimulated by three inventions: **James Hargreaves' spinning jenny**, **Samuel Crompton**'s **spinning mule**, and John Kay's (1704-1764) flying shuttle. Chronologically, the shuttle came first, and it was the use of this device that forced the spinning industry to become more productive and efficient, eventually leading to the mechanization of all spinning and weaving.

Before 1733, most weaving was done by a single weaver sitting at a large **loom**. With one hand the weaver strung horizontal *weft* thread through vertical *warp* threads, beating down each successive layer of weft with the other hand. The weft U thread was attached to a thin wooden slat called a shuttle, which the weaver passed from one hand to the other. This limited the size of the fabric, for it could only be as wide as the distance between the weaver's hands. For wider cloth, two weavers were a employed to throw the shuttle back and forth.

Kay, the twelfth child of poor farm laborers, came up with the idea for an ''automatic'' shuttle. Mounted on small **wheels**, this shuttle sat in box at one end of the loom. Using a stick called the picking peg, the weaver pulled a cord that would send the shuttle across to a box at the loom's other end, winding the weft through the warp as it went. The weaver could then send the shuttle back the other way by pulling a second cord. Thus, the weaver could create very wide cloth without leaving his seated position, all the while keeping one hand free to beat and manipulate the weft threads.

Like the inventions of Hargreaves and Crompton, Kay's flying shuttle was not welcomed by the local hand-weavers, who foresaw their imminent obsolescence. It took fourteen years for Kay to construct a reliable loom, during which time he endured the hostility of the town weavers. Finally, in 1747, he fled England for the safety of France, where he died in poverty. However, his son Robert continued to improve upon his father's invention, creating a tiered-boa system that could hold up to four shuttles; by selecting from a variety of pull-cords, the weaver could choose weft threads of several colors and textures.

Because the flying shuttle required only one weaver to quickly and easily create very wide cloth, its use drastically increased the demand for thread, putting pressure on the spinning mills. This demand led to the invention of the **water frame** spinner, the spinning jenny, and the spinning mule—which, in turn, put even greater pressure upon the weaving industry, forcing the invention of power looms. By the mid-1800s, the entire British textile industry had become mechanized.

Fodder and silage

A large percentage of crops are grown as feed for livestock. Grains, grasses, legumes, and other plants not desirable for human consumption are ideal feeds for cattle, horses, sheep and other animals. When animal eat hay or pasture grass, the ingestion of air and coarse material causes bloating. Therefore, farmers prefer to keep feed material in silos to allow it to ferment into silage, a more palatable and nutritious substance. For example, sorghums, which are poisonous when directly consumed, must be fermented. Silo storage also provides feed through months of cold or drought. The first improvement in the production of *fodder*—straw chopped small enough for livestock consumption—came with the invention of mechanical straw, or chaff, cutters during the 1800s. Perhaps the first of these was the Hotchkiss guillotine straw cutter of 1808.

The first silo was built in the United States in 1873. By the 1890s, most farmers, especially dairymen, had silos on their farms, since high quality feed increased the quality and quantity of milk production. Originally constructed of wood, which was not very air or water tight, silos were later made of stone, **brick**, or **concrete**. Corrugated metal has also become a popular building material for silos and other types of storage facilities. In the absence of silos, some farmers store fodder in water-tight pits. The ensilage chopper, which combined silage cutting with the corn picking process, was developed in 1915. An improved version was introduced in 1928. Another machine used for silage production is the tedder, developed in the 1870s, which fluffed hay for drying prior to storage. This was an important development because hay, when stored wet, rotted in the silo, making it unusable. By 1914 the side delivery rake became the preferred method for drying hay. Today, 90 percent of silage is chopped in the field. The principle silage crop of the last half of the twentieth century is corn, making up three-fourths of the total in the United States.

Food additives • See Food preservation

Food preservation

All foods begin to deteriorate, or spoil, as soon as they are harvested or slaughtered. Most spoiling is caused by microorganisms such as bacteria or mold, or by chemical changes within the food itself due to enzyme action or oxidation. The purpose of food preservation is to stop or slow down the spoilage. Ways of preserving food have been practiced since ancient times, although knowledge of what actually causes food to spoil has been gained only in modern times. Ages-old food preservation techniques include drying, salting, smoking, fermenting, pickling, cooling, and freezing. Modern methods include canning, mechanical **refrigeration** and freezing, the addition of chemicals, and irradiation.

One of the most ancient methods of food preservation is sun- or air-drying. Drying removes much of the food's water; without adequate water, microorganisms cannot multiply and chemical activities are nearly halted. Dried meat was one of the earliest staple foods of hunters and migratory peoples. Once fire was discovered, prehistoric cave dwellers heat-dried meat and fish, which probably led to the development of smoking as another way to preserve these foods. The Phoenicians air-dried quantities of fish. Ancient Egyptians stockpiled dried grains. North American natives produced a nutritious food called *pemmican* by grinding together dried meat and rendered fat.

Early northern societies quickly learned that coolness as well as freezing helped preserve foods (although they did not know why: microbial growth and chemical changes slow down at low temperatures and completely stop when water is frozen). Pre-Columbian natives in Peru and Bolivia freeze-dried potatoes, while the early Japanese and Koreans did the same to fish. Water evaporating through earthenware jars was used as a coolant in 2500 B.C. by Egyptians and Indians. Ancient Chinese, Greeks, and Romans stored ice and mountain snow in cellars or ice-houses to keep food cool.

Salting, which also inhibits microbial growth, was a preferred method of preserving fish as early as 3500 B.C. in the Mediterranean world, and was also practiced in ancient China. Other substances besides salt were found to retard food spoilage. The Chinese began using spices as preservatives around 2700 B.C. Ancient Egyptians used mustard seeds to keep fruit juice from spoiling. Jars of fruit preserved with honey have been found in the ruins of Pompeii, Italy. Melted fat—as Native Americans discovered with *pemmican*—preserved meat by sealing out air.

Fermentation was particularly useful for people in southern climates, where cooling and freezing were not options. When a food ferments, it produces acids that prevent the growth of organisms that cause spoilage. Grapes, rice, and barley were fermented into alcoholic beverages by early people. Fermentation also was used to produce **cheese** and **yogurt** from milk. Pickling—preserving foods in an acid substance like vinegar—was used during ancient times, too.

By the Middle Ages, all these ancient methods of preserving foods were widely practiced throughout Europe and Asia, often in combination. Salted fish became the staple food of poor people during this time—particularly salted herring, introduced in 1283 by Willem Beukelszoon of Holland. As the modern era approached, the Dutch navy in the mid-1700s developed a way of preserving beef in iron **cans** by packing it in hot fat and then sealing the cans. By the late 1700s, the Dutch also were preserving cooked, smoked salmon by packing it with hot butter or olive oil in sealed cans.

These traditional spoilage-prevention techniques remained adequate for most purposes until the advent of the Industrial Revolution. As populations became concentrated in ever-growing cities and towns, other methods were needed to preserve food reliably for transportation over long distances and storage pending distribution. The crucial development was the invention of canning techniques during the 1790s by the Frenchman **Nicolas François Appert**, who set up his first cannery south of Paris in 1804. Appert's method, which used **bottles**, was greatly improved by the 1810 invention of the tin can in England. First used for Arctic expeditions and by the military, canned foods came into widespread use among the general population by the mid-1800s.

Not all foods could be successfully canned; reliable methods of refrigeration were needed. Ice-houses were first used to store ice cut from frozen ponds and lakes. Among the inventors of early ice-making machines were Jacob Perkins (1766-1849) and John Gorrie (1803-1855). These machines led to the development of large-scale commercial refrigeration of foods for shipping and storage. **Clarence Birdseye** introduced tasty quick-frozen foods in the 1920s. Shortages of canned goods after World War II helped boost the popularity of **frozen foods**.

Modern methods of drying foods began in France in 1795 with a hot-air vegetable dehydrator. Dried eggs were widely sold in the United States after 1895, but dried food was not produced in volume in the U.S. until it was used by soldiers during World War I. Military need during World War II led to the development of dried skim milk, potato flakes, **instant coffee**, and soup mixes. After the war, freeze-drying was applied to items such as coffee and orange juice, and the technique continues to be applied to other foodstuffs today.

Chemical additives are now commonly used to prevent spoilage in food. Common preservatives that inhibit microbial growth include benzoic acid, sorbic acid, and sulfur dioxide. Antioxidants like BHA and ascorbic acid (vitamin C) prevent compounds in food from combining with oxygen to produce unpalatable changes. Between 1925 and 1959, the African-American chemist **Lloyd A. Hall** developed superior antioxidants and meat-curing salts that greatly improved many food products. The great proliferation of often unnecessary and sometimes harmful chemical additives to food during the late 1800s led to governmental regulation—the British Adulteration of Food and Drugs Act of 1875 and the U. S. Food and Drugs Act of 1906, among others.

Aseptic packaging, a relatively new way to keep food from spoiling, first sterilizes a food product and then seals it in a sterilized container. Aseptic packages, including **plastic**, **aluminum foil**, and **paper**, are lighter and cheaper than the traditional metal and **glass** containers used for canning. Aseptically processed foods are sterilized much more quickly than foods in cans or bottles, so their flavor is better. Aseptic packaging became commercially available in 1981. However, controversy has developed about the environmental consequences of these disposable containers.

Irradiation is another relatively new method of preservation that treats food with low doses of radiation to deactivate enzymes and kill microorganisms and insects. Although public concern about the technique persists, certain irradiated foods, such as poultry and produce, have been sold in the United States in the early 1990s. In 1997, the U.S. Food and Drug Administration approved the irradiation of beef, pork, and lamb. Worldwide, by the late 1990s, food irradiation had been approved by 37 countries for more than 40 food products. Proponents of the technique note that irradiation does not make food radioactive, nor does it change a food's nutrient content or flavor.

FOOTBALL

The vastly popular game of American football has its origins in the English sports of rugby and **soccer**. In the mid-1800s, a soccer-like game was popular in the eastern United States. The object of the game: kick a round ball across the other team's goal line. Teams often consisted of thirty or more players. As the game's popularity increased, stricter rules were created and schools organized teams. The National Collegiate Athletic Association contends that the first football game was played in 1869 between Rutgers and Princeton Universities, but the rules of that game make it seem more like a soccer match. The first official football game took place on May 14, 1874 between Harvard and McGill Universities. As more and more colleges began to play football seriously, blocking and tackling were introduced, and these skills soon became as important as running. Rules varied from game to game and teams simply agreed on them before playing. As running, blocking, and tackling became more crucial to football, so did strength and agility. Games became much more violent and, since no helmets or paddings were worn, injuries abounded. In 1905, President Theodore Roosevelt (1858-1919) threatened to ban the game from the United States after seeing a particularly gruesome photograph of an injured player. Several colleges prohibited football because of just such injuries. In 1906, when forward passing became allowed, games became less violent and more strategic, necessitating careful planning and organization.

FORCEPS, OBSTETRICAL

Forceps are a device used to help in delivering a baby during a difficult childbirth. Forceps resemble two enlarged spoons joined by a handle. The curved blades fit around an unborn baby's head, like two hands. When locked in place, they are used to gently pull the head—followed by the rest of the body—out of the birth canal. The practice of obstetrics took a giant step forward with the invention of forceps. Before forceps were available, the prospects were grim for a woman unable to deliver her baby normally: a surgeon might crush the baby's head or dismember the unborn child and then have to remove the remains from the birth canal. Forceps provided a welcome alternative and were a great improvement over the hook and loop devices previously used.

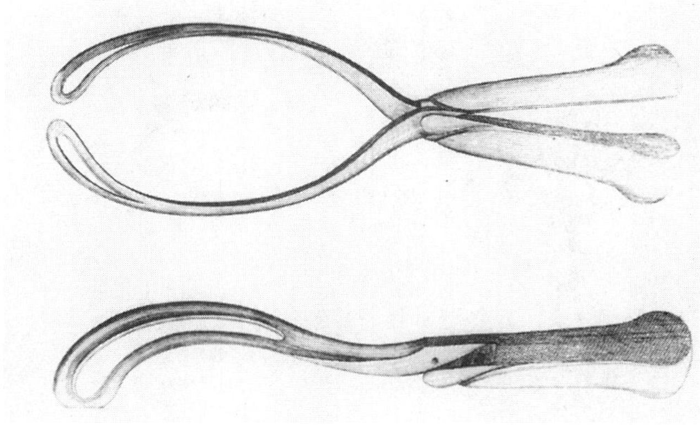

The Chamberlens, a family of barber-surgeons, invented forceps in the seventeenth century and kept the device a closely guarded secret for over 100 years.

However, forceps remained a family secret guarded for well over 100 years after their invention by a family of barber-surgeons and doctors. Dr. William Chamberlen, a French Huguenot, established the family dynasty of male midwives in London, England, in the late 1500s. His son, Peter the Elder (1560-1631), probably was the forceps' inventor. The Chamberlen family went to great lengths to surround their birth device with an aura of mystery. Two men carried it into the birth chamber in a large carved and gilded chest. The door was locked, the woman was blindfolded, and waiting relatives outside heard sounds of bells and rattling chains. The Chamberlens prospered greatly from their invention. By the 1730s, however, the last of the Chamberlens had either let the secret escape or sold it to an outsider. Thereafter, forceps were widely used until **cesarean section** became a safe choice for difficult deliveries.

While minor modifications have been made to forceps over the years, the instrument remains essentially the same as it was in the Chamberlens' time. Once forceps came into general use, they helped transform childbirth in another way. Used primarily by male midwives and doctors, the instrument was shunned by female midwives, who were later prohibited from using it. Because of this, the use of forceps promoted the shift from female-assisted to male-directed childbirth.

FORD, HENRY (1863-1947)

American engineer and industrialist

The name Henry Ford is synonymous with the **automobile** and the automobile is synonymous with Detroit. Yet Henry Ford neither invented the automobile nor built his cars in Detroit.

Henry Ford was born on the family farm in Dearborn, Michigan, a town eight miles west of Detroit, on July 30, 1863. It was in Dearborn that Henry Ford set up his manufacturing complex and built the cars that carried his name, making 'Ford' a household word the world over.

Ford was ever the tinkerer; while still a young man, he built his own **steam engine**, coupled it to an old mowing machine and built his own **tractor**. While working for the Detroit Edison Company, he built several gasoline-powered carriages

Henry Ford

in his home workshop, selling them to finance his next "horseless carriage." In 1899, he formed the Detroit Automobile Company, which was later renamed the Henry Ford Company. He left the Henry Ford Company in 1902 (this early namesake company later became the Cadillac Motor Car Company).

In 1903, Henry Ford incorporated the Ford Motor Company and began building cars, despite the fact that he was being sued for patent infringement. An industrial combine, called the Association of Licensed Automobile Manufacturers, which held the patent for gas-powered autos granted in 1895 to George Baldwin Selden (1846-1922), had taken Ford to court on the grounds that he had refused to buy a license to manufacture automobiles from them. Ford eventually won the case in court and with it the right to build autos with no regard for the Selden patent. His fight made Ford immensely popular, and his Model T was by far the world's most popular automobile. In fact, the Ford Motor Company's Model T accounted for one-half of the world's output of cars during its nineteen years of production (1908-1927). The Model T's success with the public was due to its dependability and low price, but Henry Ford's success as an industrialist was due to his innovative **mass production** techniques.

Henry Ford did not invent the principles of mass production—they were in use for nearly a century before the Model T—but Ford had the vision to apply mass production techniques to the manufacture of automobiles. Before Ford adapted mass production, assembly-line techniques to the building of Model T's, each car took twelve and one-half hours to build. After the inventor's innovations were in place, the time required to build each car was reduced to a little more than an hour and a half, and the cost per unit was lowered as well—from $950 for the first Model T's, to only $290 per auto in 1927. Ford's innovations called for the worker to stand at one place while the automobile was moved down the "assembly line" on a conveyor belt. Simultaneously, the parts the workers needed were brought to the work station on another conveyor. Bodies were built on one line; the chassis and drive train were built on another, and the two parts were bolted together at final assembly. It was an extremely efficient method of auto production, and the success of the Model T was in large part due to the low cost associated with Ford's mass production techniques.

Despite his manufacturing prowess, Ford's dictatorial management style and reluctance to alter his product to keep pace with the changing demands of the public signaled the end of the Ford Motor Company's world dominance. By 1936, the Ford Motor Company's share of the automotive market ranked the company in third place behind General Motors and the Chrysler Corporation. Henry Ford held all of the stock in the company that bore his name and kept strict control of the company until 1945, when he retired in favor of his grandson, Henry Ford II. Two years later, on April 7, 1947, this complex and innovative man died at his home in Dearborn. The stock of the company had been put under the control of the Ford Foundation in 1936, and to this day, the Ford family continues its control of the family fortune.

Forrester, Jay W. (1918-)

American engineer

Forrester was born in Anselmo, Nebraska. He received his degree in electrical engineering from the University of Nebraska and later moved to Massachusetts to work on government-related research projects at the Massachusetts Institute of Technology (MIT). One of the projects he worked on was the development of a naval **flight simulator** for training and testing new **aircraft**. In 1944, convinced that a **digital computer** would be of greater benefit than a flight simulator, Forrester decided to expand the scope of the naval assignment to create WHIRLWIND, a general-purpose digital computer. Appointed by the United States Navy as manager of WHIRLWIND, Forrester headed the project for almost twelve years.

While working on WHIRLWIND, Forrester had to overcome shortcomings in the memory storage system, which was slow, expensive and unreliable. WHIRLWIND required great speed and reliability to process air-defense information and transmit weapons instructions. Forrester realized that the available memory systems would be inadequate and began to research new storage methods, first working with vacuum glow discharge, but later dropping that research because the system was still unreliable.

In 1949 Forrester noticed that when electricity passed through wires containing magnetized iron rings, the rings retained their magnetized "on" or "off" signals. Based upon this observation, he developed magnetic core memory, which offered a low-cost, reliable method to select stored information at speeds of a few microseconds. Magnetic core memory became the dominate form of primary memory storage for two decades and was implemented on WHIRLWIND with great success. In the late 1940s, naval money became scarce and the United States Air Force stepped in with additional funding to complete the WHIRLWIND project. Finally in 1951, eight years and nearly $5 million dollars later, WHIRLWIND became operational. WHIRLWIND was responsible for several "firsts" in computer processing, including the first experiments involving teleprocessing and the first use of a conversational graphics terminal. Forrester's staff implemented the first use of marginal error checking on WHIRLWIND, which allowed the computer to identify failing components before they completely crashed. The WHIRLWIND project is considered to have been a major catalyst in the development of computers in the United States.

In 1952, following the completion of WHIRLWIND, Forrester headed the digital computer division of the Lincoln Laboratory, developing the Semiautomatic Ground Environment (SAGE) system for air defense. He also began work on WHIRLWIND II, but gave up the project in 1956 to become professor of industrial management at the Alfred P. Sloan School of Management at MIT. In 1961 Forrester published *Industrial Dynamics*, which showed new ways to understand the growth and stability of socio-economic systems.

Although it is lesser known than his other accomplishments, Forrester developed a World Dynamics computer simulation program in the 1970s, which tracked five fundamental quantities: population, pollution, food production, industrialization and consumption of resources. The program aimed to forecast the future of world civilization and resources. The results were not encouraging. According to his computer program, within less than 100 years humans would have consumed most of the earth's nonrenewable resources, at which time civilization would begin to disintegrate. This outcome caused much furor among the scientific community. However, in the years following, other researchers pointed out fundamental flaws in the data and came to regard the program as unrealistic.

See also Computer input and output devices ; Computer memory; Computer simulation

Foucault, Jean-Bernard-Léon (1819-1868)

French physicist

Jean-Bernard-Léon Foucault, born in Paris on September 19, 1819, was an extremely successful experimental physicist. He, like **Galileo**, believed experimentation and innovation were the best ways to accurately assess the properties of the natural world.

Originally trained as a physician, Foucault made his living as a highly successful science and mathematics writer, and until he was appointed to a post at the Paris Observatory later in his career, he conducted his experimental work at home. Some of Foucault's most fruitful work was done in collaboration with other prominent scientists of the day, such as French physicist Armand H. Fizeau (1819-1896).

Jean-Bernard-Léon Foucault

Using the daguerreotype developed by **Louis Daguerre**, Foucault and Fizeau produced the first photographs of the sun's surface in 1845. Since Daguerre's photos required long exposure times to produce a clear picture, Foucault had to design a device for their solar camera capable of following the sun on its daily path across the sky. (This apparent motion is a result of the earth rotating beneath the sun and stars.)

Toward this end, Foucault resurrected the concept of the siderostat, or **clock**-drive, which had originally been conceived by **Christiaan Huygens**. Foucault's device used a pendulum to regulate a clock which rotated the solar camera. As he worked with the clock-drive, Foucault noticed that regardless of how the device was turned, the pendulum tended to keep swinging in the direction in which it was started. It occurred to Foucault that this tendency could be used to demonstrate that the earth rotates. Foucault surmised that the direction in which a pendulum swung would change with respect to the earth as the earth rotated about its axis, provided the pendulum was allowed to swing freely.

After he had carried out his demonstration successfully in private, Foucault constructed a massive pendulum, consisting of a large iron ball suspended from a wire more than 200 feet (60 m) in length. Staging a dramatic exhibition with the device in 1852, Foucault set the pendulum swinging before a large crowd in the Panthéon in Paris. A spike attached to the bottom of the ball scratched a line in a plot of sand scattered on the floor. As time passed, the line shifted, proving the earth was rotating under the pendulum. Three hundred years after Copernicus proposed that the earth rotates about its axis, Foucault had provided the observational evidence.

Foucault applied the same priniciple to a heavy **wheel** by mounting it onto a shaft and spinning it rapidly. When he turned the shaft with his hands, the wheel resisted his attempts to shift the plane in which it was spinning. This wheel, called a **gyroscope**, had obvious uses. Navigators quickly found that a spinning gyroscope could accurately detect deviations from a straight course. In addition, the gyroscope could be used to regulate motors and stabilize mechanical devices. Based on his gyroscope, Foucault developed a series of devices in the 1860s that were much more effective at regulating engine velocity than **James Watt**'s **governor**. The regulators were used in **telescope** clock-drives and large **steam engine**s.

Foucault also provided an improved measurement of the velocity of light. Armand Fizeau (1819-1896) had been the first person to measure light using a terrestrial method in 1849, but his result was 5 percent too high. Foucault, working independently in 1862, built an ingenious device that worked as follows: a small **mirror** was spun at a rate of 800 times per second. A beam of light was bounced from the spinning mirror onto a stationary mirror 65 feet distant, returning to the spinning mirror. Since the mirror had moved since the beam of light had left it, the beam was deflected slightly. Foucault made calculations based on the amount of deflection, arriving at a rate of 185,058 miles (298,000 km) per second—within 1 percent of the actual speed.

Foucault also introduced the use of silvered glass in the construction of reflecting telescopes in 1857. Originally invented by Isaac Newton (1642-1727), reflecting telescopes use mirrors to gather light. In Newton's time, however, it was necessary to build the mirror from polished speculum metal which tarnished very easily. Foucault's silvered glass was lighter, cleaner and easy to re-coat when necessary. Foucault devised means of testing mirrors and lenses to be used in optical devices.

A scientist with obvious mechanical skills, Foucault inevitably made significant contributions to practical technology. He improved the **arc lamp** by developing a regulator that allowed the gas fuel to be replaced with electricity, bringing the arc lamp into theaters. He invented a photometer (a device to measure the brightness of light), and improved the performance of induction coils by creating a mercury interrupter.

During his short but productive lifetime, Foucault was also involved with a number of theoretical matters such as the conductivity in liquids, stellar spectra, and the conversion of mechanical work into heat. He died at the age of 48 on February 11, 1868.

FOUCAULT PENDULUM • See Foucault, Jean-Bernard-Léon

FOUNTAIN PEN • See Pen

FRACTURES, TREATMENTS AND DEVICES FOR

Fractures, or broken bones, are one of the most common medical problems. It is estimated that 6.8 million fractures occur in the United States each year alone, with approximately 900,000 of them requiring hospitalization. To treat a fracture so that the patient regains full use of an injured arm or leg the bone ends must first be brought back into alignment; then, the fracture must be held together until the bone ends grow back together. Closed or simple fractures, in which the bone ends do not penetrate the skin, have always been relatively easy to treat. Open or compound fractures were usually fatal prior to the advent of antiseptics in the 1860s because infection would set in.

The earliest method of holding a reduced fracture in place was to use splints—rigid strips laid parallel to each other alongside the bone. Ancient Egyptians used wood splints made of bark wrapped in **linen** to hold broken bones in place. Ancient Hindus treated fractures with bamboo splints. The writings of Hippocrates discuss management of fractures in some detail, recommending wooden splints plus exercise to prevent muscle atrophy during the immobilization.

Next, medical practitioners thought of stiffening the **bandages** that held the splints in place. The ancient Greeks used waxes and resins for this purpose. The Roman Celsus, writing in A.D. 30, describes how to use splints and bandages stiffened with starch. Arabian doctors used lime derived from sea shells. The Italian School of Salerno in the twelfth century recommended bandages hardened with a flour and egg mixture. Medieval European bonesetters used casts made of egg white, flour, and animal fat. In the sixteenth century the famous French surgeon **Ambroise Paré** made casts of wax, cardboard, cloth, and parchment that hardened as they dried.

Splints remained the basic method of immobilization until 1852, when a Dutch army surgeon, Antonius Mathijsen, introduced *roller bandages* impregnated with quick-drying **plaster** of paris (gypsum). Broken bones could be held in place while the wet bandages were applied; when dry, the bandages became a rigid cast that held the bones perfectly in place during healing. As Mathijsen himself pointed out, Arab physicians had used plaster casts for centuries, but knowledge of the technique hadn't reached the West until the end of the eighteenth century. Plaster of paris casts have remained standard treatment for fractures. Beginning in the early 1980s, casts made of **fiberglass** plaster have also come into use and are favored for their light weight and water resistance.

In addition to splints and casts, fractures have been treated with *extension* and traction. The ancient Greeks used traction—pulling on a broken limb with weights and **pulleys**—but that practice died out until after the Middle Ages. Traction was revived by the eminent French surgeon Guy de Chauliac (1300-1368) during the fourteenth century. The gifted orthopedist Hugh Owen Thomas (1833-1891) of England devised improved methods of traction as well as the Thomas splint, still used today, which allowed for extension of the limb.

Hugh Arbuthnot Lane (1856-1943) of Great Britain devised a way to hold broken bone ends together mechanically when they would not heal together naturally. In 1893 he introduced the use of steel **screws** to rejoin bones and then improved the technique around 19O5 by using steel plates screwed into the bone ends. Lane's method, of course, could only succeed after Joseph Lister (1827-1912) had introduced antisepsics to surgery, and successful treatment of compound fractures also depended on Lister's innovation. Compound fractures meant heavy contamination of the wound, which almost always led to severe infection and usually resulted in death. Since infection could not be avoided in pre-antiseptic days, the usual method of treating compound fractures until the late nineteenth century was **amputation**—which, in the case of removal of the thigh, also often resulted in fatal infection. Once Lister began the era of antiseptic surgery in 1865, infection in compound fractures could be controlled and these injuries could at last be treated successfully with surgery, casts, and plates.

FRANCIS, JOSEPH (1801-1893)

American inventor

In 1890 the United States Congress conveyed a gold medal, the largest medal ever granted by Congress, to Joseph Francis, an eighty-nine year old hero. Through his work with lifeboats, Francis had saved hundreds of lives around the world.

Born in Boston, Massachusetts, Francis grew up hearing about the frequent shipwrecks in Boston Harbor and other seaports of the northeastern United States. With the increase in **ship** travel during the nineteenth century, the occurrence of shipwrecks had also multiplied. Francis observed that the most dangerous part of an ocean crossing was the reaching the dock because the ship's navigator had to negotiate an unfamiliar and often unmarked harbor. Tragically, during heavy weather, rescue vessels were as likely to fall victim to rocks and waves as the floundering ship. The heavy wood lifeboats in use at the time could easily capsize or be smashed against the shore.

Once Francis reached adulthood, he began experimenting with different boat designs, trying to devise a lifeboat for the circumstances. He came up with the idea of using **corrugated iron** to give his boats a lighter but stronger hull. The ribbing of corrugated sheet iron gave it the rigidity needed to survive the elements. Although metal had been corrugated by hand hammering for thousands of years, Francis had to figure out how to mass produce ribbed sheet metal in the shape of boat hulls to make his boat feasible. He designed a hydraulic press to do this in 1847. The press, which stood more than twice the height of a man, consisted of a fixed upper die and a movable lower die which pressed iron sheets into the desired corrugated

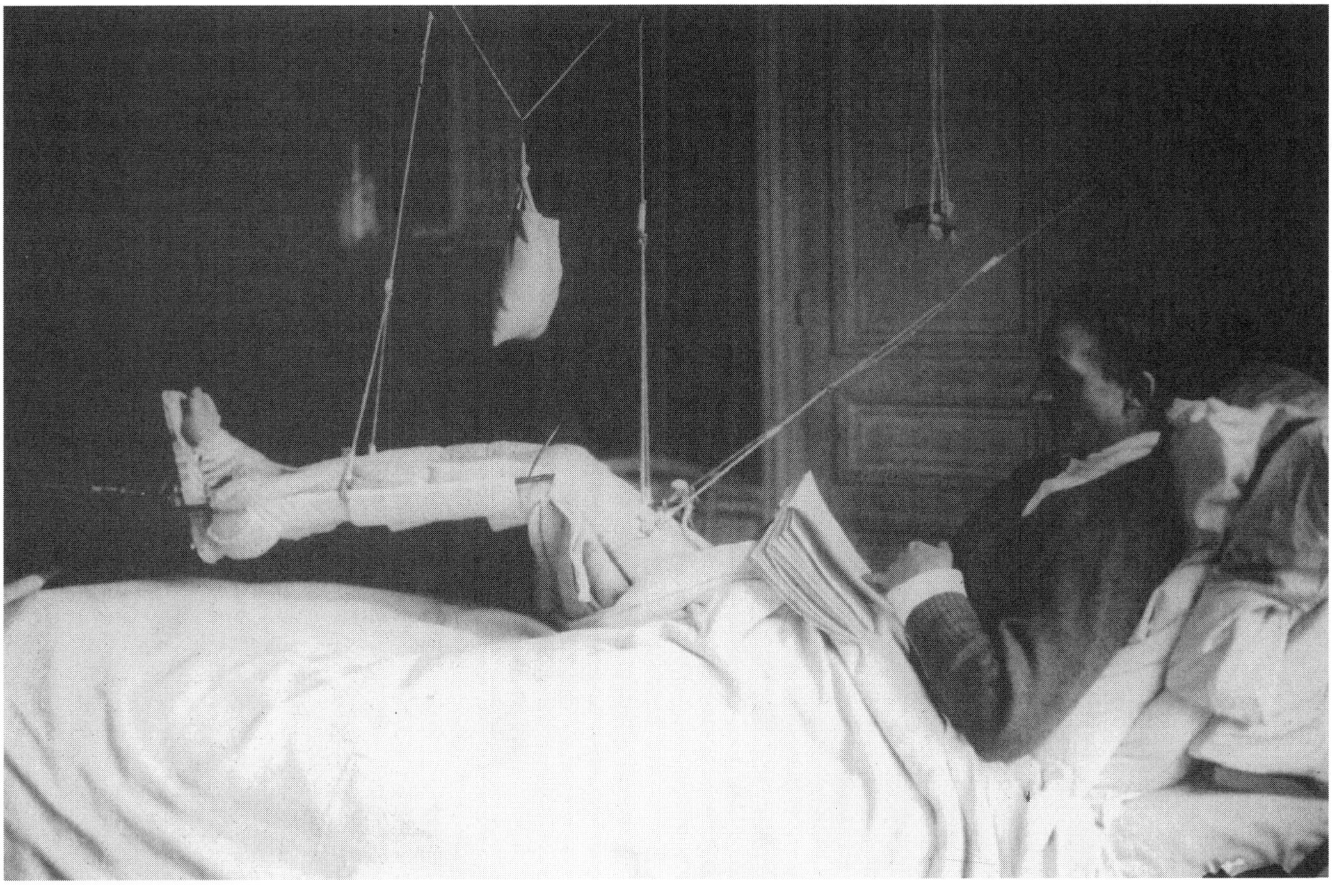

A soldier in traction recovers in a field hospital during World War I.

Joseph Francis

hull shape. Though iron is heavier than wood, the corrugated hulls were actually stronger and lighter than their wooden counterparts.

Francis opened the Metallic Life-Boat Factory at the Novelty Iron Works, to mass produce the corrugated iron hulls. His invention was recognized as highly valuable in saving lives, and he received financial support from Congress, the New York Board of Underwriters, and the Humane Society to install his boats at coastal stations along the Atlantic seaboard. Francis' first corrugated boats were open surfboats which rode high on the waves. He later invented a series of lifecars, which were enclosed lifeboats that were pulled ashore with a lifeline. The line was shot by cannon to the distressed ship, where they were attached to the lifecars carried by the vessel. The most renowned rescue involving the lifecars occurred in January of 1850. The *Ayrshire* was wrecked in a terrible snow storm off the New Jersey coast; two hundred people were safely brought ashore on the lifecars.

The life-saving vessels were also distributed around the coasts of Europe. Their noble duty earned awards for Francis from several European nations. He was also commissioned by the Russian government to build a fleet of light-weight steamers for the inland Aral Sea. Because of their corrugated hulls, they were easily transported across the mountains.

After his success with the lifeboats, Francis went on to invent a military vessel called an *amphibious duck*. Francis' lifeboats and lifecars remained in use into the late 1800s, being phased out as newer boats were invented. At the end of his career, he was awarded the Congressional gold medal in recognition of a half century of saving lives.

Besides this Congressional honor, Francis also received medals from The Franklin Institute; Ferdinand III, King of the Two Sicilies; and the European Lifesaving Society. He was also knighted by Napoleon III of France and given the knighthood of St. Stanislaus of Russia. Internationally known and recognized at home as the father of the United States Lifesaving Service, Francis is a shining example of a man using his mechanical talents for the good of mankind. Because of him, countless lives were saved, and during one short four-year period (1850-1854), more than 2,100 passengers were rescued by his invention.

Franklin, Benjamin (1706-1790)

American statesman and inventor

Born in what was then the British colony of Boston, Massachusetts, on January 17, 1706, Franklin was the fifteenth of seventeen children and received only two years of a formal education. He started working in his father's candlemaking shop at the age of ten and later became an apprentice printer, working for his brother James. As a printer he developed a love for books, from which he educated himself. He spent two years in London, where he learned more about printing, and returned to Philadelphia in 1726. There he established the *Pennsylvania Gazette and Poor Richard's Almanack*, which earned him a tidy income. Franklin's first major invention, around 1740, was the Pennsylvania fireplace, which eventually became known as the Franklin stove. Improving on an existing design, the Franklin stove had a flue around which room air could circulate. The flue acted like a radiator, increasing heating efficiency. Franklin claimed it made a room twice as warm, with one-quarter of the wood.

In 1746 Franklin had witnessed a public demonstration of electricity, and his interest was piqued. It is for his work in this subject that he became most famous. In the demonstration, a machine was used to generate static electricity which was stored in a **Leyden jar** (a water-filled bottle with a stopper through which a metal rod extended). People were instructed to join hands, forming a "circuit," and they simultaneously received a shock from the jar.

Investigating electric phenomena was all the rage, especially in Europe, but not one of the learned scientists had thought to ask the simple question Franklin framed: "How does it work?" He decided to obtain a jar of his own and find out. Franklin charged his jar, poured the water into another bottle, and found it had lost its charge. If the water did not hold the charge, it indicated that the glass of the jar did. To see if that was the case, he took a window pane, placed a thin sheet of lead on each side, and gave it a charge. He removed the sheets and tested for an electric charge. The glass sparked but no charge was indicated. Inadvertently, Franklin had just invented the electrical condenser. The condenser, also known as a **capacitor**, was destined to be one of the most important elements in **electric circuits**. Today the condenser, which received its name from **Alessandro Volta**, is used in **radios**, **televisions**, **telephones**, **radar** systems, and many other devices.

Drawing a parallel between the sparking Leyden jar and lightning, Franklin began to speculate that the sky might have an electrical charge. To ''collect'' this charge, he hit upon the idea of erecting a long metal rod on the top of Christ Church in Philadelphia. The rod would conduct electricity to a man on an insulated platform in a sentry box, who could collect the charge in a Leyden jar. Unfortunately, before the rod could be attached to the church, a French scientist named D'Alibard, who had read of Franklin's work, successfully performed the experiment himself on May 10, 1752. Franklin, always the diplomat, generously gave D'Alibard credit for being the first to ''draw lightning from the skies.''

Was D'Alibard really the first? While waiting for the installation of the **lightning rod**, Franklin thought of a quicker way to get a conductor in the sky. Yes, he really did fly a kite in a thunderstorm! He was fortunate lightning did not actually strike his kite because he would have been killed. Two other scientists who duplicated the experiment suffered this fate. Whether D'Alibard was the first to ''draw lightning'' or not is irrelevant; it is Franklin that history remembers as being the inventor of the lightning rod.

The lightning rod became indispensable for protecting buildings from the destructive force of lightning. Because he had discovered he could get the Leyden jar to spark over a greater distance with a sharply pointed rod, Franklin's lightning rods had very sharp points. (In 1776, after the unpleasantness between the Colonies and King George III had broken out, the king ordered that lightning rods with *blunt* ends be installed on his palace.) By 1782 there were four hundred lightning rods in Philadelphia.

Franklin's work with electricity led him to coin numerous terms and propose several theories. **Battery**, conductor, condenser, armature, electrician, charge, and discharge are some of the words attributed to Franklin. He came up with the idea of ''positive'' and ''negative'' electricity having ''plus'' and ''minus'' charges. He incorrectly thought electric flow was from positive to negative; the opposite is true.

To say Franklin was involved in politics is a great understatement. He established service organizations, was Postmaster of Philadelphia, and founded a college that eventually became the University of Pennsylvania. He returned to London in 1757 as an Agent of the Pennsylvania Assembly and remained there nearly 18 years. In 1775 he returned and joined the committee drafting the Declaration of Independence. While in France after the Revolutionary War, eighty-three year old Franklin came up with yet another innovation: bifocal lenses. He had become frustrated with having to continually change his glasses for reading up close or seeing at a distance and came up on the idea of mounting half of each lens in a single frame. Franklin is also credited with inventing the rocking chair, the glass harmonica, the concept of daylight savings time, and the first public library. On April 17, 1790, Franklin died in Philadelphia.

FRANKLIN STOVE • See Stove

Benjamin Franklin

FRASCH PROCESS

The Frasch process is a drilling process used to recover deeply buried sulfur deposits. It was invented by Herman Frasch (1851-1914), a young chemical engineer from Germany. Frasch came to America to seek his fortune in 1868. The Civil War had just ended, and the American economy had begun to prosper. Shortly after his arrival, Frasch established an industrial laboratory in Philadelphia, and in 1876, he patented a process for making paraffin wax from crude petroleum. This won the attention of the Standard Oil Company, which hired Frasch to work at its Cleveland, Ohio laboratories. It was there that Frasch observed that oil from many wells could not be sold because it contained sulfur compounds. This ''sour'' oil burned poorly and emitted a strong odor even after it was refined. Frasch found a way to remove these impurities. In his method, which he patented in 1887, oil was distilled in the presence of copper oxide or other metallic oxides, which extract the sulfur. Afterwards, the spent oxide is recovered and reused. This process increased the United States' usable oil supply and helped set the stage for the introduction of the **automobile**.

Frasch's next breakthough involved his idea of idea of drilling for sulfur—the mineral used to make sulfuric acid, which today is industry's most important manufactured chemical. Although sulfur is a solid material, Frasch believed that underground deposits could be melted and then pumped to the surface, much as oil is produced. At that time, the Mediterranean island of Sicily possessed a near-monopoly on world sulfur resources, where sulfur deposits were shallow and easy to mine. Additionally, Sicilian workers accepted lower wages and endured harsher conditions than would American miners. Texas and Louisiana contained vast amounts of sulfur, but it lay deep underground, covered by swamps and quicksand. In 1894 Frasch made his first attempt at drilling for sulfur in a Louisiana swamp. He adapted a method used previously for mining salt dissolved in water. To melt the sulfur, Frasch pumped water heated beyond its normal boiling point into the ground through a borehole. After overcoming many engineering problems, Frasch managed to extract a mixture of molten sulfur and water. Frasch then improved the process by using compressed air to pump the sulfur to the surface. Although much fuel was consumed to heat the water for melting sulfur, huge oil deposits were soon discovered near Frasch's first well. By 1902, the Frasch process for sulfur production had become completely practical, giving America its own supply of sulfur and sulfuric acid. This represented a major step toward making the United States less dependent of Europe for industrial chemicals. Today, the Frasch process is used to produce nearly one-third of all commercial sulfur.

See also Oil refining; Oil-drilling equipment

von Fraunhofer, Joseph (1787-1826)

German optician

Joseph von Fraunhofer was certainly the most talented lenscrafter of his time. Though he received very little in the way of formal schooling, he dedicated his life to creating optical instruments of unsurpassed quality. Through his efforts, telescopes and prisms of amazing clarity were created; in addition, he invented the device known as the **spectroscope** for observing the dark lines found in the visible spectrum, effectively founding the science of spectroscopy.

Fraunhofer was born in Bavaria in 1787. His father was a poor glass maker who could not afford to send his youngest to more than a few years of school, and so at the age of 10 Joseph began working in the family workshop. Upon his father's death in 1789 Fraunhofer became an apprentice to a Munich lensmaker. Although young Fraunhofer showed much promise, the master lensmaker used him primarily as slave labor, driving the boy to near-starvation and exhaustion.

Strangely, the luckiest day of Fraunhofer's life may have been July 21, 1801, when the run-down building in which he slept collapsed. Though unhurt, Fraunhofer was trapped beneath the wreckage, prompting a large community rescue operation; in fact, the Elector of Bavaria himself was drawn to the scene. When extricated, Fraunhofer was given a monetary award from the Elector—enough money to buy out the remainder of his miserable apprenticeship and to begin his own business.

Although he squandered much of his award on an unproductive get-rich-quick scheme, Fraunhofer retained enough money to remain in Munich, and in 1806 he was employed by the famous Munich Philosophical Instrument Company, one of the most prominent suppliers of scientific optical equipment in the world. There he was able to exercise his glassworking abilities, producing **lenses** and **prisms** of unprecedented clarity. He rapidly gained a reputation for constructing lens systems free of color flaws (such lenses were vitally important to astronomical observing). He also constructed, in 1817, a 9.5 inch lens for the Russian Dorpat Observatory; using this lens, Russian astronomers were able to discern over 2000 new double stars.

It was while he was perfecting his lenses that Fraunhofer made what was to be his most important discovery. In order to calibrate his instruments he would focus them upon a bright candle. One day, however, he chose to use the light from the sun. He was quite surprised to find a number of dark lines crossing the sun's rainbow spectrum—lines that were not present in the spectra of candlelight. Although a few dark lines had been identified years earlier by William Hyde Wollaston (1766-1828), Fraunhofer's precise equipment enabled him to discern 574 individual lines in the spectrum of the sun.

Fraunhofer began to examine the spectra of many light sources using an invention called a spectroscope. He found that different elements, when heated to incandescence, would produce lines at different positions in the spectrum. He also discovered that the light from the moon and the planets displayed the same line pattern as the sun, while other bright stars possessed individual spectral ''signatures'' (this is actually a logical conclusion, since the light from these bodies is simply reflected sunlight, but other stars generate their own light). Working tirelessly, Fraunhofer succeeded in mapping the more than five hundred lines in the solar spectrum and calculating the approximate wavelengths at which they occurred.

These dark lines—now known as *Fraunhofer lines*—were actually areas of the spectrum where light is absorbed. When the light from a star encounters the elements at the outside of its atmosphere, the elements will absorb certain frequencies. By observing any star's Fraunhofer lines, an astronomer can identify the elements in that star's atmosphere. All of this was unknown to Fraunhofer himself, and was left for Gustav Kirchhoff (1824-1887) to explain in 1859.

While conducting his research on spectral lines, Fraunhofer had been using prisms to separate light into its component colors. However, it was also common practice to obtain a spectrum by passing the light through a very thin slit. It occurred to Fraunhofer that a **wire** mesh could be assembled wherein the holes would act as hundreds of extremely thin slits, effectively splitting and re-splitting the passing light. Thus in 1821 he constructed a **diffraction grating** comprised of 260 wires, producing a widely dispersed spectrum. He later found that an even wider (and more easily examined) dispersion could be obtained if the wires were set closer together.

Fraunhofer considered his methods trade secrets and never published them; therefore, much of his work was a mystery to his peers. Also, though he was a highly skilled craftsman, he was viewed by the scientific community as nothing more than a technician. He was never truly accepted by the intelligentsia, and while he could attend scientific meetings he was never allowed to address them. He died of tuberculosis a few months before his 40th birthday.

Freeway

Freeways, also called superhighways, are roads specifically designed to allow for the free flow of traffic. Freeways typically feature two or more traffic lanes in each direction, medians to divide the opposing directions, full access control, a system of ramps to prevent merging and diverging traffic from interrupting the traffic flow, and grading to separate intersecting traffic on other roads.

The advent and eventual domination of the **automobile** created a corresponding demand for roads capable of handling the increasing traffic and loads. Increasing numbers of cars began to choke the cities with traffic. The need for linking cities to one another also became apparent, especially as the truck proved its flexibility and reliability for transporting goods, materials, and products.

The freeway was first conceived as a means for reducing the crush of traffic within the cities, and for linking the cities together. The first freeway was opened in the Grunewald Forest in Berlin, Germany in 1921. The idea for a national highway system for the United States was also developed during this time. The first U.S. freeways appeared in 1940, when California opened the Arroyo Seco Parkway between Pasadena and Los Angeles, and when Pennsylvania opened the first section of the Pennsylvania Turnpike.

The numbers of automobiles in use skyrocketed in the years after the Second World War. With this increase came an alarming increase in traffic congestion and automobile accident fatalities. In 1956, legislation was passed creating the Federal Interstate Highway System (FIHS). This network of freeways was meant to link nearly all cities in the U.S. with populations greater than fifty thousand people. Although the original plans called for exactly 40,000 mi (64,630 m) of road, by the 1990s nearly 45,000 mi (72,405 m) of road surface had been completed, carrying more than 20% of all traffic in this country. Freeways in the FIHS are constructed according to strict guidelines governing the materials and other elements of their design and construction.

Freeways dramatically changed the pattern of life in the United States. Access to the city by automobile allowed people to move beyond the traditional trolley and horse-drawn cart routes. The spread of people outside of the city created what is known as urban sprawl, in which the city extends farther and farther from its center. Meanwhile, the former centers of city life lost more and more manufacturers and other industries to the suburbs, draining the cities of vital resources and jobs. Although the freeway was originally meant to alleviate traffic, it actually increased traffic levels, by encouraging the use of the automobile over mass transportation methods such as **trains** and **buses**. The resulting increases in congestion brought problems of pollution and noise. What were once "walking cities" were now accessible only by car. Entire new communities, the suburbs, became so dependent on automobiles that most families found it necessary to have two or more. Meanwhile, the increased traffic on the roads brought corresponding increases in the number of traffic fatalities.

Nonetheless, the FIHS remains the most ambitious public works undertaking in American history. The FIHS has made nearly every part of the country accessible by car. It has brought a greater flexibility and choice of places for people to live, work, and travel, and a greater mobility over longer distances and safer roads.

Joseph von Fraunhofer

All freeways share a number of common features. A freeway has at least four lanes, two lanes in each direction. Many freeways, however, feature more than four, and as many as ten lanes, especially as they near the cities. Lanes are required to be from 11-12 ft (3.35-3.66 m) wide. Shoulder lanes provided on each side of the driving lane for each direction of the freeway allow vehicles to safely leave the traffic stream in the event of an emergency. Shoulder lanes are generally 8-10 ft (2.4-3 m) wide. A median, or center strip, separates the opposing directions of traffic. Medians may vary from 16-60 ft (4.9-18.3 m) wide. The median improves safety by preventing head-on collisions of automobiles traveling toward each other.

Freeways are called controlled access highways. This means that traffic is limited in where it may come onto or leave the freeway. These entrance and exit points are referred to as interchanges. Minor roads and driveways are diverted away from the freeway so that their traffic does not interfere with the freeway traffic flow.

Many roads, from small local roads and streets to other highways in the freeway system, intersect with a freeway. Grade separation prevents the intersection of two roads traveling crossways to each other from interrupting each others' traffic flow. Generally, one road, usually the road minor to the freeway, is raised on a grade, or slope, so that it is higher than the freeway, and allowed to cross it over a bridge. Ramps are constructed to lead the crossing road to the grade separation. Additional access ramps, often called on-ramps and off-ramps, connect the freeway to the intersecting road. They allow vehicles entering the freeway to accelerate to the proper speed before merging with the freeway traffic; the ramps allow vehicles leaving the freeway to decelerate to the slower speeds of the crossing road.

As part of the FIHS, freeways are designated by red, white, and blue signs in the shape of shields. Freeways are also numbered, with the numbering system used to indicate the direction of the road. Freeways traveling in an east-west direction are given even numbers; those traveling north-south are given odd numbers.

With the completion of the FIHS, few new freeways may be expected to be built in the United States. Existing freeways, however, will continue to be expanded and improved. In all cases, work on a freeway must be carefully planned, its route laid out, and its impact on the environment and surrounding area thoroughly investigated. Engineers design the freeways, following government specifications. In addition, geographical and geological features are examined, including the grade, or slope of the land, and the type of soil found along different sections of the proposed roadway. The type of soil will affect the nature of the pavement to be laid, so soil samples are analyzed both in the field and in the laboratory.

Many questions must be answered when designing a freeway. The expected volume of traffic must be estimated, with minimum and maximum levels established. The expected use of the freeway is another consideration, and takes into account where people live and work and how they currently travel, and also the location and type of industry in the area, the types of goods that are produced, the markets or destinations of those goods, and how those goods have been transported in the past. These questions will affect the volume of traffic on the proposed freeway; they will also affect the type of vehicles that will use it. A freeway that will serve heavy **trucks** will require different surfacing, lane widths, and bridge heights than freeways serving mostly or only automobiles.

Work begins by clearing the right-of-way, the path, of the freeway. Vegetation will be removed, and the course for the freeway will be laid out. The use of modern construction equipment, including bulldozers and other specifically designed heavy equipment, has made this process much easier and faster than in the past. At this time, hills and valleys along the freeway route may be smoothed out, to minimize the variability of the route.

At the same time, features of the water drainage system-an important part of any roadway-are formed. These include the slope of the road, and ditches and culverts alongside of the road. The drainage may be the single most costly part of constructing a freeway; yet, if the water is not properly guided away from the road, the road will quickly weaken. The cleared right-of-way, including the shoulders and drainage ditches, will next be compacted, so as to provide a firm underbed for the freeway. Any bridges to be placed along the freeway will then be constructed, before the freeway itself is paved.

Paving a freeway may actually take place in several phases, adding layer upon layer of road surface over a long period of time, even years, until the freeway has achieved its final form. This allows weaknesses, and the effects of settling, in the roadway and drainage system to be detected and corrected.

Roads, including freeways, are generally composed of three layers: the subbed, or subgrade; the bed, or base; and the surface, or pavement or wearing course. The subbed is the soil on which the freeway is built. It is prepared by leveling and compacting the soil, and may be treated with **asphalt**, tar, or other substances to provide greater firmness. Next, the base is laid, consisting of crushed stone, gravel, or **concrete** pieces in a variety of sizes ranging from dust to three inch rocks mixed in exact proportions. This allows the base to remain porous, so that moisture will not build up beneath the pavement. This course is also compacted, then sprayed with a thin, liquid layer of tar or asphalt to fill in the gaps and spaces between stones and make this surface even.

The pavement is then laid on top of the base. A layer of tar or asphalt is added, then covered with gravel or stones that are all the same size. The gravel layer is compacted into the asphalt so that they are firmly mixed together. This process, which forms the pavement, may be repeated several times, until the road surface reaches the proper thickness. Each layer is rolled with special machines until it is hard and smooth. Sudden bumps or dips in the road will make the freeway more dangerous to drive on, especially with the high speeds allowed on the freeway. The thickness of the road surface will depend on the type of traffic expected, that is, whether it is expected to be high volume, or whether it is expected to carry many heavy trucks as well as automobiles. The pavement must be watertight, because moisture can destroy the surface as it expands or contracts with temperature changes. The addition of stones or gravel in the blacktop, or surface layer, allows tires to grip the surface more easily.

Keeping the driver alert to the road is important for preventing accidents. Lighting by overhead lamps allows the driver to see the road ahead at night, even at great distances. Guardrails may be placed alongside the roadway at curves and where the land drops away suddenly beyond the shoulder. Reflectors are often placed on guardrails alongside the roadway and in the lines between lanes. Landscaping along the road and in the median helps to reduce the monotony of a long drive.

FREEZE-DRIED FOOD

Freeze-drying is a process of **food preservation** in which the food is flash frozen and placed in a refrigerated vacuum chamber, where the water is removed from the food. The vacuum causes the water to sublimate, or change directly from a frozen solid into vapor, which is whisked out of the vacuum chamber. Because the ice sublimates rather than melts, the food's tissues do not collapse. The food retains its original shape, but is lighter and spongelike.

Because freeze-dried food is extremely porous, it readily reabsorbs water and quickly returns to a condition very close to its original shape and color. The low temperatures at which the process is carried out produce fewer changes in the food than in more traditional drying methods. Freeze-dried food keeps its flavor, texture, and nutrients better than dried food. In its dried form, it is very light and can be stored indefinitely at room temperature. However, freeze-drying is an expensive process because of energy costs, so only foods of high value or for specialized uses are preserved this way.

The process of freeze-drying seems to have originated in Sweden in the 1930s as a method for drying biological materials used in the drug industry. It became important during World War II as a means of preserving plasma and penicillin. After the war, the process was applied to various foods by Americans E. W. Flosdorff and **George Speri Sperti**, among others. The process worked best on liquids, such as coffee and orange juice; thinly sliced items, such as meat and potatoes; and small vegetables, such as peas.

Freeze-drying has been most successful for **instant coffee**, since spray-drying coffee at high temperatures damages the flavor. Freeze-dried foods are highly popular among campers, backpackers, military personnel, and astronauts because they are light, compact, and retain superior flavor.

Freon

Freon is a trademark belonging to the E. I. du Pont company for a series of **chlorofluorocarbons** (CFCs). CFCs are organic compounds composed primarily of chlorine, fluorine, carbon, and hydrogen. These compounds are chemically stable, nontoxic, and nonflammable. Due to their high density, low boiling point, and low viscosity they are especially useful for use as refrigerants, aerosol propellants, solvents, and expansion agents in polyurethane foams. Since the 1970s, CFC use has declined due to concerns about their effects on the ozone layer.

The compounds that would become known as Freon were discovered in the 1930s by researchers for General Motors led by Thomas Midgley, Jr. They discovered that chlorine and fluorine atoms could be positioned along a hydrocarbon backbone to form molecules that functioned as efficient refrigerants. The first material they made was dichlorodifluoromethane, a compound that contains one carbon, two chlorine, and two fluorine atoms. This material is now referred to as Freon F-12. Other similar compounds followed shortly thereafter, including trichlorofluoromethane, dichlorotetrafluoroethane, and chlorodifluoromethane. These chemicals were given Freon names based on their structure and molecular weight. Freon molecules based on methane (CH_3) are designated with numbers below 100; those based on ethane (C_2H_5) are given numbers between 100 and 200. Freon F-11 and F-12 are the most common.

The number of applications for Freon compounds grew tremendously between the 1930s and the 1960s. Increased production of these CFCs allowed widespread manufacture of refrigerators and **air conditioner**s. Other applications for CFCs were discovered as well, for example, they are used as blowing agents in **polystyrene** foam.

In 1972, two scientists from the University of California, F. Sherwood Rowland and Mario Jose Molino, found that certain Freon materials could persist in the atmosphere long enough to reach the stratosphere where they could be broken down by ultraviolet radiation. When this occurs the CFCs release chlorine which can affect ozone production in the upper atmosphere. This discovery set the stage for fervent public contention regarding the continued use of CFCs. By the mid-1970s, the United States government banned the CFCs as aerosol propellants but it resisted a total ban for all industries. Instead, countries and industries began negotiating the process of phasing out CFCs. As CFC use is allowed in fewer and fewer applications, a black market has been growing for the chemical. In 1997, the Environmental Protection Agency, Customs Service, State Dept., Justice Dept., IRS and the FBI initiated enforcement actions to prevent (CFC) smuggling in the U.S.

Many CFCs have been replaced with related materials known as hydrochlorofluorocarbons (HCFCs) which reportedly do not interfere with the ozone layer. These are used principally in large air conditioning installations and while they are believed to be less damaging than CFCs, they are are not free from controversy. A 1997 Belgian health officials report indicate hydrochlorofluorocarbon-123 may be responsible for a rise in acute hepatitis among workers of a smelting plant. Liver damage found in nine workers is believed to be caused by a leaking refrigerant pipe in the workers' cabin. However, the U.S. Environmental Protection Agency, disputes these findings.

Meanwhile, the search for the perfect Freon replacement continues. In Europe scientists have had some success with butane mixtures as refrigerants in some European refrigerator models, but the explosive nature of the gas could create a dangerous situation if used in the frost-free models popular in the U.S.

Frequency modulation

The invention of frequency modulation (FM) was a great improvement over **amplitude modulation** (AM) **radio**.

Reginald Aubrey Fessenden made the world's first AM radio broadcast in 1906. With amplitude modulation, the amplitude (the distance between the "peak" and "trough") of the radio wave "carrier" is modulated to match the frequency of an audio signal. AM radio made the transmission of voice and music possible, but it was prone to interference by lightning storms and the earth's ionosphere.

In 1924 **Edwin Howard Armstrong** invented the way to modulate the frequency of the wavelengths, rather than the amplitude. The actual frequency of the oscillations of the carrier wave is modulated to match the frequency of the audio signal; the amplitude of the radio wave stays the same, but the frequency changes. An early way of doing this was by using a **vacuum tube**. The capacitance between the anode and cathode depends on the voltage in the grid located between them. Changing the voltage changed the oscillator frequency. Identical results are obtained today using **transistor**s in place of vacuum tubes. At the FM receiver, the process is reversed. Obviously, the **electronics** required to modulate and demodulate an FM signal are more complicated, and expensive, than those needed for AM signals.

Because it is the frequency of the signal that is being modulated, rather than the amplitude, FM is impervious to static caused by lightning. In addition, the ionosphere above the earth does not reflect and distort FM signals. Sending multiple signals (multiplexing) on the same FM carrier, allowing for high-fidelity **stereo** broadcasting, became possible. On the other hand, FM has its own set of drawbacks. The distance an FM signal travels is much more limited than that of an AM signal. FM also requires more bandwidth in which the frequencies are assigned, limiting the number of frequencies available in a given area. As mentioned earlier, the modulating and demodulation **electronics** are more complicated and expensive.

In spite of its superior quality, FM broadcasting had a very difficult time getting established. The depressed economy of the 1930s did not encourage experimentation in the new technology. The Federal Communications Commission (FCC) did not grant permission to broadcast at the FM wavelengths until 1940, at which time World War II interfered. Following the war, owners of AM radio stations saw FM broadcasting as competition to their costly investment, and they did everything they could to retard FM growth; the FCC was persuaded to arbitrarily change the frequencies allotted to FM. Armstrong found himself enmeshed in numerous lawsuits, enormous expenses, and bureaucratic red tape. ''They will stall this along until I am dead or broke,'' he asserted. On January 31, 1954, he jumped to his death from his apartment in New York City.

Fresnel lens • See Lighthouse

Frisbee

On a warm spring day on a college campus, you may need to duck in order to avoid being hit by a barrage of brightly-colored frisbees being tossed from one player to the next. However, the original frisbee was not the universally popular plastic disk often seen now. It was actually a pie tin with the name *Frisbie* stamped on it. After William Russell Frisbie's bakery opened in Bridgeport, Connecticut, in the 1870s, his pies and sugar cookies became quite popular. Years later in the 1920s, students at nearby Yale University started tossing discarded pie tins and cookie tin lids and yelling ''Frisbie.'' Just as college students of that era had popularized crazes like eating goldfish, tossing pie tins became another fad.

Twenty or so years later in California, Walter Frederick Morrison used his basement laboratory to tinker with tenite, a **plastic** used in camera parts. The son of an inventor, and himself the creator of the home popsicle maker, Morrison fashioned a plastic disk that could fly. Morrison improved the pie-tin shape by adding the familiar curled lip which keeps today's frisbee more stable in the air. He named his flying disk ''Lil' Abner'' after the popular cartoon strip character. In the late 1940s, Morrison amazed the crowds at county fairs, carnivals and beaches by tossing the Lil'Abner to his wife, claiming it was sliding along ''invisible string'' which he was willing to sell. The toy came for free.

Morrison's Lil invention become well-known enough to catch the attention of Spud Melin and Rich Knerr in 1957. This inventive pair made backyard slingshots, the sound of which gave them the idea for the name of their company— Wham-O. They bought rights to Morrison's toy and renamed it the Pluto Platter because of the nation's fascination with science fiction and flying saucers. The plastic spinning disk had the shapes of planets around the outside ring and they added the familiar grooves in the plastic. The Pluto Platter was a great success—especially on college campuses. A year later Wham-O changed the name to frisbee to pay tribute to the original. Today's frisbees come in all manner of colors and variations. There are international frisbee championships and frisbee-catching dogs. Even the military tested frisbees as a practical way of keeping flares up in the sky.

Froude, William (1810-1879)

English engineer

Froude was one of the principal players in the field of fluid dynamics. Born in Dartington, England, he was educated at Oxford, where he received a degree in engineering. His first job was as a civil engineer, working for the British railway system. It was not until a few years later that Froude became a naval engineer, and in doing so began the work for which he is best remembered.

Froude's most important work was in determining the amount of force that water exerts on a body passing through it. This was of great concern to the British navy, who wanted to maximize the speed and efficiency of their **ships**. Froude conducted a great deal of laboratory research before developing *Froude's law*, which is a tool used to determine the amount of force required to pull an object against the retarding wave that is built up in front of it as it travels. By using Froude's law, the British navy was able to tow its ships from one place to another more efficiently.

There were two important devices that were invented during the development of Froude's law: the model study tank and the water brake. In order to simulate conditions on open water, Froude had to create a water environment in his laboratory. The model study tank allowed him to perform his experiments and extrapolate their results to larger bodies of water. The water brake, a type of **dynamometer**, was actually an offshoot of Froude's research. Primarily used to measure the horsepower of **internal combustion engines**, the water brake consists of a rotor encased within a water-filled compartment; the engine turns the rotor, and the water resists that turning, slowing it down. The horsepower of any engine can thus be determined by measuring the amount of resistance exerted by the water.

See also Steam engine

Frozen food

Foods have been frozen for centuries, especially by people in cold climates like the Arctic. Freezing preserves food because low temperatures slow the growth of (but do not kill) microorganisms that cause spoilage. A method of freezing food using an ice and salt brine bath was patented by H. Benjamin in England in 1842, and Enoch Piper of Maine received a fish-freezing patent in 1861. Food freezing became widespread when mechanical **refrigeration** systems were developed in the later 1800s.

Early commercial attempts to market frozen foods focused on meat. In 1869, Dr. Henry Howard sent a shipment of frozen beef via steamboat from Texas to New Orleans, Louisiana. Two British immigrants to Australia, Thomas Mort and James Harrison, tried to capitalize on the commercial possibilities of shipping frozen beef. In 1861 Mort established the world's first meat-freezing plant in Sydney, Australia. Both Harrison and Mort gave frozen-meat lunches to the public to launch their shipments to England in 1873 and 1876, respectively. Although the shipments failed due to mechanical breakdowns aboard ship, frozen beef was transported successfully from Buenos Aires, Argentina, to Le Havre, France, aboard the *Paraguay* in 1877, and from both Australia (on the *Strathleven*) and New Zealand to London, England, in 1880.

The success of the frozen-beef trade led to the freezing and transport of various other foods. The slow-freezing method was used, in which foods were placed in very cold storage rooms until they froze solid, sometimes days later. Unfortunately, when food is frozen slowly, its fluids remain for a long time at what German physicist Max Planck (1858-1947) called the zone of maximum crystallization—a temperature just below freezing (32°F [0°C]) at which especially large ice crystals form in the food. The sharp edges of the crystals break down cell walls, and the food collapses as it is thawed. That is why frozen foods are usually mushy and/or flavorless after thawing.

Clarence Birdseye revolutionized the frozen-food industry when he discovered the process of quick-freezing, inspired by observing Eskimo methods of **food preservation** during the winter in Labrador in the early 1900s. In quick-freezing, only small ice crystals form in the food, since it passes through the zone of maximum crystallization in only a few minutes. Cell walls remain intact, and the thawed food retains its flavor, texture, and color. Birdseye perfected his process during the 1920s and began producing family-size units of quick-frozen foods in 1924, speeding up the process with his double-belt machine in 1929.

The improved frozen foods were not immediately embraced by consumers, who vividly remembered the older, slowly frozen products. Restaurants that appreciated the convenience of frozen foods carefully concealed the fact that they served these products. Birdseye worked for years at promoting frozen foods. Painstaking research by chemist Mary Engle Pennington from the 1920s through the 1940s established the best processes for freezing a wide variety of specific foods, which resulted in marked product improvement. **Frederick McKinley Jones**'s successful designs for refrigerated **trucks** and **railroad** cars made nationwide shipment of frozen foods feasible. Rationing and shortages of canned foods during World War II fostered widespread public use of frozen foods, as did the advent of home freezers after the war.

Frozen prepared meals were offered to airline passengers in 1945. U.S. food manufacturers expanded this idea by producing the much-maligned **TV dinner** in 1954. Frozen orange juice concentrate was first produced in 1947 by a U.S. citrus cooperative in Lake Wales, Florida.

Today most commercial freezing uses one of three quick-freezing methods: a blast of cold air, immersion in a very cold liquid, or contact with refrigerated hollow plates—a direct adaptation of Birdseye's idea. The widespread availability of frozen foods has added variety to the American diet and made food preparation more convenient and less time-consuming. Total retail sales of frozen foods in the United States exceeded $20 billion in 1995, according to the American Frozen Food Institute. Among the products with rising sales in the late 1990s were low-fat and all-natural products, microwaveable items, hearty helpings, single-serving and child's portions, premium items, and lunch foods.

FUEL CELL

A fuel cell is a type of **battery** which converts chemical energy into electrical energy. Its invention goes back more than 150 years, yet its potential is just being realized in the space age.

Within a few weeks of **Alessandro Volta**'s invention of his Voltaic pile in 1800, Englishmen William Nicholson (1753-1815) and Anthony Carlisle (1768-1840) used its electric current to decompose water into its two component parts, hydrogen and oxygen. This was the invention of electrolysis, and it paved the way for the fuel cell. Just as **Michael Faraday** discovered he could reverse Hans Christian Oersted's (1775-1851) discovery that electric current produces a magnetic field, William Grove (1811-1896) discovered in 1838 he could reverse electrolysis to produce an electric current, and the fuel cell was born. (Although thirty-six years earlier, **Humphry Davy** had described a fuel cell that used a carbon anode and aqueous nitric acid.)

As any chemist knows, combining hydrogen and oxygen can result in the production of prodigious amounts of energy. As early as 1903, Russian schoolteacher **Konstantin Tsiolkovsky** came up with a **rocket** design that was fueled by liquid hydrogen and liquid oxygen. Grove, in actual tests, used gaseous hydrogen and oxygen, which are less volatile than their liquid counterparts. He placed two platinum plates parallel to each other in a tank—these would be the electrodes of his "gas battery." Oxygen gas was introduced behind one plate and hydrogen behind the other. The space between the plates was filled with a liquid electrolyte (a conductor of ions). The platinum plates were porous enough to let the gas molecules penetrate, but not the liquid, so that this design prevented the two gasses from coming into direct contact with one another. As hydrogen gas mixed with the electrolyte, it produced a coating on the anode (positive electrode); oxygen gas made a coating on the cathode (negative electrode). When the electrodes were connected together, an electric current, albeit a small one, was generated. As the hydrogen and oxygen gasses were consumed, water was produced and the gasses had to be replenished.

Grove connected his fuel cells together in series to produce more power, but the fuel cell remained more of a curiosity for nearly a century. In 1932 English chemist Francis T. Bacon resurrected the concept and made improvements to it. He used nickel electrodes, which were much less expensive than those of platinum, and pressurized the gas. In 1959 his "Bacon cell" was able to produce six kilowatts of power. While six kilowatts would not jeopardize the operation of conventional electric generating plants, it was a considerable improvement and rekindled interest.

Large corporations also became involved in the technology. General Electric built an experimental **tractor** using 1,008 fuel cells, and the Chrysler Corporation constructed a fuel-cell car with four electric motors, one for each **wheel**. The first practical use of the fuel cell was in powering **satellite**s in orbit around the earth. Fuel cells were used to provide electrical power in the Gemini and Apollo **spacecraft**. On April 13, 1970, an oxygen tank aboard the *Apollo 13* command module exploded causing two of the three fuel cells to fail. The loss of oxygen, electricity, and light very nearly resulted in disaster for the astronauts, who were en route to the moon at the time.

Fuel cells are extremely efficient and can theoretically convert 100% of their fuel into electricity; those aboard the Apollo spacecraft operated at 87% efficiency, much better (and cleaner) than the 40% to 50% efficiency of a conventional power plant and the 10% to 20% percent efficiency of an **internal combustion engine**. In 1984 a 4.8-megawatt plant was put into operation in Tokyo.

Cars powered by fuel-cells would easily outperform today's electric battery-powered cars, and several companies are working to develop such technology. In 1998, Ballard Power Systems Inc. of Vancouver, British Columbia, Canada produced a fuel cell about 2 mm (0.08 in) thick which, when stacked in series, powered a automobile 250 miles. A problem with such cars is the choice of fuel: pure hydrogen, **methanol**, or **natural gas**. Prototype cars often have large tanks replacing the rear seats, and can produce some **carbon dioxide** when, for example, the methanol is converted to hydrogen. Nonetheless, the future looks bright for fuel cells to eventually provide some of the power not supplied by conventional electric generating plants, and an efficient electric **automobile** may soon make its appearance.

Fuel injection

American **automobiles** with **carburetors** were produced for the last time during the 1986 model year and were replaced by designs equipped with electronic fuel injection (EFI) systems. Clean air standards were the driving force that pushed manufacturers to reduce pollutants by re-engineering fuel delivery methods. Fuel injection, however, is not a new idea. **Rudolf Diesel** included it in his diesel engine patented in 1892 and 1893. Automotive engineers have analyzed means of maintaining car performance while meeting emissions standards; and sophisticated **electronics**, materials, and designs have made both performance and improved air quality possible through fuel-injected vehicles. Fuel injection typically improves horsepower, drivability, and fuel economy over carburetor-equipped cars.

All electronic fuel injection systems consist of sensors that feed data to an Electronic Control Unit (ECU). The ECU processes and controls the electrical operations of the engine and can adjust directions sent to the fuel injectors based on operating conditions at any given time while the automobile is running. The fuel injectors are meters for fuel flow. Engines with continuous combustion like **gas turbines** and liquid-fueled **rockets** must use injection systems because there is no suction from the motion of **pistons**.

At the inlet end of the injector, fuel flows in through a filter screen. The injector is activated by an electrical solenoid that receives its commands from the ECU. The solenoid lifts a ball off the injector seat, the injector sprays a cone-shaped spray of fuel above the throttle valve, and the fuel mixes with air and is atomized around the intake valve before the mixture is combusted. Various sensors monitor the amount of oxygen in the exhaust gas, the coolant temperature, and the manifold pressure. Inputs from these sensors tell the ECU how to adjust the air/fuel mixture. The throttle position sensor monitors throttle loads (ranging from idling while stopped to rapid acceleration), and it signals fuel needs to the ECU. Fuel temperature, air density, vehicle speed, and other characteristics are also monitored. Ports in the throttle body signal the air cleaner housing, air injector, the manifold, the spark advance, and purge systems to help the system clean itself and control air flow.

There are two types of EFI systems called throttle body injection and port injection, although all EFI systems have both throttle bodies and injectors. The throttle body system is similar to the carburetor in that the injectors spray the fuel directly into the intake manifold in this "wet manifold" method. The injectors are mounted above the throttle plates which regulate the passage of fuel through the throttle bores to the manifold where it mixes with air and then enters the combustion chambers. **Trucks** and off-road vehicles tend to have throttle body injection systems because they perform better at high altitudes, give better throttle response, and improve drivability.

In port injection, each cylinder has a dedicated injector that sprays fuel directly into the ports. A fuel rail feeds the injectors. This system is also called a "dry manifold" system because fuel is not injected into the intake manifold. Car manufacturers have their own names for variations of port injection, such as Tuned Port or Sequential Port Fuel Injection, depending on fine points of the operation. The port injection system dominates the industry because of efficiency; fuel sprayed directly into the combustion chamber burns optimally. Because more parts are required (eight injectors for a V-8 engine), however, maintenance costs may be higher.

Fuel injection systems operate best with quality **gasolines** containing additives that prevent varnish from plugging the tiny injectors and containing detergents that clean the whole system. Higher octane gasolines usually have the best combination of detergents and additives even if the engine itself doesn't need the octane level. Fuel system cleaners are also commercially available.

See also Automotive electronics

Fuller, (Richard) Buckminster (1895-1983)

American architect and inventor

Born on July 12, 1895, in Milton, Massachusetts, Richard Buckminster (Bucky) Fuller was to become one of the most creative thinkers of the twentieth century. Fuller was schooled at the Milton Academy and in 1913 entered Harvard University. After being expelled two years later for irresponsible conduct, Fuller found work as a machinist at a Sherbrooke, Quebec, textile mill. During World War I he served briefly in the U.S. Navy.

In 1917, Fuller married Anne Hewlett, the daughter of well-known architect and artist James Monroe Hewlett. Fuller and Hewlett formed a construction company, and after supervising the construction of several hundred houses, Fuller concluded that custom homes were inefficiently built.

In 1922 one of the Fullers' two daughters, Alexandra, died after a series of illnesses. Fuller blamed her death partly on a poor physical environment, and he vowed to improve that environment through comprehensive, anticipatory design. This philosophy became an important guiding force behind all of Fuller's work.

The following year, Fuller invented the stockade **brick**laying method, which consisted of bricks reinforced with concrete poured into vertical holes in the bricks. He went to Chicago, where his mother originally lived, to start a construction company, the Stockade Building System. There he had to battle strong construction workers' unions, which opposed his efficient building methods for fear that they would threaten their jobs. By 1927 he was forced to sell the business.

That year he designed his first factory-assembled Dymaxion house. ''Dymaxion'' was a term that Fuller used to refer to anything that derived maximum output from minimum material and effort. His ''house on a pole'' was a self-contained unit suspended from a central mast and included a complete recycling system.

In 1928, Fuller designed his Dymaxion car, an omnidirectional vehicle that gave minimum resistance to the wind. His two-year-old daughter, Allegra, called the teardrop-shaped car the ''zoommobile.'' While this vehicle could seat 12 passengers, run at 120 mi (190 km) per gallon, make 180 degree turns, and average 28 mi/gallon (12 km/l), it was an unprofitable venture because no automobile manufacturer would mass produce it. Although Fuller managed to pay off his debts, this enterprise bankrupted him and his family.

Although his motives were mainly altruistic, Fuller's first financial break came in 1940 with his Dymaxion Deployment Unit (DDU). This circular self-cooled living unit with pie-shaped rooms was made of corrugated steel. The British purchased DDUs for use in World War II. Before long, Fuller was shipping them all over the world.

Richard Buckminster Fuller

After the war, Fuller went on the lecture circuit and became quite a celebrity. He was appointed Dean at Black Mountain College in North Carolina in 1948, where he developed his Tensegrity Dome, which used tetrahedrons to balance the tension and integrity of the structure.

He left Black Mountain in 1949 and went to work on his **geodesic dome**. That year he constructed a model of the dome on the lawn of the Pentagon. He applied for a patent on the dome in 1951. Ford Company gave him his first major contract in 1953 to build a dome over the courtyard of the company headquarters in Detroit. The U.S. Defense Department became his largest customer, using his domes as temporary housing units and to house sensitive **radar** equipment in harsh environments.

Fuller became a research professor at Southern Illinois University in Carbondale, Illinois, in 1959. His stipend of $12,000 was insignificant, but he was given use of an entire building and staff. He went on to propose domes for whole cities, like East St. Louis, to protect them from industrial smog. By the time of his death in 1983, Fuller's domes were in place worldwide, making Fuller's name equally well known. He received even more attention after the discovery of an important new substance in 1985, two years after his death. *Fullerenes* are **carbon** atoms arranged in a sphere-like shape with many pentagonal and hexagonal faces, similar to a soccer ball. These ''Buckyballs'' can have up to 600 atoms and have unique properties.

Robert Fulton

FULTON, ROBERT (1765-1815)
American inventor

Robert Fulton, best known for his work in **steamboat** technology, was born in Little Britain, Pennsylvania, in 1765. As a child, Fulton enjoyed building mechanical devices, taking on such projects as **rocket**s and a hand-propelled **paddle wheel** boat. His interest turned to art as he matured, and by the age of seventeen, Fulton was supporting himself through his sales of portraits and technical drawings. In 1786, Fulton left the United States to study painting in England. Although he managed some success, the general response his work received was disappointing and convinced him to concentrate on his engineering skills.

The first project to capture his attention revealed his emerging interest in water transportation. His assignment involved designing a **canal** system to replace the locks that were then in use. After several years of work, Fulton came up with a double inclined plane system for which he was granted a British patent in 1794. His creative ideas continued to flow as he developed a plan for cast **iron** aqueducts and invented a digging machine; in 1796, he published a summary of his ideas on improving canal navigation in his *Treatise on the Improvement of Canal Navigation.*

In 1797 further research on canals took Fulton to Paris, France. While he was there he became fascinated with the notion of a ''plunging boat,'' or **submarine**, and began designing one based on the ideas of American inventor **David Bushnell**. Fulton approached the French government, then at war with England, with the suggestion that his submarine could be used to place powder mines on the bottom of British **warship**s. After some persuasion, the French agreed to fund the development of the boats and, in 1800, Fulton launched the first submarine, the *Nautilus*, at Rouen.

The 24 1/2 foot (7.5 m) long, oval-shaped vessel sailed above the water like a normal ship, but the mast and sail could be laid flat against the deck when the craft was submerged to a depth of twenty-five feet by filling its hollow metal keel with water. Fulton's plan was to hammer a spike from the metal conning tower into the bottom of a targeted ship. A time-released mine attached to the spike was designed to explode once the submarine was out of range. Although the system worked in the trials, British warships were much faster than the sloop used in the experiments and thus managed to elude the slower submarine. The French stopped funding the project after the failed battle attempt, but the British, who considered the technology promising, brought Fulton over to their side. Unfortunately, once again the submarine worked well in tests, but proved unsatisfactory on practical situations. After its failure in the Battle of Trafalgar (1805), the British too abandoned the project.

After these experiences, the undaunted Fulton turned to a new area of exploration—steam. Correspondence indicates that he had been aware of work on the movement of ships by steam power since at least 1793. Through his contacts in Paris, Fulton met Robert Livingston (1746-1813), the American foreign minister to France who also owned a twenty-year monopoly on steam navigation in New York State. Fulton shared some of his ideas about steam power with Livingston and, in 1802, the two decided to form a business partnership. The following year, they launched a steamboat on the Seine River that was based on the design of fellow American **John Fitch**. The vessel traveled at a speed of three miles per hour and, although some adjustments were necessary to make the craft sufficiently seaworthy, it was clear that the basic technology worked well.

Fulton returned to New York later in 1803 to continue developing his designs, conscious of the fact that his partner's monopoly was contingent on their development of a boat that could travel at least four miles per hour. After four years of work, Fulton launched the *Clermont*, a steam-powered vessel with a speed of nearly five miles per hour. The partnership between Fulton and Livingston thrived, and Fulton had at last achieved a recognized success. During the ensuing years, Fulton designed thirteen more steamboats, including the *Demologus*, a **warship**; and he established an engine works in New Jersey that produced **steam engines**.

Fulton died on February 24, 1814. His persistence and belief in his ideas helped steamboats become a major source of transportation on the rivers in the United States, and resulted in a significant reduction of domestic shipping costs.

FURFURAL

Furfural—also called furfuraldehyde, fural, 2-furaldehyde, pyromucic aldehyde, or 2-furancarboxaldehyde—is a viscous, colorless liquid with a freezing point of-37.6° F (-3.11° C) and a boiling point of 323° F (161° C). Exposed to air, it turns dark brown. A highly reactive aldehyde, furfural is derived from corn cobs, hulls of rice, cottonseeds, or oats that have been treated with hot hydrochloric acid. It is useful as an intermediate in the manufacture of many **polymers** and to dissolve impurities in petroleum compounds and vegetable and lubricating oils. Furfural is a selective solvent meaning that it dissolves only some materials.

Johann Doebereiner, (1780-1849), a German chemist accidentally discovered furfural in 1832 when he treated sugar with sulfuric acid and manganese dioxide. Large-scale manufacturing techniques were developed in the early 1920s. Furfural consists of a string of carbon atoms connected in a ring-shaped structure which contains both an aldehyde group (R-COH) and an ether bond (C-O-C). Because the bonds in furfural alternate between double and single bonds, there are several bonding sites which makes furfural highly reactive. The ring is usually opened at the ether linkage. Furfural reacts with phenols, ketones, and esters as an aldehyde; removal of the aldehyde group yields furan, which is converted to tetrahydrofuran (THF), used in the manufacture of **nylon**. Butadiene is a derivative of THF, as well. Furfural is also used in the manufacture of several synthetic resins contained in **plastic** products. Additionally, it is used in **fungicide**s, germicides, **herbicide**s, and **insecticides**, and as a catalyst in the vulcanization process. Today furfural is used to produce furfuryl alcohols which are used used in foundry sand binders.

Under the Clean Water Act, furfural is listed as a hazardous substance. Because it is a skin irritant, protective measures such as gloves, goggles, protective clothing, or engineering controls must be used to prevent contact.

FUSE • See Circuit breaker and fuse

FUZE • See Ammunition

G

Gabor, Dennis (1900-1979)
Hungarian British physicist

During the 1940s, while most American scientists were striving to perfect the military technology that would fuel the cold war, Dennis Gabor was developing a new science, one that would improve existing imaging processes by creating true three-dimensional pictures. Though essentially overlooked by the scientific community of the time, Gabor's studies would later pave the way for the invention of the **hologram**.

During his childhood in Budapest, Gabor showed an advanced aptitude for science; in their home laboratory, he and his brother would often duplicate the experiments they read about in scientific journals. He entered Budapest Technical University at age eighteen, but had only completed three years of study when he was drafted into the military. Gabor, who had already served two years in World War I, chose not to fight for Hungary's newly restored monarchy; instead, he moved to Berlin to complete his education. It was here, at Berlin Technical University, that he studied under such influential scientists as Max Planck (1858-1947), Max von Laue (1879-1960), and Albert Einstein (1879-1955). Gabor remained in Germany until the rise of Hitler, after which he fled to England, where he conducted the bulk of his research.

In England, Gabor worked primarily upon the **electron microscope**. An inherent flaw in the device frustrated him: at high magnification levels, only a very small portion of the subject could be resolved, while the rest of the information would be unreadable. As a solution, Gabor theorized that it would be possible to take a ''bad picture'' with the electron microscope, but one which contained the complete information, and then correct it later. He called this bad picture a hologram, meaning ''complete picture.''

His initial demonstrations of this process, however, were unimpressive: in order to create a true hologram, he needed a coherent light source, something that did not exist in 1947. Thus, the hologram sat idle until 1960, when **Theodore Harold Maiman** built the first working **laser**. The coherent, monochromatic light produced by the laser was ideal for Gabor's needs, and his theories were soon used to their fullest potential.

During the 1960s, the field of **holography** exploded and today has become a multimillion dollar industry. Holography has found applications in medicine, cartography, and computer information storage. Holograms can be found in advertising, on magazine covers, and on credit cards to prevent counterfeiting. Gabor was awarded numerous scientific accolades for his discovery, including the Rumford Medal, in 1968, and the Nobel Prize, in 1971.

Galilei, Galileo • See Galileo

Galileo (1564-1642)
Italian mathematician and astronomer

Galileo Galilei is credited with establishing the modern experimental method. Before Galileo, knowledge of the physical world that was advanced by scientists and thinkers was for the most part a matter of hypothesis and conjecture. In contrast, Galileo introduced the practice of proving or disproving a scientific theory by conducting tests and observing the results. His desire to increase the precision of his observations led him to develop a number of inventions, particularly in the fields of physics and astronomy.

The son of Vincenzo Galilei (c. 1520-1591), an eminent composer and music theorist, Galileo was born in Pisa. He received his early education at a monastery near Florence, and in 1581 entered the University of Pisa to study medicine. While a student he observed a hanging lamp that was swinging back and forth, and noted that the amount of time it took the lamp to complete an oscillation remained constant, even as the arc of the swing steadily decreased. He later experimented with other suspended objects and discovered that they behaved in the same way, suggesting to him the principle of the pendulum. From this discovery he was able to invent an instrument that measured time, which doctors found to be useful for mea-

Dennis Gabor

suring a patient's pulse rate, and **Christiaan Huygens** later adapted the principle of a swinging pendulum to build a pendulum **clock**.

While at the University of Pisa, Galileo listened in on a geometry lesson and afterward abandoned his medical studies to devote himself to mathematics. However, he was unable to complete a degree at the university due to lack of funds. He returned to Florence in 1585, having studied the works of Euclid (3rd century B.C.) and **Archimedes**. He expanded on Archimedes's work in hydrostatics by creating a hydrostatic **balance**, a device designed to measure the density of objects. The following year, he published an essay describing his new invention, which determined the specific gravity of objects by weighing them in water. With the hydrostatic balance, Galileo gained a scientific reputation throughout Italy.

In 1592 Galileo was appointed professor of mathematics at Padua University, where he conducted experiments with falling objects. Aristotle had stated that a heavier object should fall faster than a lighter one. It is said that Galileo tested Aristotle's assertion by climbing the leaning tower of Pisa, dropping objects of various weights, and proving conclusively that all objects, regardless of weight, fall at the same rate.

Some of Galileo's experiments did not turn out as expected. He tried to determine the speed of light by stationing an assistant on a hill while he stood on another and timed the flash of a lantern between the hills. He failed because the hilltops were much too close together to make a measurement.

In 1593 he invented one of the first measuring devices to be used in science: the **thermometer**. Galileo's thermometer employed a bulb of air that expanded or contracted as temperature changed and in so doing caused the level of a column of water to rise or fall. Though this device was inaccurate because it did not account for changes in air pressure, it was the forerunner of improved instruments.

From 1602 to 1609 Galileo studied the motion of pendulums and other objects along arcs and inclines. Using inclined planes that he built, he concluded that falling objects accelerate at a constant rate. This law of uniform acceleration later helped Isaac Newton (1642-1727) derive the law of gravity.

Galileo did not make his first contribution to astronomy until 1604, when a supernova abruptly exploded into view. Galileo postulated that this object was farther away than the planets and pointed out that this meant that Aristotle's ''perfect and unchanging heavens'' were not unchanging after all. Ironically, Galileo's best known invention, the **telescope**, was *not* his creation after all. The telescope was actually invented in 1608 by Hans Lippershey, a Danish spectacle maker. When Galileo learned of the invention in mid-1609, he quickly built one himself and made several improvements. His altered telescope could magnify objects at nine power, three times the magnification of Lippershey's model. Galileo's telescope proved to be very valuable for maritime applications, and Galileo was rewarded with a lifetime appointment to the University of Venice.

He continued his work, and by the end of the year he had built a telescope that could magnify at 30 power. The discoveries he made with this instrument revolutionized astronomy. Galileo saw jagged edges on the moon, which he realized were the tops of mountains. He assumed that the moon's large dark areas were bodies of water, which he called maria (though we now know there is no water on the moon). When he observed the Milky Way, Galileo was amazed to discover Jupiter, which resulted in his discovery of its four moons; he later called them ''satellites,'' a term suggested by the German astronomer Johannes Kepler (1571-1630). Galileo named the moons of Jupiter, Sidera Medicea (''Medicean stars'') in honor of Cosimo de Medici, the Grand Duke of Tuscany, whom Galileo served as ''first philosopher and mathematician'' after leaving the University of Pisa in 1610. Also, with repeated observation, he was able to watch the moons as they were being eclipsed by Jupiter and from this he was able to correctly estimate the period of rotation of each of the moons.

In 1610 he outlined his planetary discoveries in a small book called *Siderus Nuncius* (''The Sidereal Messenger''). Venus, seen through the telescope, exhibited phases like the Moon, and for the same reasons: Venus did not produce its own light but was illuminated by the Sun.

Saturn was a mystery: Galileo's 30-power telescope was at the limit of its ability to resolve Saturn, and the planet appeared to have three indistinct parts. When Galileo looked at the Sun, he saw dark spots on its disc. The position of the spots changed from day to day, allowing Galileo to determine the rotational rate of the Sun.

In 1613 Galileo published a book in which, for the first time, he presented evidence for and openly defended the model of the solar system earlier proposed by the Polish astronomer Nicholas Copernicus (1473-1543), who argued that the Earth, rather than being positioned at the center of the universe, as in the Ptolemaic design, was only one of several galactic bodies that orbited the Sun. While there was some support even among ecclesiastical authorities for Galileo's proof of the Copernican theory, the Roman Catholic hierarchy ultimately determined that a revision of the long-held astronomical doctrines of the church was unnecessary. Thus, in 1616 a decree was issued by the church declaring the Copernican system "false and erroneous," and Galileo was ordered not to support this system.

Following this run-in with the Catholic church and the Inquisition which forced his adherence to the Copernican theory of the solar system, Galileo focused on the problem of determining longitude at sea, which required a reliable clock. Galileo thought it possible to measure time by observing eclipses of Jupiter's moons. Unfortunately, this idea was not practical for eclipses could not be predicted with enough accuracy and observing celestial bodies from a rocking ship was nearly impossible.

Galileo wanted to have the edict against the Copernican theory revoked, and in 1624 travelled to Rome to make his appeal to the newly elected pope, Urban VIII. The pope would not revoke the edict but did give Galileo permission to write about the Copernican system, with the provision that it would not be given preference to the church-sanctioned Ptolemaic model of the universe.

With Urban's imprimatur, Galileo wrote his *Dialogue Concerning the Two Chief World Systems—Ptolemaic and Copernican*, which was published in 1632. Despite his agreement not to favor the Copernican view, the objections to it in the *Dialogue* are made to sound unconvincing and even ridiculous. Summoned to Rome to stand before the Inquisition, Galileo was accused of violating the original proscription of 1616 forbidding him to promote the Copernican theory. Put on trial for heresy, he was found guilty and ordered to recant his errors. At some point during this ordeal Galileo is supposed to have made his famous statement: "And yet it moves," referring to the Copernican doctrine of the Earth's rotation on its axis.

While the judgement against Galileo included a term of imprisonment, the pope commuted this sentence to house arrest at Galileo's home near Florence. Although he was forbidden to publish any further works, he devoted himself to his work on motions and parabolic trajectories, arriving at theories that were later refined by others and made an important impact on gunnery. He died blind at the age of 78 in 1642.

Galileo Galilei

GALVANIZED IRON • See Iron production

GALVANOMETER

The galvanometer, a device used to measure extremely small electrical currents, traces its origin back to 1820. In that year Hans Christian Oersted (1777-1851) discovered that an electric current flowing in a wire created a magnetic field around it, deflecting a magnetized needle. This effect became the basic principle behind the galvanometer. In the same year André Ampère (1775-1836) used the effect to invent a device to measure electric current. He suggested it be called the galvanometer, in honor of Luigi Galvani (1737-1798), a pioneer in the investigation of electricity.

The first practical use of the galvanometer was made by **Karl Friedrich Gauss** in 1832. Gauss built a **telegraph** that sent signals by deflecting a magnetic needle. This style is known as a *moving-magnet* galvanometer. More commonly used today is the *moving-coil* or *moving-mirror* galvanometer, sometimes called a D'Arsonval galvanometer.

The invention of the moving-coil galvanometer is credited to Johann Schweigger in 1825, three years later Italian physicist C. L. Nobilli designed an astatic type. It consists of a coil that has been wound with very fine **wire** mounted between the poles of a permanent **magnet**. Attached to the coil is a pointer. When electric current is turned on, the coil turns and the deflection angle is measured as the pointer moves along a graduated scale.

In the case of a moving-mirror galvanometer, a **mirror** is attached to the coil, and illuminated with light. When the coil moves the deflection of the light is measured along a scale. The mirror galvanometer was of major use in laying the transatlantic telegraph cable between the United States and Europe in 1866. **William Thomson**, later known as Lord Kelvin, used it to keep track of how much electric current was coursing through the cable. Thomson also invented a "siphon recorder," which was a more sensitive galvanometer. Ink was siphoned through a thin **glass** tube that was attached to the coil of wire which was mounted between the poles of a horseshoe magnet. The moving tube carried the ink onto a paper tape where it traced a line.

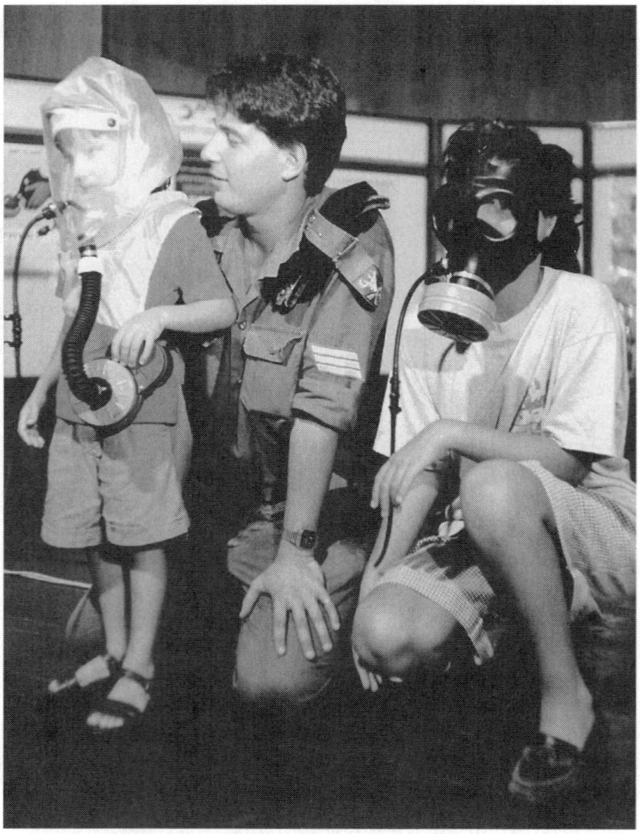

An Israeli soldier instructs children on how to wear gas masks during the Gulf War in 1990.

Galvanometers come in a variety of types. Ultraviolet recorders use light-sensitive paper and ultraviolet light in place of ink. A photoelectric galvanometer amplifies the signal using a photocell. The ballistic galvanometer is used to measure an electric pulse or burst. A cousin of the galvanometer is the direct current **ammeter**, which is a calibrated galvanometer that measures larger currents. Another cousin still is the direct-current **voltmeter**, which uses Ohm's Law to measure voltage. Digital display galvanometers, the best of which can measure a current as small as one hundred billionth (10^{-11}) of an amp, have almost entirely replaced the early analog galvanometers.

GARAND • See Rifle

GAS MASK

Anselme Payen and **Garrett Augustus Morgan** are credited with inventing the modern gas mask. Payen, a French chemist, was the son of an industrialist who established chemical factories. After his father's death, Payen took over the family business and turned his attention to a factory that refined sugar from sugar beets. In 1822 Payen used animal charcoal to remove large-molecular impurities. Eventually his use of charcoal to absorb impurities became an important feature of the gas masks used in World War I.

Morgan, an African American, was born in Paris, Tennessee in 1875. After working as a general handyman in Cincinnati, Ohio, he left for Cleveland in 1895. Morgan eventually opened a shop that sold and repaired **sewing machines**. In 1912 he came out with his most important invention, the Safety Hood or ''Breathing Device,'' as he called it. His Breathing Device was a hood placed over the head of the user. Two tubes were connected to the hood: one that provided fresh air; the other to discharge exhaled air. The fresh-air tube was lined with an absorbent material that could be moistened with water to keep out smoke and dust particles.

Although Morgan demonstrated his gas mask at several exhibits and expositions, he was not very successful in marketing his device until one crucial incident proved its value. In 1916 a violent explosion at the Cleveland Waterworks had trapped workers inside a tunnel under Lake Erie. Heavy smoke and poisonous gases prevented any rescue attempts until Morgan arrived with several of his gas masks. He and three volunteers used the masks to save the lives of more than 20 men by carrying them out of the tunnel. Morgan's invention generated much interest following the rescue, but many customers cancelled their orders after discovering that he was a black man, and in the segregated South, Morgan had to employ a white man to demonstrate his invention to potential buyers.

Meanwhile, during World War I the German army used poison gas for the first time in battle. The English, who had no access to Morgan's invention, used chemically treated cotton pads tied over their mouths and noses. Soon, advanced gas masks were devised based on Morgan's invention. These devices, which consisted of the mask itself and a large tube that connected the mask to a canister that hung in front of the soldier's body, were very cumbersome. Inside the canister was charcoal, which filtered the poison gases. Gas masks devised for World War II were lighter, better fitting, and allowed for better vision. The filter was redesigned to wear over the shoulder for easier carrying.

In the 1960s, the United States military developed the M-17 mask that proved to be a breakthrough in protecting soldiers from biological, chemical, and radiological agents. The M-17 had no hose or external canister; air is filtered through pads of flexible material enclosed in cavities molded into the facepiece of the mask. The successor to the M-17 is the M-45, developed in the late 1990s by a California firm for the U.S. Army. The M-45 used plastic parts to replace heavier metal parts in earlier models, and used an inner lining of a silicone-based compound to make the mask more comfortable than previous ones. Other advances in gas mask technology in the late 1990s included built-in microphones and drinking tubes, and interchangeable tinted lenses.

Most gas masks today are used by police and fire fighters or workers in chemical plants. However, the military and civilian need for gas masks became more marked in the 1990s, with the rise of chemical weapons. Gas masks became a commonplace of Israeli life in the 1990s, when fear of Iraq's chemical weapons arsenal was intense. First the military, then all citizens, were issued gas masks, and newborn babies were equipped with special gas-filtering cradle covers. Israel's only gas mask manufacturer had difficulty producing enough masks to cover non-citizens and visitors as well. The difficulty of protecting an entire population from chemical attack became apparent, as special masks had to be produced for heavily bearded men, and groups such as illegal aliens protested their lack of access to masks.

Gas turbine

A turbine represents a simple but effective way to harness energy by changing the force of a moving fluid into circular motion. Gas turbines are essentially a refinement of an ancient technology that underwent a rapid development in the nineteenth and twentieth centuries.

Water turbines such as water wheels, which use the energy of moving water to drive grain mills, for instance, have been in use for millennia. **Windmills**, which came in use in the Middle East in the 900s and in Europe in the 1100s, formed the next major step in the development of the turbine. The invention of machines that burned fuel to boost the power output of the turbine took place quite early—there is a description of a simple **steam turbine** dating to 120 B.C.—but remained inconsequential until the mid-19th century, when the development of the steam turbine for industrial purposes came to represent a significant share of generated power. Steam turbines are still widely used, in such technological applications as nuclear power stations and ships, because they are smaller and lighter than equally powerful piston-engines.

Attempts to develop gas turbines were first undertaken in the early 1900s, with pioneering work done in Germany. The most successful early gas turbines were built by Holzwarth, who developed a series of models between 1908 and 1933. The first industrial application of a gas turbine was installed in a steel works in Hamborn, Germany, in 1933. In 1939 a gas turbine was installed in a power plant in Neuchâtel. The basic principles of the gas turbine involve the compression of a gas, usually air, in a compressor. The compressed air then passes into a combustion chamber where it is mixed with a burning fuel. The gas expands as a result of the combustion and rushes into the turbine chamber, which is fitted out with a series of fanlike wheels. The wheels turn, driving both the compressor and an electric **generator**. The earliest gas turbines proved much less efficient than steam turbines, because sufficiently powerful compressors were not yet available. A gas turbine is most efficient at extremely high temperatures, and materials that could withstand such temperatures had yet to be developed. In addition, it turned out that most fuels were not suitable, causing too much corrosion of the turbine blades.

However, the development of gas turbines received tremendous stimulus with the discovery that such engines were vastly more efficient for high-speed, high-altitude flight than the piston-and-propeller engines then in use in **aircraft**. In addition, gas turbines are small and light, making them particularly suitable for flight applications. In Great Britain, **Frank Whittle**, an air force cadet, received a patent on an aircraft gas turbine jet propulsion engine in 1930. At the same time, German engineers, particularly Hans von Ohain, were experimenting with the technology. The first gas turbine-powered craft, a Heinkel HeS3, flew in 1939. The British first flew a gas turbine engine in 1941.

The operating principles of gas turbines in jet propulsion are slightly different, since the turbine itself is mainly used to drive the compressor, and sometimes a propeller. The bulk of the escaping hot gas serves directly to produce forward thrust, as in a balloon ''powered'' by air escaping from the inlet valve. The gas turbine revolutionized both military and commercial aviation, and today all aircraft have some form of gas turbine engine. Missiles are also powered by gas turbines. Land-based applications of gas turbine technology remain rather limited, since steam turbines have far greater fuel efficiency and can operate on many different kinds of fuel. Gas turbines are mainly used as stand-by generators to handle system overloads and in pipeline pumping stations.

Gaslight

Although gaslights might seem quaint today, the introduction of gas lighting transformed the way people lived during the 1800s. Just when books and newspapers were becoming less expensive, gas lighting made it possible for poor families to read at night. And as the industrial revolution began, factories could operate for longer hours using gaslights in place of hazardous, expensive **candle**s.

The Chinese first used gas for lighting in salt mines thousands of years ago, but the concept of indoor gas lights was not explored until the late 1700s. By then, coal was commonly used for heating, and scientists knew how to produce flammable gas from coal as well as from other solid fuels such as wood and peat. In 1792 **William Murdock**, a Scottish inventor living in Cornwall, illuminated his own house with coal-gas, even though many people feared the lights would cause a fatal explosion. The gas illumination proved safe; Murdock continued his experiments and developed methods of making, storing, and purifying coal-gas. Around the same time, French chemist **Philippe Lebon** was experimenting with gas made from sawdust. Lebon used this gas to light his home, and in 1799 he patented a wood-gas light called the Thermolamp and placed it on public display. However, the French government decided not to fund Lebon's scheme for a large-scale gas-lighting system, and his **lamp** never enjoyed widespread commercial use.

Although both Lebon and Murdock deserve credit as pioneers of the gas-lighting industry, coal-gas proved superior to wood-gas as a lighting fuel. Gas made from coal contains about 50% hydrogen; the rest is methane, carbon monoxide, and a few other gases. Compared with natural gas, coal-gas is inferior in quality, but at the time it represented a vast improvement over coal itself. Gas could be piped from place to place, which was much easier than lugging tons of coal around, and gas flames were easier to ignite and to control by simply adjusting the gas flow to the burners. When William Murdock's employers heard about Lebon's experiments in France, they began to take gas lighting seriously and asked Murdock to install gaslights at their main factory in Birmingham, England. Then in 1802, to celebrate a temporary peace treaty between England and France, all of Birmingham was illuminated by coal-gas lights. This spectacular display marked the beginning of a whirlwind of activity for the gas-lighting industry. Murdock's company began manufacturing piping and burners for gaslights, and coal-gas was used to illuminate other factories and mills. However, some still feared that gas would explode, and its use for lighting in factories was suspected of being unhealthy for workers.

When early gaslights were introduced, coal-gas was made in small vessels to supply a single installation. A German entrepreneur, **Frederick Winsor**, was responsible for changing all that. He realized that it would be more efficient and profitable to make gas in large amounts, distribute it through underground pipes to several places, and sell the leftover solid ''coke'' as a low-quality fuel. Friedrich Accum (1769-1838), a German chemist who worked for Winsor, was responsible for applying Winsor's patented gas-lighting process; he conducted the experiments that enabled Winsor's company to compete with Murdock and his employers for the British gas-lighting business. Accum later went on to write what is now considered a classic text on gas technology. To fund his idea for centralized gas distribution, Winsor advertised for partners and formed a joint venture that succeeded in gaining support from an influential committee in England's Parliament. In 1807, Winsor and his partners illuminated a garden wall for the king's birthday and then staged the first gas streetlighting display in London. This installation, now one of the oldest in the world, put to rest any remaining doubts about the safety and practicality of gaslights. Parliament finally granted Winsor a charter for the first gas company, National Heat and Light, which was founded in 1812. Soon other cities and towns began installing central gasworks and distribution systems. By 1819, London alone had nearly 300 miles (480 km) of gas mains supplying more than 50,000 burners.

Not to be left out, American cities followed in the footsteps of their European counterparts. In 1816, officials in Baltimore, Maryland authorized Rembrandt Peale (1778-1860) to install a gas distribution and lighting system for the city's streets and buildings. Earlier, Peale's father Charles Peale (1741-1827) had experimented with indoor gaslights when he established his natural history museum in Philadelphia's Independence Hall in 1802. By the late 1800s, nearly one thousand American companies were making gas from coal, mainly for lighting. The first Canadian gas company was formed in 1836 to light the streets of Montreal.

For a while, the gas-lighting industry prospered, and England in particular thrived on exports of gas-making equipment and piping. Gaslights gradually found their way from factories and streets into private homes, where they were in common use by the 1860s. Also, gas-lighting was used in tech-

Gaslights lining a street.

nical institutes, where working people could continue their education at night, and in public halls, where evening meetings encouraged people's participation in politics and other social issues. The gas industry was shocked from its complacency, however, when electric lamps were introduced in the late 1800s. Both the **arc lamp** and the electric **incandescent light bulb** threatened the industry's monopoly on the lighting business. Only **Carl Auer**'s timely invention of the gas mantle in 1885 saved gas-lighting from extinction. Auer's incandescent mantle, a cylindrical shell of metal oxides, glowed with a bright white light when placed over a gas flame. Thanks to their better quality of light, gas was able to compete with early electric lamps, which produced a relatively feeble light. In the 1890s, gas companies introduced meters that could be turned on when needed by inserting coins—a sort of pay-as-you-go lighting system. With these meters, working-class families were able to afford gas-lighting, and the number of customers was greatly increased. By comparison, electricity was expensive, and most houses did not yet have access to it. Coal-gas continued to be used for lighting well into the twentieth century. With the development of new gas-burning appliances, coal-gas became a source of heating and cooking as well. In America, coal-gas was replaced to some extent by natural gas, which could be piped at higher pressure over greater distances; however, natural gas was not discovered in Europe until much later. A few original gaslights are still in use in Europe and the United States, mainly as historical curiosities. Recently, American gas utilities have begun to look for new ways to market their product, and gas-lighting is enjoying a small resurgence in popularity as a decorative item. Researchers are developing new gaslights with more durable mantles and higher efficiency.

Gasoline

Gasoline is a mixture of liquid hydrocarbons distilled from crude oil which is used as a fuel for **internal combustion engines**. Gasoline was invented after discovery of crude oil in the late 1850s when various refining processes were developed to make the oil useable. Early types of gasoline were produced as a byproduct of the process used to make **kerosene** fuel for **oil lamps**. Since this was before the development of the internal combustion engine, much of this early gasoline was discarded because no one had any use for it.

When the automobile was invented a new market was created for gasoline. At first, automobile engines used "straight-run" gasoline—the natural gasoline fraction produced by distilling crude oil. But this process yielded less than fifteen barrels of gasoline from each barrel of oil. After the mass production of cars began in 1908, oil refiners could not keep up with the growing demand for gasoline.

In 1913, just in time for World War I, a process was invented to increase the amount of gasoline produced from crude oil. **William Burton**, who worked for Standard Oil of Indiana, developed *thermal cracking*, a process by which heavy hydrocarbons are broken down by heat and pressure into the lighter compounds used in gasoline. Since Burton's discovery, this basic process has been greatly improved. During the 1930s catalysts were introduced to promote chemical reactions during cracking. Besides increasing gasoline yields, *catalytic cracking* produces a higher quality gasoline than does thermal cracking.

During World War II catalytic cracking and other new refining processes greatly increased the United States' output of gasoline. More than eighty percent of the aviation fuel used by the Allies during the war was supplied by the United States. In Europe, gasoline became extremely scarce, and the German army had to rely on cruder types of gasoline that were produced from coal and heavy oil. The hydrogenation process for making this fuel had been developed in the 1920s by Friedrich Bergius (1884-1949), a German chemist who later fled his native country. A similar process developed in 1923 is called Fischer-Tropsch synthesis, which produces gasoline and other liquids from coal-derived synthesis gas (hydrogen and carbon monoxide). Gasoline can be made from just about any substance containing hydrogen and carbon.

Today's gasolines are blended from hundreds of hydrocarbons, and different combinations are produced to meet the needs of different engines. For example, engines vary in how hard they compress the fuel mixture of gasoline vapor and air. Although higher compression improves the engine's performance, it can also cause the gasoline to ignite too soon, creating a metallic "knocking" or pinging sound in the engine. This means that the engine is not burning fuel efficiently, and severe knocking can actually damage the engine. A gasoline's resistance to knocking is measured by its octane rating; if the gasoline's performance is ninety percent as good as that of a reference fuel (pure iso-octane), it gets an octane number of ninety.

In the early 1900s engine knock was recognized as a problem, and the auto industry began searching for a fuel that could withstand high pressures without knocking. While engineers experimented with different engine designs, chemists explored "additives"—substances that could be added to gasoline to prevent knocking. In 1921 a team of American chemists led by Thomas Midgley, Jr. and T. A. Boyd made a spectacular breakthrough at General Motors. After much trial and error, Midgley began a systematic study of promising compounds, based on the position of each compound's elements on the periodic table. "What had seemed at times a hopeless quest," he recalled, "rapidly turned into a fox hunt. Predictions began fulfilling themselves instead of fizzling." The essential compound turned out to be tetraethyl lead. When added to gasoline in minute amounts, tetraethyl lead prevents engine knock and increases the gasoline's octane rating. Between 1920 and 1950, gasoline octane numbers increased from fifty-five to eighty-five, allowing automakers to nearly double engine performance by using higher internal pressures.

Unfortunately, tetraethyl lead pollutes the air with poisonous lead compounds when the gasoline is burned, and today leaded gasoline is being phased out. Instead, new engines have been designed to run on lower octane gasoline, which is made of hydrocarbons that are resistant to knock. New additives have also been formulated to increase the octane numbers of unleaded gasoline. Other modern additives

William Gates

preserve fuel quality and prevent rust, ice, and deposits of burned solids in the engine and fueling system.

During the 1970s leaded gasoline became associated with another problem. When lead is present in exhaust fumes, it ruins the car's anti-pollution equipment. In 1970 the government required automakers to sharply reduce emissions of carbon monoxide and "unburned" hydrocarbons, which are produced when the engine is out of tune. To meet these standards, automakers introduced **catalytic converters**—devices that are attached to the exhaust system just behind the manifold. Most converters use platinum or palladium metal catalysts, which convert carbon monoxide and hydrocarbons to carbon dioxide and water vapor. The catalysts are easily poisoned by lead, however, which clogs their reactive surfaces. That is why most cars must now use unleaded gasoline.

The abundance of gasoline in the United States through most of the twentieth century has made many aspects of life convenient and pleasurable. People are able to travel greater distances to get to their jobs or to go on vacation. Farmers are able to produce more food by using gasoline-fueled machinery. But by the early 1970s, consumption of gasoline had grown so enormously that oil refiners began depending on imported oil. When foreign oil supplies were disrupted, gasoline supplies suddenly became limited. People waited for hours to fill up their tanks at service stations and gasoline prices skyrocketed from less than forty cents a gallon to more than a dollar.

Since then automakers have introduced smaller cars that use less fuel. It was the gasoline shortages of the 1970s that made compact Japanese cars popular in the United States for the first time. For a time the government also encouraged people to conserve gasoline by using public transportation, and states reduced highway speed limits. During the 1980s these conservation measures were neglected, and higher speeds are now allowed on many stretches of highway. However, the government still requires automakers to continue increasing fuel economy and reducing pollutant emissions.

In the 1990s concern about pollution is even greater and cleaner-burning gasoline formulas have been developed to meet increasingly stringent clean-air standards. These reformulated gasolines contain either corn-derived ethanol or natural-gas derivatives such as MTBE (methyl tertiary butyl ether). Both additives increase the oxygen content of the gasoline which causes it to burn more cleanly and evaporate more slowly. Another recent antipollution measure requires service stations to control the gasoline vapors released into the atmosphere during fueling operations. This involves anti-evaporation equipment which is built into the gas pumps.

See also Oil refining

Gates, William H. (1955-)

American inventor

William Gates, the son of a prominent Seattle, Washington lawyer, was an underachieving math whiz in school. He was first exposed to digital computers while in the seventh grade and worked on various computer projects for the next several years. While in high school, Gates and a friend, Paul Allen, formed a company called Traf-Data, which used the Intel 8008 **microprocessor** to help control traffic patterns in Seattle. Gates eventually sold this system to the city for $20,000 when he was only fifteen years old. He dropped out of high school for one year to work for TRW in computing, earning $30,000.

In 1974 Gates was attending Harvard University when Allen spotted an advertisement for a $350 assemble-at-home computer called the Altair 8800, manufactured by MITS, a company headquartered in Albuquerque, New Mexico. Gates and Allen worked nonstop for six weeks to devise a simple version of BASIC, a **programming language**, for the Altair. They demonstrated their finished product to the company's engineers with great success, and the following year, Gates and Allen founded Microsoft Corporation.

Microsoft had entered the personal computer (PC) industry at an opportune time. The Apple Computer Corporation had introduced its Apple II personal computer with much success. International Business Machines Corporation (IBM) decided to enter the market but needed software designed for specific tasks and an operating system that permitted hardware to interpret the software. In 1980 Microsoft won a contract with IBM to operate personal computers using Microsoft's system, commonly known as MS-DOS. By 1983 the IBM PC had become the industry standard and MS-DOS was its operating system. In 1995, Microsoft introduced Windows 95 which again revolutionized the PC market, and became an industry standard. Today some 90% of all personal computers use Microsoft Windows. In 1991 Gates was ranked number three on the *Forbes* magazine list of the wealthiest Americans; his share of Microsoft is worth approximately four billion dollars. By 1998 he had become the richest man in the world. In addition to operating systems, Microsoft produces software for word processing, spreadsheets, games, and Windows, which allows IBM PCs and their clones to use pictures and a handheld "mouse" to eliminate complicated keyboard commands, much like Apple did for its computers. Microsoft is currently the third largest company in the computer industry, trailing only IBM and Hewlett-Packard Company.

See also Computer input and output devices; Computer operating systems; Computer software; Jobs, Steven P.; Wozniak, Steve

Gatling, Richard Jordan (1818-1903)

American inventor

Born in Hertford County, North Carolina, Gatling showed signs of creativity even as a youngster. He helped his father, a plantation owner and inventor, develop machines for sowing and thinning cotton. In 1839 Gatling invented a screw propeller for ships and went on to develop agricultural machines, such as a hemp-breaking device and a steam **plow**.

When the Civil War began in 1861, Gatling focused his efforts on armaments. In 1862 he invented the weapon that has carried his name ever since, the Gatling gun. Considered the first practical **machine gun**, the Gatling gun was capable of firing 250 shots per minute. It consisted of ten breach-loading **rifle** barrels, cranked by hand, that rotated around a central axis. Each individual rifle barrel was loaded by gravity feed and fired while the entire assembly revolved. Cartridges were automatically ejected as the other barrels fired. It was operated by two persons: one who fed the **ammunition** that entered from the top, and the other who turned the crank that rotated the barrels. At first, the Union Army was uninterested in Gatling's invention, but General Benjamin Butler (1818-1893) eventually bought several Gatling guns for one of his final campaigns. They worked so well on the battlefield that the government finally agreed to adopt them in 1866, but by then the war was over.

After the war, Gatling continued to improve his gun. Eventually it was capable of firing 1,200 shots per minute at all degrees of elevation and depression. Gatling's gun was used all over the world, and remained in the U.S. military arsenal until 1911. Gatling remained active in other areas of invention: he worked on a new method for casting cannons, built a motorized plow and invented a new alloy of steel and aluminum.

Richard Jordan Gatling

Gauss, Carl Friedrich (1777-1855)

German mathematician

A true child prodigy, Carl Gauss was a mathematical genius. Born into an extremely poor family in Brunswick, Germany, on April 30, 1777, Gauss amazed his parents by learning to add numbers and make calculations before he was able to talk. He was correcting his father's addition at the age of three. Gauss taught himself to read and, when fourteen years old, received a stipend from Duke Ferdinand of Brunswick to study science. He eventually entered the University of Göttingen in 1795 and obtained his doctorate degree in 1799.

Gauss is best known for his mathematical advances. He devised the method of least squares while a teenager; calculated the orbit of the asteroid Ceres, permitting it to be rediscovered after it had been lost; calculated theories of perturbations between the planets, which led to the discovery of Neptune; constructed an equilateral polygon of seventeen sides; and established a non-Euclidean geometry. Unfortunately Gauss tended to keep some of his work secret; two others who also had devised non-Euclidean geometry received the credit when they published first. No matter; his contemporaries hailed Gauss as one of the greatest mathematicians that ever lived; up until this time only **Archimedes** and Isaac Newton (1642-1727) had been so honored.

In 1807 Gauss was appointed director of the Göttingen Observatory and he became interested in geodesy, which involves surveying large areas of the Earth. This was necessary to pinpoint the exact location of the observatory and led him to create a new, improved method of surveying. In 1821 he invented a device called the heliotrope, an instrument that could reflect sunlight over long distances. With it, parallel rays of

Carl Friedrich Gauss

light could be used to mark straight lines on the curved Earth, allowing for precise trigonometric calculations.

Next Gauss became involved in magnetism and established the first observatory to specialize in that field. Working with colleague Wilhelm Weber (1804-1891), Gauss began making a worldwide magnetic survey, the results of which allowed the accurate determination of the Earth's magnetic poles.

Gauss and Weber became interested in electromagnetism and, making use of **Michael Faraday**'s 1831 discovery of magnetic induction, invented a **telegraph**. Their version differed from that of American physicist **Joseph Henry**, who was in the process of inventing his own telegraph at the same time. Henry's receiver used a metal arm which clicked up and down; Gauss's version consisted of a large coil over a **magnet** to create the current which deflected a magnetic needle at the receiver.

In 1833 Gauss and Weber ran a **wire** across Göttingen from the physics laboratory to the observatory, but abandoned it when they were unable to gain support for additional development. Ironically, in 1837 **Charles Wheatstone** patented his telegraph in England, and in 1840 **Samuel Morse** did the same in the United States, much to the chagrin of Henry.

After 1840 Gauss's scientific work slowly began to decline. On February 23, 1855, Gauss died at Göttingen. His name was honored when the standard unit of measurement of magnetic influence, which he and Weber had established, was named the *gauss*. When asteroid number 1,001 was discovered it was christened Gaussia.

See also Surveying instruments

Gear

A gear is a component of a machine that is designed to transfer motion between rotating shafts.

A gear usually consists of a wheel with teeth around its edges, and gears operate in pairs as the teeth of one gear slip into the teeth of the other. (A tooth on a gear is also called a *cog*.) If a shaft, called the *driving shaft*, turns a wheel, this motion can be transferred to the second gear and its corresponding shaft, the *driven shaft* because the gears' teeth interlock.

The smaller (in diameter) of two engaged gears is called the ''pinion.'' If the driving shaft is the pinion, the gear pair acts to slow the rotation down, increasing the torque, or transmitted force, of the driving shaft. Vice-versa, if the driven shaft is the pinion, the pair acts to increase the rotation, decreasing the torque. Properly-designed gears can in this way be used to transfer power, and to do so efficiently, the gear teeth must be carefully designed and constructed.

Gears work best when their two shafts are not far apart, but they need not be parallel. Appropriate gears can be made whether the driving and driven shafts are parallel or nonparallel, or intersecting or nonintersecting. Different gears are designed for each case. Parallel shafts are usually connected with spurred or helical gearing, spur gears being the most familiar type of straight, toothed gear. Helical gears have twisted, screw-like teeth. Intersecting shafts require a beveled gear, which has tapered teeth around a section of a cone. Nonparallel, nonintersecting shafts are usually connected by a ''worm'' and a gear, where the worm resembles a screw.

Gears pairs can be used in a series in order to obtain larger changes (ratios) in speed and torque. Gears are found in everything from clocks to bicycles to cars to heavy machinery.

See also Wheel and axle

Geiger counter

An active area of research in physics in the early 1900s was the attempt to develop instruments to detect and count various types of radiation. Since radiation cannot be discerned by any of the human senses unaided, such devices were essential to any study of radioactivity, discovered in 1896. One of the most important among the early detection instruments invented during this time was the Geiger counter, first developed by Hans Geiger in 1908.

Born in Neustadt, Rheinland-Pfalz, on September 30, 1882, Geiger began his studies in physics at the University of Munich, and completed his doctorate at Erlangen in 1906. In the same year, he accepted a position as research assistant at the University of Manchester, where Ernest Rutherford became director in 1907. Over the next decade, Rutherford and

Geiger collaborated on a number of fundamental research studies on the nature of alpha rays and alpha particles. At the outbreak of World War I, Geiger returned to his homeland to serve in the German army. After the war, he was appointed professor of physics at the University of Kiel. He took on a post at the University of Tübingen in 1929 and, in 1936, was appointed chairman of the physics department at the Technical University at Charlottenberg-Berlin. After losing his home and all his possessions in World War II, he died in Potsdam on September 24, 1945.

Geiger constructed the earliest form of his radiation counter in 1908. The device consisted of a wire extending down the center of a sealed metal tube with a glass or mica window at one end. The wire and the metal tube were connected to a power source through an external circuit. When radiation passed through the tube, it created a track of ion pairs. The positive ions in each pair were attracted to the negatively charged metal tube, while the negatively charged electrons were attracted to the central wire.

As the ions and electrons passed through the gas, they collided with other gas molecules, producing further ionization. Because of this avalanche effect, the number of ions and electrons reaching the outside circuit was sufficient to initiate an electric current that could be recorded as a sound or observed as a flash of light.

Geiger continued to work on the design of his counter over the next two decades. In 1928, he made further modifications based on the studies of a German colleague, Walther Müller. The final design, often referred to as the Geiger-Müller counter, soon became one of the two or three most widely used radiation counters available to scientists.

A number of variations of the design have been developed, among them the ionization counter and proportional counter. These variations can be used to identify the types of radation and the characteristics of particle that pass through a counter. The Geiger-Müller counter can also be modified to detect and count non-ionizing radiation, such as neutrons. In this case, boron fluoride gas is used to fill the chamber. The reaction between neutrons and boron nuclei results in the formation of alpha particles, which can then be detected like any other alpha particle.

See also Radiation detectors

Geiger, Hans Wilhelm • See Geiger counter

Geissler, Johann Heinrich Wilhelm (1815-1879)

German inventor and glassblower

Geissler was born into a family of craftsmen who were well versed in the art of glassworking: his grandfather was an accomplished glassmaker, his father was a maker of **glass** jewelry, and his brothers were glassblowers in Berlin and Amsterdam. Geissler began his own career in this field at an early age, and by his mid-twenties he had worked for several universities constructing glass devices for scientific instruments.

In 1852 Geissler took on a permanent position as a mechanic for the University of Bonn. There he developed a working relationship with such prominent scientists as W. H. Theodor Meyer, Julius Plücker (1801-1868), and Eduard Pflüger (1829-1910). It was with Plücker that he built his reputation as a master craftsman of scientific instruments, particularly **vacuum tubes**.

The first instruments Geissler made for Plücker were **thermometers** of unprecedented precision. In order to achieve this he used very thin but strong glass, as well as a sensitive balance scale for calibration. In 1852 he constructed a "vaporimeter," a device used for measuring the alcohol content of wine.

Geissler's most important invention was an improved vacuum tube first described by Plücker in 1858, though Geissler himself claimed he had been constructing them since 1857. Scientists were eager to experiment with vacuum tubes in order to observe the behavior of electrical currents. "Geissler's tubes," as they were called, consisted of sturdy glass tubes from which most of the air had been evacuated and into which electrodes had been melted at each end. In order to create the near-vacuum within the tubes, Geissler invented a hand-cranked mercury air pump. The evacuation process was slow and tedious but provided the best results scientists had yet seen. The experiments conducted by Plücker and later by **William Crookes** eventually led to the discovery of the **cathode ray tube** and to a greater understanding of the atom. In recognition of his outstanding craftsmanship, Geissler was awarded an honorary doctorate from the University of Bonn in 1868.

Gene therapy

Gene therapy is the use of genes engineered for treating disease. The first human gene therapy was approved for clinical trial in the United States in May 1989. Because this powerful technique is still in the experimental stages, each country has its own approval process, designed to protect the patient, the health workers, and the public. In the United States, each procedure must be approved by the National Institutes of Health's Recombinant DNA Advisory Committee, by the Food and Drug Administration, and by the director of the National Institutes of Health.

Techniques

Gene therapy begins with the isolation of a gene that can have a therapeutic effect, such as producing a protein in patients who lack the protein-producing gene or who have defective genes. In the laboratory, the desired gene is cut out of a cell's deoxyribonucleic acid (DNA) with enzymes. It is then inserted into somatic (functional) cells removed from the patient, and the treated cells are returned to the patient's body. An alternative approach is to insert the gene into disabled (harmless) viral or bacterial genetic material, which serve as vectors. These molecular delivery trucks carrying the attached gene are then injected directly into the patient, where they seek and enter somatic cells. In some cases, the virus itself is altered to make many copies of the gene. Ideally the gene should be targeted to an exact location in the cell's DNA. Viral vectors are primarily retroviruses, whose genetic material is ribonucleic acid (RNA) instead of DNA. These viruses have the potential to be permanently integrated into host cells' DNA.

Techniques also exist for altering the individual's germ (reproductive) cells, which not only treat the individual but are inherited by the next generation. Such techniques are already used in plants and animals. However, this approach raises ethical dilemmas when applied to humans because permanent genetic changes (eugenics) have been attempted to harm people or eliminate groups considered inferior or undesirable. Two early extensive gene therapy trials focused on severe combined immunodeficiency (SCID) and malignant melanoma.

Severe Combined Immunodeficiency (SCID)

This rare disease inhibits immune system functioning. In a well-publicized case, a teenager named David lived for several years in a plastic bubble to protect him from infection.

Some instances of SCID result from a genetic mutation that prevents production of the protein adenosine deaminase (ADA), which protects immune system white cells called *lymphocytes*. In September 1990, Drs. R. Michael Blaese and W. French Anderson at the U.S. National Institutes of Health performed the world's first gene therapy on a four-year-old with this condition. A normal gene for ADA was inserted into a virus and allowed to enter lymphocytes that were withdrawn from her body. Then she was injected with the altered cells.

During the next year-and-a-half, she had several series of injections, along with conventional treatment. In continued studies of young SCID patients, the cells appeared to induce production of ADA, reducing infections and allowing the children to have normal lives. There were no side effects. However, because most of these children were also given a standard enzyme treatment (due to ethical considerations), scientists could not conclusively say that it was the gene therapy that caused the improvements.

Melanoma

Melanoma is a type of often-fatal skin cancer. Since 1991, the National Cancer Institute's Dr. Steven A. Rosenberg has been studying treatment of the disease using TIL (tumor-infiltrating lymphocytes) cells taken from the patient's cancerous tumor. These cells normally enter a tumor and produce the protein called *tumor necrosis factor*. But often the tumor isn't destroyed. Scientists insert a gene into the TIL cells that boosts production of tumor necrosis factor. The cells are injected into the patients, and the genes function for a short period of time. Then, as a safety feature, the injected cells die.

Other Diseases

Cystic fibrosis. A genetic treatment for this lung disease was approved in 1992. Dr. Ronald Crystal, of the National Institutes of Health inserted a needed gene into an inactive cold virus that the patients inhale. The gene then entered the lung and functioned to prevent the production of the mucus that blocks a patient's breathing. In follow-up studies, liposomes—hollow spheres of fat molecules formed in solution—were used as vectors to prevent side effects possibly caused by the viral vectors, like inflamed lungs and swollen nostrils.

Familial hypercholesterolemia. Patients with this condition lack a gene for disposing of harmful low-density lipoprotein cholesterol, allowing it to build up in their bodies. People lacking both copies of the gene usually die from a heart attack in their early teens. Someone with only one copy suffers from severe coronary disease. Scientists at several medical centers are studying insertion of the needed gene into cells from a patient's liver, then injecting the cells into the person's body.

Other. Studies are under way on genetic therapies for a host of illnesses, including hemophilia B, AIDS, liver failure, leukemia, brain tumors, several cancers, arthritis, and sickle cell anemia.

Despite ongoing advances, gene therapy remains a highly experimental procedure. Geneticists must overcome the dilemma of ensuring that a sufficient number of the therapeutic genes reach the proper cells to become functional. Most likely, viral vectors will play a key role in accomplishing this goal. Although viruses are extremely efficient at entering cells, the difficulty has been to get them to enter the proper cells. In 1998, scientists from Harvard Medical School announced a technique that can essentially build a bridge with two proteins. The bridge enables the virus to bind to and enter a specific cell. If proven effective, this approach would allow gene therapists to pinpoint cells for therapy. In addition to retroviruses as vectors, scientists have been studying adenovirus vectors to target nondividing cells, which retroviruses cannot do. However, viruses and liposomes, which adhere to cells like tumors and insert genes into them, have drawbacks as vectors. Viruses can infect many types of cells and cause immune responses or become inserted in the wrong cell. Liposomes also have a slight possibility of infecting germ cells, producing heritable changes. As a result, researchers continue to look for new vectors like plasmids (circular DNA packages) and human artificial chromosomes (HACs). These chromosomes are less likely to trigger an immune response like viruses and are not able to seep into and affect other tissues because they can only be inserted into cells that have been removed from the body. Although it is still too early to tell, HACs may be viable vectors for gene therapies in blood diseases, like hemophilia and sickle cell anemia.

An alternative approach for delivery of genes has been developed by two scientists at Northwestern University. The gene ''gun'' is powered by pressurized helium that injects microscopic gold ''bullets'' coated with genes. In animal studies focusing on skin cells around tumors, they found that the surrounding cells began producing immune cells, called cytokines, which are what the genes are encoded for. The therapy shrunk tumors and lengthened the lives of the mice. Unlike viral vectors, this approach does not seek to permanently integrate the gene into the cell's DNA. This would be a disadvantage for inherited disease because the ''mutant'' genes that cause the disease would eventually become dominant again. However, the gene gun may be useful in diseases such as cancer because of the reduced likelihood of prolonged or permanent side effects like those associated with traditional chemotherapy.

Although many obstacles must be overcome, gene therapy has shown much promise in animal studies. In 1998 researchers from Stanford University announced a gene therapy that provides proteins to inhibit inflammation in multiple sclerosis in mice. Gene therapy may also prove to be especially effective in many types of cancer, which often have a strong genetic basis, such as gene mutations that cause cancers to grow and spread. One approach uses gene-modified cancer cells as vaccines. Cancer cells are extracted from tumors removed from the body, then scientists insert the corrective genes and reinsert them into the patients' tumors, essentially immunizing them with their own tumor cells. Although good result have been obtained in animals, the therapy has yet to be proven effective in humans.

See also Genetic engineering

Generator

A generator is a machine that converts mechanical energy into electrical energy. Originally called a dynamo, it makes use of the principle of electromagnetic induction, which was discovered in 1831 by the English physicist **Michael Faraday**. He reversed the work of Hans Christian Oersted (1777-1851), who had discovered that an electric current created a magnetic field. By Faraday's reasoning, it should be possible for a magnetic field to generate an electric current. (American inventor **Joseph Henry** came to the same conclusion independently.) Faraday took a coil of **wire**, attached it to a **galvanometer** and then inserted a **magnet** into the coil. As the magnet moved, its magnetic lines of force were cut by the coil and the galvanometer registered current. The same thing happened when the magnet was removed; electricity was induced to flow in the coil, although the flow was in the opposite direction. In fact, the results were the same whether it was the magnet or the coil that was in motion. When the movement stopped, so did the electrical flow, even if the magnet was still within the coil.

Faraday's next step was to construct a device that would induce a continuous flow, not just bursts. He accomplished this by, once again, reversing the work of another person. Dominique Arago (1786-1853) had shown that a permanent magnet suspended over a copper wheel (a nonmagnetic conductor) was deflected when the wheel turned. This was because the electric current that was generated by the interaction of the permanent magnet and the moving wheel created its own magnetic field which affected the original magnet. Faraday placed a copper wheel (the *conductor*) between the poles of a permanent magnet and set it spinning. The magnetic field induced an electric current that could be drawn off the spinning wheel with a wire and put to work.

The first electrical generator, possibly the single greatest electrical discovery in history, operated on three principles. First, there must be a conductor in which to induce voltage; second, a magnetic field must be close enough to the conductor for the lines of force to be cut; and third, either the conductor or the magnetic field must be moving. Following Faraday's lead, **Hippolyte Pixii** invented a hand-driven generator in 1832 in which the magnet revolved and the conductor (coils) remained stationary, more practical than Faraday's first device. André Ampère (1775-1836) suggested using a commutator to transform the power from that generated as **alternating current** (AC) to **direct current** (DC), for which there was great interest in the scientific community. In 1846 **Ernst Werner von Siemens** devised a generator that did not need permanent magnets. His dynamo-electric principle used self-generated electricity to activate electromagnets. The residual magnetism in the electromagnets and their circuitry was enough to start the process. Siemens, along with his brother Charles (Karl Wilhelm) Siemens and J. G. Halske (1814-1890), set up a very successful business manufacturing electrical devices.

Originally, the term dynamo was applied equally to generators of AC and DC. Today the term **alternator** refers to an AC generator, and the dynamo applies to DC only. The dynamo made use of brushes in physical contact with the rotating commutator. There were some disadvantage with DC dynamos—the brushes caused arcing, and there was a limit to how far DC could be carried by power lines. Increasing the voltage to ''push'' the voltage greater distances made the arcing worse. The alternator, with its reliance on electrical induction, eliminates arcing and, with the use of transformers, produces a voltage that high tension wires can conduct to great distances.

Today, most of the world's electricity is produced by generators, which still operate under Faraday's basic conditions. Portable generators are used in the home during times of electricity blackout, in the field during times of emergencies, and for recreational purposes—anywhere electricity is needed away from the local power grid. Typically powered by gasoline engines, they can be small enough to be carried in one hand but can provide minimal electricity needs for comfortable living in the home. Automobile-based portable generators, which derive their electric power from an automobile's engine, can deliver up to 5,000 watts for as long as two days.

Genetic engineering

Genetic engineering is the altering of the genetic material of living cells in order to make them capable of producing new substances or performing new functions, like getting a microorganism to produce human insulin or a sheep to produce a human blood-clotting protein in its milk. The technique became possible during the 1950s when scientists discovered the structure of deoxyribonucleic acid (DNA) molecules and how DNA stores and transmit genetic information. Largely as the result of the pioneering work of James Watson (1928-) and Francis Crick (1916-), scientists found that the sequence of nitrogen bases that make up any specific DNA molecule codes for the manufacture of specific chemical compounds. That sequence acts, therefore, as an ''instruction manual'' that directs all cell functions.

Certain practical consequences of that discovery became almost immediately apparent. Suppose that the base sequence T-G-G-C-T-A-C-T on a DNA molecule carries the instruction ''make insulin.'' (The actual sequence for such a message would in reality be very much longer.) DNA in the cells of the islets of Langerhans in the pancreas would normally contain that base sequence since the islets are the region in which insulin is produced in mammals. But that base sequence carries the same message no matter where it is found. If a way could be found to insert the base sequence into the DNA of bacteria, for example, then those bacteria would be capable of manufacturing insulin.

Although the concept of gene transfer is relatively simple, its actual execution presents a number of difficult technical challenges. In 1973, American biochemist Paul Berg, often referred to as the father of genetic engineering, developed a method for joining the DNA from two different organisms, a monkey virus known as SV40 and a second virus known as lambda phage. The accomplishment was significant, but Berg's method was slow and laborious.

A turning point came later the same year when Stanley Cohen at Stanford and Hubert Boyer at the University of California at San Francisco discovered an enzyme that greatly increased the efficiency of the Berg process. The technique of gene transfer developed by Berg, Boyer, and Cohen is fundamentally that used in most animal genetic engineering today.

This technique requires three elements: the gene to be transferred, a host cell in which the gene is to be inserted, and a vector for transferring the gene to the body. Suppose, for example, that one wishes to insert the insulin gene into a bacterial cell. The first step is to obtain a copy of the insulin gene. This copy can be obtained from a natural source (like DNA in islets of Langerhans cells), or it can be manufactured artificially in the laboratory.

The second step is to insert the insulin gene into the vector. Viruses, liposomes (hollow spheres of fat molecules formed in solution), and plasmids (circular forms of DNA) are common vectors. Scientists have discovered enzymes that can "recognize" certain base sequences in a DNA molecule and cut the molecule open at these locations. In this case, the plasmid vector can, therefore, be cleaved at almost any point chosen by the scientist.

Once the plasmid has been cleaved, it is mixed with the insulin gene and another enzyme that has the ability to glue the DNA molecule back together. In this case, however, the insulin gene attaches itself to the plasmid before the plasmid is re-closed.

The hybrid plasmid now contains the gene whose product (insulin) is desired. It can be inserted into the host cell where it begins to function as all bacterial genes function. In this case, however, in addition to normal bacterial functions, the host cell is also producing insulin as directed by the inserted gene. This method is sometimes referred to as gene splicing. Since genes from two different sources have been combined with each other, the technique is also called recombinant DNA (rDNA) research.

The possible applications of genetic engineering are nearly limitless. For example, rDNA methods now make it possible to produce a number of natural products that were previously available in only very limited amounts. Until the 1980s, for example, the only supply of insulin available to diabetics was animals slaughtered for meat or other purposes. That supply was never adequate to treat all diabetics at moderate cost. In 1982, however, the United States Food and Drug Administration approved insulin produced by genetically altered organisms, the first such product to become available. Since 1982, a number of additional products, including human growth hormone, alpha interferon, interleukin-2, erythropoietin, tumor necrosis factor, and tissue plasminogen activator have been produced by rDNA techniques.

In addition to the production of insulin, this technique has been used to create recombinant factor VIII for the treatment of hemophilia, a hereditary blood defect that inhibits blood clotting and makes it difficult for the body to naturally control bleeding. This genetically engineered blood factor protein can help induce clotting. Available since 1993, factor VIII can, for example, be produced in hamster cell lines using bovine serum proteins. It is considered safer than similar blood-derived factors which have the potential to pass on blood viruses, such as AIDS or hepatitis. Several other similar blood factor products are under development.

The potential commercial value of genetically engineered products was not lost on entrepreneurs in the 1970s. In many cases, the founders of the first genetic engineering firms were scientists themselves, often those involved in basic research in the field. Boyer, for example, joined with venture capitalist Robert Swanson in 1976 to form Genentech (Genetic Engineering Technology). Other early firms like Cetus, Biogen, and Genex were formed similarly through the collaboration of scientists and business people.

The structure of genetic engineering (or, more generally, biotechnology) firms has been a source of controversy, with concerns about individual scientists making a profit by opening their own companies that are based on research carried out at public universities and paid for with federal funds. By the early 1990s, working relationships had, in many cases, been formalized among universities, individual researchers, and the corporations they establish. But not everyone is satisfied that the ethical issues involved in such arrangements are settled.

One of the most exciting potential applications of genetic engineering involves the treatment of genetic disorders. Medical scientists now know of more 3,000 disorders that arise because of errors in an individual's DNA and are continuously finding new links among genes and diseases. Conditions such as sickle-cell anemia, Tay-Sachs disease, Duchenne muscular dystrophy, Huntington's chorea, cystic fibrosis, and Lesch-Nyhan syndrome are the result of the loss, mistaken insertion, or change of a single nitrogen base in a DNA molecule.

Genetic engineering makes it possible for scientists to provide individuals who lack a certain gene with correct copies of that gene. If and when that correct gene begins to function, the genetic disorder may be cured. This procedure is known as human **gene therapy**.

The first approved trials of gene therapy with human patients were begun in 1989. One of the most promising sets of experiments involved a condition known as severe combined immune deficiency (SCID) or ADA deficiency. Children born with this disorder have no immune system because of the lack of a single gene. In 1990, a research team at the National Institutes of Health led by W. French Anderson attempted gene therapy with a four-year old patient with SCID. The patient received about a billion cells containing a genetically engineered copy of the ADA gene that the child's body lacked.

Human gene therapy is the source of great controversy among scientists and non-scientists alike. Few individuals would say that the technique should never be used, especially for battling life threatening diseases, like AIDS and cancer. But many critics worry about where gene therapy might lead, such as genetically engineered humans.

Genetic engineering also promises a revolution in agriculture. Recombinant DNA techniques make it possible to produce plants that are resistant to **herbicides**, that will survive freezing temperatures, that will take longer to ripen, that will convert atmospheric nitrogen to a form they can use, that will manufacture their own resistance to pests, and so on. Scientists have tested a multitude of plants engineered to have special properties such as these. In 1994, a tomato was the first genetically engineered food to appear in American supermarkets. The genetically engineered tomato was created with an "antisense" gene that allows the tomato to ripen on the vine but remain firm for shipping.

As with every other aspect of genetic engineering, however, these advances have been controversial. The development of herbicide-resistant plants, for example, only means that farmers will use still larger quantities of herbicides, critics say, not an especially desirable trend. How sure can we be, others ask, about the potential risk to the environment posed by the introduction of "unnatural" engineered plants? For example, in the case of the tomato, some people are concerned genes from the plant could spread into soil bacteria and then infect humans or animals.

Many other applications of genetic engineering have already been developed or are likely to be realized in the future. In every case, however, the glowing promises of each new technique is somewhat tempered by the social, economic and ethical questions it raises.

Genetically engineered clones (exact genetic copies of an individual), for example, became a major issue in 1997 when scientists from Scotland announced that they had successfully created the first clone of an adult mammal. The genes used to clone Dolly the sheep came from the frozen mammary tissue of a six-year-old dead sheep. The cloning was accomplished by taking the cell nucleus from an udder cell of a Finn Dorset sheep and inserting it into an egg cell of a Poll Dorset sheep from which the original nucleus had been removed. The resulting egg with the substitute nucleus was then implanted into a third Scottish Blackface sheep. Because of the three different breeds used, the researchers had ready evidence that Dolly was truly a clone. While Dolly was a major advance scientifically, she became more famous among the public because of the ethical furor that has surrounded her "creation." Using genetic engineering to clone humans could be abused to create certain "types" of people, bringing forth the specter of humans as objects instead of individuals. Both the United States and Great Britain have banned human cloning. Still, the cloning of animals has many potential benefits, including raising animals that could be genetically engineered to produce more meat or, perhaps, to produce a human gene protein that could then be used in gene therapy.

Genetic fingerprinting

Fingerprints are unique to each individual. Methods of recording and matching fingerprints have allowed police to correctly identify many criminals. Genetic scientists have recently developed another tool for identification based on the uniqueness of each person's genes.

Genetic differences account for the large variations we see among individuals. This genetic variability is expressed in obvious traits like hair color and genetic disorders such as hemophilia. However, more genetic variability is hidden from view and can only be detected by directly studying the deoxyribonucleic acid (DNA). Each human has approximately 100,000 genes in the chemical form of DNA. The genetic information coded in the genes varies greatly between individuals. Thus, no two humans, except for identical twins, have exactly the same genetic code. A description of a person's DNA that is detailed enough to distinguish it from another person's DNA is called a DNA or genetic "fingerprint."

In 1985, an English researcher named Alec Jeffreys developed a technique to visualize a person's genetic code. This direct DNA analysis revealed so much variation in the genetic code between different people that even a small section of the entire genetic code could identify an individual's special combination of traits. Jeffreys knew that human DNA had many multi-repeated segments called minisatellites, and that the number and length of minisatellite DNA varied widely from person to person. He used a special detergent to break open the human cells and release the DNA code into solution. Then a restriction enzyme called HinfI broke the chain of DNA codes at sites close to each minisatellite DNA. The fragments of DNA were then attached to a membrane and allowed to combine with a radioactive minisatellite probe. After several hours, these probe molecules located and attached to certain predefined areas of the DNA fragments. X-rays were taken of the membrane to show where the radioactive probes attached. These pictures were then used to compare bands of DNA just as fingerprints are compared.

Three years later, Henry Erlich developed a method of DNA fingerprinting so sensitive that it could be used to identify an individual from an extremely small sample of hair, blood, semen, or skin. Erlich's technique used Jeffreys' traditional method and combined it with a technique called polymerase chain reaction (PCR). First discovered by Kary Mullis, PCR was used to duplicate DNA and thus copy the genetic code. Erlich was able to duplicate and heat-separate the DNA fragments from a single human hair root many times using PCR. Ultimately, PCR multiplied the DNA from one single hair to an amount equivalent to that found in a million identical strands of hair. The amplified DNA was then used to obtain a DNA fingerprint.

Genetic fingerprinting has already proved to be a very useful tool. Initially, it was used exclusively in forensic science and law. This technique has helped to link suspects to crimes where a single drop of blood was the only clue. Maternity and paternity matters have also been settled using genetic fingerprinting. This technology's impact in the study of genetic disorders and evolutionary relationships between different animal groups is also extremely important.

While genetic fingerprinting has attained a high public profile through several sensational criminal trials, the process is not accepted by all as definitive proof of guilt. Critics argue that DNA experts may overestimate the odds of having matching DNA fingerprints; they also say that DNA samples can be contaminated when collected and that poor lab procedures can result in mistakes. These problems have been used successfully by defense attorneys to negate some DNA fingerprinting evidence presented in criminal trials. However, most courts in the U.S. allow DNA evidence, and some states have passed specific laws allowing its use.

Genetically engineered blood-clotting factor

Excessive, uncontrolled bleeding can be fatal. One well-known disease associated with this phenomenon is hemophilia. Most hemophiliacs bleed uncontrollably because a single gene on the X chromosome lacks the instructions that tell the cell how to make a specific protein. This protein, called factor VIII, is required for blood to clot normally. Purified factor VIII extracted from human blood became available around 1960, but it was very expensive. Furthermore, viral impurities in the factor VIII obtained in this manner placed many hemophiliac patients at risk of contracting serious diseases, including hepatitis and, later, AIDS.

In the early 1980s, scientists at Genentech, Inc., and Chiron Corporation in California and at the Massachusetts-based Genetics Institute began developing **genetic engineering** techniques to obtain pure, inexpensive factor VIII artificially. Genetic engineering refers to methods of rearranging genes—removing or adding them or transferring them from one organism to another.

At Genentech, Richard Lawn, Gordon Vehar, and their coworkers succeeded in isolating the normal gene for factor VIII in healthy people and inserting it into laboratory-grown hamster cells, where it joined with the DNA (deoxyribonucleic acid) of the hamsters. The hamster cells then used the genetic instructions in the DNA to make pure human factor VIII. In April 1984, after many months of work, tests showed that the genetically engineered factor VIII is able to clot hemophiliac blood. A major problem, however, with this promising method of treating hemophilia inexpensively and safely is that it is difficult to control the amount of Factor VIII that the cells produce, and too much factor VIII causes the blood to stop circulating properly. Although it will take several more years of work before the gene itself can be introduced directly into

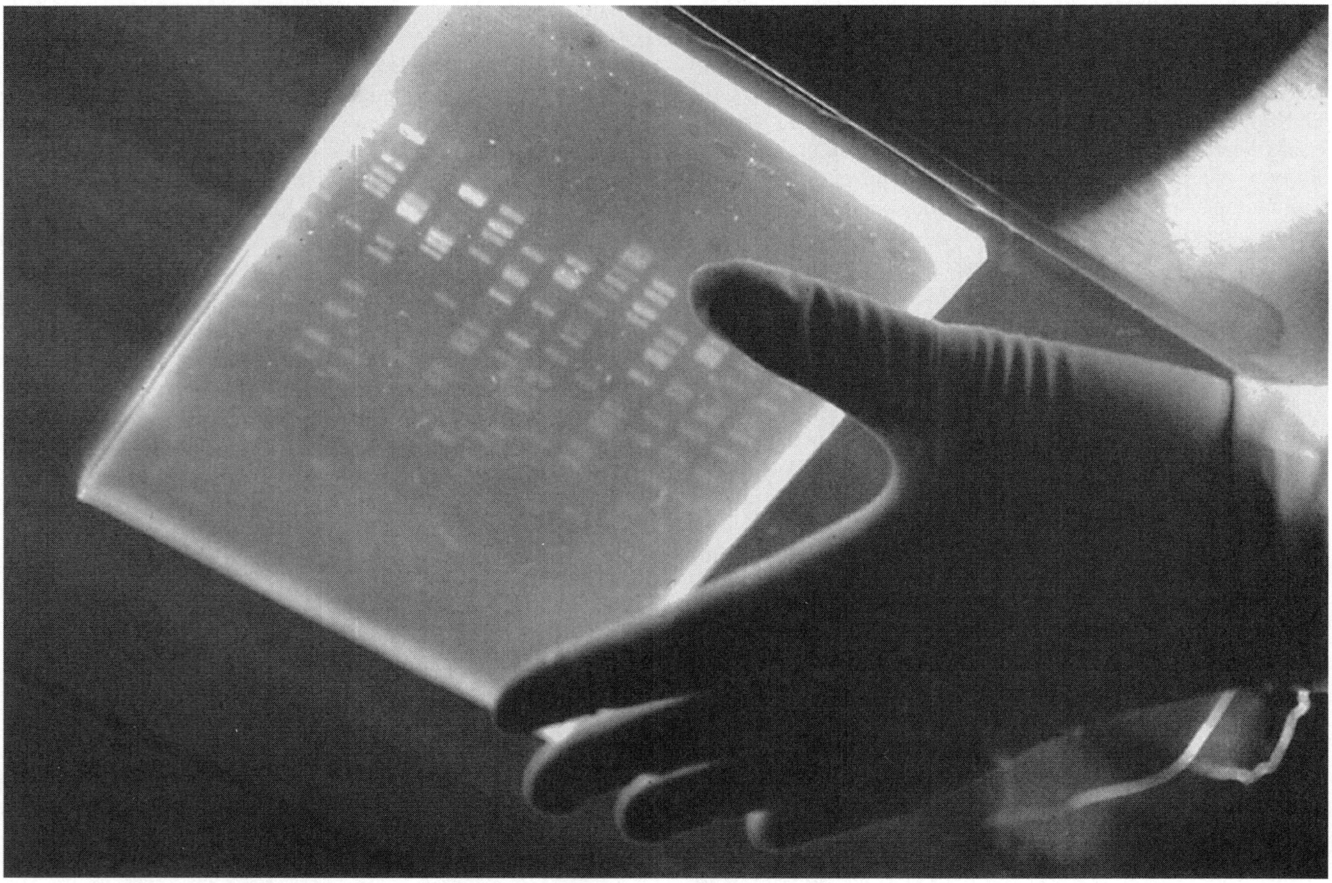

DNA fingerprint obtained by electrophoresis.

a patient, tests are currently underway to determine the best dosage of artificial factor VIII for hemophiliac patients.

See also Animal breeding

Geodesic dome

The geodesic dome was developed by **R. Buckminster Fuller** around 1949. In mathematics, "geodesic" refers to the shortest distance between two points, and the geodesic dome works on the principle of gaining maximum strength from minimal use of materials. To do this, a complex of tetrahedrons are balanced together forming the most efficient form of supporting a domed structure. The geodesic dome is the only large dome that can be set directly on the ground as a complete structure, and has no limiting size due to structural strength.

After several demonstrations, Fuller patented the dome in 1951. It was put to its first practical use in 1953 when Ford Motor Company used it to cover its headquarters rotunda in Detroit. A conventional dome would have weighed about 160 tons (145 metric tons) and was technically unfeasible. The materials on the Ford project weighed only 8.5 tons (7.7 metric tons).

In addition to being structurally sound, the geodesic dome offers virtually no resistance to the wind. For this reason, it is used to house sensitive **radar** equipment along the Defense Early Warning (DEW) line in the Arctic and in other areas with harsh environments. These domes look somewhat like large golf balls and are capable of withstanding winds exceeding 200 miles per hour (322 kph). Also, because of its lack of wind resistance, a geodesic unit can be transported by **helicopter** even under windy conditions.

Geodesic domes can be erected in a very short period of time. A symphony hall in Honolulu was erected in eighteen hours and hosted a concert an hour later. The ability of the geodesic dome to cover acres of ground make it especially favorable for exhibition buildings. It was used for the Climatron in St. Louis, Missouri, to house a tropical garden complete with fully grown palm trees. It was also used to house the United States' Exhibition at the 1967 World's Fair, the dome being twenty stories high.

Always one to take his ideas to their extreme, Fuller constructed models of his dome that were to be used to cover entire sections of cities, like East St. Louis, Illinois, to protect their citizens from smog and to offer them a controlled environment. The geodesic concept is even used in some models of backpackers' tents. The 'round' tents offer more headroom than the conventional A-frame. Many of Fuller's ideas were discounted as being extreme, and many of them died with him in 1983. However, his domes can be found throughout the world. Though they are very impersonal in design, they are an eternal memorial to one personality, that of Buckminster Fuller.

Gillete, King Camp • See Razor

Donald A. Glaser

Girdle • See Corset and girdle

Glaser, Donald A. (1926-)

American physicist

Glaser was born in Cleveland, Ohio, on September 21, 1926. After graduation from high school in Cleveland Heights, Glaser attended the Case Institute of Technology, where he earned a bachelor's degree in physics and mathematics in 1946. Three years later he was awarded a Ph.D. in the same subjects by the California Institute of Technology. His first teaching assignment was at the University of Michigan, where he later became professor of physics in 1957. In 1959 he assumed a similar position at the University of California at Berkeley.

Glaser received the Nobel Prize in physics in 1960 for his invention of the **bubble chamber**. The idea for the invention occurred to him in 1949 while carrying out research on the new "strange particles" that had recently been discovered in cosmic ray interactions. The most common device then available for photographing these particles was the **cloud chamber**.

A cloud chamber contains a vapor that has been cooled below its boiling point. When radiation passes through the chamber, it forms ions on which droplets of the vapor condense. The track formed by the droplets shows the path taken by the radiation.

The problem for particle physicists was that the cloud chamber is not a very efficient detector for high energy particles like the "strange particles" Glaser was working on. He decided that a new type of detector was needed, one that contained a medium of higher density than the vapor in a cloud chamber.

It occurred to Glaser that a superheated liquid might serve the purpose. A superheated liquid is one that has been heated to a temperature greater than its boiling point. The concept Glaser had in mind, therefore, was just the reverse of the cloud chamber: a device with a liquid ready to boil rather than a vapor ready to condense to a liquid.

Glaser has described his experience studying bottles of beer, ginger ale, and soda water before settling on a model for the bubble chamber. His first device consisted of a **glass** container filled with diethyl ether heated to a temperature of 140°C (284°F) The normal boiling point of diethyl ether is 36°C (96.8°F). When high energy radiation passed through the container, it created ions in the ether. Each time an ion formed, the ether at that location immediately boiled. The overall result was a string of tiny bubbles that showed the path of the radiation through the chamber.

Glaser next turned his attention to improvements in the bubble chamber. He varied the liquid used in the chamber and examined the effects of using containers of various shapes and sizes. Ultimately, the bubble chamber became one of the most powerful detection devices available to particle physicists.

Glass

Glass, given its durability and versatility, plays an important role in human culture. Generally manufactured by heating sand (silica), soda (alkali), lime, and other ingredients at high temperature, glass is a fundamental component of a variety of products, including tableware, **windshields**, **thermometers**, and **telescope** lenses.

Early peoples were likely to have discovered natural glass, which is created when lightning strikes sand, and were certain to have used obsidian—a dark volcanic glass—for weapons, ornaments, and money. The first manmade glass probably took the form either of glass beads or ceramic glaze and appeared around 4000 to 5000 B.C. Surviving examples of Egyptian and Mesopotamian glass objects date to around 1550 B.C.

For centuries glass, shaped by the use of molds, remained costly and difficult to produce. The invention of the **blowpipe** method of glassmaking (in which molten glass is puffed into shape with the use of a hollow tube) in about 30 B.C. made glass more commonplace. Typical uses at the time included windows as well as decorative objects.

The first four centuries A.D. are sometimes referred to as the First Golden Age of glassmaking, for during this period artisans produced a wide variety of artifacts that are now highly valued. After the decline of the Roman Empire, few developments took place in European glassmaking until the twelfth and thirteenth centuries, when **stained glass** windows (formed of pieces of colored glass outlined by lead strips and assembled into a narrative picture) began to appear in English and French churches. During the Crusades, Europeans were exposed to the accomplished glassmaking of the Near East, an influence evidenced by the growth of the craft in Italy, particularly Venice. Beginning around 1300, the Venetians ushered in the Second Golden Age of glassmaking; they became widely known for a particularly transparent glass that was worked into a number of delicate objects.

In the late 1400s and 1500s the Germans and other northern Europeans were producing containers and drinking vessels that differed markedly in their utilitarian value from those produced by the Venetians. Nonetheless, Venetian glass was immensely popular during the reign of Queen Elizabeth I (1558-1603). In 1674, George Ravenscroft brought fame to English glassmaking when he invented lead glass (now usually called lead crystal), an especially brilliant glass he produced accidentally when he added lead oxide to his mixture instead of lime. In colonial America, the glass made by this technique became known as flint glass, and was usually etched or cut into facets to lend it additional luster.

The first glass plant in the United States was built at Jamestown, Virginia, in 1608, but it survived for less than a year. Much later, in 1739, Caspar Wistar successfully launched the American glass industry with a plant in Salem City, New Jersey. Other prominent figures in early American glassmaking included Henry William "Baron" Stiegel and John F. Amelung. The renowned Sandwich glass that is now much coveted by American collectors was made by the Boston and Sandwich Glass Company; the Bakewell Company of Pittsburgh was another famous glass manufacturer.

The early 1800s saw a tremendous demand for glass windows, which were a symbol of affluence, particularly in the frontier communities of America. Window glass was originally made by spinning out a bubble of blown glass until it became flat; because of the bump or "crown" that was invariably left in its center, this was called crown glass. Around 1825, the cylinder process replaced the earlier method. Now the glass was blown into a cylinder shape that, when cooled, was cut down one side; when reheated, the cylinder flattened out to form a sheet. In 1842, John J. Adams invented a more sophisticated glass-flattening and tempering process that made not only plate glass but mirrors, showcases, and other products more widely available. During the last half of the nineteenth century, glass found wide use in medicinal containers, tableware, and kerosene lamps. Tempered glass (made exceptionally strong through a reheating process) was invented by François Royer de la Bastie in 1874, and wire glass (industrial sheet glass with metal mesh laminated into it) by Leon Appert in 1893. In 1895, Michael J. Owens invented a bottle-making machine that allowed bottled drinks to be produced inexpensively.

The great technological advances of the twentieth century broadened the range of ingredients, shapes, uses, and manufacturing processes for glass. Natural gas replaced the wood and coal that had previously been used in the glassmaking process, and huge operations were established. One of the most common forms of glass now produced is flat glass, used for windows, doors, and furniture. Formed by flattening melted glass between rollers, annealing (heat treating) in an oven called a lehr, then cutting into sheets and grinding and polishing until smooth, this category includes sheet glass and the higher quality plate glass. The best quality of all is achieved in float glass, invented in 1952 by Alistair Pilkington. Float glass is made by floating a ribbon of liquefied glass on top of molten tin so that it forms a perfectly even layer; the result is glass with a brilliant finish that requires no grinding or polishing.

In 1957, Donald Stookey of Corning Glass Works discovered that the addition of certain nucleating agents to Fotoform glass (used in electronic components and fluidic control devices) produced glass objects that could be transformed into fine-grained ceramics by heat treatment. This new Pyroceram composition became the basis of Corning Ware, first marketed in 1958. In 1964, Corning announced the invention of photochromic glass, which darkens under ultraviolet light and then clears when removed from the light source.

In 1980, Pilkington invented kappafloat glass, which features a special, energy-efficient glaze that traps thermal heat while allowing solar heat to filter through. Other modern forms of glass include the laminated safety glass used for automobile windows, which is composed of sandwiched layers of plastic and glass; fused silica optical waveguides (optical fibers used for long distance telecommunication); nonreflecting glass (invented by **Katherine Burr Blodgett** and others); structural glass, used in buildings; heat-resistant cookware such as Pyrex; and **fiberglass**.

Glass, heat-resistant

During the twentieth century, heat-resistant borosilicate **glass** replaced more fragile flint or soda-lime glass. Although German optician Carl Zeiss (1816-1888) began experiments with heat-resistant glass for use in **microscope**s by adding boric acid to silicon, the stronger material evolved from the Corning Glass Works' invention of oven-proof Nonex and Pyrex in 1915, and Vycor, an essential material in stove heating elements. Later varieties include Pyroceram, Kimax, and Corning Ware, a multipurpose **ceramic** material used to make freezer-to-table dishes for freezing, cooking, baking, microwaving, and serving as well as for laboratory equipment.

Since the 1870s, the Corning Glass Works has developed and improved other heat-resistant glass products with practical applications, particularly globes for incandescent lighting, weatherproof **lenses** for railroad signal lanterns, and colorants for lens glass.

Developed by Eugene G. Sullivan and William C. Taylor for Corning Glass Works in 1915, Corning Ware evolved from an experiment by the wife of Jesse T. Littleton, Corning's chief physicist. When she baked puddings, Mrs. Littleton substituted the bottom portion of a battery jar for breakable casseroles. From her success came Pyrex or nonexpanding glass, made from borax, alumina, sodium, and soda and fired at over 2,500°F (1,371°C). The low thermal quality reduced the danger of breakage; its reduced sodium content lessened the chance of chemical interaction.

A second application of glass technology improved the usability of Pyrex glass for baking dishes that absorb heat rather than reflecting it, as do metal pans. Also, the creation of glass pie plates, cake pans, casseroles, and custard dishes decreased the carryover of flavors, which occurs in metal pans. The perfection of Pyrex led to a successful marketing of glass bakeware. Slightly higher in price because of the cost of high temperature fuels during manufacture, Pyrex holds its own in the market because it resists breakage where earlier types of tempered glass shatter. By 1919, over 4.5 million pieces of the ovenproof bakeware were in use. The surge of interest in heat-proof glassware proved propitious for industry during World War I, when hostilities halted the supply of laboratory beakers, funnels, tubing, culture dishes, **thermometers**, and flasks from Germany.

Scientific applications for Pyrex include stronger lenses for reflector **telescopes**, as blanks for **mirrors**, and as components in construction projects, such as hulls for small boats, thermal insulation, acoustical soundproofing, and fireproof fabrics. Pyrex is also an integral part of headlight lenses for automobiles and in pipelines that carry corrosive materials such as lye. Pyrex fibers, because of their resilience and nonabsorbancy, are clean and shatterproof and can be substituted for asbestos, which is harmful to the lungs of workers. Pyrex fibers are used primarily in construction, particularly as lining for ductwork and kitchen appliances, construction of naval vessels, rockets, launch pad insulation, gas turbine silencers and mufflers, high temperature gaskets and seals, solar panels, expansion joints, high temperature filtration, coverings for furnaces and steam generators, and ports for space vehicles. They are also valuable for decorative and safety purposes, as in soundproof, noncombustible ceiling panels.

The evolution of glass as a heat-resistant material has resulted in additional products. In 1957, Donald Stookey of Corning Glass Works discovered that the addition of certain nucleating agents to Fotoform glass (used in electronic components and fluidic control devices) produced glass objects that could be transformed into fine-grained ceramics by heat treatment. This new Pyroceram composition became the basis of Corning Ware, first marketed in 1958. Pyroceram is used in bakeware, hot plates, and stove tops. The thermal endurance and stability of this and other glass-ceramics, such as Schott Glaswerkes' Robax, have made them ideal for missile nose cones and astronomical telescope mirror substrates. Transparent glass-ceramics were first introduced in the 1970s.

For comparison, a purely crystalline silicate (which is not vitreous and therefore not a glass) known as cristobalite is made at over 3,000°F (1,649°C). The raw material is turned into vapor, then condensed. Its purity makes it the most heat-resistant of all types of silicates, although a major drawback in its production is its limited malleability. Because containers for this process must be made of tungsten or graphite, the cost of crystobalite objects, such as arc tubes for lamps, crucibles for melting semiconductors, covers for solar cells, envelopes for mercury vapor lamps, optical parts, and telescope mirror, is much higher than for Pyrex.

GLASSBLOWING • See Glass

GLIDDEN, CARLOS • See Typewriter

GLIDDEN, JOSEPH FARWELL • See Barbed wire

GLIDER

There are stories of people long ago who tried to imitate the birds by attaching wings to their arms. Of course, none could truly fly this way, but undoubtedly some were able to enjoy a controlled glide over a short distance. The first successful glider flight may have occurred in England about 900 years ago. A monk fastened wings to his hands and feet, jumped off a tower and glided about 600 feet (183 m) before making a rough landing that resulted in two broken legs. In 1498, an Italian learned how to glide with wings attached, but he also suffered a crash that broke a leg. Others down through the years tried to fly, but none gave any serious thought to the mechanics and aerodynamics involved.

The first to write an extensive analysis of the theory of human flight was **Leonardo da Vinci**. He sketched many ideas for human flight, nearly all involving flapping wings. In some cases the aviator was to lie prone, like the arrangement in today's hang-gliders. In 1500, da Vinci made some intriguing sketches for a large flying machine that got most of its lift from a rigid wing. It was to be powered by flappable, hinged wingtips operated by a person suspended in a harness under the single wing. It was the first powered hang-glider that closely resembled the gliders used by **Otto Lilienthal** nearly 400 years later.

No one followed up on da Vinci's ideas until **George Cayley** in the early 1800s. In 1804, he sketched the design of a model glider based upon his ideas concerning lift and propulsion, concepts which laid the foundations of modern aerodynamics. Cayley's model possessed a wing with a semi-circular leading edge, a movable, cruciform tail unit like a modern airplane, and a sliding weight at the nose for altering the center of gravity. Much later, in 1849, he built a triplane glider which carried his coachman 900 feet (274.5 m).

In spite of the advances made by Cayley, gliders were not again actively pursued until Otto Lilienthal who put gliding on a solid theoretical and practical footing. He realized it was important to control the craft before thinking about any propulsion systems. During the 1890s he built eighteen different types of fixed-wing gliders which he flew himself. He controlled the gliders by swinging his body to shift the center of gravity, just as modern hang-gliders do today. Lilienthal also kept meticulous records of his work which helped others with their experiments. Unfortunately, he died when his glider stalled during a flight in 1898.

One American living at the same time, Octave Chanute (1832-1910), was responsible for influencing **Orville Wright** and **Wilbur Wright** through his advancement of glider design. One of the most important aeronautical experimenters of his time, Chanute refined Lilienthal's designs for better stability in the air. After publishing his findings in his work *Progress in Flying Machines*, he was contacted by the Wright brothers and began a close relationship with them. They eventually adopted Chanute's excellent braced biplane wings. Gliders played an important role as the Wright brothers systematically prepared for manned flight. From 1900 to 1902 they built three gliders which incorporated their methods of control: wing warping for turning ability, a fixed tailplane to create fore-and-aft stability, and a forward elevator for up-and-down movements.

After powered flight turned into a reality, gliders were largely abandoned. Germany, however, revived interest in them after World War I when the Treaty of Versailles banned the country from building airplanes. In 1928 the variometer was invented, a device that measures the glider's lift in thermals, the vertical currents of air rising from heated areas on the ground. After World War II, low-drag wings were shaped out of **fiberglass**, which provided a smooth and precise surface. Today gliders are used for recreation or sport. The newest sail planes have flaps, air brakes, and water ballast which can be jettisoned in flight to lower wing loading in flight.

The great expense of these sailplanes has created a new sport: hang-gliding, in which a pilot sits or lies in a harness under a lightweight wing stretched over a simple wire-braced aluminum-tube structure. Experienced hang-glider pilots can stay aloft for hours, traveling a hundred miles or more over the course of a flight. After all these years, it is interesting to note the similarity between modern hang-gliders and the early machines of da Vinci, Cayley and Lilienthal.

GLOBAL POSITIONING SYSTEM

The Global Positioning System (GPS) is a group of 24 satellites that orbit Earth at 10,907 nautical miles (20,200 km) above Earth. GPS satellites provide precise location information anywhere on Earth. Full accuracy is only available to the military, which owns and operates the satellites, but there are numerous commercial applications for GPS. By the 1990s, GPS receivers became small enough to be hand held, increasing public interest in the technology.

At a cost of $3 billion, GPS was developed by the United States military beginning in the 1970s, first in the Navy, and later by the Air Force. Their initial goal was to guide missiles and tanks. The Department of Defense launched the first GPS satellites in the 1980s, and all 24 were in orbit by 1996. GPS was first used by the military in the Persian Gulf war in 1991. Each GPS satellite has four **atomic clocks**, and continually broadcasts data concerning time, its path, and its location. A GPS receiver picks up the signal from three satellites, and compares it with its own clock, then calculates the distance to ascertain its location, within 52 ft (16 m). There are also four permanent receiver stations on Earth. When information from a mobile receiver is combined with that from a permanent receiver, a location reading can be calculated within inches. Initially, the receivers were big, unwieldy, expensive boxes, costing at least $50,000. By 1998, receivers were the size of credit cards, and the cheapest could be obtained for a few hundred dollars.

Commercial applications of GPS abound. Trucking companies use GPS to monitor the location of their fleet. Airlines use GPS to avoid mid-air collisions and land planes in fog. GPS can also help provide better routes for airplanes. GPS can monitor changes on Earth, perhaps forewarning of earthquakes. GPS can guide boaters, and help locate hikers, such as when someone is lost or injured. When GPS is combined with other technologies, such as certain computer software, the applications of GPS are endless. Computer **maps** in cars can direct drivers in unfamiliar cities. GPS satellites also have nuclear detection payload capabilities to ensure nuclear test ban treaty compliance.

There are, however, several potential problems with GPS. One concerns the Department of Defense's operation of the satellites. The military deliberately degrades the satellites' signals, making the clock incorrect by nanoseconds, so most commercial receivers are off by 109 yd (100 m). This is done in the name of national defense, though completely accurate receivers are used by the military. Because of this inaccuracy, some commercial entities have been forced to develop their own technology to correct this problem at great cost. Still, GPS is not immune to terrorist obstruction, or even solar interference from the sun. At the peak of the solar sunspot cycle, which happens about every 11 years including the year 2000, certain high-energy solar particles and ionospheric scintillation could interfere with the operation of the GPS satellites.

Technologies similar to GPS have been developed by the Russians. The Glosnass (Global'naya Navigatsionnya Sputnikovaya Sistema) also employs 24 satellites which orbit at 11,842 mi (19,100 km) above Earth. However, Glosnass operates on a different frequency.

GNOMON • Sundial

GODDARD, ROBERT (1882-1945)

American physicist

Robert Hutchings Goddard

From childhood, Goddard had been fascinated by space travel, finding inspiration in part from H. G. Wells's *War of the Worlds.* He experimented avidly as an adolescent, attempting to work out the principles and calculations for **rockets** and space travel. He began studying physics at Worcester Polytechnic Institute in 1908 and later entered Clark University where he received his Ph.D. in 1911. As a student he decided that the most effective propellant would be a combination of liquid hydrogen and liquid oxygen. Unfortunately, neither was commercially available at the time.

After his schooling, he worked briefly at Princeton as a researcher, then accepted a position in the physics department at his alma mater, Clark University. There, he speculated about travelling in space and published *A Method of Reaching Extreme Altitudes*, a now classic monograph on the topic. In 1926 he discovered an even more effective liquid fuel combination: **gasoline** and liquid oxygen. That same year he launched the world's first liquid-propelled rocket, a small (10 lb. [4.5 kg]) device that went up a grand total of 41 feet (12.5 m) and landed 184 feet (56 m) away. Despite this limited success, or perhaps because of it, reporters and fellow scientists ridiculed Goddard's efforts, dismissing him as a "crackpot" whose ideas concerning space flight were "ridiculous." This marked the beginning of his life-long struggle to be taken seriously.

Undeterred, Goddard continued his research. In July of 1929, one of his rockets exploded outside of Worcester, Massachusetts, attracting the attention of the local newspapers and the State Fire Marshal who banned Goddard from ever again testing rockets. However, the publicity the incident generated prompted the famous aviator Charles Lindbergh (1902-1974) to seek out Goddard. Impressed by the physicist's work, Lindbergh helped Goddard obtain the funding to continue his rocket research.

In 1930, Goddard moved his operations to Roswell, New Mexico, establishing the world's first professional rocket proving ground. The work to set up such a test site was incredibly difficult due to bad weather, dangerous insects, lack of tools, and the lack of any real knowledge. Goddard and his assistants had to save everything after each test in hope of repairing them for future flights. They had a terrible time developing good **parachutes** that would open long enough to keep the falling rocket from shattering on impact. One flight was especially frightening when the rocket took off, had **gyroscope** trouble, and turned around to chase Goddard and his assistant, who saved themselves by falling flat to the ground. The men also had to crouch behind sheets of galvanized steel close to the test stand in order to read their instruments, knowing that any explosion would send fragments of metal in all directions.

When World War II broke out, Goddard repeatedly offered his services to the government, but was rejected because his ideas were considered too bizarre. In addition, he had retreated into a world of secrecy because of the negative reactions he had encountered in the past. Due to the intervention of a few military men who had worked with Goddard, the United States Navy finally agreed to finance his research, but only on small boosters to help their planes lift off carrier decks. Goddard went east to set up a new site near Annapolis, but died of throat cancer soon afterwards.

The government only appreciated the importance of his work after the discovery of Germany's rocket program at the end of the war. Their devastatingly powerful V-2 rockets had been based on the same principals put forward by Goddard. In all, Goddard accomplished many things in his desire to see rockets succeed: he pioneered gas-generator-powered turbofed rockets, developed automatic launch-sequence control, set up a sequence for tank pressurization/ignition/automatic shutdown, engineered on-board control for guidance, established parachute recovery systems, pioneered gyroscopic stabilization and rocket-exhaust deflection controls, successfully used gimbal-mounted rocket motors, and set up recording and optical-telescope tracking methods. The United States government eventually awarded Goddard's estate one million dollars for all rights to the more than two hundred patents he owned.

Goldmark, Peter Carl (1906-1977)

Hungarian American inventor

Peter Goldmark was born in Budapest, Hungary, in 1906. As a student at the University of Vienna he was fascinated by the technology of **television**. While pursuing his bachelor's degree he designed a television receiver with a screen about the size of a postage stamp. After obtaining his Ph.D. in 1931, Goldmark decided to emigrate to the United States to join the Radio Corporation of America, (RCA), then the premier laboratory for television research. He was refused a position with RCA, however, and in 1936 took up with their competitors, the Columbia Broadcasting System (CBS), for whom he accrued a long and impressive list of accomplishments. In all, Goldmark received more than 180 patents for CBS.

Though he labored for several years to develop more practical television technology, it was not until 1940 that he conceived his first great idea. While attending a showing of *Gone with the Wind* (the first color film he had ever seen), he came upon an idea for color television. During the next three months he designed the system known now as field-sequential color. Basically, the system used a standard black-and-white camera and television receiver; placed in front of the camera was a spinning disk holding three colored filters—red, blue, and green. By synchronizing the camera's color disk with a second disk within the television, a very sharp color picture was achieved.

The debut of field-sequential color broadcasts was delayed by the outbreak of World War II. After the war years, CBS petitioned the Federal Communications Commission (FCC) to approve Goldmark's design. At the same time, RCA was developing its own method for color television. The picture quality of the RCA system was vastly inferior to Goldmark's; however, it was compatible with existing television sets—that is, a standard television could receive the RCA signal (albeit in black and white) but not the CBS transmission, for which a whole new television set was necessary. Since the Goldmark method would have made nine million black-and-white sets immediately obsolete, the FCC eventually ruled to make the RCA system the standard.

The field-sequential color television—which still provides the best color reproduction available—has found a niche, particularly in areas such as medical instruction that demand precise color definition. In the late 1960s, Goldmark modified his television to enable National Aeronautics and Space Administration (NASA) to clearly photograph the surface of the Moon, even from the Lunar Orbiter's altitude of 29 mi.(47 km).

Goldmark can also be considered the father of **video recording**. With the goal of producing a tool for educational media storage, he developed a system called electronic video recording (EVR). Basically, EVR consists of a cartridge wound with black-and-white film that could be slipped into a recorder/player machine. Even though it was filmed in black and white, Goldmark's ingenious design used a separate recording track to store the color signal, so that the final playback would be in color. One advantage of EVR over magnetic tape storage is that it can record still frames as well as motion; by filming a page of written information onto a single EVR frame, an entire encyclopedia can be stored on one cassette. Although video cassette recorders (VCR) have since captured the home recording market, the EVR is still considered useful in the storage of written material.

The idea for Goldmark's most successful invention came to him one evening in the early 1940s. While listening to a recording of Brahms at a party, Goldmark was annoyed by the constant clicks and interruptions in the music, since the 78 rpm records were far too short to contain the entire concert, and the host had to periodically flip or change the disks. Goldmark envisioned an album whose grooves were cut much closer together and whose turntable turned much slower, allowing more music on each side. He worked on solving this problem for three years, ultimately producing the first LP record. His microgroove recordings held the equivalent of six 78 rpm records; also, by switching to vinyl, the LP's recordings had a far superior sound quality. To combat the new CBS-manufactured LP, RCA quickly designed the 45 rpm "single." Though the single record found some success, Goldmark's LP absolutely conquered the market for more than forty years.

Golf

Golf is considered one of the world's most popular sports. The name of the game is derived from the German word for "club," but almost everything else about golf originated in

Scotland. The 500-year-old game was invented there during the early fifteenth century, most likely influenced by the Romans who occupied parts of England and Scotland and who played a game called *paganica* with bent sticks and leather balls filled with feathers.

Golf became so popular in Scotland that, in 1457, King James II (1430-1460) banned the sport because it threatened to surpass the popularity of archery, the national sport. In 1502, the ban was lifted, and golf has continued unabated ever since. Queen Mary Stuart (1542-1587) was the first woman to play the game, and she created the world's first great golf course, St. Andrews, which still exists today.

Golfers played with a leather-covered ball stuffed with feathers until 1848, when a ball of solid **gutta-percha** (''gutty'') was invented. In 1899, Coburn Haskell, an avid golfer from Cleveland, Ohio, devised a light, tightly-wound, rubber-threaded ball with a solid rubber core.

For four centuries, when golfers teed off, they scooped up clumps of dirt, molded it into small hills, and set their balls atop it. In the 1890s, rubber and paper tees began to appear but failed to gain popularity. In 1899, an African-American dentist from Boston, Massachusetts, Dr. George F. Grant, developed a wooden tee because he got ''darned tired'' of picking up dirt whenever he wanted to play golf. The tee consisted of a tapering base and a concave shoulder to hold the ball in place. With no interest in profit, Grant gave his tees away. Another American dentist, Dr. William Lowell, patented a wooden tee, and his sons marketed the ''Reddy-Tee'' in 1920, profiting enormously.

GOLF TEE • See Golf

GOODYEAR, CHARLES (1800-1860)

American inventor

Charles Goodyear

If any inventor's story epitomizes dedication to a remarkable discovery that fails to bring any financial reward to its originator, it is the life of Charles Goodyear. An acquaintance once described the inventor in these words: ''If you meet a man who has on an India-rubber cap, stock, coat, vest and shoes, with an India-rubber money purse, without a cent of money in it, it is he.''

Goodyear was born in 1800 in New Haven, Connecticut, the son of a hardware manufacturer and inventor. The frail and sickly young man began his disastrous business career as a hardware salesman in Philadelphia, Pennsylvania. Ill health brought him back to Connecticut as his father's partner. In 1826 Goodyear and his bride, Clarissa, opened the first American hardware store as an outlet for the senior Goodyear's products. Both father and son went bankrupt in 1830.

For the next few years Goodyear struggled to support his family with various small inventions while moving in and out of debtors' prison. In 1834 Goodyear bought a rubber life preserver from the Roxbury India Rubber Company in New York City and quickly invented an improved **valve** for the device. When Goodyear tried to sell his design to Roxbury, the manager sadly told him it was no use—the **rubber** itself was what needed improving, not the valve. Consumers were fed up with the way rubber melted in hot weather and hardened in cold. Goodyear's boyhood fascination with rubber instantly became his life's challenge: he would find a way to make rubber usable.

For the next five years, Goodyear dedicated himself to experimenting with rubber, in both his own and the debtors' prison kitchens. He had no idea of what to do, having no knowledge of chemistry; he had no money, and only the crudest equipment. His family lived in poverty. In 1836 Goodyear had some success in treating rubber with *nitric oxide*, but his fledgling rubber-goods company was wiped out by the economic Panic of 1837. A new venture with Nathaniel Hayward seemed secure when the United States Post Office ordered 150 mailbags made of sulphur-treated rubber. The bags, however, disintegrated in the summer heat. Moral and financial support from friends disappeared, but Goodyear persevered. The breakthrough finally came in 1839 when Goodyear accidentally discovered the *vulcanization* process—heating rubber-sulphur mixtures to yield a tough yet flexible product.

Goodyear struggled through five more years of destitution—at one point even selling his children's schoolbooks—before he was able to patent his process in 1844. Instead of profiting from his finally successful quest, Goodyear granted licenses for rubber manufacturing at ridiculously low prices, and he withdrew from manufacturing himself to contrive new uses for his product. Industrial pirates infringed on his patents, and he paid his attorney, Daniel Webster (1782-1852), more to secure his rights (successfully, in 1852) than he ever earned from his discovery. He was unable to patent his vulcanization process abroad; **Thomas Hancock** of England had already done so. He spent $30,000 for a lavish display at London's Crystal Palace Exhibition in 1851 which earned him nothing. He pooled the last of his meager resources for another extravagant display at the Paris Exhibition of 1855, which returned him once more to debtors' prison—where he was also awarded the cross of the Legion of Honor. Sick and feeble, Goodyear returned to the United States in 1858, where he found his financial affairs in disarray and his patents again infringed.

While traveling from Washington to New Haven, Connecticut, in 1860 to his dying daughter's bedside, Goodyear died in New York. While others would profit from the rubber manufacturing that Goodyear made possible, the inventor himself left $200,000 in debts.

See also Rubber, vulcanized

Gould, Gordon (1920-)
American physicist

In the late 1950s, the race was on among American scientists to build the first light-amplification machine, or **laser**. Several developers, working independently, conducted experiments based on **Charles Townes**'s early research, culminating with **Theodore Maiman**'s working model in 1960. The theories of Gordon Gould predated Maiman's work by several years, and it is Gould who, after many years, is now recognized as the true inventor of the laser.

Born in 1920 in New York City, Gould earned his undergraduate degree in physics from Union College in Schenectady, New York and later received his master's from Yale. Like many scientists of his time, Gould worked for the government during World War II, remaining in New York to participate in the Manhattan Project with the scientists developing the **atomic bomb**. After the war, he went back to school to pursue his Ph.D. at Columbia.

It was here, while studying for his doctorate in 1957, that the idea for the laser came to him, in a "flash of insight." Many scientists had been struggling with the concept of a light amplifier, ever since Townes's **maser** first achieved stimulated emission of microwaves in 1954. Indeed, the concepts essential for lasers had been around since 1917, when Albert Einstein (1879-1955) had first discussed them in depth, but nobody had developed a feasible design. Gould's model would be the first to put together all the available pieces. He quickly copied the complete design in his notebook and had it notarized, and it was in these notes that Gould coined the term "light amplification by the stimulated emission of radiation," or laser.

However, due to an unfortunate mix-up at a patent office, Gould came under the misconception that one had to exhibit a working model of a device to be awarded its patent. Thus, while he went back to the laboratory to build his machine, Townes and Arthur Schawlow (1921-) quickly patented their own remarkably similar design notes.

Gould spent the bulk of the next twenty years fighting to get the patent back, a legal struggle as much for recognition as for royalties. By the 1970s, lasers had become an integral and profitable part of several industries, and Gould's designs applied to almost ninety percent of all lasers used. His *optically pumped laser* was employed by designers and surveyors, while his *gas-discharged laser* had found use as a price scanner in supermarket checkout lanes. In 1977, the United States Patent Office finally gave Gould the patent for his initial laser design, and since then he has won major court battles with AT&T and General Motors.

Gourdine, Meredith Charles (1929-)
American physicist

Gourdine is generally recognized as one of the twentieth-century's pioneers in energy conversion, which is the science of converting fossil fuels to useful heat and electricity with as little pollution and energy loss as possible.

Gourdine was born in New Jersey and raised in New York City. He first attended a Catholic elementary school, where he became interested in math and science. After graduating from Brooklyn Tech, Gourdine entered Cornell University, where he studied math, physics, and engineering. While a college student, he participated in the 1952 Olympics in Helsinki, Finland, and won the silver medal in the long broad jump. After finishing his studies at Cornell, Gourdine went on to earn his Ph.D. in Engineering Science at the California Institute of Technology, eventually working at Jet Propulsion Laboratories in California and as chief scientist aeronaut at the Curtiss-Wright Corporation. Finding that his interests were in the theories and practical applications of electrogasdynamics (EGD), which is the conversion of gases into high-voltage electricity, he founded his own manufacturing firm, Gourdine Systems Incorporated. Gourdine Systems eventually merged with Fabricating Engineering Incorporated in 1966 and became known as Gourdine Laboratories. Gourdine also founded Energy Innovations, a research and manufacturing institution, in 1974.

Although scientists had been aware of electrogasdynamics since the 1700s, they were unable to produce a cost-effective system that could produce large amounts of energy without the need for large **steam turbines** and boilers. Gourdine developed a compact power generator that allowed very large forces to operate within a small space. The key to the generator was the development of the EGD channel. Gourdine explained: "What we contributed to the technology was a way to generate big enough forces. We had to make the generator compact, to crowd the ions into a very small space so that the charged particles could work very hard on gas moving down the channel. It's the geometry and shape of the channel which is the fundamental reason why we can get big forces and generate a lot of power in a small space."

In the 1960s Gourdine examined the possibility of setting up an electrogasdynamic power station that used coal and other fossil fuels for the United States Department of the Interior. These innovations have also been used in automotive exhaust systems and dust monitors in industrial settings. Gourdine's industries have also been responsible for *Incineraid*, a device that cleans industrial smoke of its hard contaminants as it passes through smoke stacks and is released into the air. Gourdine and his associates have also invented machinery and processes which disperse fog from airport runways and are used in industrial painting and coating processes.

Governor

Prior to the invention of the **steam engine**, governor-type devices were used to regulate the operation of grinding stones in windmills. Beyond a certain point, increased wind power hampered rather than enhanced productivity and so the idea of a regulating, or governing, device was conceived. Pendulum movement, or the centrifugal action of ball-and-rod contraptions, was the crux of such early governors.

Scottish engineer **James Watt** was the first person to fully harness and utilize the governor as a control on the power source itself. (The windmill devices only regulated the action of the millstones, not the wind.) In Watt's case, the power source was the steam engine, invented by **Thomas Newcomen** and then greatly improved by Watt during the 1770s. Around 1788 Watt added a governor to his steam engine. The device consisted of two weights attached to pivoting arms, which in turn were attached to a regulating valve. As engine speed rose, so did the weights, until the regulating valve released steam pressure and thus decreased engine speed.

Watt's invention, though subordinate to his overall work with engines, paved the way for increasingly greater factory efficiency and virtually automatic mechanical processes. A governor mounted directly on the flywheel shaft was invented in the early 1870s and an inertia governor, which magnified centrifugal forces and improved steady speed performance, was developed in 1895.

Twentieth-century governors take a variety of forms and may be found in **steam turbine**s, **electric motor**s, and gasoline engines. Automatic transmission in automobiles is an excellent example of how Watt's early work with governors continues to effect modern-day life. Working in conjunction with the oil pump, the governor in an automatic transmission system determines when throttle valve pressure crosses the threshold between low and high gears and then transfers this information to the shift valve. The governor's mechanical role has been greatly expanded beyond its original duty of simple speed regulation.

Graham, George (ca. 1673-1751)
English inventor

At age 14 George Graham became the apprentice to a London clockmaker. After serving a period of seven years in this capacity, Graham was employed by Thomas Tompion (1639-1713), the most prominent *horologist* and instrument maker of his time.

During his years with Tompion, Graham helped to construct the first machine (later known as an *orrery*) that would accurately simulate the motions of the planets. Graham assumed Tompion's business upon his mentor's death, and he concentrated his efforts upon designing **clocks** whose inner workings could prevent or compensate for certain inherent flaws, such as the effect temperature changes had upon the clocks' metal gears. By 1726 Graham had built the *mercury-compensated pendulum*, as well as the *deadbeat escapement*, a device whose application allowed clockmakers to design clocks of surprising precision. Graham also invented the *cylinder escapement*, which allowed wristwatches the same precision as clocks.

As his reputation grew, Graham began to design scientific instruments, never straying from his dedication to unparalleled accuracy. Noting that increased size would allow for greater precision in the study of stars, Graham constructed an 8-ft quadrant for the eminent astronomer Edmond Halley (1656-1742), and he also built a 24-ft zenith sector. Apart from their size, these instruments were notable for the accuracy of the graduations Graham had hand-inscribed upon their faces. His lifelong quest for precision in measurement was recognized by the British Royal Society and the French Academy in 1741 when he was asked to help develop the yard as a standard unit of measurement for both British and French scientists.

Grain elevator

The shipping and handling of grain from farm to market was once a very awkward, inefficient process. Grain had to be moved about in sacks, and for longer journeys, it had to be transferred into difficult-to-handle barrels to protect it from the elements. The development of the grain elevator and the loading and unloading methods associated with it streamlined the grain production industry.

American **Oliver Evans** patented a fully automated grain mill in 1785 that used an elevator consisting of an endless chain of buckets. This system was used to move the grain both at the storage elevator and at the mill. Mill operators did not accept Evans's elevator until after 1843 when another American, Joseph Dart, added steam power to it. Since the grain was delivered to large storage bins classified by type and quality, the grain inspectors no longer had to spend time examining every barrel.

In 1848 the first storage elevator was built in Chicago. Soon, other elevators were located at major transfer centers in the Great Lakes area and along rivers and rail lines. With the increased efficiency of grain transfer and transport, farmers could locate further from centers of population. Globally, more farming was done in regions, or belts, especially suited to grain production, such as the Great Plains, Argentina, and the Ukraine.

Today, there are four types of grain elevators, but the bucket elevator remains the most common type. It is capable of lifting grain to heights of three hundred feet at rates of two thousand tons per hour. The buckets move by either chain or conveyor belt. Similar to the bucket type is the continuous chain elevator. Paddles mounted on a chain move the grain forward inside a long steel tube either horizontally or vertically.

The pneumatic elevator literally vacuums grain from the holds of ships to dockside facilities. By using a flexible intake hose, changes in water level or the ship's own level as its load weight changes can be accommodated without adjusting the equipment. The system also loads vessels with a blower positioned above the open hatch.

Another type is the Archimedean screw, or auger, elevator. Named for the Greek mathematician who designed it, the large rotating screw mechanism pushes the grain through a shaft horizontally or at angles up to eighty degrees. It is used in farm storage facilities, often as part of a larger pneumatic system. The grain and grain dust inside an elevator are highly combustible, resulting in constant danger of explosions and fire. Newer, concrete grain elevators, somewhat lessen fire risk.

Care also must be taken to prevent spoilage and reduce invasion from pests. A forced-air treatment was developed in the 1950s in the United States to keep stored grain aerated, reducing moisture accumulation and the need to fumigate for pests. Elevators similar to those used for grain are also used extensively in the limestone industry.

With the growth of agribusinesses, farmers have begun to find alternatives to the customary grain elevator. Mobile elevators lessen the need to transfer grain to a stationary elevator, and semitrailer trucks that move the product directly to railroads or processing plants offer greater safety, speed, and allow the farmer to choose the time and place for delivery.

See also Archimedes; Farming, mass production

Gramme, Zénobe Théophile (1826-1901)

Belgian French inventor

Zénobe Gramme was a true enigma, and it is remarkable that he was able to accomplish what he did. Born on April 4, 1826, in Jehay-Bedegnée, Belgium, Gramme was the son of an educated family of modest means. With the family's interest and ability to afford an education, one might expect great things from young Zénobe. In fact, he showed no ability as a student and did not do well at school. Preferring to work with his hands, he learned only the four basic operations of simple arithmetic and remained only semi-literate throughout his life.

Gramme did, however, have a talent in tinkering, and he left school at an early age to become a joiner. At the age of 22 he moved to Liège with his family; seven years later, in 1855, he traveled to several cities before settling down in Paris as a banister-maker. Soon after, he accepted work as a model maker at a company that manufactured electrical equipment. It was here that he became interested in technology.

Back in the 1830s two physicists, **Joseph Henry** in the United States and **Michael Faraday** in England, had independently laid the foundation for the electric dynamo (**generator**), a device that produced electric current. In 1867 Gramme became interested in building an improved dynamo, but not for any scientific reasons. It was more a matter of cleanliness—he was appalled by the dirt associated with the electric batteries that were used to produce power. He not only built a dynamo for the production of **alternating current** (AC), but two years later built one that produced **direct current** (DC) that produced much higher voltages than previous dynamos.

In 1871 Gramme, who had become associated with Hippolyte Fontaine (1833-1917), opened a factory to advance the development of their machines. The business, called Société des Machines Magnéto-Électriques Gramme, manufactured the Gramme dynamo, Gramme ring, Gramme armature and other devices. In 1873 a Gramme dynamo was exhibited at the Vienna exhibition in Austria, where it was discovered that the device could be reversed and used as an **electric motor**. Today, Gramme is most remembered for building the first practical dynamos, which established the foundation of the electrical industry.

Gramophone • See Phonograph

Graphs and graphing

The use of graphs to display information became a common occurrence in the nineteenth and twentieth centuries. Newspapers, magazines, science journals, financial reports, political surveys, and, of course, the sports page, all routinely contain tables, charts, and graphs to summarize and explain current events, scientific studies, or statistical data. It was actually Florence Nightingale who became one of the first innovators in the tabulation and graphical display of descriptive statistics in her attempts to reform hospital sanitation methods. But to most people, graphs and graphing still belong primarily to the domain of high school algebra and other fields of mathematics.

One of the first graphical representations of a function occurred around the middle of the 14th century. At that time Nicole d'Oresme gave a geometrical verification of a theorem, discovered at Merton College of Oxford and hence known as the Merton rule, concerning the distance covered by a body moving with uniform acceleration. Oresme's novel idea was to draw a time-velocity graph for a body moving with uniform acceleration. He marked instants of time along a horizontal line and at each such mark he then drew a perpendicular line segment (which he called the latitude) whose length represented the velocity at that time. The end points of the latitudes lie along a straight line which we would now call the graph of the velocity as a function of time. Oresme also observed that the distances covered by the body corresponded to areas of right triangles formed by the velocity graph.

The graphical representation of functions became known as the latitude of forms. It was not studied in a systematic way, however, until the development of analytic geometry by the French philosopher and mathematician René Descartes in 1637. The introduction of a coordinate system allowed a point to be specified as a pair of numbers called its coordinates. A curve could then be graphed by plotting the points whose coordinates satisfied the equation of the curve. To Descartes, the acceptable geometric curves were those that could be expressed by a unique algebraic equation in two variables.

In 1671 Issac Newton made extensive use of the new coordinate geometry to sketch the graphs of curves from their equations. In *The Method of Fluxions and Infinite Series* he also introduced new coordinate systems, including the polar coordinate system in which points are graphed based on their distance from a fixed reference point and the angle with a fixed reference line. This work was not published, however, until 45 years after Jakob Bernoulli had devised essentially the same method for graphing special curves in polar coordinates. The main development for graphing surfaces occurred in the eighteenth century. Although Philippe de La Hire (1640-1718) had represented a surface by specifying points with three coordinates based upon a rectangular coordinate system, it was Johann Bernoulli in 1715 who introduced the three coordinate planes that we use today for graphing in three dimensions.

Modern technology has greatly increased our ability to generate accurate graphs quickly. Graphing calculators and computer programs use essentially the same technique as that developed by the 17th century mathematicians to draw the graphs of functions. A finite number of points lying on the graph are plotted and adjacent data points are connected by line segments to achieve the appearance of a smooth curve. The more points that are used, the more accurate will be the graph. Undersampling occurs when too few data points are used and important features of the graph are missed, much as a photograph may miss details if the resolution is too low. Advanced computer algebra systems often use an adaptive sampling strategy in which more points are used in regions where the graph appears to wiggle up and down over a short interval. The ability of a computer to generate graphical solutions to a differential equation has allowed mathematicians and scientists to study the qualitative behavior of these equations without having to solve them analytically. **Computer graphics** have also provided the opportunity to draw the graph of a surface or other three-dimensional object and actually rotate the object in real time to examine it from many different directions, even from the inside. This has yielded important applications to many areas such as medical diagnostics and aeronautical design.

GRAY, ELISHA (1835-1901)

American inventor

Elisha Gray

Elisha Gray was **Alexander Graham Bell**'s principle rival, first for invention of the harmonic telegraph and then of the **telephone**. He was a prolific inventor, granted some seventy patents during his lifetime. Born in Barnesville, Ohio, on August 2, 1835, and brought up on a farm, Gray had to leave school early when his father died but later continued his studies at Oberlin College, where he concentrated on physical sciences, especially electricity, and supported himself as a carpenter.

After leaving Oberlin, Gray continued his electrical experiments, concentrating on telegraphy. In 1867 he patented an improved telegraph relay, and later, a telegraph switch, an "annunciator" for hotels and large business offices, a telegraphic repeater, and a telegraph line printer. He also experimented with ways to transmit multiple, separate messages simultaneously across a single wire, a subject that was also engaging the efforts of Bell. Gray prevailed, filing his harmonic telegraph patent application in February 1875, two days before Bell's similar application.

Gray now began investigating ways to transmit voice messages, soon developing a telephone design that featured a liquid transmitter and variable resistance. In one of the most remarkable coincidences in the history of invention, Gray filed notice of his intent to patent his device on February 14, 1876—just two hours after Bell had filed his own telephone patent at the same office. Western Union Telegraph Company purchased the rights to Gray's telephone and went into the telephone business; the Bell Telephone Company launched a bitter lawsuit in return. After Western Union settled with Bell, Gray renewed his case. The Supreme Court ultimately decided in Bell's favor in 1888.

Meanwhile, Gray had been a founding partner in 1869 of Gray and Barton, an electric-equipment shop in Cleveland, Ohio. This became Western Electric Manufacturing of Chicago in 1872, which evolved into Western Electric Company, which, ironically, became the largest single component of Bell Telephone in 1881.

Despite his disappointment over the telephone patent, Gray continued to experiment with electricity. In 1888 and 1891 Gray patented his TelAutograph, which electrically transmitted handwriting or pictures, using a wide band of paper with a recording pen that moved as if hand-held. Gray demonstrated the TelAutograph at the World's Columbian Exhibition in 1893. From 1880 until his death, Gray was professor of dynamic electricity at Oberlin. Before his death on January 21, 1901 in Newtownville, Massachusetts, he was experimenting with an undersea signaling device.

See also Telegraph, electric

Greatbatch, Wilson (1919-)
American inventor

Wilson Greatbatch was born in Buffalo, New York, in 1919. When the United States entered World War II, Greatbatch left his studies at Buffalo State Teachers College and adapted his skills as an amateur **radio** operator to become a military radioman. After the war, the GI Bill gave Greatbatch the opportunity to study for an engineering degree at Cornell University, where he was distinguished for having the most children (five) of anyone in his class.

While working part-time at Cornell's animal behavior farm, Greatbatch chatted with visiting brain surgeons during lunch breaks and learned about complete heart block. In this disease, the electric impulse sent by the heart's sinus node to the heart muscles, causing them to contract and pump blood, is disrupted. Greatbatch immediately thought of designing an artificial **pacemaker** that could be implanted in the chest and deliver shocks that would cause the heart to beat. At the time, though, in the early 1950s, no components small enough to build such a device were available.

After receiving his bachelor's degree in electrical engineering from Cornell, Greatbatch earned a master's degree from the State University of New York at Buffalo in 1957. In 1958, Greatbatch met Dr. William Chardack of Buffalo's Veterans Administration Hospital and told him about the pacemaker idea. Chardack responded that Greatbatch could save 10,000 lives a year with such a device, which now could be made feasibly small because **transistors** were available. Within two weeks, Greatbatch had built a workable pacemaker; within two years he had built fifty pacemakers and the first ones were implanted successfully in humans.

In the 1970s, after more than 10 years of successful pacemaker use, Greatbatch turned to improved battery design. He found the long, reliable life pacemakers needed in the lithium **battery**, which his company began to manufacture. Next, Greatbatch delved into a biomass energy project, planting thousands of acres of poplar trees. This in turn aroused Greatbatch's interest in cloning plants and working with tissue culture and gene synthesis. His company, Greatbatch Gen-Aid, then went on to attempt the synthesis of genes that can block retroviral diseases like AIDS and T-cell leukemia. Greatbatch preferred to describe himself as an engineering executive or entrepreneur rather than an inventor, but his more than 150 patents certainly qualified him for the latter title. In 1986 Greatbatch was made a member of the National Inventors Hall of Fame.

Greek fire

Because wood was a primary building material for structures and boats in the ancient world, incendiary devices were often used in warfare. Perhaps the most famous of all these devices was *Greek fire*. It was first invented by Callinicus (c. 620-?), an Arab who fled to Constantinople, the capital of the Byzantine Empire. The composition of Greek fire remains unknown today. It is believed to have been a mixture containing a liquid derivative of petroleum; other ingredients may have included potassium nitrate to supply oxygen, and quicklime to provide additional heat through reaction with water. The liquid was forced out of a bronze tube by a pump and sprayed over the target. It even burned on water, thus making it capable of destroying wooden ships.

Greek fire was first used around 671 A.D. by the Byzantines against an Arab naval attack on its capital city. Like so many closely guarded secrets, the method of creating Greek fire was not a monopoly held by the Byzantines for long. The Arabs created their own brand of fire and used it against the Crusaders, and it may have been used in England as late as 1304 by Edward I. The advent of **gunpowder**, a more destructive invention, soon revolutionized warfare, thus making Greek fire obsolete. However modern warfare has used a variety of similar substances, such as **napalm**.

See also Chemical warfare; Flame thrower; Molotov cocktail

Greene, Catherine Littlefield • See Cotton gin

Greene, Leonard Michael (1918-)
American engineer

Leonard Michael Greene is the inventor of a number of devices that are of integral importance in modern aviation. Born in New York City in 1918, he received his bachelor's and master's degrees from the City University of New York.

Greene's most significant invention is the **aircraft** stall warning device. In aviation, a stall means something very different than it does in an automobile, where it refers to an engine shutdown. In an aircraft, a stall indicates that the airflow over the wings is no longer sufficient to provide lift — essentially, a stalled wing ceases to function. This happens at low airspeeds, such as may occur during an excessively steep climb, and the consequences can be disastrous. In the early part of the twentieth century, stalls were a leading cause of aviation deaths, and Greene himself saw a stall fatality and realized the pilot had no way of knowing what was happening. Forming a corporation in White Plains for production of aircraft equipment, he invented a device in 1946 that warns pilots of an impending stall, and no modern airplane can legally be without one. Greene's company, the Safe Flight Instrument Corporation, now produces other aircraft safety equipment in addition to the stall warning device.

In recognition of his singular contributions to aviation safety, the National Business Aircraft Association awarded Greene their 1996 Award for Meritorious Service to Aviation. Some of Greene's other major contributions to modern aviation are the development of wind shear warning devices, as well as theoretical work on supersonic travel and aircraft. He remains the president of the company he founded in 1946.

GREENWICH MERIDIAN • See Chronometer

GREGG, JOHN ROBERT • See Shorthand

GRENADE

The prototype of the modern grenade was created in the fourteenth and fifteenth centuries. They were made of bark, **glass**, clay, or earthenware pots filled with large grains of black powder that was set off by a fuse of corned powder housed in a quill or a thin tube of rolled metal. Because they looked like pomegranates with their large seeds, grenades picked up their name from the Spanish word for pomegranate: "granada." These early grenades were primarily used as incendiary devices.

Later, they were replaced by grenades with round metal bodies that could injure or kill those near its explosion. These metal grenades proved to be dangerous: the fuses were unreliable and the powder occasionally went off before the user could release it. Nevertheless, grenades maintained their popularity with the military. For example, in the 1600s, each infantry company of the British army included five "grenadiers" armed with grenades. It was dangerous to be a grenadier because of the frequency of accidents, and men who volunteered for this duty received extra pay and benefits. Grenades were thrown by hand, and also fired from musket-like tubes. But after about 1750, risky grenades were largely abandoned in favor of more accurate and powerful firearms.

Grenades became more reliable and diversified in the twentieth century. They were equipped with firing pins held up by striking levers, which in turn were secured by safety pins. They were launched from **rifle**s and guns as well as thrown by hand. They were used extensively in the Russo-Japanese war of 1904-1905, and were important again in trench warfare in World War I. Millions were manufactured in the United States for use during World War II.

Modern grenades are constructed for a variety of purposes. They can be constructed to produce anti-personnel fragments, to penetrate armor, to generate smoke or tear gas, or to fire signal and illuminating flares. Smoke-producing grenades and tear gas grenades are often used by riot police. Anti-tank grenades are essentially very large bullets. These can be shot from a shoulder-held tube called a rocket launcher or grenade launcher.

GUERICKE, OTTO VON (1602-1686)

German physicist

Otto von Guericke was born on November 20, 1602, in Magdeburg, Germany. While he studied mathematics, law, and engineering, Guericke would become famous for his experiments with a vacuum and air pressure.

Following travels to England and France, Guericke returned to Magdeburg in 1627 and became a politician. Unfortunately, this was during the Thirty Years' War, and Guericke and his family had to flee the city in 1631. After the war, he returned and helped rebuild the city, becoming mayor in 1646. Twenty years later he became a noble and added "von" to his name.

Apparatus for throwing greek fire.

During Guericke's time, scientists were involved in an argument about whether a vacuum could exist. Guericke, who believed in the Copernican theory of the solar system, was extremely interested in understanding the nature of space. He wondered whether empty space could exist. Many scientists held to the Aristotelian theory that a vacuum was impossible. Aristotle (384-322 B.C.) rightly suggested that as air became less dense an object would move faster. However, he went astray by claiming that if there wasn't any air, an object could move infinitely fast. Since he doubted the idea that infinitely increasing speed could exist, he concluded that a vacuum could not exist either.

Guericke was a rare breed of scientist who refused to accept previously held scientific facts; he believed in experimentation. He decided to try to successfully create a vacuum. In 1647 his first attempt failed, but in 1650 he built another air pump by putting a **piston** inside a cylinder that had two flap valves (a model later improved upon by **Robert Boyle**). He filled a cask with water to remove the air, sealed it, and used his pump to remove the water.

Unfortunately, the cask leaked air. He made a second attempt, placing the cask within another containing water. He hoped the water-filled cask would prevent air from entering the evacuated cask. He was right; air did not leak in, but water did.

Guericke decided to take another tack. He took a hollow copper sphere with a **valve** built in the bottom and used his pump to remove the air. The sphere promptly crumpled, proving that when the sphere was empty the external air pressure was strong enough to crush the sphere.

This is what should have happened according to René Descartes, who held that space and matter were equivalent, so

Otto von Guericke

a vacuum could not exist. Guericke did not agree with Descartes' assertion. Guericke tried again with a more substantial sphere. This time he succeeded in creating his vacuum. He then undertook a series of grandstanding experiments to demonstrate the power of air pressure. In one of the most famous, in 1657, he placed two copper hemispheres together and removed the air. Sixteen horses were unable to pull the two halves apart. Obviously, the external air had substantial pressure; so much, in fact, that the hemispheres held together when the internal air was removed.

It is interesting to note that Guericke placed the valve at the bottom of the hemispheres because he believed that air, like water, would seek the lowest level. Later he found that air was distributed evenly, since he could create a vacuum regardless of the location of the valve. This led him to think about the density of air decreasing as one's altitude increased. He studied variations in air pressure and, in 1660, invented a **barometer**, which he used to make predictions about the weather.

Another of Guericke's experiments pitted 50 men against his piston. A rope was attached to a piston in a cylinder and the men were told to pull. Guericke created a vacuum on the opposite side of the piston and the men were unable to keep the external air pressure from pushing the pistons into the cylinder. The force of air pressure became very important later in the development of the **steam engine**.

In a different vein, Guericke also invented the first mechanical static electricity **generator**, hinting at the future possibility of **Robert Van de Graaff**'s generator. Guericke rotated a sulfur sphere on a shaft. When it was rubbed, it built up an electrical charge that emitted sizeable sparks. Guericke did not give any special consideration to this electrical phenomenon, but during the next century others would continue to experiment with static electricity.

GUIDED MISSILE • See Rocket and missile

GUITAR, ELECTRIC • See Musical instruments, electric

GUN SILENCER

A silencer reduces the noise level of a discharged gun by trapping and slowing the release of gases inside the gun's muzzle. In 1908 Hiram Percy Maxim (1869-1936), the son of **Hiram Maxim**, inventor of the **machine gun**, invented such a device. Called the Maxim silencer, it consisted of a cylinder screwed onto the gun barrel. Inside the cylinder were several small chambers, separated from each other by metal rings with holes drilled in the center to allow the bullet to pass through. When the escaping gases rushed into the chambers, they expanded and slowed enough to keep from exploding from the gun's muzzle. The United States Army first used Maxim's silencer, with poor results. The military found that the velocity of the bullet, which was faster than the speed of sound, caused a loud noise by itself, thus producing its own shock wave. The only way that army snipers could make their silencers effective was to use low-velocity **ammunition** which traveled at speeds slightly less than the speed of sound. During World War II, a new variation of the silencer was introduced. Instead of using several chambers separated by metal rings, a simple tube with vent holes was attached to the muzzle of the gun. At the end were several rubber disks through which the bullet passed, and the expanding gases escaped through the holes.

Throughout the years silencers have proved to be less effective when used with most firearms, including **revolvers** and various types of semi-automatic weapons. Revolvers and semi-automatic weapons have open breeches that allow some gas to escape, and so a silencer on the muzzle does not completely eliminate the noise. Though they tend to make most firearms less effective, silencers have obvious advantages for people who want their gun activity to remain secret. Fearing that silencers contribute to illegal activity, many countries strictly regulate the sale of silencers. Some modern guns, designed specifically for clandestine military operations, are made with built-in silencers. These use vent holes and internal baffles as an integral part of the gun barrel. Like other silenced weapons, these are less accurate and use lower velocity ammunition than conventional guns.

GUNCOTTON

Guncotton, discovered in 1845 by **Christian Friedrich Schönbein**, a German chemist, represented a new phase in explosive weaponry. Previously, black **gunpowder** was used in great amounts for all weapons. The disadvantage was that it tended to give off large clouds of black smoke, obscuring the battlefield. Guncotton, on the other hand, was nearly smokeless. It was also extremely unstable: a sharp blow could make dry guncotton explode. Even wet guncotton, though more stable, could explode.

Schönbein discovered guncotton when he was experimenting with nitric and sulfuric acids in the kitchen, an activity that did not meet with his wife's approval. He spilled the acids and mopped them up with his wife's cotton apron, then hung the apron above the stove to dry. Once dry, the apron burst into flames. Schönbein researched the phenomenon and discovered that the nitric acid had bonded with the cellulose in the cotton to form nitrocellulose.

Schönbein did not reveal his method of production until 1846 when Boettger and **Nikolas August Otto** also discovered it. Schönbein placed cotton into fuming nitric and sulfuric acids, boiling the guncotton to remove impurities and any remaining acids, grounding it into a fine powder, draining it, and letting dry into slabs. Explosions in the production plants often interrupted the process; many people were killed by guncotton before it ever reached the battlefields.

Eventually, **James Dewar** and **Frederick Abel** were able to make the guncotton more stable and used it to produce the first smokeless gunpowder, called **cordite**. Cordite eventually inherited many of guncotton's uses, just as guncotton had replaced black gunpowder. Schönbein, meanwhile, developed **collodion**, a viscous mixture of guncotton and ether, which was more stable than ordinary guncotton. Collodion also had medicinal and photographic applications. Full-scale production of guncotton stopped in the early 1860s.

See also Cellulose, chemical uses of

Gundlach, Robert (1926-)

American inventor

Robert Gundlach nurtured the new science of xerography from its infancy to its adulthood. Today his well- reared child—the photocopier—is largely taken for granted. Although the son of a cosmetic chemist, Gundlach found physics to be more accessible and intriguing.

In 1949 he earned a B.S. in physics from the University of Buffalo, New York. He pursued but did not finish a M.S. degree in physics from the same institution, opting to work instead. After a short stint with the Durez Plastic Company, Gundlach joined the Haloid Company, a small photographic firm.

He quickly made his mark. While **Chester Carlson**, the inventor of xerography, and others were developing the first fully automated photocopier, the model 914, Gundlach worked on later models. Early photocopiers were cumbersome and slow, requiring three or four minutes to produce a single copy. In his first year with Haloid, the 25-year-old Gundlach had three patentable ideas. He developed a process that allowed multiple copies to be produced from the same master image, thus significantly increasing the copier's speed. He also created a process by which solid figures and shapes could be reproduced, which previously had not been possible.

Eight years after Gundlach's arrival, the Haloid Xerox 914 copier made its debut. This machine, which could produce a clear copy in less than a minute, initiated Haloid's transformation into the giant Xerox Corporation. With 15-year exclusive patent rights to xerography, Xerox—and Gundlach—refined copier machines undisturbed. By the mid-1980s, Xerox copiers could make 120 copies per minute.

In 1975 Gundlach became the first Xerox research fellow. He continues to head Xerox's EXITE Lab, a research facility, though he was granted a partial retirement in 1986 to work on his own inventions, which include a snow-making system and a backpack.

Gundlach's contributions to the Xerox Corporation, over 100 patents, set the standard for copier companies that emerged after Carlson's patents expired. Not only did Gundlach's ideas transform Xerox into one of the world's largest companies, they turned the trade name Xerox into a generic term used to designate photocopiers of all types.

See also Photocopying

Gunpowder

Gunpowder, the oldest known explosive, was probably discovered in China during the tenth century, though some sources cite a much earlier date. It consists of a mixture of saltpeter (potassium nitrate), sulphur, and charcoal, in variable proportions. Saltpeter is formed from the oxidation of organic matter containing nitrogen, and the process of making and purifying it was probably discovered in either India or China. A Chinese recipe for explosive powder dates to about 1000 A.D., and this was probably used to make incendiary bombs, **fireworks**, or terrifying explosions. It was probably not used to launch a projectile out of a gun or cannon until about 1304, the year in which the Arabs are said to have constructed the first gun, a crude device made of hollowed bamboo and reinforced with iron that was used to fire arrows. English and Italian manuscripts dating from around 1326 show the existence of guns in Europe at that time, so the technology of gunpowder seems to have spread very quickly.By the mid-14th century, gunpowder as fuel for guns, cannons, and other destructive devices was commonplace throughout Europe and much of Asia.

Prior to its first military uses, gunpowder was termed black powder. The recipe for black powder was first recorded by English philosopher Roger Bacon (c. 1214-1292) in 1242. The original formula, consisting of 7 parts saltpeter, 5 parts hazelwood (charcoal or carbon), and 5 parts sulphur, has been modified in its proportions to suit various uses, but the essential ingredients have remained the same. The mixture, though prone to rapid combustion, does not exhibit the commonly observed properties of thrust or loud explosion unless it is somehow confined. Consequently, early gun designs revolved around barrel size and strength and the manner in which gunpowder was compacted.

The proportions of essential ingredients used in gunpowder varied widely from place to place, and did not become standardized until the eighteenth century. The perfecting of gunpowder was an ongoing process. Eventually scientists realized that the powder was simply a mixture, not a new compound. It was limited by such military drawbacks as flashing and smoking. Moreover, there was the additional disadvantage of delayed reaction time. These obvious flaws, coupled with the complaints of mining and engineering companies that gunpowder was incapable of pulverizing large pieces of rock, led independent researchers in several promising directions. Both **nitroglycerine** and **TNT**, two of the most powerful explosives in existence, can trace their origins to this race in research during the nineteenth century.

The discovery that directly supplanted gunpowder for use in firearms was **guncotton**, a forerunner of **smokeless powder**. Invented by **Christian Schönbein** in 1845, guncotton is formed by a nitrating process involving the dipping of cotton (cellulose) in nitric and sulfuric acids. This nitrocellulose explosive underwent improvements until it obtained final form as a reasonably safe and clean-burning powder in 1884. This powder was ground from a hardened gelatin created by the evaporation of ether and alcohol. Its inventor was French chemist Paul Vieille (1833-1896).

Gunpowder is still employed today for a number of specialized uses but is perhaps most commonly found in fireworks. Its production involves a number of important steps. After securing ingredients of the highest quality, the manufacturer must grind and mix the material while damp, then compress it into cakes, break it into grains of varying sizes, and, finally, ensure that the grains are glazed, dried, graded, and properly packaged. These exacting requirements were arrived at after centuries of experimentation and thousands of accidental deaths. Despite vast improvements in safety guidelines and preparedness, factory and operational accidents still occur due to the volatility of the substances involved.

See also Cordite; Powdermills

Gutenberg, Johannes (1395-1468)
German inventor

Gutenberg is considered the inventor of movable metallic type that made printing many books at once practical for the first time. Movable type revolutionized printing, fostered the standardization of type size and ushered in the beginning of mass communication in fifteenth century Europe.

Printed books had been in existence long before Gutenberg. The oldest surviving book is one from China produced in 868 A.D. when wooden blocks were engraved with characters and then inked. Eleventh century Chinese books and 13th century Korean books show that printing was accomplished with some movable type. However, because Asian languages use thousands of complex characters for words rather than letters of an alphabet, movable type had a limited practicality. In Europe, perhaps twenty years before Gutenberg, a Dutch church official, Laurens Janszoon Coster, is credited with using some movable type but not with the precise system set up by Gutenberg.

Gutenberg was born sometime between 1395 and 1400 in Mainz, Germany to a wealthy, political family named Gensfleisch, but was most often known by his mother's name, Gutenberg. Little is known of his youth other than he was probably a member of the goldsmith's guild. Local political upheaval caused him to be exiled from Mainz and he moved to Strasbourg, Germany, where he worked in secret with partners on his printing inventions. Gutenberg was working on the letterpress—a mold for each individual letter of the **alphabet** cast out of equal amounts of metal alloy. The letters could be arranged into words along even lines and then locked together into a rigid form, making a page. The form was placed in a printing press and the letters covered with **inks** similar to paints used by early Flemish artists. When sheets of **paper** pressed against the letters, thousands of copies of the same page could be produced in a short time. With this invention, the type could then be separated and reused to make other pages. This process could make books faster and cheaper to manufacture.

Upon the death of one of his partners, the nature of Gutenberg's invention was brought to light in a trial at which the partner's heirs attempted to sue Gutenberg. Gutenberg won the lawsuit and later, became partners with a wealthy benefactor named Johann Fust. Fust, a practical man, wanted to print a book that would sell many copies. The two agreed to work on a Bible. Their most well-known achievement was the Gutenberg Bible printed in 1455. It is also called the 42 line Bible, referring to the fact that every page had 42 lines of type. However, Gutenberg wanted to spend time working on other creative projects and perfecting his printing process. This led Fust to sue Gutenberg—and this time Gutenberg was the big loser. Then Fust and his son-in-law Peter Schoffer, a calligrapher and former Gutenberg student, became printers of a famous hymn book in 1457. But many people believed that Gutenberg did at least part of the book because of the fine craftsmanship.

Despite the lawsuits and bickering, Gutenberg did enjoy the fruits of his labor later in his life by being named a member of the royal court and receiving a yearly allowance. By the time he died on February 3, 1468, Gutenberg's invention had already spread throughout Europe— north to Sweden, east to Poland, west to Portugal. Printing was the first **mass production** industry, allowing one person, a printer, to do the work formerly done by many scribes writing by hand. Its impact on communication was far-reaching. Although improvements were made in Gutenberg's method, no vast changes were made in printing until the twentieth century.

See also Printing technology

Gutta-percha

Natural rubber, which is essentially cis-1,4-polyisoprene, exhibits elastic properties; its trans isomer on the other hand exhibits plastic properties. Like rubber, gutta percha is an extract from tropical trees, but unlike rubber it is hard and flexible at most ambient temperatures, softening at approximately 212°F (100°C).

Gutta-percha is derived from the trunk and leaves of the Malayan gutta tree (*Palaquium gutta*), which is grown on plantations. Trees are tapped or leaves are gathered, chopped, and soaked in boiling water. The gum is then recovered and packaged. Similar to gutta-percha is balata, the inferior quality latex of the South American bully tree (*Mimusops balata*). Both gutta-percha and balata are noted for their resistance to impact and embrittlement.

Gutta-percha was first exported from the Far East in 1838. English physicist and chemist **Michael Faraday** (1791-1867) soon noted that it was an electric insulator. This property, along with its ability to resist disintegration in salt-water, made it ideal for use as insulation for underwater telegraph cables, the first one being placed between England and France in 1850. A screw extruder was developed to coat the copper conductor, thus laying the foundations for today's plastics extrusion industry. Gutta percha also found early use as a molding material. It continues to be a component in some special adhesives, and both balata and gutta-percha are used in golf ball covers.

Gyrocompass

We have all seen the principle of gyroscopic motion in a child's top, which balances as long as it continues to spin. Any spinning body, such as a wheel or a top, tends to keep its axis of spin always pointed in the same direction. **Gyroscope**s are not affected by either the downward pull of the Earth's gravity or by the presence of a magnetic field. For those two reasons, a gyroscope can be useful for navigation purposes.

In 1851 the French inventor **Jean Foucault** built the first gyroscope. It consisted of a rapidly spinning wheel within concentric rings which allowed the wheel to move freely about two axes. It demonstrated the rotation of the Earth because the spinning wheel, which was not fixed, retained its orientation in space while the Earth turned under it. Foucault also found that the force of the Earth's rotation caused the gyroscope's axis to precess gradually until it was oriented parallel to the Earth's axis in a north-south direction. Unlike traditional magnetic compasses, the gyroscope can indicate true, or geographic, north rather than magnetic north, which varies depending upon the location of the **compass**. Also, the gyroscope was unaffected by the **steel** that was increasingly replacing wood in the hulls of **ships**, which often skewed the readings of magnetic compasses.

From 1904 to 1905 Hermann Anschutz-Kaempfe, a German manufacturer, developed the first gyrocompass as a way to solve navigation problems in **submarine**s, which were unable to tell where they were by traditional means. Anschutz-Kaempfe's gyrocompass proved to be a difficult item to produce because it had to be free to turn but also strong enough to withstand the movements of a ship at sea. In addition, it needed a continuous **power supply** to keep the gyroscope rotating at high speed. At the same time, **Elmer Sperry**, a New York inventor, was working on his own gyrocompass. In 1907 he patented one that solved many of the problems found in the German design.

Gyrocompasses soon became standard equipment on submarines, battleships, and even larger commercial vessels during World War I and World War II. They also were used to detect course deviations in the German V-1 and V-2 **rocket**s and guided missiles. Conventional gyrocompasses are still extensively used in marine navigation. With the advent of the **Global Positioning System** (GPS), and its nearly pinpoint positional accuracy, though, navigational gyrocompass readings have been shown to be slightly erroneous. The combination of gyrocompasses and GPS receivers is often used in marine navigation. Gyroscopic deflection sensors are part of the standard instrumentation on modern **aircraft**.

Johannes Gutenberg

Gyroscope

Technically, a gyroscope is any body that spins on a movable axis, including a child's toy top and the Earth itself. A gyroscope maintains a fixed axis of spin in spite of forces of gravity and magnetic fields, making it useful for navigation. Today's **ships** and **aircraft** use weighted spinning wheels to stabilize and guide their vessels.

In the 1700s the British ships experimented with a spinning rotor to indicate a stable horizontal reference at sea. To avoid being jostled by rolling waters, the rotor had a pivoted support. During the 1800s, two scientists independently demonstrated the rotation of the Earth using the gyroscope's stability in space. One Scottish scientist, named Sang, was unable to construct an accurate spinning wheel, or *rotor*. **Jean Foucault** of France, however, succeeded, and in 1852, demonstrated that the axis of a spinning rotor, with its center in a fixed position but able to point in any direction, appears to revolve 360 degrees in space through the course of one day. This showed that, while the gyroscope maintained its position, the Earth revolved beneath it.

During the twentieth century, gyroscopes began to be utilized to stabilize ships at sea. German inventor Otto Schlick developed huge **gyrostablizer**s that kept a ship upright in the water despite crashing waves. A more efficient system using underwater fins and smaller gyroscopes attached to the bottom of a ship was later devised.

The invention of the **gyrocompass** saw yet another application for the gyroscope. Steel ships and onboard electrical systems created a need for a navigational system other than the magnetic **compass**, whose readings were affected by these factors. Gyroscopes, as predicted by Foucault, solved this problem. Once set in motion, they indicate true, not magnetic, north, even on a rolling ship in heavy seas. Hermann Anschutz-Kaempfe of Germany patented his gyrocompass in 1906; **Elmer Sperry** of the United States successfully deomonstrated his for the U.S. Navy in 1911.

In aircraft, gyroscopes are navigational aids. They provide artificial horizons and better control the airplane for a path true to course. Sperry invented an **automatic pilot** device that used four gyroscopes and won a prize from the French government in 1914 for an aircraft stabilization system. American **Charles Stark Draper**, during the 1940s, developed inertial navigation and guidance systems depending only on internal sensors to navigate. His method has been used on **warship**s, planes, and in **rocket and missile** guidance. Gyroscopes have become essential in space, allowing the craft to navigate from its actual position in space rather than relying on external information from **radio** signals or celestial bodies.

Gyrostabilizer

A **gyroscope** is a weighted, balanced wheel mounted in bearings and spinning at high velocity. It holds its position, no matter at what angle it is held. This has made the gyroscope valuable for several purposes.

One obvious application is to stabilize **ships** and **aircraft**. In 1868 the English inventor Matthew Watt-Boulton proposed using a gyroscope to help keep a ship steady at sea. In 1903 Otto Schlick devised the first gyrostablizer for a ship, a huge device bolted to the ship's framework. A self-contained motor drove the rotor of the gyrostablizer, which then resisted the rolling effect of waves beating against the ship. In 1913 **Elmer Sperry** began installing these gyrostabilizers aboard American **warship**s and merchant vessels.

He also applied the same principle to airplanes, inventing an **automatic pilot** that used gyroscopes to stabilize an airplane's controls. In 1914 this device won an award of 50,000 francs from the French when Sperry's son, Lawrence, flew over Paris while he took his hands off the controls and a mechanic walked out on the plane's wing.

Newer versions of the gyrostabilizer have become standard equipment on many vessels. Smaller gyroscopes are mounted to the ship's hull and connected to stabilizer fins and water tanks with gyroscopically controlled air valves. In the 1940s **Charles Stark Draper** developed a highly effective gyrostabilizer while working on his inertial navigation and guidance system for the United States Air Force. His gyrostablizer could register and correct any outside interferences that might affect the flight of aircraft and rockets. Today's sophisticated gyrostabilizers, thanks to their help in guiding and steadying various machines, free people to carry on with their duties.

Optical gyroscopes called ring laser gyros were invented in the 1960s, while fiber optic gyros were developed in the 1970s. Both operate on the principle that they are able to sense phase shifts between counterpropagating beams of light. By the 1980s, ring laser gyros began to replace mechanical or spinning gyros mainly because they are almost impervious to acceleration. By the late 1980s, ring laser gyros were used on European launch vehicles as well as on aircraft and **helicopters**. By the end of the 1990s, research on fiber optic gyroscopes had progressed to the point where it is considered to be the best technology available for inertial guidance and control.

H

HABER, FRITZ (1868-1934)

German physical chemist

Haber was born in Breslau, Germany, which is now part of Poland. Drawn to the new field of physical chemistry, which had just been established by Russian-German chemist Friedrich Wilhelm Ostwald (1853-1932) and other scientists, Haber was reluctant to stay at home and enter his father's business of selling **dye**s and other pigments. Haber obtained his doctorate from the University of Berlin in 1891 and became a professor at the University of Dahlem in 1898.

Haber's early research in electrochemistry resulted in the invention of the first **glass** electrode. This device became the most common, convenient method of measuring the acidity of solutions. Haber, who had studied under **Robert Wilhelm Bunsen**, was also interested in thermal decomposition and other chemical processes that take place in flames.

During the early 1900s, one of the problems facing scientists worldwide was the eventual depletion of a natural resource—Chile saltpeter (sodium nitrate, or $NaNO_3$), which was used to make **fertilizer**s, explosives, and other industrial products. Without nitrogen-based fertilizers to increase crop yields, it was feared that the growing world population would go hungry. Although a practically inexhaustible source of nitrogen exists in the atmosphere, which consists of nearly eighty percent nitrogen, no one knew how to convert the element into useful compounds, such as ammonia, on a large scale.

Haber began investigating the possibility of combining nitrogen from the atmosphere with hydrogen to form ammonia (NH_3). Although the chemistry of **ammonia synthesis** seemed to be simple, the technology was quite difficult. High temperatures and very high pressures were needed to produce great enough quantities of ammonia to make the process economical.

By 1908 Haber had succeeded in developing an ammonia synthesis process that worked in the laboratory. Initially the process used osmium or uranium as a catalyst, and the reactions took place at a temperature of 1,022°F (550°C) and at pressures up to two hundred times greater than normal atmospheric pressure. By 1913 **Carl Bosch** had refined the process and developed the necessary equipment for producing ammonia on an industrial scale. The industrial process, which operated at slightly lower temperatures and higher pressures, used finely divided iron as a catalyst. Essentially, the Haber-Bosch process has remained unchanged since the early 1900s and is used today to manufacture thousands of tons of ammonia worldwide. In 1911, in honor of his accomplishment, Haber was appointed director of Germany's newly created Kaiser Wilhelm Institute for Physical Chemistry and Electrochemistry.

Soon after World War I began in 1914, the Haber-Bosch process saved the German army from running out of **ammunition**. Ammonia could easily be converted into nitric acid, an essential ingredient in explosives, via the **Ostwald-Bauer process**, which had been invented by Ostwald in 1901. By synthesizing ammonia, Germany was able to manufacture ammunition even after supplies of Chile saltpeter had been cut off by Allied ships which blockaded German ports.

The outbreak of war also brought dramatic changes to Haber's life. Research at the institute he headed was redirected to support Germany's military needs and Haber became chief of the country's chemical warfare research. In 1915, Haber directed the first use of the poisonous gas chlorine, which was released into Allied trenches at Ypres, France. Haber then supervised development of the more deadly **mustard gas**, which was first used in 1917. During the war, Haber was also responsible for developing an effective **gas mask** for protecting German troops from Allied gas attacks. Germany's defeat was a bitter blow for Haber, who was a staunch patriot.

Haber was awarded the 1918 Nobel Prize in chemistry for developing an ammonia synthesis process. However, because Haber was still regarded as the villain responsible for inventing poison gas, many scientists vehemently objected to Haber's award, calling it "ill-advised" and "undiplomatic," and some prize winners protested by refusing to accept their

Fritz Haber

awards. On the other hand, many scientists defended Haber's work, calling attention to the numerous other products that can be made using the ammonia synthesis process.

Haber spent the years between the two world wars trying to help repay Germany's war debts by extracting gold from seawater. Although an experimental ship was built, the amount of gold obtained was too small, and Haber gave up the project in 1928. A few years later, Hitler's Nazis came into power, and the government planned to dismiss all Jews working at the Kaiser Wilhelm Institute. Haber was no exception, despite all of his efforts to support Germany during and after World War I. In 1933 Haber resigned his directorship, pointing out, "I have selected my collaborators on the basis of their intelligence and their character and not on the basis of their grandmothers." Although he worked briefly at Cambridge University, Haber intended to settle in Israel but died of a heart attack on the way there.

Hair care

Hair care is the process of improving the appearance of hair by cleansing, conditioning, coloring, or styling. Modern hair care is a blend of science and marketing with hundreds of companies offering thousands of products to solve every possible hair care need.

People have long used soap and water for basic hair cleansing, and this practice is still used today. However, in the last 100 years a number of products have come into fashion which were specially designed to clean hair and leave it in a manageable condition. These products are known as shampoos and instant conditioners. Shampoo, a word derived from the Hindi word for *massage*, dates to 1877, when English hairdressers boiled **soap** in soda water and added herbs for health, fragrance, and manageability. Over the last few decades shampoos have become increasingly sophisticated. Synthetic surfactants have, for the most part, replaced soap-based materials. These products are capable of gently and effectively cleansing a variety of hair types. In the 1970s and 1980s Proctor and Gamble revolutionized hair care with the introduction of the two-in-one shampoo plus conditioner. Their patented formulas are capable of cleansing and conditioning in a single step. Many other manufacturers followed with two-in-one formulas of their own and in 1997 such products accounted for approximately 20% of the shampoo market.

In addition to shampoos, instant conditioners, or cream rinses as they were once known, continue to enjoy popularity in the market place. These products are used after shampooing to smooth the hair and make it more manageable. Originally natural oils were used for this purpose. For example, during the late Victorian era, macassar oil from Indonesia was used to condition men's hair. Unfortunately, this greasy substance sparked a need for the antimacassar, a crocheted doily pinned to the back of chairs and sofas to absorb excess oil and protect upholstery. The turn of the century brought a flurry of inventions and discoveries for hairdressing., and at the Paris Exposition of 1900 E. D. Pinaud advertised a brilliantine for softening hair, beards, and moustaches. In the last 90 years, chemists have learned how to formulate non-greasy conditioning products using silicones, quaternary ammonium compounds, and fatty alcohols, which can condition hair without weighing it down.

Hair care also involves technology to reshape hair into specific styles. A number of nineteenth-and twentieth-century inventions were designed to make hair styling easier, more effective, longer lasting, and more natural looking. For instance, in 1866, **Hiram Maxim** invented the first curling iron. Four years later, two Frenchmen, Maurice Lentheric and Marcel Grateau, used hot-air drying and heated curling tongs to make deep, long-lasting Marcel waves. Twenty years later, Alexandre F. Godefroy, a French hairdresser, invented the hair dryer, composed of a bonnet attached to a flexible chimney that extended to a gas **stove**.

In 1905, **Sarah Breedlove Walker** created a **cosmetic** industry in Indianapolis, Indiana, and became the first African-American female millionaire in America by inventing a method for straightening hair, using an emollient cream and hot combs. In 1906, Charles L. Nessler, a German hairdresser working in London, applied a borax paste and curled hair with an iron to produce the first permanent waves. This costly process took twelve hours. Eight years later, Eugene Sutter adapted the method by creating a dryer containing twenty heaters to do the job of waving more efficiently. Sutter was followed by Gaston Boudou, who modified Sutter's dryer and invented an automatic roller. By 1920, Rambaud, a Paris beautician, had perfected a system of curling and drying permed hair for softer, looser curls by using an electric hot-air dryer, an innovation of the period made by the Racine Universal Motor Company of Racine, Wisconsin. Products have been made to style hair and improve grooming as well.

One of the most significant breakthroughs came in 1945, when French chemist Eugene Schueller of L'Oréal laboratories combined the action of thioglycolic acid with hydrogen peroxide to produce the first cold permanent wave, which was cheaper and faster than the earlier hot processes. By relaxing the bonds in hair protein, the process changed the configuration of the hair structure and reset it in curled form by means of oxidation. Cosmetologists learned how to control the amount of curl by varying the diameter of rods used for rolling. In related developments, Rene Lelievre and Roger Lemoine invented an electric curling iron in 1959 and the next year a Danish inventor, Aren Bybjerg Pederson, created thermal hair rollers.

Hair coloring technology has become increasingly popular in the last 50 years. Chemist Eugene Schueller invented the first ''para'' hair dye from paraphenylene-diamene. By 1927, through the addition of organic materials, hair dyes had become brighter and more natural-looking than Schueller's first attempts were. In 1953, the process of bleaching and dyeing hair was reduced to one step. Today, hair colors can be done on a temporary, semipermanent, or permanent basis.

Technology to hold hair in place was advanced in 1960 when L'Oréal laboratories introduced a **polymer** hair spray to serve as an invisible net. These polymers were originally dispersed in alcohol formulations which dried quickly to form small weld spots at the junctures where hair strands meet, locking the hair style in place. In the mid 1990s hair spray formulators were challenged by environmentally motivated legislation that mandated that alcohol and other volatile organic compounds be reduced in hair spray formulations. This requires formulators to create high quality products without the benefit of quick drying solvents such as alcohol.

Another important aspect of hair care is the treatment of hair loss. This is a key area of concern for many men and some women, and it received a boost in the 1950s from New York dermatologist Norman Orentriech's technique of transplanting plugs of from four to ten hairs each from the scalp to sections of sparse growth. The placement of plugs requires no stitching, only pressure to assure clotting and proper seating in the new location. Achieving coverage usually involves ten to fifteen once-or twice-weekly sessions in which twenty plugs are moved. The grafting process, which can be performed in a doctor's office, requires local anesthetic. Despite the pain, expense, and possibility of the hair growing in the wrong direction, the procedure is popular both in Europe and the United States. Louis Feit, a New York plastic surgeon, evolved a more drastic approach to hair replacement by grafting a single plot of tissue containing 350 to 500 hairs. His method, which spurs blood supply to the transplant, improves success rates. Another innovation, the use of antiandrogens, encourages the body to produce hair naturally. In mid 1990s the FDA approved the use of a Minoxidil, a topical over-the-counter drug solution, for use in treating male pattern baldness. In 1998, Propecia, a pill, was approved for the same purpose.

The hair care industry continues to grow by bringing new trends and technology to the marketplace. By 1997, the industry was a multi-billion dollar, worldwide market.

Hair removers

Hair removers are products designed to remove unwanted hair from areas on the body. They can work by both chemical and physical means. Formulations that chemically breakdown hair are called depilatories, and contain thioglycolate salts or sulfides. Those that remove hair by physical means are known as epilatories. These typically contain waxes which solidify on the skin and are pulled off.

Hair removers have been used for thousands of years. Evidence suggests that the earliest use of hair removers was in ancient India. These people used epilatories made of abrasive pastes and resinous plasters. In the Middle East, a lime mixture was used for a similar purpose. Antimony and arsenic compounds were also used, but their toxic nature halted their use.

Since the earliest hair removers were epilatories, their use was often a painful experience. This prompted the development of formulations that could chemically reduce hair, eliminating the need for it to be pulled out. In the early 1900s, the first depilatories were introduced. Barium sulfide was the primary active ingredient. In the 1920s, cream depilatories were introduced, and throughout the 1930s numerous patents were granted for depilatory formulations. Thioglycolate depilatories, which are now the most commonly used, were first introduced in 1938.

Epilatories were the first type of hair removers. The most common of these products is an epilating wax. A standard epilatory may be made up of beeswax and a sticky, polymeric **resin.** This product is heated and spread on the skin in the desired area. It is then allowed to cool and harden. The mass of wax is then rapidly removed, pulling with it about 80% of the hairs. The resin helps bind the wax to hair. Since it can be a painful procedure, a mild antiseptic is used to protect against skin irritation.

Depilatories are more commonly used in personal care products than epilatories. They rely on a chemical reaction between compounds in the formula with hair. The primary active ingredients in depilatories are thioglycolate salts and sulfides. The most common thioglycolate salts are calcium thioglycolate and potassium thioglycolate. Sulfides such as barium sulfide or strontium sulfide are also used. Since pH is critical for depilatories to work correctly, sodium hydroxide or calcium hydroxide are included.

When a depilatory is applied to skin, the active ingredient reacts with protein in the hair and weakens it. The hair can then be removed from the skin by gentle wiping, scraping, or rinsing. These products are effective on any hair that is above the level of the skin. Since the active ingredients in a depilatory also react with skin proteins, they cannot be left on for a long time. Therefore, depilatories are developed which have only minimal negative effects on skin.

Hair removers are available in many different forms, including creams, gels, lotions, and aerosols. These different product forms require other ingredients beyond the hair removing materials. This includes diluters, emollients, thickeners, fragrances, and colorants. Most depilatory formulas are based on water while epilatories use mineral oil.

Currently, new hair removers are being developed that are less irritating, longer lasting, and more effective. The focus of research has been on making products that are less irritating to the skin, more moisturizing, lower odor and increased effectiveness. New methods for hair removal are also being studied. Particularly drugs that can inhibit the growth of hair from follicles.

Halftone process

The halftone reproduction process is responsible for the now-familiar sets of large and small dots that comprise illustrations in today's newspapers and **comic books**. Invented by American printer **Frederic Eugene Ives** in 1880, halftone had a tremendous effect on newspaper's and other print media's ability to quickly and effectively reproduce illustrations, especially **photographs**. The invention of the halftone process arose from the need for a means of translating finely shaded pictures onto printing plates (engraved sheets of metal which are covered with **ink** and pressed to **paper** to reproduce an image). Ives devised a system for expressing the shading with large and small dots.

In halftone reproduction, the original illustration is placed behind a screen of dark, crossed lines and photographed. The effect of the grid of lines is to divide the illustration into thousands of tiny, component parts, which translate in the photograph as dots of varying size. The image thus created is projected onto a light-sensitive metal plate, creating a "photoengraved" pattern of large and small dots. After the plate has been developed, it is submerged in acid before being inked and printed onto paper. Prior to the invention of halftone, newspapers relied largely on **woodcuts** for illustrations. Unlike woodcutting, which results in an illustration composed solely of black lines on a white surface, halftone's varied dot sizes allow the reproduction of the range of shades, or "halftones," in a picture, hence its name. Though the process was soon improved and refined, through, for example, the invention of a color halftone process using multiple, overlaid screens of primary colors, the technique of photographing through a grid remained essentially unchanged until the advent of the computers, scanners, and **desktop publishing** when the process became digitalized. Halftones now can be created on a computer using such software as PhotoShop.

See also Printing technology

Hall, Charles Martin (1863-1914)

American chemist

Charles Martin Hall was born in Thompson, Ohio, in 1863, the son of a minister. He attended Oberlin College in Oberlin, Ohio. There Hall studied chemistry and was influenced by one of his professors, F. F. Jewett, himself a former student of Friedrich Wohler (1800-1882) who had developed a process of purifying small quantities of aluminum in 1845. Wohler's extraction method was so difficult that aluminum was considered a semiprecious metal. A remark by Professor Jewett that whoever could develop an efficient method of producing aluminum would become rich was taken seriously by Hall. Even before his graduation in 1885, Hall was experimenting with the process.

Only one year later, Hall succeeded in extracting a sample of pure aluminum from cryolite using an electrolytic process, instead of the thermal process then in use. In a short time, Hall had succeeded where others had been stymied for 60 years. Hall spent the next two years seeking financial support for his discovery. In 1889 he patented his extraction process. With the assistance of financier Andrew Mellon (1855-1937), Hall and a group of businessmen founded the Pittsburgh Reduction Company in 1888, which had its factory at Kensington, Pennsylvania.

In 1890 Hall became vice-president of the company, which was renamed Aluminum Company of America (Alcoa) in 1907. It produced commercially salable units of aluminum with the use of 2,000 ampere cells and required no external heating. The price of aluminum—once $100 per pound (453g)—was reduced to $0.70 per pound. By 1914 it was further reduced to $0.18 per pound.

In the same year that Hall had made his discovery (1886), Frenchman **Paul-Louis-Toussaint Héroult** developed the same process. Hall was able to get his process into commercial production faster than Heroult and won a patent dispute with Heroult in 1893. Despite this, Hall and Heroult became best friends. Hall remembered Oberlin College with a $3 million donation. Alcoa still preserves the original aluminum nodules that Hall produced in 1886.

Aluminum has become, along with steel, one of the two most important metals in the world. It is in demand for its strength, durability, and beauty. Aluminum is currently used in a variety of industries, including construction, airplane production, wiring, cookware, and packaging.

See also Aluminum production

Hall, Lloyd A. (1894-1971)

American food technologist

Hall introduced chemical innovations that helped change the way we store and preserve foods today, including the now-universal use of nitrates in place of salting to cure meats. Born in 1894 in Elgin, Illinois, Hall attended Northwestern University and the University of Chicago, where he majored in chemistry. At Northwestern, Hall befriended classmate Carroll L. Griffith, in whose Griffith Laboratories he would later spend the most productive years of his career. Hall was also one of the first African-Americans to overcome the barriers of discrimination in science and technical careers. Rejected on the first day at his first job at Western Electric Co.—after being hired over the telephone—with the blunt declaration, "We don't take niggers," Hall took a position with the Chicago Department of Health Laboratories as a chemist in 1916.

Hall soon developed an interest in the burgeoning applications of chemistry to food processing. In 1925, as a senior chemist at Griffith Labs, Hall began work on the chemical preservation of meat. Until then, salting with sodium chloride, or table salt, had been the primary, though unsatisfactory, means of curing meat. Although chemists knew that nitrate compounds, combined with salt, preserved the color and appearance of meat better, the compounds also tended to penetrate meat too fast, causing the meat to disintegrate. Hall deduced that if the nitrate could be combined with the salt at the crystal level, it would allow the two substances to penetrate at the same speed. He eventually hit upon the technique of *flash drying*. By quickly drying a solution of sodium chloride and nitrate compounds over hot rollers, the dried substance formed a salt crystal with nitrate at its center.

Hall later turned his attention to the sterilization of spices, cereals, and condiments, which, untreated, can host bacteria, yeasts, and spores, thus contaminating food. The conventional method of food sterilization—killing microorganisms with heat—would burn the spices and make them unpalatable. Hall decided to tackle the problem through chemical means, eventually discovering that ethylene oxide gas—formerly used as an **insecticide**—was effective not only in sterilizing spices, but also as a sterilizer for medical instruments and medicines, an application for which the gas is still used today.

See also Food preservation

Thomas Hancock

Halogen lamp

The invention of electric light in the late 1800s dramatically changed people's lifestyles worldwide. Most improvements to the **incandescent light bulb** were made during the early 1900s. Inert gas was used to fill the bulb, tungsten replaced carbon as the filament material, and coiled filament designs were introduced. By the 1930s researchers had developed a standard incandescentlight bulb that would last for about 1,000 hours. This bulb has since dominated the market.

Beginning in the early 1970s, as a result of sharply higher world oil prices, scientists became much more concerned with energy efficiency and conservation. Lighting equipment manufacturers began working to improve the efficiency of incandescent bulbs and extend their useful life. One development that came out of this research was a new type of incandescent light bulb called the tungsten halogen lamp or quartz lamp.

Ordinary light bulbs usually fail when their filament gradually becomes thinner and breaks. This thinning occurs when tungsten evaporates from the filament and accumulates on the inner surface of the bulb. Also, as the tungsten collects there, it gradually blackens the bulb and reduces the light output. In the halogen lamp, a small amount of iodine, bromine, or other halogen is added to the bulb's gas filling. When the tungsten evaporates from the filament, it combines with the halogen to form a gaseous compound that circulates within the bulb. Instead of blackening the bulb wall, the tungsten-halogen compound remains a gas until it comes into contact with the hot filament and breaks down. The tungsten is redeposited on the filament, and the halogen gas is free to combine with newly evaporated tungsten. However, because the tungsten is not redeposited uniformly, thin places develop on the filament and it eventually breaks.

Although halogen lamps cost more than ordinary bulbs, they last much longer (about 2,000 to 4,000 hours) and maintain their light output throughout their entire life, rather than gradually becoming dimmer. They also operate at higher temperatures, allowing them to produce a whiter, brighter light. A lower wattage halogen lamp can be substituted for a higher wattage conventional bulb; halogen lamps also require lower voltage, making them even more efficient. Instead of a glass bulb, halogen lamps use a quartz tube, which can withstand higher temperatures. Halogen lamps are commonly used in **automobile** headlights, spotlights, and floodlights and are sold for use in restaurants and private homes.

Halon • See Fire extinguisher

Hancock, Thomas (1786-1865)

English inventor

Thomas Hancock founded the rubber industry in Great Britain. He was born at Marlborough in Wiltshire, England, the third of twelve children of a lumber merchant and cabinet maker. Around 1815 Hancock went into the stagecoach business with one of his brothers in London, England. The need for an effective waterproofing agent—for coaches, drivers, and passengers—drew Hancock's attention to rubber. In April 1820 he patented India-rubber springs for various types of clothing such as gloves and suspenders. The Hancock brothers then started an "elastic works" to manufacture items using the rubber springs.

Searching for a more effective way to process his raw material and use rubber remnants, Hancock in 1820 invented his most important device: the rubber masticator. He designed a machine with revolving teeth that tore up rubber scraps. To Hancock's surprise, the shredded bits adhered into a solid mass of rubber that could then be pressed in molds into solid blocks or rolled into sheets. Hancock's masticator, which was perfected in 1821, made rubber manufacture commercially practical and gave birth to the rubber industry. (Hancock called his machine a "pickle" and kept the masticating process a secret for 10 years.)

The advantages of the Hancock rubber masticator caught the interest of **Charles Macintosh**, who in 1823 had patented a process for waterproofing fabrics with naphtha-treated rubber. Hancock in turn applied for a license in 1825 to use Macintosh's naphtha process. The two men eventually became partners in the manufacture of waterproofed items.

Although the rubber industry was prospering, rubber still suffered from the defects of becoming brittle or gummy with temperature variations. About 1840, Hancock developed a slicer, which was to play an important role in **John Wesley Hyatt's** production of celluloid sheets. Around 1842 Hancock acquired a sample of rubber that had been vulcanized by the Goodyear process. He studied the sample and discovered the effects on it of sulfur under heat. In November 1843 Hancock took out a British patent for his variation of vulcanization—a patent that blocked **Charles Goodyear's** efforts to secure a British patent for his invention of vulcanization. (Although both Hancock and Goodyear claimed to be the inventors of sulfur-modified rubber, they were both probably predated by Lüdersdorf, Bergius, Leuchs, van Geuns, and others. The addition of sulfur under heat was termed vulcanization by Brockedon, and Hancock later patented his discovery that if up to 30 percent of sulfur is added to the rubber, a hard, rigid, thermoplastically moldable product results. Termed *vulcanite*, *ebonite*, or *hard rubber*, the material would soon be used as an insulating solid in the electrical industry. In the decorative arena, vulcanite was molded to produce brooches and pipe stems.

Hancock was granted a total of 16 patents relating to rubber between 1820 and 1847. In 1822 he developed rubber tubing that could be cut into rubber bands and life-belt tubes. He devised many medical uses for rubber, and promoted the use of rubber rather than leather fire hoses. He made many improvements in the manufacture of rubber and thought of a multitude of ways in which this remarkable new product could be used. Hancock died in 1865 at Stoke Newington, London.

See also Rubber, synthetic; Rubber, vulcanized

HAND GRENADE • See Grenade

HANDCUFFS

Devices for shackling the hands, legs, and torso have existed since ancient times. Handcuffs can be distinguished from earlier forms of shackles and manacles in that they are portable. The first handcuffs took the shape of a figure eight. They were soon replaced by handcuffs that resembled two bracelets joined by a short metal chain. This design has endured with modifications to the present day.

In late eighteenth century, the Hiatt Company began manufacturing handcuffs in Birmingham, England. At the time there was a great demand for the invention for use in the slave trade. Mass civil disturbances during the French Revolution also required handcuffs. Modern police forces in the nineteenth century spurred innovations in handcuff design in Europe and the United States. As steelmaking processes developed, so did handcuffs; they became stronger and lighter than previously possible. Through most of the 1800s, the shackle mechanism had a single locking position. A bar locked into place to close each handcuff's loop. This meant handcuffs only fit one size. Captives with small wrists were able to slip free of the restraints and those with large wrists could not be handcuffed at all. Handcuffs of different sizes could be made, but carrying several pair at one time was impractical.

Around the turn of the century, handcuffs were developed that could be adjusted to fit any size wrist. The Peerless Handcuff Company of the United States introduced its version in 1914. Half of the cuff's loop was a single strand steel that locked into place between two steel strands that formed the other half of the loop. A ratchet mechanism inside the double strand portion allowed the single strand to lock into a variety of positions and ensure a snug fit. The inside edges of handcuffs are specially machined today to prevent tissue and nerve damage. Like previous handcuffs, they are unlocked with a small steel key.

Although the basic design of steel handcuffs has changed little in the twentieth century, police have begun to use handcuffs made of other materials for special situations. Nylon handcuffs are used by police when large numbers of arrests are necessary, such as during riots. Although disposable nylon handcuffs lack the durability of steel, police favor these lightweight restraints for mass arrests because they are inexpensive, easy to remove without a key, and can be carried in large numbers.

HANG GLIDER • See Glider

HARGREAVES, JAMES (ca. 1721-1770)

English inventor

Early in the eighteenth century Englishman John Kay (1704-1764) invented the **flying shuttle**, allowing weavers to produce material much faster than ever before. While this solved one problem, it created another: the spinning of yarn was still done by hand on the "Great Wheel," one thread at a time, and could not keep up with the demand brought on by Kay's new **loom**. To help increase the supply of yarn, the Royal Society of Arts offered cash prizes to anyone who invented a faster spinning machine. The first one to do so was James Hargreaves.

Hargreaves grew up in Lancashire, England, learning the trades of carpentry and weaving. He did not become an inventor until 1740, when he was employed by a local businessman to construct a better **carding machine**. A few years later, it is said, Hargreaves accidentally toppled the **spinning wheel** in his home. As it lay on its side, Hargreaves noticed the wheel and the spindle were still in motion, even though they had been tipped ninety degrees. It occurred to him that a mechanical spinner could be designed in which many spindles, set vertically and side-by-side, could spin a number of threads from just one horizontal wheel.

Hargreaves began constructing just such a machine in 1754; fourteen years and many prototypes later, the spinning jenny was complete. The **spinning jenny** was the first machine that accurately simulated the drafting motion of human fingers. This was vitally important to the success of the spinner, for it eliminated the need to draw **cotton** fibers out by hand. The jenny had one large wheel playing out cotton roving to eight different spindles, thus spinning eight threads at once.

Because the design was essentially the same as a spinning wheel (only eightfold) the yarn produced was still lumpy and uneven in places; however, it was sufficient for the weaving of many different fabrics, particularly when woven together with threads of **linen**. It was also ideal for the spinning of wool thread and yarn. Unlike many inventors that would follow him, Hargreaves did not plan to become wealthy from his invention—in fact, the first jennies were used only in his home. Soon, however, the Hargreaves family suffered some financial setbacks, and he was forced to sell a few of his machines to mills.

His neighbors feared the new machine, thinking it would soon replace them all, and in 1768 they formed a mob that gutted the Hargreaves home and destroyed his jenny. Understandably upset, Hargreaves and his family moved to Nottingham. There he entered a partnership with Thomas James, and the two men opened their own cotton mill.

In 1769, **Richard Arkwright** successfully patented his **water frame** spinning machine (along with most of the machines associated with the spinning process, not all of which were of Arkwright's design). Inspired, Hargreaves enlisted legal aid to help him patent the jenny. By that time, many Lancashire mills had copied the jenny design illegally, an infringement for which Hargreaves sought restitution. His case was dismissed, however, when the court discovered that he had sold jennies in Lancashire a few years earlier.

By 1777 the water frame had almost completely replaced the jenny as England's most popular spinning machine: the yarn it produced was stronger and smoother, much more suited to the needs of the now-dominant hosiery industry. (Both the jenny and the water frame would ultimately be replaced by **Samuel Crompton**'s spinning mule.) Hargreaves was never awarded the patent or the restitution he fought for; he died poor (compared, at least, to Arkwright) in 1778.

Harpoon

The harpoon is a hunting tool primarily used in the pursuit of whales and other sea animals. It is one of the oldest hunting tools known to man and was developed by prehistoric people as a variation of the spear.

Originally the harpoon was hand thrown like the spear, but had a double-flued head like an arrow and a line attached to the shaft so the animal could be retrieved after a chase. This remained the basic design of harpoons for hundreds of years. Its major drawback was that the head tended to slip back out of the entry wound once the animal began to thrash around.

As whaling became a more lucrative pursuit, people sought to improve the effectiveness of the harpoon. In the 1830s, a single-flued model was introduced and became the standard by 1840. The advantage of the single-flued design was that once the head entered the whale's flesh, the force of the animal pulling against the harpoon caused the unevenly weighted head to bend at a right angle to the entry wound. Thus the head anchored itself in the whale and held firmly as the snared beast struggled to become free.

In 1848, **Lewis Temple** improved the head design even further. The Massachusetts blacksmith developed a toggle head attached to the shaft of the harpoon by a rivet that allowed the head to turn ninety degrees from its upright position. The head and shaft were held in alignment by a piece of soft pine inserted through a hole in each part. Once the line attached to the harpoon was pulled taught by the whale, the wood piece snapped and the head swung into its ninety-degree position. This design proved much more reliable than its predecessors and cut down on the number of harpoon heads lost during the course of catching a whale.

Primarily, harpoons have been hand-thrown for most of their history. Although various methods of propelling the harpoon were tried at different times, the weight of the weapon and the need for accuracy demanded that it be thrown by hand at close range to the prey. In the mid-nineteenth century, however, whalers began to successfully experiment with mechanical enhancements to the harpoon. American ship captain Ebenezer Pierce developed a darting gun that was attached to the shaft of the harpoon. Although the harpoon was still thrust by hand, the darting gun helped immobilize the whale much more quickly than other methods. Upon entering the whale, the barb of the harpoon triggered the gun which released an explosive-tipped lance. The explosion normally killed or stunned the whale, allowing the whalers to deal with their quarry more easily. Pierce's invention was first used in 1865 and continues to be used by whalers in the Arctic regions.

In 1868, Norwegian whaler Svend Foyn (1809-1894) introduced a viable harpoon gun. The gun was swivel-mounted which allowed it to be raised, lowered, or rotated on its base; it was outfitted with steel springs to absorb the recoil from its discharge. Foyn also redesigned the harpoon for use with his gun. He constructed the head with a set of hinged barbs that

John Harrison

opened once they had penetrated the whale's skin. As the barbs opened, they broke a small vile of sulfuric acid, lighting a fuse that was connected to a packet of **gunpowder**. The ensuing explosion had the same effect as Pierce's darting gun, stunning or killing the animal instantly. Modern harpoons are still based on modifications of Foyn's design. Fired from a gun mounted on the bow of a whaling ship, the harpoon is composed of a shaft with a line attached, four barbs, and an exploding head at the end. Instead of using a chemical ignition system, however, these harpoons use an electric charge. Once the harpoon has been fired from the boat and is lodged in the whale's flesh, the barbs hold the head in place until it is detonated. The whale is then towed back to the ship by the line attached to the harpoon's shaft. Through minor adjustments over the course of time, the harpoon has developed into a reliable and efficient means of hunting and capturing large sea animals.

Harrison, John (1693-1776)

English horologist

John Harrison and his brother James were introduced to **clock** repair by their father, Henry. Clock making, or horology, was undergoing a developmental revolution. Mechanical clocks had existed since the fourteenth century. They remained rather primitive in their operation until **Christiaan Huygens** invented the weight-and-pendulum clock in 1656.

Self-taught individuals, John and James were also introduced to carpentry and surveying by their father. They chose, however, to become totally involved in making clocks, and almost immediately began developing their own improvements to them. One such innovation was the gridiron pendulum, which consisted of a grid of iron and brass that compensated for changes in temperature.

Perhaps the major problem with clocks in those days was that they were totally dependent upon the earth's gravity for their operation. This meant that they could not keep accurate time at sea, and could not be adapted for portability. Even moving them across a room would require adjustment.

The Harrison brothers set to work on developing a marine **chronometer** in 1728. The motivating factor was money. In 1714, the English Admiralty set up an award of £20,000 for anyone who could provide mariners with a reliable clock that, when used with celestial sightings, could keep them informed of their longitude at sea. Mariners had to rely heavily on dead reckoning to find their way, often leading to tragic results.

The Harrisons made several models of marine chronometers. The fourth model proved to be the most successful. On a nine-week voyage from England to Jamaica in 1761, the device had only a five-second error. The Harrison strategy was to design an instrument that was not only internally accurate but also externally stable.

The Board of Longitude, apparently miffed that a common artisan had achieved the coveted goal, reluctantly gave up only half of the prize. John, minus his brother, went on to get what was justly due him.

The Board subjected his invention to undue scrutiny and required him to design a fifth model. This time, Harrison outdid himself by designing a compact timepiece that resembled a modern day pocket watch. It was far more convenient than the previous models, which were heavy and bulky. The Board still refused to capitulate. It finally took a personal appeal to King George III to set things right. The King intervened, and Harrison finally received the full reward in 1773 at age seventy-nine. Harrison lived only three more years. England and the world benefitted greatly from the persistence of Harrison, whose love for his work kept him going despite the many years of frustration he suffered.

Harrow

The harrow is a mechanical device that is closely related to the **plow**. Whereas the plow does the initial breaking of the sod, the harrow further loosens the soil and also cuts field stubble. A triangular harrow, or ''A'' frame harrow, was first used in the 1600s. This harrow design remained almost unchanged until after the American Civil War. In 1869 David L. Garver of Michigan patented a spring tooth harrow. Widely used in regions such as the eastern United States where the soil is quite rocky, the spring tooth harrow uses spring steel teeth that flex when they strike an obstacle. The disk harrow was developed in the 1870s and was especially well received in the midwest and Great Plains states. Its use increased greatly in the 1890s and remained popular through World War II. Since the 1950s, use of disk harrows has decreased dramatically, with wider acceptance being given to **cultivator**s that do the same job more efficiently. Disk harrows differ from disk plows in that the plows move the soil in one direction while the harrow moves it in opposing directions. Disk plows have three to six individually mounted disks. Disk harrows have rows of blades, called gangs, mounted on axles. The single-acting disk harrow has a pair of opposing blade gangs pulled by a **tractor**. The tandem disk harrow has two sets of opposing gangs. The offset disk harrow has one gang trailing another, in which the line of pull is far to one side of the till strip, making it favorable for tilling under low tree branches, as in orchards. Another type of harrow is the spike-toothed harrow. Used to smooth the soil after plowing, it performs the same function as a rigid-tined garden rake.

Haying machines and tools

Hay is grass that has been cut and dried for use as fodder for farm animals. Numerous tools and machines have been designed over the years to process it. The earliest haying tools were homemade for manual field work. Scythes were used to cut the grass, rakes to pile it into windrows, and pitchforks to fluff the hay and toss it into stacks or onto wagons.

Efforts to mechanize haying operations began in the early 1800s with the invention of haying machines, or tedders. Their development coincided with that of mowing machines. About 1820, Englishman Robert Salmon developed a haying machine which mechanically turned the cut grass for drying. His device consisted of curved iron teeth projecting from a horizontal cylinder that turned as the machine was drawn forward. Another Englishman, Thomas Wedlake, produced an improved version of Salmon's tedder in 1843, featuring a safety function that allowed the fork tines to spring back into position when they encountered an obstruction. This basic design remained in use into the 1930s.

H. L. Emery of Albany, New York, invented a hay press, or baler, in 1853 that produced five 250-pound, bales per hour. P. K. Dederick, also of Albany, developed a more practical version of Emery's press in 1860. The press was first manufactured commercially in 1872 and was widely employed in the United States and Great Britain.

In 1866, George Ertel of Quincy, Illinois, invented a vertical hay press that operated on horsepower. Steam-powered balers began to be used in the early 1880s. The bales were automatically ejected for manual binding.

Conveyor belts also came into use in the field to deliver loose hay to the baler or the hay wagon. An American named Keystone invented a hay loader in 1875 that consisted of a cylindrical cage with teeth that picked up the hay and tossed it onto the belt as it was pulled.

During the twentieth century, a wide array of haying machines were developed. Powered hay rakes, loaders, stackers, self-tying balers (1940), and conditioners were introduced to the market by the various farm machinery companies. Hay conditioners crimp, crush, and flail the hay to speed hay curing and reduce crop loss.

A major development was the cylindrical straw bale press introduced by the American firm Allis Chalmers in 1958; this Roto-Baler could roll hay bales up to one hundred pounds. Roll balers became commercially available in 1971, and by 1975, 40% of the balers manufactured in the United States were round balers. The huge bales now produced can weigh well over a thousand pounds. Because of their weight, they are expensive to transport and are generally used on the farms that baled them.

See also Tractor; Farming, mass production; Mowing machine

HDTV (High Definition Television) •

See Television

Headlight dimmers, automatic • Se

Rabinow, Jacob

Headlights

Headlights began very simply as lanterns mounted on the earliest horseless carriages. As the versatility of **automobiles** widened, improved lighting systems were needed, and the incandescent, sealed beam headlight was devised and used almost universally for decades.

The incandescent headlight consisted of a housing or assembly, a light bulb as the light source, a reflector that bounced the light forward, a protective lens or cover that focused the light beam and protected the bulb, and connections to the car's electrical system. Double filaments in the bulb allowed for a high and a low beam. The bulbs produced a yellowish beam that caused eyestrain; they also darkened and dimmed with age. This type of headlight produced a sufficient amount of directed light for the driver to see without blinding oncoming drivers, but it was inefficient because more than 70 percent of the light produced by the bulb was either lost as heat or not collected by the reflector.

The same sealed-beam unit was improved in the 1970s when the halogen bulb was developed. Halogen bulbs don't dim or darken over time, and their light is brighter and less straining. Because the headlight was a sealed assembly, a burnt out bulb meant replacing the entire headlight. This also dictated standard shapes and sizes, and round or rectangular headlights were the only available designs until about 1980. Replaceable halogen bulbs opened new opportunities in headlight configurations, and advances in **plastics** and **fiberglass** in automotive design made model-specific headlights possible. Automotive designers could create aerodynamic masterpieces with headlights that matched the design concept. **Lenses** could also be formed from plastic instead of glass, which aided both design and weight.

Projector-beam headlights were the next technological leap. Halogen bulbs are used in devices resembling overhead projectors. The light source is projected up and out by means of a reflector and magnifier shaped like an ellipse. The reflector is so precisely shaped that the pattern of the beam is well defined. More light is made available without increasing the load on the electrical system of the car and while creating less glare for oncoming traffic. Projector-beam headlights can also be manufactured in smaller sizes to suit automobile designs. They have been used widely since 1987.

Technical advances improve components of headlights and methods of designing them. In ''Free Form'' or FF systems, engineers choose the beam pattern they want the headlight to produce, then use computer applications to design the reflector that will generate that beam. FF reflectors look like they are made of puzzle pieces because each piece reflects light to a particular point on the road. Because the reflector shapes the beam pattern, the function of the lens is less optical and more decorative or protective. More of the light from the bulb is captured by this precise design, however, and the FF beam has a 45 percent efficiency, as opposed to 30 percent for the incandescent, sealed-beam units and 35 percent for projector-beam headlights.

In 1995, the High Intensity Discharge (HID) system debuted in Europe and the United States. HID changes the bulb source itself. Electricity arcs from one electrode to another in xenon gas, rather like a fluorescent tube instead of a glowing filament inside a conventional bulb. A projector-type reflector produces an efficiency rating of 50 percent and dramatically reduces the electrical demand. The xenon gas produces a bluish light when viewed by oncoming cars, but the subject vehicle's driver sees a beam that is bright white. Fiber optic systems are also in development for automotive applications. Fiber optics will carry light from a single source in the vehicle to headlights taillights, and the dashboard. These systems have only been demonstrated in concept cars, but they may be commercially available by the year 2000.

Taillights demand less design attention because visibility and safety instead of complex beam design are the issues. Turn signals, reverse lights, and brake lights are coordinated by the car's electronics; hazard warning systems add the dimension of blinking sets of front and rear lights when the car is stopped and on the driver's command. Fog and driving lights project short, wide beams and long, intense beams, respectively. Fog lights help light the sides of the road during bad weather without causing glare, and driving lights (which are illegal unless they are used with high-beams) permit the driver to see more clearly and farther ahead. Both fog and driving lights are designed based on conventional halogen, projector-beam, or FF technology.

See also Automotive electronics; Halogen lamp; Incandescent light bulb

Headphones

Headphones are miniature **loudspeakers** that are placed over the ears and held in place by a band or wire worn over the head. They often feature cushioning to hold in sounds coming in or block outside sounds. Though airplane pilots are among the many professionals to use headphones on the job, most often headphones are thought of as auxiliary **stereo** equipment, used to listen to music without disturbing others, and in the production of music in the recording studio.

The first stereo headphones were invented in 1958 by John C. Koss, a Milwaukee-based jazz musician and audiophile. Before his time, headphones were used only in industry by telephone operators and the like. Koss's original idea consisted of a total stereo package: a small portable **phonograph** with attached speakers and his new headphones, designed with an audio engineer. When he took it to a hi-fi show in his hometown that November, the only component that drew interest was the headphones, which became an immediate hit. The idea was so successful that Japanese companies quickly designed copies of Koss's headphones. Though the original Koss headphones were crude—little more than tiny loudspeakers covered in cardboard and held in place by a military headband—they were the first headphones to provide the listener with a full amplitude of sound.

Headphones work much like a loudspeaker does. The **amplifier** sends out a signal. This signal propels a light diaphragm in the speaker. The diaphragm vibrates air in the ear canal. This describes headphones at their crudest, though, for headphones have evolved in an effort to recreate the breadth and depth of the original sound. There are several different classes of headphones. Circumaural headphones feature big soft earpads that seal the external ear, and are often bulky and heavy. Supra-aural, also known as velocity, headphones merely rest on the outer ear, with minimal foam cushioning for comfort.

Until the 1980s, headphones continued to improve technologically, but remained rather large. In the early 1980s, with the advent of the **walkman**, headphones became extremely small and lightweight. In 1990, Koss introduced its first cordless headphones which used an infrared signal to link the amplifier to the headphones. Cordless headphones allow the listener to wander the room, rather than be tethered to a stereo. There is also another kind of cordless headphone, called 900 MHZ headphones and first manufactured by Recoton. They use a tiny radio transmitter in the amplifier, and a receiver in the headphones. The 900 MHZ headphones have one significant advantage over infrared: the listener does not have to be in the same room, within a limited distance of the stereo, to listen. With 900 MHZ headphones, one can amble up to 150 ft (about 45.5 m) from the signal's source. Despite the advantages, these radio headphones were banned in Great Britain in 1997 because they work on an already assigned frequency.

In the future, digital **radio** technology will probably supplant both of these types of cordless headphones. Other burgeoning headphone technologies that will make headway include 3-D headphones (also known as surround sound headphones) which feature two or three small speakers on each side, creating the effect of space in sound. While headphones have many advantages, there are health risks associated with their use. Listening to the music too loudly can permanently damage hearing, and using headphones while engaged in potentially hazardous situations, such as riding a bike in traffic, can lead to bodily injury, if not death.

Hearing aids and implants

Hearing aids are devices that amplify sound so a person who has a hearing impairment can enjoy sounds, from a musical, symphony, to a lively conversation, to the rustle of leaves. Two types of hearing aids either conduct sound through the air or through bone. Usually a person who is hearing impaired can use an *air-conduction* hearing aid, as it amplifies sound and brings it directly to the ear. The *bone-conduction* hearing aid brings sound waves to the bony part of the head behind the ear and uses the bone to transmit sound waves to the nerves of the ear. A typical hearing aid contains a **microphone** that picks up sounds and converts them into electric signals, an **amplifier**, which increases the strength of the electric signals, and the receiver, which converts the signals back into sound waves that can be heard by the wearer. There are three main styles of hearing aids: those that fit behind the ear, those that fit into the ear, and those which slip into the ear canal. Power is provided by a small **battery**.

Devices to aid hearing have a long history. The idea of bone conduction was known in the early seventeenth century and the ear trumpet was used even before that time. The ear trumpet was shaped to gather sound and funnel it into the ear. In seventeenth century Germany, Marcus Banzer used a piece of swine bladder connected to a tube made of elk hoof to make an artificial ear drum. Later in that century the audiphone or dentiphone was invented. Made of a flexible material, the device was shaped like a fan and held at its end between the teeth as the fan was bent toward the sound. The sound vibrations captured by the fan were carried to the teeth, the jaw bones, the skull, and finally to the auditory nerve. Perhaps the largest hearing aid ever made was an imposing throne built for King John VI of Portugal in 1819. Hollow carved arms of the chair terminated with the wooden mouths of lions through which people would speak and have their voices carried by tubes to the king's ear. An artificial ear drum was devised in 1852 when an English physician, Joseph Toynbee, used a disk of vulcanized rubber attached to a rod. The Victorian era was known for some of its more elaborate concealed hearing devices, in urns, top-hats, even tiaras. In the 1870s, **Alexander Graham Bell** began experimenting with the conduction of sound through electrical devices, originally intending to help deaf children hear. His experiments led to the invention of the **telephone** instead, but his work did bring public awareness to the needs of the hearing impaired. The first electrical hearing aid was made in 1901 by Miller Reese Hutchinson and he called it the Telephone-Transmitter. During the era of **vacuum tube**s, like those used in early **radio**, new hearing aids were developed starting in 1920 with Earl Charles Hanson's Vactuphone. A 1923 model produced by the Marconi Company was the Otophone, consisting of an amplifier placed in a large case weighing 16 pounds (7 kg), making it rather bulky to use. The first "wearable" hearing aid weighed 2.5 pounds (1.1 kg) and was made by A. Edwin Steven in 1935. During the 1950s **transistor**s revolutionized electronics and Microtone introduced its compact and powerful transistor hearing aid in 1953.

For individuals whose hearing cannot be aided with a hearing aid, there is the possibility of having a cochlear implant, the first of which were done in 1973 and which involved the implantation of an electrical device which stimulated the remaining nerves in the inner ear in people suffering from nerve deafness. Although the implant did not restore normal hearing, it helped the recipient hear and interpret environmental sound. Today, multichannel electrical cochlear implants are more sophisticated and contain speech processors which allow some patients to understand speech without reading lips. Implants consist of tiny electrodes inserted in the ear and connected to a receiver implanted under the skin by an implanted wire. Robert V. Shannon of the House Ear Institute of Los Angeles has developed an auditory brainstem implant for people whose auditory nerve has been severed. The implant consists of a tiny microphone, sound processor, and transmitter all outside the ear, and an electrode that is implanted inside the head and connected to the auditory brain stem. When the microphone picks up sound, the unit converts sound energy into electric signals that are sent directly to the brain where they are interpreted as sound. Although the implant is not enough to restore hearing, it upgrades the level of environmental sound heard by the user and, in a trial situation, at least one recipient could understand limited human speech.

Computer microchips are now being used in programmable hearing aids, allowing the device to be adjusted by computer. Through programming based on the needs of the individual, signal amplification is automatically adjusted, which means softer sounds can amplified more than louder sounds. Multi-channel capability allows for programmable reception of high- and low-frequency ranges, and some devices have multi-microphone capability, which means one microphone picks up broad-range sound and the other, narrow-range sound, improving the capability of distinguishing conversation from background noise. Digital hearing aids in which a computer-controlled sound processor is contained, are also available.

Heart catheter • See Catheter, cardiac

Heart-lung machine

The heart-lung machine is an essential component in **open-heart surgery**. Blood from the veins is shunted via **catheter** to the machine, which introduces oxygen into the blood and then pumps the blood back into the patient's arteries. With the machine thus performing all the functions of the heart and lungs, the heart itself can be stopped while surgery is performed. Before the heart-lung machine, heart surgeons operated blindly, either with the heart still pumping, by slowly chilling the patient's body until circulation nearly stopped, or by connecting the patient's circulatory system to a second person's system during the operation. All of these methods were extremely risky.

While the idea of a heart-lung machine had been proposed as long ago as 1812, the device was not developed until American surgeon, John H. Gibbon, Jr. (1903-1974), decided in 1931 to build a heart-lung machine after a young female patient died of blocked lung circulation. At the time, experimental devices existed for pumping and oxygenating blood during perfusion (artificial circulation). Gibbon, who received his medical degree in 1927 from Jefferson Medical College, began his heart-lung work in 1934 at Massachusetts General Hospital during a research fellowship with the assistance of research technician Mary Hopkinson, who later became his wife. They found the action of roller pumps gentle enough to minimize both clotting and damage to blood cells, and they employed centrifugal force to spread the blood in a layer thin enough to absorb the required amounts of oxygen. In 1935 the Gibbons went to the University of Pennsylvania School of Medicine and continued their experiments, reporting successful results with animals by 1939. In 1946 John Gibbon became head of the surgical department at his alma mater and soon secured the backing of Thomas J. Watson, chairman of IBM. With the use of IBM laboratories and engineers, Gibbon's heart-lung machine was perfected after the introduction of wire-mesh screens to enhance oxygenation, filters to block air bubbles or clots, and monitoring devices. By 1952, the heart-lung bypass surgery had a ninety percent success rate on animals, and Gibbon decided to use the machine on a human patient. The first attempt failed, although the pump-oxygenator worked as required. On May 6, 1953, the second surgery using the heart-lung machine was successfully performed on Cecilia Bavolek. Despite the deaths of two subsequent patients, the era of open-heart surgery had begun. The heart-lung machine was rapidly improved: oxygenation, for example, was accomplished by more sophisticated methods. Once patients could be kept alive during heart surgery, a whole new range of operations became possible. Congenital heart defects could be repaired, diseased or damaged valves could be replaced, and coronary bypass surgery became possible through sewing in a replacement blood vessel to carry blood flow around a blocked section of artery. Thanks to Gibbon's heart-lung machine, open-heart surgery—especially coronary bypass—has become routine throughout the world.

By the late 1990s, a **laser** system being used in many protocols across the United States allowed traditional bypass surgery without the heart-lung machine. This surgery requires only a small incision between the ribs (no cutting of the chest bone as with traditional bypass surgery) to expose the heart muscle. The laser is aimed at the portion of the heart starved for oxygen and, synchronized with the heartbeat, fires automatically when the ventricle is full of blood until about 20 to 40 new channels are created through the wall of the heart muscle. The outside of these channels heals almost immediately while the inside remains open. As the heart pumps, blood is forced from the left ventricle into these new channels, providing oxygenated blood to the previously starved section. This surgery takes one to two hours as opposed to the four to six required for traditional bypass surgery.

Heating

Heating has had a major impact on human development by allowing people to reside and function far from the temperate areas of the earth. Heat warms human beings in one of three ways—conduction, convection, or radiation.

Conducted heat passes directly from a heated object into the body, as with the heated **bricks** and warming pans which chased the chill from beds in Colonial times, or modern-day **electric blanket**s and **battery**-operated socks. Convection, the main principle in forced-air furnaces, heats the air around the body. The third type, radiation, resembles the warmth of the sun in that heat moves outward in waves and maintains an even temperature. Radiation heat is commonly found in electric coil bathroom wall heaters and quartz space heaters for porches and patios.

The most primitive heating systems were cave fireplaces, which early people improved upon by cutting draft holes to allow smoke to escape. Native Americans utilized this principle in the design of tepees, which were erected around fires and channeled smoke through the opening in the top. Eventually, more sophisticated residents built chimneys to direct smoke and soot away from living spaces. However, these open fires required constant tending, removal of ash and creosote, and a screen to protect against falling logs and exploding live coals. Tile, brick, and iron **stove**s replaced fireplaces but dried out the air and posed a hazard to anyone who brushed against them. In China, families found greater comfort in sleeping on heated slabs, which they built over hearths. Around 350 B.C., a similar system in Ephesus, Greece, warmed the Great Temple through heated channels in the floor. The first centralized system of heating originated around 100 B.C. with the Roman underground hypocaust, which directed heated vapors through hollow terra cotta tubes in walls and floors of homes and public steambaths.

Until the re-emergence of central heating after the Dark Ages wiped out Roman innovations, most homes and buildings were drafty and uncomfortable because they were heated with fireplaces and stoves, which proved inefficient in that warmth stayed near the source, overheating some areas and failing to affect the temperature farther away. **Benjamin Franklin** invented an upgraded version of the stove in 1744 by regulating the draft and making the temperature easier to control. The Franklin stove remained a staple in most homes until furnaces came into common use.

An improvement on the stove was the creation of the room radiator, which was linked by ducts or pipes to a furnace. In 1777, Bonnemain, a Parisian innovator, devised a system of circulating or pumping hot water heated from coal or gas furnaces to radiators, a method which evened out temperatures in his castle by spreading heat over a wide surface. When the water cooled, it flowed back to the furnace for reheating. Bonnemain controlled this system with a crude thermostat—a lead rod in the boiler which expanded and contracted with changes in the water's temperature, thereby altering the intake of air.

In 1831, Jacob Perkins patented a high-pressure model of the heat circulator. This system, which made its appearance in the United States around 1840, proved clean and dependable. Still, it posed a fire hazard unless the furnace and flue were carefully insulated with fireproof materials and regularly cleaned of ash and soot, which impeded the draw of oxygen to the flame. **James Watt** devised a steam heating system in

1784 in his factory in Birmingham, England, employing a boiler to heat water, then directing the resulting steam through pipework and into radiators. Baron Mayer Amschel de Rothschild of England was the first to install this expensive system in a home. The system was patented in 1791 and proved clean and efficient, especially in apartment buildings, churches, schools, factories, and assembly halls. However, steam heat lacked the efficiency of later methods, which were more easily adapted to the rise or fall of temperatures. Another fault of steam heat was its inability to heat radiators at the end of the line, and the intermittent clanking in the pipes when pockets of cold air were trapped in the pipes.

But hot-air systems contained a built-in problem: they blew dust, fumes, and dirt along with the warmed air. For this reason, electric heating began edging out forced-air heating at the beginning of the twentieth century. Emanating from ceiling and floor panels or radiators, radiant electric heat offered a clean, efficient, maintenance-free alternative to earlier methods. In 1892 R. E. Bell Crompton and J. H. Dowsing invented the electric radiator, consisting of a wire protruding from an **iron** plate and coated with enamel. An American, Albert Marsh, following the lead of Englishman A. H. Barker, invented a nickel-chrome **alloy** heating coil for electric heaters in 1906; in 1908, Hugh M. Browne introduced a **governor** to control furnace dampers. Six years later, C. R. Belling devised a fireproof clay which he encircled with a wire made of the same nickel-chrome alloy, the impetus for his standard radiator.

In the 1950s, utilities engineers created the radiant ribbon, a hot water radiant system for baseboard heat. A pipe, shielded by a metal cover, followed the baseboard of a room, directing heat to the floor, which was usually the coldest point in a room. This system suited many homeowners because it heated objects rather than the air around them.

Another innovation, the **heat pump**, a much neglected design of **Lord Kelvin** in 1851, remained untapped until the twentieth century. The heat pump utilizes one cycle for heating, then reverses it for **air conditioning**. Applying the principle of refrigeration in reverse, the heat pump forces air to an evaporator, then compresses it to a higher temperature and blows it into unheated areas. This method works well in swimming pools and homes, but only where temperatures never reach extremes.

Probably the least utilized of all current heating systems is the solar collector, which absorbs energy from the sun into tiles, rocks, or water and stores it until it can be circulated or blown over cold rooms. However, the cost and bulk of solar panels and storage cells has inhibited their popularity.

Heat pump

One of the most basic laws of thermodynamics, referred to as the second law, is that heat will spontaneously travel from a warm body to a cooler body, but not the reverse. Consequently, it is impossible for a system to indefinitely transfer heat from a cooler object to a warmer object unless energy is added to the system. Such a device is called a heat pump, and is designed to make warm bodies warmer and cool bodies cooler by "pumping" heat from one place to another. Over the past decade, heat pumps have replaced traditional **air conditioning** as an affordable way to keep homes cool in the summer and warm in the winter.

The main ingredient in a heat pump is a cold fluid, such as freon or **ammonia**, called a refrigerant. The refrigerant travels between the cooler area (the inside of a house, for example, in the case of air conditioning) and the warmer area (the outside air). While inside the house, the refrigerant absorbs heat via an evaporator, reducing the inside temperature. The warmed refrigerant is pumped outside where it passes through a condenser; here, the heat collected inside is released. Before re-entering the house, the refrigerant passes through a valve that lowers its pressure and, subsequently, its temperature, and the process is repeated. (Here, the energy required to operate the valve and lower the refrigerant's pressure is the energy added to the system, in accordance with the second law of thermodynamics.) During the winter, the flow of refrigerant is reversed; the valve increases the fluid's pressure, thus warming the system and heating the house.

Heat-pump technology is used frequently in industrial applications, usually to make hot things hotter without burning fuel. They are often used in the evaporation, condensation, and distillation of liquids. Back in the home, a variation on the heat pump is the common refrigerator, used to keep cold things cold or to make them colder.

The tremendously gifted Irish scientist **William Thomson (Lord Kelvin)** is usually credited with the concept of the heat pump, though he did not have the resources to construct one. The first patent for this device was awarded in 1927 to T. G. N. Haldane, an English inventor.

Heinkel, Ernst • See Jet engine

Helicopter

The first helicopter can be traced back hundreds of years. In the early fourteenth century, the Chinese made toy flying tops that consisted of four rotor blades attached to a spindle. String was wound on the spindle, and, when pulled, sent the rotors spinning upward. In the sixteenth century, **Leonardo da Vinci** drew sketches of a flying machine with a twisting **screw**-like wing and an on-board power source. In 1784, two Frenchmen, Launoy and Bienvenu, gained widespread attention with a twin-rotor helicopter model operated by a **spring**-bow mechanism. Inventors experimented with helicopter designs throughout the nineteenth century but were unable to come up with an **engine** that was lightweight yet powerful enough to lift a full-size helicopter off the ground.

In the early 1900s, with the invention of the **gasoline** engine, new possibilities opened up for helicopter flight. In 1907, French inventor Louis Bréguet (1880-1955) achieved the first manned helicopter flight, lifting his four-rotor helicopter two feet off the ground for approximately one minute.

The foundations of the modern helicopter can be seen in the design of the autogyro, sometimes called the gyroplane.

Developed in 1923 by a Spanish engineer, Juan de la Cierva (1895-1936), the autogyro was an attempt to solve the problem of torque that helicopter designers faced: when a power-driven rotor swung overhead, it caused the fuselage to spin in the opposite direction. This forced designers to employ much complicated machinery to counteract the torque. Cierva's autogyro had a traditional engine at the front giving the craft its power, and it added an unpowered, free-rotating rotor tilted up at a slight angle so that air met the rotor blades from below and caused the blades to spin. Control in flight was handled by conventional airplane devices (ailerons, elevators, and rudder) on the autogyro's short fixed wings and tail assembly. This autorotating rotor was torqueless, so there was no need of complex machinery like other helicopter designers were forced to use. In addition, Cierva made each rotor blade hinged, allowing the blade to flap freely up or down to find its own best angle.

The autogyro was a commercial success throughout the 1920s until the Great Depression caused sales to plummet. Its limited speed, range, and cargo capacity prevented it from being used for military purposed and by the beginning of World War II, the autogyro had all but faded from the scene. Its innovative construction, however, paved the way for future helicopter designers. In the 1930s the prototypes of traditional helicopters were being built with varying results. The French had a design in 1936 with two rotors, one above the other, rotating in opposite directions to cancel out torque. This helicopter had set an altitude record of 500 feet (152.5 m) and could reach a speed of 65 miles per hour (104 kph). The Germans designed another configuration in which two rotors, again spinning opposite directions from each other, were mounted side by side on outriggers. This machine set a speed record of 76 miles per hour (122 kph), an endurance record of 1 hour and 20 minutes, and an altitude record of 10,000 feet (3050 m).

At this time **Igor Sikorsky** made a breakthrough that greatly advanced the helicopter industry. In 1909 he had built helicopters with two rotors, one above the other like the French design, but they were failures. His new idea was to use a small vertical rotor at the end of a long tail boom. This rotor would produce lateral thrust that would counteract the torque of a single main lifting rotor on top of the helicopter. Thus, he solved the problem of complex machinery used in the past to offset the torque problem. This new system, which he first tested in 1940, proved so successful that it is still the most popular configuration for all types of helicopters. United States military officers were so impressed with Sikorsky's helicopter design that they rushed it into production to be used in World War II.

Several advances in helicopter design took place after the war. The turbine engine solved a major overheating problem that occurred when the helicopter hovered, unable to use

First successful flight of Igor Sikorsky's helicopter.

moving air to cool the engine. The turbine engine also had more power, an advantage for lifting or rescue operations. Helicopters soon appeared that could transport large weights, and they therefore played crucial roles in the Korean and Vietnam wars. Their ability to take off and land vertically made them preferable to fixed-wing aircraft in the often localized and geographically remote battles of these conflicts. Today, helicopters are widely employed for commercial aviation, by news organizations (e.g., as the familiar ''eye-in-the-sky'' traffic spotters) and for rescue and emergency transport (most medium or large modern hospitals are equipped with a helipad).

See also Aircraft

HELIOGRAPH • See Photography

HELMET • See Armor

HELMHOLTZ, HERMANN VON • See Ophthalmoscope

HENRY, BEULAH LOUISE (? - ?)

American inventor

Known as the ''Lady Edison'' of the 1930s, Henry is famous for holding fifty-two patents. A native of Memphis, Tennessee, and a descendant of colonial hero Patrick Henry (1736-1799), she moved to New York City in 1919. Her prolific inventiveness is especially intriguing in the face of Henry's condition: a psychological disorder known as synesthesia, a dysfunction whereby sound is perceived as color, or taste as touch. She never quite understood how she made the complex mechanical drawings for her inventions and attributed her ability to create mechanical contraptions while knowing nothing about mechanics to the same ''inner vision'' that caused her to see a color and shape for each note of the musical scale. Her first profitable invention was an **umbrella** with snap-on cloth cover that allowed the owner to coordinate the umbrella with clothing. This success earned her enough money to employ a staff of mechanics in her laboratory. After the snap-on umbrella came numerous inventions, among them the protograph, which made four typewritten copies without carbon **paper**, the bobbinless **sewing machine**, the ''Dolly Dip'' **sponge** containing **soap** and usually used to clean **glass**es, and the ''Miss Illusion'' doll, whose eyes and hair changed color.

HENRY, JOSEPH (1797-1878)

American physicist

Joseph Henry was one of the first great American scientists after **Benjamin Franklin**. Many parallels exist between his life and work and that of the English physicist **Michael Faraday**. Like Faraday, Henry was born into a poor family. Both received little formal education and both were apprenticed at an early age, Faraday to a book binder, and Henry to a watchmaker at the age of 13. Both, finally made lasting contributions to the field of electrical research.

Joseph Henry

Henry's interest in science was sparked by an odd coincidence. He had chased his pet rabbit underneath a church. Noticing that some floorboards were missing, 16-year-old Joseph climbed into the church and found a shelf of books. He began looking through *Lectures on Experimental Philosophy* and was soon hooked on science. He entered the Albany Academy in New York and later began to teach at country schools to earn an income. He graduated from the Academy and was leaning toward studying medicine when a surveying job turned up, and that steered him toward engineering.

In 1826 Henry was back at the Albany Academy, but this time as a teacher of mathematics and science. In 1820, Danish physicist **Hans Christian Oersted** (1777-1851) had discovered that the flow of an electric current produced a magnetic field around the **wire**. This amazed scientists and many, including Henry and Faraday, began to experiment with magnetism. In 1829 Henry learned that William Sturgeon (1783-1850) had built an **electromagnet** that could lift nine pounds (4 kg). This was quite remarkable, but Henry believed he could create a magnet that was much stronger. The secret was to wrap more wire around the iron core, overlapping the levels. However, wire in that era was not insulated, and so wrapping one level over another caused a short circuit. Henry got around the problem by the laborious process of insulating the copper wire by hand, using strips of his wife's silk petticoats.

Now that he had insulated wire, Henry proceeded to experiment. By 1831 he had created an electromagnet that could lift 750 pounds (340 kg). Later that same year he gave a demonstration at Yale University and lifted more than 2,000 pounds (900 kg). For this feat he received an appointment as professor.

In addition to his large electromagnets, Henry also built small ones. In 1831 he ran a wire more than one mile (1.6 km) and attached a device consisting of an electromagnet, a movable iron arm, and a **spring**. At the other end of the wire was a **battery** and key switch. When Henry pressed the key, activating the current, the distant electromagnet engaged, attracting a metal arm with a click. Releasing the key cut the electric flow, and the spring forced the arm back to its rest position—the first **telegraph**.

There was a practical limit to the length of the wire that could be used—the longer the wire, the greater the resistance, which resulted in less current. **Georg Ohm** (1787-1854) had devised a law by which resistance could be calculated. Henry found an easy way around the resistance problem in 1835 by inventing the electric relay.

Near the close of the 1830s, Henry had a chance encounter with a man who had very little electrical knowledge, but was extremely interested in electromagnets and relays. Henry believed that the discoveries of science should be for the good of all mankind, so he never patented any of his devices. The man, **Samuel F. B. Morse**, received a great deal of advice and information from Henry. In 1840 Morse took out a patent on the electric telegraph and became a very rich man.

This was not the first time Henry had been cheated. In 1830, the teaching load at the Albany Academy monopolized Henry's time to such an extent that he had only the month of August in which to conduct research and experiment. In August of that year Henry discovered the principal of electric induction, the process in which an electric current in one coil of wire can set up a current in another coil. If the flow of electricity can produce a magnetic field, he reasoned, a magnetic field should be able to induce an electric current. But when the end of August arrived, Henry had not completed his investigation. He decided to set the work aside and return to it the following August. It was with considerable shock that Henry read Michael Faraday's announcement in 1831 of the discovery of electric induction. Rushing back to his experiments, Henry published a report of his own discovery, but he was too late; Faraday received credit for the discovery. To his own credit, Henry never argued about Faraday's priority, but privately he did resent that credit had gone elsewhere.

Henry, however, did get credit for including in his paper a discovery that Faraday had neglected to write about: the discovery of self-induction. A coil carrying electric current not only induces a flow in another coil, it can induce a current in itself. In 1834 Faraday discovered self-induction independently, but this time Henry got the credit. Estonian physicist Heinrich Lenz (1804-1865) also made the discovery independently and took it farther than both Henry and Faraday. Self-inductance became an important part in the design of electric circuits.

In 1831 Henry published a paper in which he described the working of an electric motor. Ten years earlier Faraday had built a motor, but it was little more than a toy. Henry's motor was more practical, but until Faraday developed the electric **generator**, there was no way to adequately power the motor. It is Henry's design that is used for motors in electrical appliances today.

In 1842 Henry anticipated a discovery that has been credited to **Heinrich Rudolph Hertz**. Henry discovered that he could magnetize needles in a basement with an electric spark from two floors above, correctly ascribing it to **electromagnetic wave** propagation. In another experiment, he magnetized a needle by utilizing a lightning flash 8 mi (12.8 km) away.

In addition to describing the mechanism of an electric motor, Henry was involved in many other endeavors. In 1846 he became the first secretary of the new Smithsonian Institution and encouraged the communication of scientific knowledge around the globe.Two years later he projected the image of the **sun** on a screen, made careful measurements of temperature, and discovered that the mysterious **sunspot**s were relatively cooler than the rest of the sun. He used the telegraph to obtain weather reports from across the country, initiating a system that led to the founding of the United States Weather Bureau. During the Civil War he recommended the construction of ironclad **warship**s, advice which was eventually followed. On May 13, 1878, Henry died in Washington, D.C., at the age of 80. He was finally honored in 1893 when the International Electrical Congress agreed to name the unit of inductance the Henry.

Hero of Alexandria (? - ?)

Greek scientist

Little is known about Hero, or Heron, of Alexandria except that he wrote at least thirteen books on the study and practical applications of the simple machines wedges— **lever**s, **pulley**s, and **screw**s.

Hero is thought have lived around the time of the first century. He was a teacher of mathematics, physics, pneumatics, and mechanics at the University of Alexandria. His published works—*Pneumatics*, *Mechanics*, *Dioptra*, *Automata*, *Catoptrica*, and *Belopoiika*—reveal him to be an inventor, especially an inventor of gadgets and **automata**.

In his work *Pneumatics*, Hero demonstrates that air is an actual substance, rather than the lack of substance. His simple experiment showed that water could not enter a container of air until air was removed. He further suggested that air was made up of individual particles separated by space and could be compressed into a smaller area. Amazingly, his ideas of gas compression and vacuums were valid but not studied further until fifteen centuries later.

Pneumatics also contains very basic principles of steam propulsion. In fact, one of Hero's most famous inventions was the Sphere of Aeolus, two tubes attached to opposing sides of a ball. When water was boiled in the sphere, steam rose and escaped through the bent tubes resulting in a type of ''whirligig'' that was powered by steam. Though applied to simple automata and toys, this knowledge, too, remained dormant for seventeen centuries before it was used to ease human labor. *Mechanics* is largely a textbook for architects, engineers, builders, and contractors. The book was translated and published in German as recently as 1900 and is divided into three parts. The first part studies the theory of the **wheel**, proportions and scales, centers of gravity, balance, and equilibrium, the statics of posts and beams, and the construction of a **gear** to fit an endless screw. The second part investigates the mechanical theories behind the five simple ''powers'' (winches, pulleys, levers, wedges, and screws) and their uses. The third section discusses uses of **crane**s and sledges.

In *Dioptra* Hero describes several instruments used by surveyors and provides mathematical background for measuring distances between sites. The book also details an **odometer**-like device that could measure rotations of a wheel on a level road.

Automata outlines a device that makes toys perform using an **axle**, strings, and drums. *Catoptrica* discusses **mirrors** and the theories of reflection and sight. *Belopoiika* describes crossbows and catapults.

Hero's other books briefly mention such objects as organs powered by wind or water, thermoscopes, and coin-in-the-slot dispensing machines—a predecessor to the parking meter!

Heroult, Paul Louis Touissant (1863-1914)

French metallurgist

Born on April 10, 1863, in Thury-Harcourt, Calvaedos, France, Paul-Louis-Touissant Heroult was the son of a tanner. He was influenced early in life by the writings of Henri Etienne Sainte-Claire DeVille (1818-1881) on the process of producing aluminum by sodium reduction from aluminum chloride. As a student Heroult began experiments in producing aluminum through electrolysis using a dynamo from his father's tanning business to generate a continuous electric current. In April 1886 he succeeded in making small amounts of aluminum with alumina (an oxide of aluminum) dissolved in baths of fused salt. A patent was granted to Heroult that same year, although American metallurgist **Charles Martin Hall** simultaneously made the same discovery. The minor difference was that the carbon anodes in Heroult's process were larger and less numerous than in Hall's technique. With the aid of French, German, and Swiss interests, Heroult was able to spread the use of his technique for aluminum production throughout Europe. However, Hall had the upper hand commercially since he had pursued the business aspect of his discovery more vigorously. A 15-year legal dispute between the two metallurgists ended in a compromise and an eventual friendship. Heroult also became known for his work in the development of electric furnaces for the production of steel. He died off the coast of Antibes on May 9, 1914.

See also Aluminum production; Iron production; Steel production

John Frederick William Herschel

Herschel, John Frederick William (1792-1871)

English astronomer and chemist

J. F. W. Herschel was the only child of the famous astronomer, Sir **William Herschel**. Born on March 7, 1792, he was raised in affluence and received a superior education, graduating first in his class at Cambridge. Herschel's far-reaching interests and abilities led him to make contributions in a variety of fields, including astronomy, chemistry, and photography. Celebrated during his lifetime, Herschel was a member of the premier scientific societies of his day and was knighted in 1831.

After flirting with careers in chemistry and law, Herschel in 1816 turned to astronomy. Eventually he assisted his father in making astronomical observations and constructing telescopes. He also published papers on mathematics as well as physical and geometrical optics.

Continuing their work after his father's death, Herschel observed nebulae, star clusters, and double stars in the northern hemisphere, compiling and publishing an extensive catalog that included many new sightings. From 1834 to 1838 he made detailed astronomical observations in the southern hemisphere from his base at the Cape of Good Hope in southern Africa. Herschel used his invention, the astrometer, to gauge the brightness of stars. He also developed scientific meteorology locally and used a **camera lucida** to make pictures of flowers and scenery.

Upon his return to England, Herschel analyzed his latest astronomical data and conducted photographic research, which together occupied much of his time until the mid-1840s. After learning from a note of **Louis-Jacques-Mandé Daguerre**'s photographic discoveries in France in 1839, Herschel produced his first successful photograph on glass within a week. He then made improvements to the process as he knew it, the most famous of which was the use of sodium thiosulfate, or hypo, to fix photographs. He first used gallic acid in **photography**, a chemical indispensable for printing photographs on **paper**, and is also credited with introducing the terms "positive" and "negative" into photographic vocabulary.

Herschel was more interested in theory than practice and did not seek fame as did other photographic researchers. In all, Herschel published a flurry of eleven papers describing the many chemicals and photographic processes he researched, among them the procedure for making direct positives on paper; the process used to make positive prints from glass negatives; his discovery that bromide of silver was more light-sensitive than any other silver salt; and the possibility of natural **color photography**.

Herschel spent his last years preparing his astronomical observations of the southern hemisphere. In 1849, he published the highly successful *Outlines of Astronomy*, which solidified his reputation as a premier astronomer of his day. Herschel also wrote articles for the *Encyclopedia Britannica* and translated works of classic writers. In 1850 he became master of the mint. Upon his death on May 11, 1871, Herschel was buried in Westminster Abbey near Sir **Isaac Newton**.

HIEROGLYPHICS • See Writing

HIGH DEFINITION TELEVISION

High-definition television, or HDTV, provides five times the visual information detail, 10 times the color information, and more than double the resolution as a standard **television**. In addition, HDTV features a sound quality equivalent to that provided on a digital compact disc. Development of HDTV dates as far back as 1884, when **Paul Nipkow** patented his Nipkow Disc. The Nipkow Disc produced a picture containing 18 scan lines. Corporations such as RCA, Philco, and Farnsworth latched onto his invention and eventually developed the standard television set. Even as the use of the standard television became commonplace in the 1940s, however, these and other corporations continued their research on HDTV. During this time, much of the research was dedicated to military uses for HDTV. Two airborne television systems, code-named "Block" and "Ring" were developed during World War II. After the war, the innovations, including higher resolution, advanced camera tubes and greater transmission distances, were incorporated into entertainment television.

After World War II, many Japanese companies became interested in the further development of HDTV. The most aggressive of these companies has been the Sony Corporation, which between 1975 and the early 1990s had invested $300 million toward development of its high-definition video systems (HDVS), which was intended to, and today does, rival 35mm as a film production option. HDTV has continued to evolve through the 1990s and current systems utilize 1125 scan lines. It is estimated that by the year 2008 all television broadcasts could be in HDTV format and standard television signals may become obsolete.

HIGH-SPEED FLASH PHOTOGRAPHY

High-speed **photography** requires exposure times shorter than those given by standard shutters, and at picture frequencies faster than those given by **motion picture** cameras with intermittent film movement. This makes high-speed flash photography of great value in scientific and technological studies. These photographs allow us to see the phases of movement that occur in nature when a bird flies or a cheetah runs, as well as the movements of high-speed machinery and phenomena like explosions, which normally occur with such great speed that details are not discernable to the naked eye.

The very fastest mechanical shutters give exposure times of about 1/4000th of a second. High-speed photography can require exposure times shorter than 1/1,000,000th of a second. Such extremely short exposure times are achieved with the help of magneto-optical shutters. As opposed to a mechanical shutter, which uses a physical trigger to open the shutter, this device uses a very brief electrical current to allow a small pulse of light to pass the shutter and expose the film at a much faster rate. Pulsed light sources, or quick flashes of very intense illumination, further reduce the length of time required to properly expose the film.

The earliest methods of photography were only efficient enough to record the slow- or medium-speed movements of relatively large objects because high-speed flash photography could not exceed the 1/5000th of a second exposure mark until 1851. That same year, **William Henry Fox Talbot** is said to have invented the negative/positive process of photography. Talbot's first experience with high-speed flash photography produced a successful exposure by the brief flash created from a spark during a demonstration for the Royal Institution of London. Talbot's contemporaries thought that the success of the experiment was due to the *fast exposure time* of the albumin plate (a photographic plate made of light-sensitive material derived from eggs). Ironically, as photographic plates improved, albumin came to be known as one of the least effective materials that could possibly be used. Nevertheless, Talbot's work helped lay the foundation for scientific photography.

It remained for American engineer **Harold Edgerton** to make high-speed flash photography practical. Based on his work with power **generator**s at MIT, Edgerton developed a device called the **stroboscope**, the first high-powered reusable electronic flash, in 1931. When the stroboscope was combined with a camera, Edgerton was able to photograph a particular moment in time. By holding the camera's shutter continuously open while the stroboscope produced flashes of light, exposures were created in rapid succession. Not only could the stroboscopic flash make an object appear to stand "still," but objects in cyclic motion could be made to appear slower, reversed, or could be juggled by adjusting the speed of the stroboscopic light. Edgerton continued to develop other appliances and circuits for stroboscopy, and around 1940 applied this technology to motion picture photography.

Today, the movement created by the event being photographed can actually be used to operate the camera's shutter and/or signal the flash. For example, a baseball thrown by a professional pitcher can break through a light beam to trigger the exposure. This is known as *synchronization*, and it is intended to accurately trigger the photograph at precisely the correct moment. Advances also have been made in the speeds at which film can be exposed. Today's high-speed flash photography seems limited only by the speed at which film can be mechanically advanced in a camera.

HIGHWAY INTERCHANGE • See Road building

HILLIER, JAMES • See Electron microscope

HINTON TEST • See Syphilis test

HIP, ARTIFICIAL • See Artificial limb and joint

HOCKEY

Hockey is Canada's national sport, although it's enjoyed in many countries. It's popular in Japan, Russia, the United States, and western Europe. Since 1920, hockey has been an official Olympic sport. Its roots have been traced to games of field hockey that British troops played on frozen lakes and ponds in Kingston, Ontario, and Halifax, Nova Scotia in the 1860s. In 1879, the first formal ice hockey rules were developed. Hockey is a rough, often brutal sport. Many injuries occur because of dangerous moves such as ''body checking'' (when a player throws his body against another to block his progress) and ''poke checks'' (when a player jabs opponents with his stick). In addition, the hockey puck can travel up to speeds of 115 miles (185 km) per hour.

HOE, RICHARD M. (1812-1886)

American inventor and industrialist

Born in New York City on September 12, 1812, Hoe was the son of Robert Hoe (1784-1833), a British-born printer who improved the cylinder press developed by David Napier. In 1827,

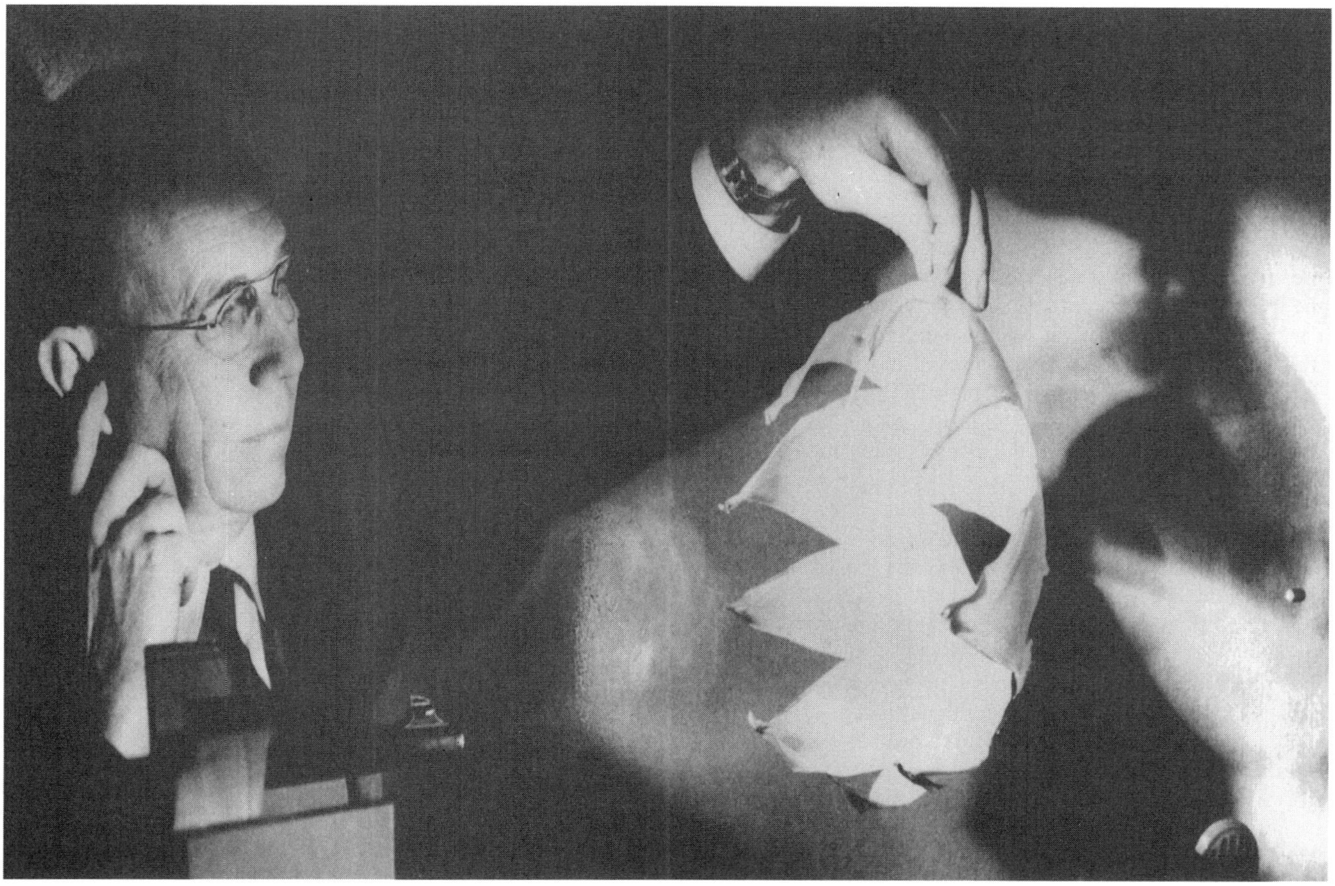

High-speed flash photograph of a 0.22 caliber bullet piercing a balloon held by Harold Edgerton.

August Wilhelm von Hofmann

at the age of fifteen, Hoe left school to join his father's printing firm, R. Hoe & Company; fifteen years later, he assumed total control of it. Hoe was among the first to realize that to increase printing speed, some alternative had to be invented. He began experimenting in 1829 with the Napier flatbed and cylinder press. In 1846, he created a new press in which he discarded the traditional flatbed and attached the type to a central cylinder, around which revolved four to ten impression cylinders. This became known as the rotary press—the first in the world. Also referred to as the "lightning press," Hoe's invention was first introduced to the printing industry in 1847 in the offices of the *Philadelphia Public Ledger*. The resulting eight thousand papers per hour revolutionized newspaper printing. Although this new speed-printing made huge daily editions possible, publishers were still limited to single sheet printing. Hoe continued to add improvements to his press and by 1871 devised a rotary web perfecting press which was fed by continuous rolls of paper, or webs, and printed on both sides in one move. The *New York Tribune* was the first newspaper printed with this process. Hoe's subsequent additions and various updates of high-speed folding apparatus virtually completed the modern newspaper press by 1875. Hoe presses were manufactured for newspapers around the globe by branch plants in Boston and London, and in New York City Hoe opened an apprentice training school for press operators. He died on June 7, 1886, while vacationing in Italy.

See also Printing technology

Hoff, Marcian E. (1937-)

American electrical engineer

Marcian Edward Hoff, Jr., known as "Ted," was born in Rochester, New York. His interest in science was encouraged by his father, who worked in railway signaling, and his uncle, a chemical engineer with Kodak. Hoff graduated in 1958 from Rochester Polytechnic Institute with a bachelor of science degree in electrical engineering, then went on to Stanford, earning his master's degree in 1959 and his Ph.D. in 1962. Hoff's inventiveness was apparent while he was still a student. Working summer jobs in Rochester, Hoff was a co-patent applicant for railway signaling and lighting devices. At Stanford, he patented an **analog computer** memory cell and a digital filter.

Hoff stayed at Stanford for six years as a research associate, then joined Intel Corporation in 1968 shortly after it was founded by **Robert Noyce**. As manager of applications research at Intel, Hoff was given the project of designing a set of integrated circuits for a line of desktop calculators. Seeking a way to simplify the complex plans, Hoff suggested putting the more complex steps into programs in memory, reserving the basic logic circuits for a single chip. Hoff—who had always wanted to build his own computer—refined his concept and produced the first microprocessor, a chip that held all the logic circuits of an entire computer central processing unit. Hoff's invention ushered in the era of the minicomputer and the "smart" computer-assisted household appliance.

Hoff worked on microprocessor development at Intel until the mid-1970s, then investigated different areas where Intel technology could be used. In 1983, he joined Atari, the video game company, as vice president of research and development, but left in 1984 when Warner Communications sold Atari. Hoff then worked as a consultant to various electronics companies in California's Silicon Valley.

See also Computer, digital

Hofmann, August Wilhelm von (1818-1892)

German chemist

Hofmann was born in Giessen. He originally planned to study law at college, but after attending chemistry lectures by **Justus von Liebig**, he became interested in chemistry. In 1841 he received his doctorate. His dissertation was on coal tar, the precursor to aniline, which in turn could be derived from phenal and ammonia. Likewise, his many hundreds of papers were on coal tars and their derivatives. In 1843, after his father's death, Hofmann joined Liebig as an assistant. He taught for a few months at the University of Bonn in 1845 and then moved on to England, where he was appointed a professor of the Royal College of Chemistry. He remained there until 1863, when he returned to Berlin, publishing a textbook entitled *An Introduction to Modern Chemistry* in 1865.

Hofmann was more of an analyst than a bench chemist—that is, he worked more comfortably analyzing and interpreting data than physically manipulating substances in experimentation. For that reason, he searched out talented assistants to carry out the laboratory techniques. Many of his assistants and students, among them **Frederick Abel**, **William Crookes**, **William Henry Perkin**, William Nicholson (1753-1815), John Newlands (1837-1898), Peter Griess (1829-1888), C. A. Martius, and Jacob Volhard (1834-1910), went on to develop synthetic **dyes** and establish the synthetic dye industries in England and Germany.

Hofmann spent many years studying nitrogen compounds like ammonia, which led to his development of polyammonias, now called triamines and diamines. Following the path started by his students, he also began to research synthetic dyes. In 1858 he reacted carbon tetrachloride with aniline which contained the impurities orthotoluidine and paratoluidine and formed a dye called rosaniline. Hofmann also succeeded in producing aniline blue, or diphenylaniline, by replacing hydrogen in rosaniline with aniline. He patented his Hofmann's violet in 1863 and successfully marketed it. He also developed new production techniques for synthetic dyes; the "Hofmann degradation" was widely used to produce amines. Hofmann also investigated methyl aldehyde, now called formaldehyde.

Hofmann was a founder of the German Chemical Society and was instrumental in standardizing the nomenclature for alkanes and alkane derivatives. He died in Berlin in 1892.

See also Ammonia synthesis; Indigo, synthetic

John Philip Holland

Holland, John Philip (1840-1914)

American inventor

Holland was born and raised in Ireland. Although employed as a schoolteacher, he read about early experiments with submarines carried out by William Bourne (d.1853), **Cornelius Drebbel**, and **David Bushnell**, and by 1870 he drew up his first plans for a **submarine**. He emigrated to the United States in 1873, resumed teaching, and continued to pursue his research on submarines. The United States Navy rejected his plans in 1875, but the American Fenian Society, a group of Irish patriots who hoped to undermine England's naval power and gain independence for Ireland, commissioned Holland to build a submarine. Holland's first model sank during testing, but the second model, the *Fenian Ram*, was successfully launched in 1881. A full-scale vessel, the *Fenian Ram* had many of the features we associate with modern submarines. The cigar-shaped sub used electrical power for propulsion under water and had an **internal combustion engine** for surface propulsion. It was fitted out with **torpedo**s and had mechanisms to steady the craft after firing. Previously designed subs submerged by sinking, but rudder planes on the *Fenian Ram* allowed the craft to dive by inclining its axis. The conning tower, however, was much smaller than has become customary for modern submarines, which are also a little sleeker.

Holland tried to interest the U.S. Navy in his work once again, and in 1895 the J.P. Holland Torpedo Boat Company was awarded a navy contract for his submarine, the *Plunger*. The vessel, in Holland's trademark fat-cigar shape, was to be 85 feet (25.5 m) long, again with a small conning tower. A great many design changes were carried out without Holland's approval during building in 1898, however. As a consequence, the sub performed so badly that the project was abandoned during the first trials. One of the main reasons for the failure of the project was that the interior became intolerably hot during submersion.

The *Plunger*'s **steam engine** for surface propulsion was also a step backward in terms of submarine design. While the *Plunger* was in production, Holland scraped together enough money to build a submarine on his own, the *Holland no. 8*. This craft, almost 56 ft. (16.8 m) long, was small and light when compared to submarines being developed by the French. It improved upon all the engineering advances of the *Fenian Ram*. The internal combustion engine for surface power also served to charge the **batteries** for the electrical engine that powered the craft under water. In this way, the submarine had a greatly increased range, not having to return to shore to recharge batteries. The *Holland* could dive and remain submerged for four hours. Surface speed was 8 knots while submerged speed was 5 knots. The craft was fitted out with a torpedo tube and dynamite gun. The Navy promptly commissioned several of the subs in 1900 when testing proved successful.

Holland's company continued to supply the U.S. Navy with many submarines over the years and has built subs for many other countries besides. Great Britain became interested

Herman Hollerith

in Holland's work in the early 1900s, commissioning several submarines. Russia and Japan also bought vessels from the Holland Boat Company. In his final years, however, Holland himself lost interest in submarines and turned his attention to aviation.

Hollerith, Herman (1860-1929)

American engineer and inventor

Herman Hollerith made a major contribution to the invention of the modern **digital computer** with his punched card tabulating machine, invented in 1890. His invention became the foundation of a company that evolved into International Business Machines (IBM). Hollerith was born in Buffalo, New York. After graduating from the Columbia University School of Mines in 1879, he went to work with the United States Census Bureau's division of vital statistics.

In 1881, at the suggestion of the division's director, Hollerith began designing a machine to tabulate census data more efficiently than hand methods. He based his machine, called a press, on electrical signals that were transmitted only when holes in paper or cards passed over the contacts. Electromechanical tabulators, like those in old-fashioned adding machines, counted the signals. He quickly changed from tape to cards that were the same size as dollar bills of that time, which let him incorporate money storage cases in his equipment. Later models could sort as well as add, and had automated card punching. Each machine could count up to 10,000 items.

Hollerith sold his first machine to the United States Army for compilation of medical statistics. He then received a contract from the Census Bureau for machines to be used in the 1890 census. He also used his invention as the basis for a dissertation and received a Ph.D. from Columbia University that same year.

Hollerith's machine saved the Census Bureau $5 million and did in one year what would have taken eight years of hand tabulating. It was the beginning of modern data processing.

In 1896, Hollerith founded the Tabulating Machine Company to sell and improve his basic machine. Patents and sales of the tabulating machine and other inventions made Hollerith a millionaire. In 1911 TMC merged with other companies, becoming the Computing-Tabulating-Recording Company. In 1914, C-T-R hired **Thomas J. Watson** as general manager. Watson later bought the company, naming it IBM.

Despite the development of computers during the twentieth century, versions of Hollerith's card tabulating machine still have a place in modern data processing. They are widely used in voting machines.

Hologram

In 1947, scientist **Dennis Gabor** was sitting by a tennis court, troubled over the poor resolving properties of the early **electron microscopes** he used. Suddenly, the solution came to him: take an electron "picture," one that was poor but that contained all the information, and correct it optically later. Gabor presented his discovery to his colleagues, who realized that while this process would undoubtedly improve the image, it would require a coherent light source—something that did not exist at that time. Thus, Gabor's solution—and the word he coined to describe it, *hologram* (meaning "complete picture")—remained merely theoretical for more than a decade.

In 1960, **Theodore Maiman** introduced the first working **laser**. His invention sent shockwaves through the scientific world and soon came to the attention of two researchers at the University of Michigan, Emmet Leith (1927-) and Juris Upatnieks (1936-), who had been working on **radar**. They saw the laser's coherent light as the final piece in Gabor's puzzle and turned their efforts toward producing the first holographic image, achieving success in 1963.

A hologram is actually a recording of *phase interference*—that is, the difference between two beams of coherent light. To make a hologram, a laser beam is split in two: one beam, called the *reference beam*, strikes a photographic plate; the second, called the *object beam*, strikes the subject and then bounces onto the plate. The difference between these two beams is the interference caused by the subject, and it is this phase difference that is recorded on the photographic plate. This process described above creates a transmission hologram, which can only be seen in laser light; another type, called a reflection hologram, can be viewed in white light, and it is with this type that most people are familiar.

A hologram produces a true three-dimensional (3-D) image: the interference pattern actually distorts the silver material of the film so that it imparts a phase difference on a beam of light reflecting off it it (in the case of a white-light hologram), creating an image with visible depth. When viewing it, the eye must re-focus to see foreground and background, and one can look ''around'' and ''behind'' the subject by tilting and turning one's head to view the image from different angles.

Besides creating a fascinating visual effect, the hologram has found a wide range of applications and has blossomed into a multimillion-dollar industry. Perhaps the most commercial application of holography is in advertising. Holograms can be found gracing the covers of magazines, books, and music recordings. In the 1970s, automakers would often debut a new car model with a cylindrical hologram: a prospective buyer could walk around the tube and view the vehicle from all angles, though the cylinder was actually empty.

The medical field was also quick to find a use for holograms. A holographic picture could be taken for research, enabling many doctors to examine a subject in three dimensions. Also, holograms can ''jump'' mediums—a hologram made using X-rays can be later viewed in white light with increased magnification and depth. Holography has also been instrumental in the development of acoustical imaging and is often used in place of X-ray spectroscopy, especially during pregnancies.

Because they are difficult to replicate, holograms have important applications in security. Many credit cards and drivers' licenses bear holograms that guarantee authenticity. Such techniques are also applicable to currency, as anticounterfeiting devices.

Perhaps the most important application of holography is in computer data storage. Magnetic tape, the most common storage device for home and small-frame computers, is two-dimensional, and thus its storage capacity is limited; because of its three-dimensional nature, a hologram can store many times more information. Imagine a cube: a two-dimensional cube is a square, offering only one ''side'' for data to be written upon; a three-dimensional cube offers six ''sides,'' all of which can be used for storage. Optical memories store large amounts of binary data on arrays of small holograms that, when viewed by the computer using coherent light, reveal a 3-D image full of information.

Scientists are now examining the possibility of using holograms to display three-dimensional images, creating true 3-D **television** and movies; in fact, Gabor himself began such research late in his life. The fruits of these efforts can be found in the appearance of holographic goggles and even a hologram-simulation video game. Ironically, the hologram never fulfilled its intended purpose, that of improving the resolution of electron microscopes. Still, after the creation of the laser, the hologram has proven an invaluable tool for scientists. For developing the basic principles of holography, Gabor was awarded the Nobel Prize in 1971.

See also Spectroscope

Holography

One of the most significant scientific breakthroughs of the twentieth century was the invention of the **laser**. Since its construction in 1960, the laser has been instrumental in the development of new techniques in numerous fields. One of the first applications of laser technology was holography, a process whose principles were laid out more than a decade before the first laser was ever built.

In the 1940s, **Dennis Gabor**, a Hungarian-born scientist working in England, was not allowed to work on the major research areas of the time; most of the scientific community was concentrating upon the war effort, especially upon the development of **radar**. A foreigner, Gabor was barred from these experiments, which were all classified as top secret. Instead, he labored to improve the budding field of electron microscopy.

At the time, **electron microscope**s were severely limited in their ability to resolve images at very high magnitudes. While sitting by a tennis court on Easter Sunday 1947, Gabor came up with the answer: take a photograph of the subject, a photograph whose quality was poor but which contained all the information, and improve the quality optically later, thus producing a clear, complete image. The name he gave this process was *holography* (from the Greek *holo*, meaning whole, and *gram*, meaning ''recording'').

Though Gabor's solution seemed practical in theory, it required a source of coherent light for best results. Thus, the holographic process remained theoretical until the first laser debuted in 1960. University of Michigan researchers Emmet Leith (1927-) and Juris Upatnieks (1936-) soon produced several **hologram**s by combining Gabor's principles with **Theodore Harold Maiman**'s laser. Though holography never did succeed in improving the resolution of electron microscopes, it has become a major industry unto itself.

There are two types of holograms: transmission and reflection. Transmission holograms, created by Leith and Upatnieks in 1963, are created splitting a laser beam into two separate beams that impinge on a photographic plate. The first beam, called the reference beam, bounces off of a **mirror** and propagates directly toward a photographic plate, unchanged. The second beam, called the object beam, contacts the subject—for example, an apple—and reflects off to travel toward the photographic plate. This object beam is no longer pristine: the lightwaves in the beam have been altered by reflecting off of the apple. The two beams interact with each other, or interfere, at the photographic plate.

The photographic plate records not the two beams but the difference between them, called the interference pattern. The only difference between the object and reference beams is the apple, and so the photographic plate essentially records the shape of the apple in three dimensions. In order to see the image, the viewer must shine a laser beam through the film, at which time a three-dimensional image of an apple can be seen.

The newer and more familiar type of hologram is the reflection hologram. It is created in essentially the same manner as the transmission hologram, except that the two beams strike opposite sides of the photographic plate, rather than the same side. The advantage of reflection holograms is that they can be viewed in white light, eliminating the need for another laser. The interference pattern recorded upon the photographic plate actually distorts the silver material within the film, creating the different depths within the image.

Not all holograms must be created by actual laser beams. Light beams can be modeled mathematically, allowing engineers to calculate the interference pattern between reference and object beam and create a computer-generated pattern. A computer-controlled electron beam etches the pattern into a substrate, and a hologram is created without a laser ever being turned on.

Unlike stereoscopic views, the image created through holography is true 3-D; a person's eyes must re-focus to adapt to different depths, and, by moving his or her head, the viewer can see "around" and "behind" the object. Gabor's discovery has found a variety of applications in numerous fields. In medicine, holography is used to create three-dimensional snapshots of specimens that could not be studied as completely through a microscope. Advertising uses holograms as intriguing and eye-catching images to attract customers to new products, and the security industry uses holograms to discourage counterfeiting in credit cards and drivers' licenses.

Another important use is in the computer industry, where holographic storage is replacing the standard **magnetic tape**: using assemblages of small holographs, vast amounts of binary data can be stored in a relatively small area and read back later by a laser scanner.

Holographic optical elements (HOEs)—specially-designed holographic patterns etched on a thin, light substrate—can redirect or redistribute light, taking the place of conventional **lenses** and **diffraction gratings**. Once the province of research labs, holographic optical elements are increasing being used by optical designers to improve performance of imaging systems or to replace one or more components in complex systems, reducing weight and size. Although production of a master HOE is costly, a large number of replicas can be made economically and rapidly, making them useful alternatives for commercial systems. For his discovery of the holographic process, Gabor was awarded the Nobel Prize in 1971.

Hooke, Robert (1635-1703)

English physicist

Robert Hooke was one of a special breed of scientist whose intellect and ingenuity spanned many different disciplines. Like his contemporaries Isaac Newton and **Christiaan Huygens**, Hooke worked in many fields, often with remarkable results.

Hooke was born in Britain, on the Isle of Wight in 1635. A sickly child who was stricken with smallpox at an early age, he was not expected to survive more than a few years. His persistent ill health forced him to remain indoors, where he found amusement in taking apart and reassembling mechanical devices. By his tenth birthday he had become adept at constructing intricate mechanical toys, including working boats and **clock**s.

After his father's death in 1648, Hooke was sent to London to attend boarding school, where the headmaster recognized his potential and placed him in a curriculum that included Latin, Greek, and mathematics. Hooke attended Oxford in 1653. Though he never completed his bachelor's degree, it was at Oxford that Hooke met some of Britain's greatest scientists, around whom the British Royal Society would later form. Among these was the physicist **Robert Boyle**, for whom Hooke served as a laboratory assistant. Under Boyle's tutelage Hooke constructed the precursor to the modern **air pump**, the first in a long line of ingenious scientific tools he would invent.

Using this new air pump, Boyle performed the research that would ultimately lead him to the finding known today as Boyle's Law. (Boyle's Law states that there is an inverse relationship between the volume and pressure of an ideal gas, at constant temperature.) In fact, some scientists have suggested that Hooke himself was the author of this law, reasoning that Hooke may have been pressured to relinquish credit for his discovery to his instructor.

Around this same time many European inventors were vying to develop the first accurate device to determine longitude on a sailing ship. Already in use, the **chronometer**, essentially a modified clock, was unreliable since the pendulum used to regulate its motion was thrown off by the ship's rocking. Sometime near 1660 Hooke introduced a chronometer design based upon a **spring** rather than a pendulum. (Today Hooke's name is associated with the law that, in many situations, the force on an object is proportional to its displacement from its resting, or equilibrium, position.) Although his design was sound, he was unable to find investors to back him, and it was not until 1674 that Christian Huygens patented his own spring-driven chronometer. Hooke immediately claimed that Huygens' invention was a derivative of his own, beginning a dispute that remains unresolved to this day.

That was not the only confrontation Hooke had with one of his peers. Perhaps the most famous was his feud with Isaac Newton, which began in the early 1670s. Newton, then a young student, had submitted a paper on light and colors to the British Royal Society. Hooke reviewed the paper and quickly dismissed it. Newton published a second paper on light in 1675, introducing a theory describing light as an undulatory wave. Hooke's reply was that Newton had stolen this wave theory outright from his own earlier publication, *Micrographia.*

Hooke later made a similar claim to Newton's theory of gravitation. The verbal battles between these two scientists were very bitter, several times driving Newton to a nervous breakdown. While his true contribution to the canon of theoretical science is unclear, Hooke was unquestionably one of society's most productive inventors of scientific equipment. Among his list of accomplishments are the universal joint, the reflecting **telescope**, the **compound microscope**, the wheel barometer, the **anemometer**, the spring-driven wristwatch, the "cross-hairs" sight for telescopes, and new standards for microscopy. The bulk of his inventions were constructed during his term as Curator of Experiments for the British Royal Society, where he was commissioned to explore new avenues and create new devices.

Though mechanics was certainly his first love, Hooke turned to architecture after a great fire burned most of London in 1666. To help with the reconstruction of the city and to aid his colleague, English architect Christopher Wren (1632-1723), Hooke designed several prominent buildings, most of which still stand.

Hoover William Henry • See Vacuum cleaner

Hopper, Grace Murray (1906-1992)

American mathematician and computer programmer

Rear Admiral Grace Hopper was a pioneer in the development of **computer languages**. She played an essential role in the development and standardization of COBOL (Common Business Oriented Language), which led to wide-spread use of computers in business, governmental, and science applications. Hopper received her Ph.D. in mathematics from Yale in 1934 and taught at Vassar College until 1943. She enlisted in the Navy and was assigned to the Bureau of Ordnance Computation Project at Harvard University. In 1944, she became one of the first programmers of the **Mark I**, the first automatic sequence digital computer. Her computer team was the first to use the term "bug" to refer to a computer glitch after a two-inch moth was found in the circuits of the Mark I. After World War II, she joined the Naval Reserve and developed software for the Mark I and Mark II while she was a research fellow at Harvard's Computation Library. She joined the Remington Rand Corporation (which later became the Sperry-Rand Corporation) in 1949 and was soon one of the first senior programmers in the computer industry. Her participation in the creation of **UNIVAC**, the first general-purpose commercial computer, led to the production of the first language compiler (also known as an assembly language, or assembler), A-0, later named the A-2. This invention led to the FLOW-MATIC, a more advanced and commercially viable English-language compiler in the mid-1950s. As a result of FLOW-MATIC, programmers could write one command, which would then set off a command sequence in machine-specific, problem-oriented machine code. This computer machine language streamlined programming, eliminating time-consuming steps and making computers more accessible to those without extensive knowledge of the machine's physical characteristics. Hopper became a proponent of the standardization of computer language, culminating in her development of COBOL in the late 1950s. COBOL utilized word commands instead of mathematical symbols, significantly simplifying required training and knowledge of specific computers. As a result, business and government began to see computers as a practical and worthwhile tool. Hopper retired from the Naval Reserve in 1966, only to be called back seven months later to further develop the standardization of COBOL. In 1973, she was promoted by a special act of Congress to the rank of captain. She retired as a rear admiral in 1986, and spent the rest of her life lecturing and writing on the further development and use of high-level computer languages.

Grace Murray Hopper

Hotchkiss, Benjamin Berkeley (1826-1885)

American inventor

Hotchkiss was born in Watertown, Connecticut, and grew up in a family involved in machinery. His father and brother, both inventors, had established a business that manufactured their patented articles. In 1856 Hotchkiss designed a **rifle** field gun that the Mexican government purchased. Along with his brother, Andrew, he came up with a projectile for rifled **artillery**. It had two sections (front and rear) made of **cast iron** that were joined by a band of lead, intended to fit into and take the shape of the grooves of the rifling in the gun. This created a much more accurate shell. Hotchkiss also helped his family to develop a better percussion fuse, and, as a result, more Hotchkiss shells were used for rifled **cannon**s during the American Civil War than any other munitions manufacturer in the country. His family was so closely associated with the Union cause that during the New York City draft riots of 1863, Hotchkiss had to ride through the streets of Manhattan concealed beneath a pile of cloth.

Success followed Hotchkiss after the war. He traveled to Paris, where he invented an improved metallic cartridge case and made many improvements in hand-held firearms and cannons. Hotchkiss designed a revolving-barrel **machine gun** and created a revolving cannon that destroyed a boat during trials, having hit the ship with 70 shots out of 119 fired. While aboard a train in 1875, he met a Romanian army officer who told him of the need for a magazine repeating rifle. In a half-hour, Hotchkiss sketched a design for such a weapon, which later proved to be superior to all others at the time.

Hotchkiss was reputed to be the world's best artillery engineer. By 1882, his B.B. Hotchkiss Company had branches throughout Europe. Hotchkiss was working on a new machine gun design when he died in 1885. His company perfected his plans after his death, and brought out a gas-operated machine gun which was used extensively in World War I. The Hotchkiss machine gun remained in use by the British army up until the late 1930s.

See also Ammunition; Gatling, Richard Jordan; Submachine gun

Hourglass

An hourglass consists of a pair of **glass** bulbs joined by a short passage, or neck. One bulb is partially filled with fine, dry sand. When the hourglass is turned to stand on the other bulb, the sand passes slowly through the neck into the empty bulb in a period of approximately one hour. This device was commonly used from the fourteenth through seventeenth centuries. Sailors continued to use them into the nineteenth century. Although marine **chronometer**s were available, not all seafarers could afford to have them; a ship's chronometer was not always accessible to all those on board a vessel. Smaller or larger glasses were made to measure smaller or larger time periods. Each hourglass was capable of measuring only a single unit of time and had to be constantly turned over to be maintained. They were also incapable of telling which hour it was. As with **sundial**s, hourglasses have passed on to novelty status. They are still used in a small way as egg timers and as board-game timers. The old poetic adage "the sands of time" refers to the hourglass. As with all things, the hourglass has become a victim of the passage of time.

Howe, Elias (1819-1867)

American inventor

Elias Howe invented the first practical **sewing machine**. Born on his parents' farm in Spencer, Massachusetts, Howe worked in his father's grist- and saw-mills as a boy. Later, he worked as an apprentice machinist in a cotton-machinery factory in Lowell, Massachusetts.

While working in Cambridge for master mechanic Ari Davis around 1841, the young Howe overheard a customer remark that the inventor of a practical sewing machine could make a fortune. Howe worked steadily from then on to perfect his design, living first with his father and then a friend, while his wife labored to support the family by taking in hand-sewing.

Howe's efforts yielded a practical sewing machine, patented in 1846, that featured an eye-pointed needle and a double-thread stitch. Howe attempted to market his machine to the clothing industry by staging and winning contests against seamstresses. When no orders resulted, however, Howe sold his invention to an English corset maker named William Thomas, who patented it in his own name and made a fortune. Nevertheless, Howe worked in England for Thomas from 1847 to 1849.

Howe sent his family back to the United States in 1849 by pawning his patent papers. After he too returned, penniless, he found his wife dying and sewing machines using his patented features selling widely. He filed a lawsuit and, after long litigation, secured the right to receive royalties on all sewing machines manufactured in the United States from 1854 until his patent expired in 1867. In the last ten years of his life, Howe was finally able to enjoy the riches he had earned through his invention.

Howitzer • See Artillery

Hubble Space Telescope

One of the unavoidable problems facing ground-based **telescope**s if a phenomenon astronomers call *seeing*: distortions of the image of a celestial object by the passage of its light through Earth's turbulent atmosphere. This the same phenomenon that causes stars to twinkle, and if the seeing is very poor, astronomical observations can be ruined. Above the atmosphere, seeing is eliminated, and the potential for top-quality observations is enormous.

The advantage of observing objects from above the atmosphere was realized by astronomers long before the launch of the Russian satellite *Sputnik 1* ushered in the space age. It was not until the 1970s, during the latter phases of the Apollo Moon program, that the Large Space Telescope, as it was then called, came under serious development by the National Aeronautics and Space Administration (NASA) in cooperation with the European Space Agency (ESA).

Integral to the deployment and servicing of the Space Telescope was the **space shuttle**. NASA engineers realized the telescope would require servicing, and the initial plan was for the space shuttle to retrieve it every five years for major servicing, with intermittent in-orbit servicing as needed. The extremely high precision of the telescope, however, coupled with the stresses it would encounter during launch and landing, led designers to scrap the idea of ever returning it to Earth. The telescope would be serviced in orbit by the Shuttle as needed, but once launched, it would remain launched for the duration of its 15-year lifetime.

With launch scheduled for 1986, NASA and ESA moved ahead rapidly on building the telescope, now renamed the Hubble Space Telescope (HST). Naming the instrument in honor of Edwin Powell Hubble, who pioneered our understanding of the expansion of the Universe, was appropriate. With its unique capability to see to the farthest reaches of the Universe unhindered by atmospheric turbulence, one of the HST's main observational goals would be to examine some of the issues Hubble and his colleagues had been unable to solve from the ground. To support HST's operations, the Space

Telescope Science Center was constructed at the campus of the Johns Hopkins University in Baltimore, Maryland, in the mid 1980s.

The Space Telescope program suffered a severe setback on January 28, 1986, with the explosion of the space shuttle *Challenger*. The ensuing shutdown of the shuttle program meant that HST was grounded until NASA decided it was safe to fly the shuttles once more. Finally, on April 25, 1990, the space shuttle *Discovery* placed the Hubble Space Telescope in orbit around Earth.

Before long, NASA was faced with one of its most embarrassing moments. The 2.4-meter (90-inch) primary **mirror** of the HST was discovered to suffer from *spherical aberration*, a defect resulting from an imperfect shape of the mirror that smeared what should have been crystal-clear images into much larger, fuzzy blurs. The telescope was still able to do some impressive research, but some of the critical observations of very distant, very faint objects were no longer feasible. Amid squeals of delight from political cartoonists (who were able to draw such things as the space shuttle deploying an enormous pair of **eyeglasses** in front of the telescope), the problem was eventually pinpointed. Making the best of an awful situation, NASA engineers designed a set of corrective optics for HST, called the Corrective Optics Space Telescope Axial Replacement (COSTAR), and in December 1993, the crew of the space shuttle *Endeavour* installed the new optical elements, fixing the problem and restoring HST to full functionality.

HST is armed with a formidable array of instruments, including its imaging camera, an **ultraviolet spectrograph**, and the Faint Object Camera (FOC), which enjoys the benefits of the COSTAR-corrected images and is able to perform the full set of original mission objectives. A large number of professional astronomers have observed with the HST, through a standard process of application for observing time and selection of the most competitive proposals.

Hula hoop

Think of American fads of the 1950s and what comes to mind? Hula hoops! And oddly enough, the hoops came from Australia. In the 1950s, Australian gym classes used 3-ft (1-m) bamboo rings for calisthenics. In 1957, this form of exercise caught on outside school gymnasiums, becoming a popular form of Aussie entertainment. The owners of the American novelty company Wham-O, Richard P. Knerr and Arthur K. (''Spud'') Melin, heard about the craze and decided to investigate. Knerr and Melin introduced the hula hoop to neighborhood kids and cocktail party guests in America and discovered that they loved playing with the toy. Immediately, Wham-O began production of the American version of the Australian ring. Christened the ''hula hoop'' and made of vividly colored **polyethylene plastics**, they cost fifty cents to make and sold for $1.98 a piece. By 1958, the hula hoop was the subject of an international hysteria. Japan's Prime Minister Kiahi received a hula hoop for his sixty-second birthday. Parisian novelists posed for photographs with them. A Belgian expedition bound for Antarctica brought along twenty of them. German world heavyweight champion Max Schmeling gyrated the hula hoop ringside. Although their popularity has waned in the decades following their initial craze, hula hoops are still in production and remain stocked in many toy stores around the world.

Elias Howe

Hunter, John (1728-1793)

Scottish surgeon and anatomist

A man with little formal education, John Hunter rose to become one of the most eminent and influential surgeons of his time. He was born in Long Calderwood, Lanarkshire, Scotland, the youngest of ten children. After his father died in 1741, Hunter left school and worked on the family farm and as a cabinetmaker. In 1748 Hunter joined his brother, William (1718-1783), an anatomist in London, England. Hunter's skill in preparing specimens for his brother's anatomy lectures was so impressive that he became William's assistant. William also arranged for Hunter to study surgery at Chelsea and St. Bartholomew's Hospitals. The younger Hunter became a master of anatomy in 1753. In 1755 William sent Hunter to Oxford to acquire a formal education; however, academics held no interest for the young man. He returned to London within two months to resume his dissection work with William and his

John Hunter

own original experimentation. He gathered his specimens from postmortem examinations, from his brother's dissecting rooms, and from ''resurrectionists''—body-snatchers who dug up freshly buried corpses from graveyards and sold them secretly to surgeons.

Suffering from ill health, Hunter became a staff surgeon with the British army in 1760, serving during the Seven Years' War. His experiences in treating wounded soldiers resulted in his important treatise on gunshot wounds, published posthumously in 1794. When the war ended in 1763, Hunter returned to London and began a private surgical practice, continuing his research at the same time. The same year, he bought two acres in Earls Court, which he filled with an overflowing and diverse collection of animals for use in his surgical studies and dissections. Hunter's practice and reputation grew rapidly. He was elected a Fellow of the Royal Society and was appointed surgeon to St. George's Hospital in 1767. He married poet Anne Home—a minor—in 1771, and was named surgeon-extraordinary to King George III (1738-1820) in 1776.

Determined to make surgical training more accessible, Hunter began lecturing on the theory and practice of surgery to students in 1773, in his own home. After a falling-out with his brother, Hunter moved to Leicester Square in 1783 and built an educational complex with lecture and conversation rooms, and a museum which housed his extensive collection of specimens and served as an invaluable teaching aid for his students, demonstrating the structure and functions of both diseased and healthy body parts in a wide range of species. One of his prized specimens was the skeleton of an eight-foot (2.4-m) tall Irishman, Charles Byrne, who had planned a burial at sea so as to remain out of Hunter's museum. Well-paid by the doctor, the resurrectionists thwarted Byrne's last wishes. Hunter's museum was purchased by the British government in 1799 but was severely damaged in the London blitz of 1941.

Hunter's contributions to anatomy and surgery were vast. His treatise on human teeth was a basic building block of modern dentistry. He made detailed studies of the structure and function of the lymph vessels and carried out early experiments on tissue transplantation, including grafting a human tooth into a cock's comb. He published important studies of venereal disease and animal hibernation, developed surgical techniques to repair the Achilles tendon and to ligate arteries in cases of aneurysm instead of amputating, and improved embalming techniques—most notably for a gentleman whose wife's body had to remain above ground in order for him to inherit her wealth. Hunter had a great influence on the development of scientific, experimentally based surgery through students who trained under him and later became famous, notably Edward Jenner (1749-1823), and Philip Syng Physick, the ''Father of American surgery.''

Hunter's health was permanently affected by an experiment he carried out on himself in 1767, through which he contracted syphilis. Increasingly troubled by angina pectoris, Hunter collapsed and died of an attack in 1793 during a meeting of the board of St. George's Hospital. His remains were reinterred at Westminster Abbey in 1859, with an inscription memorializing him as the ''Founder of Scientific Surgery.''

Huntsman, Benjamin (1704-1776)

English inventor

Until Benjamin Huntsman invented the crucible process for casting steel around 1740, steel had been produced in only very small amounts. The Huntsman process liberated slag from ''blister steel'' through cementation and carburization. From this pure and uniform castings could be made.

Born in Barton-on-Humber, Lincolnshire, England, on 1704, Huntsman apprenticed as a clockmaker and locksmith in Epworth, Lincolnshire. In 1725 he started his own business in Doncaster. Striving to create better instruments, he searched for a higher quality steel and spent a number of years experimenting with different types of moldings. When he finally succeeded, he opened a steel plant in Sheffield (1740). His steel was considered too hard by the Sheffield cutlers, and it was not accepted by them until after their counterparts in continental Europe had successfully used it.

The crucible is the vessel that holds molten steel or other metals while it is being melted or transferred. It is made of a highly heat-resistant material such as clay or porcelain. Even today, some steel mills in Eastern Europe continue to transfer molten steel through ditches in an earthen floor.

Huntsman maintained secrecy over his foundry and his steelmaking process, never patenting it. In about 1750, his secret fell into the hands of Samuel Walker, a competitor. Walker made a great profit from Huntsman's process; yet Huntsman was able to produce a higher quality product and so was able to beat his competition and even expand his business in 1770. Judged by modern standards, the crucible process was very slow and extremely costly. However, it was the best steelmaking process at the time, and it remained so until it was replaced by the open hearth and electric furnace processes of the nineteenth century. Huntsman died in Sheffield, Yorkshire on June 21, 1776.

See also Bessemer, Henry; Blast furnace; Martin, Pierre-Émile; Siemens, Charles William; Siemens, Ernst Werner von; Steel production

Hunt, Walter (1796-1859)

American inventor

Walter Hunt was a remarkably creative and prolific inventor who seldom bothered to patent or profit from his numerous devices. Born in Martinsburg, New York, Hunt invented a flax-spinning machine by the time he was 20 years old. He set up a small shop in New York City in 1826 and supported himself in the real estate business while developing dozens of practical devices. In 1827 he invented an alarm gong for **streetcar**s, fire engines, and police stations. Then he designed the Globe coal-burning **stove**, a knife sharpener, and a restaurant steam table.

Sometime between 1832 and 1834 in his Amos Street shop, Hunt invented the first modern, practical **sewing machine**. It featured an eye-pointed needle and a second thread that created a stitch. True to form, Hunt didn't patent the machine. Legend has it that his daughter said he shouldn't market the machine lest seamstresses be put out of work. When he finally did attempt to patent the machine after **Elias Howe** applied for his own sewing-machine patent in 1846, Hunt was refused on the grounds that he had abandoned the design, although he was recognized as having preceded Howe with the concept.

Hunt followed his sewing machine with many improvements to firearms, an ice-breaking boat, paraffin **candle**s, a velocipede, a machine for making nails and rivets, a street-sweeping machine, conical bullets, a self-closing inkwell, and a fountain **pen**. He even invented the Antipodean Performers-suction shoes used by circus performers to walk up walls and across ceilings.

Hunt invented the safety pin in three hours one day in 1849 to pay off a $15 debt; he did patent this device but later sold his patent outright for $400. In 1854, Hunt invented a throw-away paper collar. Again, he patented the design and this time arranged for royalty payments, but the paper collar became popular only after Hunt died. **Isaac Singer** arranged to pay Hunt $50,000 for his sewing machine design in 1858 in order to clear up the patent confusion about sewing machines, but Hunt died in New York in 1859 before Singer was able to make any payments.

Hussey, Obed (1792-1860)

American inventor

Obed Hussey was an avid inventor who rarely shied away from a challenge. He was especially intrigued by the development of new agricultural machines. Born in Maine, Hussey moved to Nantucket Island, Massachusetts, when he was very young. He probably worked as a sailor for a while. By the time he was thirty, he was in Cincinnati, Ohio, and working on his inventions.

In 1830 someone suggested to Hussey that he invent a machine for cutting grain. He had already invented, among other things, a corn-husking machine, a corn grinder, and a sugarcane press. He went to work on this latest challenge and travelled to Baltimore, Maryland, the following year to conduct his **reaper** experiments at the farm implement factory owned by Richard B. Chenaworth.

In 1832, Hussey returned to Cincinnati and built his first successful reaper. It was publicly demonstrated in July 1833 to the Hamilton County, Ohio, Agricultural Society. Hussey's machine consisted of a reciprocating sawtooth bar driven by **gears** turned by the main drive. The machine was horse-drawn. As the blade moved forward, the stalks of grain fell onto a platform where they were manually bound. This machine sold well enough between 1834 and 1838 in Illinois, New York, Pennsylvania, and Maryland that in 1838 Hussey opened a large factory in Baltimore.

Even as Hussey was developing his first reaper, **Cyrus McCormick** was working on one of his own at his father's **forge** in Virginia. McCormick's reaper was patented in 1834, and the two inventors entered into direct competition. Both Hussey and McCormick built improved versions of their machines in an effort to stay ahead of each other, and both exhibited their machines at the Crystal Palace industrial exhibition in London in 1851.

In the years following the exhibition, McCormick gradually eroded Hussey's standing in the reaper industry, shrewdly acquiring the patents of others. Hussey had an intolerance for other people and obstinately preferred to improve his machines himself. The difference between the two men gave the competitive edge to McCormick, and in 1858, Hussey sold out.

He went to work once again, however, this time on a steam-powered **plow**. His work was cut short in 1860 when he died after falling under a moving train in New England.

See also Farming, mass production; Reaper and binder

Huygens, Christiaan (1629-1695)

Dutch scientist

Though one of the most brilliant minds in history, the Dutch scientist Christiaan Huygens enjoyed relatively little fame during his lifetime. This was chiefly due to chronological factors, for Huygens worked during the period directly after the death of **Galileo** and just before the ascent of **Isaac Newton** (1642-1727), both towering figures of science. If not for the proximi-

Christiaan Huygens

ty of these two geniuses, Huygens would have certainly been considered the greatest scientist of the seventeenth century. Though his work was unregarded for many years after his death, he is today held as one of the chief contributors to the modern sciences of mechanics, physics, and astronomy.

Huygens was born in The Hague, Netherlands, in 1629. The environment in which he was raised was ideal for the nurturing of young minds: his father, Constantijn Huygens, was a diplomat and poet who understood the need for classical training, and he planned for his son a private education in mathematics, languages, literature, and music. Young Christiaan was also influenced by the mathematician-philosopher **René Descartes**, a friend of the Huygens family and frequent visitor to their home. From Descartes Huygens learned of the ''mechanistic'' philosophy of nature, and came to believe that all natural phenomena would one day be explained by science.

Huygens left his home in 1645 to study law and mathematics at the University of Leiden and, in 1647, the College of Orange in Breda. He was unsatisfied with the universities' narrow approach to learning, however, and in 1649 he returned to The Hague. There he remained until 1666, living off an allowance from his wealthy father. Given the financial freedom to study as he pleased, Huygens began to perform some of the most important research of his life. As a child, Huygens had been interested in observing the night sky; however, the optical equipment available to him allowed little more than a blurry view of the heavens. By 1650, with the help of the Dutch philosopher Benedict Spinoza (1632-1677), he had taught himself the art of **lens** grinding. Developing a new method for grinding lenses of unparalleled clarity, Huygens began to chart the sky in earnest. During the period from 1655 to 1660, Huygens made a number of landmark astronomical discoveries, including Saturn's ring and the planet's largest moon, Titan. He was the first to observe the surface features of Mars, as well as the Great Nebula in the constellation Orion.

In order to facilitate these observations, Huygens designed several new pieces of astronomical equipment. Chief among these were his lenses, which provided far greater resolution and magnification than any before. After several years he perfected an achromatic lens that corrected the ''false color'' fringes often associated with inferior lens systems; this lens, called the Huygenian eyepiece, is still used in many **telescope**s today.

In order to better view the sky, Huygens also modified the design of his telescopes. It was commonly known that longer telescopes, with their longer focal lengths, allowed for greater magnification. Huygens took this principle to its extremes, constructing telescopes up to twenty-three feet (7.0 m) long. While these remarkable instruments gave a magnificent view of the planets, Huygens remained unsatisfied. He believed that the conventional telescope design was too limiting, for it relied upon metal tubes which would bend if they were too long. His answer was a tubeless telescope called an aerial telescope. Because there was no tube to connect them, the large objective lens and the smaller eyepiece could be as far apart as practical lens construction would allow. The largest of these aerial telescopes was more than 100 feet (30 m) in length. In addition to optical aids, Huygens also invented a **micrometer** in 1658 that enabled him to measure the angular separations of objects (such as the apparent distance between Saturn and its moon) with a precision of a few seconds of arc.

While Huygens observed the heavens, the European scientific community was working toward the development of a reliable timekeeping device. Such a device was in great demand by the trading industry, who required an accurate **clock** for use in the navigation of their sailing vessels. Previously, clocks were regulated by a slowly falling weight which would turn the device's **gear**s. Unfortunately, the pace of the weight's descent was irregular, and the clocks were wildly inaccurate. Years earlier, Galileo had noted that a pendulum would swing with a precise motion, taking the same amount of time to move in one direction as it did to return. He termed this effect isochronicity (''equal time''), and suggested that it might be useful as a means for regulating timepieces; however, he was never successful in designing a working model. In 1656, Huygens found that a swinging pendulum was not truly isochronic unless the arc it described was not completely circular. Using this knowledge he devised a system combining the pendulum with a weight-driven clock—the pendulum would swing exactly once each second, precisely regulating the motion of the clock's hands. The falling weight would drive the gears, as well as give the pendulum just enough energy to overcome the slowing forces of air resistance and friction.

Before the invention of Huygens' pendulum clock there was no reliable means of measuring time. Within months of the introduction of the "grandfather clock" design, towns across the Netherlands (and, soon after, all of Europe) had large clock towers regulated by swinging pendulums. Huygens' experiments with pendulums had given him an insight into the nature of motion itself.

Using the work of John Wall as a starting point, Huygens expanded his research to include the concept of **momentum**, a property of a moving object that measures its impact should it hit another object. His theory of momentum was included in his 1673 publication *Horologux Oscillatoriux*; it is now better known as the law of conservation of momentum. This law, a precursor to Helmholtz's law of conservation of energy, states that the momentum of a moving object (its mass multiplied by its velocity) remains constant unless the object is slowed, stopped, or changes direction—that is, subject to a force. In the absence of force, momentum is conserved.

This was only the first of Huygens' forays into the field of theoretical physics. During the latter half of the 1600s, a great debate had divided the European scientific community: the nature of light. Newton, still a student but quickly gaining a reputation as Britain's most prominent scientist, had proposed the theory that light was composed of particles moving through a vacuum. His proof was the fact that, when shone through a **prism**, light would split into its individual colors—thus, white light must be a composite. This concept was contrary to Huygens' mechanistic view of the universe, a result of his childhood conversations with Descartes, which demanded that all natural phenomena be the product of other phenomena—that is, that things happen because other things make them happen. Just three years after the publication of Newton's light theory Huygens announced his own, stating that light moved in a longitudinal wave through an invisible substance called ether (in much the same was as sound waves move through air). The light waves, according to Huygens, were produced by microscopic pulses emanating from the source; as they traveled, the pulses would push the ether longitudinally (back and forth), creating a wave motion. At he time of its publication, Huygens' wave theory of light was given little consideration, chiefly because scientists were reluctant to accept the notion of the invisible ether. As Newton's reputation grew his particle theory became more and more popular, and the work of Huygens was all but forgotten. It was not until the nineteenth century that Thomas Young re-introduced the wave theory of light—this time without the dubious presence of the ether.

It is not surprising that Huygens earned little fame during his lifetime, for he published his work slowly and infrequently. He was also something of a recluse, choosing not to take on students. However, he was instrumental in the foundation of the French Academy of Sciences, and was a charter member of the British Royal Society. In addition to his clocks and his astronomical instruments, Huygens invented the **manometer** (a device used to measure the pressure of liquids and gases), as well as a prototype **internal combustion engine** using pistons.

Hyatt, John Wesley (1837-1920)
American inventor

John Wesley Hyatt went to work as a printer at age sixteen. In 1861 he patented a **knife** sharpener and went on to develop a new method for making dominoes and checkers, starting a company to produce them. His work with the playing pieces led to his search for a cheap substitute for ivory. He was also attracted by the $10,000 prize offered by Phelan and Collender for an artificial ivory to use in billiard balls.

Hyatt and his brother Isaiah had heard of **Alexander Parkes**'s (1813-1890) success using nitrated cellulose dissolved in alcohol, ether, and camphor to produce a moldable substance. The Hyatt brothers improved on Parkes's process by refining the production of nitrated cellulose and devising a variety of ingenious means by which nitrocellulose could be impregnated with camphor. The Hyatts patented their manufacturing method for billiard balls in 1869, using the substance they called celluloid. They did not win the prize, however. But the Hyatts' celluloid was marketed as rattles, collars, film, and dental plates, even though its flammability (from the nitric acid used to produce the nitrocellulose) limited its uses. Its popularity diminished when celluloid billiard balls reportedly struck each other and caught fire. Today acetic acid is used in place of nitric acid to produce the more stable cellulose acetate, which has replaced celluloid in most uses.

John Hyatt went on to develop many other inventions and established the Hyatt Roller Bearing Company in New Jersey to produce the roller bearings necessary for them. He received the Perkin Medal of Honor from the Society of Chemical Industry in 1914 for his work with celluloid. When Hyatt died in 1920 in New Jersey he held over 200 patents.

See also Cellulose, chemical uses of; Plastic

Hybrid grain • See Plant breeding

Hydraulic brake • See Braking systems

Hydraulic ram

A hydraulic ram is a pump used to raise water from one level to another by utilizing the energy of the water itself that is flowing in a pipe. The design is based on the principle that occurs when a flow of water encounters a sudden blockage. The energy generated from this sudden stop is used to force the water through pipes that raise it to another level.

In 1796 **Joseph Montgolfier**, better known for his work on early hot-air balloons, introduced the hydraulic ram. His system worked by sending water through a pipe that had a **valve** which could be shut suddenly. When the valve was closed, the pressure in the pipe immediately rose. This forced the water into a closed air chamber through a one-way valve. When the water came into the chamber, this again increased the pressure and forced the water through a delivery pipe to a higher level.

Joseph Montgolfier's son, Pierre, varied his father's design to produce what he called a *compound* or dirty water ram.

The problem with the ordinary ram was that it required a flow of water to operate that was significantly greater than the amount actually being raised. Pierre addressed the problem by using dirty water to provide the power. The dirty water was fed to the main pipe, and the rise of pressure was used to lift a **spring**-loaded plunger which pumped the clean water.

Hydraulic rams are still used, but they are not widespread because of the large quantity of water needed to operate them. They normally pump water for isolated farms. The Montgolfiers proposed to use them for **air compressors** and to help pump the bilges on **ships** through the rolling action at sea. Some have even suggested recently that the rams may be useful for driving **heat pumps**.

Hydrofoil

The hydrofoil is very similar to the hovercraft, because it moves in the boundary between air and water. It avoids drag by lifting itself out of the water, using wing-shaped structures called hydrofoils that extend into the water from the craft. These hydrofoils function like the wings on a plane, creating lift and flying the hull above the surface of the water.

The first person to work on this idea was a French priest, Ramus, in the mid-1800s. However, there was no engine that could supply sufficient thrust. In the 1890s, another Frenchman, the Count de Lambert, tried and failed to make a working model using a **gasoline** engine.

The first successful hydrofoil boats were created in the early 1900s. Enrico Forlanini, an Italian **airship** designer, built a small boat with hydrofoils in 1905. He showed **Alexander Graham Bell** a later model that impressed the famous American. Bell built one himself, based on Forlanini's patented design and set a water-based speed record of 71 miles (114 km) per hour with it in 1918. This record stood until the 1960s.

Although there were small improvements made over the next few decades, hydrofoils did not see commercial use until the 1950s, when Hans von Schertel, a German scientist, developed his designs for passenger hydrofoils. Italy created their Supramar boats, and Russia and the United States developed hydrofoils with both commercial and military applications.

There have been experiments with various types of foils and different types of engines, including the **gas turbine**, **diesel**, gasoline, and **jet engine**s.

The foils themselves have two distinct shapes. The surface-piercing models are V-shaped, so that part of the foil stays out of the water. This type is good for calm surfaces like rivers and lakes. The other foil is completely submerged. It usually consists of three foils extending straight down beneath the boat. Hydrofoils with this configuration need autopilots to keep them level. Whenever the boat shifts to one side, sensors send messages to flaps on the foils, which then adjust automatically to bring the boat back to a normal position.

Hydrofoils today are used by commuter services, fishery patrols, fire fighters, harbor control, water police, and air-sea rescues. For the military, hydrofoils can be excellent small **submarine** chasers and patrol craft.

In the mid 1990s, hydrofoils became the desired means for rapid marine transport. This was so primarily because they continued to improve both their performance and cost effectiveness. New and very large hydrofoils are being built out of light but strong marine-grade aluminum, with composite materials also being used for many parts of the craft. Propulsion systems also are becoming simpler, using direct-drive diesel/water jet systems, while new hull shapes have substantially increased speeds. The very large, Finnish-built *Stena HSS 1500* which went into ferry service in the Irish Sea in 1996 is 396 ft (120 m) long and 132 ft (40 m) wide, can reach 40 knots, and has a capacity of 1,500 passengers and 365 cars. It is the largest all-aluminum structure in the world.

Hydrogen bomb

Even as work on the first **atomic bomb** was going ahead, some scientists were thinking about an even more powerful weapon, the hydrogen, or fusion bomb. As far back as the 1920s, scientists had been exploring the possibility that small nuclei might join together—or fuse—to make larger nuclei. Ernest Rutherford (1871-1937) and his colleagues at Cambridge University's Cavendish Laboratory had observed the hydrogen fusion reaction in 1934. In 1938, the German-Austrian physicist, Hans Bethe (1906-), summarized much previous work on fusion in a theory that explained the production of energy in the stars. Bethe showed how four hydrogen nuclei—protons—might fuse to produce a single helium nucleus, with the release of enormous amounts of energy. A Japanese physicist, Tokutaro Hagiwara, had delivered a lecture in 1941 on the possibility of using uranium fission to ignite a hydrogen fusion reaction. So the theoretical groundwork for a fusion bomb had been laid even before concrete work on the atomic fission bomb began.

Many scientists realized that nuclear fusion could be used as a source of energy on Earth, for either military or peacetime applications, if a method could be found to carry out and control the reaction. But that "if" was a very large one. For one thing, nuclear fusion reactions require temperatures in the range of 40 million degrees Celsius. Because of these high temperature requirements, fusion reactions are also referred to as thermonuclear reactions. The task of working at such high temperatures was a daunting one.

One scientist, **Edward Teller**, an emigré from Hungary, began arguing for the development of a fusion weapon, or "super" atomic bomb, during the Manhattan Project. He pointed out that such a weapon would be many times more powerful than a fission (atomic) bomb and would provide the United States with an unmatched military superiority.

Teller's ideas were largely ignored, however, as the United States endeavored to complete the bombs it later dropped on Hiroshima and Nagasaki. The technical requirements for the hydrogen bomb were staggering, and production of the crude first atomic bombs was already difficult enough. For example, an important raw material needed for the hydrogen bomb, tritium (hydrogen-3) had to be made synthetically. It would have taken eighty times the effort the United States was already spending on the manufacture of plutonium for the atomic bomb to make the required tritium. There was not enough time, manpower, or equipment to work on a fusion bomb when the fission bomb had not even been developed.

After World War II ended, many scientists, appalled at the damage done in Hiroshima and Nagasaki, rejected the idea of building even more powerful weapons. Furthermore, there seemed to be no challenge to the United States' military superiority, based as it was on its possession of fission bombs. However, other scientists, particularly Edward Teller, were sure that the so-called Super bomb was possible, and they continued their research. Then in September1949, the Soviet Union exploded its first atomic bomb. The Manhattan Project reseach on the atomic bomb had been kept secret from the Soviets, though that country had been allied with Britain and the United States to defeat German and Japan. But with the help of materials stolen through espionage, Soviet scientists duplicated the atomic weapon, and let the world know they had it by testing it above ground. A few months after the Soviet test, in January 1950, President Harry S. Truman (1884-1972) ordered the Atomic Energy Commission to begin work on developing a fusion bomb.

The general concept for such a bomb is a two-part explosion, where a fission bomb then sets off a fusion reaction. The first Super design relied on the heat from the primary fission bomb to set off the secondary fusion explosion. Edward Teller, in conjunction with the mathematician Stanislaw Ulam (1909-1984), came up with a more workable idea (that became the basis for subsequent weapons) of using radiation from the primary fission bomb to generate a shock wave that would compress the thermonuclear fuel and thus set off the fusion reaction. The greatest technical problem scientists faced was finding a way to pack hydrogen isotopes together tightly enough to allow fusion to occur and to make the bomb small enough to be transportable. One of the first solutions to this problem was to surround the fission bomb with crystalline lithium hydride, made of lithium-6 and hydrogen-2. When the fission bomb explodes, tritium (hydrogen-3) is produced. This isotope then fuses with hydrogen-2.

The first thermonuclear test was carried out by the United States in May 1951. The landmark full-scale thermonuclear explosion was at Bikini Atoll on November 1, 1952. It had the destructive power of about seven million tons of TNT, roughly 500 times greater than that of the first fission bombs. Meanwhile the Soviet Union had also developed nuclear fusion bombs, the product of intensive research by the country's top scientists, supplemented by ongoing espionage. In 1948 Andrei Sakharov (1921-1989) had independently derived a fusion bomb design analogous to one Teller had described in 1946. (Teller's design was called the Alarm Clock, Sakharov's the Layer Cake.) Soviet scientists in 1953 tested the first transportable fusion bomb, one that could be air-dropped and was designed for mass production. A transportable fusion bomb was not tested by the United States until 1956.

The ability of both countries to deploy fusion bombs led to an arms race that did not effectively end until the collapse of the Soviet Union in 1989. Bombs were made small enough to be delivered from long-range missiles, and with over a thousand times the destructive power of the bomb dropped on Hiroshima. Other countries also developed thermonuclear weapons, including India and Pakistan, who aggressively tested their arsenals in May 1998. However, the drastic horror of actually using the bombs had up through that time held any country back from waging nuclear war. In the best light, the development of the hydrogen bomb had been a deterrent to conflict, an ominous stabilizing force.

Hydroponics

There are several early examples of hydroponics, or soil-free agriculture, including the hanging gardens of Babylon and the floating gardens of China and Aztec Mexico. Early Egyptian paintings also depict the growing of plants in water.

In 1600, the Belgian Jan Baptista van Helmont (1579-1644) demonstrated that a willow shoot kept in the same soil for five years with routine watering gained 160 lb (72 km) in weight as it grew into a full-sized plant while the soil in the container lost only two ounces. Clearly, the source of most of the plant's nutrition was from the water, not the soil.

During the 1860s, German scientist Julius von Sachs (1832-1897) experimented with growing plants in water-nutrient solutions, calling it *nutriculture*.

In 1929, W. F. Gericke of the University of California first coined the term hydroponics, which literally means "water labor." Gericke demonstrated commercial applications for hydroponics and became known for his 25 ft (7.5 m) tomato plants.

Hydroponics has been shown to double crop yield over that of regular soil. It can be categorized into two subdivisions: *water culture*, which uses the Sachs water-nutrient solution, with the plants being artificially supported at the base; and *gravel culture*, which uses an inert medium like sand or gravel in conjunction with wire mesh to support the plants, to which a water-nutrient solution is added.

Hydroponics was used successfully by American troops stationed on non-arable islands in the Pacific Ocean during the 1940s. It has also been practiced to produce fresh produce in arid countries like Saudi Arabia. In the 1970s, researcher J. Sholto Douglas worked on what he called the *Bengal hydroponics system*. He sought to simplify the methods and equipment involved in hydroponics so it could be offered as a partial solution for food shortages in India and other developing countries.

Successfully adopted in certain situations, hydroponics will remain in limited use as long as traditional farming methods in natural soil can support the world's population.

Hygrometer

A hygrometer gauges atmospheric humidity by measuring the reaction of certain substances to changes in the amount of water vapor present in the air. Hygrometers are used in meteorology and to regulate humidity in buildings and industrial processes.

The hygroscopic hygrometer was first built in the early 1400s by Cardinal Nicholas de Cusa, a German mathematician. This type of hygrometer used the change in the weight,

Hypatia of Alexandria

length, or twist (as with twine) of a moisture-absorbing substance to determine humidity. Although the Italian inventor **Leonardo da Vinci** was reported to have constructed a hygrometer around 1500, little attention was given to improving the instrument over the next century; it was not until the latter half of the 1600s that other methods were developed. One type of hygrometer used a **paper** strip, weighted at the middle and attached to a board. The weight would rise and fall as the strip released or absorbed moisture. Another type—the condensation hygrometer—collected dew in a cup placed under a pointed vessel filled with cold water.

In 1687, **Guillaume Amontons** developed a hygrometer that was similar to a three-liquid **barometer**. The changing volume of a mercury-filled absorbent bag at the end of a tube moved the interface level of two other immiscible (nonmixing) liquids up and down in the tube.

The eighteenth century brought more improvements for measuring humidity. Extensive studies into the properties of water vapor were made by Johann Heinrich Lambert in 1774, who built and studied hygrometers and first suggested the term "hygrometer." In 1775, Horace Benedict de Saussure (1740-1799) developed the hair hygrometer in which a suspended hair moved an indicator as the humidity changed. Jean Andre Deluc's (1727-1817) hygrometer used an ivory bulb that when damp allowed mercury to move down a tube.

The psychrometer was first developed by James Hutton (1726-1797) during the late 1700s. It measured the cooling effect of evaporating moisture on a thermometer bulb. In 1800, John Leslie (1766-1832) announced his creation of a simpler psychrometer that consisted of two thermometers mounted and attached to a handle. After rotating the thermometers—one covered with a dry cloth, the other with a wet cloth—through the air, the temperature difference was compared against a humidity chart to arrive at the final humidity measurement.

About 1819, **John Frederic Daniell** invented the dewpoint hygrometer from which the mean temperature of the appearance and disappearance of dew was calculated.

Hypatia of Alexandria (370-415)

Greek mathematician and philosopher

Although all of her work has been lost or destroyed, history regards Hypatia of Alexandria as the only famous female scholar of ancient times. She was the first woman ever known to teach and analyze highly advanced mathematics.

Hypatia probably studied mathematics and astronomy under the tutelage of her father, Theon of Alexandria (fl. c. 4th century A.D.), the last recorded member of the city's great Museum. The Museum of Alexandria in Egypt was a prominent cultural and intellectual center which resembled a large modern university. It consisted of several schools, public auditoriums, and the famous library, once one of the most comprehensive repositories of books in antiquity. Although the Museum was in Egypt, its dominant culture and a considerable portion of its population were Greek. At one time, scholars came from across the Roman Empire and even from as far away as Ethiopia and India to hear lectures on the latest scientific and philosophical ideas and to study in the city's library.

Hypatia became a teacher at Alexandria's Neoplatonic School and was appointed its director in A.D. 400. Her lively lectures won her popular esteem, and she wrote a number of books on mathematics and other subjects, as well as criticisms of philosophical and mathematical concepts which her contemporaries regarded as perceptive. She corresponded with many distinguished scholars, some of whose letters to her survive and testify to their estimation of her abilities.

Although written records are sketchy, it appears that Hypatia invented or helped to invent mechanical devices such as the plane **astrolabe**, an instrument used by Greek astronomers to determine the position of the sun and stars. This device was probably developed with Synesius of Cyrene (c. A.D. 370-413), a scholar who had attended Hypatia's classes. A letter to Hypatia from Synesius, who later became a Christian and the bishop of Ptolomais, exists in which he asks her advice on the construction of the device. Synesius also worked with Hypatia on a graduated brass hydrometer, which measured the specific gravity of liquids, and a hydroscope, which was used to observe objects submerged in water.

At the age of 45 Hypatia was brutally murdered by a group of Parabolans, fanatical monks of the Church if St. Cyril. The reasons behind her violent death are in dispute, though her personal independence and pagan beliefs seem to have created hostility among Alexandria's Christian community. Another contributing factor appears to have been her alliance with Orestes, the pagan governor of the city, and a political adversary of Cyril (c. A.D. 375-444), the Alexandrian bishop. After Hypatia was killed, her works perished, along with many other records of ancient learning, when mobs burned the library, destroying the entire collection.

HYPODERMIC NEEDLE • See Syringe

I

Ibuprofen

When ibuprofen was placed on drugstore shelves in May, 1984, it was the first new over-the-counter (OTC) pain-relief medication to enter the marketplace in a generation. Prior to its introduction, nonprescription pain relief was mainly provided by acetaminophen (introduced in 1955) and aspirin (marketed since 1899).

Ibuprofen was developed by the Boots Company, a British drug manufacturer and retailer. Early in the 1960s, researchers at Boots identified carboxylic acid as the agent that gave aspirin its anti-inflammatory property. The Boots group investigated other carboxylic acids; when they found one that was twice as strong as aspirin, they synthesized and tested more than 600 compounds from these acids. The most effective and useful of these was ibuprofen, which Boots began to sell in 1964 in the United Kingdom as the prescription medication Brufen. (Ibuprofen became available OCT in the U.K. in 1983.) Ibuprofen appeared in American pharmacies in 1974 when Boots granted a nonexclusive license to the Upjohn Company, which marketed ibuprofen as the prescription arthritis-relief drug Motrin. This drug soon became one of the most commonly prescribed drugs in the United States. A few years later, Boots began selling its own prescription-form ibuprofen, called Rufen, in the United States.

When the United States Food and Drug Administration approved OTC sales of ibuprofen at a lower dosage than in prescription form, two major drug companies immediately geared up for a product-introduction blitz. First, the Whitehall Laboratories division of American Home Products came out with Advil. This was soon followed by Nuprin, which was produced by Upjohn and marketed by Bristol-Meyers. Both operated under licenses from Boots, which held the worldwide patent for ibuprofen until May 1985 and exclusive marketing rights until September 1986. After that date, new manufacturers jumped into the lucrative market with products of their own, including Johnson & Johnson's Medipren, Thompson's Ibuprin, and a number of generic and private-label brands. Upjohn and AHP/Whitehall countered with two new ibuprofen products, Haltran and Trendar, promoted as pain relief for menstrual cramps. Sterling Drugs then introduced its own ibuprofen-based menstrual cramps product, Midol 200.

Although ibuprofen, aspirin, and acetaminophen are chemically different, all three give effective relief for minor aches and pains. Ibuprofen seems to cause fewer stomach problems than aspirin, is more effective for many women in relieving menstrual discomfort, and appears more effective for postsurgical dental pain and soft-tissue injuries. Ibuprofen should not be taken by people with certain medical conditions, including people allergic to aspirin and women in the third trimester of pregnancy.

A 1997 study shows that the likelihood of developing Alzheimer's disease may be reduced by up to 60 percent with frequent consumption of ibuprofen over two or more years, and that use for an even shorter period may reduce risk by 30 percent, according to Johns Hopkins University and the National Institute on Aging. Researchers in that study caution, however, that consistent use of ibuprofen can cause kidney failure. In a conflicting article published in the *American Journal of Medical Science* in 1990, authors believe ibuprofen itself is not an independent predictor of risk of kidney damage; however, when taken by patients 65 years or older or who have coronary artery disease, the risk factor increases.

Ice cream

No one is sure just when ice cream was first made. Water ices were popular in ancient times, especially in the East. The Roman emperor Nero (37-68 A.D.) had slaves bring down mountain snow, which was then flavored with honey and fruit pulp. Marco Polo (1254-1324) brought recipes for water ices from the Far East to Italy in 1295, and Italian water-ice recipes were brought to France by the chefs of Catherine de Medici (1519-1589) in 1533.

At some point, cream was added to the ice and ''cream ices'' were born. A Sicilian, Francesco Procopio, opened a Parisian cafe in 1670 that served ice cream and sherbet. Ice cream was probably brought to America by early English colonists. George Washington (1732-1799) and Thomas Jefferson (1743-1826) were both fond of the dish, and Dolley Madison (1768-1849) popularized it at White House dinners.

Ice cream was a rare treat until the advent of commercial ice houses, which made ice widely available, and the hand-cranked home ice cream maker invented by Nancy Johnson in New Jersey in 1846. Jacob Fussell, a Baltimore, Maryland, milk dealer, established the first large commercial ice cream factory in 1851. Rapid technical developments in the early 1900s and the invention of the continuous freezer in 1925 modernized the ice cream industry.

It is generally agreed that the ice cream soda was originated by Robert Green in 1874 at the semicentennial of the Franklin Institute in Philadelphia, Pennsylvania. He was serving a popular drink made of cream, syrup, and carbonated water. When he ran out of cream one day, he used vanilla ice cream in its place. The resulting ice cream soda was an instant hit.

The ice cream sundae seems to have been born during the 1890s either in Wisconsin or Evanston, Illinois. Since Evanston laws prohibited the sale of sodas on Sunday, soda-fountain operators simply eliminated the soda and served just ice cream and syrup. On the other hand, a Wisconsin retailer named Smithson face the problem of no ice cream deliveries on Sunday by serving less ice cream but adding syrup. Another legend names George Giffy, a Manitowoc soda-fountain operator, who started serving ice cream with chocolate syrup—a dish invented by Two Rivers retailer Ed Berner—on Sundays only. In all versions, the spelling was changed to "sundae" to avoid blaspheming the English word for Sabbath.

A number of claims are made about the origin of the ice cream cone. An Italian immigrant to the United States, Italo Marchiony, apparently made and sold cones as early as 1896; his cone mold was patented in December 1903. The cone became popular after it was invented (or reinvented) at the 1904 St. Louis Exposition in Missouri. A Syrian immigrant, Ernest Hamwi, had a stand at the fair selling zalabia, crisp pastry baked on a waffle iron and served with sugar or syrup. When an ice cream vendor next door ran out of dishes, Hamwi rolled a warm waffle into a cone shape, let it cool and harden, and gave it to his neighbor.

Ice cream served in dishes was on its way out. The ice cream bar covered with chocolate was invented by Christian Nelson in Iowa in 1919 and introduced in 1921 as the Eskimo Pie. In 1920, following Nelson's lead, Harry Burt manufactured an ice cream bar called the Good Humor Ice Cream Sucker. It was Burt's daughter who had the idea to put the messy bar on a stick, like the lollipop. The Popsicle accidentally came into being in 1923 when California lemonade manufacturer Frank Epperson inadvertently left a glass of his product with a spoon in it on a freezing New Jersey windowsill overnight. He christened his frozen discovery the Epsicle—later renaming it the Popsicle—and patented it in 1924.

Today the United States leads the world in ice-cream production and consumption, churning out over 1.5 billion gallons of ice cream and related frozen desserts in 1996. The basic mix for commercial ice cream consists of cream and other dairy ingredients plus sweeteners. The mix may also contain stabilizers, which prevent the formation of ice crystals once the ice cream is frozen. These ingredients are blended in a mixing tank. Then the mix goes to a pasteurizer, where it is heated and held at a certain temperature for a specific period of time to kill harmful bacteria. Next the hot mix goes to a homogenizer, where the milkfat globules are put under pressure and broken into smaller particles to make the ice cream smooth. After homogenization, the hot mix is quickly cooled to about 40°F. Finally the mix is frozen. During the freezing process, blades in the freezer whip and aerate the product, which keeps it from forming a hard, inedible mass. Ingredients such as fruits and nuts are sometimes added after freezing; liquid flavors are added to the mix before freezing. Once the ice cream is made, it is packed into containers. Then it goes to a hardening room, where subzero temperatures further harden the product before shipping.

Soft-serve ice cream, dispensed directly from the freezer without hardening, was invented in 1939. In recent years, other variations on ice cream and related frozen desserts have become popular. These include low-fat and reduced-fat ice cream, once called ice milk. To approximate the texture of ice cream, such products may use **fat substitute**s such as *Simplesse*, polydextrose, maltodextrin, cellulose, and various gums. Frozen yogurt is also generally lower in fat than traditional ice cream. Another category of frozen dessert is made from tofu and thus is dairy-free. Yet another category is based on fruit. Sherbet usually contains dairy products as well as fruit juice, while sorbet is usually made from whipped or frozen fruit juice without the dairy products.

Ice-cutting machine

At a Boston, Massachusetts, party in 1805, young Frederick Tudor and his friends jokingly speculated about sending ice from nearby Fresh Pond to the tropics. Within a year, Tudor had secured financial backing and sent a $10,000 shipment of 130 tons of ice to Martinique in the Caribbean. Tudor followed his cargo, educating residents about its use. Although Tudor lost money on this first venture, he persevered and traveled widely throughout the Caribbean and the southern United States, promoting iced drinks. By the mid-1820s, 3,000 tons of ice were shipped out of Boston each year, most of it by Tudor, the "Ice King." But competition was growing, and Tudor sought to lower his costs through more efficient harvesting and storage.

Meanwhile, Nat Wyeth, who had a hotel with an ice house, became interested in the ice-shipping business. In 1825, he patented a machine that cut the surface of a frozen pond into deeply grooved, uniform parts. The result was ice blocks that were easy to remove, stack, and which resisted melting as well as shifting in the hold of a ship. Many patented and unpatented improvements in cutting, shaping, moving, and storing ice followed Wyeth's invention, but ice harvesters continued to use cutters like Wyeth's throughout the 1800s.

Tudor soon purchased rights to Wyeth's invention, which greatly reduced the price of harvesting a ton of ice. He also built ice-storage houses at his southern depots and improved insulation methods by using sheepskin, sawdust, and wood shavings. In 1833, Tudor sent 180 tons of ice to Calcutta, India. In the 1840s Boston ice was being shipped around the world—to Australia, China, and the Philippines. It was also shipped to places throughout the southern United States, where it contributed to the increasing popularity of the ice box. As ice became more commonly available, its use spread to the middle class and changed what people ate and drank.

Ice-resurfacing machine

The modern ice arena, professional **hockey**, and touring ice shows all owe their great success to an ungainly machine invented in 1949 by its namesake, Frank J. Zamboni. Born in Eureka, Utah, and raised in Idaho, Zamboni joined his brother George in Paramount, California, at the age of 21. The brothers built a **refrigeration** plant and began selling ice to local farmers and householders. When the home refrigerator began destroying the market for home-delivered ice, the Zambonis decided to build the Iceland Skating Rink across the street from the ice plant.

The rink business did well enough, but Frank Zamboni was bothered by the inefficiency of the nightly cleanup. Five men starting at 10 o'clock took as much as an hour and a half to scrape the old ice, clean off the scrapings and other debris, squeegee up the dirty water, and spread a fresh layer of water with a hose. Using a Jeep he had on hand, Zamboni began experimenting in 1942 with ways to mechanize the ice clean-up process. His fourth version, a huge and lumbering contraption completed in 1949, did the job—it scraped the ice, scooped up the debris, squeegeed the surface, and spread fresh water, all in the space of 15 minutes for the entire rink.

The Zamboni machine might have remained a local phenomenon if it had not been for Sonja Henie, who rented practice time at the Iceland Skating Rink for her touring troupe. As soon as she saw the ice machine in action, she ordered two Zambonis to take along on her national tour, publicity better than any paid sales force. Ice arena managers nationwide saw the machines and began ordering them, as did the Ice Capades. International exposure came in 1960 when Zambonis were used to clear the ice at the Squaw Valley Winter Olympics. Distributorships were soon set up in Switzerland and Japan, and a secondary plant opened in Ontario, Canada.

Today Zamboni machines are used in more than 30 countries. New machines are test-driven down the streets of Paramount to Iceland and make a few turns around the rink before shipping. The Zamboni company has no competitors in the United States, though a few exist in Canada and Europe. Zambonis are so ubiquitous in ice arenas, the term Zamboni has become almost generic for all ice-resurfacing machines.

Ice skates

Ice skating, in one form or another, has existed for thousands of years. Evidence suggests that as long ago as 1000 B.C. Scandinavians were fashioning crude blades from the shank or rib bones of elk, oxen, and reindeer and strapping them onto boots. A game played on ice between teams has been recorded as early as the second century A.D. In the Netherlands, both men and women skated on the **canals** during the Middle Ages. Scottish history recounts tales of armies crossing frozen marshes on skates to attack enemy territories. Ice skating became so popular in Scotland that the first skating club was established in Edinburgh in 1742. In 1848, E.W. Bushnell invented the first all-iron ice skate that could be clipped onto a boot.

During the 1800s, the popularity of ice skating skyrocketed. Skating clubs opened in London, Vienna, and New York. Rinks were built in Toronto, Canada, and in Davos, Switzerland. In 1876, the first artificially frozen ice rink, called the Glaciarium, opened in London. During the nineteenth century, the sport of speed skating was introduced and classical dance theory was applied to create the sport of figure skating.

There are three basic types of ice skates: hockey skates, figure skates, and speed skates. Speed skates are designed for optimum swiftness in one direction, with the skater moving right foot over left. The speed skate features a straight blade up to 18 inches (46 cm) long and 0.03-0.06 inch (0.08-0.15 cm) wide. The blade is reinforced with hollow **steel** tubing. The boot is constructed of very light, thin **leather**.

Hockey skates are constructed to allow the skater to move both right foot over left and left foot over right. The blade, usually 0.06 inch (0.15 cm) wide, is also reinforced with hollow tubing. The boot is short, measuring 4-5 inches (10-13 cm) from the sole, and reinforced with **plastic** caps and extra layers of leather at the toe. This protects the skaters' feet from the blades of other skates. The original hockey skate was made of leather with a plasticized sole, a safety tip at the rear, and a hard toe. A ballistic-proof **nylon** was then introduced that provided even greater protection against cutting. The newest innovation features a plastic molded boot with plastic stanchions and plastic tubing. A heavily padded, removable liner helps to control the fit.

Figure skates are fitted with a 0.125-inch (0.32 cm) steel blade designed for spinning. The blade is hollow on the bottom so that only the outer edges touch the ice. A series of sharp angles at the front of the blade called toe picks facilitate landing from toe jumps. The figure skate has a high boot, measuring 7-8 inches (18-20 cm) from the sole to the top, completely covering the ankle.

Ideogram • See Writing

Ignition system

Although **electronics** are almost universally used to ignite the gases of an **internal combustion engine** today, these engines were first developed using much different ignition methods. The earliest internal combustion engines burned a gas called illuminating or coal gas that was widely used in city street lights. The use of **gasoline** as a fuel came much later in internal combustion engine development.

These early internal combustion engines used a burning flame to ignite the coal gas fuel. A flame was kept burning in a compartment next to the cylinder where the coal gas was induced. At precisely the right moment, a sliding **valve** would open and the flame would ignite the gas within the cylinder. The flame would be blown out, and, after combustion, another valve would have to be slid open to expose still another flame to re-ignite the internal flame used to ignite the gases.

Another crude ignition method was to insert a tube into the cylinder, the outer end of which was closed and kept red-hot by an external flame. At the moment of ignition, a valve would open and the coal gas would enter this cylinder and be exposed to the red-hot tube, igniting the gas and driving the **piston** down its cylinder. This red-hot tube system was refined by **Gottlieb Daimler**, and his modification of the system aided him in developing a lightweight, high-speed internal combustion engine.

Electricity was first used in automotive ignition systems in the early 1800s, long before inventors looked to electricity to ignite the gases that powered their internal combustion engines. Based on his discovery of electro-magnetic induction, French inventor **Hippolyte Pixii** presented the world with its first practical magneto-electric **generator** in 1831. Pixii's hand-cranked machine led the world's electric revolution. French physicists Antoine Masson and Louis Bréguet created the first induction coil in 1841—the next crucial step in the development of automotive ignition systems. The Masson and Bréguet coil produced a high-tension electrical charge that could ''jump'' a gap.

This coil was perfected in 1851 by German physicist Heinrich Ruhmkorff. The Ruhmkorff coil was adapted to ignition systems by several automotive pioneers including Herr Giesenberg (Germany, 1880), F. Forest (France, 1883), and **Nikolaus August Otto** (Germany,1884). An ignition system that used the ignition coil coupled with a storage **battery** for electrical power was first used by French inventor **Jean-Joseph-Etienne Lenoir** in 1883. Lenoir is also credited with the invention of the internal combustion engine, and, in 1885, he invented a device quite similar to the modern **spark plug**.

German inventors **Robert Bosch** and Gottlieb Honold joined forces in the Bosch company to develop the high-tension, magneto-spark plug system that was used for decades as the standard for automotive ignition systems. In the early 20th century, American inventor **Charles Kettering** made improvements to automotive ignition systems and storage batteries; these included his invention of an electric starting motor, which eliminated the need to use a crank to start an **automobile** engine. These improvements led to the gradual decline in the use of magneto ignition systems for automobiles, although magneto systems are still used in **aircraft** with internal combustion, piston engines.

The diesel engine developed by **Rudolf Diesel** does not use an electrical ignition system. In the diesel system, air drawn into the cylinder is greatly compressed—up to a ratio of 25 to 1. This compression heats the air to a temperature of up to 1,000°F (538°C). Because this super-heated air ignites fuel as it is injected into the cylinder of the diesel engine, an electrical ignition system is not necessary.

Incandescent light bulb

The history of lighting shows that no matter how many new kinds of lights are invented, the old ones are never completely replaced. So it is not surprising that the ordinary incandescent light bulb remains the most popular choice for lighting our homes, despite the proliferation of new lighting technologies that has taken place during the twentieth century. Although many improvements have made incandescent lighting more efficient than ever, the bulb that was introduced in the late 1800s and refined in the early 1900s is basically the same type of light bulb that we buy in the supermarket or hardware store today.

The development of a practical incandescent **lamp** spanned many decades of research. In the early 1800s, scientists knew that strips of platinum or other metals would glow when electrically heated to the point of incandescence. When electricity is conducted through a metal, the electric current must overcome the metal's natural resistance. This generates heat, which, in turn, excites the molecules of metal so they produce incandescent light and glow.

From 1820 to the 1840s, scientists conducted early attempts to make an incandescent bulb utilizing coils of platinum enclosed in **glass**. Warren de la Rue (1815-1889), William Grove (1811-1896), and other researchers found that the life of platinum filaments could be prolonged by removing air from the lamp's glass enclosure, thus preventing oxidation of the filament. The first patent for an incandescent bulb was granted in 1841 to British scientist Frederick de Moleyns. Instead of a single platinum coil, his lamp used two platinum filaments connected by a bridge of powdered charcoal (carbon), which would also glow when heated but would not melt as easily as platinum. However, the glass sphere enclosing the filaments quickly became blackened by soot, and the lamp had a short life.

In the late 1840s another British inventor, **Joseph Swan**, combined the idea of using carbon as the filament material with the notion of removing air from the glass enclosure. In his research, Swan had noticed a patent issued to an American, J. W. Starr, which suggested that carbon filaments could work if enclosed in a strong vacuum. Swan succeeded in making filaments with strips of carbonized paper, but incandescent lamps remained impractical for two reasons: **vacuum pump**s available at the time were unable to create a vacuum strong enough to protect the filaments (whether carbon or platinum), and electricity at the time was provided by batteries, which were expensive to operate.

But by the late 1870s, developments in other scientific fields had overcome these two problems. A more effective vacuum pump was invented by **Hermann Johann Philipp Sprengel** in 1865, and power generation machines called dynamos (**generator**s) were constantly being improved to supply cheaper, more reliable electricity than batteries. When Swan learned of these developments, he resumed his research and invented a carbon-filament incandescent lamp, which he displayed in late 1878. His own house was the first electrically illuminated private home in England. In 1881 Swan formed a company to manufacture his "glow lamps," which quickly became popular throughout England.

Meanwhile, however, the American inventor **Thomas Edison** had begun to conduct research similar to Swan's, using a carbonized cotton thread for a filament. Edison quickly caught up with Swan, and both scientists are generally credited with the simultaneous invention of the incandescent light bulb.

However, Edison had a broader vision than just the bulb itself, which may account for his more famous reputation today. Edison realized that people would not buy electric lights unless they had convenient access to cheap, reliable electricity. In the early 1880s, he introduced the idea of large power plants for generating and distributing electricity. His first power station began operating in 1881, on Pearl Street in New York, New York. Edison also focused his research on the development of accessory equipment. For example, he invented multiple circuits for incandescent lighting systems, so that the failure of one bulb would not cause others to go out. At the same time, other inventors were also working to improve incandescent lights, dynamo power generators, and other electrical equipment devices. Sir **James Swinburne**, for instance, was granted more than 100 patents, most of which were related to electricity and lighting. Swan also continued his research, patenting a process for further reducing blackening of the bulb. At first some people were suspicious of the new incandescent light bulb. They wondered if it would harm their eyesight, especially when used in factories for long working hours. Many potential manufacturers declined to invest in the new technology. Despite such doubts, the new bulb was a success, and hundreds of electric lighting systems were installed within a few years. By 1885 the United States alone was using some quarter-million incandescent lamps.

A struggle to control this new market began to take place. After fighting a battle in court, Edison and Swan merged their companies to form General Electric (GE), but other competitors such as **Hiram Maxim** in America and Sir George Lane-Fox in England continued to carve their own niches in the lighting business. Although early incandescent light bulbs had a reasonable life span, manufacturers were still looking for ways to improve the filament. In the early 1880s, the African-American inventor **Lewis Howard Latimer**, who worked at that time with Maxim, patented methods for manufacturing superior carbon filaments that resulted in inexpensive, long-lasting light bulbs. Edison also tried other filament materials, such as carbonized bamboo. Around the turn of the nineteenth century, the Nernst lamp was introduced, named after its inventor, German physicist Walther Hermann Nernst (1864-1941). The lamp's filament was made of oxides of rare-earth metals, similar to the mantles used in **gaslights** of that day. Although Nernst profited from selling his patents, the lamp was soon made obsolete by the development of the tungsten filament, which is still used in modern incandescent light bulbs. Tungsten has the highest melting point of any metal, but it is also brittle, and experimental tungsten filaments were very fragile. In 1910 William Coolidge, working at a research laboratory owned by GE, perfected a revolutionary process for making tungsten ductile, or capable of being drawn into fine wire. The Coolidge process greatly improved the durability and efficiency of incandescent light bulbs.

Thomas Edison's incandescent light bulb.

One of Coolidge's colleagues at GE, **Irving Langmuir**, made the next great advance in incandescent lighting technology. In 1913 he discovered that filling the bulb with a nonreactive gas would preserve the filament better than a vacuum would. Although nitrogen was used as the first filler gas, it was later mixed with argon, which is even more inert (less reactive). Most of today's light bulbs still use similar mixtures. The tungsten filament and the inert gas filler represent the twentieth century's major improvements to the light bulb, making it three times more efficient than the original carbon-filament version. But the design of the filament was also gradually improved. In 1918 Langmuir found that twisting the filament into a fine coil would increase its temperature and, thus, give off more light. Then in 1934 the design that is presently used was established, in which the coiled filament itself is twisted into a fine coil.

Even with all these improvements, most incandescent light bulbs today remain relatively inefficient, converting only six to eight percent of the electrical energy to light. The rest of the energy input is released as heat, which is usually not needed. Also, the average life of a standard light bulb is only about 1000 hours, and gradual blackening of the bulb reduces the light output. Since the energy crisis of the 1970s manufacturers have directed more research toward improving the efficiency of incandescent light bulbs and extending their useful life. For example, a new type of incandescent light—the quartz lamp—lasts longer than ordinary light bulbs (about 4000 hours) and maintains its light output through its entire life, rather than gradually becoming dimmer. The quartz lamp contains a small amount of halogen. The halogen eliminates bulb blackening by combining with the tungsten that is evaporated from the filament and preventing it from accumulating on the inside of the bulb. Eventually, the tungsten is redeposited on the filament, and the halogen is freed to combine with more tungsten. Most quartz lamps are small in size, compared to conventional bulbs, and tubular in shape. With today's automated manufacturing processes, incandescent light bulbs are produced at a rate of up to 5000 per hour. Their power consumption is measured in watts and their light output in "lumens." Three-way bulbs work by incorporating multiple filaments, each producing a different amount of light. Most bulbs burn out because the filament has evaporated to a point at which it breaks, which is why you usually hear a rattle when you shake a burned-out bulb.

See also Halogen lamp

Inclined plane

The inclined plane is one of the simple machines of antiquity, the physical operations of which were first theorized by **Archimedes** in the second century B.C., and then further explained by **Hero of Alexandria** in the first century A.D..

Inclined planes are often confused with wedges; while loads are moved along stationary inclined planes, wedges themselves move stationary loads. (Examples of early wedges include prehistoric **axe**s, spears, and arrowheads.) Around 2600 B.C., inclined planes in the form of ramps were used, at least in part, to raise the blocks of stone that make up the Great Pyramid. Between 1900 B.C. and 1400 B.C., inclined planes might also have been used to elevate and place large stone crosspieces at Stonehenge.

Before the advent of inclined planes, **lever**s, **pulley**s, **crane**s, **gear**s, and belts, heavy objects had to be hoisted, moved, and positioned by brute force. According to the *law of equilibrium* (which traces back to Archimedes' "law of levers"), as it applies to inclined planes, the total force required to move an object up the length of an inclined place is force = (weight × height) ÷ distance. In other words, since for an inclined plane the height is constant, the force will vary inversely with the distance (or length) of the inclined plane. Increasing the length of an inclined plane (which decreases the angle of lift) will always make a task seem easier, though the total *work* done, which equals (weight × height) is always the same no matter what the angle of the inclined plane.

Incubator

The ancient Egyptians and Chinese both devised incubators to hatch chicks from eggs without the mother hen sitting on the eggs. This enabled hens to continue laying eggs. The Egyptian incubators were large rooms heated by fires; attendants turned the eggs at regular intervals so they would warm evenly. Some Chinese incubators were warmed by fire, others by rotting manure. The Italian inventor Jean Baptiste Porta drew on the ancient Egyptian designs to build his 1588 egg incubator, but was forced to abandon his work by the Inquisition. Dutch inventor **Cornelius Drebbel** also invented an incubator to hatch eggs.

Knowledge about egg incubation was revived and introduced throughout Europe by the inventive Frenchman René-Antoine Ferchault de Réaumur (1683-1757) around 1750. Réaumur's device was warmed by a wood **stove**; temperature was controlled by a **thermometer**, also invented by Réaumur, which gave rise to the temperature scale named after the inventor. The success of Réaumur's incubator—Louis XV (1710-1774) enjoyed helping the chicks hatch—helped boost commercial production of foodstuffs at the beginning of the industrial era.

After Réaumur's death, the incubator was further developed by Abbé Jean-Antoine Nollet (1700-1770) and later by Abbé Copineau, who used alcohol **lamp**s as a source of heat. Today's incubators are electrically heated and turn the eggs automatically; large ones may hold up to 75,000 eggs.

A neonatal incubator is a device consisting of a rigid boxlike enclosure in which an infant may be kept in a controlled environment for medical care. Incubators to house prematurely-born babies unable to maintain their own body temperature were developed in the 1880s by placing pans of hot water beneath enclosures. A number of "incubators with living children" were demonstrated in Turin, Italy, at the 1898 Italian Exhibition. Today's neotatal incubators are made of **plastic** and Plexiglas. The device may include an AC-powered heater, a fan to circulate the warmed air, a container for water to add humidity, a control valve through which oxygen may be added, and access ports for nursing care. Incubators contribute greatly to ill and premature infant survival rate, even very low birth-weight infants.

Inertial guidance system

An inertial guidance system (also known as an inertial navigation system) uses **gyroscope**s, **accelerometer**s, and computers to calculate the position, velocity, and attitude of a vehicle in motion. An inertial guidance system electronically samples the current placement and acceleration of a vehicle on a continuous basis to determine its future course. Gyroscopes, often three, furnish information on fixed reference directions. Accelerometers monitors velocity changes. Computers process the data from the gyroscopes and accelerometers, ensuring the vehicle is on course. The system is self-contained and supplies navigational information without communicating with a satellite or base station. Inertial guidance systems have been used on airplanes, **submarines**, and **missiles** since the 1970s in the United States.

There are several kinds of inertial guidance systems, the gimbaled inertial system and the strap-down inertial system, with the former the more common. The gimbaled inertial system has up to four gimbals (at least one a stable platform) with three gyroscopes and accelerometers attached. When vehicle with the attached inertial guidance system is in motion, the gyroscopes drive the gimbals. The strap-down inertial system has three gyroscopes and three accelerometers mounted directly to the vehicle, and their outputs are read every one hundredth of a second. Because of the amount of data produced, computations were an encumbrance in early models. As computers have increased in speed and capabilities, strap-down systems have improved. New gyroscopes have been developed that are more precise and less prone to error, including the **laser** gyroscope and the electrostatic gyroscope. In the future, strap-down systems might replace gimbaled.

Gyroscopes and accelerometers much be precise, even small errors can put a vehicle completely off course. Gyroscopes have several potential problems. Their **bearings** are subject to friction and drift putting the vehicle off course. Errors can also come from incorrect initial route information as it is entered in the computer. There are several solutions. In airplanes flying over land, pilots can receive signals from VOR/DME transmitting stations, helping them keep on course, though the signal has strength limitations and can be subject to interference. A newer solution is using information from the **global positioning system** (GPS), a series of 24 satellites that orbit Earth. Fully functional in the 1990s, GPS allows precision information on location when a vehicle is equipped with a special receiver. Indeed, as early as 1993, combination of an inertial guidance system and a GPS receiver became available on the market. Inertial guidance systems have one advantage over GPS: because they are self-contained, they are not vulnerable to any terrorist activity or other interferences. The combination of the two technologies would seem to be the ideal guidance system for airplanes and similar vehicles.

INFRARED ASTRONOMICAL SATELLITE • See Space telescope

INFRARED PHOTOGRAPHY

Infrared **photography**, the process of recording images on film by capturing the near-infrared radiation emitted from objects, was invented in 1880 by Sir William Abney (1843- 1920). Abney is said to have produced a photograph of a boiling tea kettle by using a special **collodion** emulsion to capture the near-infrared radiation emitted by the kettle. For years many others were unable to produce the same results. It wasn't until 1903 that the first real infrared sensitizer was discovered, an emulsion known as *Dicyanine*, a chemical containing special dyes that responded to near-infrared radiation, thus producing an image of objects that emitted such radiation.

The problem with the sensitizer and the ones that followed was that the exposure time necessary to produce the image was far too long, making the use of such photography impractical. However, in 1942 Kodak solved this problem by developing an infrared photographic film sensitive enough to make quick exposures. Available in both black-and-white and color, the new film ushered in the age of infrared photography. Since 1942, infrared photography has found a variety of applications. It has been used for observing camouflaged troops and for detecting tumors and other medical ailments. Because infrared radiation is not scattered by atmospheric haze, as is light, infrared photography is used to find distant objects, including those in space. This has not only enabled astronomers to discover stars too dim to be detected visually, but also to determine the temperature and movement of certain planets and stars. Infrared photos are also used in meteorology, farming, criminology, and industry.

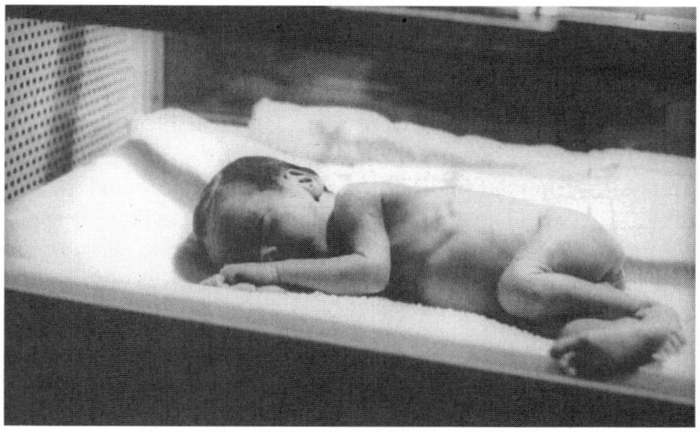

Incubators have significantly improved the survival rate of premature infants.

See also Photography

INJECTION, SUBCUTANEOUS • See Syringe

INK

Ink is a combination of a coloring agent called pigment (or **dye**) with a liquid called a *vehicle* which also contains oils, resins, and solvents. Throughout history, ink has been made from many different colored juices and extracts of plants and animals as well as synthetic materials. The use of ink for writing and printing dates back to 3200 B.C., when the Egyptians used a mixture of lampblack (a fine soot produced by the incomplete combustion of oils and other carbons) suspended in vegetable gum. Both the Egyptians and the Greeks used iron oxide to make red ink from about 2800 B.C., although this was employed in marking **linen** rather than for writing. The ancient Chinese were making both red ink (from mercury sulfate) and black ink (from iron sulfur mixed with sumac tree sap) from 2000 B.C.; like the Egyptians, they made their ink into a solid block or stick that would be mixed with water when used. The ancient Romans developed a purple ink called encaustum, from which the word *ink* is derived, and also made a dark brown ink called sepia from dried, powdered cuttlefish ink-sacs.

Iron-gall inks were first noted in Italy in the seventh century A.D. and were primarily used during the Renaissance (fourteenth through sixteenth centuries) but were prone to fading. It was not until the seventeenth century that Europeans made ink with a mixture of tannic acid from tree bark and iron salt—a recipe that formed the basis for the blue and black inks still used today. About 1450, when the printing press was first developed, printing inks were made of varnish or linseed oil with lampblack added. The first patent for colored ink was granted in England in 1772. *Invisible inks* were developed in the mid-1800s. These inks were usually milk, whey, or sugar solutions, which, when exposed to heat or light, turn brown and visible. In the nineteenth century, chemical drying agents were developed that allowed easier use of a wider range of pigments for colors.

Technological developments of the twentieth century have made the manufacture of ink increasingly complex. Aniline dyes, developed in 1926, extend the range of color printing allowing application to **plastics**, fabric, and other substances; such dyes are frequently used in indelible (irremovable) ink. Synthetic dyes are now used for printing inks in place of the natural substances formerly relied upon. During the oil crisis of the 1970s, newspaper publishers began researching possible replacements for the petroleum distillates that formed the basis of the ink they used. In 1987, General Printing Ink began marketing a **soybean**-based ink. Use of this kind of ink has increased during the 1990s, particularly because it does not produce as much hazardous waste as petroleum-based ink when it is discarded. In the late 1990s, the biodegradability of soy inks increased as the ratio of soy oil to pigment improved. (Pigments are still petroleum-based, which is the primary source of hazardous rate.) The American Soy Association claims that more than one-third of all newspapers now use soy-based ink. Other advantages of soy ink include its capacity to produce brighter colors, its extended mileage (i.e., more can be printed with less ink), its renewability, and the fact that recycling paper printed with soy ink produces a higher-than-usual quality paper. Despite its higher cost, soy ink seems to be an increasingly popular choice for environmentally concerned publishers.

In the late 1997, researchers at the Massachusetts Institute of Technology announced a new ink-related technology that could change the face of printing. Scientists lead by Joseph M. Jacobson have developed a dynamic particle ink technology. Instead of an endless supply of ink being printed on an endless supply of paper, a single sheet of reusable paper would have two layers special electronic ink inside. One would be transparent and receive page data sent via **radio** waves; the other, comprised of microencapsulated particles, would used on the imaging of the actual page. Though this technology is only in its infancy, it represents one possible future of ink.

See also Polymer and polymerization; Textiles

INSECTICIDES

Insecticides are chemical substances used to destroy or control insect populations that could otherwise do billions of dollars in damage annually. Although the primary use of insecticides is in agriculture and animal husbandry, insecticides also help to control the spread of insect-borne diseases and to protect buildings, households, and products such as clothing and books. Historically, many kinds of insecticides have been developed, as no single poison is effective against all insect pests—indeed, most insecticides do not even remain effective against a single pest indefinitely.

Insecticides can be broadly classified, based on their chemical structure, as inorganic or organic. The main ingredient of inorganic insecticides is of mineral origin, such as arsenic. Organic insecticides, whether derived from plants or made synthetically, consist of carbon in combination with one or more elements such as hydrogen, oxygen, sulfur, nitrogen, phosphorus, chlorine, or bromine. The toxicity of inorganic insecticides depends on the presence of certain key elements; the toxicity of organic insecticides is determined by their molecular structure.

Insecticide poisons can also be classified in terms of their action. Some poisons penetrate an insect's exterior membranes, whether applied directly (contact insecticides) or to surfaces that the insects will touch (residual insecticides). Stomach poisons rely on ingestion, while fumigants enter the insect through its respiratory system. Modern insecticides often combine these methods, and are applied in a variety of formulations, including dusts and sprays.

Insect control dates back two thousand years to the ancient Chinese, who used pyrethrum, extracted from the chrysanthemum flower, to kill fleas and lice. The Chinese also used arsenic sulfide to deter various plant-eating insects. In classical times, the Greeks and Romans applied sulfur and other insect-deterring substances to their crops. The Europeans discovered arsenic in the Middle Ages, and by 1681 arsenicals were being used as insecticides. In 1690, French agronomist Jean de la Quintinie discovered that tobacco was also highly effective as an insecticide. The French subsequently used tobacco against plant lice, although its active ingredient was only later identified as nicotine.

By the nineteenth century, the need for insecticides had intensified, due to the increasingly large-scale cultivation of single crops and to the exportation of insect species, on cargo ships and passenger steamers, far beyond their native habitats. Among the most important insecticides developed in the nineteenth century was Paris green (copper acetoarsenite), discovered around 1868. Paris green soon became the best-known arsenic insecticide and the first chemical insecticide to be applied on a large scale. In addition to helping combat a Colorado potato beetle infestation of the American midwest in 1872, Paris green led to the development of London purple and lead arsenate, which proved effective against the potentially devastating gypsy moth. Sulfur was also used to fight red spiders and plant lice. Though widespread application of natural organic and inorganic insecticides won some battles in the nineteenth century, insects were still winning the war. Thus, the American journalist Horace Greely reported in 1879 that the average annual loss to farmers from insect damage exceeded $100 million. Certain insects, moreover, such as the boll weevil that destroyed cotton crops in Texas, were immune to the natural insecticides available at that time. Only after years of testing was the spread of the boll weevil checked by a new pesticide, calcium arsenate.

Just prior to World War II, when organic insecticides were first created synthetically in the laboratory, scientists thought they had finally discovered the ultimate tool to prevent crop destruction. The first synthetic insecticide to be developed was of the chlorinated hydrocarbon type. In 1939, the insecticidal properties of dichloro-diphenyl-trichloroethane, known as **DDT**, were discovered by the Swiss chemist Paul Hermann Müller (1899-1965), who was awarded the Nobel Prize in 1948 for this accomplishment. This colorless and odorless pesticide was found extremely effective against both insect pests and crop-threatening insects. Furthermore, DDT was used successfully during World War II to protect American troops in the Pacific from malaria-carrying mosquitoes, and then to all but eradicate malaria domestically within the United States. DDT remains the most familiar of chlorinated hydrocarbon insecticides not only because of such benefits, however, but because of the subsequent realization of its capacity for large-scale ecological damage. DDT's extreme durability results in its transference into living organisms that are not intended targets, such as birds and fish. The ingestion of DDT, for example, has come close to exterminating the American falcon. Humans themselves are at risk when high concentrations of DDT accumulate within the body. DDT ultimately pollutes the entire food chain. In 1972, therefore, less than thirty years after its discovery, the United States Environmental Protection Agency banned the agricultural use of DDT. Other chlorinated hydrocarbon insecticides, such as chlordane, are similarly proscribed both domestically and abroad.

The destructive capabilities of the second major group of synthetic insecticides, organophosphoric compounds, became apparent even sooner in their history. Discovered by the German chemist Gerhardt Schrader in the late 1930s, these compounds affect living organisms by targeting the nervous system. Whether the victim is an insect or a warm-blooded animal, they cause violent muscular convulsions before an almost immediate death. The German government, recognizing the potential of these phosphorous compounds for military application, developed them for use during World War II, but they were not deployed. The development of organophosphoric compounds as insecticides was spurred after the war by Schrader's subsequent discovery that they could penetrate and spread through plant tissues without losing their insecticidal effect. Thousands of organic phosphorous insecticides, including parathion and malathion, were then synthesized for agricultural purposes before public opposition and government restriction gradually abated their use.

The increasing rejection of highly toxic insecticides, whether inorganic or organic, has led to a search for safer and more humane solutions. Insect repellents or attractants can thus be used (the latter lead insects away from crops and towards other food sources). Several sophisticated methods of impeding insect growth or reproduction have also been developed, and insect populations are being controlled through natural predators, through the introduction of sterile males to attract females, or through formulations that can be ingested by humans and animals but that damage insects at the larval stage. Human attempts at intervention in ecological systems have led to many disasters, however, regardless of how apparently humane or ''natural'' the intervention. Finding new ways to achieve food production and disease control without unleashing the potential ecological destructiveness of insecticides remains an important public, commercial, and governmental priority.

Instant camera (Polaroid Land camera)

The Polaroid Land camera or, as it is now known, the instant camera, is the most commonly recognized invention of **Edwin H. Land**. A chemist and physicist, he revolutionized the amateur and professional **photography** industries by inventing and producing a camera that incorporates its own processing lab, producing finished prints in less than a minute.

The inspiration for his invention came in 1943 during a family vacation, when Land snapped a photograph of his daughter. The three-year-old was impatient to see the results, and even as Land was explaining that she would have to wait, he was formulating the first plans for his instant camera.

Normally, once film is exposed, it must then be developed, rinsed, fixed, washed, dried, and printed. The instant camera eliminated several of these steps, delivering a finished print. The key to Land's instant camera was actually in the film, which featured a pod of chemicals located at the edge. When the film was squeezed between the camera's two rollers, the processing fluid in the pod spread over the picture between the exposed negative and positive sheets. In conventional photography, the unexposed portions of the image are washed away; in Land's invention, images were transferred to the print paper in contact with the film negative.

Although Land's initial instant camera was a notable success, he kept improving and producing an even better product. The first photographs rendered by the new camera in 1947 were sepia-toned—a warm, brown color—due to the chemicals involved in the process, but within ten months of the camera's debut in stores, black and white film was also available. By 1963 the Polaroid Corporation made color film available for its instant camera. All three types of film used essentially the same kind of diffusion transfer process described above.

By the 1950s Polaroid film and cameras had become extremely popular. A multitude of amateur photographers were attracted to the Polaroid's immediate results and the opportunity to ''try again'' if the results weren't satisfactory. Professional photographers also began to use Polaroid *backs* on studio cameras in order to perform test shots, evaluating lighting, composition, and decor instantly before taking a conventional photograph.

The 1972 introduction of the new SX-70 color film meant that the entire development process could be completed within the camera, eliminating the need to pull the sheet from the camera and peel the negative from the positive print. Processing time was reduced to ten seconds as a result of the introduction of ultra-fast film. However, until the 1998, black and white film was unavailable for the non-peel Polaroid camera. Two years earlier, Polaroid introduced an instant camera that shot 20 x 24 inch (about 7.9 x 9.5 cm) color pictures, though it was not generally available on the commercial market.

Instant coffee

By one account, coffee was discovered in the ninth century A.D. by an Abyssinian goatherd named Kaldi, who noticed that the berries his sheep were eating made them unusually energetic. Kaldi ate one of the berries and felt an exhilarating jolt. He soon began sharing his discovery with fellow goatherds. Drowsy monks also found that the stimulating berries helped them stay alert during prayer. Years passed before Arabs began roasting and brewing the beans, but eventually coffee evolved as a hot drink and became enormously popular throughout Arabia and Turkey. By the seventeenth century, coffee was popular throughout Europe and was even considered a cure for such ailments as gout, scurvy, headaches, constipation, and the common cold.

The first instant coffee was available in the form of a liquid extract in 1838, when the United States Congress used it as a substitute for rum in the rations of American soldiers and sailors. This liquid extract never proved practical and did not become popular among the public.

Sartori Kato, a Japanese chemist, created the first powdered instant coffee and sold it to a receptive crowd at the 1901 Pan American Exhibition in Buffalo, New York. Five years later, an American chemist, G. Washington, created a more refined water-soluble powdered coffee. This powder was better tasting and was marketed on a broad scale. Coffee connoisseurs considered instant coffee inferior to their brewed coffee and doubted that it would ever be accepted among the public. However, the United States War Department gave instant coffee sales a dramatic boost during World War II, when it purchased the entire domestic output of instant coffee.

Until the 1950s, two types of instant coffee were available: the "product type," which included carbohydrates in its mixture, and the "all coffee" blend, which has remained popular to this day. Like regular ground coffee, these soluble powders are produced by first blending, roasting, and grinding coffee beans. The bubble-like particles of instant coffee are the result of condensing moisture out of the beans and then spray-drying them. Freeze-dried instant coffee, produced by freezing the extract and placing it in a vacuum chamber where moisture is sublimed, was developed by the Nestle company in the 1960s. The solid mass left behind is processed into readily soluble granules. Freeze-dried coffee has become the most popular form of instant coffee because its excellent taste closely approximates regular ground coffee.

Today Brazil and Colombia are the world's largest coffee producers, with coffee plantations also found in other South and Central American countries, Cuba, Hawaii, Indonesia, Jamaica, and many African countries. Instant coffee is sold in a variety of forms, including decaffeinated versions and a wide range of flavors. The product is sometimes packaged in single-serving bags, much like tea bags. According to a 1997 survey by the National Coffee Association, instant coffee now accounts for about 13 percent of all the coffee consumed.

Integrated circuit

The invention of the **transistor** in 1948 eliminated the need for bulky **vacuum tubes** in computers and other electronic devices. As other components were also reduced in size, engineers were able to design miniature, and increasingly complex, electronic circuits. However, the transistors and other parts of the circuit were made separately and then had to be wired together, a tricky task that became more difficult as circuit components became tinier and more numerous. Circuit failures often occurred when the wire connections broke.

The idea of manufacturing an electronic circuit with multiple transistors as a single, solid unit arose as a way to solve this problem. The integrated circuit (IC) concept was first suggested publicly by a British radar engineer, G. W. A. Dummer, in 1952. He envisaged implanting electronic equipment in a solid layered block of semiconducting material, with connections made by cutting out areas of the layers instead of by wires. Dummer's idea got nowhere in England, but in the United States the Defense Department was distributing millions of dollars to achieve reliable miniaturization of the electronic components it wanted for new weapons systems based on transistors rather than tubes.

Success was achieved in the late 1950s by two inventors at United States electronics firms. **Jack Kilby** of Texas Instruments (TI) in Dallas, Texas, began wrestling with the circuit problem shortly after he arrived at TI in 1958, and came up with an idea similar to Dummer's. By September 1958 he had succeeded in making the first working integrated circuit consisting of tiny transistors, **resistor**s, and **capacitor**s connected by gold wires on a single chip. Kilby's 1959 patent application added a crucial element: the connections could also be made directly on the insulating layer of the semiconductor chip, without the wires.

Meanwhile, **Robert N. Noyce** of Fairchild Semiconductor in Mountain View, California, was also pursuing a solution to the miniaturization problem. Bouncing ideas off fellow Fairchild scientist Gordon Moore, Noyce—completely independently of Kilby—also thought of housing an electronic circuit and its connections on a single piece of silicon. Noyce's integrated circuit used a *planar* technique of laying down alternating layers of **semiconductor** and insulating material, with photoetching to establish the circuit. Noyce also applied for his patent in 1959.

While the ensuing patent battle wasn't settled until 1969 (in Noyce's favor), Fairchild and TI had agreed by the mid-1960s to license each other, and other companies, to manufacture the chips. Noyce and Kilby became recognized as coinventors of the integrated circuit, which completely revolutionized the electronics industry. The individual transistor, like the vacuum tube before it, was hardly ever seen—the IC was much smaller, more reliable, cheaper, and far more powerful. While at first an integrated circuit could only contain a few transistors and other components, today they can carry millions. The integrated chip has made possible the development of the **microprocessor**, and hence the personal computer, along with an array of devices such as the **pocket calculator**, **microwave oven**, and computer-guided **aircraft**.

Interferometer

An interferometer is a device constructed to split a beam of light into two perpendicular beams and then to bring the two beams together again. Any difference between the beams creates a pattern of interference that manifests itself as a group of bright spectral lines called fringes. Originally designed to measure the fluctuations in the velocity of light, interferometers are now widely used in spectroscopy, chemistry, and precision metrology and inspection.

In 1881, Albert Michelson (1852-1931) was conducting experiments to determine the speed of light. Trained in classical physics, Michelson was a firm believer in the existence of an *aether-wind*; at that time, light was imagined as an undulating wave, like ripples on a body of water. Just as those ripples need water to move through, light also required a medium for travel, which scientists at the time called the aether.

The intent of Michelson's experiment was to detect the drifting aether by sending the interferometer's perpendicular beams into it—one beam across the current, one beam against it. Michelson expected that the beam moving against the current would be slowed down by the aether wind, and his interferometer would indicate the difference in velocities by displaying an interference pattern. Much to his surprise and dismay, the results of his research were null: no evidence of an aether drift could be found.

Though he repeated his experiment with Edward Morley (1838-1923) in 1878 with the same outcome, Michelson never produced conclusive proof of an aether—in fact, the Michelson-Morley Experiment served as an epitaph for the old ether theory and paved the way for the new optical theories that ultimately led to Albert Einstein's (1879-1955) special theory of relativity in 1905.

Since its now-infamous debut, the interferometer has undergone a number of modifications and has been specialized for use in a variety of fields. In 1893, Michelson used it to measure the International Prototype Meter in Paris in units of wavelength (at that time, this metal bar was the standard for measuring length). Michelson's measurement set a new international standard—the use of the unvarying wavelength to measure length. To this day, the interferometer's primary virtue is its incredibly precise measurement of wavelength. Michelson next used his invention to calculate the velocity of light, arriving just before his death at a figure within 11 miles (17.7 km) per second of today's most accurate calculations.

The Michelson interferometer is commonly used in astronomy to determine the size and separation of distant objects such as stars. Michelson himself pioneered stellar interferometry in 1920, when he used a large interferometer mounted on a California mountainside to measure the diameter of the star Betelgeuse in Orion.

Interferometers can be found in several different incarnations, designed to perform a vast array of tasks. The *Twyman-Green interferometer* uses a point-source of light, typically a **laser** beam rather than white light, and is often used to measure surface smoothness. The most commonly used interferometer is a modified Twyman-Green model.

In 1916, the French physicist Charles Fabry (1867-1945) used an interferometer of his own construction to measure the quantity of ozone in the Earth's upper atmosphere. This interferometer differed from the others in that it used more than two beams. Fabry's findings provided a basis of research that helped establish meteorology as a legitimate science.

The *Mach-Zehnder interferometer* is an example of a very specialized version of the apparatus. By adding two cells through which the perpendicular beams are directed, this interferometer becomes ideal for observing airflow around models of **missiles**, **aircraft**, and other projectiles.

Metrology is an important application of interferometry. Optical component manufacturers use interferometers to precisely measure the shape and surface smoothness of **lenses** and **mirrors**. With the devices' unsurpassed precision, deviations of even one wavelength can be detected. Such precision is essential in the construction of sensitive optical equipment, such as **telescopes** and **microscopes**. Interferometry has applications in other fields such as semiconductor manufacturing, for example, where deviations of a fraction of a micron can significantly reduce device yield. Interferometers are used to inspect silicon wafers for smoothness and flaws prior to circuit fabrication.

One of the newest aids to the science of interferometry is the **hologram**. Designed to create a permanent three-dimensional recording of an interference pattern, a hologram is almost like a frozen interferometer. Scientists reconstruct the interference fringe stored in a hologram and cause it to react with a new comparison wave. The fact that the holographic pattern can be stored and reused makes this method more advantageous than common optical interferometry.

The interferometer was Michelson's most significant single contribution to the world of science. In 1907 he was awarded the Nobel Prize in physics for his dedication to the design of precise scientific equipment. He was the first American to receive the prize in one of the sciences.

Interferon

While the middle years of the twentieth century saw the development of antibiotics, potent new weapons against bacterial diseases, no such chemical defenses against viral diseases had yet emerged other than a few anti-viral vaccines. This was and continues to be a significant gap, since more than half of the communicable diseases that affect human beings are caused by ribonucleic acid (RNA) viruses. The first step was to find out how the body protected itself against viruses; it was known that antibodies acted only against bacteria. That step was taken by Alick Isaacs, a Scottish virologist, in 1957.

Isaacs was born in Glasgow, Scotland, in 1921, to a Russian Jewish family. He studied medicine at Glasgow University but found he preferred research to the actual practice of medicine. Accordingly, he pursued graduate studies in bacteriology at Glasgow and secured fellowships to research influenza with eminent microbiologists Stuart Harris of England and Frank MacFarlane Burnet in Australia. Returning to England,

Isaacs joined the World Influenza Centre at the National Institute for Medical Research in 1951. There, he carried out his continuing studies of viral influenza until his untimely death in 1967.

Early in his studies of influenza, Isaacs became interested in the viral interference phenomenon—the fact that an RNA virus in a cell inhibits the growth of any other viruses in that cell. He found that this interference seemed to be caused by something inside the cell. In 1957, while working with the visiting Swiss scientist Jean Lindenmann, Isaacs found that chick embryos injected with influenza virus released very small amounts of a protein that destroyed the virus and also inhibited the growth of any other viruses in the embryos. Isaacs and Lindenmann named the interfering protein *interferon.*

Further research showed that interferon was produced within hours of a viral invasion (antibodies take several days to form), and that most living things, including plants, can make the protective protein. Interferon was seen as the cell's first line of defense against viral infections, and its discovery was expected to pave the way for successful treatment of viral diseases.

Because interferon was thought to be species-specific (meaning only human interferon will work in human beings) and the body produces it in only minute amounts, interferon research inched forward at a snail's pace.

Interest in interferon was revived in the late 1960s when Ion Gresser (1928-), an American researcher in Paris, discovered that the protein stopped or slowed the growth of tumors in mice and also stimulated the production of tumor-killing lymphocytes. Gresser and the Finnish virologist Kari Cantell then developed a way to make interferon in useful amounts from human blood cells. Monoclonal antibodies, first produced in 1975, made large-scale purification of interferon possible, and the mid-1980s saw the advent of genetically engineered interferon, the first example of which was produced from bacteria by Swiss scientist Charles Weissmann in 1980. Scientists now know that there are three major types of interferon: alpha, beta, and gamma. They have also learned that interferons are not species specific but can have activity in other species.

Research of interferon's ability to kill cancer cells is active. It has been used successfully against leukemia, osteogenic sarcoma (a bone cancer), and as a therapy for delaying disease recurrence and prolonging survival in patients with melanoma (skin cancer). Research also continues on the use of interferon to treat viral diseases like rabies, hepatitis, and herpes infections. In December 1997, researchers from the Duke University Medical Center also announced study results that indicate interferon may be a way to preserve donor livers longer prior to transplantation.

Internal combustion engine

The invention of the internal combustion engine is most probably the result of the developments of several individuals. Around 1780, Dutch scientist **Christiaan Huygens** built an engine that used **gunpowder** as a fuel, but this engine was far too dangerous to be practical. His assistant, **Denis Papin**, also experimented with developing an internal combustion engine, building a simple steam-powered device around 1790. Again, this engine was not practical, and not until the early nineteenth century did the development of a practical internal combustion engine become the quest of numerous inventors.

Before then, the **steam engine** was the power plant of choice. By 1770, the steam engine had been developed to the point that the French engineer **Nicolas Cugnot** used one to successfully propel a three-wheeled vehicle, and steam power reigned supreme in industry for nearly a century.

Early in the industrial revolution, inventors struggled to develop an engine in which compressed fuel could be burned within a cylinder housing a **piston**, thus capturing a much greater amount of the potential energy of the fuel. French physicist Nicholas Carnot (1796-1832) published a book in 1824 in which he set out the principles of an internal combustion engine which would use a flammable mixture of gas vapor and air. Basing his work on Carnot's principles, another Frenchman, **Jean-Joseph-Éttien Lenoir**, presented the world with its first workable internal combustion engine in 1859. Lenoir's motor was a two-cycle, one-cylinder engine with slide **valves** and used illuminating gas (coal gas) as a fuel; it also used an electrical charge, supplied by a **battery**, to ignite the gas after it was drawn into the cylinder. Lenoir sold several hundred of his engines, and he adapted his engine to power a carriage; consequently, he is credited with inventing the first gas-powered **automobile**. Lenoir's primitive two-stroke design, however, was inefficient because each back-and-forth motion of the piston must draw in the fuel, burn it, and expel the burnt gases.

Another Frenchman, Alphonse Beau de Rochas, recognized the shortcomings of Lenoir's engine, and, in 1862, he proposed refinements to the two-stroke cycle as well as designs for a four-stroke engine that would overcome many problems associated with the gas engines of the time. The four-stroke engine doubles the motion of the piston required to accomplish intake, compression, and exhaust, and, by doing so, greatly increases the efficiency of the engine. Early French theory and practice were combined by two Germans, **Nikolaus Otto** and Eugen Langen, into a two-cycle engine that was considerably more efficient than the Lenoir engine. Introduced in 1867, Otto and Langen's engine achieved greater efficiency by compressing the gas before combustion, a refinement lacking in Lenoir's engine. The two Germans' first workable four-stroke internal combustion engine was patented in 1876 and was tremendously successful, selling over 50,000 engines in the 17 years following its introduction. The "Otto-cycle" four-stroke engine was the development that made modern automobiles, **aircraft**, **motorcycle**s, and other vehicles practical.

Gottlieb Daimler, who once worked for Otto, made several refinements to Otto's engine. He devised a cam system so that engines with more than one cylinder could be ganged together. Daimler's two-cylinder gas engine was built in a "V" shape, and he is credited with this innovation. Although supporters of **Karl Benz** and others disputed his claim, Daimler is also credited with developing the first internal combustion engine to burn **gasoline**—not coal gas—as its fuel, a development made possible by the invention of the **carburetor** by his partner, **Wilhelm Maybach**.

The internal combustion, piston engine has been developed into various configurations: "in-line" engines with 2, 4, 5, 6, or 8 cylinders; the "V"-shaped engines with 2, 4, 6, 8 or more cylinders; engines with two opposing pistons for each combustion chamber (used in **Henry Ford**'s first car); multi-cylinder opposed piston engines; and engines with an odd number of cylinders arranged radially that are used extensively in airplanes. In addition to piston-driven, gas-powered internal combustion engines, other internal combustion engines have been developed, such as the **Wankel engine** and the **gas turbine** engine. **Jet engine**s and diesel engines are also powered by internal combustion.

Future developments of the internal combustion engine are being driven by new technology, methods of reducing pollution, and the search for lower production costs. The French Variable Combustion Ratio (VCR) engine and the Peristaltic Vane engine (a rotary engine design) both show promise. Bernard Condamin developed the VCR from his perception that engines are costly because so many different kinds are needed. The VCR adjusts combustion chamber volume to load so the engine operates efficiently in a variety of different applications. The engine is small in size, and, because of its range of uses, it can be mass-produced for greater economy. The Peristaltic Vane engine, invented in the U.S. by Leo Kull, also has variable volume pressure chambers but uses sets of concentric, synchronized vanes in a rotary engine format. The rotary engine operates quietly, weighs little relative to the horsepower it produces, and is simple to construct. Technological developments on the horizon include direct injection engines that burn a lean mixture of gasoline and air and produce fewer pollutants. Ultimately, the solution to the challenges of improving the internal combustion engine may rest in hybrids or combinations of effective inventions.

See also Oil engine; Steam-powered road vehicle

INTERNET/WORLD WIDE WEB

In 1969, the U.S. Department of Defense conceived of a mechanism to safeguard communications systems in the event of a catastrophe. This mechanism was a small network of computers called ARPANET, or Advanced Research Projects Agency Network. It slowly grew into a larger network of contributors to ARPA projects. These consisted of researchers, defense contractors, and military personnel. Eventually, the military personnel felt insecure on ARPANET and broke off and created their own network called MILNET (for MILitary NETwork).

The entire idea of networks blossomed in 1986 when the National Science Foundation created their own NSFNET network. NSFNET was put in place to allow access to information resources to any school that wanted them. Once connected to this network, one was able to access all the other computers on the network. NFSNET grew faster than anyone imagined. Everyone wanted access. But instead of growing by a computer at a time, it was growing a network at a time. What it was becoming was a network of networks, dubbed an internetwork. The outcome of this growth is what is known today as the Internet. In fact, from 1969 to 1998, the Internet grew from four machines to over 29 million machines, with hundreds being added each day.

The largest use of the Internet today is **electronic mail**, or email. Email is the sending of a message from one computer user to another. This unique form of communication has taken the world by storm. Most corporate communication today is done by email, as well as much personal communication, especially over long distances. Why write someone a letter when you can send them email?

The next largest use of the internet is the World Wide Web, the web, or WWW. WWW was developed in 1991 by Tim Berners-Lee at CERN in Switzerland, the European Laboratory for Particle Physics. He desired a system whereby he could publish information so that his colleagues could see and comment on his work, and vice versa. Calling it hypertext, describing its active content, he envisioned a software program called a hypertext server which would send "pages" to a hypertext browser. The very first browser that Berners-Lee developed was text-based and showed no graphics whatsoever. The idea was that a program could display a page of information on the screen. The user, or reader, could select a portion of the page that was linked to another page and jump to that other page. These so-called "hotspots" allowed a reader to jump from link to link, drilling down to the information he or she desired. This concept was very novel at the time and created a revolution in information retrieval almost overnight.

In February of 1993, the National Center for Supercomputing Applications at the University of Illinois-Champaign Urbana developed the first graphical WWW browser. This program, called Mosaic, was written by a student named Marc Andreessen. The program was first only available for higher-end UNIX computers, but by September of 1993 it was available for Macintoshes and IBM PC compatible machines running Microsoft Windows. In March of 1994, Marc Andreessen leaves NCSA to form Mosaic Communications Corporation, and later change the name to Netscape Communications Corporation. The Mosaic Web browser becomes Netscape Navigator. With the graphical browser available, the growth of the WWW accelerated fiercely.

During the 1990s the World Wide Web grew at exponential rates. It was calculated that in mid-1998, there were over 2 million web sites available on the World Wide Web. Today it is not uncommon to see web site addresses on television commercials. The web has meshed with society, and will continue to evolve as technology marches on.

IN-VITRO FERTILIZATION

The first baby conceived by *in-vitro* fertilization (IVF) was born in England in 1978. Since then, more than 3,000 babies conceived in this way have been born. The term *in vitro* literally means "in **glass**"; the fertilization takes place outside the body in a glass petri dish. The term is used in contrast to *in-vivo*—"in the living body"—fertilization. *In vitro* fertilization is indicated when women have no functional Fallopian tubes, the male cannot produce enough sperm to fertilize an egg *in vivo*, premature menopause (which requires an egg from a donor), and other infertility problems. IVF is made up of four major steps: Development (maturation) of the egg while still in the ovary; retrieving the egg upon its release; fertilizing the egg and growing it to the embryonic stage; and transferring the embryo into the uterus.

Early attempts at IVF were made over a century ago, with successful rabbit **embryo transfer** carried out by Walter Heape in England. **Gregory Pincus** and Enzmann performed further IVF experiments in the 1930s. An editorial in the *New England Journal of Medicine* in 1937 suggested IVF as a treatment for infertility in women. Although human IVF experiments were carried out in the 1940s and 1950s by John Rock, Menkin, and Landrum Shettles, knowledge of reproductive physiology was too limited for the IVF procedures to be successful. Then it was discovered that sperm had to undergo *capacitation*—changes in the plasma membrane—after ejaculation in order to be capable of fertilization, and that the oocyte (egg) also is not ready for fertilization until just before ovulation occurs. Armed with this knowledge, researchers successfully fertilized rabbit oocytes *in vitro*. Min-Chueh Chang went a step further and implanted *in vitro* fertilized oocytes into female rabbits. IVF for many other species followed.

The final advance was made by British physicians Robert Edwards and Patrick Steptoe, who began collaborating in 1968. They developed a method of stimulating ovulation with hormone treatment, then retrieving the nearly mature ova and placing them in culture for the several hours needed for full maturation. Meanwhile, a fresh specimen of male sperm was treated so it underwent capacitation, then it was added to the oocytes in the petri dish, where fertilization took place. After undergoing initial division, the eight-celled embryo was introduced into the woman's uterus, where, in successful IVF procedures, it would implant. An English couple underwent this IVF procedure and, in 1978, Louise Brown—the first human baby conceived outside the womb—was born. Since then, IVF has become a widely used method of infertility treatment, offered by hundreds of medical centers around the world. Since the mid-1980s, **cryopreservation** (freezing) of embryos has become common as a way of saving extra fertilized eggs in the event the initial IVF attempt fails. Thorny legal and ethical questions have resulted, however, dramatized by the 1984 plane crash death of a husband and wife who left behind cyropreserved embryos in Australia. Who has the right to dispose

Louise Brown, the first child conceived through in-vitro fertilization, being held by her mother, Lesley Brown, during their appearance on the *Donahue* television show.

of cyropreserved embryos? How long can they be stored? Who owns them? Is an embryo an heir of its parents?

Several variations on IVF are also an option: One is gamete intrafallopian transfer (GIFT), in which oocytes and sperm are gathered and prepared as in IVF but are placed into the fallopian tube, uniting there rather than in the petri dish. Another is zygote intrafallopian transfer (ZIFT) in which one or more zygotes—fertilized eggs before they start to divide—are transferred to the fallopian tubes. Because they are invasive and expensive, and because higher pregnancy rates are being obtained with three-day uterine transfer, these methods have widely been abandoned.

Iron lung and other respirators

The iron lung is a mechanical respiratory device used to force air into and out of the lungs of a person unable to breathe for themselves. Invented in 1929 by Philip Drinker (1893-1977), a professor at the School of Public Health at Harvard University, the iron lung was one of the first of several inventions designed for this purpose. During the 1920s, people with respiratory failure were aided by a pulmotor, a contraption similar to fireplace bellows, which inflated and deflated the lungs by forcing air in and sucking it out. This process worked; however patients experienced chest pain because the amount of air forced into the lungs was not adjustable to the individual patient's lung capacity. Many people suffering from poliomyelitis—which often paralyzes muscles of the diaphragm, the large, dome-shaped muscle above the stomach which rises and falls to draw air into and push it out of the lungs—required the aid of these devices to keep them alive.

Drinker's idea for a breathing device came from a Swedish physician named Thunberg who had been experimenting with a vacuum to help patients breathe. Enlisting the help of his brother, Cecil, and Louis Shaw (1886-1940), Drinker built a prototype, tested it on cats, and designed a device large enough for humans. The patient's body was positioned inside an airtight metal box while the head remained outside. The box was connected to a pump which reduced the air pressure inside the box, causing a negative change in pressure in the chest cavity similar to that which occurs when the diaphragm moves downward allowing air to rush into the lungs. With this decrease in air pressure inside the box, the weight of the atmosphere outside the box forced air through the nose and mouth of the patient into the lungs. In October 1928 an eight-year-old girl unable to breathe as a result of polio was the first person to be put into the iron lung. Drinker's invention was initially known as the ''Drinker tank respirator,'' but was soon earned

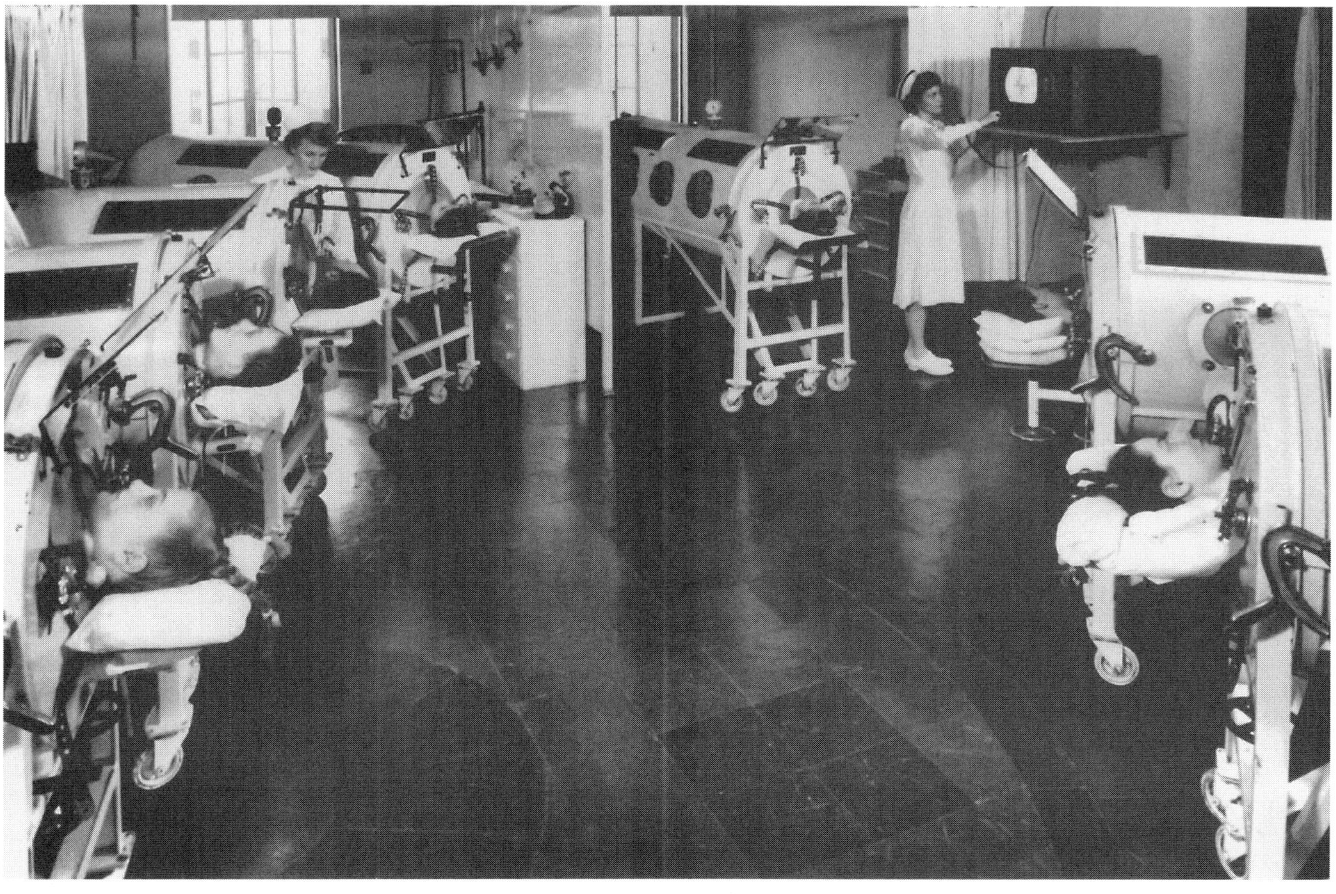

Invented in 1929 by Philip Drinker, the iron lung performed the function of the muscles that control breathing for polio victims (above) and other paralyzed patients. Modern respirators and ventilators would make these large, cumbersome machines unnecessary.

the nickname *iron lung*. Drinker and Shaw received numerous awards for their invention, which allowed many polio patients to be kept alive, and it remained the standard treatment for this purpose well into the 1950s.

Since Drinker's invention, a class of breathing machines called *ventilators* and *respirators* has been developed. A modern ventilator consists of an electrical pump connected to an air supply, a humidifier which adds moisture to the air, and a tube which is inserted into the patient's nose or mouth. Ventilators, unlike the iron lung, use positive air pressure from the pump to force air into the lungs and the patient exhales the air naturally. To adjust the oxygen-carbon dioxide ratio supplied by the ventilator, blood samples from the patient are analyzed to determine the metabolic rate. The volume of air required and the number of times per minute the patient needs to breathe in order to maintain the desired metabolic rate, is also calculated, and positive air pressure administers the correct mix and volume into the lungs, which deflate passively. Today's sophisticated hospital respiratory care units may utilize up to 15 different kinds of respirators. Ventilators assist patients unable to breathe for themselves, who suffer from degenerative muscle disease, or burns to the nose and throat. Some patients may be connected to a ventilator for months at a time, in which case a breathing tube is surgically inserted directly into the trachea (windpipe). Today, in such places as Stanford University Hospital in California, the iron lung of the 1950s is making a comeback due to its non-invasive, negative air pressure which eliminates the possibility of infections or scarring prevalent with respirators which require a tracheotomy. The iron lung can also be used at home to help a patient rest respiratory muscles either during the day or night.

Miniature ventilators are also used to help premature babies breathe and provide temporary breathing assistance to patients undergoing surgery requiring anesthesia. At University Hospital in Stony Brook, New York, mini-iron lungs are used for some patients. Nicknamed *turtles* because of their green color and shell-like shape, these miniature iron lungs can be strapped onto a patient's chest. For treating sleep apnea, a sleep disorder in which breath becomes obstructed, a small ventilator about the size of a vanity case is connected to a tiny mask placed over the nose. Intermittent positive air pressure is delivered through the nostrils and the timing can be preset to correspond with the patient's natural breathing rate.

Iron production

Since before recorded history, people have utilized iron. Evidence for the use of iron can be found as early as 4,000 B.C., but historians believe it was in use prior to that time. Early in human history, iron was expensive and precious. The technology of how to press it was a closely guarded military secret that gave those nations possessing the knowledge a strategic advantage in weapon production over neighboring nations who only knew how to produce duller and weaker bronze weapons.

Originally, iron was obtained from meteorite deposits, but as those supplies dwindled, the method of reduction of iron ore in furnaces was developed. By 1200 B.C., iron smelting techniques had made iron a viable material for all weapons and tools. Because of the relatively low temperatures attainable from furnaces prior to the Middle Ages, repeated heating and hammering in a hot charcoal fire was necessary to work out the carbon and other impurities and forge the metal into desired shapes. Around 1300, use of the **blast furnace** began to spread throughout Europe. The furnace used a steady stream of air to increase the intensity of the heat. The air was produced by bellows or by water pressure. As a result, iron production was streamlined and new industries like blacksmithing, **wire** drawing, and needle making emerged.

Cast iron was known but not widely used for a long time. It was too brittle to be worked the same way as wrought iron. But in 1709, **Abraham Darby** started using coke instead of coal as a fuel to maintain the furnace heat. The result was a product that was much stronger than previous cast iron. Darby's method also made it possible to mass produce iron cheaply and in standard shapes. Eventually, his foundry at Coalbrookdale, England, produced cast iron cylinders for **train** engines. Englishman **Henry Cort** developed a grooved rolling mill in 1783 for the production of iron bars.

The following year Cort developed a puddling process for purifying iron from pig iron. In 1856 Sir **Henry Bessemer** invented a conversion process that blew air directly through the molten iron for the efficient production of **steel** from iron ore. This was significant for the iron industry, since steel replaced iron in importance within fifty years of Bessemer's invention. The demand for iron as a building material greatly increased as architects began to adopt the use of iron building frames, partly as a fire prevention measure. In 1848 American **James Bogardus** built a sugar mill in New York entirely of cast iron. One unique feature of the building was that the beams and joists were interchangeable. Bogardus knew he had a promising product and marketed the building parts. **William Jenney**, in 1885 in Chicago, designed the first iron load-bearing frame commercial building.

Throughout the 1700s and 1800s iron was being used extensively in the railroad industry. It was used in building locomotives, **bridge**s, and railroads. It played a major role in the rapid development of the United States and in the expansion of the British Empire in India and its larger colonies. The metal was also extremely important to the shipping industry as iron-clad steamships regularly crossed the Atlantic. The discovery of large iron deposits in the Lake Superior region during the 1840s had an important influence on the American Industrial Revolution. Ores from Minnesota and Michigan were shipped to mills in Chicago, Cleveland, and Pittsburgh, where they were refined.

New techniques were developed to increase the strength and durability of iron. The process of galvanizing developed in the 1700s was later applied to iron. In galvanization, a coating of zinc bonded to the iron gave it between fifteen and thirty years of protection against the elements, especially rust. **Corrugated iron** was first mass produced by **Joseph Francis** in the 1840s. The corrugate ribbing gave rigidity to sheet iron. Francis used it to build lifeboats.

In the 1980s and 1990s, modern light-weight iron formulations were developed as the demand for lighter materials in car manufacture increased. A professor at Stockhold University, Lennart Backerud, developed a process to produce such an iron. The Backerud process samples the iron while in the molten state with a highly sensitive temperature probe and the information is process by a computer. This allows the iron to meet the customer's exact specifications. The production of iron is considered one of the greatest industrial developments in history and remains a guideline for evaluating technological advancement in developing countries.

Iron ship

By 1838, steam-powered **ships** had crossed the Atlantic Ocean. This achievement was followed by three vital engineering advances that led eventually to the truly modern ship: the advent of metal construction, the switch from paddles to propeller propulsion, and reliance on the **steam turbine** for power.

Initial experiments in the use of metal for ships started in the 1700s. **John Wilkinson**, the English iron-master, was a leading figure in the expanding iron industry and an important supplier of armaments in the middle and late eighteenth century. In 1787, he launched the first iron-hulled ship on the Severn River in England. Other English shipbuilders began to use iron, partially because good wood for ships was becoming scarce in their country. But iron had other advantages: it was stronger, safer, more economical, easier to repair, and lighter because wooden ships required huge timbers. With this final advantage, iron ships could hold more cargo. Still, many shipbuilders preferred the traditional wooden construction.

It took a disaster to prove the value of iron ships. In 1834, a violent storm drove several ships ashore along the coast of England. Among them was an all-iron vessel, *Garry Owen*. Heavy seas pounded the wooden ships to pieces, but the *Garry Owen* escaped with only a few dents and scrapes. It returned to its port under its own power, thus giving a clear demonstration of the advantages of iron in ship construction.

England took the lead in the production of iron ships, thanks in great part to an outstanding naval architect, **Isambard K. Brunel**. Brunel's ships were larger and technically superior to any that had been previously constructed. After his successful wooden-hulled *Great Western* (launched in 1837) established the first regular steamship service between England and the United States, Brunel built the first large ship made of iron, the *Great Britain*. It was 322 feet long (98 m) and weighed 3,300 tons. The value of its iron hull was made clear on the ship's maiden voyage. The *Great Britain* was beached on the southern coast of Ireland for nearly one year, but suffered very little structural damage. The *Great Britain* remained in service for 30 years as a passenger ship, sailing to San Francisco, California, journeying regularly to Australia, and even acting as a troop ship. In 1866, it was badly damaged and left to rot in South America. It was salvaged in the late 1960s, a testimony to its rugged construction.

In 1858 Brunel built an even larger ship, the *Great Eastern*. It was 692 feet long (211 m) with a displacement of 32,000 tons and was the first ship with a double iron hull made up of 30,000 iron plates and 3,000,000 rivets. Able to carry 4,000 passengers, the *Great Eastern* was considered to be the ultimate liner. However, the ship was impractical because it used far more coal than anticipated. Although a commercial failure, it made history in the 1860s by laying the first Atlantic **telegraph cable** and remained the largest ship in the world for 40 years.

The *Great Eastern*, built by Isambard K. Brunel in 1858.

The American Civil War convinced many military experts of the value of iron for their vessels. The Confederate Navy had salvaged a Union ship by replacing the old superstructure with a rectangular armored box 170 feet (52 m). The walls were over two feet thick with two layers of iron plating lying on top of wood. When this ship, the *Merrimack*, was attacked by Union vessels, cannonballs bounced harmlessly off her. Northern leaders panicked and commissioned **John Ericsson**, a Swedish-born engineer, to design and build another armored vessel. Ericcson's vessel, the *Monitor*, was 179 feet long (55 m) with a low, flat deck topped by a cylindrical, rotating metal turret housing two guns. The two ironclad ships met in battle on March 9, 1862. Neither was able to seriously damage the other, but the *Merrimack* was forced to withdraw. Naval experts realized that armor made a ship almost indestructible and the rotating turret was more effective than the fixed broadside of the past.

From 1830-1880 iron vessels dominated the shipping industry, but even then shipbuilders began experimenting with

steel because it was stronger and lighter than iron. In 1881, the *Servia*, a British ship, became the first all-steel passenger liner to cross the Atlantic. By 1900, steel had almost completely replaced iron, and newer methods of joining the units had been tried. The traditional method for joining iron or steel had been with rivets, but **welding** began to gain popularity because it was quicker and more economical. Later, prefabricated parts were used to keep costs down even further.

After World War II, fiber-reinforced **plastics** and later glass-reinforced plastics (also called composites) began to be used in place of wood for small craft of all types. These new materials were not considered appropriate for larger vessels until the 1980s and 1990s when the need to reduce both the cost and weight of ships became paramount. Research in the 1990s indicates that composites may be technically feasible for ships up to 140 meters, but they still have to contend with the economically attractive cost of heavy welded steel construction. For the future, despite the advantages of composites (light weight, stability, resistance to chemical corrosion and biological attack, and minimum maintenance), conventional steel will continue to be used.

See also Screw propeller

Irrigation

Farmers cannot always depend solely on rainfall to sustain their crops. In fact, many of the world's agricultural regions receive little or no precipitation. In these areas, water must be diverted from streams or lakes or drawn from wells. The result has been the development of irrigation systems. Mesopotamia, considered the cradle of civilization, had one of the earliest water diversion systems. Agriculture began there about ten thousand years ago. Water channels helped divert water from the Tigris and Euphrates Rivers during the dry months and also helped control the annual July floods. The civilization that developed in Egypt owed its existence to the water diversion schemes that delivered water to the fields from the Nile River and its delta. The ancient Egyptians invented the shaduf, a long pole with a bucket suspended at one end and a counterweight mounted at the other. The pole pivoted vertically to lift water in the bucket to a higher level. Shadufs are still in use. About 2000 B.C., the Chinese were diverting water from the Hwang Ho and Yangtze Rivers and building levees to keep their rice paddies flooded. In approximately 130 A.D. Native Americans built **canals**, terraces, and check **dam**s that are still visible today in the Sonoran Desert of Mexico and Arizona. The Roman Empire drew much of its stability from public works projects that included canals and **aqueducts**. The Romans had several methods of lifting water, including the treadmill-powered chain of buckets, the human-powered scoop **wheel**, the water-driven tympanum, and the **Archimedean screw**. Today, mechanical pumps raise water, while channels and sluice gates regulate its level and flow. Portable pumps are available to farmers who wish to irrigate on a short-term basis, as in a drought, and large permanent lift stations operate in larger irrigation schemes.

Irrigation networks have been constructed in many parts of the world, such as in the rice fields of eastern Arkansas, the olive groves of southern Spain, and the wheat fields of Victoria, Australia. One of the most ambitious irrigation efforts has taken place in the Sacramento, San Joaquin, and Imperial valleys of California. An important figure in the agricultural development of California was Harriet W. R. Strong. Left to fend for herself after her husband's suicide in 1883, she acquired a number of patents for irrigation and flood control measures, including, in 1887, the construction of a series of dams so that water backed up from one helped support the dam immediately upstream from it. Irrigation methods also include the use of sprinkler systems. Field-length pipes supported by wheeled legs water the corn fields of the Midwestern United States. Similarly, the potato fields of Idaho and Oregon are watered by mobile pipes that revolve in a circle. Irrigation has its environmental drawbacks, causing leaching in some areas. Water tables have been depleted from over-pumping, as in the cotton fields of west Texas. Water diversion also has fostered regional disputes over water rights, as is the case among the states of the upper and lower Missouri River Valley. In spite of its financial and environmental costs, irrigation has been and will be a vital part of man's existence for the foreseeable future.

See also Canal and canal lock; Water pump; Hydraulic ram

IUD • See Birth control

Ives, Frederic Eugene (1856-1937)
American printer and inventor

Responsible for the **halftone process** which breaks down illustrations and photographs into tiny dots for print reproduction, Ives spent most of his career studying the mechanics of **printing** and **photography**. Born near Litchfield, Connecticut, on February 17, 1856, Ives apprenticed as a young man at the *Litchfield Enquirer* newspaper, and soon after opened a photographic studio in Ithaca, New York.

After becoming the head of the Cornell University photographic laboratory in 1875, Ives began to study and experiment in the field of photographic reproduction. Newspapers and other print media at the time had not yet been able to harness the potential of photography; because printing involved pressing an engraved plate coated with **ink** onto **paper**, their challenge was to find a way to translate finely-shaded pictures such as photographs onto plates. Ives studied the process of photoengraving, through which designs are projected onto plates treated with light-sensitive substances, to create engravings. At Cornell, he developed the *swelled gelatin process*, which refined the sensitivity of the photoengraving process using a gelatin-based coating; however, the process still could only reproduce black lines on white. To translate the gradations of shading in photographs, Ives hit upon the idea of breaking down photographs into dots of various sizes to convey shades, or "halftones," which gave the name to the process that revolutionized printed illustrations.

In 1885, he developed the halftone technique which has remained largely unchanged through today: the illustration to be reproduced is placed behind a **glass** screen gridded with fine black lines and photographed. The resulting image, composed of thousands of dots of various sizes, is then photoengraved. Soon after, Ives invented a color halftone process, in which a color illustration is pictured through three screens, each of which separates out a particular primary color; the images are then overlayed to create a color reproduction.

In his lifetime, Ives patented more than 70 other inventions dealing with optics, printing, and photography. They included the photochromoscope camera and the chromogram, which, respectively, produced and projected a three-color photo negative; the latter two inventions were important in the evolution of **color photography** and printing. Ives died on May 27, 1937, in Philadelphia, Pennsylvania.

See also Printing technology

Ives, Joseph • See Clock and watch

J

Jackhammer

• See Drill

Jacquard, Joseph-Marie (1752-1834)

French inventor

During the 1700s, inventors were attempting to automate the process by which patterned **textiles** were woven—primarily due to the rising demand for fine patterned cloth. It was generally accepted that, for such a machine to work, it must satisfy two requirements. First, it must mechanically simulate the action of hand-lifting the individual warp threads, thus creating the pattern. Second, it must possess some storage medium by which the pattern is "remembered," enabling the weaver to identically duplicate the pattern again and again. Though many devices were constructed throughout the eighteenth century, none satisfied these requirements as well as the Jacquard **loom**, patented in 1804 by Joseph-Marie Jacquard.

Jacquard was born in 1752 in Lyons, France, the center of the French silk industry, with which his parents were involved. Young Jacquard, however, found little interest in the textile industry, instead apprenticing himself to a book binder and, later, a cutler. Upon the death of his parents he inherited the family home, a parcel of land, and a small sum of money, which he lost in a number of dubious entrepreneurial investments. It was not until 1790 that he came upon the idea for his automatic loom; at that time, however, he joined the revolutionists, and the loom was not actually constructed until 1801.

The result of ideas borrowed from Basile Bouchon, Falcon, and **Jacques de Vaucanson**, the Jacquard loom used a system of hooks and needles to lift the appropriate warp threads. The pattern was stored on a collection of thick paper cards perforated with rectangular holes. As the fabric was woven, the hooks were held stationary by the surface of the card. However, whenever a hole was encountered, a hook would be allowed to pass through to lift its thread. By stringing together a large number of cards, an intricate pattern could be created.

Jacquard took his loom to Paris, and in 1804 he was awarded a gold medal and a patent for its invention, as well as a handsome pension. He was not so well received by the local handweavers, however, who saw the automatic loom as a threat to their business. Several of his machines were burned, and Jacquard himself was beaten on more than one occasion. Despite this initial opposition, the Jacquard loom began to enjoy commercial success, and by 1812 thousands were in use in France alone.

The idea of using perforated cards as a method for storing information intrigued the British scientist **Charles Babbage** who, in 1823, received funding from the British government to construct an *analytical engine.* This steam-powered device would be able to perform many different mathematical functions at once, printing the result. Though it was never constructed (the technology of the time was too primitive to provide Babbage with the necessary parts), the design of the analytical engine in turn inspired American scientist **Herman Hollerith** to build a similar machine to compute the results of the 1890 census; this machine, which used punched cards as the storage medium, was the ancestor of the modern computer Hollerith's company, the Tabulating Machine Company, went on to become International Business Machines (IBM).

Jeep and Hummer

Jeep and Hummer sound like the cartoon characters of the **automobile** world, but, in fact, they may be among the most versatile, rugged, and loved vehicles ever made. The Jeep was created as a vehicle for wartime. Before the automobile, armies used horses and mules to move men and supplies. The French Army was the first to experiment with motorized vehicles by adding engines to tri- and quadricycles, but these were too fragile and awkward. **Airplanes**, tanks, **trucks**, and other motorized vehicles had significant roles in World War I, but there was no all-purpose automobile. The U.S. Army experimented with a stripped down Model T Ford equipped with balloon tires and with **motorcycles** with sidecars attached, but a vehicle that was strong but small and had the traction to cover difficult terrain was still lacking by the start of World War II.

In desperation, the Army appealed to 135 American car manufacturers to develop a ''scout car'' that was low to the ground, able to carry weapons, met strict weight and haulage requirements, and had four-wheel drive. The prototype was needed in 49 days and 70 more vehicles were needed 75 days later. The two automakers that responded were Willys-Overland Motors, Inc., and the American Bantam Car Company. The Bantam design arrived in Washington, D.C., first, but the Willys-Overland cars followed soon after. Army engineers subjected the vehicles to torture tests, and, despite breakdowns and many equipment failures, both types were considered successes. The Ford Motor Company also entered the competition late because the Army felt the other firms were too small to produce the number of vehicles needed. The engine in the Willys-Overland ''Quad'' car was so superior that this design was selected, and both Willys-Overland and Ford produced the Quad car initially with the Bantam Company assisting in later production. The three manufacturers produced 900,000 vehicles by the end of the war in August 1945.

To meet the Army's weight requirements, early jeeps were strictly bare-bones vehicles. There were no heaters, windshield wipers had to be operated by hand, seats had no cushioning, the change from two- to four-wheel drive required hand-locking the wheel hubs, and colors were camouflage green for most uses and sand brown for desert service. The jeep, however, could cover many types of terrain, and be used for command and reconnaissance. Hooks on the back could be used to hang stretchers for the wounded, and swiveling **headlights** could be turned back on the car so soldiers could work on the engines without other light sources. **Machine guns**, anti-**aircraft** guns, and mortars and **rockets** could be mounted on jeeps. An amphibious form of the jeep was invented in 1941 by Roderic Stephens, Jr., a yacht designer. Balloon tires, air pockets in the metal parts, a rudder attached to the steering wheel, and an anchor transformed the jeep into a seep.

Other armies ordered jeeps as well, and jeeps played an important diplomatic role in lend-lease shipments to American allies. The tough vehicles were credited with military victories. British commandos in jeeps swept behind the lines of the German General Erwin Rommel's famed Afrika Korps in the Sahara Desert and cut off fuel supplies and destroyed a convoy of German trucks and tanks. Jeeps plowed snow off Northern European airfields—they could be fitted with Caterpillar tracks at the rear and skis in front.

The evolution of the name ''jeep'' is unknown, varied, and almost mythical. It was nicknamed the Peep, Puddle Jumper, Beetle Bug, and Leaping Lena; but most believe the name came from the initials ''G.P.'' for general purpose vehicle. The returning veterans loved their jeeps, and Willys-Overland made improvements, such as cushioned seats, and marketed them as family cars, station wagons, and farm equipment. Even a sports model with a convertible top was designed and sold as the Jeepster Sports Phaeton. Legal hassles over the name and design rights arose between Willys-Overland and the American Bantam Car Company, but these were settled in 1950 and Jeep became a registered trademark. Kaiser Passenger Cars bought Willys-Overland, and, by the late 1960s, the Jeep Wagoneer exploded in popularity. American Motors Corporation bought Jeep, and the larger manufacturer increased production from 175 to 700 Jeeps per day in 1970. The availability of more free time, the recreational vehicle craze, comforts of civilian designs of the Jeep, and technical improvements to the engine, gas economy, and ''shift-on-the-fly'' conversion from two- to four-wheel drive accounted for Americans' love of the Jeep.

Military movers reconsidered vehicle design in the late 1970s when they realized that too many unique vehicles had evolved for specific duties. The U.S. Army announced another competition in 1979 for the design of a family of vehicles with identical mechanical features as well as adaptability to different tasks. Specifications were issued for profile, roof height, hood height, and the same mobility standards as other combat vehicles including the Bradley and the M1 tank. These included specified ability to climb grades and side slopes, ground clearance, and ability to mount vertical steps and ford streams. This new vehicle was identified as a High Mobility Multipurpose Wheeled Vehicle or HMMWV, which the military soon called the Hum-Vee.

Three defense contractors were interested in meeting the strict requirements, and prototypes and preproduction models were subjected to three phases of testing at test facilities and in the field by soldiers and Marines. AM General won the contract, and in 1983, the Army ordered 55,000 Hum-Vees in 15 different configurations. More than 100,000 additional vehicles were ordered in the next 6 years. The configurations included cargo and troop carriers, armament carriers, vehicles with TOW (Tube-launched, Optically-sighted, Wire-guided) missile launchers, shelter carriers (to house radio and other units), and ambulances. Most of these vehicles have protective armor, and all configurations are mechanically interchangeable.

By the early 1990s, greater weight and towing capacity were needed and certain configurations of the Hum-Vee were revised. The weapons carriers were given optional panels to improve their armor, and many other improvements were made to configurations and specialized equipment. The ''identical'' nature of the vehicle family was maintained by producing most of the options packages as kits.

In 1990, AM General began designing a civilian version of the Hum-Vee to be known as the Hummer. The Persian Gulf War in early 1991 brought the strong-armed appearance and versatility of the military Hum-Vee into millions of living rooms, and mass production of the civilian Hummer began in 1992. The Hummer has the low center of gravity, high ground clearance, and wide stance of its military cousin. Through brake/throttle modulation and special differential/brake design, tires that lose traction are stopped from spinning so power can be redistributed to the tires to maneuver the vehicle over obstinate terrain. The driver can inflate or deflate tires from the driver's seat using the Central Tire Inflation System (CTIS), also for better traction. The Hummer can be outfitted with equipment such as a crane, backhoe, concrete pump, firefighting apparatus, and others for industrial and civilian uses.

JEFFERSON, THOMAS (1743-1826)

American politician and inventor

Dubbed the "sage of Monticello" and the "godfather of American invention," Thomas Jefferson created a variety of practical devices, notably the dumbwaiter, lazy Susan, folding campstool, folding ladder and music stand, portable copying press, portable writing desk, revolving chair, cipher wheel for cryptography, moldboard **plow**, hemp-breaker for **threshing machines**, and improved **carding machine**. A native of Shadwell in Albemarle County, Virginia, Jefferson received a classical education from private tutors. At age 17, he entered William and Mary College and studied math and science. He was admitted to the bar in 1767, and practiced law in Williamsburg for two years.

Jefferson filled several political jobs, including county lieutenant, county surveyor, and member of the House of Burgesses and alternate to the Continental Congress. Following two terms as governor of Virginia, he retired to farm the land around his Palladian estate, Monticello. Many of his labor-saving devices date to his return to the land, notably his **nail** factory, which was intended to supplement his irregular and unpredictable income from agriculture. Inside Monticello, Jefferson's desire for privacy prompted the invention of the lazy Susan, which allowed servants to deliver food to the dining room without intruding. Dumbwaiters built into the side of the mantle allowed wine stewards to send up bottles from the wine cellar without interrupting the flow of conversation.

Jefferson's preoccupation with efficiency and convenience prompted some of his more unusual household inventions. He constructed a screened alcove bed which fit into a wall between two rooms. This gave the occupant the option of getting out of bed in whichever room he chose. At the foot of the bed was turntable with forty-eight projecting fingers which could hold ties, vests, and coats. Alongside his bed stood a revolving-top writing table he created to streamline his mechanical drawing. Opposite was a revolving chair, which also conserved motion as he reached for books and drawing supplies or referred to volumes from his sizeable library. After arthritis set in, Jefferson refined his desk chair into a chaise lounge to relieve the ache in his joints.

Jefferson was also fascinated with timekeeping and unusual **clock** designs. Over his front door at Monticello was an imposing two-faced clock which faced the outside and the foyer. Its cannonball weights also served as seven-day calendars. To reach the top of the clock for service and synchronizing, he devised a folding ladder.

To assure cooling for fresh foods, beverages, and **ice cream**, Jefferson refined the concept of the icehouse. As an aid in the removal of melted ice, he replaced the usual pump with a square tube running from the ground through the top. A servant hauled up melted ice in a bucket, which was fitted with a **leather** valve in the bottom and moved vertically up the square tube for emptying. The resulting ice water he transferred to his four cisterns, which he devised to store rainwater for domestic use and fire protection.

In his letters, Jefferson mentions his numerous and varied interest in inventions, particularly the application of steam power to navigation and manufacturing. He redesigned the moldboard plow, which farmers on both sides of the Atlantic used, and introduced rice, olives, and sheep to American farmers. Eager to standardize money, weights, and measures, he also drew up a decimal system for coins, helped regulate patent procedures, and served as the nation's first patent officer.

In 1797, Jefferson entered public office, first as vice president, then as president of the United States. On his return to Monticello, he founded the University of Virginia, one of his most satisfying accomplishments. He also devoted himself to reading the classics and helped establish the Library of Congress by donating his personal library. Jefferson tried to improve agriculture techniques at Monticello until his death in 1826.

JEFFREYS, ALEC • See Genetic fingerprinting

JENNEY, WILLIAM (1832-1907)

American architect

One of the main factors that identify modern architecture is the shift from load-bearing walls to load-bearing frameworks. William Le Baron Jenney started this architectural revolution with construction of the Home Insurance Building in Chicago in 1884. It was the first major structure to use an iron framework as its sole means of support.

Born in Fairhaven, Massachusetts, in 1832, of English Puritan ancestry, Jenney was influenced by his father, who had carried on the family's maritime tradition as agent and owner of a fleet of whaling **ships**. After the whaling industry failed in the 1850s, Jenney's father entered the insurance business.

Jenney travelled extensively during his early years. He was probably most influenced architecturally by his witnessing of the rapid rebuilding of San Francisco after fire destroyed its wood buildings in 1850, and the city was quickly rebuilt with **brick**. According to his partner, William Mundie, Jenney was also influenced by his contacts with primitive societies like the Hawaiians whose dwellings were of frame construction.

Jenney received formal engineering training at the Ecole Centrale des Arts et Manufactures in France from 1853 to 1856. During this period, the first iron beams were being produced as replacements for wood flooring for fireproofing.

Jenney married Elizabeth Cobb of Cleveland in 1867. That same year he was attracted to Chicago, and was appointed chief engineer of the West Chicago Park Engineers in 1869. Jenney was instrumental in planning Chicago's grand park scheme.

The Great Chicago Fire in October, 1871 was a disaster for most, but a boon to aspiring architects like Jenney. He designed many commercial, religious, and residential buildings during the 1870s and 1880s. His Home Insurance Building was his response for a call for improved **fireproofing techniques**.

In his later years, Jenney designed buildings for the Columbian Exhibition of 1892 and the YMCA Building in 1893. Jenney retired in 1905 and turned the business over to Mundie. Jenney loved California and its gardens, and he died in 1907 while visiting Los Angeles.

See also Cast-iron and wrought-iron construction

JEROME, CHAUNCEY • See Clock and watch

JET ENGINE

In the early part of the twentieth century, aircraft were powered by **internal combustion engines**. These engines became increasingly heavy and complex as efforts were made to create speedier planes. **Aircraft** engine designers realized the capabilities of the engines were limited. Attainable altitude and speed were capped because thin air reduced propeller and engine efficiency. As a heavier engine would only weigh down the craft, engineers turned to other ways to power aircraft.

Frank Whittle, a British engineer, worked on one possible solution: the jet engine. In 1930, Whittle patented a variation on a **gas turbine** engine already in commercial use. The engine used a turbocompressor to compress incoming air, which was mixed with fuel and ignited. The high-pressure exhaust blew out the back, producing forward thrust. The machine, however, received little attention in Britain because it was too heavy and inefficient to be mounted in a plane. After transferring its development to the United States early in World War II and making numerous modifications, Whittle saw the engine successfully power the Gloster Meteor in 1944. The Allies' first jet fighter, the Bell P-39, was not perfected in time to be used in the war.

At the time of Whittle's experiments, the Germans were independently working on a jet engine. A young engineer, Hans von Ohain, brought an idea for a jet engine to aircraft designer Ernst Heinkel (1888-1958), who was impressed enough to back Ohain. His modified engine, operating on kerosene, produced enough thrust to power the He 178, a small experimental aircraft that debuted in 1939. Later, the German military developed more powerful and reliable engines with greater fuel economy. The most successful German fighter with jet engines was the Messerschmitt Me 262, capable of flying over 550 miles per hour.

The British were the first to develop the turboprop engine, which added a propeller to the conventional jet engine. The original jet engine design, most efficient on long flights at high altitudes, wasted a great deal of power. The turboprop

Frank Whittle poses in 1962 beside one of his first jet engines.

engine had better fuel efficiency, provided a smoother ride, and needed less maintenance than piston engines. The British design, with a Rolls Royce engine, was so successful on its emergence in 1940 that it was still in production 30 years later.

While the turboprops were superior to previous propeller-driven planes, however, they did little to relieve the inherent problems of the propeller. Rolls Royce developed a new, more fuel-efficient engine in the late 1950's with a large, propeller-like fan at the front. Much of the air drawn in by the fan actually never entered the combustion chamber, but bypassed the core engine and was expelled at the back, producing additional thrust at low speeds. An even more efficient turbofan system with two compressors was later developed. For the special needs of military aircraft, engineers devised a method to burn extra fuel in the rear of the jet pipe, called reheat, or afterburning. Though the innovation wasted fuel, it provided a big thrust increase for a small increase in engine weight. In 1950, planes like the United States' F-86 Saber and the Soviet Union's Mig 17 enabled pilots to travel at the speed of sound (Mach 1). Fighters developed in the 1960s and 70s were capable of speeds of Mach 2 and Mach 3.

The introduction of jet engines to commercial aviation cut travel time dramatically. After World War II, the British produced the world's first jet-powered airliner, the de Havilland Comet, making its maiden flight in 1949. The Boeing 707, the first widely-used jet airliner, revolutionized the industry. The cooperative British and French effort that resulted in the Concorde airliner used a Rolls Royce engine and speeded travel in Europe. The latest advances in commercial airliners deal with noise reduction, the biggest problem in modern jets. So severe is the noise problem on older, so-called "Stage II" jets (such as the Boeing 727), that the Federal Aviation Administration has required all such jets to be retired by the year 2000, or, if they remain in service, to be fitted with "hushkits" that reduce engine noise. Engineers have spent much effort on modern jets to make the noise level more comfortable for those at ground level, experimenting with several factors. Turbofans proved to be quieter than other jet engines. Some noise reduction was achieved by introducing sound-absorbing material in the intakes and nozzles of the engines; multi-lobe exhaust nozzles that mix exhaust with outside air also reduce noise levels. One has only to stand near an airport and listen to an old 727 and a newer 757 takeoff to hear the dramatic improvements in noise reduction technology since the 1970s.

JET PLANE • See Aircraft

JETFOIL • See Hydrofoil

JOBS, STEVEN P. (1955-)
American computer innovator

Steven Jobs is best known as the co-developer with **Steven Wozniak** of the first widely-popular desktop computer, the Apple II in the late 1970s, and as co-founder of the Apple Computer Corporation.

Steven Jobs

Jobs was raised in the San Francisco Bay Area and became interested in computers while in junior high school. He met electronic-expert Wozniak in 1971, and the two of them began their partnership by manufacturing "blue boxes," illegal electronic devices for making long-distance telephone calls without paying for them. After spending a period in India living in a commune, Jobs again met Wozniak. The personal computer (PC) was just being developed in the mid-1970s and Wozniak had designed one, but he was unable to interest a manufacturer. Jobs, however, immediately saw its marketing possibilities. Jobs gave it the friendly name of Apple, aiming the machine at home users rather than computer enthusiasts, such as those in the influential Home Brew Computer Club, of which Wozniak was a member. The "two Steves," as they were known, sold their few valuable possessions and set up their company in the Jobs family's garage.

After sales success with the Apple I, Jobs quickly hired managerial and marketing experts, and cultivated venture capitalists to invest in their company. Together he and Wozniak designed an improved and even friendlier version, called the Apple II. By 1977, Apple Computer had moved to its own quarters in Cupertino, California, staffed in part by equally youthful experts. The company was a success from the beginning, earning almost $3 million that year. By the age of twenty-five, Jobs was a multi-millionaire, and Apple Computer eventually became one of the Fortune 500 largest companies. In the 1980s, Jobs took charge of developing the Apple Macintosh computer, the first personal computer to use a graphical user interface. However, the initial version was heavily criticized, and Jobs was forced out of the company. In 1986, he founded another firm, Next, Inc., to produce another innovative, though less popular, computer. In the 1990s, Jobs has changed Next's emphasis from hardware to software. In 1997, Jobs left Next and returned to Apple as interim CEO.

Johannsson, Carl Edvard (1864-1943)

Swedish inventor

Although denied the attention and bravura accorded upon other inventors of his time, Johannsson worked to develop a system of precise measurement so beneficial that it completely revolutionized industrial production.

As a child, Johannsson exhibited traits that would later be important in his work. He kept a careful notebook in which he recorded all the events of his life, a habit he continued until his death in 1943. He also learned to work iron and **steel** so skillfully that his help was much sought after to repair broken tools or create new ones.

In 1882, Johannsson left his home in Götlunda, Sweden, to join his brother in St. Peter, Minnesota. The two worked in lumber camps and attended Gustavus Adolphus College until 1884 when they returned to Sweden. There, in 1887, Johannsson enrolled at the Eskilstuna Technical Sunday and Evening school and later secured a position at the Carl Gastafs Stads **Rifle** Factory.

Although the concept of interchangeable parts had been pioneered by **Eli Whitney** nearly a century earlier, in Johannsson's day there was still no industrial standard for measuring equipment, and therefore no standard machine parts to be found in any two machine shops. Each production site used its own measuring gauges—which was fine only as long as all parts were made at that plant.

Recognizing this problem, Johannsson worked to create a simple, universal measuring system. First, he studied all the parts used to make a rifle, and sketched them carefully. He realized that a relatively small number of units, each a specific length, could provide, alone or in combination, all the measurements necessary to produce these parts. This revolutionary idea was called a combination gauge block set. Johannsson's first design used 102 blocks, arranged in three series, that measured all lengths between 1 and 201 mm (0.039 and 7.84 in.) rising with an increment of 0.01 mm (0.00039 in.), making altogether 20,000 measurements obtained solely by laying two or more gauge blocks together.

Ironically, the very lack of precision machining that made the blocks necessary also made them difficult to produce. They had to be perfectly flat and parallel to provide precise measurements. To ensure accurate leveling, Johannsson invented a precision instrument that measured within 0.0001 mm (0.0000039 in.); a human hair, in comparison, is 0.05 mm (0.002 in.). After the factory had produced the blocks, Johannsson finished them at home, using a converted **sewing machine** to grind the blocks with a rotating **wheel** made of cast iron. Johannsson's blocks began selling abroad in 1901, but it was not until 1907 that they were used officially in the United States.

The gauge blocks revolutionized manufacturing, particularly the American **automobile** and defense industries. This change was especially apparent during World War I, when the United States was able to arm the Allies with a flood of quickly produced weapons, a feat not possible in the days of hand-finished parts and singular measuring systems. In 1923, Johannsson went to work for **Henry Ford** at the Ford Motor Company, where he applied his gauge blocks to large-scale manufacturing. Eventually, Johannsson set up his own company which successfully manufactured standard gauge block sets and other measuring tools.

Jones, Frederick McKinley (1893-1961)

American mechanical engineer

A man with only a sixth-grade education, Frederick Jones transformed the food industry and America's eating habits with his invention of a practical **refrigeration** system for **truck**s and railroad cars.

Born in Cincinnati, Ohio, Jones was orphaned at the age of nine and was then raised by a priest in Kentucky. Jones left school after grade six and left the rectory to return to Cincinnati at age sixteen, where he got a job as an apprentice **automobile** mechanic. He boosted his natural mechanical ability and inventive mind with independent reading and study. In 1912, Jones moved to Hallock, Minnesota, where he worked as a mechanic on a 50,000-acre farm. After service with the U.S. Army in World War I, Jones returned to Hallock; while employed as a mechanic, Jones taught himself **electronics** and built a transmitter for the town's new **radio** station. He also invented a device to combine sound with **motion picture**s. This attracted the attention of Joseph A. Numero of Minneapolis, Minnesota, who hired Jones in 1930 to improve the sound equipment made by his firm, Cinema Supplies, Inc. On June 17, 1939, Jones received his first patent, for a ticket-dispensing machine for movie houses.

Around 1935, Jones designed a portable air-cooling unit for trucks carrying perishable food, and received a patent for it on July 12, 1940. Numero sold his movie sound equipment business to RCA and formed a new company in partnership with Jones, the U.S. Thermo Control Company (later the Thermo King Corporation) which became a $3 million business by 1949. Jones's air coolers for **trains**, **ships**, and **aircraft** made it possible for the first time to ship perishable food long distances during any time of the year. Portable cooling units designed by Jones were especially important during World War II, preserving blood, medicine, and food for use at army hospitals and on battlefields.

During his life, Jones was awarded sixty-one patents. Forty were for refrigeration equipment, while others went for portable **X-ray machine**s, sound equipment, and **gasoline** engines. In 1944, Jones became the first African-American to be elected into the American Society of Refrigeration Engineers, and during the 1950s he was a consultant to the U.S. Department of Defense and the Bureau of Standards. Jones died of lung cancer in Minneapolis in 1961. He was inducted into the Minnesota Inventors Hall of Fame in 1977.

JOSEPHSON, BRIAN D. • See Josephson junction

JOSEPHSON JUNCTION

The process of electron tunneling was discovered in 1957 by the Japanese physicist Leo Esaki and the Norwegian physicist Ivar Giaever. Tunneling is the process by which a charged particle is able to pass through an apparently insurmountable energy barrier. The process is possible because of the wave nature of particles.

In 1962, Brian Josephson, then a graduate student at Cambridge University, predicted tunneling could occur with *Cooper pairs* of electrons. Cooper pairs are electron pairs that form when a substance is cooled to the point where it becomes superconductive (usually, close to absolute zero). Josephson predicted that Cooper pairs could pass from one point to another even if there was no voltage drop between the two points. He said that the current would flow in one direction (DC current) if there was no external voltage applied to the system or in both directions (AC current) if there was. He also hypothesized that these currents would be very sensitive to magnetic fields in their vicinity. Josephson's predictions were confirmed by laboratory experiments completed by Anderson and Rowell for DC currents and by Shapiro for AC currents.

A *Josephson junction* consists of a pair of superconducting metals separated by a thin sheet (about 10 Å thick) of insulating material. In the absence of an external electrical source, a small DC current flows across the insulating barrier. When an external voltage is applied, a high frequency AC current develops, but with no net flow in either direction. As predicted, the presence of an external magnetic field near a Josephson junction causes very rapid changes in the current across the junction.

The sensitivity of the Josephson junction is valuable in a number of applications. Since relatively small magnetic fields induce significant current changes, the junction can be used in the design of **computer**s and scientific instruments. For example, one type of device, the **SQUID** (superconducting quantum interference device) is used as a **voltmeter** for low current measurements, as a medical device, as a magnetometer for sensitive geological measurements, and in high-speed computers.

JUDSON, WHITCOMB L. • See Zipper

JUKEBOX

The jukebox's origin dates back to the late nineteenth century. Louis Glass, an American **phonograph** manufacturer, was first to bring the public a machine that would play music as patrons inserted coins. This coin-operated phonograph, a modified version of **Thomas Edison**'s gramophone invented twelve years earlier, was introduced at the Royal Palace Saloon in San Francisco, California, in 1889.

The first phonograph to offer a selection of several tunes was the Gabel Company's *Automatic Entertainer* in 1906. This machine was dominated by a large ear trumpet speaker and was five feet (1.5 m) high. Although the *Automatic Entertainer* was more sophisticated than Glass's gramophone, it was a cumbersome machine with poor sound quality. The Gabel Company ceased production of the machine in 1908.

Dramatic advances in technology and the rise of popular culture in the early twentieth century are both reflected in the development of the jukebox. The invention of electrostatic speakers in 1926 provided the first technological boost in the jukebox's evolution. These speakers improved the sound quality of the machine and allowed greater flexibility in its design. In 1928 Justus P. Seeburg, a leading player piano manufacturer, used these new speakers in his *Audiophone*, which offered a choice of eight records. With interest in the jukebox steadily increasing during the late 1920s, more manufacturers sought to capitalize on this growing market. As Prohibition ended and the market for jukeboxes expanded along with the number of taverns, the Wurlitzer Company of Chicago, Illinois entered the field.

It was in the 1930s that the term "jukebox" actually originated. The term "juke" (*dzugu*) is a West African word that means disorder and was popularized by blacks in the southeastern United States, who coined the term "jook house" for roadhouse. The machine's technically blunt name of "coin-operated phonograph" was quickly replaced with "jukebox." With the advent of **radio**, the jukebox became a vending machine of popular music. By 1939, there were more than 225,000 jukeboxes in taverns and clubs across the United States.

Wurlitzer, Seeburg, Rock-Ola, and Rowe became the major manufacturers of jukeboxes and by the end of World War II there were over two million in operation in the United States. Ed Andrews's development of the first vertical record-changing device in 1941 increased the number of records available for selection in the machines and brought the jukebox to the brink of its peak in the 1950s. Andrews's mechanism was purchased by the Seeburg company and was first used in Seeburg's 1948 MIOOA model.

The Seeburg Company's MIOOB model, marketed in 1950, was the first jukebox to play 45's. The MIOOB offered a choice of 100 records and was characterized by a futuristic design. Throughout the 1950s the designs of jukeboxes mimicked the designs of **automobiles**. A jukebox of the 1950s characteristically had red lights and chrome rims on its facade. The Wurlitzer models were popular for the colored "bubble tubes" that ran up the sides of their jukeboxes. Jukeboxes reached the height of their popularity in the 1950s and could be found in almost every diner and truck stop.

The decline of the jukebox has been attributed to the proliferation of such competing entertainment media as portable radios, headphones, piped-in FM music and, more recently, music videos. Wurlitzer stopped its production of the jukebox

Percy Lavon Julian

in 1974, but Seeburg introduced a new 160-selection model in 1975.

Today, the jukebox often evokes a nostalgic image of the 1950s and the advent of early rock-n-roll. While jukeboxes are still being produced, the most admired and coveted models are those created in the 1950s. The Art Deco Wurlitzer Model 1015 Commercial Phonograph, which was featured in television commercials and the NBC television series ''Cheers,'' was valued for auction at $15,000. A replica of this model was priced in Germany at $4,000 in 1990. Jukeboxes found in clubs and bars today offer hundreds of song selections on recorded compact discs. Home models are available, some holding as many as 120 compact discs in a retro-jukebox cabinet. Today's machines are evidence of the incredible progress and innovation made in the field of sound reproduction during the twentieth century.

Julian, Percy Lavon (1899-1975)

American chemist

Percy Julian is best known for formulating a drug to treat glaucoma and for synthesizing sex hormones and cortisone. A native of Montgomery, Alabama and the grandson of former slaves, Julian was born first of six children on April 11, 1899. His mother, Elizabeth Lena Adams Julian, taught school and his father, James Sumner Julian, also a teacher, worked as a railroad mail clerk.

Julian graduated with the standard eight years of education from the State Normal School for Negroes in 1916 and then took additional courses to prepare for a white college. In 1920, he graduated Phi Beta Kappa and valedictorian of his class at DePauw University, where he lived in a fraternity house attic and earned tuition by waiting tables. Because doctoral programs passed over black students, he was unable to obtain a fellowship. For two years he taught chemistry at Fisk University, then received an Austin Fellowship to study at Harvard University, where he earned a master's degree in chemistry in 1923. He then worked at Harvard an additional three years as a research fellow. Returning to the classroom, Julian taught one year at West Virginia State College, then two years at Howard University. With a grant from the General Education Board, he obtained a Ph.D. in organic chemistry in 1931 from the University of Vienna, where he specialized in alkaloids.

It was under the influence of Ernst Spath that Julian first began to realize the medical applicability of soya derivatives to human illness. On his return to the United States, he headed DePauw's chemistry department from 1932 to 1936. During this period, he married sociologist Anna John and fathered two children. With the support of the Rosenwald Fund, Julian made his first great laboratory breakthrough in 1935 with the artificial creation of physostigmine, a drug that forms naturally in the adrenal glands and which lowers eye pressure in victims of glaucoma. The drug is also used in the treatment of swelling of the brain, skin and kidney disease, bronchial asthma, and leukemia.

Denied a job at the Institute of Paper Chemistry because of his race, he headed Glidden's soya research center in Chicago and later, Julian Laboratories, where he extracted sterols from soybeans. With these substances, in 1950 he contained the cost of synthetic sterols at 20 cents per gram, thereby helping arthritics obtain a derivative, cortisone, at a low price. Other aspects of his research aided the formulation of paint, printing, paper-making, and waterproofing as well as the creation of low-cost milk substitutes, livestock and poultry feed, food emulsifier, and an oxygen-impermeable fire-fighting foam used by the military in World War II.

Julian also synthesized progesterone and testosterone, both of which are essential to the body's endocrine system. His discoveries helped treat cancer and relieved problem pregnancies and menstrual disorders. He founded his own institute in Franklin Park, Illinois, in 1953, and added a satellite plant in Mexico City, where he tapped local supplies of *diosorea* (a wild yam) as a rich source of *diosgenin*, which he developed into cortexolone.

In 1964, he sold his facilities to Smith, Kline, and French and retired to Oak Park, Illinois, but continued publishing technical articles and maintained the directorship of laboratory projects until his death from cancer on April 19, 1975. Fame and utility brought a mixed reception for Julian's contributions. A devoted family man and prominent citizen of Chicago, he was named man of the year by the Jaycees and the *Chicago Sun-Times*. In spite of his acceptance by the scientific community, however, he was barred from attending some professional functions held in segregated halls, and for four consecutive years he was turned down for admittance to the Inventors Hall of Fame before his induction in 1990. When he moved to a new home, he suffered harassment from local racists but refused to be bullied. His alma mater honored his work by naming the DePauw chemistry and mathematics building for him.

JURGEN, JOHANN • See Spinning wheel

K

KAHN TEST

• See Syphilis test

KALEIDOSCOPE

Of all the inventions derived from early optical research, one of the best loved and most enduring is the kaleidoscope. Patented in 1817 by Sir **David Brewster**, the kaleidoscope became an instant success and remains to this day a popular gadget among young and old.

While conducting experiments on polarity in the early nineteenth century, Brewster noted that two **mirrors** set at thirty degree angles would create a near-infinite reflection image. He placed his mirrors in a tube with brightly colored beads in the base, and the kaleidoscope was born. Brewster's invention was immensely popular and was also easy to pirate; consequently, Brewster earned almost nothing from its sales.

Since its conception, the kaleidoscope has undergone remarkably few modifications. Polarized planes and colored foils have often replaced Brewster's beads, but the essential design is unchanged. In addition to timeless amusement, the beautiful images seen through the kaleidoscope's peephole have served as inspiration for new **carpet**, **wallpaper**, and fabric designs.

KAY, JOHN • See Flying shuttle

KECKLEY, ELIZABETH • See Dressmaking

KELLY, WILLIAM (1811-1888)

American inventor

One of the ironies of invention is that in order for the inventor to receive recognition for a new discovery, it must be made public at the risk, patent or no patent, of being copied. William Kelly was first to develop the pneumatic conversion process for making **steel**. Yet he decided to keep it a secret, and credit for its discovery was given to Sir **Henry Bessemer** a few years later.

William Kelly was born in Pittsburgh, Pennsylvania, on August 21, 1811. He was engaged in the dry goods business in Philadelphia until 1846, when he moved to Eddyville, a small town in western Kentucky, to operate a sugar kettle foundry with his brother. The foundry used the centuries-old process of heating and hammering to work out the impurities in the **iron**. A shortage of charcoal forced Kelly to search for a more efficient forging process. In 1851 he developed a method of burning off carbon from iron at a faster rate by directing blasts of air at the molten iron. By this method, he was able to develop a low grade of steel.

Instead of protecting his conversion process with a patent, Kelly chose to keep it secret and went on to build several more furnaces. In 1856 Sir Henry Bessemer patented the same process in England, completely unaware of Kelly's work. Hearing of this, Kelly finally came forward and received an American patent in 1857.

Although he could prove that he was the first to discover the process, his delay cost him the notoriety (and much of the profit) he would otherwise have earned. The process became better known as the Bessemer conversion process, and within less than a decade, Kelly's steelmaking business in the United States was overtaken by Bessemer-related interests. Kelly spent the remainder of his years in obscurity, running an axe factor in Louisville, Kentucky, while his English counterpart went on to enjoy fame and wealth. Kelly died in Louisville on February 11, 1888.

See also Blast furnace; Iron production; Steel production

William Kelly

KEMENY, JOHN G. (1926-1992)

Hungarian mathematician

Born in Budapest, Hungary, Kemeny emigrated to the United States in 1940. He received his B.S. in mathematics from Princeton University and during World War II worked in the computing center of the Manhattan Project at Los Alamos, New Mexico, assisting other pioneers in computer technology such as **John von Neumann**. After the war, Kemeny worked on his doctorate at Princeton University, helping Albert Einstein (1879-1955) with research work. With the advent of the astonishing breakthroughs in computer technology during the early 1950s, numerous computer systems were being installed across the United States. Kemeny realized such a large number of computers created a need to teach computing in an educational environment and believed that an understanding of computers would be as necessary in life as being able to read and write.

In the early 1960s, Kemeny became chairman of the department of mathematics at Dartmouth College in New Hampshire. In 1962, Thomas E. Kurtz (1928-), Dartmouth's director of computing, approached Kemeny with the idea of installing a time-share computer system at Dartmouth College. In a time-sharing system, many users are connected to one central computer, which divides processing time between users while giving them the illusion of private access. Kemeny was familiar with time sharing from research he had done earlier in his career at MIT, and gave his full approval to the project. The equipment arrived at Dartmouth in February of 1964, and Kemeny and Kurtz began the long process of installing the system and writing programs to run on it.

Although several computer languages for writing programs had been introduced during the 1950s, including FORTRAN, COBOL and LISP, Kemeny found them to be complicated and designed more for use by scientists, engineers and other technical experts. Kemeny convinced Kurtz that they needed a completely new language, simple enough for beginners to learn quickly, yet flexible enough to handle all different kinds of applications. In a few short months they developed the *Beginner's All-purpose Symbolic Instruction Code*, more commonly known as BASIC. On May 1, 1964, the Dartmouth Time-Sharing System along with BASIC language became operational.

Students were first taught BASIC programming at Dartmouth beginning in 1964. Those using the new language were able to write programs after two hours of classroom lectures, a far cry from the many weeks or months needed to master other languages in use at the time. By June of 1968, more than 80 percent of the undergraduates at Dartmouth could write BASIC programs.

Although in later years Kemeny and Kurtz wrote more powerful versions of BASIC, they always tried to maintain their original design goal of keeping BASIC simple enough for students and beginners. Although still popular in high schools and elementary schools, use of BASIC among most programmers had declined over the years in favor of other languages such as C, LISP and Pascal. But in the early 1990s, Microsoft Corporation introduced a product called Visual Basic which revitalized the BASIC programming language. By the late 1990s, Visual Basic was one of the most popular programming languages for creating computer programs for Microsoft Windows.

See also Computer network; Computer programming language; Microcomputer

KEROSENE

The invention of new **oil lamp**s in the late 1700s and early 1800s greatly improved the quality of indoor lighting, but the supply of fuel for these lamps was limited. Whale oil and other fuels were too expensive to compete with gas lighting, which was gaining popularity at the time. Then Abraham Gesner (1797-1864) produced kerosene, a pale liquid fuel that he distilled from thick crude oil. Gesner, a medical doctor from Canada who had become interested in geology, named the fuel after the Greek word *keros* (wax) and obtained United States patents for preparing it. When kerosene was successfully introduced in America in the 1850s, Gesner remarked hopefully that it might save whales from being hunted for their oil.

Around the same time, the Scottish scientist James Young (1811-1883) also began promoting the use of kerosene, which he produced by distillation from coal and oil shale. Other competitors also introduced kerosene lamp oils made from various substances, but in 1859 the discovery of huge quantities of crude oil in America made all other sources of kerosene obsolete. By the end of the century, kerosene had become the chief product of American oil refineries.

Kerosene (or kerosine, as the oil industry spells it) is a close relative of **gasoline**. Both are produced by refining crude oil, but the kerosene fraction of the oil is a little heavier. Kerosene contains a mixture of hydrocarbon compounds, most of which have 10 to 14 atoms of carbon per molecule. Because kerosene belongs to the family of hydrocarbons called alkanes or paraffins, it is sometimes referred to as paraffin oil, in addition to the nicknames coal oil, lamp oil, and illuminating oil. At room temperature, kerosene is a thin liquid that evaporates easily and smells slightly sweet. Kerosene fuel, however, is poisonous. Drinking it causes vomiting and diarrhea, and breathing it causes headaches and drowsiness.

When kerosene was first produced, the lighter gasoline fraction of the oil and the heavier fuel-oil fraction were considered useless or even dangerous waste products. American refiners were only interested in making kerosene, most of which was exported to Europe for lighting purposes. Kerosene was also widely marketed as fuel for oil **stoves**, which were introduced in the late 1800s. But around the same time, incandescent electric lights began to compete with oil lamps. Refiners were forced to market other oil products, such as fuel oil for powering ships.

Most of the kerosene produced today is used as an engine fuel for jet **aircraft** and **rocket**s. On a smaller scale, kerosene is still burned in special lamps, stoves, and space heaters, especially in rural areas that have no electricity. Kerosene appliances, however, can be dangerous; if the fuel does not burn properly, it can produce deadly carbon monoxide. Kerosene is also used as a fuel for **tractor**s and power **generator**s and as a solvent for garden chemicals such as weedkillers and **insecticide**s. In 1997 President Clinton's budget proposed a tax on kerosene which was vehemently opposed by the Petroleum Marketers Association of America (PMAA) because they said it would hurt millions of U.S. homeowners, farmers, and others that purchase tax free kerosine.

See also Gaslight; Oil refining

KEROSENE LAMP • See Oil lamp

KETTERING, CHARLES (1876-1958)
American engineer

Charles Franklin Kettering was born in Loudonville, Ohio, on August 29, 1876. In 1904, after a few years spent away from school working as a teacher and another absence from college due to his poor eyesight, Kettering graduated from Ohio State University with degrees in both mechanical and electrical engineering. Upon graduation Kettering went to work for the National Cash Register Company (NCR) in Dayton, Ohio, where he invented the first electric **cash register**.

While at NCR Kettering worked nights developing a better **ignition system** for **automobile**s. His system was so promising that he left NCR to form his own company with Edward A. Deeds, called the Dayton Engineering Laboratories Company, or Delco, as the company was soon called. It was here that Kettering developed the electric self-starter for automobiles as part of a comprehensive electrical system for automobiles that also included improvements in the ignition and lighting systems. These systems were first used in Cadillac cars in 1912. During this period Kettering also developed a small self-contained electric **generator** that revolutionized life in areas not served by electric utility power lines.

Charles Franklin Kettering

In 1916 Kettering and his associates sold Delco to the United Motor Company (which later became part of General Motors). After years of professional experience in a variety of areas, including serving as vice-president of the Dayton-Wright Aeroplane Company, Kettering agreed to transfer his interests to General Motors. He later became vice-president in charge of research, and was eventually made a director of the company. While employed at General Motors, Kettering spearheaded innovations and made significant contributions to dozens of technological advances, including tetraethyl, or leaded **gasoline** to eliminate engine knock, quick-drying lacquer finishes for cars, four-wheel **brakes**, **safety glass**, high pressure lubricants, and high-compression automobile engine. Kettering collaborated with Thomas Midgley, Jr., in the development of Freon, the gas recently used in **refrigeration** and air-conditioning units throughout the world. (Its manufacture has been prohibited since 1986 because Freon contains **chloro-**

fluorocarbons, which are linked to destruction of the Earth's **ozone** layer.) Kettering was also instrumental in refining the engine invented by **Rudolf Diesel**. These improvements led to the eventual abandonment of **steam engines** for railroad locomotion and the widespread use of diesel power on railroads and in **trucks** and **buses**.

Kettering was more than just an inventive engineer. Through his Charles F. Kettering Foundation and the Sloan-Kettering Institute for Cancer Research, he encouraged and took an active interest in various fields of medical and scientific research. He served as president of both the Society of Automotive Engineers and the American Association for the Advancement of Science, and was chairman of the National Inventors' Council and the National Patent Planning Commission. He also served on the boards of several universities and was a Fellow of the National Academy of Sciences. Kettering received more than thirty honorary degrees from universities and colleges all over the world. A much sought-after public speaker, Kettering gave hundreds of addresses and radio broadcasts. In recognition of his stature as a practical philosopher, he was elected a member of the American Philosophical Society, an organization founded over two hundred years ago by **Benjamin Franklin**. Charles Franklin Kettering died on November 25, 1958, and is buried in Dayton, Ohio.

Kevlar

Kevlar, or poly para-phenylene terephthalamide, is the product of a long search by the Du Pont Company for a high strength fiber. Several problems were encountered in its development. Because of the strength of the molecular bonds, the polymer would have a high melting temperature. Therefore it would have to be able to dissolve in solvents before forming into fibers.

In 1964, American researcher **Stephanie Louise Kwolek** began experimenting with poly-p-phenylene-terephthalate and polybenzamide. Kwolek was the first person to prepare pure monomers that could be used to synthesize polybenzamide. The intermediates required in this process proved highly sensitive to moisture and heat, and therefore also sensitive to hydrolysis and self-polymerization. Kwolek found a suitable solvent and low-temperature polymerization conditions, and succeeded in producing a fluid, cloudy liquid crystalline solution of synthetic aromatic polyamides. In 1976, Kwolek discovered the technology for spinning such solutions into fibers. The spun fibers proved to be much stronger and stiffer than any previously known.

Kwolek, a 1946 graduate of the Carnegie Institute of Technology, received the American Chemical Society Award for Creative Invention in 1980 for her patented spinning process. She retired from full-time research at Du Pont in 1986. In July 1995, she was inducted in the National Inventors Hall of Fame.

Kevlar is a polyamide fiber. It is formed by a polycondensation reaction of para-phenyldiamine and terephthalate, a chloride in solution, at low temperatures. The resulting polymer is pulverized, washed, and dried. It is then mixed with a strong acid and forced through a metal plate with small holes. Once forced into fibers, the polymer is passed through air into cold water. The fibers are washed and wound onto bobbins to dry. By altering the solvents in the first reaction with additives, different properties may be brought out.

Kevlar's strength comes from its molecular structure. Long carbon chains consisting of alternating aromatic rings and amide groups are crosslinked by hydrogen bonding. The covalent bonds between the aromatic rings and amide groups are much stronger than the hydrogen bonds, which is why there is such a great difference between the longitudinal and transverse strengths. Kevlar is highly resistant to chemicals, moisture, and heat but does show a tendency to degrade in ultraviolet light. In situations where Kevlar will be exposed to light, it is covered with UV resistant coatings. Du Pont markets Kevlar fibers under the trade names *Kevlar*, *Kevlar 29*, and *Kevlar 49*. They are used to reinforce *radial tires*, rubber products and **plastic** goods, and as ropes, cables, fabrics and tapes. Fields in which Kevlar fibers have found applications include **adhesive** and sealants (thixotropes); ballistics and defense (anti-mine boots, gloves, composite helmets, bullet-proof vests); belts and hoses (automobile); composites (**aircraft**, boats, sporting goods); **fiber optics** and electro-mechanical cables (communication and data transmission, ignition wires, **submarine** and robotic tethers); friction products and gaskets (asbestos replacement, automotive and industiral gaskets, brake pads, clutch linings); protective apparel (boots, chain saw chaps, thermal and cut protective aprons); tires (aircraft, **automobile**, off-road, race, **trucks**); ropes and cables (antennae guy **wires**, fish line, utility ropes, lifting slings, tow lines, netting and webbing, pull tapes). Kevlar's light weight and high strength makes it a highly marketable polymer fiber today.

See also Rubber, synthetic; Rubber, vulcanized; Tire, pneumatic

Key • See Lock and key

Keyless entry system

An extension of the **lock and key**, keyless entry systems were made possible in the 1970s with the advent of microchip technology. In principle, keyless entry systems are similar to combination locks because they require an alphanumeric code. This code has the advantage of being easily changed to improve security, but like the combination to a lock, must be memorized by the user to be effective. Early versions of keyless entry systems used keypads similar to those found on touch-tone **telephones**. Later models, using more advanced microchip technology, more closely resemble the keypad on a computer keyboard.

The major innovation in keyless entry systems has been the addition of remote control technology. Using a transmitter inside the remote control unit, locks can be opened from some distance with the press of a button. The remote keyless entry systems have several advantages. They eliminate the need for a keypad and alphanumeric code. In addition, it allows the holder to safely and quickly open a door without fumbling with traditional keys. Not surprisingly, remote keyless entry has found wide application in the **automobile** industry. Attached to the traditional ignition key, remote controls became virtually a standard feature in automobiles by the mid-1990s. Their use in homes is also expected to expand.

Kilby, Jack St. Clair (1923-)

American electrical engineer and inventor

When he coinvented the **integrated circuit**, or microchip, Jack Kilby also co-launched the age of modern **electronics**. He was born in Jefferson City, Missouri, and spent most of his childhood in Great Bend, Kansas. Following in his father's footsteps, Kilby earned a degree in electrical engineering from the University of Illinois in 1947, his studies having been interrupted by service in World War II. He went on to earn his master's degree in 1950 from the University of Wisconsin. At this time Kilby was also employed at the Centralab Division of Globe-Union Corporation, where he worked on ideas for manufacturing all parts of an **electrical circuit** on a single base.

After attending a seminar at Bell Laboratories in 1952 on the newly patented **transistor**, Kilby tried to develop transistor-based **hearing aids**. Dissatisfied with Centralab's commitment to the use of germanium rather than silicon transistors, Kilby joined Texas Instruments in 1958, which was already working with silicon. At TI, Kilby tackled the problem of miniaturization. While he and his fellow engineers could design ever-more-complex tiny circuits, the vast numbers of interconnections simply could not be physically made. In 1958, while working virtually alone at TI while everyone else was on vacation, Kilby came up with the idea of integrating all the parts of a circuit on a single slice of silicon. By September 1958 he had created the first working integrated circuit. He applied for a patent in 1959, and a long battle followed between Kilby/Texas Instruments and Robert Noyce/Fairchild Semiconductor, with Noyce ultimately being granted the IC patent. Nevertheless, Kilby is recognized as an independent coinventor of the microchip.

In 1965, Kilby was put in charge of directing a team at Texas Instruments to develop the world's first pocket **calculator**, made feasible by the microchip. Within a year Kilby and his colleagues had a working prototype, and a year later they filed for a patent. Kilby was recognized for his achievements by a steady series of promotions at TI which, however, increasingly distanced him from actual inventing. So, in 1970, Kilby left TI to work as a freelance inventor in North Dallas. He patented a number of diverse devices and engaged in extensive research on **photovoltaic cell**s, served as distinguished professor of electrical engineering at Texas A & M University in the 1980s, and headed the Houston Area Research Center. An unhurried, quiet man, Kilby is a classic inventor—a man who loves to take on a difficult problem and stick with it until he has found a creative, practical solution. In 1982 he was inducted into the National Inventors Hall of Fame.

Kiln, pottery • See Ceramic

Kinetophone

At the turn of the twentieth century, European and American inventors were on the brink of combining sight and sound in the **motion picture**. In 1895 **Thomas Edison** joined his **phonograph** with his **Kinetoscope**, a motion-picture viewer. As watchers looked through the Kinetoscope's eyepiece, they heard sounds from ear tubes connected to a phonograph. Although there were insurmountable problems involved in synchronizing sight and sound in the Kinetophone, as the hybrid was called, it set an important precedent for sound motion-picture viewers.

In 1899, after producing a commercial film projector, Edison resumed work on the Kinetophone. The film projector was linked to a special long-playing phonograph. The Kinetophone was developed around oversized cylinders of about 4.25 in. (11 cm) in diameter and 7.5 in. (19 cm) in length, which ran for approximately six minutes. In 1908 Edison hired an inventor, Daniel Higham, to work on the Kinetophone. Higham increased the Kinetophone's volume by using a special **valve** to improve the performance of the diaphragm that amplified the sound vibrations made by the needle on the phonograph cylinder. Edison's close associate, Miller Reese Hutchison (1876-1944), directed the Kinetophone project, and by 1910 a working Kinetophone was shown to the press. An earlier design that used an electrical **control system** had been abandoned for a mechanical control to synchronize the sight and sound.

Declaring that he had perfected talking pictures, Thomas Edison on February 17, 1913 demonstrated his new Kinetophone in a New York theater by projecting a scene from Shakespeare's *Julius Caesar* synchronized with the actors' voices for seven minutes of the one-hour silent movie. When the sound got out of synchronization with the actors' gestures or facial expressions, an operator in a projection room used a long fishing line to adjust the speed of the phonograph, which was hidden behind the viewing screen. Despite the crude methods of synchronization and the short duration of the sound, talking motion pictures had a great impact on audiences. Audible gasps were said to be heard in the theater when the audience first heard sound coming from behind the moving picture screen.

The Kinetophone was Edison's leading profit maker in 1913, exceeding the income earned from silent pictures, phonographs, and records. However, the Kinetophone's popularity was short-lived. After their initial enthusiasm, audiences became intolerant of its defects. They wanted sound to be heard throughout a film, and they noticed the synchronization problems. For another year, Edison tried to improve the Kinetophone, but after 1914, he made no further attempts to perfect talking pictures. Nevertheless, Edison had introduced the concept of sound motion pictures. Within ten years talking pictures became a reality. In 1927, *The Jazz Singer*, the first full-length sound motion picture, came to the big screen.

KINETOSCOPE

Designed as an arcade machine, the Kinetoscope was a cabinet into which a person could look through a peephole and watch pictures move. An electric motor moved a filmstrip to an eyepiece, where a slotted disc exposed the images to the viewer at a rate of approximately 40 frames per second. About 50 feet of film revolved on spools to produce a 30-second film.

Although **Thomas Edison** was involved in many pursuits, he had a strong interest in **photography** and in 1887 first began work on pictures that appeared to move. In 1889, a series of pictures could be photographed and moved rapidly using **George Eastman**'s **celluloid** camera film. Mechanic and inventor William Kennedy Laurie Dickson (1860-1937), assigned by Edison to the photography project, invented both a camera (Kinetograph) to make perforated filmstrips using celluloid roll film and a machine to view them—the Kinetoscope. Both machines were patented under Edison's name, and it was not until the 1960s that Dickson received credit for his contribution to motion picture history.

In 1893, Thomas Lombard, the promoter of Edison coin-operated **phonographs**, urged Edison to manufacture a coin operated Kinetoscope. Edison accepted the challenge and responded by manufacturing 25 Kinetoscopes, which operated for a nickel a view. These Kinetoscopes were placed in arcades next to phonographs.

For his film making, Edison erected a wooden building in the yard of his West Orange, New Jersey, laboratory. Wanting to use only natural light, he constructed the building to swing on pivots. The stage inside could follow the sun, the light of which entered the studio through a special opening in the roof. The building was covered with black tar paper and painted black inside to minimize reflections. Edison's staff soon called the studio the Black Maria.

In 1894, Edison opened the Kinetoscope Parlor in New York City, where he had two rows of peep-show machines. He played bits of vaudeville acts and some homemade moving pictures. A man sneezing, an Italian organ grinder and his monkey, a dancer, and a man smoking a cigar were among his first film subjects. In early 1893, Norman C. Raff and Frank

The earliest kinetoscope film submitted for a copyrights was of this sneeze, recorded by Thomas Edison

R. Gammon created the Kinetoscope Company and purchased a large number of his peep shows at $200 each to be exhibited in kinetoscopic parlors. The contract price included the film and the coin operation boxes. The kinetoscopic parlors enjoyed only marginal success for viewers wanted longer films than the one-minute Kinetoscope shows. The answer was to eliminate the single-viewer concept and to project longer shows on a large screen for many to enjoy. In effect, the Kinetoscope's failure was a primary contributor to the birth of the large screen **motion picture**.

See also Photographic film

Kirchhoff, Gustav Robert • See Spectroscope

Kite

The Chinese are credited with inventing the first kites, perhaps as early as 1000 B.C. No one knows what inspired their creation, but they were used for pleasure and had military purposes as well, such as measuring the distance to enemy forts in order to tunnel under them.

The use of kites eventually spread from China to the Pacific. In Japan, Korea, Myanmar (Burma), Indonesia, Melanesia, and Polynesia they attained religious and ceremonial significance. There are stories from these cultures about giant kites that could carry a person. In Japan people used kites to carry tiles and **bricks** to workmen constructing towers.

Kites eventually became known around the world, but it took a long time before any scientific use was made of them. In 1749 Alexander Wilson measured temperature variations at different altitudes by raising thermometers on six kites flying as high as 3,000 feet (915 m). Probably the most famous kite experiment of all involved **Benjamin Franklin**. In 1752 he used a kite in a thunderstorm to prove that lightning was **electricity**. At the top of his kite he attached a **wire** to attract the lightning. Near his hand he tied a **silk** ribbon and **key** to the twine. The electricity came down the wet string and out the key, as he noted when he put his hand near it.

Fifty years later, **George Cayley** used kites to explore the possibility of heavier-than-air flight. His first model **glider** used a kite as a wing unit and might have resulted in an earlier airplane if he had been able to find an adequate power source. Cayley's later gliders used kites for both wings and tail sections. One of these became the first passenger glider when it carried a servant of his across a valley in 1853.

Kites also helped save lives at sea. An Irish priest, E. J. Cordner, built a series of kites to help buoy a small boat leaving a stranded ship. In France, C. Jobart devised a lifesaving kite to fly a line from a crippled **ship** to the shore.

Kites continued to provide meteorological information. A British man in 1833 succeeded in lifting **anemometers** on kites to measure wind speeds at various altitudes. Americans established the Franklin Kite Club for further experiments, and W. R. Birt used them to study weather in 1847.

Lawrence Hargrave introduced the box kite to the world in the 1890s. This kite, the product of his search for powered flight, was extremely stable and had great lifting power. For the next thirty years it was used for meteorology. Its stability also made it a model for several early power-driven airplanes in the first decade of the twentieth century.

B. F. S. Baden-Powell was an important pioneer in man-lifting kites. He built a traditional, four-sided kite that was 36 feet (11 m) high and capable of lifting men 10 feet (3 m) off the ground. He built smaller kites that flew together, eventually carrying men as much as 100 feet (30 m) above the ground. It was with one of these kites that **Guglielmo Marconi** made his first successful transatlantic wireless reception tests from England to Newfoundland.

American Charles Lamson was the first to fly successfully inside a kite. Before this, pilots had been suspended below the kite. In 1897, using a structure that resembled future airplanes, Lamson built two pairs of biplane surfaces made of canvas stretched over wood. The pilot could move the back portion, thus adjusting the angle of incidence, and he could shift his weight left or right for further control. The device worked, and for half an hour flew at a height of 30 feet (9 m).

Meanwhile, **Otto Lilienthal** was exploring aerodynamics through kites. He flew gliders and eventually became the first person to achieve sustained controlled flight, often soaring for 750 feet (229 m). The Wright brothers used biplane kites to develop their heavier-than-air machine. In 1900 they flew this kite to confirm their theories of control through wing warping.

Samuel Cody established a remarkable system of man-lifting kites. He started with the box design, added several lifter kites to one flying cable and set up conical stops on the cable. Finally, he added a passenger-carrying kite attached to a trolley with **wheels** that ran along the cable. He hoped this arrangement could be used by observers during wartime. In 1905 he was able to send a man up 2,600 feet (793 m). Soon, however, airplanes overtook any advantages these kites might have had.

Alexander Graham Bell was also an avid kite experimenter. He created huge, multi-celled kites that looked like beehives. His Cygnet, made up of 3,393 cells and equipped with floats, carried a man for 7 minutes. He tried to add an engine for powered flight, but none were available with sufficient thrust.

As airplanes became a reality, kites were forgotten until the 1950s. But Francis Rogallo, an American, invented the flexible kite, which resulted in the famous delta wing kite used today in hang gliding. In the 1960s Domina Jolbert created a wing-shaped device made entirely of fabric that keeps its shape as the wind enters openings on its leading edge. This is now called the parafoil, the lightest, most efficient, and economical non-mechanized lifting surface yet devised.

See also Aircraft

Knife, fork, and spoon

The knife and the spoon are of prehistoric origin. Primitive peoples made knives for cutting from bones, shells, and stones, particularly flint, that flaked to form sharp edges. The Egyptians also manufactured knives from flint chipped in such a way as to form serrated edges; the flint pieces were then glued into slots formed from wooden handles. The blade was kept sharp by rubbing it in the hollow of a stone. The first spoons were probably seashells or crudely formed stone ladles that provided an ideal way to transport liquids to the mouth. Later knives and spoons were crafted from **copper** and **bronze**, clay, wood, ivory, iron, and gold. The fork is a more modern invention. For a long time food was cut into small pieces at the table with a knife, which then was used to spear the pieces and bring them to the person's mouth. It is speculated that the fork was invented to hold down meat (or other foods) while cutting it with a knife. The twelfth-century Italians were the first people known to use a two-pronged fork for eating. By the 1600s, forks and knives were commonplace eating utensils in France and in England; the first fork was brought to North America in 1630 by John Winthrop, governor of the Massachusetts colony. Today flatware for eating is made in colorful **plastics** for picnic or other casual use, or from silver, silver plate, gold plate, or stainless steel for use in homes and restaurants.

See also Swiss army knife

Knight, Margaret E. (1838-1914)
American inventor

Born in York, Maine, Margaret E. ''Mattie'' Knight, nicknamed ''Lady Edison,'' spent her childhood in Manchester, New Hampshire, where she received a meager education. Estranged from the usual pastimes of little girls, she made homemade **kites** and sleds, which were the envy of the neighborhood. At age twelve, while observing her brothers at work in a cloth factory, she saw a large metal-tipped shuttle drop from the **loom**, endangering a worker. This incident motivated her to devise a stop-motion device to prevent loom accidents.

As an adult, Knight moved to Springfield, Massachusetts, and worked for a manufacturer of **paper bag**s. After studying the operation, in 1867 she devised a mechanism that produced a bag with a square bottom, which would enable users to load the bag without holding it erect. Two years later, she took her idea to Boston and began putting the finishing touches on the device so that it could be patented. Another inventor, Charles F. Annan, copied her idea and applied for his own patent. Knight contested his claims in court and won her suit in 1870.

For most of her productive years, Knight lived in Ashland and Framingham, Massachusetts, but she also maintained a workshop in Boston, and was associated with the Knight-Davidson Motor Company of New York. She created a number of handy domestic gadgets, including various machines that cut out and sewed **shoes**, a window sash and frame, a dress and skirt shield, a **barbecue spit**, a clasp for holding robes, and a numbering device. Late in her career, she studied **rotary engine**s and evolved a sleeve-valve engine, a horizontal variation on the vertical poppet **valve**, which was posthumously patented in 1915. The return for her work, however, brought her small profit; she died of pneumonia and gallstones in 1914, leaving behind an estate valued at less than $300.

Knitting machine

During the latter half of the eighteenth century, improvements in the British textile industry were ushering in the Industrial Revolution; machines were invented to speed the production of woven fabric as well as meet the rising demand for yarn and thread. While the focus of this activity seemed to be weaving and spinning machines, similar improvements to knitting technology were being made. Today, knitting machines are just as important to the textile industry as weaving machines.

The first knitting machine was constructed around 1589 by an English reverend named William Lee (1550-1610). As the story goes, Lee was tormented by the constant clack-clacking of his wife's knitting needles. He imagined a device that, instead of producing one loop at a time, could knit an entire row of loops at once. While devices of this sort had been used by **carpet** weavers thousands of years earlier, Lee added to their design a line of hooks that would release the knitted loops, making room for the next row.

Lee soon abandoned his position in the church, traveling with his brother to the court of Queen Elizabeth I. There he presented his invention, requesting a royal grant to exploit the device, as well as sole claim to its patent (along with all the ensuing profits). The Queen refused, and Lee left England to try his luck at the court of King Henri IV of France. Though the French sovereign granted Lee the privileges he sought, before the inventor could establish himself, Henri IV was assassinated. Lee died penniless in Paris in 1610.

Though disheartened, Lee's brother continued to seek a financier for the knitting-frame. He returned to England, where he entered a partnership with a Nottingham businessman to build the world's first knitting factory. Soon the factory was so successful that the local hand-knitters appealed to the government to limit the use of the knitting-frame; by that time, several factories were in operation, manufacturing stockings so cheaply that they were no longer considered a luxury item.

The methods for knitting clothing remained unchanged until the early 1700s. About that time, improvements in the design of the knitting-frame began to appear. Machines were constructed to produce warp-knit fabric, which was less elastic and more like woven fabric than previous knits. In 1758 Jedediah Strutt designed a **loom** that could knit ribbed material, ideal for use as hosiery. The circular loom was invented by Frenchman Decroix in 1798; it created seamless tubular fabrics that were often employed as undershirts and underwear, as well as ladies' stockings; the Decroix loom was popularized by **Marc Isambard Brunel**, who patented it in 1816. Matthew Townsend improved upon the circular loom in 1847, enabling it to produce ribbed fabric. Finally, in 1855, the first water-powered knitting frame was constructed in Loughborough, England.

Arguably the most important development in knitting technology was the latch needle, patented by Townsend in 1856. This needle was essentially a half-hook with a small latch that would open and close, depending upon the needle's position. By using the latch needle the entire knitting process was made faster and more flexible. By the late 1860s, small home knitting machines were made available.

Today, knitting takes place on giant machines employing hundreds of needles per line and capable of producing more than four million stitches each second. Patterns were initially read by the machines off of punch cards; today, both home and commercial machines can be computer controlled to knit complex patterns. In many areas knitting has even superseded weaving, partially due to the fact that it can be performed almost twenty times faster.

KNOLL, MAX • See Electron microscope

KOLFF, WILLEM JOHAN (1911-)

Dutch American physician and biomedical engineer

Willem Johan Kolff was born in Leiden, the Netherlands, on February 14, 1911, the son of Adriana (de Jonge) and Jacob Kolff, a doctor who ran a tuberculosis sanatorium. After graduating from medical school at the University of Leiden in 1938, Kolff accepted a teaching post at the University of Groningen. There he was influenced by a professor who introduced Kolff to the basic concepts of dialysis—the principle whereby a solution of a high concentration passes through a semipermeable membrane to a solution with a lower concentration. He was affected as well by patients who were dying of kidney failure.

The kidneys are vital organs in the body; among other duties, the pair of organs is responsible for eliminating liquid wastes from the body, filtering over 400 gallons of blood daily. When the kidneys are not functioning correctly, uremia—a pathological condition caused by the accumulation of waste products normally removed in the urine—develops. The body swells, and if there is no intervention, death results. By the early twentieth century, doctors knew that they needed to create an external "kidney" to replace the function of damaged kidneys. As early as 1913 in the United States, doctors had managed a crude blood dialysis, or blood filtering, of dogs using celloidin as the filtering membrane. A major problem that they could not overcome, however, was preventing the coagulation of the blood once it was outside the animal. During and just after World War I, a German doctor, Georg Haas, using a new anticoagulant, heparin, successfully managed the first human dialysis, but still there were problems: Haas's apparatus could not filter enough blood quickly, and the supply of the anticoagulant was minimal.

At Groningen, Kolff was able to capitalize on these ground-breaking achievements and completed his own dialysis research with a simple sausage packing made of cellophane. He discovered that when the sausage casing was filled with a liquid, such as blood, then agitated in saline solution, the casing made a perfect membrane for filtering wastes. With the German takeover of Holland in 1940, Kolff, an outspoken anti-Nazi, left Groningen for the country town of Kampen, where he became the internist at the local hospital. There, in addition to carrying out his hospital duties and helping the resistance whenever possible, Kolff continued work on his **dialysis machine**. The anticoagulant heparin had become commercially available by that time. To solve the problem of adequate blood supply he created large rotating drums, adopting an idea automotive magnate **Henry Ford** had used to design **water pumps** in his engines. Kolff's attempts with humans initially were unsuccessful, and it was not until just after the end of the war, in September of 1945, that he successfully treated a patient, saving her life.

While still a medical student in 1937, Kolff had married Janke C. Huidekoper, and they already had four children, with another on the way, by 1950, when Kolff accepted a position in the United States at Ohio's Cleveland Clinic. Meanwhile, Kolff's dialysis machine was being improved upon by others around the world. Kolff then moved on to other medical endeavors: he was intent on creating a **heart-lung machine** that could be used to keep bodily functions operating and thus enable open-heart surgery to take place. In the 17 years he spent at the Cleveland Clinic Foundation as professor of clinical investigation, Kolff, working with C. P. Dubbelman, designed an artificial heart-lung device known as a pump-oxygenator. By 1961 another of Kolff's designs was in service: the intra-aortic balloon pump to be used in cases of circulation failure. In addition, his attention increasingly turned to designing and implanting an artificial heart. In December of 1957, Kolff and his Cleveland team removed the heart from a dog and replaced it with a pneumatic pump which kept the dog alive for 90 minutes, proving the viability of the **artificial heart**.

In 1967 Kolff moved from Cleveland to accept a position at the University of Utah as the head of both the Division of Artificial Organs and the Institute of Biomedical Engineering. As such, he led a team of surgeons, physicists, and cardiologists in designing an implantable artificial heart for humans. It was this team, using a heart designed by one of Kolff's students, Robert K. Jarvik, that implanted the first artificial heart in a human on December 2, 1982. The patient, Barney Clark, lived for 112 days, proving the viability of such a procedure. Kolff also designed an artificial ear and eye, though the latter, with its numerous devices necessary for operation, is far from a practical stage. Kolff has gone on to develop a portable kidney dialysis machine, introducing, as early as 1975, the Wearable Artificial Kidney (or WAK), which allows for home dialysis.

Kolff became a naturalized U.S. citizen in 1956 and is the author of close to 700 articles and papers on the kidney, heart, and artificial organs. Recognized worldwide for his efforts in bionics, Kolff is the recipient of more than 100 awards and honors, including the Landsteiner Silver Medal in 1942 from the Red Cross of the Netherlands, the Frances Amory Award in 1948 from the American Academy of Arts and Science, the Gairdner Prize in 1966, the Leo Harvey Prize in 1972, and the Japan Prize in 1986. He was named one of the one hundred most important Americans of the twentieth century by *Life* magazine in 1990.

König, Friedricher (1774-1833)

German printmaker and engineer

Friedricher König brought about the first major changes in mechanized printing since the innovations of **Johannes Gutenberg** in the fifteenth century. Since Gutenberg's time, there had been a number of changes in the nature and the readership of printed material, many of them caused in part by Gutenberg's development of mass printing itself. Increases in population, the spread of newspapers and periodicals, the rise of the novel as a literary form, the growth of leisure reading, and the need for print communication in the increasingly complex world of trade factors that spurred a drastic increase in the demand for sheer numbers of printed pages, a demand which became increasingly hard to meet. As late as the beginning of the nineteenth century, printers were using essentially the same techniques as their medieval counterparts—setting a matrix of moveable type, then using their own hands to feed **paper** into a printing press and to raise and lower the platen (a flat metal sheet which presses the paper against the inked type). Born on April 17, 1774, in Eisleben, Saxony, König devoted himself to the idea of using steam, which had been employed in other industries, to power the workings of the printing press. Because he could not get financial backing for his idea in Germany, König moved to London in 1806.

For eight years, König worked on a machine which moved the platen with the steam power, but was unsuccessful in synchronizing the platen's vertical motion with the horizontal motion of the type, which moved back and forth under inking rollers. Working in concert with the engineer Andreas Bauer (1783- 1860), König found a solution in 1812. Rather than simply changing the means of powering the platen, König changed the flat platen itself, replacing it with a cylinder—an idea proposed, but never developed, by English scientist William Nicholson (1753-1815) in 1790. The cylinder rolled in synchronization with the type, carrying and pressing down hand-fed sheets of paper as it moved. König refined his press to employ two cylinders, enabling two sheets to be processed with each cycle of motion. This machine was installed in 1814 to print *The Times* of London, and could print up to 1,200 copies per hour—four times the rate of a manual press. *Blumenbach's Physiology*, the first book printed with the process, was published in 1817. König and Bauer experimented further with their press, and four years later introduced the "perfecting machine," which employed a two-cylinder system to print both sides of a sheet at one pass. König died in Obserzell, Bavaria, on January 17, 1833.

See also Printing technology

Korolev, Sergei (1906-1966)

Russian engineer

Sergei Korolev was a major figure an driving force in the development of the Russian space program. Influenced by the ideas of interplanetary flight put forth by **Konstantin Tsiolkovsky**, Korolev became interested in rocketry, and in the early 1930s helped to found a **rocket** research group which was responsible for training numerous scientists and engineers who would later become the core of Russia's space program.

Korolev was responsible for many of Russia's well-known achievements in space exploration. He helped design the rocket used for **Sputnik 1**, the first artificial satellite to orbit the earth. The rocket was a modified Soviet Intercontinental Ballistic Missile (ICBM) about 100 feet (30.48 m) in length with a weight of 300 tons (272,400 kg). This became the most widely used rocket in the world. Another success associated with Korolev was *Luna 3*, which was a probe that provided the first views of the far side of the moon. In 1959 it looped around the moon, took pictures, developed them, and radioed them back to earth. This flight bolstered the prestige of the Soviet Union throughout the world. It was Korolev who led the design team responsible for *Vostok*, the first manned spacecraft that went up in 1961 with Yuri Gagarin.

Korolev was also in charge of the *Venera 3* mission, the first spacecraft to impact on another planet. It landed on Venus in 1966, and even though it failed to return any information due to loss of contact, *Venera 3* was able to relay a great deal of information about interplanetary space before it crashed. The same year, Korolev had another first when his *Luna 9* made the first successful soft landing on the moon. It sent back **television** images and showed that the feared, deep layers of lunar dust did not exist, allowing further flights and later manned missions. Korolev died in 1966 and was buried in Kremlin Wall, an honor reserved for Russians of exceptional distinction.

Korolev was also responsible for leaving behind a group of dedicated and highly trained scientists and engineers. His job entailed training large numbers of these individuals, who are still working in many space/rocket engineering research institutes and design bureaus. It was Korolev who also left behind the large "cosmodrome" where flights originated. His inexhaustible energy and talent as a researcher, his intuition regarding engineering problems, and his creative boldness in solving difficult tasks have established Sergei Korolev as a founding father of the Russian space program.

Kountz, Samuel L. (1930-1981)

American surgeon

One of the pioneering kidney **transplant** specialists in the country, Samuel Kountz was born on October 30, 1930 in Lexa, Arkansas. He graduated in 1952 from the Agricultural, Mechanical and Normal College of Arkansas and obtained a chemistry degree from the University of Arkansas. With the backing of Senator J. W. Fulbright, he won a scholarship to the University of Arkansas Medical School, becoming the first African-American to study there. While training at the Stanford Medical Center, Kountz discovered that large doses of *methylprednisolone* stopped the rejection of transplanted kidneys. In addition, Kountz performed the first kidney transplant between patients who were not identical twins in 1964. This transplant, between a mother and a daughter, made kidney transplants possible for thousands of ailing patients. In the following eight years, 5,000 kidney transplants were performed. Kountz also helped build a device to aid kidney transplants. At the University of California in 1967, Kountz and colleague Folker Belzer produced a machine capable of preserving donated human kidneys for more than two days while they were delivered to a suitable recipient. On a teaching assignment to South Africa in 1972, Kountz contracted a mysterious disease which resulted in brain damage. He never fully recovered, remaining seriously ill until he died at his home in Great Neck, New York, on December 23, 1981.

Kwoh Yik San • See Robotics

Kwolek, Stephanie Louise (1923-)

American Chemist

Stephanie Louise Kwolek is best known for her discovery of liquid crystalline solutions of synthetic aromatic polyamides and their fibers. These fibers became the first of a new generation of high performance fibers, and the basis for the commercialization of Du Pont's **Kevlar** brand fibers.

Kwolek graduated with a degree in chemistry from the Carnegie Institute of Technology in 1946, and joined Du Pont in the same year. She moved to Du Pont's Pioneering Research Laboratory at the Experimental Station when it opened four years later. There she was encouraged and supported by Hale Charch, director of the laboratory at that time, as she struggled to gain intellectual acceptance in a male-dominated research community.

Kwolek began her career scouting for new **polymers**. In 1964, Kwolek began experimenting with poly-p-Phenylene-terephthalate (PPD-T) and polybenzamide (PBA). She became the first person to prepare pure monomers that could be used to synthesize PBA.

The intermediates needed in this process proved extremely sensitive to moisture and heat, so were readily susceptible to hydrolysis and premature self-polymerization. Kwolek discovered a suitable solvent and identified the low-temperature polymerization conditions required to produce an unusual polymer solution that was fluid and cloudy, rather than clear and viscous. Upon spinning this solution, she found that tough fibers with startling properties had formed. These fibers were much stronger and stiffer than any previously made synthetic fiber. The discovery ushered in a stimulating period of research at Du Pont. The search for a candidate for commercialization finally came up with Kevlar.

Kwolek has been honored on many occasions for her pioneering studies. In 1980 she received an award for Creative Invention from the American Chemical Society. In 1995, she was inducted into the National Inventors Hall of Fame.

See also Kevlar; Polymer and polymerization; Fiber, synthetic

Sergei Korolev

L

Lacquer

Lacquers are **varnish**es made by dissolving a cellulose-based solid resin, or polymer, with modifying components (usually plasticizer but sometimes pigments) into a solvent. The resultant coating dries quickly and leaves a hard, durable, usually high-gloss finish. Because lacquers dry by the evaporation of solvents, solvents that not only dissolve the components but also evaporate quickly must be used. Lacquers are normally applied using air sprayers because of their extremely short drying times. Originally high viscosity nitrocellulose compounds were used, but today, cellulose esters and cellulose ethers are more often used.

Though crude lacquers were used in the Orient as early as 200 B.C., modern processes for making lacquer were only developed around the beginning of the twentieth century. At this time, large companies, such as Du Pont and the International Smokeless Powder and Chemical Company, saw the potential for the use of lacquers in consumer products. They purchased the rights to lacquers and began research into **plastic**-based formulas known as Duco lacquers. By 1920 plastic lacquer was being used on broom handles and toys and within five years the use of lacquers in the United States had spread to a host of products.

These Duco lacquers were particularly important for industry because prior to their development, the production of anything needing to be coated, or painted, was slowed greatly by the drying time of the coating, especially when several coats were required. At times, manufacturers had 15,000 automobiles in warehouses for a week to a month while waiting for the paint to dry. However, Duco lacquers dried in minutes and were tremendously durable. The twenty-two coats of primers and varnishes previously required for cars were replaced by three or four coats of Duco. This material could be sprayed on rather than applied with a brush as with the older more viscous varnishes. It revolutionized the way consumer products were made.

See also Cellulose, chemical uses of; Polymer and polymerization

Lacrosse

Popular in Canada and the eastern United States, lacrosse is a difficult, fast-moving game adopted from the Canadian Indians. It is widely considered the most brutal organized sport in existence; much of the game is based on the saying, ''The rules are—there ain't no rules.'' In addition, players wear only helmets and gloves for protection (although in women's lacrosse, body contact is not allowed). As the Canadian Indians played it, lacrosse was a rough, often dangerous sport in which sticks were quite often used to beat opponents. Frequently, the game involved as many as a thousand warriors and might have taken place over four or five miles (6-8 km) of territory. In 1839, Canadians began playing the game and founded the Montreal Lacrosse Club in 1856. Standard rules were established by 1867, and the National Lacrosse Association was founded that same year. In 1868, the Mohawk Club of Troy, New York, introduced the game to the United States, and as early as 1881, regular matches were being held throughout the United States.

Lamarr, Hedy (1913-)

Austrian-born American actress/inventor

Though Hedy Lamarr is best known for her work as a Hollywood acting star of the 1930s and 1940s, she also devised a radio technology that was ahead of its time and will continue to play a role in the development of telecommunications: frequency shifting.

Born into a wealthy Viennese banking family in 1913, Lamarr was born Hedwig Kiesler. As a young woman, Lamarr appeared in a Czech movie *Ecstasy*, which caught the attention of one Fritz Mandel, an Austrian arms dealer. Mandel married the 19-year-old, and Lamarr received a thorough education in arms and military technology at an endless stream of business dinners. One topic of conversation that intrigued her involved guiding radio signals while protecting them from enemy interference. She left Mandel in 1937, in part because her husband began to become more and more involved with Nazi Germany, and escaped to London, then Hollywood, where her film career took off. It was at a Hollywood dinner party in 1940 that she met George Antheil, a film scorer and avant-garde composer, and together they brainstormed the concept of frequency shifting, also known as frequency hopping.

Lamarr's and Antheil's idea is simple. To elude any interception of radio signals send between a transmitter and receiver, the signals can be divided into parts and each part is sent out separately across a frequency spectrum. The transmitter and receiver are synchronized, and a jammer could only take out a portion, if anything, of the signal, since it would be tuned to only one frequency. Antheil's idea for synchronization was based on his experience with player **pianos**, and the kind of tape they use to play music. The pair patented their idea in August 1942, and intended it to be used against the Nazis as part of the American effort in World War II. Unfortunately, the military did not use the frequency shifting idea during that war, and they first used the technology in 1962, during the Cuban Missile crisis. By then, the patent had expired, Antheil had died, and Lamarr never received any profit from her idea. Though Lamarr had other invention ideas—including one where a cube could be dropped into water to make a soda drink—none of them came to complete fruition.

However, frequency shifting has continued to grow in importance, especially with the advent of **computers** and related technologies. Though the principal remains the same, the transmitter uses a specific shared code to send a signal over a spectrum of frequencies to the receiver, instead of random shifting. With the availability of digital signal processing and fast microchips, frequency shifting has been marketed under the name Spread Spectrum and used in wireless telephones and data transmission. The principal is also used by the American government's Milstar defense communication system. Lamarr, now retired after appearing in her last film in 1958, was finally recognized for her contribution by an Electric Frontier Foundation Pioneer Award for technological innovation in March 1997.

See also Cellular telephone

Laminate

The idea of building up a complete material from a number of relatively thin layers to give a product of increased strength is an old idea. The patent for the first machine for pasting sheets of **paper** together in layers, forming **chipboard**, was granted in 1824 to an Englishman by the name of John Dickinson (1782-1869). The idea was applied again to paper in 1902, when naturally occurring resinous bonding materials were used as adhesives, still before the days of synthetic materials.

Since the time of **Leo Hendrik Baekeland** (1863-1944), laminates have come to refer to **composites** consisting of thermosetting **plastics** (phenolic, **polyester**, epoxy, or **silicone**) bonded to paper, cloth, asbestos, wood, **glass** fiber, or other material. Contemporary walls for housing, which have been made of thin aluminum sheets epoxied to **polyurethane** foam, are examples of laminar composites. The foam provides excellent thermal insulation, and the composite has a higher structural rigidity than the aluminum sheets or polyurethane foam alone. The polyurethane foam is itself a composite of air and polyurethane.

Baekeland's search for what would become the first commercial phenol-formaledhyde resin (**Bakelite**) was motivated by a desire to produce a synthetic **shellac** substitute for bonding mica in electrical apparatuses. Out of this pioneering research arose a technique for lamination.

High pressure laminates of paper and phenol-formaldehyde resins, used for electrical insulation since 1913, were an early entry into the field. By 1939, high strength weather resistant **plywood** was being made by interleaving veneers (with the grain oriented at a 90-degree angle on successive layers) with an extremely thin paper pre-impregnated with a special phenolic resin, then molding with heat and pressure. In Britain, during World War II, the de Havilland Aircraft Company built its Mosquito aircraft with this type of plywood; motor torpedo boats were also made of this material.

American designers first became aware of the potential of plastic as a modern material when the popularity of radio brought glossy black phenolic laminate into the homes in the mid-1920s. Designers shifted to Bakelite laminate initially as an imitative substitute for **lacquer** trim, and eventually as a material for surfacing entire pieces. Furniture designer Donald Deskey was particularly active in promoting phenolic laminate. In the 1930s, phenolic laminate (or Formica as it came to be known) quickly spread to every nook and cranny of public life—cafeterias, diners, even railway coaches. By the late 1930s, flawless Formica surfaces were to be found wherever one looked, seeming to contrast a futuristic dream with the real world of the Depression.

Today, paper or **textile** fabrics to be laminated are first impregnated with thermosetting compositions, then pressed with multiplaten presses. Such presses are also used for the manufacture of fiberboard (phenolic plastic with wood chips as filler). In a slight modification of this technique, hot plastic sheets are adhered to paper or fabric by pressing together while in contact, then allowing the laminate to cool. Still other techniques include the introduction of a hot filament between layers of the laminate, or the use of ultrasonic welding to join the sheets.

See also Composite materials

Lamps

From primitive torches to powerful searchlights, human progress has been marked by humankind's attempts to conquer darkness. In early huts and villages, campfires gave enough light for normal activities. Torches, made of wood or other plant material tipped with resin or bitumen, evolved from these campfires. Eventually, various holders were invented to carry torches, such as metal cups, baskets, and buckets that could also be mounted on poles or brackets. The ancient Egyptians, Greeks, and Romans all used torches, as did medieval Europeans. Even today, torches are not uncommon in less-developed societies.

Candles were invented somewhat later than torches. Primitive types of candles used by ancient civilizations in Egypt, Crete, and Italy from 3000 B.C. onward were made of animal fat, called tallow, and of beeswax, which comes from honeycombs. During the 1600s and 1700s, whale hunting became an important industry, and a type of wax derived from whales, called spermaceti, was found to be an excellent candle material. In the late 1800s, however, paraffin wax produced from crude oil began to displace spermaceti in candlemaking. In early candles, the wick had to be trimmed frequently to keep the flame burning. The modern braided candlewick, which consumes itself without trimming, was introduced in 1824.

Stone Age cave dwellers of western Europe invented the first oil lamps. These were simple hollowed-out stone cups that held a piece of moss or other spongy plant soaked with oil or tallow. Later, primitive wicks were made from dried rushes or other plant fibers. Similar lamps continued to be used as recently as 2000 B.C. by early British flint miners, and modern Eskimos still burn walrus and seal oil in lamps comparable to those of the Stone Age. Throughout the Mediterranean world, ancient civilizations began to employ other containers for oil lamps, such as shells, pottery, and metal. Some of these lamps were elaborately carved to portray deities.

The Romans developed lanterns to carry candles and oil lamps. Ancient and medieval lanterns consisted of a metal frame with sides made of some translucent material such as **glass** or oiled **paper**. Portable rushlights were also common and inexpensive; these were made by dipping clumps of dried rushes into animal fat. In Greece, lamps began to replace torches during the seventh century B.C. In fact, the name "lamp" comes from the Greek word for torch, *lampas*. The first Greek lamps were modeled after ancient Egyptian types, but the Greek lamps soon became more sophisticated, with spouts, nozzles, and handles. They were made of pottery or metal and were sometimes glazed with black or red colors. The Romans continued to introduce more complex lamps, such as elaborate metal designs representing lions and dolphins. Very large lamps were also constructed for Roman circuses and other public places, and some city streets were illuminated by hundreds of lamps.

During the Middle Ages, lighting took a step backward. Open saucers with a floating wick were most commonly used, even though this type of lamp was less developed than earlier versions. One exception to this decline in inventiveness was **Leonardo da Vinci**'s lamp, built in 1490, which enclosed the flame in a glass tube placed inside a water-filled glass globe. The Leonardo lamp not only burned more steadily than saucer lamps, but also produced more light, due to enhancement by the water.

Beginning in the 1700s lamps rapidly became much more efficient and highly developed. In the 1780s Aimé Argand, a Swiss chemist, invented an oil lamp with a round burner and circular wick, which allowed a strong draft of air to reach the flame. This design greatly intensified the light given off by the lamp. One of Argand's assistants also discovered that the flame would burn more steadily when enclosed in a glass chimney. Reflectors were also used to magnify the lamplight. Despite these improvements, however, oil lamps did not become widespread until paraffin oil, or **kerosene**, was introduced in America in the mid-1800s. Beginning in 1860, paraffin oil was exported to Europe, where it quickly replaced whale oil and other oils as a lamp fuel.

While great advancements to oil lamps were made in the early 1800s, gas lighting had begun to be installed in factories, railroad stations, and city streets. Gas made from coal was first used for lighting in the late 1700s, when Scottish inventor **William Murdock** illuminated his house with a coal-gas system. Around the same time, French chemist **Philippe Lebon** patented the Thermolamp, which burned gas made from wood. The Argand oil lamp was also quickly adapted to use gas. Soon several companies began manufacturing gas piping and burners, and cities and towns began installing central gasworks and distribution systems. Although coal-gas fuel was considered to be unsafe at first, people's fears were put to rest as more gas-lighting systems were installed and successfully operated. By the 1860s **gaslight** had found its way from factories and streets into private homes. In the 1880s the invention of the gas mantle made the light from gas lamps much brighter, enabling them to continue to compete with other artificial lighting systems.

But the greatest breakthrough in the history of lamps was electricity. As long ago as the 1600s, scientists had learned to produce light from electrical charges, and primitive electric lamps were demonstrated in the early 1700s. In 1808 Sir **Humphry Davy** invented the **arc lamp**, which produces light from a brilliant arc of electricity between two charged electrodes. Arc lamps continued to be developed throughout the 1800s (and are still used today when very bright light is needed). Around the same time, scientists began to experiment with incandescent filaments, or wires made of materials that glow when charged with electricity. During the mid-1800s, improvements to the **vacuum pump** and the dynamo **generator** made it possible for inventors to harness electric power for lighting. In 1879 the **incandescent light bulb** was introduced in the United States by **Thomas Edison** and in England by **Joseph Swan**, both of whom are credited with its invention. Better filament materials and other improvements during the early 1900s resulted in longer lasting, more efficient light bulbs. Although gaslight and oil lamps remained popular for decades, electric lighting eventually dominated the field.

Many other types of electric lamps have been invented since the 1800s. **Neon lights**, which are a type of *electric discharge lamp*, were introduced in 1910. Another kind of discharge light, the mercury vapor tube, led to the invention of fluorescent lighting in 1936. **Fluorescent lamp**s now supply two-thirds of the lighting demand worldwide. Other modern discharge lamps include sodium-vapor tubes, which are usually used for street lighting and other applications where large areas must be illuminated.

See also Halogen lamp; Miner's safety lamp; Searchlight, arc; Stage lighting

Edwin Herbert Land

LAN (LOCAL AREA NETWORK) • See Computer network

LAND, EDWIN HERBERT (1901-1991)

American inventor

While strolling down Broadway in 1926, Edwin Herbert Land was blinded by the headlights of an oncoming **automobile**. It occurred to him that there must be a way to develop a polarizing sheet that would reduce glare from light. Born on May 7, 1901, in Bridgeport, Connecticut, Land was the son of a merchant. While attending the Norwich Academy, he excelled in physics and was a member of the debate and track teams. He was attending Harvard University as a freshman when he first had his revelation about the polarizing process. He left Harvard and moved to New York, working secretly at night in a laboratory at Columbia University. In 1929 Land returned to Harvard, this time with his wife, Helen. Although Land would never graduate, he had already developed the first synthetic sheet polarizer by trapping tiny crystals in a thin sheet. By 1932 the sheet was actually being produced and would be used to make camera **lens** filters and polarizing sunglasses that cut glare from the sun. Land realized that his process could have a more far-reaching effect, but lacked the funds to do the necessary research. With the backing of Wall Street tycoons James P. Warburg, Averell Harriman (1891-1986) and Lewis Strauss, Land founded the Polaroid Corporation in 1937. One of the company's first projects was to persuade Detroit's automobile industry to install polarizing sheets in the headlamps and windshields of all its vehicles. Although Land was never able to sell his ideas to automobile manufacturers, he went on to discover many more uses for his invention.

By the 1940s, the Polaroid Corporation was making its contribution to the United States war effort. Servicemen who fought in World War II reaped the benefits of Land's inventions that resulted in polarizers for rangefinders and gunsights. The development of infrared viewers also aided night vision. By 1945 the Polaroid Corporation received millions of dollars in government contracts to develop military optics, and sales increased from the original $142,000 to $17 million. In 1943 Land got the inspiration for what was to be his most popular invention. While on vacation with his family, he snapped a photograph of his daughter. She asked to see the picture, and while Land was explaining that she would have to wait, he was already developing a plan for a camera that would process the film immediately. After the war Land began working earnestly on his **instant camera**, and in 1947 he introduced his invention to the Optical Society of America. By Christmas of 1948 the camera and film were available to deliver a finished print in 60 seconds. The first prints were sepia-toned (a warm, brown color), and black and white film went on the market less than a year later. By the 1950s Polaroid photography became a way of life.

In 1963 color film became available, and 1972 saw the introduction of the SX-70 process which eliminated the process of pullingthe film from the camera and peeling off the backing. The SX-70 simply spit out a square photo on which the image slowly appeared. Incredibly, Land even introduced instant home movies in 1977, which could be filmed, developed, and viewed without ever removing the film from its cassette. In 1980, Land founded the Rowland Institute for Science. In spite of all his discoveries and inventions, Land was personally reclusive. Although he never graduated from Harvard, he received many honorary degrees from prestigious colleges and universities. Land was granted more than 160 patents in his lifetime, and in 1963 he was given the Presidential Award of Freedom. Land died on March 1, 1991, in Cambridge, Massachusetts.

See also Photography

LAND MINE

A land mine is a hidden explosive device that is buried in the ground. There are basic one-stage mines that explode on contact, and two-stage models that fly into the air before shattering into deadly shrapnel. A few types illuminate the battlefield instead, either by burning where detonated or firing a flare into the sky.

Although land mines are a relatively recent invention, the use of underground explosives dates back hundreds of years. Besieging armies would try to capture fortified cities by putting primitive mines beneath their walls. During the American Civil War, devices called land torpedoes were used occasionally, despite universal condemnation of such ''infernal devices.'' They were usually 24-pound **artillery** shells fitted with a sensitive percussion fuse.

The Germans developed the first true land mines during World War I, and used them to disable Allied **armored vehicles**. Each mine was an artillery shell buried with the fuse facing up; when a tank ran over it, the mine exploded. During World War II, land mines became more sophisticated, and were used for specific purposes. One antitank mine, the German Teller mine, was encased in steel and could hold 12 pounds (5.4 kg) of **TNT**. It also featured an antilifting device, a fuse on the bottom that went off if the mine was unearthed. Antipersonnel mines, such as the Spreng or S-mine, were also created by the Germans. These resembled large tin cans. A trip wire projected from their tops and was attached to a nearby stick driven into the ground. If the wire was touched, a propellant fired the mine out of the ground. When it reached waist height, the anchor wire triggered the main charge (usually less than one pound of TNT), sending steel balls and shrapnel over a wide area.

As lethal as they were, steel mines had one weakness: they could be located with a metal detector. Today's land mines are housed in **plastic**, which makes finding them more difficult. They are also more brutally efficient. An English design called the bar mine has a rectangular pressure plate that rests above a much longer, bar-shaped charge. If a tank touches even the edge of the plate the charge detonates and its oblong form creates a wider explosion that ensures maximum damage. Other mines use a shaped charge to propel a high-speed jet of burning gases or to target their fragments in a specific direction. Chemical mines disperse a nonexplosive but equally deadly payload. They are especially effective when sown among conventional mines, because the explosives slow advancing troops and increase exposure to the chemical.

Some mines are designed to deactivate themselves after a set time, but these so-called ''smart'' mines are the exception, and most mines can remain active and lethal for over ten years. By 1997, it was estimated that more than 100 million mines lay scattered in over 60 countries, many in regions disturbed by civil wars. The threat of land mines continued in many areas long after the war itself was resolved. Groups in countries around the world began working to ban antipersonnel land mines in the late 1990s. The anti-land mine crusade gained momentum with the support of Diana, Princess of Wales, shortly before her death in 1997, and with the awarding of the 1997 Nobel Peace Prize to an American anti-land mine group. In December 1997, an international treaty banning land mines was signed in Ottawa, though the United States, a major land mine user, refused to sign.

Unexploded land mines about to be destroyed in Kabul, Afganistan.

Samuel Langley

While civilian protests against the use of land mines gained a victory with the signing of the treaty, land mine removal technology also advanced. Since plastic mines cannot be located with metal detectors, new methods were needed to detect them. The United States Department of Defense began researching alternate technology in 1995, and developed several methods for finding plastic mines. One of the most successful was an infrared detector that senses thermal anomalies surrounding mines. The Department of Defense also investigated sensors using ground penetrating radar, acoustic and microwave sensors, and sensors that can locate TNT by scent (as specially trained mine-detecting dogs do). A possibility for a future mine-removal system might be to genetically engineer fruit flies to locate TNT.

See also Bomb; Chemical warfare; Grenade

Lane, Samuel • See Combine harvester

Langevin, Paul (1872-1946)

French physicist

Paul Langevin was never particularly interested in fame; he preferred teaching his theories to his students rather than publishing them for the scientific community, and he rarely engaged in the risky business of invention. Nevertheless, he is best remembered for his invention, with the help of the Russian scientist Constantin Chilowski, of the echolocation system known as **sonar**.

Born in Paris in 1872, Langevin demonstrated an enthusiasm for school, especially the sciences. Always at the top of his class, he enrolled at a local engineering school, where his laboratory work was supervised by French chemist Pierre Curie (1859-1906), who, along with his wife Marie Curie and **Antoine-Henri Becquerel**, was awarded the 1903 Nobel Prize for physics. This contact strengthened Langevin's interest in physics. After graduation, he moved to Cambridge, England for a year to work with English physicist Joseph J. Thomson (1856-1940), who in 1906 received a Nobel Prize for physics. Upon his return to Paris he completed his postgraduate studies at the Sorbonne, again working with the Curies, and he received his Ph.D. in 1902. Langevin concentrated his research efforts toward the study of *paramagnetism* and *diamagnetism*.

During World War I the British fleet was being systematically decimated by German submarines, whose stealthy attacks left the British defenseless. It had been theorized for years that the distance to the bottom of a body of water could be determined by using a pulse of sound and measuring its travel time. The British army hoped that this same principle would provide a means for calculating the presence and position of submarines. Numerous scientists worked to this end, and in 1912 Chilowski built a device which was sound in principle, but was unable to generate a strong enough signal. Two years later Langevin was asked to help perfect Chilowski's design. Using piezoelectricity (a phenomenon discovered by Pierre Curie and his brother Jacques Curie) he boosted the output of ultrasonic waves to the point where echolocation was possible. Although the first working sonar system was not built until after the war, it provided an important tool for future military and civilian applications.

Langevin was a member of many scientific academies and he played a key role in bringing the theories of Albert Einstein to the French public.

Langley, Samuel (1834-1906)

American astronomer and physicist

Samuel Langley was born in Roxbury, Massachusetts, in 1834. As a child, Langley became interested in studying the stars and, despite that fact that he never attended college, went on to become a professor of astronomy and physics. Langley made many contributions to the field of astronomy, one of the most significant being the invention of the *bolometer*. Invented in 1878 while Langley was on faculty at the University of Pittsburgh, the bolometer was able to detect and measure electromagnetic radiation, thus allowing scientists to determine the energy output of the sun and stars.

While serving as secretary of the Smithsonian Institute, Langley developed an interest in aeronautics and obtained a $50,000 grant from the United States War Department to study the possibility of manned flight. He began building large, steam-powered models of an **aircraft** he named *Aerodrome*, without taking the time to first test his theories on **gliders**. By 1891 he had begun building *Aerodrome* models which were to be catapulted off the roof of a houseboat. The first five models failed, but his 1896 model flew more than half a mile. Later that year one remained airborne for nearly two minutes.

Finally, on 7 October 1903, Langley was ready to fly his first full-scale *Aerodrome* from a houseboat in the Potomac River. With the press in attendance, the machine was launched and promptly fell into the river. Langley contended that the launching mechanism was at fault, but further attempts produced the same results, and his funding was soon depleted. Just a few months later, **Orville** and **Wilbur Wright** achieved the first powered flight at Kitty Hawk, North Carolina. Throughout his life Langley maintained that if accidents had not depleted his funds, he would have achieved the fame accorded to the Wright brothers. A few years after Langley's death, experimenters did succeed in flying his *Aerodrome* after attaching a more powerful engine to it. Today, Langley Air Force Base in Virginia is named for this aviation pioneer.

Langmuir, Irving (1881-1957)

American chemist

Irving Langmuir

Irving Langmuir was born in Brooklyn, New York, with an inherent interest in science, but eyesight so poor that he could not make out even the individual leaves on trees. He had his first clear glimpse of the world of nature when he was fitted with **eyeglasses** at the age of 11, and from then on he never tired of studying nature's tiny structures—an atom of hydrogen, for example, or a film of oil only one molecule thick.

Although Langmuir's father died at an early age, he left his son enough money to continue his education in metallurgical engineering at Columbia University. For postgraduate work, Langmuir went to Germany to study chemistry under Walther Hermann Nernst, whose practical approach to science he found appealing. Langmuir's doctoral studies on the Nernst electric **lamp** laid the foundation for much of his later work. Perhaps Langmuir was also influenced toward applied science by his older brother Arthur, an industrial chemist. In 1909, after three years of teaching chemistry at the Stevens Institute of Technology in Hoboken, New Jersey, Langmuir landed a summer job at General Electric's research laboratory. At the time, GE was renowned for the freedom it gave its scientists to work on whatever interested them, even when that work was unrelated to GE business. This was a great opportunity for Langmuir who had had little time for creative research in his teaching position. He remained at GE for the rest of his career.

Almost immediately after Langmuir joined GE, his work began paying off. His improvements to GE's incandescent **light bulb** revolutionized the lighting industry and saved customers millions of dollars in electricity costs. At the time, light bulbs were made by creating a vacuum in the bulb to preserve the filament; Langmuir realized that the filament would last much longer if the bulb was filled with inert gases such as nitrogen and argon. He also found that the filament became more efficient when tightly coiled. Langmuir's work on light bulbs was but the first of a stream of successes. In the course of his career, he managed to amass 63 patents, a remarkable number, especially given that much of his work was not patentable. He discovered hydrogen in its atomic form as a film inside the bulb, and he invented an atomic hydrogen **welding** torch, which produces temperatures almost as hot as the sun's surface. His improved **vacuum pump** quickly entered widespread use in industry and laboratory research and continued to dominate the market for decades. He also invented a family of high- **vacuum tube**s that proved critical to early **radio** broadcasting. And in studying the flow of charged particles from hot metals (thermionic emission), Langmuir improved the tungsten filament by coating it with a single layer of thorium atoms.

Many modern scientific disciplines and technologies owe a debt to Langmuir's work. For example, Langmuir pioneered the study of plasma—the term he introduced to describe a complex, unstable mixture of ionized gases that exhibits unusual electrical and magnetic properties. His plasma research paved the way for developments in electron physics, astrophysics, and thermonuclear fusion. As part of this work, Langmuir invented a special probe to measure electron temperature, an entirely new concept at the time. He also modernized the theory of chemical bonding by introducing the concepts of electrovalency and covalency.

But the research that earned Langmuir the Nobel Prize for chemistry in 1932 was in the field of surface chemistry (i.e., the study of chemical forces at the boundaries between two different substances). He was the first industrial scientist to be the recipient of this high award, and as a result, industrial research labs gained new respect as havens for creative research. At GE, Langmuir had made the revolutionary discovery that gases would adsorb, or cling, onto the surface of a liquid or solid in an extremely thin layer—only one atom or molecule thick. His most extraordinary work was done with exceedingly thin oil films on water. To aid in this research he invented a device to measure the length and area of individual molecules which is still in use today. His related study of the catalytic action of gas films on platinum wires explained many mysterious surface chemistry phenomena and substantially advanced the science of catalysis.

Out of Langmuir's work in surface chemistry came the development of nonreflecting **glass**. During World War I, Langmuir was involved in developing sonic submarine detection (**sonar**); he later redirected this work toward improving the quality of **phonograph** sound recordings. Still active in World War II, he concentrated on improving **aircraft** de-icing techniques and building better smoke-screen generators. Some of this work led to his controversial attempts in the postwar years to make rain by **cloud seeding** with chemicals.

Throughout his life, Langmuir maintained a youthful love for the outdoors. He once hiked 52 mi (83.7 km) in a single day, and he climbed the Matterhorn in Europe with little preparation or conditioning. When his family vacationed at Lake George, New York, Langmuir measured the energy input and output of the lake just for his own pleasure. A competent sailor, Langmuir also learned to fly planes and became a personal friend of Charles Lindbergh. He was much concerned about public policies on wilderness conservation and atomic energy. He died of a heart attack in Falmouth, Massachusetts, at the age of 76.

See also Coolidge, William; Edison, Thomas Alva

LANGUAGE, UNIVERSAL

The idea of a universal language that could transcend national language barriers was first entertained by the ancient Greeks. During the Middle Ages and the Renaissance, Latin was considered the international language, and in the eighteenth century French was known as the language of diplomacy. Despite attempts by scholars to devise a universal language beginning in the early seventeenth century, it was not until the late nineteenth century that intellectuals—heady from a generation of advances in the physical sciences and optimistic that science could be applied with the same success to humanity's social and political barriers—began to fully formulate and widely embrace such language systems. The most successful of them were invented by European scholars, and, despite their claims to universality, each was derived in some manner from existing European languages for ease of understanding by Westerners.

The first of these languages, Volapük, was developed by German cleric Johannes Martin Schleyer (1831-1912) in 1880. Deriving its roots from Germanic and Romance languages and structurally based on German, Volapük was conceived as an attempt to develop a language free of grammatical and spelling exceptions and irregularities. During the first decade of its existence, it developed hundreds of thousands of adherents, owing to a lack of competing universal languages. However, Volapük bore two major flaws: its complexity of grammar, nearly as great as that of Latin, and its root words, which often were so far removed from their linguistic sources as to appear utterly unfamiliar to the speakers for which it was intended. It was soon supplanted by Esperanto (Latin for "hopeful"), introduced by Polish oculist Ludovic Lazarus Zamenhof (1859-1917) in 1887. With Esperanto, Zamenhof further refined the principle of removing complexities from language. Esperanto uses root words derived from, and similar to, sources in European languages; there are no exceptions to its grammatical rules and its words are spelled entirely phonetically. Because of these features, Esperanto developed an international following; an annual World Esperanto Congress was begun and persists to this day, and novels and periodicals have also been published in the language.

In 1907, French logicians Louis de Beaufront and Louis Couturat introduced Ido (a short name for Esperandido) as an attempt to revise and improve Esperanto by dropping the parent language's use of diacritical marks and altering some of its suffixes to better resemble those of European Romance languages. After a brief period of popularity, Ido waned from use. Novial, developed in 1928 by Danish philologist Otto Jespersen (1860-1943), differed from its predecessors primarily in a greater reliance on Germanic roots; it suffered much the same fate as Ido. Interlingua, a simplified form of Latin developed in 1903, was used in some scientific papers and abstracts after a brief revival in the late 1940s and early 1950s, but has largely fallen into disuse. Blissymbolics, a non-verbal language developed in the 1940s by Charles Bliss, has the potential to become a universal language. There are several hundred key symbols in blissymbolics which represent common things and ideas. Because these symbols are malleable, and work under an easily understood grammar and tense form, blissymbolics can easily express simple or complex ideas. Bliss's work went largely unnoticed until the 1980s, when it was adopted as a language for those unable to speak, especially the handicapped. Blissymbolics is not hard to learn, but has not become popular outside of limited circles.

Esperanto is to date the most successful of the spoken languages, currently boasting some 100,000 speakers. However, no artificial language has achieved the goal of weaning people from nationalistic ties to their own languages or of providing an international second language for communication in areas such as the sciences. English, in fact, is the closest of all languages to serve a universal function, particularly in the areas of science and commerce.

LAPTOP COMPUTER • See Microcomputer

LARYNGOSCOPE • See Endoscope

LASER

Arguably the most significant invention of the twentieth century, the laser has been the focus of much misconception. While many people may associate lasers with death rays and hostile aliens, most do not realize that the word *laser* (an acronym for Light Amplification by the Stimulated Emission of Radiation) describes not the beam of light, but the machine used to create that beam. The origin and invention of the laser is a subject of much debate. The elements necessary for the invention of the first laser had been around for quite some time; theories of coherence and stimulated emission were discussed in detail by Albert Einstein (1879-1955) in 1917, but they were presented only as hypotheses because, at that time, Einstein did not describe a device that could achieve stimulated emission. This was the case until 1954, when **Charles Townes** accomplished stimulated emission in microwaves. With proof that stimulated emission of light was indeed possible, scientists around the world raced to create a working laser.

In 1958 Townes and Arthur Schawlow (1921-) delivered a paper that explored the requirements for a laser radiator. At the same time, **Gordon Gould** designed what would be the working model of a laser, and it was he who coined the term laser; unfortunately, due to a misunderstanding at the patent office, Gould did not apply for a patent on his design. Townes and Schawlow eventually received the patent, but it was **Theodore Harold Maiman** who, in 1960, actually constructed the first working laser in the United States and received credit for its invention. Concurrent with these early studies in the United States, Soviet scientists F. A. Butayeva and V. A. Fabrikant had amplified light using laser technology. Failing to publish their work until years later, they received virtually no recognition in the Western world.

To understand how a laser works, we must understand how all light functions. Normal light, for example sunlight, is emitted from its source in all directions. No stimulus is required to generate this sort of light; the phenomenon is called spontaneous emission. Furthermore, the light is incoherent; that is to say, the lightwaves are disordered with respect to each other, like the groups of people walking through a shopping mall.

Laser light is generated within a medium. A large number of atoms in the medium must be excited into an energetic state, creating what is known as a population inversion3/4an avalanche of light, waiting to happen. To reach the excited state, the atoms must be energized, or pumped, generally by a beam of intense light. In Maiman's laser, a flash tube wrapped around a ruby rod generated this pump light.

Atoms can only remain in an excited state for a limited period of time. Once the atoms are excited, some of them will

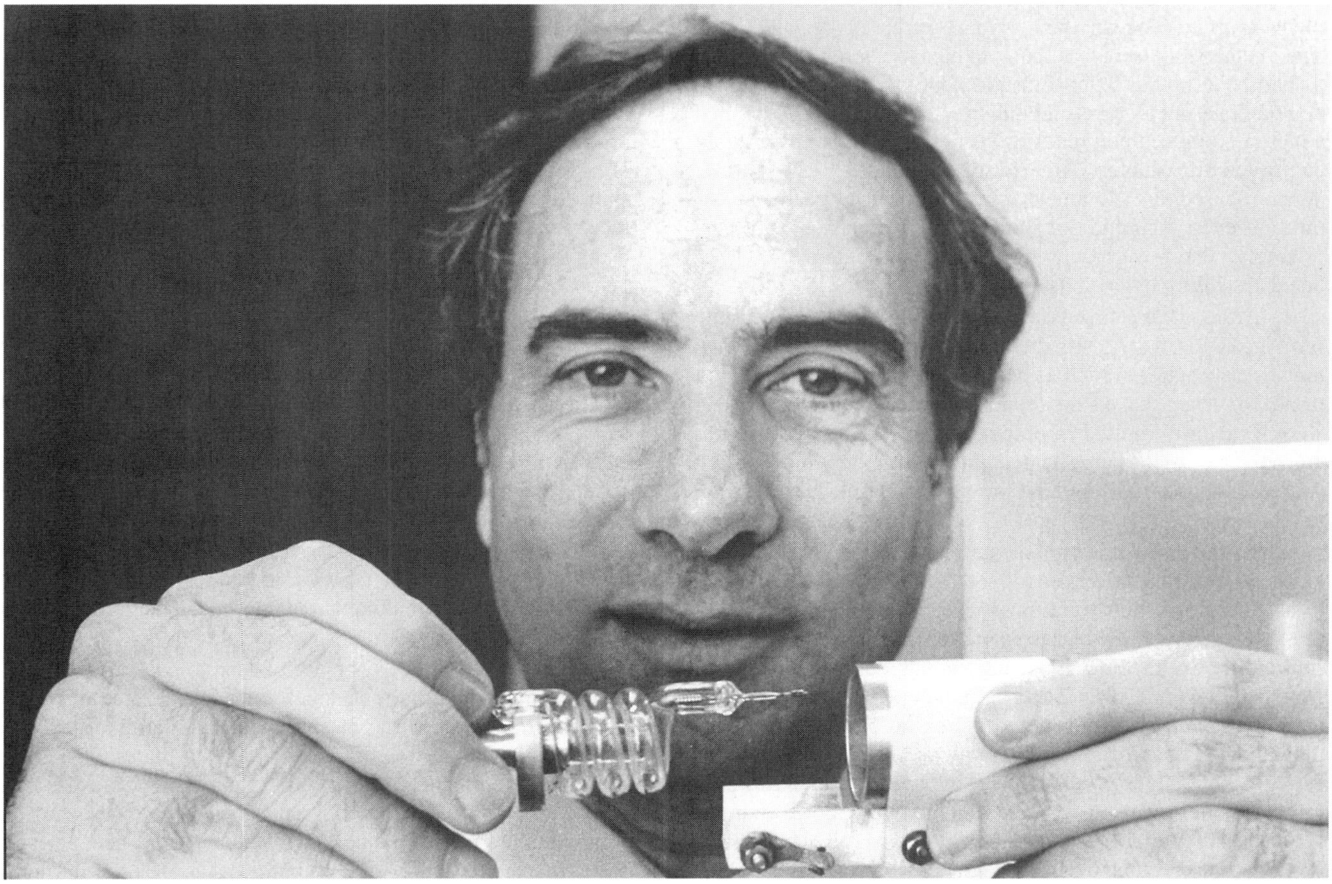

Theodore Maiman holds the first working laser, which he invented in 1960.

spontaneously emit their stored energy in the form of a lightwave, often at a different wavelength, or color, than the pump light used to excite the atoms initially. If this emitted lightwave travels to another excited atom, it will trigger that atom to also emit its stored energy as an identical lightwave. This is known as stimulated emission.

Specially aligned mirrors force the light to pass back and forth through the laser medium to extract light from as many atoms as possible, and force the lightwaves into a highly directional beam. Only a small amount of light is allowed to escape in the form of a tightly focused laser beam; most of the rest of the light continues to reflect back and forth, extracting light from the atoms, which are continually excited by the pump light. In this way, a continuous beam of light is formed.

Stimulated emission is responsible for one of the most important characteristics of laser light: its self-consistent, or coherent, nature. Because each lightwave emitted is identical to the lightwave that triggered its emission, all light in the beam is well ordered, like a platoon of soldiers marching in formation. It is this aspect of laser light that allows it to make high-precision distance measurements, create three-dimensional holograms, or beams energetic enough to cut through steel.

Maiman's laser used a flashlamp wound around a ruby rod to create a red beam of light. Commercial reaction to his invention was dramatic: within eighteen months of its debut, almost four hundred companies—as well as several government agencies—began their own research involving lasers. The intensity of lasers was found to be unhindered by distance, as had been predicted by Einstein nearly fifty years prior. In a 1962 experiment, a laser beam directed at the moon spanned a mere two miles upon reaching its destination—a remarkably controlled diffusion if compared to conventional light sources. When focused through a lens, a laser's powerful beam can cut through even diamond with unsurpassed precision. Not surprisingly, the government has maintained interest in laser beams as high power, long-range weapons.

Although the first laser demonstrated was a solid state ruby laser, for many years the most common commercial systems were gas lasers such as helium neon lasers and argon ion lasers, or lasers based on organic dyes. Helium neon lasers were frequently limited in output power, argon ion lasers required expensive, sophisticated power supplies and cooling sources, and the dyes used in dye lasers were messy and often toxic.

As the laser industry has matured, the technology trend has been toward semiconductor and solid state lasers, primarily as a result of their robustness, ease of use, and reliability. Semiconductor, or diode, lasers are based on a multiple-layer semiconductor material structure which defines the output wavelength. Diode lasers are usually pumped by electrical current, often from a power supply running off of a wall plug. Although the lasers require thermal regulation for optimal performance, this is often supplied by attaching the laser chip to a metal heatsink, in comparison to the bulky, power-hungry water-based cooling units required by many full-scale lasers. Produced by the same batch processing techniques used to fabricate computer chips, diode lasers are economical, and when configured in arrays known as diode bars, the lasers can produce tens of watts of output.

As of 1998, diode lasers operating from mid-infrared wavelengths to green wavelengths are commercially available. Several companies and research laboratories are working to make a commercially viable blue diode laser; though a number have been demonstrated, the material systems are vulnerable to small amounts of impurities which cause early failure. Common applications for diode lasers include telecommunications transceivers, CD players, and laser pointers.

Solid state lasers have gained as strong a foothold in the commercial market as diode lasers. A solid state laser is a rod or slab of laser crystal or laser glass, fitted with mirrors and a pump source. Certain types of solid state crystals can be pumped by diode lasers instead of by other lasers or by flashlamps, which is often the case. Such diode-pumped, solid state systems are reliable, economical, and easy to operate—in fact, many commercial systems are turnkey, needing only to be plugged in and turned on to operate. Solid state lasers are available from mid-infrared to ultraviolet wavelengths, and at a variety of output powers. Many solid state lasers are tunable, providing output across a range of wavelengths. Uses include medical applications, materials processing, and remote sensing.

In 1962 research began on the use of the laser in medicine, specifically its ability to perform very delicate surgery upon the eye. Initial results were promising, but the early ruby lasers did not produce a beam of the correct color or intensity for surgical purposes. It was not until the creation of the argon gas laser that laser surgery was considered viable. Now, in a relatively short procedure that requires no anesthetic, lasers are routinely used to correct myopia and astigmatism, as well to remove birthmarks, tattoos, and small varicose veins from the skin, cut tissue and seal blood vessels during surgery, or resurface skin to minimize wrinkles. Cutting-edge researchers are using visible and infrared lasers to increase bloodflow to oxygen-starved regions of the heart, to visually detect cancer cells on the skin's surface, or to image through the skin to monitor bloodflow or detect cancerous regions.

More modern applications of the laser include communications, wherein a strand of fiber optic material can be used to transmit a stream of high-speed, infrared light pulses encoded with voice discussions or computer data. Because of its coherent nature, laser light passes down the fiber with minimal loss, and multiple beams of different colors can pass down the same fiber simultaneously without interfering with each other. Thus, fiber optic lines can carry many times more information than copper wires, making them more practical for today's high-volume communications demands.

In factories, laser systems are used to measure parts, inspect them for quality, and label, cut, or resurface materials ranging from plastic film to sheet steel a quarter of an inch thick. Lasers form the basis of precision-measuring tools called **interferometers** that can measure distances less than 1/100th the thickness of a human hair, and are as useful on construction sites as in laboratories. Such instruments can be scanned over objects to create images, and are used on highways to identify vehicles automatically, or on NASA spacecraft to map the surface of the moon and asteroids. In semiconductor manufacturing, ultraviolet lasers provide the source for optical lithography, a technique used to produce computer chips with features as small as one hundred thousandth of an inch (0.25 microns).

Lasers have also found a niche in the entertainment world: concerts frequently include laser lights in a synchronized ''dance'' to the music. Laser light can also be used to form a **hologram**. In a complex process, laser light is reflected off a subject and onto a piece of film. Because of the unique properties of coherent light, the subject on film is given depth, an illusion that persists even when the holograph is viewed from many angles.

Perhaps the most familiar use of the laser is that associated with laser disk players. Disks of aluminum containing audio or visual information are encased in clear plastic; the laser beam then reads the information through the plastic without ever contacting the surface, thus providing a nearly infinite lifespan for the information.

Laser Disc

The concept of marketing items that utilized digital signal processing developed in the 1970s, spearheaded by the Sony Corporation and N.V. Philips, a Dutch company. This collaboration resulted in the release of the laser disc, for video viewing, and the audio compact disc (CD). The audio CD is viewed as a particularly revolutionary development as this item had all but replaced the record album as the standard audio unit by the 1990s. Credited with superior sound quality, decreased likelihood of scratching and the ability to play a full recording nonstop, the CD offered significant advantages over the vinyl album. CDs also offered the advantage of easier storage, given that, at 4.7 inches in diameter, they are significantly smaller than record albums.

Laser discs have revolutionized the computer industry as well. The compact-disc read-only-memory (CD-ROM) offers data storage capabilities superior to the floppy disc and allows for the creation of fully searchable databases and reference libraries as well as multimedia displays. For instance, an entire encyclopedia set can be stored on one CD-ROM. A CD-ROM can store up to 600 million characters. Audio CDs can be played on a computer CD-ROM drive. Read-write discs allow for recording sound and coding data directly to a CD.

The mid-1990s saw the development of the high-density CD (HDCD), which could store even more information (data, video or sound). Philips and Sony pioneered this development as well, along with Matsushita and JVC.

Lathe

Lathes represent one of the earliest machine tool families and were probably in use as early as 700 B.C. The invention of the cord drive for the rocking **drill**—the first mechanical cutting tool—made possible a similar drive system for the lathe, which in its simplest form is a tool used to support and rotate any material for the purpose of shaping by a cutting instrument. The many lathes that followed the ancient Etruscan, Egyptian, Syrian, and Greek rotary lathes made advances in at least one of three areas: 1) increased rigidity of the lathe spindle or axis; 2) increased power and efficiency of the drive mechanism; 3) progressive elimination of hand operations.

First came the pole lathe, around the twelfth century. The distinctive feature of this machine was a foot-operated treadle attached to a spring pole to drive the lathe spindle. The pole lathe was a significant improvement because it allowed for easy single-person use. However, it was considered ill-suited for cutting metal because the drive system could not be sustained indefinitely.

To solve this problem, the continuous band method (which relied on a combination of a large wheel and such power sources as water, horse teams, and, later, **steam engine**s and **electric motor**s) was invented in the fifteenth century. In addition to providing a reliable rotation system, the continuous band method made possible innovations in the actual cutting process, including tool holders and cross slides for enhanced precision.

The most significant developments in lathe manufacture occurred after a gradual conversion to all-metal components, which began around the mid-eighteenth century. The oldest all-metal lathe, circa 1751, was that built by Jacques de Vaucanson (1709-1782), for use in his loom machines. The Vaucanson lathe is also credited as being the first slide lathe, used expressly for cylindrical turning of metal pieces and characterized by prismatic slide bars.

In 1797 Englishman **Henry Maudslay** built a **screw**-cutting engine lathe. Precision screws had been cut by clockmakers on lathes since the late 1400s, but Maudslay's machine represented an enormous advance because of his introduction of the lead screw, geared to the lathe spindle. This lead screw allowed for controlled advancement of the cutting tool, and thus ensured the rapid machining of precision screws. By adding gears and altering the lead screw speed ratio, Maudslay was able, by the end of the century, to cut a variety of thread pitches with a single machine. For his contributions, Maudslay is regarded as the father of the industrial lathe.

During the same period, an American named David Wilkinson (1771-1852) developed a similar screw-cutting lathe that may have been based on drawings by **Leonardo da Vinci**. Wilkinson eventually produced a large general purpose lathe in 1806 that became the foundation of the American machine tool industry. Another American, **Thomas Blanchard**, first developed a lathe for producing non-symmetrical forms, such as gunstocks, and then, around 1818, a lathe that made use of a friction wheel to duplicate any of a large number of original patterns. Blanchard's machine, like Wilkinson's, was central to the development of modern manufacturing processes.

Lathe innovations, many of which related to increasing spindle speed and refining measurement accuracy, continued well into the twentieth century. Today the lathe, given its wide adaptability to automated production, is still regarded as the most widely used and most important machine tool. Shapers, planers, drilling machines, **milling machine**s, grinding machines, and a number of other important industrial inventions all trace their existence to the original machine tool, the lathe.

Latimer, Lewis Howard (1848-1928)
American inventor

Although Lewis Latimer was born before the Civil War, he was born free, unlike most African-Americans of his day. His father had escaped slavery in Virginia and moved to Boston, where abolitionists purchased his freedom. When Latimer was only ten years old, his father disappeared, leaving him to help support the family by selling copies of an anti-slavery newspaper. After the Civil War began, Latimer joined the Union Navy as soon as he reached the minimum age. At the end of the war, he found a job with a group of patent lawyers in Boston, starting out as an office boy and soon displaying a natural talent for the creating the technical drawings that illustrate patents. After buying a set of second-hand drafting tools, Latimer succeeded in getting a promotion and a raise, eventually becoming head draftsman for the firm.

During this time, Latimer met and made friends with **Alexander Graham Bell**, who was in the process of inventing the **telephone**. At Bell's request, Latimer illustrated the workings of the telephone's components, and his drawings appeared in the patent Bell obtained in 1876.

When **Thomas Edison** introduced the **incandescent light bulb** in 1879, Latimer quickly became interested in this new technology. The next year, he joined **Hiram Maxim**'s United States Electric Lighting Company as a patent draftsman. This company, the first such enterprise in America, began struggling to compete with Edison for the fast-growing electric lighting market. Central to this struggle was the development of a better filament. In 1881 and 1882, Latimer patented methods for manufacturing superior carbon filaments. With Latimer's inventions, Maxim's electric light bulbs were less expensive to make, and lasted much longer. Using Latimer's patents, his employers expanded their business and began building lighting systems for streets and railroad stations throughout the world. Latimer was responsible for setting up manufacturing plants and installing street lighting in the United States, Canada, and Britain. He also developed a new wiring scheme, using parallel circuits, for street light systems, so that when one lamp went out the others would remain lit.

Then in 1884, Latimer joined the Edison Electric Light Company in New York. After serving as an engineer for six years, he moved to the legal department where, as chief draftsman, he defended Edison's patents in court. Latimer demonstrated great skill not only in illustrating technical concepts, but also in testifying as an expert witness. Around this time, he also wrote and illustrated a pioneering book on incandescent lighting. In 1896, he was appointed to the Board of Patent Control, which had been set up by the General Electric and Westinghouse companies to oversee patent disputes. Eventually, he became an independent patent consultant. Latimer was the only black member of the Edison Pioneers, an organization of scientists who worked with Edison in the field of electricity. Upon his death, this group commended Latimer for his broad-mindedness, versatility, and genial character.

Lewis Latimer's life was not limited to scientific inventions. He wrote a book of poems, which was published for his seventy-seventh birthday celebration, and he was a talented painter and musician. He also promoted civil rights for black Americans and taught mechanical drawing to immigrants in New York.

Laundromat

The first business to provide self-service **washing machines** and **clothes dryers** for a fee opened in 1934 in Fort Worth, Texas. Later establishments also offered soap, bleach, and blueing. As time went on, improvements were made to the laundry equipment available, amenities added, and washers and dryers became coin operated. In 1984 a business featuring laundry activities on one side of the facility and a deli on the other opened in Austin, Texas. Some modern facilities serve refreshments and food, provide dance music and videos, and are decorated with everything from **neon lights** to hanging plants although they might more accurately be referred to as dollar, rather than coin operated. Many modern laundromats will wash, dry and fold customer's clothes for a fee.

Laval, Carl Gustaf de (1849-1913)
Swedish inventor

With his varied interests and his inventive talent, Carl Gustaf de Laval has been compared to **Thomas Edison**. He was involved with the dairy industry, steam turbines, electricity, and electrometallurgy in aerodynamics. At one time, de Laval had 100 engineers working to develop his devices and inventions.

De Laval's first major success was in the dairy industry. In 1877, he invented a **cream separator** whereby milk was placed in a container that was spun at 4,000 revolutions per minute by a **steam engine**. This speed caused the lighter cream to separate and remain in the center of the container while the heavier milk was pushed to the outer part and forced up through tubes to the outside, leaving only the cream in the container. Dairies all over the world used this design. He also invented a vaccuum **milking machine**.

Probably de Laval's greatest achievement was his contribution to the development of the **steam turbine**. For a long time people had known that steam could spin blades as it rushed past them. In 1887 de Laval devised a small, high-speed turbine in which jets of steam hit a single set of blades set on a rim of a wheel. To reach the speed of 42,000 revolutions per minute, de Laval improved the wheel and the nozzle that produced the jet of steam. During this period **Charles Parsons** was working on steam turbines, in which he employed several stages with different sets of blades so the steam would release its energy in small steps. On the other hand, de Laval perfected the single-stage, simple-impulse turbine that was smaller but became known for its reliability. He built a large number of these turbines from 1889-97 ranging in size from five to several hundred horsepower. He also invented special reduction gearing which allows a turbine rotating at high speed to drive a propeller or machine at comparatively slow speed, an extremely useful tool for ship propulsion.

Lawn mower

Yards have been used for centuries to enhance the beauty of large aristocratic mansions. By the time of the Revolutionary War, the open yard had become a distinctive feature of American culture. Open yards allowed colonial neighbors to keep a watch on each other's homes. In England, the open yard concept was imported from China during the Victorian era.

The most important tool of lawn maintenance is the lawn mower. Although similar in function to agricultural **mowing machines**, the lawn mower is actually an adaptation of a nineteenth-century rotary shearer that was used to remove the nap, or excess fiber, from **carpets**. Englishman Thomas Plucknett is credited with inventing a mowing machine in 1805. It consisted of a revolving blade mounted between a pair of carriage wheels and was capable of indiscriminately cutting all types of grasses, including grain.

The first mower specifically designed for keeping lawns trimmed was invented by Edwin Budding, an English textile engineer who made the adaptation from carpet shearer to grass shearer. His roller mower was patented in 1830 and was quite similar to the modern push mower. Budding's mower was particularly useful because it could cut grass while it was dry (grass cut with a scythe had to be wet). Budding designed his mower to trap and cut the blades of grass between rotating cutting blades and a fixed cutting blade. Push mowers are still popular with city homeowners and with exercise enthusiasts.

In 1832 another Englishman, Robert Ransome, secured licenses to manufacture Budding's machine. Subsequent improvements by Ransome led to the first **gasoline**-powered mower in 1899, a large riding mower designed for cutting sports fields. An electric mower was introduced in 1926, and the first electric mower with rotating blades was marketed in 1958. During the 1950s and 1960s, electric-and gasoline-powered mowers became increasingly popular as people moved into suburban homes with larger parcels of land.

The increased use of power mowers by individual homeowners brought about a call for new safety measures. The electric mower became nearly extinct during the 1970s because it was too easy to run the mower over the electric chord hidden in the grass, electrocuting the user. Also, the cutting blades on all types of power mowers would fly off when they struck a rock or other hard object, causing serious injuries. The rotors were modified to prevent this, and safety release mechanisms were also added to the handles.

Lawrence, Ernest Orlando (1901-1958)

American physicist

Ernest Orlando Lawrence was born in Canton, South Dakota, on August 8, 1901. His grandparents had emigrated from Norway in 1846. His father was an educator who eventually became superintendent of public instruction for the state of South Dakota.

Lawrence attended public schools in Canton and then Pierre, where he graduated from high school at the age of 16.

The first lawn mower, invented by British textile engineer Edwin Budding in 1830.

One of his close friends in high school was Merle A. Tuve, who later became a professional colleague of Lawrence's in the development of **particle accelerators**. Lawrence entered Saint Olaf College in Northfield, Minnesota, but, after one year, transferred to the University of South Dakota.

Although originally interested in a medical career, Lawrence soon decided to change his major to physics. He was strongly influenced in this decision by Lawrence E. Akeley (1910-), professor of electrical engineering. Lawrence graduated from the university in 1922 with high honors in physics.

After earning a master's degree at the University of Minnesota, Lawrence entered the University of Chicago to pursue a doctoral degree in physics. When his advisor, W. F. G. Swann (1884-1962), moved to Yale University, Lawrence went with him and received his Ph.D. in 1925. For the next three years, Lawrence remained at Yale as a research fellow and assistant professor of physics.

In 1928, Lawrence was offered a position at the University of California at Berkeley (UCB). Today, UCB is widely recognized as one of the world's great universities with a strong tradition of research in physics. But in 1928, the university was not well known. Lawrence's decision to leave the old and highly respected Yale institution for the uncertainty of a young and poorly regarded California university was courageous and adventurous. Two years after arriving in Berkeley, Lawrence was appointed full professor, the youngest person with that title at the university.

The invention for which Lawrence is best known is the **cyclotron**. The cyclotron is a device for accelerating protons, electrons, and other sub-atomic particles to very high velocities. The particles then collide with atomic nuclei and blow them apart. The products of such reactions reveal intimate details about the composition and structure of atoms.

Ernest Orlando Lawrence

Lawrence and two graduate students, Niels E. Edlefsen and M. Stanley Livingston (1905-1986), took nearly two years (1930-1931) to build the first cyclotron. The device was only 4.5 inches (11 cm) in diameter and operated on normal household current. But it was able to produce particles with 80,000 electron volts (80 keV) of energy and, therefore, confirmed the principle on which it had been built.

Based on the success of his first model, Lawrence was able to get a $1,000 grant from the National Research Council for further studies. With this money and donated parts and material, he built a larger cyclotron. The new machine had a diameter of 9.8 inches (25 cm) and produced particles with energies of more than one million electron volts. By the end of the decade, more than fifty cyclotrons were in use or under construction throughout the world.

During World War II, Lawrence became concerned that Nazi Germany might develop a method for making a fission weapon (an **atomic bomb**). He joined with other scientists to encourage President Roosevelt (1882-1945) to support American research in that field. When the Manhattan Project was created in response to this request, Lawrence became actively involved in research on the bomb. He converted a thirty-seven-inch (94-cm) cyclotron on the UCB campus into a device for producing uranium needed in the bomb. A friend claims that he worked ''twelve hours a day, seven days a week'' on the project.

In his spare time, Lawrence became interested in the problems of color **television**. He invented a color television picture tube that he called the chromatron. He did not have the time or financial resources, however, to see his invention through to production.

Lawrence died on August 27, 1958 during an operation for ulcerative colitis. He received many awards and honors both during his lifetime and after his death. In 1939 he was awarded the Nobel Prize in physics for his work on the cyclotron. In 1961, newly discovered element 103 was named lawrencium in his honor. Three important facilities at UCB—Lawrence Hall of Science, Lawrence Livermore Laboratory, and Lawrence Berkeley Laboratory—have been named after him.

See also Particle collider

LCD (liquid crystal display)

Liquid crystal displays (LCDs) have emerged as a popular form of electronic display, particularly since the 1980s. They use very little electricity (a thousand times less energy than a **LED [light-emitting diode]** uses) and are easily visible, even in direct sunlight because they scatter light rather than produce it. In other words, they produce dark areas that may be seen in brightly illuminated environments instead of glowing lights that could be visually overpowered by brighter light sources.

At the heart of a LCD is a material called a liquid crystal. First discovered in 1889, liquid crystals are fluid substances that have very well-ordered molecules, like those of solid crystals. When molecules of a material are aligned, they have a profound effect upon light passing through the material, often bending or splitting the individual light rays. In liquid crystals the molecules are parallel to one another and are arranged into planes, each one slightly offset from the last, creating a twisted pattern. As light passes through a liquid crystal, it is twisted and exits at a slightly different orientation from where it entered.

In a LCD, a very thin layer of liquid crystal is sandwiched between two sections of **glass**. Each piece of glass is polarized so that it will only allow light of a certain direction to pass through. Normally, because the two glasses are not aligned, light passing through one section of glass would be blocked by the other; however, the sandwiched layer of liquid crystal twists the light enough to pass through the second section of glass. Here, in the LCD's ''off'' position, the display is perfectly clear.

To turn a LCD ''on,'' an electrical charge must be applied to the liquid crystal. As the electrical field moves through the liquid, its molecules will line up with the direction of the field, ''untwisting'' the twisted planes. Light will now pass through the liquid crystal unaffected and be blocked by the second section of polarized glass. In this position the LCD appears black because it scatters all the light that enters.

In order to show numbers, letters, or pictures on the LCD, tiny metallic segments must be baked onto the surface of the polarized glass. Each segment can be turned on or off independently, forming numbers and letters. While many calculators use a simple seven-segment display format, more complex devices, for example, such as hand-held video games and laptop computers, form whole screens, called dot-matrix screens, out of tiny LCD squares. LCD screens are being developed for desktop computers as well, offering the advantages of smaller footprint, less power consumption and flat screens. Some digital cameras utilized LCD displays for viewing images as well.

While LCDs are used primarily for their legibility and low power consumption, they are essential in such dangerous industries as mining and chemistry. In these technologies LCDs are preferred for their very low voltage, current, and temperature-elements that, if too high, could start a fire or unwanted chemical reaction. Certain types of liquid crystals are also temperature sensitive and have been used in **thermometers**.

See also Calculator, pocket

Le Châtelier, Henry Louis (1850-1936)

French chemist

Henry Louis Le Châtelier grew up in a family steeped in scientific and technological traditions. Young Henry's relatives and close family friends included engineers and scientists involved in lime and **cement** production, railway construction, mining, and **aluminum** and **steel** manufacturing. France's leading chemists often visited the Le Châtelier home, and all of the Le Châtelier children pursued science-related careers. In later life, Le Châtelier said that his family had strongly influenced his research pursuits.

After interrupting his studies to serve as an army lieutenant in the Franco-Prussian War of 1870-71, Le Châtelier returned to college at the École Polytechnique and earned a degree in science and engineering in 1875. Two years later, he became a chemistry professor at the École des Mines, where he began research on cements, **ceramic**s, and **glass**. Some of his experiments with cements required the measurement of very high temperatures, for which the equipment available at the time was inadequate.

Le Châtelier invented a **thermocouple** that gave more accurate, reproducible results when measuring high temperatures. Thermocouples consist of two metal wires, of differing composition, welded together. When the junction is heated, an electric current is created, and the temperature at the junction of the thermocouple is computed by analyzing the difference in voltage between the two wires. Le Châtelier's thermocouple used one platinum wire and one platinum-rhodium alloy wire. He also introduced the use of known boiling and melting points as standards for calibrating thermocouples. Around the same time, Le Châtelier developed an optical **pyrometer**. These devices measure temperature by comparing the light emitted from hot objects against a known standard. Although other methods have replaced the pyrometer, Le Châtelier's equipment was useful at the time, and scientists continue to employ thermocouples in high-temperature research.

In the early 1880s, a series of mining disasters spurred the French government to investigate their cause and prevention. As an École des Mines professor, Le Châtelier took part in research on gas explosions. This work involved studying the ignition temperature, flame speed, and other conditions affecting explosions of methane, carbon monoxide, and hydrogen mixtures. Le Châtelier's results, applied to **acetylene** combustion, enabled other chemists to develop the **oxyacetylene welder** now used for cutting and **welding** steel. During this period, Le Châtelier also assisted in the development of safer **explosives** and improved the **miner's safety lamp**, which had been invented in the early 1800s by **Humphry Davy**.

Le Châtelier pursued further applications through his investigation of chemical reactions occurring in **blast furnace**s, which are used to manufacture steel. Engineers could not determine why carbon monoxide was present in the exhaust gases because, they believed, the compound should have reacted with iron oxides in the furnace to produce carbon dioxide. Le Châtelier realized that the iron oxides were catalyzing the reverse reaction, which created carbon monoxide. By clearing up this confusion, Le Châtelier enabled industrial engineers to develop blast furnaces that could reach higher temperatures by preheating the combustion air with hot exhaust gases.

Le Châtelier's scientific experience culminated in the discovery for which he is best known today—Le Châtelier's principle. Announced in 1884, the principle states that when a system is in equilibrium and one of the factors affecting it is changed, the system will respond by minimizing the effect of the change. Essentially, the principle predicts the direction that a chemical reaction will take when pressure, temperature, or any other condition is altered. Using *Le Châtelier's principle*, scientists were able to maximize the efficiency of chemical processes. For example, **Fritz Haber** made use of the principle to develop a practical process for **ammonia synthesis** using nitrogen and hydrogen. Le Châtelier himself had tried this, but gave up when his gas mixture exploded.

For the rest of his career, Le Châtelier continued teaching. In addition to his position at the École des Mines, he held posts at the College de France and at the Sorbonne. After working for the French government during World War I, he retired from the École des Mines in 1919 at age 69.

Leather

It probably did not take long for prehistoric hunters to realize that the skin of their prey was just as useful as its meat. Although an animal's hide would go rancid within a few hours of its death, it could be made into a strong, water-resistant material if properly cured. This process, lengthy and tiresome when done by hand, is today performed by machines. Although it was not until the 1700s that the tanning process was investigated scientifically, the process used today is almost exactly the same process that was developed by tanners thousands years ago.

There are many different kinds of leather. The skin of almost any animal can be cured and tanned—such exotic animals as ostrich, lizard, and eel have been used. However, the majority of leather is obtained from cattle and calf; sheep and lamb; goat and kid; horses and other equines; pig; and shark, seal, and alligator. The type of leather also depends upon the method by which it is tanned. In all cases, the hide is first cured—that is, it is skinned and cleaned, then dried and salted to prevent decay. After the curing process is completed, the hide can be shipped to a tannery.

In ancient times, a tannery probably consisted simply of a group of pits filled with murky solutions of water and oak bark. Other traditional tanners, such as the Eskimos and the Tartans on the steppes of Asia, have long treated (and continue to treat) hides by scraping off the fat and then working it in again with their teeth, thereby retaining the hide's natural suppleness. Still others dipped the skins in vats full of animal urine and excrement.

Modern tanneries use three different methods of tanning, each yielding a different variety of leather. The oldest method, called vegetable tanning, uses many of the same barks, roots, and seeds that were present in ancient tanning pits. Vegetable tanning is a very lengthy process (taking weeks or months) that produces strong leather which resists water; shoes are usually made from this type of leather. In the more specialized tanning method of oil tanning, after the skin has been cured and dried, it is dipped into vats of fish oil and other fats. The hide is then pounded, forcing the oil and fat into the skin; this forces the oils into the hide, where it replaces the water that has been cured out. The result is chamois leather, a very soft material that can be wetted and dried repeatedly without damage. The third method, discovered in 1852 by Swedish chemist Karl Hylte'n-Cavallius, uses mineral salts as the active tanning agent. The method has been improved over the years, making it suitable for practical and industrial use. Since the late 1800s, chrome tanning has been the principal method for tanning leather. This kind of tanning can be completed in a few hours or days. The type of leather produced is soft and pliable, and is an excellent material for gloves, coats, and other clothing, as well as for upholstery and luggage.

The bulk of leather produced today is used in the clothing industry, but leather had far more applications in those ancient cultures that pioneered its use. In addition to clothing, leather was made into ropes, sheaths for knives, harnesses, saddles, and tent coverings. Because it could be treated to resist water, buckets, bottles, and other containers were usually constructed from leather, as were small boats. When made very thin and pliable, leather vellum (parchment) proved an excellent alternative to papyrus and **paper** as a durable writing surface.

Leblanc, Nicolas (1742-1806)

French chemist

Nicolas Leblanc was born in Ivoy-le-Pré France. Emulating his guardian, a doctor, Leblanc studied medicine and, in 1780, became physician to the Duke of Orléans. Supported financially by the Orléand family, Leblanc was able to devote time to research. He studied crystallization and then developed a commercially important process to produce **soda** (sodium carbonate) out of salt (sodium chloride). The Leblanc process was crucial to industrial-chemical progress in the nineteenth century, as economical supplies of soda ash were necessary for the widespread manufacture of both **soap** and **glass**.

In 1790, Leblanc, the Duke of Orléans, a chemist named J.J. Dizé, and Henry Shée established a company to produce soda ash using the Leblanc process. Between 1791 and 1793 they built a factory at St. Denis, near Paris, France. By then, however, the French Revolution was well underway. When the Duke was guillotined in 1793, the factory was closed and later nationalized in 1794. The prize Leblanc had won for developing the process was never paid. The Committee of Public Safety forced Leblanc to publish details of his process without remuneration.

Leblanc finally regained control of his factory in 1800, but by then it needed extensive and expensive renovations. Leblanc was unable to raise the necessary capital, and the settlement of his claims against the French government in 1805 was much less than he had hoped for. Financially ruined, Leblanc killed himself in 1806. His process, however, remained in use until it was finally replaced by **Ernest Solvay**'s process in the 1860s.

Lebon, Philippe (1767-1804)

French chemist and engineer

During his brief scientific career, Philippe Lebon invented one of the first **gaslight**s and envisioned many other applications for gas that would be developed during the nineteenth century. His work was cut short in 1804, when he was murdered in the streets of Paris. Perhaps if he had lived longer, his ideas would have hastened the evolution of the gas industry. Regardless of such speculation, Lebon is recognized today as a pioneer of gas lighting.

Lebon began experimenting with gas made from sawdust when he was in his twenties, around the same time that Scottish inventor **William Murdock** was developing *coal-gas lighting* in England. In 1799 Lebon patented a method of distilling gas from wood and used it in a lighting fixture called the Thermolamp, which he exhibited as a large model for several months in 1801 in Paris. Despite the favorable attention received by this demonstration, the French government declined to finance Lebon's plans for a large-scale lighting system. However, news of Lebon's work reached **James Watt**, Murdock's employer, and the threat of competition from France spurred the British industry to complete the development of coal-gas lighting.

In addition to his *Thermolamp*, Lebon conceived the idea of running a motor on gas, using an electric fuel pump and spark ignition. This concept was far beyond its time. He also suggested that compressing the fuel charge in an engine could improve its efficiency, but Lebon did not live long enough to pursue these ideas. On the eve of Napoleon Bonaparte's coronation in 1804, Lebon was robbed and stabbed under mysterious circumstances. He died two days later, just 37 years old.

LED (Light-emitting diode)

LEDs (light-emitting diodes) can often be found in electronic toys, science kits, **calculators**, and many other household devices. LEDs are first and foremost diodes, special kinds of crystals or **transistors** that allow electrical current to pass through in one direction only.

The first **diodes** were invented by **Karl Ferdinand Braun** in 1874 as a solution to the problem of incompatible current. Specifically, Braun needed a device that could convert easy-to-produce **alternating current** (AC) into easy-to-use **direct current** (DC). The problem was that AC currents move in a sine-wave pattern, moving forward and then backward as it goes, while most electronic devices require single-direction DC. While experimenting, Braun noticed that certain crystals would allow current to pass in one direction, but hindered its passage in the opposite direction. These crystal rectifiers were used for years in **radios** and other types of communication equipment.

The next stage in diode development came in 1906 when **John Ambrose Fleming** invented the **vacuum tube** valve diode. This diode consisted of two electrodes, a cathode and an anode, encased within a vacuum-sealed **glass** tube. When a current was applied so that the cathode was negatively charged, electrons would flow to the anode and, thus, complete the circuit. However, if the current was reversed, no electrons would flow, and the diode would remain in the ''off'' position. Just like its crystal cousin, the valve diode would allow current to pass in one direction only. The modern diode is essentially a modified transistor. A piece of semiconducting material (such as **silicon** or germanium) is coated at each end with two different substances, so that one end is positively-biased and the other negatively-biased. When a current is sent through the diode negative-end first, the charge is allowed to pass; however, if the current is applied positive-end first, an electrical barrier is created within the diode, effectively blocking the current.

LEDs are of this last variety. Commonly made of gallium arsenide, the diode will give off radiation when in the ''on'' position. In order for the radiation to be within the spectrum of visible light, the diode must first be treated with phosphorus or a similar substance, otherwise it will glow in the infrared range.

See also Battery, electric; Semiconductor

Ledley, Robert S. (1926-)

American biophysicist and physiologist

Few scientists are talented enough to span more than one discipline, especially fields as diverse as medicine and mathematics. But Robert S. Ledley has done just that, with results that have revolutionized the way medical diagnosis is done.

Ledley was born in New York City in 1926, and earned a doctorate in dental surgery (D.D.S.) from New York University at the age of 22. One year later he earned a master's degree in mathematical physics from Columbia University. After brief service in the United States Army, Ledley worked at the National Bureau of Standards and at Johns Hopkins University before becoming professor of physiology and biophysics and of radiology at the Georgetown University Medical School in 1970.

Ledley is credited with establishing the field of *medical informatics*, in which **computers** and other information technologies aid physicians in the diagnosis and treatment of patients. This foundations of this field lie in the 1959 paper of Ledley and Lee B. Lusted, ''Reasoning Foundations of Medical Diagnosis,'' and in Ledley's 1965 book *Use of Computers in Biology and Medicine.*

Ledley pioneered the use of medical databases in the 1960s, compiling with Margaret O. Dayhoff a list of all known sequences of proteins and nucleic acids. He also devised methods and instruments, including a motorized **microscope** with pattern recognition capabilities, that made it possible to scan chromosomes for abnormalities that cause Down's syndrome and other disorders.

But Ledley's most important invention was that of the automatic computed transverse axial (ACTA) scanner in the early 1970s. Building on work developed independently by South African physicist Allan M. Cormack and British engineer Sir Godfrey Hounsfield, Ledley built a scanner that was capable of making cross-sectional images of any part of the human body. Such scanners have revolutionized the fields of radiology and medical imaging. The ACTA scanner set the course for modern computerized tomography (CT) machines.

Computerized tomography uses an array of detectors to collect information from beams of **x rays** (in the case of medicine) that have passed through an object, such as a portion of a human body. These beams crisscross and scatter in many directions, and the information collected is then used by a computer to reconstruct the internal structures. The three-dimensional image is then displayed on a two-dimensional display device like a television or a photographic film. CT thus presents a nonintrusive way of seeing internal structures, and it can help locate tumors, hemorrhages, and infections without the need for exploratory surgery. CT has also proved useful in geophysics, where it is an important tool for investigating the deep structure of the Earth, and in atmospheric physics, where it can be used to map the liquid contents of clouds.

Computerized tomography relies heavily on complex mathematics and sophisticated computer **algorithms** in order to process the x-ray scattering results in a reasonable amount of time. The prototype of Ledley's original ACTA scanner is now on display at the Smithsonian Institute in Washington, D.C.

Ledley holds over 60 patents, and was inducted into the National Inventors Hall of Fame in 1990. He was honored with the National Medal of Technology in 1997. Besides his professorship he is Director of the Medical and Biophysics Division in Georgetown University's department of biophysics, and is editor-in-chief of four scientific journals.

See also Computerized axial tomography; Positron emission tomography

LEE, WILLIAM • See Knitting machine

LEMELSON, JEROME (1923-1997)
American engineer

Born in New York City in 1923, Jerome Lemelson graduated from New York University in 1951 with a master's degree in aeronautical and administrative engineering. He spent the following several years working for various companies as an industrial engineer, but in 1957 Lemelson quit his job and decided to set out on his own as a full-time inventor. Out of his workshop in Metuchen, New Jersey, Lemelson began developing ideas for toys. Lemelson's first invention—a wheeled toy—was bought by the Ideal Toy Company.

Although his toy and game inventions sold well, Lemelson was far more interested in his high-tech inventions. After seeing a demonstration of a **punched card**-controlled **lathe**, he decided to begin research into the field of **automation**, particularly the pursuit to automate production machinery. In the next few years Lemelson developed a large number of very successful devices. These inventions included a system for machine vision, a universal robot capable of inspecting and measuring products, and an automatic warehousing system involving a computer or punched card-controlled stacker **crane** to pick up and deliver palletized products.

Lemelson's most important invention dates back to the early 1950s. Initially referred to as a flexible manufacturing system, it is now the backbone of **robotics**. A flexible manufacturing system consists of a series of computer-controlled machine tools with adaptable operations, based upon functions programmed into the computer. They can perform many different operations on the same unit, or several machines can be used together and connected by conveyors to move work from place to place. This flexible manufacturing system is used in automobile factories and has had a major impact in the automotive industry, particularly in the 1980s and 1990s.

The scope of Lemelson's creativity is enormous. He received his first patent in 1955 and continued to invent and develop new patents at the rate of one a month for the next twenty years. Only **Thomas Edison** and **Edwin H. Land** held more patents than the five hundred that Lemelson retains today. Unlike other inventors who create prototypes and establish companies to build and market their inventions, Lemelson is a ''professional inventor,'' living off the royalties he earns from licensing the ideas themselves. Lemelson constantly looks for new problems to solve, either from outside sources, or from magazine and newspaper articles. A lawyer who worked for Lemelson states that even while vacationing in Hawaii, Lemelson remained on the balcony of his hotel room jot-

Jerome Lemelson

ting eighty or ninety pages of notes while his wife enjoyed the beach.

In 1992, Lemelson received a 100 million dollar-settlement from a group of Japanese auto makers who had failed to pay him fees for his patents over the years. He is currently seeking about 300 million dollars from American automobile companies General Motors, Ford and Chrysler for infringement of the same patents. Lemelson recently moved from the east coast to a home in the mountains outside Reno, Nevada.

See also Computer, industrial uses of; Computer vision

Lenoir, Jean-Joseph-Étienne (1822-1900)

French engineer

The French inventor, Jean-Joseph-Étienne Lenoir was born at Mussy-la-Ville, Belgium, on January 12, 1822. He was a self-taught chemist and applied his expertise to a variety of inventions. He is best known as the first person to develop a practical **internal combustion engine** and an automobile powered by it. Before Lenoir's time, external combustion engines were used to propel road vehicles. **Nicholas-Joseph Cugnot** drove a steam carriage in 1769, and Samuel Brown of England drove a car that used hydrogen gas in an external combustion engine sometime in the early to mid-1820s. By the 1830s, steam carriages were in wide use, but steam engines had several drawbacks. They were very inefficient, cumbersome, and rather dangerous—prone to explosions and setting fires along roadsides.

Many innovators realized the advantages of an engine in which the combustion of fuel would take place directly within the cylinder and the resulting explosion would drive a **piston**. The concept of an internal combustion engine had been worked out by French physicist, Nicolas Carnot (1796-1832) and published in 1824 in his book on thermodynamics. A practical internal combustion engine was not possible until later in the century when petroleum products became widely available.

Claims for the development of the first internal combustion engine have been made by supporters of the Reverend W. Cecil and William Barnett, but it was Lenoir, in 1859, who built the first practical engine. Lenoir's model was a two-cycle, one-cylinder engine with slide valves which used ''illuminating gas'' (also called coal gas) as a fuel. This gas was used at the time in streetlights, hence its name. Lenoir's engine used an electrical charge supplied by a battery to ignite the gas after it was drawn into the cylinder.

Although Lenoir adapted an improved engine to a carriage in 1863, his two-cycle engine was too small and inefficient to power a carriage successfully. He also adapted his engine to propel a boat, but its true value was realized by the sale of hundreds to power small machinery, such as printing presses, lathes, and water pumps.

Lenoir is also credited with inventing the **spark plug** for automobile **ignition systems**. His invention is essentially the same as those used in cars today. Despite his inventions and the acclaim they gained for him during his lifetime, Lenoir died a poor man at Varenne-St. Hilaire, France, on August 4, 1900.

Leonardo da Vinci

See also Power boat

Leonardo da Vinci (1452-1519)

Italian artist, scientist, and inventor

Leonardo da Vinci is regarded as the model Renaissance Man whose endeavors reflect his accomplishment in a variety of artistic and scientific disciplines. While best known for his paintings *Mona Lisa* and *The Last Supper*, he was also a scientist who pursued an astonishing diversity of studies. His *Notebooks* contain observations and inventions that reveal a wide-ranging intellect centuries ahead of its time.

Leonardo was born in the village of Vinci in 1452, the illegitimate son of a well-to-do Florentine notary. He was educated at home, where he also received instruction in music and art. Around the year 1467 Leonardo was apprenticed to Andrea del Verrocchio, a prominent Florentine painter, sculptor, and goldsmith. Verrocchio tutored da Vinci in painting and sculpture, though more often than not, according to legend, the student corrected the master. In 1472 Leonardo was inducted into the Florentine guild of painters. During the decade that followed he became one of the most sought-after artists in Florence.

Leonardo's interests were not limited to art. Sometime between 1481 and 1483 he became a military and civil engineer for the duke of Milan, Ludovico Sforza. Leonardo flourished among prominent artists and scholars, and the two decades of his residence in Milan were the most productive of his career. He pursued studies in several disciplines and began keeping the first of his voluminous notebooks. In addition to painting such masterpieces as the *Virgin of the Rocks* and *The Last Supper*, he completed a substantial portion of a treatise on painting and laid the groundwork for texts on anatomy, architecture, and mechanics. Because most scientific and philosophical literature was not available in Italian, da Vinci taught himself Latin. This enabled him to read the works of authors, particularly **Archimedes**, whose scholastic methodology he believed paralleled his own. When Milan fell to French forces in 1499, Leonardo returned to Florence. From 1502 to 1503 he was chief architect and engineer for Cesare Borgia, designing weapons and drawing strategic maps which are early examples of modern cartography. Returning to Milan in 1506, Leonardo enjoyed a prominent position at the court of French governor Charles d'Amboise. In 1517 he was invited by Francis I to join his court at Amboise, where he lived until his death in 1519.

For over three decades da Vinci kept detailed written records and illustrations of his work. Believing the individual must ''understand how to see,'' he considered his eyes his most valuable asset and acquired knowledge chiefly through observation. The diversity of his interests, as well as his apparent passion for recording in minute detail virtually all of his perceptions, resulted in writings on a multitude of subjects. The *Notebooks* contain preliminary notes and outlines for treatises on art, architecture, and engineering, as well as studies in several branches of science, notable anatomy, zoology, physiology, geography, and astronomy.

During the early years of his career, da Vinci was chiefly interested in mechanics and engineering. His desire to understand not only how something works, but also why, led him to a number of discoveries. Included among his projects were plans for a **helicopter**, a **submarine**, and an **armored tank**. Later he concentrated almost exclusively on his scientific studies, particularly human anatomy. Leonardo performed over thirty dissections, documenting in detail the circulatory system and internal structure of the human body.

Most scholars concur that da Vinci's observations and inventions were in advance of their time, and that his scientific methodology anticipated the techniques of future generations of scientists. Today his *Notebooks* are considered among the best evidence of the transition from Renaissance to Modern scholasticism, and are highly valued for the insight they provide into both the artistic and scientific achievements of the Italian Renaissance.

LEVER

From prehistoric times levers were used for cultivation, excavation, and moving large objects. Such implements as hoes, slings, and oars were conceived and constructed to enhance human effort. The key to their operation is the relative positions of the load, the pivot called a fulcrum, and the applied effort. To maximize the applied effort, the most effective placement of the fulcrum was found to be close to the load.

As early as 5000 B.C. a simple balance scale employing a lever was used to weigh gold and other items. A Greek device called a steelyard improved on these simple scales by adding a sliding weight to enhance precision. Around 1500 B.C., the shaduf, a forerunner of the **crane**, made its appearance in Egypt and India as a device for lifting containers of water.

Scholars of this period may well have provided suitable explanations for the mechanics of the lever, but it was left to Greek theoretician **Archimedes** in the third century B.C. to document the underlying relationships between force, load, and distance from the fulcrum point. Archimedes's law of levers survives in the law of equilibrium, which states, ''effort multiplied by the length of the effort arm equals the load multiplied by the length of the load arm, where the effort arm equals the distance from the fulcrum to the point of applied effort and where the load arm equals the distance from the fulcrum to the center of the load weight.''

Simply put, this means that the longer the effort end of the lever the less force that is needed. (Today we say that the *torques* on each end of the lever are equal.) What is most important is the mechanical advantage—the reduction in pounds of effort versus pounds of load—that is gained.

Archimedes' law, which was expanded by **Hero of Alexandria**, applies to all types of levers. It should be noted that levers exist where the fulcrum rests beyond the load (e.g., the **wheelbarrow**, with the wheel serving as a fulcrum) or beyond the effort (in the case of tongs, with the elbow joint serving as a fulcrum).

In 1743 John Wyatt (1700-1766) introduced the concept of the compound lever, which consists of two or more levers working together to further reduce effort. A modest everyday example of a compound lever is a nail clipper.

The principle that underlies the movement of levers is the same as that which governs inclined planes, **pulley**s, cranes, **gear**s, and belts. Although levers are often taken for granted, they are a singularly important tool class and are incorporated in countless common objects. Crowbars, nutcrackers, seesaws, **scissors**, hammers, pliers, and bottle openers would all be unthinkable without the concept of the lever.

LEVIS • See Blue jeans

LIE DETECTOR

The origin of the modern lie detector can be traced back several centuries, to early societies who sought the truth by means of a wide variety of tests, many of them ridiculous or cruel in nature. The Bedouin of Arabia forced those who lied to them to lick a hot iron, resulting in a burnt tongue. In China, a suspect was forced to chew and spit out a handful of rice powder; if it was dry, the suspect was pronounced guilty. In England a person was asked to swallow the ''trial slice'' of cheese and bread; if he or she could not, guilt was pronounced. Despite their primitive methods, each of the aforementioned methods operated under the assumption that some form of physiological response will occur in a guilty person. The mouth will be dry and the throat will be constricted, for example.

The first scientific instrument designed to measure physiological responses was used in 1895 by an Italian criminologist, Cesare Lombroso (1836-1909). He took an existing device called a hydrosphygmograph—a small, water-filled tank into which the subject's hand was submerged; the tank was then sealed by a rubber membrane. Any changes in pulse or blood pressure would be transferred through the hand to the water, thus changing the water level. When the level changed, it caused a difference in air pressure, which was monitored by an air-filled tube connected to a revolving drum. Lombroso's modified device was now able to note any physical changes in pulse and blood pressure as a means to determine the guilt or innocence of people questioned about their activities.

The next step in the development of the modern lie detector came in 1915. After reading a book by a noted Harvard criminal psychologist that discussed the effects of emotional changes on blood pressure, respiration, galvanic skin reflex, and other physiological processes, William Marston used an ordinary **sphygmomanometer** to monitor changes in blood pressure readings during questioning. He also experimented with a **galvanometer**, which uses a gripping device to report tension in the hands and measures electric current to record skin resistance changes.

The 1920s saw continued improvements in lie detectors. John Larson, a police officer, created the first continuous-recording polygraph, which tracked changes in heart rate, blood pressure, and respiration. To help calm nervous subjects, the operator started with nonthreatening questions until a basic pattern of physiological reactions was noted. Then, key questions were periodically inserted; abrupt physiological changes in response to such questions would point to possible coverup of criminal behavior.

Leonarde Keeler, a patrolman, refined Larson's polygraph in the 1930s when he incorporated a galvanometer by fastening electrodes to the subject's fingertips, which are believed to show evidence of perspiration when a person is under emotional stress or lying. Keeler's invention is essentially the same machine used today, although advancements have been made throughout the twentieth century.

In 1950, John E. Reid devised a new format for testing subjects. Still using the polygraph, Reid changed the types of questions asked. In early polygraph testing, questions relevant to the investigation were interspersed with irrelevant questions (e.g., "What is your favorite ice cream?"). In contrast, Reid's

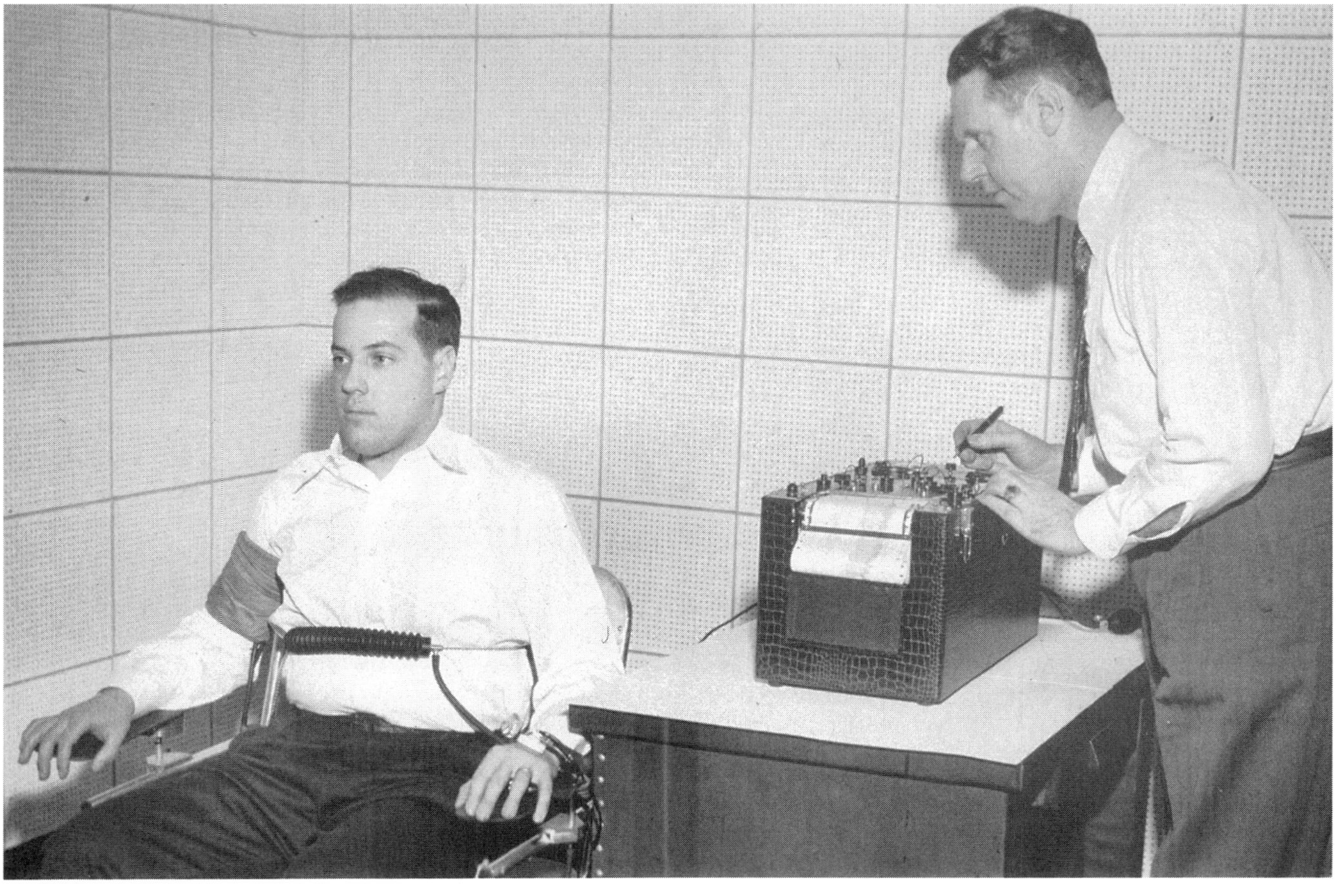

Demonstration of an early lie detector.

Justus von Liebig

system used control questions that were relevant to, but not critical to, the investigation.

The accuracy and validity of polygraph testing has been hotly contested since its invention. Some experts in the field contend that the test is accurate more than 90% of the time. Studies to confirm the reliability of polygraphs are difficult to come by, but limited research seems to support the claims of accuracy. The future of lie detecting technology in some ways depends on the resolution of the controversies over accuracy and appropriate use. While these debates are waged, however, other applications for the invention have been found. Defense attorneys, for example, will often conduct polygraph examinations on their clients to plan trial strategy.

Liebig, Justus von (1803-1873)

German chemist

Justus von Liebig did not make his reputation with a single discovery or innovation, but rather with his tremendous versatility. He conducted research in organic and inorganic chemistry, agricultural chemistry, physiology, and biochemistry and made significant contributions to the study of acids and bases, the chemistry of ether, the systematization of organic chemistry, and the production of industrial **dye**s as well as synthetic **fertilizer**s. Liebig is considered to be one of the most important chemists of the nineteenth century.

Born in Darmstadt, Germany, on May 12, 1803, Liebig was the son of a merchant who sold pharmaceuticals, dyes, and salts, so he developed a keen interest in chemistry early in his youth. By the time he was nineteen, he had earned his Ph.D. at Erlangen, and at the recommendation of Alexander von Humboldt (1769-1859) was hired to work in the laboratory of Joseph Gay-Lussac (1778-1850), where he remained for two years.

In 1825, he was appointed chairman of chemistry at the obscure University of Giessen, where he proceeded to build an excellent chemistry program. He was among the first to focus on laboratory instruction as a means of educating chemists. He remained at Giessen for 27 years before he moved to the University of Munich, staying there until his death on April 18, 1873.

Liebig is probably best known to chemistry students as the inventor of the Liebig condenser, a distillation apparatus found in almost every chemical laboratory. It consists of a **glass** tube surrounded by a glass "jacket" through which cold water can be circulated. As a substance is boiled in a flask, vapor is directed from the mouth of the flask into the tube, and it is cooled and condensed by the water flow. The condensed vapor is free of any dissolved chemicals.

He played a part in numerous key discoveries as well. For example, in the early 1820s Liebig and Friedrich Wöhler had been conducting individual research on inorganic chemicals, and Gay-Lussac noticed that a compound Liebig called silver fulminate had the same chemical formula as a compound Wöhler called silver cyanate, though the compounds were different chemically. Informed of this peculiarity by Gay-Lussac, Jöns Berzelius was inspired to derive the theory of isomers. Liebig and Wöhler began working together in organic chemistry, a field that lacked systematic theory at the time, and introduced a degree of methodical analysis to the field, including techniques to determine the content of various elements such as carbon, hydrogen, and halogens in organic chemicals. Upon discovering the benzoyl radical ($C_6 H_5 CO$-), they also attempted to find a way to define *all* organic chemicals as combinations of radicals (groups of molecules that tend to act as a unit). Though they failed, their efforts to present organizing principles in organic chemistry stimulated more successful attempts later.

Beginning in the late 1830s, Liebig pioneered the production and use of artificial **fertilizers**. At the time, it was believed that plants obtained carbon from organic chemicals in the soil and that they took in other essential nutrients in the form of organic compounds. Liebig showed that, in fact, plants receive all of their carbon in the form of carbon dioxide from the atmosphere. Furthermore, he found that in order to survive, plants required only water and minerals (such as calcium, phosphorus, and potassium) in the form of simple compounds from the soil. This meant that inorganic compounds containing minerals alone could be used to fertilize fields; organic mulch or manure was unnecessary. Following Liebig's instructions, Muspratt and Co. of Liverpool, England manufactured an experimental batch of synthetic fertilizer in 1845.

The fertilizer was not effective for two reasons. First, fearing that the mineral content of a synthetic fertilizer would be rapidly leached out of the soil by rain, Liebig locked the minerals into molecules that would not readily dissolve in water. But this same insolubility prevented the roots of the plants from absorbing the minerals. Second, he held to a false belief that all plants obtained their nitrogen directly from the atmosphere, so he did not include nitrogen in his formula. Once Liebig corrected these errors, his experiments attained their promised result of improved plant growth.

Liebig wisely argued against fertilizing with only nitrogen, as this would ultimately result in depletion of other minerals from the soil, and correctly theorized that the least abundant mineral in a given expanse of soil will limit plant growth no matter how plentiful other minerals were. This hypothesis came to be called the law of the minimum.

The insights of Liebig helped to modernize the science of plant biochemistry, and served as a catalyst in the development of the modern agricultural industry. He made noteworthy contributions for studies on fermentation and the calorific content of food.

Life vest

Recreational and professional activities bring people in contact with water everyday. Safety on and near the water is an issue for the weekend sail boater as well as for military and rescue personnel. Floatation devices are an important part of water safety. Full-body, industrial survival suits and simple water ski belts are two extremes in the family of personal floatation devices available. Although they differ tremendously in the amount of protection, the goal of all safety apparel is survival.

Life vests and life jackets are technically known as PFDs, short for personal flotation devices. They are designed to keep an individual afloat in the water in the event of an emergency and are considered life-saving equipment. The main function of a PFD is to keep a person on the water's surface in a relatively upright position to allow the person to breathe and not have to tread water to stay afloat. United States Coast Guard regulations require a boat to carry at least one Coast Guard-approved PFD per person, including one for each person water skiing.

There are five categories of Coast Guard-approved personal flotation devices. Each of the five categories provide different floatation and body positioning specifications. Types I and II PFDs are full and half-length vests designed to turn an unconscious person from a face down position to a vertical or slightly head back position floating in the water. These vests are usually big and bulky. A type III PFD, most commonly used in recreational activities, is also a buoyant vest or jacket. This type is designed to keep a conscious person afloat in a vertical or slightly head back position. This type comes in many styles and is the most comfortable. A further difference in the I, II, and III types is that the specific degree of buoyancy required increases from type III to type I. Throwable devices such as a ring buoy or the buoyant cushion typically used by boaters as a cushion for sitting are considered Type IV. Type V PFDs are special-purpose devices for aircraft pilots who fly over water, rafters, and ferryboat pilots.

Natural materials were first used to create floatation devices. Before 1900, life jackets were made from cork and balsa wood. A material called kapok was later used as the fill material in life vests. Kapok is a vegetable fiber found in tropical tree pods, resembling milkweed. The waxy coating which covers the kapok fiber provided the necessary buoyancy. The kapok fiber was sealed in vinyl plastic packets to prevent exposure to the water. One problem with the **vinyl**-sealed kapok fiber life jacket was that the packets could be punctured, causing the jacket to lose its buoyancy. Kapok is now prohibited for use in life preservers in most of Europe and in Canada.

The Coast Guard made a significant change in life vest requirements after the 1953 sinking of the ore carrier *Carl D. Bradley* in which 33 persons died. Many crewmen were found floating among their life vests, having slipped out of them after the ship had sunk. Thereafter, the Coast Guard required that life jackets be designed so that unconscious persons could not accidentally slip out of them if emersed in water.

In the 1960s, France introduced a life jacket called the flotherchoc. The flotherchoc was a light and flexible body-fitting vest. This design replaced the then popular but awkward horse-collar design. The advantage of this vest was that it was less confining and, therefore, more likely to actually be worn. The flotherchoc was made up of small, air-filled vinyl packets which were placed inside **nylon** chambers. However the flotherchoc had the same problem as the PFDs which used kapok: over time, the flotherchoc's vinyl packets could lose their buoyancy if punctured.

Plastics are now being used in the manufacture of life vests. Some vests are made from closed-cell foam or foamed plastics which are encased in nylon. Closed-cell foam has been around since the 1940s, but it wasn't until the 1970s that its use in survival wear was introduced. A closed-cell foam insert is made of tiny, individual air-filled pockets within the foam itself. The air-filled pockets are called cells. This foam structure is similar to a sponge, except that in a sponge the individual cells are connected by tunnels which run throughout the material. Closed-cell foam cells are not connected at all. It is the isolated air-filled pockets which provide the flotation. Closed-cell foam can be punctured over and over again with only minimal effect on its buoyancy. Some of the better closed-cell foam structures will not deteriorate even under tremendous compression. The air-filled pockets also provide some thermal insulation protection against hypothermia.

Light emitting diode • See LED

Lighthouse

Used for thousands of years, lighthouses function primarily to guide navigators and warn them of various hazards. The first known lighthouse was constructed on the island of Pharos, near Alexandria, Egypt, during the reign of Ptolemy II (308-246 B.C.), whose rule began in 285 B.C. One of the seven wonders of the ancient world, the lighthouse at Pharos measured about 350 ft. (110 m) high. Lit by a wood fire bright enough to be seen from sea, this architectural wonder stood until 1302, when it was finally toppled during an earthquake. The traders and conquerors of imperial Rome built lighthouses at nearly every port; after the fall of the empire, however, the use of lighthouses dwindled with the decline of commerce in the Dark Ages. As traders began to cross the seas again during the eleventh century, the old lighthouses were once again lit. However, it was the rediscovery of water-resistant **cement** (a technology which had been lost since the fall of Rome) by John Smeaton (1724-1792) that sparked the second great era of the lighthouse.

With the increased use of lighthouses, it became clear that advances needed to be made in the light itself. Simple wood fires, which were used until the eighteenth century, were not bright enough to penetrate thick fog, nor could they aim light in any specific direction. Coal fires, used after about 1550, were brighter but could consume almost a ton of coal a day; moreover, their thick smoke quickly darkened the lantern panes, obscuring the light. The English scientist William Hutchinson introduced the parabolic reflector in 1752, a device which directed the better part of a **candle**'s light along a straight path. The first lighthouse lens was designed by the eminent French optician **Augustin-Jean Fresnel** (1788-1827); this lens' face was cut into staggered planes, so that the light would be converted into a bright, straight beam. At the insistence of **David Brewster**, English lighthouses were equipped with **lenses** of Fresnel's design. Aimé Argand invented an oil-burning **lamp** in 1782 which produced a near-smokeless flame, and whose integrity and range was increased through the use of metallic reflectors. In 1862 British lighthouses began using the first forms of electric light.

Modern lighthouses are the product of centuries of design and improvements. Most use xenon **arc lamps** as a light source. The light passes through a compound lens almost identical in construction to Fresnel's. The lens, part of which is screened, revolves around the lamp, causing the light to flash intermittently. The arclamp is surrounded by **prisms**, which focus any stray light into the main lens; some of the largest lighthouses have as many as 200 of these prisms. The lens-and-prism system is housed on a revolving frame that functions electrically, either by rolling or floating devices. Many modern lighthouses also use radio beacons to warn navigators instead of the loud blasts of sound that were designed to penetrate through thick fog. Sailors recognize a specific lighthouse they are seeing by reading its radio signal. By comparing the signal to a list published by the United States Coast Guard, navigators can easily identify any lighthouse. However, recent advances in shipboard navigation, such as satellite positioning systems, have reduced the need for lighthouses; those that remain are mostly automated through the use of computers and other electronic equipment.

Lightning rod

The lightning rod was invented by **Benjamin Franklin**. In 1746 Franklin saw a public demonstration of static electricity in Europe and began to speculate whether or not lightning discharges in the sky could be related to static electricity. He decided to attempt to ''collect'' the charge by erecting a long metal rod on the top of Christ Church in Philadelphia.

A lightning stroke is the result of a buildup of static electricity in the clouds in the sky. When the charge becomes large enough it arcs out of the sky and strikes the earth. Since electricity tends to take the shortest path, it generally strikes the highest object.

Franklin's lightning rod succeeded admirably; it not only acted as a conduit for collecting the electric current, it deflected potentially damaging lightning safely into the Earth. The ''attractive range,'' the average distance over which a rod will attract lightning, is about 100 feet (30 m) horizontally when mounted 200 feet (60 m) high. The rods became indispensable for protecting buildings from lightning's destructive force.

As part of his experiments with electricity, Franklin discovered he could get a spark to travel over a greater distance if his conductor was sharply pointed, and therefore his lightning rods all had very sharp points. It is amusing to note that following the Revolutionary War, King George III ordered that the lightning rods installed on his palace have *blunt* ends. Although he acknowledged the effectiveness of lightning rods, he refused to accept an American design.

Lightpen • See Computer input and output devices

Lightwave communications • See Fiber optics

Lilienthal, Otto (1848-1896)

German aeronautical engineer

Otto Lilienthal was a major figure in the ten years preceding powered flight. His experiments with **gliders** were followed by the press and aviation enthusiasts all over the world. His exploits inspired many, including **Orville** and **Wilbur Wright**.

As a young man, Lilienthal loved the idea of flying. He and his brother studied birds, particularly storks, and built strap-on wings in the hopes of flying, but failed in their attempts. Even as a soldier in the Franco-Prussian War, he focused on flying.

As an adult, he studied the principles of flight. He soon realized a curved wing was the way to succeed. He measured the amount of lift that various rigid wings produced. He published findings in a book that became the bible for other aviation enthusiasts: *Bird-flight As the Basis of Aviation* (1889). He had tables showing calculations of lift provided by these wings, and it was this set of tables that was so admired and followed by others.

He then began experimenting with gliding as a way to prepare for powered flight. He started with short jumps, but he soon needed longer and higher areas to work. In 1894, he built an artificial hill where he achieved glides of 150 feet (45.75 m) or more. He also found hills outside Berlin where he reached a distance of 1,150 feet (350.75 m) by gliding. Lilienthal, however, probably would never have achieved successful powered flight. He maintained the only way to move through the air was by flapping wings. He devoted his energies to building machines with such flapping action, but they were so ineffective he never got to try them out.

In August of 1896, in-flight tragedy struck Lilienthal. A sudden gust of wind caused his glider to rise up swiftly. He threw his weight forward to correct the problem, but the glider stalled and crashed to the ground. He died the next day.

Limelight

The invention of limelight may not have been welcomed by all theater directors in the 1800s. Suddenly, their actors began competing for the most desirable area of the stage—the front and center, which could be brightly illuminated for the first time with Thomas Drummond's (1797-1840) new theatrical spotlight. The light was produced by pointing a hot torch at a solid block of lime (calcium oxide). When heated, the lime gave off a bright, soft white light that was easy to focus on a small area by using a **mirror** as a reflector.

Drummond, born in Scotland in 1797, invented limelight in 1816 and first used it during a survey of Ireland in 1825. Drummond's early limelight was based on a torch that burned hydrogen and oxygen, which had been developed by Robert Hare (1781-1858), an American chemist. Although limelight was considered for **lighthouse** applications, the cost of production was too high.

In 1837 limelight found its niche in the theater, where it was used not only to spotlight actors but also to create realistic special effects such as moonlight on a river or clouds moving through the sky. However, limelight had a major disadvantage; it required constant attention by a stagehand to keep turning the block of lime and tending the gas torch. In the late 1800s, limelight began to be replaced by electric arc spotlights. But its name has lived on; we still speak today of ''basking in the limelight'' of popularity or attention.

See also Stage lighting

Lincoln, Abraham (1809-1865)

American president and inventor

As a young man, Lincoln had taken flatboat trips down the Ohio and Mississippi rivers. While traveling, he saw other boats become unavoidably grounded on sandbars that were created by the river's shifting currents. His interest sparked by the sight, Lincoln began devising a plan to alleviate such difficulties. He sketched his ideas and whittled a wooden model of his device in his spare time.

Otto Lilienthal

Abraham Lincoln

Lincoln's plan utilized adjustable buoyant air chambers to elevate a **steamboat** or other river vessel; the extra buoyancy provided by these "airbags" would help ships float safely over treacherous shallows and sandbars. Lincoln's idea was patented under the title *"Buoying Vessels over Shoals."*

Lincoln was a staunch supporter of the United States patent system because he felt it "added the fuel of interest to the fire of genius in the discovery and production of new and useful things." These words are engraved above the entrance to the Department of Commerce building in Washington, D.C.

Lincoln patented his invention in 1849 (Patent No. 6469), and although he delivered a model of his system of "buoyant chambers" to the United States Patent Office himself, he did little to promote his invention. It is thought that perhaps the heavy weight of his system of ropes, pulleys, and collapsible chambers might have caused the very problem he was trying to solve. Although Lincoln's political career left him little time to invent, throughout his presidency he remained aware of the importance of new technology to American life. As President during the Civil War, Lincoln constantly pushed for such military innovations as breech-loading rifles, ironclad naval vessels, and balloons for aerial reconnaissance. Lincoln was the only president ever to own a patent. The model of his own invention is in the Smithsonian Institution, in Washington, D.C.

Linde, Karl von (1842-1934)

German engineer

Karl von Linde was born in Berndorf, Bavaria. He studied engineering at Zurich Polytechnic and began teaching at the Technische Hochschule in Munich in 1868. Gabriel Sedlmayr, a friend from the Spatenbrau brewery, who was also president of the German Brewer's Union, asked Linde to develop a **refrigeration** system that would make possible year-round beer brewing. By 1874 Linde had developed a methyl ether refrigerator, followed by an ammonia-compressor model in 1876. While Linde was not the first to use condensed ammonia as a refrigerant—the Scottish-American inventor David Boyle had patented such a system in 1872—his design was the first practical refrigerator.

By 1879 Linde had left teaching to concentrate on research, and by 1891 he had sold 12,000 domestic refrigerators in Germany and the United States. The success of the refrigeration business allowed Linde to concentrate on his next project, the removal of heat from gases and liquids at low temperatures. In 1895 he produced a machine that used the Joule-Thomson effect in a continuous process to produce large amounts of liquefied gases, which up until then could be made in only very small quantities. By 1902 Linde had developed a method for separating pure liquid oxygen and nitrogen from liquid air, again in large, commercially valuable amounts. Variations of Linde's methods are still used today.

Linde's discoveries accelerated the development of the liquid air industry, had important commercial applications in steel manufacture, and also provided an important basis for modern low-temperature, high-vacuum physics research. Linde died in Munich in 1934.

See also Liquid gas, commercial production of

Linear accelerator

The first particle accelerators to be built were linear accelerators, or linacs. In a linac, particles are introduced at one end of the machine, accelerated by a changing electrical field, and then caused to collide with a target at the opposite end of the machine.

The first machine of this type was invented by **John Cockcroft** and **E. T. S. Walton** at the Cavendish Laboratory in the late 1920s. Cockcroft and Walton developed a method for producing very high voltages in an evacuated glass tube. Protons—hydrogen atoms stripped of their electrons—were introduced at one end of the tube. They were accelerated to very high velocities by the large potential difference within the tube. As the protons traveled from one end of the tube to the opposite end, they reached energies of several hundred thousand electron-volts.

An electron-volt is the unit used to measure the energy of particles in a particle accelerator. One electron-volt is the amount of energy attained by an electron as it travels through a potential difference of one volt. The energy achieved by particle accelerators is measured in keV (kiloelectron, or thousand-electron, volts), MeV (megaelectron, or million-electron, volts), GeV (gigaelectron, or billion-electron, volts), or TeV (teraelectron, or trillion-electron, volts). The Cockcroft-Walton accelerator produced particles with energies of about 600 keV.

At about the same time that Cockcroft and Walton were building their accelerator American physicist, **Robert Van de Graaff** was constructing a somewhat similar machine at Princeton University. The Van de Graaff accelerator consists of a hollow metallic sphere on top of an insulating column which forms a large capacitor together with a conductor at the the bottom of the column. A moving belt picks up electrons at the bottom of the column and carries them into the metallic sphere where they accumulate, producing a voltage difference between the sphere and the bottom of the column.

A source of protons in the hollow sphere is injected into the column where they are accelerated by the electric field created by the voltage difference between the top and the bottom of the column. The highly energetic protons thus produced can be used to bombard nuclei in a target. The most powerful Van de Graaff generator can accelerate particles to energies of about 25 MeV.

Both the Cockcroft-Walton and Van de Graaff generators are examples of *one-stage accelerators*. The term one-stage means that particles are accelerated just once after they are introduced into the machine and before they strike a target.

As early as 1928 German physicist, Rolf Wideröe realized that particle accelerators could be made more powerful by accelerating particles a number of times. He suggested that particles be directed through a series of tubes in each of which they would experience an additional force. The cumulative effect of all these ''pushes'' would be to give particles much more energy than they could attain in a one-stage machine.

Wideröe designed a machine that consisted of a series of hollow metal cylinders. The first, third, fifth, and every odd-numbered cylinder was connected with one pole of an electric source, say the negative pole. The second, fourth, sixth, and every even-numbered cylinder was connected to the opposite pole, in this case, the positive pole. When a particle is introduced at one end of the machine, the charges are arranged so that it will be attracted to the first tube.

At the moment the particle reaches the center of the first tube, the charges on all tubes are reversed. The particle is now repelled by the first tube and attracted by the second tube. The effect is to give the particle an extra burst of energy.

The particle experiences an electric force only when it is in the space between cylinders, not while it is inside a cylinder. While it is within a cylinder, it simply drifts, which explains the name *drift tubes* given to the cylinders.

The drift tubes are progressively longer, going from source to target. Each time a particle is accelerated, it spends a little more time within the drift tube. Making the drift tubes longer each time allows the particle to stay in step with the changing electric current.

In 1946, **Luis Alvarez**, at the University of California, suggested an improvement on this basic model. In the Alvarez design, drift tubes are no longer connected to an external source of power. Instead, they are surrounded by a changing electrical and magnetic field. Bundles of particles move through the machine like a surfer traveling on a water wave. This design makes possible the construction of linacs with higher energies.

Carl Paul Gottfried von Linde

The most powerful proton linac in use today is located at the Los Alamos National Laboratory in New Mexico. It is about one half mile (0.75 km) long and accelerates protons to energies of about 800 MeV. The largest linac in the world is located at Stanford University in California. The Stanford linac is two miles (3.2 km) long and accelerates particles to energies of about 30 GeV.

See also Cyclotron; Electrostatic devices

LINEN FABRIC

Linen, derived from flax, is one of the oldest textiles used by man. Its known history begins at about 8000 B.C. in Anatolia (Turkey). Remnants found in Egypt dating to 6000 B.C. indicate that its chief uses there were for clothing and mummification. The Phoenicians carried the fiber to Greece and Rome. Its production and use eventually spread over the entire Mediterranean and temperate climatic zones of Europe and Asia, from Japan to Ireland and from the Mediterranean to Scandinavia.

In Gaul, linen was cultivated long before the Roman invasions, although Charlemagne (742?-814) was the first to promote this material there. In France, linen was in use in large quantities by the eleventh century, when it was discovered that garments made of this material promoted the healing of certain skin diseases. Priests also took to wearing it as a symbol of purity. In the thirteenth century, France's Jean-Baptiste de Cambrai wove the first fine linen canvas.

Until the eighteenth century, cultivation techniques and linen spinning and weaving remained relatively unchanged from medieval times, even though consumption had increased significantly by that time. About this time, linen woven with horsehair gave rise to crinoline. At the beginning of the nineteenth century in France, two inventions contributed to the increasing popularity of linen: **Joseph-Marie Jacquard**'s invention of the loom now bearing his name; and Philippe de Girard's (1775-1845) filing of a patent related to the spinning of linen and his invention of the first combing machine (1817).

Linen was eclipsed by cotton, however, once the latter began to be produced in large quantities in America by cheap slave labor on the Southern plantations. (Until the eighteenth century, linen and wool had been the principal fibers worn in Europe; cotton had been mainly imported from Asia as a luxury item. Before that time, practically every farm grew some flax, and the hand spinning of these fibers into yarns and the hand weaving of these yarns into cloth were cottage industries practised throughout Europe and in many parts of Asia.) The increase in cotton production led to a decrease in the price of cotton fabrics, which for the first time became cheaper than European linen. This decline in the marketability of linen continued until the 1970s and 1980s, when linen began to establish itself as a luxury fabric for men and women's clothing. (Advertisements frequently mention that the air trapped between the linen fibers makes the material a natural thermal insulator—cool in summer yet comfortable to wear in winter.)

Flax is an annual plant that grows very quickly. It can reach 3 feet (1 meter) in height. Processing consists in mechanically separating the textile fibers from the ligneous part of the plant; this is referred to as scutching. From the scutched plant can be extracted scutched flax used for house linen, clothing or furnishings; tows for rougher fabrics and rope fabric; shives for fiberboard used for carpentry and fuel; linseeds for oil, painting, varnish, and for oilcake fabrication for cattle food; and dust for organic enrichment.

Today's major producers of linen are the Soviet States, Belgium, France, Germany and Northern Ireland.

See also Cotton fabric; Textiles

Linoleum

Linoleum is a smooth-surfaced floor covering made from linseed oil, gums, and resins applied to a **felt** or burlap backing. It was invented in 1860 by Englishman Frederick Walton. Up until this time, most households had bare floors, **carpet**ing being out of the reach of all but the well-to-do.

Walton discovered that linseed oil, when oxidized (exposed to air), thickens into a **rubber**-like substance. Walton mixed this with resins and fillers and then pressed it onto a fabric backing. He called the new product linoleum, from the Latin words *linum* (''flax''—source of the linseed oil) and *oleum* (''oil'').

The original linoleum-producing process was very slow and required large amounts of space. The oil was oxidized by being spread in successive thin layers on a sheet of fabric; each layer had to dry before the next could be added. It took several weeks or months to build up a thickened oil film of an inch or more. Much space was needed for the drying sheds. After the oil-resin-gum mix was applied to the backing, it had to be dried and hardened, another time-consuming process.

Walton built the first linoleum-production factory in Staines, England, in 1864. Early in the 1900s, he developed a method for oxidizing the linseed oil in large kettles and a machine that made linoleum in different patterns. The modern process for making linoleum is very similar to Walton's original design but much faster.

Linoleum's resistance to wear, easy cleaning, and affordability made it an extremely popular floor covering for many years. Linoleum production peaked in the late 1940s. After that, **plastic** floor coverings eclipsed linoleum because of their greater resistance to moisture and chemicals, and the great variety of pattern and color they offered. Today's floors are largely, **vinyl**, which is available in tile or as planks that resemble wood. This resilient flooring often has a foam base that is comfortable to stand on and that comes in designs that include faux marble, wood, or stone. These sophisticated floorings are found in rooms throughout the house, creating an easy care, elegant environment.

Linotype

When it was introduced in 1884, *linotype* represented the first major change in the mechanics of typesetting in over four hundred years. Ever since **Johannes Gutenberg** laid out the principles of mechanical printing in the mid-fifteenth century, European typesetters—in spite of an exponential increase in the demand for and variety of printed material—used the same laborious technique of manually constructing full pages of type out of pre-made metal casts of letters. Whereas printers once had to cast type by hand, linotype allowed them to create mechanically customized type, one ''line of type'' at a time.

Previous attempts at speeding up the typesetting process involved using human hands, rather than machines, to set the full pages of type, but none of these attempts was very successful. However, **Ottmar Mergenthaler**, the inventor of linotype, discarded the idea of physically setting pre-made letters; his idea was to alter the printing process by building a machine that would custom-make letters to fit the needed document. In Mergenthaler's finished machine, the typesetter uses a typewriter-like device which has ninety keys to lift copper casts of letters and punctuation (called matrices) into place. Once the line is finished, a quick-cooling molten alloy is poured over the casts; when it cools, it forms a complete line of type, or ''slug,'' which can be set into place. Linotype drastically shortened the time needed to create a page of type for the printing press, allowing typesetters to set more than five thousand pieces of type per hour, as opposed to fifteen hundred per hour by hand. A second version of the machine was able to automatically *justify* text by inserting spaces between words so that each line is the same width. The invention was immediately and universally popular. In 1886, *The New York Tribune* became the first publication to use linotype; by Mergenthaler's death in 1899, more than three thousand of the machines were in use, produced at three factories around the world. Eventually about 100,000 linotype machines were manufactured in all. Though linotype deserves credit for making possible the tremendous boom in the publishing industry during the twentieth century, by century's end the machines were supplanted by computer technology and the advent of **desktop publishing**.

Lippmann, Gabriel Jonas (1845-1921)

French physicist

Lippmann was born in Luxembourg, but his French parents settled in Paris when he was a young boy. In 1875, he received a Ph.D. in physics from the Sorbonne, but not before inventing a sensitive **voltmeter** called the capillary electrometer.

In 1886, Lippmann, a physics professor at the Sorbonne, lectured on a means to use optical ''interference''—the same phenomenon that causes the color seen in soap bubbles, mother-of-pearl, and oil on a wet road—to induce the appearance of color on a photographic plate.

Lippmann's photographic method used a thick emulsion over a thin photographic plate coated with fine-grain silver bromide. The plate was placed in a **camera** with the emulsion side away from the **lens** in contact with mercury. When the incoming light struck the light reflected from the mercury, stationary light patterns were produced that left their impression in the emulsion. This impression reproduced the natural colors of the object photographed. In 1893, Lippmann produced a **photograph** that rendered all colors, approximating their natural brilliance. As early as 1839 others, such as J. F. W. Herschel, Edmond Becquerel and Abel Niépce De Saint-Victor, had produced color images using various methods. But none were able to keep their images from rapidly fading away. Lippmann's image was permanent.

Lippmann's color photographic method had too many drawbacks to achieve commercial success. A two- to three-hour exposure was required, the photographs were difficult to see (they had the appearance of a dense negative), and no copies of them could be made. Nevertheless, **color photography** had at last been achieved and Lippmann received the 1908 Nobel Prize for Physics for the invention. In addition to his work in color photography, Lippmann also fabricated numerous instruments, including the capillary-electrometer and the coelostat, before his death at sea on July 13, 1921.

See also Photography.

Liquid crystal display • See LCD

Liquid gases, commercial production of

The ordinary air we breathe contains several gases which are of great value to industry and science. Oxygen, for example, is essential in making steel, and helium is required by scientists who study the behavior of materials at extremely low temperatures. However, gases normally take up so much space that they must usually be liquefied, or condensed into smaller volumes, for economical transportation and distribution.

Although scientists had produced them in the laboratory for some time, liquid gases were unavailable commercially

Invented in 1884, linotype machines allowed typesetters to create entire lines of type, called slugs, for insertion into a page. Prior to this innovation, typesetters had to manually assemble pages using individual letters.

until the late nineteenth century. Oxygen and nitrogen—the two primary gases contained in air—had been liquefied, but only small amounts of the liquid gases had been produced. German chemist **Karl von Linde** changed that when he invented a continuous process for producing large quantities of liquid air in 1895. British chemical engineer William Hampson (1854-1926) is also credited with making a similar breakthough around the same time. Von Linde became an engineering professor in Munich, where he became interested in low-temperature research. In 1876, he developed the first practical refrigerator. Von Linde's commercial liquefaction process makes use of a phenomenon called the Joule-Thomson effect, which was named after its discoverers, James Joule (1818-1889) and **William Thomson** (Lord Kelvin). These scientists had shown in 1853 that a compressed gas becomes cooler when it expands, as long as it does not absorb heat from its surroundings. In von Linde's process, which is still the basis of modern production, liquid air is cooled and compressed, then allowed to expand, which cools it even more. The cold gas is constantly recycled to cool more incoming compressed gas. Because of the cumulative cooling effect, the gas gradually becomes cold enough to liquefy. Von Linde's process immediately became a commercial success and laid the foundation for today's liquid air production industry.

Von Linde also developed more economical methods of separating the liquid oxygen and liquid nitrogen, both of which have found many practical uses in research and industry. In biological research, for example, liquid nitrogen is used to freeze blood cells, sperm, tissues, and even whole small organisms. When frozen, the cell stops its normal activities, allowing scientists to examine a virtual snapshot of cell life. Nitrogen is also an important agricultural fertilizer, and in industry it is used for **refrigeration**, food processing, and metal heat treating. Although some oxygen is transported in pipelines, much of it is shipped in its liquid form in special steel cylinders. In addition to steelmaking, which consumes the largest amounts of oxygen, the gas is used in torches to produce an extremely hot flame and in rocket engine fuel.

Most of the other gases contained in air—argon, neon, krypton, and xenon—are also produced by air liquefaction and separation. Each has many uses in industry and science. Liquid helium can also be separated from air, but the process is relatively expensive because of helium's extremely low liquefaction temperature. Most commercial supplies of liquid helium are produced from natural gas and transported in railway tank cars. Larger volumes can be carried more cheaply, however, in special **truck** trailers that were recently introduced. More laboratories can now afford to use liquid helium in their low-temperature research.

Liquified Petroleum Gases (LPG) are also commercially important. Unprocessed natural gas contains about 3 to 18 percent propane (C_3H_8), as well as some butane (C_4H_{10}). These two **hydrocarbons** are extracted from the gas, then liquefied and sold as LPG. Propane and butane can also be produced from light crude oil and from gases formed when heavy crude oil is refined during the manufacture of gasoline. Liquified butane is often used in cigarette lighters and in portable stoves for camping and boating. The petroleum industry uses propane as a solvent and refrigerant during the manufacture of lubricants and as a major source of **ethylene**, an important industrial chemical. In addition to its value as a domestic fuel in rural areas, LPG can be burned in internal-combustion engines.

See also Blowtorch; Fertilizer, synthetic; Gases, liquefaction of; Refrigeration; Rocket and missile; Steel production; Train and railroad

LIQUID PAPER • See Typing correction fluid

LITHOGRAPHY

Lithography, which has developed into one of the world's most popular processes for mass-producing images, was invented not by an engineer or a trained printer, but by a writer.

The process, however, has its origin in a humble laundry list. Its inventor, Aloys Senefelder (1771-1834), was a struggling German playwright whose family had little money. Unable to afford the cost of printing his plays, he began to experiment with ways of engraving the manuscripts himself. Senefelder met with little success until one day in 1796, when, asked to take down a laundry list by his mother, he wrote the list in grease **pencil** on a polished limestone slab with which he had been experimenting. It occurred to him that he could use the water-repellent qualities of the grease coating to produce an acid etching on the slab. Over two years, he ultimately arrived at a different approach based on the same concept. In Senefelder's process of lithography (literally, "writing on stone"), the design to be reproduced is drawn directly onto a stone slab or plate. The slab is then dampened completely; the water, however, is repelled from the grease-covered areas. When a coating of **ink** is then applied to the plate, it washes away from the wet areas, but adheres to the grease marks. Senefelder found that the image created by pressing this plate to paper was exceptionally faithful in reproducing even fine lines in the original; he also found, later, that a solution of gum arabic and nitric acid was even more effective than water in completely repelling ink from the non- greased areas. Also, the stone ultimately was replaced with metal plates.

Lithography was quickly embraced by printers as a quick and effective method of mass-producing commercial illustrations and sheet music, and after Senefelder published *A Complete Course of Lithography* in 1818, it developed into a popular means of creating original art, used by painters Eugène Delacroix, Théodore Géricault, and Goya. In the mid-nineteenth century, a process was developed for using a series of different plates to produce multi-colored lithographs. The mass-market art produced by Currier and Ives, an American firm, was made possible by this technique, as were the now-famous posters of French artist Henri de Toulouse-Lautrec.

Along with the general automation of printing processes that occurred in the nineteenth century and though the turn of the twentieth century, the application of **photography** to create images on plates, through a process called photolithography, shaped modern lithography. In photolithography, the printer shines bright light through a photo negative onto a thin plate coated with light-sensitive chemicals. The areas of the plate struck by the light harden into a reproduced image, serving the same function as the grease design in early lithography. Most lithographic printing today is done through a process called offset printing, in which the thin metal plate is clamped to a cylinder and inked, and then transfers the image to a turning rubber cylinder; finally, the rubber cylinder transfers the ink onto a sheet of paper as it rolls beneath it. Like other automated, cylinder-based printing techniques, offset printing is able to print sheets repeatedly on both sides at very high speeds. Because of this, and because of the fact that photolithography allows photographic reproduction of print and photos as well as art, lithography remains today one of the most widely used printing methods.

See also Printing technology

LOCK, CANAL • See Canal and canal lock

LOCK AND KEY

Ever since people first began having private possessions, locks to keep that property out of the hands of thieves have been designed. It has been an evolutionary process: As each lock eventually yielded to thieves' attempts to pick, or open, it, ingenious locksmiths came up with new lock and key designs.

The earliest known lock is the wooden Egyptian pin-tumbler type. It was opened by a pegged wooden key, often more than a foot long, so large that it had to be carried over the shoulder. When the key was inserted into the lock, its pegs lifted pins that held the bolt in place. The ancient Chinese used a leaf-spring padlock; the key depressed the spring, allowing the bolt to slide back. The Greeks used primitive locks with large, sickle-shaped keys.

The Romans introduced metal for locks and keys and also invented the warded lock. Wards are projections inside the keyhole; only a key whose bit (flat face) is cut with projections that pass over the wards can rotate to open the lock. The Romans also independently invented the padlock, and made small keys that were worn as finger rings.

As Europe passed out of the Dark Ages into medieval times, nobles and the new merchant class acquired money and possessions that needed safeguarding. Locksmithing flourished during this period, with very intricate and precise moving parts and elaborately decorated exteriors and keys. These were still warded locks, and since most could be fairly easily bypassed, extra security was provided by hidden and dummy keyholes—even by locks with spring-loaded knives. The skeleton key emerged in this era, a long, straight key with a single tooth that could be slipped around the wards to open the lock.

Important improvements were made in lock design during the eighteenth century. The Englishman Robert Barron patented the double-action tumbler lock in 1778. Tumblers fell into place in slots in the bolt, which was held in place until the correct key raised each tumbler to the correct height, which lifted a pivoted lever and freed the bolt to move. In 1818 another English inventor, Jeremiah Chubb, improved the lever tumbler lock by adding a detector tumbler. If a tumbler was raised too high by an improper key, the detector jammed the lock, which could then only be opened by the correct key being turned backwards. Chubb was a Portsmouth ironmonger who later established a factory that produced fireproof **safes**. He was inspired to design his lock after the British government, alarmed by the high incidence of thefts and the 1817 robbing of the Portsmouth dockyard, offered a reward to the person who could invent a pick-proof lock. Chubb won the prize.

Another ingenious lock was the Bramah, patented by Englishman **Joseph Bramah** in 1778. In this lock, a tube-shaped key with a number of slots depressed a corresponding number of slides, each by a different amount depending on the length of the slot. Bramah's advertised claim that his lock was pick-proof—he offered a 200-guinea reward to the first person who could open it—held up for sixty-seven years. Finally, in 1851, an American expert, A. C. Hobbs, picked the Bramah lock, but it took him fifty-one hours to accomplish the feat.

Because Bramah's lock was complicated, he invented a number of machine tools to manufacture it, turning locksmithing away from handcrafting. This changeover was cemented by the invention of Linus Yale, Jr. (1821-1868), an American from Connecticut. Born in New York in 1921, Yale originally intended to become a portrait painter but instead emulated his father and began designing bank locks. In 1861 he patented a small, cylindrical pin-tumbler design that was an adaptation of the ancient Egyptian locks. A small, serrated key raised pins in the cylinder to proper heights, allowing the cylinder to turn. The key and pin design allowed for a vast number of possible configurations, so that no two Yale keys were alike. The small, flat keys as well as the locks could be easily mass produced and machine-made, which meant that inexpensive, secure locks were now available to the mass public. Today, Yale locks are used on most outside doors of houses and businesses. Combination locks go back to the sixteenth or early seventeenth century. A number of slotted rings, or tumblers, imprinted with numbers or letters, turn; when lined up correctly, the lock opens.

Combination locks became the favored type for tone safes because they had no keyhole in which robbers could insert explosives. James Sargeant of Rochester, New York, invented a time lock for banks in 1873; a clock controlled the hour at which the safe could be opened.

In the 1970s electronic locks were introduced. Spring-loaded buttons sound different electronic tones when pushed, like touch-tone **telephone**s. The lock opens only when the right sequence of tones is sounded. Despite the advent of combination locks, touch-tone locks, and other **keyless entry system**s, there are still many times when a key of some sort is preferred, since access to it can be more easily controlled. In addition, electronic technology has made the simple key more useful and sophisticated. Resembling a credit card, flat plastic "key cards" with a magnetic strip have become popular with hotels, businesses, and universities. When the keyholder inserts the card into (or "swipes" it through) an electronic card reader, the door's lock is opened. Inexpensive to manufacture and eas-

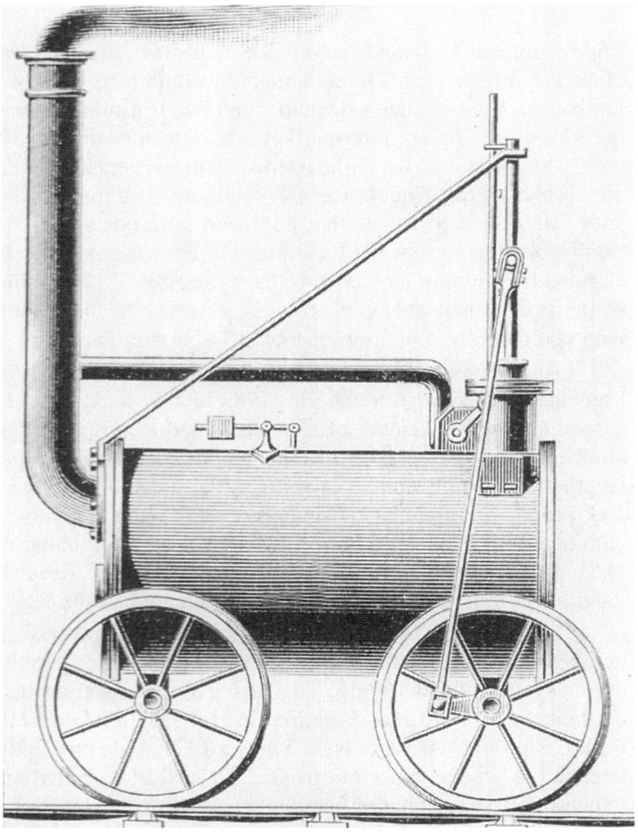

English engineer Richard Trevithick's introduced the first working steam locomotive (above) in 1804.

ily re-coded, this lock and key system can also provide information, such as the identity of the keyholder and the time of entry. In this way, the lock and key have become a form of **surveillance equipment**.

The traditional metal key has also been reinvented. In the 1990s Charles Hall, Rick Hyatt, and Doug Treat of the Medeco lock company developed a way to attach a handheld computer to a conventional key. Dubbed by its inventors the "V-Lock" and the "V-Key," the system was developed for use on pay telephones to eliminate the need for a different key for each of thousands of telephone coin boxes. The V-Key activates the V-Lock when inserted, and opens the lock when the attached computer provides the lock's identification number. Therefore, a single V-Key can open thousands of V-Locks as well as record inventory data. The continuous reinvention of the lock and key demonstrates the flexibility and genius of this ancient invention.

Locomotive

Locomotives, machines used to move trains, may be powered by steam, Diesel fuel, or electricity. The steam locomotive's energy is produced when wood, coal, or fuel oil burned in a "firebox" creates heat that turns the water in a boiler to steam; the steam is fed into cylinders, where the pressure it produces drives the piston s (steel rods) that move the locomotive's driving wheels. The Diesel locomotive's engine functions when air is compressed in its cylinders until its temperature is high enough to ignite the fuel that has been injected into the cylinder; the resulting power is transmitted to the driving wheels. The electric locomotive uses electricity transmitted from a power plant by special overhead wires or an electrical third rail. Around the middle of the twentieth century, Diesel locomotives began to replace their steam-driven predecessors. There are now more than 20,000 Diesel locomotives in the United States; less than 100 are powered by electricity, and only about 50 steam locomotives still operate (mostly for the benefit of tourists).

Although primitive wooden tramways (along which horses pulled loaded wagons) were operating in Transylvanian mines as early as the sixteenth century and in English coal mines by the seventeenth century, the concept of using a steam engine to propel vehicles along a track was not developed until the early nineteenth century. **Richard Trevithick**, an engineer employed at the Penydaren Tramway and ironworks in South Wales, is credited with inventing the locomotive, although he never received full recognition or compensation for his invention. Trevithick adapted steam engines made by **Thomas Newcomen** and **James Watt** into lighter, more compact devices that produced higher steam pressure. By 1796 he was designing steam locomotives, one of which pulled a passenger train in 1801. In 1804 he introduced what is considered the first working steam locomotive; on its maiden voyage, it carried 10 t. (9.07 t) of iron, 70 men, and 5 wagons, traveling 9.5 mi. (15.3 km) at about 5 mph (8 kph). This locomotive was never actually used because the tramways available at the time were unsuitable.

In 1808 Trevithick built a demonstration locomotive called *Catch-Me-Who-Can* and, setting it up on a circular track in London's Euston Square, offered the public rides for a shilling.

Although Trevithick is considered the first to prove that smooth metal wheels moving on smooth metal tracks could supply enough traction to pull trains, his work was overshadowed by that of the more successful **George Stephenson**. Although he did not invent the steam locomotive, George Stephenson played an important role in its development. As a teenager, Stephenson went to work in England's Killingworth Colliery, where he was in charge of the water-pumping engines. Stephenson distinguished himself by building stationary engines and trains to replace the horse-drawn sleds that had previously been used to move loads on the mine tracks. Stephenson's employers, concerned by the high cost of horse feed, commissioned him to build a steam-driven locomotive. Finished in July 1814, the *Blucher* featured an important innovation: the *flanged* wheel, equipped with a guiding rim that projected from the top of the wheel.

Stephenson established a locomotive company at Newcastle in order to make locomotives for a new railroad that was to be established between the towns of Stockton and Darlington, a distance of about 30 mi. (48 km). The railroad opened in 1825 with Stephenson's *Locomotion No. 1* at the head of its fleet. The famous Rainhill trials were held outside London in 1829 to determine the feasibility of steam locomotives for transporting goods and people. Stephenson's Rocket won the contest, impressing the onlookers with its speed of 29 mph (46.7 kph). The next year, Stephenson became chief engineer for the Liverpool-Manchester Railway, the first in the world to carry passengers. Initially operating with eight locomotives on a line that was only 30 mi. (48 km) long, the railway eventually reduced the travel time from London to Edinburgh from 12 days to a mere 50 hours. Robert Stephenson also excelled in his father's profession, producing in the *Northumbrian* and the *Planet* a standard locomotive type that would be used all over Europe and America.

Although the steam locomotive was invented by the English, American engineers and manufacturers contributed significantly to its development, responding to the challenges of their own country's terrain and transportation needs. The first full-sized locomotive to operate on a commercial railroad in the United States was the *Stourbridge Lion*, made in England and introduced by a Pennsylvania Company in 1829. Old Ironsides, built in 1827 by manufacturer and philanthropist **Matthias Baldwin**, was made of wood and iron and could travel at 28 mph (45 kph), pulling a load of 30 tons (27.2 t). The manufacturing company established by Baldwin had, by the time of his death, made over 1,500 locomotives. The South Carolina Canal and Rail Road Company's *Best Friend of Charleston*, which went into operation in December 1830, was the first American steam locomotive to be used for regular passenger and freight service. Earlier that year, Peter Cooper of New York—a prosperous manufacturer, inventor, and philanthropist—had staged a race between his diminutive locomotive, *Tom Thumb*, and a locally renowned horse in an effort to convince businessmen to use locomotives instead of horses for their transportation needs. Although the horse won, due to a minor mechanical problem with the locomotive, Cooper's enthusiasm for the steam locomotive was to prove justified.

The rugged American landscape posed challenges to those who, like Cooper, wished to promote the growth of railroads across the vast expanses of the United States. Unlike their English predecessors, American railroads were built on curved routes. The problem of negotiating sharp curves was resolved by John B. Jervis, who in 1832 designed an engine equipped with a four-wheel "truck" affixed to its underside that allowed it to swivel when the tracks curved.

The train proved a great boon to the settlement of America's frontier, and between 1830 and 1890 an average of 2,500 mi. (4,022 km) of track per year was laid across the country. Innovations added to the locomotives included flashy bells and whistles, headlights, cowcatchers, and cabs to protect the operators. American locomotive manufacturers found ways to use cheaper materials, and they designed a modified car body—featuring a long, open, single compartment with seats arranged on both sides of a center aisle, doors at each end, and platforms outside each door.

Another significant contribution to the development of the locomotive was made by Elijah McCoy (1843-1929), an African-American engineer educated in Scotland who, because of his race, was forced to take a menial job with the Michigan Central Railroad. While working as a fireman, McCoy became interested in the problem of train lubrication. In 1870 he invented a lubricating cap that eliminated the need to stop machines in order to lubricate them. This revolutionary device was soon used in many different types of machines and became so essential that users referred to it as "the real McCoy" (hence the origin of that expression). McCoy held over 50 patents for his inventions—mostly in the realm of lubrication—and his lubricating systems are now used all over the world.

At the end of its first 100 years of existence, the locomotive had doubled in size from 20 or 30 tons. (18 or 27 t) to 60 tons (54.4 t) or more. By 1938 the trip from New York to Chicago, which had taken 37 hours in 1875, had been reduced to only 16 hours. The peak popularity of trains is said to have occurred around 1916, when 77 percent of intercity freight traffic and 99 percent of intercity passenger traffic were conducted on trains. A major change was the advent of the Diesel locomotive, which was first introduced in 1923. Although Diesel locomotives were more efficient, they cost twice as much as steam locomotives. During the high-demand years of World War II, however, the efficiency of the Diesel locomotive was favored, and an improved system of mass production helped to reduce its cost. By 1960 Diesels had replaced all but about 260 steam locomotives in service in the United States.

The electric locomotive dates from about the 1830s, when attempts were made to power trains with electricity. In 1842 Robert Davidson successfully operated a battery-electric locomotive on the Edinburgh and Glasgow Railway; moving at a speed of 4 mph (6.4 kph), it hauled 6 tons (5.4 t). Ernst Werner von Siemens (1816-1892) operated an electric railway at an exhibition in Berlin, Germany in 1879 and later in Brussels, Belgium, and London, England. In the latter years of the nineteenth century, many engineers and inventors—including **Thomas Edison**—contributed to the development of the electric locomotive. Because electric trains are quiet and emit no smoke or exhaust, they are particularly well-suited for urban areas and underground use, and some proponents predict that they will eventually replace Diesel locomotives. Two notable modern electric trains are Japan's *Shinkansen* or "bullet trains," which travel at speeds up to 130 mph (209 kph) and France's *Ligne à Grande Vitesse*, which reaches speeds up to 136 mph (218.8 kph). In 1990, the French train achieved a record speed of 320 mph (515 kph). These "bullet trains" generally speed along tracks that are very straight and pull cars that are sleekly designed with curved fronts and sides to reduce drag, or wind resistance, that would slow a train down. They also usually have special disc brakes similar to those on jet airplanes.

The newest pulling technology for high-speed trains is called Maglev (for magnetic levitation). These locomotives use a set of magnets to lift or levitate the train above the ground, and although they run on rails, they actually do not touch the rails but rather float just above them by means of strong magnetic forces. Maglev technology uses an electromagnetic charge on the underside of the train which repels the opposite charge on its top. Using giant **magnets** and linear motors (electrically run with no moving parts), a Maglev train experiences none of the friction of a train running on a rail, and not only runs very fast and efficiently, but also gives an ex-

Oliver Joseph Lodge

tremely smooth and quiet ride. In 1979, a model Maglev train was tested in Japan and reached 320 mph (515 kph). Germany's *Transrapid 06* Maglev train began in 1984 as an intercity project that would run a high-speed train on a raised track. In the 1990s, plans were made in several eastern cities in the United States for commercial Maglev trains. In the future, super-Maglevs may eventually travel thousands of miles per hour.

Until this new technology becomes commonplace however, diesel and electric locomotives will continue to improve to do the jobs in which each specializes. Among the major improvements begun to be developed in the 1990s were wedge-shape ''unibody'' locomotive design; microprocessor engine control with electronic fuel injection for medium-horsepower diesels; reduced fuel consumption and emissions; electronically-monitored and controlled brakes; and in 1996, the first 6,000 horsepower diesel locomotive.

Lodge, Oliver Joseph (1851-1940)

English physicist

Although his father intended for him to become a businessman, Oliver Lodge instead became one of the pioneers of communication systems at the turn of the nineteenth century. Born on June 12, 1851, at Penkhull, Staffordshire, England, he was the oldest of nine children. His father, who was also named Oliver (and was the twenty-third of twenty-five children), was a pottery merchant.

Young Oliver went to work in his father's business at age 14, but occasional visits to London afforded him the opportunity to attend lectures given by John Tyndall (1820-1893) at the Royal Institution. These lectures turned Lodge's interests toward science, and in 1873, at age 22, he enrolled at the Royal College of Science and at University College in London, studying electricity and physics. He received his doctorate degree in 1877, and became the first professor of physics at University College in Liverpool in 1881. He spent the next nineteen years in Liverpool, where he made contributions in the theory of the ether and electromagnetic wave propagation.

The theory of light at the time required a medium through which light waves were to pass as they moved through space. Lodge helped to disprove the existence of such an ''ether,'' clearing the way for Albert Einstein's (1879-1955) new theory of matter and energy, which did not need an ether to explain the behavior of light. More importantly, Lodge studied electromagnetic radiation closely. In 1887, he discovered that traveling waves and standing waves were produced along conducting **wire**s as a result of oscillations involving the discharge of a Leyden jar. These waves were measurable and followed precisely the theory of electromagnetism established by James Clerk Maxwell (1831-1879).

Meanwhile, German physicist Heinrich Hertz (1857-1894) had generated similar invisible radiation at a wavelength of 2.2 ft. (66 cm). While Hertz just beat Lodge to the discovery, Lodge invented a better method of detecting the ''Hertzian waves.'' Lodge named his device the coherer. In 1890, French physicist Edouard Branly (1844-1940) observed that loosely connected metals would cohere when subjected to electromagnetic radiation. Lodge's coherer contained metal particles which stuck together in the presence of these waves, and made an excellent detector.

Lodge concentrated on studying long wavelengths, much longer than the wavelenghts of visible light, making him the leader in early **radio** communication. By 1894 he had used radio waves to carry **Morse code** signals half a mile (800 m).

On June 1, 1894, five month's after Hertz's unfortunate early death, Lodge gave a commemorative lecture in which he stressed the importance of syntony, resonant tuning by focusing electromagnetic radiation to obtain better results. One of his concepts involved a resonant **antenna** circuit. Lodge and his associate, Alexander Muirhead, formed a business and were farsighted enough to obtain patents covering this circuit. His lecture was printed in a book which had great influence on the development of radiotelegraphy around the world. **Guglielmo Marconi** would incorporate Lodge's devices and concepts into his radio system, which produced the first successful transatlantic **telegraph**. Lodge was knighted for his accomplishments in 1902.

In 1883, Lodge had become interested in extremely long-distance communication of another type. He began to study psychic research, telepathy, and communication with the dead. Following the death of one of his sons (Lodge had twelve children of his own) and his own retirement, his interest increased. In addition to the professional organizations to which he belonged, he was twice president of the Society of Psychical Research. On August 22, 1940, eighty-nine year old Oliver Lodge died.

LOG, SHIPS • See Ship's log

LOGIC MACHINES

The study of logic as a subject dates back to the times of Greek philosopher Aristotle (384-322 B.C.). While Francis Bacon (1561-1626), Gottfried Leibniz (1646-1716) and René Descartes (1596-1650) all made valuable contributions in the development of logical thinking during the Renaissance, it was not until the mid-nineteenth century that true breakthroughs occurred. Two mathematicians, George Boole (1815-1864) and Augustus De Morgan (1806-1871), independently proposed forms of algebra representing logical expressions. In 1854 Boole published The Laws of Thought, which expressed logic in algebraic form. Boolean algebra, as his discovery is called, has only two values, 0 and 1. These values are the basis of the binary numbering system used in **machine language**. By the nineteenth century, various logic devices were constructed which incorporated Boolean algebra to mimic logic.

In 1864 William Stanley Jevons (1835-1882), an economist and logician, published *Pure Logic: or the Logic of Quality Apart from Quantity*, which was based upon Boole's system of logic. Building on those ideas, Jevons created a logical machine which obtained conclusions mechanically from known assumptions. Consisting of letter keys and ivory result keys on the top section of the machine, Jevons' device looked somewhat like a portable piano, from which its nickname, the "logical piano" came. His machine, which he successfully demonstrated before the British Royal Society in 1870, proved the first known logic solver capable of operating faster than a human.

In 1881 Allan Marquand, a Princeton University professor, constructed a device to solve problems of formal logic. A fairly simple design based upon that of Jevons, it consisted of a wooden box containing rods and levers connected together with strings and springs. By manipulating the levers and rods, the machine displayed valid results from simple logical questions. In the next few years, Marquand designed an electrical version of the machine, which is thought to be the first electrically designed machine—though whether Marquand ever actually built it is unclear. Another, rather surprising, logic machine inventor was Charles L. Dodgson (1832-1898), perhaps better known as Lewis Carroll, author of Alice in Wonderland. Dodgson, who was also a mathematician and logician, designed a machine in 1886 which consisted of a ruled board and counters, and resembled an **abacus** in appearance.

In 1903 Annibale Pastore, a professor of philosophy at the University of Genoa, Italy, built a logic machine that provided mechanical representation of problems and their logical solutions, using belts driven by differential **gears**. Many concepts used by computer logic and **artificial intelligence** researchers today can be traced back to Boolean algebra and the early logic machines.

See also Calculating machine; Digital computer

LONGITUDE • See Chronometer

LONG-PLAYING RECORD

The LP (or long-playing) record was developed by **Peter Carl Goldmark** in 1948, changing the music industry forever. Before the LP record, the standard recorded disk was the 78 rpm (revolutions per minute) record, which had several drawbacks. The disks themselves were made of shellac compounds, which had to be imported from Southeast Asia and became very scare during World War II. Indeed the United States government banned its use in non-military items during the War. Also, shellac was hard, brittle, and easily scratched. Phonograph needles wore out quickly. The 78 rpm featured grooves of.001 in (.025 cm), and could only hold about four minutes of music per side.

Goldmark, an amateur cellist and music buff, became frustrated at the limitations of the 78 rpm record. In the middle of a recorded symphony, for example, the music would stop and the record would need to be turned over or replaced. Using the resources of his employer, CBS (Columbia), Goldmark spent three years developing a replacement for the 78. He devised the LP record, 12 inches (30.4 cm) wide, with its "microgrooves." The LP record's grooves are only.003 inches (.008 cm wide, and it plays at 33 1/3 revolutions per minute). About 25 minutes of music can fit on one side. Indeed, it would take about six 78 rpm records to fit in one LP record. And unlike the 78, the LP is made of plastic compound called PVC (polyvinyl chloride). PVC has several advantages over the shellac compound used in 78 rpm records. It is manufactured from petroleum by-products, and therefore available in the United States. PVC is also cheaper to produce, contributing to the decreased cost of LP records as compared to 78s. Most importantly, LP records sounded much better than their predecessors. LP records featured a increase in the frequency and dynamic ranges of recording, and less surface noise. Though LP records could scratch easily, they were much stronger than 78s.

Soon after the LP's introduction, RCA announced the development what it hoped to be a rival to the LP. They invented the 45 rpm record, which was only 7 in (17.8 cm) wide, but like the 78 rpm record, it could only hold several minutes of music per side. Eventually, the 45 rpm record became the standard for single, one song per side, releases.

In 1957, LP records became stereophonic. Before this date, LP records had only one channel (i.e., monophonic), and stereo sound was only available on a certain kind of tape. There were several rival formats for stereo LPs, but the Westrex 45/45 degree system eventually won out. Its groove was V-shaped, and featured two walls for the right and left channels. The Westrex system also was compatible with mono record players, which increased its appeal to the consumer market.

Original model of the loom invented by Joseph-Marie Jacquard in 1801. Jacquard's punched-card method for storing patterns was later used by Charles Babbage in the design of his analytical engine, the precursor of modern computers.

The LP record remained the dominant recording medium for consumer purchase until the 1980s, when **compact discs** were introduced. LP records relied on analog recording technology. Compact discs used digital technology and were lauded for their ability capture an even greater dynamic range than LPs. Though diehard LP record fans claimed their disks featured ''warmer'' sounds, the LP record was doomed. By 1991, many music retail stores stopped carrying them. As of the late 1990s, the LP record remains alive only as a specialty item.

See also Sound mixing and recording.

Looms

The weaving of thread into cloth has been practiced for many thousands of years. As it is today, woven fabric was created by criss-crossing threads over and over again to form a tight mesh. Fairly early on, devices were constructed to make the task of weaving easier. One of the most primitive was called a peg loom or ground loom, and was used around 5,000 B.C.

The ground loom consisted of two pegs driven into the ground, between which wooden bars were mounted. Long threads were hung over the wooden bars. The weaver sat in a shallow pit, weaving the cross-thread through the long threads. Later ground looms allowed the weaver to operate mechanisms by foot. Though this system was very crude, the quality of the fabric ultimately depended upon the skill of the weaver: an experienced weaver could produce beautiful tapestries on even the simplest of looms.

Since the creation of these first devices, the loom has evolved into one of the most complex machines in the world. Still, every loom, no matter how intricately constructed, is based on the same fundamental design. In woven fabric, there are two kinds of thread: the strong vertical *warp* thread and the weaker horizontal *weft* thread. In order for the fabric to mesh, the weft thread must be passed through the warp in an over-and-under fashion. Originally this was done by hand, but as loom technology improved mechanisms were built to lift every other warp thread, creating an opening called a *shed* through which the weft thread passed. First the even-numbered threads were lifted and the weft was passed through the shed. Then the even-numbered threads were lowered, the odd-numbered threads lifted, and the weft was passed through the other way. As each weft passage was completed the thread would be pushed, or beaten, against the cloth with a comb, creating a tightly woven fabric.

The first true frame looms were probably inspired by the silk weaving machines of China and India. These machines were equipped with foot pedals to lift the warp threads, leaving both of the weaver's hands free to pass and beat the weft thread. From this design came the vertical loom, a tall pair of vertical bars over which the warp threads were hung and kept taught by weights. This apparatus was soon found to be more efficient when laid flat, thus creating the horizontal loom, used as far back as ancient Egypt.

The horizontal loom was further improved by the addition of two devices: the *heddle* and the *shuttle*. The heddle was a foot-controlled bar that could lift an entire row of warp threads, creating a shed that extended from one side of the cloth to the other. The shuttle was a thin wooden slat upon which the weft thread was mounted. To weave the fabric, the shuttle was thrown through the shed and caught at the other side; the weaver would then beat down the weft, lift the other half of the warp threads, and throw the shuttle back through the shed. Later shuttles had spindles or bobbins built in, so that the weft would unwind behind it.

The heddle and shuttle were elegantly combined in the warping frame loom, invented by Frenchman Claude Dangon in 1606. By the early 1700s devices were added that could lift individual warp threads, allowing for the weaving of complex patterns and textures into the cloth. In 1728 another Frenchman, Falcon, suggested the use of perforated cards to ''store'' the cloth patterns; this idea would not be full explored for another seventy-five years.

The advancement of loom technology was not universally welcomed; on the contrary, each new invention was commonly received with fear and hostility, particularly on the part of the hand-weavers who saw themselves becoming rapidly obsolete. Often, the increasing use of weaving machines throughout the eighteenth and nineteenth centuries is identified with the beginnings of social unrest-particularly between the lower and upper classes. This is because weaving machines were the first mechanical devices to effectively replace large numbers of skilled workers. The word ''sabotage'' probably originates from these conflicts, for workers had been accused of throwing heavy shoes called *sabots* into the workings of the machines, destroying them.

The reaction to John Kay's (1704-1764) invention of the **flying shuttle** typifies the reception of loom technology by the hand-weavers. Kay had noticed that the primary drawback to the hand-thrown shuttle was that the width of the cloth could not be greater than the weaver's reach¾for wider cloth, two weavers had to be employed to throw the shuttle back and forth. Kay designed a wheeled shuttle that ran along a ledge called the slay. To propel the flying shuttle through the shed, the weaver simply had to pull a cord. This system allowed a single man to weave broadcloth and still keep one hand free to manipulate the thread. Though Kay's invention effectively began the British Industrial Revolution by dramatically speeding the production of woven fabrics, he was the target of violence by local workers. He fled to France in 1747, where he died in poverty.

Often, the increasing use of weaving machines throughout the eighteenth and nineteenth centuries is identified with the beginnings of social unrest-particularly between the lower and upper classes. This is because weaving machines were the first mechanical devices to effectively replace large numbers of skilled workers. The word ''sabotage'' probably originates from these conflicts, for workers had been accused of throwing heavy shoes called *sabots* into the workings of the machines, destroying them.

The next step in the evolution of the loom came in 1706 when **Edmund Cartwright** constructed the first mechanical loom. This device is reputed to have been inspired by the mechanical **water frame** spinning machines of **Richard Arkwright**, which Cartwright had observed some years earlier. The mechanical loom almost completely automated the weaving process; now the only manual task was feeding thread into the machine. In 1791 Cartwright's mill was burned to the ground. His patents disputed in court by handweavers, Cartwright, too, died penniless.

In 1801, the French inventor **Joseph-Marie Jacquard** completed his patterning loom. This was a direct descendant of Falcon's design, using **punched cards** to control the lifting of certain warp threads. He received a gold medal and a pension of several thousand francs for his invention. In addition, the punched-card method for storing patterns was copied by the English scientist **Charles Babbage**, who planned to use similar cards in his steam-powered **adding machine**. Though Babbage's ''analytical engine'' was never constructed, it became the model for early computers—most of which also used punched cards as a storage medium.

Inventors across Western Europe were busy designing looms that could be powered by water or steam. In 1822, Englishman Richard Roberts succeeded in developing a completely automatic loom; though, at first, it was able to weave only plain cloth, improvements were soon made that allowed for the weaving of complicated patterns. By the late 1800s, English-born mechanic J. H. Northrop constructed a power loom that did not need to be refilled when the weft ran out. Not only did this speed the weaving process, it also eliminated a lung affliction suffered by many loom workers who, as they sucked the weft thread through the shuttle, would inhale fabric dust. American, Erastus Bigelow, developed a power loom for the weaving of carpets, paving the way for wall-to-wall home and office carpets.

Most modern factory looms are based upon the inventions of Northrop and Bigelow. Today's looms only vaguely resemble those used even one hundred years ago. The use of a shuttle has been virtually eliminated; instead, weaving machines use pairs of gripping rapiers to push and pull the weft through. Other systems use spring-guns to fire the thread through the shed. More modern designs use jets of air or water; such machines can thread more than 1,500 wefts each minute.

See also Textiles

Loran

LORAN (Long Range Navigation) is a radio-based navigational aid first used during World War II to locate ships and planes with greater accuracy than could be achieved with conventional techniques. LORAN determines location by comparing accurately-synchronized powerful radio pulses originating from different reference transmitter sites. Pulses from nearby transmitters arrive earlier than pulses from distant transmitters since radio signals travel at a constant speed.

At least three different LORAN signals must be received to determine latitude and longitude. In practice, the distance to more than the minimum three LORAN signals increases accuracy.

The first LORAN systems were in use before **computers** were sophisticated enough to perform the complex calculations needed to process the timing comparisons. Early LORAN installations required highly-skilled operators to interpret the radio pulses. A half century later technical innovations eliminated the need for much of the skill once required to use LORAN for navigation.

LORAN has evolved through three distinct phases, LORAN A, LORAN B, and the present version, LORAN C. The A and B versions were designed for navigational assistance over relatively (short distances). LORAN A and LORAN B transmissions operated in a range of frequencies just slightly higher than the standard AM broadcast band in the United States. The present version, LORAN C, is assigned to 100 kHz, a frequency well below the AM standard broadcast band. 100 kHz, a frequency where long-distance radio propagation is very dependable. In contrast to LORAN A and LORAN B, LORAN C is reliable over distances of many hundreds of miles.

The phenomenal accuracy of LORAN is possible because radio signals travel at a constant known speed. Each coordinated LORAN transmitter sends out a continuous succession of sharp radio pulses. If the LORAN receiver is equidistant from two transmitters, the pulses will be coincident. If the pulses from one station are received earlier than the pulses from the other station, the difference in the time of arrival of the two pulses contains information about the difference in distance to the two transmitters.

Radio signals travel a distance of almost exactly 984 feet (300 m) a microsecond. If a LORAN receiver measures a 100-microsecond time delay between pulses from two identified transmitters, the receiver is somewhere along a line corresponding to all the locations that are 9,843 feet (30,000 m) closer to the station transmitting the pulses received first than to the other transmitter site. That is, the receiver is not necessarily located 9,843 feet from the closest station, but it is 9,843 feet closer to this station than to the other station. If pulses from a different pair of stations is measured, with at least one signal source not involved in the first measurement, the difference in the distance to these new transmitter sites can be determined similarly. If this second comparison reveals that one of these transmitter sites is 16,405 feet (50,000 m) closer than the other, the LORAN receiver will be along a different line where the difference in distance equals 16,405 feet. The coordinates of the point where these two lines cross satisfy both measurement pairs. A third pair of signals must be measured to remove a final ambiguity.

Early LORAN operation required the use of a previously-prepared map, covered with curved lines that corresponded to various distance differences from sets of received signal sources. The early LORAN devices indicated which map lines to use, the operator found the point on the map where the lines intersected to learn the location.

The latest versions of LORAN C receivers no longer require the use of a special map to determine location. These updated units contain a more sophisticated computer that calculates longitude and latitude directly, displaying in a format that does not need interpretation.

The LORAN C receiver automatically tunes first one then another and another of as many LORAN signals that can be received well enough to provide good data. After a short calculation delay the latitude and longitude is displayed.

As an illustration of the great locating accuracy achieved by LORAN C systems, commercial fishers sometimes use LORAN C when looking for buoys marking submerged crab traps left unattended in the open ocean.

There is a limit to the accuracy with which the relative timing of radio pulses can be measured. For greater precision, pulses need to have a very steep wavefront. That is, they must start very quickly. Pulses with steep wavefronts must have a high harmonic content, and this means that the transmitted signals will have sideband components far to either side of the assigned frequency. The LORAN signals must be confined within a fairly-narrow band of frequencies to avoid interference with other services, and this limitation blurs the definition of the start of each pulse. The result is a compromise in the accuracy of measurements of pulse timing.

There is less variation in the radio-signal propagation path taken by LORAN C signals at 100 kHz, but there are path variations that cannot be measured. The effect is to further reduce the quality of the information available to the LORAN computer. These effects are small, but they nevertheless set a limit on the available precision of navigation information that can be obtained from LORAN techniques.

A relatively new development in electronically supported navigation systems, the Global Position Satellite system, seems destined to replace LORAN C. During the years from 1978 through 1995 the United States launched more than two dozen specialized navigational satellites that each orbit the earth twice every day. These satellites transmit data that permits even portable handheld receivers and decoders to display latitude and longitude with great accuracy. The Global Positioning Satellite system provides better information than can be achieved using LORAN so it seems likely that the GPS system will soon render the LORAN system obsolete. LORAN will somebody be found only in the history of electronics-based navigational systems but it will have served the world well for better than a half century.

Lorimer, John • See Compass

Loudspeaker

A loudspeaker is a device that converts electrical signals into audible sound waves. Before electricity, amplification was obtained by using an acoustical horn of the type seen on gramophones. The invention of the speaker came with the development of **transducer**s, devices that convert mechanical energy into electrical energy. The first loudspeakers were patented in Germany in 1877 by Ernest Wermer and in 1898 by **Oliver Lodge** in England. Over the next twenty-seven years there was little advancement in loudspeaker technology.

During the early days of sending messages by **radio**, the transmitted signal was very weak. It was necessary to wear earphones and have absolute quiet in order to hear the feeble **Morse code** signals. Thanks to the work of **Lee de Forest** and **Edwin Howard Armstrong** in 1912, it became possible to amplify the weak signal considerably. (De Forest and Armstrong spent twenty years suing each other over the patent rights.) With a more powerful signal available, it was possible to construct an electric loudspeaker.

In 1924 C. W. Rice and E. W. Kelley invented the first "modern" loudspeaker in which the electric signal is sent to a coil surrounded by a magnet which is attached to a circular cone (or diaphragm) suspended from a frame. The signal, moving through the magnetic field, causes the cone to move. The movement matches the variations in the electric signal, and produces sound waves in the air. Increasing the volume is accomplished by increasing the signal passing to the coil.

As technology improved, more complicated loudspeakers were devised. In 1929 **Harold Stephen Black** invented a way to remove the distortion that had been amplified along with the signal. Different speaker cones, corresponding to a particular frequency range, were combined to produce better fidelity than the original single cone could provide. For example, a bass unit covers the 30 to 500 hertz range, a mid-range speaker goes from 500 hertz to 4 kilohertz, and a small high-frequency tweeter is used for the treble range. Electronic crossover circuits divide the audio signal into the appropriate frequency for each speaker.

In the 1950s other types of loudspeakers were invented. The ribbon loudspeaker uses a metal ribbon in a magnetic field. An electrostatic speaker uses a taut metal plate adjacent to a fixed plate. Very high voltage sets up a mechanical force between the plates and sound radiates through perforations.

In the 1990s, a very unusual looking loudspeaker was introduced. A company called NXT invented the distributed mode loudspeaker (DML) comprising little more than a perfectly flat panel. The NXT loudspeaker is simply a stiff rectangular sheet with either a transducer attached to its back with glue or a magnet and a coil clamped to it. The rippling of the panel transmits the sound. Its developers claim that this loudspeaker is more in tune with human hearing. The future of the big-sounding NXT loudspeaker probably lies in public address systems and other large places, though this may change as the NXT loudspeaker develops. Another loudspeaker that shows promise in the near future is one that relies on ultrasonic phenomenon. Although audio technology continues to evolve, allowing for more accurate sound reproduction, the operation of most loudspeakers is not much different from those of the distant past.

Lovelace, Ada Augusta, Countess of (1815-1852)

English mathematician

Lovelace was the only legitimate offspring of the poet Lord Byron (1788-1824). A quiet, studious child, she especially liked mathematics. In 1833 she attended a series of lectures on **Charles Babbage**'s difference engine and was so impressed she asked for a meeting with Babbage himself. The meeting soon developed into a lifelong friendship, and with her contacts with Scottish scientist Mary Somerville (1780-1872) and logician Augustus De Morgan (1806-1842), who instructed her in calculus, Lovelace became a respected mathematician.

In 1842 she began translating a study of Babbage's analytical engine (originally published in French by Luigi Menabrea). This work was published the following year as *The Sketch of the Analytical Engine* and included notes and commentary by Lovelace. Many scholars acknowledge Lovelace's book as an important treatise on Babbage. In addition, Lovelace used her own mathematical talents to encourage Babbage, even suggesting improvements in the design of the analytical engine.

Unfortunately, Lovelace and Babbage ran into financial problems that led to some desperate schemes. To raise money, they built machines capable of playing chess and tic-tac-toe. Their most unusual venture, however, was an attempt to create an infallible system for betting on horse races—an experiment that failed and left both of them heavily in debt. Lovelace died in 1852 of cervical cancer.

LP record • See Long-Playing record.

Lucite • See Acrylic plastic

Ada Augusta, Countess of Lovelace

Ludwig, Karl Friedrich Wilhelm (1816-1895)

German physiologist

Karl Friedrich Wilhelm Ludwig was one of the greatest researchers and teachers in the history of physiology. Born in Witzenhausen, Germany, he was the son of a former cavalry officer in the Napoleonic wars who became a civilian official at Hanau. After finishing his schooling at the Hanau gymnasium in 1834, Ludwig studied medicine at the University of Marburg, where he was expelled for dueling and political activities. He later returned to the university and earned his medical degree in 1840 after writing a dissertation on renal secretion. He then progressed through a series of anatomy and physiology professorships combined with research in Marburg in 1846; Zurich in 1849; Vienna, at the Josephinum (1855); and finally Leipzig (1865), where he remained until his death thirty years later.

Ludwig had based his investigation and teaching of physiology on chemical and physical laws, explaining all physiological processes on the basis of measurable and experimentally demonstrable phenomena—not on some speculative "vital force." He invented a number of devices and methods to carry out his scientific approach to physiology, many of them related to his interest in circulation and respiration. In 1847 he devised the kymograph to prove that blood is moved by mechanical forces, not the invisible "vital force." Ludwig's kymograph used a mercury **manometer** tube and a revolving drum to graphically record blood pressure variations and other vital signs. With later modifications, the kymograph became a standard tool for recording results of experiments.

In 1859 Ludwig and a student designed a mercury pump that separated and measured quantities of gases—oxygen and carbon dioxide—in the blood. Ludwig's *stromuhr*, or stream gauge, of 1867 measured the flow of blood. Ludwig also discovered that he could keep organs alive outside an animal by pumping blood or a saline solution through the excised part, a process called perfusion.

In addition to his important inventions, Ludwig produced a long list of major physiological observations and findings, including discoveries about salivary secretions, the mechanism of cardiac activity, respiration, blood, and blood circulation. Ludwig used his appointment in 1864 to the newly created chair of physiology at Leipzig to create a model teaching center for physiology, which became a world center for physiological study. Ludwig's textbooks of 1852 and 1856 (the first modern one on physiology), his hundreds of students, and his numerous scientific contributions have made him perhaps the most influential physiologist of the second half of the nineteenth century.

Lumière, Auguste (1862-1954)
Lumière, Louis (1864-1948)
French inventors

Auguste and Louis Lumière were the sons of Antoine Lumière, a painter and pioneer photographer who founded a factory in Lyons, France, to manufacture photographic gelatin dry-plates in 1882. The brothers were born in Besanìon, France, Auguste on October 19, 1862, and Louis on October 5, 1864. The family business gave them a natural interest in **photography** and, working as a team, they made important contributions to both still photography and **motion picture**s, as well as other scientific endeavors. But it is as the inventors of the modern cinema that the Lumières are perhaps best known.

Inspired by **Thomas Alva Edison**'s **kinetoscope**, the brothers set out to develop an improved film projection system. In February 1895, they took out a patent on their cinamatographe, an apparatus that not only took pictures but also later projected them. Using a claw movement that advanced the frames of the film 12 times per second, the cinamatographe projected still images on a screen separated by moments of blackness. The images lingered on the viewer's retina, and the viewer thus perceived a moving image. The Lumières described their invention as having "scientific curiosity, but...no commercial future whatsoever." Public opinion soon proved them wrong.

On December 28, 1895, in front of a paying audience at the Grand Cafe in Paris, the brothers staged a twenty-minute program of ten films, including one of a train as it entered a station, moving straight toward the camera. The film created panic in the audience; several women are said to have fainted. The event has gone down in history as the first public cinema performance. Within five years of the invention of the cinamatographe—during which time the Lumières promoted it throughout Europe—motion pictures were being made in every developed country in the world. The Lumières continued to innovate film technology. In Paris in 1900, they demonstrated their Photorama, which was a panoramic projector. Thirty-five years later, they introduced their stereo-cine process, which was based on the principle of anaglyphics.

As pioneers of **color photography**, the Lumières came up with the first commercially successful method of creating a color photograph on a single plate. Their autochrome process, patented in 1907, involved covering a special photographic plate, patented in 1904, with small grains of potato starch dyed green, red, and blue and then applying a thin film of panchromatic emulsion (i.e., emulsion equally sensitive to all colors). The exposure was made through the glass side of the plate through the dyed starch grains. After it was exposed, the plate produced a transparency composed of small dots of color. Since these dots were too small for the human eye to detect as separate, they gave the appearance of mixed colors.

The Lumières' process had its drawbacks. It produced transparencies rather than actual photographs, and the transparencies had to be projected onto a screen or viewed through a hand-held viewer. The density of the potato starch made the transparencies dark and grainy, and prints made from them were of poor quality. Also, the autochrome plate required a very long exposure—about 40 to 60 times longer than the best black-and-white plates. Despite its limitations, the Lumières' autochrome process paved the way for the commercial success of color photography. It remained one of the most popular methods for color photography until the 1930s, when subtractive color processes replaced it. Louis Lumière died in Bandol, France, on June 6, 1948. Auguste died in Lyon on April 10, 1954.

Auguste Lumière

M

MACHINE GUN

A few automatic weapons patented in the United States in the 1820s and 1830s were called *organ* or *machine guns*, consisting of multi-barreled weapons loaded on carts. These bear little resemblance to today's machine gun, which fires a continuous stream of projectiles as long as the trigger is depressed. One of these earlier models was invented by John W. Cochrane in 1834—apparently to no great success. **Richard Jordan Gatling**, an American inventor and the pioneer of machine guns, invented the hand-cranked, six-barreled Gatling gun in 1862. Rotating around an axis and firing .58-caliber brass cartridges, the gun, which fed ammunition from a top-mounted hopper, made its military debut at Butler's siege of Petersburg, Virginia, during the Civil War; the ten-barrel model, firing a thousand rounds per minute and covering a range of 2,400 yards (2,193.6 m), was a deciding factor in the 1898 outcome of the battle of Santiago, Cuba, during the Spanish-American War, as well as in subsequent colonial wars in Africa and Asia. The Gatling gun was rendered obsolete by the automatic reciprocating machine gun invented by **Hiram Maxim**, which operated off the energy from the erupting shell. Belt-fed and water-cooled by an attached water container, this heavier, sturdier variety of machine gun dominated the military establishment in every major country. Its familiar cartridge belt came to symbolize the fighting warrior in action.

In 1872 New York arms manufacturer **Benjamin Berkeley Hotchkiss** invented a five-barrel machine gun mounted on a solid breech block. At thirty-three rounds per minute, the Hotchkiss gun was an improvement on the Gatling gun but fired at an uneven rate and was not nearly so efficient as Hiram Maxim's gun, which could fire 666 rounds per minute. Maxim's was the first successful automatic weapon, and it presaged the coming mechanization of war, which blunted the effects of frontal infantry assault.

In 1892, Ogden, Utah, designer **John Moses Browning** (1855-1926) produced the Browning automatic **rifle** (or BAR), adopted by the United States Army in 1918 as the first machine gun working off the gas produced in firing. This multipurpose .30 or .50-caliber gun weighed only twenty pounds (9.08 kg) and could be set up with its own two-legged barrel support or fired from the shoulder. Browning followed the BAR with a short-recoil gun in 1901. Its gas-operated principle is still widely used in light automatic rifles and assault weapons.

American Army colonel and inventor Isaac Newton Lewis (1858-1931) improved on Browning's design in 1911 with the first lightweight automatic machine gun, which was efficient both in the **aircraft** and ground attacks of World War I, producing ninety-two percent of all injuries. Gas-driven and air-cooled, the thirty-six-pound (16.3 kg) Lewis gun fired rounds from a top-mounted revolving magazine, which held forty-seven cartridges. The American government, however, showed little enthusiasm for these breakthroughs. Consequently, Maxim, Hotchkiss, and Lewis looked to Europe as a market for their inventions, which became standard issue in the armies of England, France, and Germany. By World War I, the United States Coast Guard and Navy, too, had adopted Lewis's gun.

In the spring of 1915, machine guns transformed airplanes from observation crafts to a new type of weapon: a French World War I aviator, Roland Garros (1888-1918), acquired a trigger-to-cam linkage invented by French designer Raymond Saulnier (1881-1964) that allowed him to fire a machine gun through the propeller of his Morane plane. He terrorized German lines in April by firing on them from above. After Garros was forced to crash-land behind enemy lines in May, the Germans examined the gun mount, discovered its secret, and applied it to their new Parabellum, an air-cooled machine gun.

The German military hired American aircraft designer Anthony Herman Gerard Fokker to adapt Garros's device; instead, Fokker invented a new synchronizing firing mechanism similar to Franz Schneider's interrupter gear, which was invented in 1913. Fokker solved the problem of keeping bullets from destroying the propeller by letting the blade itself fire the gun. By attaching a **lever** to the cam, Fokker brought control of the device into the pilot's hand. Other adaptations included mounting machine guns on the front of pusher planes—those with propellers in the rear—and placing rear gunners in a position to fire from the tail of the fuselage. Side, wing, and swivel mounts presented additional problems of aim. The eventual synchronization of gun and plane was labeled the greatest single innovation in the evolution of aerial warfare.

Other aspects of warfare demanded heavier machine guns. German commanders equipped aircraft with 20-mm Becker guns, which were more effective against trucks and **locomotives**. The British chose the 20-mm Oerlikon machine gun to arm ships and airplanes. Browning machine guns continued to play a part in World War II and the Korean War. The United States relied on the .50-caliber Browning gun or the .37-mm cannon. During the Vietnam and Persian Gulf conflicts, American helicopters were armed with 20-mm machine guns.

An array of submachine guns—short shoulder weapons fed by box or drum magazines—was devised by the military for special purposes, such as use in border patrols and in firing from vehicles. The Thompson or ''Tommy'' model, designed by American Army Colonel John Taliaferro Thompson (1860-1940), relied on the delayed blowback principle. It could fire eight-hundred .45-caliber rounds per minute and was first used in combat by United States Marines in Nicaragua in 1925. It soon became the weapon of choice of American gangsters and the Irish Republican Army.

Inventors in many nations contributed to the evolution of the popular machine gun. Russian ordnance relied on an adaptation of the Maxim gun, while Germany developed its own MG-34 and MG-42 lightweight guns and the Erma MP 40, a simplified barrel and breech mechanism that fired five-hundred rounds per minute and could be mass-produced. In the 1950s, Russian inventors Degtyarev and Gorunov produced a series of machine guns, including the KPV, an unusual recoil-operated variety. Also, England's durable lightweight, the Bren 7.62-mm, which fed off an arc-shaped magazine, earned a reputation for dependability and cool firing.

Many of the machine gun models in use in the late 1990s are based on designs more than fifty years old, such as the Bren gun mentioned above, and the American M2, which is based on a Browning design from the 1930s. The most sophisticated and powerful modern machine gun, the Vulcan, relies on an even older source. Its design is based on one of the very earliest machine guns, the Gatling, with rotating barrels. While the Gatling was manually cranked, the Vulcan uses an outside power source, either electric or hydraulic. This gun can fire 6,000 shots per minute, and was very successful as an anti-tank and anti-aircraft gun in the Persian Gulf War.

See also Submachine gun

MACHINE LANGUAGE

In the late 1940s and early 1950s, the introduction of the electronic digital computer was greeted with great excitement. Able to perform the work of seven thousand engineers in mere seconds, these first generation computers were remarkable machines, yet the process of programming them was a difficult one. The very early computers used **vacuum tube** circuitry and the familiar decimal digits (0-9) to represent data. Computer engineers found it was very difficult to get the precise power voltages needed to represent ten different digits, and data was frequently misrepresented. In 1946 **John von Neumann**, a Hungarian mathematician working on the **EDVAC** project, decided to abandon the decimal system in favor of the binary numbering system, which has been utilized every since.

Although computers seem to perform amazing feats and accomplishments, they actually understand only two things—whether an electrical ''on'' or ''off'' condition exists in their circuits. The binary numbering system, which uses only the 0 and 1 *bi*nary digi*ts*, (later shortened to bits), lends itself ideally to represent this condition. Binary 1 represents on and binary 0 represents off. Program instructions are sent to the computer by combining bits together in groups of six or eight.

Although this process takes care of the instructional part of programs, data—in the form of decimal numbers, letters and special characters—also has to be represented. For this purpose, the EBCDIC and ASCII codes were introduced. EBCDIC (pronounced ''EB-see-dick '') stands for Extended Binary Coded Decimal Interchange Code. Developed by IBM Corporation and used in most of their computers, eight bits are used to represent a single character. A seven-bit code, ASCII (pronounced ''AS-key'') stands for American Standard Code for Information Interchange. ASCII was developed as a joint effort by several computer manufacturers to create a standard code usable on any computer, regardless of make. ASCII is used in most personal computers today and has been adopted as a standard by the U.S. government.

Although a necessary function, writing a program in machine language is a very complex, time-consuming process. It became apparent to early researchers that the ability to use symbolic codes in programming would speed up the process. Under the direction of Maurice Wilkes (1913-), the EDSAC project, which used the first symbolic coding language, was completed at Cambridge University in 1949. Around that same time, Navy Captain **Grace Hopper** put her considerable programming experience to work and developed a set of programs to translate symbols into machine language. Working as chief programmer on the **UNIVAC** I project in 1951, Hopper implemented assembly language, as it came to be called. Assembly language uses short names (known as mnemonics, or memory aids) to represent common sequences of instructions. These instructions in turn are translated back into the 0s and 1s of machine language when the program is run. Although the resultant translation slows down program execution, the time saved in initial programming makes up the difference.

With the introduction of many higher level languages in the late 1950s and 1960s, assembly language use has greatly diminished. Most operating systems, nevertheless, are still written in assembly, as it executes faster than higher level languages and allows for direct hardware interaction.

See also Computer input and output devices; Computer operating system; Computer programming language; Cybernetics

Machine tools

Machine tools are used in the construction of other tools and instruments. They are fundamental implements that change the shape, surface, or properties of an blank object made of metal, plastic, wood, or other material. In a basic sense they cut or chip away at a material, and typically they are powered by electricity. Often they require cooling in order to dissipate the heat that is generated.

The choice of machine tool is important, because the object to be constructed must frequently be to a certain accuracy, dimension, or specification. It might be required to have a certain surface finish, or a tapered edge, or grooves on a side.

There are basic machine tools in any machining shop. They include: (1) The **lathe** allows boring of a hole in an object. The object to be cut is placed on a rotating spindle, and a cutting tool is fed into the object, gouging a hole that is then shaped and made exact. (2) The **drill** is one of the most common machine tools. Drills cut cylindrical holes in objects. (3) An extruder is a machine that forces the material through a die, resulting in a specific shape. (4) Gear cutters produce a uniform series of toothed projections on the surface of the workpiece. (5) Grinding wheels remove material from an object by rubbing against it, slowly wearing the material away. The grinding wheel is typically used near the end of the machining task, to smooth a finish or to obtain extremely accurate dimensions. (6) Polishing wheels are typically built up from layers of a soft material like leather, and are used to obtain a very fine finish.

Machine tools have advanced along with technology. Where once drilling had to be done by hand, it is now powered by electricity, as in a drill press or a hand-held portable drill. Machine tools have become more powerful, and some also allow for greater and greater accuracy. Lasers are now used to cut metal. The vibrations of ultrasonic sound (about 20,000 Hz) are used to drive an abrasive, suspended in liquid, against the workpiece, which can produce impressions in **glass**, **ceramic**s, and other materials. In electric-discharge machining, electrically conductive materials (like metals) can be removed by repeated electric sparking.

A Machine Variation Analysis system developed by researchers at the Massachusetts Institute of Technology (MIT) allows designers to create machine tools on a computer, and use those virtual tools to build a computer model of a machine part. The system can predict the exact shape of the part given by any error in the machine tool. Real-life tools always have some error or imprecision, and the Machine Variation Analysis system takes some of the uncertainty from the machining process.

Macintosh, Charles (1766-1843)
Scottish chemist

Charles Macintosh was born in Glasgow, Scotland, in 1766. Although he was supposed to become a merchant, like his father, the young Macintosh had a passion for chemistry and science. By the time he was twenty, he had opened a plant in Glasgow for producing ammonia from coal-gas waste. About the same time, he introduced the manufacture of lead and aluminum acetates to Britain. He also made advances in cloth-dyeing processes, opened Scotland's first alum works, helped devise a method for making bleaching powder, and developed improved methods of iron production. Macintosh is best known, however, for the water proof garment that bears his name.

A waste product of the Glasgow gasworks was naphtha, a volatile liquid hydrocarbon mixture. In 1819 Macintosh began experimenting with the naphtha and discovered that it dissolved rubber. Applying the knowledge of textiles he had gained as a dye-maker, Macintosh had the idea of using the liquid rubber to waterproof fabrics. He painted one side of wool cloth with the rubber solution, then laid a second thickness of cloth over it. The rubber interior made the resulting sandwich of cloth waterproof.

Macintosh patented his invention in 1823. Within a year, the chemist had a flourishing factory in Manchester, England, producing rainproof cloth for the British military, the Franklin Arctic expedition, and the general public. **Thomas Hancock** joined Macintosh and his other partners around 1829. Charles Macintosh & Company became famous for its ''Macintosh Coat,'' the world's first raincoat, later known popularly as the mackintosh. (Where the *k* came from is a mystery.)

Macintosh was honored for his contributions to chemistry by his election in 1823 as a fellow of the Royal Society. He died at Dunchattan, near Glasgow, in 1843.

Magic lantern • See Slide projector

Maglev train • See Train and railroad

Magnet

The black isometric mineral, magnetite, is among the most abundant of the earth's minerals and can be found in beach sand, in meteorites, and in the natural abrasive emery. The naturally magnetic variety is known as a *lodestone*, first discovered in the Chinese desert around 3000 B.C.

The magnetic property of lodestone was first studied by the Greek philosopher Thales (624 B.C.-546 B.C.); his samples were found in the region of Magnesia, undoubtedly giving rise to the name ''magnet.'' Thales also noted the ''magnetic'' property of amber which, when rubbed, could attracted light objects. Amber, however, is not a natural magnet; its ability to attract objects is due to the build-up of static electricity.

Magnets were useful in making compasses, the invention of which is credited to the Chinese. The next advance did not occur for nearly 2,000 years when French scholar Petrus Peregrinus (1240-?) considered the possibility of using magnetic force to run a motor. In this he anticipated **Michael Faraday** and **Joseph Henry** by nearly 600 years.

In 1269 Peregrinus discovered that magnets had a north and south pole, and he was able to determine north from south (identical poles repelled each other; opposites attracted). He also showed that breaking a magnet in half did not isolate the two poles; each half became a complete magnet having both poles.

It was Peregrinus' belief that a magnetized needle pointed to the pole of the celestial sphere, which later led William Gilbert (1544-1603) to speculate about the earth having a magnetic field. Peregrinus also suggested an improvement to the compass by mounting the needle on a pivot, as opposed to mounting it on a cork floating in water.

In 1777 French physicist Charles Coulomb measured the force produced between the magnetic poles and discovered it obeyed the same inverse square law as electric charges, thereby linking the two forces. Nearly fifty years later Hans Christian Oersted (1777-1851) discovered that an electric current produced a magnetic field, paving the way for the invention of artificial magnets called electromagnets.

The first electromagnet was invented by William Sturgeon (1783-1850) in 1823. It could lift an impressive nine pounds (4 kg). American physicist Joseph Henry improved on Sturgeon's concept and produced an electromagnet that could lift more than one ton (900 kg) of metal.

Magnets and electromagnets both became essential components in **dynamos**, **alternators**, and **transformers** for the production of electricity. In 1867 Belgian-French inventor **Zénobe Gramme** improved the alternator, thereby making the production of **alternating current** (AC) electricity more practical. **Nikola Tesla** and **George Westinghouse** wired the 1893 Columbia Exposition in Chicago with AC and the world has not been the same since.

In 1916 a more powerful magnet was invented by Kotaro Honda (1870-1954). He added cobalt to tungsten steel which led to the invention of ALNICO. ALNICO was not only more strongly magnetic, but it resisted corrosion, was immune to vibration and changes in temperature, and was cheaper to manufacture than ordinary steel magnets. Not until the mid-twentieth century did the next advancement occur; an electromagnet that was wound with super-conducting **wire** and operated at liquid helium temperatures.

Magnetic core memory • See Computer memory

Magnetic recording

Magnetic recording is the technique of storing electric signals as a magnetic pattern on a moving magnetic surface. The concept of recording sound on **magnetic tape**—and thus the principle of the **tape recorder**—was worked out theoretically in an 1888 article by the English inventor Oberlin Smith (1840-1926). The article, entitled "Several Possible Developments of the Phonograph," proposed using fabric strips containing iron filings as "tapes."

It was, however, 10 years later that Smith's ideas were adapted and the first working magnetic tape recorder was introduced. In 1898, 20-year-old Danish inventor and physicist **Valdemar Poulsen** created a device that recorded and reproduced sounds by residual magnetization of a steel wire. The telegraphone, as it was called, was demonstrated at the 1900 World Fair in Paris, but the world took little notice of the new technology at the time.

The invention of magnetic recording tape is variously attributed to J. A. O'Neill (b.1909), who is said to have created a paper version in 1927 in the United States, and the German engineer Fritz Pfleumer, who in 1928 developed a tape made by bonding a thin coating of oxide to strips of either paper or film. It was Pfleumer who filed the first audiotape patent in 1929. There is no doubt, however, that audiotape was an improvement over existing methods, such as records, for recording and storing sound. The tape was easier to use, store, and edit, and less expensive to produce.

In 1935, the German electronics firm AEG produced a prototype of a record/playback machine, called a magnetophon, based on Pfleumer's idea, but using plastic tape. BASF went on to refine the tape AEG had used, presenting the first usable magnetic tape at the Berlin Fair in 1935. The first public tape recording was made by the London Philharmonic Orchestra at the BASF factory in Ludwigshafen, Germany, in 1936.

Other recording devices were also being developed concurrently. In 1937 or 1938, **Marvin Camras**, a U.S. inventor, built a magnetic wire recorder, using a variation on the early work of Poulsen. His recorder used a revolutionary magnetic recording head to record around the wire symmetrically. Early versions of his recorders were used during World War II for training and strategic purposes, such as to simulate battle sounds at noninvasion locations and therefore mislead the enemy. Camras went on to develop his recording techniques for home use. He invented the first magnetic coatings that modern recording tape is based on; these coatings are used in **videotape**, computer tape, and floppy disks for personal computers. He also discovered high-frequency bias, used on almost all tape recorders today to improve sound quality, and developed multi-track tape recording, magnetic sound for motion pictures, videotape recorders, a variety of improved recording heads, and stereophonic sound reproduction.

Thin plastic tapes have become the medium universally used in tape recorders. The tapes have magnetic coatings consisting of magnetically active particles, most commonly iron oxide and chromium dioxide. Each of these particles, in effect, is a tiny permanent magnet embedded in the coating. As the tape passes around the five magnetic heads of the tape recorder, sound is recorded, replayed, or erased according to the heads that are activated. A recording head magnetizes the passing tape in such a way that the magnetic particles on it are realigned. The resulting magnetization pattern remains on the tape, which may be rewound and replayed as often as desired, until erased or changed.

The **audiocassette**, introduced in 1963 by the Philips Company of the Netherlands, was made possible by Pfleumer's earlier development of audiotape. Audiotape was used in a reel-to-reel format, which was complicated and unwieldy, since the user had to thread tape through the machine and onto a take-up reel. Until the audiocassette format, sound recording technology had remained primarily a professional tool. Because of the ease and economy of the audiocassette, magnetic recording tape recordings could compete with long playing (LP) records (LPs). The cassette was immediately popular because it made inserting, advancing, and rewinding a tape fast and easy; it could also be stopped and ejected at any point in the tape. The cassette housing protected the tape from wear and damage due to handling and thus reduced the loss of sound quality.

Eight-track cartridges were another innovation in magnetic recording, although the audiocassette has proved more popular and long-lasting. Because the rather bulky 8-track used an endless-loop format, the tape could be played continuously, without being flipped over by the listener. But the invention of the **microchip** allowed audiotape players to be made smaller and more portable, and with the introduction of products such as Sony's compact **Walkman**, cassettes prevailed.

Although the audiocassette is economical and still widely used, digital technologies are revolutionizing the industry by enabling the recording, storage, and playback of sound in new ways. Digital audio tape (DAT) recorders became widely available in the United States in 1990. A digital system, as opposed to a standard analog system, enables a home recorder to make a tape copy that is an exact replica—not just an approximation—of the original sounds on a cassette that is half the size of the typical audiocassette. Digital technology records sound in a code of binary numbers (a series of 0s and 1s), so each subsequent recording is simply a copy of the code. Analog recordings, on the other had, record sound as a wave pattern and, like using a copy of one key as a pattern for another, each generation of recordings is subject to increasing distortion.

Magnetic tape recording is not just an audio phenomenon. Videocassettes record video through video cameras or on videocassette recorders (VCR) in much the same way as analog audiocassettes and tape recorders log sound. Standard videotapes are significantly larger than audio tapes, although the introduction of smaller casettes has allowed for the use of smaller video cameras. The video must be transferred to a standard-sized casette or placed in a standard-sized converter to be played in a VCR, however.

Magnetic resonance imaging (MRI)

Magnetic resonance imaging (MRI) allows physicians to examine tissues and organs inside the body by observing the response of atoms exposed to a strong magnetic field. In a closed MRI, the patient lies inside a machine in a 48-inch tube; in an open—or ''accessible''—MRI the patient lies on a table. More sensitive than x-ray spectroscopy, MRI does not rely upon potentially harmful radiation. Instead, a powerful electromagnet creates radio waves that cause hydrogen atoms in the body to release energy. The magnet can map this energy from almost 360 degrees, projecting images to a computer which gives an extremely high-resolution picture. Because MRI can scan through bone, it can probe the brain in search of a tumor, to assess stroke damage, or identify degeneration. It also ''sees'' under nerve coverings to help diagnose multiple sclerosis; reports joint injuries in muscle and ligaments; identifies blocked blood vessels; helps diagnose heart, liver, and kidney disease; and is among the most powerful tools for detecting tissue abnormalities such as cancer.

The series of scientific developments leading to the invention of MRI actually began in the late 1930s. Isidor Isaac Rabi (1898-1988), an American scientist, designed a process by which the magnetic strengths of atomic nuclei could be recorded. Firing a vaporized beam of silver through a magnetic field, he noted that the nuclei behaved like spinning tops, and that they wobbled at very precise frequencies; when radio signals that matched the frequencies of the wobble were applied, the nuclei reversed their spin. This phenomenon, called magnetic resonance, was very easy to observe, and much could be learned about the structure of the atom by knowing the resonance frequency.

After World War II, two other American scientists, working independently, devised improvements upon Rabi's process, making it more precise and eliminating the need to vaporize (and thus destroy) the sample. Scientists Felix Bloch (1905-1983) and Edward Purcell (1912-1997) shared the 1952 Nobel Prize in physics for *nuclear magnetic resonance (NMR) spectroscopy*, which gained an immediate place in nuclear laboratories as the most precise tool for studying molecules. Chemists had earlier discovered that a nucleus' wobble revealed information about surrounding molecules, and that each atom and molecule carried a ''signature wobble.'' By recognizing these signatures, researchers could use NMR to identify the composition of unknown chemical samples.

The most important application was forthcoming, however. Raymond V. Damadian (1916-) was the first to realize that NMR could be used on living tissue. He tested first on animals and then on humans, finding that the process was excellent for detecting areas of disease inside the body—areas that had previously required exploratory surgery to locate. NMR was especially useful for detecting cancer, since cancer cells carry their own signature resonance frequency. About this same time, the Swiss physical chemist, **Richard R. Ernst** (1933-), was working upon improving the NMR process yet again. By changing the radio signals, Ernst succeeded in making the technology more sensitive and easier to interpret—improvements that paved the way for the development of MRI, for which Ernst was awarded the 1991 Nobel Prize in chemistry.

MRI enables physicians to create three-dimensional images of large sections of molecules. It can define areas of soft tissue too thin to be picked up by x rays. Using the radio signals, the computer will search for the frequencies of specific types of atoms (such as cancer cells). Once the radio waves are turned off, the atoms emit pulses of absorbed energy; the computer reads these pulses, creating a three-dimensional image of the scanned area.

MRI scanners are generally found only in large medical research centers: the equipment is extremely expensive, and a trained radiologist must be present to supervise the procedure. The technology required to perform NMR scanning is much more affordable and easier to use; NMR units can be found in a variety of sizes at most hospitals. Both inventions have become important diagnostic tools. In fact, the Royal Swedish

Academy (while announcing Ernst's Nobel Prize) described NMR spectroscopy as "perhaps the most important instrumental measuring technique within chemistry."

See also X-ray machine

Magnetometer

A magnetometer measures the strength and direction of a magnetic field. The first magnetometers were mechanical devices that had a **spring**-loaded **magnet** which moved in relation to an external magnetic field. The greater the movement, the greater the magnetic field. Modern **electronics** have made this design obsolete, and today there are three basic designs of magnetometers: the Hall effect, fluxgate, and proton.

The principle governing the Hall effect magnetometer dates back to 1879. In that year Edwin Herbert Hall (1855-1938) discovered that an electric current flowing through a conductor in the presence of a magnetic field produced an electric potential that was perpendicular to both the current and the field. The strength of the potential was directly related to the strength of the magnetic field and created a transverse current in the conductor.

In 1966 it was discovered that the Hall effect influenced the electrons located at the interface between a **semiconductor** and an insulator and resulted in a voltage, the magnitude of which is a measure of the magnetic field strength. The Hall effect magnetometer is very well-suited for measuring strong magnetic fields.

The fluxgate magnetometer depends on two easily magnetized cores. Each core has an internal and external winding. When **alternating current** (AC) is applied the primary winding becomes magnetized and induces a current in the secondary winding. If an external magnetic field is present, the primary core becomes more magnetized. Because the outputs of both cores reinforce each other, the signal is doubled for the external field and appears as an AC voltage at twice the original frequency.

The proton magnetometer is the most accurate. It depends on a material which is rich in protons, such as paraffin or water. A coil magnetizes the paraffin to align the protons. When the coil is switched off the protons begin to return to their original random orientation. When this happens, they generate a voltage in a sensor, and the speed at which the orientation decays is a measurement of the external magnetic field. Because the rate of decay can be very accurately measured, the strength of the magnetic field can be precisely determined. Its main disadvantage is the inability to measure rapidly-changing fields.

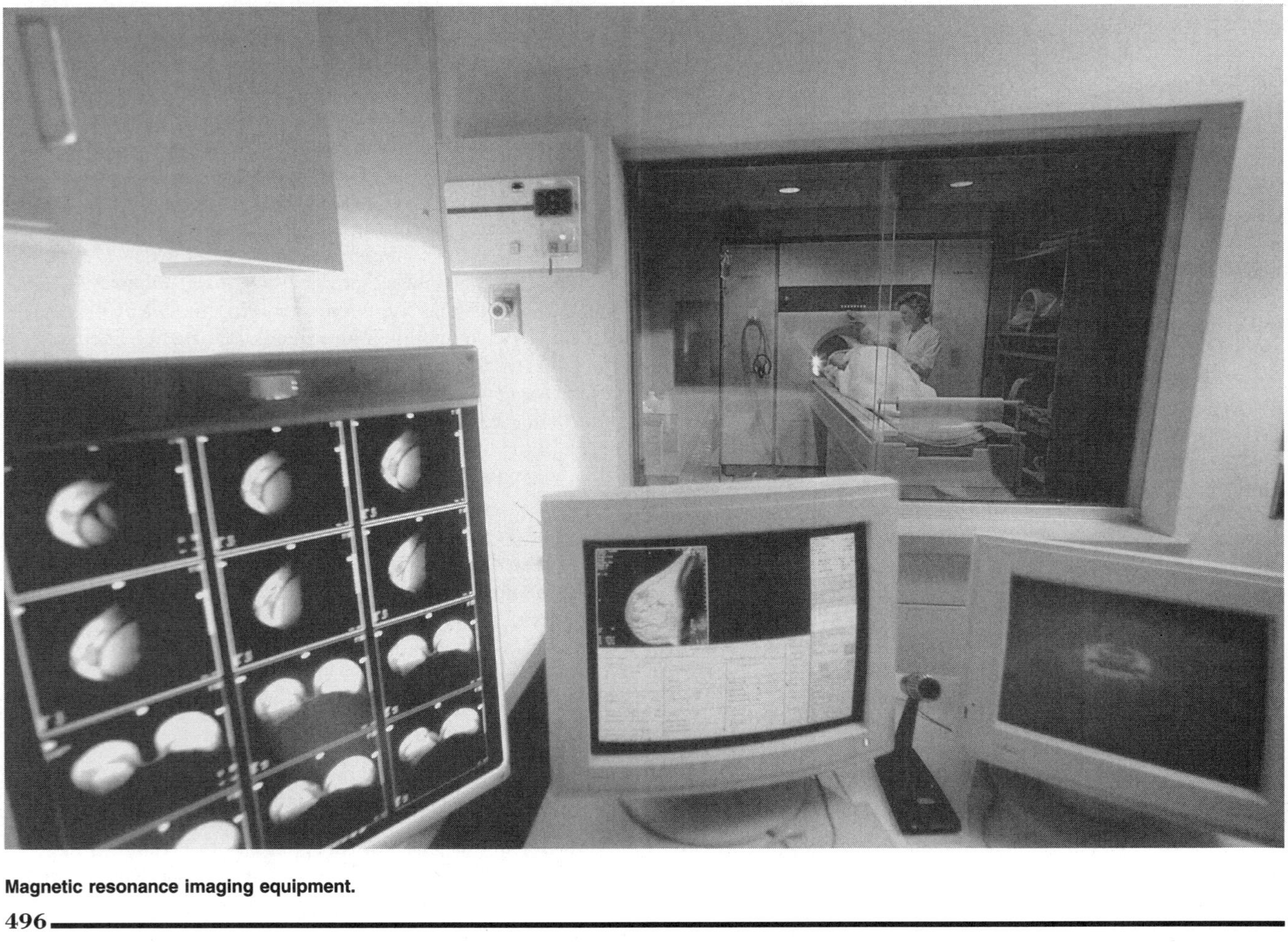

Magnetic resonance imaging equipment.

MAIMAN, THEODORE HAROLD (1927-)
American physicist

Since the late 1950s, one of the most contested issues in the science community has been that of whom to credit with the invention of the light amplifier, or **laser**. Designs for the laser were submitted by several prominent physicists, and many others contributed to the pool of knowledge from which those designs emerged. One thing, however, is certain: the very first working laser built in the United States was constructed by Theodore Maiman in 1960.

The son of an electrical engineer, Maiman worked his way through college by repairing appliances and other electrical devices. He graduated from the University of Colorado in 1949 and pursued his Ph.D. in physics at Stanford University, graduating in 1955. It was during his stay at Stanford that Maiman learned of **Charles Townes**'s work on masers (an acronym for Microwave Amplification by the Stimulated Emission of Radiation).

Shortly after the maser was constructed in 1954, Townes predicted that theories of coherence and stimulated emission might be applied to visible light, and it was to this end that Maiman turned his efforts. Townes's work was inspired by the hypotheses of Albert Einstein (1879-1955) who, in 1917, discussed the concept of stimulated emission in connection with his theory of relativity. When Townes proved Einstein's theories mechanically applicable to visible light, American scientists competed to produce the first working model of an optical maser. **Gordon Gould** is credited with designing a model as early as 1957, but he did not apply for its patent until it was too late: Townes and Arthur Schawlow (1921-) were granted the patent for a similar design in 1958.

Nobody actually constructed the device, however, until 1960, when Maiman presented his apparatus. At the heart of Maiman's laser was a ruby cylinder, the ends of which had been shorn perfectly flat and then coated with silver, so that they acted as mirrors. When excited, the atoms within the ruby rod triggered a release of energy in an internal chain reaction (called stimulated emission). Once triggered, the rod's mirrored ends would bounce the energy back and forth within the cylinder until it was strong enough to escape from one end in the form of a brief and intense red beam. Since its first appearance in 1960, the laser has found applications in medicine, communications, architecture, and numerous household devices.

Theodore Harold Maiman

MAINFRAME COMPUTER

Mainframe computers are large computers designed to be central sources of **digital computer** operations and to provide data storage for large amounts of information. The name came from the room-sized first computers, such as the **Mark I** at Harvard University and **ENIAC** at the University of Pennsylvania. The main frame was the case that held it.

The first generation of commercial computers was entirely mainframe, beginning with **UNIVAC** (UNIVersal Automatic Computer), built in 1950 by **J. Presper Eckert** and **John Mauchly** and manufactured by Remington Rand. These computers had thousands of bulky **vacuum tube**s for processing, mercury delay lines or magnetic drums for memory, punched cards for input and output, miles of wiring, and large power requirements. They were used for individual data-processing problems. UNIVAC became famous when it accurately predicted the outcome of the 1952 presidential election. Other **typewriter**, **pocket calculator**, and control equipment companies quickly entered the computer market, including Burroughs, National Cash Register, and Honeywell.

Second generation computers, such as the IBM 1401, were also mainframes. They were developed in the late 1950s and featured much faster **transistor** processors, invented by scientists at Bell Laboratories, and magnetic core memories, invented by **An Wang**. They used high-level (English-like) programming languages developed during the 1950s, principally COBOL (COmmon Business-Oriented Language), developed by **Grace Hopper** and others in the United States Navy, and FORTRAN (FORmula TRANslation), developed by John Backus and others at IBM. These computers could process large amounts of data, but they generally did not allow immediate (''real time'') interaction with their users.

The height of mainframes' popularity came with third generation computers of the mid-1960s, such as the IBM System/360. They used the much faster integrated circuits for processors and had operating systems for overall control. The S/360, for example, offered disk, tape, and card input and output, as well as a printer. Comparable mainframes were built by other companies in the United States, Japan, Germany, and elsewhere. These computers were designed for central data-processing facilities, with many terminals for users. They could run several applications programs at the same time. Mainframe systems, such as the IBM System/370 (successor to the S/360) cost one million dollars or more.

In the late 1970s, mainframes began to give way to small, but powerful computers—the fourth generation—that let users directly manage their own data. By the early 1990s, many companies were downsizing—moving from mainframes to less-expensive networks of these newer machines. Though a few companies still make them and many remain in use, the day of the mainframe has largely passed.

Mammography

Over the past century, mammography, or X-ray imaging of the breast, has become an accepted, although controversial, tool in the diagnosis of breast cancer. Breast cancer remains one of the deadliest diseases that strikes women; an estimated one out of every nine women will develop the disease in her lifetime. Mammography is considered an important screening and diagnostic tool because it can detect tumors while still small and most easily treated. In one study, women whose breast cancers were found early through mammograms had a five-year survival rate of 82 percent, while a group of women whose cancers were not found by mammograms had a five-year survival rate of just 60 percent. The American Cancer Society guidelines recommend women aged 40 and older be screened annually because the incidence of breast cancer increases with age.

Researchers began experimenting with X-ray technology in detecting cancer in the late nineteenth century. German surgeon, Albert Salomon, became the first person to use the x ray to study breast cancer. Working with breast tissue that had been removed during surgery, he used the x ray to determine the difference between cancerous and noncancerous tumors and found that the technique could successfully be used to detect breast cancer. He also discovered, based upon the differing x-ray images he obtained, that a number of different types of breast cancer existed. Salomon, who published his findings in 1913, is considered the inventor of breast radiology. However, he never used the technique in his own practice. In the 1920s, other German scientists continued Salomon's research. Out of these experiments came a study written by Leipzig researcher W. Vogel that fully described how x rays could detect the difference between cancerous and noncancerous tissues. The detailed information is still considered useful to scientists who diagnose cancer.

Stafford L. Warren, a Rochester, New York, physician, became the first to experiment with the new technology in the United States. Using a fluoroscope, he discovered images similar to those found by Salomon in breast tissue. Warren used his technique for detailed examinations prior to breast cancer surgery. Using radiology, he also discovered changes in breast tissue as a result of pregnancy, menstruation, lactation, and the beginning of breast disease. His findings were published in 1930. The first physician to advocate the wide use of x rays to screen women for breast cancer was Jacob Gershon-Cohen. In the mid-1950s, he and his colleagues began a five-year study of more than 1,300 women, screening each woman every six months. As a result, 92 were diagnosed with nonmalignant tumors and 23 with malignant tumors. A similar study conducted in Houston, Texas, found that women diagnosed early through mammography had a better recovery rate than those whose disease was discovered at a later stage by breast biopsy. By the 1960s, mammography was becoming a widely used diagnostic tool, although some critics alleged that the procedure exposed women to unnecessary levels of radiation. Through the development of more sensitive film, the amount of radiation needed to produce clear pictures of breast tissue was reduced significantly. By 1971, a quarter-million women over age thirty-five had been screened for breast cancer with mammography. The National Cancer Institute conducted a four-year study (1973-1977) of some 270,000 women throughout the United States. Some women were found to have very small benign growths, yet large numbers of these women had surgery, which some researchers felt was unnecessary. As a result of the study, the National Cancer Institute issued a set of guidelines regarding which groups of women will benefit from regularly scheduled mammograms. For women under age forty, the procedure was recommended only to those at risk of developing the disease because of family history other indicators such as palpable breast lumps.

See also X-ray machine

Manometer

Manometers, first invented in the seventeenth century, are used to measure the pressure of gases. During the 1600s, scientists tried to explain natural phenomena in logical, rational ways, instead of relying on mystical or magical explanations. At this time, most people, including the great scientist **Galileo**, believed **water pumps** demonstrated the commonly held theory that nature abhors a vacuum, but because they were unaware that air and other gases exert pressure, they could not explain why a pump could not raise water more than thirty-two feet (9.7 m). In 1643 Italian physicist **Evangelista Torricelli** tried to explain this phenomenon. In his experiments Torricelli used a vertical tube filled with mercury to measure air pressure, and, in doing so, he created the first mercury **barometer**, proving that air has weight and exerts pressure on all objects and substances on Earth. Soon afterward, British chemist **Robert Boyle** used a mercury barometer to determine the relationship between the pressure of air and its volume. In these experiments both Torricelli and Boyle used rudimentary versions of the manometer.

The simplest manometers are open-ended, U-shaped tubes that are partially filled with mercury, oil, or some other liquid. When a container of gas is connected to one end of the tube, the pressure of the gas causes the liquid in the manometer to become displaced and rise up into the arms of the tube. If the pressure of the gas introduced into the manometer is greater than the atmospheric pressure, the liquid will rise up into the open-ended arm; if the atmospheric pressure is greater, liquid rises up into the closed portion of the tube. From the amount of liquid which is displaced, scientists can determine the pressure of the gas.

Since its invention, the manometer has been altered and used for several different purposes. In some laboratory manometers, one arm of the tube is inclined at an angle instead to provide more accurate measurements. Other manometers contain a sealed-off vacuum at one end, so that changes in atmospheric pressure don't have to be accounted for in calculations. In research and industry, a special type of manometer called a McLeod gauge is used to measure extremely low pressures. Other common types of manometers include the sphygmomanometer, which is used to measure blood pressure.

See also Blood pressure measuring devices

Maps

To ask how far back mapping goes is to ask who was the first person to draw a diagram in the dust with a stick. Any attempt to schematically depict a geographic area can be considered a map, whether the purpose is to give someone directions or to define a territory. The Babylonians made maps on clay tablets as early as 2300 B.C., and the Egyptians may have had maps by 1300 B.C. It is known that diverse cultures such as Inuits and Polynesians made crude maps. Greek maps were the first to take into account a round earth and to use longitude and latitude lines. The philosopher Anaximander of Miletus (c. 611-546 B.C.) observed the revolution of the stars around the North Star and proposed that celestial objects are spheres. The historian Hecataeus (c. 550-476 B.C.) applied Anaximander's idea to his map of the earth and wrote the first book on geography in 500 B.C. Later Herodotus (c. 485-425 B.C.) expanded on Hecataeus's work, as did mathematician and astronomer Eudoxus (408-353 B.C.), who added a celestial map.

The Romans were master surveyors and builders, skills manifested in their ability to produce maps and plans. The Roman general Marcus Agrippa (63-12 B.C.) compiled a map of the Roman world based on military roads, which was completed by his sister. It is thought to have been derived from Eratosthenes' (c. 276-194 B.C.) world map. Around A.D. 150, Ptolemy (A.D. 90-168) created a world map and a set of regional maps in his eight-volume Geographica. Ptolemy lived in Alexandria and compiled his masterpiece from existing Greek maps and information brought to him by world travelers. The map was a conical projection, bowing to the notion of a round earth, and listed over 8,000 place names. Ptolemy was considered the last scientific cartographer for the next 13 centuries, meaning that no improvements were made to his map in that time. Even after the Europeans learned about the Americas, the two continents were simply added to Ptolemy's map.

Early map making by the Chinese roughly paralleled that of the Greeks. In 271 A.D. the Chinese minister P'ei Hsin designed a map of China that employed the grid system of latitude and longitude. By 801 A.D. the geographer Chia Tan drew a map 30 by 33 feet (9 by 10 m), which with its roads and the distance between towns, predated similar maps in the West.

With the advent of the **compass**, sailor's charts, called portolanos, began to emerge in the 1300s. *Portolanos* eventually had rhumb lines added to them, which crisscrossed the maps and were used as compass guidelines for mariners. The first major change to map making came in the mid-1500s when Flemish cartographer Gerhard Kramer (1512-1594), better known as **Gerardus Mercator**, developed a new map projection that made map reading easier. This map, which stretches the features from the round earth on to a flat grid, is usually the first world map used by school children. All meridians (lines of longitude) are straight and are at right angles to the parallels (lines of latitude). Although it seriously distorts the features near the poles, the Mercator made it easier to relate geographic locations to each other because navigators could plot direction with a straight line on it. Mercator was the first to refer to his map collection as an atlas. Although he was not the first to use the name America (after geographer Amerigo Vespucci) for the New World, Mercator's use of the name on his maps, which were very popular for a long time, fixed the name in the world mind. Several years after Mercator developed his map, Scottish mathematician John Napier (1550-1617) invented logarithms and used them to calculate the earth's features on Mercator's map distortions.

The next important development in map making took place near the turn of the nineteenth century in the form of specialized maps. French cartographer Philippe Buache (1700-1773) pioneered the use of contour lines to express elevation. Other isograms included Edmond Halley's (1656-1742) wind and pressure lines (isanemones and isobars) and Alexander von Humboldt's (1767-1835) lines of equal temperature (isotherms). A new equal-area projection provided the answer to Mercator's distortions. With this method, the shapes of the continents are still distorted, but their sizes are proportional and thus more accurately depict the area of land masses. Other projections include the conical and azimuthal.

In the tradition of the Romans, the United States embarked upon a surveying effort during the 1800s that coincided with its rapid westward expansions. The natural result of this was a massive mapping effort that continues today. The United States Geological Survey is aggressively producing detailed topographical maps for the entire country. A key element of this effort is photogrammetry, the compilation of maps through aerial photography. Stereo imaging is achieved by mounting two cameras side-by-side on the underside of an airplane, imitating human vision. When viewed through special **lenses**, the double images appear in three dimensions. Satellites have greatly increased the ability to survey and map the earth. Landsat satellites, or ERTS (Earth Resources Technology Satellites), photograph the earth in 115-mile (185 km) north-south strips. They are used for mineralogical and agricultural surveys, and for weather observation. A popular product of *Landsat* are the infrared photographs that reveal crop damage and effects of drought and erosion. Mapping is being taken to ever-increasing heights, so to speak, with the launching of unmanned planetary missions. Spacecraft such as the *Mariner*, *Voyager*, and *Pioneer* have delivered photographs of the other planets of the solar system and many of their moons. In some cases, complete surveys have been made of major portions of planets, such as Mercury, from which atlases have been produced.

Digital computers entered into mapping during the 1970s and are expected to be the dominant force by the year 2000. Digitized maps are increasingly taking the place of manually produced maps. The computer stores the map, and any information keyed to it such as population statistics. Computers also allow for easier updating because map information has become databased. Computers are also playing an ever-increasing role in how maps are used. Consumers can buy maps and trip-planning software for use on their home personal computer. These maps can have numerous layers of cartographic information, more than any print map could contain on one page.

See also Earth-survey satellite; Napier's bones; Surveying instruments

Marchbanks, Vance H. (1905-1973)

American physician

Vance H. Marchbanks, an African-American soldier and scientist, was "born in the service" in Wyoming where his father, a cavalry captain, was stationed. After receiving a bachelor's degree from the University of Arizona and an M.D. from Howard University, Marchbanks completed his internship and residency in Internal Medicine at Freedman's Hospital in Washington, D.C. After that he served as a medical staff member at the VA hospital in Tuskegee, Alabama, until he entered the Air Force in 1941. Marchbanks served as group surgeon in Italy and was eventually awarded the Bronze Star.

During the Korean War, Marchbanks gained further experience in aviation and medicine, accruing over 1,900 hours in prop and jet aircraft flight, gathering valuable medical data that later appeared in research publications and military manuals. In 1957 he participated in a 10,600-mile (17,066–km) nonstop flight from Florida to Argentina to New York in a B-52 jet bomber. Studying crew comfort and fatigue during the flight, Marchbanks found that the adrenal hormone content in blood and tissue was an indicator to the physical fatigue which often preceded a fatal crash. This discovery earned him an Air Force Commendation Medal. Marchbanks also received a medal for developing an oxygen mask tester, later adopted by the Air Force, which encouraged air crew members to clean their masks frequently.

With his extensive military service and his rating as a chief flight surgeon (rare because it required 1,500 flying hours and 15 years on flying status), Marchbanks was assigned to the Project Mercury space program in 1960. He prepared for this new role by sitting in on aerospace lectures, studying each astronaut's medical history, and visiting tracking stations used for the space flights. From a tracking station in Kano, Nigeria, Marchbanks monitored John Glenn (1921-) on his famous *Mercury* flight in February 1962. Marchbanks monitored Glenn's respiration, pulse, temperature, and heart reactions through electronic sensing devices attached to the astronaut and set up to relay the information to ground recorders. With his months of preparation, Marchbanks was able to compare Glenn's practice mission electrocardiograms with the actual ones recorded during the mission. He was relieved to note normal tracings throughout the flight. Marchbanks eventually headed up a medical hospital at an Air Force base in California.

See also Electrocardiograph; Space equipment; Spacecraft, manned

Marconi, Guglielmo (1874-1937)

Italian physicist

The greatest strength of Guglielmo Marconi was not his ability to innovate, but his mastery of synthesis. Assimilating the ideas and inventions of others, Marconi brilliantly fashioned a working technology. Born in Bologna, Italy, on April 25, 1874, Marconi was the son of a wealthy Italian landowner and his second wife. Educated by private tutors as a child, Marconi was later sent to the Technical Institute in Leghorn, where he studied physics and electromagnetism.

A number of key events set the stage for Marconi's experiments. British physicist James Clerk Maxwell (1831-1879) had established a theory about the existence and behavior of invisible electromagnetic radiation in the 1860s. About twenty five years later, German physicist Heinrich Hertz successfully generated such radiation, which he dubbed "Hertzian waves," using a spark-gap device. In 1894, English physicist **Oliver Lodge** invented a "coherer" capable of detecting Hertzian waves with relative efficiency, and a year later in Russia **Aleksandr Popov** had devised an **antenna** circuit capable of boosting reception and transmission.

In 1894, the year that Hertz died, Marconi came across a technical magazine that discussed some of the possibilities of Hertzian waves. Intrigued, he began to experiment with a spark-gap **generator** at his family's estate. He made a key improvement to the coherer, and devised an effective vertical antenna consisting of an elevated metal plate connected to another plate on the ground. Within a year, Marconi was successful in sending wireless **Morse code** signals a distance of more than 1.5 miles (2.4 km). Marconi also found that when he attached sheets of metal to his antenna in certain configurations, the radiated **radio** waves focused into a directional beam. When Marconi was able to transmit and receive over a hill that blocked the line of sight in September 1895, he became convinced that the potential of radio as a means of communication was far greater than anyone had anticipated.

Because the Italian government showed little interest in his work, Marconi decided to move to London in 1896. Britain was the naval power of the world, and he hoped to interest the British navy in wireless communication. Assisted by Sir William Preece, chief engineer of the British postal service, Marconi carried out a series of demonstrations on land that covered distances of up to nine miles (14.5 km), and generated an increasing amount of attention.

Marconi demonstrated better business sense than many of his contemporaries. Seeing the commercial potential of radio, he began to protect the devices he used by taking out patents with the help of his cousin, a British engineer. He received his first patent for a radio transmitting apparatus on June 2, 1896. Marconi also founded corporations both in Britain and the United States, and continued to file important patents guaranteeing his companies exclusive use of key devices.

Marconi continued to make improvements to his wireless system. In 1899 he built a wireless station to communicate with one in France, located 31 miles (50 km) across the English Channel. He also tested his system successfully on British and Italian naval vessels. Was there any limit to how far the waves would travel? Since radio waves, like light waves, seemed to move only in straight lines, many experts felt that they would travel no further than the distance to the horizon from an elevated antenna, or two hundred miles (about 300 km). But on December 12, 1901, Marconi, proved such predictions wrong, and created a major sensation when he successfully transmitted a signal 2,137 miles (3,440 km) across the Atlantic Ocean from Poldhu, Cornwall, England to St. John's, Newfoundland, Canada. It was not understood how this was accomplished until Arthur Kennelly (1861-1939) and Oliver Heaviside (1850-1925) deduced that a reflecting layer of charged particles in the upper atmosphere had to be responsible. Radio waves evidently ''bounced'' off it and back to Earth, where they were received. This layer, called the *ionosphere*, was proven to exist by Edward Appleton (1892-1965) in 1924.

Marconi's demonstration that radio signals could cross vast distances assured the future of radio as an important form of communication. By 1902, regular messages were being send across the Atlantic. In the subsequent years, Marconi helped to develop radio as a viable industry with the companies he had established, and in the process he created more important devices, including a magnetic detector and a new directional antenna. He also enlisted the help of scientists like **John Ambrose Fleming**, whose invention of the **vacuum tube** further cemented the position of radio as a practical technology.

Marconi explored the potential of shorter wavelengths for radio communication into the 1930s. Once again, Marconi's intuition proved sound, as it was found that such shortwave radio signals could carry over tremendous distances using far less power than the long waves originally used. Marconi shared the 1909 Nobel Prize with **Karl F. Braun** for innovations in radio technology. He died on July 20, 1937.

See also Armstrong, Edwin Howard; De Forest, Lee; Fessenden, Reginald

MAREY, ETIENNE-JULES (1830-1904)

French physiologist and inventor

Etienne-Jules Marey joined his interests in medicine and mechanics to invent instruments useful in the realms of physiology and cinematography. To Marey the **movie camera** was an instrument to be used in research on animal locomotion, but to the rest of the world, it was the force behind the advent of the motion picture industry.

Guglielmo Marconi

Born in Beaune, in the Burgundy region of France, on March 5, 1830, Marey was the son of a wine merchant and a schoolteacher. To please his father, Marey studied medicine at the Faculty of Medicine in Paris. He chose physiology because it was the only science field in which he could combine his interest in animals and mechanics.

Marey was particularly intrigued by blood circulation. In his first decade of research, Marey invented the sphygmograph, a device that graphically records the pulse, and the kymograph, which graphically depicts blood pressure. After 1868, when he became a professor at the Collège de France, in Paris, Marey investigated the mechanics of animal locomotion. In the course of his investigations, often involving horses or birds, Marey invented the tambour to graphically record subtle movements. This device is made up of an air-filled metal capsule encased in a rubber membrane, which is attached to a fine flexible tube. When the air pressure in the tube changed, it moved a lever on a graphical recording device.

In 1881, after seeing a photographic demonstration by **Edward Muybridge**, Marey devoted himself to animal **photography** and its mechanics, and in so doing invented the first movie camera. Unlike Muybridge, Marey decided to use one camera rather than many to produce a series of images in rapid succession. In 1882 Marey was able to take 12 pictures per second using his *fusil photographique*, or photographic gun, which looked like a rifle with a magazine made of a photographic glass plate. While creating the illusion of movement, the postage-stamp size of the photographs was too small and the frames per second too few to allow for adequate analysis of motion.

After **George Eastman** marketed, in 1885, a photographic film that used a silver bromide emulsion on a gelatin base, Marey was able to increase exposure speed. In his *chambre chronophotographique*, a paper ribbon of film that produced images 3.6 sq. in. (9 cm) was drawn along behind a shutter. There it stopped long enough to be exposed before it was moved forward. Marey was thus able to expose 60 images per second.

During the next 20 years, Marey filmed a wide variety of human and animal movements using his new apparatus. As well as slowing rapid movements by the use of high-speed cinematography, he also invented the technique of time lapse, which is used to speed up of slow movements. Marey studied his films frame by frame and published his observations in numerous articles in scientific journals.

Etienne-Jules Marey not only contributed to the study of physiology and locomotion, he opened the door to another arena—cinematography. In his quest for scientific knowledge, Marey started an industry that has entertained people worldwide. Marey died in Paris on May 15, 1904.

See also Blood pressure devices

Margarine

Margarine was originally developed and marketed as a butter substitute, but today it is considered a food in its own right. A scarcity of animal fat (a principal ingredient of butter) in France in the late 1860s prompted the government of Napoleon III (1808-1973) to offer a prize for the best ''cheap butter.'' A French chemist, Hippolyte Mège-Mouriès, conducted a series of experiments and patented his result in 1869. The product consisted of liquid beef tallow, milk, water, and chopped cow's udder, churned into solid form. Mège-Mouriès called his invention oleomargarine: from *oleo*, the French word for beef fat, and the Greek word *margarites*, ''pearl,'' because of the product's pearly white color. It was also marketed as ''butterine.''

From the time margarine was first produced commercially in 1873, the dairy industry bitterly opposed it. Excise taxes were imposed on margarine in the United States, and the sale of yellow margarine was prohibited. Margarine manufacturers fought for public acceptance with massive research on ways to improve their product. Nutritional content was boosted by the addition of vitamins A and B, and margarine was made more spreadable and creamy. To cut production costs, American manufacturers developed ways to make margarine from vegetable oils rather than animal fats in the 1930s. Cottonseed, soybean, and corn farmers dropped their support of restrictive legislation; as taxes disappeared, margarine became much cheaper than butter and secured its place as a middle-class food.

Starting in the 1960s, new research suggested that the polyunsaturated fats used in margarine were far healthier to consume than the saturated fats in butter. This further boosted margarine sales and popularity, so that by 1993, Americans were eating twice as much margarine as butter. In the mid-1990s, though, more sophisticated research raised concerns about the fat content of margarine as well-specifically, the trans fats that are created by hydrogenation. During the hydrogenation process, hydrogen molecules are added to liquid oils, such as corn and soybean oils, to make them more solid and stable. This gives margarine a texture and flavor more like butter. However, the process also rearranges some fat molecules to an unnatural, or ''trans,'' position. Newer studies have found that trans fats, like saturated fats, can raise LDL cholesterol-the ''bad'' kind. Dietitians now recommend limiting both butter and hydrogenated margarine.

The margarine industry has responded to health concerns by introducing an array of less hydrogenated products-including tub, squeeze, and spray margarine-as well as lower fat sticks. Today production of vegetable oil for export and use in margarine manufacturing is an important aspect of the economy of many developing countries.

Maria the Jewess • See Double boiler

Marine chronometer • See Chronometer

Mark I

The Automatic Sequence Control Calculator (ASCC), also known as Mark I, was the first large American digital **computer** to work from a program and produce reliable results. It was designed by **Howard H. Aiken**, a physicist at Harvard University, and built by IBM and the United States Navy.

While in graduate school during the 1930s, Aiken became dissatisfied with calculating differential equations by hand or existing calculators. He decided to build a better machine based on the Analytical Engine of the British computer pioneer **Charles Babbage**.

Aiken designed a machine to add, subtract, multiply, and divide, calculate exponents, handle probabilities, and look up trigonometric function values on tables. It would read the initial data from a punched card, perform the calculations, and punch the results on other cards. Like Babbage, Aiken organized his data in registers, and he specified fixed decimal points.

In 1937, Aiken contacted several calculating machine manufacturers about building the machine, but they turned him down. Even the president of Harvard discouraged him, saying the idea was impractical. But when Aiken presented his idea to IBM, the company's president, Thomas Watson, agreed to build it.

When construction began, in 1939, **vacuum tube** (electronic) technology was available. However, the Mark I was an electromechanical machine, making use of the standard parts and engineering expertise available at IBM. In its final form, it was a linkage of seventy-eight adding machines and calculators, over fifty feet (15.24 m) long, nine feet (2.74 m) high, and weighed five tons (4,540 kg). It used over 130 registers to hold data, seventy-two for constants and sixty hand-operated registers for intermediate processing results. Numbers were transferred from one register to another by a combination of mechanical clutches and electrical sensors.

Mark I could perform any specified sequence of operations, for numbers up to twenty-three digits long. Addition or subtraction of two twenty-three-digit numbers took one-third of a second. The program instructions were on punched paper tape. Punched cards, punched tape, or hand-set switches provided the input. The results were punched on cards or printed on an electric typewriter.

Mark I was completed in January 1943, in the midst of World War II. Aiken, who was serving as a Navy officer, moved the machine to Harvard where it performed its first calculations in May 1944. During and after the war, Mark I was used for military purposes, including calculations for ballistics trajectories and the **atomic bomb**.

When the machine was dedicated in August 1944, Aiken angered IBM's Watson by taking the entire credit for it. As a result, IBM broke its relationship with Aiken and Harvard, but the company combined its Mark I experience with other research as the foundation of its own computer line.

Mark I worked reliably at Harvard for fifteen years, but faster, entirely electronic computers were already being built. From 1945 to 1952, Aiken built three advanced versions of his machine incorporating new technology. Mark II, with electromagnetic relays, and Mark III, with magnetic drum storage, were designed for the Navy. Mark IV, built for the Air Force, used magnetic-core registers.

Marker

Markers, or felt-tip pens, serve a variety of functions. Children use them to make bright, colorful drawings. The stereotypical teacher uses a glaring, unmistakable red felt-tip to grade papers. Retail employees, roadside vendors, performers, and protesters rely on the indelible, eye-catching shades and thick inking surface of these writing and drawing utensils to announce sales, prices, and productions, or to create strongly worded posterboard signs to convey dissatisfaction. Markers are also useful for permanently marking surfaces, which is often necessary for identification purposes—putting names on clothing tags, boxes, and tape which can be adhered to almost any item.

The felt-tip pen was invented by Sidney Rosenthal in 1953. This inventor from Richmond Hill, New York, placed a felt tip on the end of a small, stout bottle of permanent **ink** and discovered that the resulting marks saturated a heavy, absorbent surface, yielding rich color and permanence.

Until 1967, the sale of margarine was illegal in the state of Wisconsin, forcing residents to buy the butter substitute in neighboring states.

The felt-tip pen had many predecessors. Inks and dyes have been used throughout human evolution for marking objects, from cave writings fashioned from natural dyes of the earth applied with sticks to graffiti applied with paint from an aerosol can. Ink is a combination of a coloring agent, or pigment, and a liquid containing oils, resins, and chemical solvents. Initially, ink was fashioned from different colored juices and plant and animal extracts. Today, synthetic materials are used in addition to these natural substances. The use of ink for writing and printing dates back to 3200 B.C., when the Egyptians used a mixture of fine soot and vegetable gum to create a substance that could be used for writing and painting. Both Egyptians and Greeks used iron oxide (or, more commonly, rust) to make red ink. Around 2000 B.C., the Chinese began making red ink from mercury sulphate and black ink from iron sulphur mixed with sumac tree sap. Like the Egyptians, they formed their ink into a solid block or stick that would be mixed with water when used. Europeans did not begin commonly using ink until the seventeenth century, using tannic acid from tree bark and iron salt to create the recipe that formed the blue and black inks still used today.

Like inks and dyes, pens have been used since antiquity. The earliest pens were made from hollow reeds and, later, hollow wing feathers of geese and swans, called quills, infused with ink. Steel pen nibs came into use in the early 1800s, and then fountain **pens**, which did not require a constant resupply of ink like previous incarnations, gained popularity.

House paint also came into popular use in the late 1800s when Edwin Binney and Harold Smith used red oxide pigments (a mixture of naturally occurring dyes and chemicals) to create a viscous coloring substance—the paint that was used to color America's first classic red barns. Binney and Smith, through their company Binney & Smith, then created a line of carbon black pigments that were used by the Goodrich Company to color its white auto **tires** black.

Next, Binney & Smith acquired a water-powered stone mill in Easton, Pennsylvania, and began fashioning slate culled from the area into **pencils**. From there, the company created dustless **chalk** in 1902. In 1903, the company fashioned a variant of its industrial wax marking crayons, which were smaller and came in a variety of colors created by colored pigments added to paraffin wax. Thus, Crayola **crayons** were born.

By the late 1950s while Binney & Smith was expanding its crayon business, Rosenthal was building up his company, Speedry Chemical Products, which manufactured and marketed his felt-tip pen invention. Rosenthal initially geared his product at the art supplies market, but soon thought to capitalize on its mass market appeal due to its suitability to poster-making, sign-lettering, and other marking purposes. Competitors threw their hats in the ring as early as 1958, when Carter's, Inc., came out with a more slender marker with an aluminum ink tube. Speedry sued Carter's for patent infringement, but lost. Other companies began marketing pens containing water-soluble inks that could be used on normal-weight paper (Rosenthal's invention required heavier paper to keep the ink from soaking straight through to the underlying surface) and with capillary flow technology, which enhanced the movement of the ink into the tip of the writing utensil. When Rosenthal changed the name of his company to Magic Marker Corporation in 1966, he was already suffering the effects of this increased competition.

Despite its name becoming synonymous with its product, regardless of the manufacturer, Magic Marker Corporation continued to lose money and filed for bankruptcy in 1980. In 1989, Binney & Smith, now a subsidiary of the Hallmark Corporation, purchased the rights to the Magic Marker name, stating that it was motivated to purchase the moniker due to remaining high consumer recognition of the brand name. The defunct Magic Marker Industries would be able to use the royalty income garnered from selling the use of its name to pay off creditors. Binney & Smith now manufacturers a wide variety of Crayola markers as well as Magic Markers.

The traditional marker is continually being enhanced. Markers are available in both indelible and washable formulas. Markers are also produced that can yield changing colors, color over colors, emit scents, and be used to create 3-D drawings. One of the most popular spin-offs of the marker is the highlighting marker, which comes in a variety of translucent, fluorescent colors and can be used for marking text. This is a popular product with students and professionals. Ink companies are also at work developing products that are more environmentally sound, such as inks using vegetable oils or water instead of traditional petroleum or other solvent ingredients.

MARTIN, ARCHER J. P. • See Chromatography

MARTIN, PIERRE-EMILE (1824-1915)
French engineer

Pierre-Emile Martin was one of the developers of the *open-hearth steelmaking* process, also known as the *Siemens-Martin* process. The open hearth, a variety of blast furnace, uses regenerated heat for more efficient production of steel. It was invented by Charles William and Friedrich Siemens in 1856. Martin improved the process shortly thereafter by using strategically placed heat-capturing chambers on either side of the furnace and by introducing a gas producing element.

Born on August 18, 1824, in Bourges, France, Martin was introduced to iron and steel production by his father, Emile, who owned a foundry in Sireuil. The increased demand for iron and steel during the nineteenth century inspired a strong competition among inventors to develop new materials and new processes. Using the Siemens heat-regeneration process, with his own modifications, Martin purified steel by dilution, using large quantities of scrap steel. After a protracted dispute with Siemens over rights to the process, a compromise was reached in 1864, recognizing both Siemens and Martin as contributors to the development of what came to be called the Siemens-Martin process.

Martin spent the remainder of his life in virtual poverty, having been drained financially by patent litigation. While he suffered, others benefited from his innovation. By the turn of the twentieth century, the Siemens-Martin process had surpassed the earlier Bessemer conversion process as the primary method of steel production, and it remained so until the late 1960s. Martin died on May 23, 1915, in Fourchambault, France.

See also Recycling; Steel production

Maser

Although the knowledge and use of **lasers** is now widespread, the laser is actually a light-emitting cousin to the maser, a device used to generate and amplify radio and light waves. The maser—an acronym for Microwave Amplification by the Stimulated Emission of Radiation —was originally designed to meet the need for an **oscillator** that could produce radiation shorter than one millimeter in wavelength. Previously, scientists had used microwave oscillators to study the basic structure of matter (the branch of physics known as microwave spectroscopy), but by the late 1940s it became increasingly clear that coherent oscillators at shorter wavelengths were essential.

Two concepts figured highly into the development of the maser: coherence and stimulated emission. Coherence describes a wave or particle whose vibratory pattern is ''in step,'' the crests and troughs aligned perfectly. Stimulated emission was first studied by Albert Einstein (1879-1955) in relation to his theory of relativity; it claimed that a group of excited atoms could be stimulated so as to emit their radiation in a controlled manner. Although these two theories laid the groundwork for maser research, actual development could not begin until after World War II. During the war, tremendous advances were achieved in **radar** technology, and it was these advances that truly opened scientists' minds to the concept of masers.

Maser research was introduced independently by Nikolai Gennadievich Basov and Alexander Mikhailovich Prokhorov in the Soviet Union, and by **Charles Hard Townes** and his students at Columbia University in the United States. The first working maser was built by Townes and his associates in 1954 using ammonia gas (NH_3). A beam of excited ammonia molecules was sent through a cylindrical ''focuser'' which allowed only those molecules in the high-energy state to pass through. This new high-energy beam was guided into a resonant cavity where, under the influence of an electrical field, amplification by stimulated emission of radiation was achieved, producing a microwave output corresponding to the resonance frequency of ammonia. In 1956, Townes and his colleagues at the University of Paris showed that maser action was possible in certain solids, leading to the development of the more common ruby maser, and the first operating ''optical maser,'' or laser (Light Amplification by the Stimulated Emission of Radiation), was built in 1960. For their efforts in this field, Townes, Basov and Prokhorov shared the 1964 Nobel Prize for physics.

The decade following 1954 saw tremendous advancement upon Townes' original ammonia-based model. One by-product was the **atomic clock**, a device whose extreme accuracy relied upon the undeviating regularity with which molecules vibrate. Ruby masers were used in radio astronomy to amplify very weak radiation from distant radio sources such as stars and planetary probes, sources that were previously undetectable among the din of background noise. Optical masers, or lasers, have found countless applications in numerous fields including medicine, engineering, and communications.

Masonite

Masonite is a synthetic hardboard used for insulation, panelling and general wall construction. It is made of compressed wood chips held together by lignin, a natural binding agent found in the wood. Masonite is produced by heating the wood chips and pressing them together to form large panels. When they are finished, the panels are smooth on one side and can be rough or smooth on the other. Masonite is a versatile, economical, and environmentally responsible product that allows wood scraps that once went to waste to be transformed into valuable building material. The product is sold under the brand names of Masonite and Abitibi.

The invention of Masonite could be considered a fortunate accident. During the 1920s, William Mason, an inventor from Mississippi, wanted to find a use for the tons of wood chips that were going to waste on sawmill floors. With financial backing from a lumber company in Wausau, Wisconsin, Mason worked on some of his ideas. Using a leaky nineteenth-century steam letter press, Mason's intended goal was to produce insulation board. One day in 1924, he accidently left the steam on while he went to lunch. He returned to find that the overdose of steam had produced a tough sheet of hardboard. Further tests proved that the product was resistant to weathering. The following year, Mason established the Mason Fibre Company, to produce his Masonite hardboard. The company later became the Masonite Corporation, which is now a subsidiary of the International Paper Company.

Mass production

Despite its obvious connection, mass production was not a corollary to the modern Industrial Revolution. Various mass production techniques had been practiced in ancient times, from ceramic production in the Orient to manufacturing in ancient Greece. The British were most likely the first modern economy to adapt water-powered, then steam-powered, machinery to industrial production methods, most notably in the **textiles** industry. Yet it is generally agreed that modern mass production techniques came into widespread use through the innovation of an assortment of Americans who substantially improved the ancient techniques. Indeed, this modern mass production was called the *American System* and its early successes are often attributed to **Eli Whitney**, who adapted mass production techniques and the interchangeability of parts to the manufacture of **muskets** for the U.S. government in the late 1790s.

While Whitney was certainly an innovator of the *American System*, others maintain that Whitney's parts were not truly interchangeable and that credit should more appropriately go to John Hall, the New England gunsmith who built flintlock **pistols** for the United States government at the Harper's Ferry armory. Hall, born in Maine in 1769, built many of the machine tools needed for precision manufacturing and instituted a system that employed accurate gauges for measuring every aspect and piece of work his factory produced. Consequently, he achieved a much higher level of interchangeability and precision than did Whitney.

Still others maintain that the credit for these modern innovations should go to a French gunsmith whose methods and results predated those of Whitney and Hall by at least a decade. In Britain, and somewhat simultaneously with Whitney, the Frenchman **Marc Isambard Brunel** adapted steam-driven machinery and assembly-line techniques to the production of 130,000 **pulley**s for the marine industry in just one year. Brunel's achievements were made possible by the design and manufacture of several machine tools by the noted British inventor, **Henry Maudslay**.

Maudslay's contribution to modern mass production was the invention of precision machine tools capable of producing the identical parts necessary for mass production techniques. It is generally conceded that the British machine tool industry was far more advanced than that of the Americans in these early stages of mass production development. Simultaneous with Whitney's innovations in the United States were those of **Oliver Evans**, whose many inventions in the flour milling process led to an automated mill that could be run by a single miller.

Samuel Colt and **Elisha King Root** were also very successful innovators in the development of industrial processes that could mass produce interchangeable parts for the assembly-line production of firearms. Colt and Root wished to advance the machining of parts so that even the most minute of tasks could be performed with the precision that they believed only machines could achieve. In these endeavors, Colt and Root were largely successful.

Eli Terry also adapted mass production methods to clockmaking in the early 1800s, and **George Eastman** made innovations to assembly-line techniques in the manufacture and the developing of photographic film later in the century. Credit for the development of large scale, assembly-line, mass production techniques is usually given to **Henry Ford** and his innovative Model T production methods. Henry Ford had his workers standing in one place while parts were brought by on conveyor belts, and the car itself moved past the workers on another conveyor belt. Bodies were built on one line and the chassis and drive train were built on another. When both were essentially complete, the body was lowered onto the chassis for final assembly. It has been said that Ford took the inspiration for his assembly line from the meat-processing and canning factories that moved carcasses along lines of overhead rails as early as the 1840s—a technique also used by **Cyrus McCormick** for building *reapers* long before Ford adapted the

Henry Ford and his innovative production methods for his Model T ushered in the modern era of large-scale, assembly-line production.

technique to the Model T. Indeed, Ransom Eli Olds (1864-1950) used assembly line techniques to build his automobiles for many years before Ford sold his first Model T.

Despite the fact that he was not the first, Ford can certainly be viewed as the most successful of these early innovators due to one simple fact—Ford envisioned and fostered *mass consumption* as a corollary to mass production. Ford's techniques lessened the time needed to build a Model T from about twelve and a half hours to an hour and a half; the price was reduced as well—from $850 for the first Model T in 1908, to only $290 in 1927 after assembly-line techniques were introduced in 1913. The automobile was no longer a luxury for the rich, the Model T fast became a necessity for nearly everyone. Indeed, Ford sold almost half of all of the automobiles bought worldwide from 1908 to 1927—the years of Model T production.

Assembly-line techniques also required that the manual skills necessary to build a product be altered. Previous to mass production techniques, as seen in the early manufacture of firearms, each workman was responsible for the complete manufacture and assembly of all of the component parts needed to build any single product. Mass production and parts interchangeability demanded that all parts be identical and the individual worker no longer be allowed the luxury of building a complete product based on his personal skills and inclinations. Machines came to dictate the production process, and each part—once created individually by hand—was now duplicated by a machine process that was merely guided by human control. The craft tradition, dominant in human endeavor for centuries, was abandoned in favor of a process that created parts by machine. Furthermore, assembly of these machine-made parts was divided into a series of small repetitive steps that required much less skill than traditional craftsmanship. Consequently, modern mass production techniques, while certainly increasing the efficiency of the manufacturing process and bringing industrial products within the reach of virtually all of humanity, also altered forever the relationship of people to their work. Mass production supplanted craft, and the repetitive assembly line is now the world's standard for all manufacturing processes.

Mass spectrograph

John Dalton's (1766-1844) atomic theory served scientists remarkably well for nearly a century after its announcement in 1803. But discoveries made in the late 1890s made it clear that Dalton's theory was incorrect or incomplete in some important details. The discovery of radioactivity by French physicist **Antoine-Henri Becquerel** (1852-1908) was among the most important of these. As scientists began to unravel the nature of this new phenomenon, they realized that atoms were far more complex than Dalton had imagined.

A particularly important discovery was made by English chemist Frederick Soddy (1877-1956) in 1913. During his investigation of naturally-occurring radioactive families, Soddy found that more than one form of an element could exist. These forms all had the same atomic number (and were, therefore, variations of a single element), but had different atomic weights. Soddy gave the name isotopes to these forms. Almost immediately, scientists began to ask whether isotopes might also be found among the stable elements. English physicist Joseph J. Thomson (1856-1940), discoverer of the electron, turned his attention to that question around 1912. After some initial discoveries, Thomson turned this research over to one of his students, Francis Aston.

Francis William Aston was born in Harborne, Birmingham, England, on September 1, 1877. He was educated first at Malvern College and later at Mason's College, Birmingham, later to become Birmingham University. He worked as a brewery chemist from 1900 to 1903 before accepting a research position at Birmingham University. After a world tour in 1909, he was invited by J. J. Thomson to join him at the Cavendish Laboratory at Cambridge. Aston spent the rest of his academic career in various positions at Cambridge. He died there on November 20, 1945.

After discovering the electron in 1897, Thomson decided to study the positively charged canal rays first observed by German physicist Eugen Goldstein (1850-1930) in 1886. He constructed an instrument that accelerated the canal rays through magnetic and electrical fields. The fields caused the rays to travel in a parabolic track whose shape was determined by the mass and velocity of the particles constituting the rays. Thomson found that canal rays produced by the excitation of neon gas formed two distinct tracks in his instrument. He was uncertain as to which of a number of explanations correctly accounted for this observation.

At this point, Thomson asked Aston to take over this line of research. Aston focused his attention at the outset on the improvement of Thomson's original instrument. Over the next decade, he made further refinements on the device until it was eventually able to resolve two tracks with an accuracy of one part in 10,000.

The mass spectrograph (also known as mass spectrometer) that Aston designed exposes a beam of positively charged particles to both an electrical and magnetic field. Aston found a way to modify the magnetic field so that particle beams are separated from each other entirely on the basis of their masses, not their velocities. With his improved instrument, Aston was able to demonstrate in 1919 that the two tracks observed by Thomson in his research were isotopes of neon with masses of 20 and 22 in a ratio of ten parts to one. Over the next two decades, Aston studied all but three of the known stable elements, identifying 212 different isotopes in the process. Eventually, Aston's mass spectrograph design was improved upon by those of A. J. Dempster (1886-1950), K. T. Bainbridge, and A. O. Nier. Today, the mass spectrograph continues to find use in a wide variety of research applications.

Masters, Sibilla (ca. 1670-1720)

American inventor

Although her two patents were in her husband's name, Sibilla, or Sybilla, Masters was among the first, if not the first, Americans to receive a patent. No record of Masters' date or place of birth exists, but it may have been in the mid to late 1670s and probably in Bermuda. Her parents, William and Sarah Righton, were Quakers. Her father was a merchant marine who may have emigrated to New Jersey from Bermuda in 1687. Sibilla married Thomas Masters, a successful merchant from Philadelphia, Pennsylvania, around 1695. After they were married, he had a large mansion built in Philadelphia overlooking the Delaware River, the fruit of overseas land investments. He held several political posts, including mayor of Philadelphia from 1707 to 1708.

Sibilla Masters travelled to London between 1712 and 1715. During this time, she designed a corn pulverizer which cleaned and cured Indian corn grown in the American colonies. Master's invention consisted of a stamp, or pestle, that, when tripped, descended through a wooden cylinder to a mortar, which held the corn. The action of the stamp turned the corn into meal, which was then transferred to bins for drying, or curing. In 1714, Thomas Masters acquired a grist mill where the corn meal was produced for sale in the Philadelphia area. It was offered as a cure for consumption, an early term for tuberculosis, though it was nothing more than a food product. In 1715 an English patent was issued to Thomas Masters for Sibilla's invention, since, by law, women were unable to receive patents.

In 1716 Masters, still on her own in England, secured another English patent in her husband's name. Her second invention involved a process by which straw and palmetto leaves were formed and stained for the adornment of women's hats and bonnets. Unfortunately, no diagram or description of the process exists. Less than a month after the patent was issued, Masters opened a shop in London, the West India Hat and Bonnet, where she sold head pieces and furniture padding made from her straw and leaf treatment. Masters' business venture was short-lived, however, and she was back in Philadelphia by mid 1717. She secured the same two patents with the colony of Pennsylvania, once again in her husband's name. It is not known if she pursued her ventures any further after this. While Sibilla Masters is recognized for her inventiveness and attempts at marketing, the real significance of her work remains her bold venture into a realm that was generally exclusive of women.

Match

The evolution of today's safe, cheap matches took hundreds of years, and many disastrous products failed along the way. Although ancient people learned long ago to carry torches as a convenient source of fire, matches were unheard of until around 1000 A.D. Possibly, the Chinese people invented the first primitive match, since Marco Polo (1254-1324) reported seeing matches on his journeys to the Orient during the late 1200s. By the late 1500s, sulfur-tipped matches were being used in England, though they bore little resemblance to modern matches. In 1681 **Robert Boyle** coated a piece of coarse **paper** with phosphorus and produced a flame by drawing a sulfur-tipped wooden splint through a fold in the paper. During the 1700s and early 1800s, several small fire-making devices were invented, including the *Ethereal Match*, the *Pocket Luminary*, and the *Instantaneous Light Box*. Most of these were **glass** tubes or bottles containing flammable chemicals, which were sold with a packet of chemically treated splinters of wood. But it was not until 1827 that the first modern match was invented. British pharmacist John Walker (1781?-1859) tipped three-inch-long wood splints with a potassium compound and a metal-sulfur compound (antimony sulfide), held together with gum arabic. Walker sold his matches, called *Congreves* after Sir William Congreve (1772-1828), an artillerist and the inventor of the Congreve rocket. *Congreves* came with a piece of striking paper coated with ground glass. Unfortunately, when Walker's customers scratched the match across the paper, they were showered with sparks from a series of small explosions, accompanied by an unpleasant smell. One of Walker's competitors began selling similar kits called *Lucifers*, which carried a warning against inhaling the match's gases. "Persons whose lungs are delicate," the warning continued, "should by no means use *Lucifers*."

Soon, matches were invented that could be lit by striking on any rough surface—a fireplace **brick**, for example. These "strike-anywhere" matches are ignited by the heat of friction. Also known as kitchen matches, they are still sold today in an improved version. The first such match was made in 1830 by French scientist Charles Sauria, who coated the wooden tip of his matches with white or yellow phosphorus. Within a year or two, the manufacture of friction matches had been well-established in Europe. But Sauria was unaware that the white and yellow forms of phosphorus are extremely poisonous. Some of his customers learned to commit murder and suicide using match tips.

In America, phosphorus matches were first patented by Alonzo Dwight Phillips in 1836. Although Phillips made his matches by hand and sold them door-to-door from a wagon, match production quickly became a factory operation. As the industry grew, so did chemical poisoning. Match factory workers developed a condition called "phossy jaw" from breathing phosphorus fumes. The disease kills the roots of the teeth and deteriorates the jawbone, creating intense pain; it also causes anemia and loss of appetite. Some patients were horribly disfigured by jaw surgery and had to live on liquid food for the rest of their lives. In the 1840s, European scientists introduced the use of the red form of phosphorus, which is less reactive than the white form. Austrian chemist Anton Ritter von Schrotter and Swedish inventor Carl Lundstrom are noted for using red phosphorus on matches. Soon, French chemists had developed a nonpoisonous chemical for match tips (phosphorus sesquisulfide), which is still used on wooden matches. Although American manufacturers purchased the French patent in 1900, these matches did not work well in American climates.

Meanwhile, Alice Hamilton (1869-1970), a medical doctor and pioneer in the study of occupational disease, publicized the terrible effects of phossy jaw disease. In the early 1900s the United States government imposed such high taxes on matches that the survival of the industry was threatened. Finally, in 1911, William Armstrong Fairburn, a young naval architect, adapted the French chemical formula for phosphorus sesquisulfide to United States climates, ending the spread of chemical poisoning among American match factory workers. Beginning in 1913, Sweden dominated the world's match industry for many years, led by the entrepreneur Ivar Kreuger (1880-1932), known as the "Match King." Kreuger's empire encompassed not only factories, but also the forests of timber and mines of chemicals needed to produce matches. As the story goes, Kreuger invented the superstition that three people using a single match is unlucky—a notion that was reinforced in the trenches of World War I, where any flame not rapidly extinguished could attract enemy gunfire. Despite his evident success, Kreuger's business collapsed, and an investigation into his affairs suggested that his dealings were not entirely legitimate. After these turns of events, he killed himself. "Safety" matches, as opposed to the strike-anywhere variety, can be ignited only on a specially treated surface that contains red phosphorus and sand. The match tip has the other chemicals needed to ignite (including a potassium compound), but neither the match nor the striking surface alone can burst into flame. This type of match was invented in 1844 by Swedish chemist Gustave E. Pasch. By the 1850s, large quantities of safety matches were being produced by John Lundstrom, a Swedish manufacturer. Today's most familiar kind of matches—those that come in "books" of folded cardboard—were invented in 1892 by a Philadelphia patent lawyer, Joshua Pusey, who made them in packs of fifty. In these early matchbooks, however, the striking surface was inside the cover, dangerously close to the match heads.

By the end of World War I, American manufacturers were making safe, practical matchbooks with external striking strips, and these became quite popular. When America entered World War II, troops fighting in tropical climates needed waterproof matches. A substance resistant to water and heat was developed by Raymond Davis Cady in 1943. With this protection, wooden matches could be submerged in water for eight hours, yet still light. Match production today is highly automated, requiring much less labor than in the days of phossy jaw. Wooden matches are manufactured at a rate of more than a million per hour in a single, continuous process by a series of huge machines. Book matches are made by two machines—one that slices the paperboard into match-sized strips and dips the tips in chemicals, and a second machine that cuts the match strips into book size and staples them into the covers.

Matzeliger, Jan Ernst (1852-1889)

American inventor

Jan Matzeliger was a nineteenth-century inventor and machinist who revolutionized the shoemaking industry and made a fortune for his financial backers. Over a period of several years, during which he sacrificed everything for the sake of his invention, Jan Matzeliger conceived, patented, built working models, and factory-tested a machine known as a shoe-lasting machine, and he eventually became a stockholder in the company that manufactured it. As revolutionary and beneficial as **Eli Whitney**'s **cotton gin** or **Elias Howe**'s **sewing machine**, Matzeliger's shoe-lasting machine could produce 150 to 700 pairs of shoes a day—compared with 50 pairs of shoes per day by hand-lasting methods.

By the 1870s, most of the steps in manufacturing shoes were already automated. In 1790, Thomas Saint, a London cabinetmaker, had invented the first sewing machine designed for use on shoe **leather**. In 1810, **Marc Isambard Brunel**, a Frenchman working in London, set up machines to mass produce nailed army shoes. In 1841, Thomas Archbold, an English machinist, applied the principle of the eyepointed needle to shoe production. A variety of other specialized machines sped the process of creating and manufacturing shoes in quantity.

It was the final step in the shoemaking process that proved to be the most difficult to automate. This final step involved connecting the upper part of the shoe to the inner sole, a process called *lasting*. Lasting, crucial to the quality of the shoe, determines its fit, walking ease, and look. A last was a wooden model of the foot, and stretching the shoe leather over the last took a great deal of skill. Tacking the finished shape into place was also difficult. When Jan Matzeliger came to work in the shoe factories, no machine had been invented that could complete the lasting process.

Jan Matzeliger was born on September 15, 1852, in the port city of Paramaribo in Dutch Guiana, now known as Surinam. His mother was a native Surinamese of African descent, and his father, a Dutch engineer who had been sent to the island colony to take charge of the government machine works, was a well-educated man and a member of a wealthy and aristocratic Dutch family.

Jan served as an apprentice in a government machine shop supervised by his father. He developed an interest in machines, eventually becoming a skilled machinist. At the age of 19 he signed on as a seaman with the Dutch East Indies Company and went to sea. He helped fix the engines on the steamship to which he was assigned. He spent two years sailing to the Far East, then came to North America with his ship. When his ship docked in Philadelphia in 1873, Matzeliger left the Dutch East Indies Company and looked for work as a machinist in Philadelphia.

In the 1870s Philadelphia was a busy center of commerce, with many factories offering opportunities to skilled machinists. Unfortunately, skilled jobs were not open to blacks in the segregated job market of the city. Matzeliger was also hindered by the fact that he spoke little English, since Dutch was his native tongue.

Eventually Matzeliger found a shoemaker's shop in Philadelphia, where he learned to use a McKay sole-sewing machine that sewed the seam of a shoe sole. He became fascinated with the shoemaking process and was advised to go to Lynn, Massachusetts, the shoe manufacturing center of North America. Matzeliger left his job in Philadelphia and arrived in Lynn on a winter day in 1877.

The social climate for African-Americans in his new home made it difficult for Matzeliger to become established in the community. It took him quite some time to find a job in the shoe factories. Finally, Harney Brothers hired him to sew shoes on the familiar McKay sole-sewing machine. While working, Matzeliger went to night school to improve his English. He was filled with a desire to learn more about machines. After some time, he managed to save enough money to buy a set of drawing instruments. He used these drawing tools to put his ideas for new kinds of machines down on paper. He observed the automated process of shoemaking in the factory in which he worked. There were specialized machines for each step of the process, except for the shoe-lasting operation. There were machines for upper work, stock fitting and bottoming, buttonholing and buffing. Each worker had his or her own part of the shoe to work on, and a machine to operate.

Matzeliger also closely observed the final step of shoelasting. Most of the time, the shoe lasters could not keep up with the machines in the factory. The lasters had a strong union and were considered kings of the shoemaking trade. It was said they often worked slowly on purpose, and they often went on strike. One day, Matzeliger said he could make a machine to do their job. His claim was greeted with skepticism.

Matzeliger was determined to learn all he could in order to enable him to invent a shoe-lasting machine. He requested a job as a millwright in the Harney Brothers factory. His new job would be to circulate through the factory and check on, and repair, all of the machines. The new position also gave him the opportunity to watch the lasters at work.

He took a room in the old West Lynn Mission to work on his plans for a shoe-lasting machine in secret, because others were also trying to develop the same machine. He worked in a cheap room at the Mission, and his quarters were not well-heated. It took six months for Matzeliger to make a model of his machine—it was made only of old cigar boxes, **wire**, nails, and scrap wood, but another inventor offered him fifty dollars for it. Wisely, he refused the offer.

Matzeliger then set out to make a working model made out of metal. Some parts he was able to salvage from junkyards, others he had to fashion himself from pieces of scrap metal. To do this, he needed a forge to heat the metal and a **lathe** to shape it. There was one shoe factory in Lynn, Beal Brothers, that had a forge Matzeliger thought he could use. So, he left Harney Brothers and went to work for Beal Brothers. His new employer gave him a workspace and the use of their forge and lathe. Both machines were old and difficult to use but his determination carried him through several years of hard work and personal expense to complete his project. In the meantime, Matzeliger also took a part-time job driving coaches that transported young people to a local park for recreation.

By 1882, Matzeliger had completed the scrap-metal model. He knew he was on the right track when another inventor offered him $1,500 for just a part of the model. He again refused the offer. Since the scrap-metal model would not stand up to factory testing, he needed to make a good working model made of new parts and with real precision. He approached businessmen in Lynn to finance his invention, but was turned down—one investor had already lost $100,000 on a shoe-lasting machine that failed.

Finally, he found two backers, C. H. Delnow and M. S. Nichols, who agreed to back him in return for two-thirds of the profits the machine would realize. Together, the three men formed the Union Lasting Machine Company, and Matzeliger set about making his third model. With a good model underway, Matzeliger could apply for a patent by submitting detailed drawings and a complete description of his invention.

In Washington, D.C., the patent officials could not understand the complicated drawings, and they didn't believe the machine could do what its inventor claimed it could. As a result, the patent office sent an examiner to Lynn to inspect the machine. Matzeliger demonstrated how the machine worked. It held the last, gripped the leather, drew the leather over the last, fitted the leather at the heel and toe, moved the last forward, fed the nails, and drove the nails. The patent official was satisfied, and on March 20, 1883, Matzeliger was granted U.S. Patent No. 274,207 for his shoe-lasting machine.

The machine fully proved itself in a factory test set for May 29, 1885, lasting seventy-five pairs of women's shoes with no trouble. Later, the machine would be able to turn out 150 to 700 pairs of shoes in one day. To begin manufacturing the machine, Delnow and Nichols needed more money. They obtained funds from George W. Brown, the northeast agent for the Wheeler Wilson Sewing Machine Company, and Sidney W. Winslow, who became known as the machinery king of New England. In exchange for their funding, Winslow and Brown took over the machine's patent from Matzeliger, who was given a block of stock in the new company, the Consolidated Lasting Machine Company.

Production of the shoe-lasting machine began in the mid-1880s and expanded rapidly—every shoe manufacturer in Lynn wanted to buy one. Shoe manufacturing boomed in New England, and exports reached a new high. Shoe prices were cut in half, and many more people now found them affordable. It was a revolution in the shoe industry. Rather than putting the shoe lasters out of work, Matzeliger's machine gave them more work to do, and it was easier work, too.

By 1897, Winslow brought together the major lasting machine manufacturers and organized a holding company, the New York Machine Company. In 1899, Winslow completed the consolidation of machine manufacturing companies to form the United Shoe Machinery Corporation with a capitalization of $20 million and himself as president. From 1899 to 1910, the United Shoe Machinery Corporation earned over $50 million and held 98% of the shoe machinery business. By 1955, the company was worth more than a billion dollars.

Demand for Matzeliger's machine was worldwide by 1889. He continued to improve the machine and received four additional patents. In spite of his success, he made few changes in his simple lifestyle. He taught oil painting classes and Sunday school, and became the leader of Christian Endeavor at the North Church. During one of their summer picnics in 1886, Matzeliger developed a cold that was later diagnosed as tuberculosis, possibly contracted from his early inventive days at the West Lynn Mission. He went into a hospital, unable to afford more expensive treatment at a sanitorium. He was bedridden for three years and died on August 24, 1889, at the age of 37.

Mauchly, John (1907-1980)
American physicist

John Mauchly, with J. Presper Eckert, designed and built several significant computers in the 1940s—**ENIAC**, EDVAC, BINAC, and **UNIVAC**. Mauchly was born in Cincinnati, Ohio, and became interested in science while very young. He studied engineering and physics at Johns Hopkins University, receiving his Ph.D. in 1932; he spent the next year there as a research assistant. In 1934, when he became a faculty member at Ursinus College, he carried out weather research.

All of this work required large amounts of computation, and he began to consider how to make calculating easier. After taking a course at the University of Pennsylvania's Moore School of Electrical Engineering in the summer of 1941, he was asked to join the faculty. There he met Eckert, who was a student.

Because of World War II, the university had a contract to develop a calculating machine for the United States Army. Mauchly and Eckert designed and patented ENIAC (Electronic Numerical Integrator And Computer), which was much faster than earlier computers and used vacuum tubes for electronic data processing. However, it required the slower punched cards for the program and intermediate processing results. Also, each processing sequence had to be set up by hand.

The school next began work on EDVAC (Electronic Discrete Variable Automatic Computer), a computer milestone featuring a **stored program**, in which the program and data are both in the computer's memory and treated alike. Mauchly and Eckert designed the memory, using mercury-filled delay lines to retain the incoming electronic signal as a much slower sound wave.

In 1948, the University wanted control of all patents for equipment produced by its faculty. Mauchly and Eckert resigned and formed their own firm, the Electronic Control Company, to produce stored-program computers based on the ENIAC patents. Their first computer, in 1949, was BINAC (BINary Automatic Computer), a fast and relatively small computer to be used on a guided missile. It was the first computer to use magnetic tape for input and output, and also featured two processing units that performed the same calculations to guarantee accuracy. At the same time, the two men were building a much larger computer called UNIVAC (UNIVersal Automatic Computer), completed in 1950. It was the first widely-available commercial computer to use stored programs.

Because they were losing money in 1950, Mauchly and Eckert sold their company and patents to the calculator and typewriter manufacturer Remington Rand. After the sale, Mauchly formed a computer-development consulting firm. He also invented the critical path method (CPM) of scheduling complex projects on the computer, which is widely used in business and research.

During the 1960s, Mauchly and Eckert were sued by Honeywell, a computer manufacturer, and John Atanasoff, a physics professor at Iowa State University. They claimed that the patents were based on Atanasoff's A-B-C computer, which he had demonstrated for Mauchly in 1941, before ENIAC was built. Mauchly and Eckert maintained that their design work was largely done before Mauchly met Atanasoff, and that the two machines were different. However, in 1973, the judge ruled in Atanasoff's favor, declaring the patents invalid.

Mauchly received numerous honors for his work, including awards from the American Federation of Information Processing Societies and the Institute of Electrical and Electronics Engineers. In addition, in 1947 he helped found the Association for Computing Machinery, the world's largest computing society.

Maudslay, Henry (1771-1831)
English engineer

Before Henry Maudslay began his work, screw threads were made by very skilled craftsmen. Large threads could be forged and filed, but smaller threads had to be cut by hand. Maudslay changed all that by designing a **lathe** that could repeatedly cut the same threads on a **screw**, and that could make identical sets of taps and dies.

Maudslay was born in Woolwich, England, on August 22, 1771. By age twelve, he was at work filling cartridges at the arsenal there, and in only two years he was promoted to the joiner's shop and apprenticed to the metal-working shop. By age eighteen his skills were already renowned, and he went to work with hydraulics and lock pioneer **Joseph Bramah**.

Maudslay rose to manager of Bramah's workshop, but left in 1797 to open his own business, where he developed his lathe to cut screw threads. He also devised the slide rest for lathe work and designed the first **micrometer** for measuring the precision of his work. His lathe was so precise that screw threads became standardized, and nuts and bolts interchangeable, a necessary step in the success of the industrial revolution.

From 1801 to 1808, Maudslay worked with **Marc Isambard Brunel** to construct a series of machines for making wooden **pulley** blocks. These machines could be operated by ten unskilled workers, and could accomplish the amount of work formerly done by 110 skilled craftsmen.

In 1807, Maudslay patented a table engine that became a compact power source for many years. Maudslay's firm proceeded to manufacture marine steam engines, first as small as 17 horsepower and later as large as 56 horsepower. Seven years after Maudslay's death, his company built the 750-horsepower transatlantic steamship *Great Western* for Isambard Branel.

Although Maudslay's main interest was in engines, he also worked on methods of purifying water and eliminating boiler scale. Maudslay died on February 15, 1831, after a trip from France. He was buried in a cast iron tomb at Woolwich.

Maxim, Hiram Stevens (1840-1916)
American British inventor

A member of a noteworthy family of mechanics and draftsmen, Hiram Maxim, the creator of the first practical automatic

Hiram Stevens Maxim

machine gun, also experimented with a steam-powered flying machine in 1894. At age twenty-four, in his uncle's engineering firm in Fitchburg, Massachusetts, he invented a curling iron. Around 1868, he created a locomotive headlight that came into common use. His later inventions included an automatic **fire extinguisher**, automatic gas **generator**, **engine** governors, steam and **vacuum pumps**, inhalers for bronchitis, a mousetrap, automatic spindle, density regulator for equalizing the illuminating value of gas, and an electrical pressure regulator, which won him a French Legion of Honor.

Born in Sangerville, Maine, Maxim was the eldest son of a farmer and woodworker. At age fourteen, Maxim was apprenticed to a carriagemaker and learned numerous other trades in New England and Canada. As a hobby, he enjoyed pugilism and toyed with the idea of becoming a professional boxer, but instead he moved to Fitchburg, Massachusetts, where he worked for an engineer. From there he migrated to Boston and worked for a maker of scientific instruments, then to New York, where he was employed at the Novelty Iron Company and secured his first patent for a curling iron. At age thirty-eight, he became the chief engineer for the United States Electric Lighting Company. His first major contributions were the graphite-rod incandescent **light bulb** in 1878, a standardized coating of carbon lamp filaments in 1880, and an electric current regulator, which won international praise in 1881 at the Paris Exhibition.

While in France, Maxim became intrigued with the possibility of creating an automatic weapon, something European designers had been struggling to produce. Maxim went to work in London, and later became a British subject. At his laboratory in Hatton Garden, London, he produced his Maxim single-barrel automated gun in 1884, which was belt-fed and water-cooled, an advancement over the hand-cranked **Gatling gun**. The device, capable of firing 666 rounds per minute, loaded, fired, then extracted and ejected the cartridge, using energy derived from the momentum of recoil. A subsequent model, dubbed the Vickers Maxim, became standard British Army issue. Later, he developed the pom-pom gun, which launched one-pound shells, and the aerial **torpedo** gun.

Maxim's contributions to the British military earned him a knighthood by Queen Victoria in 1901. He continued to work on the delayed-action fuse and **cordite**, a smokeless powder made from **nitroglycerin** and **cotton**, which he evolved with the help of Hudson Maxim, his brother and partner. Maxim's improvements in weaponry gave significant advantage to British forces during the Boxer Rebellion and the Boer War. Soldiers, grateful for the improvements, chanted: ''Whatever happens, we have got/The Maxim gun and they have not!'' Soon, the gun came into use in every major country. It was not until the perfection of **tank** warfare that the machine gun slaughter of World War I was successfully countered.

Maxim's total output included 271 patents. He tried without success to create a **steam engine** light enough to power a biplane, but recognized that he needed an **internal combustion engine** to achieve success. He died in New York on November 24,1916.

Maybach, Wilhelm (1846-1929)

German engineer

The German inventor Wilhelm Maybach was born on February 9, 1846, in Heilbronn, Württemberg, Germany. From his teenage years he was a great friend and associate of another renowned German inventor, **Gottlieb Daimler**. About 1882, Maybach and Daimler formed a partnership to develop the **internal combustion engine**. In 1885, they patented an efficient, light four-stroke engine, which was to be one of their most important engineering feats. That same year, they mounted the engine first on a **bicycle** to produce perhaps the world's first **motorcycle**, and about a year later they used the engine to propel a modified horse carriage.

At the 1889 Paris Exhibition, Daimler and Maybach exhibited a two-cylinder V-shaped engine, perhaps the first engine to use the ''V'' design. In 1890, they together formed the Daimler Motor Company, setting their factory in Cannstatt to manufacture automobiles. It was as the company's technical director that Maybach invented the float-feed **carburetor** (circa 1893), which made it possible to use **gasoline** to power internal combustion engines. (Previous to his invention, these engines were fueled by ''illuminating'' gas, which was a vapor, not a liquid.) Maybach's carburetor used screws to vary the amount of gasoline that was sent to the carburetor in a fine spray and mixed with air to form a combustible mixture. The gasoline carburetor was a revolutionary invention in the early development of the internal combustion engine. It in large part made the automobile a practical choice for transportation.

Maybach designed the world's first Mercedes automobile, which was named after Mercedes Jellinec, the daughter of an influential associate of the Daimler firm. The automobile was first run in 1901. Maybach was also responsible for the development of the internal expanding brake, and, with the possible collaboration of Daimler's son, Paul, the invention of the efficient honeycomb radiator.

In 1907, Maybach left the Daimler firm to establish his own factory, which supplied engines for Count **Ferdinand von Zeppelin**'s **airships**. After a long and inventive engineering life, Wilhelm Maybach died on December 29, 1929.

MAYNARD, EDWARD (1813-1891)

American inventor

Born in 1813 in Madison, New York, Maynard became a renowned dentist in Washington, D.C. While he is known for developing surgical procedures and inventing many dental instruments still in use today, Maynard is best known for his contributions to the firearms and munitions industries. In 1845, he patented a percussion priming system (a priming system ignites a charge to propel a bullet or ball from a **rifle**), which was later used to convert flintlock firearms. Maynard's invention consisted of a coiled, tape-like paper strip containing fifty fulminate caps spaced at regular intervals. When the hammer was cocked, one charge was automatically projected over a metal nipple. When the hammer descended, it cut off and fired the charge. The United States government bought Maynard's patent, called the Maynard tape primer, and equipped its military with new percussion priming pistols and rifles.

Maynard is also credited with making improvements to existing firearms. In 1851, he invented a percussion breechloading rifle, the Maynard rifle, in which the **ammunition** was loaded at the rear rather than from the muzzle, and improved metallic cartridges, resulting in finer precision and ruggedness. Maynard also came up with a method of converting muzzleloading arms into breechloaders in 1860. Later, he created a device that allowed two rifle or **shotgun** barrels to expand or contract endwise independently of each other to prevent warping of the barrels. In 1866, Maynard patented an invention for indicating at any time the number of cartridges in the magazine of a repeating firearm.

Maynard was considered one of the leading scientists in the world during his lifetime, and was awarded many honors. While the King of Prussia awarded him a top military honor and the King of Sweden gave him a Great Medal of Merit, Tsar Nicholas I of Russia attempted to persuade Maynard to become his Court Dentist. Maynard died in Washington D.C. on May 4, 1891.

See also Firing mechanism

MCADAM, JOHN LOUDON (1756-1836)

Scottish engineer

With the fall of the Roman Empire in the fourth century A.D., the art of road building suffered a decline for fourteen hundred years. It was the eighteenth century before engineers turned their attention to making roads safe and easy to travel. One of the leaders spearheading the road-building revival was John McAdam. Born in Ayr, Scotland, McAdam went to the American colonies at the age of fourteen, after the death of his father. There he was a salesman of property seized during military battles.

McAdam returned to Scotland in 1783 and remained in government employ for the remainder of his life. In his initial post, deputy lieutenant of Ayrshire and a road trustee, McAdam spent his own money experimenting with road construction methods. In 1789 he became a navy food supplier and in 1803 went to work for the Board of Works in Bristol, England. He was named Bristol's surveyor-general of the Roads Trust in 1815, moving in 1827 into the post of general surveyor of metropolitan roads for Great Britain.

McAdam's philosophy of road building was that the natural roadbed and its subsoil were sufficient as a road foundation. He advocated using crushed stone or granite chips, well-drained via grading and compacting. If three or four layers of broken stones were laid, water—a constant presence in the British Isles—could be absorbed into the road without affecting its load-carrying ability. He discounted a suggestion by Richard Edgeworth (1744-1817) that road surfaces could be rendered even more waterproof by using sand between the stones.

McAdam's technique was much more economical than the foundation-building methods of **Pierre Trésaguet** and **Thomas Telford**, and eventually supplanted them. The invention of the steamroller in 1866 made McAdam's method even more effective. Today, the term *macadam* refers to any road having a simple, compacted bed, considered to be the perfect type of road.

MCCORMICK, CYRUS H. (1809-1884)

American inventor

Cyrus Hall McCormick was born in Rockbridge County, Virginia, the eldest son of Robert and Mary Ann McCormick. Though he received only a limited formal education, he showed a talent for mechanics and soon learned the skills of his father, who had been experimenting with farm machinery since about 1816. The elder McCormick patented a thresher and other farm machines, but none of them were commercially successful, and his attempts to build a mechanical reaper, a device capable of cutting grain, ended in failure. In 1831, Robert abandoned work on the reaper and Cyrus took over the project.

With a new perspective on the project, Cyrus built a machine that departed radically from his father's designs. (It is suspected that McCormick's black mechanic, Joe Anderson, may have played a major role in the reaper's development without receiving recognition for it). Its workings, including

Cyrus Hall McCormick

the gearings, reciprocating knife, projecting teeth, and rotating reel became the basis for modern-day harvesting machines. McCormick gave a public demonstration of his reaper in a field outside John Steele's tavern in July 1831. He further improved the machine and exhibited it in Lexington, Virginia, the following year.

In 1833 **Obed Hussey**, who was to become McCormick's chief rival, publicly demonstrated his own reaper outside of Cincinnati, Ohio. McCormick patented his reaper in 1834 but did little with it at first. While his father's iron works business was floundering, Cyrus attempted to keep the business afloat with the limited manufacture of his machines. After the business failed in 1841, he finally had a chance to become aggressive with his own work. In 1844 McCormick made arrangements for the manufacture of his machines at factories in New York, Ohio, and other states; they sold well enough to provide McCormick with some capital to establish his own factory. McCormick, aware that the rapid opening of the Midwest and Great Plains to settlement would lead to demand for farm machines, and realizing the strategic importance of a midwestern location, opened a factory in Chicago in 1847, where he produced 800 reapers the first year.

McCormick steadily improved his reaper and supplemented his own patents by acquiring the patents of others. These acquisitions gave McCormick the competitive edge over Hussey. Following the Great Exhibition in London in 1851, at which both men displayed their machines, McCormick's business overwhelmed Hussey. In 1858, the year that Hussey sold out, McCormick was producing 4,000 machines annually.

Cyrus McCormick became a prominent public figure. He came to wield a great deal of influence among government officials, Congress, and the United States Patent Office. He left the manufacture and development of his machines to employees while he spent much of his time in court rooms and hearing chambers defending his patent rights and pursuing national and international recognition. Much of the campaign funds of a young lawyer from Illinois, **Abraham Lincoln**, came from his legal representation of McCormick's bitterest rival, John H. Manny.

In 1871 the McCormick factory burned to the ground in the great Chicago fire. With the encouragement of his wife, he not only rebuilt the business but rebuilt much of Chicago as well. For this, several Chicago landmarks are named for him.

The McCormick Harvesting Machine Company merged with Deering Harvester Company and three smaller manufacturers to become the International Harvester Corporation in 1902. In the spirit of McCormick, the company diversified, becoming a world leader in the manufacture of agricultural machines and equipment, light- and heavy-duty trucks, and lawn and garden **tractor**s. A long and bitter labor dispute in the early 1980s dealt a severe financial blow to International Harvester, and the company restructured itself and changed its name to Navistar Corporation.

See also Combine harvester; Reaper and binder

McCoy, Elijah (1843-1929)

American engineer

The prototype ''real McCoy,'' Elijah McCoy patented over fifty inventions, most relating to lubrication of **locomotive**s and earned the name ''father of lubrication.'' Born in Canada in 1843, McCoy was the third of twelve children of George and Mildred Goins McCoy, fugitive slaves who escaped bondage on a Kentucky plantation via the underground railroad. His father was a soldier who sacrificed to give him a better life. In 1859, after Elijah completed the local grammar school, his parents sent him to Edinburgh, Scotland, where he expanded his knack for tinkering into a degree in mechanical engineering. He returned to Ypsilanti, Michigan, yet found no jobs for black engineers, regardless of their expertise. He settled for a lowly position with the Michigan Central Railroad as coal-stoker and oiler.

Whiling away time, McCoy devised a method by which trains could lubricate themselves. At the age of twenty-six, he sold half interest in his patents to finance the McCoy Manufacturing Company in Detroit, where he contrived an automatic lubricator, a metal or glass drip cup which released a trickle of lubricant, an innovation which he patented two years later. The mechanism proved valuable to the industry because it allowed lubrication to take place on slides, **valve**s, **pump**s, **brakes**, **lever**s, and cylinders while an engine was in motion, thereby silencing noisy parts, stopping wasteful friction, and avoiding loss of time and money from idle machinery. Other innovations made lubrication cleaner, less wasteful, and applicable to all weights of lubricants, including graphite. These lubricating dispensers, widely used on ocean liners, steamships, locomotives, and factory engines, came to be called the ''real McCoy,'' a term that means worthy or genuine.

Not satisfied with his first creation, McCoy used the proceeds from his first patent to develop a sophisticated self-lubricator containing a stopcock, which times the application of oil to metal parts. Later modifications netted him over fifty patents for the same concept and for steam and air brakes for locomotives. He also branched out with the invention of the lawn sprinkler, vehicle tire, buggy top support, scaffold support, tire treads, and ironing table and began lecturing and consulting for the Michigan Central Railroad and other mechanical engineering firms.

Because of his race, McCoy earned little from an industry that required extensive capital. He was often the victim of ostracism particularly when the party inviting him to speak, conduct seminars, or consult discovered his color and rejected him at the door. Others cancelled orders for lubricating devices made and distributed by a black man. Focusing much of his time on family and volunteer work with troubled Detroit youth, McCoy found contentment in setting a worthy example of hard work and high ideals. An impoverished widower, his health failed in 1926, and he lived for a time in a convalescent home in Eloise, Michigan. McCoy died on October 10, 1929, of hypertension and resultant senile dementia and was buried in Detroit.

McGaffey, Ives W. • See Vacuum cleaner

McMillan, Edwin Mattison (1907-1991)

American physicist

McMillan was born in Redondo Beach, California, on September 18, 1907. He earned his bachelor's degree (1928) and master's degree (1929) at the California Institute of Technology and his doctorate in physics at Princeton in 1932. After a two-year stint with **Ernest Orlando Lawrence** at the University of California Radiation Laboratory in Berkeley, he joined the faculty at Berkeley in 1935. He became full professor of physics in 1946, associate director of the Radiation Laboratory in 1954, and director in 1958. He remained at the post until his retirement in 1973.

During World War II, McMillan played an important role in the development of microwave **radar**, **sonar**, and the atomic bomb. McMillan is best known for two accomplishments, one in the field of chemistry and one in the field of physics. In 1940, while working with Philip Abelson, McMillan discovered the first transuranium element, neptunium. For this discovery, he shared the 1951 Nobel Prize in chemistry with Glenn Seaborg. McMillan's second important achievement involved the improvement of **particle accelerators**. Lawrence's invention of the **cyclotron** in 1932 provided physicists with a powerful new tool for the study of matter.

Edwin Mattison McMillan

For a decade, larger and larger cyclotrons—''atom-smashers''—were designed, built, and utilized to probe ever deeper into the atomic nucleus. Hopes for continued progress were dampened, however, as a fundamental limitation of cyclotrons became more obvious. As particles are accelerated, they become more massive. At relatively low energies, this ''relativistic mass increase'' is not very important. In larger cyclotrons, however, with the ability to generate very high energy particles, the problem can become severe. As particles gain mass, they tend to slow down and fall out of step with the AC electrical fields that accelerate them. They become lost within the machine. A solution to this problem was devised independently in the mid-1940s by **V. I. Veksler** in the Soviet Union and Edwin McMillan in the United States. Veksler and McMillan realized that the electrical field or the magnetic field—or both—could be modified to stay in step with the slow-downed particles in a machine. For example, if the electrical field is made to change direction a bit more slowly during each revolution of the particles, it can be made to stay in phase with those particles. This theory of phase stability was soon used in the design of larger, more powerful accelerators, for which McMillan suggested the name synchrotron. For their invention of this modification of the cyclotron, Veksler and McMillan shared the 1963 Atoms for Peace Award.

Gerhardus Mercator

MICROELECTROMECHANICAL SYSTEMS

Microelectromechanical systems, or MEMS, are microscopic machines—devices only fractions of an inch in size, which can perform useful functions.

MEMS are an outgrowth of the technology used to create the integrated chip (IC). Integrated chips are small electrical circuits on tiny pieces of **silicon**. Their size and speed has revolutionized the design of electronic circuitry, which has spawned the electronic and computer revolutions. To understand MEMS it is necessary to understand the way in which ICs are constructed.

Electrical circuits—pathways of a metallic conductor—are usually carved into a piece of silicon by a technique called *photolithography.* A circuit design is first laid out, usually using computer-assisted design (CAD) software. This pattern is then duplicated on a ''mask,'' a stencil-like pattern of **chromium** on a glass plate. The piece of silicon is painted with a ''photoresist''—a **polymer** sensitive to ultraviolet light. The mask is placed in front of the painted silicon, and an ultraviolet (UV) laser is shined on it, through the glass and onto the photoresist. The UV light weakens the polymer, and it falls away when the photoresist is washed with certain chemicals. Left on top of the silicon is therefore a copy of the circuit pattern. It is possible to carve a channel in the silicon by washing this in a chemical which dissolves silicon but not the photoresist. The photoresist can then be removed by washing the chip in another chemical, one which dissolves the photoresist but not the silicon. The result is the circuit pattern carved into the silicon.

In the case of ICs, a metallic conductor is then placed in the carved channels. After the technique had been mastered for ICs, researchers began to use similar techniques to carve the silicon in other ways, creating tiny **gear**s, motors, speakers, sensors, and a host of other components. This technology provides a powerful tool for mass production and miniaturization of mechanical systems, below the size that can be reached by standard machining operations.

One of the first important uses of MEMS has been in automobile airbags, where small acceleration sensors detect large deaccelerations, assume it is a crash, and send a signal that an airbag should be inflated. Due to their batch process manufacturing, large volumes of such highly uniform MEMS devices can be created, each at a relatively low cost.

There is expected to be much greater uses of MEMS in the near future. Combined with microelectronics, MEMS components could result in a **cellular telephone** so small and light it could be worn on the wrist. Rectangular arrays of tiny mirrors, each moveable, can be coordinated to reflect light and sharpen images, much like in the field of adaptive **optics**. Thousands of MEMS sensors and actuators (devices for moving or controlling something indirectly instead of by hand) could be wrapped around the front edge of a wing on an **aircraft**, modifying the airflow and hence, the wing's lift. Model planes have shown that such a plane would be extraordinarily maneuverable. Small **microphone**s and speakers with diaphragms about 1 mm (0.04 in) across have been built, possibly to be used as **ink** controllers on ink-jet printers, or for mixing extremely small amounts of fluids for biomedical tests. MEMS sensors could be utilized as chemical sensors, and small MEMS motors could have many uses, such as in switches. In general, MEMS should enhance the capability of ''smart'' systems, allowing great advances in the gathering, processing, and control of information and the actions that are initiated from it.

The value of the worldwide MEMS market in 1995 was \$2.7 billion, with automotive applications comprising 41% or \$1.1 billion. Projections for the year 2000 estimate the worldwide MEMS market at \$12.2 billion, with the automotive sector comprising 24% or \$2.9 billion.

See also Integrated Circuit; Lithography

MERCALLI SCALE • See Earthquake measurement scale

MERCATOR, GERADRUS (1512-1594)

Flemish cartographer

One of the world's foremost figures in cartography, Gerardus Mercator developed a new projection for the **map** of the world in 1569 that still bears his name. Before Mercator's time, world maps were basically useless to navigators plotting voyages of discovery and trade. The maps may have shown landmasses correctly, but generally they did not show proportional distance and direction so navigators could not plot a compass course. In his map, Mercator drew straight, equidistant longitude lines, perpendicular to latitude lines, forming a grid which could be used to accurately determine sea routes. Mercator created and published numerous other maps, many of which were posthumously published by his son as *Atlas' or Cosmographic Meditations on the Structure of the World.* This marked the first use of the world atlas in connection with a book of maps. Mercator also introduced the use of italics to the text of maps.

Mercator was born Gerhard Kremer on March 5, 1512, in Rupelmonde, Flanders, and changed his name when he became a student at the University of Louvain in 1530. Though Mercator studied philosophy and theology, he also developed an interest in astronomy, mathematics, geography, art, and engraving. He studied the first two subjects under Gemmy Phrysius, a cartographer and mathematician. While Mercator lived in Louvain, from 1530 until 1552, Mercator made scientific instruments and worked as a surveyor, while makings his first maps and globes. His earliest globe was finished about 1536, and he published his first map in 1537. Its subject was Palestine. In 1844, Mercator was imprisoned for several months in Louvain for heresy, though he was set free due to lack of evidence. In 1552, Mercator moved to Duisberg in what is now Germany, where he was employed by the Duke of Cleves. Mercator did his most significant work under Cleves's patronage in Duisberg.

In 1569, Mercator designed his Great World Map to facilitate sea travel, inspired by his contact with sea captains and navigators. His grid based of equidistant meridians (longitude lines) and parallels (latitude lines) drawn perpendicularly is known as a graticule. Mercator's graticule allowed constant compass bearing to be plotted as a straight line. While Mercator's map was useful for navigators because it preserved constant **compass** directions, it had drawbacks. Landmasses were not depicted in their true area and proportions, except at the equator. The further from the equator the landmass is, the bigger it looks on Mercator's map. Hence, Greenland looks much larger than the continent of South America, though it is really half its size. The North and South Poles cannot be projected at all. But the relationships between these landmasses are correctly depicted. Mercator's innovation did not become widespread until 1599, when Edward Wright published corrective tables for navigator's use.

Mercator also designed, engraved, and published many maps of Europe and its different parts. In 1554, for example, he published an accurate, detailed map of Europe. Eventually, 107 of these maps appeared in the atlas published by his son in 1595. Mercator also built globes on commission, including a one made of crystal and wood for the Holy Roman Emperor, Charles V. Mercator died in Duisberg on December 2, 1594.

Mercer, John (1791-1866)

English chemist

During the mid-to-late 1800s there were a number of important advances in the **textiles** industry: the inventions of the **flying shuttle**, the **spinning jenny**, the **water frame** spinner, and the mechanical **loom** all helped to fuel the Industrial Revolution. However, none of these inventions would have had as great an impact as they did without similar advances in the textiles themselves—for example, the development of new materials and methods for processing these materials. One of the most important contributors to this field was John Mercer, the father of textile chemistry.

John Mercer

Mercer grew up in Lancashire, England—an area that would soon become the hub of the English textile industry. He first entered that industry as a boy, working as a bobbin-winder and, later, as a weaver. At the age of sixteen he became drawn to the art of dyeing. He set up a small **dye** laboratory in the Mercer home, and there experimented with new mixtures and colors. He became quite skilled at the manufacture of dyes, and that year he entered into partnership with an investor to open a dyeing shop. Their business, though small, was quite successful, and Mercer was only drawn away by an offer to become an apprentice at a print shop in nearby Oakenshaw. His time there was, unfortunately, somewhat wasted: he was prevented by a spiteful foreman from gaining any real experience, and after a year he was relieved of his apprenticeship.

Mercer spent several years as a simple weaver before once again becoming a dyer. His return to the profession was sparked by an interest in chemistry. In his home laboratory he experimented with a number of chemicals, eventually producing a new orange dye that (unlike previous dyes) was ideal for calico-printing. In 1818 he was once again employed by the

Fort brothers (who had owned the Oakenshaw print shop) as a color chemist; there he invented a number of dyes of yellow, orange, and indigo. He was made a partner in 1825.

While his employ with the Fort brothers was profitable, it took away a great deal of Mercer's free time—time that he had previously spent in his laboratory developing new chemicals for textile processing. The partnership was dissolved in 1848 and, at the age of fifty-seven, Mercer finally had both the time and financial resources to pursue the research that he had put off.

His first experiment turned out to be his most important. For years he had wondered about the effect upon **cotton fabric** of certain sodas, acids, and chlorides. He soon found that, when treated with these caustic chemicals, the material would become thicker and shorter; this made the cotton stronger, shrink-resistant, and more easily dyed. It also imparted to the material a lustrous sheen that became highly valued by textile manufacturers. Mercer called his chemical process mercerization and patented it in 1850.

Mercer himself was most interested in how mercerization aided the dyeing process. When chemically treated, the cotton fibers would swell, becoming more absorbent; mercerized fabrics require about thirty percent less dye than untreated fabrics. It soon became apparent, however, that mercerization could be applied to many other materials, including parchment and **woolen fabric**. Today, mercerization is still an important part of the cotton finishing process.

Mercer was made a Fellow of the Royal Society. He died in 1866 of a prolonged illness, brought about by falling into a reservoir of cold water.

MERCERIZED FABRIC • See Mercer, John

MERCURY LAMP • See Arc lamp

MERGENTHALER, OTTMAR (1854-1899)

German American inventor

Mergenthaler invented and perfected **linotype**, the first automated typesetting machine, which proved to be an invention largely responsible for the publishing boom in the first half of the twentieth century.

Born in Hachtel, Germany, on May 11, 1854, Mergenthaler was apprenticed to a watchmaker at the age of fourteen, but his true interest was in engineering. After emigrating to the United States in 1872 to avoid military conscription, Mergenthaler went to work in a relative's machine shop in Baltimore, Maryland; there, he became interested in **printing technology** after being approached to build a model of an automatic typesetting machine for another designer. The model failed, but the concept intrigued Mergenthaler.

There had been no real advances in the typesetting process since **Johannes Gutenberg**'s introduction of movable type; typesetters still selected cast-metal letters and punctuation from a prepared stock to piece together full pages by hand. Although earlier inventors had conceived machines that would piece together type into pages, their inventions were not efficient enough to be marketable.

Mergenthaler abandoned the basic concept behind these attempts. What typesetters needed, he reasoned, was not a machine that would take over the slow, jigsaw puzzle-like process; they needed a new process altogether. Mergenthaler conceived of a machine which would custom-make a mold of letters that could be inked and pressed onto paper. He experimented with various unsuccessful models, including a machine that would punch letters into **papier-mâché** to create a mold; then he hit upon the technique that would become linotype. He designed a machine with a keyboard that moved precast copper letter and punctuation molds into a line, over which it poured a quick-cooling alloy that became a line of finished type. Mergenthaler patented linotype in 1884; soon after, he modified it to automatically justify text, or insert spaces between words so each line is exactly the same length.

In 1886, *The New York Tribune* became the first newspaper to use linotype. Mergenthaler set up a company to market linotype but resigned in 1888 over disputes with his partners. Nonetheless, he continued to develop and patent refinements to his machine for the rest of his life. By his death on October 28, 1899, in Baltimore, more than three thousand linotype machines were in use around the world.

METAL-WORKING PROCESSES

Historians can only speculate on how metals were first discovered and worked. Since most metals must be softened by heat in order to be extracted from their ores, campfires, and hearths may have been where metals were initially observed and subjected to experimentation. Some observant individual might have noticed that rocks near the heat softened and rehardened. Eventually, someone learned to manipulate the metal ore and searched for ways to obtain more of the new material. The same act of discovery of any single metal may have been repeated in different parts of the world.

Simultaneous discoveries in metallurgy occurred in more recent times with such inventors as **William Kelly** and **Sir Henry Bessemer**, the Siemens brothers and **Pierre-Emile Martin**, and **Charles Martin Hall** and **Paul-Louis-Toussaint Héroult**. It seems reasonable to suppose similar multiple discoveries occurred even more often in prehistory or early historical periods, when communication of technology was much more limited than it has been in the last two or three centuries.

Early metalworkers had to extract, or reduce, metal from rock by repeated heating and hammering with stone or wooden mallets. Although it was laborious, people realized that metal was more useful than stone or wood for many purposes, and metallurgy became an important part of human culture. However, it took thousands of years for metals to become the everyday substances they are today. Not until the nineteenth century were metal implements mass-produced cheaply. Before that, metal was costly and used only for important tools or for items associated with wealth and prestige. The metal nails, kettles, and other implements that commoners owned were expensive items to be guarded closely and used carefully. Repeated hammering not only shaped the metal but allowed impurities to be worked out and burned off. This process, called forging, takes metal from its natural state and turns it into a useable commodity.

Around 2000 B.C., other methods of working metal, such as molding, inlaying, *soldering*, and enameling appeared. People increasingly used **alloys** of metals as well. Bronze, a mixture of copper and tin, was the first metal alloy; it made implements stronger and more workable. The period of about 3000 to 2000 B.C. is known as the Bronze Age, so named for the widespread use of bronze during that time. The making of coins probably started in Greece during the seventh century B.C. and spread throughout the Mediterranean. One of the first improvements in metalworking were the bronze dies used to strike, or mint, identical images on coins. By doing this, a ruler could spread his image throughout his domain. The Lydian king Croesus (died 546 B.C.) is supposed to have been the first to mint such standardized coins.

Throughout the first millennium A.D., knowledge of metals gradually increased. Zinc, antimony, and nickel were discovered, and monks established foundries to make bells for monasteries and iron tools for the surrounding region. In Sheffield, England, metalworking was taken to artistic heights with the manufacture of fine cutlery. During the 700s A.D. the productive output of forges increased with the development of the Catalan forge which achieved a hotter sustained heat by blowing air into the furnace. The **blast furnace**, as it is called, was gradually improved over the centuries, but working the metal still required hammering. Another advance in metalworking during the Middle Ages was **wire** drawing, the means whereby metal is elongated into wire by pulling. First developed about 1000 A.D., it was later improved by Rudolf of Nuremberg in 1350.

Although the Industrial Revolution is associated with the **steam engine** and the beginnings of automation, the advances in metalworking which occurred during the seventeenth and eighteenth centuries prepared the way for new machinery, which often required large amounts of strong, well-worked metal to withstand pressure and stress. Mechanical, or drop forging using large trip hammers was regularly practiced by the 1700s. The drop hammer had a moving die that dropped onto a stationary die, delivering a powerful blow to the molten metal while counterblow forging used converging dies. **Abraham Darby**, founder of the ironworking industry, introduced coke to the blast furnace, which mass-produced large quantities of cast iron in the early 1700s.

One of the first practical metal lathes was the screw-cutting machine **Henry Maudslay** (1771-1831) invented in 1800. It automatically cut the threads into the shaft of the screw. Other advancements were also made prior to and during the Industrial Revolution, such as the development of milling machines. These devices are similar to lathes except that the workpiece, the item being milled, is held stationary while the cutter does the moving. In the lathe, the workpiece is turned. After several prototype milling machines were introduced, the universal milling machine developed by American engineer Joseph R. Brown (1810-1876) in 1867 was adaptable to circular cutting, square cutting, boring, grooving, and grinding.

Ottmar Mergenthaler

At the turn of the twentieth century, another American engineer, B. Atha, developed the first continuous casting process for steel, in which the metal was poured into a vertical mold, however, he experienced problems with the metal clinging to the mold during cooling. In 1935 Dr. Siegfied Junghans solved Atha's problem by creating a reciprocating mold that reduced the risk of clinging. The continuous sheet mill first appeared in Teplice, Bohemia, in 1892. It was abandoned, then resumed in America, first by Charles W. Bray in 1902, then by John B. Tyrus in 1923. This final attempt led to the founding of the American Rolling Mill Company (Armco). Shell molding was invented in 1941 at a foundry in Hamburg owned by Johannes Croning (1872-1962). The mold was made of sand held together by phenolic resin. Since this was a wartime invention, it was not used outside Germany until the late 1940s, when Ford Motor Company used it to form automobile exhaust valves.

A more recent development is explosive metal-working, in which metal is shaped by impulse waves transmitted by a chemical explosion. Initial work was done by the National Aeronautics and Space Administration (NASA) in the 1960s for space-technology metals. After funding ran out, the program was picked up on an experimental basis by several international companies. The process is problematic because it has so far proven too dangerous for use in conventional factories and not adaptable to mass production.

Metal-working methods for fusing metals together include **welding** and soldering. Oxyacetylene welding, invented in 1903, can also be used to cut metal. **Electroplating** uses electrolysis to coat base metals with attractive, less corrosive metals, such as chromium, nickel, or silver. Galvanizing protects iron or steel with a protective layer of zinc. Alloys and composites combine two or more metals in order to utilize the unique characteristics of each. Alloys thoroughly combine metals, while composites introduce units of one metal into a matrix of another.

See also Composite material; Forge; Iron production; Siemens, Charles William; Siemens, Ernst Werner von; Solder and soldering iron; Steel production

Metric system

The system of weights and measures in France at the close of the eighteenth century was chaotic. There were different standard units of length in each province and there was a different method of measurement for each trade. An influential French geographer, Charles La Condamine (1701-1774), strongly advocated an international standard system of measurement but died sixteen years before anything was done to follow up on his suggestions.

On May 8, 1790, the French National Assembly finally decreed there would be reform, and the Academy of Sciences established a commission to consider what the new set of units should be. One of the committee members was Pierre-Simon Laplace, who was known as the "Newton of France." Another, Gaspard Monge, was a personal friend of Napoleon.

Committee Chairman Joseph-Louis Lagrange made a convincing argument to the commission that the new system should be based on decimals, suggesting that each unit of measurement be a multiple of ten times another, to permit easy conversion between units. For example, one kilometer would equal in length a thousand meters, and each meter in turn would be composed of one hundred centimeters. Once this was agreed, it became a matter of choosing a "natural" unit of length to use as a starting point on which to base the system. For this, the committee borrowed an idea that had been suggested by Abbé Gabriel Mouton of Lyon in 1670. When the distance from the North Pole to the equator is divided into ten million equal parts, a convenient length (roughly equal to the distance between a man's head and the end of his outstretched arm) is obtained; it was decided that the basic unit of distance would be one of these parts. Jean-Charles de Borda (1733-1799), who was a surveyor, suggested the name *metre* which came from the Greek word *metron* for measure.

The decimal-based metric system had many advantages; foremost was the ease of determining smaller and larger units given a particular quantity. Latin prefixes were added for smaller sizes and Greek prefixes were used for the larger.

The proposal was accepted on March 30, 1791, but a serious stumbling block had to be cleared first: someone had to make geographic measurements to determine exactly how long a meter was. Twelve mathematicians were assigned the project, but the most important job, that of determining the difference in latitude between Barcelona, Spain, and Dunkerque, France, was given to Jean Baptiste Delambre (1749-1822) and Pierre Méchain. This distance was great enough to establish an accurate measurement for the length of the meter. Using instruments that Borda had designed, they spent the next seven years making painstaking measurements and dodging the French Revolution. At one point they were arrested as sus-

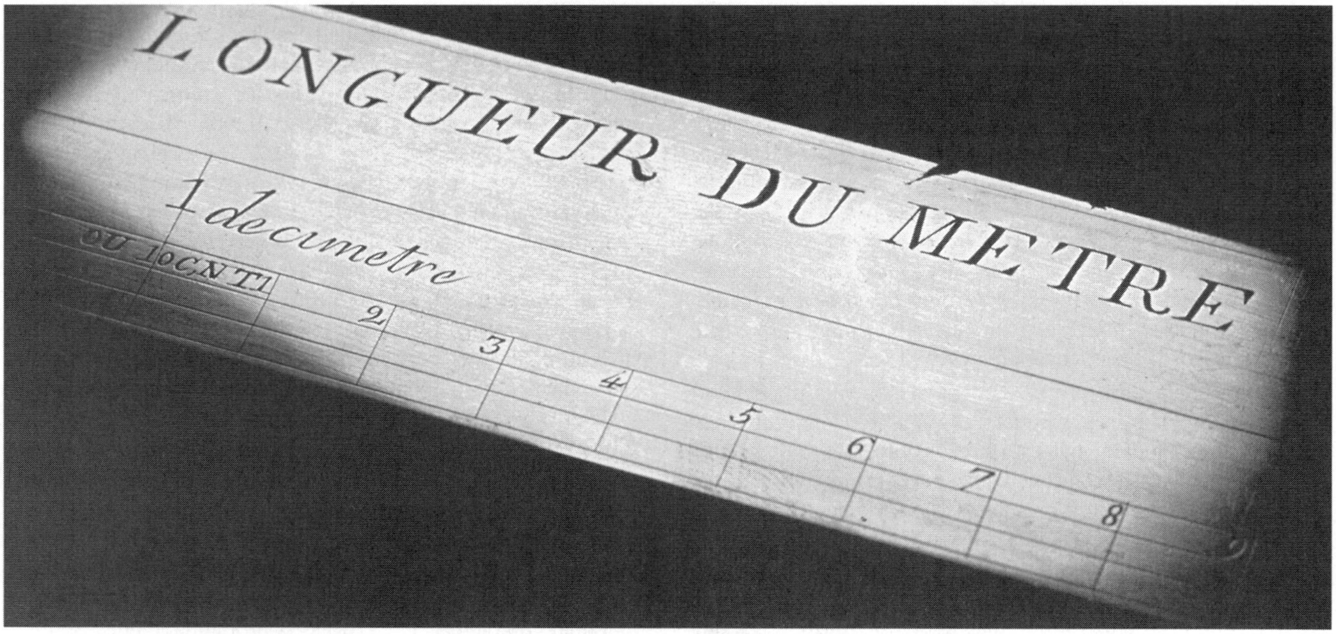

The first provisional standard scale of a meter. Made of copper in 1793, the scale is housed in the Archives Nationales de France, in Paris.

pected anti-revolutionaries because, as they carried their peculiar equipment around the countryside, they looked suspiciously like spies. Méchain died in 1804 and Delambre completed the report, which was presented to Napoleon, in 1810.

Meanwhile, mathematicians had determined other basic measurements, and the metric system was officially adopted by the French in 1799. Besides length, primary units were defined for area, volume, capacity and weight. For measurement of capacity (especially for liquids), it was agreed that a hollow cube ten centimeters square would be designated as one *liter* of water. For weight, the *gramme*, the mass of a cubic centimeter of water, would serve as the basic unit.

In 1827, French physicist Jacques Babinet (1794-1872) proposed that the standard for the meter be based upon the wavelength of light rather than a crude fraction of distance over the earth. This way any scientist could standardize measurements in a laboratory without resorting to unwieldy surveying techniques. (The idea was finally adopted when technology caught up with Babinet's vision over one hundred years later—in 1960, one meter was officially defined as 1,650,763.73 wavelengths of orange-red light from krypton-86 gas.)

Some countries were slow to adopt the metric system. Attempts to implement the use of metric system in Great Britain failed in 1819, largely through the efforts of William Hyde Wollaston, head of a royal commission looking into the matter. The British did not adopt the metric system until well into the twentieth century.

The United States has still not done so, even though Secretary of State **Thomas Jefferson** urged the United States Congress to adopt it as early as 1790. Ironically, the United States Bureau of Standards adopted the metric system in 1893, for the purpose of defining the length of the yard and pound. Legislation to implement a ten year phase-in of the metric system was considered in 1972 but was shelved because of resistance from industries and the general population.

Metronome

Usually housed in a hollow, pyramid-shaped box, metronomes visually and aurally register the tempo (speed) of music. The standard metronome consists of a pivoted pendulum with a fixed weight below and a sliding weight above. The beat to be indicated may be altered by moving the upper weight; the device ticks with the movement of the pendulum, faster or slower as the sliding weight is moved lower or higher. A number scale shows how many oscillations per minute will occur when the sliding weight is moved to various points on the pendulum.

Early "metronomes" include the seventeenth-century *chronometre*, which varied to produce up to seventy-two separate speeds; a pendulum device invented in 1756 by Robert Bremner to standardize the tempo of church music and a metronome-like instrument using a bell and clock hammer that was devised in Germany.

Although originally attributed to Austrian physician Johann Nepomuk Maelzel (1772-1838), today's metronome was actually developed by Dietrich Nikolaus Winkel (1776-1826), a German-born master organ builder living in Amsterdam. Musically and mechanically proficient, Winkel built the first compact, accurate metronome incorporating a pendulum. His device was a one-square-foot, hinged box that contained an adjustable, double-weighted pendulum kept moving by an escapement (a notched wheel that controls movement) that was itself driven by a small weight on the end of a cord wrapped around a drum.

Winkel showed his invention to Maelzel, who in turn demonstrated it to his friend Ludwig van Beethoven (1770-1827). Maelzel knew that the great composer was interested in increasing the accuracy with which his music was played. Indeed, Beethoven became the first composer to include metronome markings in his scores; the beat of one of the movements in his *Eighth Symphony* was supposedly inspired by a metronome. Maelzel, who was well-known for disreputable practices, commandeered the metronome as his own invention, acquiring a patent for it in 1816 and establishing a factory for its manufacture.

A pocket-sized metronome, Pinfold's Patent Metronome, which had a pendulum that could be reeled up like the modern tape measure, became popular during the nineteenth century. The advent of electronic technology in the twentieth century allowed for the development of extremely small, highly accurate metronomes that are capable of beating complex musical time signatures and that do not have to stand on level surfaces.

See also Balance and scale; Clock and watch; Foucault, Jean Bernard-Léon

Michell, John • See Torsion balance

Michelson, Albert • See Interferometer

Microcomputer

Microcomputers are programmable electronic information processing machines whose controlling circuits are combined into one solid state (entirely nonmechanical) component called a *central processing unit* (CPU). Microcomputers can perform complex arithmetic and logic operations faster and more reliably than is possible manually. Some machines fitting this description carry other names, such as programmable calculators and microcontrollers, but the most familiar microcomputers are the general purpose personal computers, which include desktop, laptop, notebook, and handheld models. In addition, specialized microcomputers are used in automobile ignitions, **telephone**s, **microwave oven**s, video recorders and games, robots, public safety vehicles (fire, police, and emergency medical), **cash register**s, warehouses, vehicle fleet management, heating and cooling systems, **elevator**s, weapons systems, and many other devices. Growing microcomputer use has produced new occupations and increased the educational requirements for many already existing jobs.

Like other general-purpose computers (**mainframe computers**, **minicomputers**, and **supercomputers**) most microcomputers handle information in digital format—that is, as numerical digits which can be manipulated according to rules of arithmetic and logic. Since most real-world information is in non-digital format (called linear, continuous, or analog format) digital computers must convert information from analog to digital form. For example, the keyboard converts numbers and letters to numerical codes, and most pictures and graphics images are broken down, or digitized, into millions of separate dots called pixels, all of whose characteristics can be described with numbers. Since the simplest method of representing digits electronically is with the on/off control of electrical current, digital computers use a binary (base two) number system consisting of only the digits "1" for "on" and "0" for "off." Most computers therefore are essentially assemblies of millions of switches wired together, each switch controlling the status of a single digit.

The types and speed of a microcomputer's operations are determined largely by its CPU, the random access memory available to the CPU, the programmed instructions sent to the CPU (software), and the accessories coordinated by the CPU (peripherals). Some software programs control internal system operations (operating system and utilities programs), and others apply the computer's power to specialized problems (application programs). Each type is simply a stored command list the computer executes one-by-one, a method known as sequential or serial processing. A few advanced microcomputers, some minicomputers and most supercomputers can execute many commands simultaneously. Most microcomputers are limited to serial processing because they are typically controlled by only a single CPU. For this reason, microcomputer speed is determined mainly by the processing speed of its CPU, and by the speed with which the peripherals serve the needs of the CPU.

Today's microcomputer peripherals include: *input devices* —commonly keyboards, numerical keypads, and mice, and sometimes optical scanners, musical keyboards, barcode readers, light pens, **microphones**, remote sensors, and video cameras; *output devices*—commonly printers and video monitors, and sometimes plotters, speakers, and switching devices; *storage devices*—such as hard and floppy magnetic disk drives, magnetic tape drives, optical disk drives, and solid state devices such as read only memory (ROM) chips; and *communications devices*—such as modems for telephone communication, networks for linking several computers, and even **radios** for wireless communication.

Microcomputers were made possible by a series of increasingly miniature electronic inventions using naturally occurring elements such as silicon and germanium. They were found able to function electrically as either conductors or insulators, depending on the current flowing to them. Called semiconductors, their variability enabled them to serve as the first non-mechanical, or solid state, switches. Moreover, semiconductors function regardless of their size, at microscopic—even molecular—dimensions. These and other semiconductor properties were the basis of several landmark inventions leading to the microcomputer, including:

1947: Transistors—current flow controllers much smaller and simpler than the **vacuum tubes** they replaced;

1959: **Integrated circuit** (IC) chips—miniature circuits with many transistors, **capacitors** and resistors, located on one semiconductor chip, mass produced rather than individually wired by hand;

1969: Random access memory (RAM) IC chips—specialized ICs for storing hundreds of thousands of information bits, much smaller than the magnetic core computer memories they replaced;

1971: **Microprocessor** IC chips—the "computer on a chip" containing, for the first time in history, all of the circuits needed in the central processing unit of a computer (logic, arithmetic and control circuits). When combined with other IC chips containing data and program memory, the first microprocessors enabled the almost instantaneous creation of the first general-purpose microcomputers.

In fact, in 1971 personal microcomputer kits appeared on the market for hobbyists, but established electronics companies, including IBM, bypassed their first opportunity to design microcomputers, apparently unable to see the market potential which lay just ahead. What was needed in the early 1970s, in addition to the new microprocessor technology, was a process to design, build, test, finance, mass-produce, and sell a totally new computing system—a group of new hardware and software products designed to work together to realize the potential of the microprocessor for low-cost personal computing. Today's sophisticated microcomputer systems evolved through the trials and errors of dozens of individuals and their firms.

The first widely available personal computer was the 1974 Altair from MITS (Micro Instrumentation and Telemetry Systems), for which college students **William Gates** and Paul Allen developed Microsoft BASIC, the first microcomputer operating system. The Altair was programmed with manual switches and perforated paper tape.

In 1977 many new microcomputers were developed, the most successful of which came from two California college students, **Steve Jobs** and Stephen Wozniak, who created Apple Computer Corporation to produce the Apple II computer.

Other 1977 notables were the TRS-80 Computer from Tandy/Radio Shack, the TI 99/4 from Texas Instruments, and the PET (personal electronic transactor) Computer from Commodore Computers. Commodore later produced the VIC-20 Computer (1981) and Commodore 64.

The Apple II, however, dominated the market until the IBM Personal Computer (IBM PC) appeared in 1981. Also notable in 1981 were the first portable computer (the Osborne 1) and the first computer assembled entirely in the United States which was priced under one hundred dollars (the Timex/Sinclair 1000).

While the Apple II Computer's success proved the feasibility of using microprocessor ICs for widespread, low-cost personal computing, the Apple II was not widely purchased by businesses, especially large corporations who were a potential market for literally millions of microcomputers. In July 1980 IBM's management saw this opportunity and decided to enter the microcomputer competition. With a one hundred fifty-member team based in Boca Raton, Florida, IBM designed, built and delivered the first IBM PCs thirteen months later. Within twelve more months IBM held twenty-eight percent of the personal computer market.

Thus began a rivalry between IBM and Apple Computer that continued for nearly ten years. Apple's strength was among home and school computer users. It also took the lead among graphics and publishing users with the unique, closely-guarded graphics technology of the Apple Macintosh Computer, introduced in 1984. The IBM PC (and, by 1985, hundreds of lower-priced imitations known as PC Clones), dominated the business microcomputer market.

In 1991, during a slump in most personal computer sales, IBM and Apple Computer shocked the industry when they announced an agreement to share technology and new product development. Their goals are to bridge the wide gap between IBM and Apple technologies, hoping to increase the competitiveness and sales of each company.

Today such semiconductor makers as Intel and Motorola produce millions of microprocessor ICs of all types. Each year the products have become more powerful and less expensive. Thousands of companies and individuals worldwide now assemble microcomputers, connecting ICs and other components simply by plugging them into standardized sockets on pre-wired circuit boards.

New ICs, software and peripherals have appeared frequently since 1971, permanently changing the computer business by making high-powered computers available to the average person. There is no foreseeable end to the process of invention in microcomputers. The most successful microcomputer companies have been those which were inventive at all levels—technology, product design, pricing, and marketing.

See also Computer, analog; Computer, digital; Computer disk and tape; Computer input and output devices; Computer memory; Computer network; Resistor, electric; Robotics; Video game; Video recording

Micrometer

The industrial micrometer traces its origin to an astronomical micrometer invented by William Gascoigne (1612-1644) in 1639. Gascoigne's invention was designed specifically for use with **telescope**s and consisted of a screw-adjusting device which enabled the approximate measurement of celestial bodies. Greek theoretician **Archimedes** may have been the first to propose such a device in theory.

Following Gascoigne's lead, Scottish industrialist **James Watt** developed a pocket micrometer for his personal use in 1772. The true pioneer in micrometer design, however, was **Henry Maudslay**, also known for perfecting the screw-cutting **lathe**. Around 1805 Maudslay successfully tested a bench micrometer, which he named the Lord Chancellor. The Lord Chancellor featured a gunmetal bed, one movable and one stationary anvil, an adjusting screw with 100 threads per inch, and a graduated scale. It boasted precise measurements to within 0.0001 inch.

Since Maudslay's time, micrometers have been developed to increasingly exacting standards. In 1835 British engineer **Joseph Whitworth** developed a comparator to ensure uniformity of yard lengths; the device was accurate to 0.000001 inch. One of the most recent developments in micrometers—which have come to mean all devices capable of precise measurement in small units—is a handheld micrometer that measures electronically, through diffraction grating. This device was first introduced in 1973.

Microphone

The microphone is the foundation for all voice communications, whether in the transmission or recording of sound. The device transforms acoustical energy or sound into electrical energy (a process called transduction) so it can be reproduced through a loudspeaker or recorded for later playback on a magnetic tape. In 1876, **Alexander Graham Bell** constructed what was probably the first crude microphone as a component for his **telephone**. It worked by using the vibrations produced by a sound wave to move a disk of metal, inducing a current in an adjoining magnet. The weak electrical current produced by the magnet was sent a distance over wires, where it generated pulses in a similar magnet, moving another disk of metal to reproduce the sound. But the current induced by Bell's device was too weak to produce anything more than a feeble response.

Both **Emile Berliner** and **Thomas Alva Edison** set about improving the transmission of sound almost immediately. First, Edison began to use a separate earpiece and mouthpiece (Bell had used a single microphone both to receive and transmit sound). Furthermore, Edison and Berliner simultaneously devised a more sensitive means of picking up the sound waves in 1877. They delivered the sound vibrations onto a flexible metal membrane called a diaphragm, which pressed upon a disk of carbon granules. These granules respond to pressure with changes in electrical resistance. When a current passes through the carbon, the current modulates with the changes in the disk's resistance, producing a strong and smooth analog (likeness with a similar wave pattern) of the sound that could be sent for long distances over wires. It was this device, patented in 1891 by Berliner, that became known as the first "microphone." The carbon button microphone is still used in telephones and other inexpensive devices.

Also in 1877, Charles Cuttris of the United States and **Ernst Werner Siemens** of Germany produced a moving coil microphone, using a coil attached to a diaphragm which produces an electrical current when it vibrates in the presence of a magnet. A host of other microphones has since been devised for a variety of specialized uses. In 1992, for example, the Norris Communications company introduced the Norris Ear Phone Microphone, which featured both a microphone and earphone small enough to fit in the user's ear. The microphone picks up sounds through the user's body, through his or her head or skull. It can be connected to a computer as an input device. Around the same time, the dynamic spectral phonocardiograph, or DSP, was introduced. This is an ultrasensitive microphone that can detect the sounds in the heart. Invented by

William Bennett, a professor at Yale University, and his daughter Jean, the DSP picks up the sounds and converts them into a digital code, which can be mathematically processed to display the amplitude of the sound's tone for analysis. Thus, a doctor can see what the sounds of the heart look like, allowing for an accurate diagnosis of any irregularities.

Microphones can be used in many ways today: in public address systems, CB ham, and broadcast **radio**, tape recorders, hearing aids, computers, and recording studios. In the future, all-digital microphones are possible.

See also Amplifier; Compact disk player; Intercom; Magnetic recording; Optical disk; Phonograph

Microprocessor

In 1969, after the integrated circuit had become widely available, a Japanese calculator company, Busicom, approached **Robert Noyce**'s newly formed Intel Corporation. Busicom wanted Intel to produce a set of 12 interconnected integrated circuits for a new line of desktop electronic calculators. Noyce gave the project to Marcian ''Ted" Hoff Jr., an Intel research and development engineer.

Hoff found the Busicom plan far too complex, linking it to a growing problem in the microchip industry: A different set of logic chips was needed for each individual product, making design and production of items that used integrated circuits slow, difficult, and expensive. Hoff thought of a way to solve this problem. He designed a general-purpose logic chip, one that incorporated all the logic circuitry of a computer's central processing unit (CPU). Called a microprocessor, this universal CPU chip could be combined with a separate chip for a program written for a specific device. Along with two other chips, for memory and data, the microprocessor plus program chip was a complete general-purpose computer.

After Busicom contracted with Intel to produce the microprocessor, Federico Faggin produced the detailed chip layout and circuit design. In exchange for a price reduction, Busicom gave Intel the right to market the chip. Intel put this first microprocessor, the 4004, on sale late in 1971, calling it a ''computer on a chip.'' About the size of a pencil point, the 4004 was more powerful than the original room-sized electronic computer of 1948, ENIAC. The 8008, with double the 4004's capacity, followed the next year, superseded in 1974 by the much more powerful and efficient 8080. Other manufacturers soon came out with their own microprocessors.

Sales of the microprocessor were slow at first, until engineers realized its vast potential. It made the microcomputer possible, taking the computer out of huge government and university centers and spreading it to legions of individual users across the country. Microprocessors became pervasive elements of everyday life, controlling the operations of such familiar items as cars, microwave ovens, washing machines and **clothes dryers**, **thermostats**, wristwatches, **cash registers**, **telephones**, and gas pumps.

Microscope, compound

Microscopes have been in use in various forms for more than 3,000 years. The first types were extremely simple magnifiers made of globes of water-filled **glass** or chips of transparent crystal. Ancient Romans were known to use solid, bead-like glass magnifiers; Emperor Nero (A.D. 37-68) often used a bit of cut emerald to augment his poor vision.

The first **lenses**, which were used in primitive eyeglasses, were manufactured in Europe and China in the late thirteenth century. By this time, lenscrafters realized that most clear glass or crystal could be ground into a certain shape (generally with the edges thinner than the center) to produce a magnifying effect. All of these single-lens magnifiers are called *simple microscopes.*

Until the turn of the seventeenth century, most simple microscopes could provide a magnification of 10 power (magnifying a specimen to ten times its diameter). About this time, the Dutch draper and amateur optician Antoni van Leeuwenhoek (1632-1723) began constructing magnifying lenses of his own. Though still relying upon single lenses, Leeuwenhoek's unparalleled grinding skill produced microscopes of very high power, with magnifications ranging to 500 power.

In order to achieve such results, Leeuwenhoek manufactured extremely small lenses, some as tiny as the head of a pin. Because of the very short *focal length* of these lenses, the microscope had to be held a fraction of an inch away from both the observed specimen and the observer's eye. Through his minute lenses Leeuwenhoek observed tiny ''animalcules''—what we now know as bacteria and protozoa—for the first time. His findings earned him international acclaim, and the simple microscopes he designed are still among the best-crafted. However, the limitations of the single-lens magnifier were apparent to scientists, who labored to develop a practical system to increase microscope magnification.

The next breakthrough in microscopy was the invention of the compound microscope; however, the origin of this device-as well as the identity of its inventor—is the subject of some debate. Generally, credit for the invention of the compound microscope has been given to another Dutchman, the optician Zacharias Janssen (1580-c. 1638). Around 1590 Janssen reportedly stumbled upon an idea for a multiple-lens microscope design, which he then constructed. Though he affirmed its ability, no record exists of Janssen actually *using* his invention. It is now believed that Janssen's son fabricated the story. Meanwhile, yet another Dutch-born scientist, **Cornelius Drebbel**, claimed that *he* had constructed the first compound microscope in 1619. **Galileo** also reported using a two-lens microscope to examine and describe the eye of an insect.

Regardless of its inventor, the design of the original compound microscope is very similar to those used today. Two or more lenses are housed in a long tube. Individually, none of the lenses are particularly powerful; however, the image produced by the first lens is further magnified by the second (and the third and fourth, in a multiple-lens system), producing a greatly enlarged image. In addition, multiple lenses allow for a much longer focal length, permitting both the specimen and the eye a greater distance from the lenses.

The first scientist to further improve the compound design was the Englishman **Robert Hooke** in the 1670s. Hooke was the first to use a microscope to observe the structure of plants, consisting of tiny walled "chambers" that he called *cells.* After Hooke, little advancement occurred in microscopy until the work of Carl Zeiss (1816-1888) and **Ernst Abbe** in the mid-1800s. Abbe, generally recognized as the first optical engineer, took over the design duties at the Zeiss Optical Works in 1876. The scientific instruments that resulted from Zeiss and Abbe's collaboration set new standards for optical equipment. Among their inventions were lenses that corrected blurring and color aberrations.

By the twentieth century the essential design and shape of the compound microscope had evolved into those we know today. Microscopes used in schools and small laboratories can achieve magnification of up to 400 power. More advanced microscopes used in research laboratories can magnify a specimen to almost 1,000 power. These research microscopes often have **binocular** eyepieces, relying upon a series of prisms to split the image so that it may be viewed with both eyes. Even *trinocular* microscopes—creating a third image for a camera to view—have been designed.

At its most powerful, the practical limit for any compound microscope is 2,500 power. This limited magnification capability frustrated scientists, who, in the early twentieth century, were anxious to view the world on submicroscopic and subatomic levels. In 1931 Ernst Ruska constructed the **electron microscope** to permit such investigations.

Designed much like a compound microscope, the electron microscope uses a beam of electrons focused through magnetic lenses. Since electrons have much smaller wavelengths than does visible light, the electron microscope can provide much higher magnification than light-based instruments. Through electron microscopes scientists first viewed strands of DNA. Since Ruska's invention, instruments like the **scanning tunneling microscope** and the **field ion microscope** have been developed with the ability to observe the activities and structures of individual atoms.

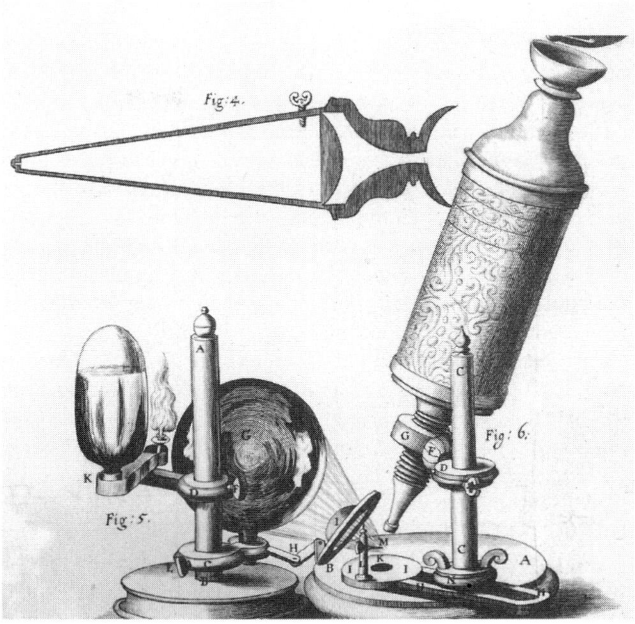

Robert Hooke's compound microscope.

MICROWAVE OVEN

If you looked into every home, apartment, lunchroom, or dormitory in the United States, you would find that more than two-thirds of them have at least one microwave oven. First introduced to the public in 1955 as the "radar range," the popularity of the microwave oven has skyrocketed to the point where it is considered a household necessity.

Like many products popularized during the 1950s, the microwave oven was an offshoot of technology perfected during World War II. In 1945 the Raytheon Manufacturing Company was experimenting with a magnetron (a tube that generates extremely short **radio** waves), trying to improve the efficiency of military **radar**. A technician at Raytheon, Percy Spencer, was standing very close to the magnetron as it operated. After a few minutes, Spencer noticed that a candy bar in his pocket had melted, even though he had not felt any heat. While this effect had been noted before, it was Spencer who first began to wonder about the domestic applications of such a device. Ten years later the "radar range" became available to consumers.

The technology behind the microwave oven is relatively simple. All microwave ovens use a small magnetron (not unlike the ones used by Raytheon) to generate microwaves. In the **electromagnet**ic spectrum, microwaves fall between radio waves and infrared rays. As the name implies, these microwaves have very short wavelengths, usually ranging from 0.039 in (1 mm) to 1.2 in (30.5 cm). Using a magnetic field, the microwaves are oscillated to change their direction 2.45 million times every second.

The microwaves are emitted from a small antenna and directed into the oven's cooking compartment through a metal tube called a **waveguide.** Just before they enter the cooking area, the microwaves are passed through a stirrer—a set of turning metal fan blades. The stirrer breaks up the microwaves, dispersing them evenly throughout the cooking compartment to ensure that the food within is neither overcooked nor undercooked.

Once inside the cooking area, the microwaves bombard the food and any stray microwaves bounce off the metal walls until they are absorbed by the food. When a microwave strikes a molecule of water, that molecule is "aligned" in the wave's direction; when the wave changes direction (due to the oscillating magnetron), the molecule is aligned in the opposite direction. Like the microwaves themselves, the water molecules are realigned several million times each second. This vibration heats the molecule. Since only the water molecules are heated by microwaves through conduction, the hot water molecules then heat the surrounding molecules, cooking the food all the way through.

When first introduced, the microwave oven had only two settings: on and off. This was because the magnetron either produced or did not produce microwaves, and there was no variance in the level of power. Modern microwave ovens are more versatile, often equipped with multi-setting controls such as high, medium, low, warm, and defrost. In fact, **computer**s in some microwave ovens allow a cook to press a pre-programmed switch that directs the oven to perform the required function. For example, pressing the word beverage will heat a cup of water to boiling. Various settings are accomplished by turning the magnetron on and off regularly during the cooking time so that more or fewer microwaves are emitted, according to the cooking level.

One popular myth about microwave ovens is that they cook food from the inside out. Microwaves can only penetrate the first few centimeters of any substance and rely upon conduction to cook the rest. However, if a relatively small object with water at its center (such as a kernel of popping corn) is placed within a microwave, the water inside the object will cook first—often causing the object to explode.

Because only water molecules are affected by microwaves, substances like **glass**, **paper**, ceramic, and many **plastics** can be placed within a microwave oven without being burnt or melted. These materials are often used to make ''microwave safe'' dishes and cookware. On the other hand, metal utensils (which also contain no water molecules) cannot be placed inside a microwave oven. Like the metal walls of the cooking compartment, a metal plate or fork will reflect microwaves. Not only does this prevent the food from heating properly, it can also create a buildup of microwaves that will eventually produce sparks and possibly damage the oven.

Microwave ovens cook food faster than conventional ovens because they are more energy efficient. In a gas oven, less than six percent of the heat energy produced actually heats the food; the rest of the energy is wasted, heating the walls of the oven and the air surrounding the food. Electric ovens are slightly more efficient at fourteen percent, but they require a substantial amount of electricity to operate. In contrast, microwave ovens direct more than fifty percent of their energy toward heating the food, and require only a small amount of electricity to do so. For many cooking processes, microwaves do not perform as well as conventional ovens. Although browning trays are available, for example, a standard oven is best for such food preparation.

MICROWAVE TRANSMISSION

Microwaves, like light rays, are a type of **electromagnet**ic radiation. In the electromagnetic spectrum, they fall somewhere between **radio** waves and infrared rays. Generally, they have wavelengths ranging from 0.039 in (1 mm) to 1.6 in (40 cm).

Microwave technology was developed during World War II in connection with secret military **radar** research. Today, microwaves are primarily used in communications. Several devices can be used to generate microwaves, the most popular being magnetrons and klystrons. Most produce microwaves of relatively low power and require the use of an amplification device, such as a **maser** (microwave amplification by stimulated emission of radiation). Like radio waves, microwaves can be modulated for communication purposes; they are more versatile than radio waves, however, and are often used in setting up communication networks and radio relay systems. Although microwaves can be easily broadcast and received via aerial **antennas**, their high frequency nature makes it difficult to transmit from one place to another once they have been received. However, in the late 1990s, at least one company (Associated Communications LLC), is planning on offering microwave transmission-based wireless telecommunications services on the Digital Electronic Message Service (or DEMS) band.

Lower frequency microwaves are occasionally transmitted using coaxial cable s. These cables consist of two conducting layers that share a single axis but are separated by a dielectric medium. The combination of two conductors allows for a greater signal load to be transferred through this type of cable. At higher frequencies, however, coaxial cables become more and more inefficient because much of the energy to be transmitted is lost. When the frequencies of microwaves are very high (as with those used by the communications industry), coaxial cable is essentially useless.

A more common variety of microwave transmission is the **waveguide**. Waveguides are hollow pipes, either circular or rectangular, that conduct microwaves along their inner walls. They are constructed from materials of very high electric conductivity and must be of an extremely precise design. Waveguides are ideal microwave conductors because they operate only at very high frequencies. In order to transmit energy of lower frequencies, the tubes would be of an impractical large width. They are best used for relatively short distances—less than one mile (1.609 km)—and, unlike water pipes, they cannot conduct microwaves through sharp turns or bends since the waveguide must be a fairly straight run. Other systems such as dielectric rods and satellite relays have been developed for the transmission of microwaves, but these are generally used only for specific tasks such as space exploration.

In addition to the familiar **microwave oven**, microwave transmission is an important part of many communication systems as well as **aircraft** radar systems. By combining these two applications, engineers have designed aircraft and other vehicles that can be operated by an ''automatic pilot.''

See also Cable television; Communication satellite; Earth-survey satellite; Navigational satellite; Space probe; Weather satellite

MILK, CONDENSED

Although a process for condensing milk was patented in England in 1835, the first successful and practical method was patented in the United States by **Gail Borden** (1801-1874) in 1856. This was a major achievement, since condensed milk was the only source of pure, wholesome milk in the days before refrigeration. Borden evaporated about 60 percent of the water from whole milk in a vacuum and sold his purified product unsealed, just as regular milk was packaged during that time. Later in 1856, he began producing a canned condensed milk sweetened with sugar, which acted as a preservative.

While condensed milk was initially rejected by a public accustomed to watered-down "swill" milk, Borden's product was used extensively during the Civil War. To keep up with military demand, Borden licensed other manufacturers to make the milk, but many sold an inferior product. Borden adopted the American Bald Eagle as a symbol for the high quality of his sweetened condensed milk. The *Eagle Brand* name is still in use today.

Unsweetened condensed milk was first canned by J. B. Meyenberg, who patented a **pressure-cooker** method for sterilizing condensed milk in Switzerland in 1884. After emigrating to the United States in 1885, Meyenberg set up and promoted a number of condensed milk factories, including the Helvetia Milk Condensing Company in Illinois.

Condensed milk was used for purposes such as infant feeding in the early 1900s. After World War I, however, many people had **refrigeration**, which brought with it the ability to store fresh milk. In the 1940s, consumers began using sweetened condensed milk in desserts, due to a shortage of sugar. This remains a major use for the product today.

MILK, PASTEURIZED

The pasteurization of milk is a process of slow heating to kill bacteria and other microorganisms in order to make milk safe for human consumption. During the process many pathogenic bacteria are destroyed including *Mycobacterium Tuberculosis* tuberculosis, one of the most heat-tolerant pathogenic bacteria that does not form protective spores. Other bacteria, encapsulated in spores, may not be destroyed and thus pasteurized milk should still be refrigerated. During the heating process (at 143.1°F [61.7°C] for a period of at least 30 minutes) some food value of certain vitamins and proteins is lost, but the possibility of spreading bacteria with the potential of causing disease is lessened substantially.

Named for its inventor, nineteenth-century French microbiologist Louis Pasteur (1822-1895), pasteurization was originally begun as a way of keeping **wine** and **beer** from turning sour. Pasteur's research led him to believe that during the fermentation of alcoholic beverages, bacteria sometimes got into the brew and produced lactic acid which soured the valuable beer and wine. By heating the beverage at a low temperature for a given length of time, the bacteria would be destroyed. Pasteur, who theorized that bacteria and other pathogenic microorganisms were responsible for causing and spreading disease, also proposed the idea that heat sterilization of everything from food to surgical instruments would cause a drastic reduction in the transmission of bacterial disease.

A champion for the pasteurization of milk in the United States was Alice Catherine Evans (1881-1975), a bacteriologist who linked the cause of human illness to milk infected with disease-causing bacteria. Evans was the first woman scientist ever hired by the Dairy Division of the Bureau of Animal Industry in Washington, D.C. In 1906 the Pure Food and Drug Act had been passed and part of her research was to find out how bacteria got in dairy products. Starting in 1917 she found a variety of bacteria in fresh milk and observed similarities between bacteria in this "certified" milk from the United States to disease-causing bacteria from other countries. Evans hypothesized that the bacteria came from cows rather than contamination that took place during storage and noted that people who had no contact with animals had contracted diseases. In many cases people who drank infected milk contracted brucellosis, a bacterial disease that was formerly called by a variety of names including Malta, Mediterranean, and undulant fever. (In her work with the pathogenic bacteria, Evans herself contracted the disease.) A related group of disease-causing bacteria was discovered and named for Sir David Bruce (1855-1931), a Scottish microbiologist who, with his wife Mary Steele Bruce, traced the cause of fever outbreaks in the Mediterranean to goat's milk. Even after the accumulation of much experimental evidence, many people in the dairy business were dubious of Evans's ideas. However, by the 1930s, pasteurizing milk had become a standard practice. Milk products that contain higher quantities of sugar or milkfat, are pasteurized at higher temperatures and for longer time periods since such products provide an environment more conducive to bacterial growth.

MILK, POWDERED

Powdered milk was first produced in France by Antoine Augustin Parmentier (1737-1813) in 1805. The first patent for producing dried milk powder was granted in England in 1855. However, because sodium carbonate was added to the fresh milk, it was not pure milk powder. After that mixture was evaporated, the resulting dough was mixed with cane sugar, dried, and ground into a powder. Milk was first evaporated in open pans, then later in a vacuum process. Malted milk was developed in 1883. A powder mixture consisting of whole milk, whole wheat, and barley malt, malted milk appeared on the market in 1887 and became quite popular.

Pure dried milk made its first appearance in 1898 in the United States. Several methods of roller-drying milk were patented around this same time. Spray dryers came next. Use of powdered milk increased gradually as improvements in drying methods and packaging gradually yielded a high-quality product. Today powdered milk is made by evaporating some of the water from fluid milk, then spraying the milk in a drying chamber.

MILKING MACHINES

Mechanization of the dairy industry coincided with the advent of **railroad**s to transport goods, which made it possible to locate dairy operations away from the cities and created dairying regions in such areas as Denmark and Wisconsin. It also allowed butter and cheese production to become factory industries. Prior to these developments, all dairy production was conducted on the farm or by city cow keepers, who kept their animals at the back of their shops. Early attempts at artificially aiding the milking process involved inserting tubes made from straw or the bones of birds' feet into the cow's udder. In 1860, American engineer, L. O. Colvin, invented a vacuum milking

Mechanization of the dairy industry and the development of railroads made it possible to locate dairy operations outside of cities and led to the development of dairying regions.

machine. The suction action drew the milk into a container set beneath the cow. However, the continuous suction irritated the cow's udder and caused bleeding. In 1889, a Scottish plumber, William Murchland, invented a machine that used a hydraulic vacuum. Requiring four people to operate it, this inefficient device also irritated the cow's udder. A machine developed in 1892 by Danish inventor Jens Neilson worked separately on the upper and lower parts of the cow's teats to simulate hand milking. Nelson's machine required hand cranking, which meant that an individual had to stay with it throughout the process, just as in hand milking. Also placing Neilson's device at a disadvantage was its improper fit, which damaged the udder. Dr. Alexander Shields of Glasgow, Scotland, invented what he called the thistle milking machine in 1895, which attempted to simulate the sucking of a calf. A very noisy machine, it could be heard two miles away. Additionally, the large amounts of air required for the sucking action caused contamination of the milk. Despite these obvious engineering flaws in milking machines, patents continued to flourish into the early 1900s. They were first marketed in the United States in 1905. In Europe, **Carl Gustaf Patrik de Laval** of Sweden marketed a machine in 1918.

Many advances have been made since these early attempts. Modern milking machines rely on the massaging action of a **rubber** liner inside the suction cups. The action aids the circulation of the blood in the teat, averting the constrictive effect of the continuous vacuum. It has a stimulative effect that "brings down" the milk from the alveoli, the areas through which the milk descends into the reservoirs, or milk cisterns, in the udder. The milking machine is but a part of a mechanized system that delivers milk to the consumer under the highest standards of cleanliness. From the milking parlors, the milk is piped directly into storage units located in separate buildings. The milk is never allowed to come into contact with air, human hands or unsanitary equipment. The milk is kept under continuous **refrigeration** to guarantee its freshness, to the moment it is consumed. Most milking machines are now computerized, as part of the trend toward agribusiness in farming. Computers track the amount and quality of the milk, and generate computerized menus for feed. Milking machines and computers combine to make dairying an exceedingly efficient industry.

Milling machine

Wheel-cutting machines, dating from the 1700s and used by clockmakers, were the precursors of industrial milling machines. Although it is unclear who invented the first actual milling machine—a device similar to the **lathe** that features a rotating cutting tool rather than a rotating workpiece—evidence supports its existence by 1818 in the United States. **Eli Whitney** is most often mentioned as the first to design and construct a milling machine that was dependable and which served as a prototype for later, improved cutting machines.

Whitney produced the machine, along with several others, with an eye toward originating **mass production** of gun parts. In 1798 he contracted with the federal government to produce a large order of **musket**s. All guns, at the time, were hand-crafted, and so gun parts were not interchangeable. Whitney remedied this with his construction of a semi-automated, machine tool factory.

In 1867, American engineer Joseph R. Brown (1805-1870) introduced his universal milling machine at the Paris Exhibition. Brown's machine arose following the testing in 1861 of one designed to solve the problem of producing spiral flutes for twist **drill**s. This machine proved amazingly versatile, and led to Brown's significant addition in 1864 of a formed cutter. Ever since, milling machines have vied with lathes as the most employed industrial machine tool. Their high adaptability is demonstrated by the numerous cutting jobs they perform, including flat surfaces, grooves, shoulders, inclined surfaces, slots, and dovetails. A turret milling machine is an example of a miller used in conjunction with a related machine tool, the drill. Another specialized miller is the hobbing machine, used to make **gear**s. Perhaps the milling machine's greatest distinction is that in 1954 it became the first machine tool to be controlled numerically, thereby representing one of the greatest industrial advances of the twentieth century.

Mine sweeper • See Warship

Miner's safety lamp

Coal mining has never been a very safe job, but it is much less risky today, thanks to an invention called the miner's safety lamp. The first lamp of this kind was invented by Sir **Humphry Davy**, an English chemist, in 1815. At that time, the industrial revolution had begun to transform England, and the demand for coal sharply increased as railroads, **steam engine**s and other new technologies dependent on the fuel were introduced. As coal mines became deeper and larger, hazards for miners increased—mainly the danger of explosions caused by a gas called "fire-damp," which seeps from coal seams. This flammable gas was ignited by the **candle**s and **lamp**s that miners carried to light their way. By 1812, explosions had become frequent and disastrous, killing hundreds of miners each year.

Davy was asked to help by a society that had been formed to prevent coal-mine accidents. After visiting mines,

Davy analyzed the composition of fire-damp gas, which he confirmed was mainly methane, and determined its flammability limits, using the results of earlier studies by German chemist Theodor Grotthuss (1785-1822).

Within three months, Davy had invented the first safety lamp which enclosed an oil-burning flame within a wire gauze cylinder, topped by a double thickness of gauze. The finely meshed metal gauze dissipated the flame's heat, preventing the lamp from reaching a high enough temperature to ignite the flammable gas. Also, the lamp's flame grew higher in the presence of fire-damp, warning miners of the danger and giving them light to escape. At the time, this lamp was regarded as nothing short of miraculous.

Although Davy refused, as a humanitarian gesture, to patent his lamp, he grew jealous when **George Stephenson** claimed its invention. Davy's claim prevailed and he was made a baronet for his service to the mining industry. Many other scientists went on to improve the Davy lamp by using **glass** enclosures, perforated brass tops, and other variations. Since then, flammable gas detectors based on electricity and light have been invented, but modern versions of Davy's lamp are still used in coal mines today.

MINICOMPUTER

Minicomputers were first developed and manufactured in the 1960s; they were smaller than the then-standard **mainframe computer** and were usually intended for a special purpose. The *minicomputer* was defined as having a 16-bit or larger processor and one or more terminals, so that it could be used by several people at the same time. With the rapid evolution of computers, the concept of the minicomputer is obsolete in the 1990s.

The first minicomputer was the Digital Equipment Corporation 's PDP (Programmed Data Processor) series. IBM, Wang, and other manufacturers also had minicomputer lines. The PDP-1 cost $120,000 in the early 1960s, and in the 1980s minicomputer prices ranged from about $20,000 to near $200,000. Even so, the price was much lower than that of mainframes, which were in the million-dollar range.

Minicomputers are part of the third computer generation, which used **integrated circuit**s for memory and computer operations. Invented in the late 1950s by scientists at Texas Instruments, Inc. and Fairchild Semiconductor, integrated circuits have large numbers of **transistor**s and other electronic circuits on a single computer chip. This miniaturization allowed computers to become smaller, yet more powerful. For example, it became possible to improve (upgrade) a computer by adding more memory chips.

Minicomputers became less attractive in the late 1970s, when the microprocessor—''processor on a chip''—resulted in computing's fourth generation machines. Chips became much more powerful and less expensive to manufacture because of what is called *large-scale integration* and very large-scale integration (VLSI). During the 1980s and 1990s, the personal computers and the desktop-sized workstations, such as those made by Compaq, Apple Computer, and Sun Microsystems, have provided greater computing power and operating flexibility than minicomputers. Though many minicomputers remain in use, their popularity has greatly diminished.

MINIÉ BALL • See Ammunition

MINING

Mining is the extraction of minerals and rocks from the earth. The history of mining stretches back to prehistoric times when **gold**, **copper**, and **silver** were apparently the first metals mined. Egyptians mined copper and **tin** as far back as 3700 B.C. A German scholar, Gergius Agricola, wrote an early comprehensive study of mining methods in the Middle Ages in his *De re metallica*. In the United States, mining did not begin until the late eighteenth century, when a Missouri lead belt began continuous production. Metal ores were found in the western part of the United States in the middle of the nineteenth century causing the California gold rush. Mining is still a major industry in the United States.

There are several methods to find potential deposits for inspection. Trenching and testpitting are used when there is only loosely packed matter covering the deposit. If the surface is harder or featured stratified layers, drills are used to take samples. Many types and sizes of drills exist for this purpose. After a sample is analyzed and a mining operation is put in place, there are two basic mining types utilized: open-pit and underground mining.

Open-pit mining, also known as surface mining or opencast mining, is employed when the material to be mined is at or near the surface. Open-pit mining is usually done in layers with materials at the edges removed. Thus, the pit is enlarged gradually. At the beginning of mining operation, if the material is slightly underground, the loose earth has to be stripped back. This is usually done by backhoes or blasting, though drilling is sometimes used. If blasting is involved, it is done with **dynamite**, bag powder, or ammonium nitrate. Flame piercing is another option. Flame piercing involves jets of burning fuel oil, or kerosene and oxygen, which dissolves the rock into a pit. As the ore or other minerals are mined, they are picked up by shovels or drag lines, though some operations use bucket-wheel excavators. One major issue in open-pit mining is keeping the pit slopes stable. However, compared to open-pit mining, underground mining is more costly, has more safety concerns, and is more complex overall.

There are several types of underground mines, each suited to a particular type of mineral deposit. A drift mine is used when the material to be mined approaches the surface at one point but the majority is located in a hill. A tunnel is inserted into the hillside. When the material to be mined is found completely underground, but not at extreme depths. A slanting entrance is built to meet the deposit, and the mined ore is taken to the surface by cars. Shaft mines are used when the mineral deposit is far underground. Two or more vertical shafts are built to the deposit area, one for hoisting the ore to the surface and the other to provide air, sometimes pushed with fans, to the miners. Because of the nature of underground mining, drilling and blasting are important to reaching deposits. Drills can be small, one-man drills (like jackhammers) or much bigger. Blasting is a quicker way to reach deposits or advance the mining area, and is accomplished with dynamite, ammonium nitrate, and bag powder. For miners to work underground, several issues have to be addressed, including ventilation, drainage, escape routes, and lighting their working area. The safety of miners is a major concern, and stability of the mine, especially the prevention of cave-ins, remains an important safety concern.

See also Miner's safety lamp

Mirror

Perhaps the most common optical instrument in the world, mirrors have been used for millennia; artifacts made of polished obsidian dating back 7,500 years have been found in Turkey. The earliest man-made mirrors, which were constructed of highly buffed copper, brass, and bronze, were manufactured in the first century A.D. Not until the thirteenth century did the Venetians develop a method for silvering glass to make it more reflective. The modern silvering process was invented in 1835 by the German chemist **Justus von Liebig**, and, with a few improvements, it is his process that we presently use to manufacture mirrors today. Technically, a mirror is any smooth surface that reflects more light than it absorbs. The mirrors with which we are familiar are actually panes of glass that have one side coated with a very thin layer of metal, which may subsequently be coated with another layer of dielectric film. The amount of light reflected by a mirror is dependent upon the kind and quality of the materials used in its construction. No matter how skillfully crafted, no mirror can reflect all the light shone upon it.

There are two basic classes of mirror: *planar* (flat) and *non-planar* (curved). In the case of a planar mirror, the light that strikes the surface will be reflected away at a precise angle; the angle at which it leaves the mirror, called the *angle of reflection*, is equal but opposite to the angle at which it arrives, called the *angle of incidence*. The image seen in a planar mirror, called the *virtual image*, is an illusion; since it appears that the subject in the mirror is *behind* the surface, at a distance exactly twice the actual distance. For example, a mirror five feet away creates a virtual image ten feet distant; actually, the light does travel ten feet—five to the mirror and five back to the observer's eye. Also, the image in a planar mirror is always reversed: the right hand becomes the left hand, and writing appears backwards. This reversal can be corrected by adding a second mirror, essentially reversing the image twice. This is the case with a periscope, which uses two mirrors to see around corners or above obstacles.

There are two major types of non-planar mirrors: concave and convex. Concave mirrors appear bowed, curving toward the observer like the outside of a sphere. The image viewed in a concave mirror appears smaller than it would in a planar mirror, but the field of view is greater. This type of mirror is often used in automobiles as side mirrors; usually bearing a warning that objects are closer than they appear, these mirrors allow the driver to see more of the road than would a simple flat mirror. In addition, drivers can now adjust automobile mirrors from inside the car, a useful safety feature. Convex mirrors are curved away from the observer, like the inside of a sphere. Mirrors with this shape can focus a path of light to a central point or expand a point of light into a wide path. Also found in automobiles, convex mirrors are used in headlights to turn the light emitted from a bright bulb into a powerful beam (a typical science class demonstration involves a device called a solar cooker; consisting primarily of a large convex mirror, the solar cooker can burn paper or even heat a hot dog by focusing a wide area of sunlight to a point). Concave and convex mirrors can also be used in place of magnifying lenses. Other curved mirrors shaped like parabolas, hyperbolas, and ellipses are manufactured for use in scientific instruments, such as reflecting **telescope**s. Mirrors are also used in homes for decorative purposes, enlarging the apparent appearance of a room.

See also Magnifying glass

Mobile telephone

Working on the 800-MHZ bandwith, Specialized Mobile Radio (known as SMR) began as a voice-only wireless communications service for private operations that required drivers and dispatchers, including delivery trucks and police offers. Radio towers connected dispatcher to their fleet, and access to a company's channel was limited to its owners. In the late 1990s, with the advent of demand for widespread wireless service and improved SMR technology, SMR has emerged a competitor to the **cellular telephone** industry.

Until recently, a SMR network was the domain of partially mobile businesses that needed to stay in touch. Beginning in the 1940s, businesses like taxi services held licenses to run their stations—be it mobile, portable, or control—through an exclusive channel paired with a specific SMR base station. By the 1990s in big cities, the demand for SMR became so great that the airwaves were nearly filled. In 1992, Motorola, then the largest SMR operator in the United States, developed a technology that increased capacity six times on already existing SMR frequencies. Essentially, Motorola brought digital technology to the SMR network with its time division multiple access (TDMA). TDMA splits what was once one analog channel into six parts so six calls can be transmitted at once. The spoken communications are converted into digital signals to accommodate the split. With TDMA, both voice and data transmissions become possible, and call privacy is increased. When SMR received data capabilities, SMR became known as "burst technology" because it can send data through a modem in rapid bursts.

The year before Motorola introduced TDMA, in 1991, the Federal Communications Commission (FCC) granted Fleet Call (now known as Nextel Communications) permission to construct a network based on digital mobile technology in its existing SMR spectrum. By upgrading to digital, the SMR system's abilities continued to increase from mobile telephone service and two-way radio dispatch, to paging, messaging, and more data services that TDMA allows. Because SMR was completely digital well before cellular telephone services were, they had an edge on the latter and became a potential player on the consumer market.

However, SMR worked on a regional basis, primarily in metropolitan markets as well as some rural areas. As of 1998, there is no complete national network. However, SMR has the potential to rival cellular in the future for several reasons. SMR receivers are more powerful and can use 12 watts compared with cellular's five. SMR is also cheaper than cellular when used as a telephone technology. While SMR telephones are bulkier than cellular's, they have more features. As SMR become more popular, though, the bulkiness might be improved. SMR systems also cover twenty-five times as much area as a cellular network.

The future of SMR is still uncertain, but looks promising. SMR technology continues to be improved. In the mid-1990s, SMR was "enhanced" (ESMR) which further boosted its capacity for both voice and data. As of 1993, there were 1.5 million SMR users.

Möbius strip

A Möbius strip is constructed by twisting one end of a rectangular strip of paper by 180 degrees (a half twist) and joining it to the other end. The result is an example of a one-sided surface. If you draw a line along the surface starting at a point A, you will pass through the point that was on the opposite side of A before the strip was twisted, and will then return to the original point A, all without ever lifting the writing instrument from the paper. In other words, it is possible to paint the entire surface with a continuous sweep of a brush. Furthermore, while an ordinary strip has two edges, a Möbius strip has only one edge. The Möbius strip was named for August Ferdinand Möbius (1790-1868), a German mathematician and astronomer, who discovered it in 1858 while studying the properties of one-sided surfaces. At almost the same time another German mathematician, Johann B. Listing (1806-1882), discovered the surface and published the first description in his book on topological investigations.

The Möbius strip has several interesting properties that makes it a favorite for mathematical demonstrations. If the surface is cut down the middle lengthwise, the result is still a single strip (though no longer a Möbius strip) that is now twice as long and half as wide, but with a double twist. However, slicing this new strip will now produce two pieces of the same length linked together like an old-fashioned linked chain. These properties play a crucial role in a short story written by William Hazlett Upson in 1949 called *Paul Bunyan versus the Conveyor Belt.* The Möbius strip has also been a favorite inspiration for artists. Maurits C. Escher (1898-1972) created three drawings based on the Möbius strip. The second, done in 1963, depicts giant ants climbing their way up one part of a Möbius strip ladder, only to find themselves descending again in an endless cycle.

See also Topology

Modem

The modem allows for the transfer of computerized information via channels such as cables, radio aerial arrays or satellites. The information is modulated into amplitude, frequency or phase, which allows for the transmission. The modem contains both a modulator for transmitting information and a demodulator for receiving information. The term modem is short for modulator-demodulator.

One of the first known modems was produced for IBM in 1954 and called the Data Transciever. A similar device was developed around the same time at the MIT Lincoln Laboratory and Bell Telephone Laboratories for the Department of Defense.

Early modems operated at a speed of 1,200 baud (bits-per-second). By the 1970s that number had been increased by 700%, to 9,600 baud, although the standard through the 1980s was only 2,400 baud. By 1991, the standard had been increased to 14.4 kilobits (14,400 baud). By the mid-1990s, modems were operating at speeds of 28.8 kilobits and higher. By this time, internal modems were standard equipment on most full-sized computers. Modems for laptop, notebook, and handheld computers were available in a credit-card sized unit; some of these units could operate using a cellular phone.

Also by this time, modems had been developed that allowed for voice as well as data transmission, although not at the same time. Simultaneous voice and data (SVD) and digital SVD (DSVD) units, which allow transmission of both voice and data at the same time, also were under development, as were integrated services digital network (ISDN) units that could transmit video.

Molotov cocktail

The Molotov cocktail was first used as a weapon during the Spanish Civil War (1936-39). This crude device consists of a bottle filled with flammable liquid—usually **gasoline**—stopped with an oily rag; the user simply has to ignite the rag and throw the bottle at his target. Named after the Soviet premier Vyacheslav M. Molotov (1890-1986), who ordered production of the device during World War II, molotov cocktails were used against advancing German tanks but were often ineffective. Because of their simplicity, molotov cocktails have become a standard projectile weapon for those without access to traditional military weapons.

See also Bomb; Chemical warfare; Napalm

Money, coin

Human civilizations have long used metals as a medium for exchange. In addition to their long-lasting properties, metals lend themselves easily to melting and casting. As early as 1000 B.C., the Chinese were using a type of metal token to represent payment. These artifacts have been labeled "spade" and "key" money because of their resemblance to a digging tool and to the modern-day Yale key. Both types bore denominations and were cast from molds. Although the ancient Egyptians did not mint coins, gold weights and rings were used to trade for products and services.

The first record of Western coins did not occur until 700 B.C., in western Asia Minor. Evidence of coins made from a naturally occurring alloy of gold and silver called electrum were found in the foundation of the temple to Artemis at Ephesus on the banks of the Aegean Sea. King Croesus of Lydia, who ruled from 560 to 546 B.C., has been credited with creating a bi-metallic system of pure gold and pure silver coins. These early coins typically carried imprints of animals, such as bulls, birds, insects, or mythical creatures. Engravings of vegetables were also popular. Imprints were stamped on one side of the coins with a tool bearing that particular design. Coin design was elevated to an art form during this period, and elaborately imprinted coins were afforded a high status. Many Greek cities vied for the distinction of having the most beautifully designed coins.

Alexander the Great built mints throughout his kingdom, from Macedonia to Babylon. He instituted uniform weights and types. It was during Alexander's reign that the coin portrait rose to popularity. Rulers, gods, and goddesses were the portraits of choice. By the fourth and fifth century A.D., engravers in Italy, and particularly in Sicily, were generally recognized as the experts in coin design. So revered was their skill that the engravers began signing their work.

Before the advent of the Industrial Age, the striking of coins was accomplished manually. A round blank of metal was placed over an anvil that had been fitted with an imprinted die. Another die was affixed to a pestle, which was then placed on top of the blank. The coin maker held the pestle in place with one hand and then brought a two-pound hammer down on top of the pestle. Remarkably, this resulted in seven tons of pressure, which forced impressions into both sides of the blank. The high relief typical of early Greek coins sometimes required two or three blows to achieve the desired effect. Heating the blank before striking often reduced the number of required strikes. This method allowed one coin to be struck every two seconds.

In the mid-1990s, the United States made preparations to join other industrialized countries in the use of a dollar coin instead of a paper bill. Although backers point to the savings that the switch would bring, and environmentalists extol the virtues of phasing out the dollar bill, traditionalists see the dollar bill as a well-entrenched symbol of the United States. Unions and trade associations representing the paper industry also voiced opposition to the new coin.

Elimination of the penny has also gained support in recent years. Ironically, the American public's view of the penny as worthless has caused millions of people to stockpile them in jars and boxes at home, to be traded in for larger denominations at a later date. This has led to a shortage of pennies in the commercial arena. Decisions about eliminating coins are intensely political, attesting to the continuing symbolic power of the metallic coin.

Money, paper

The existence of money as a means of buying or selling goods and services dates back to at least 3000 B.C., when the Sumerians began using metal coins in place of bartering with barley. The use of paper money began in China during the seventh century, but its uncertain value, as opposed to the more universally accepted value of gold or silver coins, led to widespread inflation and state bankruptcy. It was not until 1658, when Swedish financier Johann Palmstruck introduced a paper bank note for the Swedish State Bank, that paper money again entered circulation.

The first paper money in what is now the United States was issued by the Massachusetts Bay Colony in 1690. It was valued in British pounds. The first dollar bills were issued in Maryland in the 1760s. During the American Revolution, the fledgling Continental Congress issued Continental Currency to finance the war, but widespread counterfeiting by the British and general uncertainty as to the outcome of the revolution led to massive devaluation of the new paper money.

Stung by this failure, the United States government did not issue paper money again until the mid-1800s. In the interim, numerous banks, utilities, merchants, and even individuals issued their own bank notes and paper currency. By the outbreak of the Civil War there were as many as 1,600 different kinds of paper money in circulation in the United States—as much as a third of it counterfeit or otherwise worthless.

Realizing the need for a universal and stable currency, the United States Congress authorized issuing paper money in 1861. In 1865, President Lincoln established the Secret Service, whose principal task was to track down and arrest counterfeiters. This early paper currency came in several different types, designs, and denominations, but had the common characteristic of being somewhat larger in size than today's money. It was not until 1929 that the current-sized bills went into circulation. In 1945, the government stopped printing bills in denominations greater than $100, and in 1963, they stopped printing silver certificate bills with the assurance that the dollar amount was available "...in silver payable to the bearer on demand."

Starting in 1996, the Department of the Treasury began issuing $100 bills with a new front and back design and several new security features designed to make counterfeiting more difficult. Other new bills in descending denominations will be printed at the rate of one new denomination per year.

The new bills use the same paper and are the same size and color as today's bills. Multicolor images, such as are commonly found on European currency, were not used because they were too easy to duplicate with color photocopies and printers. The image of **Benjamin Franklin** still appears on the front of the new $100 bill, although his portrait is larger and is shifted to the left. A watermark, formed by reducing the thickness of the paper during manufacture, has been placed to the right of the portrait and shows a second image of Franklin when the bill is held up to the light. The imbedded security thread is also still there, although now it has been treated to glow red under ultraviolet light. The position of the thread varies depending on the denomination of the bill to prevent the counterfeiting practice of bleaching the ink off lower denomination bills and reprinting them as higher denominations.

Other new features include concentric fine lines behind Franklin's head on the front and behind the image of Independence Hall on the back. These lines are so fine that they are extremely difficult for copiers or printers to duplicate without blurring them into a solid background. Microprinting, which was introduced in 1990, is used in two places on the new $100 bill: the words "USA 100" appear within the lower lefthand numeral 100, and the words "United States of America" run down the lapel of Franklin's coat.

Perhaps the most high-tech feature is a special color-shifting ink which is used to print the numeral in the lower right-hand corner. When viewed from head on, this ink appears green, but changes to black when viewed from the side.

Lower denomination bills will have many, but not all, of the security features present in the new $100 bills. The highest level of protection was given to the $100 bill because it is the largest denomination being printed. It is also the most common bill in circulation outside the United States, and hence, is frequently counterfeited in other countries.

Some of the security features originally proposed for the new money—such as holograms, plastic films, and coded fiber optics—were not used for this latest change because they represented too great a departure from the current money or because of potential technical problems. Looking further into the future, paper money may eventually be replaced by electronic money that is downloaded onto plastic "stored value" cards from an ATM or computer. Each card would have a computer chip memory, and the money would be electronically transferred through a card reader to make purchases.

MONORAIL SYSTEM • See Train and railroad

MONTGOLFIER, JOSEPH (1740-1810)
MONTGOLFIER, JACQUES (1745-1799)
French balloonists

Joseph-Michel Montgolfier and Jacques-Ètienne Montgolfier

The sons of a successful paper manufacturer, the Montgolfier brothers were skillful mathematicians and avid experimenters (Joseph Montgolfier constructed a **parachute** in 1799 that successfully carried a sheep to safety). They read extensively, becoming familiar with **Benjamin Franklin**'s famous electrical experiment with a kite, and had read **Joseph Priestley**'s treatise on the properties of air.

In 1782 Joseph Montgolfier made an interesting observation. During a fire, he witnesses some agent—smoke, perhaps, or some sort of gas—carrying sparks high into the sky. Maybe there was a way to harness this power to lift a man. He first started with an oblong bag made of fine silk that he filled with smoke from a **paper** burning beneath an opening in the bottom of the bag. The bag sailed to the ceiling of the room: a success. He and his brother repeated the experiment outdoors where the bag rose about seventy feet before gradually losing its buoyancy and returning to the ground. They began using fire made of chopped wool and straw, creating a pungent smoke. They believed this smoke contained an unknown gas that was lighter than air, unaware that the heated air was actually responsible for the upward pressure. The brothers' first full size balloon was a great success, shooting up to 6,000 feet (1,830 m) and landing a mile (1.6 km) away.

The brothers were soon invited to Paris where they impressed scientists and the royal family with their invention. In late September of 1782, they launched the first living things to see if the upper air could sustain life. A sheep, a rooster, and a duck spent eight minutes in flight and landed without harm. The Montgolfiers next planned **balloons** for manned flight. Some suggested a criminal should be sent since the flight could prove dangerous, but others who believed that the honor should go to a more distinguished citizen prevailed. The first pilot was Jean-Francois Pilatre de Rozier, a young physician whose curiosity and enthusiasm won him the job, and on November 20, 1783, Pilatre and the Marquis d'Arlandes, an infantry major, sailed aloft for twenty-five minutes. Their flight and others like it ignited a worldwide interest in ballooning. In 1785, however, Pilatre died in a balloon crash, and confidence in the Montgolfier's invention waned. Also, their rival, physicist Jacques Charles (1746-1823), had begun perfecting his hydrogen balloon, which would soon eclipse the Montgolfier's hot-air balloon.

Garrett Augustus Morgan

Morey, Samuel (1762-1843)

American inventor

Samuel Morey was born in Hebron, Connecticut, and grew up in Orford, New Hampshire. As a youth he experimented with steam power and in 1790 harnessed the steam from a kettle to turn the spit in his fireplace. Three years later he applied the principle behind this contraption to his next creation, the first **steamboat** with paddle wheels. Capable of carrying two people to the speed of 4 miles per hour (6.4 kph), his steamboat plied nearby rivers for the following three summers as Morey demonstrated its practicality. He also modified his design, placing the paddle wheel at stern for increased efficiency.

The steamboat was of great interest at the time when traveling consisted of either rough-riding **stagecoache**s or ships dependent upon unreliable winds. Robert Livingston (1746-1813), who was to play a large role in the life of **Robert Fulton** and the development of his steamboat, offered to back Morey and his boat in a joint business venture, but Morey refused, apparently hoping to develop it himself. He later designed a crank-motion **engine** and built a steamship with side paddles in 1797. For the next few years he patented improved designs, getting ready for large-scale commercial boats.

However, his hopes for commercial success dimmed. His backers suffered financial losses, and he became discouraged when Robert Fulton succeeded with his steamboats. Morey finally sank his boat on a small lake. Its interesting to note that Fulton, often credited with the invention of the steamboat, actually owed much to Morey. It was Fulton who visited Morey to see his steamboat demonstrations before launching his own vessel.

In addition to his steamboats, Morey took on other challenges. He patented wind, water, and tide **mills**, made improvements in **steam engines** and boilers and in 1826 patented an **internal combustion engine**. He also devised a series of **locks** to aid navigation on the Connecticut River and helped construct chutes to carry logs from inaccessible heights down to ponds. Morey was so capable in so many areas of science and invention that he has been called ''the Edison of his day.''

Samuel Morey is one of the highest achieving yet least-known American inventors in history. Although he secured at least 20 patents from 1793 to 1833 and was a true pioneer of steam propulsion, he never attempted to bring suit to win a claim for any of his inventions.

See also Fitch, John

Morgan, Garrett Augustus (1875-1963)

American inventor

Garrett Morgan, an African-American inventor, is recognized for his wide range of interests and inventions. He is best known for his development of the **gas mask** and the automatic **traffic signal**, but he also gained recognition for perfecting a hair-straightening cream and a belt fastener device for **sewing machines**. Born in Paris, Kentucky, the seventh of eleven children, Morgan, like other indigents in the poverty belt, quit school at age fourteen and went to the city to look for work. Morgan moved first to Cincinnati, where he worked at odd jobs for four years. Then he moved on to Cleveland and found a job at a sewing machine repair company.

In 1907, Morgan opened his own sewing machine sales and repair shop. Two years later, he opened his own tailoring shop, and employed 32 people to make coats, suits, and dresses. Morgan encountered a problem in his sewing business: the heat and friction of the needles would often scorch wool cloth. He decided to solve the problem himself, and worked on a chemical solution to coat the needles and so reduce the friction. Morgan accidentally discovered that this solution would also straighten curly hair. After testing it on himself and a neighbor's Airedale, Morgan launched his G. A. Morgan Hair Refining Cream in 1909. The company he set up to manufacture and market his hair straightener did very well, and Morgan was alleged to be the first black man in Cleveland to own a car.

Morgan received a patent for his most important invention in 1914. This was his Safety Hood, a breathing device that evolved into the modern gas mask. It consisted of an airtight canvas hood worn over the head and connected to a breathing tubes that hung to the ground. Morgan won a gold medal at the Second International Exposition of Sanitation and Safety in New York City in 1914 for his invention, which he marketed for fire fighters, engineers, chemists, and others working with dangerous fumes. But it wasn't until Morgan proved the usefulness of his device himself that his gas mask really caught on. In 1916, Morgan and his brother Frank donned the life-saving mask and rescued more than 20 workers trapped in a tunnel 228 feet (70 m) beneath Lake Erie at the Cleveland Water Works. For this rescue, Morgan won a second gold medal, and he began to receive many more orders for the mask. His mask proved effective in World War I, and eventually became standard issue for soldiers. Morgan eventually set up his own company to manufacture gas masks. The product sold well until the public learned that the manufacturer was black. Even after he disguised himself as Big Chief Mason, a Canadian Indian, and hired white demonstrators as marketing representatives, sales of the gas mask faltered.

To occupy his talents, he turned to the development of the automatic three-stage electric traffic light. Before Morgan's invention, traffic signals indicated only Go or Stop, and they were frequently ignored. Morgan's light had arms that could be raised or lowered, and when they were in the halfway position, they indicated that drivers were to slow down. This was the precursor to the red, yellow and green lights in use today. Morgan patented his traffic light in 1923, and then sold the rights to General Electric for $40,000.

In the 1920s, he organized a newspaper, the *Cleveland Call.* A civil rights activist, he took part in the formation of the NAACP, and in 1931 ran unsuccessfully for Cleveland's city council on a platform of fair housing, employment, and representation for all people. His declining years were marred by glaucoma, but he maintained an interest in inventions until his death in 1963.

Morse code

Morse code is a series of dots and dashes that represent letters of the alphabet and numerals, used in transmitting telegraph messages. It is named for **Samuel Morse**, who invented the code.

Morse, an American portrait painter, began designing an **electric telegraph** system in 1832. In order to keep the number of transmission lines to a minimum, Morse devised a way to send messages in code, using combinations of dots and dashes; each unique combination represented a single letter of the alphabet, a numeral from 0 to 1, or a punctuation mark. To work out his code, Morse visited a typesetting shop and noted which letters were most used, which least. He assigned the simplest code symbols to the most-used letters, and the longest dot-dash combinations to the least-used characters. The letter e, for example, became a single dot in Morse's code; the letter q was dot-dot-dash-dot.

After **Alfred Vail** became Morse's partner in 1837, he helped work out the final form of the dot-and-dash code. Morse patented his "Telegraph Signs" on June 20, 1840. In order to use the code, telegraph operators punched a simple key to open and close an electric circuit; a short keystroke was a dot, while a longer keystroke was a dash. The dots and dashes were automatically marked on a moving piece of paper at the telegraph's receiving end; the receiving operator then translated these symbols into an English message. After a while, the operators found that they could decipher the message simply by listening to the clicks the receiving machine made as it recorded the dots and dashes.

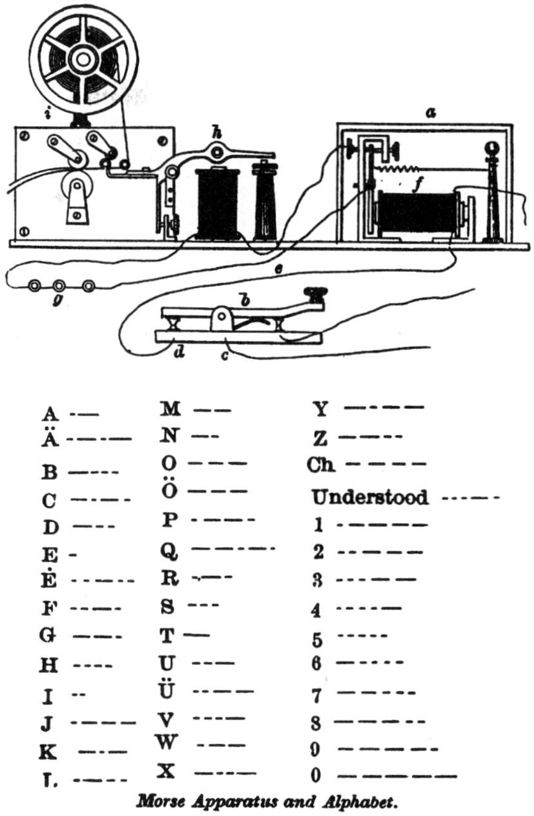

Morse code.

Because of its simplicity, Morse's code became used worldwide. To accommodate various accent marks used in languages other than English, Morse's system became the somewhat modified Continental code in Europe.

Morse, Samuel Finley Breese (1791-1872)

American artist and inventor

Samuel Finley Breese Morse was a well-known portrait painter who turned to science in mid-career and pioneered the **electric telegraph**.

Born in Charlestown, Massachusetts, on April 27, 1791, Morse was the oldest son of Jedidiah Morse, an eminent geographer and Congregational clergyman. He attended Yale, where he developed an interest in painting miniatures and attending lectures on electricity. After graduation, Morse sailed to England, where he studied painting from 1811 to 1815. On his return to Boston, Massachusetts, Morse opened a studio and soon found that portraiture was the only type of art that would sell. Within a few years, he developed a distinguished reputation as a portrait painter.

Samuel Finley Breese Morse

Burdened with financial concerns and mourning the successive deaths of his young wife, his father, and his mother, Morse returned to Europe in 1829 to continue his artistic studies. His return voyage to the United States in 1832 aboard the *Sully* became the turning point of his life. A conversation with fellow passengers—one of whom was the chemist Charles T. Jackson (1805-1880)—about experiments with electromagnetism piqued Morse's imagination. He immediately thought of sending messages over a wire via electricity, and spent the rest of the voyage sketching preliminary ideas. Morse's interest in developing the telegraph, coupled with disappointments in his artistic career, prompted him to give up painting in 1837. Morse built some prototypes of his telegraph in 1835, but his lack of background in science hobbled his efforts. At this time Morse was Professor of Arts and Design at the University of the City of New York. He turned to a fellow professor in the university's chemistry department, Leonard Gale, for help. Gale showed Morse how to improve both his electromagnet and his **battery**. Gale also introduced Morse to **Joseph Henry**, who freely shared his impressive knowledge about electromagnetism. Morse was now able to invent an electromagnetic *relay system*, which renewed the current along a line from relay to relay and made long-distance message transmission possible; he filed an intent to patent it in 1837.

In September of that year Morse met young **Alfred Vail** while demonstrating his telegraph in New York. The two men became partners, and Vail made many practical improvements to Morse's device and to the code used to transmit messages, which became known as the **Morse code**. Also in 1837, Morse applied for a grant offered by the United States Congress to construct a telegraph system. Seven long years of discouragement and poverty followed for Morse until finally, in the closing session of the 1843 Congress, he secured a $30,000 appropriation to build a telegraph line between Baltimore, Maryland, and Washington, D.C. With the aid of Vail and **Ezra Cornell**, Morse did just that, sending his famous first message "What hath God wrought!" on May 24, 1844.

Morse and Vail had intended to sell all rights in their telegraph to the United States government for $100,000, but Congress rejected their offer. The partners, along with Amos Kendall, then formed the Magnetic Telegraph Company to develop telegraph lines privately. Most of Morse's attention was taken up by prolonged and contentious litigation over patent rights to the telegraph, one of his opponents being Jackson. During this controversy, Morse unfortunately denied that Henry had ever helped him. Morse's patent rights were upheld by the United States Supreme Court in 1854.

The success of the telegraph brought Morse fame and wealth. His interests turned to politics; he supported the nativist movement, opposed abolitionism, and ran unsuccessfully for Congress in 1854. In 1857-58 Morse was an electrician for Cyrus Field's (1819-1892) attempt to lay a transatlantic **telegraph cable**. He built an estate near Poughkeepsie, New York, called Locust Grove (a historic landmark today), and enjoyed the company of his second wife, whom he had married in 1848, and his many children and grandchildren. He was a founder and trustee of Vassar College, and served in 1861 as the president of the National Academy of Design, which he had helped found in 1826 and led as president from 1826 until 1845. In his later years, Morse was a noted philanthropist. The telegraph operators of the United States honored Morse with a bronze statue in New York's Central Park in 1871. Morse died in New York on April 2, 1872.

Mortar • See Artillery

Motion detector

The development of motion detectors, also known as motion sensors, is closely related to the invention of **radar**. Like many inventions developed for military use, motion detectors found wide application in industry. In manufacturing, motion detectors proved useful as factories increasingly automated after World War II (1939-49). In addition, motion detectors became a basic component of security systems.

By the 1970s, there were essentially two varieties of motion sensors: acoustic and photoelectric. Sonic acoustic sensors use a speaker that continously emits sound waves to a receiver. Typically, the speaker and receiver are placed across a room from one another. Any movement between the speaker and receiver will interrupt the sound waves. When the interruption is detected by the receiver, an alarm sounds. The radio frequency motion detector is a variation of this invention. It uses electromagnetic waves instead of sound waves. Advancements were made to improve the sensitivity of receivers to eliminate false alarms.

Photoelectric motion detectors became much more common than the acoustic variety by the 1980s. Cheaper and more reliable, photoelectric detectors exist in several varieties. Some use an "active light beam" that passes between a transmitter and receiver. If the beam is broken, an alarm sounds much as with the acoustic detectors. Passive light detectors sense motion by detecting a change in the available light in a room. Inexpensive and reliable, this type of motion detector has become widely used in homes for turning on exterior lights, night lights, and others. Proximity detectors use an electrostatic field to detect the presence of a body near it. These detectors found application in the 1980s in automobile security systems. If the system detected a body near the car, an alarm or other warning would sound.

See also Burglar alarm; Surveillance equipment

Motion picture

The science of **photography** gave birth to modern motion pictures. This was not, however, a rapid process. The motion pictures we know today are the result of a long evolution of arts and sciences. Motion pictures are the product of a relatively complex illusion. The human brain perceives motion when pictures, taken in rapid sequence, are flashed at 15 or more frames per second—so we are able to amuse ourselves with the illusion of live action on the screen.

Sequential photographs were produced as early as 1860, but the true motion picture remained in the distant future. In 1877, **Eadweard Muybridge**, a student of animal motion, was attempting to prove that all four of a horse's hooves leave the ground when the animal gallops. A system of 24 cameras took sequential photographs of a galloping horse, and Muybridge indeed proved his theory. He also popularized this amazing illusion of motion by bringing it into homes in the form of a device called a zoetrope. His sequence of still photographs were published in strips, which were attached to the inside of the zoetrope's rotating drum. The viewer looked through a series of slots which acted as shutters as the drum spun, and the illusion of a running horse was brought to life. However, to produce an actual motion picture, even a very brief one, the exposure time necessary to take a photograph would have to be dramatically reduced.

In France, **Étienne-Jules Marey** was using paper film to shoot 120 frames with exposures as brief as 1/1000th of a second. This type of film helped lay the foundation for the science of modern cinematography. Once **George Eastman** developed strips of flexible celluloid camera film with fast exposure times in 1889, long series of pictures could be easily photographed in rapid sequence. American inventor **Thomas Alva Edison** and his collaborators were then able to combine these discoveries to produce the **kinetoscope**, a device that advanced a strip of film frame by frame in rapid succession to produce the illusion of fluid motion. Still, the images were not projected onto a screen as motion pictures are today.

The kinetoscope eventually found its way to Paris, where **Louis and Auguste Lumière** were the first to combine the flashing shutter of a camera with the bright light of projector to produce the cinematographe in 1895. The first American projectors using this technology of intermittent movement were produced by Thomas Armat (1866-1948) in 1895. Armat then struck an agreement with Thomas Edison, allowing him to produce these projectors under the name Edison Vitascope.

Though the motion picture had been achieved, attempts at the synchronization of motion pictures and sound had all but failed. Pianists tried to play music that matched the changing mood and tempo of the film but were unsuccessful, and recorded sound was played along with the film, but neither was consistent with the film's action. The quality of recorded sound was also unreliable due to the difficulty in maintaining a constant speed for both recording and playback. Fortunately, Western Electric Company was studying techniques of recording natural speech and other sounds for realistic reproduction. The company developed a wax phonographic disk and improved speakers and amplification systems for more accurate sound reproduction. The Warner Brothers studio, then a very small company, took a keen interest in these developments and formed the Vitaphone Corporation to introduce the complete sound system to the market. The integration of sound and motion pictures was complete.

The next step was to record the sound directly on film, rather than recording it on the wax phonograph disk, which made sound editing virtually impossible. **Lee De Forest** had already developed a sound-on-film process, patented as Phonofilm in 1919. Ironically, the major studios of the day considered it a mere novelty too expensive to be practical. In fact, all the major movie producers of the day rejected the idea of sound for motion pictures except Warner Brothers, which regarded it as something that could bring short-term profits. Seeing the success of Warner Brothers with its experiment in sound, Fox Film Corporation acquired the rights to a sound-on-film process based on that of De Forest. Almost overnight, sound dominated the cinema. By 1930, 95 percent of major new films had sound. It was at about the same time that photographic color was introduced.

Films had sometimes been colored by hand, but this was time consuming and impractical. As film lengths began to increase, the Pathecolor system of mechanized stenciling was developed, in which a stencil was cut for each color to be applied frame by frame. This led to faster processes of chemically tinting the film stock. However, this process interfered with the film's sound track and was abandoned. Finally, in 1922, the Technicolor Corporation introduced a new two-color method for joining two separate positive prints. This process was successfully used in the 1920s. Technicolor improved the process in 1928, making it possible to combine all the primary colors to produce the lifelike tones and hues that dominated the filmmaking industry for the next 25 years.

The combination of these new filmmaking technologies made it possible to simulate movement with an astonishing sense of reality. Smaller aperture plates in projectors and different lenses could even make the images appear wider, and wide-screen stereoscopic films were born, adding further to the illusion of reality on the screen. These processes include Cinerama, Cinemascope, and VistaVision, and rely on 70mm or modified 35mm film stock, while most films use 35mm film stock. In 1970, the largest motion picture image presentation technology to date was introduced. IMAX was originally developed for the World Expo in Osaka, Japan, that year. Shot on 65mm film stock, IMAX films are shot sideways and projected the same way, onto screens up to eight stories tall. IMAX theaters were found mostly in science museums, amusement parks, and the like, showing short documentary films about natural phenomenon until the 1990s when the company behind IMAX began expanding into the traditional motion picture market and making short fictional films. Because of the limitations of equipment and technology involved in IMAX, as of 1998, IMAX films are usually less than an hour long.

While motion picture technology has moved forward, there has been two movements in the 1980s and 1990s that look back. In the 1980s, there was a controversial trend to colorize old black and white films, to make them, as its proponents argued, more accessible to the modern consumer. At the same time, restoration of old films became widespread. Before 1951, most motion picture film stock had a nitrate base, making the film susceptible to fire and chemically unstable. Many films, especially at the beginning of motion picture history, have been lost or faded. Several academic institutions and archives, including the American Film Institute, have spearheaded a movement to find, restore, and preserve these films. When they are found, they are transferred to safety film stock, as was used in the industry after 1951. But even films produced after 1951 have to be preserved to ensure a copy is available for future viewing, showing where the motion picture industry has been.

See also Movie camera; Phonograph; Photographic film; 3-D motion picture

Motion picture, 3-D • See 3-D motion picture

Motorboat • See Powerboat

Motorcycle

The motorcycle can be classified as a motorized vehicle with less than four wheels. Using this description, the invention of the motorcycle is usually attributed to **Gottlieb Daimler** and his assistant, **Wilhelm Maybach**. During the process of developing the **internal combustion engine**, these two German engineers installed a four-stroke, one-cylinder engine on a wooden **bicycle** equipped with wooden wheels. They built and drove this motorcycle in 1885.

The claim of Daimler and Maybach, however, is disputed by the British, who maintain that the English inventor Edward Butler invented the motorcycle a year earlier. This claim arises from the fact that Butler was issued a patent for a motorized tricycle in 1884; yet, apparently, his machine was not built until 1887. Neither the Daimler-Maybach nor Butler version was put into production for various reasons.

The first mass-produced motorcycle was introduced in 1894 by two Germans, Hildebrand and Wolfmuller, who built over 1,000 models. Many improvements followed: a motorcycle with a geared transmission appeared around 1900; a four-cylinder motorcycle had its debut in 1901 (a design created by the Englishman, Colonel Holden); the sidecar appeared about 1910; an electric-starter was introduced by the American firm, Indian, in 1913 (it was a complete failure and every motorcycle sold with this feature was recalled and the electric-starter had to be removed). In 1904, Indian's rival, Harley-Davidson, went into busines and both companies produced the V-twin engine design that remains an American favorite.

While the motorbike and the motorcycle are obviously quite similar, the smaller frame and lightweight nature of the motorbike distinguishes it from the motorcycle. Credit for building the first motorbike is given to French inventors, Eugene and Michel Werner. They first attached a small **gasoline** engine to the front wheel of a bicycle and met with great success; later they moved the engine to the lower part of the frame. The original appeared in 1897; the improved model, in 1900.

There have been other variations on motorized vehicles with less than four wheels. In 1887 in France, Léon Serpollet built a steam-powered tricycle which he drove from Paris to Lyon. The motorscooter—a small, lightweight vehicle with two small wheels—was made successful shortly after World War II by the Italian firm, Vespa, and was quickly copied by others. The motorscooter has a much longer history, however. It was originally called the *autofauteuil* (auto-easychair) and was invented by Frenchman Georges Gauthier, in 1902. Several European and American firms built scooters early in the twentieth century.

Although motorcycle design and performance improved in the decades between the wars, there occurred no real revolutionary changes. After World War II and into the 1950s however, the British bike industry, led by Triumph, produced increasingly sophisticated and better performing motorcycles, but as the 1960s arrived, Japan began to emerge as the new leader. Led by Honda, Japan came to dominate the industry, and by 1969, Honda produced a four-cylinder. overhead cam engine that not only had an electric starter and front disc brakes, but could reach 120 mph (193 kph). The 1970s belonged to the Japanese, and the Kawasaki firm produced a muscular machine that reliably gave racers 130 mph (209 kph). The 1980s coined the term UJM for Universal Japanese Motorcycle to describe their aircooled large and medium cycles. Developments during that decade included fairings (aerodynamically-styled coverings), single-shock rear suspension, and radial tires, in addition to aluminum frame construction. In the 1990s, liquid-cooled engines were adopted as were antilock brakes. The motorcycle of the near-future will be cleanburning and environmentally friendly, and may contain catalytic converters, fuel injection, variable valve timing and even programmable, smart-card engine management systems, as well as forkless front suspension.

MOTOR SCOOTER • See Motorcycle

MOUSE, COMPUTER • See Computer input and output devices

MOUTHWASH

Mouthwash is an oral care product primarily designed to freshen breath and remove food debris. Certain brands are also formulated to aid in the prevention of cavities, decrease buildup of plaque, and reduce gum disease.

The first references to mouth freshening products can be found in the records of the early Christian era which indicate that liquid products designed to clean the mouth and freshen breath were in existence at least as early as the first century A.D. Not much is known about these products other than that they were probably simple mixtures of essential oils. During the Arabian Period, between the seventh and twelfth centuries A.D., the science of cosmetics began to flourish. During this time fragrance and flavor oils (such as cassia, clove, nutmeg, musk, myhrr, tamarind, and rose) were discovered. Some of these oils were eventually incorporated into early oral care products.

However, it wasn't until the early nineteenth century that truly functional modern mouthwashes were invented. The first such product was Listerine Antiseptic mouthwash, which was, and still is, marketed by Warner-Lambert. According to company history, Listerine got its start "not in the bathroom but in the operating room." In the late 1880s scientist Joseph Lister discovered that infections could be stopped by killing germs with antiseptic substances such as carbonic acid. One of Lambert's contemporaries, Dr. Joseph Lawrence, was among the many researchers also experimenting with antiseptics. In 1879 Lawrence found a unique combination of essential oils that could be safely used to kill bacteria. Lawrence formulated these oils into a product for use in surgical procedures and named it Listerine in honor of Dr. Lister. Lawrence sold the rights to his invention to in 1884 to Jordan Wheat Lambert who co-founded Warner-Lambert. Lawrence positioned Listerine as the first commercial mouthwash and it went on to become the first product of its kind to earn the American Dental Association's seal of approval. Today a variety of mouthwash products are available in a range of flavors and functions.

Most mouthwash formulations contain water, alcohol, flavor oils, antimicrobial ingredients, foaming agents, colorants, and preservatives. Water is the predominant ingredient and is used as a diluent and solvent for the other ingredients. Water makes up more than 90% of some mouthwash formulations, depending on whether or not alcohol is present. Historically, ethyl alcohol has been added to mouthwashes to give the product a certain mouth-feel, known as bite. The alcohol also helps dissolve the flavor oils and maintains product clarity.

Flavor oils such as peppermint, spearmint, wintergreen are commonly used in mouthwashes. Other popular flavoring agents are based on cinnamon or clove. Listerine uses a combination of natural oils including thymol at 0.064%, eucalyptol at 0.092%, methyl salicylate at 0.060%, and menthol at 0.042% to give it a unique flavor. This blend of ingredients also provides the product with its antimicrobial properties.

In addition to the ingredients used by Listerine, there are a number of other antimicrobial ingredients used in mouthwashes today to kill the germs that contribute to halitosis, or bad breath. These ingredients include cetylpyridinium chloride, domiphen bromide, and benzakonium chloride.

Foaming agents, or surfactants, are added to mouthwashes to help solubolize other ingredients, to aid in cleansing, and to provide a slight amount of foam. Popular foaming agents include anionic surfactants such as sodium lauryl sulfate or nonionic surfactants like Poloxamer.

Finally, a variety of colorants and preservatives are also added to these products. Colorants make the product more aesthetically pleasing while preservatives prevent spoilage from microbial growth. As with all ingredients used in oral care products these must be safe for use in the mouth.

Although mouthwashes were originally little more than flavored water, today's products can deliver efficacious therapeutic agents. For example. the FDA has approved mouthwash formulas for the delivery of fluoride to prevent cavities. In addition, many products are formulated with antiplaque and anticalculus agents. Listerine claims that it "Helps prevent and reduce supragingival plaque accumulation and gingivitis when used in a conscientiously applied program of oral hygiene and regular professional care." Furthermore, it has the American Dental Association Seal of Acceptance for helping control plaque and gingivitis.

In the 1990s mouthwashes have come under increased regulatory scrutiny. In the early 1990s, media attention was focused on a research report which indicated that high levels of alcohol in mouthwash may be linked to certain oral cancers. Despite the attention the matter has received, the data on this subject is not conclusive and the industry maintains that their products are completely safe when used as directed. Still, media pressure prompted many manufactures to eliminate alcohol from existing brands or start new brands which were alcohol free. Other areas of regulatory concern are related to the ability of mouthwashes to reduce plaque and gingivitis. While major manufacturers continue to market products claiming to treat gum disease, the FDA may one day rule that these are drug claims which are not allowable for cosmetic products.

MOVIE • See Motion picture

MOVIE CAMERA

The movie camera is a mechanical device with which a sequence of pictures is recorded in rapid succession on a roll of film. The invention of this device spawned a multi-billion dollar film industry whose movies have entertained audiences worldwide.

The history of the movie camera spans many years. In 1833, Belgian physicist Joseph Plateau (1801-1883) created the Phenakistoscope. When this apparatus, made up of a cardboard disk around which sequential drawings were attached, was rotated, the subject of the drawing appeared to move. In order to create the illusion of movement, photographs must be taken rapidly and later viewed rapidly. In 1872, Englishman **Eadweard Muybridge** used a series of cameras to take sequential pictures of a running horse. Two years later in France, astronomer Pierre Jules Janssen (1824-1907) designed a revolving camera attached to a **telescope** and used it to photograph Venus. By the late 1870s Muybridge dramatically increased camera shutter speed, and in 1880 he introduced his Zoopraxiscope, which could rapidly project series of individual photographs attached to a revolving drum.

Drawing on the work of Muybridge, Frenchman **Etienne-Jules Marey** in 1881 developed a photographic gun, with which a sequence of pictures could be quickly recorded around the circumference of a photographic plate. Later, when **George Eastman** introduced a gelatin-based film, Marey recorded images at faster speeds on rolled film. Working for **Thomas Edison** in 1889 William Kennedy Laurie Dickson (1860-1937) invented a motion picture camera, the Kinetograph. In it he used Eastman's perforated 35-mm film which set the standard for motion picture film. By 1891 Dickson had also developed the **Kinetoscope** a peephole film viewer.

Inspired by *Kinetoscope*, the team of brothers **Louis and Auguste Lumière** invented in 1894 their Cinématographe, which was both a camera and a projector. When filming, it incrementally advanced a 35-mm film behind a rotating shutter. Their first public showing of a motion picture in Paris on December 28, 1895 marked the beginning of cinema as we know it today. Frenchman Charles Pathé (1867-1957) separated the Lumière brothers' dual-purpose *Cinématographe* into a camera and an independent projector. In 1904, Pathé refined the camera to shoot film at variable speeds

Throughout the silent era of film, until about 1928, the loudness of movie cameras and their internal mechanisms did not matter during film because no sound was recorded. When sound technology was introduced, cameras had to become silent. At dawn of the sound ear, they were put in booths that drowned out their sound called blimps. However, this also limited the mobility of the camera making for very static, limited filmed images. In 1935, the first camera to contain an internal masking blimp was introduced, the Mitchell BNC (or blimped news camera). Though the Mitchell was portable, it was still heavy. Over the next forty years, movie cameras became lighter and quieter, and working on location outside of a studio became possible. Cameras become small enough to be hand held, so they could film shots as close to the subject as possible. But these hand held cameras often produced shaky, frenetic images. To produce smooth action shots, a camera would have to be mounted on a dolly, a platform on wheels that moves along a track. In 1976, the Steadicam was introduced. Invented by Garrett Brown, the Steadicam is a "hand-held" camera that shoots smooth action shots. The Steadicam operator wears a vest which has a supporting spring arm attached, and these provide the platform for the camera. Balanced by a device called a gimal, the camera can go anywhere its operator can go, and provide dolly-like motion.

Movie projector • See Motion picture

Mowing machine

Mowing machines are designed to cut grasses and other crops grown for hay and silage. They are also used to cut grass along highway rights-of-way, which may be used for feeding livestock.

Mowing machines are distinguished from **lawn mowers** in that the latter are used to keep ornamental grasses closely trimmed. They also differ from **reapers** in that reapers are used to harvest mature grain crops while mowers cut the crop while it is green, requiring a sharper cutting instrument. Some machines, however, have been designed to perform both types of cutting. Before mowing machines were invented, grasses had to be cut manually while wet—a laborious task completed by teams of men using scythes.

Peter Gaillard of Lancaster, Pennsylvania, obtained the first patent on a mowing machine in 1812. However, his attempt to recreate the cutting action of the scythe proved unsuccessful. In 1822, Jeremiah Bailey of Chester County, Pennsylvania, patented a mower which had a revolving circular scythe set horizontally a few inches above the ground.

In 1831, American William Manning invented a cutting arm with reciprocating toothed blades which was manufactured by **Obed Hussey**'s farm machinery company beginning in 1850. This cutting arm, or bar, is still used today. Another mower appearing in 1831, patented by American Cyrenus Wheeler, was designed to cut close to the ground without clogging or damaging the sod. As with many other types of machinery, the **internal combustion engine** was added to mowers in the early 1900s, making them self-propelled.

Modern mowing machines fall into three categories. First is the *reciprocating cutting bar*, a side-mounted bar added to a tractor for local cutting. The length of the bar is limited to five or six feet to reduce the chance of striking hard objects and to minimize clogging. Second, there are *flail mowers*, consisting of free-swinging knives, or flails. The grass is bent forward by an overhanging bar and is then undercut by the blades. Finally, *impact-type cutters* use a single fixed blade to cut material that is rigid enough to oppose the cutting element. After cutting, the grass is raked and conditioned, or dried. It is then delivered to feeding troughs, storage bins, and silos.

See also Fodder and silage; Haying machines and tools; Knife; Reaper and binder

Müller, Paul Hermann • See DDT

Multimeter

A multimeter is a multiple range test instrument that measures voltage, current, and electrical resistance. It was invented in the 1920s to combine the **voltmeter** and **ammeter** into one instrument.

There are two basic types of multimeters. One type has a moving coil, indicator pointer, and a graduated scale; the other has a digital display.

The moving coil (and its cousin the moving iron) multimeter makes use of a principal discovered by Hans Christian Oersted (1777-1851) in 1820; an electric current in a **wire** produces a magnetic field. The magnetic field exerts a force on a coil (or iron magnet), making it turn. This moves the attached pointer, allowing a measurement to be read off the graduated scale. Its greatest limitation is the need to use the very current being measured when it's operating as a voltmeter.

A multimeter with a digital display makes use of integrated electric circuits and semiconductors and does not have moving parts. A digital multimeter offers greater accuracy and does not draw current, but it is more expensive to build.

In either case, a multimeter has a dial, or push-buttons, on the outside and different resistors and shunts on the inside. Changing the setting allows the user to switch between the multimeter's various functions.

The primary use of the multimeter, whether it be a moving coil or digital display, is as a voltmeter to measure voltage.

Resistance is measured by the multimeter in ohms. The resistance of a conductor is one ohm when a steady current of one ampere flows through the conductor and produces a potential difference of one volt.

When used as an ammeter, the multimeter measures the flow of electrical current in amperes. Most ammeters measure **direct current** (DC). A very small current can cause a full-scale deflection of the coil, so the ammeter has to be modified (switchable) if it is used to measure high currents.

The more-intelligent digital display multimeters, which are basically digital voltmeters, have caused a decrease in the use of ammeters.

William Murdock

Murdock, William (1754-1839)

Scottish engineer

Murdock was a central figure of the English Industrial Revolution. His renowned mechanical ability probably had its origin in his early exposure to his family's milling business; his father was involved in casting the first iron-toothed gear in Scotland. In 1777, Murdock went south to look for a job with Boulton & Watt, a partnership that was selling **James Watt**'s new steam engines. During the job interview, **Matthew Boulton** noticed that Murdock was nervously handling a fine piece of woodwork that he had brought with him. Boulton recognized Murdock's talent and hired him on the spot.

Murdock entered the business as a foreman, supervising the installation of engines in tin mines in southwest England. Besides improving the engines being installed, he made a remarkable three-wheeled steam-engine ''carriage'' that reached speeds of seven miles per hour (11 kph) in 1784. Although Murdock's ideas were further developed by **Richard Trevithick**, Watt controlled the patent for steam-powered carriages, and the first **locomotive** was not developed for another 25 years.

Soon Murdock began the experiments in gas lighting for which he is famous today. Murdock was the first to realize that gas was a more convenient energy source than coal, primarily because it could be piped and controlled more easily. Despite ridicule from his peers and the danger of gas explosions, Murdock installed gaslighting in his house, using gas made from coal in his backyard and piped in through a hole in a window frame. Murdock went on to develop methods for manufacturing, storing, and purifying coal-gas.

Murdock's employers were unenthusiastic about this sideline until they heard that a similar gaslighting system, made by **Philippe Lebon**, was being used in France. Boulton and Watt then asked Murdock to install gaslighting at their main factory in Birmingham in 1802, as part of England's celebration of a temporary peace treaty with France. Soon the firm received its first commercial order to install gaslights at a cotton spinning factory. By 1806, Murdock had improved the odor of the coal-gas.

Meanwhile, Murdock had continued to apply his ingenuity to steam engine improvements, many of which were patented in 1799. He invented a new machine for boring cylinders and a better method for casting jacketed cylinders. Today, he is still known for inventing the slide valve, which injects and removes steam alternately from each end of the cylinder. Murdock also built the first model of an oscillating engine and the first free-standing steam engine.

Murdock also explored the possibilities of harnessing the power of compressed air to drive machinery. This work represented an early application of air-driven (pneumatic) systems used today in truck brakes and manufacturing equipment. Although Murdock eventually became a limited partner in

Boulton & Watt, he remained essentially a hired hand throughout his career, rather than an independent businessman. Nevertheless, he earned great respect from many rivals who were wealthier and better educated.

See also Air compressor; Braking systems; Steam-powered road vehicle

Murray, George Washington (1853-1926)

American politician and inventor

A farmer by trade, Murray led a successful career in politics and carried out a lifelong battle for voting reform in the United States. Born on September 22, 1853, near Rembert, South Carolina, Murray was the son of slaves. He went to public schools, graduating from the State Normal College at Columbia, South Carolina, in 1876. He then took up farming and teaching school.

He ran unsuccessfully for the United States House of Representatives in 1890, but was elected in 1892 when he won a last-minute challenge of a discriminatory election law engineered to keep African-Americans from the ballot box. In Congress, he fought for black rights, speaking in favor of retaining Reconstruction laws which protected black voting rights, and he highlighted African-American achievements by reading into the congressional record a list of 92 patents granted to African-Americans, eight of which happened to be his own patents for agricultural implements. When Frederick Douglass died, Murray made an unsuccessful attempt to allow the anti-slavery activist's remains to lie in state at the Capitol. Murray appeared to have lost his reelection bid in 1894, but once again he challenged voting irregularities and was awarded his seat by the House Elections Committee after a two year battle. During the next two years he was forced to spend the majority of his time trying to fight off plans in South Carolina to limit the rights of African-American voters by imposing stringent residency, literacy, and property requirements.

An unfavorable 1898 Supreme Court ruling on the question of poll taxes effectively destroyed Murray's ability to be reelected. After his electoral loss, he returned to farming and real estate. He died on April 21, 1920, in Chicago, Illinois.

Musical instruments, electric

Perhaps the most highly developed musical instrument is the human voice. It requires no special apparatus to use and can be carried anywhere. But beyond the voice, other musical instruments have been developed over the millennia because of a desire to extend the range of sounds that are possible for the listener to hear. Musical instruments are probably as old as the human race. Drums, flutes, and all manner of stringed instruments can be found among the artifacts of virtually every civilization that has ever existed. People have always used whatever materials and technology were available at the time to design and construct musical instruments. Therefore, it is only natural that as the use of electricity became increasingly widespread in the late nineteenth century, it was quickly applied to musical instruments.

One of the first successful instruments featuring an all-electric design was produced in Europe in 1920 by Lev Termin, a Russian radio engineer. The *Aetherphon*, later known as the theremin generated a sound whose pitch was controlled by two precisely positioned antennae and create a melody using carefully rehearsed movements. So startling was the effect of the instrument in its day, that it inspired numerous composers, such as Edgard Varèse, Percy Aldridge Grainger, Charles Edward Ives, and Stuart Copeland, to include the theremin in some of their compositions. The general public is probably most familiar with the theremin from science fiction films of the 1950s in which the eerie sounds of the instrument were used to evoke a mood of other-worldliness. Robert Moog, who later produced one of the first publicly marketed synthesizers, began his business career as a manufacturer of theremins.

France saw the invention of an electric, keyboard-based instrument in 1928 called the *ondes martenot*, named for its inventor Maurice Martenot. Similar in sound to the theremin, the ondes martenot has had more success in the concert hall than its rival instrument. The compositions of Darius Milhaud, Arthur Honegger, Florent Schmitt, Jacques-François Antoine Ibert, and others have featured the use of the ondes martenot. Even today, music students in some French and Canadian conservatories can receive performance instruction on the ondes martenot, just as if it were any another conventional instrument.

As interesting as the new sounds of the theremin and the ondes martenot were, there existed an unfulfilled need to create electric instruments that duplicated the sounds of traditional acoustic instruments. This need was met in 1935 with Laurens Hammond's invention of an all-electric organ. Hammond designed a clever combination of signal-generating electronic modules that could be linked together in different combinations to produce the rich ''registration'' that normally required mechanical organ technology. Since their invention, Hammond organs have been one of the staple instruments in musical settings from Broadway pit bands to church ensembles.

The rise of **radio** broadcasting and sound recording also led to the creation of new electric instruments. The impetus that led to their invention was the need for amplification, which became acute during the 1940s and 1950s when conventional acoustic instruments were found to be incapable of competing with the din of dance hall and night club audiences. One of the first attempts to add an amplifier to a guitar was made in 1929 in Los Angeles by two Czech immigrants, the Dopyera brothers. Their invention consisted of a mechanical resonator placed on the guitar, which provided a modest acoustical boost. Later the Dopyera brothers succeeded in designing an all-electric device that could pick up guitar vibrations and amplify them. But it wasn't until 1956, when Clarence Fender received a U.S. patent for his ''tremolo device for stringed instruments,'' that the modern electric guitar came into its own. The *Fender Stratocaster* was marketed to representatives of almost every stylistic genre of the music recording industry, resulting in a major change in popular music.

The conversion of the guitar to an all-electric instrument led to the creation of new guitar designs. The function of the guitar body as a resonator was no longer necessary, which made it possible to fashion guitar shapes of any size or configuration. The highly mechanized pedal steel guitar of country music does not hang from a shoulder strap, but sits flat like a piano keyboard. And the space-age body designs of guitars played by such performers as Jimi Hendrix and Eric Clapton have made the instrument into a stage prop is critical to defining the unique ''image'' sought by all rock 'n roll bands.

Any source of physical energy can be transformed into sound, either mechanically or electrically. One can only assume that as new sources of energy are discovered, the possibilities for the creation of new musical instruments will continue to increase.

Musical instruments, mechanical

The great twentieth-century composer Pierre Boulez noted that the history of music is ''littered with corpses: superfluous or over-complicated inventions, incapable of being integrated into the context demanded by the musical ideas of the age which produced them.'' With the rise of the modern, scientific age came the rise of the machine. As in the earliest ages, people used new technology to create music, particularly devices to mechanically produce music with little or no human participation. The definition of mechanical musical instrument is arbitrary at best since most musical instruments use some kind of levers (e.g., woodwind instruments) or pistons (e.g., brass instruments). However, many instruments that in some way employ mechanisms, from the marvelous to the bizarre, survive to this day.

Some of the first examples of or self-playing instruments date from the sixteenth and seventeenth centuries in the form of mechanical birds, automatic flute players, and dancing musical dolls. It was not until the 1700s that music boxes became commonplace in Europe. These devices consisted of a finely tuned comb of metal tines activated by pins that were punched onto a metal cylinder in a precise sequence designed to produce a musical phrase.

A refinement of the cylinder music box was introduced in Germany toward the end of the nineteenth century. Instead of a rotating cylinder, the German invention featured a revolving disc that contained the properly sequenced pin arrangements. The real innovation, however, was the fact that the discs could be removed and replaced with other discs, each of which produced a different tune.

Music boxes represented the application of clock-building technology to the creation of automatic musical instruments. Completely different kinds of mechanized instruments were also designed using similar clock-based technology. A document dating from 1130 describes the construction of a barrel organ in Reims, France, and by 1350, pinned barrel organ systems were used in conjunction with clock mechanisms in churches throughout much of Europe. Huge, weight driven ''super organs'' were in use by European royalty by the early seventeenth century. At this point, a change in conception and scale occurred which saw the appearance, around 1700, of the hand-cranked barrel organ, or hurdy-gurdy. Empress Maria Theresa granted barrel organ licenses to disabled soldiers, a trend that continued after the Napoleonic Wars, when licensed barrel organ players on the streets of France became a common sight.

As music boxes and barrel organs became widely available, inventors of automatic musical instruments set their sights on creating an instrument capable of mechanizing the action of several instruments simultaneously, a veritable orchestra on wheels. One of the earliest, and most famous, of these ''orchestrions,'' as they were called, was the Panharmonicon built by Johann Maelzel shortly after 1800. This noisy contraption featured automatic trumpets among its many innovations. The device is known in music history for having inspired no less a composer than Ludwig van Beethoven (1770-1827) to accept a commission to write a piece of music for it called ''Wellington's Victory.'' Smaller versions of orchestrions, such as the coin-operated nickelodeon, fared more successfully than the gargantuan attempts of Maelzel and others.

Forcing air through pipes to create musical sounds, which is the method employed in devices such as the hurdy-gurdy or orchestrion, essentially describes the mechanics of an organ. The mechanization of piano performance, defined as a hammer striking a string, was achieved in 1842 by Claude-Felix Seytre of Lyons, France. His patented invention utilized a flat music sheet that had encoded upon it the information necessary to activate the proper mechanism to strike hammers upon the appropriate **piano** strings. The innovation of the perforated player-piano ''roll'' was made by Alexander Bain in 1848. Many of these early devices were not, themselves, complete musical instruments but mechanisms that were designed to be added to an already existing piano. Completely self-contained, fully automated player-pianos were developed later in the 1800s.

In 1855 Joshua Stoddard, a Vermont native then living in Worcester, Massachusetts, received a patent for his calliope—an organ powered by a steam boiler. When the player pressed one of the brass keys of the keyboard, it opened a valve, which sent steam through a pitch pipe. Although Stoddard envisioned an instrument suitable for church use, the loud harsh tones of the calliope found their niche in the circus, where they can still be heard today.

By the end of the nineteenth century, automatic musical instruments were becoming one of the primary sources of music in the lives of the masses. An advertisement for a player-piano at the beginning of the twentieth century claimed that the player piano ''solves the problem of music in the home.... More could not be asked of mortal ingenuity...the greatest and most widely popular of musical inventions... the Royal Road to Music in the Home!'' The zenith of the player-piano's popularity was reached in the years between 1900 and 1930 when over 2 million player-pianos were produced in America.

The demise of automatic musical instruments was foreshadowed with the invention of the gramophone in 1887. Technically, the earliest **phonograph**s met the definition of an automatic musical instrument, as they reproduced sounds using a mechanical process. The storage of sound information by scratching a needle into a cylinder or disk was, in principle, the same as storing the musical information in the form of perforations on a player-piano roll. The electrification of the recording process removed phonographs from the official list of ''mechanical'' musical instruments but did not remove them from the hearts of the listening public, who turned to high quality, professional disk recordings as their primary source of musical enjoyment.

Musket

The musket may be loosely defined as a large-caliber smoothbore firearm fired from the shoulder. Early muskets weighted at least 40 pounds (18 kg) and were about 6 to 7 feet (1.8 to 2.1 m) long. These guns were so heavy that soldiers propped them in forked rests on the ground to aim and fire them. It is uncertain when the musket was first created, but Spanish accounts of musketeers—soldiers armed with muskets—date back as early as 1528.

Loading and firing muskets was tedious at best: the user dropped a charge of loose black **gunpowder**, a lead ball, and a wad of **paper** down the barrel, pushing it to the end with a ramrod. The **firing mechanism** was a matchlock, a glowing wick held in place by a moveable clamp called a serpentine. A small pan was filled with fine priming powder at the firing end, which acted much like a modern gun's hammer. When the trigger was squeezed, the serpentine swung forward to touch the glowing wick to the priming powder, which ignited and, in turn, sent a flame through a touch hole and ignited the main charge. This was a long process, and open to many mistakes. However, muskets were worth the trouble because of their powerful shot, which could easily pierce armor.

The musket soon became the most important military weapon in Europe. In 1550 the Duke of Alba, an advisor to King Philip II of Spain, introduced it to the Spanish army. Although the weapon could not shoot at great distances, it was deadly against massed troops of the period. Further improvements helped maintain the musket's popularity, including the plug **bayonet**. This was a knife blade set into a tapered shaft designed to fit into the musket's barrel, thus transforming the weapon from gun to pike. This was superceded by the socket bayonet, which allowed the gun to be fired with the blade in place. Perhaps the greatest advance for muskets was the introduction of the flintlock system. This firing method featured a spring-activated cock striking a piece of flint against a vertical plate. Under the plate was a small pan containing priming powder and a flash hole which sent the flame to the main charge inside the barrel. In the 1720s, the British military issued the *Brown Bess*, a flintlock musket which became famous for its unpainted brown walnut stock. The Brown Bess fired .75 caliber balls and was equipped with handsome brass mountings. The French countered with a lighter weapon, called the *Charleville*, with a caliber of .69 and a slender, more graceful stock.

The musket's well-known lack of accuracy contributed to its demise. With its smooth bore, the bullet flew out of the barrel heading erratically toward the target, and no matter how well aimed, a hit depended much on luck. For military use, this had long been good enough, because cavalry found it difficult to resist a massed rank of musketeers. Eventually, these smoothbore weapons gave way to rifled guns capable of sending spinning bullets toward an enemy with much better effect. Muskets continued to be used militarily through the eighteenth century. Improvements in **rifle** design in the early 1800s ultimately led to the musket's obsolescence.

See also Rifle; Shotgun

Mustard gas • See Chemical warfare

Muybridge, Eadweard James (1830-1904)

English photographer and inventor

The son of a corn merchant, Edward James Muggeridge was born in Kingston-upon-Thames, England, on April 9, 1830. While in his twenties he changed his name and emigrated to California, where he learned photography and made important contributions to that field.

In San Francisco, Muybridge worked with photographer Carleton A. Watkins. Later the U.S. government commissioned Muybridge to conduct photographic surveys of the Pacific Coast. In the late 1860s, Muybridge's photographs of Alaska and California caught the attention of Leland Stanford, a former California governor. Stanford, who was familiar with **Étienne-Jules Marey**'s photographs of running horses, hired Muybridge to confirm that a running horse's hooves were positioned under its body when they all simultaneously left the ground.

In 1872 Muybridge used a series of cameras to take sequential pictures of a running horse. As the horse passed each camera, it tripped a wire that set off the camera shutter. Though Muybridge maintained that one of his photographs demonstrated that Marey was correct, the quality of the photograph was poor and his proof inconclusive.

Muybridge's work was interrupted when in 1874 he was tried and acquitted of murdering his wife's lover. After several years on assignment abroad, he resumed his work in California, supported financially by Stanford. Muybridge improved the shutter mechanism of his camera so that it could take a 1/2000 of a second exposure.

In 1880 Muybridge produced the Zoopraxiscope to project photographs in rapid succession, thus creating a primitive form of **motion pictures**. After a three-year lecture tour in Europe, Muybridge received support for his research from the University of Pennsylvania. By 1885 he had taken more than 100,000 photographs of moving animals and humans. These images were made available in his *Animals in Motion* (1899) and *The Human Figure* (1901), which allowed scientists and artists to refine their understanding of animal movement. Muybridge died in the city of his birth on May 8, 1904.

Mylar • See Polyester

Eadweard James Muybridge

N

NAIL

The earliest nails were probably made in the Middle East about 5,000 years ago. Metal was heated and then pounded into the desired shape.

Making nails by hand, one nail at a time, continued as the method of production until the 1700s. The only slight improvement was the machine patented in 1606 by the Englishman Bevis Bulmer, which cut the iron from which nails were fashioned into rods of various thickness. Hammering out nails by hand from strips of iron was a common household task for pre-Revolutionary American colonists. About 1775, Jeremiah Wilkinson, a Rhode Island inventor, devised a machine that cut nails from a sheet of cold iron. In 1786, Ezekiel Reed of Massachusetts invented a nail-making machine, and in 1795, another Massachusetts inventor, **Jacob Perkins**, patented a nail-making machine that could cut and head nails in a single operation. This machine, which produced up to 200,000 nails per day, made mass manufacture of nails possible for the first time. Nails became widely available and affordable; by 1842, the price had dropped to 3 cents a pound, down from 25 cents a pound in 1795.

Around 1851 a New York machinist named William Hassall (or Hersell) produced the first machine for making nails out of **wire**. The availability and affordability of nails made possible a revolutionary new way to construct houses: the basket-frame or **balloon frame** method. Traditionally, the framework of a house was made of thick, heavy wooden beams fitted together at the end with notches and pegs. Such construction was slow and required skilled carpenters. The balloon or basket frame, introduced in Chicago, Illinois, in 1833, used thin sawed timbers instead, held together with nails. This type of construction could be accomplished quickly and economically, with minimally skilled labor.

Most nails today are made from wire on machines that turn out 500 a minute. Around 300 varieties of nails are available, each suited for a different purpose.

NAPALM

During World War II the United States military wanted a more efficient fuel than liquid **gasoline** for its **flame throwers**. In 1942 a team of chemical researchers at Harvard University led by Louis F. Fieser developed an inexpensive and readily available gelling agent from naphthenic acid, a **petroleum** distillate, and palmitic acid, hydrolyzed from palm oil. The new compound, called napalm, was mixed with gasoline to make a thicker, slower-burning fuel that could be projected farther and more accurately. It is a deadly chemical, clinging to whatever it touches and burning with a ferocious tenacity.

Napalm was soon adapted for other military purposes, particularly as incendiary filler for fire bombs. The Allied air forces first used it against Japanese industrial targets and at Guadalcanal. During the Korean War, napalm land mines were used. Its use continued in Vietnam, where it proved especially effective as a defoliant. Dropped from an airplane, napalm ignited large areas. The use of napalm against civilians in Vietnam contributed greatly to the unpopularity of that war, and a widely published photograph of a naked child escaping from her burning village was one of the most sinister images of the time.

Vietnam-era napalm has since been replaced by napalm-B, which consists of a polystyrene thickener, benzene, and gasoline. Napalm-B is even more lethal than its predecessor, burning hotter and longer. The United States dropped 15,000–lb (6,804 kg) bombs of ''jellied slurry,'' a napalm-like substance during the Persian Gulf War to clear areas on the ground for its helicopters. In 1998, the United States Navy was faced with the task of disposing of thousands of gallons of napalm that it had kept in storage for almost thirty years. The dangerous chemical is difficult to recycle or destroy harmlessly.

See also Chemical warfare

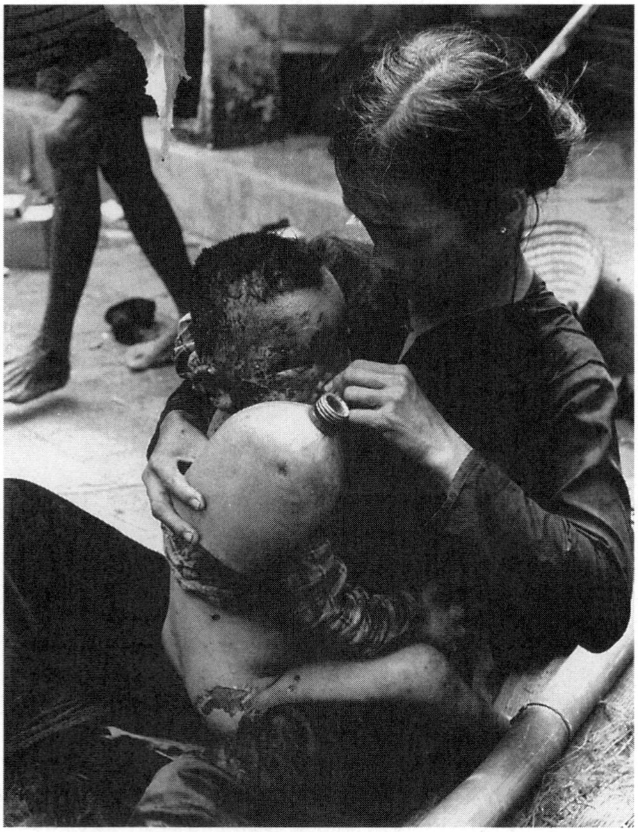

A Vietnamese mother comforts her child, burned by napalm. Use of the chemical, first developed during World War II, in the Vietnam War added to the controversy that surrounded the conflict.

Napier's bones

John Napier (1550-1617), a fifteenth century Scottish mathematician, noted that the astrologers and surveyors of his time toiled over their complex mathematical computations. He had long felt such tasks could be accomplished with less drudgery and greater accuracy. He became determined to improve the process.

Born in Edinburgh in 1550, Napier was educated in France and became increasingly interested in mathematics, which he began to study with a passion soon after his marriage in 1571. In 1594 Napier became particularly concerned with simplifying the multiplication and division processes, and the thought occurred to him that all numbers could be expressed in exponential form. Once written in such a form multiplication and division could be accomplished by adding or subtracting the exponents. He called this process of computing exponential expressions logarithms, or proportionate numbers. Exactly how Napier hit upon the use of logarithms is not clearly known, but their impact on the mathematical community of his time was highly significant.

In 1614 Napier he published a book, *Mirifici logarithmorum canonis descriptio*, in which he presented the logarithmic tables and explained their use. In the next several years, Napier took his concept one step further and invented several automatic calculating machines. One such device, a small box containing plates used to perform multiplication, was known as the *promptuary of multiplication*. Around 1617, he developed another device, which he called *rabdologiae*. Napier placed his logarithm tables on wooden cylinders, the surfaces of which contained numbers. By turning the correct cylinders (which represented the digits 0 to 9) and adding or subtracting the numbers which appeared, the correct result was displayed. As Napier's rabdologiae gained popularity, others referred to it as numbering rods, multiplying rulers and speaking rods. However, the device is best known to historians as Napier's bones—so called for the ivory or bone cylinders later used in place of the original wood. Through Napier's invention, the process of doing routine calculations was simplified to an amazing extent, causing other mathematicians to implement Napier's bones on their own calculating devices. Napier's logarithmic discovery completely transformed the mathematical calculation process.

See also Logic machine

Nasmyth, James (1808-1890)
Scottish engineer

James Nasmyth invented the steam hammer, one of the integral contributions to the Industrial Revolution in Europe.

Nasmyth was born in Edinburgh, Scotland, on August 19, 1808, the son of an artist. He left school at age twelve to make model engines and other mechanical devices. At nineteen he built a full-size steam carriage which performed with acclaim. When he was twenty-one, Nasmyth accompanied his father on a trip to London, England, where he met machinist and engineer **Henry Maudslay**. During the next two years, Nasmyth studied and worked under Maudslay, learning from him as well as making valuable contributions, such as designing hexagonal-headed nuts and a flexible shaft of coiled spring steel for drilling holes in awkward places.

In 1834, Nasmyth opened his own shop in Manchester, England, later moving to a foundry at Patricraft, England, where he became known for his craftsmanship and steam-powered tools. It was also here, in 1839, that he invented the steam hammer, a device that allowed large materials to be forged with great accuracy. The concept of the steam hammer was simple, even though the idea was totally new. A hammering block was hoisted by steam power to a vertical position above a piece of metal. Once the hammer reached an appropriate height, steam in the **piston** was released and the block fell. The pistons could be regulated not only in strength of blow, but also in frequency of strokes.

At the time, Nasmyth decided to postpone patenting, building, and marketing the new steam hammer. Two-and-a-half years later, however, while visiting a fellow machinist in France, Nasmyth was shown a steam hammer that had been built from his own rough sketches. Nasmyth quickly returned to England, patented his work, and manufactured hammers for an eager market. Soon he was making hammers with four-and five-ton blocks, and by 1843 he had improved on them by injecting steam above the piston to add force to the downward blow.

The steam hammer allowed larger forgings with heavier metals, tightened bonds, and made metals stronger and more dense. Not surprisingly, Nasmyth soon revived a previous interest and became involved in manufacturing steam **locomotives** for various railway companies. In fourteen years, he built 109 high-pressure steam engines, pumps, and hydraulic press. His steam hammer was exhibited at the Great Exhibition of 1851 alongside his prize-winning maps of the moon.

Nasmyth retired in 1856, and dedicated his last thirty years to astronomy, a life-long interest and hobby. He built a number of telescopes and charted sunspots as well as the surface of the moon. Besides his steam hammer, a direct predecessor of the pile driver, Nasmyth also devised a vertical cylinder-**boring machine** and **milling machine**s. He died a financially successful inventor (unlike many of his peers) on May 7, 1890.

James Nasmyth

NATTA, GIULIO (1903-1979)

Italian chemist

Natta was born in Imperia, Italy, on February 26, 1903. At the age of twelve, he became interested in chemistry. Natta graduated from high school in 1919 and began to study mathematics at the University of Genoa, but transferred to the Polytechnic Institute of Milan to study chemical engineering. He received his doctorate in 1924 at the age of twenty and began his science career.

Natta rose through the ranks at the Polytechnic Institute as an instructor and was named professor of chemistry in 1927. He researched the structure of inorganic compounds and industrial catalysts with X rays. In 1932, Natta learned of electron diffraction analysis from Hermann Staudinger (1881-1965) at the University of Freiburg and began to use both X-rays and electron diffraction to investigate the structure of **polymers**. Over the next six years, Natta served as professor at the University of Rome, and as professor and director at the Institute of General Chemistry at the University of Povia, the Institute of Industrial Chemistry at the Turin Polytechnic Institute, and at the Industrial Chemistry Research Center at the University of Milan.

In 1938, under the patronage of the Italian government, Natta headed the research team studying the production of **synthetic rubber**. The team was successful and production of butadiene styrene rubber began during World War II. Natta and his team went on to research catalysts and produce butadiene, formaldehyde, methanol, butyraldehyde, and various other alcohols. After World War II, Natta's research was underwritten by an Italian company named Montecatini. He began researching polymers using petroleum as his raw material.

In 1952, Natta heard a lecture given by **Karl Ziegler** in which he described his procedure of preparing **polyethylene** by using resins with aluminum or titanium atoms as catalysts. The organometallic catalysts allowed long chain polyethylene to be formed with no side chains, which greatly increased its strength. Under a licensing agreement with Ziegler's Institute, Natta's company sent some of their technicians to Ziegler for training. Making good use of their knowledge, he began to investigate the production of **polypropylene**. Propylene has one more carbon atom than ethylene and is cheaper than ethylene. In fact, polypropylene, a byproduct of refining processes, was available in large quantities.

In 1954, Natta produced a new polypropylene with great durability, heat resistance, and *tensile* strength. The compound was somewhat of a surprise because Natta had expected a linear polymer rubber with a high molecular weight. Instead he had succeeded in making polypropylene that had all of the methyl groups on the same side of the carbon chain. This arrangement was called isotactic, a name proposed by Natta's wife. This new polymer polypropylene contained a degree of crystallinity that was much higher than expected. Eventually, Natta discovered that by controlling the organometallic catalysts he could predict the three-dimensional arrangement of the atoms in the polymer. Natta shared the 1963 Nobel Prize in chemistry with Karl Ziegler for their work with the catalysts. Today polypropylene is used in molded auto parts and appliances, as well as in **textiles** for **carpet**s, ropes, and cables.

Eventually, Natta found a way to duplicate the polymerization process that occurs biologically. He was able to control *stereochemistry* so that optically active polymers could be prepared from optically inactive *monomers*. Natta's research with catalysts changed the way that polymers were developed and introduced a new class of high strength materials that outperformed metals. Natta developed Parkinson's disease later in

Giulio Natta

his life. He died in Bergamo, Italy, on May 2, 1979, from surgical complications.

See also Chemical engineering; X-ray machine

Navigational satellite

Long before the space age, people used the heavens for navigation. Besides relying on the Sun, Moon, and stars, the early travellers invented the magnetic **compass**, the **sextant**, and the seagoing **chronometer**. Eventually, radio navigation in which a position could be determined by receiving radio signals broadcast from multiple transmitters, came into existence. Improved high frequency signals gave greater accuracy of position, but they were blocked by mountains and could not bend over the horizon. This limitation was overcome by moving the transmitters into space on Earth-orbiting satellites, where high frequency signals could accurately cover wide areas.

The principle of satellite navigation is relatively simple. When a transmitter moves toward an observer, radio waves have a higher frequency, just like a train's horn sounds higher as it approaches a listener. A transmitter's signal will have a lower frequency when it moves away from an observer. If measurements of the amount of shift in frequency of a satellite radiating a fixed frequency signal with an accurately known orbit are carefully made, the observer can determine a correct position on the Earth.

The United States Navy developed such a system, called Transit, in the late 1960s and early 1970s. Transit helped **submarine**s update their on-board inertial navigation systems. After nearly ten years of perfecting the system, the Navy released it for civilian use. It is now used in surveying, fishing, private and commercial maritime activities, offshore oil exploration, and drifting **buoy**s. Transit did have some drawbacks—it was not accurate enough, a user had to wait until the satellite passed overhead, position fixes required some time to determine, and an accurate fix was difficult to obtain on a moving platform.

As a result of these shortcomings, the United States military developed another system: Navstar (Navigation Satellite for Time and Ranging) Global Positioning System (GPS). The new system can measure to within 33 feet (10.05 m), whereas Transit was accurate only to 0.1 mile (582 feet [176.5 m]). The Navstar system was fully deployed as of April 1995, with 24 satellites orbiting Earth with 12-hour periods.

Both Transit and Navstar use instantaneous satellite position data to help users travelling from one place to another. But another satellite system uses positioning data to report where users have been. This system, called Argos, is more complicated: an object on the ground sends a signal to a satellite, which then retransmits the signal to the ground. Argos can locate the object to within 0.5 mi (.80 km). It is used primarily for environmental studies. Ships, and buoys can collect and send data on weather, currents, winds, and waves. Land-based stations can send weather information, as well as information about hydrologic, volcanic, and seismic activity. Argos can be used with balloons to study weather and the physical and chemical properties of the atmosphere. In addition, the system is being perfected to track animals.

In the future, navigational satellites will continue to improve. The equipment will shrink in size and cost, while it increases in reliability. The number of people able to use the systems will also increase. Railroads are exploring ways that satellites may improve traffic management. In addition, new satellite navigation systems will be developed. In the United States, work is proceeding on Geostar, in which a user transmits a signal to three geosynchronous satellites which in turn relay the signals to a ground station. The ground station calculates the exact location and sends the information to the user or to another location. Work is also underway on an improved search and rescue operation through the use of satellites that can pinpoint emergency locator beacons.

Necktie

The necktie is thought to have its origins in the military. Soldiers of ancient Rome wore focales tied around their necks. Croatian soldiers are thought to have introduced neck scarves to France when they arrived there as mercenaries in the late seventeenth century. Worn by Croatian officers, these neck scarves were made of silk or muslin and had tassels attached to their ends. It is also believed that the French may have begun wearing neck scarves or ties even earlier, possibly during the Thirty Years War (1618-48).

During the 1700s it was common for men to sport wide, stiff whalebone stocks, fastened with straps or buckles at the back of the neck and a bow or knot in the front. In the seventeenth and eighteenth centuries, cravats, which were full, frilly, or lace-trimmed neck scarves, were a popular component of male clothing. By the middle of the nineteenth century, cravats had been replaced by string ties, knotted bow ties, and ascots.

By 1870, Western men were wearing the ''four-in-hand'' style tie that would remain popular through the twentieth century. For formal occasions, the white bow tie or black string tie was popular. The wide, soft, loosely-knotted windsor tie was worn with sporting attire. During the settlement of the American West in the nineteenth century, bandanna-patterned cotton neckerchiefs became a practical accessory of the American cowboy's wardrobe, protecting its wearer from dust during cattle drives. The twentieth century saw the arrival of the black bow tie (initially considered rather daring) and the popular striped pattern necktie.

Today's standard necktie was patented in 1920 by the American Jesse Langsdorf. Made up of four pieces of cloth cut diagonal to the grain of the cloth (to prevent twisting). Neckties remain a staple of the male wardrobe, as well as reflect fashion trends and changes in popular culture.

NEON LIGHT

Neon advertising signs from the 1920s and 1930s have become popular collectibles today, and many artists have adopted neon tubes as a medium for their creations. Such fads trace their origins to experiments by a French chemist named **Georges Claude**, who became interested in neon and other inert gases when they were first being discovered by Sir William Ramsay (1852-1916). It was Claude who invented the neon light in 1910.

Claude discovered that passing electrical current through an inert gas produced light. A **glass** tube filled with neon gas glowed an eye-catching, fire-red color when charged with electricity. Soon, Claude discovered that mixing other gases with neon would produce different colors of light, and the tubes could be twisted and shaped to make letters and pictures.

People quickly recognized the attractive value of neon lights for decorating buildings and advertising commercial establishments. With the introduction of less expensive sources of power in the late 1800s, neon signs soon transformed the advertising business. During the Roaring Twenties, neon lighting came into fashion, and the neon light's modern counterparts (which are made with **plastic** tubes instead of glass) are still effective as attention-getters.

The invention of neon lights in 1910 transformed city landscapes. Above, the Las Vegas strip in 1953.

Today, more colors can be produced by varying not only the gas filling, but also the coating inside the tube, and the tube material itself. Fluorescent lighting technology, developed in the 1930s, is a spinoff of the original neon light. Some fluorescent lights today use small amounts of neon. Even color **television** owes a debt to neon lighting. In the first practical demonstration of color television in 1928, the light source at the receiver was composed of a neon tube for red color, along with mercury-vapor and helium discharge tubes for green and blue.

Although the advertising sign is the most familiar type of neon lighting, low power neon glow lamps are widely used to indicate on/off settings and other functions on small appliances and electronic instrument panels. They can even be used to illuminate wall switches in homes. These small, lightweight neon lamps are also found in **computers**, voltage regulators, and industrial equipment.

See also Arc lamp; Fluorescent lamp; LED electricity

NEOPRENE • See Rubber, synthetic

NERVE GAS • See Chemical warfare

NEUMANN, JOHN VON (1903-1957)

American mathematician and computer scientist

John von Neumann, who was born in Budapest, Hungary, in 1903, was primarily a mathematician, and wrote numerous papers on both pure and applied math. He also made important contributions to a number of other fields of inquiry, including quantum physics, economics and computer science.

Von Neumann studied mathematics, physics and chemistry at German and Swiss universities for several years, finally receiving a Ph.D. in mathematics from the University of Budapest in 1926. He taught at Berlin and Hamburg from 1927 to 1930 and then emigrated to the United States to join the faculty at Princeton University. Three years later he took a position at the Institute of Advanced Studies at Princeton.

Until the outbreak of World War II, Von Neumann mostly did work in pure math, making important contributions to the fields of mathematical logic, set theory and operator theory. However, his work in operator theory had powerful applications in theoretical physics, and he published a book on quantum physics, *The Mathematical Foundations of Quantum Mechanics* in 1932. This work remains a standard text on the subject. During World War II, when the U.S. government called on a great many scientists to help out with the development of new technologies demanded by the war effort, Von Neumann took on numerous positions as a consultant. He was engaged in many different research projects and proved his ability as an administrator as well as a brilliant scientist. Among other consulting positions, he was involved with the development of the **atomic bomb** at the Los Alamos Scientific Laboratory. At about the same time, he caused a revolution in the social sciences with his work on game theory, *Theory of Games and Economic Behavior*, written with the economist Oskar Morgenstern and published in 1944. In those years, he also became a principal player in the development of high-speed digital computers and the stored programs used in virtually all contemporary computer applications.

While at Los Alamos, Von Neumann became impressed with the need to develop computational equipment technology that could carry out the enormously complex mathematical calculations which the scientists then had to carry out by hand. In 1944, Von Neumann became involved with efforts to develop computers, most notably **ENIAC** (Electronic Numerical Integrator and Calculator), which was then the most powerful device under construction. ENIAC could be programmed to do different tasks, but this required a partial rewiring of the machine. One of the scientists working on ENIAC, **J. Presper Eckert**, came up with the idea of a **stored program**, which would make it possible to load a computer program into computer memory from disk. The computer could then run the program without being manually reprogrammed. The idea was not used in the design of ENIAC, but a follow-up project, called EDVAC, which Von Neumann was closely associated with, did incorporate the stored program. A paper Von Neumann wrote in 1944, entitled ''First Draft of a Report on EDVAC,'' explained the revolutionary ideas that were to govern the development of computers for the next two decades. Von Neumann proposed a separation of storage, arithmetic and control functions; random-access memory (RAM); stored programs; arithmetic modification of instructions; conditional branching; a choice between binary number and decimal number representation; and a choice between serial and parallel operation. Basically, he introduced new procedures in their logical organization, the ''codes'' by which a fixed system of wiring could solve a great variety of problems. Particularly the idea of a stored program and the solutions for realizing the equipment that could deal with stored programs were revolutionary, promising great gains in speed and productivity.

Von Neumann rethought the basic design of the computer into the separate components of arithmetic function, central control (now known as the central processing unit [CPU]), memory (the hard drive) and the input and output devices. Under Von Neumann's supervision, a computer with these capabilities was developed at the Institute of Advanced Studies from 1946 to 1951. Although the machine quickly became a dinosaur, it was the first true forerunner of the contemporary high-speed digital computer.

See also Computer, digital

NEUTRON BOMB

All nuclear weapons are surrounded by controversy. Both fission and fusion bombs release such enormous amounts of energy that they are always accompanied by the massive destruction of property and human life.

The neutron bomb, also called the enhanced radiation warhead, is a small fusion bomb that can be delivered by means of conventional cannons or howitzers or in small aircraft.

The bomb is characterized by the fact that its heat and blast effects are limited to a relatively small area. These effects normally extend no more than about 327 yds (300 m) from the point of detonation.

In contrast, the bomb releases massive amounts of gamma and neutron radiation. This radiation is able to penetrate most ordinary protective shields, such as armor, **concrete**, and earth. Both gamma and neutron radiation are very destructive to living tissue. They cause painful wounds that usually cause death over a period of days.

The overall effect of a neutron bomb is that it would kill humans within its area of effectiveness without causing damage to those at greater distances. Theoretically it could be used, therefore, for destroying an attacking enemy force without endangering friendly forces or civilian populations somewhat further away. Furthermore, the bomb would cause relatively modest damage to physical structures, such as buildings, while killing people in and around those structures.

In the late 1940s American physicist Samuel Cohen (1921-), a former participant in the Manhattan Project, first made the calculations necessary to create a neutron bomb. In the 1950s scientists at the University of California's Lawrence Livermore Laboratory developed the enhanced radiation warhead concept. Since that time, it has had a checkered development history. Research went forward aggressively in the 1970s when the first underground tests were conducted. Public outcry against such a weapon was great. Production was halted in the late 1970s and then resumed a few years later. According to Cohen, a useable neutron bomb has never materialized, and with the end of the Cold War, the potential place of a such a weapon in the United States arsenal is unclear.

NEWCOMEN, THOMAS (1663-1729)

English blacksmith and inventor

Thomas Newcomen was the first person to build an effective and economical **steam engine**. His engines were based primarily on the principles of the steam pumps of English military engineer **Thomas Savery** and French physicist **Denis Papin**. The improvements and experiments begun by Newcomen, while primitive compared to **James Watt**'s engines of the late 1700s, paved the way for steam power to lead the industrial revolutions of Europe and later the United States.

Newcomen was born in Dartmouth, England, in February 1663. Little is known of his early life and training, but he was quite possibly apprenticed at an early age to an iron worker-toolsmith in Exeter. He established his own blacksmithing business and entered a partnership with a plumber by the name of John Calley.

In 1712 Newcomen and Calley unveiled their first steam engine. It is thought that they actually developed working engines before this time but had kept it quiet in order to avoid violating Savery's patent on using "the impellant force of fire" to power a machine. Nevertheless, Newcomen was forced to pay royalties to Savery.

The Newcomen engine was first used to power a pump that removed water from flooded coal mines. The pumping system consisted of a boiler, an open, brass cylinder containing a **piston** with leather sealing rings, and a wooden post that supported an unbalanced, pivoting, horizontal crossbeam that connected the piston rod to an arm of a mechanical pump. When the piston was pulled up by the downward stroke of the pump, steam from the boiler entered the cylinder at more than atmospheric pressure. The steam **valve** was closed and water was injected into the cylinder to cool the steam. The steam condensed and created a vacuum, allowing atmospheric pressure above the piston to push it to the bottom of the cylinder.

The use of air pressure rather than the force of high-pressure steam in the cycle classified this engine as an *atmospheric engine*. The engine is also designated as the second self-acting machine, clocks being the first. It remained in use, primarily at European coal and tin mines, for nearly fifty years until Watt's more efficient engines became popular. Newcomen died in London, England on August 5, 1729.

John von Neumann

NICKERSON, WILLIAM • See Razor

NICOL, WILLIAM (1768-1851)

Scottish physicist

William Nicol made two important contributions to science: he invented the first device that could conveniently polarize light, and he devised a method for preparing samples for microscopic study.

Jean Baptiste Biot (1774-1862) had first observed that a crystal of *calcite*, known as Iceland spar, had the effect of polarizing light that passed through it that is, the crystal forced the lightwave to oscillate in a particular direction. It was difficult, however, to isolate a single ray of this polarized light in order to study it.

In 1828 Nicol, then a lecturer at the University of Edinburgh, designed a polarizing **prism** from two pieces of Iceland spar. The crystals were cemented together using Canadian balsam, an adhesive also known to have light-refracting properties. A ray of light would enter the first crystal, splitting into two rays—one polarized, one not. As they hit the layer of balsam, one ray would be reflected out of the prism, while the other would pass through unchanged, eventually exiting the prism through the second calcite crystal. This device, called the Nicol prism, became an important tool for optical physicists.

A less noteworthy achievement at the time was Nicol's new method for examining crystal and rock specimens under a **microscope**. Previously, scientists were forced to study these large samples using light from a reflected source; this method only allowed them to examine the surface of the samples, not the interior. Nicol cemented the crystal to a slide plate and then ground the crystal down until it was thin enough to see through when a light was shone from below. Nicol could then observe both the surface as well as the inner structure of mineral samples. He also applied this procedure to the study of petrified woods, examining the cell structure of ancient fossilized plants. Unfortunately, the first publication of Nicol's method did not appear until 1831, and the technique was not commonly used by scientists until around 1853.

Due to the fact that he did not publish at all until he was fifty-eight, Nicol has frequently been overlooked as a pioneer in the fields of optical physics and geology.

Niepce, Joseph Nicéphore (1765-1833)

French inventor

Born in Charlon-sur-Saône, France in 1765, Joseph Nicéphore Niepce, by the age of thirty, had been a professor at an Oratorian college, a staff officer in the French army, and the Administrator of the district of Nice, France. In 1795, Niepce resigned from his position as administrator of Nice to pursue research with his brother Claude. In August, 1807, the brothers invented an **internal combustion engine**, the *pyréolophore*, which ran on powdered fuel. Claude left for Paris, and later went on to London in an attempt to generate interest in the *pyréolophore*, while Joseph stayed behind.

By 1813, Joseph Niepce, never one to stick with one pursuit for too long, had become fascinated with popular art of **lithography**. In lithography, an image is placed on a stone and treated so that some areas repel **ink** and some areas retain ink. Since Niepce himself had no artistic talent, his son Isadore would make the designs for his lithographs. Niepce would place engravings (which he made transparent) on plates coated with light-sensitive varnishes and expose them to sunlight through a process he called heliography (sun writing). When Isadore was called up for military service, Niepce decided to find a way to produce images directly from nature.

Although by the late seventeenth century the **camera obscura** projected pictures onto **paper** and in the eighteenth century the German inventor J. H. Schulze observed that silver salts darkened when exposed to light, it was more than a century later when Niepce combined these two concepts to produce **photography**. From his workroom, which overlooked the courtyard of his family's estate, Niepce made the first true attempt at photography in 1816. He used paper sensitized with silver chloride to capture a view from the camera obscura. This crude image faded away after a short time, and he could not find a means to render it permanent.

A short time later Niépce improved the same view by adding a cardboard diaphragm in front of the lens of the camera obscura. He also used nitric acid to "fix" the image briefly. Niépce continued to capture the view of the estate's courtyard, but his improvements were in vain, because he still could not make images which would last.

From 1817 to 1825, Niepce experimented with producing negative and positive images etched on metal and **glass** with light-sensitive acids. Though the processes employed were totally different from the silver chloride process which eventually became photography, he was able to produce successful and permanent copies of engravings.

In 1826 Niepce first used a professionally made camera obscura. The camera was made by Charles and Vincent Chevalier, famed Parisian opticians. On a summer day in 1826, Niepce used it to produce the first permanently fixed image from nature. The world's first photograph, a view of his courtyard on a pewter plate, had been exposed to sunlight for eight hours.

It was through the Chevalier brothers that Niépce came to know **Louis-Jacques-Mandé Daguerre**. Daguerre, who had been trying to fix images on silver chloride paper, was told by the Chevaliers of Niepce's success. He wrote the hesitant Niépce several times before Niépce began corresponding with him. Niepce and Daguerre finally met in Paris in 1827.

In London, Niepce had just discovered that his brother Claude, mentally ill, had spent much of his family's wealth on inventions which did not exist. Though Joseph Niepce did not want to reveal the details of his invention, he needed money to continue his work. On December 14, 1829, Niepce signed an agreement with Daguerre which allowed for a ten-year partnership between the two inventors. Their plan was to perfect Niepce's invention and share the profits equally. Unfortunately, Niepce died of a stroke on July 5, 1833, long before they had seen results. Daguerre's tenacity, however, assured the future of photography.

Nieuwland, Julius (1878-1936)

American chemist

Rev. Julius Nieuwland, C.S.C., was the inventor of the first synthetic rubber, neoprene, manufactured by Du Pont. At the time of his invention, Nieuwland was a professor at the University of Notre Dame and a priest of the Congregation of the Holy Cross.

Nieuwland was born of Flemish parents in Hansbeke, Belgium, and immigrated as a youngster with his family to South Bend, Indiana. He graduated from Notre Dame in 1899, studied for the priesthood and was ordained in 1903, and received his Ph.D. from Catholic University in 1904. He taught botany for a number of years at Notre Dame, and in 1918 became a professor of organic chemistry. At that time, he was working with acetylene. In the course of this work, he discovered a reaction between acetylene and arsenic trichloride that eventually led to the development of the poison gas lewisite.

Nieuwland's work with acetylene also led him into a collaboration with scientists at Du Pont. Together, they found that upon treating monovinylacetylene with hydrogen chloride to produce chloroprene and polymerizing the result, a very durable synthetic rubber, neoprene, was produced. Du Pont placed this rubber on the market in 1932 under the brand name Duprene.

Neoprene has been considered superior to rubber in terms of its resistance to sunlight, abrasion, and temperature extremes. These properties gained it popularity in many industries. Neoprene is favored for electrical cable insulation; telephone house-to-house wiring; many moulded, extruded, and sheet products; rug backings; and roofing.

See also Dielectric materials; Rubber, synthetic

Joseph Nicéphore Niepce

Night vision devices

The ability to wage battle at night is a great tactical advantage, but devices that let soldiers see well in the dark were not invented until the 1940s. Fighters illuminated land and sea battles with searchlights and flares at the beginning of the twentieth century, and star-shell **artillery** was used in the two world wars to cast light where it was needed. Then during World War II, both the Germans and the Allies developed more sophisticated night vision devices. This early night vision equipment consisted of an infrared filter placed over a searchlight. The observer looked through a **binocular**-like or **telescope**-like viewer that translated the infrared emissions into an electron flux displayed on a luminescent screen. Both the Germans and the Allies fitted their tanks with these viewers. But because both sides had infrared viewers, they could spot each other's searchlights. Even after a searchlight was turned off, it continued to emit enough infrared to make it a target. So use of this equipment was risky.

American military scientists continued to develop night vision devices after World War II, but concentrated on systems that worked by intensifying available night light—starlight and moonlight. Most of the seminal research on these so-called image intensifiers was done at Fort Belvoir, Virginia, and the technology remained top secret for a long time. American soldiers used image intensifiers called Star-Tron scopes in the Vietnam War. The scopes were almost as heavy as the rifles they were mounted on, and were extremely expensive to manufacture. They worked by focusing the dim night scene onto a photo-cathode inside a tube. The photo-cathode accelerated and multiplied the electrons from the ambient light, and focused them into a visible image. The Star-Tron scope permitted a soldier to see in the dark about 400 yards (366 m), but it had several unfortunate drawbacks. The scope would shut down if it was exposed to a sudden bright light, and the device emitted a high-pitched whine that was often loud enough to give away its user's position, thus cancelling out the stealth effect.

Further research eliminated these problems in a second generation of image intensifiers. These worked by focusing ambient light on a microchannel plate (MCP), which was made up of a thin slice of a number of hollow **glass** tubes. The tubes each emitted many electrons when struck by a single electron. Because image intensifiers using MCPs were smaller and lighter than the older devices, they could be manufactured into binoculars affixed to a helmet. These are known as night vision goggles. Night vision goggles were widely worn by military pilots for night manuevers from the 1980s on.

Other night vision devices functioned with thermal imaging. In this system, sensors detect heat, and translate temperature differences into a visual image. Thermal systems have the advantage of being able to ''see'' even through fog, smoke, and dust.

In the 1980s, night vision technology ceased to be a military secret. Several companies in the United States and Europe worked to market night vision goggles for non-military use. Some night vision devices have found their way into ordinary automobiles, to make night driving safer. In the late

1990s, night vision goggles were increasingly purchased by municipal police departments.

Nipkow, Paul Gottlieb (1860-1940)

German inventor and engineer

Nipkow, now considered the forefather of the **television** age, received little recognition for his contribution during his lifetime. Italian physicist **Guglielmo Marconi** demonstrated a device that could transmit an audio signal in the late 1890s.

Nipkow was born on August 22, 1860, in Lauenberg, Germany. Inspired by the work of Marconi, Nipkow began thinking about the challenge of transmitting a visual image while still a student in Germany. It was well known that any successful transmission device required three essential components: a device to translate the visual image into an electronic impulse, a second device to reassemble that impulse into an image, and a third device by which to transmit the impulse from the first device to the second. In 1884, even before completing his degree, Nipkow had developed and patented a transmissions system that achieved all three requirements.

Nipkow's television was based upon an ingenious device called a *Nipkow disk*, which was a metal or cardboard disk that was perforated with twenty square holes arranged in a spiral so that each hole was a little closer to the center than the last. As Nipkow spun the disk, he shined a strong light through the holes and onto the subject. Because each hole was slightly offset, the image was scanned in a series of twenty horizontal lines.

In order to translate these lines into an electrical signal, Nipkow employed a selenium **photoelectric cell** (a device with which he had previously worked with extensively). The cell recorded the light and dark areas within each of the twenty scanned lines, converting these into a transmittable signal. In order to view the signal, Nipkow essentially reversed the process. He used a light source that flashed brightly or dimly according to the incoming signal. To this he added a second scanning disk, placed in front of the flashing light and synchronized with the first disk, so that the light shone through the holes and projected twenty scanned lines onto a screen.

The main drawback to Nipkow's invention was not its design but its timing: the concept of television was so advanced that no producer or investor could envision a practical use for it. Though Nipkow used his device to transmit a visual image via telegraph wire from London to Paris, his mechanical television never quite caught on. Nipkow himself eventually abandoned electronics and spent the rest of his life as a railway engineer. He died in Berlin, Germany, on August 24, 1940.

It was not until 1929 that another scientist, the Scottish engineer **John Logie Baird**, made certain improvements upon the Nipkow design and reintroduced it to the world. While Baird's design was still essentially mechanical, it came at a time when the world was ready to embrace the concept of television. Photomechanical televisions were soon replaced by completely electronic devices, but even today's models rely upon the horizontal-scanning method first conceived of by Nipkow.

See also Telegraph cable

Nitrogen fixation

Under ordinary conditions, nitrogen does not react with other elements, so one might not expect to find its compounds in nature, at least not extensively. Yet nitrogen compounds are present in every living cell, and they can be found in all fertile soils, in many foods, in animal products such as wool and feathers, and in many other naturally occurring substances. Somehow, chemically inert nitrogen gas is "fixed," or combined with other elements, from its "free," or elemental, state. During the 1800s, scientists learned that some nitrogen fixation processes occur in nature. Although it was already known that plants absorb nitrogen compounds from the soil through their roots, most people thought that these compounds had to be restored to the soil by the addition of organic or chemical **fertilizer**s.

Then French agricultural chemist Jean Baptiste Boussingault (1802-1887) demonstrated that beans, peas, clover, and other legumes could restore nitrogen to soil by extracting, or assimilating, nitrogen from the air, although he never discovered how they did it. Boussingault also showed that other plants depend entirely on fertilizers for their nitrogen, because they cannot obtain it from the air as legumes do. Similarly, animals are unable to use atmospheric nitrogen, Boussingault found. Instead, they get their nitrogen by eating plants, or by eating animals that feed on plants. In 1862, Louis Pasteur (1822-1895) and other chemists suggested that microorganisms might be involved in nitrogen fixation. By the late 1880s, scientists had learned that certain bacteria live in a *symbiotic*, or mutually beneficial, relationship with leguminous plants. These "nitrogen-fixing" bacteria convert nitrogen from the air into ammonia (NH_3), which is then converted by "nitrifying" bacteria into nitrates—compounds that can be used by plants to make proteins and other organic nutrients. Other nitrogen-fixing bacteria live free in the soil and are also capable of assimilating nitrogen directly from the air.

Around the turn of this century, chemists began to search for an artificial nitrogen-fixation process. Demand for nitrogen-based fertilizers and explosives was growing rapidly. At the time, these products were manufactured from a natural raw material called Chile saltpeter (sodium nitrate). Scientists became concerned that this source would run out, reducing crop yields and creating food shortages, as well as depriving some countries of the materials needed to produce **ammunition**. Although a practically inexhaustible source of nitrogen exists in the atmosphere, which consists of nearly eighty percent nitrogen, no one knew how to fix the element into useful compounds such as ammonia and nitric acid. In the early 1900s, **Kristian Birkeland** and Samuel Eyde (1866-1940) developed a method of combining atmospheric nitrogen and oxygen into nitrogen oxides. Back in 1784, **Henry Cavendish** had produced nitric acid by passing sparks through a jar of air confined over water. The electricity forced the nitrogen and oxygen to combine, forming nitrogen dioxide gas, which then combined with hydrogen in the water to create nitric acid. Birkeland and Eyde succeeded in developing an electric arc process on a commercial scale, but it used such large amounts of power that it was too inefficient for most purposes.

Another nitrogen-fixation method called the cyanamide process was invented around the same time by Adolf Frank (1834-1916) and **Heinrich Caro**, but it also used too much electricity to be commercially feasible. Then in 1909, **Fritz Haber** invented one of the most important industrial processes of modern times—the fixation of nitrogen as ammonia. Initially, Haber's ammonia synthesis process used osmium or uranium as a catalyst, and the reactions took place at a temperature of 868°F (550°C) and at pressures up to two hundred times greater than normal atmospheric pressure. By 1913 the process had been translated to the industrial scale by **Carl Bosch**. Largely through Bosch's efforts, a large German chemical firm began supporting Haber's development and built two huge plants. A few years later, in 1917, French engineer **Georges Claude** developed an ammonia synthesis process based on the work by chemist **Henri Le Chatelier**. Though he worked independently, Claude's process was similar to that of Haber and Bosch. Industrially, the Haber-Bosch process was operated at slightly lower temperatures and higher pressures to maximize ammonia production, and finely divided iron was introduced as a catalyst.

The success of the **ammonia synthesis** plants built in Germany during World War I convinced scientists in other nations that the process was practical and economical, and the industry expanded rapidly. Though the Haber-Bosch process was improved in minor ways, essentially it has remained unchanged since the early 1900s and is used today to manufacture thousands of tons of ammonia worldwide. Concerns about runoff from synthetic fertilizers into streams, lakes, and rivers has encouraged biologists to develop new ways of ensuring that corn, wheat, and other grain crops can manufacture their own nitrates through genetic manipulation. This process will lessen the use of chemical fertilizers.

Nitroglycerin

A highly volatile explosive, nitroglycerin was first produced by Italian chemist Ascanio Sobrero (1812-1888) in 1847. At about the same time German-Swiss professor **Christian Schönbein** discovered **guncotton**. Both inventions, given their high degree and speed of detonation, had the potential to immediately overthrow **gunpowder** as the chief military and commercial explosive. (Nitroglycerin's speed of explosion, for example, proved to be 25 times faster than that of gunpowder.)

Sobrero, guided by scientific curiosity rather than military interests, found that by combining a commonly used skin lotion glycerol with nitric acid and sulfuric acid, a colorless, oily liquid of enormous power resulted. However, the grim potential of his discovery so frightened him that he refused to publish his findings for nearly a year and then almost in secret. Consequently, nitroglycerin went largely unnoticed for many years, while the manufacture and use of guncotton, despite several notorious accidents, spread throughout Europe.

It was not until the mid-1860s, when **Alfred Nobel** made the important discoveries, successively, of the blasting cap and **dynamite**, that nitroglycerin became a widely used and respected explosive. Since that time—and even more so since Nobel's development of nitroglycerin-based blasting gelatin in 1875—its impact on the mining and construction industries has been profound. Previously, these industries were hampered by gunpowder's inability to obliterate large sections of rock. Nitroglycerin makes up about 75% of the materials used in the manufacture of dynamite.

Sobrero's substance also found use as a key ingredient in two smokeless powders, ballistite and **cordite**, from which all modern bullets derive their construction. Lastly, and perhaps most reassuring to the memory of both Sobrero and Nobel, nitroglycerin, in minute quantities, is regularly relied upon by doctors for the treatment of heart disease. Its properties of expansion have proven ideal for dilating coronary arteries in patients suffering from hypertension, cardiac pain, and the threat of heart attack.

Alfred Bernhard Nobel

Nobel, Alfred (1833-1896)

Swedish inventor, industrialist, and philanthropist

Owner of more than 350 patented inventions during his lifetime, Nobel is best known as the discoverer of **dynamite** and the man who upon his death bequeathed much of his large estate to support the annual Nobel Prizes for accomplishments in physics, chemistry, economics, science and medicine, literature, and the promotion of peace.

Born in Stockholm, Nobel received his education from private tutors and from various apprenticeships. Like his father, a manufacturer of mines and other explosives, Nobel displayed an avid interest in engineering and chemistry and as a young man worked for a time in the laboratory of French chemist Théophile Jules Pelouze (1807-1867), who is regarded by some as the inventor of **guncotton** (most accord the honor to **Christian Schönbein**). After extensive travels, through which he acquired the sharp skills of a businessman and the distinct advantages of a multilinguist, Nobel returned to Sweden in 1863 for the singular purpose of safely manufacturing **nitroglycerin**.

Almost two decades earlier, Ascanio Sobrero (1812-1888) had invented this oily liquid, but it proved so volatile as to preclude its widespread use. Instead, **gunpowder** and **guncotton** dominated the explosives industry, despite their own shortcomings. Through his own studies and experiments, begun as early as 1859, Nobel had familiarized himself with Sobrero's compound of glycerine treated with nitric acid, and had even exploded small quantities of it under water. Sharing his interest, his father during this same time designed a method for the large-scale production of the explosive. In Nobel's mind, all that remained was to devise a special blasting charge to ensure a predictable detonation of the nitroglycerine by shock rather than heat, which he already knew to be a dangerously imperfect firing method. The result was Nobel's first important invention, the mercury fulminate cap.

A fatal factory accident the following year, in which Nobel's brother Emil was killed, led the inventor to continue his research with nitroglycerine, this time in the hope of discovering a benign substance to absorb the liquid explosive, thereby making it safe for manipulation and transportation, without seriously diminishing its eruptive characteristics. After exhaustive experimentation, Nobel found a nearly perfect substance, kieselguhr. When saturated with nitroglycerine, this porous clay became a highly desirable explosive, which Nobel termed dynamite and patented in 1867.

Virtually overnight, dynamite revolutionized the mining industry, for it was five times as powerful as gunpowder, relatively easy to produce, and reasonably safe to use. Nobel acquired a vast fortune from this invention, which spawned an intricate network of factories, sales representatives, and distributors in several industrialized countries around the world. Despite the enormous demands of his business ventures, which required that he travel almost continuously and engage repeatedly in legal battles, Nobel persevered with his scientific research. Less than satisfied with the qualities of *kieselguhr*, which occasionally leaked nitroglycerine as well as somewhat reduced the liquid's power, he began experimenting with nitroglycerine and **collodion**, a low nitrogen form of guncotton. He found that these two substances formed a gelatinous mass which, with modifications, possessed a high resistance to water and an explosive force greater than that of dynamite. The invention, perfected in 1875, became known by a variety of names, including blasting gelatin, Nobel's Extra Dynamite, saxonite, and gelignite.

One of Nobel's last significant discoveries was closely related to his work with blasting gelatin. Like a number of other inventors, Nobel was in search of a smokeless powder to replace gunpowder. In 1888 he introduced ballistite, a mixture of nitroglycerine, guncotton, and camphor which could be cut into flakes and used as a propellant; the substance was particularly valuable for its ability to burn ferociously without exploding. A year later, two British scientists invented a smokeless powder based on ballistite called **cordite**. To Nobel, the invention represented an infringement on his patent; his suit to recover damages, however, was unsuccessful.

Nobel died in 1896 but despite his long and successful career developing and manufacturing explosives he was a devoted humanitarian who wished to aid efforts that might bring about lasting peace as well as beneficial advancements in technology. To this end he composed a handwritten will which, though problematic and fiercely contested, led to the creation of the Nobel Foundation, which grants monetary prizes for contributions to scientific research and efforts toward world peace. The first prizes were awarded in 1901 and the tradition continues today.

Noise reduction systems

When **audiocassette** tapes were introduced, they quickly became more popular than open-reel and eight-track tapes because of their small size. However, their miniature size also created a new problem. On a length of tape, the narrower the recording track, the larger, relatively, are the iron oxide particles which carry the magnetized sound impulses on the surface of the tape. Thus, there is less ''room'' for the musical signal in relation to the inherent noise on the tape, caused by the random magnetization in the oxide particles. As a result, unwanted hiss may be heard on cassette tapes, especially during quiet musical passages.

There have been several systems introduced to combat this tape hiss. The first and most popular was the one created by an American, Ray Dolby, in 1967. Originally called the S/N Stretcher, the Dolby system employs complex circuitry that detects high-frequency portions of the signal coming into the tape and boosts their volume during recording. Then, it identifies these same frequencies on the tape during playback and lowers their volume. The music then sounds as it did before it was boosted, but the volume of the annoying hiss, which is a high-frequency signal, is also reduced. The Dolby circuit also has other refinements to keep hiss to a minimum. Dolby B, an improved version of the original Dolby system, was introduced in 1970, reducing tape hiss by 10 decibels. In addition to Dolby, there are other noise reduction systems, such as DBX, that operate on different principles. However DBX recorders had Dolby circuits incorporated within them, and eventually disappeared from the consumer market.

Another way to combat tape hiss is by improving the tape itself. Some tapes are produced with special oxide layers that can be more strongly magnetized than regular tapes. For instance, some ''premium'' quality tapes utilize chromium compounds in their formulation. Another innovation is the introduction of a pure metal coating, which allows for professional quality recording of compact disks.

A relatively new recording technology, digital audiotape (DAT), eliminates the problem of tape noise entirely. DAT encodes the music on the tape in the form of impulses representing binary digits (ones and zeros), the same way data is stored on a computer. These impulses are then decoded to produce music when the tape is played. Because the advanced circuitry of this system cannot mistake tape hiss for the digital data, *no* tape noise is produced upon playback. However, Dolby Laboratories introduced Dolby C in the late 1980s, then Dolby S in 1990 as analog technologies that reduce noise to a level comparable to DAT. Dolby C reduced tape hiss by 20 decibels. Based on Dolby's Spectral Recording system, commonly found in non-digital recording studios, Dolby S eventually replaced Dolby C and reduces noise by up to 24 decibels in high frequencies and 10 in lower frequencies. Dolby S works slightly differently from other Dolby systems by reacting to the signal to which it is applied, conforming to the signal's characteristics. This allows for a wider range of noise reduction. As of the late 1990s, neither DAT or Dolby S has eliminated the other, though the sale of cassettes has continued to fall in the face of the compact discs.

See also Magnetic recording; Electronics

NONSTICK PAN • See Teflon

NORTH, SIMEON (1765-1852)

American manufacturer

A native of Berlin, Connecticut, Simeon North was a successful manufacturer of scythes before he turned to gun making, eventually becoming one of the major suppliers of firearms to the United States military before the Civil War. In 1799, North secured his first contract with the army, who ordered 500 of his flintlock **pistol**s; the next year the order was increased to 1,500. Following the outbreak of the War of 1812, North expanded his plant in order to meet the government's orders. He was the first to use the principle of interchangeable parts when he filled an order for 20,000 flintlock pistols. After the war, North continued to create pistols that incorporated minor changes: an addition of a swivel ramrod, a safety catch slide on the hammer, and a brass foresight on the barrel.

North was better known for the **rifle**s he produced. In 1823, he won a government contract in which he supplied the army with standard breech-loading rifles. These initially were flintlock arms, but in 1843 North submitted a new design for governmental approval. His revamped model was christened ''North's Improvement'' for its improved method of working the breech mechanism. Until then, breech-loading rifles were flawed with loose seams between the chamber and the barrel, causing flames to spurt out the back and losing some velocity because of this gas leakage. North's design made the breech mechanism fit more closely to the barrel, thus reducing the loss of velocity. His adaptation of the Hall percussion breechloader was a success with the army, and he continued filling orders until his death in 1852.

Robert N. Noyce

NOYCE, ROBERT N. (1927-1990)

American physicist, inventor, and entrepreneur

As the co-inventor of the **integrated circuit**, Robert Noyce was a founder of the modern **electronics** industry. He was born and raised in Iowa, the son of a Congregational minister. As a physics major at Grinnell College in Grinnell, Iowa, Noyce worked with one of the first **transistors**. It had been given to his physics professor Grant Gale by one of its co-inventors, **John Bardeen**. Noyce received his Ph.D. in physics from the Massachusetts Institute of Technology in 1953 and became a research engineer at Philco Corporation, working on germanium transistor development. In 1956 Noyce began research on the physics of **silicon** transistors at the newly established Shockley Semiconductor Laboratory, started by another transistor co-inventor, William Shockley. The lab's whereabouts, near Palo Alto, California, eventually became known as Silicon Valley because so many electronics companies sprouted there.

Unhappy with Shockley's management style, he and seven fellow scientists secured financial backing from Fairchild Camera & Instrument Corporation of New York, left Shockley Lab, and founded their own firm, Fairchild Semiconductor, in 1957—establishing a pattern that would become common in Silicon Valley. As director of research and development at Fairchild, Noyce concentrated efforts on ways to improve and simplify the multiple circuit interconnections required for sophisticated electronics. Early in 1959 he combined all these elements on a single silicon chip, an invention he called an integrated circuit. Although he filed a patent application later that year, a long legal battle with Texas Instruments and **Jack Kilby**, who had also invented an integrated circuit, ensued. Noyce was ultimately awarded the patent in 1969.

Meanwhile, the electronics industry was revolutionized by integrated circuits, and Fairchild produced more of them than anyone else. In 1968, following Silicon Valley tradition, Noyce left Fairchild with fellow scientists Gordon Moore and Andrew Grove and founded Intel Corporation. The new company focused on **semiconductor** memory, developing vastly improved chips and inventing the **microprocessor**.

After 1975, Noyce increasingly worked to promote the electronics industry and strengthen it in the face of foreign competition. He founded and lobbied for the Semiconductor Industry Association and was also the leader of Sematech, a consortium of fourteen semiconductor companies he helped form in 1988. He became known as the ''Mayor of Silicon Valley,'' its elder statesman. Noyce's interests in Intel and the patent rights to the microchip made him wealthy. He was a gregarious, open man with a variety of active interests, including hang gliding, scuba diving, and piloting his own airplanes. He died suddenly after suffering a heart attack at his Austin, Texas, home at the age of sixty-two.

Nuclear reactor

One of the early discoveries relative to nuclear fission was that neutrons are one product of the fission reaction. That is, although a neutron is needed to *initiate* a fission reaction, at least one neutron is also produced *as a result of* the reaction. Some scientists realized that this fact meant that fission reactions could easily become *chain reactions*. A chain reaction is one in which the substance needed to start the reaction is also produced as a result of the reaction.

Enrico Fermi (1901-1954), **Leo Szilard**, and other physicists working in the Manhattan Project realized the practical significance of a nuclear fission chain reaction. Such a reaction would result in the release of enormous amounts of energy in a very brief period of time. They knew that they could build an **atomic bomb** employing the principle of fission which would produce a very large explosion. Slowed down in some way, the same reaction could be used to produce usable energy for the production of electrical power and other purposes.

But could a nuclear chain reaction actually be made to occur? In order to answer that question, a team under Fermi's direction built an atomic ''pile'' under the squash courts at the University of Chicago in 1942. The term ''pile'' is used because the structure consisted of blocks of uranium, uranium oxide, and graphite, stacked on top of each other. A more correct name for the structure is *nuclear reactor*.

The first atomic pile (and any other nuclear reactor) consisted of three essential parts: the fuel, a moderator, and control rods. The fuel in the original atomic pile was the uranium. Fission of uranium nuclei releases energy that can be used for destructive purposes (in a **bomb**) or peaceful uses (in a reactor).

Of the two naturally occurring isotopes of uranium, uranium-235 and uranium-238, only the former undergoes fission. The far more abundant (99.3%) uranium-238 does not. A critical problem in the construction of either a nuclear weapon or a nuclear reactor, therefore, is to increase the fraction of uranium-235 compared to that of uranium-238. That process is known as ''enrichment'' of the uranium.

In the first reactor, graphite was the moderator. The purpose of the moderator is to slow down neutrons released during fission. Fermi had discovered in 1934 that slow-moving (''thermal'') neutrons are more effective at causing fission than are fast-moving neutrons. The presence of graphite in the reactor increased the pile's efficiency.

Control rods are made of some material that absorb neutrons readily. Cadmium is one such substance. When control rods are inserted fully into a reactor, they absorb most of the neutrons released during fission. A chain reaction is not possible. As the control rods are slowly withdrawn from the reactor, more neutrons are available to initiate fission reactions.

At some point, the position of the control rods is such that one neutron is available to start a fission reaction for every neutron that is used up. At this point, a self-sustaining chain reaction can occur in the reactor. The first such reaction was observed in the University of Chicago atomic pile at 3:45 p.m. on December 2, 1942.

The immediate result of this experiment was to confirm the possibility of constructing a nuclear weapon. All of the research efforts by American and other Allied scientists in the *Manhattan Project* over the next three years were devoted to building that weapon.

After the war, scientists turned their attention to the application of nuclear fission chain reactions for peaceful purposes. By far the most important of those applications was the nuclear power plant.

At the center of a nuclear power plant is the core, essentially a nuclear reactor similar to the first atomic pile. The core consists of fuel rods that contain uranium, a moderator, and control rods. The core is surrounded by some cooling liquid or gas, such as water, liquid sodium, or carbon dioxide. The cooling material absorbs heat produced in the core and carries it to some external container. In the external container, the heat is used to boil water. Steam produced from the boiling water is then used to drive a turbine that runs a generator. Nuclear power plants of this design account for well over half of all the electricity produced in some nations (France, for example) and for a much smaller fraction in other nations (the United States, for example).

A critical discovery that made possible the nuclear power plant was made in 1939 by R. B. Roberts, L. R. Hafstad, N. C. Meyer, and P. Wang in the United States. These researchers noted that uranium continues to emit neutrons after bombardment by neutrons had stopped. That discovery means that, when a uranium nucleus fissions, all product neutrons are not released at the same time. Some are delayed. That means that, should a nuclear reactor start to go out of control, there is time to drop the control rods and stop the reaction. A potential accident can be prevented.

The earliest nuclear reactors all used some form of enriched uranium-235 as a fuel. At the time, it was the only isotope known to undergo fission. Physicists soon learned, however, that other isotopes will also undergo fission. The most important of these is plutonium-239. The nuclear bomb dropped on Hiroshima was a uranium bomb, while that used on Nagasaki was a plutonium bomb.

The association between uranium-235 and plutonium-239 in a nuclear reactor is an interesting one. When neutrons are produced during fission in a reactor core, they not only cause fission in other uranium-235 nuclei, but they also react with the far more abundant uranium-238 that is also present in the core. During that reaction, uranium-238 is converted to uranium-239, which decays by beta emission to form plutonium-239. That is, a second fissionable isotope, plutonium-239, is produced as a by-product of the fission reactions that occur in uranium.

This fact is used in the construction of a "breeder reactor," first conceived and designed by the Canadian physicist Walter Zinn in the late 1940s. In a breeder reactor, plutonium-239 produced as a result of the reactions outlined above is extracted from the reactor core. The isotope can then be used as additional fuel in a reactor core or as fuel for a nuclear bomb.

Nuclear semiconductor detectors

The invention of **transistor**s by **William Shockley** in 1948 created a revolution in many aspects of human society, from pure scientific research to a host of everyday applications. The utility of Shockley's transistor was greatly advanced by Leo Esaki's (1925-) discovery a year later of the tunnel **diode**. Tunnel diodes are able to perform many of the functions of a conventional transistor, but with much smaller size and much greater speed.

One application of the Shockley-Esaki discoveries has been in the field of particle detection. The nuclear semiconductor detector first became available in 1958. It consists of a p-n semiconductor combination separated by a thin depletion (insulating) region. Passage of radiation through the detector creates ions that initiate the movement of electrons and *holes* in the depletion region. In some cases, the electric current thus generated is read and recorded directly. In other cases, the current is amplified before being registered.

Semiconductor detectors have a number of advantages over other types of particle detectors. For example, the energy needed to produce an electron-hole pair is only about one-tenth that needed to produce ionization in a gas (as required, for example, in a **Geiger counter**). Also, since the solids of which a semiconductor detector are made are about 1,000 times more dense than a gas, these devices can be made much smaller than a Geiger counter, **spark chamber**, or cloud chamber.

Semiconductor detectors have become among the most widely used of all kinds of radiation detectors. They have found application in both basic and applied research in biology, environmental sciences, medicine, and space studies.

See also Radiation detector

Nuclear-powered submarine • See Submarine

Nutrasweet • See Artificial sweetener

Nylon

Nylon was developed at Du Pont by **Wallace Hume Carothers** and his research team. Du Pont had made a commitment to find an artificial substitute for silk. They knew that the market for silk stockings was a $70 million business and an inexpensive material with the properties of silk was bound to be successful. After several years of research, Carothers had almost given up on finding a substitute, when he and the team discovered the secret of *cold drawing*.

In 1930, Carothers and Julian Hill, his assistant, developed equipment that led to the breakthrough. They wanted to form long chain **polyester**. The polyester was formed by reacting diacids and diols. This reaction also formed water, which caused the polyester to revert to the original reactants. The equipment they invented removed the water as it was formed, resulting in a high molecular weight polyester. Hill noticed that he could stretch the material into a fiber. He and the other research members stretched the polyester down the hall of the laboratory. They realized that the fiber grew silkier and stronger as it was drawn out. By doing this they had inadvertently discovered the process of cold drawing. Cold drawing orients the molecules into a long linear chain and fosters strong bonding between molecules. Unfortunately, the polyester Julian Hill used had a low melting temperature and was unsuitable for textile applications.

Eventually, Carothers was persuaded to try the experiment using polyamides. In 1934, Donald Coffman, another assistant, drew out the first nylon fiber and by 1935, a polyamide called *Nylon 6,6* was identified as the most suitable substitute for silk. The first nylon stockings went on sale in the United States in 1938. By 1941, 60 million pairs were sold. During World War II, however, stockings became rare as nylon was diverted to the war effort and used in **parachute**s, mosquito netting, **ropes**, blood filters, and sutures.

Today, nylon is forced through small holes in a flat plate, called spinnerettes, to form fibers and then stretched by passing it through a pair of rollers rotating at different speeds. This process stretches the fiber several hundred percent and increases its strength by more than 90 percent. Nylon is stronger than steel by weight and is almost inflammable. It is used in clothing, laces, toothbrushes, strings on musical instruments, sails, fish nets, **carpet**s, and other products requiring strong, lightweight fibers.

During World War II, most supplies of nylon were diverted to the war effort, making nylon stockings a rare commodity. Above, a sailor in New York City displays his find.

O

Oberth, Hermann

• See Rocket and missile

Ocean liner

Ships have carried passengers across the oceans for centuries. Yet, it was not until the early 1800s that true ocean travel was established. Up until then, ships sailed only when they had a full load of cargo and passengers. Because international trade was expanding, the demand for better transatlantic passenger service increased. The United States took the lead in establishing regularly scheduled voyages with vessels called packet ships; they sailed regardless of weather, cargo, or passengers.

In 1818, the Black Ball Line began the first packet service. Soon, other liner companies were formed to cash in on the popularity of the service. The first packet ships were small by today's standards: the average length was about 100 ft. (30.48 m). The typical packet had three sails, the tallest of which was the middle one.

The United States lost its competitive edge in the construction of liners due to a remarkable British naval architect, **Marc Isambard Brunel**. He built ships far in advance, both technically and in size, of any before. His *Great Western*, built of wood and driven by paddles, established the first regular steamship service between the United States and England. Brunel went on to create the first liner to be made of iron and driven by a propeller: the *Great Britain*. In 1858, he built the *Great Eastern*, the first vessel with a double iron hull, which was capable of carrying 4,000 passengers.

One more advance for the liner, in addition to steam power, propellers, and iron hulls, was the **steam turbine**. A English company, the Cunard Line, gambled on the turbine to power its new passenger ship, the *Mauretania*, launched in 1905. This vessel became the first modern passenger liner with swift lines, steam turbine, steel hull, and four propellers. It was 790 ft. long, (241 m) and carried 2,335 passengers in addition to a crew of 800. It could cruise from Liverpool to New York in slightly under five days with great reliability. The now famous *Titanic* was built by the White Star Line in 1911 to compete with the Cunard ships. The largest and most luxurious ship then afloat, it was also believed to be the safest because of its advanced construction and design. When it sank on its maiden voyage after hitting an iceberg off the coast of Newfoundland, the ship became a symbol of society's excessive faith in technology.

Despite the loss of the *Titanic,* shipbuilders continued to construct more opulent and faster oceanliners. The 1930s was the golden age for ocean liners. Three of the most luxurious ships ever built were launched then: the *Normandie* (France), the *Queen Mary* (England), and the *Queen Elizabeth* (England). Each was more than 1,000 ft. long (about 305 m); the Queen Elizabeth, at 84,000 tons (76,440 t), was the largest by weight. Each was capable of crossing the Atlantic in four days. Although airplanes assumed most of the routine transoceanic travel after World War II, the liners survived as specialty cruise ships.

The last of the great steamships, *Queen Elizabeth 2*, was converted to diesel-electric in the mid-1980s. It was in that decade that the cruise ship industry experienced extraordinary growth, and by the mid-1990s, more people were traveling by ship than ever before (nearly 2.4 million in 1994). The 1900s also ushered in a period of even more growth and much bigger liners. In November 1997, the world's largest cruise ship, the 900-ft-long (273-m) *Carnival Destiny*, made its maiden voyage. As the first passenger ship to exceed 100,000 gross tons (91,000 t), it has a crew of 1,100 members who will serve as many as 3,400 passengers. That same year saw the *Splendour of the Seas* launched, and its ability to cruise 25% faster than most ships enables it to cut off a day of transit to Hawaii, giving passengers extra time to sightsee. Other new liners, like the *Grand Princess*, specialize in providing such luxurious accommodations as 750 cabins with verandas, five swimming pools, and a virtual reality theater.

"All the News That's Fit to Print."

The New York Times.

TITANIC SINKS FOUR HOURS AFTER HITTING ICEBERG; 866 RESCUED BY CARPATHIA, PROBABLY 1250 PERISH; ISMAY SAFE, MRS. ASTOR MAYBE, NOTED NAMES MISSING

Col. Astor and Bride, Isidor Straus and Wife, and Maj. Butt Aboard.

"RULE OF SEA" FOLLOWED

Biggest Liner Plunges to the Bottom at 2:20 A.M.

RESCUERS THERE TOO LATE

The Lost Titanic Being Towed Out of Belfast Harbor.

PARTIAL LIST OF THE SAVED.

CAPT. E. J. SMITH.

Front page of ***The New York Times*** **following the sinking of the** ***Titanic.***

Since there is no technical limit to how big a ship can be, a 4,320-ft long (1,310–m) vessel (five times as long as the *Carnival Destiny*) named *Freedom* is being planned, with construction to begin in 1998. Described as the world's first floating city, it is so huge that it will have to be assembled at sea. Capable of carrying as many as 115,000 people, it will circle the earth every two years and provide a 25-story high floating home for 17,000 homes and 4,000 businesses. It ''citizens'' will come and go using yachts or private jets. Its interior will be designed more like an urban area—with open spaces, parks, recreation areas, and transportation links (like a railway system)—than a traditional ship. It will be built of 520 airtight steel cells or compartments that are linked together and which act as a steel beam or keel, allowing it to ride atop waves rather than have to plow through them. Propelled by 100 3,700-hp motorized units, it will use clean-burning fuel and recycle materials as much as possible. The new ship, *Freedom*, will push the concept of an ocean liner to its limit, making it more of a moving island that is able to visit several different countries each month.

Odometer

The instrument panel displays a variety of information for the driver's use. Some dashboards contain a plethora of instruments that report aspects of an **automobile**'s performance, and others hold ''idiot'' lights that light up only to indicate hazards. Perhaps the two most common instruments are the speedometer and the odometer, and the two are really a single device that gets its information from the turning of the wheels. The speedometer tells how fast the vehicle is traveling in miles or kilometers per hour, and the odometer translates the rotations of the wheels into the distance traveled. Odometers show the total mileage logged by the vehicle, and they may also include a trip mileage recorder that can be reset at the beginning of a journey.

A vast number of inventions and improvements color the short history of the automobile, yet its odometer is one of the oldest instruments still in use. In Rome during the first century B.C., Vitruvius described the odometer in a taxi-chariot that was used to measure distance and, thus, the passenger's fare. In 1500, **Leonardo da Vinci** created an odometer that consisted of a disc geared to one wheel. This disc dislodged a stone with each complete turn, and the stone fell into a box. The number of stones at the end of the trip reflected the distance traveled.

Early automobiles recorded the total distance up to five figures. A series of dials were mounted on parallel axes. These were geared together so each turned at one tenth of the speed of the preceding dial. A window on the face of the instrument showed the numbers, except that they were hard to read if they were moving in or out of the window. A second display that could be zeroed for trip measurements was difficult to connect to the dial-type odometer.

The drum odometer solved the problems of the dial system. The drum recorder has the numbers 0 through 9 printed on the outside surface of the drum. The inner portion has two sets of gear teeth. Six of the drums are mounted on a center spindle so that the teeth of one drum mesh with those of its neighbor to form a ''train'' of gears. As the right hand drum makes one revolution, its teeth move the drum to the left by one-tenth of a revolution. The row of numbers across the driver's side of the six drums is always visible, and, by moving the spindle backwards, the drums can be reset to zero for the trip mileage indicator.

Oersted, Hans Christian • See Ammeter

Oil-cracking • See Oil refining

Oil drilling equipment

Before 1859, people drilled wells in search of water, salt, or brine (saltwater). Salt was especially important to early settlers in America because it was used to preserve food. In Pennsylvania, wells often became contaminated with oil, a nuisance to salt producers. But in the mid-1800s **kerosene**, a product of crude oil, was discovered. Kerosene proved to be an excellent fuel for **oil lamps**, which were then an important source of illumination. Samuel Kier, a salt producer whose well had been ruined by the seeping oil, began selling oil-burning lamps to provide a market for his unanticipated product. He also promoted the use of mineral oil as a medicine.

About this time, two American businessmen, George Bissell (1821-1884) and Jonathan Elvereth, became intrigued with the possibilities of producing more oil for the lighting market. In 1854 they formed a partnership, the Pennsylvania Rock Oil Company, the world's first oil company. They tried to drill for oil, but went bankrupt in 1857. In 1858, a group of investors formed the Seneca Oil Company, which hired a retired railroad conductor, "Colonel" Edwin L. Drake (1819-1880), as partner and supervisor of their first drilling project. Drake had studied the methods used for drilling brine and thought that the same equipment might be used to find oil.

At first, Drake and his crew struck water, which threatened to flood out the well. But Drake developed the idea of driving iron pipe down into the hole to shore up the sides and keep the water out. This basic concept is still used in oil drilling today. The following spring, Drake hired William ("Uncle Billy") Smith and his sons, who were experienced in drilling deep salt wells. The crew drilled almost 70 ft (21 m) down before they finally struck oil in 1859, and America's oil boom, and the world's oil industry, was launched.

Drake and other prospectors of the day used the same equipment and method—called *cable-tool* or percussion drilling—that had been used for salt wells. In percussion drilling, the hole is literally punched into the ground by a heavy cutting tool, called a *bit*, that is attached to a cable and **pulley** system. The cable hangs from the top of a four-legged framework tower called a derrick. Over and over again, the cable raises and drops the **drill** bit, shattering the rock into small pieces, or cuttings. Periodically, these cuttings have to be wet down and bailed out of the hole. By the late 1800s, **steam engines** had become available for cranking the drill bit up and down and for lowering other tools into the hole.

Although this method is sometimes still used for drilling shallow wells through hard rock, most wells today are bored into the earth by rotary drilling equipment, which works like a corkscrew or carpenter's drill. Rotary drilling originated during the early 1900s in Europe, where the ground is typically softer than in America.

In rotary drilling, a large, heavy bit is attached to a length of hollow drill pipe. As the well gets deeper, additional sections of pipe are connected at the top of the hole. The taller the derrick, the longer the sections of drill pipe that can be strung together. Although early derricks were made of wood, today's are constructed of high-strength steel. The whole length of pipe, or drillstring, is twisted by a rotating turntable that sits on the floor of the derrick. When the drill bit gets worn down, or when a different kind of bit is needed for a new layer of earth, the whole drillstring must be pulled out of the hole to change the bit. Each piece of pipe is unscrewed and stacked on the derrick. As the oil-bearing formation is reached, the hole is lined with pipe called *casing*, and finally the well is "completed," or made ready for production, with cementing material, tubing, and control **valves**.

Throughout the rotary drilling process, a stream of fluid called drilling mud is continuously forced to the bottom of the hole, through the bit, and back up to the surface. This special mud, which contains clay and chemicals mixed with water, lubricates the bit and keeps it from getting too hot. The drilling mud carries rock cuttings up out of the hole, clearing the way for the bit and allowing the drilling crew's geologists to study the rock and learn more about the formations underground. The mud also helps prevent cave-ins by shoring up the sides of the hole.

Offshore drilling processes and equipment are essentially the same as those on land, except that special types of rigs are used depending on water depth. Jackup rigs, with legs attached to the ocean floor, are used in shallow water (depths to 200 ft (61 m)). In depths up to 4,000 ft (1200 m), drilling takes place on semisubmersible rigs that float on air-filled legs and are anchored to the bottom. Drillships with very precise navigational instruments are used in deep water (depths to 8,000 ft [2400 m]). Once a promising area has been identified, a huge fixed platform is constructed that can support as many as forty-two offshore wells, along with living quarters for the drilling crew.

Many advancements have been made in oil-drilling technology in the twentieth century. In 1908, American lawyer Howard R. Hughes (1869-1924) and his partner Walter B. Sharp (1860-1912), an American drilling engineer, patented a steel, cone-shaped drill bit with rows of steel teeth. Soon the Hughes cone drill bit was being used worldwide. Through the 1920s numerous innovations were made, including the cross-roller rock bit. Hughes and his scientists also developed the first self-cleaning rock bits and introduced tungsten carbide as a bit facing hardener. Along with other improvements, this hardener resulted in a standard form of drill-bit tooth that has been used nearly continuously since 1926. The most advanced rotary cone rock bits available today can drill about 80% faster than bits from the 1920s. Depths of more than 16,500 ft (5,030 m) have recently been reached through improvements in other drilling support technologies.

OIL ENGINE • See Engine, oil

OIL LAMP

The idea of burning oil to produce artificial light probably dates backs hundreds of thousands of years, to when primitive peoples cooked meat by open fire and must have noticed the light given off by fat dripping onto flames. By 70,000 B.C., the Stone Age cave-dwellers of western Europe had invented the first oil **lamps**, which were made from hollowed-out stones that held a piece of moss or other spongy plant soaked with oil or grease. These lamps were likely used to illuminate caves and allow primitive artists to decorate the walls with paintings.

Around 2000 B.C., Iberian (Stone Age English) flint miners used similar lamps made from lumps of chalk to illuminate underground mines. Today, Eskimos still burn walrus and seal oil in lamps comparable to those of the Stone Age. Some time before 1000 B.C., people learned to make wicks from dried rushes or other plant fibers. Typically, the stone or saucer-shaped-oil container had a groove to support the wick above the oil.

Throughout the Mediterranean world, ancient civilizations began to employ other containers for oil, made of shells,

pottery, and metal. Some of these lamps were elaborately carved to portray primitive deities. The Greeks introduced more sophisticated lamps with handles and spouts or nozzles, which were used for receiving oil and holding wicks. Lamps made of pottery were sometimes glazed with black or red colors. Roman metalworkers created bronze and iron oil lamps with elaborate designs representing lions, dolphins, and other creatures.

Open saucers with a floating wick were most commonly used in medieval times, even though this type of lamp was less developed than earlier versions. One exception was **Leonardo da Vinci**'s oil lamp, invented in 1490. The flame was enclosed in a **glass** tube placed inside a water-filled glass globe. Leonardo's lamp not only burned more steadily but also produced better illumination, due to scattering of the light by the water.

Beginning in the 1700s, oil lamps became much more efficient. A lamp invented by Swiss chemist Aimé Argand (1755-1803) in the 1780s used a round burner and circular wick. This design allowed a strong draft of air to reach the flame, which greatly intensified the light it gave off. One of Argand's assistants also found that the flame would burn more steadily inside a glass chimney, and reflectors were sometimes used to magnify the lamplight. Despite these improvements, however, oil lamps did not become widespread until paraffin oil, or **kerosene**, was introduced in the mid-1800s. British scientist James Young (1811-1883) pioneered the low-temperature distillation of kerosene from coal and oil shale. Around the same time, kerosene was produced from thick asphalt rock by Abraham Gesner (1797-1864), a Canadian geologist. Beginning in 1860, kerosene produced from large crude oil deposits in America was exported to Europe, where it quickly replaced whale oil and other oils as a lamp fuel. Kerosene lamps remained the primary light source in rural America until the 1930s and 1940s. Most modern oil lamps use a cloth net, or mantle, instead of a wick.

See also Arc lamp; Gaslight

Oil Refining

Oil refining is a process used to separate crude oil into useable components. Crude oil is also known as petroleum, a term which comes from the Latin words for *rock* and *oil*. Petroleum is a complex mixture of hydrocarbon compounds, usually containing impurities such as nitrogen, oxygen, sulfur, and small quantities of trace metals. **Gasoline**, diesel fuel, heating oils, lubricants, waxes, **asphalt** and coke are obtained from refined oil.

Crude oil is the product of millions of years of decay of organic materials, such as plant life. Oil migrates upward through fissures in the ground and until it eventually reaches a solid layer of rock, where becomes trapped underneath in a reservoir. These arched layers of rock that trap oil are known as anticlines. Petroleum can also be trapped by fractured rock layers, salt formations, and stratigraphic traps. Some oil is also contained in shales and sands, but these sources are more difficult and expensive to extract than oil trapped in reservoirs.

Although ancient people were known to collect oil from surface pools and use it for a variety of medicinal and building purposes, it was not until the **kerosene** was produced by geologist Abraham Gesner (1797-1864) in the mid-1800s that any major market for petroleum developed. Kerosene soon became widely used in oil lamps, even though it gave off a foul smell when burned. Sensing the economic potential for petroleum products, two American businessmen, George Bissell (1821-1884) and Jonathan Elvereth, formed the world's first oil company. They had a sample analyzed by American chemist and geologist Benjamin Silliman (1803-1853). Silliman's report, which was the first survey of petroleum's technical and economic promise, helped launch today's oil industry.

Shortly thereafter, in 1859, the first successful oil well struck oil at a depth of almost 70 ft (21.5 m) near Titusville, Pennsylvania. By the 1860s, the oil boom had swept the country much like the gold rush had done a few years earlier. Oil production in parts of Europe such as Romania, Spain, and Italy, soon followed.

Early processes used to refine crude oil were grossly inefficient and yielded less than one barrel of gasoline from every fifteen barrels of crude oil. With the advent of the auto industry in 1908, oil refiners had trouble keeping pace with the growing demand for gasoline. In 1913, just before World War I, a process was invented to increase the efficiency of refining operations. William Burton, an employee of Standard Oil of Indiana, developed a technique called *thermal cracking*, which breaks down heavy hydrocarbons into lighter compounds through the application of heat and pressure. By the end of the war, refiners had more than doubled the amount of gasoline produced from a barrel of crude oil.

During the 1930s, the cracking process was greatly improved when catalysts were introduced to promote chemical reactions. Similarly during World War II, the oil industry proved that it could boost production quickly and create specialized new products for military needs, such as high-octane fuel for airplanes and raw materials for synthetic rubber. Over the next several decades the consumption of gasoline continued to grow until, by the early 1970s, oil refiners were highly dependant on imported oil.

Today, a variety of techniques are used to process crude oil. Distillation, for example, involves heating the oil and collecting the different vapor fractions as they boil off. These phases are then condensed and concentrated to form liquid products. Depending on the end products, atmospheric, vacuum and azeotropic, or extractive distillation techniques may be employed. Cracking is still used and includes techniques such as visbreaking (viscosity breaking) and coking. To a large extent, thermal cracking has been replaced by the more efficient catalytic processes. Hydroprocessing is a method used to treat petroleum feedstock with hydrogen at high temperature and under pressure. Hydrocracking and hydrotreating are specific types of hydroprocessing. These processes are used to create a variety of end products.

Other special processes used in refining include reforming (which increases the octane of refined gasoline), isomerization (which provides additional feedstock for alkylation units), and polymerization (which increases the molecular weight of olefin gases).

OILCLOTH • See Waterproof materials

OLEOMARGARINE • See Margarine

OMNIBUS • See Bus

OPEN-HEART SURGERY

For many years, major heart surgery—opening the chest to operate directly on an exposed heart—was considered outside the realm of possibility. The heart would cease beating during such operations. How could patients survive?

A few pioneers did perform emergency surgery directly on the open heart, one of the first being African-American surgeon **Daniel Hale Williams**, who opened the chest of a stabbing victim and sewed up the pericardium (the sac surrounding the heart) in 1893. More lengthy and complicated heart operations, however, required a way to keep the blood oxygenated and circulating while a patient's heart was undergoing the operation.

American surgeon, John H. Gibbon, Jr., devoted himself to solving this problem in the 1930s. Assisted by his wife Mary, Gibbon persisted until he had developed a workable pump-oxygenator, or **heart-lung machine**, that shunted blood from the veins through a **catheter** to a machine that supplied the blood with oxygen and then pumped the blood back into the arteries. On May 6, 1953, Gibbon connected Cecilia Bavolek, a patient suffering from heart failure, to the heart-lung machine and operated directly on her heart, closing an opening between her atria. This operation ushered in the era of open-heart surgery.

The methods and technical details of open cardiac surgery were refined throughout the 1950s by a number of surgeons and engineers, notably Owen Wangenstein at the University of Minnesota and John W. Kirklin at Minnesota's Mayo Clinic. By 1960 open-heart surgery was standard practice and began to be used not just for repair of cardiac malfunctions but also for replacement of defective heart parts—even the whole heart itself. Surgeons Albert Starr and M. L. Edwards of Portland, Oregon, designed a ball-and-cage **artificial heart valve** and successfully implanted it in a 52-year-old patient in 1961.

Rene G. Favalaro introduced coronary artery bypass surgery in 1967. The heart gets its blood supply from coronary arteries that branch off from the aorta. These arteries can narrow from accumulations of plaque, which also promote clot formation, and can thereby become blocked, causing severe

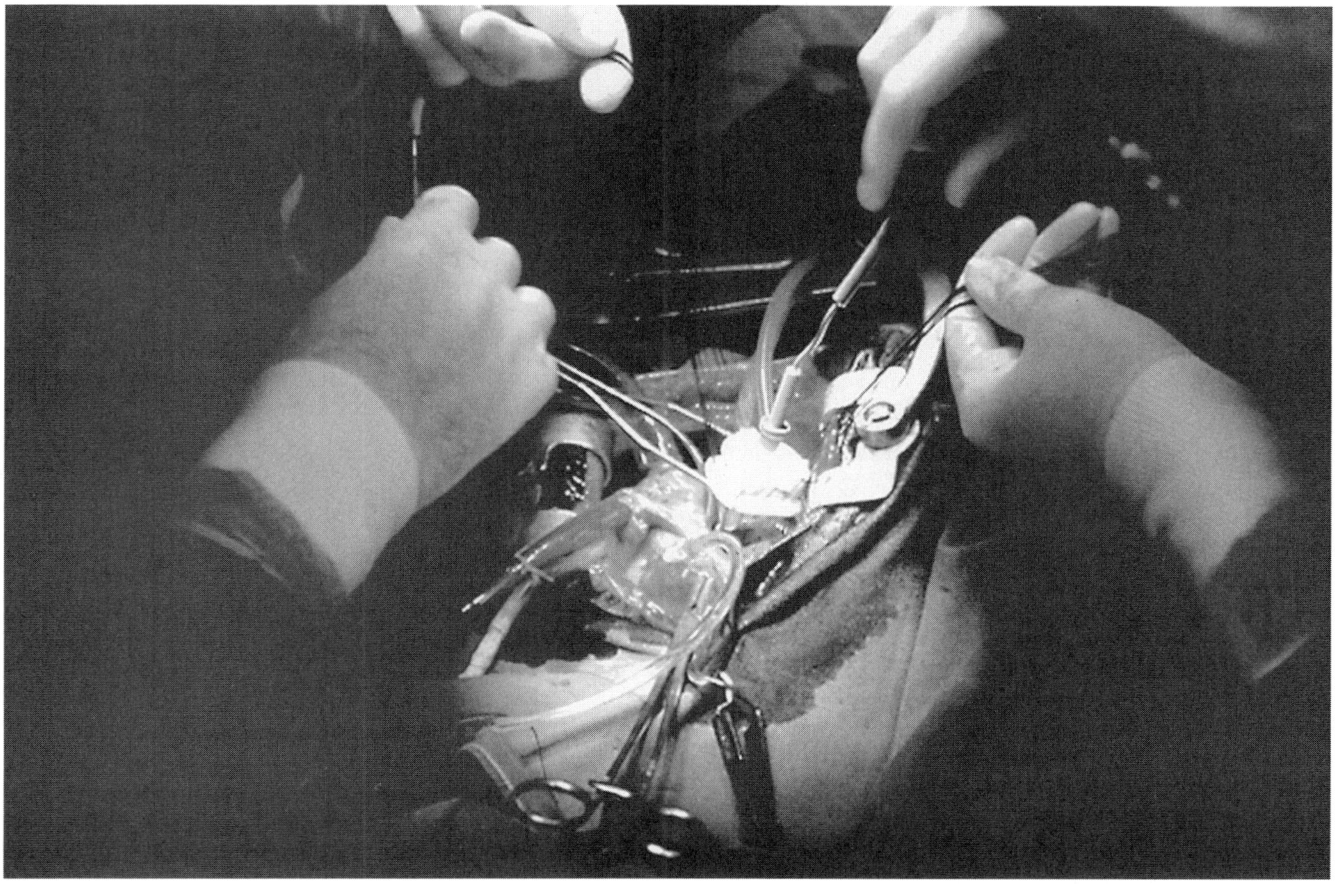

Open-heart surgery to replace a mitral heart valve. The tubes carrying blood are connected to a heart-lung machine, an invention that made such life-saving operations possible.

chest pain (angina) and, in some cases, heart attack. Favalaro and his surgical team at the Cleveland Clinic devised a technique of grafting a vein from the patient's leg around a blocked portion of a coronary artery, creating an alternate blood pathway. Favalaro's bypass surgery was made possible by the use of microsurgical techniques **arteriography** (direct images of the heart prior to open-heart surgery) and the heart-lung machine. Within three years of Favalaro's pioneering 1967 operation, coronary bypass surgery gained wide acceptance. Its use in cases of mildly clogged arteries dropped off in the late 1970s with the advent of **balloon angioplasty**.

The most dramatic development in open-heart surgery was the heart transplant, first successfully performed in Cape Town, South Africa, by Dr. **Christiaan Barnard** in 1967. Another dramatic step in open-heart surgery was revealed in 1996 when Randas Batista, an obscure cardiac surgeon from a country clinic in southern Brazil, hit the headlines in the United States. Batista cut a triangular chunk from the left ventricular muscle of an enlarged heart in an effort to save the life of patient with chronic heart failure. Researchers do not yet understand why a weak heart thickens and grows—sometimes ballooning to twice it's normal size—but they do know it no longer pumps efficiently and eventually fails completely. Contrary to popular opinion, Batista believes it is the size of the heart, not the weakened muscle, that kills people. He has used his controversial technique on more than 400 people since 1994. In 1996, prominent U.S. surgeons travelled with the ABC news team from the **televisoin** program, *20/20*, to Brazil to see Batista's technique first-hand. Since May, 1996, Patrick McCarthy of the Cleveland Clinic has performed the surgery on more than 24 patients. All but one were still living four months later.

OPEN-HEARTH PROCESS • See Steel production

OPERA GLASSES • See Binoculars

OPHTHALMOSCOPE

An ophthalmoscope enables a physician to examine the interior of the eye by directing a tiny beam of light through the pupil, the black ''window'' of the eye. Using the ophthalmoscope, the physician can look through the pupil to detect any abnormalities or pathological changes that could signal disease.

The first ophthalmoscope was invented by **Charles Babbage**, an English mathematician, in 1847. He gave the device to a physician for testing, but it was laid aside and forgotten. Four years later, German physician and physiologist Hermann von Helmholtz (1821-1894), unaware of Babbage's invention, developed his own version of the ophthalmoscope. Because he had better luck making his device known, Helmholtz is often credited as the sole inventor.

Helmholtz's instrument operated by using a **mirror** to shine a beam of light into the eye. The observer would look through a tiny aperture attached to the mirror. Helmholtz eventually found that looking through the retina into the back of the eye only produced a red reflex; consequently, he attached a condenser lens to obtain an inverted image, which was then magnified five times. He called this combination of a mirror and condenser lens an indirect ophthalmoscope. It was used regularly for eye examinations until 1920.

Helmholtz also invented the ophthalmometer, which was used to measure the curvature of the eye; studied color blindness; the speed of nervous impulses; and physiological acoustics; and wrote the classic *Handbook of Physiological Optics.* Swedish ophthalmologist, Allvar Gullstrand, who also studied physiological optics, developed another version of the ophthalmoscope and invented a slit lamp used with a microscope that enabled a physician to locate foreign bodies in the eye.

The modern ophthalmoscope is a hand-held instrument containing a small battery-powered lamp that directs the beam of light into the eye of a patient by way of a mirrored **prism**. The observer looks through a tiny hole in the prism and the instrument, which can be focused by a series of revolving lenses, magnifies the image. The lens needed to focus the image gives an approximation of the spectacle lenses needed to correct the patient's vision. A new type of ophthalmoscope that can project a **laser** beam is used in eye surgery to correct a detached retina. Another, larger type of ophthalmoscope, called the **binocular** ophthalmoscope, is used in clinical research and provides an image of the eye that is magnified fifteen times.

See also Lenses

OPPENHEIMER, J. ROBERT • See Atomic bomb

OPTICAL DISK

Optical discs are the state of the art in computer memory and storage, using **laser**s to record and retrieve information. Like magnetic discs, they can store many different kinds of information—sound, text, and pictures (both still and animated)—on the same disc. However, optical discs are far superior to their magnetic counterparts because they hold more data and, because the laser cannot damage the disc, they are more durable.

IBM began the first experiments with optical technology in the mid-1960s but made little progress until the semiconductor was developed. Sony and Philips saw the potential that the technology offered, and the two joined forces later in the decade, hoping to find a viable application for it. The result, in the 1970s, was the videodisc, which use a laser stylus to play back analog (as opposed to digital) information recorded in spiral tracks on a **plastic** disc. This spiral-track method is similar to the one used to make **phonograph** records. Videodiscs' picture quality is superior to that of videocassettes but their use is limited in this country; they enjoy wider popularity in Japan.

In the mid-1980s Sony and Philips unveiled yet another joint venture, the prerecorded compact disc, or CD. Unlike the videodisc, a CD is recorded digitally, which not only produces a better sound, it uses space on the disc more efficiently (hence the term "compact"), allowing more data (in this case, music) to be stored. Within only few years, CDs dominated the record music industry, turning **LPs** into relics.

CDs can also be used to store visual data in a form called CD-ROM (read-only memory). When used in this way they are read by CD players in a computer to reproduce text, graphics, and sound from the same disc. Because of their high density, CD-ROM discs can store incredible amounts of information; they have become the standard format for large published works like software documentation and encyclopedias.

Whether for music or text, CDs and CD-ROM discs are produced the same way. Digital data (the binary language of ones and zeroes common to all computers) are encoded onto a master disc. A laser burns small holes, or pits, into a thin metallic film sandwiched between a plastic substrate and a protective plastic or **glass** film. These pits are a rough binary equivalent of ones. Smooth areas of the disc untouched by the laser, called land, are analogous to zeroes. Thus data is translated as a series of reflections (land) and nonreflections (pits). Once transferred to the disc, they are not erasable. This master disc can be used to produce identical copies of itself.

When a CD player or **computer** reads data from an optical disc, it uses a photodetector or diode to catch low-power laser light reflected off the disc's surface. The light's intensity at any given point is determined by the arrangement of pits and land on the disc. Information received by the diode is digitized and then converted to music, text, or images.

CD-ROMs were a vast improvement over magnetic media in terms of sheer data capacity, but they could only read; users could not write new information onto the optical disc. To meet this need, IBM and other computer manufacturers came up with **WORM** (write once read many) drives and optical discs. These can both read from and write onto a computer's optical disc. Data is added using a high-power laser in the same way that CD and CD-ROM discs are encoded. Once entered, data storage is permanent; the discs cannot be erased.

WORM drives can be used in place of or in addition to magnetic discs in a personal computer. They are also manufactured in two-drive systems called jukeboxes that hold 100 discs and a potential terabyte (1 trillion bytes) of information. This is especially impressive once you realize that an entire encyclopedia set can be stored on 20% of one standard optical disc.

The next obvious evolution of optical storage technology was the erasable optical disc, and that, too, has hit the market. The most successful system to date uses magneto-optic, or magnetically assisted, recording. Unlike CD and WORM systems, data spots are heated, not burned, into the disc medium, and then magnetized. The magnetization corresponds to a zero or a one depending on its direction.

The first commercial expression of this technology, the Tahiti I disc drive, was developed by the Maxtor Corporation in 1988. Its access time was a then-breathtaking 43 milliseconds, half the rate of contemporary hard magnetic disc drives. Even more amazing was the new read/write head, which was half the weight of conventional models.

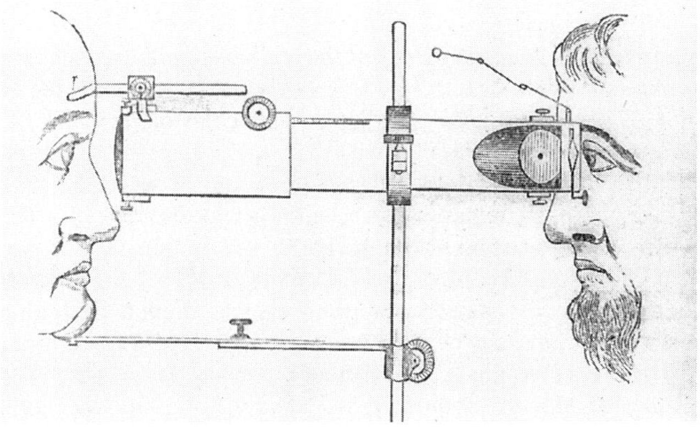

Drawing of an early ophthalmoscope from the nineteenth century.

Another type of erasable optical storage uses crystalline materials whose structure changes to an amorphous state at a certain light wavelength. This has been less successful than magneto-optical systems because the crystals eventually lose their ability to change.

Yet another erasable optical disk system hit the computer world like a thunderbolt in 1988 when the Tandy Corporation announced its THOR-CD. This hybrid system can write, retrieve, and erase data from audio OR computer systems. Although the particulars are still a trade secret, the system probably uses two lasers, one to write and one to erase data. The next set of innovations in optical disk technology were introduced in the the mid–1990s, a new and improved CD-ROM was introduced called the Digital Versatile Disk, or DVD. DVDs offered approximately seven times more storage than a CD-ROM.

ORGAN, ELECTRIC • See Musical instruments, electric

ORTHODONTICS

Orthodontics is a specialized branch of dentistry dealing with "malocclusions," or bad bite. The "bandolet," created in 1723, was the first known orthodontic apparatus; the most common modern orthodontic appliances include braces, retainers, tooth guards, and splints, all used to correct tooth and jaw problems such as crossbite, protruding or misaligned teeth, protruding or retruding jaw, tooth grinding, speech difficulties, and temporomandibular joint dislocation. Teeth-straightening and extraction to improve alignment of remaining teeth has been practiced since early times (**Leonardo Da Vinci** perhaps made the first written observation regarding teeth and bite). Today, approximately one in every 50 people (25% being adults), is being treated by an orthodontist.

Orthodontics as an independent science began developing in the l880s. The first comprehensive treatise on dentistry, *The Surgeon Dentist*, published in 1728 by Pierre Fauchard (1678-1761), devoted an entire chapter to tooth irregularities and ways to correct them. In 1757, French dentist Bourdet wrote *The Dentist's Art*, devoting a chapter to tooth alignment; in 1771, **John Hunter** wrote the first English text on the subject entitled *The Natural History of the Human Teeth*; and the term *orthodontia* was coined in 1841 by Lafoulon and appeared in a book by J. M. Alexis Schange on malocclusion. In 1858, the first article on orthodontics was written by Norman W. Kingsley. His 1880 *Treatise on Oral Deformities* served as the catalyst for the new dental science, earning him the title of ''The Father of Orthodontics.'' J. N. Farrar who published the two volume, profusely illustrated *A Treatise on the Irregularities of the Teeth and Their Corrections* and was adept at designing orthodontic appliances, suggested the use of mild force at intervals to move teeth; he ultimately became known as ''The Father of Modern Orthodontics.'' The third influential figure in orthodontics, Edward H. Angle (1855-1930), devised the first simple and logical classification system for malocclusions which is still used as the basis for orthodontic diagnosis. He contributed significantly to the design of orthodontic appliances, founded the first school and college of orthodontia, organized the American Society of Orthodontia in 1901, and founded the first orthodontic journal in 1907. His highly praised reference book, *Malocclusion of the Teeth*, went through seven editions. Other innovations in orthodontics in the late 1800s and early 1900s included the first textbook on orthodontics for students, published by J. H. Guilford in 1889; Eugene Solomon Talbot's (1847-1924) suggestion to use X rays for orthodontic diagnosis; and the use of **rubber** elastics, pioneered by Calvin S. Case (or perhaps H. A. Baker).

Developments in orthodontics include more convenient, comfortable, and less noticeable appliances such as tiny brackets instead of bands; ''space age'' wires for braces and bands developed through NASA, which require fewer replacements; braces with clear, tooth-colored, or multi-colored brackets with matching or interchangeable colored elastics; retainers with a favorite logo or photo; tiny **magnet**s attached to upper and lower molars instead of head gear to realign teeth; ''lingual'' braces connected to the back of teeth for completely invisible treatment; and computer technology to diagnose and plan treatment, and produce images of the end results.

For more extreme problems, such as a jaw too narrow to accommodate a full mouth of adult teeth, expanders—devices used to widen upper and lower jaws—prevent overcrowding. Most effective when applied before puberty, upper expanders fit over the roof of the mouth and connect to the back teeth on either side exerting outward pressure. A tiny key turned periodically further expands the frame, opening the sutures (the juncture where the bones of the mouth roof meet) allowing new bone to grow into that opening. As there is no lower suture to open, lower expanders, which fit under the tongue, are designed to upright tipped teeth.

Disorders of the *temporomandibular joint* (TMJ), located immediately in front of the ears where the jaw bone connects to the skull, affect 60 million or more Americans and has many origins—from injury to high levels of stress. Symptoms include: difficulty opening and closing the mouth—often the jaw will lock in one position; pain and noise when chewing; ear, temple, head, or face pain; debilitating headaches; loss of hearing or ringing in the ears; nausea; blurred vision; and dizziness. Diagnosis is difficult, as pain often emulates that caused by other problems. Treatments include reversible (splints, physical therapy, psychotherapy to relieve stress, chiropractic adjustments, and massage therapy), and nonreversible (crowns, surgery, and bite adjustment). Splints allow joints, muscles, and ligaments to rest, heal, and function more normally, and also reduce the damaging effects of clenching and grinding teeth, usually a stress-related activity.

Orthopedic devices • See Fractures, treatments and devices for

Oscillator

Just about any device that produces a back-and-forth or up-and-down motion—particularly if that motion is shaped like a sine wave—can be called an oscillator. One of the simplest examples is a pendulum; if the path of a pendulum were charted on a graph, that path would approximate a sine wave. The same is true of a suspended spring.

In **electronics**, an oscillator is a circuit that generates a pulsed, or periodic, signal. These signals can also resemble sine waves, though some generate square or saw-toothed pulses. Oscillators that produce sine waves are called sinusoidal, while the remainder are called non-sinusoidal.

Sinusoidal quartz oscillators are widely used as signal amplifiers. They can be found as integral parts of **radios**, **radar** systems, and just about every other electronic communication device. Because of the regularity of the pulses they generate, both sinusoidal and non-sinusoidal oscillators are used as timing devices. Such timers are used in **televisions** and digital **computers**.

Perhaps the most advanced oscillators are the **maser** and the **laser**. During the 1940s the oscillator had become an important tool for communications and the emerging television industry. However, scientists had discovered that their oscillators could not generate frequencies beyond 300,000 MHz—a limit researchers knew they would soon need to pass. American physicist **Charles Townes** developed a system that used atomic vibrations in molecules to create signals within the **microwave** frequencies. This device, called a maser (an acronym for Microwave Amplification by the Stimulated Emission of Radiation) was essentially a microwave oscillator, just as the laser (an acronym for Light Amplification by the Stimulated Emission of Radiation) was an infrared oscillator.

Oscilloscope

An oscilloscope produces a wavelike display of electrical signals. Typically the display is a **cathode ray tube**, such as that used in a personal **computer**, though usually smaller. The oscilloscope displays the value of a voltage as it changes with time. Since many physical measurements can be converted to a voltage via a **transducer**, the oscilloscope serves as a very versatile tool for a researcher who wants to analyze the behavior of his system, and they are widely used in laboratories.

An oscilloscope performs five functions. (1) It captures the electronic sample that is fed into it, (2) it provides the ability to view this sample as a function of time, (3) it allows measurement of the electronic signal, (4) it provides the ability to analyze the electronic signal, and (5) it provides the ability to store or print the results of the display screen. Typical applications for an oscilloscope are electronic circuit design, testing, and debugging, and troubleshooting faulty or intermittent circuits.

The two-dimensional display of the oscilloscope places time on the horizontal, or x axis, and the voltage to be displayed and analyzed on the vertical, or y axis. The oscilloscope has controls to allow the signal to be amplified so that small signals may be displayed, for the brightness of the screen to be increased or dimmed, for the time axis to be amplified so the rapidly varying signal may be displayed as stationary on the display, and to take a photograph of the display or record it on a paper chart. Besides showing the wavelike form of the incoming signal, an oscilloscope can be used to measure the signal's strength and frequency.

Oscilloscopes can be either analog or digital. There are a large number of analog oscilloscopes in use, but they are gradually being replaced by the superior performance of the digital oscilloscopes. Much like personal computers, the cost of a digital oscilloscope is dropping, and they are utilizing the latest electronic developments. Advances include **liquid crystal displays**, smaller "footprint" areas on a tabletop, portability, and decrease in weight.

The oscilloscope may gradually be replaced by the personal computer. A personal computer has most of the elements of an oscilloscope—in fact, of many scientific instruments: a display, a **microprocessor**, and a power supply. Circuit boards are now made that plug into a desktop or laptop computer. A cable connects the circuit boards to whatever is being tested. Software analyzes and displays the incoming signal. This "virtual oscilloscope" is cheaper than conventional ones, and such virtual instruments may be more common in the laboratory of the future.

See also Electronics; Voltmeter

Ostwald-Bauer process

The invention of the Ostwald-Bauer process enhanced the ability of chemists to produce nitric acid. Nitric acid was originally synthesized by heating saltpeter with sulfuric acids. The process was first discovered by Jābir Ibn Hayyan (c.721-c.815) in the eighth century. In 1901, Russian-born German physical chemist Friedrich Wilhelm Ostwald (1853-1932) invented a much better process for manufacturing nitric acid. He won the Nobel Prize for chemistry in 1909 for his work on catalysis, and today is regarded as the one of the founders of modern physical chemistry, the branch that investigates chemical changes by observing changes in physical properties. From a commercial point of view, however, Ostwald's greatest invention and most important contribution to the fields of science and industry was the Ostwald-Bauer process.

This method employs a platinum catalyst to promote the composition of nitric acid. Ammonia became available in economical quantities before World War I started, when German chemist **Fritz Haber** invented a process for synthesizing the element in 1908. In Ostwald's three-step procedure, ammonia is converted into nitric oxide and then changed into nitrogen dioxide, by combining with hot air and being exposed to a piece of gauze coated with platinum and rhodium. The gases that are produced and collected are then absorbed in water, thereby creating nitric acid.

Nitric acid (HNO_3) is recognized as an important industrial chemical. It is an essential ingredient in many fertilizers, and it is used in the production of **plastics**, **lacquers**, and **dyes**. One of nitric acid's most important uses, however, is in the manufacture of such explosives as **dynamite**. In the presence of sulfuric acid, nitric acid reacts with toluene to form **TNT** (trinitrotoluene).

See also Ammonia synthesis

Elisha Graves Otis

Otis, Elisha Graves (1811-1861)
American inventor

Elisha Otis, a descendant of revolutionary patriot James Otis (1725-1783), was born near Halifax, Vermont. He left his family's successful farm at the age of nineteen to pursue a number of trades, including carpentry in Troy, New York, and gristmilling, followed by carriage manufacturing and sawmilling back in Vermont. In 1845 Otis moved his family to Albany, New York, where he worked as a master mechanic at a bedstead factory and invented an automatic **lathe**.

In 1851 Otis went to Bergen, New Jersey, and then to Yonkers, New York, the following year to supervise the construction of a new bedstead factory for his employer. While in Yonkers, Otis designed a ''safety hoist'' to lift loads at the factory; his crucial innovation was a safety catch that kept the car from falling if the lifting cable or rope broke. Otis set up a small **elevator** shop in Yonkers in 1853, selling only a few for hoisting freight. To increase sales, Otis dramatically demonstrated his elevator during an exhibition at the Crystal Palace in New York City in 1854, riding in the cab high above onlookers and then having the cable cut. This did attract attention, and in 1857 Otis installed the first passenger safety elevator in a New York department store, and later his passenger elevator made the skyscraper feasible. Just before his death, Otis patented a steam-driven elevator, which was the basis for what became the Otis Elevator Company, run by Otis's two sons, Charles and Norton. Among other devices Otis patented were railroad car brakes (1852), a steam plow (1857), and a bake oven (1858).

Otto cycle

The first **gasoline**-powered engine was modeled on the **steam engine**, and was the first practical alternative to it as a power source. In the steam engine, expanding steam inside a cylinder pushed up a **piston**, which then returns to its original position when the steam is allowed to escape and the steam supply is cut off. A steam engine is said to have a *two-stroke cycle*, since the piston moves twice (through one cycle) with each blast of steam.

The two-stroke gasoline engine used a mixture of ignited gasoline and air in place of the hot steam above. The piston was attached to a crankshaft with U-shaped sections called cranks via connecting rods attached to the bends. With these cranks, the crankshaft converted the up-and-down motion of the piston into rotating mechanical energy. A two-stroke engine did not have much power, however, because it required a precisely-timed combustion of the gasoline/air mixture. Too soon or too late damped the piston and so decreased the rotary power delivered through the crankshaft.

The German engineer **Nikolaus August Otto** (1832-1891) built his first gasoline-powered engine in 1861. Three years later he formed a partnership with Eugen Langen, and together they formed the Deutz Gas Engine Factory, developing an improved engine that won a gold medal at the Paris Exposition of 1867.

In 1876 Otto built an **internal combustion engine** utilizing a four-stroke cycle. (Some attribute this invention to Otto's chief engineer, Franz Rings.) The four-stroke cycle has been patented in 1862 by the French engineer Alphonse Beau de Rochas, but since Otto was the first to build a working engine on its principle, it is commonly called the *Otto cycle.*

The piston in a four-stroke cycle moves up-and-down down twice in each cycle. The four distinct events in the four-stroke engine are commonly labeled the intake, or induction stroke, the compression stroke, the explosion, or power stroke, and the exhaust stroke: in the intake stroke, the gas-air mixture is drawn into the chamber, through the intake valve, as the piston moves down the chamber; in the compression stroke, the piston comes back up the chamber with both valves closed, compressing the gas-air mixture to a very small volume; in the explosion stroke, the compressed gas-air mixture is ignited via an electric spark, and the mixture explodes. The piston is driven back down the chamber, the stroke that sends power to the crankshaft; and in the exhaust stroke, the exhaust valve is opened and the piston comes back up the chamber, pushing the spent gasses out of the chamber. The exhaust valve closes, the intake valve opens, and the cycles begins again.

It has two valves, the intake valve, through which the gas-air fuel mixture enters the cylindrical chamber, and the exhaust valve, through which the exhaust gasses leave.

Otto's engine produced 3 horsepower and ran at 180 revolutions per minute. It was a great improvement in reliability, efficiency, and quietness, and was an immediate success. More than 30,000 of these ''Otto Silents'' were built during the next 10 years.

Otto's patent was revoked in 1886 when Beau de Rochas' earlier patent was brought to light. The Otto cycle was a starting point for later developments in the internal combustion engine, such as the **Stirling engine**, where the mixture is burned not in the piston chamber but in a separate combustion chamber.

Otto, Nikolaus August (1832-1891)

German engineer

Otto, born at Holzhausen, Hesse-Nassau, was the son of a farmer. Otto left school at 16, and later moved to Cologne where he became fascinated by the gas engines developed by Frenchman **Jean-Joseph-Étienne Lenoir** and displayed for the first time in 1859. Lenoir's engines had a two-stroke design and burned coal (or illuminating) gas; they were also notoriously inefficient. Nevertheless, in 1861, Otto built an experimental engine based on Lenoir's design. Three years later, he joined forces with industrialist Eugen Langen and together they formed the Gasmotorenfabric Company with its factory at Deutz near Cologne to build and market his engines. Otto and Langen had the able assistance of Franz Reuleaux, **Gottlieb Daimler**, and **Wilhelm Maybach** as part of their engineering team.

In 1867, Otto and Langen announced their first production engine—a noisy two-stroke engine that improved upon the Lenoir design by compressing the gas before it was ignited. This new Otto engine was a commercial success, and Otto and Langen continued their collaborative achievements by building a radically new engine in 1876—the four-stroke internal combustion engine, which was immediately dubbed the ''Otto Silent.''

The inefficiency of Lenoir's two-stroke engine was due to the fact that (in addition to not compressing the gas before ignition) the intake of the fuel, its combustion, and the exhaust of the burnt gases had to be accomplished within one back-and-forth motion of the **piston**, which was coupled with a rotary crank to convert the reciprocating motion to rotary motion.

Otto's four-stroke engine separated these various tasks. In his engine, the first downward motion of the piston drew an air and gas mixture into the cylinder; on the upward stroke, this mixture was compressed; when the piston was near the top of this motion, the mixture was ignited, driving the piston down and providing power; the fourth stroke, upward again, pushed the exhaust gases out of the cylinder. The intake and exhaust actions were controlled by **valve**s that opened at precisely the right moment by being geared directly to the crankshaft. The Otto four-cycle engine was an enormous success, so successful that the engine is quite often referred to as the "**Otto cycle**" engine. The Otto-Langen firm sold over 50,000 of these four-stroke engines in the 17 years following their introduction.

Despite Otto's commercial success, he encountered legal problems over his patent of his four-stroke engine when several others challenged his right to the invention. A Munich watchmaker, Christian Reithman, designed and built a four-stroke engine in 1872; Gustav Schmidt of Vienna set out the principles of the Otto cycle engine in a paper in 1861; and, most notably, Frenchman Alphonse Beau de Rochas described the cycles of intake, compression, power, and exhaust in a theoretical work published in 1862.

Although the courts invalidated Otto's patent in 1886 due to the claims of Beau de Rochas and others, the Otto cycle engine's superiority assured its continued success. It is the Otto cycle engine that made possible **automobile**s, **motorcycle**s, **powerboat**s, and **aircraft**, all of which require a small, light, and efficient engine. Nikolaus Otto died in Cologne on January 26, 1891.

OUGHTRED, WILLIAM • See Slide rule

OUTBOARD MOTOR • See Motorboat

OVSHINSKY, STANFORD (1922-)
American materials scientist

Stanford Ovshinsky pioneered the field of amorphous (disordered or structureless) materials that can be reversibly changed between amorphous and crystalline phases by an energy source, such as electricity or a **laser** beam. The phase change produces a change in conductivity, the amorphous structure representing zero (off) and the ordered structure representing one (on). Because the change is stable, these materials, called *ovonics*, are suitable for computer processors and memory, optical data storage, high-temperature superconductivity, and silverless photography.

Ovshinsky was born in Akron, Ohio, where he graduated from high school and from trade school as a machinist. His first invention, in the early 1940s, was an automated **lathe** that became widely used. In the early 1950s, while working as research director of an auto parts manufacturing company, he experimented with switches that worked like neurons (brain cells). Since the brain does not require ordered structures to use energy and store information, Ovshinsky reasoned that engineered energy-information systems did not need them either.

After experimenting with amorphous materials containing tellurium, arsenic, and other members of the chalcogenides, he discovered that they changed phase when a certain voltage threshold was reached. Both heat and electronic conditions are involved in the transformation.

In 1957, he built his first ovonic switch; applications have followed since then. As semiconductors, amorphous materials are very fast and less expensive to make than the crystals used in conventional semiconductors. Ovonics are being used in direct-overwrite optical data disk storage. This means new data can be written on an **optical disk** without first deleting the old data.

Ovonic materials are also being studied as superconducting materials. They retain their superconducting properties at much higher temperatures than conventional superconducting materials, which must work near a temperature of absolute zero.

Because of the novelty of his ideas and his lack of academic credentials, Ovshinsky's initial paper on ovonics was initially greeted with extreme skepticism by scientists and engineers. However, he persevered because he believed in his designs and in his ability to make products from them.

Ovshinsky is a fellow of the American Physical Society, an adjunct professor of engineering science at Wayne State University, and an adjunct professor of physics at the University of Cincinnati. In 1986, Ovshinsky was awarded the Diesel Gold Medal by the German Association of Inventors. In 1987, public television's NOVA devoted a one-hour program to his life and work. In 1988, he received the Coors American Ingenuity Award. He was awarded the Toyota Prize in 1991 for having developed the hybrid ovonic nickel-metal-hydride battery. He took first prize for the best article at the International Symposium for Electric Vehicles in 1992. In 1997, a prototype electric vehicle powered by an ovonic nickel-metal-hydride batteries traveled 216 mi (347.76 km) from Boston to New York City on a single charge at normal highway speeds, thus demonstrating that electric vehicles are both practical and capable of providing adequate driving ranges.

Ovshinsky holds over 100 ovonics-related patents and is president of Energy Conversion Devices, located in Troy, Michigan.

OWENS, MICHAEL JOSEPH • See Bottle and bottle-making machinery

OXY-ACETYLENE WELDING • See Welding

OXYGEN TENT

A French physician, Charles Michel (1850-1935), first realized the importance of oxygen to aid the healing process. It was perhaps around 1900 that he first used an oxygen chamber to improve the health of some of his patients, probably those suffering from respiratory disorders. Later the oxygen chamber was expanded around a patient's entire bed and became known as the oxygen tent. Oxygen tents then began to appear in hospitals in Europe and North America.

Oxygen tents are most often used when a medical patient suffers from pneumonia or other respiratory disease, or carbon monoxide poisoning. They are also used following an event in which the patient's body tissues have been deprived of oxygen. The gas inside the tent has a higher percentage of oxygen than found in normal air and thus the patient breathes in more oxygen per breath. The tent is a framed ''envelope'' fitted completely around the patient's hospital bed. Air inside the tent is extracted with a fan and passes through a dust filter and cooling unit where moisture in the exhaled air is condensed and removed. Oxygen-enriched air is pumped into the tent and, if necessary, humidity inside the tent can be increased by an atomizer so the patient's lungs do not dry out. Access to the patient is through a large, zippered opening.

P

Pacemaker

The rhythmic, regular beating of the heart is controlled by a natural pacemaker—a small patch of cells at the top of the right atrium called the sinus node which sends rhythmic electric impulses along specific conducting fibers to the heart muscles, stimulating them to contract and relax in a regular sequence. When the heart muscles fail to receive the pacemaker's signals, the heart ceases pumping blood. Within a few minutes, the patient faints, and within a few more minutes, dies—unless the heart muscles can be stimulated to resume contracting. An artificial pacemaker is designed to help a damaged heart beat normally—it is programmed to send an electrical impulse to stimulate the heart muscle if it does not sense a normal heart beat within a specific amount of time. Dual-chamber pacemakers, the most common type, sense and pace activity in both the atrium (upper chamber) and ventricle (lower chamber). Single-pass pacemakers use only one lead and sense only one chamber, usually the ventricle. Faster and more easily implanted than the dual lead type, this pacemaker is indicated only in certain instances.

The English surgeon W. H. Walshe first suggested using electric impulses to restart the heart in 1862. Nearly a century later, Harvard-educated American cardiologist, Paul Zoll, believed he could use the heart's responsiveness to electrical stimulation to treat cases of ''heart block.'' His first attempt, passing an electrode down the esophagus, failed, but in 1952 he developed an external pacemaker, passing an electric shock to the heart through electrodes placed on a patient's chest. In October 1952, Zoll's pacemaker was used to maintain a heartbeat in a man suffering from congestive heart failure; after two days, the patient's own heart took over again.

Zoll's machine, while effective, had inherent limitations: The shocks were painful to the chest muscles, and the machine—and thus the movement of the patient—were restricted to the nearest electrical outlet. Researchers envisioned an implantable pacemaker, and American inventor **Wilson Greatbatch** had dreamed of building one since he first heard of heart block in 1951. The restrictive size of **vacuum tubes** and storage **batteries** made it impossible, however. As **transistors** became widely available in the late 1950s, Greatbatch mentioned his idea to Dr. William Chardack of the Buffalo, New York, Veterans Administration Hospital and, with Chardack's encouragement, put together an implantable pacemaker in three weeks, working in the barn behind his home. After two years of animal testing, in 1960, Chardack and his associates implanted the first pacemakers in the chest wall of a human patient. Also in 1960, Ake Senning and his colleague, Elmqvist, also designed an implantable pacemaker with an external coil and internal receiver. Modern pacemakers are much improved over these designs. Lightweight and relatively easy to install, and their lithium batteries last up to ten years as opposed to the mercury-zinc battery's life of twenty months. The first generation of pacemakers sent signals at a preset rate; researchers Ken Anderson and Dennis Brumwell of Medtronic, Inc., in Minneapolis, Minnesota, advanced pacemaker technology immensely when they invented activity-responsive pacing in 1981 by using piezoelectric crystals that reacted to differing levels of body exertion. Medtronic's ''Activitrax,'' introduced in 1985, was the first pacemaker to adjust pacing rate to exercise level. Several even more advanced pacemakers became available in the late 1980s, among them Medtronic's ''Legend.'' These devices can be reprogrammed while they're still implanted, using **radio**-frequency signals to reset the pacemaker's **microprocessor**. They also store information about cardiac events, and some models can even transmit that information over the **telephone**, directly from the patient's chest to the doctor's office.

Two concerns surrounding the pacemaker are interference by digital wireless telephones and the ''Year 2000'' effect. Research indicates that some digital phones may interfere with some pacemakers but pose no more risk than metal detectors at airport security checkpoints or **X-rays**, and that interference is unlikely if the phone is held six in (15.24 cm) or more from the pacemaker. Research is underway to determine how this interference can be prevented. Medtronic, the world's

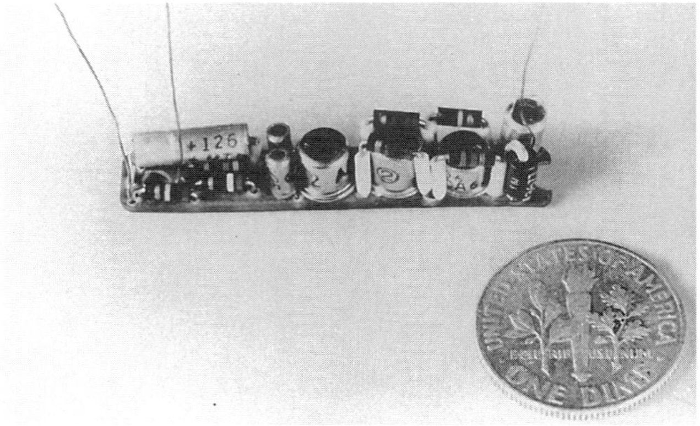

Historic cardiac pacemaker fueled by radioactive plutonium-238. This model was a prototype.

largest supplier of implantable medical devices, issued a statement in March 1998 that the anticipated ''Year 2000'' computer problems will in no way affect their computer-aided pacemakers because therapy is based on internal counters which are unrelated to any specific calendar date.

PADDLEWHEEL

As the first successful **steam engines** were developed in the 1700s, many inventors wondered how these new machines could be harnessed to drive boats. The problem was to transfer the motion of the engine to some device that would propel the boat. Several odd arrangements were tried, like mechanical oars and mechanical duck feet. It was natural for many early inventors to try paddlewheels. They took their inspiration from water wheels, which spun around under the power of running water. If paddles were attached to a boat and made to spin in such a way that they dug into the water, they might provide forward motion.

The Chinese were using paddlewheels on ships as early as the eighth century, and by the twelfth century large numbers of paddlewheel ships plied the Yangtze River. The wheels on these boats were powered by human muscle—some of the largest required as many as 200 separate wheels. Paddlewheels were first combined with a mechanical power source in the United States and Europe in the late 1700s. An American engineer, **John Fitch**, built a **steamboat** in 1788 that had stern paddles that moved like ducks' feet, and operated a ferry service with this boat between Philadelphia and Burlington, New Jersey, beginning in 1790. Fitch continued to improve the paddlewheels until they resembled a more modern configuration.

Two other American inventors, **Samuel Morey** and **Oliver Evans**, also introduced vehicles with paddlewheels in the late 1700s. While Morey designed a steamboat with paddlewheels, Evans used paddlewheels to power a strange amphibious vehicle. He built this vehicle—a floating, steam-driven dredge—at the request of the city of Philadelphia to dig and clear out the harbor area. It could move on land with wheels, switching to a paddlewheel for river travel.

Meanwhile, in Great Britain, William Symington built a steam tugboat to replace the horses that towed barges on an English canal. Symington's boat, called the *Charlotte Dundas,* was 58 feet (18 m) long, with a small steam engine driving a stern paddlewheel. Symington put his vessel into service in 1788. **Robert Fulton** built perhaps the best-known early steamboat, the *Clarmont.* As early as 1779, Fulton had experimented with paddlewheel boats, and shortly after his return from Europe in the early 1880s, he began building the *Clarmont.* Completed in 1807, it became the first commercially successful steam-driven vessel when it quickly earned back its entire initial cost of $20,000. By the 1850s, luxurious paddlewheel boats were commonplace on waterways such as the Mississippi River, where countless passengers ferried between Louisville, Kentucky and New Orleans, Louisiana on a journey of a few days.

Isambard Kingdom Brunel, a brilliant English engineer, brought the paddlewheel ship to its zenith with the *Great Western,* launched in 1838. Designed to cross the Atlantic to New York from the terminus of Brunel's Great Western Railway at Bristol, England, the *Great Western* was an enormous oaken ship with a massive hull; at the time of its launch it was the largest steamboat in the world. It made the Atlantic crossing 67 times, the quickest of these at 12 days and six hours.

But Brunel abandoned paddlewheels for his next (and larger) ship, the *Great Britain.* As more and more steamboats were built, the use of paddles came under critical fire. It was true that paddles made of wood were easy to repair and cheap to replace—all that needed to be done was to turn the wheel over until the damaged portion was above the water line. Though the paddle pulled the boat ahead as it entered the water, it also offered drag on its upward swing to clear the water. In addition, paddles did not work very well at sea, because as boats pitched from side to side, their paddles would lift completely out of the water, making them useless for propulsion. Engineers such as Brunel turned to the **screw propeller** for these reasons. It provided a powerful and efficient thrust, and remained underwater even in rough weather. Once the power and reliability offered by the screw propeller was established, the use of paddlewheel waned.

Though the paddlewheel has long since been replaced as a means of propulsion, paddlewheel boats have not entirely disappeared. The aesthetic value of these majestic ships makes them fashionable as tourist attractions in some parts of the United States.

PADLOCK • See Lock and key

PAP TEST

The Pap test is a simple and painless procedure for the early detection of the two most common and fatal forms of cancer in women: cervical and uterine. It is considered one of the most effective and significant weapons in the modern fight against cancer.

The test, also known as Papanicolaou's Smear, is named for the Greek doctor who developed it, George Nicholas Papanicolaou (1883-1962), who received his M.D. from the University of Athens in 1904. He emigrated to the United States in 1913 and was affiliated with New York Hospital and Cornell Medical College throughout his career. In 1917, Papanicolaou began a microscopic study of vaginal discharge cells in pigs. After expanding his research to humans, he observed cell abnormalities in a woman with cervical cancer, which inspired him to develop a method of detecting cancer through microscopic cell examination, or cytology. This technique had first been suggested by English physician Lionel Smith Beale (1828-1906) in 1867. Papanicolaou began publishing reports on his cytologic method of uterine and cervical cancer detection in 1928, but most of his colleagues remained committed to the standard procedures of cervical biopsy and curettage. In 1939 Papanicolaou began collaborating with gynecologist Herbert Traut. Their 1943 monograph, *Diagnosis of Uterine Cancer by the Vaginal Smear*, won wide acceptance for the procedure, and Papanicolaou began teaching it to physicians from around the world.

The significance of the Pap smear is that it allows detection of cancer in its presymptomatic stage, when the disease can best be treated. Cancer of the cervix in its earliest stages is almost 100% curable, while 80% of uterine cancer cases detected by a Pap test can also be cured. The smear technique of abnormal cell detection has been expanded to early diagnosis of cancer of many other organs.

PAPER

Derived from **cellulose** (plant) fibers that are made into pulp and felted (pressed) together, paper is one of the most common of man-made materials. There are approximately 7,000 types of paper, used not for writing or printing, but for products like **paper bag**s and cardboard boxes.

The oldest extant writing surfaces include Babylonian clay tablets and Indian palm leaves. Around 3000 B.C., the Egyptians developed a writing material using papyrus, the plant for which paper is named. This substance was composed of strips cut from stalks of the papyrus reed, which were dried, laid across each other crosswise, and glued together to form a somewhat nubby writing surface. Other early materials were parchment, made from the untanned skins of sheep or goats and vellum, or fine parchment, made from calf-, kid-, and lambskin. Used in Europe from the second century B.C., they were expensive and impractical—the equivalent of a 200-page book, for example, would require twelve animal skins.

The invention of paper similar to the kind used today is credited to Cai Lun, a Minister of Public Works under Emperor He Di during China's Han Dynasty (202 B.C. to A.D. 202). The Chinese had earlier written on bamboo, bone, and silk, but these surfaces were either too heavy and cumbersome or too expensive. In the year A.D. 105, Cai Lun devised a way to make the bark of the mulberry tree into paper. The process he used contained the basic elements still found in paper mills today: the raw material was chopped and mixed with water to form a pulp, which was spread thin on porous screens, drained, and dried in sheets. The Chinese also used hemp, rag, and fish nets to make paper. At first, paper was used for clothing, household articles, decorative arts, and for wrapping. By the fourth century, papermaking was firmly established in China, although the technology would not reach Europe for another 1,000 years.

Depiction of Chinese papermaking from the seventeenth century. Here, paper sheets are dried on a wall heated by a stove (right).

The Chinese managed to keep papermaking a secret for 500 years, and it was not until the seventh century that is spread to Korea and Japan. Chinese prisoners (from eighth-century battles fought in what is now Turkistan) taught their Arab captors to make paper, which led to the establishment of papermaking workshops at Baghdad, Damascus, and Samarkind. The Spanish Moors finally brought paper to Europe in the twelfth century; they used essentially the same process as the Chinese, producing a very absorbent product similar to modern plotting paper. Europeans initially referred to paper as *cloth parchment*, and used linen rags (sometimes with cotton or straw added) to produce it. By the year 1276, a paper mill had been established in Italy's Appenine region. The invention of the **printing** press in 1450 created an increased demand for paper. During the sixteenth century, sizing (treating paper with starch to stiffen it and prevent smudging) was introduced, and papers began to be coated with opaque mineral powders (such as **lime**) to improve quality.

The first American paper mill was established near Philadelphia in 1690; by 1810 there were 185 paper mills in the United States. The papermaking process remained quite slow and tedious until 1750, when a Dutch inventor designed a machine that reduced the time it took to break rags down to fiber.

In 1798, Nicholas Louis Robert, a Frenchman, invented a machine that made paper in continuous rolls. The English Foudrinier brothers, Henry and Sealy, financed his idea and produced a successful model in 1805. The device, which bears their name, is still used in modern papermaking. The Foudrinier machine was first used in the United States in 1816, when Thomas Gilpin and his brother Joshua built America's first mechanized paper mill. They "borrowed" a state-of-the-art design from English inventor John Dickinson and even persuaded Dickinson's foreman to defect to their operation. The Gilpins' monopoly on American papermaking was broken only when improved technology became widely available.

By the early nineteenth century, wood and other plant pulps began to replace the rags that had been used earlier, particularly in Europe. In 1840, a method for reducing logs to chips was invented, thus facilitating pulp production. In 1867 an American inventor named Tilghman discovered that wood fibers could be easily separated if the chips were soaked in a sulfurous acid solution. In 1883, German inventor Carl Dahl incorporated sodium sulfate into his papermaking process, producing a very strong pulp that became known as kraft (which means "strength" in German). Kraft pulp is still used, in bleached and unbleached forms, for such products as paper bags, milk cartons, and paper cups. Most paper now produced is made from wood, either harvested particularly for that purpose or from lumber or woodworking waste. Wood pulp is created by both *mechanical* and *chemical* processes designed to break down the wood fibers. Once the pulp has been produced, it is washed to remove any chemicals; foreign matter is filtered out, and the pulp is drained, bleached, washed, and formed into sheets that are pressed by heavy rollers and dried by air suction. Cylindrical colander presses put a smooth finish on the paper, which is finally wound into large rolls. Some paper, however, is made from such plants as cotton, rice, wheat, cornstalks, hemp, and jute; very high quality "rag" paper is still derived from cotton rags.

By 1991, Americans were producing nearly 70 million tons (64 million metric tons) of paper per year and consuming 660 pounds (300 kilograms) per person each year. With increasing concern about the environmental implications of forest depletion and waste disposal, many communities have begun paper recycling programs, chopping newspaper, magazines, and other waste papers into a pulp that is then treated to remove ink. The end product, called secondary fiber, is made into paperboard and some forms of printed paper, napkins, and towels. Despite the continued promise of a "paperless office" because of computer technology, paper consumption continues to grow. The future of paper might be affected by such technology, however. In 1994, Xerox corporation announced its development of "smart paper." Such paper will have dataglyphs, which can microencode digital information, embedded within it. Smart paper can count copies made from it, assure authenticity, and become part of the interoffice routing system.

Paper bag

Although the paper bag is a common household article, no one knows who first invented it. Many inventors, however, have improved on the original product. The first, and probably most significant improvement made in modern times came from American inventor **Margaret Knight**. In 1869, while working in a shop that produced paper bags, she devised a machine that would mechanically fold square-bottomed paper bags. This was a big improvement since it meant the bag would stand on its own. Although Knight came up with the idea in 1867, it took two years for the machine to be installed. In February 1872, Bostonian Luther Childs Croswell improved on Knight's square-bottomed design by adding two longitudinal, inward folds. This made the bag easier to store and better able to stand. Between 1884-1894 William Purvis, an African-American inventor, devised 10 different machines for making paper bags. He also invented a **fountain pen**, hand stamp and had three patents on an electrical railway. He later licensed his paper bag inventions to Union Paper Bag Co. of New York.

Beginning in the early 1970s, an effort was made by retailers to offer only **plastic** bags, which had been designed in Sweden and introduced in the United States at that time. With the advent of money-saving automated production in the early 80s plastic bags became a much less expensive alternative for supermarkets and other vendors, and customers had to request paper bags. However, questions about the **recycling** aspects of plastic bags have led many commercial establishments to offer shoppers a choice of paper or plastic packaging. Nonetheless, the vast majority of bags used today are plastic.

Paper chromatography • See Chromatography

Paper clip

Invented in 1900 by Norwegian Johann Waaler, the now common item was actually patented in Germany. Originally a metallic clip, the invention allowed several pieces of paper to be joined together, but then easily separated when necessary. Now Waaler's invention is distributed by countless companies in materials ranging from **plastic** to **steel** to goldplate and in a vast range of sizes. Specialty clips are sold for resealing food bags such as **potato chips** or cereals.

Pâpier maché

A mixture of paper pulp, glue, chalk, and sand, pâpier maché first became popular in Europe in the mid-eighteenth century for making ornamental items such as trays, boxes, and mirror frames; it was later used in architectural decoration. The pulp mixture was pressed by hand into oiled wood molds and then baked. Since architects could supervise the making of the mold, they had greater control over the finished product. Previously, decorative work was completed on-site by a plasterer, a process which made changes or repairs difficult and costly.

Henry Clay of Birmingham patented a new and stronger version of pâpier maché in 1772. Starting with a sheet of metal or wood, he applied the pâpier maché mixture to both sides so that the finished product would not warp or bend. When enough layers had been applied to obtain the strength desired, the sheet could be cut at the edges to obtain a smooth, straight surface. It was then oven-dried to remove excess moisture and hand-oiled or varnished. The finished product could be screwed or dovetailed much like wood.

In 1842 Benjamin Cook even patented a means of using pâpier maché for casting metal. The use of pâpier maché reached its zenith at the Great Exhibition of 1851, where, in addition to decorative items, chairs, couches, and tables made of pâpier maché were featured. Its use quickly declined, however, perhaps due to the advent of **plywood**, which was stronger. Pâpier maché is still used today to produce and decorate small items and for the amusement of children.

Papin, Denis (1647-1712)

French physicist

Denis Papin was an early pioneer in the study of steam pressure. In fact, Papin is credited with making the first real developments with steam since the time of **Hero of Alexandria** 1,500 years earlier.

Papin was born in France on August 22, 1647. He studied medicine, mathematics, and physics before assisting **Christiaan Huygens** and **Robert Boyle** with their work on air pumps. Papin's best-known and most influential invention was his 1679 steam "digester," a direct forerunner of modern **pressure cooker**s and hospital steam sterilizers.

Papin evidently realized the connection between steam pressure and the possibility of using it to move objects. His experiments showed that when small amounts of water were heated at one end of a container, the resulting steam would move an object at the other end. In 1690 he developed an atmospheric engine using this principle. He boiled water in a three-inch (7.6 cm) diameter vertical tube. The tube was sealed at one end with a movable **piston.** The piston rose as the water turned to steam. As the steam cooled and condensed, atmospheric pressure forced the piston back to its original position. Papin tried to use this concept to build a pump for removing water from mines and joined efforts with the English engineer **Thomas Savery**. He studied Savery's **water pump** and incorporated a piston, instead of Savery's vacuum chamber, to provide suction.

In 1707 Papin again built upon Savery's work to create a boat propelled by side paddles. While Savery proposed to power his boat with human muscle, Papin propelled his with paddles turned by a ratchet mechanism that was powered by steam. Unfortunately, the vessel was destroyed by river boatmen while enroute to its demonstration on the Thames River in London, England, and Papin could not afford to build another model.

In spite of the work done by Papin and Savery, it was **Thomas Newcomen** who actually used their ideas to create the first successful **steam engine** in 1712, the same year Papin died in poverty and obscurity.

Denis Papin

Parachute

The earliest known parachute design is that of **Leonardo da Vinci**, who in 1485 devised what he called a "tent roof." It had rigid frame at the base on the assumption it wouldn't stay open otherwise. The canopy, over twenty feet wide, was to be made of linen.

The first jumps, however, were not made until the eighteenth century. The French physicist Louis-Sébastien Lenormand (1757-1839), used two parasols to jump from a tree in 1783. André-Jacques Garnerin (1769-1832) made the first jump from an aerial machine on October 22, 1797, when he descended from a **balloon** at an altitude of about 3,200 feet (976 m). He went on to make many more exhibition jumps including one from 8,000 feet (2440 m) over England in 1802.

During the nineteenth century, parachutes were only used by entertainers to thrill crowds that had grown used to balloon flights. With the invention of **aircraft** in the twentieth century, parachute use increased dramatically, particularly among the military. Capt. Albert Berry of the United States Army made the first successful parachute jump from an airplane in 1912. However, during most of World War I, the only personnel given parachutes were observers in balloons. Parachutes were considered impractical for pilots and were disdained by those who thought the life-saving device contradicted the fighter pilot's daredevil image.

With the advent of World War II, parachutes became an integral part of military strategy, particularly for German forces who used them for landing troops, supplying ground corps, and saving fliers who ejected from their aircraft. Since the war, parachutes have been used to recover capsules from space, slow down jet planes, and to drop food and medicine in rough terrain.

Parachute design and materials have gone through many changes. Originally made from canvas, they were later fashioned from **silk.** Today, they are made from **nylon** because it is much stronger and cheaper. The canopy is made out of many smaller panels, each of which has even smaller sections sewn in various directions for added strength. **Parafoils** have changed the old picture of parachutists helpless in determining where to land. Now they can use the combination parachute/sail to glide to their intended landing site. These parafoils are often prominently in the public eye, sometimes used by the military by demonstration maneuvers, or for dramatic (and not always completely successful) entrances into stadiums at major sporting events.

PARAFFIN • See Kerosene

PARÉ, AMBROISE (1510-1590)

French surgeon

The uneducated son of a country artisan, Ambroise Paré became the greatest surgeon of the sixteenth century. Renowned as much for his compassion as his surgical skill, Paré guided his life with a humble credo of patient care: "I dressed him, God cured him."

Paré was born during an era when physicians considered surgery beneath their dignity; they left all cutting to the "lowly barber-surgeons." He served as an apprentice to a barber in his youth and, at the age of 19, went to Paris where he became a surgical student at the Hôtel Dieu hospital. After attaining the rank of master barber-surgeon in 1536, he joined the army as a regimental surgeon. Over the next 30 years, Paré developed a flourishing surgical practice while continuing to serve occasionally in the army. His medical writings and his fair treatment of soldiers regardless of their rank earned him considerable fame and, before his career ended, he had served as surgeon to four French kings.

During the siege of Turin in 1536-1537, Paré made his first great medical discovery. Gunshot wounds, a new medical condition, were considered to be poisonous and were routinely treated by cauterization with boiling oil. When Paré ran out of oil during the siege, he turned instead to simple dressings and

Parachutes first became an integral part of military strategy during World War II.

soothing ointment, and immediately noted the improved condition of his patients. Paré popularized this revolutionary treatment in his *Method of Treating Wounds* in 1545.

Paré's second critical contribution to medicine was his promotion of ligature of blood vessels to prevent hemorrhage during **amputation**s. Paré's classic *Treatise on Surgery*, written in 1564, disseminated knowledge of this life-saving technique. In this book, Paré also included several sections of *De corporis fabrica*, the authoritative work on anatomy written by Belgian anatomist Andreas Vesalius (1514-1564), translated from the original Latin into French. This dramatically improved the training, development, and effectiveness of surgeons, most of whom could not read Latin and thus had no previous access to the seminal work.

The innovative Paré often developed new surgical techniques by departing from established practices. He advocated massage and designed a number of **artificial limbs** as well as an artificial eye. He advanced obstetrics by reintroducing podalic version (the *in utero* turning of a fetus to a birth position) and inducing premature labor in cases of uterine hemorrhage. As always, he spread knowledge of these discoveries through his writings in French.

Paré's role in disseminating surgical knowledge to the barber-surgeons of his time and his efforts to elevate the status of surgery to a level of prestige and professionalism led him to be generally recognized as the Father of Modern Surgery.

See also Artificial limb and artificial joint

Ambroise Paré

Parker, John P. (1827-1900)

American inventor

John P. Parker started life in slavery and rose to the role of innovator and businessman. The son of a black slave mother and a white father, he was born at Norfolk, Virginia, and was sold into slavery at the age of eight. Sent to Mobile, Alabama, he was purchased by a physician in 1843. While under the doctor's ownership, Parker learned to read and write. He was also apprenticed as a molder at an iron foundry. Twice his high productivity was a source of irritation to his coworkers, at the foundry in Mobile and at a foundry in New Orleans. At a third foundry, Parker stayed for two years and accumulated $1,800 in savings, the amount he needed to purchase his freedom.

In 1845, a free man, he went first to Indiana, where he was involved in abolitionist activities and the Underground Railroad, which rescued people from slavery in the South. Parker married Miranda Boulden of Cincinnati in 1848. Two years later, Parker and his family moved to Ripley, Ohio, where he again became involved in abolitionism, aiding in the freeing of over a thousand slaves.

He established a small foundry in Ripley, the Ripley Foundry and Machine Company, in 1854 to manufacture castings. In 1863 he became a recruiter for the 27th Regiment of the United States Army, a black unit, to serve in the Civil War. His foundry also made castings for the war effort.

In 1884, Parker obtained a patent for a **screw** for tobacco presses and a year later patented a type of **harrow** called the Parker Pulverizer. Both items were produced in his foundry. These patents were issued at a time when patents were rarely awarded to black inventors. Parker's foundry remained in operation until 1918, well after his death.

See also Forge; Iron production

Parkes, Alexander (1813-1890)

English metallurgist and chemist

Born on December 29, 1813, in Birmingham, England, Parkes began his career as an apprentice in the art metal trade and then moved on to an **electroplating** firm, where he silver-plated diverse objects such as spider webs and plants. His work with silver solutions and the chemicals used to produce them—namely phosphorus and carbon disulfide—led him to investigate solutions of **rubber** and cellulose nitrate. In 1841 he patented a method of waterproofing fabrics by coating them with rubber. He received a second patent in 1843 for an electroplating process.

In 1855 Parkes patented the first **plastic**. By dissolving cellulose nitrate in alcohol and camphor containing ether, he produced a hard solid which could be molded when heated, which he called Parkesine (later known as **celluloid**). Unfortunately, Parkes could find no market for the material. **John Wesley Hyatt**, an American chemist, would rediscover celluloid and market it successfully as a replacement for ivory in the 1860s. Parkes died in London on June 29, 1890.

See also Cellulose, chemical uses of; Waterproofing material

Parking meter

In ancient Greece, **Hero of Alexandria** used simple machines to create many ingenious devices, among them a device that raised a **lever** when a coin or token was dropped against the opposite side of the fulcrum. The lever opened a **valve** and delivered a cup of holy water to anyone who dropped a coin. Centuries later, two men in Oklahoma used the same principle to create the parking meter, one of the earliest traffic control devices.

As early as the 1920s, **automobiles** were causing congestion in larger cities. Not only were narrow streets filled with moving traffic, but cars parked along curbs and in front of stores added to the congestion. Shoppers often had to walk many blocks to get to a store because parking spaces were so scarce.

Carlton C. Magee, editor of the Oklahoma City *Daily News* and member of the chamber of commerce traffic committee, became concerned with this parking problem. He proposed a device to "charge" people for parking spaces. Magee presented his idea to mechanical engineering professor Gerald A. Hale at Oklahoma Agricultural and Mechanical College. Hale became so interested in the project that he quit his job to perfect the mechanism and enter a partnership with Magee. In December 1932, Magee applied for a patent on the parking meter.

On July 16, 1935, the citizens of Oklahoma City found meters installed on one side of the street. Spaces on the free side of the block quickly filled up as usual. Spaces with meters were still available when stores opened but they too filled up. Unlike the unmetered side, however, these spaces had a constant turnover. After three days, merchants on the free side of the street petitioned city council for meters. Other cities were not so open to the idea of paying for parking spaces. In Mobile, Alabama, and Carthage, Texas, vigilante committees tried to destroy the parking meters.

Following World War II, the number of automobiles increased and so did the demand for parking meters. These machines now help solve traffic and parking problems in major cities throughout the world. Convenient parking is now available for handicapped citizens, to allow them easier access to facilities. While most of these meters still resemble Magee's original design, newer electronic models are also available.

Parsons, Charles (1854-1931)

Irish English engineer

Charles Algernon Parsons was born into a wealthy and talented family. His father, the third Earl of Rosse, was president of the Royal Society and a distinguished astronomer, and his mother was a pioneer photographer. Privately tutored, Parsons spent much of his time as a child in his father's workshop, listening to the many scientists who frequented his parents' home, Birr Castle, Ireland. At the age of twelve, he constructed, with the aid of his brothers, a steam carriage that could travel at speeds of up to ten miles an hour (16 kph).

After enrolling at Trinity College, Dublin, Parsons transferred to Cambridge, where he earned a degree in mathematics. In 1884 he went to work for Clarke, Chapman, and Company, manufacturers who specialized in building electric dynamos. Like all dynamos of the time, these machines were driven by a belt connected to the flywheel of a **steam engine.** Realizing that a significant amount of energy was being lost between the engine and the dynamo, Parsons set about designing a machine that would directly use the steam's energy.

Later that same year Parsons produced the first **steam turbine**. Like a **waterwheel** or a windmill, Parson's machine operated on the turbine principle that mechanical energy can be derived from the movement of a fluid or gas. It consisted of a rotor in which several vaned **wheels** were attached to a shaft. Steam entered and expanded, causing the shaft wheels to rotate. Other stationary blades forced the steam against those that rotated, thus making use of as much energy as possible. The steam then continued until it encountered another set of turbine blades designed to work with the same steam at a slightly lower pressure. Because the steam expanded as it continued on its way through the turbine, Parsons made the second set of blades larger and the exhaust end wider than the intake end.

Parsons saw the advantage of his device immediately. His turbine achieved a speed of 18,000 revolutions per minute, compared to a previous maximum of 1,500 rpm. Parsons left Clark, Chapman, and Company in 1889 to form his own company. When, one year later, two of his devices were installed at a power station in Newcastle, England, it became the world's first station of its kind. Turbines later became the most widely used method of providing electricity for large-scale processes.

Parsons also wanted to apply this technology to marine uses. He built the *Turbinia*, a 100-foot (30 m) vessel capable of speeds near 20 knots (37 kph). He redesigned the propulsion system until he had three shafts, each with three propellers. In 1887 he put on a public demonstration of the vessel at a naval review for Queen Victoria's Diamond Jubilee. Spectators were amazed as the *Turbinia* sped along at 34 knots (63 kph), seven knots faster than any other ship then in existence. Parsons soon saw his steam turbine used for all sorts of commercial and military ships, because it provided high speeds with less vibration than the traditional reciprocating steam engines. Engines with a capacity of 70,000 horsepower provided the propulsion for some famous ships, including the *Mauretania* and the *Lusitania.*

Parsons was also successful in other engineering endeavors. He worked on searchlights and optical instruments, founding companies that specialized in producing lenses for scientific instruments. However, he will continue to be best known for developing the most convenient, useful, and efficient means of converting power into motion—the steam turbine.

Particle accelerator

By the late 1890s, it had become obvious to scientists that John Dalton's (1766-1844) concept of a solid, indivisible atom was incorrect. In 1897 Joseph J. Thomson (1856-1940) showed that even smaller particles—electrons—could be removed from an atom. A decade later, Ernest Rutherford (1871-1937) identified another structure within the atom, a central body that he called the *nucleus*. The next stage in learning about the composition of matter seemed to be to explore this new structure, the atomic nucleus.

During the late 1910s, Rutherford himself carried out some of the earliest experiments in this area. He directed alpha particles from naturally radioactive materials at gases confined within a **glass** tube. He found that the alpha particles collided with and broke apart the nuclei of the gaseous atoms. These experiments constituted the first example of atom smashing, the process by which an atomic nucleus is broken apart in order to find out what is inside of it.

Atom-smashing presents some difficult practical problems for a researcher. The atomic nucleus is itself positively charged. It is surrounded by a cloud of electrons that is negatively charged. Any charged particle directed at the atom will, therefore, experience a force of repulsion either from the electron cloud (if the incident particle is negatively charged) or from the nucleus itself (if the particle is positively charged). Rutherford's experiments were successful only because the alpha particles produced by radioactive materials had sufficient energy to overcome these forces of repulsion.

The energy of alpha particles is, however, relatively modest. It was obvious that some device was needed to accelerate particles to higher speeds and to give them enough energy to overcome the electrical repulsion of atomic nuclei and their surrounding electron clouds. That device was the *particle accelerator*.

Over the past eighty years, two types of particle accelerators have been developed, **linear accelerators** and **cyclotrons.** In a linear accelerator, particles are introduced at one end of a machine, accelerated by a changing electric field, and directed at a target at the opposite end of the machine. When the accelerator particles strike the target, they break apart some of the atomic nuclei in the target.

The first linear accelerators were those built by **John Cockcroft** and **E. T. S. Walton** in England in the late 1920s and by **Robert Van de Graaff** in the United States in the early 1930s. In a cyclotron, particles are accelerated along a curved path. In the simplest version of a cyclotron, the particles are introduced into the center of the machine. Then they are accelerated in an ever-widening spiral path by a changing electrical field. The path taken by the particles is controlled by a magnetic field imposed above and below the machine. When the particles have attained some maximum energy, they are directed out of the cyclotron and made to collide with some external target. The first cyclotron was designed and built by **E. O. Lawrence** at the University of California in 1932.

In later versions of the cyclotron, particles are maintained in a circular path by means of carefully controlled electrical and magnetic fields. The discovery by **Edwin McMillan** and **Vladimir I. Veksler** of the principle of phase stability in the 1940s made possible the synchrotron, a far more powerful machine than had been possible with Lawrence's original design.

Charles Algernon Parsons

In the most powerful accelerators available today, two beams of particles are accelerated in opposite directions. At some point, the two beams are caused to collide with each other, releasing very large amounts of energy.

One of the most successful versions of this **particle collider** is the modification of the Super Proton Synchrotron at the European Center for Nuclear Research (CERN). This modification was made at the suggestion of Simon van der Meer (1925-) and Carlo Rubbia (1934-) in the late 1970s and early 1980s. In the Rubbia-van der Meer design, a beam of protons revolving in one direction through the machine is caused to collide with a second beam of antiprotons revolving in the opposite direction. The energy released as a result of this collision has been sufficient to result in the formation of W and Z bosons, particles predicted by the electroweak theories of Abdus Salam (1926-), Steven Weinberg (1933-), and Sheldon Glashow (1932-).

Particle accelerators have a number of uses in basic research. When highly energetic particles pass through an atom, they interact with particles in the nucleus of that atom. The nature of that interaction—that is, the way the particles are scattered—provides information about the nature of the particles within the nucleus.

E. O. Lawrence standing before the particle accelerator at Berkeley Laboratory.

The energy released when particles collide with nuclei or with each other may also be converted into new kinds of subatomic particles. The discovery of these subatomic particles has made possible a comprehensive theory about the structure of matter, the *standard model*.

Accelerators also have many practical applications. They are used in the medical profession for both diagnosis and therapy. Particles produced in accelerators can, for example, be used in carrying out CAT (**computerized axial tomography**) scans. Accelerators are also used in industry for a variety of purposes, such as the production of radioactive isotopes, lithographic etching, and the analysis of stress and damage in materials.

Particle collider

The original concept behind the construction of **particle accelerator**s was to build machines that could accelerate protons, electrons, and other small particles to very high energies. These particles would then be allowed to collide with stationary targets outside the machine. These collisions would provide information about the fundamental structure of matter.

The earliest **linear accelerator**s and **cyclotron**s were all built on this design, and they were very successful in producing new data on the composition of matter. But they all possessed one major handicap.

When high energy particles collide with a stationary target, only a relatively small fraction of their energy is available to take part in the reaction. By far the greatest fraction of their energy is used simply to move pieces of the target around. When a proton moving with one TeV (teraelectron, or trillion-electron, volts) of energy strikes a stationary proton, for example, only 41 Gev (gigaelectron, or billion-electron, volts), or 4.1% is actually available for an interaction between the protons.

One solution to this problem was envisioned by Donald Kerst (1911-) at the University of Illinois in 1956. Kerst suggested that two particle beams be accelerated in opposite directions at the same time. For example, a synchrotron might consist of two rings arranged next to each other. Particles might be accelerated in a clockwise direction in one ring and in a counter-clockwise direction in the other ring. At some particular moment, the two particle beams could be allowed to collide with each other. When that happens, all the energy possessed by both beams could be utilized in particle interactions. None would be wasted in lost motion, as is the case with stationary targets. The collision of two one-TeV proton beams, for example, would release two TeV of energy for interactions.

The first successful demonstration of this principle was achieved in 1963 at the Italian National Laboratory in Frascati. A 200-MeV beam of electrons was allowed to collide with a 200-MeV beam of positrons (positive electrons).

The most powerful particle accelerators today all make use of the colliding-beam principle. Such accelerators also make use of a *storage ring*. A storage ring looks like any other synchrotron ring. But the magnetic field around the storage ring is controlled so that particles within it are maintained at constant energies for long periods of time—sometimes hours—until they are needed. When an experiment is about to be carried out, the particles are released from the storage ring and allowed to collide with a second beam traveling in the opposite direction.

In 1972, the **linear accelerator** at Stanford University was modified to make possible colliding-beam experiments. Particles leaving the 3.2 kilometer-long accelerator are divided into two beams. The particles in one beam are converted to positrons and then allowed to collide with electrons from the other beam. This arrangement (known as SPEAR) produces the most energetic electron-positron collisions available anywhere in the world.

The largest and most powerful circular colliders in existence are located at the Fermi National Accelerator Laboratory (Fermilab) near Chicago, Illinois, at the Centre Européen pour la Recherche Nucléair (CERN) in Geneva, Switzerland, and at the Deutsches Electronen-Synchrotron (DESY) in Hamburg, Germany. The Fermilab accelerator is also known as the tevatron because of its ability to accelerate particles to energies of one-TeV. All three machines have been used to make important discoveries of new particles.

For many years, scientists have been eager to build an even larger particle accelerator. All the major discoveries that are possible with one TeV or lower energy particles appear to have been made. New frontiers will be crossed, scientists believe, only when a larger machine is available. For theoretical reasons, one with an energy output of about 20-TeV is especially desirable.

The most serious technical problem involved in building such a machine is the size of **magnets** needed to control particle beams. By one estimate, the magnets needed for a 20 TeV machine would require four billion watts of power, more than the combined output of the three largest nuclear power plants in the United States.

Fortunately, an alternative solution for this problem is available. In 1911, the Dutch physicist, Heike Kamerlingh Onnes (1853-1926), discovered the phenomenon of superconductivity, the tendency of some materials to lose their resistance to the flow of electrical current at very low temperatures. Magnets made of superconducting materials can be operated at about one percent the cost of conventional magnets. An accelerator that used superconducting magnets appeared to be a realistic possibility. Thus was born the idea of the superconducting super collider (SSC).

For more than a decade, planning for the SSC has gone on under the auspices of the Universities Research Association, a group of 56 universities in 26 states and the province of Ontario. In June 1987, President Ronald Reagan (1911-) announced his support for the construction of the SSC. A year later, he approved a site near Waxahachie, Texas, for the location of the machine.

The SSC will consist of a storage ring with a circumference of 51.54 mi (82.94 km) buried 18 ft (6 m) below the earth's surface. The ring will consist of a concrete tunnel 10 ft (3 m) in diameter. Particles will travel in two metal tubes a few centimeters in diameter and 28 in (70 cm) apart. Protons will be accelerated to energies of 20 TeV, resulting in a release of 40 TeV when two opposing beams collide with each other. Before entering the main ring, protons will be accelerated by four smaller machines, a linear accelerator and three smaller synchrotrons.

Plans for construction of the SSC were surrounded by vigorous controversy. Critics claimed that the projected cost of the machine (originally, $4.5 billion) was more than the nation should spend on basic research which will provide few practical benefits. Many scientists worried that the project would drain money away from other research priorities. One Congressman called the scheme a ''quark-barrel project,'' after one of the subatomic particles that would be studied with the SSC.

In spite of these objections, construction of the SSC began in the early 1990s. The project has not gone smoothly, however. Prototype magnets have not worked as well as anticipated, requiring a rethinking of many aspects of the SSC's design. This has driven up the estimated cost of the SSC at least fifty percent, and the United States government has had to interest other nations in supporting the project. These obstacles once again brought the machine into the center of political debate, and President Bill Clinton (1946-) has cancelled funding for the project.

Pascal, Blaise • See Pascaline

Pascaline

The pascaline, one of the earliest mechanical devices that performed addition and subtraction, was invented in 1642 by the eighteen-year-old French prodigy **Blaise Pascal**. Although other inventors, such as **William Schickard**, produced prototypes of **calculating machines** before Pascal, the pascaline was the first device that was manufactured and sold for general use. Pascal was granted a monopoly on the manufacture and sale of his invention by royal decree in 1649. Unfortunately its high price made it available only to the few who could afford it. Today, seven copies of the device still exist in museums and private collections.

Pascal's main motivation for designing a calculating machine was to ease the work of his father who, as president of the Board of Customs and Excise at Clermont Ferrand, was charged with the laborious work of reorganizing finances and taxation. As an accountant's tool, the pascaline was able to add and subtract numbers quickly as well as convert exchange rates for different currencies. The mechanism of the pascaline was similar in principle to the modern **odometer** in the sense that it contained a row of **wheels** with numbers upon them connected by gears. The cogs and **gears** connecting the wheels were arranged so that one complete revolution of the dial furthest to the right would cause the adjacent dial to turn by only 1/10 of a revolution, and so on. Later scientists and inventors, such as Gottfried Wilhelm von Leibniz, would produce calculating machines including mechanisms that made it possible to perform multiplication and division. Later still, calculating devices would adopt a binary system of mathematics, using ones and zeroes, which is easier for machines to handle.

After he had built over 50 variations of his device, Pascal patented the final version of his invention in 1649 and, in 1652, presented a copy of it to Queen Christina of Sweden, hoping that the calculating machine would yield a profit for him. It did not, and Pascal left it behind, directing his prodigious creativity toward matters of physics, mathematics, and religion, which occupied him until his untimely death at the age of thirty-nine in 1662.

Pasta

Though Italy comes to mind when we think of pasta, the earliest pasta was prepared in China. It is a popular belief that Marco Polo (1254-1324) introduced pasta to Italy when he returned from the Orient. While many researchers agree that Marco Polo may have increased the popularity of noodles, they note that noodles were being eaten by Arabs and Indians at least half a century before Marco Polo visited the Orient. Trade between Arab lands and Italy was fairly common in the early Middle Ages, and many commentators note that pasta may have been introduced to Italy through trade as early as the eleventh century.

Indian noodles were called *sevika* and the Arabic word for noodle is *rishta*; both words translate roughly as thread. In the same vein, the Italian word for noodles, *spago*, means string. Stemming from the word paste, pasta is made from a starchy dough extracted from cereal grains. This paste is processed into one of a variety of shapes and boiled until it softens. Gluten is the distinctive ingredient that helps pasta maintain its shape when boiled.

There is a bit of confusion surrounding the origin of the various shapes of pasta. The most common name for pasta in the late Middle Ages was "macaroni." However, unlike the curved tubular macaroni of today, the early macaroni noodle was flat. The first "macaroni and cheese" dish was described in a fourteenth-century English cookbook, which suggested serving the noodle with butter and grated **cheese.**

For many centuries, pasta was made by hand. Noodle-making was such strenuous work that travelers visiting Italy were often startled by the "sturdy" arms of the women who made pasta. A wooden screw press invented in Naples, Italy, in the 1800s made **mass production** of pasta possible.

Today American dry pasta is made from semolina, which is often produced by grinding kernels of durum wheat. The semolina is mixed with water to form a dough. Other ingredients, such as eggs, spinach, or tomato, may be added as well. The dough is kneaded, then pushed through a metal disc with holes in it. The shape of the holes determines the shape of the pasta. When the pasta is the right length, it is cut with sharp blades. Then it is slowly dried in special machines that circulate hot, moist air. Finally the dried pasta is packed in boxes or bags.

Inexpensive and easy to prepare, pasta is one of America's favorite foods. Popular pasta dishes include spaghetti, macaroni and cheese, ravioli, and lasagna. Supermarkets now boast dried, frozen, and fresh pasta in a wide range of shapes and flavors.

Pasta is high in complex carbohydrates. Since January 1998, all commercial pasta made in the United States has also been fortified with folic acid, a nutrient that reduces the risk of neurological birth defects when consumed by pregnant women.

PASTEURIZED MILK • See Milk, pasteurized

PATENT LEATHER • See Leather

PATENT MEDICINE

Patent medicines first appeared in England in the 1600s. When a medication was patented, its formula was owned by the patent holder, and no one else could duplicate and sell it. In order to qualify for a patent, a medicine only had to be original; no proof of effectiveness or safety was required. Because the ingredients of patented remedies had to be listed, many sellers of proprietary medications never applied for patents. Instead, they registered distinctive trade names in order to market their nostrum—ingredients unspecified—under a unique brand name. In time, all of these medicines being promoted for public sale became known as "patent" medicines, whether they were in fact patented or not. Most were promoted as astonishingly effective cures for an equally astonishing range of maladies. For example, this 1800s advertisement for Dr. Jayne's Alternative claims the following cures:

For the cure of Scrofula, King's Evil, White Swellings, Ulcers, Tumours, Mercurial and Syphilitic Affections, Rheumatism, Gout, Scurvy, Neuralgia, Cancer, Goitre, Enlargements of the Bones, Joints, Glands, or Ligaments or of the Ovaries, Liver, Spleen, Kidneys, etc. All the various Diseases of the Skin, Dyspepsia and Liver Complaint, Jaundice and Nervous Diseases, Dropsical Swelling, Constitutional Disorders, and diseases originating from a depraved or impure state of the Blood or other fluids of the body.

Among the earliest British patent medicines were Anderson's Pills, Daffy's Elixir, and Lockyear's Pills, all of which date from the 1600s. In the 1700s came Dr. Batemen's Pectoral Drops, Dr. Hooper's Female Pills, and Robert Turlington's Balsam of Life.

Patent medicines followed the British colonists to America. The first American-made medicine to be patented (in Great Britain, in 1715) was Tuscorora Rice (actually made from Indian corn), a tuberculosis remedy invented by Mrs. **Sibilla Masters** of Pennsylvania. Another early American nostrum was Widow Read's Ointment for the ITCH, advertised for the widow by her son-in-law, **Benjamin Franklin,** in his *Pennsylvania Gazette* in 1731; heavy national advertising would later be the prime means of promoting patent medicine sales in America. In 1796, the first patent issued in the United States for a medicine was granted to Samuel Lee, Jr. of Windham, Connecticut, for his Bilious Pills, which promised to cure a score of ailments.

A popular nostrum of the 1800s was Swaim's Panacea, a syrup of sarsaparilla introduced by William Swaim, a New York bookbinder (or perhaps a Philadelphia harness-maker) in 1820. When Swaim ran a six-page advertisement for his Panacea in the 1832 *Farmers and Mechanics Almanac*, he pointed the way toward a new advertising gimmick for patent medicines: the free annual almanac devoted to promoting of an individual remedy. Dr. David Jayne launched his series *Medical Almanac and Guide to Health* in the 1840s to push such medicines as Jayne's Sanative Pills, Jayne's Vermifuge, and Jayne's Alternative. The first *Hostetter's American Almanac* was published in 1860; it promoted the sale of Doctor Hostetter's Celebrated Stomach Bitters, which had indeed been formulated by a doctor and then commercialized by the doctor's son in 1853.

While patent medicines flourished in the United States from their inception, they were most popular in the latter part of the nineteenth century. Perhaps the best-known and most enduring among these was a home remedy devised by Mrs. Lydia Estes Pinkham (1819-1883) of Lynn, Massachusetts, to cure "female complaints." Mrs. Pinkham had been preparing her herbal concoction on her kitchen stove for years, taken from a recipe supposedly given to her by a machinist named George Clarkson Todd in payment of a debt to Mr. Pinkham. When the family fell into near poverty following the panic of 1873, Pinkham's son Dan suggested selling the compound. She produced the remedy—now called Lydia E. Pinkham's Vegetable Compound—and wrote advertising copy, along with a four-page brochure titled "Guide for Women." Dan and his brother Will distributed handbills and the brochures. When they began advertising in the *Boston Herald* in 1876, a successful mail-order business flourished. When Pinkham's portrait was added to the Compound's label in 1879, sales escalated, and she became the most recognized woman in America. Letters poured in from women all over the country seeking medical advice; their queries were answered by a staff of women supervised by Pinkham. Will and Dan both died of tuberculosis in 1881, and Pinkham herself suffered a stroke and died in 1883. The business remained in the Pinkham family until 1968, and the Compound was still being marketed in the 1980s.

Ironically, **Louis Pasteur**'s scientifically-based germ theory of disease was brought to the American public by the very unscientific claims of patent medicine peddlers. Chief among these was William Radam, a Prussian emigre residing in Texas. Inspired by Pasteur's discovery of the microbe, Radam determined to develop a medication to fight the microscopic entities within the human body. The result was Radam's Microbe Killer, patented in 1886. Its popularity was unshaken by analysis revealing it to be 99 percent water.

While patent medicine makers always advertised and promoted their products heavily, the most colorful promotions were the traveling medicine shows and entertainments. These shows existed in colonial times, continued to grow in size and scope during the 1800s, and reached a climax in the 1880s and 1890s. The shows offered a variety of entertainment—drama, vaudeville, circus, minstrels, and magic—to pitch the product, but perhaps the biggest and best-known were the Kickapoo Indian or Wild West shows staged by John E. Healy of New Haven, Connecticut, and ''Texas Charley'' Bigelow. As many as 75 Kickapoo shows toured the country at a time, each staffed with half-a-dozen Native Americans, plus a ''scout'' and several other whites. An Indian ''medicine man'' would impressively describe the virtues of the particular remedy in his native language, while the ''scout'' interpreted his speech. Remedies included the popular Kickapoo Indian Sagwa, plus Kickapoo Indian Salve, Kickapoo Indian Worm Killer, and Kickapoo Cough Cure.

Another cure falsely attributed to Native Americans was Clark Stanley's Snake Oil Liniment, which Stanley claimed originated with the medicine men of the Moki Pueblo in Wolpi, Arizona. Stanley started marketing his remedy in 1886 and promoted it most colorfully by killing hundreds of rattlesnakes before audiences at the Chicago World's Fair of 1893. Analysis, however, showed that Stanley's liniment contained no rattlesnake oil.

During the nineteenth century, makers of patent medicines often staged lavish travelling medicine shows to pitch their products, the most famous of which was the Kikapoo Indian or Wild West shows. During the entertainment, salesmen dressed as "medicine men" would circulate through the crowd. Above, a patent-medicine salesman from the early twentieth century hawks his product.

Although patent medicines were very popular, concerns began to grow about their ingredients. Many had high levels of alcohol—Lydia Pinkham's, for example, was 19 percent alcohol; the widely sold Dr. Kilmer's Swamp Root was 12 percent; Hostetter's Bitters was a dizzying 32 percent. Other products, including medicines for children, were laced with such addictive drugs as heroin, opium, and cocaine. These concerns led to the passage of the Pure Food and Drug Act of 1906. Patent medicines now had to list their ingredients on packaging labels. A supplementary law passed in 1938 required manufacturers to test their products for safety before marketing them; tests for effectiveness were required as of 1962.

Not all patent medicines were of the ''snake oil'' variety. Some of the most familiar legitimate patent medicines originated in the late 1800s and early 1900s. Nineteenth-century chemists knew that salicylic acid had pain-relieving qualities, but the acid burned throats and upset stomachs. In 1853 French chemist Charles F. Gerhardt synthesized a primitive form of acetylsalicylic acid, or aspirin. In 1897 Felix Hoffmann of the Bayer Company found a better method to synthesize acetylsalicylic acid and discovered that the drug overcame the unpleasant side effects while maintaining the therapeutic effects of the acid. In 1899 Bayer began marketing the new product as Aspirin, a trade name. Bayer lost the use of the trade name in 1919 as part of Germany's concessions to the Allies at the end of World War I, and the name aspirin passed into generic usage in the United States and United Kingdom. Today, however, aspirin is still a registered trademark of Bayer in many countries. One aspirin-based product, Anacin, originated in the United States in 1918, the invention of a Wisconsin dentist.

Listerine, an antiseptic and disinfectant, was developed in 1879 by Jordan W. Lambert, cofounder of the Warner (later Warner-Lambert) pharmaceutical firm, and marketed to physicians. The product was named after Joseph Lister (1827-1912), the English physician who pioneered antiseptic surgery. Lister was said not to be pleased with the name. Lambert's son, Gerald, introduced Listerine to the mass market in 1921; advertising played heavily on the product's effectiveness in saving the user from the social scourge of halitosis.

New York chemist Charles Henry Phillips coined the name Milk of Magnesia in 1880 for his antacid, a white suspension of magnesium hydroxide in water. Lunsford Richardson, a North Carolina pharmacist, developed an external cold remedy in the 1890s that he called Richardson's Croup and Pneumonia Cure Salve. When he renamed his product for his brother-in-law, Dr. Joshua Vick, sales of Vick's Salve, later Vick's VapoRub, skyrocketed. Ex-Lax, the laxative with a chocolate flavor, was the 1905 invention of a Hungarian-born New York scientist; he originally called his remedy Bo-Bo, for bon-bon, to stress its candy-like flavor.

Among more modern patent medicines, Alka-Seltzer, a tablet composed of an antacid, aspirin, and an agent formulated to bubble when immersed in a glass of water, was introduced by Miles Laboratories in 1931. In 1955 the Johnson & Johnson company marketed Tylenol, which uses acetaminophen to relieve pain and reduce fever just as aspirin does but which does not cause the side effects aspirin produces in some people. At first a prescription medication, Tylenol became an over-the-counter product in 1960. The deaths of seven people in the Chicago area in 1982 from poisoned Tylenol capsules laced with cyanide purchased at local stores led to the widespread use of tamper-resistant packaging.

See also Barnes, Albert C.; Cough drops; Sperti, George Speri

Payen, Anselme (1795-1871)

French chemist

Payen was the son of an entrepreneur who had started several chemical production factories. Payen studied chemistry first with his father and later with Nicolas Louis Vauquelin (1763-1829) and Michel-Eugène Chevreul (1786-1889).

In 1815 Payen was promoted to head of his father's borax production plant. While there Payen discovered a method of preparing borax from boric acid which was readily and cheaply available from Italy. Because production costs for this method were very low, Payen was able to undersell his competitors, the Dutch, who until that time had a monopoly on borax.

Five years later Payan began investigating the production of sugar from sugar beets. He developed a process in which charcoal was used to decolor the sugar. Charcoal filters were later used in **gas mask**s since carbon filters absorbed dangerous organic gases. Payan continued his research and in 1833 developed a chemical from malt extract that catalyzed the starch-to-sugar conversion. He named the organic catalyst diastase, which was the first enzyme produced in concentrated form. Enzymes discovered since then are named with the *ase* suffix, the pattern started by Payen.

Payen went on to study woods and during his research began to separate out a substance resembling starch. He named this substance, which he found in abundance in the cell walls of all the plants he studied, cellulose. Again, Payen established a naming system: **carbohydrates** end with *ose*. Although Payen was the first to isolate cellulose, it was not until the late 1930s through the work of Wanda Farr that it was discovered how plants produce this carbohydrate.

Cellulose, a natural **polymer,** was the building block of many other inventions. Treated with various acids and additives it became the main ingredient in **nitrocellulose**, **guncotton**, **collodion**, **rayon**, **celluloid**, **cellophane**, and many other products.

In 1835 Payen accepted a professorship at the Ecole Centrale des Arts et Manufactures and concentrated exclusively on research. He died in Paris during the Franco-Prussian War in 1871 after refusing to leave the city as the Prussian army advanced.

See also Cellulose, chemical uses of; Polymer and polymerization

PC (Personal computer) • See Microcomputer

Peake, George (1722-1827)
American inventor

Despite prejudice and slavery, some black people managed to overcome the odds in early America and become financially successful. George Peake was one such individual. He was born in the colony of Maryland and spent his early life in Pennsylvania. He participated as a British soldier in the battle of Quebec in September 1754 during the French and Indian War, but he later deserted.

In 1809, Peake took up residence in Cleveland, Ohio, only a small frontier town at the time. He and his family, Cleveland's first black settlers, acquired a one-hundred-acre farm on the western outskirts of the town. Peake's wife was apparently a women of means and helped support the family. Peake is credited with inventing a labor-saving device—a stone hand **mill**. Settlers had learned from the Indians to grind corn with a pestle—a pounding device—and a mortar, or vessel. This method required a degree of strength to grind corn meal.

Peake's mill consisted of two **wheel**-like stones about 19 inches (483 mm) in diameter. Corn placed between the rotating millstones was quickly ground and produced a better quality meal than did the pestle and mortar. The mill became popular in Peake's community and contributed to the prosperity of his family. They became well-known among their neighbors and were apparently well liked. George Peake died in 1827 at the age of 105.

Peanut butter • See Peanut products

Peanut products

The peanut, or *Arachis hypogaea,* is a member of the legume family commonly called the groundpea or goober. It produces two to four edible subterranean seeds in separate pods on viney plants. Peanuts are made up of a shell, skin, and a kernel. It is composed of 26.2% protein, 48.7% oil, 1.8% water, and 20.6% carbohydrates. They are typically grown in the sandy or friable loamy soil of South America, Africa, Asia, and the southern and western United States.

Introduced by explorers and used by colonial farmers to fatten hogs, the peanut was not cultivated for human consumption until the end of the nineteenth century. This was primarily because it required intense hand labor to harvest and process, and was subject to mold, rat and worm infestation, desiccation, and other storage problems. However, this portable food was used by troops during the United States' Civil War. When the war was over, demand for the legume was high. This demand coupled with improvements in harvesting led to the birth of the peanut industry.

In 1890, a St. Louis doctor created a nutritional and digestible food for patients by crushing peanuts and reducing them to a cream he called peanut butter. In modern times, the peanut has become a worldwide staple—an inexpensive source of protein, minerals, vitamins, fiber, fat, and food energy for undernourished people in Third World countries. Chief importers of peanuts are Canada, Japan, and Europe.

Because of the peanut's versatility and nutritional value, botanist **George Washington Carver** devoted a major portion of his career to developing peanut products, including lipstick, shampoo, **soap**, shoe polish, **paint**, shaving cream, **paper**, creosote, imitation marble, **dye**, **plastic**, wood stain, flour, and **cheese** substitutes. To bolster peanut cultivation, he published a bulletin, the Peanut Promoter, touting peanut milk, flavoring for **ice cream**, **instant coffee**, Worcestershire sauce, and peanut bread.

Today, the peanut, with 50 percent of its weight comprised of fat, produces valuable cooking, salad, and canning oil, and **margarine** and shortening, as well as pulp for peanut butter. As food, peanuts are eaten raw, boiled, roasted, or salted and fried. Their usual dressing is plain salt or a blend of salt and sugar. They form a major component in confections, including candy bars, cakes, cookies, and other baked goods. By-products of peanuts are used as food additives, animal feed, fertilizer, forage, and conditioner for compacted soil.

In industry, the peanut yields **varnish**, paint, lubricant, soap, **cosmetics**, textiles, **plastic,** wallboard, abrasive, cat litter, absorbent cleaners, abrasives for **toothpaste**, **adhesive**s, sizing for cloth and paper, waterproof moldings for plastic, firefighting foam, synthetic fibers, propagators of antibiotics, **leather** dressing, **insecticide**, furniture polish, and **nitroglycerine**. In addition, the peanut shells may be compressed into synthetic firelogs or briquettes or rendered for xylose, furfural, cellulose, or mucilage.

Pelham, Robert • See Calculating machine

Pen

The pen, an instrument for writing with **ink**, has been used since antiquity. The earliest pens were made from hollow reeds that held a small amount of ink. Reed pens gave way to quill pens, made from the hollow wing feathers of geese and swans. The quills were sharpened with an instrument called a penknife, the term still used for small knives. Quill pens were the standard writing implement from the seventh century until the nineteenth century, except in the Far East, where brushes were used.

Steel pen nibs came into use in the early 1800s, thus eliminating the necessity of constantly resharpening the pen point, but the problem of having continually to resupply the pen with ink remained. Fountain, or reservoir, pens that held their own ink supply were developed to overcome this problem. A number of fountain pens appeared during the 1800s—the first British patent was awarded in 1809—but these pens all had severe problems with clogging and leaking.

The first practical fountain pen was finally patented by Lewis Edson Waterman (1837-1901) in 1884. Waterman, a native of Decatur, New York, had abandoned teaching for selling insurance in his mid-twenties. After losing a large sale when the fountain pen he owned flooded the about-to-be signed contract, Waterman was determined to improve the implement, eventually devising a pen that used capillary action to control ink flow. Waterman made his fortune from this invention; he ran his very successful pen company until his death.

Waterman's fountain pen had to be filled with an eyedropper. Self-filling pens appeared in the early twentieth century; a **piston** was used to draw up ink. **Levers** were introduced in 1908. Ink-cartridge and more sophisticated capillary action fountain pens were developed in the 1950s.

Although Waterman's pen was a great improvement over its predecessors, such problems as leaking and refilling remained. A revolutionary approach to pen design was patented in 1888 by a Weymouth, Massachusetts inventor named John J. Loud. Loud's pen replaced the steel nib with a rotating ball that was constantly bathed in ink from a reservoir. The ink flow was uncontrolled, however, so the pen leaked and was of little practical use.

It was not until the 1930s that an effective ballpoint pen was designed by two Hungarian brothers. Ladislao Biro, a journalist, hypnotist, painter, sculptor, and proofreader, would no longer tolerate ink splotches and refilling his pen. With his older brother Georg Biro, a chemist, Ladislao developed a workable pen with a rolling ball in 1938. The Biros emigrated to Argentina during World War I—Ladislao arrived in Buenos Aires with ten dollars in his pocket—and began to produce pens there in 1943. Henry Martin, an English financier, began manufacturing Biro pens in England for use by Royal Air Force pilots; the ballpoint was the only pen that would not leak at high altitudes. By war's end, the "biro" was in wide use in Europe.

The Biros' ballpoint pen came to the United States via two routes. The Shaeffer Company bought marketing rights from Biro for half a million dollars in 1944 and launched a heavy advertising campaign. An American entrepreneur, Milton Reynolds, bought a few Biro pens in Argentina in 1944, brought them back to the United States, changed the design enough to satisfy patent requirements, and sold hundreds of thousands of his new product in 1946 and 1947. One New York department store sold 10,000 units at $12.50 each in one day.

However, a problem remained—the ink used in ballpoint pens was unsatisfactory. It clogged, it skipped, it washed off when moistened, and it disappeared when exposed to sunlight. As a result of these flaws, some banks refused to accept checks signed with a ballpoint pen. This dilemma was solved in 1949 by an Austrian chemist living in California, Franz Seech. Working in his home kitchen, Seech developed an even-flowing ink that dried quickly and cleanly.

Meanwhile, Baron Marcel Bich (pronounced "beek") began manufacturing ever-cheaper ballpoint pens in a shed in Clichy, a suburb of Paris, in 1945. Soon, seven million Bic pens were being produced daily at Bich factories. Bich entered the American market by buying out the original Waterman company.

Since 1950, several other types of pens have been introduced. The rolling-ball pen, first seen in the 1960s, combines the rotating ball of the ballpoint with the free-flowing ink of the felt-tip. Pentel, a Japanese firm, introduced the felt-tip pen in 1963, which uses a soft, absorbent tip fed by free-flowing, often colorful ink. Pentel followed with the felt-nibbed ballpoint pen in 1973 and the ceramic nib in 1981. The Gillette Company introduced an erasable ballpoint pen, the Eraser Mate, in 1979. In 1984, another Japanese manufacturer, Nakayaina, began marketing a luminous ballpoint pen for nighttime use. Despite some nostalgic purchase of fountain pens by aficionados the ballpoint pen remains the writing instrument of choice for most people although felt tip markers, really made of synthetic fibers are available in many colors for specialized uses as both water-based and permanent markers.

Pencil

A pencil is a thin rod of a solid marking material enclosed in a tube of wood, metal, or **plastic**, used for writing or drawing. The marking material today is usually composed of a mixture of graphite and clay. Pencil-like marking devices have been used since earliest times. Prehistoric peoples used sticks of charred wood. The ancient Greeks and Romans used pieces of lead. In 1564, the discovery of a large deposit of pure graphite (thought to be a type of lead) in Cumberland, England, made the modern lead pencil possible. Pieces of graphite were wrapped with string and used for writing.

By 1565, the German-Swiss scientist Konrad von Gesner (1516-1565) described the first lead pencil—a piece of graphite enclosed in a wood holder. In 1779 the "lead" used in pencils was proved to be a form of carbon. It was then named graphite, from the Greek *graphein*, meaning "to write." The first pencil-making factory was established in 1761 in Nuremburg, Germany, by Kasper Faber. Faber's brother, John Eberhard Faber (1822-1879) immigrated to the United States in 1848 and built the first American pencil factory in 1861 in New York City.

Joseph Dixon (1799-1869) introduced the rounded pencil and developed the method of gluing grooved halves of cedar cylinders together to form pencils. The French chemist Nicholas Jacques Contè (1755-1805) developed a much-improved pencil in 1795 by mixing powdered graphite and clay with water and baking it; he found he could adjust blackness and hardness by varying the proportions of these substances. Contè's manufacturing method remains the basis of today's pencil industry. Natural graphite was replaced by graphite made from coke after this process was invented by Edward G. Acheson (1856-1931), a former assistant to **Thomas Alva Edison**, in 1896.

The mechanical pencil, in which the lead is automatically sharpened, appeared as early as 1822; the Eagle Pencil Company patented its design in 1879. Mechanical pencil sharpeners for common pencils were not introduced until the 1890s. The American Hyman Lipman patented the idea of attaching an **eraser** to the pencil's end in 1858. In the 1990s, the Faber Castell Company in New Jersey introduced the EcoWriter. The barrel of the EcoWriter is made out of recycled newspaper and cardboard.

Pennington, Mary Engle (1872-1952)
American chemist

Born in Nashville, Tennessee, Mary Pennington grew up in Philadelphia, where she often joined her father, a business executive, in his backyard gardening hobby. A library book on medical chemistry revealed Pennington's life direction when she was twelve years old.

In 1892 Pennington completed the requirements for a B.S. degree at the University of Pennsylvania but was not allowed to receive her degree because of her sex. She later earned her Ph.D. from that same university in 1895. After several years of postgraduate study, Pennington founded the Philadelphia Clinical Laboratory, where she performed bacteriological analyses for physicians. Soon she became director of the bacteriological laboratory of the Philadelphia Department of Health. There she developed techniques and standards for milk inspection and dairy product preservation that came into use throughout the country.

In 1907 Pennington joined the staff of the Bureau of Chemistry of the United States Department of Agriculture after taking the civil service examination as ''M. E. Pennington'' so it would not be apparent that she was a woman. In the following year, Pennington was named director of the Food Research Laboratory, a new division within the Bureau of Chemistry. She continued working out of her own lab in Philadelphia, where she investigated and devised methods of storing and shipping poultry, eggs, and other perishable foods without spoilage. She established national standards for ice-cooled refrigerator cars and solved the problem of humidity control in cold storage.

In 1922 Pennington became a private consultant in New York City. During the next thirty years, she contributed to advances in the design and construction of home and commercial **refrigeration** units and conducted research that revealed the best ways to freeze many different foods. Pennington's life work contributed greatly to the successful storage and transportation of perishable foods of all varieties.

See also Food preservation; Frozen food ; Refrigerated trucks and railway cars

Percussion caps • See Ammunition

Perfume

Since the beginning of recorded history, humans have attempted to mask or enhance their own odor by using perfume which emulates nature's pleasant smells. Many natural and man-made materials have been used to make perfume to apply to the skin and clothing, to put in cleaners and cosmetics, or to scent the air. Because of differences in body chemistry, temperature, and body odors, no perfume will smell exactly the same on any two people.

Perfume comes from the Latin ''per'' meaning ''through'' and ''fumum,'' or ''smoke.'' Many ancient perfumes were made by extracting natural oils from plants through pressing and steaming. The oil was then burned to scent the air. Today, most perfume is used to scent bar **soaps.** Some products are even perfumed with industrial odorants to mask unpleasant smells or to appear ''unscented.''

While fragrant liquids used for the body are often considered perfume, true perfumes are defined as extracts or essences and contain a percentage of oil distilled in alcohol. Water is also used. The United States is the world's largest perfume market with annual sales totalling several billions of dollars.

According to the Bible, Three Wise Men visited the baby Jesus carrying myrrh and frankincense. Ancient Egyptians burned incense called *kyphi*—made of henna, myrrh, cinnamon, and juniper—as religious offerings. They soaked aromatic wood, gum, and resins in water and oil and used the liquid as a fragrant body lotion. The early Egyptians also perfumed their dead and often assigned specific fragrances to deities. Their word for perfume has been translated as ''fragrance of the gods.'' It is said that the Moslem prophet Mohammed wrote, ''Perfumes are foods that reawaken the spirit.''

Eventually Egyptian perfumery influenced the Greeks and the Romans. For hundreds of years after the fall of Rome, perfume was primarily an Oriental art. It spread to Europe when thirteenth century Crusaders brought back samples from Palestine to England, France, and Italy. Europeans discovered the healing properties of fragrance during the seventeenth century. Doctors treating plague victims covered their mouths and noses with **leather** pouches holding pungent cloves, cinnamon, and spices which they thought would protect them from disease.

Perfume then came into widespread use among the monarchy. France's King Louis XIV used it so much that he was called the ''perfume king.'' His court contained a floral pavilion filled with fragrances, and dried flowers were placed in bowls throughout the palace to freshen the air. Royal guests bathed in goat's milk and rose petals. Visitors were often doused with perfume, which also was sprayed on clothing, furniture, walls, and tableware. It was at this time that Grasse, a region of southern France where many flowering plant varieties grow, became a leading producer of perfumes.

Meanwhile, in England, aromatics were contained in lockets and the hollow heads of canes to be sniffed by the owner. It was not until the late 1800s, when synthetic chemicals were used, that perfumes could be mass marketed. The first synthetic perfume was nitrobenzene, made from nitric acid and benzene. This synthetic mixture gave off an almond smell and was often used to scent. In 1868, Englishman William Perkin synthesized coumarin from the South American tonka bean to create a fragrance that smelled like freshly sown hay. Ferdinand Tiemann of the University of Berlin created synthetic violet and vanilla. In the United States, Francis Despard Dodge created citronellol—an alcohol with rose-like odor—by experimenting with citronella, which is derived from citronella oil and has a lemon-like odor. In different variations, this synthetic compound gives off the scents of sweet pea, lily of the valley, narcissus, and hyacinth.

Just as the art of perfumery progressed through the centuries, so did the art of the perfume **bottle.** Perfume bottles were often as elaborate and exotic as the oils they contained. The earliest specimens date back to about 1000 B.C. In ancient Egypt, newly invented glass bottles were made largely to hold perfumes. The crafting of perfume bottles spread into Europe and reached its peak in Venice in the eighteenth century, when **glass** containers assumed the shape of small animals or had pastoral scenes painted on them. Today perfume bottles are designed by the manufacturer to reflect the character of the fragrance inside, whether light and flowery or dark and musky.

Perfumes today are being made and used in different ways than in previous centuries. Perfumes are being manufactured more and more frequently with synthetic chemicals rather than natural oils. Less concentrated forms of perfume are also becoming increasingly popular. Combined, these factors decrease the cost of the scents, encouraging more widespread and frequent, often daily, use.

Using perfume to heal, make us feel good, and improve relationships between the sexes are the new frontiers being explored by the industry. The sense of smell is considered a right brain activity, which rules emotions, memory, and creativity. Aromatherapy—smelling oils and fragrances to cure physical and emotional problems—is being revived to help balance hormonal and body energy. The theory behind aromatherapy states that using essential oils helps bolster the immune system when inhaled or applied topically. Smelling sweet smells also affects one's mood and can be used as a form of psychotherapy.

Like aromatherapy, more research is being conducted to synthesize human perfume—that is, the body scents we produce to attract or repel other humans. Humans, like other mammals, release pheromones to attract the opposite sex. New perfumes are being created to duplicate the effect of pheromones and stimulate sexual arousal receptors in the brain. Not only may the perfumes of the future help people cover up ''bad'' smells, they could improve their physical and emotional well-being as well as their sex lives.

Perkin, William Henry (1838-1907)

English chemist

William Perkin is considered to be the father of the synthetic **dye** and **perfume** industries.

Perkin was born in London, England, and as a child attended the City of London School. There he came into contact with **Michael Faraday** who fostered his fledgling fascination with chemistry. In 1853, Perkin entered the Royal College of Chemistry where, at seventeen, he was named an assistant to the school's director, a renowned German chemist named **August Wilhelm von Hofmann**. Although Hofmann was a brilliant chemist, he was awkward with laboratory work and depended on talented assistants to help him in his research on coal tar and its derivatives.

It was under Hofmann's tutelage in 1856 that Perkin experienced his first major success. That year, Perkin spent his Easter vacation attempting to synthesize quinine from aniline, a coal-tar derivative. Although he failed to produce artificial quinine, the results of his experiment determined the course of his career. As part of his process, Perkin mixed aniline with potassium dichromate and alcohol, which yielded a purple liquid. Thinking it might be useful as a dye, Perkin named the liquid *aniline purple* and sent a sample to a silk dyeing firm. When the company sent back for more dye, it became clear that this was a lucrative business opportunity so Perkin convinced his father and brother to invest in a company to produce the new dye. Soon the company began marketing aniline purple, which became known as mauve (from the French word for the plant previously used to make violet).

While his family tended to the practical aspect of the business, Perkin headed up the company's research department. His experiments led to the development of more dyes, including violets and rosanilines. Over the next few years, he introduced several other colors into his company line: aniline red (1859), aniline black (1863), and alkalate magenta (1864). In 1868, Perkin used the work of two German chemists, Carl Graebe (1841-1927) and Carl Liebermann (1842-1914), as a basis for synthesizing alizarin, the chemical component of the madder plant essential in dye making. While Graebe and Liebermann had developed a workable synthesis process, it was too expensive to be of practical use. Perkin came up with a cost-effective production version of his fellow chemists' process, and by 1871, his company was producing two hundred twenty tons of alizarin annually. Within a short time, Perkin's curiosity and drive paid off as his synthetic dyes replaced natural dyes all over the world.

Perkin's further experimentation led to his discovery of a method for changing the structure of organic compounds on a molecular level. Using this process, known as the ''Perkin synthesis,'' he produced a coumarin, a synthetic perfume which has been described as smelling like fresh hay or vanilla. Although he technically retired at age thirty-six, he launched a second career in the synthetic perfume business. He later teamed up with B.F. Duppa to research and develop other aspects of the synthetic fragrance field. Their accomplishments include the development of a process for producing glycine, racemic acid, and tartaric acid, as well as significant research into the similarities between tartaric acid and maleic acids. In 1889, Perkin received the Davy medal of the Royal Society, and the British government recognized Perkin's contribution to science, industry, and his country by knighting him in 1906. He died one year later, on July 14, in Sudbury, England.

Perkins, Jacob (1766-1849)

American inventor

Born in 1766 in Newburyport, Massachusetts, a descendant of an Englishman who came to America in 1631, Perkins was a gifted and prolific inventor who had very little formal schooling. He was apprenticed to a goldsmith at age thirteen, took over his master's business at age fifteen, and went on to invent a method of plating **shoe** buckles. By age twenty-one, Perkins had been hired by the State of Massachusetts to make dies for copper coins.

In 1790, Perkins made what is perhaps his most important invention, a machine for cutting and heading **nails** in one operation. As a result, nails could be mass-produced and sold cheaply. Perkins patented his machine in 1795 and set up a nail-manufacturing company, but a subsequent lawsuit about the invention ruined the business.

Perkins next invented **steel** (rather than copper) plates for bank-note engraving that made counterfeiting money nearly impossible. He pursued his interest in engraving in Boston, then New York, and finally Philadelphia. With his partner, Gideon Fairman, Perkins sailed for England in 1818 to establish their engraving process in that country. They set up a factory in 1819 that for many years made plates and printed notes for local banks. In 1840 their firm was awarded the contract to produce England's first penny **postage stamp**s; the plant printed many millions of stamps over the next 40 years.

With an ever-curious mind, Perkins continued to produce creative inventions of incredible diversity. Around 1823 he began conducting numerous experiments with high-pressure **steam engine**s and boilers, and eventually introduced many innovations and improvements. Perkins unveiled an improved **paddlewheel** in 1829; in 1831 he invented a way to achieve free circulation of water in boilers; his 1834 description of the vapor compression cycle for **refrigeration** and ice-making was revolutionary.

Other Perkins inventions included the steam-gun; a pleometer, to measure the speed of a vessel moving through water, a bathometer, to measure the depth of water, a process of transferring **engravings** from one steel plate to another, a ship's pump, and a method of ventilating rooms and ships' holds. Perkins was recognized in several countries for his work, and received various awards. He died in London in 1849.

Personal computer (PC) • See Microcomputer

Pesticide

The word ''pesticide'' is a broad term that refers to any device, method, or chemical that kills plants or animals that compete for humanity's food supply or are otherwise undesirable. Pesticides include **insecticides,** fungicides, **herbicides,** nematocides (used to kill nematodes, elongated cylindrical worms), and rodenticides. Of these various pesticides, insecticides have a longer and more noteworthy history, perhaps because the number of insects labeled ''pests'' greatly exceeds the number of all other plant and animal ''pests'' combined. Hence, this article focuses on the use of agricultural insecticides.

Since they first began cultivating crops (around 7000 B.C.) if not before, humans have devised methods to prevent insects from eating or otherwise destroying precious crops. Some cultures relied on the practice of planting during certain phases of the moon. Other early agricultural practices that indirectly kept insect populations low were rotating crops; planting small, varied crops; and selecting naturally resistant plants. People picked bugs off plants by hand and made noise to ward off grasshoppers. Chemicals were also used early on. The crushed petals of the pyrethrum (a type of chrysanthemum), sulfur, and arsenic were used in the Middle East, Rome, and China, respectively. The Chinese also used natural predators such as ants to eat undesirable insects.

All attempts at pest control were pretty much individual affairs until the 1840s, when a North American fungus called *powdery mildew* invaded Britain, and the epidemic was controlled with large-scale applications of sulfur. The Colorado beetle in the western United States was the next target: by 1877 western settlers had learned to protect their potato crop by using water-insoluble chemicals such as paris green. Other pesticides such as derria, quassia, and tar oil followed, but nineteenth-century pesticides were weak. They had to be supplemented by introducing natural predators, or, in some cases, by grafting threatened plants onto more resistant rootstock.

By World War II, only about 30 pesticides existed. Research during the war yielded **DDT** (dichloro-diphenyl-trichloro-ethane), which had been synthesized in 1874 but wasn't recognized as an insecticide until 1942. Other strong pesticides soon followed, such as chlordane in 1945 and endrin in 1951. Poison gas research in Germany yielded the organophosphorus compounds, the best known of which is parathion. These new pesticides were very strong. Further research yielded hundreds of organophosphorus compounds, the most noteworthy being malathion, which was recently used in California against the medfly.

Until the 1800s, when people began to spray personal gardens using fairly large machines, pesticides were generally applied by hand. Airplanes were not used until the 1920s, and slow, well-controlled, low-level flights were not implemented until the 1950s. The first aerial spraying of synthetic pesticides used large amounts of inert materials, 4,000 liters per hectare (a hectare equals 2.47 acres). This quantity was rapidly reduced to 100 to 200 liters/hectare, and by the 1970s the amount had been reduced (in some cases) to.3 liters per hectare of the ingredient itself (for example, malathion) applied directly to the fields.

Today, some 900 active chemical pesticides are used to manufacture 40,000 commercial preparations. The Environmental Protection Agency (EPA) estimates that the use of pesticides doubled between 1960 and 1980. Currently, over 372 million kilograms a year are used in the United States, with over 1.8 billion kilograms a year used worldwide.

Petersen scale

The Petersen Scale, a chart designed to help sailors assess weather conditions on the water, was devised by a German ship captain by that name. Adopted at the 1939 International Meteorological Conference in Berlin, Petersen's scale drew together the criteria used in the parallel **Beaufort Wind Scale** and **Douglas Sea Scale.**

By 1947, it was generally recognized that the state of the sea was all that really mattered to the ship masters. The hybrid scale was deemed unnecessary. The World Meteorological Conference in Washington replaced it that year with a single wave code.

Petroleum • See Oil Refining

Phase • See Alternating current

Phase-contrast microscope

For hundreds of years scientists have observed the structure of once-living cells through microscopes, whether they be simple **lenses** (such as those used by **Robert Hooke**) or **compound microscopes** (like those found in most school laboratories) or large **electron microscopes** (such as those found in most hospitals today). All of these devices have one major fault: the inability to observe living cells. The phase-contrast microscope was invented to make observation of living cells possible.

When an object is placed under a normal microscope and a light is placed below it the object will block a good deal of the light and create shadows. A viewer looking through the eyepiece sees the dark (shadowed) areas compared to the light (non-shadowed) areas. This comparison between light and dark is called contrast.

Because living cells are largely transparent and allow most of the light to pass through unaffected, they cast very few shadows; thus, the cell appears to the viewer to have very little contrast, and it becomes difficult to make out one part of the cell from another. The simple solution is to add a bit of **dye** to the object, greatly enhancing the image's contrast but killing the cell in the process.

The phase-contrast microscope increases the image's contrast without dyes; instead, it works on the principle of *refraction.* When light passes through a transparent material it is slowed down a bit, but speeds up again after exiting the material. The amount it is slowed depends upon the material's refractive index. When compared to light that does not pass through the material, the refracted light will seem to ''arrive'' later than the direct light. The amount of difference between their two arrivals is called the *phase difference.*

In a phase-contrast microscope, both the refracted light (having passed through the cell) and the direct light (from the background) are collected. Using a series of optical devices called phase plates, the two lights are overlapped. If the two are overlapped just right so that the phase difference is one-quarter of a wavelength, the direct light will be amplified and the refracted light darkened, thus producing an image of very high contrast. This type of image is called *positive phase contrast.* By changing the phase difference slightly, the object will appear bright against a dark background (*negative phase contrast*).

There are two drawbacks to the phase-contrast microscope. First, even in perfectly adjusted microscopes, a halo can be seen around the object image—dark in negative phase contrast, light in positive phase contrast. This halo is caused by the phase plates in between the object and the eyepiece and is considered a permanent and acceptable defect in the microscope's design. The second flaw is known as the *zone of action* effect: when a cell is viewed through a phase-contrast microscope, only the edges will exhibit very high contrast. This contrast decreases toward the middle of the cell, and in its center very little contrast is observed. While slightly annoying, this too is viewed as an acceptable defect.

The inventor of the phase-contrast microscope was the Dutch physicist Fritz Zernicke (1888–1966), who constructed the first model in 1934, though its introduction was delayed until after World War II. He was awarded the 1953 Nobel Prize for Physics

Microscope, compound

Phillips, William D. (1948-)

American physicist

William Daniel Phillips was born on November 5, 1948, in Wilkes-Barre, Pennsylvania. He grew up in Camp Hill, outside Harrisburg, and attended Juniata College. Phillips was awarded his B.S. degree from Juniata in 1970. He then continued his studies at the Massachusetts Institute of Technology (MIT), from which he earned a Ph.D. in physics in 1976. He then spent two years as a postdoctoral Chaim Weizmann Fellow at MIT. In 1978, Phillips accepted an appointment as a physicist at NIST, a post he retained for the next two decades.

Phillips has been interested in the problem of cooling and trapping atoms since his years at MIT. This line is of growing interest among physicists because it provides a way of studying the structure and properties of individual molecules and atoms. Traditionally, scientists have been able to study only the bulk properties of matter, that is, the average properties of, at best, trillions and trillions of atoms and molecules. Yet, it is known that these average properties mask the properties of individual particles that make up a sample. The study of atoms and molecules is of interest not only for theoretical reasons, but also because of its potential practical applications on a macroscopic scale. Engineers are beginning to learn how to construct particles that consist of a relatively small number of particles, called nanostructures, with potentially revolutionary applications. In order to advance in this field of research, it has become necessary to understand how individual atoms and molecules behave.

The key to these studies is to find ways to cool atoms and molecules, that is, to reduce the speed at which they travel. At room temperature, the particles in a gas are typically traveling at speeds of a few thousand miles (kilometers) per hour. At these speeds, particles escape from a field of view so quickly that no meaningful measurements can be made. In order to keep the particles within a field of view, their velocity must be reduced. Simply cooling a gas by conventional methods (such as putting it in a refrigerator) does not solve this problem. Under those conditions, the gas particles do lose speed, but they eventually condense to form a liquid or solid. In both liquids and solids, particles are so close to each other that individual properties cannot be determined. The objective, then, is to cool down a gas without permitting it to condense.

Theoreticians were proposing methods for solving this problem since the 1950s. One of the most popular approaches was to bombard a moving particle with a photon of energy moving towards that particle. Imagine, for example, a stream of ping pong balls fired from a gun. Those ping pong balls might be compared to the particles in a gas. Then imagine that short puffs of air are directed at the stream of ping pong balls. The puffs of air correspond to photons of energy. As each puff of air strikes a ping pong ball, it transfers energy to the ball and reduces the speed with which the ball is moving in the forward direction. If this process is repeated over and over again with any given ping pong ball, the ball eventually slows down and comes to a stop.

This analogy is greatly oversimplified, however. Real particles that are struck by a photon quickly give off the energy they have absorbed and recoil in a direction that cannot be predicted. Another photon aimed at the same particle will miss contact because the particle has changed its line of flight.

Another factor to be considered is that atoms and molecules absorb only certain quantized levels of energy. A photon with the wrong energy will have no effect on any particle that it strikes. Furthermore, the energy needed to excite a particle changes as the particle moves toward or away from the source of energy. This effect is known as the Doppler effect and is familiar to anyone who has heard the pitch of a train whistle increase or decrease as it moves toward or away from an observer.

Still, given all these conditions, an experimenter should be able to design an apparatus by which bursts of energy can be used to slow down the movement of atoms and molecules. In the mid-1980s, **Steven Chu** and his co-workers at the Bell Laboratories in Holmdel, New Jersey, designed an ''atom trap'' in which particles are bombarded by three pairs of **laser** guns arranged at right angles to each other. This arrangement was able to slow atoms down to a speed of about 12 in per second (30 cm per second), equivalent to a temperature of about 240 microkelvin. Theoretical calculations had shown that this temperature was probably the lowest that could be reached by the methods described above.

In 1988, Phillips attained a remarkable breakthrough by repeating the kinds of experiments carried out earlier by Chu and his colleagues. Phillips' group found that they were able to attain a minimum temperature of 40 microkelvin, six times lower than that predicted by theory and observed by Chu. Phillips later wrote that his results were ''at first difficult to believe, especially considering the attractive simplicity of the Doppler cooling theory (and the generally held belief that experiments never work better than one expects). This turn of events,'' he went on, ''was both welcome and unsettling.'' The theoretical explanation for these observations was later provided by **Claude Cohen-Tannoudji** at the École Normale Supérieure in Paris, an accomplishment for which he was awarded a share of the 1997 Nobel Prize in Physics.

Phillips produced other breakthroughs in the study of low temperature behavior. In 1985, for example, he found a way to trap atoms in a maze of laser beams. The beams were arranged so as to produce constructive and destructive interference patterns. When placed into this kind of maze, gaseous atoms at very low temperatures fell into low-energy pockets in somewhat the manner that eggs fall into the depressions of the cartons in which they are sold.

Using devices such as these, Phillips has been able to create entirely new forms of matter that exist only under low temperature conditions and only for very brief periods of time. For example, in one experiment, he brought together a collection of sodium atoms that formed a molecular-like substance but that was much larger than an ordinary molecule. The collection of atoms broke apart after about 10 nanoseconds.

PHONOGRAPH

The phonograph, a device that makes audible recorded sounds imprinted within the narrow grooves of phonograph records, is made of several parts: a tone arm, a revolving turntable on which a record spins, a pickup cartridge that sends vibrations in the form of electric waves to an **amplifier** that boosts the volume of these waves, and a **loudspeaker** that broadcasts them as sounds. On the tone arm is a needle, usually made of diamond or sapphire, which must be replaced periodically. The phonograph developed fairly quickly after its invention in the late 1870s, growing in sophistication and popularity until it became by the early twentieth century a common accessory in American households.

The American inventor **Thomas Alva Edison** is credited with inventing the phonograph, which was reportedly his favorite creation. Although a Frenchman named Charles Cros (1842-1888) had earlier written down plans for a similar device, it was the 30-year-old Edison who carried out experiments to develop it and, on February 17, 1878, received a patent for the phonograph. In late 1877, Edison had been working in his Menlo Park, New Jersey research laboratory—known as the ''invention factory''—on improvements to the **telephone** (which had recently been invented by **Alexander Graham Bell**). Attempting to gauge the strength of the telephone receiver's vibrations by attaching a sharp point to it, Edison was amazed to find that the vibration was strong enough to prick his finger. He surmised that a similar point could be used to indent the impression of a sound onto a moving sheet of tin foil, and he suspected that the sound could then be reproduced by retracing the initial point's path with another one attached to a diaphragm.

Edison shared this idea with his friend Edward H. Johnson, who mentioned it in several subsequent lectures and generated public interest in a ''talking machine.'' Thus Edison was inspired to sketch a plan for a such a machine, which he gave to John Kruesi, the Swiss-born foreman of Edison's machine shop, with a scrawled directive to ''Make this.'' The device Kruesi built consisted of a brass cylinder inscribed with spiral grooves and wrapped with a sheet of tin foil; when turned by a hand **crank**, the cylinder simultaneously rotated and moved lengthwise. At each side was situated a diaphragm equipped with a stylus (needle). A receiver would carry sound waves to one needle, which would be applied to the tin foil as the crank was turned and would follow the cylinder's grooves. Then the cylinder would be reset to the beginning and the other point—attached to the device's amplifier—would turn into sound the vibrations etched into the tin foil.

On December 6, 1877, Edison tested his device by reciting the nursery rhyme ''Mary Had a Little Lamb.'' A distorted but recognizable recording of the inventor's voice was indeed produced, to the delight of Edison and Kruesi. News of the ingenious talking machine spread rapidly, interesting not only the National Academy of Sciences and the Smithsonian Institution but President Rutherford B. Hayes, who is said to have sat up until 3 A.M. listening to the device. Edison later replaced the hand crank with a steam-driven belt that ran through the floor of his laboratory to the machine shop below; this improvement stabilized the pitch of the sound produced, which was nevertheless still of poor quality.

Once Edison had proved it was possible to record and reproduce sound mechanically, other inventors began to further develop this fledgling technology. In 1885, Chichester A. Bell and Charles S. Tainter (1854-1940) invented the graphophone, which featured a cardboard cylinder coated with wax instead of Edison's metal cylinder with its covering of tin foil. Two years later, **Emile Berliner**—a German engineer who had become chief inspector for Bell's Telephone after emigrating to the United States—produced the gramophone, which introduced the flat disk that would permanently replace the cylinder. Berliner's version of the phonograph featured a needle, mounted to a horizontal arm, that vibrated from side to side rather than up and down, thus improving the sound quality somewhat. Berliner began marketing his invention in 1898, founding a company called Deutsche Gramophone Gesellschaft. By 1904 he had developed a method for **mass production** of recordings, using a metal-plated master disk to imprint durable, **shellac**-coated disks.

One of the most important contributors to the development of the phonograph was Eldridge R. Johnson (1867-1945), a Camden, New Jersey inventor and machine shop owner. Assigned a contract to make **spring** motors for Berliner's gramophone, Johnson began working to improve the quality of the device's sound reproduction. In 1897 he received some master disks from Berliner. Closely examining the microscopic grooves, Johnson determined that engraving the sound waves into them would be more effective than the etching method that had previously been used. Instead of a needle, he used a tiny, sharp cutting tool, similar to a **lathe**, that carved deeper grooves, ultimately producing a more distinct sound and a longer lasting impression. Johnson's other improvements included making the master disc of wax rather than the zinc Berliner had used, producing a more sophisticated recording apparatus (albeit featuring a **rubber** hose), and attaching an electric motor to rotate the disk at a regular speed to ensure even pitch.

Johnson founded the Victor Talking Machine Company in 1901 and began to manufacture gramophones and recorded disks, quickly designing a machine to automatically stamp out disks duplicated from a metal-plated master. The next year, Johnson's company produced more than one million records, which were enthusiastically purchased by a public eager to bring recorded music into their own homes. The celebrated Italian tenor Enrico Caruso (1873-1921) made a series of re-

Thomas Edison and his phonograph.

cordings for Victor in 1903, which generated even more popularity for this technology and made both Caruso and Johnson wealthy. The very commercially minded Johnson is credited with making phonographs and records affordable, accessible commodities for American consumers. The electronic recording methods introduced in the mid-1920s made vast improvements in sound quality, but the basic method Johnson devised for producing records is still used today.

Until the late 1940s, records—called "78s" because they made that many revolutions per minute—were constructed of a shellac-and-clay mixture and (while certainly preferable to the cylinder Edison had used) were easily broken; in addition, they carried only about five minutes worth of sound per side. The latter problem in particular annoyed Peter Goldmark (1906-1977), a Hungarian-born engineer who worked at Columbia Broadcasting Systems (CBS) Laboratories. An accomplished cellist and pianist, Goldmark was listening to a recording of Brahms at a party when the music abruptly stopped—the record had ended in the middle of a movement and had to be turned over. Goldmark was thus spurred to invent the first long-playing **(LP) record,** which was made of **plastic** rather than the unwieldy shellac of the 78. The LP disk rotated at 33-and-one-third revolutions per minute, and accommodated 250 grooves per inch rather than 80 so that it played for 20 minutes or more per side.

The first recordings made on the new LP disks were *Violin Concerto* by Felix Mendelssohn (1809-1847), *Fourth Symphony* by Piotr Ilyich Tchaikovsky (1840-1893), and the score to the musical comedy *South Pacific* by Richard Rodgers (1902-1979). CBS offered to share this new technology with competitors, but RCA—a company that had rejected Goldmark's application for employment soon after his arrival in the United States—chose instead to announce the development of a seven-inch, 45 rpm record, which would never prove as successful as the LP. The first stereophonic recordings, which feature sound vibrations drawn from two different directions (as opposed to the previous monaural recordings) into left and right **stereo** channels and imprinted as two sets of sound waves in the record's microscopic grooves, were produced in 1933 on 78 rpm records by the British electronic company EMI (Electric and Musical Industries). Engineer Alan Dower Blumlein conducted experimental work on stereophonic records during subsequent years, and in 1958 the product was marketed by the Audio Fidelity Company in the United States and by the Pye and Decca firms in Britain. Almost all records were stereophonic by the late 1960s, and their fine sound quality fueled public demand for high-fidelity phonographs. With the development and increasing popularity in the 1980s and 1990s of such advanced technology as the **compact disk**—which offered exceptionally high sound quality at a higher but still fairly reasonable consumer cost than records—and the **laser disk**, the role of the traditional flat plastic disk in recording is all but over.

See also Magnetic recording

Photocopying

The term photocopying refers to the duplication of an existing document by using light. There are three major types of photocopying: photographic copying, which uses photographic techniques to develop copies on photosensitive materials; thermography, which is a dry-copying process that uses the action of heat on heat-sensitive chemicals; and xerography, which uses electrostatic charges, dry **ink,** and heat to fuse an image onto **paper**.

Although the process of photocopying is commonly thought to be a recent invention, the foundation for modern photocopying was laid before the nineteenth century. In 1780 the Scottish engineer and inventor **James Watt** took out a patent on a letter-copying machine after becoming frustrated with the inefficiency of making hand-written copies of letters. The device consisted of a roller press which created a reverse impression on the copy paper. The paper which was used in this process was thin enough that the document could be read from left to right on the other side. The first models of the copying press were in great demand until the last quarter of the nineteenth century. In 1877 **Thomas Alva Edison** developed a stencil machine that squeezed ink onto to a type of wax paper. The only improvement that Edison made on Watt's copying machine was that the document was copied on the correct side and no longer had to be read through the paper. Edison's model was improved on by David Gestetner, who introduced the Cyclostyle pen which consisted of a spiked wheel **pen** that created perforations in the stencil. Next came the mimeograph; developed by A. B. Dick in 1884, this is commonly referred to as the *ditto* method.

In a parallel development to the early stencil devices, copy cameras were used to duplicate documents beginning in the early 1800s. Contact photocopying, one of the first camera-based copying devices, used lights and developers such as ammonia vapor or water to produce copies. Contact photocopying is still used to create blueprints. Microfilm was developed in 1839 by John Danzer. Danzer substituted a microscope **lens** for the regular lens in a camera. When microfilming, documents are imaged on a very small area of silver halide film. Fine-grain films and special lenses allow documents to be filmed at reduction ratios. The films are then contained in a retrieval magazine. If paper documents are discarded after microfilming, users can obtain large file space savings. Microfilm is used mostly for the archival and retrieval of documents. It is the second most common photocopying technique used today, and is a very efficient method for libraries and large companies.

Carl Miller invented the dry-copying method of thermography in 1944. He was inspired when he observed that snow melted under fallen leaves more quickly than in open air. This concept of heat absorption by a dark surface led to research in the 3M labs in Minnesota. 3M called its thermographic copier the Thermofax. Thermography allows infrared rays to produce a copy directly onto specially treated, thermosensitive paper without using a negative.

Xerography is the most efficient and a more advanced method of photocopying. The word *xerox* comes from the

Greek *xeros*, meaning dry, and *graphein* meaning to write. Today, photocopying is sometimes called xeroxing in the business world. Xerography depends upon the use of photoconducting metals, such as selenium, which conduct electricity only when exposed to light. A xerographic copying machine has a metal sheet coated with a photoconducting material that has a negative charge at the beginning of the copying cycle. Light is reflected off the white areas of the original document onto the coated metal. The electric charge disappears where light from the reflection strikes the coated metal surface, so only dark parts of the image remain charged. Positively charged particles of toner powder (dry ink) are applied to the metal surface. The charged parts of the metal surface attract the dark powder, which is transferred to paper. High temperatures seal the powder to the paper, and a xerographic copy is produced.

Xerography was developed by the American inventor **Chester Carlson** in 1938. Carlson developed the basic principles of the first electrostatic copier while doing experiments in the kitchen of his New York apartment. For a number of years he encountered difficulty in finding a company interested in his invention. In 1947 a small photo company, Haloid, agreed to develop the copier. The company eventually changed its name to Xerox and marketed its first machine, the Xerox 914, in 1959. Since Carlson's model, the xerox machine (photocopier) has been significantly improved. The copiers have increased speed and the ability to produce two-sided copies from single-sided originals, and reduced or enlarged copies. They can even collate the papers into packets and staple them. Color photocopiers became common in the 1980s and 1990s. They worked essentially the same way as black and white copiers, only the exposure and toner steps are repeated three times, and special filters and toners are involved to produce full-color.

In the 1990s, digital photocopiers became widely available. Photocopiers also became enhanced; multifunction photocopiers also worked as scanners, fax machines and printers, all in the same unit. In 1997, the Xerox company announced the development of digital copiers that can be linked to personal computers, and with special software, can output documents from anywhere on a network, and in the future, from the Internet. Thus, an original is not needed to make a copy; the photocopier handles both. The future of photocopiers also includes such items as the Recycle Copier, which can photocopy and reverse the process and strip a photocopy of its text, allowing a piece of paper to be used up to ten times. Another enhanced photocopier can automatically translate a document from English to Japanese. In the twenty-first century, photocopiers will continue to merge with printing and computer technologies into do-it-all machines.

Photoelectric cell

During the latter half of the nineteenth century many scientists and engineers were simultaneously observing a strange phenomenon: electrical devices constructed from certain metals seemed to conduct electricity more efficiently in the daytime than at night. This phenomenon, called the photoelectric effect, had been noted years earlier by the French physicist A. E. Becquerel (1820-1891), who had invented a very primitive device for measuring the intensity of light by measuring the elecrical current produced by photochemical reactions. It was becoming evident that one metal in particular—selenium—was far more reactive when exposed to light than any other substance. Using selenium as a base, several scientists set out to develop a practical device for measuring light intensity.

A number of them succeeded. In 1883 the American inventor Charles Fritts created a working photoelectric cell; that same year a German engineer, **Paul Nipkow**, used a photoelectric cell in his ''Nipkow's disk''—a device which could take a picture by measuring the lighter and darker areas on an object and translate them into electrical impulses. The precursor to the modern photoelectric cell was invented by the German physicists Hans Geitel (1855-1923) and Julius Elster (1859-1920) by modifying a **cathode-ray tube**.

Strangely, the explanation for why selenium and other metals produced electrical current did not come until 1902, when **Phillip Lenard** showed that radiation within the visible spectrum caused these metals to release electrons. This was not particularly surprising, since it had been known that both longer **radio** waves and shorter X-rays affected electrons. In 1905 Albert Einstein (1879-1955) applied the quantum theory to show that the current produced in photoelectric cells depended upon the intensity of light, not the wavelength; this proved the cell to be an ideal tool for measuring light.

The affordable Elster-Geitel photoelectric cell made it possible for many industries to develop photoelectrical technology. Probably the most important was the invention of transmittable pictures, or **television**. Employing a concept similar to that used in Nipkow's scanning disk, a television **camera** translates the light and dark areas within its view (and, later, the colors within) into a signal that can be sent and decoded into a picture.

Another interesting application of photoelectric cells was the invention of **motion pictures**. As a film is being shot, the sound is picked up by a **microphone** and converted into electrical impulses. These impulses are used to drive a **lamp** or **neon light** tube that causes a flash, and this flash is recorded on the side of the film as a sound track. Later, when the film is played back, a photoelectric cell is used to measure the changes in intensity within the soundtrack and turn them back into electrical impulses that, when sent through a speaker, become sound. This method replaced the old practice of playing a **gramophone** recording of the actors' voices along with the film, which was very difficult to time to the action on the screen. Stored on the same film, a soundtrack is always perfectly synchronized with the action.

The photoelectric cell has since proven useful in many different applications. In factories items on a conveyor belt pass between a beam of light and a photoelectric cell; when each item passes it interrupts the beam and is recorded by a computer, so that the exact number of items leaving a factory can be known simply by adding up these interruptions. Small light meters are installed in streetlights to turn them on automatically when darkness falls, while more precise light meters are used daily by professional photographers. Alarm systems have been designed using photoelectric cells that are sensitive to ultraviolet light and are activated when movement passes a path of invisible light. Cousin to the photoelectric cell is the **photovoltaic cell** which, when exposed to light, can produce electricity. Photovoltaic cells form the basis for solar **batteries** and other **solar-powered machines**.

PHOTOGRAPHIC FILM • See Photography

PHOTOGRAPHY

Photography is the process whereby light produces an image on a sensitized surface. The precursor of photography and the modern camera is the **camera obscura** (Latin for "dark chamber"). In its basic form, as developed by tenth-century Islamic scientists, the camera obscura was a darkened enclosure with a small aperture to admit light. The light rays would cast an inverted image of external objects onto a flat surface opposite the aperture. This image could be studied and traced by someone working inside the camera obscura, or viewed from the outside through a peep-hole. In the sixteenth century, the Italian scientist Giambattista della Porta (1538-1615) published his studies on fitting the aperture with a **lens** to strengthen or enlarge the image projected. Made increasingly versatile through additional improvements, the camera obscura become popular among seventeenth and eighteenth-century European artists—including, perhaps most notably, the Dutch painter Jan Vermeer (1632-1675). The essential challenge in the development of photography was to capture the image produced by a camera obscura in a permanent form. In 1727, the German professor of anatomy Johann Heinrich Schulze took a step in this direction by discovering that silver salts darkened when exposed to sunlight. Schulze "printed" words on a **bottle** by coating it with silver salts, attaching templates cut out of **paper,** and exposing the bottle to sunlight. Thomas Wedgwood (1771-1805), of the English pottery-works family, made some of the first attempts to actually "fix" such images and make them permanent. Wedgwood, however, only managed to produce transient silhouettes.

It was the French inventor **Joseph-Nicéphore Niepce** who first achieved Wedgwood's goal. One area of Niepce's experimentation derived from his interest in reproducing **engravings** lithographically. In **lithography,** an image is printed from a porous plate whose **ink**-receptive surface is partially drawn over with a hard, ink-repellent substance. Niepce's son, who was charged with drawing the lithographic plates, eventually left for military service. Therefore, Niepce tried to devise a way to "draw" the lithographic plates photographically. Niepce experimented with oiling **engravings** to make them transparent, placing them across plates that he had treated with a light-sensitive **varnish**, and then exposing the engraving-covered plates to sunlight.

In 1822, Niepce succeeded in this process. Niepce used bitumen of Judea, a form of **asphalt,** as the basis for the light-sensitive **varnish.** Sunlight would solidify the varnish only in areas under light portions of the engraving. After a period of exposure, therefore, Niepce could rinse the plate to obtain a photographic reproduction that could be used in lithographic printing. Using improvised cameras, Niepce also conducted more conventional photographic experiments. As early as 1816, Niepce captured an image of Paris on paper treated with silver chloride. The exposure time required was at least an hour, and Niepce could not adequately fix the image. By 1826, however, Niepce had achieved the first permanent photograph taken from nature, a view of his country estate. Niepce had fitted a professionally made camera obscura with a lens. Inside, he placed a pewter plate treated with the bitumen solution used in the lithographic experiments. To take the photograph, Niepce exposed the plate for approximately eight hours. After rinsing the plate, the lights were represented by solidified bitumen and the darks by the pewter background. Niepce termed his discovery heliography (Greek for "sunwriting").

The world's first photograph, taken by Joseph Nicéphore Niépce in 1826. It is of the courtyard of his family's estate in France.

In 1826, the French scene painter **Louis-Jacques-Mande Daguerre**, who wanted to exploit photographic images in creating theatrical backdrops, began corresponding with Niepce about heliography. This led to a commercial partnership that lasted from 1829 until Niepce's death in 1833. Working with copper plates coated with silver iodide, a technique he owed to Niepce, Daguerre discovered that a latent image, resulting from a relatively short thirty-minute exposure, developed fully when exposed to mercury vapor. Daguerre then used salt or sodium thiosulfate to fix the fully-formed image. Ignoring his debt to Niepce, Daguerre marketed this discovery as the daguerreotype. Although the daguerreotype produced a reversed left-to-right image, and the exposure time initially required precluded its use for portraiture, these impressively detailed and burnished photographs were an immediate commercial sensation in Paris.

During this same period, the English scientist **William Henry Fox Talbot** was attempting to capture a permanent photographic image on paper. To do so, Talbot first created a negative, in which light and dark tones are reversed. Talbot formed his negatives by placing patterned objects, such as twigs and lace, across transparent sheets of paper saturated with silver chloride. Talbot then exposed this ensemble to light. The silver chloride would darken except where covered by the objects, creating a reverse silhouette. After fixing the negative image with a salt or sodium thiosulfate solution, Talbot would place it against a fresh sheet of photosensitive paper. Exposing the pair of sheets to light would again reverse the image, yielding a positive photographic print. Talbot's 1835 image of his family estate (Lacock Abbey, Wiltshire) was the first negative-based photograph taken from nature in England. Talbot subsequently developed the calotype. Like the earlier daguerreotype, Talbot's breakthrough was based on developing a latent image though a separate chemical process, and thus reducing the exposure time required. With the calotype, an exposure time of a minute or less left an imperceptible negative image on light-sensitized, paper. This image was then developed in a wash containing gallic acid. Early cameras for Talbot's calotype consisted of wooden boxes fitted with a lens and partially lined with the photosensitive paper.

In 1839, fully aware of Daguerre's success, Talbot presented his calotype process to the Royal Society, Britain's premier scientific body. By 1841, Talbot had perfected his discovery and patented it under the name of talbotype. The talbotype method enjoyed two significant advantages over the daguerreotype: it allowed several positive prints to be obtained from a single negative, and the final image was not reversed. However, the talbotype photograph remained somewhat grainy. This soft-grained quality could be exploited for romantic artistic effects, but the daguerreotype's bright, finely-detailed image was generally considered superior.

Thus, it was not until the early 1850s, when the English portrait sculptor Frederick Scott Archer (1813-1857) promoted the **collodion** process, that the daguerreotype was superseded as the most popular form of photography. With Archer's process, glass plates were treated with a preparation of collodion (powdered cotton dissolved in ether). Once the solvents evaporated, a moist photosensitive film was left on the plate. This allowed for a finely-detailed negative to develop very rapidly, and positive paper prints could be taken from the negative for as long as it remained damp. The collodion process, which was widely used up through the 1880s, did have a significant limitation. As the glass plates had to be treated, exposed, and developed immediately, photographs could not be taken unless darkroom facilities, whether permanent or portable, were at hand. This drawback was partially balanced by the fact that the collodion process was never patented—an economically unwise self-effacement very much in contrast to the policies of Talbot and Daguerre.

The American inventor **George Eastman** benefitted far more than any of these predecessors from the mass-merchandising of photography and cameras. Eastman, who left school at the age of fourteen, developed a passion for photography while working as a bank clerk. Experimenting in his mother's kitchen, Eastman devised a way to coat glass photographic plates with a gelatin emulsion containing silver chloride. This emulsion, once solidified, left a light-sensitive ''dry plate'' that (unlike Archer's collodion-coated plates) would last indefinitely and could be developed at leisure. In 1879, Eastman developed a machine to mass produce his invention. In 1881, to market it, he founded the Eastman Dry Plate Company of Rochester, New York. Eastman's Dry Plate Company marked a new era in photography. Through his company, Eastman introduced several breakthrough products and procedures. The dry plate, sixty times more sensitive than those treated with collodion, meant that tripods were no longer mandatory to anchor the camera during exposure. In 1884, Eastman patented a **photographic film**, packaging it in a small **plastic** roll. Replacing plates with compact rolls of film enabled a welcome reduction to the size of cameras. In 1889, Eastman began producing a **celluloid** film that improved on his original paper-and-gelatin version. Eastman (who launched the trade name Kodak in 1888) ultimately marketed a highly innovative series of ''folding,'' ''hand,'' and ''pocket'' cameras that made photography accessible to even the casual amateur. Eastman-Kodak also exploited contemporary developments in **color photography**. The 35mm film format became the norm for non-instant photograophy.

In 1937, the American inventor **Edwin Herbert Land** founded the Polaroid company in Cambridge, Massachusetts. Land sponsored the next significant breakthrough in photography: the **instant camera**. During the 1940s, Land experimented with film systems that would rapidly and automatically generate positive photographic prints. In 1948, Land introduced the first such camera, the Polaroid Model 95. The film in the Model 95 consisted of sealed, two-sided packets. After a picture was taken, the camera would automatically eject a single two-sided packet though a set of rollers. As the packet passed through these rollers, a pod of developing chemicals would burst open. After one minute, during which the photograph developed, the packet could be separated into its two halves. The positive side displayed a sepia-toned image that quickly hardened and became permanent. The side coated with the negative emulsion was disposable.

Photography continues to grow in the second half of the twentieth century with the development of infra-red, ultraviolet, thermal, **laser**, electronic, and electromagnetic photographies, though these are often used in sophisticated technological applications. In the 1990s, the consumer photography market saw rapid, if not revolutionary, changes from traditional 35mm photography. In 1996, the Advanced Photo System (APS) was introduced. Developed jointly by Kodak and Fuji, two of the largest photography companies in the world, APS features smaller cameras and smaller 24 mm film. Though smaller, the film is more sophisticated, containing a magnetic strip which registers data about the conditions under which the photo was shot. This information is used in processing to correct exposure as necessary. APS cameras also allow three different sized pictures to be shot on the small role of film, from standard prints up to panoramic shots. APS has several drawbacks, however, including the system's incompatibility with 35mm cameras (which means the purchase of a new camera) and the lack of black and white film. However, the APS film itself might be part of the future of 35mm. APS film is made of a polyethylene napthalate base, which is thinner, stronger and hardier than the cellulose triacetate base used in 35mm. Some authorities predict that the base will eventually be used in 35mm film. Whether APS will replace 35mm altogether has yet to be determined.

There are two non-film-based technologies that threaten traditional photography. In the early 1990s, the still video camera was introduced. It combines camcorder and camera, using a floppy disk or an attached computer instead of videotape to record images. The information is stored in analog form. Still video cameras do not produce sharp pictures, no better than a camcorder, but because the images are stored electronically, they can be viewed immediately on a computer screen. There is no photoprocessing involved. Still video cameras have not caught on, though by 1997, digital still video cameras were available that produced sharper images and had more memory available on disk. What might replace film photography all together is **digital camera**s. Available on the consumer market as earlier as 1994, digital cameras take pictures electronically and store them as computer files, to be edited with computer software such as Adobe's PhotoShop. No photoprocessing is involved. Though the images produced by digital cameras are sharper than still-video cameras, they do not yet match 35mm pictures. As digital cameras improve, they look to be the future of photography in the 21st century.

PHOTOGRAPHY, INFRARED • See Infrared photography

PHOTOGRAPHY, UNDERWATER • See Underwater photography

PHOTOTYPESETTING

In modern printing systems, **ink** is pressed directly onto **paper** from plates of photographic paper or film, which carry an image of a finished page; the ink adheres only to the parts of the photographic plate which will show up in print, as was earlier done with raised type.

Since the turn of the twentieth century, the lettered parts of plates were made by photographing actual metal plates of finished type—the exact type of plates which, in earlier days, would have been inked themselves. In **linotype** machines that cast type with metal, the typesetter inputs text at a keyboard. The machine sets up a row of pre-made models (or ''matrices'') of letters. When the row is finished, it pours a quick-setting alloy over the row, which cools to form a metal row of type; this row is then dropped into place on the plate.

Obviously, the process of creating an entire finished page of metal simply to be photographed wasted considerable time and material. In the mid-twentieth century, inventors marshaled new photographic and computer technology to create a method of setting type directly onto photographic plates, thus bypassing the metal casting stage entirely. Such systems of typesetting are called *phototypesetting*, or photocomposition.

Early phototypesetting machines, such as the Fotosetter, developed in 1947, were designed as photographic versions of hot-metal casting machines. In these, the matrices are simply pictures of letters, and rather than casting them in metal, the machine photographs them one row at a time. These machines, however, were not considerably faster than the hot-metal machines they were meant to replace. It was in 1949 that two French inventors, René Higonnet and Louis Moyroud, revolutionized phototypsetting with their Lithomat (later called the Lumitype). In the Lithomat, the letter matrices are on a disk revolving at a steady, high speed; a high-speed **lens** catches the disk at exactly the right point in its rotation to photograph the correct letter directly. An early version of the machine could produce 28,000 characters per hour; by 1957, through refinements in the process, the speed reached 80,000 characters per hour. The speed was ultimately limited, though, by physical limits on how fast a disk could spin. In 1959, the Lumizip 900 was introduced; in this machine, only the lens moved, allowing a speed of up to 2,000,000 characters per hour. In the 1960s, electronic typesetters eliminated even the lens, using electron beams to directly draw characters based on matrices stored in a computer's memory.

Perhaps the most important aspect of phototypesetting machines is that the typesetting is done in a machine separate from the keyboard. This allows numerous typists to key in a text and store it on paper tape, *magnetic tape*, or **computer disk**. With the actual keying divided among limitless typists, computerized phototypesetters can set volumes of text in hours, whereas a hot-metal typesetter might have taken a year; furthermore, encyclopedias and indices can now be updated with minimal effort and reprinted rapidly. Phototypesetting might become obsolete as personal computers give their users the ability to typeset and do page layouts on their own. Personal computers can be directly linked to their printing source without need for intermediate retyping.

See also Printing technology

PHOTOVOLTAIC CELL

As far back as the 1800s scientists realized that, through certain chemical reactions, sunlight could be converted into electricity. The first experimenter to successfully accomplish this feat was A. E. Becquerel (1820-1891), who built a device that could measure the intensity of light by observing the strength of an electric current produced between two metal plates. Later scientists discovered that the metal selenium was particularly sensitive to sunlight, and during the 1880s Charles Fritts constructed the first selenium solar cell. Fritts's device was woefully inefficient, however, converting less than one percent of the received light into usable electricity.

The Fritts selenium solar cell was mostly forgotten until the 1960s when the drive to produce an efficient solar cell was renewed. It was known that the key to the photovoltaic cell was in creating a semiconductor that would release electrons when exposed to radiation within the visible spectrum. During this time researchers at the Bell Telephone Laboratories had been developing similar semiconductors to be used in communications systems. Quite by accident, Bell scientists Calvin Fuller and Daryl Chapin (who had been working independently on projects unrelated to solar cells) found the perfect semiconductor: a hybridized crystal called a ''doped'' cell made of phosphorous and boron. The first solar cells using these new crystals debuted in 1954 and yielded a conversion efficiency of nearly six percent. Later improvements in the design increased the efficiency to almost fifteen percent, a high mark by even today's standards.

In 1957 Bell Telephone used a silicon solar cell to power a **telephone** repeater station in Georgia. The process was considered a success, though it was still too inefficient to penetrate the general marketplace. The first real application of silicon solar cells came in 1958, when a solar array was used to provide electricity for the **radio** transmitter of *Vanguard 1,* the second American satellite to orbit the earth. Solar cells have been used on almost every satellite launched since.

Once space exploration had proven their efficacy, photovoltaic cells began to appear more and more frequently in homes. Just about any small appliance can be adapted to run on solar power, but by far the most successful have been watches and **pocket calculators**. These devices generally use a solar cell to charge a nickel-cadmium battery, so that they can be used even during periods of dim light.

Even in the early days of solar technology scientists envisioned vast photovoltaic arrays that could power entire cities. Those early dreams have been realized to a small extent by many homeowners who have installed solar panels in their homes. The usefulness of these panels is limited, however, by their relatively low rate of efficiency, as well as by the limited hours of available sunlight. It is unlikely that solar power will replace that gleaned from fossil fuels even in the distant future; still, the production of solar energy has increased steadily, approximately doubling each year. At that rate, it is conceivable that, by the year 2000, ten percent of the energy produced in the world will be produced by solar cells.

See also Battery, electric; Communication satellite ; Earth-survey satellite; Navigational satellite ; Space probe; Weather satellites

PIANO

The piano may be the best known and loved of all musical instruments. It also has the broadest range of any instrument, so music for all other instruments can be composed on it. It can be played solo, but most other instruments, including the voice, use the piano for accompaniment. Technically, the piano may also be the most complicated musical instrument with over 2,500 parts.

The piano is a stringed instrument. Its many parts are organized into five general structural and mechanical areas of ei-

A building sided with photovoltaic cells, which capture the energy of the sun and convert it to electrical energy.

ther grand or vertical pianos. These are: the case of the wing-shaped grand piano (or the cabinet of the vertical or upright piano); the soundboard and the ribs and bridges that are its components; the cast iron plate; the strings; and, collectively, the keys, hammers, and piano action or mechanism. The case has many structural parts for attaching legs and tuning pins, but perhaps the rim and the keybed or shelf where the keys and piano action will be installed are most important. The soundboard amplifies the vibrations of the strings which are transmitted through bridges.

The cast iron plate is installed over the soundboard and pinblock (part of the case), and it provides the strength to anchor the strings under tension. Nose bolts and perimeter bolts anchor the plate to the braces and inner rim of the case. The 220-240 strings of the piano are attached to hitch pins along the curved edge of the cast iron plate and to tuning pins across the front of the piano, roughly parallel to the keyboard. The piano action is still more complicated and includes the keys, hammers, and mechanism or action.

Names for pianos usually indicate their sizes. Grand (wing-shaped) pianos range in length from 4 ft 7 in-9 ft 6 in (1.4-2.9 m) from the front of the keyboard to end of the bend. The "baby" grand is 5 ft-5 ft 2 in (1.52-1.57 m) in length; smaller grand pianos are called "apartment size," and the larger sizes are the medium grand and concert grand. Modern vertical piano design has changed little since 1935. Verticals range in height from 12-17 yd (10.9-15.7 m) with small variations in width and depth. The five standard sizes from smallest to tallest are the spinet, consolette, console, studio, and professional pianos. Pianos are frequently chosen for appearance, and cabinets are available in most furniture styles and finishes.

The piano's ancestors are the first stringed instruments. Plucking, striking, and bowing of strings was known among all ancient civilizations; the harp is mentioned in the Book of Genesis in the Bible. The psaltery was an ancient box-type instrument with strings that were plucked with a pick. Keys were added to stringed instruments to make the family of instruments led by the harpsichord, the most popular instrument of the seventeenth century. The harpsicord created sound by using keys to pluck the strings. However, the dulcimer, another instrument developed at the same time, created sound by using the keys to strike the strings. Keys and strings were paired in a striking instrument in the clavichord, which led directly to the invention of the pianoforte or fortepiano.

Bartolomeo de Francesco Cristofori made harpsichords in Padua, Italy. He is credited with having invented the piano in 1700. Cristofori's piano had hammers that struck the strings by falling by momentum, after having been moved by the action parts linking the hammers to the keys. The hammers were caught by back checks or hammer checks to keep them from bouncing up and down on the strings after the initial strike. This method of fall allowed the strings to continue to vibrate and make sound and for them to be struck loudly or softly unlike the sameness of tone in the harpsichord. Johann Andreas Silbermann of Strasbourg, France, continued Cristofori's interest in the pianoforte, and the instrument became popular in Germany after Frederick the Great purchased several and Johann Sebastian Bach approved of it in 1747.

The piano had replaced the harpsichord in importance by the end of the eighteenth century. Cabinetmakers built beautiful cases for them. The square piano was built mid-century, and more musicians began writing music specific to the piano, rather than borrowing harpsichord tunes. Piano building began in America in 1775, and changes to the design of the hammers and to the playing mechanism or action improved the sound and responsiveness of the instrument. Jean Henri Pape of Paris patented 137 improvements for the piano during his life (1789-1875). In England, John Broadwood developed machines to manufacture pianos and reduce their cost.

Improvements continued from 1825 to 1851 with over 1,000 patents in Europe and the United States for stronger, more deft pianos with greater control and repetitive motion. By the mid-nineteenth century, the modern piano had emerged based on the development of the cast iron plate for structural strength and cross-stringing by fanning bass strings over trebles. By 1870, Steinway & Sons had developed this fanning method called the over-strung scale, so that the strings crossed most closely in the center of the soundboard where the best sound is produced.

C. F. Theodore Steinway also developed the continuous bent rim for the case, which enhanced sound transmission by using the acoustic properties of long wood fibers. These improvements were adapted to all styles of pianos including grand, upright, and square pianos. By 1911, there were 301 piano builders in the United States. Production peaked in the 1920s and declined greatly because of the Great Depression in the 1930s. Today, there are approximately 15 piano manufacturers in the United States, and Japan is the world's largest producer of pianos.

The design of the piano has not changed appreciably since the late 1800s, although manufacturers may use different materials or approaches to the manufacturing process. The manufacturing process for the grand piano is described below; there are some differences in manufacturing the vertical or upright piano and in operation methods, particularly the angle at which the hammers strike the strings.

PIANO, ELECTRIC • See Musical instruments, electric

PIANO, PLAYER • See Musical instruments, mechanical

PICCARD, AUGUSTE (1884-1962)

Swiss physicist

Auguste Piccard was a scientist and inventor whose exploring instincts took him to record heights and depths. He and his twin brother, Jean Felix, initially achieved prominence together, but Auguste's accomplishments were more sensational in nature and gained him a greater degree of popularity. The two brothers were influenced by their father, Jules, a chemistry professor, and their uncle, Paul, who was commercially involved with hydroelectric projects. Auguste and Jean Felix

Auguste Piccard

earned doctorates at the Zurich Polytechnic School. Both went on to professorships at various institutions. While Jean Felix moved on to the University of Chicago and the Massachusetts Institute of Technology in the United States, Auguste stayed in Europe to work on his famous projects.

Even as a student at the Zurich School, Auguste was laying plans for his ventures. The Piccards began making **balloon** ascents in 1913 and participated in the balloon section of the Swiss Army. Auguste developed an airtight, pressurized gondola in 1930. On May 27, 1931, he made his first record-setting ascent in the gondola at Augsburg, Germany, to an altitude of 51,775 feet (15,781 m). The seventeen-hour flight landed him world attention. Another record flight from Zurich took him to 53,153 feet (16,201 m) in 1932. He continued his stratospheric activities for five more years, making his last flight in 1937.

In the late 1930s, Piccard began heading in the opposite direction, working on an idea for an independent deep-sea diving craft some think he first visualized when he was a student. From 1930, William Beebe and Otis Barton were making record dives in their **bathysphere**. The sphere was limited to vertical dives since it depended upon a tether and air supply from a surface support ship. Auguste Piccard's **bathyscaphe**, even in its earliest stages of conception, never required direct support. Auguste developed two craft, the French Navy *FNRS-2* and the *Trieste*. On both, he took the lighter-than-air concept of stratospheric penetration and developed a compartmentalized envelope containing lighter-than-water **gasoline** weighted with ballast. He added to this a propeller system and a spherical pressure-resistant compartment. This combination gave him the free-moving craft that he was after. After years of research and test dives, Piccard's vessels started setting records. In 1954, a French team dove to 13,287 feet (4,050 m) off West Africa. Piccard's son, Jacques, took the *Trieste* to 35,800 feet (10,912 m) in the Marianas Trench near Guam in 1960 on the same day as Jean Felix's death.

Auguste Piccard was neither a meteorologist nor an oceanographer, yet his technical achievements were of great importance to both sciences. His inventions and explorations took scientific research to new levels.

Pin

The pin has been used since ancient times to fasten pieces of clothing or **paper** together, and as jewelry. Bronze pins with gold heads have been found in Egyptian tombs. The Greeks and Romans fastened their garments with bronze pins that resembled our modern safety pin. Pins made of wood, bone, and various metals were widely used in medieval Europe. The manufacture of wire pins was well established in France by the late 1500s; England's first large pin factory was established by John Tilsby in 1625 in Gloucester.

In early pin manufacturing, the heads were made separately; attaching and securing them was difficult and produced unreliable results. In 1817 American inventor Seth Hunt patented a machine for making one-piece pins. Similar English machines were designed in 1824 by Samuel Wright and in 1838 by Daniel Foote-Taylor and Henry Shuttleworth. In 1832, American inventor John Ireland Howe patented the first practical and highly effective machine for making one-piece hatpins.

Once the manufacture of one-piece pins was successfully established, packaging the pins efficiently remained problematic. Howe patented a machine in 1843 that solved this problem by inserting pins into sheets of paper. The ''paper of pins'' soon became a household necessity and continues so to this day.

The modern safety pin was invented in three hours by Walter Hunt (1796-1859), a gifted and prolific New York inventor. Seeking a way to pay off a $15.00 debt, Hunt twisted a piece of brass wire into various shapes until he came up with a coil spring at one end that held the point securely into a clasp at the other end. The problem of being pricked by the sharp end of the pin was eliminated, and so was Hunt's debt: the inventor patented the design and sold the rights to it for $400.00 to his creditor.

Pincus, Gregory (1903-1967)

American endocrinologist

Gregory Pincus is best known for his central role in developing ''the pill''—the oral contraceptive or **birth control** pill. He also investigated the biochemistry of aging, arthritis, cancer, and the adrenal system's response to stress. Pincus was born in

Woodbine, New Jersey. Both of his parents had interests in agriculture and the arts, and his father taught at an agricultural school. In 1924, Pincus graduated from Cornell University where he not only studied science but founded a literary magazine. In 1927 he received master's and doctoral degrees from Harvard University and, following further study in Europe, joined Harvard's biology faculty. In 1938 he joined the faculty at Clark University, in Worcester, Massachusetts, as an experimental zoologist and, in 1944, co-founded the independent Worcester Foundation for Experimental Biology where he continued his earlier research on the way the reproductive system and female hormones functioned.

Since the discovery of the sex hormones, scientists had searched for a natural, safe, and foolproof method of using female hormones to treat infertility and prevent pregnancy. Several scientists beginning in the 1920s showed that progesterone inhibited ovulation. In the 1940s, British chemist Robert Robinson tried unsuccessfully to synthesize various female hormones. Pincus's work attracted the attention of Margaret Sanger (1879-1966), the United States' best-known advocate of birth control. Financed by Sanger's friend, philanthropist Katherine Dexter McCormick (1875-1967), Pincus led a group of scientists in the early 1950s who began developing a hormone-based substance to make the body mimic pregnancy—the one time when a woman is almost certain not to become pregnant. The biologist **Min-Chueh Chang** carried out experiments on laboratory animals with various compounds of progestin, a synthetic progesterone developed in Mexico by the American chemist Carl Djerassi. Another collaborator, physician **John Rock**, had already been experimenting with progesterone to cure infertility. Tests of the new substance on women took place in Massachusetts, Puerto Rico, Haiti, Mexico, and California, supervised by Pincus, Rock, Celso-Ramon Garcia and Edris Rice-Wray. Because contraception was illegal in Massachusetts and due to objections from religious groups—principally the Roman Catholic Church—the initial tests were to treat infertility rather than prevent pregnancy. However, these tests showed that the compound prevented ovulation and in 1960, progesterone was approved by the U.S. Food and Drug Administration as the first contraceptive pill.

Gregory Pincus continues to be hailed as the primary force behind the oral contraceptive. Among the many honors he received during his lifetime was membership in the National Academy of Sciences of the U.S.A.

Pinkham, Lydia • See Patent medicine

Pistol

A pistol is a small gun that can be held and fired with one hand. Firearms had been in use for centuries before the handheld gun became practical. Until the 16th century, soldiers had to ignite the **gunpowder** by hand, a time-consuming and dangerous undertaking even with the largest cannon. In the early 1500s, however, the wheellock, an ignition mechanism that could be installed inside the firearm, was invented. Similar to the **ignition system** found in cigarette lighters today, the wheellock consisted of a **steel** wheel revolving against a flint, producing a spark to set off the powder. Handguns were soon in production and have remained popular ever since because they can be carried easily and leave one hand free to wield another weapon. Although it is not clear who was the first to develop a pistol, the first designs of about 1540 are commonly credited to Camilleo Vitelli of Pistoia. Pistols with different firing mechanisms, such as the flintlock and the matchlock, were brought on the market in quick succession. The **revolver**, a kind of pistol with a revolving magazine tripped by the trigger action, soon followed.

Gregory Pincus

A major drawback associated with early pistols—excepting revolvers—was that they could be fired only once before requiring manual reloading. A major advance in the design of pistols seemed in the offing when **Hiram Maxim** introduced a recoil-operated **machine gun** in 1884, a gun that used the energy released in firing a shot to reload itself. Within a decade the makers of firearms found ways to apply the technique to handguns. The first repeating pistol, with a **lever** that activated a bolt action, was brought on the market by Laumann, an Austro-Hungarian inventor in 1890. Another Laumann design, created two years later, used the so-called "blowback" system, which used the energy produced by the cartridge case as it attempted to leave the chamber when the gun was fired. This gun, called the Schönberger, is widely considered the prototype of the modern automatic pistol.

A more practical design was developed in 1893 by a German gunmaker, Hugo Borchardt, who came up with a toggle system and recoil mechanism similar to those in Maxim's machine gun. The recoil principle is generally favored over the blowback system by the military because it produces more power. Borchardt was also the first to propose a removable box magazine to be loaded inside the pistol butt.

The next design improvement came in 1895 with the German Mauser pistol, produced by Peter Mauser (1838-1914) and his brother Wilhelm, previously successful makers of bolt-action **rifle**s, who had tried, unsuccessfully, to sell a revolver design to the German military. The Mauser pistol employed the same principles as the Borchardt gun, but had a simpler design and was much stronger. It was assembled without a single **screw** or bolt in the firing mechanism, its interlocking elements tightly held in place by the pistol frame. However, these and other automatic pistols available at the time were expensive and considered unsuitable for military purposes. The automatic pistol brought out by Georg Luger (1848-1922), who interested Borchardt in the project for a while, signaled a change in military attitudes. The Borchardt-Luger pistol, smaller than previous models, had an angled grip and an improved trigger mechanism. The internal **spring** in the recoil mechanism was also improved, and a safety device prevented accidental firing. Perhaps the most significant improvement was the use of a smaller cartridge, which reduced the recoil and gave the weapon greater accuracy. The Swiss military adopted the Borchardt-Luger.

New designs were being developed at the same time in many different places in Europe. The American **John Browning** worked with a Belgian gunmaker, bringing out his first design using the blowback system in 1898. A greatly improved design, marketed three years later, became tremendously popular in Europe, not least because it had only three major components: the barrel, frame and the slide. Browning then went on to develop a pistol for Colt, which became the famous .45 automatic. One of the oldest European companies producing firearms, the Beretta firm, produced a pistol with an external hammer, which showed immediately whether the gun was cocked or not. The next improvement came with Carl Walther's 1929 blowback automatic with a fixed barrel, which had a protruding pin indicating whether there was a cartridge in the chamber. Subsequent improvements included safety catches easily operable by the left or right hand and a pistol that could fire three-round bursts.

By and large pistols have changed little over the last half century. They are used primarily by law enforcement agents and for personal defense. Since they are small and light, pistols are easily stolen or used by people other than their owners. This problem was addressed in the late 1990s by several American pistol manufacturers, who developed so-called "smart" guns that could only be fired by their owners. This was to prevent accidental deaths, especially by children playing with parents' guns. The smart gun is fitted with a **radio** transmitter that emits a signal. The gun's user has a transponder (radio receiver/transmitter), worn in some designs embedded in a finger ring, in others in a small badge affixed to the user's clothing. When the transponder picks up the weapon's signal, it sends a code back to the gun, and the gun automatically removes a pin blocking the trigger mechanism, allowing the gun to be fired. Because the transponder has a range of only a few inches, the gun should not fire in anyone but the legitimate user's hand, even if it is snatched away in a close struggle. Though personalizing a pistol in this way added greatly to its cost, it seemed able to alleviate much concern about handgun safety.

PISTON

A piston is a device that is forced up and down within a cylinder, creating motion in engines and other similar mechanisms.

The use of pistons can be traced to the first century A.D., when Greek engineers developed pneumatic tools (utilizing compressed air) containing pistons. Later, around 1650, **Otto von Guericke** built an air pump that utilized a piston.

Dutch scientist **Christiaan Huygens** built a dangerous **gunpowder** engine around 1680 that used the power from an explosion to drive a piston through a cylinder. Huygens's assistant, French physicist **Denis Papin**, suggested that the condensation of steam would create a vacuum in a cylinder into which a piston could be drawn, and in 1690 he built an engine based on this concept. Huygens and Papin were among the earliest contributors to the eventual development of an **internal combustion engine**.

Of course, it is within the internal combustion engine that the piston was put to its most practical use. The piston slides up within the cylinder to compress fuel vapor. When this is ignited, it explodes and drives the piston down, creating usable power. In any of these piston engines, the reciprocating power of the piston is converted into rotary power through the use of the connecting rod and crankshaft. This internal combustion engine process is the same whether the engine is fueled by **gasoline**, diesel fuel, or some other combustible fuel.

PITCH

Pitch, a derivative of bituminous coal from which the oil has been removed, is popular as a road-paving material for several reasons: it is easily applied to the road after heating, it forms a good seal against moisture, and it is pliable enough when hardened to allow for a smooth ride for the traveler.

Excavations show evidence that pitch was used at Jericho circa 2500 B.C. to seal the spaces between the stones of its infamous wall. The Babylonians, Vikings, and others also used pitch in shipbuilding to seal hulls against water. Road-building pioneers **Pierre Trésaguet** and **Thomas Telford** used it as a filler between paving stones, a method rejected by **John McAdam**, who preferred to allow surface rock to absorb moisture. Pitch is now used worldwide for road surfacing, often mixed with rock or cement to strengthen it. Roads surfaced in this way are referred to as bitumen roads in some countries, while the terms **asphalt** or blacktop are commonly used in North America. Pitch is also used in the manufacturing of electrodes.

See also Road building

Pitman, Isaac • See Shorthand

Pixii, Hippolyte (1808-1835)

French engineer

Pixii was born in France and followed in his father's footsteps, becoming a builder of scientific instruments. He had a short life, dying at the age of 27 in 1835. Pixii's claim to fame came as a result of English physicist **Michael Faraday**'s discovery of electromagnetic induction. In 1831 Faraday, experimenting with his discovery, had taken a copper disk and spun it between the poles of a permanent **magnet**. This created an electric current which could be drawn off the disk with a **wire** and put to work.

Faraday announced his discovery to the Royal Society of England and included a description of a simple dynamo, a machine that produces electricity. Pixii, guided by Faraday's description, proceeded to build his own **generator**, more efficient than Faraday's. His machine was hand-driven and had stationary coils around which revolved a field magnet. Pixii's generator produced **alternating current** (AC) which was of little interest at the time. Following a suggestion by French physicist **André Ampère** (1775-1836), Pixii installed a commutator which converted the AC into **direct current** (DC). Alternating current would not be utilized until 160 years later, thanks largely to the efforts of **Nikola Tesla**.

The device that Pixii built was essentially a working model, but it was the first practical generator built on the principle Faraday had discovered. Later, **Zénobe Gramme** established a very profitable business building electric generators.

Pizza

Pizza, in some form, has been eaten by Mediterranean people since the earliest days of their civilizations. Roman soldiers disliked the flat *matzo* **bread** of the Jews in Israel and attempted to make the bread more like the *focaccia* bread of their home. They sprinkled herbs, oil, and crushed **cheese** on it, creating the first pizza pie. The Italian word, *pizza*, is roughly translated "flat pie."

The style of pizza popular today began in pre-Renaissance Naples, Italy. Because the ingredients needed for pizza were few and inexpensive, pizza was a peasant food that became popular in the Neapolitan slums. Women of Naples combined flour, olive oil, lard, herbs, and cheese in creative ways and perfected the pies until all of Italy praised the pizza pies of Naples. Neapolitans continued to experiment with the pies and began adding some of the mysterious yellow berries Christopher Columbus (1451-1506) brought back from the New World, creating the first *pizza al pomidoro* (pizza with tomatoes).

The stature of pizza received a boost from Queen Maria Carolina of Naples (1752-1814) in the early eighteenth century. She persuaded her husband, King Ferdinand IV (1751-1825), to allow their chef to prepare pizza in the royal oven. The popularity and variety of pizza pies grew rapidly throughout Italy, and the *pizzaiolos* (pizza chefs) were given the respect often accorded to sushi chefs today. In 1889, the famous *pizzaiolo*, Raffael Esposito, received a royal command from Queen Margherita (1851-1925) to make pizzas for a royal feast. Esposito created one pizza for the feast that resembled the Italian flag by using tomatoes for the red, basil for green, and mozzarella cheese for the white field of the flag. This pizza, named Pizza Margherita, is the most popular pizza in America today.

Variations on pizza include the *calzone*, which is a pizza folded in half. The calzone is called *mezzaluna* in Italy. A deep-fried *calzone* is called a *panzerotti*. Another dish called *torta rustica*, uncommon in the United States, is a deep pizza stuffed with a variety of vegetables and cheese and is served on special occasions in Italy.

Pizza's popularity in America has grown steadily since Gennaro Lombardi opened the first pizzeria at 53 1/2 Spring Street in New York City in 1905. Though much of the fast food pizza in America pales in comparison to the original pies, many restaurants offer delicious variations of the original dish. Types of pizza popular in America today vary dramatically from one region of the country to the next. Neapolitan pizza, also called New York pizza, has a very thin and crisp crust with just a light layer of toppings. Chicago-style, or deep dish pizza, is characterized by a thick, bready crust and a heavy sauce. California-style pizza reflects the area's interest in gourmet health food and has a light crust topped with a wide variety of exotic cheeses and mushrooms.

In addition to fresh pizza, frozen pizza is also common. By the late 1990s, according to the American Frozen Food Institute, over 1.8 billion slices of frozen pizza were sold each year. More than one-third of these pizzas were microwavable.

Plane

Hand planes used by carpenters date back at least to the time of the Roman Empire. Made of iron, these planes were used to shape, edge, and smooth wooden surfaces. The origin of industrial planing machines, for cutting metal, may be traced to these tools as well as to the first sophisticated cutting machine, the **lathe**.

A direct forerunner of metal-cutting planes was constructed by Nicolas Focq in 1751. Operated by hand crank, Focq's machine allowed the cutting tool to travel across the surface to be planed—later planing machines reversed this process. Despite Focq's early experimentation, planes did not begin to make their mark in the **machine tool** industry until the 1810s and 1820s. Among the most prominent inventors associated with the plane are Matthew Murray, Joseph Clement, James Fox, and Richard Roberts, each of whom had by 1820 produced at least one machine that significantly advanced the precise manufacture of other specialized machines, particularly the lathe, as well as engine parts. Like their successors, these planing machines featured a reciprocating table, upon which the workpiece could be clamped and then passed beneath a stationary cutting tool.

One outstanding exception of a plane featuring a movable tool was that produced by **Joseph Whitworth** in 1835. The

Luther Burbank was one of the first horticulturists to successfully apply the principles of plant breeding to large-scale farming.

tool rest on Whitworth's machine pivoted automatically at the end of each cutting stroke to enable the tool to work in both directions. However, the demand for greater cutting power and increased stability led ultimately to the standard design in which the cutting tool is kept stationary.

Today planing machines can cut surfaces up to 50 feet long. Shaping machines represent a continuation of the movable tool concept and are employed for finer work, accommodating pieces up to 36 inches long. Hand-held metal planes complement these two industrial machines as a standard carpentry tool. Leonard Bailey, who sold his patent rights to the Stanley Rule and Level Company, is credited with the basic design, featuring an adjustable cutting angle and sturdy planing surface.

Plant breeding

Plant breeding involves the development of new plant varieties that possess traits human beings consider desirable. Breeding improves the quantity and quality of crop production. Man's earliest influence on plant selection was haphazard. The very act of organizing plants into garden plots and fields brought them into close proximity with each other and natural breeding within and among crops took place. Similarities among plants were strengthened while differences were reduced, and new varieties were created naturally among genetically similar species.

The research of German botanist Rudolph Camerarius (1665-1721) in 1694 marked the first serious attempt to understand the sexual roles played by different parts of a flower and the relationship of insects and wind to the pollination process. Since then, humanity has learned to manipulate plant reproduction in significant ways. The research of Austrian botanist Gregor Mendel (1822-1884) during the 1860s on garden peas and other plants established the basic laws of heredity and proved the existence of chromosomes, the paired rod-like units within a cell nucleus that carry genetic material.

In 1870, **Luther Burbank**, a horticulturist from Massachusetts, began his work on the development of hybrid plants. After five years of research, he bred the successful Burbank potato, introduced in Ireland to reestablish the potato crop lost in the blight of the 1840s. Burbank moved to California in 1875, where he went on to develop over a hundred new varieties of fruits, vegetables, and ornamental plants during the next fifty years, including about sixty varieties of plums. He took plant breeding out of the laboratory and brought it into the field and to the grocer's shelf.

Hybridization varies according to the way different types of plants reproduce. Some plants cross-pollinate while others self-pollinate; some reproduce asexually. Self-pollinating hybrids are achieved by producing a generation of plants that is genetically varied. Variations that possess the traits desired for the end hybrid are carefully selected, while the remainder are discarded. It may take two or more generations to produce the desired hybrid. A genetically pure hybrid can be maintained for many generations, allowing for a consistent crop. In the hybridization among cross-pollinating plants, genetic purity and uniformity are achieved through several generations of inbreeding. Corn, the best example of this, is inbred to produce vigor and resistance to pests. Commercial seed companies conduct complex series of tests to evaluate yields of various hybrid combinations to develop hardy seed stocks and to eliminate variation within crops.

Some crops reproduce both sexually and asexually. The potato, for example, can reproduce sexually by going to seed and asexually by sprouting from the "eyes." These kinds of plants are improved vegetatively—the individuals with desired traits are selected out and planted. Recent research has led to a detailed understanding of the composition of protoplasts, structures which together compose the cell. Increased knowledge of gene structures has opened the way for scientists to directly alter genes. Through **genetic engineering**, traits may be developed that might never have been achieved through selection. Gene alteration in plants can be achieved by blasting cells with a gene gun. For instance, a frost-resistant gene from an apple tree can be injected into the cells of a citrus tree, the result being individual citrus trees that remain unaffected by freezing temperatures. Genetic engineering has been successful enough in plants that altered tomatoes and potatoes are currently on the market. Research is also being done on altering bacteria which might be used in the elimination of gypsy moths, locusts, and other pests.

Past experience has shown that every technological achievement in agriculture may have environmental costs. With each step it must be determined if the gain is worth the cost. Although progress is being made in current research in genetic engineering and **cloning** of plants such as tomatoes, potatoes and peanuts, organizations such as Greenpeace and the Union of Concerned Scientists have voiced concerns about harm to consumers and damage to the environment resulting from these experiments.

See also Plant breeding; Farming, mass-production

PLANT REPRODUCTION

An essential requirement of agriculture is to understand how plants grow and reproduce.The earliest practitioners of crop production some ten thousand years ago had to learn to plant seeds and to keep them alive through watering and cultivation. Man eventually invented tools and machines that made farming more efficient and learned about plants themselves, leading to the development of crops that were of greater human benefit. The science of botany evolved from humankind's desire to know how plants live and reproduce. Their reproduction systems are functionally similar to those of animals. Most plants grown for agriculture have their sexual organs in their flowers. Whether the plants cross-pollinate or self-pollinate, it is essential that fertilization takes place for the production of seeds.

In the 1500s, German botanist Valerius Cordus (1515-1544) used the word pollen, Latin for powder or fine flour, to describe flower dust. In 1592, a Venetian named Prospero Alpini (1553-1616 or 1617) studied Egyptian flora while serving as physician to his government's consul in Cairo and published observations on the fertilization of date palms. He also disproved classical plant correlations. In the 1600s, an Englishman, John Ray (1627-1705), attempted a grand systematic description of all known plants and animals. Out of this came, in 1660, a three-volume work that listed and classified the plants in the area surrounding Cambridge. That effort was significant in that it sought to organize plants along lines of similarity and discussed, for the first time, sexuality among plants. Another Englishman, Nehemiah Grew (1641-1712), wrote in 1682 of the possibility of plant sexuality, suspecting that the parts of the flower played separate roles in plant reproduction. Rudolph Camerarius (1665-1721), a German botanist, conducted extensive research at the Botanic Garden in his home town of Tubingen. In 1694, he noted that the flower anthers, the pollen-bearing parts of the stamen, were the plant's male organs, while the style and stigma were the female organs. He not only confirmed what Ray, Grew, and others had suspected earlier, but also described the pollination process.

English botanist Stephen Hales (1677-1761) applied his interest in both plants and animals to describe the similarities in their circulation systems: that sap is to a tree what blood is to a horse or dog. His experiments led him to recognize plants' ability to absorb water and air and the importance of light in their growth. He summarized his conclusions in his book *Vegetable Staticks* (1727). The German botanist Joseph Gottlieb Koelreuter (1733-1806) published accounts of his research on plant pollen during a five-year period beginning in 1761. His successful cross-pollination of tobacco plants was his effort to prove the conclusions that Camerarius had reached earlier. Koelreuter also noted the uniformity and sterility of most hybrid plants, and observed the roles of nectar, wind, and insects in the transmittal of plant pollen. He is considered the father of plant hybridization. Another German botanist, Christian Konrad Sprengel (1750-1816), began research in 1787 on the pollination of geraniums, publishing his findings in 1793. He went further than Koelreuter by describing in greater detail flower structures and the role of insects. He discovered that the ability of some flowers to self-pollinate was attributable to the different maturation times of the anther and stigma.

German botanist Wilhelm Hofmeister (1824-1877) studied simple plants such as ferns and mosses in the late 1800s, discovering that these plants alternated generations, sexually reproducing in one generation while reproducing asexually in the next. He looked at plant cell division in microscopic detail and observed that plant ovules developed into embryos. He seems to have come close to discovering chromosomes in plant cells and is considered the father of modern botany. Gregor Mendel's discoveries of the laws of heredity in the late 1800s and the practical application of plant cross breeding by Luther Burbank in the early 1900s led to a new era in agriculture. For the first time in the long history of farming, agronomists could directly manipulate the reproduction of their crops. **Plant breeding** became an important part of mass crop production during the twentieth century. Further discoveries of gene structure—in particular the discovery of DNA (deoxyribonucleic acid)—has opened a new branch of agricultural research called **genetic engineering**.

Scientists hope that crop quality, quantity, and consistency can be achieved in the laboratory by altering gene structures and through the use of cloning techniques. Although much progress has been made in these areas of research, many individuals and groups voice concerns about the environmental impact and the effect on food safety of these processes. As man becomes more deeply involved with reproductive processes, utmost care must be taken to prevent environmental catastrophes that could affect both man and nature.

See also Plant breeding; Animal breeding; Farming, mass-production

PLANTÉ, GASTON • See Battery, electric

PLASTER

Plaster is most commonly used to coat the surface of the interior walls of buildings. Also known as *plaster of paris*, it is a mixture of solids and water that sets to a hard solid.

Plastering is one of the oldest building techniques. Early plaster was mostly mud, used to strengthen buildings against weather and animals. Over time, other materials came to replace mud. The Great Pyramids of Egypt contain plasterwork that originated over 4,000 years ago and is still durable, made of gypsum and identical to plaster of paris. Even the plastering procedures used by the Egyptians were similar to those used today.

Plaster consists of water, sand, and usually gypsum. Heating the calcium-based mineral gypsum to temperatures up to 482°F (250°C) produces a fine, white powder. When mixed with a material like sand (to provide bonding), moistened and allowed to dry, it produces the hard material with which we are familiar. (Gypsum near Paris, France was early used to make such plaster, hence the name "plaster of paris.")

Such plaster has been applied to the interior of building walls, where it provides structure, a base for decoration and painting, soundproofing, is easily cleaned, and also works as a fire retardant. The ornamental ceilings of the English monarchs Henry VIII, Elizabeth I, and James I are still admired for their artistry. When applied to the outside of buildings plaster is known as "stucco," a architectural technique first used by the Greeks. Plaster is also used in casts and molds—once poured, it hardens and expands slightly, and has been used for ornamental work on ceilings and walls, sculpture, and casts for broken arms and legs.

See also Ceramic; Concrete and cement

Plastics

Plastics are a type of **polymer** characterized by the fact that they can be molded with heat. Thermoset polymers are extremely rigid and once molded and hardened cannot be remelted. Thermoplastic materials are softer, more flexible and may be remelted. They are also recyclable.

The first plastics were often based on cellulose. One of the first plastics was developed to take the place of a material in short supply. In 1836, the Phelan and Collender company, makers of ivory billiard balls, sponsored a contest with a $10,000 prize for the inventor of synthetic ivory. **John Wesley Hyatt** and his brother set out to win the prize. Building on the research of Parkes and Schonbein they formed **celluloid** out of colloidian in 1875. Although they did not win the prize, the Hyatt brothers founded a company and successfully marketed celluloid. Alexander Parkes (1813-1890) patented his **cellulose** nitrate material called *Parkesine* in 1855. The patent was the culmination of years of investigation into rubber and cellulose nitrate solutions. Unfortunately, there was no market for it and Parkes sold the patent to the Hyatts.

Christian Friedrich Schönbein, a German chemist, announced his discovery of **guncotton**, an explosive, in 1846. He produced the guncotton by treating cotton with fuming nitric and sulfuric acids, a process simultaneously discovered by Bettger and **Nikolas August Otto**. Another product developed by Schönbein was derived from a solution of guncotton in ether. It was called **collodion** and was used in medicine and **photography**.

The first man-made fiber, **rayon**, was developed by **Comte Louis de Chardonnet**, a French chemist who devised artificial silk from cellulose nitrate. He began searching for an alternative to silk because an epidemic threatened to wipe out the silkworms in France. In 1884, Chardonnet applied for a patent for extruding cellulose nitrate through very fine **glass** capillaries. He developed the technique by making a close study of silkworms.

Edward John Bevan, **Charles Frederick Cross** and Beadle created viscous rayon when they dissolved cellulose in carbon disulfide and aqueous sodium xanthate. The mixture was passed through small holes (or spinerettes) into a dilute acid solution to regenerate the cellulose. The fibers were then spun and woven into fabrics. **Jacques E. Brandenberger** later developed a process for making films from the mixture, creating a new market—**cellophane** packaging.

In 1907, **Leo Baekeland**, a Belgian chemist, created a substitute for **shellac.** He had learned that **Adolf von Baeyer**, when mixing phenol and formaldehyde, had produced an insoluble mass. Baekeland thought that if he could find a solvent to dissolve the material he could use it as shellac. He failed to find the solvent but did end up with a castable, moldable material. He called it **Bakelite.** It is still used today for automotive applications. Bakelite was the first totally synthetic plastic in that it was not created to replace a natural product.

Around 1907, **Frederic Stanley Kipping**, a British chemist, prepared the first silicone in an attempt to produce double-bonded silicon compounds from simple carbon compounds. Nearly forty years after Ripping's research, Rochow discovered the method for obtaining **silicones** rapidly. Further research by Patnode of General Electric and Hyde of Dow Corning showed that treating silicones with acid produced an intermediate which could be polymerized using acid or basic catalysts. Silicon oils, resins, and elastomers were produced in great quantity during World War II.

Hermann Staudinger (1881-1965), a German chemist, began his study of macromolecules with **rubber**, a natural elastomer. He was one of the first to express the belief that covalent bonds were the integral bonds in large molecules. Staudinger and his research group began a systematic investigation into the formation of polymers. They succeeded in proving that polymers are chain molecules with normal valance bonds, terminated by characteristic groups. These terminating groups may be used to determine the length of the chain. Staudinger's research into condensation reactions proved that the polymerization requires the activation of a monomer, which in turn attracts other monomers. Staudinger laid much of the foundation for modern polymer science and he was awarded the Nobel Prize in chemistry in 1953.

Wallace Carothers, an American chemist conducting research at DuPont, confirmed Staudinger's view of polymers and distinguished between addition and condensation reactions. Carothers went on to develop **nylon** in 1938. His research group also developed cold drawing—a technique that straightened the polymer chains and greatly increased the strength. Dickson and **John Rex Whinfield**, British chemists, picked up the research where Carothers left off and went on to develop **polyesters** in 1941. Polyesters also benefit from cold drawing.

Research conducted by **Karl Ziegler** and **Giulio Natta** resulted in the production of high density **polyethylene** and **polypropylene**. They also discovered the organometallic and stereospecific catalysts that revolutionized the plastics industry by making room temperature and low pressure polymerization possible.

Plastics are often formed into powders or pellets before they are processed. A few may be in liquid or sheet form. Injection molding, blow molding, casting, foaming, thermoforming, laminating, extrusion and cold drawing are just a few of the ways plastics are processed. The choice of processing technique depends on the composition and state of the plastic and the type of end product being manufactured.

While plastics have greatly simplified some production processes and led to inexpensive substitutes for natural products, they have also created other problems. Plastics do not biodegrade quickly—some apparently do not degrade at all. This quality helps them last, but makes them difficult to dispose of once they have outlived their usefulness. To some extent, recycling has helped reduce this problem. Additionally, scientists have developed new polymers and additives that help the plastics breakdown more readily when they are discarded.

Player piano • See Musical instruments, mechanical

Plexiglas • See Acrylic plastic

Plimsoll mark

In the mid-1800s, Samuel Plimsoll (1824-1898), an unemployed British coal merchant, became acquainted with the unjust treatment of the impoverished, particularly with regard to the poor working conditions on British **cargo ship**s. Nicknamed ''coffin ships,'' British vessels were largely unseaworthy due to their shoddy construction and merchants' tendencies to overload them. Many lives were lost at sea as a result of these factors. Shipowners heavily insured their vessels, and so were indifferent to any loss they might suffer, whether it was a loss of merchandise or human lives. In 1873, Plimsoll expressed his outrage by publishing *Our Seamen,* a popular book that exposed and heavily criticized the gross injustices of the shipping industry. Primarily as a result of Plimsoll's actions, the Merchant Shipping Act was passed in 1876. The Act mandated strict inspection of all vessels and required the painting of a special mark, called the ''Plimsoll mark,'' on the side of each ship to indicate the depth to which that ship could be loaded.

This Plimsoll mark was actually a series of marks. It showed six loading levels that reflected different seasons and locations: tropical fresh water, fresh water, tropical sea water, summer sea water, winter sea water, and winter North Atlantic. Tropical fresh water levels were indicated by the highest line, as a ship could be more heavily loaded in calmer water. The lowest mark was the North Atlantic mark, which reflected the more dangerous water found in that area. In addition to these loading marks, there was also a circle bisected by a horizontal line with large, capital letters that indicated the registration society with which the ship was associated. The horizontal line on the registration mark also served as an indicator of the summer freeboard, the distance between the uppermost deck considered watertight and the official load line corresponding to summer sea water.

The Plimsoll mark prevented shippers from overloading their vessels, considerably decreasing the likelihood of death by drowning for crew members. Other countries soon established similar sets of markings. The United States established the Load Line Act in 1929, imposing strict limits on cargo weights and guidelines on shipping procedures. An international treaty was signed in 1968, incorporating new rules and limits for ship cargo, and reflecting the better design and construction of ships since the creation of the Plimsoll mark.

One of the ironies of the huge shipping losses during Plimsoll's time (nearly 1,000 sailors were drowned every year) was that such losses of life were occurring at a time when cargo ship technology was improving. **Iron** hulls and steam power made **ships** more maneuverable and sturdy, while both lighting and **buoys** also had been improved considerably. A century after Plimsoll's achievements however, the same irony has been allowed to prevail. During one modern 20-year period (1959-1979) when the state of technology should never have allowed such a thing to happen, losses of world shipping had doubled. Further, in the 1980s, a number of disastrous car ferry capsizings and a series of sinkings of giant ore carriers (including the British *Derbyshire* that disappeared with all hands in the Pacific), revealed a system that had allowed standards like the Plimsoll Mark to be ignored.

Illustration of a plow from the fifteenth century.

Plows

The plow—or plough—one of the oldest types of agricultural implements, is used to break up soil in preparation for planting. Modern plows are also used to bury, or plow under, crop remains at the end of the growing season.

The earliest plow was the swing plow (a plow without **wheels**) used in Egypt and Mesopotamia by about 4000 B.C. Civilizations that did not have the plow had to do their planting either by broadcasting the seed onto soil without burying it—which wastes a great deal of seed—or by planting seeds one at a time with a stick or sharp piece of antler. The first wheeled plow was developed in what is now Switzerland during the first century B.C. The wheels gave the plow weight and stability and also gave the furrow a more even depth.

The first **iron** plowshare—the share is the cutting edge of the plow—appeared among the Celts in Britain around 400 B.C. Plows made completely of wood broke easily, especially in rocky soil, and a heavier iron plow was developed in central and northern Europe to accommodate the stiffer soils in those areas. This led to greater food production and eventually a significant population increase around 600 A.D.

Innumerable plow designs were developed in England, the American colonies, and other countries according to the individual farmer's particular needs and resources. **Thomas Jefferson** is credited with the invention of several plows for use on his Virginia estate.

In 1785 Englishman Robert Ransome patented a plow with a cast-iron share. Later, in 1808, he designed cast-iron plows with detachable pieces, while in 1797 American Charles Newbold invented a single-piece all-iron plow. The problem with Newbold's design was that if any part broke, the whole plow was ruined.

In 1813, Richard B. Chenaworth of Baltimore invented a cast-iron plow with separate share, moldboard, and landside. The landside is the vertical part of the plow that guides it and absorbs the side pressure. The moldboard receives the soil and turns it over. In America, Jethro Wood of Massachusetts invented a plow similar to Chenaworth's. John Lane of Chicago patented a **steel** plow in 1833, but was unable to recover profits due to patent infringements.

In 1837, American **John Deere** (1804-1886) developed a steel plow fashioned from a circular saw blade. He was responding to the need of farmers in the midwestern United States whose iron plows could not cut through the difficult prairie soils. Deere began manufacturing the plow the following year, and his company became one of the largest farm machinery manufacturers in the world. His plow played a major role in opening the western states to settlement and agriculture.

The disc plow was first developed in the United States in 1847 but was not actually put to use until 1893. In Australia disc plows were first used in 1877. Similar to a disc **harrow**, the disc plow more readily cuts through crop stubble and rolls over field obstructions.

Gang plows with twelve to fourteen blades were invented before the American Civil War. They required teams of horses rather than single animals to maintain the necessary momentum for cutting through the soil. American F. S. Davenport invented the sulky plow, which gave the farmer a seat to ride on, in 1864.

A radical change came in 1927 with the introduction of the carried plow. This innovation allowed the plow apparatus to be mounted to the rear of a **tractor** and carried out to the field, then lowered for plowing. In 1960, the semi-mounted plow was developed to prevent the tractor from tipping. This plow was pulled to the field like a trailer with the gang raised. A rotary plow mixes surface growth with soil, while the chisel plow, has a gang of rigid shanks with shovel points and is used to loosen hard, dry soil prior to regular plowing.

See also Farming, mass-production

PLUNKETT, ROY • See Teflon

PLYWOOD

Sheets of plywood are made from thin layers of wood joined with glue. They are used to make a wide variety of products, including wall paneling; truck, airplane, and trailer bodies; furniture; cabinets; and counters.

Plywood is always made from an odd number of layers or plies (also called veneers) that are usually arranged so that the grain of one layer is at a right angle to the grain of the layer next to it. The outside layers are called ''faces'' and ''backs,'' the ones directly next to those are ''crossbands,'' and the inner layers make up the ''core.'' These layers give equal strength to both the length and width, lessen expansion and contraction, and prevent splitting.

Today plywood is made from a log that is attached to a large **lathe** that spins against a cutter blade. This process produces a large, continuous sheet, which is then cut to size. The plies are glued together with various bonding agents, including synthetic **plastics** and resins. The wood is either softwood (mostly Douglas fir) or hardwood. (oak, poplar, birch, cherry, and walnut are among the more than 80 varieties used). The final sheets are usually 4 × 8 ft (1.22 × 2.44 m) and start at one quarter inch (6.35 mm) thick.

Though its use was not widespread before the introduction of **mass production** methods, plywood was employed in Ancient China and Egypt, and in Europe around 1830 by cabinetmaker Michael Thonet, who built chairs of veneer strips glued together. Plywood was introduced in the United States in 1865 and was first manufactured in Portland, Oregon, in 1905 by U.S. Plywood Corporation.

Plywood can be variously treated during its manufacture to instill it with several desired properties, a kind of wood-made-to-order. Exterior-type plywood is made with waterproof glue—it can also be used interiorly anywhere water is present. Plywood whose glue is only moisture-resistant can withstand an occasional soaking but not repeated soakings or continuous high humidity. Fire-retardant and rot-retardant plywood is also manufactured and is used for roofing.

PNEUMATIC DRILL • See Drill

PNEUMATIC TIRE • See Tire, pneumatic

Pneumatics

Pneumatics rely on the power of compressed air. One of the best known pneumatic devices is the pneumatic hammer, or "jackhammer." A pneumatic hammer is powered by high-pressure air. Such air is delivered via a hose from an attached **air compressor**, which is a machine that increases the pressure of air from its normal atmospheric pressure to a higher pressure, by reducing its **volume**. The high pressure air drives a **piston** inside the hammer, much like the piston of an **internal combustion engine** but without the electric spark and internal explosion—an important safety factor in some explosive or wet environments.

Pneumatic devices are usually supplied with compressed air at about 90 psi (pounds per square inch), which equals 6.2×10^5 Pascals; atmospheric pressure is 14.7 psi (1.01×10^5 Pa). The jackhammer strikes the ground about 2,000 times per minute, delivering about 40,000 foot-pounds (54,000 J) of energy each minute.

An ordinary hand bellows was one of the first air compressors, used by blacksmiths for working iron and other metals. In the 17th century compressor technology was improved greatly by the German **Otto von Guericke**, and in the United States the first compressor used for large-scale work was for construction of the Hoosac Tunnell in North Adams, Massachusetts in 1866.

The use of pneumatic devices increased greatly in the 20th century. Improved engine technologies, such as the **jet engine**, stimulated use of certain types of compressors. The further development of automatic machinery, labor-saving devices, and automatic-control systems led to an increase in the use of pneumatics. The use of digital logic in the 1960s spawned a new class of pneumatic-controlled components which were used in various power and control systems.

Other pneumatic devices besides the jackhammer include rock **drills**, riveters, forging presses, paint sprayers, blast cleaners, paint mixers, atomizers, and **brakes** on railroad cars, **buses**, and large **trucks**. Hovercraft ride on a cushion of air, and some air-cushioned trains are under development in a number of countries.

Pneumatic dispatch is a method of sending closed containers through tubes by means of air pressure, The containers carrying small objects like letters, packages, and money. In 1853 J. Latimer Clark installed a 220 yard long pneumatic tube connecting the London Stock Exchange with Central Station, and a network of 3-inch tubes existed in Paris in the 19th century with railroad-like switches to route a carrier—to a small degree the network is still in use today. Such systems are also used by banks, department stores, and, increasingly, hospitals, who use it as a way to transport vital samples or specimens. Kaiser-Permanente Medical Center in Los Angeles has 15 miles of tubing that it uses to transport medical records throughout its 11-building complex. "Tube trains" have been proposed for the New York city subway system, and for a 500-mile-per-hour shuttle from Boston to Washington, D.C.

Poison gas • See Chemical warfare

Engraving of Robert Boyle working with Denis Papin on the development of the pneumatic machine.

Polar fleece

In 1979, when Malden Mills first sold their lightweight, hydrophobic **polyester** fleece under the trademark Polarfleece, they did not realize that they would be coining a term, i.e., polar fleece, that would eventually come to describe a type of fabric rather than a brand.

Malden Mills' original fabric was revolutionary because it was both lightweight and warm. It picked up less than 1% of its weight in moisture, and even when completely wet, it maintained its insulating properties. It made an ideal outerwear fabric because it actively wicked moisture away from the body.

Today the term polar fleece is applied to a class of high technology, high performance products that offer tremendous warmth relative to their weight, that are soft to the touch, and that are able to wick moisture away from the body so that they feel dry even when soaking wet. This class of fabric includes both fleece and pile fabrics, even though the two fabric types are constructed differently.

Pile fabrics emerged in the 1960s and 1970s as a result of the introduction of finer gauge, high speed, computerized knitting machines with sophisticated jacquard capabilities accompanied by the introduction of microfibers—high-crimp polyesters and fine denier **polypropylenes**. Pile fabrics are slightly lighter in weight than fleeces. The loose fibers are dyed, mixed and combed into a long fluffy rope that is guided into a sliver knitting machine. The machine permanently locks the fibers into a fine denier backing. A pile results from the individual fibers standing on end. Finally, the fabric is sheared and finished.

In contrast, fleece is made by first twisting fibres into yarns that are then knitted into a fabric. The fabric is then brushed vigorously to loosen some of the fibers to create a fuzzy (fleece) surface that is then sheared and finished.

In 1998 there were a half dozen or so North American manufacturers of polar fleece; these included Malden Mills, Dyersburg Corporation, Menra Mills, Huntington Mills, Siltex Mills, and Glenoit Mills, in addition to numerous offshore producers.

The history of one leader in the fleece market, Dyersburg Corporation, parallels that of the industry. Begun in 1929 as a family-owned business, Dyersburg originally made **cotton** sweaters, long johns and cotton fleece gloves. In the 1930s, they introduced the first knitted fleece fabric, a wool blend. By the end of the 1960s, they had expanded into pile fabrics. In 1972, they followed up with the introduction of the first flame-retardant fleece made of Monsanto's SEF Modacrylic. By 1976 they had become the leader in acrylic fleece production. During the 1980s they emerged as a fleece specialist. Responding to changing market trends, they put their expertise in surface finishing to work, creating new cotton-rich fleece products. In 1993 they introduced the world's first post-consumer recycled polyester fleece made from recycled plastic soda bottles.

See also Dacron; Soda pop; Textiles; Waterproof materials; Wyeth, Nathaniel

Polychlorinated biphenyl (PCBs)

Polychlorinated biphenyls (PCBs) are organo-halogens, a class of chemical compounds that contain two or more chlorine atoms attached to a bonded pair of benzene rings. When improperly disposed or burned, PCBs are highly *carcinogenic* (cancer causing) and *lipophilic*. This means they persist indefinitely in the body fat of living organisms and are easily passed on to other creatures through the food chain. First developed in 1881, PCBs were introduced into American industry on a large scale in 1929, and since then their commercial use has been widespread. They are nonflammable, chemically stable, heat resistant, and have low electrical conductivity. PCBs are also thick, odorless, colorless, and inert. These properties made PCBs very attractive to industry.

PCBs were first used to insulate electrical transformers and as hydraulic fluids. Their applications multiplied as electricity became the primary industrial power source, and eventually included **transformers** (especially those used on railway locomotives and self-propelled cars), **capacitors**, **circuit breakers**, voltage regulators, and switches. At one time, they were even used in newsprint ink. Unfortunately, their great stability, so helpful to industry, is also a drawback because it prevents PCBs from deteriorating.

These dangerous chemicals are pervasive in the environment. They have been measured as far away as the polar ice caps, and can be detected in practically every living organism from plants to animals; trace amounts have even been found in mother's milk. Although they can be absorbed through the lungs or the skin, most PCBs enter the human body through the food chain in contaminated plant and animal products. Scientists believe that people can tolerate small amounts of PCBs in their fatty tissues without damaging effects. A 1981 study revealed that virtually 100% of those tested had detectable levels of PCBs. In large concentrations, their buildup can cause such toxic reactions as digestive-tract disturbance, liver dysfunction, numbness in the hands and feet, decreases in reproductive capacity, and changes in blood chemistry.

Previously, the only known way to destroy PCBs is by incinerating them. This had to be done at extremely high temperatures however, or the PCBs would mutate and form even more toxic compounds, the most infamous of which are dioxins. Ineffectively burned PCBs have been known to contaminate entire buildings to such an extent that people could not enter them without wearing protective clothing. Because of their potential hazards, PCBs have been banned in the United States since 1979 and in Great Britain since 1980.

A new method which is being studied uses a device called a Destructor. This device may one day be used to clean-up existing PCB found in household appliances all over the industrialized world. The Destructor is a mobile device that uses a closed-loop chemical process to reduce PCBs. It works by heating the PCBs to a gaseous state and then soaking the vapor with hydrogen. The hydrogen displaces the chlorine an bonds to the carbon on the molecules. When sodium bicarbonate is added, an environmentally-safe salt is produced. Other byproducts include methane and carbon dioxide.

Polyester

Commercial polyesters are long chain molecules, or polymers, of which eighty percent consist of an ester derived from a dihydric alcohol and terephalic acid. A great deal of the groundwork for the development of polyesters as **textile** fibers was prepared by **Wallace Carothers** and his research staff. Carothers went to work at DuPont as the head of a basic research program in 1928. One of the goals of the program was to develop a synthetic fiber. He began to study diamines and carboxylic acids. Through this search, he eventually discovered neoprene, a synthetic rubber, and **nylon**.

Carothers and his staff also investigated polyesters, which are polymers resulting from the interaction of dialcohols and dihydric acids. The researchers eventually abandoned polyesters as useless synthetic fibers—finding them lacking in strength and possessing melting points that were too low. However, Carothers's group had not tried the polyester formed from the combination of a dihydric alcohol and terephalic acid.

John Whinfield and J. T. Dickson, two British chemists, began their research where Carothers left off. They discovered that by making a polyester with terephalic acid and by using a fiber-forming method (pioneered by Carothers's group) called *cold drawing*, a fiber suitable for use in textiles was formed. They called the new fiber Terylene.

In cold drawing, the polymer is heated and forced through a metal plate with small holes. The polymer fibers are solidified with cool air as they emerge from the holes. They are then passed through a pair of rollers that rotate at different speeds. The greater the difference in the rollers' speeds the stronger the fiber becomes. The drawing process aligns the molecular chains in the polymer. The fiber was marketed by the Imperial Chemical Industries in Great Britain and by DuPont in the United States under the names Fibre V and **Dacron**. Polyesters are used mostly in textile applications, both alone and in blends. Like many discoveries, polyester was developed by one group building upon the research of another.

See also Fiber, synthetic; Plastics ; Polymer and polymerization; Rubber, synthetic

Polyester leisure suit from the 1970s.

Polyethylene

Polyethylene is a long chain **polymer** produced by the polymerization reaction of liquid ethylene at high temperatures and pressures. It is a solid **plastic** which has a milky transparent appearance.

The discovery of polyethylene in the 1930s greatly aided the Allied war effort in World War II. Polyethylene was used as insulation for cables vital to the Allied information network. Polyethylene was accidently discovered by J. Swallow and M. Perrin at Imperial Chemical Industries in Britain. In 1933, while researching the effects of high pressure on chemical reactants, a fellow scientist, R. Gibson, managed to produce a waxy solid from ethylene and benzaldehyde. He repeated the experiment with ethylene alone with no success. The experiment had taken place at 280°F (138°C) with ethylene at a pressure of 1,400 atmospheres. Theorizing that a higher pressure was needed, the three colleagues set to work designing and building improved laboratory equipment.

Nearly three years later, Swallow and his companions tried the experiment at a higher temperature. In the course of the polymerization process the pressure dropped and they added more ethylene to compensate. The procedure yielded approximately eight grams of polyethylene. After inspecting the equipment they discovered that there had been a leak. The polyethylene they had added contained oxygen. Further investigation revealed that the oxygen was vital to the polymerization process. The polyethylene, like polyesters, benefitted from the cold drawing technique developed by **Wallace Carothers** at DuPont.

Polyethylene, or polythene as it was marketed in Britain, began as a polymer with little practical use. It would have remained so if an employee had not noticed that the mechanical properties of polyethylene were comparable to those of *gutta percha*, a natural product used to insulate **telegraph cables**. Polyethylene was used to insulate the cables laid between France and Britain, providing a crucial line of communication towards the end of World War II. After the war, polyethylene film was used for packaging, liners, tank and pool covers, and drop cloths.

The early polyethylene was a low density branched polymer. This means that there was a good deal of empty space in each molecule, and the molecules were formed in a branching pattern. These characteristics accounted for the fact that this polyethylene was not a very strong material. However, in 1953, German scientist **Karl Ziegler** and his staff discovered a method of producing high density, linear polyethylene (whose molecules had less empty space and were arranged in rope-like strands, both of which give a substance greater tensile strength). Ziegler was researching organometallic compounds (carbon compounds which contain metals) and their effects on polymerization reaction. He attempted to polymerize ethylene using catalysts at a much lower pressure than that used by the British process.

At first, Ziegler was puzzled when an experiment yielded a dimer (a compound of two radicals, rather than the several radicals which compose a polymer) of polyethylene rather than the expected low molecular weight polymer. He discovered trace amounts of nickel on the laboratory equipment had inhibited the reaction. As a result of his investigation of this occurrence, Ziegler eventually succeeded in producing a very high molecular weight polyethylene with a very high melting point using metal chloride and organoaluminum compounds as catalysts. Polyethylene could now be carried out at low temperatures and normal pressure, which greatly simplified the industrial production process.

High density polyethylene is currently used in dishes, squeezable bottles, and other soft plastics products. It is recyclable. Linear low density polyethylene (LLPE) now substitutes for the older branched low density polyethylene (LDPE). LLPE is formed using the Ziegler process using hydrogen to regulate molecular weight. All of the various polyethylenes may be processed by injection molding or extrusion.

See also Polymer and polymerization

POLYGRAPH • See Lie detector

POLYMER AND POLYMERIZATION

A polymer is a high molecular-weight material, either synthetic or natural, that is composed of repeating units called monomers. In some cases a natural polymer like rubber has been replaced on the market by a synthetic polymer that closely resembles, or even duplicates, the molecular structure of the natural model. The synthetic copies are usually altered to emphasize the desired properties. Types of synthetic polymers include **plastic**s (thermosets and thermoplastics), elastomers, fibers, coatings, and adhesives.

Thermoplastic materials can be formed and then remelted due to their long flexible carbon chains. Examples include ABS (acrylonitrile-butadienestyrene), cellulose acetate, **nylon**, **polyethylene**, **polypropylene**, **polystyrene**, PVC (polyvinyl chloride), and polyvinylidene chloride. These materials are also the most easily recyclable plastics.

Thermoset plastics are those polymers whose long carbon chains are bonded together like a rigid net. Well known thermoset plastics include allylic, epoxy, melamine and urea-formaldehyde, **polyester**s, **polyurethane**, and **silicone**s.

Elastomers are rubber-like materials characterized by their tendency to return to their original shapes. They are flexible to varying degrees and are used in gaskets, tires, and tank liners. Examples include neoprene, isoprene, SBR, ethylene and propylene polymers, silicon rubbers and styrene block copolymers. Examples of fibers include **rayon**, nylon, polyesters, spandex, and acetates, while polymer coatings have been based on synthetic resins— **vinyl** acetate acrylic resins are the most popular. **Adhesives** have made use of nearly every type of polymer, and many polymers developed for coatings or elastomers were eventually developed into adhesives.

Copolymers are polymers formed from two or more different monomers. The monomer groups may be randomly interspersed or grouped in blocks. In some polymers—urethanes for example—the blocks form soft and hard segments in the polymer chain, the mixture of which imparts strength as well as flexibility. Polymers may also be grafted onto one another.

Polymers may also be atactic, isotactic, or syndiotactic (stereoregular). These terms refer to the position of the pendant groups in relation to the main carbon chain. Atactic polymers have randomly arranged pendant groups. The groups alternate in syndiotactic polymers, while isotactic polymers have all pendant groups on the same side. **Giulio Natta** and **Karl Ziegler** were instrumental in discovering the catalysts that could control the positioning of the pendant groups. Their positions affect bonding, reactivity, and melting point temperatures.

Polymerization may take place by chain (addition) or step (condensation) reactions. Chain polymerization is very rapid, and once a polymer is terminated, it will not react with monomers or other polymers. On the other hand, in step polymerization each polymer will continue to react with monomers or polymers. Chain polymerization starts with an initiation reaction that forms a free radical, carbonium, or a carbonium ion. The next step, called propagation, is very rapid. A monomer is added to the end of the polymer. This is repeated many times until termination. Polymers may be processed through injection molding, extrusion, blow molding, thermoforming, rotational molding, compression and transfer molding, or foaming.

POLYPROPYLENE

Polypropylene is a polymeric material produced by the polymerization of propene. It is a solid material that typically has a molecular weight of over forty thousand. It is a thermoplastic polymer and is softened by solvents as well as by heat, chlorine, fuming nitric acid, and strong oxidizing agents. Ultraviolet light can degrade it. The polymer is easily dyed and resists fungi and mold.

Although **Giulio Natta** was the man who actually developed polypropylene, he drew largely upon the breakthroughs of **Karl Ziegler**. Ziegler and his collaborators discovered that organometallic compounds (carbon compounds containing a transition metal) were catalysts for **polymer** reaction. In 1952 Ziegler succeeded in preparing a high molecular weight **polyethylene** at room temperature.

Natta, working for the Italian company Montecatini, set out to develop a high molecular weight polypropylene, choosing propylene because it was much cheaper and more readily available than ethylene. Natta sent workers to Ziegler's research lab to learn how to work with catalysts. By 1954 he and his research group created a polypropylene with high durability, heat resistance, and tensile strength. It also had another property that they had not anticipated.

All of the methyl groups along the carbon chain of the polypropylene were on the same side. This kind of polymer is now called an *isotatic* polymer. Natta and his group continued their research until they were able to use catalysts to place the pendant groups of the carbon chains where they wanted them. This catalyzed polypropylene turned out to be the first of many stereoregular or syndiotactic polymers (polymers whose pendant groups attach themselves to alternate sides of the molecule).

Polypropylene is used in films, automotive parts, appliances, housewares, **carpet** fibers, crates, labware, fish nets and surgical supplies. Its discovery led to a change in the perception of polymers. Until that time, there had been substances whose structures could not be predicted; the discovery that stereoregularity could be controlled, however, ushered in a new era of ''designer'' **polymers**.

See also Fiber, synthetic ; Plastic

Polystyrene

Polystyrene is a **polymer** consisting of styrene molecules covalently bonded in a long carbon chain. It is a thermoplastic polymer which means that is flexible and may be deformed by heat. It can be colorless to yellow and it has a penetrating odor. It has excellent electrical insulation properties, a relatively high resistance to water and a high refractive index.

Polystyrene was first synthesized by E. Simons of Berlin in 1839. Simons, an apothecary, distilled storax resin with a solution of sodium carbonate. The product was an oily residue that Simons called styrol because he assumed that that he had made styrene oxide. After Simons's first discovery there was a flurry of activity among the German scientific community as they tried to pin down the structure of styrol. J. Blyth and A. W. Hofman confirmed that the carbon-to-hydrogen ratio of the compound was the same as in benzene and a compound called cinnamol. This suggested that cinnamol and styrol, or styrene, were the same compound. This hypothesis was lent credibility by H. Kopp's work showing the two compounds had very similar physical properties. The final barrier in confirming the two substances were the same was that styrene prepared by Simons' method was optically active whereas cinnamol was not. In 1874 Jacobus Van't Hoff (1852-1911) finally proved that the optical activity of styrene was due to impurities.

The full scale production of polystyrene would not have been possible without a crucial discovery by Charles Friedal (1832-1899) and James Crafts (1839-1917). In 1877, while attempting to produce amyl iodide by treating amyl chloride with aluminum and iodide, Friedal and Crafts discovered that the reaction produced large amounts of hydrocarbons. Until that time hydrocarbons could not be produced quickly. Their discovery showed that aluminum chloride is a catalyst that facilitates the production of ethylbenzene with fewer impurities. With fewer impurities the reaction proceeds quickly, making the production of styrene—and therefore polystyrene—more economically feasible.

Polystyrene is a very versatile material and it may be formed into sheets, beads, foamed, injection molded or extruded. People encounter it nearly every day in the form of coffee cups, coolers, **insulation** and, until recently, fast food packaging. Foamed polystyrene has long been a staple for cheap thermal insulation. Polystyrene is also a frequent additive to **plastic** housewares, toys, and furniture. Its monomer (single unit) form, styrene, is a major component of *acrylonitrile-styrene-butadiene rubber* (ASB), which is used in many automotive applications.

Polystyrene foam has fallen out of favor in the past few years. About 90 percent of the **chlorofluorocarbon**s (CFCs) used to make the resin foam are released into the atmosphere during production thus posing a significant threat to the ozone layer. The foam's disposal poses even more problems. Over 25 billion foam cups are thrown away every year in the United States alone. Incineration of these produces toxic air pollution and hazardous ash. Burying them uses precious landfill space. To solve these problems scientists have investigated the possibility of recycling the material. This has led to the use of the foam for a variety of plastic products.

Polytetrafluorethylene (PTFE) • See Teflon

Polyurethane

Polyurethane is an elastic polymer formed by a condensation reaction between diisocyanates and dihydroxy compounds, usually alcohols. This results in a molecule consisting of a rigid urethane group alternating with a soft chemical group. These alternating soft and hard segments are responsible for the elastic properties of the urethanes. Urethane elastomers—substances resembling rubber—are made by using an ether or ester type constituent in one of the prepolymers. Urethane elastomers resist ozone but are easily degraded by acids, alkalies, and steam.

Polyurethane might never have been discovered if not for the scarcity of rubber during World War II and the success of Du Pont's **nylon**. During the 1940's, Du Pont started a project to develop alternatives to rubber. In 1947, **Adolf von Baeyer** published a paper detailing the research he and his team had accumulated on diisocyanates and dihydroxy compounds used to form polyurethanes and polyureas from diisocyanates and diaminos. Von Baeyer stated that the original purpose of the research was to develop a substitute for nylon without violating any of Du Pont's patents. When scientists found that these polyurethanes could be made into fine threads the usefulness of these materials were soon realized.

Polyurethanes are used to produce **spandex** clothing which is lightweight and form-fitting. They are also used in solvent-resistant coatings, abrasive resistant **rubbers**, **fiber**s and foams. The latter currently represents the biggest market for polyurethanes; the foams vary from very flexible to extremely rigid and are used in furniture, **carpet** backing, and insulation. The rigidity of the foam is determined by the degree of crosslinking that occurs. Lower molecular weight reactants tend to produce more rigid foams.

See also Polymer and polymerization

Polyvinyl chloride (PVC) • See Vinyl

Popcorn

Popcorn was discovered by Native Americans over five thousand years ago and predates corn eaten off the cob. Archaeologists have discovered 5600-year-old popcorn in New Mexico's Bat Cave and a thousand-year-old popped kernel in a dry cave in Utah. This unusual food fascinated the early explorers. Christopher Columbus (1451-1506) brought a popcorn lei from the Caribbean Indians to the court of Ferdinand (1452-1516) and Isabella (1451-1504). Hernando Cortes (1485-1547) noted in 1510 that the Aztec Indians of Mexico used popcorn in religious rituals. The brother of Chief Massasoit (1580-1661) brought popcorn to the Pilgrims at the first Thanksgiving dinner in 1621, and Pilgrim women began serving popcorn with warm cream for breakfast.

Methods of popping corn have varied over the centuries. Indians held ears of corn over fire, then tried removing the kernels from the ears when they popped. Later Indians covered the kernels with hot sand and dug them out when they were done. Some Indians believed that tiny demons inhabited the kernels of corn and caused them to explode when heated.

Eventually, a less exciting explanation was determined as the cause: The kernels pop when the water within the kernel is heated and turns to steam. When the steam builds up enough pressure, the hard shell of the kernel gives way and explodes into a white starchy mass. A kernel must contain at least fourteen percent moisture in order to pop. Kernels that do not pop are often referred to as old maids or duds.

In 1885, Charlie Cretors invented a popcorn machine that could be pushed on foot, pulled by a horse, or mounted on a truck. This made it easier for vendors to reach their customers. Today much of the popcorn sold at movies and fairs is still made in poppers manufactured by the Cretors family. Home poppers were first introduced around 1925. Two decades later, Percy Spencer discovered that popcorn would pop when placed under microwave energy. This led to experiments with other foods and the invention of the **microwave oven**. Microwave popcorn remains a popular use for the appliance.

In 1952, agronomists Charles Bowman and Orville Redenbacher developed a hybrid corn that produced a low percentage of unpopped kernels. Popcorn companies rejected this new popcorn, claiming that consumers would not pay more for this "gourmet" popcorn. Redenbacher persevered and proved the large manufacturers wrong when his popcorn became the leading brand in America.

The idea of selling popcorn in movie theaters is attributed to Jacob Beresin of Philadelphia, Pennsylvania. Beresin worked at Philadelphia's Metropolitan Opera House and augmented his salary by selling popcorn and other snacks during intermissions. He worked so diligently that he and his partner, Edward Loeb, were soon selling snacks in all nine of the city's theaters. With the advent of silent movies and then "talkies," Beresin and Loeb's business grew steadily.

Today over a million pounds of popcorn are eaten in American movie theaters and living rooms each day, and the average popcorn eater consumes about two pounds of popcorn annually. By 1998, according to the Popcorn Institute, microwave popcorn alone accounted for $240 million in yearly U.S. popcorn sales.

In the 1990s, health concerns were raised about the fat content of popcorn that is popped in oil and coated in butter. In response, some movie theaters switched from oils that are high in cholesterol-raising saturated and trans fats, such as coconut oil and hydrogenated canola oil, to healthier oils, such as liquid corn oil. Health-conscious consumers can now make air-popped popcorn at home, and microwave popcorn is sold in reduced-fat versions as well as many flavors.

Popov, Aleksandr Stepanovich (1859-1906)

Russian physicist

Had Aleksandr Stepanovich Popov stayed with his initial vocation, he would never have been involved with **radio**.

Born on March 16, 1859, in Turinsk, Russia, Popov followed in his father's footsteps and studied for the priesthood. During his seminary years, he became interested in engineering, physics, and mathematics. When he graduated from the seminary, he decided to attend the St. Petersburg University's Faculty of Physics and Mathematics.

When Popov graduated in 1882, he was offered a teaching position at the university. He refused because there was no opportunity to become involved in electrical engineering research. Instead, he became an assistant at the Torpedo School in Kronstadt, Russia, teaching electricity and magnetism to naval specialists.

In 1888, Heinrich Hertz (1857-1894) discovered invisible electromagnetic radiation. Two years later Édouard Branly (1844-1940) discovered that electrical resistance in metallic powders decreased when subjected to an electromagnetic discharge; in other words, the powders stuck together. **Oliver Lodge** used this discovery to invent a device to detect electromagnetic waves. Called a coherer, the device contained metal filings that cohered under the influence of electromagnetic waves.

Popov further improved the coherer. By attaching an **antenna** to the device, he showed it was possible to receive "Hertzian waves" that had been transmitted a distance of 260 ft (80 m). Obviously, this was a short distance, but it was only the beginning.

On May 7, 1895, Popov gave a public lecture in which he made one of the first demonstrations of electromagnetic wave reception. The following January he published a paper on the subject and included a detailed diagram of his tuning circuit. He also wrote of "the hope that my apparatus...may be used for the transmission of signals over a distance...." Lo and behold, in the autumn of 1896 the first printed notice of **Guglielmo Marconi**'s radiotelegraphy invention appeared. Marconi's 1897 application for a patent included a diagram of his apparatus, which nearly coincided exactly with Popov's January 1896 description. The Russian Physicochemical Society claimed, perhaps justifiably so, that Popov should be considered the inventor instead of Marconi. Though he is considered, along with Lodge and Marconi, to have independently devised the principles of radio, Marconi's later achievements eclipsed Popov's initial success.

In any case, Popov was still the first to use an antenna to send long distance signals. In 1898, he transmitted a **ship**-to-shore message nearly six miles (10 km). This was three years before Marconi would send his famous trans-Atlantic communication. Unfortunately, Popov became sidetracked at this point. He was able to get the Russian navy and army to install radio equipment, but he began to spend more of his time using his receiver to study the physics of thunderstorms by detecting lightning strikes. This gave Marconi the time he needed to establish himself.

Popov went on to become a professor at the St. Petersburg Institute of Electrical Engineering in 1901, eventually becoming its director in 1905. His health began to fail, and he died on January 13, 1906. His "hope" of long-distance communication has been realized far beyond his expectations.

Aleksandr Stepanovich Popov

PORCELAIN • See Ceramic

POSITRON EMISSION TOMOGRAPHY

Before about 1970, physicians had few techniques with which to examine the internal structure of the living human body. Wilhelm Conrad Röntgen in 1895, had been utilized for decades, but could only show bones (which absorb x rays) and some soft tissue around them. Viewing other organs was difficult.

In the early 1970s the field of *medical informatics* was developed by **Robert Ledley** and others, in which **computers** and other information technologies aid physicians in diagnosis. South African physicist Allan M. Cormack and British engineer Sir Godfrey Hounsfield independently developed the field of x-ray computed tomography, referred to as simply CT, or CAT. Computed tomography uses the fact that different tissues absorb varying amounts of x-ray energy—the denser the tissue, the more energy absorbed.

An x-ray beam sent through tissue will therefore exit with a reduced energy, depending on what tissue and organs it went through. The information obtained from many x rays passed through the body at many different angles can provide a picture of the body section. The CAT scan was superior to x rays, but still suffered from some soft tissue x-ray absorption. However, its development did provide scientists with the mathematical framework and computer algorithms necessary to reconstruct images from scattered beams.

Scientists next developed scanning methods using biological *tracers*. A tracer is a compound made radioactive by replacing one of its atoms with a radioactive atom. When the tracer is placed in the body through a blood vein, it will accumulate in an organ (such as the brain), and when it gives off a gamma ray (a high energy photon), that particle can be detected outside the body. Much like the CAT scan, multiple photon scans are performed and an image is reconstructed. This single photon emission computerized tomography (SPECT) has several drawbacks, as many of the photons either emitted are missed by the detectors or located only within a cone roughly centered around the location of their creation. Also, there is a limit to the amount of radioactive tracer that can be safely injected into the body.

Positron emission tomography (PET) combines the principles of CT and SPECT together. PET uses tracers too, but these tracers emit positrons, a particle with all the properties of the electron except for its positive charge. Common positron-emitting isotopes used as PET tracers such as fluorine-18, carbon-11 and a form of water, hydrogen combined with oxygen-15. When these tracers are injected into the body, the emitted positron comes to rest, and, attracted to a charge of the opposite sign, annihilates with an electron. This event produces two photons which fly away in opposite directions (so as to conserve momentum), and the PET detectors detect both, which arrive at nearly the same time but from opposite directions. CT methods are then used to make an image. This detection along a line makes PET about 10 to 50 times more sensitive than SPECT.

Positron emission tomography has proved especially useful for studying brain and heart functions. PET has been used to show which regions of the brain are active at any one time, and has provided real-time images of the physical changes associated with thought processes—a mapping of the neural systems responsible for thought. A typical PET can locate changes in activity to within just a few millimeters.

Since about 1997 scientists have been using yet another imaging stragegy, functional magnetic resonance imaging (fMRI) to map brain functions, mostly replacing PET for this particular application.

See also Computerized axial tomography; Magnetic resonance imaging

POSTAGE STAMP

Before the invention of the postage stamp, letters were usually paid for by the person receiving them. If the sender did prepay the postage, the letter would be marked "Paid" by hand or by

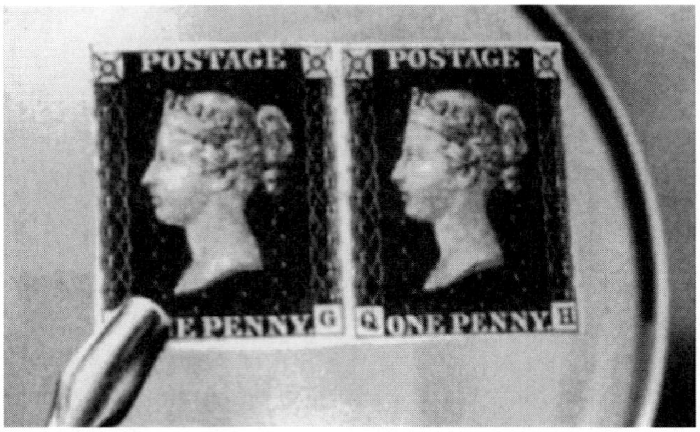

Introduced in 1840, Great Britain's "Penny Black" was the first postage stamp.

a hand stamp. Sometimes, wrappers preprinted with a stamp were used. Postage rates were high, and varied depending on how far a letter had to travel and how many sheets the letter contained.

A retired English schoolteacher, Rowland Hill (1795-1879), published a pamphlet in 1837 titled *Post Office Reform: Its Importance and Practicability,* in which he demonstrated that the cost of delivering a letter in Great Britain had less to do with the distance sent than with the amount of time spent for handling and calculating and collecting postage. He proposed an inexpensive, universal postage rate based on the weight of the letter. Hill also suggested that postage should be prepaid by the use of envelopes stamped as such, and "by using a bit of paper...covered at the back with a glutinous wash, which the bringer might, by the application of a little moisture, attach to the back of the letter." An Irishman, James Chalmers of Dundee, Scotland, had in fact produced some experimental **adhesive**-backed postage stamps in 1834.

The British government soon adopted Hill's proposals. Penny post became effective in January 1840. The first postage stamps went on sale on May 1, 1840, printed by the American firm Perkins, Bacon, and Petch, using a process invented by Jacob Perkins (1766-1849). The one-pence stamp carried a portrait of young Queen Victoria (1819-1901) on a black background; it became popularly known as the Penny Black. Its mate was called the Twopenny Blue. Pre-stamped envelopes also went on sale, but the individual postage stamp was far more popular.

The use of prepaid postage stamps rapidly spread worldwide. Brazil and several Swiss cantons issued postage stamps in 1843, followed by Mauritius and Trinidad in 1847. Privately issued postage stamps appeared in the United States in 1842, and the federal government issued the first official United States postage stamps on July 1, 1847—a five-cent stamp bearing a portrait of **Benjamin Franklin** (the first Postmaster General of the United States) and a ten-cent stamp featuring George Washington.

The first postage stamps were not scored for easy separating; purchasers had to cut them apart with **scissors** or a **knife**. The Englishman Henry Archer invented a machine that, once perfected, produced the first perforated stamps in 1854. Today, postage stamps are sold with an adhesive already attached so there is no need to provide moisture, one simply peels the stamp from its paper backing and attaches it to the envelope. People collect early postage stamps as well as special-issue commemorative stamps. Businesses that have vast amounts of correspondence usually acquire special permits for postage meters that permit them to stamp and cancel their mail.

Post-it note

Post-It Notes were originally yellow pieces of **paper** with a sticky edge used for stick-on messages in offices and homes everywhere.

Post-Its were the inspiration of Arthur Fry, a chemist at the Minnesota Mining and Manufacturing Company (3M). One Sunday in 1974 at the North Presbyterian Church, Fry began considering better ways to mark his place in his hymnal; the slips of paper he used kept falling out, and he could not find the right hymn when it came time to sing. Suddenly he thought of a failed experiment of a 3M colleague, Spencer Silver. Trying to develop a super-strong adhesive, Silver had produced a super-weak one instead—just the thing for his purposes, Fry reasoned.

Fry experimented with the glue, adjusting the formula so the notes would not leave a residue when peeled off, and passed out samples to coworkers until he convinced 3M management to promote the product as notepaper. "Press & Peel" was test-marketed in 1977. The notes did not become popular until 3M promoted their use by distributing free samples. A flood of samples and advertising in Boise, Idaho, in 1978 produced heavy sales, and national marketing began in 1979.

Today, Post-It Notes come in a variety of colors and sizes, are a best-seller for 3M, and seem indispensable to legions of devoted users.

Potato chip

The potato chip was created by a frustrated chef. George Crum, an Adirondak Indian chief and chef of the Moon Lake House Hotel in Saratoga Springs, New York, was exasperated by a customer who complained repeatedly that his french fried potatoes were sliced too thick. Crum decided to respond with an exaggeratedly thin and unappetizing potato. He sliced the potatoes paper thin, boiled them in fat, triple salted them, and then watched as Commodore Cornelius Vanderbilt (1794-1877) sampled the chips. Much to Crum's chagrin, Vanderbilt loved them. In fact, customers began requesting them, and "Chief George Crum's Saratoga Chips" appeared on the menu the next day.

Crum's Saratoga chips became popularly known as potato chips around the turn of the century. In 1925, A. A. Walter and Company built the first potato chip production plant in Albany, New York.

In the 1960s, Proctor and Gamble initiated research aimed at producing better potato chips and introduced *Pringle's Newfangled Potato Chips* in 1969. These chips differed dramatically from earlier types, as they were made from potato granules that were moistened, rolled in sheets, cut into uniform bits, and fried. Because these chips were identical in size and shape, they could be stacked and packaged in cylinders. The Potato Chip Institute International complained to the Food and Drug Administration (FDA) that these new chips should not be identified as potato chips, and in 1975 the FDA ruled the processed potato product had to be identified as "potato chips made from dried potatoes."

Chips are also popular in Europe where people enjoy paprika-, beef-, and chicken-flavored chips. In England they are called potato crisps to avoid confusion with the English chips, which more closely resemble the American french fried potato.

Though there is an ever-growing number of snack foods available in America today, potato chips are still a favorite. In fact, almost 90 percent of U.S. households buy potato chips each year, and 11 percent of America's potato crop is used for the production of chips.

To satisfy an increasingly health-conscious public, many varieties of baked, unskinned, and unsalted chips now appear on supermarket shelves. In 1998, potato and tortilla chips became the first foods on the national market made with *Olean*, a **fat substitute** from Proctor & Gamble. Potato chips fried in *Olean* cooking oil contain no fat and only 75 calories per one-ounce serving, compared to 10 grams of fat and 150 calories in regular chips.

Potentiometer

A potentiometer is used for precise measurements of voltage, and sometimes current. It also can be used as a three-terminal potential (voltage) divider, which is a type of rheostat. Nearly every type of electronic instrument uses a potentiometer. The most common use is for volume control for **radio**s, **television**s and other **amplifier**s.

Potentiometers make use of *rheostats,* which are variable **resistor**s. A rheostat is a coil of **wire** that has been wrapped around an insulating tube and attached to two terminal points. Next to the coil are metal "brushes" that make contact with the coil. When a knob (or "slider") is moved, it causes the position of the brushes on the coil to change, resulting in the resistance between them, and so the terminals of the coil, to change. Heat is produced by resistance, but a rheostat can handle large amounts of current because the coil easily radiates the heat away.

The potentiometer works by balancing one voltage against another. If two voltages are connected together so that one opposes the other, the electromotive force (emf or "driving force") is the difference in the two voltages. A **galvanometer** internal to the potentiometer can be added to the circuit, and a reading of the current flow and the inbalance between the voltages can then be made. If the galvanometer shows no deflection, the voltages are equal and completely balanced. In practice one of the voltages is known, usually by a "standard cell" (a cell whose emf is known). A standard cell has an absolute value which remains stable for years.

The *sliding-wire potentiometer* makes use of a taut wire, which has uniform resistance, that is attached between two terminals. The unknown voltage located at any point on the wire is proportional to the distance from the terminal. Two leads are taken from the source of the voltage. One lead is attached to the end terminal of the wire; the other end is attached to a hand-held contact after passing through a galvanometer. When the contact touches the wire, the galvanometer registers the imbalance. The contact is moved along the wire until the galvanometer reads zero. This procedure is repeated substituting the standard cell for the unknown voltage and a new distance is measured. With the voltage of the standard cell known the two distance measurements are made, and it is possible to calculate the unknown voltage. This is essentially how the **Wheatstone bridge** measures resistance.

A "voltage divider" is used when large voltages are to be measured. The divider is comprised of a series of resistances which are tapped at various points along its length. After the voltage is reduced the potentiometer can be used to measure the remaining voltage.

See also Electric circuit

Potter's wheel • See Ceramic

Pottery • See Ceramic

Poulsen, Valdemar (1869-1942)

Danish engineer

Valdemar Poulsen was a Danish engineer who invented the magnetic recorder in 1898. His ideas were the basis of **magnetic recording** and led to crucial developments in communications and computer technology. Born in Copenhagen on November 23, 1869, Poulsen exhibited an early interest in science. He later studied various aspects of science in depth, though he never earned a degree. Around 1890, Poulsen began working in local machine shops, finally joining the Copenhagen Telephone Company as an assistant engineer in its technical section in 1893. It was while working there, at the age of 20, that he developed the telegraphone, a machine that recorded sounds magnetically on a **wire**.

The concept of recording sounds magnetically had originated with the British inventor Oberlin Smith (1840-1926) in 1888, but Poulsen put that concept into practice ten years later. Smith's ideas had involved using "tapes" of fabric containing **iron** filings, but Poulsen's device used magnetized **steel piano** wire to record and reproduce sounds. The wire, running between spools, could record continuously for 30 minutes, moving at a speed of 84 inches per second. Poulsen conceived the telegraphone initially as a message-taking machine for use with telephones; messages could be erased at will. He applied for a patent on the device in 1898.

While a working model created great interest at the Paris Exhibition in 1900, its impact did not reach beyond that initial presentation. Since Poulsen could not find financial backers in Europe, he came to the United States in 1903 and obtained benefactors who founded the American Telegraphone Company (in Springfield, Massachusetts) to produce and sell his machine. The telegraphone was promoted as an office dictation apparatus and an automatic **telephone** message recorder, but it did not find wide application due to its drawbacks: poor amplification via earphones and unwieldy bulk. Inept company management also affected the product's success. However, Poulsen's system was sound in principle and laid the basis for later inventors (such as **Marvin Camras**) to develop commercially successful magnetic recorders.

Poulsen had also described methods to magnetize steel or **paper** tape coated with magnetic powder. With these methods, the recording medium could be erased and reused many times without loss of sound quality. Thus Poulsen's helped pave the way for the development of modern recording tape. German engineer Fritz Pfleumer refined Poulsen's ideas and created the first magnetic recording tape in 1928. Initially made of paper and then of plastic, the tape was coated with magnetic materials as Poulsen had described. While plastic tape is lighter and more flexible than Poulsen's original wire and allows multiple tracks on the same tape, wire is still used for flight recorders on **aircraft**, where durability is the most important consideration. Poulsen later turned his attention to **radio** communications.

In 1903 he aided the development of radio broadcasting when he invented a device for generating continuous radio waves. He then went on to patent an adaptation of an arc oscillator, or "singing arc," for radio transmitting purposes which helped make long-wave radio broadcasting possible by 1920. The U.S. Navy adopted his arc **generator** as standard equipment in 1912. However, subsequent devices by other inventors in the 1920s would rapidly make Poulsen's arc generator obsolete. Poulsen continued his research into radio communications until his death in Copenhagen in July 1942.

Powdermills

The mechanized manufacture of **gunpowder** first began about 1435 near Nuremberg, Germany. Until that time the primary ingredients of gunpowder (charcoal, sulfur, and potassium nitrate) were mixed dry and crushed by hand, often during the midst of war. The performance of gunpowder on the battlefield was by no means predictable, and accidents or misfires regularly occurred due to wet powder, too much powder, too tightly compacted powder, or poorly mixed powder.

Technology was greatly improved, however, when it was discovered that by moistening the primary substances, cakes could be formed, dried, and then broken so that grains of differing shapes and sizes would result. This technique was developed in the controlled environment of the powdermill. Not only did the new process reduce the possibility of accidental explosion while mixing, it also helped ensure that proper combustion and propulsion occurred when the powder was packed into guns or cannons and fired.

Through the centuries, the manufacture of gunpowder has become a highly systematized operation. One of the most important developments in production was the added step of glazing, in which the powder is tumbled in large cylinders to remove rough edges. Graphite, a versatile type of carbon, is added at this time to ensure that the individual grains are perfectly sealed and moisture resistant. The grains are then graded by size and repackaged; occasionally they are formed into prisms. The most important gunpowder manufacturer in the United States is E. I. du Pont de Nemours & Company, which began operations in 1802 in Wilmington, Delaware, under the guidance of French-born Eleuthère-Irénée du Pont (1771-1834). Now, a greatly diversified corporate giant, the Du Pont Company traces its fortunes to its ability to supply, both cheaply and efficiently, gunpowder in vast quantities for the purposes of mining and war.

Among important discoveries made by the company during the nineteenth century were soda powder, used for blasting, and **cocoa** powder, used for added propulsion in large caliber guns. During World War I, Du Pont served as a key supplier of powder, **bombs,** and other munitions for the Allied forces and also conducted extremely important chemical research. Today the name Du Pont is most often associated with the synthetic material **nylon**, first developed during the 1930s.

Power steering

Power steering was once considered an optional luxury, but changes in **automobile** configurations made this feature essential to most drivers. Steering originally depended only on the driver's muscular exertions, but cars became heavier as equipment was added and tires became softer to cushion the ride. The softer tires hugged the road, but friction also increased so more force was needed especially during turning. The passenger compartment and the engine moved forward as improvements were made to road handling and comfort, but this shifted more of the load to the front **wheels.** Front-wheel drive also added efficiencies to car operations but complicated the average driver's ability to steer.

At first, steering was adapted to these changes by mechanical improvements to the bearings, linkages to the front wheels, and the gears that transmit the movement of the steering wheel to the front wheels. The "average" driver was changing too, and the increase in the number of women drivers also pushed the industry to provide steering assistance.

Power-assisted steering was actually used with some steam-powered vehicles in the early 1800s. It was rediscovered by the military and adapted to heavy **trucks** and military vehicles in the 1930s, but it was not revived and applied to passenger cars with **internal combustion engines** until the 1950s. During power-assisted steering, as the driver applies turning moment or torque to the steering wheel, a hydraulic device driven by the engine amplifies this torque at the wheels to overcome the forces of weight and friction. Chrysler introduced power steering by adapting hydraulic power cylinders and jacks that had been developed by the **aircraft** industry. A pressure-sensitive **valve** responds to steering force by supplying fluid under pressure from a pump to the jack cylinder connected to the chassis. As the car begins to move and turn, less steering force is needed, the fluid supply and resulting boost is reduced, the steering wheel returns to the forward direction, and the process reverses until the wheels are moving straight ahead.

Other power steering systems use track rods connected to the control valve or, instead of a hydraulic jack, a ball nut inside the steering **gear** casing forms a similar **piston** inside a cylinder. Improvements to power steering allow the system to recognize the relationship between speed and the need for powered assistance. The need for power steering diminishes with increased speed, but the system remains sensitive to flat tires or other sudden slow-downs and keeps the wheels straight despite these apparently abrupt turns.

Power supply

All power supplies provide electrical energy to do work, but how the feat has been accomplished over the centuries has varied considerably. The first power supplier was the Leyden jar. Invented in 1745, it could store sizeable electric charges created by **electrostatic devices**, such as **Alessandro Volta**'s electrophorus. The charge could be drawn from the jar and put to work. In 1800, Volta created the first **battery**, the ''Voltaic pile.'' This reliable source of power produced electricity by means of a chemical reaction.

The first mechanical electrical **generator** was invented by **Michael Faraday** in 1831. **Joseph Henry** and Faraday had independently discovered that a moving **magnetic field** could create the flow of electricity in a conductor. Faraday used this process of electrical induction to create an **alternating current** power supply. Soon after, **Hippolyte Pixii** built a hand-driven generator that produced alternating current (AC), and added a commutator to convert the power into **direct current** (DC).

In 1867, inventor **Zénobe Gramme**, using the principles established by Henry and Faraday, built an improved dynamo for producing AC, and two years later he improved the DC dynamo. The two methods of power production divided scientists, with factions led by **Thomas Edison** and **Nikola Tesla**. Edison's invention of the **incandescent light bulb** in 1879 had created a demand for electricity, so he established a DC-generated power supply company in New York City. Direct current power supplies had two main disadvantages: power production was limited by arcing from the brushes that drew electricity from the dynamo's rotor, and long-distance transmission was prevented by resistance in electrical **wires** carrying the current. In 1884 Edison's plant was supplying power to over 11,000 electric lights in 500 buildings, with another 60,000 buildings receiving power from individual generators Edison had supplied.

Tesla, who had worked with Edison, tried to convince him that AC was the wave of the future. It could be produced in abundance without arcing, and could be carried very efficiently over power lines. Unable to persuade Edison, Tesla took his ideas to **George Westinghouse**. In 1893, Tesla and Westinghouse successfully used AC power supplies to provide electricity for the World Columbia Exposition in Chicago, ending the monopoly of DC dynamos. In 1895 the first hydroelectric AC power generator began operation at Niagara Falls and, with the use of power **transformers**, supplied electricity over great distances.

The first mechanical electrical generator, Michael Faraday's magneto-electric induction apparatus.

Alternating current power is created by rotating conductors (turbines) within a magnetic field. Water is heated (by burning wood or coal, or through atomic decay) and steam pressure is produced, rotating the turbines. In hydroelectric plant, the force of moving water is used to spin turbines. **Gas turbines** and tidal power are also used. The standard AC power systems in use today are either three-phase with three wires, mainly for use in transmission of high voltages, or three-phase with four wires, where low voltage power is needed. They operate at a frequency of either 50 or 60 *Hertz* (cycles per second), with each phase synchronized to the entire system. Voltage going into a system can be controlled by changing the ''tapping'' on transformers to alter voltage ratios, varying the rotor field strength, or switching in and out shunt reactors. The frequency of the system is controlled by adjusting the steam supply going to the turbines. Power is created at remotely located generating stations, stepped up by transformers, and distributed to substations where step-down transformers, switches (busbars), isolaters and **circuit breakers** are used to reduce the voltage and protect equipment.

The transformed power is sent out over various distribution networks which handle the appropriate voltage to supply the demand. Duplication of power supplies, transformers and distribution networks back up the system in case of a failure. Power is transmitted over long distances by underground cables or overhead lines, each with its advantages. Underground cables are free from the danger of storms and falling trees, but difficult to access when problems arise. Above-ground lines are cheaper to install, but vulnerable and unsightly. Some have claimed that high electric and magnetic fields produced by high-tension wires could be a health risk to people living in their vicinity, but most studies now seem to refute this idea, except for the possibility of some childhood leukemias being caused in part from high-current power lines.

Powerboat

The powerboat is defined as any kind of boat powered by any engine except a **steam engine**. While steam-powered boats date from as early as the late 1790s, boats with other power sources did not come into existence until nearly a century later. Perhaps the first powerboat designed for pleasure was developed by F. W. Ofeldt of the United States in 1885. His 21-foot (6.4 m) launch used a two-horsepower engine that was fueled by naphtha gas, a type of liquid fuel related to gasoline.

German inventor **Gottlieb Daimler** is usually credited with building the first boat powered by a **gasoline** engine—a project that he soon abandoned in favor of his work on the **automobile**. A boat powered by an **electric motor** and **battery** was exhibited at the Paris Exposition in 1889. Despite these early developments, boating did not become popular until certain improvements to the gasoline engine made powerboating practical in the 1900s.

All of these early powerboats were inboards; that is, the motor was located within the hull. These engines were most often automobile engines that had been adapted to power a boat. The invention of the first outboard motor—a motor that was specifically designed for a boat—is credited to **Ole Evinrude**, who built his first outboard engine in 1906. The invention of the outboard engine, which made powerboating more economical and easier, greatly advanced the popularity of pleasure boating. After 1940, engines made of lightweight **alloy**s were developed, allowing for outboard engines of greatly increased horsepower. This too gave rise to a surge in boating popularity in the 1940s.

These early boats were generally built of wood, but later powerboats were built of metals or of **fiberglass**. The design was altered to allow for greater speed, and the power sources varied as well—ranging from multi-cylinder gasoline engines to **jet engine**s to diesels.

Racing boats such as hydroplanes and unlimited hydroplanes are considered to be powerboats, as are yachts. Looking ahead, the powerboat engine industry will soon enter a new age. By 1997, the United States Environmental Protection Agency had already mandated that by the year 2006, all two-stroke outboard engines must reduce their hydrocarbon emission levels by 75 percent. To achieve such a large reduction in emission levels and smoke, the industry will create a new generation of computer-controlled engines. These new engines will not only improve fuel economy by 30 to 40 percent, but will result in smoother running and easier starting engines that also have increased reliability and more power. Just as automakers were able to give drivers generally trouble-free, reliable cars that could go much longer between service intervals, so makers of powerboat engines will soon do the same. While all of the major engine producers are racing to meet both government requirements and consumer expectations, it should be noted that all of the planned two-stroke engines are based on **fuel-injection** systems. Also, they all will require sophisticated electronic systems to control engine operation. These systems' sensors will tell a small computer how long to activate the fuel injector based on such variables as altitude, temperature, and engine load. They will also automatically adjust the ignition timing to the best setting relative to any working conditions. Like modern car engines, they will be both highly reliable and highly efficient.

Prenatal diagnostic techniques

Prenatal diagnosis, assessing the condition of a fetus before it is born, has become an important—although controversial—part of pregnancy care. Increasingly sophisticated technology adds to the variety and accuracy of prenatal tests.

The earliest prenatal testing was simple. The mother noted activity inside her womb; the physicians felt it from the outside. Machines were developed for listening to the fetal heartbeat and diagnose fetal ''stress;'' the 1950s saw development of **amniocentesis** and the ability to culture cells from amniotic fluid in the 1960s allowed identification of chromosomal disorders—specifically Down's syndrome—and today, a wide range of metabolic and chromosomal disorders can be detected through tests such as ultrasound, chronic villus sampling (CVS), Bart's triple test, maternal serum screen (MSS)/alpha-fetoprotein (AFP), umbilical vein sampling/cordoncentesis, fetoscopy, and genetic counselling.

Ultrasound scanning has had the most comprehensive effect on prenatal testing. Sound waves produce a picture of the developing fetus, and the procedure has no known harmful effect. The development of real-time ultrasound (1977) gave detailed images of the moving fetus, permitting invasive diagnostic techniques through visual guiding of needles into the womb for samples blood or tissue while avoiding damage to the fetus, placenta, or other vital structures.

Nuclear **magnetic resonance imaging** (MRI) and spectroscopy (NMR) reveal biochemical information about fetal soft-tissue and organ structure and function. Amniocentesis, normally recommended at 16-18 weeks' gestation, is sometimes conducted earlier (11-12 weeks); however, at this point fetal loss is increased by 7.6%, and the risk of related negative consequences is higher.

Chorionic villus sampling involves collecting a small sample of the developing placenta for chromosomal, biochemical, and DNA content. Obtaining chorionic villi through the cervix was described by Jan Mohr in 1968 and by other researchers in the early 1970s. In 1982, a Soviet group reported CVS with ultrasound guidance; and the University College Hospital of London developed the now widely used technique of transcervical CVS guided by real-time ultrasound which can be performed much earlier than amniocentesis (at 9-14 weeks). Higher incidence of limb defects is reported when performed earlier than nine weeks, and risk of miscarriage is one in 100.

Maternal serum alpha-fetoprotein (MSAFP) measures the levels of AFP—a protein produced by the fetal liver—in the mother's blood. High levels can indicate a longer pregnancy than calculated, twins, spina bifida, or anencephaly (incomplete development of the skull or brain). Low levels can indicate a shorter time pregnant than calculated, or other problems such as Down's syndrome. These varying indications dictate further tests—such as ultrasound or amniocentesis.

Fetal blood sampling for diagnostic purposes has been done since 1972. A fiber optic **endoscope** inserted through an incision in the uterine wall which allows observation of the placenta, fetus, and amniotic fluid, as well as sampling of fetal, placental, or umbilical cord blood or tissue for diagnosis of hereditary blood or skin disorders.

This radical, high-risk procedure is used only in exceptional circumstances. DNA analysis for specific gene disorders was introduced to prenatal testing in 1976. Fetal samples from amniocentesis, CVS, and maternal and paternal blood samples are examined. More than 260 gene defects can now be diagnosed through DNA analysis, and the list grows. In 1982, Fernand Daffos and his team used real-time ultrasound imaging for cordoncentesis, or percutaneous umbilical blood sampling (PUBS), in which a hollow needle passes through the uterine wall and collects blood from a vessel in the umbilical chord, allowing chromosome count; measurement of oxygen, carbon dioxide, and bicarbonate if fetus growth retardation is suspected; and detection of fetal anemia and infections such as rubella and herpes.

In August, 1997, *The Lancet* reported that researchers from Britain, Italy, and Hong Kong discovered samples of fetal DNA in maternal blood plasma. This non-invasive test holds promise of a safer, more effective method of genetic testing. Bart's triple test, a new test which also screens maternal blood for AFP, human chorionic gonadotropin, and unconjugated estriol, helps determine the need for amniocentesis.

Pre-implantation genetic diagnosis (PGD), in which a woman's eggs are fertilized *in vitro* with her partner's sperm, allows earliest detection of fetal deformity thus far. One cell removed from the 8-cell fetus in the petri dish is checked for genetic defect before implantation into the uterus. The first PGD baby was born in 1989 and, by 1997, more than 30 babies had been born worldwide using this technique. Investigational as of 1998, a specialized test—the InSight Prenatal Test—uses flourescent DNA probes on amniotic fluid to detect, with a high degree of accuracy, the most common chromosome abnormalities. Prenatal diagnosis is now moving toward prenatal treatment: A few conditions—including respiratory distress syndrome, a prime killer of prematurely born infants, and a rare vitamin metabolism defect called methylamalonic acidemia—can be treated in the womb by medicating the mother, and **prenatal surgery** on the fetus is in its infancy.

See also Ultrasound devices

Prenatal surgery

The first human fetal surgery, performed in the 1950s, were fetal blood transfusions for Rh-positive babies whose red blood cells were under attack by the Rh-negative mother's immune system. Attempts to improve this technique via direct access to the fetus through an open uterus were unsuccessful, and open-womb fetal surgery was abandoned. The development of real-time **ultrasound** imaging in 1977 made intrauterine fetal surgery feasible: visual imaging permitted surgeons to accurately guide needles similar to those used for **amniocentesis** without the risk of damage to placenta or fetus. In the early 1980s, surgeons at several medical centers attempted closed-womb surgery to relieve fetal hydrocephalus (abnormal accumulation of fluid around the brain) by draining excess fluid. Although fluid could be drained, most babies were born with severe neurological defects. In 1981, pediatric surgeon Michael Harrison of the University of California at San Francisco developed a closed-womb shunting procedure to correct a urinary tract obstruction which afflicts one in every 2,000 male fetuses. In this defect, a blocked bladder damages or destroys the unborn baby's kidneys and, because the presence of the baby's urine in the amniotic fluid is vital for its lung development, the absence of urine is fatal. Although this technique is now widely used, it remains high-risk.

Harrison performed the first successful major open-womb surgery on a fetus on June 15, 1989, when he repaired a hole in the diaphragm of a 24-week-old fetus. In this pioneering operation, the unborn baby's arm was lifted out of his mother's womb through an incision in her abdomen and uterus, exposing the left side. The baby's misplaced internal organs were repositioned and the diaphragm and abdominal incisions patched with Gortex. The baby was born seven weeks later, premature but healthy. Harrison and his team successfully performed the same operation on a second fetus soon after. These successes were preceded by minor surgery on fetuses either inside or partially removed from the womb, although six previous procedures on humans had failed. Harrison prepared for human prenatal surgery by performing more than a thousand intrauterine operations on animals—the first successful animal fetal operations being carried out in the 1920s.

Prenatal surgery is still in its earliest stages and the success rate is not yet high, even though it is higher for some closed-womb procedures such as the bladder shunt. Risk remains high for both fetus and mother, and prenatal surgery is recommended for only a few conditions. One technique that has exciting possibilities is fetal stem-cell transplantation in which stem cells, the source of all blood cells, from a dead fetus are injected into a living fetus, providing healthy genetic material in an effort to overcome defective genes in the recipient fetus.

Pressure cooker

A pressure cooker is a container with an airtight lid that traps steam from boiling water; the steam increases pressure inside the cooker, which raises the water's boiling point. The higher temperature cooks food much faster than usual. The pressure cooker includes a safety **valve** to prevent explosion if the steam pressure gets too high.

Both the device and the valve were invented in 1679 by the French physician **Denis Papin**, who called it a "steam digester." He demonstrated his invention to the Royal Society by reducing bones to an edible jelly. Papin described his "New Digester" in a 1681 pamphlet, emphasizing the fact that food cooked under pressure in this way retains flavor, color, and nutritive value that otherwise would be lost. In spite of these advantages, Papin's "digester" was used less for cooking than as an autoclave—a vessel that uses the high temperatures produced by steam under pressure to rapidly sterilize medical and scientific materials, an application in standard use today.

After the United States Department of Agriculture announced in 1917 that pressure cooking was the only safe way to preserve low-acid foods, home pressure cookers came on the market, but these were at first large and cumbersome to use. After World War II, many manufacturers began producing home devices, and pressure cooking became much more common. Today, pressure cooking methods are also used to cook wood chips so they become wood pulp that can be used to manufacture **paper**.

See also Can and canned food

Pretzel

While **popcorn** and **potato chip**s are indigenous American foods, the pretzel has European roots. The packaged and processed pretzels we are familiar with today are less popular than the early pretzels that were soft, bready, and served warm.

The Italian word for pretzel, *bracciatelli*, means folded arms. According to legend, the first pretzels were made in 1610 by an Italian monk who rewarded his best students with a bready snack that was shaped to resemble arms folded in prayer. The pretzel has no other religious significance but is often eaten during Lent in many parts of Europe.

Americans today often think of the ''Bavarian'' pretzel and assume that pretzels originated in Germany. Though the pretzel came first from Italy, its popularity soon spread throughout Europe. In the early 1600s, Turkish soldiers, who tried to penetrate the walled fort in Vienna, Austria, by digging a tunnel beneath it, were foiled by pretzel vendors who heard them and warned the Viennese troops. From that point on, the pretzel became the recognized symbol for baker throughout Europe.

When the Europeans immigrated to America, they brought the art of pretzel baking with them. In 1861, Julius Sturgis opened America's first commercial pretzel bakery in Lititz, Pennsylvania, a small town about ten miles from Lancaster. As early as 1879, a couple of Lancaster men invented a machine that twisted pretzels, a tedious task when done by hand. Today about 90 percent of all pretzels are made by fully mechanized bakeries.

In modern bakeries, after the pretzel dough is made, it is placed in the hopper of a twisting machine. A small bit of dough is pushed out and rolled into the desired length. Then it is twisted and dropped onto a long belt, which carries it to the cooker, where the pretzel passes through a hot solution that gives it a golden brown color. Next the pretzel is salted and carried through an oven, where the baking process is completed.

Though pretzel vendors still appear on the streets of American cities, the popularity of the pretzel lagged behind the overwhelming number of competing snack foods. But pretzels remain a popular item in bars, as they are often served with beer. By the late 1990s, according to the National Pretzel Bakers Institute, over 300 million pounds of pretzels and pretzel products were produced each year in the United States.

Priestley, Joseph • See Soda water

Printing technology

If we define printing as the process of transferring repeatable designs onto a surface, then the first known printing was done by the Mesopotamians, who as early as 3000 B.C. used stamps to impress designs onto wet clay. Printing on **paper** developed much later; Chinese inventor Ts'ai Lun (50?-118? A.D.) produced the first paper in 105 A.D. Chinese books printed with inked wood blocks survive from the T'ang Dynasty (618 A.D.-907 A.D.), and it was the Chinese—not German printer **Johannes Gutenberg**, as is widely believed—who developed movable type, allowing printers to compose a master page from permanent, raised characters. Pi Sheng invented movable type around 1045; he fashioned his characters of heat-fired clay. Later printers improved on his design by using characters made from lead. However, movable-type printing did not achieve the popularity in medieval China that it later would in Europe; as the Chinese language has some 80,000 characters, their printers ultimately found it more convenient to use carved blocks.

Thus Johannes Gutenberg, who built his first printing press around 1440-1450, could be said to be the first inventor of a lasting system of movable-type printing. Printing technology had only a brief existence in Europe before Gutenberg. The Mongol invasions of the fourteenth century introduced wood-block printing to Europeans, who had relied on copyists, particularly monks, to produce laborious, handwritten manuscripts. The design of Gutenberg's press, which employed metal characters, was based on machines used to press cheese and grapes. The set type, or *matrix*, was placed in a sliding track, moistened by **ink**-coated leather balls, and pressed to paper by a plate, or *platen*, which was moved by hand. The printing of Gutenberg's famous Bible in 1455-56 prefigured an explosion in European printing. The crude machine (it could print only 300 sheets per day) was nonetheless enormously faster than hand copying; the improvement was so staggering that many believed it the work of Satan.

Before the printing press, books were luxuries even among the educated; a small shelf of books was the hallmark of wealth. By 1500, several million books had been printed throughout Europe. Mechanical printing swelled the readership and variety of printed material, making possible magazines, newspapers, and the rise of the novel as a literary form. Yet, almost four hundred years later, printers were still using essentially the same machine as Gutenberg's—setting type, inserting paper, and pressing the platen all by hand (a few innovators refined the machine's design, replacing the original wood with stone or metal). Printing shops had difficulty in meeting the public demand for print products.

Finally, in 1811, German inventor **Friedrich König** discovered a way to harness steam power to drive a press in which the flat platen was replaced by a revolving cylinder; his machine could print over a thousand pages an hour. The constant motion offered by a cylinder is far faster and more efficient than moving a flat surface in a vertical or horizontal plane, and for this reason American inventor **Richard M. Hoe** invented a press in 1846 that also used a cylinder to carry the type. An 1857 version of Hoe's rotary press could print 20,000 individual sheets an hour. In 1865, William Bullock developed the first web press in 1865, in which rolls of paper replaced single sheets, further speeding the process. Although printing presses were now capable of producing material in large volumes and at high speeds, printing technology was still constrained by the typesetting process.

Typesetters still employed the method of setting pre-cast letters by hand, or by operating machines that were a little more efficient. In 1886, German-born **Ottmar Mergenthaler** invented the **linotype** machine, which operated on an entirely different principle. An operator typing at the linotype keyboard moved letter forms into place; the machine automatically spaced the line, then poured over it a metal **alloy,** which cooled into a full "line of type." Its speed was a great benefit to newspapers, which often need to be reset as news breaks. Linotype and its rival, monotype, introduced in 1897 by American inventor Tolbert Lanston (1844-1913), dominated typesetting until the mid-twentieth century.

The rise of **photography** in the late 19th century allowed printers to apply methods used for reproducing art—such as **lithography** and **engraving**—to text, and it eventually allowed them to make text matrices almost instantaneously. In each of these processes, including photoengraving and photolithography, the object to be reproduced—whether a piece of art or a sheet of type—is photographed. Light is then shone through the clear parts of the negative (those which would show up dark in print) onto a plate coated with a light-sensitive gel. The exposed areas of gel harden and the rest is washed away, resulting in a reproduction; the plate can then be inked (with an ink that adheres only to the gel) or etched (with an acid that eats only the exposed portions), and the thin plate is affixed to a cylinder on the printing press. Printers were enabled to reproduce finely-shaded material, like photographs, through the **halftone process** invented by **Frederic Eugene Ives**.

The major printing innovations of the late twentieth century center around electronic methods of typesetting. Printing in the early twentieth century involved an awkward marriage of modern and old-fashioned technology, in which entire metal plates of text had to be produced simply to be photographed and then destroyed. Photocomposition, or **phototypesetting**, developed after World War II, is the process of printing text directly on film (bypassing the hot-metal typesetting stage), using "master" characters stored on film or in computer memory. Just as nineteenth-century printing innovations made use of technology, phototypesetters employed computer technology; text is input not by a single operator, but is fed in by tape or **computer disk**, so that any number of typists can key in parts of the same text. These phototypesetting systems can set type at speeds of tens of millions of characters per hour, uninhibited by the physical limits of their operators, and massive works, such as encyclopedias and indices, can be stored on disk and revised without being completely retyped.

In the 1990s, the film step was being replaced by computer to plate printing technology (known as CTP), usually a digital process. Instead of converting files to film, plates are made directly from computer files. There are computer programs, such as PROSE (production order specification/EDD), which facilitate CTP. It is predicted that the printing process will soon become all digital, and film will disappear entirely. Another innovation is waterless printing, which involves different kinds of plates. Instead of metal, usually aluminum, plates are made of silicon or plastic. Parts of these plates are not affected by ink, so there is no need for water. (Water was used to protect non-imaging parts of the plate from ink.) Computer technology will also effect binding equipment, so advertising and articles can vary to target specific customers.

Gutenberg based the design of his printing press on that of the wine and cheese presses of his day. The set type was placed on a sliding track, moistened with ink, and pressed to paper by a plate, which was moved by hand.

A common problem found with newspapers since the inception of the press was ink that smudged, leaving a residue on the reader's hands. In the 1980s and 1990s, low-smudge inks were developed to combat this problem. The low-quality ink used on newspapers smudged because it was made from carbon black and mineral oils, which never fully dry on paper. There are several kinds of low-smudge ink, made with water instead of oil bases, or with agents that force the ink to fully dry. Other, less expensive, low-smudge options include ink that dyes the paper, invented by Rodger Gamblin of Dayton Tinker Corp. in Ohio.

Today, electronic miniaturization, and "pressureless" printing methods (those used by photocopying machines and laser printers) allow many individuals to produce printer-quality documents at home.

Prism

A prism is a clear solid used to manipulate the path of a light ray, either by bending it, splitting it, or polarizing it. Generally, the light ray is reflected one or more times within the body of the prism, an effect known as *total internal reflection.*

Prisms can be found in a wide variety of shapes and sizes and can possess a number of sides. Any prism whose lateral faces are perpendicular to the base is called a *right* prism; other prisms are called *oblique* prisms. Most are made of **glass** or quartz and are at least partially transparent, allowing light to pass through the body.

One of the best known prismatic effects is the rainbow. When a rainbow occurs, thousands of tiny drops of water suspended in the air act as prisms, splitting sunlight into its color spectrum. In addition to separating colors, prisms, such as those used in **binocular**s and **telescope**s, help to bend and direct light, as well as to correct color aberrations.

One particular prism, the *Nicol prism*, is used to convert simple light into polarized light. Invented by the Scottish scientist **William Nicol**, this device is actually composed of two prisms cemented together with *Canadian balsam*; the completed system has the effect of polarizing light along a certain plane.

Programming languages

A digital **computer** language is a group of words or numbers that people use to issue commands to computers. Like human languages, computer languages use their own vocabulary and syntax (grammar) to convey specific meanings (semantics). These include performance of actions, such as arithmetic, logical operations (using *and* and *or*), reading data from a disk or storing it on a disk, and communication of data to a printer or other device. If necessary, the language also provides a method of translation so that it can be understood directly by the computer.

A low-level language is closely related to what the computer can execute directly, and is usually created for a specific computer. Examples are **machine languages** (sometimes called first generation languages) and second generation assembly languages.

A machine language consists of sets of numbers, letters, or symbols that the machine recognizes as an operation to be carried out and the rules for doing so. Typically each instruction carries a symbol, called a label or tag, that identifies it and a memory address (location) for the data that it will use.

A computer operates on the principle of *binary* (base 2) numbers, with zero and one representing the two electrical states. To make it easier for programmers to write in machine language, a base 16 (hexadecimal or hex) system is used. Each hex number represents a set of four binary digits.

Because machine language can be difficult to remember, assembly language was created. It uses some abbreviations of human language as symbols, such as MPY for multiply. It also uses symbolic names for memory addresses and data. The symbols are translated into machine language for execution. Some of the earliest assembly language work was done in the late 1940s by the British scientist Maurice Wilkes on the EDSAC computer. The first English-language assembly program was FLOW-MATIC, written in the early 1950s by the American naval officer **Grace Hopper** for the **UNIVAC** computer.

A high-level language uses very carefully defined human words as instructions. This allows use of the language on computers with different architectures (designs). The instructions must be translated into machine language before execution.

High-level, procedure-oriented languages, such as FORTRAN, COBOL, and BASIC, are procedure-oriented, meaning that they implement procedures or algorithms. They are sometimes called *third-generation languages.*

Translation into machine language is performed with either a compiler or an interpreter. A compiler translates all the high-level commands into machine language before the program is executed. Recompiling is not necessary if the program is executed more than once, because an interpreter translates each high-level command into machine language and executes it before translating the next command. This makes it easier to find mistakes (bugs) in the program, but the program must be reinterpreted each time it is used.

The first widely used high-level languages were FORTRAN and COBOL, written in the 1950s. They are still popular for programming on mainframe computers.

FORTRAN (FORmula TRANslator) was developed in the mid-1950s by John Backus (1924-) and others at IBM for number-intensive problems in mathematics and science. Many dialects of FORTRAN were written, but they all have many common features. FORTRAN allows straightforward expression of mathematical calculations. It also allows repeated calculations (iteration), conditional (IF) statements, and branching (going to a specified place in the program). Provision for subroutines allows a specific set of commands to be compiled once and called (executed) when needed.

Backus, with Peter Naur (1928-), also codeveloped a standard method of describing a programming language's grammar, called Backus-Naur Form.

COBOL (COmmon Business Oriented Language) was written in 1959 by Grace Hopper, one of the first programmers of the Harvard University **Mark I** computer. Hopper, with a group of naval and civilian programmers, developed COBOL as an uncopyrighted data processing language that business managers and administrators could use comfortably. COBOL allows programmers to use entire English language words, phrases, and statements, such as Purchase Order and Add Interest. Each program begins by identifying the program. It then specifies the environment, or computer equipment. The data section describes the data format, and the procedure section defines how the operations are to be carried out. Despite its age, COBOL is still used for accounting, inventory, billing, ordering, personnel management, and other data-processing uses.

PL/I (Programming Language One) was also developed in the 1960s by IBM. Originally intending to write an extension of FORTRAN, the scientists developed a new language instead. PL/I was written for the IBM System/360 mainframe computer and was intended for use in both scientific and business computing.

BASIC (Beginner's All-purpose Symbols Instruction Code) was developed by the mathematicians **John G. Kemeny** and Thomas E. Kurtz (1928-) at Dartmouth College in the 1960s (where Kemeny later served as president). It was developed when *time-sharing*—the use of a mainframe by many users at separate terminals—was first developed. Along with a specially-written operating system, BASIC made it possible for all students at the college to use the computer as a routine part of their college courses. Today BASIC is taught in elementary and high school.

BASIC uses simple commands; consequently, it was the first language widely used by people other than computer professionals. Its popularity grew rapidly in the mid-1970s, with the invention of the personal computer. In 1975, an interpreted version of BASIC was the first product of Harvard dropout **William Gates**'s Microsoft Corporation.

C was written by Dennis Ritchie at Bell Telephone Laboratories in the early 1970s. Though it is a high-level language, it has some characteristics of assembly language, and can be directly translated into machine language. It is closely related to the Unix operating system and is also used for compilers and communications applications.

Ada was developed by Jean Ichbeah in the late 1970s for the United States Department of Defense, for embedded (built-in) systems in computerized weapons and communications systems, as well as general programming. It consists of a kernel of essential commands, called KAPSE (Kernel Ada Program Support Environment); commands or tools for developing programs, called MAPSE (Minimal Ada Program Support Environment); and specialized routines called APSEs (Ada Program Support Environment).

Ada was named after the nineteenth century English computer pioneer, **Augusta Ada, Countess of Lovelace**, considered to be the first computer programmer.

Pascal was developed in the 1970s by Nicklaus Wirth at ETH (the Zurich Technical School), in Zurich, Switzerland, for use in teaching programming. It has been popular in colleges for that purpose. Pascal was named for Blaise Pascal (1623-1662), the French mathematician and **calculating machine** inventer.

Some high-level languages describe functional relationships of non-numeric data, rather than linear mathematical processing. An example is LISP (LISt Processing language), developed by John McCarthy and others at MIT during the 1950s and 1960s. LISP was designed to process lists of words and logical and mathematical statements. A list consists of elements, and each element can be a single item or another list. This flexibility makes it appropriate for building and decomposing data structures. LISP is often used in writing the **artificial intelligence** programs known as **expert systems.**

Object-oriented programming uses a modular approach to writing a program in order to make programming more efficient. An object contains both data variables and the procedures and functions for carrying out operations. Object-oriented languages include an extension of C called *C++*, Java, and versions of Pascal, Ada, and LISP.

Smalltalk, developed in the 1970s at the Xerox Palo Alto Research Center, was the first object-oriented language. Its user interface was the basis for today's graphical user interfaces, such as those used in the Apple Macintosh computer and Microsoft Windows for IBM-compatible computers.

PROLOG (PROgramming in LOGic) uses the rules of logic for solving problems in mathematics, logic, biochemistry, automation, and artificial intelligence. It was developed in the early 1970s by Alain Colmerauer and others at the University of Marseilles, France.

Fourth-generation languages are non-procedural and use English-like words to specify results, rather than the steps leading to the results. They are usually written for a specific application, such as for the "friendly" user interfaces that allow non-experts to use complex databases without having to learn the database language SQL (Structured Query Language).

PROKHOROV, ALEKSANDR • See Maser

PRONY BRAKE • See Dynamometer

PROPELLER, SCREW • See Screw propeller

PROSTHESIS • See Artificial limb and joint

PTFE (POLYTETRAFLUORETHYLENE) • See Teflon

PULLEY

The first pulley was likely part of a simple **crane** dating back to around 1000 B.C., though there is pictorial evidence that pulleys may have been in use since 8000 B.C.

A pulley may be defined as any **wheel** that supports a **rope**, belt, cable, cord, or chain for the purpose of transferring motion and energy. Pulleys are affixed to frames or blocks and an entire pulley system is usually referred to as a **block and**

tackle. The principle governing the mechanics of pulleys, outlined by **Archimedes**, is virtually the same as that for **lever**s. Consequently, pulleys are employed to gain a mechanical advantage through a reduction in the effort required to move or lift an object.

In the case of single pulleys, there is no real mechanical advantage and the main benefit consists only of changing the direction of the force. Compound pulleys—pulleys containing several wheels—date back to about 400 B.C. A five-pulley, two-axle arrangement of this period, the pentapaston, produced a 5:1 mechanical advantage. (Ratios for compound pulleys may be calculated by simply counting the number of pulley wheels on both the upper and lower axles. The magnification of effort is equal to this number. However, it must be kept in mind that the magnification will always be slightly less than this ideal number, given the presence of friction.) Interestingly, Greek dramatists may have been among the first to avail themselves of compound pulleys, in order to create such special effects as gods descending from the heavens.

Pulley systems, for centuries, have been used in construction and manufacturing for centuries. However, **elevator**s, **escalator**s, and a large number of other commonly encountered devices are also founded on the operation of the pulley.

PULLMAN CAR

The first railroad sleeping cars appeared in America in the late 1830s. They were used between Baltimore and Philadelphia and provided berths for two dozen passengers. In 1856 Theodore T. Woodruff, a master car builder with the Terre Haute and Alton Railroad, patented a railroad car with three tiers: the lowest was made by moving seat bottoms together, the middle by folding up the backs of seats, and the top by lowering panels attached above the windows. These cars were quite popular until George Pullman (1831-1897) introduced his brand of railroad cars in 1865.

Pullman was born March 3, 1831, and grew up in Brocton, New York, but moved to Chicago in 1855. In 1858, Pullman arranged with the Chicago and Alton Railroad Company to remodel two of their boxcar-like coaches. The first few experiments were not successful, but in 1861 he designed a model called the *Pioneer*.

Pullman improved upon Woodruff's earlier design by eliminating the middle tier. He added interior decoration and improved riding comfort by reinforcing each spring with rubber cushions. He added a foot to the car's width and 2.5 feet (76.2 cm)to its height. This extra size, while more comfortable and popular, however, meant that existing bridges, platforms, and other standards would have to be altered to accommodate the cars. This could have become a problem except that the United States government chose one of Pullman's new cars to carry President Abraham Lincoln's body from Chicago to Springfield in May 1865. Every station and bridge between the cities was modified to accommodate the oversized car. People along the route praised the luxury and elegance of the car. By 1867, forty-eight of these luxury cars were in operation in Illinois, Michigan, Canada, and New York. Pullman soon founded the Pullman Palace Car Company. He also introduced the first combined sleeping and restaurant car. The first dining car followed in 1868 and the first chair car in 1875.

Some historians believe that Pullman's most important legacy was not the invention of Pullman cars, but the innovation of railroad comfort. His company also made contributions towards railroad safety by introducing such features as the vestibule in 1887 which allowed riders safe passage between cars. The Pullman Company also invented the ventilator, a fine screen that allowed air to circulate into cars while filtering out cinders and soot.

George Pullman died on October 19, 1897, three years after a major strike at his Pullman, Illinois, manufacturing plant. The strike sparked a sympathy boycott of his cars across the country and disrupted rail traffic for a time. The strikes ended bitterly when President Cleveland sent an army regiment to protect the flow of mail after strikers derailed a postal **train** made up of Pullman cars.

After the strike, the company eventually reestablished itself as an integral part of train travel, and by the 1920s there were nearly 10,000 Pullman passenger cars in operation in the United States, Canada, and Mexico, with a growing freight-car business as well. After Pullman's death, the company continued his business practice of controlling all aspects of first-class travel, and sought to limit the railroads' involvement with any of its operations. Thus the Pullman Company and not the railroads were responsible for the sleeping, dining, and parlor cars to the extent that it provided all maintenance, staffing, linens, and other luxuries. This worked well until 1947 when the government brought antitrust action against what it considered a monopoly. Pullman eventually had to separate its car-building function from actual operations, and with the emergence of the **automobile** and the commercial airliner, the Pullman sleeping car business began a steady fall-off. Pullman maintained its freight-car building business however, and at the end of the twentieth century it is still an on-going concern.

See also Train and railroad

PUNCHED CARD

During the eighteenth century, the demand for fine woven and patterned fabrics increased dramatically; while weaving was facilitated by the rising use of mechanical spinners and **looms**, there seemed to be no suitable method for simulating the hand-lifting of individual warp threads that was involved in patterning. In order to mechanize this process, two requirements had to be met: a memory medium in which to store the desired pattern, and a system for lifting the appropriate threads.

In 1725, French inventor Basile Bouchon constructed a device that used hooks and needles to lift the warp threads. The pattern was stored as a series of holes punched into a long sheet of paper; as the fabric was woven, the holes would allow certain hooks to pass through and lift the pattern threads. To change the pattern, one had only to change the paper sheet. (This system closely resembled that of a **player piano**.) While Bouchon's apparatus was very clever it was also very small, and thus found little application in the textile industry. The hook-and-paper system was improved upon by another Frenchman, named Falcon, who in 1728 replaced the sheet of paper with a series of punched cards, each one controlling a single hook or needle. This design allowed for much larger patterns to be woven but required two men to operate.

It was not until 1801 that a practical punched-card loom was constructed by **Joseph-Marie Jacquard**. His device—whose design was essentially a combination of Falcon's, Bouchon's, and **Jacques de Vaucanson**'s—employed a foot pedal to allow a single weaver to create patterned textiles. Some improvements were made in 1845, and the device was quickly adapted to fit power looms.

One man impressed by the success of punched cards was the British scientist **Charles Babbage**. He envisioned an **adding machine** (which he called the ''analytical engine'') that could perform many simultaneous calculations and print the result onto a long tape of punched paper. While Babbage received government funding in 1823 to build his machine, the technology of the time was not sophisticated enough to match his design, and the project was never completed. Almost seventy years later American inventor **Herman Hollerith** succeeded in constructing a punched-card machine to calculate the results of the 1890 census; in 1896 Hollerith opened his own business, the Tabulation Machine Company, which, in 1924, became International Business Machines (IBM). Using punched cards as their principal storage medium, IBM led the way in the development of computers, eventually becoming the leading computer manufacturer in the world.

See also Computer, digital

PVC (POLYVINYL CHLORIDE) • See Vinyl

PYREX • See Glass, heat-resistant

PYROMETER

There are many types of devices used to measure temperature. **Thermometer**s can reliably reflect temperatures from 32-212°F (0-100°C). For higher temperatures, modified thermometers such as the platinum resistance thermometer are best suited to the task. At several hundred degrees celsius, the most precise thermometer is a **thermocouple**, an electronic device that measures temperature by observing voltage changes. However, at temperatures greater than 2,400°F (1,300°C) most thermometers are rendered unusable. For these extreme temperatures one must use a pyrometer.

There are two basic kinds of pyrometers: electrical and clay. Electrical pyrometers use a **wire** filament to measure very high temperatures. When a current is run through the filament it will heat up and glow. The temperature of the wire can be regulated by adjusting the current. Scientists then compare the color and brightness of the filament to the inside of a **furnace**. When the furnace matches the glowing wire the temperatures are known to be the same. This type of device is called an optical pyrometer. Other electrical pyrometers have no visible filaments, but rather an internal sensor that can be used to raise or lower the furnace's temperature.

Clay pyrometers are placed within a hot furnace. When they reach a certain sustained temperature the clay-salt cones will break down or melt. This type of pyrometer is used exclusively to measure, rather than regulate, temperature, and is often found in kilns.

The first pyrometer was invented by Josiah Wedgwood (1730-1795) in 1780. Further improvements on his design came in 1821 from John Frederic Daniell and later from **Henri Louis Le Châtelier**. French physicist Antoine-César Becquerel (1788-1878) was the first to introduce a practical electrical pyrometer.

PYROXYLINE • See Collodion

R

Rabinow, Jacob (1910-)
American inventor

Jacob Rabinow was born in the Russian Ukraine as Yakov Aaronovich Rabinovich in 1910. The Russian Revolution forced his family out of Russia and into China, where his father died. In 1921 Rabinow, with his mother and brother, settled in Brooklyn, New York.

Rabinow received a B.S. in electrical engineering from the City College of New York in 1933, and in 1938 he began work as a mechanical engineer at the U.S. government Bureau of Standards. He left the Bureau of Standards in 1954 and started the to develop his reading machines. He sold the company to Control Data Corporation in 1964, staying on as the company's vice president from 1964 to 1972. In 1972 he returned to the government Bureau of Standards, where he is still employed.

Rabinow has received numerous awards and honors, including a Certificate of Merit from President Truman in 1948 for his war-time efforts. In 1980 he was named Scientist of the Year by the editors of *Industrial Research and Development* magazine.

A prolific inventor, Rabinow holds patents on over 200 inventions. He has, by his own estimate, thousands of unpatented "inventions in progress" filling a shelf full of notebooks in his Bethesda, Maryland, home.

Rabinow began his career as an inventor during World War II while working for the U.S. military, designing bomb and **rocket** fuses. His first patent—for a camera designed to record airplane flight paths—was granted in 1947 and was a direct result of his ordnance work.

Among Rabinow's most important inventions is the optical character recognition machine—the machine that reads numbers like those that appear on the bottom of bank checks and other documents. First developed in 1954 while working for the Bureau of Standards, Rabinow's sophisticated machines do not actually "read" the numbers but work more like the human eye. That is, they take strings of numbers—even badly smudged or marred ones—and compare them to a set of "best guess" samples. Variants of these optical scanners are widely used in business, industry and by the U.S. Postal Service. Rabinow's first reading machine is on permanent display in the National Museum of American History, which is a component of the Smithsonian Institution.

Rabinow often invents things on a dare (such as his pick-proof lock or his self-justifying typewriter) or merely for the challenge (tired of constantly adjusting mechanical watches, he invented the self-regulating **clock**). He holds patents in a wide range of technological areas: automated mail-handling equipment (sixteen patents); **phonograph** tone arms, turntables, and other audio equipment (eighteen patents); and various timekeeping mechanisms (seven patents). He has also patented roadway reflectors that lie flush with the road, a pressurized can to keep tennis balls bouncy, the first magnetic disk file for computers, an auto-focusing **camera**, a device to indicate whether the owner of a telephone has received a call in his absence, a camera tripod with legs that can be adjusted from a single point, and a sophisticated automatic automobile headlight dimmer. Rabinow's magnetic particle clutch is being used on some Subaru automobiles, on the drive mechanisms of some tape recorders, and to control the flaps in Lear jets.

One of Rabinow's latest patents, issued in July 1986, is for a pick-proof lock. It is the result of nearly a half-century of pondering and tinkering with various models. Inspired by the remark of a lock company engineer who claimed it could not be done, Rabinow blended the mechanisms of both the combination lock and a standard key mechanism to create his pick-proof model. Rabinow is quoted as having said: "A really good invention, in fact, is a work of art....[A] good invention has the same qualities [as] a good piece of poetry." He has also claimed that "inventions enrich both the natural wealth and the cultural life of a nation and the world....[T]hey are an art form and should be supported as such."

See also Computer pattern and character recognition

Radar

For the last century, much of the world's scientific research has been conducted at the request of governments, and most often under the supervision of the military. The technological advances resulting from such research—from the development of the computer chip to the invention of flame-retardant fabrics—have affected nearly every sector of society. During the 1930s one such government-sponsored project was Britain's top-secret study of *radio detection and ranging*, now familiarly abbreviated as radar.

The properties of **radio** waves had been studied for many years before the idea of radio ranging emerged. In 1904, Christian Hullsmeyer, a German engineer, used radio waves in his telomobiloskop, a crude radar device. The telomobiloskop was invented to prevent collisions between ships. In 1922 Italian physicist **Guglielmo Marconi** noted that metallic objects reflect these waves, and in 1924 English physicist Edward Appleton (1892-1965) and Miles Barnett used radio signals to find the ''mirror in the sky'' known as the ionosphere (a layer of ionized particles that often reflects radio waves). The British government became intrigued with the military potential of radio in 1925, when Gregory Breit and American physicist Merle Tuve (1901-1982) introduced the pulse signal. The British Air Ministry subsequently funded a major project to develop a radio ''death ray'' that would immobilize enemy aircraft and would send them tumbling to earth. This concept was dismissed when Sir Robert Watson Watt took over the project; he instead suggested the use of radio for echolocation.

To find an object with echolocation, a signal must be transmitted. Since the velocity of radio waves (equivalent to the speed of light) is a constant, it is easy to calculate the distance to an object by measuring the time it takes for a radio signal to travel that distance. When one of the transmitted signals returns to the source, the back-and-forth time is recorded, as well as the direction from which it is reflected. This information is collated and displayed on a *cathode-ray screen*, pinpointing the exact position of the object.

The concept of echolocation was not new at the time of early radar research; **Paul Langevin** had just perfected his **sonar** system in the years following World War I. However,

Radar room on an aircraft carrier during World War II. The development of radar technology contributed significantly to the Allied victory.

the rise of Nazism in Germany and the attendant threat of invasion by air motivated the British government to accelerate their research. In order to adapt sonar technology for use as radar, a chain of radiolocation stations had to be constructed which would throw an invisible net of radio waves over Britain with the ability to detect the approach and precise location of any aircraft. By 1938 Britain had built a network of radio stations along their eastern border—the direction from which Nazi bombers would arrive. When Harry Boot and John Randall invented the *multicavity magnetron* (the first practical microwave transmitter), the system was complete, just in time for its most important test.

In 1940 Hitler ordered a massive air strike against England. The Nazi air force, which greatly outnumbered the Royal Air Force, would certainly have defeated the British had it not been for the radar shield which foiled the Nazis' usual stealthy tactics. Soundly beaten in the daytime bombings of the Battle of Britain, Hitler's forces concentrated their efforts upon night bombings. This strategy failed as well; the radar's eye could easily locate the Nazi planes, even in darkness and poor weather. The military use of radar continued throughout World War II. Compact transmitters were developed that could be mounted on the underside of a plane to scan the ground far below for targets. **Bomb**s and shells equipped with radar tracking systems were designed that could "look" for their targets, exploding at just the right moment.

Though radar use was confined to the military for years, radar devices began to trickle into everyday use soon after the end of World War II. In 1947 a young engineer named John Barker, while attempting to use radar to regulate traffic lights, noticed that a passing automobile would reflect a radio pulse, and that the velocity of the vehicle could then be determined by examining the returning signal. Much to the dismay of chronic speeders, Barker had devised the first radar *speed-gun*, now used by police and highway patrolmen worldwide. Indeed, Robert Watson-Watt himself was caught speeding in Canada in 1954 with a radar speed gun. Subsequently radar detectors for drivers, often called fuzzbusters, were invented in 1968 and became widely used in the 1980s. They detect when a police radar transmitter is working within a certain area, and communicate the distance in a series of beeps. Other radar detector, such as the Safety Warning System, can also inform drivers of any highway hazards such as accidents or construction.

The civilian use of radar is also widespread: marine navigators, surveyors, meteorologists, and astronomers all have found use for radiolocation technology. A continuous-wave version called Doppler radar is often used to track storms and hurricanes. Microwave Early Warning radar is used by air traffic controllers to help guide airplanes. Probes launched into space have also used radar to map the surfaces of other planets including Venus in the early 1960s. Radar has also been used to map Earth from space. On a space shuttle mission in 1994, an experimental radar used microwave frequencies to map a segment of Earth's vegetation in color. This radar uses three different frequencies to end up with one color picture.

The future of radar lies in the micropower impulse radar, or MIR. Invented by Tom McEwan, an electronics engineer at California's Lawrence Livermore National Laboratory, the battery-operated MIR can fit on a 1.5 inch (3.81 centimeters) small circuit board. Working within a 200 feet (about 610 meters) radius, the MIR uses a rapid fire sampler to analyze very short radio pulses. It has amazing accuracy at low power. Because it is also unaffected by noise or temperature, and is inexpensive, the MIR has potentially numerous uses. It can be placed in a car to detect if anything in its path when it backs out of a parking space or change lanes. The signal from MIR is nearly undetectable making it useful in such items as burglar alarms. The MIR also has potential applications in medical diagnostics and, of course, the military.

See also Cathode-ray tube; Traffic signal

Radial keratotomy

Radial keratotomy (RK) is a surgical procedure in which precisely placed micro incisions are made in a patient's cornea—the clear matter which covers the eyeball—to permanently correct near-sightedness, or myopia. In myopia, the curvature of the cornea causes light rays entering the eye to fall short of the *retina*. Radial keratotomy flattens the eyeball minutely, allowing light rays to focus at the retina. First performed in Japan in 1955 by T. Sato, Soviet ophthalmologist Svyatoslav Fyodorov developed the procedure in the 1970s after removing glass splinters from a patient whose vision actually improved once the eye healed. The procedure was brought to the United States in 1977 and further refined. Although once considered a risky procedure—and there are still risks involved—more than one million people worldwide have undergone treatment. In most instances, RK can improve myopic vision to 20/40 or better, even though some cases the patient still needs lenses due to under-correction and in others because of over-correction which causes far-sightedness.

Other permanent procedures as alternatives to RK include photorefractive keratectomy (PRK) which uses an excimer (ultraviolet) **laser** to vaporize corneal tissue; automated lamellar keratectomy (ALK), in which a microkeratome is used to remove portion of the cornea which is frozen, reshaped by grinding, and sewn back into the eye; laser-assisted intrastromal keratomileus (LASIK), which combines portions of ALK and LASIK to make a tiny flap on the cornea thrugh which the inside layers are reshaped and the flap then closed; and orthokeratology—flattening the cornea by the rubbing of hard contact lenses which exert pressure on the eye.

Radiation detectors

Nuclear radiation cannot be detected by any of the five human senses directly. For example, radiation from naturally-occurring sources such as carbon-14 or potassium-40 is present in the environment around us at all times. But without special instruments to detect its presence, humans are totally unaware of its existence. It was necessary, therefore, to develop instruments for the detection of radiation as soon as the existence of that radiation was recognized.

Detection devices can be classified in a variety of ways. For example, some instruments are *counters*. They record the presence and the number of particles present in a situation. The **Geiger counter** is an example of such a device.

Other instruments record the track of a particle. A **cloud chamber**, for example, allows one to take a picture of a particle's movement through an electrical and/or magnetic field. The picture provides information about the particle's mass, charge, velocity, and other properties.

The first radiation detection device ever used was the **photograph film**. In fact, Antoine Becquerel's (1852-1908) discovery of radioactivity in 1896 occurred when radiation from a uranium ore fogged a photographic plate. That principle is still used in one of the most common of all radiation detection devices: the film badge. A film badge consists of nothing other than an employee identification badge that also contains a piece of photographic film wrapped in black paper. Nuclear radiation to which the employee is exposed penetrates the badge and fogs the photographic film. The amount of fogging is directly proportional to the radiation received by the film.

The two general categories of detection devices were developed early in the history of nuclear research. In 1899, for example, Becquerel observed that various forms of radiation produce luminescence in certain substances, including zinc sulfide, barium platinocyanide, and diamond. That discovery was put to practical use in **William Crookes**' invention of the spinthariscope in 1903, a device for observing and counting the flashes of light produced when radiation strikes a zinc sulfide screen.

The major drawback of **scintillation counter**s, such as the spinthariscope, is the practical difficulty of counting the individual pulses of light associated with each particle contact. The invention of the photomultiplier tube in the 1930s provided a solution for that problem. Each pulse of light is converted into an electrical current, multiplied, and then electronically recorded. Photomultiplier tubes made scintillation counters relatively simple and efficient to use so that they find application today in nearly every aspect of research on radioactivity.

Another method for counting particles depends on the fact that most forms of radiation have the tendency to ionize an atom. When an alpha particle passes through a gas, for example, it tears loose one or more electrons from the atoms near its path. The result is the formation of free negative electrons and positively-charged ions. If the event takes place within an electrical field, the positive ions and negative electrons then travel to opposite poles, producing an electrical current that can be measured. This principle was first recognized by Marie Curie (1867-1934) in 1898 and by **Ernest Rutherford** in 1900. It forms the basis of the Geiger counter first invented by Hans Geiger (1882-1945) in 1913.

The general principle of the Geiger counter has been modified and used in a number of other detection devices. For example, a **spark chamber** consists of a stack of thin metallic plates separated from each other by narrow gaps. The plates are highly charged and insulated from each other. When an ionizing particle passes through the chamber, the ions it produces are attracted to the charged plates, producing sparks as they do so. The spark pattern can be photographed or otherwise recorded, indicating both the number and path of the particles that pass through the chamber.

A device similar to the spark chamber is the multiwire proportional counter (MPC). In the MPC, an array of charged thin wires and plates allows the tracing of particles that move through the device.

Spark chambers and MPCs illustrate the point that many detectors can be used to record both particle numbers and particle paths. Another early example of particle detector, the Wilson cloud chamber, has the same property. The cloud chamber was developed by **Carl Wilson** between 1896 and 1911. The cloud chamber contains a vapor that can be instantaneously cooled below its boiling point. When a particle passes through the chamber, it produces ions that act as nuclei on which the supercooled vapor can condense. The particle's path is clearly distinguishable by means of the line of droplets formed in the chamber.

A more efficient form of the cloud chamber, the bubble chamber, was developed by **Donald A. Glaser** in 1952. The bubble chamber employs a superheated liquid rather than a supercooled vapor to record the track of a particle. When a particle passes through the bubble chamber, the ions it forms act as nuclei for the instantaneous boiling of the liquid. The path of the particle is mapped by the tiny bubbles formed in the chamber.

Physicists are continually looking for new ways of identifying particles. In some important cases, the discovery of a new physical phenomenon has been put to use in the development of a new particle detector. The analysis of radiation produced by particles traveling faster than the speed of light by Pavel Cherenkov (b.1904) in the 1930s is an example. Cherenkov found that the angle at which such radiation is emitted is a function of its velocity. This discovery led to the development of the Cherenkov detector, used to measure the velocity of particles that pass through it.

Similarly, research on semiconductors led to the development in 1958 of nuclear **semiconductor** detectors. In these devices, the presence of a particle is detected by the creation of an electron-hole pair within the semiconductor.

Radiation detection devices have a wide variety of applications today. Some are highly practical, as in the measurement of radiation to which workers are exposed or in the detection of radiation used for medical or industrial purposes. The devices are also critical in pure research, however, as in particle research carried out in accelerators. In such cases, very large banks, often containing a number of different types of detectors, are employed.

RADIATOR, AUTOMOBILE • See Engine cooling systems

RADIO

A radio is much more than a device that provides communication and entertainment. Its invention led to a revolution in technology.

Although turning on a radio produces sound, radio waves themselves can not be ''heard'' and have nothing to do with sound waves. Sound waves are a vibration of the air; radio waves are a part of the electromagnetic spectrum. They have long wavelengths and are located below the red and infrared portion of the spectrum. As history has shown, radio waves have the excellent property to act as a ''carrier,'' carrying the information which a radio detector (receiver) then decodes and amplifies.

The existence of radio waves was first predicted by James Clerk Maxwell (1831-1879). Maxwell, who had worked out the laws concerning electromagnetism in 1865, had predicted that an oscillating current would produce radiation with an extremely long wavelength.

In 1890 **Elihu Thomson** discovered the principal of **alternating current** repulsion which eventually led to the invention of AC **motors**. During his work with high-frequency **generators** and **transformers**, he passed electricity through a spark induction coil and created the electromagnetic waves which Maxwell had predicted.

The person who made real progress with radio waves was Heinrich Hertz (1857-1894). He used an **oscillator** and a spark-gap to generate radio waves. He also used a simple ''Hertzian dipole **antenna**'' to detect and measure the shape and intensity of his ''Hertzian waves.''

In 1890 Edouard Branly (1844-1940) discovered that loose metal powders cohered together when they were subjected to electromagnetic radiation. **Oliver Lodge** incorporated this discovery in his coherer, which made for a very efficient radio wave detector. Lodge was a proponent of ''focusing'' radio waves by tuning them, thus obtaining better results in their detection. In 1894 he used radio waves to transmit Morse code one-half mile (800 m). In the year following Lodge's success, Russian physicist Aleksandr Stepanovich Popov (1859-1906) added a wire to the coherer, inventing the first modern antenna.

Meanwhile, in 1893, the brilliant inventor **Nikola Tesla** (1856-1943) devised a method of sending information via wireless **telegraph**. In 1897 he gave a demonstration over a distance of 25 miles (40 km), and the following year he built a radio-controlled model ship. He predicted that radio would become a means of ''enlightening the masses,'' but instead of developing radio telegraphy into a means of communication, Tesla allowed himself to be sidetracked, devoting himself to the establishment of a power distribution system using wireless transmission towers.

Karl F. Braun, attempted to increase the range of spark-gap generated radio signals by increasing the power of the transmitter, but found it was effective only up to a certain limit. In 1899 he invented a method of using magnetic induction to attach the transmitter directly to the antenna, which

The receiver Guglielmo Marconi used when he transmitted the first radio signal across the Atlantic Ocean on December 12, 1901.

greatly improved the range. Magnetic induction has since become the standard for all similar forms of transmission. Braun also experimented with **crystal rectifiers**, having discovered (in 1874) that some mineral metal sulfides conducted electricity in one direction only. Early crystal radio sets used his "cat's whisker" until the crystal was replaced by the **vacuum tube**.

The man credited with the invention of radio was **Guglielmo Marconi**. Marconi amazed the world on December 11, 1901, when he successfully transmitted a **Morse code** signal 2,137 miles (3,440 km) from England to Newfoundland. With his invention, it was possible for ships to communicate over great distances. (Radio played a role in signalling for help when the *Titanic* struck an iceberg and then sunk in 1912.)

According to the experts, it should not have been possible for Marconi to transmit a signal such a distance. Radio waves should behave like light waves and not bend around the curvature of the earth. Arthur Kennelly (1861-1939) and Oliver Heaviside (1850-1925) theorized that a layer of charged particles in the upper atmosphere might be responsible for reflecting radio waves. The Kennelly-Heaviside layer was discovered by Edward Appleton (1892-1965) in 1924, and is now known as the ionosphere.

In the meantime, **Reginald Aubrey Fessenden** made a great advancement in 1901 with the invention of a high-frequency **alternator** (a generator that produces alternating current). It produced a continuous radio wave instead of the intermittent spark-generated pulse with which Marconi was working. Fessenden's liquid barretter detector would eventually replace the coherer as a radio detector.

In addition, Fessenden discovered the way to modulate the amplitude (the distance from a "peak" to a "trough") of radio waves. He used a **microphone** to convert sound waves into an electric signal which he then superimposed on the continuous radio waves. The **amplitude modulated** (AM) radio waves matched the amplitude of the electric signal, and were sent out over the transmitting antenna. On Christmas Eve, 1906, Fessenden used his invention to make the first vocal broadcast in history, sending his voice, live violin, and recorded music over the radio.

The invention of the vacuum tube made for a revolution in radio technology. While working with his electric light bulb in 1884, **Thomas Edison** placed a metal plate between the glowing filaments and unwittingly invented the **diode**. This device allowed electricity to flow from the positive side of the filament to the plate, but not from the negative. This "Edison Effect" was not understood and was treated as a curiosity.

Later, a former assistant of Edison's, **John Ambrose Fleming** of England, became involved in designing a radio transmitter for Marconi. In 1904 Fleming, realizing that the diode could convert alternating current into **direct current**, incorporated it into his very efficient radio wave detector. Fleming's diode was named the thermionic valve because it used heat to control the flow of electricity just as a valve controls the flow of water. In the United States the valve was called a vacuum tube.

Lee de Forest (1873-1961) improved on Fleming's diode in 1906. He added a third element, creating a triode, which he called an audion. The audion was a more sensitive receiver, but it had implications beyond what de Forest realized.

Edwin H. Armstrong realized the full potential of the audion in 1912. Using the triode to create a "regenerative circuit," Armstrong found he could amplify radio signals to such a degree, they could be sent to a loudspeaker and heard without the use of headphones. In 1924 he invented another innovation, **frequency modulation** (FM). With FM, radio signals were impervious to the interference that lightning discharges wrought on AM signals. Also, the ionosphere does not reflect and distort the FM signal, as it does an AM signal. Frequency modulation also permitted multiplex broadcasting, opening the door for **stereo** sound.

Armstrong's triode had an unfortunate side-effect; it amplified distortion as well as the original signal. In 1923 Harold Stephen Black invented a "feedback-feedforward" system that suppressed, but did not eliminate, distortion. Four years later he came up with the negative-feedback **amplifier** that had the ability to filter out any amount of distortion.

The vacuum tube, the device that made all these advancements possible, became an essential component of other forms of communication as well. Photo tubes were used to record and retrieve sound for **motion pictures**, the **cathode-ray tube** led to the development of **television**, **microwave** tubes were used in the early days of space communication, and storage tubes made the invention of the **computer** possible. The vacuum tube reigned supreme until the invention of the **transistor** by **Walter H. Brattain**, **John Bardeen**, and **William B. Shockley** in 1948. The transistor did everything that its more fragile, larger, power-hungry predecessor did. Transistors became commercially available after 1960, and the Sony Corporation introduced the first small, portable transistor radio. The invention of complete **integrated circuits** permitted even further miniaturization.

In the early days of radio it had become accepted that the wavelengths in the Very Low Frequency, Low Frequency, and Medium Frequency ranges were the best for communication. "Short" waves (SW), which were in the High Frequency range, were ignored as being of no commercial value. However, amateurs soon found a use for them. Very High Frequency (VHF) and Ultra High Frequency (UHF) waves were set aside for television broadcasting.

The range on either side of a frequency is known as the sideband. In areas where there is a lot of broadcasting, the available frequencies become congested and the sidebands create a lot of wasted space. To overcome this, single-sideband transmission is used to filter out one of the sidebands. Superimposing two single-sidebands onto a common carrier wave permits stereo broadcasting.

Today's technology has made radio communication available to the masses. A walkie-talkie is a hand-held transceiver (transmitters and receivers) which contains its own source of power and antenna. A mobile station, also known as a Citizen's Band radio, is installed in **trucks** and **automobiles**. A base station, having a large fixed antenna and greater power, is used by shortwave radio amateurs ("hams"), police, military, and businesses. New technology and direct **satellite** broadcasting has made possible the abundance of **cellular tele-**

phones and personal pocket pagers (''beepers''). Each of these devices operates on the same tried-and-true principle.

Tesla's 1897 prediction has been realized far beyond his wildest expectations.

Radio interferometer

A radio interferometer was invented by merging **radio** astronomy with **computer** technology.

Radio astronomy began in the 1930s when Karl Jansky (1905-1950) built a radio receiver designed to locate interference that was plaguing long distance **telephone** communications. His discovery of radio signals coming from space created a sensation in 1933.

The first actual **radio telescope** was invented when **Grote Reber** built a 29.5–ft (9–m) parabolic dish receiver in 1937. With it he was able to create a map of numerous radio sources in the sky.

In the case of optical **telescope**s, the greater the aperture, the more light can be gathered. Likewise, the greater the area of the radio telescope dish, the more radio waves detected. Unfortunately, radio waves are at the low end of the electromagnetic spectrum. These wavelengths are long and stretched out. A really gigantic dish is needed to produce sharp resolution of radio objects, and that does not come inexpensively.ingenious alternative. Instead of building a massive radio dish with a huge aperture, he planned to link up smaller radio telescopes to synthesize an aperture that would achieve the same result.

In 1955, Ryle used a combination of twelve radio telescopes to simultaneously observe the same object. The data was recorded and then transmitted to a single receiver. A computer was used to synchronize and analyze the information to produce a single image that was much more detailed than any single radio telescope could have created.

This process became known as radio interferometry with aperture synthesis. In 1964 Ryle used three radio telescopes that were 1 mi (1.6 km) apart, creating a combined resolution that was essentially equivalent to a single dish of the same diameter.

The radio interferometer greatly expanded the potential of radio astronomy. Interferometers have been established at radio observatories throughout the world. A three-telescope interferometer was built by the National Radio Astronomy Observatory (NRAO) at Green Bank, West Virginia, with each dish having a diameter of 85 ft (26 m).

In the late 1970s, NRAO built the Very Large Array (VLA) radio interferometer in New Mexico. It includes 27 telescopes, each with a diameter of 82 ft (25 m), mounted on rails in a ''Y'' shape. Each arm of the ''Y'' has a maximum length of about 12 mi (19 km). This is currently the world's largest single unit, but even bigger interferometers have been synthesized.

For instance, Very Long Baseline Interferometry (VLBI) creates a radio telescope with an effective diameter up to several thousand kilometers. The Merlin system of the United Kingdom and the United States VLBA are examples. When finished, VLBA will be able to resolve an arc in the sky the thickness of a pea in San Francisco, California, as seen from New York City. Supernova 1987A, in the Large Magellanic Cloud, was studied with radio telescopes in Australia, South America, and South Africa in a VLBI array. Two telescopes on opposite sides of the earth could produce the resolution of a single dish nearly 8,000 mi (12,756 km) across. Not even the sky's the limit, since plans for orbiting radio interferometers would create an effective diameter larger than the earth.

Radio telescope

Radio waves are a part of the electromagnetic spectrum. Just as an optical **telescope** gathers visible light and magnifies it, a radio telescope is a large dish-shaped device that gathers and amplifies radio frequencies from space. Radio telescopes do not ''see'' the way an optical telescope does; the radio signals the telescopes gather are used to create diagrams showing where the strongest radio sources are located.

Cosmic radio waves were discovered accidentally by Karl Jansky (1905-1950) in 1932. He had been using a receiver to search for the source of radio noise that was interfering with long distance radio-**telephone** conversations. That the source could be out in the cosmos was a big surprise, but there was no follow-up to the discovery because radio astronomy did not exist as a science.

One person, an amateur ham-radio enthusiast named **Grote Reber**, was electrified by Jansky's discovery. Reber built the world's first radio telescope in 1937, spending $1,300 of his own money to build a receiver in his backyard from lumber and galvanized sheet metal. The dish—the portion of the telescope that collects radio waves—had a diameter of 31 feet (9.5 m). The receiver—the portion of the telescope that amplifies the waves—was mounted at the focus of the dish.

Reber's device was similar in design to the backyard satellite dishes of today, though at a considerably larger scale. Unlike backyard dishes, which are intended only to follow man-made satellites in a narrow band of sky, however, a radio telescope needs to be moved so it can point at different parts of the sky. To keep costs at a minimum, Reber used a meridian-transit mount; the dish could be moved in a north-south line only. The rotation of the earth provided movement from west to east.

The bigger the dish, the more data that can be collected, and the greater the resolution and sensitivity of the telescope. In 1947 Sir Bernard Lovell (1913-) directed the construction of a parabolic radio reflector 216 feet (66 m) in diameter at Jodrell Bank, England. It was also here that a 250-foot (76 m) dish went into operation in August, 1957. A gear-and-rack mechanism once used to turn gun turrets aboard battleships was used to aim the telescope. It was the world's largest steerable telescope, but only for five years.

Although dish diameter determines telescope performance, the size cannot be scaled up indefinitely. Past a certain size, structural issues come into play—how do you support such a massive dish? How do you turn it? Consider, for example, the 300-foot (91 m) diameter dish at Green Bank, West

Virginia, built by the National Radio Astronomy Observatory in 1962. It was the largest steerable radio telescope in the world, until it collapsed from metal fatigue in 1989.

The largest individual dish currently in existence is in Arecibo, Puerto Rico. Built by the engineer Thomas Kavanagh (1912-), the radio telescope was created out of a natural depression within a ring of mountains in 1963. With a diameter of 1,000 feet (305 m), the telescope's reflecting surface is too large to be steerable. Instead, a movable feed suspended above it the dish provides access to most of the sky.

Arecibo pushed the limits on the size of a single-dish telescope, but astronomers developed other approaches that circumvented the structural issues involved in building large-aperture radio telescopes. In 1955, Sir Martin Ryle (1918-1984) invented the **radio interferometer**, a device in which several telescopes are linked to synthesize the performance of a single telescope with an extremely large aperture. As with optical telescopes, the enhanced signal-gathering ability provided by the increased aperture size greatly increased the resolution of these telescopes, permitting the detection of more distant objects.

The Very Large Array (VLA) is one of the best examples of this technique. Built in New Mexico in the 1970s and dedicated in 1980, the VLA consists of an array of radio telescopes that can operate with the resolving power of a single antenna as large as 36 km in diameter. The 27 telescopes are mounted in a ''Y'' configuration, with nine telescopes on each arm of the ''Y''. Weighing approximately 360 tons, each of the 82 foot (25 m) dishes can be moved from place via a system of railroad transporters to one of four standard configurations; each configuration yields a different resolving power. The maximum antenna separation ranges from 0.62–22.3 mi (1–36 km), and changing from one configuration to another requires an average of two weeks.

During observing projects at the VLA, which generally take from half an hour to several days, data from the antennas is acquired in real time in a process called correlation. This technique is feasible for even an array as large as the VLA. Past a certain distance, however, the antennas in a synthesis array are separated widely enough that this level of data communications is no longer possible. To move beyond the size and resolving power of the VLA, then, scientists were forced to develop Very Long Baseline Interferometry (VLBI), in which the signals from widely-spaced antennas are recorded on magnetic tape and correlated later by computer.

In 1986, the National Science Foundation began construction of the VLBA, an array of ten radio telescopes designed for VLBI observation.

The Very Large Array (VLA) radio telescope in Socorro, New Mexico. Constructed in 1977, it is the most powerful radio telescope in operation to date.

RADIO TUBE • See Vacuum tube

RADIOIMMUNOASSAY (RIA)

Radioimmunoassay is an extremely sensitive method of measuring very small amounts of a substance in the blood. The isotopic method was developed in 1959 by the Americans, biophysicist **Rosalyn Yalow** and physician Solomon A. Berson (1918-1972) at the Bronx (New York) Veterans Administration Hospital.

Yalow had established the hospital's radioisotope laboratory in 1947. In 1950, she was joined by Berson, a resident in internal medicine who became interested in her work. Berson was born in New York City in 1918, and received his B.S. in 1938 from the City University of New York and his M.D. in 1945 from New York University.

Yalow and Berson developed their first radioisotopic technique to study blood volume and iodine metabolism, then adapted the method to study hormones. Working with very pure insulin, they became the first to discover that small molecules can induce production of antibodies. They were also able to show that, contrary to theory, Type II (adult onset) diabetes is caused by inefficient use of insulin, not by lack of it in the body.

In 1959 they perfected their method, naming it radioimmunoassay (RIA). Its extreme sensitivity—it measures one thousand billionths of a gram of material per milliliter of blood—quickly made it a standard laboratory tool. To measure insulin, the first step is mixing known amounts of radioisotope-tagged insulin and antibodies. These combine chemically. Next, a small amount of the patient's blood is added, the insulin it contains displaces some of the tagged insulin. The free tagged insulin is then measured with isotope detectors and the patient's insulin level is calculated.

RIA has been used for measuring peptide and steroid hormones, as well as other substances such as morphine, viruses, vitamins, cyclic adenosine monophosphate (cAMP), and messenger RNA (ribonucleic acid). It has been applied to narcotics detection, blood bank screening for hepatitis virus, early cancer detection, measurement of growth hormone levels in shorter-than-normal children, tracking of leukemia virus, diagnosis and treatment of peptic ulcers, and research with brain chemicals called neurotransmitters.

In 1968 Berson became chair of the Department of Medicine, Mt. Sinai School of Medicine, still continuing to work with Yalow. His death in 1972 made him ineligible to share the portion of the Nobel Prize in physiology or medicine that Yalow received in 1977 for their work.

RADIOTHERAPY

Since the development of x-ray technology, medical professionals have embraced the concept of radiation as a diagnostic and therapeutic tool. This nuclear science, called radiotherapy, utilizes radioactive particles and elements for diagnosis and treatment of a variety of illnesses. Because x-ray radiation can only image dense materials such as bone, it has limited in application and overexposure can be harmful. Diagnosing certain disorders in soft tissue is now possible through the use of radionuclides, or **radioactive tracers** which are introduced into the body by way of an injection called a **radiopharmaceutical**. According to the type of radionuclide, the tracer will collect in one or more areas of the body. As the isotope emits radiation it is easily detected by a **Geiger counter** or scanning device so physicians are able to follow its path, checking the organs under surveillance. The radionuclide is flushed out of the body like any ordinary waste product. Radioactive trace elements are becoming a preferred method of diagnosis for two reasons. First, they can be used to target individual organs, such as the kidneys or the thyroid. Second, they emit radiation from the body while x-ray radiation is absorbed by the body; thus, the amount of radiation to which the patient is exposed is much lower. Trace elements are also often used to replace riskier and more traumatic procedures such as contrast **arteriography**.

Advances in radiotherapy are increasing its application in disease detection, diagnosis, and treatment, particularly of certain types of cancers. *3-dimensional conformal radiotherapy* uses CAT (CT) or **magnetic resonance imaging** (MRI) scans to detect disease and permit precise treatment targeting of tumors. Treatments may consist of either a well-defined beam of radiation from a high-voltage ''linear accelerator,'' neuron or proton beam therapy, or implants of radioactive ''seeds'' of iodine or palladium. For example, treatment of early prostate cancer may involve the insertion of Iodine-125, a low-energy, gamma-emitting isotope, directly into the prostate gland. Implanted under ultrasound guidance, this one-day, cost-effective, and convenient form of treatment—called prostate *brachytherapy*—appears to control local disease as effectively as external beam therapy. Brachytherapy is also used for gynecological, head and neck, and soft tissue cancers.

Stereotactic radiosurgery, an extremely complicated therapy, delivers multiple converging beams of high-dose radiation directly to the target lesion minimally affecting surrounding tissue. Used primarily to treat brain tumors, it is also used for vascular malformations, certain recurring head and neck tumors, and lesions at the base of the skull. Total skin electron irradiation (TSEI), a complex treatment for certain types of cutaneous lymphomas, uses an electron (as opposed to photon) beam to treat the entire skin surface. Because penetration of the electron beam is superficial, underlying tissue remains relatively unaffected. *Intraoperative radiotherapy* involves the insertion of an electron cone into the bed of a surgically removed tumor which has a high risk of recurrence. Radiation is therefore delivered to the tumor bed only. In a July 1997 article in *The Lancet*, British researcher Dr. Michele Saunders and colleagues released a study of a radiotherapy regimen called *continuous hyperfractionated accelerated radiotherapy* (CHART), found to increase the survival rate of patients with non-small-cell-lung cancer. Still in its experimental stages, this unique radiotherapy schedule delivers more than one treatment each day, delivering the same total dosage in approximately 12 days as compared to the normal treatment time of four to six weeks. This accelerated dose schedule reduces the problem of radiation resistance encountered in traditional therapy.

The person most responsible for the acceptance of nuclear medical procedures is American radiologist Edith Quimby. Quimby was the first to precisely measure the amount of radiation necessary for tracing within the body, and later determined exact dosages required for therapy. In addition to diagnostic and therapeutic applications, nuclear radiation is used to sterilize medical equipment, making the old method of sterilization through steam practically obsolete. Because it can be administered at low temperatures, radiation can be used to sterilize plastic instruments that would be destroyed by steam and reaches small crevices that steam would often miss.

RAILROAD • See Train and railroad

RAILWAY CAR COUPLER

Railroads have played a very important part in American history. After the Civil War, the railroad industry became one of the nation's largest employers and provided many inventors with an ongoing source of inspiration.

In 1897, a young African-American railroad worker, Andrew Beard, filed a patent for an improved coupling mechanism. He had witnessed many of his co-workers become injured or maimed by the traditional ''pin'' system, in which a coupling pin had to be placed into eyes on the ends of each car to join the cars together. This method of coupling also caused jarring lurches and jolts when the train started and stopped.

While many inventors turned their attention to this problem, Beard devised an automatic system to prevent the risk to the lives of railway workers when train cars were attached or disconnected. His first-hand experience with the system and his clear ingenuity seemed to offset a lack of formal training in either engineering or metal working, and Beard successfully introduced the Jenny Coupler. The device consisted of a large, curved head that pivoted to lock onto a swiveling head from another train car. His system was such a success that Beard received $50,000 for his invention before the end of 1897. Although some improvements were made during the next century, most modern couplers were based on the same construction and action as Beard's design.

However, by the mid 1990s, an entirely new railcar called the Flexiliner had been produced in Denmark that employed a revolutionary coupling system. Using state-of-the-art computer technology, each Flexiliner car has a rubber-rimmed front that, upon command, automatically deflates its rubber diaphragm and rolls forward slowly until it couples with the car ahead of it. Once joined in the traditional manner, the lead unit's rubber shrouding reinflates to provide an air-and-water-tight seal. Since all the air and electrical control circuits are contained within the coupler, there is no need for the ages-old ritual of connecting hoses and wires. Further, uncoupling is even faster and more efficient since Flexiliner cars can uncouple while moving. This is achieved by having the front car continue at speed while the second car applies its brakes. In 1996, these new cars and their new coupling systems made successful demonstration runs in Las Vegas, Nevada.

RAM • See Computer memory

RAMSEY, NORMAN • See Atomic clock

RANGE • See Stove

RAYON

Rayon is a generic word for fibers developed from cellulose. The first patent for synthetic fibers was granted in 1855 to a Swiss chemist, George Audemars. In 1880 Sir **Joseph Swan**, an English chemist succeeded in using nitrocellulose to produce fibers. Although these two men started the cellulose-based fiber industry, it is Count **Louis Comte de Chardonnet** who accelerated its production.

Rayon is produced from nitrocellulose, a substance that, due to its nitro groups, is highly flammable. Rayon was often denitrated with a nitric acid bath. Unfortunately, this process weakened the fibers. Chardonnet used an acid sulfide solution to denitrate the fibers and preserve their strength, which is comparable to that of silk.

A twist on the production of rayon was introduced in 1897 by Hermann Pauly, who dissolved cellulose with an ammoniacal solution of copper hydroxide. The rayon produced this way is called cuprammonium rayon, Pauly silk, or Bemberg rayon. Very little of the rayon made in the United States (about five percent in 1950) was made by using Pauly's process. Today most rayon is produced through the viscose process.

The viscose process begins when cellulose from wood pulp is mercerized in sodium hydroxide, aged for a few days, and then xanthated with carbon disulfide. The cellulose xanthate is dissolved in sodium hydroxide which lowers the viscosity. After three to six hours the solution is thick and orange, and known as viscose. Pigments are added and the viscose is forced through spinnerettes—small metal plates covered with fine holes-into an acid bath where it is regenerated. The rayon is then stretched and twisted slightly, washed, and dried.

The second major type of rayon is cellulose acetate rayon. Viscose and cuprammonium rayons are made of regenerated cellulose. Acetate rayon is an ester of cellulose. **Charles Frederick Cross** and **Edward John Bevan**, both chemists from England, developed the industrial production process for cellulose acetate in 1894. Originally cellulose acetate was dissolved in solvents and used as a coating on airplane wings, cloth, and metals. Eventually it was formed into fibers.

The cellulose for acetate rayon comes from cotton rather than wood pulp. Cotton fibers are impregnated with glacial acetic acid and sulfuric acid. This causes the fibers to swell. Acetic anhydride is added and the solution is allowed to age for several hours. More acetic acid is added and the solution is aged again this time for twelve to sixteen hours. The solution is run under water, the acetate precipitates as white flakes, which are then dissolved in acetone, filtered, and deaerated. Pigments are added and the solution is forced through spinnerettes, stretched, and dried.

Because it is based on an organic material, rayon "breathes" more than most synthetic fibers, making it as comfortable as cotton or linen in hot weather. The material itself tends to be heavy, even for very thin fabrics; consequently, rayon tends to offer a great deal of movement, making it excellent for draping and flowing designs. Rayon was popular in the 1930s and 1940s, but in the 1950s it was supplanted by the other synthetic fibers and fell out of favor. In the 1980s, consumers rediscovered the fabric, and it regained its popularity. In the late 1980s, a distressed form of rayon called sandwashed rayon provided a new twist on the material, offering a texture and appearance similar to sandwashed silk for a significantly lower price.

Just as the discovery of rayon, the first commonly used manufactured fiber, revolutionized the **textile** industry, the investigation of its production radically changed worker safety laws. The production of rayon was not without risk. Carbon disulfide attacks the central nervous system and can cause mental disorders, blindness, and paralysis. In addition, the hydrogen sulfide given off during the reaction can also cause mental disorders and paralysis. Workers in early rayon plants were obviously in danger. Still, nothing might have been done to alleviate these industrial hazards if not for the efforts of Alice Hamilton(1869-1970).

Hamilton was a graduate of the University of Michigan medical school and a noted pathologist. She was appointed director of Occupational Disease Commission of Illinois in 1910 and dedicated her time to making factories safe for workers. Hamilton received a telegram from a company nurse at a rayon factory asking for help with an insanity epidemic. Knowing that the workers were probably suffering from carbon disulfide poisoning, Hamilton replied quickly and offered to help.

Plant management started an information blackout, refusing Hamilton's request for information. Since government officials were not interested in getting involved, Hamilton took matters into her own hands. She began examining the workers in secret, with the help of specialists at the University of Pennsylvania. Eventually Hamilton and the others published their findings. Their efforts resulted in the passage of laws in Pennsylvania granting compensation for occupational diseases and in the development of engineering controls to protect workers.

See also Fiber, synthetic

Razor

A razor is a device used to cut hair from the face or other areas of the body. Men have been shaving off their beards with sharp implements since ancient times. In fact, cave paintings show shells, shark's teeth, and sharpened flint were used as razors. Bronze was the favored substance for razors during the Bronze Age while gold and copper razors were found in ancient Egyptian tombs. The steel straightedge razor was created in the 1700s in Sheffield, England.

However, shaving with this sharp, unprotected blade, was a dangerous procedure. The French cutler Jean-Jacques Perret, therefore, invented a safety razor in the 1760s that guarded all but the edge of the blade. Another edge-guard razor was made in Sheffield in 1828, and a similar design was produced in the United States. The modern T-shape of the safety razor was invented by William Henson, an Englishman. Nevertheless, the old straightedge razor remained in universal use until the advent of King Camp Gillette's disposable blade in 1901.

Gillette (1855-1932) was born in Fond du Lac, Wisconsin, and grew up in Chicago, Illinois. When the fire of 1871 left his family destitute, Gillette became a traveling salesman. In 1891 he went to work for William Painter, the inventor who had earned a fortune with his disposable crown cork bottle cap. Painter noted Gillette's interest in inventing and advised him to develop a product that people would use once and throw away, thus ensuring a steady demand. Struck by the idea, Gillette went through the alphabet searching for a logical disposable product. While shaving one morning in 1895, he saw the answer in his mirror—a thin, double-edge blade that could be thrown away when it got dull, fastened to a special guarded holder.

Technical experts assured Gillette that it was impossible to produce steel hard, thin, and cheap enough to make the disposable blades. A young graduate of the Massachusetts Institute of Technology, William Nickerson, agreed to attempt the project. Nickerson had already invented push-button controls for elevators as well as automatic machinery for packaging and weighing cereal. By 1903, Nickerson had developed the necessary design and machinery and Gillette had secured the needed financial backing. The Gillette Safety Razor Company began producing razors with disposable blades in South Boston in 1901.

Sales steadily increased, and an entire generation was converted to safety razor use when the United States government issued Gillette razors to its troops during World War I. While other disposable blades eventually appeared on the market, Gillette dominated the field. He retired from active participation in his company in 1913 and devoted himself to promoting his cherished "World Corporation," a version of utopian socialism. He died in California.

Another notable change in shaving products came with the advent of the electric razor. Early models were patented in America as early as 1900 and in England shortly before World War I, but the first successful electric razor was patented in 1928 and marketed in 1931 by Jacob Schick, a retired United States Army colonel and committed inventor. During a mining expedition in British Columbia in the winter of 1910, Schick was inspired to develop a razor that worked without soap or water. After World War I, Schick devoted himself to inventing an electric razor, and his wife mortgaged their Connecticut home to finance the venture. The patented design that resulted used a series of slots to hold the hairs while a series of moving blades cut the hairs off. Competing models followed, but all used the basic Schick principle. By the time of Schick's death in 1937, sales of electric razors had become significant.

The completely disposable razor was introduced in 1975 by Baron Marcel Bich (pronounced "beek"), who had invented the Bic pen thirty years earlier. Expanding upon Gillette's disposable concept, Bich conceived of a half-blade razor with a plastic handle that could be produced cheaply enough to be thrown away after use.

Disposable razor technology continues to dominate the market in the 1990s. Razors have been developed which employ multiple blades, lubricating strips, and other special technology to maximize the removal of beard hair.

Reaming machine • See Boring machine

Reaper and binder

Reapers are machines designed to cut, or harvest, grain. As opposed to **mowing machine**s, which are designed to cut green grass for hay, reapers are used in fields of mature grain whose stalks require a heavier serrated blade.

An early corn reaper built in southern Gaul during the first century A.D. consisted of a case with teeth mounted on a pair of **wheel**s and pushed by an ox or donkey; Pliny the Elder (23-79 A.D.), the Roman naturalist, described it. However widespread its use might have been, reaper technology was lost over the centuries and was not retrieved until the early 1800s. Instead, sickles and shears were used for manual grain harvesting.

In 1805 Thomas Plucknett, an implement maker from London, England, patented a machine with a very sharp circular steel plate positioned parallel to the ground. The problem was that the grain fell backwards on top of the operator. In 1811 an individual named Smith of Denton, England, invented a horse-drawn version of the circular-blade reaper, later adapted for use as grain gatherer, using a rake with twenty-four prongs. In 1822 the English team of Ogle and Brown developed a machine that effectively reaped wheat and barley. It used a serrated horizontal bar that moved from side to side. The felled grain was automatically delivered to a binding platform.

Patrick Bell of England designed a reaper with a reciprocating blade in 1826. The bar had triangular plates, or "fingers," that clipped the grain with a scissor-like motion. The grain was pushed downward with six cross-bars, called sails, that rotated above the blade, holding the stalks firmly so the blades could cut them. So many reapers were being invented during this time that none could be effectively marketed. Most were handmade and were good for only one harvest season.

In 1831 American inventor **Cyrus McCormick** developed a reaper with a serrated reciprocating cutting bar later manufactured on a large scale, while another American, **Obed Hussey**, designed his own reaper in 1833. It had a saw-toothed scissor-like cutting blade. McCormick and Hussey became stiff competitors. During the 1850s, McCormick overtook Hussey and went on to become the modern-day International Harvester corporation.

One of the major problems with the early reaping machines was that the time gained in the cutting operation was lost binding the sheaves, which was still done manually. In 1871 American Walter Wood invented a binder that used twisted wire to bind the grain.

This and other wire binders left flecks of metal in the grain that damaged machinery and made farm animals sick when they ate the grain. American **John Appleby** developed the twine binder in 1874 from an idea he had come up with as a teenager. The Appleby Knotter could be used in the field in tandem with the reaper. The development of combines in the mid-1800s gradually rendered the binder obsolete. Combines harvested and threshed the grain in the field, making binding unnecessary. Mechanical corn pickers gradually replaced the corn reaper-binder beginning in the 1920s. Reapers and binders, eclipsed by more modern farm machinery, are no longer in use today

See also Combine harvester; Farming, mass production; Haying machines and tools

Réaumur, Réne-A.F. de • See Thermometer

Reber, Grote (1911-)

American radio engineer

Grote Reber was enthralled by radio. He built his own radio set and was a "ham" operator by the age of fifteen. When Karl Jansky (1905-1950) announced his discovery of radio noise from outer space in 1933, Reber became excited enough to further the work on his own. He tried to reflect radio signals off the Moon, but was unsuccessful. He was ahead of his time; this would not be accomplished until after World War II by the Army Signal Corps.

In 1937, Reber built what can be considered the first **radio telescope** in his backyard in Wheaton, Illinois, using rafters from the local lumberyard, galvanized sheet metal, and old **automobile** parts. The receiving dish was 31 feet (9.5 m) in diameter. The entire apparatus weighed nearly two tons (18 t) and cost $1300, which Reber paid for entirely out of his own pocket.

Although radio astronomy observations can be made during the day as well as at night, Reber worked from midnight to dawn. There was less traffic on the road at that time and electronic "noise" from automobile ignitions would be minimized. His curious apparatus made him the object of wild rumors among neighbors. His first attempts at detecting radio frequencies from space were disappointing. Reber had been looking at the 9 cm range, which would provide the best angular resolution, but he found nothing. He upgraded his equipment in 1938 and tuned for 33 cm wavelengths and still found nothing. It wasn't until the spring of 1939 that Reber found success at the longer 1.87-m (160 megahertz) wavelength.

Reber began making a complete survey of the sky in 1941. He used a chart recorder to collect data and plotted contour lines on a chart, creating the first radio map of the sky. There was a peak in intensity in Sagittarius which lies in the direction of the center of our galaxy. Other peaks were found and later matched to visible objects like the Crab nebula in Taurus. This, and another peak in Cassiopeia, turned out to be the remnants of a supernova explosion. Reber remained the only radio astronomer until after World War II when other scientists embraced radio astronomy following the adaption of radar tracking dishes for use as radio telescopes.

Record, phonograph • See Phonograph

Recycling

Recycling is a way to conserve natural resources by salvaging metals, **paper**, **plastic**, **glass**, and other materials used in packaging and industry, and turning them into new and usable products. Successful recycling involves a three-step process: collecting the material; processing it, which can include sorting, compacting, and transporting it to the buyer; and finally, remanufacturing the material into new items.

One of recycling's many benefits is its conservation of energy and resources; its popularity is increasing with public awareness of environmental issues. Also, as disposal costs grow and landfill space shrinks, recycling becomes a way to recover materials from these landfills, decreasing the amount of new raw materials used, and reserving land for purposes other than dumping.

The energy-saving benefits of recycling were first recognized by the glass and aluminum industries, which require tremendous amounts of energy to produce containers from raw materials. There, it was discovered that far less energy was needed to produce aluminum cans and glass bottles when recycled materials were used. For example, recycling aluminum requires 95% less energy than mining and processing aluminum ore to make new soft drink cans. The total energy saved per can is enough to keep a 100-watt light bulb burning for 100 hours. In addition, recycling aluminum produces 95% less air pollution and 97% less water pollution than processing raw materials to make new cans. Because of these benefits, the recycling industry was willing to pay for collected cans. Thus, the recycling rate for aluminum in the United States, at about 55% in 1989, was higher than that of all other recyclable materials. (By 1995, however the recycling rate for aluminum had fallen to about 50%, slightly behind the rates for paper and steel.)

Communities that recycle use one of two collection methods. With low-tech or manual services, residents separate such recyclable materials as cans, **bottles**, and paper, which are put into a specially designed collection truck with different compartments for each material. Automated collection services can vary, but typically involve a truck that will pick up loads of mixed recyclables. After the truck delivers the comingled materials to a processing center, they are sorted manually as they move along a conveyor belt.

Studies have shown that when recycling is made convenient, people are more apt to participate. Sorting by residents is valuable because it helps educate people about recycling, but programs with comingled collection show higher partici-

Grote Reber

pation and recovery rates. The sorting issue will continue to be an important determinant in any recycling program, especially as communities seek to recycle additional materials. In many cities, junk mail, phone books, cardboard, and certain plastics are being included in collection programs.

By the end of 1995, approximately 50% of all aluminum, steel, and paper packaging in the U.S. was being recycled, compared to about 28% of the glass and 10% of the plastics packaging.

A 1997 report published by the Environmental Defense Fund concluded that from 1990 to 1996, over 13 times more virgin plastic packaging was produced than was recycled. As suggested by these statistics, plastics, which make up about 20 volume percent of current landfill input, pose particular difficulty in spite of the fact that research on the recycling of plastics has been underway since about 1970. This slow progress is due in part because not all plastics can be recycled together, and because there are at least eight different types of popular plastics. Success in recycling depends very much on the chemistry of the plastic involved and on the purity of the scrap, and reprocessing of comingled waste leads to poor physical poperties. Considerable effort, therefore, has been devoted to developing automated processes for segregation according to plastic type. One bright spot is the case of polyethylene terephthalate, where recycling can be achieved by chemical depolymerization to polyethylene terephthalate and other products. Such processes are used for products that come in contact with food.

Until materials like plastics are made easier to recycle, some communities are considering advanced disposal fees, or small surcharges that the consumer would pay to finance the disposal cost of these items.

Reese, Charles L. (1862-1940)

American chemist

The developer of nonfreezing **dynamite** (safe for use in cold-weather situations), permissible explosives (for use in underground coal mines), and the industrial explosive sodatol (created from **TNT** and picric acid), Reese is an important American chemist of the early twentieth century.

Born in Baltimore, Maryland, Reese received his doctorate in Heidelberg, Germany, while studying under **Robert Bunsen**, inventor of the **spectroscope** and **Bunsen burner**. In 1902, after a brief but distinguished career in academics, he assumed the position of chemical director at the Du Pont Company, which he held until his retirement in 1924.

Virtually all of the achievements for which Reese is known came through his work at Du Pont, where he oversaw the rapid expansion of one of the largest and most advanced research departments in the world. He is less well known for his involvement with efforts during World War I to secretly escort two German scientists, who possessed valuable knowledge regarding dyestuff and synthetic-nitrates, to the United States. This successful mission may well have been one of the important turning points for both the war and the field of chemistry, which had long been dominated by the Germans. He died in 1940.

Refrigerated trucks and railway cars

Refrigerated trucks and railroad cars have had a great impact on the economy and eating habits of Americans. As the United States became more urbanized, the demand for fresh food shipped over long distances increased. Meat products were especially in demand.

In the mid-1800s, cattle raised in Texas were shipped by rail to Chicago, Illinois. Although it was more efficient to slaughter the cattle in Chicago and ship the carcasses to the East, rather than send live cattle east by rail, carcasses could only be shipped during the cold winter months. The first refrigerator car patent was issued in 1867 for a crude design developed by William Davis for meat-packer George Hammond. While Hammond was able to ship meat to Boston by 1872, the cars had to be reloaded with ice once a day, and the meat arrived discolored from contact with the ice.

The first successful refrigerator car was patented in 1877 by Joel Tiffany of Chicago. A similar design was developed the same year by meat packer Gustavus Franklin Swift (1839-1903) and his engineer, Andrew Chase. Ice stored on the car's roof dropped cold air down through the car; warm air was ventilated out through the floor. Once meat could be reliably shipped east, the Chicago slaughterhouse industry boomed, and such meat-packing companies as Swift and Armour made fortunes. Refrigeration with ice is still used in railroad cars as well as in **truck**s and ships, with powerful **fan**s circulating the cooled air.

An obvious problem with iced refrigeration of transported perishable foods is that the food may spoil if the ice melts before the shipment reaches the market. In the late 1930s, at the request of the Werner Transportation Company, Minneapolis engineer **Frederick McKinley Jones** (1892-1961) sought ways to build an automatic, ice-free air-cooling unit for long-distance trucking. He designed a compact, shock-proof air conditioner that could withstand the vibrations and jolting of overland travel. Jones's first **air conditioning** device, which was installed under the truck, failed when it was clogged with mud. A unit mounted in front of the truck, above the cab, was a success.

Jones patented his truck air conditioner in 1940. The system was later adapted for use on railroad cars and ships. Jones's invention changed the food industry. For the first time, perishable foods could be reliably transported over long distances at any time of the year. In turn, food production facilities could be located anywhere; foods could be marketed anywhere. A much greater variety of fresh and frozen foods was now available to millions of people.

After World War II, carbon dioxide and ammonia refrigeration systems became standard for trucks and rail cars. By the late 1980s, truck cargo containers were developed to replace refrigerated containers called "reefers" that used chlorinated fluorocarbons (chemicals that damaged the earth's ozone layer). One of these new systems for trucks was an adaptation of the thermos bottle concept. This new system fits a vacuum vessel with slightly curved, airtight double walls inside a standard truck cargo container. The perishable products are loaded cold or frozen and will remain that way for long periods, with no need for further, active refrigeration. For rail systems, cryogenic boxcars that refrigerate with carbon dioxide worked extremely well for frozen products. By the mid-1990s, another system consisting of an insulated boxcar made of high strength composite materials was introduced. Using insulation only, the car had minimal temperature loss over a seven-day run when loaded with cold beer. If fabrication costs come down, it may become competitive with carbon dioxide rail cars and eventually with conventional insulated steel cars as well (since composites are both larger and lighter).

Refrigeration

Refrigeration is the process of removing heat from a substance to produce a low temperature. Heat always flows from a warmer to a cooler body or substance. Refrigeration works by placing something to be cooled near a refrigerant, something cooler that will absorb heat. Since ancient times, people have used refrigeration to help preserve food. Storage at or near 32° F (0° C) inhibits the growth of food-spoiling organisms and also decreases enzymes that change the texture, color, and flavor of food.

As early as 1000 B.C., the Chinese cut and stored ice. The Greeks and Romans filled cellars with mountain snow. Natural-ice refrigeration became a large-scale industry in nineteenth-century America. Entrepreneurs like Frederic Tudor harvested and shipped tons of northern pond ice to southern states and the tropics.

As cities and towns grew in both Europe and the United States during the 1800s, however, the need for mechanized refrigeration became critical. Ever-expanding amounts of food had to be transported over long distances without spoiling. Natural-ice refrigeration could not meet the need. It was also discovered that ice harvested from lakes and ponds contained microbes that caused disease. It had been previously thought that freezing killed microorganisms in the same manner as boiling.

Inventors worldwide rose to the task—scores of ice-making and cooling machines were patented throughout the nineteenth century. These operated on the principle that liquids absorb heat when they change into a gas. Both compression and absorption systems were used to change a liquid refrigerant into a gas and then back into a liquid again in a continuing, self-contained cycle.

The first known artificial refrigeration devices were developed in Scotland. In 1758, William Cullen evaporated ethyl ether in a partial vacuum, obtaining a small amount of ice. In 1777, Gerald Nairne accelerated Cullen's process by using sulfuric acid to absorb the water. In 1805, the American inventor

The first frigidaire refrigerator shipped from the Delco Light Company's Ohio plant.

Oliver Evans designed a compressed-ether refrigeration machine with a closed cycle; however, the Evans machine never advanced beyond the prototype stage. Two early refrigeration machines were patented in 1834. Jacob Perkins (1766-1849), a Massachusetts inventor living in London, England, designed a compression machine with a closed cycle that used ether as the refrigerant. L. W. Wright patented an ice-making machine that used an air-compression process.

In 1844, Florida doctor John Gorrie (1803-1855), in the mistaken belief that malaria was caused by hot, humid air, designed a compressed-air refrigerating machine—an air conditioner and ice-maker—to cure his patients. Gorrie patented his ice-making machine in 1851, but the idea was ridiculed by New York newspapers. James Harrison, an immigrant from Glasgow, Scotland, improved the Perkins ether-compression system and installed the world's first commercial refrigerating machinery in an Australian brewery in 1851. Harrison and fellow British immigrant Thomas Mort later attempted to ship frozen beef to England from Australia.

Ferdinand Carré of France patented the first absorption refrigeration system in 1859, with ammonia as the refrigerant and water as the absorbent. Refrigerators using Carré's design came into wide use industrially. Ferdinand's younger brother, Edmond, put Cullen and Nairne's ideas into practical use with his 1866 water vapor refrigerating machine. Swiss physicist Raoul-Pierre Pictet (1846-1929) developed a compression refrigeration system in 1874 that used sulfur dioxide as the refrigerant. Pictet's machine made possible the world's first artificial skating rink, installed in London in 1876. The first successful compression system using ammonia was designed by **Karl Paul Gottfried von Linde** in 1876 for a German brewery, although an earlier patent for such a system was obtained in 1872 by David Boyle, a Scottish-American inventor. After some years of research, Linde converted his industrial design into a unit suitable for home use.

Today's domestic refrigerators use essentially the same cooling system that Linde had devised in the 1870s. A serious problem with the use of ammonia as a refrigerant is that ammonia is highly toxic; therefore, leaks are very hazardous. In the early twentieth century, refrigeration engineers searched for a viable ammonia substitute. The most well known of these was **freon** (dichlorodifluoromethane), a synthetic substance developed by Thomas Midgley and Albert Henne of General Motors. Artificial refrigerants are now used in systems worldwide.

As artificial ice and mechanical refrigeration became widely available, cold storage and transportation of perishable foods expanded greatly. As early as 1870, chilled beef was being shipped successfully from the United States to England, using an ice-salt coolant. The refrigerator boxcar and, later, the **refrigerated truck** distributed fresh foods throughout the United States.

Home storage of perishables became much more feasible. The use of ice and iceboxes expanded greatly after 1880. Linde marketed some of the first domestic mechanical refrigerators in the 1880s. The Kelvinator, designed by Nathaniel Wales, appeared in the United States in 1918, followed by the Frigidaire in 1919. Mass production began in 1931 with the Electrolux in Sweden and the Servel in America.

Such technical developments as smaller and more efficient equipment, safe artificial refrigerants, and the electric motor contributed to an expansion of the refrigeration industry from the 1930s to the present. Mechanical refrigeration is important today for many uses beyond **food preservation**-for example, **air conditioning**, storage of medical supplies and drugs, medical and surgical techniques, and manufacturing and industrial processes of all kinds, including liquid-air applications.

See also Refrigerated trucks and railway cars

Reinforced concrete • See Concrete and cement

Remington, Eliphalet • See Typewriter

Remote Control

Remote control was first developed for military use. The concept was used by the German Navy during World War I. A remote apparatus, operated from a distance, was used to steer radio-controlled motorboats that sabotaged enemy ships. By World War II, military forces had developed radio-controlled bombs and other weapons operated by remote apparatus.

One of the first post-war uses for the remote control was as an automatic garage door opener, a concept developed by American manufacturers in the 1940s. Remote control for the **television** was first introduced in the mid-1950s and today is considered standard equipment with any television set. Most stereo systems today feature a remote control apparatus as well.

The television remote control operates by way of a beam of infrared rays, which are sent from the control apparatus to the television set. A device called a photodetector converts the rays to an electronic signal. The signal is sent to a remote control unit, which controls the various functions of the television set. This is known as infrared remote control.

Like the first television sets, the first remote controls were very basic. Some changed channels only and could not control volume. Also, many remote controls only allowed to change channel by moving up or down, not by punching in the precise channel numbers. Today, television remote control systems are as complex as the televisions they operate, and can be used to perform all the various functions that today's televisions offer.

Remote control is used in many other realms also. **Radio** remote control continues to be used today and outside of military uses are used to steer model airplanes and other toys, as well as position antennas and satellites. In the military realm, remote control works in conjunction with the **computer** to steer guided missiles and other objects of destruction. Ultrasonic remote control operates telephone answering machines and also some television sets, among other devices. In such devises, a **microphone** converts the sound waves into electronic signals that flip electronic switches which, in turn, control operation of the device. Electrical remote control can be used over a telephone line to transfer money in a bank account and conduct other similar transactions. Mechanical remote control allows for the operation of a machine from a remote point. Such a concept was applied at Three Mile Island, where robots performed much of the cleanup after the nuclear disaster there.

RENAULT, LOUIS (1877-1944)

French automobile designer

In 1898, the gifted and innovative Frenchman Louis Renault built his first car. Only twenty-one years old, he converted a motorized tricycle into a ''minicar'' by adding another road wheel and a transmission system. The car he designed had a three-speed gearbox plus reverse. His gearbox provided a direct drive, with no **gear** wheels transmitting power on the high gear. This was an improvement on earlier models that transmitted each speed indirectly through a pair of gear wheels, thus adding more noise and creating frictional losses. His innovation of putting universal joints in the drive shaft to enable the axle to rise and fall on springs was also a boon to the industry. The vehicle was quiet, flexible, and lighter than most other cars in those days. The car was such a success that he and his brothers, Fernand and Marcel, united in 1898 to found the *Société Renault Frères*.

Renault cars were extremely popular on the racing circuit and brought the company a great deal of success. Unfortunately, the personal lives of the founding brothers were often tragic. In 1903, Louis's brother Marcel was killed in a racing accident. At that point, Louis gave up racing to concentrate on building his automobile enterprise. He was able to use knowledge gained from his racing days to design a number of vehicles. Renault produced five vehicle models in 1905; he introduced a city passenger **bus** in Paris in 1906; and his company's tanks escorted French troops during World War I. After the war, the company branched into production of farm equipment and industrial and marine machinery. The *Société Renault Frères* grew to become France's largest manufacturer and exporter of motor vehicles.

Renault continued his innovative work in the industry. During World War II, his company produced military equipment under German occupation. After France's liberation, he was jailed on charges of collaboration with the Nazis. He died awaiting trial in 1944, and his company was subsequently nationalized by the French government.

See also Armored vehicle; Automobile, gasoline; Transmission, automobile; Wheel and axle

RENNIE, JOHN (1761-1821)

Scottish engineer

During the eighteenth and nineteenth centuries, Great Britain generated spectacular wealth from its colonies. Among the benefits of this wealth was a renewed attention to civil projects—roads, **bridge**s, harbors and, later, railroads. New names to the field of engineering spurred these projects with new ideas.

John Rennie

John Rennie was born in the midst of this renaissance. He became one of several Scottish engineers of the period whose influence was felt throughout Britain and Europe. Rennie was the son of a farmer, and received his education at the University of Edinburgh. When he was only nineteen, he moved to London to begin his career.

Rennie's first significant project was the Albion Flour Mills in London. Completed in 1784, it was the first factory in Britain to be constructed entirely of cast iron. He also made innovative use of machinery in the factory's construction and operation.

From 1804, Rennie worked on the construction of docks in London, Liverpool, Holyhead, Hull, Dublin, and several other ports using masonry and cast iron. He improved and employed steam-powered **crane**s and **dredger**s in the construction process. He engineered several major canal projects, including the 78-mile Lancaster Canal at Liverpool, the Rochdale Canal at Manchester, the Kennet, and Avon Canal in the south of England, and the Crinan Canal in western Scotland.

Rennie was best known for his bridges. One of his most famous bridges, the New London Bridge, was composed of multiple masonary **arch**es. The bridge was completed by his son John in 1831, ten years after Rennie's death. It was dismantled in 1970 and reassembled at Lake Havasu City, Arizona. Many of Rennie's bridge projects were closely tied to the new road network of fellow Scottish engineer **Thomas Telford**. Other bridges of note are the Waterloo Bridge (1817), constructed of granite, and the Southwick Bridge (1819), with its inverted cast iron arches.

Rennie's sons, George and John, went on to follow in their father's footsteps, becoming notable engineers in their own right, George as a mechanical engineer and John as a civil engineer.

See also Road building

Reservoir • See Water supply and drainage

Resins

Resins are solid or semisolid complex amorphous mixtures of organic compounds that cannot be characterized by definite melting points and which show no tendency to crystallize. Historically, however, the term resin has been applied to a group of substances obtained as gums from trees or manufactured synthetically. But strictly speaking, resins are complex mixtures, whereas gums are compounds that can be represented by a single chemical formula.

The word gum was originally applied to any soft sticky product derived from trees; for example, the latex obtained from Hevea trees, which is the source of natural or gum rubber. Natural rubber, i.e, chemically unsaturated polyisoprene, is a polymeric material that can also be produced synthetically. Thus, although the term resin when applied to **polymers** actually antedates the understanding of the chemistry of polymers and originally referred to the resemblance of polymer liquids to the pitch on trees, it has by association also come to refer to synthetic polymers.

The term natural resin usually refers to plant products consisting of amorphous mixtures of carboxylic acids, essential oils, and isoprene-based hydrocarbons. These materials occur as tacky residues on the bark of many varieties of trees and shrubs. In addition, natural resins have also come to describe **shellac**, which is a natural, alcohol-soluble, flammable material made from deposits on tree twigs left by the lac insect in India and other parts of Asia; amber, which is a fossilized polymeric material derived from a coniferous tree; and natural liquid substances such as linseed and similar drying oils.

A varnish is a solution of a hard resin, drying oil, and a solvent. It is transparent as it does not contain the pigments present in paint. Shellac varnish is a solution of shellac, in denatured alcohol. As a point of reference, by 1000 B.C., the Egyptians are known to have used a varnish of the gum of the acacia tree.

Vegetable-derived natural resins generally fall in one of four categories: rosins, which are resinous products obtained from the pitch of pine trees, and are used in varnishes, **adhesives**, and various compounds; oleoresins, which are natural resins containing essential oils of plants; fossil resins, which are natural resins from ancient trees that have been chemically altered by long exposure such as amber and copal and gum resins, which are natural mixtures of true gums and resins. Examples of gum resins along with their traditional or modern uses are galbanum (obtained from a perennial herb of western Asia and having medicinal uses); myrrh obtained from small trees of India, Arabia, and northeast Africa and used for incense and perfumes; medicinal tonics, stimulants, and antiseptics); asafetida (obtained from a perennial herb and used as an Asian food flavoring; and for medicines and perfumes in the United States; creosote bush resin (obtained from the leaves of the greasewood bush or creosote bush of the desert regions of Mexico and the southwestern United States and used in adhesives, insecticides, core binders, insulating compounds, and pharmaceuticals); okra gum (obtained from the pods of a plant native to Africa but now grown in many countries, and used for foodstuffs, pharmaceuticals; also as an antioxidant and chemical stabilizer, and as a gelation agent); ammoniac resin (obtained from the stems of a desert perennial plant of Persia and India and used in adhesives, perfumes, and medicinal stimulants).

Synthetic resins are polymeric materials, which are better known as polyesters, **polyurethanes**, and amino resins. Although the first polyester resin was made by the Swedish chemist Jacob Berzelius (1779-1848) in 1847, the first thermosetting unsaturated polyester resins did not appear on the market until nearly 100 years later in 1946 when they found use in the construction and transportation industries. American **Leo Hendrik Baekeland** (1863-1944) prepared the first phenolic resin in 1909, but did not succeed in commercializing it until 1932 when it found use in electrical products such as ovens and toasters, wiring devices, switch gears, pulleys, pot and cutlery handles.

Thermoplastic resins are polymeric materials that can be softened and resoftened indefinitely by the application of heat and pressure, provided that the heat that is applied does not chemically decompose the resin. They include **polyethylene** (first synthesized in 1933; commercialized in 1939) used in packaging and non-packaging films; **polystyrene** (synthesized 1839; commercialized 1930) used in molded products such as cassettes, audio equipment cabinets, packaging film, and foodstock trays; acrylonitrile/butadiene/styrene (ABS) (synthesized 1953; commercialized 1954) used in injection-molded automotive components; polyethylene terephthalate (synthesized 1941; commercialized 1954) used in food packaging, soda pop bottles, textiles; polyvinyl chloride (synthesized 1912; commercialized 1927) used in flooring, pipes and conduits, siding; polycarbonate (synthesized 1953; commercialized 1959) used in compact discs and optical memory discs; nylon (synthesized 1935; commercialized 1940) used in transportation industry products; polyacetal (synthesized 1956; commercialized 1959) used in transportation industry products; silicone resins (synthesized in the 1930s; commercialized 1943); **Teflon** (synthesized 1938; commercialized 1949) used in coatings; and poly(methyl methacrylate) (synthesized 1933; commercialized 1936) used as a substitute for silica **glass**.

See also Bakelite; Gutta-percha; Plunket, Roy; Rubber, synthetic; Rubber, vulcanized; Wyeth, Nathaniel

Resistor

A resistor is a component of an **electric circuit** that resists the flow of current and protects delicate electronic equipment from surges. The resistor allows a specific amount of current to pass, but inhibits a high voltage spike that could cause damage to other components. Most resistors are of a fixed value, but others, called a *rheostat*, are tunable.

In 1827, Georg Simon Ohm (1789-1854) experimented with resistance and found that the amount of current which was transmitted through a conductor was inversely proportional to the length and directly proportional to the thickness of the conductor. This relationship between the amount of current I, the electric potential V, and resistance R became known as Ohm's Law, famously written as V=IR. Today, in Ohm's honor, the value of resistance is measured in *ohms*. (Henry Cavendish (1731-1810) discovered this same relationship nearly 50 years earlier, but he failed to publicize it and Ohm received the credit.)

As the electric current passes through a resistor, the energy is reduced and dissipated as heat, as the charge-carrying electrons collide with small imperfections in the atomic structure of the metallic conductor. Power levels are varied to prevent the heat from burning up the resistor. They range from fractions of a watt, for use in transistor **radios**, to kilowatts, needed in electrical power plants.

In 1850 temperature sensitive nonlinear resistors, known as thermistors, were invented. In thermistors, changes in resistance are induced by a change in temperature. Today, because of the high demand for resistors in electronic circuits, they are mass produced from simple ingredients at very low cost. Carbon composition resistors, made of a mixture of powdered carbon black and an inert filler, are among the most common. The ratio of the two ingredients determine the value of the resistor. A liquid resin is added and the mixture is pressed into rods and molded with a copper conductor at each end. After the resistor is tested, bands of paint, corresponding to a standard color code indicating value, are applied.

While carbon composition resistors are adequate for most tasks, they have a poor tolerance and are not suited for high precision applications. For greater precision, a thin film of resistive material, such as carbon or tin oxide, is deposited onto a ceramic base. For very high precision, an extremely thin film of nickel and chromium is used. Because of its thermal properties, the thin film resistor can dissipate more power than a carbon resistor of the same size. These more expensive resistors can also be adjusted to a very small tolerance by cutting a grove in their surface or adding a thin resistance **wire**, giving stability even at high temperatures.

Tunable resistors, such as rheostats or potentiometers (or pots) make use of a sliding contact. Turning a screw causes the contact to move along the resistive film or wire, changing its value.

In modern electronics individual resistors have been completely replaced by the **integrated circuit** (IC). Integrated circuits are inexpensive to make, easy to mass produce, and can incorporate many components on a single board. Tiny resistive elements on an IC are created using very small amounts of metals or semiconductors, carefully controlled.

RETINOGRAPHY

Biometric devices identify people by their physical characteristics. Some examples include fingerprint scanners and other devices that can recognize a particular voice, hand, or signature. The subject of science fiction for decades, the retinal scanner is a recent addition to this family of machines. It is a shoebox-size device that can recognize the unique pattern of blood vessels in the back of each person's eye. The pattern is so complex that even identical twins do not have the same blood vessel configuration. Its proponents claim it has an error rate of only one in a million.

A retinal scanner uses infrared light for its mapping. As a person looks into the eyepiece, an invisible beam of low-energy infrared light traces a circular path on the retina, located at the back of the eye. Because blood-filled capillaries absorb more of the infrared light than the surrounding tissue, there is a variation in the intensity of the reflection. The scanner measures this reflection at 320 points along the beam path and assigns an intensity grade between 0-4,095. The resulting numbers are compressed into an 80-byte computer code that can then be compared with patterns that have already been entered in the computer's database.

These scanners are already found in the Pentagon, national laboratories, and companies with large computer rooms, and there are plans to use them in other areas soon. For example, Congress passed a law in 1988 requiring truck drivers and bus drivers to produce some sort of *unique identifier* when they seek a commercial driver's license, starting in 1992. This bill is designed to prevent inept drivers from holding licenses in several states to hide bad driving records.

Retinal scanners have several advantages over other means of identification. They do not carry the negative, criminal connotations of the fingerprint system. Also, far fewer bytes of computer memory are required to do a retinal scan than to do a fingerprint check, which translates to faster processing of information. In fact, some suggest a retinal scan could be done in three seconds compared to a fingerprint search that takes several hours to complete.

In the late 1990s, corporations developed new optical biometric devices that scan the iris instead of the retina. One such device uses a video camera to take a picture of the subject's eye and then compares it with an image from a database. Iris scanning is may replace retinal scanning because many consider the process less invasive and more accurate.

See also Fingerprinting

REVOLVING DOOR

Revolving doors originated in Europe sometime in the nineteenth century, and first appeared in the United States in Kansas City, Missouri. There they kept out the dust and smell from cattle drives through town streets while still offering patrons simple access to business establishments.

First made of wood and glass, revolving doors today are primarily made from brass, bronze, or **aluminum**. They consist of four door wings and a permanent barrier between the inside and outside of a building, to which the wings are attached. The foremost characteristic of revolving doors is that they are "always open, always closed." They allow people to enter and exit a building while maintaining the interior environment of a building, such as heat and **air conditioning** (or, in the above case in Kansas City, air quality and odor).

Standard hinged doors often do not function properly in tall structures. In such buildings, heating, ventilation, and air conditioning systems create an internal pressure, called "stack pressure.'' Stack pressure affects the opening and closing of standard side-hinged doors. Depending on the direction of door swing, stack pressure will either make doors very difficult to open or very difficult to close, and they cannot be properly closed and latched until the internal systems are turned off. Revolving doors have an advantage because the seal they provide to the building is never broken, and stack pressure does affect their operation.

Revolving doors also allow control of the pedestrian traffic entering a building. By controlling the rotation rate of the door, architects can control how many people might congregate in a building's lobby or wait for an elevator. Revolving doors will revolve reasonably unrestricted up to a preset rate, when a centrifugal brake comes into play and restricts excess acceleration.

A disastrous fire at the Coconut Grove Nightclub in Boston in 1942 lead to the redesign of revolving doors. 492 people died in the fire, partly because many were unable to exit the burning building because revolving doors became jammed with people pushing on both sides, so the doors would not rotate. (Other factors contributing to the number of deaths were an excess of flammable drapes and other decorations, and overcrowding.) Revolving doors in the United States are now required have an emergency mechanism whereby three of the four wings ''breakaway'' and stack flat against one another, allowing a open exit into and out of the building. Hinged doors are also found on the side of revolving doors for emergency exit.

Today revolving doors are much used for security purposes. For a traditional door entrance to a secured building, an authorized person can, in principle, momentarily take control and allow any number of unauthorized people to enter the building. Even in nonthreatening situations, politeness often causes people to hold such a door open for the next person, providing a potential for unauthorized access. Security revolving doors tightly control each person who enters a building, and can be instantly frozen in place should a security problem arise. Sensors can detect two people in the same revolving door section, and lock or reverse direction to gently back them out the door. Identification cards can provide further security, so that once a person has been permitted to enter a building they will only be permitted to exit. This prevents the sharing or unwanted duplication of cards. Equivalent control with traditional, side-hinged doors must often be provide manually or at greater cost.

Revolver

Small handguns were convenient weapons because they could be easily carried on horseback, and a person armed with a pair of pistols could shoot off one and then the other, getting two shots off before needing to reload. Attempts were made as early as the sixteenth century in Europe to manufacture handguns which could let off several shots consecutively. Some used a cluster of barrels, and others, like the modern revolver, rotated a series of loaded cylinders around a central barrel. A revolver is considered any gun with a revolving series of barrels or chambers that are turned to align them with the firing mechanism. A London lawyer, James Puckle (1667-1724) is credited with inventing an early revolver that met with limited success. Puckle advertised that his gun could fire 63 times in seven minutes, an extremely high rate of fire. The revolving cylinders were turned by hand, and a crank screwed each cylinder tightly to the barrel. Puckle advertised his gun widely, but it was apparently used only twice in battle, and Puckle's gun manufacturing company soon failed.

Puckle patented his revolver in 1718, and it was not until a hundred years later that others improved on his idea. A Massachusetts inventor, Artemus Wheeler, took out a patent in the United States on a handgun with an automatic mechanism for rotating cylinders around a central barrel in 1818. Apparently Wheeler gave up on selling his revolver when he could not interest the Navy in it. But another inventor, Elisha Collier, patented what might have been the same gun in England the same year. Another inventor took out a patent in France for a similar weapon. Collier claimed to have made a lot of money of his revolver in England and Europe. But the first person to produce a really successful revolver was American inventor **Samuel Colt**, who patented his weapon in 1836.

Colt's revolver was of the ''cap-and-ball'' type—that is, it was loaded from the front of the revolving cylinder with separate charges of powder and a ball. Along with the fact of the revolving cylinders themselves, Colt's weapon was unique in that the action of cocking the hammer would also bring the next cylinder, or charge, into line with the percussion hammer and barrel. Other early revolvers required the shooter to line up the cylinder and the barrel and cock the hammer in separate actions. And Colt's gun was simple, strong, and easy to use. The war with Mexico that broke out in 1847 provided Colt with a ready market for his weapons, and he dominated the revolver market until his patent expired in 1857. At that time, two other Americans, Horace Smith and Daniel Wesson, marketed a revolver with rim-fire copper cartridges that could be easily and quickly loaded from the rear of the weapon. When the Smith and Wesson patent expired in 1872, new revolvers appeared. An important innovation was the addition of a mechanism to eject the spent cartridge casings. Another innovation saw the trigger action serving to not only fire the gun, but to first advance the cylinder so as to bring the next cartridge into line for firing.

Among other notable innovators in revolver manufacture was the Austrian George Luger (1848-1922), whose semiautomatic weapon became the standard for military use for a number of countries both in Europe and elsewhere after its introduction in 1900. The American **John Moses Browning** also met with considerable success with his.32 caliber **pistol**—250,000 pistols were sold in the six years after its introduction in 1900. The famous Colt 45 is said to have been based on the design of the Browning pistol. Revolvers are used today by military and police all over the world. The basic design has changed little in the past 50 years.

RIFLE

The modern rifle is considered a descendant of the **musket**, a large caliber smoothbore weapon for infantry use. By 1550, the musket had become the most important weapon in Europe, because the ammunition was easier to load and the destructive power was greater than any alternative weapons. About the same time, gunsmiths in central Europe had created a better way to reach a target: rifling. The term, which came from the German *riefeln*, referred to a method of cutting spiral grooves in the gun barrel to impart spin to the ball. Such spin would help stabilize the ball in flight and give it greater accuracy over a longer range. At first rifling was not a success because it was expensive to cut the grooves and the ball often got stuck on the grooves when it was loaded.

In the 1720s German and Swiss gunmakers who had settled in the Pennsylvania colony addressed these problems. Over a period of 50 years they worked to finally produce the famous Kentucky rifle, probably named for its most famous user, Kentuckian Daniel Boone (1734-1820). The caliber of the ball was reduced to .40 inches (1.02 cm), which allowed a lighter weight and increased velocity. The stock was slimmed down and the barrel was extended to as much as 48 inches (122 cm). The ball was made slightly smaller than the inside diameter; for a tight fit, it was wrapped in a patch of greased linen or very thin buckskin. For the next 100 years the Kentucky rifle was the most accurate weapon in the world.

Shortly before the American Revolution, the demand for a gun that could be loaded more quickly than the standard muzzle loaders prompted British army major Patrick Ferguson (1744-1870) to create a breechloader of simple design. He made a vertical screw-plug which could be lowered by turning a lever to expose a chamber in the breech of the gun. The ball was dropped into this vertical hole and rolled forward until stopped by the raised tracks of the rifle barrel. Powder was added and the plug was screwed back to seal the breech. It could be fired at a much faster rate than its counterparts. In 1819 John Hall of Maine invented a variation of this gun: his had a hinged breech block arranged so that its front end could be raised above the top of the barrel, thus exposing a chamber which could be quickly loaded and the block snapped down again to line up with the bore of the gun.

About the same time Hall was working on his breechloader, the Scottish minister and inventor Alexander Forsyth (1769-1843) began experimenting with **firing mechanisms**, and he eventually created the first percussion lock. Joshua Shaw (1777-1860), an English artist living in Philadelphia, Pennsylvania, came up with a small percussion copper cap that fit over a metal nipple. When the hammer hit the cap containing a small chemical charge, it exploded and fired the main charge. The United States Army used Shaw's cap in the rifle it produced for its forces in 1842. This weapon was the best in the world for years to come: it was accurate and could penetrate 8 inches of pine at 100 yards. **Eli Whitney**, known more for his work with the **cotton gin**, created the concept of interchangeable parts for these rifles, thus paving the way for the mass production of firearms.

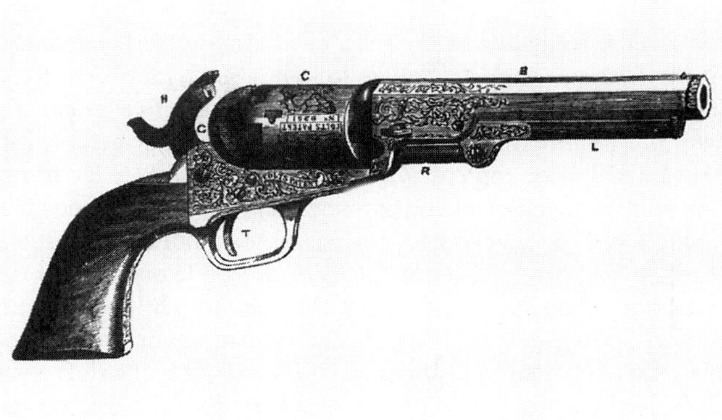

Original colt revolver.

Improvements in **ammunition** for rifles came next. The existing spherical balls were not aerodynamically sound, primarily because they usually had dimples from the process used to mold them. The French army captain Claude-Étienne Minié (1804-1879) invented a bullet (known as the Minié Ball) which was shaped much like modern bullets with a flat base and a pointed nose. The flat base was hollowed out slightly, because when the powder exploded, it expanded the bullet's diameter enough so it would fit the bore tightly. The shape gave the bullet greater accuracy and range. With its reduced size it could be dropped in as easily as a musket ball, thus making the rifle as quick to fire as a smoothbore musket. This development eliminated the musket's last advantage and gave predominance to the rifle. The Minié ball and the rifle were used extensively in the American Civil War with devastating effects.

In 1848, **Christian Sharps**, John Hall's former apprentice, created a reliable, single-shot rifle with a breechblock that lowered when the trigger-guard lever had been dropped. The cartridge was inserted, and, as the breech was closed, a metal edge cut the end off the paper cartridge, exposing the powder charge to the action of the percussion primer. This weapon, used by many snipers during the Civil War, gave birth to the term ''sharpshooter.''

When metal cartridges appeared with built-in primers, the rifle was ready for another major improvement. Gunsmen noticed that metallic self-contained cartridges could be fed mechanically into a barrel repeatedly instead of being fed one at a time. The Henry rifle, named for American B. Tyler Henry, came out in 1860. It carried 15 cartridges in a tube under the barrel, from which the cartridges were fed into the breech by a mechanism operated by swinging the trigger guard forward and back. Oliver Winchester (1810-1880), who had no knowledge of firearms, purchased a Connecticut arms-manufacturing company and hired Henry as supervisor. Henry's system became the basis of the famous *Winchester repeating rifle*.

A German design, the Mauser M1898, served as a model for further United States Army rifle development. In 1903 the Springfield Armory created a gun that served well in both World War I and II. The Springfield was a bolt-action rifle: it had a steel cylinder bolt containing most of the mechanism needed to make the gun shoot. A knobbed handle stuck out on the right side of the bolt, enabling the soldier to move it forward and back as well as to lock upon a cartridge in the chamber. When the bolt was all the way back, a five-round clip was put into the magazine from the top. Pushing the bolt forward carried the top cartridge into the chamber. After firing, the bolt was opened, which extracted and ejected the fired shell and cocked the gun for the next shot. Most hunting rifles still use the bolt-action technology. World War II occasioned the adoption of an even better rifle, the Garand. Unlike the bolt-action Springfield, the Garand was semiautomatic: it removed its own fired shell and put a new one into the chamber by a little gas spurting from a hole in the barrel near the muzzle and moving a **piston** in a cylinder under the barrel. As a semiautomatic, the trigger had to be pulled by hand and released after every shot.

Since World War II, the rifle has been refined, but the basic design and mechanism have changed little. The United States military replaced the Garand with the M-1 and, later, the M-16, capable of firing 20 bullets in 6 seconds. Many rifles in use around the world are based on designs from the 1950s or before, such as the well-known AK-47. Some more recent rifles are designed with plastic parts, and though light weight, are extremely durable. Some rifles popular for military use in the late 1990s are more more compact than their predecessors, with the magazine and breech located behind the trigger. An example is the French FA MAS assault rifle.

Rillieux, Norbert (1806-1894)

American engineer and inventor

Norbert Rillieux was a brilliant student of thermodynamics who became famous for devising evaporators for sugar cane, revolutionizing the sugar-refining industry and easing the labor of slaves.

Born free on March 17, 1806, on a New Orleans plantation to Vincent Rillieux, a prosperous engineer and inventor of a steam-operated cotton baler, and his slave wife, Constance Vivant, Norbert was baptized at the St. Louis Cathedral in the Latin Quarter. Exceptionally privileged for a Southern negro of his day, he was educated at Catholic Schools, then at L'Ecole Centrale in Paris.

In 1830, Rillieux's skill in engineering brought him a teaching post in applied mechanics at his Paris alma mater. That same year he published his findings on the applicability of steam economy to industry, and began working on the problem of evaporating moisture from cane juice while lowering heat to produce a whiter, more refined, sugar crystal. At the same time that he evolved the basic machinery, he created lunettes, which are glass chambers through which the technician could observe the process, a catchall for preventing sugar from escaping from one pan to another, and cast-iron vessels to replace costlier copper containers.

Ten years after beginning work, Rillieux tested his multi-effect vacuum evaporating chamber, a bulky locomotive-sized apparatus containing a network of condensing coils for evaporating raw cane juice. A secondary advantage to the internal coils was the use of vapor from the first stage of the process as the heat source for the rest of the procedure. By removing intense human labor and increasing fuel economy, the device improved the product, increased the rate of production, and cut expenses and the cost of sugar. He patented the device in 1843, but for two years he found no investor for his system. Rillieux ultimately found a prospective client and tinkered with his system for over two years before turning out a suitable product. After the system was permanently installed in 1846, he obtained a patent on modifications.

Rillieux's evaporators, which quickly gained popularity in the sugar industry, were used at Myrtle Grove Plantation, Louisiana, and thousands of other plantations throughout the southeastern United States, Mexico, and the Caribbean. Eventually, refined sugar crystals, which were a specialty item, became an ordinary commodity for which refiners found increasing markets. As a result, the demand for slave labor increased the price of field hands to $5,000 each. Rillieux received an offer to head Edmund Forstall's New Orleans sugar factory, but upon Rillieux's return to the United States, he resigned over a quarrel between Forstall and Rillieux's father.

After Rillieux's evaporation system reached European markets, he began applying the concept to sugar beets, thereby lessening the cost factor in sugar production. Ultimately, the process was applied to all industrial evaporation processes, including the making of condensed milk, gelatin, soap, glue, whisky, and other products, and the recycling of wastes from paper mills.

As a means of reducing yellow fever from the mosquitoes breeding in Louisiana's lowlands and swamps, Rillieux also studied New Orleans's sewage disposal system, but his proposal was rejected because of his race. Subsequent systems resembling his were later instituted. Rillieux grew depressed and bitter, believing Southerners were allowing racism to override progress.

In 1854, as the racial climate of Louisiana became more restrictive, Rillieux, in revolt against having to carry a pass, became one of many black expatriates to settle in Paris. He returned to his old teaching job and was advanced to headmaster. He gained a scholarship, then studied engineering, wrote articles for scientific journals, and worked on translations of Egyptian hieroglyphics. He died October 8, 1894, in Paris and was buried among France's illustrious dead in Pere Lachaise Cemetery, leaving behind considerable wealth to his wife Emily Rillieux. Around 1930, the Dutch began a push to honor Norbert Rillieux. From their effort came a bronze plaque at the Louisiana State Museum.

Road building

The Mesopotamians may have built the first roads, and India's Mauryan civilization had a sizable network of roads by 400 B.C. The Persians built the first highways circa 500 B.C., running from Asia Minor to India, primarily to serve military and administrative needs rather than trade or routine travel. The ancient Chinese also boasted a highway system, while the Greeks, wary of disturbing the natural landscape, which they believed was the habitat of supernatural entities, constructed few major roadways.

The Romans had no such reservations, and their highway system—some of which is still is use today, so well was it constructed—was fashioned of stone blocks covered with sand or broken stones and embraced 50,000 mi (80,000 km) of roads, complete with drainage systems and stair step mounting stations. Rome's decline in the fourth century A.D. heralded the onset of fourteen hundred years of neglect of the western world's road systems, and mud roads and wooden bridges that lasted less than a year were the standard. When progress did resume in the 1600s, it began in the cities of Europe with the introduction of street lighting, ferry service, and municipal regulations. European governments, especially in France and England, assumed increasing responsibility for the planning and maintenance of roads, which led to opportunities for several road-building pioneers. One of the first was **Pierre Trésaguet**, who headed the French École des Ponts et Chaussées during the mid-1700s. His department controlled 25,000 mi (40,000 km) of roads, many of which followed the original Roman routes. His surfacing methods, using layers of paving stones and drainage gutters, were similar to those of the Romans.

England's road system was transformed at the hands of two Scotsmen during the 1800s. **Thomas Telford** built the 194-mi (312 km) London-to-Holyhead road using a costly technique which involved the complete reconstruction of the roadbed. **John McAdam**, on the other hand, relied on the simpler method of using the compacted natural roadbed as a base. His roads were cheaper to build. Today, the term *macadam road* refers to any road using this simplified paving method. Road development reached a standstill with the coming of the and **train and railroad** during the mid-1800s. Even though there were few improvements during this period, the extension of farm-to-market roads from rail centers became the basis for the United States' secondary road system.

The nineteenth century brought with it steam-driven machines. For road building, the most important of these was the steamroller, more accurately called the road roller, invented by Frenchman Louis Lemoine. By 1803 a French company was manufacturing rollers, and a roller designed by W. F. Batho, of England, was being used in Calcutta, India. Its steam power was replaced by **gasoline** in the early 1900s, and the roller was used to smooth all types of road surfaces, including stones. Rollers continue to play a fundamental part in road building as part of a team of men and equipment that includes spreaders, graders, paving machines, and trucks. Portland cement, first used for paving in Scotland in 1865, has become the favored material for city streets while bitumen, or **pitch**, and concrete are used for primary cross-country routes.

Remote rural roads are often dirt and gravel. The 1925 construction of the Bronx River Parkway ushered in the era of the super highway, and road building reached its peak during a thirty-year period following World War II. Inspired by Germany's autobahns, the United States' Interstate Highway System was launched in 1956 and resulted in the construction of 40,000 mi (64,000 km) of multi-lane highways. They have limited access, allow nonstop automobile travel throughout the country, and feature intricate interchange systems among intersecting highways. Today's highway design requires careful study of soil type, topography, and drainage options. The Interstate System has had a profound effect on the social structure of the United States, contributing to unprecedented population and business migrations, dependency on the automobile, and the decay of city centers and small towns. Yet, as with the Roman Empire, it has served as a vital part of the economic growth of the United States. Most of the industrialized world boasts similar systems.

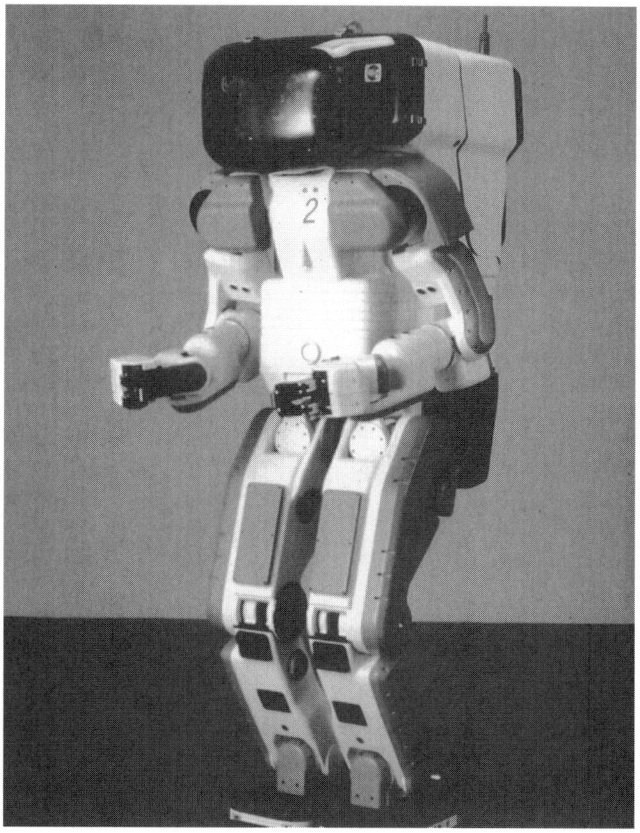

Honda Motor's humanoid robot is the first machine with two "arms" and two "legs" that can change direction under various conditions and climb stairs or slopes.

See also Concrete and cement

ROBOHELP • See Robotics

ROBOTICS

Robotics and its ''offspring''—robots—have long been a source of fascination to human beings, both in and out of the scientific community. The term robot was first used in a 1920s play written by Czech author Karel Capek (1890-1938) and comes from the Czechoslovakian word *robota*, meaning work. Today, robots have come to mean programmable machines whose output copies the function of human beings in one respect or another. Robotics, a term coined in 1942 by science fiction writer Isaac Asimov (1920-1992), is the study of the construction, maintenance and use of robots.

The origins of robotics can be traced back to early Egypt, where priests used steam-activated mechanisms to open temple doors, thus showing their "mystical" powers. Ancient Greeks, Chinese, and Ethiopians built water and steam-powered statues and experimented with other water powered devices. By the 1700s, automation was utilized to a much greater extent, particularly following the invention of the automatic control mechanism, first used on **James Watt**'s ball **governor** in 1788. Around the same period, two Swiss brothers, Pierre and Henri Jacquet-Droz developed an interesting invention. Known as *jaquemarts*, these spring-powered mannequins could play musical instruments, write or draw pictures and strike the hours on **clock** bells.

By the late 1800s industrial **automation** devices were avidly employed, particularly the 1892 invention of Seward Babbitt of the United States, a motorized **crane** with a hinged gripper which was capable of reaching into a furnace, grasping a hot ingot of steel and placing it where desired. Although none of these devices were true robots as we know them today, they represent the first steps of automation and robotics technology.

With the arrival of the 1940s and 1950s, many exciting technological breakthroughs in computers and **cybernetics** (automatically adjusted machines) appeared. Searching for a safe method to handle radioactive nuclear materials, researchers developed a mechanized manipulator arm—a basic component of modern-day industrial robots. In 1946, George Devol developed a magnetic process machine capable of storing and replaying program instructions. While the programmed machine itself was an innovation, Devol realized it would be even more useful if equipped with the manipulator arm. In 1954 he combined the two devices together to create the first programmable industrial robot.

Another American inventor, Jerome Lemelson, began experimenting with automated manufacturing systems and industrial robots in the early 1950s. By 1954 Lemelson had developed and patented a design for a universal robot to perform production and measurement functions. In 1961, the first commercial industrial robots were offered for sale by Unimation Inc. The most popular model was the Unimate 2000, a heavy-duty hydraulically-powered robot that gained wide use, first in Japan, later in the United States. By the late 1970s and 1980s, industrial robots, or "steel collar workers" as they are sometimes called, became commonplace in automotive and manufacturing plants. In 1980, there were estimated to be 5,000 industrial robots employed in the United States, by the year 2000, forecasters suggest the figure will be half a million.

Less widespread, but none the less fascinating, is non-industrial robotics research. Still in the fairly early stages of implementation, general-purpose robots are being employed and used for a host of duties, the scope of which is limitless.

Equipped with **microprocessors** and specialized sensors to see, feel, move and in some cases, think—robots have been invaluable in exploring unknown or dangerous realms. Lunkahod, a robot sent to explore the moon in 1970 is one such example. While a mission to Mars is still considered risky for human explorers, in 1976, a robot on the *Viking I* probe landed on Mars. Controlled from the earth, the robot provided much valuable information about the planet and conducted various test projects.

In the field of medicine, robots have made surprising achievements. Dr. Yik San Kwoh, director of CAT Scan research at the Long Beach Medical Center in California, observed his colleagues' tedious work and was intrigued by the possibility of training a robot to perform the work. After receiving donations from private individuals and a robot arm compliments of Unimation Inc., Dr. Kwoh proceeded to build Ole, a mechanical neurosurgeon. Ole began performing human surgeries in the late 1980s and is still "practicing" in California with great success. The use of robots has also been considered as a method of providing long-term patient care, in view of the current nursing shortage. In 1983 just such a nursing robot was created by Japanese Professor Hiroyasu Funakubo. Dubbed Melkong, (Medical Electric King Kong) the robot can hold a patient in its "arms," wash him or her, put him or her back in bed and tuck in the sheets.

Since they were first depicted in early science fiction movies of the 1940s, walking, talking robots to do all our household work have been eagerly anticipated. Although some household robots are now available, they are not quite so advanced as those portrayed in Hollywood. In 1982, Heath Co., a division of Zenith Radio Corp., began marketing HERO I, a self-contained mobile robot. Designed mainly as a teaching aid for industrial robotics, HERO can be programmed to turn lights on or off, stand guard at the front door, and make idle chitchat. Personal robots of the future propose to be complete entertainment centers, able to sing, dance, and tell jokes, as well as control all your electronic equipment—**television**, **radio**, **stereo**, computer games, and telephone.

While some researchers have suggested future robotics research may combine microprocessor and biogenetic technology to create "super-robots" based on living cells instead of silicon, exactly what the future holds remains unclear. With the rapid advances in robotics and artificial intelligence research, the concept of a personal robot to handle all of our work may be closer than ever.

See also Computer, digital; Computer, industrial uses of; Microcomputer; Space probe

ROBOTS • See Robotics

ROCK, JOHN • See Birth control

ROCKET AND MISSILE

Rocket technology has its origins in ancient and medieval civilizations. Around 160 A.D. the Greek mathematician **Hero of Alexandria** invented the Sphere of Aeolus, the first known device to demonstrate the principles of propulsion. It consisted of two tubes attached to opposing sides of a sphere. When water was boiled in the sphere, steam escaping from the tubes produced enough pressure to spin the sphere. Hero's invention, however, and the principles it demonstrated elicited no more than idle curiosity from his contemporaries.

In China, rocketry advanced more quickly, closely following the invention of **gunpowder**. A Chinese text from the thirteenth century describes "arrows of flying fire" that could be launched without the aid of a bow. Essentially handheld Roman candles, these rockets were also used for religious ceremonies and celebrations.

Eventually knowledge of these weapons spread from China to Mongolia and the Arab world. Crusaders introduced rockets to Europe where they became a part of military strategy during the Middle Ages. For example, the forces of Joan of Arc used them at the Battle of Orleans in 1429. However, advances in artillery eventually eclipsed rockets, which were then only used for entertainment. Interest in the military potential of rockets was revived in Europe some three hundred years later when, in the late eighteenth century, Indian troops effectively used rockets to defeat British forces in a number of battles.

In 1804 William Congreve (1772-1828), a British army officer, developed a metal-cased, stick-guided rocket capable of being fired in large barrages against any enemy position up to 2,000 yards (1,828 m) away. Congreve's rockets were used effectively in military engagements throughout the nineteenth century, including the War of 1812, as attested to in Francis Scott Key's "Star Spangled Banner."

Several improvements were made to Congreve's original design as the century progressed. Englishman William Hale developed a three-finned tail that replaced the original stick, giving the rocket greater accuracy. In the 1890s Swede Wilhelm Teodor Unge introduced more powerful propellants and increased the rocket's spin rate. As a result he could match the accuracy and distance of most artillery then available. However, most armies continued to prefer artillery and later the **airplane**.

At the beginning of the twentieth century, scientists began to contemplate the broad implications of rocketry, including the possibility of space flight. Russian theorist **Konstantin Tsiolkovsky** formulated the first modern theories of rocketry and is considered the father of space travel. In 1903 he published his first article on the subject, which focused on the value of liquid propellants. Later, he improved his drawings and explanations to include methods of cooling the combustion chamber, methods of guidance, and the possibility of developing a multistage rocket. He then proposed that rockets would be the ideal for space travel due to the fact that their propulsion systems were completely self-contained and did not rely on atmospheric oxygen for combustion.

American **Robert Goddard** took Tsiolkovsky's ideas further, actually building and flying the first liquid-fueled rockets. His first successful launch, using gasoline and liquid oxygen, took place in 1926 with a flight of 184 feet (56 m). Goddard continued work on rocket power plants, pumps, fuel systems, and control mechanisms that included gyrostabilization and steering jet vanes in the rocket exhaust. He made so many advances that one rocket expert recently said, "Every liquid-fueled rocket that flies is a Goddard rocket."

Hermann Oberth (1894-), a German schoolteacher, pioneered the science of rocketry in Germany. Inspired as a child by the futuristic works of Jules Verne, Oberth abandoned plans for a medical career to study mathematics, physics, and astronomy. In 1923 he published basic mathematical formulas crucial to rocket space flight. He also proposed designs for manned spacecraft and high-altitude research rockets and advanced the idea of orbital rendezvous for refueling and resupply through the use of a space station. He also provided a practical design for a complex liquid-propellant rocket to explore the upper atmosphere. His research prompted the formation of rocket clubs which would later play a significant role in the advancement of rocket technology.

Robert Goddard

One such club, Verein für Raumschiffahrt (Society for Spaceship Travel), was recruited by the Nazi government to develop rockets for military use. **Wernher von Braun**, still in his mid-twenties, became the technical director of the operation. By 1942 von Braun and his team had created the A-4, a 46-foot (14.02 m) rocket with a range of 170 miles (273.5 km). In 1944 the first of more than 4,000 A-4s were launched under the name V-2. Capable of flying at 3,600 miles per hour (5790 km/h) and carrying 2000 pounds (908 kg) of explosives, these missiles inflicted heavy damage upon targets in England. Near the end of the war, von Braun and his associates surrendered to the Americans, and eventually played a major part in the success of America's space program.

After the war both the United States and the Soviet Union continued work on rockets. The Soviets, under the leadership of **Sergei Korolev**, made tremendous progress on work that had started before the war. In 1955 the Soviet government built a major launch facility called Baikonur where the crew tested the Soviet Union's first ICBM (intercontinental ballistic missile), fueled with kerosene and liquid oxygen and comprised of a central core with four strap-on boosters. In the United States, von Braun developed a medium-range ballistic missile, the Redstone, and an intermediate-range missile, the Jupiter.

The Soviet Union's successful launch of **Sputnik** in 1957 initiated a race for superiority between the two super powers. Von Braun was able to send up America's first satellite, **Explorer 1**, with a modified *Jupiter* rocket. It was obvious the Soviets were using larger rockets to send heavier payloads into orbit, so the United States developed larger and more powerful boosters over the next decade: Thor, Atlas, Delta, and Titan. The ultimate rocket was the Saturn, designed for a manned trip to the moon. The final configuration that sent three men to the moon in 1969 was the *Saturn 5*: three stages, a height of 363 feet (111 m), a weight of 3,000 tons (2,721 t), and five main engines that developed 7.75 million pounds (3.52 million kg) of thrust (20 times that of the Atlas which launched the early one-man Mercury capsules). Newer rocket models continue to appear, including the Delta Clipper, which underwent successful flight tests in 1994-1995. The Delta Clipper incorporates many of the technologies necessary for the development of a space plane, a vehicle that can fly inro orbit, rather than being launched vertically as conventional rockets are.

Since the moon missions ended, both the Soviet Union and the United States, as well as other nations, have pursued rocketry even if not at the same hectic pace. The Soviets have introduced a new, heavy-lift rocket, the Energia, which is two-thirds as powerful as the *Saturn* rocket and can carry its own **space shuttle** into orbit. The United States developed its space shuttle, the world's first reusable manned space vehicle, in the 1980s. Two solid-fuel rocket boosters provide the main thrust of 5.3 million pounds (2.4 million kg), while three engines on the shuttle itself provide an additional 1.1 million pounds (500,000 kg) of power. They release energy equivalent to that of 23 Hoover Dams and consume 500,000 gallons (1.892 million l) of liquid hydrogen fuel in 8.5 minutes.

Both the American and Soviet space programs received serious set backs around the same time. In January 1986, the American space shuttle *Challenger* exploded on launch, killing all seven astronauts aboard, and the American government was forced to spend many months reassessing the effectiveness of its rockets and space programs. By 1990, shuttle launches had been resumed.

In the early 1990s, the political breakup of the Soviet Union and the ensuing economic crisis put a temporary end to Russia's space program. The situation was worsened by the failure of Russia's Mars96 mission, an unmanned **space probe**, carrying numerous experiments, that was to explore Mars. The probe crashed into the Pacific shortly after launch.

ROEBLING, JOHN (1806-1869)

German American engineer

John Augustus Roebling was a pioneer in American **bridge** building, not only for his designs, but for the materials and methods he used in their construction. His crowning achievement was New York's Brooklyn Bridge.

Roebling was born in Muhlhausen, Germany, the son of a tobacco shop owner. While visiting Bamberg, Bavaria, Roebling saw a suspension bridge for the first time, fell in love with it, and made a sketch of it. At that moment he decided on a career in bridge building.

Roebling studied engineering and philosophy at the Royal Polytechnic Institute in Berlin, Germany, and graduated in 1826. Four years later he emigrated to the United States to take advantage of the great opportunities that awaited engineers. Upon his arrival, Roebling bought a farm in Pennsylvania, where he waited out the residency requirement for U.S. citizenship. A year after becoming a citizen in 1837, he went to work for the State of Pennsylvania as a canal engineer. It was obvious that Roebling had bridge building on his mind, for it was not long before he invested his earnings as a canal builder into the production of steel cable. Wire rope, or cable, was a recent invention, and Roebling apparently saw its potential as a bridge building material before anyone else. Roebling invented his own machinery to twist steel wire into cable. In 1841 he established a factory in Saxonburg, Pennsylvania, to mass-produce cable.

At first, his new product was used in the canal business for hoisting and pulling canal boats. Roebling employed it in bridges at the earliest opportunity, however, to replace the chains and rigid supporting rods then used. His first bridge project was a suspension bridge over the Monongahela River at Pittsburgh in 1846. He then opened a new cable factory in Trenton, New Jersey, and went on to design and build suspension bridges at Niagara Falls (1854), over the Allegheny River, at Pittsburgh (1860), and across the Ohio River at Cincinnati (1867).

The Brooklyn Bridge design was originally submitted in 1857, but it was not approved until 1869. As with the Cincinnati bridge, the Civil War was at least partly to blame for its delay, but once approved, the work was immediately started. On June 28, 1869, Roebling was standing on a set of wood pilings at the edge of the East River taking observations for his new bridge. A boat moved against the pilings, crushing several of Roebling's toes. The injury itself was not life threatening, but tetanus set in, and the infection took Roebling's life on July 22nd. His son, Washington Augustus Roebling, took over the Brooklyn Bridge project and saw to its completion in 1883. The Brooklyn Bridge was unique in that its steel cables supported a span of 1,595 feet (486.2 m) from its 274-foot (83.5 m) twin towers, allowing ships to pass freely underneath. Had it not been for Roebling's untimely death, his mark on American bridge architecture might well have been far greater than it was.

See also Canal and canal lock

ROHRER, HEINRICH (1933-)

Swiss physicist

Heinrich Rohrer was born in 1933, in Buchs, St. Gallen, Switzerland, the son of Hans Heinrich Rohrer, a distributor of manufactured goods, and Katharina Ganpenbein Rohrer. He received his Ph.D. in physics in from the Federal Institute of Technology in Zurich. His doctoral research involved superconductivity.

In 1963, Rohrer joined the staff of the research laboratory of International Business Machines (IBM) in Zurich, eventually becoming manager of the physics department as well as an IBM Fellow. Rohrer has remained at the IBM lab throughout his career. In 1961, Rohrer married Rose-Marie Eggar; the couple had two daughters.

After joining IBM, Rohrer expanded his research in physics beyond superconductivity, investigating magnetic fields and critical phenomena. He became interested in the little-understood and complex atomic structures of the surfaces of materials. While electron microscopes had been developed to probe the internal arrangements of atoms in materials, attempts to uncover the very different characteristics of surface atoms had been decidedly unsuccessful. In 1978 **Gerd Binnig**, a young German who had just received his Ph.D., joined Rohrer's research team in Zurich. Together, Rohrer and Binnig began to explore oxide layers on metal surfaces. They decided to develop a spectroscopic probe and in the process invented an entirely new type of microscope, the **scanning tunneling microscope** (STM).

Rohrer and Binnig began with the phenomenon called tunneling. As revealed through quantum mechanics, electrons behave in a wavelike manner that causes them to produce a diffuse cloud as they leak out from the surface of a sample. When electron clouds from two adjacent surfaces overlap, electrons tunnel from one surface or cloud to the other. Tunneling through an insulating layer had been used often to reveal information about the materials on either side of the insulation. Rohrer and Binnig decided to tunnel through a vacuum and then use a sharp, needlelike probe within the vacuum to scan the sample's surface. As the scanning tip closely approached the sample, the electron clouds of each overlapped and a tunneling current began to flow. A feedback mechanism used the tunneling current to keep the tip at a constant height above the sample's surface. In this way, the tip followed the contours of the individual atoms of the scanned surface, and a computer processed the tip's motion to produce a three-dimensional, high-resolution image of that surface.

One large problem was the sensitivity of the scanning tip to disturbances from vibration and noise. Here Rohrer's background in superconductors was helpful, because transducers too are extremely sensitive to vibration. Rohrer and Binnig solved the problem by shielding their scanner from disturbances with magnets and a heavy stone table set on inflated **rubber** tires. They successfully tested their new device in 1981 and then worked to refine it technologically. By the mid-1980s the STM could fit in the palm of the and could show some details as tiny as 0.1 angstrom (with 1 angstrom being about the diameter of a single atom, or 2.5 billionths of an inch). STMs were also developed that worked in water, air, and cryogenic fluids as well as vacuums. By 1987 Rohrer's group at IBM had developed an STM the size of a fingertip.

Rohrer and Binnig shared the Nobel Prize in 1986 for their STM, which opened up entirely new fields for the study of the structure of matter. Rohrer shared other international awards with Binnig for his work on the STM. He is a member of many important scientific societies and has been awarded honorary doctorates by several universities.

See also Scanning tunneling microscope

ROLLER BEARING • See Bearings

ROLLING MILL • See Metal-working processes

ROOFING MATERIALS

Roofs are fundamental to all shelters. They protect the inhabitants of a building from rain, snow, wind, sun, and heat or freezing. Types of roof, and the materials used to cover them, have varied over the centuries, changing with geography, innovation, and technology. A wide range of roofs exists around the world even today.

Primitive huts probably had roofs made of straw rushes, reeds, palm leaves, or a similar material, depending on what grew locally. Such "thatched roofs" had to be constructed at a steep angle, to allow rainwater to run off. Later mud was mixed with these materials, to provide strength, and eventually the reeds and rushes were replaced with branches from trees and larger timbers. Mud, clay, or other materials could then be used to create a mostly watertight roof.

In areas with little rainfall, such as the Middle East and the American southwest, roofs could be flat. In these hot, dry climates roofs were often made of local materials like clay or mud. Later, in the nineteenth century, Europeans and Americans made widespread use of flat roofs with the development of waterproof roofing materials and steel and **concrete** for structural strength.

With the development of brick and cut stone as building materials, roofs became more elaborate, such as domes and vaults. (Domes are hemispheres—a sphere cut in half, such as that over the U.S. Capital building in Washington, D.C. Vaults are a series of arches, like half a cylinder cut lengthwise, first used by the Romans around 1 A.D.) Roofs became not just a protective structure, but often elaborate shapes meant to delight and inspire. Trusses, structural beams inside a dome or vault, were used to strengthen such roofs. They were first made of wood, which had the disadvantage of being heavy. With the development of steel and reinforced concrete more elaborate roofs became possible, such as the immense roofs that cover large areas such as stadiums.

The material used to cover a roof must stop rainwater from penetrating a building. As mentioned earlier, mud and clay were used on early roofs. Tar was later used (and still is today). Rocks or other materials are often put on top the tar to prevent wear and tear. Layers of asphalt or tar-saturated fabric sheets can be used to additional protection and insulation.

Shingles are also commonly used to cover roofs. These are small, waterproof pieces of wood, slate, tar-paper, or other materials that overlap one another so that water rolls over them and down the roof unimpeded. Such shingles are usually nailed into place. Larger pieces of corrugated sheets of steel, aluminum, or other materials also are used in the same way.

Root, Elisha King (1808-1865)
American mechanic and inventor

Elisha K. Root was born in 1808 near Springfield, Massachusetts, and became one of New England's most accomplished machinists. He began his industrial apprenticeship at the age of ten working as a bobbin boy in the textile industry. At fifteen Root served as an apprentice at a machine shop in Ware, also in Massachusetts. He then honed his industrial skills in a machine shop in Chicopee, Massachusetts.

Root made the acquaintance of **Samuel Colt**, who, in 1829 at the age of 15, detonated one of his inventions—an electrically-ignited **submarine** mine—which doused the crowd with muddy water. Root was among the onlookers, and, as the story goes, it was Root who helped Samuel Colt escape the crowd's wrath.

In 1849, when Colt was looking for a superintendent to run his munitions factory, he naturally turned to Root, who was by then known as the best mechanic in the Connecticut valley. At the time, Root was working for the Collins Company, which made axes, and Root was busy building machines (along with his reputation) to produce the axeheads more efficiently. Colt lured Root from the Collins factory by doubling his previous salary, an action that reportedly made Root the highest paid ''mechanic'' in New England.

Root immediately set about transforming the Colt firearms factory, building nearly 400 machines to produce the interchangeable parts that made the Colt factory the model of industrial efficiency. It was Root's mechanical genius that is credited with making the Colt firearms company a success. Among Root's most successful machine tools was the ''universal'' miller. His miller could be easily adapted to perform a great variety of metal-cutting operations, each of which previously required a separate machine. The machine was made for the Colt factory at the Lincoln iron works in Hartford, Connecticut, and became known as the ''Lincoln Miller.'' With more than 150,000 sold, it became the most commonly used machine tool in America.

Elisha Root continued to apply his successful machine tool inventiveness to all aspects of the Colt manufacturing process. When Samuel Colt died in 1862, Root took over as president of the Colt armories. He ably managed this now very large and prosperous business until his death three years later.

See also Mass production

Rope

Ropes have been used for more than 10,000 years and were originally constructed from vines and strips of animal hide. About 3,500 B.C., the Egyptians began using papyrus, hemp, and other fibrous plants.

Almost any fiber can be used to make rope, including flax, cotton, manila, and sisal. Today, many types of rope are made from synthetic materials, which can better withstand the effects of time and weather. Almost all ropes are made from three or more strands, which are themselves composed of many yarns. The yarns are arranged in concentric layers (to make them compact) and are twisted slightly, forming a strand. Three such strands are then twisted together in the opposite direction—this keeps the rope from unraveling, even when stretched. The rope can also be plaited or braided, making it even stronger. A hawser or cable-laid rope is made from three complete ropes that are twisted together.

During the time of ancient Egypt ropes were used primarily as equipment for seagoing vessels. Then, long buildings were constructed wherein rope could be manufactured—the longer the better, since the rope could only be as long as the building it was made in. These buildings could be up to 400 yards (365 m) long, and came to be known as ropewalks due to the tremendous amount of footwork required of rope makers. Though mechanized, many modern rope machines are still called ropewalks.

As technology has advanced, new synthetic materials have begun to replace natural fibers in the manufacturing of rope; polymer fibers can be made longer and stronger, and are better suited for certain applications. Also, because the fibers are manmade, they can be any length, so that a single polymer fiber can run the entire length of rope.

Nylon rope is known for its ability to stretch without breaking. It is often used for towing as well as climbing, where the weight from a sudden drop can sometimes snap ropes made from natural fibers.

On boats, manila and hemp ropes have been almost entirely replaced by ropes made from **polyester**, and **dacron** and **polypropylene**. Unlike natural ropes, these synthetic ropes will not stretch when wet, nor will they rot. They are also easier to grip without abrasion.

Rotary engine • See Wankel engine

RU 486

French biochemist Etienne-Emile Baulieu (1926-), who has spent more than thirty years researching hormones and co-founded the International Society for Research in Biology and Reproduction, soared to international notoriety in the 1980s with one of the most controversial medical breakthroughs of recent times—the invention of RU 486, known as the ''abortion pill.'' RU 486—now known as Mifepristone—is an ''anti-hormone'' which blocks the effects of progesterone, the hormone that allows an embryo to develop in the uterus. Once this happens, the fertilized egg and uterine lining separate from the uterine wall and are expelled through vaginal bleeding.

After developing the steroid in 1980, Baulieu and his colleagues at the French pharmaceutical company Roussel

Uclaf performed numerous tests on animals and found the drug to be nontoxic. In 1985, extensive tests on human volunteers were carried out in France, Great Britain, Holland, Sweden, and China. RU 486 was found to be safe and to have an 85% success rate when used in the first eight weeks of pregnancy. When a low dose of the hormone-like substance, prostaglandin, was given following RU 486, the success rate rose to 96.9%. In September 1988, RU 486 was approved for use in France, where abortion has been legal since 1975. One month later, however, the manufacturer withdrew the drug from the market when anti-abortion activists threatened a boycott of the company's products. Responding to rising protests from members of the World Congress of Obstetrics and Gynecology, the French government ordered the company to release the drug to state-controlled clinics. Within a year, about one-fourth to one-third of all early abortions in France were conducted with RU 486. In September 1989, Baulieu was given the Albert Lasker Award for his work. Following the death of a French woman who had used the drug in April 1991, he made refinements that he says make the drug safer.

Baulieu staunchly defends his invention as a safe option for women who choose to end a pregnancy. He calls it a "contragestion" medication and lists among its benefits the fact that it is private and poses far less risk than other abortion procedures that kill up to 200,000 women each year when they are performed improperly. Opponents of RU 486 believe that it tacitly sanctions abortions by making them easier, thereby encouraging sexual behavior that results in unwanted pregnancies.

In early 1992, the newly-elected president of the United States, Bill Clinton, began lobbying to lift the ban on the import of RU 486 into the United States. By 1995, clinical trials involving 2,100 were concluded, and the Food and Drug Administration granted *conditional* approval in September 1996, with final approval still pending in early 1998. France, Sweden, and Great Britian are the only European countries in which RU 486 is legally available outside clinical trials. In 1998, Abortion Rights Mobilization announced introduction of their RU 486 clone to their clinical trials; 1600 women at 12 different sites received the drug.

Although RU 486 was not originally intended for other purposes, Baulieu and other researchers suspected that, as a steroid, it might have additional applications. Indeed, it has shown promise in treating a number of diseases, including Cushing's syndrome, benign tumors of the brain and spine, endometriosis, and breast cancer. In 1995, some evidence suggested it may prevent cell infection and subsequent replication of the HIV virus. Early tests have even indicated that RU 486 may function as a male **birth control** drug.

RUBBER BANDS • See Rubber, vulcanized

RUBBER, SYNTHETIC

Natural rubbers, before vulcanization, tend to be sticky and soft at high temperatures, while at low temperatures they are brittle and stiff, making them difficult to process. Because of these properties as well as the difficulties associated with obtaining adequate and affordable supplies of natural rubber, the search for natural rubber substitutes began.

In 1906 Farbenfabriken of Elderfeld began to search for a viable production process. Fritz Hofmann was appointed head of the research group. He attempted to polymerize isoprene, which was known to be a component of rubber. Convinced that the purity of the isoprene was paramount, he spent two years researching methods of producing pure isoprene. Finally he developed a six-step process. He next attempted to polymerize the pure isoprene. None of the techniques discussed in the scientific journals of the time were successful. Ultimately Hofmann simply heated the isoprene in an autoclave. Various viscous liquids were used in the heating process to emulsify the rubber. The product was comparable to natural rubber in toughness and elasticity and was much less sticky. But the mixture had to be shaken at 140°F (60°C) for several weeks. Because of the long production time and high cost, the process was abandoned as impractical.

Once World War I started and German access to natural rubber supplies was cut off, Farbenfabriken developed two grades of synthetic rubber, both made from 2,3-dimethyl butadiene. The softer grade was formed by pretreating the *monomers* (single units) with oxygen and polymerizing them at 149°F (65°C). The firmer grade used an initiator (usually preformed rubber) and was polymerized at 86°F (30°C). Despite the initiators both reactions took several weeks to complete.

After the war, the price of natural rubber fell below the cost of production for synthetic rubber. It was not until natural rubber producers raised the price from $.17/lb to $1.21/lb that interest in synthetic rubbers resumed. Buna rubber was made of butadiene (Bu) by sodium polymerization. Buna S was a copolymer of butadiene and styrene; Buna N used butadiene and acrylonitrile as copolymers. Both were developed by Hermann Staudinger (1881-1965) and his research team, and patented in 1929. Both were more resistant to oil, gasoline, and aging than natural rubber.

Neoprene is another synthetic rubber discovered in the pre-World War II years. Father Julius Nieuwland (1878-1936) of Notre Dame University shared his research with Wallace Carothers (1896-1937) of Du Pont. Father Nieuwland was researching the polymerization of acetylene, and his research showed strong similarities between the structure of polymerized acetylene and natural rubber.

Arnold Collins, a research chemist on Carother's team, purified a sample of Nieuwland's acetylene and produced a small amount of a liquid which, when left out over a weekend, formed an *elastomeric* (rubber-like) solid. After additional research, a reaction converting acetylene to vinyl acetylene, using copper chloride as a catalyst and hydrogen chloride as an additive, was patented. The product of the reaction was 2-chlorobutadiene, called chloroprene because of its similarity to isoprene. Du Pont renamed it neoprene and began to market it in 1930. It was more expensive than natural rubber but had an even greater resistance to oil, gasoline, and ozone.

Thiokol Chemical Corporation produced a synthetic rubber from ethylene dichloride and sodium polysulfide. It was developed by J. C. Patrick, Thiokol's president, and marketed in 1929. Although it emitted an unpleasant odor, it found use in fuel tank linings for aircraft and for windshield seals for cars.

GR-S (government rubber, styrene type) came on the market during World War II, when there were again restricted supplies of natural rubber.

Studies by the German scientist **Karl Ziegler** (1898-1973) and by the Italian **Giulio Natta** (1903-1979) showed that organometallic compounds greatly increased the speed of some reactions. By using organometallic catalysts, scientists were able to prepare copolymers of styrene and butadiene very quickly, with reactions completing in less than twenty-four hours. GR-S rubbers are now commonly called SBRs (styrene-butadiene rubbers). Ziegler's catalyst system led to the development of ethylene-propylene rubber in 1960. Du Pont succeeded in crosslinking the ethylene-propylene rubber by adding small amounts of a third monomer (termonomer) in 1961; by 1967, Uniroyal, Copolymer Corporation, and Jersey Standard were producing ethylene-propylene rubber using a termonomer patented by Du Pont. Ethylene-propylene polymers (EPMs) are most often used in abrasion-resistant applications.

Today there are many synthetic rubbers on the market. **Silicone** rubbers are linear polymers (polymers whose molecules are arranged in long chain-like structures) derived from dimethyl silicone. They are difficult to process but are stable over a wide range of temperatures -130-601°F (-90-316°C). They are used in **wire** and cable insulation and gasket applications.

Besides these synthetic rubbers, many styrene-based polymers have been used in sporting goods. Thermoplastic elastomers are currently the fastest reacting and most economical synthetic rubbers available.

Most developments in the rubber industry since 1955 have been more characterized by technology transfer rather than innovation. The major West German synthetic rubber producer Chemische Werk Hüls decided to abandon its own obsolete technology and purchase a GR-S plant from Firestone in 1954. Synthetic rubber production began in Great Britain and Italy in 1957, in France and Japan two years later, and in Brazil in 1962. By the early 1970s, Japan had become a major producer of SBR as well as of all-cis polybutadiene and polyisoprene.

See also Plastic; Polyethylene; Polymer and polymerization; Polypropylene; Rubber, vulcanized

RUBBER, VULCANIZED

When rubber was first introduced to Europe from the New World, it was considered a marvelous novelty. Early explorers were amazed to find Caribbean natives playing games with bouncing balls made from the milky white juice (latex) of certain trees. A French explorer, Charles Marie de la Condamine, brought back samples of Indian-made rubber from the Amazon Valley in 1738 and set about promoting interest in the new substance. It was variously called *caoutchouc* (from the South American Indian word for it), *gum elastic*, and *India rubber*—''India'' because the substance came from the West Indies, which Columbus thought were India, and ''rubber'' from the observation of British chemist **Joseph Priestley** (1733-1804) that the substance rubbed out lead **pencil** marks.

Europeans were fascinated with rubber's attributes: it was elastic, waterproof, strong, springy, and moldable. Since latex coagulates quickly, rubber always arrived in Europe in solid form, usually as bottles. Manufacturers like **Thomas Hancock** (1786-1865) in England sliced up the bottles to make rubber novelties like shoe lasts, tobacco pouches, and rings that were used as garters and wristbands—the Western world's first rubber bands. Experimenters soon found that the hardened gum could be dissolved in turpentine and then reshaped. Cloth soaked in the liquid became waterproof, but it also smelled like turpentine. Hancock and Charles Macintosh solved that problem in the early 1820s by dissolving rubber in naphtha instead. Hancock also designed commercially successful rubber manufacturing machines.

A great craze for India-rubber products ensued, both in England and the United States. Five hundred pairs of India rubber boots were imported to Boston, Massachusetts, in 1823. In England, **Charles Macintosh** (1766-1843) began producing rubber-lined waterproof coats. Manufacturers vied with one another to produce rubber overshoes, coats, caps, wagon covers, and suspenders. Hancock developed rubber tubing from which he cut rubber bands and hoses. But natural rubber's most serious flaw soon showed itself: it is unstable over varying temperatures. People soon discovered that their overshoes became stiff and brittle in cold weather, and that in the heat their raincoats dissolved into a stinking, gummy mess that had to be disposed of by burial.

In 1839, **Charles Goodyear** (1800-1860), a Yankee inventor who had devoted himself to improving the usefulness of rubber, discovered the answer. Goodyear, with absolutely no knowledge of chemistry, had spent five years mixing crude gum elastic with every possible substance, with the idea that sooner or later something would work. He had no success with salt, sugar, castor oil, ink, soap, or even cottage cheese. Magnesia, quicklime, and nitric acid all seemed promising for a time, but failed. Goodyear then experimented with Nathaniel Hayward's process of mixing rubber with sulphur. While doing so, Goodyear had a famous accident: he spilled some of his rubber-sulphur compound on a hot stove and was amazed to find that instead of melting, as natural rubber would have, it ''charred like leather'' and lost its stickiness. Goodyear noticed a tiny line of perfectly cured rubber on the edge of the piece. Further tests revealed that the cured rubber remained flexible even when left outdoors overnight in intense cold.

Goodyear spent the next five years perfecting his rubber-sulphur curing-by-heat process, securing a patent for it in 1844. The process became known as vulcanization—named after Vulcan, the Roman god of fire. A sample of Goodyear's vulcanized rubber found its way across the Atlantic to Hancock in England. Hancock studied and experimented with the sample and soon took out a British patent for his version of rubber-sulphur vulcanizing. Other variations of the vulcanizing process were developed, one by **Alexander Parkes** (1813-

1890) in 1846 which used sulphur monochloride vapor and another by S. J. Peachy using sulphur dioxide gas.

Vulcanizing made rubber a practical, eminently usable product. The rubber industry flourished, spewing out hundreds of everyday items (including waterproof clothing and footwear, fire hoses, rubber bands, mattresses, combs, and balloons), and contributed greatly to the process of industrialization. Rubber provided the electrical and communications industries with the effective insulation they badly needed. Rubber seals perfected industrial machinery. Perhaps most importantly, the use of rubber for **pneumatic tires** helped to expand both the automotive and rubber industries.

Today, natural rubber, which is essentially cis-1,4-polyisoprene, remains the strongest of all rubbers, with its excellent dynamic properties such as resistance to fatigue. In some products the rubber of choice is determined solely by properties (e.g., aircraft tires which require 100 percent natural rubber) but in many products the choice between between natural and synthetic rubbers hinges on price and properties. World consumption of natural rubber in 1996 was 6.13 million metric tons, corresponding to 39% of world consumption of all elastomers (i.e. natural plus synthetic rubber).

See also Eraser; Waterproof materials

Rubik's cube

A puzzle that confounded people worldwide during the late 1970s and early 1980s was patented by Hungarian inventor Erno Rubik in 1977. The puzzle was a cube, assembled around a sphere, so that all six sides moved in all directions. When the puzzle was purchased, each side of the cube was all one color. After mixing up the colors, the frustrated puzzle solver would try to get the cube back to its original position. With millions of combinations possible, many of the puzzles were never solved.

Although the Ideal Toy Company, which purchased the American rights, made $75 million on the cube between 1980 and 1983, a successful 1984 patent infringement suit by American game inventor Larry D. Nichols cut into those profits significantly. Nichols claimed he had invented a similar puzzle in 1969.

Rumsey, James (1743-1792)

American inventor

James Rumsey, a self-taught Virginia blacksmith and builder, held patents for a jet-propelled **steamboat**, as well as improvements to the **steam boiler**, the grist mill, the saw mill, and the steam pump. Although his ideas received little recognition during his lifetime, they provided a basis for future inventors.

Native Brazilians manufacturing rubber shoes.

After the American Revolution, in which he fought as a patriot, Rumsey helped build George Washington's house in Virginia. His ideas about mechanically-propelled boats captured Washington's imagination, and the nation's new leader helped Rumsey secure a monopoly on that type of sea craft in the state of Virginia. However, Rumsey was unable to interest other inventors in the project and had to direct his efforts elsewhere.

His interest remained in the area of water transport, and he was inspired by the work of **John Fitch** on steamboat technology. In 1787, after several years of work, Rumsey introduced the world's first jet-propelled boat. The vessel was equipped with a **steam engine** to power a pump that took water in at the helm and expelled it from the stern in a jet stream, thus moving the boat forward. Although the design was highly advanced, it needed a more powerful engine to provide the speed necessary to appeal to potential investors.

In 1788, as he continued to pursue support, Rumsey presented his ideas to the American Philosophical Society. To his delight, a group of its members responded enthusiastically to his presentation and formed the Rumsian Society to back him. They funded his trip to England where he was able to obtain patents for his boat and other inventions. While he was there, Rumsey met English inventor and entrepreneur **Matthew Boulton** who proposed that they enter into a partnership. Rumsey declined out of loyalty to his American supporters.

When he returned to the United States, Rumsey hoped investors would be more interested in his projects because he had obtained patents for them. Instead he met with disappointment again, losing even the support of the Rumsian Society. He died in 1792, before the completion of his second boat, the *Columbia Maid.*

The role played by **Benjamin Franklin** in Rumsey's choice of jet propulsion over paddle wheels is not known by many. While Franklin was in France, he had the opportunity to observe some disappointing experiments by the French with paddle wheels, and concluded that they were in no way the best way to use steam power to propel a boat. Upon his return to the United States, Franklin's opinion became known among those interested, and it was Franklin's disenchantment with paddle wheels that led Rumsey to build a boat that moved forward by reaction as water was forced out of its stern (the jet propulsion principle).

Ruska, Ernst (1906-1988)

German physicist

Ruska was born in Heidelberg, Germany, on December 25, 1906. He wasthe fifth child of Julius Ferdinand Ruska, an Asian studies professor, and Elisabeth (Merx) Ruska. After receiving his undergraduate education in the physical sciences from the Technical University of Munich and the Technical University of Berlin, he was certified as an electrical engineer in 1931. He then went on to study under Max Knoll at Berlin, and received his doctorate in electrical engineering in 1933. During this period he and Knoll created an early version of the electron microscope, and Ruska concurrently was employed by the Fernseh Corporation in Berlin, where he worked to develop television tube technology. He left Fernseh to join Siemens as an electrical engineer, and at the same time accepted aposition as a lecturer at the Technical University of Berlin.

His ability to work in both academic and corporate milieus continued through his time at Siemens, and expanded when in 1954 he became a member of the Max Planck Society. In 1957 he was appointed director of the Society's Institute of Electron Microscopy, and in 1959, he accepted the TechnicalUniversity of Berlin's invitation to become professor of electron optics and electron microscopy. He remained an active contributor to his field until his retirement in 1972.

Prior to Ruska's invention of the **electron microscope** in 1931, the field of microscopy was limited by the inability of existing microscopes to see features smaller than the wavelength of visible light. The conception that it was possible to construct a microscope that used electrons instead of light was realized in the late 1920s when Ruska was able to build a short-focus magnetic lens using a magnetic coil. A prototype of the electron microscope was then developed in 1931 by Ruska and Max Knoll at the Technical University in Berlin. Although it was less powerful than contemporary optical microscopes, the prototype laid the groundwork for a more powerful version, which Ruska developed in 1933. That version was ten times stronger than existing light microscopes. Ruska subsequently worked with the Siemens Company to produce for the commercial market an electron microscope with a resolution to one hundred angstroms (by contrast, modern electron microscopes have are solution to one angstrom, or one ten-billionth of a meter).

Ruska's microscope—called a transmission microscope—captures on a fluorescent screen an image made by a focused beam of electrons passing through a thin slice of metalized material. The image can be photographed. In 1981 **Gerd Binnig** and **Heinrich Rohrer** took Ruska's concept further by using a beam of electrons to scan the surface of a specimen (rather than to penetrate it). A recording of the current generated by the intermingling of electrons emitted from both the beam and specimen is used to build a contour map of the surface. The function of this ''**scanning electron microscope**'' complements, rather than competes against, the transmission microscope, and its inventors shared the 1986 Nobel Prize in physics with Ruska.

In addition to the Nobel Prize, Ruska's work was honored with the Senckenberg Prize of the University of Frankfurt am Main in 1939, the Lasker Award in 1960, and the Duddell Medal and Prize of the Institute of Physics in London in 1975, among other awards. He also held honorary doctorates from the University of Kiev, the University of Modena, the Free University of Berlin, and the University of Toronto. Ruska died in West Berlin on May 30, 1988.

See also Electron microscope

Rutherford, Ernest • See Geiger counter

Ryle, Martin • See Radio interferometer

S

SAFE

Early safes, used primarily by banks, were crude wooden chests with padlocks or boxes covered with sheet **iron** as a slight protection against fire. Neither posed any serious problem to determined thieves and robbers.

In 1844, Frenchman Alexander Fichet created the first modern safe, a box made of solid iron that was too heavy to haul away and too strong to be pried open with a crowbar. He also devised a ''burglarproof'' **lock** for the safe; such locks, unfortunately, were little match against **gunpowder**, **dynamite**, and other **explosives**.

The keyhole, the most vulnerable part of the safe, was eliminated with the development of the combination lock. Invented by American Linus Yale in 1863, this new lock featured a revolving dial imprinted with numbers. When the correct sequence of numbers were dialed (the ''combination''), the corresponding tumblers would be properly positioned for the lock to open. One problem arose when James Sargent, a skilled locksmith, invented a **micrometer** to be attached to the lock knob. This device was so sensitive that it could measure the movements of the internal tumblers that controlled the opening of the safe. While Sargent had created this device to help locksmiths open safes for their owners, there was an immediate danger that criminals could also use it to ''crack'' a safe. As a result, combination locks were created that functioned on magnetic principles and could not be defeated by micrometers.

Better explosives, powerful drills, and cutting torches used by thieves led to many advances in safe design and construction throughout the twentieth century. In the 1920s, sophisticated alloys were developed to make safes drill-proof and fireproof. In addition, massive room-sized safes called vaults were built into the foundations of banks and other buildings. Vaults can feature steel doors several feet thick and fit their openings to precise specifications. The time lock, another invention by Sargent, will only allow the vault to open during designated hours of the day.

Since modern safes and vaults are virtually impossible to break open, electronic innovations have included computer controlled **surveillance equipment** and monitoring devices to record the all activity around the safe.

See also Lock and key

SAFETY BELT

Safety devices to protect occupants against injury and death have been a primary focus for the automotive industry and governmental agencies. First, it was important to maintain ''passenger-compartment integrity'' so that occupants remain in the compartment (thanks to door-latches and safety glass) and impacts are deflected. Within the compartment, energy-absorbing devices are in the forms of padded dashboards and visors, energy-absorbing steering columns, cushioned steering wheels, and hinged steering columns. Other energy absorbers have been built into the chassis and body of the car; many of these are stress risers in the frame that buckle on impact. The final group of devices consists of active and passive restraints for the driver and passengers, of which the safety or seat belt may be the best known.

In automotive design, the seat belt may be the perfect proof that everything old is new again. The motor cars of 1903 vintage used tillers to steer and were fitted with rigid wheels or tires. Small bumps would eject the driver. Babies of the day were held secure in their carriages, and G. Pounce suggested a similar harness for drivers. As early as 1885, E. J. Claghorn was awarded a patent for a lap belt. Richard Radtke, in 1908, designed a sophisticated lap belt, shoulder harness, and chest harness that was fastened to the frame with springs to allow some comfort and flexing with small motions. Similar belts and harnesses were adapted to early airplanes and have been used consistently in aircraft. As better seats, windshields, tires, and steering improved the comfort and apparent security of the automobile, seat belts were forgotten by the public and manufacturers from approximately 1910 to 1950.

Tests in the 1940s showed that severe head injuries could be reduced if car occupants were restrained by lab belts. During the period from 1956 through 1970, some companies borrowed the reel-type lap belt from a design made for aircraft by Pacific Scientific Products. The reel allowed the strap to be slowly extended but it would lock with sharp movements. Another design had a locking ball that would roll into place only with rapid acceleration or deceleration. Shoulder harnesses were added when experiments using **crash test dummies** showed that riders often jackknifed into the dashboard. Lap and shoulder belts are made of reinforced fabric that can withstand loads of 6,000 lb (2,700 k). Latches that can be attached and released quickly were designed to encourage passenger use while withstanding the same forces.

In 1972, all cars manufactured in the United States were required to have seat belts and reminder systems, but 60% of car occupants admitted to not using lap belts, and less than 3% used shoulder harnesses regularly. Many states have enacted fines if drivers and passengers are not wearing seat belts when stopped. Other efforts have been directed toward inventing passive systems that don't require occupants to act to be buckled up. The self-applying seat belt was developed by Ford in 1966 with two ends that attached to the rear of the door with the third between the seats. Movement of the door and the passenger in the seat secure the belt.

A child seat and harness were invented in 1962 by Jean Ames in England. Straps pin the padded seat to rear seat of the car; within the safety seat, a full harness extending over the shoulders and between the legs restrain the child. A series of designs to adapt to children as they grew soon followed. The **airbag** as a passive restraint system has also been developed and tested but remains controversial because of fatalities caused by the force of the airbag itself. Accident avoidance systems that rely heavily on electronics are also projected for use soon after the year 2000 in conjunction with the **automated highway system**, and elimination of accidents altogether may be the ultimate passive restraint.

Sail

A sail catches the power of the wind and uses it to propel a boat or ship across the water. The earliest boats were simple rafts made of logs or reeds lashed together, and it was to these that the first sails were added by the ancient Egyptians around 3500 B.C. In its most essential form, a sailboat consists of the sail itself (made of papyrus or linen by the Egyptians), the mast (one or two poles attached vertically to the front half of the boat), and a long spar called a yard attached horizontally across the top of the mast (to which the sail is attached, and raised and lowered, by lines called halyards). These first Egyptian sails were square and harnessed the wind's power to travel upstream on the Nile River, since that was the way the winds usually blew. The boats were paddled rather than sailed on the return trip downstream into the wind. By about 2000 B.C., a single pole mast was used and the bottom of the sail was attached to a second horizontal spar. Boats were also made of wood and were large enough to be called ships. Although the Egyptians still did not venture far out into the ocean, their sail technology spread to Crete, Phoenicia, and even Greece. A typical large Phoenician trading ship of about 1000 B.C. derived nearly all of its propulsion from its single, square sail.

Some believe however, that the sail was invented in Indonesia rather than in Egypt, and there is some evidence for this argument. These Pacific Island societies used an upside-down triangular sail attached to a single vertical pole. It is known that they influenced the Polynesians whose ocean-going ''tainui''—which look like modern catamarans—used the inverted Indonesian sail as one of its two sails.

In the West however, square and then rectangular sails prevailed for some time, despite the fact that they were of little use if the wind was not blowing in the direction that the ship had to go. It was not until after 200 B.C. that triangular ''lateen'' sails were adopted in the Mediterranean region. Pioneered by the Arabs possibly as early as 200 B.C., these triangle-shaped sails were slung across the mast to a long, sloping crossbar, and were the earliest example of fore-and-aft sails. Angling from high above the mast down nearly to the deck, their maneuverability permitted a ship to sail almost into the wind. This ability to ''tack'' into the wind greatly increased the potential and range of a sailing ship, and proved to be a major breakthrough when combined with the square sail. Together, they produced a vessel capable of conquering the oceans.

In the East, China became a major sea power around 1000 A.D. because of its development of the junk. Their use of the lugsail, which was shaped like a trapezoid and strengthened by bamboo strips, combined with their invention of the rudder, made these ships one of the most efficient vessels ever to move under sail power. Junks were often five-masted, and the stiffness of their lugsails allowed them to make use of even the slightest gusts of wind.

Around the 1450s, western Europeans combined the stern-hung rudder with the use of square and lateen sails to produce the deep-hulled carrack—the first ship capable of making long sea voyages without interruption and thus capable of making voyages of discovery. Sail technology continued to improve and reached its apex with the clipper ships of the 1840s. These long, low ships were sleekly designed for speed and had masts nearly as tall as the keel was long, with every inch covered by sails. By this time, the end was in sight for these as well as the massive, two-masted ''brig'' which had more than 20 separate square or triangular sails, since steam and iron were soon taking over. Steam and other types of engines signalled the permanent decline of the sail as a primary means of moving a ship, and increasingly sails were used only for pleasure and competition. It was not until 1980 that a ship of any size employed any version of sail. That year, Japan introduced its tanker *Shin Aitoku Maru* whose computer-controlled sails lowered its fuel consumption by 10 percent (since the main diesel engines could reduce power when the sails were in use).

Sandglass • See Hourglass

Sandwich

Now an integral part of nearly all Western cuisines, the sandwich was named after John Montagu, the Fourth Earl of Sandwich. A prominent eighteenth-century English statesman known both for corruption and administrative savvy, Montagu was also apparently devoted to gambling. It was during an epic, 24-hour stint at the gaming tables in 1762 that he—or the cook who brought him his snack—invented the sandwich by calling for bread and meat so that he could keep playing. Word of this prestigious gentleman's unusual meal spread, and it was soon being copied by England's fashionable aristocrats. During the nineteenth century the sandwich gained popularity elsewhere on the European continent, eventually entering the French language.

Still admired for its simplicity, portability, and limitless variety, the sandwich may incorporate any type of bread and filling and may be served hot or cold. Versions from around the world include the British tea sandwich, the Scandinavian open-faced smorre-brod, and those American favorites, the hamburger and the peanut-butter-and-jelly. One popular modern variation is the wrap, in which a flour tortilla or flat bread is topped with an array of ingredients, then rolled into a burrito-shaped package. Another is the pocket, in which the filling is placed inside hollow pita bread.

John Montagu, Fourth Earl of Sandwich

Saran

Saran is derived from the polymerization of vinylidene chloride or the copolymerization of vinylidene chloride with lesser amounts of other unsaturated monomers, most often **vinyl** chloride or acrylonitrile. This thermoplastic polymer was first polymerized in 1920 by the Dow Chemical Company. Today, it is most often used in copolymers to form packing film, rigid pipe, and fibers. As a film, it acts as a barrier to oxygen, water vapor, grease, and liquid. About 85% of the polyvinylidene chloride produced is used as a thin layer between **cellophane**, **paper**, and **plastic** packaging to improve barrier performance. Saran Wrap, well known as the cling wrap designed for household and commercial use that Dow Chemical first introduced in 1949, revolutionized the food packaging industry. Today, saran films are used as a barrier package for cheese, baked goods, processed meats, and seafood.

Saran is produced by polymerizing vinylidene chloride with other monomers to form long stereoregular chains. These chains are susceptible to crystallization, giving rise to the superior clarity and excellent barrier properties of saran.

See also Polymer and polymerization

Savery, Thomas (ca. 1650-1715)

English military engineer

As an English army officer, Thomas Savery was once ejected from the Lord of the Admiralty's office as a lunatic because he proposed a ship that could be propelled by side-mounted wheels rather than by wind or oars. Fortunately, some of his other ideas were better received, among them a **water pump**, the first working steam-powered machine.

Savery was born in Devonshire, England, about 1650, but little else is known about his first four decades of life. Some sources even question his career as a military officer. He was an inventor and owned several patents, such as one for a 1696 machine for polishing plate **glass**. He was also a published writer, and his writings include *Navigation Improved* (1698), and a translation from the Dutch *Treatise on Fortifications* by Baron van Coehoorn (1641-1704).

His most important contribution was his 1698 water pump. An atmospheric engine by definition, Savery's pump did not produce any motion. The pump used the expansion and contraction of steam to create vacuums that raised water from flooded mines. This ''Miner's Friend'' consisted basically of a boiler, a separate collecting chamber, and two pipes. Water in the boiler was heated until it turned to steam. **Valve**s opened to allow this steam to pass into the closed storage chamber. This hot chamber was then cooled from the outside to condense the steam and create a vacuum. Water from the mine was then sucked up to replace the vacuum and forcibly expelled through the drain pipe.

However, the pump did not prove to be very practical since it could not be sealed tightly enough to raise water much higher than 25 ft (7.6 m). Nonetheless, it was used for decades to supply water to towns and country homes and to recycle water back into millraces.

Savery seemed to have given up making improvements on the water pump, but others like **Denis Papin** and **Thomas**

Newcomen were quick to see other possibilities of steam power, and went on to create the **steam engine** and other important inventions.

SCALE • See Balance and scale

SCALPEL • See Surgical instruments

SCANNING ELECTRON MICROSCOPE • See Electron microscope

SCANNING PROBE MICROSCOPE

A scanning probe microscope (SPM) is any type of microscope that can see surfaces on a very fine scale—so fine, in fact, that individual atoms may be discerned. There are several types of SPMs, all about the size of a fingernail, but all rely on a small, extremely sharp probe that scans across a surface, recording the **force**s it encounters and transmitting this information back to a sensor. With such microscopes scientists have studied surfaces at the level of individual atoms.

There are three basic types of scanning probe microscopes, all of which work slightly differently: (1) the atomic force microscope (AFM), (2) the scanning tunneling microscope (STM), and (3) the near-field scanning optical microscope (NSOM). The first scanning probe microscope was an STM, invented in 1982 by Heinrich Rohrer (1933-) and **Gerd Binnig** of the IBM Zurich Research Center in Switzerland. In 1986 Binnig and Rohrer shared the Nobel Prize in Physics for their design of the scanning tunneling microscope.

The atomic force microscope is something like a **phonograph**ic needle. Using **lithography**, a very small tip is made of silicon nitride (similar to the way computer chips are made). The tip is tapered as narrow as possible, possibly to a single atom. Hung from a **cantilever**, it then moves across the surface to be investigated, touching it, bending and flexing in response to the extremely small electromagnetic and mechanical forces it feels from the atoms on the surface. A small **laser** beam reflected off the back of the tip is detected, and the small changes in the location of the reflected beam are used to reconstruct the surface over which the tip is moving. The AFM can make an image down to atomic dimensions, about 0.1 nanometers (nm), somewhat like a topographical map, and works best over relatively smooth surfaces.

The scanning tunneling microscope was built before the AFM, and, since the only difference is in their tips, the STM invention provided all the image reconstruction techniques needed by other SPMs. A very fine wire tip is brought to within a few tenths of a nanometer of a surface. (The surface must be an electrical conductor for the STM to work, a limitation that the AFM does not have.) At this distance **electron**s can tunnel across the gap, from the surface up to the tip, generating a current that is detected. This current is also fed back into the tip-controller, a piezoelectric crystal that expands or contracts slightly when it encounters current, and the controller then tells the tip where to move next—up, down, left, right, back, or forward. The STM was a major advance in the study of materials, but it was limited to analyzing only conductors, usually metals.

The near-field scanning optical microscope runs a fiber optic probe over the sample. A small laser beam shines out the end of the probe, which is reflected off the surface for most materials. For opaque samples it is transmitted all or partially through, and some samples fluoresce under the influence of the laser. The light is then gathered and used to form the image. The detail that can be seen by the NSOM is limited by the wavelength of light used, in practice about 400-500 nm, although techniques exist to reduce this to about 40-50 nm—still not nearly as accurate as the AFM or STM, but useful for larger structures.

Since AFM can analyze electrical insulators as well as conductors, it has become more widely used, especially among biologists studying the surfaces and molecules of the biological world. The placement of a small magnetic coating on the AFM tip allows it to measure magnetic properties of a sample, and other schemes work for other properties.

With the SPM, scientists have taken pictures of individual atoms and crystal surfaces, they have made movies of fundamental biological molecules like DNA and RNA bind and move together, the molecular basis of phenomena like adhesion and lubrication, and a host of other new and creative applications. Using the SPMs, scientists are also construction s mall machines like switches and batteries, a new field called nanotechnology.

SCANNING TUNNELING MICROSCOPE

The latter half of the twentieth century has opened to scientists an entirely new world: the world of atomic and subatomic particles. With the inventions of the **electron microscope** and the **field ion microscope**, scientists have been able to observe the microcosm as never before.

Until the early 1980s, however, one mystery that eluded researches was the nature of the surfaces of substances. Since the arrangement of atoms in the surface of a substance differs greatly from that of its bulk, it requires other methods of analysis. Scientists had lacked a mechanism for studying the intricacies of surfaces until 1981, when German physicists **Gerd Binnig** and Heinrich Rohrer (1933-) invented the scanning tunneling microscope (STM).

The word *tunneling* used here describes an effect of quantum mechanics theorized upon for years and first verified in the laboratory in 1960. It was known that an electron orbits about the nucleus of an atom, its motion random and diffuse, behaving as a cloud. When an element is placed very close to another, the cloud-orbits of the surface atoms will overlap slightly. The resulting diffusion of electrons is called tunneling.

When Binnig and Rohrer met in 1978 they were both working at IBM research laboratories in different cities, each studying the atomic structures of surfaces. They decided to combine their efforts toward using tunneling to explore these structures. By 1980 they had constructed a prototype STM, and in the spring of 1981 they succeeded in obtaining microscopic images using electron tunneling.

The STM that Binnig and Rohrer had built was actually based upon the field ion microscope invented by **Erwin Wilhelm Müller**. The field ion microscope uses a tiny sharpened needle placed within a **cathode ray tube**; as an electrical field is applied, metal ions are emitted from the tip of the needle, creating an image of the metal's atomic structure upon the cathode screen.

In Binnig and Rohrer's device, a similar needle is placed in a vacuum, above a specimen to be scanned at a height of less than one nanometer. The sharper the needle, the more precise is the STM reading; the best needle tips are only one or two atoms wide. A very low voltage is applied, causing the overlapping clouds of electrons to tunnel from the needle to the specimen. This electron flow is called a tunneling current, and by measuring this flow irregularities in the surface of the specimen can be determined. The scanning tip is swept over the sample, so that the entire surface may be mapped.

Even the very first tests of the STM showed it to be extremely powerful. When scanning crystals of calcium-iridium, Binnig and Rohrer resolved surface hills only one atom high. Maintaining the very small distance between the needle tip and the specimen proved to be difficult, however, since noises as unobtrusive as a footstep would jar the instrument. Sinnig and Rohrer used **magnet**s to suspend the microscope over a table equipped with shock absorbers to solve this problem. Other improvements increased the magnification of the STM, and today's tunneling microscopes can resolve features as small as one hundred-billionth of a meter, or about one-tenth the width of a hydrogen atom. It has also been discovered that STMs are equally useful in air, water, and cryogenic fluid media.

The incredibly precise three-dimensional images provided by STMs have found varied applications in a number of industries. They are used for quality control in manufacturing digital recording heads as well as in the construction of compact audio disk stampers. In 1991 the STM was used to move and place 35 atoms of xenon in a predetermined pattern; this ability to manipulate matter at the level of a single atom may allow scientists to customize molecules, possibly creating ultramicroscopic data storage chips.

Because it is effective in many media and uses a very low voltage, the STM can be used to study the atomic structure of living and biologic matter if that matter readily conducts electrons. Most biologic matter does not, however, and must be coated with a thin layer of a conducting substance. The STM is standard equipment in most atomic research laboratories. It is undeniably the most powerful optical tool yet invented, and for its invention Gerd Binnig and Heinrich Rohrer shared the 1986 Nobel Prize for Physics, along with Ernst Ruska, the inventor of the electron microscope.

Schick, Jacob • See Razor

Schick test

During the late 1800s, diphtheria swept through western Europe and the United States killing thousands of children. Diphtheria is a serious contagious disease of the respiratory system that causes membranes to grow on the inner throat. The infection can spread through the blood to the heart and kidneys.

During the epidemic, an enormous effort was launched to find effective treatments and immunizations for the disease. One of the first findings that came out of this period was the diphtheria test developed by Béla Schick (1877-1967), a Hungarian pediatrician who specialized in childhood diseases. The *Schick test*, as the procedure came to be known, was based on the toxin-antitoxin research of Emil von Behring (1854-1917). Behring's research revealed that when a toxin (produced, for example by an invading bacteria) enters the body, the body naturally produces substances called *antitoxins* or *antibodies* capable of neutralizing the invading substance.

The test works by injecting a small amount of specially-prepared diphtheria toxin beneath the skin. This causes a red swollen rash to appear around the injection if the person is susceptible to the disease, in which case the person should receive the diphtheria vaccine, a serum containing diphtheria toxoids which stimulate the production of diphtheria antibodies to ward off disease.

Used in concert with diphtheria vaccine, Schick's test dramatically reduced the incidence of diphtheria worldwide. Until the 1920s, there were 150,000 to 200,000 cases of diphtheria in the United States each year. The number dropped to less than 10 cases per year by the 1970s. Other scientists using Schick's approach developed similar tests for diseases such as measles, tuberculosis, pertussis, gonorrhea, and syphilis, saving thousands of lives in the process.

Schickard, Wilhelm (1592-1635)

German mathematician

Born in Herrenberg, Germany, Schickard was a brilliant student. By the age seventeen, he had already received his B.A. in theology and Oriental languages from the University of Tübingen. Continuing his studies, Schickard received his M.A. in 1611. In 1617 he met astronomer Johannes Kepler (1571-1630), who inspired Schickard's interest in mathematics and astronomy. While toiling over the many tedious calculations necessary in astronomy work of that time, Schickard's thoughts turned to the notion of mechanically performing mathematical calculations. Although the discovery of logarithms and logarithmic tables by John Napier (1550-1617) several years earlier had greatly simplified the process of multiplication and division, Schickard sought to develop a calculating machine to completely automate these functions.

In 1623 Schickard wrote a letter to his friend Johannes Kepler, describing his progress in developing such a machine. Schickard placed **Napier's bones**, the logarithmic calculating device invented by Napier, on cylinders that were selected by the turn of a dial. Employing the use of accumulator wheels allowed six digit calculations to be performed. Results were generated by turning large knobs, while the answers were viewed through small windows. Dubbed a calculator-clock, the machine was capable of performing all four mathematical functions, but before accounts of the invention were publicized to the world, the machine became lost in the chaos of the Thirty Years War. Schickard set about building a second machine; unfortunately, it, too, was destroyed by fire while still under construction. It appeared that all of Schickard's designs and records were lost as well—a great misfortune as it would take fifty years before the advancements achieved by Schickard could be duplicated by Gottfried Leibniz (1646-1716) and later inventors.

In 1935 while researching a book on Kepler, a scholar found a letter from Schickard and a sketch of his calculator, but did not immediately recognize the designs or their great importance. Another twenty years passed before the book's editor, Franz Hammer, found additional drawings and instructions for Schickard's second machine and released them to the scientific community in 1955. A professor at Schickard's old university, Tübingen, reconstructed the calculator based upon Schickard's original plans; it is still on display there today.

Schickard died of bubonic plague in 1635. Unfortunately, the significance of his work was not fully appreciated until 320 years later.

See also Leibniz calculator; Calculating machine

SCHMIDT CAMERA • See Telescope

SCHÖNBEIN, CHRISTIAN (1799-1868)
German chemist

Schönbein was a chemist who is best known for his discovery and research on ozone (O_3).

Schönbein's family was financially unable to continue his formal education past the age of fourteen, so he was apprenticed to a chemical and pharmaceutical company in Boeblingen. For the next seven years, Schöenbein worked at the plant and studied chemistry, philosophy, mathematics and several languages in his spare time. In 1820 he finished his apprenticeship and moved to Augsburg where, for a short time, he translated scientific papers from French to German. He soon found a position with another chemical company where he remained for three years.

In 1823, Schönbein took a post under Friedrich Froebel (1782-1852) teaching chemistry, physics, and mineralogy. After three years with Froebel, Schönbein moved to England to teach mathematics and natural history. He stayed in England only one year, then moved again in 1827, this time to France. There he was awarded an honorary Ph.D. and appointed a professor of physics and chemistry at the University of Basel. It was during these years that Schönbein conducted his most significant research.

Schönbein began his studies by exploring the passivity of iron. Although he reached several interesting conclusions, the methods he used to conduct his studies were qualitative and relied heavily on analogies rather than empirical data. Thus, most of his research into passive iron has since been disproved. Schönbein's major accomplishment came almost coincidentally. Because his laboratory at Basel was poorly ventilated, he was able to notice an unusual odor that was produced during the electrolytic decomposition of water. He observed that this odor was similar to the smell produced around large electrical apparatus machines. He theorized that in both cases this smell was due to a new type of gas. This same gas, he discovered, was also produced when phosphorus oxidized. He also discovered the gas was positively charged and that it resembled chlorine and bromine in its chemical properties. Schönbein named the gas ozone (from the Greek *ozon*, meaning ''odor'') because of its peculiar smell.

Another of Schönbein's major discoveries was also a fortunate accident. He discovered **guncotton** while experimenting with nitric and sulfuric acids in his kitchen. He spilled some acid and, not wanting his wife to know that he had been using the kitchen as a laboratory, quickly cleaned the liquid up with an apron and hung it on the stove to dry. Much to Schönbein's surprise, once the apron dried, it burst into flames. He then realized that the nitric acid had reacted with the **cellulose** in the cotton apron to form nitrocellulose. Because the guncotton, as Schönbein named it, would burn without creating a great deal of smoke, it briefly flourished as a replacement for black **gunpowder**. Schönbein died in 1868.

SCHREYER, HELMUT • See Zuse, Konrad

SCHWARZ, DAVID (1845-1897)
Austrian engineer

In 1893 David Schwarz, an Austrian engineer, drew up plans for an **airship** which was constructed of an inside framework that was strong enough to support a gas envelope. By 1897, Schwarz, who was working in Berlin, Germany, had built the first practical airship with a metal envelope. The structure was sheathed in **aluminum foil** (a new metal at this time) and shaped like a cannon shell, with measurements of 135 ft (41 m) in length, 46 ft (14 m) in diameter, and a volume of 130,000 ft^3 (3,640 m^3). It had a twelve horsepower engine turning three propellers. The craft was not tested until after Schwarz's death in 1897 (which is said to have been caused by a heart attack when the Prussian Ministry of War offered to buy his invention). The test, however, ended disastrously. The propeller belts broke, and the pilot valved gas too quickly, causing the ship to bounce to a crash landing which destroyed it.

One man who watched this crash was determined that no such failure would ever affect his plans. Count **Ferdinand von Zeppelin** became interested in the airship, bought the patent rights from Schwarz's widow, and went on to perfect the rigid airship.

SCINTILLATION COUNTER

The ability of radiation to produce luminescence in certain types of materials was recognized by **Antoine-Henri Becquerel**

(1852-1908) in 1899. The first application of the principle in a detecting device was made by **William Crookes** in 1903 with the invention of the spinthariscope. Crookes' spinthariscope consisted of a brass tube with a zincsulfide screen at one end and a magnifying lens at the other end. A tiny crystal of a radioactive salt was mounted on a pin about a millimeter from the zinc sulfide screen. Radiation from the salt struck the screen, producing tiny flashes of light that could be viewed through the lens.

The potential for using such a device as a radiation detector and counter was noted by a number of investigators. **Ernest Rutherford**wrote in 1904 that the Crookes device "would offer a very convenient means of actually counting the number of particles...if each particle gave rise to a flash of light." Working with Hans Geiger (1882-1945), Rutherford showed that such a method could be used for counting alpha particles and was as accurate as was the **Geiger counter**. The first working scintillation counter was apparently made at about this time, however, by the German physicist Ernst Regener (1881-1955), who used diamond rather than zinc sulfide as the scintillation source.

A number of early researchers used primitive scintillation counters in their work on nuclear reactions. A fundamental drawback of the instrument, however, was the difficulty of observing and counting individual flashes of light. As a result, scintillation counters became less popular than Geiger counters, **cloud chamber**s, and other instruments that were easier to use.

By the 1930s, a solution to the problem of observing flashes—the photomultiplier tube—had been invented. In the photomultiplier tube, a single flash of light falls on a metal plate, causing the emission of a number of electrons. These electrons, in turn, are accelerated towards other metal plates, resulting in the release of even larger quantity of electrons. After a number of repetitions, the single flash of light has been multiplied enough times to produce an electrical current that can be easily read and recorded.

Scintillation counters are now used in a wide variety of applications in nearly every field of science. They can be produced with very small dimensions, making them useful in situations where other detectors might be too bulky. Where conditions demand, they can be made very large in order to scan a broad area of an experiment. Since scintillations occur very quickly (as fast as 10^{-9} second), they are sometimes more efficient at recording all the events that occur in a particular experiment, a characteristic not shared by some other types of detectors.

Scissors

Scissors were an early development in hand tools, probably originating about 1500 B.C., using blades connected by a C-shaped **spring** at the handle end. In ancient Egypt, people used scissors to cut silhouettes, and turned this into high art. Various blades and **bronze** or iron scissors were produced in India, Europe, parts of the Orient, as well as in the ancient Roman world. Methods of **metal** forging continued to improve and, as a result, scissors and other implements became more widely used.

Christian Schönbein

It was not until 1761, however, that true mass production of scissors occurred. In Sheffield, England (a city known for its cutlery since the 1300s), Robert Hinchliffe used crucible-cast **steel** (a highly refined type made in clay vessels) to manufacture the blades. By the 1800s, beautifully ornate hand-filed scissors became popular in Europe.

Today, machines are used to produce scissors. Red hot steel bars are turned into blanks (unfinished pieces of steel), which are then forced between the dies of drop hammers. These hammers form the rough shape of the blade. The final product is made by trimming the forgings from the blade; the blade is finally hardened (through a heating process) and tempered. The steel used in scissors contains varying amounts of carbon, depending on the quality of scissors. Surgical and other specialized scissors are made of stainless steel; cheaper scissors are made with softer steel that is cold pressed. Scissors come in many sizes and their blades are shaped and sharpened for specific functions that range from cutting fabric to cutting steel.

Screw

The winch, **pulley**, **lever**, wedge, and screw are considered the five simple machines of antiquity. The discovery of the screw is usually credited to **Archimedes**, who used an enclosed screw or auger to lift and move water.

In his book, *Machanica*, **Hero of Alexandria** mentioned a screw-type machine that was used in the first century A.D. One such machine, an olive press, has been excavated in Pompeii, Italy, lending support for the theory that screws were in existence around 79 A.D. In the Middle Ages, screw-like battering rams were used to break through castle gates. In the thirteenth century, screw hoists were being used to raise loads, and towards the end of the 1400s screw jacks were used to lift wagon **wheels** for repairs.

By the 1500s, gun and **armor** manufacturers were using tools much like turnscrews to form and adjust their products. About the same time, craftsmen discovered that they could get a better seal between seams and joints if they gave their tacks and **nails** a twist. Once they were hammered in, they remained difficult, if not impossible, to remove. Soon, slots were cut into the heads so that they could be removed with a "screwpuller" or "unscrewer." Around 1780, toolmakers in London, England, introduced a screwdriver that also increased the everyday use of the screw.

Screw propulsion systems for boats and ships began appearing in the late 1700s, but did not become commonplace until 1838 when Swedish engineer **John Ericsson** successfully demonstrated propellers on a small ship named *Archimedes*. Around the same time, **Cyrus H. McCormick** built the first **combine harvester** (1835). Thought by many to have been the most important farming tool since the **plow**, the combine employs a system of enclosed screws and augers, much like the Archimedean screw, to harvest all types of crops.

Screws have now become so ordinary that they are hardly noticed. In some parts of the world, however, people still rely on screws for such basic roles as transporting water.

See also Inclined plane; Water pump

Screw propeller

When steam power became practical in the early 1800s, shipbuilders turned to three methods in an attempt to harness this improved power source: **paddle wheel**s, jet propulsion, and the screw propeller. Jet propulsion was attempted but with little practical success. Paddles were often used at the sides or backs of ships navigating on western rivers because they could take a great deal of punishment and could be easily repaired since half of the paddle wheel sticks out above the waterline. But early steam-powered, oceangoing ships equipped with paddle wheels experienced difficulties in rough seas. The ships would roll, and the wheel would often come completely out of the water, wasting power and making the vessel difficult to handle.

There was a third alternative for transforming power into thrust: the screw. As early as 1785 the Englishman **Joseph Bramah** patented a sixteen-blade propeller to drive boats. But the first to actually experiment with the concept was American engineer John Stevens (1749-1838), who built a small **steamboat** in 1804 powered by a high-pressure engine and propelled up to 4 mph (6.4 kph) by a pair of four-bladed screw propellers. Unfortunately, America did not have the tools or craftsmen to machine the parts the engine required. Progress was made in the 1830s. An English inventor, Francis Pettit Smith (1808-1874), and **John Ericsson**, who later built the Union's ironclad *Monitor*, constructed screw-propelled craft that aroused a great deal of interest in England. Smith's propellers drove a large steamship, and Ericsson's device was applied to a tugboat. These vessels impressed **Isambard K. Brunel**, who was building the *Great Britain*, the first all-iron oceanliner. He scrapped his plans for paddle wheels and substituted a six-bladed propeller 15.5 feet (4.7 m) in diameter. In 1843 the *Great Britain* sailed from Liverpool to New York in 14 days.

Two years later, Brunel convinced the British Admiralty to conduct tests to determine the best means of propulsion. One test involved two ships of equal horsepower and weight being attached by their sterns, the only difference being that one used paddle wheels and the other used a propeller designed by Francis Smith. After the tug-of-war commenced, the paddle wheels churned the water, but the propeller-driven vessel ended up towing the other behind it at 2.5 knots (4.6 kph).

This and the other tests helped convince people of the value of propellers, but it was the experience of the British Navy in the Crimean War (1854-1856) that persuaded naval authorities to use only propellers on warships. Russian gunners had an easy time disabling British ships with exposed pa wheels. Soon after the war, screw propulsion was used by nearly all the world's navies and merchant vessels. Paddle wheels were used only for ships sailing coastal or inland waters.

Ever since, there have been modifications in the screw to give it better performance. During the 1970s, for example, the American Navy developed super-cavity propellers, which because of their shape and action in the water, achieved greater speed.

By the mid-1990s, propellers for large commercial vessels as well as for pleasure runabouts were no longer made of once-popular cast iron. Starting in the 1960s, the use of high-tensile brass accounted for the majority of propellers, but by the 1980s, some combination of nickel-aluminum-bronze accounted for the bulk of all propellers. By the 1990s, heavy cruisers, yachts, and sport-fishing boats were usually outfitted with either manganese bronze propellers that were malleable, easy to repair, and cost-efficient, or a new material called Nibral (Nickel, Bronze, and Aluminum). Although more expensive, Nibral has proven to be an especially good combination for high-power use. It was also in the 1990s that the three-blade propeller finally became the one most commonly used.

A propeller is a deceptively simple device that is in fact, a highly complex piece of equipment when it is in action. In addition to a propeller's seven characteristics (diameter, pitch, bore, number of blades, rotation, style, and material), its performance can be influenced by such variables as hull shape and even water temperature. Propellers of the twenty-first century will be computer-machined to exact specifications, and their manufacture will be digitally controlled, improving their precision and specialization of design and use.

Screwdriver • See Screw

Scuba diving

The underwater exploration of the sea has been a desire and a fascination for humankind since the earliest times. Divers are mentioned in Homer's *Illiad*; Xerxes of Persia used divers to recover treasure in 475 B.C., and divers breathing through tubes are mentioned by Aristotle in 355 B.C. Early attempts at underwater exploration made use of the **diving bell** and special rigid suits supplied with air from surface pumps. In 1889, Jules Verne (1828-1905), in *Twenty Thousand Leagues Under the Sea*, describes divers moving freely to hunt in underwater forests while breathing air from iron tanks "fastened to their backs with straps." Verne claimed that the equipment had been invented by two French men.

In 1943, two French men, **Jacques Cousteau** and Emile Gagnan did, in fact, invent the aqualung, a *S*elf-*C*ontained *U* nderwater *B*reathing *A*pparatus (SCUBA) and gave birth to a new sport and a new profession. The key element of the invention was the demand regulator. The regulator delivered air to the diver's breathing mouthpiece at the same pressure as the diver's underwater surroundings allowing the diver to breath in a normal and comfortable manner. It is called a demand regulator because air only flows in response to the diver's inhalation effort, rather than free flowing. The scuba diver now was free to move underwater without dependence on a supply of air from the surface.

Since 1943, innovations in scuba equipment design have improved the safety and comfort of the sport. In addition to the mask, fins and snorkel used in skin diving, a scuba diver utilizes a broad array of equipment specialized for the underwater environment. Compressed air is carried in a steel or aluminum tank and two pressure regulators (the first and second stages) are mounted on the tank. The first stage reduces the air pressure to an intermediate level and the second includes the breathing mouthpiece and reduces the pressure further so that it matches the diver's surrounding pressure. An additional second stage regulator is frequently carried as a backup against equipment failure. A weight belt and an inflatable vest called a buoyancy compensator are worn to control underwater buoyancy. In water that is below about 75°F (24°C), most divers wear a protective body suit to keep warm.

Two types of suits, wet and dry, are in common use. The wet suit is made of neoprene foam rubber and includes jacket, pants, hood, gloves and boots. The neoprene is a thermal insulator as is the thin layer of water trapped between the suit and the skin. The dry suit has water-tight seals at the neck and wrists and uses a heavy insulating undergarment to keep the diver warm. While somewhat more difficult to use and prone to the occasional leak, the dry suit is substantially more effective than the wet suit.

A diver carries a watch and a depth gauge to monitor the duration and depth of the dive. Underwater computers are available that continuously monitor depth and calculate how long and how deep the diver may safely remain underwater. The amount of air remaining in the tank is monitored with a pressure gauge and the diver carries a knife to free himself from underwater entanglements.

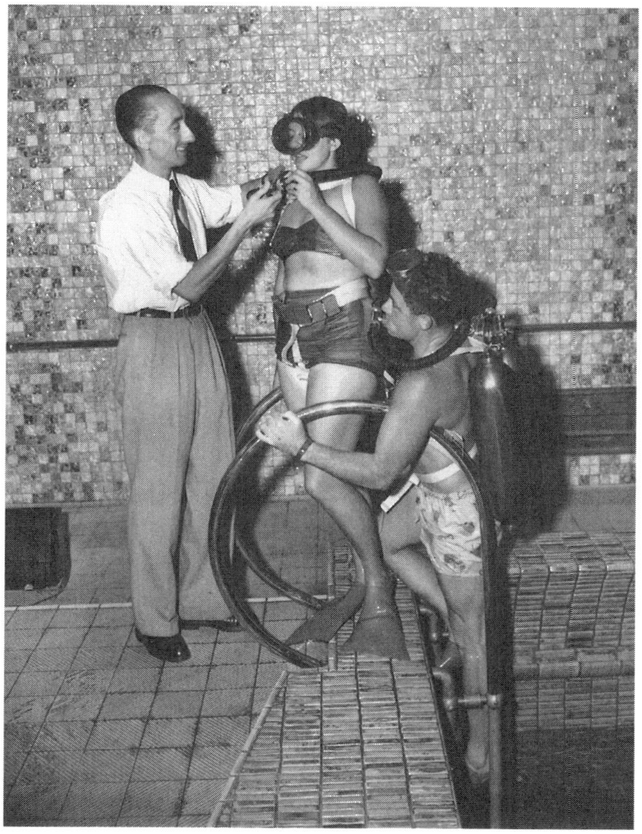

Jacques Cousteau, co-inventor of the aqualung, demonstrates how to use the apparatus in 1950.

Knowledge and training are required to avoid the risks associated with the underwater environment. While underwater, nitrogen gas accumulates in the diver's body tissues. If a diver remains too deep for too long and surfaces too quickly, the nitrogen gas will form bubbles within the body. This condition is called decompression sickness (DCS) or the *bends*. Depending on their location, these bubbles may cause joint pain, choking, blindness, seizures, unconsciousness, or death. The cause of these effects was first identified in 1878 by Paul Bert (1833-1886) and in 1907, John Scott Haldane (1860-1936) developed the first decompression tables to limit time under pressure and, thus, avoid the *bends*. The most effective treatment of decompression sickness is immediate recompression in a recompression chamber. This chamber was invented by **Edwin Link** and uses elevated pressure to reduce the size of the nitrogen bubbles. The diver can then be decompressed very slowly without the formation of new bubbles.

Nitrogen is normally an inert gas but, at depth, the increased pressure of the nitrogen causes it to become a narcotic and effect the thinking process. Nitrogen narcosis or *rapture of the deep* can cause divers to take illogical and unsafe actions. This effect may occur at depths as shallow as 60 ft (18.2 m) and is treated by simply moving into shallower water.

Oxygen toxicity is another potential diving problem related to the gases breathed. Toxic effects may begin to appear in as little as 33 ft (10 m) of water when breathing pure oxygen. Onset of toxicity may occur with air at depths of 265 ft (80 m) or so. To avoid or reduce the risk of decompression sickness, nitrogen narcosis, and oxygen toxicity, many commercial divers, as well as a few recreational divers, breath mixed gases when deep diving. These consist of various mixtures of nitrogen, oxygen and in some cases, helium, designed to circumvent these diving maladies.

The increased mobility and safety of scuba diving permits a wide range of new underwater activities for hobbyists, scientists, and entrepreneurs. These include activites as varied as nautical archaeology, shipwreck exploration, studies of marine and freshwater ecology, dredging for gold, and treasure hunting.

See also Diving apparatus; Cousteau, Jacques

Seaplane

It was obvious from the very start of heavier-than-air flights that a craft capable of taking off and landing on water would be desirable. As early as 1910 a Frenchman, Henri Fabre (1883-?), built the first seaplane by attaching floats rather than wheels to his **aircraft**. The next year Glenn Curtiss (1878-1930) designed a plane that had a light boat for its body. Curtiss demonstrated the military potential of his seaplanes by landing near a United States Navy ship and allowing the ship to hoist the craft on board via a crane.

Seaplanes were used during World War I. Although they did not receive the publicity or acclaim that land-based fighters did, they were widely used by both sides for reconnaissance, bombing, and **torpedo** attacks. It was a seaplane that warned the British about the movements of the German ships that led to the battle of Jutland. By the end of the war, the British had developed long-range seaplanes that could cross the North Sea to attack German **airships** patrolling the coast.

After the war, the Schneider Trophy Races helped accelerate the development of seaplanes. The first race was won in 1923 by an American whose craft averaged 47.75 miles per hour (76.8 kph). Later races were noted for design and engine innovations. The British came out with Supermarine aircraft—capable of over 407 miles per hour (655 kph)—that anticipated the great Spitfire fighters of World War II.

The seaplane distinguished itself in another field: commercial aviation. A German aircraft designer, Claudius Dornier (1884-1969), had already designed several successful seaplanes when he tackled the problem of transporting passengers across the Atlantic profitably, comfortably, and safely. In 1929 he unveiled the Dornier Do.X, the largest airplane of its time with twelve engines, a wingspan of 157 feet (47.88 m), and a length of 131 feet (39.95 m). Although it never flew, Dornier's craft served as the model for succeeding commercial seaplanes, including the Boeing 314 Clippers, which established the first scheduled routes across the Atlantic. Introduced in 1939, these huge flying boats provided luxurious accommodations for their passengers on Atlantic and Pacific ocean routes. During World War II, seaplanes were used extensively as patrol craft by the Allied forces. One famous descendant of the Clipper was the Catalina, which was responsible for helping the British find and destroy the German battleship *Bismarck*.

After the war, improvements in runways led to the abandonment of seaplanes. Airline companies also needed planes capable of transporting a large number of passengers at maximum speed but at minimum expense and the luxury flying boats could not compete with the new jet aircraft. Today, seaplanes are still used in rugged backwoods areas where there

Howard Hughes' *Spruce Goose*, the largest seaplane ever built.

are few runways, and in fire fighting where they can scoop up water from lakes.

SEEBERG, J.P. • See Jukebox

SEEDING DEVICES

Seeding devices such as the seed **drill** were important developments in agricultural history. Before seeding devices were developed, seeds were planted by hand or by scattering them over a field. Once seeding machines were invented, farmers could use less seed to yield higher crops in a shorter amount of time. As early as the third millennium B.C., Babylonians were presumably using seed drills along with ox-drawn plows. The seed drills depicted on Babylonian seals were simple vertical tubes through which seeds were dropped into newly opened furrows, thus making it easier to plow for irrigation and weed control. Seed drills like these were in use in India and China by 100 A.D., but they remained unknown in Europe until the 1700s when they were possibly introduced by Jesuit missionaries who had traveled in the Orient.

Once introduced to Europe, mechanical improvements were attempted, but seed drills were not practical until **Jethro Tull**, an English musician and "gentleman farmer," perfected a mechanical drill in 1701. Born in 1674 in Berkshire, England, Tull was educated at St. John's College, Oxford, to be a lawyer. He was called to the bar in 1699, but ill health prevented him from practicing or following a political career. Instead, he began experimental farming at Howberry, an estate near Wallingford, England, and later, in 1709, at Prosperous Farm near Hungerford, England. Tull's design used the grooved sound board of an organ, a brass cover and **spring**, and a toothed **wheel**. The machine sowed three rows at once and was pulled by one horse. Though described in his 1731 publication *Horse Hoeing Husbandry*, the drill did not become commonly used until the 1800s. In addition to the advantages of row planting, it has been theorized that the mechanical drill saved a great deal of seed. While early colonists rejected Tull's successful seed drill in favor of scattering, or broadcasting, seeds, later Americans were eager to use the new tool. Even so, as late as 1850, it is believed that fewer than twenty-five percent of the farmers in the northeast used modern methods and equipment such as the replaceable steel plows and seed drills. Today, multi-row grain seed drills are common **tractor** attachments.

SEISMOGRAPH

Seismographs are instruments used to measure and record ground motion, or seismic waves, caused by explosions and earthquakes. In the late 1800s, John Milne, an English mining engineer, developed the first precise seismometer, the sensor in a seismograph that detects and measures motion. Since then, seismograms, the data recorded by a seismograph, have helped seismologists predict much more than ground movement. These devices have also led to discoveries about the nature of the Earth's interior.

The process of using a tool to detect ground motion dates back to the ancient Han Dynasty when Chang Heng (78-139 A.D.), a Chinese astronomer and mathematician, invented the first seismometer in 132 A.D. Although the internal mechanics of this device are unknown, it probably used a pendulum that, when moved during an earthquake, would dislodge a bronze ball from the mouth of one of eight ornamental dragon's heads that adorned the exterior. While this did not provide clues of an earthquake's force, it gave Heng an idea of the direction of the shock waves and their source. Since then Heng's concept has since been refined considerably.

Later seismographs employed a heavy pendulum with a *stylus*, or needle, suspended above a revolving drum. The drum was covered with paper or film on which the etchings from the needle were recorded. During an earthquake, the pendulum and needle remained stationary while the drum on the base moved, recording the ground motion. As much as these later pendulum seismographs improved upon the ancient Chinese method, they still failed to solve many of the problems that arose with more precise readings. For example, once a strong motion set off a seismograph's pendulum, the pendulum would swing indefinitely, failing to record aftershocks that followed the initial disturbance. Also, the seismographs of the late 1800s recorded only a limited range of wave sizes and numbers.

The inverted pendulum, invented by German seismologist Emil Weichert in 1899, helped overcome many of these limitations. Weichert employed a system of mechanical levers that linked the pendulum movement more closely to the ground motion. In 1906 Boris Golitsyn, a Soviet physicist and seismologist, devised the first electromagnetic seismograph; for the first time, a seismograph could be operated without mechanical levers. To record the magnitude of an earthquake, electromagnetic seismograms convert ground motion to electrical currents that are then read by a **voltmeter**. They can therefore register very small waves with great precision. During the 1930s Hugo Benioff designed a vertical component seismometer capable of providing much higher amplification of vertical ground motion and therefore much more precise records.

Although many of the modern seismographs are complicated technical devices, all such instruments contain five basic parts. The *clock* records the exact time that the event takes place and marks the arrival time of each specific wave. The *support structure*, which is always securely attached to the bedrock, transfers the ground motion to the recording device during the earthquake or explosion. The *inertial mass* is the component that remains stationary as the ground and the support structure oscillate around it. The *pivot* holds the inertial mass in place, that is, it allows the inertial mass to remain stationary in spite of the ground motion. The vibrations are registered through the *recording device*; in simple systems, this is a pen attached to the inertial mass and a roll of paper. The paper moves along with the support structure while the pen remains stationary. Most modern units now use a digital recorder connected to a **computer**. In either case, this records the pattern of seismic waves as a set of wavy lines of varying amplitude and period, revealing the strength of the various waves as well as the frequency with which they occurred.

After the first modern seismograph was installed in the United States at the University of California at Berkeley, it recorded the 1906 earthquake that devastated San Francisco. The same year, Weichert and fellow scientist Richard Oldham (1858-1936) determined the existence of the Earth's core through precise recordings of seismic waves. In 1909 the use of a seismograph helped Yugoslavian seismologist and meteorologist Andrija Mohorovicic (1857-1936) discover the depth at which the Earth's crust meets the upper mantle. That discovery was followed in 1914 by German-American seismologist Beno Gutenberg's (1889-1960), determination of the depth to the core-mantle boundary, now known as the Gutenberg discontinuity. Then in 1936, Danish seismologist Inge Lehmann's work suggested the existence of an inner core. These important findings revealed the existence of all of the Earth's major layers: the inner core, the outer core, the mantle, and the crust.

By the late 1950s, the range of motions detectable with seismographs made possible for the first time very precise determinations of earthquake magnitude and depth, rock density and heterogeneity, etc. However, as the sensitivity of seismographs has increased, the need for filtering—removal of background vibrations from local activities—has had to keep pace. Without computers, recent improvements in sensitivity and filtering would not be possible. Likewise, the techniques for artificially inducing seismic waves, by explosions or surface concussion, along with greater instrument sensitivity, enable seismologists to greatly expand the abilities and utility of seismic studies. For example, seismology was critical in the development, and in the continued refinement, of the theory of plate tectonics. It provides evidence for the character of many important features of plates and plate boundaries, including mid oceanic ridge and rift systems, deep ocean trenches, transform faults, and volcanic island arcs.

In addition to helping scientists understand the phenomena of natural ground movement, seismology also includes the study of manmade earth motion. Seismographs provide miners with tools to effectively monitor how much dynamite should be used for quarry blasts and other explosions. They are also widely used today to detect variations in the speed of seismic waves traveling in near-surface rocks, clues which may help locate economic deposits, such as oil and coal. Since the development of the **atomic bomb** and the heightened threat of nuclear war, seismologists are in greater demand. The need to quickly and accurately decipher the difference between natural events and manmade explosions has brought about the estab-

The primary tool of seismologists, seismograph detects, measures, and records ground vibrations. Above, a seismologist takes readings following the 1991 explosion of Mt. Pinatubo in the Philippines.

lishment of cooperating seismic stations in sixty countries, called the World-Wide Standardized Seismograph Network (WWSSN). Seismologists can now routinely establish the time, location, depth, and explosive yield of nuclear detonations with great accuracy. For this reason, non-WWSSN member countries such as Russia, France, and Canada have modernized their earthquake observation systems as well.

Since the early 1980s, major advances have been made in linking seismometers to computer networks, enabling seismologists to work with increasingly large sets of data and allowing new types of analyses and graphical presentations. For example, seismic tomography provides two- or three-dimensional views of the Earth's interior by computer-synthesizing hundreds or thousands of seismic records into an image. Seismic tomography works somewhat like a **computerized axial tomography** or CAT scan. CAT scans are interior views of a person's body produced by computers that combine hundreds of digital x-ray cross sections of the body into a coherent picture. The density of the different tissues provides the contrast that allows recognition of medical conditions. In seismic tomography, differences in density, primarily due to temperature differences, reveal details about the structure of the Earth's interior. This technology only became possible in the mid to late 1990s as computers grew capable of processing huge data sets at extremely rapid speeds. Seismic tomography has shown that the base of the mantle has "mountains" and "valleys" much larger than anything at the Earth's surface. At these locations, low areas probably result from the cooler (denser) mantle sinking downward somewhat into the core and highs result as molten core material convects upwards into the mantle. Seismic tomography also reveals convection currents within the mantle itself, which may be the driving force behind plate tectonics.

Seltzer water • See Soda water

Semaphore

To meet the need for swift communication between Napoleon's (1769-1821) far-flung armies in the 1790s, the Frenchman Claude Chappe (1763-1805) invented an optical-relay system of visual **telegraphs** that he called the semaphore. His system consisted of tall vertical posts, each of which supported a movable crossbar with movable arms at each end. Using a system of **pulleys** and ropes, the semaphore operator would move the crossbeam and its indicator arms into a variety of positions that represented numbers and the different letters of the alphabet. Chappe installed his signalling mechanisms on hilltop towers which allowed an unobstructed view. While such a system would have been possible many centuries earlier, Chappe's made use of both a signalling code and the recently invented **telescope** to read signals across distances much greater than would previously have been possible.

Giving the name *telegraph* to his signalling system, Chappe secured a commission from the French government as *Ingénieur-Télégraphe* in 1793 and installed a line of semaphore towers from Paris, France, to Lille, France. With each tower was approximately 6 miles (10 km) apart, the distance covered totaled 144 miles (231.7 km). Messages could be sent 90 times faster with the Chappe semaphore than by horse-riding couriers. The new telegraphy achieved wide popularity when it was used to bring news of victories on the Napoleonic battlefields in 1794. Soon Chappe semaphores covered most of France, and were only supplanted by the **electric telegraph** in the 1850s.

In England, George Murray and John Gamble had built an optical signaling system in 1795, consisting of a tower-mounted box with six movable shutters that opened and closed to produce coded messages. Chains of these shutter telegraphs were built in Britain in the early 1800s, but they were gradually replaced between 1811 and 1814 with the Chappe semaphore system. In the United States, Jonathan Grout built the first commercial semaphore system in 1800 between Boston, Massachusetts, and Martha's Vineyard to send news about ship and cargo arrivals. Various telegraph and signal hills can still be found on maps of Cape Cod. Proposals in the United States Congress to award $30,000 to build a semaphore telegraph system all along the Atlantic Coast led to **Samuel Morse**'s successful application for government support of his electric telegraph. Once the electric telegraph was perfected, the semaphore was doomed by its inherent limitations: it was unusable at night or when bad weather reduced visibility, and was restricted in its speed of transmission by the manual demands it put on the operator.

Semon, Waldo L. (1898-)

American chemical engineer

In 1912, the European scientist Klattes had discovered that hydrochloric acid combined with acetylene gives **vinyl** chloride. Another European, Ostromislenski, in the same year patented the production of a rubber-like material from vinyl chloride. However, it was not until 1930, when the American Waldo L. Semon advanced the use of high-boiling organic esters, especially phosphates, to plasticize polyvinyl chloride, yielding rubber-like masses, that polyvinyl chloride began to find widespread use. Plasticized polyvinyl chloride soon found applications as cable insulation, substitute leather cloth, and in packaging.

Pressed to find additional uses for rubber, Semon came up with the idea of bubble gum, a form of rubber. He patented his idea in October of 1931. Semon's expectation was that bubble gum would replace conventional **chewing gum**. His employer, B.F. Goodrich, however, looked upon the bubbles as a product defect, and declined to develop and commercialize Semon's invention. After the time had run out on the patent, Bazooka trademarked the product, and Semon ended up having no financial stake in the commercialization of his invention.

In the course of his 37-year career with B.F. Goodrich, Semon made significant discoveries in the areas of diarylamine rubber antioxidants, **synthetic rubbers** and oil-resistant nitrile rubber. From 1954 to 1961, he was Director of Polymer Research at B. F. Goodrich where he provided the technical leadership that fueled the discovery of three major new families of polymeric materials: thermoplastic polyurethane, synthetic 'natural' rubber, and the first oil-resistant synthetic rubbers.

Semon was born in Demopolis, Alabama, in 1898. He moved to the Pacific Northwest when he was seven. In 1916 he entered the University of Washington determined to be a chemist. He graduated cum laude in 1920 then continued his studies at the University of Washington, where he received his Ph.D. in chemistry in 1923. After a short period as an instructor at the university, Semon joined B.F. Goodrich. Following his retirement from that company in 1963, he served as a research professor at Kent State University. In the course of his career Semon was awarded 116 patents. In recognition of his contributions to polymer science, Semon received the ACS Charles Goodyear Medal, the Edward W. Morely Medal, and the Thomas Midgely Award; he was also inducted into the National Inventors Hall of Fame.

See also Plastics; Polymer and polymerization

Senefelder, Aloys • See Lithography

Servomechanisms

A servomechanism is a device that automatically controls an object by means of feedback.

A simple example is the driver of a car. His eyes tell him where he is on the road, and compare it to where he should be, and this information makes its way to his brain. The brain decides what action should be taken in order to move the car from where it is to where it should be, and sends a signal to the muscles in the arm, turning the steering wheel to realign the car.

Servomechanisms are generally more mechanical than this example, but the principle remains the same. All servomechanisms have the following parts: (1) a way to measure what is desired and what is being accomplished, (2) a way to transport this information, (3) a way to determine the difference between the actual condition and the desired condition, and (4) a means to amplify this difference (which is often small) and use it to move the actual condition towards the desired condition. In the example of the car, 1 are eyes, 2 is the optic nerve and pathways to the brain, 3 is the brain, and 4 are the arms and steering wheel. A small turn of the wheel translates into a major turn for the car.

Servomechanisms are useful to control motion without human attendants, or to maintain the accuracy of an environment like a power plant, and to control action from a remote isolated station. The controller typically uses (and has) much less power than that of what is being controlled. Almost always it is the position or velocity which is being controlled.

Servomechanisms were first used in military applications, such as an antiaircraft gun that tracks a **plane** via **radar**. As the plane moves the **radar** gives the plane's position information to the gun, which computes the new position of the plane and realigns. This process can go indefinitely. Some other applications are satellite tracking and satellite antenna alignment systems, automatic **machine tools**, star-tracking systems on **telescope**s (since the stars' position changes as the earth rotates), and navigation systems.

Sewage systems

''Out of sight, out of mind'' has been the traditional philosophy of waste disposal. Technologically, it may well be the last thing scientists want to be associated with, so it has received relatively little attention until recent decades, when the negative results of lax policies have shown themselves. Improper sewage disposal results in pollution-related problems—besides being foul and unpleasant, pollution can damage plant and animal communities, contaminate food and water supplies, and spread serious human diseases.

In the 1870s, **Louis Pasteur** (1822-1895) discovered the existence of microbes, and made the association between pollution and diseases such as cholera and typhoid. Yet it took the better part of a century for practice to come into line with knowledge, and many parts of the world still do not have adequate sanitation.

Sewer systems have been part of the urban scene since ancient times. Latrines and drains were used in the Indus Valley circa 2500 B.C., and Rome boasted sewers as early as 500 B.C.. Water closets and toilets began to appear in homes during the 1700s. In cities, the effluent had to be carried in buckets to gutters and cesspools, while it was buried or used as fertilizer in rural areas. In both cases, water and food supplies were easily contaminated, either directly or by bacteria carried by rats, mosquitoes, and flies. The first sewer system to resemble modern underground sewer plans was built in Hamburg in 1842. In 1854, London rebuilt its sewer system after a cholera outbreak. Though Pasteur had yet to make his discovery, suspicions about the link between disease and improper disposal of human waste had surfaced.

The new London sewer system made extensive use of Portland **cement**, a highly impervious material invented by Joseph Aspden in 1824. Yet nothing was done to treat the effluent before it was sent into the Thames River. Waste treatment began about 1890 when filters and chemicals started to be used in some cities. In 1916 two United States innovators, Lockett and Ardern, devised the activated sludge method, relying on biological organisms to purify waste, but modern waste treatment evolved slowly during the twentieth century.

About 80% of waste has industrial origins. The special nature of each factory and its pollutants requires individualized attention at the site before the waste is sent into the public system. Incineration, radiation treatment, and **composting** are among those treatments. Modern sewage treatment is comprised of three stages. The primary stage separates out larger solids, or sludge. The second stage includes further filtering and allows bacteria to break down the particles in aeration tanks. The third stage introduces lime to aggregate the remaining suspended particles and allow it to settle in ponds. In rural areas, where homes and businesses are widely scattered, sewage systems are expensive to build. Therefore, septic tanks are located near each building. Invented in 1896, they collect the effluent, break it down with bacteria, and gradually introduce it to the surrounding soil.

One method for disposal of sewage wastes, the spreading of *sludge,* is used by many countries, and has had renewed, and controversial, interest in the United States in recent years. Sludge, sometimes referred to as *biosolids*, is processed sewage waste, left over after attempts to remove debris (e.g. gum wrappers), heavy metals (e.g. lead from old city pipes), and pathogens (**virus**es, molds and **bacteria**). Some sludge, deemed to be cleaner, is heat-treated and sold as **fertilizer**. The rest, if not considered toxic, is used to feed the soil, and provide a cheap method of sewage disposal. The waste industry, and some state and federal government agencies, advocate the spreading of sludge as beneficial to growing and grazing land, and as the best way to handle the nation's sewage disposal problems. But citizen and environmental groups have protested vigorously that the sludge frequently contains toxic heavy metals, fecal disease-causing fecal organisms, or radioactive compounds. Scientists have discovered that the practice of spreading only thin layers of sludge increases the amount of mercury—an environmental contaminant—that evaporates into the atmosphere and which, ultimately, condenses out in cold Artic air. There is also concern that processed sludge sold as fertilizer may contain disease-causing fecal organisms such as *Salmonella.* The real problem, some scientists point out, is that the vast array of pollutants in sludge cannot all be treated by one, single disposal method. As population increases, the spreading of sludge as a means of sewage disposal is sure to remain contentious.

Promising future directions in sewage handling include the use of reed beds to remove toxins, reliance on peat or lignite to absorb harmful bacteria, and the immobilization of bacteria in a solid medium like carbon powder.

Sewing machine

The sewing machine stitches materials together by using a needle and thread. It is important both as a home and an industrial appliance. Because it made possible the mass manufacture of reasonably priced ready-made clothing, the sewing machine relieved women of one of their most demanding, time-consuming domestic chores—manually sewing all of the family's clothing.

The first sewing machine patent was issued in 1790 to Thomas Saint, an English cabinetmaker. This machine used a notched needle, awl, and chain stitch to sew leather and heavy canvas, but the device was not practical. Various inventors made attempts in the early 1800s to develop a sewing machine focused on duplicating the motions of hand sewing. A machine patented in England in 1807 by Edward and William Chapman made the first crucial innovation: it used an ''eye-pointed'' needle, a needle with the eye at the point rather than at the top.

One of the first successful sewing machines was developed in 1829 by Barthelémy Thimmonier (1793-1859), a French tailor. This machine, patented in 1830, used a hooked needle to produce a chain stitch. By 1841 Thimmonier had eighty machines at work turning out uniforms for the French army. Infuriated at Thimmonier's threat to their livelihood, an angry mob led by Parisian tailors stormed Thimmonier's factory, destroyed the machines, and nearly killing the inventor. Thimmonier fled for his life and later died in poverty.

An improvement over Thimmonier's design and the first successful sewing machine made in the United States was the creation of **Walter Hunt**, a gifted and prolific New York inventor. Hunt's 1834 machine used an eye-pointed needle and a second thread that produced a lock stitch—a great improvement over the chain stitch, which would completely unravel if a thread were pulled in the wrong direction. Hunt, however, never applied for a patent and showed no further interest in his design.

The first patented and practical American sewing machine was designed in 1846 by **Elias Howe**. Howe, a Boston, Massachusetts machine-shop employee struggling to provide for a growing family, was supported financially by a friend while he developed his machine. Like Hunt's invention, it used an eye-pointed needle with a bobbin for a second thread to make a lock stitch. Although Howe's machine operated successfully, it did not sell.

Howe went to England in 1847 to market his invention but instead practically gave away his rights to it. Howe's design was being widely copied by other American manufacturers, as the penniless Howe discovered when he returned from England in 1849. Howe gathered financial backers and sued a number of sewing machine companies for patent infringement, chief among them the I.M. Singer Company.

By 1854 these lawsuits had been successfully settled or decided in Howe's favor. The proliferation of sewing machine patents by now threatened to hobble all manufacturers, since a practical, successful machine needed to incorporate a number of competing patented features. In 1856, four manufacturers created a ''combination,'' or patent pool, where rights to the pooled patents could be purchased. Howe received a royalty on each sewing machine sold.

The most important advances in sewing machine design during the 1850s were made by Allen B. Wilson (1824-1888) and Isaac Singer. Wilson, an American cabinetmaker, invented both a rotary hook shuttle (1851) and four-motion feed (1854), features that are still used today. Singer, another Boston machine-shop worker, had entered into the sewing machine industry in 1850 when a machine was brought into his shop for repairs. Singer developed a new design that corrected the drawbacks of Howe's machine, introducing vertical (rather than horizontal) movement of the needle, a yielding presser foot, a foot treadle to replace the hand crank, and—most importantly—a continuous feed mechanism (the patent for which Singer purchased from its inventor, John Bachelder).

Singer's other, equally important, contribution to the sewing machine industry was his company's masterful use of merchandising, which brought the machine into thousands of households around the world. Singer also introduced such pioneering sales tactics as installment payments, trade-ins, and a repair service. Soon, Singer was selling one thousand machines a week.

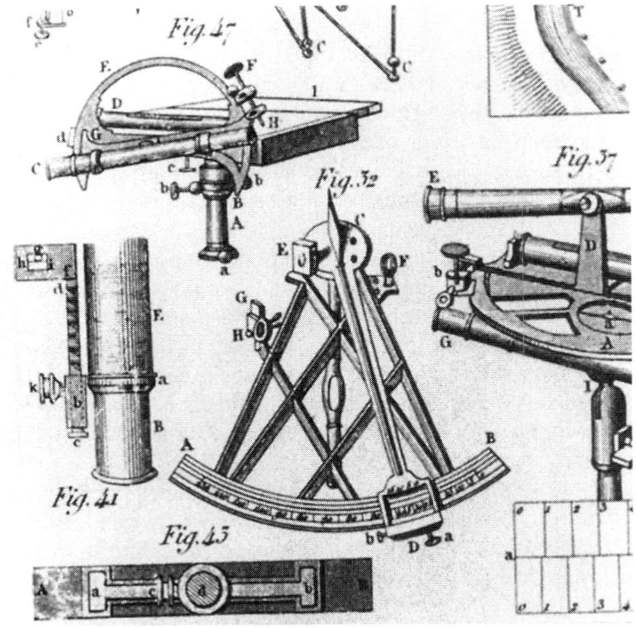

Edmond Halley's sextant.

The sewing machine revolutionized the manufacture of clothing by creating the new industry of ready-made clothes, and it improved innumerable other industries, including boot and shoe making, carpeting, bookbinding, hosiery, and upholstery. Today, specialized sewing machines serve a host of industrial functions. Sergers, for example, sew a seam, trim it back, and wrap thread around the raw edge all in one step to create a durable, finished seam commonly seen in factory-produced garments.

Singer introduced the first electric sewing machine for the home in 1889. The first zigzag (swing-needle) machine was used industrially before 1900 but home models were not available in America until after World War II. Contemporary sewing machines have seen tremendous advances courtesy of the computer. A microprocessor automatically sets the machine parameters, adjusting needle motion and stitch length for various stitch types, and thread tension and presser foot pressure for a variety of fabrics. Computerized stepping motors also improve stitch accuracy and uniformity. Top-of-the-line machines can even be hooked up to home computers to automatically produce complicated embroidery from patterns that can be downloaded off of the Internet.

Sewing pattern • See Dressmaking

Sextant

The sextant is an instrument that measures angular distance of the sun and stars from the horizon to determine latitude. It is used with a **chronometer** and nautical charts by navigators of ships at sea to determine location. The sextant consists of a **telescope** rigidly mounted to a bar and aligned with the horizon. A pair of **mirror**s, one fixed in line with the horizon, the other mounted on an adjustable arm, reflect the celestial object. When the object aligns with the horizon in the fixed mirror, the angular reading is taken from the sextant's arc.

Mariners originally took sightings with an **astrolabe**, a flat circular instrument that had a **map** on its face and an adjustable ring that represented the local horizon. They were compact, but complicated to use. Many seafarers preferred the simpler cross-staff. It had a staff topped with a peep-sight and a graduated slide-rod from which the apparent altitude of the sun could be determined. The *quadrant*, or "hog-yoke," was the immediate predecessor to the sextant. It was invented simultaneously in 1730 by Englishman John Hadley (1682-1744) and American, Thomas Godfrey (1704-1749).

Hadley was an instrument maker and friend of astronomer Edmond Halley (1656-1742). His was a double-reflecting quadrant, with a 90° arc, or 1/4 of total arc. At the suggestion of Captain John Campbell (c. 1720-1790), Hadley extended the arc to 120°. From then on it was referred to as a sextant. Hadley also added a spirit level for readings in foggy conditions. Thomas Godfrey was a glazier; he installed the windows at Independence Hall in Philadelphia. Encouraged by the Governor of Pennsylvania, James Logan, and influenced by his landlord, **Benjamin Franklin**, Godfrey entered into mathematics and science. He completed work on his quadrant in the same year as Hadley, but never received recognition for it from the Royal Observatory.

Sextants have become more sophisticated, though their basic working principle has remained unchanged. Any sextant has a set of mirrors or **prism**s through which the observer gazes at the objects he wishes to measure. As it is adjusted, the sextant brings the objects together into a single image, and the observer reads the angle between them on a graduated scale. Both in the air and at sea, sextants are often mounted to the frame of the craft for greater stability and more accurate readings. The mounts are often gimballed for easy adjustment. With the advent of electronic navigation instruments, and more recently, satellite-linked computers, like the **Global Positioning System**, traditional instruments like the sextant has generally fallen out of use. Indeed, as of 1998, the United States Naval Academy ceased to teach celestial navigation.

Sharps, Christian (1811-1874)

American inventor

Sharps, a native of Washington, New Jersey, was employed under gunmaker John Hall (1788-1841) at Harpers Ferry, West Virginia, before relocating to Cincinnati, Ohio, in 1844. In 1848 he patented his designs for a basic breechloading system and created a single-shot **rifle** that used a linen or paper wrapped cartridge. After first lowering the breechblock, the user then inserted the cartridge. As the breech was closed, a metal piece was sheared off the end of the cartridge, exposing the powder charge to the action of the percussion primer. Sharps' gun was both simple and strong. He formed the Sharps Rifle Manufacturing Company in 1851 to market his arms. Though Sharps himself resigned from the company in 1853, guns bearing his name went on to become very well known. By the end of the American Civil War, the Union Army had purchased approximately 100,000 Sharps rifles. Later, the Sharps rifle became the preferred gun of buffalo hunters because of its power and long-range accuracy. The Sharps system remains the basis for many single-shot rifles manufactured today.

But Sharps left his rifle company and soon turned his attention to **pistols**, succeeding in creating a 0.22 and 0.32 caliber rim-fire metallic cartridge four-barrel pistol. Sharps' invention featured a hammer with a ratchet that revolved the firing pin by cocking and firing the four barrels in rotation. Before he died in 1874, he had secured patents on more than fifty other inventions.

Shaving cream & aftershave

Shaving creams and lotions are personal care products used to make the process of shaving body hair easier and less painful. They are typically applied prior to shaving and removed with a **razor**. Aftershaves are lotions that are applied after shaving to improve skin condition, relieve irritation, and provide a fresh, clean scent.

Shaving has been an important facet of society for centuries. Shaving instruments have been discovered dating back to the Neolithic and Stone ages. These devices were constructed from things such as shark teeth, clam shells, or flint and were ground to a sharp razors edge. Discoveries of ancient Egyptians artifacts show that these people used solid gold and copper blades. To these early societies, shaving was not done primarily for personal grooming, but for religious and political reasons. In Egypt priests were known to have their entire bodies shaved every three days. People in the ancient Greek and Roman societies were also known to shave. The practice was popularized among the Greeks by Alexander the Great. It is suggested that he ordered his soldiers be clean shaven so enemies could not use beard hair as something to grab during battle.

Over the years, improvements were made in both the shaving instruments and the associated creams. These improvements focused on making shaving a more pleasant, convenient, and comfortable experience. Blades were more refined and could give a closer, longer lasting shave. Preshaving products were developed for similar reasons. As an outgrowth of these technological advances, **hair removers** like depilatories and epilatories, were developed as alternatives to shaving.

From the middle of the fourteenth century up to World War II, shaving preparations were primarily shaving soaps. These were bars, cakes or powders which would foam when applied. A shift away from these solid products to aerosols was a result of adaptation of insecticidal, high pressure spray systems. By 1946, aerosol lather shave creams became the standard product. Since then many minor ingredient changes have been introduced to shaving creams and aftershaves.

Shaving creams improve the shaving process in a variety of ways. One of the primary ways in which they work is by lubrication. The compounds in shaving creams help reduce friction between the cutting blade and the skin. This helps to significantly reduce the occurrences of nicks and cuts. This also helps to eliminate the pain and razor rash associated with the pulling of hair that happens when shaving dry.

Another way in which shaving creams improve the shaving process is by reducing the strength of hair. The hair absorbs water from the shaving cream thereby lowering the hair's tensile strength. During shaving, weaker hair gives less resistance to cutting which makes the whole process less painful. Additionally, shaving creams function by positioning the hairs in the optimal cutting position. When the lather is applied to skin, tiny air bubbles surround each hair. This causes the hairs to be separated from each other and maintain an upright position.

While shaving creams prepare the skin for shaving, there may still be a significant degree of skin trauma. For this reason, aftershave lotions are used. These products are designed to relieve the discomfort, irritation and physical damage that may be caused during shaving. They work by refreshing and cooling the skin and soothing minor irritations.

Shaving creams are sold in both foaming and nonfoaming variants. The first types of shaving creams required a brush for application. These products required mixing with the brush to create a foam which could then be applied to the skin. These were the standard product until the advent of the aerosol container. Today, the most common type of shaving product is a foaming aerosol shaving cream. This product is sold in a high pressure container and is dispensed as a foam when a button is pushed. The foam can be applied to the skin immediately after dispensing. Another popular type of shaving cream is a post foaming shaving gel. These products are also aerosols but they are dispensed in the form of a thick gel. The rubbing motion used when they are applied to skin, helps create foam. These products are generally easier to spread than traditional shaving creams. Powdered shaving creams are still available but rarely used. Aftershaves are typically clear, water-thin lotions that are splashed on after shaving. They can be provided in both glass and plastic containers.

There are numerous ingredients used in making shaving creams including such things as surfactants, emollients, humectants, and diluents. During manufacture, each of these ingredients are mixed together in a large container to produce the final formula.

Foam is one of the key characteristics of shaving creams and surfactants are used for this reason. The original formulas were based on soap surfactant. These materials foamed well but they had a drying effect on skin. Today, synthetic surfactants such as stearic acid are used to produce a quick, rich lather that is non-irritating, adheres to the face, has good wetting properties, and is easily rinsed. Coconut oil may also be used to boost foam.

Lubricity and feel are other important characteristics. Therefore, oils and emollients are included in shaving cream formulas. These materials help improve the condition of skin and reduce friction between the razor and skin. Superfatting agents such as beeswax or petrolatum are examples. In addition, these materials also inhibit rust formation on the razor.

Humectants are added to impart smoothness and softness to the products. They have the added benefit of preventing excessive water loss caused by evaporation which would reduce the stability of the foam. Glycerin is a common example.

A variety of other ingredients are used. Water, which makes up the largest weight percentage of the formula, is included as a diluent. To control the thickness of the product viscosity stabilizers like cellulose derivatives are employed. Preservatives are used to prevent spoilage. Fragrance and colors are also included for aesthetic reasons. For aerosol products, propellants are required.

While shaving creams are water based formulas, aftershaves are based on alcohol. This makes them evaporate more quickly and gives them an astringent characteristic. Additional ingredients may include fragrances, colorants, and UV absorbers. Moisturizers and emollients such as witch hazel are also added.

Shearing machine

Wool, one of the most important materials for making warm-weather clothing, blankets, and yarn, is produced from the thick coats of domesticated sheep.

Early shearing was done with sharpened stone or metal cutting implements. Sometimes wool was simply ripped off the sheep's skin. Hand shears, developed later, were large **scissors** that resembled garden shears. With development of the shearing machine, use of hand shears declined considerably during the twentieth century.

S. G. Dorr of Albany, New York, invented the first mechanical shearing machine in 1792. It was referred to as the wheel of knives and was not commercially successful. Australian James Higham of Melbourne invented a shearing machine in 1868, but it, too, failed to sell well.

In 1872 Frederick Wolsely, a British inventor who immigrated to Australia, created a prototype of a shearing machine. He and his assistants continued to work on improving the device until 1885, when another British immigrant, John Howard, a mechanic, helped achieve the final breakthrough.

The new mechanical shears were demonstrated at shearing competitions. The machines not only sheared faster, but left less wool on the sheep. However, skillful hand-shearing required years to perfect and was a source of pride among the workers who composed the shearing teams. It was not until 1888, when the Dunlop sheep station of New South Wales equipped its shearing sheds with the machines, that commercial success began for Wolsely and Howard.

Sheep stations in Australia, New Zealand, America, and other parts of the world gradually adopted the machines. By 1900, their use was widespread. However, in areas where spring came late, like Wyoming, hand-shearing continued until about 1940 because some wool had to be left on the sheep to protect them from the cold. Eventually, the high cost of shearing forced the total adoption of the machines in the United States by the 1940s and worldwide acceptance by the 1960s.

See also Farming, mass production; Woolen fabric

Shell • See Ammunition

Shellac

Shellac is a natural, alcohol-soluble, flammable **resin** (based on laccaic acid) that is made from deposits on tree twigs left by the lac insect in southern Asia and India. Shellac has been known in Asia for nearly 4,000 years. It was introduced to Europe about 300 years ago, and was described in *The Plictho*, published in 1540 in Venice. The word shellac is most probably derived from the Indian word *laksha* (one hundred thousand).

Shellac is the only commercially used natural resin of animal origin and it is quite different from all other natural resins. Lac, a resinous substance secreted by the lac insect (such species as *Lakshadia chimensis*, *Lakshadia communis*, *Kerria lacca*, *Tacchardia lacca*, or *Coccus lacca*) is collected from the twigs and young branches of certain trees in southern Asia, including fig species in northern India.

Lac females produce the resinous secretion as a protective coating for their larvae. This substance, known as sticklac or seedlac, forms a hard, resinous layer on branches and twigs. After collecting and crushing this *sticklac*, the resin and wood are separated by mechanical means. The remaining resin is ground, washed, and finally dried. After further treatment, the material is drawn into thin sheets of finished shellac.

In the nineteenth century and early twentieth century, shellac found use in Europe and the United States (records of shellac's use there appear as early as 1815 in *The Dyer's Companion*) as a wood polish for music instruments and furniture, and later in antifouling paints for ships, **phonograph** records, wood and wallpaper paints, printing **ink**s, resins for electrical applications, floor polishes, **fireworks**, sealing wax, etc. Over the years the use of shellac in some of these applications disappeared.

By the 1850s, photography had gained considerable popularity. Early photograhs were presented in a folding case, which cast a shadow over the image and enhanced its appearance. In the United States, the case was made of tooled wood covered with a composition of shellac reinforced with wood flour. The shellac coating allowed fine detail to be reproduced; this ability to reproduce fine detail was the principal reason that shellac was used in 78 r.p.m. gramophone records until after the Second World War.

At the turn of the century, **Leo Hendrik Baekeland** (1863-1944) was searching for a solvent to substitute for shellac in bonding mica in electrical apparatuses when he discovered the first commercial phenol-formaledhyde resin (**Bakelite**).

Modern applications for shellac have included coatings for fruits and vegetables, food and confections, pills, tablets, and vitamins in the pharmaceutical industry. Shellac continues to be used in woodcoatings, printing inks, leather finishes, and stone polishing. There has been renewed interest in shellac resins for waterbased coatings and water-based printing inks as people become aware of the environment and the toxicity of chemical raw materials. Even after 4,000 years of recorded use, shellac's nontoxic and physiologically harmless characteristics as well as its watersolubility, biodegradability, film-forming properties, excellent adhesion, hardness, high gloss, superior electrical properties, compatibility with other resins, and crosslinking properties continue to suggest new applications.

Ship

Large ocean-going vessels are called ships, while all other water craft are called boats. The difference between ships and boats therefore is primarily a matter of size, with boats being the smaller vessels. If the earliest boat was probably a log or a bundle of reeds lashed together, then the first ship could be said to be large log rafts or the long, spoon-shaped reed vessels made by the Egyptians. Despite their size however, such craft were in no way capable of venturing out to the open sea, and remained largely river vessels or coastal craft.

By about 4000 B.C., the Egyptians had learned how to build long, narrow boats called galleys that were powered by rows of paddlers. As another 500 years passed, they developed the first sail and soon learned how to build ships out of planks of wood. By about 3000 B.C., they were able to sail north from the mouth of the Nile River and made sea voyages that hugged the coastline. With the invention of the sail and the use of planks, shipbuilding for the next few thousand years consisted of building larger, more seaworthy wooden ships and improving and adding to the rig or the system of sails, masts, and ropes.

By about 1200 B.C., the Phoenicians and the Greeks constructed both large, roomy cargo ships with sails and warship galleys powered by oars. The Greeks and later the Romans used a system of wooden ribs to stiffen and strengthen the hull. Between the Romans and early Medieval times, Viking ships dominated the waters of northern Europe from about 700 A.D. to the late 1000s. Their open longships were clinker-built (overlapping planks held together by iron rivets) and carried these intrepid sailors across the Atlantic Ocean to North America. The other Viking craft, called the knarr, was a trading vessel built wide and tubby to accommodate more goods. This shape was modified by shipbuilders of the north by about 1200 to create the "cog" which was used in northern Europe as both a trading vessel and a warship. These ships usually had one large sail and were characterized by a high platform structure called the "castle." In the south of Europe about the same time, shipbuilders in the Mediterranean were developing their use of the triangular or lateen sail on very light vessels, and these two schools of shipbuilding eventually came to combine the best features of each around the middle of the fifteenth century. This produced a sailing ship called the "carrack" that was capable of making long, uninterrupted ocean voyages and which would become standard in Europe for the next 300 years. It was on such a ship that Columbus discovered the New World.

By the eighteenth century, this ship had grown into the floating fortress called the "galleon" which was capable of carrying 100 cannons and a crew of 850. The cargo version of the galleon was designed to hold as much goods as possible. By the beginning of the nineteenth century, trade concerns were predominant, and in America, shipbuilders sought something that would give them a seafaring edge over British dominance. What American designers decided upon around 1840 were extremely speedy vessels rather than ships that could carry enormous loads. With this in mind, they made their new ship hulls sharp and narrow which, combined with a forward-slanting bow, would enable them to slice through waves and "clip" off miles from their travel time. The largest clippers had as many as six tiers of sails, and the American ship, *Great Republic*, was the largest wooden sailing ship ever built. These clippers carried passengers from New York to California during the Gold Rush and raced from China with tea and wool from Australia. Despite their success, the expansion of the railroad and the coming of steam propulsion put an end to the brief career of these sleek merchant vessels.

By 1838, steam had already begun to replace sails, and in that year the British steamship *Great Western* crossed the Atlantic in only 16 days. This lead to the precursor of the modern ship, the *Great Britain*. Although it carried six masts for sails in case its engines failed, it was made of iron instead of wood, and in 1845 became the first propeller-driven ship to cross the Atlantic. Steam power did not fully replace sailing ships until a completely new type of marine engine, the steam turbine, was developed in the 1890s. The turbine-powered *Mauretania* was one of the first modern liners, and it held the Atlantic crossing speed record for over two decades. By 1910, diesel engines that used heavy oil as fuel proved more efficient, and by 1920, oil began to replace coal as fuel for steam turbines.

The great age of the ocean liner reached its height in the 1930s with the launching of several speedy, luxurious ships, but as airplane travel became more common and proficient after World War II, it became apparent that ships could no longer compete with airplanes in crossing the Atlantic quickly. By the mid-1950s, new passenger ships were designed as cruise ships or floating resorts whose actual destination was often incidental to the trip itself. Postwar developments in cargo ships also saw an amazing degree of specialization as each ship was designed and built to carry a particular type of cargo. One of these was the container ship that was built to carry standard-size metal containers and looked like a huge barge with a pointed bow. These made loading, storage, and off-loading a routine, efficient, and very rapid exercise. The largest container ship is about 700 feet (210 meters) long. Tanker ships, which have been carrying liquid petroleum products since around 1900, attained enormous lengths, with a mid-1990 supertanker reaching 1,500 feet (457 meters). Future ships promise to be even more specialized as well as more efficient and cheaper to operate. Increasingly, tomorrow's ships will be more automated, with computers actually operating all major systems. Crews will also be smaller and ships will be built of aluminum and other materials that do not rust.

Ship's log

On land a vehicle's speed is measured by a **speedometer** that reacts to the revolutions of the **wheels**. At sea, **ships** use a device called a log for the same purpose. In its earliest form, the log literally consisted of a log or other wooden float attached to a line. The log was let out at the stern of the ship and allowed to drift away from the ship for an interval of time. Then it was reeled in. The length of the line told the mariners how far they had traveled during that time interval. The original log was not an accurate method of measuring speed, however, for the log itself moved about in the water as the ship moved forward.

An accurate reading would have required that the log remain stationary. In the sixteenth century, a triangular chip weighted with lead was let out in the same manner as the log. Lowering the chip to a moderate depth made it less susceptible to surface drift. In 1801 Edward Massey, an Englishman, created a mechanical log that automatically measured the line as it was let out. This was followed by the taffrail log, which consisted of a rotator at the end of a tow line connected to a recording device. Because it measured distance traveled, a mariner had to take two readings to establish the ship's speed.

The modern pitometer log uses a device called a Pitot tube that also projects from the ship's bottom. The tube has one forward-facing orifice and two side-facing orifices. The difference in pressure caused by the ship's forward movement is calculated to establish the ship's speed. When the ship is at rest, the difference is zero. The best modern method of measuring a ship's speed is by counting the revolutions of the propeller. Since no two ships are exactly alike, the standard speed of a ship relative to its propeller revolutions must be calculated in a series of trial runs shortly after the ship is launched. Still, such factors such as weather and aging of the ship's hull can affectthe accuracy of this method.

The book in which readings are recorded is called the logbook, and in mariner's jargon it, too, is often shortened to ''log.'' All of the events of the ship's journeys, both routine and unusual, are noted in the logbook.

See also Accelerometer; Screw propeller

Shockley, William (1910-1989)

American physicist

During his youth, Shockley was strongly influenced to pursue a scientific career by his father, a mining engineer, and by his mother, a mineral surveyor, both of whom encouraged his inclination toward science and mathematics. Shockley, furthermore, grew up in Palo Alto, California, the home of Stanford University, where he became acquainted with many people who were involved in scientific education and research.

After spending his first year of college at the University of California at Los Angeles, Shockley transferred to the California Institute of Technology, where he graduated in 1932 with a degree in physics. In 1936, Shockley earned a Ph.D. in physics from the Massachusetts Institute of Technology and went to work at Bell Telephone Laboratories. For several years, Shockley conducted research on **vacuum tubes**, which were used to control and amplify electrical currents.

Mervin J. Kelly, the research director at Bell Laboratories, realized during the 1930s that the drawbacks of vacuum tubes—such as their bulk and their tendency to overheat—were restricting technological advances in **telephone** switching equipment. In 1939, prompted by the search for a smaller, more efficient replacement for vacuum tubes, Shockley developed a plan for producing a solid-state **amplifier**. However, the materials available at the time made the project unfeasible, and Shockley's research was interrupted by World War II, during which he directed the United States Navy's anti-**submarine** research.

When Shockley was reunited with his colleagues at Bell Laboratories in 1945, a research team was established to investigate semiconducting materials, the most likely replacements for vacuum tubes. **Semiconductors**, such as germanium and silicon, can either conduct or resist an electric current. Although semiconductors had been used to control the direction of electrical currents, the researchers wanted to control amplification it as well.

In 1947, Shockley and two of his team members, **John Bardeen** and **Walter Houser Brattain**, demonstrated the world's first ''transfer resistor,'' or **transistor**, which used a germanium semiconductor placed between metallic contacts. With this device, the team was able to amplify an electric current. For inventing the transistor, which revolutionized electronic technology, Shockley shared the 1956 Nobel Prize in Physics with Bardeen and Brattain. Shockley also invented a more versatile type of transistor that was easier to mass-produce. In Shockley's junction, or bipolar, transistor, tiny amounts of elements such as phosphorous or boron are added to the semiconductor to alter its electrical properties.

Shockley left Bell Laboratories in 1954 to start his own company. At the same time, he began directing weapons research for the U.S. Defense Department. In 1963, Shockley became a professor of engineering science at Stanford University, where he remained until 1975.

During the early 1970s, Shockley generated much controversy when he expounded the theory that intelligence was determined genetically and therefore represented an inherited trait. Citing culturally-biased IQ tests and test scores, he concluded that since the disadvantaged social position of African-Americans was caused more by heredity than by environment, they could never be as intelligent as whites. Although his theories were soon discredited by the scientific community, by 1973 Shockley's views became so controversial he was labeled a racist and was prevented from speaking at several college campuses.

Shoe and shoemaking

Shoes are as old as clothing itself. Primitive people wrapped animal skins around their feet, and found that hides were warm and comfortable. Not long after, they began to color and preserve the pelts with vegetable dyes and natural oils. In ancient civilizations, shoemaking was a well-established craft. Egyptians wore sandals, and Mesopotamians wore soft, simple shoes. Even the Neolithic people of early Switzerland made shoes, as evidenced by a wooden model of a foot, called a *last*, that has been found.

The Greeks were adept shoemakers. Urns and vases show craftsmen with a variety of tools. They not only improved the craft of shoemaking, they changed the technique as well. Ancient Asian and Egyptian cobblers worked on the ground, and either squatted or knelt above their task. The Greeks worked at a bench, and because of this were able to develop more sophisticated leather working techniques.

Under the Romans, shoemakers banded into guilds, and practiced many techniques of modern shoemaking. They used iron lasts to shape shoes, and fitted them for right and left feet a practice that would be lost with the fall of the empire. Roman cobblers bound leather pieces together with thongs or animal tendons, and sometimes used rivets or nails for decoration and better wear. Soldier's sandals (*solea*), for example, had nailed soles that improved traction. Open-toed laced boots (remember old gladiator movies?) called *caliga* were sturdy and comfortable. They also made *gallica*, or ''Gaulish shoes'' that had wooden soles and leather uppers for protection from rain and mud. We know them today as rubber galoshes. Another form of Roman footwear, *calcei*, were short closed boots with thongs that tied around the ankle. This style was still being worn in Europe during the Middle Ages.

In the colder climates of northern Europe, boots were a necessity. Viking boots left the fur on the inside, and tied around the ankle. Russian tribes were wearing boots as early as the fifth century B.C. In the Middle Ages, shoes were a mark of one's social station. Peasants often went barefoot or wore simple wooden shoes. The nobility expressed their gentility with shoes of velvet, silk, and cloth of gold. But these luxurious fabrics were no match for the grime of medieval life, and wooden *pattens*, or clogs, were often attached. These were refined a few hundred years later when, as legend claims, Queen Elizabeth I had the first pair of heels put on her shoes in an attempt to improve upon her short stature.

Shoemaking began in the New World in 1629 when Thomas Beard and his apprentice, Isaac Rickman, landed in the Massachusetts Bay Colony. Before their arrival, all shoes were imported from England, although early settlers had learned how to make moccasins from the Indians. These eventually became so popular that they were exported to England.

In the mid-1700s, John Adams Dagyr, a Welshman, established America's first shoe factory in Lynn, Massachusetts. He divided the shoemaking process into discrete operations, and set each worker to one task. This enabled them to produce shoes in quantity, rather than waiting for an order from a specific customer. Progressive as it was, however, Dagyr's shop still made shoes entirely by hand. In 1845, a rolling machine was invented to soften and strengthen the leather; the following year saw the invention of Elias Howe's sewing machine, which was quickly adapted for leather. With **mass production** of shoes came the standardization of shoe sizes, and shoes were made once again for both right and left feet.

During the nineteenth century the rest of the shoemaking process was gradually mechanized. The first shoe manufacturing device was the *McKay stitching machine*, developed in 1858, which stitched the upper to the insole. In 1874, **Charles Goodyear, Jr.**, invented a welt-stitching machine that joined the welt, insole, lining, and upper. But the modern shoe industry really began with a revolutionary invention by a young black American, **Jan Ernest Matzeliger**.

William Shockley

After working as an apprentice for a cobbler, in 1878 Matzeliger eventually settled in Lynn, Massachusetts, which by then had become the center of shoemaking in the United States. Matzeliger worked to conquer the seemingly impossible task of making shoes completely by machine. The problem lay in the fact that there was no machine that could join the uppers to the soles of the shoe, a process called *lasting*. This was done by hand, adding time and expense to the final product. Matzeliger was determined to succeed, and experimented nightly with all sorts of odds and ends: scraps of wood, old cigar boxes, and other improvised tools. He finally created a lasting machine, patented in 1883, that held the last in position and moved it forward while other parts punched the leather and drew it over the last, fitting the leather at the difficult points of the toe and heel. Nails were fed into position automatically and driven into place. The machine could make 75 pairs of womens' shoes a day, and cut their price in half. Technology has improved the machinery, but modern shoes are still made in basically the same way today, with each step performed on a different machine. Most soles are now made of plastic and rubber; humanmade fibers and plastics are used al-

Cobblers made shoes by hand before the development of such innovations as Jan Matzeliger's lasting machine allowed for mass production.

most exclusively for heels and other parts. Leather still predominates for uppers, but synthetics and coated fabrics are gaining ground. The leading shoe-producing state is still Massachusetts.

In the first half of the twentieth century, specialized sports shoes were costly and available only to elite athletes; most other people merely wore canvas sneakers for sports activities. In the 1960s, however, more widespread general interest in exercise led to the commercialization of athletic shoes. The running boom in the 1970s saw a number of manufacturers produce shoes with specialized support and shock-absorption structures. In the 1990s, athletic shoes incorporate sophisticated design and robust synthetic materials suitable for industrial applications, and consumers can buy shoes designed specifically for every type of sports and exercise activity from aerobics to riding exercise bicycles.

Sholes, Christopher Latham (1819-1890)

American printer and journalist

Christopher Sholes is known as the father of the **typewriter**. He was born on a farm in Mooresburg, Pennsylvania, on February 14, 1989. As a young man, Sholes served a four-year printing apprenticeship before moving with his parents to Wisconsin. He went to work as a printer, then became editor of his brother's newspaper in Madison, Wisconsin, the *Wisconsin Enquirer*. After marrying in 1841, Sholes moved to Southport (now Kenosha), Wisconsin, and served as editor of the *Southport Telegraph* for the next four years. He became a local postmaster and in the 1850s served in the state legislature. In 1860 he moved to Milwaukee, Wiscinsin, and returned to newspaper editing, which he quit once again when President Lincoln (1809-1865) appointed him collector of the port of Milwaukee. Sholes found enough spare time to pursue his interest in inventing.

Having already designed a way to address newspapers mechanically for mailing, he began to work with a machinist friend, Samuel W. Soulé, on a paging machine, which they patented in 1864, and then on a machine that would automatically number the blank pages of a book. Alongside Soulé and Sholes in the same machine shop was Carlos Glidden, who was tinkering with his own design for a spading machine intended to replace the **plow**. Glidden struck upon the idea of adapting the numbering machine to print letters of the alphabet as well, and referred Sholes to a magazine article about a mechanical writer recently invented in London, England.The suggestion fired Sholes's imagination, and he worked on typewriter designs for the rest of his life.

Initially, the three inventors built a working model of the first practical typewriter and patented it in 1868. Sholes worked on improvements to this early design for the next five years, securing several more patents, and sought financial backing to market the typewriter. He consulted with **Thomas Edison**, but Edison was interested only in a telegraphic printer. A businessman and former newspaper publisher, James Densmore, stepped in, offering to pay expenses in return for a share in the future profits. He also demanded specific improvements in the machine, wanting it lighter and easier to use. Meanwhile, Soulé and Glidden gave up their patent rights.

Finally, unable to market his invention, Sholes sold his patent rights to the Remington Arms company for $12,000. This proved to be all he would earn for his invention of the typewriter. He continued to experiment with typewriter improvements, however, with the help of his two sons, and he shared his results with Remington. Sholes received his last typewriter patent in 1878. Having suffered with a delicate constitution all his life, Sholes developed tuberculosis in 1881 and died nine years later on February 17, 1890, in Milwaukee.

Shorthand

Greek inscriptions indicate that as early as the fourth century B.C. people have used abbreviated script in an attempt to keep pace with the speed of speech (140 words per minute, on average). The first known organized system of shorthand writing, or stenography, emerged around 50 B.C. and is attributed to the Roman freedman Marcus Tullius Tiro (born ca. first century B.C.), a secretary to the orator Cicero (106-43 B.C.). The Tironian method, which employed brief strokes to represent the characters of the alphabet, was taught in schools and used to record speeches throughout the Roman era and the early Christian period; another system was devised by Roman writer and philosopher Seneca (ca. 4 B.C.-ca. 65 A.D.). During the early Middle Ages, however, shorthand became associated with witchcraft and fell into disuse.

It was not until 1588 that shorthand was revived by English clergyman Timothy Bright (1551?-1615), in his *Characterie: An Art of Short, Swift, and Secret Writing by Character.* Bright's system was followed through the seventeenth and eighteenth centuries by a number of shorthand methods, none of which established widespread dominance. Seventeenth-century English author Samuel Pepys (1633-1703) wrote his famous literary diary in shorthand; the secrecy of the method is thought to have helped him to record personal anecdotes that he might otherwise have excluded for fear of discovery. The systems introduced during this period, while undeniably faster than longhand writing, were flawed by their complexity and overabundance of symbols, which made them relatively slow and difficult to learn.

Phonetic stenography, introduced by William Tiffin in the mid-eighteenth century, was a significant step in making stenography faster. Earlier systems were alphabetic, employing symbols to stand for letters; Tiffin's system employed symbols that stand for sounds, enabling the stenographer to transcribe a word using fewer strokes. A number of competing variations again followed on Tiffin's innovation; it was not until 1837, when Sir Isaac Pitman (1813-1897), an English educator, published Stenographic Sound—Hand, that a single dominant English stenographic system arose. Pitman, who had learned a phonetic system developed by stenographer Samuel Taylor, was the first to develop a shorthand system based on scientific analysis of the sounds that comprise speech.

In the Pitman system, each consonant sound has its own symbol, though similar sounds, such as *p* and *b*, are distinguished only by shading; vowel sounds are indicated by dots and dashes placed near the consonant strokes. The system, like most major shorthand systems, also employs short forms, symbols for commonly used words and phrases; the Pitman system contains 214 such symbols. Its relative speed and accessibility quickly made it the preeminent shorthand system in the English-speaking world (it remains the standard system in Great Britain), and it was adapted for many European languages as well. In 1888 American educator John Robert Gregg (1867-1948) introduced a rival system in his book. Also a phonetic system, Gregg shorthand distinguishes similar-sounding consonants by length rather than shading; the result is a script that flows much like longhand. Gregg's system eventually became the dominant method in the United States, where Pitman shorthand had not caught on as quickly as in Europe.

In 1949 the Gregg system was revised to reduce the number of short forms used from several hundred to 184. Owing to the phonetic approaches of the Gregg and Pitman methods, today's stenographers can match and even surpass the rate of speech; speeds of over 280 words per minute have been recorded. Systems introduced in the twentieth century are generally simpler and easier to learn than the Gregg and Pitman methods, but they are also slower. Despite competition from typewriter-like stenography machines now used in recording court proceedings, the two nineteenth-century systems still dominate manual stenography. However, with the advent of the personal computer and voice-activated input systems, many managers and executives are finding it easier to do their own typing, and shorthand may be a dying art, used primarily by reporters when taking notes during interviews.

During World War II, civilian shortwave radio operators, such as these NBC employees, aided the war effort by listening into broadcasts from Asia.

See also Alphabet; Writing

Shortwave radio

On December 11, 1901, **Guglielmo Marconi** amazed the world when he successfully transmitted a **Morse code** signal 2,137 miles (3,440 km) from England to Canada using **radio** waves. The feat was not only remarkable but, according to the experts, impossible.

Scientists had been dabbling with radio waves ever since 1885, when German physicist Heinrich Rudolf Hertz (1857-1894) used a spark-gap to create them. Nine years later English physicist **Oliver Lodge** invented the coherer, a device to detect *Hertzian waves*, and in 1895 Russian physicist **Aleksandr Popov** invented an **antenna** to send and receive signals.

It was known that radio waves were a part of the electromagnetic spectrum, located below the wavelengths of visible light. Although radio waves cannot be detected with the eye, they were expected to behave in the same manner as light waves; that is, they should move in straight lines and not bend around corners. It was believed that the curvature of the earth would limit radio transmissions to a distance of 200 miles (300 km). Marconi's feat sent the scientists scrambling for an explanation.

Arthur Kennelly (1861-1939) in the United States and Oliver Heaviside (1850-1925) in England independently suggested there might be a layer of charged particles in the upper atmosphere that were responsible for reflecting the radio waves. This layer became known as the *Kennelly-Heaviside layer.* In 1924 Edward Appleton (1892-1965) discovered the existence of the Kennelly-Heaviside layer, now called the ionosphere, about 60 miles (100 km) high. In 1926 Breit and Tuve discovered there were four individual layers which were named (from bottom to top) D, E, F_1 and F_2.

Meanwhile scientists had been studying radio waves from the turn of the nineteenth century. It had become accepted that those radio wavelengths that were in the Very Low Frequency, Low Frequency, and Medium Frequency ranges were best suited for communication. The High Frequency waves, know as "short waves" were of no commercial value and were ignored.

In addition to the scientists, many amateurs had become interested in radio communication. There were no government regulations and anyone with the interest (and money) could build their own radio sets. With the start of World War I, however, amateurs in Europe and the United States were abruptly prohibited from using their equipment. Restrictions in the United States were finally lifted on October 1, 1919, thanks to the efforts of amateurs such as **Hiram Maxim**. The amateurs discovered that as the radio wavelengths became shorter (increased in frequency), they traveled farther, even using low power. Obviously the ionosphere was very efficient at reflecting these wavelengths, making communication on a global scale possible. By the 1930s, worldwide shortwave communication had been established.

Shortwave radio networks, such as the United Statesò Voice of America, air in numerous languages, reaching areas not able to be served by **amplitude modulation** (AM) **frequency modulation** (FM) radio broadcasts. Though shortwave is basically limited to voice (and not music) transmissions due to its low-fidelity nature, it performs an important international information dissemination service. For example, during the events surrounding the end of communism in the Soviet Union in the early 1990s, shortwave was one way of hearing news from the country itself. Sometimes news can be learned before other media picked up on the story. In the 1990s, shortwave radio's future seemed threatened by digital satellite delivery of radio on the L band, but if such a network develops, its transmissions can be easily blocked, while shortwave radio cannot. Between 10 and 20 million shortwave radio sets were in use in the United States in 1995.

Shotgun

A shotgun is a smoothbore firearm that has no spiral grooved cut on the inner surface of the gun barrel like the **rifle**. Designed to fire a number of pellets simultaneously, it is used primarily in hunting small game, particularly birds. The first shotguns, called fowling pieces, appeared in sixteenth-century Europe. In the following century, the prevailing technological theory was that longer gun barrels allowed a greater charge of **gunpowder** to be used which would result in a longer range. This led gunsmiths to make firearms with barrels up to 7 feet (2.13 m) long and weights up to 12 pounds (5.44 kg). In addition to their unwieldy design, the slow flintlock ignition system was a drawback because it made hitting a moving target nearly impossible.

Several significant changes brought the shotgun into the modern age of weaponry. The invention of the percussion cap created a more dependable system of ignition so moving game could be hit more consistently. In addition, English gunsmiths realized the cumbersome nature of shotguns made them difficult to use, so they brought out shorter and lighter weight guns for better use out in the fields. In 1787, Henry Nock (1741-1804) patented a new kind of breech. It made the gun fire more rapidly and efficiently by igniting a small amount of gunpowder in a chamber behind the main charge. This meant the weapon had more power and distance, so gunmakers could shorten barrels to as little as 30 inches (76 cm). One other result of this lighter gun was the increased practicality of a side-by-side double-barreled construction. Guns of this design had been made in the past but they were heavy and difficult to use. Joseph Manton (1760-1835), an English gunsmith, made the double-barrel even more popular when in 1806 he patented an elevated sighting rib, which made aiming easier. Muzzle-loading shotguns faded out of use in the 1800s. Johannes Pauly (1766-1820) introduced a break-open breechloader and made a special cardboard-wrapped cartridge with a metal head in 1812.

Another improvement in shotgun design occurred following the American Civil War when several people began experimenting with the barrel's shape. W. R. Pape, an Englishman, patented a design in 1866 by which the barrel was constricted near the muzzle, making the shot go further. Unfortunately, accurate aim was still difficult. Pape's design, now called choke boring, was perfected by Fred Kimble, an American hunter. Kimble experimented with various shaped barrels and was able to reshape the constriction at the muzzle until he could keep all the shot load within a 30-inch (76 cm) circle at 40 yards (37 m).

The creation of fixed **ammunition** during the late 1870s brought about a successful repeating shotgun. Up until then, shells had been manufactured and sold empty to be loaded by the gunner. With fixed ammunition now available, the demand for faster repeating shotguns grew. **John Moses Browning** created the Winchester Model 1887, a lever-action, five-shot gun that sold extremely well. He also designed a pump shotgun that was a big success. Browning then sketched out plans for an autoloader, which came out on the market in 1905 and survived for over 50 years. Several manufacturers brought out semiautomatic shotguns using a gas-operated system in the mid-1950s. These guns used the gas released when the shot is fired to push a piston, which ejects the fired cartridge and brings up the next round. Few important changes have been made in these basic shotgun designs since then. Shotguns are used primarily for hunting and sport shooting. However, they were also used by American troops during World War I, primarily by guards. The barrels were shortened for convenience, leading to the term ''sawed-off shotgun.'' Shotguns were also used in training troops during World War II, and they are still often used by police forces worldwide.

SHRAPNEL • See Ammunition

SHULL, CLIFFORD GLENWOOD (1915-)
American physicist

Clifford Glenwood Shull was born on September 23, 1915, in the Glenwood section of Pittsburgh, Pennsylvania. Along with his older brother and sister, Shull attended neighborhood schools, then enrolled in Schenley High School. At Schenley, a dynamic instructor sparked Shull's interest in physics, causing him to reconsider his initial ambition to be an aeronautical engineer. After graduating, he won a partial scholarship to the Carnegie Institute of Technology, now Carnegie Mellon University, and immersed himself in his studies. During Shull's freshman year, however, his father died suddenly, precipitating not only an emotional, but a financial crisis for the family. As a result, Shull's older brother, Perry, quit his job as an art teacher and ran their father's business until his younger sibling had graduated from college.

After receiving his B.A. from Carnegie, Shull entered the graduate physics program at New York University. Early on he became associated with the nuclear physics research group and took part in several different experiments using **particle accelerators**. For his Ph.D. thesis he used the department's newly constructed Van de Graaff accelerator to determine whether or not electrons have a spin or polarization.

A month after receiving his Ph.D. from New York University, Shull accepted a research position at The Texas Company, now Texaco, in Beacon, New York. There, he used gas absorption as well as x-ray diffraction and scattering to analyze catalysts used in the production of high-performance aviation fuel. This research took on great importance following the entry of the United States into World War II. However, Shull, like many other young physicists of the time, was very interested in the Manhattan Project. He had seen several of his former university friends and professors join in the race to produce the first nuclear weapon and longed to take part as well. Yet when an attempt was made to recruit him, The Texas Company blocked his way, eventually convincing a government manpower board that his research for them was more crucial to the wartime effort.

At the end of the war Shull was free to pursue his interest in nuclear physics, and in 1946 he left The Texas Company for the Clinton Laboratory in Oak Ridge, Tennessee. Eventually renamed the Oak Ridge National Laboratory, it had been one of the top-secret sites of the Manhattan Project, and its **nuclear reactor** had supplied a significant portion of the plutonium used in the **atomic bombs** dropped on Hiroshima and Nagasaki. At the time Shull arrived, the laboratory's administrators had begun to shift its focus from military to civilian uses, and he was assigned to work with Ernest Wollan. Wollan, who had been with the laboratory since its inception, had begun to investigate how the neutron beams produced by the reactor could be used to analyze the structure of matter. Over the next decade, the two scientists developed a method of neutron diffraction that eventually became the foundation of an entirely new branch of physics research.

Shull and Wollan discovered that neutrons, upon hitting the atoms within a fluid or solid sample, ricochet in a characteristic fashion. Specifically, the angle at which the neutrons are deflected shows how the atoms are arranged. In other words, neutron scattering effectively determines the atomic structure of a sample. Neutron scattering was a truly revolutionary discovery, opening new avenues of research as it shed new light on important substances, such as the element hydrogen, that had been resistant to earlier methods of analysis. It wasn't until 1994 that the Nobel committee recognized the work of Shull and Wollan, awarding the Nobel Prize in physics to Shull (and **Bertram Brockhouse**). Sadly, Wollan had died in 1984.

Shull left the Oak Ridge National Laboratory in 1955 to become a professor of physics at the Massachusetts Institute of Technology, retiring as emeritus professor in 1986.

SIEGE WEAPONS

Sieges were the neutron bombs of ancient warfare—a total concentration of military force on a fixed defensive, usually civilian, position. Whenever a city retreated behind its walls and a standoff loomed, the attacking army surrounded it and cut off its supplies. Capitulation often took months or even years, depending on how well provisioned the town was. The besieging army needed a method to hasten surrender, a way to bring the walls down and allow them to move in for the kill. The Assyrians, one of the oldest-known civilizations (2000-612 B.C.), are thought to have invented siege warfare. One of their most effective weapons was the battering ram, a tool that was used and improved by both the Greeks and the Romans. The first rams were rather crude, consisting of little more than a tree trunk about 15 feet (4.5 m) long hung by a rope from a turret (about 18 feet [5 m] high) on a wheeled base. The business end was sharpened like an axe, and the log was swung repeatedly, usually at the gates. The soldiers that worked these devices were obviously sitting ducks for anyone with a bow and arrow or a pot of boiling oil, so they were covered by archery fire from their own troops while they worked. A battering ram shown on a seventh-century B.C. clay tablet was roofed with hides to protect its operators. While those inside the walls were preoccupied with the ram pounding at the gates and the constant hail of arrows, attackers with spears climbed scaling ladders, hoping to find a few holes in the defense and bring the whole episode to a victorious conclusion.

The ancient Greeks added other weapons to their siege arsenal, most notably several types of **catapult**, which had originally been developed by the Phoenicians. The smallest version, the *katapeltes*, could throw arrows or spears or even a small boulder with some accuracy to a range of about 250 yards (229 m). The *petrobolos* was larger and could handle a bigger payload—stones up to 55 pounds (25 kg) or so. Both models used twisted sinew or lengths of human hair as a spring to propel the missile. Another device adopted from the Phoenicians was the *siege tower*, a tall wheeled structure with a roof and a drawbridge that could be brought close to the walls, eventually disgorging its occupants onto the ramparts. As destructive as these machines were, the sturdy walls of most cities generally withstood the assaults made on them. The town of Plataea resisted the Athenians' fiery missiles during the Peloponnesian War by erecting barriers made of hides around its walls. The inhabitants of Syracuse stopped Athenian battering rams by pouring a fiery combination of sulfur and pitch on their attackers. At this point in history, the combined weapons of starvation and treachery carried the day far more often than military technology.

The best example of this is the famous Trojan horse, which, according to the *Iliad*, the Greeks used to capture Troy. In 398 B.C., however, the city of Motya fell after a siege by Greek forces. Dionysius I of Syracuse (405-367 B.C.) used towers, rams, and catapults to bombard the town. Later, men with scaling ladders climbed the walls and overran the defense. Alexander the Great (356-323 B.C.), the most successful Greek commander in history, also used towers, rams, and catapults in his conquests.

Few towns stood a chance against a Roman siege. Their first step was to surround it with a bulwark constructed just out of the range of fire; if a long siege was expected, the soldiers built a double-walled, roofed structure. Whatever the fortification, the object was to prevent anyone from getting out or bringing food in. The Romans too relied heavily on the battering ram. Their version used a tree trunk tipped with an iron ram's head, swung by as many as fifteen hundred men. Those closest to the walls were covered by a roof called a testudo, or tortoise, which shielded them from the various deadly objects dropped on them by the besieged.

Like the Greeks, the Romans also used various forms of catapult. The two heaviest models were the ballista and the onager, which could hurl rocks as large as 60 pounds (27 kg) as far as 500 yards (457 m). This is rather impressive when you consider that naval cannon in use at the time of the American Revolution had only twice the range for a shot that was half the weight—and both were about equally accurate. The onager, which means wild ass, got its name because it kicked up its hind end when fired. It had a shorter range than the heavier ballista, and was used at closer targets. Catapults, the smallest of the Roman hurling engines, could pivot from side to side, and were poised so that their elevation could be adjusted as well. They usually fired short but substantial spears from their 4-foot (122 cm) arms, and had a range equal to the ballistas. Mining was another favored tactic, and tunnels were dug to try to get under city walls. The occupants of the castle often retaliated by digging a countermine, usually an inescapably deep hole the miners would fall into, or a mass of straw soaked in oil and set alight. If enough water was on hand, the tunnel could be flooded. The cover of darkness was a useful tool as well. Roman besiegers would often build an earthen ramp gradually closer to the city each night, the idea being to bring the siege towers in to do some damage and eventually breach the walls. But the townspeople could, if they had the chance, build their wall correspondingly higher at that point, nullifying any advantage the ramp provided.

When the Roman Empire finally fell in the fifth century, the West began to struggle with invasions from ''barbarian'' tribes like the Vikings and the Magyars. Around this threat the feudal military and social system developed, and the castle became the surest means of defense. The old Roman siege weapons were still used, but the lack of military technique that accompanied the loss of civilization rendered them far less effective, even if later technology did make them more powerful. Medieval siege weapons were familiar variations of their Roman counterparts, although they were given new names and sometimes new purposes. Battering rams were still used, but were largely useless against a 10-foot (3 m) thick castle wall; their chief advantage was against flimsier city gates. Siege towers were called *beffrois* or cats, although their use was stymied by the water-filled moat that surrounded most castles. Scaling ladders were known as *escalades*, but moats presented problems here too, and it was difficult to assault a fortified castle from below while armed defenders rained all manner of opposition on their attackers. For these reasons, medieval armies relied heavily on those old standbys, missile-throwing engines, or *gyns*. The *springal* was a piece of light artillery that usually threw flaming darts. (As castle design evolved, roofs were made of lead to protect against this kind of attack.) The ballista became a kind of giant crossbow, hurling massive spears. Catapults came in two categories, the *mangonel*, or nag, which threw stones and bucked like the Roman onager, but was really a much clumsier machine. The *trebuchet* was the largest of the gyns, using an entire tree trunk as its base. Its size was so formidable that it was usually built on site during the siege. The design was unchanged until well into the sixteenth century, although it was eventually equipped with metal bearings. In addition to boulders and javelins, besieging armies would use almost anything else they could find as ammunition. Dead animals were especially spectacular, and for real emphasis nothing beat a live prisoner if the attackers could capture one.

Mining was another possibility, and one that was arguably more successful. The tunnel was supported with timbers, then, when the appropriate spot had been reached, the beams were doused with fuel and set afire. As the wood burned away, the ground collapsed and the walls tumbled. Later armies would use this technique to plant land mines and other explosives underground. The invaders could also try to ignite wooden gates by soaking them with pitch, but slits in the wall above the gates were often built to quash any such attempts. As cannonballs and gunpowder replaced the ballista and battering ram, the castle's inviolability began to crumble, and sieges became decidedly more one-sided. The siege of Constantinople in 1453 was a turning point in military history, for it marked the end of the medieval era. The strength of the city's walls had been legendary, but the Turks' relentless bombardment destroyed them in six weeks.

Siemens, (Karl) Wilhelm (1823-1883)
German English engineer

Karl Wilhelm Siemens was born into a family of scientists and inventors at Lenthe, near Hanover, Prussia, on April 4, 1823. After his father's death in 1840, he was encouraged by his brother **Ernst Werner von Siemens** to enter technical school at Magdeburg. Siemens also apprenticed at a **steam engine** factory there.

In 1843 he traveled to England to promote Werner's inventions, in particular an **electroplating** process. For the same reason, Siemens's younger brother, Friedrich (1826-1904), joined him in 1847.

The two brothers collaborated on their own inventions. Karl Wilhelm's first success was a water meter, from which he lived comfortably from the royalties. Financially secure, the brothers set about working on a new smelting process, the open hearth furnace, which would become the foremost steel-making process of the twentieth century.

The Siemens brothers' process used heat regeneration. Heat that otherwise would have been lost to the atmosphere was recaptured and used to heat the furnace more efficiently. Their initial experiments were only marginally successful. An encounter with **stove** designer E. A. Cooper at Smethwick, England, gave them the technical solution to their furnace design, and they were finally able to apply the process to making steel.

The Siemens brothers received a patent on their regenerative metal processing chamber in 1856. This technique, which became known as the open-hearth process, was applied to steelmaking two years later, with that patent awarded to Karl Wilhelm. A shallow hearth containing molten metal was exposed to flames from above, while a pair of chambers at either end of the furnace captured heat and returned it for more efficient processing.

In 1864 Frenchman Pierre-Émile Martin took the Siemens brothers' process further by modifying the location of the chambers and introducing scrap steel to the process. Some legal battles were waged between the Siemens and Martin interests. In the end, a compromise was reached: so that both parties would be recognized for their contributions, the process was named the Siemens-Martin process. By 1900 this technique had surpassed the Bessemer process as the most efficient method of processing steel. Karl Wilhelm Siemens tried unsuccessfully to establish his own steel manufacturing plant at Landore, England. While the business foundered in the 1880s, Siemens enjoyed success in the field of electric telegraphy. This later business effort was, again, the initiative of his brother Werner.

Werner's firm of Siemens and Halske was involved in producing **telegraph cable** and laid the first transatlantic cable in 1875. From 1850 Wilhelm served as an agent for his brother's firm and made improvements to lighting in Britain. He became a naturalized British subject in 1859. In 1878 he created an electric arc furnace that allowed the current to flow through loosely packed material that was high in resistance and low in conductivity, producing an iron-melting temperature. One of his final achievements was the creation of an electric passenger railway in Portrush, Ireland, in 1883, and he was also knighted that year. Wilhelm died in London on November 18, 1883.

Although Sir William, or Wilhelm, Siemens was the catalyst in the industrial collaborations of the Siemens brother, all three made substantial contributions.

See also Bessemer, Henry; Blast furnace; Steel production

Ernst Werner von Siemens

Siemens, Ernst Werner von (1816-1892)
German engineer

Ernst Werner von Siemens was born at Lenthe, Hanover, Germany, in December 1816, the oldest of four brothers. After receiving an education at Lübeck, he went into the army and became an officer. However, he was eventually sent to prison for dueling. Making good use of his time, Siemens set up a laboratory in his cell where he experimented on gold and silver plating, selling the rights to his process which he patented in 1842. He also experimented with nickel plating, independently of the work being done by John Wright.

Siemens established his reputation in 1846 when he modified the dial telegraph which had been invented by **Charles Wheatstone** in England. Siemens used ''make-and-break'' **circuits** to make the telegraph self-acting. Instead of using permanent **magnets**, he used self-generated electricity to activate magnetic field coils.

In 1847 Siemens, along with mechanic Johann Georg Halske (1814-1890), established Siemens & Halske, a company that manufactured and repaired telegraphs. They ran telegraph wires in Germany as well as Russia. Siemens & Halske eventually became one of the major electrical manufacturing companies in Europe.

Siemens also made numerous technical advances. In 1847 he made the first use of **gutta-percha** to insulate electric cables; in 1866 he independently discovered the dynamo-electrical principle and became interested in the development of the self-excited dynamo and electric-traction. He cofounded the Physical Society in Berlin, and delivered a paper on another of his interests, hot-air engines. He invented the ozone tube, an alcohol meter, electrolytic refining processes, and established an electrical standard of resistance based on mercury.

Siemens had sent his brother **Karl Wilhelm Siemens** (later known as Sir Charles William Siemens) to England to promote his projects and head the British branch of the company. There Carl invented and improved a water meter. In 1861 he worked with another brother, Friedrich Siemens (1826-1904), who had joined him in England in 1848, to devise a regenerative furnace which became the basis of the Siemens-Martin, or open-hearth, process used in the **glass** and **steel** industries. Carl tried to apply his regenerative principle to conserve heat that was wasted by **steam engines**, but was unsuccessful. In 1867 he delivered an important paper on electric generators before the Royal Society.

On December 13, 1892, one week before his seventy-sixth birthday, Ernst Siemens died a very rich man at Charlottenburg, Germany.

Sign language

The most familiar sign language systems today are those developed for the deaf and hearing-impaired. However, non-verbal communication through bodily movements is much older than these relatively recent systems. The need for secrecy, vows of silence, and language barriers between people of different cultures have spurred people to invent non-verbal language systems. Members of religious orders sworn to silence often rely on simple gestures rather than a coded system to communicate with one another; the English historian and cleric Venerable Bede (673-735) devised a system in which manually signed numbers, representing letters, were used to spell out words. Chinese and Japanese, whose languages use the same written characters but pronounce them differently, will sometimes trace characters onto another's palms to communicate. In the nineteenth century, North American Plains Indians developed a system of signs bridging tribal language barriers. The system did not match words directly with signs, but instead used gestures to stand for common concepts; for example, one would sign a circle in the air to represent the moon.

However, only sign languages conceived for the deaf have attained the complexity, versatility, and depth of vocabulary of spoken languages. Deaf and mute people have long devised systems of signing, but the first such language to be fully schematized was devised by Charles-Michel, Abbé de l'Epée, a French teacher of deaf children, in the mid-eighteenth century. His system combined a manual alphabet with a vocabulary of gestures that stood for whole concepts. L'Epée's system evolved into French Sign Language (FSL), which in 1816 was brought to the United States by Thomas Gallaudet (1787-1851), who founded the American School for the Deaf in Hartford, Connecticut. Various elements of rudimentary systems used in the United States were incorporated into the language to form American Sign Language (ASL). In addition to hand gestures, ASL uses facial expressions and body posturing for inflection and grammar; signers use these to construct complex sentences just as punctuation and word order is used in written language. A change in the signer's head position and expression can change two sentences (''Your keys are in the living room. You lost them last week.'') into one, complex sentence (''Your keys, which you lost last week, are in the living room.''). Changes in inflection can change the meaning of signs, so a single sign can represent multiple phrases and concepts. Various national sign languages have been adapted from ASL, but there is considerable overlap between them, as signs represent concepts, not words.

Prior to the invention of these language systems, deaf children were simply never educated, with the rare exception of the children of the wealthy. However, as sign language gained in popularity and teachers of the deaf grew in number, many of these teachers resisted its use. Opponents of sign language hoped that deaf children eventually could be taught to speak and lip-read effectively, and that sign language would further isolate the deaf from hearing society. In 1880, an international congress of teachers for the deaf passed a resolution against its use, keeping it out of most schools for decades; in many American states the language was banned from schools by law. However, deaf people and their families continued to use the language, and it maintained a flourishing, *underground* existence until the 1960s, when linguists began to accept the utility of sign language.

In 1965, *A Dictionary of American Sign Language* was published. With the increased scientific respectability of sign language, more schools began to accommodate demands for signing teachers, and sign-language interpreters became increasingly visible on television and at public events. Deaf artists began to use the special properties of signing creatively, notably in the drama *Children of a Lesser God.* It is now widely acknowledged, despite the well-intentioned objections of early educators, that sign language has been indispensable in improving educational and social opportunities for the deaf.

Sikorsky, Igor (1889-1972)

Russian engineer

As a child in Kiev, Russia, Igor Sikorsky studied and dreamed about flying. First studying the designs of **Leonardo da Vinci** and later the work of **Louis Blériot** and Count **Ferdinand von Zeppelin**, he designed and built his first **helicopter** in 1909 at the age of twenty. After some setbacks, he realized that several problems had to be overcome: inadequate lift, destructive vibration, and the need for better, lighter materials. He temporarily abandoned helicopters to became Russia's chief **aircraft** designer in World War I, but he left the country following the Bolshevik Revolution.

He settled in the United States and started his aircraft company, Sikorsky Aero Engineering Company, in 1923. Airplane designs took up most of his time, but he still dreamed of successful helicopter flights. He patented the basic configuration: a single lifting rotor with a small vertical rotor at the tail to offset torque. In 1939, he built the VS-300 helicopter. It was flimsy-looking, with a skeleton of metal tubes and no enclosed space for the pilot. After many trials and errors, he re-designed it by putting two small horizontal rotors at the tail to help with stability.

On May 13, 1940 the revamped VS-300 flew free. It stayed up for fifteen minutes, but had trouble flying straight ahead. Eventually, he gave up on the extra rotors at the rear and settled for the one main rotor on top and one rotor to the rear. Sikorsky took the lead in helicopter design and manufacturing, and by the end of World War II, more than 400 of his helicopters came off the assembly lines. Sikorsky's major contribution to practical helicopter flight was his tail rotor configuration. His idea of using a small tail rotor at the end of a long tail boom proved successful and is still the most popular design for all types of helicopters.

Igor Ivan Sikorsky

Silicones

Silicone is a generic term denoting compounds which have a backbone of alternating silicon and oxygen atoms with additives hanging off the silicon atoms as pendant groups. Silicon and oxygen are the two elements in greatest supply on the Earth's crust.

Silicon was first isolated by Jons J. Berzelius (1779-1848) in its amorphous state in the 1820s. In 1854 Henri-Etienne Sainte-Claire Deville (1818-1881) obtained it in crystalline form. Charles Friedel (1832-1899) researched silicon compounds as part of his search for synthetic diamonds. By 1863 Friedel and his partner, James Crafts (1839-1917), had produced the first organo-silicon compound. Alfred Stock (1876-1946), Albert Ladenburg (1842-1911), and A. Polis all conducted research into carbon-silicon bonds. It was not until Victor Grignard (1871-1935) had discovered Grignard reagents, organo-magnesium compounds, that Frederic Stanley Kipping prepared his first silicone using the Grignard reagents. While there is some doubt as to whether he was the very first to form a silicone, he was the first to use the term.

Kipping was an English chemist born in Manchester, England. He attended Owens College in Manchester and graduated with a chemistry degree. He then moved on to the University of Munich and worked in **Adolf von Baeyer**'s (1835-1917) laboratory as a graduate student, receiving his doctorate degrees from the University of Munich and the University of London in 1887.

Kipping's use of Grignard reagents to form silicones led him to further research the compounds. He conducted many experiments trying to synthesize a double-bonded silicone compound. He never succeeded, but he published more than 50 papers on silicones that laid the foundation for the industry, which today manufactures over 500,000 tons of silicone for use in a wide range of products.

It was not until the 1940s that the silicone industry was able to make use of Kipping's research. In 1940 E.G. Rochow, a chemist with the General Electric Company, discovered an easy way to form silicones. He combined methyl chloride gas and heated silicon and copper to form compounds with silicon-carbon bonds. Earlier research by Kipping proved that these compounds could be reacted with water to form silicones, which owe their great strength and stability to the unique characteristics of the silicon-oxygen bond.

Rochow's discovery and the increased need of certain products during World War II led to a silicone boom. Silicones played important role in the war effort. Electronic equipment, **radios**, and **aircraft spark plugs** were waterproofed with silicones. Silicone rubber was used in gaskets for searchlights and for superchargers in aircraft engines.

General Electric, Dow-Corning, Union Carbide, Stauffer Chemical, Wacher-Chemie, and Farbenfabriken Bayer A.G. all began large-scale silicone production within ten years of Rochow's discovery. Other companies soon followed. Further research by James F. Hyde of Dow-Corning led to the production of silicone rubber.

Silicone rubber material is particularly valuable because its useful temperature range is -178°F (-117°C) to 600°F

(315°C), while the range of natural rubber is -60°F (-51°C) to 180°F (82°C). Silicones today have thousands of applications. They are used primarily as release agents, lubricants, and sealants. Silicone-treated fabrics, bricks, and concrete walls resist water. Silicones in paints give improved weatherability. They are used as electrical insulation and in oven and aircraft door gaskets. Silicone rubber is used in high-temperature applications and space technology, including astronaut boots. The majority of the silicone produced is used in paper and automobile, and food industries, as well as being used in office and medical equipment.

Silicone has been used in plastic surgery to build nose, cheeks, chin, and other bony structures of the face. Its use for cosmetic and replacement breast implants has been clouded by concern that these implants might cause breast cancer and other health problems. In 1995, a jury awarded $10 million in damages to a woman who alleged that a leaky breast implant manufactured by Dow Corning had caused skin disorders, muscle pain, tremors, incontinence, and other medical problems. By 1997, about 100,000 women had settled lawsuits with manufacturers of silicone implants, and another 500,000 had reserved the right to sue. But in September 1997, a review of scores of medical studies that appeared in *The Journal of the National Cancer Institute* concluded that there was no evidence that silicone breast implants do indeed breast cancer, and that there is only a slight statistical suggestion that breast implants result in some types of connective tissue disease. All the same, by 1997, vegetable oil and other viscous materials were being studied as implant materials to replace silicone.

Silk

Silk has set the standard in luxury fabrics for several millennia. The origins of silk date back to Ancient China. Legend has it that a Chinese princess was sipping tea in her garden when a cocoon fell into her cup, and the hot tea loosened the long strand of silk. Ancient literature, however, attributes the popularization of silk to the Chinese Empress Si-Ling, to around 2600 B.C. Called the Goddess of the Silkworm, Si-Ling apparently raised silkworms and designed a loom for making silk fabrics.

The Chinese used silk fabrics for arts and decorations as well as for clothing. Silk became an integral part of the Chinese economy and an important means of exchange for trading with neighboring countries. Caravans traded the prized silk fabrics along the famed Silk Road into the Near East. By the fourth century B.C., Alexander the Great is said to have introduced silk to Europe. The popularity of silk was influenced by Christian prelates who donned the rich fabrics and adorned their altars with them. Gradually the nobility began to have their own clothing fashioned from silk fabrics as well.

Initially, the Chinese were highly protective of their secret to making silk. Indeed, the reigning powers decreed death by torture to anyone who divulged the secret of the silkworm. Eventually, the mystery of the silk-making process was smuggled into neighboring regions, reaching Japan about A.D. 300 and India around A.D. 400. By the eighth century, Spain began producing silk, and 400 years later Italy became quite successful at making silk, with several towns giving their names to particular types of silk.

The first country to apply scientific techniques to raising silkworms was Japan, which produces some of the world's finest silk fabrics. Other countries which also produce quality silks are China, Italy, India, Spain, and France. China was the largest exporter of raw silk in the early 1990s, accounting for about 85% of the world's raw silk, worth about $800 million. Exports of China's finished silk products were about half of the world's total at about $3 billion.

Silk is highly valued because it possesses many excellent properties. Not only does it look lustrous and feel luxurious, but it is also lightweight, resilient, and extremely strong—one filament of silk is stronger then a comparable filament of steel! Although fabric manufacturers have created less costly alternatives to silk, such as **nylon** and **polyester**, silk is still in a class by itself.

The secret to silk production is the tiny creature known as the silkworm, which is the caterpillar of the silk moth *Bombyx mori.* It feeds solely on the leaves of mulberry trees. Only one other species of moth, the *Antheraea mylitta,* also produces silk fiber. This is a wild creature, and its silk filament is about three times heavier than that of the cultivated silkworm. Its coarser fiber is called *tussah.*

The life cycle of the *Bombyx mori* begins with eggs laid by the adult moth. The larvae emerge from the eggs and feed on mulberry leaves. In the larval stage, the Bombyx is the caterpillar known as the silkworm. The silkworm spins a protective cocoon around itself so it can safely transform into a chrysalis. In nature, the chrysalis breaks through the cocoon and emerges as a moth. The moths mate and the female lays 300-400 eggs. A few days after emerging from the cocoon, the moths die and the life cycle continues.

The cultivation of silkworms for the purpose of producing silk is called sericulture. Over the centuries, sericulture has been developed and refined to a precise science. Sericulture involves raising healthy eggs through the chrysalis stage when the worm is encased in its silky cocoon. The chrysalis inside is destroyed before it can break out of the cocoon so that the precious silk filament remains intact. The healthiest moths are selected for breeding, and they are allowed to reach maturity, mate, and produce more eggs.

Generally, one cocoon produces between 1,000-2,000 ft (303-606 m) of silk filament, made essentially of two elements. The fiber, called fibroin, makes up between 75 and 90%, and sericin, the gum secreted by the caterpillar to glue the fiber into a cocoon, comprises about 10-25% of silk. Other elements include fats, salts, and wax. To make one yard of silk material, about 3,000 cocoons are used.

Sericulture is an ancient science, and the modern age has not brought great changes to silk manufacture. Rather, man-made fibers such as polyester, nylon, and acetate have replaced silk in many instances. But many of the qualities of silk cannot be reproduced. For example, silk is stronger than an equivalent strand of steel. Some recent research has focused on the molecular structure of silk as it emerges from the silkworm, in order to better understand how new, stronger artificial fibers might be constructed. Silk spun by the silkworm starts out as a liquid secretion. The liquid passes through a brief interim state with a semi-ordered molecular structure known as nematic liquid crystal, before it solidifies into a fiber. Materials scientists have been able to manufacture durable fibers using liquid crystal source material, but only at high temperatures or under extreme pressure. Researcher are continuing to study the silkworm to determine how liquid crystal is transformed into fiber at ordinary temperatures and pressures.

Silk-screen printing

Silk-screen printing dates back to at least 500 A.D., when it was used in both China and Japan to print works of art. In Europe, it first appeared in the fifteenth century, when it was used to manufacture religious images and playing cards.

The process of silk-screen printing is relatively simple. The printed stencil (usually made of silk) is stretched over a frame. A viscous ink is then forced through the stencil onto the subject—which can include cloth, **paper**, wood, plastic, metal, or **glass**—with a ''squeegee,'' a soft-bladed tool which scrapes across the surface.

The versatility of the method made it a popular means of decorating household furnishings in the eighteenth and nineteenth centuries. With the 1907 patenting of an automated process by American Samuel Simon, the commercial applications of silk-screen flourished. Today, artificial fabrics have, for the most part, replaced silk as the stencil medium and the technique is primarily used to impress inked patterns on mass-market clothing. Light-sensitive screens are used to reproduce **photograph**s and illustrations, such as those often seen on T-shirts. The ability of silk-screen printing to apply inks of various thicknesses makes it possible to print weather-resistant images on outdoor signs and billboards.

Silly putty

Silly putty was invented accidently by James Wright, an engineer with General Electric, in the early 1940s when he was asked to develop a low-cost **synthetic rubber** substitute for the military. The United States War Production Board hoped that GE engineers could invent a chemically synthesized rubber which could be used in the mass production of **gas mask**s, military gear, and jet and airplane tires.

Wright's experimentations with silicone oil and boric acid resulted in a rubber-like compound with bizarre properties. You could roll it into a ball and bounce it extremely high; you could stretch it to enormous length; you could lift images

Depiction of ancient Chinese silkmaking.

from newspaper print and **comic strips**. It was an amazing new product but useless to the war effort. There were no industrial advantages to the "bouncing putty," only psychological ones: GE engineers often entertained themselves for hours with it.

In 1949, Wright brought his bouncing putty to a party. Ruth Fallgatter, a New Haven toy store owner, was intrigued by the putty's huge popularity with the partygoers. At the time, she was designing a toy catalogue with the help of Peter Hodgson, an advertising copywriter. They included Wright's putty in their issue, and it quickly outsold all other items in the catalogue.

Despite the putty's amazing success, Fallgatter was unimpressed. Hodgson, however, foresaw the market value of the putty. For $147.00, he bought a huge quantity of the putty from GE and hired a Yale student to cut it into one-ounce balls. Hodgson then packaged the putty in colored, **plastic** eggs and sold it for one dollar a piece. It was now called *Silly Putty*, but there was nothing silly about its sales: it became an overnight sensation and racked up sales of millions of dollars, becoming one of the most successful toys ever marketed in the United States. Hodgson became a wealthy man and upon his death, his estate was worth almost $140 million.

SILO • See Fodder and silage

SIMPLESSE • See Fat substitute

SINGER, ISAAC M. (1811-1875)

American inventor

Isaac Singer developed the first practical home **sewing machine** and brought it into general use. Born in Pittstown, New York, to German-Jewish immigrants, Singer left home at age twelve and roamed the Northeast for many years, working variously in carnivals, as an actor, and a mechanic. In 1839 he patented an excavator, and in the 1840s, a metal and woodcarving machine.

In 1850 Singer was working in a Boston, Massachusetts, machine shop when he was asked to analyze a Blodgett & Lerow sewing machine that had been brought in for repair. Singer developed a new design based on that machine, patented it in 1851, and along with Edward Clark, cofounded the I.M. Singer Company to market the machine.

Although Singer's machine was a great improvement over existing models, partly because of its continuous-feed feature, he was successfully sued three years later for patent infringement by **Elias Howe**, who had registered his own sewing machine design in 1846. However, the advent of patent pooling and licensing agreements in 1856 allowed the manufacture of Singer machines to continue with constant improvements. By 1860 the Singer Manufacturing Company had become the world's largest maker of sewing machines, and by 1863 Singer had received twenty patents for the machines.

Singer earned millions of dollars from his company and lived flamboyantly, enjoying rides through New York City's Central Park in his yellow coach with his mistresses—not a proper image for a company trying to sell sewing machines to middle-class housewives. Singer retired from the business in 1863, traveling throughout Europe before settling in Torquay, England, where he built a mansion and encouraged his twenty-four children (legitimate and illegitimate) to visit. Upon his death Singer left behind an estate of $13 million.

See also Howe, Elias; Sewing machine

SKATEBOARD

Skateboards have their origin in the scooter of the 1950s, a popular pre-teen toy constructed from an orange crate, a two-by-four board, wooden steering handles, and rollerskate wheels. In order to practice in bad weather, Californian surfers reduced these four pieces to two elements—the board and the wheels—so that they could "sidewalk surf."

Skateboarding became fairly popular in the 1960s, especially when rock stars Jan and Dean rhapsodized about the sport. However, enthusiasm waned quickly when people realized the skateboard's wheels—made of steel, clay, and rubber—wore out fast, snagged easily, and permitted few stunts. Frank Nasworthy, however, introduced a new type of wheel made of durable urethane. In 1973, he formed a production company after discovering that, while these wheels were useless for rollerskaters, they were perfect for skateboarders. On asphalt or concrete surfaces, these wheels offered much better traction and maneuverability. Within one year, Nasworthy's company sold ten thousand sets of the "Cadillac Wheels" and they were trademarked in twelve countries.

Other subsequent innovations helped improve the skateboard. Henry Larrucea researched the skateboard's shape and increased its flexibility by extending the wheel out beyond the axle nut. Bearings were refined, the wheel assembly was improved, and the skateboard became more streamlined.

Once associated only with children, skateboarding is now considered a challenging, highly skilled sport for people of all ages.

SKELETON KEY • See Lock and key

SKIN, ARTIFICIAL • See Artificial skin

SKIN CARE PRODUCTS

Skin care products are used to cleanse and condition the body and face. These include soaps, bubble baths and shower gels, moisturizing creams and lotions, and a variety of special treatment products like facial masks, blemish control products, and sunscreens.

The earliest skin care products were natural oils and fats applied to soften skin. Galen (circa A.D. 130-200), who is considered to be one of the greatest physicians of antiquity, discovered to how mix oils and water to form the first cold cream emulsion which he called ceratum refrigerans, or cooling ointment. Through the ages, Galen's discovery has been used to create a variety of skin care products. As recently as the 1700s lard, mutton tallow, and white and yellow wax were still key ingredients in pomades and ointments. In the last few hundred years, emulsions and other skin care products have become increasingly sophisticated. For example, in the early 1820s coconut oils were first used to make **soaps** which offered improved foam characteristics compared to lye soaps based on animal fats. Similar advances were made in the production and refinement of natural oils. Over the next few decades, breakthroughs in colloid chemistry and glycerol purification techniques aided in the formulation of improved cosmetic creams, lotions, and cleansers. Despite these advances, many popular skin care products in the late 1800s were still based on relatively simple natural ingredients such as almond paste, beef tallow, vinegar, lemon juice, and rice and talcum powders.

The early twentieth century saw an unprecedented expansion of skin care products and technology. One of the first products of the new century was Pompein Massage Cream which was introduced in 1901. This product and others like it helped prompt a boom in the beauty care business that stretched into 1920s. The advent of World War I caused shortages of many natural raw materials and forced the chemical industry to develop synthetic substitutes. For example, glycerine was a main ingredient in soap making, but its war time shortage drove the development of the synthetic detergents which dominate the market today.

A large number of new detergents, fragrance oils, and other cosmetic raw materials were developed in the first part of the century. Shaving products and other specialty skin products derived from olive oil also became widely available. Massage oils and other skin treatment products continued to grow in popularity in the 1920s and beauty salons flourished. Well-known names of the period include Elizabeth Arden, Harriet Hubbard Ayer, Jane Curran, Primrose House, and Helena Rubenstein. In the decades that followed, the skin care industry continued to grow exponentially. Today skin care involves hundreds, if not thousands, of products with annual sales in the billions.

Most skin care products are based on the scientific principle of surfactancy. Surfactants are molecules which have an affinity for both oil and water. This property gives surfactants the ability to form stable mixtures of oils and water, which are known as emulsions. Skin creams and lotions are examples of emulsions. When surfactants are used as detergents they allow water to dissolve oils and dirt and wash them away. This same technology is used in pharmaceuticals, paints and coatings, detergents and agricultural chemicals.

Chemists blend surfactants with other cleansing and conditioning agents to create cosmetics. These products must effectively cleanse or condition without drying out the surface of skin or without leaving it feeling excessively greasy. It is the challenge of the cosmetic formulator to ensure they are efficacious, aesthetically appealing, and safe to use. In addition to cosmetics which improve the appearance of skin, formulators also develop products which interact with the physiology of the skin. Such products are regulated as drugs because of the way they affect the structure and function of skin. Examples of skin care products which are over-the-counter drugs include anti-acne treatments, wrinkle control creams, and **sunscreens**.

In the 1990s the skin care industry continues to seek increasingly innovative technology and new highly effective ingredients are constantly being introduced. Recent examples include improved **silicone**-based molecules which have superior skin feel and moisturizing ability; milder and more biodegradable surfactants; and a multitude of cosmetically elegant oils and waxes. Special delivery systems have also become popular as a way to more effectively carry ingredients to the skin. Liposomes, tiny capsules which penetrate the skin, are one example. These and other new developments continue to fuel the growth of the skin care industry. Future advances will be made and more effective products will be created as knowledge of skin biology is expanded.

Isaac Merrit Singer

Skis and ski bindings

Like many winter sports, skiing is Nordic in origin. For centuries, it was viewed solely as a means of transportation. With heavy snowfall common in everyday life, it was developed simply to survive.

Originally, skis were made from the sharpened bones of large animals. In Stockholm, Sweden, a 5,000-year-old pair of "bone" skis are on display. Besides everyday use, skis were often used for military purposes. Ski troops took part in the Battle of Oslo in 1200, and Sweden used them in 1521 during conflicts with Denmark. United States ski troops fought in the Alps during World War II.

Skiing as a sport is believed to have begun in the early 1800s. In 1877, the first ski-jumping contests were held in Norway. In 1860, Norwegian Sondre Nordheim invented a swifter ski by bending up the front end. An Austrian inventor named Mathias Zdarsky shortened the skis and equipped them with metal bindings. These bindings, called *ski bindings*, were a crucial addition to skis.

The purpose of ski bindings is to keep the skier's boot firmly in place so that the ski can be controlled. The earliest bindings were found on prehistoric skis preserved in Scandinavian peat bogs and were simply leather straps. By the early 1900s, standard bindings consisted of two upright toe irons linked by a leather strap and a loop of spring cable clipped to the ski in front of the toe.

However, these bindings were not too efficient, as they jammed easily in the snow. Modern bindings are of the "step-in" type. The toe of the boot is inserted under the flange of the toe-piece and the heel lowered on the bottom flange of the heel-piece. The binding clicks firmly shut, and the snugness helps many skiers.

Skyscrapers

Until one hundred years ago, the skylines of the world's cities were characterized by silhouettes of trees and church towers, but as urban populations have grown, urban land has become extremely valuable. To keep up with the demands of the population but still provide affordable retail, office, and living space, builders have developed methods of constructing vertical rather than horizontal buildings.

The urban skyline began to experience change in 1883 when **William Le Baron Jenney** designed the first skeletal frame building, the Home Office Building in Chicago. Jenny's innovations opened the way for many other architects to literally soar to enormous heights. Among those moving in the direction Jenny had established were Louis Sullivan (1856-1924), William Holabird (1854-1923), and Daniel Burnham (1846-1912).

Many innovations occurred around the end of the nineteenth century that contributed to the success of the skyscraper. Rolled steel beams were introduced by the Butterfly Company in 1880 and provided building material for the structures. The first commercial **elevator** was designed based on drawings by **Elisha Graves Otis** and installed in the Demarest Building in New York City in 1889. (Steam-powered elevators had been introduced as early as 1857.) Subsequently, cities like Chicago and New York experienced a flurry of competition among industrial corporations. The new buildings filled many city blocks with large, steel-framed buildings that had facades of brick and masonry. Many famous buildings, including the 102-story, 1,250 ft (381 m) Empire State Building, were built during this time.

Although construction virtually came to a halt during the Great Depression and World War II, it resumed with a vengence after the war. However, the ornate facades of earlier designs were replaced by streamlined **glass** and steel. Although the new materials represented progress and sophistication, they also became symbols of the insensitivity and impersonality of modern living. For the first time, psychological studies became a part of the building design process.

During the 1970s and 1980s, architects began to move away the cold, utilitarian designs of the previous decades, once again adding variety to their work. Postmodern architects such as Robert Venturi and Michael Graves returned to the use of color and shape so that a building could once again be identified by its appearance, rather than simply by its address. Stress-skin construction, in which the load-bearing frame is incorporated into the external wall, began with the 1982 completion of the One Mellon Bank Center in Pittsburgh. The new method allows the architect more freedom in the placement of floor supports rather than having support columns dictate design.

Some cities have rejected high-rise construction. Many European cities like Milan and Paris have opted to retain their historical character by keeping their older structures and expanding outward instead of upward.

See also Balloon frame

Skywriting

Skywriting is the art of using an airplane to create messages in the sky with chemically produced smoke. There are two basic kinds of skywriting: standard and skytyping.

In standard skywriting one airplane produces a continuous stream of smoke and it maneuvers in the sky to create a message that can be seen for 15-20 miles (24-32 km) in all directions. The pilot, who has a diagram of the message in the cockpit, must plan the mission carefully because the writing is done upside down and backwards in relation to the pilot's position. He or she flies up to at least ten thousand feet (3,050 m) to avoid most of the convection currents that could quickly destroy the message. The pilot maintains a speed of 150-175 miles per hour (240-280 kph), flying into the wind to keep all the letters in the same spatial relationship. Some letters are more difficult to skywrite than others because they require precise timing and maneuvering: N, M, W, and S are especially demanding.

Skytyping uses a formation of airplanes that release short, individual puffs of smoke to create letters. It resembles dot-matrix printing in this regard. On the ground a punched piece of paper tape is prepared with the message. The tape, on a reel, goes into a machine in the lead plane. This machine reads the tape and translates it into **radio** signals and sends them to receivers in the other planes (usually four) to govern the formation of letters that can be five hundred feet (152 m) tall and a finished message that can stretch five miles (8 km) in length. One flight can cover a whole state with several skytyped messages.

In 1986, a new skytyping method was introduced that only requires one plane: the plane tows a 300-foot (91 m) cable with seven cylinders attached. These cylinders, which containing a fogging substance, are turned on and off by a computer in the cockpit. All types of skywriting usually utilize the same kind of smoke. Specially formulated petroleum oils containing paraffin additives are sprayed onto the aircraft's hot exhaust pipe. Up to sixty gallons (227 l) of oil can be used in one day of skywriting. An innovation in skywriting is colored smoke, created with the addition of light foam **plastic** or oil-soluble dye.

See also Aircraft; Computer graphics; Computer input and output devices

Slater, Samuel (1768-1835)

English American manufacturer

In the late 1700s, England controlled the **textile** industry. This was largely due to the spinning machines of **Samuel Crompton** and **James Hargreaves**, as well as the water-powered machines of Sir **Richard Arkwright**. American industrialists were eager to dip into this profit pool; English businessmen, however, were not about to share the secrets of their trade. Several American textile companies began to offer rewards and bounties to mill workers who would emigrate from England—bringing their knowledge of textile machinery with them, of course. One of the men lured across the ocean in this way was Samuel Slater.

Slater was born in Derbyshire, England, in 1768. His father, a farmer, died when he was fourteen, at which time he was apprenticed to a neighbor, Jebediah Strutt. Just a few years earlier, Strutt had entered into partnership with Arkwright to construct the first **water frame** spinning machine. Strutt employed Slater as the supervisor of one of his textile mills. For six and a half years Slater learned all about the process of manufacturing cotton yarn. At the end of his apprenticeship he decided to travel to the United States.

Since it was illegal to export textile technology (such as parts, designs, and sketches) Slater memorized the construction plans for the Arkwright factory. He did not tell anyone of his decision to leave the country; even his family did not learn of his departure until receiving a letter days later. At the docks, Slater told authorities that he was a farm laborer.

He landed in Philadelphia in 1789, moving shortly afterward to New York City where he took a position at the New York Manufacturing Company. His stay there was brief, however, for within a few months he made contact with Moses Brown of the Almy and Brown textile firm. Brown made Slater a generous offer: if he would come to Providence, Rhode Island, and set up an English-style textile factory, he would be allowed to keep all of the profits. Slater, of course, accepted.

Slater and Brown met at an old mill in Pawtucket, Rhode Island. The idea was to adapt the existing technology to operate as Arkwright's factories did. It was immediately apparent to Slater, though, that the mill's machinery was insufficient. Working from the designs he had memorized, Slater oversaw the construction of completely new machinery, almost identical to that found in English mills. It was nearly a year before the machines were complete, and another two before the factory was put into operation. In 1793 the mill began producing high-quality cotton yarn.

The first skyscraper ever built, the Home Office Building in Chicago was designed by William Le Baron Jenney.

Although the little factory in Pawtucket enjoyed relative success, it took six years for the venture to be profitable enough to consider expansion. In 1789 Almy, Brown, and Slater opened a second mill, then a third, and eventually controlled the production of cotton yarn in much of Rhode Island, Massachusetts, and New Hampshire.

One of the keys to the success of Slater's mills was the use of children—particularly those between the ages of four and ten. Unlike most sweat-shops employing children, the conditions within Slater's factories were quite comfortable, with the children receiving good food and working relatively short hours. In a society where the same children would be put to

Samuel Slater

work on the family farm as soon as they could stand, employment in a textile factory was highly coveted.

In 1798, Slater and his brothers (who had emigrated to the U.S. a few years after him) started their own corporation, Samuel Slater & Company. The primary goal of this company was the production and marketing of a new sewing thread made from very fine cotton yarn, which replaced the more expensive linen thread that had been commonly used. This thread had been invented by none other than Slater's wife, Hannah. She received a patent for her thread—probably the first American patent issued to a woman.

Sled

The sled may have been the first land vehicle used by mankind. Also called a sledge, it is a vehicle with parallel runners instead of wheels that is dragged or pushed and is used to convey a load, especially over snow and ice. Probably the most primitive sled was made of animal skin onto which early man placed his load, usually a dead animal, and dragged it using vines or ropes of some kind. Dragging a heavy load is always easier than carrying it any distance. The true sled with some type of runners is thought to have evolved from the primitive ''travois'' which was composed of either a single branch of a tree or two wooden poles, one end of which dragged along the ground while the other end was supported or tied to the person or animal pulling it. The load being carried or dragged by a travois was usually placed and tied across the two branches or poles at about midlength. Another early and very simple precursor of the sled was made from a forked or V-shaped branch with cross pieces tied to the wider parts of the V, and which was dragged by its single end. These early dragging devices inevitably became a flat platform that was dragged along the ground, to which runners of some sort eventually were added. It is known that the skilled forest dwellers of Neolithic Europe made wooden runners for their sleds some 8,000 years ago, since some still survive having been preserved in the peat swamps of Denmark. These sleds are thought to have been pulled with thongs made of hide. In the north, sleds ran easily over ice and snow, but they were also employed in the south on dry land, with water or even milk poured on the ground in front of the runners when necessary.

With the advance of civilization and the growth of organized communities, mankind was required to move ever heavier objects. In ancient Egypt, heavy stone and timber loads to build the pyramids were hauled along the ground by teams of straining slaves. Many images still exist of massive sleds with huge loads being pulled by ropes. Others also show highly decorated sleds being used as a hearse in Egyptian royal funeral processions. It is significant to note that the Egyptians used sleds over sand in which wheels would have sunk. They also used sleds without runners that were aided by the insertion of log rollers underneath the sled's floor. As each roller was in turn left behind as the sled rolled slowly forward, it was taken back to the front to be reused and rolled over once again. From this roller system, the wheel eventually evolved.

Sleds were most commonly used in northern climates because of their ease in running over ice and snow, and the excavation at Oseberg, Norway of a tenth century Viking burial ship contained several sleds among other things. Besides one hand-drawn sled that had obviously received everyday use, the remaining sleds were all very ornate and used only for ceremonial purposes. Still, sleds were used extensively in the frozen north. Also, in northern Asia and North America which is covered with snow much of the time, native peoples perfected the use of harnessed dogs or reindeer to pull light sleds. As late as the beginning of the nineteenth century, it was customary in parts of England to load hay and wheat that had been cut onto light sleds and haul them from the fields to the barn. Sleds were especially useful over terrain that could not accommodate a wheeled vehicle. Sleds naturally played a major role in polar expedition, and the dog-sled was the only method of transporting men and supplies across the vast wastes of ice that mid-nineteenth and early twentieth century Arctic explorers had to cross. It is partly this exploring tradition that later Canadian dog-sled ''Derbys'' would commemorate. Today's annual Iditarod race in Alaska is a continuation of this old winter pastime.

Sleds are also used for the simple fun of gliding down a snowy hill, and in nineteenth-century America, sledding became a regular winter pastime for children. Early American sleds were made of hardwood and often had turned up runners. However, sleds with runners that thrust to a point were called ''Clippers'' and used only by boys. Girls used ''Cutters'' whose runners were elegantly curled upward. Boys rode downhill headfirst, while girls sat down. The first ''Flexible Flyer'' was the invention of the Pennsylvania farmer, Samuel Leeds Allen. In 1889, his patent described a T-shaped runner with slatted seat that was steerable. Over fifty years later, the sled had become such an integral part of growing up in America that it appeared in the classic film, ''Miracle on 34th Street.'' By the end of the twentieth century, sleds have made the nearly complete transition from transport vehicle to recreation vehicle, and the use of molded plastic sleds that coast easily without runners has transformed sled design.

SLEEPING CAR, RAILROAD • See Pullman car

SLIDE PROJECTOR

In our world of animated film, the slide projector, which projects still images onto a viewing screen, occupies a small niche, mostly in education and business.

The earliest written record of a slide projector, then known as a magic lantern, dates from 1646, when Athanasius Kircher, a German Jesuit scholar, depicted a candle-lit device in his publication *Ars magna lucis et umbra.*

Magic lanterns that projected images from hand-painted glass plates were popular in the eighteenth and nineteenth centuries. The first public shows, Gaspard Robert's Fantasmagorias, took place in Paris in 1798. After projectors and slides were massed produced beginning in 1845, they became available to ever wider audiences.

In the 1870s and 1880s, J. A. R. Rudge developed a number of magic lantern projectors, and in 1880 **Eadweard Muybridge** introduced his Zoopraxiscope, the first projector to show actual photographs.

When color film came on the market in the early twentieth century, allowing amateurs to take high quality color slides, slide projectors came into great demand. Using the same basic process as the magic lantern, the modern projector shines a beam of light, which is concentrated by mirrors and lenses, through processed film (positive transparency) held in a carousel or tray. A straight tray called the Slidex, in which the slides are laid flat, had been popular in Europe before its introduction in the United States in 1987. An audience can view the projected photos can on a wall or screen.

In addition to enhancing business presentations, families could view photos together. As rival technologies, like videotape, developed in the late twentieth century, slide projectors became more sophisticated. For example, the Arion Mirage 901, introduced in 1995, allows slide shows to have special effects like fades, dissolves, and a synchronized soundtrack.

Computer engineers also developed a projector-computer hybrid that allows the user to project the contents of the computer cathode ray tube onto a screen for group viewing. Though the slide projector has proven its utility, the emergence of the computer and the video monitor have taken over many of the slide projectorôs functions both in the business and domestic arenas.

See also Motion picture; Video recording

SLIDE REST • See Lathe

SLIDE RULE

With the publication of logarithmic tables by John Napier (1550-1617) in 1614, astronomers and mathematicians were freed from much of the drudgery long associated with their professions. Some scholars realized that logarithms could be used in mechanical devices to take automatic calculation one step further. In 1620 Edmund Gunter (1581-1626), a professor at Gresham College, London, created a forerunner of the slide rule, which he described as his ''logarithmic line of numbers.''

Around that same time, English mathematician William Oughtred (1574-1660), began developing his own device employing Napier's logarithms. Although he was an ordained minister, Oughtred was most interested in mathematics and he spent every spare minute studying and planning his slide rule. In 1621 Oughtred introduced the device he had been perfecting—the first linear (or straight) slide rule. His device consisted of two sliding rulers marked with graduated scales representing logarithms, allowing calculations to be performed mechanically by sliding one ruler against the other. The invention caused quite a quarrel between Oughtred and his former pupil, Richard Delamain, because in 1630 Delamain described plans for his own design, a circular slide rule. Delays in publishing caused accounts of Delamain's device to appear in print before write-ups of Oughtred's slide rule appeared. While Oughtred accused Delamain of stealing his ideas, it has been generally concluded that the devices were both developed independently. In addition to his slide rule, Oughtred was also responsible for the introduction of the multiplication sign ''x'' and the trigonometric abbreviations still commonly used today—sin, cos and tan, for sine, cosine, and tangent.

Few innovations were made to slide rule design until 1902 when Norwegian Carl George Lange Barth copyrighted a design for a circular slide rule. An advanced model, it was intended to speed up mathematical calculations and offered several advantages, perhaps the most important of which was it did not ''run off the scale,'' as the more traditional straight line device was known to do.

While early slide rules were commonly used by tax collectors, masons and carpenters, slide rules today are most often used in engineering. Simple, yet versatile instruments, they perform calculations quickly and with reasonable accuracy.

See also Napier's bones

SLINKY

The Slinky is a toy made of a coil of steel that slithers up and down steps, one half propelling the other demonstrating the physical laws of inertia (matter's propensity to keep moving in the same direction) and Hooke's law (its tendency to return to its original shape after being stretched).

The Slinky was invented by Richard James during World War II. While working at the Cramp Shipyard, the young Philadelphia engineer experimented with coil **springs** that he hoped to make into shock absorbers for fragile instruments. While he was sitting at his desk one day, a torsion spring fell down from a shelf above and ''walked'' across a pile of books. Thus, James later claimed that he had not really invented the Slinky but that it had ''practically walked into my life.'' Shrewdly anticipating the public's fascination for this curiously self-propelling object, James and his wife Betty took out a $500 loan and made 400 springs, which they named ''Slinky'' (which implies stealthiness and sinuousness) after Betty searched the dictionary for an appropriate appellation.

In November of 1946, the couple set up their wares in a corner of Gimbel's New York department store. They sold all 400 Slinkies in ninety minutes, and many millions in following years. Slinkies are now made of both metal and **plastics**, with crimped edges to protect the user from injury. A Slinky is on exhibit in the collection at the Smithsonian Institution's National Museum of American History.

Slot machine and vending machine

The slot machine or one-armed bandit is a gambling device that, when a coin is dropped into its slot and its handle pulled down, activates three or more horizontal reels marked with various symbols such as stars, card suits, numbers, fruit (especially plums, cherries, lemons, oranges, and watermelons), or the word *jackpot*. The operator receives a payoff if, when the reels come to rest, matching symbols are shown in the display windows; the amount of the payoff may range from several coins to all of the coins in the machine, depending on how the symbols line up. The slot machine pays the winner by dropping coins into a cup or trough.

The first slot machines, which appeared in the United States toward the end of the nineteenth century, were novelty devices usually found in taverns. They did not reward users with coins, but offered wagering opportunities to would-be gamblers. For example, such a machine might have two automated horses attached to its top that would race. By around 1892, paying machines had begun to appear, usually featuring a circular display window and a spinning indicator that would come to rest with its pointer on a particular number, color, or picture. In the early twentieth century, a three-reel (later expanded to five) device became standard.

One popular legend attributes the invention of the slot machine to H. S. Mills, a former Chicago newsboy who became a lemonade stand entrepreneur and equipped his operations with a device called a *Kalamazoo* that had a slot and three

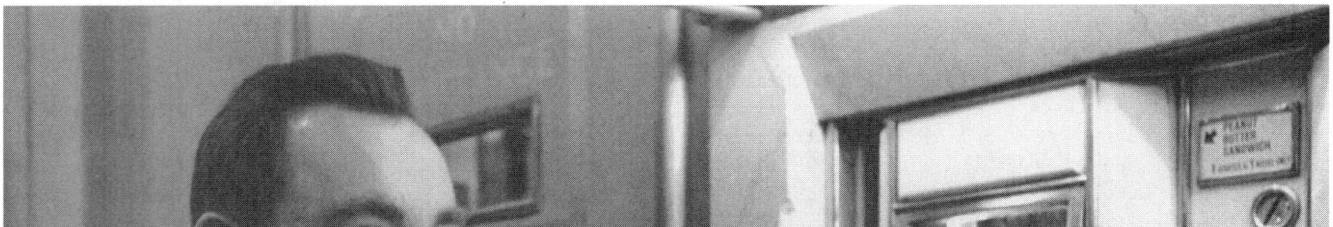

Automats, cafeteria-like establishments where customers make their selections from coin-operated vending machines, were a fixture in New York City until the rise of fast food restaurants in the 1970s. Above, French chef Pierre Charlot receives special attention froman automat manager in 1969.

tubes. A penny inserted into the slot would, when it emerged from two of the tubes, bring two or three more pennies with it, but it would not emerge at all from the third tube. Another legendary figure in the development of the slot machine is Charles Fey of San Francisco, who rented his famous *Liberty Bell* machines to saloons and kept half of the profits for himself. This device was decorated with art nouveau designs and stood on lions' paw feet. Fey set up the Mills Novelty Company in 1889 to manufacture slot machines, and by 1932 the factory was producing seventy thousand per year.

During the 1920s and 1930s, the slot machine became an immensely popular entertainment, particularly at American resorts. In response to mounting disapproval from those who found gambling immoral, some slot machines disguised their payoffs by dispensing gum, candy, or music along with coins. A 1932 article in *Fortune* entitled "Plums, Cherries, and Murder" claimed that racketeers bought slot machines from legitimate manufacturers and then exploited them for huge profits—as much as $150 million in the United States alone. The belief that organized crime controlled slot machines led to legislation restricting their sale, transportation, and use (except in private social clubs). By 1951, slot machines were prohibited in all states except Nevada, where gambling remained legal. Over the next several decades, however, other states and countries were attracted by the possibility of revenue from the machines and began to permit their use.

Early slot machines tended toward patterns that could be divined to maximize payoffs, but manufacturers were eventually able to randomize the timing of the reels. Another problem was solved with the 1931 invention of the coin tester, which rejected slugs and bad coins. Modern versions can check the diameter, weight, and metal content of coins inserted into the slot machine. The term *one-armed bandit* referred not only to the machine's single handle but to the fact that owners adjusted the payoff rate according to high-and low-volume play periods (with higher payoffs allowed during the latter). By the late twentieth century, the record jackpot—on a five-reel, $1 machine—was $1 million. Despite the growing popularity of electronic machines offering poker, keno, and blackjack, gamblers' enthusiasm for the slot machine has continued unabated.

Closely related to the slot machine is the vending machine, which responds similarly to the insertion of coins in a slot. Vending machines dispense items as diverse as candy, cigarettes, hot and cold drinks, food, air travel insurance, and suntan lotion. Coin-operated washers and dryers are also examples of vending machines.

The earliest vending machine, dating to 215 B.C., dispensed holy water in a Greek temple in Egypt. In 1857, an Englishman named Denham patented a stamp vending machine; about thirty years later, R. W. Brownhill patented an effective coin-operated gas meter that allowed consumers to purchase varying amounts of gas within their own homes. The first vending machines in the United States were chewing gum dispensers installed on New York City train platforms in 1888; candy and gum machines appeared during the 1920s. Vending machines became increasingly popular with a convenience-minded public, and by the mid-1980s the industry was doing $16 billion a year in business.

John Smeaton

One interesting and particularly famous use of the vending machine was in the automat, an eatery that featured an array of vending machines rather than the conventional kitchen and waitpersons. The first automat was opened by Horn and Hardart in Philadelphia in 1906, using German-made vending machines. Automats subsequently grew popular in New York City, where their art deco exteriors and the convenience they offered were both much appreciated by the public. The automat has, in fact, entered the lexicon of American popular culture and is remembered with nostalgia—perhaps most famously by pop artist Andy Warhol, who dreamed of establishing "Andymats." The automat was eventually overshadowed by the fast-food restaurant, and the last one (located on 42nd Street in Manhattan and operating primarily as a cafeteria during its final days) closed in 1991.

See also Parking meter

SMEATON, JOHN (1724-1792)
English civil engineer

John Smeaton, like so many of his contemporaries, made a name for himself in several fields. He can best be described as a professional inventor. Considered to be the founder of instrument making in Britain, he is best remembered for his work as a civil engineer.

Smeaton was born in England of Scottish ancestry—many of the engineers that followed in his footsteps were

Scotsmen. Scotland is a rocky country and stone masonry was a part of daily life there. His father was an attorney and gave his son a start in his law firm. This profession did not appeal to Smeaton, and he soon left it to pursue a profession of a more mechanical nature. He began writing papers on instrument-making and by 1753 was elected to the Royal Society. He was selected in 1755 to design a new **lighthouse** for the Eddystone Rocks near Plymouth, England, and the project was completed in 1759. It included stone blocks with interlocking dovetails. He also developed a cement made of limestone and clay that could be set under water.

Smeaton next went to work on water mills, which were the main source of power at that time. His experiments led to a major paper written in 1759. His tests proved that overshot **waterwheels** were much more efficient than undershot. The advantage was the result of gravity, which is greater in falling water than in flowing water. His work earned him the Copley Award in 1759. Smeaton went on to build forty-three mills, the most significant being the mill at the Carron Company ironworks in 1769. Smeaton invented a tidal pump at London Bridge for the supply of water to subscribers in 1767. It was more compact and efficient than another pump installed earlier at the same site. In the late 1760s, he developed a water pressure engine that pumped water to a nobleman's residence. It was superseded a few years later by **James Watt**'s condensing **steam engine**. Smeaton also made improvements to Edmund Halley's **diving bell** by adding an air pump to it, and used it in the construction of Ramsgate harbor in 1774. Smeaton engineered several **canal** and **bridge** projects from 1757 onward. These included the Forth and Clyde Canal in Scotland. He also invented a metal **boring machine** in 1769.

In his later years, Smeaton experimented with the steam engine and made improvements to it. He died in 1792 at Austhorpe, Leeds, the place of his birth. His influence changed the thinking of many of his peers to such a degree that the Institute of Civil Engineers, founded in 1771 and of which Smeaton was a founding member, was renamed the Smeatonian Society. In his wake came many noted British civil engineers, most of them from Scotland, who built roads, bridges, harbors and canals all over the British Isles and the European Continent. Among these were **Thomas Telford**, John McAdam, **John Rennie**, **Marc Isambard Brunel**, and **Isambard K. Brunel**.

SMITH, FRANCIS PETTIT • See Screw propeller

SMITH, HORACE (1808-1893) WESSON, DANIEL BAIRD (1825-1906)

American inventors and manufacturers

Horace Smith and Daniel Baird Wesson formed a firearms manufacturing partnership in Springfield, Massachusetts, that held many patents in the mid-1800s. They produced a variety of **rifles**, **pistols**, and **revolvers**. The Smith & Wesson .22 caliber revolver became a worldwide success.

Smith, born in Cheshire, Massachusetts, moved to Springfield at the age of four, and upon completing his public school education, joined his father working in the United States armory as an apprentice. Through his early career, he worked his own gun-manufacturing business, as well as producing pistols and tools for the manufacture of rifles for other companies. He patented an improvement on the breech-loading rifle in 1851, a year before he met Wesson through their work at Allen, Brown, and Luther rifle barrel manufacturers.

Wesson was born in Worcester, Massachusetts, and was the son of a farmer and manufacturer of wooden plows. He attended public school until he was eighteen years old, and then apprenticed himself to his eldest brother, a gunsmith. After his brother died, Wesson owned the business for a short time, then became employed by Allen, Brown, and Luther. In his off-hours, Wesson experimented on an improved metallic cartridge, and in 1853, convinced Smith to go into business with him. The following year, the pair patented a pistol that used a cartridge with powder and lubricant, and had a new repeating action, which became the basis for the famous Winchester rifle.

Smith retired from the business in 1855, when Wesson became superintendent of the Volcanic Arms Co., and they sold their patents to the company. Under Wesson, the Volcanic Arms Co. produced the self-primed metallic cartridge used throughout the Civil War. In 1857, the two men rejoined to produce the Smith & Wesson revolver. Manufactured with interchangeable parts, it used repeating action, metallic cartridges, and an open cylinder. It was the only one of its type, and it became an enormous success. It was adopted by U.S. military authorities and also secured foreign markets in European and South American countries, as well in China and Japan. In 1860 Smith & Wesson expanded their 25-member workforce to 600 and built a factory in Springfield, Massachusetts.

Smith and Wesson improved their inventions with patents of their own and those purchased from other inventors. In 1869, they bought a design by William C. Dodge that emptied shells from the gun. Wesson patented a safety revolver that prevented unintentional firing in 1887.

In 1883, Smith sold his interest in the company to Wesson and retired, involving himself in local business activities. Wesson took on his two sons as partners ten years later. Smith & Wesson is still one of the largest gun manufacturers in America.

SMOKELESS POWDER

Smokeless powder was the long-sought successor to **gunpowder**, which had reigned as the chief military and commercial explosive for nearly 500 years. Advances in scientific knowledge and experimental methods created an environment ripe for new inventions by the mid-nineteenth century. Ironically, Ascanio Sobrero's (1812-1888) discovery of **nitroglycerin**—which would later become a key ingredient of smokeless powders—was largely overlooked when first introduced in 1846. Instead, it was **Christian Schönbein**'s formation of **guncotton** the same year which attracted widespread excitement for its powerful explosive properties.

To create guncotton, Schönbein, through a process termed nitrating, dipped cotton (cellulose) in **nitric acid** and **sulfuric acid**, and then removed the acids by washing the cotton with water. This explosive underwent a series of improvements culminating in French chemist Paul Vieille's (1833-1896) derivation of smokeless powder, or Poudre B, in 1884. **Alfred Nobel**'s ballistite and **Frederick A. Abel**'s **cordite**, discovered in 1888 and 1889, respectively, were other popular powders founded on the same principles of nitrating the mixture to a specific ratio, which encouraged rapid burning without explosion. Other early experimenters with smokeless powder include Prussian officer J. F. E. Schultze (1825-1874), who combined nitrated wood with potassium nitrate, and chemists at the Explosives Company in England, who treated guncotton with potassium nitrate, alcohol, and ether. Gunpowder's major drawbacks—its tendency to flash, smoke, and corrode, as well as its limited power—sparked all of these discoveries.

Unlike gunpowder which is a mixture of three commonly found **elements**, this new class of smokeless powders was based on an organic compound called nitrocellulose. Upon combustion, this compound produces a swift and predominantly gas reaction which enhances bullet propulsion. When **nitroglycerin** is added to nitrocellulose a hard gelatin may be formed, allowing for the precise formation of a wide range of charges. Perhaps most significantly, the invention of smokeless powder engendered related inventions in bullet and rifle technology such as longer, slimmer cartridges, magazines and repeating rifles, each of which took advantage of the powder's promotion of higher velocities, lower trajectories, and increased accuracy.

SNAPS • See Buttons and other fasteners

SNOWMOBILE

The first patents for vehicles designed to travel over snow were issued in the late 1800s. Some of these machines used rotating tracks for propulsion, while others sat up high on runners and were powered by propellers. Most of these early inventions were built for several passengers, so they were large and slow.

The first successful small motorized ski sled was developed in 1927. Carl Eliason, a Wisconsin storekeeper, built a long, wooden, toboggan-like sled whose rear portion housed a two-chain track system. The **engine**, mounted in front, was a small outboard design with a **radiator** from a Model T Ford. Over the next thirty years, Eliason made about forty of these vehicles, selling them for $360 each. He gradually adapted the design to include a cowl and windshield for passenger protection, a metal chassis, and a rear-mounted engine.

About the same time that Eliason patented his invention, Joseph-Armand Bombardier, a Canadian teenager, developed his own motorized sled. His design combined a Model T engine with a hand-carved propeller attached to a frame with four runners. He operated the engine controls while his brother steered the runners with his feet. Their father, fearful of the dangers presented by the machine, ordered them to disassemble it.

Despite his father's disapproval, Bombardier continued his work. In 1928 he created a snowmobile from an **automobile** frame. It used skis in place of front tires and had four **wheel**s at the back that spun a track system similar to a **tank**'s. The cab was enclosed for passenger warmth. In the early 1930s he experimented again with his propeller design, using a lightweight two-cycle engine. He finally achieved commercial success in 1940 with his B-12, a full-track (no skis) snowmobile driven by a Ford VS engine that could carry twelve passengers.

In 1957 Bombardier began work on the machine that would make him famous. He created a small sled, which he called the Ski-Dog, because he believed that it would replace the dog sled. The Ski-Dog had a 7.5-horsepower, two-cycle engine in the front with a cowling that surrounded the vehicle. It ran on a **rubber** track that rotated on a rubber sprocket or toothed wheel. In addition, the track had **pneumatic tire**s along its length for shock absorption and suspension. The big advantage of the Ski-Dog was that it weighed only 500 lb (225 kg). The Ski-Dog sold well and was eventually redesigned as the Ski-Doo in 1959. Since then other machines have been introduced, but Bombardier's basic design has remained successful.

SOAP AND DETERGENT

Soap is a cleansing agent which, when dissolved in water, removes dirt from soiled surfaces. Made from animal fat and wood ashes, soap is one of the earliest chemical inventions and was first used as a salve or ointment. Many ancient cultures made and used soap, the Sumerians as early as 3000 B.C. The ancient Egyptians used a soap-like paste to treat skin disease.

The Romans learned about soap from conquered peoples, perhaps the Celts or the Gauls. In his *Historia naturalis* the Roman naturalist Pliny the Elder (ca. 23 A.D.-79 A.D.) mentions soap used by the Gauls as a hair dye and salve. The Greco-Roman physician Galen (ca.130-c.200) wrote about soap as a medication and—for the first time—as a cleanser in one of his treatises published in the second century A.D. The Arab alchemist Jābir ibn Hāyyan (Geber) also mentioned soap as a cleanser in the eighth century.

By then, it was known throughout southern Europe. Marseilles, France, and Genoa and Venice, Italy, became the leading manufacturers of the substance because of their abundant supplies of olive oil and crude soda. Soapmaking in England began around 1200 in Bristol. Refinements to the soapmaking process in the sixteenth century produced a purer soap. Since English authorities considered soap a luxury, it was heavily taxed until 1853, and soap-boiling pans were locked each night by local tax collectors.

Soapmakers plied their trade in America at Jamestown, Virginia, beginning in the early 1600s. Although manufactured soap was available in America by 1800, rural Americans continued to make their own soap at home from wood ashes and animal fat, just as the Sumerians and Phoenicians had, into the early 1900s.

The modern soapmaking industry was launched in 1791 when French chemist **Nicolas Leblanc** developed a method of obtaining sodium carbonate from common salt rather than from plant ashes. Methods were also developed to recover glycerin during soapmaking, which could then be reused in place of purchased vegetable oils. These improvements made soapmaking easier and cheaper, which in turn made soap widely available at affordable prices.

In 1823 another French chemist, Michel-Eugène Chevreul (1786-1889), uncovered the chemical constitution of fats in soap, putting soapmaking on a sound scientific basis. Ingredients were gradually refined so that soap cleaned more effectively and also became mild, fragrant, and attractively colored.

Making soap at home was a major—and tedious—domestic chore. Even after manufactured soap became available, it was sold only in bar form, and shaving or slicing it constituted yet another household chore. Lever Brothers alleviated this problem in 1906 with its introduction of Lux Flakes. A proliferation of soap chips, flakes, powders, and beads followed rapidly, prompting such intense advertising competition among soap manufacturers that the radio serials on washday Mondays became known as ''soap operas.''

Unfortunately, soap as a home laundering agent had a serious disadvantage: it reacted with substances in hard water to produce an insoluble scum—the ring around the bathtub and the dull grayness on freshly washed clothes. In 1916 a German scientist, Fritz Gunther (1877-1957), developed a synthetic surfactant that kept oil and grease suspended in water so that it was rinsed away rather than deposited as soap scum. In 1933 Proctor & Gamble (P & G) introduced Dreft, the first synthetic detergent using surfactants for household use. P&G's Tide, which debuted in 1947, was the first effective synthetic detergent and ushered in the detergent era. By 1955, detergents were outselling soaps. Much soap is sold in bar form today for personal bathing, although liquid soap dispensers for bathroom and kitchen use are increasing in popularity.

Until about 1940, soap was made in huge kettles; most soap today is made by a continuous process using stainless steel tubes called hydrolizers. After the soap is produced, it is pumped into giant mixers called crutchers, where other ingredients such as perfumes, colors, and builders—substances that help loosen dirt—are added. The soap is then hardened into bars, rolled into flakes, or spray-dried into powder.

Initially, synthetic detergents posed an environmental threat because their surfactants did not break down under bacterial action in the soil or during sewage treatment. The use of biodegradable surfactants in detergent formulas has helped relieve that hazard, although concern continues over the use of phosphates in detergents. Any of a range of compounds from phosphoric acid and a natural plant nutrient as well, phosphates appear in detergent formulas as a brightening agent. Unfortunately, their natural fertilizing capabilities can alter the ecological balance of lakes and streams by stimulating algae and aquatic plant overgrowth, or eutrophication. The subsequent decay of that overgrowth robs the water of oxygen and in severe instances leads to massive fish kills. The trend today is toward elimination of these compounds.

Soccer

Soccer probably originated in China as early as 400 B.C. with a game called tsu-chin. In 200 A.D., Romans and Greeks played a game in which two teams tried to move a ball across a line on a field. During the tenth century, London children played forms of soccer in the streets.

Two English towns, Derby and Chester, claim to have organized the first soccer game around 1220. The annual game, played on Shrove Tuesday, remained an English tradition until 1314. That year, King Edward II banned the sport because ''too much noise is created in the towns by all the shoving and pushing around of the big balls, which may give rise to any number of the many evils that God forbids.'' Happily for soccer fans, the edict was never applied.

Soccer, as we know it today, was created in 1863, when the London Football Association distinguished it from **football** rugby and set up different rules for it. In 1871, the Football Association Challenge Cup Competition was introduced and remains one of soccer's most famous contests. By 1885, professional soccer had begun in England.

Soccer soon spread to other parts of the world. By 1900, Belgium, Chile, Denmark, and Switzerland all had soccer associations. In 1930, Frenchman Jules Rimet founded the World Cup—an international competition still held every four years.

Until the mid-1800s, soccer was the only form of football played in the United States. American football, however, gained in popularity and overtook soccer as the favored sport. Many American schools and colleges, however, continue to have soccer teams.

Soda fountain • See Soda water

Soda pop

Soda pop (also known as soda and pop) is a soft drink made from carbonated water and syrup. The water is carbonated by the addition of carbon dioxide gas. The syrup consists of sweeteners and a concentrate that is a mixture of flavoring and acid; sometimes coloring is also added. Typically, a soda pop company manufactures the syrup or concentrate and then sells it to a franchised bottler, who adds carbonated water (plus sweetener, if just concentrate is being used) and then bottles and sells the finished drink. Soda pop got its name from the term ''**soda water**,'' used for carbonated water, and from the popping sound made by the cork stoppers when pulled out of the bottles.

Soda pop made its first appearance in Philadelphia, Pennsylvania, during the 1830s, but it did not become widely popular until the last quarter of the century. Most soda pops were at first mixed and sold at drugstore counters, called soda fountains, as healthful tonics. Individual pharmacists developed and sold their own special syrups, dispensing a dollop of syrup and then filling the rest of the glass with carbonated water from the taps at the fountain.

When bottled soda pop grew in popularity during the 1890s, national brands began to appear. The world's most popular soda pop, Coca-Cola, was the invention of Dr. John Stith Pemberton (1850-1888), an Atlanta, Georgia, druggist, who experimented with various ingredients until he came up with his distinctive and stimulating thirst-quencher in 1886. Pemberton's bookkeeper, Frank Robinson, devised its name from two of the drink's ingredients, the coca leaf and cola nut, and designed its trademark flowing script used to spell the product's name. First-year sales totaled 25 gal (94.5 l).

Pemberton died in 1888, and another Atlanta druggist, Asa G. Candler, became the sole owner of Coca-Cola in 1891. Candler aggressively marketed his product, and granted the first bottling rights to Joseph Biedenharn of Vicksburg, Mississippi, in 1894. Benjamin Thomas and Joseph Whitehead of Chattanooga, Tennessee, also secured bottling rights, in 1899. The famous trademarked bottle, inspired by the shape of the cola nut, appeared in 1916, designed by Alex Samuelson of the Root Glass Company in Terre Haute, Indiana. The Coca-Cola Company at first opposed the public's spontaneous use of the nickname Coke for its product, but eventually yielded to the inevitable and trademarked that name as well.

Other familiar soda pops also trace their origins back to the 1880s and 1890s. Root beer became the first nationally popular soda pop when Charles F. Hires (1851-1937) exhibited his mixture of dried roots, barks, and herbs at the 1876 Philadelphia Centennial Exhibition. Originally designed to be brewed into a drink at home, Hires Root Beer became available in bottles around 1893.

The second-oldest soda pop was concocted by a lovestruck soda jerk in Virginia around 1880. Fired by the pharmacist for whom he worked because he was in love with the druggist's daughter, this anonymous young man moved to Waco, Texas, where he began dispensing his drink at the Old Corner Drug Store. Patrons of the soda fountain dubbed the invention Dr. Pepper, after the Virginia druggist. A Waco bottler, R. S. Lazenby, perfected the drink's formula and began marketing it in bottled form in 1885. Dr. Pepper remained a regional specialty until the 1920s; today, Dr. Pepper is one of the largest American soda manufacturers.

Coca-Cola's major rival, Pepsi-Cola, is the invention of pharmacist Caleb D. Bradham of New Bern, North Carolina.

Early advertisement for Coca-cola.

While local customers called the soda ''Brad's Drink,'' the inventor chose the name Pepsi-Cola in 1898. Pepsi almost disappeared in 1920, when Bradham bought up vast quantities of sugar at 22 cents a pound but was wiped out when sugar prices fell precipitately to 3 cents a pound. Pepsi bounced back in 1933 during the Depression, when the company doubled the size of its bottle but kept the price at five cents.

Moxie, once one of the most popular sodas in the United States, began as a concentrated medicinal tonic marketed by the Moxie Nerve Food Company of Lowell, Massachusetts, in 1885; it was soon sold in bottled, carbonated form. Ginger ale was bottled in the 1880s but became popular only after pharmacist and chemist John J. McLaughlin of Toronto, Ontario, improved the old-style product in 1904 by making it pale and less gingery. Canada Dry Pale Dry Ginger Ale rapidly became popular in Canada, and then in the United States beginning in 1919. Orange Crush, one of the first orange soda pops, was introduced in 1916 in Chicago, Illinois.

St. Louis, Missouri, was the birthplace of 7-Up, developed in 1929 by bottler C. L. Grigg who, with his partner E. G. Ridgway, had been selling an orange soda called Howdy since 1920. When citrus growers successfully lobbied for legislation requiring orange soda to contain real orange flavoring, Grigg concocted a new lemon-lime mix. Its unwieldy name, Bib-Label Lithiated Lemon-Lime Soda, was soon changed to 7-Up, inspired by its seven-ounce container. 7-Up rapidly came to dominate the 600-odd other lemon-lime sodas on the market.

Today average per capita consumption of soda pop in the United States is 49 gal (185 l) per year. In fact, according to the National Soft Drink Association, one out of every four beverages consumed by Americans, from water to wine, is a soft drink. There are now about 450 soft drinks on the market in the United States, and the beverages are consumed in over 195 countries. The most popular flavor is cola, followed by, in order of preference, lemon-lime, orange, ginger ale, root beer, and grape. Diet sodas first appeared on the market around 1960. Two **artificial sweetener**s, aspartame and saccharin, have been approved for use in U.S. diet sodas, and other sweeteners are now under development.

SODA, SYNTHESIZED COMMON

Soda ash was one of the earliest manufactured chemicals. Ancient peoples produced soda ash by pouring water through burned plants and then evaporating the water from the resulting mix. This soda was used for making both **glass** and **soap**.

The chemical industry expanded during the 1700s, producing ever-increasing amounts of glass, soaps, and **textiles**, all of which required common soda. Soda derived from plants was expensive and could not keep up with the demand, especially when sources of supply were disrupted by frequent wars. A number of chemists experimented with ways to produce soda from alternative sources. The French Academy of Sciences, intent on spurring industrial growth, offered a monetary prize around 1775 for the best method of producing sodium carbonate from common salt (sodium chloride).

In response to the Academy's competition, the French physician and chemist **Nicolas Leblanc** developed a usable method of treating salt from seawater with sulfuric acid to produce sodium sulfate, which was then heated with limestone and charcoal to produce soda ash and calcium sulfide. Leblanc received a patent for his process in 1791 and won the Academy's prize which, however, was never paid. His efforts to establish a factory to produce soda ash were destroyed by the French revolutionary government. The Leblanc process was finally used commercially in England in the 1820s and soon became the industrial standard worldwide.

Although Leblanc's process did produce artificial soda in large quantities for industry, it had its drawbacks. Sulfuric acid was expensive at first, and the byproducts of the process were both pollutants and difficult to find uses for. Chemists turned their attention to the possibilities of extracting soda from sodium chloride with ammonia. They noted that when ammonia and carbon dioxide gas act on sodium chloride, sodium bicarbonate and ammonium chloride are formed. Simple heating of the bicarbonate produces sodium carbonate—common soda—and lime decomposes the ammonium chloride to ammonia, which can then be used to start the reaction again.

Around 1810, chemists from France, Germany, and England were attempting to find a practical means of exploiting the ammonia-soda cycle. Unaware of these chemists' efforts, **Ernest Solvay** of Belgium devised a feasible way to use the ammonia-soda cycle on a commercial scale in 1861. He designed metallic towers that allowed large amounts of gas to mix with the salt and ammonia solution, producing large amounts of soda; this method was able to recover 99 percent of the expensive ammonia for reuse. Solvay's factory in Couillet, Belgium, began producing commercial amounts of synthetic soda ash in 1865. Ludwig Mond (1839-1909), a German-born English chemist, improved Solvay's process in the 1870s. Today, the Solvay process remains in use to produce most of the world supply of soda ash.

Natural soda ash is mined and purified from deposits in ancient lakes and seas. Salt electrolysis—the breaking down of sodium chloride into its elements by electric current—is being developed as an alternative to the Solvay process for producing synthetic soda.

Soda ash is used in the production of a variety of products including soap, glass, and textiles, as well as wood pulp, **paper**, and water softeners. In 1996 demand for soda ash in the United States was at 7.1 million tons.

SODA WATER

European scientists became interested in the natural effervescence of mineral waters in the sixteenth century. Mineral-water springs were very popular at the time because of their supposed therapeutic properties. Chemists and scientists studied the nature of the bubbles in the water and considered ways to reproduce them artificially.

The Irish scientist **Robert Boyle** wrote about mineral waters and the possibility of imitating them chemically in 1685. Jan Baptista van Helmont (1579-1644) was the first to use the word *gas* to describe the vapor in the bubbles (which was carbon dioxide), in the early part of the seventeenth century. Joseph Black (1728-1799) used the term *fixed air.* Many investigations of the chemistry of mineral waters were reported in the bulletin of the Royal Society of London during the 1700s. American scientists who became interested in mineral waters beginning in the 1770s included Dr. Benjamin Rush (1745-1813) of Philadelphia, Pennsylvania, **Benjamin Franklin**, **Thomas Jefferson**, and James Madison.

An English chemist named Joseph Priestley (1733-1804) discovered a way to introduce carbon dioxide into still water in 1772. Priestley, a failed preacher, lived next door to a brewery, which produced abundant supplies of "fixed air" in layers above the vats of fermenting beer. Experimenting with the gas, Priestley found that if he introduced it into plain water, the water absorbed some of it and became fizzy. He discovered that this new mixture had a pleasant taste, and began to market it. Since the addition of flavoring and sugar to Priestley's soda water would later create **soda pop**, Priestley is acknowledged as the "father of the soft drink industry."

Priestley demonstrated a small carbonating apparatus in 1772. Both he and Antoine-Laurent Lavoisier (1743-1794), a French chemist, suggested using a pump to increase carbonation. The term "soda water" was soon applied to carbonated water because sodium carbonate was used in manufacturing it. A druggist named Thomas Henry (1734-1816) first produced soda water commercially in Manchester, England. A Swiss jeweler, Jacob Schweppe, also adapted Priestley and Lavoisier's ideas and was selling a highly carbonated soda water in Geneva, Switzerland, by 1794. He later expanded his business to London. In the United States, scientist **Benjamin Silliman** founded the first soda water manufacturing facility in New Haven, Connecticut, in 1807. The first American patent for making artificial mineral water was granted to Joseph Hawkins of Philadelphia in 1809.

Mineral salts and flavors were soon added to soda waters, and by 1865, fifteen soda-water flavors were advertised. As more flavorings (plus sweeteners) were added, soda water evolved into soda pop. In 1850, 64 plants were making bottled soda water in the United States; in 1870 the number had climbed to 387, and it reached an all time high of 8,220 plants before the stock market crash of 1929.

SODIUM LAMP • See Arc lamp

SOLAR CELL • See Photovoltaic cell

SOLAR SAIL

Scientists are looking for ways to propel **spacecraft** with less fuel than a **rocket** uses. The solar sail intrigues many because it requires no fuel at all.

In the late 1860s the theory of **electromagnet**ism, first described by **James Clerk Maxwell**, defined light as a packet of energy called a *photon.* These photons have characteristics both of a wave and a particle, meaning at times they can behave like either, and they have energy and momentum when they move. The Russian space pioneer **Konstantin Tsiolkovsky** imagined that some day large thin sheets could be spread out in space to capture photons, which would push the reflective sail along as they bounced off it.

The National Aeronautics and Space Administration (NASA) has studied solar sailing. They launched small metal needles into the atmosphere high above the earth and found that sunlight altered their orbit. Using this discovery, they built the *Mariner 10* spacecraft, which traveled to Venus and Mercury. It had small panels that turned the craft when they were tilted at different angles to the sun. NASA also explored the possibility of using a sail to propel a craft to rendezvous with Halley's comet when it swept through the inner solar system in 1985-86. Scientists were unable to devise such a technique by the time of the comet's arrival, so the plan was dropped.

The principles of solar sailing are very simple: The photons hit a reflective surface and propel the craft. To turn, the sail is slanted, much like a sail on a sailboat when a person tacks with the wind. The craft can increase up to 13 miles (21 km) per hour and continue this acceleration as long the sail is extended to catch photons; in just one day the vessel could gain a speed of 225 miles (362 km) per hour; in 12 days it could cover 1/2 mile (.8 km) per second. At that rate of acceleration, it could reach Mars in 400 days. A traditional rocket could beat the time to Mars, but the solar sail could haul far greater payloads.

The design and construction of solar sails is also relatively simple. The material used to reflect photons would probably be a thin layer of **aluminum** sprayed onto **plastic**, which would give it strength and form. There are only three basic configurations for a solar sail: square, disk, or heliogyro. The square design is similar to that of a kite, with stiff support pieces to give it shape. Both the disk and heliogyro spin to keep the material stable and rigid. The only difference between the two is that the disk is solid, while the heliogyro is made of rotating strips, much like the blades on a helicopter.

The size of these projected configurations is extraordinary. A square sail, to achieve the speed indicated above, would require an area of at least one square mile (2.6 square km). The heliogyro would need blades 20 miles (32 km) long and 30 feet (9 m) wide. Such a craft would be able to carry a payload about the size of the early Apollo command module.

Other variations in solar-sail design are being considered, such as **laser**s to provide propulsion, or beamed microwaves to provide great acceleration and speed.

In the late 1990s, solar sails became the subject of much research. The World Space Federation in the United States has built a prototype square sail with 90-foot booms. The French are also pursuing this idea: hoping to encourage more research and development in solar sailing, they are offering a prize for the winner of a solar sail race to the moon.

SOLAR TELESCOPE

Before the late 1800s, little was known about the three layers of the solar atmosphere: the photosphere, the chromosphere, and the corona. It was only during total eclipses of the Sun—a time period amounting to only a few minutes a year—that astronomers were able to even begin examining the details of solar phenomena. The chromosphere and the corona comprise the outermost layers of the Sun's total atmosphere. Activities and conditions of scientific importance, such as such as sunspots, prominences, spicules, and flares, occur in both areas. These conditions can be studied with a solar telescope, a reflecting (or refracting) **telescope**, modified with special spectral instruments.

Creation of the first solar telescope took place in two separate areas of the world: the United States and France. American astronomer George Ellery Hale (1868-1938) began developing the spectrohelioscope in 1889. Combining a telescope with a **spectroscope**, the spectrohelioscope separates the Sun's spectral rays into individual wavelengths. Chromospheric hydrogen, for example, shows up as red and ionized calcium as ultraviolet. By 1889 Hale began work on the spectroheliograph, an instrument to photograph solar activity.

About the same time in France, astrophysicist Henri-Alexandre Deslandres (1853-1948), worked to develop theories on and instruments for studying the Sun and its atmosphere. He built upon Hale's concepts and also created his spectroheliograph in 1894 to permanently record solar activity. Unfortunately, it was still nearly impossible to study the Sun's atmosphere during times other than an eclipse.

For this reason, in 1930, French astronomer Bernard Lyot built the coronagraph. This instrument caused the Sun to be artificially eclipsed by a blackened disk. The image of the corona was formed on a black screen slightly larger than the disk's image. The coronagraph also scattered the bright light of the sun so it wouldn't overwhelm the faint image of the corona. This coronagraph permitted far more precise measurements of the sun as well as the study of stellar composition and dynamics, including temperature and luminosity.

Lyot devised other methods of solar study. He combined the coronagraph with the spectroheliograph to photograph the Sun's two brightest spectral lines clearly over the entire solar disk; he developed the quartz-polaroid monochromatic interference filter system that filtered polarized light reflected from planet surfaces to allow rapid observations of prominences over the entire Sun (he also used this to take moving films of various wavelengths); and finally he developed the coronagraphic photometer, which allowed astronomers to trace weak spectral lines even with ordinary telescopes.

Since Lyot's time, other advances in solar study have been made. Modern solar telescopes, for example, include long-focus telescopes locked into position, allowing sunlight to hit a mirror and to then be directed to a measuring instrument or spectrograph; larger solar telescopes are frequently built as solar towers, in which the sun's rays are passed down the tower shaft and reflected into subterranean chambers housing measuring instruments. This cools and protects the instruments from the intense solar heat. One type of coronagraph was an integral piece of equipment on the 1974 mission of the United States' *Skylab* and provided researchers with 1,000 hours of data.

In December of 1995, NASA, in partnership with the European Space Agency, launched the Solar and Heliospheric Observatory (SOHO). Unlike previous solar telescopes, which orbited the Earth and thus suffered period interruptions in their view of the sun, SOHO has a 24-hour-a-day view of the solar disk. The telescope orbits a Lagrangian point in the Earth-Sun system, a region in space nearly a million miles from Earth, at which the gravitational attraction and centrifugal forces of the bodies cancel one another out. In this position, SOHO will maintain its position with respect to the Earth, observing the sun for at least two years.

SOHO carries a group of 12 instruments, designed to study the solar corona, the solar disk, and the solar wind. Developed and maintained by European and American scientists, the payload includes spectrometers, particle analyzers, telescopes, coronagraphs, oscillation measurement instruments, and Doppler imagers. Detailed and controlled studies done with these instruments will provide scientists with much valuable information for use in better understanding the intricacies of solar activity.

Solder and soldering irons

Soldering is one of the most accurate methods of fusing pieces of metal together. It is used in extremely localized operations and when visual appeal is more important than strength. Bolting and riveting metal do not create a complete seal and may spoil an object's appearance. Direct **welding** may damage the metal and alter the dimensions of the metal component by melting its edges. Soldering fuses pieces of metal by introducing a separate substance, or flux, that can be melted at a temperature lower than the host metal. The solder material may be of the same type of metal as the host metal with an **alloy** of a second metal such as tin to reduce the melting point. Because of its accuracy, soldering is popular among artisans in the creation of jewelry, dinnerware and small statuary. For larger works of art, such as bronze statues, where strength is required, welding is acceptable. Welding scars, undetectable from a distance, can be seen upon close examination of the joints.

Soldering was in use by 2500 B.C. when the Chaldeans at Ur soldered gold sheets together. It was also used in Egypt. In 1838 the Frenchman Desbassaynes de Richemond made the suggestion that a flame from a stream of oxygen could produce the localized source of high heat necessary for accurate fusing. This insight eventually led to the development of the electric soldering iron, which supplies the necessary heat without a flame.

For soldering to be carried out properly, the surface of the metal must first be cleaned. The flux material comes in rods, and the metallurgist holds one end while the other is applied to the joint. The alignment gap has a tolerance of no more than .005 inch. Heat is applied by electric current at the end of the soldering iron as the flux is fed into the joint. Finally, the joint is allowed to cool and the excess flux removed. Tin-lead solders are used in plumbing and electrical work. Zinc-chloride flux is used on copper alloys. Precious metal alloys using gold, silver and copper are used in jewelry, **clocks**, and other detailed work. Dip soldering is used in the **automobile** industry. The metal is fused by dipping it into molten flux.

As the toxicity of lead became more commonly known in the late twentieth century, different alloy combinations were experimented with to replace tin-lead solders, especially in plumbing and water supplies. In 1993, AT&T Bell labs found that when one percent zinc was added to a tin-silver alloy, the solder was as strong as a tin-lead solder. Another potential water line replacement was a tin-antimony alloy. A tin-silver-copper alloy, which is more durable than tin-lead solder, was developed for use in the automotive and electronic industries. More environmentally conscious technology was also being developed for fluxes. Fluxes were often cleaned with chloroflurocarbons (CFCs), which contribute to the destruction of the ozone layer. New technologies were developed that employed water-based fluxes, low-residue solder pastes, low-solids fluxes, and laser soldering.

Solenoid

A solenoid is device that uses the principle of an **electromagnet** to convert electrical current into mechanical movement. Most typically it is a coil of insulted **wire**, which, when electric current flows through the wire, has an induced magnetic field inside the coil. This field is used to magnetize a metallic armature or plunger inside the coil—an electromagnet.

Hans Christian Oersted's (1775-1851) discovery in 1820 that an electric current in a **wire** produced a magnetic field set off a flurry of activity among scientists.

In the forefront was **André Ampère** (1775-1836), whose experiments with electricity and magnetism showed that two parallel wires with electric currents flowing in the same direction attracted each other. When the current flowed in opposite directions, the wires repelled one other. He also experimented with magnetic fields that were created by current flowing in a circular wire and theorized that a helix of wires, or a wire coiled like a bedspring, would behave like a bar magnet, and consequently have a north and south pole. He called his helix a solenoid, but did not actually build the device.

The person who first did build a solenoid was William Sturgeon (1783-1850), in 1823. He took an iron bar, wrapped a wire around it eighteen times, and sent electric current through it. The result was remarkable; he had built an electromagnet that could lift nine pounds (four kilograms), or nine times its own weight.

Joseph Henry then improved on Sturgeon's design. By painstakingly hand-wrapping his wire in silk for insulation, Henry was able to increase the number of turns of wire in his coil. His electromagnet could lift 750 pounds (340 kg); a later design doubled that amount.

Henry also used his device to attract a small metal bar that used a spring to snap it back into place when the current was turned off— a relay. He set up a "relay" of devices, so one relay could operate another, thus remotely controlling a series of electromagnets. (This was the forerunner of the **telegraph**.)

When an iron bar is placed within a tube that has been wrapped with a coil of wire, it will move *within* the coil. When the current is turned on, the bar is pulled inward; when the current is switched off, a spring attached to the bar pulls it back out. The strength of the magnetic field within the coil is fairly constant, but at the ends it drops off to about half of the strength at the center.

This movement gives the solenoid the capability of controlling other circuits. It can operate relays, switches, shutters, **valves**, and numerous other spring-loaded devices. The starter of an automobile is actuated by a solenoid; turning the key energizes the coil, which pulls the solenoid in to complete the starting circuit, and the engine starts. The most common use of a solenoid in the home is found in doorbell chimes.

Solvay, Ernest (1838-1922)

Belgian chemist and industrialist

Ernest Solvay was born in Rebecq-Rognon, near Brussels, Belgium, to a family of industrial chemists. Although his only formal education was at local schools, young Solvay read and experimented widely in chemistry and electricity. He joined his father's salt-making business, and then, at age twenty-one, went to work with his uncle at a gasworks factory near Brussels.

Solvay began experimenting with the ammonia-soda reaction as a means of producing soda ash, which was then in great demand by the **glass** and **soap** industries. Most soda ash at the time was produced by the Leblanc process of converting common salt, a method which was expensive and created unusable byproducts. Solvay solved the practical problems of conducting the ammonia-soda process on a large scale, unaware that many chemists had tried and failed to do this over the past fifty years. He devised carbonating towers, which allowed large amounts of ammonia, salt solution, and carbon dioxide to be mixed; his process also allowed the recovery of expensive ammonia, which could then be reused.

Solvay patented his process in 1801 and founded a company with his brother, Alfred. They built a factory in 1863 and began commercial production of soda ash, using the Solvay process in 1865. Solvay protected his process with many patents, but offered other manufacturers licensing rights to use his technology. By 1890 he had established plants in many foreign countries.

By the end of the 1800s the Solvay process produced most of the world's soda ash. Solvay became very wealthy as a result, and he used his money to endow schools and, in particular, to found he Solvay International Institutes of Chemistry, of Physics, and of Sociology. These institutes held well respected international conferences; those on physics were particularly noted for their inquiries into atomic structure and quantum mechanics. Solvay died in Brussels in 1922.

Sonar

The concept of using sound to determine the depth of a lake or ocean was first proposed in the early nineteenth century by the French physicist Francois Arago (1786-1853); however, for nearly a century little research was conducted on the subject. Interest in underwater ranging was renewed in 1912, when the luxury vessel Titanic collided with an iceberg and ruptured. Experimenters then began to use *supersonic waves* in an attempt to detect submerged obstacles, but the majority of the scientific world was not committed to this research until two years later. On September 22, 1914, in less than an hour, a single German **submarine** sank three British cruisers carrying more than 1,200 men. In order to defend themselves against this new weapon, the British government funded a tremendous effort toward the creation of an underwater detection system.

At the time, the best idea concerning such a system came from the Russian engineer Constantin Chilowski. He designed a model utilizing *ultrasonic waves* that, though theoretically effective, did not produce a strong enough signal. In 1915 **Paul Langevin** helped Chilowski perfect his apparatus; he suggested using the principles of piezoelectricity that Langevin had learned while working with Pierre Curie (1859-1906), wherein a sample of quartz, if compressed, produces an electric charge, while a charge applied to the quartz makes it expand. With the quartz sample acting as both a source and receiver, a signal was emitted that could be received at a range of nearly two miles (almost 3 km).

Though the first working model did not debut until after the end of World War I, the entire operation was conducted in complete secrecy; in fact, the project operated under the code name *asdic* (which stood for Allied Submarine Detection Investigating Committee). The device retained the name asdic in Britain until the late 1950s, when the American term *sonar* (for Sound Navigation and Ranging) was adopted.

The principle behind sonar is simple: a pulse of ultrasonic waves is sent into the water, where it bounces off a target and comes back to the source. By observing the returning signal, the distance and location of the target can be calculated.

There are two types of sonar ranging: active and passive. The active method (also called echolocation) is as described above, and it is this type that Langevin and Chilowski first achieved. The passive method does not send out a pulse signal; instead, the searcher listens for the characteristic sound of another boat or submarine. This stealthy technique identifies the target without revealing the nature or location of the searcher, and it is this method that is most often used during wartime. However, since a submarine is usually completely submerged, it *must* use active sonar at times, generally to navigate past obstacles.

In 1915 Professor R. W. Boyle and the Royal Navy tested the new system, successfully detecting icebergs, reefs, and underwater topography. The perfected system was installed on all World War II submarines, usually with just one hydrophone listening for signals. In the 1980s, passive sonar became more sensitive. Scientists found that an array of hydrophones pulled by a surface ship gives sonar operators more directional control. With the advent and continual improvement of computers, sonar has become digitized, and therefore more easier to analyze. Early sonar was analyzed by listening; modern sonar is almost completely visual.

Since 1915, sonar devices have become standard equipment for most commercial and many recreational ships. Fishing boats use active sonar to locate schools of fish. The device is instrumental in the science of *acoustic fathometry*, wherein echo pulses are used to map the undersea terrain, and it is vital in navigating submerged vessels under the Arctic Ocean ice sheets. A newer technology, synthetic aperture sonar (SAS) can show images with 20 times the clarity of acoustic fathometry, by using a number of highly sensitive transmitters/receivers. Treasure hunters also use sonar to search for and within shipwrecks. A cane for the blind has been developed that uses sonar to help steer users clear of obstacles.

The future of sonar might not be passive nor active. In 1998 researchers from the National University of Singapore introduced a new type of sonar based on a dolphin's sonar, which identifies an object based on how it affects the ocean's natural ''white noise,'' such as waves, rain, and passing ships. Microphones are placed in different directions, monitoring a small area. The pick up a pattern of a target, and when processed on a computer, show the object. Other technologies that might be in sonar's future are **fiber optics**, sophisticated computers, and **artificial intelligence** systems that mimic the sonar of other animals such as bats. The sensitivity of traditional active and passive sonar will also increase.

SOUND MIXING AND RECORDING

Sound mixing and recording is the process by which music is recorded and prepared for release. Pre-1925 sound recordings were made acoustically with sound waves driving the process. Sounds captured by collecting horns directed a stylus or similar device to cut grooves directly into a disc. The disc, usually made of wax, was used as a master to make copies. Recording studios were cramped and heavily curtained, making them uncomfortable for musicians. In 1925, electronic recording was developed, and the first **microphone**s were introduced. Styluses could cut a wider range of frequencies into grooves when driven by electric current, but until the 1940s, when magnetic tape became widely available, recordings could not be edited or mixed. The recording process was also limited by the time available on each disk. It also took an enormous amount of time to set up a studio, and each master disk had to be checked for quality before another was recorded. An error would result in the whole process starting over again. Sound quality was also not great. Indeed, until the introduction of stereo in the 1950s, recordings could not reflect the depth and breadth of sound.

Before magnetic tape, there were efforts as early as 1898 to make **magnetic recordings** on steel **wire** (though no reliable machines existed until 1912), and steel tape. Magnetic wires found limited use through World War II, but by war's end, magnetic tape technology had come of age. A resilient tape, one coated with magnetic oxides, became available. Magnetic tape recordings are made by storing sound wave patterns as analogous variations of magnetic flux. Magnetic tape proved much more flexible than discs in the recording process. The tape length far exceeded anything possible with discs. Tape could be easily started and stopped. Mistakes were easily erased and redone. Tapes could be edited. The best part of a performance take could be saved, and joined to other superior takes. Indeed, all the musicians did not have to be in the same studio at the same time to record as discs required. Tapes could be recorded one place and edited together at another.

Early magnetic tape recordings were single track affairs. In the late 1950s, multitrack tape recording was introduced. Multitrack recording involved the creation of individual tracks on different tapes. Individual instruments could have their own microphone, resulting in their own track. With individual tracks available, mixing became possible. Tapes could be mixed on mixing console. In mixing, the relationship between sounds can be changed. Sounds could be optimized, and texture, balance, and spatial relationships could be composed. Musicians had the means to make their intentions clearer. Effects became possible, as well as overdubbing and other studio wonders. By 1970, 24-track recording was possible. One major problem of magnetic tape recording was solved in this time period: tape noise. Noise reduction systems were introduced in the mid-1960s, and the most successful ones were invented by Ray Dolby.

In the mid-1970s, digital recording came to fruition. Digital recordings are made by regular sampling of sound waves, and converting them into a number sequence. This data can be stored on magnetic tape or computer. When digital recordings are stored on **computers**, recordings can be instantly, easily played back without problems of tape speed fluctuation or background noise. Indeed, when digital recordings are stored on discs, moments can be more easily and precisely pinpointed than on tape. Digital recording also offers more sound effects and sound modification possibilities, as well as more mixing possibilities. With the advent of the compact disc, a totally digital means of hearing music, on the consumer market in 1980s, digital recording became dominant.

As technologies like DAT (digital audio tape) players and recorders become more inexpensive and portable available, musicians can record and mix tapes as good in quality as those made in a recording studio. Personal computers also have recording and mixing capabilities via sound cards and special software. They also can be used in home recording. In the future, computer's role in sound mixing and recording will continue to grow.

See also Compact disc player; Computer input and output devices

Soybean products

The soybean, or Glycine max, is a cultivated summer annual not found in the wild, and a member of the legume family. A staple food that is grown around the world, the plant has been labeled the world's most valuable bean.

Asian cultures in particular have developed the use of soybeans. In fact, cultivation of the soybean probably began in China before recorded history. There it was revered as a healing and purifying agent and counted among the five sacred grains alongside rice, wheat, millet, and barley.

Many foods are created by soaking, cooking, grinding, creaming, aging, and extruding the soybean. The Japanese, for example, use soybeans as the base of shoyu or soy sauce to flavor meat and vegetables. The salty, aromatic liquid is made from fermented soy paste and cereal. A similar emulsion, miso, combines soybeans and rice for a soy paste essential in pickling and in soup base. Other emulsified, fermented soybean products from the Orient include *tempeh*, *ontjam*, *natto*, *hamanatto*, *tao tjo*, *kochu chang*, and *ketjap*.

Not all soybean products are fermented. Soybean milk, a vegetable milk made by soaking, grinding, and heating the beans, then filtering them through a mesh, is sold as a flavorful, nourishing beverage. It has been used in the United States as a substitute for cow's milk in lactose-intolerant infants and adults.

Tofu, which is steamed and formed into blocks on a wedge press, resembles cheese and is used in vegetarian recipes as a meat substitute. It retains its shape and can be chopped and sauteed like meat. The soybean eventually reached the Western world by the eighteenth century, when European farmers grew them as a cheap source of protein, vitamins, and minerals.

Benjamin Franklin brought the bean to the United States from the Jardin des Plantes in Paris sometime before 1804, but because of its disagreeable odor and taste, it remained a novelty crop despite its benefit. It was not until the end of the nineteenth century that the soybean was first seriously introduced to American farmers by the Department of Agriculture. At that point the bean developed into a major agricultural commodity for Nebraska, Iowa, Kansas, Arkansas, Illinois, Minnesota, Indiana, Missouri, Ohio, Mississippi, Tennessee, Louisiana, South Carolina, and Delaware, thus earning its title as America's "Cinderella Crop."

Adopted by makers of vegetable oil as a stable, unsaturated source of raw oil, it was subjected to hydraulic and screw presses in oil mills. By the 1930s, these primitive devices gave place to countercurrent solvent extraction, a German innovation, as a more streamlined method of processing the bean.

Another boon to the soybean's popularity in the United States was the boll weevil. In the 1930s, the cotton-based economy of the South was devastated by the animal's fondness for its chief product. The soybean provided an alternative crop to cotton. In fact, soybeans became such a part of Southern life that developers named new strains after Confederate heroes, including Lee, Davis, Pickett, Jackson, Rebel, Forrest, and Bragg.

The soybean grew in importance as American and Canadian researchers advanced its usability. Many uses were explored. In the 1930s, while metal was scarce, **Henry Ford** attempted to mold a car frame from reinforced phenolic resin, which was synthesized from soybean oil. As a result of this research and development, by the late 1930s, the United States became the world's top producer of soybeans. Over a fifty-year span, soybean cultivation increased from 13.9 million bushels in 1930 to over 2 billion bushels in 1980. Other countries, particularly Central and South America and the Caribbean, observed the United States' success with the soybean and sought similar results. Thus cultivation of the crop spread even further. Part of the reason for the popularity of soybeans is the crop's versatility.

Before processing, soybeans can be used as hay, forage, and fertilizer. The food industry uses soybeans in meat, coffee, milk and cheese substitutes, flour, **margarine**, whipping agents, shortening, paste, confections, and sauces. They are served as snack food, sprouts, baked beans, curd, and dietetic foods. Many packaged foods, including grits, flakes, formula and **baby food**, **pasta**, frozen desert, sardine packing, ground meat casings, seasonings, soups, mayonnaise, cheese, and flavorings are based on soybeans.

Soybeans are also useful in processed feed for cattle, swine, and poultry, and as a green manure or cover crop for soil enrichment. While 90 percent of the soybean crop is used for food, industry adapts soybeans as a raw material in many items, including **linoleum**, rubber substitutes, glycerin, **soap**, **plastic**, **paint**, enamel, **varnish**, oilcloth, stain sealer, caulking compound, waterproofing, **insecticide**, pharmaceuticals, lubricants, disinfectant, adhesive, textile fiber, and waxing agents. It is an essential ingredient in plywood glue, coatings, fire-fighting foam, **cosmetics**, emulsifier, and printing ink. The diversity of products that are made from the soybean have clearly made the crop one of the world's most valuable beans.

In recent years, soybeans have been **genetically engineered** and concerns have surfaced among public interest groups and some scientists about the safety of these products for public consumption and their environmental impact. As always, scientists and the larger community must weigh genetic improvements against environmental ills.

Space equipment

When humans go into space, they face conditions far different from those on the earth. There is no air, and the gravitational forces experienced by the astronauts will be quite different than those they experience on Earth. All food and water must be carried along. Temperatures vary greatly. Astronauts have to rely on sophisticated equipment to enable them to survive this hostile environment.

From the first *Mercury* flight on February 20, 1962 to recent **space shuttle** missions, scientists have endeavored to design **space equipment** that makes orbital flight possible for human beings. During the Mercury missions, astronauts depended on a dual life support system, one subsystem to maintain temperature and pressure for the spacecraft cabin and the other for within the pilot's suit.

The early spacesuit was bulky, its many layers prohibiting easy movement by the astronauts. Air was propelled through the suit, exiting the helmet and passing through a charcoal bed and a lithium hydroxide-carbon dioxide scrubber. The gas was then cooled through evaporation and fed back into the suit. In case of sudden leakage or overconsumption, oxygen from a pair of high-pressure bottles could be utilized.

For the *Gemini* spacecraft, scientists were determined to improve the flow of water vapor and their control of the temperature. To this end, astronauts used **fuel cell**s which produced water as a by-product. This helped to prevent dehydration, which had been a major problem on the Mercury missions. In addition, temperature was more efficiently controlled through liquid coolants, electronic equipment and fuel cells.

The *Gemini* spacesuit was more advanced, sporting a thinner protective layer, multiple visors and greater mobility for the astronauts. Oxygen entered at the waist, circulated, and then exited at the waist; the oxygen supply originated from a cryogenic container, much lighter than pressurized bottles. Because improvement in cabin life-support was significant on the *Gemini* missions, crews could remove their suits for a period of time and work in long underwear.

The *Gemini* crew was the first to leave the spacecraft and do extravehicular work, and an umbilical hose was set up in order to transport oxygen to the astronauts. Unfortunately, this system was easily overtaxed and proved inadequate in the long run.

The life-support systems for the Apollo missions were similar to those of the *Gemini*. Fuel cells provided water, lithium hydroxide removed carbon dioxide, and oxygen was transported in cryogenic canisters. Two types of spacesuits were designed, one for inside the spacecraft and one for lunar exploration. The best-known suit was the one worn on the moon by the Apollo astronauts. It had twenty-one layers: the outer one was made of **fiberglass** fabric, while the next was a rubber pressure-tight layer. Air pressure was kept up by tubes that carried oxygen for breathing as well. Water tubes threaded through the underwear to cool the person. Both sets of tubes were connected to the backpack, which stored water, oxygen, and control equipment. Under it all the astronaut wore a urine-collection assembly. The helmet had a thin gold coating to act as a sun visor. These suits weighed 57 pounds (26 k) on Earth, but only 1/6 as much on the Moon, due to its weaker gravity. It took one hour just to get into the complicated device. The suit worn inside the craft was only six layers, with an outer layer made of fireproof **Teflon**.

The suits worn by the space shuttle crew are the safest and the least constrictive of any spacesuit design: liquid-cooled underwear, pants and boots with five layers of flexible metallic **mylar** for heat reflection and micrometeorite protection, pleated fabric in places where a person needs to bend or walk, a hard-shelled torso unit that connects via a metal ring to the pants that allows for limited turning at the waist, helmet and gloves attached through intricate locking devices to prevent them from accidentally coming loose. They are the result of years of research and testing, and are significantly advanced from the bulky, oppressive suits of the 1960s. Inside the spacecraft is equipment necessary for drinking, eating, sleeping, going to the bathroom, and exercising.

Before the early Mercury flights, there was much of worry about whether eating was affected by zero-gravity. Scientists didn't know whether people would be able to swallow solid food, or if it would perhaps get stuck in their throats. As a result, the early astronauts ate pureed foods, such as applesauce, out of squeeze tubes. In later Mercury flights, after scientists realized the only danger was keeping food on the spoon, astronauts ate from plastic sacks of freeze-dried foods that could be injected with water. As interior space increased, the crew ate better and more varied meals. For example, shuttle crews now have eggs and bacon for breakfast, spaghetti and meatballs for lunch, and barbecued beef for dinner.

The early astronauts had to use bags for urinating and defecating. Now, at the cost of three million dollars, the shuttle has a space toilet that looks a little like those on airliners. The differences include foot restraints, handholds, and a waist restraint which helps maintain a seal between the user and the seat. It has been designed for both male and female crew members. Located in a central tube under the seat, a fan with blades provides a downward air flow that forces the fluids and solid wastes into collection receptacles.

Skylab was equipped with a shower in a folding stall that relied on a vacuum system to dispose of the water, but it was difficult to find uses for. The shuttle, since it flies for a much shorter time, has no shower facilities, so crew members take sponge-baths instead, reminiscent of the Mercury, Apollo and Gemini missions.

Today's larger spacecraft have enclosures much like sleeping bags for the crew to use so they can sleep unstrapped without floating. Exercise on long missions is accomplished with treadmills, stationary **bicycle**s and special exercises that stimulate the circulation of blood through the legs. Today, equipment onboard the space shuttle is much more sophisticated and conducive to human inhabitation than on the early Mercury missions.

As space technology advances and flights travel further and last for longer periods of time, scientists will have to design devices and equipment that protect the astronauts as well as enhance the space experience for human beings. This long-term goal has led to a number of orbiting space laboratories, used both for scientific research as well as to study the effects of long-term existence in the space environment on humans. The first such station was NASA's *Skylab*, launched in 1973. It provided scientists with their first glimpse of the effects of prolonged weightlessness. In 1986 the Russians launched their space station *Mir*, which in Russian means both "world" and "peace." This proved to be an extremely durable station, remaining operation well past its design lifetime into the late 1990s. American astronauts were able to participate in the Mir program through several dockings with the Space Shuttle.

In the mid- to late-1990s, increasing effort was devoted to the initiation of the International Space Station (ISS), spearheaded by the United States and Russia, perhaps the most complex piece of space equipment yet devised.

Space probe

The term *space probe* refers here to those spacecraft that have escaped earth-orbit. The former Soviet Union was first in this area, as they had been with other space shots. In 1959 they launched *Luna 1*, which missed the Moon by 3,728 miles (5,998 km). They followed that flight with a spectacular circumlunar orbit by *Luna 3*, which gave us our first pictures of the far side of the Moon.

The development of probes in the United States also revolved around the Moon at this time. After several unsuccessful attempts to reach the Moon with the Pioneer series, the National Aeronautics and Space Administration (NASA) launched the Ranger series. It planned to crash-land the spacecraft onto the moon's surface, taking photos up until impact. The first few probes were unsuccessful, but the last three—*Ranger 7*, *Ranger 8*, and *Ranger 9*—took over 17,000 pictures beginning in 1963.

The former Soviet Union scored another first in 1966 when it soft-landed *Luna 9* on the Moon's surface, which carried television pictures for three days. The Americans followed suit with their *Surveyor* series, soft-landing a capsule four months after *Luna 9* had landed. The *Surveyor* boasted more sophisticated landing ability and sent back over 11,000 pictures.

During this time the two nations launched missions to Venus and Mars. One great success was the United States' *Mariner 2*, which came within 21,594 miles (34,745 km) of Venus after its launch in 1962. *Mariner 4* returned the first close-up photographs of Mars when it approached within 6,118 miles (9,844 km) of the red planet in 1965. The former Soviet Union focused more of its missions on Venus, parachuting capsules into the harsh atmosphere where temperatures were hot enough to melt lead and obtaining some readings before the pressure destroyed the crafts. *Venera 7* actually landed on Venus in 1970 and became the first probe to send back information from a planet. *Venera 9* and *Venera 10* transmitted television cameras that sent back startling photos of a rocky, harsh world. In 1973 the United States launched *Mariner 10*, which photographed the clouds of Venus before sending back the first pictures of the moon-like Mercury.

Mars was the prime target for many probes. *Mariner 6* and *Mariner 7* flew within 3,400 miles (5,470 km) of Mars in 1969, showing a surface of craters. One of the most successful missions was *Mariner 9*, which settled into orbit around Mars in 1971. Unfortunately, a severe dust storm initially hid the surface, but the probe eventually sent over 7,000 pictures of the planet, which revealed huge canyons that dwarfed the Grand Canyon, monstrous volcanoes, and what appeared to be dried-out beds of ancient rivers.

The Viking program launched by the United States was a highly successful Mars mission. Consisting of an orbiter and a lander, *Viking 1* and *Viking 2* reached the red planet in 1976, sending back thousands of high-quality photographs in color, back-and-white, and stereo. The landers scooped up Martian soil to analyze for any biochemical reactions that would indicate the existence of life (there was no clear evidence of organisms, past or present). They made daily weather reports, allowing scientists to understand weather patterns better. Overhead, the *Viking* orbiters mapped much of the Martian surface and probed its surface with sensitive instruments.

Two other American probes headed for deep space at this time. *Pioneer 10* and its sister craft *Pioneer 11* took close-up photos of Jupiter and its famous Red Spot. *Pioneer 11* went on to Saturn, where in 1979 it was able to help determine the small size of the particles making up the famous rings surrounding the giant planet. After its encounter with Jupiter, *Pioneer 10* continued into interstellar space, the first man-made object to leave the solar system.

Probably the most famous space probes were the two *Voyagers* launched by the United States in 1977. The crafts had several highly sophisticated scientific instruments and two television cameras. *Voyager 1* passed Saturn, then left the solar system for interstellar space, while *Voyager 2* continued on, exploring Uranus and Neptune. The *Voyagers* sent back the first clear photos of Jupiter's Great Red Spot and of Saturn's rings. *Voyager 2* also sent back new information regarding the motion, composition, and origin of Uranus and its moons. *Voyager 2* visited Neptune, the final encounter, in 1989, twelve years after its launch, discovered new moons, rings, arcs within rings, intriguing cloud features, and an unexpectedly skewed magnetic field.

To get a better idea of what Venus was really like, the United States launched *Magellan* from the space shuttle Atlantis in 1989. *Magellan* was able to penetrate the thick clouds of Venus because it had a radar mapper rather than a visual camera. It returned pictures of impact craters and volcanoes, improving our understanding of geologic history of the planet. *Magellan* ended its mission in a blaze of glory, as mission controllers braked it and plunged it into the Venusian atmosphere, returning on last set of data about Venus's atmosphere before being destroyed by the intense heat and pressure of Venus's hostile environment.

After many delays, the American probe Galileo was launched to explore the asteroid Gaspra, and then to fly on to its main target, Jupiter. Because the United States had to cancel a powerful upper stage for the Centaur rocket, *Galileo* left the Earth in 1989 on an odd trajectory. First, it swung by Venus for a gravity-assisted velocity boost, then passed Earth in 1990. *Galileo* encountered Gaspra in late 1991 and will swung once more past earth in late 1992, at which time it will have gained enough velocity to slingshot out to Jupiter. It arrived near Jupiter in December 1995, where it dropped a probe into Jupiter's clouds and continued to circle the giant planet, providing revolutionary images of the giant planet and many of its moons.

Mars became a favorite target for space probes again in the 1990s, although success was a while in coming. In 1993, NASA's billion-dollar *Mars Observer* made the trip to the red planet—and disappeared from radio contact, never to be heard from again. In 1996, the Russian-lef *Mars96* ended even more disastrously, crashing into the Pacific Ocean shortly after launch. Finally, in 1997, NASA's *Mars Pathfinder* landed on Mars, and deployed the robot crawler *Sojourner*. *Sojourner* was the first mobile craft ever deployed to another planet, and for several publicity-crazed weeks in July 1997, it implanted itself almost as firmly in the American popular culture as the fictional R2-D2 of *Star Wars*.

Launched in 1977, the interplanetary probe *Voyager 2* explored the regions of space around Jupiter, Saturn, and Uranus

In the mid-1990s the Moon also became the target of unmanned space probes. In 1994, the *Clementine* probe carried out a topographic survey of the entire Moon, and in early 1998 the *Lunar Prospector* probe made the startling discovery of water on the Moon, located in the shadowed regions of some craters near the lunar poles. The presence of water on the Moon is an enormous logistical aid for future permanent colonies there.

Unmanned spacecraft continue to explore the outer solar system. On October 17, 1997, the *Cassini* spacecraft was launched, scheduled to arrive at Saturn in 2004. It carries a formidable array of experiments and deployable probes, and is expected to revolutionize our understanding of Saturn in the way that *Galileo* did with Jupiter. Some controversy surrounded the launch of *Cassini*, due to its use of plutonium as a power source for its time in the outer solar system (plutonium was not, and could not, have been used as a launch fuel), but other craft, such as the Voyagers, have also employed plutonium, and great care was taken to ensure that the toxic element was safely packaged, even in the even of a lauch failure.

A vigorous program of unmanned space probes remains in place as the twentieth century draws to a close. Missions to Jupiter's moon Europa, as well as the most distant planet, Pluto, are planned. As the Voyagers and Pioneers head outward into interstellar space, the next generation of craft with follow them, furthering our understanding of the small corner of the Universe we inhabit.

SPACE SHUTTLE

About the size of a DC-9 commercial airliner, the shuttle is a high-tech, **rocket**-boosted space craft. The shuttle is launched vertically like any other rocket, using its own engines firing along with two attached solid rocket boosters that are jettisoned and recovered for re-use. While in orbit, the shuttle can use its own rocket motors for changes in direction. For reentry, the shuttle turns around, brakes with its engines, and descends as a glider landing on special three-mile long runways in Florida or California like an **airplane**.

The shuttle was envisioned as a cheaper way to transport material and satellites into orbit for military, scientific, and large-scale commercial operations in space as well as the construction of a permanent space station. The concept of a craft with the characteristics of both a **rocket** and an airplane has roots in World War II. The Germans had planned a rocket-boosted, winged bomber that could strike America in World War II, but lost the war before such a vessel could be built.

The technological competition between the Soviet Union and the United States that began in the wake of the war fueled the innovations that would eventually lead to the production of a hybrid airplane-spacecraft. In the late 1960s and early 1970s NASA and the U.S. Air Force worked on sending aloft reduced-wing or wingless aircraft designed to generate less friction than conventional airplanes as they traveled high in the atmosphere. The United States gained much information about high-altitude flight at supersonic speed with its X-series of planes, starting with the X-1, the first plane to break the sound barrier, and continuing with the X-15, which reached speeds of over 4,000 miles per hour (6,436 kph) and a height of over 300,000 feet (91,500 m).

In 1969, a NASA group headed by **Maxim Faget** began planning a reusable spacecraft. Since the rockets used at the time could be launched only once, every space flight involved a tremendous waste of resources and materials. An associate of Faget's, Robert Gilruth, suggested the idea of a shuttle to Faget, a flexible and innovative designer. Within two weeks, Faget had fashioned a crude prototype model. With Faget's matter-of-fact announcement, "we are going to build the next generation spacecraft," a team of 20 top scientists began fleshing out the space shuttle's design. Faget preferred a straight-wing design which could land at any major airport, but secretive military officials rejected his idea, favoring the use of special military airfields. The group settled on a delta-shaped wing for greater efficiency high in the atmosphere. Due to budget constraints, an expendable tank and solid-propellant boosters were chosen, which would later cause the *Challenger* disaster. The result of the scientists' work was a craft resembling a stocky airplane 122 ft (36.9 m) long, with a wingspan of 78 ft (23.6 m).

NASA had originally envisioned the shuttle as a way to ferry personnel and supplies to a space station, but America's love of space travel was waning due to a weak economy, so a number of other advantages were touted. The shuttle was sold as a money-saving, reusable replacement for NASA's major launch vehicles. In addition, NASA said, a successful shuttle could fly missions under contract to industries and other nations for a fee, as well as helping to establish industrial operations in space where zero-gravity conditions allow precision manufacturing.

The first shuttle, named *Enterprise* for the ship of the same name in *Star Trek*, started tests in February 1977. Never intended to reach orbit, the *Enterprise* proved that a shuttle could glide safely to a landing on Earth. It also became the heaviest glider ever flown at 150,000 pounds (68,040 kg).

In April 1981, the *Columbia* became the first shuttle to orbit the earth. The only problem the shuttle encountered was the loss of some of the silica fiber tiles used to protect against the heat of re-entry. None were crucial, and the *Columbia* returned to earth safely 54 hours after liftoff.

From April 1981 to January 1986 the shuttle fleet—*Challenger*, *Discovery* and *Atlantis*—flew 24 consecutive successful missions. During these flights the shuttle delivered 28 non-military satellites to orbit, 24 of which were commercial communications satellites. *Spacelab*, a European built module containing equipment for scientific experiments, was carried in the cargo bay on four flights. The shuttles made use of a remote arm that could manipulate large objects either to lift them out of the cargo bay or to retrieve them from space. The astronauts were able to use a manned maneuvering unit (MMU) which allowed them to fly unattached into space near the shuttle. With these devices, crews were able to retrieve four disabled satellites, repair two, and retrieve two more for ground repair and future re-launch.

However, the program encountered disaster on the cold morning of January 28, 1986. *Challenger* exploded 73 seconds after launch due to a faulty seal in its solid rocket booster. All seven crew members died as a result. The fleet of shuttles was grounded for 32 months while over 400 changes in the shuttle's construction were made. NASA then put the shuttles back in flight. A replacement for *Challenger*, called *Endeavor*, was built, but no additional shuttles are planned.

In the 1990s the shuttles continued regaular operations, with over 90 missions flown by 1998. A significant international effort in 1997 and early 1998 involved the docking of several shuttles with the Russian space station *Mir*. In December 1998, the first female commander of a mission, Eileen Collins, is expected to lead the crew of the 93rd shuttle flight. Although the shuttles never achieved the financial benefit touted by NASA in the 1970s, they nevertheless hold a unique place in the history of aviation and space flight, and represent a significant development in mankind's presence in space.

Space station

In the wake of the successful American moon landings of the late 1960s and early 1970s, the Soviet Union moved to pioneer manned space stations ahead of the Americans. They launched the first such station, *Salyut 1*, in April 1971, followed by its first crew—Georgi T. Dobrovolsky (1928-1971), Viktor I. Patsayev (1933-1971) and Vladislav N. Volkov (1935-1971)—aboard the *Soyuz 24* in June.

Salyut was constructed from four cylinders of different diameters. At the forward end was a docking port and transfer compartment linking the station with the *Soyuz 24* ferry craft. *Soyuz* had a set of solar panels to collect energy for the station's power. Its length was 47 feet (14.3 m), and its largest diameter was 13 feet (4 m). The crew of the *Salyut* carried out astronomical research, plant growth experiments, and earth observations. Unfortunately, the mission ended in tragedy. As the crew headed back to earth, the seal of their *Soyuz* capsule broke and, without pressurized space suits, the three cosmonauts died instantly.

Satellite view of the space shuttle *Challenger* with its cargo doors open.

The United States had also been planning a space station and, following the initial flight of *Salyut*, hurried the construction of their vessel, called *Skylab*. Perhaps wisely, engineers decided not to start from scratch but rather to use the shell of an Apollo rocket tank as a ''skeleton.'' *Skylab* was a two-story complex with enough room for three men. The bottom section consisted of a ward room, sleep compartments, and a washroom or toilet. The upper deck contained a spacious workshop for the astronauts to carry out their assignments. It had about the same space as a small three-bedroom house. Large solar panels similar to those found on *Salyut* powered the station, and a thin aluminum shield protected against micrometeorites and excessive solar heat.

Soon after launch on May 14, 1973, *Skylab* encountered trouble. Part of the shield tore off, carrying with it one of the solar panels. The station was now seriously underpowered and was vulnerable to overheating from the sun. Eleven days later, a three-man crew was launched to meet the station. They first had to work outside the station to repair the damage to the solar panel and to deploy a hastily engineered sun shade. They were then able to carry out their original mission. For the next 28 days they observed the sun, photographed the earth, and carried out biomedical experiments.

The second mission to *Skylab*, launched in July 1973, was also successful. The crew conducted a large group of experiments in biology, space medicine, solar physics, and astrophysics over a period of 59 days.

The final crew, which boarded *Skylab* in November the same year, broke the records of the previous two flights. The astronauts spent a great deal of time on research (including observation of a solar flare and a comet) and performed physical exercise on a stationary bicycle and portable treadmill. After 84 days in space, the astronauts remained fairly healthy.

In 1979, *Skylab* met its fate. As its orbit decayed, plans were made for the United States' new **space shuttle** to attach a booster rocket to the *Skylab*, carrying it to a higher orbit. But before the shuttle was ready, *Skylab* plunged back into the atmosphere in July and burned up.

Undaunted by the initial American success, the Soviets persisted. After two more disheartening failures, the Soviets launched Salyut 3 in June 1974. A crew visited it for 16 days, but due to docking problems with a later Soyuz, the station was not visited by any more cosmonauts. The Soviets hit their stride beginning in 1974 with a number of increasingly ambitious Salyut missions. The later Salyut stations had more solar power for further experiments and longer missions. In addition, the Soviets perfected the resupply of Salyut by unmanned, automatic vehicles. These robot ships brought fresh supplies of air, food, water, and other consumables, as well as scientific equipment. The *Salyuts* also had re-startable engines

***Mir* in 1997.**

to push them into higher orbits when necessary. These improvements allowed the cosmonauts to establish a string of duration records. One crew stayed in the station for 237 days in 1984.

A new space station, *Mir*, was launched in February 1986. It had more docking ports and improved crew accommodations: separate compartments with table, chair, and intercom. The same year, the Soviets carried out the first successful transfer of cosmonauts between stations, from its *Salyut 7* to the *Mir* space station. In three years, the *Mir* cosmonauts completed nine EVAs or "space walks" and carried out a host of scientific experiments. Future modules that can be added to *Mir* are anticipated, but the breakup of the Soviet Union has put a halt to manned Russian space flights for the immediate future.

Mir lasted well beyond its design lifetime, with crews still inhabiting it in 1998. American astronauts joined the Russians through a series of *Mir*-Space Shuttle dockings. This entire period was puncuated with moments of nervousness, as the old space station began showing signs of its age. Equipment failers, power losses, and computer crashes began to plague the station, and in June 1998 the United States retrieved its last astronaut from *Mir*.

The latest, most ambitious space station yet conceived was still on the drawing board as *Mir* neared the end of its service life. The International Space Station, scheduled for initial in-orbit construction in late 1998, with completion in 2004.

Space telescope

The Earth's atmosphere provides a layer of protection and promotes life on the planet. Unfortunately, it also acts as a barrier, preventing us from easily studying and observing the universe. The images gathered by **mirror**s or **lens**es ripple and shake due to atmospheric turbulence. In addition, our atmosphere shuts out most radiation (other than visible light and radio waves), so many wavelengths cannot be measured from Earth itself. For example, scientists have learned that a majority of objects in space—from cold planets and stars to extremely hot galaxies and quasars-emit infrared rays. Yet this radiation is absorbed by water vapor in the atmosphere, and most ground **telescope**s are unable to gather these important data.

To minimize atmospheric distortion, telescopes have traditionally been located on mountains where the atmosphere is thinnest. This solution, however, remains extremely limiting. To observe different celestial radiation types and to overcome atmospheric interference, scientists have developed space telescopes. In 1983 Great Britain, the Netherlands, and the United States shared in the development of the Infrared Astronomical Satellite (IRAS). Launched on January 25, 1983, IRAS operated for only ten months, but it surveyed almost the entire sky. It discovered many galaxies glowing with infrared radiation barely detectable with Earth-based optical telescopes. The **satellite** also collected an immense catalog of data about the infrared universe. The European Space Agency (ESA) hopes to initiate additional studies by launching the Infrared Space Observatory (ISO) during this decade.

Since ultraviolet, x-ray, and gamma ray radiation is readily absorbed by the atmosphere, observations must be made from space. Astronomers are keenly interested in ultraviolet radiation. In the 1970s Apollo and Skylab carried small ultraviolet telescopes into space. The International Ultraviolet Explorer (IUE) was launched in 1979, and it conducted some of the best ultraviolet work to date. It was used to study supernovas as well as to produce very high quality photographs of spectra.

X rays and gamma rays are also of major interest to astronomers. X rays provide information about the violent processes responsible for the creation and destruction of stars. Several small satellites launched during the 1970s looked closely at x-ray and gamma ray radiation, revealing hundreds of previously unknown sources, including at least one black hole. Beginning in 1977 and throughout the rest of that decade, several High Energy Astrophysical Observatories (HEAO) were launched, carrying an array of x-ray and gamma ray detectors which provided readings on even the weakest sources of radiation. The Small Astronomy Satellite (SAS) was also launched in the late 1970s to detect gamma ray sources in the Milky Way (primarily along the outer edge). Some of these sources can be identified with mature supernovas.

The 1990s have seen increased activity in exploration of the universe. In 1990 the ESA launched ROSAT, a satellite designed to capture sharp x-ray images and to detect undiscovered ultraviolet wavelengths. Since its mission began, ROSAT has uncovered what could be a new class of very bright stars and supernova remnants. The National Aeronautics and Space Administration (NASA) has also initiated its own programs: the Gamma Ray Observatory (GRO); the Advanced X-Ray Astrophysics Facility (AXAF); the Space Infrared Telescope Facility (SIRTF); and, perhaps the most well-known, the **Hubble Space Telescope (HST)**.

The HST was deployed by a **space shuttle**. Developed by NASA, it has a mirror nearly 8 ft (2.4 m) in diameter and is able to probe more deeply into the universe than the strongest ground-based telescope. It can detect stars 50 times dimmer than those that can be photographed from Earth. Unfortunately, a slight imperfection in the shape of the primary mirror has resulted in a decrease in resolution level. The HST also experiences instabilities in each orbit (it circles the Earth every 96 minutes) which jar and shift its delicate instruments. Both of these problems will be corrected in future shuttle service missions. Regardless, HST has been an invaluable source of information for astronomers—it is providing information on galaxy collisions and newly-discovered black holes, and is photographing infant stars.

In early 1997, astronauts from the space shuttle *Discovery* installed two new instruments in the HST, the Near-Infrared Camera and Multi-Object Spectrometer (NICMOS), and the Space Telescope Imaging Spectrograph (STIS). The NICMOS is a camera capable of studying and imaging the universe at discrete infrared wavelengths. The NICMOS allows the Hubble to see within the dust clouds that surround objects near the center of the galaxy, and has already enabled astronomers to discover the most massive star ever observed.

The STIS captures an infrared scene and images it at a series of discrete wavelengths, creating a series of images, each at a different infrared "color." Such data sets can allow scientists to extract details of chemistry, density, and velocity of stellar objects.

Even as the HST is finishing out its useful lifetime (the telescope was designed to be operational through 2003), scientists are already designing its replacement. According to current plans, the Next Generation Space Telescope (NGST) will be a near-infrared instrument designed to allow scientists to study the birth of the first stars and galaxies. Fitted with a 4.36-8.72 yd (4-8 m) primary mirror and a wide-field camera and spectrometer, the NGST will offer an angular resolution on the order of the HST's, with high-sensitivity, low-noise operation over a wide field of view.

Spacecraft, manned

During the late 1950s the Soviet Union and the United States confined their Cold War competition for dominance in space to unmanned satellites. The Soviets successfully launched the first satellite, **Sputnik 1**, on October 4, 1957, and the U.S. quickly answered the challenge with their own *Explorer 1.* President Dwight D. Eisenhower, impressed by the success of *Explorer 1* and determined that this new technology should remain outside the control of the military, sponsored the creation of the National Aeronautics and Space Administration (NASA). However, he remained unconvinced that humans belonged in space and, despite the arguments of such eminent scientists as **Wernher von Braun** to the contrary, believed that unmanned space flight should be NASA's primary pursuit.

Yuri Gagarin

Unlike the United States, the Soviet Union actively pursued manned flight. In 1959 and 1960 a design team headed by **Sergei Kolorev** constructed several prototypes of the first manned spacecraft, Vostok ("East"). Soviet scientists also launched several live dogs in Sputnik satellites to test the possible stresses that humans might undergo in space. The most famous of these canine travellers was Laika, the first living creature to ascend into space. Finally, on April 12, 1961, *Vostok 1* lifted off from the Baikonur Cosmodrome carrying Yuri Gagarin, a Soviet Air Force test pilot and the first astronaut. He circled the planet for 108 minutes then reentered Earth's atmosphere. At about 4 miles (6.4 km) above ground, his ejection seat propelled him from the capsule, and he parachuted to the ground.

Project Mercury

Gagarin's flight galvanized the American public and the new administration of John F. Kennedy. Concerned that the United States' leadership in technology had been compromised, President Kennedy announced the goal of landing an American on the Moon within eight years. As a result, the manned Mercury space program, approved in 1958 after much controversy, was accelerated. Commissioned by NASA prior to Gagarin's flight, the Space Task Group at Langley Research Center had modified the military's *Redstone* and *Atlas* **rocket**s for the launching of manned flights and developed *Mercury's* unique bell-shaped capsule that could withstand the incredible heat caused by the friction of reentry into the atmosphere. Possible failure on board the *Mercury* capsule was compensated for by backup systems and, as a last resort, the ability of the astronaut to control flight manually. Each of the original seven astronauts mastered one aspect of the launch, and all united together on certain key design issues. For example, they lobbied to have a window installed on the capsule despite the resistance of engineers.

Liftoff of *Apollo 11* from the Kennedy Space Center on July 16, 1969.

On May 5, 1961, Alan Shepard (1923-) ascended into suborbital flight on board *Freedom 7*, spending 15 minutes and 22 seconds in flight before descending to Earth. Unlike Gagarin, he remained in the spacecraft until it splashed down in the Atlantic and a Navy ship rendezvoused with the capsule. Virgil ("Gus") Grissom (1926-1967) went on another such flight in July of that year. John Glenn (1921-) became the first American to orbit the Earth on February 20, 1962, circling the planet three times in less than five hours. As a result of this accomplishment, Glenn became one of the most celebrated national heroes since Charles Lindbergh, and the U.S. manned space program gained valuable public and media support.

Gemini and Voskhod

Project Mercury gave way to the Gemini program in 1964, the goal of which was to design a two-man spacecraft and develop the skills and technology necessary for achieving a lunar landing. However, the Soviet Union had once again overtaken the Americans, launching *Voskhod* (''Sunrise''), its new three-man spacecraft in October 1964, well before the first *Gemini* flight. In March 1965 Soviet cosmonaut Alexei Leonov (1934-) also took the first space walk, spending 10 minutes outside the *Voskhod* capsule connected to the craft by only telephone and telemetry cables. The United States Gemini spacecraft consisted of two sections: a manned capsule capable of carrying two astronauts and an adapter section. Between March 1965 and November 1966 ten manned *Gemini* spacecraft were sent into orbit in order to extend mission times, perfect and practice orbital maneuvering, rendezvous, and docking techniques, and train astronauts in extra-vehicular activity. There were several highlights: on *Gemini 4* Ed White (1930-1967) took a 21-minute spacewalk, *Gemini 6* and *7* rendezvoused together in space, and *Gemini 10* docked with an Agena rocket and fired its engine for a new orbit.

Apollo

NASA applied the advances made during the Gemini missions to the Apollo spacecraft, the first manned vehicle to land on the Moon. Developed by North American Aviation, the craft consisted of three parts: the command module where three astronauts would travel; the service module that carried fuel, oxygen, water, the electrical system, and communications equipment; and the lunar module, which would make the actual descent and lift-off from the Moon's surface.

The first Apollo missions would consist of testing the command and service modules as well as testing the ability of the command and lunar modules to rendezvous and dock. In 1967, however, disaster struck when, on a ground test of *Apollo 1*, three astronauts—Gus Grissom, Ed White, and Roger Chaffee (1935-1967)—died after an electrical spark ignited the pure oxygen of their cabin, causing fire and toxic fumes to spread in seconds. As a result of this tragedy the program was delayed for over two years as questions arose concerning the spacecraft's safety (Grissom at one point had hung a lemon on the craft), and the pressures placed on designers by government officials eager to surpass the Soviets. During that period over 1,500 modifications were made to the command module to ensure the astronauts' safety.

The *Apollo* spacecraft was ready for flight in October 1968, and, after several ''rehearsal'' flights, *Apollo 11* was launched on July 16, 1969, with astronauts Neil Armstrong (1930-), Michael Collins (1930-), and Edwin ''Buzz'' Aldrin (1930-) on board. Four days later, Aldrin and Armstrong descended into the Sea of Tranquility in the lunar module Eagle. As Armstrong stepped down onto the Moon's surface he uttered the now-famous words ''that's one small step for man, one giant leap for mankind.'' The culmination of nearly a decade of technological advances, *Apollo 11* has been described as the greatest achievement of the modern world.

Soyuz

While the United States had been developing the *Apollo* spacecraft, the Soviet Union had begun work on *Soyuz* (the Russian word meaning ''Union''). Like *Apollo*, *Soyuz* consisted of three modules for orbit, descent, and storage. Also, unlike earlier Soviet spacecraft, cosmonauts could control its navigation and guidance systems. Problems, however, plagued the program and the spacecraft's maiden flight was repeatedly delayed. Finally in April 1967—three months after the *Apollo 1* fire—*Soyuz 1* was launched. Yet little over 24 hours later, the spacecraft crashed to Earth, with cosmonaut Vladimir Komarov (1927-1967) on board. Although Soviet officials denied that design flaws were to blame for the crash, Western analysts concluded that the controls of the craft had failed, causing it to tumble wildly and become tangled in its parachute lines. This disaster, coupled with the death of the visionary Kolorev a year earlier, effectively halted the Soviet space program.

By the time of their first flight following *Soyuz 1* some 18 months later, the Soviets had unofficially conceded defeat in the race for the Moon. They focused instead on establishing the first orbiting **space station**, *Salyut* (''Salute'').

Space Stations

The Russians launched their first space station, *Salyut 1*, in April 1971. Orbiting approximately 200 miles (321 km) above the Earth, the station was powered by two solar panels and divided into several different modules, three of which were pressurized. While cosmonauts onboard *Soyuz 10* were unable to enter the station (Soviet officials claimed otherwise), the three-man crew of *Soyuz 11* successfully entered *Salyut 1* on June 7, 1971.

On board, the cosmonauts used a **telescope** and **spectrograph** for observations and conducted experiments in plant growth. Their highly successful three-week stay set a new record for human endurance in space. Yet during their reentry into Earth's atmosphere, a cabin seal released prematurely and the spacecraft lost air pressure. The three crew members—Georgi T. Dobrovolsky (1928-1971), Viktor I. Patsayev (1933-1971), and Vladislav N. Volkov (1935-1971)—had not been issued pressure suits and suffocated instantly. As a result of this disaster, the Soviets were forced to allow *Salyut 1* to fall out of its orbit and burn up in reentry, having lost the means of visiting and refueling the station.

Despite this major setback, they were eventually able to launch other Salyut stations as the decade progressed. By the early 1980s, the Soviet Union had begun to establish a significant lead in the duration of manned space flight using their *Salyut* stations. As an example of the Soviets' ability to keep people in space for extended periods of time, three cosmonauts stayed aloft in 1984 for 237 days, nearly three times longer than the American record of 84 days.

In 1986 the Soviets launched a new space station, *Mir* (''Peace''), which had less scientific equipment than Salyut but had better accommodations for its cosmonauts as well as improved life support and ventilation systems. Over the next three years, *Mir* cosmonauts carried out astronomical observations, materials processing and medical experiments, and extensive **photography** of the Earth.

Following the initial flight of *Salyut*, the United States stepped up production of their own space station, *Skylab*. Unlike Soviet engineers, U.S. engineers chose not to start from scratch but rather to use the shell of an *Apollo* rocket tank as a ''skeleton.'' The size of a small two-bedroom house, the station boasted comforts unheard of in spacecraft, including a shower, toilet, and sleep compartments.

Launched into orbit in May 1973, *Skylab* was powered by large solar panels similar to those found on *Salyut*, and a thin aluminum shield protected the station against micrometeorites and excessive solar heat. Soon after launch on May 14, 1973, scientists discovered that part of the shield had torn off, carrying with it one of the solar panels. Realizing the station was now seriously underpowered and was vulnerable to overheating from the Sun, NASA sent a three-man crew to meet the station. Working in space suits, the astronauts repaired the damage to the solar panel and deployed a hastily engineered sun shade. Now able to carry out their original mission, the crew conducted astronomical observations and biomedical experiments for 28 days. Two more missions took place before 1979 when *Skylab* reentered the Earth's atmosphere and disintegrated despite the efforts of NASA to save it.

The Space Shuttle

When the Apollo program drew to a close in the mid-1970s, NASA presented Congress with three possible long range goals for manned spaceflight: a lunar-orbiting space station, a manned voyage to Mars, and a 50-person Earth-orbiting space station serviced by a fleet of reusable **space shuttle**s. Although Congress rejected all three as too expensive, NASA officials were determined to at least win approval for the space shuttle. They approached the Department of Defense (DOD), and an agreement was struck which stipulated that the DOD help fund the project in return for use of the shuttle for military projects.

The idea of a reusable spacecraft dated back to the early 1950s with the development of the X series of high-altitude **aircraft**, and the actual design process for the space shuttle commenced in 1969 when engineer Robert Gilruth brought his idea of a reusable spacecraft to NASA engineer **Maxime Faget**. By 1977 Faget and his team had completed the development phase of the shuttle, and on April 12, 1981, the *Columbia* was successfully launched and returned to Earth for its second voyage on November 4, 1981.

About the size of a DC-9 commercial airliner, the *Columbia* and its sister shuttles were intended to serve as an economical transport for the development of manufacturing and large-scale commercial operations in space and the construction of a permanent space station. Although the shuttle program failed to meet the ambitious flight schedule set for it by government officials, the first 24 missions were flown successfully and the American public soon grew used to seeing shuttle launches on **television**. However, on January 28, 1986, the seemingly routine nature of space flight was shattered when the *Challenger* exploded a minute into the flight, killing all seven crew members. An investigation of the disaster revealed that the solid rocket's O-rings had become brittle in the cold weather, then had broken in flight, resulting in the explosion. The next shuttle did not fly again until nearly three years later, after some 400 modifications had been made to its design and its payload schedule revised. A replacement for the *Challenger*, the *Endeavor*, saw its first service in 1992.

Despite the resumption of shuttle flights, the future of manned spaceflight remains uncertain. In the United States the *Challenger* tragedy lost NASA its near-mythic reputation with the American public, and budget constraints have caused officials to seriously question the funding of multi-billion dollar space programs. And with the collapse of the Soviet Union, the ''space race''—which drove the achievements of the 1960s—has ended as Russia struggles to achieve economic stability. However, several nations, including Russia, are forging ahead with plans to construct the so-called International Space Station (ISS), with initial in-orbit construction slated to begin in November 1998, with assembly to be completed by 2004.

SPAETH, MARY (1938-)
American physicist

Theodore Maiman introduced the first working **laser** to the world in 1960, initiating one of the most intense periods of scientific activity in modern history. Every major think tank in the United States turned its attention toward perfecting and modifying Maiman's machine. Immersed within this flood of optical research was Mary Spaeth who, just six years after Maiman's invention, developed the tunable dye laser.

Spaeth graduated from Valparaiso University and received her master's degree in physics from Wayne State University. During her studies, Spaeth (along with most other scientists at that time) had become frustrated with the monochromatic nature of lasers: lasers could be built to emit different colors, but, once chosen, that color was fixed. It became clear that a tunable laser was needed—a laser whose color could be changed in mid-stream. Peter Sorokin and J. R. Lankard had, in 1966, shown how certain **dyes** affected the coloration of a laser, and it was upon these findings that Spaeth based her design for a tunable laser. Like **Charles Townes**'s **maser** and **Gordon Gould**'s laser, the puzzle pieces that formed the tunable dye laser had been around for some time; before Spaeth, however, nobody had successfully assembled the pieces.

Spaeth's tunable laser is used primarily to separate isotopes of certain elements, especially uranium and plutonium. Scientists had long known of these elements' potential, but only the enriched isotopes were useful. For years they had employed tedious physical means to obtain these enriched isotopes. The tunable dye laser provided a relatively simple and inexpensive means to separate the natural isotopes from the more desirable ones.

With this laser scientists can change the molecular structure of isotopes, that is, atoms of the same element with a different number of neutrons in the nucleus. Due to their unique structure, each isotope absorbs light differently. A colored light can be absorbed by one kind of isotope without affecting another. When an isotope absorbs light, the extra energy changes the shape and size of the isotope—and with enough energy, its electrical charge. Changing the electrical charge of only one of two very similar isotopes makes it easier to separate them. Atoms with a negative charge will be attracted to a positively charged plate and vice versa.

Initially the production of enriched uranium and plutonium isotopes was of primary interest to the military, which was trying to mass produce nuclear weapons; the patent for the tunable dye laser, in fact, belongs to the United States Army, for whom Spaeth worked at the time. Since then, the tunable laser has become the primary source for deriving the isotopes used in nuclear reactors. Work on Spaeth's invention has continued, and she herself was named deputy associate director of Lawrence Livermore National Laboratory's Laser Isotope Separation program.

Spam

Spam luncheon meat was introduced by Hormel Foods in 1937. The previous year, the company had devised a recipe for a canned loaf of spiced ham. Because this product was quickly copied by imitators, though, Jay C. Hormel was determined to give it a distinctive name that would set it apart from competitors. He offered a $100 prize, and the winning name was submitted by Kenneth Daigneau, the brother of a Hormel vice president.

In 1941, the United States Congress passed the Lend-Lease Act, providing for aid to Allied forces. Soon many millions of cans of *Spam* were being shipped to Great Britain and Russia. Before long, the ubiquitous luncheon meat had become notorious as "the ham that didn't pass its physical" and "a meatball without basic training." Nevertheless the product played an important role in helping sustain Allied forces during World War II.

After the war, *Spam* remained a popular food. By 1994, the five-billionth can of the luncheon meat had been produced. Today about 150 million cans of *Spam* are sold worldwide every year. In the United States alone, a can of *Spam* is eaten about every four seconds. Yet despite the product's proven popularity, it continues to be the butt of jokes. Internet users have even adopted the term "spam" to refer to electronic junk mail and junk newsgroup postings.

Spandex

Rubber threads, known as Lastex and Lactron, have been known since 1925, but it is only since the late 1970s and early 1980s that synthetic **plastic** threads based on elastomeric long-chain polyester polyurethanes with some crosslinking have had a major impact on the market. Such fibers are known collectively in the United States as spandex fibers, defined as polymers containing at least 85% of **polyurethane**.

E. I. du Pont de Nemours & Company, Inc. produced the first spandex fiber in 1958. By 1959 there was sufficient demand that Du Pont decided to trademark the fiber, giving it the name Lycra. U.S. Rubber introduced an elastomeric fiber called Vyrene in the 1960s. During the 1960s there were on the market a number of other spandex fibers, including Cleerspun, Glospan, Numa, and Unel. Foreign manufacturers were also producing spandex under such tradenames as Roica, Fujibo, Kanebo, Mobilon, Opelon, and Espa, some of which found its way into the U.S. marketplace. By 1998, the U.S. spandex market was supplied by three manufacturers: Du Pont (Lycra), Globe Manufacturing Co. (Cleerspun, Glospan), and Bayer (Dorlastan).

Spandex is commonly used in such clothing articles as swimwear, foundation garments, **brassieres**, lingerie straps, sock tops, hosiery, and medical products requiring elasticity. Spandex is never used alone; it is always part of a blend of fibers, most frequently with **nylon**. It is more flexible than conventional elastic threads, and has two to three times the restraining power. It can be repeatedly stretched to over 500% without breaking and still recover instantly to its original length.

See also Polymer and polymerization; Textiles

Spark chamber

Imagine an electrical system consisting of a brass plate and a **wire** parallel to it, separated by a gap of about 1 millimeter. Then, establish a potential difference of at least 1,000 volts across the gap. In 1931, the German physicist Greinacher found that the passage of an alpha particle through such a system generated a spark between the wire and the plate. With that discovery was born the principle of the spark chamber.

Since Greinacher's initial discovery, the spark chamber has gone through a number of modifications. In one form, the device consists of a number of parallel metallic plates, all charged to a high voltage. The plates are surrounded by some gas, originally argon, but more recently, air, neon, an argon-neon mixture or a helium-neon mixture. Credit for the first working model of a spark chamber of this design is usually given to two Japanese physicists, Fukui and Miyamoto, who announced their design in 1959.

When a charged particle passes through a spark chamber, it creates a series of gas ions. These ions act as conductors between the charged plates, setting off a line of sparks that corresponds to the path of the particle. After sparking has occurred, the chamber must be cleared of ions before it can be activated again. The recovery time of the most efficient spark chambers is less than 1 millisecond.

One of the most desirable properties of a spark chamber is the ability to design the system so that it is activated only when an important event occurs. That is, the spark chamber can be wired to a "preview chamber" that detects the presence of an incoming particle and prepares the main chamber to record its pathway.

Spark chambers are highly adaptable instruments. It is possible to change the material, thickness, number, and spacing of plates to deal with the special requirements of an experiment. An external magnetic field can also be used to determine the charge, mass, and velocity of particles passing through the chamber. Particle paths in a spark chamber can be recorded by stereo cameras, electronic recording devices, and acoustical systems.

Spark plug

In 1769, the first self-powered carriage was produced by **Nicholas-Joseph Cugnot**, a French military engineer. The machine's power was steam-generated and required so much cumbersome equipment that it was unmanageable and inefficient. It quickly became obvious that a more effective machine had to be developed.

John Barber was the first to act on the theory of using a combination of highly condensed air and fuel to drive these self-powered carriages. He used matches and candles to produce the spark necessary to ignite the fuel and create the drive force. He patented his **gas turbine** in 1791.

It was not until 1860, however, that Frenchman, **Jean-Joseph-Étienne Lenoir** developed the earliest version of the **internal combustion engine** that used an electrical spark from a spark plug to ignite an air-fuel mixture. Lenoir's spark plug consisted of a hollow brass bolt, porcelain, and two platinum **wire**s. Later, in 1885, Lenoir refined his spark plug, and the later version closely resembles what is used today.

Timing of combustion is controlled by the **ignition system**, but the spark plug's task is to produce the spark. Voltage from the ignition system ionizes (removes electrons from) air molecules around the electrodes of the spark plug. The voltage concentrates along a sharp edge of the electrode, causing a spark to arc across the electrode and ignite the exposed fuel. Sparking wears away the edges of the electrodes, so spark plugs have to be replaced periodically.

Modifications to spark plug design have been part of the search for internal combustion engines that burn leaner air-fuel mixtures and run cleaner. Metal alloys improve thermal conductivity; gold-palladium, copper-core and platinum-tipped plugs are claimed by spark plug makers to be more durable and to fire more reliably. The shape and exposure of the electrode improves combustion efficiency, and the two-pronged side electrode has been proven to improve fuel economy and engine performance. Specialized plugs like the Halo Disc are made of tungsten, nickel, and iron and have specialized insulators, electrodes, and inductors to minimize fouling, quenching, and electromagnetic interference. Such design features account for the wide assortment of spark plugs on the market.

SPECTACLES • See Eyeglasses

SPECTROSCOPE

Any instrument used to examine the component colors of light is called a spectroscope. The effect of light separated into its individual colors is often seen in the form of a rainbow, which results when rays of sunlight pass through tiny droplets of water that remain in the atmosphere after a rainstorm. The tiny water droplets function as natural spectroscopes, revealing a spectrum of colors.

The light displayed by a spectroscope is called an *emission pattern*, or spectrum. If the observed light is white, the spectrum will contain all the primary colors; however, if the light itself is colored, the emission spectrum will be incomplete. A **prism** simple type of spectroscope. Sir Isaac Newton (1647-1727), using a glass prism, was the first to realize that white light was actually composed of all the colors within the spectrum mixed in equal proportions.

The first spectroscope precise enough for laboratory use was invented in 1814 by **Joseph von Fraunhofer**. His device used a **telescope** to magnify a distant light source, and a **lens** to focus the light from the telescope onto a prism. Through this spectroscope Fraunhofer observed the spectrum of sunlight, finding hundreds of tiny dark lines (now called Fraunhofer lines) within the colored pattern. Such a pattern is called an absorption spectrum; the dark lines indicated the presence of specific elements in the sun that absorbed light at these wavelengths. Fraunhofer also used his invention to examine the different spectra of distant stars. Revealing information carried by the light from these stellar objects, Fraunhofer's spectroscope was the first scientific instrument that could be used to perform an analysis of objects billions of miles away.

Probably the most important use of the spectroscope came in 1859, when **Robert Wilhelm Bunsen** and **Gustav Robert Kirchhoff** teamed to learn about the nature of chemicals through the observation of their spectra. The pair were particularly interested in the phenomenon of incandescence, wherein a substance is heated (or burned), emitting a glow (or flame) of a certain color. They knew that, by examining the substance's incandescent light through a spectroscope, fundamental information about the substance could be derived. The research, however, was thwarted because Fraunhofer's device caused the colors within spectra to overlap and blur.

To correct this, Bunsen and Kirchhoff passed the light through a narrow slit before it entered the prism. The light that entered the spectroscope was thus refined to a thin beam that, when split, revealed a spectrum consisting of many thin, colored lines. Through extensive research, Bunsen and Kirchhoff found that every known element possessed a spectrum as individual as a fingerprint, and that by searching for unfamiliar spectra one could identify previously unknown elements.

Using this method the two scientists discovered cesium and rubidium. Kirchhoff also defined two types of spectra: emission patterns and *absorption patterns*. When an element that has been heated or burned emits light, a bright line appears in the spectrum; when placed in the path of light, a cold element will absorb the light corresponding to that same line. The relationship between emitted and absorbed frequencies of light formed the basis for much of Kirchhoff's research.

Bunsen and Kirchhoff's improvements utilizing incandescence have made the spectroscope an invaluable tool for scientists. Chemists, in particular, have used the device to analyze and identify the component parts of unknown substances. Kirchhoff also used his spectroscope to study stars. Comparing the spectrum of sunlight to the catalog of known elements, Kirchhoff discovered that numerous elements (such as gold) are present within the sun. The Italian astronomer Pietro Secchi (1818-1878) was the first to conduct systematic research of the heavens using spectroscopy, studying more than four thousand stars and their spectra.

The next great advance in spectroscopy came in 1912, when H. G. J. Moseley (1887-1915) developed the x-ray spectroscope. This design uses a diffraction grating rather than a prism. Based upon English physicist Charles Barkla's (1877-1944) theories of radiation absorption, the x-ray spectroscope used in x-ray fluoroscopy soon became of great use to physicians who could detect impurities in blood and other bodily fluids through spectrochemical analysis.

Elmer Ambrose Sperry

Modern computer technology has led to an improved form of spectroscope called a spectrometer. A spectrometer is a spectroscope that can provide quantitative data. Rather than simply producing a spectrum of bright lines and dark lines, a spectrometer can produce a spectrum of peaks whose height corresponds to the amount of a substance present in a sample. Spectrometers can be used to perform chemical analysis, match paint samples, or inspect food and products on manufacturing lines. Equipped with sophisticated scanning equipment and flown in an airplane, spectrometers are used to produce geological maps, to detect pollutants in shallow water, or to monitor crops and forest for damage.

Raman spectroscopy is a highly sensitive, sophisticated variation on the basic instrument and technique. In Raman spectroscopy, **laser** light is projected on a sample. The atoms that form the molecules of the sample are bound together, but oscillate to varying degrees along those bonds. When a photon of light from the laser impinges upon a moving atom, the velocity imparted by the atom shifts the wavelength of the light as it scatters the light back to the spectrometer. This is known as a Raman shift. The spectrum of Raman-shifted peaks provides information about the atomic structure and dynamics of the sample.

The simple yet powerful designs of Fraunhofer, Bunsen and Kirchhoff, have been hybridized and specialized for other types of applications. Astronomical spectroscopes observe the composition, temperature, density, and magnetic fields of distant stars, and atomic absorption spectroscopes analyze elements in incredibly small quantities. Microwave and gamma-ray spectroscopes have also been constructed.

Sperry, Elmer (1860-1930)

American inventor

Sperry was always interested in machinery, inventing at age six a horseradish grater for his aunt. Later he studied machinery at various shops in his hometown of Cortland, New York, and made informal arrangements to sit in on lectures at nearby Cornell University. In recognition of his talents, the local YMCA collected money to send him to the Centennial Exposition at Philadelphia, where he was so awestruck with the mechanical marvels there that he said it influenced the rest of his life.

While attending Cornell, Sperry made suggestions concerning the construction of a dynamo at the university that led to his being placed in charge of its building, as well as the development of a new type of arc light. At the age of twenty Sperry founded the Sperry Electric Company, which manufactured dynamos, **arc lamp**s, and other electrical equipment. The company was a success and was responsible for lighting systems in a number of cities, including Omaha, Nebraska, Kansas City, Missouri, and St. Louis, Missouri. Sperry also attracted great publicity by building and installing the world's highest beacon on the 390-foot (119 m) Board of Trade Tower in Chicago.

Sperry soon became interested in other mechanical devices. While visiting a coal mine, he observed the inefficiency of the labor used and decided to do something about it. He organized the Sperry Electric Mining Machine Company in 1888, which designed and manufactured electrically driven mining equipment—electric chain cutters, electric **generator**s, and electric **locomotive**s for use in mines. Sperry later set up another company to manufacture streetcars for Cleveland. In addition, Sperry patented and built electric cars, devising a different kind of storage battery capable of taking a car 100 miles (161 km) on a single charge. He established yet another company to manufacture caustic soda, hydrogen, and chlorine compounds from salt. He later introduced a high-intensity **searchlight**, five times brighter than others available, which many countries had adopted it for anti-**aircraft** use in the military.

Despite these achievements, Sperry will always be best known for his invention of the **gyrocompass**. At the heart of this device is the **gyroscope**, a weighted, balanced wheel mounted in bearings and spinning at high speed. A key feature of a gyroscope is its ability to maintain its position no matter at what angle it is held. As early as 1851 the French scientist **Jean-Bernard-Léon Foucault** had predicted that the gyroscope might someday be used as a **compass**. Unlike the magnetic compass, which is confused by the steel ship hulls and electric systems on board these ships, the gyroscope remains undisturbed. In Sperry's gyrocompass, the gyroscope's wheel is part of the rotor of an electric motor, so that it spins at high speed. A weight on one side of the innermost ring responds to the

Earth's rotation and forces the gyroscope to align itself with the Earth until gravitational attraction holds it in place. Thus, the gyroscope assumes a north-south position, valuable for a compass, especially since the gyroscope points true north, not magnetic north as traditional compasses do (true north being the axis around which the Earth rotates).

Sperry used the principle of the gyroscope for two other key applications. First, he introduced a **gyrostabilizer** in 1913, a huge gyroscope mounted in the hull of a ship that resisted rolling from side to side. A destroyer outfitted with one reduced a total roll of thirty degrees down to six degrees. In the 1920s Sperry invented the first **automatic pilot** control for ships, named Metal Mike, which also used a gyroscope.

Sperry then moved to aircraft stabilization. At first, gyroscopes were used to control roll and pitch. When they sensed movement, the gyros activated compressed air, which acted on pistons to activate the aircraft controls. In later aircraft cockpits, pilots could use turn indicators, bank indicators, artificial horizons, and other devices controlled by gyroscopes. Today, gyroscopes form the basis for the inertial guidance systems prominent on **submarine**s, **aircraft**, and **rocket**s. During his lifetime Sperry held over 400 patents and founded eight manufacturing companies. Upon his death, he left the YMCA one million dollars to repay the organization for sending him to the exposition that gave direction to his life.

Sperti, George Speri (1900-1991)

American inventor and engineer

A prolific inventor of many different products, George Speri Sperti was born in Covington, Kentucky. While studying electrical engineering at the University of Cincinnati, Sperti invented the K-vas meter, a device for measuring large-scale consumption of electricity. The Westinghouse Electric Corporation bought Sperti's apparatus for $30,000.

After receiving his degree in 1923, Sperti worked in the research lab of the Union and Duncan electric companies in Cincinnati, Ohio, and Lafayette, Indiana, respectively. He then returned to the University of Cincinnati, where he co-founded the Basic Science Research Laboratory and served as its director for ten years. During this time, Sperti developed a process of increasing the vitamin D content of milk by means of irradiation. General Foods bought the process for $300,000, and Sperti put the money back into the laboratory.

In 1935 Sperti founded the Institutum Divi Thomae in cooperation with the Catholic Archdiocese of Cincinnati. Named after St. Thomas Aquinas, the institute was a scientific research and graduate studies center later called the St. Thomas Institute for Advanced Studies. Sperti directed the Institute until 1988 and supported it with profits from his many inventions.

Sperti and his laboratory held more than 120 patents. Among Sperti's inventions are: Preparation H, the well-known hemorrhoid treatment; Aspercreme, a medication for arthritis relief; the Sperti **sunlamp**; a meat tenderizer; a process for freeze-drying orange juice concentrate; ultraviolet germicidal devices; burn ointment; yeast enrichment of animal feed; and many cosmetics and medical creams. Sperti died in Cincinnati at the age of ninety-one.

Sphygmomanometer • See Blood pressure measuring devices

Spinning jenny

John Kay's (1704-1764) invention of the flying shuttle in 1761 made it possible for one man to weave material wider than the length of his extended arms—material so wide that a normal **loom** would require two or even three men to cast the shuttle back and forth. As use of the flying shuttle increased, a new shortage of thread became apparent, for one man, using a **spinning wheel**, could only produce one thread at a time. A reward was offered to the man who invented a machine to increase the spinner's productivity. While it is not known if he ever received that reward, the man who ultimately invented the machine was **James Hargreaves**, and the machine was the spinning jenny.

The origin of the jenny's name is unclear. Some say that it was named after Hargreaves' wife; others, that it is a colloquialism for engine; still others believe that it has no origin at all. What is known is that Hargreaves formulated the idea for the spinning jenny when he saw a spinning wheel that had been knocked over onto its side. Though it was tipped ninety degrees, both the wheel (now horizontal) and the spindle (now vertical) were still spinning. It occurred to Hargreaves that several spindles could be arranged this way, side-by-side, all taking their roving from a single wheel.

Hargreaves began development of the spinning jenny in 1754, and after fourteen years of experimentation the first jenny was completed. Through the use of a moving bar that was operated by a foot pedal, the cotton fibers were drawn out; this was the first successful mechanical simulation of how a spinner would draw the fibers out by hand. The fibers were then twisted lightly and spun onto eight separate spindles. Like the yarn produced on a hand-spun wheel, the jenny's yarn was weak and sometimes lumpy—usable as *weft* but unsuitable as *warp* (for which much stronger linen thread was used).

Within a few years jennies had been constructed that supported 80 to 120 spindles. Soon after its invention in 1768, Hargreaves was forced to sell several jennies in order to support his family. The local spinners feared, correctly, that the new machine would replace them; they organized a vigilante group to vandalize Hargreaves' home and destroy his spinning machines. Though Hargreaves and his family were driven to Nottingham he did not cease production of the jenny. He opened a small mill and, in 1770, sought to obtain a patent for the spinning jenny; by that time, unfortunately, the machines he had sold years earlier had been copied, and jennies were operating in mills across Britain. Hargreaves was never awarded a patent.

Spinning mule

Invented by **James Hargreaves**, the **spinning jenny** was the first machine to successfully simulate the action of spinning yarn by hand. The next, Sir **Richard Arkwright**'s water frame machine, almost completely automated a process for producing strong, smooth yarn. The final step in the evolution of the spinning machine was the invention in 1779 of the spinning mule by another Englishman, **Samuel Crompton**.

Crompton, who was in charge of many of the household chores, was frustrated with the spinning jenny: though it allowed for the spinning of much finer yarn, the yarn was often too fine and would break while being spun. He decided to build his own spinning machine—a hybrid, combining the best aspects of both the jenny and the water frame.

Crompton spent five years and all of his money to develop the spinning mule (so called because of its hybrid nature). Unfortunately, Crompton was a very poor businessman. He could not afford to purchase the patent for his inventions, and instead sold it to the textile industry on a handshake deal. The unscrupulous textile owners reneged on their promise almost immediately, and Crompton spent the rest of his life seeking compensation for his stolen invention.

Though the spinning mule was one of the most intricate machines of its time, its basic design is fairly simple. A sliver or roving of cotton is passed between a set of rollers, arranged in pairs; these rollers (borrowed from Arkwright's water frame) draw apart the cotton fibers, producing a finer, smoother sliver. The roving is then spun on bare spindles that are mounted on a moving carriage, such as that used in the spinning jenny. The moving spindles draw and spin the yarn without tension, so that strong yarn can be spun without being broken in the process. More than one spindle can be mounted on the carriage; Crompton's held 48, while some held as many as 150.

The spinning mule was able to spin yarn so quickly and cheaply it forced the textile industry to improve its weaving machines just to keep up. A steam-powered mule was soon developed, though it still required two people to operate. Later, Richard Roberts added a quadrant, camshaft, and winding **chain** to the mule's design to produce the first self-acting mule.

Operating without human assistance, the self-acting spinning mule could be more than 130 feet (40 m) long with 1,400 spindles and thirty thousand moving parts. The self-acting mule became the dominant machine of the textile industry. It could spin all types of yarn, from very fine to very coarse. Though its design underwent very few modifications after 1830, the use of the mule continued for almost a century and a half; it has only recently been replaced by the winder and ring spinner. The last spinning mule was shut down in 1974, and today most exist only in museums.

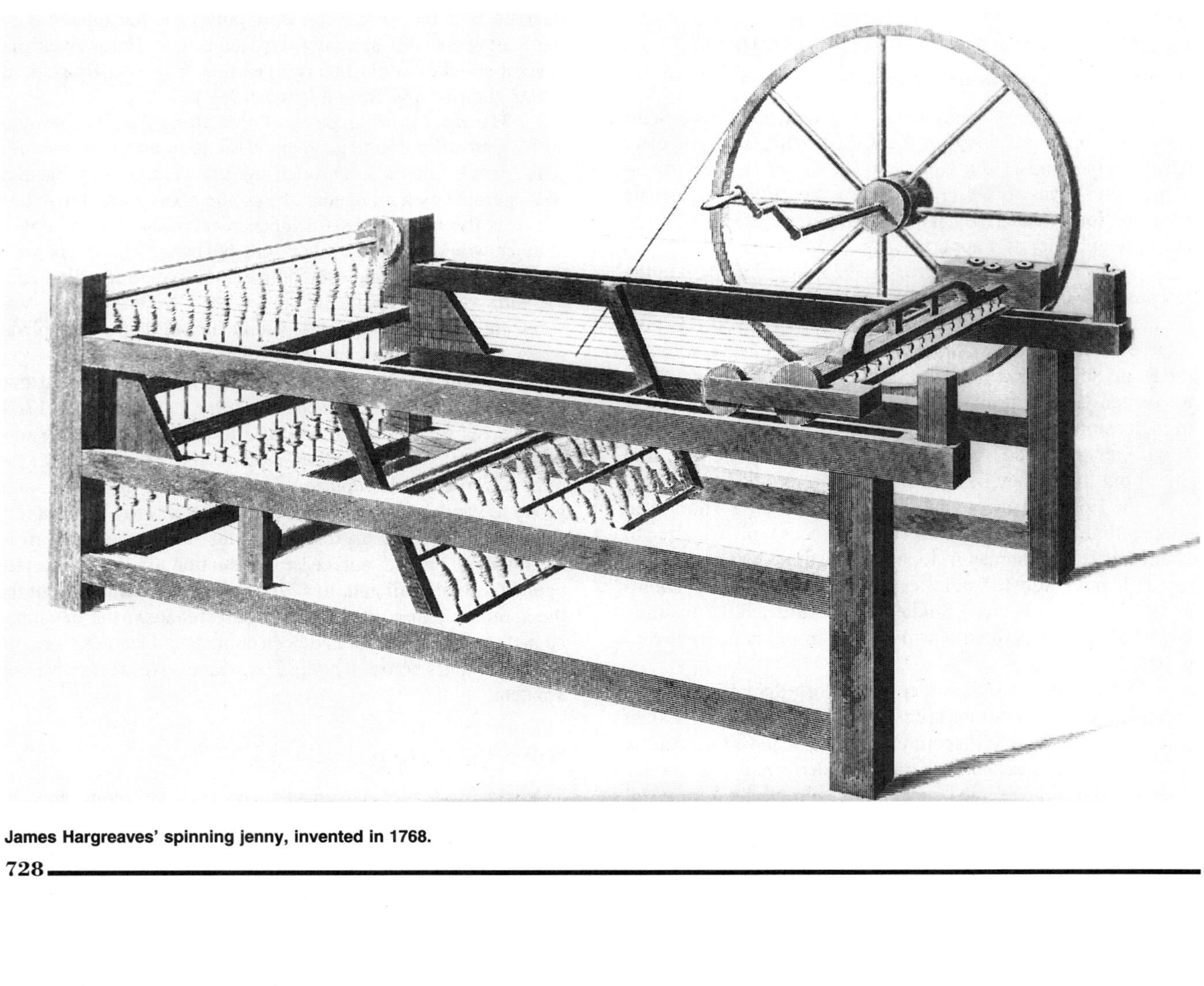

James Hargreaves' spinning jenny, invented in 1768.

Spinning wheel

Clothing made from fabric (rather than animal hides) was not invented until about 6500 B.C., when the first spinning devices were developed. Before the appearance of these devices—which were almost certainly invented by women—there had been no way to spin fibers for **woolen fabric** or other materials into yarn (which could be woven into cloth) and thread (to sew a garment together). However, these early devices did not actually spin. They simply held the fibers taut so that the spinner could twist them by hand. It was not until much later that a true spinning wheel was developed.

The first spinning wheels were probably inspired by the silk spinning machines used in India around 700 B.C. They were used throughout Europe and the British Isles, and most had an almost identical design. The most prominent feature was a large hand-turned wheel that was attached by a belt to a smaller spindle; this design caused the spindle to turn very fast, even though the large wheel rotated slowly. In her left hand the spinner held a sliver of wool that was fed onto the spindle; with her right she turned the great wheel. As the spindle spun it pulled and twisted the wool into coils of yarn; at the same time, the spinner would draw the sliver away from the spindle, playing it out with her fingers so that the thickness of the yarn would be even. After a short time the spindle would collect too much yarn, threatening to snap or snarl the coils. The spinner would then have to reverse the motion of the wheel, allowing some of the yarn to slip off, thus making room for more.

Though the spinning wheel was a relatively simple device, it took skill and experience to simultaneously execute all of the drawing, turning, and winding necessary to produce a consistent length of yarn. The art of spinning was made easier in the fifteenth century A.D. by the invention of the Saxony wheel. In this spinning wheel the yarn, once spun, was pulled through a small hole and onto a turning bobbin. The bobbin collected the yarn, eliminating the reversing step and allowing for continuous spinning. Soon after, foot pedals were added to turn the wheel; this device freed both the spinner's hands for drawing and pulling the sliver, yielding yarn of a much higher quality.

Although by the eighteenth century the perfected spinning wheel was the keystone of textile manufacturing, it would not be long before **James Hargreaves** would invent his **spinning jenny**, a device that would almost instantly render the old wheels obsolete. By the turn of the nineteenth century the jenny-along with **Richard Arkwright**'s **water frame** and **Samuel Crompton**'s **spinning mule**—had ushered in Britain's Industrial Revolution, and established that country's dominance of the **textiles** industry.

Spoilage retardants • See Food preservation

Sponge, synthetic

Synthetic sponges are **polymer** foams. Depending on the intended use, the foam may be *open* or *closed-cell.* The most familiar sponge is the open cell which readily absorbs liquid through capillary action. Closed-cell foams do not readily absorb liquids and are used in **life preserver**s and buoys.

The sponge is made from a polymer and a chemical agent that introduces cells, or bubbles, into the polymer. The polymers used in sponges may be cellulose acetate or one of the phenolic chemicals; both can absorb over ten percent of their volume in water. Cellulose acetates are cheaper and are used more often.

Open-celled foams are made during free expansion, that is, with no external pressure so that the cells may reach their maximum size, while closed-cell foams are made while pressure is applied, increasing their density. Inorganic salts which give off carbon dioxide are used to form the cells in the polymer; the gas then dissipates.

See also Cellulose, chemical uses of

Spoon • See Knife, fork, and spoon

Sprengel, Hermann Johann Philipp (1834-1906)

German English inventor

Like so many of his contemporaries, Hermann Sprengel worked in Germany at the famous laboratory of **Robert Wilhelm Bunsen**. Born and educated in Hanover, Germany, Sprengel moved to England in 1859, where he began research in chemistry at Oxford and continued his work at several laboratories in London.

Sprengel's invention of a more efficient vacuum pump in 1865 had far-reaching implications for technological advances in many fields. For example, the **vacuum pump** was critical to the development of the **incandescent light bulb**, because early bulbs required a high vacuum to keep the carbon filament from being eaten away by oxygen in the air. As early as 1860, Joseph Swan had developed an incandescent bulb, but the filament would not last long enough for practical use. Years later, Swan learned of Sprengel's vacuum pump from the work of **William Crookes**, who had found the device essential to his analysis of certain gases. Encouraged by this news, Swan resumed his research, but by that time, **Thomas Edison** had begun similar work. The two scientists are credited with introducing the incandescent bulb at about the same time, in 1879.

Sprengel's pump made these inventions practical by providing an effective, economic means of producing high vacuums. Sprengel's pump was derived from **Johann Heinrich Wilhelm Geissler**'s earlier version, which Sprengel improved by mechanizing the slow, tedious evacuation process. The invention of the **radio**, the **television**, and the **digital computer** also depended on vacuum technology. Early radios and televisions were constructed with **vacuum tube**s, also known as thermionic valves. Likewise, the first generation of computers used vacuum tubes, which have since been replaced by electronic components. Sprengel's technology also allowed more rapid progress in the field of atomic physics, which requires high vacuums for the study of small particles.

In addition to the vacuum pump, Sprengel developed a type of **explosive** that could be mixed at the blasting site, thus reducing the danger of transporting live explosives over long distances.

SPRING

Nearly any solid material possesses a certain amount of elasticity, or resistance to bending. A substance that has enough elasticity to cause it to recover its initial shape when bent is called a spring.

Probably the first springs were branches and young saplings used by primitive hunters for traps, such as noose-traps. More advanced peoples developed bows using two springs—a bent frame of wood and a stretched piece of sinew, or string. Soon, these simple springs became the basis for a series of new tools, such as **looms**, **lathe**s, and **drill**s. Some time around 4 B.C., it was observed that these flexible materials resisted twisting as well as bending, thus inspiring the *torsion* spring. The torsion spring quickly found use in weapons, such as the crossbow and the **catapult**, as well as in new domestic devices.

Since these early times, the spring has become one of society's simplest and most frequently used tools, appearing in many forms and adapted to a wide variety of tasks. Probably the most familiar is the *coil* spring, but many others—*helical*, *leaf*, *balance*, and *mainspring* springs—have been developed.

Within a spring's molecular structure are two forces: the attracting force and the repelling force, derived ultimately from the electrical forces between the spring's material's atoms and molecules. When the spring is at rest, these two forces balance. However, when a spring is compressed, the repelling force increases, pushing the molecules outward; thus, when released, the spring expands. The converse is true for a stretched spring, wherein the attracting force escalates in an attempt to contract the spring.

Today, springs can be found just about everywhere, from the coils in a bed's mattress, to the helical spring in a mousetrap, to the leaf spring in an automobile's suspension. One of the most important springs ever invented is the mainspring that drives all wind-up clocks. A very special device, it is a coiled length of thin metal designed to uncoil itself at a specific and unvarying rate, so that the inner workings of the clock can be regulated. A similar spring, called the *balance* spring, was invented by the Dutch scientist **Christiaan Huygens**.

SPRINGFIELD • See Rifle

SPUTNIK 1

Sputnik 1 was the world's first space satellite. It was successfully launched into space by a team of Russian scientists on October 4, 1957, from the town of Baikonur, Kazakhstan. The name of the satellite, *Sputnik*, is the Russian word for "traveler." *Sputnik 1*'s successful launch shocked American scientists and engineers, who believed that America would be the first to send a satellite into space. *Sputnik 1*'s reported weight of 184 lbs. (83.6 kg) further astonished and intimidated American scientists, who were at work on America's first satellite that weighed only twenty pounds.

Sputnik 1, created by the design team led by **Sergei Korolov**, was a moon-shaped, spherical apparatus with a diameter of 22.83 in. (58 cm). It had four flexible whip antennae that ranged in length from approximately 2.2 to 2.6 yards (2.4 to 2.9 m). Two of the antennae were **radio** transmitters that sent signals back to Earth on two different frequencies as the satellite traveled through space at approximately 17,360 miles (28,000 km) per hour. As the satellite circled the Earth once every ninety-five minutes, it gathered valuable information about the ionosphere and space temperatures. *Sputnik 1* used **battery** power to relay this data back to Earth via its radio antennae.

Sputnik 1 fell back to Earth on January 4, 1958. The success of *Sputnik 1* signalled the beginning of the increasingly tense competition that would characterize the relationship between the scientific communities and governments of Russia and the United States for the next thirty years.

SPY SATELLITE

The Soviet Union's launch of the *Sputnik 1* satellite in 1957 caused an international uproar. The satellite was harmless, but thre was significant concern in the United States that the Soviets were gaining a tremendous upper hand in space operations. The advantages of satellites for communications, weather monitoring, and navigation were obvious—and equally obvious was their utility for spying on enemies. Spy **aircraft** were well-established, a famous example being the United States's U-2. As the U-2 eventually demonstrated, however, it was not invulnerable to attack, even at the rarefied altitudes at which it cruised. Satellites, orbiting far above the atmosphere, could observe the enemy with impunity, however, and it was not long before the space-capable countries, at the time the United States and the Soviet Union, were developing ambitious spy satellite programs.

The United States delivered the first of these new contenders in the cold war, with its 1960 satellite *Discovery*. Not intended for an extended mission, the satellite never achieved a stable orbit in space, but it did complete three passes over the Soviet Union, photographing everything it could. *Discovery* had a rather ignominious end, splashing into the Pacific, where it was recovered by Navy vessels. The next round of satellites went in to bona fide orbit, and delivered their photographs either by physically dropping canisters containing them, or (in obviously a more advanced technique) by transmitting images of them back to Earth,

In the early 1960s, the United States launched the most useful orbital spies then devised, the MIDAS (Missile Defense Alarm System) satellites. These took advantage of a fact pointed out by the science fiction writer Arthur Clarke: satellites orbiting about 22,300 mi (35,903 km) above Earth's surface would have an orbital period of 24 hours, and thus could constantly observe a single part of the Earth. By the end of the twentieth century, a host of these so-called *geosynchronous* satellites would be in place, performing a variety of functions including spying. Of course, the Soviet Union was kept pace with the United States, and within ten years of the launch of little *Sputnik 1*, both superpowers had an never-sleeping eye on each other.

Since the 1960s, the United States and the former Soviet Union have launched thousands of spy satellites, and they tend to fall into three categories. From the start of the space age, both sides have used *photoreconnaissance* satellites. These are generally short-lived (typically a year or less for American satellites, and only a matter of days for Russian satellites) craft that with modern technology can provide extremely high-resolution photographs of missile sites, airports, city blocks — whatever is deemed important to observe.

More advanced are the so-called *ferrets*. These are permanent satellites, some of them geosynchronous satellites, some of them closer to Earth. The MIDAS satellites were some of the earliest ferrets. Their successors were the Vela satellites, designed to detect nuclear tests. Where the MIDAS craft were equipped with infrared detectors (to detect the heat associated with missile launches), the much more advanced Vela craft carried a variety of detectors, and had much more sophisticated methods of identifying events on the Earth below as nuclear tests.

The photoreconnaissance satellites' capabilities are augmented by a third class of satellite that uses microwaves to observe enemy actions even through clouds. The latest generation of such satellites is capable of seeing individual people on a city street.

Perhaps the most ambitious spy satellite network of all was the Strategic Defense Initiative (SDI), proposed in 1983 by President Ronald Reagan. Significant SDI research has been carried out, but it seems unlikely that the network of satellites envisaged by Reagan will ever be put in place.

After all the worry surrounding the launch of *Sputnik 1*, spy satellites could be argued to have significantly improved relations between the superpowers. In the early 1960s, the United States and the Soviet Union operated in a state of fearful ignorance, neither side knowing exactly what the other was doing, and either ready to pull the nuclear trigger at the first—and possibly erroneous—sign of trouble. Spy satellites made it effectively impossible for nations to hide anything from one another. It has been said that the first approach toward peace is communication, and spy satellites, albeit in a slightly unorthodox way, opened those lines, setting the stage for the easing of tensions that rode so high during the Cold War.

***Sputnik 1*, the first artificial satellite.**

SQUIDs

SQUIDs is an acronym for Superconducting Quantum Interference Devices. In superconductivity, a low temperature causes metal to lose its electrical resistance: *extremely* low temperatures or temperatures on the order of 1–10 K above absolute zero (Absolute zero is defined as the temperature where all molecular motion ceases. It is equivalent to 0 K, which is -273°C, or -459°F). Electric current (super-current) can flow indefinitely in a ring made of superconducting material, so long as the temperature remains below the conductor's *transition zone* (the zone at which it is a SQUID).

The **superconductor** is believed to be a very modern invention, but it originated in 1907 when Heike Kamerlingh Onnes (1853-1926) was able to liquify helium gas for a coolant. In 1911 superconductivity was first performed with **mercury** as the conductor. Most all metallic elements, as well as hundreds of alloys and intermetallic compounds, can be superconductors.

In 1933, German physicist Walther Meissner discovered that magnetic fields which were below a specific strength could not penetrate superconductors. Instead, they were expelled from them. Type II superconductors, however, are immune to this *Meissner Effect*. Magnetic solenoids made of Type II superconductors can create high fields in relatively large volumes of material with nearly no loss of power. Nearly any material can have Type II characteristics if it is alloyed with other metals.

The BCS theory, named for **John Bardeen**, Leon N. Cooper (1930-), and John R. Schrieffer (1931-), was presented in 1958. This deals with superconductivity at the microscopic level. Their theory suggests that the electrons which are conducting the current condense into a state in which pairs of electrons become correlated with each other. When they are in this condition, the electrons can take advantage of *quantum-mechanical zero-point* motion of the positive ions and lower their energy.

In 1962 Brian D. Josephson (1940-) theorized that if two SQUIDs were separated by just a thin oxide layer of insulation, it is possible for the electron pairs to pass through the energy gap. This *Josephson junction* would make it possible to maintain a weak super-current in a ring of superconducting material.

The Josephson current and its two other related quantum mechanical effects, the direct current and alternating current Josephson effects, have significant practical applications. They can be used to make measurements of extremely small electrical currents and magnetic fields, such as those found in computers, the brain, and heart. They can also help in locating oil, water and mineral deposits that evade discovery by traditional methods.

A tremendous advance was made in 1986 when J. George Bednorz and K. Alex Muller discovered materials that were able to become SQUIDs at relatively much higher temperatures than previously believed possible. Scientists have been investigating various combinations of *copper oxides* since then, hoping to discover still higher transition temperatures. Ceramic copper oxide materials have been fabricated into SQUIDs that can superconduct at 77 K, which is the temperature of liquid nitrogen, and some materials have found that work at 120 K.

William A. Goddard III devised a new theory for superconductivity in 1988, predicting a transition temperature limit of about 200 K for the copper-oxide materials. That temperature is easily reached by using frozen carbon dioxide (dry ice). Should this prove to be the case, an explosion in commercial applications would occur, especially in the electronics industry.

Unless materials can be found that are superconductive at room temperatures, power storage facilities, electrical transmission lines and magnetic levitation trains will be prohibitively expensive to build. The discovery, in the early 1990s, that some forms of carbon, called fullerenes, can be made into SQUIDs holds great promise for the future.

SST • See Aircraft

Stabilizer • See Gyrostabilizer

Stage decompression • See Scuba diving

Stage lighting

In ancient Greece and Rome, plays were staged outdoors in daylight, usually on a hillside. This tradition continued through the Middle Ages until the 1500s, when theater first moved indoors. Everyday sources of artificial light, such as **candle**s and **oil lamp**s, were first used to illuminate the indoor stage, although their flickering light lacked color and intensity. The systematic development of stage lighting began in Italy during the sixteenth century, when flasks of blue- or amber-tinted water were placed in front of candles and torches to create special effects. Lighting remained uneven and primitive for the next 200 years, and of necessity both the auditorium and stage were lit during performances.

The advent of gas lighting in the early 1800s was a major advancement, since gas flames could be controlled more easily and smoothly than candle or torch flames. Unfortunately, they were also hot, malodorous, and caused many disastrous theater fires. A further refinement was made in 1816 with the invention of **limelight**, a **gaslight** with a lime filament that cast a brilliant white light. With it, actors could be sharply spotlighted for the first time. Even better results were obtained with the **arc lamp**, which was first introduced at the Paris Opera in 1846. It, too, was used for spotlighting, as well as for special effects like rainbows and fountains. The use of these bright controllable lights naturally led to changes in theaters and performances. During the 1860s the practice of darkening the theater began, leaving only the stage brightly illuminated.

The modern era of stage lighting began with the invention of the **incandescent light bulb** in the late 1800s. Incandescent lamps gradually replaced gas lights, and even the arc spotlight became obsolete. In 1882 London's Savoy Theater was the first to use electric lights. Although by modern standards they were weak and difficult to use, they gave producers more options that ever before; they also complemented the more natural and realistic stagings that began to evolve. By the turn of the century, electric lights were used exclusively in most theaters.

During the twentieth century, stage-lighting design became an art in its own right, emerging from the obscurity of props, set design, and costumes. Great efforts were made to bring the subtlety and drama of affective light to the stage, most notably by American playwright and producer David Belasco (1853-1931) and his assistant Louis Hartman, who developed many lighting instruments. Jean Rosenthal, another pioneer of American stage lighting, invented a system for recording a particular lighting sequence so that it could be faithfully repeated. Other technical advances included special **lenses**, reflectors, projectors, and new plastic materials for color filters. In 1948, mechanical dimmer switches were replaced by electronic dimmers capable of instantaneous remote control. Today, the lighting designer programs and operates a computerized control board that can recreate any instant of a lighting performance.

Stained glass

The oldest extant stained **glass** works date from the eleventh century, and a depiction of stained glass art appears in a tenth century manuscript. The design of these early examples, however, resemble that of Byzantine art, suggesting that the craft may have originated in the Near East. In the Middle Ages, stained glass was used primarily as a religious medium, employed in churches and cathedrals to illustrate Biblical passages, a famous example being the Cathedral of Chartres, France, completed in the mid-1100s.

The basic steps involved in stained-glass making have remained essentially unchanged throughout its history. The artist draws a full-sized *cartoon*, or pattern, for the finished piece on **paper**, based on an enlargement from a scale drawing. Pieces of colored glass are created to match the cartoon, either by suspending colored particles in the glass at its manufacture, or by coating glass with dyes—usually iron oxides—and oven-firing it at extremely high temperatures (approximately 1,150°F or 620°C), so that the **dye** finish becomes permanent. The stained pieces, cut to size, are joined into sections by lead strips; finally, the entire work is assembled on-site, section by section. Changes in technology and style, rather than basic technique, have changed the art over the centuries.

Early medieval stained-glass artists had access to only a limited number of dyes; hence, their figures tended to be expressionistic, rather than strictly realistic, and deep in color, dominated by reds and blues. The crudely fired glass used in these early works contained a number of imperfections and bubbles, which, however, had the pleasing effect of refracting light so that it appeared to shimmer and jump. Medieval craftsmen were more interested in illustrating an idea than creating natural or realistic images. They employed rich, jewel colors played of against milky, dull neutrals. A dark brown enamel, grisaille, was matted to the glass surface to delineate features. By the fourteenth century, artisans had begun to develop a more naturalistic, detailed style of design.

In the fifteenth century, the apex of the Gothic style of architecture, paint work became more sophisticated, and the rediscovery of silver stain as a lightening agent allowed the artist to realistically depict yellow hair and golden garments. Craftsmen of that century adopted the use of paler colors that admitted more light, and worked with larger figures. Thus stained glass came to be seen as more in terms of the picture and less as atmosphere. The trend continued through the 1500s, when stained-glass makers began to strive toward the realism of oil painting; the result, however, was generally inferior work, and the art declined in quality and popularity.

By the seventeenth century, the discovery of enamel painting was allowing artists to paint a single pane of glass in different colors rather than joining together cut pieces of colored glass. This technique sacrificed the original symbolism and innate beauty of stained glass. The eighteenth century saw the removal of many medieval stained glass windows, which were destroyed as old-fashioned and replaced by painted glass.

Critical and commercial interest in stained glass did not revive until the late nineteenth century, when artists, notably Americans Louis Comfort Tiffany (1848-1933) and John LaFarge (1835-1910), began experimenting with stained glass design. LaFarge developed opalescent glass in 1879, and Tiffany employed it not only in windows, but in decorative objects such as lamps. This was made possible by using thin strips of copper as a substitute for lead, allowing for intricate sections in windows. Artists of this period began using hand-blown or *muff* glass of high technical quality. In the twentieth century, French artists such as Henri Matisse (1869-1954) and Fernand Léger (1881-1955), and Russian-born Marc Chagall (1887-1985) created commissioned stained glass works that reflected the progressive style of their paintings.

The years since 1970 have seen a resurgence of interest in the possibilities of this art form. Homes have been embellished with beveled stained glass entryways, stained glass bathroom windows, and Tiffany-style lampshades; and decorative panels have come to be purchased just to hang in a sunny window, with hot-formed glass pieces not infrequently adorning tables, walls, shelves, and fill windows.

STAINLESS STEEL • See Steel alloys

STANLEY, FRANCIS EDGAR (1849-1918)
STANLEY, FREELAN OSCAR (1849-1940)
American inventors

The Stanley twins were born on June 1, 1849, in Kingfield, Maine, and showed mechanical inclination even as young boys. In 1883, Francis developed a machine that coated dry photographic plates. After receiving a patent for their process, the brothers set up a factory in Newton, Massachusetts, to manufacture the plates. In the summer of 1897, they attended a local fair where they witnessed a French inventor demonstrate his steam-driven car. Apparently impelled by his wife's inability to ride a **bicycle**, Francis vowed to build something that his wife could ride. The French inventor's steam car was the impetus Francis needed. After the fair, the brothers began to develop a steam car of their own.

The stained glass at the Chartres Cathedral in France was completed in the mid-1100s.

The brothers formed a car company in 1898 and produced their first steam car, which was dubbed The Flying Teapot. An instant success, the car was easy to run and achieved a top speed of 35 miles per hour (56 kph), quite fast for the turn of the century. Its major drawback was the need to stop every 10 miles (16 km) or so to refill the boiler. The brothers sold their company after only a few months, but they returned to the business of making cars in 1902 when they formed the Stanley Motor Carriage Company. They staged various events to publicize their steam cars, including racing up mountains and racing against gas-powered cars. Eventually the Stanleys sold their photographic plate business to **George Eastman** and concentrated on the manufacture of their steam cars, which came to be known unofficially as Stanley Steamers.

The brothers continued to build race-winning, steam-powered cars. In 1906, one of their cars—The Rocket, driven by Stanley employee Fred Marriott—set the world's record for the fastest mile: 28.2 seconds, which is a speed of more than 127 miles per hour (204 kph). In 1918, Francis was killed while driving one of his automobiles. He swerved to avoid an obstruction in a mountain road and plunged down an embankment near Ipswich, Massachusetts. At the time of his death, the Stanley Motor Company had been suspended automobile production to manufacture engines to pump out Allied trenches during World War I. After the war, **Henry Ford**'s Model T soon came to dominate the American automobile industry. Developments in gas-powered engines, and the limitations of steam cars, signalled the end of the steam-auto era. The Stanley Motor Carriage Company ceased production in 1924. Freelan Oscar Stanley died in 1940 in Boston.

See also Automobile, gasoline; Steam-powered road vehicle

Stapler

The stapler, a necessity for any office, is credited to Charles H. Gouldin 1868. There is little difference between Gould's stapler and the desk model we use today. Both have U-shaped **wires** fed in a channel (raceway) and forced into paper by a blade that breaks off individual wires from the strip. A small metal anvil under the papers bends the wire flat.

Due to high-volume office and industrial needs, more precise automatic versions of the stapler have been developed. The basic principle remains the same, but each new incarnation is able to complete tasks in a different manner. For example, industrial staplers are used in furniture making, upholstering, packaging, and magazine binding. Some stapling machines, called tackers, use the power of compressed air to drive staples into material without bending them. A related variation is the hog ring machine, a device that bends staple legs into a tight ring around the twisted neck of a bag. This is used to secure filled sacks and bags without piercing the material in the container.An upright stapler, designed to be used with one hand, is also available. Staples are used in some medical procedures, some types being absorbed by the body, making it unnecessary for a surgeon to remove them.

Stealth aircraft

The term *stealth aircraft* was first introduced to the public in the 1980s and 1990s when the United States Air Force allowed a glimpse of the highly classified F-117A fighter and B-1 bomber. However, these **aircraft** and the cutting-edge technology involved in their design are the culmination of a process that began during World War I.

That war's top fliers understood the importance of spotting the enemy first and attacking before being detected. The most successful strategy involved maneuvering into the universal blind spot—silhouetted against the sun—before attacking. To make their craft less visible in that position, the Germans covered two of their planes—the Fokker E.1 and the Gotha Bomber—with heavy **cellophane** skins. Unlike conventional canvas coverings, the cellophane allowed sunlight to pass through the plane. This new material, however, failed to hold up under the stress of flying and was soon abandoned.

By the 1930s, aircraft could fly high enough and fast enough to arrive at a target unseen by the human eye, at that time the only means of detection. As a result, only nominal emphasis was placed on stealth design. The one significant exception, however, was the American P-38 *Lightning*, designed by Lockheed, which was difficult to see, with its smaller side and front profiles.

The development of **radar** at the beginning of World War II significantly changed air strategy. Airplanes previously invisible to observers could now be sought out and destroyed. To counteract land-based radar, aircraft designers equipped planes with radar of their own and devices to jam or distort enemy signals. Some bombers even dropped chaff, thin metal strips that confused radar and disguised their location. Electronic warfare, as it came to be known, escalated dramatically during the war and following the advent of the **jet engine** in the 1950s. As a result, aircraft, particularly bombers, grew larger and more complicated. Also, as radar grew more sophisticated and diversified, strike missions became more dangerous and often involved a wide range of aircraft, several of which were exclusively dedicated to creating an electronic smoke screen. By the end of the Vietnam War, the danger and expense of relying on electronic warfare compelled the United States military to seek an alternative. It eventually turned to Lockheed, the company that had designed some of the military's most successful reconnaissance aircraft.

During World War II the need for covert flights had become apparent as spy networks proved increasingly inadequate. In 1952, Air Force Major John Seaberg, an aeronautical engineer, developed a design concept for a spy plane which required that consideration be given not just to performance but to minimizing detectability. With the help of the Central Intelligence Agency, the chief engineer at Lockheed, Clarence L. "Kelly" Johnson, and his design team set about realizing Seaberg's concept, focusing on the radar cross-section (RCS) or detectability of their prototype by radar. The "Skunk Works," as the group became known, eventually produced the U-2. First flown in 1955, the U-2, with its small size, slender profile, and radar-absorbent black paint, proved the value of designing with stealth in mind. Lockheed followed the U-2 with the A-12, which eventually evolved into the SR-71 *Blackbird*, one of the most distinctive airplanes ever built. Its wing-mounted engines and slim fuselage created a slender side profile, and its smooth body, boasting few straight angles, made it difficult to detect with radar. Designers also used a radar-absorbent **plastic** material on the wing's leading edges and control surfaces. Despite these advances, however, manned reconnaissance was largely abandoned after the now famous incident in which a U-2 was shot down over the Soviet Union in 1960.

The United States military then turned to spy satellites and unmanned drones. These drones, first developed by the Ryan Aeronautical Company, utilized much of the stealth technology generated by the manned flights. In designing its own drone, the D-21, Lockheed made more extensive use of radar-absorbent and radar-transparent materials and created an extended tail pipe that masked possible infrared trails. Lockheed also developed the manned QT-2 and the YO-3A, both of which produced only minimal engine noise and were used successfully during the Vietnam War.

In an effort to attract the attention of the United States Air Force, Lockheed began to develop a fighter jet that incorporated the technology of the A-12 and the D-21. The Air Force had already allocated some funds to similar research. Leo Windecker, an aircraft designer from Texas, had approached the military as early as 1963 with a proposal for a plane made of a composite material transparent to radar. By the early 1970s Windecker's *Eagle* had become the Air Force's prototype for stealth research into non-metallic airframes. Lockheed eventually persuaded the military to fund their prototype as well. Officially designated *Have Blue*, the project was also unofficially referred to as *Project Harvey*, after the invisible rabbit of the 1950s movie.

The members of *Have Blue* had to overcome a wide range of obstacles. When in flight, an aircraft is betrayed by many different signals: it puts out infrared radiation due to its warmth, its noise and vibrations send out waves, it absorbs and reflects natural light, and it generates a radar signature whenever a radar beam hits it. Designers countered many of these telltale signs by having the weapons carried internally, burying the engines inside long inlet and exhaust ducts, and avoiding steep vertical slopes, large flat surfaces, and straight lines. To eliminate hot spots picked up by infrared detectors, the exhaust was deflected upward through a series of ports stretching along both sides of the rear fuselage. In addition, more advanced radar-absorbent materials were manufactured and used as the skin of the prototype.

These efforts culminated in the F-117A, finally revealed to the public in 1990, thirteen years after its first flight as a prototype. The F-117A proved its worth in the Persian Gulf War of 1991, when it successfully completed missions against Iraq's most heavily defended targets without suffering damage. The Stealth fighters have also had less fortunate moments in a well-publicized incident, an F-117A broke apart and crashed during an airshow in Maryland in early 1998.

Along with the F-117A, the most advanced stealth aircraft yet designed is the B-2 bomber. Although the government originally considered using Lockheed or Rockwell,

Production of the B-2 "stealth" bomber in Palmdale, California.

which had manufactured the B-1 bomber, it ultimately awarded the contract to Northrop Corporation in 1981, due in large part to that company's *Flying Wing* project. In designing the *Flying Wing*, engineers eliminated the fuselage and tail of the craft, leaving a large wing over which the weight of the craft is distributed. Northrop adapted this design for the B-2, effectively hiding the cockpit and engines in the wing. As a result, the almost perfectly flat bomber with its radar-absorbing skin diffracts most incoming radar waves. Its ability to hide from radar becomes obvious in a comparison with the B-52 bomber. The older aircraft echoes about as much radar energy as a minivan, while the B-2 appears more like a butterfly. Although the Air Force requested more than 100 B-2s initially, only 20 were authorized by Congress for construction. The first B-2s entered service in 1993, with several hundred flights logged by 1996. However, the cost of the B-2s (about $2 billion each) has kept the program under constant scrutiny and skepticism since its inception, and it remains unclear if authorization for additional bombers will be granted.

STEAM BOILER • See Steam engine

STEAM BRAKE • See Braking systems

STEAM ENGINE

Simple machines (such as **levers**, **pulleys**, **inclined planes**, **screws**, wheels and axles, and wedges) have been in use since antiquity; complex machines, though more recent inventions, rely largely on combinations and refinements of these basic machines. Since machines change the direction of work or the force required to perform it, machines need energy to function, whether it derived from animal or human muscle, wind or water currents, or heat-generated energy, such as steam. Steam engines were one of the earliest and most important inventions to convert one form of energy to another. Some historians even credit the steam engine with enabling Europe's industrial revolution, which occurred during the late eighteenth and nineteenth centuries.

The use of steam to power machines initially depended on the ability to create a vacuum. Experiments with vacuums and air pumps began in the 1600s, when the German scientist **Otto von Guericke** developed a hand-pump that could remove air from sealed containers. Experiments showing that candles would not burn and that a bell's ring could not be heard within such containers proved that a vacuum had been created. In 1654, Guericke more spectacularly demonstrated before the Emperor Ferdinand III that a team of eight horses could nor separate two half-spheres joined together through the force of an internal vacuum.

One of the first steam-powered machines was built in 1698 by the English military engineer **Thomas Savery**. His invention, designed to pump water out of coal mines and known as the Miner's Friend, was not precisely a steam engine but rather a water pump that used steam and condensation to create a vacuum. Savery's machine, which had no moving parts, consisted of a simple boiler, a steam chamber whose valves were located on the surface, and a pipe leading to the water in the flooded levels of the mine below. The Miner's Friend was based on expansion and suction effects achieved by generating and then condensing steam. Water was heated in the boiler until its steam filled the chamber. This created an expansion that forced out any water or air inside the chamber. A **valve** was then closed, after which cold water was sprayed over the chamber; this chilled and condensed the steam inside to form the vacuum. When the valves were reopened, water was forced up from the mine, and the process could then be repeated. However, the suction range of this apparatus was only twenty-five feet, and the pump never truly became practical for its intended use in mines.

The French-born British physicist **Denis Papin**, who was familiar with Savery's pump, also experimented with exploiting the properties of steam to create a mine pump. In 1707, he designed a pump with a moving **piston**, two receiving chambers, and the first safety valve. Even with these improvements, Papin's design remained inadequate for the large-scale requirements of mining operations.

Further advancements on the Savery pump were made in 1712 by the British engineer **Thomas Newcomen**, who, with his assistant John Calley, became the first to harness power by setting a moving piston inside a cylinder, a technique still in use today. Newcomen's steam engine modified earlier pumps in two important ways: a closed cylinder replaced Savery's chamber, and the piston, instead of a vacuum, was used to create motion. Newcomen's design was generally significant in its capacity to produce motion to power a machine. The motion created allowed the engine to sustain its own movement; earlier steam-powered machines required that their water and steam valves be constantly monitored and manipulated. Newcomen's engine is called an *atmospheric engine* because its motion relies on atmospheric pressure rather than on counterweights or the force of steam itself. Steam approximately two pounds per square inch (psi) above atmospheric pressure was admitted to the underside of a piston. At a determined point in the cycle, a jet of water was injected to the same space to condense the steam, lowering the pressure underneath the piston to a near-vacuum. The atmospheric pressure in the open end of the cylinder then forced the piston back down, producing a power stroke. The replacement of brass cylinders with iron cylinders after 1724 was an additional step toward the higher-pressure steam engines to come.

In the 1720s, the German inventor Jacob Leupold designed a remarkably advanced engine that eliminated the condensation and vacuum steps of this process. Leupold's design called for two symmetrical cylinders that moved in opposition to create a continuous power source. The engine was apparently never built, however, probably because of limitations in the materials available and the current levels of craftsmanship.

The Scottish engineer **James Watt** who, contrary to popular belief, did not single-handedly invent the steam engine, still faced these technological limitations nearly fifty years later. Nonetheless, his design improvements of the 1770s and 1780s contributed most to the development of the modern steam engine. The first such design modification was to seal the engine's cylinder and install steam valves on both ends. Earlier engines had injected steam only on one side of the cylinder and employed counterweights or atmospheric pressure and vacuums to reset the piston in position. These earlier engines suffered in efficiency because of the heat dissipated during the procedure, which made reheating necessary at the beginning of each cycle. Watt found it difficult to seal the cylinders because those available were still being bored as they had been in Newcomen's time, and would not seal tightly enough for Watt's purpose. Watt tried to use jackets around his cylinders and a patented ''stuffing box'' to prevent steam from escaping when the piston rod moved through the cylinders, but these techniques proved inadequate.

To address this problem, Watt entered into a partnership in 1773 with businessman **Matthew Boulton**, who financed and organized a search for techniques to manufacture a well-sealed cylinder. **John Wilkinson**'s boring machine, introduced in 1775, provided the solution. The refined design that resulted used one-third less fuel than a comparable Newcomen engine.

Watt also was responsible for an even more impressive innovation: attaching a flywheel to the engine. Flywheels accomplished two tasks. First, the inertia of a large flywheel allowed the engine to run more smoothly by creating a more constant load. Second, flywheels converted the conventional back-and-forth power stroke into a circular motion that could be adapted more readily to run machinery. Watt initially connected the engine and flywheel with a series of **gear**s, but began using a crankshaft for this purpose when its earlier patent expired. Watt continued to experiment with, and improve upon, steam engine technology. In 1784, he patented a steam jet condenser and a parallel-motion, double-acting engine that admitted steam to both sides of the piston.

The next advance in steam engine technology involved the realization that steam itself, rather than the condensing of steam to create a vacuum, could power an engine. The American inventor **Oliver Evans** designed the first high-pressure, non-condensing engine by 1804. This engine forced steam into a cylinder at pressures of up to fifty psi. Evans's engine, which was stationary, operated at thirty revolutions per minute (rpm) and was used to power a marble-cutting saw. The British mechanical engineer and inventor **Richard Trevithick** was experimenting at this same time with similar engines. The high-pressure engines of both men retained the use of large cylindrical tanks of water, heated from beneath, to produce steam.

The first practical alternative to these large boilers was introduced by **John Stevens**, a pioneer of the American **steamboat**. This new boiling system used rows of long, narrow pipes to carry water through flames. Water recirculated until it was converted to high-pressure steam, which then forced itself out of the pipes. The Stevens boiler system, however, frequently exploded when its unmonitored and uncontrolled pressure grew too great for the boiler and pipes to contain. Several mid-nineteenth century devices attempted to solve this problem though the use of stronger metals and improved seams—or to at least contain the damage of an explosion, as was the case with Isaiah Jenning's design for an outer casing for these boilers.

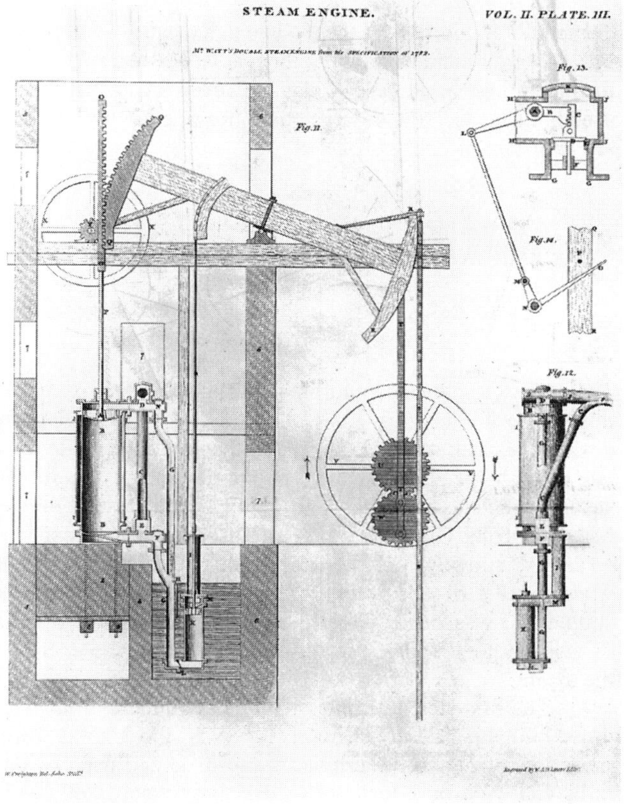

James Watt's steam engine was the primary power source of the Industrial Revolution.

In this same period, a somewhat peripheral debate persisted over whether a cylinder should lie vertically or horizontally. Each option seemed to have its own advantages and applications. For example, boats were best served by a horizontal cylinder that ran the length of the vessel. In buildings such as mills, however, where floor space rather than a height limitation was at issue, vertical cylinders that often extended through several floors were employed. Neither position seemed more efficient, safer, or easier to access than the other.

But this debate did trigger some new ideas. Some of these innovations involved resorting to alternated positions, as was the case with the oscillation or pendulum engine. In this engine, the cylinder swung back and forth, relieving stress in many parts of the engine and allowing for smoother operation. Alternate types of cylinders were also proposed, including the rotary cylinder, which eliminated altogether the flywheel and its interconnected moving parts. While such designs brought their own set of disadvantages—they often required more fuel or broke down more frequently—they continued the process of improvement and innovation on which technological advances depend.

In the 1860s and 1870s, **George Corliss** essentially perfected the steam engine. His large engines operated so smoothly that they could power textile mills in Scotland without breaking delicate threads, yet were large enough to power all the exhibits at the Machinery Hall during Philadelphia's 1876

Centennial Exhibition. Since the late nineteenth-century, gasoline-powered engines have largely replaced steam engines in the industrial market. Even today, however, steam turbines are used in the production of large quantities of energy for secondary distribution.

STEAM ROLLER • See Road building

STEAM TURBINE

The earliest known steam turbine was built by **Hero of Alexandria** in the first century A.D. Called the *Sphere of Aeolus*, the device consisted of a hollow sphere that could be made to rotate at high speed by feeding the interior of the sphere with steam, which escaped through small spouts resembling elbows. In 1629 Giovanni Branca told how a **wheel** could be turned by allowing jets of steam to hit vanes attached to the wheel's outer rim. However, the significant drawback to this simple idea was the need for the wheel to revolve with unheard of velocity to absorb even a useful fraction of the energy of the steam.

The first steam turbines with any commercial use were those built in 1831 by William Avery in the United States. About 50 of his crude turbines were made and used in sawmills and even on a **locomotive**. But they were finally abandoned because they needed constant repair and would not maintain a constant speed.

In 1883, Gustav de Laval constructed a simple steam turbine, consisting of a wheel with blades on its perimeter, that he used to power small machines such as his **cream separator**s. Laval's impulse turbine spun too fast (up to 30,000 rpm) to be used without step-down gears, and since it did not capture much of the energy from the steam that passed through it, it couldn't compete with other efficient engines in use at the time.

Seeing that a more efficient turbine would have tremendous commercial potential, Charles Parsons designed a practical one. Parsons's turbine had a rotor consisting of a drum with several sets of vanes attached to a shaft. Moving steam caused the drum to rotate as it pushed the vanes on its way through the turbine. Other stationary blades forced the steam against those that rotated, making use of as much energy as possible. The steam then continued until it encountered another set of turbine blades designed to work with the same steam at a slightly lower pressure. Because the steam expanded as it passed through the turbine, Parsons made the second set of blades larger and the exhaust end wider than the intake end. His turbines, while they had the ability to operate at slower speeds, could also be designed to spin as fast as 18,000 revolutions per minute with a high degree of efficiency. They were, however, prone to leakage between the moving blades and the casing of the turbine.

Parsons set up two of his turbines at a power station in England, thereby providing the first electricity by steam turbine. He also applied his technology to ships. He built a 100-foot (30 m) vessel, the *Turbinia*, which reached speeds near 20 knots (37 km/hr). Eventually, with three propellers this ship achieved the distinction of the fastest afloat when it reached 34 knots (63 km/hr). His engines provided high speeds with less vibration than the traditional reciprocating steam engines. Later, much larger steam engines (70,000 horsepower) drove famous liners like the *Mauritania* and *Lusitania*.

Around the turn of the century, two other engineers built upon the foundations laid by de Laval and Parsons. Auguste Rateau, a Frenchman, and Charles Curtis, an American, introduced modifications to turbine designs which vastly increased efficiency. The Curtis turbine, a kind of Laval turbine with many rows of blades, could produce the output of a piston engine at one-third its cost and one-tenth its size. In 1897, Curtis sold the rights for his design to the General Electric Company, which produced a working model in 1902. Rateau's multistage impulse turbine, built around 1897, corrected some of the leakage problems encountered by Parsons's turbine.

Within ten years, steam turbines had become an essential power source in naval vessels and electric generation plants. Today's steam turbines still rank among the most important and powerful machines in the world. Just one turbine powering a generator can supply all the electricity needed by 3,000,000 people.

STEAMBOAT

Like many inventions, a number of people can be credited with the development of the steamboat. As early as 1690 the French inventor **Denis Papin** used a **steam engine** similar to that of **Thomas Newcomen** to drive a **paddle wheel** boat. When he sailed a successful model of his vessel down the Fulda River in 1707, Papin was attacked by boatmen who, fearing for their livelihood, destroyed his craft. Thirty years later Englishman Jonathan Hulls (1699-1758) patented a design for a steam-driven boat, but his proposed engine was so large that the boat would have sunk under its weight. In 1783 the Marquis Claude de Jouffroy d'Abbans designed and built a 150-foot (45 m) ship that used a more efficient steam engine designed by **James Watt**. It was successfully piloted on a river near Lyons, France, for over a year but the Marquis could not generate enough financial interest in the project and was eventually forced to abandon it.

At the end of the eighteenth century, work on steamboats shifted to colonial America because such great rivers as the Hudson and the Mississippi were ideally suited for steamboat travel. Although several Americans had designed and built steamboats by the late 1780s, most financiers dismissed the work of these visionaries as unprofitable.

James Rumsey, one of these farsighted individuals, produced a jet-powered boat based upon **Benjamin Franklin**'s suggestion that if water were pumped from the front of the vessel and ejected at high speed, the expelled water would shove the boat ahead. Rumsey eventually demonstrated his boat for George Washington, who invested some money for further development. Rumsey continued but met with lack of adequate funding, a tale common to so many of these early inventors.

John Fitch was another whose attempts met with defeat. An early steamboat of his was 34 feet (10 m) long with 12 vertical oars connected to a steam engine in such a way that they alternately dipped in the water for a stroke and then were lifted out and carried forward for the next stroke. The mechanism was obviously too elaborate to work well, and the boat suffered many breakdowns. Fitch did succeed in building a small steam launch that traveled over 2,000 miles (3,218 km) as a ferry on the Delaware River, but he had no luck getting enough money behind his efforts.

Oliver Evans (1755-1819) refined the steam engine used for boats, which at that time used condensed steam and atmospheric pressure to move the **piston**. By 1804 he had invented a high-pressure, noncondensing engine in which the steam itself did the pushing under a pressure many times higher than the atmosphere. Naturally, this required a steam boiler, made of copper, which he made of copper, that could withstand a pressure of 50 pounds per square inch. Unfortunately, the combination of high pressure and nineteenth-century technology made for terrible explosions that plagued steamships for years.

One man who came close to success was **Samuel Morey**. In the late 1790s he built several steamboats. His first, while capable of carrying only two people, used its paddle wheel to navigate on rivers for three years. He tinkered with it by moving the paddle wheel from the side to the stern where it increased his speed to 5 miles per hour (8 kph). Robert Livingston (1746-1813), who later helped finance **Robert Fulton**'s successful venture in steamboats, offered to back Morey and his boat in a commercial venture, but Morey declined, apparently out of a desire to control any such lucrative operation himself. Livingston, however, eventually teamed up with Fulton, and Morey lost his financial support. Although Fulton visited Morey and must have been inspired by his steamboat, very little credit came to Morey for his achievements.

Meanwhile in Great Britain, Scottish inventor William Symington (1763-1831) installed a steam engine that powered paddle wheels mounted between the twin hulls of a catamaran. The British Secretary for War saw the ship and commissioned a steam tugboat to tow barges on a **canal**. The resulting sternwheel ship, the *Charlotte Dundas*, was 38 feet (11 m) long and is considered the first functional steamship. It was in regular service for only a month, however, because its wash damaged the canal banks.

Fulton, who rode on the *Charlotte Dundas*, created the first commercially successful steamboat. He had been commissioned by Livingston, then the American ambassador to France, to build a steamboat. Although Fulton had no direct experience with these vessels, he had worked for seven years as an inventor and engineer in Europe. While building a **submarine**, Fulton had in fact designed a small steamer, but he miscalculated stresses on the hull and broke the vessel. He learned a great deal from this disaster, and as a result, was able to proceed without the tedious process of trial and error that hampered others. His steamboat was a success in France, and Livingston urged him to build another in America. Fulton did so and in 1807 launched the Clermont in the United States.

Once steamships could operate on rivers, the next challenge was ocean travel. The problem was storing enough coal to fuel the engine. One temporary solution was to combine sail and steam. In 1819 a full-rigged American ship, the Savannah, crossed the Atlantic in 21 days, but used her engine for only a few hours. In 1838 the Sirius made the journey with only engines to power her. The Great Western, launched the same year, was the first profitable steamship to cross the Atlantic.

Also during this time a better means of propulsion, the **screw propeller**, won out over the paddle wheel, which was ill-suited to rough ocean conditions. In 1838 Swedish inventor **John Ericsson** used a screw propeller to drive a small ship named the Archimedes. This impressed the creator of the Great Western, **Isambard K. Brunel**, who was designing the

Robert Fulton's *Clermont*, the first commercially successful steamboat.

Great Britain, the world's first all-metal liner. After seeing the *Archimedes*, he scrapped his plans for paddle wheel and substituted a six-bladed propeller that proved a success.

In 1894 steamboat design changed again when **Charles Parsons** built the first successful **steam turbine** to increase power and speed. He used steam to drive a vaned wheel much like wind does a pinwheel. His boat, the *Turbinia*, equipped with three propellers, astounded British naval experts when it reached speeds of 34 knots (63 kph). The first of the giant steam liners, the Mauretania, was launched with turbine power in 1905.

One final change in steamships was their fuel. Coal was used originally, but it was a bulky material that required a great deal of handling to load and unload. Oil eventually replaced coal because it could be pumped aboard quickly, stored in otherwise useless tank space, and also because it boosted the ship's cruising range. Another advantage of switching from coal to oil for the furnaces of steamships was the cleanliness of the engine room, since every aspect of handling, storing, and burning coal made for filthy conditions.

The oil-fired boiler/steam turbine combination remained standard until after World War II, when the gas turbine and the diesel engine slowly replaced it. The gas turbine offered very high power for its very small size and light weight, while the diesel offered reliability and economy. During the 1970s, the energy crisis and the steep rise in oil prices effectively eliminated the steam turbine as a major source of power for merchant ships. It simply could not compete with the efficiency of the diesel. Steam remains in use at the close of the twentieth century on only two types of vessels: very large ships like aircraft carriers that use steam turbines powered by nuclear fuel (a nuclear reactor is used to produce steam, which then turns a turbine), and the liquefied natural gas carrier. As a means of providing propulsive power to a ship, steam really has no future, which now belongs to the diesel.

Steam-powered road vehicles

Credit for the first steam-powered road vehicle is given to **Nicholas-Joseph Cugnot**, a French military engineer. The French minister of war, the Duc de Choiseul, commissioned Cugnot to build a steam carriage capable of transporting large **artillery** pieces. Cugnot's response was to build a heavy three-wheeled vehicle that carried no reserves of fuel or water. Consequently, it was very limited in its use, although it did perform as a carriage by carrying four passengers in Paris in 1770. It moved at only a walking pace and was considered impractical because, after the pressure diminished, the copper steam tank had to be refilled with water and a fire had to be built under it to develop more steam.

Steam engines use the heat of combustion to boil water, converting it into steam vapor. By this action, its volume is expanded greatly, creating pressure. The resulting pressurized steam is then transferred to an engine, where it is used to push a **piston**; a crankshaft is used to convert the reciprocating motion of the moving piston into rotary power that can be used to turn lathes or other machine tools, or, as in Cugnot's application, to propel a vehicle. The greatest drawback of the steam engine is its inefficiency. Much of the potential energy of the burning fuel is converted to heat, and more is lost in the transfer of the steam to the engine. Also, other than building an engine roughly to the size considered appropriate to its proposed use, it is extremely difficult to equate the amount of fuel burned to the amount of energy needed; obviously, this adds to its inefficiency.

Cugnot developed at least two subsequent models, but his mentor, the Duc de Choiseul, was removed from office in about 1771. As a result, Cugnot's steam-powered carriages were abandoned and his work languished until the British developed steam carriages at the turn of the century.

Credit for the first practical steam carriage is often given to Englishman **Richard Trevithick**, who drove his carriage in 1801. His invention was essentially a four-wheel stagecoach equipped with a steam engine. Austrian Joseph Bozek and several British inventors followed Trevithick's lead. In the 1820s and 1830s in England, several carriages ran a service between Cheltenham and Gloucester, attaining speeds of up to 14 miles per hour (22 kph) and carrying up to 22 passengers. Among the notable Britishers was **James Nasmyth** who, at age 19, built a steam-powered carriage that was the model of efficiency for its time. The most successful of these early British steam-carriage makers was Goldsworthy Gurney, whose carriages carried 3,000 passengers over 3,500 miles (5,631 km) without a serious accident.

Despite these successes, steam carriages were noisy, tended to scare horses, dirtied the air with smoke, and scattered hot sparks that set fire to crops and wooden **bridge**s. These early British carriages were also opposed by those who wished to maintain the supremacy of horse-powered haulage. Due to the problems mentioned above and faced with active opposition from various other competitors, the British government passed several laws that made it difficult for steam carriages to compete. One of these laws required that each carriage be preceded by a man waving a red flag; another law restricted their speed to 4 miles per hour (6 kph). These laws, often modified but never repealed, stifled English efforts to develop steam carriages and remained in effect in England until 1896. By this time, engineers in other countries—notably France and Germany—had developed carriages powered by gas engines, supplanting the steam engine as the power plant of choice for self-propelled vehicles.

In the United States, credit for the first steam-powered vehicle is given to **Oliver Evans**, who built a machine powered by a steam engine to dredge the Philadelphia harbor in 1805. Evans built the dredge on the shore and coupled its steam engine to power not only its paddle-wheel, but the wheels that moved the machine over land as well. Evans's Orukter Amphibolos, as he called his creation, propelled itself to the river's edge and supplied its own power on the water, becoming the first vehicle known to travel on both land and water.

Although many Americans, such as J. N. Carhart, Richard Dudgeon, Sylvester Roper, and Ransom Eli Olds (1864-1950), developed steam-powered **automobile**s, the most successful U.S. manufacturers were the twins **Francis Stanley** and **Freelan Stanley**, who built their first steam-powered car in

1898. In 1906, the brothers built a steam-powered car that set a world's record for the fastest mile—28.2 seconds, which is more than 127 miles per hour (204 kph). Despite this success, the public did not wholeheartedly accept the steam automobile. Steam cars were hard to start and hard to operate. They were also impractical for long-distance travel, and the public was reluctant to embrace an automobile that required an open fire and hot steam for propulsion.

Francis Stanley was killed in an automobile accident in 1918; his brother was ill at the time, and the Stanley Motor Company built only a few automobiles after Francis's death. Instead, the company's engines were shipped to Europe where they were used extensively to pump water from the trenches at the front during the World War I. While the Stanley Steamer was a relatively successful automobile, continuing developments in **gasoline**-powered engines led to the abandonment of steam as a practical source of automotive power. After the war, **Henry Ford**'s "Tin Lizzie" dominated the automobile market in the United States. The public's aversion to steam-powered cars led to the bankruptcy of the Stanley Motor Company, which ceased production in 1924.

Steel alloys

Alloy steels are made of steel combined with other metals such as nickel, chromium, or vanadium. The result of combining the metals is a steel product that has increased hardness, strength, durability, malleability, and resistance to corrosion.

The first alloy was developed by the English physicist **Michael Faraday** in the 1830s. His experimentation in the area of electromagnetic induction and electrolysis resulted in an iron-chromium alloy. In 1883, English metallurgist Sir Robert Hadfield (1858-1940) invented manganese steel at his father's steelmaking firm. Manganese gave the steel strength and resistance to wear. He also invented silicon steel and other alloy steels. Hadfield erroneously assumed that chromium impaired corrosion resistance, otherwise he might have discovered stainless steel.

An American, Elwood Haynes (1857-1925), developed several steel alloys. In 1881 he invented tungsten chrome steel which retains its strength at high temperatures. A chromium/nickel alloy followed in 1897; in 1900 he introduced one of cobalt/chromium; and in 1911 came stainless steel. Building on Haynes's work, other scientists developed stainless steels. German scientists P. Monnartz and W. Borchers were among the first to realize the usefulness of stainless steels. However, the English metallurgist, Harry Brearly, receives the most

The first American steam-engine wagon, built by Oliver Evans in 1804.

credit for developing stainless steel. He accidentally discovered that nickel-chromium steel made a good anti-corrosion alloy when he found a sample of it among a pile of discarded experimental scraps with its shine still intact. In 1912 Brearly proposed that the new alloy be used for naval guns. By 1914 he had proved its usefulness as a material for cutlery and obtained an American patent on it in 1915.

Frederick Mark Becket (1875-1942), a Canadian-American metallurgist, developed the silicon reduction process for the mass production of low-carbon ferroalloys and stainless steels. John A. Mathews of Washington, D. C., participated in the development of permanent **magnet** and corrosion-resistant steels, inoculated iron and high speed vanadium steels in the early 1900s. Mathews received the Andrew Carnegie Gold Medal for Research for his work. In 1912, German researchers Edward Maurer and Benno Strauss developed the austenitic group of stainless steels which are heat-resistant and shock-resistant, making them attractive for use in cookware and chemical equipment.

All steel and steel alloys are classified according to their carbon and alloyed metal content. A higher carbon content increases the strength of the metal. Low-carbon steels containing carbon of 0.08-0.15% are used in automobile bodies, containers and tubing. Medium-carbon steels having 0.15-0.35% carbon is used in railroad equipment, bridge and building frames, ship structures and machine parts. The high-carbon steels, 0.65-1.2%, are used in steel rails, drill bits and cutting tools. In alloy steels, the carbon content generally is very low and acts mainly as an alloying agent. The exception is in superplastic steel alloys developed in 1975. Superplastic steel alloys have 1-2.1% carbon are hard yet not as brittle as such a high carbon count would indicate. Because they have an ultrafine grain size, about 2 microns, superplastic alloys can be stretched to 11 times their size, and display high wear-resistance.

The characteristics of each alloy depend upon the metal that is added to the steel. Some, like nickel and chromium, impart toughness, heat and corrosion resistance. This alloy is used in axles, gears and bearings, and in most stainless steels. Other metals, like manganese, which is added to all steel, improve its workability, and in higher quantities, increases the alloy's strength. Molybdenum is added for its hardening ability and heat resistance. Tungsten is also used as a heat resistant alloy. Boron adds strength to steel at amounts smaller than other alloys. Only 0.0005% is enough to achieve the desired level of strength and toughness. Steel alloys have allowed many important advancements in modern society and are an essential part of the complex **steel production** industry.

STEEL CABLE • See Wire

STEEL PRODUCTION

For more than a century, steel has been the most important metal in industry. By the turn of the twentieth century, steel had replaced its parent metal, iron, as the primary metal for heavy industries such as construction, railroads, and, later, the **automobile** and airline industries. The invention of **Henry Bessemer**'s converter in 1856 was the true beginning of steel as a commercially viable material.

Steel is a malleable alloy of iron with carbon and other trace elements. It is made by carefully removing excess amounts of carbon from pig iron, which has about four or five percent carbon content, and adding the other trace elements. Carbon steel contains less than one per cent carbon. Steel has been known for about two thousand years. Even earlier, smiths worked meteoric iron, which usually contains a large portion of nickel and resembles stainless steel in its composition. In Sri Lanka, in the seventh to the eleventh centuries, for example, they produced high-quality steel using furnaces that used monsoon winds fan flames for the furnace. These furnaces might have produced the legendary Damascus swords. There is evidence of co-fusion production of steel in Merv (located in Western Asia) in the ninth century.

However, steel was never produced in large quantities until the late 1800s. New methods in the manufacture and use of iron during the 1700s and early 1800s made iron more practical and affordable than ever, which helped keep steel in the background for more than a century. **Blast furnace**s and rolling mills were producing wrought and cast iron in affordable quantities. Both the Industrial Revolution and the architectural revolution that utilized load-bearing iron frames, were founded on the iron industry.

Bessemer's converter, or pneumatic conversion process, was the first efficient method of removing carbon from pig iron in amounts necessary for mass production of steel. The method applied blasts of cool air directly to the molten iron for rapid carbon burn-off. The same process credited to Bessemer was actually invented about five years earlier by the American **William Kelly**. Kelly chose to keep his invention secret rather than patent it. Bessemer, unaware of Kelly's work, made the same discovery and gained the commercial recognition by going public with it right away. Later, in 1875, Englishman **Sidney Gilchrist Thomas** took the Bessemer process further by introducing burned limestone to remove the phosphorus from iron ore. Bessemer had used phosphorous-free ores. Thomas' improvement widened the range of ores suitable for steel production.

Bernard Lauth of Pittsburgh inadvertently discovered cold rolling when a pair of tongs was reshaped after an accidental trip between the rollers. After developing the process, he opened the first cold-rolling mill in 1860. Lauth's method processed the metal by rolling iron and steel at room temperature at high speed. Since hot rolled steel required additional processing, the cold rolled method soon replaced it.

The next improvement to steel production was the open-hearth process. The open hearth used regenerated heat for a more efficiently sustained heat. The process is called the Siemens-Martin Process. Sir **Charles William Siemens** invented the process in 1856. The following year, **Pierre-Émile Martin** made improvements in the arrangement of Siemens' heat-capture chambers and also introduced scrap steel to further cheapen the steel production process.

In 1878, Siemens developed an **electric arc** furnace which produced an iron-melting temperature with an electric current. In 1898, Paul-Louis-Touussaint Heroult (1863-1914) incorporated the electric arc for commercial steel production.

The Siemens-Martin Process surpassed the Bessemer Process as the leading steelmaking process by the beginning of the twentieth century. It remained unchallenged until the 1950s and the development of the basic oxygen process. This process is also known as the L-D process, for Lin-Donawitz, the Austrian town where it was first developed. The basic oxygen process is similar to the Bessemer conversion process except that air is blown in from above instead of below and at supersonic speed. Molten iron is poured into the furnace to melt scrap steel faster. By 1969 the basic oxygen process had become the leading steelmaking process in the United States. In the 1980s, bottom-stirred furnaces were introduced, which reduced the carbon levels in steel, and became industry standards.

Steel is produced in units called ingots which are quenched, or rapidly cooled by dipping them in water or oil. The ingots are then delivered to semi-finishing mills where they are further reduced to blooms (rolled ingots), bars, and billets (short bars). The steel is then sent on to finishing mills where it is shaped to fit its intended use, as plates, sheets, **wire**, pipe or more specialized shapes. Some of the finished steel is delivered to warehouses or construction sites, while other forms are sent to assembly plants to go into **automobiles**, **washing machines**, **trains**, **airplanes** and a multitude of other products.

Steel is often alloyed with other metals to combine the qualities of both metals. Metals such as chromium, tungsten, nickel and vanadium are added to increase hardness, resiliency, durability and resistance to heat and corrosion under various conditions. Stainless steel has chromium as its chief alloy. Chromium resists rust and renders the metal more visually attractive.

Steel and its related industries has become central to the economies of many regions. In the United States, production is centered around the Great Lakes and the Ohio River valley. These areas are strategically situated to take advantage of water and rail transportation routes, which allow raw materials to be brought in and finished products to be shipped out. The coal reserves of the Appalachian mountains and central Illinois are near by to provide fuel for the furnaces. Finally, the steel mills require large amounts of water which is provided by the lake or river. Other world steel production areas are Japan, the

Henry Bessemer's invention of the pneumatic conversion process, or bessemer process, in 1856 made steel a commercially viable material. His method, which involved blasting molten iron with cold air, was the first efficient mean of removing carbon from pig iron in amounts necessary for mass production.

Charles Proteus Steinmetz

Ruhr Valley of Germany, central England and the Donetsk Valley of Russia.

International competition has caused a decline in steel production in some areas that were once major industrial centers. Many mills closed during the 1970s due to inefficient employment structures and failure to modernize older plants, causing hardships for many communities in which the mills were once their lifeblood. However, several new methods of steel production were in development in the 1990s. American's developed direct steel-making, which reduces steelmaking to a one-step process and eliminates the need for coke. Untreated coal, iron ore and scrap metal are mixed together and injected with oxygen, treated, and purified. The coal's function is two-fold. It works as a heat-generating fuel and as a reducing agent. When the molten metal is ready, it is poured into a ladle and further refined into partially finished products like sheets. Direct steel-making cuts costs and also releases fewer toxins into the atmosphere. With such innovations, steel production will remain viable in the United States for the foreseeable future.

Steinmetz, Charles Proteus (1865-1923)

German American electrical engineer

An **electronics** genius, Charles Proteus Steinmetz founded the General Electric laboratory and refined and standardized the study and notation of **alternating current** circuitry. In all, he patented some 200 electrical inventions. A public-spirited visionary, he also predicted problems with air and water quality, especially from the burning of soft coal.

Born in Breslau, Germany, on April 6, 1865, Steinmetz inherited an oversized head, twisted spine, and hunchback, deformities passed down by his father and grandfather. His father sent him to the University of Breslau, where he compensated for his disability by becoming politically and socially involved in campus activities. Extending his education in Berlin and Zurich, he studied language, science, math, and medicine. Because Steinmetz edited a socialist newspaper and wrote an inflammatory article, he was forced to flee Germany or face arrest. Abandoning the final stages of a doctoral degree, he escaped first to the Polytechnic Institute in Zurich, Switzerland, and then on to America, where he became a citizen in 1894 and Americanized his name from Karl August Rudolf Steinmetz to Charles Proteus Steinmetz, Proteus being his university nickname.

Rejected by **Thomas Edison**'s factory, Steinmetz drafted plans for streetcar motors at Eickemeyer and Osterheld Manufacturing, a small Manhattan electrical company. In Yonkers, New York, Steinmetz maintained an experimental laboratory and studied alternating current, which varied from the mainstream **direct current** system developed by Thomas and Charles Edison. Steinmetz delivered addresses on **transformer** design to the American Institute of Electrical Engineers, and explained how to generate electrical power cheaper and more efficiently. As a consultant for General Electric, the first major American electrical company, he moved to Lynn, Massachusetts, to oversee the calculations department. From experiments conducted at the Schenectady plant, he patented the magnetite arc street lamp and the aluminum lightning arrester and developed **turbine**s for the Niagara Falls power station.

Advancing studies begun by **Nikola Tesla**, Steinmetz's research on hysteresis (a magnetic phenomenon that caused power loss in motors) was applied to the design of AC circuits, and resulted in precise calculations of magnetic resistance. Steinmetz also originated standard symbolic notation for schematic drawings of ciruits. Steinmetz was always far ahead of his colleagues in mathematical skills, and few in his lecture's audience understood the brilliance of his ideas about magnetism and **alternating current** circuits. Steinmetz published his theories in articles and textbooks, including educational books to teach the mathematics required to understand his ideas. At General Electric Steinmetz staffed the company lab with bright, creative people working under his tutelage.

Steinmetz served as president of the American Institute of Electrical Engineers, head of Schenectady's board of education, and president of the common council. From 1902 to 1914, he taught electrical engineering at Union College. In this capacity, he studied lightning and its effects on power relay systems. From his studies of artificial lightning in 1921, he created measures to protect high-tension power equipment from destruction by lightning bolts. Steinmetz maintained productive friendships with noted inventors and scientists, including **Henry Ford**, **Guglielmo Marconi**, Tesla, and **Albert Einstein**. He also published an impressive list of nine scientific volumes, including *Theory and Calculation of Alternating Current Phenomena* (1897). He died of heart failure on October 26, 1923.

See also Arc lamp

Stellarator

The stellarator is a device for containing the very hot gases that are used in experiments on the production of nuclear fusion for commercial purposes. Nuclear fusion is the process by which small atomic nuclei, such as protons and deuterium atoms, combine with each other to form larger nuclei. Theoretically, this reaction can be used to generate enormously large energies. The **hydrogen bomb** is a practical example of the way in which fusion energy can be utilized, albeit in a very destructive way.

Scientists would like to find a way to harness the energy produced by fusion for peaceful uses. The problem is that the combination of nuclei does not take place very easily. Two protons, for example, carry a common positive charge and experience a strong electrical force of repulsion. Only if the two nuclei are given very large kinetic energies can this force of repulsion be overcome. In practical terms, that means the nuclei must be heated to temperatures ranging into the hundreds of millions of degrees.

The technical problem, then, is how to contain a reaction that occurs at, say, 100,000,000K (99,999,727°C). Obviously, no ordinary construction material can withstand this temperature. The most common solution to this problem is to trap the hot gases inside a magnetic field. In the 1950s, the American astrophysicist, Lyman Spitzer, designed a magnetic field that could be used to contain a fusion reaction. The magnetic field takes the shape of a figure eight, wrapped around the hot gases. The field acts like a kind of magnetic "bottle" holding the fusion reaction inside.

Spitzer arrived at his design as a result of his earlier interest in the formation of stars. He tried to understand how stars can form out of cool hydrogen gas in the presence of weak magnetic fields. The results of his speculation form the basis of our present understanding of star formation in galaxies.

Spitzer was born in Toledo, Ohio, on June 26, 1914. He received his doctorate from Princeton University in 1938. He was on the faculty at Yale University from 1939 until 1942. Then, during World War II, he worked on undersea warfare. After the war, he returned to Yale. In 1947, he was appointed chairman of the astronomy department at Princeton. Spitzer was one of the first American scientists to express an interest in space research using **rocket**s and artificial satellites.

Stephenson, George (1781-1848)

English inventor and industrialist

George Stephenson was exposed to steam power at a very early age. He was born near Newcastle, England, on June 9, 1781. By the age of fourteen, George was assisting his father, who was a fireman for a steam-powered pump at a coal mine. Within three years, George too was performing as a fireman and learning more about steam technology.

George Stephenson

Stephenson attended night school so he could learn to read and study the writings of **James Watt** and **Richard Trevithick**. Later he assisted his own son, **Robert Stephenson**, with homework in order to broaden his own education. Studying and reading evidently helped him, for soon after 1812 George had built thirty-nine stationary **steam engines** and had begun replacing horse-drawn coal sleds with cars on wooden rails pulled by stationary engines.

Stephenson was also trying to design a traveling steam engine that could carry itself, and cargo, over land. Stephenson's employers commissioned him to build a **locomotive** to carry coal out of their mines. His *Blucher* began operating on July 25, 1814. His design of flanged wheels, as well as his perfection of other innovations such as Trevithick's steam-blast technique (tested on the *Wellington* and *My Lord*) led to his selection as engineer of a railway between Stockton and Darlington, England. The steam-blast technique directed waste steam up the chimney of the boiler fire. This technique increased the draft of the fire and raised both temperature and pressure in the boiler. This innovation, based on Trevithick's early designs, was perhaps the single most important contribution to the history of the locomotive because it eventually led to faster running engines.

George persuaded Edward Pease (1767-1858) to help him establish a locomotive works at Newcastle where he could manufacture his designs. The 12-mi (19 km) Stockton and Darlington Railway opened on September 27, 1825, with Ste-

Robert Stephenson

phenson's *Locomotion* pulling the first train. Though designed primarily as a freight line, the S&L also carried people, becoming the first railway to carry passengers. Thirty-eight cars were pulled at speeds averaging 12-16 mi (19-26 km) per hour.

In 1827 Robert Stephenson joined his father's business. In 1829 their *Rocket* made its famous run at the Rainhill Trials, reaching a top speed of 36 mi (58 km) per hour. The next year George opened his Liverpool & Manchester Railway, the world's first railway built to provide passenger service. Opening day trains carried 600 persons.

The success of the L&M line triggered the rapid development of rail transportation in Europe and the United States. George retired from engineering in 1840, leaving his son to carry on the family business and make additional contributions to the steam industry. Still, he continued to play a major role in the development of rail transportation in England until his death near Chesterfield on August 12, 1848. Over the years, he has become recognized as the father of railway transportation.

Stephenson, Robert (1803-1859)

English engineer

Robert Stephenson was born near Newcastle, England, on October 16, 1803. He was the only son of **George Stephenson** and was sent to the best schools and served a challenging apprenticeship with his father. He served as a mine supervisor in Colombia from 1824 to 1827. Upon his return he managed Robert Stephenson and Company, manufacturers of **locomotives**, which was founded in 1823 by his father with the financial support of Edward Pease (1767-1858).

The firm's first engine was the 1828 *Lancashire Witch.* The *Witch* had inclined cylinders that were connected directly to crank pins on the **wheels**. This innovation was intended to lessen the jumping motion created by vertical cylinders and to reduce the frequent derailments associated with vertical movement. The *Witch* was a direct predecessor to the infamous *Rocket.*

The *Rocket* was completed in 1829 and began the century of the steam locomotive in both England and the United States. In the late 1820s, railroads were still facing many challenges: financial supporters, for example, were skeptical that steam locomotives could ever completely replace horse-drawn transportation.

Robert was determined that the new Liverpool and Manchester Railroad would rely on steam power. The L&M announced a contest, with a 500-pound prize, to find a suitable engine for the line. The *Rocket* met the requirements and was announced the winner. As small as it was, the *Rocket* was able to perform consistently at an average speed of 14 miles (22.5 km) per hour on the 60-mile (96.5 km) Rainhill track and hit a peak speed of 36 miles (58 km) per hour with a cargo of more than two dozen passengers.

The *Rocket* incorporated many new technological advancements. First, like its forerunner, the *Lancashire Witch*, its **piston** was connected directly to its wheels by a crankshaft rather than by the previously used sets of rocking arms, chains, **levers**, and counterweights. Second, exhaust steam from the *Rocket*'s boilers was successfully sent up the smoke stack to increase the fire's draft and temperature. This innovation became known as the steam-blast technique, which was introduced by **Richard Trevithick** around the turn of the century as part of his high-pressure steam applications.

The Rainhill Trials proved that rail transportation could indeed be powered by steam. When the 31-mile (50 km) long L&M opened in 1830, it became the first railway in the world to rely on steam locomotives and was known as the fastest and most regular line in operation. Robert and his father continued to build locomotives such as the *Meteor*, *Comet*, *Dart*, and *Arrow*. The firm's design reached near perfection with the 1830 *Planet* which utilized all of Stephenson's innovations to that date. The company soon shipped locomotives to France, Germany, Italy, Belgium, and the United States.

In 1833 Robert became engineer-in-chief of the new London-Birmingham Railway. He moved to London, England, to serve on Parliamentary committees and to solve problems related to the new railway. On June 24, 1838, the first of its trains traveled through tunnels, over subterranean quicksand, and over bridges.

Robert's work with the railway led him to another area of interest: bridge engineering. Although his first structure collapsed with a train on it, he soon developed a tubular construction which eventually led to the building of the Britannia Bridge over the Menai Straits with two 459-foot (140 m) spans (the longest wrought-iron span prior to this was 31 feet [9.5 m]). Similar structures were built at Conway, over the St. Lawrence at Montreal, Canada, at Newcastle and at Berwick, England, and in Egypt.

The Stephensons became known as the parents of steam-powered rail transportation and were honored throughout the world. When Robert died in London on October 12, 1859, he was buried in Westminster Abbey.

Stereo

On August 30, 1881, Clement Ader (1841-1926) received a patent in Germany for the first stereophonic system. Two groups of **microphone**s were placed on either side of a stage. These microphones relayed the sound of a play or musical directly to two **telephone** receivers, which paying subscribers held to their ears. He displayed this invention the same year at the Paris Exposition.

During World War I a similar system was used. What was referred to as "binaural receiving trumpets," were set up to locate enemy **aircraft**. Two large horns, like those on early **phonograph** systems, were connected by rubber tubes from their thin ends to the ears of operators. By using these horns, the operator could more readily hear from what direction planes were coming.

In the 1930s, further progress was made in stereophonic systems. The Bell Telephone Laboratories, under the leadership of Harvey Fletcher (1884-1981) and others, including the renowned conductor Leopold Stokowski (1882-1977), was the prime investigator in the area. Researchers there set up a tailor's dummy, nicknamed Oscar, who had microphones built into his ears in an attempt to reproduce the conditions of human hearing as closely as possible. If the electrical signals were sent through separate **amplifier**s to spaced **loudspeaker**s, the listener was able to tell where the original instrument had been located in the room where it was recorded. Headphones improved on this system: when left and right signals were fed to the left and right earpiece, the stereo effect was very good. In Germany today this idea is still used; it can reproduce a 360-degree sound field that seemingly surrounds the listener.

However, this system of using a dummy head was not very flexible. The British engineer Alan Dower Blumlein was responsible for using ordinary microphone arrangements for stereo recording. He found that good stereo could be created by either of two methods that depended on the intensity of the signal or the time of the signal's arrival. For the intensity method, he set up two microphones next to each other but turned in such a way that each aimed at the side of the stage. Thus, a sound located in the center would not register as intense, while a sound either to the left or right would be picked up more readily by whichever microphone was aimed that way. He also found that stereo could be created by placing two microphones far apart; they picked up signals at a slightly different time depending on where the sound originated. In 1933, Blumlein created the first stereo recording for the British firm EMI (Electric and Musical Industries). The record contained two sets of information: the left-hand signal was engraved on the inside of the groove, while the right-hand signal appeared on the outside of the groove.

Since the 1970s, stereo has become far more sophisticated. **Magnetic recording** tapes and **compact discs** are among the foremost developments affecting stereo sound. Broadcasters learned how to transmit stereo through multiplex systems that first put the two channels into a composite signal that was later restored to true stereo through a matrix circuit in a stereo receiver. Sound engineers have also created four-channel stereo (quadraphony), often called surround sound which consists of four microphones, four amplifiers, and four loudspeakers that together provide remarkably realistic sound reproduction. In the 1990s, a 5.1 channel surround sound system became popular. Consisting of a left, right, center, two surrounds and a subwoofer, this multichannel system provides realistic sound space. Though first embraced by home theater enthusiasts, the 5.1 system will inevitably take over home stereo use as well.

See also Noise reduction system

Stereoscope

The stereoscope was the first device to show **photograph**s with the illusion of three dimensions. Ordinarily, a **camera** only shows images in two dimensions, but two different cameras used simultaneously can simulate two eyes, if the **lenses** of the cameras are the right distance apart. Thus, when the resulting photographs are viewed together, the image appears to have depth.

The first stereoscopes were developed in France in the early nineteenth century as an offshoot of the intense research on light and optics that was happening. The stereoscope was modified in 1849 by Sir **David Brewster**, who had invented the **kaleidoscope** some thirty-five years earlier. The improved stereoscope featured a system of lenses and **prism**s that would overlap the two photographs without straining the viewer's eyes.

Like the kaleidoscope, the stereoscope was a popular success; marketed as "refined amusement," it found its way into the drawing rooms of high-minded audiences everywhere, offering views of famous landmarks and faraway vistas. Brewster even advertised its properties as an educational tool, claiming its three-dimensional images showed the nuances of art and nature that textbooks and lectures could not.

Modern inventors have taken Brewster's stereoscope-with its handle, slide, and screened-lens prisms—and have incorporated several improvements. The most common modern stereoscope is housed in a **plastic** casing with two holes for viewing, and is most familiar as a child's toy that uses slides on a **paper** disc to display three-dimensional vignettes from popular cartoons and **television** shows.

The first stereoscopic camera was invented in 1859 by noted physician and essayist Oliver Wendell Holmes (1841-1935). Dubbed the stereopticon, Holmes's device replaced the two cameras previously required with one camera bearing two lenses and yielding two negatives. Though the stereo camera saw brief periods of popularity among amateur photographers, its use dwindled until the early 1950s, when filmmakers became intrigued by the idea of three-dimensional movies. In 1952 producers unveiled the first **3-D motion picture** projector, which utilized a system not unlike Holmes' stereopticon. Moviegoers donned special glasses, often with red and green lenses, that would overlap the two images on the screen to create the illusion of depth.

The IMAX format of ultralarge-frame film projected on a 80-ft-high (24.2-m) screen has been adapted to 3-D. The IMAX camera contains two objective lenses separated by the approximate interocular distance of the human eye. Each lens records on a separate film strip, one for the right eye, one for the left. When the film is shown, both films are screened simultaneously.

To isolate the images so that the left eye sees the left-eye film strip and the right eye sees the right-eye film strip, IMAX 3-D designers use the phenomenon of polarization. Polarization controls the oscillation properties of a lightwave; certain types of filters will admit light polarized in one direction, but will block light polarized in the other. When the movie is shown, filters apply a distinct polarization to the left eye and to the right eye film strips. Viewers in turn wear special glasses with polarizing filters matched to the projector such that the left lens of the glasses only admits the images for the left eye, and the right lens only admits images for the right eye. The result is an amazingly realistic illusion of three-dimensionality, without the headaches associated with the red and green glasses of the 1950s.

In addition to its entertainment applications, stereoscopic technology is frequently used in medical research, as well as topographical surveys.

STEREOSCOPIC PHOTOGRAPHY • See Stereoscope

STERILIZER • See Pressure cooker

STETHOSCOPE

The stethoscope is an instrument for listening to sounds inside the human body for diagnostic purposes. Until the stethoscope was invented, clinical examination of patients was largely limited to external observations.

Medical knowledge about the inside of a patient's body took its first important step forward when Leopold Auenbrugger (1722-1809), a Viennese doctor, developed a technique he called percussion. Auenbrugger tapped on his patient's chest and then analyzed the different sounds to tell what conditions existed inside the chest. He published his findings in a 1761 pamphlet, which was ignored by the medical profession.

In the early 1800s, Jean Nicholas Covisart (1755-1821), Napoleon Bonaparte's personal physician, espoused Auenbrugger's percussion technique and translated the doctor's pamphlet into French. Covisart encouraged one of his students, René Theophile Laënnec (1781-1826), to study acoustic diagnosis.

Laënnec invented the stethoscope in 1816 during an examination of a young woman with a heart affliction. Due to both the patient's stoutness and prevailing standards of modesty, Laënnec was unable to put his ear to the woman's chest. In a burst of inspiration, Laënnec rolled a sheaf of paper tightly into a tube, placed one end of the tube over the patient's heart, and listened from the other end. The doctor later wrote, "I was both surprised and gratified at being able to hear the beating of the heart with much greater clarity and distinctness than I had ever done before by direct application of my ear."

Later, Laënnec developed a wooden stethoscope. When his book describing his instrument and the diagnoses to be made with it appeared in 1819, the publisher gave a stethoscope to each purchaser of the book.

As the stethoscope came into standard use, promoted especially by the Austrian doctor Joseph Skoda (1805-1881), some modifications were made. Pliable tubing was introduced in 1850, the American doctor George P. Cammann developed a binaural stethoscope in 1852, and the electronic stethoscope appeared in 1980. Although advanced diagnostic tools such as CAT scans have reduced the importance of the stethoscope, it remains a valuable and widely used instrument.

STEVENS, JOHN (1749-1838)

American engineer

John Stevens provided the impetus for the **steamboat** industry that revolutionized the United States. The son of a merchant and ship owner and member of a very wealthy family, Stevens was born in New York City and graduated from King's College in 1768. Though he became an attorney three years later, he never went into practice and instead followed his father into politics, becoming treasurer of New Jersey during the Revolutionary War, obtaining the rank of colonel. He later used his political connections in Congress to petition for the passage of the first U.S. patent laws in 1790.

In 1784, Stevens bought a huge estate on the west side of the Hudson River (most of which is now Hoboken, New Jersey) and later, bought the ferry service between Hoboken and New York. His interest in improving the service with steam-driven boats was inspired by the work of **John Fitch** and **James Rumsey**.

After patenting an improved steam boiler and **steam engine**, Stevens interested Robert Livingston (1746-1813), his college friend and brother-in-law, in building a steamboat. Together, they joined with mechanic Nicholas Roosevelt and, after several experimental models, Stevens produced the *Little Juliana*. The boat used a new high-pressure steam engine and two **screw propellers** to cross the Hudson in 1804.

Livingston, who had purchased a temporary exclusive charter for steamboats on the Hudson River, meanwhile was in France convincing American inventor **Robert Fulton** to produce his five-mile-an-hour steamboat in America. On his return, to the U.S., Livingston was not impressed with the *Little Juliana*, which failed to meet the contract's speed requirements. He offered Stevens a partnership in Fulton's future steamboat, but Stevens refused, feeling that Livingston had broken his word.

Livingston fulfilled the charter when Fulton's *Clermont* made its successful first voyage from New York to Albany in

1807. A short time later, Stevens launched his 100-foot (30 m) *Phoenix*, the first ocean-going steamboat. He used the steamer to set up a steam-powered ferry service on the Delaware river while the Livingston-Fulton line ran on the Hudson.

About 1810, Stevens focused his attention on adapting steam technology to the **railroad** and argued the advantages of rail transportation over canals in Congress. His efforts resulted in the passage of the first American railway act, which created a company to build a railroad from the Delaware to the Raritan river. In 1825, he constructed the first steam locomotive in the United States, operating it on a track on his estate grounds. Before his death, he designed a **bridge** and underwater tunnel from Hoboken to New York, as well as an elevated railroad system for New York City.

Stevens, Robert Livingston (1787-1856)

American inventor

Robert Livingston Stevens regarded himself as a naval architect. Indeed he designed and built more that twenty ferries and steamboats in his life, continuously trying to increase both safety and speed in his vessels. Stevens was born in Hoboken, New Jersey on October 18, 1787. He was the son of **John Stevens**, an early proponent of steam navigation and later rail transportation, and one of the first people to receive a patent under the 1790 United States patent law.

Robert assisted his father in the construction of *Little Juliana*, a small **steamboat** driven by **screw propellers** that crossed the Hudson River in 1804. In June 1809 he captained their *Phoenix*, a 100-foot paddle wheel steamboat, on its maiden voyage to Philadelphia. This trip is recorded as the world's first sea voyage made by a steamship. After this trip Stevens piloted the *Phoenix* as a ferry along the Delaware River. The War of 1812 led Stevens to ideas of developing metal-clad ships but navy officials showed no interest until the 1840s and Stevens was unable to construct one before his death.

By 1830 Robert joined his father's rail transportation company. He traveled to England to study **locomotives** and bring back some iron rails. As a result, he designed the T-shaped type of rail which is still used. He also discovered that iron rails set on wooden cross ties over a gravel bed provided a safer and more comfortable ride than other methods of the day.

That same year the Stevens family organized the Camden and Ambay Railroad and Transportation Company with Robert as president and chief engineer. Using the *John Bull* locomotive from England, the first steam railroad service in New Jersey began in 1831. A Stevens innovation that became important because it helped to reduce the number of derailments on sharp curves was a two-wheeled guide called a *pilot* that he attached to the front of the *John Bull*. This pilot assisted in leading locomotives around the numerous curves in American railways and became a necessity for that reason for many decades.

In 1844 Stevens returned to his naval roots and designed the *Maria*, a yacht that was known as the fastest sailing ship for the next twenty years. He died in Hoboken on April 20, 1856.

Caricature of René Laënnec wearing his invention, the stethoscope, from 1824.

See also Train and railroad

Stevin, Simon (1548-1620)

Flemish mathematician and engineer

Simon Stevin was born in Bruges, Flanders, the son of Antheunis Stevin and Cathelijne van de Poort. As a young man, he began work as a bookkeeper, but soon found work as an itinerant engineer. He traveled widely, visiting France, Germany, and perhaps even Poland and Norway before returning to Bruges in 1577 to become a tax collector.

In 1581 Stevin traveled north to enter the University of Leiden, where the first classes for engineers were being held. The works of antiquity had been newly rediscovered, and those of Euclid, Apollonius of Perga, and Archimedes would play a large part in Stevin's own scientific accomplishments. (Stevin added to this cultural revival by translating the works of Diophantus of Alexandria for the first time.)

Stevin is best known as a contributor to two branches of physics: *statics* (the science of forces producing equilibrium) and *hydrostatics*.

Stevin, like René Descartes and **Galileo**, wrote almost all his works in the vernacular, marking the decline of Latin as the European scholarly language. He considered Dutch perfect for scientific discourse, and he coined many new terms which survive in that language today.

A true Renaissance man, Stevin published books on a variety of scientific and humanistic topics, but devoted most of his studies to mathematics and engineering. His works are characterized by his uncommon skill in combining theory and practice, and his uncanny ability to foreshadow later discoveries. They reflect the contemporary belief in the prevalence of reason, and the ultimate comprehensibility of nature.

His first publication, *Tafelen van interest* (*Tables of interest*) (1582), listed rules for computing interest and tables for calculating discounts and annuities. This information had been closely guarded by banks, primarily because there were few people with the skill to perform such computations, but perhaps it preserved a financial advantage as well. After Stevin's work was published, interest tables were available to anyone who could read.

Stevin's principal work in statics is *De Beghinselen der weeghconst*, published in 1586. In it, Stevin described his most famous discovery, the *law of inclined planes*, which he proved by drawing an imaginary circle of connected, equal weights called a *clootcrans*, or wreath of spheres. The basic premise of the law is that less weight on a steep slope can balance more weight on a gentler slope. Stevin was so delighted with his find that beneath the illustration he wrote *Wonder en is gheen wonder*—'' what seems mysterious can be understood.'' The phrase became his motto, and the seal, appearing on his letters, his instruments, and the title pages of his books.

Stevin's other famous publication, *De Beghinselen des waterwichts*, was the first since antiquity to study Archimedes's principle of displacement. Stevin added many new ideas of his own, including one that is the fundamental principle of hydraulics: the pressure exerted by a liquid depends only on its height, and not on the shape of its container. This meant that a small amount of fluid could produce a large amount of pressure if it were held in a long, narrow tube.

In another branch of science, Stevin was upstaged by his contemporary Galileo. Although credit has historically been given to the Italian, it was Stevin who first refuted Aristotle's mistaken belief that heavier bodies fall faster that light ones. He dropped two lead balls, one 10 times heavier than the other, from a height of 30 feet and found that they hit the ground simultaneously. He published his findings years before Galileo, but never attained the same degree of fame.

In *De thiende* (The tenth) and De disme (The decimal), both published in 1585, Stevin discussed decimal fractions and described their usefulness in everyday life. Until this time, decimals had been used only in the rarefied realm of trigonometry, and then only occasionally. Although awkward by modern standards, Stevin's notation was a vast improvement over sixteenth-century methods, and decimal fractions were soon in wide use.

Stevin parted company with his fellow mathematicians on a few points. He believed, for example, that all numbers, even irrational or imaginary numbers, were basically alike, a view not widely held until the development of algebra. He also challenged the centuries-old Greek method of mathematical proof, *reductio ad absurdum*. He introduced a different means, which, although unwieldy, hinted at improvements made later in calculus. Neither was he afraid of controversy. In *De Hemelloop* (1608), an astronomical treatise, Stevin explained and supported the Copernican theory, in which the Earth and other planets orbit the sun. This book was published several years before Galileo's famous clash with the pope over the same topic, and predated most other scientists' acceptance of a sun-centered cosmos.

Stevin also proposed a decimal system for weights, measures, and currency, ideas that were not adopted until hundreds of years later. The true inventor of decimal notation is not known, although the idea may have originated in China or India. It is known that the fifteenth-century Arabic mathematician Muhammad al-Qushchi (died c. 1430) used decimals to calculate the value of pi to 16 places. In Europe, C. Rudolff (fl. 1500s) described decimal notation and its manipulation 50 years before Stevin advocated its general use.

Stevin was a practical man, and he brought his engineering skills to the burgeoning commercial and industrial world of the Netherlands. He was a prolific inventor, solving problems wherever he found them. Perhaps his best-known achievement was a system of sluices and locks that used tides to flush **canals**; the **valve**s could also be opened to flood the country in case of an invasion. He invented a winch to lift boats out of the water, and a mechanical spit for use in cooking. A rather whimsical design published in 1599 describes a 26-passenger sail-propelled carriage that was intended for use along the seashore. Stevin also patented embanking and drainage inventions (a constant priority in the waterlogged Dutch countryside), including notable improvements in windmill design. Using mathematics, Stevin analyzed both the scoop wheels that mills used to remove water and the gears that drove them. In *Van de Molens* (*About windmills*), he proposed slowing the wheels' revolution and changing the gears' construction to engage the teeth on the face, not the rim. Several mills were altered to these new specifications; unfortunately, technology sorely lagged behind theory, and the craftsmen of the day were not able to execute Stevin's design as perfectly as he envisioned it.

Stevin eventually entered the service of Prince Maurice of Orange, who commanded the Dutch army in the country's struggle for independence from Spain. Stevin served as the Prince's tutor and as a trusted consultant on matters of defense and navigation. He became the army's Quartermaster-General in 1604, responsible for providing accommodations for the troops. Stevin's military experiences influenced his later inventions, although his ideas on the art of fortification, too expensive for their time, were used successfully by armies a century later. They did have one practical expression—a combination pick, shovel, and axe that Stevin devised for use in building fortifications.

In his later years, Stevin organized a school for engineers at Leiden. He settled down with a young woman named Catherine Cray, by whom he had four children. One of them, his son Hendrick, became a scientist in his own right.

Stibitz, George R. (1904-1995)

American mathematician and computer scientist

George Robert Stibitz was born on April 20, 1904, in York, Pennsylvania. Stibitz's childhood was spent in Dayton, Ohio, where his father taught at a local college. Because of the interest in and aptitude for science and engineering that he had exhibited, Stibitz was enrolled at an experimental high school in Dayton established by **Charles Kettering**, inventor of the first automobile ignition system.

Stibitz enrolled at Denison University in Granville, Ohio. After earning his bachelor's degree there in 1926, he went on to Union College in Schenectady, New York, where he was awarded his M.S. degree in 1927. After graduating from Union, he worked as a technician at General Electric in Schenectady for one year before returning to Cornell University to begin his doctoral program. Stibitz received his Ph.D. in mathematical physics from Cornell in 1930. Stibitz's first job after graduation was as a research mathematician at the Bell Telephone Laboratories in New York City.

In the fall of 1937 Stibitz made the discovery for which he is now best known, the use of relays for automated computing. A relay is a metallic device that can assume one of two positions—open or closed—when an electrical current passes through it. The relay acts as a kind of gate, therefore, that will control the flow of electrical current, and was a common device used to regulate **telephone** circuits.

In November 1937 Stibitz decided to see if relays could be used to perform simple mathematical functions. He borrowed a few of the metal devices from the Bell stockroom, took them home, and assembled a simple computing system on his kitchen table. The system consisted of the relays, a dry cell, flashlight bulbs, and metal strips cut from a tobacco can. He soon had a device in which a lighted bulb represented the binary digit "1" and an unlighted bulb, the binary digit "0." The device was also able to use binary mathematics to add and subtract decimal numbers. Stibitz's colleagues later gave the name "K-Model" to this primitive **computer** because it was built on his kitchen table.

When Stibitz first demonstrated his K-model computer for company executives, they were not very impressed. Less than a year later, however, Bell executives had changed their minds about the Stibitz invention. An important factor in that decision was the increasing pressure on Bell to find a way of solving its increasingly complex mathematical problems. The company agreed to finance construction of a large experimental model of Stibitz's invention. Construction on that machine began in April 1939, and the final product was first put into operation on January 8, 1940. Called the Complex Number Calculator (CNC), the machine had the capacity to add, subtract, multiply, and divide complex numbers—just the kinds of problems that were particularly troublesome for engineers at Bell.

Nine months later, Stibitz recorded another milestone in the history of computer science. At a meeting of the American Mathematical Society at Dartmouth College, he hooked up the new CNC in New York City with a **telegraph** system. He then sent problems from Dartmouth to the CNC in New York, which solved the problems and sent the answers back to Dartmouth by means of the telegraph. This type of data transmission has now become commonplace in a modern day society of modems and **fax machine**s.

During World War II, Bell Labs permitted Stibitz to join the National Defense Research Council. There the demands of modern military artillery convinced Stibitz even more of the need for improved computer hardware, and he spent most of the war working on improved versions of the CNC, also known as the Model 1. The Model 2 computer, for instance, used punched tapes to store programs that would give the computer instructions; in this manner the computer could perform the same complex calculations many times on different sets of numbers. This proved useful in calculating weapons trajectories.

At the end of World War II, Stibitz moved with his family to Vermont where he became a consultant in applied mathematics. After two decades, Stibitz was offered a job at Dartmouth's Medical School, where he was asked to show how computers can be used to deal with biomedical problems. He accepted and was appointed professor of physiology; in that capacity he investigated the motion of oxygen in the lungs and the rate at which drugs and nutrients are spread throughout the body. In 1972 he retired from his position and was made professor emeritus; nevertheless, he continued to contribute his knowledge to the department. He died in his Hanover, New Hampshire, home on January 31, 1995. He was 90.

Stirling engine

As water, wind, and animal power gave way to the Industrial Revolution, **steam engines** provided the cheap power necessary for **mass production**. They were notoriously dangerous and unreliable, often shoddily built of unsuitable materials, and they exploded with tragic frequency. Moved by such calamities, Scottish clergyman Robert Stirling (1790-1878) decided to find a safer alternative, one without a pressurized boiler. The result was an external-combustion engine that alternately heats, compresses, and cools a gas (Stirling used air) to produce energy. Although it never replaced the steam engine as Stirling hoped, it proved to be a quiet and efficient machine that still intrigues scientists today.

The Stirling engine has five major components: a *heater*, which burns fuel in an external-combustion chamber, is connected to a gas-filled cylinder. Inside, the *power piston* compresses the gas as it is heated. The hot gas expands, and the force generated is transmitted to a drive mechanism. At this point the *displacer piston* moves down, pushing the gas through the *regenerator* and the *cooler*. Then the cycle starts again as the gas enters the heater. The regenerator, or heat exchanger, stores the thermal energy produced as the gas is compressed and returns it to the system. This process, known as a thermodynamic cycle, was discovered by Nicolas Carnot (1796-1832) in his search for an efficient engine. Stirling was the first to put the concept to practical use.

Stirling's engines were patented in 1816 and manufactured through the 1920s. Small and reliable, they performed limited tasks like pumping water. A factory model was built in 1843, but it proved too expensive for commercial use. Stirling engines powered portable generators during World War II in the Netherlands. In the 1970s, Ford and Philips tried without success to develop a Stirling engine for automobiles.

Today the Stirling engine uses hydrogen or helium instead of air, but has found application only as a cryogenic refrigerator. There have been sporadic efforts to adapt it for wider industry, but results have been disappointing, although its theoretical efficiency continues to tantalize researchers.

Stored program

A **digital computer** program contains instructions for processing data. A stored computer program is one that is held in the computer along with the data—the method used by today's computers, developed during the 1930s and 1940s using a nineteenth-century design.

Stored program use is a two-step process. First the program is loaded from **disk** or **tape** into the computer's memory or provided on a special computer chip. Then the processor copies the program into its own unit and carries out the instructions.

The nineteenth-century English mathematician **Charles Babbage** was the first to use the word *store* to describe a computer program location. Babbage's Analytical Engine design, a forerunner of the modern computer, was based on a cotton mill. The *mill* performed the computer operations, and the *store* was the location of the original numbers and the results of the mill's operations.

The term *storage* carried over into twentieth-century computer development and can have two different meanings. Program and data held in the computer for processing are said to be in memory or working storage. Long-term storage is on disks, tape, or another permanent medium.

The idea of stored programs in modern computer design came from the British mathematician **Alan Turing** in the 1930s. Turing's "machine" concept, developed while he was doing graduate work at Princeton University, depended on the device's ability to store programs in the computer's memory. However, Turing did not design a computer, and none could have been built from his description.

EDVAC (Electronic Discrete Variable Automatic Computer), designed for the United States Army during 1944 and 1945, was the first computer built as what is now called a von Neumann machine. It is the standard architecture that most of today's computers still use. Both program and data are kept in the computer's memory, and the program's instructions are used to process the data sequentially.

EDVAC was designed by the American mathematician **John von Neumann**, from the Institute for Advanced Studies, in Princeton, New Jersey, the mathematician Herman Goldstine (1913-), and the philosopher-engineer Arthur Burks (1915-) of the University of Pennsylvania. Other design contributors included **ENIAC** creators **J. Presper Eckert** and **John Mauchly** also at the University of Pennsylvania.

EDVAC's most advanced feature was its stored program. ENIAC's program and intermediate processing results were kept on punched cards, which greatly slowed its speed. EDVAC treated the program and the data the same way (as data objects), holding them in a mercury delay line memory. The incoming electronic signal was converted to a much slower sound wave in the mercury, delaying its reconversion. This process could be repeated as often as needed while the computer was in use.

Because of engineering problems, EDVAC didn't start operating until 1952. In the meantime, some of its stored-program features were added to ENIAC, which had already been built. The mercury relay line memory was later used in commercial computers, such as Eckert and Mauchly's **UNIVAC**.

The first functioning computer with stored programs was EDSAC (Electronic Delay Storage Automatic Calculator). It was principally designed and built by the British mathematician, physicist, and engineer Maurice Wilkes (1913-). Wilkes built EDSAC at the University of Cambridge, United Kingdom, as part of his efforts to develop computer programs for practical applications. Based on Eckert and Mauchly's work, EDSAC had a mercury delay line memory that could hold 512 34-bit words. Input and output were on paper tape. EDSAC went into operation in May 1949, and was used routinely until 1958.

In 1949, Eckert and Mauchly formed their own firm to produce computers with stored programs. After building the relatively small and fast BINAC (BINary Automatic Computer) for military use on a guided missile, they built a much larger computer intended for the civilian market. UNIVAC (UNIVersal Automatic Computer), completed in 1950 and manufactured by Remington Rand, was the first widely-available commercial computer to use stored programs. All general-purpose computers since then have had stored programs.

Today the principle of stored programs is unchanged from the early years. However, there have been vast improvements in the size and speed of **computer memory** and processors, including the use of transistors in the 1950s and the constantly-improved integrated circuit s (computer chips) used since the 1960s. Also, architecture for some computers uses parallel processing, in which several processors work on different parts of the same problem.

Some programs are too large to be loaded entirely into the computer's memory. Instead, part of the program is loaded while the rest of it remains on the disk or other permanent storage device. As computer operations proceed, needed program segments are moved into and out of memory.

These innovations have allowed use of larger and more complex programs that interact constantly with the data. The first stored programs were written in machine language, which is directly usable by a digital computer for conversion into electronic signals as zero or one. Assembly language programs are more flexible, but are still very close to machine language. In contrast to these low-level languages, high-level languages are written in human-language words, then translated into machine language for execution. Two of the first high-level languages were written in the 1950s and are still in use— COBOL (COmmon Business-Oriented Language), developed by **Grace**

Hopper and others in the United States Navy, and FORTRAN (FORmula TRANslation), developed by John Backus and others at IBM.

Many languages are now available for processing numerical and non-numerical data. Some of them are designed for general use and others are created for specific purposes, such as equipment design, tool control, and **artificial intelligence**.

Stove

Trying to harness heat from the fireplace for cooking purposes, people in fifteenth-century Europe invented the first stoves. These stoves used wood and were made of **brick** and tile. Unfortunately, these stoves had no storage areas for the ash they produced and proved so inefficient that people continued to cook food in their fireplaces for the next two hundred years.

In the 1630s, John Sibthorpe of England developed an oven that used coal as its fuel. A decade later, the first cast iron stoves were manufactured in Lynn, Massachusetts. The design was simple: a cast iron box with a lid into which fuel would be added. Trying to capture heat from the fireplace for room heating purposes, **Benjamin Franklin** developed his cast iron stove in the early 1740s. The Franklin stove extended from the fireplace so that three sides of the cast iron could radiate heat throughout the room. Franklin marketed his stove shrewdly, claiming that it reduced house drafts created by fireplaces and prevented premature aging of women's faces caused by exposure to the harsh heat of fireplaces.

The Industrial Revolution brought about an abundance of stove designs suited for two main purposes—heating and cooking. In the early 1800s, Isaac Orr of Philadelphia, Pennsylvania, invented the first round-bodied heating stoves. Another popular model was the Oberlin stove created by Philo Stewart (1798-1868) in 1834. These stoves produced great heat and were intended for heating the room. Perhaps the hottest of these stoves was Jordan L. Mott's pot-bellied baseburner stove of 1833. The baseburner was the distinctive feature of the pot-bellied stove. Until the development of the baseburner, wood was the most common stove fuel. The baseburner was filled with coal from the top and fed small amounts of coal to the fire as required. The baseburner was also first to utilize the principles of radiation and *convection* to spread its heat over a wide area. With their ornate cast-iron designs, these stoves replaced the fireplace as the social center of the home and were often symbols of prestige. Commentators have noted that the typical room of the nineteenth century was as overheated as the room of the twentieth century is overcooled. The arrival of central heating prompted the demise of the stove as a home's centerpiece, and it was relegated to its more common place as a kitchen appliance.

The coal range developed when the round stove made it necessary to have another stove for cooking purposes. In 1780, Thomas Robinson invented the first kitchen stove. From their inception these stoves typically included an oven for baking, cooking plates on which pots were placed, and reservoirs for warming rain water. Oil was another fuel source for the early cooking stoves. Oil stoves resembled large **lamps** and had large wicks which held the flame beneath a single metal plate. Later oil stoves had three or more plates and larger fuel tanks but never proved as popular as coal-burning stoves.

Cartoon from 1800 depicting "The Comforts of a Rumford Stove."

Gas stoves became common in United States in 1860. Natural gas, coal gas, or a mixture was fed into a burner ring with tiny holes. Gas flame burned cleaner and more efficiently than coal. The pilot light was an improvement in the design of gas stoves that made ignition of the stove easier and safer.

Electric stoves began to appear in 1890. An early model was the centerpiece of the first fully electric kitchen installed at a New York restaurant in 1894. Electric ranges were introduced to American homes in the 1930s. An electrically charged coil is the source of heat on electric ranges. The first electric ranges had poor temperature controls and cooked slowly compared to the gas-flame ranges, but improvements in their dial mechanisms now allow for faster cooking.

Even the roles of these modern stoves have become diminished in recent years. The toaster oven and the microwave oven now provide alternatives to gas and electric ranges and can cook food in a fraction of the time necessary in conventional ovens. A self cleaning capacity for conventional and some toaster ovens has become standard on most ranges sold today.

Florence W. Van Straten

Straten, Florence W. Van (1913-)
American meteorologist

Florence Van Straten made significant contributions to the collection of meteorological information.

The daughter of Dutch immigrants, Van Straten received a Ph.D. in physical chemistry at the Massachusetts Institute of Technology in 1933. During World War II, she enlisted in the United States Navy where she became a weather forecaster. She helped develop methods of using weather phenomena, such as storms, in the planning of ship maneuvers and carrier-based airplane flights.

In 1946, she became a civilian adviser to the Chief of Naval Operations. During the decade that followed, she went on to make the contributions for which she is known.

She was instrumental in the development of the rocketsonde, which launched a data-collecting package, called a sonde, into the upper atmosphere, which was then parachuted back to earth. It was at her suggestion that meteorological data be used in planning the trajectory of rocket launchings.

She also developed the constant-altitude balloon. The deflated **weather balloon** was carried aloft with its sonde by a bubble of helium. The balloon inflated as the atmosphere became thinner until it was full, at which point its altitude remained constant.

Also to her credit was the floating weather station, called NOMAD, the National Oceanographic Meteorological Automatic Device, which could be anchored to 3,667 yd (3,337 m)

She invented the weather instrument shelter, that protects sensitive instruments from the elements, and the tipping bucket rain gauge which tipped with every 1/100 inch of precipitation and automatically recorded the action at the station.

Florence Van Straten was responsible for bringing the weather station to its modern day standard.

Strauss, Levi • See Blue jeans

Streetcar and trolley

Trams, also called streetcars, were four-wheeled vehicles that traveled over rail systems called tramways, or street railways. Horse-drawn trams quickly replaced horse-drawn buses, or omnibuses, for a simple reason: steel wheels on steel rails gave a smoother ride with less friction. Horses could also pull twice as much weight on wheels rolling on steel rails as they could on dirt roads.

The first street railway was built in New York City in 1832. The New York and Harlem Railroad route was originally a mile long and used two cars to transport thirty people each. Within two years the route had grown to four miles and ran cars every fifteen minutes. The popularity of the streetcars soon spread to Europe, first to Paris in 1853, then to England in 1860.

Steam-powered trams were devised in Cincinnati, Ohio, and in London, England, but existing tracks built for horse-drawn vehicles could not support the weight of an engine. The next step in the development of self-propelled cars was the cable car. Cable cars could still be powered by a **steam engine**, but the engine did not have to be mounted on the vehicle. Instead, an arm suspended from the vehicle engaged a moving cable buried beneath the street. The cable was moved by a steam engine located somewhere along the route. Andrew S. Hallidies's 1873 San Francisco system was the first to use the cable car, and it grew to more than 500 miles of cable tramways by 1890.

German engineer **Ernst Werner von Siemens** built the first electric-powered trams in Berlin in 1881. He used one rail to supply current and the other to return current. This method was soon found to be dangerous, and he quickly introduced an overhead contact wire. In 1886 a tram route with electric motors operated in Mobile, Alabama. This early **direct current** system, based on Siemens's updated technology, received power through an overhead wire and completed its circuit through one of the rails in the street. The device that collected electricity from the overhead wire was known as a *troller*. Electric-powered trams, or streetcars, therefore became known as trolleys. Once Frank J. Sprague (1857-1934) perfected the swivel trolley pole, electric streetcar lines expanded to nearly 22,000 miles (35,398 km) in barely a dozen years. In fact, by World War I, most towns and cities had a trolley route of some kind.

Though electric-powered trolleys operate safely and quietly, their range of movement is limited by the location of guideways. Trolleybuses, or trackless trolleys, allowed varied routes and steering flexibility and became popular during the 1920s and 1930s when much of the rails of the older systems began to deteriorate. Rather than laying new track, transit companies invested in the new rubber-tire vehicles. Unfortunately, they were also too limited in range when compared to newer gasoline-powered **bus**es, and in most places, they declined in popularity during the 1950s. In recent years, however, electric transit, in the form of trolleybuses, has made a dramatic return in cities plagued by heavy traffic and air pollution.

During the 1970s and 1980s, several surviving rail systems in the United States were rebuilt, modernized, and even extended, and by the late-1980s, eight new lines were built in Canada and the United States. No longer called trolleys or streetcars, these new systems are part of the modern ''light rail'' concept which has proven to be an excellent middle ground between bus operations and heavy rail rapid transit. These new light rail systems are characterized by simplicity and economy, as exemplified by the adoption of such systems as a self-service, barrier free fare collection system. The standard vehicle for these new urban electric railways is the articulated light rail vehicle. By the beginning of the twenty-first century, the old streetcar and trolley promises to be reborn in the light rail boom, as more than a dozen American cities have projects planned or proposed.

Beyond this near future, light rail technology may evolve into a fixed-guideway concept that has been under development for some time. Called System 21, this transit system consists of streamlined, pod-like cars that are cantilevered from either side of a single, elevated beam. These cars are built of a rugged fiberglass/plastic composite and have steel wheels that travel in a contoured rail truck that is recessed in the beam. They are powered by standard commercial DC power delivered by the beam. The cars of this fixed guideway system will appear to hang over the edge of the beam, but will be able to carry large numbers of passengers at speeds of up to 60 mph (96.6 kph). Finally, unlike monorail systems that require expensive dual tracks or work only in a closed loop, this design offers simultaneous two-way traffic on a single, 6-ft (1.8 m) wide steel beam.

Stroboscope

A stroboscope is a light that *freezes* motion by flashing at regular, controlled intervals. People dancing under such a light appear to be almost immobile. This illusion of immobility or slow motion stems from the persistence of vision—the eye's ability to retain an image for a brief moment after it has vanished. This phenomenon also underlies the principle of **motion picture**s.

Joseph Plateau (1801-1883), a Belgian physicist, built the phénakistiscope, a device with flashing lights akin to the stroboscope, in 1832. In the same year, an Austrian named Simon von Stampfer devised the first actual stroboscope. In the 1920s, Harold Edgerton, an American engineer, pioneered the use of modern stroboscopes in **photography**. These strobes produce extremely short flashes of intense light from a gas-discharge lamp. Strobe photography can capture multiple images of bullets and other fast-moving objects with extreme clarity. Stroboscopes are used in industry to study the movement of machinery and in science to analyze objects as they undergo rapid, cyclic motion, making the objects appear stationary or nearly so by matching the stroboscope's rate of flashing to the object's frequency of motion. Strobe lights are used in this way in automotive timing instruments.

See also Arc lamp

Photograph of an early San Francisco cable car.

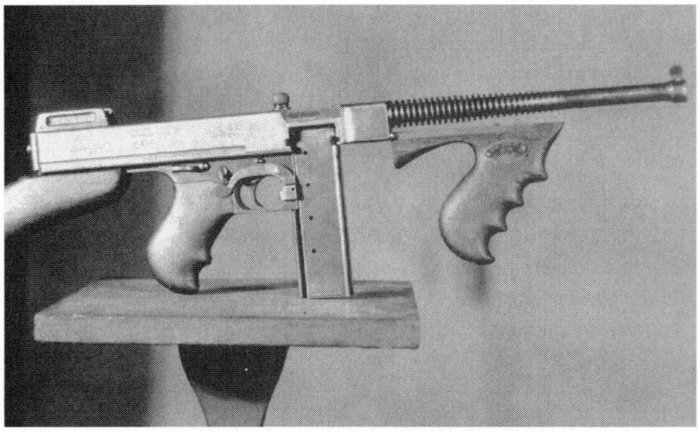

A Thompson "Tommy" gun.

STYROFOAM • See Polystyrene

SUBMACHINE GUN

Submachine guns are small, portable machine guns intended for use at close range. The first submachine gun was made by Hugo Schmeisser, a German firearm designer, toward the end of World War I. He experimented with semiautomatic **pistols**, which fired rounds at lower velocities than standard guns. His weapon, the MP 1918, had an 8-inch (20.32cm) barrel and fired 9 mm rounds, which were originally used in pistols. The MP 1918 operated using blowback, a simple system during which spent cartridge cases were blown backward out of the chamber due to the gases created when the weapon was fired. This action forced the bolt back against a spring and tripped the mechanism that ejected the old cartridge. After the gases dissipated, the spring pressed the bolt forward as a new cartridge was introduced to the chamber. This cycle continued as long as the trigger was held. Schmeisser used a heavy bolt to minimize the spring's motion so the gun would not operate faster than the mechanism could handle. As a result, this first submachine gun fired about 400 rounds per minute.

Others soon followed with variations on Schmeisser's design. Vasily Degtyarev (1890-1959) of the former Soviet Union built a submachine gun capable of firing 900 rounds per minute, a speed too great for accuracy. In the 1920s, John Taliaferro Thompson (1860-1940), an American army officer and armament consultant, developed and patented a submachine gun that was later called the "Tommy gun"; it had a large drum magazine, which was later replaced by a box magazine, and fired the same .45 caliber cartridges used in the Colt pistol. This gun was closely connected to the gangland wars in American cities during the 1920s and 1930s.

World War II saw the creation of many more submachine guns. The Germans came out with a version of the Schmeisser, often called the "burp gun" by the Allies. It was made of inexpensive sheet metal and fired at a rate of 500 rounds per minute. It used a box magazine that did not jam as easily as the drum variety did. The United States introduced the M3, often called the grease gun because it looked like a car mechanic's device. The M3 used a piece of heavy wire with a bend in it as a shoulder stock. It was also made of inexpensive sheet metal with a long, narrow magazine that fit into the bottom of the gun. It could fire effectively up to 100 yards (91.44 m).

Since World War II, there have been mostly minor modifications to these weapons. The British and German submachine guns were reconfigured to take the NATO standard 9-mm cartridge. A big change occurred in 1948, when Vaclav Holec created a telescoping bolt in a Czechoslovakian gun. This bolt was hollowed out and slid partially over the barrel when a round was brought into the chamber. With this innovation, guns could have a much shorter barrel. Perhaps the best example of this configuration is the Uzi, designed by an Israeli army major, Uziel Gal, following the Arab-Israeli war of 1948. The Uzi is approximately 25 inches (65 cm) long and weighs about 8 pounds (3.5 kg). While the submachine gun is now less prevalently used by the military, it has found its niche as a police and anti-terrorist weapon.

See also Machine gun; Rifle

SUBMARINE

The first known treatise on submarines was written in 1578. Published by William Bourne in his *Inventions or Devices*, the document describes a ship with two hulls, the outer made of wood and the inner made of **leather**. While no record exists concerning its manufacture, the ship, according to Bourne, could be submerged or raised by taking in or expelling water from between the double hulls. The first known submarine to be built was that of Dutch inventor **Cornelius Drebbel**, which consisted of greased leather over a wooden framework. It was propelled either on or beneath the surface by eight oars sealed through the sides with leather flaps. During a demonstration for James I in 1620, this vessel was successfully piloted just under the surface of the Thames River. It was unable, however, to make deep descents.

During the American Revolution **David Bushnell** built a one-man submarine called the *Turtle*. It resembled an egg squashed thin with a height of six feet (2 m), and had two hand-cranked **screw propellers**, a hand-operated control lever connected to the rudder, foot-operated pumps to let water in or send it out (to submerge or surface), and a crudely-lit control panel. As if it was not dangerous enough simply to get in the water while sealed inside this device, the *Turtle* also had a large explosive attached to it in the hopes the operator could maneuver under an enemy ship, screw the explosive into the ship's hull, and depart before the explosive's timing device discharged it. Unfortunately, the *Turtle* failed to sink any ship. On its only test mission, the *Turtle* was assigned the task of bombing the British HMS Eagle in New York, but its pilot was unable to screw the explosive into the *Eagle*'s copper hull.

Others, such as English carpenters Symons and Day, included ballast systems on their submarines to permit descents. Day's submarine resembled a sloop, and had two large bags of stones hanging from its bottom to serve as ballast. Day would sink, then jettison the rocks to return to the surface. After two successful tests, Day confidently decided he would test his vessel off Plymouth Sound, a site with a depth of 900 feet (273 m). Apparently his ship was crushed by high water pressure, for when he and his crew descended, a crowd of onlookers waited in vain for his return. Day and his crew had become the first victims of a submarine mishap.

Perhaps the most successful early submarine was designed by **Robert Fulton**. In an age of naval battles, Fulton, who detested war, felt that a device capable of neutralizing the effectiveness of **warship**s would end war altogether. While living in France in 1767, he outlined plans to build a sub called the **Nautilus** and unsuccessfully attempted to interest the French government in his idea. By 1801, however, he had managed to complete a submarine on his own. A 21-foot (6 m) vessel with a two-bladed propeller, the *Nautilus* performed well in tests, even sinking a ship with an explosive charge. But he was once again rejected by the French government, so he moved to England, hoping for a better reception there.

It soon became clear that the English did not want his submarine either. In fact, Fulton had failed not because his vessel did not work, but because major naval powers feared his vessel and did not want to participate in developing a weapon that could negate their military strength. Fulton went on to produce his famous **steamboat**s in the United States.

After the American Civil War, designers, spurred on by the invention of the self-propelled **torpedo** in 1866, increasingly sought alternatives to human-powered propulsion for submarines. Several systems proved unsuitable—**steam engine**s made the craft unbearably hot and an electric **battery** could not be recharged at sea. In the late 1890s, however, Irish-born American **John Holland** solved the problem with the use of a new power source, the gasoline engine. Because it needed oxygen, the gasoline engine could not be used while a submarine was underwater, but on the surface it could not only provide propulsion but also charge the batteries used while submerged. Holland's vessels incorporated many of the features we associate with modern subs: a powerful engine, advanced control and balancing systems, and a circular-shaped hull to withstand pressure. The United States Navy accepted his submarine, the *Holland*, in 1900.

Around this time, two other improvements were introduced. Simon Lake (1866-1945), who also built an early gasoline-powered submarine, created the first **periscope** specifically for submarines: it provided a magnified view and a wide angle of vision. In the 1890s **Rudolf Diesel** invented an engine that was fired by compression rather than an electric spark. The **diesel engine** was more economical than the gasoline engine and its fumes were much less toxic and volatile. This new engine became the mainstay of all submarines until nuclear power was introduced as a means of propulsion in the 1950s.

Launched in 1955, the *U.S.S. Nautilus* (above) was the first nuclear submarine ever built.

Germany made good use of diesel propulsion. Unlike Britain's small, coastal subs, Germany's vessels, displacing up to 3,200 tons, were capable of crossing the Atlantic. Their U-boats (short for *Unterseeboot*) sent more than 11 million tons of Allied shipping to the bottom and, in the process, created a new, terrifying type of warfare.

In World War II submarines played an even larger role in Germany's repeated attacks on Allied shipping, eventually destroying 14 million tons of shipping. meanwhile, American submarines crippled the Japanese by sinking nearly 1,400 merchant and naval ships.

The greatest improvement came through the development of the snorkel, a set of two fixed air pipes that projected from the sub's topside. One tube brought fresh air into the vessel, and the other vented engine exhaust fumes. Now a sub could stay hidden below the surface when running on its diesel engine and recharging its batteries. The greatest advance in submarine technology was the advent of nuclear power. With the encouragement of U.S. Navy Captain Hyman Rickover, American inventors Ross Gunn and Philip Abelson designed the *U.S.S. Nautilus*, the first nuclear-powered submarine. Launched in 1955, the *Nautilus* carried a reactor in which controlled nuclear fission provided the heat that converted water into steam for turbines. With this new power source, the submarine could remain under water indefinitely and cruise at top speed for any length of time required.

For a submarine able to remain under water for longer distances at higher speeds, a needle-like shape proved inefficient. The Davis Taylor Model Basin in the United States developed a new teardrop design, first tested on its Albacore submarine. Vessels with this improved shape easily drove through the water at speeds of 35-40 knots per hour. The U.S. Navy later adopted the *Albacore*'s shape for its submarines. Submarines have also benefitted from advances in navigation equipment. Inertial navigation systems, relying on **gyroscope**s, now fix their position with extreme accuracy. The *U.S.S. Skate* used this system to navigate under the polar ice cap at the North Pole in 1959.

In 1960, the United States launched the first of its large, George Washington-class vessels. These were the first nuclear-powered submarines to carry long-range ballistic nuclear missiles. By the late 1960s, all the major navies of the world had well-established nuclear-powered submarines, although not all carried missiles. The cold war eventually necessitated a changed role for submarines, as their traditional mission of preying only on enemy surface ships expanded greatly to make them mobile, undersea carriers of nuclear missiles. These new submarines effectively became potential underwater launching pads for ballistic missiles and soon became a major factor in the strategic policy of nuclear deterrence.

In 1976, America built the first of sixty-two Los Angeles-class submarines, which would become the sophisticated workhorses of its modern undersea navy. Setting new standards for speed and quietness, they were built with an extremely strong single hull and with their ballast tanks and other equipment located at the ends. Running as quietly as possible became the mandatory design and construction goal for all modern submarines, and this meant changes in fin design, propeller, and hull shape. Quiet propulsion gear was also essential. At 360 feet (109.2 m) long, the Los Angeles-class submarine can cruise at more than 30 knots (about 30 mph) and dive to almost 1,500 feet (455 m).

In 1977, the Soviet Union produced the world's largest submarine. Called appropriately the Typhoon class, it was 200 feet (60.6 m) longer than the Los Angeles-class submarine. In 1997, the United States built the entirely new Seawolf-class submarine. At 353 feet (117.6 m), it is only slightly longer than its Los Angeles predecessor, but its beam (the width of its hull) is 40 feet (12.1 m) compared to 33 feet (10 m). As the Navy's most computerized submarine, it has forty-four multifunction display terminals and is equipped with torpedoes, antiship missiles, cruise missiles, mines, and decoys. Each new Seawolf submarine is expected to cost over $2 billion.

See also Ocean research vessel

Subsurface detection

Subsurface detection involves determining the characteristics of subsurface materials indirectly, that is without direct physical contact. Surveys can be run on land, at sea, or from the air. Seismic reflection and refraction studies using seismographs have allowed scientists to explore the **geology** of the interior of the Earth. Shallow seismic studies are usually done with an explosion or air blast, especially at sea, or in some cases, a special truck that makes the ground vibrate. Deep **seismology** surveys make use of **earthquake**s or nuclear explosions to explore the deepest interior of the Earth.

Seismic reflection involves interpretation of rock or sediment depth and character based upon the amount of time required for a seismic wave to travel from the Earth's surface down to a reflective layer and back again to the surface. The waves are generally created by a small concussion below ground or, if at sea, the water surface. The waves bounce off of multiple surfaces, called reflectors, below ground giving a detailed reflection profile, a computer-generated graph of the travel time to subsurface layers. Travel time is influenced by the nature of the subsurface materials. The denser the materials, the faster the travel time. With this in mind, the profile is interpreted with respect to depth and rock character. Generally only very abrupt changes in rock type make good reflectors.

Seismic refraction is similar to the reflection technique except the waves of interest are those that are refracted (bent) as they pass through layers of varying density or from one layer to another. Again based on travel times, the depth and character of the subsurface materials can be interpreted.

Seismic tomography is a technique involving computer analysis of hundreds or thousands of seismic profiles, and computer-generation of maps and three-dimensional diagrams based on these profiles. It provides a much clearer picture of the complex form of the Earth's interior.

Sonar, an acronym for sound navigation and ranging, is closely related to seismic analysis. In sonar surveys, a ship uses an ''air gun'' to produce sound shock waves that move downward through the water. Then the waves hit bottom and echo back up to the surface where they are detected by a sensor towed behind the ship. Analysis of travel time for the shock waves indicates depth, or bathymetry. Some sonar systems, called echo sounders, can penetrate shallow thicknesses of bottom sediments and produce a profile of sediment layers, like a seismic survey.

The Earth has two natural phenomena that can be used for subsurface detection: gravity and magnetism. Variations in these fields allow interpretation of the volume and character of subsurface rocks surveyed. The Earth's gravity field increases in the presence of rocks of high density, such as iron-rich igneous rocks, and decreases in areas of low density rocks, such as salt domes. The gravity field potential is measured using a gravimeter, a highly sensitive scale that measures the gravitational ''pull'' on a suspended mass. Ground-based surveys generally cover a small area and are rather detailed. Large-scale surveys are usually done from an airplane or a ship. The end product is a map of the Earth's gravity field showing the distribution of high and low density rock, which can be interpreted with respect to the character and origin of the rocks involved.

Recently a remote sensing technique that is closely related to gravity field detection—satellite-based altimetry—has been used to map the seafloor. The satellite, using radar, precisely determines sea level. Sea level is influenced by the mass (gravity) of features on the seafloor. More massive features, due to their effect on gravity, ''pull'' water towards them increasing sea level while less massive features cause a relative decrease in sea level. The resulting survey indirectly measures gravity and determines seafloor topography, or bathymetry.

Earth's magnetic field varies from place to place based on the presence of minerals that influence the field—iron-bearing minerals, for instance. Few sedimentary rocks strongly influence the field, whereas many igneous and metamorphic rocks do. For magnetic surveys, a magnetometer is used to measure Earth's magnetic field potential and the results are plotted in map form. Care must be taken when performing magnetic surveys since pipelines, electric cables and the like influence magnetic measurements as well. The results can be used to determine the thickness of sedimentary rock cover over igneous and metamorphic rocks, known as the ''depth to basement,'' to trace economically important rock units within the subsurface, and to verify findings from gravimeter surveys.

Ground penetrating radar is the most recently developed ground-based, subsurface detection technique. The user moves a cart- or sled-mounted radar unit over the ground in a series of transects. The radar waves reflect off objects or layers in the subsurface. Based on travel time, a profile of the shallow subsurface is constructed. Generally, the greater the density of the object the stronger the signal. Electrical fields from buried cables confuse the signal and must be avoided. Ground penetrating radar has been successfully used to explore for buried materials such as gasoline storage tanks and steel drums, and so has seen wide use in environmental geology and engineering.

Electrical resistivity has been used extensively in both surface surveys and downhole (well) surveys. The basic premise is that different Earth materials conduct (or resist) induced electrical currents at different rates, so that the subsurface geology can be determined by measuring the resistance of the materials to the passage of the current. In practice, surface surveys are performed by deploying detector electrodes at specified distances from a source electrode. The depth of signal penetration can be varied based on the distance from the source to the detector. When the current is turned on, the strength of the electrical signal received at each electrode is an indication of the subsurface geology at a particular depth. Similar to magnetic surveys, metal objects interfere with electrical resistivity surveys. Downhole surveys work on the same principle, except the instrument, called a sonde, is lowered down the well on a wire line. Spacing of the electrodes is controlled by the length of the sonde. The nature of the fluid in the rock or sediment pores significantly influences the resistivity, so electrical resistivity has been used by petroleum companies to search for oil and gas and by the environmental industry to detect pollution.

Radioactivity has also been used for surface and downhole surveys. Both natural radioactivity and induced radioactivity—produced by bombarding rock materials with gamma rays or neutrons from a radioactive source—have been used for subsurface detection surveys. In either case, the rock or sediment type is determined based on the measured nuclear decay rate of the subsurface materials. This has been used extensively in prospecting for uranium deposits and searching for oil and gas.

Subway

As cities have become increasingly congested with vehicular traffic, mass transit systems have been developed to relieve traffic jams, taking passengers to their destinations without the need for **automobiles**. Subways serve as the nucleus of many transit systems and are generally underground rail lines that offer commuters a relatively speedy ride to and from urban centers. In most cases, the subway forms the nucleus of a larger network of ground-level and above-ground **train** and **bus** lines.

Two types of rail subways exist, the open-cut and the tubular. The open-cut subway is constructed by digging a trench, then covering it after the tunnel structure has been put into place. Open-cut subways are rectangular in shape.

The tubular subway, or tube, is constructed by boring a round linear hole parallel to the ground surface. The surface is penetrated only for passenger stations and ventilation shafts.

The first tunneling shield—a means of supporting the ground as workers tunnel through the earth—was patented in Great Britain in 1818 by its inventor, **Marc Isambard Brunel**. Later shields also were devised by Peter William Barlow and Englishman James Henry Greathead (1844-1896) in the nineteenth century, all of them comprising technology crucial to the development of the subway.

Subways are usually built in alignment with existing streets since the sublevels of office buildings can obstruct any alternative. Beyond the city center, the trains may go in any direction and run beneath rivers and harbors.

Some subways, like Montreal's *Metro*, run several levels below the city center, with the upper levels used for shops, offices, museums, and hotels. Construction costs, which can be millions of dollars per mile, make subways impractical in less densely developed areas.

The first subway, suggested by city solicitor Charles Pearson, opened in London in 1863; that initial 3.75 mi. (6.033 km) was constructed by the open-cut method and carried 9,500,000 passengers in its first year. Passengers were first carried by steam **locomotives** that burned coke and coal, both major pollutants, but electricity was in use by 1890. All subways built since then have used electric trains, which are quieter and do not foul the subway passages with pollution. In 1866 work began on what became known as "The Tube" after tunneling methods and shields were perfected. American financier Charles Tyson Yerkes (1837-1905) came to London in 1900 to construct additional tube rail lines and electrify the open-cut segments.

In 1896, Budapest, Hungary, became the first country in continental Europe to open a subway; the Paris *Metro* opened four years later. Boston opened the first subway in the United States in 1897, and the New York subway system was begun in 1904 and is now the largest system in the world. Dozens of other subway systems have been built in cities worldwide. San Francisco's *Bay Area Rapid Transit System* (BART), built in 1971, was the first fully automated subway system, eliminating the use of human drivers.

In many cities in the United States, the migration of major employers to suburban areas has posed a new challenge to transit planners. Despite this, the subway will remain a permanent part of the urban scene.

This is evidenced not only by the 1990s completion of a subway that extends throughout the Washington, D.C., metropolitan area and the beginning of one in Los Angeles, but by the 1997 commitment of the nation's largest rapid transit system (serving metropolitan New York) to purchase a fleet of 1,080 state-of-the-art subway cars that will bring their system into the twenty-first century. In 1998, New York began ordering the most technologically advanced subway cars ever built. These sleek, rider-friendly cars have improved interior space, seating, and information systems, as well as more and better safety features and a smoother ride. Called the R142, the new car is equipped with computer-based communications and diagnostics systems that will interact with an onboard monitoring system. Passengers in these new cars will enjoy ample, bench-style seats and use wider doors. Electronic signs and prerecorded announcements will keep passengers informed. Roof-mounted air conditioning units also will be more reliable, and energy absorbing "crush zones" at each car end will keep travellers safe. Even the operator's cab is ergonomically designed. Tomorrow's subways will run faster, closer together, safer, and with less breakdowns.

Opened in 1866, London's "Tube" was the first successful subway system in the world.

Sundback, Gideon • See Zipper

Sundial

One of the earliest timekeeping devices is the sundial, which consists of a pillar (called a style) or other object that casts its shadow across a dial with hour markings on it. The style is aligned parallel to the Earth's axis.

As early as 3500 B.C., sundials existed as a stick, pillar, or pyramid that cast a shadow but did not have a dial. The Egyptians developed the first complete sundial in the eighth century B.C., and in 300 B.C. Babylonian astronomer Berossus devised a hemispherical sundial. these early sundials divided the day into 6 or 12 segments, which, because the orbit of the earth around the sun is elliptical, varied in duration during the year.

The Greeks, among them astronomer Ptolemy (ca.90-168 A.D.) and mathematician Appolonius of Perga (fl. 250-220 B.C.), developed complex sundials. One of the most famous of these is the Tower of Winds at Athens, which still survives. Built about 100 B.C., the octagonal tower has eight dials that allow for the changing arc of the sun during the year. The Romans used hemispherical sundials and developed portable versions.

Although mechanical **clocks** were developed in the fourteenth century, sundials, which were often works of art as well as timepieces, were used throughout the Renaissance period. In response to the equal-hour marking system used on clocks, sundials were made to also measure in equal hours. However, after the invention of the weight-and-pendulum clock in 1656 by **Christiaan Huygens**, sundials were relegated to ornamental status.

Sunscreen

The image of a healthy person today does not necessarily include a deep, dark tan. Research has linked exposure to the sun's ultraviolet rays to skin cancer, premature wrinkles, and other skin problems. While protective clothing such as hats, pants, and long sleeve shirts are the most effective blocks against these damaging rays, sunscreen lotions also provide remarkable protection for the skin.

The light emitted by the sun consists of three frequency bands of radiation: infrared, visible, and ultraviolet. Of the three, only the ultraviolet is harmful to most humans. Ultraviolet (UV) radiation is further divided into three categories: UVA, UVB, and UVC. UVA radiation penetrates the skin without burning the surface layers. These rays can penetrate to a depth of 0.04 in (1.0 mm) and cause damage to cell membranes and the immune system. UVA radiation has been linked to skin cancers and premature aging and wrinkling of the skin. UVB radiation is responsible for the painful, red burn people get after prolonged exposure to the sun. UVB rays also cause skin cancer and can damage the cornea and lens of the eye. The third category, UVC radiation, is generally absorbed by the earth's atmosphere and is not considered harmful. Sunscreen lotion which provide protection against both UVA and UVB radiation are known as having a broad spectrum of protection.

There are two basic types of sunscreen lotions on the market: products that penetrate the outermost layer of skin to absorb ultraviolet rays, and products which coat the surface of the skin to act as physical barriers to ultraviolet rays. Both of these types are rated with a sun protection factor (SPF), which lets the consumer know how much protection against UVB rays the product provides. The SPF of a product is the ratio of the time required for a person's protected skin to redden after being exposed to sunlight compared to the time required for the same person's unprotected skin to redden. For example, a product with SPF15 means that a person whose unprotected skin would redden in ten minutes can apply the product and stay in the sun 15 times longer, or 150 minutes, before they get a sunburn.

Researchers believed for a long time that UVB rays—the rays that actually cause a sunburn—were solely responsible for all forms of skin cancer. However, recent studies prove that UVA rays are also responsible. Although many sunscreens now contain UVA protectors, there are currently no standards set by the Food and Drug Administration (FDA) for protection against UVA rays. The SPF rating on a product applies only to protection against UVB rays. The FDA requires strict regulations and testing prior to the manufacture of any new sunscreen lotion. Sunscreen producers go through an expensive and lengthy process to get FDA approval. This approval authorizes the manufacturer to produce the exact formulation applied for and is limited to only one SPF rating and one specific usage.

Today's target market for sunscreens are highly specialized. Sunscreen products are continually being redeveloped to meet the changing needs of specific consumers. For instance, formulations for athletes may contain ingredients which are more waterproof and sweatproof to provide protection for up to eight hours. Athletes may also desire a lotion that feels dry so as not to affect their grip. Children's skins are more sensitive than adults because the outermost layer is thinner. This supports the observation that most sun damage to the skin occurs during childhood and the early teenage years. Sunscreens developed for the children's market tend to contain natural ingredients such as aloe vera and vitamin E.

In the development phase for a new sunscreen, a team of chemists and lab technicians develop the sunscreen formulation from synthetic and natural ingredients. Initial formulations are made in quantities of 10 gal (38 l) and are stored in stainless steel vats. These initial formulations are tested and finalized before applying to the FDA for approval. FDA approval requires further testing which may be done in-house or by an outside laboratory. Examples of the kind of testing required for FDA approval include tests to measure the effective sun protection factor according to FDA guidelines, tests to determine how safe the product is to use on the skin, and tests to measure the waterproof tolerance of a lotion.

Many combinations of synthetic and natural ingredients may go into the formulation of a single sunscreen. A formulation is generally geared towards a specific SPF rating or the needs of a specific consumer group. Perhaps the best-known synthetic material used for protection against UVA rays is avobenzone, or Parsol 1789, which is used in products worldwide.

Broad spectrum protection is provided by other synthetic ingredients such as benzophenone and oxybenzone, which protect by absorbing UV light. PABA (para-aminobenzoic acid) was once a popular UV-absorbing sunscreen ingredient, but it can cause skin irritation in some people and is now replaced by Padimate-O, a derivative of PABA. Other broad spectrum synthetic ingredients are octyl methoxycinnamate and menthyl anthranilate.

Titanium dioxide is a natural mineral and a popular ingredient for broad spectrum protection. Titanium dioxide works by scattering UV light instead of absorbing it. Although not as opaque as zinc oxide, it has a similar whitening effect in the higher SPF ratings. Antioxidants are often combined with titanium dioxide to slow down the oxidation of oils and thereby delay the deterioration of the lotion. Some examples of natural antioxidants are vitamins E and C, rice bran oil and sesame seed oil. Another popular antioxidant in the natural category is green tea. Many newer sunscreen products also contain skin soothing and moisturizing additives such as aloe and chamomile.

Researchers look to nature in search of the next wave in sunscreen development. Some plants have natural defenses against the damaging rays of the sun. For example the single-cell alga called *Dunaliella Bardawil* that thrives in the Dead Sea and the Sinai desert makes its own sunscreen. Scientists at the Weizmann Institute of Science in Rehovot, Israel, isolated the protein that this plant produces when sunlight gets too intense. The protein acts as a solar deflector by funnelling light down to where photosynthesis takes place. Excess light which could interfere with photosynthesis is shunned by a yellow-orange pigment produced by the algae.

The human body also has a natural defense called melanin. Melanin is the brownish-black pigment found in skin and hair. It reflects and absorbs ultraviolet rays to provide a broad spectrum of protection. Dark-skinned people have a higher concentration of melanin and, as a result, have a lower incidence of skin cancer as well as less physical and medical signs of aging skin. Melanin was once painstakingly collected by extraction from exotic sources such as cuttle fish and cost about $3,000 per ounce ($101 per ml). However it can now be made in an inexpensive procedure using fermentation jars. One technology being used as a method of incorporating melanin into sunscreen lotions is to encapsulate it into microsponges which hold the melanin on the surface of the skin where it is most effective. The microsponges are invisible to the eye, and can only be seen under a microscope. Researchers continue to gain approval for the many uses of natural and synthetic melanin as an ingredient in sunscreen formulations.

Supercomputer

The introduction of the microprocessor **integrated circuit** (IC) in 1971 gave birth to two new categories of computer, the **microcomputer** and the **supercomputer**. The microcomputer developed first, drawing on the sudden progress in miniaturization represented by the **microprocessor**. Shortly thereafter, as further scientific breakthroughs doubled the microprocessor's speed and complexity each year, the term ''supercomputer'' came to be used to designate whatever computer currently contained the largest number of ICs, had the highest speed, and (usually) sold for the highest price.

The *Cray 1 Computer*, designed by Seymour Cray, founder of Cray Research of Chippewa Falls, Wisconsin, is generally regarded as the first supercomputer. Containing two hundred thousand ICs, it operated at one hundred fifty megaflops, which means one hundred fifty million floating point calculations per second. A floating point calculation is the addition, subtraction, multiplication or division of two very large or small numbers of the type described in scientific notation, such as 1.345 x 1011. Such calculations require much more work than fixed-point or integer operations.

Supercomputers are used for the most calculation-intensive tasks such as oil and mineral prospecting, analysis of subatomic activity, studying the Earth's changing ozone layer, and many computer design and simulation projects, including design of **aircraft**, microprocessors, and ships; and simulation of weather systems, the birth of the universe, nuclear reactions, DNA, and many other tasks.

Cray, who had previously worked for Univac and later helped found Control Data Corporation, made his goal the production of the world's most powerful computers. Until the late 1980s, Cray fulfilled that goal nearly all of the time, but with increasing competition in the U.S. and Japan. Cray and others achieved very powerful computing in several ways, some of which include:

1. Adding more circuits to single ICs—As scientists found more ways to create smaller and smaller transistors, capacitors, resistors and diodes within semiconductors, the complexity and speed of each single IC quickly increased.
2. Adding parallel processing—Scientists also realized that the fastest, most powerful computing device, the human brain, achieved its power not from ultrafast processors, but from having many processors operating simultaneously, or in parallel. This has led most supercomputer firms to design computers having two or more microprocessors. For instance, the Cray-2 uses four microprocessors and has a peak speed of about 2.2 gigaflops (2.2 billion floating point calculations per second).
3. Conversion to massively parallel processing—Based on the success of their initial efforts in parallel processing, most supercomputer firms, including Cray, Control Data, Ncube, Inc., NEC, Fujitsu, Hitachi, Thinking Machines, Intel, IBM and others, are designing supercomputers with thousands or even millions of microprocessor ICs operating simultaneously. Most have set a goal of producing a computer by the year 2000 capable of operating at one teraflop (one trillion floating point operations per second).
4. Use of gallium arsenide instead of silicon chips—Cray's new company, Cray Computer, Inc., of Colorado Springs, is attempting to build its Cray 3 Supercomputer using gallium arsenide chips, which operate at lower

voltages, higher speeds, and produce less heat. The Cray 3 is designed to have a peak speed between twenty and thirty gigaflops.

In 1996, Cray merged with workstation giant Silicon Graphics, Inc. to jointly produce supercomputers.

See also Minicomputer

Superconducting supercollider • See Particle collider

Superheterodyne • See Radio

Supersonic aircraft • See Aircraft

Surfboard

Early polynesian peoples were the first to ride ocean waves on surfboards, but no one knows which particular group of islanders invented them. At the time, there were two types of boards: the *alaia* and the *olo*.

The smaller alaia boards, which weighed only about 11 lbs. (4.98 kg) and were made from the breadfruit tree, were typically used by women and children. They measured approximately 6 ft. (1.83 m) long, 14 in. (35.56 cm) wide, and 1/2 in. (1.27 cm) thick. The larger olo boards measured 15 ft. (4.57 m) long and were 6 in. (15.24 cm) thick. At 160 lbs. (72.43 kg), these boards were considerably heavier and were carved with tools made of stone or bone and then smoothed and polished by hand.

Construction of the boards remained virtually unchanged until early in the twentieth century. Duke Paoa Kahanamoku (1890-1968), a Hawaiian Olympic champion swimmer, was among those who worked on improving surfboard construction and design. Tom Blake, one of the earliest Caucasian inventors of surfboards, built the first paddleboard—a big, hollow board with straight rails (sides), a semi-pointed tail, and laminated wood for the deck (top surface).

After World War II, surfboard builders in Hawaii and California accelerated their experimentation. Single and double skegs (fins) were tried along with various convex and concave shapes. Boards during this period were made of solid wood, laminated wood, or **plastic.** One innovation that largely changed this was the introduction of **foam rubber** and **fiberglass**, materials which rendered wood, at least as a primary material, obsolete.

Ensuing design improvements included the creation of scoop noses that allowed the tips of surfboards to stay on the surface rather than dip under the waves. One person closely associated with such improvements was Robert Simmons, who lost his life in a surfing accident.

See also Windsurfer

Surge suppressor

Power line disturbances occur on an average of four times a day, according to studies by the Institute of Electrical and Electronic Engineers (IEEE). These disturbances—increases in current that can damage electronic devices plugged into outlets—may be caused by lightning or other weather-related incidents; traffic accidents affecting power lines; the use of electrical products such as motors, compressors, and fluorescent lights (these all cause power-spikes); high-powered electrical equipment and voltage fluctuations initiated by a power company (both cause power surges) and high-frequency noise.

Such disturbances can interrupt or wipe out power service and damage electrical equipment. Both surges and spikes can be prevented by using a surge suppressor, an apparatus that serves to ground the interference or, in other words, run the interference into the ground rather than into the electrical equipment, and/or absorb the excess flow of current throughout an electrical system.

Spikes and surges can damage electronic equipment even when it is not turned on. One of the first surge suppressors was developed by the General Electric Company in the 1950s. A similar device was being developed in Japan around the same time. Initial surge suppressors utilized selenium rectifiers (components used to convert direct current to **alternating current**) and, later, carbon piles (disk-shaped components used to carry current). Most surge suppressors commonly used today utilize spark-gap technology—a system whereby the suppressor breaks down, or shorts out, currents as the voltage applied to an electronic device exceeds the maximum tolerance, or rating, of the device.

There are three types of spark-gap surge suppressors: gas tube suppressors, metal oxide varistor (MOV) suppressors, and silicon avalanche suppressors (a specific type of transient voltage suppressor [TVS]). The type names refer to the materials that are the main component of each type of surge suppressor.

A TVS utilizes a process called ''reverse bias voltage clamping.'' In this process, the excess current flowing through a TVS **diode** (a tubular semiconductor device that is similar to a resistor) breaks down and becomes a short circuit when the voltage applied to a circuit exceeds that circuit's rated avalanche, or maximum, level. A TVS is only used with **direct current** (DC) circuits, where the current only flows one way (such as in an automobile battery) but two TVSs placed back to back will protect an alternating current (AC) system, where current flows both ways. A gas tube suppressor is commonly used with communications and voltage power supply lines.

The gas tube suppressor shorts out a power surge and carries any associated high current to the ground, bypassing the circuit which would otherwise be affected, by providing a low-voltage path for the excess current through a ceramic or metal tube. Gas in the tube ionizes during a surge and creates a conducting state within the suppressor. The surge is transformed in this conducting atmosphere into a discharge arc, which shorts out the suppressor, and any high current is grounded. The gas subsequently de-ionizes and the suppressor is reset. A MOV uses variable resistors, or varistors, and is thus most functional with an AC system.

An MOV suppressor absorbs extra voltage during both positive and negative swings in an AC system through a process known as clamping. When the voltage applied to the circuit exceeds the MOV rating the load above which the MOV is supposed to be triggered a short circuit is caused and the voltage is transferred into the absorbent body of the MOV suppressor, thereby bypassing the circuit that would have otherwise been disrupted.

Because surge suppressor design is fairly simple, new innovations center more on enhancing the original product rather than changing its makeup. While MOV surge suppressors remain the most common apparatus for residential and commercial surge protection, many models of MOV surge suppressors are available. Some models incorporate resistance to phone line interference and state-of-the-art methods of eliminating noise interference. Many changes are package-oriented: the outlet configuration is altered to allow for more appliances to be connected to the suppressor and to accomodate plug-in **transformers**, the casing is made of more durable materials, like aluminum, diagnostic **LED** indicators (lights indicating the status of the line and whether there is any interference) have been added.

As residential and commercial electronics need continue to grow at a quick pace, and protection of electronic equipment especially **computer** and telecommunications equipment becomes paramount, this small, simple electronic devise is fast becoming a necessary component of any electronic setup.

SURGICAL INSTRUMENTS

The earliest surgical operations were probably circumcision (removal of the foreskin of the penis) and trepanation (making a hole in the skull, for release of pressure and/or spirits). Primitive surgical instruments consisted of flint or obsidian knives and saws. Stone Age skulls from around the world have been found with holes from trepanning. Primitive people used knives to remove fingers, and the ancient Mesopotamian cultures practiced surgery to some degree. Small copper Sumerian knives of about 3000 B.C. are believed to be surgical instruments. The Babylonian *Code of Hammurabi* of about 1700 B.C. mentions bronze lancets—sharp-pointed two-edged instruments used to make small incisions. The Code, however, provided harsh penalties for poor treatment outcomes, so surgery was practiced only sparingly. Likewise, ancient Chinese and Japanese cultures were opposed to cutting into the human body, so surgical instruments were used very little.

By contrast, the ancient Egyptians recorded surgical procedures as early as 2500 B.C. They fashioned sharper instruments with metal and copper, and designed special tools to remove the brain when preparing bodies for mummification.

The ancient Hindus excelled at surgery. The great surgical textbook, *Sushruta Samhita*, probably dates back to the final centuries B.C. This work described 20 sharp and 101 blunt surgical instruments, including **forceps**, pincers, trocars (sharp-pointed instruments fitted with a small tube), and cauteries (irons to heat and sear tissue), mostly made of steel. The ancient Hindus used lancets for **cataract surgery**, scalpels to restore amputated noses via plastic surgery, and sharp knives to remove bladder stones. Ancient Peruvians performed trepanation and left behind various obsidian surgical instruments, including scalpels and chisels.

The Greeks practiced surgery, mostly on external parts, using forceps, knives, and probes, among other instruments. Bronze Roman surgical instruments found at Pompeii include a scalpel with a steel blade, **spring** and **scissor** forceps, a sharp hook, and shears. Celsus in the first century A.D. described the use of ligatures to tie off blood vessels and reduce bleeding during operations. Galen (130-200) in the second century A.D. gave detailed instructions on the use of surgical instruments.

After ancient times, medical knowledge atrophied, and surgeons fell to a lowly status. In the absence of knowledge about antiseptics, surgery was highly risky, so only the simplest and most urgent operations, such as **amputation**s, were performed. A few physicians sought to spread knowledge of surgical procedures and published texts that illustrated surgical instruments. Among these were the Muslim Spaniard Albucasis (eleventh century; his favorite instrument was the cautery), the Germans Fabricius and Scultetus (1600s), and the Britons William Clowes (1591), Peter Lowe (1596), and John Woodall (1639). Most important was the Frenchman **Ambroise Paré** (1517-1590), the great surgeon of the Renaissance, who revived use of the ligature and invented many surgical procedures and instruments, among the latter being the ''crow's beak'' to hold blood vessels while tying them off. He also perfected an instrument for cataract removal.

The era of modern surgery began with the introduction of anesthesia and antiseptics/antisepsis in the mid-1800s. Louis Pasteur (1822-1895) suggested sterilizing surgical instruments in 1878, and the American doctor William Halsted (1852-1922) introduced sterile rubber gloves to surgery in the 1890s. The discovery of **X ray**s in 1895 gave surgeons an invaluable diagnostic tool. Refinements in surgery were made possible by the introduction of the operating **microscope** (microsurgery) in the mid-twentieth century and **laser** surgery in the 1970s.

New uses for or processing of existing materials continually enhances surgical and diagnostic processess: **glass** in the form of optical fibre is used in the manufacture of the flexible endoscope, a thin tube with light source, camera, and tiny surgical tools that provides visual examination and allows non-invasive surgery through a natural body opening such as the throat or through a tiny incision; NiTi, bonded nickel and titanium, provides strong, biocompatible, and corrosive resistant metal for implantable devices such as the Mitek anchor used in orthopedic surgery; non-magnetic surgical tools utilizing titanium are durable and lightweight, and compatible with **magnetic resonance imaging**; and bendable handles on surgical instruments—which straighten with the heat of sterilization—allow precision in **open-heart surgery**.

Increasingly sophisticated technology makes ever-more-precise surgical tools possible, among them voice-activated operating microscopes and robotic surgical hands: NASA's Technology Commercialized Program unveiled a robotic "brain surgeon" in June 1997; their "Smart Probe," robotically guided through the brain, uses minute sensors to measure tissue density and blood flow and an endoscope to transmit real-time images and information to the surgeon via adaptable software that literally learns from experience. Also in 1997, the University of California at Berkeley developed a "millirobot" prototype—not yet used by a surgeon—which enters the body and differentiates between healthy and cancerous tissue. Looking much like a fountain pen, this little robot transmits data to the surgeon who directs the operation from a remote location using a television console and controls like joy sticks. Researchers predict that by the end of the first decade in the 2000s, surgeons miles away from an accident victim or battlefield casualty will be directing robotic surgery.

SURVEILLANCE EQUIPMENT

Surveillance equipment allows for the continous monitoring of people and places, the desire for which has existed forever. In some cases, the observer wishes to be hidden from the people being watched or listened to; examples include hidden **cameras** and "bugs," or hidden listening devices. Other times, surveillance equipment is made as visible as possible through signs, conspicuous camera and monitor placement, and other means. In all cases, however, surveillance equipment enhances the observers natural ability to see and listen.

Strategic locations, either manmade or natural, were early forms of visual, or optical, surveillance equipment. A castle tower, the crow's nest of a ship, or a tall tree all provided the "bird's-eye view" essential for surveillance. Combined with optical inventions such as the **telescope**, early surveillance relied primarily on the human eye.

Modern visual surveillance equipment began with the development of photography. During the American Civil War (1861-65), still photographs were used to gather information. **Motion pictures**, invented in the 1890s, were quickly adapted for use as surveillance equipment. In World War I (1914-18), motion picture cameras were mounted in balloons floating above the battlefield to monitor enemy troops. As with most surveillance equipment, films were adapted to civilian security use during peacetime. To provide continuous monitoring, movie cameras were set up to run automatically without a human operator. To save film, they used time phase photography, which slowed down the exposure rate. Motion picture cameras were used for surveillance until the 1960s until they were finally replaced by videotape.

Surveillance equipment for listening was also developed for military use. During World War II (1939-45), advances in magnetic tape recording and solid state electronic components were used for espionage. Eavesdropping on telephone communications, known as "wiretapping" became a useful tool for military intelligence and law enforcement. "Bugging" refers to the recording of sounds, usually conversations, without the subject's knowledge. Audial surveillance has relied heavily on advances in miniaturization. As microphones and tape recorders were made smaller and smaller, they became more useful. Concerns about abuses of these inventions eventually led to

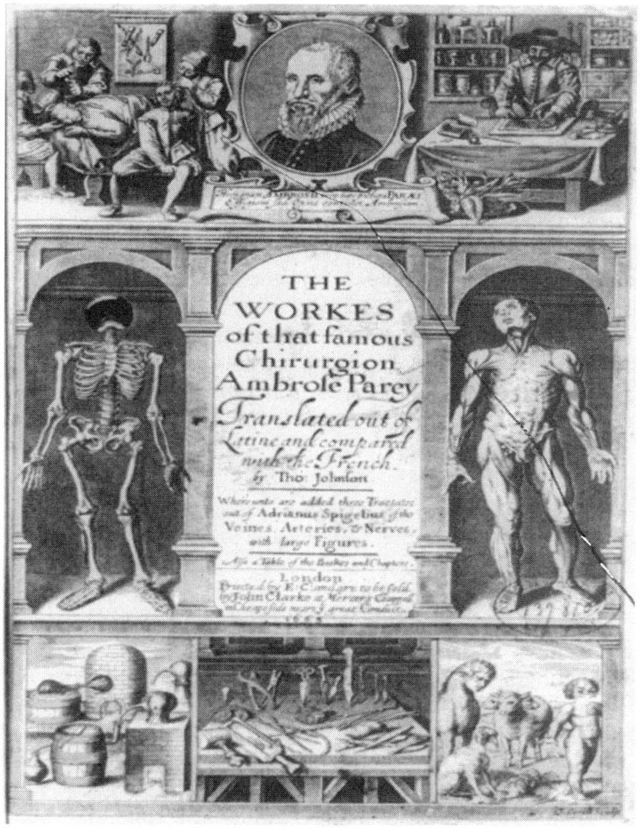

Frontpiece of *The Workes of that famous Chirurgion Ambrose Parey.* Published in 1665, the text outlined for Renaissance doctors many of the advancements made by the pioneering french surgeon, including the use of ligatures and the "crow's beak" for clamping blood vessels.

the Electronic Surveillance Law of 1968, which put limits on the use of such equipment.

The major breakthrough in visual surveillance came with the invention of television, and specifically, closed-circuit **television** (CCTV). Unlike broadcast television, closed-circuit television connects the camera to designated television monitors by cables. The major advantage of CCTV is the ability to conduct "remote monitoring," whereby numerous cameras are placed at various locations but all the monitors are centrally located. As the cost of television equipment dropped in the 1950s, such systems became a popular means of security for businesses and institutions.

In the 1960s, another breakthrough in surveillance equipment occurred when the emerging technology of videotape recording was combined with CCTV. Magnetic videotape offered many advantages over film, including the ability to be erased and reused. Advances in television cameras made them cheaper, smaller, more durable, and better able to capture images in low light. Videotape recording was standardized with the innovation of **videocassette** recording (VCR). Through these advances, video surveillance systems became a $1 billion per year industry by the early 1980s.

The video surveillance revolution has continued well into the 1990s. The trends of falling prices and miniaturization have opened vast new markets for video surveillance equipment. No longer the exclusive domain of law enforcement, spies, and security companies, consumers began to use use surveillance equipment to monitor domestic help, babysitters, and even their own children.

In the 1990s, digital technology became the state of the art for surveillance equipment. Images from CCTV and videotape could be digitized and stored on **computer disk**. Also, software was developed to eliminate the need for human monitoring of video images. Digitized images are compared to a series of ''profiles'' stored in the computer and the appropriate action is taken. If, for example, the image captured matches the profile for a dog, the system identifies it as such and ignores it. If a human is identified, the system can sound an alarm or call for human intervention. Digital technology facilitates around-the-clock surveillance

''Wireless,'' or cellular technology, also changed surveillance equipment in the 1990s. Previously, remote monitoring required monitor and camera to be physically connected via cable. Wireless technology allowed remote monitoring dozens, or even hundreds, of miles away. This advancement has tremendous appeal for security companies. From a small number of locations, they can monitor hundreds of cameras located all over the nation.

Although most citizens accept the ubiquitous presence of surveillance equipment, the prospect of being watched and recorded around-the-clock by computers located hundreds of miles away is of concern to many. In the late 1990s, however, the debate over privacy issues has not slowed the development of surveillance equipment.

Surveying instruments

Surveying is the apportionment of land by measuring and mapping. It is employed to determine boundaries and property lines, and to plan construction projects. Any civilization that had any degree of sophistication in construction methods required surveys to ensure that work came out according to plan.

Surveying is thought to have originated in ancient Egypt as early as 2700 B.C., with the construction of the Great Pyramid of Khufu at Giza, though the first recorded evidence of boundary surveying dates from 1400 B.C. in the Nile River valley. The classic surveyors were the Romans. In order to forge an Empire that stretched from the Scottish border to the Persian Gulf, a large system of roads, **bridge**s, aqueducts, and canals was built, binding the country economically and militarily. Surveying was a major part of Roman public works projects. It also was used to divide the land among the citizens. Roman land surveying was referred to as centuriation after the *century*, which was a common rectangular unit of land area. These land parcels can still be seen in aerial photographs taken over France and other parts of Europe, the work of the Roman agrimensores, or ''measurers of land.'' The property lines were usually marked by stone walls and boundary markers.

With the advent of new methods of **trigonometry** and calculus, new surveying instruments emerged. The theodolite was invented in the sixteenth century. Its precise origin is unclear, but one version was invented by English mathematician Leonard Digges in 1571, who gave it its name. A great theodolite was invented by Jesse Ramsden more than 200 years later in 1787, and its use led to the establishment of the British Ordnance Survey. Made of a telescope mounted on a **compass** or of a quadrant plus a circle and a compass, the theodolite is used to measure horizontal and vertical angles. The modern theodolite is usually equipped with a **micrometer**, which gives magnified readings up to 1/3600°, or one second of arc. The micrometer is derived from the **vernier** scale, which was invented by French engineer and soldier Pierre Vernier (1584-1638) in 1631 to measure in fractions.

The transit is a theodolite capable of measuring horizontal and vertical angles, as well as prolonging a straight line or determining a level sight line. A telescope atop a tripod assembly is clamped in position to measure either horizontal or vertical angles. The transit employs a plumb bob hanging from the center of the tripod to mark the exact location of the surveyor.

The practice of triangulation was introduced by Gemma Frisius in 1533. By measuring the distance of two sides of a triangle in a ground survey, the third side and the triangle's area can be calculated. Triangulation was aided by the inventions of the prismatic **astrolabe** and the heliotrope. The latter was invented by German mathematician Johann Gauss (1777-1855), who is considered the father of geodesy, the science of earth measurement. Both instruments used a beam of sunlight to signal the positions of distant participants in a land survey.

Other survey instruments include the surveyor's compass, which is used for less precise surveying, the surveyor's level used to measure heights of points above sea level or above local base points, and metal tapes, first introduced by English mathematician Edmund Gunter in 1620, used for shorter measurements.

In the late twentieth century, surveying has been aided greatly by remote sensing. Photogrammetry employs aerial photography to survey large areas for topographic mapping and land assessment purposes. Artificial satellites and global positioning techniques enable surveryors to make quick and accurate measurements of the distance between two points, and global positioning is also used in mapping utility lines and monitoring shipping.

Laser technology has increased the precision of survey sightings. Laser-based speed and distance measurement instruments, or *laser rangefinders,* are used in a wide variety of applications ranging from traffic speed enforcement to forestry, from surveying and mapping to golfing and hunting, and from docking ships to docking the Space Shuttle. Lasers have even been used to measure the distance between the Earth and the **moon**. During the first landing on the moon, Apollo 11 astronauts set up a laser reflector that is still in operation. By measuring how long it takes a laser pulse to return to Earth (about 2.5 seconds), the Earth-moon distance has been measured to an astonishing accuracy of about 1 in (2.5 cm).

Suspension bridge • See Bridge

Suspension systems, automobile

Automobile suspension systems combine the principles of two key elements: pneumatic **spring**s and shock absorbers. These not only make the ride more comfortable, but also protect the mechanical components of the vehicle. Early attempts to soften carriage travel appeared in the sixteenth century. The carriage body was hung like an upside-down table, suspended by leather straps from the tops of the chassis ''legs.'' Unfortunately, this solution simply created more motion. By the mid-1600s, carriage makers had developed large C-shaped springs, but this system was little better.

Early self-powered vehicles used modified versions of the basic carriage C-springs, called leaf springs. They were strengthened by adding **steel** strips that tapered lengthwise from the fixed end. Their shape soon became elliptical; and they were bolted between the axles, the chassis, and body. Other springs included helical coils and straight torsion bars (coiled springs that had been straightened out).

In 1896, J. R. Heath applied the principles of pneumatics to vehicular suspension. He put air bags around the ends of the axles and held them in place with metal loops attached to the chassis. This idea was furthered in 1901 by M. A. Yeakley, who built a car using air-operated springs on all four axles. His creation allowed the driver to adjust the suspension height (based on the load carried) through manual control of an engine-driven air compressor.

To help eliminate the bouncing that occurred after every bump, Emile Mors created a friction-type shock absorber in 1899. This was improved by C. L. Horock in 1901. His design, called the telescopic shock absorber, is still used today. It had a **piston** and cylinder inside a metal sleeve; it also included a one-way **valve** in the piston. As air (or oil) flowed into the cylinder through the valve, the piston moved freely. In the other direction, the piston encountered friction as it pushed against air or oil, thus cushioning the vehicle's ride.

Like other automobile systems, today's suspension systems use computer-controlled adjustments to respond to ride height, road conditions, the speed of the vehicle, weight and load, pressure from the power-steering system, acceleration, and braking. The main components of the modern system are adjustable shock absorbers, the electronic suspension (which may operate by air or hydraulic pressure or both), air or fluid lines, actuators and actuator motors (to move the rod or piston inside the shock absorber cylinder), sets of connectors for the shocks or struts, and sensors that relay information to the onboard computer. Designers of suspension components seek improvements in ride-control systems that add durability, safety (for both the vehicle and its occupants), comfort, and economy.

Sutures, vascular • See Carrel, Alexis

Svedberg, Theodor (1884-1971)
Swedish physical chemist

Svedberg was raised by a civil engineer who shared his love of nature with his son. He studied chemistry in school because he believed many unsolved problems in biology could be explained as chemical events. As a scientist he became interested in the chemistry of colloids which led him to develop the ultracentrifuge a device which is useful in separating suspended particles.

Theodor Svedberg

A colloid is a mixture in which tiny particles of one or several substances are suspended in liquid such as water. These colloidal particles are larger than molecules or ions, but they are usually too small to settle under the force of gravity or to be seen with an ordinary microscope. Some important biological colloid particles include proteins, DNA, and viruses.

Svedberg wanted to identify the exact size of the colloids he studied, but this was difficult because the only way to make such measurements was to note how the particles settled through sedimentation. However, only the largest colloid particles settled out a rate fast enough to be measured. Other, smaller particles proved so small that the incessant movement of water molecules was enough to keep them suspended. Svedberg attempted to make the small particles sink faster by spinning the liquid in a centrifuge. Centrifuges were already being used to separate milk from cream and blood corpuscles from blood plasma. Yet, to force much smaller colloids out of solution, a much more powerful centrifuge was needed. In 1923 Svedberg developed the ultracentrifuge, a device that was capable of rotating fast enough to create a force hundreds of thousands of times that of gravity.

By 1936 he had produced an ultracentrifuge that spun at 120,000 times per minute and created a centrifugal force of 525,000 times the earth's gravitational force. With these centrifuges scientists could discern both the size and shape of colloid particles. This machine had far-reaching effects, enabling biologists, biochemists, physicians, and other life scientists to examine viruses and how they attack cells, the nucleus of a cell, cell functions, and individual protein and nucleic acid molecules, as well as explore **genetic engineering** possibilities.

Svedberg was awarded the Nobel Prize in Chemistry in 1926 for his contributions to colloid chemistry. In later years, Svedberg studied the electric synthesis of metals suspended in a colloid solvent, carried out research in radioactive processes, and helped develop **synthetic rubber** during World War II.

Swan, Joseph (1828-1914)

English chemist and physicist

After serving as a pharmacist's apprentice, Swan became a partner in a chemical firm that manufactured photographic plates. At that time, one of the most difficult steps in the production of photographic plates involved adding chemicals in a liquid state to the plates. However, Swan discovered that the fluids used on the plates became more sensitive when carefully heated. This meant that the plate could actually work better when dry. By 1871, Swan had developed the much simpler "dry-plate" photographic method. His innovation initiated a new era in **photography**, leading to rapid improvements over the next 20 years. Swan also invented the first bromide **paper**, which is still widely used for printing photos.

Swan's primary interest, however, was electric light. Recalling the days of his youth, when indoor lighting was dim and expensive, Swan observed that "as a rule, the common people went to bed soon after sunset." Swan began to experiment with incandescent filaments while in his twenties. He came across a patent on an incandescent **lamp** in the name of J. W. Starr (1822-1847), an American who had died at age twenty-five before he could fully develop his idea. Although Starr's lamp used a platinum filament, he had noted that in a vacuum, carbon could be used. By 1860, Swan succeeded in making filaments with strips of carbonized paper, but his lamps remained impractical for two reasons: **vacuum pumps** at the time were unable to create a high enough vacuum to protect the carbon filament and electricity supplied by **batteries** was expensive and unstable.

Swan abandoned this research for 17 years. Then in the late 1870s, he read about the work of **William Crookes**, who was analyzing new gases, and learned that **Hermann Johann Philipp Sprengel** had invented a more effective vacuum pump. Also, power **generators** called dynamos had been improved to the point where they could supply cheaper, more reliable electricity than batteries. Soon Swan had invented a practical incandescent filament lamp, which he displayed in December 1878. His own house was the first electrically illuminated private home in England. In 1881, Swan formed a company to manufacture his "glow lamps," and they quickly became popular in England. By the early 1880s, they were installed in the Savoy Theater, the House of Commons, and the British Museum. At one point, Swan's factory received an order from the United States for 25,000 lamps.

Meanwhile, **Thomas Alva Edison** had begun similar research and had aggressively protected his work with patents. In 1882, Edison sued Swan for patent infringement. When Edison's lawsuit failed to stop work at Swan's factory, the two inventors settled out of court, joining forces in 1883 to form the United Electric Light Company. Swan continued to improve incandescent filaments and, as an offshoot of his research, created what was to become the first synthetic fiber when he developed a method for making fine threads out of nitrocellulose, which could be converted to less flammable cellulosefibers. The French chemist Louis Comte de Chardonnet adapted Swan's process to produce artificial silk, later named **rayon**. In recognition of his scientific contributions, Swan was knighted in 1904.

See also Light bulb, incandescent

Swinburne, James (1858-1958)

Scottish inventor

During his lifetime, Sir James Swinburne was granted more than one hundred patents, mostly related to **electricity** and lighting. Born into a titled seafaring family in Scotland, Swinburne started out as an apprentice engineer at a locomotive factory in England. In 1880, he was sent by Sir **Joseph Swan** to set up electric lamp factories in France and America.

Swinburne's next job was with Crompton & Co., a British manufacturer of dynamos, machines that generate direct-current electric power. By improving dynamo design, Swinburne helped spur the growth of the electric lighting industry. Previously, electricity was supplied mainly by batteries, which were relatively expensive to operate. Dynamos provided a cheaper source of power that could be widely distributed. Swinburne also invented many significant improvements to the **incandescent light bulb**.

In 1899, Swinburne established his own engineering consulting firm in London, complete with laboratory and workshop, and his ingenuity and dexterity made him a success. Swinburne worked to patent an important new **plastic** made by mixing **phenol** and **formaldehyde**. The dark, resinous product had first been formed some thirty years earlier by German chemist Adolf von Bayer. But Swinburne, who was attempting to find a better material for electric cable insulation, realized that the resin's properties could be commercially useful. When the liquid sludge was heated, it could be molded and solidified into a permanent shape (*thermoset*); once hardened, liquid sludge was resistant to water, heat, and solvents, and could not conduct electricity. Swinburne considerably improved the method of preparing the resin, using a catalyst to accelerate production. But unknown to him, a chemist working independently in America, Belgian-born **Leo Hendrik Baekeland** (1863-1944), also recognized the resin's value. He patented the material in 1907, one day earlier than Swinburne, and named it **bakelite**. Thus began the modern plastics industry.

Bakelite production grew by leaps and bounds along with the automotive and electrical equipment industries, since the plastic was ideal for making small switches and other components. Meanwhile, Swinburne went on to develop a related material, a synthetic lacquer that he named "damard." After World War I, Swinburne and Baekeland agreed to merge their businesses into Bakelite Ltd., which eventually became part of Union Carbide Corp. Swinburne remained Bakelite's chairman until 1948. Swinburne's industrial career was exceptionally long-lived. He held a record seventy-three year membership in the Institution of Electrical Engineers. In addition, he was also a gifted composer who set some of Alfred Lord Tennyson's poetry to music.

See also Generator

Swiss army knife

The Swiss army knife was first produced for the Swiss military by Karl Elsener in 1891. It was a wood-handled device with a newly designed **spring** mechanism that allowed such utensils as a **bottle** opener, **screwdriver**, and fingernail file, to fold alongside the blade. Elsener also designed a knife for civilian use; this one had a red handle so that it could be found easily in snow.

Today, the Victorinox Cutlery Corporation, whose president is the inventor's great-grandson, continues to make the swiss army knife. The largest version, called the Swiss Champ, weighs approximately 7 ounces (198 g) and has 29 attachments, including a **ballpoint pen**, **magnifying glass**, and a leather-stitching needle. Numerous companies make similar knives that sport a variety of attachments for men and women.

See also Knife

Sword

Human beings have presumably always carried weapons, and developed varieties of spears, bows and arrows, axes, daggers and flails very early in our history. Swords, which are essentially daggers with lengthened blades, are relatively late arrivals in the arsenal. They were first manufactured during the Bronze Age, when techniques were developed to make sufficiently strong blades out of bronze with a 10% tin content. Other improvements included the addition of extra longitudinal ribbing for greater rigidity of the blade and the addition of "horns" to protect the hand of the swordsman. The earliest known uses of swords occur among the Assyrians, Gauls, and Greeks. Swords may be divided into weapons designed for thrusting and those designed for cutting. Both varieties have been found among the ancient Greeks, although they seem to have favored the thrusting sword. Swords of both varieties were developed in Western Europe, while all over Asia swords with a curved blade best suited for cutting were used almost exclusively. The dreaded Eastern scimitar is the best known example of these.

With the introduction of iron, swords were again greatly improved. The discovery that iron could be hardened into a mild kind of **steel** by heating it with small quantities of carbon gave the Romans a superior weapon that aided them in many military victories. The Middle Ages brought new defenses against lethal weaponry in the form of plate armor. In response, swords were developed that could pierce an opponent's armor. They became longer, eventually reaching a length of almost five feet (1.5 m), and heavier, so that they had to be wielded with both hands. The medieval sword had a straight, two-edged blade.

The dawn of the modern age in Western Europe saw a proliferation in types of swords. Rapiers first came into use in the early sixteenth century with the abandonment of plate armor. Since the fighter was now more vulnerable, a weapon was developed suitable for parrying attacks. The rapier is a long, narrow, straight, two-edged sword, usually with elaborate basket-guards as much for decorative purposes as to protect the hand of the swordsman. At about the same time, swords came into vogue for other than military uses. The rapier became a part of the civilian dress of wealthy men, divorcing the sword from its military context and often serving more as a status symbol than a weapon. In addition, rapier and dagger play became a popular sport, eventually resulting in the highly elaborate art of fencing, at times put to lethal use in duels. Fencing schools sprang up all over Europe. Other straight swords in use in the sixteenth and seventeenth centuries were the broadsword, a lighter, shorter version of the medieval sword, and the backsword, which could be described as a broadsword with a single-edged blade. Swords with curved blades, inspired by the Eastern scimitar, were also commonly used.

The all-purpose hanger, which had a slightly convex cutting edge and was introduced in the sixteenth century, inspired a variety of swords that remained in use for many centuries. Both the hunting sword and the naval cutlass were modeled on the hanger. The last descendent of the hanger was the **bayonet**, which came into use in the seventeenth century with the development of firearms. Some bayonets were permanently attached to **rifle**s by a hinge, so they could be folded back along the barrel when not needed. Others were detachable. The saber, adopted by the United States military in the 19th century, is a hybrid sword suitable for both thrusting and cutting with its slightly convex cutting edge and sharp point. Twentieth century advances in military technology have virtually eliminated hand-to-hand combat, making the sword obsolete as a weapon. It is still used for decorative purposes and as a symbol of power as a part of full military regalia in many countries the world over.

The only exception to this might be Japan, where swords figure in a large percentage of attacks and murders in the late 1990s. Medieval Japanese smiths produced some of the finest swords ever made. The blades were made with a special clay process that produced a sword that was both sharp and flexible. Fine blades were passed down through families, and Japanese soldiers as recently as World War II sometimes wielded weapons that were hundreds of years old. Excellent blades are still made today by Japanese Shinto priests. The Japanese reverence for the sword has been compared to Americans' fascination with the gunslingers of the Old West. Swords are reputed to be the weapon of choice for Japanese organized crime groups, and they are also favored by some far-right political splinters. Perhaps only in Japan has the sword remained the dangerous and desirable weapon it once was.

SYNCHROCYCLOTRON • See Cyclotron

SYNCROTRON • See Cyclotron

SYNGE, RICHARD L.M. • See Chromatography

SYNTHETIC DIAMOND • See Synthetic gemstones

SYNTHETIC FIBER • See Fiber, synthetic

SYNTHETIC GEMSTONES

For centuries, humans have been fascinated by gems because of their rarity, durability, and beauty. Efforts to produce imitation gemstones go far back into history also. Charlatans have long tried to pass off as the real thing minerals that only looked like true gemstones.

The first attempts to produce gemstones in the laboratory occurred in the 1830s. French chemist Marc Gaudin (1804-1880), tried melting chemical compounds that would decompose to produce aluminum oxide, the compound of which rubies are made. Gaudin added a small amount of chromium to impart the characteristic color of the ruby. Although he did not realize it at the time, Gaudin was successful in producing a few tiny crystals of ruby.

In 1904 French scientist Auguste Verneuil (1823-1895) was more successful. Verneuil invented a special furnace for first melting a raw material, such as aluminum oxide, and then allowing it to cool slowly. When an impurity such as chromium, titanium, or iron was added to the molten material, Verneuil's process yielded synthetic rubies or sapphires. When magnesium oxide was used in place of aluminum oxide, spinel, emerald, aquamarine, blue zircon, green tourmaline, or milky moonstone was the product. Verneuil's process is still in use today. In all, about one-hundred different synthetic minerals and gemstones can be produced by the method.

The Verneuil process, however, cannot be used to produce some gemstones, such as opal. Opal can be synthesized by producing tiny spheres of silica in a chemical reaction and then bonding the spheres together by compression.

The greatest amount of effort to produce synthetic gemstones has probably been devoted to the manufacture of synthetic diamonds. Diamonds are an important material for both jewelry and industrial cutting applications. Neither of the methods described above can be used in the production of synthetic diamonds. The technique that is used mimics the process by which diamonds are produced naturally in the earth.

Scientists believe that diamonds (a form of pure carbon) are produced when other forms of carbon are subjected to very high temperatures and pressures at depths of 90-120 mi (145-200 km) beneath the Earth's surface. Based on laboratory research, they think that temperatures of at least 2,700°F (1,482°C) and pressures of 975,000 lb per sq in (66,300 atmospheres) are involved in the creation of natural diamonds.

As far back as the 1850s, scientists attempted to produce temperatures and pressures in the laboratory that would be sufficient to convert graphite to diamond. Scottish chemist James Hanney, French chemists Charles Friedel (1832-1899) and Ferdinand Moissan (1852-1907), British engineer **Charles Parsons**, and American inventor **Edward G. Acheson**, were only a few of those who tried—and failed—to find a method for synthesizing diamonds.

That breakthrough finally occurred on December 16, 1954. Working with a high-pressure device invented by **Percy Bridgman**, a research team at General Electric's Research Laboratories, under H. Tracy Hall, was able to convert graphite powder into tiny diamond crystals. The team used Bridgman's **diamond anvil cell** to produce a temperature of more than 4,800°F (2,649°C) and a pressure of more than 1.5 million lb per sq in (100,000 atmospheres) to achieve success.

The General Electric process is still used to produce tens of thousands of pounds of diamonds with little gemstone value, but with great industrial importance. In 1970, the General Electric process was improved and refined to allow the production of carat-size diamonds. The process is too expensive, however, to make these diamonds competitive with natural diamonds.

In 1984, the diamond anvil cell was employed to manufacture yet another gemstone, jadeite. Robert C. De Vries and James F. Fleischer, at the General Electric Research and Development Center, used a pressure of 440,000 lb per sq in (30,000 atmospheres) and a temperature of 2,700°F (1,482°C) to convert a mixture of sodium, aluminum, and silicon oxides to jadeite. They were able to produce crystals 0.25 in (6.35 mm) thick and 0.50 in (12.7 mm) long colored green (when chromium was added), lavender (when manganese was added), or banded with the two colors.

Not all methods of producing synthetic gemstones are safe. One method involves bombarding stones with neutrons in a nuclear reactor. In 1997, a number of chrysoberyl cat's eye gems found in Hong Kong were made by this method. The procedure converted the initial stones from yellow to brown and made them worth thousands of dollars more per carat. Although this procedure increased the value of the gems, it also caused them to emit radioactivity at dangerous levels. Authorities found most of these radioactive stones and confiscated them.

SYNTHETIC JADE • See Synthetic gemstones

SYNTHETIC RUBBER • See Rubber, synthetic

SYNTHESIZED SPEECH • See Voice synthesizer

Synthesizer, music

The virtuoso demands that composers placed on musicians at the end of the 1800s were but a foretaste of things to come in the twentieth century. Members of the orchestra were complaining that the music of contemporary composers was unplayable because of the enormous difficulty of complex orchestral writing styles. With the Paris premiere of Igor Stravinsky's *Le Sacre Du Printemps* in 1913, it seemed that the limits of performability had been reached and that the music world was about to go over the brink. After a break in compositional flow during World War I, composers explored new, uncharted musical domains. In the 1920s and 1930s, some composers created intricate and complex avant garde music, demonstrating the ultimate limitations of human musicians. They did not know it at the time, but these pioneers were attempting to write electronic music before the needed technology had been invented.

After World War II, European composers began to experiment with a new invention, the tape recorder. Here was a medium in which an artist could actually hold sounds in her own hands. Chopping up a tape recording and reassembling the pieces in a different order opened up a new world of sounds for composers to explore. It also required artists to come to grips with the phenomenon of sound itself, and questions like what it was made of and how sounds differed from each other. These problems were eventually solved on paper but a real tool was required to give composers the ability to actually manipulate the building blocks of sound. In the 1950s, **electronics** technology had been developed to the point where it was finally able to meet this demand, leading ultimately to the development of the first music synthesizer by Harry Olson and Herbert Belar in the laboratories and studios of RCA.

The synthesizer is a device that creates sounds electronically and allows the user to change and manipulate the sounds. All sounds in nature consist of waves of varying air pressure. A synthesizer creates sounds by generating an electrical signal in the form of a waveform, usually either a sine wave or other simple mathematical wave, which is amplified and used to drive an acoustic speaker. Unfortunately, the sound quality of a simple waveform is somewhat "raw" and unmusical, at least to most people. The waveform is usually altered in numerous ways, using filters to create the interesting *timbres*, or colors, of sound that are usually associated with certain acoustical instruments. Changing the frequency of the waveform raises or lowers the pitch of the sound. A synthesizer can control and change the beginning attack of a sound, its duration, and its decay, in addition to controlling the waveform itself.

Synthesizers can receive information from numerous sources about how to set the different parameters of its output sound. Any electronic device, such as a computer program, or person can control the synthesizer. An obvious way to accomplish this is to build the synthesizer in such a manner that it resembles an already-existing musical instrument, such as a piano. A piano-like keyboard is often used to generate signals that control the pitch of the synthesizer, although a keyboard is not required, or even necessarily desirable, to do the job. One of the first commercially available keyboard-based synthesizers marketed to the general public was built by Robert Moog in the 1960s. Other early competitors of the Moog Synthesizer were built by Don Buchla and Al Perlemon.

All of the early synthesizers were built using **analog computer** technology. In the late 1970s, however, digital synthesis became the premiere technology in synthesizer design. In the process of digitally recording a sound, called sampling, any sound recording can be converted into a series of numbers that a computer can analyze. The computer takes "snapshots" of the sampled sound in very short increments, about forty thousand times a second. Mathematical techniques, such as Fourier analysis, are then used to calculate the complex waveform of the original sound. The sound can then be easily reproduced in real-time from a synthesizer keyboard. This technique for creating sounds, and others, form the design basis of most digital synthesizers such as the New England Digital Synclavier and the Kurzweil music synthesizer. The same technique can be applied to synthesize drums, voices or any other kind of sound. Digital instruments can also receive input not just from a keyboard, but from the actual breath of the performer, for instance. Digital flutes and other wind-instrument synthesizers convert the force of the musicians breath into a signal that can modify any desired parameter of the output sound. Synthesizers have shown themselves capable of creating a wide variety of new and interesting sounds.

With its introduction in the 1980s, MIDI (Musical Instrument Digital Interface), a computer language, has made synthesizer technology available to those who were not professional musicians while surpassing the output quality of digital audio. MIDI goes beyond recording the sound: it records all actions related to performance on any MIDI-friendly instrument, including information like how hard keys are pressed when played. The resulting computer files can be edited with a flexibility previously unavailable and used in capacities beyond traditional recording such as musical backdrops for web sites and other multimedia uses. Many of the latest developments in music instrument technology, including so-called hyperinstruments, rely on the capabilities of MIDI.

The one limitation of synthesizers is that they sound only as good as the speaker that amplifies their signal. Most humans can hear sounds far beyond the range that even the best speakers can reproduce, sounds that acoustical instruments have always been capable of generating. Because of this limitation, and others, synthesizers are not commonly viewed as replacements of traditional instruments, but rather as a creative tool that enhances the musician's already rich palette of musical possibilities.

Syphilis test

Once a ravaging disease that rose to epidemic proportions, syphilis is effectively treated today with penicillin and other antibiotics. Because there is no known inoculation, accurate testing has become a key determinant of quick and successful treatment.

In 1903, Russian biologist Elie Metchnikoff (1845-1916) and French scientist Pierre-Paul-Emile Roux demonstrated that syphilis could be transmitted to monkeys and, consequently, could be studied in the laboratory. Two years later, German zoologist Fritz Schaudinn and his assistant Erich Hoffmann isolated the bacterium that causes syphilis, a spiral-shaped spirochete called *Treponema pallidum*.

The first effective test for syphilis was developed in 1906 by German physician and bacteriologist August von Wassermann (1866-1925). Wassermann was influenced by the work of German researcher Paul Ehrlich (1854-1915), who established basic theories of blood immunity. Wassermann's test consisted of taking a blood sample from the patient, and testing it for the antibody to the syphilis bacterium. If antibodies were present, the test was positive. After treatment, if the antibodies disappeared, the test was negative. The Wassermann test was useful in diagnosing syphilis in 95% of cases.

Unfortunately, the Wassermann test required a two-day incubation period. Reuben Leon Kahn (1887-1979), a Russian-born American immunologist, developed a new syphilis test in 1923 that was faster and simpler. This modified test used an extract from beef heart to detect syphilis antibodies. More sensitive than the Wassermann test, the Kahn test could be completed in a matter of minutes. However, the test could be inaccurate, showing false positive or false negative reports. For many years, Kahn studied the reaction of the test animals' immune systems. His research focused on tissue immunity and he found this could vary with the age of the animal, growing stronger with maturity. Kahn conducted his research at the University of Michigan and Howard University.

Another effective syphilis test was developed by William A. Hinton (1883-1959), an African-American physician who became a leading expert on venereal disease. Hinton, who worked out of Harvard Medical School, also collaborated with J. A. V. Davies on the Davies-Hinton test. Several other syphilis tests have been developed. The VDRL test, which came out of the Venereal Disease Research Laboratory, is performed on either cerebrospinal fluid or blood; however, this test produces false negative results about one-third of the time. The rapid plasma reagin (RPR) test is very sensitive but false positives are common, especially in anyone with a rheumatologic disorder. All positive results should be confirmed with one of the treponemal-specific tests such as the fluorescent treponema antibody absorption test (FTA-ABS).

Because syphilis still occurs in epidemic proportions in developing countries and parts of the United States, a quick, low-cost, easy-to-read, and accurate "IC strip test" has been developed by Omega Diagnostics in the United Kingdom with funding from USAID, the World Bank, and others. In 1998, field trials to validate effectiveness of the strip were being planned.

Syringe

A pumplike device used for hypodermic or subcutaneous (beneath the skin) injections or to remove liquids by suction, the syringe consists of a tube—usually made of **plastic**—that is tapered at one end and has a plunger at the other that either creates suction when pulled back or forces out fluid when pushed forward. The syringe may be used for intravenous (going into the veins), intramuscular (into the muscle), or intradermal (between skin layers) injections to administer drugs or vaccines.

The conceptualization of the syringe is thought to have originated in fifteenth-century Italy, but it was not developed for practical use until several centuries later. In 1657, experiments were conducted on syringe-like devices by two Englishmen, Christopher Wren (1632-1723) and **Robert Boyle**. Dominique Anel, a surgeon to the seventeenth-century army of French King Louis XIV (1638-1715), is usually credited with the invention of the kind of syringe used today, which he devised to clean wounds with suction. The first true hypodermic syringe was created by Charles Pravaz, a French physician, in 1853; it was made entirely of silver and held one cubic centimeter of liquid. Around the same time, Scotsman Alexander Wood devised a subcutaneous injection method, allowing physicians to administer intravenous anesthesia for the first time. An Englishman named Fergusson used a syringe made partially of **glass**, thus allowing visual monitoring of injections. The all-glass syringe developed by a man named Luer in France in 1869 further reduced the risk of infection.

Today, increased public awareness about acquired immune deficiency syndrome (AIDS) and concern about halting the progress of the deadly human immunodeficiency virus (HIV) has led to the widespread use of disposable syringes, which are used only once and then discarded. In some cities, these syringes have been distributed free to users of illegal, injectable drugs, whose sharing of unsanitary needles containing traces of HIV-infected blood has made them among the most common victims of AIDS. Although many health practitioners and others believe this practice is in the public's best interest, it has met with resistance from those who feel it encourages, or at least communicates, tolerance of illegal drug use.

Szilard, Leo (1898-1964)

Hungarian American physicist

Szilard was born in Budapest on February 11, 1898. He had studied only a year at the Technical Academy in Budapest before he was drafted into the army. His military experience was so unpleasant that he became a life-long pacifist.

After returning to civilian life, Szilard left Hungary to enroll first at the Technical Academy at Charlottenburg and then at the University of Berlin, where he received a Ph.D. in nuclear physics in 1922. The rise of Hitler convinced Szilard, as it did so many other Jewish scientists at the time, to flee Germany. He traveled first to London in 1934 and then, four years later, to the United States. He became an American citizen in 1943.

While in London, Szilard conceived of the notion of a nuclear chain reaction. "It occurred to me," he wrote, "that a chain reaction might be set up if an element could be found that would emit two neutrons when it swallowed one neutron." Szilard first considered using beryllium for such a reaction, but, as he notes, "for some reason or other the crucial experiment was never carried out."

Only a few years later, the discovery of nuclear fission by Hahn and Strassman presented another possibility. Szilard realized that the fission of uranium nuclei, with its release of neutrons, could be used to produce a chain reaction. Further, he saw the potential for using such a reaction in an **atomic bomb**.

Horrified by the thought of such a weapon in German hands, Szilard convinced his American colleagues not to publish their research on fission. He saw this decision as the only way of preventing the Germans from keeping up to date with fission research and constructing their own weapon. Along with two other Hungarian emigrés, **Edward Teller** and Eugene Wigner, Szilard convinced Albert Einstein (1879-1955) to write President Franklin D. Roosevelt about the potential of nuclear fission and the need for the United States to build a fission bomb.

Szilard's role in the ensuing Manhattan Project was the development of the first self-sustaining chain reaction. He worked with Enrico Fermi (1901-1954) on this project at the University of Chicago. Szilard and Fermi found that graphite could be used as a moderator to slow down neutrons produced during the fission of uranium. These slowed-down neutrons were more efficient at producing other fission reactions.

This research was critical in the development of the first atomic bombs. However, Szilard was among those bomb researchers who argued against the weapon's use over populated areas, a position that was not adopted by the United States government.

After the war, Szilard decided to leave the field of nuclear research. In 1946, he became professor of biophysics at the University of Chicago. He devoted much of his time and energy on efforts to ban nuclear weapons testing and nuclear warfare. In 1959, he received the Atoms for Peace Award. He joined the Salk Institute for Biological Studies in 1956 and died in La Jolla, California, on May 30, 1964.

See also Hydrogen bomb

Leo Szilard

T

Talbot, William Henry Fox (1800-1877)

English inventor

William Henry Fox Talbot, the son of an army officer, received his master's degree from Cambridge University in 1825. In 1831, he was made a member of the British scientific society, the Royal Society, for his expertise in mathematics. He came to be quite well known for his studies in calculus and in the translation of ancient cuneiform tablets though his greatest fame would come from his work in photographic science.

In 1833 he entered Parliament, but retired the next year and began experimenting with **photography**. Unbeknownst to Talbot, **Louis-Jacques-Mandé Daguerre** was performing similar experiments in photography at the same time in France. While using a **camera lucida** to capture natural images from the shores of Italy's Lake Como, Talbot realized that he lacked the drawing talent required to sketch the image on **paper**. Instead Talbot was inspired to find a means to record images without drawing. From 1834 to 1835, Talbot used wooden **camera obscura**s, dubbed ''mousetraps,'' to produce miniature negative imprints of objects on paper sensitized with silver salts. The exposure time for these images was about one half-hour. He fixed the images with a solution of cooking salt.

In 1839, Talbot devised an improved process to form a negative picture on transparent paper coated with silver salts. Such an image still required an exposure of at least a minute, but many positives could be made from a single negative image by exposing a second light-sensitive sheet through the negative image itself. Talbot called these images *calotypes*, from the Greek for ''beautiful.'' This method of converting a negative to a positive is the basis for modern photography, and for this reason Talbot is considered (along with **Joseph-Nicéphore Niépce**) to be one of the two founders of photography.

The announcement of Daguerre's process in France moved Talbot to present his ''photogenic drawing'' to the Royal Society in 1839. By the autumn of 1840, Talbot had discovered that gallic acid sped up the development of latent images, and in 1841 he patented his perfected process, now calling it the *Talbotype*. Unlike the Talbotype, Daguerre's process could not produce multiple prints of an image. But Talbot's photographs had a grainy quality and could not compare with the brilliant, minutely detailed silvered plates produced by Daguerre. Nevertheless, Talbot's process, once refined, would hold the key to the future of photography.

Talbot published the first book illustrated with photographs in 1844, and pioneered flash photography of fast-moving objects in 1851. He developed a way to speed up the exposure of images by coating the photographic paper repeatedly with alternate washes of salt and silver, then exposing the paper in a moist state. This vastly reduced the exposure time required for portraits, and the agonizing effort required to remain still during a sitting was eliminated.

Talbot guarded his patent closely and worried about infringement of his rights to his invention. Without license from Talbot, British photographers were prohibited from using his new photographic process. In France, advances in photography were published freely. As a result, photography grew much faster in France than it did in England. In addition, Talbot often passed off other's ideas as his own. He incorporated, for example, **J. F. W. Herschel**'s idea for a better fixative into his patent. His claims continued to retard the growth of photography in Britain until his patent infringement suit against Frederick Scott Archer's (1813-1857) **collodion** process failed in court.

Tanker • See Cargo ship

Tanning • See Leather

Tape recorder • See Magnetic recording

TARMAC • See Road building

TAXICAB

A taxicab is a vehicle hired to transport passengers, usually within a city and its suburbs. Its name comes from the taximeter, invented by Wilhelm Bruhn in 1891, an instrument that records both distance traveled and time elapsed, thus computing an accurate fare. The suffix-*cab* refers to the cabriolet, a type of carriage often used for passenger travel. A Parisian coachman, Nicolas Sauvage, started the first taxi service in 1640. He eventually had twenty coaches running different routes through the city. In 1703 the French police drafted laws for their use and assigned a number to each, the first such vehicle registration known. Horse-drawn carriages remained the standard for taxi service through most of the next two centuries. The first automobiles to be used as cabs were electric; they made their debut in the late 1890s. A few years later, **Louis Renault** launched a fleet of small, specially built cars with two-cylinder **gasoline** engines, and the modern taxi was born. During World War I a desperate French general used 700 of these to carry his troops to the front. In other parts of the world different vehicles have been used as taxis. The ricksha, invented in Japan in 1870, was a two-wheeled covered cart pulled by a man. It was eventually replaced by the pedicab, another human-powered cart that resembled a tricycle, before it too was replaced by the **automobile**. Most American taxis are specially modified passenger cars, with reinforced frames and stronger suspension systems that enable them to withstand of the rigors of heavy use. In some places, taxis must meet special requirements in order to be licensed. The distinctive British cabs are one example of this. Another example are the over 12,000 taxis in New York City that, until 1995, had to meet the requirements specified for police cars.

Taxicabs of the very near future will no longer be only sedans like police cars, but will most likely be sport utility vehicles or mini-vans. Compared to even large sedans, these new vehicles are roomier for passengers and luggage, and offer far more comfort with headrests in the back and rear air conditioning vents and controls. They also get better gas mileage than sedans. Another piece of new equipment will be a satellite tracking device that not only gives the vehicle's position and the best route to any destination, but also will tell the driver the trip distance and can alert him to wrong turns.

Since gasoline-powered taxicabs are such a large part of any major city's transit system, they necessarily are pumping tons of contaminants into the air annually. In 1998, New York City took the first step toward its goal of replacing its entire taxi fleet with cleaner vehicles that burn natural gas instead of gasoline. Besides the initial higher cost, running natural gas taxicabs is cheaper, since it costs about 20% less than gasoline. The city has also increased the number of natural gas refueling stations and removed its restrictions barring such vehicles from **bridges** and **tunnels**.

TAYLOR, HARDEN • See Frozen food

TEA AND TEA BAG

Tea has existed as a beverage since 2000 B.C. The brewing, serving, and drinking of tea are time-honored rituals throughout the world. While there is general agreement that the tea trade began in China, both China and India lay claim to discovering the dietary properties of tea leaves. The Chinese tell the story of a mythical emperor named Shen Nung who was so particular about his nutrition that he boiled his drinking water before he drank it. One day, the story goes, the wind caught some of the leaves on the tree branches that he had used to build a fire. The leaves floated into his boiling water and tea—as a beverage—was created.

In India, the discovery is attributed to Bodhidharma, an actual person who founded the Ch-an School of Buddhism. In A.D. 527, after four years of a self-imposed nine-year meditation, Bodhidharma grew sleepy. In an attempt to stay awake, he began to chew on the twigs of a nearby tree and suddenly found himself wide awake; he had discovered tea.

The tea bush is a white-flowered evergreen in the Camellia family. Chinese documents record it as indigenous to the Hunan province in southwest China. In modern times, it is generally accepted that the original tea bush grew in India and was brought to China. It thrives in a rocky terrain. In approximately A.D. 350, tea cultivation was also reported in the Szechwan province along the Yangtze River.

During the T'ang Dynasty in the eighth century, tea drinking achieved the status of an art form. Tea merchants hired a man named Lu Yu to compile the first written record of Chinese tea ceremonies. Entitled *Ch'a Ching* (*The Class of Tea*), the three-volume work revolutionized the tea industry. The second volume includes an exhaustive list of the equipment necessary to brew tea correctly; 24 items total are listed.

In the twelfth century, monks from the Zen sect of Buddhists brought the tea-brewing process home to Japan from their travels in China. Yeisei, a Buddhist abbot, is credited with writing Japan's first tea book: *Kitcha-Yojoki* (*Book of Tea Sanitation*). Since then, Zen Buddhism and tea have been synonymous.

Several Europeans mentioned bringing tea back home from their visits to the Orient. One of them was Giambattista Ramusio, an editor of travel books and a diplomatic representative of the Venetian government. Gaspar da Cruz, a Portuguese Jesuit priest and missionary, also brought tea when he returned. In 1520, Ferdinand Magellan discovered the straits that now bear his name around the southern tip of South America and opened the door to a what would become a well-traveled westward trade route between Europe and the Orient.

Holland was the first to record the purchase of tea in 1607. The tea was first sold at apothecary shops, then in stores where spices and sugar were sold. By the eighteenth century, stores devoted entirely to the sale of tea and coffee had opened. The first tea sold to the English public occurred in 1657 at a coffee house called Exchange Alley. When King Charles II married the Portuguese princess Catherine of Braganza in 1661, her dowry included tea. Fifty-five years later, on October 12, 1712, Thomas Twining opened England's first tea shop.

England's East India Company grew prosperous from its opium trade to China, and although the English addiction to tea was not as detrimental as the Chinese addiction to opium, the company also profited by the sale of tea it brought back in exchange. The English government ignored the questionable nature of the company's business so that it could levy a high import tax on tea. In spite of the tariff, or perhaps because of it, a great deal of tea was smuggled into the country. The tea tax also figured prominently in the American Revolution, as witnessed by the infamous Boston Tea Party when rebellious colonists pitched a shipment of East India Company tea into the harbor.

Until the early twentieth century, tea was sold loose. Tea bags were invented quite by accident in 1904 when Thomas Sullivan, an enterprising merchant, wrapped samples of tea leaves in **silk** bags and sent them to prospective customers, some of whom dipped the bags directly into boiling water. The silk bags gave way to gauze pouches and eventually to specially treated filter **paper**. The filter paper is made primarily of abaca, the leafstalk of Philippine bananas also known as Manila hemp.

Herb teas, caffeine-free and cultivated from a variety of plant leaves, flowers, roots, bark, and seeds, have become enormously popular over the last 20 years. Although they were probably brewed as early as prehistoric times, herb teas were primarily consumed for medicinal purposes.

Tea bags are composed of two main ingredients: processed tea leaves and filter-paper bags. The top tea leaves and leaf buds are hand-picked from the plant. The leaves are then subjected to several processes, including withering, rolling, drying, cutting, and blending. The intensity and duration of each process differs according to the type of tea.

Tea continues to be a fashionable beverage, in spite of coffee's popularity. The Tea Council of the United States estimates that Americans alone consume 122 million cups of tea each day. Worldwide, it is the second most popular beverage, preceded only by water.

Recently, the scientific community has become interested in the potential health benefits of tea, particularly in those properties that could lower blood pressure and blood cholesterol levels, stabilize blood sugar, prevent tooth decay, and inhibit the growth of cancerous tumors.

Teaching aids

Educators use teaching aids to impart and emphasize information, stimulate interest, and facilitate the learning process. They range from simple to sophisticated and can be aural, visual, or increasingly more frequently, computerized. One of the earliest known teaching aids is the hornbook, which was used in English schools from the mid-1400s (and later in colonial America) through the early nineteenth century. It was a flat board on which a sheet of paper printed with the alphabet, the Lord's prayer, and several simple words were pasted. A transparent piece of horn covered the paper to protect it. The blackboard probably evolved from the hornbook, and has traditionally been one of the most widely-used teaching aids. It was patented in 1823 by Samuel Read Hall of Concord, Vermont, a Congregational minister who founded the Concord Academy to train teachers in "school keeping." Hall's version was made of pine board, planed smooth and painted black. Today, the blackboard is often called a chalkboard; it may be green instead of black and made of slate, **glass**, or wood.

The globe is another long-popular teaching aid. It provides a representation of the earth and other terrestrial (and sometimes celestial) bodies that is true to scale, without the unavoidable distortions of flat maps. Globes are hollow spheres usually made of plastic or metal, although sometimes translucent or inflatable, on which a map, cut in triangular or tapering strips called gores, is pasted or printed. A physical globe shows the natural characteristics of the body it represents, while a political globe shows countries, cities, and other governmental features. The ancient Greeks realized that the earth was spherical and are known to have used globes as early as 150 B.C. The oldest existing celestial globe (showing the constellations) is the Farnese globe, made in A.D. 25. Another historic globe was made in 1492 by Martin Behaim of Nuremburg, Germany, who is thought to have influenced Christopher Columbus's decision to try to reach Asia by sailing west. In 1810, a Vermont farmer and copper engraver named James Wilson made the first American globes. He drew highly accurate maps that fit perfectly on wooden spheres; many of his globe-making techniques are still used.

These early teaching aids were, of necessity, strictly visual media. This type of instruction, however, remains a vital part of education today. Modern teachers still use blackboards and globes. Until quite recently, additional contemporary visual aids included bulletin, flannel, and magnetic boards; models; overhead transparencies and slides (shown with special projectors); **photograph**s; drawings; and specimens. Audio aids include **tape recordings** (used, for instance, to teach foreign languages) and **radio** or **television**, with some students establishing their own stations. During World War II, educators used **motion pictures** to train large numbers of troops quickly. Films came into widespread classroom use in the 1940s and 1950s. Students today can watch educational broadcasts to closed-circuit lectures on cable or satellite television; they can also view instructional **videotape**s and videodiscs.

In the past, audio-visual materials that were presented in an interactive format, were part of a class of instructional media called teaching machines. These allowed students to teach themselves, usually as they progressed through a series of increasingly difficult questions. The first such device was invented by Sydney L. Pressey in the 1920s. It tested students' knowledge by requiring them to press the correct levers before they could proceed to another question or problem.

Computers have complemented, supplemented, and in many cases supplanted earlier teaching aids. Educators have moved far from the drill and practice lessons and tutorial help of early computers in schools. With the advent of Internet access, and development of the World Wide Web for communication and information, students, who in many cases know more than their teachers about technology and how to manipulate it, require a different manner of guidance, consequently, a different style of teaching aid. The teacher's changing role from knowledge provider to learning coach and facilitator, requires the ability to guide students in making the best educational and informational use of technology. Thus, new learning aids that encourage students to think critically and creatively must be devised. The not too long ago, but nonetheless outdated, era of computer-assisted instruction (CAI) is gone, to be replaced by a period in which young people can communicate globally under the guidance of teachers who encourage them to become lifelong learners, intelligently using the vast information resources available to them.

TEAR GAS • See Chemical warfare

TECHNICOLOR • See Motion picture

TEDDY BEAR

The story of the teddy bear begins in November 1902, when President Theodore ''Teddy'' Roosevelt traveled to Mississippi to mediate a border dispute between that state and Louisiana. It is told that he took some time out from his work to go on a hunting expedition and that his hosts trapped a bear cub to ensure the President's venture would be successful. When Roosevelt realized this scheme, he refused to shoot the cub. Hearing of the incident, cartoonist Clifford Berryman drew a sketch of the President holding a **rifle** and standing with his back to a terrified bear cub; the caption read, ''Drawing the line in Mississippi.'' The cartoon caught the attention of Brooklyn candy store owner Morris Michtom, a Russian immigrant, who fashioned a brown plush toy bear with button eyes and movable limbs. He placed the bear in the front window of his store, along with a copy of Berryman's cartoon and a sign that said, ''Teddy's Bear.'' Michtom's customers loved his bear and asked him to make more and sell them. Before doing so, the cautious inventor wrote to Roosevelt seeking permission to use the President's nickname for his product. Roosevelt obliged, commenting that he doubted his name was ''worth much to the toy bear cub business.'' He may have been wrong about that, for by 1906 the teddy bear had become America's top-selling toy. Production had reached almost a million in 1907, and bears fashioned from mohair in various colors were available in sizes from 3.5-40 in (9-102 cm). Pirated versions appeared all over the world. Children so loved their toy bears that some people feared they would transfer to the plush creatures their affection for their mothers. Michtom was able to abandon his candy business and found the Ideal Novelty and Toy Company, which later became the Ideal Toy Company, the world's largest dollmaker.

At about the same time Ideal began marketing the teddy bear in the United States, the German company—Steiff—began distributing a similar toy in Europe. The Steiff bear featured a metal rod skeleton (rather than the usual cardboard disk and pin joints) that gave the toy movable joints. Developed from sketches the company's founder, Richard Steiff, made while watching bears at the zoo, the bear looked quite realistic when made to stand on all four legs. By 1910, many companies in Europe and the United States were producing teddy bears, including models that incorporated clockwork mechanisms to make the bears tumble and those that came equipped with muzzles and leads. **Glass** began to replace the **buttons** previously used for eyes. During the 1920s, pastel-colored bears gained popularity. Teddy bears produced during the early years of the twentieth century are much coveted by modern collectors.

Named for President Theodore "Teddy" Roosevelt, teddy bears became the most popular toys of the twentieth century.

TEFLON

Teflon is a fluorocarbon **polymer** known by the chemical name polytetrafluoroethylene. It is a slippery material that is chiefly known for its use in nonstick cookware and heart valves. It was first used in the gaskets and **valves** needed to concentrate Uranium 235 and resist the highly corrosive uranium hexaflouride gas.

Roy J. Plunkett of New Carlisle, Ohio, accidentally discovered Teflon. Plunkett graduated from Manchester College in 1932 with a B.A., and received a Ph.D from Ohio State University in 1936. There he became friends with future Nobel Prize winner Paul Flory (1910-1985). He then went to work at Du Pont's laboratories to research refrigerants, namely flourochlorohydrocarbons. To facilitate that research, he needed about 100 lbs. (45 kg) of tetraflouroethylene.

He developed a small pilot plant to produce the chemical, which was then stored in cylinders in a cold box. During a subsequent experiment on April 6, 1938, Plunkett and his assistant Jack Rebok realized that a full cylinder apparently contained no gaseous tetraflouroethylene. Instead of simply getting another tank, Plunkett began to investigate. Plunkett checked the valve on the tank and found a white powder. When he sawed the tank open it was full of a slippery, white, waxy substance. He theorized that the gas had polymerized.

Teflon, as it came to be known, is resistant to strong acids, bases, heat, and solvents. Initially, it was extremely expensive to produce and was unavailable to the general public until 1960. It played an important role in the war effort, and in peacetime has come to be used in cookware, electrical insulators, space suits, and as nose cones, heat shields, and fuel tanks for space vehicles. Because Teflon is essentially inert, the body does not reject it, so it is used in **artificial limb and joint** replacements, sutures, heart valves, and **pacemakers**.

See also Polymer and polymerization; PTFE (polytetrafluorethylene)

Advertisement for the first successful transatlantic telegraph cable.

TELEGRAPH CABLE, UNDERSEA

Almost as soon as the electric telegraph came into widespread use, people began thinking of laying telegraph cables underwater for intercontinental message transmission. Successful installation of **submarine** cables was not possible at first because insulating materials were inadequate. Early lines run by **Samuel Morse** in New York Harbor in 1842 and by **Ezra Cornell** across the Hudson River in 1845, for example, both failed. Then the insulating properties of Malaysian **gutta-percha** were discovered, and machines for coating telegraph cables with gutta-percha were invented by **Werner von Siemens** (1847) and **Thomas Hancock** (1848).

The English brothers John and Jacob Brett laid the first submarine telegraph cable between England and France in 1850, but it was not strong enough, and broke. In 1851 the Bretts laid a second cross-Channel cable. The success of this cable led to the laying of many more underwater lines which, inevitably, led to thoughts of a transatlantic cable. The project turned out to be a herculean task, accomplished thanks to the unstinting efforts of American entrepreneur Cyrus Field (1819-1892).

Approached by a Canadian engineer about laying a cable from New York to Newfoundland, Canada, Field consulted noted oceanographer Matthew Maury and telegraph inventor Morse about the feasibility of the project. Encouraged by both men, Field went forward, forming his first submarine telegraph cable company and securing technical advice from the English physicist **William Thomson (Lord Kelvin)**. Installation began in 1855, and in a portent of things to come, the cable was cut loose in the midst of a gale. This line was finally finished in 1856, setting the stage for a cable to be laid between Newfoundland and Ireland.

Field organized a new company, and in 1857 the United States Navy ship *Niagara* and the Royal Navy ship *Agamemnon* began laying a cable. It broke. The same ships tried again in 1858, but more breaks ended this attempt as well. Field persevered, and guided a third attempt, which ended successfully in August 1858. Queen Victoria (1819-1901) cabled her greetings to President Buchanan (1791-1868); the public went wild with enthusiasm. Just a few weeks later, however, a telegraph operator applied too much voltage and ruined the cable.

Although he remained passionately committed to the transatlantic cable, Field was unable to proceed with any other attempts during the Civil War years. In 1865 he was able to put together another try at the cable. The huge steamboat *Great Eastern* laid a working cable from Ireland almost to Newfoundland before the cable parted, a heartbreaking failure that might have been final if Field had not pushed on. He organized yet another company in 1866, and the *Great Eastern* set out again. Success finally came on July 27, 1866.

Field was not finished, however. He immediately returned to sea with the *Great Eastern* and recovered and successfully spliced the lost 1865 cable. Amazingly, after all the effort and failure, two transatlantic telegraph cables came into operation at nearly the same time.

Telegraph, electric

The need to communicate rapidly over long distances reached critical proportions once the Western world became heavily industrialized and urbanized. Visual telegraphy systems such as the **semaphore** had insuperable physical and speed limitations. Advances in the science of electricity permitted the crucial breakthrough, finally making it possible to send messages via electrical impulses over long distances through **wire**s.

Experimenters began investigating the possibility of sending messages alphabetically by electric current during the early 1700s. As early as 1727 an experimenter in London managed to send an electrical impulse one sixth of a mile along thread. A writer identified only as C.M. in a *Scots Magazine* of 1753 described a static electricity telegraph that would spell out messages over 26 wires, one for each letter of the **alphabet**. Claude Chappe (1763-1805), developer of the semaphore, experimented with a synchronized clockwork electric telegraph in 1790, and a system using this principle was developed by the Englishman Francis Ronalds (1788-1873) in 1816.

It was Chappe who coined the term telegraph (for his semaphore), from the Greek *tele*, ''far,'' and *graphien*, ''to write.'' The practical development of electrical telegraphy was made possible by the invention of the **electric battery**, or voltaic pile, by **Alessandro Volta** in 1800. A steady, low-pressure form of electricity was now available. In 1804 Francisco Salva, who had earlier developed a static-electricity telegraph, now devised a battery-powered system that could send messages a short distance over wire. S. T. von Sömmering (1755-1830) demonstrated a similar device in 1809 in Germany, but its many wires made this telegraph impractical.

The next crucial step was the discovery by Hans Christian Oersted (1777-1851) of Denmark in 1819 that a magnetic needle was deflected by an electric current passing by it. The Frenchman André Ampère (1775-1836) suggested using the deflection of the needle for an electric telegraph. Baron Schilling, a Russian diplomat stationed in Germany, who had seen a demonstration of von Sömmering's device in 1810, followed up on the magnetic-needle idea. By 1832 Schilling had designed an electromagnetic needle telegraph with an alarm that required significantly fewer wires because the letters of the alphabet were transmitted by a code. The German Carl-August von Steinheil (1801-1870) developed a single-circuit needle telegraph in 1827. None of these systems, however, was put into practical commercial use.

That final development was the accomplishment of two Englishmen, William F. Cooke (1806-1879) and **Charles Wheatstone**, and an American, **Samuel Morse**. Cooke and Wheatstone formed a partnership and in 1837 patented a five-wire, five-needle telegraph. Electric current moved magnetized needles, which pointed singly or in pairs to a background panel imprinted with letters and numerals. This system was demonstrated on a short line between Euston and Camden in London in 1837. It was then installed for use in 1838 by the Great Western Railway Company on its line between Paddington (London) and West Drayton, about 13 miles (208 km), and extended to Slough, about 18 miles (288 km), in 1842. Cooke and Wheatstone simplified their telegraph to a two-needle design in 1842 and patented a single-needle instrument in 1845. While these telegraph lines were very successful, they attracted little public attention until 1845, when a suspected murderer was spotted boarding a train to London. A message telegraphed to the destination station of Paddington resulted in the man's arrest and later execution, as well as much public excitement about the possibilities of telegraphy.

In the United States the well-known American artist Morse had become intensely interested in 1832 in developing a way to transmit messages via electricity. His first crude device, built in 1835, consisted of a typesetting sender that made and broke an electric circuit, transmitting a coded message to an electromagnetic receiver that recorded the message as dots and dashes on a moving tape. Morse substituted a simple operator key to open and close the circuit in 1837. Since Morse had no background in science, he was stymied by technical problems, especially insufficient current in his battery and electromagnet. Chemistry professor Leonard Gale stepped in with valuable assistance in 1836, and also introduced Morse to **Joseph Henry** and his work on electromagnets. Henry had actually constructed a working electromagnetic telegraph, or relay, in 1831, and shared his knowlege with Morse.

Using what he now knew, Morse made one of the greatest practical advances in the telegraph. He designed a relay system using a series of electromagnets to open and close circuits along the telegraph wire, which kept the current strong enough to travel long distances. Morse demonstrated his telegraph in 1837. In 1838 he took on a partner, **Alfred Vail**, who brought mechanical expertise and his father's financial backing to Morse's enterprise. Vail made many improvements to the telegraph's design and helped refine the message-transmitting system into what became known as the **Morse code**. In 1843 the partners were granted $30,000 by Congress to build an experimental telegraph line between Baltimore, Maryland, and Washington, D.C. Initial attempts to bury the line met with failure due to inadequate insulation. Young **Ezra Cornell**, who had invented a **plow** to bury the cable, suggested instead stringing it overhead.

Successful installation of the line was dramatically demonstrated by Morse on May 24, 1844, when he sent the famous message ''What hath God wrought!'' from the Supreme Court room in the Capitol to a waiting Vail in Baltimore. After this, telegraph lines proliferated rapidly across the United States. The telegraph became the standard way of communicating within cities until the widespread availability of the **telephone** around 1880. Many competing telegraph companies were formed amid contentious litigation over the Morse patents.

Mergers soon consolidated these disparate businesses, with Western Union Telegraph Company, the largest, forming in 1856. By 1861 a telegraph line ran across the continent, and by 1866 it was possible to send messages across the Atlantic via a submarine cable, installed after numerous failures and herculean effort.

Refinements were continually made. The Morse system's receiver replaced inked messages with a simple "sounder" in 1856; operators could "read" the clicks they heard and transcribe them into written messages. Improved insulating methods, especially the use of India **rubber** and **gutta-percha**, reduced overhead as well as underground line failure. A duplex circuit developed in Germany allowed messages to travel simultaneously in opposite directions along a single line. **Thomas Alva Edison** designed a quadruplex in 1874, making two messages in each direction possible. Jean-Maurice Baudot invented a multiplex in 1872 that could handle five messages at once.

Soon after telegraph systems came into use, attempts were made to develop printing telegraphs to eliminate the need for skilled operators and to improve reception time. Vail invented a printing system in 1837; Alexander Bain (1818-1903) designed one in 1843, followed by Royal House (1814-1895) in 1846. Wheatstone developed a model in 1841, as did fellow Englishman John W. Brett (1805-1863) in 1846. David Hughes (1831-1900) of Kentucky designed a continuous-movement printing system in 1855 which successfully synchronized sending and receiving apparatus. With improvements, the Hughes machine became the industry standard. In 1878 the **typewriter** was adapted for use as a telegraphic printer.

The impact of worldwide telegraphy was enormous. Instant communication over long distances, never before possible, became commonplace. Journalism was transformed; events were reported almost as soon as they occurred. Large, far-flung corporate structures became feasible. The United States could be treated as one large national market. Also, individuals, through news reports or personal communication, felt closely connected with people and places in far distant parts of the continent and globe.

See also Telegraph cable, undersea

Teleconferencing

Teleconferencing is the use of **telephone** or other means to conduct meetings between two or more people who are not in the same place. With teleconferencing—or its newer counterparts, videoconferencing (uses television or other visual means) and computer teleconferencing—people can exchange information and discuss related matters without the expense or hassle of traveling. Teleconferencing can also be used as a means for families separated by distance as well as an educational tool. Institutions of higher learning can conduct seminars with a teacher in one location and students in others, via videoconferencing. The only limitations of teleconferencing, videoconferencing, and computer teleconferencing are technological.

Teleconferencing, in the strictest sense, is the use of the telephone as a conferencing medium for people at different locations. Teleconferencing was prohibitively expensive until the late 1960s and early 1970s because of the amount of work telephone operators did to set up conference calls. With direct dial, teleconferencing became more feasible and economical. Multiple teleconferencing locations require special bridging equipment, and up to 80 locations can call a prearranged number to telephone conference. The drawback to teleconferencing via the telephone is the lack of visual contact. This can be overcome by use of a **fax machine** or other means of information transfer. This exchange of information, however, does not make up for the lack of eye contact.

Early attempts at videoconferencing were tried in the 1960s, but proved cumbersome and expensive. The first years of videoconferences were usually closed circuit links for large groups in only two locations. One way conferences were offered beginning the 1970s, usually with one location speaking and the other receiving. By the early 1980s, AT&T offered a videoconferencing service. Customers would have to come to studio locations at prearranged times to meet. This idea has become more sophisticated throughout the 1990s. Studio rooms were enhanced for optimal visual quality, fitted with equipment to aid presentations. With communications satellites and digital compression, transmission capabilities have increased and videoconferencing has decreased in cost. One advantage to videoconferencing is the ability to tape the meeting for future reference.

Computer conferencing threatens to take over both teleconferencing and videoconferencing, because it is more flexible and less costly, using normal telephone lines. Computer conferencing can take place between a few or many people; in real time or not; face to face or with e-mail and electronic bulletin boards. If a computer is fitted with a sound card and a desktop video camera, visuals can be sent electronically. While picture quality is not yet as sharp as found in most videoconferencing studios, the technology is rapidly improving. With a cybercast, for example, participants call into a service, can access both video and sound, and the proceedings can be recorded for future reference. Indeed, all computer conferencing can be saved by electronic means.

See also Computer applications; Computer input and output devices

Telemetry

The original definition of telemetry encompassed any form of remote measuring, but the term has become more narrowly defined to describe the observation of variables and physical measurements from a remote point. Early uses of telemetry were primarily for military purposes. The first documented use of telemetry was in Russia in 1812 by a man named Shilling, who applied the concept to the firing of mines. In 1845, a man named Takobi devised a military data-transmission system using telemetry. That same year, Konstantinov and Pouli created a telemeter that could recorde and analyze the flight of cannonballs. By 1857, **warships** used **telegraph** signals between the bridge and engine and gun locations.

Soon telemetry began to be applied in other fields. Olland developed a meteorological telemeter which he used to measure weather variables in 1874, and in 1906 seismic telemetering came into use. The Weather Bureau began to use the **radio** meteorological telemeter in 1913. Telemetering was used extensively to measure variables and conditions during the construction of the Panama Canal in 1913 and 1914.

By 1941, Princeton and Johns Hopkins universities were researching means to apply telemetry to aircraft and missile technology, and the first guided missiles, utilizing a telemetric guidance system, were tested at Eglin Field in Florida in 1946. In 1947, the first missile was fired into the atmosphere for the purpose of conducting upper-atmosphere research. The successes of these projects prompted research into the use of artificial satellites guided by telemetric systems and in 1957, Russia launched its exploratory satellites *Sputnick I* and *II*, while the United States launched *Explorer I* and *Vanguard I*.

Industrial telemetry was also being developed during this time, building toward the creation of **digital computers**. One of the first such machines was put into use by the Louisiana Power and Light Company in 1958. The first **television** communications relay via satellite occured in 1962. Today, telemetry is applied to a variety of remote applictions, including operation of satellites and antennas. Telemetry is also used to operate land-based mechanical apparati; robots perform a variety of automated functions in various industries and sometimes carry out functions too dangerous for humans. Robots performed much of the cleanup after the nuclear disaster at Three Mile Island in Pennsylvania, for instance.

Telephone

The telephone is a device for electrically transmitting voice communications over **wire** and is based on the fact that the human voice vibrates air. In the phone, this air vibrates a diaphragm in turn, which produces a varying electric current. This current is transmitted through wires to a receiver, where it causes fluctuations in the power of an **electromagnet** which, in turn, causes a diaphragm to vibrate that moves the air so as to reproduce the original sound.

Simple string telephones, like the children's toy, were known by the early eighteenth century. The word (from Greek words for ''far sound'') was originally used for a variety of devices for transmitting sounds and voices. A number of researchers described theoretical electrical voice-transmission systems, but the first practical, working one seems to have been invented by Johann Philipp Reis (1834-1874), a German, in 1863. In the Reis machine, which he called a telephone, a vibrating membrane opened and closed an **electric circuit** to reproduce musical sounds. Reis used his device simply as a scientific toy, to demonstrate the nature of sound.

The first operational telephone was patented and produced in the United States in 1876 by **Alexander Graham Bell**. An intent to patent a very similar design was filed simultaneously by **Elisha Gray**. Both men had begun by attempting to invent a ''harmonic telegraph''—a device that simultaneously would send multiple messages through a single **telegraph** wire. This interest led both men to investigate the possibility of voice transmission. Gray built a steel-diaphragm receiver in 1874 and designed a variable-resistance transmitter in 1875. Gray apparently did not test his completed device; instead, he filed notice of intent to patent it, on February 14, 1876. He then discovered that Bell had filed his own patent two hours earlier at the same patent office.

While Gray came to his interest in telegraphy and telephony from his background in electricity, Bell's interest arose from his studies of vocal physiology and his teaching of speech to the deaf. Bell's patent design used a skin membrane and only as an afterthought mentioned variable resistance. The critical breakthrough in Bell's experiments came on June 2, 1875, while he and his assistant, Thomas A. Watson (1834-1954), were working on their harmonic telegraph. A stuck reed on Watson's transmitter changed an intermittent current into a continuous current, and Bell's extraordinarily sharp hearing picked up the sound on his receiver in another room. Bell immediately grasped the significance of this, which he had previously suspected: only continuous, varying electric current could transmit and reconvert continuously varying sound waves (the electric telegraph, by contrast, operated with pulses of current). Bell immediately gave Watson directions for building a telephone apparatus, which was the basic design that was granted a patent on March 7, 1876. Three days later, Bell tested a new transmitter; Watson, in an adjoining room, clearly heard Bell's summons on the receiver: ''Mr. Watson, come here, I want you!''

Bell's new invention was exhibited at the Philadelphia Centennial Exposition in June 1876, impressing the scientific judges and astonishing Dom Pedro (1825-1891), the emperor of Brazil, who exclaimed, ''My God—it talks!'' Watson and Bell spent the rest of the year perfecting their new device and then threw themselves into a series of public demonstrations to promote the commercial potential of the telephone. Together with partners Gardiner Hubbard and Thomas Sanders, they formed the Bell Telephone Company in 1877. That same year, after declining to buy Bell's patent for $100,000, Western Union Telegraph Company entered the telephone business by purchasing Gray's patents and using a carbon **microphone** invented by **Thomas Alva Edison**. The Bell company countered with improved transmitters designed by **Emile Berliner** and Francis Blake (1850-1913), and then sued Western Union for patent infringement. This suit was settled in 1879; it was only the first of hundreds of lawsuits over Bell's patents, which were finally upheld by the United States Supreme Court in 1888.

Use of the telephone spread rapidly. A switchboard system was soon needed to connect multiplying subscribers. The first commercial exchange served 21 customers in New Haven, Connecticut, beginning in January 1878. These early switchboards were worked manually by operators, who were initially young men. The ''telephone lads,'' however, proved too boisterous and were swiftly replaced by young women, whom company supervisors considered more cooperative. The first automatic number-calling mechanism was devised by a Kansas City, Missouri, undertaker named Almon B. Strowger. He suspected a switchboard operator was diverting calls intended for his business to a rival establishment, and patented his automatic switching system to bypass operators in 1891. Strowger's design used push buttons and eventually evolved into the dial telephone. The first automatic commercial switching system, based on Strowger's patent, went into operation in La Porte, Indiana, in November 1892. The first coin-operated public telephone was installed in the Hartford (Connecticut) Bank in 1889. The first telephones used a single microphone; the user spoke into it, then transferred it to the ear to listen. The resulting user confusion was eliminated by Edison's 1877 design for separate ear-and mouth-pieces. The handheld unit incorporating both earpiece and mouthpiece came into standard use in the 1930s.

Long-distance service began almost as soon as the telephone was invented, with experimental lines used by Bell and Watson. The first public long-distance line went into service between Boston, Massachusetts, and Providence, Rhode Island, in January 1881. After the original iron wires were replaced with much more satisfactory copper **alloy**s, long-distance lines proliferated. Still, long distance calls relied on shouting and careful listening. A great advance was made in 1900 when Michael Pupin (1858-1935) of Columbia University and George Campbell (1870-1954) of Bell System both used the theories of English physicist Oliver Heaviside (1850-1925) to design loading coils that minimized distortion. The **vacuum tube** invented by **Lee De Forest** in 1906 was adapted as a telephone amplifier in 1912, which solved the other great problem of long-distance telephone transmission, loss of signal strength over distance. A transcontinental line was now possible; it was officially demonstrated on January 25, 1915, with a call from Bell in New York City to Watson in San Francisco.

Attention next turned, logically, to transatlantic telephone transmissions. Shortwave **radio** was used for this purpose beginning in 1927, but it was plagued by atmospheric interference. A **submarine** cable across the Atlantic was finally laid in 1956, made possible by long-lived amplifiers and a new **plastic** insulating material, **polyethylene**; many more undersea cables followed. Direct dialing by customers for domestic long-distance calls was introduced in Englewood, New Jersey, in November 1951. International direct dialing (IDDD) by customers was introduced in March 1970 between New York and London. IDDD is now available in over 100 countries. Beginning in 1962, orbiting **communication satellite**s were used to relay and amplify telephone transmissions. **Fiber-optic** transmitting systems were introduced in 1980 in Atlanta, and a transatlantic **fiber-optic** cable went into operation in 1988. Mobile telephones using cellular radio technology came into use in 1982. Cordless phones were introduced in 1980, and by 1998, voice-activated cordless telephones were available.

In the mid-1980s, telephone companies installed a new system for routing calls that was based on an international standard called the Signaling System 7 (SS7), which allowed phone companies to route calls easier. SS7 sent information about call setup and the actual voice transmission separately, and both phone numbers were included. New Jersey Bell was the first to realize that this data could be sold, and in 1987, caller ID was introduced. Caller ID allows the call receiver to see the name and phone number of the caller, with a special box or telephone. By the early 1990s, the service, which required a separate fee, was available nation-wide. Other new technologies related to SS7 include: call blocking—blocks certain preprogrammed numbers; call trace—when prompted by a code entered by a call receiver, traces the call; priority call or custom ringing—programmed numbers that ring through have a distinctive ring pattern; return call—when a special code is entered, calls the last number that rang through back; and repeat call—dials a busy number for up to 30 minutes. By 1995, 40% of households in the United States had callwaiting.

Since the late 1920s and 1930s, videophones—which transmit an image of the speaker with he voice—have been experimented with to limited success. They have worked, but never had caught on because of technological limitations and the high cost. A 1992 version transmitted color pictures, but because the video was sent only at two to ten frames a second, the picture appeared jerky. Only 40,000 of these phones were sold world-wide by 1996. In 1992, screenphones (also known as smart phones) were also introduced and became increasingly sophisticated throughout the 1990s. Screenphones are telephones with some computer-like capabilities. They feature a small **LCD** screen (usually with touch screen features), a keyboard, and sometimes, a printer. In addition to placing phone calls and handling all the services offered by telephone companies, screenphones can be used to send e-mail and search the World Wide Web as well as conduct some business transactions, perhaps using smart cards, and store phone numbers and other information. By 1995, 250,000 screenphones had been sold in the United States. In 1994, a telephone handset that fits in your ear was developed. The Jabra 1000 ear-bud phone fits inside the useròs ear, like the speakers found on a **Walkman**. It also transmits the useròs voice as well, picking up the sound vibrations created by speaking from the useròs bones.

By 1998, more than 50% of all telephone lines were used to transmit data rather than voice calls. Telephone lines were increasingly used to send faxes, access the **Internet**, and videoconferencing. With that in mind, Sprint, a national telephone company, announced the development of the Integrated On-Demand Network (ION) which allows one phone line to be used simultaneously for voice, video, and data, and at a faster rate than previously possible. Previously, a separate, dedicated line was needed for each to be used at the same time.

See also Cellular telephone; Telephone cable, trans-atlantic

TELEPHONE ANSWERING DEVICE

The **telephone** answering machine was introduced by the American manufacturer Code-A-Phone in 1958. These early machines were large, clunky, expensive devices that were mostly used by small businesses. They consisted of two separate tape players housed in a single cabinet. One played the outgoing message on a continuous-loop cartridge or cassette. The other was usually a reel-to-reel tape that recorded the incoming message. The phone company required these machines to be connected to the phone line through a coupler, owned and installed by the phone company. Installation costs ranged from ten to fifty dollars, and the customer also had to pay a monthly rental fee of three dollars to eight dollars for the coupler.

During the mid-1980s demand for telephone answering machines took off, attributed mainly to the large increase in single-person households and families with two wage earners who left homes and phones untended for significant periods of time. Dramatic price drops also fueled their new-found popularity; costs plummeted from $750 and higher to $100 and less, thanks mostly to cheaper production in the Far East. During 1985, sales rose 40% and machines were in short supply. The new generation of answering machines was much smaller than their predecessors, using just two small cassettes. Owners won the right to install the machines themselves, directly, without the coupler.

Today's machines offer many features unknown in 1958. To retrieve messages when one is not at home, one enters a personal code into a touch-tone phone to receive a playback of recorded messages. Some machines allow one to change an outgoing message over the remote phone. Others announce the time and date of each message with a **voice synthesizer**. Some manufacturers now use an electronic chip in place of the outgoing message tape; since the chip has no moving parts, it seldom breaks down. PhoneMate's ADAM, available in 1990, was an All-Digital Answering Machine, the first phone answering device to use digital signal processing, which converts incoming messages into signals that are stored in memory rather than on tape. Throughout the 1990s, digital answering machines have continued to improve. The time available for recording messages increased from several minutes to nearly 30 (the equivalent of tape machines). Other features available on digital machines included multiple mailboxes for messages and multiple outgoing messages. With the right software and connections, computers also have the ability to act as answering machines, using the same principles as **voice mail**.

Voice mail is an alternative to answering machines. Used in offices since the early 1980s, it became available for home use through phone companies in 1988. With voice mail, the phone itself becomes an electronic mailbox, via digital signal processing. Callers, directed by a voice prompt, may leave a message directed to a specific family member or business associate; each individual can retrieve his or her own messages by entering a personal code.

Telephone booth

The first **telephone** provided for public use was installed at the New Haven office of the Connecticut Telephone Company on June 1, 1880. In Great Britain, "Call Office Suites" were established in 1882 by the National Telephone Company at the Stock Exchange and Baltic Exchange in London and at the Wool Exchange in Bradford. Users paid an attendant, who allowed them access to the locked booths. Regular users had subscriber's keys and could use the phones at any time.

The first coin-operated public phone was installed at the Hartford Bank in Connecticut by the Southern New England Telephone Company in 1889. In 1891, William Gray—the inventor of the coin mechanism—formed Gray Telephone Pay Station Company and began renting coin-operated phones to storekeepers. England's first coin-operated telephone was installed in April of 1906 at the Ludgate Circus Post Office by the Western Electric Company. The first telephone kiosk (booth) appeared in England in 1908, featuring a door that opened after coins were deposited and, from 1912 until the beginning of the first World War, doodle pads to discourage users from defacing the walls.

Most early public telephones were operated on a post-payment system, that is, when a call was finished, the user was told by the operator how many coins to deposit. Multi-coin prepayment boxes first appeared in 1925. With the advent of **cellular telephones**, the need for public phone booths has somewhat decreased, although they are still actively used by many people, particularly at airports and hotels. No longer the bi-fold door, wooden booths of yesterday, phone booths are often made of metal. Some even offer phones that provide **Internet** access, for a fee.

Telephone cable, trans-Atlantic

Cross-Atlantic **telephone** transmissions started in 1927, via **radio**, but weather conditions frequently disrupted communications. An undersea cable seemed to be the logical answer, but it took nearly a hundred years to follow the first successful trans-Atlantic **telegraph** cable with a similar telephone cable.

Undersea telephony presented several difficult problems. Repeaters, to boost the signal as it weakened over distance, did not last long, and failed repeaters in an undersea line could not be replaced economically. Also, the interaction of two conductors, **wire** and water, caused distortion of the signal that could not be overcome with available insulating materials.

Bell Labs tackled the amplification problem in the late 1940s, developing a repeater that would last for twenty years. At the same time, **plastic polyethylene** was developed and found to have excellent insulating properties. A test line was tried between Miami, Florida, and Havana, Cuba, in 1950. It was successful enough for AT&T to plan a trans-Atlantic cable. In June 1955, the ship *Monarch* set out from Newfoundland, Canada, laying cable and adding a repeater every 10–40 mi (16–64.4 km). The *Monarch* reached Scotland in September. The following summer the ship made the return cable-laying voyage. The resulting trans-Atlantic telephone line went into service in September 1956. On the first full day of service, 75% more cross-Atlantic telephone calls were made by cable than had been made by radio in the previous ten days.

The world community responded enthusiastically to the reality of live person-to-person communication across the ocean. Overseas lines proliferated. The first cable included 64 channels; the next one had 128 two-way circuits; within 15 years an SC cable carried 4,200 circuits.

Satellite transmission of telephone communications challenged **submarine** cables beginning in 1962. Satellite signals, however, were often hindered by electromagnetic interference, which blocked high-speed data transmission. As telephone usage increasingly involved data as opposed to voice messages, this interference became more troublesome. Also, satellite transmissions could be easily intercepted. **Fiber optic** transatlantic cables, beginning in 1988, are an answer to these problems, offering very secure transmissions with little interference.

Teleprinter and teletype

The teleprinter, also called a teletypewriter, is a typewriter-printer that sends messages over **telegraph** or other data communication lines to a similar receiving printer, which automatically prints the message. The transmitter inputs the message on a typewriter-like keyboard, which converts each keystroke to a coded electrical impulse and sends it to the receiver, which interprets the impulses to reproduce the original keystrokes. The teleprinter was first developed by Frederick G. Creed (1871-1957), a Nova Scotian who was inspired by his work as a telegrapher to find a way to send printed messages without using **Morse code**. Creed immigrated to Scotland in 1897, where he experimented with his **typewriter** (which he always kept with him) to produce the first teleprinter. Although the British Post Office bought a dozen of Creed's machines in 1902, their use did not catch on.

In the United States, Charles L. Krumm designed the prototype of the modern teleprinter in 1907. A refined version of this machine was widely marketed in the United States in the 1920s, where it was most often referred to by its American Telephone and Telegraph trade name, Teletype. A similar machine was sold at the same time in Germany by Siemens-Halske. Both used a code originally devised by Emile Baudot (1845-1903) for his multiplex telegraph and adapted by Donald Murray of New Zealand in 1903 for his multiplex page-form-telegram system. Other teleprinters were developed that used a different keyboard code, the American Standard Code for Information Interchange (ASCII). Commercial teleprinter exchange services, called TRX and **Telex**, were developed in the 1930s. News **wire** services became heavy users of teleprinter communications.

Early teletypewriters printed on strips of gummed tape, which was cut to length and pasted onto telegram forms. Later machines printed messages in page form. At first, teleprinters produced up to 500 characters per minute. By the mid-1960s these machines had upped their output to 900 words per minute. Beginning in the 1980s, teleprinters were being replaced by direct transmissions from **communication satellite**s to computers, electronic computer delivery, and **fax machine**s.

Telescope

The principle of the telescope was first developed by a Dutch **spectacle**-maker, Hans Lippershey (1570-1619). He used his first telescope, made in 1608, for observing grounded objects from a distance, rather than astronomy. His invention was not openly embraced by the scientific community; he was, in fact, unable to patent it.

In 1609, not far away, Italian mathematics professor **Galileo** developed his own *refractor* telescope, without seeing even a model of Lippershey's work. His creation had an object **glass** that bent light rays to a focus near the eye. There a second lens, an eyepiece, magnified the image. His invention grew to be quite popular, as glass was relatively cheap and **mirror**s of the day were of very poor quality. Galileo's first telescope was small by today's standards and its object glass was only one inch (2.54 cm) in diameter.

This simple instrument allowed Galileo to make astonishing discoveries. He saw that the Milky Way comprised thousands of stars, he identified darkened blemishes on the moon's surface as craters, and he also noted changes (phases) of the moon's shape. Galileo's instrument was soon enhanced by that of Johannes Kepler (1571-1630), the German astronomer whose creation increased the field-of-view as well as the he discovered Uranus in 1781.

Even today, reflectors are perhaps the most prominent type of telescope. Some of the largest reflective telescopes are located on Palomar Mountain near San Diego, California and on Mount Pastukhov in Russia. Reflectors are cost-effective, produce little false color, and maintain a high resolution. The mirrors used in larger reflectors, however, often cause distortion due to the weight on the instrument. Newer reflectors are incorporating mirrors of varying shapes (hexagonal glass segments, for example) and are produced using lighter, more durable materials (such as **Pyrex**). Another approach to eliminate sag is to build several large mirrors, mount them separately on a common base, and link them via computer into one central unit.

The latter part of the nineteenth century saw the resurgence of refracting telescopes. What began as a hobby for American astronomer **Alvan Clark** ended as a very notable enterprise. For years, Clark had made his own mirrors and **lenses**. Realizing his were superior to any made in Europe, he set out to manufacture the highest quality lenses possible for sale worldwide. He was successful enough to eventually be given the task of building what was then world's largest refractor: the 36-inch (91 cm) Lick Observatory telescope in California.

After the telescope's completion in 1887, Clark worked with the University of Southern California in developing an even larger telescope—the university was to buy the lenses and the land was to come from private donation. Unfortunately, due to problems in the real estate market, the project was cancelled. Upon hearing of this, George Ellery Hale (inventor of the *spectroheliograph* used in solar studies) formulated a plan to complete what Clark and the university had begun. Hale worked to secure funding for the project. He struck an agreement with Charles Yerkes, a Chicago railroad millionaire, who fronted over $300,000 for the construction of the world's largest refractor—Yerkes Observatory was built in 1897 near Chicago, Illinois.

It soon became obvious to Hale that because of resolution flaws and inconsistencies in the resulting images, the size was close to the maximum for a refracting telescope. Realizing the need for sharper, more distant images the scientific community turned again to the reflector telescope and a new instrument developed by German optician Bernhard Schmidt—the *refractor-reflector telescope*. Schmidt invented the first combination refractor-reflector telescope in 1931 for wide-angle astro **photography**. It had a thin, specially shaped lens at the end of a tube and a regular mirror at the other end; a photographic plate was also included.

The largest Schmidt telescope is located at Palomar Observatory (California). It can photograph an area of the sky more than 300 times as large as that seen by other reflectors; this is key in mapping the skies and closely studying objects at a significant distance. The multiple atmospheric layers have hampered telescope observations from the Earth's surface, absorbing infrared, ultraviolet, and x-ray radiation. Scientists, therefore, have developed other more efficient means for gathering data on our solar system. These instruments include the **space telescope** and the **solar telescope**.

In an attempt to raise telescopes above the atmosphere without launching them into space, scientists have flown instruments in airplanes, performing observations at high altitudes. One such example was the Kuiper Airborne Observatory, a C-141 cargo plane fitted with a 450-lb (203 kg) infrared telescope. The missions generally took place at 41,000 feet (12,300 m), where the telescope could be above 99% of the atmospheric water vapor responsible for absorbing infrared radiation. Commissioned in 1974, the Kuiper discovered the rings of Uranus in 1977, and atmosphere around Pluto in 1988. Plans call for the now-retired Kuiper to be replaced by the Stratosphic Observatory for Infrared Astronomy (SOFIA).

In an extension of the Kuiper Observatory principle, in 1995 scientists flew a trio of ultraviolet telescopes in the space shuttle *Endeavor*. Known collectively as Astro-2, the telescopes remained aloft for 16 days, capturing image after image of the universe. Because ultraviolet radiation is emitted by hot young stars or very old stars, the Astro-2 images showed a very different picture of the universe, revealing structures within galaxies, and even whole galaxies themselves, not seen in visible-light images.

See also Eyeglasses; Heat-resistant glass; Navigational satellite; Space probe; Spy satellite; Weather forecasting methods

TELETYPE • See Teleprinter and teletype

TELEVISION

About the same time that **Guglielmo Marconi** was experimenting with **radio** transmissions in the late 1890s, other scientists were exploring the possibility of transmitting visual images. The first inventor to do so was Abbe Caselli, an Italian-born priest who succeeded in sending very elemental shadow pictures via the French telegraph lines in 1866. Almost twenty years later an Englishman named Shelford Bidwell developed a device that he called an electric distant vision apparatus. This machine used a selenium cell mounted on a box that moved up and down to scan an image.

While still a student, the German scientist **Paul Nipkow** designed an electric **telescope** that divided its target into scanned lines. Using this as a base he developed a photomechanical image scanner in 1884 that he called a *Nipkow disk.* The device was comprised of a metal or cardboard disk perforated with a series of square holes in a spiral pattern, so that each hole was slightly closer to the disk's center than the last. As the disk spun, a light was shone through the holes and onto the target; by looking through the holes one could see the target revealed as a series of horizontal lines. Though the Nipkow disk became the basis for later photomechanical televisions, the technology in the mid-1880s was not advanced enough for Nipkow himself to pursue his work further.

It was a Scottish engineer, **John Logie Baird**, who eventually continued the research Nipkow had begun. In 1923 Baird designed a television system utilizing the Nipkow disk. By adding a **photoelectric cell**, Baird's system was able to read the areas of dim and bright light that made up each scanned line and convert them into an electrical signal. To play back the image a second Nipkow disk was used, along with a flashing light bulb. By using the incoming signal to make the bulb flash brightly or dimly, and then by synchronizing that flashing to the second spinning disk, a crude but recognizable image was formed. Baird's design was used to send bicoastal and transatlantic transmissions during the late 1920s, and it was his device that was first used by the British Broadcasting Company (BBC).

However, it was becoming very clear to scientists that the photomechanical systems had their limits, and that an all-electrical device would be far more advantageous. This line of thinking was not new: both the American physicist Charles F. Jenkins and the Scottish-born engineer A. A. Campbell-Swinton had suggested designs for an all electrical television by the turn of the twentieth century. Decades ahead of its time, Campbell-Swinton's design was outstanding in that it was the first to suggest the use of an electron scanner to convert the images.

The basic model for modern television was designed simultaneously yet independently by two scientists: American **Philo Farnsworth** and Russian immigrant **Vladimir Zworykin**. Farnsworth got the idea for his "image dissector" as a sophomore at Brigham Young University. He showed his design to several investors who gave him enough money to continue his research. In 1927 he demonstrated a working model of his television, which, though not practical for mass-market production, it showed potential. Unfortunately, Farnsworth's investors soon backed out, and at the onset of World War II, the entire project was abandoned.

Zworykin fared much better. After immigrating to the United States in 1919, he found a job with the Westinghouse Electric Corporation. It was there that he developed his iconoscope, an all-electrical television **camera**. The iconoscope used a vast array of tiny drops of selenium, each acting as an individual photoelectric cell. As the drops were exposed to light, they stored the amount of light they had "seen" as a very small electric pulse. An electron gun was then used to convert the stored pulses into an electric signal that could be transmitted. With the drops' pulses discharged, the process could be begun anew and was repeated many times each second. In 1924, just a year after applying for the patent for his iconoscope, Zworykin developed the kinescope, the precursor to the modern television receiver. Zworykin's tremendous advances in the field attracted the attention of the Radio Corporation of America (RCA), and in 1929 he began to work for them full-time. It was Zworykin's iconoscope/kinescope system that evolved into the television systems of today.

Television operates because of a principle known as persistence of vision—that is, the human eye is slower than the brain, so that if a series of similar images are played rapidly, the brain will blur them together, creating the illusion of animation. This is the same principle by which **motion pictures** work. In a standard American television set each picture contains 525 lines. These lines are scanned the way a book is, starting at the top left and going from left to right and from top to bottom. In the first pass, the scanner shows the odd-numbered lines (1, 3, 5) and then goes back to scan the even-numbered lines (2, 4, 6); this practice helps to decrease screen flicker. In all it takes just 1/30th of a second to scan all 525 lines. This method was adopted in 1951 by the National Television System Committee (NTSC) in order to standardize all American television broadcasts.

In the early years of television, all broadcasts were in black and white. It was not until the late 1950s that a viable color system was approved, and not until after a bitter battle between the two largest television companies, RCA and Columbia Broadcasting System (CBS). CBS was the first to offer a practical color image based upon the work of **Peter Goldmark**. Goldmark's field-sequential color system delivered color quality unsurpassed even today; however, the mechanical nature of the Goldmark design would have made it necessary for all television owners to exchange their existing sets for new field-sequential sets. RCA offered a color system designed by **Ernst Alexanderson**; this design was not as clear as the one developed by CBS, but it was compatible with existing black-and white sets. The Federal Communications Commission (FCC) eventually chose to make the RCA color system the industry standard in 1954. The color and scanning standards adopted in other countries (SECAM in France and PAL in Great Britain, for example) are not necessarily compatible with American television sets, and digital translators are required in order to receive these images.

In the late 1990s, cathode ray technology continued to develop. The trend in televisions moved towards thinner and flatter—the opposite of the bulk inherent in one big **cathode ray tube** television, Philips Research Laboratories have devised cathode-ray tube panels (CRP), which are thinner and flatter. Researchers have produced early prototypes that are one centimeter (.4 in) thick. Japan's Matsushita Television Research Laboratory reduced cathode ray bulk with Flat Vision which has 10,000 tiny tubes, managing the image with an electron beam sweep. The resulting television is merely .25 in (.6 cm) thick, and might become commercially available in the

Experimental broadcast of television in the early 1930s.

United States by the year 2000. Another television technology in development is 3-D television, being developed in the Computer Laboratory at Cambridge University in England. Called the Cambridge Autostereo Display, the 3-D is created by eight or sixteen scene views being displayed then updated in rapid succession, about thirty times per second. The viewers' eyes see the projected images in 3-D. However, this technology will take years to come to fruition and be available on the consumer market.

What has hit the consumer market in 1998 after 17 years of promise is high-definition television, also known as HDTV. The HDTV system offers a much sharper image, comparable to that seen at a movie theater. This is accomplished by transmitting each picture in at least 720 lines and up to 1080 lines—twice as many as in a standard television. Because HDTV can require the transmission of so much information, **cable television** and direct broadcasting satellite are the preferred HDTV broadcast mediums, though transmission via standard broadcast is possible because of digital multicasting.

DirecTV began broadcasting two channels of HDTV programming in 1998. Available to consumers in 1994, DirecTV is a digital television programming broadcast medium that relies on satellites circling the Earth at 22,300 miles (236,380 km) above the earth. The signals sent by the satellites are picked up by small dishes owned or rented by a consumer that are only 18 inches (45 cm) across. DirecTV, and similar products like Primestar, promoted themselves as an alternative to cable, with up to 150 channels available and sharp, clear pictures. It could reach consumers in rural areas where cable in unavailable. There were several drawbacks to the DirecTV system. Remote controls are unavailable, so channel changing requires going back to one on-screen grid. The dish and television receiver can be costly. As of 1996, there were 1.3 million DirecTV household subscribers. In 1996, DirecTV partnered with Microsoft to develop multi-media satellite broadcasts with computer links, providing web content related to television. In 1997, the first commercially available link between computer and television was announced: WebTV. A set-top box, WebTV allows viewers to explore the Internet using the television and telephone lines. WebTV Plus, introduced in 1998, allows viewers to watch television while they use the Internet and featured a hard drive to store data and the capabilities for printer hook-up. WebTV and WebTV Plus were priced much less than a computer, and such the future of both television and computers might lie in this kind of convergence.

See also Animated film; Incandescent light bulb; Telegraph cable

Telex machine

Telex is an international message-transfer service for subscribers using **teleprinters**. A subscriber attaches its Telex-compatible teleprinter to a Telex line, then dials (much like a telephone call) the fellow subscriber to whom the message is being sent. The message travels as a low-bit-rate electrical signal. The recipient's Telex-compatible teleprinter converts the signal, then prints out the message.

Telex originated in Germany in the early 1930s and soon expanded to other European countries. In the United States, American Telephone & Telegraph Company (AT&T) started a manually-switched teletypewriter exchange service called TWX in 1931. Police departments were one of TWX's prime early users. TWX became automatically switched in 1962 and began serving Canada in 1963. Western Union Telegraph Company initiated its Telex system in the United States in 1962, registering the name as a trademark. (In Europe it had been a generic term.)

After Western Union acquired the TWX network in 1970, Western Union linked Telex and TWX, which became respectively known as Telex I and Telex II. Because Telex I teleprinters use the Baudot code and Telex II (TWX) teleprinters use the ASCII code, and transmit at different speeds, exchange of messages between the two systems is somewhat delayed while a processing computer does the necessary conversions. Other companies in the United States, such as MCI and ITT, also provide switched teleprinter exchange services.

By 1990 more than 190 countries used Telex, and in that year, AT&T bought Western Union's Telex network. Sometimes called the first electronic mail service, Telex was, at one time, the most-used written telecommunication system in the business world. Though Telex is still available as of 1998, the widespread use of **fax machine**s and computers, especially electronic mail and the **Internet**, have lead to Telex's severe decline in popularity.

Telford, Thomas (1757-1834)

Scottish civil engineer

Thomas Telford, the son of a Scottish shepherd, worked on farms and apprenticed as a stonemason during his teen years. He moved to London in 1782, initially working as a stonemason. He was named surveyor of public works for the County of Shropshire in 1787, and in 1793 was appointed chief engineer of the Ellesmere Canal on the English-Welsh border, a project noted for its two transport aqueducts.

Telford went to work for the Scottish Highland Roads Commission in 1802, a position which placed him in charge of the construction of over 900 mi. (1,400 km) of roads, 1,200 **bridge**s (some made of cast iron), and several harbors. His major project during this period was the Caledonian Canal connecting Scotland's two coasts at Glasgow. An expensive undertaking, it was not completed until 1847. Telford also supervised construction of the Gotha Canal, which traversed southern Sweden; for that effort he was named to the Swedish order of knighthood.

In the 1820s, Telford supervised construction of the 194-mi. (312 km) road from London to Holyhead, Wales. He used a foundation of large stones laid on end, a technique used in France by **Pierre Trésaguet**, and used **pitch** as a paving agent. Crowning this project was the Menai Strait suspension bridge connecting the Welsh mainland with the Isle of Anglèsey. Like the Caledonian project, this effort proved quite costly. Telford's construction methods may have been too good—

the reason, perhaps, that **John McAdam** supplanted them with his simpler and less costly road foundations.

See also Canal and canal lock; Water supply and drainage

TELLER, EDWARD (1908-)
Hungarian American physicist

Teller was born in Budapest on January 15, 1908. He attended the Karlsruhe Technische Hochschule, the University of Munich, and the University of Leipzig. He earned his Ph.D. degree in theoretical physics at Leipzig in 1930.

Teller's earliest research interests involved the application of quantum theory to physical chemistry. His first paper, "Hydrogen Molecular Ions" outlines a theory of the hydrogen molecule that is still widely accepted. From 1931 to 1933, Teller continued his research at the University of Göttingen. Following that period, he spent a year working with Niels Bohr at the University of Copenhagen.

By the mid-1930s, it had become clear to many European scientists that they could not continue to work in their homelands. The rise of fascist governments in Germany and Italy created a serious threat not only to their scientific careers, but also to their very survival. Teller was one of dozens of researchers who decided to leave the continent for the United States. From 1935 to 1941, he held the post of professor of physics at George Washington University in the District of Columbia.

During this time, Teller's research interests shifted to the newly discovered phenomenon of nuclear fission. In 1941, he accepted an assignment to work on the Manhattan Project developing the first **atomic bomb**.

Part of Teller's notoriety today has its origin in the interpersonal struggles that arose during the bomb research. Teller was especially enthusiastic about moving beyond the fission bomb to a larger and more powerful "superbomb," one that made use of fusion reactions. For a variety of reasons, he was unable to convince other scientists or the United States government to go forward on this project.

After the war, Teller became embroiled in the controversy about J. Robert Oppenheimer's security clearance. Many scientists believe that Teller's testimony in the case was critical in having Oppenheimer branded as a security risk and denying him further security clearance.

Teller has long been an active and outspoken advocate of U.S. military preparedness. He has written, spoken, and otherwise campaigned for a strong defense program. He regarded the election of like-minded Ronald Reagan as president in 1980 as a "miracle" for the nation. Only three years later, he had convinced Reagan of the need for and feasibility of a massive national defense program, the Strategic Defense Initiative (SDI). Nicknamed "Star Wars" for its futuristic plans, SDI continues to consume billions of dollars in research and development funds each year. In spite of the demise of the Soviet Union, the program is expected to cost an additional $90 billion between the years 1992 and 2007. Critics continue to attack the program, however, claiming it is based on overly optimistic projections, faulty science, and unrealistic expectations.

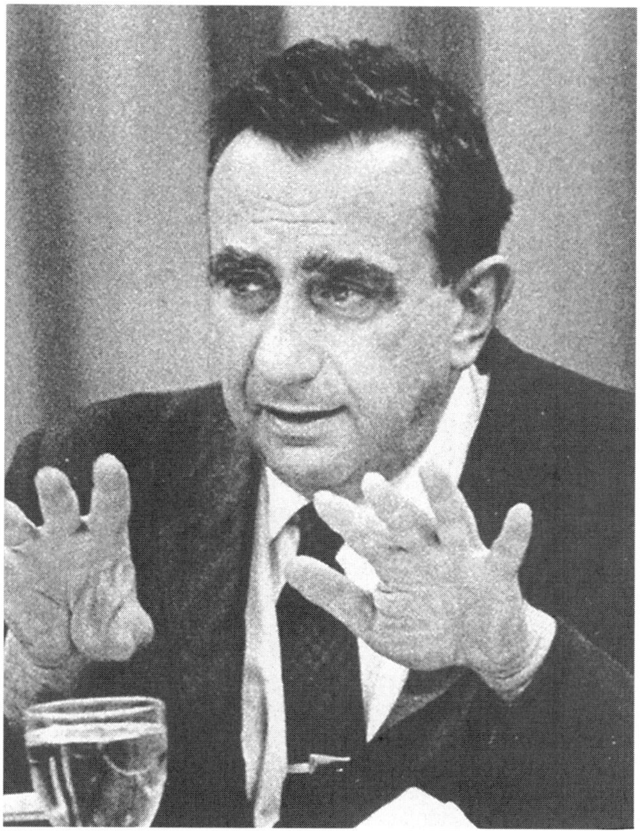

Edward Teller

The Soviet Union's detonation of its first atomic bomb in 1949 rekindled United States' interest in Teller's "superbomb." The following year, President Harry S. Truman gave approval for the development of that weapon. Although the accuracy of Teller's actual computations on the bomb have been questioned, he is widely acknowledged to be the "father of the **hydrogen bomb**."

Since 1952, Teller has been associated with the Lawrence Livermore Laboratory at the University of California at Berkeley. He has received the Enrico Fermi Award (1962), the Albert Einstein Award (1977), and the National Medal of Science (1982).

TEMPLE, LEWIS (1800-1854)
American inventor

By upgrading the original arrowhead **harpoon** with a locking toggle device, Lewis Temple, an African-American inventor, created "Temple's Iron" and revolutionized the whaling industry. A native of Richmond, Virginia, Temple never received formal schooling. In 1829, he married Mary Clark and established a family in New Bedford, Massachusetts, a major whaling center. As a metal worker at Coffin's Wharf, he studied the primitive design of the whaling harpoon, which whalers threw at the animal, then approached in small boats to complete the kill with lances.

In 1845, Temple replaced the less sophisticated harpoon model with a pivoting device set at right angles to the shaft. Held in place by a wooden pin, the mechanism set in the whale's jaw when tugs on the harpoon broke the pin and set the barb. This innovation secured the animal on the whaler's line. Because of his ingenuity, harpoon makers copied the device, and whalers increased their catch during an upsurge in the industry. Temple made no claim on the mechanism, but from 1848 to 1868 he earned a steady income from its manufacture.

An active abolitionist and temperance worker, Temple took an interest in improving life in New Bedford. In 1854, he built a brick blacksmith shop near Steamboat Wharf, but he did not survive its completion after suffering a fall over a plank left in the street by city sewer workers. Because he received permanent injuries to his arms and internal organs, he successfully sued the city for $2,000. However, Temple died before he could collect his compensation, and his wife and son were left in poverty, selling his property and goods to cover his debts.

TENNANT, CHARLES • See Bleach

TENNIS

The worldwide popularity of tennis could be attributed to the various stars on the international circuit. Tennis appears to be very glamorous, but it is in fact a grueling game that requires extreme strength and stamina.

Tennis has its roots in many countries. The ancient Greeks played a game called *a la phoeninde* in which one player hits a ball to another player. In France, an indoor game called "royal tennis" was developed around 1050. Native Americans used to play a game called *poona*.

In 1873, Walter C. Wingfield (1833-1912), an English army major stationed in India, combined rules from *poona* with a grass playing field and thus invented tennis as we know it today. He called the game *spharistike*, after *spherique*, the Greek words for "games with balls." The name "tennis" is derived from the French. When players were ready to serve, they would shout "*Tenez*!" ("Hold!") to warn their opponent.

Many years later, a friend of Major Wingfield met Mary Ewing Outerbridge, a young American woman who expressed great enthusiasm in the sport. In 1874, she introduced the game to the United States.

TERRY, ELI (1772-1852)

American manufacturer

Born in East Windsor, Connecticut, Terry was apprenticed to a master clockmaker at the age of fourteen. At that time, clocks were considered to be a status item of little practical value, for the rigors of daily life were regulated by the sun or by the amount of oil in an oil lamp. Thus, clocks were custom built by hand, mostly of wooden parts.

In 1793 Terry had opened his own clock-shop in Plymouth, Connecticut. He had heard of **Eli Whitney**'s methods and saw the potential in his own business. Terry adapted his machines to be powered by water and hired several workmen to cut the individual wheels, cogs, and other clock parts which he later assembled to make the finished clocks.

Terry put his system to a test. In 1807 he formed a partnership with Seth Thomas (1785-1859) and Silas Hoadley to build 4,000 clock-works for Connecticut businessmen Edward Porter and Levi Porter. This huge order was considered impossible for any one man to complete in his entire lifetime, but Terry promised to do it in three years. He made good on his promise, and his success showed the feasibility of adapting mass production techniques to the building of clocks. It was also a very profitable venture, and after this initial run of 4,000 clocks, Thomas and Hoadley bought out Terry's interest in the business, and became very successful in their own clockmaking ventures.

Terry went on to invent the wooden shelf clock (or mantle clock) which featured a system of weights and **pulley**s suspended within the clock on either side of the face and placed within an attractive case. His innovative production methods and interchangeable parts allowed Terry to build as many as 12,000 clocks per year. An extensive network of "Yankee Peddlers" hawked these popular wooden clocks, mass-produced by Terry and Thomas, door-to-door all over New England, and both men made their fortunes in the venture.

Terry retired from the clockmaking business in 1833, turning the venture over to his sons and other family members. By the time of his death in 1852, Terry's innovations had earned him ten patents for improvements to clocks.

See also Clock and watch

TESLA, NIKOLA (1856-1943)

Croatian American electrical engineer

Perhaps more than any other individual, Nikola Tesla was responsible for developing the AC (**alternating current**) system of power supply that provides the world with electricity. A prolific inventor of keen intelligence and exceptional insight, Tesla patented more than 700 inventions in his lifetime.

Born on July 10, 1856, in Smijan, Croatia, Tesla was the son of a clergyman about whom little is known. Tesla may have acquired his knack for inventing from his mother. Although illiterate, she was a clever person who invented numerous implements for use at home and on the farm.

Tesla studied mathematics and engineering at the University of Graz and philosophy at the University of Prague. From the start, he was especially interested in electricity, and in 1881 he took a job as an electrical engineer in Paris, where he designed equipment and repaired power systems. It was around this time that Tesla first developed the polyphase synchronous motor, a "brushless" motor which would revolutionize the electrical industry. At the time, power was supplied by **direct current**, which was reliable but could not be delivered over great distances like AC could. In addition, the DC motors then in use contained brushes which sparked as the motors spun, limiting their maximum voltage. Not only could Tesla's motor convert the zig-zag impulses of an alternating current into a spinning motion in one direction, but it could run at high voltages with no danger of sparking.

Though Tesla was convinced that a system of AC using his motors should be used to supply power, he could not persuade his employers to consider his system, and he decided to leave for the United States to try his luck there. He was hired to work in **Thomas Alva Edison**'s laboratory, and although Edison developed enormous respect for Tesla's brilliant work, he was no more receptive to Tesla's ideas than the Europeans had been. Edison and Tesla hotly denounced one another, and following an argument over wages, Tesla quit in 1885.

Tesla was approached by a group of promoters and formed his first company, the Tesla Electric Light Company. He went into competition with Edison, developing an **arc lamp** for use in lighting streets and factories. This venture proved to be quite successful, earning Tesla seven patents. But Tesla was swindled by his financiers and had only a worthless stock certificate in his possession when he was forced out of the company.

During this low point in Tesla's life, he worked as a ditch-digger. In 1887 Tesla's sympathetic foreman introduced him to an official at the Western Union Telegraph Company, who provided him with the backing to start a company once again. Soon Tesla was devising a flurry of inventions in the laboratory of his Tesla Electric Company: split-phase, induction and synchronous motors, **generator**s, **transformer**s, all manner of electrical equipment. Soon his work came to the attention of one of the most important businessmen of the day, **George Westinghouse**.

Shrewdly anticipating the future, Westinghouse offered Tesla a large sum for his patent rights to the polyphase induction motor as well as a royalty on power generated. Westinghouse also gave the inventor a job as a consultant. Outbidding Edison, Westinghouse and Tesla arranged to provide the Columbian Exposition of 1893 with AC power. This victory paved the way for an even greater triumph late that year when Westinghouse was offered the first contract to harness Niagara Falls for electrical power. By 1895 a generator was in place and two years later the Tesla generators were supplying electricity to Buffalo, New York, 22 miles (35 km) away. The Edison Company scrambled to catch up, and began installing AC transmission equipment at all of its power stations.

Around this time, Tesla was bursting with creativity. He became captivated by the generation of high-frequency currents and built a machine that produced frequencies of up to 25,000 cycles per second. Machines of this type would play an important part in the development of **radio**—and, in fact,

Nikola Tesla in his laboratory.

Tesla had demonstrated the principles of wireless transmission of signals as early as 1893. In 1898, when **Guglielmo Marconi** was just beginning to send crude **Morse code** messages over long distances, Tesla was demonstrating a radio-controlled model boat at Madison Square Garden. This early forerunner of robots was so far ahead of its time that the baffled United States Navy could not see any use for such a device.

Tesla also began work on higher-frequency devices, the most important of which was Tesla coil, or resonant air-core transformer. The Tesla coil could produce current of almost any frequency and magnitude.

A mesmerizing speaker, Tesla traveled throughout Europe and the United States exhibiting his devices and lecturing. Often his lectures included spectacular trickery. For example, he astounded audiences by illuminating fluorescent tubes without any electrical connections using his Tesla coil; he magically spun a metal egg on a table, using an induction coil hidden below; he shot ''artificial lightning'' bolts through the room; and he sent one-million volts of AC current harmlessly through his body.

At times, Tesla's insight into the future of technology was uncanny. He intended his experiments with his **robotic**, remote-controlled boat to demonstrate that a day would come in which wars would be fought with such devices. He experimented with creating a particle beam, and attempted to devise a high-speed turbine, but failed because metal **alloys** strong enough to hold it together did not yet exist. Tesla was one of the first to advocate the use of solar and geothermal power to obtain energy, and he even described the principles of **radar** and electronic **television** before either existed.

Unfortunately, Tesla's enthusiasm for innovation began to get the best of him, and his ideas, though often visionary, became more and more impractical. He became obsessed with the possibility of broadcasting electric power through the air, and lost millions on a pair of grandiose towers he built in Colorado and New Jersey designed for this purpose. The second of these towers was intended to broadcast power and radio worldwide, but Tesla went bankrupt before it could be completed, and it was demolished during World War I as a potential security risk.

Such disastrous projects had lost Tesla the support of his old friend Westinghouse. When another of Tesla's patrons, J. P. Morgan, died in 1913, Tesla faced financial ruin. Over the years, his hot temper, flamboyant spending habits, and opinionated manner had damaged his reputation among the businessmen he depended upon, and Tesla was left nearly penniless. He spent the last years of his life in increasing seclusion, feeding the pigeons daily on the steps of the New York Public Library. Only occasionally did he break his isolation to speak to the press about a new invention he was working on, or to make eccentric claims. He died quietly on January 8, 1943.

TEST-TUBE BABY • See In-vitro fertilization

TETRAETHYL • See Gasoline

TEVATRO • See Cyclotron

TEXTILES

Animal hides, such as **leather**, provided the first material for the making of clothing; they came in one large, supple piece that could be cut to any shape and sewn together with thin leather strips. However, hides were not ideal for all purposes or all climates. After thousands of years, early civilizations began to construct clothing from natural fibers (both animal and vegetable), thus creating the first textiles.

It is difficult to pinpoint the emergence of textiles, since the materials used decay easily. However, it is probable that the weaving of textiles was inspired by the weaving of mats, baskets, and other vessels. While these objects—dating as far back as 6500 B.C.—were made primarily from reeds and grasses, there soon began the use of other natural fibers for the creation of clothing.

The four most important clothing fibers were flax, wool, cotton, and silk. Flax fibers are the basis for **linen** cloth. They were first used in Ancient Egypt; the lightweight fabric was ideal for the hot, dry climate. Flax is a very difficult material to prepare for weaving, since the most useful fibers are found deep within the stalk. The flax stem must first be *retted*, a process originally accomplished by soaking it in stagnant water for several weeks. The now-pliable stalk is then dried and beaten to break the core and loosen the fibers. Next, the broken flax must be *hackled*, drawing out the light *bast* fibers from which the linen can be woven. Lastly, the fibers are spun into a yarn or thread thick enough for weaving.

The next important natural fiber was wool. This was a readily available material, since sheep have been domesticated since 4000 B.C. There are two types of fleece follicles: primary and secondary. Wool from the primary follicles is usually too coarse for use, and this type of fleece has been bred out of most modern animals. The secondary follicles produced fleece fibers that are fine, short, and somewhat scaly. These scales allow the fibers to interlock, so that little twisting is required to make wool yarn. Because they lock so readily, wool fibers also trap a great deal of air between them; this acts as an insulator, making wool an excellent material for colder climates.

While sheep are the most popular source of wool, the fleece of goats, camels, alpacas, llamas, and vicunas is also used. Another fabric, felt, can be easily made from most wools. If a quantity of wool is moistened, heated, and compressed, the fibers will lock tightly to create a thick, somewhat rigid material-the more wool that is used, the more rigid the felt will be. Since it is produce through compression rather than weaving, felt can be made into any shape: hats, **shoes**, **carpet**s, and other products can be easily made from felt. Felt is also a very inexpensive material, since it is created through a one-step process.

Cotton, probably the most important natural fiber used in textile production, was discovered independently by the ancient Egyptian and Peruvian civilizations. Both civilizations found that cotton, once cleaned, was easily separated into fibers that could be spun into a strong yarn. Cotton fabrics are absorbent, washable, and smooth against the skin. Also, because it burns without odor, cotton is an ideal material for **candle** and **lamp** wicks.

The most recently developed natural fiber is silk. It is very strong and stretchable, and is valued for its sheen. The fiber was first harvested in China and India around 3000 B.C.. The silk moth, whose worm produces natural silk, thrives upon the leaf of the white mulberry tree. Silk manufacturers kept acres of mulberry trees, upon which the moth would lay its eggs. When the silk grub builds its cocoon it excretes a long, thin strand of *fibroin*, then coats it with *sericin* (known as silk gum). To obtain the silk, the cocoon must be collected before the worm emerges, for it uses an acid to burn through the cocoon. While the worm lies dormant, the cocoon is plucked from the tree and immersed in boiling water; this kills the worm and causes its cocoon to unravel.

Once uncoiled, a single strand of silk can be up to 1,093 yards (1,000 m) in length. Because the silk thread is so thin, it is usually combined with four or five other strands. As they are dried, the traces of silk gum still clinging to them help to form one stronger, thicker strand. The result is a natural yarn that is very easy to weave and sew.

In the late 1800s, chemists began experimenting with **synthetic fibers**. These had been anticipated much earlier by such great scientists as **Robert Hooke** and René Reaumur (1683-1757). None, however, had successfully produced such a fiber.

In 1883 **Joseph Swan** patented the first artificial fiber, made by forcing nitrocellulose through small holes. A few years later, **Louis Comte de Chardonnet** succeeded in creating an artificial silk called **rayon**; though recognized as an important discovery, rayon was not widely used by the textile industry until the 1930s.

In 1938, the next true textile revolution began in the laboratory of Dr. **Wallace Hume Carothers**, for in that year he invented **nylon**. Unlike previous man-made fibers, which were produced from wood pulp and other natural materials, nylon was completely synthesized—produced chemically from benzene, nitrogen, oxygen, and hydrogen. Marketed by Du Pont, nylon was found to have applications almost everywhere fabric was used. It was strong, waterproof, and stretchable. Being synthetic, it could be formed into any shape, and when so formed would always return to that shape (such an ability is called memory pressing). Since the invention of nylon other synthetics such as terylene, **acrylics**, polyolefin, and bonded fibers have played important roles in the textile industry.

THEODOLITE • See Surveying instruments

THERMIONIC RADIO VALVE • See Vacuum tube

THERMISTOR

There are many different **thermometers** for many applications—the common mercury thermometer is used for measuring to within a degree, the **thermocouple** is used to electrically measure temperatures within a very wide range, and the **pyrometer** is used to observe and regulate temperatures above 1,300°C (2,400°F). One of the most sensitive devices for measuring temperature is the *thermistor*.

A type of **semiconductor**, the thermistor is usually constructed of metal oxides, such as nickel oxide or copper oxide, and then coated in epoxy to protect it. It works on the principle that electrical resistance within an **electric circuit** will increase as the temperature increases. Because the resistance within a circuit can be measured very precisely, the thermistor is able to detect temperature variations as small as 0.001°C. This makes it among the most accurate thermometers available. Thermistors are generally found in hospitals, coupled with a digital readout. They are also used in other biological environments, particularly those that require careful measurement.

THERMOCOUPLE

In 1821 the German scientist Thomas Seebeck (1770-1831) was experimenting with the thermal properties of certain metals. He constructed a loop of two different metals and applied heat to one of them. To his surprise a magnetic field was formed, as if a current had been generated between the two metals. This phenomenon was named the Seebeck effect, even though Seebeck himself did not fully understand it.

Nearly a century later Seebeck's work became the basis for the field of thermoelectricity, and his metal loop was recognized as the first *thermocouple.* The Seebeck effect can be observed when an **electric circuit** is formed from **wires** of two different metals and the two ends are at different temperatures. A simple **galvanometer** will then detect the voltage difference between the two metal ends, and the temperature of one end can be determined if the temperature of the other is known. This kind of circuit is called a thermocouple, and for many years after Seebeck's discovery it was used only as a sensitive **thermometer**.

Thermometers of unsurpassed accuracy and range have been developed using thermocouples. Used primarily in industry, these thermometers have a range from a few degrees above absolute zero to several thousand degrees Fahrenheit. Beginning in the early twentieth century, scientists began to imagine a new use for thermocouples. If a small amount of heat created a small electric current, they thought, then more heat could generate a more powerful current, possibly powerful enough to run machinery. For several years scientists in Europe, the former Soviet Union, and the United States worked to find the best combination of **alloys** to maximize the output of thermocouples. They developed such devices as an emergency **radio** that ran on the heat from a kerosene lamp. After World War II, the beginning of the atomic age, physicists experimented with thermocouples that ran on the heat from decaying isotopes. These nuclear thermocouples are used to power satellites and deep **space probes**, devices that must run unattended for many years. Such devices have become controversial, as with the 1997 launch of the *Cassini* mission to Saturn, because of the small but nonzero chance of a launch accident of a craft that is carrying radioactive material.

There is another thermoelectric effect, called the Peltier effect, that has found some usefulness in recent years. Essentially the opposite of the Seebeck effect, the Peltier effect (named after French physicist Jean-Charles-Athanase Peltier) shows that when a current is sent through a thermocouple in a certain direction the thermocouple will heat up, and it will cool off when the current is sent in the other direction. The primary application of the Peltier effect is as a refrigerant, since a sizable current can cool a thermocouple to a temperature low enough to liquify nitrogen and helium.

Thermometer

In the earliest days of the medical profession, there was no device available to measure a patient's temperature. Patients were either too hot, normal, or too cold, depending upon the doctor's personal observation methods. It was not until nearly the seventeenth century that scientists devised an instrument that could detect the changes in air temperature, and many years later that a medical thermometer was constructed.

The first thermometers were created to measure the changes in atmospheric temperature. The most famous of these was invented by **Galileo** in 1592; called an air-thermoscope (or air thermometer). It consisted of a long **glass** tube with a wide bulb at one end. The tube was heated, causing the air within to expand and some to be expelled. While still warm, the open end of the tube was placed into a flask of water; as the tube cooled the warm air would contract, drawing water into the tube as its volume decreased. Once the tube-and-water system reached a steady state, any change in air temperature would cause the level of the water within the tube to rise or fall.

While the principles behind the air-thermoscope were scientifically valid, there were two major hindrances to its acceptance. First, the varying heights and widths of the tubes made it very difficult to graduate the device—that is, to establish a degree scale. The first scientist to do so was Italian doctor Santorio Santorre, who created a scaled thermometer in 1612. Second, it was soon discovered that the air-thermoscope was unreliable, giving widely varying readings for apparently identical temperatures. Scientists puzzled over this phenomenon until the 1660s, when it was realized that an open-ended system would react to air pressure as well as temperature (in other words, the air thermometer acted as a **barometer** as well). This realization came several years after the solution, for in 1654 Duke Ferdinand II of Tuscany had constructed a sealed liquid-in-glass thermometer that was not prey to the changes in air pressure.

With this as a starting point, European scientists began to perfect the design of the thermometer. One of the issues they strove to address was the need for an instrument that could travel by sea; ordinary liquid thermometers were rendered unreliable by the ships' rocking motions. In 1695 a French physicist, Guillaume Amontons, designed a thermometer using an unshaped tube filled with compressed air and capped with a layer of mercury. As the temperature increased the air would expand, causing the mercury level to rise; as the temperature decreased, the mercury would fall. Another Frenchman, René de Réaumur (1683-1757), sought to improve upon Amontons' design by replacing the air-and-mercury system with a mixture of alcohol and water. Réaumur's thermometer was remarkable in that he devised an 80-degree temperature scale based upon the freezing and boiling points of water—the same points that, years later, would become the basis for the more widely accepted scales of Celsius and Fahrenheit.

At the turn of the eighteenth century the most discouraging issue for scientists and instrument makers was the lack of a standard temperature scale. The level of glass-blowing technology was too poor to make thermometer tubes that were exactly alike, and so every scientist's temperature scale was different. All that changed in 1717 when a Dutch instrument maker, **Daniel Fahrenheit**, introduced a line of mercury-filled thermometers of nearly identical proportions. His use of mercury in very thin tubes allowed him to graduate the scale into many degrees; using the boiling and freezing points of water as fiducial points, Fahrenheit developed the first scale to be accepted as a worldwide standard, with water's boiling point at 212° and its freezing point at 32°.

The Fahrenheit scale enjoyed global popularity for many years, until the introduction of the hundred-degree scale by **Anders Celsius** in 1746. Several scientists had attempted to popularize a hundred point scale, but Celsius was the first to also utilize water's freezing and boiling points as the zero and hundred-degree marks. Originally, Celsius placed the freezing point at 100° and the boiling point at 0°; this was reversed in 1747, at which time the centigrade (meaning "five hundred steps") scale began to rise in popularity. In 1946 the Celsius scale was adopted by most of the world as the official temperature scale.

Probably the most familiar thermometer is that found in a doctor's office, the clinical thermometer. It was invented in 1866 by Sir Thomas Clifford Allbutt, an English physician. The important features of this thermometer were that it was relatively short, usually no longer than 6 in (15 cm), and it responded quickly to the patient's temperature. Previous instruments required nearly twenty minutes to get an accurate reading, while Allbutt's thermometer could reach equilibrium in less than five minutes. This made it easier for doctors to follow the course of a fever, since temperatures could be taken more quickly and, therefore, more often.

Modern thermometers come in many different varieties. Mercury **thermostat**s are used to regulate temperatures in ovens and household air systems. Digital thermometers have been designed to measure temperature changes using electronic equipment. New thermometers are being designed that can read a patient's temperature using infrared technology; these devices can determine a person's temperature in about one minute, and can take a reading from inside the ear, rather than the mouth.

Thermonuclear weapon • See Hydrogen bomb

Thermos • See Vacuum bottle

Thermostat

For many years, heat was an insoluble mystery. The first step toward understanding the mystery came in the mid-1600s, when scientists began constructing the first **thermometers**. Once they began to realize that temperature was measureable, they determined that anything that could be measured could be controlled. So began the effort to design a device to govern the temperature of a system.

That effort culminated around 1660 with the invention of the thermostat. Named from the Greek words for ''constant temperature,'' the thermostat was invented by Dutch scientist **Cornelius Drebbel**. His device, used to regulate the temperature within a duck-and chicken-egg **incubator**, was actually an elegant combination of several different technologies. Placed above a furnace, the incubator was surrounded by a water-filled jacket. Inside the incubator was a container of alcohol that would expand as it was heated—this expansion pushed down on a U-shaped tube filled with mercury, and as one end of the U was pushed down, the other end would rise up, raising with it a rod. The rising rod would raise a lever connected to the furnace's flue, so that the airway would be partially closed and the temperature would begin to fall. When the temperature fell beyond a certain point the alcohol once again contracted, thus releasing the mercury, lowering the lever, and re-opening the furnaces airway. This type of thermostat underwent a series of modifications over the next two centuries before it found its way into industry.

Thermostats were an important part of the Industrial Revolution, and today are essential to the heating and **air conditioning** systems of homes, **automobiles**, and heavy machinery. Modern thermostats fall into two general categories: metallic and electronic. The metallic thermostat has at its heart a bimetallic strip—a thin strip of two metals fused together. The two metals have different coefficients of thermal expansion; that is, they expand or contract by different amounts as the temperature rises or falls. When a bimetallic strip is heated, it will bend to one side. In a thermostat, that bending can be used to push a **lever** or a **valve**, or to complete an electrical circuit. Many household thermostats use a bimetallic strip. The thermostat that is becoming more common is the electrical thermostat. In electronics, the resistance of most metals will increase as the temperature does. By running an electrical current through a circuit, temperature can be measured by observing the change in the circuit's resistance. This type of thermostat is particularly useful when very precise temperatures must be maintained, such as in scientific laboratories and hospitals. Modern electronics are further combined with thermostats to produce *programmable* thermostats, essentially small **microcomputers** that control the temperature. Such units keep full track of the time and date, have a digital display readout, and can be preset to change to certain temperatures, such as an automatic lowering of a house temperature at night. It can then be programmed to raise the temperature of the house a hour or so before waking, saving money overnight that would have otherwise been spent heating an essentially empty house. Such units can save between 5% and 15% of a household's energy usage.

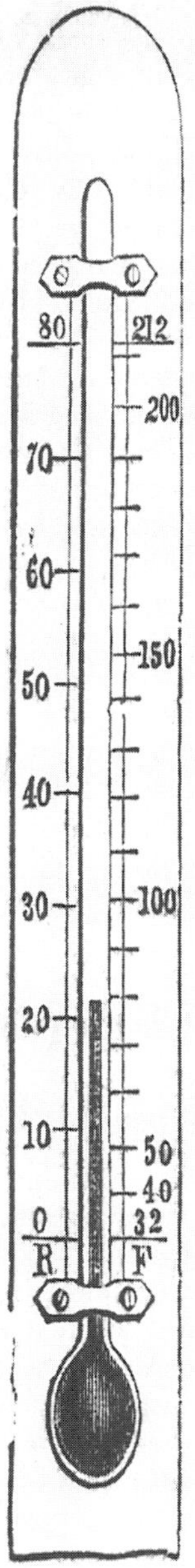

Engraving of Daniel Fahrenheit's thermometer. Introduced in 1717, his scale was the first to be accepted as a worldwide standard.

Thimonnier, Barthélemy • See Sewing machine

Thomas, Sidney Gilchrist (1850-1885)

English metallurgist

One of the major impurities often found in iron ore is phosphorus. During the 1850s, **Henry Bessemer** developed a conversion process whereby carbonic impurities could be burned away, but his method worked only in phosphorus-free ores. In 1875, Sidney Gilchrist Thomas devised a process of removing the phosphorus using the Bessemer converter. His method was based on the introduction of a basic compound such as burned limestone to the converter. The phosphorous adhered to the compound and could then be removed from the iron ore.

Thomas's livelihood hardly reflected his stature in the history of technology. For twelve years (1867-1879) he held a menial job as police court clerk, continuing even after making his discovery. He had learned of the need for eliminating phosphorous from iron ore during lectures that he attended at Birkbeck College in England. He later studied chemistry and metallurgy at the Royal School of Mines.

Thomas's cousin, Percy Carlyle Gilchrist (1851-1935), organized the initial trials of the new process at the steel works at Blaenavon, where he was employed as a chemist. The process was hence named the Thomas-Gilchrist Process. This and the precipitate fertilizer known as Thomas Slag were patented and protected.

Thomas was forced to leave his court position to keep pace with demands for his new process. His health was failing, however, and an attempt to travel to a healthier climate came too late. He died at the age of thirty-four.

See also Iron production; Steel production; Fertilizer, synthetic

Thompson, Benjamin (Count Rumford) (1753-1814)

American English inventor

Benjamin Thompson was something of a soldier-of-fortune: he was a spy for the British during the Revolutionary War; he used his position of influence in the government to take bribes, and he worked for several countries to advance the science of armaments, through which he first earned international acclaim. However, he was also a shrewd and intuitive scientist, and was almost solely responsible for the acceptance of the concept of heat as a form of motion rather than a fluid. He is better known in the annals of science as ''Count Rumford.''

Benjamin Thompson was born in Woburn, Massachusetts, in 1753. As a teen he worked as an apprentice to a Salem storekeeper. He was apparently a very poor apprentice, viewing himself as destined for greater achievements; in fact, by the age of 17 he had taught himself French, philosophy, and the sport of fencing, all in anticipation of his future position. When he was 19 he became a schoolmaster, moving to nearby Rumford (now Concord, New Hampshire). There he met and married a wealthy older widow, and they had a child.

About this time Thompson was the center of local controversy. Members of the growing anti-British movement accused him of selling secrets to the British army. Thompson soon fled Rumford, leaving his wife and child behind, and moved to England. It was later discovered that he had indeed been a spy, having always considered himself an Englishman at heart. Thompson continued his career of espionage until the end of the Revolutionary war; upon the defeat of the King's army it became clear that he could never return to the land of his birth.

He began his life of exile in the employ of King George III, eventually holding the titles of Minister of War, Minister of the Interior, and Royal Scientist to the King.

In the years shortly after the war Thompson dedicated himself to studying and improving upon the science of weaponry and, in particular, **gunpowder**. He devised a new type of **mortar** that could be used to determine the explosive potential of gunpowder. In 1781 he was elected to the Royal Society for his work with **explosives**; many years later he would establish the gunpowder standard. The same year Thompson was accepted into the exclusive Royal Society, he was banished from England for selling naval secrets to the French. However, many of his duties for the crown were top secret, and instead of an execution he was given an appointment as a diplomat to Bavaria, where he would again serve as a spy for Britain. Once he was safely in Germany, though, Thompson severed his ties with King George and entered the service of Elector Karl Theodor of Bavaria.

Bavaria in the early 1780s was beset with beggars. As one of his first duties as an administrator, Thompson was given the task of finding some use for them. He came up with the ingenious idea of developing workhouses for the poor, enabling them to earn money while providing a source of cheap labor for the government.

Thompson's workhouses began producing clothing—first army uniforms and then winter coats for civilians. The coats were of Thompson's own design, based upon research he had conducted into the properties of convection and insulation. In order to feed this new workforce, Thompson sought an inexpensive yet nutritious foodstuff; he was thus responsible for bringing the potato to continental Europe.

The Bavarian workhouses became Thompson's own little laboratory: he envisioned and designed new devices and assigned their construction to the state's laborers. In this way he was responsible for the invention of the **double boiler**, the percolating coffee pot, the range **stove**, an improved **pressure cooker**, a meat roaster, and the **thermos** bottle, all with the purpose of making food easier to prepare and store. In order to keep his workhouses well-lit Thompson invented what is now called the Rumford **oil lamp**. His experiments with illumination also led to the construction of a shadow photometer that also bears his name.

Thompson's success earned him the gratitude of the state and its citizens, and the Elector made him a Count of the Holy Roman Empire. He chose the name Rumford, after the small town in which he had met his wife.

While studying the insulating properties of certain materials, Rumford became intrigued by the nature of heat itself. During the late 1700s, the common conception was that heat was the manifestation of an invisible liquid called *caloric*. As you heated a substance, caloric flowed into it; as it cooled, caloric seeped out. (This was probably based upon the fact that metal expands when heated—apparently filling up with some invisible substance.)

One day in 1798 Rumford was observing the drilling of a **cannon**. He noticed that, as the **drill** bored into the *brass* shaft of the cannon, the metal became very hot—so hot that the workers continuously doused the casing with water to avoid being scorched. Contemporary science held that the spinning drill reduced the brass to shavings, allowing the caloric to escape. However, it seemed to Rumford that far too much caloric was being released—that such a quantity of the fluid would have melted the brass it was supposedly contained within. He began to theorize that heat was a form of motion, generated by the spinning drill bit against the cannon casing.

Rumford's theory of heat was not popular among his peers, but it began a revolution in thermodynamics that would culminate fifty years later with James Clerk Maxwell's (1831-1879) proof of the motive nature of heat. While attempting to support his theory, Rumford invented a device called a **calorimeter** that could accurately measure the amount of heat released by a substance. He later modified his calorimeter so that it could compare the heat of combustion of different fuels.

In 1799 the Elector of Bavaria died; Rumford, whose abrasive personality had won him few friends, chose this time to return to England. He stayed in London long enough to establish the Royal Institution with Sir Joseph Banks. The Institution soon attracted such brilliant young scientists as Thomas Young (1773-1829) and **Humphry Davy** as lecturers. They continued Rumford's research into the nature of heat, though it would be nearly fifty years before James Clerk Maxwell would finally put the caloric theory to rest.

Meanwhile, Rumford himself grew restless in London, and in 1804 he moved to Paris, where he would remain for the rest of his life. There he met and married a wealthy older widow who, coincidentally, had been married to Antoine Lavoisier (1743-1794), the originator of the caloric theory.

Perhaps to atone for the dishonorable actions of his youth, Rumford made arrangements in his will for his substantial wealth to be used for the advancement of science. He established the Rumford medals at the Royal Society and the Academy of Arts and Science at Boston. The remainder of his fortune he gave to Harvard University for the creation of a Rumford professorship. All of these accolades exist today, and are among the most prestigious in their fields.

Benjamin Thompson (Count Rumford)

THOMPSON, JOHN TALIAFERRO • See Submachine gun

THOMSON, ELIHU (1853-1937)
American electrical engineer

Scarcely known today, Elihu Thomson was arguably the most important early contributor to the development of electricity as a power and light source. An outstanding scientist as well as a brilliant engineer, Thomson applied his insights in theoretical science to practical devices in a way that few others have.

Thomson was born in Manchester, England, on March 29, 1853, and his family moved to Philadelphia five years later. The son of an engineer and machinist, Thomson was tinkering and inventing before he had even reached his teens. Keenly interested in electricity, Thomson constructed a crude **electrostatic device** from a wine bottle at the age of eleven, as well as experimented with telegraphy, **electromagnets**, and other electrical devices.

Shortly after graduating from Central High School, Thomson became a faculty member there, and began working with another teacher, Edwin J. Houston. Appointed professor of chemistry and mechanics in 1876, Thomson spent his time lecturing on electricity at the Franklin Institute, building **lenses** and optical devices, and constructing new electrical motors. It was around this time that Thomson made what was almost certainly the earliest demonstration of the existence and behavior of radio waves. Just a few years before, James Clerk Maxwell had established a theory that finally showed the relationship of electricity to magnetism. One of the predictions of Maxwell's theory was that the oscillation of an electric charge would produce an electromagnetic pulse, composed of long waves, that traveled out from its source. Thomson generated such waves, passing them through brick walls and floors, ten years before Heinrich Hertz's (1857-1894) independent discovery of radio waves would lead to the invention of the **radio** receiver. Apparently not seeing the potential of his radio experiments (the kind of blunder he seldom made), the prolific Thomson moved on, more interested in projects to make electrical power generation and lighting practical.

Toward this end, Thomson collaborated with Houston on an improved arc lighting system. Thomson constructed a number of devices, and with the help of financial backers, Thomson and Houston opened a factory in 1879. Their successful arc lighting predated Thomas Edison's **incandescent light bulb** and launched Thomson on a commercial career. Resigning his position as professor, Thomson began to devote all of his time to his new enterprise.

Before long, Thomson expanded their lighting company, known as Thomson-Houston Electric, into the manufacture of incandescent lights, motors, dynamos, and even electric trains. In 1892 the Thomson-Houston Electric Company merged with the Edison General Electric Company to become General Electric, becoming the largest manufacturer of electrical devices in the world. Thomson stayed with the company as an engineer and consultant for the rest of his life, while pouring out invention after invention. By the time he died in 1937 he had accumulated a remarkable 700 patents.

Thomson's impressive list of inventions includes the watt-hour meter (used to measure electricity consumption), the first method of **welding** metals with electricity, a high-frequency **generator**, a type of objective lens for a refracting **telescope**, a constant-current **transformer**, an electrically-operated pipe organ, a **cream separator**, a type of **lighting rod**, and an **alternating current** repulsion motor which helped introduce the use of AC as a source of power.

After x-rays were discovered in 1895, Thomson worked on improving the design of x-ray devices, devising a means of producing stereoscopic x-ray photographs. He is even credited with first suggesting that, in order to avoid suffering the bends, divers working at great depths (or workers deep in **caisson**s) should use a helium-oxygen air supply rather than one of nitrogen-oxygen.

Thomson, Robert William (1822-1873)

Scottish engineer

Born at Stonehaven, Kincardineshire, Scotland, in 1822, Robert Thomson was sent to Charleston, South Carolina, in 1836 to train as a merchant. Thomson's interests lay elsewhere, however, and when he returned to Scotland he began a program of self-education in engineering and science. During a stay in London in 1841, **Michael Faraday** encouraged young Thomson, who worked in blasting operations and as a railway engineer.

The problem of cushioning horse-drawn carriages from shock and vibration began to interest Thomson. In 1845 he patented his improvement over the steel tires of the day: a rubber tube filled with air inside a **leather** protective cover. This was the world's first **pneumatic tire**. A set of Thomson's "Aerial Wheels" was fitted to a carriage and tested very successfully, running 1,200 miles (1,930 km). Thomson's tires did not catch on, however, and the idea died until **John Boyd Dunlop** reinvented pneumatics in 1888.

Thomson went on to design and invent many other items. He patented a fountain **pen** in 1849. He greatly improved sugar-manufacturing machinery during his years in Java from 1852 to 1862, after which he resettled in Edinburgh, Scotland. In 1865 he patented improvements to steam boilers and in 1866 to steam gauges. From 1867 through 1873, Thomson was granted various patents for a very successful traction-engined road-steamer. He died in Edinburgh in 1873.

Thomson, William [Lord Kelvin] (1824-1907)

Irish English scientist

William (Lord Kelvin) Thomson is recognized as the premier scientific mind of the nineteenth century, and perhaps the greatest thinker since Isaac Newton (1642-1727). He originated new schools of thought in physics, thermodynamics, **electronics**, and mathematics. He was knighted in 1866 for his work in salvaging the first **telegraph** cable to span the Atlantic, became wealthy enough to own a 126-ton yacht and an estate, and in 1892 was made Baron Kelvin of Largs.

Thomson was born in Belfast, Ireland, in 1824. The son of a respected mathematician, he was quickly recognized as a child prodigy. When his father was given a professorship at the University of Glasgow, William (then a mere eight years old) would attend lectures. At age 10 he entered the University of Glasgow (as did his 11-year-old brother, James), finishing second in his class. While still in his teens he pursued his graduate degrees at Cambridge and Paris—his first published paper appeared when he was 16 years old. In 1846, the 22-year-old Thomson was appointed (with some help from his father) Professor of Natural Philosophy at the University of Glasgow. There, he was responsible for the construction of Britain's first physics laboratory.

It did not take long for the young professor to stir up the European scientific community. Less than a year after his appointment, Thomson announced his findings regarding the age of Earth. Beginning with the assumption that Sun and Earth were once the same temperature, he determined that the time it would take for our planet to cool to its present temperature was somewhere between 20 million and 400 million years, and probably very close to 100 million years. This number was much lower than those previously determined by geologists (through by completely different means). Biologists in particular were dismayed, for it meant that all of the evolution of life must now fit within the span of a million years. They began to consider the possibility of evolutionary "leaps"—shortcuts that would dramatically decrease the amount of time needed for life to evolve. Though Thomson's figures were later shown to be incorrect, the furor among biologists led to the theories of evolutionary mutation.

During his studies of the age of our planet, Thomson became intrigued by the relationship between heat and energy. In 1847 he met James Joule (1818-1889), the author of some of the most innovative heat theories of the time; at that juncture, however, Joule's work was relatively unknown, particularly in England. Using his tremendous influence Thomson introduced Joule's theories to the Royal Society, giving Joule the recognition he deserved. During the next few years Thomson and Joule worked together, experimenting with the heat and energy of certain gases. One phenomenon they observed was that as a gas was introduced into a vacuum its temperature would drop, and if that drop were enough the gas could be converted to a liquid. Called the Joule-Thomson effect, this phenomenon became the basis for the liquification of most gases and, much later, the science of cryogenics.

Thomson's work with Joule continued to feed his curiosity about the nature of heat. He was especially interested in the work of the French physicist Jacques Charles (1746-1823), who had found that, for every degree celsius below zero a gas was cooled, the volume would decrease by a factor 1/273.16. The implication of this theory was that, at -273.16° C; the volume of the gas would be zero. Scientists were unable to explain exactly why this would happen, just as they were unable to prove Charles's law false. In 1848 Thomson explained the effect in this way: when the temperature of a gas is reduced, so is the energy level of the atoms; as the atoms move less they take up less room, therefore decreasing the volume. At -273.16° C the energy level of the atoms reaches zero—they stop moving, take up almost no space, and their temperature cannot be lowered any further. Because this theory should be true for any substance, Thomson called -273.16° C the absolute zero of temperature.

The most famous application of this idea was Thomson's 1848 invention of an absolute scale of temperature. This scale essentially drops the centigrade scale by 273.16° C, so that zero and absolute zero coincide. Thomson called his scale the absolute scale, but after his death it was renamed the Kelvin scale. Zero kelvin (O K) is the lowest possible temperature, or absolute zero. In celsius it equals -273.16° C, and in fahrenheit it equals -459.69° F.

The absolute scale was much easier for scientists to use than centigrade, since it directly took into account energy differences. James Clerk Maxwell utilized this scale in his formulation of his kinetic gas theory. In 1851 Thomson used the scale to show that all energy is eventually converted into unusable heat, and that heat is dissipated into the atmosphere. This concept of degradation was later explored by Rudolf Clausius and was reintroduced in a much clearer and more explicit form as the second law of thermodynamics, stated using the concept of entropy.

During the late 1800s much of the British scientific community was busy working on the first transatlantic cable. Thomson lent his immense knowledge of electrical theory to this effort, inventing a number of ultra-sensitive **galvanometers**. His theory was that only very low voltages could transmit the telegraph signals at a sufficient rate over such a long cable. This clashed with the views of E. O. W. Whitehouse, an electrician who had been placed in charge of the project. Using a system of his own design, Whitehouse completed construction of a high-voltage telegraph cable. When tested in 1856, the cable was a complete failure. After several years of litigation, the transmitters were replaced with Thomson's low-voltage system, and beginning in 1865 the underwater cable provided instant communication across the Atlantic. Thomson was recognized as the man who had rescued a giant financial investment, and was knighted for his work.

During the 1880s Thomson lectured on the virtues of Victorian science, a school of thought that believed that all of the important discoveries in physics had been completed. So committed was he to this idea that he completely rejected the theories of radioactivity, ignoring and thus missing the onset of the next great scientific age. He died in 1907 and was buried next to Isaac Newton in Westminster Abbey.

William Thomson (Lord Kelvin)

See also Telegraph cable, undersea

3-D MOTION PICTURE

As early as 1894 attempts were made to enhance cinema with 3-D special effects. It was not until the early 1900s, however, after Edwin S. Porter (1870-1941) and W. E. Waddell developed the anaglyphic process that 3-D cinema became a reality. When twin images (one red, the other green) were superimposed and viewed with special red and green glasses, they appeared as one 3-D image.

In 1915 the first commercial 3-D film was shown publicly in New York and consisted of three black-and-white shorts. Movie makers have flirted with 3-D sporadically ever since.

After the development of the *Polaroid* system of filming by **Edwin Land**, the first color movie with sound to use 3-D was released in Germany in 1937. Fifteen years later producer Arch Oboler used the Polaroid system to shoot *Bwana Devil*, the first feature film to employ 3-D effects. Oboler employed two cameras whose reels had to be specially synchronized. To visualize the 3-D effect, spectators had to wear Polaroid glasses. Polaroid's improved process sparked a short-lived 3-D infatuation. Over one hundred films were produced in this way.

In 1966, Oboler shot *The Bubble* with the use of a special **lens**, the polarisator. The invention of the polarisator did away with the inconveniences of the double projection method, but the shaky camera movements and other problems with how the film looked remained.

In the 1980s, the Canadian Film Board and ISTEC, a private manufacturer, worked together to invent the 3-D Rig, a camera set-up which steadied the movements. By 1990, the IMAX Corporation, whose films are usually documentaries shot on special film that is larger and sharper than conventional 35mm filmstock and shown on huge screens, became interested in combining IMAX and 3-D technologies. They invented a new camera which had two lens and two film movement mechanisms inside the film was synced internally. IMAX produced its first IMAX 3-D short feature film in 1994, *Wings of Courage*, and planned to do others as well as a full-length feature film. Despite these kinds of innovations, many viewers dislike wearing the special glasses needed, and 3-D films seems destined to remain novelties.

See also Motion picture

Threshing machine

Threshing, the process whereby ripened grain is removed from the chaff—the husks and straw—is done after grain has been harvested. The earliest method of threshing was to pound the stalks in order to loosen the grains. Grain could be flailed on barn floors with flails made of wooden bars, or swingles, fastened loosely to a handle. The grain was manually beaten loose. An early labor-saving threshing device was the Egyptian charantz, a sled-like device that threshed the grain with spiked cylinders that turned as it was pulled. The Hebrews had a similar device called a moreg. The early Japanese used flails equipped with iron or wooded teeth that stripped the grain from the stalk.

Another early threshing method was treading. Horses, and sometimes oxen, were forced to walk through the grain. The force of their weight loosened the grain. Although a great deal of grain was missed this way, the increased volume in threshed grain justified the method.

Michael Menzies, a Scot, invented one of the first mechanical threshers in 1732. It consisted of flails attached to a hydraulically operated **wheel** and delivered 1,320 strokes per minute. Since the grain had to be brought in from the fields to the stationary machine, and since the flails were prone to breakage, the machine was not successful.

Another Scot, Andrew Meikle (1719-1811), invented a successful thresher in 1786 that consisted of a rotating drum with four vertical blades and a stationary shield through which the grain was fed. A fan blew the chaff away from the heavier kernels of grain, a process called winnowing.

Around 1816 Robert McCormick, father of **Cyrus McCormick**, inventor of the **reaper**, experimented with threshers and other farm machines, but was not successful. In 1828 Samuel Lane of Maine invented a portable thresher that could also harvest—the first **combine harvester**. Hiram A. Pitts and John A. Pitts of Winthrop, Massachusetts, developed a threshing machine in 1837 that had an ''endless belt'' with pins that threshed and cleaned the grain. It was efficient and inexpensive to operate.

By the mid 1800s steam power was adopted for threshers and other farm machinery, and methods for stacking and bagging the straw were added to the process. Other functions, like binding and weighing, were in operation by the early 1900s. Because of the popularity of the combine harvester, beginning in the 1940s, the use of threshers as a separate operation has declined greatly during the twentieth century.

See also Tractor

Audience wearing special polarizing glasses to view a 3-D motion picture.

THROTTLE VALVE • See Carburetor

TILE • See Ceramic

TIME ZONES

As Earth rotates on its axis from west to east, night and day arrive at different parts of the earth at different times. Until the late 1800s, every town, county, or isolated group of islands observed its own time and set **clocks** according to the local sunrise and sunset. Time differences between locations were practically unnoticeable, however, because it took days, weeks, or months to travel, and instant modes of communication did not exist.

Demand for a unified time system evolved as a result of two technological advances: the **telegraph** and the **locomotive**. In the 1830s, the telegraph made possible instantaneous communication between distant points, and the first locomotives developed in England and America made rapid travel possible. With the development of telegraph and railroad networks on a continental scale in North America, Europe, and elsewhere, local times came into conflict for it was nearly impossible to create schedules or to relay messages effectively. Railroads wrote timetables and set station clocks accordingly; yet, these schedules were not coordinated among the railroad companies or among political areas.

American meteorologist **Cleveland Abbe**, who helped found the United States Weather Bureau in 1870, pioneered a system of weather reporting and forecasting using the telegraph to collect and disseminate information. As accurate timekeeping was important to accurate weather forecasting, Abbe persuaded North American railroads to adopt time zones by 1883.

The following year an international system of 24 time zones was adopted. The line of 0° longitude, which runs through the Greenwich Observatory at London, England, is the prime, or starting, *meridian*. The zones extending east from Greenwich increase one hour each for a total of 12 hours. The zones extending west decrease by a total of 12 hours. The 24th zone is divided by the International Date Line (IDL), at 180° longitude. The time difference on either side of the line is 24 hours, one day greater west of the line than east of it.

Major deviations from the ideal zone system exist. For instance, Newfoundland and the Cook Islands are located in half-hour zones. In the former Soviet Union, all time zones are one hour greater than the zones adopted in 1884. Time zone boundaries, especially in populated areas, are deliberately defined along political boundaries or physical features. Even the IDL, which runs down the center of the pacific Ocean, has been flexed through the Bering Strait so that Siberia and Alaska are entirely within their own time units.

TIRE, PNEUMATIC

The pneumatic tire was invented in 1845 by the Scotsman **Robert Thomson**. Vehicles of that time had wooden **wheel**s with steel tires, which wore well but produced a jolting, vibrating ride were subject to skidding. Thomson designed a great improvement, a non-stretchable cover over a rubber inner tube pumped up with air. Although Thomson's ''Aerial Wheels'' did give a much smoother ride, they attracted little interest and were soon forgotten.

Solid rubber tires began replacing steel tires in 1870 in England, but they didn't solve the vibration problem. As the **bicycle** craze grew in the 1880s, the need to improve the ride of these ''boneshakers'' became more pressing. A Scottish veterinarian practicing in Ireland, **John Boyd Dunlop**, found the solution for his son's tricycle. He put together a rubber tube with a one-way **valve** for inflating, covered it with a rubber casing, and attached it to the trike's rear wheels with wrappings of tape. Dunlop patented his new pneumatic tire in 1888, unaware of Thomson's similar patent design of 1845.

These fat ''mummy'' tires were ridiculed at first, but they rapidly gained favor with racing cyclists, who appreciated both the smoother ride and the easier pedaling pneumatics produced. Dunlop-type tires soon became standard for bicycles. In 1890 Charles K. Welch of Tottenham patented an improved method of fastening the tire to the rim using **wire** embedded in the casing, and William Erskine Bartlett, an American expatriate living in England, invented the bead on the edge of the casing. The French brothers, Edouard and André Michelin, introduced the removable pneumatic tire in 1891, which allowed the driver rather than a mechanic to fix a blowout quickly and efficiently.

The greatest application and importance of the pneumatic tire was for something Dunlop had not envisioned. It made possible the **automobile** age—cars couldn't run successfully without pneumatic tires. The Michelins surprised competitors by using pneumatics on an automobile for the first time in 1895, in the great Paris to Bordeaux race. Later that year pneumatics made their first appearance on an American car in the Chicago-Evanston race. The United States tire industry soon established itself at Akron, Ohio: Benjamin F. Goodrich (1841-1888) had moved there in 1870 and made its first pneumatic tires in 1896; Goodyear Tire Company was incorporated there in 1898; Harvey Firestone (1868-1938) started tire manufacturing in Akron in 1903.

Two-part high-pressure pneumatic tires appeared in the early 1900s, being a casing with a flexible rubber inner tube inside. Low-pressure balloon tires were introduced in 1922, pioneered by Firestone. The tubeless tire with an airtight casing came in 1948 and by the mid-1950s became the industry standard. Recently, tires have been developed that seal themselves and can continue to run when punctured.

For many years, bias-ply tires were the standard type of pneumatic tire. In a bias-ply tire, fabric cords in the inner lining of the tire are laid at angles to the wheel axle and run in layers over each other. In radial-ply tires, the cords run from one bead to another, parallel to the axle. Conventional bias-ply tires have fabric cords; belted bias-ply tires have **fiberglass** belts that circle the tire, over the cords; belted radial-ply tires have steel mesh belts. A belted radial ply tire design was patented in 1914 by Christain H. Gray and Thomas Sloper of England, but the tire was not marketed. The Michelin company applied for the patent on today's modern radial-ply belted tire in 1946; the tires themselves went into production early in 1948. They soon became standard in Europe, but were not commonly used in the United States until the 1970s.

Radial tires proved to give roughly 20% less rolling resistance (or friction) than bias-ply tires, and therefore provided a significant reduction in fuel consumption. In the United States from 1970 to 1992, radial tires were responsible for saving approximately 77 billion gallons (293 billlon L) of gasoline. Once radials were generally adopted, tire engineers in the 1980s concentrated on further decreasing their rolling resistance by eliminating any excess mass in the shape of the tire itself. Tires of the 1980s thus had a slimmer look that replaced the wide-shouldered ''macho'' tires of the 1970s. In the 1990s, engineers concentrated on increasing a tire's efficiency, and eventually turned their attention to composite materials.

Tires have long been made out of a combination of many different natural and synthetic materials, one of which is called carbon black. This necessary oil by-product stiffens or hardens a tire's tread, but it also creates unnecessarily high friction. In the late 1980s, engineers began experimenting with different materials and tried silica—the basic component of ordinary beach sand—to replace carbon black. It was found to reduce friction remarkably well. Eventually, once the problems of lower durability, decreased traction in wet weather, high cost, and complicated manufacture were solved by the Michelin tire company, it began to produce a new silica-blend tire by the end of the 1990s. These new tires reduce friction with no sacrifice of any other factor, and are only slightly more expensive. It is estimated that once America converts to silica-blend tires, 2.4 billion (9 billion L) gallons of fuel will be saved each year, with a related reduction of carbon dioxide emissions.

See also Goodyear, Charles

Tiselius tube

In order for the science of biochemistry to advance significantly, methods needed to be developed to separate individual molecules so their characteristics could be known. This was made possible by the invention of the Tiselius tube.

A brilliant Swedish physical biochemist, Arne Tiselius (1902-1971), turned his attention to this problem at the beginning of his career in the 1920s, and made many contributions to the field throughout his research career. Born in Stockholm, Sweden, in 1902, Tiselius earned his M.A. in chemistry, physics, and mathematics from the University of Uppsala in 1924. The following year he became an assistant in the research laboratory of the eminent physical chemist Theodor Svedberg (1884-1971) in Uppsala and joined the university faculty in 1930 after receiving his doctorate. He later became director of Uppsala's Institute of Biochemistry.

When Tiselius joined Svedberg's lab, Svedberg was developing the high-speed **centrifuge** in order to study the size and shape of protein molecules. Svedberg assigned Tiselius the project of using electrophoresis to further study the nature of proteins. Electrophoresis is a method of separating protein molecules in a mixture influenced by an electric field. Molecules of different sizes and electric charges move at different rates and in different directions within the field, separating out from each other. Tiselius was able to separate blood plasma proteins using electrophoresis; this work was the basis of his doctoral dissertation of 1930.

However, discouraged by the limitations of electrophoresis—it failed to separate molecules in many substances—Tiselius turned to other studies. Contact with prominent chemists while studying at Princeton University's Frick Chemical Laboratory in 1934-35 refocused Tiselius's interest in improving electrophoresis methods. Back in Uppsala, Tiselius redesigned his electrophoresis apparatus and method. He performed his experiments at 104°F (40°C), thereby eliminating variations in temperature and solution thickness which had seriously distorted molecule movement. He devised a rectangular U-shaped tube with portions carefully fitted together that could be divided to yield samples of individual proteins from separated parts of a solution. Tiselius also used special refractive optics that showed the wavelike bands of different proteins as they separated.

Using his improved electrophoresis apparatus, Tiselius was able to separate four protein bands in horse serum, making it possible for the first time to study the different elements of blood serum. He and his colleagues went on to separate and purify many more biologically and medically important proteins. Tiselius continued to improve electrophoresis methods and also developed techniques for adsorption **chromatography**, another approach to molecular separation. For these important contributions to the development of biochemistry, Tiselius was awarded the Nobel Prize for chemistry in 1948.

TNT (trinitrotoluene)

Trinitrotoluene (TNT) is a substance that traces its roots to the **dye** industry and later research by German chemist **Adolf von Baeyer**. It is the most powerful nonatomic military explosive of the twentieth century. Sources credit J. Wilbrand with its initial discovery in 1863. Although probably first employed in the Russo-Japanese War of 1904-05, TNT was not mass-produced or regularly used until its refinement by the Germans during World War I. Fired by long-range guns, TNT shells encased in steel exploded with a force of 2,250,000 pounds per square inch.

In both world wars, new forms of TNT were introduced. These included TNT in combination with such similarly volatile compounds as TNX, PETN, and RDX. One mixture, RDX-TNT, or cyclonite, with a detonation pressure of 4,000,000 pounds per square inch, is regarded as the most powerful of this new class of weaponry. It is especially forceful when combined with aluminum in the form of torpex. TNT itself is composed of nitrogen, hydrogen, carbon, oxygen. In its most rudimentary form, achieved by nitrating (adding nitric and sulfuric acid to) the colorless petroleum-based liquid toluene, it appears as pale yellow or brownish crystals. Despite its violent potential when detonated, it is extremely safe to cast into shells and handle and is thus a preferred high explosive.

See also Abel, Frederick A.; Cordite; Dynamite; Guncotton; Machine gun; Nitroglycerine; Pistol; Revolver; Rifle

Toaster

The rapid automation of the American kitchen during the twentieth century is due to a greater knowledge and manipulation of energy sources such as gas and electricity. The electric toaster is one of the most common appliances in American households.

While a few prototypes appeared before 1900, the first models marketed to the public were those invented by the General Electric Company of Schenectady, New York, in 1909. These early toasters were simply strips of bare **wire** wound around mica strips. The wire was heated by electrical current, and the mica strips spread the heat evenly over the bread.

In 1927, a mechanic from Stillwater, Minnesota, invented a toaster that could heat both sides of the bread. He also installed a clock-like mechanism and spring that would cut off the toaster's power and release the toasted bread. In the early 1930s further enhancements were made by adding a **thermostat** that would shut off the toaster's power when the bread reached a certain temperature. The spring mechanism of the toaster was perfected in 1932, and the first pop-up toaster began to appear in kitchens across America. Toasters are now available in under-the-counter models, and in models that have wider slots for bagels or other larger rolls.

Tobacco products

When Christopher Columbus (1451-1506) reached America, he found that the Native Americans who had occupied the land for thousands of years were already using tobacco in much the same ways it is used today. The tobacco plant is native to America, primarily to the Caribbean area. The Indians believed that tobacco had medicinal value and passed this idea on to European colonists. The natives also used tobacco in official ceremonies in which the traditional peace pipe was shared by all. They mainly used wild tobacco, though some groups did cultivate it.

The colonists pursued tobacco cultivation as an industry. The first known instance of colonial tobacco growing was in 1531 at Santo Domingo. The French ambassador to Portugal, Jean Nicot (c. 1530-1600) first introduced tobacco to Europe in 1556. The tobacco genus Nicotiana was named for him, and the term nicotine was derived from his last name. All of Europe had tobacco by 1600. Early tobacco items in use among the Europeans and colonists included rolled cigars, pipes, chewing tobacco, and snuff. Snuff consists of finely grated tobacco scented with jasmine, cloves, or other fragrances. It is sniffed or dipped—applied between the lip and gums. By 1600, tobacco was popular in England, and its popularity increased in Europe generally during the seventeenth and eighteenth centuries. Elaborate pipes and finely ornamented snuffboxes became the fashion rage.

By the 1630s, tobacco was being grown in colonies from Brazil to Virginia and eventually became the most important colonial commodity traded with Europe. After tobacco is harvested, it must be carefully dried. This process is called curing. At first, tobacco was air-cured in barns, which took several weeks. Smoke-curing was adopted later. This method was faster, but the smoke affected the tobacco's flavor. To overcome this, charcoal-curing was developed in 1825. In 1839 Stephen Slade, the slave of a prosperous North Carolina tobacco farmer, accidentally discovered that tobacco turned bright yellow and became more flavorful when cured slowly over hot coals. The widespread adoption of Slade's method greatly increased the demand for tobacco and its value as a cash crop.

Development of a light form of tobacco called White Burley from green-deficient tobacco plants began in Ohio in 1864. It was used in the manufacture of the American tobacco-blended cigarette. Paper-wrapped tobacco was first used in Seville, Spain, in the 1600s and gradually spread to all parts of Europe. It had been considered offensive to be offered a cigarette when a fine cigar or snuff was so much more preferable, but widespread acceptance of the cigarette began in America during the economic crisis of 1873 and with the introduction of White Burley. Tobacco is produced in North America in Virginia, North Carolina, Maryland, Kentucky, Tennessee, Ohio, Missouri, and southern Ontario. It is grown and processed worldwide in temperate and tropical regions. Mechanical harvesting of tobacco was not introduced until the 1970s. As with cotton, human labor was plentiful for tobacco harvesting and selecting high quality leaves requires human judgement.

Extensive research into the health risks of tobacco use has inspired extensive anti-smoking campaigns. Contrary to early beliefs concerning tobacco's healthful benefits, its use causes cancer, emphysema, heart disease, and painful, premature death, not only for the user, but for those exposed to secondhand tobacco smoke. Despite the ominous warnings, tobacco use has increased in some segments of society, particularly among women and teenagers. Regional economies and livelihoods depend on the tobacco industry, while users defend their right of choice. However, a number of recent court cases, and efforts to impose government regulation on tobacco companies has resulted in massive negative publicity for tobacco companies, and in some government imposed fines on tobacco companies.

See also Farming, mass production

Toilet

The earliest known flush toilet dates from 1800 B.C. Featuring a wooden seat, it was installed in the royal palace at Knossos in ancient Crete and used a drainage system with venting air shafts. Water-flushed latrines were in use in the Indus Valley circa 2500-1500 B.C., and by the fourth century A.D. the Romans used them as well. The flush toilet then disappeared for many centuries, and chamber pots, emptied into public streets, became the norm in Europe during the Middle Ages.

The concept of the flush toilet was finally revived around 1590 by Queen Elizabeth's godson Sir John Harington (1561-1612). He designed and had installed in his home a water closet that featured an overhead water tank with a valve that released the water on demand. Although the queen had one of her godson's inventions installed in her palace, the water closet did not catch on. Drainage and venting problems made these early toilets smelly and unsanitary, and plumbing and **sewage systems** were primitive or nonexistent. Amazingly, even the lavish French palace of Versailles, built in 1661, had no toilets; residents and guests were burdened with the indignity of relieving themselves outdoors among the statuary and shrubbery.

Water closet design languished until English watchmaker Alexander Cummings patented a version with an improved **valve** in 1775. British cabinetmaker **Joseph Bramah** patented a design with an even better hinged valve in 1778. Toilets then began to come into common use in England. A major problem remained, however: disposal of their contents. Water closets typically drained into cesspools, which, once filled, invariably resulted in leaking pipes that fouled the surrounding soil. Drainage into sewer systems usually resulted in discharge into a nearby river, polluting the drinking water supply.

These problems were alleviated with the invention of the septic system in the mid-1800s and the development of predischarge sewage treatment. London had a modern sewage system by the 1860s. That year, Bramah's water closet design was improved by London plumber Thomas Crapper, who added an automatic flush shutoff, his last name becoming a fixture in English slang.

Flush toilets only came into use in the United States after 1870, and then slowly, because so many homes lacked running water. The outhouse or privy, backed up by the chamber pot, remained standard in both rural America and tenements well into the twentieth century. Newly popular today are low flush toilets which use less water and a strong vacuum system as an environmentally sound process. Early water saver toilets did not work well, although recent ones operate well. Most campgrounds and many camper vehicles use composting toilets that process human waste naturally over time.

Gayetty's Medicated Paper was the first modern—that is, soft—toilet paper, introduced by New Yorker Joseph C. Gayetty in 1857. Toilet paper in rolls was the contribution of Philadelphia brothers E. Irvin and Clarence Scott in 1879.

Tokamak

Scientists have known for approximately seven decades that nuclear fusion reactions are a potentially important source of energy. A fusion reaction is one in which two small nuclei, such as protons or deuterons, combine with each other to form one large nucleus. In the fusion process, large amounts of energy are released.

A 1747 engraving of an Englishman accepting tobacco pipes from a Native American woman.

Advertisement for Dr. Scott's Electric Toothbrush.

The primary obstacle to achieving fusion is the fact that nuclei are all positively charged and, therefore, repel each other strongly. In order to overcome this force of repulsion, the nuclei must be given very large amounts of kinetic energy. In practical terms, that means heating the nuclei to temperatures in excess of 20,000,000 K. Such conditions exist within stars, where fusion is the primary source of energy, but are unknown in any natural condition on the earth.

The first instance in which scientists were able to initiate fusion reactions artificially was the hydrogen (fusion) bomb. The **hydrogen bomb** is a device that is ignited by the explosion of an atom (fission) bomb. The **atom bomb** briefly produces temperatures of 20,000,000-40,000,000 K or more, sufficient to initiate fusion reactions in hydrogen isotopes that surround the atom bomb.

The development of controlled nuclear fusion has been a far more challenging task. The problem has been to find a way to contain the very hot gasses required for fusion to occur. Obviously, no conventional construction material can survive temperatures of a few million degrees. The most successful method discovered so far is to confine the hot gases within a magnetic field.

A critical problem with this approach is to design a magnetic field with exactly the correct geometric shape. Simply surrounding the reacting gases with a magnetic field is not sufficient since they tend to leak out of the field at its extremities. Designing a controlled fusion reaction becomes a problem in geometry, therefore, as much as it is a problem in chemistry and physics.

The effectiveness of any design is determined by its ability to confine a large number of particles at a high enough temperature for a sufficiently long period of time. The method has to be able to result in the release of more energy than the large quantities used to initiate the reaction.

Perhaps the most promising technique yet developed is called the *tokamak*, developed largely as the result of research by the Russian physicist, Lev Artsimovich (1909-1973), in the 1950s. The name *tokamak* is an acronym for "toroid camera with magnetic field." In a tokamak, nuclei are trapped in the middle of a magnetic field that has the shape of a *torus*, a hollow, doughnut-shaped figure. The torus prevents particles from escaping from the field of reaction, turning them back onto themselves.

At the high temperatures required for fusion reactions, neutral atoms are completely broken apart into positively-charged nuclei and negative charged electrons. The swarm of nuclei and electrons is known as plasma, a state of matter with properties different from those of solids, liquids, and gasses. Some of these special properties of plasma present additional technical problems that must be solved in containing the hot material.

Artsimovich, the tokamak's inventor, was born in Moscow on February 25, 1909. He graduated from the Belorussian State University in Minsk at the age of 19 and took a position at the Leningrad (now St. Petersburg) Physical-Technical Institute. Some of Artsimovich's earliest research dealt with the properties of the recently discovered neutron. He made critical discoveries concerning the ability of nuclei to capture slow-moving neutrons. His later work dealt with methods for accelerating electrons in **particle accelerator**s and for the electromagnetic separation of isotopes.

Artsimovich began teaching at the Leningrad Polytechnical Institute in 1930 and, later, at Leningrad University also. He received a number of honors and awards from the Soviet Union and international scientific organizations. He died in Moscow on March 1, 1973.

TOMMY GUN • See Submachine gun

TOMOGRAPHY • See Computerized axial tomography

TOOTHBRUSH AND TOOTHPASTE

The earliest toothbrushes were simply small sticks mashed and frayed at one end to increase their cleaning surface. Ancient Roman patricians employed special slaves to clean their teeth, and toothbrushing formed part of some ancient religious rituals.

The bristle brush was probably invented by the Chinese; it came to Europe during the seventeenth century and soon was widely used. French dentists, who were the most advanced in Europe at the time, advocated the use of toothbrushes in the seventeenth and early eighteenth centuries. In pre-

Revolutionary America, dentists urged the use bristle toothbrushes. **Nylon** has replaced natural bristles in modern brushes. Hard bristles, once recommended, are now thought to be too abrasive, and soft nylon bristles with rounded ends are preferred. Special brushes have also been designed to remove plaque and debris from relatively large spaces between teeth, or spaces between caps and bridges.

Dr. Scott's Electric Toothbrush was marketed in 1880; its manufacturer claimed the brush was "permanently charged with electro-magnetic current." The first real electric toothbrush was developed in Switzerland after World War II. This corded model was introduced to the United States market in 1960 by Squibb under the name Broxodent. General Electric followed in 1961 with its rechargeable cordless model. Although it seemed like an odd idea to many people, the electric toothbrush was an immediate success.

Like toothbrushes, compounds for cleaning teeth (and freshening breath) have been used since ancient times. Early Egyptian, Chinese, Greek, and Roman writings describe numerous mixtures for both pastes and powders. The more palatable ingredients included powdered fruit, burnt shells, talc, honey, ground shells, and dried flowers. The less appetizing ingredients included mice, the head of a hare, lizard livers, and urine. Powder and paste formulas continued to proliferate through the Middle Ages. Unfortunately, many of these recipes used agents that corroded or abraded the non-replaceable tooth enamel.

Modern toothpastes began to appear in the 1800s. Peabody suggested adding soap to tooth cleaners in 1824, chalk was popularized by John Harris in the 1850s, and soon the well-known S. S. White Company introduced a paste in a collapsible tube. Dr. Washington W. Sheffield, a Connecticut dentist, put his popular Dr. Sheffield's Creme Dentifrice, in its collapsible tube, on the market in 1892. The toothpaste tube reigned supreme until 1984, when the pump dispenser—which originated in Europe—was introduced to the U.S. market.

Fluoride was added to toothpaste in 1956 when Proctor & Gamble's launched its Crest product. Home whitening solutions, which utilize hydrogen peroxide, are available to remove discoloring stains from tooth enamel. In 1996, the federal government approved the use of **lasers** for teeth whitening. This procedure uses laser energy to "excite" the hydrogen peroxide, causing it to expel free radical oxygen which "attacks" the organic composition of the stain. Still controversial because of its expense, and the lack of knowledge concerning long-term effect on tooth enamel, this procedure is fast and even has some success at lightening teeth discolored by antibiotic use during pregnancy.

See also Fluoride treatment

TOOTH-EXTRACTION DEVICES

In primitive societies teeth were extracted with a chisel-shaped piece of wood held against the tooth and pounded with a mallet. Early Chinese tooth-pullers used their fingers, strengthening them for the task by spending hours pulling nails out of planks; ancient Greeks used double-lever **forceps** around 300 B.C. Romans used forceps of various designs, including a thin-root forceps, and pliers to remove small pieces; and Abulcasis (963-1013), an Arab surgeon from Spain, illustrated a number of dental extraction devices in his eleventh-century *Treatise on Medicine and Surgery*, including elevator chairs, forceps, and lancets for loosening the gum.

Johann Schrenk of Germany used and described an instrument called a pelican—a form of forceps—in 1481, and a similar device was illustrated by famed surgeon Guy de Chauliac (c.1300-1370). In fourteenth-century England, barber-surgeons regularly extracted teeth, and their familiar red-and-white barber poles—advertisements they used to indicate they would also bleed the sick—were sometimes adorned with teeth they had pulled.

Renowned French surgeon **Ambroise Paré** used a three-instrument approach to tooth extraction: a root-exposer to loosen the gum, a pusher to ease the tooth out of its socket, and a pelican to lift the tooth out. In the late 1500s Fabricius (1537-1619) described nine different pairs of forceps, most named for the mouth or beak of the animal or bird it resembled. Thomaseus devised a heavy-toothed forceps in 1525; Dutch surgeon Anton Nuck advocated anatomical dental forceps designed to fit the shape of the teeth they were to extract in the 1600s; and American dentist Cyrus Fay practicing in London built the first anatomical forceps in 1822. John Tomes (1815-1895) of London also designed anatomically based forceps, in 1841; and French emigre toolmaker Evrard (1800-1882) manufactured Tomes' instruments.

The dental key was first described in 1725 by Parisian J. C. de Garengeot, who improved but did not invent the device, the origins of which are unknown. Unlike the horizontal extractors used up to that time, the key featured a solid handle set at right angles to its long shaft and was turned until the tooth popped out. Unfortunately, the key-extracted tooth often broke, leaving the root behind. John Aitkins of London further refined that device in 1771. Between 1790 and 1840 the battle of the tooth extractors raged; contestants were the horizontal key and the new vertical tooth-extracting devices which pulled the tooth straight out. Thomas Bruff patented one of the latter devices in 1797. Modern tooth extraction tools include fine "root extractors" which overcome that problem. One company offers 14 different models with special features including very fine working tips with graduated widths in the form of a gouge providing easy access to tooth roots. Curved root extractors allow access even to difficult areas such as maxillary molars.

An improved elevator chair introduced by the French dentist Lecluse in 1750 was still widely used well into the twentieth century. J. J. J. Serre of Vienna designed a screw for removing root remnants in 1790 which also, with modifications, continued to be used into the 1900s. Americans Horace Wells and W. T. G. Morton used the first general anesthesia for tooth extraction—nitrous oxide and ether—in the mid-1840s, and James Robinson of London was the first to use general anesthesia—ether—in England in 1846.

TOOTHPASTE • See Toothbrush and toothpaste

TOP • See Gyroscope

TORPEDO

The torpedo, named after an electric shock-giving fish, was originally applied to explosive charges moored at sea (now called mines). The same name was also applied to explosive charges attached to the ends of spars projecting from the bows of small boats or attached to lines and towed by boats. This type of torpedo was used in the American Civil War by the Confederate submersible *H.L. Hunley* to sink the *U.S.S. Housatonic* in 1864.

In 1865 the modern torpedo appeared. An Austrian captain named Luppis devised a clockwork-driven, boat-shaped craft guided by lines attached to its rudder. It was, in effect, a small electric boat with an explosive bow steered from shore. This proved to be too difficult to operate effectively, so Luppis took his idea to Robert Whitehead (1823-1905), the English manager of a marine engine factory in Austria.

Whitehead developed a much better torpedo. His design was a spindle-shaped underwater missile driven by a compressed-air engine at six knots for a few hundred yards and carrying an explosive charge of 18 pounds (8.2 kg) of dynamite in its head. Whitehead's device could stay at a set depth by a valve actuated by water pressure and linked to the horizontal rudders. In 1868 he added a balance chamber with a pendulum connected to horizontal rudders that automatically corrected any fore-and-aft tilt. Back in England, Whitehead demonstrated his invention. The British navy was impressed with his 16-inch (41 cm) diameter torpedo. Driven by two contra-rotating propellers, the device had a range of 1,000 yards (914 m) at 7 knots or 300 yards (274 m) at 12 knots. A British ship used one of these torpedoes in action in an 1877 battle.

The torpedo was improved in many ways after this. An Austrian invented a gyro-controlled steering mechanism in 1878. The British increased the range and speed by using a mixture of steam and oil vapor with compressed air to give a hot run instead of the original cold run under compressed air only. Torpedoes stayed much the same until World War II, when the Germans developed an electrically-driven one that proved impossible to track back to the **submarine**. The steam-powered ones had left tell-tale bubbles in their path. The Germans also developed torpedoes that acoustically homed in on ship propellers.

Modern torpedoes are impressive weapons. The United States Navy has a 21-inch (53.4 cm) diameter design that is driven at high speeds by machinery using hydrogen peroxide for its fuel. The U.S. Navy also has Subroc, a device fired as a torpedo deep below the surface. It emerges from the water, flies like a guided missile, moves on an interception course toward the target, submerges again, acts as a homing torpedo, and finally explodes near the enemy like a depth charge. Many torpedoes incorporate passive or active homing systems or wire guidance in cooperation with electronic sensors. Due to the tremendous progress in miniaturization, there have been many small torpedoes developed for launching from **helicopters**. Another torpedo in use in the late 1990s is an antisubmarine rocket (ASROC) that launches a rocket-propelled ballistic missile containing an acoustic-homing torpedo.

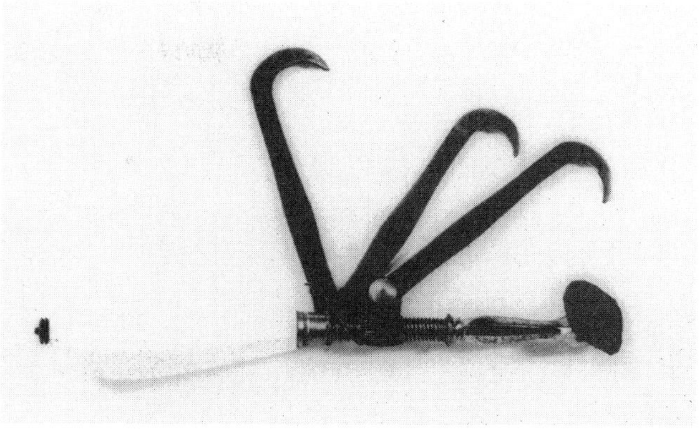

Dental key from the nineteenth century.

See also Rocket and missile

TORPEDO BOAT • See Submarine

TORRICELLI, EVANGELISTA (1608-1647)
Italian physicist

Evangelista Torricelli was first educated in Jesuit schools in his native Faenza, near Ravenna. His abilities as a physicist and mathematician were so great that he was sent to Rome for further education under the direction of Benedetto Castelli (1578-1643), a student of. Through Castelli, Torricelli was introduced to the aging **Galileo**, and he became Galileo's secretary and assistant for the last few months of Galileo's life. After Galileo's death in January 1642, the Grand Duke of Tuscany offered Torricelli Galileo's old position as court mathematician and philosopher. Torricelli held this position until his death from a sudden fever shortly before reaching his 40th birthday.

As a scientist Torricelli became well known for his study of the motion of fluids and was declared the father of hydrodynamics by **Ernst Mach**. Torricelli also conducted experiments on what we now call gases, though the term was not then in use. This led to his most important invention, the mercury **barometer**. The barometer was developed directly from his experimental work on air pressure and his effort to settle a dispute about the nature of the vacuum. The debate was a very old one. The Greek philosopher-scientist Aristotle believed that a vacuum could not exist, stating that "nature abhors a vacuum" Though Galileo felt that a vacuum could exist, he contended that the action of suction (in a water pump, for example) was produced by a vacuum itself and not by the pressure of the air pushing on the liquid being pumped. In fact, Galileo felt that air was weightless.

Participants in the debate noted that, for reasons not understood, suction pumps in mines could not raise water more than eighteen *bracci* (about 30 feet or 9 m) regardless of their size or power. If nature "abhorred a vacuum" why didn't

Evangelista Torricelli

water flow all the way to the top of these pumps? Torricelli made sense of this phenomenon and, in the process, invented the mercury **barometer**, an instrument that launched a flurry of study into the physics of the atmosphere and the behavior of gases.

Torricelli also made a contribution to meteorology with his suggestion that wind was not caused by the "exhalations" of vapors from a damp earth, but by differences in the density of air which in turn were caused by differences in the air temperature.

To replicate the action of a suction pump in a small tube, Torricelli began to substitute heavier liquids such as sea water, honey, and finally mercury for pure water. Using mercury, Torricelli could observe the effect of a vacuum using relatively short tubes sealed at one end. Torricelli filled such a tube of about a meter in length with mercury, sealed the open end with his finger, and then, inverting the tube, immersed the open end into a dish of mercury. The column of mercury dropped part way down the tube leaving an empty space at the top and a column of mercury in the tube about one and one-third *bracci* (about 30 inches or 0.75 m) in height.

Torricelli settled the debate about the nature of the vacuum by interpreting his experiment in the following way: a vacuum was not pulling mercury up the tube, but rather the weight of air pushing down on the dish of mercury prevented the mercury in the tube from falling out completely. This weight was enough to retain some thirty inches of mercury in the tube.

Applying his interpretation to the action of suction pumps in mines, Torricelli observed that such pumps, by evacuating the air pressure above a column of water, could cause the water to move upwards, but that the water would move up only as far as the air pressure below pushed it up. Once the weight of the water exceeded the power of the air pressure below, the water came to a stop, no matter how hard the pump was working.

Torricelli also noticed that the height of mercury in his tubes varied slightly from day to day. He concluded that this was due to changes in the air pressure over time, a principle which has come to have great importance in meteorology and geophysics.

In 1644 the French scientist Marin Mersenne (1588-1648) visited Torricelli and took back to his friend Blaise Pascal (1623-1662) the idea of the mercury barometer. If, Pascal thought, air was indeed pressing downward upon us as Torricelli contended, the total weight of the air, and hence its pressure, should decrease as altitude increases. With the help of his brother-in-law, Pascal showed that barometric pressure did indeed decrease as one ascended a mountain. Pascal's observational evidence showed beyond any doubt that Torricelli's theory was correct.

Writing some years later, Vincenzo Antinori stated that just as the telescope had transformed astronomy, Torricelli's invention of the barometer had changed physics. Torricelli was also active in the improvement of the telescope, an instrument that had first been used in astronomy by Galileo. Torricelli was able to grind **lenses** with such accuracy that he was able to produce some of the finest telescopes of his day.

Torricelli's investigations in mathematics played an important role in scientific history as well. Based on Francesco Cavalieri's "of indivisibles," Torricelli worked out equations upon curves, solids, and their rotations, helping to bridge the gap between Greek geometry and calculus. Along with the work of René Descartes, Pierre de Fermat, Gilles Personne de Roberval, and others, these works enabled Isaac Newton and Gottfried Wilhelm Leibniz to give calculus its first complete formulation. Though not as great a scientist as his older contemporary, Galileo, Torricelli continued the tradition of Italian scientific pioneering. This tradition was not to last long after his own death in the middle of the seventeenth century, however; by the beginning of the next century the center of scientific progress had shifted to northern Europe.

Torsion balance

The torsion balance is a device used to measure extremely small forces. Its invention is credited to Charles Augustin Coulomb (1736-1806), but John Michell (1724?-1793) and Henry Cavendish (1731-1810) also deserve acknowledgement for their work with the device.

Michell, born in Nottinghamshire, England, was a church rector who dabbled in science. He invented a torsion balance in 1790, intending to use it to measure gravitational attraction. Using that value in Isaac Newton's (1642-1727) gravitational formula would enable him to calculate the density of the earth. Michell died before he could accomplish this, but he did make two great accomplishments in the field of astronomy which made him famous. He was the first person to make a realistic estimate as to the distance of the stars, and he discovered the existence of physical double stars.

Born in Nice, France, the independently wealthy Cavendish was part of an aristocratic English family. The eccentric scientist was especially interested in electricity and gravitation. Unlike Michell, he was well educated in science, having spent four years at Cambridge. Cavendish had learned of Michell's work with the torsion balance and built one of his own in 1798. He suspended a rod by a **wire** at its center and attached a light lead ball to each end of the rod. When he brought two large balls close to the light ones, the gravitational attraction between them caused the rod to pivot, providing him with a measurement. Knowing the distance between the centers of the balls and the value of their attraction, he solved for Newton's gravitational constant and determined the earth's mass and density.

Cavendish's accomplishment went unnoticed because his work was not published until 1879, due to James Clerk Maxwell (1831-1879). In the eighty-one year interim between Cavendish's work and its publication, Coulomb had been experimenting.

Born at Angoulême, Charente, France, Coulomb was a military engineer. After an injury sustained in service, he returned to Paris and began experimenting in science and, in 1777, invented a torsion balance. His invention, made independently and without knowledge of the work of Michell and Cavendish, so impressed scientists that he was elected to the French Academy in 1781.

Experimenting with electrically charged spheres, Coulomb was able to use his torsion balance to measure the force of attraction and repulsion. He found that the force was proportional to the product of the charge of each ball, and inversely proportional to the square of the distance between the center of the balls. This relationship, similar to the rule governing gravitational forces that Newton had determined, became known as Coulomb's Law. The unit measuring electrical charge was dubbed the *coulomb* in his honor.

TOWNES, CHARLES H. (1915-)

American physicist

Charles Townes is generally considered the American inventor of the **maser** (an acronym for microwave amplification by the stimulated emission of radiation), an honor he shares with two Russian scientists, Aleksandr Prokhorov and Nikolai Basov. The microwave theories he introduced and pursued throughout the 1960s paved the way for such advances as the modern **laser**.

Charles H. Townes

The son of a South Carolina attorney, Townes studied modern languages and physics at Furman University in his home town of Greenville, S.C., graduating *summa cum laude* in 1935. He obtained his master's degree in one year at Duke University and his Ph.D. at the California Institute of Technology in 1939. After completing his education, Townes worked for Bell Telephone Laboratories on the design of **radar bombing systems**, in support of the U.S. effort in World War II. It was during these years at Bell that Townes' interest in microwave technology bloomed; in 1948, when he was asked to join the physics department at Columbia University, he gladly accepted.

Townes concentrated his research on microwave spectroscopy, the study of the basic structure of matter. Though some advancement was made, it rapidly became clear that an **oscillator** was needed which could produce radiation of very short wavelengths. Unfortunately, Townes also knew that existing technology was incapable of constructing a device small enough to produce such radiation.

In 1951, while sitting on a park bench in Washington, D.C., it struck Townes that an extremely small device might be found if he concentrated not upon electrical **circuits** but rather upon molecules. Since molecules had very specific rates of vibration, and microwaves had very specific wavelengths, molecular vibrations, if somehow converted into radiation, would be equivalent to the essential short-wavelength microwaves.

Frantically writing on the back of an envelope, Townes calculated that it was possible to produce microwaves if ammonia molecules were "excited" by pumping energy into them and then were stimulated to emit that energy in a controlled, or coherent, pattern. These theories were by no means new; coherence had been studied for years, and stimulated emission was first discussed in depth in Albert Einstein's (1879-1955) theory of relativity. But the application of those two theories for microwaves was unprecedented.

By 1953, Townes, A. L. Schawlow, and Townes' students at Columbia had finally constructed a working ammonia maser. One of its first applications was in timekeeping, for the steady, undeviating frequency of the microwaves was far more accurate than any timepiece prior, and was dubbed the "**atomic clock**." But Townes and others felt that a more versatile device could be developed if the gaseous ammonia were replaced with a solid crystal.

The resulting solid-state masers were instrumental in the field of radio astronomy, where they were used to amplify very weak signals from distant radio sources. The faint reflected signals from the *Echo I* **satellite** were successfully amplified in this manner, as well as radar scannings of the planet Venus.

Perhaps the most important advance in maser technology began in 1957, when Townes and his brother-in-law began speculating on the possibility of creating an "optical maser," delivering infrared or visible light rather than microwaves. Such light would be coherent; rather than spreading out like normal light, the maser light would maintain a tight beam almost indefinitely. Such a beam could also focus its energy to an extremely fine point, making it a cutting tool unsurpassed in power and precision.

In 1960 the first ruby maser, or **laser** (light amplification by the stimulated emission of radiation), was constructed, paving the way for a technology that would revolutionize engineering, medicine, and communications. For these advances Townes was awarded the 1964 Nobel Prize for physics, sharing it with Prokhorov and Basov (who worked independently in Russia).

In 1959, Townes took a two year leave of absence from Columbia to serve as vice-president and director of research of the Institute for Defense Analyses in Washington, D.C. In 1961, he became provost and professor of physics at the Massachusetts Institute of Technology. He was elected to the National Academy of Sciences in 1956.

Tracheotomy

A tracheotomy is a life-saving surgical procedure in which an opening is made in a patient's windpipe (trachea) and tube is inserted into the opening in the throat to allow breathing to continue in the event of airway obstruction. Once the emergency situation passes, the tube can be removed and the opening closed.

The first tracheotomy was performed in 1825 by French physician Pierre Bretonneau (1778-1862) on a four-year-old girl whose throat had become obstructed with the scar tissue that forms in the throats of diphtheria victims. Bretonneau had attempted two tracheotomies previously and failed, but his determination, skill, and dexterity finally paid off, saving the girl's life. Bretonneau, the son of a surgeon, became a physician at the hospital in Tours. Practicing medicine among the poor, he was the first to study such diseases as typhoid fever and diphtheria in detail and was the first to use the term "diphtheria." Also a skilled craftsman, Bretonneau made hydraulic hammers, **barometers**, and **thermometers**.

Tracheotomies can be used for people who need long-term artificial airway support—such as poliomyelitis victims or people paralyzed from the neck down, people with respiratory infections, cancer, airway burns, upper airway obstruction, and even extreme cases of sleep apnea.

Tractor

Early in the history of agriculture, draft animals replaced humans for such difficult tasks as plowing and pulling heavy loads. More recently, machines have replaced animals in many parts of the world. The most familiar farm machine is the tractor.

The steam tractor was developed in response to the increase in the size of farms during the westward expansion of the United States and Canada. The Case company built the first steam tractor in 1829. Several other companies soon followed with their own machines. Steam tractors (then called portable traction motors or agricultural motors) were extremely large and heavy. The boiler had to be made of cast-iron to contain the heat, and the **wheel**s had to be large and heavy to carry the weight. These tractors were often owned by contractors who offered their services over a certain territory. The tractor would be driven to the edge of the field, where it would remain stationary; it would then reel a **plow** or other implement across the field along a guide line.

A few mobile tractors were later developed, the first being the French *Albaret locotractor* of 1856. Despite improved mobility, these machines remained hot, heavy, and difficult to operate. If they broke down in the field, they were nearly impossible to move to a repair site. There was also a constant danger of setting the field ablaze by sparks from the engine.

In 1889, the *Burger tractor*, the name given to it by its manufacturer, Charter Engine Company of Chicago, had an **internal combustion engine** affixed to a steam tractor chassis. It was the first gasoline-powered vehicle in America. However, it had so little power that it could not even pull a plow.

The first practical gasoline-powered tractor was built in 1892 by the Van Duze Gas and Gasoline Engine Company of Cincinnati at the request of Iowa farmer John M. Froelich. Its two-cylinder engine had to be constantly pumped by the operator. Weighing nine thousand pounds, it had only thirty horsepower; however, this was an improvement over steam tractors. The first industrial-scale manufacture of gasoline tractors took place at the Hart-Parr tractor company of Charles City, Iowa. Charles Hart and Charles Parr developed their machine in 1901 and started the company two years later. In 1906, they became the first to use the term tractor.

In 1904, American Benjamin Holt developed the caterpillar tractor for the purpose of carrying steam engines more efficiently. At that time some five thousand steam tractors were still being built annually. The track tractor was first used for farm work in 1914 and became the inspiration for the military tank. The *Wallace Cub* was a smaller, lighter model built in 1913 by the Bull Tractor Company of Minneapolis. It had three wheels—one in front and two in back—for driving in the crop furrows.

With the increase in mechanical implements available to the farmer, the all-purpose Farmall tractor built by International Harvester in 1924 answered the need for versatility and was the first of the light modern tractors.

In 1928, citrus farmers in Florida began to add rubber to their tractor wheels to protect tree roots from damage. The B. F. Goodrich tire company responded in 1931 with solid rubber tires. The next year, Firestone tire company offered inflatable tractor tires; these tended to puncture and did not provide adequate traction. The company switched to water-filled tires.

During the late 1930s, the Irish-English farm machine manufacturer, Harry G. Ferguson (1884-1960), developed the *Ferguson tractor system*, which integrated the tractor with interchangeable implements. This system was efficient, dependable, and affordable. They were manufactured by the Ford company throughout World War II, after which Ferguson manufactured them on his own.

Four wheel drive and air conditioned cabs added greatly to the comfort and effectiveness of tractors. The gasoline-powered farm tractor played a major role in the transformation of the farm. The number of tractors on farms increased steadily from the 1940s onward, while the number of farms decreased. The development of the tractor was not only a response to change, but also spurred change.

See also Armored vehicle; Farming, mass production; Steam-powered road vehicle; Tire, pneumatic

Traffic signal

Ever since Roman times, society has tried to control traffic. Even the fabled Roman road system created a conflict between pedestrian and equine travellers. However, a practical solution was not developed until the mid-nineteenth century, when J. P. Knight, a railway signaling engineer, created the first traffic signal, which was installed near Westminster Abbey in London, England in 1868. Unfortunately, the device exploded, killing a police officer, and its use was discontinued after being in operation for only a short time.

The modern traffic light was invented in America. New York had a three-color system in 1918 that was operated manually from a tower in the middle of the street. Other cities soon adopted the idea of having someone on the scene to control the lights. **Garrett Morgan**, inventor of the **gas mask**, also developed traffic signalling devices. Having witnessed an accident between a car and a carriage, Morgan felt compelled to devise a system to prevent such collisions at street intersections. In 1923 he patented an electric traffic light system using a pole with a cross section on which the words STOP and GO were illuminated.

These basic designs were soon improved. In 1926 the first automatic signals were installed in London; they depended on a timer to activate them. In the 1930s vehicle-activated lights were created in which cars rolled over half-buried rubber tubes. Air in the tubes was displaced by the weight of the car rolling over them, and the increased pressure operated an electric contact, activating the lights. But these tubes wore out quickly. A better idea was the inductive-loop device: a loop of wire was imbedded in the road itself and connected to a box controlling the lights; a current of electricity passed through the loop, and when the steel body of a car passed overhead, it produced a signal that activated the light.

Today, traffic is automatically routed onto limited access highways courtesy of a **computer** activated guidance system that determines traffic volume on the highway. **Global positioning satellite systems** (GPS) are installed in many cars. These systems connect with a satellite and inform drivers where they are and possible routes to their destination. Such systems will eventually enable a drive to determine the best route to a destination given prevailing traffic conditions.

Train and railroad

The use of rails, or other types of guides, for roads dates to antiquity, when the Greeks and Romans carved trails in stone streets to guide cart wheels. By the third century B.C. this system of sunken guide ways was used in some parts of China. The first raised rail roads were likely used in mining and quarrying industries during the sixteenth century.

These early rails were made of wood, which wore and rotted quickly. Over the years engineers tried to protect the wooden rails with various materials, such as layers of hardwood and later metals. Finally, in 1738, the first wholly iron rails were introduced in Britain. Still the power source used on these rail roads was a team of horses.

A number of inventors, including **James Watt** and John Stevens (1749-1838) and his son Robert (1787-1856), aided in the development of steam engines. Experiments with and demonstrations of steam-powered locomotives reached a pivotal point in 1804 when **Richard Trevithick** introduced the first viable steam **locomotive**. Still others were more concerned with the track aspect of land transportation.

In 1811 English engineer John Blenkinsop designed a cog rail system to overcome fears that the traction between metal wheels and metal rails might be insufficient to climb hills. Prior to this time, horses were still used to pull loads uphill. Blenkinsop's device utilized an extra, toothed rail laid inside the regular tracks. A gear mounted on the locomotive meshed into the teeth and provided the extra traction needed to replace horse teams.

In 1821 John Birkinshaw perfected a method of rolling wrought-iron rails in 15 ft. (4.6 m) sections. These rolled rails withstood extreme weight and required fewer joints. Rail promoter **George Stephenson** admitted the superiority of Birkinshaw's rails over his own and laid them on England's Stockton and Darlington Railway (S&D). The S&D railway was originally planned as a horse-powered, wooden rail road, but Stephenson's appointment as engineer led to the laying of iron track and the use of some steam locomotives. Stephenson's Locomotion eventually demonstrated some of the advantages of steam power over horses.

While serving as engineer on that line, Stephenson also began drafting improvements for his new Liverpool and Manchester branch (L&M). Unlike the S&D, he wanted the L&M to operate entirely with steam locomotion. He not only had many technological obstacles to overcome, he also had to convince his proprietors that it was a practical idea.

At the Stephenson Works in Newcastle, George and his son Robert began designing and building a locomotive so improved that even his strongest skeptics would be convinced. The result of their work was the 1829 Rocket—and success. When the 31 mi. (50 km) long L&M opened the next year, it was the first railway in the world to rely exclusively on steam locomotion. It was also the fastest and most reliable line to that date.

Railway mania had begun in Europe. English rail projects had proven the feasibility and profitability of land transport by steam. After years of battling skepticism, money for development and expansion suddenly became readily available. In the mid-1800s, England's rail companies were employing 250,000 construction workers, nearly 10,000 mi. (16,090 km) of railroads had been laid, and capital investments exceeded 250-million pounds.

Although slowed by a vast lobby of canal supporters and enthusiasts, railway mania was not slow in arriving in America. The Delaware and Hudson Railroad opened in 1829 with the imported locomotive, *Stourbridge Lion*. Another foreign locomotive, the *John Bull*, served the Camden and Amboy Rail-Road and Transportation Company (C&A) in New Jersey. It had been built in England by the firm of George and Robert Stephenson. Shipped disassembled, it arrived in 1831 and was put together by a teenage boy named Isaac Dripps, who added a distinctly American feature: the cowcatcher.

Robert Stevens was in charge of laying the track for the C&A. During this period, rails were anchored directly onto stone blocks. When a delivery of blocks was delayed to the construction site, Stevens laid a bed of crushed rocks and placed on top a series of wooden cross ties to which the rails were anchored. To his surprise, his design provided a smoother ride for passengers and helped absorb and distribute the weight the locomotive's weight. It has been used for virtually every rail placed since.

The "Golden Spike" ceremony held at Promontory Point, Utah, on May 10, 1869, officially linked the Central Pacific Railroad and the Union Pacific Railroad, creating North America's first transcontinental railroad.

Henry Campbell invented a locomotive in 1836 that became an American prototype. Traction was a concern of many at the time. Europeans solved the problem with cog railways and by making bigger and heavier locomotives. Physical limitations on American railways prevented this solution, so Campbell proposed using two sets of driving wheels placed close together on each side of the locomotive. Despite its rougher ride, the new arrangement worked and became standard on American locomotives.

The Philadelphia and Columbia Railroad opened in 1831 with horse-drawn vehicles, but by 1834 it had acquired locomotives. Other early rail companies included the Mohawk and Hudson Company and the South Carolina Canal and Rail-Road Company.

The Baltimore and Ohio Railway (B&O) used Peter Cooper's small *Tom Thumb* (1830) to prove that steamlocomotives could compete with canals and horse teams even on curvy and hilly track.

The *Tom Thumb*—the first to pull a load of passengers in America—was a short, wheeled platform that supported an engine with a vertical cylinder and a vertical boiler with tubes made from gun barrels. The vertical layout was important because of the many curves and sudden grade changes common to early American railways. In 1831 the B&O announced a contest for designing a lightweight locomotive that could reach a speed of 15 mph (24 kph) and pull a load of 15 t. (13.6 t) on level ground.

Phineas Davis's *York*, which won the contest, used a vertical boiler and vertical cylinders. Two problems persisted over the many years railways used vertical layouts. The locomotives were top heavy, and the up-and-down motion of the pistons made the engines bounce and derail. Robert Stevens's pilot, an attachment of two wheels connected in front of a locomotive to act as guides, partly solved this problem. The B&O chief locomotive designer, Ross Winans, improved engines, some of which operated so well that the line continued using vertical engines into the 1890s after most companies had adopted horizontal layouts. The switch became possible when curves and grades were improved as companies acquired more direct rights-of-way and dug tunnels, built bridges, and eased turns. Horizontal layouts both solved problems and allowed space to construct longer and larger boilers to generate more power.

As rail lines spread and grew in importance, demand increased for higher speeds and greater efficiency. Belgian engineer Egide Walschaerts improved the valve gear in 1844, and by 1847 Britain's Great Western Railway ran some routes at 60 mph (90.5 kph).

In 1859 greater rail adhesion was achieved with coupled driving wheels. By 1860 steel rails began to replace wrought iron, but the joint between rails proved to be a weakness until ''fish plates'' for actually connecting end-to-end rails were introduced in 1847. Between the Civil War and 1885 the United States eased travel between systems by standardizing the gauge, or width between rails, at 4 ft. 8.5 in. (1.42 m). And in 1898, Wilhelm Schmidt from Germany introduced the use of superheated steam to reduce condensation losses.

In the 1860s comfort became another consideration. Railways had originally been envisioned as a method for moving freight; yet, in spite of many early catastrophic accidents, people soon began to seek passage on them. At first, passengers sat on wooden benches in open wagons. Those who could afford to travel at a higher level of comfort sat in their own carriages, which had been securely attached to a flat bed car. In the mid-1800s, special cars were commissioned for Ludwig II of Bavaria and Pope Pius IX, which consisted of three room suites and ornate throne rooms.

In America George Pullman began making sleeping and dining cars for common use about this time. Though his were not the first, they were the earliest to be designed for safety and comfort. Early cars still resembled stage coaches: They were crowded, filled with smoke and ash from both the engine and car heaters, and tended to jump and rock. Pullman's 1864 Pioneer led the change in attitude and design, and before long his cars were in such demand, even in Europe, that he had built large manufacturing facilities near Chicago.

Safety devices also became more necessary with staggering increases in rail traffic. Derailments caused by locomotives and cars jumping the track, especially on sharp curves, were largely solved by the similar and controversial invention of swiveling wheel trucks by Horatio Allen (on the *South Carolina*), John B. Jervis in 1831, and Ross Winans a few years later. Joining cars, a dangerous job for railroad workers into the middle of the nineteenth century, involved dropping a pin into a link at the ends of cars until adequate coupling mechanisms and systems were introduced. Claude Chappe's system of **military semaphores** became widely adopted for manual signaling. Once the **telegraph** became common, it was widely used for sending location messages. Braking systems were nearly always of primary importance, but none were as reliable as George Westinghouse's pressure (1868) and later vacuum (1872) brakes.

Before the Civil War began, the American rail network covered 30,000 mi. (48,270 km). By the end of the century, more than 200,000 mi. (321,800 km) of track were considered largely responsible for opening the western half of the country to settlement. While developments may not be as dramatic or important as they once were in the realm of rail transportation, some advancements continue to be made.

In the track department, after little change in track technology for many decades, continuous welded rails became common after 1950, providing improved safety, reduced noise, and lower maintenance. Since the 1960s several studies have focused on lower-friction and more maintenance-free types of track and bases using concrete. Current technology allows for eliminating rails altogether.

In 1958 engineers began experimenting with TACVs (Tracked Air-Cushion Vehicles). Based on the technology of hovercrafts, TACVs would not need rail maintenance at all and could operate faster and more smoothly since they ride on a cushion of air rather than a guideway. Another related high-speed option is magnetic levitation, or MAGLEV technology, in which an electromagnetic charge on the underside of a vehicle repels the opposite charge on the top of a support. Traveling about 4 in. (10.2 cm) above its guides, one prototype has demonstrated a speed of 321 mph (516.5 kph), a world ''rail'' speed record. Both of these technologies have been tested in Europe, Japan, and the United States, and by the mid-1990s, Germany committed itself to building a Transrapid maglev system.

Proponents of Maglevs argue they are environmentally friendly, consume less energy than conventional track trains, and also are safer in that they eliminate the possibility of derailment. Critics say they are too costly and that their extremely high speeds could never be attained in short, between-city runs. Still, besides the German's definite commitment, Switzerland has proposed an underground Maglev and New York State is considering a Maglev link between New York City and Albany. Fuels also are a research focus. While diesel engines are still the most common power source in the United States, trains, buses, and other electric-powered vehicles are increasingly used in cities.

While duorail, or two-rail, systems have predominated the railway scene since the beginning, monorail, or single-rail, systems have interested some inventors. The first monorail system was patented in 1821 by English engineer Henry Palmer and drawn by horses. In 1869 another English engineer, J. L. Hadden, used steam power for a system in Syria. In 1889 E. Moody Boynton of Portland, Maine, unveiled a bicycle monorail locomotive. In 1894 a monorail line based on French inventor C. Lartique's design (c. 1850) was built and electrified in France. Gyroscopes kept the 1903 system of Irish engineer Louis Brennon upright while carrying 50 passengers. Yet not until the later part of the twentieth century were commercially safe and successful monorail systems built. Modern systems usually operate in urban or commercial areas and frequently use rubber tires for smoother, quieter rides.

As the 21st century begins, the next major change for trains and railroads will be the use of advanced technology to run trains. By 1995, radio remote control systems were available and in use in which transmitters were adapted to existing freight locomotives, resulting in increased productivity and reductions in crew size. Safety is an issue that divides many, but Canada's years of remote control experience argue that remote control systems are actually safer than conventional systems. By 1997, communications-based train control was considered by most to be at the center of the next revolution in train control. In fact, most would agree that such systems are not only inevitable and welcome by the railroad industry, but that the only things keeping it from happening sooner are issues like standardization and interoperability.

TRAM • See Streetcar and trolley

TRAMPOLINE

George Nissen, who invented the trampoline, was not only a creative inventor, but also knew how to promote and sell his products.

Nissen first got the idea for the trampoline by watching circus high wire and trapeze artists. He was enthralled by the antics of the performers as they bounced in the safety nets strung above the circus floor. Instead of merely trying to get out of the nets as quickly as possible, some of them showed off with twists, spins and somersaults before alighting to the ground.

In 1926 Nissen took over his family's garage in Cedar Rapids, Iowa, and began work on his version of the bouncing table. As a young high school student he scrounged for the materials needed—**springs**, rubber inner tubes and scraps of iron—from the town dump, and anywhere else he could find them. Somehow he found an industrial **sewing machine** and adapted it to his specific needs. Nissen's resourcefulness meant he was able to move forward with his project even though he had little money to spend on it.

Although the idea of a bouncing table was not new, Nissen wanted to improve on the design of those that already existed. It was important that the table be simple to transport, set up and store. It had to be large and strong enough to withstand various types of jumping safely, but not use any more floor space than necessary.

After graduating from college in 1938 Nissen began to work in earnest, perfecting and promoting his invention. Nissen also invented the machines necessary to mass produce it, then strapped a trampoline to the top of his car, touring the country and demonstrating his invention wherever a crowd could gather. He used the profits from those first sales to purchase advertising in various sports magazines.

Nissen enlisted in the United States Navy during World War II and convinced both the Army and Navy to use his invention in their pilot training programs. After the war he expanded his promotional touring, including Europe. He even brought a kangaroo with him to prove he could outjump it using his invention. By 1948 the NCAA and AAU added trampolining to their gymnastics competitions. And the feeling of weightlessness that the trampoline simulates was useful to America's training astronauts. By 1964 the trampoline volleyball game ''Spaceball'' had swept the country. Although no longer in existence, Nissen's company branched out into production of all sorts of gymnastics, **basketball** and scoreboard equipment, sold primarily to schools and colleges.

TRANSDUCER

Just about any device that is used to convert energy from one form into another can be called a transducer. Most electronic measuring devices use some type of transducer in order to display their measurements.

An example of one type of transducer is an automobile's fuel gauge. In this system a float within the fuel tank is allowed to rise and fall according to the amount of fuel beneath it. The rise and fall is measured by a **potentiometer** as relative displacement, and the electrical signal produced by the potentiometer is used to control the position of the dashboard's fuel gauge. Like most transducers, this system utilizes a two-step process by which first a mechanical and then an electrical element are used to produce an electrical signal.

The transducer described above would be called a potentiometric transducer, since it uses a potentiometer as its electrical element. Other types are the inductive and capacitive transducers, strain gauges, ultrasonic transducers, as well as **anemometers**, flowmeters, and **accelerometers**.

One of the first kinds of transducers was the **microphone**, developed in the late 1800s by the American inventor **Thomas Edison**. The microphone converts sound waves by allowing them to vibrate a sensitive cone, later changing that vibration into an electrical signal. In order to reproduce the sound, one must use a speaker (another type of transducer) to reverse the process; here, the same signal is used to vibrate a similar cone, thus creating sound waves.

Another, more recent variety of transducer is the **photoelectric cell**. These *solar* cells consist of certain elements—often selenium or **silicon**—that readily lose electrons when exposed to bright light, particularly sunlight. The liberated electrons will produce an electric current across the cell. Photoelectric cells are often employed as light meters, measuring the intensity of light by recording the strength of the electric current it produces. A cousin to the photoelectric cell, the **photovoltaic cell**, is also used to convert sunlight into usable energy; though usually encountered in watches and pocket calculators. Photovoltaic cells have shown promise as a potential replacement for fossil fuels when assembled into large solar arrays.

See also Automobile, gasoline; Calculator, pocket; Ultrasound devices

Transformer

The transformer makes modern power transmission possible, changing voltages up or down with very little loss of power. Its development dates back to the early nineteenth century, when, in 1820, **Hans Christian Oersted** (1775-1851) discovered that an electric current in a conductor created a magnetic field and started a flurry of research into the area of **electromagnetism**. **Michael Faraday** decided to reverse Oersted's discovery by using a magnetic field to create an electric current. In 1831, he discovered that the current sent through a wire coil set up ''magnetic lines of force'' that expanded outward and affected a nearby second coil. By alternating the expansion and collapse of the lines, he could induce an **alternating current** (AC) of electricity in the second coil. Experimenting with the number of turns of wire in his coil, Faraday discovered that if the secondary coil had twice as many turns as the primary, the electric current was doubled; if the secondary coil had half as many turns, the current was cut in half. Thus the transformer was born, and with great efficiency was able to ''step up'' low-generated current for high-current transmission and step it down again at its destination.

The scientific community took little notice of Faraday's work for half a century, because their attention had been fixed on direct current (DC) since the invention of the electric battery by **Alessandro Volta** in 1800. By the 1880s, Thomas Edison was lighting electric street lamps with DC dynamos. However, there were disadvantages to DC, including the power loss caused by arcing in the dynamos and a limited transmission distance due to resistance in the **wires**. In England, Lucien H. Gaulard and John D. Gibbs used a transformer (which they called a secondary **generator**) to power incandescent lamps in an AC arc-lighting system in 1882. Though impractical, the Gaulard-Gibbs equipment did encourage others to experiment.

The prototype for today's lighting systems was designed by three Hungarian engineers, Max Déri, Otto T. Bláthy and Karl Zipernowski, who demonstrated their transformer in 1885 at the Hungarian National Exhibition in Budapest. In 1884, Croatian inventor **Nikola Tesla** began working for Edison, and repeatedly tried to sway him to the advantages of AC. Failing in his attempts, Tesla took his ideas to the industrialist **George Westinghouse**. The ultimate success of Tesla and Westinghouse at the 1893 Columbia Exposition in Chicago ended the DC monopoly on electric power supply. In 1895 the first hydroelectric AC power generator was build at Niagara Falls, using a *generation* (step-up) transformer to conduct the current many miles over high tension wires to a *transmission* (step-down) transformer for distribution to customers.

The quality of transformers has continued to improve with better construction materials. A good generator has a core with high permeability—that is, it produces a high number of lines of magnetic force. Resistivity minimizes energy lost to eddy currents. A high saturation level insures the ability of the material to amplify the magnetic force to a high plateau. A quality generator also minimizes loss of efficiency due to hysteresis, a ''magnetic memory'' that wastes energy from a lag in the AC cycle, due to the build up of a magnetic influence.

Transistor

Since World War II, no invention has made a larger impact on the communications field than the transistor, which replaced old-fashioned **vacuum tubes** in electronic equipment. The transistor revolutionized electronic technology and created a billion dollar industry that sells everything from popular consumer items—such as home computer games, **pocket calculatos**, and portable **stereos**—to the complex electronic systems used by business and industry.

In the 1930s researchers were aware of the limitations of vacuum tubes, which were used to increase electric current, a process called amplification, and to force it to flow in only one direction, which is known as *rectification*. Vacuum tubes were bulky and fragile, and they consumed a lot of power. The tubes also had a tendency to overheat, so they needed large, reliable cooling systems, and when **radar** was developed during World War II, vacuum tubes were unable to provide the range of frequencies, power, and voltages needed for radar circuits. After the war, when scientists began trying to develop bigger and better computers, they were frustrated by the huge, complex arrays of vacuum tubes and cooling systems that were required.

In 1945 Bell Telephone Laboratories established a research team to develop a ''solid-state'' electronic device that could amplify and control electrical current without using moving parts or vacuum tubes. The most promising replacements were **semiconductors**—materials such as silicon and

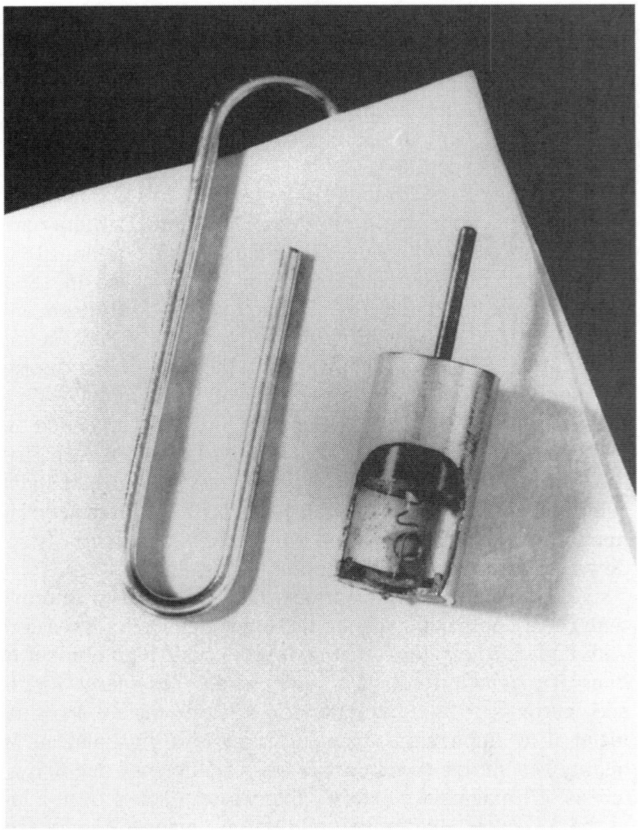

A 1950 photo of the then-new miniature transistor. The invention of the transistor made it possible to hod more electronics in one's hand than could be installed in a building using bulky electron tubes.

germanium that can strengthen an electrical current, create a current, and either conduct or resist a current, which means they can *semi-conduct* a current by switching it on and off. At that time, semiconductors had been used to rectify electrical current, but no one knew how to make them amplify it.

Bell's research team, which included American physicists **William Shockley**, **John Bardeen**, and **Walter Houser Brattain**, began studying semiconducting materials with the hope of applying them to **electric circuits**. Shockley theorized that electrical current could be amplified by using semiconductors and metallic conductors. When Shockley's initial experiments failed, Bardeen explained that what was needed was a connection between the two materials to enable current to flow.

Bardeen, Brattain, and Shockley then performed several ground-breaking experiments. They mounted a germanium oxide semiconductor on a metal base and then placed the semiconductor between two fine metal wires, which served as metallic contacts through which the current flowed. By 1948 the team had successfully amplified electrical current by a factor of fifty using the world's first point contact resistor, or transistor. Tiny changes in the current induced great variations in the power output. In 1956 the three scientists were awarded the Nobel Prize in physics for their work in transistors and semiconductors.

Still, the operation of the transistor was not well understood. Shockley suggested an important improvement: placing a positively charged region between two negatively charged regions, all contained within a single crystal of semiconducting material. In the negative, or *n-type* region, current is carried by electrons, which are negatively charged. In the positive *p-type* region, current flows through ''holes''—spaces left behind by vacated electrons. These holes act as positive charges by accepting electrons from the negative region. Shockley's transistor proved easier to manufacture than the point-contact type as well as more versatile in electronic applications. The positive and negative regions were created by adding tiny amounts of such chemicals as arsenic, borium, and gallium to the semiconductor to alter its electrical properties. Today this process is called *doping* the semiconductor, and Shockley's improved device is known as the *junction* or bipolar transistor.

With financial support from the United States military, Bell Labs redesigned electronic equipment to accommodate transistors and soon found many valuable markets for the device. During the 1950s transistors began to be used in **telephone** switching equipment, **hearing aid**s, **radios**, and computers. IBM introduced a computer that used only five percent as much power as vacuum-tube models in 1955. Then, in 1958, scientists learned how to integrate many transistors on a single piece of semiconducting material. These electronic devices are called **integrated circuits**, or silicon chips. By the early 1970s mass-production of computer chips had led to the commercial introduction of small, powerful **microprocessor**s.

See also Calculator, pocket; Computer, analog; Computer, digital

TRANSIT • See Surveying instruments

TRANSMISSION, AUTOMOBILE

The windmill and the motor car may seem unlikely cousins, but both transmit power through rotating shafts. Methods of connecting vertical and horizontal shafts, equipment of different weights, and shafts turning at different speeds were crucial obstacles to automotive pioneers. Steam carriages moved forward by simply opening a throttle, but the **internal combustion engine** must not be connected to the rear wheels (where the load is exerted), it must continue running while it is stationary, and it must accelerate from a low initial speed to a higher one. Consequently, a device is needed to apply loads gradually after the engine is started.

In the 1890s, **Emile Levassor** adapted a cone-shaped clutch and sliding gears he observed on woodworking machinery to the transmission of automotive power. **Louis Renault** followed in 1899 and created the general layout of the transmission that was to be adopted by the fledgling automotive industry; he furnished a propeller shaft with universal joints that drove a pinion and crown wheel linked to the **differential** on the rear axle. This speed-changing mechanism allows the power to be controlled at low speeds and increased at high speeds for maximum engine efficiency. A ratio or relationship exists between engine torque (turning capacity) and wheel speed which is the same factor as the increase of engine speed.

To select the best gear ratio (in the simplest type of transmission), the driver moves a shift lever to snag a spur gear that engages the forward speed gears or the reverse gear. This sliding gear transmission used clutches and a blocker ring to prevent the gear teeth from grinding. The driver has to select the appropriate gear for the speed of the car (and other conditions) and manipulate the shift lever while operating the accelerator and clutch pedals. The automatic transmission eliminated this complex action.

Clutches progressed through several evolutionary steps from the cone clutch to those constructed of combinations of interleaves and discs or multiple plates. Moulded friction materials improved simultaneously, and, by 1921, the complication of sticking plates was eliminated and replaced with a single plate, the operation of which was smoothed by helical springs and diaphragms.

The automatic transmission was invented by **Gaston Fleischel** of France in 1933. Automatic transmissions may use a fluid coupling, hydraulic torque converter, or mechanical linkage to adjust the torque over the gears without the driver's intervention. The driver still manually selects a low drive gear or reverse, but the automatic transmission then adjusts for grades on hills and the change of gears, although at a lower gear ratio than the manual method. By 1970, the hydraulic torque converter predominated. A governor in this transmission makes shifts depending on engine speed, and the driver can control the gear changes to a certain extent by adjusting the car's speed. Multiple-disk clutches and bands in oil are hydraulically operated by the governor. Automatic transmissions tend to produce the best performance because power is continuously applied. The opportunity for error and inefficiency exists when the driver can interrupt this power flow.

Hydraulic valves used to control transmissions have been replaced as of the mid-1970s by electronic systems. Sensors that can measure road and driving conditions improve the efficiency of the automatic transmission by adjusting the shift speed for hills, for example. While electronics may ''read'' conditions, they still actuate fluid pressure to engage brakes and clutches within the transmission.

Transmission electron microscope • See Electron microscope

Transplant, surgical

Stories of transplanted tissue and body parts go far back in myth and legend. Chinese folklore tells of organ transplants; in the sixth century A.D., the Western patron saints of medicine, Cosmos and Damian, supposedly replaced the cancerous leg of a white man with the healthy leg of a recently-deceased black man. Records of transplants begin in India in the sixth century A.D. with skin grafts to replace amputated noses (the penalty for adultery).

This practice was introduced to Western medicine in the sixteenth century by Italian surgeon Gaspare Tagliacozzi who grafted to a patient's nose a flap of skin from the forearm. Portion of this flap was left in tact on the arm until the graft took several weeks later when the flap was severed and the arm released. Tagliacozzi used a patient's own skin because he felt that an individual's ''force and power'' would reject foreign tissue—thus anticipating the problem of rejection which remains a major impediment to successful transplantation to this day.

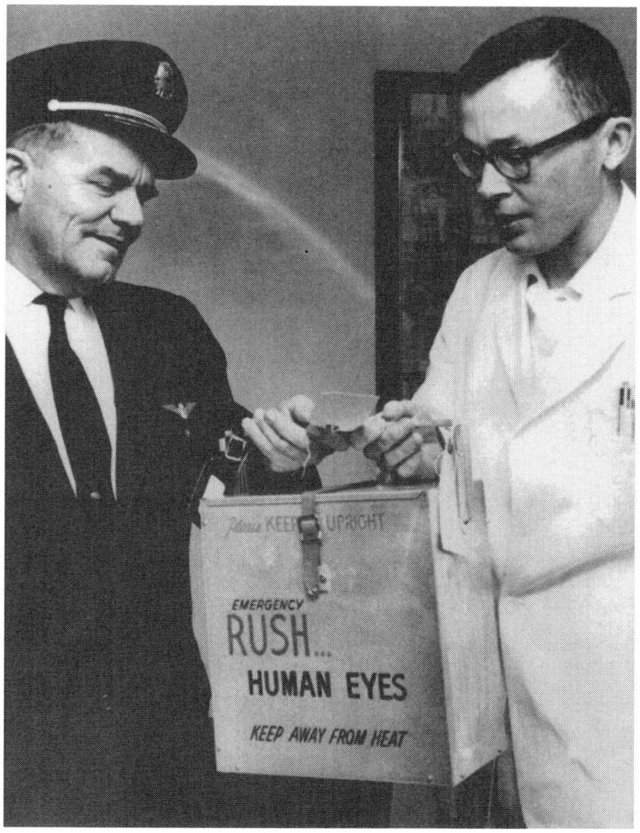

A box of human eyes, intended for corneal transplants, being transported in the 1950s.

Scottish anatomist **John Hunter** revived some interest in tissue transplantation in the eighteenth century and, during the nineteenth century, skin grafting was reintroduced with increasingly successful results when, in 1869, a young French surgeon, Jacques Louis Reverdin, discovered grafts needed to be thin. Scottish surgeon William MacEwen reported success with bone allografts (transplants from one person to another) in children in 1881, and von Hippel carried out a corneal transplant in 1877.

In 1902 the French surgeon **Alexis Carrel** developed a method of sewing together small blood vessels with tiny needles and fine thread, making transplantation technically feasible. Testing these methods with Dr. Charles Guthrie of the University of Chicago, Carrel performed a series of organ transplants on animals. Organs functioned well only for a short time. Other researchers experienced similar results and transplantation came to a standstill.

Researchers, including Guthrie, began to suspect rejection of the implanted organ was an immune system response to foreign tissue. British biologist Peter Medawar became interested in skin graft problems while working with severely burned soldiers during World War II. He found that a second set of skin grafts between the same two subjects was rejected twice as quickly as the first set. To Medawar, this was clearly an immune response. In 1948 he found that grafts between twins were successful and, in 1954, showed that immune tolerance was acquired during embryo development and that injection of foreign substances into embryos or newborn mice would produce permanent tolerance to those substances later in life.

Meanwhile, researchers began focusing on kidney transplantation because of the kidney's relatively simple blood-supply system. Russian surgeon Yuri Voronoy performed the first human kidney transplant in Kiev, Ukraine in 1933, and several more during the 1940s. Ten kidney transplants were carried out in Paris, France from 1950 to 1953, one in Chicago, Illinois in 1950, and several at Peter Bent Brigham Hospital in Boston, Massachusetts. All failed except one—a 26-year-old Boston doctor who lived for six months. Discouragingly, matching donor and recipient blood types didn't improve outcomes. Drs. Joseph Murray and John Merrill and The Peter Bent Brigham team finally achieved success in 1954 when they transplanted a kidney donated by 24-year-old Ronald Herrick for his dying twin brother, Richard, and the Boston team performed 23 identical-twin kidney transplants between 1954 and 1966.

Transplants between non-twins still resulted in rejection, however, and unsuccessful attempts were made to suppress immune response using whole-body **X-rays** and X-rays of the organ to be transplanted. Two major breakthroughs occurred in the early 1960s. The first began with the 1931 observation by Karl Landsteiner (1868-1943), that blood could be classified by types, indicating that human tissue types might also exist. Also after George Snell found the major histocompatibility complex (MHC), a system of tissue-compatibility genes in mice in the 1940s, Jean Dausset identified a similar genetic set of human white blood cell antigens called human lymphocyte antigens (HLA). Beginning in Paris in 1962, HLA characteristics were increasingly used to closely match donor and recipient tissue markedly decreasing the likelihood of rejection. The second breakthrough began in 1959 when Robert Schwartz and William Dameshek of Tufts University in Boston showed that a drug called 6-mercaptopurine was a strong suppressor of the immune response in animals; this response was given the name immunosuppression. The Murray transplant team at Peter Bent Brigham, and the visiting British surgeon **Roy Calne**, worked with a derivative of 6-MP developed by Dr. George Hitchings at Burroughs-Wellcome. After us of this improved immunosuppressant, called Imuran (azathioprine), became standard procedure in 1962, kidney transplantation—with extended survival—was possible. Dr. Thomas Starzl of the University of Colorado developed the current procedure of administering steroids along with azathioprine in 1963. Immunosuppressant therapy was greatly improved, with a dramatic shift in transplantation success, by the discovery of cyclosporin, a substance isolated from soil samples. Microbiologist, Jean-Francois Borel, with the Swiss drug firm Sandoz, discovered in 1972 that cyclosporin inhibited the activity of the T lymphocytes, the immune system components that attack and destroy transplanted tissue. Trials carried out by British physicians Calne and David White in 1976-78 demonstrated cyclosporin's effectiveness in preventing rejection, and its widespread use ushered in the era of organ transplantation. Again, Starzl showed cyclosporin to be more effective when used with steroids.

Until cyclosporin, organ transplants other than kidneys fared poorly. This included the early heart transplants, the first of which electrified the world. South African doctor, **Christiaan Barnard**, drawing on techniques developed by Drs. Norman Shumway and Richard Lower of the United States, transplanted the heart of a young woman into Louis Washansky, a 55-year-old grocer. Washansky survived only 18 days, but Barnard's second patient, dentist Philip Blaiberg, lived for 17 months. Heart transplantation rapidly mushroomed, with Dr. Denton Cooley of Houston, Texas, performing more than 20 in 1968. Outcomes, however, remained poor because of rejection, and heart transplantation virtually died out until the use of cyclosporin in 1983 when heart transplants began again with much greater success, and other organ transplants have also become more common. Starzl and Calne pioneered the field of liver transplantation when Starzl performed the first successful liver transplant at the University of Colorado in 1967. Human pancreas transplantation was first performed by Drs. Richard Lillehei and William Kelly of the University of Minnesota in 1966, and the first successful combined heart-lung transplant was carried out in 1981 at Stanford University Medical Center by Drs. Bruce Reitz and Shumway.

Other body parts are often transplanted, however problems remain: Many grafts do not survive permanently; cyclosporin and tacrolimus (another major immunosuppressant) are expensive and have serious side effects; and graft-vs.-host reaction, discovered by Simonsen in 1957, in which lymphocytes in the transplanted tissue attack the "foreign" host tissue, is difficult to control. In the 1990s, investigations by Starzl at the Starzl Transplantation Institute at the University of Pittsburgh into *tolerance induction* and the state of *chimerism* (the coexistence of donor and recipient cells), achieved particularly through simultaneous donor bone marrow infusion with the organ transplant, hold promise for reducing or eliminating the need for immunosuppressive drugs.

With the success rate of organ transplantation increasing, surgeons and patients alike face a new dilemma— shortage of donor organs. As of April 1, 1998, more than 60,000 patients in the United States alone awaited donor organs; in 1996, almost 4,000 Americans died awaiting organs. An area of serious research to help solve this dilemma is *xenotransplantation*— transplanting the organ from another species into a human. In 1992, Dr. John Fung, chief of transplant surgery and his team at the Starzl Transplantation Institute, performed the world's first liver transplant from a baboon to a man dying of hepatitis B. Although the man died of a brain hemorrhage shortly after the transplant was completed, researchers learned from the attempt that that rejection could be managed with immunosuppressants. A second failed xenotransplant brought a temporary halt to the practice until further research, along with the decision to use organs from non-primates. Pigs were chosen for several reasons; however, human antibodies trigger hyperacute rejection of the xenograft organ. To overcome this obstacle, research is now focusing on three fascinating possibilities: the combination of chimerism and **genetic engineering** to breed "transgenic" animals which express human "inhibitory" proteins and prevent host destruction of the transplanted organ; infusion of bone marrow cells into a pig fetus in order to modify its immune system as an adult and allow cohabitation of organ and recipient cells; and genetic manipulation of dendritic cells (DCs)—specialized white blood cells which control the activity of other immune cells—to either "teach" host T cells not to attack the transplanted organ, or by altering donor DCs (which enter the host with the transplanted organ) to produce their own supply of immunosuppressants, thus protecting the transplanted organ from rejection by the host.

Trésaguet, Pierre (1716-1796)

French civil engineer

Pierre Marie Jerome Trésaguet launched an eighteenth-century **road building** revival. Not since the decline of the Roman Empire in the fourth century had roads received such attention from planners. He began his road-building career at the École des Ponts et Chaussées in 1747, which put him in control of 25,000 miles (40,000 km) of roads and **bridges**.

In 1764, he began working on the road at Limoges, introducing a road foundation consisting of large base stones over which a shallow layer of smaller stones was laid. Trésaguet also introduced the use of **pitch** as a binding material.

In 1775, Trésaguet was named France's inspector general of public works. Under his direction, the country developed the finest road system of the time, one which used the same routes the Romans had surveyed thirteen hundred years earlier.

Trésaguet's reliance on foundation stones and pitch was adopted in England by **Thomas Telford** on the Holyhead Road in the 1820s. However, the roads of Tresaguet and Telford were expensive to build and were later abandoned for the simpler methods of **John McAdam**.

Trevithick, Richard (1771-1833)

English engineer

Contrary to the beliefs of the time, Richard Trevithick proved that pressures higher than the atmosphere's could be safely used to power machinery.

Trevithick was born in Cornwall, England, on April 13, 1771. His father managed a coal mine near there, and Richard had an early exposure to the **steam engines** made by **James Watt**. When Watt's patents expired in 1800 Trevithick built a double-acting engine, which was used in sugar mills, corn mills, ironworks, and pumping stations throughout South Wales. With pressures above 145 pounds per square inch, Trevithick's high-pressure engines could safely use a smaller cylinder to produce more power. Additional power was tapped by releasing waste steam into the smoke stack to increase the fire's draft and temperature.

Trevithick applied many of the same principles towards building a **locomotive**, which, on Christmas Eve 1801, carried passengers a short distance over a poor road. This demonstration proved that smooth metal **wheels** meeting a smooth metal track could indeed create enough friction to move weight. An 1804 model, the *New Castle*, was tested on an existing horse-tram route with cast-iron rails. Unfortunately, this locomotive broke the rails. Many of Trevithick's discoveries and principles were used by **George Stephenson** and **Robert Stephenson**, who eventually were called the "parents" of steam-powered railway transportation.

In his later years, Trevithick spent time designing dredging and **threshing machines**. He attempted to dig a **tunnel** under the Thames River using a **dredger** he created, but was unsuccessful, and the venture bankrupted him. Trevithick left for Peru and Costa Rica in 1816 where he planned to introduce his improvements to the steam engine, and his plan for linking the Atlantic and Pacific Oceans by rails. He was again unsuccessful in this venture and after borrowing money from Robert Stephenson, who was now rich from profits gained through railways that relied on many of Trevithick's ideas, returned to England. He died in poverty in 1833, and was buried in an unmarked grave only after the men in his workshop contributed enough money for his funeral.

Tricycle • See Bicycle

Trolley • See Streetcar and trolley

Truck

One of the major concerns of industrialized nations is the transportation of raw material and manufactured goods from one point to another. Prior to the introduction of the truck, the primary means of transport were rail and water. These methods limited the locations that could be serviced by the transportation system. Clearly, waterways were arranged by nature, and their routes were difficult to alter. The rail system was limited by the expense of building rail lines, especially in undeveloped regions of the country.

When the **automobile** was introduced, industrialists quickly realized the advantages of using this technology to meet their transportation needs. Truck designs began to circulate in the market place in 1898, when the German company Daimler presented a vehicle in which the driving compartment was positioned over the **engine**, allowing more room for cargo in the back. Although this met the need for space, the vibrations from the engine caused discomfort for the driver. It was not until the 1930s that engine vibrations were reduced enough to make this design practical.

Meanwhile, Charles Martin, an American inventor, introduced a tractor-trailer combination in which cargo compartments of various sizes could be attached to a standard tractor unit. Martin's design, developed for the Knox company, featured a "fifth wheel" turntable which was mounted on the truck's rear axle, causing the trailer's weight to be distributed to the truck's wheels, rather than the trailer's frame. This increased the efficiency of the truck because the frame could be constructed of lighter materials because it did not have to bear the full weight of the goods inside. Perhaps the best known model of this design was the 1920 British Scammel which boasted a forty-seven horsepower engine.

The separate tractor-trailer design was not alone in its success. A three-axle rigid chassis with combined cab and cargo space was used successfully by the Goodyear Tire Company to promote its innovative line of large, air-filled **tires**. The truck was eventually designed to distribute drive power to both rear axles, which contributed to its success. With the 1908 invention of the universal joint by Wisconsin engineers Otto Zachow and William Besserdich and its application to trucks in 1912, front-wheel drive was made available in cargo-carrying vehicles. The United States Army put this technology to use shortly thereafter, ordering 15,000 trucks with front-wheel drive for use on battlefields during World War I.

All of the variations in truck style from 1898 to the end of World War I had at least one characteristic in common: they were all primarily designed for the transportation of dry goods. The need to transport liquids had been overlooked by the trucking industry until 1926, when the frameless tanker made its debut. This model was composed simply of a large tank mounted on a "fifth wheel" turntable in the front and an axle in the back—it needed no other frame for support. Its capacity for hauling all sorts of liquids and gases was a welcome development, and the design was so successful that it remained virtually unchanged for thirty years.

Perhaps the most important invention for the trucking industry was the diesel engine. First developed for the automobile by the Benz company in 1923, this engine was not immediately successful because it was expensive and difficult to start. Its fuel economy and greater pulling power, however, made the diesel engine attractive to the commercial transportation industry. Orders for trucks with diesel engines began to flood the Mercedes-Benz company, and, by 1931, they no longer offered trucks with **gasoline** engines.

Throughout the twentieth century, many changes have taken place in commercial transit. **Ships** and **trains** started out as the major means of conveyance and remain an important part of the system. The airplane has increased the speed with which certain commodities can be moved. But the truck above all has risen to become the kingpin of commercial transportation.

Tsiolkovsky, Konstantin (1857-1935)

Russian engineer

At the age of ten, Tsiolkovsky became almost totally deaf due to scarlet fever. Consequently, he spent much time alone, but his father, recognizing his son's potential, encouraged his interests in math and physics. At the age of twenty-two Tsiolkovsky became a teacher, but he also continued to pursue his own experiments in physics and invention. His interests extended to **airships** and he developed the first **wind tunnel** in Russia to experiment on how much friction a metallic plane could generate at a certain speed.

In 1895, he first mentioned space travel in an article that he fully expected never to be published. It was, however, and thoughts of travel in space began to dominate him. By 1898, he completed a preliminary study of space travel and outlined many of the basic concepts now taken for granted by scientists. He proposed that humans could only enter space in a sealed cabin with oxygen reserves and air purification devices. He also knew that **rocket**s could be the only means of propulsion in empty space because, unlike conventional engines, they did not rely on oxygen for combustion. Realizing that a rocket powerful enough to carry humans into space needed fuel with a higher exhaust speed, he advocated liquid fuels such as **kerosene** as opposed to the solid fuels then being tested by engineers. His theories in this regard predated the research of **Robert Goddard**.

Tsiolkovsky's ideas were initially dismissed by the Russian scientific community. However, following the October Revolution, the Communist government began to look more closely at his work and in 1921 awarded him a pension, allowing him to concentrate fully on his studies. Dr. Jakov Perelman, a writer and editor, helped to popularize Tsiolkovsky's ideas and by the mid-1920s Tsiolkovsky had garnered international attention. He also based a novel, *Outside of Earth*, on his research, which told of a journey through space, furthering his reputation among the general public.

In his later life, Tsiolkovsky was given many honors. His seventy-fifth birthday prompted glowing tributes in Soviet papers and honors from the Communist government. Following his death, the launching of **Sputnik 1** was timed to coincide with the one hundredth anniversary of his birthday, but missed by twenty-nine days. Nevertheless, their efforts affirmed Tsiolkovsky's belief that "the Earth is the cradle of the mind, but one cannot live forever in the cradle."

Tsvett, Mikhail • See Chromatography

Tuberculin test

Tuberculin tests are used to detect the presence tubercle bacilli, the pathogen which causes tuberculosis (TB), a serious infectious disease that attacks the lungs. Once well under control in the Western World, more than 400,000 new cases of this reemerging disease are reported in industrial countries every year, and it kills more than three million people world-wide annually.

Tubercle bacilli was discovered by German bacteriologist Robert Koch (1843-1910) in 1882. Although unable to find a cure, Koch did find a way of diagnosing the disease. He developed a test for sensitivity to the tubercle bacilli in 1890, which gained him the 1905 Nobel Prize for physiology or medicine. Today there are several ways of testing for the presence of TB. For non-pulmonary TB, specimens are taken for biopsy or microscopy and culture. Pulmonary TB tests consist of chest X-rays and a sputum smear test to detect tubercle bacilli. The tuberculin skin test—administered in one of two ways—tests for tuberculolipoprotein hypersensitivity from previous infection only. The substance used in this test is a purified protein derivative (PPD) which comes from the filtrates of laboratory cultures of tubercle bacilli. The *Mantoux* test is an injection of tuberculin PPD into the skin of the forearm. The *Heaf* test, (or tine test), used more for mass screening of low-risk populations, involves smearing undiluted strength of tuberculin PPD on the skin of the forearm which is then punctured with a six-pin Heaf apparatus. In each case, a hardening of the skin within 48-72 hours causing inflammation over 10 mm in size indicates that the person was at some time infected with the TB bacteria. However, a positive test does not indicate active disease; previous immunization against TB will likely show a positive test.

The skin test currently used was developed by Dr. Florence B. Seibert, a biochemist, who first isolated the active protein in tuberculin and used it to make TB skin tests more accurate. If a person shows an inflammatory reaction, a physician may recommend a chest X-ray to see if there is a spot on the lung.

TULL, JETHRO • See Seeding devices

TUNNEL

Tunnels are used for transportation, mining, and drainage purposes, as well as for installation of power sources. Although tunneling is difficult and dangerous, the benefits are often worth the effort. Early civilizations relied on tunnels to transport water for drinking and irrigation. The Egyptians tunneled into cliffs to construct temples, and in 2100 B.C., the Babylonians went so far as to build a pedestrian tunnel under the Euphrates River by diverting the river during the tunnel's construction. The Greeks and Romans practiced tunneling extensively, the Romans constructing **aqueduct** tunnels through mountains.

Tunnel technology went through a long hiatus until about the 1700s, when tunnels were included in the development of **canal** networks in Europe and North America. The use of **gunpowder** blasting in excavations of solid rock was a major advance; different explosives are used today, although the blasting method is still similar. As railroads gained prominence beginning in the 1830s, tunnels became crucial elements of the rail system. Since **trains** had to operate on the lowest grade slope possible, tunnels were built in such great numbers that tunnels and trains became closely associated with each other. The father-and-son team of **Marc Isambard Brunel** and **Isambard K. Brunel** engineered the Thames River Tunnel, the world's first underwater tunnel, in the early 1800s. Several cave-ins led them to develop a cast-iron tunneling shield to protect workers in 1825. Peter Barlow improved this shield in 1869, and James Greathead (1844-1896) used compressed air to reduce seepage of water and mud into the work space.

Compressed air was also used in construction of the London subway in 1886. In 1855, the 4.5 mi. (7.2 km) Hoosac Tunnel, begun in the Berkshire Mountains of Massachussetts, became the first railroad tunnel in the United States. It took eighteen years to complete, but many new procedures were developed in the process. While hand-drilling had been standard practice before the Hoosac project, **dynamite** and compressed-air **drills** were used for the first time during its construction.

The twentieth century has seen the development of several tunneling machines. The rotary excavator, or *mole*, forces its way forward through the bedrock. The loosened rock is then carried back through the tunnel in muck cars. *Rock mechanics* has become a major part of tunnel planning, since a number of disasters have occurred because the rock structure was either not evaluated or was misjudged. Accordingly, the mass of the rock must be analyzed and geological stress predicted, since stress from earthquakes and other past geologic events can affect the tunnel. Another important consideration in tunnel construction is *stand-up time*—the amount of time, whether seconds or hours, that an excavated stretch will stand without support. The tunneling crew has to reinforce the tunnel at its *heading* or *excavation face*.

Most modern tunnels have been built to accommodate **automobile** traffic. Since the grade of the roadbed is less critical a factor than it is with railroads, road cuts can suffice for a highway where a tunnel might have been necessary for railroad tracks. Many major automobile tunnels run under rivers and harbors, and *trench* tunnels are preferred for these situations. In this method a trench is dug at the bottom of the body of water, and the tunnel sections, often double-barreled for opposite traffic flow, are lowered into the trench, then buried and

Konstantin Tsiolkovsky

drained of water. *Pipe-jacking* is used for smaller tunnels; this involves lowering pipe segments into a vertical shaft, then moving them into place as the tunnel boring progresses. Mining shafts and tunnel networks can extend miles into the earth, both vertically and horizontally.

Despite all precautions, underground mining is quite dangerous. Gases can accumulate in pockets of a mine, causing explosions; the noise from the machinery can damage hearing and make communication difficult; and cave-ins are a constant threat. Construction on the English Channel Tunnel between England and France, an engineer's dream for centuries that was envisioned and encouraged by Napoleon, was begun in 1987. Referred to as *the Chunnel*, it was completed in 1994 at a cost of $13 billion. The two rail tunnels (one for northbound and one for southbound traffic) and one service tunnel are each 38 km (24 mi) in length and have an average depth of 40 m (131 m) under the seabed. It is the first physical link between Britain and the European Continent. Passenger rail service is provided, as well as the ferrying of automobiles and trucks. Travel times from London to Paris have been reduced from more than five hours (over sea) to three hours via the Chunnel.

The Seikan Tunnel in Japan was placed in service in 1988. The 53.85 km (33.44 mi)-long tunnel connects the northern tip of Japan's main island of Honshu with the island of Hokkaido, passing under the Tsugaru Strait. The Seikan Tunnel is the world's longest submarine tunnel, involving excavation 100 m (330 ft) below the seabed across a strait where the sea is up to 140 m (460 ft) in depth.

Other such monumental tunneling projects will likely be attempted. It may be more realistic to consider a tunnel or submerged conduit for the Strait of Gibraltar, for example, than to attempt to bridge it.

See also Train and railroad

TURBOJET ENGINE • See Jet engine

TURBOPROP ENGINE • See Jet engine

TURING, ALAN M. (1912-1954)

British mathematician and computer scientist

Alan Turing was one of the leading theoreticians of **digital computer** science during the 1930s, 1940s, and 1950s. Throughout his life, he used mathematics as a way to explore

In 1991, British workers welcomed their French counterpart after a drilling machine broke through the last section of the "chunnel," the 30 mi (48 km) tunnel that now spans the English Channel.

whether nature, including human thought, worked like a machine. He is generally considered to be the founder of the field of **artificial intelligence**.

Turing was born in London, England. His father was a colonial civil servant and his mother came from a family of scholarly men and women. Turing received his undergraduate degree in mathematics at Cambridge University in 1935. He spent the next three years doing graduate work at Princeton University, where he worked with the logician Alonzo Church.

While at Princeton, Turing defined a theoretical machine that used a binary code to solve mathematical problems or carry out a series of instructions one step at a time. The machine could work with only a specified number of conditions, and it used a tape divided into squares, each of which carried a symbol. The machine scanned one square at a time, remembering all the symbols. The machine's behavior at any one time depended on the number of conditions and the number of remembered symbols. It could change its behavior by changing either the number of conditions or by shifting from one square to another. Furthermore, the machine could be employed for a special purpose or for universal use on any special problem, as long as the tape carried the necessary instructions.

Turing also proved logically that some problems could not be solved by such a machine. He soon learned that several other theorists had devised their own systems to produce an equally strong proof. One was Church's lambda (λ) calculus, which is now used in **artificial intelligence** programming. The Turing machine became a starting point for modern automata theory, a formal mathematical study of computers and other logical machines.

During World War II, Turing helped design computer-like devices, such as the **Colossus** series, that broke German military codes produced with its Enigma cipher machines. While the war was in progress, Turing again spent some time in the United States, possibly conferring with the mathematician **John Von Neumann** of the Institute for Advanced Studies in Princeton, New Jersey, who was working on computer designs.

After 1945, when the war ended, Turing designed the Automatic Computing Engine (ACE) at Britain's National Physical Laboratory. However, the existing technology was not adequate and computer construction was very slow.

Frustrated, he joined the faculty at Manchester University in 1948, where he turned his efforts to the nature of intelligence and information. Like the American Norbert Wiener (1894-1964), the founder of **cybernetics**, Turing believed that intelligence was strictly a matter of processing information, and that human thought and the operations of an intelligent machine differed only by degree. He was convinced that just as artificial sensors performed the same operations as human ones, such as cameras and eyes, computers could perform the functions of the human brain.

In 1950, Turing invented a test for an intelligent machine: If a knowledgeable person cannot tell whether a problem is being solved by a computer or a person, the computer that solved the problem is ''intelligent.'' Today some people are still trying to design computers and programs that can pass Turing's test. However, not all students of artificial intelligence support Turing's definition. Some believe that human thought involves more than computation alone.

Turing also investigated the mathematical aspects of biological interactions. He was named a Fellow of the Royal Society in 1951. He died in 1954, probably by suicide and possibly because of concerns that his homosexuality would become publicly known.

TURN SIGNALS • See Directional signals, automobile

TV DINNER

The original TV dinner was introduced nationally by C.A. Swanson & Sons in 1954. Clarke and Gilbert Swanson, whose father had founded the company, were responding to a new social force: **television**. At the time, television programming was only available three or four hours a day, so if you had a set, you watched it during those hours. Since most of the programming was on in the late afternoon and early evening, that's when people gathered around their sets. The Swansons' idea was to create a portable meal that people could carry into their living rooms to eat.

The Swanson TV dinner was the first widely available frozen dinner. The original dinner contained roast turkey, cornbread dressing and gravy, peas, and sweet potatoes in a three-compartment aluminum tray. It was packaged in a box designed to look like a television, with the food pictured inside a ''screen'' surrounded by what looked like a wood-grain console complete with knobs at the bottom. In 1987, Swanson's original aluminum dinner tray took its place in the Smithsonian Institution.

Swanson soon added fried chicken and roast beef dinners as well. The TV dinners were an instant success. In their first year of national distribution, 10 million dinners were sold. Today frozen dinners and entrees still comprise the largest category of **frozen food**, with over $4 billion in annual U.S. sales. In 1994, 382 new dinner and entree products were introduced, up nearly 50 percent from the previous year.

TYPEWRITER

The typewriter is a machine that prints characters one after the other on sheets of paper. Many attempts to design a character-printing machine were made in the eighteenth and nineteenth centuries, especially to create raised characters for reading by the blind. The first recorded patent for a typewriter was issued by Queen Anne (1665-1714) in January 1714 to an English engineer named Henry Mill. Mill described ''An Artificial Machine or Method for the Impressing or Transcribing of Letters Singly or Progressively one after the other,'' but no drawing or model of the device exists.

The first United States patent for a typewriter was issued in 1829 to **William A. Burt** of Detroit, Michigan, for a ''typographer.'' The letters on this table-size printer were set around a circular carriage, which was rotated by hand—a very slow process. When Burt failed to attract financial backing, he in-

The daughter of Christopher Latham Sholes pictured in 1872 using one of her father's early typewriters.

vented a very successful solar surveying **compass** instead. An improved circular-carriage machine was patented by Charles Thurber (1803-1886) in 1845; it rotated automatically and featured an inked roller. The ancestor of the type-bar machine was invented by Xavier Projean, a printer in Marseilles, France. His *machine cryptographique* had each character mounted on a single, separate bar. Projean proudly claimed that his machine would write almost as fast as a **pen**.

The invention of the modern commercial typewriter is credited to **Christopher Latham Sholes**, an American printer and editor, with the help of Carlos Glidden and Samuel W. Soulé. Sholes and Soulé had been working to develop a machine to print book page numbers; Glidden suggested designing the machine to print letters of the alphabet as well. Glidden also pointed out to Sholes a *Scientific American* article about a typewriting machine recently invented by John Pratt of London. With help from Glidden and Soulé, Sholes designed a type-bar machine with a carriage that automatically moved one space to the left when a letter was printed, and keys that worked with ''pianoforte'' action—they all struck the platen at exactly the same point. Sholes built a working model of his machine in 1867. Glidden, Soulé, and Sholes patented their design in 1868. For the next five years, Sholes worked to improve his machine, driven by his financial backer, James Densmore. In particular, he struggled with the problem of colliding type bars when fast typing was attempted. He finally solved this by devising a unique keyboard arrangement, with the most-used letters spaced apart from each other. This QWERTY keyboard remains in standard use today, even though more efficient arrangements have been devised.

In 1873, Densmore and Sholes interested the Remington Arms company in their machine. Remington was struggling to find new products, since the arms manufacturing business had collapsed following the end of the Civil War. The company bought Sholes's patents for $12,000 and put the Remington Model 1 on the market in 1876. One of the first was purchased by Mark Twain (1835-1910), who used it to produce the first typed manuscript for a publisher. The typewriter did not catch on immediately. Remington gave up on it, selling their rights to it in 1886. By the early 1890s, however, business offices had discovered the new machine. The mass market in typewriters boomed, and with it came profound social change: women by the millions had a new, respectable form of employment.

Many improvements were soon made to the early Remington model. A shift-key mechanism was added in 1878 so lowercase as well as capital letters could be typed. Double keyboards, featuring separate keys for capital and small letters, also appeared, but they didn't work well for touch typing and eventually disappeared. American inventor John Williams developed the front-stroke machine in 1890 so typing could be seen as it was being produced. The first portable typewriter, the Blick, was available in the early 1890s. In the 1920s electricity was added to the typewriter, for more uniform, effortless, and faster operation; portable electrics were introduced in 1956. IBM came out with its spherical type element and stationary paper carriage in 1961. Typewriters with correction tape appeared on the market in 1973, and rotating print wheels were added in 1978. IBM introduced the first electric typewriter with a memory in 1965; Olivetti offered the portable electronic typewriter in 1980. In the late 1980s and early 1990s, voice-activated typewriters were being developed, but they could not create text nearly as quickly as a moderately speedy typist. By 1993, with personal computers becoming less expensive and more sophisticated word processing software available, typewriters' usefulness became increasingly limited. Only 800,000 typewriters were sold in the United States in 1992. By 1995, the once-preeminent American typewriter manufacturer, Smith Corona, declared bankruptcy. At the dawn of the twenty-first century, typewriters seem destined for oblivion.

Typing correction fluid

A white fluid used to correct typing mistakes was invented by Bette Nesmith Graham who herself was a rather poor typist.

Born Bette Claire McMurray in 1924 in Dallas, Texas, Graham graduated from high school when the firm where she was a secretary sent her to night school. By 1951 she was working as an executive secretary at Texas Bank & Trust in Dallas. Since her typing skills had not significantly improved, her new **typewriter** with its carbon film ribbon presented a problem. When she tried to use her **eraser** to correct errors, a smudged black splotch was left on the **paper**.

Graham reasoned that she should be able to cover her mistakes with white, tempera waterbase paint and began bringing a small bottle and brush to work with her. Her co-workers began to request bottles of their own, and by 1956 she had a thriving business going. In 1957 Graham and her son Michael filled more than 100 bottles per month of ''Mistake Out'' in their garage, and Graham decided it was time to patent her product and trademark the name. She also changed the product's name to Liquid Paper. An article about her invention appeared in a national magazine and orders increased.

It took more than 10 years, but by 1968 Liquid Paper produced more than 10,000 bottles a day and sales totaled $1 million. By the middle of the next decade, the company was producing 25 million bottles a year, and in 1979 the Gillette Corporation bought Liquid Paper for more than $47 million and agreed to pay royalties on every bottle sold until the end of this century. In addition to liquid paper, correction tapes are used to correct errors. These tapes are available in rolls, or as an integral part of ribbons on electric typewriters.

Graham died in 1980 leaving half of her $50 million estate to her favorite charities and the other half to her son, Michael Nesmith, who gained fame as a member of the musical group, The Monkees.

U

ULTRAMICROSCOPE

The ultramicroscope is a tool used for viewing very small particles or objects that are too minute to be seen through a conventional microscope. When first invented it was the most powerful device available for observing such specimens, but it has since been surpassed in effectiveness by the **electron microscope**.

The ultramicroscope was invented by the German chemist **Richard Zsigmondy**, an industrial researcher at the Schott Glass Manufacturing Company, which provided specialized **glass** for the optical equipment built at the renowned Zeiss Optical Works. Zsigmondy had become interested during college in colored glass: it was suspected that tiny particles suspended in glass produced glass tints, but the extremely small size of the particles made them impossible to examine under a microscope. Conventional **compound microscopes** view beams of light reflected off or through a sample; thus, in order to be resolved by a microscope, the object *must* be at least the size of one wavelength of visible light. Since most particles in colloid compounds—such as colored glass—are smaller than a wavelength, they are beyond the resolving power of any conventional microscope.

Zsigmondy found that colloid particles, when placed within the path of a beam of light, would scatter that light, and that the particles could be viewed as tiny bright dots against a dark background (like motes of dust swirling in a beam of sunlight). This method, called *dark-field illumination*, is the basis of ultramicroscopy. The ultramicroscope shines a strong horizontal beam (such as that from an arc lamp) upon the subject particles. The rays scattered by the particles are brought together by a series of condensing **lenses** and are then focused through a final eyepiece lens. Particles as small as 10 millionths of a millimeter can be observed through an ultramicroscope.

ULTRASOUND DEVICES

Before World War II, there was no way to accurately diagnose the condition of a fetus without surgery—a procedure that was extremely risky for both the mother and the unborn child. An alternate method was the use of **X-ray** radiation. However, a 1958 study showed a much higher rate of leukemia among children exposed to radiation *in utero*, and that same year, a new science officially began—ultrasonography. Using high-frequency sound waves, physicians could harmlessly observe the condition of an unborn child. By the 1990s ultrasound had become the second most widely used method for diagnostic imaging.

The man most responsible for the development of ultrasound technology was English physician Ian Donald. Donald grew up during the years when British scientists were developing **sonar** systems in response to the sinking of the *Titanic* in 1912 and to protect ships from German U-boats. While Donald was serving in the Royal Air Force, the Allied forces were experimenting with **radar** and sonar devices. Although initially classified as top secret, information on this technology was released to the scientific community following the end of World War II—about the same time Donald began his medical career. Donald was certain that ultrasonics could be applied in medicine, particularly in his field of obstetrics, and began collaborating with engineers in the early 1950s in an attempt to modify current ultrasonic devices to the task of observing the human body. After several years of failure, Donald appealed to the Kelvin Hughes Company, then the largest producer of industrial ultrasonic equipment. At this time, ultrasonics were used chiefly in the testing of machine parts, using beams of high frequency sound waves to detect cracks, flaws, and bubbles in the metal. With the financial and scientific support of Kelvin Hughes, Donald succeeded in constructing an ultrasonic device for medical diagnosis. It was first tested in 1957, when Donald used sound waves to correctly diagnose the na-

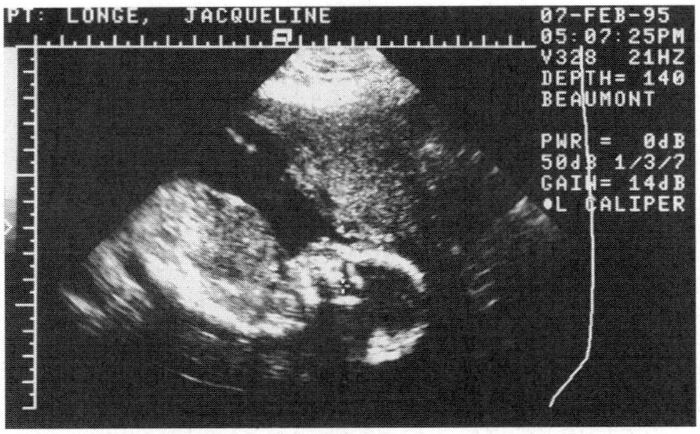

An ultrasound image of Anabelle Ashlyn Longe at 20 weeks.

ture of a patient's heart condition. A year later the procedure called ultrasonography was being used on pregnant women.

Ultrasonics works because sound waves of very high frequencies (from one to ten megahertz) easily and harmlessly penetrate human flesh. As they encounter different materials such as bone and internal organs, the waves reflect back to the source allowing physicians to identify types of tissue because different tissues reflect waves differently. A machine assimilates this information, forming a picture of the inside of the body. Since the 1960s, improvements to ultrasonographic technology have made it the most common procedure for observing the fetus and, to date, no evidence of harmful effects has been found. Ultrasonics is also used to examine other delicate organs such as the heart, lungs, and kidneys, and researchers are continuously developing devices to provide faster, cost-effective, and higher frequency image resolution diagnostics. For example, ''Sequoia'' high-speed echocardiography allows doctors to view and measure blood flow in the small arteries around the apex—or tip—of the heart for the first time; Digital Image Optimizers, Tissue Doppler Imaging, and Directional Color Angio are new technologies which provide detailed pictures of tissue previously difficult to image accurately; in 1996, the Food and Drug Administration (FDA) approved a High Definition Imaging (HDI) ultrasound device—already in use for general purpose diagnostics—for use in conjunction with mammography to help determine whether a biopsy is needed if a breast lump is found; in 1998, the FDA approved pre-market application for SAHARA, a clinical bone sonometer which estimates bone density of the heel for diagnosing osteoporosis; and NASA engineers have successfully tested ultrasonic sound for removing ice from—and preventing icing of—**aircraft**.

Along with diagnostic applications, ultrasonics is becoming a useful tool in the treatment of certain conditions: the Cavitron, invented in 1980, which uses a narrow beam of sound waves to destroy a tumor without removing it from the body, is particularly useful in the treatment of brain tumors. A similar method pulverizes gall stones, making their passage much less painful; and, in 1994, the FDA approved the Sonic Accelerated Fracture Healing System (SAFHS), a portable ultrasound device to accelerate the healing of bone fractures in the lower leg and forearm. In 1996, and Australian research scientist, Dr. John Falconer, pioneered a powerful Digital Alpha workstation enabling scanned images of a fetus to be sent via ''tele ultra-sonography'' from remote, rural hospitals to specialist in city-based medical centers; and a group in the United States is in the process of developing an all-digital ultrasound system small enough to fit into a clinician's hand or lab coat pocket.

See also Submarine; X-ray machine

Ultraviolet camera/spectrograph

Unlike so many inventions that are the result of fortunate accident, the ultraviolet camera/spectrograph was intentionally designed for a specific purpose: to record data yielded by space explorations. On its first mission aboard the *Apollo 16* **spacecraft**, which touched down on the moon in April 1972, it fulfilled its purpose admirably. Among the images it provided were the first photographs of the ultraviolet equatorial bands of atomic oxygen that surround the earth. Four years later, aboard *Skylab* 4, it tracked the movements of the comet Kohoutek.

The instrument was essentially a miniature observatory that acquired imagery and spectra at far-ultraviolet wavelengths. A Schmidt camera captured images of the scene. When the instrument was in spectrographic mode, the light captured by the Schmidt camera was instead routed through a dispersing element that split the light into distinct ultraviolet wavelengths. These wavelengths formed spectra that yielded information about the lunar surface, the Earth's atmosphere, the solar corona, and the stars and galaxy.

Developed at the Naval Research Laboratory in Washington, D.C., the ultraviolet camera/spectrograph was the work of two men: William Conway, who adapted the instrument for the *Apollo* 16 mission, and George Carruthers, who designed it. Carruthers, an African-American born in Chicago in 1940 and raised on that city's south side, constructed his first **telescope** at the age of 10. After receiving his Ph.D. in physics from the University of Illinois in 1964, he began a long career at the Naval Research Laboratory in the Rocket-Astronomy Program. For his development of the ultraviolet camera/spectrograph, the National Aeronautics and Space Administration (NASA) awarded him its Exceptional Scientific Achievement medal.

See also Photography; Spectroscope

Umbrella

A collapsible device that forms a canopy to protect the bearer from the sun or rain has been in use for more than 3,000 years. In many countries it has been used as a ceremonial item and to show high rank. The word comes from the Italian word *ombrella*, which means ''little shade.''

The umbrella is known to have first appeared in China in the eleventh century B.C., and was reportedly invented by a woman. Members of the royal family and other aristocrats were allowed to use one and it was usually carried by a servant who trailed several steps behind. It is also known to have existed in ancient Egypt, sometimes even used in battle and to this day is favored by tribal societies in Africa. In ancient Greece and Rome, it was carried only by women.

The umbrella was introduced to Europe by the Greeks, probably about 2,000 years ago. During the twelfth century it became ceremonial garb for the Roman Catholic Church, and in the sixteenth century the Pope determined it to be a symbol of honor and was seen under one whenever he appeared in public. By the middle of the 1700s Englishmen began to use them daily as protection from frequent rain showers, although its first English enthusiast, Jonas Hanway, was teased mercilessly. In fact, cabbies who saw the umbrella as a threat to their business, would purposely spray mud and insults in Hanway's direction.

By the nineteenth century Americans were also in the swing, and the parasol, first cousin to the umbrella, was seen in Paris and London, too. Elegant ladies carried parasols decorated with fringe, lace and beads to protect them from the slightest hint of sun. Umbrellas replaced swords as an emblem of aristocracy among gentlemen.

The first umbrellas were heavy and clumsy, usually made of wood. Whalebone became the next material of choice, and by 1850 **steel** was used. Covers ranged from silk and linen to alpaca. The devices were finally mass produced by the English by the middle 1800s. Modern umbrellas use a lightweight metal or **plastic** frame with a **nylon** or **polyester** cover. Some of them fold into such a small bundle that they can be tucked into a purse or pocket.

UNDERGROUND RAILWAY • See Subway

UNDERWATER PHOTOGRAPHY

English engineer and amateur photographer William Thompson devised the first method of taking underwater photographs in 1856. As he watched a swollen river sweep beneath a bridge during a rainstorm, he wondered how **photography** might be used to assess the damage if the bridge actually collapsed and divers descended to examine it. Accordingly, he and a friend constructed a watertight container in which to place a camera, covering one end of the box with plate **glass.** Over this they fit a wooden shutter raised and lowered by a string, focused the **lens,** and loaded a dry **collodion** plate into the camera. Thompson then rowed into Weymouth Bay and lowered the device to a depth of 18 ft. (5.5 m), pulled the shutter, and left his creation underwater for ten minutes. Unfortunately, the container yielded to pressure, and the camera filled with water. Thompson did obtain a faint image of the bay bottom, submitting it to the Society of Arts in London as the first underwater photograph.

In 1893 Louis Boutan, a French professor of zoology, went underwater in a diving suit and used a magazine camera adapted with a fixed-focus lens and holding six gelatin dry plates; he fitted his modification into a watertight metal container. By means of levers he could release the shutter and change the plates, but the device was cumbersome. Boutan also developed the precursor of the magnesium flash bulb, later supplanted by more sophisticated underwater lighting methods.

American **Charles William Beebe** (1877-1962) took the first underwater photographs at extreme depths in 1935 when he descended nearly 3,000 ft. (900 m) to take photographs through the porthole of a **bathyscaphe**—a submersible ship used in deep-sea exploration and featuring a watertight cabin. Ten years later American geophysicist Maurice Ewing (1906-1974) devised the first automatic underwater camera, used chiefly for photographing deep-water geological structures.

In 1989, Kodak introduced the Weekend 35, a disposable camera housed in a clear plastic body that made it waterproof. The camera could be submerged up to 12 feet (3.7 m). Two years later, Nikon introduced the first 35mm single-lens reflex underwater camera which featured an aluminum alloy body, the Nikonos RS. This camera could be submerged up to 328 feet (1000 m), without having to be housed in a bulky sealed package. With these innovations, anyone could take photographs under water.

See also Photography

UNICYCLE • See Bicycle

UNIVAC

UNIVAC (UNIVersal Automatic Computer), invented in 1950 by the physicist **J. Presper Eckert** and the engineer **John Mauchly**, was the first commercial computer to be widely available. The German engineer **Konrad Zuse**'s Z machines were available several years earlier in Europe, but not in quantity, and they were unknown in the United States. UNIVAC was the first commercial computer to use **stored programs**, in which the computer reads the program from storage into its memory and follows its instructions to process data.

UNIVAC grew out of two computers Eckert and Mauchly worked on for the United States Army at the University of Pennsylvania during World War II, ENIAC and EDVAC. **ENIAC** (Electronic Numerical Integrator And Calculator), which they designed and patented, featured **vacuum-tube** processors and could add and subtract twenty-digit decimal numbers. It also performed multiplication. Though faster than earlier computers, ENIAC required considerable human interaction to perform a particular series of operations. Its program and intermediate processing results were held outside the computer on **punched cards.**

EDVAC (Electronic Discrete Variable Automatic Computer) was more like today's computers. Its memory stored a program as well as data, and its processors handled the program and the data the same way. The computer's overall design was developed by the mathematicians **John von Neumann**, of the Institute for Advanced Studies in Princeton, New Jersey, and Herman Goldstine (1913-), and the logician

Early UNIVAC mainframe computer. Interest in computers soared when the first UNIVAC correctly predicted Eisenhower's victory in the 1952 presidential campaign.

and engineer Arthur Burks (1915-), of the University of Pennsylvania. Eckert and Mauchly designed the EDVAC memory, which consisted of mercury-filled delay lines, in which the incoming electrical signal was converted to a sound wave in the mercury, delaying its output. This process could be repeated as often as necessary to retain the signal.

In 1948, the University of Pennsylvania changed its policy, deciding to take control of patents for equipment designed by faculty members. Mauchly and Eckert left the school and formed the Electronic Control Company in order to build commercial computers based on their ENIAC patents and EDVAC experience. The first UNIVAC model, completed in 1950, had vacuum tube processors. The mercury delay-line memory held one thousand twelve-digit numbers. Permanent storage was magnetic tape, which could hold up to one million characters. Computer instructions were written in **machine language** that the computer could understand directly.

Because of financial difficulties, in 1950 Mauchly and Eckert sold their company, including the UNIVAC and ENIAC patents, to the **typewriter** and calculator manufacturer Remington Rand Company. Eckert stayed with Remington Rand, eventually becoming a corporate vice-president for UNIVAC development. Mauchly became an independent computer developer and consultant.

Despite the computing ability of the new machines, businesses were reluctant to purchase them. The United States Census Bureau, always interested in more efficient data processing, bought a UNIVAC in 1951. But the machine's real popularity began in 1952, when it was used to predict that Dwight D. Eisenhower would win the presidential election by 438 electoral votes over his opponent, Adlai E. Stevenson. Eisenhower won by 432 votes. The first large American corporation to purchase one of the machines was General Electric Co., in 1954, for its accounting department. Eventually, almost fifty UNIVACs were sold, with prices in the million-dollar range.

By the late 1950s, UNIVAC's success brought competition from IBM and other companies, and it soon became one of many mainframe brands on the market. In the 1970s the ENIAC patents were overturned by another computer manufacturer, Honeywell, and by the physicist **John Atanasoff**, who proved to a court that Atanasoff's A-B-C computer was the basis for ENIAC's design.

Computers named UNIVAC were manufactured into the 1970s, as Remington Rand merged with Sperry Corporation to become Sperry Rand, then Sperry Univac. By the mid-1980s, even the name Univac disappeared, as the company eventually merged with Burroughs and was renamed Unisys.

V

Vacuum bottle

The vacuum bottle was one of those devices that had applications far beyond what its inventor had envisioned.

Scottish chemist and physicist **James Dewar** was interested in studying the nature of liquid gases. The biggest hurdle in his work was keeping the gasses cool enough to remain liquid while they were stored in flasks.

In 1872 Dewar hit upon the idea of insulating a flask from the surrounding air by enclosing it in a larger flask and creating a vacuum between the two. The vacuum would prevent the transfer of heat that normally occurred through conduction and convection. To further insulate the flasks, he silvered them. The silver coating would prevent the absorption of radiant energy from the outside or the escape of cold from the inside by reflecting them.

Dewar's invention was a great success, especially when he successfully liquefied oxygen in 1885 and became the first person to liquefy hydrogen in 1898. His original vacuum bottle was made of **glass,** but later models were made of metal, which was stronger and made larger models possible.

Surprisingly, Dewar did not patent his vacuum flask. It was not until 1904 that Reinhold Burger realized its domestic potential: not only would it keep cold liquids cold, it could keep hot liquids hot. Burger offered a prize for a name and, ironically, thermos, the Greek word for hot, won.

Modern vacuum bottles are constructed with walls of glass or **steel;** the stopper and internal support are made of cork, which helps to insulate as well. If dropped, the result in the glass lined bottles is disastrous; one or both of the walls of the container could shatter. This destroys the vacuum, allowing air to enter and eliminating the flask's ability to insulate. Today, many vacuum bottles have stainless steel lines, rendering them unbreakable. Their use has extended to decorative carafes and serving pieces useful for entertaining. Vacuum bottles do not "make" the contents objects hot or cold, they merely maintain the original temperature of the contents when put into the flask.

Vacuum cleaner

As mass-produced **carpet**s became widely available and affordable during the 1800s, the need for an effective method of cleaning carpets grew pressing. The **carpet sweeper** was one solution. Devices for blowing or sucking dirt and dust out of carpets were another approach.

Before the advent of electricity, machines that used the vacuum principle to suck up dirt were hand- or foot-powered, with bellows for pumping air. Most of these required two people for operation, one to power the bellows and the other to run the nozzle. Some cleaners ran on compressed air; but since they failed to recapture much of what they blew out of carpets, they served mainly to redistribute dust and dirt.

The first practical suction vacuum cleaner usable in homes was patented in 1901 by Herbert C. Booth of England. While attending a demonstration of a compressed-air machine that blew dust out, Booth became convinced that suction was the correct method to use. He is supposed to have tested his theory by applying his mouth to the upholstered back of his chair in a fancy London restaurant. Sucking in, Booth extracted a chestful of dust. He promptly designed a vaccuum cleaner with an effective dust filter, powered by a five-horsepower **piston** motor. While Booth's machine worked well, it was extremely heavy, large, and awkward. It worked best when mounted on a wagon, with long hoses run through the windows of a building from the outside.

At the same time across the ocean, two Americans invented their own versions of the vacuum cleaner. Corrine Dufour of Georgia developed an "Electric Sweeper and Dust Gatherer" that used an electrically-powered **fan** to suck dust into a wet **sponge**. A New Jersey plumber, David E. Kenney, devised a very large suction machine that could be installed in the basement of a building with pipes leading to outlets in each room.

Once the small electric motor was developed, the small, convenient home vacuum cleaner became possible. A down-

Advertisement for the Hoover vacuum cleaner from the 1920s.

on-his-luck Ohio inventor named James Murray Spangler was toiling as a department-store janitor in 1907. Plagued with a cough caused by the dust stirred up by the inefficient carpet sweepers he was forced to use, Spangler decided to invent a lighter, more efficient, easier to use machine. His prototype was patched together from a soap box, an old electric fan, goat bristles, a broom handle, and a pillowcase as the bag. Spangler patented his invention in 1908, but his Electric Suction Sweeper Company was not a financial success.

Spangler's luck changed when he sold one of his cleaners to a cousin. The cousin's husband, William H. Hoover, became president of the new Hoover Company, with Spangler as superintendent. Hoover soon noticed that his product sold only when it was demonstrated by someone who could show a potential customer how to use it. The door-to-door vacuum cleaner salesperson was born and secured the Hoover vacuum cleaner's position as the industry leader.

The Hoover machine, like all others of the time, was an upright. The first cylinder or tank style vacuum machine originated in Sweden and was brought to the United States in 1924 by Gustaf Sahlin. This Electrolux glided along the floor and featured a flexible hose that made it possible for the first time to vacuum hard-to-reach places plus furniture and draperies. The portable, cordless, handheld vacuum cleaner with rechargeable power handle was introduced in the mid-1970s by Black and Decker. Its improved version, the Dustbuster, came out in 1979 and became instantly popular.

Vacuum pump

A *vacuum* is a space completely devoid of matter. A pure vacuum does not exist—even outer space has many particles per cubic meter. In practice one uses a *vacuum pump* to partially exhaust a space to the highest degree possible, and this partially-evacuated space is what is commonly, if not completely accurately, referred to as a "vacuum."

That air could be pumped like water and create a vacuum in the process was something **Aristotle** (384-322 B.C.) would never have believed, as he thought that the elements formed a continuum of mass and therefore a vacuum could not exist. His erroneous belief endured for centuries.

However, that changed in 1650, when **Otto von Guericke** decided to determine by experiment whether a vacuum could, or could not, exist. Von Guericke created an airtight mechanism that was able to pump air out of an enclosed container. The resulting vacuum, while of course not perfect, nonetheless dampened sound, extinguished candle flames, and took the life of experimental animals. Guericke initially thought air would seek the lowest level, like water, so he placed his **valve** at the bottom of his container. He later discovered the location of the valve was irrelevant.

Robert Boyle learned of Guericke's invention, realized its scientific potential, and built his own air pump. With the help of his assistant, **Robert Hooke**, Boyle designed a pump that was a considerable improvement over Guericke's. With it they verified Guericke's experiments, and confirmed **Evangelista Torricelli**'s findings on air pressure. This led to the establishment of Boyle's law: there is an inverse relationship between the volume and pressure of an ideal gas, at constant temperature. Boyle was the first to make use of a vacuum-sealed **thermometer**, which was a great improvement over one made by **Galileo**. He also proved Galileo's theory that all objects fall at the same velocity in a vacuum.

Hooke, meanwhile, was giving serious thought to air pumps and vacuums, and how they could perform mechanical work. He was not alone in his musings. **Denis Papin**, who invented the **pressure cooker**, had built a vacuum pump he called the atmospheric engine in 1690. It used the condensation of steam to create a partial vacuum which moved a cylinder.

In 1698, **Thomas Savery** became the first to pump water using a vacuum. Savery poured cold water over a sealed tank to condense steam and create a vacuum which drew the water. Savery received a patent on his device, which he called the "Miner's Friend," since it was used primarily to remove water from flooded mines.

The atmospheric engine was improved by **Thomas Newcomen**. Newcomen's first engine, which used a **piston** and a cylinder, was set up in 1712. Because Savery held the patent for the atmospheric engine, Newcomen was obligated to make him a partner in the venture. Newcomen engines were used extensively throughout the eighteenth century. It was soon discovered that using steam pressure to push the piston directly worked better than condensing the steam to create a vacuum, and so the **steam engine** was born.

Two hundred years after Guericke's innovation and Torricelli's **barometer**, **Johann Heinrich Wilhelm Geissler** invented an air pump that had no moving mechanical parts. Moving a column of mercury up and down within a tube created a vacuum, used to suck the air out of an enclosed vessel. Geissler created better vacuums than anyone else, and his tubes became known as Geissler tubes. Scientists were later able to pass electrical discharges through his evacuated vessels; their research would eventually lead to the discovery of the electron.

Today there are two main types of pumps used to create a vacuum: the oil-sealed rotary pump and the vapor diffusion pump. The first type compresses a volume of air until the pressure is high enough to force the air out of an exhaust valve. The second type, which is more efficient, is used in conjunction with a rotary pump to provide the primary vacuum. A boiler then injects a pressurized oil vapor that condenses on the walls. This process is repeated and results in the "pumping" of a vacuum. Other specialty vacuum pumps exist, and a vacuum as low as 10^{-15} millimeters of mercury, or one billionth of a billionth of atmospheric pressure, has been reached.

VACUUM TUBE

The invention of the vacuum tube (or *valve*, as it is known in Britain) led to the electronic revolution. Its history begins in 1884.

Thomas Edison, while working on his **incandescent light bulb**, inserted a metal plate between glowing filaments. He observed that electricity would flow from the positive side of the filament to the plate, but not from the negative. He didn't understand why this was so and treated this Edison Effect as a curiosity. Unwittingly, he had created the first **diode**.

Later, **John Ambrose Fleming** of England, one of Edison's former assistants, became involved in designing a **radio** transmitter for **Guglielmo Marconi**. In 1904 Fleming realized that the diode had the ability to convert **alternating current** (AC) into **direct current** (DC), and incorporated it into his very efficient radio wave detector. Fleming called his device the thermionic valve because it used heat to control the flow of electricity just as a **valve** controls the flow of water. In the United States the invention became known as a vacuum tube.

In Germany, Arthur Wehnelt, who also worked with thermionic emission, had applied for a patent in January 1904 for a tube that converted AC into DC. However, he neglected to mention the use of the device in radio wave detection and was unable to sell his invention for that purpose after Fleming applied for his own patent.

Lee De Forest (1873-1961) improved on Fleming's valve by adding a third element in 1906, thus inventing the triode. This made an even better radio wave detector but, like Edison, he didn't realize the full potential of his invention; his *audion* created an electrical current that could be amplified considerably.

In 1912 **Edwin Howard Armstrong** realized what De Forest had wrought. He used the triode to invent a regenerative circuit that not only received radio signals, it amplified them to such a degree they could be sent to a **loudspeaker** and heard without the use of headphones. Armstrong immediately found himself involved in a four-way lawsuit. De Forest, **Irving Langmuir** at General Electric, and Alexander Meissner in Germany all claimed the invention as their own.

Diodes were usually made of two concentric cylinders, one inside the other. The cathode emitted electrons and the anode collected them. Fleming's thermionic valve operated at a temperature of 4500° F (2500° C), generating a considerable amount of heat. Deforest placed a grid between the cathode and anode. The electrons passed through the triode's grid, inducing a larger current to flow.

These early vacuum tubes were called "soft valves." The vacuum was not the best and some air remained within the tube, shortening its lifespan. Langmuir devised a more efficient **vacuum pump** in 1915; with a better vacuum, the tubes lasted longer and were more stable. The improved tubes were called hard valves and their operating temperature dropped to 3600° F (2000° C). In 1922 the temperature was reduced yet again, to 1800° F (1000° C), with the introduction of new elements. Indirect heating improved tube efficiency.

Triodes were limited to low frequencies of less than one megahertz. In 1927 American physicist Albert Wallace Hull (1880-1966) invented the tetrode to eliminate high-frequency oscillations and improve the frequency range. A year later the pentode, which improved performance at low voltage, was developed and became the most commonly used valve.

Over the course of years, a variety of vacuum tubes came into use. Low-voltage/low-power tubes were used in radio receivers as well as early **digital computer**s. Photo tubes were used in sound equipment, making it possible to record and retrieve audio from **motion picture** film. The **cathode-ray tube** focused an electron beam, leading to the invention of **oscilloscopes**, **televisions**, and **cameras**. Microwave tubes were used in **radar**, early space communication, and **microwave ovens.** Storage tubes, which could store and retrieve data, were essential in the advancement of computers.

Despite its numerous advantages, the vacuum tube had many drawbacks. It was extremely fragile, had a limited life, was fairly large, and required a lot of power to operate its heating element. The successor to the vacuum tube, the **transistor**, invented by **Walter Houser Brattain**, **John Bardeen**, and **William Shockley** in 1948, had none of these drawbacks. After 1960 the small, lightweight, low-voltage transistors became commercially available and replaced vacuum tubes in most applications.

VAIL, ALFRED (1807-1859)

American inventor

Alfred Vail was born on September 25, 1807, in Morristown, New Jersey, where his father, Stephen, operated the Speedwell Iron Works. After completing his early schooling, Vail worked for his father and became a skilled mechanic. He then returned to school, earning his degree from the University of the City of New York in 1836. At the university in 1837, Vail observed a demonstration by **Samuel Morse** of his **electric telegraph**. Vail immediately persuaded Morse to take him on as a partner, which was a great stroke of fortune for Morse. Vail brought with him his mechanical expertise, a practical inventiveness, and his father's financial resources. Vail agreed to construct a full set of telegraph equipment (at his father's shop) and finance the American and foreign patent-application process, in return for an interest in Morse's rights to the telegraph.

During 1838, Vail helped make many improvements to Morse's original design, including a simple sending key and a much more compact size. He also worked out the final form of the Morse code, and invented a printing telegraph in 1837. By 1838 the partners were demonstrating their perfected telegraph; the superb mechanics of the system were largely Vail's contribution. In the years between 1838 and 1843, while the partners' application for a Congressional grant languished in political limbo, Vail lost interest in the telegraph and began working for his father's firm in Philadelphia, Pennsylvania. However, when Congress approved funds to construct an experimental telegraph line between Baltimore, Maryland, and Washington D.C. in 1843, Vail returned to become Morse's assistant while the line was built. Vail was the recipient, in Baltimore, of Morse's famous first message telegraphed along the line, "What hath God wrought!"

Vail remained with Morse for the next four years, publishing *The American Electro Magnetic Telegraph* in 1845. He retired to Morristown in 1848, intending to manufacture telegraphic equipment, but his plans were never realized. His first wife, whom he had married in 1839, died in 1852, leaving their three sons; Vail remarried in 1855. He died in poverty in January 18, 1859. Because the terms of Vail's partnership agreement specified that all patents would be in Morse's name, it is difficult to tell precisely which telegraphic innovations were Vail's invention, and to what extent.

VALVE

A valve is a device used to control the flow of a substance, usually a fluid, in an enclosed area. Valves date from antiquity and are used today in such varied mechanical systems as faucets, **automobile** engines, dishwashers, and aqualungs.

Early metalworkers used a check valve in bellows of their forges. Check valves are made with a hinged flap or a ball held on a seat by gravity and allow the air to pass in only one direction. When the Greek physicist Ctesibus (second century B.C.) invented a hydraulic water organ that used slide valves to control air intake to the organ pipes, valves found applications in machinery. During the Renaissance, **Leonardo da Vinci** designed a conical poppete valve, or safety valve. This valve is held in place by a spring or weight and opens when the fluid behind it reaches a certain pressure. His contemporary, Aleotti, used a butterfly valve: two hinged plates were fashioned so that they flipped up or down in a pipe. Three hundred years later **Carl Benz** used the butterfly valve with great success to regulate the fuel-air mixture in his automobile **carburetor**.

As scientists developed new materials, such as **steel alloys**, and refined metalworking processes, valves could be used in a greater variety of machines, such as **internal combustion engines**. **James Watt**'s nineteenth-century **steam engine** employed valves to admit low-pressure steam, which forced the **piston** down. Other valves released the air that accumulated in the cylinder. Watt's innovation is now called a slide valve: the moving parts of the valve slide back and forth across the openings made in the cylinder, thus opening and closing them as needed.

For turning on or shutting off the flow of liquids, globe, gate, and needle valves are used. Operated by an outside force, these valves have a plug, disk, or needle on the inside that is screwed down across the flow of water, thus cutting off the flow. To allow the liquid to flow, the **wheel** on top is twisted in the opposite direction, which raises the stem that had been blocking the tube where the liquid was contained.

VAN ALLEN, JAMES (1914-)

American physicist

From his earliest years, Van Allen was fascinated by electricity and the laws of physics. At the age of twelve he and a friend built a high-voltage machine that produced its own "lightning." In high school, his physics teacher regularly chased Van Allen from the laboratory at the end of the day so he could lock up and go home.

As a student at Iowa Wesleyan College and the University of Iowa, Van Allen proved a better scientist than did many professors. He helped develop and check the instrumentation used during Admiral Richard Byrd's (1888-1957) second Antarctic expedition. While testing this equipment, he made cosmic ray measurements, foreshadowing future interest in cosmic rays.

During World War II, Van Allen helped design and build a radio-proximity fuse, a tough assignment because the delicate instrumentation of a radio transmitter and receiver had to withstand the shock of being fired from large-caliber guns. He learned an important lesson for the future: how to get a tremendous amount of scientific instrumentation into a tiny space and to make it so rugged it would withstand unbelievable shock. Many believe the resulting device was the United States Navy's most important World War II achievement.

After the war, Van Allen was asked to join a team of scientists testing German V-2 **rockets** captured during the war. His job was first to determine what experiments should be done, carry out those experiments, and then explain the results. Van Allen served as chairman of the V-2 Rocket Panel from 1947 to 1958. While V-2 testing continued, Van Allen was charged with the task of finding replacement rockets for the program after the V-2s were expended. His research resulted in the Aerobee, a rocket capable of carrying 150 pounds (68 kg) of instrumentation to an altitude of 300,000 feet (91,500 m). To counteract fears concerning the rocket's safety, Van Allen designed a tiltable launch tower to compensate for the winds at the White Sands Proving Ground in New Mexico and installed a **radio** command to cut off the rocket's fuel if it should go off course. Projects studied with the Aerobee included solar radiation, sky brightness, and atmospheric composition.

In 1952 Van Allen introduced the Rockoon. As early as 1949 he had considered the possibility of somehow combining rockets with **balloons**. If a rocket could be lifted above most of the atmosphere and then ignited, it would be able to travel much higher, thus delivering more information to Earth-bound scientists. The rockoons, lifted into the stratosphere by balloon, would then fire off on signal from the ground. These rockets were able to travel an additional 50 to 70 miles (80 to 112 km) closer to space.

During the early 1950s, Van Allen became chairman of the satellite project for the International Geophysical Year, an ambitious program during which sixty-six countries carried out joint research between July 1957 and December 1958. Through Van Allen's efforts, the participants agreed to allow their scientific instruments to be placed aboard the United States' first satellite. Luckily, he also decided to make the package fit on the Army's Jupiter-C rocket even though the government had picked the Navy's Vanguard to be the designated satellite launcher. Under his exacting direction, the instruments were designed to survive the violence of the launch and function far from earth in a hostile environment, and, after laborious refinement, achieved their goal of a package six inches (15 cm) around that weighed no more than 20 pounds (9 kg).

On January 31, 1958, *Explorer 1* was successfully launched with Van Allen's instruments on board. Among the most exiting data to be produced by the satellite's flight was that concerning the cosmic ray activity in space. To better study this radiation, Van Allen refined his instruments for future satellites. Eventually, instruments aboard *Pioneer 3* confirmed that a belt of electrically charged particles surrounded the Earth, trapped by the planet's magnetic field. The radiation belts were named after Van Allen because the discovery had been made by his instruments. Thanks to his use of artificial satellites and their miniaturized instrumentation, Van Allen was able to gain much knowledge about the earth's environment.

For his many contributions, Van Allen has been honored with several prizes, including the prestigious National Medal of Science, awarded to him by President Reagan in 1987.

VAN DE GRAAFF, ROBERT (1901-1967)
American physicist

Robert Jenison van de Graaff was born in Tuscaloosa, Alabama, on December 20, 1901. He earned bachelors and masters degrees in mechanical engineering at the University of Alabama in 1922 and 1923, respectively. He worked briefly for the Alabama Power Company before enrolling at the Sorbonne in Paris. In 1925, he moved to Oxford University as a Rhodes Scholar and earned both a bachelors (1926) and Ph.D. (1928) degree.

Robert van de Graaff (left) demonstrates his generator for the president of MIT in 1931.

While at Oxford, van de Graaff became interested in the difficulties of accelerating particles to high velocities for use in nuclear research. It occurred to him that one technique that could be used would be to deposit electrical charges on a moving belt, and then to accumulate those charges inside a hollow metal sphere. In 1929, he began work on the construction of such a device. The result of this research was the machine that now bears his name, the van de Graaff **generator**.

The van de Graaff generator consists of a long vertical hollow column containing a moving belt. At the top of the column is a dumbbell-shaped sphere that holds the electrical charge swept off the belt. The earliest models of the machine were able to generate a potential of 80,000 volts, although later machines produced particles with an excess of 5,000,000 volts. Van de Graaff envisioned a number of applications for his machine. He was particularly interested in bombarding heavy atoms, like uranium and thorium, with protons. He thought that such bombardment might result in the disintegration of these already unstable atoms, or, if the heavy nuclei caught protons, they might transform into elements with atomic numbers greater than 92.

Later in his career, van de Graaff devised methods for accelerating uranium atoms to be used in the bombardment of other uranium atoms. In the process, he was able to create uranium ions in which up to fifty electrons had been removed from the corresponding atoms. Van de Graaff worked at the Massachusetts Institute of Technology (MIT) from 1931 to 1960. He then resigned to cofound the High Voltage Engineering Corporation (HVEC). He continued his research at HVEC for the remainder of his career, and died in Boston on January 16, 1967.

See also Electrostatic devices

Varnish

Varnishes are coatings that are applied as liquids and dry to hard film finishes. They are used to protect wood, metal, and masonry from water, abrasion, and other damage. They consist of a resin, either natural or synthetic, and a drying oil, which *polymerizes* as it absorbs oxygen from the air. Spirit varnishes are dissolved in a thinner, which evaporates when the varnish is applied, leaving a transparent finish behind. Moisture-cure varnishes are suspended in a water base and dry upon reaction with air.

The earliest varnishes were produced by cooking plant resins with natural oils (such as linseed oil, still used in varnishes), then diluting the mixture with turpentine. By 1000 B.C. the Egyptians had developed a varnish based on the gum of the acacia tree. The Japanese used the sap of a tree known as the varnish tree to produce **lacquer**, a highly polished coating with a durable, water-resistant finish. **Shellac** is a spirit varnish made from a resin secreted by the lac insect. First used in Asia, the resin was collected from the trees, washed, melted, dried into sheets, and then dissolved in alcohol. Shellacs dry quickly and have a light color. They are not moisture resistant, but still find use in sealing wax, hair sprays, and furniture finishes.

Natural resins were eventually displaced by synthetic varieties. Phenolic resins similar to **Bakelite** were first developed for use in varnishes by chemist **Leo Baekeland.** Spar varnishes, an example of this type, are extremely weather resistant and are used for finishes on marine vessels. Phenolics in turn were deposed by alkyds, which have become the most widely used type of varnish today. These resins are combined with an alcohol, an acid, and an oil such as castor, coconut, or soybean, to produce a versatile coating that can also be mixed with colors. **Polyurethane** resins are used in moisture-cure varnishes that require good abrasion resistance and a high gloss. The varnish industry is a growing market which contributes a significant amount to the approximately $16 billion paint, varnish and coatings market.

See also Polymer and polymerization

Vaucanson, Jacques de (1709-1782)

French inventor

Jacques de Vaucanson was a prolific inventor who made many contributions to the Industrial Revolution. Born at Grenoble, France, in 1709, he showed an early interest in machinery. He was educated at the Jesuit College of Grenoble and then went to Paris, France, to study mechanics.

In Paris he became interested in building **automata**, life-like devices driven by **clock**-like mechanisms. Vaucanson's first automation, built in 1738, was "The Flute Player." The next, in 1739, was "The Tambourine Player." His most famous automaton was "The Duck," a robot bird that quacked, swam, flapped its wings, ate, drank, and even digested food. The Vaucanson Duck entertained audiences across Europe. In order to build his automata, Vaucanson designed various precision **machine tools** such as **lathe**s and **drills**.

After being appointed inspector of silk factories in 1741, Vaucanson began designing improved weaving machines. He developed an automated **loom** in 1745 that used perforated cards to select the warp threads. This loom was later improved on by **Joseph-Marie Jacquard** and became one of the most important machines of the Industrial Revolution. During the 1770s Vaucanson built an unconventional **iron**-framed slide lathe that permitted a much higher degree of precision than the old wooden-bed ornamental lathes.

For all of his inventions, Vaucanson continued to design improved machine tools that became widely used. Late in his life, he collected many of his tools and inventions, plus those of others, and this collection was incorporated into the Conservatoire des Arts et Metiers in Paris in 1794. There, Vaucanson's slide lathe has been exhibited ever since, the oldest engineering lathe in existence. Also at the Conservatoire, Jacquard discovered the loom that he then redeveloped. Vaucanson died in Paris in 1782.

Vaulting

Vaulting refers to the space created by an **arch**ed ceiling. Just as a horse vaults across a barricade, arched roofs literally vault across enclosed spaces and were devised to span far more area than could a simple **column** and lintel structure. While a vaulted roof can be a **dome**, domes comprise a class of their own, so a vaulted ceiling usually refers to all other arched ceilings.

Vaults are three-dimensional arched structures and fall into several categories. The simplest is the *barrel* vault, a continuous arch. The *groined* vault consists of two or more *barrel* vaults joined at right angles, and the *ribbed* or *Gothic* vault features pointed arches and decorative ribs.

While vaulting was used in ancient Egypt and the Middle East, it came into wide usage with the Romans, who used the simpler *barrel* vault in drainage systems and developed the *groin* vault for use in buildings. Romanesque columns and ceilings were massive in scale and the space between them small.

The Gothic style that began in the 1100s and continued through the 1500s took vaulting to its peak. The Gothic pointed arch made it possible to take walls and roof to airy heights. Towering Gothic arches were vaulted to create a powerfully spiritual atmosphere in medieval churches.

Vaulting prevailed well into the twentieth century with the construction of churches, courthouses, and monuments. During a Gothic revival between 1870 and 1910, arches and vaults also found their way into residential architecture. American Gothic, with its high roofs and detailed appointments, became a symbol of westward expansion in the United States, but the architecture of the later twentieth century reverted to function and practicality. The Industrial Revolution, two world wars, and an economic depression meant there was little time or money for frills, and the vaulted ceiling and its kindred features have been put aside until the next revival.

VCR • See Video recording

Veksler, Vladmir Iosifovich (1907-1966)
Ukrainian physicist

Veksler was born in Zhitomir, Ukraine, on March 4, 1907. He graduated from the Moscow Institute of Energetics in 1931. He was employed at the All-Union Electrotechnical Institute from 1930 to 1936 and at the Institute of Physics of the U.S.S.R. from 1936 to 1956. In the latter year, he moved to the Joint Institute of Nuclear Research.

Most of Veksler's research involved nuclear and particle physics. He studied the use of Geiger-Müller proportional counters to detect the effects of x-rays and cosmic rays on matter. His most famous accomplishment was the development of the theory of phase stability of particles.

When particles are accelerated in a **cyclotron**, they gain mass. At low velocities, this mass increase is relatively insignificant. But at velocities close to the speed of light—common in most cyclotrons—mass increase becomes so great that particles slow down and fall ''out of synch'' with the electrical field that accelerates them.

In 1944, Veksler demonstrated that one way to deal with this problem was to slow down the frequency of the accelerating electrical field. When that happens, particles are able to keep in step with the electrical field and continue to gain energy, even as they gain mass. This principle, developed independently at about the same time by **Edwin McMillan**, led to the development of a new type of cyclotron known as the synchrocyclotron, the frequency-modulated (FM) cyclotron or, in the then Soviet Union, the phasotron.

For his contribution to the development of **particle accelerator**s, Veksler shared the Atoms for Peace Award with McMillan in 1963. He died in Moscow on September 23, 1966.

See also Geiger counter

Velcro

Velcro is a fabric-strip fastener that takes the place of **zipper**s, **button**s, hooks, and snaps. It was invented by a Swiss engineer named Georges de Mestral. In 1948, on returning from a day's hunting in the Alps, de Mestral wondered what made burrs stick to his clothing and his dog. He examined a burr under the **microscope** and discovered that it consisted of hundreds of tiny hooks that would catch and hold onto anything with loops. De Mestral decided to design a fastener that worked on the burr principle.

It took de Mestral eight years to perfect his invention: two strips of woven **nylon**, one containing thousands of minute hooks and the other thousands of tiny loops. When pressed together, the strips quickly and strongly cling to one another, just like burrs to clothing, and could be meshed and separated thousands of times. De Mestral called his invention Velcro, from the French *velours*, ''velvet,'' and *crochet*, ''hook.'' Velcro was patented worldwide in 1957.

This basic patent ran out in 1978, so competing brands are now manufactured. However, velcro remains a registered trademark, so only the original bears the name. It is used in an amazing variety of applications, from the United States space program to shoelace replacement. The shear strength of Velcro (the force applied to parallel pieces pulled in opposite directions) is 10 to 15 pounds per square inch. This means that the more area of Velcro you have, the more shear strength exists—which accounts for the feasibility of Velcro jumping (leaping off a **trampoline** wearing Velcro-hook coveralls and sticking to a Velcro-loop-covered wall) as demonstrated on **television** by David Letterman.

Velocipede • See Bicycle

Venturi tube

The venturi tube is a flowmeter that allows scientists to measure the drop in air pressure that occurs as a fluid's velocity increases. It was designed by the Italian physicist Giovanni Venturi (1746-1822), but the principles it demonstrates are those of the Swiss mathematician Daniel Bernoulli (1700-1782) as outlined in his book *Hydromachia*.

Venturi's flowmeter is a **glass** tube with a sloped constriction in the middle. As fluid passes into the tube, it does so at a constant velocity; however, as it reaches the constriction, the fluid's velocity increases. When the fluid reaches the end of the tube the velocity drops again to its initial value. Attached to the tube at three points are **manometers** (devices for measuring pressure)—one at the tube's entrance, one at the point of constriction, and one at the exit. During the experiment, the manometers will show a drop in pressure at the constriction, caused by the increased velocity at that point. The venturi tube must be designed precisely so that the fluid passes through without becoming turbulent, especially at the constriction point.

The principles involved in the venturi flowmeter form the foundation of hydrodynamics (the movement of water) and aerodynamics (the movement of air). As air passes around the wing of an airplane, the air moving above the wing moves faster than that below. This unequal velocity causes a pressure drop at the top of the wing, creating the lift necessary to become and remain airborne. Fluid-motion pressure drops are also used in the laboratory to simulate vacuums.

See also Aircraft

Vernier

A vernier is a two-part scale that measures angles and lengths in small divisions. It consists of a large stationary scale used to measure in whole numbers and an attached small movable scale used to measure in fractions, often to .001 in (.025 mm).

This device bears the name of its inventor, French engineer and soldier Pierre Vernier (1584-1638). The son of a lawyer who was probably also an engineer, Vernier early learned to survey with the measuring devices of his time, such as the **astrolabe** of Portuguese Nunez Salaciense (1492-1577). Made up of fixed concentric scales, astrolabes were difficult to contruct and use. Instrument makers had to measure and engrave accurately the degrees. The user had to calculate to arrive at the correct measurement. Invented in 1631, the vernier was much simpler in concept because it allowed the user to measure in small increments without calculating.

The vernier was ahead of its time for in the seventeenth century survey instruments were not precise enough to require the detail that the vernier could provide. It was not until the mid-1700s that it came into regular use. Today the vernier is a standard part of surveying equipment and is built into such implements as microscopes, **micrometers**, **theodolites**, and transits. The vernier is valuable wherever extremely small measurements are need, as in surveying where error multiplies with distance.

See also Maps; Surveying instruments

Vertical takeoff and landing vehicles

Commonly referred to as Vertical/Short Takeoff and Landing (V/STOL) **aircraft**, V/STOL planes arose from aviators' desire to combine the takeoff and landing abilities of **helicopters** with the fast, long-range flight abilities of airplanes. V/STOL technology has been of particular interest to the military, since it both enhances the ability of aircraft to operate from platforms other than **aircraft carriers**, as well as to operate in remote areas where airports with adequate runways are not available.

V/STOL development began shortly after the end of World War II, and after a number of failures (an inevitable part of experimentation with new technology) the first V/STOL prototype, the Hawker P1127, made its initial test flights in 1960. This prototype eventually led to production of the most successful V/STOL aircraft yet produced, the AV-8 Harrier "Jump jet."

Although proven in battle, the Harrier has all the drawbacks of V/STOL aircraft. The dual requirements of helicopter-like and plane-like performance lead to extremely heavy aircraft, and the Harrier can neither carry a large payload nor fly supersonically. To alleviate this, experiments are underway regarding aircraft that can land vertically but take off on a significantly shortened runway. The relaxation of the takeoff requirements results in a greatly increased payload, while still not requiring a full runway or full-size aircraft carrier for safe operations.

A different type of V/STOL aircraft is the helicopter-like V-22 Osprey, which employs two swiveling tilt-rotors that allow it to take off and land very much like a helicopter, while flying with performance comparable to a turboprop airplane. Although the Osprey is considerably slower than the Harrier, managing only 300 knots in cruise mode, it can carry a much larger payload and answer some of the objections about the Harrier's limited transport ability. An outgrowth of the Bell XV-15 tiltrotor aircraft program, the Osprey was developed in the 1980s and made its first flight in 1989, after which funding for the program was cancelled. A renewal occurred in 1992, however, and a lighter, faster version of the Osprey was delivered by Bell Boeing in March of 1997.

Future developments in V/STOL technology may fuse the two concepts above. Planes like the Harrier employ **jet engines** that can provide vertical as well as horizontal thrust, and in general they look more like airplanes than helicopters. The Osprey, on the other hand, looks more like a mutant helicopter, employing tiltrotors than can propel it forward more efficiently than a true helicopter's. Investigations are proceeding that may combine these concepts, employing a tiltrotor that is used for hovering takeoff and landings, and which then literally stops and turns into a wing for fixed-wing flight. The technical difficulties with such a craft are manifest, and remain under development.

Video • See video recording

Video camera • See Video recording

Video games

While many of the high-technology breakthroughs in **electronics** have been focused on advances in the fields of science and business, the video game industry has also grown rapidly, due to sophisticated electronic devices.

The concept of video games had its beginnings on September 1, 1966, with Ralph Baer, a consumer products engineer at Sanders, a defense contracting firm located in Nashua, New Hampshire. While sitting in a New York bus terminal, Baer thought of the sixty million **television** sets in the United States and struck on the idea that games using the sets could be a lucrative proposition. He immediately set to work developing a game in his spare time, and three months later produced a basic game similar in which a spot chasing another around the screen. After showing his prototypes for other games to Sanders executives, Baer and two other engineers, William T. Rusch and William L. Harrison, devised a basic paddle and ball game. By the end of 1967, they had also come up with a video hockey game. On April 25, 1972, Baer and his associates were issued a patent for their creations. Shortly thereafter, Baer licensed his video game technology to Magnavox and the video game industry took off.

Another early pioneer was Noland Buschnel, an American engineer who introduced the game Pong in 1972. An electronic tennis game, Pong became very popular, particularly in arcades. Buschnel formed his own company, known as Atari and in 1976, Warner Communications bought Atari for $28 million and began offering video game cartridges for the home television.

PacMan, a game similar in concept to Baer's first prototype, was introduced in 1983, the product of a joint effort by the Namco Company of Japan and Midway of the United States. With hundreds of thousands of games already sold, PacMan remains one of the most popular video games ever created; however, that claim to fame may be upstaged with the introduction of home video games from Nintendo. The Nintendo Entertainment System (NES) offers players a wide array of game choices, joysticks, wireless remote controls, and many other options. As computer technology continues to advance, the future direction video games will take is unclear. The use of three-dimensional graphic displays, video and laser disks have already begun to appear in the latest games. With the rapid increases in computer chip speeds and capacities, as well as speech recognition/synthesis technology, future video games are certain to bear little resemblance to their crude counterparts of the 1960s.

See also Computer graphics; Computer input and output devices; Computer speech recognition; Microcomputers; Optical disks; Voice synthesizer

Video recording

The very first device used to record a video signal was developed by **John Logie Baird** in 1928. A ''video album'' that stored video information, Laird's invention used a 78 r.p.m. disc identical to those used in audio recording. Soon, the increasing demand for practical video storage made this design obsolete, and scientists began a sustained effort to create an affordable machine for the recording and playback of video signals.

The first machines to feature recording and playback debuted in 1948. Their main drawback was that they used a very high unwinding speed and, thus, required a vast supply of thick, heavy **magnetic tape**. Another problem was the machine's inability to keep the tape speed constant, which caused a great deal of interference and signal loss.

In 1951 Alexander M. Poniatoff assembled a team of scientists at his California laboratory to design what would eventually become the prototype video tape recorder. Among his team were Charles Ginsberg, Charles Anderson, and Ray Dolby. By 1954, the team had developed a recorder with a stationary head, eliminating the need for large quantities of tape. They demonstrated their machine to their parent company, Ampex, in 1955, and on November 30, 1956, it was used in the first retransmission of a recorded **television** program, allowing a California television station to rebroadcast that day's showing of ''Douglas Edwards and the News.''

For more than a decade, Ampex became the premier designer of video recording technology. They debuted the first color video recorder in 1958.

Another American company, 3M, was responsible for the first mass production of video magnetic tape. Mel Sater and Joe Mazzitello knew of Ampex's plans to introduce their recording machine into the general market in 1956. Working long, exhausting hours, they managed to present their own invention on the exact day Ampex's recorder hit the stores.

In 1966 Ampex created the Videofile system, which used a video tape recorder to store up to 200,000 documents on a reel of tape. In 1967 Ampex created a short-term, quick access recorder, which was able to record only thirty seconds at a time, but allowed an operator to locate and display segments in just four seconds. This machine has found a niche as an instant-replay recorder. Ampex also designed the cartridge-loading recorders used to play very short recordings, such as commercials and news briefs.

Since the 1960s video technology has been dominated by Japanese producers, a domination which has its roots in the late 1950s. In 1958 Toshiba produced the first singlehead recorder. This was followed the next year by the two-head machine manufactured by Japan Victor Company (JVC) and in 1964 by the Sony Corporation's marketing of the very first video tape recorder for home use.

With the market expanding, it became necessary for the wide array of video tape recorder producers to agree upon certain standards, the most important being the style of videotape. In 1970 the Philips Company finished its newest project: a video cassette recorder, or VCR. The enclosed cassette design replaced the old, bulky reel-to-reel systems. Within two years JVC and Matsushita, working jointly, developed a cassette using a three-quarter-inch tape that quickly became the industry standard.

As more and more homeowners purchased VCRs they began asking for a new type of machine, one that could be used to film home movies as well as television broadcasts. The first home video cameras came in two varieties: those with separate recorders and those with built-in recording machines. The latter, despite its compact size, was not well received by the public, primarily because it was necessary to use small ''micro cassettes'' that could not be played on a home VCR. Later models have since been developed that use a standard cassette for both recording and playback.

The next generation of video recording devices will most likely employ laser discs in place of magnetic tapes. Not unlike the concept of Baird's video album, the videodisc player uses a weak laser to read microscopic bumps on the disc's surface. A typical videodisc can hold several hours of video information, as well as thousands of pages of documentation—all of which can be accessed immediately. Current models that allow playback but without the ability to record are available for home use.

See also Audiocassette; Magnetic recording; Optical disk; Phonograph

VIDEOCASSETTE RECORDER • See Video recording

VINCI, LEONARDO DA • See Leonardo da Vinci

VINYL

The term *vinyl* refers to the monomers that are the building blocks of this class of **polymer,** vinyl monomers being small molecules containing carbon-carbon double bonds. **Polyethylene**, **polystyrene**, **polypropylene**, polymethyl methacrylate, polytetrafluoroethylene, polyvinyl chloride, polyvinyl acetate, polyvinylidene chloride, polyvinyl alcohol, polyvinylacetals, polyvinyl fluoride, polyvinyl pyrrolidone, polyvinylcarbazole, and polyvinyl ethers are all examples of compounds referred to generically as vinyl polymers. Polyethylene, the simplest vinyl polymer, is made from the monomer ethylene. When polymerized, the ethylene molecules are joined along the axes of their double bonds to form a long chain of many thousands of carbon atoms containing only single bonds between atoms. By adding HCl, HF, CH_3OOH, or CH_3OH, to acetylene one can form vinyl chloride, vinyl fluoride, vinyl acetate, and vinyl methyl ether, respectively. These monomers are polymerized through various methods, and the resulting thermoplastics are processed through injection molding and extrusion.

Henry Victor Regnault (1810-1878) first produced polyvinyl chloride in 1835 by combining ethylene dichloride with potassium hydroxide in alcohol. Ostromislen researched it again in 1912. It was patented in 1913 by Klatte, a German chemist. Klatte also discovered vinyl acetate and vinyl esters. Reppe discovered vinyl ethers in 1927. The first vinyl polymer produced in large amounts was poly(vinyl chloroacetate), used as a **lacquer** during World War I. Poly(vinyl acetate) **adhesives** were introduced in 1926.

Probably the most familiar of these vinyls, however, is polyvinyl chloride (PVC). It was soon discovered that a plasticizer mixed with PVC yields a flexible plastic. By 1931 IG Farbenindustrie had developed an industrial production process using Klatte's methods. Production of PVC, which IG *Farbenindustrie* called PC, began in Germany in 1932 and moved to the United States shortly after that. Production of polyvinyl chloride began in France in 1941 under the name Rhoval. PVC is weather-resistant, is an excellent electrical insulator, and is nonflammable. These characteristics lent themselves to applications on military hardware during World War II.

Today both rigid and plasticized PVC are in great demand. Both are often used in copolymers, most commonly copolymers of vinyl chloride and vinyl acetate. These have been used in everything from records to vinyl floor tiles. Vinyl chloride is also blended with ethylene, propylene, vinylidene chloride, ethyl, n-butyl, or 2-ethyl-hexyl acrylate. PVC is chemically inert, making rigid PVC ideal for piping. Plasticized PVC is used in garden hoses, imitation **leather**, raincoats, and electric plugs.

The Carbide and Carbon Chemicals Corporation developed vinyon filaments and fibers and licensed the American Viscose Corporation to produce them in 1939. Vinyon is a copolymer of 88 percent vinyl chloride and 12 percent vinyl acetate. It was the first **plastic** fiber produced on a large scale in the United States. The copolymer is dissolved in acetone and forced through spinnerets to form fibers. The fibers are subjected to hot air to evaporate the solvents and are then stretched in a process similar to cold drawing. The stretching increases the strength of the fiber but lowers its elasticity. The fiber does not take **dyes** and becomes sticky if heated to over 152°F (67°C). At 170°F (77°C); garments made of the fiber will shrink. Further research led to the development of the fiber Vinyon N, a copolymer of vinyl chloride and acrylonitrile patented in 1947.

Dates of commercialization of important vinyl polymers are as follows: polyvinyl chloride (1927; B. F. Goodrich), polystyrene (1930; I. G. Farben), polymethyl methacrylate (1936; Rohm and Haas), polyethylene (1939; ICI), and polytetrafluoroethylene (1946; Du Pont).

See also Waterproof materials

VIRTUAL REALITY

Virtual reality combines state-of-the-art imaging with computer technology to allow users to experience a three-dimensional simulated environment called cyberspace (a term coined by science-fiction writer William Gibson). This interactive medium incorporates powerful **computers** with video displays, sensors, electronic headsets, and gloves. With these tools, users can both see and manipulate a phantom environment that appears real. Virtual reality tools under development include a whole body suit, which, like diving into water, would totally immerse the user in a virtual world. Although virtual reality has been popularized as a new form of entertainment, it has applications in business, industry, and medicine.

The concept of virtual reality dates back to World War II and the use of **flight simulators** to train pilots, but the computerized version of this phenomenal technology can be traced to the mid-1960s. In 1966, Ivan Sutherland conducted experiments with the first head-mounted three-dimensional displays at the Massachusetts Institute of Technology's Lincoln Laboratory. Although the headset was extremely cumbersome, the user was able to view a computer-generated three-dimensional cube floating in space and, by moving his or her head, inspect various aspects of the cube and determine its dimensions. Sutherland moved to the University of Utah, where he built the first fully functional head-mounted display unit in 1970.

Myron Krueger also worked on the infant science of virtual reality, first at the University of Utah and later at the University of Connecticut. His ''artificial realities'' used both computers and video systems. VIDEOPLACE was first exhibited in 1975 at the Milwaukee Art Center. Using video displays, **computer graphics**, and position-sensing technologies, Kreuger was able to create a virtual environment in darkened rooms containing large video screens. People in the room could see their own computer-generated silhouettes and follow their movements in the virtual world projected onto the screen. In addition, people in two different rooms could see each others' silhouettes and interact in the same virtual world.

Much of the initial development of virtual reality was funded by the military. By 1972, the General Electric Corporation had built one of the first computerized flight simulators, using three screens surrounding the training cockpit to provide a 180-degree field of view that simulated flying conditions. In 1979, virtual reality technology was incorporated into a head-mounted display developed by the McDonnell-Douglas Corporation. Three years later, Thomas Furness III, who had created visual displays for the military since 1966, developed the prototype Visually Coupled Airborne Systems Simulator. Donning a specialized oversized helmet, pilots were presented for the first time with an abstract view of flying conditions instead of a reality-based image. Since they were unable to see anything but the computerized cockpit's field of view, pilots became totally immersed in the graphic representation.

While scientists like Sutherland and Furness concentrated on the visual components of virtual reality, Frederick Brooks began experimenting with tactile feedback, or the sense of touch, in the early 1970s at the University of North Carolina. However, it was not until 1986 that the computer industry developed the tools to simulate tactile experience. Brooks was able to develop his GROPE-III system, which used a specialized remote manipulator based on a device that mimicked arm motions to handle radioactive substances. Specifically, the GROPE-III system generated stereoscopic images of molecules and protein structures that could be felt and manipulated as though they existed in the physical world.

In its most basic rendering, virtual reality systems consist of the computer and software—known as the reality engine-and input and output sensors. In standard computer technology, input devices are the familiar keyboard, mouse, knobs, and joysticks; output devices include the printer and video display. Virtual reality input/output devices include head- and ear-mounted equipment for hearing and seeing, and gloves for controlling the virtual world. The fourth ''component'' is the user, who both directs and reacts to the chosen environment.

The reality engine employs both computer hardware and software to create the virtual world. Reality engines are based largely on the same components that make up a personal computer (PC), although much more computing power is required for the reality engine than what is available in a standard PC.

One key to virtual reality is creating a world that appears real. The images created by the computer and software are extremely complex compared to the relatively simple line-based graphics associated with computer games. Virtual reality images are made with pixels, also known as picture elements, which use hundreds of thousands of dots to make up the images. The more pixels there are per inch, the better or more realistic the image will be. Creating realistic images that can be manipulated is known as ''realization.'' These images can be either opaque, in which all the viewer sees is the virtual world, or see-through, in which the virtual image is projected or superimposed onto the outer world.

Sound enriches the virtual world. The experience of soaring through the air in a simulated cockpit is more realistic if the user hears the roar of the engines. Sound also enhances participation in the virtual world by providing the user with audio cues. For example, the user may be directed to look for the virtual airplane flying overhead.

To incorporate the total experience, the reality engine may use haptic enhancement, which engages the participant's other senses of touch and pressure into the virtual world. Haptic cues however, are complex and expensive and have been used primarily for military and research applications.

Head-mounted display (HMD) units use a small screen or screens (one for each eye) that are worn in a helmet or a pair glasses. Unlike a movie, where the director controls what the viewer sees, including the angle and field of view, the HMD allows viewers to look at an image from various angles or change their field of view by simply moving their heads.

HMD units usually employ **cathode-ray**-tube (CRT) or **liquid-crystal-display (LCD)** technology. CRTs incorporate optic systems that reflect an image onto the viewer's eye. Although more bulky and heavy than LCD displays, CRT systems create images that have extremely high resolutions. In addition, CRT images can be semi-reflective, allowing the viewer to see the outside world as well. Such units have practical applications since the user can operate a machine or other device while viewing the virtual world.

Although LCD technology has lagged behind CRT in picture quality, LCD systems are slimmer, lighter, and less expensive, making them better suited for home use. These units use liquid crystal monitors to display two slightly different images that the brain processes into a single three-dimensional view. Initial efforts to market this technology to home users failed because of poor LCD image quality. But rapid advances in LCD technology have improved the images, and higher quality LCD-based units have become available for home use.

Sound effects in virtual reality rely on a prerecorded sound set that is difficult to alter once the reality engine begins to generate audio. The audio portion of virtual reality is transmitted through small speakers placed over each ear. Audio cues may include voices, singing, the sound of bubbling water, thud-like noises of colliding objects—in short, any sound that can be recorded.

Three-dimensional, or omnidirectional, sound further enhances the virtual reality experience. Sound that seems to come from above, below, or either side provides audio cues that mimic how sounds, like footsteps approaching or a plane flying overhead, are heard in the real world. Three-dimensional sound is achieved through the use of highly complex filtering devices. This technology must take into account factors like interaural time difference (which ear hears the sound first) and interaural amplitude difference (which ear hears the sound louder). The most complex human hearing dynamic is called head-related transfer functions (HRTF). HRTF accounts for how the ear drum and inner ear process sound waves, taking into consideration the various frequencies at which these waves travel as well as how waves are absorbed and reflected by other objects. HRTF audio processing enables the listener not only to locate a sound source but also to focus in on a specific sound out of a multitude of sounds, like distinguishing the call of a peanut vendor out of a crowd at a baseball game.

Gloves in virtual reality allow the user to interact with the virtual world; for example, the user may pick up a virtual block, turn it over in the virtual hand, and set it on a virtual table. Wired with thin **fiber-optic** cables, some gloves use **light-emitting diodes** (**LED**s) to detect the amount of light passing through the cable in relation to the movement of the hand or joint. The computer then analyzes the corresponding information and projects this moving hand into the virtual reality. Magnetic tracking systems are also used to determine where the hand is in space in relation to the virtual scene.

Some gloves use haptic enhancement to provide a sense of touch and feel. In haptic enhancement, the reality engine outputs the tactile experience, which may include force, heat, and texture. Tactile experiences are created by remeasuring a pattern of forces, which is programmed into the reality engine and then relayed back to the user when the appropriate object is touched. Virtual reality gloves may use either air pressure (such as strategically placed, inflated air pockets in the glove) or vibrating transducers placed next to the skin (such as a voice coil from a stereo speaker or alloys which change shape through the conduction of electrical currents) to simulate tactile experience.

Many other virtual reality tools are in the phases of research and development. Remote control robotic or manipulator haptic devices are being tested for industry and medicine. Special wands with sensors, joysticks, and finger sensors such as picks and rings will eventually be as common to virtual reality technology as microwaves are to cooking. The technology to control the virtual world through voice commands is also rapidly advancing.

Perhaps the most impressive technology under development is the whole body suit. These suits would function similarly to the gloves, creating a virtual body that could take a stroll through a virtual world and feel a virtual windstorm.

The potential for virtual reality as an entertainment medium is apparent. Instead of manipulating computerized images of two boxers or a car race, the virtual playground allows the user to experience the event. Disney World's Epcot Center houses a virtual reality system that propels the user on a magic carpet ride like the one featured in the popular animated feature movie *Aladdin.* Although most entertainment applications are primarily visually based, virtual reality players of the future may also experience a variety of tactile events. For example, in a simulated boxing match, virtual reality users would bob and weave, throw and land punches, and—unless they were very adept—take a few punches themselves.

Virtual reality also has practical applications in the realms of business, manufacturing, and medicine. NASA has developed a virtual **wind tunnel** to test aerodynamics shape. Virtual reality holds promise for discovering the most efficient manufacturing conditions by allowing planners to evaluate the actual physical motions and strength needed to complete a job. The McDonnell-Douglas Corporation is using virtual reality to explore the use of different materials and tools in building the F-18 E/F **aircraft**. The study of people in relation to their environments, also known as ergonomics, may also be revolutionized by trials in cyberspace. Engineers at the Volvo car company use virtual reality to test various designs for the dashboard configuration from the perspective of the user. In medicine, virtual reality systems are being developed to help surgeons plan and practice delicate surgical procedures. Philip Green, a researcher at SRI International, is developing a telemanipulator, a special remote-controlled **robot**, to be used in surgery. Using instruments connected to a computer, doctors will be able to perform the operation in cyberspace, while the computer sends signals to direct the telemanipulator. Virtual reality may even have applications in psychiatry; for example, someone with acrophobia (a fear of heights) may be treated by having the patient stand atop virtual skyscrapers or soar through the air like a bird.

Considering the rapid advances in computer technology, virtual reality machines of one decade will be viewed as a crude devices by the next decade. For example, technology is being developed to use the retina of the eye as a screen for images that could be transmitted directly to the brain through the optic nerve. Developing the technology to facilitate the remote operation of **aircraft** and other machines as though the user were actually in the machine is also under way. Traveling to France or Greece, including the experience of climbing the Eiffel Tower or basking on a sun-drenched beach, may be as easy as donning a headset and body suit and plugging in.

Like most technological advances, virtual reality has social and psychological ramifications. Critics argue that virtual reality could cause some people to forego emotions and interpersonal relationships for the safe, controllable virtual world. But proponents say advances from the proper applications of this technology—both as a means of interacting with the real world and as an end of facilitating training and entertainment-far outweigh the potential for antisocial abuse.

VLA (Very Large Array) • See Radio interferometer

Voice mail system

A voice mail system uses computer technology similar to electronic mail to answer **telephone** calls, and route and store resulting messages. When a call is answered by a voice mail system, the voice message is digitized (converted to electronic signals) and stored on magnetic disks attached to computers or computers themselves. The call receiver enters his ''mailbox'' by punching in his or her security code to retrieve his or her messages. After listening to the messages, callers can save or delete the messages. A voice mail system can save many more messages than a stand-alone answering machine can. Time and date information can also be recorded and retrieved. Voice mail can do more than answer calls to one line. Recorded prompts can route callers to the proper mailbox to leave messages. Messages can also be forwarded and appended, and one recorded message can be sent to multiple callers automatically.

Though **telephone answering devices** have a long history, voice mail systems were not possible until the development in the 1960s of touch tone, a telephone technology. Touch tone allows computers and telephones to communicate. Computers also had to develop to the point where computer digitized voices sounded nearly normal, a process that did not come to fruition until the 1970s. The earliest voice mail systems were available in 1972 from Sudbury Systems, and were sold mostly to hospitals. Early systems were problematic for short message allowances (as little as 90 seconds allowed to record a messages) and other shortcomings. By the early 1980s, many of these problems were solved and voice mail systems became common in the offices of the United States.

When a call is made to a voice mail system, it is answered automatically and a pre-recorded greeting is played. If the caller chooses, the call can be transferred to a specific extension. This transfer is called the automated receptionist or operator. The caller can leave a message, or if the prompt is offered, the call can be transferred to another line. Human receptionists and operators can do the same thing. More complex voice mail systems offer its users the ability to leave several different outgoing messages to separate messages. In many voice mail systems found in businesses, messages can be forwarded and appended. Distribution lists can be created to forward messages, appended or not, within the company, and with some systems, to outside numbers as well. Voice mail can also be used to screen calls. When this option is chosen, a caller is prompted to say his or her name, and the receiver can hear the name before choosing to accept or reject the call. Many businesses use voice mail as informational bulletin boards to present data to customers. These mailboxes cannot receive messages, but merely give out information as the customer goes through prompts. For example, to a movie theater could give out times and synopses of films being screened by having callers press 1 for the movie on screen one, pressing 2 for the movie on screen 2, and so on.

By 1988, voice mail systems were promoted for use in private residences by phone companies as a replacement for answering machine. Many of these systems have features similar to ones found in the office, but not all. One common use for voice mail in the home is a multi-mailbox system. In this system, each member of a household has his or her own mailbox with its own access code, so one line can be shared yet phone messages remain private. The outgoing message prompts callers to chose a mailbox to ensure each person gets his or her messages. Voice mail can also record messages when another call is being answered, and because it does not rely on electricity like an answering machine, will continue to record messages even if electricity becomes unavailable. Though there are advantages to voice mail in residences, there is usually a monthly fee involved, while an answering machine represents a one-time cost. Most voice mail systems also do not have a notification light to indicated messages are waiting. In the future, voice mail might become integrated with e-mail. When connected to a computer with special software like Visual Mailbox, the source of the call, and the text of the call itself, becomes accessible by computer.

See also Computer applications; Computer, digital; Computer speech recognition

VOICE SYNTHESIZER

The earliest known ''talking machine'' was developed in 1778 by Wolfgang von Kempelen. Eyewitnesses reported that it could speak several words in a timid, childlike voice. While the talking machine's success appears genuine, Baron von Kempelen's accomplishments are not above suspicion. Nine years earlier, he had built a chess-playing machine, which defeated many players, including Napoleon (who, incidentally, made several unsuccessful attempts to cheat). Eventually, it was discovered that the machine was a fraud—its cabinet concealed a hidden, human chess player, who controlled the game. In 1830, Professor Joseph Faber, of Vienna, Austria, produced his own speaking **automaton**. Faber's machine, dubbed Euphonia, had taken twenty-five years to construct. Designed to look like a bearded Turk, the creation could recite the alphabet, whisper, laugh, and ask ''How do you do?'' Speech was produced by its inner workings—double **bellows**, **levers**, **gears**, and **keys** located inside the mannequin. Strangely enough, Euphonia spoke English with a German accent.

The first talking machines employing electronic technology were developed in the 1930s. The Voice Operation Demonstrator, or Voder, invented by Dudley in 1933, could imitate human speech using electrical analogues of a human vocal tract via electric circuits. It could utter complete sentences as its operator pressed keys on a board. Speech-synthesis technology evolved further with the rapid development of computer technology in the 1950s. During the late 1960s, the MITalk System was developed at the Massachusetts Institute of Technology. Although originally designed as a reading machine for the blind, once completed, the system could convert virtually any type of text into speech-synthesized output. Raymond Kurzweil also developed speech-synthesis technology to aid the blind. In 1976, he produced the Kurzweil reading machine which could read everything from a phone bill to a full-length novel and provided unlimited-vocabulary synthesized output. Sometimes called a set of eyes for the blind, the reading machine has proved very popular. Another use of a voice synthesizer for the visually impaired was a navigational system developed by a student in Ottawa, Canada, in 1986. Charles LaPierre used **global positional systems** satellites and a self-created database in a portable computer to determine his position when walking around a city. The voice synthesizer relays

Alessandro Volta

location, including street signs, so someone could navigate around the city. In 1986, another voice synthesizer was developed by Bloomfield R & D, for use in toys and for patients recovering from surgery. Called P.O. Vox—for the post-operative synthesized voice—the **battery**-operated device featured a tube that could be placed in the patient's mouth. When the patient touched the case, he or she could speak, no matter how softly, through the tube and be heard via a synthesized voice. This allows the patient to communicated clearly with doctors after surgery or when too weak to speak clearly.

Today, speech synthesis is a useful way to convey information in public places. Cars, appliances, and even games are being equipped with voice-synthesizer chips. In the early 1990s, ''talking chessboard'' was introduced that played an amateurish game of chess and announced each move it makes. Macintosh computers came with a Speech Manager, which was a text-to-speech system extension, throughout the 1990s.

See also Communication Satellites; Computer game; Computer input and output devices; Video game

Volta, Alessandro (1745-1827)

Italian physicist

Alessandro Volta knew what he wanted his life's work to be at a fairly early age. Volta was born in Como, Lombardy, Italy, on February 18, 1745. His family believed he was developmentally-impaired because he did not begin to talk until the age of four. When he turned seven, he caught up with, and began to surpass, his classmates. At age fourteen, Volta decided to become a physicist after reading The History and Present State of Electricity by Joseph Priestley (1733-1804).

In 1774 Volta was appointed professor of Physics in the Como high school. In the following year he made his first major invention, a device called the electrophorous that used to generate static electricity. The electrophorous consisted of two metal plates. One plate was covered in ebonite which, when rubbed with a dry cloth, built up a negative static electric charge. The second plate, which had an insulated handle, was brought near the negatively charged ebonite plate. Since opposite charges attract, the ebonite attracted a positive charge in the second plate, which collected on the side nearest the ebonite. The side of the plate opposite the ebonite was left with a negative charge. By attaching a **wire** to the negative side of the second plate, the charge was drained away into the earth. By repeating the process, Volta was able to build, and store, a considerable positive charge on his plate. The electrophorous replaced the Leyden jar which, up until that time, was the standard device used for storing charges. Volta's invention became the forerunner of the modern **capacitor** which stores electricity in **electric circuit**s.

Volta was duly proud of his invention, which he first described to Priestley. Volta became famous and in 1779 was appointed a professor at the University of Pavia. He continued to work with electricity, modifying the electrophorous so it could be used to measure weak electrical charges. He also became involved in a controversy which, eventually, led him to the invention of a device for which he became even more famous.

In 1791 Italian anatomist Luigi Galvani (1737-1798) made public a conclusion he had drawn through his experimentation with electricity and muscles. He announced that an animal electricity existed. When two different metal probes touched the muscle in a frog leg, the animal electricity was released, Galvani said, causing the muscle to twitch. Galvani's theory received a large following, but not everyone accepted it. Volta strongly disagreed as did the illustrious scientist **Charles Augustin de Coulomb** (1736-1806). Volta began his own investigation into animal electricity in 1794 and found muscle tissue was completely irrelevant to creating the current. Only the presence of two different metals produced electricity; the fluid in the muscle merely enhanced the charge. The controversy raged on until 1800, when Volta built a device that produced a large flow of electricity. He filled bowls with a saline solution and ''connected'' them with strips of different metals. One end of the strip was copper; the other end was tin or zinc. By bending his strips from one bowl into another, Volta was able to create a constant flow of electrical current—the world's first electric **battery**. Volta had proven that the metal was the source of the electricity, and animal electricity did not exist.

In the interest of making his battery smaller, Volta used round discs of copper, zinc, and cardboard that had been soaked in a saline solution. He stacked his discs one on top of the other; attaching a wire to the top and bottom of his pile allowed the electric-current to flow. The invention of this ''Voltaic pile'' marked the apex of Volta's career. Volta died in Como on March 5, 1827. His greatest honor was bestowed when the force that moves electric current was named the Volt.

VOLTAIC PILE • See Battery, electric

VOLTMETER

A voltmeter is a device that measures the voltage difference between two points. Voltmeters come in four basic types: moving coil, moving iron, electrostatic, and digital.

The principal behind the voltmeter was established by Danish physicist **Hans Christian Oersted** (1777-1851) in 1820 when he discovered that an electric current in a **wire** produced a magnetic field around it. The moving coil and moving iron voltmeters, both of which are similar to **ammeter**s, make use of this discovery. The moving coil voltmeter has a rectifier within its circuitry to ensure that **direct current** (DC) flows through it. When current flows, it sets up a magnetic field that causes the coil to turn; an indicator needle, attached to the coil, moves over a graduated scale, allowing the voltage to be read. In the moving **iron** version, it is a small **magnet** with an attached needle that moves.

The problem with the moving coil and moving iron voltmeters is they require current from the very circuit they are measuring in order to operate, so any measurement will miss at least part of the current it is trying to measure. This problem can be lessened if an **amplifier** is used as part of the circuit.

An ammeter can be turned into a voltmeter by including a high resistance in series, (in accordance with Ohm's law), which increases accuracy. A highly sensitive ammeter is known as a **galvanometer**, which was used by, André Ampère (1775-1836) in 1820 to measure electric current.

An electrostatic voltmeter operates on the principal that unlike charges attract each other. When an electric current flows through the voltmeter, moving vanes are drawn in toward fixed plates. The rotating parts are controlled by a **spring**, and so the electrostatic voltmeter is able to measure voltage without drawing any current.

Modern **electronics**, **semiconductor**s, and **integrated circuit**s have made the digital voltmeter possible. This instrument compares the measured voltage with its own internal ''reference'' voltage. The difference between the two voltages controls a digital display. While the digital voltmeter is more expensive to build than its moving coil and moving iron cousins, it offers greater accuracy, sensitivity, and resolution. Digital voltmeters are now the preferred instrument for almost all AC and DC measurements. A voltmeter is an integral part of a **multimeter**, a multi-range test instrument that also measures current and resistance.

See also Electric circuit

VON GUERICKE, OTTO • See Guericke, Otto von

VON NEUMANN, JOHN • See Neumann, John von

VON SIEMENS, ERNST WERNER • See Siemens, Ernst Werner von

VON ZEPPELIN, COUNT FERDINAND • See Zeppelin, Count Ferdinand von

VULCANIZED RUBBER • See Rubber, vulcanized

VYCOR • See Glass, heat-resistant

W

WALKER, SARAH BREEDLOVE (1867-1919)

American inventor

Sarah Breedlove Walker, a St. Louis washerwoman, created a cosmetic empire by inventing a system of hair straightening. Born to a poor farm family on December 23, 1867, in Delta, Louisiana, Sarah received very little education and, by the age of ten, had moved away from home to look for work. By age nineteen Sarah was a widow with a young daughter to support, and she moved to St. Louis to work as a hotel washerwoman. In 1906, Sarah married Denver newspaperman Charles Joseph Walker and changed her name to Madame C. J. Walker.

Around 1910, Sarah came up with idea of straightening hair with a hot iron comb and an ointment to add softness and shine. This was an important development because for generations before her revolutionary process, blacks had straightened tightly curled hair on ironing boards which endangered the scalp and face and broke the hair. Sarah made up batches of her pomade in a tub, then packed the mixture in jars and sold them to customers. She developed a variety of products to serve a range of hair care needs.

To increase sales of her pomade, hair growing tonic, strengtheners, hot combs, toiletries, fragrances, and facial treatments, she peddled them door to door, then organized agents in ''Walker Clubs.'' She opened a shop, trained assistants, then later added mail-order sales followed by a beauty school that taught the Walker Method of hair straightening and hair growing.

In 1910, she moved the operation to Indianapolis and built her first factory. By 1917, Sarah Walker was employing 3,000 workers in America's largest black-owned business and was drawing sizeable return from sales of equipment and supplies and from a chain of beauty schools. She became a social leader among the black middle class and opened Lelia College, a hair care laboratory, and a chain of beauty salons in Harlem.

Sarah Walker was a generous contributor to the National Association for the Advancement of Colored People (NAACP) as well as to St. Louis and Indianapolis charities. She bankrolled scholarships for women at Tuskegee Institute, Bethune-Cookman College, and Palmer Memorial Institute and endowed black YWCA chapters and orphanages. She died from nephritis on May 24, 1919.

WALKMAN

When **audiocassette**s replaced **eight-track tapes** as the preferred tape format of consumers in the 1970s, small cassette recorders with built-in speakers (and poor monophonic audio quality) were already on the market. However, high-quality mobile **stereo** systems with headphones were not yet available.

In 1979 Akio Morita (1920?-), president of Sony Corporation, created a pocket-sized cassette player using the miniaturized **electronics** pioneered, in part, by his own company. This pocket-sized player, dubbed the ''Walkman,'' featured tiny motors and a miniature circuit board housed in a plastic casing, and equipped with lightweight headphones which allowed listeners almost unlimited portability. The Walkman proved so popular that other varieties appeared, including models with AM-FM **radio**s, recording capability, and, recently, models capable of playing compact disks.

See also Amplitude modulation; Frequency modulation

WALLIS, BARNES (1887-1979)

English engineer

Wallis was born in Derbyshire, England, and educated at Christ's Hospital school. He began his career building **ships**, but soon left to join the Vickers Company, which designed many British airplanes during the First World War.

At Vickers Wallis first worked on rigid **airship**s, whose form later inspired his lattice-like geodetic structures for **aircraft**. When the airship program was temporarily halted by World War I in 1915, Wallis joined the Artists' Rifles, an army regiment consisting of artists and designers. After the war, he returned to head Vickers newly reinstated airship program. He also returned to school in 1920, earning his engineering degree only four months later. Wallis designed several superior airships during this period, including the R80 and the R100. However, the crash of a separately designed sister ship effectively ended airship manufacture at Vickers and ultimately in England.

Wallis was eventually transferred to the company's airplane works in 1930. One of his first achievements was a lighter wing structure for the *Viastra 2*, an Australian commercial airplane. Research for this project and others led finally to Wallis's *geodesic construction* in which the members of a frame are formed into a curved latticework system so that the load each member bears is offset by tension loads in crossing members. In 1932 Wallis designed an aircraft with a geodesic fuselage that could handle incredible stress without collapsing. The British army, however, believing that such construction was too difficult for **mass production**, refused to order the planes from Vickers. Yet the company, confident in Wallis, decided to manufacture the plane without the army's financial backing. Wallis and his team then designed tools and techniques so that a complete *Wellington* **bomber** airframe could be assembled in 24 hours using semiskilled labor. The army quickly revised its opinion and the bomber became a successful part of the Allied arsenal during World War II with over 11,000 being built.

Wallis also applied aerodynamic streamlining to **bombs**. He invented the 1,200-pound *Tallboy* bomb and the 22,000 pound *Grand Slam* bomb, but is best known for the famous bouncing bomb. Designed to destroy Germany's massive **dams**, the bouncing bomb was hung crosswise in the bay of an aircraft. Just before being dropped, the bomb was given a backward spin by an **electric motor**. This spin caused the bomb to skip across the surface of the water until it hit the dam. Then, instead of detonating or bouncing off, the bomb, because of its spin, would plow under the water to a depth of 30 feet where it would explode, causing extensive damage. The bombs were a great success, depriving many vital German industries of their power sources.

For his achievements Wallis was widely honored by fellow engineers and by his government. He continued to design aircraft, including the supersonic Swallow, until his death in 1979.

WALLPAPER

Wallpaper is a non-woven (**paper**) or woven (fabric) backing, decoratively printed for application to walls of a residence or business. Wallpaper is not considered essential to the decoration of a structure; however, it has become a primary method by which to impart style, atmosphere, or color into a room.

The wallpaper industry divides the manufacture of wallpaper into those used in residences and those hung in businesses or other public buildings. The two categories of paper differ in weight, serviceability, and quality standards. Residential-use wallpapers are made from various materials and can be purchased prepasted or unpasted. There are no mandated serviceability tests for residential papers. The commercial-grade wallpapers are divided into categories based on weight, backing composition, and **laminate**/coating thickness. All commercial-use wallpapers must have a **vinyl** surface and pass rigorous physical and visual tests as mandated by the Chemical Fabrics and Film Association.

There are also four popular methods used to print wallpapers and designers choose the printing technique based on cost as well as aesthetics.

The earliest wallpapers used in Europe as early as the thirteenth century were painted with images of popular religious icons. These "domino papers" were pasted within homes of the devout; however, they also enlivened the bleak homes of the poor. Within the next few centuries, papers were hand block-printed, but only remained popular with the poor.

By the sixteenth century, however, more expensive wall covering, depicting tapestries hung in homes of nobility, became popular with the middle class. Small sheets either carried a repeating image, or several blocks produced a pattern spread across many sheets. Fashionable eighteenth century Americans puchased wallpapers from France and England—"paper stainers" were producing wallpapers in this country by the early nineteenth century if not before.

Two problems plagued wallpaper stainers until the mid-nineteenth century. One was the problem of producing long sheets of paper for printing, the other was printing attractive wallpaper inexpensively. Until the mid-1700s, rag-based paper was individually printed in sheets, then applied to walls. Then, wallpaper manufacturers were pasting the pieces together, ground coating them, then printing. In the late nineteenth century, the paper industry developed "endless" paper, or paper made in very long strips. By 1870, wood pulp had supplanted rag stock, resulting in a very cheap backing for wallcovering.

In the nineteenth century, printing costs were greatly reduced by abandoning labor-intensive block printing in favor of cylinder printing. Wood-block printers applied each color by hand using a separate block for each color in the pattern. Thus, each block had to be inked with the right color, pressed down on the paper, tapped to ensure a quality imprint, lifted up, and re-inked as the printer moved down the paper roll—an expensive process. Wood blocks were supplanted by copper cylinders, which carried the design below the surface of the roll, each roll printing a single color. The cylinders were mounted within one machine and the paper was mechanically fed between cylinders until the paper was completely printed—no hand printing involved. Thus, by about 1885 wood pulp paper printed with cylinders so greatly reduced wallpaper costs that it was cheaper to wallpaper a house in the United States than to paint it.

More recent advances include development of additional printing methods, new **inks** and solvents, and use of latex and vinyl as coatings or laminates.

WALTON, ERNEST T. S. (1903-)

Irish physicist

Walton was born in Dungarvan, County Waterford, Ireland, on October 6, 1903. He earned a bachelor of arts degree from Trinity College, University of Dublin, in 1926, and master of science (1928) and master of arts (1934) degrees from the same institution. He received a scholarship to Cambridge University, from which he received a Ph.D. degree in 1931. He returned to Trinity College, Dublin, in 1934, where he was elected Erasmus Smith Professor of Natural and Experimental Philosophy and Head of the Department of Physics in 1947.

While a graduate student at Cambridge, Walton completed the research that led to a share of the 1951 Nobel Prize for physics (with **John Cockcroft**). The prize was awarded for Cockroft and Walton's development of a particle accelerating device that now carries their names.

Walton's interest in **particle accelerator**s went back to the late 1920s. Like many other scientists, he recognized the inherent limitations of using particles from naturally occurring radioactive materials to induce changes in atomic nuclei. Rutherford had been successful in producing the first nuclear transformations using alpha particles from naturally-radioactive isotopes. But alpha particles from such sources are too few in number and have too little energy to be used for most of the transformations that scientists want to study.

In 1928, Walton experimented with efforts to accelerate electrons in straight lines and in circles. Both methods failed because the high frequency sources needed for the machines were not available. Walton's methods were, however, to form the basis for a more successful **linear accelerator** and bevatron constructed in the 1930s and 1940s.

While working at the Cavendish Laboratory in Cambridge in 1929, Walton and Cockroft heard about speculations by G. Gamow, E. U. Condon, and R. W. Gurney that high energy particles have a small, but significant, probability of overcoming the electrical repulsion of an atom and entering its nucleus. In response to this hypothesis, the two scientists constructed a machine that could accelerate hydrogen ions (protons) to energies of 700,000 electron volts.

When these protons were used to bombard a lithium target, Walton and Cockroft found that large numbers of alpha particles were emitted. They concluded that protons had collided with lithium nuclei, causing them to break apart into two helium nuclei (alpha particles) as follows: ${}_3Li^7 + {}_1H^1 \rightarrow 2\ {}_2He^2$.

The 1951 Nobel Prize recognized both the development of the particle accelerator and the discoveries of nuclear reactions Walton and Cockroft made with it.

Ernest T. S. Walton

WANG, AN (1920-1990)

American computer scientist and industrialist

An Wang was one of the pioneers in developing advanced computer memories and also founded Wang Laboratories, a leading manufacturer of word-processing equipment in the 1970s. Wang was born in Shanghai, China; his father taught school in a small town in the vicinity. He was interested in mathematics even as a small child, and he learned English from his parents at home. He earned a B.S. in electrical engineering from Chiao-Tung University in Shanghai in 1940 and spent the next five years, during World War II, building **radio** equipment for the Chinese army.

In 1945, Wang came to the United States on what was supposed to be an advanced engineering apprenticeship. When the position was not available, Wang became a graduate student at Harvard, earning his M.S. in applied physics in one year and a Ph.D. specializing in nonlinear mechanics.

In 1948, he began postdoctoral studies with **Howard H. Aiken** at the Harvard Computation Laboratory. There he quickly developed his first invention, the magnetic core memory, which became widely used in computers during the 1950s and 1960s. It consists of tiny ferrite ''doughnuts'' (iron-based and magnetic) that take and hold the charges placed by current-carrying conductors. Each doughnut or ring represents one bit, and zeros and ones are represented by different magnetic field directions.

When linked in a matrix (an improvement devised by **Jay Forrester** of MIT), the memories proved to operate quickly. Early magnetic core memories were tiny by modern standards, for example, 256 bits, but they were a great improvement over memories in computers of that day. Though superseded by silicon-based **semiconductor**s in the late 1960s, the magnetic core

An Wang

memory continued to have some applications because it did not rely on electricity to retain its data.

In 1951, Wang patented the magnetic core memory and established his firm, Wang Laboratories. After selling his patent to IBM in 1950s, he manufactured electronic calculators, computers, and in the 1970s computer-based word-processing equipment.

Wang was known for his charitable contributions and he received many professional honors, including being named a fellow of the Institute of Electrical and Electronic Engineers. He was awarded the United States Medal of Freedom in 1986.

Wankel engine

Soon after the invention of the conventional **internal combustion engine**, engineers realized that a tremendous savings in both weight and energy could be realized if the combustion of **gasoline** could be used to produce rotary motion directly instead of via the reciprocating action of the typical **piston** engine. The engineer who transformed this idea into a working rotary engine was Felix Wankel.

Felix Wankel was born in Germany at Lahr, a town near the French border, in 1902. During World War II he worked in the German Aeronautical Research Establishment researching rotary piston technology. After the war, he went to work for a German automotive firm. By 1957 he had built a prototype of his ''Wankel'' engine and received numerous patents, including several for the special tools needed to build the rotary engines.

The rotary-piston engine uses a triangular-shaped piston which revolves within an oval-shaped chamber that is slightly constricted in the middle. (This shape is called an *epitrochoid.*) One of the early major drawbacks of the rotary engine was the necessity of maintaining a perfect seal between the engine walls and the leading edges of the triangular rotor. Wankel designed a series of sprung sealing plates to solve this problem. The rotating triangular piston also serves to seal off the intake and exhaust ports, thus eliminating the need for **valves**.

The power of the ignited gasoline is transferred directly to the driveshaft through a series of **gears** in a three-to-one ratio. This design resulted in a motor that produces three power ''strokes'' per revolution, compared to only one power stroke per two revolutions of the typical four-stroke gasoline engine. Further efficiency and smoothness of operation can be achieved by building engines with more than one rotor. A three-rotor engine corresponds in power to an eight cylinder reciprocating engine.

The engine's major advantages when compared to conventional piston engines are its light weight, short length, efficient energy transfer, and insensitivity to grades of fuel burned. Also due to the rotary engine's lack of a crankshaft, camshaft, valves, and valve-train, it is much easier to maintain. However, its fuel consumption, manufacturing costs, and pollutant emission rates are above those of reciprocating piston engines.

The early dreams of seeing eight out of ten cars powered by the Wankel engine by the 1980s were not realized, yet the Wankel engine has been used by General Motors in the United States, Rolls-Royce in Great Britain, Alfa Romeo in Italy, Citroën in France, and Toyota in Japan. The Japanese Mazda Motor Corporation was the first to use the rotary engine in a production car in 1967, and Mazda continues to build automobiles with the rotary engine. General Motors, however, abandoned its plans for a rotary-powered vehicle in 1976 due to concerns over the engine's emissions problems. Several companies are actively researching the rotary engine's potential for use in the **aircraft** industry, where its small size and easy serviceability are decided advantages over bigger and more complicated gasoline engines.

Throughout his life, Felix Wankel continued to refine the engine that bears his name, heading his own research company at Lindau in Germany. He died in 1988.

Wankel, Felix • See Wankel engine

Warping frame • See Looms

Warship

Warships are armed naval fighting ships that range from small vessels to giant **ships**. The earliest known warships—called galleys by the Egyptians and Greeks—sailed nearly 3,000 years ago. They were long open warships that used square shaped sails and 40-foot (12.2 m) oars to move through the water. Extended from the ship's bow was a long sharp point for ramming enemy ships. Shields made of animal hide or other material covered the sides to protect the slaves who manned the oars. The Assyrians, Greeks, and Romans (who developed three-tiered versions called triremes) used these vessels successfully in their battles.

In the Middle Ages the Vikings developed the drakkar, or long boat. These ships had a tall stern and bow and their sides were protected by shields. They were not only much lighter than the galleys, but were also stronger and more maneuverable. By 1000 A.D., the drakkars ruled their part of the world. In the late 1400s and 1500s, explorations of the New World required heavier vessels. King Henry VIII of England developed a galleon from an existing Italian armed merchant vessel; its distinguishing feature was a large, raised deck at the rear, with holes in its sides for several large guns. The Spanish galleon was used to transport gold from the Americas, but it was extremely poor at maneuvering, a fact that led to England's destruction of the Spanish Armada in 1588. The design of warships of the era were based on cargo vessels; the British built the first war frigate in 1652, the *Constant Warwick*, a ship 90 feet (27.4 m) long carrying 42 guns.

Three inventions in the 1800s revolutionized warfare at sea. The first was the **steam engine**, which freed ships from their dependence on wind and increased maneuverability. The second was the introduction of iron and **steel** as armor to protect the ship's hull. These ''ironclads'' were used in 1862 during the American Civil War, when the Confederate Navy's *Merrimack* fought the Union's *Monitor*. The third invention was **John Ericsson**'s pivoting turret, used aboard the *Monitor*, a device that allowed cannons and other large **artillery** to rotate in all directions.

In 1906, England launched a warship that became the prototype for battleships of the twentieth century. This was the *Dreadnought*—a very heavily armored vessel guarded by ten 12-inch (30.5 cm) guns, and powered for the first time by **steam turbine**s, which made her the fastest battleship afloat. By 1915, battleships could fire 15-inch (38.1 cm) shells weighing nearly 1900 pounds (861.8 kg) a distance of 20 miles (32.2 km). Other types of warships were created to support these massive craft. The battle cruiser, for example, was smaller and had less armor than battleships, but it was faster and had a greater cruising range. The destroyer was developed to protect the big ships against small **torpedo** boats and, later, against the **submarine**s.

During World War II, naval engineers came up with amphibious boats. These were landing craft with shallow drafts and large hinged doors that could put **armored vehicle**s, **trucks**, supplies, and soldiers ashore. Some of these were given chain treads much like those on tanks so they could crawl over reefs

The first battle between iron war ships occurred during the Civil War when between the Confederacy's *Merrimac* and the Union's *Monitor*.

and even move up the beach after landing. The **aircraft carrier** also debuted in World War II. This type of warship had actually been around since 1919, when the English built flight decks on an old battleship and a steamship.

Following World War II, the traditional multi-ton battleship gave away to sleek nuclear-powered vessels. Less glamorous ships are also part of modern naval warfare. Minesweepers are used to clear waters for friendly vessels. **Missile** boats are speedy craft that can attack much larger ships with guided missiles. Patrol boats, small and maneuverable, are used to carry torpedoes against enemy ships.

Perhaps the most technologically advanced warships of the late 1990s are manufactured in England. England's Sea Wraith ships are designed with a slew of features that allow the ship to hide from and confuse enemy **radar**. Sea Wraiths can change the size and shape of their radar images by raising or lowering retractable reflectors. To baffle heat-seeking missiles, these ships can surround themselves with an artificial fog bank which cancels out their heat radiation. And to foil infrared sensors which identify heat emissions from a ship's engines, the Sea Wraith uses a low-heat diesel-electric fuel drive.

A new warship propulsion system was under development by American shipbuilders in the late 1990s. It is an electric system that uses a magnet-driven motor. A series of high-density, high-powered magnets mounted in the motor spin the shaft and screw. The magnets are lighter and cleaner than conventional or nuclear power sources, and they take up less room. With a smaller motor, ship builders have more options in designing other parts of the warship, so that ships using this propulsion system could carry more cargo or more weapons.

WASHING MACHINE

The invention of the washing machine relieved householders of an age-old drudgery—for centuries, clothes had been cleaned by soaking them in stream water and pounding them with rocks. In 1797, the invention of the washboard for scrubbing eliminated the need for rocks.

Hundreds of mechanical washing machines were designed in the first half of the nineteenth century, but they were hand powered. The earliest models rubbed clothes to clean them; later designs featured mechanisms that moved the clothes through the water. The user either turned a handle to rotate or rock the washing box or pumped a dolly to agitate the clothes.

Steam power was applied to commercial washing machines in the 1850s. An enterprising California miner washed shirts with his machine in exchange for gold dust in 1851. Brother David Parker of the Canterbury, New Hampshire, Shaker community patented his ''Wash-Mill'' for hotel use in 1858.

Most home washing machines remained hand powered. James King's 1851 model used the rotating cylinder principle. Hamilton E. Smith of Philadelphia, Pennsylvania, patented a reversible-action machine in 1863. An underwater agitator design was patented in 1869. The Blackstone washer of 1874 featured a handle-and-gear device and sold for $2.50. By 1880, 4,000 to 5,000 washing machine designs had been patented.

In 1847, a design featuring a single wringer to remove water from laundered clothes appeared; the two-roller wringer appeared in 1861. A washing machine using a spinning basket to extract water was patented in 1873.

Electric washing machines first appeared in the early 1900s. A chain-driven electric washer was produced in 1906. Both the Automatic Electric Washer Company and the Hurley Machine Corporation offered electric washers in 1907. Electric wringer washers appeared in 1910. In 1911 the Maytag Company, which had produced its first hand-powered machine in 1907, introduced an electric Hired Girl wringer washer. These early electric washers were adaptable to **gasoline** power and could be hand operated as well. Maytag added the vaned agitator to its machines in 1922. Although a wringerless model was marketed as early as 1926 by the Easy Washing Machine Company, wringer washers continued to be the industry standard.

The automatic washing machine was introduced in 1937 by Bendix. This Model S required only two settings of the dial during its cycles, although the machine vibrated so ferociously it had to be bolted to the floor. A fully automatic Bendix appeared in 1947, along with a flood of other machines to satisfy the postwar demand for consumer products. Spin-dry machines overtook the old wringer types in popularity by 1953. Most machines that were manufactured in the mid-1950s featured lint filters, **bleach** and **soap** dispensers, and speed selection. By 1958, ninety percent of all American households used electric washing machines.

The contemporary washing machine offers many options so that the user can tailor the type of washing to the type of garment being washed. For example, the user can select cycles for permanent press, synthetic, or delicate fabrics; a range of temperatures for both wash and spin cycles; length of cycles; and prewash and soak options. Modern washing machines are either front loading or top loading—front loaders wash by tumbling the clothes inside a revolving washbasket, while top loaders agitate the clothes within the washbasket.

WASSERMAN TEST • See Syphilis test

WATCH • See Clock and watch

WATER CLOCK • See Clock and watch

WATER CLOSET • See Toilet

WATER FRAME

The water frame was the first spinning machine that performed constant spinning of **cotton** yarn. Through the use of a series of rollers, the water frame was able to produce very strong, smooth yarn that paved the way for the weaving of all-cotton fabrics.

The water frame was invented in England by **Richard Arkwright** and was patented in 1769. At that time, **James Har-**

greaves' spinning jenny was enjoying great success. However, the spinning jenny simply mechanized the actions of a hand-turned spinning wheel; it could not improve upon the actual quality of yarn produced.

Arkwright came up with a design that would produce stronger yarn by drawing the cotton fibers apart further and more evenly. The cotton was pulled through a set of eight rollers arranged in four pairs. In each pair, the top roller was covered in **leather**, a material that easily gripped the rough cotton. The bottom rollers were made of either wood or metal, and had flutes cut into them that allowed the cotton fibers to pass through. The first set of rollers was designed to turn slowly, while each successive pair turned slightly faster; this pulled the fibers further and further apart, producing a roving that was even and free of lumps. As the cotton emerged from the final set of rollers it was twisted tightly into a strong yarn.

Arkwright's prototype spinner was powered by horses, but in 1771 he perfected a water-powered spinner, hence the name "water frame." For years, cotton yarn had been used in the weaving of fabrics. Unfortunately, the yarn spun using the "great wheel" or the spinning jenny was too weak for many applications, including hosiery. In **looms**, cotton yarn could be used as weft, but was not strong enough to serve as warp; in most cases, stronger threads of linen were used for this purpose. Arkwright's water frame produced the first yarn suitable for use as both warp and weft. For the first time in England, all-cotton materials could be woven.

Like many inventors, Arkwright possessed the inspiration but not the experience to construct his water frame. In 1768 he enlisted the help of two machinists, John Kay (1704-1764) and Thomas Highs, who built for him the important parts for his spinning machine. Years later, this partnership was the source of legal trouble, as Arkwright sought to enforce his patents upon a number of competitors in the yarn industry. He eventually lost the patent for the water frame, but remained a rich and powerful player in that industry. He was knighted in 1786.

Water pump

The Mesopotamians invented the first pump about 3000 B.C.; they had previously used buckets alone to water their crops in the Nile River valley. Their *shaduf,* or swipe, was a long wooden **lever** that pivoted on two upright posts situated on the riverbank. A pole with a bucket attached to it was suspended from one end of the lever, with a counterweight fixed to the other end. To operate the water pump, the user pushed the pole down so that the bucket filled with water; the counterweight then helped to raise the bucket, which was emptied into a trough that led to the irrigation ditch constructed between the river and the fields. The *shaduf* became popular throughout the Middle East, and was the only form of water pump used in that region for the next two thousand years.

Around 500 B.C., three new water-lifting devices came into use. The *saqiya,* or **waterwheel**, which had pots attached around its circumference, was mounted over the water so that the lowest pots filled. As the wheel turned, the filled pots would rise to the top and dump their water into a chute that led to the **irrigation** ditch. Eventually the pots were replaced by troughs. Both the *saqiya* and the tympanum—a wheel comprised of radiating, watertight compartments with openings to allow water in and out—were mentioned by the Roman engineer Vetruvius in his writings from the first century B.C.

Another ancient water pump was the bucket chain, a continuous loop of buckets that passed over a pulley-wheel, thought to have been used to irrigate the Hanging Gardens of Babylon around the year 600 B.C. Perhaps the most famous of early water pumps was the Archimedean screw. The Greek **Archimedes** (287-212 B.C.) was a famous mathematician and inventor who, in about 250 B.C., devised a pump made of a metal pipe in a corkscrew or helix shape that drew water upward as it revolved. This device proved particularly useful for removing water from the hold of a ship. (Although the invention of the screw-type pump is usually attributed to Archimedes, many researchers believe he merely popularized an invention he had seen in Egypt.) The modern screw pump features helices rotating in open, inclined troughs; it is particularly well-suited to pumping sewage in wastewater treatment plants, because its design allows debris to pass without clogging.

The ancestor of the modern force pump—which features a cylinder with a plunger or **piston** at the top that creates a vacuum and draws water upward through **valves** at the bottom—was reputedly designed by Ctesibus of Alexandria, Egypt. Vetruvius mentioned this device, as he did the *saqiya,* in his first century B.C. chronicles. Remains of this pump, which was usually made of bronze, have been found in many buildings dating from the days of the Roman Empire. The valves and plungers were particularly valuable inventions that were incorporated into other kinds of machinery, including military equipment. The Romans, for instance, used this kind of pump to hurl flammable liquids at invading Arabs.

One of the most common water pumps in use today is the centrifugal pump, comprising a motor-driven, propeller-like impeller contained in a housing—when turned, the impeller creates suction that draws a continuous flow of fluid. The positive displacement pump, on the other hand, traps individual portions of fluid in an enclosed area, then moves them along. French inventor **Denis Papin** invented the centrifugal pump in the late 1600s. His impeller had straight vanes, whereas that developed by the British inventor, John G. Appold, in 1851 had the curved vanes still preferred today.

Other modern pumps include axial-flow pumps used as compressors in **jet engines** since the 1940s; jet pumps, which send a jet of steam or water through the fluid to be moved and which are used to raise water from wells deeper than 200 feet (60 m); and electromagnetic pumps, used in **nuclear reactors**.

Water supply and drainage

Ancient hunters and gatherers remained near water supplies when they could, devising means to carry water with them when necessary and tapping water sources below the surface or from plants. The Tigris and Euphrates, Nile, and Indus rivers supplied ample water for the ancient civilizations near them, and wells dating back to 3000 B.C. are the earliest man-made water sources and served small communities. Caves functioned as early storage chambers—*cisterns*—for water diverted for later use, as did crevices and fissures in rocky surfaces. Prized for its purity, rainwater often was caught in rooftop containers in the Middle East and the Mediterranean, and as construction methods became more sophisticated, cisterns were built underground. Istanbul's Hall of a Thousand and One Columns, built in the sixth century B.C., is one of the largest such cisterns.

The modern-day equivalent of the cistern is the familiar water tower see on the landscape of so many cities and towns today. As growing populations demanded increasing water supplies, deeper wells became necessary, as did transporting water from distant locations. Reservoirs and systems of **canals** and **aqueducts** evolved as a result, culminating in the highly advanced water distribution system of ancient Rome, which encompassed eleven aqueducts 359 mi. (578 km) in length and carried over 50 million gal. (189 million l) each day. Lead pipes relayed the water from cisterns to city fountains and baths. A modern parallel to Rome is Southern California, which uses a system of aqueducts to bring water to its cities and crops.

Dams, too, were in use as an early method of water containment. Rome's attention to furnishing citizens with an adequate supply of clean water did not prevail in the Middle Ages, when contamination and the disease that went with it prevailed. By the seventeenth and eighteenth centuries, however, water-distribution systems of cast-iron pipes, aqueducts, and pumps were operating in London and Paris, and the Chelsea Water Company introduced sand-filtration to treat water from the Thames River in 1828. Both *sedimentation*—the process of allowing matter in water to settle out—and *filtration* through sand and stones date from antiquity, and today water treatment has evolved to encompass a range of methods.

Water collection occurs on a large scale in semi-arid regions, which rely on reservoirs to collect and store water for large populations, as in California. Even in areas with heavier rainfall, such as New York City, reservoirs provide emergency supplies for summer dry spells. Coastal desert areas lacking ready access to fresh water supplies sometimes turn to **desalination**—a means of removing saline from salt water. The first such plant opened in Kuwait in 1949. Water diversion rather than water collection is used mainly where large quantities are in demand, and greater numbers of people are served through this method than any other. Water is piped from source to user, or from storage point to user.

Every water tap represents the user end of a pump-and-pipe supply system. The water is brought from its initial source, purified, then delivered directly or stored for later use. Treatment is a crucial stage, with water first stored to allow sedimentation to occur, and visible and ultraviolet light to kill bacteria. Care must be taken to counter algae growth, and filtration systems using chemicals like aluminum sulphate coagulate suspended solids so that they can be removed. Water is sometimes disinfected with chlorine, and further refinements include softening water by removing calcium and magnesium and aerating it—infusing the water with small air bubbles—to mitigate odor or unpleasant taste.

Just as an adequate supply of water is essential, so is proper drainage, which keeps large amounts of water from collecting in roadways, buildings, and farmland; prevents stagnant accumulations of water that can breed disease-carrying mosquitoes and other pests; and facilitates flood and erosion control and land reclamation. Land drainage methods include open-channel drains, networks of underground pipes, and mole drainage—a system of earthen drainage channels in the

The Pont du Gard, a Roman aqueduct in Nimes, France, dates from before the fifth century.

subsoil. In towns and cities, drainage entails both **sewage systems** and systems that address surface water drainage. Street sewers collect rainwater and deliver it to an elevation lower than that of the local population, while sanitary sewers are a part of the purification system and carry waste-containing water to the plant for treatment.

Water turbine

The **waterwheel** is an ancient technology that converts the energy of rushing water to human use. Mostly used for milling grain in ancient times, the simplest waterwheel has a series of submerged paddles fanning out from a vertical shaft. The moving water rotates the paddles and shaft, which passes through a fixed millstone and is fastened to a movable millstone on top. The more familiar traditional form employs a vertical **wheel** and horizontal shaft.

With the growing mechanization of the production process in the 1800s came an increasing demand for energy. **Steam engines** were feasible only where coal was cheap, and the old waterwheels did not make efficient use of available hydrological power. An urgent need to improve the efficiency of waterwheels was felt particularly in countries that had little or no coal. The first improvements were made by the Frenchman Jean Victor Poncelet (1788-1867), who proposed curbed blades and the use of iron instead of wood for the wheel itself. Wheelpits were built of masonry to form a tight enclosure around the wheel and force all the water coming through the millrace through the wheel.

In 1828, another Frenchman, Claude Burdin, coined the word *turbine*, from the Latin *turbo*, or spinning top. Burdin was a teacher at St. Etienne and taught a course on waterwheels. One of his students, Benoit Fourneyron, introduced the most momentous design changes in 1833. Fourneyron developed an enclosed water turbine that used two wheels. The wheels were completely submerged, and the inner, fixed wheel had gates that channeled the water to the blades of the outer rotating wheel, greatly improving the efficiency of the machine. By 1837, he had developed a design that was four times more efficient than the traditional waterwheel.

The design pioneered by Fourneyron was what has now come to be called a reaction turbine and was most efficient with fast-flowing water. Reaction turbines in use today are of several types, according to the available water pressure. The Francis type, developed by James Francis and Uriah Boyden in the 1850s, has an outside casing with guide vanes that simultaneously directed water against all the spirally curved vanes on the inner rotor. It is best suited for medium water pressures. The Kaplan type reaction turbine, produced by V. Kaplan in 1920, has four very large blades on an outer rotor in a design very similar to Fourneyron's. However, the pitch of the blades can be altered during the actual running to get the best results from varying water pressures. This latter type is most commonly used with low pressure. Finally, the Deriaz type combines features of the other two.

Where a small volume of water falls a great distance, the Pelton wheel, or impulse turbine, is most commonly used. It was invented by Lester Pelton, an American engineer who learned of waterwheels during the California gold rush. His device involves high-pressure water jets which are directed against the rim of a wheel fitted with large twin buckets.

Water turbines, an important feature of the Industrial Revolution, gradually made way for the more powerful and efficient steam turbines of the late 19th century. Turbine technology has also come to be used in jet propulsion, and has revolutionized aviation in the second half of the 20th century. Today water turbines no longer serve to drive industrial machinery directly, except in developing nations. However, they are still in use in hydroelectric plants. Natural waterfalls and **dams** provide falling water to power arrays of water turbines which are hooked up with electric **generators**. The energy is then distributed across vast electrical networks. A significant share of global energy consumption is contributed by this clean and cheap form of power.

Waterbed

Although an early mention of a waterbed was made in a 1961 science fiction novel by Robert A. Heinlein—*Stranger in a Strange Land* —the concept was not practically applied until 1965, when two Tufts University physicians, Dr. James Weinstein and Dr. Barry Davidson, designed a water-filled mattress for use with hospital patients.

During the 1960s and 1970s, the waterbed gained popularity, beginning in the young, ''hippy'' culture and gradually working its way into mainstream American society. The bed was proclaimed ideal for both sleeping and sex, although its detractors claimed it a hazard because of its weight (about 1800 lb [817 kg]) and the possibility of leakage or—worse yet—electrocution. Nevertheless, such devotees as Playboy magnate Hugh Hefner (who covered his with Tasmanian opossum fur) continued to promote the advantages of the waterbed, and in 1984 U.S. sales reached $3.4 million. Waterbeds are now available with a layer of mylar sheeting which absorbs body heat and makes use of an electric heater unnecessary.

Waterman, Lewis E. • See Pen

Waterproof materials

Ever since people began wearing fabric clothing and going out in the rain, efforts have been made to keep water from soaking through the garments. Early waterproofing methods relied on natural properties of the fabrics. Wool was treated with lanolin emulsions, boosting the water repellency of the naturally occurring body oil in wool. Cotton was given a wax emulsion. These treatments were inexpensive, but were not durable when the fabric was cleaned.

In 1823 Charles Macintosh (1766-1843), a Scottish chemist, patented his method of producing rainproof cloth using rubber. He coated one side of wool cloth with a liquid rubber solution, then covered it with another piece of wool cloth. Since rain could not penetrate the **rubber**, the inner layer of garments made with Macintosh's material remained dry. The rainproof coats made by Macintosh's company became popularly known as mackintoshes.

An American inventor named Otis Ferrin patented a waterproof "painting cloth" in the mid-1800s. His process applied heavy oil to heavy material; the resulting waterproof fabric became known as oilcloth and was used worldwide for table coverings and raingear for well over a hundred years before being supplanted by **plastics**. Advances in chemistry gave rise to new water-repellent finishes. **Silicones**, which are more expensive than the waxes but more durable to cleaning, played a strategic part in keeping U.S. **aircraft** airborne during World War II. Beginning in 1942, when American pilots began flying B-17 bombers from factories in the U.S. to airbases in Great Britain, pilots sometimes noticed a blue glow around the leading edge of engine compartments. The engine would begin misfiring, and then die altogether as more **pistons** quit. Loss of more than one engine and the crew would have to abandon the craft. The explanation turned out to be that the high-voltage electricity of the aircraft engine system was ionizing moisture from the ocean that had worked its way into tiny pores in the rubber **wires** insulating the **ignition system**. A silicone **polymer** that had been developed for waterproofing electrical equipment aboard **submarines** provided the solution. By applying this coating to the **spark plug** wires, the silicone grease kept the wires dry and the bombers flying. It was used for all U.S. bombers throughout the war.

Another type of repellent finish is the fluorine-based finish, the best known of which is Scotchguard, an accidental invention of the ever-productive Minnesota Mining and Manufacturing Company (3M) research lab. An assistant in the lab accidentally spilled an experimental fluid on her canvas sneaker and found that while she could not remove the substance, the sneaker stayed cleaner than its mate. Lab chemists Patsy Sherman and Samuel Smith sensed possibilities, and launched experiments that produced Scotchguard, first marketed in 1956. Applied to upholstery, rainwear, and clothing—either by the manufacturer or by the consumer—Scotchguard makes a fabric stain-resistant by making the material shed rather than absorb liquids. Today, Scotchguard is the world's most-sold protective finish for **leather**, fabrics, and **carpeting**; it is also now being used on wood.

The 1980s saw the birth of a new type of waterproof fabric that uses its microscopic structure rather than a chemical finish to repel water. The best-known of these, Gore-tex, was invented by W. L. and Bob Gore. W. L. Gore left the Du Pont company in 1958 to devote himself full-time to developing the company's **Teflon** product into insulation for computer wiring. Gore's son, Bob, earned his Ph.D. in chemical engineering and joined his father's lab. One day in 1969, Bob yanked a rod of Teflon instead of gently and slowly pulling it; the rod stretched into a porous, strong filament. This expanded Teflon was christened Gore-tex. It was ideal for transmitting computer signals, for use in surgical implants—and to produce a waterproof material consisting of a Teflon film bonded to a synthetic fabric. It is now widely used in outdoor activity garments and raingear.

Waterwheel

The waterwheel is considered the first rotor mechanism in which an outside force created power to spin a shaft. The Greeks are said to have first developed the waterwheel, using it to raise water from rivers. Polls or pots were attached around the circumference of a large **wheel**; then oxen would walk in a circle around a vertical shaft connected through a simple **gear** to the horizontal shaft of the waterwheel. When possible, the current of a fast-flowing river would do the work of the oxen.

By Roman times, ancient engineers realized that the spinning shaft could be used as a power source to turn millstones. However, few were built for this purpose due to the abundance of slaves to grind grain into flour. After the fall of the Roman Empire waterwheels spread throughout Medieval Europe largely through the efforts of monks who introduced the technology to landowners, many of whom had lost workers to disease and war. The first factories—textile mills—were powered by water. Other early uses of waterwheels included sawmills for sawing lumber and gristmills for grinding grain. Many large cities owe their existence to power drawn from nearby rivers, but this power source was not very dependable due to floods and droughts that changed the amount of water flowing in the river.

Several types of waterwheels evolved over the centuries. The two most used were the overshot wheel, which involved water falling on the paddles from above, and the undershot wheel, which was placed directly in the water and worked especially well in swift streams. A more unusual and less frequently used type was the reaction wheel in which water was dropped into a hollow tube from a great height. The tube had hollow spouts attached to it, and the water would spray out the spouts, causing the tube to rotate much like today's lawn sprinklers. The tidal wheel required an enclosed pond with a gate. As the tide rose, water flowed into the pond until the gate was closed, effectively trapping the water. Later, the gate was opened, allowing water to flow out and turn a wheel for power. By the late eighteenth century, more efficient and powerful **steam engines** began to replace waterwheels as Europe's primary source of power. However, the principle of the waterwheel lives on in today's **water turbine**.

Watson-Watt, Sir Robert (1892-1973)

Scottish physicist

In the early days of World War II, British scientists had just completed what would be the most advanced defense system of their day: a network of **radio**-signal stations that could detect the approach of Nazi bombers, day or night, in fair or foul weather. This network was given the ungainly title "Radio Detection and Ranging" but is most commonly known as **radar**. Though the list of scientists who contributed to its creation is long, the name at the top would certainly be Sir Robert Watson-Watt, whose work made the creation of radar possible. Watson-Watt was born on April 13, 1892, and as a student in

Brechin, Scotland, he developed an interest in radio telegraphy; in 1915 he began working for the London Meteorological Office using radio waves to monitor thunderstorm activity, an essential task to aid the safe flight of early fragile **aircraft**. He continued to work for the government for much of his life, first by supervising two radio research stations and later by researching the use of radar for navigation. All this time, Watson-Watt knew that radar could eventually be used for echolocation—in fact, he patented this concept in 1919, anticipating the future construction of a radar location device.

In the early 1930s, Watson-Watt was appointed scientific adviser to the Air Ministry and was given the task of designing just such a device. Several important discoveries were instrumental to the invention of the radar system; for example, in 1922 the **cathode-ray tube**, which would be used for visualizing the returning radar signal, became available. In 1936 pulsed radar replaced continuous-wave emitters; the old system could detect only the presence of an object, while the pulsed signal could also pinpoint its location. Finally, in 1939 the practical microwave transmitter was constructed, allowing the radar locater to operate through cloud and fog. The construction of a radar defense shield, which was completed in secrecy, passed its first test during the Battle of Britain, where the early warning system made possible the Allied victory over the Nazi air force. Ironically, German scientists had reportedly also been working on a radar location system. However, Adolf Hitler (1889-1945) and Hermann Göring (1893-1946) saw it as useful in a defensive situation only; since Germany would never be on the defensive end of an air war, they concluded, the development of radar was of low priority. Watson-Watt was invited to the United States in 1941 to help Americans develop their own radar system. Since World War II, the use of radar for nonmilitary application has become widespread. Watson-Watt was knighted in 1942. He died on December 5, 1973.

See also Microwave transmission

Robert Watson-Watt

WATT, JAMES (1736-1819)
Scottish engineer and inventor

The Industrial Revolution in Europe could not have taken place without the work of James Watt, who is commonly credited with inventing the **steam engine**.

Watt was born in Greenock, Scotland, on January 19, 1736. At an early age he helped his father build **ships**, and he was exposed to the various technology of the time. In 1755, he left for London to study the craft of instrument-making. He began working with steam in 1764 when Glasgow University, for which he worked, brought him a Newcomen engine for repair.

Watt not only repaired the engine, but he also began improving on it. He noted, for example, that the engine wasted time cooling the piston chamber during every cycle. Within five years Watt built a demonstration model that he patented as "a new method of lessening the consumption of steam and fuel in fire engines." This model introduced a second chamber where the condensing cycle took place, leaving the steam chamber hot and ready for a new batch of steam.

Watt's work with steam caught the attention of English engineer **Matthew Boulton**. In 1775 Watt entered a partnership with Boulton, whose factory was widely respected for its quality metalwork, plating, and silversmithing. Watt was continually searching for ways to improve existing technology. One of his innovations was the 1772 **micrometer**. This invention helped **John Wilkinson** perfect a **boring machine** that could drill cylinders with unprecedented uniformity, a necessary step towards Watt's next development—introduction of steam on both sides of the **piston**. This innovation forced the piston in both directions rather than allowing atmospheric pressure to complete the condensation/vacuum cycle in its own time. These were the engines that he and Boulton began to market in 1775.

In 1781 Watt devised mechanical attachments to convert the rocking motion of the piston to a rotary movement that could more easily be used to power machinery. It was this step perhaps more than any other that led the steam engine to the forefront of Europe's Industrial Revolution. It was used to pump bellows for **blast furnaces**, to power huge hammers for shaping and strengthening forged metals, and to turn machinery at textile mills. For the first time, mills and factories were not limited to locations near streams or windy plains.

In 1788 Watt unveiled another major contribution to the realm of instrumentation and machine control by introducing the centrifugal **governor**. Since a similar device had been in use in **windmills**, Watt made no effort to apply for a patent;

James Watt

still, it became commonly known as the Watt governor. Its importance centered around its automatic control of steam output. The governor consisted of two metal spheres mounted on a vertical rod that was spun by the engine's output of steam. The faster the rod spun, the farther the two spheres were thrown outward by centrifugal force. The farther the spheres were thrown, the more they choked off the steam outlet. As steam output decreased, so did intake as well as the engine's power output. As power and steam output decreased, the slower the spheres rotated, re-opening the steam outlet and beginning the reverse of the process. Engine output and speed could be controlled and adjusted between these two limits.

By 1790 Watt's engines had nearly replaced Newcomen's engines which were, for the most part, at least fifty years old. By this time Watt had added many other improvements such as a new type of condenser that used a system of tubes instead of one large chamber, an air pump to maintain a vacuum in the condenser, a jacket and "stuffing box" to prevent steam from escaping where the piston rod passed through the cylinder, a *sun and planet gear* that introduced rotary power to a generation of mills in Soho, and the later crankshaft that replaced this gear.

Watt's other ideas include the steam radiator heat, an office copying press, and a device that could reproduce sculptured busts.

Watt retired in 1800. He received an honorary doctorate degree from Glasgow University and was elected to the Royal Society. He died near Birmingham, England, on August 19, 1819, a successful and respected inventor whose contributions are used to this day, and he has been honored by the use of his name for the unit of power, the watt (one joule per second). He is buried beside his former business partner, Matthew Boulton.

See also Newcomen, Thomas

WATT'S GOVERNOR • See Governor

WAVEGUIDES

A waveguide is a device used primarily in the conduction and transmission of microwaves. It is a hollow tube or pipe that takes the place of **wire** or coaxial cable to transmit a microwave signal a certain, usually short, distance. Because of the nature of microwaves, an efficient waveguide can carry more than 100,000 transmissions simultaneously.

It did not take scientists long to realize that simple cables were excellent for transmitting **radio** signals, but were inefficient in the transmission of high frequency signals. They had experimented with several tube-design conductors, finding that they, too, were impractical—unless the frequency of the signal was very high, much higher than any radio waves. As their knowledge of microwaves increased, scientists realized that pipe-design conductors were ideal for microwave transmission.

A waveguide can actually be a dielectric rod as well as a tube conduit, but the latter is used more frequently. Usually made from drawn brass, the waveguide is constructed in small circular or rectangular segments that are later assembled into a long pipe. The inside of the pipe, along which the signal travels, is plated with cadmium or nickel, although some special applications demand that the inside be made of gold or silver. As a microwave signal is sent through the pipe it bounces along the inside walls, taking the path described by the waveguide. If a waveguide is constructed properly, little loss of signal will occur throughout the pipe's "run."

Though it somewhat resembles a water pipe, the waveguide cannot conduct microwave signals around sharp curves or corners, so the run must be fairly straight. Also, waveguides lose efficiency over long distances, and many are restricted to less than 10 ft (3 m).

Almost every microwave application uses some kind of waveguide to conduct the signal from one place to another. The most common application is in a **microwave oven**, where a waveguide is used to direct the heating waves toward the cooking compartment. Microwave **antennas** use waveguides to carry microwaves to transmitting arrays and to collect them from receiving arrays. They are also used in **radio telescopes**, mass spectrometry, radio surveying, and **burglar alarms**.

The most recent evolution of the waveguide is a tube of light-carrying material called **fiber optic** cable. Utilizing the principle of total internal reflection, fiber optic cable bounces a beam of light over a long distance while retaining most of its original intensity. Fiber optics have formed the foundation of many new communication and information storage systems, particularly those that rely upon digital coding.

See also Fiberscope; Microwave transmission; Prism; Spectroscope

WEATHER BALLOON

The invention of the weather balloon inaugurated the age of remote sensing, the ability to collect information from unmanned sources.

The first observation **balloon** was launched immediately before the first manned balloon flight by Frenchmen Jean-François de Rozier and the Marquis d'Aalandes on November 21, 1783, for a pre-flight wind reading.

Later, French meteorologist Leon Teisserenc de Bort (1855-1913) pioneered the use of weather **balloons**, handily proving their utility. With balloon-acquired data, he determined the existence of a lower level of the atmosphere, which he termed the troposphere or "sphere of change," where weather takes place.

Since the 1930s, when **radio** tracking systems were invented, balloons have been used as complete floating weather stations, employing such instruments as **thermometers**, **barometers**, **hygrometers**, cameras, and **telescopes**. They are used widely to collect such atmospheric information as temperature, pressure, and humidity.

Since their inception, the elongated bags of helium have been carrying aloft more sophisticated observation devices, taking the science of weather observation to the edges of outer space.

With the invention of radio tracking systems in the 1930s, weather balloons became complete floating weather stations, employing such instruments as thermometers, barometers, and cameras. Above, a balloon 128 ft (39 m) in diameter, is being filled with helium prior to its 90,000 ft (27,430 m) ascent into the atmosphere.

WEATHER FORECASTING METHODS

For as long as they have observed the sky, humans have been keenly interested in predicting what the weather may bring. Most of the world's indigenous peoples came to understand the climate and weather patterns in their area, and could often predict the weather for the next day or so. However, weather forecasting in the modern sense had to await a more detailed understanding of the atmosphere and the development of many weather instruments, such as the **barometer**, **hygrometer**, **weather balloon**, and **radar**. Three major technological developments in particular have led weather forecasting from its days of inception to its current status: the development of instant communications beginning in the late 1800s, remote sensing devices starting in the early 1900s, and computers in the 1940s and 50s.

Weather recording instruments date from the fifteenth century when **Leonardo da Vinci** invented the hygrometer, an instrument to measure atmospheric humidity. About 1643 **Evangelista Torricelli** created the barometer to measure air pressure differences. These instruments were improved upon in the eighteenth century by Frenchman Jean Andre Deluc (1727-1817), and have been refined numerous times since then.

Once surface weather observations began to be collected, it became necessary to organize this data in a meaningful form. One obvious way is to draw up a weather map. Weather information has long been displayed in map form. In 1686 English astronomer Edmond Halley (1656-1742) drafted a map to explain regular winds, trade winds, and monsoons. Over 200 years later, in 1863, French astronomer Edme Hippolyte Marie-Davy (1820-1893) published the first isobar maps, which depicted barometric pressure differences.

The availability of weather data allowed scientists to try to forecast weather. The United States Weather Service, established in 1870 under the supervision of **Cleveland Abbe**, unified communications and forecasting. **Telegraph** networks made it possible to collect and disseminate weather reports and predictions. By the turn of the twentieth century, the **telephone** and **radio** further increased meteorologists' ability to collect and exchange information. This was the period of *synoptic* forecasting: by collecting surface weather data and plotting it on a map, skilled meteorologists could often recognize large scale patterns and make reasonable forecasts, especially if they were familiar with the region.

In many cases however, synoptic forecasting failed because surface observations failed to tell the whole story. What was missing was information on the atmosphere throughout its depth, the temperature, pressures, and winds aloft. Remote sensing, the ability to collect information from unmanned sources, was developed to fill in this gap. Remote sensing originated with the invention of the weather balloon by Frenchman Leon Teisserenc de Bort (1855-1913). Designed to make simple preflight tests of wind patterns, balloons were eventually used as complete floating weather stations with the addition of a **radio** transmitter to the balloon's instruments. The development of radar, **rockets**, and satellites greatly improved remote data collection. Weather radar first came into use in the United States in 1949 with the efforts of Horace Byers (1906-) and R. R. Braham. **Weather satellites**, the first launched in 1960, can produce sequence **photography** showing cloud and frontal movements, water-vapor concentrations, and temperature changes.

By the close of the nineteenth century, armed with mounting observations and better communications, many scientists were adding to the pool of meteorological knowledge. Among them was Englishman Ralph Abercromby, who in his 1887 book, *Weather*, depicted a model of a depression (low pressure area) that was used for many years. During World War I, the father-son team of Vilhelm Bjerknes (1862-1951) and Jacob Bjerknes (1897-1975) organized a nationwide weather-observing system in their native Norway. With the available data they formulated the theory of polar fronts: The atmosphere is made up of cold air masses near the poles and warm tropical air masses, and fronts exist where these air masses meet. In the 1940s, Englishman R. C. Sutcliffe and Swede S. Peterssen developed three-dimensional analysis and forecasting methods. The science of *dynamic* meteorology, which focuses on describing the motion of the atmosphere in terms of mathematic equations, continued to develop throughout the early twentieth century. This mathematical approach to predicting the atmosphere held great promise, except for one glaring problem: to make a prediction, the dynamic equations had to solved for each point in the atmosphere, and had to be solved repeated to recreate the time development of the weather. Attempting to use dynamics therefore required some simplifications: for example, solving the equations for widely separated geographic points, and stepping forward in time by large jumps. Even so, the calculations required to produce a forecast were so complicated and time consuming that dynamic weather forecasting had to await the invention of electronic computers.

In the late 1940s, a group of American computer scientists and meteorologists began to develop computer weather ''models,'' collections of dynamical equations modified so that computers can solve them, along with a set of initial conditions obtained from observations. The models divided the atmosphere into layers, and the layers into a set of points making up a grid, so that the dynamics equations could be solved at each point. The only computer available at the time was so slow that it took 24 hours to produce a 24-hour forecast. In time, as digital computers increased in speed and reliability, it became possible to model the weather more accurately. By the mid 1950s, American and Swedish meteorologists were producing simple but fairly accurate computer forecasts on a regular basis.

The rapid development of computers in the 1960s and 70s allowed great leaps in the size and complexity of computer models. Today complex computerized models provide many types of weather forecasts for the National Weather Service, major airlines, the military, and virtually everyone with an interest in the weather. However, experience learned over years of forecasting often allows a local forecaster to fine tune the computer model predictions for a particular area. Hence, weather forecasting still remains very much a human activity.

Weather radar

Radio Detection And Ranging systems, known as **radar**, were developed in Britain in the 1930s as a defense against German bombing raids. While their military use flourished during World War II, radar was not used commercially until the 1950s. Today, radar has become commonplace. Flight crews routinely use radar tracking features to navigate **aircraft** to their destinations safely. Radar is also commonly used by meteorologists to track weather patterns. For most **television** viewers of the weather forecast, the image of a green, circular radar screen—complete with a sweeping arm of light—is a familiar one. Using a high-intensity **microwave transmission**, meteorologists can detect and follow large masses of precipitation, whether they be rain, snow, or cloud. A beam of pulsed microwaves travels until it hits an obstacle (for meteorological purposes, a cloud or band of precipitation). It is then reflected back to the source, where it is received by a radar **antenna**. By measuring the time taken for the signal to reach the obstacle and return, its distance can be easily calculated. With thousands of pulses emitting and returning, a two-dimensional image of the weather formation is displayed on a **cathode-ray tube**, showing its precise position.

The first use of radar to study the atmosphere occurred in 1945, when Major Harry Wexler of the U.S. Air Weather Service published the first radar image of a hurricane. After World War II, the U.S. Navy gave the civilian Weather Bureau 25 surplus radars, and the Bureau began to establish a radar network for tracking tornado movements in the Midwest. Later in the 1950s, radar networks were extended to the hurricane prone eastern seaboard of the United States. In the mid fifties, the Weather Bureau began to install a few Doppler radars. A more elaborate version of radar tracking, Doppler radar uses a continuous signal rather than a pulsed wave. Doppler radar can determine both the direction and velocity of wind patterns, as well as areas of precipitation. Doppler radar measures the shift in frequency caused by a moving particle. If the returning frequency is higher than when transmitted, the particle is moving toward the source; if it is lower, the particle is moving away. However, the system only works when a particle is approaching or receding from the transmitter; Doppler radar cannot detect the velocity of a particle moving perpendicular to the radar signal. For this reason, signals from more than one radar source must be combined to produce an image free of gaps. Although standard radar can often detect tornadoes by a characteristic hook shaped echo, a Doppler system can detect the presence of funnel clouds and tornadoes much more reliably, and is now used quite commonly by weather forecasters, as well as **radio** and television stations, to monitor thunderstorms for the presence of strong winds and tornadoes. Doppler radar forms an important part of the Next Generation Radar (NEXRAD) network, an extensive upgrade to the radar facilities of the National Weather Service put in place in the 1980s and 90s. With NEXRAD, forecasters are now able to continuously track severe storms in any region of the country, providing potentially life-saving weather information.

WEATHER SATELLITE

The first attempt to look at the earth's weather from space occurred early in the history of the United States space program. In 1959 *Vanguard II* was launched with light-sensitive cells able to provide information about the earth's cloud cover. Unfortunately, the satellite tumbled in orbit and was unable to return any information. *Explorer VI*, also launched in 1959, was more successful and transmitted the first photographs of the earth's atmosphere from space.

In 1960 the United States launched the first experimental weather satellite, *TIROS 1*. The acronym for Television and Infra Red Observation Satellite, *TIROS 1* televised over 22,000 photos before it failed six weeks later. It detected potential hurricanes days before they could have been spotted by any other means. It watched the spring breakup of the ice in the St. Lawrence River and helped forecast weather for the supplying of Antarctic bases. *TIROS 1* also used infrared detectors to measure the amount of heat radiated by the earth's surface and the clouds.

Later versions of TIROS improved upon the original with television cameras that provided direct, real-time readouts of pictures to simple stations around the world. In 1970 *ITQS-l* was launched with the capability of not only direct-readout, automatic picture transmission but also the ability to store global images for later transmission and processing. Another successful series was called NOAA after the National Oceanic and Atmospheric Administration. Some of these satellites were placed in geostationary orbit (moving at the same speed as the Earth) and thus were able to continuously observe one area. This helped in the detection of severe storms and tor-

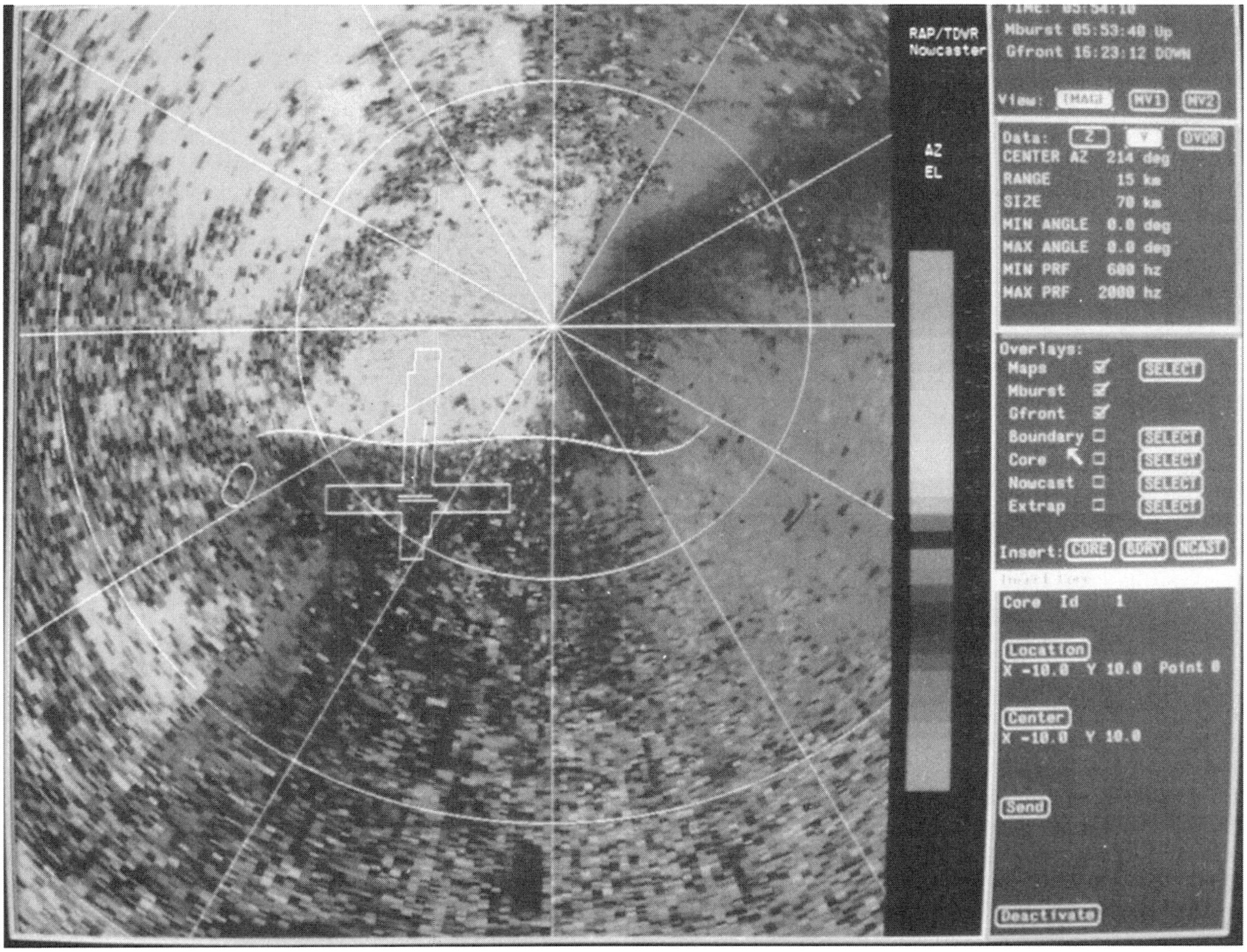

Winds shown on a Doppler weather radar.

nadoes and provided real-time coverage at an earlier stage of cloud and frontal weather movements.

Other TIROS-type satellites, such as *NIMBUS* (1960s) and *NOAA-9* (1980s-1990s), are in polar orbit, where their infrared sensors measure temperatures and water vapor over the entire globe. Several GOES (Geostationary Operational Environmental Satellites) also cover the western and eastern hemispheres. The GOES satellites form part of an international global weather forecasting system administered by the World Meteorological Organization (WMO). In addition to two U.S. GOES satellites (*GOES East* and *GOES West*, covering the two U.S. coasts), the eastern Pacific ocean is monitored by a Japanese geostationary satellite, India's *INSAT* covers the Indian Ocean, Russia maintains a geostationary satellite over western Asia, and the European Space Agency's *Meteosat* observes the eastern Atlantic Ocean, Europe and Africa. These satellites are able to provide weather reports for places that have not been covered very well in the past: ocean regions, deserts, and polar areas. They also trace hurricanes, typhoons, and tropical storms, in the process saving many lives. Their data are used to produce state-of-the-art charts showing sea-surface temperatures, information useful to the shipping and fishing industries.

New satellites that probe the earth's atmosphere by day and night in all weather are being developed in many countries. Since the weather satellite is now an established tool of meteorologists all over the world, both developed and developing nations will continue to rely on these crafts.

WEDGE • See Inclined plane

WELDING

There are many methods for fusing or welding metal together. Depending upon the type of metal involved, the nature of the task, and the desired strength of the bond, there are welds to fit every need. The forge, or fire, method is generally used in blacksmithing. This method is the oldest welding process and establishes fusion by hammering the heated pieces of metal together. Blacksmithing fell into disuse at the end of the nineteenth century when more accurate welding techniques became available.

In the late 1860s, *electric welding* was introduced in America and *oxyacetylene welding* followed in 1903. Oxyacetylene welding has become the most common type of welding in non-industrial settings. In high-pressure oxyacetylene systems, which are most often used for small jobs, oxygen and acetylene are fed by hoses from separate cylinders to a torch mechanism where the two gases are mixed and burned at a high temperature. Low-pressure oxyacetylene systems generate acetylene from calcium carbide. This method is more attractive to users of large quantities of acetylene, since it eliminates the need to store a large number of acetylene tanks. Oxyacetylene welding is inexpensive, easy to use, and applicable to many types of metal.

Metal-inert gas (MIG) welding uses an inert-gas shielded arc of electricity. The arc is struck between the object being welded and a filler metal electrode. During the welding, the electrode continuously melts and is fed by a motor drive. In *tungsten-inert gas* (TIG) welding, the arc is struck between the work and a tungsten electrode, which is not consumed by the process. A filler metal is fed separately. The advantage of MIG welding is that it requires no *flux* (a substance applied to the metal to aid its bonding). The most common inert gas used for welding in the United States is argon, while helium is preferred in the United Kingdom and many other countries. *Thermit welding* uses a chemical reaction between aluminum and iron oxide. The resulting flame burns at 5000° F—twice the temperature of liquid **steel**.

An energy-efficient microwave welding was developed in the early 1990s. It ionises a gas, argon for example, using microwave energy which is maximized by a tuned cavity that centers the microwave welder's electric field. It can reach temperatures of 4000° F. Ultrasonic welding was first developed in the 1960s, but refined in the 1990s to be clean and fast and regulated by a **microprocessor**. An ultrasonic welder transforms electrical energy to ultrasonic frequency mechanical vibratory energy. This energy is transferred to a horn via a booster; the horn transfers the energy to the objects being welded. The actual weld is made by friction. Another welding technique developed in the 1990s is fiery foils. The foils are ultrathin (no thicker than a strand of human hair), highly reactive metallic films made of two layers of elements that have a strong desire to bond with each other, like aluminum and nickel. The foil is placed over the two surfaces, and a one small spark, from a match for example, causes the foil to ignite itself, giving off enough heat in mere milliseconds to weld the surfaces without damaging the rest of the objects. Extremely portable, fiery foils can be used anywhere, including places traditional welding equipment could never be carried. Because no oxygen is required, fiery foils can also be used in outer space or under water.

Welding is used extensively in construction and manufacture, serving as a substitute for or in addition to bolts and rivets. Welding is often utilized by metal sculptors as well.

See also Electric arc; Solder and soldering iron; Blowtorch; Liquid gas, commercial production of

WELL • See Water supply and drainage

WESTINGHOUSE, GEORGE (1846-1914)

American inventor and manufacturer

George Westinghouse was born on October 6, 1846 in New York. He worked in his father's factory, where he learned about **gears**, **pistons**, and cams. In 1865, after serving in the Union army and navy, Westinghouse received his first of patent (one of over 400 patents he would eventually receive), for a small rotary engine. By the age of 30, he had already been president of his first corporation, and had exhibited his air **brakes**, **air compressors**, speed indicators, engines, and engine **governors** at Philadelphia's Centennial Exhibition World's

Fair of 1876, alongside inventions of **Alexander Graham Bell** and **Thomas Edison**. But he had yet to enter the industry, or found the company, that made his name a household word.

As business returned to normal following the Civil War, Westinghouse was one of the few men who became appalled at the lack of safety on the world's railroads. In fact, after nearly four decades of *modern* railroading in America, adequate braking and signaling systems were yet to be developed. At this time, railroad cars were still stopped by brakemen who were stationed along the length of a **train** to turn hand brakes on each car after hearing whistle signals from the engineer. Westinghouse saw the need to stop the whole train using one braking system.

Following experiments with pressure-and steam-activated brakes, Westinghouse discovered a magazine article that described a **tunnel** in Switzerland that was being excavated using compressed air. In 1868 he finished a model of a train brake that consisted of an air pump powered by the train's engine, a single control **valve** for use by the engineer or brakeman, pipes and flexible connections that ran the length of the train, and mechanical devices on each car to activate brakes.

Using Westinghouse's new brake invention, an engine and four cars were outfitted for a demonstration in front of railroad officials. Ironically, as the train was arriving for its show, a man was thrown unconscious onto the tracks. The engineer reached for the new brake valve, turned it, and watched as the man's body drew near. Air rushed from the reservoir, through the pipes, and into the cylinders that forced the brakes against the wheels. According to the story, the train stopped four feet from the man's body. At age 23, Westinghouse founded the Westinghouse Air Brake Company.

Further testing revealed two problems with the Westinghouse braking system. First, if part of the train separated from the engine, the air lines broke and the brakes became useless. Second, a leak or malfunction of the brakes in one car sometimes led to the failure of all the brakes. To rectify these problems, Westinghouse tried applying another theory: instead of activating the brakes with air, why not activate them with a drop in air pressure? The idea worked and, with only minor changes, Westinghouse's 1872 air brake principles continue to be used today. Westinghouse also developed and patented signaling systems and other safety devices for railways, which led to the founding of the Union Switch and Signal Company in 1880.

In 1883 Westinghouse drilled a small gas well for himself on the grounds of his Pittsburgh estate, *Solitude*. Two months later, another well was drilled, but this one exploded with such volume that it blew out of control for several weeks. The gas was lighted and became known as Westinghouse's Torch. As a result of this mishap, Westinghouse became interested in natural gas as a power source, and devoted himself to engineering a successful transmission network. He soon founded the Philadelphia Company to provide gas service to the Pittsburgh area and received 38 patents, including one for a gas meter, an automatic cutoff regulator, and a leak-proof piping system.

Westinghouse was 39 years old when he began exploring the field that would make his name legendary. Though not a new theory, Westinghouse was perhaps the first to foresee the importance and versatility of **alternating current** (AC) over **direct current** (DC) for electrical **power supply**. Westinghouse proceeded to buy the patent for the first modern electrical **transformer** of Frenchman Gaulard and Englishman Gibbs. With the help of engineers such as **Nikola Tesla**, he improved their experimental designs and proceeded to promote AC power.

George Westinghouse

During 1866 experimental AC distribution networks were built in Great Barrington, Massachusetts and Lawrenceville, Pennsylvania. The Westinghouse Electric Company began producing dynamos, transformers, and **electric motor**s that same year. In 1893 Westinghouse received contracts to build the Niagara Falls generating station and to light the World's Columbian Exposition in Chicago, Illinois.

Westinghouse lost most of his fortune in the economic panic of 1907. He died on March 12, 1914, almost 20 years after bringing electricity to New York City.

See also Braking systems; Railway signaling systems; Train and railroad

Wheatstone, Charles (1802-1875)
English physicist

Charles Wheatstone was a physicist whose fertile, questioning mind produced numerous discoveries and inventions in the fields of optics, acoustics, electricity, and telegraphy.

Charles Wheatstone

Wheatstone was born in Gloucester, England, to a family of musical instrument makers and dealers. He had no formal education in science. He was apprenticed to an uncle in 1816 as a musical instrument maker, and invented the concertina in 1829. His musical background led Wheatstone to experimental studies of acoustics, which in turn led to his appointment as professor of experimental physics at King's College, London, in 1834. Wheatstone held this position for the rest of his life, although he seldom lectured, concentrating instead on research in electricity and optics.

Wheatstone's contributions to acoustics included the kaleidophone, a device that produced a visual demonstration of sound vibration. He also made important discoveries about plate vibrations and vibrations in columns of air. In the field of optics, Wheatstone invented the **stereoscope**, which demonstrated how two pictures could be visually combined to create the illusion of depth and three dimensions. He anticipated the development of spectroscopy with his 1835 paper showing that the spectra of spark discharges differed according to the metal used for the electrode. Wheatstone also invented a polar **clock**, which used the angle of polarization of light to determine the time of day.

In the 1830s Wheatstone turned to the study of electricity. He used a rotating **mirror** in 1834 to make the first measurement of the velocity of an electrical current through a **wire**. (The same apparatus, by Wheatstone's suggestion, was later used to measure the speed of light.) These experiments, combined with his early interest in acoustics, led Wheatstone to speculate on the possibilities of sending messages along the wire using the electric current. At this point, Wheatstone was approached by William F. Cooke (1806-1879), an Englishman who had joined the East India army at the age of 20 in 1826 but who had been forced to resign his commission because of ill health. In 1835 Cooke had observed a demonstration of an electric **telegraph**, and, his imagination fired, returned to England in 1836 to devote himself to telegraph design. He turned to Wheatstone for advice, and the two formed a partnership. They patented their five-needle telegraph in 1837, constructed a demonstration line the following month, and installed the first working commercial electric telegraph for the Great Western Railway in 1838.

Although Cooke and Wheatstone persistently quarreled about claims of who actually invented their telegraph, they remained partners. Cooke concentrated on business affairs, while Wheatstone continued development work on the telegraph for many years. In 1840 he brought out the first of his many types of letter-showing dial telegraphs; in 1841 came the type-printing telegraph; then he introduced an automatic transmitting and receiving system. Wheatstone also studied submarine telegraphy, experimenting with an actual line in 1844 in Swansea Bay.

Wheatstone made important contributions in the field of electicity as well. He improved early versions of the **dynamo**. He recognized the great theoretical and practical importance of Ohm's law, and spread knowledge of it in England, where it was little known at that time. He invented the rheostat (an adjustable resistor), and popularized a method of accurately measuring electrical resistance that had been invented by Samuel Christie (1784-1865) but became known as the Wheatstone bridge.

Wheatstone was also an accomplished cryptographer, and invented a machine called the Playfair cipher to create indeciperable secret messages. Wheatstone received numerous honors for his scientific accomplishments. He married in 1847 and had five children. Wheatstone was knighted in 1868 and died on a visit to Paris in 1875.

WHEEL AND AXLE

The wheel has undoubtedly been among the most significant of human inventions. Prior to its invention, population growth was limited, because people could not move easily to find more food and water once their resources had been diminished. The population in any area had to remain small enough for the land to support that group indefinitely. Only small quantities of goods could be transported, so trade was not common. Many other tasks which are basic to our way of life today could not be accomplished without the wheel.

The wheel has had many uses throughout human history. One of the earliest was the potter's wheel, which first appeared in Egypt around 3000 B.C. Wheels have also been used as winches to draw water up from wells or anchors up onto a **ship**, as **waterwheels** and **windmills** to generate power, as grinding mills, as **pulleys**, and as **spinning wheels** for textile workers.

Perhaps the most common use of the wheel, however, has been in transportation. Sometime before 3500 B.C., people in Mesopotamia began to use sleighs-like vehicles to increase the load their beasts of burden could carry. By 3500 B.C., the runners of the sleighs had been replaced with disk wheels fixed to an axle to which a cart body was attached. The earliest examples of these inventions are two- and four-wheeled Mesopotamian chariots. At first, wheels were made of wood planks fastened together, then cut into circles and often bound with cross-struts. Natural knot-holes were frequently placed in the center to serve as holes for the axle.

Around 3000 B.C., rims, made of wood studded with **nails** or metal strips, were added to the wheel. Between 2000 and 1000 B.C., spoked wheels came into use in Mesopotamia, Egypt, China, and Scandinavia. Spoked wheels were much lighter and could be made with a larger diameter, allowing chariots to be driven at high speeds over rough ground—which made them well-suited for use in battle. The axles and rawhide **bearings** of these vehicles were lubricated with animal fat, vegetable fat, or mineral oil.

Vehicle wheel design experienced only slight changes until the sixteenth century A.D., when dished wheels were invented. The dish wheel spokes were arranged in a flattened cone, so that the tops of the wheels angled out to accommodate wider, swaying loads on narrow tracks. Wheels continued to be made from wood with metal rims until the 1870s, when **wire** wheels were developed for the **bicycle**.

The pneumatic tire, a rubber casing inflated with compressed air, became more popular beginning in the late 1880s. The wheels on early **automobiles** were wooden and spoked or made of wire, although some had **artillery** wheels (one-piece, cast iron copies of wooden wheels). During the 1930s, light, strong, and inexpensive pressed **steel** wheels replaced other varieties for use on cars. Cast wheels, similar to the old artillery wheels but made of light, strong magnesium or **aluminum** alloy are now found on some sports cars.

WHEELBARROW

The wheelbarrow, a single-wheeled vehicle with a pair of handles at the rear, is designed to transport small, heavy loads over short distances. The person operating it can pick up the rear and push the load forward; the **wheel** takes the place of a second person who might otherwise be required to assist in carrying a heavy or cumbersome load.

Modern wheelbarrows have a pair of legs under the rear to keep the load level when the wheelbarrow is not in motion, but some earlier varieties had four legs, one under each corner of the bed. Wheelbarrows are distinguished from carts, which have two wheels and can be either manually pushed or pulled by hitches, and from wagons, which generally have four wheels and are pulled by hitches. A hitch apparatus does not work on a wheelbarrow since there must be at least three points of support for it to remain upright, and the hitch and wheel would supply only two.

The origins of the wheelbarrow can be traced to the first century A.D. in China. Evidence of them has been found in ancient Chinese tombs. Speculation is that wheelbarrows may have been invented not long after the wheel, probably around 4000 B.C. Europeans did not begin using the wheelbarrow until about 1200 A.D., however. It is thought that the European version was created from verbal descriptions brought over the silk routes of Asia.

The Chinese wheelbarrow had its wheel directly under the load, while the European wheelbarrow had its wheel far out in front of the load. The Chinese wheelbarrow had the advantage of bearing more of the load, but the European version was less prone to tipping over. Some Chinese wheelbarrows had sails mounted in front. Obviously the wind had to be in one's favor for this arrangement to be of help.

Wheelbarrows were an important part of the construction of the great cathedrals of Europe. Today, they continue to be a common feature at construction and excavation sites and are also widely used by gardeners and farmers.

See also Wagon and coach; Wheel and axle

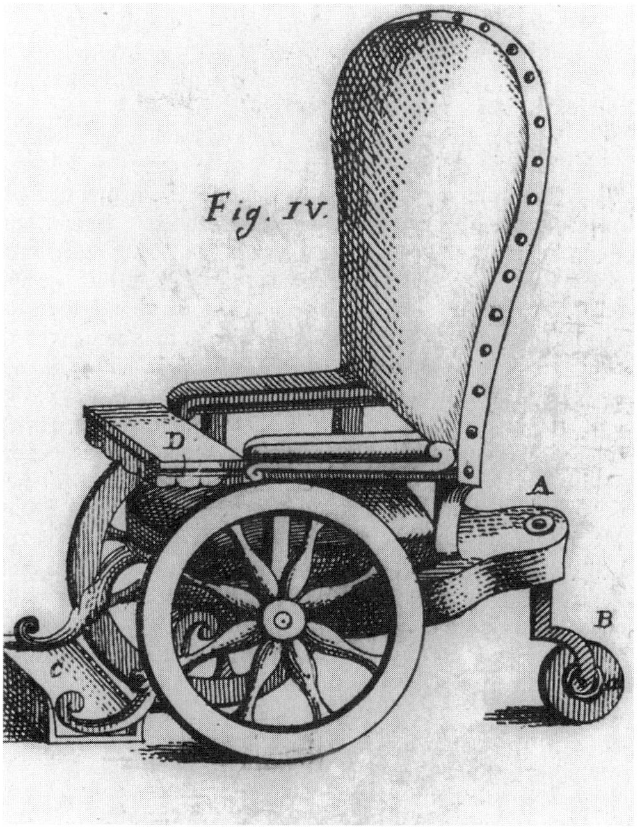

Engraving of a wheelchair published in 1766.

WHEELCHAIR

No one knows where or when the first wheelchair was invented, but a few one-of-a-kind wheeled chairs apparently were built during the 1600s. Wooden wheelchairs were in use in the United States by the 1860s, and the first modern wheelchair—a lightweight folding model—was designed and marketed in

1932 simultaneously by Sam Duke and by the team of Harry C. Jennings, a mechanical engineer, and Herbert Everest, an injured mining engineer.

Although the electrically powered wheelchair first appeared around the time of World War I, manually powered wheelchairs remained the norm until the rising number of quadriplegics in the 1960s created sufficient demand for wheelchairs propelled by other means. Electric wheelchairs can now be activated by hand, head movement, tongue, and breath. In 1982, astronomy student Martine Kempf, a native of France, developed a computer program that responded to voice commands—the *Katalavox*—which is now used to power both wheelchairs and **microscopes**.

The growing interest in wheelchair sports has benefited from improvements in wheelchair durability and maneuverability. Tom Houston, a pipefitter paralyzed by a fall in 1979, designed the *HiRider* with the help of fellow pipefitter Ray Metzger. This revolutionary wheelchair, which went on the market in 1989, allows the user to maintain a standing position while moving about. Wheelchairs capable of climbing stairs and curbs also are manufactured now. In 1987, West Germany issued the *Rollsteiger*, and in 1992 in the United States a high-priced computer-and-sonar-equipped *ACCESS Mobility System* was offered in 1992. On flat surfaces, both models utilize the standard four **wheels**, but convert to tank-like treads to climb over obstacles.

The stair-climbing ACCESS chair uses its track mode to adapt to curbs and even misaligned sidewalk slabs. Controlled by an armrest-mounted joystick, ACCESS can maintain a 6 mph pace for 12 miles or ascend and descend staircases as steep as 36 degrees. Also in the early 1990s, the first compact, elevating, powered wheelchair was produced. Called the *Mangar Freestyle*, it is able to raise its seat using pressurized air like a bellows. This allows the user to reach objects above the normal range of ordinary wheelchairs. Many of the newest chairs use the same ''ultralight'' materials that go into **aircraft**. For races and wheelchair competition of all sorts, tough stainless **steel** chairs are designed that offer both speed and endurance.

Whinfield, John Rex (1901-1966)

English chemist

Whinfield was born in Sutton, England and attended the Merchant Taylor's School and Caius College, studying chemistry. He worked as an assistant for **Charles F. Cross** and **Edward John Bevan** and became interested in textile fibers, perhaps inspired by Cross's and Bevan's earlier work with viscose **rayon** in 1892.

When **Wallace Carothers** introduced **nylon** in 1935, Whinfield, working as a research chemist at the Calico Printer's Association, began researching other substances that might be viable fiber material. Carothers and his research team had investigated **polyesters**, formed by reacting diacids and dialcohols, but were unable to develop the polyesters into fibers of significant strength. They turned to polyamides research.

In 1941 Whinfield and J. T. Dickinson reacted terephthalic acid and ethylene glycol in a condensation reaction; ironically, terephthalic acid was the sole diacid Carothers and his group did not try. Whinfield and Dickinson named the terephthalic-acid-based product Terylene. In 1946 they were granted a patent for their polyester, which was marketed by Imperial Chemical Industries in Great Britain and by Du Pont, under the names Fibre V and Dacron, respectively, in the United States.

Terylene fibers are made by a process known as melt spinning, in which Terylene is heated and forced through a metal plate with small holes. The polymer fibers emerging from the holes solidify in the cool air and are then passed through godet wheels, or rollers, that rotate at different speeds. This draws the fibers out and orients the molecules into a long linear chain.

Terylene revolutionized the British textile industry. It was used alone and in wool blends, and is still a staple of the clothing industry. Whinfield published few papers and received little recognition for his achievement, but by the time he died Terylene/Dacron was being produced at the rate of hundreds of tons a year.

See also Plastics

Whitney, Eli (1765-1825)

American inventor

Born in Westboro, Massachusetts, on December 8, 1765, Eli Whitney showed unusual mechanical ability at an early age. In fact, Whitney's mechanical skills kept him employed making and fixing various machines and paid his way through Yale University. Upon his graduation in 1792, he traveled to Savannah, Georgia, where he planned to teach while studying law.

In Georgia Whitney met Phineas Miller, another Yale graduate close to his age who managed the plantation owned by the widow of the American Revolutionary War general Nathanael Greene (1742-1786). Catherine Littlefield Greene soon employed Whitney to attack several mechanical problems attendant to running a large plantation. Foremost among them was the slow and tedious work of removing the seeds from the short cotton grown in the Savannah area.

Stories of Whitney's invention of the **cotton gin** often attribute his invention to significant help from both Catherine Greene and the slaves who worked the plantation. The slaves used a simple comblike device to clean the cotton, and it is probable that Whitney simply mechanized this manual process. It is also maintained by some that Whitney was not the first to develop a cotton gin; gins of various designs had been in use in the British colonies from the seventeenth century, notably one designed by Joseph Eve (1760-1835) for use in the West Indies.

Despite arguments about the origin and invention of the cotton gin, there is no question that Whitney built and patented a rather simple device that revolutionized the cotton industry. His cotton gin received a U.S. patent in 1807—about fourteen years after his original gin was developed. The design was quite simple and very easy to duplicate. Due to widespread pirating of his design and constant court fights to protect his patent, Whitney (in partnership with Miller) never profited from his invention.

Discouraged with the cotton gin business, Whitney turned his mechanical genius to the manufacture of firearms, a business that he also revolutionized. At the time, skilled craftsmen made **musket**s one at a time; consequently each firearm was unique. If a musket broke down, its replacement parts had to be individually made to fit that particular musket. Whitney transformed his arms factory in New Haven, Connecticut, to produce musket parts that were precisely machined so that they were identical and thus interchangeable.

He received a contract in about 1797 to supply the U.S. government with 10,000 muskets in two years. This contract should be compared with the productivity of other musketmakers at this time: two national armories had produced only one thousand muskets in the previous three years. Whitney's revolutionary production methods proved to be successful. Yet, due to epidemics and supply problems, requiring extensions from the government, it took his arms factory over ten years to fulfill the contract.

For one of his extension requests, Whitney mounted a public demonstration for government officials. He dumped the parts necessary to build ten muskets into a pile and challenged the officials to build ten muskets. The officials completed the task and were instantly convinced of the value of Whitney's production methods. His extension was granted, his success assured, and arms manufacture was never to be the same.

Due to the success of his musket-manufacturing methods, Whitney is credited with pioneering the use of precision interchangeable parts assembled to a final product on a production line. This method of dividing the labor necessary to build a musket among several workmen was also revolutionary for the times. Whitney also helped develop the **machine tool** industry by inventing many of the machines required by his new production methods.

Unlike his cotton gin, Whitney's arms manufactures proved to be financially successful. After his death on January 8, 1825, in New Haven, Whitney's arms plants were placed under the control of his son, Eli Whitney, Jr. Whitney's methods were the foundation for later assembly line factory production.

See also Mass production

Eli Whitney

Whittle, Frank (1907-1996)

English engineer

Born at Coventry in 1907, Frank Whittle became a Royal Air Force (RAF) apprentice in 1923. During the next several years he advanced to pilot, flight instructor, and RAF officer.

Between World War I and World War II, **aircraft** designers realized that propeller-driven planes were limited both in terms of velocity and optimal altitude. To reach speeds in excess of 350 miles (560 km) per hour, planes required larger **internal combustion engine**s. Of course such engines meant additional weight, which would, ironically, slow down the aircraft and reduce propeller efficiency. This technical problem was aggravated by an atmospheric problem. Attempts during World War I to fly at extremely high altitudes—a definite tactical advantage—had been thwarted by the thinness of the air itself, which likewise reduced propeller as well as engine efficiency.

Whittle had been contemplating a solution to the matter for some time. In 1928 he learned that a French inventor had suggested using the **gas turbine** as an airplane's power source. Unlike the **piston**-driven engine then in use, the new engine would, according to Whittle, become more efficient at higher speeds due to its tunnel design and compression of air by a supercharger, or **turbocompressor**.

Whittle patented his design in 1930 but was unable to convince the Air Ministry or outside interests to pursue his idea with further research. Advised by colleagues that the gas turbines used in industry were too heavy and inefficient to serve his purpose, Whittle set out to prove the critics wrong. Those who believed in him helped set up funding in 1936 to launch Power Jets Ltd., a company which just a year later succeeded in producing a demonstration turbo **jet engine**. There were, however, numerous initial problems. The engine wore out quickly and had to be rebuilt; it lost several turbine blades that required replacing; combustion chambers had to be redesigned; **bearings** burned out; and fuel vaporizers malfunctioned. After all of these problems were corrected, Whittle's

Frank Whittle

engine was placed in an experimental aircraft named the Gloster E28/39 in 1941. With later modifications, it was incorporated into the Gloster Meteor in 1944. Even though jet power and thrust have increased dramatically since the first flights of these airplanes, Whittle is remembered as a major force in the development of modern aeronautics.

In honor of his achievements, Whittle was knighted in 1948, and spent his later years as a research professor at the United States Naval Academy. He died on August 8, 1996.

Whitworth, Joseph (1803-1887)
English engineer

Joseph Whitworth brought the practice of **machine tool** manufacture to new levels of accuracy and precision. He was born in Stockport, Cheshire, England, in 1803, the son of a schoolmaster and Congregational minister. His formal schooling ended when, at the age of 14, Whitworth was apprenticed to his uncle, a Derbyshire cotton-spinner. Fascinated with the mill machinery, Whitworth left Stockport in 1821 and worked for four years in Manchester, England, as a mechanic. In 1825 he moved with his bride to London, and joined **Henry Maudslay**'s workshops. There, he perfected a scraping technique for making true metal plane surfaces. He returned to Manchester in 1833 and set up his own machine shop.

Whitworth soon became interested in precision manufacturing. At the time, each workshop made its own machine parts to its own standards; even **screw** threads were individual. In 1841 Whitworth proposed a uniform system of screw threads, and by 1860 his specifications were standard throughout Great Britain. He also developed ways to measure close tolerances with unprecedented accuracy, and he designed cutting, shaping, drilling, slotting, turning, and planing machines capable of achieving these very precise measurements. He invented a **knitting machine** in 1835 and, in 1842, a mechanical street-sweeper. He also promoted the advantages of the decimal system.

At the Great Exhibition of 1851, Whitworth machines were universally recognized for their high quality and precision. When the Crimean War began in 1854, Whitworth became interested in the manufacture of armaments. He performed many experiments on improved rifling and boring, and he developed a method of casting steel under pressure for heavy **artillery** that minimized metal-weakening air bubbles.

Whitworth's enterprises were highly successful, and he applied his considerable wealth to promoting the training of engineers. He established 30 Whitworth scholarships for university engineering students, and donated large sums to several technical colleges and universities. Whitworth received many honors, including a knighthood in 1869 and admission to the French Legion of Honor. He died in Monte Carlo in 1887 after a long illness.

Widere, Rolf • See Cyclotron

Wigs and hairpieces

Wigs are worn for either prosthetic, cosmetic, or convenience reasons. People who have lost all or part of their own hair due to illness or natural baldness can disguise the condition. For strictly cosmetic reasons (or perhaps to alter their appearance), people might wear a wig to quickly achieve a longer or fuller hairstyle or a different color. In an article in *Vogue* magazine, the wife of a prominent politician was described as using a wardrobe of wigs to avoid $8,400 and 160 or more hours spent with professional hairdressers each year, in addition to the complicated task of finding appropriate hair care while traveling.

Based on an ivory carving of a woman's head found in southwestern France, anthropologists speculate that wigs may have been used as long as 100,000 years ago. Wigs were quite popular among ancient Egyptians, who cut their hair short or shaved their heads in the interests of cleanliness and comfort (i.e., relief from the desert heat). While the poor wore felt caps to protect their heads from the sun, those who could afford them wore wigs of human hair, sheep's wool, or palm-leaf fiber mounted on a porous fabric. An Egyptian clay figure that dates to about 2500 B.C. wears a removable wig of black clay. The British Museum holds a beautifully made wig at least 3,000 years old that was found in the Temple of Isis at Thebes; its hundreds of tiny curls still retain their carefully arranged shape.

Wigs were popular in ancient Greece, both for personal use and in the theater (the color and style of wigs disclosed the nature of individual characters). In Imperial Rome, fashionable women wore blond or red-haired wigs made from the heads of Germanic captives, and Caesar used a wig and a laurel wreath to hide his baldness. Both Hannibal and Nero wore wigs as disguises. A portrait bust of Plautilla (ca. 210 A.D.) was made without hair so wigs of current fashion could always adorn this image of Emperor Caracalla's wife.

During the reign of Stephen in the middle third of the twelfth century, wigs were introduced in England; they became increasingly common, and women began to wear them in the late sixteenth century. Italian wigs of that time were made of either human hair or silk thread. In 1630, embarrassed by his baldness, Louis XIII began wearing a wig made of hair sewn onto a linen foundation. Wigs became fashionable, increasing in popularity during the reign of Louis XIV, who not only wore them to hide his baldness but also to make himself seem taller by means of towering hair. During the Plague of 1665, hair was in such short supply that there were persistent rumors of the hair of disease victims being used to manufacture wigs. This shortage of hair was partially remedied by using wool or the hair of goats or horses to make lower grades of wigs (in fact, horsehair proved useful since it retained curls effectively). For several decades around 1700, men were warned to be watchful as they walked the streets of London, lest their wigs be snatched right off their heads by daring thieves.

The enormous popularity of wigs in England declined markedly during the reign of George III, except for individuals who continued to wear them as a symbols of their professions (e.g., judges, doctors, and clergymen). In fact, so many wigmakers were facing financial ruin that they marched through London in February 1765 to present George III with a petition for relief. Bystanders were infuriated, noticing that few of the wigmakers were wearing wigs although they wanted to protect their jobs by forcing others to wear them. A riot ensued, during which the wigmakers were forcibly shorn.

During the late eighteenth century, Louis XVI wore wigs to hide his baldness, and wigs were very fashionable throughout France. The modern technique of ventilating (attaching hairs to a net foundation) was invented in this environment. By 1784, springs were being sewn into French wigs to make them fit securely. In 1805, a Frenchman invented the flesh-colored hair net for use in wigmaking. A series of other improvements followed rapidly, including knotting techniques, fitting methods, and the use of silk net foundations. These matters were so important that a major lawsuit arose, and one inventor committed suicide after selling his patent cheaply and watching others become rich using his technique. One of the manufacturing processes that was tried at this time was based on the use of pig or sheep bladders to simulate bald heads on actors. In the mid-1800s, some wigs and toupees were made by implanting hairs in such bladders using an embroidery needle. In the late nineteenth century, children and apprentices of wigmakers amused themselves by playing the "wig game," in which each participant accumulated points by throwing an old wig up to touch the ceiling and catching it on the head as it fell.

By the early 1900s, jute fibers were being used as imitation hair in theatrical wigs. Today, a favorite material for theatrical wigs, particularly those worn by clowns, is yak hair from Tibet. The hair of this ox species holds a set well, is easily dyed, and withstands food and shaving cream assaults.

Wigs of synthetic (e.g., acrylic, modacrylic, **nylon**, or **polyester**) hair are popular for several reasons. They are comparatively inexpensive (costing one-fifth to one-twentieth as much as a human hair wig). During the past decade, significant improvements in materials have made synthetic hair look and feel more like natural hair. In addition, synthetic wigs weigh noticeably less than human hair versions. They hold a style well—so well, in fact, that they can be difficult to restyle. On the other hand, synthetic fibers tend not to move as naturally as human hairs, and they tend to frizz from friction along collar lines. Synthetic hair is also sensitive to heat and can easily be damaged (e.g., from an open oven, a candle flame, or a cigarette glow).

Human hair remains a popular choice for wigs, particularly because it looks and feels natural. It is easily styled; unlike synthetic hair, it can be permed or colored. During periods of scarcity of cut human hair for wigs, manufacturers have used combings (hairs that fall out naturally at the end of their life cycle). However, actively growing hair that is cut for wigmaking is preferred. United States wigmakers import most of their hair. Italy is known as a prime source of hair with desirable characteristics; other colors and textures of hair are purchased in Spain, France, Germany, India, China, and Japan. Women contract with hair merchants to grow and sell their hair. After cutting, the hair is treated to strip the outer cuticle layer, making the hair more manageable. Wigmakers pay $80 or more per ounce for virgin hair, which has never been dyed or permed; a wig requires at least 4 oz (113.4 g) of hair.

Some manufacturers blend synthetic and human hair for wigs that have both the style-retaining qualities of synthetic hair and the natural movement of human hair. However, this can complicate maintenance, since the different types of hair require different kinds of care.

WILKINSON, JOHN (1728-1808)

English industrialist and inventor

John Wilkinson is perhaps best known as the man who perfected a **boring machine** that, among other things, made possible the successful manufacture and operation of **James Watt**'s **steam engine**.

Born in 1728 at Clifton, Cumberland, England, Wilkinson was exposed to **iron**-furnace work at an early age through his father, an ironmaster, farmer, and part-time inventor. By 1763 he had become operator of his own furnace, and soon became a leading supplier of castings and armaments. His first boring machine, patented in 1774, was constructed for cannonmaking. Because of its rigid design that enabled precise cutting, the machine, with a few adaptations, proved instrumental in boring engine cylinders to Watt's exacting standards. The essential components of Wilkinson's landmark machine were a fixed hollow cylinder and a rotating, double-supported boring bar that passed through the inside of it to advance the cutting head.

Wilkinson incorporated the Watt engine into his factory work, and, from it, produced the first steam hammer. His other important invention is a machine which could produce quality lead pipe through an extrusion process that reduced the pipes wall dimensions. Wilkinson's continuing and pioneering work with iron, which included the construction of an iron-hulled barge, several innovations in iron smelting, and the supply of castings for an iron **bridge** across the Severn River, earned him a lasting reputation as one of the great industrialists of the eighteenth century. Wilkinson died in 1808 in Bradley, Staffordshire, England.

WILLARD, SIMON • See Clock and watch

WILLIAMS, DANIEL HALE (1858-1931)

American physician

Daniel Williams, a meticulous, knowledgeable surgeon and founder of the first interracial hospital in the United States, advanced further in medicine than any other African-American doctor of his time. Among his many achievements, Williams is credited with having performed the first emergency **open-heart surgery**.

Williams was born the fifth of seven children of Daniel and Sarah Ann Price Williams on January 18, 1858, in Hollidaysburg, Pennsylvania. At the age of eleven, he was left an orphan after his father, a prosperous barber, died of tuberculosis and his mother deserted him. First apprenticed to a cobbler, he rebelled against repetitive, menial labor and moved to Edgerton, Wisconsin, to live with his sister Sally. He boarded with a foster family and found work as a barber and guitarist in a string band so that he could attend Haire's Classical Academy in Janesville, from which he graduated in 1877.

Following the example of his older brother, he studied law for a time. After a year, intrigued by the work of the town doctor, he requested a position as doctor's assistant. To increase his knowledge of medicine, he read journals and texts. After two years apprenticeship, working as a laborer on a lake steamer, and borrowing money from friends and family to pay his tuition, he completed a medical degree from Chicago Medical College in 1883. Because African-American doctors were denied privileges at white hospitals, he worked as a surgeon at the South Side Dispensary in a ghetto area until 1892.

Under primitive conditions—either in his office or in the patient's home—Williams performed necessary surgery without an anesthesiologist or X-rays, either in his office or in the patient's home. To improve surgical standards and his patients' chances for survival, he founded Provident Hospital, a twelve-bed facility which accepted patients of all races, augmenting his meager funds by organizing donations of linen, beds, cleaning supplies, and kitchenware. Fundraisers, including abolitionist Frederick Douglass, provided the money for medical instruments and drugs.

Williams took a personal interest in his community hospital by working lengthy stints and ended his day by disinfecting floors and instruments. He evolved high standards for employees, including his nurse's training program, which accepted only qualified applicants. By maintaining strict standards and because of his skills as a surgeon, he achieved a better mortality ratio than other hospitals. Because of his reputation, patients from surrounding states requested his services.

In 1893, under the supervision of six associates, Williams saved a stockyard laborer suffering from a knife wound to the pericardium by administering only local anesthesia, cleansing with a saline solution, and performing open-chest surgery, a dramatic departure from standard protocol. The success of the procedure rated headlines in the *Chicago Daily Inter-Ocean*, although many people doubted that a African-American doctor could evolve such an innovation.

From this major breakthrough came offers of positions at other institutions. He chose to serve as chief surgeon of Washington's Freedmen's Hospital, revamping its surgical program with the same antiseptic controls and nurse's training he had instituted at Provident. To modernize Freemen's further, he staffed the hospital with qualified specialists and created departments for each specialty. So successful were his innovations that he held training sessions at Howard University.

After a shift in the political climate reduced Williams's effectiveness, he opted to return to Chicago. Having married teacher Alice Johnson Williams, he settled in as staff associate of Mercy and St. Luke's hospitals, taking time to serve as the first African-American member of the Illinois board of health in 1889 and again in 1891 and as surgeon for the City Railway Company, a rare opportunity for a African-American man. He also helped organize the National Medical Association for Black Doctors and taught at Meharry Medical College and Howard University.

Williams received an LL.D. degree from Wilberforce University in 1908 and was named a fellow of the American College of Surgeons and a member of the Chicago Surgical Society. Always interested in the training of African-American medical professionals, he immersed himself in the work of the NAACP as well as in the creation of schools that gave African-Americans an opportunity to develop medical skills. He retired from medicine in 1920. After his wife's death, he attended a few private patients, gardened and swam in his spare time. Already a victim of diabetes, he suffered a paralyzing stroke in 1925 and died at his summer home in Idlewild, Michigan, on August 4, 1931.

WILLIAMS, O. S. (OZZIE) (1921-)

American engineer

Ozzie Williams, an aeronautical engineer and designer of small **rocket** engines, developed an airborne **radar** device for locating downed **aircraft**. A graduate of the College of Engineering at New York University, Williams was the first African-American to attain the post of engineer at Republic Aviation. He moved on to serve as projects engineer at Greer Hydraulics, where he invented a radar beacon for air searches of wrecked planes. In 1961, Williams accepted a post as manager of rocket systems for Grumman International. In this position, he developed a control rocket for the Apollo space program to guide lunar modules during landings.

Wilson, Charles Thomas Rees (1869-1959)

Scottish physicist

Wilson was born in Glencorse, Scotland, on February 14, 1869. However, his family moved to Manchester in 1873 when his father died. Wilson entered Owens College in Manchester at the age of fifteen and graduated with a bachelor of science degree in biology three years later. He then enrolled at Sidney Sussex College, Cambridge, and earned his doctorate in physics in 1892.

An important event in Wilson's life occurred in 1894. He spent a few weeks at the observatory on top of Ben Nevis, the highest mountain in Great Britain. Wilson was fascinated by the process of cloud formation that he observed every day at the observatory. He decided to find ways of imitating the process in the laboratory.

Wilson's method was to allow moist air within a **glass** container to expand quickly. As the air expanded, it became cooler and tiny droplets of moisture condensed as artificial clouds. Wilson observed that the amount of condensation was greatly reduced if all dust was removed from the container. Yet, some condensation occurred even in a completely dust-free environment.

Wilson concluded that ions within the container served as nuclei for the condensation of water droplets. He tested this idea in the late 1890s by exposing the container to X-rays and nuclear radiation. He found that the amount of condensation greatly increased in the presence of either type of radiation.

Thus was born the principle of the **cloud chamber**. By 1911, Wilson had refined the concept. Particles applied to the cloud chamber formed ions that served as nuclei for the condensation of water droplets. Magnetic fields surrounding the cloud chamber caused charged particles to travel in curved paths. The shape of such paths reflect the mass and charge of particles that pass through the container. Using this device, nuclear physicists could identify a host of otherwise invisible subatomic particles by their "vapor trails." J. J. Thomson called the cloud chamber a device of "inestimable value to the progress of science." For his invention, Wilson was awarded the 1927 Nobel Prize in physics.

Wilson made a second important discovery. He observed that an electroscope completely shielded from surrounding electrical fields still loses its charge slowly. He predicted that this loss of charge was due to some kind of radiation from "outside our atmosphere." This idea was the first hypothesis of the existence of cosmic radiation.

Deeply attached to his Scottish homeland, Wilson lived the last twenty-five years of his life there. He died in Carlops, Peeblesshire, on November 15, 1959.

See also Electrostatic devices

Wimshurst machine • See Electrostatic devices

Wind tunnel

Amateur nineteenth-century aviators typically studied the flight behavior of birds and then built flying machines accordingly. The resulting bird-like craft failed miserably because the builders had no knowledge of aerodynamics and aeronautics, particularly of lift and drag forces acting on surfaces cutting through the air.

The earliest invention for testing flight characteristics was a whirling arm. Benjamin Robbins, an English mathematician, first employed such a device in the eighteenth century. He mounted such shapes as pyramids and oblong plates on the arm tip and spun them in different orientations. He found that no simple theory would account for the complex forces acting on moving objects. **George Cayley** also used a whirling arm to measure drag and lift in the early nineteenth century. The major drawback to whirling arms was the disturbance of air created; the **aircraft** models flew into their own wakes, thus precluding any clear findings.

Frank Wenham of England is credited with designing and operating the first wind tunnel in 1871. It was 12 ft. (4 m) long and 18 in. square. A fan driven by a **steam engine** propelled air down the tube to the model. Wenham soon discovered that wings could support much heavier loads than had been thought previously; thus flight was seen as a real possibility.

Orville and Wilbur Wright used a wind tunnel to improve their planes. The first two gliders they built were quirky and unpredictable in the air. They realized that all scientific information they were using for flight data, gathered from previous experimenters, was incorrect and that they would have to conduct their own investigations. The Wright brothers first built a square tube for channeling the air, a driving **fan**, and a two-element balance mounted in the airstream. They attached various wings to the balance and, as the contraption revolved, observed the relative lifting forces. They went on to build a larger and more sophisticated tunnel, in which they obtained the critical data they needed for their first manned, powered aircraft.

By the time World War I began, leadership in aerodynamic research had shifted to Europe. Many wind tunnels were built there while facilities in the United States were almost nonexistent. In the 1920s, however, aviation's growth spurred renewed interest in American research. The National Advisory Committee for Aeronautics (NACA) built new wind tunnels, one of which was constructed vertically to test aircraft spinning.

The 1930s saw increased speed in the tunnels; by the end of the decade NACA had created tunnels that enabled engineers to study forces acting on an airfoil near Mach 1 (or, one times the speed of sound) speed. In 1939 NACA built a 19 ft. (6 m) in diameter, high-pressure tunnel that used an 800 horsepower **electric motor** to drive a 34.5 ft. (11 m) propeller, which created a wind speed of 300 miles (48 km) per hour. After World War II, there was a real need for supersonic wind tunnels. They were complicated, involving narrowed tunnel walls and compressor fans.

In 1948 a 4 ft by 4 ft supersonic tunnel began providing data which led to the development of the B-58 bomber, the X-2 research plane, and several fighters, including the F-102 and F-105. Beginning in the 1950s, hypersonic tunnels were created to test flight at Mach 5 (five times the speed of sound). They required unusually high temperatures and speeds, and, because of power restrictions, the runs had to be of short duration.

In the last 30 years there have been several refinements in wind tunnels. Hypersonic tunnels can now operate much longer (one can provide Mach 10 flow continuously); helium has been used to create a Mach-50 blast of wind; hydrogen and oxygen have been exploded to provide a blast wave of air; and electric explosions have created similar shock waves. With extensive computerization, self-streamlining walls, **laser**-based instrumentation, cryogenic operations, and magnetic model suspension, wind tunnel technology is unquestionably keeping pace with the rapid advances in scientific research.

Wind vane

The wind vane, or weather vane, is the simplest and oldest of weather instruments. As far back as 100 B.C., the Greeks mounted wind vanes atop statues of their gods. These functioned as ornaments until the sixteenth century.

Leonardo da Vinci made illustrations of many mechanical devices during the early 1500s, including the wind vane. Although often credited as being an inventor, many of his sketches were of devices already in existence.

In 1578, Egnation Dante, an astronomer from Bologna, was the first to use a directional indicator with a vane. His vane had a vertical dial. The first splayed dial, familiar today, was created in 1797 by G. R. Parrot of Paris. The first practical recording vane was developed in 1868 by F. Pfeiffer. It used a recording drum rotated by the vane's own power. Today, wind direction and velocity are usually recorded electronically at stations distant from the vane.

Windmill and wind turbine

The windmill is a simple device used over the years for many valuable purposes: grinding grain, driving saws, pumping water, driving electric **generators**, and charging storage **batteries**. The windmill harnesses natural wind energy for human use—the whirling propeller shaft transmits its rotary motion through **gears**, converting it into power that people can use.

Windmills go far back in history. Nearly 4,000 years ago the Babylonian emperor Hammurabi (1792-1750 B.C.) used them for **irrigation**. About 200 years before Christ, **Hero**

Langley Laboratory's first wind tunnel, which became operational in June 1920.

of Alexandria described a windmill in his writings. By 1000 A.D. these machines were in practical usage throughout Europe. In 1105 a French convent built both a water mill and a windmill.

It was the Dutch who made the most extensive use of windmills. Inhabitants of the Netherlands faced a drainage problem for centuries. Their land was actually a series of marshes separated from the sea by a belt of dunes, mostly below sea level. Sections of land called poders were surrounded by high earthen walls, or dikes, to protect them from the sea, and people led precarious lives huddled on top of earthen mounds. In the fifteenth century this situation changed. Using powerful windmills, the Dutch were able to pump unwanted water from the polders. For example, by 1608 a 10-foot (3 m) deep lake had been emptied by 26 windmills.

At the height of their use, over 9,000 windmills lined the dikes and **canals** which covered the land. These windmills had sails made of 30-to 50-foot (9-15 m) long wooden frames with canvas covering their lattice or framework. The sails were attached to a central wooden axle, or shaft, that was connected to large, toothed gears which could provide power for pumping. The older types of windmills were made so the entire mill house, with its sails, could be turned around its shaft to face the wind. The Dutch also used the configuration of the windmill's wings to send messages—if the wings were positioned straight up and down in a cross formation, this indicated the birth of a baby. To announce weddings, the sails formed an X with garlands of flowers woven into the framework. During World War II, windmill sails signaled German troop movements to the Dutch resistance fighters. With the invention of the **steam engine** in England in the nineteenth century, the Dutch, too, began to shift to the new form of power. But many windmills still exist there, either as homes or as functioning windmills.

In the United States there was little interest in the European-style windmills because they were costly to build and cumbersome to operate. In 1854, Daniel Halliday revolutionized the entire concept of wind power when he invented an easily constructed tower, topped by a light **wheel** fitted with automatic mechanisms to break its speed when the wind increased or when a tank of water was full. This type of windmill was later used to generate enough electricity to meet the needs of a small house. Some historians believe that this windmill was a major factor in the rapid settling of the West.

As fossil fuel prices skyrocket and supplies are depleted, wind power is becoming an attractive alternative energy source. Much thought and effort has gone into designing better windmills. A German design uses two blades and an electric generator mounted on the axis between the two rotors. It does away with gears, and is able to convert more of the wind's en-

Wind generators in Whitewater, California.

ergy to use than conventional windmills, at a cost very competitive with fossil fuels. Simple *egg beater* windmills with a vertical axis and two thin, curved blades attached at the top and bottom are used in many locations. Of course, windmills have the further advantage that they emit no pollution.

The federal government and several universities have sponsored wind power research projects. One interesting possibility is the Chalk windmill, an airfoil that looks like a **bicycle** wheel, with 48 blades capable of spinning in breezes which will not budge other windmills. Another wind visionary, William Heronemus of the University of Massachusetts, has called for a network of gigantic windmills off the coastline of the United States. The energy produced by them would be used to convert ocean water to hydrogen and oxygen, and the hydrogen would be shipped ashore to be combined with the oxygen in fuel cells to produce electricity without pollution.

Many believe that wind power will be the power of the future. It is estimated that, with today's technology, wind power could supply about 20% of the United States' electricity. For the twenty-first century, it is thought that installation costs for wind turbines, and ultimately the electricity they produce, will be cheaper than more conventional means like hydroelectric plants or nuclear reactors—without concerns about a nuclear accident or environmental damage. To date Europe has been ahead in the exploration and acceptance of windmills and wind turbines to generate power needs.

See also Dam and dike

Windshield and windshield wiper

Because early autos traveled at fairly slow speeds, little, if any, wind-breaking protection needed to be provided in the car design itself. As speeds increased, this type of protection became more important. The earliest ''windshields'' were nothing more than vertical sheets of plate **glass** in wooden frames.

The plate glass was an obvious source of injury to drivers and passengers. Hoping to strengthen and reinforce windshield glass, John C. Wood (in 1905) glued two sheets of glass together with a central layer of **celluloid** to create a type of laminated glass. Unfortunately, in addition to being difficult and expensive to produce, the celluloid tended to yellow over time, causing yet another safety hazard.

Work continued on laminated glass in efforts to eliminate the yellowing and to increase the ease of production. Although laminated glass could be curved to suit specific manufacturing needs, it most often required a frame, and this made it less desirable. In 1928, the car maker Hillman was the first to use laminated **safety glass** for use in windshields as well as in side and rear windows.

Toughened glass was developed as an alternative. The French glassmaking firm St. Gobain developed toughened glass for car windshields in 1929 based on previous findings of Owens-Corning, the U.S.-based company. This type of glass is super-heated and rapidly cooled in single sheets. Toughened glass was found to be more pliable than laminated glass, making it ideal for the curved slopes of the windshield. Unlike laminated glass, it could be fitted directly into the bodywork and did not need a frame. Unfortunately, the finished product was so rigid that it was hazardous. It did not yield upon impact and could severely injure individuals in the vehicle.

Further development of laminated glass continued. The Triplex company created curved laminated glass for use in streetcars. In 1956, a British car, the Vauxhall Victor, was fitted with the first wraparound, laminated windshield.

Additional features developed after World War II include window defoggers and tinting. In 1948, Rolls Royce first embedded tungsten **wire**s in the laminations of the rear window for defogging and de-icing. Tinted glass (under the brand name Sundym), for cutting down on heat and glare, was introduced in American cars years later. Safety in window glass has also been a concern. Car makers created differential zoning, which allows windshields to crack and break in such a way that larger, undamaged sections will be left in front of the driver.

Another feature created to improve the driver's field of vision is the windshield wiper. In 1903, J. H. Apjohn invented a system of two brushes that he could move up and down the glass. Other systems based on the same principles were later developed using rubber strips rather than brushes. Until 1917, drivers had to use one hand to manipulate the wipers and the other to steer and shift. Ormand Wall, a Hawaiian dentist, recognized the danger in this practice and devised an **electric motor** in the middle of the windshield to oscillate a long rubber blade. Other wiper designs featured double blades with rubber on the outside to clear the windshield and a tandem, felt-lined squeegee to defog the inside. In 1929, another American, William Folberth, designed the vacuum-powered wiper that was run by suction from the engine's intake manifold. Similar versions of this system evolved, but the timing of the mechanisms was problematic. The wipers would speed up and slow down with the speed of the car.

By the 1930s, wipers propelled by an electric motor were refined and became the standard. Blower vents were added to the dashboard to defrost windshields. With the introduction of a curved windshield, the British car company Vauxhall, in 1933, produced a blade made of a number of short blades all controlled by a main curved arm. By the late 1930's, the American car manufacturer Nash introduced automobile **air conditioning**, and defrosting and defogging methods were incorporated in this system.The intermittent wiper was invented in 1962 by Bob Kearns, and microsensors in windshields of the 1990s can sense rain on the windshield and adjust intermittent wipers for showers or downpours.

Windsurfer

In the 1960s, two California friends carried on a debate regarding the merits of surfing compared to sailing. Hoyle Schweitzer, vice president of a computer firm, argued that sailing was a better sport because surfers had to waste so much time waiting for the right waves and because the shore area was too crowded with other surfers. Jim Drake, an aeronautical engineer, believed just the opposite: surfers had the edge because their sport was so simple and much less time consuming than sailing.

Both men were creative thinkers who began thinking about some way to combine the positive elements of both sports into one new package—a sailing **surfboard**. Both men set to work. Drake exchanged new sailing ideas with another scientist, Fred Payne. Several ideas were discarded for one reason or another. In 1967, Drake and Schweitzer tried an enlarged surfboard, but they had problems steering it without a rudder like other **boats** had. They finally attached the mast to the board with a universal joint, which allowed the sail to turn freely in all directions. No other sailing craft can be used like a windsurfer because of this universal joint: it makes it possible for the mast and sail to lie flat in the water or to tilt and turn in any direction. It lets the boardsailor shift the sail easily to maneuver with the wind. They also created a wishbone-shaped boom that allowed a standing sailor to keep the sail extended; it also provided support for the individual on the board. The two men called their creation a "free sail system." The board itself was first made of **fiberglass**, like surfboards of the time. But they came up with **polyethylene** because it was lighter, longer lasting, and less expensive. Today's boards are slightly longer than the traditional surfboard with straps on the top to anchor feet into and a skeg on the bottom to add stability. There is also a daggerboard, a small rectangular board, that plunges down through a slot into the water below to keep the board moving straight ahead.

Public reaction was positive; experts called it the first really original sailing idea in 100 years. Hoyle Schweitzer quit his job and devoted himself to turning out more of these windsurfers because the demand was high. One man who saw a windsurfer near the coast, stopped his car, ran down to the beach, and ordered six of them on the spot. Today, the windsurfer is very popular because it does combine the advantages of both sailing and surfing.

WINEMAKING

Since ancient times, people have been drinking wine. Wine is the fermented juice of fruit, most notably grapes, which have naturally occurring yeast in their skins. After the fruit juice has been squeezed from the fruit, yeast reacts with the natural sugar in the juice and converts it into alcohol through the process of fermentation.

The oldest wine yet discovered was detected in the residue in two clay jars found when archeologists excavated a 7,000-year-old farming village in the Zagros Mountains of northern Iraq. Although the jars were uncovered in 1968, the residue was not identified until 1996. That's when an archeological chemist at the University of Pennsylvania analyzed the residue and detected tartaric acid, found in large quantities only in grapes, and terebinth tree resin, known as an age-old wine preservative.

Winemaking was originally a practical matter: with little pure water to drink and no refrigeration, the fermented juice of fruits was safer to drink than ordinary fruit juice, which spoiled easily. In the ancient Greek and Roman cultures, the intoxicating effect of wine played a very important role. The Greek god Dionysus and the Roman god Bacchus symbolized not only wine, but good living and civilization in general. Wine became a part of many religious celebrations and rituals.

Today wine is made in many parts of the world but is especially concentrated in Italy, France, Spain, Russia, Argentina, the United States, Portugal, and Germany. Most wines are made from grapes, but there are also apple, pear, cherry, and blackberry wines.

In traditional times, workers would tread on the grapes (usually the species *Vitis vinifera*) to break the skins; today mechanical crushers are used to crush the grapes without breaking the seeds. The juice from the crushed grapes, known as *must*, runs into vats; additional yeast is sometimes added, and the mixture is left to ferment. When the fermentation has progressed to the desired stage, the wine is drawn from the vat. Sometimes a wine is then clarified to remove yeast or other undesirable particles. The wine is placed into casks where it remains until it has properly aged, developing its *bouquet*—its desired flavor and aroma.

A wine suitable for drinking that still contains some of its grape sugar is known as a *sweet wine*; if most of the sugar has been converted into alcohol, it is a *dry wine*. A wine's *vintage* is the year in which the grapes were harvested.

Wines vary in a number of ways. Color is dependent upon the variety of grape used and the inclusion of the grape

Ancient Egyptian depiction of winemaking from the tomb of Nakht at Thebes.

skins in the must. Green grapes usually produce white wines and purple grapes produce red wines. However, if the skins of purple grapes are removed from the must, a white wine can be produced. Wine runs a spectrum of color from ruby red to pink to white, and its alcohol content can vary from 7 to 14 percent.

During fermentation, carbon dioxide gas is released. If this gas is allowed to escape, a *still wine* is produced, but if the gas is trapped when the wine is bottled, a *sparkling wine* is produced containing hundreds of bubbles of carbon dioxide. The most famous sparkling wine is *champagne*, named after the French province in which it was first made—a pale, amber-colored wine with a 12 percent alcohol content.

Some wines, known as *fortified wines*, contain added distilled spirits, usually brandy, resulting in an alcohol content of 18 to 24 percent. Example of fortified wines include *sherry*, a golden brown wine originally made in Spain; *Marsala*, which originated in Sicily; and *Tokay*, from Hungary. When herbs and spices are added to the wine, what results is a *flavored wine* such as *vermouth*. Because of differences in bouquet, wines are served at varying temperatures and used to accompany different foods.

Most wines are classified by the geographic region of their origin and by the species of grapes that grow well in that location. *Bordeaux* in southwestern France is famous for its red wines made with *Cabernet Sauvignon* grapes. The white wines of *Burgundy* in east-central France are usually made from *Chardonnay* grapes. The *Sangiovese* grapes native to Tuscany in central Italy produce *Chianti* wine. Along the Rhine River in Germany, *Reisling* grapes are used to produce a distinctive wine. In the United States, *California* wines—notably from the grape-growing regions of the Napa and Sonoma valleys—produce many varieties of wines.

Winsor (Winzer), Frederick Albert (1763-1830)

German businessman

On trips to France and England, Winsor searched for inventions in need of financing and development and, on one occasion, found gas lighting. After trying without success to buy the technology for **Philippe Lebon**'s thermolamp, Winsor built his own **gaslight** and demonstrated it at London's Lyceum Theatre in 1804. Winsor's chemist did much of the initial work, which was done in competition with **William Murdock** and his employers, **Matthew Boulton** and **James Watt**.

Winsor realized it would be more efficient and profitable to make gas in large amounts, distribute it through underground pipes to several places, and sell the leftover solid ''coke'' as a low-quality fuel. Such an undertaking would require more money than any single person could provide, so Winsor mounted an advertising campaign that challenged anyone to join him in this bold venture. Although many people made fun of his approach, Winsor won the support of an influential committee in the English government, and in 1807 Winsor and his partners illuminated a garden wall for the king's birthday. They went on to stage the first gas streetlighting display in London, England. Parliament finally granted Winsor a charter for the first gas company, National Heat and Light, which was founded in 1812. Ironically, Winsor was ousted from the company's management later that year, and its name was changed to the Gas Light and Coke Company, better known today as British Gas.

See also Lamps

Wire

Wire and wire rope were developed as stronger substitutes for their fiber counterparts. String, twine and **rope** made from fibrous materials have their own uses. Metal wire and rope, on the other hand, are unsurpassed in their ability to support heavy loads. They are also invaluable for the transmission of electricity and sound. Wire may be made from **iron**, **steel**, copper, **aluminum** and many other metals, depending on the nature of its intended use. Wire was originally formed by forging and hammering strips of metal into long strands. A method called drawing was developed in A. D. 1000 in which wire is elongated by pulling, or drawing out. This process produces a stronger and more uniform wire than does forging. Rudolf of Nuremberg is credited with having used water power to draw wire about 1350. During the 1800s, wire drawing was first done using **steam engines**. Ichabod Washburn is considered the father of the steel wire industry. He founded a wire mill in Worcester, Massachusetts in 1831.

Today, nearly all wire is made by machine. Ingots are rolled into bars, called billets. The billets are rolled into smaller rods, which are baked to reduce brittleness, coated with a lubricant, then, finally, passed through a series of dies which reduce the rods to wire. To produce greater thicknesses, wire is twisted into rope rather than producing a single thick strand which might break due to a single weak spot. This is done by twisting wires around a center wire to create a strand of moderate thickness. Such strands, in turn, are twisted around a center core. The core can consist of either hemp or wire, the former being cheaper, the latter being stronger. Wire rope was first made in the 1830s, and in 1840 a patent for wire rope was granted to Englishman Robert Newell. Using this twisting method, **John Roebling** pioneered the manufacture of massive steel cables for the construction of suspension **bridges**, including the Brooklyn Bridge in New York in 1883.Vast amounts of wire cable are used today for telecommunications and computer applications, wiring individual components within a machine, and connecting machines in both local and wide area networks.

See also Clock and watch; Concrete and cement

Wire glass • See Glass

Wire rope • See Wire

Wood-burning stove • See Stove

Woodcut

Although relief, or raised designs cut into wood, was used by the Mesopotamians to impress images into clay as early as 3000 B.C., the first examples of wood carvings being used as repeatable stamps date only to the sixth century A.D., when the practice was used by Egyptians, Chinese, and Europeans. Woodcut stamps featuring images of the Buddha found in Turkestan have been dated before 800 A.D. In Europe, the large-scale use of woodcut as a means of printing and artistic expression only began with advances in **paper** production in the fourteenth and fifteenth centuries. **Playing cards** made from woodcuts first appeared in the late fourteenth century. In the fifteenth century, a number of artists helped to make woodcut one of the most widespread and distinctive forms of medieval art. Typically, the medieval artist drew a full paper sketch of the artwork, which was then actually cut into the wood not by the artist but by a trained craftsman; the carver would cut the white spaces into the wood, leaving a finely detailed raised design that would be inked and pressed to paper. Therefore, the success of a finished woodcut, atypical of most art forms, depended on the skills of two individuals.

Because woodcuts could be reproduced and because paper was the cheapest of art media, woodcuts were affordable by a much greater segment of the population than paintings and thus were commissioned in high numbers. The period in which woodcut flourished was an intensely religious one, and one made somber by the aftermath of the Black Plague in the fourteenth century. Hence, woodcut has become associated more so than most art forms with religious subjects. Perhaps the most famous woodcut artist, the German Albrecht Dürer (1471-1528), created one of the most famous sets of woodcuts in 1498, his *Apocalypse* series, inspired by the widespread belief among Europeans that the world would end in the year 1500. The distinctive appearance of woodcut art, which used the stark contrast of fine dark lines on a white background to create an almost three-dimensional visual effect, lent itself well to the often grim religious themes of the era. This is also exemplified in a number of medieval *ars morendi*—books of print and text that literally instructed Christians in the ''art of dying.'' Woodcuts were used for other such ''block books'' (crude books consisting of a few bound woodcuts and named for the carved blocks used in the process) but they eventually waned in popularity with the spread of movable type in Europe during the fifteenth and sixteenth centuries. Superseded in speed, versatility, and accuracy of reproduction by processes such as **lithography**, woodcut is today used almost exclusively by artists exploiting its distinctive black and white visual effects.

See also Printing technology

Woods, Granville T. (1856-1910)

American inventor

A multi-talented inventor, Granville Woods, dubbed the ''black Edison,'' created the railroad **telegraph**, a device which transmitted messages between moving **train**s via static electricity. This invention was an important advancement in railroad safety, reducing the number of train wrecks by enabling engineers to communicate with each other and monitor track and weather conditions. Woods was born in Columbus, Ohio, on April 23, 1856, the son of Tailer and Martha Woods. At the age of ten, he had to end his public education to take a job in a machine shop. However, he continued to seek tutoring and private classes in the evening and borrowed books from the public library and friends to augment his studies.

Woods developed his mechanical skills by going to Missouri to work as a railroad engineer and fireman on the Iron Mountain Railroad. At the age of twenty, Woods began studying mechanical and electrical engineering, earning his tuition by working in a machine shop. In 1878, because he could find no jobs to suit his talents, he served as engineer of the British steamer *Ironsides* for two years. By 1880, he advanced to **locomotive** engineer for the Danville and Southern Railway Company, with which he served until 1884. During this era, he patented a steam boiler **furnace** which improved combustion and fuel economy.

While working in his shop, called Woods Electric Company, in Cincinnati with Lyates Woods, his brother and fellow inventor, Woods made a major breakthrough with the invention of a synchronized railway telegraph, which enabled railroad companies to keep tabs on trains, thereby preventing crashes. In contrast to devices invented by **Alexander Graham Bell**, the unique quality of his innovation was that it could send both voice or **Morse code** messages. He named this hybrid **telephone** and telegraph the ''telegraphony'' and sold it to the Bell Telephone Company. Pleased by his success, he abandoned manufacturing and concentrated solely on invention.

Woods created other practical electronic products: an electric telegraph transmitter, relay instrument, telegraph alarm, electromagnetic **brake** and brake housing, galvanic **battery**, safety devices to prevent shock from **electric circuits**, **incubator**, and **circuit breaker**, all of which he evolved between 1879 and 1899. Additional inventions included an electrical translator and regulator, railway brakes, electrical **thermostat**, **tunnel** construction for **streetcars**, and dynamotor to avert fires on streetcars. From his design for the troller or ''third rail,'' a grooved **wheel** which reduced friction while receiving electrical signals, the English language gained the word ''trolley.''

Woods's career was beset by difficulties. He successfully challenged both Lucius Phelps and **Thomas Alva Edison** for rights to a similar telegraph system. However, after he moved to New York in 1890, he spent much of his earnings on litigation resulting from his charge that the American Engineering Company had stolen his electric streetcar concept. He suffered a stroke and died at Harlem Hospital in New York City on January 30, 1910, and was buried in St. Michael's Cemetery in Queens. Eulogizers honored him for modernizing public transportation. In 1969, a public school in Brooklyn was named for him.

Wool

As with many discoveries of early man, anthropologists believe the use of wool came out of the challenge to survive. In seeking means of protection and warmth, humans in the Neolithic Age wore animal pelts as clothing. Finding the pelts not only warm and comfortable but also durable, they soon began to develop the basic processes and primitive tools for making wool. By 4000 B.C., Babylonians were wearing clothing of crudely woven fabric.

People soon began to develop and maintain herds of wool-bearing animals. The wool of sheep was soon recognized as one of the most practical to use. During the eleventh and twelfth centuries, wool trade prospered. The English had become proficient in the raising of sheep, while the Flemish had developed the skills for processing. As a result, the British began to sell their wool to the Flemish, who processed the raw material and then sold it back to the English.

The ambitious British soon realized the advantages of both producing and processing their own wool. As Britain began to prosper, it sought to enhance its position by enacting laws and embargoes that would stimulate its domestic production. Some laws, for example, required that judges, professors, and students wear robes made of English wool. Another law required that the dead be buried in native wool. When the American colonies began to compete with the motherland, the English passed a series of laws in an attempt to protect their "golden fleece." One law even threatened the amputation of the hand of any colonist caught trying to improve the blood line of American sheep.

Today, wool is a global industry, with Australia, Argentina, the United States, and New Zealand serving as the major suppliers of raw wool. While the United States is the largest consumer of wool fabric, Australia is the leading supplier. Australian wool accounts for approximately one-fourth of the world's production.

What for centuries was a small home-based craft has grown into a major industry. The annual global output is now estimated at 5.5 billion pounds. Though cotton is the number one plant used for fabrics and the number one fiber overall, the number one source for animal fiber is still wool.

The current widespread use and demand for wool is so great that there is little doubt that wool will continue to maintain its position of importance in the fabric industry. Only a major innovation that encompasses the many attributes of wool—including it warmth, durability, and value—could threaten the prominence of this natural fiber.

WORM optical drives

WORM optical drives are backup storage and data retrieval devices for personal computers. One of the earliest WORM systems was introduced about 1985 by the Colorado-based company Information Storage, Inc. Unlike magnetic disk drives which utilize mechanically operated read/write heads, WORM drives employ **laser** beam optics to read and write disk information. The term WORM is an anagram for Write Once, Read Many—the method in which information on optical drives is used and the only type of drive available up until mid-1987. Since that time, Sony Corporation and other drive manufacturers have begun introducing erasable **optical disks**, which have now surpassed the non-erasable disks in popularity and usage.

Although optical disk drives are much more expensive and slower in operation than magnetic disks, their advantages are their extremely large capacities and increased security over magnetic disks, which are prone to erasures, crashes, etc. Optical disks can hold the equivalent of several hundred books, in a fraction of the space required by standard magnetic disks or tape storage systems. Optical drives are especially suited for information searches in huge databases where speed is not a particularly important consideration. As the drive technology advances, the speed and cost factors of optical drives can be expected to improve vastly.

Wozniak, Steve (1950-)

American inventor

Born and raised in San Jose, California, Wozniak designed his first computer at age thirteen. He credited his father, an electrical engineer; Tom Swift books (a series about a young inventor); and local science fairs with encouraging his interests in **electronics**.

In high school, Wozniak met **Steve Jobs**, who shared his enthusiasm for electronics, and they soon became friends. In 1971, they assembled and sold "blue boxes," an illegal device used to make long-distance **telephone** calls without being charged. After dropping out of college, Wozniak began designing calculators at Hewlett-Packard, and Jobs gained employment at Atari, the **video game** company. They stayed in touch, though, and Wozniak, at Jobs's request, designed a new game for Atari called "Breakout," which earned him $700.

In 1975, the two friends joined the Homebrew Computer Club (an informal information exchange group for computer enthusiasts based in Menlo Park, California). When the club displayed an Altair 8800, an assemble-at-home personal computer (PC) manufactured by MITS, Wozniak believed he could improve on its design, and had a chance to prove his theory when MOS Technology brought out a new **microprocessor** chip. He wrote a version of BASIC programming language for the chip and, in a matter of only a few weeks, designed a circuit board that using that chip, along with interfaces that connected it to a keyboard and a video monitor. Although Wozniak said he designed this computer to show off to his friends, he underestimated the Jobs's marketing skills, who convinced Wozniak to join him in manufacturing personal computers. Jobs, who had been working at an apple orchard at the time, called their new company Apple Computers. To finance their venture, Jobs sold his car and Wozniak sold his two Hewlett-Packard calculators. At first they built their computers in Jobs's garage. In 1976, however, a local retailer ordered 50 Apple I machines, as they were known, and Apple Computers soon became a major force in the personal computer industry. With Jobs in charge of marketing, Wozniak set about improving the Apple I. After attending a Homebrew Computer Club meeting where he watched a color television displaying computer graphics, Wozniak decided to incorporate this technology into the company's next project. The result, the

Apple II, was more sophisticated than any other personal computer available at that time.

In 1978 Wozniak created a floppy disk system for the Apple II to replace the cassette tape used by personal computers of that time and wrote the program for the accompanying software. He later designed the Apple IIe, III, IIc, Lisa, and Macintosh computers. He left Apple temporarily in 1981 to complete his education and returned a few years later. He resigned in 1985 to begin another company, but remains an engineering consultant for Apple.

See also Computer, digital

Steve Wozniak

Wright, Wilbur (1867-1912)
Wright, Orville (1871-1948)

American inventors

Wilbur and Orville were born in Indiana and Ohio, respectively; neither sought education beyond high school, though both showed an aptitude for mechanics and independent study from an early age. In 1878 the Wright brothers first became interested in flight when their father brought a toy whirligig home for them to play with. They tinkered with the basic design and built their own models, displaying an inventiveness that would stay with them throughout their lives.

When the brothers heard of **Otto Lilienthal**'s fatal **glider** crash in 1896, they began to seriously consider the problem of flight. They meticulously studied contemporary aeronautical research—including that of American engineer Octave Chanute (1832-1910), a pioneer of the **biplane**—and observed the flights of soaring birds.

They first tackled the problem of controlling the machine in all three axes of movement (up and down, side to side, forward and backward) without resorting to the pilot twisting or shifting his body weight as Lilienthal had. After observing buzzards control their flight by twisting their wing tips, they settled on wing warping as the best method of maintaining balance.

To test their theories, the Wrights traveled to Kitty Hawk, North Carolina, an ideal site for wind and terrain. Following experiments with a biplane **kite**, they built several gliders, each more sophisticated than its predecessor. During the process they discovered errors in the mathematical tables that Lilienthal had created to explain lift. To achieve the correct calculations, they built their own **wind tunnel** with a homemade pressure-testing device to check the amount of lift in various wing configurations.

By the time they returned to Kitty Hawk in 1902 with their last glider, the Wright brothers had solved the basic problems of control. They gave this glider narrower wings which curved much less than the previous ones. They also mounted a tail assembly on the back that acted as a rudder, with control **wire**s linking it to the wing-warping mechanism. Before returning to Dayton, Wilbur Wright flew the glider 622 feet (189.71 m) in just 26 seconds.

The Wrights immediately prepared for powered flight the following year at Kitty Hawk. They needed a **gasoline engine** to provide the power, but none in existence was light enough. Consequently, the resourceful brothers designed and built their own four-cylinder, 12-horsepower engine.

They also worked on propellers for the craft, only to discover that little was known about how they worked. They soon realized that a propeller was nothing more than a wing moving in a circular course and finally designed and built two blades that would be mounted at the rear of the plane so the craft would be undisturbed by propeller turbulence. In addition, the blades would spin in opposite directions to prevent torque from pulling the craft to one side.

The Wrights returned to Kitty Hawk in September 1903, ready to try their powered machine. Storms, engine problems, and the need to devise a takeoff system slowed progress. Then, on December 14, 1903, the brothers were ready and Wilbur, who had won a coin toss, took the first ride. Unaccustomed to the engine's power, he rose too steeply, stalled, and then plowed into the sand, slightly damaging the craft. On December 17, Orville took his turn. He rose into the air, climbing to about 10 feet (3 m), and landed 12 seconds later, some 120 feet (37 m) beyond the takeoff point. The Wrights flew three more times that day, with Wilbur covering 852 feet (259 m) in a 59-second flight. In the next two years, the brothers built improved models of their **aircraft** and, by 1905, they were staying aloft for as long as 38 minutes and covering a distance of 24 miles.

Wilbur and Orville Wright

In 1908, they sold the first military airplane to the U.S. Department of War; this same year Orville established several aloft records in excess of one hour before crashing and injuring himself. Fortunately, he fully recovered and the Wrights formed both a German and an American company to manufacture airplanes. Although they never became wealthy due to patent infringements and lawsuits, the brothers were universally hailed as the first to unlock the secrets of flight through their determination, experimentation, and inventiveness.

Writing

Considering that historians have traditionally identified the beginning of civilization with the emergence of written language, writing could well be considered humankind's most important invention. The various writing systems in existence today have followed different paths of development; all, however, have their origin in the use of pictures to represent objects. The earliest extant cave drawings are at least 15,000 years old. It appears that the most ancient writing systems emerged, changing from idiosyncratic drawings into organized representation, when primitive cultures developed into settled agricultural societies; anthropologists and historians speculate that the increased need for record-keeping spurred writing's development as trade and government became more complex.

The oldest known form of writing is probably cuneiform, developed by the Sumerians in the Near East as early as 4000 B.C. Cuneiform derives its name from the Latin *cuneus* (wedge), referring to the wedges that were used to inscribe such writing onto soft clay tablets. Cuneiform symbols, like those of other ancient languages, originated as pictograms, simple drawings which stood for the spoken names of the objects they represented; pictograms are generally considered the first step in the evolution of pictures into words. For example, an early hunter-gatherer might have drawn a picture of a bird to represent that animal. Eventually, this picture would become recognized as the equivalent of the spoken word "bird"; once this connection was made, the picture would be simplified with repeated use, becoming a pictogram. Another well-known example of such stylized picture-writing is the Egyptian system of hieroglyphics, which is believed to have originated around 3000 B.C. and continued in use until around 300 A.D. In the case of cuneiform writing, as people became accustomed to the concept of pictograms, they began to draw them in increasingly abstract, simplified ways which bore less and less resemblance to the original picture. The cuneiform for "bird," to continue the example, eventually evolved into two crossed wedge marks next to a vertical wedge mark, a symbol which no longer looked anything like a bird, but which still was equated with the spoken word "bird" and which could be written much faster than a picture could be drawn. Such abstract signs are called logograms.

Logograms had limitations, however. Firstly, they necessitated a writing system in which one had to memorize thousands of symbols (cuneiform had more than 20,000). Secondly, they expressed inadequately, if at all, abstract concepts, such as "life." In response to these and other problems, the Sumerians began to use cuneiform symbols to stand for the *sounds* of the spoken words as well as the words themselves, so that two logograms could be strung together to produce the sound of a third word that could not be pictured. The technique is similar to that of modern-day *rebus* puzzles, in which, for example, a picture of a beer bottle next to the numeral "8" would spell the word "berate" (beer + eight = berate). This stage in the development of writing, in which a symbol can stand for a word or a sound, is called *word-syllabic* writing.

The Semites of the Mediterranean coast, neighbors of the Sumerians, took the Sumerian word-sign system and, between 1500 and 1000 B.C., simplified it radically, retaining only a few of the original symbols and using them only to represent sounds, from which they strung together the words of their spoken language. One Semitic group, the Phoenicians, developed a 22-character system around 1000 B.C., in which each character stood for a one-syllable sound beginning in a consonant and ending in a vowel. This system is considered by many the first **alphabet**; other scholars contend that the first true alphabet was that developed by the Greeks, who added vowel symbols to the consonant-syllables of the Phoenicians. The Greek alphabet was adopted by the Etruscans, from whom it evolved into the Latin alphabet, which was spread throughout the Roman Empire and is today used in Romance languages. The Arabic and Cyrillic alphabet, used in eastern Europe, also derive ultimately from Phoenician.

Not all writing systems in use today have gone through all these developmental stages. The Chinese system, for example, has more than 80,000 symbols that can stand for words or phonetic sounds, as in Sumerian word-syllabic writing; used by twenty percent of the world's population, it has remained essentially unchanged since the second millennium B.C.

See also Ink; Language, universal; Paper; Pen; Pencil; Printing technology

WROUGHT IRON • See Iron production

WYETH, NATHANIEL (1911-1990)
American engineer and inventor

A member of the famous artistic family, Nathaniel Wyeth stood out by becoming an engineer and inventor. Born in Chadds Ford, Pennsylvania, Nat was the son of famed illustrator N. C. Wyeth, who constantly encouraged his children to develop and express their imaginations. While brother Andrew and three sisters flourished artistically and musically, Nat dismantled **clocks** and transformed them into toy speedboat engines. Far from disapproving, N. C. early on suspected his son's talents would lie in engineering and changed the boy's given name from Newell Convers (N. C.'s own name) to Nathaniel, the name of an uncle who was an engineer. N. C. recruited another uncle to survey colleges where Nat could receive good technical training. Nat attended the uncle's choice, the University of Pennsylvania, graduating with a B.S. in 1936. During his college years, Wyeth designed and built a 20-foot hydroplane.

Upon graduating, Wyeth worked briefly for the Delco company in Ohio and then joined Du Pont as a field engineer. He remained at Du Pont for the rest of his career, and was eventually named the company's first senior engineering fellow. After designing a new **valve** for a production machine, Wyeth was transferred to the mechanical development laboratory, where he was able to devote himself to engineering design and invention. One of his first projects was an automatic **dynamite**-cartridging machine; a later one was a bonding machine to produce a **polypropylene** fabric called Typar used as a backing material on rugs.

Wyeth's most famous invention was the **plastic soda pop bottle**. Wyeth became curious about why plastic was not being used for carbonated beverages, and found out that when plastic was molded into bottles, certain parts of the bottle were weaker than others and could not withstand the carbonation pressure. Adopting a stretching technique used to strengthen **nylon** fibers, Wyeth started experimenting with cold-stretching plastic molecules. He devised a way to extrude the plastic with crisscrossed flow lines, which reinforced themselves and created a uniformly strong bottle. The plastic he was using, however, was not self-balancing—it kept stretching in one direction until it burst. He switched to polyethylene terephthalate, or PET, which stretched in one place when the limit was reached somewhere else. The result was the first successful plastic soda pop bottle, patented in 1973, and now the industry standard. Wyeth retired from Du Pont in 1976. He died in 1990 in Glen Cove, Maine, the holder of 25 patents.

The Rosetta Stone provided modern scholars with a key to deciphering Egyptian hieroglyphics.

X

X-RAY CRYSTALLOGRAPHY

X-ray crystallography is a process by which the extremely fine atomic structure of many crystals can be examined and recorded. It was first developed not as a research tool but as a means of determining the nature of X-rays themselves.

X-rays were discovered—quite accidentally—in 1895 by Wilhelm Röntgen (1845-1923). Although his intensive research revealed much about the properties of these new rays, such as their ability to penetrate certain substances, Röntgen could not ascertain whether X-rays consisted of particles or longitudinal waves. This question puzzled scientists until 1912, when German physicist Max von Laue (1879-1960) directed an X-ray beam through a crystal. As the X-ray struck the lattice-like pattern of atoms within the crystal, an interference (or diffraction) pattern was formed—an effect that could only occur if X-rays were waves, like light.

Laue's experiment proved to his fellow scientists the longitudinal nature of X-rays. However, it was an Australian professor, **William Henry Bragg**, and his son **William Lawrence Bragg** who realized the significance of Laue's discovery. They surmised that the structure of the crystal on a molecular level could be deduced from a study of the interference pattern. In order to prove their theory they also designed an X-ray spectrometer to measure the specific wavelengths of X-rays, and devised a mathematical system for analyzing the information. In 1915 the father-son team shared the Nobel Prize for Physics for the establishment of a new scientific method, X-ray crystallography.

In crystallography, X-rays are used to probe the structure of a variety of crystals. The pattern of diffracted X-rays is analogous to an atomic "shadow"—by examining where the X-rays are blocked by the crystal's atoms, scientists can define the structure of those atoms. This was first seen by the Braggs, who found that crystals consist not of molecules, but rather of groups of layered ions; for example, a sodium chloride crystal is formed from sodium ions and chlorine ions. X-ray crystallography quickly became an important tool for validating many of Danish physicist Niels Bohr's (1885-1962) theories of atomic structure.

Perhaps the most important application of X-ray crystallography is its use in synthesizing substances, particularly in medicine. Many of the medicinal chemicals that have been discovered by scientists are very difficult to produce naturally in large amounts. In this case, it becomes necessary to create the chemicals in the laboratory through synthesis. However, before a chemist can synthesize a substance, a very specific map of its atomic structure must be obtained, a map that can only be drawn by using X-ray crystallography. Few scientists have been more successful at this than the British chemist Dorothy Hodgkin (1910-). During World War II, Hodgkin and her colleagues determined the structure of penicillin, whose synthesis was necessary to supply army hospitals. Since then, Hodgkin's team has worked on the crystallographic cartography of vitamin B_{12} (prescribed to prevent pernicious anemia) and insulin (used in the treatment of diabetes). Other researchers have used X-ray technologies to record the structures of proteins, hemoglobin, and the now-familiar double-helix of DNA (deoxyribonucleic acid).

The development of X-ray crystallography also created the science of mineralogy. Once they were able to examine in detail the inner structure of many minerals, mineralogists were able to define the major mineral groups. The understanding that stems from crystallography has also allowed scientists to construct the man-made minerals used in industry.

X-RAY MACHINE

The very first X-ray device was discovered accidentally by the German scientist Wilhelm Röntgen (1845-1923) in 1895. He found that a **cathode-ray tube** emitted certain invisible rays that could penetrate paper and wood and, the first person in the world to see through human flesh, even saw a perfectly clear outline of the bones in his own hand. Röntgen studied these new rays—which he called X-rays—for several weeks before publishing his findings in December of 1895. For his great discovery, he was given the honorary title of Doctor of Medicine and awarded the 1901 Nobel Prize for physics. Adamant his discovery was free for the benefit of humankind, Röntgen refused to patent it.

X-rays are waves of *electromagnetic energy* which behave in much the same way as light rays, but at wavelengths approximately 1,000 times shorter than the wavelength of light. X-rays can pass uninterrupted through low-density substances such as tissue, whereas higher-density targets reflect or absorb the X-rays because there is less space between the atoms for the short waves to pass through. Thus, an X-ray image shows dark areas where the rays traveled completely through the target (such as with flesh) and light areas where the rays were blocked by dense material (such as bone). Following X-ray's discovery in 1895, this scientific wonder was seized upon by sideshow entertainers who allowed patrons to view their own skeleton and gave them pictures of their own bony hands wearing silhouetted jewelry.

The most important application of the X-ray, however, was in medicine, an importance recognized almost immediately after Röntgen's findings were published. Within weeks of its first demonstration, an X-ray machine was used in America to diagnose bone fractures. **Thomas Alva Edison** invented an X-ray fluoroscope in 1896, which was used by American physiologist Walter Cannon (1871-1945) to observe the movement of barium sulfate through the digestive system of animals and, eventually, humans. In 1913 the first X-ray tube designed specifically for medical purposes was developed by American chemist **William Coolidge**. X-rays have since become the most reliable method for internal diagnosis.

At the same time, a new science was being founded on the principles introduced by German physicist Max von Laue (1879-1960), who theorized that crystals could be to X-rays what diffraction gratings were to visible light. He conducted experiments in which the interference pattern of X-rays passing through a crystal were examined; these patterns revealed a great deal about the internal structure of the crystal. **William Henry Bragg** and his son **William Lawrence Bragg** took this field even farther, developing a system of mathematics that could be used to interpret the interference patterns. This method, known as **X-ray crystallography**, allowed scientists to study the structures of crystals with unsurpassed precision and is an important tool for scientists, particularly those striving to synthesize chemicals. By analyzing the information within a crystal's interference pattern, enough can be learned about that substance to create it artificially in a laboratory, and in large quantities. This technique was used to isolate the molecular structures of penicillin, insulin, and DNA.

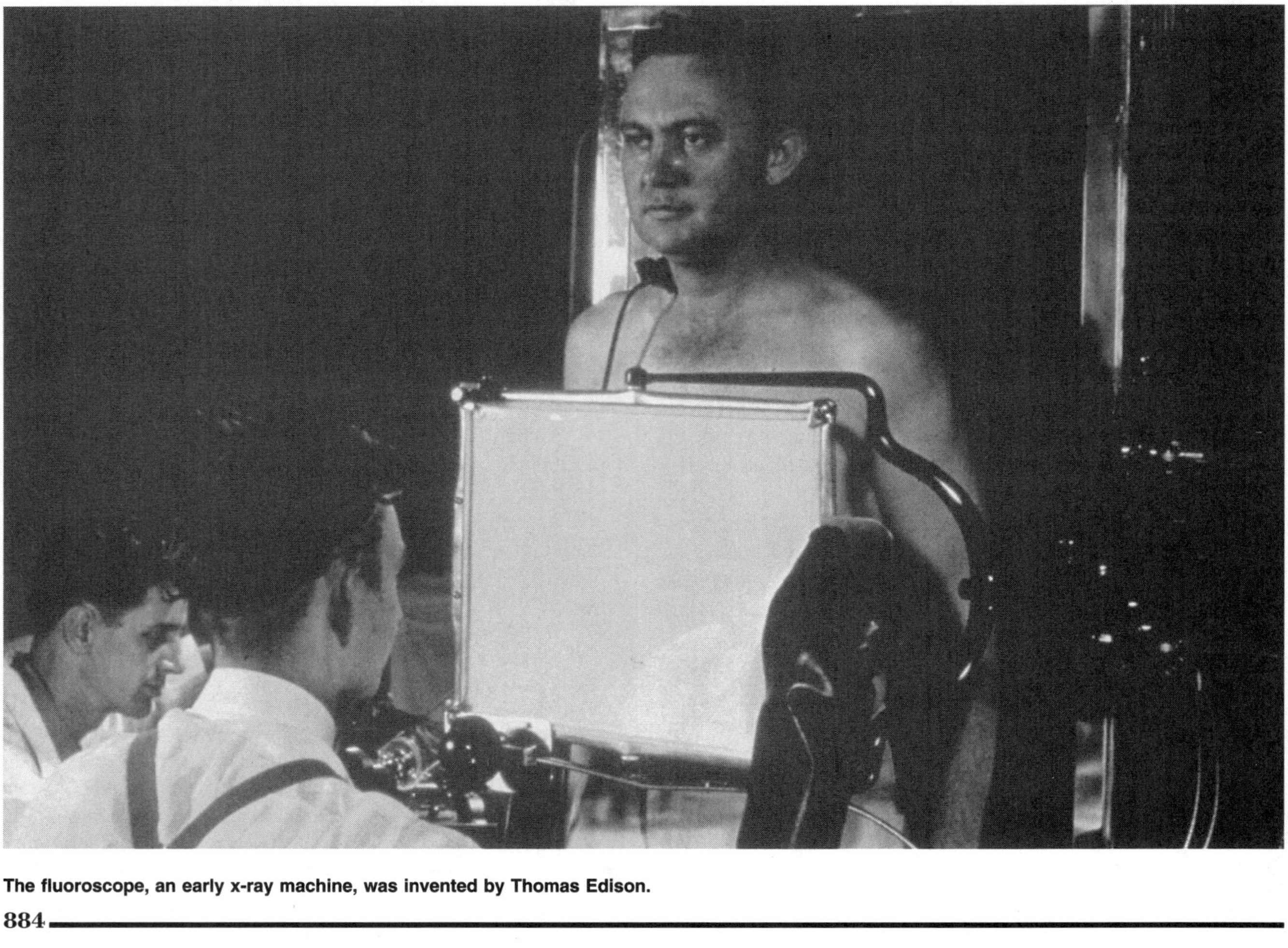

The fluoroscope, an early x-ray machine, was invented by Thomas Edison.

Modern medical X-ray machines are grouped into two categories: ''hard'' or ''soft'' X-rays. Soft X-rays, which operate at a relatively low frequency, are used to image bones and internal organs and, unless repeated excessively, cause little tissue damage. Hard X-rays, very high frequencies designed to destroy molecules within specific cells thus destroying tissue, are used in **radiotherapy**, particularly in the treatment of cancer. The high voltage necessary to generate hard X-rays is usually produced using **cyclotron**s or synchrotrons (variations of **particle accelerators**, or atom smashers).

In 1996, *Amorphous silicon X-ray detectors* were introduced which produce real-time, high resolution images by converting X-rays into light, the light into electrical signals which are interpreted by a computer, which produces digital data displayed as digital images, which can be enlarged to target a specific area. Images are filmless and instantly available, formatted for electronic storage and/or transmission. First applied to **mammography**, this technology reduces radiation, cost of film and storage, and can be used in industrial applications. Also in 1996, researchers at NASA's Marshall Space Flight Center developed the *high resolution* or *high brilliance* X-ray which generates beams 100 times more intense than conventional X-rays. These beams can be controlled and focused by reflecting them through tens of thousands of tiny curved capillaries, much as light is directed through **fiberoptics**. NASA is using this instrument to define the atomic structure of proteins for use as blueprints in designing drugs. It may also initiate smaller, less expensive, and safer X-ray sources.

A familiar use for X-rays is the security scanner for examining baggage at airports, while a new application for the Advanced Photon Source (APS)—the worlds largest X-ray device valued at $800 million—is in archaeology. The University of Chicago plans to utilize the machine—designed to provide biologists, chemists, physicists, and materials scientists with information into molecular structures—to analyze ancient tools. Not only harmless to these ancient artifacts, it is 10,000,000 times more accurate than other methods of determining chemical content. Meanwhile, the timber industry hopes to save millions of dollars with new technology known as the *Glass Log Technique* developed at Monash University in Australia. This X-ray tube produces cross-section images of logs and, combined with **computerized tomography**, construct three-dimensional images of the wood inside log, revealing its quality without having to saw the log open.

Y

Yale, Linus

• See Lock and key

Yalow, Rosalyn S. (1921-)

American biophysicist

Rosalyn S. Yalow is the co-developer of **radioimmunoassay** (RIA), an extremely sensitive isotopic method of measuring **hormones** and other substances in blood. Her work earned her part of the 1977 Nobel Prize in physiology or medicine.

Rosalyn Sussman was born in New York City where her father owned a small paper and twine business. Neither of her parents had attended high school, but they encouraged her studies. Interested in mathematics from childhood, she graduated from Hunter College in 1941, the first woman to receive a degree in their recently-established physics department, and was the only woman in her entering class of 400 at the University of Illinois College of Engineering, earning her Ph.D. in nuclear physics there in 1945. There, also, she met her future husband, fellow physics student Aaron Yalow.

In 1947, Yalow established the radioisotope laboratory at the Bronx (New York) Veterans Administration Hospital, where she spent her entire career. Until 1950, she also taught physics full-time at Hunter College. That year, Solomon A. Berson, a resident in internal medicine at the hospital, became interested in her work. They worked together for twenty-two years, until his untimely death in 1972 (making him ineligible to share the Nobel Prize). Their first research was using isotopes to study blood volume and diagnose thyroid diseases by measuring iodine metabolism. Yalow and Berson then adapted the same method to hormones, including insulin, which was widely available. By 1959 they had perfected their method, which they called **radioimmunoassay (RIA)**, so that it was sensitive enough to detect one thousand billionths of a gram of material per milliliter of blood. In the process, they discovered that hormones bind with antibodies and also that, contrary to the prevailing theory, Type II (adult onset) diabetes is caused by the body's inefficient use of insulin, not by failure to produce the hormone.

RIA rapidly became a standard laboratory technique. Medical uses include diagnosis of cancer and measurement of blood levels of hormones, vitamins, and other substances. It is also used in forensic work, for example, to determine narcotics and poison levels in blood.

Besides being a pioneer in her own field, Yalow is a strong supporter of women in science. Among her honors is the 1976 Albert Lasker Basic Medical Research Award, of which she was the first woman recipient.

Yogurt

Yogurt is made by curdling milk with purified bacteria cultures that cause the milk sugar, or lactose, to turn into lactic acid. The result is a high-calcium, high-protein dairy food. From early on in its history, yogurt has been linked with longevity and has even been used for medicinal purposes.

The exact origin of yogurt is uncertain, but legend has it that the Biblical patriarch Abraham offered yogurt to the angels when they told him of the birth of his son Isaac. Over the years, yogurt has been credited as the reason for the longevity of Abraham, who lived to the advanced age of 175, and some Biblical scholars believe that yogurt is the milk of the phrase "milk and honey."

In the early 1500s, a Turkish doctor was credited with saving the life of the French monarch François I (1494-1547) by prescribing a diet of yogurt for the king. From that point on, yogurt was dubbed *le lait de la vie eternelle*, or the milk of eternal life, by the French.

In the early 1900s, the Russian bacteriologist Elie Metchnikoff (1845-1916), went to observe a group of Bulgarian peasants who were said to be thriving and even bearing children past the age of one hundred. Metchnikoff noted that the men and women would frequently pause from their work and eat large bowls of yogurt mixed with onions, nuts, and vegetables. Metchnikoff was convinced that yogurt was re-

Rosalyn S. Yalow

sponsible for their incredible strength and longevity and began to study the food. He isolated the two strains of bacteria that were present in yogurt: *Streptococcus thermophilus* and *lactobacillus bulgaricus*, named for the country in which Metchnikoff discovered it. Metchnikoff discovered that these bacteria are rich in B vitamins and are able to combat a common intestinal virus, which he believed produced toxins that hasten the aging process. In addition to killing this intestinal virus, yogurt is easily digested and aids in the digestion of other foods.

In 1929, a Parisian, Isaac Carasso, began selling yogurt commercially, naming his product *Dannone* after his son Daniel. By the 1950s, Carasso had the largest yogurt-producing factory in the world. Carasso introduced his product as *Dannon* in the United States, and by the mid-1960s, yogurt had established itself in America's burgeoning ''health food'' market.

No longer a specialized product found only in health food stores, yogurt is now a staple item on supermarket shelves and is available in a wide range of flavors. Like milk, it comes in whole, low-fat, and non-fat varieties. In the 1980s, frozen yogurt stores, offering a healthy alternative to **ice cream**, were among the fastest growing franchises in America.

Z

Zeppelin, Count Ferdinand von (1838-1917)

German inventor

The son of German nobility, Zeppelin entered the military and, as was expected of aristocrats of the time, served in the American Civil War. While in the United States, he became interested in air travel after observing **balloons** being used by the military for intelligence gathering. He returned to Germany and rose to the rank of lieutenant-general in the German army. He then left the army in 1891 to devote himself to aviation.

While hospitalized for a riding injury in 1874, he outlined plans for an **airship** as big as an ocean liner, capable of carrying cargo, mail, and twenty passengers. He already knew how the dirigible should be constructed: a rigid metal structure of vertical rings held in place by long rows of girders; separate gas cells between the rings; a fabric skin covering the metal structure. However, no power source was yet available to drive this airship through the skies, so Zeppelin had to wait until the **internal combustion engine** was refined.

In June 1898, construction began on his first airship with money raised through public donations and his own fortune. Completed in 1900, the LZ-1 (Luftschiff Zeppelin 1) proved a failure because, despite its length of 400 feet (122 m) and diameter of 38 feet (11.59 m), it could lift a payload of only 660 pounds (299.64 kg) and was not very maneuverable. His money gone, Zeppelin dismantled the airship and dismissed his workers, except for his chief engineer, Ludwig Durr.

Zeppelin tried again in 1904. The King of Wurttemberg held a public lottery and Zeppelin mortgaged his wife's country estates in Latvia. In January 1906, the new ship, LZ-2, was destroyed during a storm, but the German government, already convinced of the airship's military potential, funded the construction of the LZ-3. Built with the addition of horizontal fins for stability, this airship proved a success, flying for over two hours at a speed of 24 miles per hour (38.6 kph) carrying 5,500 pounds (2,497 kg) of water, ballast, and eleven people.

The LZ-3, however, could not meet the military's requirement that it remain aloft for twenty-four hours, so Zeppelin created the LZ-4. On August 4, 1908, this airship took off in an attempt to meet those endurance standards. It not only surpassed the requirements, but it also achieved the speed of forty miles per hour (64 kph). The zeppelin set down with an engine problem, an unexpected squall came up, and the craft lurched into the air. A crew member brought it down into some trees where a branch tore the outer cover and a gas cell. No one knew that the cell's material, rubberized cloth, would shoot off sparks of static electricity if it rubbed on itself. As a result, the entire airship went up with a whoosh of flame and a gigantic explosion when the flammable hydrogen gas ignited.

The German people, who had come to know and love the determination of Count von Zeppelin, immediately contributed to a fund established to help him build another. When LZ-5 flew the longest voyage yet by any powered craft (thirty-eight hours in 1909), it looked like a success for the Count. But the airship was too slow for the military, which refused to buy it and the already-built LZ-6.

Zeppelin realized he needed some other market for his creation, so he turned to public transportation. He established the German Airship Transport Company (Deutsche Luftschiffahrts Aktien Gesellschaft—DELAG). It began operations between German cities in 1910, but all three of its dirigibles either crashed or burned. The next airship, LZ-10, had new and better engines as well as other modifications which allowed it to be more reliable and maneuverable. It was a success when it first flew in 1911 allowing DELAG to build three more zeppelins. When World War I started in 1914, these airships had carried over 10,000 passengers without any problems.

In his last years, the Count finally saw the military accept his airships for patrol and bombing purposes. His determination had captivated a nation, and his designs had done much to establish the value of lighter-than-air craft.

Zhang Heng • See Seismograph

Ferdinand Adolf August Heinrich von Zeppelin

Ziegler, Karl (1898-1973)

German chemist

Karl Ziegler was born in Helsa, Germany, on November 26, 1898. At age eighteen he entered the University of Marburg/Lahn to begin advanced courses in chemistry. In 1920 he received his doctorate in organic chemistry. Following a brief stint at the University of Frankfurt/Main, he became a professor at the University of Heidelberg, where he remained for 10 years. He married Maria Kurz in 1922.

Ziegler became interested in trivalent carbon compounds—carbon molecules with three functional groups. Because these compounds are *free radicals* with an unbonded electrons, they tend to be highly reactive. Ziegler was specifically interested in carbons with *aromatic rings* as substituents because they tend to be more stable. He studied the steric and electronic effects of substituents on reactions, proving through systematic studies that stable free radicals could be formed even if the ring substituents were not present, and that larger substituents led to more stable free radicals. For the next thirty years Ziegler and his colleagues studied the effects of functional groups on reactions and the formation of free radicals. Their research showed that *organometallic compounds*—organic compounds with carbon-metal bonds—are instrumental in forming free radicals.

In 1928, while Professor of Chemistry at the University of Heidelberg, he discovered that by combining an organometallic compound containing potassium with an olefin, he could insert the olefin between the compound and the potassium, thereby lengthening the carbon chain. When an excess of olefin was used, a *polymer* would result as the addition reaction repeated itself. Ziegler went on to research the formation of carbon rings, succeeding in creating rings that contained up to 30 carbon atoms with very high yields. He was appointed Professor of Chemistry and Director at the University of Halle/Saale in 1936. From 1943 until 1969 he was Director of the Max-Planck-Institut für Kohlenforschung (formerly the Kaiser-Wilhelm-Institut für Kohlenforschung. In 1953, while doing research at the institute, Ziegler discovered, mostly through serendipity, a process for generating high-molecular weight polyethylene using metallic catalysts. While trying to obtain it, Ziegler and Heinz Martin, ended up with a dimer—a short-chain polymer composed of two molecules of the same chemical composition—of ethylene. After some investigation they surmised that the reaction vessel they had used had contained a small amount of nickel. Realizing that other metals could also affect the reaction and inhibit *polymerization*, they began to investigate. To their surprise they found that while some metals were found to inhibit polymerization, others enhanced it. They also found that a metal chloride used with organoaluminum compounds caused the rapid low-temperature polymerization of ethylene into a very high molecular weight linear chain. The discovery of catalysts that made possible low-temperature polymerization revolutionized the chemical industry.

Karl Ziegler was unique in that he contributed both to theoretical knowledge and practical industrial chemistry. He shared the Nobel Prize in chemistry in 1963 with the Italian chemist **Giulio Natta** (1903-1979) and went on to research electrolytic processes. He died in 1973 after a short illness.

See also Fiber, synthetic; Polymer and polymerization; Plastic

Zipper

A zipper is a slide fastener with two edges of teeth attached to a fabric tape. The teeth lock into a snug fit when they are drawn together by a slide. When the slide is pulled back, the teeth separate.

The zipper was invented by Whitcomb L. Judson, a Chicago, Illinois, mechanical engineer, to relieve the tedium of fastening by hand the fashionable high-buttoned boots of the time. Judson's fastener, called the Clasp Locker and patented in 1893, consisted of a movable guide that meshed together two sets of hooks and eyes. Judson also invented a machine to mass produce his fasteners cheaply. However, the machine broke down frequently and the fastener itself had a way of spontaneously unfastening.

In 1905 Judson invented an improved fastener, the C-curity, but like its predecessor, it tended to break open unexpectedly. Perhaps for this reason, clothing manufacturers showed no interest in the device. It was Gideon Sundback (1880-1956), a Swedish engineer employed by Judson, who developed the first really practical and successful slide fastener. Sundback's 1913 invention used small, interlocking teeth that were flexible and remained locked together. (Catharina Kuhn-Moos patented a similar fastener in Europe the same year.) Sundback also invented efficient machinery to produce his improved fastener cheaply.

Although the slide fastener was now ready to be mass-produced for wide spread usage, clothing manufacturers continued to ignore it—except for one, which contracted in 1918 to supply the United States Navy with flying suits equipped with the device. Judson's company suddenly had an order for thousands of the fasteners.

Manufacturers began to realize how useful the fastener could be. Soon it appeared in gloves and tobacco pouches. In 1923 the B.F. Goodrich Company added the slide fasteners to their **rubber** galoshes, calling this new footwear design "zippers," which from then on became the popular name for the fastener itself. Zippers finally appeared on clothing, first in men's pants, and, in the late 1920s, in women's garments. Today's design is little changed from Sundback's original.

Zsigmondy, Richard (1865-1929)

German chemist

The son of a very successful dentist, Richard Zsigmondy was educated at the finest Austrian and German institutions. After completing his undergraduate studies in Vienna, Austria, he enrolled at the University of Munich, where he received his Ph.D. in 1890. An organic chemistry student in Vienna, Zsigmondy turned to the study of colored **glass** during his postgraduate years. Thus, after graduating from the University of Munich and lecturing there for a few years, he began work in Jena, Germany, as an industrial chemist at the Schott Glass Manufacturing Company, whose relationship with the renowned Zeiss Optical Works had earned it international recognition.

At Schott, Zsigmondy studied *colloidal compounds* in tinted glass, inventing a popular white variety called milk glass. It was this research that led him to the invention of the **ultramicroscope**. Colloids are substances composed of very tiny particles suspended in a fluid. The particles within colloids are so fine as to remain stably dispersed within the fluid—that is, they do not settle. They are far too small to be seen, even in a **compound microscope**, but their presence affects the properties of the fluid; for example, tiny particles of gold suspended in glass (such as the glass at the Schott Company) give it a deep red or purple color. Other examples of colloids are plasma and egg white.

Though Zsigmondy knew that colloid particles could not be seen under a microscope's lens, and were thus impossible to study directly, he hoped that they might be observed indirectly. He reasoned that, if placed in the path of a beam of bright light, the tiny particles would scatter that light, revealing their presence and position (just as particles of dust can be seen dancing in a shaft of sunlight). Working with H. F. W. Siedentopf (1872-1940), a physicist at Zeiss Optical, Zsigmondy designed the ultramicroscope, a device allowing him to view the motion of particles within a colloid. Though it has since been made obsolete by the **ultracentrifuge** and the **electron microscope**, the ultramicroscope was the first breakthrough in the very specialized field of colloidal chemistry. For his work in this area, Zsigmondy won the 1925 Nobel Prize in chemistry.

Karl Waldemar Ziegler

Zuse, Konrad (1910-1995)

German engineer and computer designer

Konrad Zuse was the designer of the first operational, fully electronic, program-controlled, general-purpose computer. He created the computer in Germany in 1941, but the machine was largely unknown until the 1960s.

Zuse was born in Berlin and studied engineering at the Berlin Technical School. Interested in speeding up mathematical calculations, Zuse designed a device similar to that developed by the nineteenth century English computer pioneer, **Charles Babbage**, whose work was unknown to him. After graduating in 1935, he went to work for an aircraft company, spending his weekends building a computer (the Z1) in his parents' living room. From the beginning, Zuse's design used binary numbers and he wrote his instructions in logical form, using *and*, *or*, and *not*. Instructions were punched on movie film. For memory, the Z1 used slotted metal plates with pin positions in the slots representing one or zero.

Zuse's employer provided the resources for his Z3 model, which he completed in December 1941. The Z3 featured floating point arithmetic (for varying decimals), was

Richard Adolf Zsigmondy

fully operational, and used electric telephone relays (on-off switches) for computing operations. It used 1,800 relays for memory, 600 for computation, and 200 for input-output. The Z3 was destroyed in a 1945 air raid during World War II.

In 1949, Zuse set up Z4 at the Technical School (ETH) in Zurich, Switzerland, later moving it to Basel where it continued to operate until 1959. He then founded his own computer-manufacturing company, Zuse AG, which designed and produced a series of innovative specialized computers. In 1958 Zuse designed a parallel processor (able to perform several operations at once) for differential equations. In 1969, Zuse sold his company to the large computer manufacturer Siemens AG.

Zuse also devised a computer language, Plankalkul, that could be used for both numerical and non-numerical problems, including programming chess games. In 1966 he became a professor at the University of Gottingen, Germany. Among his numerous honors is a medal from the American Federation of Information Processing Societies.

Zworykin, Vladimir Kosma (1889-1982)

Russian American inventor

Zworykin is often called the father of **television**. During the 1920s he invented the iconoscope, a transmission device that became the precursor to the modern television **camera**, as well as the kinescope, a reception device almost identical to the television tubes used today.

Born in Mourom, Russia, Zworykin was the son of a prominent river boat merchant. Though his father had hoped to give the family business to his son, he did not balk at sending Zworykin to the St. Petersburg Institute of Technology to study electrical engineering. After receiving his degree in 1912, Zworykin moved to France to study under physicist **Paul Langevin**. While focusing primarily upon the phenomenon of **x rays**, Zworykin began to explore the possibility of transmitting a visual signal via a wireless system. His graduate work was interrupted after just two years when, in 1914, Russia entered World War I. Serving as a **radio** operator and technician, Zworykin was able to study the inner workings of radio systems, knowledge that would prove essential for his future endeavors.

After the war, Zworykin remained in Russia only briefly, electing to emigrate to the United States at the onset of the Russian Revolution. Thus, in 1919, he came to America and, after a year spent learning English, got a job in the laboratories of the Westinghouse Electric Corporation. Curious as to its possible applications, he studied the **cathode-ray tube** there, and within five years he had developed the iconoscope.

Previous television systems were *photomechanical*—that is, they used a large perforated spinning disk to scan an image. The advantage of Zworykin's iconoscope was that it was completely electronic. Within the device's tube was a plate covered with thousands of tiny droplets of cesium-silver, each acting as a microscopic **photoelectric cell**. As light passed through a lens and fell upon the droplets each one produced a small electric charge, with each droplet's charge being different from the others. For example, if a droplet saw a bright spot, it would produce a greater charge than if it had been exposed to a dim spot. These droplets retained their charge until being scanned by an electron gun within the iconoscope; as the electron beam passed over the droplets the charges were converted into an electronic signal that could be sent to a receiver. Since the electron beam scans the photoelectric plate sixty times every second, it is possible to collect a series of images and run them together, creating motion.

In 1925, the year after he had invented the iconoscope, Zworykin developed the kinescope. Using a cathode-ray tube and another electron gun, the kinescope receives the signal from an iconoscope and scans the image back onto a fluorescent screen, converting it once again into a visual image. Zworykin was awarded the patents for his iconoscope and his kinescope in 1968.

Though his electronic television was certainly a landmark achievement, it did not impress his superiors at Westinghouse. However, it did attract the attention of David Sarnoff of the Radio Corporation of America (RCA). In 1929 Zworykin became the director of electronic research at RCA, rising to vice-president in 1947.

While at RCA Zworykin spent several years and close to four million dollars improving and refining his television system, and by the 1950s it had blossomed into a nationwide industry. In later years he began to turn his attention toward using his vast knowledge for other applications. He took his basic iconoscope design and modified it, making it possible to magnetically focus a stream of electrons in order to magnify an image. Using this as a springboard, James Hillier (b.1915) developed a practical and affordable **electron microscope**, a device that was used by medical professionals to examine viruses and human tissue in unprecedented detail. Zworykin also invented a device called the **scintillation counter**, one of the most precise tools for measuring radioactivity. In all, Zworykin held more than eighty patents.

For his work, Zworykin received numerous accolades, including the Rumford Medal in 1941 and the United States National Medal of Science in 1967.

Sources Consulted

Aaseng, Nathan. *Better Mousetraps*. Minneapolis: Lerner Publications, 1990.

Abbot, Charles Greeley. *Great Inventions*. Washington, D.C.: Smithsonian Institution, 1949.

Abbott, David, ed. *Astronomers*. New York: Peter Bedrick Books, 1984.

——. *Chemists*. New York: Peter Bedrick Books, 1983.

——. *Biologists*. New York: Peter Bedrick Books, 1983.

——. *Mathematicians*. New York: Peter Bedrick Books, 1985.

——. *Physicists*. New York: Peter Bedrick Books, 1984.

——. *Engineers and Inventors*. London: Fredrick Muller, Ltd., 1985.

Abell, George. *Exploration of the Universe*. 6th ed. Philadelphia: Saunders College Press, 1991.

Albert, Lillian Smith, and Kent, Cathryn. *The Complete Button Book*. Garden City, NY: Doubleday, 1949.

Allen, Hugh. *The House of Goodyear: Fifty Years of Men and Industry*. Cleveland: Corday & Gross, 1949.

Alper, Joseph, and Nelson, Gordon. *Polymeric Materials*. Washington, D.C.: American Chemical Society, 1989.

American Men and Women of Science. 17th ed., 8 Vols. New York: R. R. Bowker, 1989.

Angellucci, Erzo and Bellucci, Alberso. *The Automobile: From Steam to Gasoline*. New York: McGraw- Hill Book Company, 1975.

Arnold, Oren. *What's in a Name? Famous Brand Names*. New York: Julian Messner, 1979.

Arya, S. Pal. *Introduction of Micrometeorology*. New York: Harcourt Brace, 1988.

Ashworth, William. *The Encyclopedia of Environmental Studies*. New York: Facts on File, 1991.

Asimov, Isaac. *Asimov's Biographical Encyclopedia of Science and Technology*. Garden City: Doubleday and Co., 1972.

Augarten, Stan. *Bit by Bit: An Illustrated History of Computers*. New York: Ticknor & Fields, 1984.

Bailey, Joseph H. *Small Inventions That Make a Big Difference*. Washington, D.C.: National Geographic Society, 1984.

Baldwin, T. O., et. al., eds. *Chemical Aspects of Enzyme Biotechnology: Fundamentals*. New York: Plenum, 1990.

Ballantine, Richard and Grant, Richard. *Richards' Ultimate Bicycle Book*. New York: Dorling Kindersley, Inc., 1992.

Barach, Arnold B. *Famous American Trademarks*. Washington, D.C.: Public Affairs Press, 1971.

Barry, R. G., and Chorely, R. J. *Atmospheric Weather and Climate*. New York: Holt, Rinehart & Winston, 1970.

Beard, Charles A., ed. *A Century of Progress*. Freeport, NY: Books for Libraries Press, 1932, 1960.

Behling, Robert. *Computers and Information Processing: An Introduction*. Boston: Kent Publishing Co., 1986.

Belloc, Hilaire. *The Road*. New York: Harper & Bros., 1840.

Bensançon, Robert M., ed. *The Encyclopedia of Physics*. 2nd ed. New York: Van Nostrand Reinhold Co., 1974.

Berkow, M.D., Robert. *Merck Manual of Medical Information—Home Edition*. Whitehouse Station, NJ: Merck Research Laboratories, 1997.

Berry, Arthur. *A Short History of Astronomy*. New York: Dover, 1961 (reprint of 1898 edition).

Bettmann, Otto L. *A Pictorial History of Medicine*. Springfield, IL: Charles C. Thomas, 1956.

Billings, Henry. *Bridges*. New York: Viking Press, 1958.

Bishop, Chris, ed. *The Vital Guide to Combat Guns and Infantry Weapons*. Shrewsbury, England: Airlife Publishing Ltd., 1996.

Blohm, Hans, et. al. *Pebbles to Computers: The Thread*. Oxford: Oxford University Press, 1986.

Boettinger, H. M. *The Telephone Book*. Croton-on-Hudson, NY: Riverwood Publishers, 1977.
Bond, Peter. *Heroes in Space: From Gagarin to Challenger*. New York: Basil Blackwell, 1987.
Bordley, James. *Two Centuries of American Medicine*. Philadelphia: W.B. Saunders Company, 1976.
Bowers, Horace R. *General Meteorology*. New York: McGraw Hill, 1959.
Bremner, M. D. K. *The Story of Dentistry*. Brooklyn, NY: Dental Items of Interest Publishing Co., 1959.
Brennan, Richard P. *Levitating Trains and Kamikaze Genes*. New York: Wiley, 1990.
Bridges, T. C. *The Book of Invention*. London: George G. Harrap & Co., 1925.
Brooks, John. *Telephone: The First Hundred Years*. New York: Harper & Row, 1976.
Brumbaugh, Robert S. *Ancient Greek Gadgets and Machines*. New York: Crowell, 1966.
Burke, James. *Connections*. Boston: Little, Brown, 1978.
Burlingame, Roger. *Machines That Built America*. New York: Harcourt Brace, 1953.
Burt, McKinley, Jr. *Black Inventors of America*. Portland: National Book Company, 1969, 1989.
Bynum, W. F., et. al., eds. *Dictionary of the History of Science*. Princeton: Princeton University Press, 1981.
Campbell, Bernard G. *Humankind Emerging*. Boston: Little, Brown and Company, 1982.
Campbell, Hannah. *Why Did They Name It...?* New York: Fleet Publishing, 1964.
Campbell, Neil A. *Biology*. Menlo Park, CA: The Benjamin/Cummings Publishing Company, 1987.
Caney, Steven. *Steve Caney's Invention Book*. New York: Workman Publishing, 1985.
Capron, H.L., and Williams, Brian K. *Computers and Data Processing*. 2nd ed. Menlo Park, CA: The Benjamin/Cummings Publishing Co., 1984.
Carson, Rachel. *Silent Spring*. New York: Houghton, 1962.
Carwell, Hattie. *Blacks in Science*. Hicksville, NY: Exposition Press, 1977.
Cheney, Margaret. *Tesla: Man Out of Time*. Laurel, 1981.
Coe, Sophie D., and Coe, Michael D. *The True History of Chocolate*. London: Thames and Hudson, 1996.
Cohen, Nancy Wainer, and Estner, Lois J. *Silent Knife*. South Hadley, MA: Bergin and Garvey, 1983.
Cole, Franklyn W. *Introduction to Meteorology*. John Wiley & Sons Inc., 1970.
Condit, Carl W. *American Building*. Chicago: University of Chicago Press, 1968.
Cone, Robert J. *How the New Technology Works*. Pheonix: The Oryx Press, 1991.
Considine, Douglas M., and Considine, Glenn D., eds. *Van Nostrand's Scientific Encyclopedia*. 6th ed. New York: Van Nostrand, 1983.
Cordata, James W. *Historical Dictionary of Data Processing: Biographies*. New York: Greenwood Press, 1987.
Cornwell, Nancy. *Adventures with Polarfleece: A Sewing Expedition*. Iola, WI: Krause Publications, 1997.
Coyle, L. Patrick, Jr. *The World Encyclopedia of Food*. New York: Facts on File, 1982.
Crabbe, M. J. C., ed. *Enzyme Biotechnology: Protein Engineering, Structure, Prediction and Fermentation*. New York: Ellis Horwood, 1990.
Daintith, John, et. al. *A Biographical Encyclopedia of Scientists*. 2 Vols. New York: Facts on File, 1981.
Daumas, Maurice, ed. *A History of Technology and Invention*. 3 vols. New York: Crown, 1979.
Day, John. *The Bosch Book of the Motor Car*. New York: St. Martin's, 1976.
De Bono, Edward, ed. *Eureka! An Illustrated History of Inventions*. New York: Holt, Rinehart & Winston, 1974.
De Camp, L. Sprague. *The Heroic Age of American Invention*. Garden City, New York: Doubleday and Co., 1961.
De Vries, Leonard. *Victorian Inventors*. American Heritage Press, 1971.
Dennis, Landt. *Catch the Wind: A Book of Windmills and Windpower*. New York: Four Winds Press, 1976.
Diagram Group. *Weapons: an International Encyclopedia*. New York: St. Martin's Press, 1990.
Dilke, O. A. W. *The Roman Land Surveyors*. New York: Barnes & Noble Inc., 1971.
Downs, Robert B. *Landmarks in Science: Hippocrates to Carson*. Littleton, CO: Libraries Unlimited, Inc., 1982.
Downs, Robert B., et.al. *More Memorable Americans: 1750-1950*. Littleton, CO: Libraries Unlimited, Inc., 1985.
Drucker, Peter F. *Innovation and Entrepreneurship*. New York: Harper & Row, 1985.
Duffy, John. *The Healers: The Rise of the Medical Establishment*. New York: McGraw-Hill, 1976.
Duke, Marc. *Acupuncture*. New York: Pyramid House, 1972.
Dummer, G. W. A. *Electronic Inventions and Discoveries*. 3rd ed. Oxford: Pergamon Press, 1983.
Dutton, William S. *Du Pont: One Hundred Forty Years*. New York: Charles Scribner's Sons, 1951.
Eco, Umberto, and Zorzoli, G. B. *The Picture History of Inventions*. New York: Macmillan, 1963.
Edwards, Perry, and Broadwell, Bruce. *Data Processing, Computers in Action*. Belmont, CA: Wadsworth Publications, 1979.
Elliot, Clark A. *Biographical Dictionary of American Science: The Seventeenth Through the Nineteenth Centuries*. Westport: Greenwood Press, 1979.
Ewing, Elizabeth. *Underwear: A History*. New York: Theatre Arts Books, 1972.
Fagen, Brian M. *Staying Put! The Art of Anchoring*. Santa Barbara, CA: The Caractacus Corp., 1993.
Faith, Nicholas. *Classic Ships: Romance and Reality*. Osceola, WI: Motorbooks International Publishers & Wholesalers, 1995.
Fang, Irving E. *The Computer Story*. St. Paul: Rada Press, 1988.
Feldman, Anthony, and Ford, Peter. *Scientists and Inventors*. New York: Facts on File, 1979.
Findlay, A. *A Hundred Years of Chemistry*. London: Gerald Duckworth and Co., 1955.

Finley, John M. *Practical Wound Management*. Chicago: Year Book Medical Publishers, 1981.

Flinn, Richard A., and Trojan, Paul K. *Engineering Materials and Their Applications*. 3rd ed. Dallas: Houghton Mifflin Company, 1986.

Forbes, R.J. *Man the Maker: A History of Technology and Engineering*. London: Abelard-Schuman, 1958.

Ford, Grace Horney. *The Button Collector's History*. Springfield, MA: Pond-Ekberg, 1943.

Frances A. *The Unseen Minority: A Social History of Blindness in the United States*. New York: David McKay Company, 1976.

Frisinger, H. Howard. *The History of Meteorology: to 1800*. New York: Science Histoy Publications, 1977.

Fuller, Edmund. *Tinkers and Genius*. New York: Hastings House, 1955.

Fuller, R. Buckminster. *Inventions: The Patented Works of R. Buckminster Fuller*. New York: St. Martins Press, 1983.

Garrison, Fielding H. *An Introduction to the History of Medicine*. Philadelphia: W.B. Saunders, 1929.

Garrison, Webb. *Why Didn't I Think of That?* Englewood Cliffs, NJ: Prentice-Hall, 1977.

Gedzelman, Stanley David. *The Science and Wonders of the Atmosphere*. New York: John Wiley & Sons, 1980.

Gerhartz, W., ed. *Enzymes in Industry: Production and Applications*. New York: VCH, 1990.

Gernsheim, Helmut. *The History of Photography from the Camera Obscura to the Beginning of the Modern Era*. New York: McGraw-Hill, 1969.

Gillispie, Charles Coulston, ed. *Dictionary of Scientific Biography*. 14 Vols. New York: Charles Scribner's Sons, 1970.

Giscard d'Estaing, Valérie-Anne. *The World Almanac Book of Inventions*. New York: World Almanac Publications, 1985.

Glasscheib, H. S. *The March of Medicine*. New York: G. P. Putnam's Sons, 1964.

Goldsmith, Edward and Hildyard, Nicholas *The Earth Report*. Los Angeles: Price, Stern, Sloan, 1988.

Golob, Richard and Eric Brus, eds. *The Almanac of Science and Technology: What's New and What's Known*. New York: Harcourt Brace Jovanovich, 1990.

Graf, Rudolf F., and Whalen, George J. *How It Works Illustrated: Everyday Devices and Mechanisms*. New York: Harper & Row, 1974.

Graham, Ada, and Graham, Frank. *The Big Stretch: The Complete Book of the Amazing Rubber Band*. New York: Knopf, 1985.

Greenwood, David. *Mapping*. Chicago: University of Chicago Press, 1964.

Gregory, Richard L., ed. *The Oxford Companion to the Mind*. Oxford: Oxford University Press, 1987.

Grossinger, Tania. *The Book of Gadgets*. New York: David McKay, 1974.

Guthrie, Douglas. *A History of Medicine*. Philadelphia: J.B. Lippincott, 1946.

Gutkind, Lee. *Many Sleepless Nights: The World of Organ Transplantation*. New York: W. W. Norton, 1988.

Haggard, Howard W. *Devils, Drugs, and Doctors*. New York: Blue Ribbon Books, 1929.

Halprin, Lawrence. *Freeways*. New York: Reinhold, 1966.

Harris, Harry. *Good Old-Fashioned Yankee Ingenuity*. Chelsea, MI: Scarborough House, 1990.

Harrison, Michael R., et al., eds. *The Unborn Patient: Prenatal Diagnosis and Treatment*. 2nd ed. Philadelphia: W.B. Saunders, 1990.

Hartsuch, B. *Introduction to Textile Chemistry*. New York: John Wiley and Sons, 1950.

Haser, L. *Women Pioneers in Science*. New York: Harcourt Brace Jovanovich, 1979.

Haskins, Jim. *Outward Dreams: Black Inventors and Their Inventions*. New York: Walker and Co., 1991.

Hatch, Alden, *Buckminster Fuller: At Home in the Universe*. New York: Crowe Publishers, Inc., 1974.

Hawke, David Freeman. *Nuts and Bolts of the Past*. New York: Harper & Row., 1988.

Hawthorne, Peter. *The Transplanted Heart*. Chicago: New York: Rand McNally, 1968.

Hazelton, Nika Standen. *Chocolate!* New York: Simon & Schuster, 1967.

Helfman, Elizabeth S. *Blissymbolics: Speaking without Speech*. New York: Elsevier/Nelson Books, 1987.

Heyn, Ernest. *Fire of Genius: Inventors of the Past Century*. Garden City, NY: Anchor Press/Doubleday, 1976.

Hinz, Earl R. *The Complete Book of Anchoring and Mooring*. Centreville, MD: Cornell Maritime Press, 1994.

Hitchcock, H. Wiley, ed. *The New Grove Dictionary of American Music*. New York: Macmillan Press Limited, 1986.

Hogg, Ian. *Military Pistols and Revolvers*. Dorset: Arms and Armor, 1987.

Hooper, Meredith. *Everyday Inventions*. New York: Taplinger Publishing, 1976.

Hoyt, Edwin P. *One Penny Black: The Story of Stamp Collecting*. New York: Duell, Sloan & Pierce, 1965.

Hughes, Patrick. *A Century of Weather Service: A History of the Birth and Growth of the National Weather Service 1870-1970*. New York: Gordon & Breach, 1970.

I'll Buy That! Mt. Vernon, NY: Consumers Union (Consumer Reports), 1986.

Illingworth, Valerie. *The Facts on File Dictionary of Astronomy.* New York: Facts on File, 1979.

The Illustrated Science and Invention Encyclopedia. Westport: H.S. Stuttman, 1983.

Inglis, Brian. *A History of Medicine*. Cleveland: World Publishing, 1965.

International Dairy Foods Association. *The Latest Scoop*. Washington, D.C.: International Dairy Foods Association, 1997.

Jackson, John M., and Shinn, Byron M. *Fundamentals of Food Canning Technology*. Westport, CT: Avi Publishing, 1979.

James, Portia P. *The Real McCoy: African-American Invention and Innovation, 1619-1930*. Washington, D.C.: Smithsonian Institution Press, 1989.

Jesperson, James, and Fitz-Randolph, Jane. *Mercury's Web:*

The Story of Telecommunications. New York: Atheneum, 1981.

Jones, Arthur. *Cellulose, Lacquers, Finishes and Cements*. New York: J.B. Lipponcott Company, 1938.

Kahanamoku, Duke. *Duke Kahanamoku's World of Surfing*. New York: Grosset & Dunlap, 1968.

Katcher, Brian S. *Prescription Drugs*. New York: Atheneum, 1988.

Kemp, Peter. *The History of Ships*. London: Orbis Publishing, 1978.

Koller, Larry. *Fireside Book of Guns*. New York: Simon & Schuster, 1959.

Kranzberg, Melvin, and Pursell, Carroll W., Jr., eds. *Technology in Western Civilization*. New York: Oxford University Press, 1967.

Lambert, Samuel W., and Goodwin, George M. *Medical Leaders From Hippocrates to Osler*. Indianapolis: The Bobbs-Merrill Company, 1929.

Laver, James. *Costume*. New York: Hawthorn Books, 1963.

LaWall, Charles H. *Four Thousand Years of Pharmacy*. Philadelphia: J.B. Lippincott, 1927.

Leatham, G. F., and Himmel, M. E., eds. *Enzymes in Biomass Conversion*. Washington, D.C.: American Chemical Society, 1991.

Lefevre, Herve. *The Fiber-Optic Gyroscope*. Boston: Artech House, 1993.

Leitch, Michael. *The Romance of the Sail*. London: The Hamlyn Publishing Group Limited, 1975.

Lifshey, Earl. *The Housewares Story*. Chicago: National Housewares Manufacturers Association, 1973.

Logan, Rayford W., and Winston, Michael R., eds. *Dictionary of American Negro Biography*. New York: W. W. Norton & Co., 1982.

Logsdon, Tom. *Computers Today and Tommorrow: The Microcomputer Explosion*. Rockville, MD: Computer Science Press, 1985.

Lurie, Alison. *The Language of Clothes*. New York: Random House, 1981.

Magill, Frank N., ed. *The Nobel Prize Winners: Chemistry*. Pasadena: Salem Press, 1990.

——. *Great Events from History II: Science and Technology Series*. 5 vols. Pasadena: Salem Press, 1991.

——. *Great Lives from History: British and Commonwealth Series*. Pasadena: Salem Press, 1987.

Magnusson, Magnus. *Cambridge Biographical Dictionary*. Cambridge University Press, 1990.

Mahoney, Tom. *The Merchants of Life*. New York: Harper & Bros., 1959.

Mandell, Steven L. *Computers and Data Processing: Concepts and Applications*. St. Paul: West Publishing Company. 1979.

Margen, Sheldon. *The Wellness Encyclopedia of Food and Nutrition*. New York: Health Letter Associates, 1992.

Martin, George. *Making Music: The Guide to Writing, Performing & Recording*. New York: Quill, 1983.

Massett, Larry, and Sutherland, Earl W., III. *Everyman's Guide to Drugs and Medicines*. Washington, D.C.: Robert B. Luce, 1975.

Maymie R. *All About the Months*. New York: Harper & Row, 1966.

McAleer, Neil. *The Omni Space Almanac*. New York: World Almanac, 1987.

McCorduck, Pamela A. *Machines Who Think*. San Francisco: W. H. Freeman & Company, 1979.

McGraw-Hill Modern Men of Science. New York: McGraw Hill, 1986.

McGraw-Hill Yearbook of Science and Technology. New York: McGraw-Hill, 1982-1993.

McGrew, Roderick E. *Encyclopedia of Medical History*. New York: McGraw-Hill, 1985.

Mez-Mangold, Lydia. *A History of Drugs*. Totowa, NJ: Barnes & Noble Books/Rarthenon Publishing, 1986.

Morawetz, Herbert. *Polymers: The Origins and Growth of a Science*. New York: John Wiley and Sons, 1985.

Morgan, Hal. *Symbols of America*. New York: Viking, 1986.

Morton, Maurice, ed. *Rubber Technology*. New York: Van Nostrand, 1959.

Mossman, S.T.I., and Morris, P.J.T., eds. *The Development of Plastics*. Cambridge, England: The Royal Society of Chemistry, 1994.

Mount, Ellis, and List, Barbara A. *Milestones in Science and Technology*. Pheonix: The Oryx Press, 1987.

Neill, C. R., ed. *Guide to Bridge Hydraulics*. Toronto: University of Toronto Press, 1973.

The New Oxford Companion to Music. Oxford: Oxford University Press, 1983.

Newhouse, Elizabeth, ed. *Inventors and Discoverers: Changing Our World*. Washington, D.C.: National Geographic Society, 1988.

Newsholme, Arthur. *The Story of Modern Preventive Medicine*. Baltimore: The Williams and Wilkens Co., 1929.

Newton, David. *Encyclopedia of Crytology*. Santa Barbara, CA: ABC-CLIO, 1997.

Norman, Bruce. *The Inventing of America*. New York: Taplinger Publishing, 1976.

Ogden, Tom. *Two Hundred Years of the American Circus: From Aba-Daba to the Zoppe-Zavatta Troupe*. New York: Facts on File, Inc., 1993.

Ogilvie, Marilyn Bailey. *Women in Science*. Cambridge: The MIT Press, 1986.

Oliver, John W. *History of American Technology*. New York: Ronald Press, 1956.

Oxford Companion to Ships and the Sea. Oxford: Oxford University Press, 1976.

Palicia, Joan. *Glexible Flyer and Other Great Sleds for Collecting*. Atglen, PA: Schiffer Publishing, 1997.

Parker, Sybil P. *McGraw-Hill Concise Encyclopedia of Science and Technology*. 2nd ed. New York: McGraw-Hill, 1989.

Partingen, J. R. *A History of Chemistry*. New York: Macmillan and Co., 1964.

Patton, Phil. *Made in the USA*. New York: Grove Weidenfeld, 1992.

Peters, Thomas J., and Waterman, Robert H., Jr. *In Search of Excellence*. New York: Harper & Row, 1982.

Peterson, Harold. *Encyclopedia of Firearms*. New York: E.P. Dutton, 1964.
——. *The Great Guns*. NY: Grosset & Dunlap, 1971.
Pfeiffer, Michael and Gassebner, Jurgen. *Custom Motorcycles*. Oxford: MetroBooks, 1997.
Polymer Science and Engineering: The Shifting Research Frontiers. Washington, D.C.: National Academy Press, 1994.
Prinz, Hermann. *Dental Chronology*. Philadelphia: Lea & Febiger, 1945.
Queisser, Hans. *The Conquest of the Microchip*. Cambridge, MA: Harvard University Press, 1988.
Radford, Thomas. *The Construction of Roads and Pavements*. New York: McGraw- Hill, 1940.
Refrigeration and Air Conditioning. Englewood Cliffs, NJ: Prentice-Hall, 1979.
Reichardt, Jasia. *Robots: Fact, Fiction and Prediction*. New York: Penguin, 1978.
Reid, T. R. *The Chip: How Two Americans Invented the Microchip and Launched a Revolution.* New York: Simon & Schuster, 1984.
Reid, William. *Arms Through the Ages*. New York: Harper & Row, 1976.
Reidman, Sarah R. *Masters of the Scalpel*. Chicago: Rand McNally, 1962.
Richardson, Robert G. *Surgery: Old and New Frontiers*. New York: Scribner, 1968.
Ries, Estelle H. *The Ingenuity of Man*. Westport, CT: Greenwood Press, 1962.
Robertson, Patrick. *The Book of Firsts*. New York: Clarkson N. Potter, Inc., 1974.
Robinson, Victor. *Pathfinders in Medicine*. New York: Medical Life Press, 1929.
Rodriquez, Ferdinand. *Principles of Polymer Systems*. New York: Hemisphere Publishing, 1982.
Rolt, L. T. C. *A Short History of Machine Tools*. Cambridge: The M.I.T. Press, 1965.
Roth, Leland, M., ed. *America Builds*. New York: Harper & Row, 1983.
Rowland, K. T. *Eighteenth Century Inventions*. New York: Barnes and Noble, 1974.
Sadie, Stanley. *The Norton/Grove Concise Encyclopedia of Music*. New York: W.W. Norton & Company, 1988.
Samuel P. Wilson. *Pyroxylin, Enamels and Lacquers*. 2nd ed. New York: Van Nostrand, 1927.
Scheele, Carl H. *A Short History of the Mail Service*. Washington, D. C.: Smithsonian Institution Press, 1970.
Schlebecker, John T. *Whereby We Thrive: A History of American Farming, 1607-1972*. Ames, IA: The Iowa State University Press, 1975.
Schuck, H., et. al. *Nobel: The Man and His Prizes*. New York: Elsevier, 1962.
Schuyler, Hamilton. *The Roeblings: A Century of Bridge Builders and Industrialists*. Princeton: Princeton University Press, 1931.
Seagrave, Sterling. *Yellow Rain: A Journey Through the Terror of Chemical Warfare*. New York: M. Evans and Company, Inc., 1981.
Seinfeld, John H. *Atmospheric Chemistry and Physics of Air Pollution*. New York: John Wiley & Sons, 1986.
Serway, Raymond A. *Physics for Scientists and Engineers*. 3rd ed. Philadelphia: Saunders College Publishing, 1990.
Shurkin, Joel. *Engines of the Mind: A History of the Computer.* New York: Norton, 1984.
Singer, Charles, et al., eds. *A History of Technology*. 8 vols. Clarendon Press, 1954-84.
Sittig, Marshall. *Vinyl Chloride and PVC Manufacture*. Park Ridge, NJ: Noyes Data Corporation 1978.
Smith, Merritt Roe. *Harper's Ferry Armory and the New Technology: The Challenge of Change*. Ithaca: Cornell University Press, 1977.
Sobol, Robert, and Sicilia, David B. *The Entrepreneurs: An American Adventure*. Boston: Houghton Mifflin, 1986.
Stewart, George F., and Amerine, Maynard A. *Introduction to Food Science and Technology*. New York: Academic Press, 1973.
Talbott, John H. *A Biographical History of Medicine*. New York: Grune and Stratton, 1970.
Tames, Richard. *Isambard Kingdom Brunel*. London: Shire Publications, Ltd., 1972. Reprinted 1992.
Tannahill, Reay. *Food in History*. New York: Stein & Day, 1973.
Tapley, M.D., Donald F., et al.*Columbia University College of Physicians and Surgeons: Complete Home Medical Guide*. 3rd revised ed. New York: Crown Publishers Inc., 1995.
Taylor, L. B. *Chemical & Biological Warfare*. New York: F. Watts, 1985.
Thomas, Shirley. *Men of Space*. 5 Vols. Philadephia: Chilton Co., 1960-62.
Thompson, Claudia G. *Recycled Papers: The Essential Guide*. Cambridge, MA: The MIT Press, 1992.
Thompson, Robert Luther. *Wiring a Continent: The History of the Telegraph Industry in the United States*. Princeton: Princeton University Press, 1947.
Thorwald, Jurgen. *The Triumph of Surgery*. New York: Pantheon, 1960.
Tolmazin, David. *Elements of Dynamic Oceanography*. London: Allen & Unwin Inc., 1985.
Towler, Jean, and Bramall, Joan. *Midwives in History and Society*. London: Croom Helm, 1986.
Trewartha, Glenn T. *An Introduction to Weather and Climate*. New York: McGraw Hill, 1943.
Tunis, Edwin. *Weapons*. New York: World, 1954.
Tuplin, W. A. *Great Western Saints and Sinners*. London: Allen & Unwin, 1972.
Turak, Theordore. *William LeBaron Jenney: A Pioneer of Modern Architecture*. Ann Arbor: UMI Research Press, 1967, 1986.
Turner, Roland, and Goulden, Steven L., eds. *Great Engineers and Pioneers in Technology*. New York: St. Martins Press, 1981.
Van Doren, Charles, ed. *Webster's American Biographies*. Sprinfield, MA: G. & C. Merriam Co., 1974.

Vare, Ethlie Ann. *Mothers of Invention*. New York: William Morrow and Co., 1988.

Veglahn, Nancy J. *Women Scientists*. New York: Facts on File, 1991.

Wallace, John M. and Hobbs, Peter V. *Atmospheric Science: An Introductory Survey*. Orlando, FL: Academic Press, Inc., 1977.

Walton, John, et. al., eds. *The Oxford Companion to Medicine*. Oxford: Oxford University Press, 1986.

Wangensteen, Owen H., and Wangensteen, Sarah D. *The Rise of Surgery*. Minneapolis: University of Minnesota Press, 1978.

Wasson, Tyler, ed. *Nobel Prize Winners*. New York: H. W. Wilson Co., 1987. Wertz, Richard W. and Wertz, Dorothy C. *Lying-In: A History of Childbirth in America*. New York: Schocken Books, 1977.

Whitman, Joan, ed. *Craig Claiborne's The New York Times Food Encyclopedia*. New York: The New York Times Company, 1985.

Whitnah, Donald R. *A History of the United States Weather Bureau*. Champaign-Urbana: University of Illinois Press, 1961.

Williams, Trevor I. *A Biographical Dictionary of Scientists*. London: Adam and Charles Black, 1969.

Wilson, Charles Morrow. *Trees & Test Tubes: The Story of Rubber*. New York: Henry Holt, 1943.

Wilson, Hugo. *The Encyclopedia of the Motorcycle*. London: Dorling Kindersley, 1995.

Wilson, James Grant, and Fiske, John. *Appleton's Cyclopedia of American Biography*. New York: D. Appleton & Co., 1887-89.

Wilson, Mitchell. *American Science and Invention: A Pictoral History*. New York, Simon and Schuster, 1954.

Wilson, P.W. *The Romance of the Calendar*. New York: W.W. Norton, 1937.

Woodroof, Jasper Guy, and Phillips, G. Frank. *Beverages: Carbonated and Noncarbonated*, rev. ed. Westport, CT: Avi Publishing, 1981.

Wright, Lawrence. *Clean and Decent*. New York: Viking Press, 1960.

Wriston, Walter B. *The Twilight of Sovereignty*. New York: Charles Scribner's Sons, 1992.

Wulffson, Don L. *Extraordinary Stories Behind the Invention of Ordinary Things*. New York: Lothrop, Lee & Shepard, 1981.

Yarwood, Doreen. *Encyclopedia of Architecture*. New York: Facts on File, 1986.

Yost, Edna. *Modern Americans in Science and Invention*. New York: Frederick A. Stokes Co., 1941.

Zaret, M.D., Barry L., sr. ed. *The Patient's Guide to Medical Tests*. New York: Houghton Mifflin Company, 1997.

Subject Index

Agriculture

Automotive engineering

Aviation and Aerospace

Chemistry

Civil engineering and Construction

Clothing, Textiles, and their Manufacture

Communications and Graphic Arts

Computer science and Mathematical devices

Electrical engineering and Electrical devices

Electronics

Everyday items

Food and Food science

Geology

Household appliances

Lighting and illumination

Materials

Mechanical engineering

Medicine and Dentistry

Metallurgy

Meteorology

Musical instruments

Navigation

Oceanography

Optics

Personal care items

Physics

Security systems and Related items

Sports, Games, Toys and Fads

Timepieces, Measuring devices, and Related Items

Transportation

Weapons and Related Items

General Index

A

B

C

D

G

H

I

K

L

M

N

O

P

Q

R

T

U

V

W

X